MW01075191

Seismic Waves and Sources

second edition

Ari Ben-Menahem
Sam and Ayala Zacks Professor of Geophysics
The Weizmann Institute of Science
Rehovot 76100, Israel

and

Sarva Jit Singh
Professor of Mathematics
Maharshi Dayanand University
Rohtak, India

DOVER PUBLICATIONS, INC.
Mineola, New York

Copyright

Published in Canada by General Publishing Company, Ltd., 30 Lesmill Road, Don Mills, Toronto, Ontario.

Bibliographical Note

This Dover edition, first published in 2000, is the unabridged second, corrected edition of the text of *Seismic Waves and Sources,* originally published in 1981 by Springer-Verlag New York Inc. It includes 307 illustrations and 81 tables. The Author Index provided in the original edition is not reproduced in the Dover edition.

Library of Congress Cataloging-in-Publication Data

Ben-Menahem, Ari.
Seismic waves and sources / by Ari Ben-Menahem and Sarva Jit Singh. — 2nd, corr. ed.
p. cm.
Includes bibliographical references and index.
ISBN 0-486-40461-7 (pbk.)
1. Seismic waves. 2. Seismology—Mathematics. I. Singh, Sarva Jit. II. Title.

QE538.5.B46 2000
551.22—dc21

98-40854
CIP

Manufactured in the United States of America
Dover Publications, Inc., 31 East 2nd Street, Mineola, N.Y. 11501

In memory of my father, Moshe Ben-Menahem

(A.B-M.)

To my parents

(S.J.S.)

Preface to the First Edition

Earthquakes come and go as they please, leaving behind them trails of destruction and casualties. Although their occurrence is little affected by what we do or think, it is the task of earth scientists to keep studying them from all possible angles until ways and means are found to divert, forecast, and eventually control them.

In ancient times people were awestruck by singular geophysical events, which were attributed to supernatural powers. It was recognized only in 1760 that earthquakes originated within the earth. A hundred years later, first systematic attempts were made to apply physical principles to study them. During the next century scientists accumulated knowledge about the effects of earthquakes, their geographic patterns, the waves emitted by them, and the internal constitution of the earth.

During the past 20 years, seismology has made a tremendous progress, mainly because of the advent of modern computers and improvements in data acquisition systems, which are now capable of digital and analog recording of ground motion over a frequency range of five orders of magnitude. These technologic developments have enabled seismologists to make measurements with far greater precision and sophistication than was previously possible. Advanced computational analyses have been applied to high-quality data and elaborate theoretical models have been devised to interpret them. As a result, far-reaching advances in our knowledge of the earth's structure and the nature of earthquake sources have occurred.

The primary objective of this book is to give the reader a comprehensive account of the generation of elastic waves by realistic earthquake sources and their propagation through realistic earth models. There has been a wide gap between the levels of the textbooks available to seismologists and the work appearing in research journals. No previous modern work has bridged this gap, even partially. We hope that our treatise indeed fulfills this objective.

We seek to represent earthquake seismology as a science that stands firm on its own theoretical foundations and is able to render a satisfactory quantitative account of seismic observations over the entire spectral range of recorded wave phenomena.

The hard core and the general framework of the material presented here are based on lectures given by one of us (A.B-M.) to students of the Feinberg

Graduate School of the Weizmann Institute of Science through the years 1966–1975. The uniform representation of sources and fields, which is the theoretical backbone of the treatise, was developed mostly by the authors in their joint work during the period 1967–1974. This unifying formalism has enabled us to develop the mathematical theory of seismic fields from first principles and to take the reader to the latest developments in the subject. Functional both as a textbook and a handbook, this work should prove useful to university students and research workers in the various branches of earth sciences, applied mathematics, theoretical physics, and engineering.

The material covers some 160 years in the history of seismology, starting with the equations of motion derived by Louis Navier in 1821. In the *pre-seismograph era* (1821–1891), intensive theoretical work was done by mathematicians and physicists, who laid the foundations to the mathematical theories of infinitesimal elasticity and seismic fields. The *precomputer era* (1892–1950) was characterized by the availability of instrumental data, which motivated theoretical research. Simple models for earth structure and seismic sources were established and tested against the data. The fundamental properties of seismic body and surface waves were discovered. Concurrently, mathematicians and theoretical physicists discovered new methods for tackling problems of wave propagation. Among these were the method of steepest descents, integral relations among plane, spherical and cylindrical waves, operational methods, variational techniques, and asymptotic solutions of differential equations. These methods were applied by seismologists to solve problems of wave propagation in the earth. The third era (1951 onward) has been marked by two outstanding features: the development of sensitive long-period seismographs and the increasing influence of the computer, both on the choice of problems and on the methods of attack.

In selecting material out of this vast literature, we were guided by two principles: First, only well-established theoretical results were reported, ones that had been verified by repeated observations and careful data analyses. Second, the scope of the book forced us to concentrate our efforts on the topics that belong to the mainstream of contemporary seismology. For example, a discussion of dynamic theories of fracture was deleted for the first reason and a discussion of scattering and diffraction of seismic waves, as well as the theory of leaking modes, were excluded for the second reason.

The book gives an up-to-date, comprehensive, rigorous, and lucid account of the propagation of elastic waves in the earth. Although the main emphasis is on earthquake waves, the theories of gravity waves in water and acoustic-gravity waves in the atmosphere are also included. The book is well illustrated with figures, tables, and solved examples and is made self-sufficient with the addition of several appendices, which introduce most of the mathematical tools employed. The dyadic approach is used for elegance and brevity. The material is fully indexed and a fairly comprehensive list of references is given at the end of each chapter and at the end of the appendices. We make no pretensions as to completeness or historical balance. The references are not included to give credit for results but to help the reader find more complete discussions of various topics.

Chapter 1 is concerned with a brief but lucid account of those results of classical continuum dynamics which are essential to seismology. The fundamental concepts of stress and strain are presented and the basic field equations are derived. It includes the Lagrangian formulation, which is found to be a useful tool in the amplitude theory of surface waves in elastic and anelastic media.

In Chapter 2 we deal mainly with the eigenvector solutions of the Navier equation for infinite space and the plane-wave solutions of this equation in Cartesian, circular cylinder, and spherical coordinates. Emphasis is laid on the integral interrelations between plane, cylindrical, and spherical waves. The concepts of primary (P), vertical shear (SV), and horizontal shear (SH) waves are introduced.

Chapter 3 concerns the interaction of vector plane waves with planar discontinuities. The basic laws of reflection and refraction are derived. The properties of surface waves are exhibited, first for simple structures and eventually for a multilayered half-space with the aid of the matrix propagator algorithm. Later in the chapter, we introduce numerical methods that enable one to derive crustal and upper-mantle structure of selected geological provinces from observed dispersion data of seismic surface waves over that region.

Chapter 4 renders a comprehensive and systematic study of seismic sources and the application of elasticity theory of dislocations to seismology. Starting from the fundamental Stokes–Love solution of the inhomogeneous Navier equation, the theory of dipolar point sources is developed and is finally linked with the key concept of displacement dislocation via the Volterra relation. The discussion goes on to incorporate other relevant aspects of modern seismology which bear on the theory of earthquake sources, such as fault-plane geometry, theoretical seismograms in an infinite medium, displacements in the near and the far fields, explosions in pre-stressed media and finally the theory of radiation of body waves from finite moving sources.

Chapter 5 contains a detailed account of one of the greatest triumphs of modern seismology, namely the ability to account for the observed amplitudes of surface waves from earthquakes in terms of the kinematic parameters of the source. The new field of "terrestrial interferometry" by which the fault length and rupture velocity can be routinely calculated from the seismic "Doppler effect" via the directivity function, is explained and illustrated with several examples. The material includes details of numerical algorithms and data reduction routines that are widely used in the analysis of earthquake signals.

One of the noteworthy achievements of seismology since the late fifties is the discovery that earthquakes can excite the earth's free oscillations to a level that can be measured by most long-period seismographs. This new field is known as "terrestrial spectroscopy." Chapter 6 gives a detailed account of the theory of excitation of the free oscillations of the earth by earthquakes. It includes numerical methods for the calculation of the eigenfrequencies and spectral amplitudes for realistic structural models. The effects of the rotation of the earth and the source's finiteness and motion on the line spectra are also given.

Chapter 7 is about rays. While the free oscillation data renders important information concerning the gross features of the earth and surface waves are useful in deriving average crustal structures and the parameters of seismic sources, our main tool for the investigation of the core and deep mantle comes chiefly from the study of body waves that travel everywhere inside the earth in the form of rays. Starting again from the equation of motion, the amplitude theory of body waves in a radially inhomogeneous earth is developed. The useful concept of a generalized ray is introduced and numerical methods are developed by which ground motion can be represented as a sum of generalized rays. The theory of initial motions and the phenomenon of caustics are discussed.

The duality of the normal modes and rays is explored in Chapter 8. Here, the asymptotic theory of earth's normal modes is studied in great detail. It is shown how the exact normal-mode solution yields the partial fields of body waves, diffracted waves in the shadow zone and the field on a caustic. Finally, the "rainbow expansion" is used to generate generalized rays in spherical earth models and to derive amplitudes of isolated body wave signals. Topics such as Fresnel diffraction, tunnelling, and earth-flattening transformation are also presented.

Atmospheric and water waves associated with earthquakes and explosions are discussed in Chapter 9. The chapter opens with a brief summary of the basic principles of hydrodynamics, including the equations of sound waves and long gravity waves in liquids. The theoretical results are then applied to waves and oscillations excited by earthquakes such as tsunamis, seiches, air waves, pressure-induced surface waves, and coupled air–sea waves. Rayleigh waves and acoustic–gravity waves excited by nuclear explosions are also discussed. It is demonstrated that many of the concepts and numerical algorithms developed earlier for layered elastic media are valid also for waves in fluid media.

In Chapter 10 we return to seismic waves and examine their propagation in low-loss anelastic substances of which the real earth is made. After introducing the basic concepts of the theory of viscoelastic solids, we discuss the propagation of seismic pulses in unbounded anelastic media and apply the results to the attenuation of seismic waves in the earth. The causal dispersion of attenuated surface waves is explained.

The appendices furnish ample information on the various mathematical techniques used in the book. There should be no need for the reader to look for the material in other sources, where his comprehension may be unnecessarily hampered by alien notation and applications to fields foreign to seismology. The information given in the appendices is brief, concentrated and with seismological applications in mind.

Parts of this book were written while one of us (A.B-M.) was a visiting professor at the department of Geophysics of Stanford University (1975–1977) and the Institute of Geophysics and Planetary Physics, UCLA (1979). The hospitality and technical assistance afforded there are gratefully acknowledged. Thanks are due especially to Kathleen Hart, Linda and Bob Kovach, and Prof. George A. Thompson of Stanford University, and Dr. A. K. Chatterjee of UCLA. Among those who made useful suggestions we must single out Dane Brooke of our Geophysical Laboratory and Dr. Shahar Ben-Menahem of the Stanford Linear Accelerator Center. We are thankful to the Department of Applied Mathematics of the Weizmann Institute for grants which enabled one of the authors (S.J.S.) to travel several times to Rehovot and work at the Institute. We are much indebted to Professor Robert L. Kovach of Stanford University for his help in writing Sections 3.9, 6.5, and 6.7.1. The technical production of the book has been accomplished with the devoted assistance of Ms. Sara Fligelman, who prepared the typescript. Mr. Yehuda Barbut drew most of the figures and Mr. A. Silberberg of the Wix Library assisted us in the photographic reproductions. Special acknowledgment must be made to Adolpho Bloch for his generous support of seismological research at the Weizmann Institute of Science. Finally, we would like to thank the staff of Springer-Verlag for their unfailing efforts in producing this book.

Rehovot
November 4, 1980

Ari Ben-Menahem
Sarva Jit Singh

Preface to the Second Edition

This book was originally published in 1981 by Springer-Verlag, New York, and for the past 19 years has served as a basic reference book for the mathematical theory of seismology. A generation of graduate students, researchers, and professional seismologists have used this text as an authoritative starting point for their theories and computations.

Indeed, *Seismic Waves and Sources* is a single-volume encyclopedia of the accumulated lore of modern seismological theory since its nascence, almost two hundred years ago. It is still the most comprehensive treatise that presents earthquake seismology as a science firmly resting on its own foundations. It provides in-depth quantitative assessment of seismic observations over the entire range of recorded wave phenomena. Although the new technology of the past decade has motivated a shift of emphasis toward computer-based methods, the fundamental theory that links seismic sources to the global field of observation withstood the test of time and needs no modifications. For this reason, we found no need to introduce changes in the text, except for minor corrections and modifications intended to make the notation consistent throughout this vast text and to enhance the clarity of certain derivations.

Thus, in chapter 1 the derivation of the maximum shearing stress was shortened. In (4.4.2), the interpretation of the decomposition of the field of a general shear dislocation was simplified. Certain missing phase-factors in Eqs. (7.206)–(7.214) were restored. Our treatment of the energy of plane viscoelastic waves in (10.3.2) was deemed unsatisfactory, and appropriate changes were made. Eq. (10.312) was simplified. Finally, the list of abbreviations (p. 1084) was expanded.

On April 14, 1992, an unpredicted earthquake of magnitude six occurred in the heart of Europe, amidst hundreds of seismographs, computers, and professors of seismology. It was the strongest shock in central Europe since the Lisbon earthquake (1755), which marked the dawn of modern seismology. Such "unpredictions" have by now become routine in Japan, California, Italy, Turkey, and elsewhere. While seismic research centers are flooded with enormous amounts of data and connected to supercomputers and satellites, "mother earth" continues to surprise us almost daily. The message is clear and unequivocal: there is too much reliance on computer simulation, and insufficient effort to forge new mathematical weapons and novel theoretical ideas.

We hope that the knowledge acquired by creative readers of this book will inspire them to build upon the rich legacy of the theory expounded here, to advance the frontiers of seismic wave and source theory.

A. B–M.
S. J. S.

Seismology—Milestones of Progress

Science is the knowledge of many,
orderly and methodically digested and arranged,
so as to become attainable by one.

(J. F. W. Herschel)

1660	Robert *Hooke* (England) stated his law: "Ut tensio sic vis."
1760	John *Michell* (England) recognized that earthquakes originate within the earth and send out elastic waves through the earth's interior.
1821	Louis *Navier* (France) derived the differential equations of the theory of elasticity.
1828	Simeon-Denis *Poisson* (France) predicted theoretically the existence of longitudinal and transverse elastic waves.
1849	George Gabriel *Stokes* (England) conceived the first mathematical model of an earthquake source.
1857	First systematic attempt to apply physical principles to earthquake effects by Robert *Mallet* (Ireland).
1883	*Rossi–Forel* scale for earthquake effects published.
1885	C. *Somigliana* (Italy) produced formal solutions to Navier equations for a wide class of sources and boundary conditions. Lord *Rayleigh* (England) predicted the existence of elastic surface waves.
1892	John *Milne* (England) constructed in Japan a seismograph suitable for world-wide use. Seismological observatories were set up on global basis to measure ground movements.
1897	Emil *Wiechert* (Germany) conjectured on the presence of a central fluid core in the earth. R. D. *Oldham* (England) identified in seismograms the three main types of seismic waves.
1899	C. G. *Knott* (England) derived the general equations for the reflection and refraction of plane seismic waves at plane boundaries.
1901	First Geophysical Institute founded in Göttingen, Germany.

1903 A. E. H. *Love* (England) developed the fundamental theory of point-sources in an infinite elastic space. International seismological association founded.

1904 Horace *Lamb* (England) layed the theoretical foundation for propagation of seismic waves in layered media.

1906 R. D. *Oldham* (England) established Wiechert's conjecture from seismic data. Electromagnetic seismograph developed by Count B. *Galitzin* (Russia).

1907 Vito *Volterra* (Italy) published his theory of dislocations based on Somigliana's solution.

1909 A. *Mohorovičić* (Zagreb) discovered the discontinuity between the crust and the mantle and demonstrated that the structure of the earth's outer layers can be deduced from travel times of reflected seismic waves.

K. *Zoeppritz* and L. *Geiger* (Germany) calculated velocities of longitudinal waves in the earth's mantle.

1914 Beno *Gutenberg* (Germany) measured the radius of the earth's core.

1935 H. *Benioff* (U.S.A.) invented the linear strain seismograph.

1936 I. *Lehmann* (Denmark) suggested the existence of a solid inner core.

1940 Sir Harold *Jeffreys* (England) and K. E. *Bullen* (Australia) published travel-time tables for seismic waves in the earth.

1952 M. *Ewing* and F. *Press* (U.S.A.) developed a sensitive long-period seismograph.

1959 Ari *Ben-Menahem* (Israel) discovered that energy release in earthquakes takes place through a propagating rupture over the causative fault. He then derived the fault-length and average rupture-velocity of the Chilean earthquake of May 22, 1960 from the spectra of its surface waves.

1960 Existence of the *free oscillations of the earth* first established from analyses of records of the Chilean earthquake of May 22, 1960. The rotational splitting of the free oscillations determined by C. L. *Pekeris* (Israel) and G. *Backus* and F. *Gilbert* (U.S.A.).

1967 Global seismicity patterns and earthquake generation linked to *Plate motions*.

1970 NASA (U.S.A.) put a seismograph on the moon.

Acknowledgments

Numerous research papers and review articles were consulted in the writing of this book. We have included them in the bibliography at the end of the relevant chapters. Omissions, if any, are regretted. We mention below the sources from which some of the tables and figures have been taken. Most of these figures have been redrawn and modified to suit our needs. It is a pleasure to thank the authors of these papers:

Alsop (1964; Table 8.1)*; Alsop, Sutton, and Ewing (1961; Fig. 10.21); Alterman, Jarosch, and Pekeris (1961; Table 8.4); Alterman and Kornfeld (1966; Figs. 8.23–8.27); Arkhangel'skaya (1964; Table 5.6e—crustal and upper-mantle structure only); Brune (1964; Fig. 8.19); Brune and Dorman (1963; Table 5.6b—crustal and upper-mantle structure only); Brune, Nafe, and Alsop (1961; Fig. 8.4); Chapman (1969; Fig. 8.43, Table 8.11); Engdahl (1968; Fig. 8.39); Gilbert and MacDonald (1960; Fig. 8.7); Gupta and Narain (1967; Table 5.6f—crustal and upper-mantle structure only); Harkrider (1968; Fig. 3.10, Table 3.2); Harkrider, Newton, and Flinn (1974; Fig. 9.23); Harkrider and Press (1967; Figs. 9.11, 9.12, and 9.16–9.22); Haskell (1964; Figs. 4.39–4.42); Herrin (1968; Table 7.1); Hill (1973; Figs. 7.64–7.66); Hirasawa and Stauder (1965; Fig. 4.35); Jeffreys and Lapwood (1957; Fig. 8.9); Kanamori (1977; Fig. 10.15b); Knopoff and Fonda (1975; Table 5.6h—crustal and upper-mantle structures only); Kovach (1965; Tables 5.6a,d,g—crustal and upper-mantle structures only); Kovach (personal communication; Figs. 3.9 and 7.51–7.56); Landisman, Satô, and Nafe (1965; Tables 8.2, 8.3, and L.1); Liu and Kanamori (1976; Figs. 10.15a, 10.20); Pekeris (1965; Table 8.5); Pekeris, Alterman, and Abramovici (1963; Figs. 7.67–7.69); Press and Harkrider (1962; Figs. 9.7–9.9); Rial (1978; Fig. 5.25b); Richards (1973; Figs. 8.32, 8.41, and 8.42); Richards (1976; Fig. 8.36); Saito and Takeuchi (1966; Table 5.6c—crustal and upper-mantle structure only); Sato (1969; Fig. 8.18); Satô, Usami, Landisman, and Ewing (1963; Figs. 8.20–8.22); Sterling and Smets (1971; Fig. 10.17); Takeuchi, Dorman, and Saito (1964; Tables 6.6–6.10 and L.6).

* All Figure and Table numbers are as they appear in this volume, and do not reflect the numbers as they appear in the original works.

Contents

List of Tables

CHAPTER 1

Classical Continuum Dynamics

Man can only conquer nature by obeying her.

(Francis Bacon)

1.1. The Stress Dyadic and Its Properties

If a deformable body is subjected to the action of forces distributed over its surface, these forces are transmitted to every point in the interior of the body. As a result, the relative positions of its physical elements change. Unlike *body forces*, which act on each volume element within the body (e.g., gravitational forces), the *surface forces* act at an internal point P across a surface in which P is embedded. Consider a vectorial surface element through P, $\Delta\mathbf{S} = \mathbf{n}\Delta S$, where the normal $\mathbf{n}$ is given with respect to some fixed Cartesian coordinate system. We may define the resultant force $\Delta\mathbf{F}$, acting on $\Delta\mathbf{S}$, as the measure of the action of the continuum I (on the side of ΔS toward which $\mathbf{n}$ is directed) on continuum II (Fig. 1.1). The ratio $\Delta\mathbf{F}/\Delta S$ is the average stress vector on ΔS. If the limit of this quotient exists as $\Delta S \to 0$ ($\mathbf{n}$ fixed), we write

$$\lim_{\Delta S \to 0} \frac{\Delta\mathbf{F}}{\Delta S} = \frac{d\mathbf{F}}{dS} = \mathbf{T}(\mathbf{n}) \tag{1.1}$$

and call $\mathbf{T}$ the *stress vector* (or *traction*) at P associated with the normal $\mathbf{n}$. In this connection, the following may be noted:

1. By Newton's law of action and reaction, side II exerts upon side I an equal and opposite force; i.e.,

$$\mathbf{T}(-\mathbf{n}) = -\mathbf{T}(\mathbf{n}). \tag{1.2}$$

2. We assume that the stress vector at P associated with the normal $\mathbf{n}$ is independent of the surface used for its definition (Cauchy's stress principle). This is a fundamental hypothesis. It allows us to replace the unknown actual intermolecular forces by a single force that depends upon two geometric entities alone, viz., the coordinates of the point relative to the applied surface forces and the orientation of the normal.

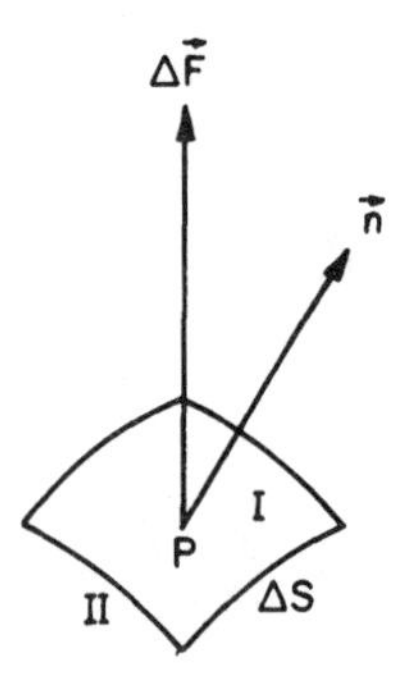

Figure 1.1. $\Delta\mathbf{F}$, the action of continuum II upon continuum I. In general, $\Delta\mathbf{F}$ is not in the direction of the normal $\mathbf{n}$ to ΔS.

3. Given a finite surface S within the body, the total force acting across it is determined by the integral

$$\int_S \mathbf{T}(\mathbf{n})dS. \tag{1.3}$$

This integral will vanish if S is a closed surface, provided no body forces (including inertial forces) are acting on the volume bounded by S. If the body forces are present, we shall assume that the surface forces transmitted into the continuum are, at each moment, in equilibrium with the body forces.

4. Equation (1.1) does not specify the nature of the functional relationship $\mathbf{T}(\mathbf{n})$. However, this can be found by solving a *canonical problem* that relates to the surface force distribution alone, with no body forces. If the whole body is in equilibrium under the action of the surface forces, so is an infinitesimal volume element around P. For the sake of simplicity, we choose it to be a tetrahedron $PABC$ three of whose faces dS_i $(i = 1, 2, 3)$ coincide with the coordinate planes. Because the sides of the tetrahedron are small, Eqs. (1.2) and (1.3) yield (Fig. 1.2)

$$\begin{aligned}\int_S \mathbf{T}(\mathbf{n})dS &= \mathbf{T}(\mathbf{n})dS_n + \mathbf{T}(-\mathbf{e}_1)dS_1 + \mathbf{T}(-\mathbf{e}_2)dS_2 + \mathbf{T}(-\mathbf{e}_3)dS_3 \\ &= \mathbf{T}(\mathbf{n})dS_n - \mathbf{T}(\mathbf{e}_1)dS_1 - \mathbf{T}(\mathbf{e}_2)dS_2 - \mathbf{T}(\mathbf{e}_3)dS_3,\end{aligned} \tag{1.4}$$

where $\mathbf{e}_i$ $(i = 1, 2, 3)$ is the unit vector along the x_i direction and dS_n is the area of the face ABC. Because $dS_i = n_i\, dS_n = (\mathbf{n}\cdot\mathbf{e}_i)dS_n$ and the matter inside $PABC$ is in equilibrium, we may write Eq. (1.4) formally as

$$\mathbf{T}(\mathbf{n}) = \mathbf{n}\cdot[\mathbf{e}_1\mathbf{T}(\mathbf{e}_1) + \mathbf{e}_2\mathbf{T}(\mathbf{e}_2) + \mathbf{e}_3\mathbf{T}(\mathbf{e}_3)]. \tag{1.5}$$

Here, for example, $\mathbf{T}(\mathbf{e}_1)$ denotes the stress vector across the x_1-plane with normal $\mathbf{e}_1$. As the volume of the tetrahedron is allowed to shrink to zero, the plane ABC contains the point P and the vector $\mathbf{T}(\mathbf{n})$ becomes the stress vector at P associated with the normal $\mathbf{n}$. Equation (1.5) shows that it is sufficient to know the value of the stress on three mutually perpendicular planes passing through the point P of the deformed body in order to evaluate the stress on an arbitrary plane passing through that point.

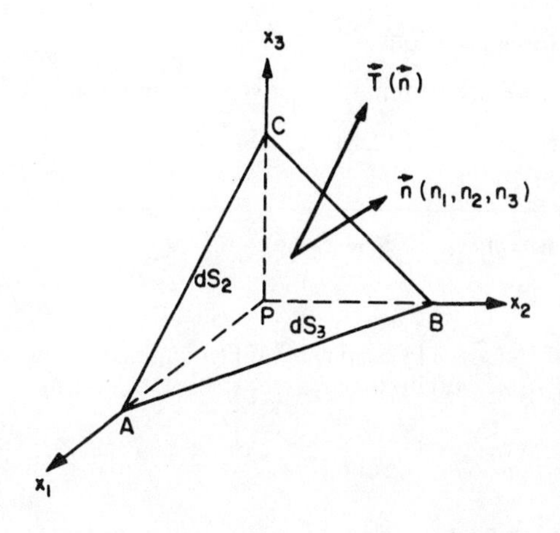

Figure 1.2. A volume element in the form of a tetrahedron. **T(n)** is the stress vector across the plane ABC having normal **n**. dS_3 is the area of the face PAB.

The entity in the brackets of Eq. (1.5) is a dyadic (App. A). We denote it by $\mathfrak{T}$ and call it the *stress dyadic* (or *stress tensor*). Because $\mathbf{T}(\mathbf{e}_i)$ is a vector, it is expressible as a linear combination of the base vectors $\mathbf{e}_i$:

$$\begin{aligned} \mathbf{T}(\mathbf{e}_1) &= \tau_{11}\mathbf{e}_1 + \tau_{12}\mathbf{e}_2 + \tau_{13}\mathbf{e}_3, \\ \mathbf{T}(\mathbf{e}_2) &= \tau_{21}\mathbf{e}_1 + \tau_{22}\mathbf{e}_2 + \tau_{23}\mathbf{e}_3, \\ \mathbf{T}(\mathbf{e}_3) &= \tau_{31}\mathbf{e}_1 + \tau_{32}\mathbf{e}_2 + \tau_{33}\mathbf{e}_3. \end{aligned} \tag{1.6}$$

The coefficient τ_{ij} is the x_j component of the stress vector on a plane whose normal points in the x_i direction (Fig. 1.3).

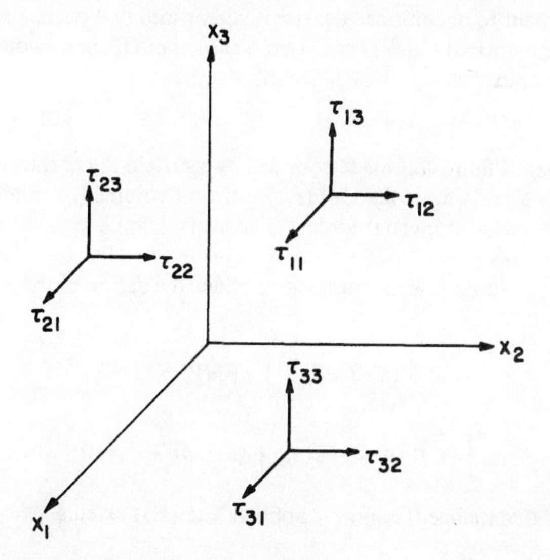

Figure 1.3. The nine components of the stress tensor. τ_{ij} is the component of the stress vector parallel to x_j on a plane having $\mathbf{e}_i$ as normal.

Equation (1.5) can be written in three notations:

1. *Indicial notation.* In this notation we use summation convention and write

$$T_i = n_j \tau_{ji} \qquad (i = 1, 2, 3), \tag{1.7}$$

where

$$\mathbf{T} = T_i \mathbf{e}_i, \qquad \mathbf{n} = n_j \mathbf{e}_j, \qquad \mathfrak{T} = \tau_{ji} \mathbf{e}_j \mathbf{e}_i.$$

2. *Matrix notation.*

$$[T_1, T_2, T_3] = [n_1, n_2, n_3] \begin{bmatrix} \tau_{11} & \tau_{12} & \tau_{13} \\ \tau_{21} & \tau_{22} & \tau_{23} \\ \tau_{31} & \tau_{32} & \tau_{33} \end{bmatrix}. \tag{1.8}$$

3. *Vector notation.*

$$\mathbf{T}(\mathbf{n}) = \mathbf{n} \cdot \mathfrak{T}, \tag{1.9}$$

where

$$\begin{aligned} \mathbf{T} &= T_1 \mathbf{e}_1 + T_2 \mathbf{e}_2 + T_3 \mathbf{e}_3, \\ \mathbf{n} &= n_1 \mathbf{e}_1 + n_2 \mathbf{e}_2 + n_3 \mathbf{e}_3, \\ \mathfrak{T} &= \tau_{11} \mathbf{e}_1 \mathbf{e}_1 + \tau_{12} \mathbf{e}_1 \mathbf{e}_2 + \tau_{13} \mathbf{e}_1 \mathbf{e}_3 + \tau_{21} \mathbf{e}_2 \mathbf{e}_1 + \tau_{22} \mathbf{e}_2 \mathbf{e}_2 + \tau_{23} \mathbf{e}_2 \mathbf{e}_3 \\ &\quad + \tau_{31} \mathbf{e}_3 \mathbf{e}_1 + \tau_{32} \mathbf{e}_3 \mathbf{e}_2 + \tau_{33} \mathbf{e}_3 \mathbf{e}_3. \end{aligned}$$

The components of $\mathfrak{T}$, τ_{ij}, depend on the coordinate system and, like the components of a vector, change with the change in the orientation of the base vectors. However, the law $\mathbf{T} = \mathbf{n} \cdot \mathfrak{T}$ is independent of the reference frame.

It is convenient to decompose the stress vector into two parts, one, $\mathbf{T}_n$, in the direction of the normal (*normal stress*) and the other, $\mathbf{T}_s$, perpendicular to it in the tangent plane (*shearing stress*). Clearly,

$$\mathbf{T}_n = \mathbf{n}(\mathbf{T} \cdot \mathbf{n}) = \mathbf{n}(\mathbf{n} \cdot \mathfrak{T} \cdot \mathbf{n}), \qquad \mathbf{T}_s = \mathbf{T} - \mathbf{T}_n = \mathbf{n} \times \mathbf{T} \times \mathbf{n}. \tag{1.10}$$

Consider next a finite volume V bounded by a surface S. Let the matter in V be acted upon by a body force field, $\rho\mathbf{F}(\mathbf{r}, t)$, per unit volume ($\rho$ is the density and the force field includes inertial forces). The surface force transmitted across a surface element $d\mathbf{S}$ is $(\mathbf{n} \cdot \mathfrak{T})dS$, where $\mathfrak{T}(\mathbf{r}, t)$ is the dyadic stress field. The fact that the entire volume V is in equilibrium under the action of these forces at all times implies that

$$\int_S (\mathbf{n} \cdot \mathfrak{T}) dS + \int_V \rho \mathbf{F} \, dV = 0, \tag{1.11}$$

$$\int_S [\mathbf{r} \times (\mathbf{n} \cdot \mathfrak{T})] dS + \int_V (\mathbf{r} \times \rho \mathbf{F}) dV = 0. \tag{1.12}$$

When Gauss' divergence theorem is applied, Eq. (1.11) yields

$$\int_V (\operatorname{div} \mathfrak{T} + \rho \mathbf{F}) dV = 0.$$

Because the volume of integration is arbitrary and the integrand is continuous, we have the *equation of equilibrium*

$$\operatorname{div} \mathfrak{T} + \rho \mathbf{F} = 0, \tag{1.13}$$

valid at each point of V. Applying Gauss' theorem and using Eq. (1.13), Eq. (1.12) can be recast in the form

$$\int_V [\operatorname{div}(\mathfrak{T} \times \mathbf{r}) - (\operatorname{div} \mathfrak{T}) \times \mathbf{r}] dV = 0. \tag{1.14}$$

However, every dyadic $\mathfrak{T}$ satisfies the identity

$$\operatorname{div}(\mathfrak{T} \times \mathbf{r}) - (\operatorname{div} \mathfrak{T}) \times \mathbf{r} = \langle \mathfrak{T} \rangle,$$

where $\langle \mathfrak{T} \rangle$ is the vector of the dyadic (App. A). A dyadic is symmetric if and only if its vector is identically zero. Equation (1.14), therefore, shows that the stress dyadic is symmetric; i.e.,

$$\tau_{ij} = \tau_{ji} \qquad (i, j = 1, 2, 3). \tag{1.15}$$

Because of this symmetry, the stress at each point is specified by six quantities and the stress field in the body is given by six functions of position. These are referred to as *components of stress.*

The components of the stress dyadic at any given point depend on the orientation of the base vectors $\mathbf{e}_i$. If we rotate the base vectors in such a way that they always form an orthonormal system, the components of the stress dyadic will change according to a known transformation law.

Because any real symmetric matrix has at least three mutually perpendicular eigenvectors, $\mathbf{a}_i$, there is at least one position of the axes in which the stress dyadic can attain the *canonical form*

$$\mathfrak{T} = \tau_1 \mathbf{a}_1 \mathbf{a}_1 + \tau_2 \mathbf{a}_2 \mathbf{a}_2 + \tau_3 \mathbf{a}_3 \mathbf{a}_3. \tag{1.16}$$

In Eq. (1.16), τ_i are the eigenvalues of the matrix of $\mathfrak{T}$ corresponding to the eigenvectors $\mathbf{a}_i$ and are known as *principal stresses.* In the coordinate system $(\mathbf{a}_1, \mathbf{a}_2, \mathbf{a}_3)$, the stress vectors on coordinate planes are normal to the coordinate planes and the shearing stresses vanish on these planes. The directions $(\mathbf{a}_1, \mathbf{a}_2, \mathbf{a}_3)$ are known as the *principal directions of stress* and the planes normal to the principal directions of stress are known as the *principal planes of stress.*

Let

$$\mathbf{n} = \alpha \mathbf{a}_1 + \beta \mathbf{a}_2 + \gamma \mathbf{a}_3 \tag{1.17}$$

be the unit vector normal to a plane through P. From Eq. (1.16), the normal stress is given by

$$N = \mathbf{n} \cdot \mathfrak{T} \cdot \mathbf{n} = \tau_1 \alpha^2 + \tau_2 \beta^2 + \tau_3 \gamma^2. \tag{1.18}$$

Because $\alpha^2 + \beta^2 + \gamma^2 = 1$, we have

$$N = \tau_1 (1 - \beta^2 - \gamma^2) + \tau_2 \beta^2 + \tau_3 \gamma^2.$$

Stationary values of N are obtained by putting $\partial N/\partial\beta = 0, \partial N/\partial\gamma = 0$. This yields

$$\alpha = \pm 1, \qquad \beta = 0, \qquad \gamma = 0, \qquad N = \tau_1.$$

Similarly, two other stationary values of N can be obtained. They are

$$\alpha = 0, \qquad \beta = \pm 1, \qquad \gamma = 0, \qquad N = \tau_2,$$

and

$$\alpha = 0, \qquad \beta = 0, \qquad \gamma = \pm 1, \qquad N = \tau_3.$$

This shows that the principal directions of stress are those directions in which the normal stress is stationary.

Consider next the shearing stress. From Eqs. (1.16)–(1.18), we find

$$\begin{aligned} S^2 = |\mathbf{T}_s|^2 &= |\mathbf{n}\cdot\mathfrak{T}|^2 - N^2 \\ &= \tau_1^2\alpha^2 + \tau_2^2\beta^2 + \tau_3^2\gamma^2 - (\tau_1\alpha^2 + \tau_2\beta^2 + \tau_3\gamma^2)^2. \end{aligned} \tag{1.19}$$

Applying the relation $\alpha^2 + \beta^2 + \gamma^2 = 1$, Eq. (1.19) can be made to yield

$$S^2 = \tau_1^2(1 - \beta^2 - \gamma^2) + \tau_2^2\beta^2 + \tau_3^2\gamma^2 - [\tau_1(1 - \beta^2 - \gamma^2) + \tau_2\beta^2 + \tau_3\gamma^2]^2.$$

Stationary values of S^2 can be obtained by putting $\partial S^2/\partial\beta = 0$, $\partial S^2/\partial\gamma = 0$. Assuming $\tau_1 > \tau_2 > \tau_3$, these conditions yield

$$\alpha = \pm 1, \beta = 0, \gamma = 0, S = 0 \tag{1.20a}$$

or

$$\alpha = \pm 1/\sqrt{2}, \beta = 0, \gamma = \pm 1/\sqrt{2}, S = \tfrac{1}{2}(\tau_1 - \tau_3), \tag{1.20b}$$

or

$$\alpha = \pm 1/\sqrt{2}, \beta = \pm 1/\sqrt{2}, \gamma = 0, S = \tfrac{1}{2}(\tau_1 - \tau_2). \tag{1.20c}$$

Three more stationary values of S can be obtained by eliminating β and γ, in turn, and following the above procedure. We find,

$$\alpha = 0, \beta = \pm 1, \gamma = 0, S = 0, \tag{1.21a}$$

$$\alpha = 0, \beta = 0, \gamma = \pm 1, S = 0, \tag{1.21b}$$

$$\alpha = 0, \beta = \pm 1/\sqrt{2}, \gamma = \pm 1/\sqrt{2}, S = \tfrac{1}{2}(\tau_2 - \tau_3). \tag{1.21c}$$

Equations (1.20a) and (1.21a, b) give the minimum (zero) values of S. Since $\tau_1 > \tau_2 > \tau_3$, the greatest shearing stress is given by Eq. (1.20b). Therefore, the maximum shearing stress is equal to one-half the difference between the greatest and the least principal stresses. It acts in a plane that bisects the angle between the directions of the largest and smallest of the principal stresses.

In addition to the specification of the stress dyadic in terms of its six independent components τ_{ij}, it is sometimes useful to emphasize the simultaneous dilatation and shearing effected by the stress dyadic at a point. This is achieved by the formal decomposition

$$\underset{}{\mathfrak{T}} = \underbrace{p\mathfrak{J}}_{\text{Mean stress}} + \underbrace{(\mathfrak{T} - p\mathfrak{J})}_{\text{Stress deviator}}, \tag{1.22}$$

where

$$p = \tfrac{1}{3}(\tau_{11} + \tau_{22} + \tau_{33}) = \tfrac{1}{3}(\tau_1 + \tau_2 + \tau_3), \tag{1.23}$$

is also known as the *isotropic* or the *spherical part* of the stress tensor. The stress deviator is a symmetric dyadic. Its principal directions coincide with the principal directions of $\mathfrak{T}$ and its eigenvalues are $\frac{1}{3}(2\tau_1 - \tau_2 - \tau_3)$, $\frac{1}{3}(2\tau_2 - \tau_3 - \tau_1)$, and $\frac{1}{3}(2\tau_3 - \tau_1 - \tau_2)$, respectively. In particular, in eigenvector coordinates, we have

$$\begin{aligned} \mathfrak{T} &= \tau_1\mathbf{a}_1\mathbf{a}_1 + \tau_2\mathbf{a}_2\mathbf{a}_2 + \tau_3\mathbf{a}_3\mathbf{a}_3 \\ &= p\mathfrak{J} + \tfrac{1}{3}(2\tau_1 - \tau_2 - \tau_3)\mathbf{a}_1\mathbf{a}_1 + \tfrac{1}{3}(2\tau_2 - \tau_3 - \tau_1)\mathbf{a}_2\mathbf{a}_2 \\ &\quad + \tfrac{1}{3}(2\tau_3 - \tau_1 - \tau_2)\mathbf{a}_3\mathbf{a}_3. \end{aligned}$$

This may be written in the following useful form

$$\begin{aligned} \mathfrak{T} \equiv \begin{bmatrix} \tau_1 & & \\ & \tau_2 & \\ & & \tau_3 \end{bmatrix} &= p\begin{bmatrix} 1 & & \\ & 1 & \\ & & 1 \end{bmatrix} + \tfrac{1}{3}(\tau_1 - \tau_2)\begin{bmatrix} 1 & & \\ & -1 & \\ & & 0 \end{bmatrix} \\ &+ \tfrac{1}{3}(\tau_2 - \tau_3)\begin{bmatrix} 0 & & \\ & 1 & \\ & & -1 \end{bmatrix} + \tfrac{1}{3}(\tau_3 - \tau_1)\begin{bmatrix} -1 & & \\ & 0 & \\ & & 1 \end{bmatrix}. \end{aligned} \tag{1.24}$$

Note that $\frac{1}{3}|\tau_3 - \tau_1|$ is $\frac{2}{3} \times$ (maximum shearing stress). The dyadic $(\mathbf{a}_1\mathbf{a}_1 - \mathbf{a}_2\mathbf{a}_2)$ is known as *simple shear* stress and is obtained when tension is applied across one plane together with an equal compression across a perpendicular plane. Similar interpretation can be given to the remaining constituents in Eq. (1.24).

EXAMPLE 1.1

Let the stress at any point P be given by the matrix

$$\begin{bmatrix} 2 & 0 & 0 \\ 0 & 3 & 4 \\ 0 & 4 & -3 \end{bmatrix}. \tag{1.1.1}$$

The eigenvalues of this matrix are the roots of the equation

$$\|\mathfrak{T} - \tau\mathfrak{J}\| = (2 - \tau)(\tau^2 - 25) = 0,$$

i.e.,

$$\tau_1 = 2, \qquad \tau_2 = 5, \qquad \tau_3 = -5.$$

Let $\mathbf{a}_1 = \alpha_1\mathbf{e}_1 + \alpha_2\mathbf{e}_2 + \alpha_3\mathbf{e}_3$ be the eigenvector corresponding to the eigenvalue τ_1. The equations for α_i are

$$(\tau_{ij} - \tau_1\delta_{ij})\alpha_j = 0, \qquad \alpha_j\alpha_j = 1 \qquad (i = 1, 2, 3).$$

Therefore, for $\tau_1 = 2$, we have

$$\alpha_2 + 4\alpha_3 = 0, \qquad 4\alpha_2 - 5\alpha_3 = 0, \qquad \alpha_1^2 + \alpha_2^2 + \alpha_3^2 = 1,$$

yielding $\alpha_1 = \pm 1$, $\alpha_2 = \alpha_3 = 0$. We repeat this procedure for the other eigenvalues, obtaining the right-handed system of eigenvectors

$$\mathbf{a}_1 = \mathbf{e}_1,$$

$$\mathbf{a}_2 = \frac{1}{\sqrt{5}}(2\mathbf{e}_2 + \mathbf{e}_3),$$

$$\mathbf{a}_3 = \frac{1}{\sqrt{5}}(-\mathbf{e}_2 + 2\mathbf{e}_3).$$

The transformation from the old system $(\mathbf{e}_1, \mathbf{e}_2, \mathbf{e}_3)$ to the new system $(\mathbf{a}_1, \mathbf{a}_2, \mathbf{a}_3)$ can be represented with the help of the dyadic whose matrix is

$$\begin{bmatrix} 1 & 0 & 0 \\ 0 & 2/\sqrt{5} & 1/\sqrt{5} \\ 0 & -1/\sqrt{5} & 2/\sqrt{5} \end{bmatrix}. \tag{1.1.2}$$

The stress dyadic in the new system is

$$\mathfrak{T} = 2\mathbf{a}_1\mathbf{a}_1 + 5\mathbf{a}_2\mathbf{a}_2 - 5\mathbf{a}_3\mathbf{a}_3. \tag{1.1.3}$$

The matrix of the stress dyadic in the new system can be obtained by multiplying the stress matrix (1.1.1) from the left by the matrix (1.1.2) and then from right by the transpose of the matrix (1.1.2):

$$\begin{bmatrix} 1 & 0 & 0 \\ 0 & 2/\sqrt{5} & 1/\sqrt{5} \\ 0 & -1/\sqrt{5} & 2/\sqrt{5} \end{bmatrix} \begin{bmatrix} 2 & 0 & 0 \\ 0 & 3 & 4 \\ 0 & 4 & -3 \end{bmatrix} \begin{bmatrix} 1 & 0 & 0 \\ 0 & 2/\sqrt{5} & -1/\sqrt{5} \\ 0 & 1/\sqrt{5} & 2/\sqrt{5} \end{bmatrix}$$

$$= \begin{bmatrix} 2 & 0 & 0 \\ 0 & 5 & 0 \\ 0 & 0 & -5 \end{bmatrix}.$$

1.2. Geometry of Small Deformations

According to classical mechanics, the action of forces upon a rigid body results in its translation and rotation as a whole. If the body is *elastic*, however, the applied body or surface forces can also change the relative positions of the internal parts of the body. The body returns to its initial state after the forces are removed. Unless otherwise stated, we shall restrict our treatment to *isotropic* materials in which the mechanical properties are independent of direction at any given point. We shall also ignore the motion of the body as a whole and consider only its deformation.

Consider two neighboring points $P(\mathbf{r})$ and $Q(\mathbf{r} + d\mathbf{r})$, prior to deformation, relative to a fixed coordinate system O (Fig. 1.4). After the deformation, these points are displaced to new positions, P' and Q', respectively. If we denote by

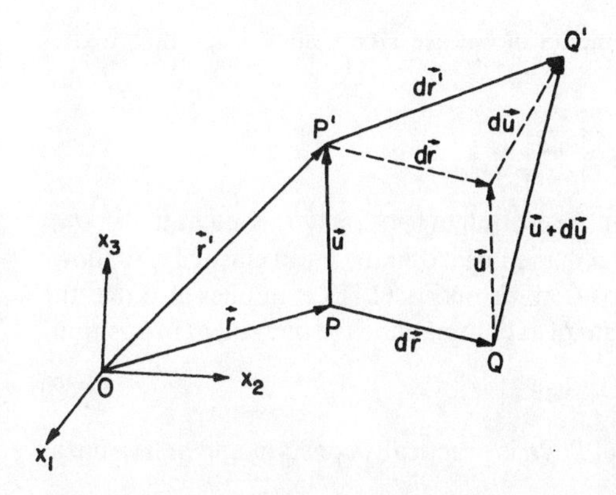

Figure 1.4. The displacement **u** and the relative displacement *d***u**.

$\mathbf{u}(\mathbf{r}) = \mathbf{r}' - \mathbf{r}$, the *displacement* of the point P to its new position P'; the corresponding displacement of the neighboring point Q to its new position Q' will be given by $\mathbf{u}(\mathbf{r} + d\mathbf{r}) = \mathbf{u} + d\mathbf{u}$, where $d\mathbf{u}$ is the displacement of Q relative to P. Expanding the scalar function $u_1(\mathbf{r} + d\mathbf{r})$ in Taylor series, we have

$$u_1(\mathbf{r} + d\mathbf{r}) = u_1(\mathbf{r}) + d\mathbf{r} \cdot \nabla u_1(\mathbf{r}) + \cdots$$

or

$$du_1 = u_1(\mathbf{r} + d\mathbf{r}) - u_1(\mathbf{r}) = d\mathbf{r} \cdot \nabla u_1,$$

discarding $|d\mathbf{r}|^2$, etc. We repeat this process with the other two components of the displacement and write the combined result as

$$d\mathbf{r}' - d\mathbf{r} = d\mathbf{u} = d\mathbf{r} \cdot \nabla\mathbf{u}, \tag{1.25}$$

or, equivalently,

$$d\mathbf{r}' = d\mathbf{r} \cdot (\mathfrak{I} + \nabla\mathbf{u}) = d\mathbf{r} \cdot \boldsymbol{\psi}. \tag{1.26}$$

The dyadic $\nabla\mathbf{u}$ is known as the *displacement gradient*, whereas $\boldsymbol{\psi}$ is known as the *deformation dyadic*. The matrix of $\nabla\mathbf{u}$ is

$$\begin{bmatrix} \dfrac{\partial u_1}{\partial x_1} & \dfrac{\partial u_2}{\partial x_1} & \dfrac{\partial u_3}{\partial x_1} \\[2ex] \dfrac{\partial u_1}{\partial x_2} & \dfrac{\partial u_2}{\partial x_2} & \dfrac{\partial u_3}{\partial x_2} \\[2ex] \dfrac{\partial u_1}{\partial x_3} & \dfrac{\partial u_2}{\partial x_3} & \dfrac{\partial u_3}{\partial x_3} \end{bmatrix}, \tag{1.27}$$

i.e.,

$$(\nabla\mathbf{u})_{ij} = \frac{\partial u_j}{\partial x_i}. \tag{1.28}$$

We will assume that the components of $\nabla \mathbf{u}$ are small, such that their higher powers may be neglected

$$\left|\frac{\partial u_j}{\partial x_i}\right| \ll 1. \tag{1.29}$$

This means that the displacement vector changes slowly throughout the continuum. This is a basic assumption in the linearized theory of elasticity. It allows us to use the principle of superposition. From Eq. (1.25), it implies also that the increment in the displacement is small as compared to the increment in position

$$|d\mathbf{u}| \ll |d\mathbf{r}|. \tag{1.30}$$

Moreover, in the neighborhood of P, $\nabla \mathbf{u}$ is practically constant and $d\mathbf{r}'$ is a linear vector function of $d\mathbf{r}$.

We introduce the notation

$$\mathfrak{E}(\mathbf{u}) = \tfrac{1}{2}(\nabla \mathbf{u} + \mathbf{u}\nabla), \qquad \varepsilon_{ij} = \frac{1}{2}\left(\frac{\partial u_i}{\partial x_j} + \frac{\partial u_j}{\partial x_i}\right), \tag{1.31}$$

$$\mathfrak{R}(\mathbf{u}) = \mathfrak{J} + \tfrac{1}{2}(\nabla \mathbf{u} - \mathbf{u}\nabla) = \mathfrak{J} - \mathfrak{J} \times (\tfrac{1}{2}\,\mathrm{curl}\,\mathbf{u}). \tag{1.32}$$

The deformation dyadic therefore represents the sum of two linear transformations:

$$\psi = \mathfrak{J} + \nabla \mathbf{u} = \mathfrak{E}(\mathbf{u}) + \mathfrak{R}(\mathbf{u}).$$

Equation (1.25) now yields

$$d\mathbf{u} = \mathfrak{E}(\mathbf{u}) \cdot d\mathbf{r} + (\tfrac{1}{2}\,\mathrm{curl}\,\mathbf{u}) \times d\mathbf{r}. \tag{1.33}$$

The dyadic $\mathfrak{R}$ is called the *infinitesimal spin dyadic* for the following reason. A rotation of the coordinate axes by an angle θ about the z axis is given by the transformation

$$\begin{bmatrix} x' \\ y' \\ z' \end{bmatrix} = \begin{bmatrix} \cos\theta & \sin\theta & 0 \\ -\sin\theta & \cos\theta & 0 \\ 0 & 0 & 1 \end{bmatrix} \begin{bmatrix} x \\ y \\ z \end{bmatrix}.$$

This can be written as $\mathbf{r}' = \mathfrak{R} \cdot \mathbf{r}$, where

$$\mathfrak{R}(\theta, \mathbf{e}_z) = \mathbf{e}_z\mathbf{e}_z(1 - \cos\theta) + \mathfrak{J}\cos\theta - (\mathfrak{J} \times \mathbf{e}_z)\sin\theta.$$

For small angles, to the first order in θ, we have

$$\mathfrak{R}(\theta, \mathbf{e}_z) = \mathfrak{J} - (\mathfrak{J} \times \mathbf{e}_z)\theta.$$

This result remains valid for an arbitrary axis of rotation $\mathbf{e}$:

$$\mathfrak{R}(\theta, \mathbf{e}) = \mathfrak{J} - (\mathfrak{J} \times \mathbf{e})\theta. \tag{1.34}$$

When Eqs. (1.32) and (1.34) are compared it can be seen that $\mathfrak{R}(\mathbf{u})$ represents a rigid rotation of a small volume element, containing the point P, through the small angle $\theta = \frac{1}{2}|\text{curl } \mathbf{u}|$ about an axis directed parallel to curl $\mathbf{u}$. This is a local rotation and should not be confused with the rigid rotation of the whole body, which has been excluded from $\mathbf{u}$ *ab initio*.

The symmetric dyadic $\mathfrak{E}(\mathbf{u})$ is the *infinitesimal strain dyadic*. Its matrix is given by

$$\begin{bmatrix} \frac{\partial u_1}{\partial x_1} & \frac{1}{2}\left(\frac{\partial u_2}{\partial x_1} + \frac{\partial u_1}{\partial x_2}\right) & \frac{1}{2}\left(\frac{\partial u_3}{\partial x_1} + \frac{\partial u_1}{\partial x_3}\right) \\ \frac{1}{2}\left(\frac{\partial u_1}{\partial x_2} + \frac{\partial u_2}{\partial x_1}\right) & \frac{\partial u_2}{\partial x_2} & \frac{1}{2}\left(\frac{\partial u_3}{\partial x_2} + \frac{\partial u_2}{\partial x_3}\right) \\ \frac{1}{2}\left(\frac{\partial u_1}{\partial x_3} + \frac{\partial u_3}{\partial x_1}\right) & \frac{1}{2}\left(\frac{\partial u_2}{\partial x_3} + \frac{\partial u_3}{\partial x_2}\right) & \frac{\partial u_3}{\partial x_3} \end{bmatrix}. \tag{1.35}$$

The components ε_{ij} have a simple geometric interpretation. Indeed, from Eq. (1.26), we have

$$|d\mathbf{r}'|^2 = \{d\mathbf{r}\cdot(\mathfrak{I} + \nabla\mathbf{u})\}\cdot\{(\mathfrak{I} + \mathbf{u}\nabla)\cdot d\mathbf{r}\} = d\mathbf{r}\cdot(\mathfrak{I} + 2\mathfrak{E})\cdot d\mathbf{r}, \tag{1.36}$$

neglecting the product $\nabla\mathbf{u}\cdot\mathbf{u}\nabla$.[1] Hence

$$|d\mathbf{r}'|^2 - |d\mathbf{r}|^2 = d\mathbf{r}\cdot 2\mathfrak{E}\cdot d\mathbf{r}. \tag{1.37}$$

Let $d\mathbf{r} = ds\mathbf{a}$ denote a line element of length ds and direction $\mathbf{a} = \mathbf{a}(l, m, n)$. It then follows from Eq. (1.37) that

$$\mathbf{a}\cdot\mathfrak{E}\cdot\mathbf{a} = \mathbf{aa}\colon\mathfrak{E} = \frac{ds'^2 - ds^2}{2\,ds^2} \simeq \frac{ds' - ds}{ds} = \text{change in length per unit length}$$
$$= \text{extension in the direction } (l, m, n). \tag{1.38}$$

The explicit expression for $\mathbf{aa}$ is

$$\mathbf{aa} = (l\mathbf{e}_1 + m\mathbf{e}_2 + n\mathbf{e}_3)(l\mathbf{e}_1 + m\mathbf{e}_2 + n\mathbf{e}_3)$$
$$\equiv \begin{bmatrix} l^2 & lm & ln \\ ml & m^2 & mn \\ nl & nm & n^2 \end{bmatrix}. \tag{1.38a}$$

Therefore, from Eq. (1.38), the extension in the direction (l, m, n) is given by

$$\varepsilon = l^2\varepsilon_{11} + m^2\varepsilon_{22} + n^2\varepsilon_{33} + 2mn\varepsilon_{23} + 2nl\varepsilon_{31} + 2lm\varepsilon_{12}. \tag{1.38b}$$

Taking $l = 1$, $m = n = 0$, this relation shows that ε_{11} is the extension of a line element that, in the unstrained state, is parallel to the x_1 direction. The entities ε_{22} and ε_{33} have similar interpretations. The components ε_{11}, ε_{22}, and ε_{33} are

[1] Mathematically, the term $(\nabla\mathbf{u}\cdot\mathbf{u}\nabla)$ cannot be neglected if the contributions of $\nabla\mathbf{u}$ and $\mathbf{u}\nabla$ are absent. This is the case, for instance, in pure shear, in which $u_x = sy$, $u_y = 0$, $u_z = 0$. Here the components of $\nabla\mathbf{u}$ and $\mathbf{u}\nabla$ in the yy direction are zero. Consequently, we have $[\nabla\mathbf{u} + \mathbf{u}\nabla + 2\nabla\mathbf{u}\cdot\mathbf{u}\nabla]_{yy} = 2s^2$, which must not be neglected. In seismology, however, these second-order effects are usually too small to be recorded by conventional seismographs.

known as *normal strains*, and the components ε_{23}, ε_{31}, and ε_{12} are known as *shearing strains*. It can be shown that $2\varepsilon_{12}$ is equal to the decrease, caused by deformation, in the angle between two line elements that were parallel to the x_1 and x_2 directions, respectively, before the deformation.

The components of the strain dyadic at a given point depend upon the orientation of the base vectors, $\mathbf{e}_i$. Because the matrix of the strain dyadic is real and symmetrical, there exist three mutually orthogonal unit vectors $(\mathbf{a}_1, \mathbf{a}_2, \mathbf{a}_3)$ such that

$$\mathfrak{E} = \varepsilon_1 \mathbf{a}_1 \mathbf{a}_1 + \varepsilon_2 \mathbf{a}_2 \mathbf{a}_2 + \varepsilon_3 \mathbf{a}_3 \mathbf{a}_3, \tag{1.39}$$

where $(\varepsilon_1, \varepsilon_2, \varepsilon_3)$ are the eigenvalues corresponding to the eigenvectors $(\mathbf{a}_1, \mathbf{a}_2, \mathbf{a}_3)$ of the matrix of $\mathfrak{E}$. We call $(\varepsilon_1, \varepsilon_2, \varepsilon_3)$ the *principal strains* and the directions $(\mathbf{a}_1, \mathbf{a}_2, \mathbf{a}_3)$ the *principal directions of strain.*

To the first order in the components of strain Eq. (1.26) yields

$$|d\mathbf{r}|^2 = d\mathbf{r}' \cdot (\mathfrak{I} - \nabla \mathbf{u} - \mathbf{u}\nabla) \cdot d\mathbf{r}'. \tag{1.40}$$

Therefore, the particles on a sphere centered at P and of small radius R prior to deformation are displaced to lie on the surface

$$R^2 = d\mathbf{r}' \cdot (\mathfrak{I} - 2\mathfrak{E}) \cdot d\mathbf{r}'. \tag{1.41}$$

If we put $d\mathbf{r}' = x'\mathbf{a}_1 + y'\mathbf{a}_2 + z'\mathbf{a}_3$ and use Eq. (1.39), Eq. (1.41) becomes

$$(1 - 2\varepsilon_1)x'^2 + (1 - 2\varepsilon_2)y'^2 + (1 - 2\varepsilon_3)z'^2 = R^2, \tag{1.42}$$

which is an ellipsoid. This is known as the *strain ellipsoid*. The principal axes of the strain ellipsoid coincide with the principal axes of strain. The volume of a sphere of radius R centered at P prior to deformation is

$$dV = \tfrac{4}{3}\pi R^3. \tag{1.43}$$

The volume of the corresponding ellipsoid, Eq. (1.42), is

$$\begin{aligned} dV' &= \tfrac{4}{3}\pi[(1 - 2\varepsilon_1)(1 - 2\varepsilon_2)(1 - 2\varepsilon_3)]^{-1/2} R^3 \\ &= \tfrac{4}{3}\pi(1 + \varepsilon_1 + \varepsilon_2 + \varepsilon_3)R^3, \end{aligned} \tag{1.44}$$

with second-order terms in ε_i neglected. Therefore, the increase in volume per unit volume, known as *cubical dilatation*, is given by

$$\begin{aligned} \frac{dV' - dV}{dV} &= \varepsilon_1 + \varepsilon_2 + \varepsilon_3 = \varepsilon_{11} + \varepsilon_{22} + \varepsilon_{33} \\ &= \frac{\partial u_1}{\partial x_1} + \frac{\partial u_2}{\partial x_2} + \frac{\partial u_3}{\partial x_3} = \operatorname{div} \mathbf{u}. \end{aligned} \tag{1.45}$$

Because for a given mass the density is inversely proportional to the volume, the relative change in density is given by

$$\frac{\delta\rho}{\rho} = \frac{\rho' - \rho}{\rho} = -\operatorname{div} \mathbf{u}. \tag{1.46}$$

If the deformation is independent of the coordinates, the corresponding strain is known as *homogeneous strain.* We shall now describe the three common homogeneous strain fields.

1. Simple expansion.

$$\mathbf{u} = \varepsilon \mathbf{r}, \qquad \mathfrak{E} = \varepsilon \mathfrak{J}, \qquad \operatorname{div} \mathbf{u} = 3\varepsilon, \qquad \operatorname{curl} \mathbf{u} = 0, \qquad \varepsilon > 0. \tag{1.47}$$

2. Simple shear.

$$\begin{gathered} \mathbf{u} = \tfrac{1}{2}\varepsilon(x\mathbf{e}_x - y\mathbf{e}_y), \qquad \mathfrak{E} = \tfrac{1}{2}\varepsilon(\mathbf{e}_x\mathbf{e}_x - \mathbf{e}_y\mathbf{e}_y), \\ \operatorname{div} \mathbf{u} = 0, \qquad \operatorname{curl} \mathbf{u} = 0. \end{gathered} \tag{1.48}$$

This deformation is accompanied by neither dilatation nor rotation, hence the name. If we rotate the axes by 45° about the z axis, the new unit vectors $(\mathbf{e}_{x'}, \mathbf{e}_{y'})$ are related to $(\mathbf{e}_x, \mathbf{e}_y)$ via the relations

$$\mathbf{e}_x = \frac{1}{\sqrt{2}}(\mathbf{e}_{x'} - \mathbf{e}_{y'}), \qquad \mathbf{e}_y = \frac{1}{\sqrt{2}}(\mathbf{e}_{x'} + \mathbf{e}_{y'}),$$

with similar relations for the coordinates. Hence, we have

$$\mathbf{u} = -\tfrac{1}{2}\varepsilon(x'\mathbf{e}_{y'} + y'\mathbf{e}_{x'}), \qquad \mathfrak{E} = -\tfrac{1}{2}\varepsilon(\mathbf{e}_{x'}\mathbf{e}_{y'} + \mathbf{e}_{y'}\mathbf{e}_{x'}). \tag{1.49}$$

Note that $(\mathbf{e}_x, \mathbf{e}_y, \mathbf{e}_z)$ are the principal directions of strain.

3. Pure shear.

$$\mathbf{u} = \varepsilon y \mathbf{e}_x, \qquad \mathfrak{E} = \tfrac{1}{2}\varepsilon(\mathbf{e}_x\mathbf{e}_y + \mathbf{e}_y\mathbf{e}_x), \qquad \operatorname{div} \mathbf{u} = 0, \qquad \operatorname{curl} \mathbf{u} = -\varepsilon\mathbf{e}_z. \tag{1.50}$$

It is shown in Appendix B that in orthogonal curvilinear coordinates (q_1, q_2, q_3), the components of the gradient of a vector $\mathbf{u}$ are given by

$$(\nabla \mathbf{u})_{\alpha\beta} = \frac{1}{h_\alpha}\frac{\partial u_\beta}{\partial q_\alpha} + \sum_{\gamma=1}^{3} \frac{u_\gamma}{h_\alpha}\left(_{\gamma}{}^{\beta}{}_{\alpha}\right). \tag{1.51[2]}$$

Adding to this expression its transpose, $(\mathbf{u}\nabla)_{\alpha\beta}$, and dividing by 2, we obtain

$$\varepsilon_{\alpha\beta} = [\mathfrak{E}(\mathbf{u})]_{\alpha\beta} = \frac{1}{2}\left[\frac{1}{h_\alpha}\frac{\partial u_\beta}{\partial q_\alpha} + \frac{1}{h_\beta}\frac{\partial u_\alpha}{\partial q_\beta} + \sum_{\gamma=1}^{3} u_\gamma\left\{\frac{1}{h_\alpha}\left(_{\gamma}{}^{\beta}{}_{\alpha}\right) + \frac{1}{h_\beta}\left(_{\gamma}{}^{\alpha}{}_{\beta}\right)\right\}\right]. \tag{1.52}$$

Hence

$$\varepsilon_{kk} = \frac{1}{h_k}\left[\frac{\partial u_k}{\partial q_k} + \sum_{\gamma=1}^{3} u_\gamma\left(_{\gamma}{}^{k}{}_{k}\right)\right], \tag{1.53}$$

$$\varepsilon_{\alpha\beta} = \frac{1}{2}\left[\frac{h_\beta}{h_\alpha}\frac{\partial(u_\beta/h_\beta)}{\partial q_\alpha} + \frac{h_\alpha}{h_\beta}\frac{\partial(u_\alpha/h_\alpha)}{\partial q_\beta}\right], \qquad (\alpha \neq \beta). \tag{1.54}$$

[2] Summation convention not used in Eqs. (1.51)–(1.54).

In cylindrical coordinates (Δ, ϕ, z), we obtain

$$\begin{aligned}
&\varepsilon_{\Delta\Delta} = \frac{\partial u_\Delta}{\partial \Delta}, \qquad \varepsilon_{\phi\phi} = \frac{1}{\Delta}\left(\frac{\partial u_\phi}{\partial \phi} + u_\Delta\right), \qquad \varepsilon_{zz} = \frac{\partial u_z}{\partial z}, \\
&\varepsilon_{\phi z} = \varepsilon_{z\phi} = \frac{1}{2}\left(\frac{1}{\Delta}\frac{\partial u_z}{\partial \phi} + \frac{\partial u_\phi}{\partial z}\right), \\
&\varepsilon_{z\Delta} = \varepsilon_{\Delta z} = \frac{1}{2}\left(\frac{\partial u_\Delta}{\partial z} + \frac{\partial u_z}{\partial \Delta}\right), \\
&\varepsilon_{\Delta\phi} = \varepsilon_{\phi\Delta} = \frac{1}{2}\left(\frac{1}{\Delta}\frac{\partial u_\Delta}{\partial \phi} + \frac{\partial u_\phi}{\partial \Delta} - \frac{u_\phi}{\Delta}\right).
\end{aligned} \tag{1.55}$$

In spherical coordinates (r, θ, ϕ), we find

$$\begin{aligned}
&\varepsilon_{rr} = \frac{\partial u_r}{\partial r}, \qquad \varepsilon_{\theta\theta} = \frac{1}{r}\left(\frac{\partial u_\theta}{\partial \theta} + u_r\right), \\
&\varepsilon_{\phi\phi} = \frac{1}{r}\left(\frac{1}{\sin\theta}\frac{\partial u_\phi}{\partial \phi} + u_r + \cot\theta\, u_\theta\right), \\
&\varepsilon_{\theta\phi} = \varepsilon_{\phi\theta} = \frac{1}{2r}\left(\frac{\partial u_\phi}{\partial \theta} - \cot\theta\, u_\phi + \frac{1}{\sin\theta}\frac{\partial u_\theta}{\partial \phi}\right), \\
&\varepsilon_{\phi r} = \varepsilon_{r\phi} = \frac{1}{2}\left(\frac{1}{r\sin\theta}\frac{\partial u_r}{\partial \phi} + \frac{\partial u_\phi}{\partial r} - \frac{u_\phi}{r}\right), \\
&\varepsilon_{r\theta} = \varepsilon_{\theta r} = \frac{1}{2}\left(\frac{\partial u_\theta}{\partial r} - \frac{u_\theta}{r} + \frac{1}{r}\frac{\partial u_r}{\partial \theta}\right).
\end{aligned} \tag{1.56}$$

EXAMPLE 1.2: Geometric Representation of the State of Strain at a Point

Consider a point P in a strained state E in a body relative to the axes (x, y, z) such that the z axis is taken along one of the principal axes at P. Therefore, we have

$$E = \begin{bmatrix} \varepsilon_{xx} & \varepsilon_{xy} & 0 \\ \varepsilon_{yx} & \varepsilon_{yy} & 0 \\ 0 & 0 & \varepsilon_{zz} \end{bmatrix}. \tag{1.2.1}$$

We rotate the axes through an angle θ about the z axis (Fig. 1.5a). The state of strain E' with respect to the new axes is given by

$$E' = \Pi E \tilde{\Pi},$$

where

$$\Pi = \begin{bmatrix} \cos\theta & \sin\theta & 0 \\ -\sin\theta & \cos\theta & 0 \\ 0 & 0 & 1 \end{bmatrix}, \qquad \tilde{\Pi} = \begin{bmatrix} \cos\theta & -\sin\theta & 0 \\ \sin\theta & \cos\theta & 0 \\ 0 & 0 & 1 \end{bmatrix}. \tag{1.2.2}$$

Carrying out the matrix multiplication, we get

$$E' = \begin{bmatrix} \varepsilon_{x'x'} & \varepsilon_{x'y'} & 0 \\ \varepsilon_{y'x'} & \varepsilon_{y'y'} & 0 \\ 0 & 0 & \varepsilon_{zz} \end{bmatrix}, \tag{1.2.3}$$

where

$$\begin{aligned} \varepsilon_{x'x'} &= a + b \cos 2\theta + c \sin 2\theta, \\ \varepsilon_{y'y'} &= a - b \cos 2\theta - c \sin 2\theta, \\ \varepsilon_{x'y'} &= -b \sin 2\theta + c \cos 2\theta; \end{aligned} \tag{1.2.4}$$

$$a = \frac{\varepsilon_{xx} + \varepsilon_{yy}}{2}, \qquad b = \frac{\varepsilon_{xx} - \varepsilon_{yy}}{2}, \qquad c = \varepsilon_{xy}. \tag{1.2.5}$$

Note that Eq. (1.2.4) yields

$$(\varepsilon_{x'x'} - a)^2 + \varepsilon_{x'y'}^2 = b^2 + c^2, \tag{1.2.6}$$

which is the equation of a circle in the XY plane with $X = \varepsilon_{x'x'}$, $Y = \varepsilon_{x'y'}$. The transformation of the elements of E is given in Fig. 1.5b.

Let us now consider θ as a parameter. Then, the directions $\theta = \phi$ for which $\varepsilon_{x'y'} = 0$ are the principal directions. From Eq. (1.2.4),

$$\phi = \frac{1}{2} \operatorname{tg}^{-1}\left[\frac{2\varepsilon_{xy}}{\varepsilon_{xx} - \varepsilon_{yy}}\right] \quad \text{or} \quad \frac{1}{2} \operatorname{tg}^{-1}\left[\frac{2\varepsilon_{xy}}{\varepsilon_{xx} - \varepsilon_{yy}}\right] + \frac{\pi}{2}. \tag{1.2.7}$$

At $\theta = \phi$, E' is represented by the diagonal matrix

$$E'(\phi) = \begin{bmatrix} \varepsilon_1 & 0 & 0 \\ 0 & \varepsilon_2 & 0 \\ 0 & 0 & \varepsilon_{zz} \end{bmatrix}, \tag{1.2.8}$$

where

$$\varepsilon_{1,2} = a \pm \sqrt{(b^2 + c^2)}. \tag{1.2.9}$$

Suppose next that we choose the axes (x, y) along the principal axes at P, i.e., $\varepsilon_{xy} = 0$. The corresponding strains in the rotated system are [cf. Eq. (1.2.4)]

$$\varepsilon_{x'x'} = a + b \cos 2\theta, \qquad \varepsilon_{y'y'} = a - b \cos 2\theta, \qquad \varepsilon_{x'y'} = -b \sin 2\theta.$$

Eliminating θ between $\varepsilon_{x'x'}$ and $\varepsilon_{x'y'}$ we obtain

$$(\varepsilon_{x'x'} - a)^2 + (\varepsilon_{x'y'})^2 = b^2$$

which is a special case of Eq. (1.2.6) with $c = 0$. It represents a circle centered at $(a, 0)$ with radius b (Fig. 1.5c) and is known as *Mohr's circle*. It is essentially a graphical representation of the normal and shear strains corresponding to any direction relative to the principal strains.

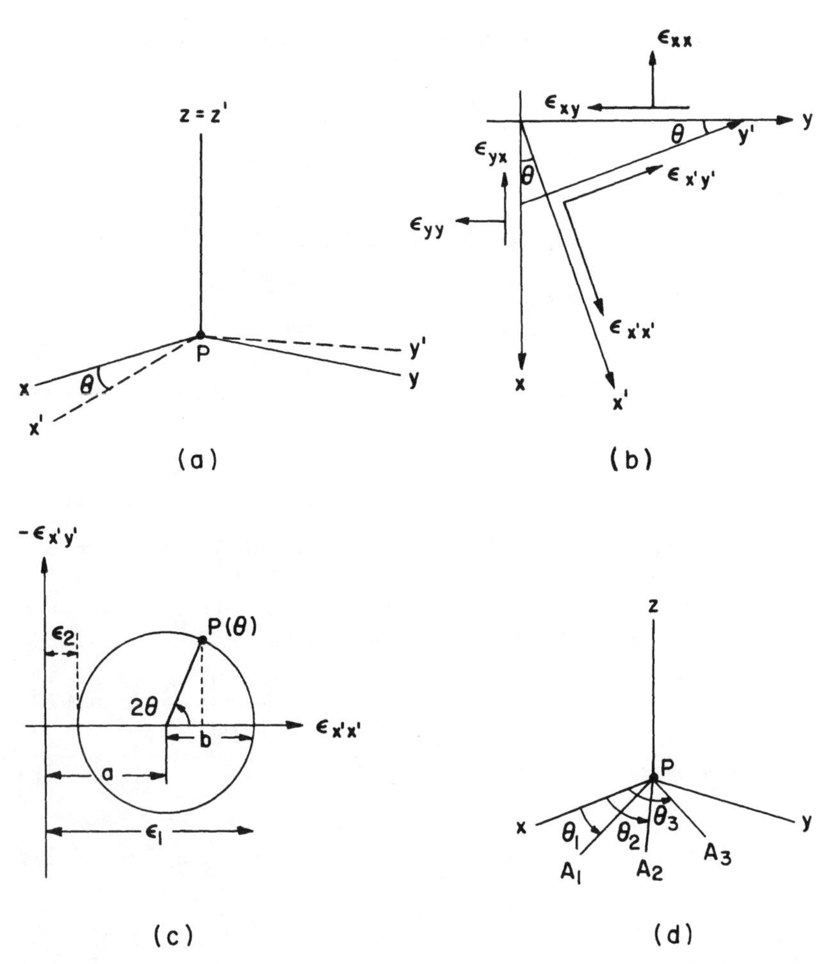

Figure 1.5. Geometric representation of the state of strain at a point. (a) Rotation of the axes about z at P; (b) transformation of the strain components as a result of a rotation of the axes; (c) Mohr's circle; (d) strains at a point in terms of three elongations along three arbitrary directions.

Returning to the initial coordinate system, we measure elongations along three arbitrary directions θ_1, θ_2, and θ_3 in the xy plane (Fig. 1.5d), which we can imagine to coincide, say, with the earth's surface. According to Eq. (1.2.4), we can write

$$\begin{aligned}
\varepsilon_1 &= \varepsilon(\theta_1) = a + b\cos 2\theta_1 + c\sin 2\theta_1,\\
\varepsilon_2 &= \varepsilon(\theta_2) = a + b\cos 2\theta_2 + c\sin 2\theta_2, \qquad (1.2.10)\\
\varepsilon_3 &= \varepsilon(\theta_3) = a + b\cos 2\theta_3 + c\sin 2\theta_3.
\end{aligned}$$

Solving for ε_{xx}, ε_{yy}, and ε_{xy}, we have

$$\varepsilon_{xx} = 4\left[\frac{\varepsilon_1 \sin(\theta_3 - \theta_2)\sin\theta_3 \sin\theta_2 + \varepsilon_2 \sin(\theta_1 - \theta_3)\sin\theta_1 \sin\theta_3 + \varepsilon_3 \sin(\theta_2 - \theta_1)\sin\theta_1 \sin\theta_2}{\sin 2(\theta_3 - \theta_2) + \sin 2(\theta_1 - \theta_3) + \sin 2(\theta_2 - \theta_1)}\right],$$

$$\varepsilon_{yy} = 4\left[\frac{\varepsilon_1 \sin(\theta_3 - \theta_2)\cos\theta_3 \cos\theta_2 + \varepsilon_2 \sin(\theta_1 - \theta_3)\cos\theta_1 \cos\theta_3 + \varepsilon_3 \sin(\theta_2 - \theta_1)\cos\theta_1 \cos\theta_2}{\sin 2(\theta_3 - \theta_2) + \sin 2(\theta_1 - \theta_3) + \sin 2(\theta_2 - \theta_1)}\right],$$

$$\varepsilon_{xy} = -\left[\frac{(\varepsilon_3 - \varepsilon_2)\cos 2\theta_1 + (\varepsilon_1 - \varepsilon_3)\cos 2\theta_2 + (\varepsilon_2 - \varepsilon_1)\cos 2\theta_3}{\sin 2(\theta_3 - \theta_2) + \sin 2(\theta_1 - \theta_3) + \sin 2(\theta_2 - \theta_1)}\right]. \quad (1.2.11)$$

1.3. Linear Elastic Solid

In a deformable body the deformation represented by the dyadic $\mathfrak{E}$ is caused by the forces applied to the body. Hence, certain relations must exist between $\mathfrak{E}$ and $\mathfrak{T}$ that depend on the physical constitution of the body in question. These relations must be obtained from experiments. We assume that there is no flux of heat across any boundary in the body, i.e., the deforming process will be assumed to take place without changes in temperature.

To establish the basic relations between the strain and stress, we introduce the concept of *potential strain energy*. Suppose that the external forces deform the body at a sufficiently slow rate such that the kinetic energy of the particles can be neglected. If we superimpose on the already deformed body an additional displacement $\delta\mathbf{u}$, then the work done by the surface tractions and body forces is given by

$$\delta A = \int_S (d\mathbf{S}\cdot\mathfrak{T})\cdot\delta\mathbf{u} + \int_V \rho(\mathbf{F}\cdot\delta\mathbf{u})dV = \int_V [\operatorname{div}(\mathfrak{T}\cdot\delta\mathbf{u}) + \rho\mathbf{F}\cdot\delta\mathbf{u}]dV. \quad (1.57)$$

However, because

$$\operatorname{div}\mathfrak{T} = -\rho\mathbf{F}, \quad (1.58)$$

$$\operatorname{div}(\mathfrak{T}\cdot\delta\mathbf{u}) - (\operatorname{div}\mathfrak{T})\cdot\delta\mathbf{u} = \mathfrak{T}\colon\nabla(\delta\mathbf{u}) = \mathfrak{T}\colon\delta\mathfrak{E}, \quad (1.59)$$

we have

$$\delta A = \int_V (\mathfrak{T}\colon\delta\mathfrak{E})dV. \quad (1.60)$$

Here $\delta\mathfrak{E}$ is the increment in the strain resulting from the additional displacement $\delta\mathbf{u}$.

Let us now confine our discussion to a small volume element dV about the point P and let us denote by dA the work done per unit volume. Then, Eq. (1.60) yields

$$dA = \mathfrak{T}\colon d\mathfrak{E}. \quad (1.61)$$

The total work done per unit volume is therefore given by

$$A = \int (\mathfrak{T} : d\mathfrak{E}). \tag{1.62}$$

Because the body is perfectly elastic, the integral should not depend on the path of the deformation and, consequently, the integrand must be a total differential. In other words, there must exist a potential function $W = W(\varepsilon_{ij})$, known as the *strain energy density*, such that

$$dW = \mathfrak{T} : d\mathfrak{E} = \tau_{ij}\, d\varepsilon_{ij}, \qquad \tau_{ij} = \frac{\partial W}{\partial \varepsilon_{ij}}, \qquad \delta A = \delta W. \tag{1.63}$$

We have assumed that W depends explicitly only on ε_{ij}. Now, because ε_{ij} are small, W can be expanded in a power series:

$$W = a_0 + b_{ij}\varepsilon_{ij} + \tfrac{1}{2}C_{ijkl}\varepsilon_{ij}\varepsilon_{kl} + \cdots .$$

We normalize the energy in such a way that the undeformed state, $\mathfrak{E} = 0$, corresponds to zero energy. Hence $a_0 = 0$. Also, the energy must have a minimum value at $\mathfrak{E} = 0$, or else the material will strain itself to reach that level. Hence, $\tau_{ij} = 0$ for $\varepsilon_{ij} = 0$. Equation (1.63) then leads to $b_{ij} = 0$. Neglecting third and higher order terms in the components of strain, W becomes a quadratic function of ε_{ij}:

$$W = \tfrac{1}{2}C_{ijkl}\varepsilon_{ij}\varepsilon_{kl}, \qquad C_{ijkl} = C_{klij}. \tag{1.64}$$

Equations (1.63) and (1.64) now yield

$$\tau_{ij} = C_{ijkl}\varepsilon_{kl}. \tag{1.65}$$

Equation (1.65) states the well-known experimental result that under given conditions of small displacements and small strains, the components of the stress and strain in an ideal elastic solid are related by linear homogeneous equations (*generalized Hooke's law*). Homogeneity ensures that the deformation is measured from the state of zero stress.

The array of coefficients C_{ijkl}, which are $3^4 = 81$ in number in the most general case, constitutes a tensor of the fourth order, known as the *tensor of elastic moduli*. Its components depend upon the reference frame. For uniform materials, C_{ijkl} are constant throughout the body. Because $\tau_{ij} = \tau_{ji}$ and $\varepsilon_{kl} = \varepsilon_{lk}$, it follows from Eq. (1.65) that

$$C_{ijkl} = C_{jikl} = C_{ijlk}. \tag{1.66}$$

This reduces the number of independent constants to 36. This number further reduces to 21 because of the relation $C_{ijkl} = C_{klij}$.

In mechanically isotropic materials the elastic properties at any given point are the same in all directions and, consequently, C_{ijkl} is the general isotropic tensor of the fourth order. (An isotropic tensor is defined as a tensor whose components retain the same value however the axes are rotated.) It can be shown that the general isotropic tensor of order 4 may be expressed in the form

$$C_{ijkl} = \lambda\delta_{ij}\delta_{kl} + \mu(\delta_{ik}\delta_{jl} + \delta_{il}\delta_{jk}) + \nu(\delta_{ik}\delta_{jl} - \delta_{il}\delta_{jk}).$$

Equation (1.66) shows that in our case $\nu = 0$ and, therefore, for an isotropic material, we have

$$C_{ijkl} = \lambda\delta_{ij}\delta_{kl} + \mu(\delta_{ik}\delta_{jl} + \delta_{il}\delta_{jk}). \tag{1.67}$$

This tensor has 21 nonvanishing components, of which only two are independent:

$$\begin{aligned} &C_{1111} = C_{2222} = C_{3333} = \lambda + 2\mu, \\ &C_{1122} = C_{1133} = C_{2211} = C_{2233} = C_{3311} = C_{3322} = \lambda, \\ &C_{2323} = C_{3223} = C_{3232} = C_{2332} = C_{3131} = C_{1331} = C_{1313} \\ &\quad = C_{3113} = C_{1212} = C_{2112} = C_{2121} = C_{1221} = \mu. \end{aligned}$$

The polyadic form of Eq. (1.67) is

$$\mathbf{C} = \lambda\mathfrak{I}\mathfrak{I} + \mu(\mathbf{\Pi}' + \mathbf{\Pi}), \tag{1.67a}$$

where $\mathbf{\Pi}$ denotes the unit tetradic defined by the property that, for an arbitrary dyadic $\mathfrak{A}$,

$$\mathfrak{A}:\mathbf{\Pi} = \mathbf{\Pi}:\mathfrak{A} = \mathfrak{A}.$$

Similarly, the tetradic $\mathbf{\Pi}'$ is defined as

$$\mathfrak{A}:\mathbf{\Pi}' = \mathbf{\Pi}':\mathfrak{A} = \tilde{\mathfrak{A}}.$$

Equations (1.65) and (1.67) give the following relationship valid for isotropic solids

$$\tau_{ij} = \lambda\delta_{ij}\varepsilon_{kk} + 2\mu\varepsilon_{ij} = \lambda\delta_{ij}\frac{\partial u_k}{\partial x_k} + \mu\left(\frac{\partial u_j}{\partial x_i} + \frac{\partial u_i}{\partial x_j}\right). \tag{1.68}$$

This may be written in the form

$$\mathfrak{T} = \mathbf{C}:\mathfrak{E} = \lambda\mathfrak{I}\operatorname{div}\mathbf{u} + 2\mu\mathfrak{E} = \lambda\mathfrak{I}\operatorname{div}\mathbf{u} + \mu(\nabla\mathbf{u} + \mathbf{u}\nabla). \tag{1.69}$$

Also, from Eqs. (1.64) and (1.69)

$$W = \tfrac{1}{2}\mathfrak{E}:\mathbf{C}:\mathfrak{E} = \tfrac{1}{2}(\mathfrak{E}:\mathfrak{T}) = \tfrac{1}{2}(\lambda\varepsilon_{kk}^2 + 2\mu\varepsilon_{ij}\varepsilon_{ij}). \tag{1.70}$$

From Eq. (1.68) we have

$$\tau_{kk} = (3\lambda + 2\mu)\varepsilon_{kk}. \tag{1.71}$$

Hence, the relationship inverse to Eq. (1.69) is

$$\mathfrak{E} = \frac{1}{2\mu}\left[\mathfrak{T} - \frac{\lambda}{3\lambda + 2\mu}\tau_{kk}\mathfrak{I}\right]. \tag{1.72}$$

The coefficients λ and μ are known as *Lamé parameters.* They are determined by experiment for any given material. The following three experiments give the relations between various measures of elasticity:

1. Uniaxial tension. A long, thin, cylindrical wire is stretched. If the x_1 axis is chosen along the wire, only τ_{11} differs from zero:

$$\mathfrak{T} = \tau\mathbf{e}_1\mathbf{e}_1. \tag{1.73}$$

Then, from Eq. (1.72),we have

$$\mathfrak{E} = \frac{\tau}{2\mu(3\lambda + 2\mu)}[2(\lambda + \mu)\mathbf{e}_1\mathbf{e}_1 - \lambda(\mathbf{e}_2\mathbf{e}_2 + \mathbf{e}_3\mathbf{e}_3)]. \tag{1.74}$$

We notice that Hooke's law predicts an extension along the direction of the tension and, in addition, a contraction in the perpendicular direction. The ratio of the tension to the elongation in the direction of the tension is known as *Young's modulus*:

$$\frac{\tau_{11}}{\varepsilon_{11}} = \frac{\mu(3\lambda + 2\mu)}{\lambda + \mu} = Y. \tag{1.75}$$

The ratio of the contraction in a direction perpendicular to the tension to the elongation in the direction of the tension is known as *Poisson's ratio*:

$$-\frac{\varepsilon_{22}}{\varepsilon_{11}} = -\frac{\varepsilon_{33}}{\varepsilon_{11}} = \frac{\lambda}{2(\lambda + \mu)} = \sigma. \tag{1.76}$$

2. Pure shear. A bar with a rectangular cross section is in equilibrium under a system of shearing forces in the x_1x_2 plane:

$$\mathfrak{T} = \tau(\mathbf{e}_1\mathbf{e}_2 + \mathbf{e}_2\mathbf{e}_1). \tag{1.77}$$

The corresponding strain, found from Eq. (1.72), is

$$\mathfrak{E} = \frac{\tau}{2\mu}(\mathbf{e}_1\mathbf{e}_2 + \mathbf{e}_2\mathbf{e}_1).$$

Recalling our earlier interpretation of $2\varepsilon_{12}$ as the decrease in the angle between two line elements that were parallel to the x_1 and x_2 axes, respectively, before the deformation, we may interpret μ as

$$\mu = \frac{\tau_{12}}{2\varepsilon_{12}} = \frac{\text{shearing stress}}{\text{decrease in angle}}. \tag{1.78}$$

We call μ the *rigidity* or *shear modulus.*

3. Hydrostatic pressure. A small mass of material is put in a large vessel containing a liquid and a pressure p is exerted on the liquid. By Pascal's law, the body experiences only normal traction, $\mathbf{T} = -p\mathbf{n}$, and the stress dyadic is therefore of the form $\mathfrak{T} = -p\mathfrak{I}$. From Eq. (1.72), the resulting strain is

$$\mathfrak{E} = -\frac{p}{3\lambda + 2\mu}\mathfrak{I}.$$

However, from Eq. (1.45), the increase in volume per unit volume is given by

$$\operatorname{div} \mathbf{u} = \varepsilon_{kk} = -\frac{p}{\lambda + (\frac{2}{3})\mu}.$$

This leads us to the following definition of the *bulk modulus*, κ,

$$\kappa = \lambda + \frac{2}{3}\mu = \frac{p}{(-\operatorname{div} \mathbf{u})} = \frac{\text{hydrostatic pressure}}{\text{decrease in volume per unit volume}}. \tag{1.79}$$

Table 1.1. Elastic Moduli of Some Common Substances

	σ	λ (10^{11} dyne/cm^2)	μ (10^{11} dyne/cm^2)	κ (10^{11} dyne/cm^2)	Y (10^{11} dyne/cm^2)
Steel	0.26	8.84	8.19	14.3	20.64
Gold	0.42	14.73	2.80	16.6	7.95
Copper	0.33	8.65	4.58	11.7	12.18
Glass	0.25	2.69	2.72	4.50	6.80
Fluids	0.50	Large	0	Large	0
Earth (crust)	0.28	4.5	3.6	6.0	9.2
Incompressible materials	0.50	∞	μ	∞	3μ
Poisson's case	0.25	λ	λ	$\frac{5}{3}\lambda$	$\frac{5}{2}\lambda$
General	$-1 \leq \sigma \leq \frac{1}{2}$	—	≥ 0	≥ 0	≥ 0

Of the five parameters λ, μ, Y, σ, and κ, only two are independent. Using Eqs. (1.75), (1.76), and (1.79), we derive the following relations between the Lamé parameters, λ and μ; Poisson's ratio, σ; Young's modulus, Y; and the bulk modulus, κ:

$$\begin{aligned}
\lambda &= \frac{2\sigma}{1-2\sigma}\mu = \kappa - \frac{2}{3}\mu = \frac{\sigma}{(1+\sigma)(1-2\sigma)}Y, \\
\mu &= \frac{1-2\sigma}{2\sigma}\lambda = \frac{1}{2(1+\sigma)}Y, \\
\sigma &= \frac{\lambda}{2(\lambda+\mu)} = \frac{Y}{2\mu} - 1, \\
Y &= \frac{3\lambda+2\mu}{\lambda+\mu}\mu = \frac{(1+\sigma)(1-2\sigma)}{\sigma}\lambda = \frac{9\kappa\mu}{3\kappa+\mu} = 2(1+\sigma)\mu, \\
\kappa &= \lambda + \frac{2}{3}\mu = \frac{1+\sigma}{3\sigma}\lambda = \frac{2(1+\sigma)}{3(1-2\sigma)}\mu = \frac{1}{3(1-2\sigma)}Y.
\end{aligned} \tag{1.80}$$

Table 1.1 gives the values of σ, λ, μ, κ, and Y for some common materials.

Hooke's law can be expressed in an alternative form that is sometimes more useful. We recall our decomposition, Eq. (1.22), of the stress tensor into an isotropic part and a stress deviator:

$$\tau_{ij} = \tfrac{1}{3}\tau_{kk}\delta_{ij} + \tau'_{ij}. \tag{1.81}$$

Similarly, for the strain tensor,

$$\varepsilon_{ij} = \tfrac{1}{3}\varepsilon_{kk}\delta_{ij} + \varepsilon'_{ij}. \tag{1.82}$$

It is easy to show that Hooke's law, Eq. (1.68), is equivalent to the following relations:

$$\tau_{ii} = 3\kappa\varepsilon_{ii}, \qquad \tau'_{ij} = 2\mu\varepsilon'_{ij}. \tag{1.83}$$

Because $\varepsilon'_{ii} = 0$, $\kappa = \lambda + (\frac{2}{3})\mu$, Eqs. (1.70), (1.82), and (1.83) now yield

$$W = \tfrac{1}{2}\kappa\varepsilon_{ii}^2 + \mu\varepsilon'_{ij}\varepsilon'_{ij} = \frac{1}{18\kappa}\tau_{ii}^2 + \frac{1}{4\mu}\tau'_{ij}\tau'_{ij}. \tag{1.84}$$

Equation (1.84) exhibits the dilatational and deviatoric strain energies separately. In pure shear, for example, $\tau_{ii} = 0$ and the strain energy is purely deviatoric:

$$W = \frac{1}{4\mu}\tau'_{ij}\tau'_{ij} = \frac{1}{2\mu}\tau^2. \tag{1.85}$$

In the case of hydrostatic pressure, the stress deviator $\tau'_{ij} = 0$, and the strain energy is purely dilatational:

$$W = \frac{1}{18\kappa}\tau_{ii}^2 = \frac{1}{2\kappa}p^2. \tag{1.86}$$

1.4. The Field Equations

Consider a continuum in a state of motion relative to an inertial frame of reference. An infinitesimal volume of mass $(\rho\, dV)$ will be referred to as a particle. The density field $\rho(\mathbf{r}, t)$ and the velocity field $\mathbf{v}(\mathbf{r}, t)$ are defined throughout a volume V with boundary S. The volume V is assumed to contain the same material particles at all times. If the total mass in V is m, the total linear momentum is $\mathbf{P}$, and the total moment of momentum is $\mathbf{M}$, then

$$m = \int_V \rho\, dV, \qquad \mathbf{P}(t) = \int_V \mathbf{v}\rho\, dV, \qquad \mathbf{M}(t) = \int_V (\mathbf{r} \times \mathbf{v})\rho\, dV. \tag{1.87}$$

Let $\boldsymbol{\Lambda}$ be the total applied force and $\boldsymbol{\Gamma}$ the total applied torque about the origin of the reference frame. The following fundamental principles are assumed to hold in V:

1. $\dfrac{Dm}{Dt} = 0$ — Conservation of mass (1.88)

2. $\dfrac{D}{Dt}\mathbf{P}(t) = \boldsymbol{\Lambda}$ — Newton–Euler principle of linear momentum (1.89)

3. $\dfrac{D}{Dt}\mathbf{M}(t) = \boldsymbol{\Gamma}$ — Newton–Euler principle of angular momentum. (1.90)

It is shown in Appendix C that the material derivative D/Dt is given by

$$\frac{D}{Dt} = \frac{\partial}{\partial t} + \mathbf{v}\cdot\nabla \tag{1.91}$$

and, if $A(\mathbf{r}, t)$ is an arbitrary differentiable function,

$$\frac{D}{Dt}\int_V A(\mathbf{r}, t)dV = \int_V \left[\frac{DA}{Dt} + A(\operatorname{div} \mathbf{v})\right]d\dot{V}. \tag{1.92}$$

The force $\mathbf{\Lambda}$ includes body forces ($\rho\mathbf{F}$) acting throughout V and surface forces ($\mathbf{n} \cdot \mathfrak{T}$) acting on S. The total force and the total moment acting on the material occupying the region V are

$$\mathbf{\Lambda} = \int_S (d\mathbf{S} \cdot \mathfrak{T}) + \int_V \rho\mathbf{F}\, dV = \int_V (\operatorname{div} \mathfrak{T} + \rho\mathbf{F})dV, \tag{1.93}$$

$$\mathbf{\Gamma} = \int_S \mathbf{r} \times (d\mathbf{S} \cdot \mathfrak{T}) + \int_V (\mathbf{r} \times \rho\mathbf{F})dV. \tag{1.94}$$

Applying Eq. (1.92) with $A = \rho(\mathbf{r}, t)$, Eqs. (1.87), (1.88), and (1.91) yield

$$\frac{Dm}{Dt} = \frac{D}{Dt}\int_V \rho\, dV = \int_V \left[\frac{D\rho}{Dt} + \rho \operatorname{div} \mathbf{v}\right]dV = \int_V \left[\frac{\partial\rho}{\partial t} + \operatorname{div}(\rho\mathbf{v})\right]dV = 0. \tag{1.95}$$

Because the equation is valid for an arbitrary V, the integrand must vanish at each point. We therefore arrive at the *equation of continuity*:

$$\frac{D\rho}{Dt} + \rho \operatorname{div} \mathbf{v} = 0 \quad \text{or} \quad \frac{\partial\rho}{\partial t} + \operatorname{div}(\rho\mathbf{v}) = 0. \tag{1.96}$$

However, if we apply Eq. (1.92) with $\mathbf{A} = \rho\mathbf{v}$, we have from Eqs. (1.87), (1.89), and (1.96),

$$\begin{aligned}\mathbf{\Lambda} &= \frac{D}{Dt}\int_V \rho\mathbf{v}\, dV = \int_V \left[\frac{D(\rho\mathbf{v})}{Dt} + \rho\mathbf{v} \operatorname{div} \mathbf{v}\right]dV \\ &= \int_V \left[\rho\frac{D\mathbf{v}}{Dt} + \mathbf{v}\left(\frac{D\rho}{Dt} + \rho \operatorname{div} \mathbf{v}\right)\right]dV = \int_V \left(\rho\frac{D\mathbf{v}}{Dt}\right)dV.\end{aligned} \tag{1.97}$$

From Eqs. (1.93) and (1.97), we arrive at *Euler's equation of motion*

$$\operatorname{div} \mathfrak{T} + \rho\mathbf{F} = \rho\frac{D\mathbf{v}}{Dt}. \tag{1.98}$$

The momentum equation, Eq. (1.90), puts no further restriction on the motion. It just yields the symmetry of $\mathfrak{T}$. The proof is similar to the one given in Eq. (1.12) for an elastic solid.

When Eqs. (1.96) and (1.98) are applied to a linear elastic solid, the basic assumptions underlying the linearization conditions used in Eq. (1.29) must be invoked. Therefore, we make the approximations

$$\begin{aligned}\mathbf{v} &= \frac{D\mathbf{u}}{Dt} = \frac{\partial\mathbf{u}}{\partial t} + \mathbf{v}\cdot\nabla\mathbf{u} \simeq \frac{\partial\mathbf{u}}{\partial t}, \\ \frac{D\mathbf{v}}{Dt} &= \frac{\partial\mathbf{v}}{\partial t} + \mathbf{v}\cdot\nabla\mathbf{v} \simeq \frac{\partial^2\mathbf{u}}{\partial t^2}.\end{aligned} \tag{1.99}$$

Under these conditions, Eq. (1.98) reduces to *Cauchy's equation of motion* for the infinitesimal theory of elasticity:

$$\operatorname{div} \mathfrak{T} + \rho \mathbf{F} = \rho \frac{\partial^2 \mathbf{u}}{\partial t^2}. \tag{1.100}$$

Equation (1.100) can be obtained directly from Eq. (1.13) if we include the inertial terms as additional body forces. The equation of continuity, Eq. (1.96), leads us to Eq. (1.46):

$$\frac{\delta\rho}{\rho} = -\operatorname{div} \mathbf{u}. \tag{1.101}$$

In problems pertaining to elastic wave propagation in the solid earth, the density changes occurring because of the wave motions are small.

Equations (1.100) and (1.101) constitute a system of four equations in 10 unknown functions, ρ, u_i, τ_{ij}, of time and position. However, when the generalized Hooke's law is added, the set is complete. If the density is given, we have a set of nine equations in nine unknowns, u_i and τ_{ij}:

$$\operatorname{div} \mathfrak{T} + \rho \mathbf{F} = \rho \frac{\partial^2 \mathbf{u}}{\partial t^2}, \qquad \mathfrak{T} = \mathbf{C} : \mathfrak{E}. \tag{1.102}$$

The solution of Eqs. (1.102) is facilitated by expressing $\mathfrak{T}$ in terms of the displacements:

$$\operatorname{div}\{\mathbf{C} : \mathfrak{E}(\mathbf{u})\} + \rho \mathbf{F} = \rho \frac{\partial^2 \mathbf{u}}{\partial t^2}. \tag{1.103}$$

Solving for $\mathbf{u}$ when ρ, $\mathbf{F}$, and $\mathbf{C}$ are known, we may then evaluate the strain and stress everywhere by differentiation and applying Hooke's law.

When the medium under consideration is isotropic, we have

$$\begin{aligned} \operatorname{div} \mathfrak{T} &= \operatorname{div}[\lambda \mathfrak{J} \operatorname{div} \mathbf{u} + \mu(\nabla \mathbf{u} + \mathbf{u}\nabla)] = \nabla(\lambda \operatorname{div} \mathbf{u}) + \mu \operatorname{div}(\nabla \mathbf{u} + \mathbf{u}\nabla) \\ &\quad + \nabla\mu \cdot (\nabla \mathbf{u} + \mathbf{u}\nabla) \\ &= \lambda \operatorname{grad} \operatorname{div} \mathbf{u} + (\nabla\lambda)\operatorname{div} \mathbf{u} + \mu(\nabla^2 \mathbf{u} + \operatorname{grad} \operatorname{div} \mathbf{u}) + \nabla\mu \cdot (\nabla \mathbf{u} + \mathbf{u}\nabla). \end{aligned} \tag{1.104}$$

Using this in Eq. (1.100), we find

$$\mu\nabla^2 \mathbf{u} + (\lambda + \mu)\operatorname{grad} \operatorname{div} \mathbf{u} + (\nabla\lambda)\operatorname{div} \mathbf{u} + \nabla\mu \cdot (\nabla \mathbf{u} + \mathbf{u}\nabla) + \rho \mathbf{F} = \rho \frac{\partial^2 \mathbf{u}}{\partial t^2}. \tag{1.105}$$

This is the equation of small motion of an isotropic perfectly elastic continuous medium.

If $\mathbf{n}$ is a unit vector, we have

$$\mathbf{n} \cdot (\nabla \mathbf{u} + \mathbf{u}\nabla) = \mathbf{n} \cdot (2\nabla \mathbf{u} + \operatorname{curl} \mathbf{u} \times \mathfrak{J}) = 2\frac{\partial \mathbf{u}}{\partial n} + \mathbf{n} \times \operatorname{curl} \mathbf{u}. \tag{1.106}$$

We define a *vertically heterogeneous medium* as one in which the density and the Lamé parameters are functions of the z coordinate only. Let $\rho = \rho(z)$, $\lambda = \lambda(z)$, and $\mu = \mu(z)$. Then, using Eq. (1.106), Eq. (1.105) becomes

$$\mu\nabla^2\mathbf{u} + (\lambda + \mu)\text{grad div }\mathbf{u} + \mathbf{e}_z\frac{d\lambda}{dz}\text{div }\mathbf{u} + \frac{d\mu}{dz}\left(2\frac{\partial\mathbf{u}}{\partial z} + \mathbf{e}_z \times \text{curl }\mathbf{u}\right) + \rho\mathbf{F}$$
$$= \rho\frac{\partial^2\mathbf{u}}{\partial t^2}. \tag{1.107}$$

Similarly, for a *radially heterogeneous medium*, $\rho = \rho(r)$, $\lambda = \lambda(r)$, $\mu = \mu(r)$, and the equation of motion becomes

$$\mu\nabla^2\mathbf{u} + (\lambda + \mu)\text{grad div }\mathbf{u} + \mathbf{e}_r\frac{d\lambda}{dr}\text{div }\mathbf{u} + \frac{d\mu}{dr}\left(2\frac{\partial\mathbf{u}}{\partial r} + \mathbf{e}_r \times \text{curl }\mathbf{u}\right) + \rho\mathbf{F}$$
$$= \rho\frac{\partial^2\mathbf{u}}{\partial t^2}. \tag{1.108}$$

It may be noted that Eqs. (1.107) and (1.108) remain valid even if the density is an arbitrary function of the coordinates.

In a homogeneous medium, the equation of motion, Eq. (1.105), takes the following simplified form

$$\mu\nabla^2\mathbf{u} + (\lambda + \mu)\text{grad div }\mathbf{u} + \rho\mathbf{F} = \rho\frac{\partial^2\mathbf{u}}{\partial t^2}, \tag{1.109}$$

or, equivalently,

$$(\lambda + 2\mu)\text{grad div }\mathbf{u} - \mu\text{ curl curl }\mathbf{u} + \rho\mathbf{F} = \rho\frac{\partial^2\mathbf{u}}{\partial t^2}. \tag{1.110}$$

When the constants α and β are introduced through the relations

$$\alpha = \left(\frac{\lambda + 2\mu}{\rho}\right)^{1/2}, \qquad \beta = \left(\frac{\mu}{\rho}\right)^{1/2}, \tag{1.111}$$

Eq. (1.110) may be written as

$$\alpha^2\text{ grad div }\mathbf{u} - \beta^2\text{ curl curl }\mathbf{u} + \mathbf{F} = \frac{\partial^2\mathbf{u}}{\partial t^2}. \tag{1.112}$$

Taking the divergence of the last equation, we find

$$\alpha^2\nabla^2(\text{div }\mathbf{u}) + \text{div }\mathbf{F} = \frac{\partial^2}{\partial t^2}(\text{div }\mathbf{u}). \tag{1.113}$$

Similarly, taking the curl of Eq. (1.112), we have

$$\beta^2\nabla^2(\text{curl }\mathbf{u}) + \text{curl }\mathbf{F} = \frac{\partial^2}{\partial t^2}(\text{curl }\mathbf{u}). \tag{1.114}$$

Therefore, the volume dilatation and the rotation in an isotropic homogeneous medium obey similar equations.

We have discussed in Section 1.3 the total conversion of the work done by the external forces into potential strain energy of the elastic body under static conditions. However, if the volume elements are able to move, part of the external work will be converted into kinetic energy. Proceeding as in the parallel case of discrete particle dynamics, we obtain from the Cauchy equation of motion, Eq. (1.100),

$$(\text{div}\, \mathfrak{T}) \cdot \dot{\mathbf{u}} + \rho \mathbf{F} \cdot \dot{\mathbf{u}} = \rho \ddot{\mathbf{u}} \cdot \dot{\mathbf{u}} = \frac{1}{2} \rho \frac{\partial}{\partial t} \dot{\mathbf{u}}^2, \tag{1.115}$$

where a dot signifies differentiation with respect to time, t. Using the relation

$$(\text{div}\, \mathfrak{T}) \cdot \dot{\mathbf{u}} = \text{div}(\mathfrak{T} \cdot \dot{\mathbf{u}}) - \mathfrak{T} : \nabla \dot{\mathbf{u}} \tag{1.116}$$

Eq. (1.115) can be made to yield

$$\frac{1}{2} \frac{\partial}{\partial t} (\mathfrak{T} : \mathfrak{E}) + \frac{1}{2} \rho \frac{\partial}{\partial t} \dot{\mathbf{u}}^2 = \text{div}(\mathfrak{T} \cdot \dot{\mathbf{u}}) + \rho \mathbf{F} \cdot \dot{\mathbf{u}}. \tag{1.117}$$

Integrating over the volume V, we get

$$\begin{aligned} \frac{\partial}{\partial t}\left[\int_V \frac{1}{2} (\mathfrak{T} : \mathfrak{E} + \rho \dot{\mathbf{u}}^2) dV\right] &= \int_V \text{div}(\mathfrak{T} \cdot \dot{\mathbf{u}}) dV + \int_V \rho (\mathbf{F} \cdot \dot{\mathbf{u}}) dV \\ &= \int_S (\mathfrak{T} \cdot \dot{\mathbf{u}}) \cdot d\mathbf{S} + \int_V \rho (\mathbf{F} \cdot \dot{\mathbf{u}}) dV. \end{aligned} \tag{1.118}$$

This is the *energy equation*, showing that the rate of change of total energy (strain + kinetic) contained in the volume V is equal to the rate at which work is being done by the surface tractions and body forces.

We introduce the *Hamiltonian density* (energy per unit volume), H; the total *mechanical energy*, U; the *canonical momentum density* (momentum per unit volume), $\mathbf{P}$; and the *energy-flux density*, $\mathbf{\Sigma}$, through the relations

$$H = \tfrac{1}{2}(\mathfrak{T} : \mathfrak{E} + \rho \dot{\mathbf{u}}^2), \tag{1.119}$$

$$U = \int_V H \, dV, \tag{1.120}$$

$$\mathbf{P} = \rho \dot{\mathbf{u}}, \tag{1.121}$$

$$\mathbf{\Sigma} = -\mathfrak{T} \cdot \dot{\mathbf{u}}. \tag{1.122}$$

Equation (1.118) may now be written as

$$\frac{\partial U}{\partial t} = -\int_S \mathbf{\Sigma} \cdot d\mathbf{S} + \int_V (\mathbf{F} \cdot \mathbf{P}) dV = \int_V (-\text{div}\, \mathbf{\Sigma} + \mathbf{F} \cdot \mathbf{P}) dV. \tag{1.123}$$

However, Eq. (1.117) yields the *equation of continuity of energy density*

$$\text{div}\, \mathbf{\Sigma} + \frac{\partial H}{\partial t} = \mathbf{F} \cdot \mathbf{P} \tag{1.124}$$

valid at every point of the body. Because $\mathbf{\Sigma}$ determines the magnitude as well as the direction of energy flow across a boundary, Eq. (1.123) states that the rate of change of energy in V, plus the rate of flow of energy across S, equals the power pumped into the volume by the body forces. This is the law of conservation of energy.

1.5. Lagrangian Formulation

The dynamic behavior of an elastic medium can be specified by a single function, *a Lagrangian density* $\mathscr{L}$, which is a function of the displacements and their spatial and temporal derivatives. Therefore, for the linearized theory of elasticity, we have

$$\mathscr{L} = \frac{1}{2}\rho\left(\frac{\partial \mathbf{u}}{\partial t}\right)^2 - \frac{1}{2}(\mathfrak{T}:\mathfrak{E}). \tag{1.125}$$

If ρ is considered to be a constant, we have

$$\frac{\partial}{\partial t}\left(\frac{\partial \mathscr{L}}{\partial \dot{\mathbf{u}}}\right) = \rho\frac{\partial^2 \mathbf{u}}{\partial t^2}, \tag{1.126}$$

$$\frac{\partial \mathscr{L}}{\partial(\nabla \mathbf{u})} = -\mathfrak{T}. \tag{1.127}$$

The Cauchy equation of motion, Eq. (1.100), in the Lagrangian form is then

$$\frac{\partial}{\partial t}\left(\frac{\partial \mathscr{L}}{\partial \dot{\mathbf{u}}}\right) = \rho\mathbf{F} - \frac{\partial}{\partial \mathbf{r}}\cdot\frac{\partial \mathscr{L}}{\partial(\nabla \mathbf{u})}. \tag{1.128}$$

Note that if we define the four-gradient operator

$$\partial_\mu = \frac{\partial}{\partial x_\mu} = \left(\nabla, \frac{\partial}{\partial t}\right) \qquad (\mu = 1, 2, 3, 4), \tag{1.129}$$

the Lagrange equation, Eq. (1.128), can be recast in a more compact form

$$\frac{\partial}{\partial x_\mu}\frac{\partial \mathscr{L}}{\partial(u_{k,\mu})} = \rho F_k \qquad (k = 1, 2, 3). \tag{1.130}$$

1.5.1. Hamilton's Principle

Hamilton's principle states that under the influence of conservative forces, the actual motion of each particle of a body is such that the action integral over time and space is extremal. However, for applications to regimes that involve forces not derivable from a potential-energy function, such as an elastic body subjected to surface tractions, the statement of Hamilton's principle must be modified.

In Eqs. (1.57)–(1.63) we have derived the relation between the displacement $\delta\mathbf{u}$ and the virtual work done on the body by the external forces and surface tractions under conditions of static equilibrium. In the dynamic case, we must add the contribution of the inertial forces. Therefore, in place of Eq. (1.60), we now have

$$\delta A = \int_V \rho(\mathbf{F}\cdot\delta\mathbf{u})dV + \int_{S'}(d\mathbf{S}\cdot\mathfrak{T})\cdot\delta\mathbf{u} = \int_V(\rho\ddot{\mathbf{u}}\cdot\delta\mathbf{u})dV + \int_V(\mathfrak{T}:\delta\mathfrak{E})dV, \tag{1.131}$$

where S' is that part of the body over which the surface tractions are specified, whereas $\delta\mathbf{u}$ is arbitrary. Over the rest of the surface, the displacements are prescribed.

Equation (1.131) is, in fact, a restatement of the energy balance principle. To obtain the action integral, we integrate Eq. (1.131) over an arbitrary time interval (t_1, t_2). The inertial term yields to the following transformations

$$\begin{aligned}
\int_{t_1}^{t_2} dt \int_V \left(\rho\frac{\partial^2\mathbf{u}}{\partial t^2}\cdot\delta\mathbf{u}\right)dV &= \int_V \rho\, dV\left[\int_{t_1}^{t_2}\left(\frac{\partial^2\mathbf{u}}{\partial t^2}\cdot\delta\mathbf{u}\right)dt\right] \\
&= \int_V \rho\, dV\left[\frac{\partial\mathbf{u}}{\partial t}\cdot\delta\mathbf{u}\right]_{t_1}^{t_2} - \int_V \rho\, dV\int_{t_1}^{t_2}\left(\frac{\partial\mathbf{u}}{\partial t}\cdot\delta\frac{\partial\mathbf{u}}{\partial t}\right)dt \\
&= -\int_{t_1}^{t_2}\delta\left[\frac{1}{2}\int_V\rho\left(\frac{\partial\mathbf{u}}{\partial t}\right)^2 dV\right]dt,
\end{aligned} \tag{1.132}$$

where $\delta\mathbf{u}$ is chosen to vanish everywhere at $t = t_1$ and $t = t_2$. Equation (1.131) can now be recast in the variational form

$$\delta\int_{t_1}^{t_2}\left[\int_V(\mathscr{L} + \rho\mathbf{F}\cdot\mathbf{u})dV + \int_{S'}(\mathfrak{T}\cdot\mathbf{u})\cdot d\mathbf{S}\right]dt = 0, \tag{1.133}$$

assuming that $\mathbf{F}$ is independent of $\mathbf{u}$. Note that the surface tractions enter explicitly in the surface integral over S' and also implicitly in $\mathscr{L}$ because the effect of the surface tractions is transmitted into the medium via $\mathfrak{E}$ by Newton's second law and via $\mathfrak{T}$ by Newton's third law (action and reaction). True, the internal stresses in a finite body are caused by the surface tractions, but both forces must be included separately. Omitting the surface integral term would be like writing the dynamical equations taking into account the reaction forces alone.

The variation of the action integral is done by the transformations $\mathbf{u} \to \mathbf{u} + \delta\mathbf{u}$, $\dot{\mathbf{u}} \to \dot{\mathbf{u}} + \delta\dot{\mathbf{u}}$ and $\nabla\mathbf{u} \to \nabla\mathbf{u} + \nabla(\delta\mathbf{u})$. When this process is carried out, the volume integral in Eq. (1.133) will render the Lagrange equation of motion (or the Cauchy equation of motion), whereas the surface integral will give the boundary condition over S'. Hamilton's extended principle is therefore equivalent to both the equations of motion and the boundary conditions.

1.5.2. Stress–Energy Tensor

We define the four-vectors

$$(x_1, x_2, x_3, x_4 = t), \qquad \mathbf{f} = (\rho F_1, \rho F_2, \rho F_3, \mathbf{P} \cdot \mathbf{F}),$$

and the nonsymmetric tensor

$$\mathfrak{R} \equiv \begin{bmatrix} -\tau_{11} & -\tau_{12} & -\tau_{13} & P_1 \\ -\tau_{21} & -\tau_{22} & -\tau_{23} & P_2 \\ -\tau_{31} & -\tau_{32} & -\tau_{33} & P_3 \\ \Sigma_1 & \Sigma_2 & \Sigma_3 & H \end{bmatrix}. \tag{1.134}$$

It may be verified easily that

$$\operatorname{div} \mathfrak{R} = \mathbf{f} \tag{1.135}$$

incorporates both the principle of energy conservation, Eq. (1.124), and the principle of conservation of linear momentum (Cauchy's equation of motion)

$$-\operatorname{div} \mathfrak{T} + \frac{\partial \mathbf{P}}{\partial t} = \rho \mathbf{F}.$$

We call $\mathfrak{R}$ the *matter stress–energy tensor*, because $\mathfrak{T}$ and $\mathbf{P}$ are essentially forces and momenta associated with a volume element of matter, respectively.

We construct a second stress–energy tensor, $\mathfrak{B}$, whose corresponding elements are the forces acting on the elastic waves propagating through the matter and the momenta of the waves. This tensor is derived from the Lagrangian formulation in the following way: We put

$$B_{44} = H = \dot{\mathbf{u}} \cdot \frac{\partial \mathscr{L}}{\partial \dot{\mathbf{u}}} - \mathscr{L} = \rho \dot{\mathbf{u}}^2 - \mathscr{L}, \tag{1.136}$$

and construct the fourth column as a generalization of B_{44} in the sense that $\partial/\partial t$ in $\dot{\mathbf{u}}$ is replaced by a spatial operator $\partial/\partial \mathbf{r} = \nabla$,

$$B_{i4} = \left(\nabla \mathbf{u} \cdot \frac{\partial \mathscr{L}}{\partial \dot{\mathbf{u}}}\right)_i - \mathscr{L}\delta_{i4} = \left(\rho \nabla \mathbf{u} \cdot \frac{\partial \mathbf{u}}{\partial t}\right)_i = Q_i \qquad (i = 1, 2, 3), \tag{1.137}$$

where $\mathbf{Q}$ is the *wave momentum density* (momentum/volume). The generalization of the spatial part of B_{ij} is then

$$W_{ij} \equiv \nabla \mathbf{u} \cdot \frac{\partial \mathscr{L}}{\partial (\nabla \mathbf{u})} - \mathscr{L}\mathfrak{J} = -(\nabla \mathbf{u} \cdot \mathfrak{T}) - \mathscr{L}\mathfrak{J} \qquad (i, j = 1, 2, 3), \tag{1.138}$$

where W_{ij} is the *wave force dyadic* (force/area). The complete wave stress–energy tensor can then be written in the form

$$B_{ij} = \sum_{k=1}^{3} u_{k,i} \frac{\partial \mathscr{L}}{\partial u_{k,j}} - \mathscr{L}\delta_{ij} \qquad (i, j = 1, 2, 3, 4), \tag{1.139}$$

or, in the matrix form,

$$\mathfrak{B} = \begin{bmatrix} W_{11} & W_{12} & W_{13} & Q_1 \\ W_{21} & W_{22} & W_{23} & Q_2 \\ W_{31} & W_{32} & W_{33} & Q_3 \\ \Sigma_1 & \Sigma_2 & \Sigma_3 & H \end{bmatrix}.$$

We next note that

$$\mathfrak{B} \cdot \nabla \equiv \frac{\partial B_{ij}}{\partial x_j} = \frac{\partial^2 u_k}{\partial x_j \partial x_i} \frac{\partial \mathscr{L}}{\partial u_{k,j}} + \frac{\partial u_k}{\partial x_i} \frac{\partial}{\partial x_j} \frac{\partial \mathscr{L}}{\partial u_{k,j}} - \frac{\partial \mathscr{L}}{\partial x_i} \tag{1.140}$$

$$(i, j = 1, 2, 3, 4;\ k = 1, 2, 3).$$

However, because $\mathscr{L}$ depends on x_j only through the time and space derivatives of u_k,

$$\frac{\partial \mathscr{L}}{\partial x_i} = \frac{\partial \mathscr{L}}{\partial u_{k,j}} \frac{\partial^2 u_k}{\partial x_j \partial x_i}.$$

Also, from Eqs. (1.127) and (1.130) we find that Eq. (1.140) shrinks to

$$\mathfrak{B} \cdot \nabla \equiv \rho \frac{\partial u_k}{\partial x_i} F_k = \rho(\partial_i u_k) F_k \qquad (k = 1, 2, 3;\ i = 1, 2, 3, 4). \tag{1.141}$$

For $i = 4$,

$$\partial_4 u_k \equiv \frac{\partial \mathbf{u}}{\partial t}, \qquad \operatorname{div}(B_{41}, B_{42}, \mathrm{B}_{43}, B_{44}) = \operatorname{div} \boldsymbol{\Sigma} + \frac{\partial H}{\partial t},$$

which is the energy conservation equation, Eq. (1.124). For $i = 1, 2, 3$,

$$\partial_i u_k \equiv \nabla \mathbf{u}, \qquad \mathfrak{B} \cdot \nabla = \mathfrak{W} \cdot \nabla + \frac{\partial \mathbf{Q}}{\partial t},$$

namely

$$\mathfrak{W} \cdot \nabla + \frac{\partial \mathbf{Q}}{\partial t} = \rho(\nabla \mathbf{u}) \cdot \mathbf{F}. \tag{1.142}$$

Equation (1.142) is related to the Cauchy equation of motion in the following sense: The equation of motion renders the rate of change of momentum of an elementary material volume (particle) caused by the external forces and the forces that the wave exerts on the elastic medium; the relation in Eq. (1.142) gives the rate of change of momentum of the wave caused by the forces that the medium exerts on the wave (e.g., scattering) and the external forces.

EXAMPLE 1.3: Complex Notation of the Lagrangian and Field Tensors in Lossless Media

The energy conservation laws and the subsequent expressions that we have derived for the Lagrangian of a lossless elastic solid, Eqs. (1.119)–(1.124), are expressed in terms of instantaneous values of real entities. When the displacements are given as complex quantities, however, we always substitute the real part of the displacements into the energy integrals. This procedure may sometimes be rather tedious, especially when double dot products of tensors are involved. It can be greatly simplified for *time-harmonic displacements*, where calculations are made using the complex conjugate symbol. In order to gain full advantage of the simplifications provided by complex notation, therefore, it is important to recast Eqs. (1.119)–(1.124) without reconverting to real variables.

Consider, for example, two complex quantities z_1 and z_2 of the form

$$z_1 = Z_1 e^{i\omega t}, \qquad z_2 = Z_2 e^{i\omega t}, \tag{1.3.1}$$

where Z_1, Z_2 are independent of t. Then

$$\begin{aligned} \operatorname{Re} z_1 &= \tfrac{1}{2}(Z_1 e^{i\omega t} + Z_1^* e^{-i\omega t}), \\ \operatorname{Re} z_2 &= \tfrac{1}{2}(Z_2 e^{i\omega t} + Z_2^* e^{-i\omega t}). \end{aligned}$$

Integrating over a cycle, we get

$$\frac{1}{T}\int_0^T (\operatorname{Re} z_1)(\operatorname{Re} z_2)dt = \tfrac{1}{4}(Z_1 Z_2^* + Z_1^* Z_2) = \tfrac{1}{2}\operatorname{Re}(Z_1 Z_2^*). \tag{1.3.2}$$

Therefore the average value of the kinetic energy of a mass m associated with the displacement $z(t) = Ze^{i\omega t}$ is given by

$$\tfrac{1}{2}m\left[\frac{1}{T}\int_0^T (\operatorname{Re} \dot{z})^2\, dt\right] = \tfrac{1}{4}m\omega^2 zz^*. \tag{1.3.3}$$

We are thus led to the convention that the square of the magnitude of the displacement is twice the time average of the square of its real part. Assuming a harmonic time dependence and taking a time average, Eqs. (1.119) and (1.122) become

$$\langle H\rangle = \rho\frac{\omega^2}{4}(\mathbf{u}\cdot\mathbf{u}^*) + \tfrac{1}{4}(\mathfrak{T}:\mathfrak{E}^*), \tag{1.3.4}$$

$$\langle \boldsymbol{\Sigma}\rangle = \frac{i\omega}{4}[\mathbf{u}^*\cdot\mathfrak{T} - \mathbf{u}\cdot\mathfrak{T}^*], \tag{1.3.5}$$

$$\langle \mathscr{L}\rangle = \tfrac{1}{2}\rho\omega^2(\mathbf{u}\cdot\mathbf{u}^*) - \langle H\rangle. \tag{1.3.6}$$

If the time dependence is arbitrary and the spatial part of $\mathbf{u}$ is still complex,

$$\mathscr{L} = \rho\left(\frac{\partial \mathbf{u}}{\partial t}\cdot\frac{\partial \mathbf{u}^*}{\partial t}\right) - (\mathfrak{T}:\mathfrak{E}^*). \tag{1.3.7}$$

Because the real and imaginary parts of $\dot{\mathbf{u}}$ will be varied independently in the variational process used to obtain the Lagrange equations of motion, we can consider $\dot{\mathbf{u}}$ and $\dot{\mathbf{u}}^*$ as being varied independently. Therefore, the variation of the integral of $\mathscr{L}$ gives rise to two equations, one for $\dot{\mathbf{u}}$ and one for $\dot{\mathbf{u}}^*$

$$\frac{\partial}{\partial x_\mu}\frac{\partial \mathscr{L}}{\partial(u_{k,\mu})} = \rho F_k, \qquad \frac{\partial}{\partial x_\mu}\frac{\partial \mathscr{L}}{\partial(u^*_{k,\mu})} = \rho F_k, \qquad k = 1, 2, 3, \qquad \mu = 1, 2, 3, 4. \tag{1.3.8}$$

Because of the symmetry of $\mathscr{L}$ in $\dot{\mathbf{u}}$ and $\dot{\mathbf{u}}^*$, these two equations are identical.

The form of the energy–momentum tensor ought to be modified because second-order terms must be real. Therefore

$$B_{44} = \dot{\mathbf{u}} \cdot \frac{\partial \mathscr{L}}{\partial \dot{\mathbf{u}}} + \dot{\mathbf{u}}^* \cdot \frac{\partial \mathscr{L}}{\partial \dot{\mathbf{u}}^*} - \mathscr{L} = H \tag{1.3.9}$$

and in general

$$B_{ij} = \sum_{k=1}^{3} u_{k,i}\frac{\partial \mathscr{L}}{\partial u_{k,j}} + \sum_{k=1}^{3} u^*_{k,i}\frac{\partial \mathscr{L}}{\partial u^*_{k,j}} - \mathscr{L}\delta_{ij}, \qquad i, j = 1, 2, 3, 4. \tag{1.3.10}$$

Note that the continuity equations in a force-free medium

$$\frac{\partial H}{\partial t} + \operatorname{div} \mathbf{\Sigma} = 0, \qquad \frac{\partial \mathbf{Q}}{\partial t} + \mathfrak{W} \cdot \nabla = 0$$

remain the same as before.

EXAMPLE 1.4: Anisotropic Solid

Consider the general case where the tensor of elastic moduli $\mathbf{C}$ has 21 independent elements. The only allowed symmetries are

$$C_{ijkl} = C_{klij} = C_{ijlk} = C_{jikl}. \tag{1.4.1}$$

It then follows that the equation of motion is

$$\rho\frac{\partial^2 \mathbf{u}}{\partial t^2} = \operatorname{div}[\mathbf{C} : \nabla \mathbf{u}] \tag{1.4.2}$$

which is equivalent to Eq. (1.103). Note that, in contradistinction to the isotropic case, it is not always possible to separate pure transverse and pure compressional waves.

Parallel to the isotropic case, however, we may define the basic field tensors in the following way:

$$\text{Lagrangian density } \mathscr{L} = \tfrac{1}{2}\rho\left(\frac{\partial \mathbf{u}}{\partial t}\right)^2 - \tfrac{1}{2}(\nabla \mathbf{u} : \mathbf{C} : \nabla \mathbf{u}), \tag{1.4.3}$$

$$\text{Energy-flux density } \mathbf{\Sigma} = -\dot{\mathbf{u}} \cdot \{\mathbf{C} : \nabla \mathbf{u}\}, \tag{1.4.4}$$

$$\text{Wave force dyadic } \mathfrak{W} = -(\nabla \mathbf{u}) \cdot \{\mathbf{C} : \nabla \mathbf{u}\} - \mathscr{L}\mathfrak{I}, \tag{1.4.5}$$

$$\text{Wave momentum density } \mathbf{Q} = \rho(\nabla \mathbf{u}) \cdot \left(\frac{\partial \mathbf{u}}{\partial t}\right), \tag{1.4.6}$$

$$B_{44} = \tfrac{1}{2}\rho\left(\frac{\partial \mathbf{u}}{\partial t}\right)^2 + \tfrac{1}{2}(\nabla \mathbf{u} : \mathbf{C} : \nabla \mathbf{u}). \tag{1.4.7}$$

1.6. One-Dimensional Approximations

Although we are mostly interested in three-dimensional boundary value problems, it is sometimes desirable to reduce the number of dependent and independent variables of the deformation field by making certain simplifying assumptions.

The simplest physical regime is that in which body forces and stresses depend upon a single Cartesian coordinate, say x_1. We then have

$$\varepsilon_{ij} = \begin{bmatrix} \dfrac{\partial u_1}{\partial x_1} & \dfrac{1}{2}\dfrac{\partial u_2}{\partial x_1} & \dfrac{1}{2}\dfrac{\partial u_3}{\partial x_1} \\ \dfrac{1}{2}\dfrac{\partial u_2}{\partial x_1} & 0 & 0 \\ \dfrac{1}{2}\dfrac{\partial u_3}{\partial x_1} & 0 & 0 \end{bmatrix},$$

$$\tau_{ij} = \begin{bmatrix} (\lambda + 2\mu)\dfrac{\partial u_1}{\partial x_1} & \mu\dfrac{\partial u_2}{\partial x_1} & \mu\dfrac{\partial u_3}{\partial x_1} \\ \mu\dfrac{\partial u_2}{\partial x_1} & \lambda\dfrac{\partial u_1}{\partial x_1} & 0 \\ \mu\dfrac{\partial u_3}{\partial x_1} & 0 & \lambda\dfrac{\partial u_1}{\partial x_1} \end{bmatrix}. \tag{1.143}$$

The Cauchy equation of motion, Eq. (1.100), can now be written in the form

$$\frac{\partial \tau_{11}}{\partial x_1} + \rho F_1 = \rho\frac{\partial^2 u_1}{\partial t^2},$$
$$\frac{\partial \tau_{12}}{\partial x_1} + \rho F_2 = \rho\frac{\partial^2 u_2}{\partial t^2}, \tag{1.144}$$
$$\frac{\partial \tau_{13}}{\partial x_1} + \rho F_3 = \rho\frac{\partial^2 u_3}{\partial t^2}.$$

If, in addition, $u_2 = u_3 = 0$, we have the case of *longitudinal strain* in which

$$\mathfrak{E} = \mathbf{e}_1\mathbf{e}_1\frac{\partial u_1}{\partial x_1}, \qquad \mathfrak{T} = (\lambda\mathfrak{J} + 2\mu\mathbf{e}_1\mathbf{e}_1)\frac{\partial u_1}{\partial x_1},$$
$$\frac{\partial}{\partial x_1}\left[(\lambda + 2\mu)\frac{\partial u_1}{\partial x_1}\right] + \rho F_1 = \rho\frac{\partial^2 u_1}{\partial t^2}, \qquad F_2 = F_3 = 0. \tag{1.145}$$

For a homogeneous medium, the equation of motion reduces to

$$\alpha^2\frac{\partial^2 u_1}{\partial x_1^2} + F_1 = \frac{\partial^2 u_1}{\partial t^2}, \qquad \alpha^2 = \frac{\lambda + 2\mu}{\rho}. \tag{1.146}$$

If, in contrast, we require that

$$\mathfrak{T} = \tau_{11}\mathbf{e}_1\mathbf{e}_1, \tag{1.147}$$

we obtain from Eq. (1.72)

$$\mathfrak{E} = \frac{\tau_{11}}{2\mu}\left(\mathbf{e}_1\mathbf{e}_1 - \frac{\lambda}{3\lambda + 2\mu}\mathfrak{J}\right). \tag{1.148}$$

This is called a state of *longitudinal stress*. The corresponding equation of motion may be written in the form

$$\gamma^2 \frac{\partial^2 u_1}{\partial x_1^2} + F_1 = \frac{\partial^2 u_1}{\partial t^2}, \tag{1.149}$$

where

$$\gamma^2 = \frac{Y}{\rho}, \quad Y = \frac{\mu(3\lambda + 2\mu)}{\lambda + \mu}.$$

Finally, if $\mathbf{u} = u_2\mathbf{e}_2 + u_3\mathbf{e}_3$, we have

$$\mathfrak{T} = \mu(\mathbf{e}_1\mathbf{e}_2 + \mathbf{e}_2\mathbf{e}_1)\frac{\partial u_2}{\partial x_1} + \mu(\mathbf{e}_1\mathbf{e}_3 + \mathbf{e}_3\mathbf{e}_1)\frac{\partial u_3}{\partial x_1},$$

$$F_1 = 0, \qquad \beta^2 \frac{\partial^2 u_2}{\partial x_1^2} + F_2 = \frac{\partial^2 u_2}{\partial t^2}, \qquad \beta^2 \frac{\partial^2 u_3}{\partial x_1^2} + F_3 = \frac{\partial^2 u_3}{\partial t^2}, \qquad \beta^2 = \frac{\mu}{\rho}. \tag{1.150}$$

The entire motion is a *shear* confined to a plane normal to the x_1 axis and the two components of the motion are uncoupled.

1.7. Two-Dimensional Approximations

It is sometimes permissible (e.g., in cases of plane symmetry) to assume that the components of the displacements are independent of a single Cartesian coordinate, say x_3. In that case

$$\begin{aligned}
\tau_{11} &= \lambda\left(\frac{\partial u_1}{\partial x_1} + \frac{\partial u_2}{\partial x_2}\right) + 2\mu\frac{\partial u_1}{\partial x_1}, \\
\tau_{22} &= \lambda\left(\frac{\partial u_1}{\partial x_1} + \frac{\partial u_2}{\partial x_2}\right) + 2\mu\frac{\partial u_2}{\partial x_2}, \\
\tau_{12} &= \mu\left(\frac{\partial u_1}{\partial x_2} + \frac{\partial u_2}{\partial x_1}\right), \\
\tau_{13} &= \mu\frac{\partial u_3}{\partial x_1}, \qquad \tau_{23} = \mu\frac{\partial u_3}{\partial x_2}, \qquad \tau_{33} = \lambda\left(\frac{\partial u_1}{\partial x_1} + \frac{\partial u_2}{\partial x_2}\right) = \sigma(\tau_{11} + \tau_{22}).
\end{aligned} \tag{1.151}$$

The equation of motion, Eq. (1.100), then yields

$$\frac{\partial \tau_{11}}{\partial x_1} + \frac{\partial \tau_{21}}{\partial x_2} + \rho F_1 = \rho \frac{\partial^2 u_1}{\partial t^2}, \tag{1.152}$$

$$\frac{\partial \tau_{12}}{\partial x_1} + \frac{\partial \tau_{22}}{\partial x_2} + \rho F_2 = \rho \frac{\partial^2 u_2}{\partial t^2}, \tag{1.153}$$

$$\frac{\partial \tau_{13}}{\partial x_1} + \frac{\partial \tau_{23}}{\partial x_2} + \rho F_3 = \rho \frac{\partial^2 u_3}{\partial t^2}. \tag{1.154}$$

On substitution from Eq. (1.151), Eq. (1.154) becomes an equation in u_3 alone

$$\beta^2 \nabla_1^2 u_3 + F_3 = \frac{\partial^2 u_3}{\partial t^2}, \tag{1.155}$$

$$u_3 = u_3(x_1, x_2; t), \qquad \nabla_1^2 = \frac{\partial^2}{\partial x_1^2} + \frac{\partial^2}{\partial x_2^2}. \tag{1.156}$$

Similarly, Eqs. (1.152) and (1.153) lead us to two simultaneous equations in u_1 and u_2 which do not involve u_3. The motion represented by u_3 is known as *antiplane strain* (or *antiplane shear*), whereas the motion represented by (u_1, u_2) is known as *plane strain*. It is obvious that in both cases, the body forces and stresses are independent of x_3.

1.7.1. Plane Strain

For a body in plane strain parallel to the $x_1 x_2$ plane

$$u_1 = u_1(x_1, x_2; t), \qquad u_2 = u_2(x_1, x_2; t), \qquad u_3 = 0. \tag{1.157}$$

Equations (1.152)–(1.154) yield

$$F_1 = F_1(x_1, x_2), \qquad F_2 = F_2(x_1, x_2), \qquad F_3 = 0. \tag{1.158}$$

Therefore, the entire deformation field is truly two dimensional

$$\varepsilon_{ij} = \begin{bmatrix} \varepsilon_{11} & \varepsilon_{12} & 0 \\ \varepsilon_{21} & \varepsilon_{22} & 0 \\ 0 & 0 & 0 \end{bmatrix}, \qquad \tau_{ij} = \begin{bmatrix} \tau_{11} & \tau_{12} & 0 \\ \tau_{21} & \tau_{22} & 0 \\ 0 & 0 & \sigma(\tau_{11} + \tau_{22}) \end{bmatrix},$$

$$\tau_{11} = (\lambda + 2\mu)\varepsilon_{11} + \lambda\varepsilon_{22}, \qquad \tau_{12} = 2\mu\varepsilon_{12},$$

$$\tau_{22} = \lambda\varepsilon_{11} + (\lambda + 2\mu)\varepsilon_{22},$$

$$\tau_{33} = \lambda(\varepsilon_{11} + \varepsilon_{22}) = \sigma(\tau_{11} + \tau_{22}).$$

The equation of motion, Eq. (1.109), may be written in the form

$$\mu \nabla_1^2 u_i + (\lambda + \mu) \frac{\partial}{\partial x_i}\left(\frac{\partial u_1}{\partial x_1} + \frac{\partial u_2}{\partial x_2}\right) + \rho F_i = \rho \frac{\partial^2 u_i}{\partial t^2} \qquad (i = 1, 2). \tag{1.159}$$

If we put

$$u_1 = \frac{\partial \Phi}{\partial x_1} + \frac{\partial \Psi}{\partial x_2}, \qquad u_2 = \frac{\partial \Phi}{\partial x_2} - \frac{\partial \Psi}{\partial x_1},$$

Eqs. (1.159) show that Φ and Ψ satisfy the equations

$$\alpha^2 \nabla_1^2 \Phi = \frac{\partial^2 \Phi}{\partial t^2}, \qquad \beta^2 \nabla_1^2 \Psi = \frac{\partial^2 \Psi}{\partial t^2}, \tag{1.160}$$

assuming $F_1 = F_2 = 0$.

Physically, plane strain is applicable whenever the dimensions of the body in the x_3 direction are large compared to its dimensions in the x_1, x_2 directions and when such a body is under the action of *body and surface forces* perpendicular to the x_3 direction and independent of it. Under these conditions, it is safe to assume that there is no motion in the x_3 direction and that the displacement at each cross section perpendicular to x_3 will not depend on it. This is the *plane-strain hypothesis*, which leads to a truly two-dimensional theory.

1.7.2. Plane Stress

A body is said to be in a state of *plane stress* parallel to the x_1x_2 plane if

$$\tau_{13} = \tau_{23} = \tau_{33} = 0. \tag{1.161}$$

If we put $\tau_{33} = 0$, Eq. (1.68) yields

$$\frac{\partial u_3}{\partial x_3} = -\frac{\lambda}{\lambda + 2\mu}\left(\frac{\partial u_1}{\partial x_1} + \frac{\partial u_2}{\partial x_2}\right).$$

Therefore

$$\operatorname{div} \mathbf{u} = \frac{2\mu}{\lambda + 2\mu}\left(\frac{\partial u_1}{\partial x_1} + \frac{\partial u_2}{\partial x_2}\right)$$

and the stress–displacement relations may be written in the form

$$\begin{aligned} \tau_{11} &= \bar{\lambda}\left(\frac{\partial u_1}{\partial x_1} + \frac{\partial u_2}{\partial x_2}\right) + 2\mu \frac{\partial u_1}{\partial x_1}, \\ \tau_{22} &= \bar{\lambda}\left(\frac{\partial u_1}{\partial x_1} + \frac{\partial u_2}{\partial x_2}\right) + 2\mu \frac{\partial u_2}{\partial x_2}, \\ \tau_{12} &= \mu\left(\frac{\partial u_1}{\partial x_2} + \frac{\partial u_2}{\partial x_1}\right), \\ \bar{\lambda} &= \frac{2\lambda\mu}{\lambda + 2\mu}. \end{aligned} \tag{1.162}$$

The equations of motion, Eqs. (1.152) and (1.153), now yield

$$\mu \nabla_1^2 u_i + (\bar{\lambda} + \mu)\frac{\partial}{\partial x_i}\left(\frac{\partial u_1}{\partial x_1} + \frac{\partial u_2}{\partial x_2}\right) + \rho F_i = \rho \frac{\partial^2 u_i}{\partial t^2} \qquad (i = 1, 2). \tag{1.163}$$

Equations (1.162) and (1.163) coincide with the corresponding equations for plane strain if we replace $\bar{\lambda}$ by λ.

It may be noted that in the case of plane stress, the components u_1 and u_2 of the displacement may depend on x_3 and ε_{33} is not in general zero. Hence the problem is not truly two dimensional. We will show next how the problem can be made two dimensional.

1.7.3. Generalized Plane Stress

Consider a cylinder whose height $2h$ is small compared with the linear dimensions of its cross section (Fig. 1.6). Let the generators of the cylinder be parallel to the x_3 axis and let the plane ends of the cylinder be in the planes $x_3 = \pm h$. To simplify the field equations and at the same time render the problem two dimensional, we make the following assumptions:

1. The plane ends of the cylinder are free from the applied forces.
2. The forces acting on the curved surface lie in planes parallel to the middle plane and are symmetrically distributed with respect to this plane.
3. The component F_3 of the body forces vanishes and the components F_1 and F_2 are symmetrically distributed with respect to the middle plane.

It is obvious that the points of the middle plane will undergo no displacement in the direction of the x_3 axis and the mean value of u_3 with respect to the thickness of the cylinder will vanish; i.e.,

$$\bar{u}_3(x_1, x_2) = \frac{1}{2h} \int_{-h}^{h} u_3(x_1, x_2, x_3) dx_3 = 0. \tag{1.164}$$

Moreover, the mean values of u_1 and u_2 will be independent of x_3; i.e.,

$$\bar{u}_1 = \bar{u}_1(x_1, x_2), \qquad \bar{u}_2 = \bar{u}_2(x_1, x_2).$$

Because $F_3 = 0$ and because $\partial\tau_{31}/\partial x_1 = \partial\tau_{32}/\partial x_2 = 0$ at $x_3 = \pm h$, the third equilibrium equation renders $\partial\tau_{33}/\partial x_3 = 0$ at $x_3 = \pm h$. It means, therefore, that both τ_{33} and its normal derivative vanish on these boundaries.

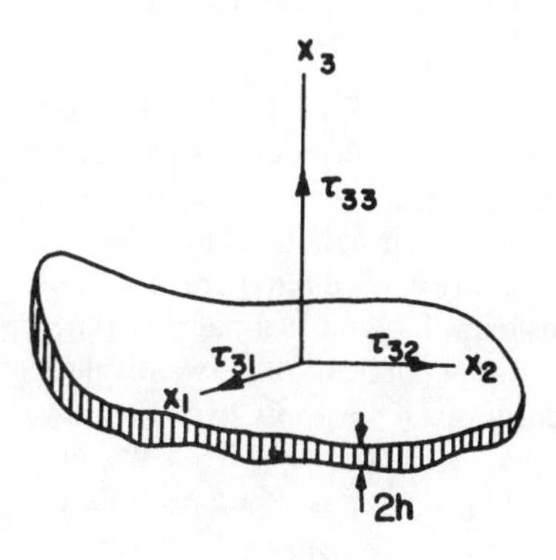

Figure 1.6. Generalized plane stress.

Because the cylinder is thin, we can take $\tau_{33} = 0$ everywhere. This makes the problem two dimensional. The equations of elasticity may now be expressed in terms of the mean values of the displacements and stresses:

$$\bar{\varepsilon}_{ij} = \begin{bmatrix} \bar{\varepsilon}_{11} & \bar{\varepsilon}_{12} & 0 \\ \bar{\varepsilon}_{21} & \bar{\varepsilon}_{22} & 0 \\ 0 & 0 & -\dfrac{\sigma}{1-\sigma}(\bar{\varepsilon}_{11} + \bar{\varepsilon}_{22}) \end{bmatrix}, \qquad \bar{\tau}_{ij} = \begin{bmatrix} \bar{\tau}_{11} & \bar{\tau}_{12} & 0 \\ \bar{\tau}_{21} & \bar{\tau}_{22} & 0 \\ 0 & 0 & 0 \end{bmatrix},$$

$$\begin{aligned} \bar{\tau}_{11} &= \bar{\lambda}\left(\frac{\partial \bar{u}_1}{\partial x_1} + \frac{\partial \bar{u}_2}{\partial x_2}\right) + 2\mu \frac{\partial \bar{u}_1}{\partial x_1}, \\ \bar{\tau}_{22} &= \bar{\lambda}\left(\frac{\partial \bar{u}_1}{\partial x_1} + \frac{\partial \bar{u}_2}{\partial x_2}\right) + 2\mu \frac{\partial \bar{u}_2}{\partial x_2}, \\ \bar{\tau}_{12} &= \mu\left(\frac{\partial \bar{u}_1}{\partial x_2} + \frac{\partial \bar{u}_2}{\partial x_1}\right), \end{aligned} \tag{1.165}$$

$$\mu \nabla_1^2 \bar{u}_i + (\bar{\lambda} + \mu)\frac{\partial}{\partial x_i}\left(\frac{\partial \bar{u}_1}{\partial x_1} + \frac{\partial \bar{u}_2}{\partial x_2}\right) + \rho \bar{F}_i = \rho \frac{\partial^2 \bar{u}_i}{\partial t^2} \qquad (i = 1, 2).$$

It may be noted that the mean values of the displacements and stresses for the generalized plane stress problem satisfy the same equations that govern the problem of plane strain, provided that the parameter λ is replaced by $\bar{\lambda} = 2\lambda\mu/(\lambda + 2\mu)$.

1.8. Representation of Finite Strains

In the theory of infinitesimal strain (Sect. 1.2) we have made two basic assumptions:

1. Second and higher powers of $|\delta \mathbf{r}|$ in the Taylor expansion of $\delta \mathbf{u}$ are neglected; i.e., the analysis is confined to a small neighborhood of a point.
2. The first derivatives of the displacements with respect to the coordinates are small as compared to unity. Therefore, products of these derivatives are negligible and deformations are considered to be infinitesimal.

The first assumption implies that at a given point P, $\nabla \mathbf{u}$ is fixed and $\delta \mathbf{r}'$ is a linear vector function of $\delta \mathbf{r}$. Viewed as a linear transformation, small plane segments in the neighborhood of P are carried into small plane segments near the image point P'. Straight lines are carried into straight lines, parallel lines are carried into parallel lines, and parallel planes are carried into parallel planes. Also, a small rectangular parallelepiped transforms into another parallelepiped.

The second assumption implies that the order of application of two displacement fields has no effect on the final configuration; if we apply two successive deformations, $d\mathbf{r}' = d\mathbf{r} \cdot (\mathfrak{I} + \nabla \mathbf{u})$ followed by $d\mathbf{r}'' = d\mathbf{r}' \cdot (\mathfrak{I} + \nabla \mathbf{u}')$, the final result will be $d\mathbf{r}'' = d\mathbf{r} \cdot [\mathfrak{I} + \nabla(\mathbf{u} + \mathbf{u}')]$. This property is recognized as the *principle of superposition* and is the underlying principle of the linear theory of elasticity.

We shall now remove the second restriction, allowing for larger deformations at the cost of abandoning the principle of superposition. Equation (1.37) will then read

$$ds'^2 - ds^2 = |d\mathbf{r}'|^2 - |d\mathbf{r}|^2 = d\mathbf{r} \cdot 2\mathfrak{D}^L \cdot d\mathbf{r}, \tag{1.166}$$

where

$$2\mathfrak{D}^L = 2\mathfrak{E} + \nabla\mathbf{u} \cdot \mathbf{u}\nabla \tag{1.167}$$

is the symmetric *Lagrange* (*or Green*) *deformation tensor.* The Cartesian components of $[\nabla\mathbf{u} \cdot \mathbf{u}\nabla]$ are

$$\begin{aligned}
(11) &= \left(\frac{\partial u}{\partial x_1}\right)^2 + \left(\frac{\partial v}{\partial x_1}\right)^2 + \left(\frac{\partial w}{\partial x_1}\right)^2,\\
(22) &= \left(\frac{\partial u}{\partial x_2}\right)^2 + \left(\frac{\partial v}{\partial x_2}\right)^2 + \left(\frac{\partial w}{\partial x_2}\right)^2,\\
(33) &= \left(\frac{\partial u}{\partial x_3}\right)^2 + \left(\frac{\partial v}{\partial x_3}\right)^2 + \left(\frac{\partial w}{\partial x_3}\right)^2,
\end{aligned} \tag{1.168}$$

$$\begin{aligned}
(12) = (21) &= \left(\frac{\partial u}{\partial x_1}\right)\left(\frac{\partial u}{\partial x_2}\right) + \left(\frac{\partial v}{\partial x_1}\right)\left(\frac{\partial v}{\partial x_2}\right) + \left(\frac{\partial w}{\partial x_1}\right)\left(\frac{\partial w}{\partial x_2}\right),\\
(13) = (31) &= \left(\frac{\partial u}{\partial x_1}\right)\left(\frac{\partial u}{\partial x_3}\right) + \left(\frac{\partial v}{\partial x_1}\right)\left(\frac{\partial v}{\partial x_3}\right) + \left(\frac{\partial w}{\partial x_1}\right)\left(\frac{\partial w}{\partial x_3}\right),\\
(23) = (32) &= \left(\frac{\partial u}{\partial x_2}\right)\left(\frac{\partial u}{\partial x_3}\right) + \left(\frac{\partial v}{\partial x_2}\right)\left(\frac{\partial v}{\partial x_3}\right) + \left(\frac{\partial w}{\partial x_2}\right)\left(\frac{\partial w}{\partial x_3}\right).
\end{aligned} \tag{1.169}$$

To obtain the geometric interpretation of the components of $\mathfrak{D}^L$ we follow the method used earlier for infinitesimal strains. From Eq. (1.166) it follows that

$$\left(\frac{ds'}{ds}\right)^2 = 1 + 2\mathfrak{D}^L : \frac{d\mathbf{r}}{ds}\frac{d\mathbf{r}}{ds}.$$

When a unit vector is defined as $\mathbf{a}(l, m, n) = d\mathbf{r}/ds$, the *elongation in the* (l, m, n) *direction* is

$$e = \frac{ds' - ds}{ds} = \frac{ds'}{ds} - 1 = [1 + 2\mathfrak{D}^L : \mathbf{aa}]^{1/2} - 1. \tag{1.170}$$

For small $\mathfrak{D}^L$, a Taylor expansion of the square root in Eq. (1.170) yields $e = \mathfrak{E} : \mathbf{aa}$ as in Eq. (1.38).

We next consider the change of angle between two given directions in the unstrained state at P (Fig. 1.7a). Let these directions be $\mathbf{a}_1(l_1, m_1, n_1)$ and $\mathbf{a}_2(l_2, m_2, n_2)$ with an angle θ between them. After the deformation, the directions change to $\mathbf{a}_1'(l_1', m_1', n_1')$ and $\mathbf{a}_2'(l_2', m_2', n_2')$ and θ changes to θ'. In complete analogy to the linear case, we define four increments $d\mathbf{r}_1$, $d\mathbf{r}_2$, $d\mathbf{r}_1'$, $d\mathbf{r}_2'$, in the

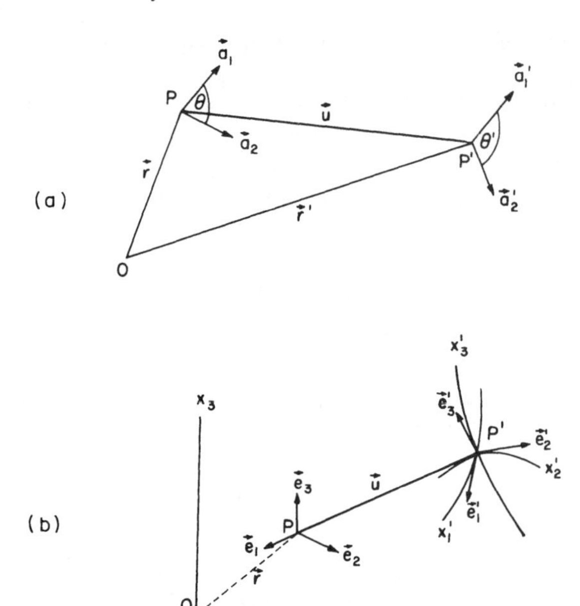

Figure 1.7. Finite deformation. (a) A change of angle between two given directions in the unstrained state vs. the strained state; (b) the deformation as a transformation of the coordinate axes.

corresponding directions $\mathbf{a}_1 = d\mathbf{r}_1/ds$, $\mathbf{a}_2 = d\mathbf{r}_2/ds$, $\mathbf{a}'_1 = d\mathbf{r}'_1/ds'$, and $\mathbf{a}'_2 = d\mathbf{r}'_2/ds'$. Then

$$d\mathbf{r}'_1 \cdot d\mathbf{r}'_2 = d\mathbf{r}_1 d\mathbf{r}_2 : (\mathfrak{I} + 2\mathfrak{D}^L). \tag{1.171}$$

However, because $|d\mathbf{r}'_1| = (1 + e_1)|d\mathbf{r}_1|$, $|d\mathbf{r}'_2| = (1 + e_2)|d\mathbf{r}_2|$, we have

$$\cos\theta' = \frac{\mathbf{a}'_1 \cdot \mathbf{a}'_2}{|\mathbf{a}'_1||\mathbf{a}'_2|} = \frac{\mathbf{a}_1\mathbf{a}_2 : (\mathfrak{I} + 2\mathfrak{D}^L)}{(1 + e_1)(1 + e_2)} = \frac{\cos\theta + (2\mathfrak{D}^L : \mathbf{a}_1\mathbf{a}_2)}{(1 + e_1)(1 + e_2)}$$

$$= \frac{\mathbf{a}_1\mathbf{a}_2 : (\mathfrak{I} + 2\mathfrak{D}^L)}{[1 + 2\mathfrak{D}^L : \mathbf{a}_1\mathbf{a}_1]^{1/2}[1 + 2\mathfrak{D}^L : \mathbf{a}_2\mathbf{a}_2]^{1/2}}. \tag{1.172}$$

Consider the special case $\mathbf{a}_1 = (1, 0, 0)$, $\mathbf{a}_2 = (0, 1, 0)$ and define $\phi = \pi/2 - \theta'$. Because $\theta = \pi/2$, the change in angle between two orthogonal directions in the undeformed state will be given by the angle $\phi = \sin^{-1}\gamma_{12}$, where

$$\gamma_{12} = \frac{2D_{12}}{(1 + 2D_{11})^{1/2}(1 + 2D_{22})^{1/2}}. \tag{1.173}$$

For small strains, $\gamma_{12} \simeq \phi$. Similar results are obtained for γ_{13} and γ_{23}.

Having determined the change of angle between a pair of orthogonal axes, we wish to consider the transformation of an orthogonal base of unit vectors $(\mathbf{e}_1, \mathbf{e}_2, \mathbf{e}_3)$ as a result of the deformation. In this sense the deformation may be viewed as transformation of the coordinate axes. A point P in a fixed Cartesian system $O(x_1, x_2, x_3)$ is displaced to a point P' (Fig. 1.7b). Prior to the deformation at P, the unit vectors at P $(\mathbf{e}_1, \mathbf{e}_2, \mathbf{e}_3)$ are parallel to the unit vectors of the coordinate axes at O. Because of the deformation, straight coordinate lines at P are transformed into curved coordinate lines at P'. Drawing tangents to these lines at P', we define a base of unit vectors that in general need not be orthogonal. We wish to relate the "new" unit vectors at P' to the "old" ones at P. To this end we equate two different representations of $d\mathbf{r}'$, namely,

$$d\mathbf{r}' = \frac{\partial \mathbf{r}'}{\partial x_1} dx_1 + \frac{\partial \mathbf{r}'}{\partial x_2} dx_2 + \frac{\partial \mathbf{r}'}{\partial x_3} dx_3 \tag{1.174}$$

and [cf. Eq. (1.32)]

$$d\mathbf{r}' = d\mathbf{r} \cdot (\mathfrak{I} + \nabla \mathbf{u}) = d\mathbf{r} \cdot [\mathfrak{I} + \mathfrak{E} - \mathfrak{I} \times \boldsymbol{\omega}], \qquad \boldsymbol{\omega} = \tfrac{1}{2} \operatorname{curl} \mathbf{u}. \tag{1.175}$$

Hence

$$\begin{aligned} \frac{\partial \mathbf{r}'}{\partial x_1} &= \{1 + \varepsilon_{11}, \varepsilon_{12} + \omega_3, \varepsilon_{13} - \omega_2\}, \\ \frac{\partial \mathbf{r}'}{\partial x_2} &= \{\varepsilon_{12} - \omega_3, 1 + \varepsilon_{22}, \varepsilon_{23} + \omega_1\}, \\ \frac{\partial \mathbf{r}'}{\partial x_3} &= \{\varepsilon_{13} + \omega_2, \varepsilon_{23} - \omega_1, 1 + \varepsilon_{33}\}. \end{aligned} \tag{1.176}$$

The scale factors of this transformation (App. B) are then found to be

$$\begin{aligned} h_1 &= \left| \frac{\partial \mathbf{r}'}{\partial x_1} \right| = (1 + 2D_{11})^{1/2}, \\ h_2 &= \left| \frac{\partial \mathbf{r}'}{\partial x_2} \right| = (1 + 2D_{22})^{1/2}, \\ h_3 &= \left| \frac{\partial \mathbf{r}'}{\partial x_3} \right| = (1 + 2D_{33})^{1/2}. \end{aligned} \tag{1.177}$$

In accordance with these results

$$\begin{aligned} \mathbf{e}_1' &= \frac{1 + \varepsilon_{11}}{(1 + 2D_{11})^{1/2}} \mathbf{e}_1 + \frac{\varepsilon_{12} + \omega_3}{(1 + 2D_{11})^{1/2}} \mathbf{e}_2 + \frac{\varepsilon_{13} - \omega_2}{(1 + 2D_{11})^{1/2}} \mathbf{e}_3, \\ \mathbf{e}_2' &= \frac{\varepsilon_{12} - \omega_3}{(1 + 2D_{22})^{1/2}} \mathbf{e}_1 + \frac{1 + \varepsilon_{22}}{(1 + 2D_{22})^{1/2}} \mathbf{e}_2 + \frac{\varepsilon_{23} + \omega_1}{(1 + 2D_{22})^{1/2}} \mathbf{e}_3, \\ \mathbf{e}_3' &= \frac{\varepsilon_{13} + \omega_2}{(1 + 2D_{33})^{1/2}} \mathbf{e}_1 + \frac{\varepsilon_{23} - \omega_1}{(1 + 2D_{33})^{1/2}} \mathbf{e}_2 + \frac{1 + \varepsilon_{33}}{(1 + 2D_{33})^{1/2}} \mathbf{e}_3. \end{aligned} \tag{1.178}$$

We notice that both rotation and shear partake in the deformation of the axes.

Finally, we evaluate the *cubical dilatation*. To this end we consider an element of volume in the form of rectangular parallelepiped around P in the undeformed state. Its edges are the vectors $dx_1\mathbf{e}_1$, $dx_2\mathbf{e}_2$ and $dx_3\mathbf{e}_3$ and its volume is $dv = dx_1dx_2dx_3$, where $\mathbf{e}_1 = \partial\mathbf{r}/\partial x_1$, $\mathbf{e}_2 = \partial\mathbf{r}/\partial x_2$, $\mathbf{e}_3 = \partial\mathbf{r}/\partial x_3$ (see App. B). The element of volume in the deformed body is a parallelepiped with edges $dx_1(\partial\mathbf{r}'/\partial x_1)$, $dx_2(\partial\mathbf{r}'/\partial x_2)$, $dx_3(\partial\mathbf{r}'/\partial x_3)$. Using the triple scalar product, we find the deformed element of volume to be

$$dv' = \left(\frac{\partial\mathbf{r}'}{\partial x_1}dx_1\right)\cdot\left(\frac{\partial\mathbf{r}'}{\partial x_2}dx_2\right)\times\left(\frac{\partial\mathbf{r}'}{\partial x_3}dx_3\right). \tag{1.179}$$

Hence

$$\left(\frac{dv'}{dv}\right)^2 = \left\|\frac{\partial\mathbf{r}'}{\partial\mathbf{r}}\cdot\left(\frac{\partial\mathbf{r}'}{\partial\mathbf{r}}\right)^{\sim}\right\| = \|(\mathfrak{I} + \nabla\mathbf{u})\cdot(\mathfrak{I} + \mathbf{u}\nabla)\| = \|\mathfrak{I} + 2\mathfrak{D}^L\|. \tag{1.180}$$

Let J_1, J_2, J_3 denote the invariants of the dyadic $\mathfrak{D}^L$ (see App. A). It can then be shown that

$$\|\mathfrak{I} + 2\mathfrak{D}^L\| = 1 + 2J_1 + 4J_2 + 8J_3, \tag{1.181}$$

where

$$\begin{aligned} J_1 &= D_{11} + D_{22} + D_{33}, \\ J_2 &= (D_{11}D_{22} - D_{12}^2) + (D_{11}D_{33} - D_{13}^2) + (D_{22}D_{33} - D_{23}^2), \\ J_3 &= \|\mathfrak{D}^L\|. \end{aligned} \tag{1.182}$$

Therefore

$$\text{Cubical dilatation} = \frac{dv' - dv}{dv} = \left(\frac{dv'}{dv}\right) - 1 = [1 + 2J_1 + 4J_2 + 8J_3]^{1/2} - 1. \tag{1.183}$$

For small strains, the invariants J_2 and J_3 are of the orders D_{11}^2 and D_{11}^3, respectively. These invariants can be neglected with respect to J_1, which is of the order D_{11} only.

Bibliography

Cauchy AL (1823) Recherches sur l'équilibre et le mouvement intérieur des corps solides et fluides, élastiques ou non élastiques. Presented to the Paris Academy, 30 September, 1822. Bull Philomat, January 1823: 9–13.

Cauchy AL (1827a) De la pression dans les fluides. Exercises de Mathématiques 2: 23–24.

Cauchy AL (1827b) De la pression ou tension dans un corps solide. Exercices de Mathématiques 2: 41–56.

Cauchy AL (1827c) Sur la condensation et la dilatation des corps solides. Exercices de Mathématiques 2: 60–69.

Cauchy AL (1827d) Sur les relations qui existent, dans l'état d'équilibre d'un corps solide ou fluide, entre les pressions ou tensions et les forces accélératrices. Exercices de Mathématiques 2: 108–111.

Cauchy AL (1827e) Sur les équations qui expriment les conditions d'equilibre, ou les lois du mouvement interier d'un corps solide, élastique, ou non élastique. Exercices de Mathématiques 2: 160–187.

Lamé MG (1852) Leçons sur la Théorie Mathématique de l'Elasticité des Corps Solides. p. 335.

Lamé MG, Clapeyron BPE (1833) Mémoire sur l'équilibre de corps solides homogénes. Reported to the Paris Academy by Poinsot and Navier, 29 September, 1828. Mémoires Présentées par divers Savans 4: 465–562.

Navier L (1827) Mémoire sur les lois d'équilibre et du mouvement des corps solides elastiques. Read to the Paris Academy, 14 May, 1821. Mémoires de l'Institut 3: 375–393. (This memoir is the foundation of the modern theory of elastodynamics.)

Poisson SD (1827) Note sur les vibrations des corps sonores. Ann. Chim. Phys. 36: 86–93.

Poisson SD (1829) Memoire sur l'equilibre et le mouvement des corps elastiques. Read to the Paris Academy, 14 April, 1828. Mémoires de l'Academe 8: 357–570.

Todhunter I, Pearson K (1960) A History of the Theory of Elasticity and the Strength of Materials, Vol I. Dover, New York.

CHAPTER 2

Waves in Infinite Media

The enormous and the minute are interchangeable manifestations of the eternal.

(*William Blake*)

2.1. Elementary Solutions of the Wave Equation

A *wave* is a disturbance, usually periodic, that travels with finite velocity through a medium. Sound waves, water waves, and electromagnetic waves are some examples. All wave motions have two important characteristics in common: First, energy is propagated to distant points and, second, the disturbance travels through the medium without giving the medium as a whole any permanent displacement. Each successive particle of the medium performs a motion similar to its predecessor's but later in time, and returns to its origin. Whatever the nature of the medium that transmits the waves, be it air, a stretched string, a liquid, or an electrical cable, these two properties enable us to relate all wave motions together. Indeed, many types of waves are governed by a second-order linear partial differential equation

$$\nabla^2 \Psi = \frac{1}{c^2} \frac{\partial^2 \Psi}{\partial t^2}, \tag{2.1}$$

where $\Psi(\mathbf{r}, t)$ represents the disturbance traveling with the velocity c. Equation (2.1) is known as the *wave equation.*

Some elementary solutions of Eq. (2.1) can be easily derived. Let us change the independent variables in Eq. (2.1) through the relations

$$\begin{aligned} u &= t - \frac{1}{c}(lx + my + nz), \\ v &= t + \frac{1}{c}(lx + my + nz), \end{aligned} \qquad (l^2 + m^2 + n^2 = 1). \tag{2.2}$$

Noting that

$$\frac{\partial}{\partial t} = \left(\frac{\partial}{\partial v} + \frac{\partial}{\partial u}\right), \qquad \frac{\partial}{\partial x} = \frac{l}{c}\left(\frac{\partial}{\partial v} - \frac{\partial}{\partial u}\right), \text{ etc.,}$$

we can transform Eq. (2.1) to

$$4\frac{\partial^2\Psi}{\partial u \partial v} = 0. \tag{2.3}$$

Integrating, we find

$$\Psi = f(u) + g(v) = f\left(t - \frac{lx + my + nz}{c}\right) + g\left(t + \frac{lx + my + nz}{c}\right), \tag{2.4}$$

where f and g are arbitrary, twice-differentiable functions. This is known as *D'Alembert's solution* of the wave equation. In Eq. (2.4), at a given time t, Ψ is constant for all x, y, z satisfying $lx + my + nz =$ const., which is the equation of a plane with (l, m, n) as the direction cosines of the normal. Hence these waves are known as *plane waves* and the planes $lx + my + nz =$ const. as *wave fronts.* When $\mathbf{p} = l\mathbf{e}_x + m\mathbf{e}_y + n\mathbf{e}_z$, Eq. (2.4) may be written as

$$\Psi = f\left(t - \frac{\mathbf{p}\cdot\mathbf{r}}{c}\right) + g\left(t + \frac{\mathbf{p}\cdot\mathbf{r}}{c}\right).$$

If we consider the one-dimensional counterpart of Eq. (2.1)

$$\frac{\partial^2\Psi}{\partial x^2} = \frac{1}{c^2}\frac{\partial^2\Psi}{\partial t^2}, \tag{2.5}$$

D'Alembert's solution becomes

$$\Psi = f\left(t - \frac{x}{c}\right) + g\left(t + \frac{x}{c}\right). \tag{2.6}$$

Let us first assume that

$$\Psi(x, t) = f\left(t - \frac{x}{c}\right).$$

Because

$$\Psi(x + c\tau, t + \tau) = f\left(t + \tau - \frac{x + c\tau}{c}\right) = f\left(t - \frac{x}{c}\right) = \Psi(x, t),$$

the value of Ψ at distance x and time t is equal to its value at distance $x + c\tau$ at time $t + \tau$. Therefore $\Psi = f(t - x/c)$ represents a disturbance moving in the positive x direction with velocity c, its profile remaining unchanged. Similarly, $\Psi = g(t + x/c)$ represents a disturbance moving in the negative x direction. Therefore, the solutions, Eqs. (2.4) and (2.6), are known as the *progressive solutions* of the wave equation.

In a spherical coordinate system with spherical symmetry, Eq. (2.1) can be reduced to

$$\frac{\partial^2\Psi_1}{\partial r^2} = \frac{1}{c^2}\frac{\partial^2\Psi_1}{\partial t^2}, \qquad \Psi_1 = r\Psi. \tag{2.7}$$

From Eqs. (2.5)–(2.7), we find

$$\Psi = \frac{1}{r} f\left(t - \frac{r}{c}\right) + \frac{1}{r} g\left(t + \frac{r}{c}\right) \qquad (r \neq 0). \tag{2.8}$$

Therefore the wavefronts are concentric spherical surfaces; i.e., at any given time, Ψ is constant for all points that are at the same distance from the origin. Therefore, these waves are known as *spherical waves*. The first term on the right-hand side of Eq. (2.8) represents waves moving away from the origin, whereas the second term represents those moving toward the origin.

Next, we apply the Fourier transform to both sides of Eq. (2.1). Defining

$$S(\mathbf{r}, \omega) = \int_{-\infty}^{\infty} \Psi(\mathbf{r}, t) e^{-i\omega t}\, dt, \tag{2.9}$$

Eq. (2.1) transforms to the time-independent wave equation

$$\nabla^2 S + k_c^2 S = 0, \qquad k_c = \frac{\omega}{c}. \tag{2.10}$$

Equation (2.10) is known as the *Helmholtz equation*. The solutions of Eq. (2.10) are related to those of Eq. (2.1) through the relation

$$\Psi(\mathbf{r}, t) = \frac{1}{2\pi} \int_{-\infty}^{\infty} S(\mathbf{r}, \omega) e^{i\omega t}\, d\omega. \tag{2.11}$$

Because $\Psi(\mathbf{r}, t)$ is real, we find, from Eq. (2.9),

$$S(\mathbf{r}, -\omega) = S^*(\mathbf{r}, \omega), \tag{2.12}$$

where an asterisk signifies a complex conjugate. Equation (2.11) now yields

$$\Psi(\mathbf{r}, t) = \frac{1}{2\pi} \int_{0}^{\infty} S(\mathbf{r}, -\omega) e^{-i\omega t}\, d\omega + \frac{1}{2\pi} \int_{0}^{\infty} S(\mathbf{r}, \omega) e^{i\omega t}\, d\omega$$

$$= \frac{1}{\pi} \operatorname{Re} \int_{0}^{\infty} S(\mathbf{r}, \omega) e^{i\omega t}\, d\omega. \tag{2.13}$$

We define a unit vector $\mathbf{p}$ and the *propagation vector* $\mathbf{k}$ by the relations

$$\mathbf{p} = l\mathbf{e}_x + m\mathbf{e}_y + n\mathbf{e}_z, \qquad \mathbf{k} = k_c \mathbf{p}. \tag{2.14}$$

Considering only the f part of Ψ in Eq. (2.4) and using Eq. (2.9), we get

$$S(\mathbf{r}, \omega) = \int_{-\infty}^{\infty} f\left(t - \frac{\mathbf{p}\cdot\mathbf{r}}{c}\right) e^{-i\omega t}\, dt = \left\{\int_{-\infty}^{\infty} f(t) e^{-i\omega t}\, dt\right\} e^{-i\mathbf{k}\cdot\mathbf{r}}$$

$$= S(0, \omega) e^{-i\mathbf{k}\cdot\mathbf{r}}. \tag{2.15}$$

Assuming

$$S(0, \omega) = A(\omega) e^{-i\chi_0(\omega)} \qquad (A, \chi_0 \text{ real}),$$

we obtain, from Eq. (2.13), the *spectral representation*

$$\Psi(\mathbf{r}, t) = \frac{1}{\pi} \operatorname{Re} \int_0^\infty A(\omega) e^{i[\omega t - \mathbf{k}\cdot\mathbf{r} - \chi_0(\omega)]}\, d\omega. \tag{2.16}$$

Equation (2.16) expresses the wave function $\Psi(\mathbf{r}, t)$ as an integral of plane waves over the entire frequency range. This is the *Fourier principle of superposition.* A *spectral component* of $\Psi(\mathbf{r}, t)$ is therefore of the form

$$\Psi(\mathbf{r}, t, \omega_0) = A(\omega_0)\cos\left[\omega_0\left(t - \frac{\mathbf{p}\cdot\mathbf{r}}{c}\right) - \chi_0\right], \tag{2.17}$$

where ω_0 is a particular value of ω. The amplitude factor $A(\omega_0)$ is the *spectral amplitude* and the argument of the cosine term is the *phase.* The value of Ψ in Eq. (2.17) remains unchanged if we replace t by $(t + T)$, where $T = 2\pi/\omega_0$ is the *period.* We then say that Ψ is harmonic with respect to time. Similarly, Ψ is unaffected if we replace $(\mathbf{p}\cdot\mathbf{r})$ by $(\mathbf{p}\cdot\mathbf{r}) + \Lambda$, where $\Lambda = 2\pi c/\omega_0$ is the *wavelength.* Hence Ψ is harmonic in space as well.

The angular frequency ω, the period T, the wavelength Λ, and the wave number k_c are related by the relations

$$k_c = \frac{\omega}{c} = \frac{2\pi}{cT} = \frac{2\pi}{\Lambda}. \tag{2.18}$$

Because of the periodicity of $\Psi(\mathbf{r}, t, \omega_0)$ in both t and $(\mathbf{p}\cdot\mathbf{r})$, a still photograph of it at a fixed time t ($\mathbf{r}$ variable) will be identical to a cinematographic picture of it at a fixed location $\mathbf{r}$ (t variable).

Solutions of the wave equation exist that are periodic in time and space but still do not represent progressive waves. These occur if we take $f = g$ in Eq. (2.4), i.e., if we superpose two progressive waves propagating in opposite directions. Equation (2.17) was obtained by taking only f into account. However, if we take both f and g into account and, moreover, put $f = g$, Eq. (2.17) becomes

$$\begin{aligned}\Psi(\mathbf{r}, t, \omega_0) &= A(\omega_0)\left\{\cos\left[\omega_0\left(t - \frac{\mathbf{p}\cdot\mathbf{r}}{c}\right) - \chi_0\right] + \cos\left[\omega_0\left(t + \frac{\mathbf{p}\cdot\mathbf{r}}{c}\right) - \chi_0\right]\right\} \\ &= 2A(\omega_0)\cos\left[\frac{\omega_0}{c}(\mathbf{p}\cdot\mathbf{r})\right]\cos(\omega_0 t - \chi_0).\end{aligned} \tag{2.19}$$

This is known as a *standing wave.* For such waves the wave profile does not move forward but Ψ vanishes on the planes $(\mathbf{p}\cdot\mathbf{r}) = \pm(\Lambda/4), \pm(3\Lambda/4), \ldots$ for all times. These planes are known as *nodal planes.*

2.2. Separability of the Scalar Helmholtz Equation

A common mode of obtaining solutions of the Helmholtz equation is through the method of the *separation of variables.* We shall discuss these solutions in Cartesian, cylindrical, and spherical coordinate systems.

2.2.1. Cartesian Coordinates (x, y, z)

Let

$$S = X(x)Y(y)Z(z) \tag{2.20}$$

be a solution of the Helmholtz equation. Substituting this expression in Eq. (2.10) and dividing by XYZ, we find

$$\frac{X''}{X} + \frac{Y''}{Y} + \frac{Z''}{Z} + k_c^2 = 0, \tag{2.21}$$

where a prime denotes differentiation with respect to the argument. We put

$$k_c^2 = k_x^2 + k_y^2 + k_z^2, \tag{2.22}$$

where k_x, k_y, k_z are known as separation constants. Clearly, Eq. (2.21) is satisfied if X, Y, Z satisfy the equations

$$X'' + k_x^2 X = 0, \qquad Y'' + k_y^2 Y = 0, \qquad Z'' + k_z^2 Z = 0. \tag{2.23}$$

The solutions of Eqs. (2.23) are proportional to

$$e^{\pm ik_x x}, \qquad e^{\pm ik_y y}, \qquad e^{\pm ik_z z},$$

respectively. Introducing the notation $k_x = lk_c$, $k_y = mk_c$, and $k_z = nk_c$ and choosing the negative exponentials, we have

$$S \propto e^{-ik_c(lx + my + nz)} = e^{-i\mathbf{k}\cdot\mathbf{r}}, \qquad l^2 + m^2 + n^2 = 1, \tag{2.24}$$

where $\mathbf{k}$ is given by Eq. (2.14). Therefore, for every value of k_c, there are an infinite number of solutions of Eq. (2.10) such that Eq. (2.21) is satisfied. The parameters (l, m, n) may take complex values as well subject to the condition $l^2 + m^2 + n^2 = 1$. Hence the integral

$$S(\mathbf{r}, \omega) = \iint A(l, m, \omega) e^{-ik_c[lx + my + z\sqrt{(1 - l^2 - m^2)}]}\, dl\, dm, \tag{2.25}$$

where $A(l, m, \omega)$ is an arbitrary function of l and m, is also a solution of Eqs. (2.10).

From Eq. (2.25), we have

$$S(\mathbf{r}, \omega) = \frac{1}{(2\pi)^2}\int_{-\infty}^{\infty}\int_{-\infty}^{\infty} A(k_x, k_y, \omega) e^{-i[k_x x + k_y y + z\sqrt{(k_c^2 - k_x^2 - k_y^2)}]}\, dk_x\, dk_y, \tag{2.26}$$

where now the variables k_x and k_y are allowed to take real values only and the factor $(2\pi)^{-2}$ is introduced for the sake of uniformity in notation. The function $A(k_x, k_y, \omega)\exp[-iz\sqrt{(k_c^2 - k_x^2 - k_y^2)}]$ in Eq. (2.26) may be considered as the double Fourier transform of $S(\mathbf{r}, \omega)$,

$$A(k_x, k_y, \omega) = \int_{-\infty}^{\infty}\int_{-\infty}^{\infty} S(\mathbf{r}, \omega) e^{i[k_x x + k_y y + z\sqrt{(k_c^2 - k_x^2 - k_y^2)}]}\, dx\, dy. \tag{2.27}$$

From Eqs. (2.9), (2.11), (2.26), and (2.27), we have

$$A(k_x, k_y, \omega) = \int_{-\infty}^{\infty}\int_{-\infty}^{\infty}\int_{-\infty}^{\infty} \Psi(\mathbf{r}, t)e^{-i(\omega t - \mathbf{k}\cdot\mathbf{r})}\, dx\, dy\, dt, \tag{2.28}$$

$$\Psi(\mathbf{r}, t) = \frac{1}{(2\pi)^3}\int_{-\infty}^{\infty}\int_{-\infty}^{\infty}\int_{-\infty}^{\infty} A(k_x, k_y, \omega)e^{i(\omega t - \mathbf{k}\cdot\mathbf{r})}\, dk_x\, dk_y\, d\omega. \tag{2.29}$$

Equation (2.29) represents the *general principle of superposition.*

2.2.2. Cylindrical Coordinates (Δ, ϕ, z)

In this system, Eq. (2.10) takes the form

$$\frac{\partial^2 S}{\partial \Delta^2} + \frac{1}{\Delta}\frac{\partial S}{\partial \Delta} + \frac{1}{\Delta^2}\frac{\partial^2 S}{\partial \phi^2} + \frac{\partial^2 S}{\partial z^2} + k_c^2 S = 0. \tag{2.30}$$

Applying the method of the separation of variables, we find that Eq. (2.30) is satisfied by

$$S = Z(\Delta)\Phi(\phi)H(z) \tag{2.31}$$

if Z, Φ, H satisfy the equations

$$\frac{d^2 Z}{d\Delta^2} + \frac{1}{\Delta}\frac{dZ}{d\Delta} + \left(k^2 - \frac{m^2}{\Delta^2}\right)Z = 0, \tag{2.32}$$

$$\frac{d^2\Phi}{d\phi^2} + m^2\Phi = 0, \tag{2.33}$$

$$\frac{d^2 H}{dz^2} - (k^2 - k_c^2)H = 0, \tag{2.34}$$

where k and m are separation constants.

Equation (2.32) is the *Bessel equation* of order m, a solution of which is denoted by $Z_m(k\Delta)$ (App. D).

The solutions of Eq. (2.33) are

$$\Phi = \begin{matrix}\cos\\ \sin\end{matrix}(m\phi) \quad \text{or} \quad e^{\pm im\phi}.$$

In most of the applications, we require that S be a single-valued function. Consequently, $\Phi(2\pi + \phi) = \Phi(\phi)$. This restricts m to be an integer. This also justifies the choice of m^2 in Eq. (2.33). If we were to take $-m^2$ instead of m^2, Φ would be of the form $\exp(\pm m\phi)$ and then S would not be single valued.

If $v_c^2 = k^2 - k_c^2$, Eq. (2.34) yields

$$H = e^{\pm v_c z}.$$

Therefore, by taking $k > k_c$ and taking the lower sign in the exponential, it is possible to make H tend to zero as z tends to infinity. This is sometimes desired in physical problems.

Collecting the various results, we find that

$$Z_m(k\Delta)\begin{Bmatrix}\cos\\ \sin\end{Bmatrix}(m\phi)\Big\} e^{\pm \nu_c z} \tag{2.35}$$

is an appropriate solution of Eq. (2.30). Consequently, we may take

$$S(\mathbf{r}, \omega) = \sum_m \begin{Bmatrix}\cos\\ \sin\end{Bmatrix}(m\phi) \int A(m, k) Z_m(k\Delta) e^{\pm \nu_c z}\, dk,$$

where $A(m, k)$ is an arbitrary function of m and k. If we require that the solution be finite at $\Delta = 0$, then $Z_m(k\Delta) = J_m(k\Delta)$. In that case, we have

$$S^{\pm} = \sum_{\sigma = c, s} \sum_{m=0}^{\infty} \int_0^{\infty} A(m, k) \Phi_m^{\sigma, \pm}(k_c \Delta) dk,$$

or, alternatively,

$$S^{\pm} = \sum_{m=-\infty}^{\infty} \int_0^{\infty} A(m, k) \Phi_m^{\pm}(k_c \Delta) dk,$$

where

$$\begin{aligned} \Phi_m^{\pm}(k_c\Delta) &= e^{\pm \nu_c z} Y_m(k\Delta, \phi), & \Phi_m^{\sigma, \pm}(k_c\Delta) &= e^{\pm \nu_c z} Y_m^{\sigma}(k\Delta, \phi), \\ Y_m(k\Delta, \phi) &= J_m(k\Delta) e^{im\phi}, & Y_m^{c, s}(k\Delta, \phi) &= J_m(k\Delta)(\cos m\phi, \sin m\phi). \end{aligned} \tag{2.36}$$

2.2.3. Spherical Coordinates (r, θ, ϕ)

For spherical coordinates the Helmholtz equation is

$$\frac{\partial^2 S}{\partial r^2} + \frac{2}{r}\frac{\partial S}{\partial r} + \frac{1}{r^2 \sin\theta}\frac{\partial}{\partial\theta}\left(\sin\theta \frac{\partial S}{\partial\theta}\right) + \frac{1}{r^2 \sin^2\theta}\frac{\partial^2 S}{\partial\phi^2} + k_c^2 S = 0. \tag{2.37}$$

Substituting

$$S = R(r)\Theta(\theta)\Phi(\phi),$$

we find that Eq. (2.37) is identically satisfied provided R, Θ, Φ satisfy the equations

$$\frac{d^2 R}{dr^2} + \frac{2}{r}\frac{dR}{dr} + \left[k_c^2 - \frac{l(l+1)}{r^2}\right] R = 0, \tag{2.38}$$

$$\frac{1}{\sin\theta}\frac{d}{d\theta}\left(\sin\theta \frac{d\Theta}{d\theta}\right) + \left[l(l+1) - \frac{m^2}{\sin^2\theta}\right]\Theta = 0, \tag{2.39}$$

$$\frac{d^2\Phi}{d\phi^2} + m^2\Phi = 0, \tag{2.40}$$

where l and m are separation constants. As in the case of cylindrical coordinates, m must be an integer if we require $S(2\pi + \phi) = S(\phi)$.

The solutions of Eq. (2.39) are the associated Legendre functions (App. D). For most applications we shall need the function $P_l^m(\cos\theta)$ which, for integral l, is finite throughout the range $0 \leq \theta \leq \pi$. Equation (2.38) is solved by the spherical Bessel functions $z_l(k_c r)$ (App. D). Therefore,

$$z_l(k_c r)P_l^m(\cos\theta)\begin{Bmatrix}\cos\\ \sin\end{Bmatrix}(m\phi), \tag{2.41}$$

where l and m are integers, is an appropriate solution of Eq. (2.37). More generally, we may take

$$S^{\pm} = \sum_{\sigma=c,\,s}\sum_{l=0}^{\infty}\sum_{m=0}^{l} A(l, m)\Phi_{ml}^{\sigma,\,\pm}(k_c r), \tag{2.42}$$

or, alternatively,

$$S^{\pm} = \sum_{l=0}^{\infty}\sum_{m=-l}^{l} A(l, m)\Phi_{ml}^{\pm}(k_c r), \tag{2.43}$$

where

$$\Phi_{ml}^{\sigma,\,\pm}(k_c r) = \begin{Bmatrix} j_l(k_c r) \\ h_l^{(2)}(k_c r)\end{Bmatrix} Y_{ml}^{\sigma}(\theta, \phi), \tag{2.44}$$

$$\Phi_{ml}^{\pm}(k_c r) = \begin{Bmatrix} j_l(k_c r) \\ h_l^{(2)}(k_c r)\end{Bmatrix} Y_{ml}(\theta, \phi), \tag{2.45}$$

$$Y_{ml}^{c,\,s}(\theta, \phi) = P_l^m(\cos\theta)(\cos m\phi, \sin m\phi), \tag{2.46}$$

$$Y_{ml}(\theta, \phi) = P_l^m(\cos\theta)e^{im\phi}. \tag{2.47}$$

Having studied the solutions of the scalar Helmholtz equation in three coordinate systems we now proceed to investigate its separability in different coordinate systems.

Separability of a scalar partial differential equation in three-dimensional space (3-space) means that the equation can be reduced to three ordinary differential equations that are related only by two constants, known as separation constants. Whether separation is possible or not depends on both the equation and the coordinate system.

In the curvilinear coordinate system (q_1, q_2, q_3), the Helmholtz equation, Eq. (2.10), becomes

$$\frac{1}{h_1 h_2 h_3}\left[\frac{\partial}{\partial q_1}\left(\frac{h_2 h_3}{h_1}\frac{\partial S}{\partial q_1}\right) + \frac{\partial}{\partial q_2}\left(\frac{h_3 h_1}{h_2}\frac{\partial S}{\partial q_2}\right) + \frac{\partial}{\partial q_3}\left(\frac{h_1 h_2}{h_3}\frac{\partial S}{\partial q_3}\right)\right] + k_c^2 S = 0. \tag{2.48}$$

We have seen that Eq. (2.48) is separable in Cartesian, circular cylinder, and spherical coordinate systems. In general, the Helmholtz equation can be said to be separable in the curvilinear coordinate system (q_1, q_2, q_3) if the substitution $S = Q_1(q_1)Q_2(q_2)Q_3(q_3)$ leads to the separation of Eq. (2.48) into a set of three ordinary differential equations $L_i(Q_i) = 0, (i = 1, 2, 3)$, where L_i are second-order differential operators. It can be shown that the Helmholtz equation, Eq. (2.48), separates in only 11 coordinate systems, of which four are cylindrical, four are rotational, and the remaining three are of general nature (Table 2.1).

The separability of the cylindrical and the rotational systems can be demonstrated with the aid of the theory of analytic functions. The analytic function $\zeta = F(w)$ transforms a rectangular map in the w plane into a map containing the desired shape of boundary in the ζ plane. The resulting plane map can then be translated in a direction perpendicular to itself to give a cylindrical system in 3-space.

Let $\zeta = x_1 + ix_2$ and $w = q_1 + iq_2$; then $x_1 = f_1(q_1, q_2)$, $x_2 = f_2(q_1, q_2)$ constitute two families of plane curves that are mutually orthogonal [because $F(w)$ is analytic]. A translation of the ζ plane in a direction normal to itself (x_3 direction) generates a cylindrical system given by

$$x_1 = f_1(q_1, q_2), \tag{2.49}$$

$$x_2 = f_2(q_1, q_2), \tag{2.50}$$

$$x_3 = q_3. \tag{2.51}$$

The scale factors of this system are easily shown to be

$$h_1 = h_2 = \left[\left(\frac{\partial f_1}{\partial q_1}\right)^2 + \left(\frac{\partial f_1}{\partial q_2}\right)^2\right]^{1/2}, \qquad h_3 = 1. \tag{2.52}$$

If we assume $S = Q_1(q_1)Q_2(q_2)\exp\{\pm\sqrt{(k^2 - k_c^2)}q_3\}$, the Helmholtz equation, Eq. (2.48), for curvilinear cylindrical systems reduces to

$$\frac{1}{Q_1}\frac{d^2Q_1}{dq_1^2} + \frac{1}{Q_2}\frac{d^2Q_2}{dq_2^2} + h_1^2k^2 = 0. \tag{2.53}$$

It can be shown that the four cylindrical systems listed in Table 2.1 are the only separable systems for this equation. Some confusion may arise with regard to the circular cylinder system. The physical definition of this system requires $f_1 = q_1 \cos q_2$, $f_2 = q_1 \sin q_2$. The corresponding complex function is not analytic. To achieve analyticity, we may take $q_1 = \ln \Delta$ which yields $h_1 = h_2 = \Delta$, $F(w) = e^w$. However, in practice, it is hardly ever necessary to do this.

The general form of the group of curvilinear rotational systems is

$$x_1 = f_1(q_1, q_2)\cos q_3, \tag{2.54}$$

$$x_2 = f_1(q_1, q_2)\sin q_3, \tag{2.55}$$

$$x_3 = f_2(q_1, q_2), \tag{2.56}$$

with the constraint that $f_1 + if_2$ is an analytic function of the complex variable $q_1 + iq_2$. To prove the orthogonality of this system it is sufficient to show that the off-diagonal elements of the metric tensor, g_{ij}, vanish. Indeed

$$\begin{aligned} g_{12} &= \frac{\partial x_1}{\partial q_1}\frac{\partial x_1}{\partial q_2} + \frac{\partial x_2}{\partial q_1}\frac{\partial x_2}{\partial q_2} + \frac{\partial x_3}{\partial q_1}\frac{\partial x_3}{\partial q_2} \\ &= \frac{\partial f_1}{\partial q_1}\frac{\partial f_1}{\partial q_2}(\cos^2 q_3 + \sin^2 q_3) + \frac{\partial f_2}{\partial q_1}\frac{\partial f_2}{\partial q_2} = 0, \end{aligned} \tag{2.57}$$

Table 2.1. Separability of the Scalar Helmholtz Equation

Coordinate System	Variables q_1	q_2	q_3	Scale Factors h_1	h_2	h_3	Coordinate Surfaces	x_1	x_2	x_3	$\mathbf{a}$	Eigen-Functions	Generating Analytic Function $F(w)$
CYLINDRICAL													
Rectangular	x	y	z	1	1	1	Planes	x	y	z	$\mathbf{e}_z$	Exponential	w
Circular cylinder	$\Delta(\ln \Delta)$	ϕ	z	$1(\Delta)$	Δ	1	Circular cylinder, planes	$\Delta \cos \phi$	$\Delta \sin \phi$	z	$\mathbf{e}_z$	Bessel, exponential	e^w
Elliptic cylinder	u	v	z	$a\sqrt{(\text{sh}^2 u + \sin^2 v)}$	$a\sqrt{(\text{sh}^2 u + \sin^2 v)}$	1	Elliptic cylinder, hyperbolic cylinder, plane	$a \text{ ch } u \cos v$	$a \text{ sh } u \sin v$	z	$\mathbf{e}_z$	Mathieu, exponential	$a \text{ ch } w$
Parabolic cylinder	ξ	η	z	$\sqrt{(\xi^2 + \eta^2)}$	$\sqrt{(\xi^2 + \eta^2)}$	1	Parabolic cylinders, plane	$\frac{1}{2}(\xi^2 - \eta^2)$	$\xi\eta$	z	$\mathbf{e}_z$	Weber, exponential	$\frac{1}{2}w^2$
ROTATIONAL													
Spherical	r	θ	ϕ	1	r	$r \sin \theta$	Sphere, cone, plane	$(r \sin \theta)\cos \phi$	$(r \sin \theta)\sin \phi$	$r \cos \theta$	$\mathbf{r}$	Bessel, Legendre, exponential	e^w
Prolate spheroidal	u	v	ϕ	$a\sqrt{(\text{sh}^2 u + \sin^2 v)}$	$a\sqrt{(\text{sh}^2 u + \sin^2 v)}$	$a \text{ sh } u \sin v$	Spheroid, hyperboloid, plane	$(a \text{ sh } u \sin v)\cos \phi$	$(a \text{ sh } u \sin v)\sin \phi$	$a \text{ ch } u \cos v$	—	Lamé, exponential	$a \text{ ch } w$
Oblate spheroidal	u	v	ϕ	$a\sqrt{(\text{sh}^2 u + \cos^2 v)}$	$a\sqrt{(\text{sh}^2 u + \cos^2 v)}$	$a \text{ ch } u \sin v$	Spheroid, hyperboloid, plane	$(a \text{ ch } u \sin v)\cos \phi$	$(a \text{ ch } u \sin v)\sin \phi$	$a \text{ sh } u \cos v$	—	Lamé, exponential	$a \text{ sh } w$
Parabolic	ξ	η	ϕ	$\sqrt{(\xi^2 + \eta^2)}$	$\sqrt{(\xi^2 + \eta^2)}$	$\xi\eta$	Paraboloids, plane	$(\xi\eta)\cos \phi$	$(\xi\eta)\sin \phi$	$\frac{1}{2}(\xi^2 - \eta^2)$	—	Whittaker, exponential	$\frac{1}{2}w^2$
GENERAL													
Conical	r	λ	μ	1	*	*	Sphere, elliptic cones	*	*	*	$\mathbf{r}$	Bessel, Lamé	—
Ellipsoidal	λ	μ	ν	*	*	*	Ellipsoid, hyperboloids	*	*	*	—	Lamé	—
Paraboloidal	λ	μ	ν	*	*	*	Paraboloids	*	*	*	—	Baer	—

* Expression is too lengthy to be given here.

using the Cauchy–Riemann relations. Similarly, it can be shown that $g_{13} = g_{23} = 0$, whereas

$$h_1 = h_2 = \left[\left(\frac{\partial f_1}{\partial q_1}\right)^2 + \left(\frac{\partial f_1}{\partial q_2}\right)^2\right]^{1/2}, \qquad h_3 = f_1. \tag{2.58}$$

Assuming the separation $S = F(q_1, q_2)e^{imq_3}$, the Helmholtz equation, Eq. (2.10), reduces to

$$\left(\nabla^2 F + k_c^2 F - \frac{m^2}{f_1^2} F\right)e^{imq_3} = 0. \tag{2.59}$$

Here again it can be shown that the four rotational systems listed in Table 2.1 are the only systems that lead to the separation of this equation, plus, of course, the circular cylinder system.

2.3. Separability of the Vector Helmholtz Equation

In the Section 2.2 we saw that the scalar Helmholtz equation in 3-space is separable in 11 coordinate systems; i.e., in 11 systems the partial differential equation can be reduced to a set of three ordinary differential equations of the second order, one in each independent variable.

When we treat the vector equation

$$\nabla^2 \mathbf{u} + k_c^2 \mathbf{u} = 0, \tag{2.60}$$

the question of separability becomes more complex. Let us first consider rectangular coordinates. There are three independent variables, namely, the coordinates x_1, x_2, x_3 and three dependent variables u_1, u_2, u_3. Because $\nabla^2 \mathbf{e}_1 = \nabla^2 \mathbf{e}_2 = \nabla^2 \mathbf{e}_3 = 0$, Eq. (2.60) separates into three scalar equations

$$\nabla^2 u_1 + k_c^2 u_1 = 0, \qquad \nabla^2 u_2 + k_c^2 u_2 = 0, \qquad \nabla^2 u_3 + k_c^2 u_3 = 0. \tag{2.61}$$

Each equation contains only a single dependent variable. However, if we consider the situation in spherical coordinates, then the vector equation, Eq. (2.60), splits into the scalar equations

$$\nabla^2 u_r + \left(k_c^2 - \frac{2}{r^2}\right)u_r - \frac{2}{r^2 \sin\theta}\frac{\partial}{\partial\theta}(u_\theta \sin\theta) - \frac{2}{r^2 \sin\theta}\frac{\partial u_\phi}{\partial\phi} = 0, \tag{2.62}$$

$$\nabla^2 u_\theta + \left(k_c^2 - \frac{1}{r^2 \sin^2\theta}\right)u_\theta + \frac{2}{r^2}\frac{\partial u_r}{\partial\theta} - \frac{2}{r^2}\frac{\cos\theta}{\sin^2\theta}\frac{\partial u_\phi}{\partial\phi} = 0, \tag{2.63}$$

$$\nabla^2 u_\phi + \left(k_c^2 - \frac{1}{r^2 \sin^2\theta}\right)u_\phi + \frac{2}{r^2 \sin\theta}\frac{\partial u_r}{\partial\phi} + \frac{2}{r^2}\frac{\cos\theta}{\sin^2\theta}\frac{\partial u_\theta}{\partial\phi} = 0. \tag{2.64}$$

This is a system of three simultaneous, second-order differential equations in three unknown functions, u_r, u_θ, u_ϕ, that leads eventually to a sixth-order differential equation in each of the variables. A similar situation arises in the rest of the nine orthogonal curvilinear systems for which the scalar Helmholtz equation is separable. To avoid this difficulty and to simplify the solution of the vector equation, we must look for ways of breaking **u** into three independent vectors such that each of them leads to a second-order partial differential equation in a single scalar potential. When this is achieved it is called vector separability of the vector Helmholtz equation. Furthermore, each of the resulting scalar equations must be separable in its independent variables. If such vectors can be found, they will constitute a complete solution of the vector equation.

For a solution of this kind to be useful, three conditions must be met:

1. The vector solutions must be orthogonal in some sense.
2. The three scalar equations must be separable in their independent variables.
3. One of the vector solutions must be tangential to a coordinate surface and a second solution must be normal to it.

A vector field that satisfies these requirements is said to be *vector separable.*

It now remains for us to choose the three scalars with which to express the vector field. We have seen that the choice of the vector components along the curvilinear axes leads to difficulties. Another way is suggested by the Helmholtz decomposition theorem

$$\mathbf{u} = \text{grad}\,\Phi + \text{curl}\,\mathbf{A}; \qquad \text{div}\,\mathbf{A} = 0. \tag{2.65}$$

The gradient term is known as the *longitudinal component*, because it points in the direction of greatest rate of change of the scalar potential. The curl term is known as the *transverse component.* This splitting of the field enables us to simplify the procedure of applying the boundary conditions in accord with requirement 3. Because requirement 1 is satisfied by the two vectors it remains for us to choose the transverse solution in such a way that it becomes either normal or tangential to one of the coordinate surfaces. Note that the gauge condition, div $\mathbf{A} = 0$, reduces the number of independent scalar potentials to three.

Consider the curvilinear coordinates (q_1, q_2, q_3) with scale factors (h_1, h_2, h_3). Let **a** be a vector normal to the surface $q_1 = $ const. and $\Psi(q_i)$ be a solution of $\nabla^2\Psi + k_c^2\Psi = 0$. Because $\mathbf{a}\Psi$ is normal to $q_1 = $ const., the vector $\mathbf{M} = \text{curl}(\mathbf{a}\Psi)$ will be tangential to this surface. Applying the operator $(\nabla^2 + k_c^2)$ to **M** and using $\nabla^2\Psi = -k_c^2\Psi$, we find

$$\nabla^2\mathbf{M} + k_c^2\mathbf{M} = \text{curl}[\Psi\nabla^2\mathbf{a} + 2(\text{grad}\,\Psi \cdot \text{grad}\,\mathbf{a})]. \tag{2.66}$$

If **M** is to satisfy the vector Helmholtz equation it is sufficient that both $\nabla^2\mathbf{a}$ and curl(grad $\Psi \cdot$ grad **a**) vanish. These conditions are met when either **a** is a constant vector in the q_1 direction or $\mathbf{a} = \mathbf{r}$. These restrictions on **a** reduce the 11 scalar-separable coordinate systems to just six orthogonal curvilinear coordinate systems in which the vector Helmholtz equation is separable. They are: rectangular, circular cylinder, elliptic cylinder, parabolic cylinder, spherical, and conical. The vector **a** is $\mathbf{e}_z$ for the first four systems and **r** for the last two.

To obtain a second transverse solution that generates a field normal to a coordinate surface we try

$$\mathbf{N} = \frac{1}{k_c} \operatorname{curl} \operatorname{curl}(\mathbf{a}\chi), \qquad \nabla^2\chi + k_c^2\chi = 0.$$

Obviously, $\nabla^2\mathbf{N} + k_c^2\mathbf{N} = 0$ because the operator $(\nabla^2 + k_c^2)$ commutes with curl. The factor $1/k_c$ is introduced to equalize the dimensions of the two transverse solutions.

We have therefore defined three independent vector fields that depend on three potentials such that their linear combination can be used to generate the general solution of the vector Helmholtz equation in a form that allows a simple application of physical boundary conditions. They are:

$$\begin{aligned} \mathbf{M} &= \operatorname{curl}(\mathbf{a}\Psi) = (\operatorname{grad} \Psi) \times \mathbf{a}, \\ \mathbf{N} &= \frac{1}{k_c} \operatorname{curl} \operatorname{curl}(\mathbf{a}\chi), \\ \mathbf{L} &= \frac{1}{k_c} \operatorname{grad} \Phi, \end{aligned} \tag{2.67}$$

where Φ, Ψ, and χ are solutions of the scalar Helmholtz equation. The vectors **L, M, N** are known as *Hansen vectors.* The reader may easily verify that they are linearly independent. He or she can also verify without much effort the truth of the following relations:

$$\begin{aligned} &\operatorname{div} \mathbf{M} = 0, \qquad \operatorname{div} \mathbf{N} = 0, \qquad \operatorname{div} \mathbf{L} = -k_c\Phi, \\ &\operatorname{curl} \mathbf{L} = 0, \qquad \mathbf{N} = k_c\,\mathbf{a}\chi + \frac{1}{k_c} \operatorname{grad} \frac{\partial(w\chi)}{\partial q_1}, \end{aligned} \tag{2.68}$$

where $\mathbf{a} = w\mathbf{e}_1$ and $\mathbf{e}_1$ is a unit vector in the q_1 direction. Moreover, if we do away with the distinction among the potentials Φ, Ψ, and χ, we have

$$\mathbf{M} = k_c \mathbf{L} \times \mathbf{a} = \frac{1}{k_c} \operatorname{curl} \mathbf{N}, \qquad \mathbf{N} = \frac{1}{k_c} \operatorname{curl} \mathbf{M}.$$

Substituting into the expressions of Eq. (2.67) for the eigenvectors the separated form of a scalar eigenfunction, $S = f(q_1)Y(q_2, q_3)$, we find

$$\begin{aligned} \mathbf{M} &= f\mathbf{C}, \\ k_c\mathbf{N} &= \left[k_c^2(wf) + \frac{1}{h_1}\frac{\partial^2}{\partial q_1^2}(wf)\right]\mathbf{P} + \frac{1}{w}\frac{\partial(fw)}{\partial q_1}\mathbf{B}, \\ k_c\mathbf{L} &= \left(\frac{1}{h_1}\frac{\partial f}{\partial q_1}\right)\mathbf{P} + \frac{1}{w}f\mathbf{B}. \end{aligned} \tag{2.69}$$

The vectors **P**, **B**, **C** depend only on the coordinates q_2 and q_3 and are known as *vector surface harmonics.* We have

$$\begin{aligned}\mathbf{P} &= Y\mathbf{e}_1,\\ \mathbf{B} &= w \operatorname{grad} Y = \mathbf{e}_1 \times \mathbf{C},\\ \mathbf{C} &= \operatorname{curl}(\mathbf{a}Y) = (\operatorname{grad} Y) \times \mathbf{a} = \mathbf{B} \times \mathbf{e}_1.\end{aligned} \tag{2.70}$$

In spherical coordinates these are known as *vector spherical harmonics.* Their explicit form is

$$\begin{aligned}\mathbf{P}_{ml}(\theta, \phi) &= \mathbf{e}_r Y_{ml}(\theta, \phi) = \mathbf{e}_r P_l^m(\cos\theta)e^{im\phi},\\ \sqrt{[l(l+1)]}\mathbf{B}_{ml}(\theta, \phi) &= r \operatorname{grad} Y_{ml}(\theta, \phi) = \left(\mathbf{e}_\theta \frac{\partial}{\partial\theta} + \mathbf{e}_\phi \frac{1}{\sin\theta}\frac{\partial}{\partial\phi}\right) Y_{ml}(\theta, \phi),\\ \sqrt{[l(l+1)]}\mathbf{C}_{ml}(\theta, \phi) &= \operatorname{curl}[\mathbf{r}Y_{ml}(\theta, \phi)] = \left(\mathbf{e}_\theta \frac{1}{\sin\theta}\frac{\partial}{\partial\phi} - \mathbf{e}_\phi \frac{\partial}{\partial\theta}\right) Y_{ml}(\theta, \phi).\end{aligned} \tag{2.71}$$

The vectors $\mathbf{P}_{ml}$, $\mathbf{B}_{ml}$, and $\mathbf{C}_{ml}$ satisfy the following orthogonality relations:

$$\mathbf{P}_{ml} \cdot \mathbf{B}_{m'l'} = \mathbf{P}_{ml} \cdot \mathbf{C}_{m'l'} = \mathbf{B}_{ml} \cdot \mathbf{C}_{m'l'} = 0, \tag{2.72}$$

$$\int_0^{2\pi}\int_0^{\pi} \mathbf{B}_{ml} \cdot \overset{*}{\mathbf{C}}_{m'l'} \sin\theta \, d\theta \, d\phi = 0, \tag{2.73}$$

$$\begin{aligned}\int_0^{2\pi}\int_0^{\pi} \mathbf{P}_{ml} \cdot \overset{*}{\mathbf{P}}_{m'l'} \sin\theta \, d\theta \, d\phi &= \int_0^{2\pi}\int_0^{\pi} \mathbf{B}_{ml} \cdot \overset{*}{\mathbf{B}}_{m'l'} \sin\theta \, d\theta \, d\phi\\ &= \int_0^{2\pi}\int_0^{\pi} \mathbf{C}_{ml} \cdot \overset{*}{\mathbf{C}}_{m'l'} \sin\theta \, d\theta \, d\phi = \delta_{mm'}\delta_{ll'}\Omega_{ml},\end{aligned} \tag{2.74}$$

where

$$\Omega_{ml} = \frac{4\pi}{2l+1}\frac{(l+m)!}{(l-m)!} \tag{2.75}$$

and the asterisk stands for a complex conjugate.

The relationship in Eq. (2.72) follows directly from the definition of the vectors. Let us next prove Eq. (2.73): We have

$$\begin{aligned}&\int_0^{2\pi}\int_0^{\pi} \sqrt{[l(l+1)]}\mathbf{B}_{ml} \cdot \sqrt{[l'(l'+1)]}\overset{*}{\mathbf{C}}_{m'l'} \sin\theta \, d\theta \, d\phi\\ &\quad = -2\pi im\delta_{mm'} \int_0^{\pi} \left[\frac{dP_l^m}{d\theta} P_{l'}^m + P_l^m \frac{dP_{l'}^m}{d\theta}\right] d\theta\\ &\quad = -2\pi im\delta_{mm'}[P_l^m(\cos\theta)P_{l'}^m(\cos\theta)]_0^\pi = 0,\end{aligned} \tag{2.76}$$

because $P_l^m(\pm 1)$ is zero for nonzero values of m.

Again,

$$\int_0^{2\pi}\int_0^{\pi} \mathbf{P}_{ml}\cdot\overset{*}{\mathbf{P}}_{m'l'}\sin\theta\,d\theta\,d\phi = \int_0^{2\pi}\int_0^{\pi} Y_{ml}\overset{*}{Y}_{m'l'}\sin\theta\,d\theta\,d\phi$$

$$= 2\pi\delta_{mm'}\int_0^{\pi} P_l^m(\cos\theta)P_{l'}^m(\cos\theta)\sin\theta\,d\theta$$

$$= \delta_{mm'}\delta_{ll'}\Omega_{ml}, \qquad (2.77)$$

using the orthogonality relation of the Legendre functions.

Similarly,

$$\int_0^{2\pi}\int_0^{\pi}\sqrt{[l(l+1)]}\mathbf{B}_{ml}\cdot\sqrt{[l'(l'+1)]}\overset{*}{\mathbf{B}}_{m'l'}\sin\theta\,d\theta\,d\phi$$

$$= 2\pi\delta_{mm'}\int_0^{\pi}\left(\frac{dP_l^m}{d\theta}\frac{dP_{l'}^m}{d\theta} + \frac{m^2}{\sin^2\theta}P_l^mP_{l'}^m\right)\sin\theta\,d\theta. \qquad (2.78)$$

The last integral can be evaluated as follows. We know that

$$\frac{1}{\sin\theta}\frac{d}{d\theta}\left(\sin\theta\frac{dP_l^m}{d\theta}\right) + \left[l(l+1) - \frac{m^2}{\sin^2\theta}\right]P_l^m = 0.$$

Hence

$$\frac{1}{\sin\theta}\frac{d}{d\theta}\left(\sin\theta P_{l'}^m\frac{dP_l^m}{d\theta}\right) + l(l+1)P_l^mP_{l'}^m = \frac{dP_l^m}{d\theta}\frac{dP_{l'}^m}{d\theta} + \frac{m^2}{\sin^2\theta}P_l^mP_{l'}^m.$$

Multiplying both sides by $\sin\theta$ and integrating over θ between the limit $(0, \pi)$, we obtain

$$\int_0^{\pi}\left(\frac{dP_l^m}{d\theta}\frac{dP_{l'}^m}{d\theta} + \frac{m^2}{\sin^2\theta}P_l^mP_{l'}^m\right)\sin\theta\,d\theta = l(l+1)\int_0^{\pi}P_l^mP_{l'}^m\sin\theta\,d\theta$$

$$= \frac{1}{2\pi}l(l+1)\Omega_{ml}\delta_{ll'}. \qquad (2.79)$$

The use of Eq. (2.79) in Eq. (2.78) completes the proof of the relation, Eq. (2.74), because the same integral occurs for the $\mathbf{C}_{ml}\cdot\overset{*}{\mathbf{C}}_{m'l'}$ case.

In the vector spherical harmonics, Eq. (2.71), $Y_{ml}(\theta, \phi)$ may be replaced by

$$Y_{ml}^{c,s}(\theta, \phi) = P_l^m(\cos\theta)(\cos m\phi, \sin m\phi),$$

and denoted as $\mathbf{P}_{ml}^{c,s}$, etc. If this is done, the orthogonality relation, Eq. (2.74), becomes

$$\int_0^{2\pi}\int_0^{\pi}\mathbf{P}_{ml}^{\sigma}\cdot\mathbf{P}_{m'l'}^{\sigma'}\sin\theta\,d\theta\,d\phi = \int_0^{2\pi}\int_0^{\pi}\mathbf{B}_{ml}^{\sigma}\cdot\mathbf{B}_{m'l'}^{\sigma'}\sin\theta\,d\theta\,d\phi$$

$$= \int_0^{2\pi}\int_0^{\pi}\mathbf{C}_{ml}^{\sigma}\cdot\mathbf{C}_{m'l'}^{\sigma'}\sin\theta\,d\theta\,d\phi$$

$$= \frac{1}{\varepsilon_m}\delta_{mm'}\delta_{ll'}\delta_{\sigma\sigma'}\Omega_{ml}, \qquad (2.80)$$

where σ, σ' can be c or s and ε_m is the Neumann factor:

$$\varepsilon_m = \begin{cases} 1 & \text{for } m = 0, \\ 2 & \text{for } m > 0. \end{cases} \tag{2.81}$$

Certain linear combinations of the vector spherical harmonics are useful in deriving the orthogonality relations of the eigenvectors. They are defined as follows:

$$\begin{aligned} \mathbf{A}^1_{ml} &= l\mathbf{P}_{ml} + \sqrt{[l(l+1)]}\mathbf{B}_{ml}, \\ \mathbf{A}^2_{ml} &= -(l+1)\mathbf{P}_{ml} + \sqrt{[l(l+1)]}\mathbf{B}_{ml}, \\ \mathbf{A}^3_{ml} &= \sqrt{[l(l+1)]}\mathbf{C}_{ml}. \end{aligned} \tag{2.82}$$

Using Eqs. (2.72)–(2.74), we can easily show that

$$\int_0^{2\pi}\int_0^{\pi} \mathbf{A}^i_{ml} \cdot \overset{*}{\mathbf{A}}{}^j_{m'l'} \sin\theta \, d\theta \, d\phi = 0 \qquad (i \neq j;\, i, j = 1, 2, 3), \tag{2.83}$$

and

$$\int_0^{2\pi}\int_0^{\pi} \mathbf{A}^k_{ml} \cdot \overset{*}{\mathbf{A}}{}^k_{m'l'} \sin\theta \, d\theta \, d\phi = \gamma_k(l)\delta_{mm'}\delta_{ll'}\Omega_{ml}, \tag{2.84}$$

where

$$\gamma_1 = l(2l+1), \qquad \gamma_2 = (l+1)(2l+1), \qquad \gamma_3 = l(l+1).$$

Returning to Eq. (2.69), we note that in spherical coordinates $w = q_1 = r$ and $f = f_l^{\pm}(k_c r)$, where

$f_l^+ = j_l$, spherical Bessel function of the first kind,

$f_l^- = h_l^{(2)}$, spherical Hankel function of the second kind.

Consequently, the eigenvectors become

$$\begin{aligned} \mathbf{M}^{\pm}_{ml}(k_c r) &= \sqrt{[l(l+1)]}\mathbf{C}_{ml} f_l^{\pm}(k_c r), \\ \mathbf{N}^{\pm}_{ml}(k_c r) &= \mathbf{P}_{ml} l(l+1) f_l^{\pm}(k_c r)/(k_c r) + \sqrt{[l(l+1)]}\mathbf{B}_{ml}[f_l'^{\pm}(k_c r) \\ &\quad + f_l^{\pm}(k_c r)/(k_c r)], \\ \mathbf{L}^{\pm}_{ml}(k_c r) &= \mathbf{P}_{ml} f_l'^{\pm}(k_c r) + \sqrt{[l(l+1)]}\mathbf{B}_{ml} f_l^{\pm}(k_c r)/(k_c r), \end{aligned} \tag{2.85}$$

a prime denoting differentiation with respect to the argument. The superscript + indicates an interior solution, i.e., a solution that is finite at the origin. The superscript −, in contrast, indicates an exterior solution that is singular at the origin and is bounded at infinity.

Expressing the spherical eigenvectors in terms of the vectors $\mathbf{A}_{ml}^{i}$ and using Eqs. (2.83) and (2.84), we obtain the following orthogonality relations:

$$
\begin{aligned}
&\int_0^{2\pi}\int_0^{\pi} \mathbf{L}_{ml}\cdot\overset{*}{\mathbf{M}}_{m'l'} \sin\theta\, d\theta\, d\phi = \int_0^{2\pi}\int_0^{\pi} \mathbf{N}_{ml}\cdot\overset{*}{\mathbf{M}}_{m'l'} \sin\theta\, d\theta\, d\phi = 0,\\
&\int_0^{2\pi}\int_0^{\pi} \mathbf{L}_{ml}\cdot\overset{*}{\mathbf{N}}_{m'l'} \sin\theta\, d\theta\, d\phi = \frac{l(l+1)}{2l+1}\Omega_{ml}(f_{l-1}^2 - f_{l+1}^2)\delta_{ll'}\delta_{mm'},\\
&\int_0^{2\pi}\int_0^{\pi} \mathbf{M}_{ml}\cdot\overset{*}{\mathbf{M}}_{m'l'} \sin\theta\, d\theta\, d\phi = l(l+1)\Omega_{ml} f_l^2 \delta_{ll'}\delta_{mm'},\\
&\int_0^{2\pi}\int_0^{\pi} \mathbf{N}_{ml}\cdot\overset{*}{\mathbf{N}}_{m'l'} \sin\theta\, d\theta\, d\phi = \frac{l(l+1)}{2l+1}\Omega_{ml}[(l+1)f_{l-1}^2 + l f_{l+1}^2]\delta_{ll'}\delta_{mm'},\\
&\int_0^{2\pi}\int_0^{\pi} \mathbf{L}_{ml}\cdot\overset{*}{\mathbf{L}}_{m'l'} \sin\theta\, d\theta\, d\phi = \frac{1}{2l+1}\Omega_{ml}[l f_{l-1}^2 + (l+1) f_{l+1}^2]\delta_{ll'}\delta_{mm'}.
\end{aligned}
\tag{2.86}
$$

Note that **L** and **N** are not always orthogonal. This is of little importance because the orthogonality of **P**, **B**, **C** is sufficient for most applications.

In cylindrical coordinates, we define *vector cylindrical harmonics* as follows:

$$
\begin{aligned}
\mathbf{P}_m(k\Delta, \phi) &= \mathbf{e}_z Y_m(k\Delta, \phi) = \mathbf{e}_z J_m(k\Delta)e^{im\phi},\\
\mathbf{B}_m(k\Delta, \phi) &= \frac{1}{k}\operatorname{grad} Y_m(k\Delta, \phi) = \left(\mathbf{e}_\Delta \frac{\partial}{\partial k\Delta} + \mathbf{e}_\phi \frac{1}{k\Delta}\frac{\partial}{\partial\phi}\right) Y_m(k\Delta, \phi),\\
\mathbf{C}_m(k\Delta, \phi) &= \frac{1}{k}\operatorname{curl}[\mathbf{e}_z Y_m(k\Delta, \phi)] = \left(\mathbf{e}_\Delta \frac{1}{k\Delta}\frac{\partial}{\partial\phi} - \mathbf{e}_\phi \frac{\partial}{\partial k\Delta}\right) Y_m(k\Delta, \phi).
\end{aligned}
\tag{2.87}
$$

The vectors $\mathbf{P}_m$, $\mathbf{B}_m$, and $\mathbf{C}_m$ satisfy the following orthogonality relations:

$$
\begin{aligned}
\mathbf{P}_m(k\Delta, \phi)\cdot\mathbf{B}_{m'}(k'\Delta, \phi) &= \mathbf{P}_m(k\Delta, \phi)\cdot\mathbf{C}_{m'}(k'\Delta, \phi)\\
&= \mathbf{B}_m(k\Delta, \phi)\cdot\mathbf{C}_m(k\Delta, \phi) = 0,
\end{aligned}
\tag{2.88}
$$

$$
\int_0^{2\pi}\int_0^{\infty} \mathbf{B}_m(k\Delta, \phi)\cdot\overset{*}{\mathbf{C}}_{m'}(k'\Delta, \phi)\Delta\, d\Delta\, d\phi = 0,
\tag{2.89}
$$

$$
\begin{aligned}
&\int_0^{2\pi}\int_0^{\infty} \mathbf{P}_m(k\Delta, \phi)\cdot\overset{*}{\mathbf{P}}_{m'}(k'\Delta, \phi)\Delta\, d\Delta\, d\phi\\
&\quad= \int_0^{2\pi}\int_0^{\infty} \mathbf{B}_m(k\Delta, \phi)\cdot\overset{*}{\mathbf{B}}_{m'}(k'\Delta, \phi)\Delta\, d\Delta\, d\phi\\
&\quad= \int_0^{2\pi}\int_0^{\infty} \mathbf{C}_m(k\Delta, \phi)\cdot\overset{*}{\mathbf{C}}_{m'}(k'\Delta, \phi)\Delta\, d\Delta\, d\phi\\
&\quad= \frac{2\pi}{\sqrt{kk'}}\delta_{mm'}\delta(k-k').
\end{aligned}
\tag{2.90}
$$

Relations (2.88) are obvious from the definitions. We now prove Eq. (2.89). We have

$$\int_0^{2\pi}\int_0^{\infty} \mathbf{B}_m(k\Delta, \phi)\cdot \overset{*}{\mathbf{C}}_{m'}(k'\Delta, \phi)\Delta\, d\Delta\, d\phi$$

$$= -2\pi i m \delta_{mm'} \int_0^{\infty} \left[\frac{d}{dk\Delta} J_m(k\Delta) \frac{1}{k'\Delta} J_m(k'\Delta) + \frac{1}{k\Delta} J_m(k\Delta) \frac{d}{dk'\Delta} J_m(k'\Delta)\right]\Delta\, d\Delta$$

$$= -2\pi i m \delta_{mm'} \frac{1}{kk'} [J_m(k\Delta) J_m(k'\Delta)]_0^{\infty} = 0, \tag{2.91}$$

because $J_m(x)$ tends to zero as x tends to infinity and $J_m(0)$ is zero for nonzero values of m.

Next, we consider the integral

$$\int_0^{2\pi}\int_0^{\infty} \mathbf{P}_m(k\Delta, \phi)\cdot \overset{*}{\mathbf{P}}_{m'}(k'\Delta, \phi)\Delta\, d\Delta\, d\phi$$

$$= \int_0^{2\pi}\int_0^{\infty} Y_m(k\Delta, \phi)\overset{*}{Y}_{m'}(k'\Delta, \phi)\Delta\, d\Delta\, d\phi$$

$$= 2\pi\delta_{mm'} \int_0^{\infty} J_m(k\Delta) J_m(k'\Delta)\Delta\, d\Delta = \frac{2\pi}{\sqrt{kk'}} \delta_{mm'}\, \delta(k - k'). \tag{2.92}$$

Similarly,

$$\int_0^{2\pi}\int_0^{\infty} \mathbf{B}_m(k\Delta, \phi)\cdot \overset{*}{\mathbf{B}}_{m'}(k'\Delta, \phi)\Delta\, d\Delta\, d\phi$$

$$= \frac{2\pi}{kk'} \delta_{mm'} \int_0^{\infty} \left[\frac{dJ_m(k\Delta)}{d\Delta}\frac{dJ_m(k'\Delta)}{d\Delta} + \frac{m^2}{\Delta^2} J_m(k\Delta) J_m(k'\Delta)\right]\Delta\, d\Delta$$

$$= \frac{2\pi}{\sqrt{kk'}} \delta_{mm'}\, \delta(k - k'). \tag{2.93}$$

As in the case of vector spherical harmonics, we can define vector cylindrical harmonics $\mathbf{P}_m^{c,\,s}$, etc., with $Y_m(k\Delta, \phi)$ in Eq. (2.87) replaced by

$$Y_m^{c,\,s}(k\Delta, \phi) = J_m(k\Delta)(\cos m\phi, \sin m\phi).$$

Then, the orthogonality relation, Eq. (2.90), must be modified to

$$\int_0^{2\pi}\int_0^{\infty} \mathbf{P}_m^{\sigma}(k\Delta, \phi)\cdot \mathbf{P}_{m'}^{\sigma'}(k'\Delta, \phi)\Delta\, d\Delta\, d\phi$$

$$= \int_0^{2\pi}\int_0^{\infty} \mathbf{B}_m^{\sigma}(k\Delta, \phi)\cdot \mathbf{B}_{m'}^{\sigma'}(k'\Delta, \phi)\Delta\, d\Delta\, d\phi$$

$$= \int_0^{2\pi}\int_0^{\infty} \mathbf{C}_m^{\sigma}(k\Delta, \phi)\cdot \mathbf{C}_{m'}^{\sigma'}(k'\Delta, \phi)\Delta\, d\Delta\, d\phi = \frac{2\pi}{\sqrt{kk'}\,\varepsilon_m} \delta_{mm'}\, \delta_{\sigma\sigma'}\, \delta(k - k'). \tag{2.94}$$

Having defined vector cylindrical harmonics, we are ready to consider the corresponding eigenvectors. With reference to Eq. (2.69), we note that in a cylindrical system $w = 1$ and $f = \exp(\pm \nu_c z)$, where $\nu_c = (k^2 - k_c^2)^{1/2}$. Consequently, the eigenvectors become

$$\mathbf{M}_m^{\pm} = \mathbf{C}_m e^{\pm \nu_c z},$$

$$\mathbf{N}_m^{\pm} = \frac{1}{k_c}(k\mathbf{P}_m \pm \nu_c \mathbf{B}_m)e^{\pm \nu_c z}, \tag{2.95}$$

$$\mathbf{L}_m^{\pm} = \frac{1}{k_c}(\pm \nu_c \mathbf{P}_m + k\mathbf{B}_m)e^{\pm \nu_c z}.$$

The set of solutions in Eqs. (2.95) is suitable only when the boundaries are of the form $z =$ const. This set of solutions cannot be used for solving problems concerning circular cylinders. For such cases, we can put $\mathbf{a} = \mathbf{\Delta} = \Delta \mathbf{e}_\Delta$ in Eq. (2.67). Therefore, for the two-dimensional problems when there is no dependence on z, we find

$$\mathbf{M} = \operatorname{curl}(\mathbf{\Delta}\Psi) = -\mathbf{e}_z\left(\frac{\partial \Psi}{\partial \phi}\right),$$

$$\mathbf{N} = \frac{1}{k_c}\operatorname{curl}\operatorname{curl}(\mathbf{\Delta}\Psi) = -\frac{1}{k_c}\left(\mathbf{e}_\Delta \frac{1}{\Delta}\frac{\partial}{\partial \phi} - \mathbf{e}_\phi \frac{\partial}{\partial \Delta}\right)\frac{\partial \Psi}{\partial \phi},$$

$$\mathbf{L} = \frac{1}{k_c}\operatorname{grad}\Psi = \frac{1}{k_c}\left(\mathbf{e}_\Delta \frac{\partial}{\partial \Delta} + \mathbf{e}_\phi \frac{1}{\Delta}\frac{\partial}{\partial \phi}\right)\Psi,$$

$$\Psi = Z_m(k_c\Delta)e^{\pm im\phi}, \qquad \left(k_c = \frac{\omega}{c}\right),$$

where Z_m denotes a Bessel function. The parameter m must be a positive integer if the condition $\Psi(2\pi + \phi) = \Psi(\phi)$ is to be satisfied. Otherwise, it can be nonintegral or even complex. Therefore, if $m = in$, where n is real, we have

$$\Psi = Z_{in}(k_c\Delta)e^{\pm n\phi}.$$

2.4. Eigenvector Solutions of the Navier Equation

We return to the equation of motion, Eq. (1.112), of a homogeneous isotropic elastic solid under the action of a body force distribution $\mathbf{F}(\mathbf{r}, t)$ per unit mass

$$\alpha^2 \operatorname{grad}\operatorname{div}\mathbf{u} - \beta^2 \operatorname{curl}\operatorname{curl}\mathbf{u} + \mathbf{F} = \frac{\partial^2 \mathbf{u}}{\partial t^2}. \tag{2.96}$$

With definitions of the Fourier transforms

$$\mathbf{u}(\mathbf{r}, \omega) = \int_{-\infty}^{\infty} \mathbf{u}(\mathbf{r}, t)e^{-i\omega t}\,dt, \tag{2.97}$$

$$\mathbf{F}(\mathbf{r}, \omega) = \int_{-\infty}^{\infty} \mathbf{F}(\mathbf{r}, t)e^{-i\omega t}\,dt, \tag{2.98}$$

Eq. (2.96) is transformed to the spectral Navier equation

$$\alpha^2 \text{ grad div } \mathbf{u} - \beta^2 \text{ curl curl } \mathbf{u} + \mathbf{F} + \omega^2 \mathbf{u} = 0. \tag{2.99}$$

Let us assume that [Eq. (2.65)]

$$\mathbf{u} = \mathbf{u}_\alpha + \mathbf{u}_\beta, \qquad \text{curl } \mathbf{u}_\alpha = 0, \qquad \text{div } \mathbf{u}_\beta = 0, \tag{2.100}$$

so that $\mathbf{u}_\alpha$ is the irrotational part and $\mathbf{u}_\beta$ the solenoidal part of the displacement vector. Then, for $\mathbf{F} = 0$, Eq. (2.99) may be written as

$$\alpha^2 \text{ grad div } \mathbf{u}_\alpha - \beta^2 \text{ curl curl } \mathbf{u}_\beta + \omega^2(\mathbf{u}_\alpha + \mathbf{u}_\beta) = 0,$$

i.e.,

$$\alpha^2(\nabla^2 + k_\alpha^2)\mathbf{u}_\alpha + \beta^2(\nabla^2 + k_\beta^2)\mathbf{u}_\beta = 0, \tag{2.101}$$

with $k_\alpha = \omega/\alpha$, $k_\beta = \omega/\beta$. Equation (2.101) will be identically satisfied if $\mathbf{u}_\alpha$ and $\mathbf{u}_\beta$ satisfy the equations

$$(\nabla^2 + k_\alpha^2)\mathbf{u}_\alpha = 0, \qquad (\nabla^2 + k_\beta^2)\mathbf{u}_\beta = 0. \tag{2.102}$$

Because curl $\mathbf{L} = 0$, div $\mathbf{M} = 0$, div $\mathbf{N} = 0$, we see that the three independent eigenvector solutions of the force-free Navier equation, Eq. (2.99), are given by

$$\begin{aligned} \mathbf{L} &= \frac{1}{k_\alpha} \text{ grad } \Phi_\alpha, \\ \mathbf{M} &= \text{curl}(\mathbf{a}\Phi_\beta) = (\text{grad } \Phi_\beta) \times \mathbf{a} = \frac{1}{k_\beta} \text{ curl } \mathbf{N}, \\ \mathbf{N} &= \frac{1}{k_\beta} \text{ curl curl}(\mathbf{a}\Phi_\beta) = \frac{1}{k_\beta} \text{ curl } \mathbf{M}, \end{aligned} \tag{2.103}$$

where $\mathbf{a} = \mathbf{e}_z$ in Cartesian and cylindrical coordinates and $\mathbf{a} = \mathbf{r}$ in spherical coordinates. The potentials $\Phi_{\alpha,\beta}$ satisfy the Helmholtz equations

$$(\nabla^2 + k_\alpha^2)\Phi_\alpha = 0, \qquad (\nabla^2 + k_\beta^2)\Phi_\beta = 0. \tag{2.104}$$

The eigenfunctions $\Phi_{\alpha,\beta}$ and the Hansen vectors $\mathbf{L}$, $\mathbf{M}$, $\mathbf{N}$ for the cylindrical and spherical coordinates are given in Table 2.2. In this table, $\mathbf{L}_{ml}$ and $\mathbf{N}_{ml}$ are given in terms of the spherical Bessel functions $f_{l-1}^{\pm}$ and $f_{l+1}^{\pm}$. Alternatively, these vectors may be expressed in terms of $f_l^{\pm}$ and $f_l'^{\pm}$ as was done in the case of the eigenvectors for the vector Helmholtz equation [Eq. (2.85)].

In general, the displacement field will be a linear combination of the three Hansen vectors. Therefore, $\mathbf{u}$ may be expressed in the form

$$\mathbf{u}(\mathbf{r}) = U(r)\mathbf{P}_{ml}(\theta, \phi) + V(r)\sqrt{[l(l+1)]}\mathbf{B}_{ml}(\theta, \phi) + W(r)\sqrt{[l(l+1)]}\mathbf{C}_{ml}(\theta, \phi).$$

The corresponding stress dyadic is given by

$$\mathfrak{T} = \lambda \mathfrak{I} \text{ div } \mathbf{u} + \mu(\nabla \mathbf{u} + \mathbf{u}\nabla),$$

Table 2.2. Eigenvector Solutions of the Navier Equation

Spherical (r, θ, ϕ)		Cylindrical $(\Delta\ \phi, z)$	
Interior	Exterior	Interior	Exterior
$\mathbf{M}_{ml}^{+}(k_\beta r) = \operatorname{curl}[\mathbf{r}\Phi_{ml}^{+}(k_\beta r)]$	$\mathbf{M}_{ml}^{-}(k_\beta r) = \operatorname{curl}[\mathbf{r}\Phi_{ml}^{-}(k_\beta r)]$	$\mathbf{M}_{m}^{+}(k_\beta \Delta) = \frac{1}{k}\operatorname{curl}[\mathbf{e}_z\Phi_{m}^{+}(k_\beta\Delta)]$	$\mathbf{M}_{m}^{-}(k_\beta \Delta) = \frac{1}{k}\operatorname{curl}[\mathbf{e}_z\Phi_{m}^{-}(k_\beta\Delta)]$
$= j_l(k_\beta r)\sqrt{[l(l+1)]}\mathbf{C}_{ml}$	$= h_l^{(2)}(k_\beta r)\sqrt{[l(l+1)]}\mathbf{C}_{ml}$	$= e^{\nu_\beta z}\mathbf{C}_m$	$= e^{-\nu_\beta z}\mathbf{C}_m$
$\mathbf{N}_{ml}^{+}(k_\beta r) = \frac{1}{k_\beta}\operatorname{curl}\operatorname{curl}[\mathbf{r}\Phi_{ml}^{+}(k_\beta r)]$	$\mathbf{N}_{ml}^{-}(k_\beta r) = \frac{1}{k_\beta}\operatorname{curl}\operatorname{curl}[\mathbf{r}\Phi_{ml}^{-}(k_\beta r)]$	$\mathbf{N}_{m}^{+}(k_\beta \Delta) = \frac{1}{kk_\beta}\operatorname{curl}\operatorname{curl}[\mathbf{e}_z\Phi_{m}^{+}(k_\beta\Delta)]$	$\mathbf{N}_{m}^{-}(k_\beta \Delta) = \frac{1}{kk_\beta}\operatorname{curl}\operatorname{curl}[\mathbf{e}_z\Phi_{m}^{-}(k_\beta\Delta)]$
$= \frac{1}{2l+1}[(l+1)j_{l-1}(k_\beta r)\mathbf{\Lambda}_{ml}^{1}$	$= \frac{1}{2l+1}[(l+1)h_{l-1}^{(2)}(k_\beta r)\mathbf{\Lambda}_{ml}^{1}$	$= \frac{1}{k_\beta}e^{\nu_\beta z}[k\mathbf{P}_m + \nu_\beta\mathbf{B}_m]$	$= \frac{1}{k_\beta}e^{-\nu_\beta z}[k\mathbf{P}_m - \nu_\beta\mathbf{B}_m]$
$- lj_{l+1}(k_\beta r)\mathbf{\Lambda}_{ml}^{2}]$	$- lh_{l+1}^{(2)}(k_\beta r)\mathbf{\Lambda}_{ml}^{2}]$		
$\mathbf{L}_{ml}^{+}(k_\sigma r) = \frac{1}{k_\sigma}\operatorname{grad}\Phi_{ml}^{+}(k_\sigma r)$	$\mathbf{L}_{ml}^{-}(k_\alpha r) = \frac{1}{k_\sigma}\operatorname{grad}\Phi_{ml}^{-}(k_\sigma r)$	$\mathbf{L}_{m}^{+}(k_\sigma \Delta) = \frac{1}{k_\sigma}\operatorname{grad}\Phi_{m}^{+}(k_\sigma\Delta)$	$\mathbf{L}_{m}^{-}(k_\sigma \Delta) = \frac{1}{k_\alpha}\operatorname{grad}\Phi_{m}^{-}(k_\sigma\Delta)$
$= \frac{1}{2l+1}[j_{l-1}(k_\alpha r)\mathbf{\Lambda}_{ml}^{1}$	$= \frac{1}{2l+1}[h_{l-1}^{(2)}(k_\alpha r)\mathbf{\Lambda}_{ml}^{1}$	$= \frac{1}{k_\sigma}e^{\nu_\alpha z}[\nu_\sigma\mathbf{P}_m + k\mathbf{B}_m]$	$= \frac{1}{k_\sigma}e^{-\nu_\alpha z}[-\nu_\sigma\mathbf{P}_m + k\mathbf{B}_m]$
$+ j_{l+1}(k_\sigma r)\mathbf{\Lambda}_{ml}^{2}]$	$+ h_{l+1}^{(2)}(k_\sigma r)\mathbf{\Lambda}_{ml}^{2}]$		
$\Phi_{ml}^{+}(k_c r) = j_l(k_c r)Y_{ml}(\theta, \phi)$	$\Phi_{ml}^{-}(k_c r) = h_l^{(2)}(k_c r)Y_{ml}(\theta, \phi)$	$\Phi_{m}^{+}(k_c\Delta) = e^{\nu_c z}Y_m(k\Delta, \phi)$	$\Phi_{m}^{-}(k_c\Delta) = e^{-\nu_c z}Y_m(k\Delta, \phi)$
$\mathbf{\Lambda}_{ml}^{1} = l\mathbf{P}_{ml} + \sqrt{[l(l+1)]}\mathbf{B}_{ml}$	$\mathbf{\Lambda}_{ml}^{2} = -(l+1)\mathbf{P}_{ml} + \sqrt{[l(l+1)]}\mathbf{B}_{ml}$	$k_c = \omega/c, \quad \nu_c = (k^2 - k_c^2)^{1/2}, \quad c = \alpha \text{ or } \beta$	

where

$$\operatorname{div} \mathbf{u} = \left[\frac{dU}{dr} + 2\frac{U}{r} - l(l+1)\frac{V}{r}\right] Y_{ml}(\theta, \phi),$$

$$\begin{aligned}\nabla\mathbf{u} + \mathbf{u}\nabla = {} & \frac{2}{r}\left[\left\{U + V\frac{\partial^2}{\partial\theta^2} + W\frac{1}{\sin\theta}\left(\frac{\partial}{\partial\theta} - \cot\theta\right)\frac{\partial}{\partial\phi}\right\} Y_{ml}(\theta, \phi)\right]\mathbf{e}_\theta\mathbf{e}_\theta \\
& + \frac{2}{r}\left[\left\{U - V\frac{\partial^2}{\partial\theta^2} - l(l+1)V\right.\right. \\
& \left.\left. - W\frac{1}{\sin\theta}\left(\frac{\partial}{\partial\theta} - \cot\theta\right)\frac{\partial}{\partial\phi}\right\} Y_{ml}(\theta, \phi)\right]\mathbf{e}_\phi\mathbf{e}_\phi \\
& + 2\frac{dU}{dr} Y_{ml}(\theta, \phi)\mathbf{e}_r\mathbf{e}_r + \frac{1}{r}\left[\left\{2V\frac{1}{\sin\theta}\left(\frac{\partial}{\partial\theta} - \cot\theta\right)\frac{\partial}{\partial\phi}\right.\right. \\
& \left.\left. - W\left(2\frac{\partial^2}{\partial\theta^2} + l^2 + l\right)\right\} Y_{ml}(\theta, \phi)\right](\mathbf{e}_\theta\mathbf{e}_\phi + \mathbf{e}_\phi\mathbf{e}_\theta) \\
& + \left[\left\{\left(\frac{dV}{dr} + \frac{U - V}{r}\right)\frac{1}{\sin\theta}\frac{\partial}{\partial\phi}\right.\right. \\
& \left.\left. - \left(\frac{dW}{dr} - \frac{W}{r}\right)\frac{\partial}{\partial\theta}\right\} Y_{ml}(\theta, \phi)\right](\mathbf{e}_\phi\mathbf{e}_r + \mathbf{e}_r\mathbf{e}_\phi) + \left[\left\{\left(\frac{dV}{dr} + \frac{U - V}{r}\right)\frac{\partial}{\partial\theta}\right.\right. \\
& \left.\left. + \left(\frac{dW}{dr} - \frac{W}{r}\right)\frac{1}{\sin\theta}\frac{\partial}{\partial\phi}\right\} Y_{ml}(\theta, \phi)\right](\mathbf{e}_r\mathbf{e}_\theta + \mathbf{e}_\theta\mathbf{e}_r). \qquad (2.105)\end{aligned}$$

The stress vector across a spherical surface $r =$ const. can now be easily calculated as

$$\begin{aligned}\mathbf{T}(\mathbf{e}_r) = {} & \mathbf{e}_r \cdot \mathfrak{T} = \lambda\mathbf{e}_r \operatorname{div}\mathbf{u} + \mu\mathbf{e}_r \cdot (\nabla\mathbf{u} + \mathbf{u}\nabla) \\
= {} & \lambda\mathbf{e}_r\left[\frac{dU}{dr} + 2\frac{U}{r} - l(l+1)\frac{V}{r}\right] Y_{ml}(\theta, \phi) \\
& + \mu\left[2\mathbf{e}_r\frac{dU}{dr} + \mathbf{e}_\theta\left\{\left(\frac{dV}{dr} + \frac{U - V}{r}\right)\frac{\partial}{\partial\theta}\right.\right. \\
& \left. + \left(\frac{dW}{dr} - \frac{W}{r}\right)\frac{1}{\sin\theta}\frac{\partial}{\partial\phi}\right\} + \mathbf{e}_\phi\left\{\left(\frac{dV}{dr} + \frac{U - V}{r}\right)\frac{1}{\sin\theta}\frac{\partial}{\partial\phi}\right. \\
& \left.\left. - \left(\frac{dW}{dr} - \frac{W}{r}\right)\frac{\partial}{\partial\theta}\right\}\right] Y_{ml}(\theta, \phi). \qquad (2.106)\end{aligned}$$

EXAMPLE 2.1: Two-dimensional Fields

We shall show that if the motion is independent of the y coordinate, the displacement vector (u, v, w) may be expressed in the form

$$u = \frac{\partial\Phi}{\partial x} + \frac{\partial\Psi}{\partial z}, \qquad v = v, \qquad w = \frac{\partial\Phi}{\partial z} - \frac{\partial\Psi}{\partial x}, \qquad (2.1.1)$$

and the corresponding stress vector across a plane parallel to the xy plane is given by

$$\begin{aligned}\tau_{zx} &= \mu\left(2\frac{\partial^2\Phi}{\partial x\partial z} - \frac{\partial^2\Psi}{\partial x^2} + \frac{\partial^2\Psi}{\partial z^2}\right),\\ \tau_{zy} &= \mu\frac{\partial v}{\partial z},\\ \tau_{zz} &= \lambda\nabla^2\Phi + 2\mu\left(\frac{\partial^2\Phi}{\partial z^2} - \frac{\partial^2\Psi}{\partial x\partial z}\right).\end{aligned} \tag{2.1.2}$$

Let

$$\mathbf{u} = \mathbf{L} + \mathbf{M} + \mathbf{N}, \tag{2.1.3}$$

where

$$\mathbf{L} = \text{grad}\,\Phi, \qquad \mathbf{M} = \text{curl}(\mathbf{e}_z\Psi_1), \qquad \mathbf{N} = \text{curl curl}(\mathbf{e}_z\Psi_2). \tag{2.1.4}$$

The potentials Φ, Ψ_1, and Ψ_2 satisfy the equations

$$\nabla^2\Phi = \frac{1}{\alpha^2}\frac{\partial^2\Phi}{\partial t^2}, \qquad \nabla^2\Psi_{1,2} = \frac{1}{\beta^2}\frac{\partial^2}{\partial t^2}\Psi_{1,2}. \tag{2.1.5}$$

Because the motion is independent of y, $\partial/\partial y \equiv 0$, and Eqs. (2.1.4) yield

$$\begin{aligned}\mathbf{L} &= \mathbf{e}_x\frac{\partial\Phi}{\partial x} + \mathbf{e}_z\frac{\partial\Phi}{\partial z},\\ \mathbf{M} &= -\mathbf{e}_y\frac{\partial\Psi_1}{\partial x},\\ \mathbf{N} &= \mathbf{e}_x\frac{\partial^2\Psi_2}{\partial x\partial z} - \mathbf{e}_z\frac{\partial^2\Psi_2}{\partial x^2}.\end{aligned} \tag{2.1.6}$$

From Eqs. (2.1.3) and (2.1.6), we find

$$u = \frac{\partial\Phi}{\partial x} + \frac{\partial\Psi}{\partial z}, \qquad w = \frac{\partial\Phi}{\partial z} - \frac{\partial\Psi}{\partial x}, \tag{2.1.7}$$

where $\Psi = \partial\Psi_2/\partial x$. Putting $\partial/\partial y \equiv 0$ in Eq. (1.68), we get

$$\tau_{zx} = \mu\left(\frac{\partial u}{\partial z} + \frac{\partial w}{\partial x}\right), \qquad \tau_{zy} = \mu\frac{\partial v}{\partial z}, \qquad \tau_{zz} = \lambda\left(\frac{\partial u}{\partial x} + \frac{\partial w}{\partial z}\right) + 2\mu\frac{\partial w}{\partial z}, \tag{2.1.8}$$

which, on using Eq. (2.1.7), yields Eq. (2.1.2).

It may be noted that Eqs. (2.1.1) and (2.1.2) express the displacement and stress vectors in terms of three functions: Φ, Ψ, and v. The function v is uncoupled from the functions Φ and Ψ, although Φ and Ψ themselves are coupled together. Therefore, two independent problems can be discussed. In the first problem $\Phi = \Psi = 0$ and, therefore, $u = w = \tau_{zx} = \tau_{zz} = 0$. In the second problem $v = 0$, which yields $\tau_{zy} = 0$. This separation of the original problem into two independent problems is of great practical value.

EXAMPLE 2.2: Axially symmetric Fields

Here we show that if the motion is symmetrical about the z axis, the displacement vector (u_Δ, u_ϕ, u_z) may be expressed in the form

$$u_\Delta = \frac{\partial \Phi}{\partial \Delta} + \frac{\partial \Psi}{\partial z}, \qquad u_\phi = v, \qquad u_z = \frac{\partial \Phi}{\partial z} - \frac{\partial \Psi}{\partial \Delta} - \frac{\Psi}{\Delta}, \tag{2.2.1}$$

and the corresponding stress vector across a plane $z = \text{const.}$ is given by

$$\begin{aligned} \tau_{z\Delta} &= \mu\left(2\frac{\partial^2 \Phi}{\partial \Delta \partial z} + \frac{\partial^2 \Psi}{\partial z^2} - \frac{\partial^2 \Psi}{\partial \Delta^2} - \frac{1}{\Delta}\frac{\partial \Psi}{\partial \Delta} + \frac{\Psi}{\Delta^2}\right), \\ \tau_{z\phi} &= \mu \frac{\partial v}{\partial z}, \\ \tau_{zz} &= \lambda \nabla^2 \Phi + 2\mu\left(\frac{\partial^2 \Phi}{\partial z^2} - \frac{\partial^2 \Psi}{\partial \Delta \partial z} - \frac{1}{\Delta}\frac{\partial \Psi}{\partial z}\right). \end{aligned} \tag{2.2.2}$$

Equations (2.1.3)–(2.1.5) of Example 2.1 are valid even in the present case. On putting $\partial/\partial\phi \equiv 0$, Eq. (2.1.4) can be made to yield

$$\begin{aligned} \mathbf{L} &= \mathbf{e}_\Delta \frac{\partial \Phi}{\partial \Delta} + \mathbf{e}_z \frac{\partial \Phi}{\partial z}, \\ \mathbf{M} &= -\mathbf{e}_\phi \frac{\partial \Psi_1}{\partial \Delta}, \\ \mathbf{N} &= \mathbf{e}_\Delta \frac{\partial \Psi}{\partial z} - \mathbf{e}_z\left(\frac{\partial \Psi}{\partial \Delta} + \frac{\Psi}{\Delta}\right), \end{aligned} \tag{2.2.3}$$

where $\Psi = \partial \Psi_2/\partial \Delta$. Because Ψ_2 satisfies

$$\nabla^2 \Psi_2 = \frac{1}{\beta^2}\frac{\partial^2 \Psi_2}{\partial t^2}, \tag{2.2.4}$$

it is easily seen that Ψ satisfies the equation

$$\nabla^2 \Psi - \frac{1}{\Delta^2}\Psi = \frac{1}{\beta^2}\frac{\partial^2 \Psi}{\partial t^2}. \tag{2.2.5}$$

From Eqs. (2.1.3) and (2.2.3), we get Eq. (2.2.1). Next, putting $\partial/\partial\phi \equiv 0$ in Eq. (1.55) and using

$$\mathbf{e}_z \cdot \mathfrak{T} = \lambda \mathbf{e}_z \operatorname{div} \mathbf{u} + 2\mu \mathbf{e}_z \cdot \mathfrak{E},$$

we find

$$\begin{aligned} \tau_{z\Delta} &= \mu\left(\frac{\partial u_\Delta}{\partial z} + \frac{\partial u_z}{\partial \Delta}\right), \\ \tau_{z\phi} &= \mu \frac{\partial u_\phi}{\partial z} = \mu \frac{\partial v}{\partial z}, \\ \tau_{zz} &= \lambda\left(\frac{\partial u_\Delta}{\partial \Delta} + \frac{u_\Delta}{\Delta}\right) + (\lambda + 2\mu)\frac{\partial u_z}{\partial z}. \end{aligned} \tag{2.2.6}$$

Using Eqs. (2.2.1) in Eqs. (2.2.6) we get Eqs. (2.2.2).

Once again we note that the original problem splits into two independent problems. The first one has $u_\Delta = u_z = \tau_{z\Delta} = \tau_{zz} = 0$ and the second $u_\phi = \tau_{z\phi} = 0$. The former is solved in terms of v and the latter in terms of Φ and Ψ. Note that although Φ satisfies the wave equation, Ψ does not.

2.5. Plane Waves

We have seen in Section 2.1 that the three-dimensional scalar wave equation, Eq. (2.1), admits solutions of the form

$$\Psi_c(\mathbf{r}, t) = f\left(t - \frac{\mathbf{p}\cdot\mathbf{r}}{c}\right), \tag{2.107}$$

where f is an arbitrary, twice-differentiable function; $\mathbf{p}$ is a constant unit vector; $\mathbf{r}$ denotes the position vector; and c is the constant velocity of propagation. At any given time t, the equation $\mathbf{p}\cdot\mathbf{r} =$ const. describes a plane normal to the unit vector $\mathbf{p}$. The solution Ψ_c therefore represents a *progressive plane wave* whose plane of constant phase is normal to $\mathbf{p}$ and propagates with velocity c. The corresponding eigenvector solutions of the source-free Navier equation are obtained from Eq. (2.103)

$$\begin{aligned}
\mathbf{L} &= \frac{1}{k_\alpha}\operatorname{grad}\Psi_\alpha = -\frac{1}{\alpha k_\alpha}\Psi'_\alpha\mathbf{p},\\
\mathbf{M} &= \operatorname{curl}(\mathbf{e}_3\Psi_\beta) = -\frac{1}{\beta}\Psi'_\beta(\mathbf{p}\times\mathbf{e}_3),\\
\mathbf{N} &= \frac{1}{k_\beta}\operatorname{curl}\mathbf{M} = -\frac{1}{\beta^2 k_\beta}\Psi''_\beta[(\mathbf{p}\times\mathbf{e}_3)\times\mathbf{p}].
\end{aligned}\tag{2.108}$$

Here a prime denotes differentiation with respect to the argument.

The particular choice $f(t) = e^{i\omega t}$ renders *a harmonic plane wave*

$$\Psi_c(\mathbf{r}, \omega) = e^{i(\omega t - \mathbf{k}_c\cdot\mathbf{r})} = e^{i[\omega t - k_c(lx_1 + mx_2 + nx_3)]}, \tag{2.109}$$

where

$$\mathbf{k}_c = k_c\mathbf{p}, \qquad k_c = \frac{\omega}{c}, \qquad c = \alpha \text{ or } \beta, \qquad \mathbf{p} = l\mathbf{e}_1 + m\mathbf{e}_2 + n\mathbf{e}_3,$$

$$l^2 + m^2 + n^2 = 1. \tag{2.110}$$

The vector $\mathbf{k}_c$ is the *propagation vector* and its magnitude k_c is the *wave number*. Equations (2.108) become

$$\begin{aligned}
\mathbf{L} &= -i\Psi_\alpha\mathbf{p} = -i(l\mathbf{e}_1 + m\mathbf{e}_2 + n\mathbf{e}_3)\Psi_\alpha,\\
\mathbf{M} &= -ik_\beta\Psi_\beta(\mathbf{p}\times\mathbf{e}_3) = -i(m\mathbf{e}_1 - l\mathbf{e}_2)k_\beta\Psi_\beta,\\
\mathbf{N} &= k_\beta\Psi_\beta[(\mathbf{p}\times\mathbf{e}_3)\times\mathbf{p}] = [-ln\mathbf{e}_1 - mn\mathbf{e}_2 + (l^2 + m^2)\mathbf{e}_3]k_\beta\Psi_\beta.
\end{aligned}\tag{2.111}$$

Therefore, the three eigenvectors are mutually perpendicular. The vector **L** describes a motion with velocity α in which the displacement is in the direction of propagation. This partial field is therefore purely longitudinal and the corresponding waves are known as *longitudinal* or *P waves* (*P* for primary). In contrast, each of the displacement vectors **M** and **N** is perpendicular to the direction of propagation, the corresponding disturbance propagating with velocity β. Therefore, these waves are known as *transverse* or *S waves* (*S* for secondary or shear).

To gain a better understanding of the overall motion implied by these solutions we set up a Cartesian coordinate system (x_1, x_2, x_3) in which a plane wave propagates away from the origin in the direction of the vector $\mathbf{p}$.

Figure 2.1 shows the geometric parameters that constitute a plane wave. $Q(\mathbf{r})$ is a point on the wavefront Π at which the field is observed. Because $\Psi_{\alpha,\beta}$ are constants all over the plane at a given frequency and time, so are **L**, **M**, and **N** at every point over Π. By rotating the system about the x_3 axis (through an angle λ), the normal $\mathbf{p}$ can be made to lie in the x_1x_3 plane (Fig. 2.2). Equations (2.109) and (2.111) then yield ($m = 0$)

$$\mathbf{L} = (l\mathbf{e}_1 + n\mathbf{e}_3)\exp\left[i\omega\left(t - \frac{lx_1 + nx_3}{\alpha}\right) - \frac{i\pi}{2}\right],$$

$$\mathbf{N} = lk_\beta(n\mathbf{e}_1 - l\mathbf{e}_3)\exp\left[i\omega\left(t - \frac{lx_1 + nx_3}{\beta}\right) + i\pi\right], \qquad (2.112)$$

$$\mathbf{M} = lk_\beta\mathbf{e}_2 \exp\left[i\omega\left(t - \frac{lx_1 + nx_3}{\beta}\right) + \frac{i\pi}{2}\right],$$

with $l^2 + n^2 = 1$. We take the x_1x_3 plane to be vertical and the x_3 axis as vertically upwards. From Eq. (2.112) it is clear that both **L** and **N** lie in the vertical plane x_1x_3, while **M** is horizontal and perpendicular to this plane. The vector **N** then describes a shear-wave motion that is linearly polarized in the vertical plane. This is designated in seismology as *SV motion* (*SV* for vertical shear). The plane of **L** and **N** is called the *plane of incidence.* Clearly, **M** describes a shear motion polarized linearly in a horizontal plane. It is called *SH* motion (*SH* for horizontal shear). The angle between the vectors **M** and (**M** + **N**), ε, is the *polarization angle.* The plane of **L** and (**M** + **N**) is sometimes called the *plane of polarization.* Note that **M** has no vertical component, whereas both **L** and **N** have vertical as well as horizontal components.

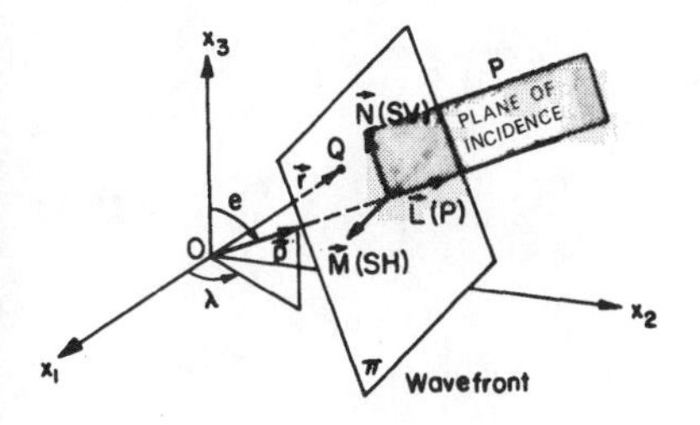

Figure 2.1. A plane wavefront and the **L**, **M**, **N** vectors representing the *P*, *SH* and *SV* motions, respectively. The vector **L** is perpendicular to the wavefront. According to our convention (**L**, **N**, **M**) form a right-handed triad.

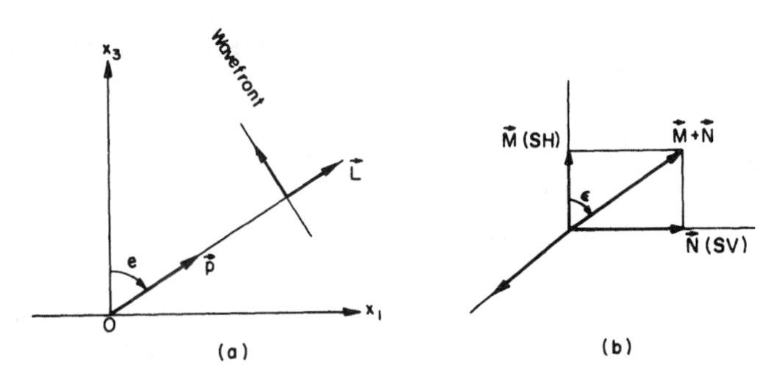

Figure 2.2. The polarization angle ε and the plane of incidence $(x_1 0 x_3)$. (a) Vectors associated with a longitudinal plane wave; (b) vectors associated with a shear plane wave.

Writing $\mathbf{p} = l\mathbf{e}_1 + n\mathbf{e}_3$ in Eq. (2.112), we can write the displacement fields induced by plane P, SV, and SH waves in the form:

$$\begin{aligned} \mathbf{u}_{\mathrm{P}} &= A\mathbf{p} \exp\left[i\omega\left(t - \frac{\mathbf{p}\cdot\mathbf{r}}{\alpha}\right)\right], \\ \mathbf{u}_{\mathrm{SV}} &= B(\mathbf{e}_2 \times \mathbf{p})\exp\left[i\omega\left(t - \frac{\mathbf{p}\cdot\mathbf{r}}{\beta}\right)\right], \\ \mathbf{u}_{\mathrm{SH}} &= C\mathbf{e}_2 \exp\left[i\omega\left(t - \frac{\mathbf{p}\cdot\mathbf{r}}{\beta}\right)\right]. \end{aligned} \tag{2.113}$$

Assuming A, B, and C to be real and considering the real parts only, we have

$$\begin{aligned} \mathbf{u}_{\mathrm{P}} &= A\mathbf{p} \cos \omega\left(t - \frac{\mathbf{p}\cdot\mathbf{r}}{\alpha}\right), \\ \mathbf{u}_{\mathrm{SV}} &= B(\mathbf{e}_2 \times \mathbf{p})\cos \omega\left(t - \frac{\mathbf{p}\cdot\mathbf{r}}{\beta}\right), \\ \mathbf{u}_{\mathrm{SH}} &= (C\mathbf{e}_2)\cos \omega\left(t - \frac{\mathbf{p}\cdot\mathbf{r}}{\beta}\right). \end{aligned} \tag{2.114}$$

The corresponding expressions for the stress dyadics are

$$\begin{aligned} \mathfrak{T}(\mathbf{u}_{\mathrm{P}}) &= \frac{A}{\alpha}\,\omega(\lambda\mathfrak{I} + 2\mu\mathbf{pp})\sin \omega\left(t - \frac{\mathbf{p}\cdot\mathbf{r}}{\alpha}\right), \\ \mathfrak{T}(\mathbf{u}_{\mathrm{SV}}) &= \mu\frac{B}{\beta}\,\omega[\mathbf{p}(\mathbf{e}_2 \times \mathbf{p}) + (\mathbf{e}_2 \times \mathbf{p})\mathbf{p}]\sin \omega\left(t - \frac{\mathbf{p}\cdot\mathbf{r}}{\beta}\right), \\ \mathfrak{T}(\mathbf{u}_{\mathrm{SH}}) &= \mu\frac{C}{\beta}\,\omega(\mathbf{p}\mathbf{e}_2 + \mathbf{e}_2\mathbf{p})\sin \omega\left(t - \frac{\mathbf{p}\cdot\mathbf{r}}{\beta}\right). \end{aligned} \tag{2.114a}$$

Because the energy-flux density is given by $\mathbf{\Sigma} = -\mathfrak{T}(\mathbf{u}) \cdot \dot{\mathbf{u}}$, we find

$$\begin{aligned}
\mathbf{\Sigma}_{\mathrm{P}} &= A^2\left(\frac{\omega^2}{\alpha}\right)(\lambda + 2\mu)\mathbf{p} \sin^2 \omega\left(t - \frac{\mathbf{p}\cdot\mathbf{r}}{\alpha}\right), \\
\mathbf{\Sigma}_{\mathrm{SV}} &= B^2\left(\frac{\omega^2}{\beta}\right)\mu\mathbf{p} \sin^2 \omega\left(t - \frac{\mathbf{p}\cdot\mathbf{r}}{\beta}\right), \\
\mathbf{\Sigma}_{\mathrm{SH}} &= C^2\left(\frac{\omega^2}{\beta}\right)\mu\mathbf{p} \sin^2 \omega\left(t - \frac{\mathbf{p}\cdot\mathbf{r}}{\beta}\right).
\end{aligned} \tag{2.115}$$

We will be interested in the amount of energy that flows across a unit area of the plane $x_3 = 0$ in a unit time. This is given by

$$\begin{aligned}
\mathbf{e}_3 \cdot \mathbf{\Sigma}_{\mathrm{P}} &= \rho\alpha\omega^2 A^2 n \sin^2 \omega\left(t - \frac{l x_1}{\alpha}\right), \\
\mathbf{e}_3 \cdot \mathbf{\Sigma}_{\mathrm{SV}} &= \rho\beta\omega^2 B^2 n \sin^2 \omega\left(t - \frac{l x_1}{\beta}\right), \\
\mathbf{e}_3 \cdot \mathbf{\Sigma}_{\mathrm{SH}} &= \rho\beta\omega^2 C^2 n \sin^2 \omega\left(t - \frac{l x_1}{\beta}\right).
\end{aligned} \tag{2.116}$$

Averaging over a cycle, we finally get

$$\begin{aligned}
\mathbf{e}_3 \cdot \mathbf{\Sigma}_{\mathrm{P}} &= \tfrac{1}{2}\rho\alpha n\omega^2 A^2, \\
\mathbf{e}_3 \cdot \mathbf{\Sigma}_{\mathrm{SV}} &= \tfrac{1}{2}\rho\beta n\omega^2 B^2, \\
\mathbf{e}_3 \cdot \mathbf{\Sigma}_{\mathrm{SH}} &= \tfrac{1}{2}\rho\beta n\omega^2 C^2.
\end{aligned} \tag{2.117}$$

EXAMPLE 2.3: Stress–Energy Tensors for Plane Waves

For P waves:

$$\mathbf{p} = p_1\mathbf{e}_x + p_2\mathbf{e}_y + p_3\mathbf{e}_z, \qquad \phi_P = \omega\left(t - \frac{\mathbf{p}\cdot\mathbf{r}}{\alpha}\right),$$

$$\mathbf{u}_P = A\mathbf{p}\cos\phi_P, \qquad |\mathbf{p}| = 1,$$

$$B_P = \rho\omega^2 A^2 \sin^2\phi_P \begin{bmatrix} -p_1^2 & -p_1p_2 & -p_1p_3 & -p_1/\alpha \\ -p_1p_2 & -p_2^2 & -p_2p_3 & -p_2/\alpha \\ -p_1p_3 & -p_2p_3 & -p_3^2 & -p_3/\alpha \\ p_1\alpha & p_2\alpha & p_3\alpha & 1 \end{bmatrix}.$$

For S waves:

$$\phi_S = \omega\left(t - \frac{\mathbf{p}\cdot\mathbf{r}}{\beta}\right), \qquad \mathbf{u}_S = B\mathbf{q}\cos\phi_S, \qquad |\mathbf{q}| = 1, \qquad \mathbf{p}\cdot\mathbf{q} = 0,$$

$$B_S = \rho\omega^2 B^2 \sin^2\phi_S \begin{bmatrix} -p_1^2 & -p_1p_2 & -p_1p_3 & -p_1/\beta \\ -p_1p_2 & -p_2^2 & -p_2p_3 & -p_2/\beta \\ -p_1p_3 & -p_2p_3 & -p_3^2 & -p_3/\beta \\ p_1\beta & p_2\beta & p_3\beta & 1 \end{bmatrix}.$$

2.6. Interrelations Among Plane, Cylindrical, and Spherical Waves

The principle of superposition, a direct result of the linearity of the wave equation, can be used as a powerful tool to generate new solutions by integration or summation. Let $\Psi = \Psi_n(x_1, x_2, \ldots, x_n; t)$ satisfy the n-dimensional wave equation

$$\left(\frac{\partial}{\partial x_1^2} + \frac{\partial}{\partial x_2^2} + \cdots + \frac{\partial^2}{\partial x_n^2}\right)\Psi = \frac{1}{c^2}\frac{\partial^2 \Psi}{\partial t^2}. \tag{2.118}$$

Clearly $\Psi = \Psi_n(x_1 + \xi_1, x_2, \ldots, x_n; t)$ is also a solution of Eq. (2.118). Hence

$$\Psi_{n-1} = \int_{\xi_1 = -\infty}^{\xi_1 = +\infty} \Psi_n(x_1 + \xi_1, x_2, \ldots, x_n; t)d\xi_1 = \int_{\xi = -\infty}^{\xi = +\infty} \Psi_n(\xi, x_2, \ldots, x_n; t)d\xi \tag{2.119}$$

satisfies the wave equation in $(n - 1)$ dimensions, provided the integral converges. Therefore, starting with a solution Ψ_n in n dimensions, solutions in $(n - 1)$, $(n - 2)$, etc., dimensions can be obtained. The method essentially consists of superposing wave solutions in the direction of one axis.

However, if we superpose solutions corresponding to different values of an angle, a new solution in $(n + 1)$ dimensions is obtained from a given solution in n dimensions. This method is particularly useful in three-dimensional cylindrical and spherical coordinates. Hence, when the two-dimensional solution $\Psi_2 = \Psi(z, x)$ is written as $\Psi_2 = \Psi(z, \Delta \cos \phi)$, we get a three-dimensional solution of the form

$$\Psi_3(z, \Delta, \phi) = \int_{\lambda_1}^{\lambda_2} \Psi[z, \Delta \cos(\phi - \lambda)]g(\lambda)d\lambda, \tag{2.120}$$

where $g(\lambda)$ is an arbitrary weighting function.

We now turn to the Helmholtz equation, Eq. (2.10). Starting with the one-dimensional solution $S_1 = e^{ik_c x} = e^{ik_c \Delta \cos \phi}$ and noting that $\Delta \cos(\phi - \lambda) = x \cos \lambda + y \sin \lambda$, we generate the two-dimensional solution

$$S_2 = \int_{\lambda_1}^{\lambda_2} e^{ik_c(x \cos \lambda + y \sin \lambda)} g(\lambda)d\lambda. \tag{2.121}$$

An elementary wave function in cylindrical coordinates is [Eq. (2.36)]

$$e^{i\omega t}\Phi_m^{\pm}(k_c \Delta) = e^{im\phi} J_m(k\Delta)e^{i\omega t \pm \nu_c z}, \qquad \nu_c = \sqrt{(k^2 - k_c^2)}. \tag{2.122}$$

This represents standing cylindrical waves. It can also represent plane waves in which planes of constant phase are propagated along the z axis with the velocity $\omega/[\text{Re}\sqrt{(k_c^2 - k^2)}]$, whereas the amplitude over these planes is a function of Δ

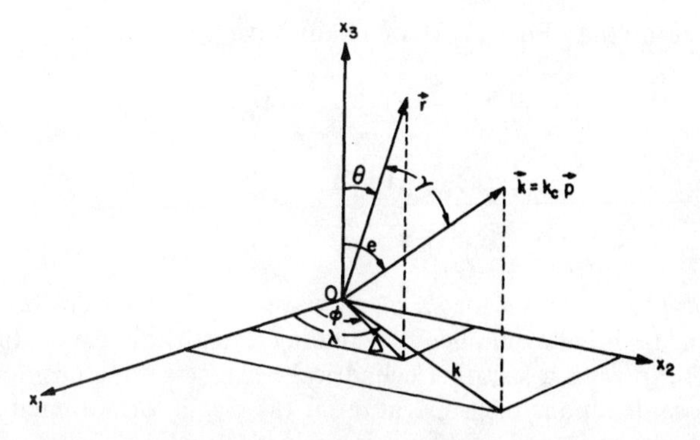

Figure 2.3. The spherical coordinates (r, θ, ϕ) and the corresponding coordinates in the wave-number space (k_c, e, λ). $(\Delta = r \sin\theta,\ k = k_c \sin e.)$

and ϕ. Nevertheless, Eq. (2.122) can be resolved into homogeneous plane waves. To show this we first note that

$$e^{-i\mathbf{k}\cdot\mathbf{r}} = e^{-i[k_c\Delta \sin e \cos(\phi-\lambda) + k_c z \cos e]} = e^{-ik\Delta\cos(\phi-\lambda) \pm v_c z}, \tag{2.123}$$

where (Fig. 2.3)

$$\mathbf{k} = k_c\mathbf{p} = k_c(\sin e \cos\lambda \mathbf{e}_1 + \sin e \sin\lambda \mathbf{e}_2 + \cos e \mathbf{e}_3),$$

$$\sin e = \frac{k}{k_c}, \qquad \cos e = \frac{1}{k_c}\sqrt{(k_c^2 - k^2)} = \pm\frac{iv_c}{k_c}, \qquad v_c = \sqrt{(k^2 - k_c^2)}. \tag{2.124}$$

Further, $J_m(\zeta)$ is the coefficient of $\exp(im\alpha)$ in the Fourier expansion of $\exp(i\zeta \sin\alpha)$; i.e.,

$$e^{i\zeta\sin\alpha} = \sum_{m=-\infty}^{\infty} J_m(\zeta)e^{im\alpha}, \tag{2.125}$$

$$J_m(\zeta) = \frac{1}{2\pi}\int_0^{2\pi} e^{i(\zeta\sin\alpha - m\alpha)}\,d\alpha. \tag{2.126}$$

Hence, we have

$$e^{ik\Delta\sin(\phi-\lambda-\pi/2)} = \sum_{m=-\infty}^{\infty} J_m(k\Delta)e^{im(\phi-\lambda-\pi/2)}, \tag{2.127}$$

$$J_m(k\Delta) = \frac{1}{2\pi}\int_0^{2\pi} e^{i[k\Delta\sin(\phi-\lambda-\pi/2) - m(\phi-\lambda-\pi/2)]}\,d\lambda. \tag{2.128}$$

Equations (2.123), (2.127), and (2.128) now yield

$$e^{-i\mathbf{k}\cdot\mathbf{r}} = \sum_{m=-\infty}^{\infty} J_m(k\Delta)e^{im(\phi-\lambda-\pi/2)\pm v_c z}, \tag{2.129}$$

$$J_m(k\Delta)e^{im\phi \pm v_c z} = \frac{1}{2\pi}\int_0^{2\pi} e^{i(-\mathbf{k}\cdot\mathbf{r} + m\pi/2 + m\lambda)}\,d\lambda. \tag{2.130}$$

Using the notation in Eq. (2.122), we finally have

$$e^{-i\mathbf{k}\cdot\mathbf{r}} = \sum_{m=-\infty}^{\infty} \frac{1}{g_m(\lambda)} \Phi_m^{\pm}(k_c\Delta), \tag{2.131}$$

$$\Phi_m^{\pm}(k_c\Delta) = \frac{1}{2\pi}\int_0^{2\pi} g_m(\lambda)e^{-i\mathbf{k}\cdot\mathbf{r}}\,d\lambda, \tag{2.132}$$

where $g_m(\lambda) = \exp[im(\lambda + \pi/2)]$.

Equation (2.131) shows that a plane wave may be considered as a superposition of an infinite number of standing cylindrical waves. In contrast, Eq. (2.132) expresses a standing cylindrical wave as a superposition of plane waves whose directions of propagation at the origin form a circular cone of angle $2e$ about the x_3 axis.

To find the interrelationship between the plane and the spherical waves, we first note that [Eq. (2.124)]

$$e^{-i\mathbf{k}\cdot\mathbf{r}} = e^{-ik_c r\cos\gamma}, \tag{2.133}$$

where γ, the angle between the propagation vector and the radius vector (Fig. 2.3), is given by

$$\cos\gamma = \cos\theta\cos e + \sin\theta\sin e\cos(\phi - \lambda). \tag{2.134}$$

For given k_c and $\mathbf{r}$, $\exp(-i\mathbf{k}\cdot\mathbf{r})$ is a function of the two angles, e and λ. Therefore, it should be possible to expand it in a convergent series of surface harmonics. To achieve this end, we start from the series exapnsion (App. D)

$$e^{-ik_c r\cos\gamma} = \sum_{l=0}^{\infty}(-i)^l(2l+1)j_l(k_c r)P_l(\cos\gamma). \tag{2.135}$$

Using the addition theorem

$$P_l(\cos\gamma) = \sum_{m=-l}^{l}\frac{(l-m)!}{(l+m)!}Y_{ml}(\theta,\phi)\overset{*}{Y}_{ml}(e,\lambda) \tag{2.136}$$

in Eq. (2.135), we get the desired expansion

$$e^{-ik_c r\cos\gamma} = \sum_{l=0}^{\infty}\sum_{m=-l}^{l}(-i)^l(2l+1)\frac{(l-m)!}{(l+m)!}\Phi_{ml}^{+}(k_c r)\overset{*}{Y}_{ml}(e,\lambda), \tag{2.137}$$

where $\Phi_{ml}^{+}(k_c r)$ is a spherical wave function

$$\Phi_{ml}^{+}(k_c r) = j_l(k_c r)P_l^m(\cos\theta)e^{im\phi}. \tag{2.138}$$

Multiplying both sides of Eq. (2.137) by $\sin e\,Y_{m'l'}(e,\lambda)$, integrating over e $(0 \le e \le \pi)$ and λ $(0 \le \lambda \le 2\pi)$, and using the orthogonality relation

$$\int_0^{2\pi}\int_0^{\pi} Y_{m'l'}(e,\lambda)\overset{*}{Y}_{ml}(e,\lambda)\sin e\,de\,d\lambda = \delta_{mm'}\delta_{ll'}\Omega_{ml}, \tag{2.139}$$

we get

$$4\pi(-i)^l\Phi_{ml}^{+}(k_c r) = \int_0^{2\pi}\int_0^{\pi} e^{-ik_c r\cos\gamma}Y_{ml}(e,\lambda)\sin e\,de\,d\lambda. \tag{2.140}$$

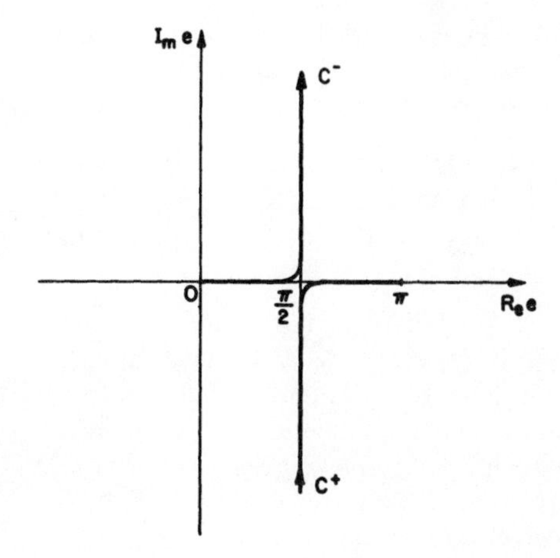

Figure 2.4. The contours C^+ and C^- in the e plane.

Combining Eqs. (2.133), (2.137), and (2.140), we have

$$e^{-i\mathbf{k}\cdot\mathbf{r}} = \sum_{l=0}^{\infty} \sum_{m=-l}^{l} \left[(-i)^l(2l+1)\frac{(l-m)!}{(l+m)!}\overset{*}{Y}_{ml}(e, \lambda)\right]\Phi_{ml}^{+}(k_c r), \quad (2.141)$$

$$\Phi_{ml}^{+}(k_c r) = \frac{i^l}{4\pi}\int_0^{2\pi}\int_0^{\pi} Y_{ml}(e, \lambda)e^{-i\mathbf{k}\cdot\mathbf{r}} \sin e \, de \, d\lambda. \quad (2.142)$$

Equation (2.141) shows that a plane wave may also be considered as a superposition of standing spherical waves, whereas Eq. (2.142) represents standing spherical waves as a superposition of plane waves propagating in all possible directions in space.

If the spherical Bessel function in Eq. (2.138) is replaced by the spherical Hankel function of the second kind, Eq. (2.142) is modified as follows

$$\Phi_{ml}^{-}(k_c r) = \frac{i^l}{2\pi}\int_0^{2\pi}\int_{C^{\pm}} Y_{ml}(e, \lambda)e^{-i\mathbf{k}\cdot\mathbf{r}} \sin e \, de \, d\lambda, \quad (2.143)$$

where the contours $C^{\pm}$ are shown in Fig. 2.4. The contour C^+ begins at $\pi/2 - i\infty$, goes to $\pi/2$ along a line parallel to the imaginary axis, and then goes to π along the real axis. In contrast, the contour C^- starts at the origin, moves to $\pi/2$ along the real axis, and then goes to $\pi/2 + i\infty$ along a line parallel to the imaginary axis. Putting $e = \pi/2 \mp i\alpha \, (0 < \alpha < \infty)$ (for parts of $C^{\pm}$, respectively, parallel to the imaginary axis) and using Eq. (2.124), we have

$$\begin{aligned}\mathbf{k} &= k_c(\sin e \cos \lambda \mathbf{e}_1 + \sin e \sin \lambda \mathbf{e}_2 + \cos e \mathbf{e}_3) \\ &= k_c \cosh \alpha(\cos \lambda \mathbf{e}_1 + \sin \lambda \mathbf{e}_2) \pm i k_c \sinh \alpha \mathbf{e}_3.\end{aligned} \quad (2.144)$$

This yields

$$e^{-i\mathbf{k}\cdot\mathbf{r}} = e^{-ik_c \cosh \alpha(x \cos \lambda + y \sin \lambda) \pm k_c \sinh \alpha z}. \quad (2.145)$$

Therefore, the integrand in Eq. (2.143) remains bounded as $\alpha \to \infty$ provided we choose the contour C^- for $z > 0$ and the contour C^+ for $z < 0$. Therefore, we have

$$\cos e = -i\varepsilon \sinh \alpha, \qquad (\varepsilon = 1 \text{ for } z > 0, = -1 \text{ for } z < 0), \tag{2.146}$$

$$e^{i(\omega t - \mathbf{k}\cdot\mathbf{r})} = \exp\left[i\omega\left\{t - \frac{\cosh \alpha}{c}(x \cos \lambda + y \sin \lambda)\right\} - k_c \sinh \alpha |z|\right]. \tag{2.147}$$

Equation (2.147) represents a wave propagating in the direction ($\cos \lambda$, $\sin \lambda$, 0) with velocity $c/\cosh \alpha$, which is always less than c. At any given moment, the planes of constant phase $x \cos \lambda + y \sin \lambda =$ const. are orthogonal to the planes of constant amplitude $z =$ const. The amplitudes decay exponentially with z. A wave of this type is known as an *inhomogeneous plane wave* or *surface wave.* It may be noted that the attenuation of the waves is not caused by the physical absorption in the medium, which will be considered later.

To recapitulate, whereas a standing spherical wave $\Phi^+_{ml}(k_c r)$ can be regarded as a superposition of homogeneous plane waves only, a diverging spherical wave can only be obtained by a superposition of both homogeneous and inhomogeneous plane waves.

When Eqs. (2.124) and (2.146) are compared, it is apparent that, in the relation $\cos e = \pm i\nu_c/k_c$, we must choose the upper sign for $z < 0$ and the lower sign for $z > 0$; i.e.,

$$\cos e = \frac{-i\varepsilon\nu_c}{k_c}. \tag{2.148}$$

Equations (2.130) and (2.143) now yield

$$\begin{aligned}\Phi^-_{ml}(k_c r) &= i^{l-m}\int_{C^\pm} J_m(k\Delta)e^{im\phi - \nu_c|z|}P_l^m(\cos e)\sin e\, de \\ &= \frac{i^{m-l+1}}{k_c}\int_0^\infty J_m(k\Delta)e^{im\phi - \nu_c|z|}P_l^m\left(\frac{i\varepsilon\nu_c}{k_c}\right)\frac{k\,dk}{\nu_c} \\ &= \frac{i^{m-l+1}}{k_c}\int_0^\infty \Phi_m^{-\varepsilon}(k\Delta)P_l^m\left(\frac{i\varepsilon\nu_c}{k_c}\right)\frac{k\,dk}{\nu_c}.\end{aligned} \tag{2.149}$$

It may be noted that although the integration in the e plane is over a complex contour, the integration in the k plane is along the positive real axis only. Equation (2.149) may be written in the form of the *Erdélyi integral*

$$h_l^{(2)}(k_c r)P_l^m(\cos\theta) = \frac{i^{m-l+1}}{k_c}\int_0^\infty J_m(k\Delta)e^{-\nu_c|z|}P_l^m\left(\frac{i\varepsilon\nu_c}{k_c}\right)\frac{k\,dk}{\nu_c}. \tag{2.150}$$

The integration path is such that the indentation at $k = k_c$ passes *above* the k-axis. If the indentation passes *below* the k-axis, the corresponding result for $h_l^{(1)}(k_c r)P_l^m(\cos\theta)$ will ensue (see, for example, Fig. 5.2).

Equation (2.150) expresses a spherical eigenfunction in terms of cylindrical eigenfunctions. The inverse operation of expressing the cylindrical functions in terms of the spherical functions is rather straightforward. Substituting the expression (2.141) for $\exp(-i\mathbf{k}\cdot\mathbf{r})$ in Eq. (2.132) and evaluating the integral over λ, we get

$$\Phi_m^{\pm}(k_c\Delta) = \sum_{l=0}^{\infty} i^{m-l}(2l+1)\frac{(l-m)!}{(l+m)!}P_l^m\left(\pm\frac{i\nu_c}{k_c}\right)\Phi_{ml}^{+}(k_c r). \tag{2.151}$$

On putting $l = 0$, $m = 0$, Eq. (2.150) yields the *Sommerfeld integral*

$$\frac{e^{-ik_c r}}{r} = \int_0^{\infty} J_0(k\Delta)e^{-\nu_c|z|}\frac{k\,dk}{\nu_c}. \tag{2.152}$$

Replacing $J_0(k\Delta)$ by $\frac{1}{2}[H_0^{(1)}(k\Delta) + H_0^{(2)}(k\Delta)]$ and using the relation $H_0^{(1)}(k\Delta e^{i\pi}) = -H_0^{(2)}(k\Delta)$, Eq. (2.152) becomes

$$\frac{e^{-ik_c r}}{r} = \frac{1}{2}\int_{-\infty}^{\infty} H_0^{(2)}(k\Delta)e^{-\nu_c|z|}\frac{k\,dk}{\nu_c}. \tag{2.153}$$

The Erdélyi integral, Eq. (2.150), is of great use in solving boundary-value problems with plane-parallel boundaries. The source in an unbounded medium is, in general, represented in terms of spherical wave functions. With this representation, it is not possible to satisfy boundary conditions on a plane boundary of the type z = const. However, when we use the Erdélyi integral to express the source in terms of the cylindrical wave functions, boundary conditions can be easily satisfied.

Equation (2.143) expresses the spherical eigenfunctions as a superposition of plane waves. Starting from this equation, one can derive the corresponding expansions for the spherical Hansen vectors given below (see Example 2.5)

$$\mathbf{L}_{ml}^{-}(k_c r) = \frac{i^{l-1}}{2\pi}\int_0^{2\pi}\int_{C^{\pm}} \mathbf{P}_{ml}(e, \lambda)e^{-i\mathbf{k}\cdot\mathbf{r}}\sin e\,de\,d\lambda, \tag{2.154}$$

$$\mathbf{N}_{ml}^{-}(k_c r) = \frac{i^{l-1}}{2\pi}\int_0^{2\pi}\int_{C^{\pm}} \sqrt{[l(l+1)]}\,\mathbf{B}_{ml}(e, \lambda)e^{-i\mathbf{k}\cdot\mathbf{r}}\sin e\,de\,d\lambda, \tag{2.155}$$

$$\mathbf{M}_{ml}^{-}(k_c r) = \frac{i^{l}}{2\pi}\int_0^{2\pi}\int_{C^{\pm}} \sqrt{[l(l+1)]}\,\mathbf{C}_{ml}(e, \lambda)e^{-i\mathbf{k}\cdot\mathbf{r}}\sin e\,de\,d\lambda, \tag{2.156}$$

where

$$\begin{aligned}
\mathbf{P}_{ml}(e, \lambda) &= \mathbf{p}Y_{ml}(e, \lambda), \\
\sqrt{[l(l+1)]}\,\mathbf{B}_{ml}(e, \lambda) &= \left(\mathbf{e}_e\frac{\partial}{\partial e} + \mathbf{e}_\lambda\frac{1}{\sin e}\frac{\partial}{\partial \lambda}\right)Y_{ml}(e, \lambda), \\
\sqrt{[l(l+1)]}\,\mathbf{C}_{ml}(e, \lambda) &= \left(\mathbf{e}_e\frac{1}{\sin e}\frac{\partial}{\partial \lambda} - \mathbf{e}_\lambda\frac{\partial}{\partial e}\right)Y_{ml}(e, \lambda).
\end{aligned} \tag{2.157}$$

In the case of the vectors $\mathbf{L}^{+}_{ml}(k_c r)$, etc., we begin with Eq. (2.142) and obtain

$$
\begin{aligned}
\mathbf{L}^{+}_{ml}(k_c r) &= \frac{i^l}{4\pi i}\int_0^{2\pi}\int_0^{\pi}\mathbf{P}_{ml}(e,\lambda)e^{-i\mathbf{k}\cdot\mathbf{r}}\sin e\,de\,d\lambda \\
&= \frac{i}{4\pi i^l}\int_0^{2\pi}\int_0^{\pi}\mathbf{P}_{ml}(e,\lambda)e^{i\mathbf{k}\cdot\mathbf{r}}\sin e\,de\,d\lambda, \\
\mathbf{N}^{+}_{ml}(k_c r) &= \frac{i}{4\pi i^l}\int_0^{2\pi}\int_0^{\pi}\sqrt{[l(l+1)]}\mathbf{B}_{ml}(e,\lambda)e^{i\mathbf{k}\cdot\mathbf{r}}\sin e\,de\,d\lambda, \\
\mathbf{M}^{+}_{ml}(k_c r) &= \frac{1}{4\pi i^l}\int_0^{2\pi}\int_0^{\pi}\sqrt{[l(l+1)]}\mathbf{C}_{ml}(e,\lambda)e^{i\mathbf{k}\cdot\mathbf{r}}\sin e\,de\,d\lambda.
\end{aligned}
\tag{2.158}
$$

Similarly, Eq. (2.149) expresses spherical wave functions in terms of cylindrical wave functions. Starting from this equation, we obtain the following expansions of the spherical Hansen vectors in terms of cylindrical Hansen vectors

$$
\mathbf{L}^{-}_{ml}(k_c r) = \frac{i^{m-l+1}}{k_c}\int_0^{\infty}\mathbf{L}^{(0)}_m(k_c\Delta)P^m_l(\eta)\frac{k\,dk}{v_c}, \tag{2.159}
$$

$$
\begin{aligned}
\mathbf{N}^{-}_{ml}(k_c r) = &-\frac{i^{m-l}}{k_c^2}\int_0^{\infty}\mathbf{N}^{(0)}_m(k_c\Delta)\left[\frac{dP^m_l(\eta)}{d\eta}\right]\frac{k^2\,dk}{v_c} \\
&- i^{m-l+1}\int_0^{\infty}\hat{\mathbf{M}}^{(0)}_m(k_c\Delta)P^m_l(\eta)\frac{dk}{v_c},
\end{aligned}
\tag{2.160}
$$

$$
\begin{aligned}
\mathbf{M}^{-}_{ml}(k_c r) = &-\frac{i^{m-l}}{k_c^2}\int_0^{\infty}\mathbf{M}^{(0)}_m(k_c\Delta)\left[\frac{dP^m_l(\eta)}{d\eta}\right]\frac{k^2\,dk}{v_c} \\
&- i^{m-l+1}\int_0^{\infty}\hat{\mathbf{N}}^{(0)}_m(k_c\Delta)P^m_l(\eta)\frac{dk}{v_c}.
\end{aligned}
\tag{2.161}
$$

Here $\eta = i\varepsilon v_c/k_c$,

$$
\begin{aligned}
\mathbf{L}^{(0)}_m(k_c\Delta) &= \frac{1}{k_c}(-\varepsilon v_c\mathbf{P}_m + k\mathbf{B}_m)e^{-v_c|z|}, \\
\mathbf{N}^{(0)}_m(k_c\Delta) &= \frac{1}{k_c}(k\mathbf{P}_m - \varepsilon v_c\mathbf{B}_m)e^{-v_c|z|}, \\
\mathbf{M}^{(0)}_m(k_c\Delta) &= \mathbf{C}_m e^{-v_c|z|},
\end{aligned}
\tag{2.162}
$$

and $\hat{\mathbf{M}}^{(0)}$, $\hat{\mathbf{N}}^{(0)}$ are obtained from $\mathbf{M}^{(0)}$, $\mathbf{N}^{(0)}$ on replacing $Y_m(k\Delta,\phi)$ in the expressions for $\mathbf{P}_m$, etc., by $\partial/\partial\phi\, Y_m(k\Delta,\phi)$. Equations (2.159)–(2.161) are also valid when $Y_{ml}(\theta,\phi)$ in $\mathbf{L}^{-}_{ml}$, etc., is replaced by $Y^{c,s}_{ml}(\theta,\phi)$ and $Y_m(k\Delta,\phi)$ in $\mathbf{L}^{(0)}_m$, etc., is replaced by $Y^{c,s}_m(k\Delta,\phi)$.

Relations (2.159)–(2.161) are useful in wave propagation problems because they immediately give the representation of a source in cylindrical coordinates if its representation in spherical coordinates is known.

EXAMPLE 2.4

Consider the dyadic defined by the expansion

$$\mathfrak{S} = \sum_{m=-\infty}^{\infty} \int_0^{\infty} [\mathbf{M}_m^-(k_c\Delta)\overset{*}{\mathbf{M}}{}_m^+(k_c\Delta_0) + \mathbf{N}_m^-(k_c\Delta)\overset{*}{\mathbf{N}}{}_m^+(k_c\Delta_0)$$

$$+ \mathbf{L}_m^-(k_c\Delta)\overset{*}{\mathbf{L}}{}_m^+(k_c\Delta_0)] \frac{k\,dk}{v_c}, \qquad (z > z_0).$$

For $z < z_0$, the roles of Δ and Δ_0 are interchanged. Here, Δ stands for (Δ, ϕ, z) and Δ_0 that for (Δ_0, ϕ_0, z_0). Using the value of the eigenvectors $\mathbf{M}_m$, etc., in terms of the vector cylindrical harmonics from Eq. (2.95) and writing $\mathbf{P}$ for $\mathbf{P}_m(k_c\Delta)$ and $\mathbf{P}_0$ for $\mathbf{P}_m(k_c\Delta_0)$, with similar notation for $\mathbf{B}$ and $\mathbf{C}$ vectors, we have

$$\mathfrak{S} = \sum_{m=-\infty}^{\infty} \int_0^{\infty} \left[\mathbf{C}\overset{*}{\mathbf{C}}_0 + \frac{1}{k_c^2}(k\mathbf{P} - v_c\mathbf{B})(k\overset{*}{\mathbf{P}}_0 + v_c\overset{*}{\mathbf{B}}_0) \right.$$

$$\left. + \frac{1}{k_c^2}(-v_c\mathbf{P} + k\mathbf{B})(v_c\overset{*}{\mathbf{P}}_0 + k\overset{*}{\mathbf{B}}_0) \right] e^{-v_c|z-z_0|} \frac{k\,dk}{v_c}$$

$$= \sum_{m=-\infty}^{\infty} \int_0^{\infty} (\mathbf{P}\overset{*}{\mathbf{P}}_0 + \mathbf{B}\overset{*}{\mathbf{B}}_0 + \mathbf{C}\overset{*}{\mathbf{C}}_0) e^{-v_c|z-z_0|} \frac{k\,dk}{v_c}$$

$$= \sum_{m=-\infty}^{\infty} \int_0^{\infty} \mathfrak{J} J_m(k\Delta) J_m(k\Delta_0) e^{im(\phi-\phi_0)} e^{-v_c|z-z_0|} \frac{k\,dk}{v_c}$$

$$= \mathfrak{J} \int_0^{\infty} J_0(kD) e^{-v_c|z-z_0|} \frac{k\,dk}{v_c}$$

$$= \mathfrak{J} \frac{1}{R} \exp(-ik_c R),$$

by the Sommerfeld formula, Eq. (2.153), where

$$R^2 = D^2 + (z - z_0)^2, \qquad D^2 = \Delta^2 + \Delta_0^2 - 2\Delta\Delta_0 \cos(\phi - \phi_0).$$

EXAMPLE 2.5: Integral Representation of the Spherical Eigenvectors

The relation in Eq. (2.154) is obtained simply by invoking the definition of $\mathbf{L}$ and the integral representation of the potential given in Eq. (2.143)

$$\mathbf{L}_{ml}^-(k_c r) = \frac{1}{k_c} \nabla_r[\Phi_{ml}^-(k_c r)] = \frac{i^l}{2\pi} \int_0^{2\pi} \int_{C^\pm} \frac{1}{k_c} \nabla_r[e^{-i\mathbf{k}\cdot\mathbf{r}}] Y_{ml}(e, \lambda) \sin e \, de \, d\lambda. \tag{2.5.1}$$

Because

$$\frac{1}{k_c} \nabla_r[e^{-i\mathbf{k}\cdot\mathbf{r}}] = \frac{1}{i} \frac{\mathbf{k}}{k_c} e^{-i\mathbf{k}\cdot\mathbf{r}},$$

the result in Eq. (2.154) follows. The derivation of the other two relations is somewhat more "tricky." We start with the vector identity

$$\mathbf{k} \times \nabla_k[e^{-i\mathbf{k}\cdot\mathbf{r}} Y_{ml}(e, \lambda)] = e^{-i\mathbf{k}\cdot\mathbf{r}}[\mathbf{k} \times \nabla_k Y_{ml}(e, \lambda)] - i(\mathbf{k} \times \mathbf{r})e^{-i\mathbf{k}\cdot\mathbf{r}} Y_{ml}(e, \lambda). \tag{2.5.2}$$

Next, we multiply the relation in Eq. (2.154) vectorially by $\mathbf{r}$ to obtain

$$k_c \mathbf{L}_{ml}^-(k_c r) \times \mathbf{r} = \mathbf{M}_{ml}^-(k_c r) = \frac{i^{l-1}}{2\pi} \int_0^{2\pi} \int_{C^\pm} (\mathbf{k} \times \mathbf{r}) Y_{ml}(e, \lambda) e^{-i\mathbf{k}\cdot\mathbf{r}} \sin e \, de \, d\lambda. \tag{2.5.3}$$

Using Eq. (2.5.2) and the identity

$$\mathbf{k} \times \nabla_k Y_{ml}(e, \lambda) = -\sqrt{[l(l+1)]}\mathbf{C}_{ml}(e, \lambda)$$

together with the fact that $\int_S (\mathbf{k} \times \nabla_k Y_{ml}) dS \equiv \mathbf{0}$ for $\mathbf{k}$ normal to S, the last relation in Eq. (2.156) follows. The integral representation of $\mathbf{N}$ follows from the relation

$$\mathbf{N} = \frac{1}{k_c} \operatorname{curl} \mathbf{M}.$$

In order to derive Eqs. (2.159)–(2.161) from Eqs. (2.154)–(2.156), we take notice that the vector harmonics in this representation are defined in the wave-number space (k, e, λ) whose unit vectors are related to the cylindrical unit vectors through the equations (Fig. 2.3)

$$\begin{aligned} \mathbf{e}_k &= \mathbf{e}_z \cos e + \mathbf{e}_\phi \sin e \sin(\lambda - \phi) + \mathbf{e}_\Delta \sin e \cos(\lambda - \phi), \\ \mathbf{e}_e &= -\mathbf{e}_z \sin e + \mathbf{e}_\phi \cos e \sin(\lambda - \phi) + \mathbf{e}_\Delta \cos e \cos(\lambda - \phi), \\ \mathbf{e}_\lambda &= \mathbf{e}_\phi \cos(\lambda - \phi) - \mathbf{e}_\Delta \sin(\lambda - \phi). \end{aligned} \tag{2.5.4}$$

We express the vector harmonics of Eq. (2.154) in terms of these unit vectors and carry out the integration over λ. We then appeal to Eq. (2.149) and two other equations obtained from it by differentiating it with respect to ϕ and Δ. Finally, we make use of the Erdélyi integral, Eq. (2.150), to achieve our goal.

EXAMPLE 2.6: Operational Representation of Wave Functions

1. Spherical coordinates. Let $\mathbf{a}$ be a constant vector and $\Phi(\mathbf{r}, \omega)$ a solution of the Helmholtz wave equation $\nabla^2\Phi + k_c^2\Phi = 0$. Because

$$\nabla^2(\mathbf{a} \cdot \operatorname{grad} \Phi) = \mathbf{a} \cdot \nabla^2 \operatorname{grad} \Phi = \mathbf{a} \cdot \operatorname{grad}(\nabla^2\Phi) = -k_c^2(\mathbf{a} \cdot \operatorname{grad} \Phi), \tag{2.6.1}$$

the scalar function $(\mathbf{a} \cdot \nabla\Phi)$ is also a solution of the same wave equation. Choosing

$$\mathbf{a} = \mathbf{e}_z, \qquad \Phi = \Phi_{ml} = i^l f_l(k_c r) P_l^m(\cos\theta) e^{im\phi},$$
$$f_l = \text{spherical Bessel function},$$

we see that $(\partial/\partial z)\Phi_{ml}$ is a wave function. Noting that

$$\frac{\partial}{\partial z} = \cos\theta \frac{\partial}{\partial r} - \frac{\sin\theta}{r} \frac{\partial}{\partial \theta}$$

we have

$$\frac{\partial}{\partial(k_c z)}\Phi_{ml} = i^l\left\{\frac{\partial}{\partial(k_c r)} f_l(k_c r)\cos\theta P_l^m(\cos\theta) - \frac{f_l(k_c r)}{k_c r}\left[\sin\theta\frac{\partial}{\partial\theta}P_l^m(\cos\theta)\right]\right\}e^{im\phi}. \tag{2.6.2}$$

Using the recurrence relations (App. D)

$$\begin{aligned}
(2l+1)\frac{f_l(k_c r)}{k_c r} &= f_{l-1} + f_{l+1},\\
(2l+1)\frac{d}{dk_c r}f_l(k_c r) &= lf_{l-1} - (l+1)f_{l+1},\\
(2l+1)\cos\theta P_l^m(\cos\theta) &= (l+m)P_{l-1}^m + (l-m+1)P_{l+1}^m,\\
(2l+1)\sin\theta\frac{\partial}{\partial\theta}P_l^m(\cos\theta) &= l(l-m+1)P_{l+1}^m - (l+1)(l+m)P_{l-1}^m,
\end{aligned} \tag{2.6.3}$$

we can express Eq. (2.6.2) as

$$(2l+1)\frac{\partial}{\partial(ik_c z)}\Phi_{ml} = (l+m)\Phi_{m(l-1)} + (l-m+1)\Phi_{m(l+1)}. \tag{2.6.4}$$

However, Eq. (2.6.4) is similar in form to the recurrence relation for the associated Legendre functions (App. D)

$$(2l+1)\mu P_l^m(\mu) = (l+m)P_{l-1}^m(\mu) + (l-m+1)P_{l+1}^m(\mu), \qquad \mu = \cos\theta. \tag{2.6.5}$$

The analogy will be complete if we set the correspondence

$$\mu \Leftrightarrow \frac{\partial}{\partial(ik_c z)}.$$

Hence, it is natural to try the relationship

$$i^l f_l(k_c r)P_l(\cos\theta) = P_l\left(\frac{\partial}{\partial ik_c z}\right)f_0(k_c r), \tag{2.6.6}$$

which, on test, is confirmed. In general, it can be shown by the method of induction that

$$\Phi_{ml}^{\pm} = i^l f_l(k_c r)P_l^m(\cos\theta)e^{\pm im\phi} = (D^{\pm})^m P_l^{(m)}\left(\frac{\partial}{\partial ik_c z}\right)f_0(k_c r), \tag{2.6.7}$$

where

$$D^{\pm} = \frac{\partial}{\partial ik_c x} \pm i\frac{\partial}{\partial ik_c y} = e^{\pm i\phi}\left[\sin\theta\frac{\partial}{\partial ik_c r} + \frac{\cos\theta}{ik_c r}\frac{\partial}{\partial\theta} \pm \frac{i}{ik_c r\sin\theta}\frac{\partial}{\partial\phi}\right], \tag{2.6.8}$$

$$P_l^{(m)}(\mu) = \frac{d^m}{d\mu^m}P_l(\mu). \tag{2.6.9}$$

The special case $m = 0, \theta = 0$, yields

$$f_l(k_c z) = i^{-l} P_l\left(\frac{\partial}{\partial i k_c z}\right) f_0(k_c z). \tag{2.6.10}$$

2. Cylindrical coordinates. Consider next the cylindrical wave function in two dimensions

$$W_n = i^n Z_n(k\Delta)\cos n\phi, \qquad x = \Delta\cos\phi, \qquad y = \Delta\sin\phi. \tag{2.6.11}$$

Applying the operator

$$\frac{\partial}{\partial x} = \cos\phi\,\frac{\partial}{\partial\Delta} - \frac{\sin\phi}{\Delta}\frac{\partial}{\partial\phi} \tag{2.6.12}$$

together with the recurrence relations

$$\begin{aligned} 2\frac{n}{k\Delta} Z_n(k\Delta) &= Z_{n-1}(k\Delta) + Z_{n+1}(k\Delta), \\ 2\frac{\partial Z_n(k\Delta)}{\partial(k\Delta)} &= Z_{n-1}(k\Delta) - Z_{n+1}(k\Delta), \end{aligned} \tag{2.6.13}$$

we derive the relation

$$2\left(\frac{\partial}{\partial ikx}\right)W_n = W_{n-1} + W_{n+1}.$$

A comparison with the recurrence relation for the *Chebyshev polynomials* $T_n(\mu)$

$$2\mu T_n(\mu) = T_{n-1}(\mu) + T_{n+1}(\mu)$$

suggests the correspondence $\mu \Leftrightarrow \partial/\partial(ikx)$.

Hence

$$i^n Z_n(k\Delta)\cos n\phi = T_n\left[\frac{\partial}{\partial ikx}\right] Z_0(k\Delta). \tag{2.6.14}$$

The special case $\phi = 0$ yields

$$Z_n(kx) = i^{-n} T_n\left[\frac{\partial}{\partial ikx}\right] Z_0(kx). \tag{2.6.15}$$

A useful generalization of Eq. (2.6.14) to wave functions in $2\nu + 2$ dimensions is feasible. Let

$$W_n^\nu = i^n \frac{Z_{n+\nu}(k\Delta)}{(k\Delta)^\nu} C_n^\nu(\cos\phi), \tag{2.6.16}$$

where C_n^ν is the Gegenbauer function.

Using some recurrence relations of the Bessel functions (App. D), we find

$$\begin{aligned} \frac{\partial}{\partial k\Delta}\left[\frac{Z_{n+\nu}(k\Delta)}{(k\Delta)^\nu}\right] &= \frac{1}{2(k\Delta)^\nu}\left[\frac{nZ_{n+\nu-1} - (n+2\nu)Z_{n+\nu+1}}{n+\nu}\right], \\ 2\frac{Z_{n+\nu}(k\Delta)}{(k\Delta)} &= \frac{Z_{n+\nu-1} + Z_{n+\nu+1}}{n+\nu}. \end{aligned} \tag{2.6.17}$$

We next apply the operator $\partial/\partial x$ of Eq. (2.6.12) to Eq. (2.6.16) together with the recurrence relations

$$(n+1)C_{n+1}^{\nu}(\mu) + (n-1+2\nu)C_{n-1}^{\nu}(\mu) = 2(n+\nu)\mu C_n^{\nu}(\mu), \tag{2.6.18}$$

$$-(1-\mu^2)\frac{\partial C_n^{\nu}(\mu)}{\partial \mu} + (n+2\nu)\mu C_n^{\nu}(\mu) = (n+1)C_{n+1}^{\nu}(\mu), \tag{2.6.19}$$

$$n\mu C_n^{\nu}(\mu) + (1-\mu^2)\frac{\partial C_n^{\nu}(\mu)}{\partial \mu} = (n-1+2\nu)C_{n-1}^{\nu}(\mu), \tag{2.6.20}$$

$$\frac{\partial C_n^{\nu}}{\partial \mu} = 2\nu C_{n-1}^{\nu+1}(\mu).$$

The result is

$$(n+1)W_{n+1}^{\nu} + (n-1+2\nu)W_{n-1}^{\nu} = 2(n+\nu)\frac{\partial}{\partial ikx}W_n^{\nu}. \tag{2.6.21}$$

Comparing Eq. (2.6.21) with Eq. (2.6.18), we obtain the symbolic relation

$$i^n \frac{Z_{n+\nu}(k\Delta)}{(k\Delta)^{\nu}} C_n^{\nu}(\cos\phi) = C_n^{\nu}\left[\frac{\partial}{\partial ikx}\right]\frac{Z_{\nu}(k\Delta)}{(k\Delta)^{\nu}}. \tag{2.6.22}$$

The special case $\phi = 0$, $k = 1$, $\nu = m + \frac{1}{2}$ yields

$$i^n \frac{f_{n+m}(x)}{x^m}\frac{(2m+n)!}{(2m)!n!} = C_n^{m+1/2}\left[\frac{\partial}{\partial ix}\right]\left[\frac{f_m(x)}{x^m}\right]. \tag{2.6.23}$$

For $m = 0$, Eq. (2.6.23) reduces to Eq. (2.6.10).

EXAMPLE 2.7

Let $\nabla^2\Phi_{ml} + k_c^2\Phi_{ml} = 0$, $\Phi_{ml} = f_l(k_c r)Y_{ml}(\theta, \phi)$. It is required to write the vector $\operatorname{curl}(\mathbf{e}_z\Phi_{ml})$ in terms of the spherical eigenvectors. We shall need the following relations:

$$-\left\{\frac{\partial^2 Y_{ml}}{\partial\theta^2} + \cot\theta\frac{\partial Y_{ml}}{\partial\theta}\right\} = \frac{1}{\sin^2\theta}\frac{\partial^2 Y_{ml}}{\partial\phi^2} + l(l+1)Y_{ml}, \tag{2.7.1}$$

$$\mathbf{e}_z = \mathbf{e}_r\cos\theta - \mathbf{e}_\theta\sin\theta, \tag{2.7.2}$$

$$\mathbf{M}_{ml} = \operatorname{curl}(\mathbf{r}\Phi_{ml}) = \left(\mathbf{e}_\theta\frac{1}{\sin\theta}\frac{\partial}{\partial\phi} - \mathbf{e}_\phi\frac{\partial}{\partial\theta}\right)\Phi_{ml}, \tag{2.7.3}$$

$$k_c\mathbf{N}_{ml} = \operatorname{curl}\operatorname{curl}(\mathbf{r}\Phi_{ml}) = \frac{l(l+1)\Phi_{ml}\mathbf{e}_r}{r} + \left(\mathbf{e}_\theta\frac{\partial}{\partial\theta} + \mathbf{e}_\phi\frac{1}{\sin\theta}\frac{\partial}{\partial\phi}\right)\frac{1}{r}\frac{\partial}{\partial r}[r\Phi_{ml}], \tag{2.7.4}$$

$$\operatorname{curl}(\mathbf{e}_z\Phi) = (\operatorname{grad}\Phi)\times\mathbf{e}_z. \tag{2.7.5}$$

1. First step. Equations (2.7.2) and (2.7.5) yield

$$\operatorname{curl}(\mathbf{e}_z\Phi_{ml}) = \frac{\partial\Phi}{\partial\phi}\frac{\mathbf{e}_r}{r} + \left(\mathbf{e}_\theta\frac{1}{\sin\theta}\frac{\partial}{\partial\phi} - \mathbf{e}_\phi\frac{\partial}{\partial\theta}\right)\frac{\Phi}{r}\cos\theta - \mathbf{e}_\phi\sin\theta\left(\frac{\Phi}{r} + \frac{\partial\Phi}{\partial r}\right). \tag{2.7.6}$$

2. Second step. From Eq. (2.7.1) follows the interesting identity

$$\left(\mathbf{e}_\theta \frac{1}{\sin\theta}\frac{\partial}{\partial\phi} - \mathbf{e}_\phi \frac{\partial}{\partial\theta}\right)\sin\theta \frac{\partial Y_{ml}}{\partial\theta} = \left(\mathbf{e}_\theta \frac{\partial}{\partial\theta} + \mathbf{e}_\phi \frac{1}{\sin\theta}\frac{\partial}{\partial\phi}\right)\frac{\partial Y_{ml}}{\partial\phi} + l(l+1)Y_{ml}\sin\theta\,\mathbf{e}_\phi. \tag{2.7.7}$$

Hence, from Eqs. (2.7.6) and (2.7.7), we have

$$\begin{aligned}\operatorname{curl}(\mathbf{e}_z\Phi_{ml}) = {} & \left\{\frac{\mathbf{e}_r}{r}\frac{\partial\Phi_{ml}}{\partial\phi} + \frac{1}{l(l+1)}\left(\mathbf{e}_\theta\frac{\partial}{\partial\theta} + \mathbf{e}_\phi\frac{1}{\sin\theta}\frac{\partial}{\partial\phi}\right)\frac{1}{r}\frac{\partial}{\partial r}\left(r\frac{\partial\Phi_{ml}}{\partial\phi}\right)\right\} \\ & + \left(\mathbf{e}_\theta\frac{1}{\sin\theta}\frac{\partial}{\partial\phi} - \mathbf{e}_\phi\frac{\partial}{\partial\theta}\right)\left[\frac{f_l}{r}Y_{ml}\cos\theta \right. \\ & \left. - \frac{1}{l(l+1)}\sin\theta\frac{\partial Y_{ml}}{\partial\theta}\left(\frac{f_l}{r} + \frac{\partial f_l}{\partial r}\right)\right].\end{aligned} \tag{2.7.8}$$

3. Third step. From Eq. (2.6.3), we find

$$\begin{aligned}\frac{f_l}{r}(Y_{ml}\cos\theta) - \sin\theta\frac{\partial Y_{ml}}{\partial\theta}\left\{\frac{1}{l(l+1)}\left(\frac{f_l}{r} + \frac{\partial f_l}{\partial r}\right)\right\} \\ = \frac{k_c(l-m+1)}{(l+1)(2l+1)}\Phi_{m(l+1)} + \frac{k_c(l+m)}{l(2l+1)}\Phi_{m(l-1)}.\end{aligned} \tag{2.7.9}$$

Therefore, Eq. (2.7.8) becomes

$$\begin{aligned}\operatorname{curl}(\mathbf{e}_z\Phi_{ml}) = {} & \frac{1}{l(l+1)}\operatorname{curl}\operatorname{curl}\left(\mathbf{r}\frac{\partial\Phi_{ml}}{\partial\phi}\right) + k_c\frac{(l-m+1)}{(l+1)(2l+1)}\operatorname{curl}[\mathbf{r}\Phi_{m(l+1)}] \\ & + k_c\frac{l+m}{l(2l+1)}\operatorname{curl}[\mathbf{r}\Phi_{m(l-1)}].\end{aligned} \tag{2.7.10}$$

Taking the curl of this equation we get

$$\begin{aligned}\operatorname{curl}\operatorname{curl}(\mathbf{e}_z\Phi_{ml}) = {} & \frac{k_c^2}{l(l+1)}\operatorname{curl}\left(\mathbf{r}\frac{\partial\Phi_{ml}}{\partial\phi}\right) \\ & + k_c\frac{(l-m+1)}{(l+1)(2l+1)}\operatorname{curl}\operatorname{curl}[\mathbf{r}\Phi_{m(l+1)}] \\ & + k_c\frac{l+m}{l(2l+1)}\operatorname{curl}\operatorname{curl}[\mathbf{r}\Phi_{m(l-1)}].\end{aligned} \tag{2.7.11}$$

2.7. Dyadic Plane Waves

Monochromatic vector plane-wave solutions are formed by a linear superposition of the eigenvectors **L**, **M**, and **N**. Denote by **D** a linear combination of the *transverse* vectors **M** and **N**:

$$\mathbf{D} = \operatorname{Re}\{\mathbf{d}e^{i(\omega t - \mathbf{k}\cdot\mathbf{r})}\}, \tag{2.163}$$

where

$$\mathbf{d} = \mathbf{d}_1 - i\mathbf{d}_2, \qquad (\mathbf{d}_1 \cdot \mathbf{d}_2) = 0 \tag{2.164}$$

in which $\mathbf{d}_1$ and $\mathbf{d}_2$ are real vectors, constant in space and time along **N** and **M**, respectively. We choose a fixed right Cartesian reference system (ξ, η, ζ) such that the ξ and η axes are parallel to $\mathbf{d}_1$ and $\mathbf{d}_2$, respectively, and the ζ axis is along the direction of propagation of the wave, that is, in the **k** direction (Fig. 2.5a). The vectors **d** and **D** rotate but remain always parallel to a fixed plane in space, which is the $\xi\eta$ plane. We say that these vectors are *plane polarized* and call the plane that includes **k** and **D** the *plane of polarization.* In the (ξ, η, ζ) system, the vector **D** has the form:

$$\mathbf{D} = \operatorname{Re}\{\mathbf{d}e^{i(\omega t - k_c\zeta)}\}, \tag{2.165}$$

$$D_\xi = d_1 \cos(\omega t - k_c\zeta), \qquad D_\eta = d_2 \sin(\omega t - k_c\zeta). \tag{2.166}$$

Eliminating the phase from these expressions, we obtain

$$\left(\frac{D_\xi}{d_1}\right)^2 + \left(\frac{D_\eta}{d_2}\right)^2 = 1 \tag{2.167}$$

which is the curve (ellipse) that the end point of **D** decribes at a given point in space on the phase plane $\xi\eta$. The polarization is right handed if the curve is

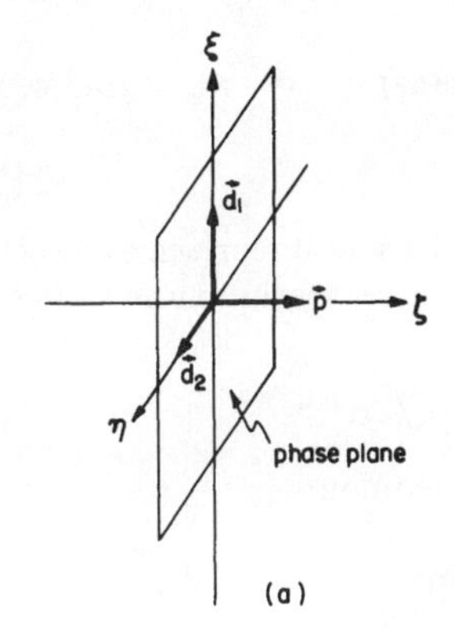

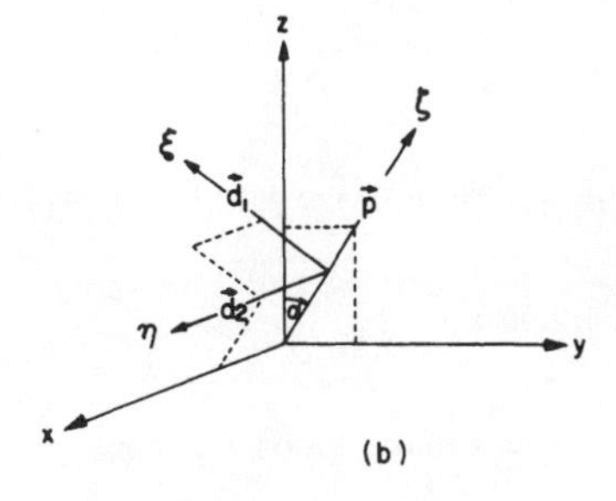

Figure 2.5. Geometry of polarized waves. (a) Geometry of the phase–plane; (b) vector plane waves.

described in a clockwise direction with respect to an observer looking in the direction from which the wave is coming. The polarization is left handed in the counterclockwise case.

We say that $\mathbf{D}$ is *elliptically polarized* if $d_1 \neq d_2$, *circularly polarized* if $d_1 = d_2$, and *linearly polarized* if either $d_1 = 0$ or $d_2 = 0$. Note that $\mathbf{M}$ and $\mathbf{N}$ are each linearly polarized. The combination $(\mathbf{M} - \mathbf{N})$, however, is circularly polarized. Let us examine this latter case in more detail. We may, without loss of generality, take $\mathbf{p} = \mathbf{k}/k_c$ to be in the yz plane (Fig. 2.5b). Writing the components (l, m, n) of $\mathbf{p}$ as $(0, \sin\alpha, \cos\alpha)$, we have

$$\mathbf{M} - \mathbf{N} = \sqrt{2}\boldsymbol{\varepsilon} e^{i(\omega t - \mathbf{k}\cdot\mathbf{r})}, \qquad \mathbf{d}_1 = (0, -\cos\alpha, \sin\alpha), \qquad \mathbf{d}_2 = (1, 0, 0),$$

where

$$\boldsymbol{\varepsilon} = \frac{1}{\sqrt{2}}(\mathbf{d}_1 - i\mathbf{d}_2) = \frac{1}{\sqrt{2}}[(-\cos\alpha\mathbf{e}_y + \sin\alpha\mathbf{e}_z) - i\mathbf{e}_x] \tag{2.168}$$

and $\boldsymbol{\varepsilon}$ is the unit *polarization vector.*

A *stress wave*, such as given by Eq. (2.114a), is a *dyadic plane wave.* It is a solution of the *dyadic wave equation*

$$\nabla^2\mathfrak{G} + k_c^2\mathfrak{G} = 0. \tag{2.169}$$

It is easily verified that there are nine independent basic solutions of Eq. (2.169)

$$\mathfrak{G}_{1,2,3} = \operatorname{curl}(\mathbf{au}), \qquad \mathfrak{G}_{4,5,6} = \frac{1}{k_c}\operatorname{curl}\operatorname{curl}(\mathbf{av}), \qquad \mathfrak{G}_{7,8,9} = \operatorname{grad}\mathbf{w}, \tag{2.170}$$

where $\mathbf{a}$ has the same meaning as in Eq. (2.67) and $\mathbf{u}$, $\mathbf{v}$, or $\mathbf{w}$ represents one of the eigenvectors $\mathbf{L}$, $\mathbf{M}$, $\mathbf{N}$. Restricting ourselves to transverse solutions in Cartesian coordinates we put

$$\mathbf{a} = \mathbf{e}_x, \qquad \mathbf{u} = \mathbf{v} = \mathbf{M} - \mathbf{N} = \sqrt{2}\boldsymbol{\varepsilon} e^{i(\omega t - \mathbf{k}\cdot\mathbf{r})}$$

and align $\mathbf{p}$ to be in the yz plane. Under these conditions

$$\operatorname{curl}(\mathbf{e}_x\mathbf{v}) = -\mathbf{e}_x \times \nabla\mathbf{v} = ik_c(\mathbf{e}_x \times \mathbf{p})\mathbf{v},$$

$$\frac{1}{k_c}\operatorname{curl}\operatorname{curl}(\mathbf{e}_x\mathbf{v}) = -k_c[(\mathbf{e}_x \times \mathbf{p}) \times \mathbf{p}]\mathbf{v}$$

and

$$\mathfrak{G} = \operatorname{curl}(\mathbf{e}_x\mathbf{v}) + \frac{1}{k_c}\operatorname{curl}\operatorname{curl}(\mathbf{e}_x\mathbf{v}) = (2ik_c)\boldsymbol{\varepsilon\varepsilon} e^{i\omega[t - (z\cos\alpha + y\sin\alpha)/c]}. \tag{2.171}$$

Here, $\boldsymbol{\varepsilon\varepsilon}$ is the *polarization dyad* or *polarization tensor.*

Because

$$\mathbf{e}_\xi = -\cos\alpha\mathbf{e}_y + \sin\alpha\mathbf{e}_z, \qquad \mathbf{e}_\eta = \mathbf{e}_x, \qquad \mathbf{e}_\zeta = \sin\alpha\mathbf{e}_y + \cos\alpha\mathbf{e}_z = \mathbf{p}$$

it follows that $\boldsymbol{\varepsilon} = (\mathbf{e}_\xi - i\mathbf{e}_\eta)/\sqrt{2}$. Consequently,

$$2\boldsymbol{\varepsilon}\boldsymbol{\varepsilon} = (\mathbf{e}_\xi\mathbf{e}_\xi - \mathbf{e}_\eta\mathbf{e}_\eta) - i(\mathbf{e}_\xi\mathbf{e}_\eta + \mathbf{e}_\eta\mathbf{e}_\xi) = \mathfrak{D}_1 - i\mathfrak{D}_2. \tag{2.172}$$

We also notice that

$$2\boldsymbol{\varepsilon}\boldsymbol{\varepsilon} = \underset{\text{Simple shear}}{\begin{bmatrix} 1 & 0 & 0 \\ 0 & -1 & 0 \\ 0 & 0 & 0 \end{bmatrix}} - i \underset{\text{Pure shear}}{\begin{bmatrix} 0 & 1 & 0 \\ 1 & 0 & 0 \\ 0 & 0 & 0 \end{bmatrix}} = \begin{bmatrix} 1 & -i & 0 \\ -i & -1 & 0 \\ 0 & 0 & 0 \end{bmatrix}. \tag{2.173}$$

This shows that the state of tensor polarization is composed of two parts: First, a tension along the ξ direction accompanied by a compression in the η direction (simple shear); second, the same state rotated about the ζ axis by 45° (pure shear). Each of these may be considered as a linear polarization tensor analogous to the linear polarization vectors $\mathbf{d}_1$ and $\mathbf{d}_2$. The combination $\mathfrak{D}_1 - i\mathfrak{D}_2$ is a left-handed circular polarization tensor. Moreover, because $\mathfrak{D}_1 : \mathfrak{D}_2 \equiv 0$, we may stretch the above argument further and say that $\mathfrak{D}_1$ is orthogonal to $\mathfrak{D}_2$ in analogy to the orthogonality of $\mathbf{d}_1$ and $\mathbf{d}_2$. We have to accustom ourselves, however, to the odd fact that although $\mathfrak{D}_2$ is obtained from $\mathfrak{D}_1$ by a rotation of 45°, it is nevertheless orthogonal to it.

If we rotate the axes (ξ, η, ζ) in a counterclockwise sense about the ζ axis by an angle θ, the components of the right polarization tensor $\mathfrak{D}_1 + i\mathfrak{D}_2$ will rotate by an angle of 2θ:

$$\begin{bmatrix} \cos\theta & \sin\theta & 0 \\ -\sin\theta & \cos\theta & 0 \\ 0 & 0 & 1 \end{bmatrix} \begin{bmatrix} 1 & i & 0 \\ i & -1 & 0 \\ 0 & 0 & 0 \end{bmatrix} \begin{bmatrix} \cos\theta & -\sin\theta & 0 \\ \sin\theta & \cos\theta & 0 \\ 0 & 0 & 1 \end{bmatrix}$$
$$= e^{2i\theta} \begin{bmatrix} 1 & i & 0 \\ i & -1 & 0 \\ 0 & 0 & 0 \end{bmatrix}. \tag{2.174}$$

Bibliography

Ben-Menahem A, Singh SJ (1968a) Eigenvector expansions of Green's dyads with applications to geophysical theory. Geophys Jour Roy Astron Soc (London) 16: 417–452.

Ben-Menahem A, Singh SJ (1968b) Multipolar elastic fields in a layered half-space. Bull Seismol Soc Amer 58: 1519–1572.

Eisenhart LP (1934a) Separable systems of Stäckel. Ann Math 35: 284–305.

Eisenhart LP (1934b) Separable systems in Euclidean 3-space. Phys Rev 45: 427–428.

Erdélyi A (1937) Zur Theorie der Kugelwellen. Physica 4: 107–120.

Hansen WW (1935) A new type of expansion in radiation problems. Phys Rev 47: 139–143.

Mie G (1908) Beiträge zur Optik trüber Medien, speziell kolloidaler Metallösungen. Ann Phys 25: 377–445.

Moon P, Spencer DE (1961) Field Theory Handbook. Springer, New York.

Morse PM, Feshbach H (1953) Methods of Theoretical Physics, Parts I and II. McGraw-Hill, New York.

Sommerfeld A (1909) Über die Ausbreitung der Wellen in der drahtlosen Telegraphie. Ann Phys 28: 665–736.

Sommerfeld A (1964) Partial Differential Equations in Physics. Academic Press, New York, 335 pp.

Sternberg E, Eubanks RA (1957) On stress functions for elastokinetics and the integration of the repeated wave equation. Quart Appl Math 15: 149–153.

Stratton JA (1941) Electromagnetic Theory. McGraw-Hill, New York.

Van der Pol B (1936) A generalization of Maxwell's definition of solid harmonics to waves in n dimensions. Physica 3: 393–397.

Watson GN (1966) A Treatise on the Theory of Bessel Functions. Cambridge University Press, Cambridge, 804 pp.

Weyl H (1919) Ausbreitung elektromagnetischer Wellen über einen ebenen Leiter. Ann Phys 60: 481–500.

Wheeler LT, Sternberg E (1968). Some theorems in classical elastodynamics. Archive for Rational Mechanics and Analysis 31: 51–90; Corrigendum 31: 402.

CHAPTER 3

Seismic Plane Waves in a Layered Half-Space

Here is to pure mathematics, may it never find an application.

(*G. H. Hardy*)

3.1. Reflection and Refraction of Plane Waves—General Considerations

When seismic waves impinge on the earth's surface or on an internal surface of discontinuity, the laws of reflection and refraction of a plane wave at a plane surface of discontinuity can be applied only approximately because both the wavefront and the surface are, in fact, curved. However, in many applications this approximation is quite satisfactory.

In Section 2.5, we saw that in the case of plane waves, the P, SV, and SH motions are described by the eigenvectors **L**, **N**, and **M**, respectively. If we denote the angle of incidence of P waves by e and that of S waves by f (Figs. 3.1 and 3.2), then from Eq. (2.112) the displacement caused by incident P, SV, and SH waves may be expressed in the form

$$\mathbf{u}_{\mathrm{P}} = A(\mathbf{a}_1 \sin e - \mathbf{a}_3 \cos e)\exp\left[i\omega\left(t - \frac{x_1 \sin e - x_3 \cos e}{\alpha}\right)\right],$$

$$\mathbf{u}_{\mathrm{SV}} = B(\mathbf{a}_1 \cos f + \mathbf{a}_3 \sin f)\exp\left[i\omega\left(t - \frac{x_1 \sin f - x_3 \cos f}{\beta}\right)\right], \tag{3.1}$$

$$\mathbf{u}_{\mathrm{SH}} = C\mathbf{a}_2 \exp\left[i\omega\left(t - \frac{x_1 \sin f - x_3 \cos f}{\beta}\right)\right],$$

where $\mathbf{a}_i$ denotes the unit vector in the x_i direction. Similarly, if the angle of reflection of P waves is e_1 and that of S waves is f_1, then the displacement field

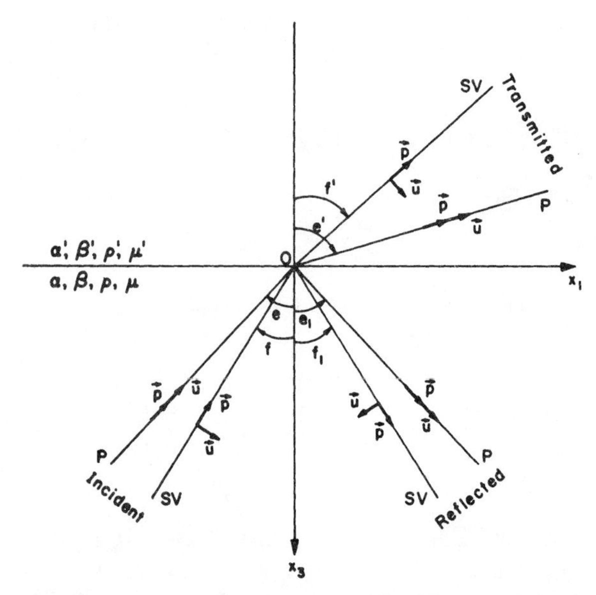

Figure 3.1. The angle e corresponds to P waves and f to S waves. $\mathbf{p}$ is a unit vector in the direction of propagation. For P waves, the positive direction of the particle displacement is along $\mathbf{p}$. For S waves, it is in the direction $(\mathbf{p} \times \mathbf{a}_2)$, where $\mathbf{a}_2$ is drawn perpendicular to the plane of the paper toward the reader.

caused by reflected waves is given by

$$\begin{aligned}
\mathbf{u}_P &= A_1(\mathbf{a}_1 \sin e_1 + \mathbf{a}_3 \cos e_1)\exp\left[i\omega\left(t - \frac{x_1 \sin e_1 + x_3 \cos e_1}{\alpha}\right)\right], \\
\mathbf{u}_{SV} &= B_1(-\mathbf{a}_1 \cos f_1 + \mathbf{a}_3 \sin f_1)\exp\left[i\omega\left(t - \frac{x_1 \sin f_1 + x_3 \cos f_1}{\beta}\right)\right], \\
\mathbf{u}_{SH} &= C_1\mathbf{a}_2 \exp\left[i\omega\left(t - \frac{x_1 \sin f_1 + x_3 \cos f_1}{\beta}\right)\right].
\end{aligned} \tag{3.2}$$

Denoting the angles of transmission of P and S waves by e' and f', respectively, the displacements caused by transmitted waves are:

$$\begin{aligned}
\mathbf{u}_P &= A'(\mathbf{a}_1 \sin e' - \mathbf{a}_3 \cos e')\exp\left[i\omega\left(t - \frac{x_1 \sin e' - x_3 \cos e'}{\alpha'}\right)\right], \\
\mathbf{u}_{SV} &= B'(\mathbf{a}_1 \cos f' + \mathbf{a}_3 \sin f')\exp\left[i\omega\left(t - \frac{x_1 \sin f' - x_3 \cos f'}{\beta'}\right)\right], \\
\mathbf{u}_{SH} &= C'\mathbf{a}_2 \exp\left[i\omega\left(t - \frac{x_1 \sin f' - x_3 \cos f'}{\beta'}\right)\right].
\end{aligned} \tag{3.3}$$

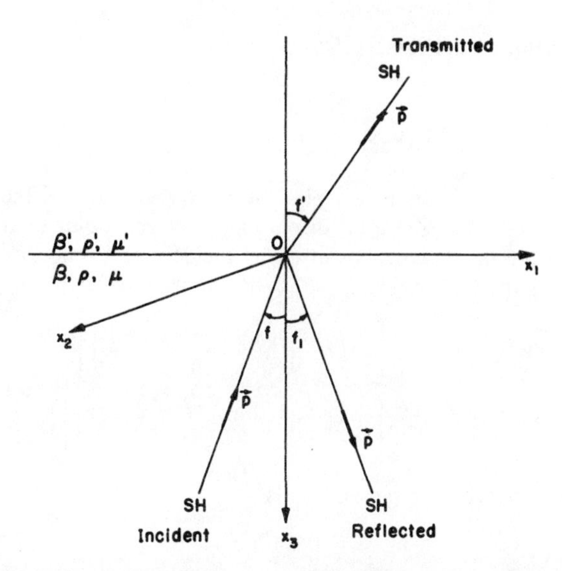

Figure 3.2. The angle of incidence (f), the angle of reflection (f_1), and the angle of transmission (f') for an incident *SH* wave. The waves are polarized in a direction perpendicular to the plane of the paper.

In Figs. 3.1 and 3.2, **p** is a unit vector in the direction of propagation and **u** gives the positive direction of the displacement at a given moment. We take **u** along **p** for *P* waves, in the direction of the positive x_2 axis for *SH* waves, and in the direction of the unit vector $\mathbf{p} \times \mathbf{a}_2$ for *SV* waves.

Our next task is to calculate the stress vector $\mathbf{a}_3 \cdot \mathfrak{T}$. If **a** is a constant vector, the stress associated with the displacement $\mathbf{a}e^{ik(vt-\mathbf{p}\cdot\mathbf{r})}$ is

$$\mathfrak{T} = -ik\{\lambda(\mathbf{p}\cdot\mathbf{a})\mathfrak{I} + \mu(\mathbf{pa} + \mathbf{ap})\}e^{ik(vt-\mathbf{p}\cdot\mathbf{r})}.$$

This yields

$$\mathbf{a}_3 \cdot \mathfrak{T}(\mathbf{u}) = \begin{cases} \mathbf{a}_1\left[2\mu\dfrac{\partial u_1}{\partial x_3}\right] + \mathbf{a}_3\left[\lambda\dfrac{\partial u_1}{\partial x_1} + (\lambda+2\mu)\dfrac{\partial u_3}{\partial x_3}\right] & \text{for } P \text{ waves,} \\ \mathbf{a}_1\left[\mu\left(\dfrac{\partial u_1}{\partial x_3} + \dfrac{\partial u_3}{\partial x_1}\right)\right] + \mathbf{a}_3\left[2\mu\dfrac{\partial u_3}{\partial x_3}\right] & \text{for } SV \text{ waves,} \\ \mathbf{a}_2\left[\mu\dfrac{\partial u_2}{\partial x_3}\right] & \text{for } SH \text{ waves.} \end{cases} \quad (3.4)$$

Equations (3.1)–(3.4) immediately yield the stress vector $\mathbf{a}_3 \cdot \mathfrak{T}$ for incident, reflected, and transmitted waves. From these equations it is clear that whereas an incident *SH* wave gives rise to a reflected *SH* wave and a transmitted *SH* wave, an incident *P* or *SV* wave, in general, gives four waves: a reflected *P*, a reflected *SV*, a transmitted *P*, and a transmitted *SV*. In the case of a free surface, the transmitted waves are absent.

3.2. Reflection at a Free Surface

3.2.1. Incident *SH* Waves

We assume that the medium occupies the region $x_3 > 0$ and the boundary $x_3 = 0$ is traction free. The total displacement at any point of the medium is the sum of the displacements caused by incident and reflected *SH* waves. Equations (3.1) and (3.2) yield

$$\mathbf{u} = \mathbf{a}_2 \left[C \exp\left[i\omega\left(t - \frac{x_1 \sin f - x_3 \cos f}{\beta} \right) \right] \right.$$
$$\left. + C_1 \exp\left[i\omega\left(t - \frac{x_1 \sin f_1 + x_3 \cos f_1}{\beta} \right) \right] \right]. \tag{3.5}$$

The boundary condition to be satisfied is

$$\mathbf{a}_3 \cdot \mathfrak{T} = 0 \quad \text{at} \quad x_3 = 0. \tag{3.6}$$

Equations (3.4)–(3.6) give

$$C \cos f e^{-i\omega x_1 \sin f/\beta} - C_1 \cos f_1 e^{-i\omega x_1 \sin f_1/\beta} = 0. \tag{3.7}$$

This can be satisfied for all values of x_1 only if $f_1 = f$, in which case Eq. (3.7) yields $C_1 = C$. Hence, the surface amplitude is twice the amplitude of the incident wave alone.

3.2.2. Incident *P* Waves

In the case of incident *P* waves we have reflected *P* as well as reflected *SV* waves. The displacement vector therefore is given by

$$\mathbf{u} = A(\mathbf{a}_1 \sin e - \mathbf{a}_3 \cos e) \exp\left[i\omega\left(t - \frac{x_1 \sin e - x_3 \cos e}{\alpha} \right) \right]$$
$$+ A_1(\mathbf{a}_1 \sin e_1 + \mathbf{a}_3 \cos e_1) \exp\left[i\omega\left(t - \frac{x_1 \sin e_1 + x_3 \cos e_1}{\alpha} \right) \right]$$
$$+ B_1(-\mathbf{a}_1 \cos f_1 + \mathbf{a}_3 \sin f_1) \exp\left[i\omega\left(t - \frac{x_1 \sin f_1 + x_3 \cos f_1}{\beta} \right) \right]. \tag{3.8}$$

In order that the boundary condition, Eq. (3.6), be satisfied for all values of x_1 and t, we must have $e_1 = e$ and

$$\frac{\sin e}{\alpha} = \frac{\sin f}{\beta}, \tag{3.9}$$

where f has been written for f_1. This is known as *Snell's law*. Equations (3.4), (3.6), (3.8), and (3.9) yield two homogeneous, simultaneous equations in A, A_1, and B_1. Solving these, we find the amplitude ratios:

$$\frac{A_1}{A} = \frac{\sin 2e \sin 2f - (\alpha/\beta)^2 \cos^2 2f}{\sin 2e \sin 2f + (\alpha/\beta)^2 \cos^2 2f},$$
$$\frac{B_1}{A} = \frac{-2(\alpha/\beta)\sin 2e \cos 2f}{\sin 2e \sin 2f + (\alpha/\beta)^2 \cos^2 2f}. \tag{3.10}$$

It may be noted that, for reflected P, the amplitude ratio for the horizontal component of the displacements is A_1/A, whereas for the vertical component, it is $-A_1/A$. Similarly, for reflected SV, the amplitude ratio for the horizontal component of the displacements is $(-B_1 \cos f)/(A \sin e)$, whereas for the vertical component it is $(-B_1 \sin f)/(A \cos e)$.

The total surface displacement is given by

$$\mathbf{u}_0 = A(\mathbf{a}_1 \sin 2f - \mathbf{a}_3 \cos 2f)\frac{2(\alpha/\beta)^2 \cos e}{\sin 2e \sin 2f + (\alpha/\beta)^2 \cos^2 2f} e^{i\omega(t - x_1 \sin e/\alpha)}. \tag{3.11}$$

This gives $(u_1/u_3)_0 = \tan(\pi - 2f)$; i.e., the displacement vector at the free surface makes an angle $2f$ with the vertical upward direction (Fig. 3.3). Using Snell's law as in Eq. (3.9), it is useful to recast the denominator in Eq. (3.11) in another form in which the angle f does not appear

$$\sin 2e \sin 2f + \left(\frac{\alpha}{\beta}\right)^2 \cos^2 2f = \left(\frac{\beta}{\alpha}\right)^2 \Delta_p(e),$$

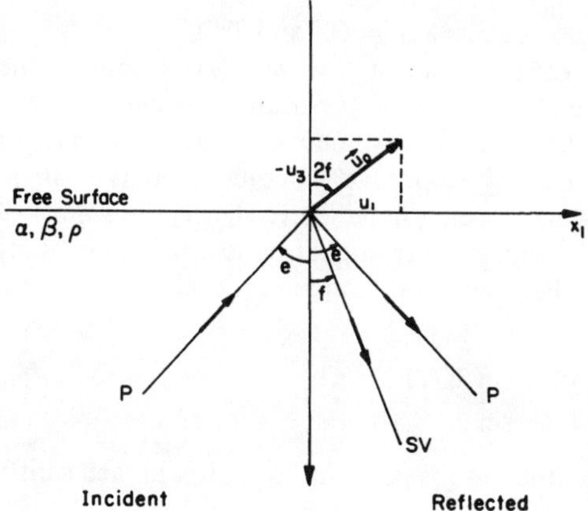

Figure 3.3. The surface displacement ($\mathbf{u}_0$) for P waves incident at a free surface.

where

$$\Delta_p(\Omega) = \left[\left(\frac{\alpha}{\beta}\right)^2 - 2\sin^2\Omega\right]^2 + 4\sin^2\Omega|\cos\Omega|\sqrt{\left[\left(\frac{\alpha}{\beta}\right)^2 - \sin^2\Omega\right]}. \quad (3.11a)$$

Because $\alpha > \beta$, Snell's law, Eq. (3.9), assures real values for f corresponding to each value of e in the range $0 \le e \le \pi/2$. Therefore, reflected SV waves always exist. The following are some of the important features of the problem.

1. Normal incidence: $e = 0, f = 0, A_1 = -A, B_1 = 0,$

$$\mathbf{u}_0 = -2A\mathbf{a}_3 \exp^{i\omega t}$$

   There is no reflected SV wave; the whole energy is reflected in the form of P waves. Further, there is no horizontal surface displacement, whereas the vertical surface displacement is doubled.
2. Grazing incidence: $e = \pi/2$, $f = \arcsin(\beta/\alpha)$, $A_1 = -A$, $B_1 = 0$. The surface displacement is zero. There is no reflected SV, whereas at the free surface, the reflected P annihilates the incident P.
3. Mode conversion. The reflected P will be absent for the angles of incidence given by

$$\sin 2e \sin 2f = \left(\frac{\alpha}{\beta}\right)^2 \cos^2 2f.$$

   Using Snell's law, this is made to yield

$$2\sin e \sin 2e\left(\frac{\alpha^2}{\beta^2} - \sin^2 e\right)^{1/2} = \left(\frac{\alpha^2}{\beta^2} - 2\sin^2 e\right)^2. \quad (3.12)$$

   In this case

$$\frac{B_1}{A} = -\frac{\alpha}{\beta}\cot 2f = \frac{\alpha}{\beta}\left(\frac{u_3}{u_1}\right)_0.$$

   For $\sigma = \frac{1}{4}$, the mode conversion occurs at $e = 60°$ and $77°13'$.
4. Energy equation: From Eq. (2.117), incident P waves carry energy to the free surface at an average rate of $\frac{1}{2}\rho\alpha(\cos e)\omega^2 A^2$ per unit area per unit time. Similarly, reflected P and reflected SV waves remove energy at the rates of $\frac{1}{2}\rho\alpha(\cos e)\omega^2 A_1^2$ and $\frac{1}{2}\rho\beta(\cos f)\omega^2 B_1^2$, respectively. Because there is no stress across the free surface, no work is being done there. Hence, the supply of energy to the surface by the incident wave should be just balanced by its removal by reflected waves. When Snell's law is applied, the above statement leads to the energy equation

$$\left(\frac{A_1}{A}\right)^2 + \frac{\sin 2f}{\sin 2e}\left(\frac{B_1}{A}\right)^2 = 1. \quad (3.13)$$

   It is easily seen that the amplitude ratios given in Eq. (3.10), in fact, satisfy this equation.

3.2.3. Incident *SV* Waves

With incident *SV* waves, the displacement at any point results from incident *SV* waves, reflected *P* waves, and reflected *SV* waves. By Snell's law, displacement may be expressed in the form

$$\mathbf{u} = [B(\mathbf{a}_1 \cos f + \mathbf{a}_3 \sin f)e^{i\omega x_3 \cos f/\beta} + A_1(\mathbf{a}_1 \sin e + \mathbf{a}_3 \cos e)e^{-i\omega x_3 \cos e/\alpha} + B_1(-\mathbf{a}_1 \cos f + \mathbf{a}_3 \sin f)e^{-i\omega x_3 \cos f/\beta}]e^{i\omega(t - x_1 \sin f/\beta)}. \tag{3.14}$$

The amplitude ratios are found to be

$$\frac{A_1}{B} = \frac{(\alpha/\beta)\sin 4f}{\sin 2e \sin 2f + (\alpha/\beta)^2 \cos^2 2f}, \tag{3.15}$$

$$\frac{B_1}{B} = \frac{\sin 2e \sin 2f - (\alpha/\beta)^2 \cos^2 2f}{\sin 2e \sin 2f + (\alpha/\beta)^2 \cos^2 2f}. \tag{3.16}$$

The following are some of the salient features of this case.

1. Normal incidence: $f = 0$, $e = 0$, $A_1 = 0$, $B_1 = -B$,

$$\mathbf{u}_0 = 2B\mathbf{a}_1 \exp^{i\omega t}$$

   There is no reflected *P*, the whole energy going to reflected *SV*. There is no vertical surface displacement and the horizontal surface displacement is doubled.
2. Mode conversion: Reflected *SV* waves will be absent for the angles of incidence given by

$$\sin 2e \sin 2f - \left(\frac{\alpha}{\beta}\right)^2 \cos^2 2f = 0.$$

   When Snell's law is applied, this becomes

$$2 \sin f \sin 2f \left(\frac{\beta^2}{\alpha^2} - \sin^2 f\right)^{1/2} = \cos^2 2f. \tag{3.17}$$

   For $\sigma = \frac{1}{4}$, this yields $f = 30°$ and $34°16'$.
3. Inhomogeneous waves: From Snell's law

$$\sin e = \left(\frac{\alpha}{\beta}\right) \sin f.$$

   Therefore, e is real if $0 \le f \le f_1 = \arcsin(\beta/\alpha)$. The angle f_1 (not to be confused with the angle of reflection) always exists and is known as the *critical angle*. For $f = f_1$, we have $e = \pi/2$ and

$$\frac{A_1}{B} = \frac{4(1 - \beta^2/\alpha^2)^{1/2}}{(\alpha^2/\beta^2 - 2)}, \qquad \frac{B_1}{B} = -1. \tag{3.18}$$

For $\sigma = \frac{1}{4}$, $f_1 = \arcsin(1/\sqrt{3}) = 35°16'$, approximately. When f is increased beyond f_1, e ceases to be real, because $\sin e > 1$. In fact, e is then of the form $\frac{1}{2}\pi + i\gamma$ (γ real) and ($\cos e$) becomes a pure imaginary number. We define

$$\cos e = -i(\sin^2 e - 1)^{1/2} = -i\frac{\alpha}{\beta}(\sin^2 f - \sin^2 f_1)^{1/2}, \qquad (f > f_1). \quad (3.19)$$

The above choice of the sign of the radical makes the P-wave amplitude decay exponentially as we move away from the boundary into the half-space [Eq. (3.14)]. It is assumed here that $\mathrm{Re}\,\omega > 0$. However, if $\mathrm{Re}\,\omega < 0$, we must take $\cos e = i(\sin^2 e - 1)^{1/2}$.

Equations (3.15), (3.16), and (3.19) give

$$\frac{A_1}{B} = \frac{(\alpha/\beta)\sin 4f}{(\alpha/\beta)^2\cos^2 2f - 2i\sin e(\sin^2 e - 1)^{1/2}\sin 2f},$$
$$\frac{B_1}{B} = -\frac{(\alpha/\beta)^2\cos^2 2f + 2i\sin e(\sin^2 e - 1)^{1/2}\sin 2f}{(\alpha/\beta)^2\cos^2 2f - 2i\sin e(\sin^2 e - 1)^{1/2}\sin 2f}. \quad (3.20)$$

Therefore, for $f > f_1$,

$$\frac{A_1}{B} = 2\left(\frac{\beta}{\alpha}\right)\tan 2f \sin\chi e^{i(\pi/2-\chi)}, \qquad \frac{B_1}{B} = e^{-2i\chi}, \quad (3.21)$$

where

$$\tan\chi = \frac{\cos 2f \cot 2f}{2\sin f(\sin^2 f - \beta^2/\alpha^2)^{1/2}}. \quad (3.22)$$

Equation (3.21) is valid for $\mathrm{Re}\,\omega > 0$. If $\mathrm{Re}\,\omega < 0$, however, we must replace i by $-i$ in Eq. (3.21).

Equations (3.14) and (3.21) lead us to the following expressions for the reflected field when $f > f_1$:

$$\mathbf{u}_P = 2B\sin f\tan 2f\sin\chi\left[i\mathbf{a}_1 + \mathbf{a}_3\left(1 - \frac{c^2}{\alpha^2}\right)^{1/2}\right]$$
$$\times\exp\left[-\left(\frac{\omega}{\beta}\right)x_3\left(\sin^2 f - \frac{\beta^2}{\alpha^2}\right)^{1/2} + i\omega\left(t - \frac{x_1}{c}\right) - i\chi\right], \quad (3.23)$$
$$\mathbf{u}_{SV} = B(-\mathbf{a}_1\cos f + \mathbf{a}_3\sin f)\exp\left[i\omega\left(t - \frac{x_1\sin f + x_3\cos f}{\beta}\right) - 2i\chi\right],$$

where

$$c = \frac{\alpha}{\sin e} = \frac{\beta}{\sin f}. \quad (3.24)$$

Beyond the critical angle, therefore, the reflected P wave decays exponentially as we move away from the boundary into the half-space. In other words, the reflected P wave becomes an inhomogeneous plane wave, or simply a *surface wave*, that travels in the x_1 direction with the *phase velocity* $c = \beta/\sin f$, which is greater than β. Both the amplitude and the phase of the reflected P wave are distorted. There is also a phase delay of $\pi/2$ of the vertical component of the displacement relative to the horizontal component. In contrast, the reflected SV wave has the same amplitude as incident SV but suffers a phase delay of 2χ, which is independent of the frequency. The phase shift χ is equal to $\pi/2$ for $f = f_1$ and for $f = \pi/2$. The phase shift is zero at $f = \pi/4$ when $A_1/B = 0$.

The surface displacement is given by

$$(u_1)_0 = B \begin{cases} \dfrac{2(\alpha/\beta)^2 \cos f \cos 2f}{\sin 2e \sin 2f + (\alpha/\beta)^2 \cos^2 2f}, & (f \leq f_1) \\ 2 \cos f \sec 2f \sin \chi e^{i(\pi/2-\chi)}, & (f > f_1), \end{cases} \tag{3.25}$$

$$(u_3)_0 = B \begin{cases} \dfrac{2(\alpha/\beta)\cos e \sin 2f}{\sin 2e \sin 2f + (\alpha/\beta)^2 \cos^2 2f}, & (f \leq f_1) \\ \operatorname{cosec} f \cos \chi e^{-i\chi}, & (f > f_1). \end{cases} \tag{3.26}$$

Therefore

$$\left(\frac{u_1}{u_3}\right)_0 = \begin{cases} \tan e \cot f \cot 2f, & (f \leq f_1) \\ i \tan 2f \tan \chi, & (f > f_1). \end{cases} \tag{3.27}$$

We consider next the displacements in the *time domain*. With reference to the horizontal component of the SV displacement [Eq. (3.14)] we let

$$b(t - t_0) = \frac{\cos f}{2\pi} \int_{-\infty}^{\infty} B(\omega) e^{i\omega(t-t_0)}\, d\omega, \tag{3.28}$$

where $t_0 = (x_1 \sin f - x_3 \cos f)/\beta$. Then, for the reflected P and SV, we have

$$\begin{aligned} a_1(t - t_P) &= \frac{\sin e}{2\pi} \int_{-\infty}^{\infty} A_1(\omega) e^{i\omega(t-t_P)}\, d\omega, \\ b_1(t - t_S) &= -\frac{\cos f}{2\pi} \int_{-\infty}^{\infty} B_1(\omega) e^{i\omega(t-t_S)}\, d\omega, \end{aligned} \tag{3.29}$$

where $t_P = (x_1 \sin e + x_3 \cos e)/\alpha$, $t_S = (x_1 \sin f + x_3 \cos f)/\beta$. For $f \leq f_1$, we simply have

$$a_1(t - t_P) = \frac{A_1}{B}\left(\frac{\sin e}{\cos f}\right) b(t - t_P), \qquad b_1(t - t_S) = -\frac{B_1}{B} b(t - t_S), \tag{3.30}$$

where the amplitude ratios A_1/B and B_1/B are given by Eqs. (3.15) and (3.16), respectively. However, for $f > f_1$, we have from Eq. (3.21),

$$a_1(t - t_P) = \frac{1}{2\pi}\int_{-\infty}^{\infty} \frac{A_1 \sin e}{B \cos f} B \cos f e^{i\omega(t-t_P)}\, d\omega$$

$$= 2 \tan f \tan 2f \sin \chi \frac{\cos f}{2\pi}\int_{-\infty}^{\infty} B(\omega) e^{i\omega(t-t_P)+i(\pi/2-\chi)\operatorname{sgn}\omega}\, d\omega,$$

$$b_1(t - t_S) = -\frac{1}{2\pi}\int_{-\infty}^{\infty} \frac{B_1}{B} B \cos f e^{i\omega(t-t_S)}\, d\omega \tag{3.31}$$

$$= -\frac{\cos f}{2\pi}\int_{-\infty}^{\infty} B(\omega) e^{i\omega(t-t_S)-2i\chi\operatorname{sgn}\omega}\, d\omega.$$

Because $(\cos f)B(\omega)$ is the Fourier transform of $b(t)$ and $\{-i\pi \operatorname{sgn} \omega\}$ is the Fourier transform of $1/t$, the convolution theorem yields

$$\frac{\cos f}{2\pi}\int_{-\infty}^{\infty} B(\omega)(i \operatorname{sgn} \omega) e^{i\omega t}\, d\omega = \hat{b}(t), \tag{3.32}$$

where $\hat{b}(t)$ is the *Hilbert transform* of $b(t)$

$$\hat{b}(t) = -\frac{1}{\pi} \mathrm{P}\int_{-\infty}^{\infty} b(t - \tau)\frac{1}{\tau}\, d\tau = \frac{1}{\pi} \mathrm{P}\int_{-\infty}^{\infty} \frac{b(\tau)}{\tau - t}\, d\tau. \tag{3.33}$$

In Eq. (3.33) P denotes the Cauchy principal value of the integral. The function $\hat{b}(t)$ is known as *allied* to $b(t)$. Equations (3.31) and (3.32) yield

$$\begin{aligned} a_1(t - t_P) &= 2 \tan f \tan 2f \sin \chi[(\sin \chi)b(t - t_P) + (\cos \chi)\hat{b}(t - t_P)], \\ b_1(t - t_S) &= -(\cos 2\chi)b(t - t_S) + (\sin 2\chi)\hat{b}(t - t_S). \end{aligned} \tag{3.34}$$

If we choose the time variation of the incident pulse to be a unit step function, we get

$$b(t - t_0) = b_0 H(t - t_0),$$

$$\hat{b}(t - t_0) = -\frac{b_0}{\pi} \ln|t - t_0|.$$

Equation (3.34) then becomes

$$a_1(t - t_P) = 2b_0 \tan f \tan 2f \sin \chi\left[(\sin \chi)H(t - t_P) - (\cos \chi)\frac{1}{\pi} \ln |t - t_P|\right],$$

$$b_1(t - t_S) = -b_0\left[(\cos 2\chi)H(t - t_S) + (\sin 2\chi)\frac{1}{\pi} \ln |t - t_S|\right]. \tag{3.35}$$

It may be noted that in Eq. (3.34) the reflected pulse yields a disturbance prior to the time of the application of the incident pulse. The reason for this is that an elastic half-space is an ideal linear system whose impulse response function is not causal.

4. Energy equation. When $f < f_1$, it can be shown, just as in the case of incident P waves, that the energy equation is

$$\frac{\sin 2e}{\sin 2f}\left(\frac{A_1}{B}\right)^2 + \left(\frac{B_1}{B}\right)^2 = 1. \tag{3.36}$$

However, when $f > f_1$, this needs modification. From Eq. (3.21), $|B_1/B| = 1$. Therefore, the above equation will be satisfied only if we ignore the first term in the left-hand side and replace B_1/B by $|B_1/B|$. By proceeding as in Section 2.5, it can be shown that the average of $\mathbf{a}_3 \cdot \boldsymbol{\Sigma} = -\mathbf{a}_3 \cdot \mathfrak{T}(\mathbf{u}_P) \cdot \dot{\mathbf{u}}_P$, with $\mathbf{u}_P$ given by Eq. (3.23), is indeed zero. Detailed calculations are given in Section 3.3.1 for SH waves.

3.2.4. Phase Change of P Waves

We note from Eq. (3.10) that A_1/A changes sign when

$$\sin 2e = q^2 \cos 2f \cot 2f, \qquad q = \frac{\alpha}{\beta}.$$

Squaring both sides, setting $T = \sin^2 e$, and using Eq. (3.9), we obtain

$$16(q^2 - 1)T^3 - 8q^2(3q^2 - 2)T^2 + 8q^6 T - q^8 = 0.$$

The cubic in T has one or three real roots, depending on whether the discriminant $\Delta = 11q^6 - 62q^4 + 107q^2 - 64$ is greater or less than zero. For $\sigma = \frac{1}{4}$, we have the roots $T = 0.750, 0.951, 3.549$. The angles for the first two roots are $e_1 = 60°$, $e_2 = 77.2°$. The third root is the square of the ratio of the P velocity to that of the Rayleigh-wave velocity along the surface. Hence, for $\sigma = \frac{1}{4}$, there is no change of phase upon reflection when the angle of incidence lies between 60° and 77.2°. In the general case, an examination of the discriminant shows that Δ is less than, equal to, or greater than 0 according as q is less than, equal to, or greater than 1.764. The corresponding value of the Poisson ratio σ is 0.2631. Further, σ is a monotonic increasing function of q in the range $\sqrt{2} < q < \infty$. Therefore, for $\sigma > 0.2631$, there is always a phase change on reflection. For $\sigma < 0.2631$, there are certain angles of incidence for which there is no phase change on reflection. These results are applicable to the problem of change of polarity of the seismic signals pP and PP upon reflection at the earth's surface (see Chap. 8).

3.3. Reflection and Refraction at a Solid–Solid Interface

3.3.1. Incident SH Waves

Let two half-spaces, H and H', be in welded contact (Fig. 3.2). The interface between H and H' is taken as $x_3 = 0$, H occupying the region $x_3 > 0$ and H' the region $x_3 < 0$. Plane SH waves are incident on the boundary between these two half-spaces from H. From Eqs. (3.1)–(3.3), the displacement field is given by

$$\mathbf{u} = \mathbf{a}_2\left[C \exp\left[i\omega\left(t - \frac{x_1 \sin f - x_3 \cos f}{\beta}\right)\right]\right.$$
$$\left. + C_1 \exp\left[i\omega\left(t - \frac{x_1 \sin f_1 + x_3 \cos f_1}{\beta}\right)\right]\right], \qquad (x_3 > 0) \quad (3.37)$$
$$\mathbf{u}' = \mathbf{a}_2 C' \exp\left[i\omega\left(t - \frac{x_1 \sin f' - x_3 \cos f'}{\beta'}\right)\right], \qquad (x_3 < 0),$$

where f_1 is the angle of reflection and f' is the angle of transmission. The boundary conditions are

$$\mathbf{u} = \mathbf{u}' \quad \text{and} \quad \mathbf{e}_3 \cdot \mathfrak{T}(\mathbf{u}) = \mathbf{e}_3 \cdot \mathfrak{T}(\mathbf{u}') \quad \text{at} \quad x_3 = 0. \tag{3.38}$$

These must be satisfied for all values of x_1. Consequently,

$$f_1 = f, \qquad \frac{\sin f}{\beta} = \frac{\sin f'}{\beta'}. \tag{3.39}$$

From Eqs. (3.4) and (3.37)–(3.39), we find

$$C + C_1 = C', \qquad \frac{\mu \cos f}{\beta}(C - C_1) = \frac{\mu' \cos f'}{\beta'} C'. \tag{3.40}$$

On solving these equations, we get the amplitude ratios:

$$\frac{C_1}{C} = \frac{\sin 2f - (\rho'/\rho)\sin 2f'}{\sin 2f + (\rho'/\rho)\sin 2f'}, \qquad \frac{C'}{C} = \frac{2 \sin 2f}{\sin 2f + (\rho'/\rho)\sin 2f'}. \tag{3.41}$$

The energy equation can be derived with the help of Eq. (2.117). We find

$$\left(\frac{C_1}{C}\right)^2 + \frac{\rho' \sin 2f'}{\rho \sin 2f}\left(\frac{C'}{C}\right)^2 = 1. \tag{3.42}$$

If $\beta' < \beta$, we see from Eq. (3.39) that f' is real for each value of f in the range $(0, \pi/2)$. However, if $\beta' > \beta$, f' is complex for $f > f_1 = \arcsin(\beta/\beta')$, where f_1 now denotes the critical angle. We then define

$$\cos f' = -i(\sin^2 f' - 1)^{1/2}, \qquad (f > f_1). \tag{3.43}$$

The above choice of the sign of the radical makes the amplitude of the transmitted SH wave tend to zero as we move away from the interface into H', i.e., as $x_3 \to -\infty$. Therefore, for $f > f_1$,

$$\frac{C_1}{C} = -e^{-2i\chi}, \tag{3.44}$$

$$\frac{C'}{C} = 2 \sin \chi e^{i(\pi/2 - \chi)}, \tag{3.45}$$

where

$$\tan \chi = \frac{(\mu/\mu')\cos f}{(\sin^2 f - \beta^2/\beta'^2)^{1/2}}. \tag{3.46}$$

Therefore, for $f > f_1$, we have from Eq. (3.37),

$$\mathbf{u} = 2\mathbf{a}_2 C \sin\left(\frac{\omega x_3 \cos f}{\beta} + \chi\right) e^{i\omega(t - x_1 \sin f/\beta) + i(\pi/2 - \chi)}, \tag{3.47}$$

$$\mathbf{u}' = 2\mathbf{a}_2 C \sin \chi e^{\omega x_3 \beta^{-1}(\sin^2 f - \beta^2/\beta'^2)^{1/2} + i\omega(t - x_1/c) + i(\pi/2 - \chi)}, \tag{3.48}$$

where $c = \beta/\sin f$ is the phase velocity along x_1. In this case, therefore, the transmitted SH becomes an inhomogeneous plane wave traveling along the x_1 direction with velocity c and decaying exponentially as it moves away from the interface. Let us next calculate the energy for the inhomogeneous wave. Taking the real part of Eq. (3.48), we have

$$\mathbf{u}' = -2\mathbf{a}_2 C \sin \chi \sin\left[\omega\left(t - \frac{x_1}{c}\right) - \chi\right] e^{\omega x_3 \beta^{-1}(\sin^2 f - \beta^2/\beta'^2)^{1/2}}. \tag{3.49}$$

This yields, at $x_3 = 0$,

$$\begin{aligned}(-\mathbf{a}_3)\cdot\boldsymbol{\Sigma} &= \mathbf{a}_3 \cdot \mathfrak{T}(\mathbf{u}')\cdot\dot{\mathbf{u}}' \\ &= 2\frac{\mu'}{\beta}\omega^2 C^2 \sin^2 \chi \left(\sin^2 f - \frac{\beta^2}{\beta'^2}\right)^{1/2} \sin 2\left[\omega\left(t - \frac{x_1}{c}\right) - \chi\right].\end{aligned}$$

Averaging over a cycle, we find

$$\frac{1}{T}\int_0^T (-\mathbf{a}_3)\cdot\boldsymbol{\Sigma}\, dt = 0. \tag{3.50}$$

This shows that the inhomogeneous wave may be assumed to carry no energy. The energy equation, Eq. (3.42), is satisfied if we ignore the term corresponding to the inhomogeneous wave and replace C_1/C by its modulus which is unity by Eq. (3.44).

3.3.2. Incident P Waves

The geometry of the problem of incident P waves is shown in Fig. 3.1. The displacement in H is caused by incident P, reflected P, and reflected SV and is given by

$$\begin{aligned}\mathbf{u} = [&A(\mathbf{a}_1 \sin e - \mathbf{a}_3 \cos e)e^{i\omega x_3 \cos e/\alpha} \\ &+ A_1(\mathbf{a}_1 \sin e + \mathbf{a}_3 \cos e)e^{-i\omega x_3 \cos e/\alpha} \\ &+ B_1(-\mathbf{a}_1 \cos f + \mathbf{a}_3 \sin f)e^{-i\omega x_3 \cos f/\beta}]e^{i\omega(t - x_1/c)}.\end{aligned} \tag{3.51}$$

In contrast, in H' we have

$$\begin{aligned}\mathbf{u}' = [&A'(\mathbf{a}_1 \sin e' - \mathbf{a}_3 \cos e')e^{i\omega x_3 \cos e'/\alpha'} \\ &+ B'(\mathbf{a}_1 \cos f' + \mathbf{a}_3 \sin f')e^{i\omega x_3 \cos f'/\beta'}]e^{i\omega(t - x_1/c)},\end{aligned} \tag{3.52}$$

where

$$c = \frac{\alpha}{\sin e} = \frac{\beta}{\sin f} = \frac{\alpha'}{\sin e'} = \frac{\beta'}{\sin f'}. \tag{3.53}$$

Using Eq. (3.4), the boundary conditions, Eq. (3.38), yield the following equations for the amplitude ratios

$$\begin{bmatrix} -\sin e & \cos f & \sin e' & \cos f' \\ \cos e & \sin f & \cos e' & -\sin f' \\ \sin 2e & -\frac{\alpha}{\beta}\cos 2f & \frac{\rho'\alpha}{\rho\alpha'}\left(\frac{\beta'}{\beta}\right)^2 \sin 2e' & \frac{\rho'\alpha}{\rho\beta'}\left(\frac{\beta'}{\beta}\right)^2 \cos 2f' \\ -\cos 2f & -\frac{\beta}{\alpha}\sin 2f & \frac{\rho'\alpha'}{\rho\alpha}\cos 2f' & -\frac{\rho'\beta'}{\rho\alpha}\sin 2f' \end{bmatrix} \begin{bmatrix} A_1 \\ B_1 \\ A' \\ B' \end{bmatrix}$$

$$= A\begin{bmatrix} \sin e \\ \cos e \\ \sin 2e \\ \cos 2f \end{bmatrix}. \tag{3.54}$$

It is clear from Eq. (3.53) that the angle f is always real; i.e., there is no critical angle corresponding to reflected SV waves. If $\alpha' > \alpha$, then the angle e' becomes complex for $e > e_1$, where e_1 is the critical angle corresponding to the transmitted P waves given by $\sin e_1 = \alpha/\alpha'$. Guided by the requirement that the amplitude of the transmitted P should remain bounded as the point of observation moves to infinity and noting that $x_3 < 0$ in H', we define

$$\cos e' = -i(\sin^2 e' - 1)^{1/2}, \qquad (e > e_1). \tag{3.55}$$

If $\beta' > \alpha$, there is a critical angle e_2 for transmitted SV also, such that $\sin e_2 = \alpha/\beta'$. Obviously $e_2 > e_1$. We define

$$\cos f' = -i(\sin^2 f' - 1)^{1/2}, \qquad (e > e_2).$$

The energy equation is

$$\left(\frac{A_1}{A}\right)^2 + \frac{\sin 2f}{\sin 2e}\left(\frac{B_1}{A}\right)^2 + \frac{\rho' \sin 2e'}{\rho \sin 2e}\left(\frac{A'}{A}\right)^2 + \frac{\rho' \sin 2f'}{\rho \sin 2e}\left(\frac{B'}{A}\right)^2 = 1. \tag{3.56}$$

3.3.3. Incident SV Waves

In the case of incident SV waves, we have

$$\begin{aligned} \mathbf{u} = [&B(\mathbf{a}_1 \cos f + \mathbf{a}_3 \sin f)e^{i\omega x_3 \cos f/\beta} + A_1(\mathbf{a}_1 \sin e + \mathbf{a}_3 \cos e)e^{-i\omega x_3 \cos e/\alpha} \\ &+ B_1(-\mathbf{a}_1 \cos f + \mathbf{a}_3 \sin f)e^{-i\omega x_3 \cos f/\beta}]e^{i\omega(t - x_1/c)} \end{aligned} \tag{3.57}$$

in H. For the upper half-space, H', the displacement is given by Eq. (3.52). The amplitude ratios are governed by the equations

$$\begin{bmatrix} -\cos e & -\sin f & -\cos e' & \sin f' \\ -\sin e & \cos f & \sin e' & \cos f' \\ \frac{\beta}{\alpha}\sin 2e & -\cos 2f & \frac{\rho'\beta'^2}{\rho\alpha'\beta}\sin 2e' & \frac{\rho'\beta'}{\rho\beta}\cos 2f' \\ \frac{\alpha}{\beta}\cos 2f & \sin 2f & -\frac{\rho'\alpha'}{\rho\beta}\cos 2f' & \frac{\rho'\beta'}{\rho\beta}\sin 2f' \end{bmatrix} \begin{bmatrix} A_1 \\ B_1 \\ A' \\ B' \end{bmatrix} = B \begin{bmatrix} \sin f \\ \cos f \\ \cos 2f \\ \sin 2f \end{bmatrix}. \quad (3.58)$$

In this case, there is always at least one critical angle corresponding to reflected P. There will be two critical angles if $\alpha' > \beta > \beta'$ and three critical angles if $\beta' > \beta$. We define

$$\cos e = -i(\sin^2 e - 1)^{1/2}, \qquad (f > f_1), \quad (3.59)$$

where $\sin f_1 = \beta/\alpha$;

$$\cos e' = -i(\sin^2 e' - 1)^{1/2}, \qquad (f > f_2, \alpha' > \beta),$$

where $\sin f_2 = \beta/\alpha'$; and, last, (3.60)

$$\cos f' = -i(\sin^2 f' - 1)^{1/2}, \qquad (f > f_3, \beta' > \beta),$$

where $\sin f_3 = \beta/\beta'$.

The principle of energy conservation yields

$$\frac{\sin 2e}{\sin 2f}\left(\frac{A_1}{B}\right)^2 + \left(\frac{B_1}{B}\right)^2 + \frac{\rho' \sin 2e'}{\rho \sin 2f}\left(\frac{A'}{B}\right)^2 + \frac{\rho' \sin 2f'}{\rho \sin 2f}\left(\frac{B'}{B}\right)^2 = 1. \quad (3.61)$$

The explicit expressions for the amplitude ratios are given in Section 7.6.2.

3.4. Reflection and Refraction at a Solid–Liquid Interface

3.4.1. Incident P Waves

Consider a solid half-space $H(\alpha, \beta, \rho, \mu)$ in contact with a liquid half-space $H'(\alpha', \rho')$. We assume that P waves are incident at the boundary between H and H' from H. In this case there will be no transmitted SV waves; i.e., $B' = 0$. Further, because at a solid–liquid interface there is a possibility of slip, the boundary conditions require that $u_3 = u_3'$ instead of $\mathbf{u} = \mathbf{u}'$ as mentioned in Eq. (3.38) for the solid–solid case. These considerations lead us to the following system of equations for the amplitude ratios:

$$\begin{bmatrix} \cos e & \sin f & \cos e' \\ \sin 2e & -\frac{\alpha}{\beta}\cos 2f & 0 \\ -\cos 2f & -\frac{\beta}{\alpha}\sin 2f & \frac{\rho'\alpha'}{\rho\alpha} \end{bmatrix} \begin{bmatrix} A_1 \\ B_1 \\ A' \end{bmatrix} = A \begin{bmatrix} \cos e \\ \sin 2e \\ \cos 2f \end{bmatrix}. \tag{3.62}$$

The displacements are given by Eq. (3.51) for H and Eq. (3.52) for H' with $B' = 0$. Equation (3.62) easily can be solved for the amplitude ratios A_1/A, B_1/A, and A'/A. The explicit expressions for these ratios are given in Section 7.2.5.

3.4.2. Incident *SV* Waves

In the case of incident SV waves, we assume the displacement vector, Eq. (3.57), for H and Eq. (3.52) for H' with $B' = 0$ and get

$$\begin{bmatrix} -\cos e & -\sin f & -\cos e' \\ \frac{\beta}{\alpha}\sin 2e & -\cos 2f & 0 \\ \frac{\alpha}{\beta}\cos 2f & \sin 2f & -\frac{\rho'\alpha'}{\rho\beta} \end{bmatrix} \begin{bmatrix} A_1 \\ B_1 \\ A' \end{bmatrix} = B \begin{bmatrix} \sin f \\ \cos 2f \\ \sin 2f \end{bmatrix}. \tag{3.63}$$

3.5. Reflection and Refraction at a Liquid–Solid Interface

We suppose that a liquid half-space $H(\alpha, \rho)$ is in contact with a solid half-space $H'(\alpha', \beta', \rho', \mu')$ and P waves are incident from H. Hence, there will be no reflected SV waves; i.e., $B_1 = 0$. The displacement $\mathbf{u}$ at any point of H is given by Eq. (3.51) with $B_1 = 0$ and $\mathbf{u}'$ at any point of H' is given by Eq. (3.52). On applying the boundary conditions, we get

$$\begin{bmatrix} \cos e & \cos e' & -\sin f' \\ 0 & \sin 2e' & \frac{\alpha'}{\beta'}\cos 2f' \\ -1 & \frac{\rho'\alpha'}{\rho\alpha}\cos 2f' & -\frac{\rho'\beta'}{\rho\alpha}\sin 2f' \end{bmatrix} \begin{bmatrix} A_1 \\ A' \\ B' \end{bmatrix} = A \begin{bmatrix} \cos e \\ 0 \\ 1 \end{bmatrix}. \tag{3.64}$$

The explicit expressions for the amplitude ratios are given, with a slight change in notation, in Section 7.2.5.

3.6. Surface Waves

3.6.1. General Considerations

We have seen in Section 2.5 that the displacement field induced by plane waves may be expressed in the form

$$\begin{aligned}
\mathbf{u}_{\mathrm{P}} &= A_{\mathrm{P}}(l\mathbf{a}_x + n\mathbf{a}_z)\exp\left[i\omega\left(t - \frac{lx + nz}{\alpha}\right)\right],\\
\mathbf{u}_{\mathrm{SV}} &= A_{\mathrm{SV}}(-n\mathbf{a}_x + l\mathbf{a}_z)\exp\left[i\omega\left(t - \frac{lx + nz}{\beta}\right)\right],\\
\mathbf{u}_{\mathrm{SH}} &= A_{\mathrm{SH}}\mathbf{a}_y \exp\left[i\omega\left(t - \frac{lx + nz}{\beta}\right)\right],
\end{aligned} \tag{3.65}$$

where $l^2 + n^2 = 1$. When l and n are real, these equations represent homogeneous plane waves propagating in the direction $(l, 0, n)$. The following form of the displacements will be found to be more convenient in discussing surface waves

$$\begin{aligned}
\mathbf{u}_{\mathrm{P}} &= A(\mathbf{a}_x \pm \eta_\alpha \mathbf{a}_z)e^{ik(ct - x \mp \eta_\alpha z)},\\
\mathbf{u}_{\mathrm{SV}} &= B(\mp\eta_\beta \mathbf{a}_x + \mathbf{a}_z)e^{ik(ct - x \mp \eta_\beta z)},\\
\mathbf{u}_{\mathrm{SH}} &= C\mathbf{a}_y e^{ik(ct - x \mp \eta_\beta z)},
\end{aligned} \tag{3.66}$$

where

$$\omega = ck, \qquad \eta_\alpha = \left(\frac{c^2}{\alpha^2} - 1\right)^{1/2}, \qquad \eta_\beta = \left(\frac{c^2}{\beta^2} - 1\right)^{1/2}. \tag{3.67}$$

Moreover, the upper sign corresponds to the waves moving in the positive z direction, and the lower sign to the waves moving in the negative z direction.

However, we need not restrict l and n to being real. The only condition is that they satisfy the relation $l^2 + n^2 = 1$. Therefore, if $l > 1$, we have

$$\begin{aligned}
\mathbf{u}_{\mathrm{P}} &= A(\mathbf{a}_x \mp i\gamma_\alpha \mathbf{a}_z)e^{\mp\gamma_\alpha kz + ik(ct - x)},\\
\mathbf{u}_{\mathrm{SV}} &= B(\pm i\gamma_\beta \mathbf{a}_x + \mathbf{a}_z)e^{\mp\gamma_\beta kz + ik(ct - x)},\\
\mathbf{u}_{\mathrm{SH}} &= C\mathbf{a}_y e^{\mp\gamma_\beta kz + ik(ct - x)},
\end{aligned} \tag{3.68}$$

where

$$\gamma_\alpha = \left(1 - \frac{c^2}{\alpha^2}\right)^{1/2}, \qquad \gamma_\beta = \left(1 - \frac{c^2}{\beta^2}\right)^{1/2}. \tag{3.69}$$

Equation (3.68) represents surface waves (inhomogeneous plane waves) propagating in the positive x direction with their amplitudes decaying exponentially with $\pm z$.

Equations (3.66) and (3.68) represent a two-dimensional displacement field. We have seen before that in the case of two-dimensional wave propagation, the *SH* motion is decoupled from the *P–SV* motion. Surface waves of the *SH* type are known as *Love waves*, whereas surface waves of the *P–SV* type are known as *Rayleigh waves*.

3.6.2. Love Waves

Let us consider first the possibility of the propagation of Love waves in a homogeneous, semiinfinite, elastic medium occupying the region $z \geq 0$. From Eq. (3.68), the displacement at any point of the medium is given by

$$\mathbf{u} = C\mathbf{a}_y e^{-\gamma_\beta kz + ik(ct - x)}, \tag{3.70}$$

where $c < \beta$. The boundary condition that the plane $z = 0$ be stress free yields

$$\mu C(-\gamma_\beta k) = 0.$$

Therefore, $C = 0$, thereby implying that in the case of a homogeneous semi-infinite medium Love waves do not exist.

Consider next a homogeneous half-space, (β_2, ρ_2), covered with a homogeneous layer, (β_1, ρ_1), of thickness H. The free surface is taken as the plane $z = 0$ and the z axis is drawn into the medium (Fig. 3.4). From Eqs. (3.66) and (3.68), we have

$$\mathbf{u} = \begin{cases} \mathbf{a}_y(Ae^{-i\eta_1 kz} + Be^{i\eta_1 kz})e^{ik(ct - x)}, & (0 < z < H), \\ C\mathbf{a}_y e^{-\gamma_2 kz + ik(ct - x)}, & (z > H), \end{cases} \tag{3.71}$$

where

$$\eta_1 = \left(\frac{c^2}{\beta_1^2} - 1\right)^{1/2}, \qquad \gamma_2 = \left(1 - \frac{c^2}{\beta_2^2}\right)^{1/2}. \tag{3.72}$$

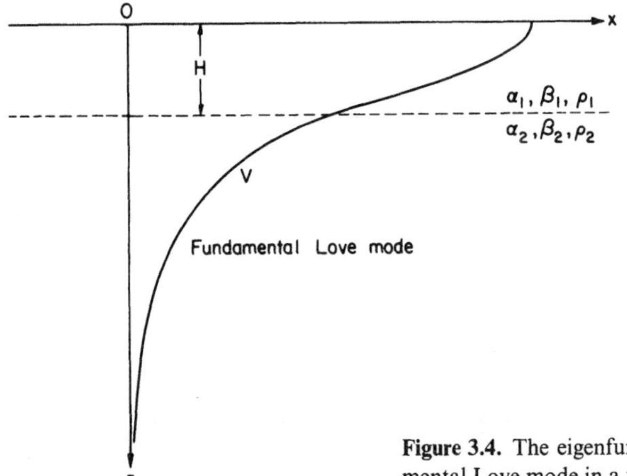

Figure 3.4. The eigenfunction V for the fundamental Love mode in a two-layered half-space.

The traction across the free surface must be zero and the displacements and the tractions must be continuous across the interface. These conditions yield

$$\begin{aligned} A - B &= 0, \\ Ae^{-i\eta_1 kH} + Be^{i\eta_1 kH} - Ce^{-\gamma_2 kH} &= 0, \\ Ae^{-i\eta_1 kH} - Be^{i\eta_1 kH} + iC\frac{\mu_2\gamma_2}{\mu_1\eta_1}e^{-\gamma_2 kH} &= 0. \end{aligned} \tag{3.73}$$

The above equations in A, B, and C will be consistent if their determinant vanishes. This leads us to the *period equation* (*frequency equation*) for Love waves

$$\tan(\eta_1 kH) = \frac{\mu_2\gamma_2}{\mu_1\eta_1}. \tag{3.74}$$

It is clear from Eq. (3.72) that $c < \beta_2$, because otherwise, $\mathbf{u}$ will not tend to zero as $z \to \infty$. Further, if η_1 is purely imaginary, say $-i\gamma_1$, Eq. (3.74) becomes

$$-\gamma_1 \tanh(\gamma_1 kH) = \left(\frac{\mu_2}{\mu_1}\right)\gamma_2, \tag{3.75}$$

which has no relevant solution because the two sides are of opposite signs for positive values of k. Therefore, η_1 is real; i.e., $\beta_1 < c$. We therefore see that $\beta_1 < c < \beta_2$, which implies that the shear-wave velocity in the layer has to be less than the shear-wave velocity in the half-space.

Equation (3.74) is a transcendental equation. For any value of c in the interval $\beta_1 < c < \beta_2$, it determines a set of possible values of $\eta_1 kH$, the first in the interval $(0, \pi/2)$, the second in the interval $(\pi, 3\pi/2)$, and so on. Let us denote the corresponding values of k by k_0, k_1, etc. The corresponding periods are then obtained from the relation $T_n = 2\pi/\omega_n = 2\pi/[ck_n(c)]$, $n = 0, 1, 2, \ldots$. Here, T_0 corresponds to what is known as the *fundamental Love mode* and $T_n\,(n \geq 1)$ corresponds to the nth Love mode.

From Eqs. (3.71) and (3.73), we have

$$\mathbf{u} = 2A\mathbf{a}_y V(z)e^{ik(ct-x)}, \tag{3.76}$$

where

$$V(z) = \begin{cases} \cos(\eta_1 kz), & 0 < z < H, \\ \cos(\eta_1 kH)e^{-\gamma_2 k(z-H)}, & z > H. \end{cases} \tag{3.77}$$

Therefore, V is unity at the free surface, is continuous across the interface, and goes to zero as $z \to \infty$. Figure 3.4 shows the variation of V with z for the fundamental Love mode, assuming $\mu_2/\mu_1 = 1.8$, $\beta_1 = 3.6$ km/s, $\beta_2 = 4.6$ km/s, and $c = 4.0$ km/s. V does not vanish anywhere in the region $z > H$. However, the displacement in the layer vanishes for all values of x and t when $\eta_1 kz$ takes any of the particular values $\frac{1}{2}\pi, \frac{3}{2}\pi, \frac{5}{2}\pi, \ldots$. The corresponding planes, which are parallel to the free surface, are known as *nodal planes*. It is obvious that for values of k for which $\eta_1 kH < \frac{1}{2}\pi$, there will be no nodal planes; for values of k for which $\frac{1}{2}\pi < \eta_1 kH < \frac{3}{2}\pi$, there will be one nodal plane; and so on. We have seen before that $0 < \eta_1 k_0 H < \frac{1}{2}\pi$, $\pi < \eta_1 k_1 H < \frac{3}{2}\pi$, $2\pi < \eta_1 k_2 H < \frac{5}{2}\pi$, etc. Hence, the fundamental Love mode has no nodal plane, whereas the nth Love mode has n nodal planes.

The period equation, Eq. (3.74), can be recast in the form

$$\tan(\eta_1 kH) = \frac{\rho_2 \beta_2}{\rho_1 \beta_1}\left(\frac{\eta_0^2}{\eta_1^2} - 1\right)^{1/2}, \tag{3.78}$$

where $\eta_0^2 = \beta_2^2/\beta_1^2 - 1$. For a given value of kH, this can be solved for η_1 from which c may then be determined using Eq. (3.72).

Consider next the possibility of the propagation of Love waves in a vertically heterogeneous medium in which $\mu = \mu(z)$ and $\rho = \rho(z)$. For SH-type motion, we know that $u_z = 0$, div $\mathbf{u} = 0$. Therefore, Eq. (1.107), becomes

$$\nabla^2 \mathbf{u} + \frac{\mu'}{\mu}\left[2\frac{\partial \mathbf{u}}{\partial z} + (\mathbf{e}_z \times \text{curl}\, \mathbf{u})\right] = \frac{1}{\beta^2}\frac{\partial^2 \mathbf{u}}{\partial t^2}, \tag{3.79}$$

where a prime denotes differentiation with respect to z and $\beta(z) = \sqrt{(\mu/\rho)}$.

For Love waves propagating in the positive x direction and polarized in the y direction, we may take

$$\mathbf{u} = \mathbf{a}_y V(z) e^{ik(ct - x)}. \tag{3.80}$$

Equation (3.79) will be satisfied if V is an integral of the equation

$$\frac{d}{dz}\left(\mu \frac{dV}{dz}\right) + (\rho\omega^2 - \mu k^2)V = 0. \tag{3.81}$$

Substituting $V(z) = F(z)/\sqrt{\mu}$, Eq. (3.81) becomes

$$\frac{d^2 F}{dz^2} + \left[\frac{\omega^2}{\beta^2} - k^2 - \frac{\mu''}{2\mu} + \left(\frac{\mu'}{2\mu}\right)^2\right]F = 0. \tag{3.82}$$

The boundary condition yields

$$\frac{dF}{dz} - \frac{F}{2\mu}\frac{d\mu}{dz} = 0 \quad \text{at} \quad z = 0.$$

If Eq. (3.82) can be solved for a given variation of μ and β with z, the period equation can then be obtained from the boundary condition. Table 3.1 gives $F(z)$ for some particular functional forms of μ and β.

When μ and ρ are not given as explicit functions of z, Eq. (3.81) must be solved numerically. However, this equation is not suitable for numerical integration, because it is necessary to evaluate the derivative of the empirically determined function $\mu(z)$. This difficulty can be obviated by transforming

Table 3.1. Solutions of Eq. (3.82) for Given $\mu(z)$ and $\beta(z)$ ($k_0 = \omega/\beta_0$; μ_0 and β_0 Constants)

Model	$\mu(z)/\mu_0$	$\beta(z)/\beta_0$	$F(z)$	Remarks
I	$(1+\delta z)^2$	$(1+\delta z)(1+\varepsilon z)^{-1/2}$	$W_{\pm l,m}(\pm\xi)$ (Whittaker functions)	$\xi = 2k(z+1/\delta)$; $l = \varepsilon k_0^2/(2k\delta^2)$; $m^2 = (\varepsilon-\delta)k_0^2/\delta^3 + \frac{1}{4}$
II	$(1+\delta z)^2$	$1+\varepsilon z$	$\xi^{1/2}[K_m(\theta), I_m(\theta)]$ (Modified Bessel functions)	$\theta = k(z+1/\varepsilon)$; $\xi = 1+\varepsilon z$; $m^2 = \frac{1}{4} - k_0^2/\varepsilon^2$
III	$(1+\delta z)^2$	$(1+\gamma e^{\varepsilon z})^{-1/2}$	$H_m^{(1,2)}(le^{\xi})$ (Hankel functions)	$\xi = \frac{1}{2}\varepsilon z$; $l = 2k_0\sqrt{\gamma}/\varepsilon$; $m^2 = 4(k^2-k_0^2)/\varepsilon^2$
IV	$e^{\delta z}$	$(1+\varepsilon z)^{1/2}$	$W_{\pm l,1/2}(\pm\xi)$	$\xi = (4k^2+\delta^2)^{1/2}(z+1/\varepsilon)$; $l = (k_0^2/\varepsilon)(4k^2+\delta^2)^{-1/2}$
V	$(1+\delta z)^p$	$1+\delta z$	$\xi^{1/2}[K_m(\theta), I_m(\theta)]$	$\xi = 1+\delta z$; $m^2 = \frac{1}{4}(p-1)^2 - k_0^2/\delta^2$; $\theta = k(z+1/\delta)$
VI	$(1+\delta z)^{\varepsilon}$	1	$\xi^{1/2}[K_m(l\xi), I_m(l\xi)]$	$\xi = 1+\delta z$; $m = \frac{1}{2}(\varepsilon-1)$; $l = (k^2-k_0^2)^{1/2}/\delta$
VII	$e^{\delta z}$	$e^{(1/2)(\delta-\varepsilon)z}$	$H_m^{(1,2)}(le^{\xi})$	$\xi = \frac{1}{2}(\varepsilon-\delta)z$; $l = 2k_0/(\varepsilon-\delta)$; $m^2 = (4k^2+\delta^2)(\varepsilon-\delta)^{-2}$

Eq. (3.81) into the following system of two simultaneous differential equations

$$\frac{dy_1}{dz} = \frac{1}{\mu} y_2, \qquad \frac{dy_2}{dz} = (\mu k^2 - \rho\omega^2)y_1, \tag{3.83}$$

where $y_1 = V$. From the boundary conditions, y_2 vanishes at the free surface and both y_1 and y_2 are continuous across a parallel internal discontinuity.

If (y_1, y_2) and (x_1, x_2) are two possible solutions of Eq. (3.83) subject to some arbitrary boundary conditions, we have

$$\frac{d}{dz}(x_1 y_2) = (\mu k^2 - \rho\omega^2)x_1 y_1 + \mu\frac{dx_1}{dz}\frac{dy_1}{dz}.$$

Therefore, on account of the symmetry of the right-hand side in x_1, y_1, we have

$$\frac{d}{dz}(x_1 y_2 - y_1 x_2) = 0; \tag{3.84}$$

i.e., $x_1 y_2 - y_1 x_2$ is independent of z.

3.6.3. Dispersion

Equation (3.74) shows that the phase velocity, c, of Love waves is not a fixed constant but depends on the particular value of $k = 2\pi/\Lambda$, i.e., on the wavelength of the Fourier component under consideration. Waves of different wavelengths (or frequency) will, in general, propagate with different phase velocities. This phenomenon is known as *dispersion.* It is caused by the inhomogeneity of the medium. The inhomogeneity may result either from a continuous change of the elastic parameters or from abrupt discontinuities within the medium. We have seen earlier that waves in an unbounded homogeneous medium are undispersed, the velocity of propagation of the waves being independent of their wavelengths.

A plane monochromatic wave, traveling all by itself in a homogeneous space, is an idealization that is never strictly realized in nature. Most sources emit signals with a continuous spectrum over a limited frequency band. As such waves propagate through inhomogeneous media, they exhibit dispersion. The linear relation $\omega = ck$ is then replaced by $\omega = \omega(k)$, where the functional relationship depends on the characteristics of the medium and the boundary conditions of the problem at hand.

Consider two adjacent Fourier components within the signal with frequencies $\omega \pm \Delta\omega$ and wave numbers $k \pm \Delta k$. Assuming that the amplitudes of the two components are close to each other, the result of superposition is

$$e^{i[(\omega-\Delta\omega)t-(k-\Delta k)x]} + e^{i[(\omega+\Delta\omega)t-(k+\Delta k)x]} = 2\cos\left[\Delta k\left(\frac{\Delta\omega}{\Delta k}t - x\right)\right] \times \exp\left[ik\left(\frac{\omega}{k}t - x\right)\right].$$

The wave profile therefore has the form of a *carrier* moving with the phase velocity $c = \omega/k$ and a *modulator* (envelope) moving with the velocity $U_g = \Delta\omega/\Delta k = d\omega/dk$, in the limit. The velocity U_g is the velocity of an *interference pattern* within the wave profile and is known as the *group velocity.* It follows immediately that

$$U_g = \frac{d\omega}{dk} = c^2\frac{dT}{d\Lambda} = c + k\frac{dc}{dk} = c - \Lambda\frac{dc}{d\Lambda},$$
$$\frac{1}{U_g} = \frac{1}{c} + \frac{T}{c^2}\frac{dc}{dT} = \frac{1}{c} - \frac{\omega}{c^2}\frac{dc}{d\omega}. \tag{3.85}$$

Consider, for example, the case of Love waves in a layer over a half-space. The period equation is given by Eq. (3.74). On taking its logarithmic derivative, we find that the group velocity is given by

$$U_g = c + k\frac{dc}{dk} = \frac{\beta_1^2}{c}\left(\frac{c^2/\beta_1^2 + \Omega}{1 + \Omega}\right), \tag{3.85a}$$

where

$$\Omega = kH\gamma_2\left[\frac{\rho_1}{\rho_2}\left(\frac{c^2-\beta_1^2}{\beta_2^2-\beta_1^2}\right)+\frac{\mu_2}{\mu_1}\left(\frac{\beta_2^2-c^2}{\beta_2^2-\beta_1^2}\right)\right].$$

It then follows that as $c \to \beta_1$, $U_g \to \beta_1$, and as $c \to \beta_2$, $U_g \to \beta_2$. Therefore, the phase-velocity curves and the group-velocity curves meet at both ends.

Equation (3.85) can be used to compute U_g from c. However, sometimes it is necessary to compute c from U_g. To this end, we note first that

$$\frac{d(cT)}{(cT)^2} = \frac{dT}{T^2 U_g}.$$

Integrating between the limits (T, T_1), we find

$$\frac{1}{c} = T\int_T^{T_1}\frac{dT}{T^2 U_g(T)} + \frac{1}{c_1}\left(\frac{T}{T_1}\right), \tag{3.86}$$

where $c_1 = c(T_1)$. This is a very useful formula for calculating $c(T)$, if $U_g(T)$ is known from observations.

If U_g can be regarded as constant, we find

$$c = \frac{U_g}{1-\varepsilon T},$$

where ε is a constant. Therefore, if we represent a signal $\Psi(x, t)$ as a Fourier integral, we have

$$\begin{aligned}\Psi(x,t) &= \frac{1}{2\pi}\,\mathrm{Re}\int_{-\infty}^{\infty} S(\omega)e^{i\omega(t-x/c)}\,d\omega \\ &= \mathrm{Re}\left\{e^{2\pi i\varepsilon x/U_g}\left[\frac{1}{2\pi}\int_{-\infty}^{\infty} S(\omega)e^{i\omega(t-x/U_g)}\,d\omega\right]\right\}.\end{aligned}$$

Because U_g is assumed to be constant, the signal will propagate with constant velocity U_g, with its profile remaining unaltered. However, the polarity of the signal will be reversed at regular intervals as the epicentral distance is increased by the critical distance, $x_c = U_g/(2\varepsilon)$.

Physical signals can be represented formally as a sum of plane harmonic waves with different wave numbers in the form of a Fourier integral. Therefore, in the one-dimensional case, we have

$$\Psi(x,t) = \frac{1}{2\pi}\int_{-\infty}^{\infty} F(k)e^{i[\omega(k)t-kx]}\,dk. \tag{3.87}$$

We will apply the Kelvin's method of stationary phase to evaluate the integral in Eq. (3.87) approximately. This method asserts that, for large values of t, the main contribution to the integral arises from the neighborhood of the points at which the phase $\phi(k) = \omega t - kx$ is stationary. Let one such point be denoted by k_0: Then

$$\left(\frac{\partial \phi}{\partial k}\right)_0 = \left(\frac{d\omega}{dk}\right)_0 t - x = 0,$$

i.e.,

$$(U_g)_0 = \left(\frac{d\omega}{dk}\right)_0 = \frac{x}{t}. \tag{3.88}$$

The frequency $\omega_0 = ck_0$ is known as the *predominant frequency* at the location x for the time t. It may be noted that the relation $U_g = x/t$ is valid only at $\omega = \omega_0$. The contribution of k_0 to the value of the integral in Eq. (3.87) obtained by the Kelvin's method of stationary phase is given by [cf. Eq. (E.12), in App. E]

$$\Psi(x, t) = \frac{F(k_0)}{\sqrt{(2\pi t|\omega_0''|)}} \exp\left[i\left(\omega_0 t - k_0 x + \frac{\pi}{4} \operatorname{sgn} \omega_0''\right)\right]. \tag{3.89}$$

However, if $\Psi(x, t)$ is real, then $F(-k_0) = F(k_0)$, $\omega(-k_0) = -\omega(k_0)$, and $U_g(-k_0) = U_g(k_0)$. Consequently, for each given x and t, both k_0 and $-k_0$ satisfy Eq. (3.88). The sum of the contributions of $\pm k_0$ then yields

$$\Psi(x, t) = \frac{2F(k_0)}{\sqrt{(2\pi t|\omega_0''|)}} \cos\left(\omega_0 t - k_0 x + \frac{\pi}{4} \operatorname{sgn} \omega_0''\right). \tag{3.90}$$

In this representation

$$k_0 = k_0(t, x), \qquad \omega_0 = \omega_0(k_0), \qquad \omega_0'' \neq 0.$$

Because

$$\omega'' = \frac{dU_g}{dk} = U_g \frac{dU_g}{d\omega} = -\frac{T^2}{2\pi} U_g \frac{dU_g}{dT}, \tag{3.91}$$

we finally have

$$\begin{aligned}\Psi(x, t) &= \frac{1}{2\pi} \int_{-\infty}^{\infty} e^{i(\omega t - kx)} F(k) dk \\ &= \left\{\frac{2F(k)}{T\sqrt{(x|dU_g/dT|)}}\right\}_{T=T_0} \cos\left(\omega_0 t - k_0 x - \frac{\pi}{4} \operatorname{sgn} \frac{dU_g}{dT}\right),\end{aligned} \tag{3.92}$$

provided that $(dU_g/dk)_{k_0} \neq 0$. We shall use this formula later as an asymptotic approximation for the Fourier-integral representation of certain waveforms.

If $d^2\omega/dk^2 = 0$ at $k = k_0$ but $d^3\omega/dk^3 \neq 0$ there, the right-hand side of Eq. (3.92) is replaced by the expression

$$\begin{aligned}&\frac{1}{(2\pi)^{2/3}} CF(k_0) \left(t \left|\frac{d^3\omega}{dk^3}\right|_0\right)^{-1/3} \cos(\omega_0 t - k_0 x) \\ &= C\left\{\frac{F(k_0)}{\sqrt[3]{(xT^4 U_g |d^2 U_g/dT^2|)}}\right\}_{T=T_0} \cos(\omega_0 t - k_0 x),\end{aligned} \tag{3.93}$$

where

$$C = 2\sqrt{3}\,\frac{\Gamma(\frac{4}{3})}{(\pi/3)^{1/3}} = 3.0462. \tag{3.94}$$

In the derivation of Eq. (3.93), we have used the following relations obtained from Eq. (3.88)

$$\begin{aligned} \frac{d^3\omega}{dk^3} &= \frac{d\omega}{dk}\frac{d}{d\omega}\left(U_g\frac{dU_g}{d\omega}\right) = U_g^2\frac{d^2U_g}{d\omega^2} + U_g\left(\frac{dU_g}{d\omega}\right)^2, \\ \left(\frac{d^3\omega}{dk^3}\right)_0 &= \left(U_g^2\frac{d^2U_g}{d\omega^2}\right)_0 = \frac{1}{4\pi^2}\left(T^4U_g^2\frac{d^2U_g}{dT^2}\right)_0. \end{aligned} \tag{3.95}$$

If the integral is over ω, we put $d\omega = U_g\,dk$ and obtain

$$\frac{1}{2\pi}\int_{-\infty}^{\infty} e^{i(\omega t - kx)}G(\omega)d\omega$$

$$= \begin{cases} \left[\dfrac{2G(T)}{(T/U_g)\sqrt{(x|dU_g/dT|)}}\right]_{T=T_0}\cos\left(\omega_0 t - k_0 x - \dfrac{\pi}{4}\operatorname{sgn}\dfrac{dU_g}{dT}\right), & \left(\dfrac{dU_g}{dk}\right)_0 \neq 0, \\ \left[\dfrac{CG(T)}{\sqrt[3]{(x(T^4/U_g^2)|d^2U_g/dT^2|)}}\right]_{T=T_0}\cos(\omega_0 t - k_0 x); & \left(\dfrac{dU_g}{dk}\right)_0 = 0,\ \left(\dfrac{d^2U_g}{dk^2}\right)_0 \neq 0. \end{cases} \tag{3.96}$$

Writing $\phi_0 = \omega_0 t - k_0 x$ and using Eq. (3.88), we have

$$\begin{aligned} \frac{\partial\phi_0}{\partial x} &= t\frac{d\omega_0}{dk_0}\frac{\partial k_0}{\partial x} - k_0 - x\frac{\partial k_0}{\partial x} = -k_0, \\ \frac{\partial\phi_0}{\partial t} &= \omega_0 + t\frac{d\omega_0}{dk_0}\frac{\partial k_0}{\partial t} - x\frac{\partial k_0}{\partial t} = \omega_0. \end{aligned} \tag{3.97}$$

Consequently, for values of x near $U_g t$, the motion is approximately harmonic with wavelength $2\pi/k_0$, whereas for values of t near x/U_g, it is approximately harmonic with period $2\pi/\omega_0$. The function $\Psi(x, t)$ therefore represents an individual harmonic component that passes the coordinate x at time t with a local phase velocity $c = c(x, t) = \omega_0/k_0$. If, however, we fix the value of x and consider a time interval $t_1 \leq t \leq t_2$, a correspondence can be set up between the values of t in this interval and the values of $\omega_0(t)$. In this sense, Eq. (3.92) is in effect another spectral representation of $\Psi(x, t)$ with $t = t(\omega_0)$.

If the dispersion equation is given in the implicit form $F(\omega, k) = 0$, the relation $dF = (\partial F/\partial\omega)d\omega + (\partial F/\partial k)dk = 0$ implies

$$U_g = -\frac{\partial F/\partial k}{\partial F/\partial\omega}. \tag{3.98}$$

3.6.4. Rayleigh Waves

We have seen in Section 3.6.2 that surface waves of the *SH* type do not exist in a homogeneous semiinfinite medium. We will now show that in such a configuration, surface waves of the *P–SV* type do exist.

Taking the free surface as the $z = 0$ plane and assuming that the medium occupies the region $z > 0$, the displacement field follows from Eq. (3.68)

$$\mathbf{u} = [A(\mathbf{a}_x - i\gamma_\alpha \mathbf{a}_z)e^{-\gamma_\alpha kz} + B(i\gamma_\beta \mathbf{a}_x + \mathbf{a}_z)e^{-\gamma_\beta kz}]e^{ik(ct-x)}. \qquad (3.99)$$

This represents waves traveling in the positive x direction with phase velocity c and decaying exponentially with z, provided $\gamma_{\alpha,\beta}$ are real; i.e., $c < \beta$.

The boundary condition that the traction $\mathbf{e}_z \cdot \mathfrak{T}$ vanishes at the free surface yields the following two equations in A and B:

$$2\gamma_\alpha A + i\left(2 - \frac{c^2}{\beta^2}\right)B = 0, \qquad -i\left(2 - \frac{c^2}{\beta^2}\right)A + 2\gamma_\beta B = 0. \qquad (3.100)$$

Therefore, we must have

$$\left(2 - \frac{c^2}{\beta^2}\right)^2 - 4\gamma_\alpha\gamma_\beta = 0. \qquad (3.101)$$

This is the period equation for Rayleigh waves and is known as the *Rayleigh equation.* It shows that c is independent of ω. In other words, in a semiinfinite homogeneous medium, Rayleigh waves are nondispersive.

Substituting for γ_α and γ_β from Eq. (3.69), Eq. (3.101) becomes

$$f(\xi) \equiv \xi^3 - 8\xi^2 + 8\xi\left(3 - \frac{2\beta^2}{\alpha^2}\right) - 16\left(1 - \frac{\beta^2}{\alpha^2}\right) = 0, \qquad (3.102)$$

where $\xi = c^2/\beta^2$. Because $f(0) = -16(1 - \beta^2/\alpha^2) < 0$ and $f(1) = 1 > 0$, it is apparent that $f(\xi) = 0$ has a root between 0 and 1; i.e., the Rayleigh equation has a root c_R such that $0 < c_R < \beta$, for all values of β/α. For $\sigma = \frac{1}{4}$, $\alpha^2/\beta^2 = 3$, and we get

$$3\xi^3 - 24\xi^2 + 56\xi - 32 = 0, \qquad (3.103)$$

yielding

$$\xi = \frac{c^2}{\beta^2} = 4, \qquad 2 + \frac{2}{\sqrt{3}}, \qquad 2 - \frac{2}{\sqrt{3}}.$$

Because c must satisfy the inequality $0 < c < \beta$, we have

$$\frac{c_R}{\beta} = \sqrt{\left(2 - \frac{2}{\sqrt{3}}\right)} \simeq 0.9195,$$

$$\gamma_\alpha = \frac{1}{3}\sqrt{(3 + 2\sqrt{3})} \simeq 0.8475, \qquad \gamma_\beta = \frac{1}{\sqrt{3}}\sqrt{(2\sqrt{3} - 3)} \simeq 0.3933. \qquad (3.104)$$

Using Eq. (3.100) in Eq. (3.99) and taking the real part, we get

$$
\begin{aligned}
u_x &= Q\left[e^{-\gamma_\alpha kz} - \left(1 - \frac{c^2}{2\beta^2}\right)e^{-\gamma_\beta kz}\right]\sin(\omega t - kx) \\
&= Q(k)U(z)\sin(\omega t - kx), \\
u_z &= Q\left(1 - \frac{c^2}{\alpha^2}\right)^{1/2}\left[-e^{-\gamma_\alpha kz} + \left(1 - \frac{c^2}{2\beta^2}\right)^{-1} e^{-\gamma_\beta kz}\right]\cos(\omega t - kx) \\
&= Q(k)W(z)\cos(\omega t - kx).
\end{aligned} \tag{3.105}
$$

The variation of U and W with z for $\sigma = \frac{1}{4}$ is shown in Fig. 3.5. W is always positive, whereas U is positive at the free surface but changes sign at a depth h given by

$$
e^{-(\gamma_\alpha - \gamma_\beta)kh} = 1 - \frac{c^2}{2\beta^2}. \tag{3.106}
$$

For $\sigma = \frac{1}{4}$, this gives $kh \simeq 1.209$, whereas Eq. (3.105) yields

$$
\begin{aligned}
u_x &= Q(e^{-0.8475kz} - 0.5773e^{-0.3933kz})\sin(\omega t - kx), \\
u_z &= Q(-0.8475e^{-0.8475kz} + 1.4679e^{-0.3933kz})\cos(\omega t - kx).
\end{aligned} \tag{3.107}
$$

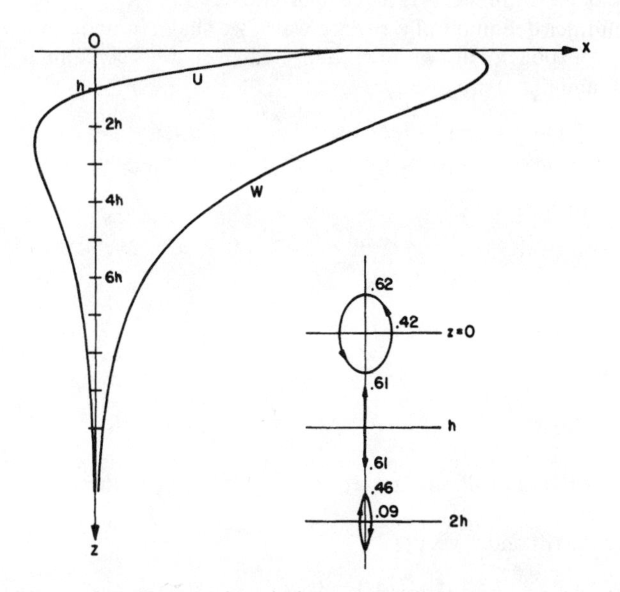

Figure 3.5. The horizontal (U) and vertical (W, down) displacements for Rayleigh waves in a homogeneous half-space. U vanishes at a depth h. The path of the particles is elliptic retrograde for $z < h$ and elliptic direct for $z > h$ [see Eq. (3.107)].

At $z = 0$, we have

$$u_x = 0.4227Q \sin(\omega t - kx),$$
$$u_z = 0.6204Q \cos(\omega t - kx). \tag{3.108}$$

At any point of the elastic medium the particle motion resulting from Rayleigh waves is elliptic. In the depth range $0 < z < h$ in which $U(z) > 0$, $W(z) > 0$, the particle motion on the ellipse is counterclockwise (*retrograde*). This becomes obvious by noting that $\dot{u}_x = QU\omega \cos(\omega t - kx)$, $\dot{u}_z = -QW\omega \sin(\omega t - kx)$. However, when $z > h$, $U(z) < 0$ and $W(z) > 0$, the particle motion is clockwise (*direct*). The ratio of horizontal to vertical axes at $z = 0$ is known as *surface ellipticity* and is given by

$$\frac{U(0)}{W(0)} = \frac{1 - c^2/2\beta^2}{(1 - c^2/\alpha^2)^{1/2}}, \tag{3.109}$$

which is approximately equal to 0.6812 when $\sigma = \frac{1}{4}$.

3.6.5. Dispersed Rayleigh Waves

We have established the existence of nondispersive Rayleigh waves in a homogeneous simiinfinite elastic solid. The simplest earth model (Fig. 3.4) that exhibits dispersion of Rayleigh waves is that of a layer overlying a half-space. To comply with our former definition of a surface wave, we shall require *a priori* that the waves decay exponentially with depth in the half-space. Therefore, in line with Eqs. (3.66) and (3.68), we assume

$$\begin{aligned}\mathbf{u}_1 = [&A_1'(\mathbf{a}_x + \eta_{\alpha 1}\mathbf{a}_z)e^{-ik\eta_{\alpha 1}z} + A_1''(\mathbf{a}_x - \eta_{\alpha 1}\mathbf{a}_z)e^{ik\eta_{\alpha 1}z} \\ &+ B_1'(-\eta_{\beta 1}\mathbf{a}_x + \mathbf{a}_z)e^{-ik\eta_{\beta 1}z} + B_1''(\eta_{\beta 1}\mathbf{a}_x + \mathbf{a}_z)e^{ik\eta_{\beta 1}z}]e^{ik(ct-x)}, \\ &\qquad 0 < z < H; \\ \mathbf{u}_2 = [&A_2(\mathbf{a}_x - i\gamma_{\alpha 2}\mathbf{a}_z)e^{-k\gamma_{\alpha 2}z} + B_2(i\gamma_{\beta 2}\mathbf{a}_x + \mathbf{a}_z)e^{-k\gamma_{\beta 2}z}]e^{ik(ct-x)}, \\ &\qquad z > H;\end{aligned} \tag{3.110}$$

where

$$\eta_{\alpha 1} = \left(\frac{c^2}{\alpha_1^2} - 1\right)^{1/2}, \qquad \gamma_{\alpha 2} = \left(1 - \frac{c^2}{\alpha_2^2}\right)^{1/2}, \quad \text{etc.} \tag{3.111}$$

It is obvious that $\mathbf{u}_2$ will tend to zero as $z \to \infty$ provided that $\gamma_{\alpha 2}$ and $\gamma_{\beta 2}$ are real; i.e., $c < \beta_2$.

The boundary conditions are

$$\begin{aligned}\mathbf{a}_z \cdot \mathfrak{T}(\mathbf{u}_1) = 0 \quad &\text{at} \quad z = 0, \\ \mathbf{u}_1 = \mathbf{u}_2, \qquad \mathbf{a}_z \cdot \mathfrak{T}(\mathbf{u}_1) = \mathbf{a}_z \cdot \mathfrak{T}(\mathbf{u}_2) \quad &\text{at} \quad z = H.\end{aligned} \tag{3.112}$$

Equations (3.110) and (3.112) yield a set of six homogeneous equations in six unknown coefficients A_1', A_1'', B_1', B_1'', A_2, and B_2. The condition of consistency of these equations leads us to the following dispersion equation

$$\Delta_R \equiv A_0 + B_0 \cos \hat{k}\eta_{\alpha 1} \cos \hat{k}\eta_{\beta 1} + C_0 \sin \hat{k}\eta_{\alpha 1} \sin \hat{k}\eta_{\beta 1} + D_0 \cos \hat{k}\eta_{\alpha 1} \sin \hat{k}\eta_{\beta 1} + E_0 \sin \hat{k}\eta_{\alpha 1} \cos \hat{k}\eta_{\beta 1} = 0, \tag{3.113}$$

where $\hat{k} = kH$ and

$$A_0 = 4k^2(2k^2 - k_{\beta_1}^2)\left[\left(2k^2 f - \frac{\mu_2}{\mu_1} k_{\beta_2}^2\right)\left(2k^2 f - \frac{\mu_2}{\mu_1} k_{\beta_2}^2 + k_{\beta_1}^2\right) - 2f k^2 \gamma_{\alpha 2}\gamma_{\beta 2}(2k^2 f + k_{\beta_1}^2)\right],$$

$$B_0 = 4k^4\gamma_{\alpha 2}\gamma_{\beta 2}[(2k^2 f + k_{\beta_1}^2)^2 + f^2(2k^2 - k_{\beta_1}^2)^2] - 4k^4\left(2k^2 f - \frac{\mu_2}{\mu_1} k_{\beta_2}^2 + k_{\beta_1}^2\right)^2 - (2k^2 - k_{\beta_1}^2)^2\left(2k^2 f - \frac{\mu_2}{\mu_1} k_{\beta_2}^2\right)^2,$$

$$C_0 = 4k^4\eta_{\alpha 1}\eta_{\beta 1}\left[\left(2k^2 f - \frac{\mu_2}{\mu_1} k_{\beta_2}^2\right)^2 - 4f^2k^4\gamma_{\alpha 2}\gamma_{\beta 2}\right] + \frac{(2k^2 - k_{\beta 1}^2)^2}{\eta_{\alpha 1}\eta_{\beta 1}} \times \left[\left(2k^2 f - \frac{\mu_2}{\mu_1} k_{\beta_2}^2 + k_{\beta_1}^2\right)^2 - (2k^2 f + k_{\beta_1}^2)^2\gamma_{\alpha 2}\gamma_{\beta 2}\right], \tag{3.114}$$

$$D_0 = -\frac{\mu_2}{\mu_1} k_{\beta_1}^2 k_{\beta_2}^2\left[4k^4\gamma_{\alpha 2}\eta_{\beta 1} + (2k^2 - k_{\beta_1}^2)^2\frac{\gamma_{\beta 2}}{\eta_{\beta 1}}\right],$$

$$E_0 = -\frac{\mu_2}{\mu_1} k_{\beta_1}^2 k_{\beta_2}^2\left[4k^4\eta_{\alpha 1}\gamma_{\beta 2} + (2k^2 - k_{\beta_1}^2)^2\frac{\gamma_{\alpha 2}}{\eta_{\alpha 1}}\right],$$

$$f = \frac{\mu_2}{\mu_1} - 1, \qquad k_{\beta_1} = \frac{\omega}{\beta_1}, \qquad k_{\beta_2} = \frac{\omega}{\beta_2}, \qquad k_{\alpha_1} = \frac{\omega}{\alpha_1}, \qquad k_{\alpha_2} = \frac{\omega}{\alpha_2}.$$

The requirement that $c > \alpha_1$ is not essential for the existence of Rayleigh waves because, in the layer, z is bounded. However, $c > \alpha_1$ will imply an oscillatory dependence on z in the layer with possible nodal planes, whereas $c < \alpha_1$ will render nonoscillatory motion in the layer.

Clearly, the shape of the dispersion curves for different modes will depend on the four velocities α_1, α_2, β_1, β_2 and the rigidity contrast μ_2/μ_1. It will be useful therefore, to investigate the dependence of the roots of the period equation upon these structural parameters. It is easier, however, to reverse the process, i.e., to assume upper and lower bounds for the phase velocity and study the consequences. We will consider two cases.

3.6.5.1. Case 1: $\alpha_1 < c < \beta_2$. The coefficients $A_0 \cdots E_0$ are real quantities. Equation (3.113) can be trigonometrically manipulated to yield the equivalent form

$$2A_0 + \sqrt{[(B_0 - C_0)^2 + (E_0 + D_0)^2]}\cos[kH(\eta_{\beta 1} + \eta_{\alpha 1}) - \theta_2] + \sqrt{[(B_0 + C_0)^2 + (E_0 - D_0)^2]}\cos[kH(\eta_{\beta 1} - \eta_{\alpha 1}) + \theta_1] = 0, \tag{3.115}$$

where

$$\tan\theta_1 = \frac{E_0 - D_0}{B_0 + C_0}, \qquad \tan\theta_2 = \frac{E_0 + D_0}{B_0 - C_0}.$$

From Eq. (3.115), it is apparent that

$$2|A_0| < [(B_0 - C_0)^2 + (E_0 + D_0)^2]^{1/2} + [(B_0 + C_0)^2 + (E_0 - D_0)^2]^{1/2}. \tag{3.116}$$

Equation (3.113) can be recast into the form of a quadratic in $\tan(kH\eta_{\beta 1}/2)$ or $\tan(kH\eta_{\alpha 1}/2)$. When it is solved, considering the former as the unknown variable, the dispersion relation is expressible in the form

$$\tan\left(\frac{kH\eta_{\beta 1}}{2}\right) = \frac{-b \pm \sqrt{(b^2 - ad)}}{a}, \tag{3.117}$$

where a, b, d are functions of $\tan(kH\eta_{\alpha 1}/2)$ and the real coefficients $A_0 \cdots E_0$. In fact,

$$\begin{aligned} a &= A_0 - B_0 \cos kH\eta_{\alpha 1} - E_0 \sin kH\eta_{\alpha 1}, \\ b &= C_0 \sin kH\eta_{\alpha 1} + D_0 \cos kH\eta_{\alpha 1}, \\ d &= A_0 + B_0 \cos kH\eta_{\alpha 1} + E_0 \sin kH\eta_{\alpha 1}. \end{aligned} \tag{3.118}$$

When $H = 0$, $a = A_0 - B_0$, $b = D_0$, and $d = A_0 + B_0$ and Eq. (3.117) yields

$$0 = -D_0 \pm \sqrt{D_0^2 - A_0^2 + B_0^2}, \qquad (A_0 \neq B_0).$$

Because, for $\alpha_1 < c < \beta_2$, D_0 is negative [Eq. (3.114)], the above equation will possess a real root only if $D_0^2 > A_0^2 - B_0^2$ and only if we choose the lower sign before the radical. If we do that, the dispersion relation becomes $A_0 + B_0 = 0$, which, on using Eq. (3.114), yields

$$4k^4\gamma_{\alpha 2}\gamma_{\beta 2} - (2k^2 - k_{\beta_2}^2)^2 = 0.$$

A comparison with Eq. (3.101) reveals that this is the period equation for the Rayleigh waves in the lower medium.

The roots of the dispersion relation Eq. (3.117) will be always real except for those values of kH for which $b^2 < ad$. If we exclude this possibility, for each value of kH, we shall have two values of c. The minus sign before the radical in Eq. (3.117) corresponds to what is known as the M_1 *branch*, whereas the plus sign corresponds to the M_2 *branch*. Each branch has infinite modes of propagation. The nomenclature is $M_{11}, M_{12}, \ldots$ for the M_1 branch and $M_{21}, M_{22}, \ldots$ for the M_2 branch. It can be shown that the M_2 branch corresponds to the *antisymmetrical* modes of a free plate, whereas the M_1 branch corresponds to the *symmetrical* modes. The mode M_{11} is known as the *fundamental Rayleigh mode.* We have shown above that in the limit $kH \to 0$, the M_2 branch has no real root and the phase velocity corresponding to the M_1 branch equals the Rayleigh-wave velocity in the lower medium.

3.6.5.2. Case 2: $\beta_1 < c < \alpha_1$, $c < \beta_2$. A_0, B_0, and D_0 are real, whereas C_0 and E_0 are purely imaginary. Because $\eta_{\alpha 1} = -i\gamma_{\alpha 1}$ ($\gamma_{\alpha 1}$ real), the dispersion relation is reduced to the form

$$A_0 + \sqrt{|C_0|^2 + |E_0|^2}\,\mathrm{sh}(kH\gamma_{\alpha 1})\sin(kH\eta_{\beta 1} + \theta_3) + \sqrt{B_0^2 + D_0^2}\,\mathrm{ch}(kH\gamma_{\alpha 1})\cos(kH\eta_{\beta 1} - \theta_4) = 0, \tag{3.119}$$

where

$$\tan\theta_3 = \frac{(-iE_0)}{(-iC_0)}, \qquad \tan\theta_4 = \frac{D_0}{B_0}.$$

Here, $\cos\theta_{3,4}$ has the same sign as the denominator of the fraction in the expression for $\tan\theta_{3,4}$.

Once the dispersion equation is solved, the surface displacements can be evaluated in terms of one of the six unknown coefficients. The amplitudes and velocities of Rayleigh waves have the following properties:

1. The particle orbital motion at the surface is retrograde elliptic for the M_1 branch and direct elliptic for the M_2 branch.
2. The energy of the M_{21} branch is concentrated mainly in the vicinity of the interface $z = H$, whereas the energy of M_{11} is concentrated mainly near the free surface $z = 0$.
3. At the long-wave limit ($kH \to 0$) the phase velocity of the fundamental mode M_{11} approaches the velocity of Rayleigh waves in the half-space ($= 0.9195\,\beta_2$ for $\lambda_2 = \mu_2$).
4. All modes, except for the fundamental, have threshold periods above which no surface motion exists. This is known as the *cutoff period.* The cutoff period decreases with the mode number. The phase velocity at the cutoff period equals the shear-wave velocity in the lower medium, β_2.
5. At the short-wave limit ($kH \to \infty$) the phase velocity for M_{11} approaches the Rayleigh-wave velocity in the layer. Under some very stringent conditions, there would be an additional mode in the M_2 branch for which the phase velocity approaches the velocity of interface waves between two half-spaces as $kH \to \infty$. For all other modes, the phase velocity approaches the shear-wave velocity in the layer, β_1.

Let us next consider the possibility of the propagation of Rayleigh waves in a vertically heterogeneous half-space, $\rho = \rho(z)$, $\lambda = \lambda(z)$, $\mu = \mu(z)$. In the absence of external body forces, the equation of motion, Eq. (1.107), may be written in the form

$$\mu\nabla^2\mathbf{u} + (\lambda + \mu)\,\mathrm{grad}\,\mathrm{div}\,\mathbf{u} + \mathbf{a}_z\frac{d\lambda}{dz}\,\mathrm{div}\,\mathbf{u} + \frac{d\mu}{dz}\left(2\frac{\partial\mathbf{u}}{\partial z} + \mathbf{a}_z \times \mathrm{curl}\,\mathbf{u}\right) = \rho\frac{\partial^2\mathbf{u}}{\partial t^2}. \tag{3.120}$$

Assuming

$$\mathbf{u} = [-iU(z)\mathbf{a}_x + W(z)\mathbf{a}_z]e^{ik(ct-x)}, \tag{3.120a}$$

we find that Eq. (3.120) will be satisfied if U and W satisfy the equations

$$\begin{aligned} &\frac{d}{dz}\left[\mu\left(\frac{dU}{dz} + kW\right)\right] + [\rho\omega^2 - (\lambda + 2\mu)k^2]U + \lambda k\frac{dW}{dz} = 0, \\ &\frac{d}{dz}\left[(\lambda + 2\mu)\frac{dW}{dz} - k\lambda U\right] + (\rho\omega^2 - \mu k^2)W - \mu k\frac{dU}{dz} = 0. \end{aligned} \tag{3.121}$$

These equations are to be solved subject to the radiation condition that $U \to 0$, $W \to 0$ as $z \to \infty$ and the boundary conditions

$$(\lambda + 2\mu)\frac{dW}{dz} - \lambda kU = 0, \qquad \frac{dU}{dz} + kW = 0 \quad \text{at} \quad z = 0.$$

In general, Eqs. (3.121) must be integrated numerically. To circumvent the determination of the derivative of empirically known quantities $\lambda(z)$ and $\mu(z)$, we transform Eqs. (3.121) into an equivalent set of four simultaneous linear differential equations. To this end, we define

$$\begin{aligned} &y_1 = W, \qquad y_2 = (\lambda + 2\mu)\frac{dW}{dz} - k\lambda U, \\ &y_3 = U, \qquad y_4 = \mu\left(\frac{dU}{dz} + kW\right). \end{aligned} \tag{3.122}$$

From Eqs. (3.121) and (3.122), we find

$$\begin{aligned} \frac{dy_1}{dz} &= \frac{1}{\lambda + 2\mu}(y_2 + k\lambda y_3), \\ \frac{dy_2}{dz} &= -\rho\omega^2 y_1 + ky_4, \\ \frac{dy_3}{dz} &= -ky_1 + \frac{1}{\mu}y_4, \\ \frac{dy_4}{dz} &= -\frac{k\lambda}{\lambda + 2\mu}y_2 + \left(-\rho\omega^2 + 4k^2\mu\frac{\lambda + \mu}{\lambda + 2\mu}\right)y_3. \end{aligned} \tag{3.123}$$

The boundary conditions reduce to:

1. $y_2 = 0$, $y_4 = 0$ at the free surface;
2. y_1, y_2, y_3 and y_4 are continuous across an internal surface of discontinuity, $z = \text{const}$;
3. y_i $(i = 1, 2, 3, 4) \to 0$ as $z \to \infty$.

3.6.6. Energy Integrals

In the following we shall derive certain relations involving energy integrals. To this end we consider the basic energy integral, Eq. (1.120), and consider the total energy in an arbitrary volume N bounded by a surface S averaged over a cycle

$$\langle U \rangle = \frac{1}{T}\int_0^T dt \int_{\mathrm{N}} \frac{1}{2}(\mathfrak{T}:\mathfrak{E} + \rho\dot{\mathbf{u}}^2)d\mathrm{N}. \tag{3.124}$$

We now apply this integral to the propagation of Love waves in a semi-infinite half-space in which the structural parameters depend only on the coordinate z. The explicit expressions for the field entities appearing in the integrand of the energy integral can be obtained from Eq. (3.80), taking, of course, the real parts only

$$\mathbf{u} = \mathbf{a}_y V(z)\cos(\omega t - kx),$$

$$\mathfrak{E} = \frac{1}{2}\left[(\mathbf{a}_y\mathbf{a}_z + \mathbf{a}_z\mathbf{a}_y)\frac{dV}{dz}\cos(\omega t - kx) + k(\mathbf{a}_y\mathbf{a}_x + \mathbf{a}_x\mathbf{a}_y)V\sin(\omega t - kx)\right],$$

$$\mathfrak{T} = 2\mu\mathfrak{E}. \tag{3.125}$$

Substituting in Eq. (3.124) and considering the energy per unit surface area, we are left only with integrals over the vertical coordinate

$$\langle U \rangle = \frac{1}{4}\omega^2\int_0^\infty \rho(z)V^2(z)dz + \frac{1}{4}\int_0^\infty \mu(z)\left\{\left[\frac{dV(z)}{dz}\right]^2 + k^2V^2(z)\right\}dz. \tag{3.126}$$

It may be noted that the first term on the right-hand side of Eq. (3.126) represents the kinetic energy, whereas the rest of the terms represent the potential energy. We now apply the *virial theorem*, which states that for a conservative system in which the potential energy is a quadratic function of the coordinates, the time average of the potential and kinetic energies are equal. Hence

$$\omega^2 I_0^L = k^2 I_1^L + I_2^L, \tag{3.127}$$

where

$$I_0^L = \int_0^\infty \rho V^2\,dz, \qquad I_1^L = \int_0^\infty \mu V^2\,dz, \qquad I_2^L = \int_0^\infty \mu\left(\frac{dV}{dz}\right)^2 dz. \tag{3.128}$$

Equation (3.127) is basically a dispersion relation because it is an implicit relation between ω and k, V being a function of k. Indeed, the reader may verify this by integrating the eigenfunction given by Eq. (3.77) for the case of a single layer over a half-space. Equation (3.127) will then lead directly to Eq. (3.74). Equation (3.127), however, can also be used to evaluate the group velocity without the differentiation implied by the defining Eq. (3.85). To this end, we invoke the *Rayleigh principle*, which asserts that a small perturbation of the eigenvalue $k = k_L$ of a vibrating system by the amount δk_L will induce a corresponding change in the eigenfunction $V(k_L)$ that is of second order of smallness in δk_L. Applying this principle to Eq. (3.127), we perturb k_L to $k_L + \delta k_L$ and consider the corresponding change of ω to $\omega + \delta\omega$. Because of the Rayleigh principle, the change $V(z, k_L + \delta k_L) - V(z, k_L)$ will be of second order in δk_L. Keeping only first-order quantities in δk_L, we have, from Eq. (3.127),

$$\omega(\delta\omega)I_0^L = k_L(\delta k_L)I_1^L + O(\delta k_L)^2.$$

In the limit as $\delta k_L \to 0$, we have

$$c_L U_L = \frac{I_1^L}{I_0^L} = \frac{\int_0^\infty \mu V^2\,dz}{\int_0^\infty \rho V^2\,dz}. \tag{3.129}$$

We shall now use Eq. (3.129) to demonstrate the equivalence of group velocity and the velocity of energy transport in a lossless, isotropic, elastic solid. From Eqs. (3.126) and (3.127), we have

$$\langle U \rangle = \frac{1}{2}\omega^2 \int_0^\infty \rho V^2\,dz. \tag{3.130}$$

However, from Eqs. (1.122) and (3.125), the average energy-flux density is

$$\langle \boldsymbol{\Sigma} \rangle = \frac{1}{2}\mathbf{a}_x \omega k \int_0^\infty \mu V^2\,dz. \tag{3.131}$$

Hence, the velocity of energy transport is given by

$$\frac{|\langle \boldsymbol{\Sigma} \rangle|}{\langle U \rangle} = \frac{1}{c_L}\frac{\int_0^\infty \mu V^2\,dz}{\int_0^\infty \rho V^2\,dz} = U_L, \tag{3.132}$$

using Eq. (3.129).

Relation (3.127) can be obtained directly from the equation of motion, Eq. (3.81). Let us denote the eigenfunctions y_1 and y_2 corresponding to the eigenvalue k_i by y_{1i} and y_{2i}. We then have

$$\frac{dy_{21}}{dz} = (\mu k_1^2 - \rho\omega_1^2)y_{11},$$

which further yields

$$\frac{d}{dz}(y_{12}y_{21}) = (\mu k_1^2 - \rho\omega_1^2)y_{11}y_{12} + \mu\frac{dy_{11}}{dz}\frac{dy_{12}}{dz}. \tag{3.133}$$

Replacing y_{12} by y_{11} and integrating with respect to z, we find

$$y_{11}y_{21}\Big|_0^\infty = k_1^2\int_0^\infty \mu y_{11}^2\,dz - \omega_1^2\int_0^\infty \rho y_{11}^2\,dz + \int_0^\infty \mu\left(\frac{dy_{11}}{dz}\right)^2 dz.$$

Noting that $y_{21}(0) = 0, y_{21}(\infty) = 0$, we get Eq. (3.127).

Next, interchanging the roles of y_{11} and y_{12} in Eq. (3.133), we obtain

$$\frac{d}{dz}(y_{11}y_{22}) = (\mu k_2^2 - \rho\omega_2^2)y_{12}y_{11} + \mu\frac{dy_{12}}{dz}\frac{dy_{11}}{dz}. \tag{3.134}$$

Equations (3.133) and (3.134) yield

$$(y_{12}y_{21} - y_{11}y_{22})\Big|_0^\infty = (k_1^2 - k_2^2)\int_0^\infty \mu y_{11}y_{12}\,dz - (\omega_1^2 - \omega_2^2)\int_0^\infty \rho y_{11}y_{12}\,dz.$$

Once again, because $y_{21}(0) = y_{21}(\infty) = y_{22}(0) = y_{22}(\infty) = 0$, we obtain

$$(k_1^2 - k_2^2)\int_0^\infty \mu y_{11}y_{12}\,dz = (\omega_1^2 - \omega_2^2)\int_0^\infty \rho y_{11}y_{12}\,dz. \tag{3.135}$$

Consequently, if y_{11} and y_{12} are two eigenfunctions corresponding to two different eigenvalues k_1 and k_2 of the same frequency ω, they are orthogonal with a weighting function $\mu(z)$. However, if $\omega_1 \neq \omega_2$ but $k_1 = k_2$, they are orthogonal with a weighting function $\rho(z)$.

Another interesting relation is obtained from Eq. (3.135) by setting $y_{12} = y_{11}^*$ (complex conjugate). It is evident from the equation of motion that if $y_1(z, \omega, k)$ is an eigenfunction, so is $y_1^*(z, \omega^*, k^*)$. Then, Eq. (3.135) yields

$$\mathrm{Re}(k)\mathrm{Im}(k)\int_0^\infty \mu|y_1|^2\,dz = \mathrm{Re}(\omega)\mathrm{Im}(\omega)\int_0^\infty \rho|y_1|^2\,dz, \tag{3.136}$$

where Re and Im denote the real and imaginary parts, respectively. Therefore, $\{\mathrm{Re}(\omega)\mathrm{Im}(\omega)\}$ is always of the same sign as $\{\mathrm{Re}(k)\mathrm{Im}(k)\}$. If we assume $\mathrm{Im}(\omega)$ to be negative, the eigenvalue k lies in the second or fourth quadrant of the complex k-plane when $\mathrm{Re}(\omega) > 0$, and in the first or third quadrant when $\mathrm{Re}(\omega) < 0$.

For Rayleigh waves, the displacements in a vertically heterogeneous medium can be represented in the form [see Eq. (3.120a)]

$$\mathbf{u} = \mathbf{a}_x U(z)\sin(\omega t - kx) + \mathbf{a}_z W(z)\cos(\omega t - kx). \tag{3.137}$$

Proceeding as in the case of Love waves and taking the average over a cycle, we find

$$\text{Kinetic energy per unit surface area} = \frac{\omega^2}{4} I_0^R,$$

$$\text{Potential energy per unit surface area} = \tfrac{1}{4}k^2 I_1^R + \tfrac{1}{2}k I_2^R + \tfrac{1}{4}I_3^R, \tag{3.138}$$

where

$$\begin{aligned}
I_0^R &= \int_0^\infty \rho(U^2 + W^2)dz,\\
I_1^R &= \int_0^\infty \mu\left\{\frac{\alpha^2}{\beta^2}U^2 + W^2\right\}dz,\\
I_2^R &= \int_0^\infty \mu\left\{-\frac{\lambda}{\mu}U\frac{dW}{dz} + \frac{dU}{dz}W\right\}dz,\\
I_3^R &= \int_0^\infty \mu\left\{\frac{\alpha^2}{\beta^2}\left(\frac{dW}{dz}\right)^2 + \left(\frac{dU}{dz}\right)^2\right\}dz.
\end{aligned} \tag{3.139}$$

Equating the mean kinetic and potential energies over a cycle of the motion, we obtain

$$\omega^2 I_0^R = k^2 I_1^R + 2k I_2^R + I_3^R. \tag{3.140}$$

The Rayleigh principle gives an integral expression for the group velocity

$$U_R = \frac{I_1^R + I_2^R/k_R}{I_0^R c_R}. \tag{3.141}$$

Here again it is useful to derive the relations between the energy integrals in an alternative way. Let y_{i1} and y_{i2} ($i = 1, 2, 3, 4$) represent two sets of independent solutions of Eq. (3.123) bounded at infinity. Clearly, any solution that is bounded at infinity may be expressed as a linear combination of these two solutions in the form

$$y_i(z) = \theta_1 y_{i1}(z) + \theta_2 y_{i2}(z), \tag{3.142}$$

where θ_1 and θ_2 are arbitrary constants of integration determined by the boundary conditions. Putting $z = 0$, and remembering that $y_2(0) = y_4(0) = 0$, we get

$$\begin{aligned} \theta_1 y_{21}(0) + \theta_2 y_{22}(0) &= 0, \\ \theta_1 y_{41}(0) + \theta_2 y_{42}(0) &= 0. \end{aligned} \tag{3.143}$$

These equations have a nontrivial solution if, and only if,

$$\Delta_R(\omega, k) \equiv y_{21}(0) y_{42}(0) - y_{22}(0) y_{41}(0) = 0. \tag{3.144}$$

This is the *dispersion relation* for Rayleigh waves in a medium with vertical heterogeneity.

We next multiply the second and the fourth equations in Eqs. (3.123) by arbitrary functions $x_1(z)$ and $x_3(z)$, respectively. Adding the results and using Eq. (3.123), we have

$$\begin{aligned} \frac{d}{dz}(x_1 y_2 + x_3 y_4) = &-\rho\omega^2(x_1 y_1 + x_3 y_3) + (\lambda + 2\mu)\dot{x}_1 \dot{y}_1 - k\lambda(\dot{x}_1 y_3 + x_3 \dot{y}_1) \\ &+ \mu \dot{x}_3 \dot{y}_3 + k\mu(\dot{x}_3 y_1 + x_1 \dot{y}_3) \\ &+ k^2[\mu x_1 y_1 + (\lambda + 2\mu) x_3 y_3], \end{aligned} \tag{3.145}$$

where a dot signifies differentiation with respect to z. Integration from $z = 0$ to $z = \infty$ yields

$$\begin{aligned} (x_1 y_2 + x_3 y_4)\Big|_{z=0}^{\infty} = &-\omega^2 \int_0^\infty \rho(x_1 y_1 + x_3 y_3) dz \\ &+ k^2 \int_0^\infty \mu\left[\frac{\alpha^2}{\beta^2} x_3 y_3 + x_1 y_1\right] dz \\ &+ \int_0^\infty \mu\left\{\frac{\alpha^2}{\beta^2} \dot{x}_1 \dot{y}_1 + \dot{x}_3 \dot{y}_3\right\} dz \\ &+ k \int_0^\infty \mu\left\{(\dot{x}_3 y_1 + x_1 \dot{y}_3) - \frac{\lambda}{\mu}(\dot{x}_1 y_3 + x_3 \dot{y}_1)\right\} dz. \end{aligned} \tag{3.146}$$

Taking $x_1 = y_1$, $x_3 = y_3$ and using the boundary conditions $y_2(0) = y_4(0) = y_2(\infty) = y_4(\infty) = 0$, we get

$$\omega^2 \int_0^\infty \rho(y_1^2 + y_3^2)dz = k^2 \int_0^\infty \mu\left(\frac{\alpha^2}{\beta^2} y_3^2 + y_1^2\right)dz + \int_0^\infty \mu\left(\frac{\alpha^2}{\beta^2} \dot{y}_1^2 + \dot{y}_3^2\right)dz$$
$$+ 2k \int_0^\infty \mu\left(y_1\dot{y}_3 - \frac{\lambda}{\mu}\dot{y}_1 y_3\right)dz,$$

which coincides with Eq. (3.140).

The orthogonality relation for the Rayleigh eigenfunctions is derived in a manner similar to the Love-wave case. Defining

$$x_2 = (\lambda + 2\mu)\dot{x}_1 - k\lambda x_3, \qquad x_4 = \mu(\dot{x}_3 + kx_1),$$

and assuming that y_i correspond to the eigenvalues (ω_1, k_1) and x_i correspond to the eigenvalues (ω_2, k_2), we find

$$(\omega_1^2 - \omega_2^2) \int_0^\infty \rho(x_1 y_1 + x_3 y_3)dz = 4(k_1^2 - k_2^2) \int_0^\infty \frac{\mu(\lambda + \mu)}{\lambda + 2\mu} x_3 y_3\, dz$$
$$+ (k_1 - k_2) \int_0^\infty \left[(x_1 y_4 + x_4 y_1) - \frac{\lambda}{\lambda + 2\mu}(x_3 y_2 + x_2 y_3)\right] dz. \quad (3.147)$$

Hence, if $k_1 = k_2$ but $\omega_1 \neq \omega_2$, the vector $\{y_1(\omega_1), y_3(\omega_1)\}$ is orthogonal to the vector $\{x_1(\omega_2), x_3(\omega_2)\}$ with ρ as the weighting function.

3.7. Spectral Response of a Multilayered Crust

The crust of the earth, which extends on the average to some 10 km under the oceans and 35 km under the continents, plays a dominant role in shaping the form of the seismic wave as it travels on its way from the source to the receiving instrument on the earth's surface. The influence of the crust on the spectrum of body waves therefore must be taken into account when P and S waves are used for source studies. However, we must make the simplifying assumption that the curvature of the spherical wavefronts is small for waves whose wavelengths are much smaller than the distance traveled.

In Section 3.2, we derived certain relations between the amplitudes of the incident and reflected plane body waves of longitudinal and transverse types at the free surface of a homogeneous medium. These relations are valid only for wavelengths that are very large compared with the total thickness of the crustal layers. For periods commonly observed in seismic body waves, the thickness of the crustal layers is far from a negligible fraction of a wavelength, and we may expect that for an incident wave of a given type, the surface amplitude will be strongly dependent on the period as well as on the angle of incidence.

Consider a semiinfinite elastic medium consisting of $N - 1$ parallel, homogeneous layers overlying a homogeneous half-space. The layers are numbered serially, the layer at the top being layer 1 and the half-space being layer N. There are altogether N boundaries, including the free surface. We place the origin of a right-handed Cartesian coordinate system (x, y, z) at the free surface with the z axis drawn into the medium and the y axis pointing positively from the plane of the paper toward the reader. The nth layer, D_n, is of thickness d_n and is bounded by the plane $z = z_{n-1}$ from above and $z = z_n$ from below. The parameters of this layer are λ_n, μ_n, ρ_n, α_n, and β_n (Fig. 3.6).

Let a plane harmonic wave (SH, SV, or P) be incident on the $(N - 1)$th interface from below at a known angle of incidence. Part of the wave is reflected back into the half-space and part is transmitted into the stratified medium above. Each layer then sustains an upgoing field and a downgoing field such that the following boundary conditions are met:

1. The tractions on the free surface $z = 0$ vanish.
2. The displacements and stresses at each interface are continuous.

In the case of SH waves there are $2N - 1$ unknown field coefficients to be determined by $2N - 1$ boundary-condition equations. When the incident wave is SV or P, there are $4N - 2$ equations in the $4N - 2$ unknown coefficients. In order to derive an explicit expression for the surface displacements in terms of the incident amplitude and angle of incidence, we will have to face the formidable computational labor involved in inverting a determinant of a high order.

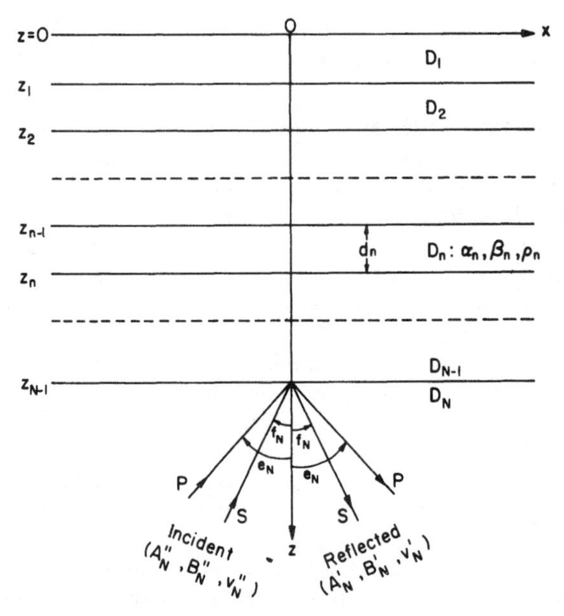

Figure 3.6. Enumeration of layers and interfaces for a multilayered model of the earth.

We shall present here, however, an alternative method in which the problem is formulated with the aid of (2×2) and (4×4) matrices. We begin with the simpler case of incident *SH* waves.

3.7.1. Incident *SH* Waves

From Eqs. (3.1) and (3.2), the displacement field in the nth layer may be written in the form

$$\mathbf{u}_n = \mathbf{a}_y v_n, \tag{3.148}$$

where

$$v_n = \underset{\text{Upgoing}}{v_n'' e^{i(\omega t - k_n x \sin f_n + k_n z \cos f_n)}} + \underset{\text{Downgoing}}{v_n' e^{i(\omega t - k_n x \sin f_n - k_n z \cos f_n)}}, \tag{3.149}$$

and $k_n = \omega/\beta_n$. This solution is interpreted geometrically as a sum of two plane waves, each of which is moving in the direction of the positive x axis with a *horizontal phase velocity* (also known as the *apparent velocity*) $c = \beta_n/\sin f_n$. One of these waves propagates upward (toward the free surface) with vertical velocity $\beta_n/\cos f_n$ and the other propagates downward with the same velocity. Because of the boundary conditions, the phases of all the waves in all the layers must be equal for all values of x and t. This implies the constancy of c,

$$c = \frac{\beta_1}{\sin f_1} = \frac{\beta_2}{\sin f_2} = \cdots = \frac{\beta_n}{\sin f_n} = \cdots = \frac{\beta_N}{\sin f_N}. \tag{3.150}$$

This is known as Snell's law. It is exact for plane waves in homogeneous media at all frequencies but will later be shown to hold for spherical wavefronts in nonhomogeneous media only at high frequencies.

Introducing the notation

$$k = k_n \sin f_n,$$
$$\eta_n = \begin{cases} \sqrt{[(c/\beta_n)^2 - 1]}, & c > \beta_n \\ -i\sqrt{[1 - (c/\beta_n)^2]}, & c < \beta_n, \end{cases} \tag{3.151}$$

we find the following expressions for the transverse displacement and shear stress in the nth layer $(z_{n-1} \le z \le z_n)$:

$$\begin{aligned} v_n(z) &= v_n'' e^{ik\eta_n z} + v_n' e^{-ik\eta_n z}, \\ \tau_n(z) &= [\tau_{zy}(z)]_n = \mu_n \frac{\partial v_n}{\partial z} = ik\mu_n \eta_n (v_n'' e^{ik\eta_n z} - v_n' e^{-ik\eta_n z}). \end{aligned} \tag{3.152}$$

The common factor $\exp\{i(\omega t - kx)\}$ has been temporarily deleted. It is convenient to treat the normalized velocity $\dot{v}/c$ rather than v itself, where c is the same for all the layers. Therefore, we have for the $(n-1)$th interface, the dimensionless entities:

$$\frac{\dot{v}_n(z_{n-1})}{c} = ik(v_n'' e^{ik\eta_n z_{n-1}} + v_n' e^{-ik\eta_n z_{n-1}}),$$
$$\frac{1}{\mu_n \eta_n} \tau_n(z_{n-1}) = ik(v_n'' e^{ik\eta_n z_{n-1}} - v_n' e^{-ik\eta_n z_{n-1}}). \tag{3.153}$$

Similarly, for the nth interface,

$$\frac{\dot{v}_n(z_n)}{c} = ik(v_n'' e^{ik\eta_n(z_{n-1}+d_n)} + v_n' e^{-ik\eta_n(z_{n-1}+d_n)}),$$
$$\frac{\tau_n(z_n)}{\mu_n \eta_n} = ik(v_n'' e^{ik\eta_n(z_{n-1}+d_n)} - v_n' e^{-ik\eta_n(z_{n-1}+d_n)}). \tag{3.154}$$

Equations (3.153) and (3.154) may be expressed in the form

$$\frac{\dot{v}_n(z_{n-1})}{c} = i\Omega_n^+, \qquad \frac{\tau_n(z_{n-1})}{\mu_n \eta_n} = i\Omega_n^-,$$
$$\frac{\dot{v}_n(z_n)}{c} = i\Omega_n^+ \cos\theta_n - \Omega_n^- \sin\theta_n, \quad \frac{\tau_n(z_n)}{\mu_n \eta_n} = -\Omega_n^+ \sin\theta_n + i\Omega_n^- \cos\theta_n, \tag{3.155}$$

where

$$\theta_n = k\eta_n d_n,$$
$$\Omega_n^+ = k(v_n'' e^{ik\eta_n z_{n-1}} + v_n' e^{-ik\eta_n z_{n-1}}), \tag{3.156}$$
$$\Omega_n^- = k(v_n'' e^{ik\eta_n z_{n-1}} - v_n' e^{-ik\eta_n z_{n-1}}).$$

Eliminating Ω_n^+ and Ω_n^- from Eqs. (3.155), we obtain a matrix relation that ties up the values of the displacement and stress at the top and bottom of the nth layer,

$$\underset{\text{Bottom of } D_n}{\begin{bmatrix} \dot{v}_n(z_n)/c \\ \tau_n(z_n) \end{bmatrix}} = \begin{bmatrix} \cos\theta_n & i\mu_n^{-1}\eta_n^{-1}\sin\theta_n \\ i\mu_n\eta_n \sin\theta_n & \cos\theta_n \end{bmatrix} \underset{\text{Top of } D_n}{\begin{bmatrix} \dot{v}_n(z_{n-1})/c \\ \tau_n(z_{n-1}) \end{bmatrix}}. \tag{3.157}$$

When the notation

$$a_n = \begin{bmatrix} \cos\theta_n & i\mu_n^{-1}\eta_n^{-1}\sin\theta_n \\ i\mu_n\eta_n \sin\theta_n & \cos\theta_n \end{bmatrix}, \tag{3.157a}$$

is introduced, Eq. (3.157) becomes

$$\begin{bmatrix} \dot{v}_n(z_n)/c \\ \tau_n(z_n) \end{bmatrix} = a_n \begin{bmatrix} \dot{v}_n(z_{n-1})/c \\ \tau_n(z_{n-1}) \end{bmatrix}. \tag{3.158}$$

Similarly, for the $(n-1)$th layer, we have

$$\begin{bmatrix} \dot{v}_{n-1}(z_{n-1})/c \\ \tau_{n-1}(z_{n-1}) \end{bmatrix} = a_{n-1} \begin{bmatrix} \dot{v}_{n-1}(z_{n-2})/c \\ \tau_{n-1}(z_{n-2}) \end{bmatrix}. \tag{3.159}$$

Using the boundary conditions that the values of $\dot{v}$ and τ at the top of the nth layer are equal to the corresponding values at the bottom of the $(n-1)$th layer, Eqs. (3.158) and (3.159) yield

$$\begin{bmatrix} \dot{v}_n/c \\ \tau_n \end{bmatrix} = a_n a_{n-1} \begin{bmatrix} \dot{v}_{n-2}/c \\ \tau_{n-2} \end{bmatrix}, \tag{3.160}$$

where

$$\begin{aligned} \dot{v}_n &= \dot{v}_n(z_n), \\ \tau_n &= \tau_n(z_n), \qquad (n = 1, 2, \ldots, N-1). \end{aligned} \tag{3.161}$$

It is now obvious that

$$\begin{bmatrix} \dot{v}_{N-1}/c \\ \tau_{N-1} \end{bmatrix} = a_{N-1} a_{N-2} \cdots a_1 \begin{bmatrix} \dot{v}_0/c \\ \tau_0 \end{bmatrix}, \tag{3.162}$$

where $\dot{v}_0 = \dot{v}_1(0)$, $\tau_0 = \tau_1(0)$.

Setting $a_{N-1}a_{N-2} \cdots a_1 = A^L$, Eq. (3.162) yields

$$\begin{aligned} \frac{\dot{v}_N}{c} &= \frac{\dot{v}_{N-1}}{c} = A_{11}^L \frac{\dot{v}_0}{c} + A_{12}^L \tau_0, \\ \tau_N &= \tau_{N-1} = A_{21}^L \frac{\dot{v}_0}{c} + A_{22}^L \tau_0, \end{aligned} \tag{3.163}$$

where A_{ij}^L are the elements of the matrix A^L and $\dot{v}_N = \dot{v}_N(z_{N-1})$, $\tau_N = \tau_N(z_{N-1})$. Also for $n = N$, Eqs. (3.155) and (3.163) yield

$$i\Omega_N^+ = A_{11}^L \frac{\dot{v}_0}{c} + A_{12}^L \tau_0, \qquad i\mu_N \eta_N \Omega_N^- = A_{21}^L \frac{\dot{v}_0}{c} + A_{22}^L \tau_0. \tag{3.164}$$

Invoking at this point the boundary condition that $\tau_0 = 0$ and using the relation $\dot{v}_0/c = ikv_0$, we obtain from Eqs. (3.156) and (3.164)

$$\frac{v_N'}{v_N''} = \frac{A_{11}^L - A_{21}^L/(\mu_N \eta_N)}{A_{11}^L + A_{21}^L/(\mu_N \eta_N)} e^{2ik\eta_N z_{N-1}}, \quad \text{(reflected } SH) \tag{3.165}$$

$$\frac{v_0}{v_N''} = \frac{2}{A_{11}^L + A_{21}^L/(\mu_N \eta_N)} e^{ik\eta_N z_{N-1}}, \quad \text{(free surface } SH) \tag{3.166}$$

where the amplitude of the incident wave v_N'' is known.

From Eq. (3.158), it is obvious that whether η_n is real or imaginary, the matrix a_n is of the form

$$\begin{bmatrix} R & I \\ I & R \end{bmatrix},$$

where R stands for a real quantity and I stands for an imaginary quantity. Consequently, $A^L = a_{N-1}a_{N-2} \cdots a_1$ is also of the above form; i.e., A^L_{11} is real and A^L_{21} is imaginary. If the angle of incidence at the interface $z = z_{N-1}$ is denoted by f_N, then $\eta_N = \cot f_N$ is real ($c > \beta_n$). Therefore, the numerator and denominator on the right side of Eq. (3.165) are complex conjugate quantities; i.e., the amplitudes of the incident and reflected waves are equal in magnitude but differ in phase.

The results for a single layer of thickness H overlying a homogeneous half-space ($N = 2$) follow immediately:

$$\frac{v_2'}{v_2''} = \frac{\cos\theta_1 - i(\mu_1/\mu_2)(\eta_1/\eta_2)\sin\theta_1}{\cos\theta_1 + i(\mu_1/\mu_2)(\eta_1/\eta_2)\sin\theta_1} e^{2ik\eta_2 H},$$
$$\frac{v_0}{v_2''} = \frac{2e^{ik\eta_2 H}}{\cos\theta_1 + i(\mu_1/\mu_2)(\eta_1/\eta_2)\sin\theta_1}. \tag{3.167}$$

If we add an additional condition that the displacement and stress fields in the half-space must tend to zero as z tends to infinity (that is, no wave is incident from infinity), we obtain the dispersion relation for Love waves. Indeed, inserting $v_N'' = 0$ into Eq. (3.164) and using Eq. (3.156), we find

$$A^L_{21} = -\mu_N \eta_N A^L_{11}. \tag{3.168}$$

In the two-layer case, $A^L = a_1$, and therefore Eq. (3.168) reduces to

$$\tan\theta_1 = \frac{i\mu_2\eta_2}{\mu_1\eta_1},$$

which coincides with Eq. (3.74).

3.7.2. Incident P and SV Waves

In the case of incident P or SV waves, we have four partial fields: upgoing P, downgoing P, upgoing SV, and downgoing SV. Keeping our previous notation and using Eqs. (3.1) and (3.2), we may write the total field in the nth layer ($z_{n-1} \le z \le z_n$) as

$$\begin{aligned}
\mathbf{u}_n = {} & A_n''(\mathbf{a}_x \sin e_n - \mathbf{a}_z \cos e_n)e^{i(\omega t - k_{\alpha n}x \sin e_n + k_{\alpha n}z \cos e_n)} && \text{(upgoing } P) \\
& + A_n'(\mathbf{a}_x \sin e_n + \mathbf{a}_z \cos e_n)e^{i(\omega t - k_{\alpha n}x \sin e_n - k_{\alpha n}z \cos e_n)} && \text{(downgoing } P) \\
& + B_n''(\mathbf{a}_x \cos f_n + \mathbf{a}_z \sin f_n)e^{i(\omega t - k_{\beta n}x \sin f_n + k_{\beta n}z \cos f_n)} && \text{(upgoing } SV) \\
& + B_n'(-\mathbf{a}_x \cos f_n + \mathbf{a}_z \sin f_n)e^{i(\omega t - k_{\beta n}x \sin f_n - k_{\beta n}z \cos f_n)} && \text{(downgoing } SV).
\end{aligned} \tag{3.169}$$

As in the case of SH waves, we define the horizontal phase velocity via Snell's law

$$c = \frac{\beta_n}{\sin f_n} = \frac{\alpha_n}{\sin e_n} \qquad (n = 1, 2, \ldots, N). \tag{3.170}$$

Introducing the notation

$$k = k_{\alpha n} \sin e_n = k_{\beta n} \sin f_n = \frac{\omega}{c}, \qquad \frac{\dot{\mathbf{u}}_n}{c} = \left(\frac{\dot{u}_n}{c}, 0, \frac{\dot{w}_n}{c}\right), \qquad \gamma_n = \frac{2\beta_n^2}{c^2},$$

$$\eta_{\alpha n} = \begin{cases} \sqrt{[(c/\alpha_n)^2 - 1]}, & c > \alpha_n \\ -i\sqrt{[1 - (c/\alpha_n)^2]}, & c < \alpha_n \end{cases} \qquad \eta_{\beta n} = \begin{cases} \sqrt{[(c/\beta_n)^2 - 1]}, & c > \beta_n \\ -i\sqrt{[1 - (c/\beta_n)^2]}, & c < \beta_n, \end{cases} \tag{3.171}$$

and deleting temporarily the common propagator $e^{i(\omega t - kx)}$, we may rewrite Eq. (3.169) in the form

$$\begin{aligned} \frac{\dot{u}_n}{c} = {} & ik_{\alpha n}\left(\frac{\alpha_n}{c}\right)^2 [(A'_n + A''_n)\cos(k\eta_{\alpha n}z) - i(A'_n - A''_n)\sin(k\eta_{\alpha n}z)] \\ & - ik_{\beta n}\left(\frac{\beta_n}{c}\right)^2 \eta_{\beta n}[(B'_n - B''_n)\cos(k\eta_{\beta n}z) - i(B'_n + B''_n)\sin(k\eta_{\beta n}z)], \\ \frac{\dot{w}_n}{c} = {} & ik_{\alpha n}\left(\frac{\alpha_n}{c}\right)^2 \eta_{\alpha n}[(A'_n - A''_n)\cos(k\eta_{\alpha n}z) - i(A'_n + A''_n)\sin(k\eta_{\alpha n}z)] \\ & + ik_{\beta n}\left(\frac{\beta_n}{c}\right)^2 [(B'_n + B''_n)\cos(k\eta_{\beta n}z) - i(B'_n - B''_n)\sin(k\eta_{\beta n}z)]. \end{aligned} \tag{3.172}$$

The corresponding normal stress $\sigma = \tau_{zz}$ and the shear stress $\tau = \tau_{zx}$ are obtained with the aid of Eq. (1.68)

$$\begin{aligned} \sigma_n = {} & ik_{\alpha n}\rho_n\alpha_n^2(\gamma_n - 1)[(A'_n + A''_n)\cos(k\eta_{\alpha n}z) - i(A'_n - A''_n)\sin(k\eta_{\alpha n}z)] \\ & - ik_{\beta n}\rho_n\frac{c^2}{2}\gamma_n^2\eta_{\beta n}[(B'_n - B''_n)\cos(k\eta_{\beta n}z) - i(B'_n + B''_n)\sin(k\eta_{\beta n}z)], \\ \tau_n = {} & -ik_{\alpha n}\rho_n\alpha_n^2\eta_{\alpha n}\gamma_n[(A'_n - A''_n)\cos(k\eta_{\alpha n}z) - i(A'_n + A''_n)\sin(k\eta_{\alpha n}z)] \\ & - ik_{\beta n}\rho_n\frac{c^2}{2}(\gamma_n - 1)\gamma_n[(B'_n + B''_n)\cos(k\eta_{\beta n}z) - i(B'_n - B''_n)\sin(k\eta_{\beta n}z)]. \end{aligned} \tag{3.173}$$

Defining

$$\begin{aligned} \Omega_{\alpha n}^{+}(z) &= -ik_{\alpha n}(A''_n e^{ik\eta_{\alpha n}z} + A'_n e^{-ik\eta_{\alpha n}z}), \\ \Omega_{\alpha n}^{-}(z) &= ik_{\alpha n}(A''_n e^{ik\eta_{\alpha n}z} - A'_n e^{-ik\eta_{\alpha n}z}), \\ 2\Omega_{\beta n}^{+}(z) &= ik_{\beta n}(B''_n e^{ik\eta_{\beta n}z} + B'_n e^{-ik\eta_{\beta n}z}), \\ 2\Omega_{\beta n}^{-}(z) &= -ik_{\beta n}(B''_n e^{ik\eta_{\beta n}z} - B'_n e^{-ik\eta_{\beta n}z}), \end{aligned} \tag{3.174}$$

we can write the values of the displacements and stresses in the nth layer in the form

$$\left[\frac{\dot{u}_n}{c}, \frac{\dot{w}_n}{c}, \sigma_n, \tau_n\right]_{z_{n-1}} = E_n[\Omega_{\alpha n}^{+}, \Omega_{\alpha n}^{-}, \Omega_{\beta n}^{-}, \Omega_{\beta n}^{+}]_{z_{n-1}}, \tag{3.175}$$

$$\left[\frac{\dot{u}_n}{c}, \frac{\dot{w}_n}{c}, \sigma_n, \tau_n\right]_{z_n} = E_n[\Omega_{\alpha n}^{+}, \Omega_{\alpha n}^{-}, \Omega_{\beta n}^{-}, \Omega_{\beta n}^{+}]_{z_n}, \tag{3.176}$$

where E_n denotes the 4×4 matrix

$$E_n = \begin{bmatrix} -(\alpha_n/c)^2 & 0 & -\gamma_n\eta_{\beta n} & 0 \\ 0 & -(\alpha_n/c)^2\eta_{\alpha n} & 0 & \gamma_n \\ -\rho_n\alpha_n^2(\gamma_n - 1) & 0 & -\rho_n c^2\gamma_n^2\eta_{\beta n} & 0 \\ 0 & \rho_n\alpha_n^2\gamma_n\eta_{\alpha n} & 0 & -\rho_n c^2\gamma_n(\gamma_n - 1) \end{bmatrix}. \tag{3.177}$$

It may be noted that the two matrices in Eqs. (3.175) and (3.176), although written as row matrices, are in fact column matrices.

It follows directly from Eqs. (3.174) that

$$\begin{bmatrix} \Omega_{\alpha n}^+ \\ \Omega_{\alpha n}^- \\ \Omega_{\beta n}^- \\ \Omega_{\beta n}^+ \end{bmatrix}_{z_{n-1}+d_n} = \begin{bmatrix} \cos P_n & -i\sin P_n & 0 & 0 \\ -i\sin P_n & \cos P_n & 0 & 0 \\ 0 & 0 & \cos Q_n & -i\sin Q_n \\ 0 & 0 & -i\sin Q_n & \cos Q_n \end{bmatrix} \begin{bmatrix} \Omega_{\alpha n}^+ \\ \Omega_{\alpha n}^- \\ \Omega_{\beta n}^- \\ \Omega_{\beta n}^+ \end{bmatrix}_{z_{n-1}} \tag{3.178}$$

where

$$P_n = k\eta_{\alpha n}d_n, \qquad Q_n = k\eta_{\beta n}d_n. \tag{3.179}$$

Equations (3.176)–(3.178) yield

$$\left[\frac{\dot{u}_n}{c}, \frac{\dot{w}_n}{c}, \sigma_n, \tau_n\right]_{z_n} = D_n[\Omega_{\alpha n}^+, \Omega_{\alpha n}^-, \Omega_{\beta n}^-, \Omega_{\beta n}^+]_{z_{n-1}}, \tag{3.180}$$

where D_n is the matrix

$$D_n = \begin{bmatrix} -(\alpha_n/c)^2\cos P_n & i(\alpha_n/c)^2\sin P_n & -\gamma_n\eta_{\beta n}\cos Q_n & i\gamma_n\eta_{\beta n}\sin Q_n \\ i(\alpha_n/c)^2\eta_{\alpha n}\sin P_n & -(\alpha_n/c)^2\eta_{\alpha n}\cos P_n & -i\gamma_n\sin Q_n & \gamma_n\cos Q_n \\ -\rho_n\alpha_n^2(\gamma_n - 1)\cos P_n & i\rho_n\alpha_n^2(\gamma_n - 1)\sin P_n & -\rho_n c^2\gamma_n^2\eta_{\beta n}\cos Q_n & i\rho_n c^2\gamma_n^2\eta_{\beta n}\sin Q_n \\ -i\rho_n\alpha_n^2\gamma_n\eta_{\alpha n}\sin P_n & \rho_n\alpha_n^2\gamma_n\eta_{\alpha n}\cos P_n & i\rho_n c^2\gamma_n(\gamma_n - 1)\sin Q_n & -\rho_n c^2\gamma_n(\gamma_n - 1)\cos Q_n \end{bmatrix}. \tag{3.181}$$

Eliminating $\Omega_{\alpha n}^+(z_{n-1})$, etc., from Eqs. (3.175) and (3.180), we get a linear relationship between the values of $\dot{u}/c$, $\dot{w}/c$, σ, and τ at the top and the bottom of the nth layer

$$\left[\frac{\dot{u}_n}{c}, \frac{\dot{w}_n}{c}, \sigma_n, \tau_n\right]_{z_n} = a_n\left[\frac{\dot{u}_n}{c}, \frac{\dot{w}_n}{c}, \sigma_n, \tau_n\right]_{z_{n-1}}, \tag{3.182}$$

where

$$a_n = D_n E_n^{-1}. \tag{3.183}$$

The matrix E_n^{-1} is the inverse of E_n and is given by

$$E_n^{-1} = \begin{bmatrix} -2(\beta_n/\alpha_n)^2 & 0 & (\rho_n\alpha_n^2)^{-1} & 0 \\ 0 & c^2(\gamma_n - 1)(\alpha_n^2\eta_{\alpha n})^{-1} & 0 & (\rho_n\alpha_n^2\eta_{\alpha n})^{-1} \\ (\gamma_n - 1)(\gamma_n\eta_{\beta n})^{-1} & 0 & -(\rho_n c^2\gamma_n\eta_{\beta n})^{-1} & 0 \\ 0 & 1 & 0 & (\rho_n c^2\gamma_n)^{-1} \end{bmatrix}. \tag{3.184}$$

From Eqs. (3.181) and (3.184), the elements of the product matrix a_n may be computed as follows

$$\begin{aligned}
(a_n)_{11} &= \gamma_n \cos P_n - (\gamma_n - 1)\cos Q_n, \\
(a_n)_{12} &= i[(\gamma_n - 1)\eta_{\alpha n}^{-1} \sin P_n + \gamma_n\eta_{\beta n} \sin Q_n], \\
(a_n)_{13} &= -(\rho_n c^2)^{-1}(\cos P_n - \cos Q_n), \\
(a_n)_{14} &= i(\rho_n c^2)^{-1}(\eta_{\alpha n}^{-1} \sin P_n + \eta_{\beta n} \sin Q_n), \\
(a_n)_{21} &= -i[\gamma_n\eta_{\alpha n} \sin P_n + (\gamma_n - 1)\eta_{\beta n}^{-1} \sin Q_n], \\
(a_n)_{22} &= -(\gamma_n - 1)\cos P_n + \gamma_n \cos Q_n, \\
(a_n)_{23} &= i(\rho_n c^2)^{-1}(\eta_{\alpha n} \sin P_n + \eta_{\beta n}^{-1} \sin Q_n), \\
(a_n)_{24} &= (a_n)_{13}, \\
(a_n)_{31} &= \rho_n c^2\gamma_n(\gamma_n - 1)(\cos P_n - \cos Q_n), \\
(a_n)_{32} &= i\rho_n c^2[(\gamma_n - 1)^2\eta_{\alpha n}^{-1} \sin P_n + \gamma_n^2\eta_{\beta n} \sin Q_n], \\
(a_n)_{33} &= (a_n)_{22}, \\
(a_n)_{34} &= (a_n)_{12}, \\
(a_n)_{41} &= i\rho_n c^2[\gamma_n^2\eta_{\alpha n} \sin P_n + (\gamma_n - 1)^2\eta_{\beta n}^{-1} \sin Q_n], \\
(a_n)_{42} &= (a_n)_{31}, \\
(a_n)_{43} &= (a_n)_{21}, \\
(a_n)_{44} &= (a_n)_{11}.
\end{aligned} \tag{3.185}$$

The boundary conditions require that the values of $\dot{u}/c$, $\dot{w}/c$, σ, and τ evaluated at the top of the nth layer be the same as the values computed at the bottom of the $(n - 1)$th layer. This implies

$$\left[\frac{\dot{u}_n}{c}, \frac{\dot{w}_n}{c}, \sigma_n, \tau_n\right]_{z_{n-1}} = \left[\frac{\dot{u}_{n-1}}{c}, \frac{\dot{w}_{n-1}}{c}, \sigma_{n-1}, \tau_{n-1}\right]_{z_{n-1}}. \tag{3.186}$$

If we write

$$\dot{u}_n = \dot{u}_n(z_n), \qquad \dot{w}_n = \dot{w}_n(z_n), \qquad \sigma_n = \sigma_n(z_n), \qquad \tau_n = \tau_n(z_n) \qquad (n = 1, 2, \ldots, N-1),$$

Eqs. (3.182) and (3.186) yield

$$\left[\frac{\dot{u}_n}{c}, \frac{\dot{w}_n}{c}, \sigma_n, \tau_n\right] = a_n\left[\frac{\dot{u}_{n-1}}{c}, \frac{\dot{w}_{n-1}}{c}, \sigma_{n-1}, \tau_{n-1}\right]. \tag{3.187}$$

By a repeated application of Eq. (3.187), we have

$$\left[\frac{\dot{u}_{N-1}}{c}, \frac{\dot{w}_{N-1}}{c}, \sigma_{N-1}, \tau_{N-1}\right] = a_{N-1}a_{N-2}\cdots a_1\left[\frac{\dot{u}_0}{c}, \frac{\dot{w}_0}{c}, \sigma_0, \tau_0\right], \tag{3.188}$$

where $\dot{u}_0 = \dot{u}_1(0)$, etc. Equations (3.175), (3.186), and (3.188) now yield

$$[\Omega_{\alpha N}^+, \Omega_{\alpha N}^-, \Omega_{\beta N}^-, \Omega_{\beta N}^+] = J\left[\frac{\dot{u}_0}{c}, \frac{\dot{w}_0}{c}, \sigma_0, \tau_0\right], \tag{3.189}$$

where

$$\begin{aligned} \Omega_{\alpha N}^+ &= \Omega_{\alpha N}^+(z_{N-1}), \quad \text{etc.,} \\ J &= E_N^{-1}A^R, \\ A^R &= a_{N-1}a_{N-2}\cdots a_1. \end{aligned} \tag{3.190}$$

Setting the normal and tangential stresses at the free surface, σ_0 and τ_0, equal to zero in Eq. (3.189), we have

$$\begin{aligned} \Omega_{\alpha N}^+ &= J_{11}\frac{\dot{u}_0}{c} + J_{12}\frac{\dot{w}_0}{c}, \\ \Omega_{\alpha N}^- &= J_{21}\frac{\dot{u}_0}{c} + J_{22}\frac{\dot{w}_0}{c}, \\ \Omega_{\beta N}^- &= J_{31}\frac{\dot{u}_0}{c} + J_{32}\frac{\dot{w}_0}{c}, \\ \Omega_{\beta N}^+ &= J_{41}\frac{\dot{u}_0}{c} + J_{42}\frac{\dot{w}_0}{c}. \end{aligned} \tag{3.191}$$

We now deal with the problems of incident P and SV waves separately.

3.7.2.1. Incident P Waves. For an incident P wave with given amplitude A_N'', we set $B_N'' = 0$ in Eqs. (3.174). Equations (3.191) then constitute a set of four equations in the four unknowns, A_N' (reflected P into the half-space),

B'_N (reflected SV into the half-space), and the two components of the surface displacement, u_0 and w_0. Their solutions are

$$\frac{A'_N}{A''_N} = \frac{1}{D}[(J_{11} + J_{21})(J_{32} - J_{42}) - (J_{12} + J_{22})(J_{31} - J_{41})]e^{2ik\eta_{\alpha N}z_{N-1}},$$

$$\frac{B'_N}{A''_N} = -\frac{4}{D}\frac{\beta_N}{\alpha_N}(J_{32}J_{41} - J_{31}J_{42})e^{ikz_{N-1}(\eta_{\alpha N}+\eta_{\beta N})}; \tag{3.192}$$

$$\frac{u_0}{A''_N} = \frac{2c}{D\alpha_N}(J_{42} - J_{32})e^{ik\eta_{\alpha N}z_{N-1}},$$

$$\frac{w_0}{A''_N} = \frac{2c}{D\alpha_N}(J_{31} - J_{41})e^{ik\eta_{\alpha N}z_{N-1}}; \tag{3.193}$$

$$D = (J_{11} - J_{21})(J_{32} - J_{42}) - (J_{12} - J_{22})(J_{31} - J_{41}). \tag{3.194}$$

Equations (3.193) give the *crustal transfer functions* relating the output (surface displacements) to the input (incident P wave amplitude) and Eqs. (3.192) give the reflection coefficients.

3.7.2.2. Incident SV Waves. Setting $A''_N = 0$ in Eq. (3.174), we obtain, from Eqs. (3.191),

$$\frac{A'_N}{B''_N} = -\frac{\alpha_N}{D\beta_N}(J_{11}J_{22} - J_{12}J_{21})e^{ikz_{N-1}(\eta_{\alpha N}+\eta_{\beta N})}, \tag{3.195}$$

$$\frac{B'_N}{B''_N} = -\frac{1}{D}[(J_{12} - J_{22})(J_{31} + J_{41}) - (J_{11} - J_{21})(J_{32} + J_{42})]e^{2ik\eta_{\beta N}z_{N-1}};$$

$$\frac{u_0}{B''_N} = -\frac{c}{\beta_N D}(J_{12} - J_{22})e^{ik\eta_{\beta N}z_{N-1}},$$

$$\frac{w_0}{B''_N} = -\frac{c}{\beta_N D}(J_{21} - J_{11})e^{ik\eta_{\beta N}z_{N-1}}. \tag{3.196}$$

3.7.2.3. Rayleigh Waves. If we add an additional condition that the displacement and stress field in the half-space must tend to zero as z tends to infinity (no sources at infinity), we obtain the dispersion relation for Rayleigh waves. Taking

$$A''_N = 0, \qquad B''_N = 0$$

in Eqs. (3.174) and eliminating A'_N and B'_N, we obtain

$$\frac{\dot{u}_0}{\dot{w}_0} = \frac{u_0}{w_0} = \frac{J_{22} - J_{12}}{J_{11} - J_{21}} = \frac{J_{42} - J_{32}}{J_{31} - J_{41}}. \tag{3.197}$$

This gives the dispersion relation for Rayleigh waves. We note from Eqs. (3.193) and (3.196) that

$$u_0/w_0 = \begin{cases} (J_{42} - J_{32})/(J_{31} - J_{41}) & \text{for incident } P \text{ waves,} \\ (J_{22} - J_{12})/(J_{11} - J_{21}) & \text{for incident } SV \text{ waves.} \end{cases} \tag{3.198}$$

Using the explicit form, Eq. (3.184), of the matrix E_n^{-1}, we may express the amplitude ratio $\dot{u}_0/\dot{w}_0$ directly in terms of the elements of A^R. We find

$$-\frac{\dot{u}_0}{\dot{w}_0} = \frac{K}{L} = \frac{M}{N}, \tag{3.199}$$

where (suppressing the superscript R of A_{ij}^R)

$$\begin{aligned}
K &= \gamma_N \eta_{\alpha N} A_{12} + (\gamma_N - 1)A_{22} - \frac{\eta_{\alpha N} A_{32}}{\rho_N c^2} + \frac{A_{42}}{\rho_N c^2},\\
L &= \gamma_N \eta_{\alpha N} A_{11} + (\gamma_N - 1)A_{21} - \frac{\eta_{\alpha N} A_{31}}{\rho_N c^2} + \frac{A_{41}}{\rho_N c^2},\\
M &= -(\gamma_N - 1)A_{12} + \gamma_N \eta_{\beta N} A_{22} + \frac{A_{32}}{\rho_N c^2} + \frac{\eta_{\beta N} A_{42}}{\rho_N c^2},\\
N &= -(\gamma_N - 1)A_{11} + \gamma_N \eta_{\beta N} A_{21} + \frac{A_{31}}{\rho_N c^2} + \frac{\eta_{\beta N} A_{41}}{\rho_N c^2}.
\end{aligned} \tag{3.200}$$

In the expressions of Eq. (3.185) for the elements of the matrix a_n, it will be observed that the quantities $\sin P_n$, $\sin Q_n$, $\eta_{\beta n}$, and $\eta_{\alpha n}$, which may be either real or imaginary depending upon the value of c, occur only in the combinations $\eta_{\alpha n}^{\pm 1} \sin P_n$ and $\eta_{\beta n}^{\pm 1} \sin Q_n$. Because $\sin P_n$ is real or imaginary according as $\eta_{\alpha n}$ is real or imaginary and $\sin Q_n$ is similarly related to $\eta_{\beta n}$, these combinations are always real for real values of c. With regard to the real or imaginary properties of its elements, the matrix a_n then has the form

$$a_n = \begin{bmatrix} R & I & R & I \\ I & R & I & R \\ R & I & R & I \\ I & R & I & R \end{bmatrix},$$

where an R indicates a real quantity and an I indicates an imaginary quantity. The product of any two matrices of this form is also a matrix of the same form. Hence, of the elements of A^R occurring in Eqs. (3.200), A_{11}, A_{22}, A_{31}, and A_{42} are real; A_{12}, A_{21}, A_{32}, and A_{41} are imaginary. By definition, a surface wave is one whose amplitude tends to zero as $z \to \infty$. This means that in our case $\eta_{\alpha N}$ and $\eta_{\beta N}$ must be imaginary; that is, we are concerned only with values of $c < \beta_N$. Then, referring to Eqs. (3.200), all terms of K and N are real and all terms of L and M are imaginary. Therefore, the ratio $\dot{u}_0/\dot{w}_0$ will always be imaginary, which implies a phase difference of 90° between the horizontal and vertical displacements at the free surface. The particle motion is therefore an ellipse whose axes are vertical and horizontal. The phase difference, however, may be of either sign, and hence the sense of the motion around the ellipse is not necessarily retrograde with respect to the direction of propagation at all frequencies, as is the case with Rayleigh waves in a homogeneous half-space.

Figure 3.7 shows the crustal amplitude response for the vertical component of the displacement for a single-layer crustal model when P waves are incident at the base of the crust. The five curves correspond to the five angles of incidence (THN): 10°, 20°, 30°, 40°, and 50°. The effect of the crustal structure on the

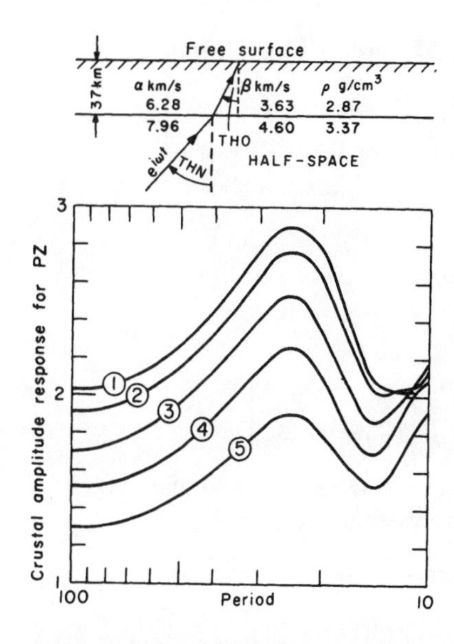

Figure 3.7. Variation of the crustal amplitude response (vertical component) with the angle of incidence (THN) of P waves of unit amplitude incident at the base of a single-layer crust for (1) THN = 10°, THO = 7.88°; (2) THN = 20°, THO = 15.66°; (3) THN = 30°, THO = 23.20°; (4) THN = 40°, THO = 30.50°; (5) THN = 50°, THO = 37.20°. THO is the angle of incidence of the wave at the free surface.

spectral amplitude of an incident P wave is shown in Fig. 3.8. Note that, up to periods of 20 s the variations of the structure do not affect the gross features of the spectra. However, below 10 s, as the wavelength becomes aware of the structural details, there is a strong dependence on the geologic province at which the interaction takes place.

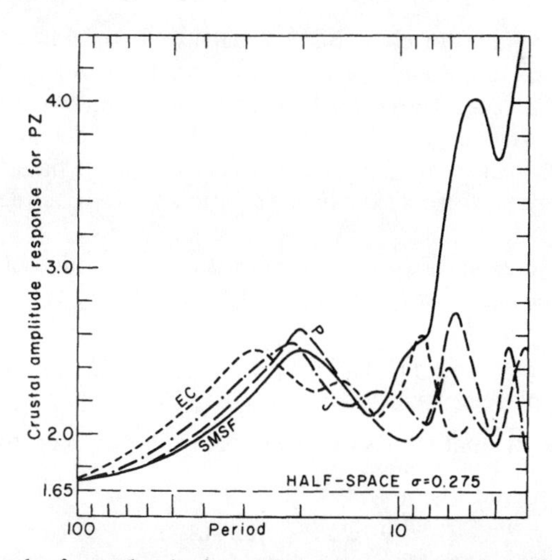

Figure 3.8. The role of crustal and subcrustal structure on spectral amplitude response for incident P wave of unit amplitude and of phase velocity 15 km/s. SMSF, Santa Monica–San Francisco; EC, eastern Colorado; J, Japan; P, Pacific. Period of wave is measured in seconds of time.

3.8. Generalization of the Matrix Method—The Matrix Propagator

The matrix method, used to calculate phase velocities of surface waves in a multilayered half-space, can be viewed from a different angle. We rewrite the basic Love-wave Eqs. (3.83) in the form

$$\frac{dy}{dz} = P(z)y, \tag{3.201}$$

where

$$y = \begin{bmatrix} y_1 \\ y_2 \end{bmatrix}, \qquad P = \begin{bmatrix} 0 & \mu^{-1} \\ \mu k^2 - \rho\omega^2 & 0 \end{bmatrix}.$$

Let us consider two sets of independent solutions (y_{11}, y_{21}) and (y_{12}, y_{22}). An arbitrary solution (y_1, y_2) may be expressed in the form

$$\begin{aligned} y_1 &= A_1 y_{11} + A_2 y_{12}, \\ y_2 &= A_1 y_{21} + A_2 y_{22}, \end{aligned} \tag{3.202}$$

where A_1 and A_2 are arbitrary constants of integration determined by the surface boundary conditions. On account of the linearity of the governing differential equations, we have

$$\frac{dY}{dz} = P(z)Y, \qquad Y = \begin{bmatrix} y_{11} & y_{12} \\ y_{21} & y_{22} \end{bmatrix}. \tag{3.203}$$

We need not confine ourselves to 2×2 matrices. In the following discussion, it will be assumed that P and Y are $n \times n$ matrices. The matrix Y is called the *integral matrix* of Eq. (3.201). Each column of this matrix is a solution of Eq. (3.201). If Y is nonsingular for every z in its domain of definition, it is known as a *fundamental matrix* of Eq. (3.201). It can be shown that, for any $n \times n$ matrix B and any z_0, there is exactly one solution $Y(z)$ of Eqs. (3.203) such that $Y(z_0) = B$.

We wish to integrate Eqs. (3.203), given the initial value of Y at $z = z_0$. We introduce the *matrizant* (*propagator*) of P from $z = z_0$ by

$$\Omega(z, z_0) = I + \int_{z_0}^{z} P(\zeta_1)d\zeta_1 + \int_{z_0}^{z} P(\zeta_1) \int_{z_0}^{\zeta_1} P(\zeta_2)d\zeta_2\, d\zeta_1 + \cdots, \tag{3.204}$$

where I denotes the unit matrix of the nth order. It is evident from Eq. (3.204) that

$$\frac{d}{dz}\Omega(z, z_0) = P(z)\Omega(z, z_0) \tag{3.205}$$

and, for $z = z_0$,

$$\Omega(z, z_0) = I. \tag{3.206}$$

Hence

$$Y(z) = \Omega(z, z_0)Y(z_0) \tag{3.207}$$

satisfies the given system of differential equations and the initial condition at $z = z_0$. The solution is valid unless the path of integration in Eq. (3.204) encounters any singularity of $P(z)$. The matrices $\Omega(z, z_0)$ and $\Omega(z, z_1)\Omega(z_1, z_0)$ are both solutions of Eqs. (3.203) and are equal when $z = z_1$. Therefore, the uniqueness theorem assures their equality for all z,

$$\Omega(z, z_0) = \Omega(z, z_1)\Omega(z_1, z_0). \tag{3.208}$$

Because $\Omega(z_0, z_0)$ is the unit matrix, Eq. (3.208) shows that $\Omega(z_1, z_0)$ is the inverse matrix of $\Omega(z_0, z_1)$. In general, if $z_1, z_2, \ldots, z_{m-1}$ are intermediate points in the range (z_0, z), then the exact solution of Eq. (3.203) follows from Eqs. (3.207) and (3.208)

$$Y(z) = \Omega(z, z_{m-1})\Omega(z_{m-1}, z_{m-2}) \cdots \Omega(z_1, z_0)Y(z_0). \tag{3.209}$$

It is now apparent why the matrix Ω is sometimes known as the propagator matrix. It is also clear from the foregoing discussion that for any z, $\Omega(z, z_0)$ has an inverse and hence that any propagator is a fundamental matrix. Equation (3.207) yields

$$\Omega(z, z_0) = Y(z)Y^{-1}(z_0). \tag{3.210}$$

Therefore, the propagator from any point z_0 can be calculated immediately by matrix inversion, once n linearly independent $n \times 1$ column matrix solutions of Eq. (3.201) are known, namely,

$$y(z) = \Omega(z, z_0)y(z_0). \tag{3.211}$$

The propagator can be used to solve the inhomogeneous system of linear differential equations

$$\frac{dy}{dz} = P(z)y + g(z), \tag{3.212}$$

where $g(z)$ is a given $(n \times 1)$ column matrix. The solution, which may be verified by direct substitution, is

$$\begin{aligned} y(z) &= Y(z)\left[\int_{z_0}^{z} Y^{-1}(\zeta)g(\zeta)d\zeta + Y^{-1}(z_0)y(z_0)\right] \\ &= \int_{z_0}^{z} \Omega(z, \zeta)g(\zeta)d\zeta + \Omega(z, z_0)y(z_0). \end{aligned} \tag{3.213}$$

The reader can easily show that

$$\|Y(z)\| = \|Y(z_0)\| \exp \int_{z_0}^{z} \mathrm{trace}\{P(\zeta)\}d\zeta. \tag{3.214}$$

In particular,

$$\|\Omega(z, z_0)\| = \exp \int_{z_0}^{z} \mathrm{trace}\{P(\zeta)\}d\zeta. \tag{3.215}$$

The calculation of the matrizant is usually difficult except when (1) $P(z)$ commutes with $\int_{z_0}^{z} P(\zeta)d\zeta$ for every z, or (2) $P(z)$ is a constant matrix. In the first case, the solution of Eqs. (3.203) can be written as

$$Y(z) = \exp\left[\int_{z_0}^{z} P(\zeta)d\zeta\right] Y(z_0). \tag{3.216}$$

In the second case, we find from Eq. (3.204) that

$$\Omega(z, z_0) = I + (z - z_0)P + \tfrac{1}{2}(z - z_0)^2 P^2 + \cdots = e^{(z - z_0)P}. \tag{3.217}$$

Let the eigenvalues of P be denoted by ν_i $(i = 1, 2, \ldots, n)$. Then, from Sylvester's theorem, we have

$$E(d) = e^{dP} = \sum_{i=1}^{n} e^{\nu_i d} \prod_{j \neq i} \frac{\nu_j I - P}{\nu_j - \nu_i}. \tag{3.218}$$

When n is even (2 for Love waves and 4 for Rayleigh waves), the eigenvalues are written as $\nu_0 \pm \nu_i$ $(i = 1, 2, \ldots, n/2)$. Hence we have

$$E(d) = e^{dP} = e^{\nu_0 d} \sum_{i=1}^{n/2} \left[I \operatorname{ch} \nu_i d + (P - \nu_0 I)\frac{\operatorname{sh} \nu_i d}{\nu_i}\right] \prod_{j \neq i} \frac{\nu_j^2 I - (P - \nu_0 I)^2}{(\nu_j^2 - \nu_i^2)}. \tag{3.219}$$

In particular, for $n = 2$ (Love waves), we have $\nu_0 = 0$, $\nu_1 = \pm\sqrt{(k^2 - k_\beta^2)} = \pm\nu$. Consequently, the layer matrix is given by

$$E(d) = I \operatorname{ch} \nu d + P\frac{\operatorname{sh} \nu d}{\nu} = \begin{bmatrix} \operatorname{ch} \nu d & \dfrac{1}{\mu}\dfrac{\operatorname{sh} \nu d}{\nu} \\ \mu\nu \operatorname{sh} \nu d & \operatorname{ch} \nu d \end{bmatrix}. \tag{3.220}$$

The elements of $E(d)$ are always real whether ν is real or imaginary. It may be noted that $E(d)$ is related to the layer matrix a_n defined earlier in Eq. (3.158).

If in Eq. (3.209) the interval (z_{k-1}, z_k) is taken small enough, we may substitute for P in Eq. (3.204), its value at the middle of the interval

$$\bar{P}^{(k)} = P\left(\frac{z_{k-1} + z_k}{2}\right), \qquad z_{k-1} < z < z_k. \tag{3.221}$$

Then, an approximation to the propagator

$$\Omega(z_k, z_{k-1}) \simeq e^{(z_k - z_{k-1})\bar{P}^{(k)}} = E^{(k)}(z_k - z_{k-1}) \tag{3.222}$$

is computed from Eq. (3.219). After the propagator in each interval is found, a solution will be obtained through matrix multiplications as required by Eq. (3.209),

$$Y(z) = \{E^{(m)}(z - z_{m-1})E^{(m-1)}(d_{m-1}) \cdots E^{(1)}(d_1)\} Y(z_0), \tag{3.223}$$

where $d_k = z_k - z_{k-1}$. The above solution is exact if P is constant in each interval. The averaging process in Eq. (3.221) is of course not unique. We note that the solution given in Eq. (3.223) corresponds to the one given in Eq. (3.162) for a stack of $N - 1$ homogeneous layers.

The matrizant method has the advantage of being able to supply us with higher order approximations. For example, we may improve the approximation given in Eq. (3.223) by the substitution

$$Y(z) = \{e^{(z-z_0)\bar{P}^{(1)}}\}\hat{Y}(z) = E(z - z_0)\hat{Y}(z) \qquad (z_0 \le z \le z_1), \quad (3.224)$$

where

$$\bar{P}^{(1)} = P\left(\frac{z_0 + z_1}{2}\right).$$

Then, Eqs. (3.203) reduce to

$$\frac{d\hat{Y}}{dz} = \hat{P}(z)\hat{Y}, \qquad \hat{Y}(z_0) = Y(z_0), \quad (3.225)$$

where

$$\hat{P}(z) = e^{-(z-z_0)\bar{P}^{(1)}}[P(z) - \bar{P}^{(1)}]e^{(z-z_0)\bar{P}^{(1)}}. \quad (3.226)$$

In particular,

$$\hat{P}\left(\frac{z_0 + z_1}{2}\right) = 0. \quad (3.227)$$

Because Eq. (3.225) is of the same form as Eqs. (3.203), the matrizant solution at $z = z_1$ according to Eq. (3.209) is

$$\hat{Y}(z_1) = \Omega(z_1, z_0)Y(z_0). \quad (3.228)$$

An approximate solution that avoids the direct evaluation of the matrizant $\Omega(z_1, z_0)$ can be obtained by applying the fourth-order Runge–Kutta approximation to the differential equation, Eq. (3.225), at $z = z_1$. Using Eq. (3.227), we have

$$\begin{aligned}\hat{Y}(z_1) = \hat{Y}[z_0 + (z_1 - z_0)] &= \hat{Y}(z_0) + \frac{z_1 - z_0}{6}\{\hat{P}(z_0)\hat{Y}(z_0) + \hat{P}(z_1)\hat{Y}(z_0)\} \\ &= [I + \tfrac{1}{6}(z_1 - z_0)\{\hat{P}(z_0) + \hat{P}(z_1)\}]Y(z_0)\end{aligned} \quad (3.229)$$

which, together with Eq. (3.224), renders

$$Y(z_1) = \bar{\Omega}(z_1, z_0)Y(z_0),$$

$$\begin{aligned}\bar{\Omega}(z_1, z_0) = E(z_1 - z_0) + \tfrac{1}{6}(z_1 - z_0)E(z_1 - z_0)[P(z_0) - \bar{P}^{(1)}] + \tfrac{1}{6}(z_1 - z_0) \\ \times [P(z_1) - \bar{P}^{(1)}]E(z_1 - z_0).\end{aligned} \quad (3.230)$$

This formula is accurate to the fourth power of $(z_1 - z_0)$. Once $\bar{\Omega}(z_1, z_0)$ is known, the propagator from z_0 to an arbitrary point z is realized by means of Eq. (3.209).

If we approximate the earth's structure by a series of flat homogeneous layers, the exact solution for surface and body waves is given by the theory given in Section 3.7. However, if we prefer to obtain an approximate solution for a more exact earth model, we may use solutions of the type given in Eqs. (3.230). The two approaches should be compared on the basis of accuracy and speed of computation.

3.9. The Inverse Surface-Wave Problem

Much of the difficulty in comparing theoretical surface-wave dispersion curves with observed data can be minimized with the use of partial derivatives of phase and group velocities. Partial derivatives specify the change in the phase or group velocity with respect to elastic parameter changes in a layered wave guide and make it possible to compute surface-wave dispersion for any desired earth model. The calculation of the partial derivatives is commonly based on the application of Rayleigh's principle.

Consider the development for Love waves. We are interested in examining the variation in the phase velocity with the perturbation of a specific elastic parameter holding the wave number k fixed. Application of Rayleigh's principle to Eq. (3.127) leads to the relation

$$2\omega(\delta\omega)I_0^L + \omega^2(\delta I_0^L) = \delta I_2^L + k^2(\delta I_1^L). \tag{3.231}$$

Therefore, if we vary the density in a particular depth interval $(z - \varepsilon, z + \varepsilon)$ holding the rigidity fixed, the change in the phase velocity is governed by the relation

$$-\frac{\rho}{c}\left(\frac{\partial c}{\partial \rho}\right)_{k,\mu} = \frac{1}{2I_0^L}\int_{z-\varepsilon}^{z+\varepsilon} \rho V^2\, dz. \tag{3.232}$$

The right-hand side of Eq. (3.232) can be recognized as the ratio of the kinetic energy over the perturbed interval to the total energy in the system. Similarly, we can show that

$$\frac{\mu}{c}\left(\frac{\partial c}{\partial \mu}\right)_{k,\rho} = \frac{1}{2\omega^2 I_0^L}\int_{z-\varepsilon}^{z+\varepsilon} \mu\left[k^2V^2 + \left(\frac{dV}{dz}\right)^2\right]dz. \tag{3.233}$$

The Rayleigh-wave can be treated along similar lines. From Eq. (3.140), we find

$$\begin{aligned}
-\frac{\rho}{c}\left(\frac{\partial c}{\partial \rho}\right)_{\lambda,\mu} &= \frac{1}{2I_0^R}\int_{z-\varepsilon}^{z+\varepsilon} \rho(U^2 + W^2)dz,\\
\frac{\lambda}{c}\left(\frac{\partial c}{\partial \lambda}\right)_{\mu,\rho} &= \frac{1}{2\omega^2 I_0^R}\int_{z-\varepsilon}^{z+\varepsilon} \lambda\left[k^2U^2 - 2kU\frac{dW}{dz} + \left(\frac{dW}{dz}\right)^2\right]dz,\\
\frac{\mu}{c}\left(\frac{\partial c}{\partial \mu}\right)_{\rho,\lambda} &= \frac{1}{2\omega^2 I_0^R}\int_{z-\varepsilon}^{z+\varepsilon} \mu\left[k^2(2U^2 + W^2) + 2kW\frac{dU}{dz}\right.\\
&\qquad\left. + \left(\frac{dU}{dz}\right)^2 + 2\left(\frac{dW}{dz}\right)^2\right]dz.
\end{aligned} \tag{3.234}$$

It is more convenient to take (α, β, ρ) as the independent variables instead of (λ, μ, ρ), because the former set is determined directly from the observations. Using the relations

$$\lambda = \rho(\alpha^2 - 2\beta^2), \qquad \mu = \rho\beta^2,$$

we find

$$\begin{aligned} \left(\frac{\partial c}{\partial \alpha}\right)_{\beta,\rho} &= 2\rho\alpha\left(\frac{\partial c}{\partial \lambda}\right)_{\mu,\rho}, \\ \left(\frac{\partial c}{\partial \beta}\right)_{\rho,\alpha} &= 2\rho\beta\left\{-2\left(\frac{\partial c}{\partial \lambda}\right)_{\mu,\rho} + \left(\frac{\partial c}{\partial \mu}\right)_{\rho,\lambda}\right\}, \\ \left(\frac{\partial c}{\partial \rho}\right)_{\alpha,\beta} &= \left(\frac{\partial c}{\partial \rho}\right)_{\lambda,\mu} + (\alpha^2 - 2\beta^2)\left(\frac{\partial c}{\partial \lambda}\right)_{\mu,\rho} + \beta^2\left(\frac{\partial c}{\partial \mu}\right)_{\rho,\lambda}. \end{aligned} \tag{3.235}$$

An examination of the partial derivative curves as a function of period gives insight into the relative importance of the various parameters at different depths. Figure 3.9a shows the Love-wave phase-velocity partial derivatives for a continental-earth model (App. L). The density derivatives are small, emphasizing sensitivity to variations in shear-wave velocity. In this model, an increase in any layer's shear velocity will produce an increase in the phase velocity, but the density partials can be either positive or negative. For Rayleigh waves (Fig. 3.9b) the shear velocity is again the controlling parameter, although for certain period ranges the density is an important variable.

3.9.1. Perturbation of Love-Wave Spectra

We begin with the equation of motion for the horizontal transverse displacement [Eq. (3.81)]

$$\frac{d}{dz}\left(\mu\frac{dV}{dz}\right) - k^2\mu V + \omega^2\rho V = 0 \tag{3.236}$$

in which ρ and μ are functions of z only. Multiplying Eq. (3.236) by V, integrating from $z = 0$ to $z = \infty$, and making use of the continuity of V and $\mu(dV/dz)$ at the discontinuities of the elastic parameters, we obtain

$$\mathscr{L} \equiv \omega^2 I_0 - k^2 I_1 - I_2 = -\left(V\mu\frac{dV}{dz}\right)\bigg|_0^\infty, \tag{3.237}$$

where

$$I_0 = \int_0^\infty \rho V^2\,dz, \qquad I_1 = \int_0^\infty \mu V^2\,dz, \qquad I_2 = \int_0^\infty \mu\left(\frac{dV}{dz}\right)^2 dz \tag{3.238}$$

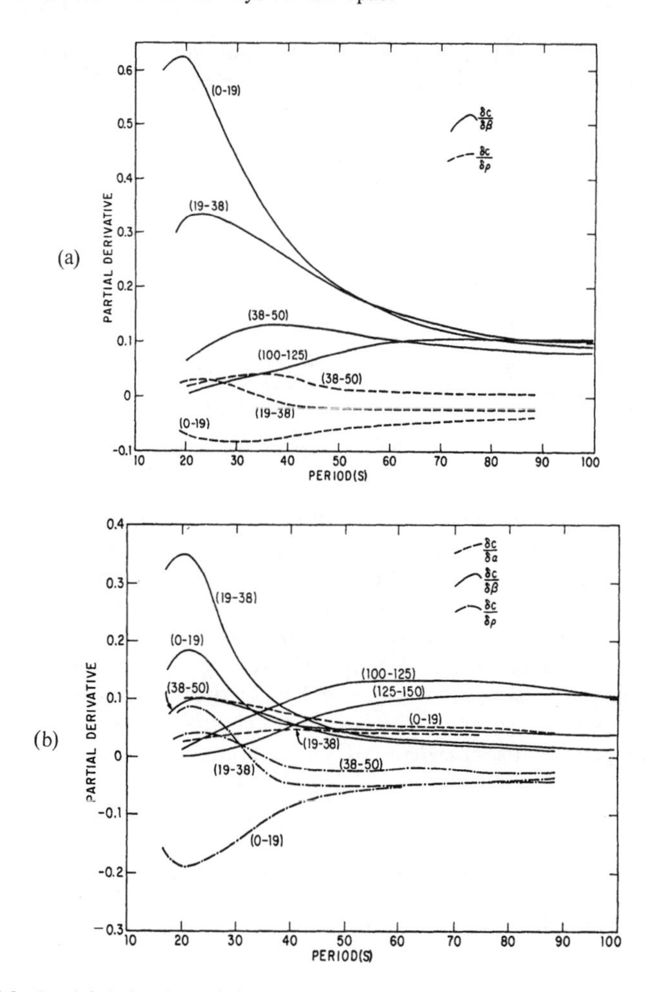

Figure 3.9. Partial derivatives of the phase velocity relative to the elastic parameters for fixed values of k. (a) Love waves (b) Rayleigh waves. The numbers labeling the curves indicate the depth ranges.

and $\mathscr{L}$ is the *Lagrangian function.* For Love waves, we impose the boundary conditions of a free surface at $z = 0$, i.e.,

$$\mu \frac{dV}{dz}\bigg|_{z=0} = 0;$$

and require that the motion vanishes at infinity. Therefore, for Love waves in a vertically inhomogeneous half-space, the Lagrangian vanishes,

$$\mathscr{L}(\omega, k, \rho, \mu) = 0. \tag{3.239}$$

Let p be one of the variables ω, k, ρ, μ. We multiply Eq. (3.236) by $\partial V/\partial p$ and integrate from $z = 0$ to $z = \infty$, keeping the other three parameters constant. Applying the continuity of $\partial V/\partial p$ across interfaces, together with the boundary conditions at $z = 0$ and $z = \infty$, we obtain

$$\omega^2 \int_0^\infty \rho \frac{\partial(V^2)}{\partial p} dz = k^2 \int_0^\infty \mu \frac{\partial(V^2)}{\partial p} dz + \int_0^\infty \mu \frac{\partial}{\partial p}\left[\left(\frac{\partial V}{dz}\right)^2\right] dz \qquad (p = \omega, k, \rho, \mu). \tag{3.240}$$

Equation (3.239) may be written in the form

$$\omega^2 I_0 = k^2 I_1 + I_2. \tag{3.241}$$

After small perturbations $\delta\rho$ in density and $\delta\mu$ in rigidity, this equation becomes

$$(\omega + \delta\omega)^2(I_0 + \delta I_0) = (k + \delta k)^2(I_1 + \delta I_1) + I_2 + \delta I_2. \tag{3.242}$$

From Eqs. (3.241) and (3.242), we find

$$\omega^2 \delta I_0 + 2\omega(\delta\omega)I_0 = k^2 \delta I_1 + 2k(\delta k)I_1 + \delta I_2, \tag{3.243}$$

where small quantities of the second and third orders have been neglected. Hence, if ω and ρ are held fixed, we have

$$\omega^2(\delta I_0)_{\omega,\rho} = k^2(\delta I_1)_{\omega,\rho} + 2k\delta k I_1 + (\delta I_2)_{\omega,\rho}. \tag{3.244}$$

We next use Eqs. (3.238) to derive

$$(\delta I_0)_{\omega,\rho} = \int_0^\infty \rho(\delta V^2)_{\omega,\rho}\, dz, \tag{3.245}$$

$$(\delta I_1)_{\omega,\rho} = \int_0^\infty (\delta\mu)V^2\, dz + \int_0^\infty \mu(\delta V^2)_{\omega,\rho}\, dz, \tag{3.246}$$

$$(\delta I_2)_{\omega,\rho} = \int_0^\infty (\delta\mu)\left(\frac{dV}{dz}\right)^2 dz + \int_0^\infty \mu\left[\delta\left(\frac{dV}{dz}\right)^2\right]_{\omega,\rho} dz. \tag{3.247}$$

In contrast, Eq. (3.240) gives

$$\omega^2 \int_0^\infty \rho(\delta V^2)_{\omega,\rho}\, dz = k^2 \int_0^\infty \mu(\delta V^2)_{\omega,\rho}\, dz + \int_0^\infty \mu\left[\delta\left(\frac{dV}{dz}\right)^2\right]_{\omega,\rho} dz. \tag{3.248}$$

Equations (3.244)–(3.248) yield

$$k^2 \int_0^\infty (\delta\mu)V^2\, dz + 2k(\delta k)I_1 + \int_0^\infty (\delta\mu)\left(\frac{dV}{dz}\right)^2 dz = 0. \tag{3.249}$$

Using the relation $\omega = ck$, we find

$$\frac{1}{c}(\delta c)_{\omega,\rho} = -\frac{1}{k}(\delta k)_{\omega,\rho} = \frac{1}{2I_1}\left[\int_0^\infty (\delta\mu)V^2\, dz + \frac{1}{k^2}\int_0^\infty (\delta\mu)\left(\frac{dV}{dz}\right)^2 dz\right]. \tag{3.250}$$

Similarly, it can be shown that

$$\frac{1}{c}(\delta c)_{\omega,\mu} = -\frac{c^2}{2I_1}\int_0^\infty (\delta\rho)V^2\,dz \tag{3.251}$$

$$= -\frac{c}{2UI_0}\int_0^\infty (\delta\rho)V^2\,dz, \tag{3.252}$$

$$\frac{1}{c}(\delta c)_{k,\rho} = \frac{U}{2cI_1}\left[\int_0^\infty (\delta\mu)V^2\,dz + \frac{1}{k^2}\int_0^\infty (\delta\mu)\left(\frac{dV}{dz}\right)^2 dz\right], \tag{3.253}$$

$$\frac{1}{c}(\delta c)_{k,\mu} = -\frac{1}{2I_0}\int_0^\infty (\delta\rho)V^2\,dz. \tag{3.254}$$

3.9.2. Layered Media

In a multilayered structure of homogeneous layers, μ and ρ are constants in each layer, whereas V varies. The energy integrals therefore assume the form

$$I_0 = \sum_{j=1}^n \rho_j D_j, \qquad I_1 = \sum_{j=1}^n \mu_j D_j, \qquad I_2 = \sum_{j=1}^n \mu_j S_j,$$
$$D_j = \int_{z_{j-1}}^{z_j} V^2\,dz, \qquad S_j = \int_{z_{j-1}}^{z_j} \left(\frac{dV}{dz}\right)^2 dz. \tag{3.255}$$

If we perturb the parameters only in the mth layer, Eq. (3.240) becomes

$$\omega^2 \sum_{j=1}^n \rho_j \left(\frac{\partial D_j}{\partial k}\right)_{\omega,\mu,\rho} = k^2 \sum_{j=1}^n \mu_j \left(\frac{\partial D_j}{\partial k}\right)_{\omega,\mu,\rho} + \sum_{j=1}^n \mu_j \left(\frac{\partial S_j}{\partial k}\right)_{\omega,\mu,\rho},$$
$$\omega^2 \sum_{j=1}^n \rho_j \left(\frac{\partial D_j}{\partial \mu_m}\right)_{\omega,k,\rho} = k^2 \sum_{j=1}^n \mu_j \left(\frac{\partial D_j}{\partial \mu_m}\right)_{\omega,k,\rho} + \sum_{j=1}^n \mu_j \left(\frac{\partial S_j}{\partial \mu_m}\right)_{\omega,k,\rho}, \tag{3.256}$$

etc. The reader may then easily verify the following relations:

$$\left(\frac{\partial c}{\partial \mu_m}\right)_{\omega,\rho} = \frac{c}{k^2}\frac{(k^2 D_m + S_m)}{2I_1} = \frac{c}{U}\left(\frac{\partial c}{\partial \mu_m}\right)_{k,\rho},$$

$$\left(\frac{\partial c}{\partial \rho_m}\right)_{\omega,\mu} = -\frac{c^3 D_m}{2I_1} = \frac{c}{U}\left(\frac{\partial c}{\partial \rho_m}\right)_{k,\mu},$$

$$\left(\frac{\partial c}{\partial \beta_m}\right)_{\omega,\rho} = \frac{c}{2I_1}\left[2\beta_m\rho_m D_m + 2\frac{\beta_m\rho_m}{k^2}S_m\right],$$

$$\left(\frac{\partial c}{\partial \beta_m}\right)_{\omega,\mu} = \frac{c}{2I_1}\left[\beta_m^2 D_m + \frac{\beta_m^2}{k^2}S_m - c^2 D_m\right],$$

$$\left(\frac{\partial U}{\partial \mu_m}\right)_{\omega,\rho} = \frac{1}{cI_0}\left[D_m + \sum_{j=1}^n \mu_j\left(\frac{\partial D_j}{\partial \mu_m}\right)_{\omega,\rho}\right] - \frac{U}{I_0}\sum_{j=1}^n \rho_j\left(\frac{\partial D_j}{\partial \mu_m}\right)_{\omega,\rho} - \frac{U}{c}\left(\frac{\partial c}{\partial \mu_m}\right)_{\omega,\rho},$$

$$\left(\frac{\partial U}{\partial \rho_m}\right)_{\omega,\mu} = \frac{1}{cI_0}\sum_{j=1}^{n}\mu_j\left(\frac{\partial D_j}{\partial \rho_m}\right)_{\omega,\mu} - \frac{U}{I_0}\left[D_m + \sum_{j=1}^{n}\rho_j\left(\frac{\partial D_j}{\partial \rho_m}\right)_{\omega,\mu}\right] - \frac{U}{c}\left(\frac{\partial c}{\partial \rho_m}\right)_{\omega,\mu},$$

$$\left(\frac{\partial U}{\partial \beta_m}\right)_{\omega,\rho} = \frac{2\beta_m\rho_m}{cI_0}\left[D_m + \sum_{j=1}^{n}\mu_j\left(\frac{\partial D_j}{\partial \mu_m}\right)_{\omega,\rho}\right] - \frac{2\beta_m\rho_m U}{I_0}\sum_{j=1}^{n}\rho_j\left(\frac{\partial D_j}{\partial \mu_m}\right)_{\omega,\rho} - \frac{U}{c}\left(\frac{\partial c}{\partial \beta_m}\right)_{\omega,\rho},$$

$$\left(\frac{\partial U}{\partial \rho_m}\right)_{\omega,\beta} = \frac{1}{cI_0}\left\{\beta_m^2 D_m + \sum_{j=1}^{n}\mu_j\left[\left(\frac{\partial D_j}{\partial \rho_m}\right)_{\omega,\mu} + \beta_m^2\left(\frac{\partial D_j}{\partial \mu_m}\right)_{\omega,\rho}\right]\right\} - \frac{U}{I_0}\left\{D_m + \sum_{j=1}^{n}\rho_j\left[\left(\frac{\partial D_j}{\partial \rho_m}\right)_{\omega,\mu} + \beta_m^2\left(\frac{\partial D_j}{\partial \mu_m}\right)_{\omega,\rho}\right]\right\} - \frac{U}{c}\left(\frac{\partial c}{\partial \rho_m}\right)_{\omega,\beta},$$

$$\left(\frac{\partial A}{\partial \mu_m}\right)_{\omega,\rho} = -\frac{A}{I_0}\left[D_m + \sum_{j=1}^{n}\mu_j\left(\frac{\partial D_j}{\partial \mu_m}\right)_{\omega,\rho}\right],$$

$$\left(\frac{\partial A}{\partial \rho_m}\right)_{\omega,\mu} = -\frac{A}{I_1}\sum_{j=1}^{n}\mu_j\left(\frac{\partial D_j}{\partial \rho_m}\right)_{\omega,\mu},$$

$$\left(\frac{\partial A}{\partial \beta_m}\right)_{\omega,\rho} = -\frac{2\beta_m\rho_m A}{I_1}\left[D_m + \sum_{j=1}^{n}\mu_j\left(\frac{\partial D_j}{\partial \mu_m}\right)_{\omega,\rho}\right],$$

$$\left(\frac{\partial A}{\partial \rho_m}\right)_{\omega,\beta} = -\frac{A}{I_1}\left\{\beta_m^2\left[D_m + \sum_{j=1}^{n}\mu_j\left(\frac{\partial D_j}{\partial \mu_m}\right)_{\omega,\rho}\right] + \sum_{j=1}^{n}\mu_j\left(\frac{\partial D_j}{\partial \rho_m}\right)_{\omega,\mu}\right\}, \tag{3.257}$$

where A is the amplitude response.

EXAMPLE 3.1: Perturbation in the Case of a Layer Over a Half-Space

The Love-wave energy integrals for this case are

$$I_0 = \frac{1}{2}\rho_1 H\left[1 + \cos\theta_1\frac{\sin\theta_1}{\theta_1}\right] + \frac{1}{2}\rho_2 H\frac{\cos^2\theta_1}{\theta_2}, \tag{3.1.1}$$

$$I_1 = \frac{1}{2}\mu_1 H\left[1 + \cos\theta_1\frac{\sin\theta_1}{\theta_1}\right] + \frac{1}{2}\mu_2 H\frac{\cos^2\theta_1}{\theta_2}, \tag{3.1.2}$$

$$I_2 = \frac{1}{2}\mu_1\frac{\theta_1^2}{H}\left[1 - \cos\theta_1\frac{\sin\theta_1}{\theta_1}\right] + \frac{1}{2}\frac{\mu_2}{H}\theta_2\cos^2\theta_1, \tag{3.1.3}$$

$$\theta_1 = kH\left[\frac{c_L^2}{\beta_1^2} - 1\right]^{1/2}, \qquad \theta_2 = kH\left[1 - \frac{c_L^2}{\beta_2^2}\right]^{1/2}. \tag{3.1.4}$$

Using the period equation $\tan\theta_1 = (\mu_2/\mu_1)(\theta_2/\theta_1)$, the expressions for I_0 and I_1 can be simplified to

$$I_0 = \frac{H}{2}\left[\rho_1 + \frac{\cos^2\theta_1}{\theta_1}\left(\rho_1\frac{\mu_2}{\mu_1}\frac{\theta_2}{\theta_1} + \rho_2\frac{\theta_1}{\theta_2}\right)\right], \tag{3.1.5}$$

$$I_1 = \frac{H}{2}\left[\mu_1 + \frac{\cos^2\theta_1}{\theta_1}\mu_2\left(\frac{\theta_2}{\theta_1} + \frac{\theta_1}{\theta_2}\right)\right]. \tag{3.1.6}$$

Then by Eqs. (3.257)

$$\frac{\mu_1}{c}\left(\frac{\partial c}{\partial \mu_1}\right)_\omega = \frac{\left(1 + \dfrac{\cos^2\theta_1}{\theta_1}\dfrac{\mu_2}{\mu_1}\dfrac{\theta_2}{\theta_1}\right) + \dfrac{\theta_1^2}{k^2H^2}\left(1 - \dfrac{\cos^2\theta_1}{\theta_1}\dfrac{\mu_1}{\mu_2}\dfrac{\theta_2}{\theta_1}\right)}{2\left[1 + \dfrac{\cos^2\theta_1}{\theta_1}\dfrac{\mu_2}{\mu_1}\left(\dfrac{\theta_2}{\theta_1} + \dfrac{\theta_1}{\theta_2}\right)\right]}, \tag{3.1.7}$$

$$\frac{\rho_1}{c}\left(\frac{\partial c}{\partial \rho_1}\right)_\omega = -\left(\frac{c^2}{\beta_1^2}\right)\frac{\left(1 + \dfrac{\cos^2\theta_1}{\theta_1}\dfrac{\mu_2}{\mu_1}\dfrac{\theta_2}{\theta_1}\right)}{2\left[1 + \dfrac{\cos^2\theta_1}{\theta_1}\dfrac{\mu_2}{\mu_1}\left(\dfrac{\theta_2}{\theta_1} + \dfrac{\theta_1}{\theta_2}\right)\right]}. \tag{3.1.8}$$

These equations and Table 3.2 have been used to obtain Fig. 3.10.

Table 3.2. Perturbation of Love-Wave Spectra in a Two-Layered Half-Space

H (km)	β (km/s)	ρ (g/cm^3)	μ (10^{10} dyne/cm^2)
Standard			
40	3.6	2.8	36.288
∞	4.5	3.3	66.825
Perturbed			
40	3.8	3.0	43.320
∞	4.5	3.3	66.825

T (s)	c (km/s)	U (km/s)	A, 10^{-15} (μm/dyne)
120.13	4.4550	4.3677	0.2591
113.90	4.4500	4.3534	0.2896
103.83	4.4400	4.3250	0.3514
95.99	4.4300	4.2970	0.4145
89.65	4.4200	4.2694	0.4788
79.90	4.4000	4.2154	0.6108
69.64	4.3700	4.1377	0.8167
64.51	4.3500	4.0883	0.9587
55.06	4.3000	3.9733	1.3276
48.35	4.2500	3.8713	1.7119
43.16	4.2000	3.7825	2.1059
38.90	4.1500	3.7068	2.5043
35.24	4.1000	3.6440	2.9024
28.99	4.0000	3.5544	3.6820
23.45	3.9000	3.5089	4.4228
18.01	3.8000	3.5027	5.1204
11.96	3.7000	3.5324	5.8013
6.70	3.6350	3.5711	6.3047
3.45	3.6100	3.5909	6.5887

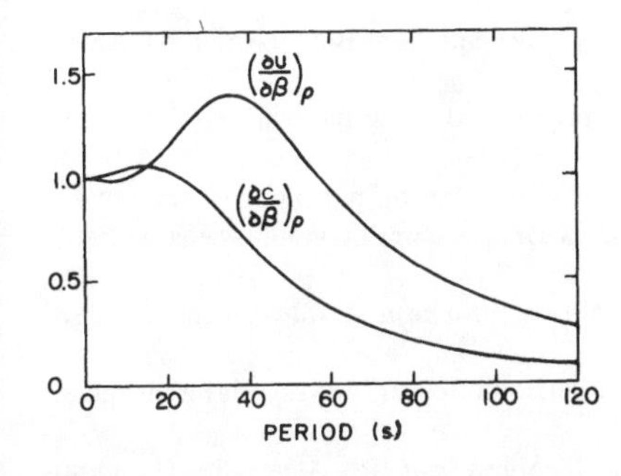

Figure 3.10. First derivatives of the phase, c, and group, U, velocities of the fundamental Love-wave mode with respect to the surface-layer shear velocity.

Bibliography

Alsop LE (1970) The leaky-mode period equation—a plane-wave approach. Bull Seismol Soc Amer 60: 1989–1998.

Alsop LE, Goodman AS, Gregersen S (1974) Reflection and transmission of inhomogeneous waves with particular application to Rayleigh waves. Bull Seismol Soc Amer 64: 1635–1652.

Båth M (1954) The density ratio at the boundary of the earth's core. Tellus 6: 408–414.

Biot, MA (1957) General theorems on the equivalence of group velocity and energy transport. Phys Rev 105: 1129–1137.

Costain JK, Cook KL, Algermissen ST (1963) Amplitude, energy and phase angles of plane *SV* waves and their application to earth crustal studies. Bull Seismol Soc Amer 53: 1039–1074; Corrigendum 55: 567–575, 1965.

Friedlander FG (1948) On the total reflection of plane waves. Quart Jour Mech Appl Math 1: 376–384.

Gilbert F, Backus GE (1966) Propagator matrices in elastic wave and vibration problems. Geophysics 31: 326–332.

Harkrider DG (1968) The perturbation of Love wave spectra. Bull Seismol Soc Amer 58: 861–880.

Haskell NA (1953) The dispersion of surface waves on multilayered media. Bull Seismol Soc Amer 43: 17–34.

Haskell NA (1962) Crustal reflection of plane *P*- and *SV*-waves. Jour Geophys Res 67: 4751–4767.

Havelock TH (1972) The Propagation of Disturbances in Dispersive Media. New York, N.Y., Hafner Publishing Co. p. 87.

Jeffreys H (1961) Small corrections in the theory of surface waves. Geophys Jour Roy Astron Soc (London) 6: 115–117.

Knopoff L, Fredricks RW, Gangi AF, Porter LD (1957) Surface amplitudes of reflected body waves. Geophysics 22: 842–847.

Knott CG (1899) Reflection and refraction of elastic waves with seismological applications. Phil Mag 48: 64–97.

Love AEH (1911) Some Problems of Geodynamics. Cambridge, Engl., Cambridge, at the University Press.

McCamy K, Meyer RP, Smith TJ (1962) Generally applicable solutions of Zoeppritz' amplitude equations. Bull Seismol Soc Amer 52: 923–955.

Meissner E (1926) Elastische Oberflächen-Querwellen. Proc Second Internatl Congr Appl Mech, Zurich, 3–11.

Pilant WL (1972) Complex roots of the Stoneley-wave equation. Bull Seismol Soc Amer 62: 285–299.

Rayleigh (Lord), (JW Strutt) (1885) On waves propagated along the plane surface of an elastic solid. Proc Lond Math Soc (Ser 1) 17: 4–11.

Singh SJ, Ben-Menahem A, Shimshoni M (1970) Comments on papers by Costain *et al.* [1963] and McCamy *et al.* [1962] on the solution of Zoeppritz' amplitude equations. Bull Seismol Soc Amer 60: 277–280.

Thomson WT (1950) Transmission of elastic waves through a stratified medium. Jour Appl Phys 21: 89–93.

Thrower EN (1965) The computation of the dispersion of elastic waves in layered media. Jour Sound Vib 2: 210–226.

Woodhouse JH (1976) On Rayleigh's principle. Geophys Jour Roy Astron Soc (London) 46: 11–22.

Zoeppritz K (1919) Über Reflection und Durchgang seismischer Wellen durch Unstetigkeitsflächen. Erdbebenwellen VIIb, Göttingen, Nachrichten 1: 66–84.

CHAPTER 4

Representation of Seismic Sources

If, for the study of earth physics, it becomes necessary to resort to complex theoretical devices, then the earth is to blame, not the theoretician.

(*Harold Jeffreys*)

On 26 November, 1849 *G. G. Stokes*, then Fellow of Pembroke College and Lucasian Professor of Mathematics at the University of Cambridge, read his paper on the dynamical theory of diffraction. As a model for a light source in the luminiferous ether, he chose a tangential force in an infinite elastic solid. Drawing on earlier results of *Poisson* (see Bibliography, Chapter 1) he obtained an exact solution for the displacement field caused by a single force in an infinite elastic medium. Without knowing it, he had conceived the first mathematical model of an earthquake.

For half a century no follow-up occurred, but the great shocks that jolted the globe with repeated fervor at the turn of the century also shook the dust off Stokes' old manuscript, and on 11 June, 1903, *H. Lamb* presented his "problem" before the Royal Society of London "On the Propagation of Tremors Over the Surface of an Elastic Solid." Soon afterwards, on 12 November, 1903, *A. E. H. Love* read his paper before the Mathematical Society of London on "The Propagation of Wave-Motion in an Isotropic Elastic Solid Medium," therein extending Stokes' results to an arbitrary initial disturbance and a wider class of body forces.

4.1. A Concentrated Force in a Homogeneous Medium

4.1.1. The Stokes–Love Solution

Having found the solution of the force-free Navier equation, we shall proceed to derive its solution for an arbitrary force distribution. To this end, we return to the Navier equation

$$\alpha^2 \operatorname{grad} \operatorname{div} \mathbf{u} - \beta^2 \operatorname{curl} \operatorname{curl} \mathbf{u} + \mathbf{F} = \frac{\partial^2 \mathbf{u}}{\partial t^2}, \tag{4.1}$$

where $\mathbf{F}$ represents the body force per unit mass. In the case of a force of magnitude F_0 concentrated at the point $\mathbf{r} = \mathbf{r}_0$, we may take

$$\rho \mathbf{F} = \mathbf{a} F_0 g(t) \delta(\mathbf{r} - \mathbf{r}_0), \tag{4.2}$$

where $\mathbf{a}$ is a unit vector, the function $g(t)$ is dimensionless, and $\delta(\mathbf{r} - \mathbf{r}_0)$ is the three-dimensional delta function. The singularity defined by Eq. (4.2) is known as a *single force* or a *concentrated force*. Substituting in Eq. (4.1) and then taking its Fourier transform, we get

$$\alpha^2 \text{ grad div } \mathbf{u} - \beta^2 \text{ curl curl } \mathbf{u} + \omega^2 \mathbf{u} = -\mathbf{a} \frac{F_0}{\rho} g(\omega) \delta(\mathbf{r} - \mathbf{r}_0), \tag{4.3}$$

where $g(\omega)$ is the Fourier transform[1] of $g(t)$. Noting that

$$\mathbf{a}\delta(\mathbf{r} - \mathbf{r}_0) = -\frac{\mathbf{a}}{4\pi} \nabla^2 \frac{1}{R} = -\frac{1}{4\pi}\left[\text{grad div}\left(\frac{\mathbf{a}}{R}\right) - \text{curl curl}\left(\frac{\mathbf{a}}{R}\right)\right], \tag{4.4}$$

and assuming

$$\mathbf{u} = F_0 g(\omega)[\text{grad div}(\mathbf{a} S_\alpha) - \text{curl curl}(\mathbf{a} S_\beta)], \tag{4.5}$$

we see that Eq. (4.3) is satisfied if S_α and S_β satisfy the equations

$$\nabla^2 S_\alpha + k_\alpha^2 S_\alpha = \frac{1}{4\pi\alpha^2 \rho R}, \qquad \nabla^2 S_\beta + k_\beta^2 S_\beta = \frac{1}{4\pi\beta^2 \rho R}, \tag{4.6}$$

where

$$k_\alpha = \frac{\omega}{\alpha}, \qquad k_\beta = \frac{\omega}{\beta}, \qquad \mathbf{R} = \mathbf{r} - \mathbf{r}_0.$$

Particular solutions of Eq. (4.6) such that $S_\alpha(\infty, \omega) = 0$, $S_\beta(\infty, \omega) = 0$, are

$$S_\alpha(R, \omega) = \frac{1}{4\pi\omega^2 \rho}\left(\frac{1 - e^{-ik_\alpha R}}{R}\right), \qquad S_\beta(R, \omega) = \frac{1}{4\pi\omega^2 \rho}\left(\frac{1 - e^{-ik_\beta R}}{R}\right). \tag{4.7}$$

Substituting in Eq. (4.5) and using the identity

$$(\nabla^2 + k_c^2)\left(\frac{e^{-ik_c R}}{R}\right) = -4\pi\delta(\mathbf{r} - \mathbf{r}_0),$$

we obtain

$$\mathbf{u} = \frac{F_0 g(\omega)}{4\pi\mu}\left[\frac{e^{-ik_\beta R}}{R}\mathbf{a} + \text{grad.div}\left(\frac{e^{-ik_\beta R} - e^{-ik_\alpha R}}{k_\beta^2 R}\mathbf{a}\right)\right] \tag{4.8a}$$

$$= \frac{F_0 g(\omega)}{4\pi\mu}\left[\mathfrak{I}\frac{e^{-ik_\beta R}}{R} + \text{grad grad}\left(\frac{e^{-ik_\beta R} - e^{-ik_\alpha R}}{k_\beta^2 R}\right)\right]\cdot \mathbf{a} \tag{4.8b}$$

$$= \frac{F_0 g(\omega)}{4\pi\mu}\left[\mathfrak{I}\frac{e^{-ik_\beta R}}{R} + \text{grad div}\left(\mathfrak{I}\frac{e^{-ik_\beta R} - e^{-ik_\alpha R}}{k_\beta^2 R}\right)\right]\cdot \mathbf{a} \tag{4.8c}$$

where $\mathfrak{I} \equiv \delta_{ij}$ is the unit dyadic.

[1] The letter g is used here for two *different* functions.

The expression

$$\mathfrak{G}(\mathbf{r}|\mathbf{r}_0;\omega) = \left[\mathfrak{J}\frac{e^{-ik_\beta R}}{R} + \text{grad grad}\left(\frac{e^{-ik_\beta R} - e^{-ik_\alpha R}}{k_\beta^2 R}\right)\right] \tag{4.9}$$

is known as the *spectral Green's dyadic*. Because

$$\mathbf{u} = F_0 g(\omega)\mathfrak{G}(\mathbf{r}|\mathbf{r}_0;\omega)\cdot\mathbf{a}, \tag{4.10}$$

and $\mathbf{u}$ satisfies Eq. (4.3), the spectral Green's dyadic must satisfy the equation

$$\rho\alpha^2 \text{ grad div } \mathfrak{G} - \rho\beta^2 \text{ curl curl } \mathfrak{G} + \rho\omega^2\mathfrak{G} = -\mathfrak{J}\delta(\mathbf{r} - \mathbf{r}_0). \tag{4.11}$$

With the aid of the identities $\text{div}(\mathfrak{J}\phi) = \text{grad } \phi$, $\text{curl}(\mathfrak{J}\phi) = (\text{grad } \phi) \times \mathfrak{J}$ it is immediately seen from Eq. (4.9) that

$$\text{div } \mathfrak{G} = \frac{1}{4\pi(\lambda + 2\mu)} \text{grad}\left(\frac{e^{-ik_\alpha R}}{R}\right), \qquad \text{curl } \mathfrak{G} = \frac{1}{4\pi\mu}\left[\text{grad}\left(\frac{e^{-ik_\beta R}}{R}\right) \times \mathfrak{J}\right].$$

The expression, Eq. (4.9), for the Green's dyadic may be written in the form

$$\begin{aligned}\mathfrak{G} = &-\frac{ik_\alpha}{12\pi(\lambda + 2\mu)}\left[\mathfrak{J}h_0^{(2)}(k_\alpha R) + \left(\mathfrak{J} - \frac{3}{R^2}\mathbf{RR}\right)h_2^{(2)}(k_\alpha R)\right]\\ &+ \frac{ik_\beta}{12\pi\mu}\left[-2\mathfrak{J}h_0^{(2)}(k_\beta R) + \left(\mathfrak{J} - \frac{3}{R^2}\mathbf{RR}\right)h_2^{(2)}(k_\beta R)\right] = \mathfrak{G}_\alpha + \mathfrak{G}_\beta.\end{aligned} \tag{4.12}$$

Note that $\mathfrak{G}$ has the dimensions of (displacement $\times$ time)/force.

It is sometimes advantageous to work in terms of cylindrical coordinates. Let us suppose that the force acts at the point $(0, 0, h)$ and the x_3 axis is drawn vertically upward. We then have

$$x_1 = \Delta\cos\phi, \qquad x_2 = \Delta\sin\phi, \qquad x_3 = z, \qquad R^2 = \Delta^2 + (z - h)^2. \tag{4.13}$$

If the force is acting along the x_3 axis, as shown in Fig. 4.1, there is then axial symmetry, and we obtain from Eq. (4.8a)

$$\begin{aligned}u_\Delta^3 &= \frac{F_0 g(\omega)}{4\pi\mu}\left[\frac{1}{k_\beta^2}\frac{\partial^2}{\partial\Delta\,\partial x_3}\left\{\frac{1}{R}(e^{-ik_\beta R} - e^{-ik_\alpha R})\right\}\right],\\ u_\phi^3 &= 0,\\ u_z^3 &= \frac{F_0 g(\omega)}{4\pi\mu}\left[\frac{1}{k_\beta^2}\frac{\partial^2}{\partial x_3^2}\left\{\frac{1}{R}(e^{-ik_\beta R} - e^{-ik_\alpha R})\right\} + \frac{1}{R}e^{-ik_\beta R}\right].\end{aligned} \tag{4.14}$$

Similarly, if the force is acting at $(0, 0, h)$ parallel to the x_1 axis, we have

$$u_\Delta^1 = \frac{F_0 g(\omega)}{4\pi\mu}\cos\phi\left[\frac{1}{k_\beta^2}\frac{\partial^2}{\partial\Delta^2}\left\{\frac{1}{R}(e^{-ik_\beta R} - e^{-ik_\alpha R})\right\} + \frac{1}{R}e^{-ik_\beta R}\right], \tag{4.15}$$

$$u_\phi^1 = -\frac{F_0 g(\omega)}{4\pi\mu}\sin\phi\left[\frac{1}{k_\beta^2}\frac{1}{\Delta}\frac{\partial}{\partial\Delta}\left\{\frac{1}{R}(e^{-ik_\beta R} - e^{-ik_\alpha R})\right\} + \frac{1}{R}e^{-ik_\beta R}\right], \tag{4.16}$$

$$u_z^1 = \frac{F_0 g(\omega)}{4\pi\mu}\cos\phi\left[\frac{1}{k_\beta^2}\frac{\partial^2}{\partial\Delta\,\partial z}\left\{\frac{1}{R}(e^{-ik_\beta R} - e^{-ik_\alpha R})\right\}\right]. \tag{4.17}$$

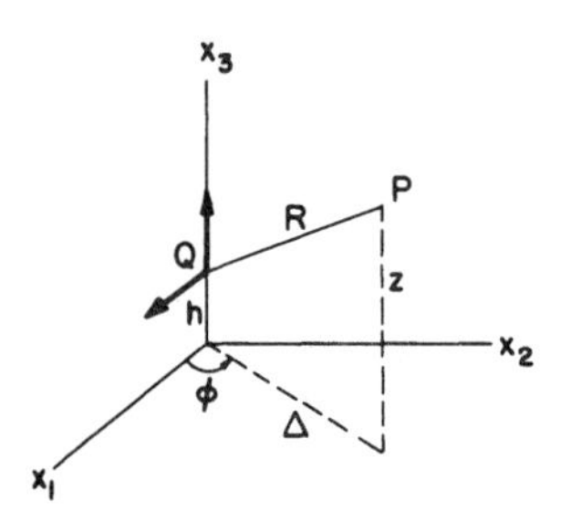

Figure 4.1. A vertical and a horizontal concentrated force.

Equations (4.15)–(4.17) are obtained immediately by observing that

$$\text{grad div}[F(R)\mathbf{e}_1] = \text{grad}\left[\frac{\partial F(R)}{\partial x_1}\right] = \text{grad}\left[\cos\phi\,\frac{\partial F(R)}{\partial \Delta}\right]$$

$$= \cos\phi\,\frac{\partial^2 F}{\partial \Delta^2}\,\mathbf{e}_\Delta - \sin\phi\,\frac{1}{\Delta}\frac{\partial F}{\partial \Delta}\,\mathbf{e}_\phi + \cos\phi\,\frac{\partial^2 F}{\partial z\,\partial \Delta}\,\mathbf{e}_z. \quad (4.18)$$

EXAMPLE 4.1

Prove that for a concentrated force $Fe^{i\omega t}$ acting at the point $Q(y_i)$ in the x_j direction, the x_i component of the displacement at the point $P(x_i)$ is given by

$$u_i^j(P, Q; \omega) = -\frac{iFk_\beta}{12\pi\mu}\,e^{i\omega t}\left[\left(\frac{\beta}{\alpha}\right)^3\left\{\delta_{ij}h_0^{(2)}(k_\alpha R) + \left(\delta_{ij} - 3\,\frac{X_iX_j}{R^2}\right)h_2^{(2)}(k_\alpha R)\right\}\right.$$

$$\left. + 2\delta_{ij}h_0^{(2)}(k_\beta R) - \left(\delta_{ij} - 3\,\frac{X_iX_j}{R^2}\right)h_2^{(2)}(k_\beta R)\right],$$

where $R^2 = X_1^2 + X_2^2 + X_3^2$, $X_i = x_i - y_i$.

4.1.2. Time-Domain Solution

To find the response in the time domain, we make use of the convolution theorem for the Fourier transform. According to this theorem, if $f(\omega)$ and $f_1(\omega)$ are the Fourier transforms of $f(t)$ and $f_1(t)$, respectively, then $f(\omega)f_1(\omega)$ is the Fourier transform of the convolution integral

$$\int_{-\infty}^{\infty} f(t')f_1(t - t')dt'.$$

If $f_1(t)$ vanishes for $t < 0$, the above integral becomes

$$\int_{-\infty}^{t} f(t')f_1(t - t')dt'. \quad (4.19)$$

We take $f_1(\omega) = \omega^{-2}$, so that $f_1(t) = -tH(t)$. Therefore, from Eq. (4.19), $\omega^{-2}f(\omega)$ is the Fourier transform of

$$-\int_{-\infty}^{t} f(t')(t - t')dt'. \tag{4.20}$$

The inverse Fourier transform of $g(\omega)\exp(-ik_\alpha R)$ is $g(t - R/\alpha)$. Therefore, from Eqs. (4.8d) and (4.20), we obtain the following expression for the field caused by a single force in the time domain, $R \neq 0$

$$\begin{aligned}\mathbf{u}(\mathbf{r}, t) &= \frac{F_0}{4\pi\rho}\left[\text{grad div}\left\{\frac{\mathbf{a}}{R}\,\hat{g}\left(t - \frac{R}{\alpha}\right)\right\} - \text{curl curl}\left\{\frac{\mathbf{a}}{R}\,\hat{g}\left(t - \frac{R}{\beta}\right)\right\}\right] \\ &= \frac{F_0}{4\pi\rho}\left[\text{grad div}\left\{\frac{\mathbf{a}}{R}\,\hat{g}\left(t - \frac{R}{\alpha}\right) - \frac{\mathbf{a}}{R}\,\hat{g}\left(t - \frac{R}{\beta}\right)\right\} + \nabla^2\left\{\frac{\mathbf{a}}{R}\,\hat{g}\left(t - \frac{R}{\beta}\right)\right\}\right],\end{aligned} \tag{4.21}$$

where

$$\hat{g}(t) = \int_{-\infty}^{t} g(t')(t - t')dt'. \tag{4.22}$$

We can show that

$$\begin{aligned}\text{grad div}\left\{\frac{\mathbf{a}}{R}\,g\left(t - \frac{R}{\alpha}\right)\right\} &= \frac{\mathbf{R}(\mathbf{R}\cdot\mathbf{a})}{\alpha^2 R^3}\,g\left(t - \frac{R}{\alpha}\right) + \left[3\,\frac{\mathbf{R}(\mathbf{R}\cdot\mathbf{a})}{R^5} - \frac{\mathbf{a}}{R^3}\right]\left[\hat{g}\left(t - \frac{R}{\alpha}\right)\right. \\ &\quad \left. + \frac{R}{\alpha}\,\hat{g}'\left(t - \frac{R}{\alpha}\right)\right],\end{aligned} \tag{4.23}$$

where the prime denotes differentiation with respect to the argument. There is also the relation, for $R \neq 0$,

$$\nabla^2\left\{\frac{\mathbf{a}}{R}\,\hat{g}\left(t - \frac{R}{\beta}\right)\right\} = \frac{\mathbf{a}}{\beta^2 R}\frac{\partial^2}{\partial t^2}\,\hat{g}\left(t - \frac{R}{\beta}\right) = \frac{\mathbf{a}}{\beta^2 R}\,g\left(t - \frac{R}{\beta}\right). \tag{4.24}$$

Equations (4.21)–(4.24) yield

$$\begin{aligned}\mathbf{u} = \frac{F_0}{4\pi\rho}&\left[\frac{\mathbf{R}(\mathbf{R}\cdot\mathbf{a})}{R^3}\left\{\frac{1}{\alpha^2}\,g\left(t - \frac{R}{\alpha}\right) - \frac{1}{\beta^2}\,g\left(t - \frac{R}{\beta}\right)\right\} + \frac{\mathbf{a}}{\beta^2 R}\,g\left(t - \frac{R}{\beta}\right)\right. \\ &+ \left\{3\,\frac{\mathbf{R}(\mathbf{R}\cdot\mathbf{a})}{R^5} - \frac{\mathbf{a}}{R^3}\right\}\left\{\hat{g}\left(t - \frac{R}{\alpha}\right) + \frac{R}{\alpha}\,\hat{g}'\left(t - \frac{R}{\alpha}\right) - \hat{g}\left(t - \frac{R}{\beta}\right)\right. \\ &\left.\left. - \frac{R}{\beta}\,\hat{g}'\left(t - \frac{R}{\beta}\right)\right\}\right].\end{aligned} \tag{4.25}$$

It may be seen from Eq. (4.22) that

$$\hat{g}\left(t - \frac{R}{\alpha}\right) + \frac{R}{\alpha}\,\hat{g}'\left(t - \frac{R}{\alpha}\right) = \int_{R/\alpha}^{\infty} g(t - t')t'\,dt'. \tag{4.26}$$

Equation (4.25) now becomes

$$\mathbf{u} = \frac{F_0}{4\pi\rho R}\left[\frac{\mathbf{R}(\mathbf{R}\cdot\mathbf{a})}{R^2}\left\{\frac{1}{\alpha^2}g\left(t-\frac{R}{\alpha}\right)-\frac{1}{\beta^2}g\left(t-\frac{R}{\beta}\right)\right\}+\frac{\mathbf{a}}{\beta^2}g\left(t-\frac{R}{\beta}\right)\right.$$
$$\left.+\left\{3\frac{\mathbf{R}(\mathbf{R}\cdot\mathbf{a})}{R^4}-\frac{\mathbf{a}}{R^2}\right\}\int_{R/\alpha}^{R/\beta} g(t-t')t'\,dt'\right]. \tag{4.27}$$

This is the infinite space "seismogram" for a single force $F_0 g(t)$. Equation (4.27) expresses the displacement as the sum of the displacement in the longitudinal and transverse waves.

Equation (4.27) gives

$$\mathbf{u} = F_0\mathfrak{G}(\mathbf{r}|\mathbf{r}_0;t)\cdot\mathbf{a}, \tag{4.28}$$

where $\mathfrak{G}(\mathbf{r}|\mathbf{r}_0;t)$, the Green's dyadic in the time domain, is given by

$$\mathfrak{G}(\mathbf{r}|\mathbf{r}_0;t) = \frac{1}{4\pi\rho R}\left[\frac{\mathbf{RR}}{R^2}\left\{\frac{1}{\alpha^2}g\left(t-\frac{R}{\alpha}\right)-\frac{1}{\beta^2}g\left(t-\frac{R}{\beta}\right)\right\}+\frac{\mathfrak{I}}{\beta^2}g\left(t-\frac{R}{\beta}\right)\right.$$
$$\left.+\left\{3\frac{\mathbf{RR}}{R^4}-\frac{\mathfrak{I}}{R^2}\right\}\int_{R/\alpha}^{R/\beta} g(t-t')t'\,dt'\right]. \tag{4.29}$$

The physical significance of the Green's dyadic is best revealed when we write its components in a Cartesian frame of reference. Denoting (Fig. 4.2)

$$\mathbf{r} = (x_1, x_2, x_3), \qquad \mathbf{r}_0 = (y_1, y_2, y_3), \qquad X_i = x_i - y_i,$$

Eq. (4.29) yields

$$4\pi\rho G_{ij}(\mathbf{r}|\mathbf{r}_0;t) = \frac{1}{R^3}\begin{bmatrix} X_1^2 & X_1X_2 & X_1X_3 \\ X_2X_1 & X_2^2 & X_2X_3 \\ X_3X_1 & X_3X_2 & X_3^2 \end{bmatrix}\left\{\frac{1}{\alpha^2}g\left(t-\frac{R}{\alpha}\right)-\frac{1}{\beta^2}g\left(t-\frac{R}{\beta}\right)\right.$$
$$\left.+\frac{3}{R^2}\int_{R/\alpha}^{R/\beta} g(t-t')t'\,dt'\right\}+\frac{1}{R}\begin{bmatrix} 1 & 0 & 0 \\ 0 & 1 & 0 \\ 0 & 0 & 1 \end{bmatrix}\left\{\frac{1}{\beta^2}g\left(t-\frac{R}{\beta}\right)\right.$$
$$\left.-\frac{1}{R^2}\int_{R/\alpha}^{R/\beta} g(t-t')t'\,dt'\right\}. \tag{4.30}$$

From Eq. (4.28), it is obvious that G_{ij} represents the ith displacement component per unit force caused by a concentrated force in the jth direction. It may be noted that the dyadic $\mathfrak{G}$ satisfies the reciprocity relations

$$G_{ij}(P, Q) = G_{ji}(P, Q) = G_{ij}(Q, P). \tag{4.31}$$

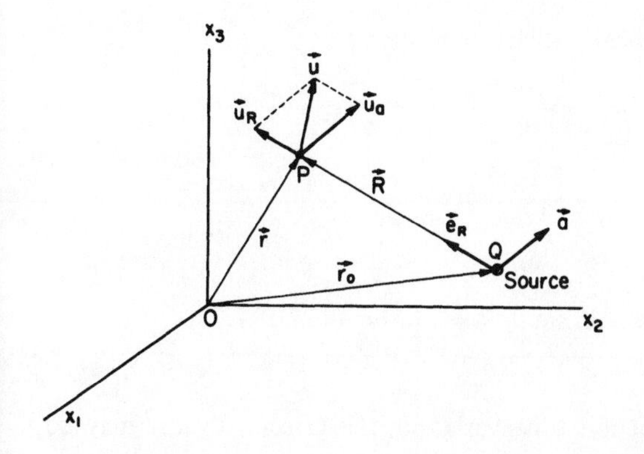

Figure 4.2. The source–observer system.

EXAMPLE 4.2

Prove that for a concentrated force $Fg(t)$ acting at the point $Q(y_i)$ in the x_j direction, the x_i component of the displacement at the point $P(x_i)$ is given by

$$u_i^j(P, Q; t) = \frac{F}{4\pi\rho}\left[\frac{1}{R}\frac{\partial R}{\partial x_i}\frac{\partial R}{\partial x_j}\left\{\frac{1}{\alpha^2}g\left(t - \frac{R}{\alpha}\right) - \frac{1}{\beta^2}g\left(t - \frac{R}{\beta}\right)\right\}\right.$$

$$\left. + \delta_{ij}\frac{1}{\beta^2 R}g\left(t - \frac{R}{\beta}\right) + R^2\frac{\partial^2 R^{-1}}{\partial x_i\,\partial x_j}\int_{1/\alpha}^{1/\beta} g(t - \xi R)\xi\,d\xi\right].$$

Hint: Substitute $\mathbf{a} = \mathbf{e}_j$ in Eq. (4.27) and use the relations

$$R \text{ grad } R = \mathbf{R}, \qquad \text{grad grad}\,\frac{1}{R} = 3\,\frac{\mathbf{RR}}{R^5} - \frac{\mathfrak{I}}{R^3}.$$

The result of the problem follows immediately on making the change of variable $t' = \xi R$ in the integral.

EXAMPLE 4.3

A force of magnitude F and time dependence $g(t)$ acts in the x_3 direction at the origin. Prove that the spherical components of the displacement are given by

$$u_r = \frac{F\cos\theta}{4\pi\rho r}\left[\frac{2}{r^2}\int_{r/\alpha}^{r/\beta} t' g(t - t')dt' + \frac{1}{\alpha^2}g\left(t - \frac{r}{\alpha}\right)\right],$$

$$u_\theta = \frac{F\sin\theta}{4\pi\rho r}\left[\frac{1}{r^2}\int_{r/\alpha}^{r/\beta} t' g(t - t')dt' - \frac{1}{\beta^2}g\left(t - \frac{r}{\beta}\right)\right],$$

$$u_\phi = 0.$$

Hence show that the far-field displacements are:

$$u_r = \frac{F\cos\theta}{4\pi(\lambda+2\mu)r}\,g\left(t-\frac{r}{\alpha}\right),$$

$$u_\theta = -\frac{F\sin\theta}{4\pi\mu r}\,g\left(t-\frac{r}{\beta}\right),$$

$$u_\phi = 0.$$

Hint: Put $\mathbf{a} = \mathbf{e}_3 = \mathbf{e}_r\cos\theta - \mathbf{e}_\theta\sin\theta$, $R = r$ in Eq. (4.27).

EXAMPLE 4.4

Prove that for delta-function time variation, the Green's dyadic may be expressed in the form

$$\mathfrak{G} = \frac{1}{4\pi}\left(\frac{1}{\lambda+2\mu}\mathfrak{G}_\alpha + \frac{1}{\mu}\mathfrak{G}_\beta\right),$$

where

$$\mathfrak{G}_\alpha = \frac{\mathbf{RR}}{R^3}\left[\delta\left(t-\frac{R}{\alpha}\right) + 3\frac{\alpha}{R}H\left(t-\frac{R}{\alpha}\right) + 3\left(\frac{\alpha}{R}\right)^2 F\left(t-\frac{R}{\alpha}\right)\right] - \frac{\mathfrak{I}}{R}\left[\frac{\alpha}{R}H\left(t-\frac{R}{\alpha}\right) + \left(\frac{\alpha}{R}\right)^2 F\left(t-\frac{R}{\alpha}\right)\right],$$

$$\mathfrak{G}_\beta = -\frac{\mathbf{RR}}{R^3}\left[\delta\left(t-\frac{R}{\beta}\right) + 3\frac{\beta}{R}H\left(t-\frac{R}{\beta}\right) + 3\left(\frac{\beta}{R}\right)^2 F\left(t-\frac{R}{\beta}\right)\right] + \frac{\mathfrak{I}}{R}\left[\delta\left(t-\frac{R}{\beta}\right) + \frac{\beta}{R}H\left(t-\frac{R}{\beta}\right) + \left(\frac{\beta}{R}\right)^2 F\left(t-\frac{R}{\beta}\right)\right],$$

and where $\delta(t)$, $H(t)$, and $F(t)$ represent, respectively, the delta function, the Heaviside unit function, and the ramp function $tH(t)$.

4.1.3. Theoretical Seismograms Resulting from a Single Force

We shall use the term "theoretical seismogram" to indicate the explicit dependence of the displacement on time. We can imagine an ideal recorder, placed at a certain point of observation, that records this variation of the displacement with time faithfully, without distortion.

According to Eq. (4.27), the displacement field induced by a single force can be written as a sum of two vectors, one pointing in the direction of the force and the other pointing in the direction of the unit vector $\mathbf{e}_R$ (Fig. 4.2). Therefore,

$$\mathbf{u} = \mathbf{a}u_a + \mathbf{e}_R\,u_R, \tag{4.32}$$

where

$$u_a = \frac{F_0}{4\pi\mu R}\left[g\left(t - \frac{R}{\beta}\right) - \frac{\beta^2}{R^2}\int_{R/\alpha}^{R/\beta} g(t - t')t'\,dt'\right],$$

$$u_R = \frac{F_0(\mathbf{e}_R \cdot \mathbf{a})}{4\pi\mu R}\left[\frac{\beta^2}{\alpha^2} g\left(t - \frac{R}{\alpha}\right) - g\left(t - \frac{R}{\beta}\right) + \frac{3\beta^2}{R^2}\int_{R/\alpha}^{R/\beta} g(t - t')t'\,dt'\right]. \tag{4.33}$$

The asymptotic value as $t \to \infty$ is given by

$$u_a = \frac{F_0\bar{g}}{8\pi\mu R}\left(1 + \frac{\beta^2}{\alpha^2}\right), \tag{4.34}$$

$$u_R = \frac{F_0\bar{g}(\mathbf{e}_R \cdot \mathbf{a})}{8\pi\mu R}\left(1 - \frac{\beta^2}{\alpha^2}\right), \tag{4.35}$$

where $g(t) \to \bar{g}$ as $t \to \infty$.

When $g(t) = H(t)$, we find from Eq. (4.33),

$$u_a = \begin{cases} 0, & t < \dfrac{R}{\alpha}, \\[2ex] -\dfrac{F_0}{8\pi\rho R^3}\left(t^2 - \dfrac{R^2}{\alpha^2}\right), & \dfrac{R}{\alpha} < t < \dfrac{R}{\beta}, \\[2ex] \dfrac{F_0}{8\pi\mu R}\left(1 + \dfrac{\beta^2}{\alpha^2}\right), & t > \dfrac{R}{\beta}; \end{cases} \tag{4.36}$$

$$u_R = \begin{cases} 0, & t < \dfrac{R}{\alpha}, \\[2ex] \dfrac{F_0(\mathbf{e}_R \cdot \mathbf{a})}{4\pi\rho R}\left[\dfrac{1}{\alpha^2} + \dfrac{3}{2R^2}\left(t^2 - \dfrac{R^2}{\alpha^2}\right)\right], & \dfrac{R}{\alpha} < t < \dfrac{R}{\beta}, \\[2ex] \dfrac{F_0(\mathbf{e}_R \cdot \mathbf{a})}{8\pi\mu R}\left(1 - \dfrac{\beta^2}{\alpha^2}\right), & t > \dfrac{R}{\beta}. \end{cases} \tag{4.37}$$

The variation of the dimensionless displacements $(4\pi\mu R/F_0)u_a$ and $[4\pi\mu R/(F_0 a_R)]u_R$ with dimensionless time $(\beta/R)t$ is shown in Figs. 4.3 and 4.4, respectively, for a fixed value of R. As R increases, the bulge ABC decreases, and it practically disappears at long ranges, leaving behind a step.

In the case of a vertical force acting at the origin in the direction of the z axis (for an observer in the xy plane), Eq. (4.36) gives the vertical component (u_z) of the displacement, whereas Eq. (4.37) shows that the horizontal component of the displacement vanishes. Therefore, Fig. (4.3) also represents the variation of the vertical component of the displacement for a point of observation in the xy plane when a vertical force acts at the origin along the z axis.

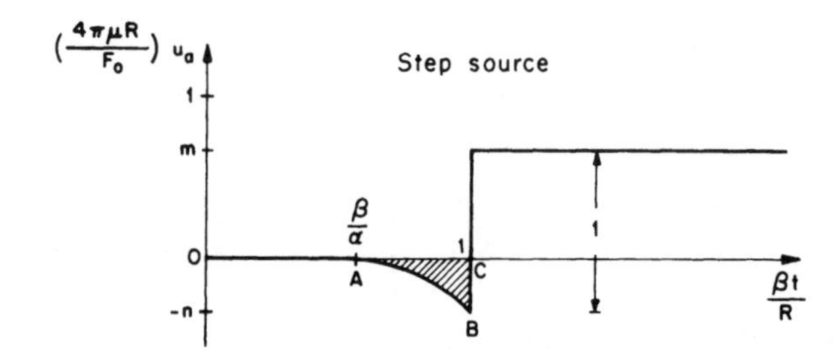

Figure 4.3. The variation of the dimensionless displacement, $(4\pi\mu R/F_0)u_a$, parallel to the concentrated force $F_0H(t)$ with dimensionless time, $(\beta/R)t$, for fixed R. [$m = \frac{1}{2}(1 + \beta^2/\alpha^2)$, $n = \frac{1}{2}(1 - \beta^2/\alpha^2)$.]

Consider next the case of a single force acting at the origin along the x axis. Putting $\mathbf{a} = \mathbf{e}_x$, $R = \Delta$, we find from Eqs. (4.36) and (4.37) the following components of the displacement at a point on the xy plane

$$u_x = \begin{cases} 0, & t < \dfrac{\Delta}{\alpha}, \\[2ex] \dfrac{F_0}{4\pi\rho\Delta}\left[\dfrac{(2 - 3\sin^2\phi)}{2\Delta^2}\left(t^2 - \dfrac{\Delta^2}{\alpha^2}\right) + \dfrac{\cos^2\phi}{\alpha^2}\right], & \dfrac{\Delta}{\alpha} < t < \dfrac{\Delta}{\beta}, \\[2ex] \dfrac{F_0}{4\pi\mu\Delta}\left[1 - \dfrac{1}{2}\sin^2\phi\left(1 - \dfrac{\beta^2}{\alpha^2}\right)\right], & t > \dfrac{\Delta}{\beta}; \end{cases}$$

$$u_y = \begin{cases} 0, & t < \dfrac{\Delta}{\alpha}, \\[2ex] \dfrac{F_0}{8\pi\rho\Delta}\sin 2\phi\left[\dfrac{1}{\alpha^2} + \dfrac{3}{2\Delta^2}\left(t^2 - \dfrac{\Delta^2}{\alpha^2}\right)\right], & \dfrac{\Delta}{\alpha} < t < \dfrac{\Delta}{\beta}, \\[2ex] \dfrac{F_0}{16\pi\mu\Delta}\sin 2\phi\left(1 - \dfrac{\beta^2}{\alpha^2}\right), & t > \dfrac{\Delta}{\beta}; \end{cases} \tag{4.38}$$

$$u_z = 0.$$

The variation of the dimensionless displacement $(4\pi\mu\Delta/F_0)u_x$ with dimensionless time $(\beta/\Delta)t$ is shown in Fig. 4.5a for $\sin^2\phi < \frac{2}{3}$ and in Fig. 4.5b for $\sin^2\phi > \frac{2}{3}$. In the latter case, the curve crosses the time axis at the point $(\beta/\alpha)\sin\phi(3\sin^2\phi - 2)^{-1/2}$ provided $\sin^2\phi > \frac{3}{4}$. When Eq. (4.38) is compared with Eq. (4.37), we find that the behavior of u_y is similar to that of u_R as depicted in Fig. 4.4. The variation of the displacements with time when $g(t) = \delta(t)$ is shown in Fig. 4.6.

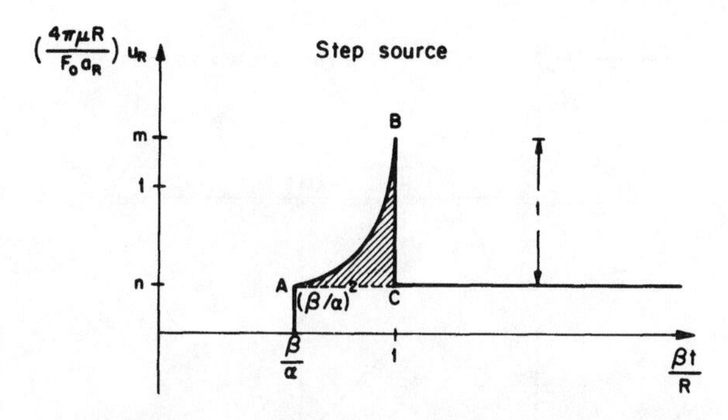

Figure 4.4. The variation of the dimensionless radial displacement, $[4\pi\mu R/(F_0 a_R)]u_R$, caused by the concentrated force $F_0 H(t)$ with dimensionless time, $(\beta/R)t$, for fixed R. $[m = \frac{1}{2}(3 - \beta^2/\alpha^2), n = \frac{1}{2}(1 - \beta^2/\alpha^2).]$ Here, $a_R = \mathbf{e}_R \cdot \mathbf{a}$.

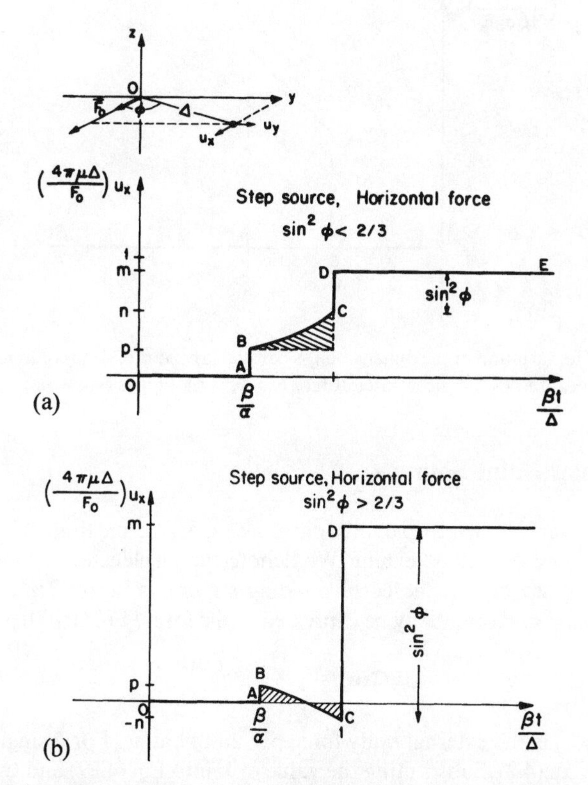

Figure 4.5. The variation of the dimensionless displacement, $(4\pi\mu\Delta/F_0)u_x$, parallel to a concentrated force $F_0 H(t)$ with dimensionless time, $(\beta/\Delta)t$, for fixed Δ and for (a) $\sin^2\phi < 2/3$ and (b) $\sin^2\phi > 2/3$. $[m = 1 - \frac{1}{2}\sin^2\phi(1 - \beta^2/\alpha^2),\ n = 1 - \frac{1}{2}\sin^2\phi(3 - \beta^2/\alpha^2),$ $p = (\beta^2/\alpha^2)\cos^2\phi.]$

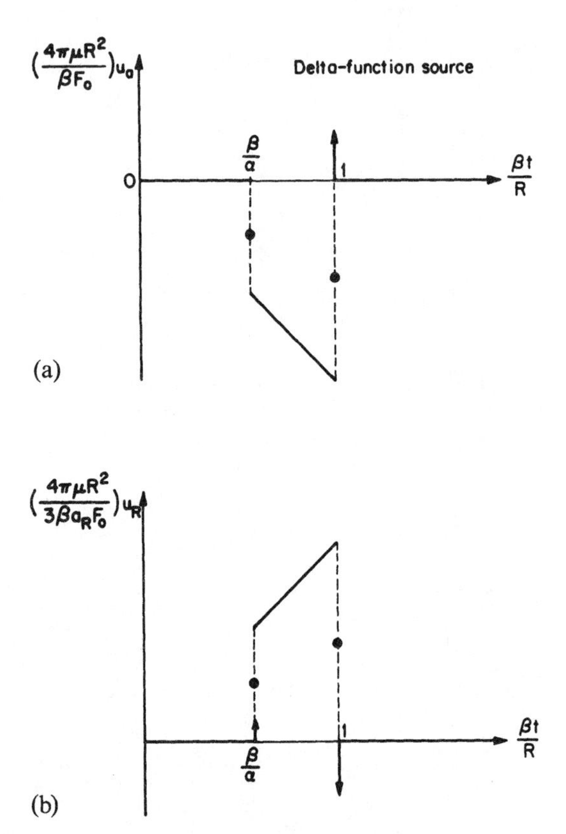

Figure 4.6. The variation of the dimensionless parallel (a) and radial (b) components of the displacement caused by a concentrated force, $F_0\delta(t)$, with dimensionless time.

4.2. Dipolar Point Sources

Let us consider an anisotropic, inhomogeneous, elastic medium. The medium may be of finite or infinite extent. We denote the displacement vector at any point $P(\mathbf{r})$ by $\mathbf{u}(\mathbf{r}, t)$ and the corresponding stress dyadic by $\mathfrak{T}(\mathbf{u})$. The first-order equation of motion may be expressed in the form [Eq. (1.100)]

$$\operatorname{div} \mathfrak{T}(\mathbf{u}) + \rho\mathbf{F} = \rho \frac{\partial^2 \mathbf{u}}{\partial t^2}, \tag{4.39}$$

where $\rho\mathbf{F}(\mathbf{r}, t)$ is the external body force per unit volume. For a single force, $\mathbf{F}$ is given by Eq. (4.2). Substituting the value of $\mathbf{F}$ into Eq. (4.39) and then taking the Fourier transform of both sides, we obtain the equation of motion in the frequency domain:

$$\operatorname{div} \mathfrak{T}(\mathbf{u}) + \rho\omega^2\mathbf{u} = -F_0 g(\omega)\delta(\mathbf{r} - \mathbf{r}_0)\mathbf{a}. \tag{4.40}$$

The solution of the equation of motion, Eq. (4.40), with the prescribed boundary conditions is the displacement field induced by a concentrated force acting at the point $\mathbf{r}_0$ of the medium. Then the *Green's dyadic* for a medium with the same boundary conditions is defined by Eq. (4.10). We note that the Green's dyadic satisfies the equation

$$\operatorname{div} \mathfrak{T}(\mathfrak{G}) + \rho\omega^2 \mathfrak{G} = -\delta(\mathbf{r} - \mathbf{r}_0)\mathfrak{I}. \tag{4.41}$$

Writing $G_{ij} = G_i^j$, it follows that $G_i^j(P, Q)$ is the x_i component of the displacement at P caused by a concentrated force acting at Q in the x_j direction. Note that Eq. (4.31) does not hold in the case of heterogeneous media, i.e., in general,

$$G_i^j(P, Q) \neq G_j^i(P, Q),$$

$$G_i^j(P, Q) \neq G_i^j(Q, P).$$

However, it will be proved subsequently that

$$G_i^j(P, Q) = G_j^i(Q, P),$$

i.e.,

$$\mathfrak{G}(\mathbf{r}\,|\,\mathbf{r}_0) = \tilde{\mathfrak{G}}(\mathbf{r}_0\,|\,\mathbf{r}), \tag{4.42}$$

where $\tilde{\mathfrak{G}}$ is the transpose of $\mathfrak{G}$. This is known as the *seismic reciprocity relation.*

The displacement at the point $P(\mathbf{r})$ induced by a concentrated force $F_0\,\delta(t)$ acting at the point $Q(\mathbf{r}_0)$ in the direction of the unit vector $\mathbf{a}$ is given by

$$\mathbf{u}(\mathbf{r}) = F_0\,\mathfrak{G}(\mathbf{r}\,|\,\mathbf{r}_0)\cdot\mathbf{a}(\mathbf{r}_0).$$

Let us consider next two forces of the same magnitude F_0 acting at the points $Q_1(\mathbf{r}_0 + (\varepsilon/2)\mathbf{v})$ and $Q_2(\mathbf{r}_0 - (\varepsilon/2)\mathbf{v})$ in opposite directions, as shown in Fig. 4.7, $\mathbf{v}(\mathbf{r}_0)$ being a unit vector. The resultant displacement at the point P resulting from these forces is

$$\begin{aligned}\mathbf{u}(\mathbf{r}) &= F_0\left[\mathfrak{G}\left(\mathbf{r}\,|\,\mathbf{r}_0 + \frac{\varepsilon}{2}\mathbf{v}\right) - \mathfrak{G}\left(\mathbf{r}\,|\,\mathbf{r}_0 - \frac{\varepsilon}{2}\mathbf{v}\right)\right]\cdot\mathbf{a}(\mathbf{r}_0)\\ &= F_0\,\varepsilon[(\mathbf{v}\cdot\operatorname{grad}_0)\mathfrak{G}(\mathbf{r}\,|\,\mathbf{r}_0)\cdot\mathbf{a} + O(\varepsilon)]\\ &= F_0\,\varepsilon[\mathbf{a}(\mathbf{r}_0)\mathbf{v}(\mathbf{r}_0):\operatorname{grad}_0\tilde{\mathfrak{G}}(\mathbf{r}\,|\,\mathbf{r}_0) + O(\varepsilon)],\end{aligned} \tag{4.43}$$

where the operator grad_0 operates on the source coordinates $(\mathbf{r}_0)$ only and the colon (:) signifies the double dot product (App. A).

Taking the limit as $F_0 \to \infty$, $\varepsilon \to 0$ in such a manner that the product $F_0\varepsilon \to M$, we obtain

$$\mathbf{u}(\mathbf{r}) = M\mathbf{a}\mathbf{v}:\operatorname{grad}_0\tilde{\mathfrak{G}}(\mathbf{r}\,|\,\mathbf{r}_0). \tag{4.44}$$

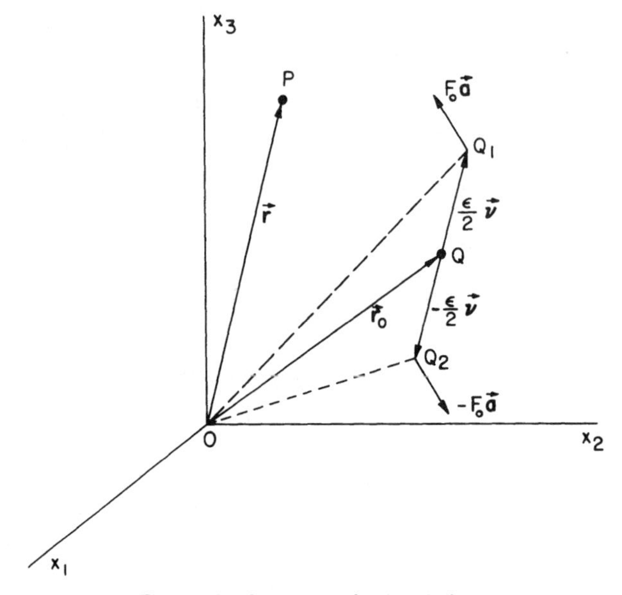

Figure 4.7. Geometry of a double force.

The quantity M has the dimensions of force $\times$ length and is called the *moment* of the dipolar source. Equation (4.44) gives the displacement field induced at $P(\mathbf{r})$ by a *double force* at $Q(\mathbf{r}_0)$ of moment M. The entity

$$\Gamma_i^{jk}(P, Q) = [\text{grad}_0\, \tilde{\mathfrak{G}}(\mathbf{r}\,|\,\mathbf{r}_0)]_{kji} = \frac{\partial}{\partial y_k} G_j^i(P, Q) \tag{4.45}$$

is a tensor of the third order. Here, $\mathfrak{G} \equiv G_{ij} = G_i^j$ and $\mathbf{r}_0 = \mathbf{e}_i y_i$.

Two cases are of fundamental importance:

1. $\mathbf{v}$ coincides with $\mathbf{a}$, which gives a *double force without moment* or a *dipole.*
2. $\mathbf{v}$ is perpendicular to $\mathbf{a}$, which gives a *single couple.*

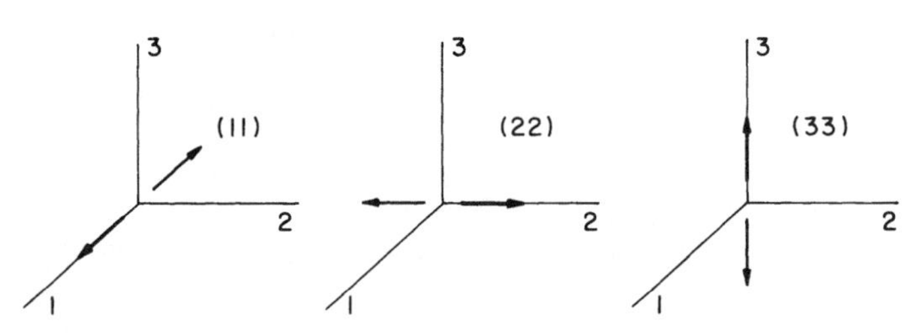

Figure 4.8. Three fundamental dipoles corresponding to Γ^{11}, Γ^{22}, Γ^{33}.

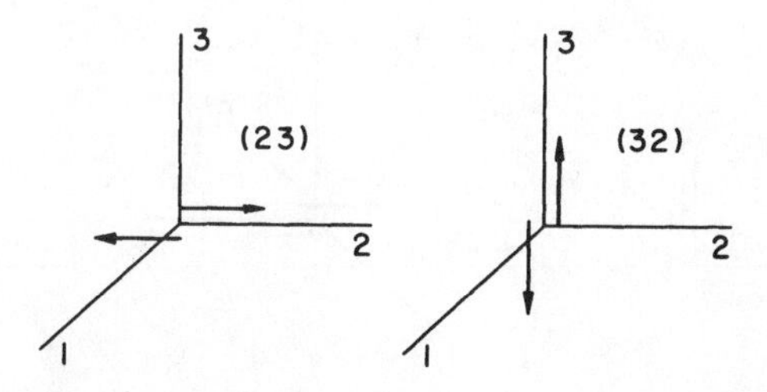

Figure 4.9. The single couples represented by Γ^{23} and Γ^{32}. The first index corresponds to the direction of the force.

In the latter case, M is the moment of the couple. When $j = k$, $M\Gamma_i^{jk}$ is the x_i component of the displacement at P caused by a dipole in the x_j direction at Q (Fig. 4.8). In contrast, *when $j \neq k$, Γ_i^{jk} is the x_i component of the displacement per unit moment at P resulting from a single couple at Q whose forces are parallel to the x_j direction and whose arm is in the x_k direction* (Fig. 4.9). It may be noted that, in general, Γ_i^{jk} is not equal to Γ_i^{kj}.

Let us consider three dipoles of the same strength directed along three arbitrary mutually perpendicular directions as shown in Fig. 4.10. From Eq. (4.44), the combined effect of the three dipoles is

$$\mathbf{u}(\mathbf{r}) = M\mathbf{e}_i\mathbf{e}_i : \operatorname{grad}_0 \tilde{\mathfrak{G}} = M\mathfrak{I} : \operatorname{grad}_0 \tilde{\mathfrak{G}} = M \operatorname{div}_0 \tilde{\mathfrak{G}}. \tag{4.46}$$

It is obvious that the individual directions of the dipoles are of no importance as long as the three dipoles are of the same strength and are acting along mutually perpendicular directions. This singularity is known as a *center of compression* or an *explosion* and is symmetrical about the point Q. However, the generated displacement field will not be symmetrical about Q if the properties of the medium are not symmetrical about this point.

Let us return to Eq. (4.44). Assuming that $\mathbf{v}$ is perpendicular to $\mathbf{a}$, the displacement field resulting from a single couple whose forces are parallel to the vector $\mathbf{v}$ is given by

$$\mathbf{u}(\mathbf{r}) = M\mathbf{va} : \operatorname{grad}_0 \tilde{\mathfrak{G}}(\mathbf{r}|\mathbf{r}_0) \qquad (\mathbf{a}\cdot\mathbf{v} = 0). \tag{4.47}$$

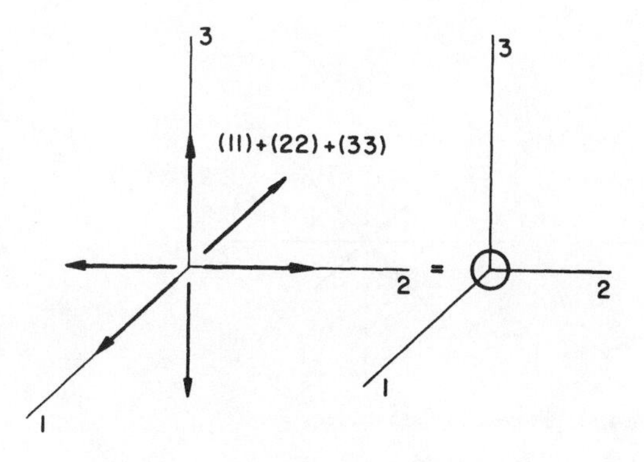

Figure 4.10. Center of compression $\Gamma^{11} + \Gamma^{22} + \Gamma^{33}$. It is schematically represented by a sphere.

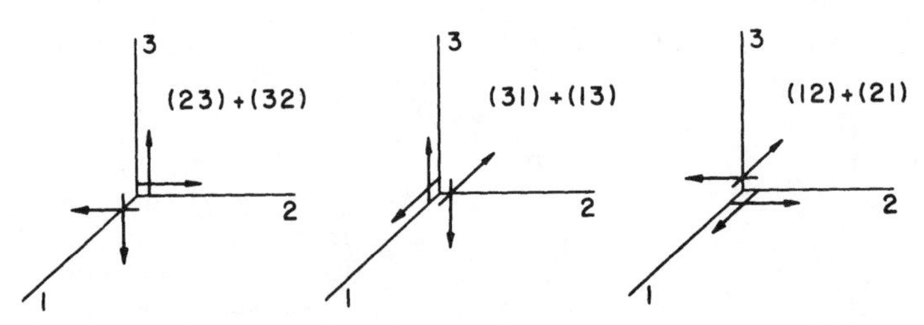

Figure 4.11. Three fundamental double couples $\Gamma^{23} + \Gamma^{32}$, $\Gamma^{31} + \Gamma^{13}$, and $\Gamma^{12} + \Gamma^{21}$.

The moment of this single couple is equal and opposite to the moment of the single couple represented by Eq. (4.44). Furthermore, the forces of one couple are perpendicular to the forces of the other couple. Consequently,

$$\begin{aligned} \mathbf{u}(\mathbf{r}) &= M(\mathbf{av} + \mathbf{va}) : \text{grad}_0 \,\tilde{\mathfrak{G}}(\mathbf{r}\,|\,\mathbf{r}_0) \\ &= M\mathbf{av} : [\text{grad}_0 \,\tilde{\mathfrak{G}}(\mathbf{r}\,|\,\mathbf{r}_0) + \{\text{grad}_0 \,\tilde{\mathfrak{G}}(\mathbf{r}\,|\,\mathbf{r}_0)\}^{213}] \qquad (\mathbf{a}\cdot\mathbf{v} = 0) \end{aligned} \quad (4.48)$$

represents the field caused by a *double couple* of zero net moment and zero net force. In Eq. (4.48), $\{\text{grad}_0 \,\tilde{\mathfrak{G}}\}^{213}$ is the left transpose of $\{\text{grad}_0 \,\tilde{\mathfrak{G}}\}$, i.e., obtained from it by interchanging the first two vectors in each of its triads. In the indicial notation, Eq. (4.48) implies that

$$u_i = M(\Gamma_i^{jk} + \Gamma_i^{kj}), \qquad (j \neq k)$$

is the displacement field induced by a double couple in the $x_j x_k$ plane, M representing the moment of either of the two single couples (see Fig. 4.11).

Let us define two mutually perpendicular unit vectors through the relations

$$\mathbf{a}' = \frac{1}{\sqrt{2}}(\mathbf{v} + \mathbf{a}), \qquad \mathbf{v}' = \frac{1}{\sqrt{2}}(\mathbf{v} - \mathbf{a});$$

so that the vector $\mathbf{a}'$ makes an angle of 45° with both $\mathbf{a}$ and $\mathbf{v}$, as shown in Fig. 4.12 for the case when $\mathbf{a} = \mathbf{e}_2$, $\mathbf{v} = \mathbf{e}_3$. We note that

$$\mathbf{av} + \mathbf{va} = \mathbf{a}'\mathbf{a}' - \mathbf{v}'\mathbf{v}'. \quad (4.49)$$

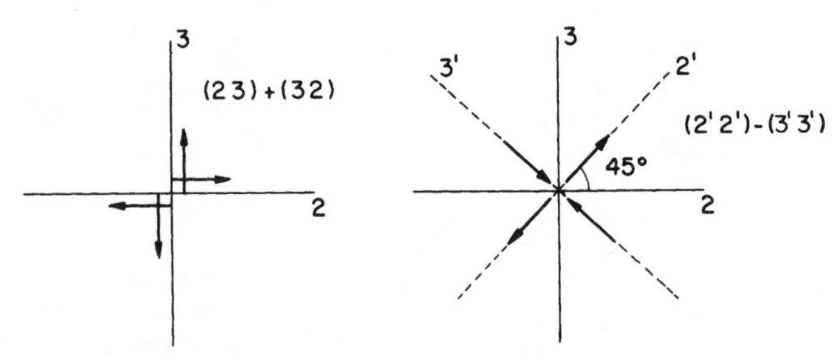

Figure 4.12. Two equivalent sources.

Using this relation in Eq. (4.48), it is seen that the field induced by a double couple with forces in the **a** and **v** directions is given by

$$\mathbf{u}(\mathbf{r}) = M\mathbf{a}'\mathbf{a}' : \text{grad}_0\, \tilde{\mathfrak{G}}(\mathbf{r}\,|\,\mathbf{r}_0) - M\mathbf{v}'\mathbf{v}' : \text{grad}_0\, \tilde{\mathfrak{G}}(\mathbf{r}\,|\,\mathbf{r}_0).$$

Comparing the last equation with Eq. (4.44), we find that the displacement field induced by a double couple is identical with the displacement field induced by two mutually perpendicular dipoles of equal strengths, one pressure dipole and one tension dipole, acting along the bisectors to the angles between the directions of the forces of the double couple.

A double couple is obtained by combining two coplaner single couples of equal but opposite moments with mutually perpendicular force directions. However, if we combine two single couples of the type described but with moments equal in magnitude and sense, we get a *torque* or a *center of rotation*, in which the net force is zero, whereas the net moment is not. The displacement field induced by a center of rotation is

$$\mathbf{u}(\mathbf{r}) = M(\mathbf{a}\mathbf{v} - \mathbf{v}\mathbf{a}) : \text{grad}_0\, \tilde{\mathfrak{G}}(\mathbf{r}\,|\,\mathbf{r}_0) \qquad (\mathbf{a}\cdot\mathbf{v} = 0), \tag{4.50}$$

the rotation being from **v** to **a**. In indicial notation

$$u_i = M(\Gamma_i^{kj} - \Gamma_i^{jk}), \qquad (j \neq k)$$

gives the x_i component of the displacement at P resulting from a center of rotation in the $x_j x_k$ plane, the rotation being in the $x_j \rightarrow x_k$ sense. The three centers of rotation with moments about three directions parallel to the coordinate axes are shown in Fig. 4.13.

In the case of a concentrated force $F_0 g(\omega)$, the equation of motion is Eq. (4.40). The corresponding Green's dyadic satisfies Eq. (4.41). Beginning with this dyadic, we derived displacement fields induced by various dipolar sources by performing suitable differential operations on the dyadic. By performing the same operations on the right-hand side of Eq. (4.41), we shall get the equations of motion for various dipolar sources. We therefore write

$$\text{div}\, \mathfrak{T}(\mathbf{u}) + \rho\omega^2\mathbf{u} = -g(\omega)\mathbf{Q}, \tag{4.51}$$

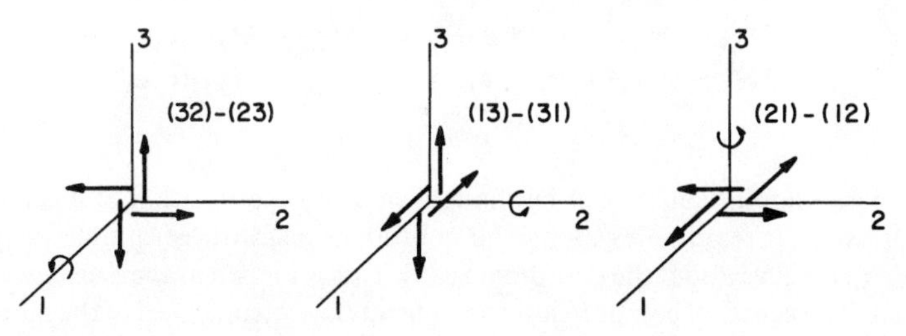

Figure 4.13. Three fundamental torques, $\Gamma^{32} - \Gamma^{23}$, $\Gamma^{13} - \Gamma^{31}$, and $\Gamma^{21} - \Gamma^{12}$, with rotations about the x_1, x_2, and x_3 axes, respectively.

where the source vector $\mathbf{Q}$ plays the same role as the vector $F_0\delta(\mathbf{r}-\mathbf{r}_0)\mathbf{a}$ does in the case of a concentrated force. From Eqs. (4.44), (4.46), (4.47), and (4.50), we obtain

$$
\begin{aligned}
\mathbf{Q} &= M\mathbf{av}:\mathrm{grad}_0[\delta(\mathbf{r}-\mathbf{r}_0)\mathfrak{I}], && \text{double force;}\\
&= M\,\mathrm{div}_0[\delta(\mathbf{r}-\mathbf{r}_0)\mathfrak{I}], && \text{center of compression;}\\
&= M(\mathbf{av}+\mathbf{va}):\mathrm{grad}_0[\delta(\mathbf{r}-\mathbf{r}_0)\mathfrak{I}], && \text{double couple;}\\
&= M(\mathbf{av}-\mathbf{va}):\mathrm{grad}_0[\delta(\mathbf{r}-\mathbf{r}_0)\mathfrak{I}], && \text{center of rotation.}
\end{aligned}
$$

Because $\mathrm{grad}(u\mathfrak{I}) = (\mathrm{grad}\,u)\mathfrak{I}$, $\mathrm{div}(u\mathfrak{I}) = \mathrm{grad}\,u$, and $\mathrm{grad}_0\,\delta(\mathbf{r}-\mathbf{r}_0) = -\mathrm{grad}\,\delta(\mathbf{r}-\mathbf{r}_0)$, the above equations may be simplified to

$$
\begin{aligned}
\mathbf{Q} &= -M\mathbf{a}[\mathbf{v}\cdot\mathrm{grad}\,\delta(\mathbf{r}-\mathbf{r}_0)], && \text{double force;}\\
&= -M\,\mathrm{grad}\,\delta(\mathbf{r}-\mathbf{r}_0), && \text{center of compression;}\\
&= -M(\mathbf{av}+\mathbf{va})\cdot\mathrm{grad}\,\delta(\mathbf{r}-\mathbf{r}_0), && \text{double couple;}\\
&= -M(\mathbf{av}-\mathbf{va})\cdot\mathrm{grad}\,\delta(\mathbf{r}-\mathbf{r}_0), && \text{center of rotation.}
\end{aligned}
\tag{4.52}
$$

Table 4.1 gives explicit expressions for $\mathbf{Q}$ for various sources. The point source is situated at the point $Q(y_1, y_2, y_3)$ and the observer at $P(x_1, x_2, x_3)$. The notation used conforms with the notation of Figs. (4.8)–(4.13).

Beginning with Eq. (4.44) for the displacement field generated by a double force, we introduce a *general dipolar source* as one whose displacement field is given by

$$\mathbf{u}(\mathbf{r}) = \mathfrak{M}:\mathrm{grad}_0[\tilde{\mathfrak{G}}(\mathbf{r}\,|\,\mathbf{r}_0)],$$

i.e.,

$$u_i(P) = M_{jk}(Q)[\mathrm{grad}_0\,\tilde{\mathfrak{G}}(P, Q)]_{kji}, \tag{4.53}$$

where $\mathfrak{M} \equiv M_{jk}$ is known as the *source moment tensor*. Because

$$
\begin{aligned}
\mathfrak{M} = M_{jk}\mathbf{e}_j\mathbf{e}_k &= \tfrac{1}{3}(M_{11}+M_{22}+M_{33})(\mathbf{e}_1\mathbf{e}_1+\mathbf{e}_2\mathbf{e}_2+\mathbf{e}_3\mathbf{e}_3)\\
&+\tfrac{1}{3}(2M_{11}-M_{22}-M_{33})\mathbf{e}_1\mathbf{e}_1\\
&+\tfrac{1}{3}(2M_{22}-M_{33}-M_{11})\mathbf{e}_2\mathbf{e}_2+\tfrac{1}{3}(2M_{33}-M_{11}-M_{22})\mathbf{e}_3\mathbf{e}_3\\
&+\tfrac{1}{2}(M_{32}+M_{23})(\mathbf{e}_3\mathbf{e}_2+\mathbf{e}_2\mathbf{e}_3)+\tfrac{1}{2}(M_{32}-M_{23})(\mathbf{e}_3\mathbf{e}_2-\mathbf{e}_2\mathbf{e}_3)\\
&+\tfrac{1}{2}(M_{13}+M_{31})(\mathbf{e}_1\mathbf{e}_3+\mathbf{e}_3\mathbf{e}_1)+\tfrac{1}{2}(M_{13}-M_{31})(\mathbf{e}_1\mathbf{e}_3-\mathbf{e}_3\mathbf{e}_1)\\
&+\tfrac{1}{2}(M_{21}+M_{12})(\mathbf{e}_2\mathbf{e}_1+\mathbf{e}_1\mathbf{e}_2)+\tfrac{1}{2}(M_{21}-M_{12})(\mathbf{e}_2\mathbf{e}_1-\mathbf{e}_1\mathbf{e}_2),
\end{aligned}
\tag{4.54}
$$

the general dipolar source can be thought of as a superposition of a center of compression, three dipoles along the coordinate axes, three double couples, and three torques about the coordinate axes. This is known as the *decomposition theorem.* The center of compression comes from the isotropic part of the moment tensor and the remaining nine sources from the deviatoric part.

If the moment tensor is symmetric, then $M_{ij} = M_{ji}$ and the torques in Eq. (4.54) will vanish, leaving behind seven sources. We know that the eigenvalues of a symmetrical second-order tensor are all real and its eigenvectors are mutually orthogonal. Assuming $\mathfrak{M}$ to be symmetrical, let its eigenvalues be M_1, M_2, and M_3 and let the corresponding eigenvectors be $\mathbf{a}_1$, $\mathbf{a}_2$, and $\mathbf{a}_3$, respectively. Then we have [see also Eq. (1.24)]

$$\begin{aligned}\mathfrak{M} &= \tfrac{1}{3}(M_1 + M_2 + M_3)\mathfrak{I} + \tfrac{1}{3}(2M_1 - M_2 - M_3)\mathbf{a}_1\mathbf{a}_1 \\ &\quad + \tfrac{1}{3}(2M_2 - M_3 - M_1)\mathbf{a}_2\mathbf{a}_2 + \tfrac{1}{3}(2M_3 - M_1 - M_2)\mathbf{a}_3\mathbf{a}_3. \qquad (4.55)\end{aligned}$$

In other words, a general dipolar source with $M_{ij} = M_{ji}$ can be expressed as a superposition of a center of compression and three dipoles along three mutually orthogonal directions.

Equation (4.55) may be written in the form

$$\begin{aligned}\mathfrak{M} &= \tfrac{1}{3}(M_1 + M_2 + M_3)\mathfrak{I} + \tfrac{1}{3}M_1(2\mathbf{a}_1\mathbf{a}_1 - \mathbf{a}_2\mathbf{a}_2 - \mathbf{a}_3\mathbf{a}_3) \\ &\quad + \tfrac{1}{3}M_2(2\mathbf{a}_2\mathbf{a}_2 - \mathbf{a}_3\mathbf{a}_3 - \mathbf{a}_1\mathbf{a}_1) + \tfrac{1}{3}M_3(2\mathbf{a}_3\mathbf{a}_3 - \mathbf{a}_1\mathbf{a}_1 - \mathbf{a}_2\mathbf{a}_2). \qquad (4.56)\end{aligned}$$

In the above equation, $2\mathbf{a}_1\mathbf{a}_1 - \mathbf{a}_2\mathbf{a}_2 - \mathbf{a}_3\mathbf{a}_3$ represents a compressional dipole of moment 2 along $\mathbf{a}_1$ axis and two dilatational dipoles each of unit moment along the $\mathbf{a}_2$ and $\mathbf{a}_3$ axes. This type of source is known as a *compensated linear vector dipole*. Therefore, a general dipolar source with $M_{ij} = M_{ji}$ is equivalent to a center of compression plus three mutually orthogonal compensated linear vector dipoles. Because

$$2\mathbf{a}_1\mathbf{a}_1 - \mathbf{a}_2\mathbf{a}_2 - \mathbf{a}_3\mathbf{a}_3 = (\mathbf{a}_1\mathbf{a}_1 - \mathbf{a}_2\mathbf{a}_2) + (\mathbf{a}_1\mathbf{a}_1 - \mathbf{a}_3\mathbf{a}_3),$$

a compensated linear vector dipole is equivalent to two double couples.

Equation (4.55) can also be expressed as

$$\begin{aligned}\mathfrak{M} &= \tfrac{1}{3}(M_1 + M_2 + M_3)\mathfrak{I} + \tfrac{1}{3}(M_1 - M_2)(\mathbf{a}_1\mathbf{a}_1 - \mathbf{a}_2\mathbf{a}_2) \\ &\quad + \tfrac{1}{3}(M_2 - M_3)(\mathbf{a}_2\mathbf{a}_2 - \mathbf{a}_3\mathbf{a}_3) + \tfrac{1}{3}(M_3 - M_1)(\mathbf{a}_3\mathbf{a}_3 - \mathbf{a}_1\mathbf{a}_1). \qquad (4.57)\end{aligned}$$

Because, for example, $\mathbf{a}_1\mathbf{a}_1 - \mathbf{a}_2\mathbf{a}_2$ gives a double couple [see Eq. (4.49)], we note that a symmetrical moment tensor represents a center of compression plus three double couples. If the isotropic part of the moment tensor vanishes, we are left with just three double couples, acting in three mutually orthogonal planes.

If $M_2 = M_3$, Eq. (4.56) becomes

$$\mathfrak{M} = \tfrac{1}{3}(M_1 + 2M_2)\mathfrak{I} + \tfrac{1}{3}(M_1 - M_2)(2\mathbf{a}_1\mathbf{a}_1 - \mathbf{a}_2\mathbf{a}_2 - \mathbf{a}_3\mathbf{a}_3),$$

and we simply have a center of compression plus a compensated linear vector dipole.

Table 4.1. The Source Vector **Q** for Various Sources

Source	**Q**
Single force of magnitude F_0	
1. In the x_1 direction	$F_0\mathbf{e}_1\delta(x_1 - y_1)\delta(x_2 - y_2)\delta(x_3 - y_3)$
2. In the x_2 direction	$F_0\mathbf{e}_2\delta(x_1 - y_1)\delta(x_2 - y_2)\delta(x_3 - y_3)$
3. In the x_3 direction	$F_0\mathbf{e}_3\delta(x_1 - y_1)\delta(x_2 - y_2)\delta(x_3 - y_3)$
Dipole (Fig. 4.8)	
1. In the x_1 direction	$-M\mathbf{e}_1\frac{\partial}{\partial x_1}\delta(x_1 - y_1)\delta(x_2 - y_2)\delta(x_3 - y_3)$
2. In the x_2 direction	$-M\mathbf{e}_2\delta(x_1 - y_1)\frac{\partial}{\partial x_2}\delta(x_2 - y_2)\delta(x_3 - y_3)$
3. In the x_3 direction	$-M\mathbf{e}_3\delta(x_1 - y_1)\delta(x_2 - y_2)\frac{\partial}{\partial x_3}\delta(x_3 - y_3)$
Single couple (Fig. 4.9)	
1. (12)	$-M\mathbf{e}_1\delta(x_1 - y_1)\frac{\partial}{\partial x_2}\delta(x_2 - y_2)\delta(x_3 - y_3)$
2. (21)	$-M\mathbf{e}_2\frac{\partial}{\partial x_1}\delta(x_1 - y_1)\delta(x_2 - y_2)\delta(x_3 - y_3)$
3. (23)	$-M\mathbf{e}_2\delta(x_1 - y_1)\delta(x_2 - y_2)\frac{\partial}{\partial x_3}\delta(x_3 - y_3)$
4. (32)	$-M\mathbf{e}_3\delta(x_1 - y_1)\frac{\partial}{\partial x_2}\delta(x_2 - y_2)\delta(x_3 - y_3)$

5. (31)	$-M\mathbf{e}_3 \frac{\partial}{\partial x_1} \delta(x_1 - y_1)\delta(x_2 - y_2)\delta(x_3 - y_3)$
6. (13)	$-M\mathbf{e}_1 \delta(x_1 - y_1)\delta(x_2 - y_2) \frac{\partial}{\partial x_3} \delta(x_3 - y_3)$
Double couple (Fig. 4.11)	
1. (23) + (32)	$-M\left[\mathbf{e}_2 \delta(x_2 - y_2) \frac{\partial}{\partial x_3} \delta(x_3 - y_3) + \mathbf{e}_3 \frac{\partial}{\partial x_2} \delta(x_2 - y_2)\delta(x_3 - y_3)\right]\delta(x_1 - y_1)$
2. (31) + (13)	$-M\left[\mathbf{e}_3 \frac{\partial}{\partial x_1} \delta(x_1 - y_1)\delta(x_3 - y_3) + \mathbf{e}_1 \delta(x_1 - y_1) \frac{\partial}{\partial x_3} \delta(x_3 - y_3)\right]\delta(x_2 - y_2)$
3. (12) + (21)	$-M\left[\mathbf{e}_1 \delta(x_1 - y_1) \frac{\partial}{\partial x_2} \delta(x_2 - y_2) + \mathbf{e}_2 \frac{\partial}{\partial x_1} \delta(x_1 - y_1)\delta(x_2 - y_2)\right]\delta(x_3 - y_3)$
Center of rotation (Fig. 4.13)	
1. (32) − (23)	$-M\left[\mathbf{e}_3 \frac{\partial}{\partial x_2} \delta(x_2 - y_2)\delta(x_3 - y_3) - \mathbf{e}_2 \delta(x_2 - y_2) \frac{\partial}{\partial x_3} \delta(x_3 - y_3)\right]\delta(x_1 - y_1)$
2. (13) − (31)	$-M\left[\mathbf{e}_1 \delta(x_1 - y_1) \frac{\partial}{\partial x_3} \delta(x_3 - y_3) - \mathbf{e}_3 \frac{\partial}{\partial x_1} \delta(x_1 - y_1)\delta(x_3 - y_3)\right]\delta(x_2 - y_2)$
3. (21) − (12)	$-M\left[\mathbf{e}_2 \frac{\partial}{\partial x_1} \delta(x_1 - y_1)\delta(x_2 - y_2) - \mathbf{e}_1 \delta(x_1 - y_1) \frac{\partial}{\partial x_2} \delta(x_2 - y_2)\right]\delta(x_3 - y_3)$
Center of compression (Fig. 4.10)	$-M\left[\mathbf{e}_1 \frac{\partial}{\partial x_1} \delta(x_1 - y_1)\delta(x_2 - y_2)\delta(x_3 - y_3) + \mathbf{e}_2 \delta(x_1 - y_1) \frac{\partial}{\partial x_2} \delta(x_2 - y_2) \times \delta(x_3 - y_3) + \mathbf{e}_3 \delta(x_1 - y_1)\delta(x_2 - y_2) \frac{\partial}{\partial x_3} \delta(x_3 - y_3)\right]$

4.3. Relations of Betti, Somigliana, and Volterra

4.3.1. Betti's Relation and Reciprocity

Consider a simply connected volume V bounded by a surface S. Let $\mathbf{u}_1$ and $\mathbf{u}_2$ be two single-valued, twice continuously differentiable vector fields in $V + S$, representing two possible displacement fields in the region. The corresponding stress and strain dyadics are denoted by $\mathfrak{T}(\mathbf{u}_1)$, $\mathfrak{E}(\mathbf{u}_1)$ and $\mathfrak{T}(\mathbf{u}_2)$, $\mathfrak{E}(\mathbf{u}_2)$. Because $\mathfrak{T}(\mathbf{u})$ is a symmetrical dyadic, it follows that

$$\operatorname{div}[\mathfrak{T}(\mathbf{u}_1) \cdot \mathbf{u}_2] = \mathbf{u}_2 \cdot \operatorname{div} \mathfrak{T}(\mathbf{u}_1) + \mathfrak{T}(\mathbf{u}_1) : \mathfrak{E}(\mathbf{u}_2). \tag{4.58}$$

Integrating both sides of Eq. (4.58) over the volume V and using Gauss' theorem, we get

$$\int_V \mathbf{u}_2 \cdot \operatorname{div} \mathfrak{T}(\mathbf{u}_1) dV = \int_S \mathbf{n} \cdot [\mathfrak{T}(\mathbf{u}_1) \cdot \mathbf{u}_2] dS - \int_V \mathfrak{T}(\mathbf{u}_1) : \mathfrak{E}(\mathbf{u}_2) dV, \tag{4.59}$$

where $\mathbf{n}$ is the unit outward drawn normal to the element dS of S (Fig. 4.14). Using Hooke's law, Eq. (1.65), it may be verified that

$$\mathfrak{T}(\mathbf{u}_1) : \mathfrak{E}(\mathbf{u}_2) = \mathfrak{T}(\mathbf{u}_2) : \mathfrak{E}(\mathbf{u}_1). \tag{4.60}$$

Therefore, antisymmetrizing Eq. (4.59) with respect to $\mathbf{u}_1$ and $\mathbf{u}_2$, we obtain

$$\int_V [\mathbf{u}_1 \cdot \operatorname{div} \mathfrak{T}(\mathbf{u}_2) - \mathbf{u}_2 \cdot \operatorname{div} \mathfrak{T}(\mathbf{u}_1)] dV = \int_S \mathbf{n} \cdot [\mathfrak{T}(\mathbf{u}_2) \cdot \mathbf{u}_1 - \mathfrak{T}(\mathbf{u}_1) \cdot \mathbf{u}_2] dS. \tag{4.61}$$

Let us next suppose that $\mathbf{u}_1$ and $\mathbf{u}_2$ are caused by the body force distributions $\mathbf{F}_1$ and $\mathbf{F}_2$, respectively, per unit mass. Then, we have the equations

$$\operatorname{div} \mathfrak{T}(\mathbf{u}_1) + \rho\omega^2 \mathbf{u}_1 = -\rho \mathbf{F}_1, \qquad \operatorname{div} \mathfrak{T}(\mathbf{u}_2) + \rho\omega^2 \mathbf{u}_2 = -\rho \mathbf{F}_2. \tag{4.62}$$

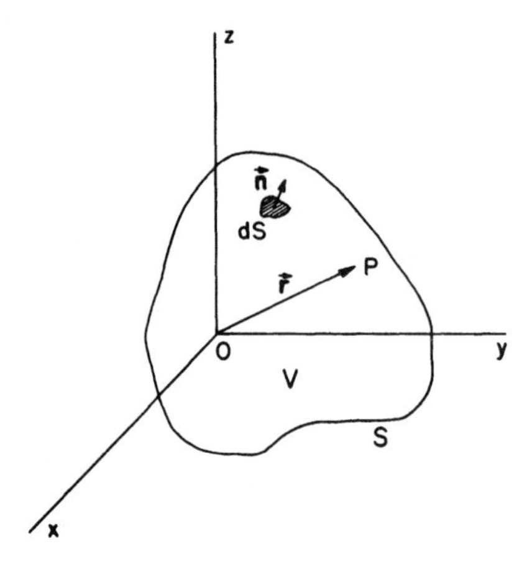

Figure 4.14. The closed surface S bounds a volume V. $\mathbf{n}$ is a unit outward drawn normal to the element dS of S.

Equations (4.61) and (4.62) yield

$$\int_V \rho(\mathbf{u}_1 \cdot \mathbf{F}_2 - \mathbf{u}_2 \cdot \mathbf{F}_1) dV = \int_S \mathbf{n} \cdot [\mathfrak{T}(\mathbf{u}_1) \cdot \mathbf{u}_2 - \mathfrak{T}(\mathbf{u}_2) \cdot \mathbf{u}_1] dS. \tag{4.63}$$

This is known as *Betti's relation.* Starting from this relation the reciprocity relation, Eq. (4.42), can be easily proved. If we assume homogeneous boundary conditions on S, the surface integral in Eq. (4.63) vanishes. For example, if the surface S is traction free, $\mathbf{n} \cdot \mathfrak{T}(\mathbf{u}_1) = 0$, $\mathbf{n} \cdot \mathfrak{T}(\mathbf{u}_2) = 0$ on S. Equation (4.63) then becomes

$$\int_V \rho \mathbf{u}_1 \cdot \mathbf{F}_2 \, dV = \int_V \rho \mathbf{u}_2 \cdot \mathbf{F}_1 \, dV. \tag{4.64}$$

Next, let $\mathbf{u}_1$ and $\mathbf{u}_2$ be the displacements induced by localized forces $\mathbf{F}_1$ and $\mathbf{F}_2$, acting at $Q_1(\mathbf{r}_1)$ and $Q_2(\mathbf{r}_2)$, respectively. Then

$$\begin{aligned} \rho \mathbf{F}_1 &= g(\omega)\delta(\mathbf{r} - \mathbf{r}_1)\mathbf{a}_1(\mathbf{r}_1), \\ \rho \mathbf{F}_2 &= g(\omega)\delta(\mathbf{r} - \mathbf{r}_2)\mathbf{a}_2(\mathbf{r}_2), \end{aligned} \tag{4.65}$$

where $g(\omega)$ is the Fourier transform of $g(t)$ and $\mathbf{a}_1, \mathbf{a}_2$ are unit vectors. Using the properties of the Dirac delta function, Eqs. (4.64) and (4.65) give

$$\mathbf{u}_1(\mathbf{r}_2) \cdot \mathbf{a}_2 = \mathbf{u}_2(\mathbf{r}_1) \cdot \mathbf{a}_1. \tag{4.66}$$

Define Green's dyadics $\mathfrak{G}(\mathbf{r}|\mathbf{r}_1)$ and $\mathfrak{G}(\mathbf{r}|\mathbf{r}_2)$ as in Eq. (4.10). Equation (4.66) then yields

$$[\mathfrak{G}(\mathbf{r}_2|\mathbf{r}_1) \cdot \mathbf{a}_1(\mathbf{r}_1)] \cdot \mathbf{a}_2(\mathbf{r}_2) = [\mathfrak{G}(\mathbf{r}_1|\mathbf{r}_2) \cdot \mathbf{a}_2(\mathbf{r}_2)] \cdot \mathbf{a}_1(\mathbf{r}_1). \tag{4.67}$$

The last equation implies that

$$\mathfrak{G}(\mathbf{r}_1|\mathbf{r}_2) = \tilde{\mathfrak{G}}(\mathbf{r}_2|\mathbf{r}_1). \tag{4.68}$$

Equation (4.68) holds provided $\mathbf{u}_1(\mathbf{r}_2) \cdot \mathbf{a}_2 \neq 0$. Relation (4.68) is the required *reciprocity relation.* Equation (4.66) also expresses the reciprocity and implies the following:

> If, in an anisotropic, inhomogeneous, elastic medium of finite or infinite extent, a concentrated force applied in a particular direction $\mathbf{a}_1$ at a point Q_1 produces a displacement field whose component in a direction $\mathbf{a}_2$ at a point Q_2 is v, then the same force when applied at the point Q_2 in the $\mathbf{a}_2$-direction will produce a displacement field whose component in the $\mathbf{a}_1$-direction at the point Q_1 will be v.

4.3.2. Somigliana's Relation and the Representation Theorem

Let us return to Betti's relation, Eq. (4.63), and assume that $\mathbf{u}_2$ is the displacement field induced in an infinite medium by a concentrated unit force having delta-function time variation and acting at a point $\mathbf{r} = \mathbf{r}_0$ in the direction of a constant unit vector $\mathbf{a}$. The infinite medium is such that its properties inside V are identical to the properties of the finite volume V occurring in Eq. (4.63). It does not matter how the properties of the infinite medium outside V are chosen. Substituting $\rho \mathbf{F}_2 = \delta(\mathbf{r} - \mathbf{r}_0)\mathbf{a}$ and omitting the subscript 1 of $\mathbf{u}_1$ and $\mathbf{F}_1$ in Eq. (4.63), we obtain

$$\mathbf{u}(\mathbf{r}_0) \cdot \mathbf{a}(\mathbf{r}_0) = \int_V \rho \mathbf{F}(\mathbf{r}) \cdot \mathbf{u}_2(\mathbf{r}) dV + \int_S \mathbf{n}(\mathbf{r}) \cdot [\mathfrak{T}\{\mathbf{u}(\mathbf{r})\} \cdot \mathbf{u}_2(\mathbf{r}) - \mathfrak{T}\{\mathbf{u}_2(\mathbf{r})\} \cdot \mathbf{u}(\mathbf{r})] dS. \tag{4.69}$$

This is known as *Somigliana's relation.* We have assumed above that the point of application of the force, $\mathbf{r}_0$, lies within V. However, if $\mathbf{r}_0$ lies outside V, one must put zero on the left-hand side of Eq. (4.69).

Let us next take $\mathbf{F}_2 = 0$ in Eq. (4.63) and denote the corresponding displacement by $\hat{\mathbf{u}}_2$. Proceeding as above, we find

$$0 = \int_V \rho \mathbf{F}(\mathbf{r}) \cdot \hat{\mathbf{u}}_2(\mathbf{r}) dV + \int_S \mathbf{n}(\mathbf{r}) \cdot [\mathfrak{T}\{\mathbf{u}(\mathbf{r})\} \cdot \hat{\mathbf{u}}_2(\mathbf{r}) - \mathfrak{T}\{\hat{\mathbf{u}}_2(\mathbf{r})\} \cdot \mathbf{u}(\mathbf{r})] dS. \tag{4.70}$$

Adding Eqs. (4.69) and (4.70), we obtain

$$\mathbf{u}(\mathbf{r}_0) \cdot \mathbf{a}(\mathbf{r}_0) = \int_V \rho \mathbf{F} \cdot (\mathbf{u}_2 + \hat{\mathbf{u}}_2) dV + \int_S \mathbf{n} \cdot [\mathfrak{T}(\mathbf{u}) \cdot (\mathbf{u}_2 + \hat{\mathbf{u}}_2) - \mathfrak{T}(\mathbf{u}_2 + \hat{\mathbf{u}}_2) \cdot \mathbf{u}] dS. \tag{4.71}$$

We define Green's dyadics $\mathfrak{G}_\infty$ and $\hat{\mathfrak{G}}$ through the relations

$$\begin{aligned} \mathbf{u}_2(\mathbf{r}) &= \mathfrak{G}_\infty(\mathbf{r} \mid \mathbf{r}_0) \cdot \mathbf{a}(\mathbf{r}_0), \\ \hat{\mathbf{u}}_2(\mathbf{r}) &= \hat{\mathfrak{G}}(\mathbf{r} \mid \mathbf{r}_0) \cdot \mathbf{a}(\mathbf{r}_0), \end{aligned} \tag{4.72}$$

and put $\mathfrak{G} = \mathfrak{G}_\infty + \hat{\mathfrak{G}}$. Equations (4.71) and (4.72) then yield

$$\mathbf{u}(\mathbf{r}_0) = \int_V \rho \mathbf{F} \cdot \mathfrak{G}\, dV + \int_S \mathbf{n} \cdot \mathfrak{T}(\mathbf{u}) \cdot \mathfrak{G}\, dS - \int_S \mathbf{u} \cdot [\mathbf{n} \cdot \mathfrak{T}(\mathfrak{G})] dS. \tag{4.73}$$

It may be noted that $\hat{\mathbf{u}}_2$ is a displacement field satisfying the homogeneous equation of motion and $\hat{\mathfrak{G}}$ is the corresponding Green's dyadic. We may choose $\hat{\mathbf{u}}_2$ in such a manner that $\mathbf{u}_2 + \hat{\mathbf{u}}_2$ represents the field in V satisfying required boundary conditions on S. We call $\mathfrak{G}_\infty$ the Green's dyadic for the infinite medium and $\mathfrak{G}$ the Green's dyadic for the volume V bounded by S, which satisfies required boundary conditions on the surface S.

Equation (4.73) is an *elastodynamic representation theorem.* The displacement field at a point $\mathbf{r}_0$ of the volume V bounded by S consists of three parts:

1. The volume integral of the body force distribution in V weighted by the Green's dyadic $\mathfrak{G}$.
2. The surface integral of the stress vector distribution over S weighted by the same Green's dyadic.
3. The surface integral of the displacement distribution over S weighted by $\mathbf{n} \cdot \mathfrak{T}(\mathfrak{G})$.

In Eq. (4.73), we have introduced a new entity, namely, $\mathfrak{T}[\mathfrak{G}(\mathbf{r}|\mathbf{r}_0)]$, which is obtained from $\mathfrak{G}$ in the same manner as we obtain $\mathfrak{T}(\mathbf{u})$ from $\mathbf{u}$. If $\mathfrak{T}$ is considered as an operator, then in $\mathfrak{T}[\mathfrak{G}(\mathbf{r}|\mathbf{r}_0)]$, it operates on the $\mathbf{r}$ dependence of $\mathfrak{G}$ only. In other words, in the differentiations implied in $\mathfrak{T}(\mathfrak{G})$, the source coordinates ($\mathbf{r}_0$) behave as constants. Remembering the definition, Eq. (4.72), of the Green's dyadic, it is apparent that $\mathfrak{T}[\mathfrak{G}(\mathbf{r}|\mathbf{r}_0)] \cdot \mathbf{a}(\mathbf{r}_0)$ is the stress dyadic at $P(\mathbf{r})$ resulting from a concentrated unit force acting at $Q(\mathbf{r}_0)$ in the $\mathbf{a}$ direction. For an isotropic medium, we have

$$\mathfrak{T}[\mathfrak{G}(\mathbf{r}|\mathbf{r}_0)] = \lambda\mathfrak{I} \operatorname{div} \mathfrak{G} + \mu[(\operatorname{grad} \mathfrak{G}) + (\operatorname{grad} \mathfrak{G})^{213}], \tag{4.74}$$

where $(\operatorname{grad} \mathfrak{G})^{213}$ is the left transpose of $(\operatorname{grad} \mathfrak{G})^{123}$.

On the right-hand side of Eq. (4.73), the argument of $\mathbf{u}$, $\mathbf{F}$, and $\mathbf{n}$ is $\mathbf{r}$, whereas the argument of $\mathfrak{G}$ is $\mathbf{r}|\mathbf{r}_0$. Interchanging the roles of $\mathbf{r}$ and $\mathbf{r}_0$ (P still being at $\mathbf{r}$ and Q at $\mathbf{r}_0$) we find that, if $\mathbf{r}$ lies within V,

$$\mathbf{u}(\mathbf{r}) = \int_V \rho\mathfrak{G} \cdot \mathbf{F}\, dV(\mathbf{r}_0) + \int_S \mathfrak{G} \cdot [\mathbf{n} \cdot \mathfrak{T}_0(\mathbf{u})] dS(\mathbf{r}_0) - \int_S \mathbf{u} \cdot [\mathbf{n} \cdot \mathfrak{T}_0(\tilde{\mathfrak{G}})] dS(\mathbf{r}_0), \tag{4.75}$$

where $\mathbf{u} = \mathbf{u}(\mathbf{r}_0)$, $\mathbf{F} = \mathbf{F}(\mathbf{r}_0)$, $\mathbf{n} = \mathbf{n}(\mathbf{r}_0)$, and $\mathfrak{G} = \mathfrak{G}(\mathbf{r}|\mathbf{r}_0)$. Use has been made of the reciprocity relation $\mathfrak{G}(\mathbf{r}_0|\mathbf{r}) = \tilde{\mathfrak{G}}(\mathbf{r}|\mathbf{r}_0)$.

If the medium under consideration is isotropic,

$$\mathfrak{T}_0(\tilde{\mathfrak{G}}) = \lambda\mathfrak{I} \operatorname{div}_0(\tilde{\mathfrak{G}}) + \mu[(\operatorname{grad}_0 \tilde{\mathfrak{G}}) + (\operatorname{grad}_0 \tilde{\mathfrak{G}})^{213}]. \tag{4.76}$$

In the indicial notation, if $\mathfrak{G} \equiv G_{km} = G_k^m$,

$$\begin{aligned} T_k^{ij}(P, Q) = [\mathfrak{T}_0(\tilde{\mathfrak{G}})]_{ijk} &= \lambda\delta_{ij} G_m^{k,m} + \mu(G_j^{k,i} + G_i^{k,j}) \\ &= \lambda\delta_{ij}\Gamma_k^{mm} + \mu(\Gamma_k^{ji} + \Gamma_k^{ij}), \end{aligned} \tag{4.77}$$

where, for example,

$$\Gamma_k^{ij} = G_i^{k,j} = \frac{\partial}{\partial y_j} G_i^k, \qquad \mathbf{r}_0 = \mathbf{e}_i y_i. \tag{4.78}$$

Obviously,

$$T_k^{ij} = T_k^{ji}. \tag{4.79}$$

We are now in a position to give an interpretation to the tensor T_k^{ij}. If $i \neq j$, Eq. (4.77) shows that

$$T_k^{ij}(P, Q) = \mu(\Gamma_k^{ji} + \Gamma_k^{ij}). \tag{4.80}$$

Comparing Eqs. (4.48) and (4.80), we note that T_k^{ij} ($i \neq j$) represents, up to a constant, the x_k component of the displacement at $P(\mathbf{r})$ resulting from a double couple at $Q(\mathbf{r}_0)$ in the $x_i x_j$ plane. Similarly, if $i = j$,

$$T_k^{ij}(P, Q) = \lambda\Gamma_k^{mm} + 2\mu\Gamma_k^{ij}. \tag{4.81}$$

Noting that $\Gamma_k^{mm} = [\operatorname{div}_0(\tilde{\mathfrak{G}})]_k$ and comparing Eqs. (4.45) and (4.46) with Eq. (4.81), we see that T_k^{ij} ($i = j$) represents, up to a constant, the x_k component of

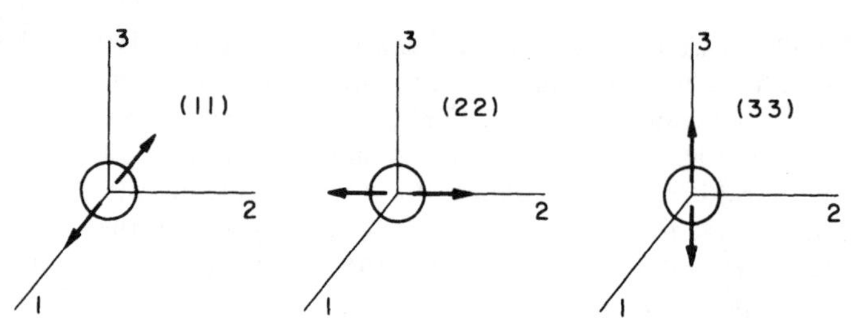

Figure 4.15. The sources T^{11}, T^{22}, and T^{33}. Each consists of a center of compression plus a dipole.

the displacement at P caused by a center of compression plus a dipole in the x_i direction at Q, of suitable strengths. Alternatively, the equivalent force system is a dipole of strength proportional to $\lambda + 2\mu$ in the x_i direction plus two more dipoles of identical strengths, proportional to λ, parallel to the other two coordinate axes. If a center of compression is represented by a sphere then the three sources corresponding to T_k^{ij} $(i = j)$ may be shown as in Fig. 4.15.[2]

4.3.3. Volterra's Relation and Dislocations

Returning to Eq. (4.75), let us suppose that the volume V is bounded internally by a surface S and externally by a surface S'. Consequently, the surface integral in Eq. (4.75) should be taken over $S + S'$. Let us further assume that the surface S shrinks, such that in the limit it consists of the two sides of an open surface. We denote the two sides of S as S^+ and S^-. Let the displacement vector $\mathbf{u}(\mathbf{r}_0)$ and the stress vector $\mathbf{n}(\mathbf{r}_0) \cdot \mathfrak{T}_0[\mathbf{u}(\mathbf{r}_0)]$ be discontinuous across S. We put

$$\begin{aligned} \Delta\mathbf{u}_0(\mathbf{r}_0) &= \mathbf{u}^+(\mathbf{r}_0) - \mathbf{u}^-(\mathbf{r}_0), \\ -\Delta\boldsymbol{\tau}_0(\mathbf{r}_0) &= \mathbf{n}^+(\mathbf{r}_0) \cdot \mathfrak{T}_0[\mathbf{u}(\mathbf{r}_0)]^+ + \mathbf{n}^-(\mathbf{r}_0) \cdot \mathfrak{T}_0[\mathbf{u}(\mathbf{r}_0)]^-, \end{aligned} \tag{4.82}$$

where the superscripts + and − are used to indicate that the corresponding quantities are measured at neighboring points on S^+ and S^-, respectively. Remembering that $\mathbf{n}$ denotes a unit outward (with respect to V) drawn normal, $\mathbf{n}^+$ $(\mathbf{n}^-)$ denotes a unit normal drawn in the + to − (− to +) sense (see Fig. 4.16). Equation (4.75) then yields

$$\begin{aligned} \mathbf{u}(\mathbf{r}) = {} & \int_V \rho\mathfrak{G} \cdot \mathbf{F}\, dV + \int_{S'} \mathfrak{G} \cdot [\mathbf{n} \cdot \mathfrak{T}_0(\mathbf{u})]dS' - \int_{S'} \mathbf{u} \cdot [\mathbf{n} \cdot \mathfrak{T}_0(\tilde{\mathfrak{G}})]dS' \\ & - \int_{S^+} \mathfrak{G} \cdot \Delta\boldsymbol{\tau}_0\, dS^+ - \int_{S^+} \Delta\mathbf{u}_0 \cdot [\mathbf{n}^+ \cdot \mathfrak{T}_0(\tilde{\mathfrak{G}})]dS^+. \end{aligned} \tag{4.83}$$

[2] It may be noted that the (11) of Fig. 4.8 is different from the (11) of Fig. 4.15. The former corresponds to Γ_i^{11}, whereas the latter corresponds to $T_i^{11} = \lambda(\Gamma_i^{11} + \Gamma_i^{22} + \Gamma_i^{33}) + 2\mu\Gamma_i^{11}$.

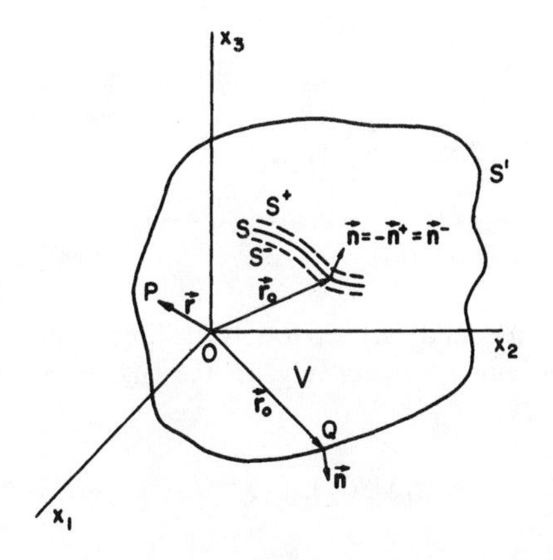

Figure 4.16. The volume V is bounded externally by a surface S' and internally by a surface S, which in the limit consists of the two sides of an open surface. **n** denotes a unit outward drawn normal with respect to V.

In deriving the above equation, use has been made of the fact that $\mathfrak{G}$ and $\mathbf{n}\cdot\mathfrak{T}_0(\tilde{\mathfrak{G}})$ are continuous across S.

If we assume homogeneous boundary conditions on S', the corresponding surface integrals in Eq. (4.83) vanish. For example, if S' is traction free,

$$\mathbf{n}\cdot\mathfrak{T}_0(\mathbf{u}) = 0, \tag{4.84}$$

$$\mathbf{n}\cdot\mathfrak{T}_0(\tilde{\mathfrak{G}}) = 0, \tag{4.85}$$

on S'. Writing S for S^+, Eq. (4.83) becomes

$$\mathbf{u}(\mathbf{r}) = \int_V \rho\mathfrak{G}(\mathbf{r}|\mathbf{r}_1)\cdot\mathbf{F}(\mathbf{r}_1)dV(\mathbf{r}_1) - \int_S \mathfrak{G}(\mathbf{r}|\mathbf{r}_0)\cdot\Delta\boldsymbol{\tau}_0(\mathbf{r}_0)dS(\mathbf{r}_0)$$
$$+ \int_S \Delta\mathbf{u}_0(\mathbf{r}_0)\cdot[\mathbf{n}(\mathbf{r}_0)\cdot\mathfrak{T}_0\{\tilde{\mathfrak{G}}(\mathbf{r}|\mathbf{r}_0)\}]dS(\mathbf{r}_0), \tag{4.86}$$

where $\mathbf{n} = -\mathbf{n}^+ = \mathbf{n}^-$.

From the properties of the delta function, we note that

$$\mathfrak{G}(\mathbf{r}|\mathbf{r}_0) = \int_V \delta(\mathbf{r}_1 - \mathbf{r}_0)\mathfrak{G}(\mathbf{r}|\mathbf{r}_1)dV(\mathbf{r}_1), \tag{4.87}$$

$$\mathbf{n}(\mathbf{r}_0)\cdot\mathfrak{T}_0\{\tilde{\mathfrak{G}}(\mathbf{r}|\mathbf{r}_0)\} = \int_V [\mathbf{n}(\mathbf{r}_0)\cdot\mathfrak{T}_0\{\mathfrak{I}\delta(\mathbf{r}_1 - \mathbf{r}_0)\}]\cdot\tilde{\mathfrak{G}}(\mathbf{r}|\mathbf{r}_1)dV(\mathbf{r}_1). \tag{4.88}$$

Equations (4.86)–(4.88) yield

$$\mathbf{u}(\mathbf{r}) = \int_V \mathfrak{G}(\mathbf{r}|\mathbf{r}_1)\cdot[\rho\mathbf{F}(\mathbf{r}_1) - \int_S \{\delta(\mathbf{r}_1 - \mathbf{r}_0)\Delta\boldsymbol{\tau}_0(\mathbf{r}_0)$$
$$- \Delta\mathbf{u}_0(\mathbf{r}_0)\mathbf{n}(\mathbf{r}_0) : \mathfrak{T}_0[\mathfrak{I}\delta(\mathbf{r}_1 - \mathbf{r}_0)]\}dS(\mathbf{r}_0)]dV(\mathbf{r}_1). \tag{4.89}$$

Consider the expression within the outer brackets in Eq. (4.89). Because the surface integral is involved in the same manner as $\rho\mathbf{F}(\mathbf{r}_1)$ and $\rho\mathbf{F}(\mathbf{r}_1)$ represents a body force per unit volume, we conclude that the effect of the prescribed discontinuities $\Delta\boldsymbol{\tau}_0$ and $\Delta\mathbf{u}_0$ across S is the same as the effect of introducing extra body forces $\mathbf{Q}(\mathbf{r}_1)$, per unit volume, into the unfaulted medium, given by

$$\mathbf{Q}(\mathbf{r}_1) = \int_S \Delta\mathbf{u}_0(\mathbf{r}_0)\mathbf{n}(\mathbf{r}_0) : \mathfrak{T}_0[\mathfrak{I}\delta(\mathbf{r}_1 - \mathbf{r}_0)]dS(\mathbf{r}_0)$$
$$- \int_S \delta(\mathbf{r}_1 - \mathbf{r}_0)\Delta\boldsymbol{\tau}_0(\mathbf{r}_0)dS(\mathbf{r}_0). \tag{4.90}$$

We shall deal with the discontinuities $\Delta\boldsymbol{\tau}_0$ and $\Delta\mathbf{u}_0$ separately, calling them *stress dislocation* and *displacement dislocation*, respectively. In the case of a stress dislocation, we have

$$\mathbf{Q}(\mathbf{r}_1) = -\int_S \delta(\mathbf{r}_1 - \mathbf{r}_0)\Delta\boldsymbol{\tau}_0(\mathbf{r}_0)dS(\mathbf{r}_0). \tag{4.91}$$

It is apparent from Eq. (4.91) that a stress dislocation is equivalent to a distribution of single forces over the dislocation surface. The strength of the single force at a point of the dislocation surface is proportional to the magnitude of the dislocation at that point and its direction is opposite to the direction of the dislocation. For a point stress dislocation at $\mathbf{r} = \mathbf{r}_0$,

$$\mathbf{Q}(\mathbf{r}_0) = -dS\,\Delta\boldsymbol{\tau}_0\,\delta(\mathbf{r} - \mathbf{r}_0), \tag{4.92}$$

which is a concentrated force of magnitude $dS|\Delta\boldsymbol{\tau}_0|$ acting at $\mathbf{r}_0$ in the direction opposite to the direction of the discontinuity in the traction.

However, by the law of action and reaction, one may assume that for earthquake faults $\Delta\boldsymbol{\tau}_0 = 0$. In the fault zone, stresses build up within the earth until the strength of the rock in some region is exceeded, and fracture occurs. The fault is a surface across which there is a discontinuity in the displacement vector.

For a displacement dislocation, we have, from Eq. (4.90),

$$\mathbf{Q}(\mathbf{r}_1) = \int_S \Delta\mathbf{u}_0\,\mathbf{n} : \mathfrak{T}_0[\mathfrak{I}\delta(\mathbf{r}_1 - \mathbf{r}_0)]dS(\mathbf{r}_0). \tag{4.93}$$

The resultant equivalent force is obtained by integrating over V. Clearly

$$\int_V \mathbf{Q}(\mathbf{r}_1)dV(\mathbf{r}_1) = \int_S \Delta\mathbf{u}_0\,\mathbf{n} : \mathfrak{T}_0\left[\mathfrak{I}\int_V \delta(\mathbf{r}_1 - \mathbf{r}_0)dV(\mathbf{r}_1)\right]dS(\mathbf{r}_0) = 0;$$

because the volume integral is 1 and $\mathfrak{T}_0(\mathfrak{I}) = 0$. Further,

$$\int_V \mathbf{r}_1 \times \mathbf{Q}(\mathbf{r}_1)dV(\mathbf{r}_1) = \int_S \Delta\mathbf{u}_0\,\mathbf{n} : \mathfrak{T}_0\left[\int_V \delta(\mathbf{r}_1 - \mathbf{r}_0)(\mathbf{r}_1 \times \mathfrak{I})dV(\mathbf{r}_1)\right]dS(\mathbf{r}_0)$$
$$= \int_S \Delta\mathbf{u}_0\,\mathbf{n} : \mathfrak{T}_0(\mathbf{r}_0 \times \mathfrak{I})dS(\mathbf{r}_0) = 0,$$

because $\mathfrak{T}_0(\mathbf{r}_0 \times \mathfrak{I}) = 0$. Hence, in an inhomogeneous, anisotropic medium, a displacement dislocation is equivalent to a force system that has zero net force and zero net moment.

For a point displacement dislocation at $\mathbf{r}_0$, Eq. (4.93) yields

$$\mathbf{Q}(\mathbf{r}_0) = U_0\, dS\, \mathbf{en} : \mathfrak{T}_0[\mathfrak{I}\delta(\mathbf{r} - \mathbf{r}_0)], \tag{4.94}$$

where $\Delta\mathbf{u}_0 = U_0\mathbf{e}$ and $\mathbf{e}$ is a unit vector that will be referred to as *unit-slip vector*. The direction of the unit vector $\mathbf{e}$ is known as the *direction of motion* and U_0 as *slip*. The quantity $U_0\, dS$ is a measure of the strength of the source and is called the *potency* of the dislocation source. If we define the dyadic $\mathfrak{q} = \mathfrak{I}\delta(\mathbf{r} - \mathbf{r}_0)$ then Eq. (4.51) can be put in the form

$$\operatorname{div} \mathfrak{T}(\mathbf{u}) + \rho\omega^2\mathbf{u} = U_0 g(\omega) dS[\mathbf{en} : \mathfrak{T}(\mathfrak{q})]. \tag{4.94a}$$

If the medium in the immediate neighborhood of the source may be regarded as isotropic, we may use definition (4.76) for the operator $\mathfrak{T}_0$. Equation (4.94) then becomes

$$\mathbf{Q}(\mathbf{r}_0) = -U_0\, dS[\lambda\mathbf{e}\cdot\mathbf{n} \operatorname{grad} \delta(\mathbf{r} - \mathbf{r}_0) + \mu(\mathbf{en} + \mathbf{ne})\cdot \operatorname{grad} \delta(\mathbf{r} - \mathbf{r}_0)]. \tag{4.95}$$

The following results have been used in deriving Eq. (4.95):

$$\mathbf{en} : \mathfrak{I} = \mathbf{e}\cdot\mathbf{n}, \tag{4.96}$$

$$\mathbf{en} : \operatorname{grad}(\mathfrak{I}u) = \mathbf{en} : [(\operatorname{grad} u)\mathfrak{I}] = \mathbf{en}\cdot \operatorname{grad} u, \tag{4.97}$$

$$\mathbf{en} : [\operatorname{grad}(\mathfrak{I}u)]^{213} = \mathbf{ne} : \operatorname{grad}(\mathfrak{I}u). \tag{4.98}$$

Two types of displacement dislocations are of particular interest: *tangential (or shear) displacement dislocation* and *tensile displacement dislocation*. As the names suggest, in a tangential dislocation $\mathbf{e}$ is perpendicular to $\mathbf{n}$ and in a tensile dislocation $\mathbf{e}$ coincides with $\mathbf{n}$. For a tangential dislocation, Eq. (4.95) gives

$$\mathbf{Q}(\mathbf{r}_0) = -\mu U_0\, dS(\mathbf{en} + \mathbf{ne})\cdot \operatorname{grad} \delta(\mathbf{r} - \mathbf{r}_0), \qquad (\mathbf{e}\cdot\mathbf{n} = 0). \tag{4.99}$$

Comparing Eqs. (4.52) and (4.99), we note that a point shear dislocation is equivalent to a double couple of moment

$$M = \mu U_0\, dS. \tag{4.100}$$

Equation (4.100) is known as the *equivalence theorem* for shear dislocations. The directions of the forces of the double couple are the directions of the vectors $\mathbf{e}$ and $\mathbf{n}$ for the shear dislocation.

In the case of a tensile dislocation, Eq. (4.95) yields

$$\mathbf{Q}(\mathbf{r}_0) = -U_0\, dS[\lambda \operatorname{grad} \delta(\mathbf{r} - \mathbf{r}_0) + 2\mu\mathbf{nn}\cdot \operatorname{grad} \delta(\mathbf{r} - \mathbf{r}_0)]. \tag{4.101}$$

Comparing Eqs. (4.52) and (4.101), we find that a point tensile dislocation with normal in the direction of a unit vector $\mathbf{n}$ is equivalent to a dipole of moment

$$M = 2\mu U_0\, dS \tag{4.102}$$

in the direction of the vector $\mathbf{n}$ plus a center of compression, which, in turn, is equivalent to three mutually perpendicular dipoles each having the moment

$$M = \lambda U_0\, dS. \tag{4.103}$$

From Eq. (4.93), it may be concluded that an arbitrary displacement dislocation is equivalent to a distribution of dipoles and double couples of suitable strengths over the dislocation surface S.[3]

Let us return to Eq. (4.86). In the absence of external body forces, this equation becomes, with $g(\omega) = 1$,

$$\mathbf{u}(\mathbf{r}) = \int_S U_0 \, \mathbf{en} : \mathfrak{T}_0(\tilde{\mathfrak{G}}) dS - \int_S \mathfrak{G} \cdot \Delta\tau_0 \, dS. \tag{4.104}$$

This is known as *Volterra's relation.* The Green's dyadic $\mathfrak{G}$ satisfies the same boundary conditions on the outer surface S' as $\mathbf{u}$. It may be recalled that $\mathfrak{G} = \mathfrak{G}_\infty + \hat{\mathfrak{G}}$, where $\mathfrak{G}_\infty$ is the Green's dyadic for an infinite space. The unknown dyadic $\hat{\mathfrak{G}}$ is chosen in such a manner that $\mathfrak{G}$ satisfies the prescribed boundary conditions on S'.

Equation (4.104) gives the displacement field throughout the volume V bounded by S' caused by jumps in the displacement and stress vectors across an internal surface S. These jumps are given by Eq. (4.82). For a displacement dislocation, Eq. (4.104) reduces to

$$\mathbf{u}(\mathbf{r}) = \int_S U_0 \, \mathbf{en} : \mathfrak{T}_0(\tilde{\mathfrak{G}}) dS. \tag{4.105}$$

If the medium is unbounded, we have

$$\mathbf{u}(\mathbf{r}) = \int_S U_0 \, \mathbf{en} : \mathfrak{T}_0(\tilde{\mathfrak{G}}_\infty) dS. \tag{4.106}$$

We now consider the special case when S is a small planar surface of area dS. In that case Eq. (4.105) may be approximated as:

$$\mathbf{u}(\mathbf{r}) = U_0 \, dS \, \mathbf{en} : \mathfrak{T}_0(\tilde{\mathfrak{G}}). \tag{4.107}$$

Because the operator $\mathfrak{T}_0$ is symmetrical, we can interchange the vectors $\mathbf{e}$ and $\mathbf{n}$ in Eq. (4.107). Therefore,

$$\mathbf{u}(\mathbf{e}, \mathbf{n}) = \mathbf{u}(\mathbf{n}, \mathbf{e}). \tag{4.108}$$

4.4. Fault-Plane Geometry

4.4.1. Nomenclature

If the source is placed in an infinite medium, we can choose the origin at the source and any three mutually perpendicular lines through the source as the three coordinate axes. However, because we intend to solve the problem of a

[3] We have shown here the equivalence of a shear displacement dislocation to a double couple and of a tensile displacement dislocation to a center of compression plus a dipole, in the case of an isotropic, inhomogeneous medium only. The generalization to the case of an anisotropic medium is immediate.

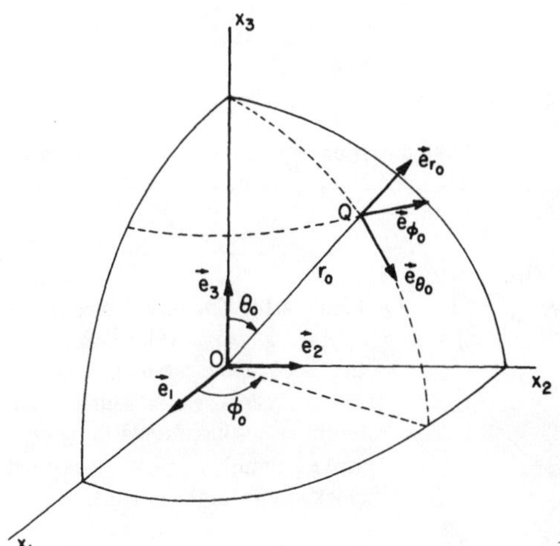

Figure 4.17. The geographic system ($\mathbf{e}_1, \mathbf{e}_2, \mathbf{e}_3$) and the epicentral system ($\mathbf{e}_{\theta_0}, \mathbf{e}_{\phi_0}, \mathbf{e}_{r_0}$).

source inside the earth, we define a set of coordinate systems that will be useful for this purpose.

The first system of coordinates, which we shall call the *geographic system of coordinates*, has the center of the earth O as its origin. The x_1 axis is drawn in the Greenwich meridian and the x_3 axis through the North Pole. The third axis is determined from the condition that the system of coordinates is right handed (Fig. 4.17). The unit vectors along x_i are denoted by $\mathbf{e}_i$. We also have a spherical coordinate system r, θ, ϕ with unit vectors ($\mathbf{e}_r, \mathbf{e}_\theta, \mathbf{e}_\phi$) centered at O. The azimuth angle ϕ is measured counterclockwise from Ox_1 and $0 \leq \theta \leq \pi$, $0 \leq \phi < 2\pi$. The point source is situated at the point $Q(r_0, \theta_0, \phi_0)$. The three unit vectors at Q in the directions of (r_0, θ_0, ϕ_0) are ($\mathbf{e}_{r_0}, \mathbf{e}_{\theta_0}, \mathbf{e}_{\phi_0}$).

We now execute the following orthogonal transformations on the geographic system:

1. Rotate it about Ox_3 through an angle ϕ_0. This leaves x_3 axis undisturbed while the x_1 axis is brought to the source meridian.
2. Rotate the system of coordinates just obtained about the new position of the x_2 axis through an angle θ_0. This makes the x_3 axis pass through the source.

The system of coordinates thus obtained is called the *epicentral system of coordinates* and is denoted by ($\mathbf{e}_{\theta_0}, \mathbf{e}_{\phi_0}, \mathbf{e}_{r_0}$). In this system the three axes are in the directions of south, east and up, respectively. The geographic system and the epicentral system are connected through the relations

	$\mathbf{e}_1$	$\mathbf{e}_2$	$\mathbf{e}_3$
$\mathbf{e}_{\theta_0}$	$\cos\theta_0 \cos\phi_0$	$\cos\theta_0 \sin\phi_0$	$-\sin\theta_0$
$\mathbf{e}_{\phi_0}$	$-\sin\phi_0$	$\cos\phi_0$	0
$\mathbf{e}_{r_0}$	$\sin\theta_0 \cos\phi_0$	$\sin\theta_0 \sin\phi_0$	$\cos\theta_0$

(4.109)

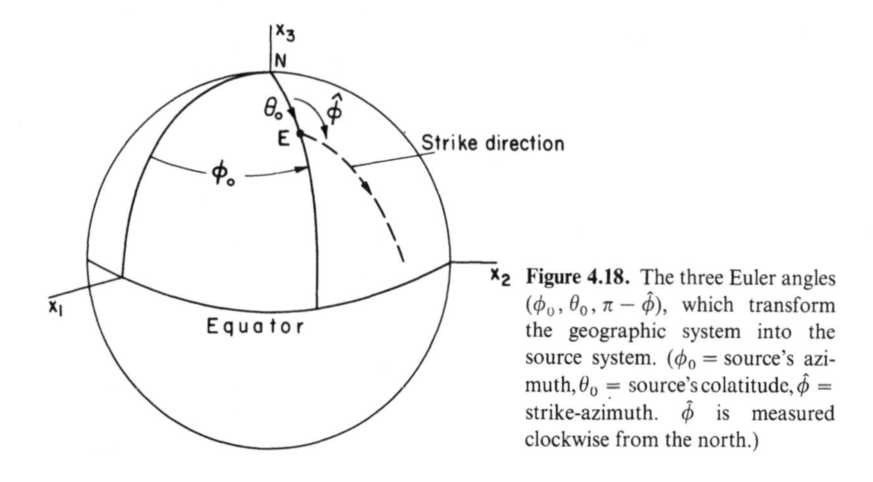

Figure 4.18. The three Euler angles $(\phi_0, \theta_0, \pi - \hat{\phi})$, which transform the geographic system into the source system. (ϕ_0 = source's azimuth, θ_0 = source's colatitude, $\hat{\phi}$ = strike-azimuth. $\hat{\phi}$ is measured clockwise from the north.)

If the source lies on the x_3 axis, the geographic system and the epicentral system are identical.

Let a point displacement dislocation be situated at Q. Then the plane through the point Q with normal **n** is called the *fault plane* or *dislocation plane.* This is the surface of the fracture simulated by the displacement dislocation. The plane through Q with normal **e** is the *auxiliary plane.* There exists an ambiguity between the fault plane and the auxiliary plane because of relation (4.108), and, therefore, these two planes are known as *conjugate planes.* The line of intersection of the fault plane with the surface of the earth is known as the *strike direction* or the *strike* of the fault. Its orientation is usually expressed in terms of an angle $\hat{\phi}(0 \le \hat{\phi} < 2\pi$; see Fig. 4.18), measured clockwise from the north. The angle $\hat{\phi}$ so defined is known as the *fault's azimuth* or the *strike azimuth.*

The mass of the rock below an inclined fault plane is known as the *foot wall,* and the mass of the rock above it as the *hanging wall.* A line on the earth's surface perpendicular to the strike drawn in the direction in which the fault plane is dipping is known as the *dip direction.* A line in the fault plane perpendicular to the strike drawn downward is the *dip of the fault* (Fig. 4.19). We take the positive direction of the strike to the right of an observer facing the foot wall.

By rotating the epicentral system of coordinates $(\mathbf{e}_{\theta_0}, \mathbf{e}_{\phi_0}, \mathbf{e}_{r_0})$ about $\mathbf{e}_{r_0}$ through an angle $\pi - \hat{\phi}$, we obtain yet another system of coordinates $(\mathbf{e}_1^\circ, \mathbf{e}_2^\circ\ \mathbf{e}_3^\circ)$, known as the *source system* (Figs. 4.20 and 4.21). We have the transformation

	$\mathbf{e}_{\theta_0}$	$\mathbf{e}_{\phi_0}$	$\mathbf{e}_{r_0}$
$\mathbf{e}_1^\circ$	$-\cos\hat{\phi}$	$\sin\hat{\phi}$	0
$\mathbf{e}_2^\circ$	$-\sin\hat{\phi}$	$-\cos\hat{\phi}$	0
$\mathbf{e}_3^\circ$	0	0	1

(4.110)

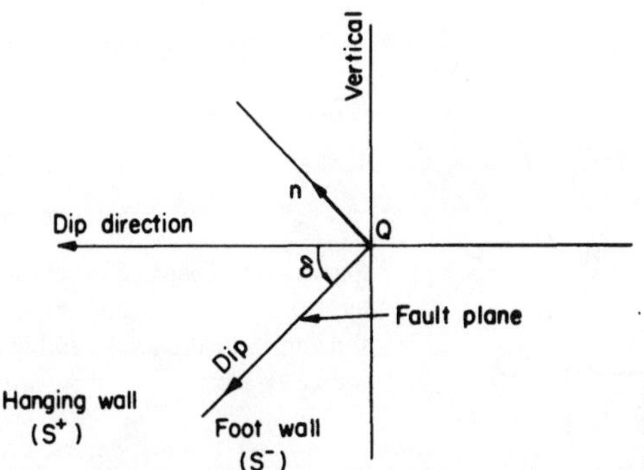

Figure 4.19. Displacement dislocation: side view showing the dip direction, the dip, and the dip angle.

It may be noted that $(\phi_0, \theta_0, \pi - \hat{\phi})$ are the Euler angles, which transform the geographic system into the source system.

Let us study the properties of the source system. The x_3° axis passes through the source and is drawn vertically upward. The x_1° axis is in the strike direction. The choice of the direction of the unit vector $\mathbf{n}$ normal to the fault plane is at our disposal. We choose the direction of $\mathbf{n}$ to be such that the angle between $\mathbf{n}$ and $\mathbf{e}_3^\circ$ is acute. The unit vectors $\mathbf{e}$, $\mathbf{n}$, and $\mathbf{e}_i^\circ$, all drawn at the point Q, are shown in Fig. 4.21. The vector $\mathbf{n}$ lies in the $x_2^\circ x_3^\circ$ plane. We define a unit vector $\mathbf{b}$ as $\mathbf{b} = \mathbf{n} \times \mathbf{e}$ and call it the *null vector*. The vectors $(\mathbf{e}, \mathbf{b}, \mathbf{n})$ form a right-handed system. Let λ be the angle that the vector $\mathbf{e}$ makes with the strike direction, measured counterclockwise when viewed from the hanging wall side of the fault plane. Further, let δ be the angle that the direction of the vector $\mathbf{n}$ makes with the x_3° axis (vertical). Obviously, δ is also the angle that the fault plane makes with the horizontal plane through the source. We adhere to the convention $0 \le \lambda < 2\pi$, $0 \le \delta \le \pi/2$. The angle λ is known as the *slip angle* and the angle δ is known as the *dip angle*.

If $\lambda = 0$ or $180°$, the fault is known as *strike-slip*, *lateral*, *transcurrent*, or *wrench fault*. In contrast, if $\lambda = 90°$ or $270°$, the fault is known as *dip-slip fault*.

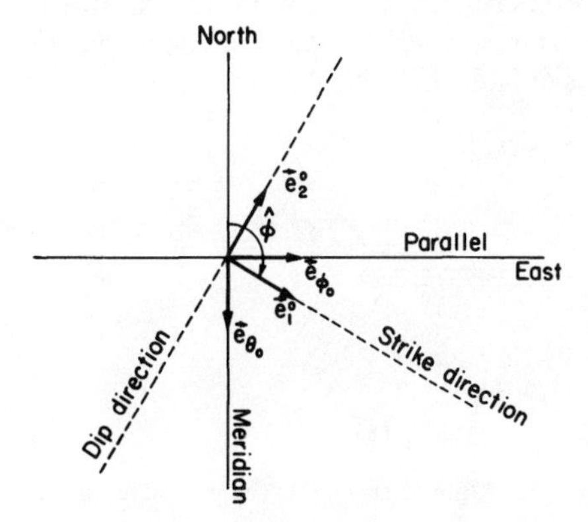

Figure 4.20. Displacement dislocation: top view showing the dip direction and the strike direction; also the epicentral system $(\mathbf{e}_{\theta_0}, \mathbf{e}_{\phi_0}, \mathbf{e}_{r_0})$ and the source system $(\mathbf{e}_1^0, \mathbf{e}_2^0, \mathbf{e}_3^0 = \mathbf{e}_{r_0})$. $\mathbf{e}_3^0$ is perpendicular to the plane of the paper, toward the reader.

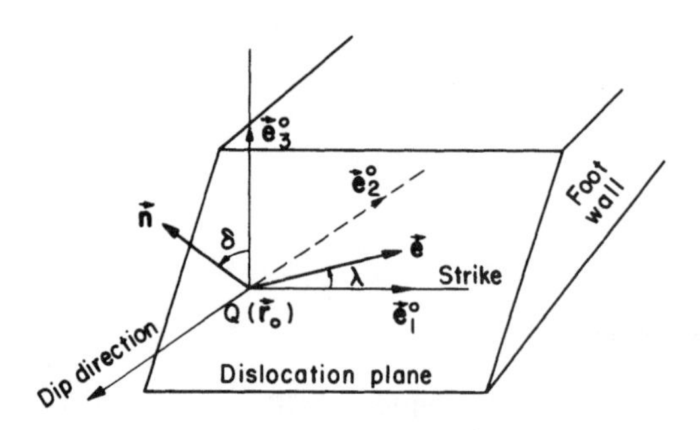

Figure 4.21. Displacement dislocation: front view showing the slip angle, the dip angle, and the source system.

It may be noted that Fig. 4.21 shows the foot wall. However, the displacement $U_0\mathbf{e}$ is the displacement of the hanging wall (not shown in Fig. 4.21) relative to the foot wall. In the notation of Fig. 4.16 and Eq. (4.82), therefore, the hanging wall has been chosen as S^+ and the foot wall as S^-. In the case of a strike-slip fault, the relative displacement is along the strike direction. In the case of a dip-slip fault, however, the relative displacement is perpendicular to the strike direction, in the fault plane. In general, the fault will not be purely of a strike-slip or a dip-slip type. A more general fault is equivalent to the superposition of a strike-slip fault of slip $U_0 \cos \lambda$ and a dip-slip fault of slip $U_0 \sin \lambda$, and is known as *oblique-slip fault*.

In geology, it is common to choose λ and δ in the first quadrant and to define two kinds of strike-slip faults, *dextral* and *sinistral*, and two kinds of dip-slip faults, *normal* and *reverse*. In a dextral strike-slip fault, which is also known as *right-lateral fault*, the direction of the relative displacement of the side of the fault opposite the observer who is facing the fault is to the right (see Fig. 4.22a). In a sinistral strike-slip fault (*left-lateral fault*), the direction of the relative displacement of the opposite side is to the left (Fig. 4.22b). A *normal fault* is an inclined fracture along which the hanging wall moves down relative to the foot wall (see Fig. 4.23a). A *reverse fault* is an inclined fracture along which the hanging wall moves up relative to the foot wall (Fig. 4.23b). It is obvious that in normal faults the horizontal extent increases, whereas in reverse faults the slip shortens the horizontal extent. Sometimes, a reverse fault with $\delta < 45°$ is also known as *thrust fault*, and with $\delta < 10°$ as *overthrust fault*. We have summarized the geologic terminology of various kinds of faults in Table 4.2.

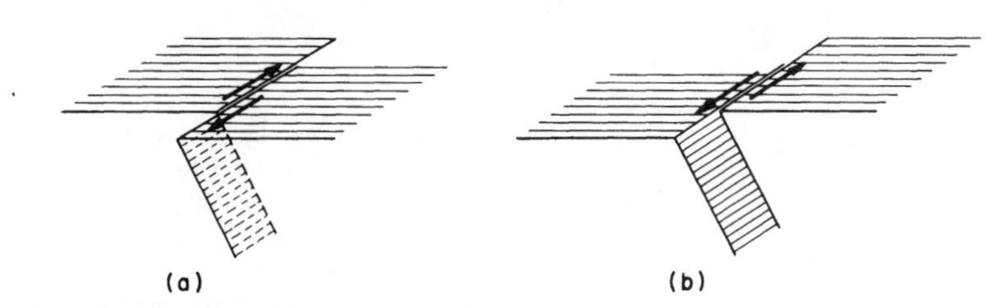

Figure 4.22. Two kinds of strike-slip faults. (a) Right-lateral fault; (b) left-lateral fault.

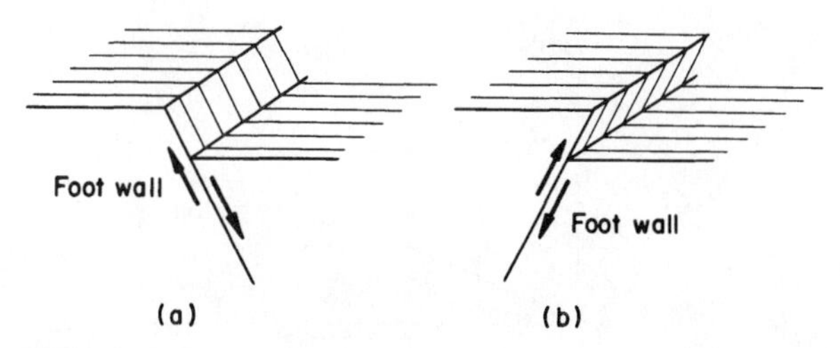

Figure 4.23. Two kinds of dip-slip faults. (a) Normal; (b) reverse.

According to our convention, the three displacement components along $(\mathbf{e}_1^\circ, \mathbf{e}_2^\circ, \mathbf{e}_3^\circ)$ at the source are positive on the hanging wall side of a reverse left-lateral fault.

The motion on the fault can be described alternatively in terms of the *trend of motion* and the *plunge of motion* (also known as *rake* or *pitch*). The trend of motion (t) is the azimuth of the projection of the unit-slip vector $\mathbf{e}$ on the earth's surface, measured from strike, counterclockwise. The plunge of motion (p) is the angle that the vector $\mathbf{e}$ makes with the horizontal plane through the source, measured upward. These angles are shown in Fig. 4.24.

The breakdown of the total motion U_0 on the fault is given below (Fig. 4.24):

$$\begin{aligned} &\text{Motion along the strike} && (QA) = U_0 \cos\lambda, \\ &\text{Motion along the dip} && (AC) = U_0 \sin\lambda, \\ &\text{Motion in vertical direction} && (BC) = U_0 \sin\lambda \sin\delta, \\ &\text{Motion in dip direction} && (AB) = U_0 \sin\lambda \cos\delta, \\ &\text{Total horizontal motion} && (QB) = U_0(1 - \sin^2\lambda \sin^2\delta)^{1/2}. \end{aligned} \tag{4.111}$$

We further have

$$\begin{aligned} \sin t &= \frac{\sin\lambda\cos\delta}{\sqrt{(1 - \sin^2\lambda\sin^2\delta)}}, \\ \sin p &= \sin\lambda\sin\delta. \end{aligned} \tag{4.112}$$

le 4.2. Geologic Terminology of a Shear Fault

	λ					
δ	0°–90°	90°–180°	180°–270°	270°–360°	0° or 180°	90° or 270°
45°	Left-lateral thrust	Right-lateral thrust	Right-lateral normal	Left-lateral normal	Strike slip	Dip slip
-90°	Left-lateral reverse	Right-lateral reverse	Right-lateral normal	Left-lateral normal	Strike slip	Dip slip

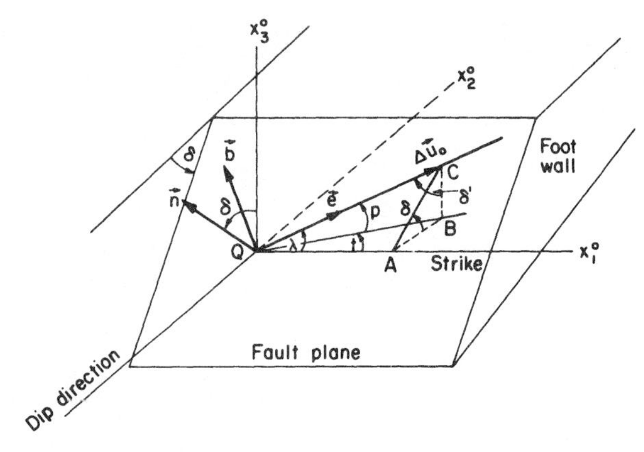

Figure 4.24. The angles λ (slip), δ (dip), p (plunge), and t (trend) and the unit vectors $\mathbf{e}$ (unit-slip vector), $\mathbf{n}$ (normal), and $\mathbf{b}$ (null vector). The angle δ' is defined by Eq. (4.126).

Remembering that $\mathbf{b} = \mathbf{n} \times \mathbf{e}$, it is easily seen that the coordinate system having unit vectors $(\mathbf{e}, \mathbf{b}, \mathbf{n})$ as base vectors is obtained from the source system x_i° through the following transformations (see Fig. 4.24):

1. Rotating the source system about x_1° axis through an angle δ.
2. Rotating the system just obtained about the new position of the x_3° axis through an angle λ.

We therefore have

	$\mathbf{e}_1^\circ$	$\mathbf{e}_2^\circ$	$\mathbf{e}_3^\circ$
$\mathbf{e}$	$\cos\lambda$	$\sin\lambda\cos\delta$	$\sin\lambda\sin\delta$
$\mathbf{b}$	$-\sin\lambda$	$\cos\lambda\cos\delta$	$\cos\lambda\sin\delta$
$\mathbf{n}$	0	$-\sin\delta$	$\cos\delta$

(4.113)

4.4.2. Fundamental Shear Dislocations

In the case of a shear dislocation, $\mathbf{e}\cdot\mathbf{n} = 0$ and, therefore, Eq. (4.107) may be written as $[g(\omega) = 1]$

$$\mathbf{u}(\mathbf{r}) = \mu U_0\, dS(\mathbf{en} + \mathbf{ne}) : \mathrm{grad}_0\, \tilde{\mathfrak{G}}. \tag{4.114}$$

When Eqs. (4.53) and (4.114) are compared, it is obvious that for a point shear dislocation, the moment tensor is $\mathfrak{M} = M(\mathbf{en} + \mathbf{ne})$, where $M = \mu U_0\, dS$. In terms of the angles λ and δ, we have, from Eq. (4.113),

$$\begin{aligned}\mathbf{en} + \mathbf{ne} = &- \sin\lambda\sin 2\delta(\mathbf{e}_2^\circ\mathbf{e}_2^\circ - \mathbf{e}_3^\circ\mathbf{e}_3^\circ) + \sin\lambda\cos 2\delta(\mathbf{e}_2^\circ\mathbf{e}_3^\circ + \mathbf{e}_3^\circ\mathbf{e}_2^\circ)\\ &+ \cos\lambda\cos\delta(\mathbf{e}_3^\circ\mathbf{e}_1^\circ + \mathbf{e}_1^\circ\mathbf{e}_3^\circ) - \cos\lambda\sin\delta(\mathbf{e}_1^\circ\mathbf{e}_2^\circ + \mathbf{e}_2^\circ\mathbf{e}_1^\circ).\end{aligned} \tag{4.115a}$$

Sometimes it is necessary to express $\mathfrak{M}$ in terms of the epicentral system. Equations (4.110) and (4.115a) yield

$$\begin{aligned}\mathbf{en}+\mathbf{ne} = &-(p_2+p_5)\mathbf{e}_{\theta_0}\mathbf{e}_{\theta_0}+(p_2-p_5)\mathbf{e}_{\phi_0}\mathbf{e}_{\phi_0}+2p_5\,\mathbf{e}_{r_0}\mathbf{e}_{r_0}\\ &-p_3(\mathbf{e}_{\phi_0}\mathbf{e}_{r_0}+\mathbf{e}_{r_0}\mathbf{e}_{\phi_0})-p_4(\mathbf{e}_{r_0}\mathbf{e}_{\theta_0}+\mathbf{e}_{\theta_0}\mathbf{e}_{r_0})-p_1(\mathbf{e}_{\theta_0}\mathbf{e}_{\phi_0}+\mathbf{e}_{\phi_0}\mathbf{e}_{\theta_0}),\end{aligned} \tag{4.115b}$$

where

$$\begin{aligned}p_1 &= (\sin\lambda\sin\delta\cos\delta)\sin 2\hat{\phi}+(\cos\lambda\sin\delta)\cos 2\hat{\phi},\\ p_2 &= (\cos\lambda\sin\delta)\sin 2\hat{\phi}-(\sin\lambda\sin\delta\cos\delta)\cos 2\hat{\phi},\\ p_3 &= -(\cos\lambda\cos\delta)\sin\hat{\phi}+(\sin\lambda\cos 2\delta)\cos\hat{\phi},\\ p_4 &= (\sin\lambda\cos 2\delta)\sin\hat{\phi}+(\cos\lambda\cos\delta)\cos\hat{\phi},\\ p_5 &= \sin\lambda\sin\delta\cos\delta.\end{aligned} \tag{4.116}$$

From Eqs. (4.115b) and (4.116), it is found that

$$\begin{aligned}\mathfrak{M}(\lambda,\delta,\hat{\phi}) = {}& p_1\mathfrak{M}\left(0,\frac{\pi}{2},0\right)-p_3\,\mathfrak{M}\left(\frac{\pi}{2},\frac{\pi}{2},0\right)+2p_5\,\mathfrak{M}\left(\frac{\pi}{2},\frac{\pi}{4},0\right)\\ &+(p_2+p_5)\mathfrak{M}\left(0,\frac{\pi}{2},\frac{\pi}{4}\right)-p_4\,\mathfrak{M}\left(\frac{\pi}{2},\frac{\pi}{2},\frac{\pi}{2}\right).\end{aligned} \tag{4.117}$$

In Eq. (4.117),

$$\mathfrak{M}\left(0,\tfrac{\pi}{2},0\right) = -M(\mathbf{e}_{\theta_0}\mathbf{e}_{\varphi_0}+\mathbf{e}_{\varphi_0}\mathbf{e}_{\theta_0}), \tag{4.118a}$$

$$\mathfrak{M}\left(0,\tfrac{\pi}{2},\tfrac{\pi}{4}\right) = M(\mathbf{e}_{\varphi_0}\mathbf{e}_{\varphi_0}-\mathbf{e}_{\theta_0}\mathbf{e}_{\theta_0}), \tag{4.118b}$$

$$\mathfrak{M}\left(\tfrac{\pi}{2},\tfrac{\pi}{2},0\right) = M(\mathbf{e}_{\varphi_0}\mathbf{e}_{r_0}+\mathbf{e}_{r_0}\mathbf{e}_{\varphi_0}), \tag{4.119a}$$

$$\mathfrak{M}\left(\tfrac{\pi}{2},\tfrac{\pi}{2},\tfrac{\pi}{2}\right) = M(\mathbf{e}_{r_0}\mathbf{e}_{\theta_0}+\mathbf{e}_{\theta_0}\mathbf{e}_{r_0}), \tag{4.119b}$$

$$\mathfrak{M}\left(\tfrac{\pi}{2},\tfrac{\pi}{4},0\right) = M(\mathbf{e}_{r_0}\mathbf{e}_{r_0}-\mathbf{e}_{\varphi_0}\mathbf{e}_{\varphi_0}), \tag{4.120}$$

Equation (4.117) is of fundamental importance. It shows that the displacement field induced by an arbitrary shear dislocation can be expressed in terms of the displacement fields induced by five double couples. The double couples

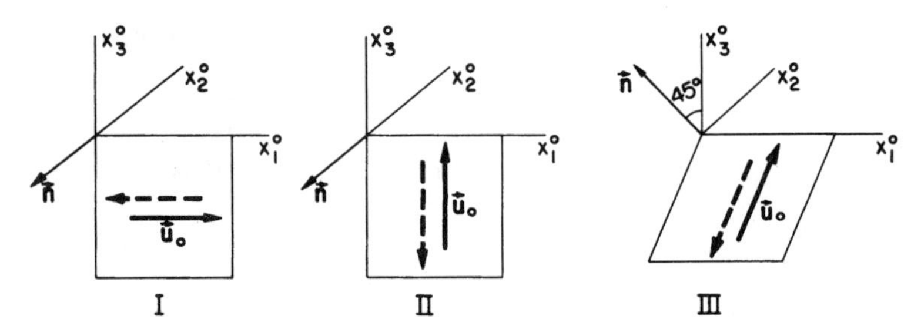

Figure 4.25. The three fundamental shear dislocations: I, vertical strike slip; II, vertical dip slip; III, dip slip on a 45° dipping plane.

represented by Eqs. (4.118a, b) correspond to vertical strike-slip faults, the double couples represented by Eqs. (4.119a, b) correspond to vertical dip-slip faults, and the double couple represented by Eq. (4.120) corresponds to dip-slip on a 45° dipping fault (see Fig. 4.25 and Table 4.3).

When treating dislocation sources in infinite space it is sometimes more convenient to define a *fault coordinate system* (η_1, η_2, η_3) with unit vectors $(\mathbf{i}_1, \mathbf{i}_2, \mathbf{i}_3)$ such that the fault plane itself constitutes one of the coordinate planes. In this system, shown in Fig. 4.26, the η_1 axis is taken in the strike direction and the η_3 axis is taken along $(-\mathbf{n})$. Hence, we have

$$\begin{array}{c|ccc} & \mathbf{e} & \mathbf{b} & \mathbf{n} \\ \hline \mathbf{i}_1 & \cos\lambda & -\sin\lambda & 0 \\ \mathbf{i}_2 & -\sin\lambda & -\cos\lambda & 0 \\ \mathbf{i}_3 & 0 & 0 & -1 \end{array} \tag{4.121a}$$

$$\begin{array}{c|ccc} & \mathbf{e}_1^\circ & \mathbf{e}_2^\circ & \mathbf{e}_3^\circ \\ \hline \mathbf{i}_1 & 1 & 0 & 0 \\ \mathbf{i}_2 & 0 & -\cos\delta & -\sin\delta \\ \mathbf{i}_3 & 0 & \sin\delta & -\cos\delta \end{array} \tag{4.121b}$$

$$\mathbf{en} = \cos\lambda(-\mathbf{i}_1\mathbf{i}_3) - \sin\lambda(-\mathbf{i}_2\mathbf{i}_3). \tag{4.121c}$$

Table 4.3. Fundamental Shear Dislocations

Source	λ	δ	$\mathbf{e}$	$\mathbf{n}$	Geologic terminology
I	0	$\pi/2$	$\mathbf{e}_1^\circ$	$-\mathbf{e}_2^\circ$	Vertical strike slip
II	$\pi/2$	$\pi/2$	$\mathbf{e}_3^\circ$	$-\mathbf{e}_2^\circ$	Vertical dip slip
III	$\pi/2$	$\pi/4$	$(\mathbf{e}_3^\circ + \mathbf{e}_2^\circ)/\sqrt{2}$	$(\mathbf{e}_3^\circ - \mathbf{e}_2^\circ)/\sqrt{2}$	Dip slip on a 45° dipping fault

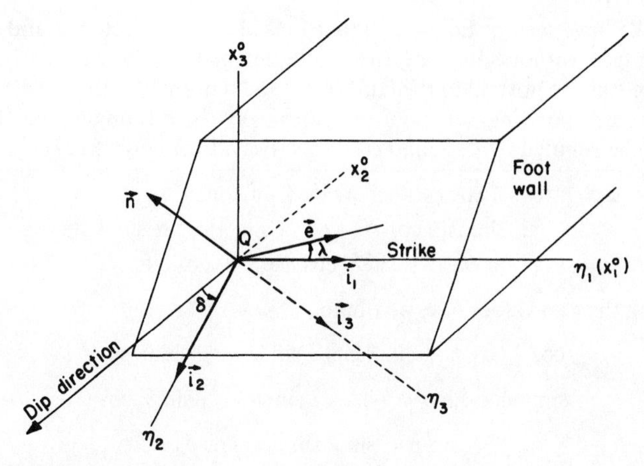

Figure 4.26. The fault system (η_1, η_2, η_3) and the source system (x_1^0, x_2^0, x_3^0). $(\mathbf{i}_3 = -\mathbf{n}.)$

Therefore, the whole displacement field splits into two parts: a strike-slip field resulting from a *longitudinal shear fault* $(-\mathbf{i}_1\mathbf{i}_3)$ and a dip-slip field resulting from a *transverse shear fault* $(-\mathbf{i}_2\mathbf{i}_3)$, as depicted in Fig. 4.27. The terms longitudinal and transverse are related to the direction of rupture of a finite fault treated in Section 4.7.

4.4.3. Fault-Plane—Auxiliary-Plane Ambiguity

From Eqs. (4.110) and (4.113), we get $(\mathbf{e}, \mathbf{b}, \mathbf{n})$ in terms of $(\mathbf{e}_{\theta_0}, \mathbf{e}_{\phi_0}, \mathbf{e}_{r_0})$:

	$\mathbf{e}_{\theta_0}$	$\mathbf{e}_{\phi_0}$	$\mathbf{e}_{r_0}$
e	$-\cos\lambda\cos\hat{\phi}$ $-\sin\lambda\cos\delta\sin\hat{\phi}$	$-\sin\lambda\cos\delta\cos\hat{\phi}$ $+\cos\lambda\sin\hat{\phi}$	$\sin\lambda\sin\delta$
b	$\sin\lambda\cos\hat{\phi}$ $-\cos\lambda\cos\delta\sin\hat{\phi}$	$-\cos\lambda\cos\delta\cos\hat{\phi}$ $-\sin\lambda\sin\hat{\phi}$	$\cos\lambda\sin\delta$
n	$\sin\delta\sin\hat{\phi}$	$\sin\delta\cos\hat{\phi}$	$\cos\delta$

(4.122)

In this equation, λ, δ, $\hat{\phi}$ are, respectively, the slip angle, the dip angle, and the strike azimuth, $\mathbf{n}$ being normal to the fault plane and $\mathbf{e}$ normal to the auxiliary

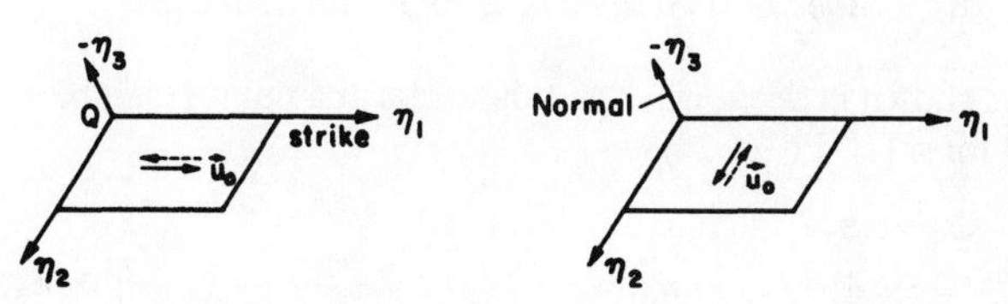

Figure 4.27. Two kinds of shear dislocations. (a) Longitudinal shear; (b) transverse shear.

plane. We have seen in Eq. (4.108) that the roles of the vectors **e** and **n** can be interchanged without affecting the displacement field. Consequently, we can also take **e** as the normal to the fault plane and **n** normal to the auxiliary plane. Let the corresponding values of the slip angle, the dip angle, and the strike azimuth be denoted by λ', δ', and $\hat{\phi}'$, respectively. From Eq. (4.122)

$$\begin{aligned}\mathbf{e} &= -(\cos\lambda\cos\hat{\phi} + \sin\lambda\cos\delta\sin\hat{\phi})\mathbf{e}_{\theta_0} \\ &\quad - (\sin\lambda\cos\delta\cos\hat{\phi} - \cos\lambda\sin\hat{\phi})\mathbf{e}_{\phi_0} + \sin\lambda\sin\delta\mathbf{e}_{r_0} \\ &= \sin\delta'\sin\hat{\phi}'\mathbf{e}_{\theta_0} + \sin\delta'\cos\hat{\phi}'\mathbf{e}_{\phi_0} + \cos\delta'\mathbf{e}_{r_0}.\end{aligned} \tag{4.123}$$

Equating the two values of **e**, we obtain

$$\cos\lambda\cos\hat{\phi} + \sin\lambda\cos\delta\sin\hat{\phi} = -\sin\delta'\sin\hat{\phi}', \tag{4.124}$$

$$\sin\lambda\cos\delta\cos\hat{\phi} - \cos\lambda\sin\hat{\phi} = -\sin\delta'\cos\hat{\phi}', \tag{4.125}$$

$$\sin p = \sin\lambda\sin\delta = \cos\delta'. \tag{4.126}$$

Eliminating λ from Eqs. (4.124)–(4.126), we get

$$\tan\delta\tan\delta'\cos(\hat{\phi} - \hat{\phi}') = -1. \tag{4.127}$$

Further, from Eqs. (4.124), (4.126), and (4.127), we obtain

$$\cos\lambda = \sin\delta'\sin(\hat{\phi} - \hat{\phi}'). \tag{4.128}$$

Relations (4.126)–(4.128) are obtained by equating the direction cosines of the vector **e**. However, if we repeat the process for the vector **n**, we get relation (4.127) and

$$\sin\lambda'\sin\delta' = \cos\delta, \tag{4.129}$$

$$\cos\lambda' = \sin\delta\sin(\hat{\phi}' - \hat{\phi}). \tag{4.130}$$

The above analysis shows that for a given problem of point shear dislocation, there exist two solutions: $\lambda, \delta, \hat{\phi}$ and $\lambda', \delta', \hat{\phi}'$. The parameters of the two solutions are connected through the relations (4.126)–(4.130). Given one solution, these relations can be used to get the other solution. Beginning with $(\lambda, \delta, \hat{\phi})$ solution, δ' can be found from Eq. (4.126), using the condition $0 \le \delta' \le \pi/2$. Next, Eq. (4.127) gives $\cos(\hat{\phi} - \hat{\phi}')$ and Eq. (4.128) gives $\sin(\hat{\phi} - \hat{\phi}')$. Hence $\hat{\phi}'$ is known uniquely. Last, Eq. (4.129) gives $\sin\lambda'$ and Eq. (4.130) gives $\cos\lambda'$. Hence λ' is determined. For example, if $\lambda = \pi/2$, $\delta = \pi/4$, then, following the method just described, it may be seen that $\lambda' = \pi/2$, $\delta' = \pi/4$, and $\hat{\phi}' = \hat{\phi} + \pi$ when $0 \le \hat{\phi} < \pi$, and $\hat{\phi}' = \hat{\phi} - \pi$ when $\pi \le \hat{\phi} < 2\pi$.

4.4.4. Faults, Triaxial Stresses, and Fracture

It has been shown in Section 1.1 that the stress at a point P can be written in the canonical form [cf. Eq. (1.24)]

$$\begin{aligned}\mathfrak{T} &= \tau_1\mathbf{a}_1\mathbf{a}_1 + \tau_2\mathbf{a}_2\mathbf{a}_2 + \tau_3\mathbf{a}_3\mathbf{a}_3 \\ &= p\mathfrak{J} + (\tau_1 - p)\mathbf{a}_1\mathbf{a}_1 + (\tau_2 - p)\mathbf{a}_2\mathbf{a}_2 + (\tau_3 - p)\mathbf{a}_3\mathbf{a}_3,\end{aligned}$$

where $\mathbf{a}_1$, $\mathbf{a}_2$, $\mathbf{a}_3$ are the eigenvectors of the stress dyadic at P, τ_1, τ_2, τ_3 are the principal stresses (eigenvalues) at that point, and $p = \frac{1}{3}(\tau_1 + \tau_2 + \tau_3)$. Moreover, we have proved that:

1. The principal stresses include the greatest (τ_{max}) and the least (τ_{min}) normal stresses at P.
2. The greatest shearing stress at P is equal to one-half the difference between the maximum and the minimum principal stresses, $\frac{1}{2}|\tau_{max} - \tau_{min}|$, and acts on a plane that bisects the angle between the directions of the said principal stresses.
3. The normal stress on this plane is equal to $\frac{1}{2}(\tau_{max} + \tau_{min})$.

If we accept the theory that the fracture takes place in a plane of maximum shearing stress, then the fault vectors $\mathbf{n}$ and $\mathbf{e}$ will be related in a simple manner to the directions of the principal stresses at the fault. Assuming that $\tau_1 > \tau_2 > \tau_3$, we have the relations

$$\begin{aligned} \tau_1' &= \tau_1 - p = \tfrac{1}{3}[(\tau_1 - \tau_2) + (\tau_1 - \tau_3)] > 0 \qquad &\text{(compression);} \\ \tau_3' &= \tau_3 - p = -\tfrac{1}{3}[(\tau_1 - \tau_3) + (\tau_2 - \tau_3)] < 0 \qquad &\text{(tension);} \\ \tau_2' &= \tau_2 - p = \tfrac{1}{3}[(\tau_2 - \tau_3) - (\tau_1 - \tau_2)]. & \end{aligned} \tag{4.131}$$

The isotropic part of $\mathfrak{T}$ does not contribute to shear fracture except for its indirect influence on the normal stress, which increases the friction on the fault

$[(\tau_1 + \tau_3)/2 = \frac{3}{2}p - \frac{1}{2}\tau_2]$. (We assume, in this context, that the coefficient of friction on the fault is unity. Otherwise, the greatest shearing stress does not act on a plane that bisects the angle between the directions of extremal stresses.) The acting stresses are therefore τ_1' (compression), τ_2', and τ_3' (tension). Fracture then takes place on a plane that passes through the direction of τ_2' and is equally inclined to the directions of τ_1' and τ_3'. Because no shearing stress acts along the earth–air interface, one of the principal stresses must be normal to the earth's topographic surface. This principal stress is considered to be approximately vertical. There are then three cases of interest, shown in Figs. 4.28 and 4.29:

1. Normal fault: Faulting is produced by a vertical compression (τ_1') accompanied by a horizontal tension ($-\tau_3'$), and τ_2' is horizontal.
2. Strike-slip fault: Because failure takes place on a vertical plane, the directions of τ_1' and τ_3' are horizontal and τ_2' is vertical.
3. Reverse fault: The stress τ_3' is vertical, whereas τ_1' is horizontal; τ_2' is horizontal.

Note that τ_2' is always in the direction of the null vector (intersection of the conjugate planes). It is an experimental fact that τ_2' has no influence on shear fracture, and therefore its magnitude and sign are not important.

Therefore, our previous representation of a shear dislocation by an equivalent set of two orthogonal dipoles with opposite polarities finds here an important application, in that we can specify the directions of the active principal stresses utilizing knowledge of the dip of the fault, its slip, and its azimuth.

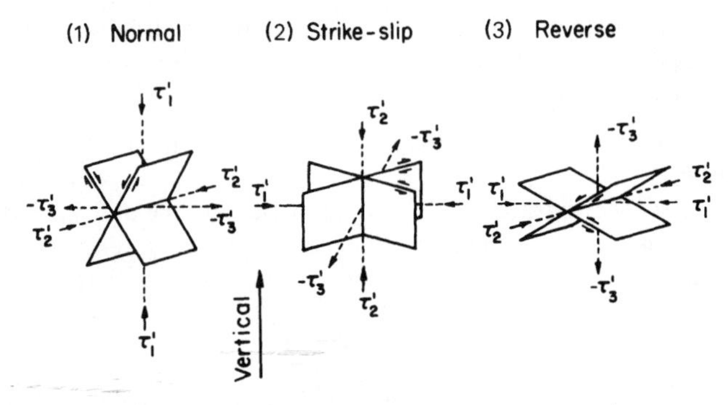

Figure 4.28. Principal triaxial stresses and conjugate planes.

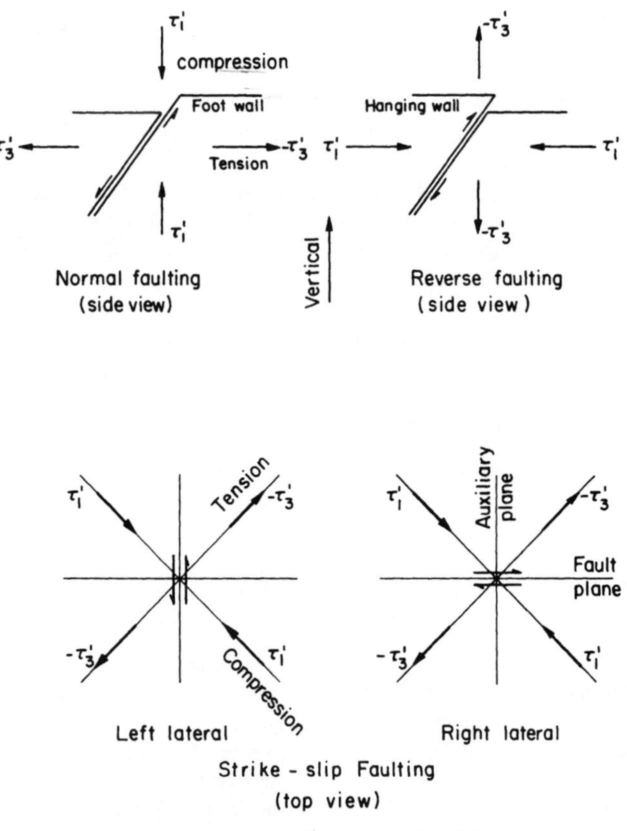

Figure 4.29. Principal stresses and faulting.

We define a unit vector **t** in the direction of τ'_3 and another unit vector **p** in the direction of τ'_1. Then (**p**, **b**, **t**) form a right-handed system that can be obtained from the (**e**, **b**, **n**) system by rotation about the **b** axis through 45°. Consequently, we have

	e	**b**	**n**
p	$1/\sqrt{2}$	0	$-1/\sqrt{2}$
b	0	1	0
t	$1/\sqrt{2}$	0	$1/\sqrt{2}$

(4.132)

It has been known almost since the beginning of instrumental seismology that during any earthquake certain stations record an impulse away from the epicenter (which we call a *compression*), whereas other stations record an impulse toward the epicenter (a *dilatation*). Further, the areas of compression and dilatation are arranged in a pattern. For a shear dislocation, the quadrant in the **en** plane in which **t** lies will yield compressions, and the quadrant in which **p** lies will yield dilatations. Opposite quadrants have similar patterns. In Fig. 4.12, therefore, the first and the third quadrants will have compressions, whereas the second and the fourth quadrants will have dilatations. This property is of great use in source mechanism studies using first-motion observations. The boundaries that separate those stations which record compressional impulses from those which record dilatational impulses are the conjugate planes, also known as *nodal planes*. Equations (4.113) and (4.122) can also be put in a convenient matrix form. Defining

$$E_1 = \begin{bmatrix} \cos\lambda & \sin\lambda & 0 \\ -\sin\lambda & \cos\lambda & 0 \\ 0 & 0 & 1 \end{bmatrix},$$

$$E_2 = \begin{bmatrix} 1 & 0 & 0 \\ 0 & \cos\delta & \sin\delta \\ 0 & -\sin\delta & \cos\delta \end{bmatrix},$$

$$E_3 = \begin{bmatrix} -\cos\hat{\phi} & \sin\hat{\phi} & 0 \\ -\sin\hat{\phi} & -\cos\hat{\phi} & 0 \\ 0 & 0 & 1 \end{bmatrix},$$

we obtain

$$\begin{bmatrix} \mathbf{e} \\ \mathbf{b} \\ \mathbf{n} \end{bmatrix} = E_1 E_2 \begin{bmatrix} \mathbf{e}_1^\circ \\ \mathbf{e}_2^\circ \\ \mathbf{e}_3^\circ \end{bmatrix} = E_1 E_2 E_3 \begin{bmatrix} \mathbf{e}_{\theta_0} \\ \mathbf{e}_{\phi_0} \\ \mathbf{e}_{r_0} \end{bmatrix}.$$

Combining Eqs. (4.132) with Eqs. (4.110) and (4.113), we obtain the directions of the active principal stresses in terms of the fault's dip, slip, and azimuth:

$$\begin{bmatrix}\mathbf{t}\\ \mathbf{b}\\ \mathbf{p}\end{bmatrix} = \begin{bmatrix} -\frac{1}{\sqrt{2}}[\cos\lambda\cos\hat{\phi} + (\sin\lambda\cos\delta - \sin\delta)\sin\hat{\phi}] & \frac{1}{\sqrt{2}}[\cos\lambda\sin\hat{\phi} - (\sin\lambda\cos\delta - \sin\delta)\cos\hat{\phi}] & \frac{1}{\sqrt{2}}(\sin\lambda\sin\delta + \cos\delta) \\ \sin\lambda\cos\hat{\phi} - \cos\lambda\cos\delta\sin\hat{\phi} & -(\sin\lambda\sin\hat{\phi} + \cos\lambda\cos\delta\cos\hat{\phi}) & \cos\lambda\sin\delta \\ -\frac{1}{\sqrt{2}}[\cos\lambda\cos\hat{\phi} + (\sin\lambda\cos\delta + \sin\delta)\sin\hat{\phi}] & \frac{1}{\sqrt{2}}[\cos\lambda\sin\hat{\phi} - (\sin\lambda\cos\delta + \sin\delta)\cos\hat{\phi}] & \frac{1}{\sqrt{2}}(\sin\lambda\sin\delta - \cos\delta) \end{bmatrix}\begin{bmatrix}\mathbf{e}_{\theta_0}\\ \mathbf{e}_{\phi_0}\\ \mathbf{e}_{r_0}\end{bmatrix}.$$

Another useful transformation is obtained from Eqs. (4.109) and (4.110):

$$\begin{bmatrix}\mathbf{e}_1^\circ\\ \mathbf{e}_2^\circ\\ \mathbf{e}_3^\circ\end{bmatrix} = \begin{bmatrix} -\cos\hat{\phi} & \sin\hat{\phi} & 0\\ -\sin\hat{\phi} & -\cos\hat{\phi} & 0\\ 0 & 0 & 1\end{bmatrix}\begin{bmatrix}\cos\theta_0 & 0 & -\sin\theta_0\\ 0 & 1 & 0\\ \sin\theta_0 & 0 & \cos\theta_0\end{bmatrix} \times \begin{bmatrix}\cos\phi_0 & \sin\phi_0 & 0\\ -\sin\phi_0 & \cos\phi_0 & 0\\ 0 & 0 & 1\end{bmatrix}\begin{bmatrix}\mathbf{e}_1\\ \mathbf{e}_2\\ \mathbf{e}_3\end{bmatrix}$$

$$= \begin{bmatrix} -\cos\theta_0\cos\phi_0\cos\hat{\phi} - \sin\phi_0\sin\hat{\phi} & -\cos\theta_0\sin\phi_0\cos\hat{\phi} + \cos\phi_0\sin\hat{\phi} & \sin\theta_0\cos\hat{\phi}\\ -\cos\theta_0\cos\phi_0\sin\hat{\phi} + \sin\phi_0\cos\hat{\phi} & -\cos\theta_0\sin\phi_0\sin\hat{\phi} - \cos\phi_0\cos\hat{\phi} & \sin\theta_0\sin\hat{\phi}\\ \sin\theta_0\cos\phi_0 & \sin\theta_0\sin\phi_0 & \cos\theta_0\end{bmatrix}\begin{bmatrix}\mathbf{e}_1\\ \mathbf{e}_2\\ \mathbf{e}_3\end{bmatrix}.$$

These equations link the orientation of the source axes with the geographic axes through the strike azimuth and the geographic latitude and longitude.

EXAMPLE 4.5

Assuming

$$\mathbf{u}(\mathbf{r}_0) = U(r_0)\overset{*}{\mathbf{P}}_{ml}(\theta_0, \phi_0) + V(r_0)\sqrt{[l(l+1)]}\overset{*}{\mathbf{B}}_{ml}(\theta_0, \phi_0) + W(r_0)\sqrt{[l(l+1)]}\overset{*}{\mathbf{C}}_{ml}(\theta_0, \phi_0), \tag{4.5.1}$$

and (**en** + **ne**) given by Eq. (4.115b), we have

$$\mathbf{en} : \mathfrak{T}_0(\mathbf{u}) = \mu_0 \, \mathbf{en} : (\nabla_0 \mathbf{u} + \mathbf{u}\nabla_0), \tag{4.5.2}$$

$$\begin{aligned}
\mathbf{en} : (\nabla_0 \mathbf{u} + \mathbf{u}\nabla_0) &= \frac{1}{2}(\mathbf{en} + \mathbf{ne}) : (\nabla_0 \mathbf{u} + \mathbf{u}\nabla_0) \\
&= p_5\left[2\left(\frac{dU}{dr_0} - \frac{U}{r_0}\right) + l(l+1)\frac{V}{r_0}\right]\overset{*}{Y}_{ml}(\theta_0, \phi_0) \\
&\quad + \left[p_3\left(\frac{dW}{dr_0} - \frac{W}{r_0}\right) - p_4\left(\frac{dV}{dr_0} + \frac{U - V}{r_0}\right)\right]\frac{\partial \overset{*}{Y}_{ml}(\theta_0, \phi_0)}{\partial \theta_0} \\
&\quad - \left[p_4\left(\frac{dW}{dr_0} - \frac{W}{r_0}\right) + p_3\left(\frac{dV}{dr_0} + \frac{U - V}{r_0}\right)\right]\frac{1}{\sin\theta_0}\frac{\partial \overset{*}{Y}_{ml}(\theta_0, \phi_0)}{\partial \phi_0} \\
&\quad - 2\left[p_2\frac{W}{r_0} + p_1\frac{V}{r_0}\right]\frac{\partial}{\partial \theta_0}\left\{\frac{1}{\sin\theta_0}\frac{\partial \overset{*}{Y}_{ml}(\theta_0, \phi_0)}{\partial \phi_0}\right\} \\
&\quad + \left[p_1\frac{W}{r_0} - p_2\frac{V}{r_0}\right]\left\{2\frac{\partial^2}{\partial \theta_0^2} + l(l+1)\right\}\overset{*}{Y}_{ml}(\theta_0, \phi_0).
\end{aligned} \tag{4.5.3}$$

When the source lies on the x_3 axis, it is necessary to calculate the limit of $\mathbf{en} : (\nabla_0 \mathbf{u} + \mathbf{u}\nabla_0)$ as $\theta_0 \to 0$, $\phi_0 \to 0$. To this end, we use the integral

$$P_l^m(\cos\theta_0) = \frac{1}{(l-m)!}\int_0^\infty \exp(-t\cos\theta_0) J_m(t\sin\theta_0) t^l \, dt,$$

$$0 < \theta < \frac{\pi}{2}, \qquad (l + m + 1) > 0. \tag{4.5.4}$$

Replacing the associated Legendre function by the integral in Eq. (4.5.4) and using the relationship $J_m(0) = \delta_{0m}$, we find that, as $\theta_0 \to 0$, $\phi_0 \to 0$,

$$\begin{aligned}
&\lim[\overset{*}{Y}_{ml}(\theta_0, \phi_0)] = \delta_{0m}, \\
&\lim\left[\frac{\partial \overset{*}{Y}_{ml}(\theta_0, \phi_0)}{\partial \theta_0}\right] = \frac{1}{2}[l(l+1)\delta_{1m} - \delta_{(-1)m}], \\
&\lim\left[\frac{1}{\sin\theta_0}\frac{\partial \overset{*}{Y}_{ml}(\theta_0, \phi_0)}{\partial \phi_0}\right] = -\frac{i}{2}[l(l+1)\delta_{1m} + \delta_{(-1)m}], \\
&\lim\left[\frac{\partial}{\partial \theta_0}\left\{\frac{1}{\sin\theta_0}\frac{\partial \overset{*}{Y}_{ml}(\theta_0, \phi_0)}{\partial \phi_0}\right\}\right] = -\frac{i}{4}\left[\frac{(l+2)!}{(l-2)!}\delta_{2m} - \delta_{(-2)m}\right], \\
&\lim\left[\left\{2\frac{\partial^2}{\partial \theta_0^2} + l(l+1)\right\}\overset{*}{Y}_{ml}(\theta_0, \phi_0)\right] = \frac{1}{2}\left[\frac{(l+2)!}{(l-2)!}\delta_{2m} + \delta_{(-2)m}\right].
\end{aligned} \tag{4.5.5}$$

4.5. Dipolar Sources in a Homogeneous Medium

The field induced by a concentrated force in a homogeneous medium has been derived in Section 4.1. We now use these results and the theory of dipolar sources and dislocations developed in Sections 4.2 and 4.3 to obtain the field induced by dipolar sources and dislocations in a homogeneous medium.

4.5.1. The Spectral Field

Because in the case of a homogeneous isotropic medium of infinite extent the Green's dyadic is a function of $R = |\mathbf{r} - \mathbf{r}_0|$ only, we have

$$\begin{aligned}
\tilde{\mathfrak{G}}(\mathbf{r}|\mathbf{r}_0) &= \mathfrak{G}(\mathbf{r}|\mathbf{r}_0),\\
\mathfrak{G}(\mathbf{r}|\mathbf{r}_0) &= \mathfrak{G}(\mathbf{r}_0|\mathbf{r}),\\
\operatorname{grad}_0 \tilde{\mathfrak{G}}(\mathbf{r}|\mathbf{r}_0) &= -\operatorname{grad} \mathfrak{G}(\mathbf{r}|\mathbf{r}_0),\\
\operatorname{div}_0 \tilde{\mathfrak{G}}(\mathbf{r}|\mathbf{r}_0) &= -\operatorname{div} \mathfrak{G}(\mathbf{r}|\mathbf{r}_0),\\
\mathfrak{T}_0[\tilde{\mathfrak{G}}(\mathbf{r}|\mathbf{r}_0)] &= -\mathfrak{T}[\mathfrak{G}(\mathbf{r}|\mathbf{r}_0)].
\end{aligned} \tag{4.133}$$

The first two identities are simply the reciprocity relations. Equation (4.133) helps us in computing the displacement field for dipolar sources from expression (4.12) for the Green's dyadic, by using the results derived in Section 4.2. We find

$$\begin{aligned}
\operatorname{grad}_0 \mathfrak{G}(\mathbf{r}|\mathbf{r}_0) = {}& \frac{ik_\alpha^2}{20\pi(\lambda + 2\mu)}\Bigg[-\frac{1}{R}\{h_1^{(2)}(k_\alpha R) + h_3^{(2)}(k_\alpha R)\}\\
&\times\{\mathbf{R}\mathfrak{I} + \mathfrak{I}\mathbf{R} + (\mathfrak{I}\mathbf{R})^{132}\} + \frac{5}{R^3}h_3^{(2)}(k_\alpha R)\mathbf{RRR}\Bigg]\\
&-\frac{ik_\beta^2}{20\pi\mu}\Bigg[\frac{5}{R}h_1^{(2)}(k_\beta R)\mathbf{R}\mathfrak{I} - \frac{1}{R}\{h_1^{(2)}(k_\beta R) + h_3^{(2)}(k_\beta R)\}\\
&\times\{\mathbf{R}\mathfrak{I} + \mathfrak{I}\mathbf{R} + (\mathfrak{I}\mathbf{R})^{132}\} + \frac{5}{R^3}h_3^{(2)}(k_\beta R)\mathbf{RRR}\Bigg],
\end{aligned} \tag{4.134}$$

with

$$(\mathfrak{I}\mathbf{R})^{132} = (\mathbf{R}\mathfrak{I})^{213} = \mathbf{e}_1\mathbf{R}\mathbf{e}_1 + \mathbf{e}_2\mathbf{R}\mathbf{e}_2 + \mathbf{e}_3\mathbf{R}\mathbf{e}_3. \tag{4.135}$$

If we take $\mathbf{a} = \mathbf{e}_j, \mathbf{v} = \mathbf{e}_k$, Eqs. (4.44) and (4.134) give the following expression for the displacement caused by the double force $\mathbf{e}_j\mathbf{e}_k$

$$\begin{aligned}
\mathbf{u} &= M\mathbf{e}_j\mathbf{e}_k : \operatorname{grad}_0 \mathfrak{G}(\mathbf{r}|\mathbf{r}_0)\\
&= -\frac{iMk_\alpha^2}{20\pi(\lambda + 2\mu)}\Bigg[\frac{1}{R}\{h_1^{(2)}(k_\alpha R) + h_3^{(2)}(k_\alpha R)\}(X_k\mathbf{e}_j + X_j\mathbf{e}_k + \delta_{jk}\mathbf{R})\\
&\quad - \frac{5}{R^3}h_3^{(2)}(k_\alpha R)X_jX_k\mathbf{R}\Bigg] - \frac{iMk_\beta^2}{20\pi\mu}\Bigg[\frac{5}{R}h_1^{(2)}(k_\beta R)X_k\mathbf{e}_j - \frac{1}{R}\{h_1^{(2)}(k_\beta R)\\
&\quad + h_3^{(2)}(k_\beta R)\}(X_k\mathbf{e}_j + X_j\mathbf{e}_k + \delta_{jk}\mathbf{R}) + \frac{5}{R^3}h_3^{(2)}(k_\beta R)X_jX_k\mathbf{R}\Bigg],
\end{aligned} \tag{4.136}$$

where

$$\mathbf{r} = x_i \mathbf{e}_i, \qquad \mathbf{r}_0 = y_i \mathbf{e}_i, \qquad X_i = x_i - y_i. \tag{4.137}$$

From Eq. (4.136), the field induced by a dipole in the x_j direction can be obtained by taking $j = k$ and that induced by a single couple in the $x_j x_k$ plane with the forces of the couple parallel to the x_j axis by taking $\delta_{jk} = 0$. The field induced by a double couple or a center of rotation in the $x_j x_k$ plane can be immediately calculated from the field induced by the single couple.

The displacement induced by a center of compression can be found by using Eq. (4.46)

$$\begin{aligned} \mathbf{u}(\mathbf{r}) = M \operatorname{div}_0 \mathfrak{G} &= -\frac{M}{4\pi(\lambda + 2\mu)} \operatorname{grad}\left(\frac{e^{-ik_\alpha R}}{R}\right) \\ &= -\frac{M}{4\pi(\lambda + 2\mu)} \left[\frac{1}{R}\frac{d}{dR}\left(\frac{e^{-ik_\alpha R}}{R}\right)\right]\mathbf{R}. \end{aligned} \tag{4.138}$$

The radial stress vector at any point can be calculated through the relation

$$\mathbf{e}_R \cdot \mathfrak{T}(\mathbf{u}) = (\lambda \operatorname{div} \mathbf{u})\mathbf{e}_R + \mu\left(2\frac{\partial \mathbf{u}}{\partial R} + \mathbf{e}_R \times \operatorname{curl} \mathbf{u}\right). \tag{4.139}$$

Substituting for $\mathbf{u}$, we get

$$\mathbf{e}_R \cdot \mathfrak{T}(\mathbf{u}) = -\frac{M\mathbf{e}_R}{4\pi(\lambda + 2\mu)}\left[(\lambda + 2\mu)\frac{d^2}{dR^2} + \frac{2\lambda}{R}\frac{d}{dR}\right]\left(\frac{e^{-ik_\alpha R}}{R}\right). \tag{4.140}$$

The displacement field generated by a displacement dislocation is given by the Volterra relation, Eq. (4.105). In order to get the displacement field, we must first compute the triadic $\mathfrak{T}_0(\tilde{\mathfrak{G}})$. From Eqs. (4.76) and (4.134), we obtain, for a homogeneous medium of infinite extent,

$$\mathfrak{T}_0(\mathfrak{G}) = \mathsf{T}_\alpha + \mathsf{T}_\beta, \tag{4.141}$$

where

$$\mathsf{T}_\alpha = -\frac{ik_\alpha^2}{20\pi}\left(\frac{1 - 2\sigma}{1 - \sigma}\right)[h_1^{(2)}(k_\alpha R)\mathsf{A} + h_3^{(2)}(k_\alpha R)\mathsf{B}], \tag{4.142}$$

$$\mathsf{T}_\beta = \frac{ik_\beta^2}{20\pi}[h_1^{(2)}(k_\beta R)\mathsf{C} + 2h_3^{(2)}(k_\beta R)\mathsf{B}], \tag{4.143}$$

$$\mathsf{A} = \frac{1}{R}\left[\frac{1 + 3\sigma}{1 - 2\sigma}\mathfrak{I}\mathbf{R} + \mathbf{R}\mathfrak{I} + (\mathbf{R}\mathfrak{I})^{213}\right], \tag{4.144}$$

$$\mathsf{B} = \frac{1}{R}\left[-\frac{5}{R^2}\mathbf{RRR} + \mathfrak{I}\mathbf{R} + \mathbf{R}\mathfrak{I} + (\mathbf{R}\mathfrak{I})^{213}\right], \tag{4.145}$$

$$\mathsf{C} = \frac{2}{R}\mathfrak{I}\mathbf{R} - \frac{3}{R}[\mathbf{R}\mathfrak{I} + (\mathbf{R}\mathfrak{I})^{213}], \tag{4.146}$$

and σ is the Poisson ratio. Inserting expression (4.76) for the triadic $\mathfrak{T}_0(\mathfrak{G})$ in Eq. (4.105) [or Eq. (4.107) for a point source], we get the displacement field in an infinite medium. However, if the medium under consideration is of finite extent, we must first find a dyadic satisfying the prescribed boundary conditions and then apply the above method.

In the case of a homogeneous medium, we can assign an alternative meaning to the triadic $\mathfrak{T}_0[\mathfrak{G}(\mathbf{r}|\mathbf{r}_0)]$. Remembering that $F_0\mathfrak{G}(\mathbf{r}|\mathbf{r}_0)\cdot\mathbf{a}$ is the displacement vector at $P(\mathbf{r})$ resulting from a single force F_0 acting at $Q(\mathbf{r}_0)$ in the direction of the unit vector $\mathbf{a}$, it is obvious that $F_0\mathfrak{T}[\mathfrak{G}(\mathbf{r}|\mathbf{r}_0)]\cdot\mathbf{a}$ represents the stress dyadic at P resulting from the single force at Q. Because, for a homogeneous medium,

$$-\mathfrak{T}_0[\mathfrak{G}(\mathbf{r}|\mathbf{r}_0)] = \mathfrak{T}[\mathfrak{G}(\mathbf{r}|\mathbf{r}_0)],$$

we note that $-F_0\mathfrak{T}_0[\mathfrak{G}(\mathbf{r}|\mathbf{r}_0)]\cdot\mathbf{a}$ is the stress dyadic at P resulting from a single force F_0 acting at Q in the $\mathbf{a}$ direction. Such an interpretation cannot be given in the case of a heterogeneous medium.

We next put expression (4.141) for the triadic $\mathfrak{T}_0(\mathfrak{G})$ into the Volterra relation, Eq. (4.107), to obtain the displacement field induced by a displacement point dislocation in a homogeneous, isotropic, unbounded medium. We find

$$\mathbf{u} = \mathbf{u}_\alpha + \mathbf{u}_\beta, \tag{4.147}$$

where

$$\begin{aligned}\mathbf{u}_\alpha = &-\frac{ik_\alpha^2}{20\pi}\left(\frac{1-2\sigma}{1-\sigma}\right)U_0\,dS\Big[h_1^{(2)}(k_\alpha R)\left\{\frac{1+3\sigma}{1-2\sigma}(\mathbf{e}\cdot\mathbf{n})\mathbf{e}_R + (\mathbf{n}\cdot\mathbf{e}_R)\mathbf{e} + (\mathbf{e}\cdot\mathbf{e}_R)\mathbf{n}\right\}\\ &+ h_3^{(2)}(k_\alpha R)\{-5(\mathbf{e}\cdot\mathbf{e}_R)(\mathbf{n}\cdot\mathbf{e}_R)\mathbf{e}_R + (\mathbf{e}\cdot\mathbf{n})\mathbf{e}_R + (\mathbf{n}\cdot\mathbf{e}_R)\mathbf{e} + (\mathbf{e}\cdot\mathbf{e}_R)\mathbf{n}\}\Big],\\ \mathbf{u}_\beta = &\frac{ik_\beta^2}{20\pi}U_0\,dS\Big[h_1^{(2)}(k_\beta R)\{2(\mathbf{e}\cdot\mathbf{n})\mathbf{e}_R - 3(\mathbf{n}\cdot\mathbf{e}_R)\mathbf{e} - 3(\mathbf{e}\cdot\mathbf{e}_R)\mathbf{n}\}\\ &+ 2h_3^{(2)}(k_\beta R)\{-5(\mathbf{e}\cdot\mathbf{e}_R)(\mathbf{n}\cdot\mathbf{e}_R)\mathbf{e}_R + (\mathbf{e}\cdot\mathbf{n})\mathbf{e}_R + (\mathbf{n}\cdot\mathbf{e}_R)\mathbf{e} + (\mathbf{e}\cdot\mathbf{e}_R)\mathbf{n}\}\Big].\end{aligned} \tag{4.148}$$

In the case of a shear dislocation, $\mathbf{e}\cdot\mathbf{n} = 0$, and we obtain

$$\begin{aligned}\mathbf{u}_\alpha = &-\frac{ik_\alpha^2}{10\pi}\left(\frac{\beta}{\alpha}\right)^2 U_0\,dS\Big[-5(\mathbf{e}\cdot\mathbf{e}_R)(\mathbf{n}\cdot\mathbf{e}_R)\mathbf{e}_R h_3^{(2)}(k_\alpha R) + \{(\mathbf{n}\cdot\mathbf{e}_R)\mathbf{e} + (\mathbf{e}\cdot\mathbf{e}_R)\mathbf{n}\}\\ &\times\{h_1^{(2)}(k_\alpha R) + h_3^{(2)}(k_\alpha R)\}\Big],\\ \mathbf{u}_\beta = &\frac{ik_\beta^2}{20\pi}U_0\,dS\Big[-10(\mathbf{e}\cdot\mathbf{e}_R)(\mathbf{n}\cdot\mathbf{e}_R)\mathbf{e}_R h_3^{(2)}(k_\beta R) + \{(\mathbf{n}\cdot\mathbf{e}_R)\mathbf{e}\\ &+ (\mathbf{e}\cdot\mathbf{e}_R)\mathbf{n}\}\{-3h_1^{(2)}(k_\beta R) + 2h_3^{(2)}(k_\beta R)\}\Big].\end{aligned} \tag{4.149}$$

The Volterra relation, Eq. (4.105), may be written in the form

$$u_k(P) = \int_S \Delta u_i(Q)T_k^{ij}(P, Q)n_j(Q)dS,$$

where the tensor T_k^{ij} is given by Eq. (4.141). From this relation, one can find the stress tensor with the aid of Hooke's law. Therefore, the kl component of the stress tensor at $P(x_i^\circ)$ induced by a displacement dislocation over S is found to be

$$\tau_{kl}(P) = \int_S \Delta u_i(Q) H_{kl}^{ij}(P, Q) n_j(Q) dS,$$

where

$$H_{kl}^{ij} = \lambda \delta_{kl} \frac{\partial}{\partial x_m^\circ} T_m^{ij} + \mu\left(\frac{\partial}{\partial x_l^\circ} T_k^{ij} + \frac{\partial}{\partial x_k^\circ} T_l^{ij}\right) \tag{4.150}$$

is the ij stress component resulting from a (kl) dipolar source. After a lengthy but straightforward analysis, Eqs. (4.141) and (4.150) yield

$$\begin{aligned} H_{kl}^{ij} = &-\frac{i\mu k_\alpha^2}{20\pi}\left(\frac{1-2\sigma}{1-\sigma}\right)[C_{1\alpha} h_1^{(2)}(k_\alpha R) + C_{2\alpha} h_2^{(2)}(k_\alpha R) + C_{3\alpha} h_3^{(2)}(k_\alpha R)] \\ &+ \frac{i\mu k_\beta^2}{20\pi}[C_{1\beta} h_1^{(2)}(k_\beta R) + C_{2\beta} h_2^{(2)}(k_\beta R) + C_{3\beta} h_3^{(2)}(k_\beta R)], \end{aligned} \tag{4.151}$$

where

$$\begin{aligned} C_{1\alpha} &= \frac{2}{R}\left[\Sigma\delta_{ij}\delta_{kl} + 5\frac{\sigma(2-\sigma)}{(1-2\sigma)^2}\delta_{ij}\delta_{kl}\right], \\ C_{2\alpha} &= -5k_\alpha\left[2\left(\frac{\sigma}{1-2\sigma}\right)^2\delta_{ij}\delta_{kl} + \frac{2\sigma}{1-2\sigma}\frac{1}{R^2}(\delta_{ij}X_kX_l + \delta_{kl}X_iX_j) \right. \\ &\quad \left. + \frac{2}{R^4}X_iX_jX_kX_l\right], \\ C_{3\alpha} &= \frac{2}{R}\left[\Sigma\delta_{ij}\delta_{kl} - \frac{5}{R^2}\Sigma\delta_{ij}X_kX_l + \frac{35}{R^4}X_iX_jX_kX_l\right], \\ C_{1\beta} &= \frac{2}{R}[5\delta_{ij}\delta_{kl} - 3\Sigma\delta_{ij}\delta_{kl}], \\ C_{2\beta} &= \frac{5}{R^2}k_\beta\left[\delta_{ik}X_jX_l + \delta_{il}X_jX_k + \delta_{jk}X_iX_l + \delta_{jl}X_iX_k - \frac{4}{R^2}X_iX_jX_kX_l\right], \\ C_{3\beta} &= 2C_{3\alpha}, \\ X_i &= \mathbf{R}\cdot\mathbf{e}_i^\circ, \\ \Sigma\delta_{ij}\delta_{kl} &= \delta_{ij}\delta_{kl} + \delta_{ik}\delta_{jl} + \delta_{il}\delta_{jk}, \\ \Sigma\delta_{ij}X_kX_l &= \delta_{ij}X_kX_l + \delta_{ik}X_jX_l + \delta_{il}X_jX_k + \delta_{jk}X_iX_l + \delta_{jl}X_iX_k + \delta_{kl}X_iX_j. \end{aligned} \tag{4.151a}$$

Therefore, for a homogeneous, isotropic, infinite medium

$$\begin{aligned} H_{ij}^{kl}(P, Q) &= H_{ji}^{kl}(P, Q) = H_{kl}^{ij}(P, Q), \\ H_{ij}^{kl}(P, Q) &= H_{ij}^{kl}(Q, P). \end{aligned}$$

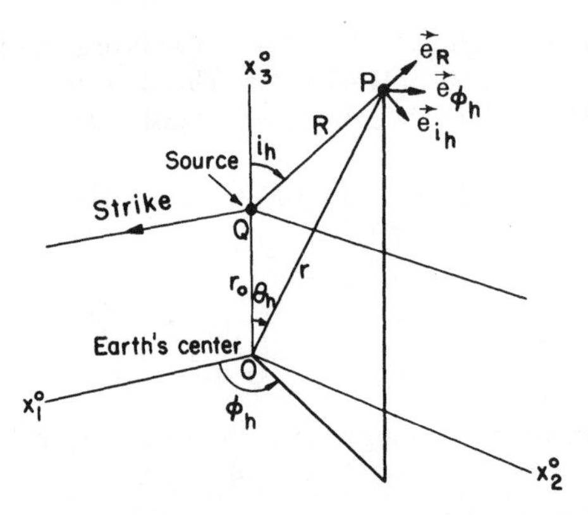

Figure 4.30. The spherical system (R, i_h, ϕ_h) at the source, the spherical system (r, θ_h, ϕ_h) at the center of the earth, and the source system (x_1^0, x_2^0, x_3^0).

We next set up a spherical coordinate system (R, i_h, ϕ_h) at the point Q as shown in Fig. 4.30. This figure also shows the source system x_i° with unit vectors $\mathbf{e}_i^\circ$. The azimuthal angle is measured counterclockwise from the x_1° axis. The transformation is expressed through the relations

	$\mathbf{e}_1^\circ$	$\mathbf{e}_2^\circ$	$\mathbf{e}_3^\circ$
$\mathbf{e}_R$	$\sin i_h \cos\phi_h$	$\sin i_h \sin\phi_h$	$\cos i_h$
$\mathbf{e}_{i_h}$	$\cos i_h \cos\phi_h$	$\cos i_h \sin\phi_h$	$-\sin i_h$
$\mathbf{e}_{\phi_h}$	$-\sin\phi_h$	$\cos\phi_h$	0

(4.152)

From Eqs. (4.113) and (4.152), it follows that

$$(\mathbf{e}\cdot\mathbf{e}_R)(\mathbf{n}\cdot\mathbf{e}_R) = \frac{1}{6}F,$$

$$(\mathbf{e}\cdot\mathbf{e}_R)(\mathbf{n}\cdot\mathbf{e}_{i_h}) + (\mathbf{e}\cdot\mathbf{e}_{i_h})(\mathbf{n}\cdot\mathbf{e}_R) = \frac{1}{6}\frac{\partial F}{\partial i_h}, \tag{4.153}$$

$$(\mathbf{e}\cdot\mathbf{e}_R)(\mathbf{n}\cdot\mathbf{e}_{\phi_h}) + (\mathbf{e}\cdot\mathbf{e}_{\phi_h})(\mathbf{n}\cdot\mathbf{e}_R) = \frac{1}{6\sin i_h}\frac{\partial F}{\partial \phi_h},$$

where

$$F(\lambda, \delta; i_h, \phi_h) = a_0 P_2(\cos i_h) + (a_1\cos\phi_h + b_1\sin\phi_h)P_2^1(\cos i_h) + (a_2\cos 2\phi_h + b_2\sin 2\phi_h)P_2^2(\cos i_h),$$

$$a_0 = 3\sin\lambda\sin 2\delta,$$

$$a_1 = 2\cos\lambda\cos\delta, \qquad b_1 = 2\sin\lambda\cos 2\delta, \tag{4.154}$$

$$a_2 = \tfrac{1}{2}\sin\lambda\sin 2\delta, \qquad b_2 = -\cos\lambda\sin\delta.$$

Equations (4.147), (4.149), (4.153), and (4.154) now yield

$$\mathbf{u} = -\frac{iU_0\,dSk_\beta^2}{24\pi}\sum_{m=0}^{2}\left[2\left(\frac{\beta}{\alpha}\right)^4\{a_m\mathbf{L}_{m,2}^c(k_\alpha R) + b_m\mathbf{L}_{m,2}^s(k_\alpha R)\}\right.$$

$$\left. + a_m\mathbf{N}_{m,2}^c(k_\beta R) + b_m\mathbf{N}_{m,2}^s(k_\beta R)\right], \tag{4.155}$$

where

$$\mathbf{L}_{ml}^{c,s}(k_\alpha R) = \frac{1}{k_\alpha}\,\mathrm{grad}[h_l^{(2)}(k_\alpha R)P_l^m(\cos i_h)(\cos m\phi_h, \sin m\phi_h)], \tag{4.155a}$$

etc. (see Table 2.2). It is interesting to note that the displacement field in an infinite medium induced by a shear dislocation at the origin can be expressed in terms of **L** and **N** vectors; the vector **M** is not needed.

The field induced by a concentrated force can also be expressed in terms of the vectors **L** and **N**. To this end, we first note that if S is a function of R only and **a** is a constant vector, then

$$\begin{aligned}\mathrm{grad\,div}(S\mathbf{a}) &= \mathrm{grad}\left[\frac{1}{R}\frac{dS}{dR}(\mathbf{R}\cdot\mathbf{a})\right],\\ \mathrm{curl\,curl}(S\mathbf{a}) &= -\mathrm{curl\,curl}\left[\frac{1}{R}\frac{dS}{dR}\mathbf{R}(\mathbf{R}\cdot\mathbf{a})\right].\end{aligned} \tag{4.156}$$

Equations (4.5), (4.152) and (4.156) now yield

$$\mathbf{u} = -\frac{ik_\beta F_0}{4\pi\mu}\left[\left(\frac{\beta}{\alpha}\right)^3\sum_{m=0}^{1}\{a_m\mathbf{L}_{m,1}^c(k_\alpha R) + b_m\mathbf{L}_{m,1}^s(k_\alpha R)\}\right.$$

$$\left. + \sum_{m=0}^{1}\{a_m\mathbf{N}_{m,1}^c(k_\beta R) + b_m\mathbf{N}_{m,1}^s(k_\beta R)\}\right], \tag{4.157}$$

where

$$\mathbf{a} = a_1\mathbf{e}_1^\circ + b_1\mathbf{e}_2^\circ + a_0\mathbf{e}_3^\circ.$$

Equation (4.157) expresses the displacement field generated by a single force of magnitude F_0 acting in an infinite medium at the origin along the direction of the unit vector **a** in terms of spherical eigenvectors.

In a similar manner, the spectral displacement fields induced by other sources can be expressed in terms of the spherical eigenvectors **L**, **N**, and **M**. The results have been summarized in Table 4.4. It may be noted that all the spherical vectors in this table are in the (R, i_h, ϕ_h) coordinate system and involve $h_l^{(2)}(k_{\alpha,\beta}R)$ as the radial function. Once we know the field in terms of the spherical eigenvectors, Eqs. (2.159)–(2.161) immediately express the spectral displacement field in terms of the eigenvectors in cylindrical coordinates. These results have also been included in Table 4.4. The cylindrical eigenvectors $\mathbf{L}_m^\sigma$, $\mathbf{N}_m^\sigma$ and $\mathbf{M}_m^\sigma$ used in this table are defined as follows

$$\mathbf{L}_m^{\sigma,(0)}(k_\alpha\Delta) = \frac{1}{k_\alpha}(-\varepsilon\nu_\alpha\mathbf{P}_m^\sigma + k\mathbf{B}_m^\sigma)e^{-\nu_\alpha|z|},$$

$$\mathbf{N}_m^{\sigma,(0)}(k_\beta\Delta) = \frac{1}{k_\beta}(k\mathbf{P}_m^\sigma - \varepsilon\nu_\beta\mathbf{B}_m^\sigma)e^{-\nu_\beta|z|}, \tag{4.158}$$

$$\mathbf{M}_m^{\sigma,(0)}(k_\beta\Delta) = \mathbf{C}_m^\sigma e^{-\nu_\beta|z|},$$

where $z = x_3^0 - r_0$, $\Delta = R\sin i_h$ and

$$\varepsilon = \begin{cases} +1, & z > 0 \\ -1, & z < 0, \end{cases}$$

$$\sigma = c\text{ (for cos)} \quad \text{or} \quad s\text{ (for sin)}.$$

The vectors $\mathbf{P}_m^{c,s}$, etc., are defined as in Eq. (2.87) with $\exp(im\phi)$ replaced by $(\cos m\phi_h, \sin m\phi_h)$.

From the expressions given in Table 4.4, we immediately get the representation of sources in terms of the potentials in spherical and cylindrical coordinates with the help of the Eq. (2.67).

4.5.2. Expansion of the Field Tensors in Terms of the Spherical Eigenfunctions

We have introduced three tensors $\mathfrak{G}$, T, and $\mathbf{H}$. The second-order tensor $\mathfrak{G}$ gives the displacement field resulting from a concentrated force. The third-order tensor T represents the displacement field resulting from a point displacement dislocation and some other dipolar sources. Alternatively, it can be considered as giving the stress field induced by a concentrated force. Last, the fourth-order tensor $\mathbf{H}$ yields the stress field induced by a point displacement dislocation and some other dipolar sources. The solution of boundary-value problems in spherical coordinates requires that these tensors be expressed in terms of spherical eigenfunctions.

We will consider the spherical coordinate system (R, i_h, ϕ_h) with its origin at the point source Q and the focal radius as the polar axis (Fig. 4.30). This system is related to the source system through Eq. (4.152), which shows that

$$X_1 = \mathbf{R}\cdot\mathbf{e}_1^\circ = R\sin i_h\cos\phi_h,$$

$$X_2 = \mathbf{R}\cdot\mathbf{e}_2^\circ = R\sin i_h\sin\phi_h,$$

$$X_3 = \mathbf{R}\cdot\mathbf{e}_3^\circ = R\cos i_h. \tag{4.159}$$

We define the spherical eigenfunctions in the (R, i_h, ϕ_h) system as

$$\Phi_m^{c,s}(k_\alpha R) = h_l^{(2)}(k_\alpha R)P_l^m(\cos i_h)(\cos m\phi_h, \sin m\phi_h), \tag{4.160}$$

with a similar expression with α replaced by β.

Table 4.4. Representation of the Displacement Field Induced by Various Sources in Terms of the Spherical Eigenvectors $\mathbf{L}_{ml}^{\sigma}$, $\mathbf{N}_{ml}^{\sigma}$, and $\mathbf{M}_{ml}^{\sigma}$, and in Terms of the Cylindrical Eigenvectors $\mathbf{L}_{m}^{\sigma,(0)}$, $\mathbf{N}_{m}^{\sigma,(0)}$, and $\mathbf{M}_{m}^{\sigma,(0)}$; in the Latter Case, the Integral Operator $\int_0^\infty$ ---$k\,dk$ Is Understood[a]

Source	$\frac{1}{K_1}$ Displacement	K_1	$\frac{1}{K_2}$ Displacement	K_2	Indices m	l
Center of compression (Fig. 4.10)	$\mathbf{L}_{00}^{c}$	$\frac{iMk_\alpha^2}{4\pi(\lambda+2\mu)}$	$\mathbf{L}_0^{c}$	$-\frac{Mk_\alpha}{4\pi(\lambda+2\mu)\nu_\alpha}$	0	0
Center of rotation (Fig. 4.13)						
1. (32) − (23)	$\mathbf{M}_{11}^{c}$	$-\frac{iMk_\beta^2}{4\pi\mu}$	$\mathbf{N}_1^{s} + \varepsilon\frac{\nu_\beta}{k_\beta}\mathbf{M}_1^{c}$	$\frac{Mk_\beta}{4\pi\mu\nu_\beta}$	1	1
2. (13) − (31)	$\mathbf{M}_{11}^{s}$	$-\frac{iMk_\beta^2}{4\pi\mu}$	$-\mathbf{N}_1^{c} + \varepsilon\frac{\nu_\beta}{k_\beta}\mathbf{M}_1^{s}$	$\frac{Mk_\beta}{4\pi\mu\nu_\beta}$	1	1
3. (21) − (12)	$\mathbf{M}_{01}^{c}$	$-\frac{iMk_\beta^2}{4\pi\mu}$	$\frac{k}{k_\beta}\mathbf{M}_0^{c}$	$\frac{Mk_\beta}{4\pi\mu\nu_\beta}$	0	1
Concentrated force						
1. In the x_1 direction	$\gamma_1\mathbf{L}_{11}^{c} + \mathbf{N}_{11}^{c}$	$-\frac{iF_0k_\beta}{4\pi\mu}$	$-\frac{k}{\nu_\sigma}\left(\frac{\beta}{\alpha}\right)\mathbf{L}_1^{c} - \varepsilon\mathbf{N}_1^{c} - \frac{k_\beta}{\nu_\beta}\mathbf{M}_1^{s}$	$-\frac{F_0}{4\pi\mu k_\beta}$	1	1
2. In the x_2 direction	$\gamma_1\mathbf{L}_{11}^{s} + \mathbf{N}_{11}^{s}$	$-\frac{iF_0k_\beta}{4\pi\mu}$	$-\frac{k}{\nu_\sigma}\left(\frac{\beta}{\alpha}\right)\mathbf{L}_1^{s} - \varepsilon\mathbf{N}_1^{s} + \frac{k_\beta}{\nu_\beta}\mathbf{M}_1^{c}$	$-\frac{F_0}{4\pi\mu k_\beta}$	1	1
3. In the x_3 direction	$\gamma_1\mathbf{L}_{01}^{c} + \mathbf{N}_{01}^{c}$	$-\frac{iF_0k_\beta}{4\pi\mu}$	$-\varepsilon\left(\frac{\beta}{\alpha}\right)\mathbf{L}_0^{c} - \frac{k}{\nu_\beta}\mathbf{N}_0^{c}$	$-\frac{F_0}{4\pi\mu k_\beta}$	0	1

(continued)

Table 4.4—(*continued*)

Source	$\frac{1}{K_1}$ Displacement	K_1	$\frac{1}{K_2}$ Displacement	K_2	Indices m	Indices l
Single couple (Fig. 4.9)						
1. (12)	$\gamma_2 \mathbf{L}^s_{22} + \mathbf{N}^s_{22} - 6\mathbf{M}^c_{01}$	$-\frac{iMk_\beta^2}{48\pi\mu}$	$\frac{k}{v_\alpha}\left(\frac{\beta}{\alpha}\right)\mathbf{L}^s_2 + \varepsilon\mathbf{N}^s_2 - \frac{k_\beta}{v_\beta}(\mathbf{M}^c_0 + \mathbf{M}^c_2)$	$\frac{Mk}{8\pi\mu k_\beta}$	0, 2	1, 2
2. (21)	$\gamma_2 \mathbf{L}^s_{22} + \mathbf{N}^s_{22} + 6\mathbf{M}^c_{01}$	$-\frac{iMk_\beta^2}{48\pi\mu}$	$\frac{k}{v_\alpha}\left(\frac{\beta}{\alpha}\right)\mathbf{L}^s_2 + \varepsilon\mathbf{N}^s_2 + \frac{k_\beta}{v_\beta}(\mathbf{M}^c_0 - \mathbf{M}^c_2)$	$\frac{Mk}{8\pi\mu k_\beta}$	0, 2	1, 2
3. (23)	$2\gamma_2 \mathbf{L}^s_{12} + 2\mathbf{N}^s_{12} - 6\mathbf{M}^c_{11}$	$-\frac{iMk_\beta^2}{48\pi\mu}$	$2\varepsilon\left(\frac{\beta}{\alpha}\right)\mathbf{L}^s_1 + 2\frac{v_\beta}{k}\mathbf{N}^s_1 - 2\varepsilon\frac{k_\beta}{k}\mathbf{M}^c_1$	$\frac{Mk}{8\pi\mu k_\beta}$	1	1, 2
4. (32)	$2\gamma_2 \mathbf{L}^s_{12} + 2\mathbf{N}^s_{12} + 6\mathbf{M}^c_{11}$	$-\frac{iMk_\beta^2}{48\pi\mu}$	$2\varepsilon\left(\frac{\beta}{\alpha}\right)\mathbf{L}^s_1 + 2\frac{k}{v_\beta}\mathbf{N}^s_1$	$\frac{Mk}{8\pi\mu k_\beta}$	1	1, 2
5. (31)	$2\gamma_2 \mathbf{L}^c_{12} + 2\mathbf{N}^c_{12} - 6\mathbf{M}^s_{11}$	$-\frac{iMk_\beta^2}{48\pi\mu}$	$2\varepsilon\left(\frac{\beta}{\alpha}\right)\mathbf{L}^c_1 + 2\frac{k}{v_\beta}\mathbf{N}^c_1$	$\frac{Mk}{8\pi\mu k_\beta}$	1	1, 2
6. (13)	$2\gamma_2 \mathbf{L}^c_{12} + 2\mathbf{N}^c_{12} + 6\mathbf{M}^s_{11}$	$-\frac{iMk_\beta^2}{48\pi\mu}$	$2\varepsilon\left(\frac{\beta}{\alpha}\right)\mathbf{L}^c_1 + 2\frac{v_\beta}{k}\mathbf{N}^c_1 + 2\varepsilon\frac{k_\beta}{k}\mathbf{M}^s_1$	$\frac{Mk}{8\pi\mu k_\beta}$	1	1, 2
Double couple (Fig. 4.11)						
1. (23) + (32)	$2\gamma_2 \mathbf{L}^s_{12} + 2\mathbf{N}^s_{12}$	$-\frac{iMk_\beta^2}{24\pi\mu}$	$2\varepsilon\left(\frac{\beta}{\alpha}\right)\mathbf{L}^s_1 + 2\frac{\Omega}{kv_\beta}\mathbf{N}^s_1 - \varepsilon\frac{k_\beta}{k}\mathbf{M}^c_1$	$\frac{Mk}{4\pi\mu k_\beta}$	1	2
2. (31) + (13)	$2\gamma_2 \mathbf{L}^c_{12} + 2\mathbf{N}^c_{12}$	$-\frac{iMk_\beta^2}{24\pi\mu}$	$2\varepsilon\left(\frac{\beta}{\alpha}\right)\mathbf{L}^c_1 + 2\frac{\Omega}{kv_\beta}\mathbf{N}^c_1 + \varepsilon\frac{k_\beta}{k}\mathbf{M}^s_1$	$\frac{Mk}{4\pi\mu k_\beta}$	1	2
3. (12) + (21)	$\gamma_2 \mathbf{L}^s_{22} + \mathbf{N}^s_{22}$	$-\frac{iMk_\beta^2}{24\pi\mu}$	$\frac{k}{v_\alpha}\left(\frac{\beta}{\alpha}\right)\mathbf{L}^s_2 + \varepsilon\mathbf{N}^s_2 - \frac{k_\beta}{v_\beta}\mathbf{M}^c_2$	$\frac{Mk}{4\pi\mu k_\beta}$	2	2

Dipole (Fig. 4.8)						
1. (11)	$\gamma_2(2\mathbf{L}^c_{00} + 2\mathbf{L}^c_{02} - \mathbf{L}^c_{22}) + 2\mathbf{N}^c_{02} - \mathbf{N}^c_{22}$	$\frac{iMk_\beta^2}{48\pi\mu}$	$\frac{k}{v_\sigma}\left(\frac{\beta}{\alpha}\right)(\mathbf{L}^c_0 - \mathbf{L}^c_2) + \varepsilon(\mathbf{N}^c_0 - \mathbf{N}^c_2) - \frac{k_\beta}{v_\beta}\mathbf{M}^s_2$	$-\frac{Mk}{8\pi\mu k_\beta}$	0, 2	0, 2
2. (22)	$\gamma_2(2\mathbf{L}^c_{00} + 2\mathbf{L}^c_{02} + \mathbf{L}^c_{22}) + 2\mathbf{N}^c_{02} + \mathbf{N}^c_{22}$	$\frac{iMk_\beta^2}{48\pi\mu}$	$\frac{k}{v_\alpha}\left(\frac{\beta}{\alpha}\right)(\mathbf{L}^c_0 + \mathbf{L}^c_2) + \varepsilon(\mathbf{N}^c_0 + \mathbf{N}^c_2) + \frac{k_\beta}{v_\beta}\mathbf{M}^s_2$	$-\frac{Mk}{8\pi\mu k_\beta}$	0, 2	0, 2
3. (33)	$\gamma_2(2\mathbf{L}^c_{00} - 4\mathbf{L}^c_{02}) - 4\mathbf{N}^c_{02}$	$\frac{iMk_\beta^2}{48\pi\mu}$	$-2\frac{v_\alpha}{k}\left(\frac{\beta}{\alpha}\right)\mathbf{L}^c_0 - 2\varepsilon\mathbf{N}^c_0$	$-\frac{Mk}{8\pi\mu k_\beta}$	0	0, 2
Tangential dislocation (Fig. 4.25)						
1. Case I ($\lambda = 0°, \delta = 90°$)	$\gamma_2\mathbf{L}^s_{22} + \mathbf{N}^s_{22}$	$\frac{iU_0\,dSk_\beta^2}{24\pi}$	$\frac{k}{v_\alpha}\left(\frac{\beta}{\alpha}\right)\mathbf{L}^s_2 + \varepsilon\mathbf{N}^s_2 - \frac{k_\beta}{v_\beta}\mathbf{M}^c_2$	$-\frac{U_0\,dSk}{4\pi k_\beta}$	2	2
2. Case II ($\lambda = 90°, \delta = 90°$)	$2\gamma_2\mathbf{L}^s_{12} + 2\mathbf{N}^s_{12}$	$\frac{iU_0\,dSk_\beta^2}{24\pi}$	$2\varepsilon\left(\frac{\beta}{\alpha}\right)\mathbf{L}^s_1 + \frac{2\Omega}{kv_\beta}\mathbf{N}^s_1 - \varepsilon\frac{k_\beta}{k}\mathbf{M}^c_1$	$-\frac{U_0\,dSk}{4\pi k_\beta}$	1	2
3. Case III ($\lambda = 90°, \delta = 45°$)	$-\gamma_2(3\mathbf{L}^c_{02} + \frac{1}{2}\mathbf{L}^c_{22}) - 3\mathbf{N}^c_{02} - \frac{1}{2}\mathbf{N}^c_{22}$	$\frac{iU_0\,dSk_\beta^2}{24\pi}$	$-\frac{3k^2 - 2k_\alpha^2}{2kv_\alpha}\left(\frac{\beta}{\alpha}\right)\mathbf{L}^c_0 - \frac{1}{2}\frac{k}{v_\alpha}\left(\frac{\beta}{\alpha}\right)\mathbf{L}^c_2 - \varepsilon\frac{1}{2}(3\mathbf{N}^c_0 + \mathbf{N}^c_2) - \frac{1}{2}\frac{k_\beta}{v_\beta}\mathbf{M}^s_2$	$-\frac{U_0\,dSk}{4\pi k_\beta}$	0, 2	2
Tensile dislocation						
1. In the x_1 direction	$\gamma_2(\gamma_3\mathbf{L}^c_{00} + 2\mathbf{L}^c_{02} - \mathbf{L}^c_{22}) + 2\mathbf{N}^c_{02} - \mathbf{N}^c_{22}$	$\frac{iU_0\,dSk_\beta^2}{24\pi}$	$\frac{2v_\alpha^2 - v_\beta^2}{kv_\alpha}\left(\frac{\beta}{\alpha}\right)\mathbf{L}^c_0 - \frac{k}{v_\alpha}\left(\frac{\beta}{\alpha}\right)\mathbf{L}^c_2 + \varepsilon(\mathbf{N}^c_0 - \mathbf{N}^c_2) - \frac{k_\beta}{v_\beta}\mathbf{M}^s_2$	$-\frac{U_0\,dSk}{4\pi k_\beta}$	0, 2	0, 2
2. In the x_2 direction	$\gamma_2(\gamma_3\mathbf{L}^c_{00} + 2\mathbf{L}^c_{02} + \mathbf{L}^c_{22}) + 2\mathbf{N}^c_{02} + \mathbf{N}^c_{22}$	$\frac{iU_0\,dSk_\beta^2}{24\pi}$	$\frac{2v_\alpha^2 - v_\beta^2}{kv_\alpha}\left(\frac{\beta}{\alpha}\right)\mathbf{L}^c_0 + \frac{k}{v_\alpha}\left(\frac{\beta}{\alpha}\right)\mathbf{L}^c_2 + \varepsilon(\mathbf{N}^c_0 + \mathbf{N}^c_2) + \frac{k_\beta}{v_\beta}\mathbf{M}^s_2$	$-\frac{U_0\,dSk}{4\pi k_\beta}$	0, 2	0, 2
3. In the x_3 direction	$\gamma_2(\gamma_3\mathbf{L}^c_{00} - 4\mathbf{L}^c_{02}) - 4\mathbf{N}^c_{02}$	$\frac{iU_0\,dSk_\beta^2}{24\pi}$	$-\frac{2\Omega}{kv_\alpha}\left(\frac{\beta}{\alpha}\right)\mathbf{L}^c_0 - 2\varepsilon\mathbf{N}^c_0$	$-\frac{U_0\,dSk}{4\pi k_\beta}$	0	0, 2

[a] $\gamma_1 = (\beta/\alpha)^3$, $\gamma_2 = 2(\beta/\alpha)^4$, $\gamma_3 = 2 + 3\lambda/\mu$, $\Omega = k^2 - \frac{1}{2}k_\beta^2$, $\varepsilon = +1(z > 0), -1(z < 0)$.

Expressing $\mathfrak{G}$ of Eq. (4.12) in the indicial notation and using Eq. (4.159), we find

$$G_{kl} = -\frac{i}{24\pi\mu}\sum_{m,n}\left[k_\alpha\left(\frac{\beta}{\alpha}\right)^2 a_{mn}(k,l)\Phi^c_{mn}(k_\alpha R) + k_\beta b_{mn}(k,l)\Phi^c_{mn}(k_\beta R)\right] \quad (k+l = \text{even});$$

$$= -\frac{i}{24\pi\mu}\sum_{m,n}\left[k_\alpha\left(\frac{\beta}{\alpha}\right)^2 a_{mn}(k,l)\Phi^s_{mn}(k_\alpha R) + k_\beta b_{mn}(k,l)\Phi^s_{mn}(k_\beta R)\right] \quad (k+l = \text{odd});$$

$$(mn) = (00), (02), (12), (22). \tag{4.161}$$

The values of the coefficients a_{mn} and b_{mn} are given in Table 4.5.

Next, expressing the triadic T given in Eq. (4.141) in indicial notation and proceeding as above we find, for $m + k + l = \text{odd}$,

$$T^m_{kl} = \frac{i}{20\pi}\sum_{n,p}\left[k_\alpha^2\left(\frac{\beta}{\alpha}\right)^2 A_{np}(k,l,m)\Phi^c_{np}(k_\alpha R) + k_\beta^2 B_{np}(k,l,m)\Phi^c_{np}(k_\beta R)\right],$$

$$(np) = (01), (11), (03), (13), (23), (33). \tag{4.162}$$

Similarly, when $m + k + l = \text{even}$, we have

$$T^m_{kl} = \frac{i}{20\pi}\sum_{n,p}\left[k_\alpha^2\left(\frac{\beta}{\alpha}\right)^2 A_{np}(k,l,m)\Phi^s_{np}(k_\alpha R) + k_\beta^2 B_{np}(k,l,m)\Phi^s_{np}(k_\beta R)\right]. \tag{4.163}$$

The coefficients A_{np} and B_{np} are given in Table 4.6. In this table $\gamma = 1/[2(1-\sigma)]$.

In the case of the tensor $\mathbf{H}$ given in Eq. (4.151), we need only H^{33}_{kl} in most of the applications. We find

$$H^{33}_{kl} = \frac{i\mu}{420\pi}\sum_{mn}[k_\alpha^3 C_{mn}(k,l)\Phi^c_{mn}(k_\alpha R) + k_\beta^3 D_{mn}(k,l)\Phi^c_{mn}(k_\beta R)] \quad (k+l = \text{even})$$

$$= \frac{i\mu}{420\pi}\sum_{mn}[k_\alpha^3 C_{mn}(k,l)\Phi^s_{mn}(k_\alpha R) + k_\beta^3 D_{mn}(k,l)\Phi^s_{mn}(k_\beta R)] \quad (k+l = \text{odd}),$$

$$(mn) = (00), (02), (04), (12), (14), (22), (24), \tag{4.164}$$

where the coefficients C_{mn} and D_{mn} are given in Table 4.7.

For problems involving a displacement dislocation in a multilayered half-space, the tensor T^m_{kl} should be expressed in terms of cylindrical eigenvectors.

Table 4.5. Numerical Values of the Coefficients a_{mn} and b_{mn}

	a_{mn}				b_{mn}			
kl	00	02	12	22	00	02	12	22
11	2	2	0	−1	4	−2	0	1
22	2	2	0	1	4	−2	0	−1
33	2	−4	0	0	4	4	0	0
23	0	0	−2	0	0	0	2	0
31	0	0	−2	0	0	0	2	0
12	0	0	0	−1	0	0	0	1

Table 4.6. Numerical Values of the Coefficients A_{np} and B_{np}

			A_{np}						B_{np}					
m	k	l	01	11	03	13	23	33	01	11	03	13	23	33
1	1	1	0	$-\frac{1+4\gamma}{1-\gamma}$	0	-1	0	$\frac{1}{6}$	0	-4	0	1	0	$-\frac{1}{6}$
2	1	1	0	$\frac{3-8\gamma}{1-\gamma}$	0	$-\frac{1}{3}$	0	$\frac{1}{6}$	0	2	0	$\frac{1}{3}$	0	$-\frac{1}{6}$
3	1	1	$\frac{3-8\gamma}{1-\gamma}$	0	-2	0	$\frac{1}{3}$	0	2	0	2	0	$-\frac{1}{3}$	0
1	2	2	0	$\frac{3-8\gamma}{1-\gamma}$	0	$-\frac{1}{3}$	0	$-\frac{1}{6}$	0	2	0	$\frac{1}{3}$	0	$\frac{1}{6}$
2	2	2	0	$-\frac{1+4\gamma}{1-\gamma}$	0	-1	0	$-\frac{1}{6}$	0	-4	0	1	0	$\frac{1}{6}$
3	2	2	$\frac{3-8\gamma}{1-\gamma}$	0	-2	0	$-\frac{1}{3}$	0	2	0	2	0	$\frac{1}{3}$	0
1	3	3	0	$\frac{3-8\gamma}{1-\gamma}$	0	$\frac{4}{3}$	0	0	0	2	0	$-\frac{4}{3}$	0	0
2	3	3	0	$\frac{3-8\gamma}{1-\gamma}$	0	$\frac{4}{3}$	0	0	0	2	0	$-\frac{4}{3}$	0	0
3	3	3	$-\frac{1+4\gamma}{1-\gamma}$	0	4	0	0	0	-4	0	-4	0	0	0
1	2	3	0	0	0	0	$\frac{1}{3}$	0	0	0	0	0	$-\frac{1}{3}$	0
2	2	3	-2	0	-2	0	$-\frac{1}{3}$	0	-3	0	2	0	$\frac{1}{3}$	0
3	2	3	0	-2	0	$\frac{4}{3}$	0	0	0	-3	0	$-\frac{4}{3}$	0	0
1	3	1	-2	0	-2	0	$\frac{1}{3}$	0	-3	0	2	0	$-\frac{1}{3}$	0
2	3	1	0	0	0	0	$\frac{1}{3}$	0	0	0	0	0	$-\frac{1}{3}$	0
3	3	1	0	-2	0	$\frac{4}{3}$	0	0	0	-3	0	$-\frac{4}{3}$	0	0
1	1	2	0	-2	0	$-\frac{1}{3}$	0	$\frac{1}{6}$	0	-3	0	$\frac{1}{3}$	0	$-\frac{1}{6}$
2	1	2	0	-2	0	$-\frac{1}{3}$	0	$-\frac{1}{6}$	0	-3	0	$\frac{1}{3}$	0	$\frac{1}{6}$
3	1	2	0	0	0	0	$\frac{1}{3}$	0	0	0	0	0	$-\frac{1}{3}$	0

Table 4.7. Numerical Values of the Coefficients C_{mn} and D_{mn}

		C_{mn}							D_{mn}						
k	l	00	02	04	12	14	22	24	00	02	04	12	14	22	24
1	1	$7\frac{1+8\gamma-24\gamma^2}{1-\gamma}$	$10(12\gamma-5)$	$48(1-\gamma)$	0	0	$5(12\gamma-5)$	$-4(1-\gamma)$	28	−20	−48	0	0	−10	4
2	2	$7\frac{1+8\gamma-24\gamma^2}{1-\gamma}$	$10(12\gamma-5)$	$48(1-\gamma)$	0	0	$-5(12\gamma-5)$	$4(1-\gamma)$	28	−20	−48	0	0	10	−4
3	3	$7\frac{-7+24\gamma-32\gamma^2}{1-\gamma}$	$40(8\gamma-1)$	$-96(1-\gamma)$	0	0	0	0	−56	40	96	0	0	0	0
2	3	0	0	0	$10(8\gamma-1)$	$-24(1-\gamma)$	0	0	0	0	0	10	24	0	0
3	1	0	0	0	$10(8\gamma-1)$	$-24(1-\gamma)$	0	0	0	0	0	10	24	0	0
1	2	0	0	0	0	0	$5(12\gamma-5)$	$-4(1-\gamma)$	0	0	0	0	0	−10	4

In this connection, the Erdélyi formula, Eq. (2.150), proves very useful. Equations (4.162) and (4.163) then yield

$$T^1_{11} = \cos\phi_h \int_0^\infty J_1(k\hat{r})\Delta^1_{11}\,dk + \cos 3\phi_h \int_0^\infty J_3(k\hat{r})\Theta^1_{11}\,dk,$$

$$T^2_{11} = \sin\phi_h \int_0^\infty J_1(k\hat{r})\Delta^2_{11}\,dk + \sin 3\phi_h \int_0^\infty J_3(k\hat{r})\Theta^1_{11}\,dk,$$

$$T^3_{11} = \varepsilon\left[\int_0^\infty J_0(k\hat{r})\Delta^3_{11}\,dk + \cos 2\phi_h \int_0^\infty J_2(k\hat{r})\Theta^3_{11}\,dk\right],$$

$$T^1_{22} = \cos\phi_h \int_0^\infty J_1(k\hat{r})\Delta^2_{11}\,dk - \cos 3\phi_h \int_0^\infty J_3(k\hat{r})\Theta^1_{11}\,dk,$$

$$T^2_{22} = \sin\phi_h \int_0^\infty J_1(k\hat{r})\Delta^1_{11}\,dk - \sin 3\phi_h \int_0^\infty J_3(k\hat{r})\Theta^1_{11}\,dk,$$

$$T^3_{22} = \varepsilon\left[\int_0^\infty J_0(k\hat{r})\Delta^3_{11}\,dk - \cos 2\phi_h \int_0^\infty J_2(k\hat{r})\Theta^3_{11}\,dk\right],$$

$$T^1_{33} = \cos\phi_h \int_0^\infty J_1(k\hat{r})\Delta^1_{33}\,dk,$$

$$T^2_{33} = \sin\phi_h \int_0^\infty J_1(k\hat{r})\Delta^1_{33}\,dk,$$

$$T^3_{33} = \varepsilon \int_0^\infty J_0(k\hat{r})\Delta^3_{33}\,dk,$$

$$T^1_{23} = \varepsilon \sin 2\phi_h \int_0^\infty J_2(k\hat{r})\Theta^3_{11}\,dk,$$

$$T^2_{23} = \varepsilon\left[\int_0^\infty J_0(k\hat{r})\Delta^2_{23}\,dk - \cos 2\phi_h \int_0^\infty J_2(k\hat{r})\Theta^3_{11}\,dk\right],$$

$$T^3_{23} = \sin\phi_h \int_0^\infty J_1(k\hat{r})\Delta^3_{23}\,dk,$$

$$T^1_{31} = \varepsilon\left[\int_0^\infty J_0(k\hat{r})\Delta^2_{23}\,dk + \cos 2\phi_h \int_0^\infty J_2(k\hat{r})\Theta^3_{11}\,dk\right],$$

$$T^2_{31} = \varepsilon \sin 2\phi_h \int_0^\infty J_2(k\hat{r})\Theta^3_{11}\,dk,$$

$$T^3_{31} = \cos\phi_h \int_0^\infty J_1(k\hat{r})\Delta^3_{23}\,dk,$$

$$T^1_{12} = \sin\phi_h \int_0^\infty J_1(k\hat{r})\Delta^1_{12}\,dk + \sin 3\phi_h \int_0^\infty J_3(k\hat{r})\Theta^1_{11}\,dk,$$

$$T^2_{12} = \cos\phi_h \int_0^\infty J_1(k\hat{r})\Delta^1_{12}\,dk - \cos 3\phi_h \int_0^\infty J_3(k\hat{r})\Theta^1_{11}\,dk,$$

$$T^3_{12} = \varepsilon \sin 2\phi_h \int_0^\infty J_2(k\hat{r})\Theta^3_{11}\,dk, \tag{4.165}$$

where

$$
\begin{aligned}
\hat{r} &= R \sin i_h, \qquad x_3^\circ - r_0 = R \cos i_h, \\
X &= -\frac{k^2}{8\pi k_\beta^2} e^{-\nu_\alpha |x_3^\circ - r_0|}, \qquad Y = -\frac{k^2}{8\pi k_\beta^2} e^{-\nu_\beta |x_3^\circ - r_0|}, \\
\nu_\alpha &= \sqrt{(k^2 - k_\alpha^2)}, \qquad \nu_\beta = \sqrt{(k^2 - k_\beta^2)}, \qquad \gamma = \frac{1}{2(1-\sigma)}, \\
\varepsilon &= \begin{cases} +1, & x_3^\circ > r_0 \\ -1, & x_3^\circ < r_0, \end{cases} \\
\Delta_{11}^1 &= \frac{X}{\nu_\alpha}[2(1-2\gamma)k_\beta^2 - 3k^2] + \frac{Y}{\nu_\beta}(3k^2 - 4k_\beta^2), \\
\Theta_{11}^1 &= k^2\left[\frac{X}{\nu_\alpha} - \frac{Y}{\nu_\beta}\right], \\
\Delta_{11}^2 &= \frac{X}{\nu_\alpha}[2(1-2\gamma)k_\beta^2 - k^2] + \frac{Y}{\nu_\beta}k^2, \\
\Delta_{11}^3 &= \frac{2X}{k}[(1-2\gamma)k_\beta^2 - k^2] + 2kY, \\
\Theta_{11}^3 &= 2k[X - Y], \\
\Delta_{33}^1 &= 4\left[\frac{k^2 - \frac{1}{2}k_\beta^2}{\nu_\alpha} X - \nu_\beta Y\right], \\
\Delta_{33}^3 &= 4\left[\frac{k^2 - \frac{1}{2}k_\beta^2}{k} X - kY\right], \\
\Delta_{23}^2 &= 2\left[-kX + \frac{\nu_\beta^2}{k} Y\right], \\
\Delta_{23}^3 &= 4\left[\nu_\alpha X - \frac{k^2 - \frac{1}{2}k_\beta^2}{\nu_\beta} Y\right], \\
\Delta_{12}^1 &= -\frac{X}{\nu_\alpha}k^2 + \frac{Y}{\nu_\beta}(k^2 - 2k_\beta^2).
\end{aligned}
\tag{4.166}
$$

4.5.3. Eigenvector Expansion of Green's Dyadic

We begin with the expansion (App. F)

$$
\frac{\delta(e - e')\delta(\lambda - \lambda')}{\sin e'} \mathfrak{I} = \sum_{l=0,1,1}^{\infty} \sum_{m=-l}^{l} \frac{1}{\Omega_{ml}} [\mathbf{P}_{ml}(e, \lambda)\overset{*}{\mathbf{P}}_{ml}(e', \lambda') + \mathbf{B}_{ml}(e, \lambda)\overset{*}{\mathbf{B}}_{ml}(e', \lambda') + \mathbf{C}_{ml}(e, \lambda)\overset{*}{\mathbf{C}}_{ml}(e', \lambda')]. \tag{4.167}
$$

The vectors **P**, **B**, and **C** are the vector spherical harmonics defined in **k** (k_c, e, λ) space [Eq. (2.157) and Fig. 2.3] in which

$$\mathbf{k} = k_c\mathbf{p} = k_c(\sin e \cos \lambda \mathbf{e}_1 + \sin e \sin \lambda \mathbf{e}_2 + \cos e\mathbf{e}_3).$$

An asterisk signifies a complex conjugate and

$$\Omega_{ml} = \frac{4\pi}{2l+1}\frac{(l+m)!}{(l-m)!}.$$

Furthermore, the summation over l begins with $l = 0$ for $\mathbf{P}\overset{*}{\mathbf{P}}$, whereas it begins with $l = 1$ for $\mathbf{B}\overset{*}{\mathbf{B}}$ and $\mathbf{C}\overset{*}{\mathbf{C}}$. Multiplying both sides of Eq. (4.167) with

$$\sin e' \exp(i\mathbf{k}' \cdot \mathbf{r}_0),$$

integrating over $e'(0, \pi)$ and $\lambda'(0, 2\pi)$, and using Eqs. (2.157) and (2.158), we get

$$e^{i\mathbf{k}\cdot\mathbf{r}_0}\frac{\mathfrak{I}}{4\pi} = \sum_{l=0,1,1}^{\infty}\sum_{m=-l}^{l}\frac{i^{l-1}}{\Omega_{ml}}\Bigg[\mathbf{P}_{ml}(e, \lambda)\overset{*}{\mathbf{L}}{}^{+}_{ml}(k_c r_0) + \frac{1}{\sqrt{[l(l+1)]}}\mathbf{B}_{ml}(e, \lambda)\overset{*}{\mathbf{N}}{}^{+}_{ml}(k_c r_0) + \frac{i}{\sqrt{[l(l+1)]}}\mathbf{C}_{ml}(e, \lambda)\overset{*}{\mathbf{M}}{}^{+}_{ml}(k_c r_0)\Bigg]. \tag{4.168}$$

We next multiply both sides of Eq. (4.168) with $(\sin e)\exp(-i\mathbf{k}\cdot\mathbf{r})$, integrate over e (along $C^{\pm}$, Fig. 2.4) and $\lambda(0, 2\pi)$, and use Eq. (2.143) with $m = l = 0$ and Eq. (2.154) to get, for $r > r_0$,

$$\frac{\mathfrak{I}}{4\pi R}e^{-ik_cR} = -ik_c\sum_{l=0,1,1}^{\infty}\sum_{m=-l}^{l}\frac{1}{\Omega_{ml}}\Bigg[\mathbf{L}^{-}_{ml}(k_c r)\overset{*}{\mathbf{L}}{}^{+}_{ml}(k_c r_0) + \frac{1}{l(l+1)}\mathbf{N}^{-}_{ml}(k_c r)\overset{*}{\mathbf{N}}{}^{+}_{ml}(k_c r_0) + \frac{1}{l(l+1)}\mathbf{M}^{-}_{ml}(k_c r)\overset{*}{\mathbf{M}}{}^{+}_{ml}(k_c r_0)\Bigg]. \tag{4.169}$$

It may be noted that the dyadic $\mathfrak{I}e^{-ik_cR}/(4\pi R)$ satisfies the dyadic equation

$$(\nabla^2 + k_c^2)\mathfrak{G}(\mathbf{r}\,|\,\mathbf{r}_0) = -\mathfrak{I}\delta(\mathbf{r} - \mathbf{r}_0). \tag{4.170}$$

The expansion in Eq. (4.169) is valid for $r > r_0$. However, if $r < r_0$, we should interchange the roles of r and r_0. Because the dyadic on the left-hand side of Eq. (4.169) is symmetric, we can interchange the order of the dyads in its right-hand side.

The expansion of the Green's dyadic for the Navier equation is now straightforward. We have simply to use the expansion in Eq. (4.169) in expression (4.9) for the Green's dyadic. We note that when the right-hand side of Eq. (4.169) is differentiated with respect to the observer's coordinates, the source vectors behave as constant. Therefore, using the definitions of the **L**, **M**, and **N** vectors, we have

$$\text{curl curl}[\mathbf{M}(k_\beta r)\mathbf{M}(k_\beta r_0)] = k_\beta^2 \mathbf{M}(k_\beta r)\mathbf{M}(k_\beta r_0),$$
$$\text{curl curl}[\mathbf{N}(k_\beta r)\mathbf{N}(k_\beta r_0)] = k_\beta^2 \mathbf{N}(k_\beta r)\mathbf{N}(k_\beta r_0),$$
$$\text{grad div}[\mathbf{L}(k_\alpha r)\mathbf{L}(k_\gamma r_0)] = -k_\alpha^2 \mathbf{L}(k_\alpha r)\mathbf{L}(k_\alpha r_0). \tag{4.171}$$

Inserting the expansion (4.169) in Eq. (4.9) and using Eq. (4.171), we obtain $(r > r_0)$

$$\mathfrak{G}(\mathbf{r}\,|\,\mathbf{r}_0;\omega) = -\frac{i}{\mu}k_\beta \sum_{l=1,1,0}^{\infty}\sum_{m=-l}^{l}\frac{1}{\Omega_{ml}}\left[\frac{1}{l(l+1)}\mathbf{M}_{ml}^{-}(k_\beta r)\overset{*}{\mathbf{M}}{}_{ml}^{+}(k_\beta r_0)\right.$$
$$\left.+\frac{1}{l(l+1)}\mathbf{N}_{ml}^{-}(k_\beta r)\overset{*}{\mathbf{N}}{}_{ml}^{+}(k_\beta r_0)+\left(\frac{\beta}{\alpha}\right)^3\mathbf{L}_{ml}^{-}(k_\alpha r)\overset{*}{\mathbf{L}}{}_{ml}^{+}(k_\alpha r_0)\right]. \tag{4.172}$$

Expansions corresponding to Eqs. (4.169) and (4.172) in cylindrical coordinates are

$$\frac{\mathfrak{I}}{R}e^{-ik_cR} = \sum_{m=-\infty}^{\infty}\int_0^{\infty}[\mathbf{M}_m^{-}(k_c\Delta)\overset{*}{\mathbf{M}}{}_m^{+}(k_c\Delta_0)+\mathbf{N}_m^{-}(k_c\Delta)\overset{*}{\mathbf{N}}{}_m^{+}(k_c\Delta_0)$$
$$+\mathbf{L}_m^{-}(k_c\Delta)\overset{*}{\mathbf{L}}{}_m^{+}(k_c\Delta_0)]\frac{k\,dk}{v_c}, \tag{4.173}$$

$$\mathfrak{G}(\mathbf{r}\,|\,\mathbf{r}_0;\omega) = \frac{1}{4\pi\mu}\sum_{m=-\infty}^{\infty}\int_0^{\infty}\left[\frac{1}{v_\beta}\{\mathbf{M}_m^{-}(k_\beta\Delta)\overset{*}{\mathbf{M}}{}_m^{+}(k_\beta\Delta_0)\right.$$
$$\left.+\mathbf{N}_m^{-}(k_\beta\Delta)\overset{*}{\mathbf{N}}{}_m^{+}(k_\beta\Delta_0)\}+\frac{1}{v_\alpha}\left(\frac{\beta}{\alpha}\right)^2\mathbf{L}_m^{-}(k_\alpha\Delta)\overset{*}{\mathbf{L}}{}_m^{+}(k_\alpha\Delta_0)\right]k\,dk, \tag{4.174}$$

valid for $z > z_0$. If $z < z_0$, we should interchange the roles of Δ and Δ_0 in the above expansions.

EXAMPLE 4.6: Eigenvector Expansion of Dyadic Plane Waves

In Eq. (4.168) we put $e = 0$. This makes **k** point in the z direction and reduces all the vectors **P**, **B**, and **C** to zero except at $m = 0$ for **P** and $m = \pm 1$ for **B** and **C**. To prove this statement, we need the limits of $P_l^m(\cos\theta)$, $[P_l^m(\cos\theta)]/\sin\theta$, and $(\partial/\partial\theta)P_l^m(\cos\theta)$ as $\theta \to 0$. Indeed, using the integral representation given in Eq. (4.5.4), it easily can be proved that

$$\lim_{\theta\to 0} P_l^m(\cos\theta) = \delta_{0m}, \tag{4.6.1}$$

$$\lim_{\theta\to 0}\frac{P_l^m(\cos\theta)}{\sin\theta} = \lim_{\theta\to 0}\frac{\partial P_l^m(\cos\theta)}{\partial\theta} = \frac{1}{2}[l(l+1)\delta_{1m} - \delta_{(-1)m}]. \tag{4.6.2}$$

Equation (2.71) then yields

$$\lim_{\theta\to 0} \mathbf{P}_{ml}(\theta, \phi) = \delta_{0m}\mathbf{e}_r = \delta_{0m}\mathbf{e}_z, \tag{4.6.3}$$

$$\lim_{\theta\to 0} \sqrt{[l(l+1)]}\mathbf{B}_{ml}(\theta, \phi) = \frac{1}{2}[l(l+1)e^{i\phi}\delta_{1m} - e^{-i\phi}\delta_{(-1)m}]\mathbf{e}_\theta + \frac{i}{2}[l(l+1)e^{i\phi}\delta_{1m} + e^{-i\phi}\delta_{(-1)m}]\mathbf{e}_\phi, \tag{4.6.4}$$

$$\lim_{\theta\to 0} \sqrt{[l(l+1)]}\mathbf{C}_{ml}(\theta, \phi) = \frac{i}{2}[l(l+1)e^{i\phi}\delta_{1m} + e^{-i\phi}\delta_{(-1)m}]\mathbf{e}_\theta - \frac{1}{2}[l(l+1)e^{i\phi}\delta_{1m} - e^{-i\phi}\delta_{(-1)m}]\mathbf{e}_\phi. \tag{4.6.5}$$

These can also be written as

$$\frac{2\mathbf{B}_{ml}(0, \phi)}{\sqrt{[l(l+1)]}} = (\mathbf{e}_x + i\mathbf{e}_y)\delta_{1m} + \frac{(-\mathbf{e}_x + i\mathbf{e}_y)}{l(l+1)}\delta_{(-1)m},$$

$$\frac{2\mathbf{C}_{ml}(0, \phi)}{\sqrt{[l(l+1)]}} = (i\mathbf{e}_x - \mathbf{e}_y)\delta_{1m} + \frac{(i\mathbf{e}_x + \mathbf{e}_y)}{l(l+1)}\delta_{(-1)m}.$$

We now put these expressions in Eq. (4.168) and change $\mathbf{r}_0$ to $\mathbf{r}$. Taking notice that $\Omega_{1l} = [4\pi/(2l+1)]l(l+1)$, $\Omega_{(-1)l} = [4\pi/(2l+1)]\{1/[l(l+1)]\}$, the sum over $m = \pm 1$ eventually yields

$$\mathfrak{I}e^{ik_c z} = \sum_{l=0,1,1}^{\infty} \frac{2l+1}{l(l+1)} i^l\{-il(l+1)\mathbf{e}_z\mathbf{L}_{0l}^{+} + \mathbf{e}_x[\mathbf{M}_{1l}^{s,+} - i\mathbf{N}_{1l}^{c,+}] - \mathbf{e}_y[\mathbf{M}_{1l}^{c,+} + i\mathbf{N}_{1l}^{s,+}]\}, \tag{4.6.6}$$

where (s) and (c) refer to the respective ϕ dependence $\sin\phi$ and $\cos\phi$ in the expressions for the eigenvectors.

Multiplying Eq. (4.6.6), in turn, scalarly by $\mathbf{e}_x$, $\mathbf{e}_y$ and $\mathbf{e}_z$ we obtain the respective expansions for plane waves polarized in the directions of the Cartesian axes,

$$\mathbf{e}_x e^{ik_c z} = \sum_{l=1}^{\infty} \frac{2l+1}{l(l+1)} i^l[\mathbf{M}_{1l}^{s,+} - i\mathbf{N}_{1l}^{c,+}],$$

$$\mathbf{e}_y e^{ik_c z} = -\sum_{l=1}^{\infty} \frac{2l+1}{l(l+1)} i^l[\mathbf{M}_{1l}^{c,+} + i\mathbf{N}_{1l}^{s,+}],$$

$$\mathbf{e}_z e^{ik_c z} = \sum_{l=0}^{\infty}(2l+1)i^{l-1}\mathbf{L}_{0l}^{+}. \tag{4.6.7}$$

These formulas are useful in solving the problem of scattering and diffraction of plane waves by spherical obstacles. Note that the longitudinal wave $\mathbf{e}_z e^{ik_c z}$

is expressible in terms of irrotational vectors only, whereas the transverse waves $\mathbf{e}_x e^{ik_c z}$ and $\mathbf{e}_y e^{ik_c z}$ are represented by both solenoidal vector sets (*SH* and *SV* motion).

EXAMPLE 4.7: Expansions of Green's Dyadics in Terms of the Hansen Vectors

Consider the integral

$$\int_0^{2\pi}\int_0^{\pi} \overset{*}{\mathbf{M}}{}_{ml}^{-\varepsilon}(\mathbf{r}) \frac{e^{-ik_c R}}{R} \sin\theta\, d\theta\, d\phi, \qquad \varepsilon = \operatorname{sgn}(r - r_0), \qquad R = |\mathbf{r} - \mathbf{r}_0|. \tag{4.7.1}$$

To evaluate it we first express the spherical unit vectors appearing in $\mathbf{M}$ in terms of Cartesian unit vectors. We obtain

$$\begin{aligned} \overset{*}{\mathbf{M}}_{ml} &= \mathbf{a}_{-1}\overset{*}{\Psi}_{(m-1)l} + \mathbf{a}_0 \overset{*}{\Psi}_{ml} + \mathbf{a}_1 \overset{*}{\Psi}_{(m+1)l}, \\ \Psi_{ml}^{\pm}(\mathbf{r}) &= Y_{ml}(\theta, \phi) f_l^{\pm}(k_c r), \end{aligned} \tag{4.7.2}$$

where $\mathbf{a}_i$ are constant vectors that depend on m and l but whose explicit form is not needed for the evaluation of the integral. Applying the known results

$$\int_0^{2\pi}\int_0^{\pi} \Psi_{m'l'}^{-\varepsilon}(\mathbf{r}) \overset{*}{\Psi}{}_{ml}^{-\varepsilon}(\mathbf{r}) \sin\theta\, d\theta\, d\phi = \delta_{mm'}\delta_{ll'}\Omega_{m'l'}[f_{l'}^{-\varepsilon}(k_c r)]^2, \tag{4.7.3}$$

$$\frac{e^{-ik_c R}}{4\pi R} = -ik_c \sum_{l'=0}^{\infty} \sum_{m'=-l'}^{l'} \frac{1}{\Omega_{m'l'}} \overset{*}{\Psi}{}_{m'l'}^{\varepsilon}(\mathbf{r}_0) \Psi_{m'l'}^{-\varepsilon}(\mathbf{r}), \tag{4.7.4}$$

we obtain

$$\begin{aligned} &\int_0^{2\pi}\int_0^{\pi} \overset{*}{\mathbf{M}}{}_{ml}^{-\varepsilon}(\mathbf{r}) \frac{e^{-ik_c R}}{4\pi R} \sin\theta\, d\theta\, d\phi \\ &= -ik_c \sum_{l',m'} \frac{1}{\Omega_{m'l'}} \overset{*}{\Psi}{}_{m'l'}^{\varepsilon}(\mathbf{r}_0) \int_0^{2\pi}\int_0^{\pi} \Psi_{m'l'}^{-\varepsilon}(\mathbf{r}) \overset{*}{\mathbf{M}}{}_{ml}^{-\varepsilon}(\mathbf{r}) \sin\theta\, d\theta\, d\phi \\ &= -ik_c \sum_{l',m'} \frac{1}{\Omega_{m'l'}} \overset{*}{\Psi}{}_{m'l'}^{\varepsilon}(\mathbf{r}_0) \int_0^{2\pi}\int_0^{\pi} \Psi_{m'l'}^{-\varepsilon}(\mathbf{r}) \{\mathbf{a}_{-1} \overset{*}{\Psi}{}_{(m-1)l}^{-\varepsilon}(\mathbf{r}) + \mathbf{a}_0 \overset{*}{\Psi}{}_{ml}^{-\varepsilon}(\mathbf{r}) \\ &\quad + \mathbf{a}_1 \overset{*}{\Psi}{}_{(m+1)l}^{-\varepsilon}(\mathbf{r})\} \sin\theta\, d\theta\, d\phi \\ &= -ik_c \sum_{l',m'} \frac{1}{\Omega_{m'l'}} \overset{*}{\Psi}{}_{m'l'}^{\varepsilon}(\mathbf{r}_0) \{\mathbf{a}_{-1}\delta_{m'(m-1)} + \mathbf{a}_0 \delta_{m'm} + \mathbf{a}_1 \delta_{m'(m+1)}\} \delta_{ll'} \Omega_{m'l'} \\ &\quad \times [f_{l'}^{-\varepsilon}(k_c r)]^2, \\ &= -ik_c [f_l^{-\varepsilon}(k_c r)]^2 [\mathbf{a}_{-1} \overset{*}{\Psi}{}_{(m-1)l}^{\varepsilon}(\mathbf{r}_0) + \mathbf{a}_0 \overset{*}{\Psi}{}_{ml}^{\varepsilon}(\mathbf{r}_0) + \mathbf{a}_{-1} \overset{*}{\Psi}{}_{(m+1)l}^{\varepsilon}(\mathbf{r}_0)] \\ &= -ik_c [f_l^{-\varepsilon}(k_c r)]^2 \overset{*}{\mathbf{M}}{}_{ml}^{\varepsilon}(\mathbf{r}_0). \end{aligned}$$

In the same way we derive the corresponding integrals for the other two eigenvectors. We thus obtain the following Hansen integrals:

$$\int_0^{2\pi}\int_0^{\pi} \overset{*}{\mathbf{M}}{}_{ml}^{-\varepsilon}(\mathbf{r}) \frac{e^{-ik_cR}}{R} \sin\theta\, d\theta\, d\phi = -4\pi i k_c [f_l^{-\varepsilon}(k_c r)]^2 \overset{*}{\mathbf{M}}{}_{ml}^{\varepsilon}(\mathbf{r}_0),$$

$$\begin{aligned}\int_0^{2\pi}\int_0^{\pi} \overset{*}{\mathbf{N}}{}_{ml}^{-\varepsilon}(\mathbf{r}) \frac{e^{-ik_cR}}{R} \sin\theta\, d\theta\, d\phi = -\frac{4\pi i k_c}{2l+1} [l(l+1)\{&[f_{l-1}^{-\varepsilon}(k_c r)]^2 \\ &- [f_{l+1}^{-\varepsilon}(k_c r)]^2\}\overset{*}{\mathbf{L}}{}_{ml}^{\varepsilon}(\mathbf{r}_0) \\ &+ \{(l+1)[f_{l-1}^{-\varepsilon}(k_c r)]^2 \\ &+ l[f_{l+1}^{-\varepsilon}(k_c r)]^2\}\overset{*}{\mathbf{N}}{}_{ml}^{\varepsilon}(\mathbf{r}_0)],\end{aligned}$$

$$\begin{aligned}\int_0^{2\pi}\int_0^{\pi} \overset{*}{\mathbf{L}}{}_{ml}^{-\varepsilon}(\mathbf{r}) \frac{e^{-ik_cR}}{R} \sin\theta\, d\theta\, d\phi = -\frac{4\pi i k_c}{2l+1} [\{&l[f_{l-1}^{-\varepsilon}(k_c r)]^2 \\ &+ (l+1)[f_{l+1}^{-\varepsilon}(k_c r)]^2\}\overset{*}{\mathbf{L}}{}_{ml}^{\varepsilon}(\mathbf{r}_0) \\ &+ \{[f_{l-1}^{-\varepsilon}(k_c r)]^2 - [f_{l+1}^{-\varepsilon}(k_c r)]^2\}\overset{*}{\mathbf{N}}{}_{ml}^{\varepsilon}(\mathbf{r}_0)].\end{aligned} \tag{4.7.5}$$

We next assume that

$$\mathfrak{I}\frac{e^{-ik_cR}}{4\pi R} = \sum_{l,m} [\mathbf{M}_{ml}^{-\varepsilon}(\mathbf{r})\mathbf{X}_{ml}(\mathbf{r}_0) + \mathbf{N}_{ml}^{-\varepsilon}(\mathbf{r})\mathbf{Y}_{ml}(\mathbf{r}_0) + \mathbf{L}_{ml}^{-\varepsilon}(\mathbf{r})\mathbf{Z}_{ml}(\mathbf{r}_0)], \tag{4.7.6}$$

where **X**, **Y**, and **Z** are unknown vectors in the $\mathbf{r}_0$ system (source coordinates). Multiplying scalarly both sides of Eq. (4.7.6) with $\overset{*}{\mathbf{M}}{}_{m'l'}^{-\varepsilon}(\mathbf{r})$, integrating over the surface of a unit sphere, and using Eqs. (4.7.5) and (2.86), we obtain

$$\mathbf{X}_{ml}(\mathbf{r}_0) = -\frac{ik_c}{l(l+1)\Omega_{ml}} \overset{*}{\mathbf{M}}{}_{ml}^{\varepsilon}(\mathbf{r}_0).$$

Similarly, multiplying in turn with $\overset{*}{\mathbf{N}}{}_{m'l'}^{-\varepsilon}(\mathbf{r})$ and $\overset{*}{\mathbf{L}}{}_{m'l'}^{-\varepsilon}(\mathbf{r})$ and repeating the procedure just used, we get two simultaneous equations in **Y** and **Z**. On solving, we find

$$\mathbf{Y}_{ml}(\mathbf{r}_0) = -\frac{ik_c}{l(l+1)\Omega_{ml}} \overset{*}{\mathbf{N}}{}_{ml}^{\varepsilon}(\mathbf{r}_0),$$

$$\mathbf{Z}_{ml}(\mathbf{r}_0) = -\frac{ik_c}{\Omega_{ml}} \overset{*}{\mathbf{L}}{}_{ml}^{\varepsilon}(\mathbf{r}_0).$$

Inserting the expressions for **X**, **Y** and **Z** in Eq. (4.7.6), we get

$$\begin{aligned}\mathfrak{I}\frac{e^{-ik_cR}}{R} = -ik_c \sum_{l=0,1,1}^{\infty} \sum_{m=-l}^{l} (2l+1)\frac{(l-m)!}{(l+m)!} \Big[&\mathbf{L}_{ml}^{-\varepsilon}(\mathbf{r})\overset{*}{\mathbf{L}}{}_{ml}^{\varepsilon}(\mathbf{r}_0) \\ &+ \frac{1}{l(l+1)}\mathbf{M}_{ml}^{-\varepsilon}(\mathbf{r})\overset{*}{\mathbf{M}}{}_{ml}^{\varepsilon}(\mathbf{r}_0) + \frac{1}{l(l+1)}\mathbf{N}_{ml}^{-\varepsilon}(\mathbf{r})\overset{*}{\mathbf{N}}{}_{ml}^{\varepsilon}(\mathbf{r}_0)\Big].\end{aligned} \tag{4.7.7}$$

4.5.4. Representation of a Shear Dislocation in Terms of Jumps in the Displacement and Stress Vectors

The expressions for the displacements resulting from various sources given in Table 4.4 are with respect to the coordinate system (R, i_h, ϕ_h), having its origin at the source. In seismology, however, we have to take into account the interaction of elastic waves with the earth's free surface as well as with its interior discontinuities. For this reason, we must refer the displacements to a coordinate system having its origin at the center of the earth.

The spectral displacement field in a homogeneous boundless space is given by the Volterra relation

$$\mathbf{u}(\mathbf{r}, \omega) = U_0\, dS\, \mathbf{en} : \mathfrak{T}_0(\mathfrak{G}), \qquad g(\omega) = 1, \tag{4.175}$$

where $\mathfrak{G}(\mathbf{r}\,|\,\mathbf{r}_0; \omega)$ is given by Eq. (4.172). There is obviously a discontinuity in both the displacement and the radial stress vectors at the source's level $r = r_0$. For homogeneous media, this discontinuity is incorporated into the structure of the eigenvector expansion of $\mathfrak{G}(\mathbf{r}\,|\,\mathbf{r}_0; \omega)$ which is different for $r \lessgtr r_0$. In real-earth models, however, the explicit analytic expression for $\mathfrak{G}$ is usually unattainable in simple form. Therefore, we first solve a canonical problem. We use Eq. (4.172) to obtain the displacement and stress jumps at the source's level. These jumps will then serve as a unique characterization of the source. The resulting expressions bear the properties of the source, not that of the medium, and therefore are valid even when the source is placed in an inhomogeneous bounded medium. We therefore propose to evaluate

$$\begin{aligned} \delta\mathbf{u}(r_0) &= \mathbf{u}(r_0 + 0) - \mathbf{u}(r_0 - 0), \\ \delta\{\mathbf{e}_r \cdot \mathfrak{T}(\mathbf{u})\}_{r_0} &= \{\mathbf{e}_r \cdot \mathfrak{T}(\mathbf{u})\}_{r_0+0} - \{\mathbf{e}_r \cdot \mathfrak{T}(\mathbf{u})\}_{r_0-0}. \end{aligned} \tag{4.176}$$

From Eqs. (4.172), (4.175), and (4.176), we find

$$\delta\mathbf{u}(r_0) = \frac{iU_0\, dS\, k_\beta}{4\pi\mu} \sum_{l=1,1,0}^{\infty} \sum_{m=-l}^{l} \frac{2l+1}{l(l+1)} \frac{(l-m)!}{(l+m)!} \times \left[\mathbf{X}_{m,l} + \mathbf{Y}_{m,l} + l(l+1)\left(\frac{\beta}{\alpha}\right)^3 \mathbf{Z}_{m,l}\right],$$

where

$$\begin{aligned} \mathbf{X}_{m,l} &= \mathbf{en} : [\{\mathfrak{T}(\overset{*}{\mathbf{M}}{}^+)\}\mathbf{M}^- - \{\mathfrak{T}(\overset{*}{\mathbf{M}}{}^-)\}\mathbf{M}^+]_{r=r_0}, \\ \mathbf{Y}_{m,l} &= \mathbf{en} : [\{\mathfrak{T}(\overset{*}{\mathbf{N}}{}^+)\}\mathbf{N}^- - \{\mathfrak{T}(\overset{*}{\mathbf{N}}{}^-)\}\mathbf{N}^+]_{r=r_0}, \\ \mathbf{Z}_{m,l} &= \mathbf{en} : [\{\mathfrak{T}(\overset{*}{\mathbf{L}}{}^+)\}\mathbf{L}^- - \{\mathfrak{T}(\overset{*}{\mathbf{L}}{}^-)\}\mathbf{L}^+]_{r=r_0}. \end{aligned} \tag{4.177}$$

Carrying out the indicated differentiations and using the Wronskian relations

$$\begin{aligned} j_l(x)h_{l-1}^{(2)}(x) - h_l^{(2)}(x)j_{l-1}(x) &= -ix^{-2}, \\ j_{l+1}(x)h_{l-1}^{(2)}(x) - h_{l+1}^{(2)}(x)j_{l-1}(x) &= -(2l+1)ix^{-3}, \\ j_{l+2}(x)h_{l-1}^{(2)}(x) - h_{l+2}^{(2)}(x)j_{l-1}(x) &= ix^{-2}[1 - (2l+1)(2l+3)x^{-2}], \end{aligned} \tag{4.178}$$

we obtain

$$\delta \mathbf{u}(r_0) = \frac{U_0\, dS}{4\pi r_0^2} \sum_{\sigma=c,s} \sum_{m=0}^{2} \sum_{l=m}^{\infty} (2l+1)[p_{ml}^{\sigma} \mathbf{P}_{ml}^{\sigma}(\theta, \phi) + b_{ml}^{\sigma} \sqrt{\{l(l+1)\}} \mathbf{B}_{ml}^{\sigma}(\theta, \phi) + c_{ml}^{\sigma} \sqrt{\{l(l+1)\}} \mathbf{C}_{ml}^{\sigma}(\theta, \phi)]. \tag{4.179}$$

The corresponding jumps in the radial stress vector across a spherical surface passing through the source are given by

$$\delta[\mathbf{e}_r \cdot \mathfrak{T}(\mathbf{u})]_{r_0} = \frac{\mu U_0\, dS}{4\pi r_0^3} \sum_{\sigma=c,s} \sum_{m=0}^{2} \sum_{l=m}^{\infty} (2l+1)[\bar{p}_{ml}^{\sigma} \mathbf{P}_{ml}^{\sigma}(\theta, \phi) + \bar{b}_{ml}^{\sigma} \sqrt{\{l(l+1)\}} \mathbf{B}_{ml}^{\sigma}(\theta, \phi) + \bar{c}_{ml}^{\sigma} \sqrt{\{l(l+1)\}} \mathbf{C}_{ml}^{\sigma}(\theta, \phi)]. \tag{4.180}$$

The values of the nonzero source coefficients are

$$\begin{aligned}
p_{0l}^{c} &= \left(\frac{\beta}{\alpha}\right)^2 \sin\lambda \sin 2\delta, & \bar{p}_{0l}^{c} &= \sin\lambda \sin 2\delta\left(3 - 4\frac{\beta^2}{\alpha^2}\right), \\
b_{1l}^{s} &= \frac{1}{l(l+1)} \sin\lambda \cos 2\delta, & \bar{b}_{0l}^{c} &= -\frac{1}{2}\bar{p}_{0l}^{c}, \\
b_{1l}^{c} &= \frac{1}{l(l+1)} \cos\lambda \cos\delta, & \bar{b}_{2l}^{c} &= -\frac{1}{2l(l+1)} \sin\lambda \sin 2\delta, \\
c_{1l}^{s} &= b_{1l}^{c}, & \bar{b}_{2l}^{s} &= \frac{1}{l(l+1)} \cos\lambda \sin\delta, \\
c_{1l}^{c} &= -b_{1l}^{s}, & \bar{c}_{2l}^{c} &= -\bar{b}_{2l}^{s}, \\
& & \bar{c}_{2l}^{s} &= \bar{b}_{2l}^{c}.
\end{aligned} \tag{4.181}$$

The jumps in cylindrical coordinates are obtained in a similar way by using Eq. (4.174). These can also be obtained from the results given in Table 4.4. We find

$$\delta \mathbf{u}(0) = \mathbf{u}(z = +0) - \mathbf{u}(z = -0) = \frac{U_0\, dS}{2\pi} \sum_{\sigma=c,s} \sum_{m=0}^{2} \int_0^{\infty} [p_m^{\sigma} \mathbf{P}_m^{\sigma} + b_m^{\sigma} \mathbf{B}_m^{\sigma} + c_m^{\sigma} \mathbf{C}_m^{\sigma}] k\, dk, \tag{4.182}$$

$$\delta[\mathbf{e}_z \cdot \mathfrak{T}(\mathbf{u})]_{z=0} = [\mathbf{e}_z \cdot \mathfrak{T}(\mathbf{u})]_{z=+0} - [\mathbf{e}_z \cdot \mathfrak{T}(\mathbf{u})]_{z=-0} = \frac{\mu U_0\, dS}{2\pi} \sum_{\sigma=c,s} \sum_{m=0}^{2} \int_0^{\infty} [\bar{p}_m^{\sigma} \mathbf{P}_m^{\sigma} + \bar{b}_m^{\sigma} \mathbf{B}_m^{\sigma} + \bar{c}_m^{\sigma} \mathbf{C}_m^{\sigma}] k^2\, dk, \tag{4.183}$$

where the nonzero coefficients are

$$
\begin{aligned}
p_0^c &= \left(\frac{\beta}{\alpha}\right)^2 \sin\lambda \sin 2\delta, & \bar{b}_0^c &= -\frac{1}{2}\left(3 - 4\frac{\beta^2}{\alpha^2}\right)\sin\lambda\sin 2\delta,\\
b_1^c &= \cos\lambda\cos\delta, & \bar{b}_2^s &= \cos\lambda \sin\delta,\\
b_1^s &= \sin\lambda\cos 2\delta, & \bar{b}_2^c &= -\frac{1}{2}\sin\lambda\sin 2\delta,\\
c_1^c &= -b_1^s, & \bar{c}_2^c &= -\bar{b}_2^s,\\
c_1^s &= b_1^c, & \bar{c}_2^s &= \bar{b}_2^c.
\end{aligned}
\tag{4.184}
$$

In particular, the discontinuities in the displacement and stress vectors for the three fundamental shear dislocations are as follows:

1. Vertical strike slip ($m = 2$, $\lambda = 0°$, $\delta = 90°$)

$$
\begin{aligned}
\delta\mathbf{u}(r_0) &= 0,\\
(\delta\mathbf{T})_{r_0} &= \frac{\mu U_0\,dS}{4\pi r_0^3}\sum_{l=2}^{\infty}\frac{2l+1}{l(l+1)}\sqrt{[l(l+1)]}[\mathbf{B}_{2,l}^s(\theta,\phi) - \mathbf{C}_{2,l}^c(\theta,\phi)],\\
(\delta\mathbf{u})_{z=0} &= 0,\\
(\delta\mathbf{T})_{z=0} &= \frac{\mu U_0\,dS}{2\pi}\int_0^\infty (\mathbf{B}_2^s - \mathbf{C}_2^c)k^2\,dk;
\end{aligned}
\tag{4.185}
$$

2. Vertical dip slip ($m = 1$, $\lambda = 90°$, $\delta = 90°$)

$$
\begin{aligned}
\delta\mathbf{u}(r_0) &= \frac{U_0\,dS}{4\pi r_0^2}\sum_{l=1}^{\infty}\frac{2l+1}{l(l+1)}\sqrt{[l(l+1)]}[-\mathbf{B}_{1,l}^s(\theta,\phi) + \mathbf{C}_{1,l}^c(\theta,\phi)],\\
\delta\mathbf{T}(r_0) &= 0,\\
(\delta\mathbf{u})_{z=0} &= \frac{U_0\,dS}{2\pi}\int_0^\infty (-\mathbf{B}_1^s + \mathbf{C}_1^c)k\,dk,\\
(\delta\mathbf{T})_{z=0} &= 0;
\end{aligned}
\tag{4.186}
$$

3. 45° dip slip ($m = 0, 2$, $\lambda = 90°$, $\delta = 45°$)

$$
\begin{aligned}
\delta\mathbf{u}(r_0) &= \frac{U_0\,dS}{4\pi r_0^2}\left(\frac{\beta}{\alpha}\right)^2\sum_{l=0}^{\infty}(2l+1)\mathbf{P}_{0,l}^c(\theta,\phi),\\
(\delta\mathbf{T})_{r_0} &= \frac{\mu U_0\,dS}{4\pi r_0^3}\Bigg[\left(3 - 4\frac{\beta^2}{\alpha^2}\right)\sum_{l=0}^{\infty}(2l+1)\{\mathbf{P}_{0,l}^c(\theta,\phi)\\
&\quad - \tfrac{1}{2}\sqrt{[l(l+1)]}\mathbf{B}_{0,l}^c(\theta,\phi)\}\\
&\quad - \frac{1}{2}\sum_{l=2}^{\infty}\frac{2l+1}{l(l+1)}\sqrt{[l(l+1)]}\{\mathbf{B}_{2,l}^c(\theta,\phi) + \mathbf{C}_{2,l}^s(\theta,\phi)\}\Bigg],\\
(\delta\mathbf{u})_{z=0} &= \frac{U_0\,dS}{2\pi}\left(\frac{\beta}{\alpha}\right)^2\int_0^\infty \mathbf{P}_0^c k\,dk,\\
(\delta\mathbf{T})_{z=0} &= -\frac{\mu U_0\,dS}{4\pi}\int_0^\infty\left[\left(3 - 4\frac{\beta^2}{\alpha^2}\right)\mathbf{B}_0^c + \mathbf{B}_2^c + \mathbf{C}_2^s\right]k^2\,dk.
\end{aligned}
\tag{4.187}
$$

4.5.5. Displacement Field in the Time Domain

From Eqs. (4.29) and (4.76), we get the following expression for the tensor $\mathfrak{T}_0(\mathfrak{G})$ in the time domain. The corresponding expression in the frequency domain has been given in Eq. (4.141). We find

$$\begin{aligned}4\pi\mathfrak{T}_0(\mathfrak{G}) = &-\frac{6\beta^2}{R^4}\mathsf{B}\int_{R/\alpha}^{R/\beta} g(t-t')t'\,dt' - \frac{2\beta^2}{R^2}\mathsf{B}\left[\frac{1}{\alpha^2}g\left(t-\frac{R}{\alpha}\right) - \frac{1}{\beta^2}g\left(t-\frac{R}{\beta}\right)\right]\\ &+\frac{1}{R^3}\left[\mathbf{R}\mathfrak{J} + (\mathbf{R}\mathfrak{J})^{213} - \frac{2}{R^2}\mathbf{RRR}\right]\left[g\left(t-\frac{R}{\beta}\right) + \frac{R}{\beta}\dot{g}\left(t-\frac{R}{\beta}\right)\right]\\ &+\frac{2}{R^3}\left(\frac{\beta}{\alpha}\right)^2\left[\frac{\sigma}{1-2\sigma}\mathfrak{J}\mathbf{R} + \frac{1}{R^2}\mathbf{RRR}\right]\left[g\left(t-\frac{R}{\alpha}\right) + \frac{R}{\alpha}\dot{g}\left(t-\frac{R}{\alpha}\right)\right],\end{aligned} \tag{4.188}$$

where the triadic B is given in Eq. (4.145) and the dot signifies differentiation with respect to t.

The displacement field induced by a point shear dislocation now follows from the Volterra relation, Eq. (4.107). We find

$$\mathbf{u}(\mathbf{r};t) = \frac{U_0\,dS}{4\pi}\,\mathbf{en}:\left\{\frac{1}{R}\mathsf{D}_1 + \frac{1}{R^2}\mathsf{D}_2\right\}, \tag{4.189}$$

where

$$\begin{aligned}\mathsf{D}_1 = &\frac{1}{\beta}\dot{g}\left(t-\frac{R}{\beta}\right)[\mathbf{e}_R\mathfrak{J} + (\mathbf{e}_R\mathfrak{J})^{213} - 2\mathbf{e}_R\mathbf{e}_R\mathbf{e}_R] + \frac{2}{\alpha}\left(\frac{\beta}{\alpha}\right)^2\dot{g}\left(t-\frac{R}{\alpha}\right)\\ &\times\left[\frac{\sigma}{1-2\sigma}\mathfrak{J}\mathbf{e}_R + \mathbf{e}_R\mathbf{e}_R\mathbf{e}_R\right],\\ \mathsf{D}_2 = &-6\mathsf{B}\int_{\beta/\alpha}^{1} g\left(t-\frac{sR}{\beta}\right)s\,ds - 2\beta^2\mathsf{B}\left[\frac{1}{\alpha^2}g\left(t-\frac{R}{\alpha}\right) - \frac{1}{\beta^2}g\left(t-\frac{R}{\beta}\right)\right]\\ &+[\mathbf{e}_R\mathfrak{J} + (\mathbf{e}_R\mathfrak{J})^{213} - 2\mathbf{e}_R\mathbf{e}_R\mathbf{e}_R]g\left(t-\frac{R}{\beta}\right) + 2\left(\frac{\beta}{\alpha}\right)^2\\ &\times\left[\frac{\sigma}{1-2\sigma}\mathfrak{J}\mathbf{e}_R + \mathbf{e}_R\mathbf{e}_R\mathbf{e}_R\right]g\left(t-\frac{R}{\alpha}\right).\end{aligned} \tag{4.190}$$

4.5.5.1. Far Field. From Eqs. (4.189) and (4.190), the far field is given by

$$\begin{aligned}u_R &= \frac{U_0\,dS}{2\pi\beta R}\left(\frac{\beta}{\alpha}\right)^3\dot{g}\left(t-\frac{R}{\alpha}\right)(\mathbf{e}\cdot\mathbf{e}_R)(\mathbf{n}\cdot\mathbf{e}_R) + O\left(\frac{1}{R^2}\right),\\ u_{i_h} &= \frac{U_0\,dS}{4\pi\beta R}\dot{g}\left(t-\frac{R}{\beta}\right)[(\mathbf{e}\cdot\mathbf{e}_{i_h})(\mathbf{n}\cdot\mathbf{e}_R) + (\mathbf{e}\cdot\mathbf{e}_R)(\mathbf{n}\cdot\mathbf{e}_{i_h})] + O\left(\frac{1}{R^2}\right),\\ u_{\phi_h} &= \frac{U_0\,dS}{4\pi\beta R}\dot{g}\left(t-\frac{R}{\beta}\right)[(\mathbf{e}\cdot\mathbf{e}_{\phi_h})(\mathbf{n}\cdot\mathbf{e}_R) + (\mathbf{e}\cdot\mathbf{e}_R)(\mathbf{n}\cdot\mathbf{e}_{\phi_h})] + O\left(\frac{1}{R^2}\right).\end{aligned} \tag{4.191}$$

Using Eq. (4.153), this yields

$$
\begin{aligned}
u_R &= \frac{U_0\,dS}{12\pi\beta}\left(\frac{\beta}{\alpha}\right)^3 F\,\frac{\dot{g}(t - R/\alpha)}{R} + O\left(\frac{1}{R^2}\right),\\
u_{i_h} &= \frac{U_0\,dS}{24\pi\beta}\frac{\partial F}{\partial i_h}\frac{\dot{g}(t - R/\beta)}{R} + O\left(\frac{1}{R^2}\right),\\
u_{\phi_h} &= \frac{U_0\,dS}{24\pi\beta}\frac{1}{\sin i_h}\frac{\partial F}{\partial \phi_h}\frac{\dot{g}(t - R/\beta)}{R} + O\left(\frac{1}{R^2}\right),
\end{aligned}
\tag{4.192}
$$

where the function $F(\lambda, \delta; i_h, \phi_h)$ is given by Eq. (4.154).

The radial motion propagates with the velocity of longitudinal waves. These behave as acoustical waves in the sense that the particle motion at each point coincides with the direction of propagation. In contrast, u_{i_h} and u_{ϕ_h} are shear waves, in which the particle motion is normal to the direction of propagation.

A dislocation source in an infinite elastic solid emits both shear and longitudinal waves. The ratio of the amplitudes and energies of these two wave types is diagnostic of the source and it is therefore used to obtain some theoretical estimates. Assuming that the average values of $\ddot{g}(t - R/\alpha)$ and $\ddot{g}(t - R/\beta)$ are the same, we integrate over a unit sphere to obtain from Eq. (4.192) the following expression for the energy ratio

$$
\begin{aligned}
\frac{E_\alpha}{E_\beta} &= \frac{\int_0^{2\pi}\int_0^{\pi} \dot{u}_R^2 \sin i_h\, di_h\, d\phi_h}{\int_0^{2\pi}\int_0^{\pi} (\dot{u}_{i_h}^2 + \dot{u}_{\phi_h}^2)\sin i_h\, di_h\, d\phi_h}\\
&= \frac{\int_0^{2\pi}\int_0^{\pi} (F)^2 \sin i_h\, di_h\, d\phi_h}{\int_0^{2\pi}\int_0^{\pi} [(\partial F/\partial i_h)^2 + (1/\sin^2 i_h)(\partial F/\partial \phi_h)^2]\sin i_h\, di_h\, d\phi_h}\left[4\left(\frac{\beta}{\alpha}\right)^6\right].
\end{aligned}
\tag{4.193}
$$

For a vertical strike-slip source, this yields

$$
\frac{E_\alpha}{E_\beta} = \frac{2}{3}\left(\frac{\beta}{\alpha}\right)^6 \tag{4.194}
$$

which is $\frac{2}{81}$ for $\sigma = \frac{1}{4}$.

4.5.5.2. Residual Field. Equations (4.189) and (4.190) give the following displacement components determined by the triadic D_2:

$$
\begin{aligned}
u_R &= \frac{U_0\,dS}{2\pi R^2}\left[3\int_{\beta/\alpha}^{1} g\left(t - \frac{sR}{\beta}\right)s\,ds + \frac{4}{3}\left(\frac{\beta}{\alpha}\right)^2 g\left(t - \frac{R}{\alpha}\right) - g\left(t - \frac{R}{\beta}\right)\right]\\
&\quad \times 3(\mathbf{e}\cdot\mathbf{e}_R)(\mathbf{n}\cdot\mathbf{e}_R),\\
u_{i_h} &= \frac{U_0\,dS}{4\pi R^2}\left[-6\int_{\beta/\alpha}^{1} g\left(t - \frac{sR}{\beta}\right)s\,ds - 2\left(\frac{\beta}{\alpha}\right)^2 g\left(t - \frac{R}{\alpha}\right) + 3g\left(t - \frac{R}{\beta}\right)\right]\\
&\quad \times \{(\mathbf{e}\cdot\mathbf{e}_R)(\mathbf{n}\cdot\mathbf{e}_{i_h}) + (\mathbf{e}\cdot\mathbf{e}_{i_h})(\mathbf{n}\cdot\mathbf{e}_R)\},\\
u_{\phi_h} &= \frac{U_0\,dS}{4\pi R^2}\left[-6\int_{\beta/\alpha}^{1} g\left(t - \frac{sR}{\beta}\right)s\,ds - 2\left(\frac{\beta}{\alpha}\right)^2 g\left(t - \frac{R}{\alpha}\right) + 3g\left(t - \frac{R}{\beta}\right)\right]\\
&\quad \times \{(\mathbf{n}\cdot\mathbf{e}_R)(\mathbf{e}\cdot\mathbf{e}_{\phi_h}) + (\mathbf{n}\cdot\mathbf{e}_{\phi_h})(\mathbf{e}\cdot\mathbf{e}_R)\}.
\end{aligned}
\tag{4.195}
$$

A repeated use of Eq. (4.153) renders

$$u_R = \frac{U_0\, dS}{4\pi R^2} F(\lambda, \delta; i_h, \phi_h)$$

$$\times \left[3 \int_{\beta/\alpha}^{1} g\left(t - \frac{sR}{\beta}\right) s\, ds + \frac{4}{3}\left(\frac{\beta}{\alpha}\right)^2 g\left(t - \frac{R}{\alpha}\right) - g\left(t - \frac{R}{\beta}\right)\right],$$

$$u_{i_h} = \frac{U_0\, dS}{24\pi R^2} \frac{\partial F}{\partial i_h}$$

$$\times \left[-6 \int_{\beta/\alpha}^{1} g\left(t - \frac{sR}{\beta}\right) s\, ds - 2\left(\frac{\beta}{\alpha}\right)^2 g\left(t - \frac{R}{\alpha}\right) + 3g\left(t - \frac{R}{\beta}\right)\right], \tag{4.196}$$

$$u_{\phi_h} = \frac{U_0\, dS}{24\pi R^2} \frac{1}{\sin i_h} \frac{\partial F}{\partial \phi_h}$$

$$\times \left[-6 \int_{\beta/\alpha}^{1} g\left(t - \frac{sR}{\beta}\right) s\, ds - 2\left(\frac{\beta}{\alpha}\right)^2 g\left(t - \frac{R}{\alpha}\right) + 3g\left(t - \frac{R}{\beta}\right)\right].$$

4.6. Stress Distributions on a Spherical Cavity and Their Equivalent Sources

Tractions are applied on the surface of a spherical cavity in an infinite elastic solid of rigidity μ and velocities α and β. Let the cavity be of radius a and let its center be the origin of a spherical coordinate system (r, θ, ϕ). Following the formalism established earlier, we expand the spectral displacement field as an infinite sum of eigenvectors

$$\mathbf{u}(\mathbf{r}) = \sum_{m,l} [\alpha_{ml}\mathbf{M}_{ml}^{-}(k_\beta r) + \beta_{ml}\mathbf{N}_{ml}^{-}(k_\beta r) + \gamma_{ml}\mathbf{L}_{ml}^{-}(k_\alpha r)], \tag{4.197}$$

where $k_\alpha = \omega/\alpha$, $k_\beta = \omega/\beta$ and ω is the angular frequency. The coefficients α_{ml}, etc. are constants. Let the boundary conditions at the surface of the cavity be

$$\mathbf{e}_r \cdot \mathfrak{T}(\mathbf{u}) = \mathbf{F}(\theta, \phi) \quad \text{at} \quad r = a, \tag{4.198}$$

where $\mathbf{e}_r$ is a unit vector in the direction of increasing r, $\mathfrak{T}(\mathbf{u}) = \lambda\mathfrak{J}\, \text{div}\, \mathbf{u} + \mu(\nabla\mathbf{u} + \mathbf{u}\nabla)$ is the stress dyadic, and $\mathbf{F}$ is a known distribution. The expressions for the vectors $\mathbf{e}_r \cdot \mathfrak{T}(\mathbf{M})$, etc., can be obtained directly from Eq. (2.106) and Table 2.2. Expanding $\mathbf{F}(\theta, \phi)$ into a series of vector spherical harmonics,

$$\mathbf{F}(\theta, \phi) = \sum_{m,l} [\alpha_{ml}^0 \sqrt{\{l(l+1)\}}\mathbf{C}_{ml} + \beta_{ml}^0 \mathbf{P}_{ml} + \gamma_{ml}^0 \sqrt{\{l(l+1)\}}\mathbf{B}_{ml}], \tag{4.199}$$

we obtain, after some straightforward calculations,

$$\alpha_{ml} = \frac{\alpha_{ml}^0}{\mu k_\beta} \frac{1}{\chi F_{l,1}(\chi)}, \qquad \beta_{ml} = \frac{1}{\mu k_\beta} \frac{\beta_{ml}^0 F_{l,1}(\zeta) - \gamma_{ml}^0 F_{l,3}(\zeta)}{\Delta_l},$$
$$\gamma_{ml} = \frac{1}{2\mu k_\alpha} \frac{-\beta_{ml}^0 F_{l,2}(\chi) + 2l(l+1)\gamma_{ml}^0 F_{l,1}(\chi)}{\Delta_l}, \tag{4.200}$$

where

$$\chi = ak_\beta, \qquad \zeta = ak_\alpha,$$
$$\Delta_l = 2l(l+1)F_{l,1}(\zeta)F_{l,1}(\chi) - F_{l,2}(\chi)F_{l,3}(\zeta),$$
$$F_{l,1}(x) = \frac{(l-1)}{x^2} f_l(x) - \frac{1}{x} f_{l+1}(x),$$
$$F_{l,2}(x) = \left[\frac{2}{x^2}(l^2-1) - 1\right] f_l(x) + \frac{2}{x} f_{l+1}(x), \tag{4.201}$$
$$F_{l,3}(x) = \left[\frac{1}{x^2} l(l-1) - \frac{1}{2}\left(\frac{\alpha}{\beta}\right)^2\right] f_l(x) + \frac{2}{x} f_{l+1}(x),$$
$$f_l(x) = h_l^{(2)}(x) = \text{spherical Hankel function of the second kind.}$$

Four particular cases will be considered.

4.6.1. Pressure in a Spherical Cavity

In this section, we describe a source that emits compressional waves only. A pressure $P(t) = p_0 g(t)$ is applied to the inner walls of the cavity. Because the boundary condition requires that the normal stress at the surface of the cavity be equal to the negative of the applied pressure, we write

$$\mathbf{F}(\theta, \phi) = -\mathbf{e}_r p_0 g(\omega) = -p_0 g(\omega)\mathbf{P}_{00}, \tag{4.202}$$

where $g(\omega)$ is the Fourier transform of $g(t)$. Comparing Eq. (4.202) with Eq. (4.199) we note that $\beta_{00}^0 = -p_0 g(\omega)$, and all the other coefficients in expansion (4.199) are zero. Therefore, $m = l = 0$. Equations (4.200) and (4.201) then yield the only relevant nonzero coefficient

$$\gamma_{00} = -\frac{p_0 g(\omega)}{2\mu k_\alpha} \frac{1}{F_{0,3}(ak_\alpha)} = -\frac{ia^3 p_0 k_\alpha^2 g(\omega)}{\mu(a^2 k_\beta^2 - 4iak_\alpha - 4)} e^{iak_\alpha}. \tag{4.203}$$

Because $\mathbf{N}_{00} = 0$, the displacement is given by

$$\mathbf{u}(\mathbf{r}; \omega) = \gamma_{00} \mathbf{L}_{00}^-(k_\alpha r), \qquad \mathbf{L}_{00}^- = \left(\frac{i}{k_\alpha^2}\right) \nabla \left[\frac{1}{r} e^{i\omega(t - r/\alpha)}\right]. \tag{4.204}$$

The corresponding displacement in the time domain is obtained by the application of the inverse Fourier transform. We denote $\mathbf{u} = \nabla S$ and choose the source-time function to be $H(t)e^{-\sigma t}$, $\sigma > 0$. Then

$$S = \frac{-p_0 a^3}{8\pi\mu r} \int_{-\infty}^{\infty} \frac{e^{i\omega\tau}\, d\omega}{[1 - a^2\omega^2/(4\beta^2) + ia\omega/\alpha](\sigma + i\omega)}, \qquad \tau = t - \frac{r - a}{\alpha}. \tag{4.205}$$

The integrand has three simple poles at $\omega_1 = (2\beta/a)e^{i\varepsilon}$, $\omega_2 = (2\beta/a)e^{i(\pi-\varepsilon)}$, $\omega_3 = i\sigma$, ($\sin \varepsilon = \beta/\alpha$). Taking the residues at these poles and then substituting $\sigma = 0$ in the result, we get the explicit expression for a step source

$$S = -\frac{p_0 a^3}{4\mu r}[1 - (\sec \varepsilon)e^{-b\tau} \cos(\omega_0 \tau - \varepsilon)],$$

$$b = \frac{2\beta^2}{a\alpha}, \qquad \omega_0 = \left(\frac{2\beta}{a}\right)\cos \varepsilon. \tag{4.206}$$

The displacement field (radial component only) has the form

$$u_r(\mathbf{r}, t) = \frac{p_0 a^3}{4\mu r^2}[1 - (\sec \varepsilon)e^{-b\tau}\cos(\omega_0 \tau - \varepsilon)] + \frac{p_0 a^2}{2\mu r\sqrt{(\alpha^2/\beta^2 - 1)}} e^{-b\tau} \sin \omega_0 \tau. \tag{4.207}$$

The most interesting feature of the motion is its exponential decay with time. This occurs because an initial finite energy is spreading into an infinite space and this must decay with time at each point or else the energy integrals over time will not converge. The static limit of u_r is $p_0 a^3/(4\mu r^2)$. The variation of the dimensionless displacement $U = [4\mu/(ap_0)]u_r$ against dimensionless time $\omega_0 \tau$ is shown in Fig. 4.31.

A center of compression is obtained as a limiting process when the radius, a, tends to zero while the pressure, p_0, tends to infinity in such a way that the entity $E_0 = (4\pi/3)a^3 p_0$ remains finite. For an arbitrary pressure pulse $g(t)$,

$$\lim_{\substack{a \to 0 \\ p_0 \to \infty}} \gamma_{00} = \frac{3iE_0 k_\alpha^2 g(\omega)}{16\pi\mu}.$$

Therefore, from Eq. (4.204), we have

$$\mathbf{u}(\mathbf{r}, t) = -\frac{3E_0}{16\pi\mu} \operatorname{grad}\left\{\frac{g(t - r/\alpha)}{r}\right\} = \frac{3E_0}{16\pi\mu}\left[\frac{g(t - r/\alpha)}{r^2} + \frac{g'(t - r/\alpha)}{r\alpha}\right]\mathbf{e}_r. \tag{4.208}$$

At the near field, the shape of the initial pulse is preserved. As we move far away from the origin, the waveform is the derivative of the pressure pulse.

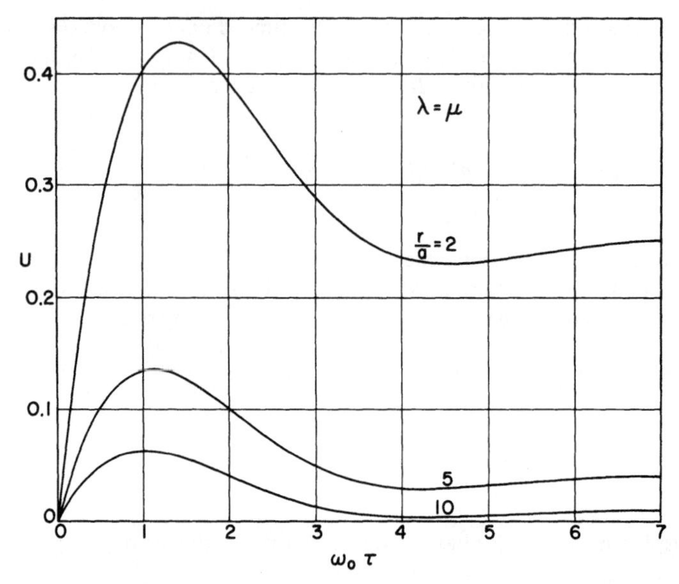

Figure 4.31. The variation of the dimensionless radial displacement against dimensionless time for three locations of the observer (see Eq. (4.207)).

4.6.2. Shear on a Spherical Cavity

We next discuss a source that emits shear waves only. Suppose that a shearing stress $\tau_{r\phi} = \tau_0 \sin\theta g(t)$ is applied at the inner surface of a spherical cavity of radius a centered at the origin of a spherical coordinate system (r, θ, ϕ). The boundary condition is $\mathbf{e}_r \cdot \mathfrak{T}(\mathbf{u}) = \tau_0 \sin\theta g(\omega)\mathbf{e}_\phi$. In the notation of Eq. (4.199), we have

$$\mathbf{F}(\theta, \phi) = -\tau_0 \frac{\partial}{\partial\theta} P_1(\cos\theta) g(\omega)\mathbf{e}_\phi = \sqrt{2}\,\tau_0 \mathbf{C}_{0,1} g(\omega). \tag{4.209}$$

Comparing Eqs. (4.199) and (4.209), we find that $\alpha_{01}^0 = \tau_0 g(\omega)$ and all other coefficients in expansion (4.199) are zero. Equation (4.200) now yields the coefficient α_{01}. The spectral displacement is found to be

$$\begin{aligned} \mathbf{u}(\mathbf{r}, \omega) &= \alpha_{01}\mathbf{M}_{01}^{-} = \sin\theta\mathbf{e}_\phi \frac{ds(\omega)}{dr}, \\ s(\omega) &= \frac{\tau_0 a^3}{3\mu r}\left[\frac{g(\omega)e^{-ik_\beta(r-a)}}{1 - a^2\omega^2/(3\beta^2) + ia\omega/\beta}\right]. \end{aligned} \tag{4.210}$$

The corresponding displacement in the time domain is determined by the inverse Fourier integral

$$S(t) = \frac{\tau_0 a^3}{6\pi\mu r}\int_{-\infty}^{\infty} \frac{g(\omega)e^{i\omega\tau}\, d\omega}{1 - a^2\omega^2/(3\beta^2) + ia\omega/\beta}, \qquad \tau = t - \frac{r-a}{\beta}. \tag{4.211}$$

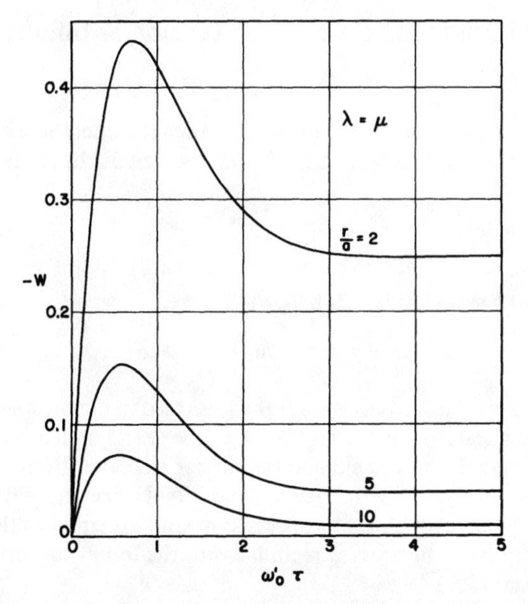

Figure 4.32. The variation of the dimensionless azimuthal component of the displacement against dimensionless time for three locations of the observer (see Eqs. (4.210) and (4.212)).

From this point on, the discussion closely follows our treatment of the pressure source (Section 4.6.1). For a step-function time dependence, we find

$$S(t) = \frac{\tau_0 a^3}{3\mu r}\left[1 - 2e^{-b'\tau}\cos\left(\omega_0'\tau - \frac{\pi}{3}\right)\right], \tag{4.212}$$

where

$$b' = \frac{3\beta}{2a}, \qquad \omega_0' = \frac{\beta\sqrt{3}}{2a}. \tag{4.213}$$

The variation of the dimensionless displacement $W = 3\mu u_\phi/(\tau_0 a \sin\theta)$ with dimensionless time $\omega_0'\tau$ is shown in Fig. 4.32. The displacement W begins at zero and settles down to its static value—$(a/r)^2$.

As a tends to zero while τ_0 tends to infinity such that $(4\pi/3)a^3\tau_0 = M_0$, *a point torque source* about $\mathbf{e}_3$ is created. The displacement vector in the time domain for this limiting case is

$$\begin{aligned}\mathbf{u}(\mathbf{r}, t) &= \frac{M_0}{4\pi\mu}\sin\theta\,\mathbf{e}_\phi\frac{\partial}{\partial r}\left[\frac{1}{r}g\left(t - \frac{r}{\beta}\right)\right]\\ &= -\frac{M_0}{4\pi\mu}\sin\theta\,\mathbf{e}_\phi\left[\frac{g(t - r/\beta)}{r^2} + \frac{g'(t - r/\beta)}{r\beta}\right].\end{aligned} \tag{4.214}$$

4.6.3. Explosion in the Presence of Tension at Infinity in a Given Direction

A fixed Cartesian coordinate system (x, y, z) is set at O. Let there exist an initial stress $\tau_{zz} = \tau_0$ in the z direction. It can be represented by the stress dyadic

$$\mathfrak{T}_0 = \tau_0 \mathbf{e}_z \mathbf{e}_z. \tag{4.215}$$

Because $\mathbf{e}_z = \mathbf{e}_r \cos\theta - \mathbf{e}_\theta \sin\theta$, the stress vector at any point on the surface of a sphere having its center at the origin, is given by

$$\mathbf{e}_r \cdot \mathfrak{T}_0 = \tau_0(\cos^2\theta \mathbf{e}_r - \sin\theta\cos\theta \mathbf{e}_\theta). \tag{4.216}$$

Suppose that a spherical cavity of radius a is formed in the prestressed medium, at O. The field generated by the formation of the cavity is such that the surface tractions resulting from this field and the surface tractions given by Eq. (4.216), taken together, vanish on the surface of the cavity. Hence the field generated by the creation of the cavity in the prestressed medium is equal to the field of the spherical cavity in an unstressed medium with the following surface tractions over its boundary:

$$\begin{aligned}\mathbf{F}(\theta, \phi) &= -\tau_0(\cos^2\theta \mathbf{e}_r - \sin\theta\cos\theta \mathbf{e}_\theta) \\ &= -\frac{\tau_0}{3}(\mathbf{P}_{00} + 2\mathbf{P}_{02} + \sqrt{6}\mathbf{B}_{02}), \quad \text{at} \quad r = a.\end{aligned} \tag{4.217}$$

Therefore, in Eq. (4.199), the only surviving coefficients are

$$\beta_{00}^0 = -\tfrac{1}{3}\tau_0, \qquad \beta_{02}^0 = -\tfrac{2}{3}\tau_0, \qquad \gamma_{02}^0 = -\tfrac{1}{3}\tau_0. \tag{4.218}$$

Equation (4.200) then yields

$$\begin{aligned}\beta_{00} &= -\frac{\tau_0}{3\mu k_\beta}\left[\frac{F_{0,1}(ak_\alpha)}{\Delta_0}\right], \qquad \gamma_{00} = \frac{\tau_0}{6\mu k_\alpha}\left[\frac{F_{0,2}(ak_\beta)}{\Delta_0}\right], \\ \beta_{02} &= \left[\frac{\tau_0}{3\mu k_\beta}\right]\frac{\{-2F_{2,1}(ak_\alpha) + F_{2,3}(ak_\alpha)\}}{\Delta_2}, \\ \gamma_{02} &= \left[\frac{\tau_0}{3\mu k_\alpha}\right]\frac{\{-6F_{2,1}(ak_\beta) + F_{2,2}(ak_\beta)\}}{\Delta_2}.\end{aligned} \tag{4.218a}$$

The rest of the coefficients are zero. Substituting in Eq. (4.197), we have

$$\mathbf{u}(\mathbf{r}) = \beta_{02}\mathbf{N}_{02}^- + \gamma_{00}\mathbf{L}_{00}^- + \gamma_{02}\mathbf{L}_{02}^-. \tag{4.219}$$

This is the displacement field generated by the creation of a traction-free cavity in a prestressed medium.

When the cavity shrinks to a point ($a \to 0$), we take the limit of the functions $F_{l,j}(x)$ in Eq. (4.201) as $x \to 0$. We find

$$\begin{aligned}
F_{0,1}(x) &= -2ix^{-3} - x^{-2} + \cdots, \\
F_{0,2}(x) &= -2x^{-2} + ix^{-1} + \cdots, \\
F_{0,3}(x) &= 2ix^{-3} - \left[\frac{i\sigma}{1-2\sigma}\right]x^{-1} + \cdots, \\
F_{2,1}(x) &= -12ix^{-5} - ix^{-3} + \cdots, \\
F_{2,2}(x) &= 48ix^{-5} + 3ix^{-3} + \cdots, \\
F_{2,3}(x) &= 36ix^{-5} + i\left[\frac{1-5\sigma}{1-2\sigma}\right]x^{-3} + \cdots,
\end{aligned} \tag{4.220}$$

where σ is the Poisson's ratio. Using these results in Eq. (4.201), we obtain

$$\Delta_0 = \frac{4i}{k_\alpha^3 k_\beta^2 a^5} + \cdots, \qquad \Delta_2 = -\frac{12}{k_\alpha^5 k_\beta^3 a^8}\frac{7-5\sigma}{1-\sigma} + \cdots, \tag{4.221}$$

and hence

$$\beta_{02} = -\frac{5ik_\beta^2}{4\pi\mu}\left(\frac{1-\sigma}{7-5\sigma}\right)M_0, \qquad \gamma_{02} = -\frac{5ik_\alpha^2}{2\pi\mu}\left(\frac{\beta}{\alpha}\right)^2\left[\frac{1-\sigma}{7-5\sigma}\right]M_0,$$

$$\gamma_{00} = \frac{ik_\alpha^2}{16\pi\mu}M_0, \qquad M_0 = \lim_{\substack{a\to 0\\ \tau_0\to\infty}}\left(\frac{4\pi}{3}\right)a^3\tau_0.$$

The displacement field is then written as

$$\mathbf{u}(\mathbf{r}, \omega) = \mathbf{u}_1 + \mathbf{u}_2,$$

$$\mathbf{u}_1 = \left[\frac{ik_\alpha^2}{16\pi\mu}M_0\right]\mathbf{L}_{00}^-, \quad \mathbf{u}_2 = -\left[\frac{5ik_\beta^2}{4\pi\mu}\frac{1-\sigma}{7-5\sigma}M_0\right]\left\{\mathbf{N}_{02}^- + 2\left(\frac{\beta}{\alpha}\right)^4\mathbf{L}_{02}^-\right\}. \tag{4.222}$$

It is clear from our previous results (Table 4.4) that $\mathbf{u}_1$ is a field resulting from a center of compression, whereas $\mathbf{u}_2$ is a field resulting from a dipole in the z direction with moment

$$M = 15\frac{1-\sigma}{7-5\sigma}M_0. \tag{4.223}$$

If the cavity is formed by an explosion, the field induced by this explosion, namely,

$$\mathbf{u}_0 = \frac{ia^3 p_0}{4\mu}k_\alpha^2\mathbf{L}_{00}^-$$

must be added to the former results.

4.6.4. Explosion in the Presence of Pure Shear at Infinity

Assume a state of pure shear defined by the dyadic

$$\mathfrak{T}_0 = \tau_0(\mathbf{e}_x\mathbf{e}_y + \mathbf{e}_y\mathbf{e}_x). \tag{4.224}$$

Substituting for $\mathbf{e}_x$ and $\mathbf{e}_y$ in terms of $\mathbf{e}_r$, $\mathbf{e}_\theta$ and $\mathbf{e}_\phi$, we find

$$\mathbf{e}_r \cdot \mathfrak{T}_0 = \frac{\tau_0}{6}[2\mathbf{P}_{22}^s + \sqrt{6}\mathbf{B}_{22}^s]. \tag{4.225}$$

As in Section 4.6.3, if a spherical cavity with radius a is suddenly formed in a medium with initial stress as given by Eq. (4.224), there arises an additional displacement field caused by the creation of the spherical cavity in an unstressed medium with the boundary conditions

$$\mathbf{e}_r \cdot \mathfrak{T}(\mathbf{u}) = -\frac{\tau_0}{6}[2\mathbf{P}_{22}^s + \sqrt{6}\mathbf{B}_{22}^s], \quad \text{at} \quad r = a. \tag{4.226}$$

A comparison of Eq. (4.226) with Eq. (4.199) yields

$$\beta_{22}^{0,s} = -\frac{\tau_0}{3}, \qquad \gamma_{22}^{0,s} = -\frac{\tau_0}{6},$$

and, therefore,

$$\beta_{22}^s = -\frac{\tau_0}{6\mu k_\beta}\frac{[2F_{2,1}(ak_\alpha) - F_{2,3}(ak_\alpha)]}{\Delta_2},$$

$$\gamma_{22}^s = -\frac{\tau_0}{6\mu k_\alpha}\frac{[6F_{2,1}(ak_\beta) - F_{2,2}(ak_\beta)]}{\Delta_2},$$

$$\mathbf{u}(\mathbf{r}, \omega) = \beta_{22}^s\mathbf{N}_{22}^{-;s} + \gamma_{22}^s\mathbf{L}_{22}^{-;s}. \tag{4.227}$$

Using Eqs. (4.220) and (4.221), we obtain in the limit as $a \to 0$, $\tau_0 \to \infty$,

$$\mathbf{u} = -\frac{5ik_\beta^2}{8\pi\mu}M_0\frac{1-\sigma}{7-5\sigma}\left[\mathbf{N}_{22}^{-;s}(k_\beta r) + 2\left(\frac{\beta}{\alpha}\right)^4\mathbf{L}_{22}^{-;s}(k_\alpha r)\right], \qquad M_0 = \frac{4\pi}{3}a^3\tau_0. \tag{4.228}$$

We note from Table 4.4 that Eq. (4.228) gives the field of an equivalent strike-slip displacement dislocation for which

$$U_0\,dS = -15\left[\frac{1-\sigma}{7-5\sigma}\right]\left(\frac{M_0}{\mu}\right). \tag{4.229}$$

Equation (4.229) allows us to represent the elastodynamic field of an explosion in a medium with shear at infinity by means of a shear dislocation plus an explosion.

If the cavity is formed by an explosion, **u** of Eq. (4.204) must be added to Eq. (4.228). In that case, the spectral ratio of the S wave amplitude resulting from the excited dislocation to the P wave amplitude of the exciting explosion is

$$\left[\frac{10}{3}\frac{1-\sigma}{7-5\sigma}\frac{\alpha^2}{\beta^2}\right]\frac{\tau_0}{p_0} \simeq 1.30\frac{\tau_0}{p_0}, \qquad (\text{for } \sigma = \tfrac{1}{4}).$$

If the initial shear stress is not necessarily in the z direction, we introduce the general *prestress dyadic*

$$\mathfrak{T}_0 = \tau_0(\mathbf{ne} + \mathbf{en}), \tag{4.230}$$

where **e** and **n** are two orthogonal unit vectors. In this case, we get a shear dislocation (**e**, **n**), with a potency $U_0\,dS$ given by Eq. (4.229).

4.7. Radiations from a Finite Moving Source

The sources of seismic waves in the earth are not always localized in time and space. Sufficient evidence has accumulated in recent years to show that fracture zones at the origin of major earthquakes may be as long as 800 km. Moreover, in many cases, the rupture propagates along the fault with an average velocity of 3–3.5 km/s. In order to account for the observed radiation field from such sources, we must build suitable theoretical models that take into account the finiteness of the source and nonzero *rupture velocity.* The simplest way is to start with the expressions for the field of a point dislocation source and integrate over a finite area with proper time delays so as to simulate a source moving with a uniform velocity. This is known as the *kinematic source model.*

4.7.1. Deterministic Kinematic Source Model

Let us consider the Volterra relation, Eq. (4.105), for a finite source

$$\mathbf{u}(\mathbf{r}) = \int_S U_0\,\mathbf{en} : \mathfrak{T}_0(\tilde{\mathfrak{G}})dS. \tag{4.231}$$

For the following discussion, it is more convenient to work in the source system $(x_1^\circ, x_2^\circ, x_3^\circ)$ shown in Fig. 4.24. Let us suppose that the dislocation U_0 is a function of time and space, so that $U_0 = U_0(\xi, \eta, t)$, where $\xi = \eta_1$, $\eta = \eta_2$ and (η_1, η_2, η_3) is the fault system shown in Fig. 4.26. We further suppose that the fault is of length L and width W and the rupture velocity is V_f along ξ.

A general point on the fault is S, with coordinates $(\xi, \eta, 0)$ in the fault system (Fig. 4.33). The observer is at P (R, i_h, ϕ_h). Assuming that $R \gg \xi, \eta$, we have, on using Eqs. (4.121b) and (4.152)

$$\begin{aligned} D &= [(R\sin i_h\cos\phi_h - \xi)^2 + (R\sin i_h\sin\phi_h + \eta\cos\delta)^2 \\ &\quad + (R\cos i_h + \eta\sin\delta)^2]^{1/2} \\ &\simeq R - \xi\sin i_h\cos\phi_h + \eta(\sin\delta\cos i_h + \cos\delta\sin i_h\sin\phi_h). \end{aligned} \tag{4.232}$$

Next, we assume that the dislocation is independent of η. Replacing R in Eq. (4.192) by D and t by $t - \xi/V_f$ (to take into account a finite velocity of

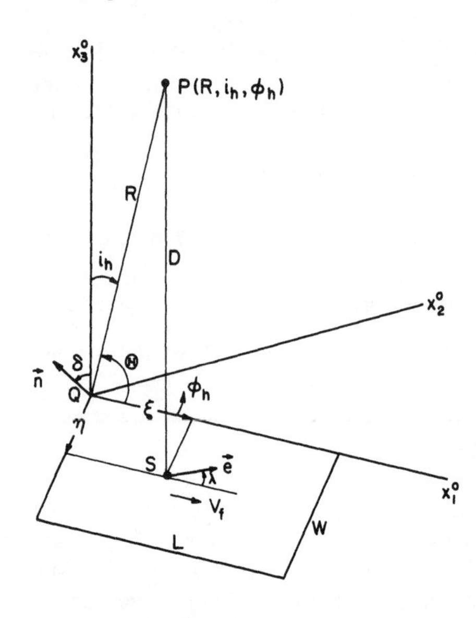

Figure 4.33. Model of a moving fault in an infinite medium. Θ is the angle that the line joining the observer with the point of fracture initiation makes with the direction of the fault propagation.

rupture), making the approximation $D \simeq R - \xi \sin i_h \cos \phi_h$, and integrating over $\xi(0, L)$, Eq. (4.192) yields

$$
\begin{aligned}
u_R &= \frac{W}{12\pi\beta}\left(\frac{\beta}{\alpha}\right)^3 \frac{F}{R}\int_0^L \dot{U}_0\left[\xi; t - \frac{R}{\alpha} - \frac{\xi}{V_f}\left(1 - \frac{V_f}{\alpha}\cos\Theta\right)\right]d\xi + O\left(\frac{1}{R^2}\right),\\
u_{i_h} &= \frac{W}{24\pi\beta R}\frac{\partial F}{\partial i_h}\int_0^L \dot{U}_0\left[\xi; t - \frac{R}{\beta} - \frac{\xi}{V_f}\left(1 - \frac{V_f}{\beta}\cos\Theta\right)\right]d\xi + O\left(\frac{1}{R^2}\right),\\
u_{\phi_h} &= \frac{W}{24\pi\beta R \sin i_h}\frac{\partial F}{\partial \phi_h}\int_0^L \dot{U}_0\left[\xi; t - \frac{R}{\beta} - \frac{\xi}{V_f}\left(1 - \frac{V_f}{\beta}\cos\Theta\right)\right]d\xi + O\left(\frac{1}{R^2}\right),
\end{aligned}
\tag{4.233}
$$

where $F(\lambda, \delta; i_h, \phi_h)$ is given by Eq. (4.154) and $\cos\Theta = \sin i_h \cos \phi_h$, Θ being the angle between QP and Qx_1° (Fig. 4.33). Knowing $U_0(\xi, t)$, we can calculate the far-field displacements.

We next assume that $U_0(\xi, t) = g(t)s(\xi)$. Then, the integration in Eq. (4.233) is immediate in the following two cases.

4.7.1.1. Case I. $s(\xi) = \text{const.} = \overline{U}$. In this case,

$$
\begin{aligned}
&\dot{U}_0\left[\xi; t - \frac{R}{\alpha} - \frac{\xi}{V_f}\left(1 - \frac{V_f}{\alpha}\cos\Theta\right)\right]\\
&\quad = \overline{U}\dot{g}\left[t - \frac{R}{\alpha} - \frac{\xi}{V_f}\left(1 - \frac{V_f}{\alpha}\cos\Theta\right)\right]\\
&\quad = -\frac{V_f\overline{U}}{1 - (V_f/\alpha)\cos\Theta}\frac{\partial}{\partial\xi}g\left[t - \frac{R}{\alpha} - \frac{\xi}{V_f}\left(1 - \frac{V_f}{\alpha}\cos\Theta\right)\right],
\end{aligned}
\tag{4.234}
$$

etc. Therefore, Eq. (4.233) becomes

$$u_R = \frac{P_0}{12\pi\beta}\left(\frac{\beta}{\alpha}\right)^3 \frac{F}{R}\left[\frac{g(t_\alpha) - g(t_\alpha - t_{d_\alpha})}{t_{d_\alpha}}\right],$$

$$u_{i_h} = \frac{P_0}{24\pi\beta R}\frac{\partial F}{\partial i_h}\left[\frac{g(t_\beta) - g(t_\beta - t_{d_\beta})}{t_{d_\beta}}\right], \tag{4.235}$$

$$u_{\phi_h} = \frac{P_0}{24\pi\beta R \sin i_h}\frac{\partial F}{\partial \phi_h}\left[\frac{g(t_\beta) - g(t_\beta - t_{d_\beta})}{t_{d_\beta}}\right],$$

where

$$P_0 = \bar{U}LW = \text{source potency},$$

$$t_\alpha = t - \frac{R}{\alpha}, \tag{4.236}$$

$$t_{d_\alpha} = \frac{L}{V_f}\left(1 - \frac{V_f}{\alpha}\cos\Theta\right),$$

with similar expressions for t_β and t_{d_β}. The time t_{d_α} is known as the *duration* of the P signal at the observation point. We assume that $V_f < \beta$ (*subshear*), so that t_{d_α} and t_{d_β} are always positive. The entity $(1 - (V_f/\alpha)\cos\Theta)$ is known as the *propagation factor* for P waves.

The finite source field generated by a constant dislocation along a fault can be interpreted as the difference of the fields of two point dislocations at Q, each of potency $P_0/t_{d_{\alpha,\beta}}$.

From Eq. (4.236), we note that for $i_h = \pi/2$, $\phi_h = 0$, $t_{d_\beta} \to 0$ as $V_f \to \beta$. Therefore, using Eq. (4.154), we have

$$u_{i_h} \to -\frac{P_0}{4\pi\beta R}\cos\lambda\cos\delta\dot{g}(t_\beta),$$

$$u_{\phi_h} \to -\frac{P_0}{4\pi\beta R}\cos\lambda\sin\delta\dot{g}(t_\beta).$$

Therefore, the shear waveform in the direction of rupture is shaped like the time derivative of the source-time function.

For earthquake sources, it has been found that V_f is less than β but is very close to it. Therefore, the pulse shapes of P and S waves will be quite different. The P pulse will be of a shorter duration. Furthermore,

$$t_{d_\alpha} - t_{d_\beta} = \frac{L}{\beta}\left(1 - \frac{\beta}{\alpha}\right)\cos\Theta = \left(\frac{0.42L}{\beta}\right)\cos\Theta.$$

For major earthquakes, this difference could be as much as 1 min.

Let us suppose that $g(t \geq \tau) = 1$ and $g(t) = 0$ for $t < 0$. Then, we may write

$$g(t) = Y(t)[H(t) - H(t-\tau)] + H(t-\tau), \quad \tau > 0, \qquad Y(0) = 0, \qquad Y(\tau) = 1.$$

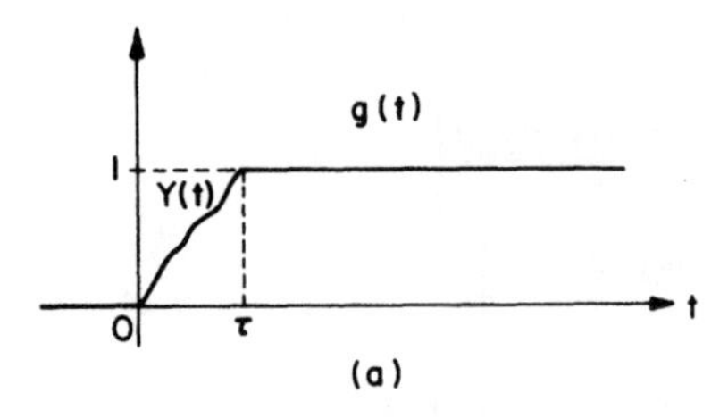

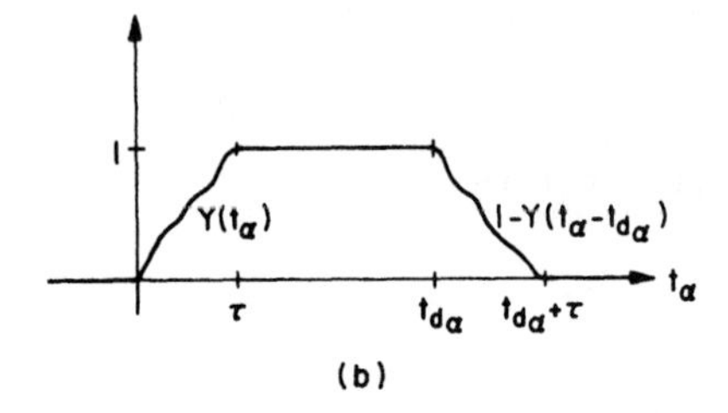

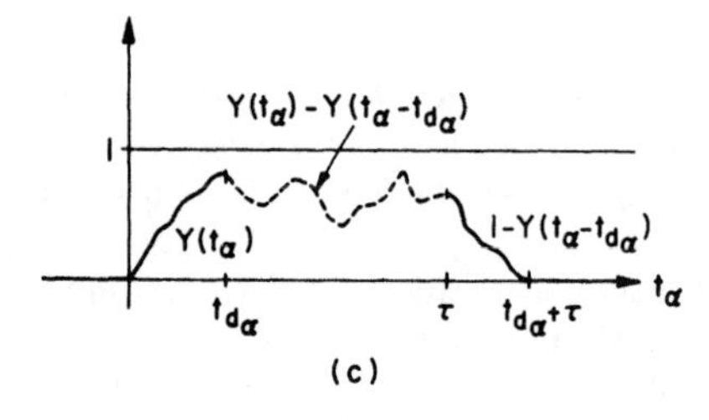

Figure 4.34. An example of the source-time function. (a) $g(t) = 0$ $(t < 0)$, $Y(t)$ $[0 < t < \tau,\ Y(0) = 0,\ Y(\tau) = 1]$, 1 $(t > \tau)$; (b) $g(t_\alpha) - g(t_\alpha - t_{d_\alpha})$ for $\tau < t_{d_\alpha}$; (c) $g(t_\alpha) - g(t_\alpha - t_{d_\alpha})$ for $\tau > t_{d_\alpha}$.

The parameter τ is known as the *rise time* or *time constant* of the source. It is obvious from Eq. (4.235) that the P signal will always terminate at $t = R/\alpha + t_{d_\alpha} + \tau$ and the S signal at $t = R/\beta + t_{d_\beta} + \tau$. Figure 4.34a shows $g(t)$, Fig. 4.34b shows $g(t_\alpha) - g(t_\alpha - t_{d_\alpha})$ when $\tau < t_{d_\alpha}$, and Fig. 4.34c shows $g(t_\alpha) - g(t_\alpha - t_{d_\alpha})$ when $\tau > t_{d_\alpha}$.

When $g(t)$ is known, Eqs. (4.235) yield the radiated field. We give u_R for three particular cases in the following list. The expressions for u_{i_h} and u_{ϕ_h} are similar and can be obtained from the corresponding expressions for u_R according to Eqs. (4.235).

1. $g(t) = (1 - e^{-t/\tau})H(t)$

$$u_R = \frac{P_0}{12\pi\beta\tau}\left(\frac{\beta}{\alpha}\right)^3 \frac{F}{R} \begin{cases} 0, & t_\alpha < 0 \\ \dfrac{1}{\gamma}(1 - e^{-t_\alpha/\tau}), & 0 < t_\alpha < t_{d_\alpha} \\ \dfrac{1}{\gamma}(1 - e^{-\gamma})e^{-(t_\alpha - t_{d_\alpha})/\tau}, & t_\alpha > t_{d_\alpha} \end{cases} \tag{4.237}$$

where

$$\gamma = \frac{1}{\tau} t_{d_\alpha} = \frac{L}{V_f \tau}\left(1 - \frac{V_f}{\alpha}\cos\Theta\right).$$

2. $Y(t) = \frac{1}{2}\left(1 - \cos\frac{\pi t}{\tau}\right), \tau < t_{d_\alpha}$

$$u_R = \frac{P_0}{12\pi\beta t_{d_\alpha}}\left(\frac{\beta}{\alpha}\right)^3 \frac{F}{R}\begin{cases} 0, & t_\alpha < 0 \\ \sin^2\left(\frac{\pi t_\alpha}{2\tau}\right), & 0 < t_\alpha < \tau \\ 1, & \tau < t_\alpha < t_{d_\alpha} \\ \cos^2\left[\frac{\pi}{2\tau}(t_\alpha - t_{d_\alpha})\right], & t_{d_\alpha} < t_\alpha < t_{d_\alpha} + \tau \\ 0, & t_\alpha > t_{d_\alpha} + \tau \end{cases} \quad (4.238)$$

3. $Y(t) = \frac{t}{\tau}\left(1 - \frac{\sin\omega_n t}{\omega_n t}\right), \quad \omega_n = \frac{2\pi n}{\tau}, \quad \tau < t_{d_\alpha}$

$$u_R = \frac{P_0}{12\pi\beta t_{d_\alpha}}\left(\frac{\beta}{\alpha}\right)^3 \frac{F}{R}\begin{cases} 0, & t_\alpha < 0 \\ \frac{t_\alpha}{\tau}\left(1 - \frac{\sin\omega_n t_\alpha}{\omega_n t_\alpha}\right), & 0 < t_\alpha < \tau \\ 1, & \tau < t_\alpha < t_{d_\alpha} \\ 1 - \frac{t_\alpha - t_{d_\alpha}}{\tau}\left[1 - \frac{\sin\omega_n(t_\alpha - t_{d_\alpha})}{\omega_n(t_\alpha - t_{d_\alpha})}\right], & t_{d_\alpha} < t_\alpha < t_{d_\alpha} + \tau \\ 0, & t_\alpha > t_{d_\alpha} + \tau. \end{cases}$$

(4.239)

To examine the effect of the source motion on the observed field, we consider two diametrically opposite points: $P(R, i_h, \phi_h)$ and $P'(R, \pi - i_h, \pi + \phi_h)$. Then, the P wave amplitude of the forward field (at P) is bigger than that of the backward field (at P') by the factor

$$\frac{u_R(R, i_h, \phi_h)}{u_R(R, \pi - i_h, \pi + \phi_h)} = \frac{t'_{d_\alpha}}{t_{d_\alpha}} = \frac{1 + (V_f/\alpha)\cos\Theta}{1 - (V_f/\alpha)\cos\Theta},$$

assuming $t_\alpha < t_{d_\alpha}$. The duration of the forward field is less than the duration of the backward field by an amount $t'_{d_\alpha} - t_{d_\alpha} = (2L/\alpha)\cos\Theta$, which is independent of V_f. Therefore, the forward field will be richer in high frequencies, whereas the backward field will be richer in low frequencies.

In the case of a vertical strike-slip source ($\lambda = 0$, $\delta = \pi/2$), we find from Eq. (4.235) for points on the plane $x_3^0 = 0$ ($i_h = \pi/2$) and $g(t) = H(t)$,

$$u_R = -\frac{P_0 V_f}{4\pi\beta LR}\left(\frac{\beta}{\alpha}\right)^3 \frac{\sin 2\phi_h}{1 - (V_f/\alpha)\cos\phi_h}[H(t_\alpha) - H(t_\alpha - t_{d_\alpha})],$$

$$u_{i_h} = 0,$$

$$u_{\phi_h} = -\frac{P_0 V_f}{4\pi\beta LR}\frac{\cos 2\phi_h}{1 - (V_f/\beta)\cos\phi_h}[H(t_\beta) - H(t_\beta - t_{d_\beta})].$$

Therefore, the radiation pattern of P waves is given by the factor $\sin 2\phi_h/(1 - (V_f/\alpha)\cos\phi_h)$ and that for S waves by the factor

$$\frac{\cos 2\phi_h}{1 - (V_f/\beta)\cos\phi_h}.$$

These are shown in Fig. 4.35.

4.7.1.2. Case II. $s(\xi) = \bar{U}\Lambda(\xi), g(t) = H(t), V_f = V_f(\xi)$. In writing Eq. (4.233), it was assumed that V_f was constant. When V_f is a function of ξ, we replace V_f in Eq. (4.233) by $\bar{V}_f(\xi)$ such that the time of rupture up to the point $(\xi, \eta, 0)$ on the fault remains the same, i.e.,

$$\text{Time of rupture to the point } (\xi, \eta, 0) = \int_0^\xi \frac{d\xi}{V_f(\xi)} = \frac{\xi}{\bar{V}_f(\xi)}.$$

With this modification, Eq. (4.233) yields

$$u_R = \frac{\bar{U}W}{12\pi\beta}\left(\frac{\beta}{\alpha}\right)^3 \frac{F}{R}\int_0^L \delta[f_\alpha(\xi)]\Lambda(\xi)d\xi, \tag{4.240}$$

with

$$f_\alpha(\xi) = t - \frac{R}{\alpha} - \frac{\xi}{\bar{V}_f(\xi)}\left[1 - \frac{\bar{V}_f(\xi)}{\alpha}\cos\Theta\right].$$

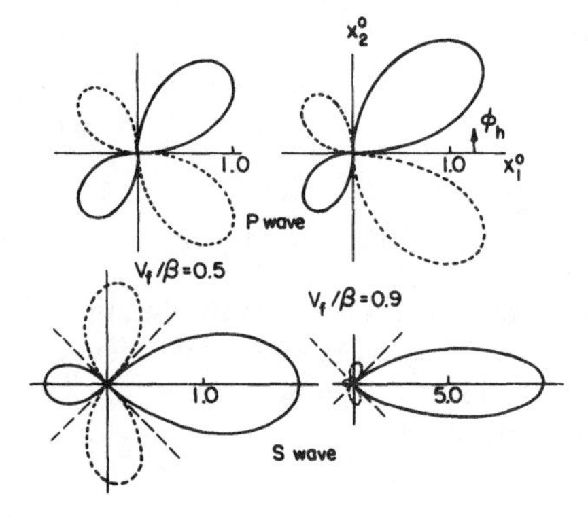

Figure 4.35. Horizontal radiation patterns of the P and S waves from a finite moving source of the strike-slip type. Solid lines indicate positive quantities, dashed lines indicate negative ones.

We assume that $V_f(\xi)$ is such that there is only one zero of $f_\alpha(\xi)$ at $\xi = \xi_{\alpha_0}$ in the range $0 < \xi < L$. Using the sifting property of the delta function and the result

$$\delta[f_\alpha(\xi)] = \frac{\delta(\xi - \xi_{\alpha 0})}{|f'_\alpha(\xi_{\alpha 0})|}, \tag{4.241}$$

Eq. (4.240) becomes

$$\begin{aligned} u_R &= \frac{\bar{U}W}{12\pi\beta}\left(\frac{\beta}{\alpha}\right)^3 \frac{F}{R}\frac{\Lambda(\xi_{\alpha 0})}{|f'_\alpha(\xi_{\alpha 0})|}[H(\xi_{\alpha_0}) - H(\xi_{\alpha_0} - L)] \\ &= \frac{P_0}{12\pi\beta}\left(\frac{\beta}{\alpha}\right)^3 \frac{F}{R}\left[\frac{\Lambda(\xi_{\alpha 0})}{t_{d_\alpha}(\xi_{\alpha 0})}\right][H(\xi_{\alpha_0}) - H(\xi_{\alpha_0} - L)], \end{aligned} \tag{4.242}$$

where

$$P_0 = \bar{U}LW, \qquad t_{d_\alpha}(\xi_{\alpha 0}) = \frac{L}{\bar{V}_f(\xi_{\alpha 0})}\left[1 - \frac{\bar{V}_f(\xi_{\alpha 0})}{\alpha}\cos\Theta - \frac{\xi_{\alpha 0}\bar{V}'_f(\xi_{\alpha 0})}{\bar{V}_f(\xi_{\alpha 0})}\right]. \tag{4.243}$$

Similarly, for S waves, we find

$$\begin{aligned} u_{i_h} &= \frac{P_0}{24\pi\beta R}\frac{\partial F}{\partial i_h}\left[\frac{\Lambda(\xi_{\beta 0})}{t_{d_\beta}(\xi_{\beta 0})}\right][H(\xi_{\beta_0}) - H(\xi_{\beta_0} - L)], \\ u_{\phi_h} &= \frac{P_0}{24\pi\beta R \sin i_h}\frac{\partial F}{\partial \phi_h}\left[\frac{\Lambda(\xi_{\beta 0})}{t_{d_\beta}(\xi_{\beta 0})}\right][H(\xi_{\beta_0}) - H(\xi_{\beta_0} - L)]. \end{aligned} \tag{4.244}$$

In the particular case when V_f is a constant, $\xi_{\alpha_0} = Lt_\alpha/t_{d_\alpha}$ and Eq. (4.242) becomes

$$u_R = \frac{P_0}{12\pi\beta}\left(\frac{\beta}{\alpha}\right)^3 \frac{F}{R}\left[\frac{\Lambda(Lt_\alpha/t_{d_\alpha})}{t_{d_\alpha}}\right][H(t_\alpha) - H(t_\alpha - t_{d_\alpha})]. \tag{4.245}$$

Similar expressions can be written for S waves.

Let us now suppose that the rupture velocity diminishes because of the energy exhaustion of the fracture process according to the law

$$V_f(\xi) = V_0\left[1 - (1 - K)\left(\frac{\xi}{L}\right)^2\right], \qquad (K < 1). \tag{4.246}$$

We then find

$$\begin{aligned} \frac{1}{\bar{V}_f(\xi)} &= \frac{\xi_0}{V_0\xi}\,\mathrm{th}^{-1}\left(\frac{\xi}{\xi_0}\right), \qquad \xi_0 = L(1 - K)^{-1/2}, \\ f_\beta(\xi) &= t_\beta - \frac{\xi_0}{V_0}\,\mathrm{th}^{-1}\left(\frac{\xi}{\xi_0}\right) + \frac{\xi}{\beta}\cos\Theta. \end{aligned} \tag{4.247}$$

Therefore, ξ_{β_0} is a root of the equation

$$a + bx = \mathrm{th}^{-1}x, \tag{4.248}$$

where

$$a = \frac{V_0 t_\beta}{\xi_0}, \qquad b = \frac{V_0}{\beta} \cos \Theta, \qquad x = \frac{\xi}{\xi_0}. \tag{4.249}$$

4.7.2. Supershear Rupture

Although it is believed that rupture on earthquake faults propagates with velocities less than the shear-wave velocity in the surrounding medium, it is interesting to examine the theoretical implications of rupture velocity greater than the shear-wave velocity. For simplicity, let us consider the case when $g(t) = H(t)$, $s(\xi) = \text{const.} = \bar{U}$, and $V_f = \text{const.}$ Equation (4.244) then yields

$$u_{i_h} = \frac{P_0}{24\pi\beta R} \frac{\partial F}{\partial i_h} \frac{1}{t_{d_\beta}} [H(\xi_\beta) - H(\xi_\beta - L)], \tag{4.250}$$

where

$$\xi_\beta = \frac{L t_\beta}{t_{d_\beta}} = \frac{V_f(t - R/\beta)}{1 - (V_f/\beta)\cos \Theta}. \tag{4.251}$$

A nonzero contribution to u_{i_h} is secured only if $0 < \xi_\beta < L$. Let us consider the subshear and supershear situations separately. All the forthcoming conclusions are valid in the far field only.

1. Subshear ($V_f < \beta$): The condition $\xi_\beta > 0$ implies that $t > R/\beta$. Hence there is no signal for $t < R/\beta$, and, for $t > R/\beta$, the signal reaches all points of space with amplitude proportional to $1/t_{d_\beta}$.
2. Supershear ($V_f > \beta$): Let $\cos \Theta_0 = \beta/V_f$. Then $\xi_\beta > 0$ will be satisfied if $t > R/\beta$ for $\Theta > \Theta_0$ and if $t < R/\beta$ for $\Theta < \Theta_0$. The cone $\Theta = \Theta_0$ is known as the *Mach cone*. In the supershear condition, therefore, the signal arrives at points inside the Mach cone at times $t < R/\beta$ and arrives at points outside this cone at times $t > R/\beta$. The amplitude of the signal is inversely proportional to the pulse width

$$|t_{d_\beta}| = \frac{L}{\beta} |\cos \Theta_0 - \cos \Theta|,$$

which tends to zero as $\Theta \to \Theta_0$. It is obvious from Eq. (4.250) that, as $\Theta \to \Theta_0$,

$$u_{i_h} \to \frac{P_0}{24\pi L R} \frac{\partial F}{\partial i_h} \delta\left(t - \frac{R}{\beta}\right). \tag{4.252}$$

Similarly,

$$u_{\phi_h} \to \frac{P_0}{24\pi L R \sin i_h} \frac{\partial F}{\partial \phi_h} \delta\left(t - \frac{R}{\beta}\right). \tag{4.253}$$

4.7.3. Spectral Displacements

We will write the explicit expressions for u_R only. The results for u_{i_h} and u_{ϕ_h} can be obtained by adjusting the multiplying factors suitably.

Assuming $U_0(\xi, t) = s(\xi)g(t)$ and applying the Fourier transform, we obtain, from Eq. (4.233),

$$u_R(\mathbf{R}; \omega) = \frac{W}{12\pi\beta}\left(\frac{\beta}{\alpha}\right)^3 \frac{F}{R} e^{-ik_\alpha R}[i\omega g(\omega)]$$

$$\times \int_0^L s(\xi)\exp\left[-i\omega\frac{\xi}{V_f}\left(1 - \frac{V_f}{\alpha}\cos\Theta\right)\right]d\xi, \tag{4.254}$$

where $k_\alpha = \omega/\alpha$ and $g(\omega)$ is the Fourier transform of $g(t)$. If $s(\xi) = \text{const.} = \overline{U}$, we get

$$u_R = \frac{P_0}{12\pi\beta}\left(\frac{\beta}{\alpha}\right)^3 \frac{F}{R}[i\omega g(\omega)]\left(\frac{\sin X_\alpha}{X_\alpha}\right)e^{-ik_\alpha R - iX_\alpha}, \tag{4.255}$$

where

$$X_\alpha = \frac{L}{2V_f}\omega_\alpha, \qquad \omega_\alpha = \omega\left(1 - \frac{V_f}{\alpha}\cos\Theta\right). \tag{4.256}$$

The factor $(\sin X_\alpha/X_\alpha)\exp(-iX_\alpha)$ is a first-order correction for the finiteness of the source and its motion. The *Doppler frequency shift* affects both the amplitude and the phase. An observer at (R, i_h, ϕ_h) will notice a phase retardation and an amplitude modulation, which will be functions of the frequency, ω; the polar angle, Θ; the rupture duration, L/V_f; and the *Mach numbers*, V_f/α and V_f/β. The *P*-wave amplitude vanishes for the values of ω that make X_α an integral multiple of π; i.e., whenever $L(\alpha/V_f - \cos\Theta)$ is an integral multiple of the wavelength. This is the condition for the *destructive interference* of *P* waves at the point of observation.

If $s(\xi)$ is not constant, we may write Eq. (4.254) in the form

$$u_R(\mathbf{R}; \omega) = \frac{W}{12\pi\beta}\left(\frac{\beta}{\alpha}\right)^3 \frac{F}{R}[i\omega g(\omega)]I(p_\alpha)e^{-ik_\alpha R}, \tag{4.257}$$

where $I(p_\alpha)$ is the *finiteness transform* of $s(\xi)$ defined by

$$I(p_\alpha) = \int_0^L s(\xi)e^{-ip_\alpha\xi}\,d\xi, \qquad p_\alpha = \frac{\omega}{V_f}\left(1 - \frac{V_f}{\alpha}\cos\Theta\right). \tag{4.258}$$

Table 4.8 gives $I(p)$ for several particular values of $s(\xi)$ of interest in seismology. Similarly, Table 4.9 lists a number of common time functions and their Fourier transforms.

The dependence of the spectral displacements on ω is diagnostic of the spatial and temporal behavior of the source function. The frequency dependence of the displacement $u_R(\mathbf{R}; \omega)$ is embodied in the factor

$$[i\omega g(\omega)]\int_0^L s(\xi)\exp\left[-i\frac{\omega}{V_f}\xi\left(1 - \frac{V_f}{\alpha}\cos\Theta\right)\right]d\xi.$$

Table 4.8. Finiteness Transform[a] s

	$s(\xi)$	ξ_m	$I(p) = \int_0^L s(\xi) e^{-ip\xi}\, d\xi$	Remarks
1	$\overline{U}$	—	$L\overline{U}\dfrac{\sin X}{X} e^{-iX}$	
2	$\overline{U}\left[\sin\left(\dfrac{\pi\xi}{L}\right)\right]^{2m}$	$\dfrac{L}{2}$	$\dfrac{L\overline{U}}{N}\left[\dfrac{\sin X}{X \prod_{k=1}^{m} (1 - (X^2/\pi^2 k^2))}\right] e^{-iX}$	$N = 2^{2m}\dfrac{(m!)^2}{(2m)!}$
3	$\overline{U}\left[\sin\left(\dfrac{\pi\xi}{L}\right)\right]^{2m-1}$	$\dfrac{L}{2}$	$\dfrac{L\overline{U}}{N}\left[\dfrac{\cos X}{\prod_{k=1}^{m} \{1 - (X^2/\pi^2(k - \frac{1}{2})^2)\}}\right] e^{-iX}$	$N = \dfrac{\pi m(2m)!}{2^{2m}(m!)^2}$
4	$\overline{U}\left[4\dfrac{\xi}{L}\left(1 - \dfrac{\xi}{L}\right)\right]^{m}$	$\dfrac{L}{2}$	$\dfrac{L\overline{U}}{N}\left[\dfrac{(2m+1)!!\, j_m(X)}{X^m}\right] e^{-iX}$	$N = \dfrac{(2m+1)!}{2^{2m}(m!)^2}$
5	$\overline{U} A(\mu, \nu)\left(\dfrac{\xi}{L}\right)^{2\mu-1}\left(1 - \dfrac{\xi}{L}\right)^{2\nu-1}$	$\dfrac{(2\mu - 1)L}{2(\mu + \nu - 1)}$	$\dfrac{L\overline{U}}{N}\left[\dfrac{M_{\mu-\nu, \mu+\nu-1/2}(2iX)}{(2iX)^{\mu+\nu}}\right] e^{-iX}$	$M_{x,y}(z)$ = Whittaker function $B(m, n)$ = Beta function $A(\mu, \nu) = \left[\dfrac{2(\mu + \nu - 1)}{2\mu - 1}\right]^{2\mu-1}\left[\dfrac{2(\mu + \nu - 1)}{2\nu - 1}\right]^{2\nu-1}$ $N = [A(\mu, \nu) B(2\mu, 2\nu)]^{-1}$
6	$\overline{U}\left(\dfrac{\lambda e}{n}\xi\right)^{n} e^{-\lambda\xi}, (n < \lambda L)$	$\dfrac{n}{\lambda}$	$\dfrac{L\overline{U}}{N}\left[\dfrac{M_{n/2,(n+1)/2}(2iZ)}{(2iZ)^{(n+2)/2}}\right] e^{-iZ}$	$N = (n + 1)\left(\dfrac{n}{\lambda e L}\right)^{n}$ $Z = X - \dfrac{iL\lambda}{2}$
7	$\overline{U} e^{-\gamma(\xi - L/2)^2}$	$\dfrac{L}{2}$	$\dfrac{L\overline{U}}{N} e^{-X^2/(\pi N^2)} e^{-iX}$	$N = L\sqrt{\left(\dfrac{\gamma}{\pi}\right)}$

[a] $p = \dfrac{\omega}{V_f}\left(1 - \dfrac{V_f}{c}\cos\Theta\right)$, $c = \alpha$ or β, $X = \frac{1}{2}Lp$, $s_{max}(\xi, 0 \le \xi \le L) = \overline{U}$ at $\xi = \xi_m$.

Table 4.9. Source Time-Functions and Their Spectra

Name	$g(t) = \frac{1}{2\pi}\int_{-\infty}^{\infty} g(\omega)e^{i\omega t}\,d\omega$	$g(\omega) = \int_{-\infty}^{\infty} g(t)e^{-i\omega t}\,dt$	Behavior at high frequencies
1. Delta function	$\delta(t - t_0)$	$e^{-i\omega t_0}$	1
2. Step function	$H(t - t_0)$	$\pi\delta(\omega) + \frac{1}{i\omega}e^{-i\omega t_0}$	ω^{-1}
3. Exponential decay	$H(t)e^{-t/\tau}, \quad (\tau > 0)$	$\frac{\tau}{1 + i\omega\tau}$	ω^{-1}
4. Exponential buildup	$H(t)(1 - e^{-t/\tau}), \quad (\tau > 0)$	$\frac{1}{i\omega(1 + i\omega\tau)}$	ω^{-2}
5. Gamma kernel	$H(t)t^{\beta}e^{-t/\tau}, \quad (\tau > 0, \beta > -1)$	$\frac{\tau^{\beta+1}\Gamma(\beta + 1)}{(1 + i\omega\tau)^{\beta+1}}$	$\omega^{-(\beta+1)}$
6. Boxcar function	$H(t) - H(t - T)$	$T\left[\frac{\sin(\omega T/2)}{(\omega T/2)}\right]e^{-i(\omega T/2)}$	ω^{-1}
7. Ramp function	$R(t) = \begin{cases} 0, & t < 0 \\ t/T, & 0 \le t \le T \\ 1, & t > T \end{cases}$	$\frac{1}{\omega}\left[\frac{\sin(\omega T/2)}{(\omega T/2)}\right]e^{-i(\omega T/2) - \pi i/2}$	ω^{-2}

(continued)

Table 4.9. *(continued)*

Name	$g(t) = \frac{1}{2\pi}\int_{-\infty}^{\infty} g(\omega)e^{i\omega t}\,d\omega$	$g(\omega) = \int_{-\infty}^{\infty} g(t)e^{-i\omega t}\,dt$	Behavior at high frequencies
8. Modulated ramp	$R_n(t) = \begin{cases} 0, & t < 0 \\ \frac{t}{T}\left(1 - \frac{\sin \omega_n t}{\omega_n t}\right), & 0 \le t \le T \\ 1, & t > T \end{cases} \left(\omega_n = \frac{2\pi n}{T}\right)$	$\frac{1}{\omega}\left[\frac{\sin(\omega T/2)}{(\omega T/2)}\right]\frac{e^{-i(\omega T/2) - \pi i/2}}{1 - (\omega/\omega_n)^2}$	ω^{-4}
First derivative	$\dot{R}_n(t) = \begin{cases} \frac{1 - \cos \omega_n t}{T}, & 0 \le t \le T \\ 0, & \text{otherwise} \end{cases}$	$\left[\frac{\sin(\omega T/2)}{(\omega T/2)}\right]\frac{e^{-i(\omega T/2)}}{1 - (\omega/\omega_n)^2}$	ω^{-3}
Second derivative	$\ddot{R}_n(t) = \begin{cases} \frac{\omega_n}{T}\sin \omega_n t, & 0 \le t \le T \\ 0, & \text{otherwise} \end{cases}$	$\frac{2}{T}\left[\frac{\sin(\omega T/2)}{1 - (\omega/\omega_n)^2}\right]e^{-i(\omega T/2) + \pi i/2}$	ω^{-2}
9. Half-sine function	$H_s(t) = \begin{cases} 0, & t < 0 \\ \frac{1}{2}\left(1 - \cos\frac{\pi t}{T}\right), & 0 \le t \le T \\ 1, & t > T \end{cases}$	$\frac{1}{\omega}\left[\frac{\cos(\omega T/2)}{1 - (\omega T/\pi)^2}\right]e^{-i(\omega T/2) - \pi i/2}$	ω^{-3}
10. Delta comb	$\sum_{n=-N}^{N} \delta(t - nT)$	$\frac{\sin(N + \frac{1}{2})\omega T}{\sin(\omega T/2)}$	—
11.	$\sum_{n=-N}^{N} g(t - nT)$	$g(\omega)\frac{\sin(N + \frac{1}{2})\omega T}{\sin(\omega T/2)}$	—

12. Power law	$H(t)t^{\alpha}$	$\frac{\Gamma(\alpha+1)}{\lvert\omega\rvert^{\alpha+1}} e^{-(\pi i/2)(\alpha+1)\operatorname{sgn}\omega}$	$\omega^{-(\alpha+1)}$
13.	$[1+e^{-(t-t_0)/\tau}]^{-1}$	$\frac{1}{i\omega}\left[\frac{\pi\omega\tau}{\sinh(\pi\omega\tau)}\right]e^{-i\omega t_0}$	$e^{-\pi\omega\tau}$
14. Attenuated cosine	$e^{-t/\tau}\cos\omega_0 t H(t) \quad (\tau>0)$	$\frac{\tau(1+i\omega\tau)}{(1+i\omega\tau)^2+\omega_0^2\tau^2}$	ω^{-1}
15. Attenuated sine	$e^{-t/\tau}\sin\omega_0 t H(t) \quad (\tau>0)$	$\frac{\omega_0\tau^2}{(1+i\omega\tau)^2+\omega_0^2\tau^2}$	ω^{-2}
16. Bessel function	$J_0(\omega_0 t)H(t)$	$(\omega_0^2-\omega^2)^{-1/2}, \quad \omega_0>\omega$ $i(\omega^2-\omega_0^2)^{-1/2}, \quad \omega>\omega_0$	ω^{-1}
17.	$(1-e^{-t/\tau})^{\nu-1}H(t) \quad (\operatorname{Re}\nu>0, \tau>0)$	$\tau B(\nu, i\omega\tau)$	$\omega^{-\nu}$
18.	$\left[1-\left(1+\frac{t}{\tau}\right)e^{-t/\tau}\right]H(t)$	$\frac{1}{i\omega(1+i\omega\tau)^2}$	ω^{-3}
19. Sampled causal function	$g(nT) \quad (n=0,1,2,\cdots)$ $nT<t<(n+1)T$	$\frac{1-e^{-i\omega T}}{i\omega}\sum_{k=0}^{\infty} g(kT)e^{-i\omega\kappa T}$	
20. Sampled buildup function	$(1-e^{-nT/\tau})H(t)$ $(n=0,1,2,\cdots)$	$\left[1-\frac{1-e^{-i\omega T}}{e^{-(T/\tau)(1+i\omega\tau)}}\right]\frac{1}{i\omega}$	

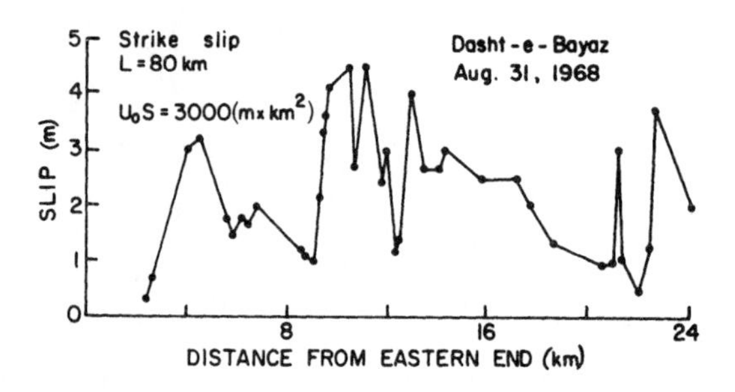

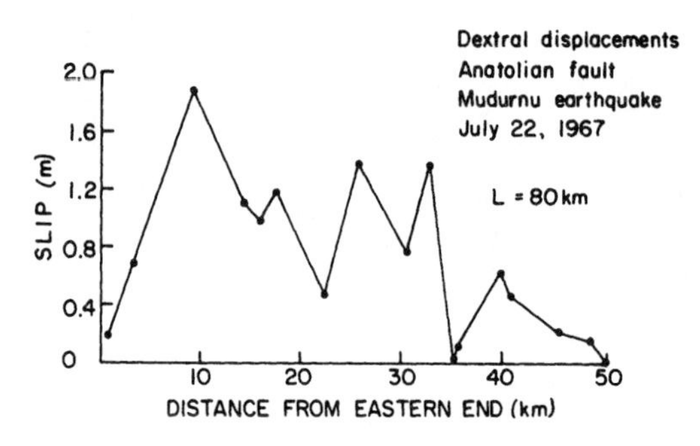

Figure 4.36. Examples of the observed variation of the slip $s(\xi)$ along the length of the fault.

For a constant slip along the fault, this factor reduces to

$$[i\omega g(\omega)]\left[L\bar{U}\frac{\sin X_\alpha}{X_\alpha}e^{-iX_\alpha}\right]$$

with the modulus $\omega|g(\omega)||\sin X_\alpha/X_\alpha|$ multiplied by a constant factor. At the high-frequency end of the spectrum, the dependence on ω is roughly that of $|g(\omega)|$. At low frequencies, in contrast, $\sin X/X \simeq 1$ and the dependence is close to $\omega|g(\omega)|$. We usually require that $g(t) = 0$ for $t < 0$ and $g(t) = 1$ for $t > T$. We then put

$$g(t) = Y(t)[H(t) - H(t-T)] + H(t-T), \tag{4.259}$$

$Y(0) = 0$, $Y(T) = 1$, and find

$$i\omega g(\omega) = i\omega\int_0^T Y(t)e^{-i\omega t}\,dt + e^{-i\omega T} = \int_0^T \dot{Y}(t)e^{-i\omega t}\,dt.$$

Figures 4.36 and 4.37 show the observed variation of $s(\xi)$ along the fault for four earthquakes.

In writing Eq. (4.254), we have made the approximation

$$D \simeq R - \xi\sin i_h\cos\phi_h$$

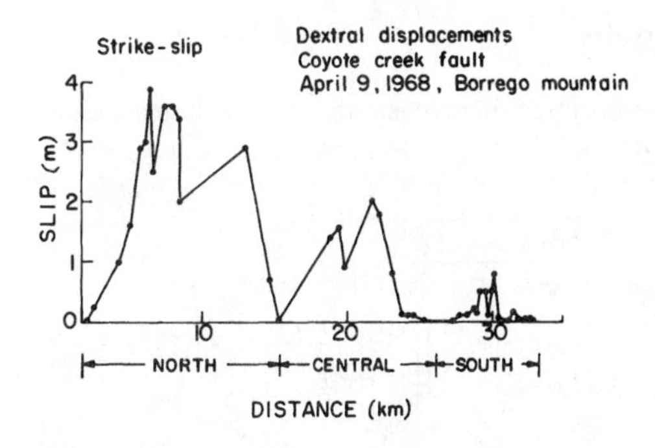

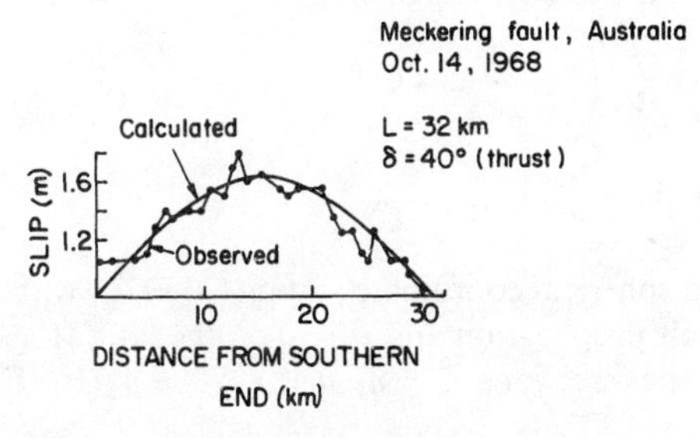

Figure 4.37. Examples of the observed variation of the slip $s(\xi)$ along the fault. In the Meckering fault, the calculated values for a theoretical model are also shown.

(Fig. 4.33). This is valid if W is small as compared to L. In general, we must use the approximation, Eq. (4.232). In that case Eq. (4.254) becomes

$$u_R(\mathbf{R};\omega) = \frac{1}{12\pi\beta}\left(\frac{\beta}{\alpha}\right)^3 \frac{F}{R} e^{-ik_\alpha R}[i\omega g(\omega)]$$
$$\times \int_0^L s(\xi)\exp\left[-i\omega\frac{\xi}{V_f}\left(1 - \frac{V_f}{\alpha}\cos\Theta\right)\right]d\xi$$
$$\times \int_0^W \exp\left[-i\omega\frac{\eta}{\alpha}(\sin\delta\cos i_h + \cos\delta\sin i_h\sin\phi_h)\right]d\eta.$$

Assuming $s(\xi) = \text{const.} = \overline{U}$, this yields

$$u_R(\mathbf{R};\omega) = \frac{P_0}{12\pi\beta}\left(\frac{\beta}{\alpha}\right)^3 \frac{F}{R}[i\omega g(\omega)]\left(\frac{\sin X_\alpha}{X_\alpha}\right)\left(\frac{\sin Y_\alpha}{Y_\alpha}\right)e^{-ik_\alpha R - iX_\alpha - iY_\alpha},$$

where

$$Y_\alpha = \frac{\omega W}{2\alpha}(\sin\delta\cos i_h + \cos\delta\sin i_h\sin\phi_h).$$

4.7.4. Directivity

The ratio of the spectral displacements at two diametrically opposite points relative to the origin, is known as the *directivity*. Equation (4.255) shows that if $s(\xi) = \overline{U}$, the directivity is given by

$$D_\alpha(i_h, \phi_h) = \frac{u_R(R, i_h, \phi_h; \omega)}{u_R(R, \pi - i_h, \pi + \phi_h; \omega)}$$
$$= \left[\frac{1 + (V_f/\alpha)\cos\Theta}{1 - (V_f/\alpha)\cos\Theta}\right]\left[\frac{\sin\{\omega L/(2V_f)[1 - (V_f/\alpha)\cos\Theta]\}}{\sin\{\omega L/(2V_f)[1 + (V_f/\alpha)\cos\Theta]\}}\right] e^{i\omega(L/\alpha)\cos\Theta}. \quad (4.260)$$

The zeros of the directivity function, D_α, are located at the frequencies $2\pi n/t_{d_\alpha}$, and the poles at the frequencies $2\pi n/t'_{d_\alpha}$, where $n = 1, 2, 3, \ldots$ and

$$t_{d_\alpha} = \frac{L}{V_f}\left(1 - \frac{V_f}{\alpha}\cos\Theta\right), \qquad t'_{d_\alpha} = \frac{L}{V_f}\left(1 + \frac{V_f}{\alpha}\cos\Theta\right). \quad (4.261)$$

4.7.5. Total Radiated Energy

We introduce a spherical coordinate system (R, Θ, ψ) with the polar axis in the direction of fault propagation and $\psi = 0$ on the fault (Fig. 4.38). Denoting the unit vectors in the directions $(\xi = \eta_1, \eta = \eta_2, \zeta = \eta_3)$ by $(\mathbf{i}_1, \mathbf{i}_2, \mathbf{i}_3)$, respectively, we have

	$\mathbf{i}_1$	$\mathbf{i}_2$	$\mathbf{i}_3$
$\mathbf{e}_R$	$\cos\Theta$	$\sin\Theta\cos\psi$	$\sin\Theta\sin\psi$
$\mathbf{e}_\Theta$	$-\sin\Theta$	$\cos\Theta\cos\psi$	$\cos\Theta\sin\psi$
$\mathbf{e}_\psi$	0	$-\sin\psi$	$\cos\psi$

(4.262)

Further, from Eq. (4.121a) and Fig. 4.38, we note that

$$\mathbf{e} = \cos\lambda\,\mathbf{i}_1 - \sin\lambda\,\mathbf{i}_2, \qquad \mathbf{n} = -\mathbf{i}_3. \quad (4.263)$$

Equations (4.262) and (4.263) yield

$$(\mathbf{e}\cdot\mathbf{e}_R)(\mathbf{n}\cdot\mathbf{e}_R) = \frac{1}{6}E,$$
$$(\mathbf{e}\cdot\mathbf{e}_R)(\mathbf{n}\cdot\mathbf{e}_\Theta) + (\mathbf{e}\cdot\mathbf{e}_\Theta)(\mathbf{n}\cdot\mathbf{e}_R) = \frac{1}{6}\frac{\partial E}{\partial\Theta}, \quad (4.264)$$
$$(\mathbf{e}\cdot\mathbf{e}_R)(\mathbf{n}\cdot\mathbf{e}_\psi) + (\mathbf{e}\cdot\mathbf{e}_\psi)(\mathbf{n}\cdot\mathbf{e}_R) = \frac{1}{6\sin\Theta}\frac{\partial E}{\partial\psi},$$

where

$$E(\lambda, \Theta, \psi) = 3(\sin\lambda\sin^2\Theta\sin 2\psi - \cos\lambda\sin 2\Theta\sin\psi). \quad (4.265)$$

Therefore, the (R, Θ, ψ) components of the displacement are obtained from the (R, i_h, ϕ_h) components of the displacement on replacing in Eq. (4.233)

$$\left[F, \frac{\partial F}{\partial i_h}, \frac{1}{\sin i_h}\frac{\partial F}{\partial\phi_h}\right]$$

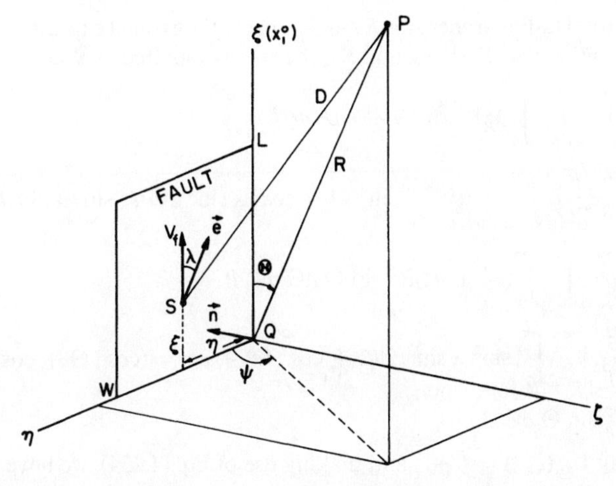

Figure 4.38. The fault coordinates (ξ, η, ζ) for the calculation of the total energy and the spherical coordinate system (R, Θ, ψ) having the direction of the fault propagation as the polar axis.

by $[E, \partial E/\partial\Theta, 1/(\sin\Theta)\partial E/\partial\psi]$, respectively. Equation (4.233) then gives the following expressions for particle velocities

$$\dot{u}_R = \frac{W}{12\pi\beta}\left(\frac{\beta}{\alpha}\right)^3 \frac{E}{R} I_\alpha,$$

$$\dot{u}_\Theta = \frac{W}{24\pi\beta R}\frac{\partial E}{\partial\Theta} I_\beta,$$

$$\dot{u}_\psi = \frac{W}{24\pi\beta R\sin\Theta}\frac{\partial E}{\partial\psi} I_\beta, \tag{4.266}$$

where

$$I_\alpha = \int_0^L \ddot{U}_0\left[\xi; t_\alpha - \frac{\xi}{V_f}\left(1 - \frac{V_f}{\alpha}\cos\Theta\right)\right] d\xi, \tag{4.267}$$

$$I_\beta = \int_0^L \ddot{U}_0\left[\xi; t_\beta - \frac{\xi}{V_f}\left(1 - \frac{V_f}{\beta}\cos\Theta\right)\right] d\xi. \tag{4.268}$$

If $\boldsymbol{\Sigma}$ denotes the energy-flux density, we have

$$\begin{aligned}
\mathbf{e}_R \cdot \boldsymbol{\Sigma} &= -\mathbf{e}_R \cdot [\mathfrak{T}(\mathbf{u}) \cdot \dot{\mathbf{u}}] \\
&= -(\tau_{RR}\dot{u}_R + \tau_{R\Theta}\dot{u}_\Theta + \tau_{R\psi}\dot{u}_\psi) \\
&= -\rho\left[\alpha^2 \dot{u}_R \frac{\partial u_R}{\partial R} + \beta^2\left(\dot{u}_\Theta \frac{\partial u_\Theta}{\partial R} + \dot{u}_\psi \frac{\partial u_\psi}{\partial R}\right)\right] + O\left(\frac{1}{R^3}\right) \\
&= \rho\alpha\dot{u}_R^2 + \rho\beta(\dot{u}_\Theta^2 + \dot{u}_\psi^2) + O\left(\frac{1}{R^3}\right).
\end{aligned} \tag{4.269}$$

Hence, the total radiated energy is $E_\alpha + E_\beta$, where the explicit expressions for E_α, the energy radiated as P waves, and E_β, the energy radiated as S waves, are

$$E_\alpha = \rho\alpha \int_{-\infty}^{\infty}\int_0^{2\pi}\int_0^{\pi} \dot{u}_R^2 R^2 \sin\Theta \, d\Theta \, d\psi \, dt$$

$$= \frac{\rho W^2}{16\pi\beta}\left(\frac{\beta}{\alpha}\right)^5 \int_{-\infty}^{\infty}\int_0^{\pi} (\sin^2\lambda \sin^4\Theta + \cos^2\lambda \sin^2 2\Theta) I_\alpha^2 \sin\Theta \, d\Theta \, dt,$$

$$E_\beta = \rho\beta \int_{-\infty}^{\infty}\int_0^{2\pi}\int_0^{\pi} (\dot{u}_\Theta^2 + \dot{u}_\psi^2) R^2 \sin\Theta \, d\Theta \, d\psi \, dt \tag{4.270}$$

$$= \frac{\rho W^2}{16\pi\beta} \int_{-\infty}^{\infty}\int_0^{\pi} [\sin^2\lambda \sin^2\Theta(1 + \cos^2\Theta) + \cos^2\lambda(\cos^2\Theta + \cos^2 2\Theta)]$$

$$\times I_\beta^2 \sin\Theta \, d\Theta \, dt.$$

Assuming $U_0(\xi, t) = \bar{U}g(t)$ and making use of Eq. (4.234), we have

$$I_\alpha = \frac{L\bar{U}}{t_{d_\alpha}}[\dot{g}(t_\alpha) - \dot{g}(t_\alpha - t_{d_\alpha})]$$

and, therefore,

$$\int_{-\infty}^{\infty} I_\alpha^2 \, dt = 2\left(\frac{L\bar{U}}{t_{d_\alpha}}\right)^2 \int_{-\infty}^{\infty} [\dot{g}^2(t) - \dot{g}(t)\dot{g}(t - t_{d_\alpha})]dt. \tag{4.271}$$

Equation (4.271) expresses the radiated energy in terms of the autocorrelation of $g(t)$. Similar results hold for S waves.

In terms of the function $Y(t)$ of Eq. (4.259), we find

$$\int_{-\infty}^{\infty} I_\alpha^2 \, dt = 2\left(\frac{L\bar{U}}{t_{d_\alpha}}\right)^2 \left[\int_0^T \dot{Y}^2(t)dt - \int_{t_{d_\alpha}}^T \dot{Y}(t)\dot{Y}(t - t_{d_\alpha})dt\right], \qquad (t_{d_\alpha} < T)$$

$$= 2\left(\frac{L\bar{U}}{t_{d_\alpha}}\right)^2 \int_0^T \dot{Y}^2(t)dt \qquad (t_{d_\alpha} > T). \tag{4.272}$$

The integral in Eq. (4.271) may be evaluated for some given forms of the time function. For example, if

$$g(t) = (1 - e^{-t/\tau})H(t), \qquad (\tau > 0)$$

we find

$$\int_{-\infty}^{\infty} I_\alpha^2 \, dt = \frac{1}{\tau}\left(\frac{L\bar{U}}{t_{d_\alpha}}\right)^2 (1 - e^{-t_{d_\alpha}/\tau}). \tag{4.273}$$

4.7.6. Spectral Energy Density

Let $F(\omega)$ be the Fourier transform of a real time function $f(t)$. Then

$$F(\omega) = \int_{-\infty}^{\infty} f(t)e^{-i\omega t} \, dt,$$

$$f(t) = \frac{1}{2\pi}\int_{-\infty}^{\infty} F(\omega)e^{i\omega t} \, dt, \tag{4.274}$$

and $F(-\omega) = F^*(\omega)$. Writing

$$F(\omega) = |F(\omega)|e^{i\chi(\omega)},$$

we have $|F(-\omega)| = |F(\omega)|$, $\chi(-\omega) = -\chi(\omega)$. Parseval's theorem now yields

$$\int_{-\infty}^{\infty} [f(t)]^2\, dt = \frac{1}{2\pi}\int_{-\infty}^{\infty} |F(\omega)|^2\, d\omega = \frac{1}{\pi}\int_0^{\infty} |F(\omega)|^2\, d\omega = 2\int_0^{\infty} |F(\omega)|^2\, d\nu, \tag{4.275}$$

where $\nu = \omega/2\pi$ is the frequency. The function $2|F(\omega)|^2$ is called the *spectral energy density*.

Denoting the Fourier transform of the displacement vector $[u_R(t), u_\Theta(t), u_\psi(t)]$ by $[u_R(\omega), u_\Theta(\omega), u_\psi(\omega)]$, Eqs. (4.270) and (4.275) yield

$$E_\alpha = \int_0^{\infty} \varepsilon_\alpha(\nu) d\nu, \qquad E_\beta = \int_0^{\infty} \varepsilon_\beta(\nu) d\nu, \tag{4.276}$$

where

$$\begin{aligned}
\varepsilon_\alpha(\nu) &= 2\rho\alpha \int_0^{2\pi}\int_0^{\pi} |\dot{u}_R(\omega)|^2 R^2 \sin\Theta\, d\Theta\, d\psi,\\
\varepsilon_\beta(\nu) &= 2\rho\beta \int_0^{2\pi}\int_0^{\pi} [|\dot{u}_\Theta(\omega)|^2 + |\dot{u}_\psi(\omega)|^2] R^2 \sin\Theta\, d\Theta\, d\psi.
\end{aligned} \tag{4.277}$$

The functions $\varepsilon_\alpha(\nu)$ and $\varepsilon_\beta(\nu)$ are the spectral energy densities for P and S waves, respectively. Assuming $U_0(\xi, t) = \bar{U}g(t)$ and proceeding as in Section 4.7.3, Eqs. (4.270) and (4.275) yield

$$\begin{aligned}
\varepsilon_\alpha(\nu) &= \frac{\rho P_0^2}{2\pi\beta}\left(\frac{\beta}{\alpha}\right)^5 \omega^2 |g(\omega)|^2 [\cos^2\lambda B_1(\omega) + \sin^2\lambda B_3(\omega)],\\
\varepsilon_\beta(\nu) &= \frac{\rho P_0^2}{2\pi\beta}\omega^2 |g(\omega)|^2 [\cos^2\lambda B_2(\omega) + \sin^2\lambda B_4(\omega)],
\end{aligned} \tag{4.278}$$

where

$$\begin{aligned}
B_1(\omega) &= \frac{1}{4}\int_0^{\pi}\left(\frac{\sin X_\alpha}{X_\alpha}\right)^2 \sin^2 2\Theta \sin\Theta\, d\Theta,\\
B_2(\omega) &= \frac{1}{4}\int_0^{\pi}\left(\frac{\sin X_\beta}{X_\beta}\right)^2 (\cos^2\Theta + \cos^2 2\Theta)\sin\Theta\, d\Theta,\\
B_3(\omega) &= \frac{1}{4}\int_0^{\pi}\left(\frac{\sin X_\alpha}{X_\alpha}\right)^2 \sin^5\Theta\, d\Theta,\\
B_4(\omega) &= \frac{1}{4}\int_0^{\pi}\left(\frac{\sin X_\beta}{X_\beta}\right)^2 (1 + \cos^2\Theta)\sin^3\Theta\, d\Theta,
\end{aligned} \tag{4.279}$$

$$P_0 = \bar{U}LW, \qquad X_\alpha = \frac{\omega L}{2\alpha}\left(\frac{\alpha}{V_f} - \cos\Theta\right), \qquad X_\beta = \frac{\omega L}{2\beta}\left(\frac{\beta}{V_f} - \cos\Theta\right).$$

Therefore, B_1 and B_2 correspond to a longitudinal shear fault and B_3 and B_4, to a transverse shear fault.

The integrals $B_i(\omega)$, $(i = 1, 2, 3, 4)$, can be expressed in terms of certain elementary functions and their integrals. Let us consider, for example, the B_2 integral. Making the change of variable, $y = \beta/V_f - \cos\Theta$, we find

$$B_2(\omega) = \frac{2}{q_\beta^2}[K(b) - J(b, q_\beta)], \tag{4.280}$$

where

$$b = \frac{\beta}{V_f}, \qquad q_\beta = \frac{\omega L}{\beta},$$

$$4K(b) = \int_{b-1}^{b+1} (a_0 y^{-2} + a_1 y^{-1} + a_2 + a_3 y + a_4 y^2)dy$$

$$= \frac{2a_0}{b^2 - 1} + a_1 \ln\frac{b+1}{b-1} + 2a_2 + 2ba_3 + \tfrac{2}{3}(1 + 3b^2)a_4,$$

$$4J(b, q) = \int_{b-1}^{b+1} (a_0 + a_1 y + a_2 y^2 + a_3 y^3 + a_4 y^4)\frac{\cos qy}{y^2}dy$$

$$= A_0^+ \cos[(b+1)q] + A_0^- \cos[(b-1)q] + A_1^+ \sin[(b+1)q]$$

$$+ A_1^- \sin[(b-1)q] + a_1\{Ci[(b+1)q] - Ci[(b-1)q]\}$$

$$- a_0 q\{Si[(b+1)q] - Si[(b-1)q]\},$$

$$Si(x) = \int_0^x \frac{\sin t}{t}dt, \qquad Ci(x) = \gamma + \ln x + \int_0^x \frac{\cos t - 1}{t}dt, \tag{4.281}$$

$$\gamma = 0.5772 \text{ (Euler's constant)},$$

$$A_0^+ = -\frac{a_0}{b+1} + \frac{1}{q^2}[a_3 + 2(b+1)a_4],$$

$$A_0^- = \frac{a_0}{b-2} - \frac{1}{q^2}[a_3 + 2(b-1)a_4],$$

$$A_1^+ = \frac{1}{q}\left[a_2 + (b+1)a_3 + \left\{(b+1)^2 - \frac{2}{q^2}\right\}a_4\right],$$

$$A_1^- = -\frac{1}{q}\left[a_2 + (b-1)a_3 + \left\{(b-1)^2 - \frac{2}{q^2}\right\}a_4\right].$$

The coefficients a_0, etc., are given in Table 4.10.

Table 4.10. Finite-Fault Energy Density Coefficients[a]

	B_1	B_2	B_3	B_4
a_0	$-4a^2(a^2-1)$	$4b^4-3b^2+1$	$(a^2-1)^2$	$-(b^4-1)$
a_1	$8a(2a^2-1)$	$-2b(8b^2-3)$	$-4a(a^2-1)$	$4b^3$
a_2	$-4(6a^2-1)$	$3(8b^2-1)$	$2(3a^2-1)$	$-6b^2$
a_3	$16a$	$-16b$	$-4a$	$4b$
a_4	-4	4	1	-1

[a] $b = \beta/V_f$, $a = \alpha/V_f$.

Although $K(b)$ and $J(b, q)$ are singular at $b = 1$, $B_2(\omega)$ is not singular there. In fact, we have

$$B_2(\omega; b=1) = \frac{1}{q_\beta^2}\left[5\{Ci(2q_\beta) - \ln(2q_\beta) - \gamma\} + q_\beta Si(2q_\beta) + \frac{1}{2}\cos 2q_\beta + \left(\frac{8}{q_\beta^2} - 5\right)\frac{\sin 2q_\beta}{2q_\beta} + \frac{59}{6} - \frac{8}{q_\beta^2}\right]. \quad (4.282)$$

The other B integrals can be evaluated in a similar way. Formula (4.280) is then valid with the appropriate a_i coefficients given in Table 4.10. In the cases of B_1 and B_3, the parameter b in the expressions for K and J is to be replaced by $a = \alpha/V_f$. For $B_4(\omega)$, we also have

$$B_4(\omega; b=1) = \frac{2}{q_\beta^2} \times \left[-Ci(2q_\beta) + \ln(2q_\beta) + \gamma + \left(1 - \frac{1}{q_\beta^2}\right)\frac{\sin 2q_\beta}{2q_\beta} + \frac{1}{q_\beta^2} - \frac{5}{3}\right]. \quad (4.283)$$

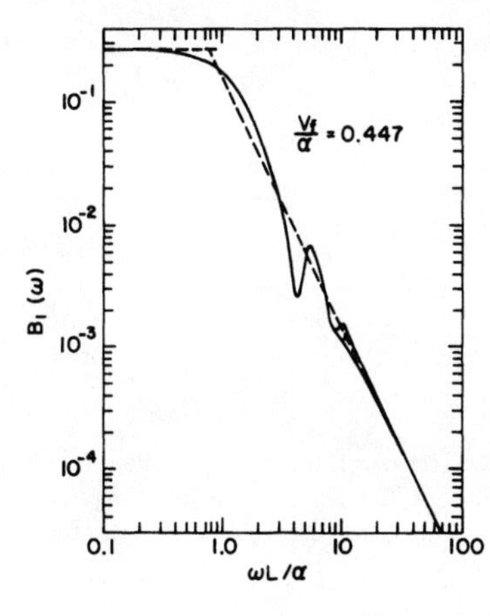

Figure 4.39. The function $B_1(\omega)$ vs. the dimensionless frequency $\omega L/\alpha$.

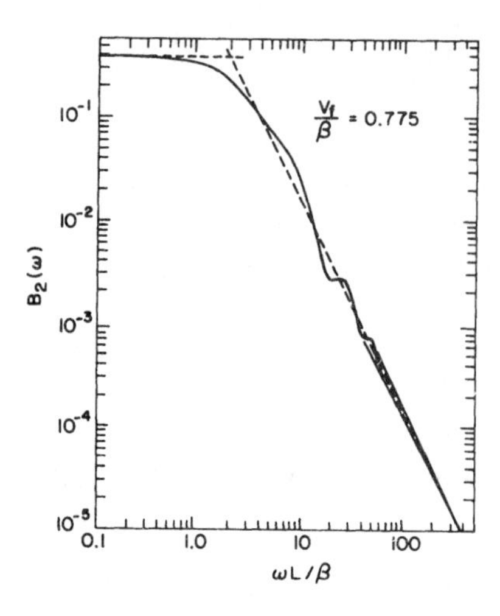

Figure 4.40. The function $B_2(\omega)$ vs. the dimensionless frequency $\omega L/\beta$.

The function $B_1(\omega)$ is plotted against the dimensionless frequency, $\omega L/\alpha$, in Fig. 4.39 on a log–log scale. Similarly, Fig. 4.40 shows the variation of $B_2(\omega)$ with $\omega L/\beta$. For the purpose of calculating spectral energy density, we have assumed that

$$g(t) = \begin{cases} 0, & t < 0 \\ t/T, & 0 \le t \le T \\ 1, & t > T \end{cases}$$

so that

$$g(\omega) = \frac{1}{\omega}\frac{\sin(\omega T/2)}{\omega T/2} e^{-i(\omega T/2) - i\pi/2}.$$

We have also assumed that

$$\frac{\beta T}{L} = 0.2, \qquad \alpha = \sqrt{3}\beta, \qquad \frac{V_f}{\beta} = 0.775.$$

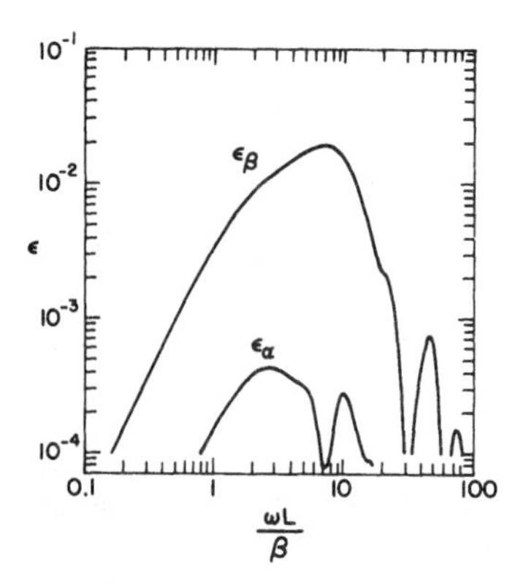

Figure 4.41. The spectral energy densities for P and S waves for the Ramp time function, with $\beta T/L = 0.2$ and $\alpha = \beta\sqrt{3}$ for a longitudinal shear fault ($\lambda = 0$).

The spectral energy density can now be calculated from Eq. (4.278). Figure 4.41 shows on a log–log scale ε_α and ε_β in units of $2\rho P_0^2/(\pi\beta T^2)$ against $\omega L/\beta$ for a longitudinal shear fault ($\lambda = 0$). The total spectral energy density is shown in Fig. 4.42 on a linear scale. The maximum in the spectral energy density occurs, approximately, at $\omega L/\beta = 7$. With $L = 700$ km and $\beta = 4.619$ km/s, the corresponding period would be about 135 s.

The asymptotic behavior of the energy density with respect to the frequency is important in the study of earthquake sources, because it can be inferred from the analysis of seismic waves. For $g(t) = [1 - \exp(-t/\tau)]H(t)$, we find that, as $\omega \to \infty$

$$
\begin{aligned}
&B_2(\omega) \to \frac{\pi}{2q}, && (b = 1) \\
&B_2(\omega) \to \frac{2K(b)}{q^2}, && (b > 1) \\
&\varepsilon_\beta(\omega) \to \left(\frac{\rho\bar{U}^2 L W^2}{4\tau^2}\right)\omega^{-1}, && (b = 1) \\
&\varepsilon_\beta(\omega) \to \left[\frac{\rho\beta\bar{U}^2 W^2 K(b)}{\pi\tau^2}\right]\omega^{-2}, && (b > 1) \\
&\varepsilon_\alpha(\omega) \to \left[\frac{\rho\alpha\bar{U}^2 W^2 K(a)}{\pi\tau^2}\right]\left(\frac{\beta}{\alpha}\right)^5 \omega^{-2}.
\end{aligned}
\qquad (4.284)
$$

Furthermore, as $\omega \to 0$

$$B_2(\omega) \to \tfrac{2}{5}, \qquad \varepsilon_\beta \to 0.$$

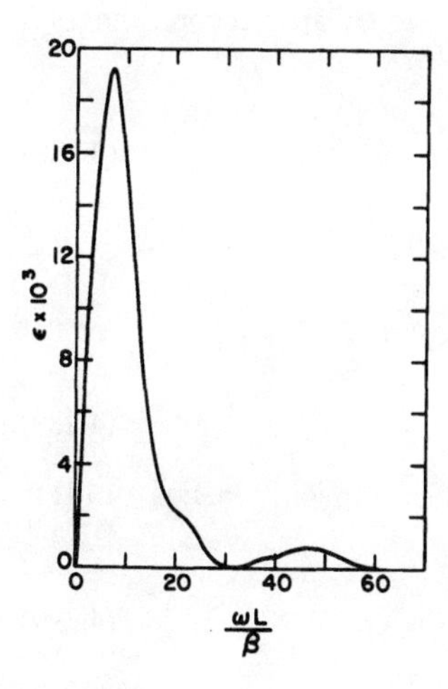

Figure 4.42. Total spectral energy density for the Ramp time function, with $\beta T/L = 0.2$ and $\alpha = \beta\sqrt{3}$ for a longitudinal shear fault.

In general, if $g(\omega) \to \omega^{-m}$ as $\omega \to \infty$, we have

$$\varepsilon_\beta \to \left(\frac{\rho \bar{U}^2 L W^2}{4\tau^{2m-2}}\right)\omega^{-(2m-3)}, \qquad (b = 1)$$
$$\varepsilon_\beta \to \left(\frac{\rho\beta \bar{U}^2 W^2 K(b)}{\pi\tau^{2m-2}}\right)\omega^{-(2m-2)}, \qquad (b > 1). \tag{4.285}$$

4.8. Radiation of Elastic Waves by Volume Sources

A region bounded by a closed surface S in an isotropic elastic medium undergoes a change of form resulting either from internal nuclei of strain (e.g., heat sources) or from external agents (e.g., tectonic forces). A local strain field $\mathfrak{E}^{(0)}$ is generated with the corresponding stress field

$$\mathfrak{T}^{(0)} = \lambda(\text{trace } \mathfrak{E}^{(0)})\mathfrak{I} + 2\mu\mathfrak{E}^{(0)}. \tag{4.286}$$

Let, at a certain time, a *stress relaxation* take place inside S. Let us assume that as far as an observer outside S is concerned, this relaxation is equivalent to the placing of a layer of body forces $(\mathfrak{T}^{(0)} \cdot \mathbf{n})$ on each element of S, where $\mathbf{n}$ denotes a unit normal drawn outwardly with respect to the inclusion. This force distribution generates a deformation field outside S (called henceforth "the medium"), characterized by displacements u_i^c, strains ε_{ij}^c, and stresses τ_{ij}^c, which we now wish to calculate.

Let $\mathbf{r}(x_1, x_2, x_3)$ be any point of the medium and $\mathbf{r}'(x_1', x_2', x_3')$ any point on S. Then the displacement field at $\mathbf{r}$ is given by the surface integral [cf. Eq. (4.73)]

$$\mathbf{u}^c(\mathbf{r}) = \int_S \mathbf{n}(\mathbf{r}') \cdot \mathfrak{T}^{(0)}(\mathbf{u}') \cdot \tilde{\mathfrak{G}}(\mathbf{r}|\mathbf{r}')dS(\mathbf{r}'). \tag{4.287}$$

Assuming that $\mathfrak{T}^{(0)}$ is *uniform* throughout the inclusion, Gauss' theorem enables one to recast Eq. (4.287) in the form

$$\mathbf{u}^c(\mathbf{r}) = \mathfrak{T}^{(0)} : \int_V (\nabla_{r'} \tilde{\mathfrak{G}})dV, \tag{4.288}$$

where V is the volume of the inclusion. However

$$\begin{aligned}\mathfrak{T}^{(0)} : \nabla_{r'} \tilde{\mathfrak{G}} &= [\lambda(\text{trace } \mathfrak{E}^{(0)})\mathfrak{I} + 2\mu\mathfrak{E}^{(0)}] : \nabla_{r'} \tilde{\mathfrak{G}}(\mathbf{r}|\mathbf{r}') \\ &= \mathfrak{E}^{(0)} : \mathfrak{T}_{r'}(\tilde{\mathfrak{G}}),\end{aligned} \tag{4.289}$$

where, for a homogeneous medium, $\mathfrak{T}_{r'}(\tilde{\mathfrak{G}})$ is given explicitly in Eqs. (4.141)–(4.146). Hence

$$\mathbf{u}^c(\mathbf{r}) = \mathfrak{E}^{(0)} : \int_V \mathfrak{T}_{r'}(\tilde{\mathfrak{G}})dV(\mathbf{r}'). \tag{4.290}$$

Clearly, for a point **r** far away from the inclusion,

$$\mathbf{u}^c(\mathbf{r}) \simeq [\mathfrak{E}^{(0)} : \mathfrak{T}_{r'}(\mathfrak{G})]V. \tag{4.291}$$

Now, the displacement at large distances from a Volterra dislocation of area S, normal **n**, and unit-slip vector **e** [Eq. (4.175)] is given by

$$\mathbf{u}(\mathbf{r}) \simeq U_0 S[\mathbf{en} : \mathfrak{T}_{r'}(\mathfrak{G})]. \tag{4.292}$$

Comparing Eqs. (4.291) and (4.292), we note that the far field induced by a Volterra dislocation is the same as the far field induced by an inclusion of arbitrary shape whose volume and "strain drop," $\mathfrak{E}^{(0)}$, satisfy the relation

$$V\mathfrak{E}^{(0)} = \tfrac{1}{2}(\mathbf{ne} + \mathbf{en})U_0 S. \tag{4.293}$$

The stress field induced outside the inclusion is derived from Eqs. (4.150) and (4.290)

$$\tau_{ij}^c(\mathbf{r}) = \varepsilon_{kl}^{(0)} \int_V H_{kl}^{ij}(\mathbf{r}, \mathbf{r}')dV(\mathbf{r}'). \tag{4.294}$$

In the case of a homogeneous medium, we use the value of $\mathfrak{G}(\mathbf{r}|\mathbf{r}')$ from Eq. (4.9). Equations (4.288) and (4.133) then yield

$$u_i^c(\mathbf{r}) = \frac{1}{4\pi\mu}\tau_{jk}^{(0)}\left[-\delta_{ij}\frac{\partial}{\partial x_k}\Psi_\beta + \frac{1}{k_\beta^2}\frac{\partial^3}{\partial x_i\,\partial x_j\,\partial x_k}(\Psi_\alpha - \Psi_\beta)\right], \tag{4.295}$$

where

$$\Psi_\alpha = \int_V \frac{e^{-ik_\alpha R}}{R}\,dV(\mathbf{r}'), \qquad \Psi_\beta = \int_V \frac{e^{-ik_\beta R}}{R}\,dV(\mathbf{r}'), \tag{4.296}$$

$$\frac{\partial\Psi}{\partial x_i} = -\frac{\partial\Psi}{\partial x_i'} = \Psi_{,i}, \qquad R = |\mathbf{r} - \mathbf{r}'|. \tag{4.297}$$

It may be noted that

$$(\nabla^2 + k_\alpha^2)\Psi_\alpha = 0 \quad \text{or} \quad -4\pi \tag{4.298}$$

according as the point **r** lies outside or inside V. Equations (4.286), (4.295), and (4.298) yield

$$u_i^c = \frac{1}{2\pi}\varepsilon_{jk}^{(0)}\left[-\frac{\sigma}{2(1-\sigma)}\delta_{jk}\Psi_{\alpha,i} - \delta_{ij}\Psi_{\beta,k} + \frac{1}{k_\beta^2}(\Psi_\alpha - \Psi_\beta)_{,ijk}\right]. \tag{4.299}$$

The corresponding equation for the stress is

$$\tau_{ij}^c = \frac{\mu}{2\pi}\varepsilon_{kl}^{(0)}\left[-\frac{\sigma}{1-\sigma}\delta_{ij}\left\{\frac{\sigma}{1-2\sigma}\delta_{kl}\Psi_{\alpha,nn} + \Psi_{\alpha,kl}\right\} - \frac{\sigma}{1-\sigma}\delta_{kl}\Psi_{\alpha,ij}\right.$$
$$\left. - (\delta_{il}\Psi_{\beta,jk} + \delta_{jl}\Psi_{\beta,ik}) + \frac{2}{k_\beta^2}(\Psi_\alpha - \Psi_\beta)_{,ijkl}\right]. \tag{4.300}$$

From Eq. (4.288), the dilatation and rotation are given by

$$(\text{div } \mathbf{u})^c = u^c_{i,i} = -\frac{1}{4\pi(1-\sigma)}\varepsilon^{(0)}_{jk}[\sigma\delta_{jk}\Psi_{\alpha,ii} + (1-2\sigma)\Psi_{\alpha,jk}], \quad (4.301)$$

$$(\text{curl } \mathbf{u})^c \equiv \tfrac{1}{2}(u^c_{i,j} - u^c_{j,i}) = \frac{1}{4\pi}[\varepsilon^{(0)}_{jk}\Psi_{\beta,ik} - \varepsilon^{(0)}_{ik}\Psi_{\beta,jk}]. \quad (4.302)$$

If $\mathfrak{E}^{(0)}$ happens to be a pure dilatation so that

$$\mathfrak{E}^{(0)} = \tfrac{1}{3}(\text{div } \mathbf{u})^{(0)}\mathfrak{I}, \quad (4.303)$$

we find from Eqs. (4.299)–(4.301)

$$u^c_i = -\frac{1}{12\pi}\frac{1+\sigma}{1-\sigma}(\text{div } \mathbf{u})^{(0)}\Psi_{\alpha,i}, \quad (4.304)$$

$$\tau^c_{ij} = -\frac{1}{12\pi}\frac{1+\sigma}{1-\sigma}(\text{div } \mathbf{u})^{(0)}[\lambda\delta_{ij}\Psi_{\alpha,mm} + 2\mu\Psi_{\alpha,ij}], \quad (4.305)$$

$$(\text{div } \mathbf{u})^c = -\frac{1}{12\pi}\frac{1+\sigma}{1-\sigma}(\text{div } \mathbf{u})^{(0)}\nabla^2\Psi_\alpha. \quad (4.306)$$

When $\mathfrak{T}^{(0)}$ is not uniform throughout V, Eq. (4.287) yields

$$u^c_i(\mathbf{r}) = \int_V \left[\tau^{(0)}_{jk}(\mathbf{r}')\frac{\partial}{\partial x'_k}G_{ij}(\mathbf{r}|\mathbf{r}') - \left\{\frac{\partial}{\partial x'_k}\tau^{(0)}_{jk}(\mathbf{r}')\right\}G_{ij}(\mathbf{r}|\mathbf{r}')\right]dV(\mathbf{r}'). \quad (4.307)$$

The elastic energy of the inclusion and the medium is evaluated by means of Eq. (1.62)

$$A = \frac{1}{2}\int_V [\mathfrak{E}^{(0)} : (\mathfrak{T}^{(0)} - \mathfrak{T}^c)]dV(\mathbf{r}), \quad (4.308)$$

where the integration is carried over the volume of the inclusion alone. If $\mathfrak{E}^{(0)}$ is a pure dilatation, we have, from Eqs. (4.298), (4.303), and (4.305),

$$A = \frac{2}{9}\mu\frac{1+\sigma}{1-\sigma}[(\text{div } \mathbf{u})^{(0)}]^2\left\{V - \frac{1}{8\pi}\frac{1+\sigma}{1-\sigma}k^2_\alpha\int_V \Psi_\alpha\, dV(\mathbf{r})\right\}. \quad (4.309)$$

Although the energy of the medium in the static case [$k_\alpha = 0$, compare with Eq. (1.86)] does not depend upon the shape of the inclusion, it does depend upon it in the dynamic case.

Bibliography

Bakun WH, Stewart RM, Bufe CG (1978) Directivity in the high-frequency radiation of small earthquakes. Bull Seismol Soc Amer 68: 1253–1263.

Ben-Menahem A (1961a) Radiation of seismic waves from propagating faults and the determination of earthquake fault-length and rupture velocity from the observed waves spectrums. PhD Thesis, California Institute of Technology, Pasadena.

Ben-Menahem A (1961b) Radiation of seismic surface waves from finite moving sources. Bull Seismol Soc Amer 51: 401–435.

Ben-Menahem A (1962) Radiation of seismic body waves from a finite moving source in the earth. Jour Geophys Res 67: 345–350.

Ben-Menahem A, Singh SJ (1968a) Eigenvector expansions of Green's dyads with applications to geophysical theory. Geophys Jour Roy Astron Soc (London) 16: 417–452.

Ben-Menahem A, Singh SJ (1968b) Multipolar elastic fields in a layered half-space. Bull Seismol Soc Amer 58: 1519–1572.

Ben-Menahem A, Singh SJ (1972) Computation of models of elastic dislocations in the earth. In: Bolt BA (ed) Methods in Computational Physics, Vol 12, p. 299–375. Academic Press, New York.

Berckhemer H (1962) Die Ausdehnung der Bruchfläche im Erdbebenherd und ihr Einfluss auf das seismische Wellenspektrum. Gerl Beit Geophys 71: 5–26.

Blake FG Jr (1952) Spherical wave propagation in solid media. Jour Acoust Soc Amer 24: 211–215.

Burridge R, Knopoff L (1964) Body force equivalents for seismic dislocations. Bull Seismol Soc Amer 54: 1875–1888.

Chinnery MA (1961) The deformation of the ground around surface faults. Bull Seismol Soc Amer 51: 355–372.

Chinnery MA (1963) The stress changes that accompany strike-slip faulting. Bull Seismol Soc Amer 53: 921–932.

De Hoop AT (1958) Representation theorems for the displacement in an elastic solid and their application to elastodynamic diffraction theory. DSc Thesis, Technische Hogeschool, Delft, The Netherlands, 84 pp.

Gangi AF (1970) A derivation of the seismic representation theorem using seismic reciprocity. Jour Geophys Res 75: 2088–2095.

Haskell NA (1964) Total energy and energy spectral density of elastic wave radiation from propagating faults. Bull Seismol Soc Amer 54: 1811–1841.

Hirasawa T, Stauder W (1965) On the seismic body waves from a finite moving source. Bull Seismol Soc Amer 55: 237–262.

Jarosch H, Aboodi E (1970) Towards a unified notation of source parameters. Geophys Jour Roy Astron Soc (London) 21: 513–529.

Knopoff L, Gangi AF (1959) Seismic reciprocity. Geophysics 24: 681–691.

Lamb H (1904) On the propagation of tremors over the surface of an elastic solid. Phil Trans Roy Soc (London) A. 203: 1–42.

Love AEH (1903) The propagation of wave-motion in an isotropic elastic solid medium. Proc Lond Math Soc (Ser 2) 1: 291–344.

Love, AEH (1944) A Treatise on the Mathematical Theory of Elasticity, 4th ed. Dover, New York.

Maruyama T (1963) On the force equivalents of dynamical elastic dislocations with reference to the earthquake mechanism. Bull Earthquake Res Inst (Tokyo) 41: 467–486.

Nakano H (1923) Notes on the nature of the forces which give rise to the earthquake motions. Seismol Bull Ctr Meteorol Obser Japan, Tokyo 1: 92–120.

Reid HF (1910) Elastic rebound theory. Univ Calif Publ, Bull Dept Geol Sci 6: 413–433.

Savage JC, Hasegawa HS (1965) A two-dimensional model study of the directivity function. Bull Seismol Soc Amer 55: 27–45.

Sharpe JA (1942) The production of elastic waves by explosion pressures. I. Theory and empirical field observations. II. Results of observations near an exploding charge. Geophysics 7: 144–154, 311–321.

Singh SJ (1973) Generation of *SH*-type motion by torsion-free sources. Bull Seismol Soc Amer 63: 1189–1200.

Singh SJ, Ben-Menahem A, Vered M (1973) A unified approach to the representation of seismic sources. Proc Roy Soc (London) A331: 525–551.

Starr AT (1928) Slip in a crystal and rupture in a solid due to shear. Proc Camb Phil Soc 24: 489–500.

Steketee JA (1958a) On Volterra's dislocations in a semi-infinite medium. Can Jour Phys 36: 192–205.

Steketee JA (1958b) Some geophysical applications of the elasticity theory of dislocations. Can Jour Phys 36: 1168–1198.

Stokes GG (1849) On the dynamical theory of diffraction. Trans Camb Phil Soc 9: 1. (Reprinted in Stokes' Math Phys Papers 2: 243–328, Cambridge, 1883.)

Stokes GG (1880) Propagation of an arbitrary disturbance in an elastic medium. Camb Math Papers 2: 257–280.

Volterra V (1907) Sur l'equilibre des corps elastiques multiplement connexes. Ann Sci Ecol Norm Super Paris 24: 401–517.

CHAPTER 5

Surface-Wave Amplitude Theory

I had a feeling once about Mathematics — that I saw it all. Depth beyond Depth was revealed to me — the Byss and the Abyss. I saw a quantity passing through infinity and changing its sign from plus to minus. I saw exactly how it happened and why the tergiversation was inevitable — but it was after dinner and I let it go.

(*Winston Churchill*)

5.1. Surface-Wave Amplitudes in Simple Configurations

In Section 3.6, we demonstrated the existence of Rayleigh waves in a homogeneous half-space and Love waves in a two-layered half-space. We also derived the general form of the dispersion relations for both Love and Rayleigh waves in a multilayered half-space. In this chapter we intend to obtain the amplitudes of surface waves in multilayered media excited by point dislocations.

We shall begin with the simpler case of the Rayleigh-wave field induced by a shear dislocation in a uniform half-space. We introduce a cylindrical coordinate system $(\mathbf{e}_\Delta, \mathbf{e}_\phi, \mathbf{e}_z)$ with the z axis drawn out of the medium (vertically upward) and a source at depth h. If we should prefer to take the z axis drawn into the medium (vertically downward), we would change z, $\mathbf{e}_z$, ϕ, and $\mathbf{e}_\phi$ to $-z$, $-\mathbf{e}_z$, $-\phi$, and $-\mathbf{e}_\phi$, respectively, in all the equations (Fig. 5.1).

Let $\mathbf{u}$ denote the spectral displacement at any point of the half-space, $z < 0$, and $(\mathbf{e}_z \cdot \mathfrak{T})$ the corresponding stress vector across a plane $z =$ const. in the absence of the source. We assume

$$\mathbf{u} = \int_0^\infty \sum_m \mathbf{u}_m(k) k\, dk,$$
$$\mathbf{e}_z \cdot \mathfrak{T}(\mathbf{u}) = \int_0^\infty \sum_m \mathbf{T}_m(k) k\, dk, \tag{5.1}$$

where the values that m can take depend upon the source. For dipolar sources, $m = 0, 1, 2$. Because $\mathbf{u}_m$ is a linear combination of the Hansen eigenvectors

$$\mathbf{u}_m = A_m \mathbf{L}_m^+ + B_m \mathbf{N}_m^+ + C_m \mathbf{M}_m^+, \tag{5.2}$$

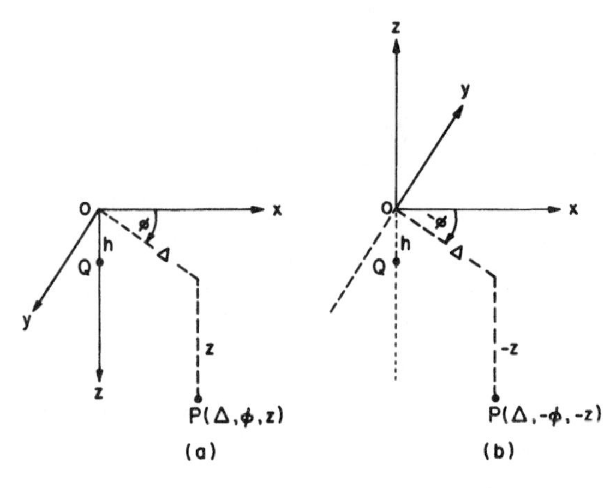

Figure 5.1. Cylindrical coordinate system with the $\bar{z}$ axis drawn vertically (a) downward and (b) upward.

Table 2.2 enables us to write

$$\mathbf{u}_m = \bar{x}_m \mathbf{P}_m + \bar{y}_m \mathbf{B}_m + \bar{z}_m \mathbf{C}_m,$$
$$\mathbf{T}_m = \bar{X}_m \mathbf{P}_m + \bar{Y}_m \mathbf{B}_m + \bar{Z}_m \mathbf{C}_m, \tag{5.3}$$

where

$$\bar{x}_m = A_m \nu_\alpha e^{\nu_\alpha z} + B_m k e^{\nu_\beta z}, \qquad \bar{X}_m = \mu(2k^2 - k_\beta^2) A_m e^{\nu_\alpha z} + 2\mu k \nu_\beta B_m e^{\nu_\beta z},$$
$$\bar{y}_m = A_m k e^{\nu_\alpha z} + B_m \nu_\beta e^{\nu_\beta z}, \qquad \bar{Y}_m = 2\mu k \nu_\alpha A_m e^{\nu_\alpha z} + \mu(2k^2 - k_\beta^2) B_m e^{\nu_\beta z},$$
$$\bar{z}_m = C_m e^{\nu_\beta z}, \qquad \bar{Z}_m = \mu \nu_\beta C_m e^{\nu_\beta z}, \tag{5.4}$$
$$\nu_\alpha = (k^2 - k_\alpha^2)^{1/2}, \qquad \nu_\beta = (k^2 - k_\beta^2)^{1/2}, \qquad k_\alpha = \frac{\omega}{\alpha}, \qquad k_\beta = \frac{\omega}{\beta}.$$

In Eqs. (5.2) and (5.3), A_m, B_m and C_m are arbitrary constants to be determined from the boundary and source conditions.

The displacement field resulting from a source situated in an infinite medium is expressible in terms of the Hansen eigenvectors. These results have been given in Table 4.4. Beginning with this representation, we can express the displacement and stress vectors, $\mathbf{u}^{(0)}$ and $\mathbf{e}_z \cdot \mathfrak{T}[\mathbf{u}^{(0)}]$, resulting from a source situated at the point $z = -h$ on the z axis in an infinite medium in terms of the vectors **P**, **B**, and **C**. We assume

$$\mathbf{u}^{(0)} = \int_0^\infty \sum_m \mathbf{u}_m^{(0)}(k) k \, dk,$$

$$\mathbf{e}_z \cdot \mathfrak{T}[\mathbf{u}^{(0)}] = \int_0^\infty \sum_m \mathbf{T}_m^{(0)}(k) k \, dk,$$

where

$$\mathbf{u}_m^{(0)} = \bar{x}_m^{(0)}\mathbf{P}_m + \bar{y}_m^{(0)}\mathbf{B}_m + \bar{z}_m^{(0)}\mathbf{C}_m,$$
$$\mathbf{T}_m^{(0)} = \bar{X}_m^{(0)}\mathbf{P}_m + \bar{Y}_m^{(0)}\mathbf{B}_m + \bar{Z}_m^{(0)}\mathbf{C}_m. \tag{5.5}$$

We are mainly interested in the field induced by a shear dislocation. It has been shown earlier [Eq. (4.120)] that the field induced by an arbitrary shear dislocation can be expressed as a linear combination of the fields resulting from three fundamental shear dislocations, namely, case I (vertical strike slip), case II (vertical dip slip), and case III (45° dip slip). From Table 4.4, we find

$$\sum_m \mathbf{u}_m^{(0)} = -\frac{k^2}{\nu_\alpha}\left(\frac{\beta}{\alpha}\right)\mathbf{L}_2^{s,(0)} - \varepsilon k\mathbf{N}_2^{s,(0)} + \frac{kk_\beta}{\nu_\beta}\mathbf{M}_2^{c,(0)}, \qquad \text{(case I)}$$

$$= -2\varepsilon k\left(\frac{\beta}{\alpha}\right)\mathbf{L}_1^{s,(0)} - \frac{2k^2 - k_\beta^2}{\nu_\beta}\mathbf{N}_1^{s,(0)} + \varepsilon k_\beta\mathbf{M}_1^{c,(0)}, \qquad \text{(case II)}$$

$$= \frac{3k^2 - 2k_\alpha^2}{2\nu_\alpha}\left(\frac{\beta}{\alpha}\right)\mathbf{L}_0^{c,(0)} + \frac{k^2}{2\nu_\alpha}\left(\frac{\beta}{\alpha}\right)\mathbf{L}_2^{c,(0)}$$

$$+ \frac{\varepsilon}{2}k[3\mathbf{N}_0^{c,(0)} + \mathbf{N}_2^{c,(0)}] + \frac{kk_\beta}{2\nu_\beta}\mathbf{M}_2^{s,(0)}, \qquad \text{(case III)} \tag{5.6}$$

using the notation of Eq. (4.158) with the modification that $|z|$ is replaced by $|z + h|$. Moreover,

$$\varepsilon = \begin{cases} +1, & z + h > 0 \\ -1, & z + h < 0. \end{cases}$$

In writing Eq. (5.6), a multiplying factor $[U_0\, dS/(4\pi k_\beta)]$ has been withheld. It will be reintroduced whenever necessary.

Sometimes, it is necessary to compare the field induced by a shear dislocation with the corresponding field induced by a center of compression. To this end, we find from Table 4.4 the following expression for the displacement field induced by a center of compression in an infinite medium

$$\sum_m \mathbf{u}_m^{(0)} = -\frac{Mk_\alpha}{4\pi(\lambda + 2\mu)\nu_\alpha}\mathbf{L}_0^{c,(0)}, \tag{5.7a}$$

where M is related to the pressure p_0 inside a spherical cavity of small radius, a, through the relation (Sect. 4.6.1)

$$M = \frac{3}{4}\left(\frac{\alpha}{\beta}\right)^2 E_0, \qquad E_0 = \frac{4\pi}{3}a^3 p_0. \tag{5.7b}$$

Expressing $\mathbf{L}_m^{\sigma,(0)}$, etc., in terms of $\mathbf{P}_m$, $\mathbf{B}_m$, and $\mathbf{C}_m$ with the help of Eq. (4.158), Eqs. (5.4)–(5.7) yield

$$
\begin{aligned}
\bar{x}_m^{(0)} &= -i_m \nu_\alpha \varepsilon^{m+1} e^{-\nu_\alpha |z+h|} + j_m k \varepsilon^{m+1} e^{-\nu_\beta |z+h|},\\
\bar{y}_m^{(0)} &= i_m k \varepsilon^m e^{-\nu_\alpha |z+h|} - j_m \nu_\beta \varepsilon^m e^{-\nu_\beta |z+h|},\\
\bar{z}_m^{(0)} &= k_m k_\beta \varepsilon^m e^{-\nu_\beta |z+h|},\\
\bar{X}_m^{(0)} &= \mu \varepsilon^m [i_m (2k^2 - k_\beta^2) e^{-\nu_\alpha |z+h|} - 2 j_m k \nu_\beta e^{-\nu_\beta |z+h|}],\\
\bar{Y}_m^{(0)} &= \mu \varepsilon^{m+1} [-2 i_m k \nu_\alpha e^{-\nu_\alpha |z+h|} + j_m (2k^2 - k_\beta^2) e^{-\nu_\beta |z+h|}],\\
\bar{Z}_m^{(0)} &= -\mu \varepsilon^{m+1} k_m k_\beta \nu_\beta e^{-\nu_\beta |z+h|}, \quad (m = 0, 1, 2).
\end{aligned}
\tag{5.8}
$$

The quantities i_m, j_m, and k_m are given in Table 5.1 for a shear dislocation. In the case of a center of compression, the only nonzero coefficient is

$$i_0 = -\frac{M}{4\pi(\lambda + 2\mu)\nu_\alpha}. \tag{5.9}$$

In Table 5.1, each value of i_m, j_m, and k_m is followed by either cos or sin. This corresponds to cos $m\phi$ and sin $m\phi$ in the definitions of the vectors $\mathbf{P}_m$, etc.:

$$\mathbf{P}_m^{c,s}(k\Delta, \phi) = \mathbf{e}_z J_m(k\Delta)(\cos m\phi, \sin m\phi).$$

The total field is $\mathbf{u} + \mathbf{u}^{(0)}$. Applying the condition of a free surface at $z = 0$

$$\mathbf{e}_z \cdot \mathfrak{T}[\mathbf{u} + \mathbf{u}^{(0)}] = 0 \quad \text{at} \quad z = 0, \tag{5.10}$$

we get

$$\bar{X}_m^{(0)} + \bar{X}_m = 0, \qquad \bar{Y}_m^{(0)} + \bar{Y}_m = 0, \qquad \bar{Z}_m^{(0)} + \bar{Z}_m = 0 \quad \text{at} \quad z = 0. \tag{5.11}$$

Table 5.1. Source Coefficients for Shear Dislocations[a]

Source	m	i_m		j_m		k_m	
Case I	2	$-\dfrac{k^2}{k_\beta \nu_\alpha}$	sin	$-\dfrac{k}{k_\beta}$	sin	$\dfrac{k}{\nu_\beta}$	cos
Case II	1	$-\dfrac{2k}{k_\beta}$	sin	$-\dfrac{2k^2 - k_\beta^2}{k_\beta \nu_\beta}$	sin	1	cos
Case III	0	$\dfrac{3k^2 - 2k_\alpha^2}{2k_\beta \nu_\alpha}$	cos	$\dfrac{3k}{2k_\beta}$	cos	0	—
	2	$\dfrac{k^2}{2k_\beta \nu_\alpha}$	cos	$\dfrac{k}{2k_\beta}$	cos	$\dfrac{k}{2\nu_\beta}$	sin

[a] A common factor $U_0\, dS/(4\pi k_\beta)$ has been withheld and the delta-function time dependence is assumed.

Solving for the coefficients A_m, B_m, and C_m and substituting the results in the expressions for the total displacement field, we find

$$\mathbf{u}^{\text{total}} = \int_0^\infty \sum_m \mathbf{u}_m^{\text{total}} k\, dk, \tag{5.12}$$

where

$$\mathbf{u}_m^{\text{total}} = U_P \mathbf{P}_m + U_B \mathbf{B}_m + U_C \mathbf{C}_m,$$

$$U_P = \nu_\alpha i_m \left\{ -(\varepsilon)^{m+1} e^{-\nu_\alpha |z+h|} - \frac{R^+(k)}{R(k)} e^{-\nu_\alpha (h-z)} + \frac{4k^2(2k^2 - k_\beta^2)}{R(k)} e^{-\nu_\alpha h + \nu_\beta z} \right\}$$

$$+ k j_m \left\{ (\varepsilon)^{m+1} e^{-\nu_\beta |z+h|} - \frac{R^+(k)}{R(k)} e^{-\nu_\beta (h-z)} + \frac{4\nu_\alpha \nu_\beta (2k^2 - k_\beta^2)}{R(k)} e^{-\nu_\beta h + \nu_\alpha z} \right\},$$

$$U_B = k i_m \left\{ (\varepsilon)^m e^{-\nu_\alpha |z+h|} - \frac{R^+(k)}{R(k)} e^{-\nu_\alpha (h-z)} + \frac{4\nu_\alpha \nu_\beta (2k^2 - k_\beta^2)}{R(k)} e^{-\nu_\alpha h + \nu_\beta z} \right\} \tag{5.13}$$

$$+ \nu_\beta j_m \left\{ -(\varepsilon)^m e^{-\nu_\beta |z+h|} - \frac{R^+(k)}{R(k)} e^{-\nu_\beta (h-z)} + \frac{4k^2(2k^2 - k_\beta^2)}{R(k)} e^{-\nu_\beta h + \nu_\alpha z} \right\},$$

$$U_C = k_\beta k_m [(\varepsilon)^m e^{-\nu_\beta |z+h|} + e^{-\nu_\beta (h-z)}],$$

$$R(k) = (2k^2 - k_\beta^2)^2 - 4k^2 \nu_\alpha \nu_\beta, \qquad R^+(k) = (2k^2 - k_\beta^2)^2 + 4k^2 \nu_\alpha \nu_\beta.$$

Each term on the right-hand side of the expressions for U_P, U_B, and U_C lends itself to a simple geometric interpretation. Substituting $k = k_\alpha \sin e = k_\beta \sin f$ [as in Eqs. (2.124) and (3.9)], we have $\nu_\alpha = ik \cot e$, $\nu_\beta = ik \cot f$. This yields

$$-\frac{(2k^2 - k_\beta^2)^2 + 4k^2 \nu_\alpha \nu_\beta}{(2k^2 - k_\beta^2)^2 - 4k^2 \nu_\alpha \nu_\beta} = \frac{\sin 2e \sin 2f - (\alpha/\beta)^2 \cos^2 2f}{\sin 2e \sin 2f + (\alpha/\beta)^2 \cos^2 2f} = R_{PP} = R_{SS},$$

$$\frac{4k^2(2k^2 - k_\beta^2)}{(2k^2 - k_\beta^2)^2 - 4k^2 \nu_\alpha \nu_\beta} = \frac{-2(\alpha/\beta) \sin 2e \cos 2f \sin f / \cos e}{\sin 2e \sin 2f + (\alpha/\beta)^2 \cos^2 2f} = \frac{\sin f}{\cos e} R_{PS}, \tag{5.14}$$

$$\frac{4\nu_\alpha \nu_\beta (2k^2 - k_\beta^2)}{(2k^2 - k_\beta^2)^2 - 4k^2 \nu_\alpha \nu_\beta} = \frac{(\alpha/\beta) \sin 4f \cos e / \sin f}{\sin 2e \sin 2f + (\alpha/\beta)^2 \cos^2 2f} = \frac{\cos e}{\sin f} R_{SP},$$

where R_{PP} and R_{PS} are the amplitude ratios of an incident P wave at a free boundary [see Eq. (3.10)] and R_{SP} and R_{SS} are the amplitude ratios of an incident SV wave at this boundary [see Eqs. (3.15) and (3.16)]. Therefore, for a fixed value of k the coefficients in the expressions for U_P and U_B are simply the plane-wave reflection coefficients for P and SV waves. These results are physically expected because it has been shown in Section 2.6 that plane waves can be superposed to form cylindrical as well as spherical waves.

The integral expressions for the displacements at $z = 0$ for the three fundamental shear dislocations are given in the following list.

1. Case I ($\lambda = 0°, \delta = 90°$)

$$u_z = \frac{U_0\,dS}{2\pi} \sin 2\phi \int_0^\infty \frac{k}{R(k)} J_2(k\Delta) f_2(k) dk,$$

$$u_\Delta = -\frac{U_0\,dS}{2\pi} \sin 2\phi \int_0^\infty \frac{k^2}{R(k)} \left[f_4(k) \frac{\partial J_2(k\Delta)}{\partial k\Delta} + \left\{ \frac{2R(k)}{\nu_\beta} \frac{J_2(k\Delta)}{k\Delta} \right\} e^{-h\nu_\beta} \right] dk,$$

$$u_\phi = -\frac{U_0\,dS}{2\pi} \cos 2\phi \int_0^\infty \frac{k^2}{R(k)} \left[2f_4(k) \frac{J_2(k\Delta)}{k\Delta} + \left\{ \frac{R(k)}{\nu_\beta} \frac{\partial J_2(k\Delta)}{\partial k\Delta} \right\} e^{-h\nu_\beta} \right] dk. \tag{5.15}$$

From Eq. (5.13), we also derive the vertical motion at any depth originating from a surface source ($h = 0$, $\varepsilon = -1$)

$$u_z = \frac{U_0\,dS}{2\pi} \sin 2\phi \int_0^\infty \frac{k^3}{R(k)} J_2(k\Delta) \{2\nu_\alpha \nu_\beta e^{z\nu_\alpha} - (2k^2 - k_\beta^2) e^{z\nu_\beta}\} dk. \tag{5.16}$$

2. Case II ($\lambda = 90°, \delta = 90°$)

$$u_z = \frac{-U_0\,dS}{\pi} \sin \phi \int_0^\infty \frac{k^2}{R(k)} \nu_\alpha (2k^2 - k_\beta^2) J_1(k\Delta) \{e^{-h\nu_\alpha} - e^{-h\nu_\beta}\} dk,$$

$$u_\Delta = \frac{U_0\,dS}{2\pi} \sin \phi \int_0^\infty \frac{k^2}{R(k)} \left[4\nu_\alpha \nu_\beta k \frac{\partial J_1(k\Delta)}{\partial k\Delta} e^{-h\nu_\alpha} - \left\{ \frac{(2k^2 - k_\beta^2)^2}{k} \frac{\partial J_1(k\Delta)}{\partial k\Delta} + \frac{J_1(k\Delta)}{k\Delta} \frac{R(k)}{k} \right\} e^{-h\nu_\beta} \right] dk, \tag{5.17}$$

$$u_\phi = \frac{U_0\,dS}{2\pi} \cos \phi \int_0^\infty \frac{k^2}{R(k)} \left[4\nu_\alpha \nu_\beta k \frac{J_1(k\Delta)}{k\Delta} e^{-h\nu_\alpha} - \left\{ \frac{(2k^2 - k_\beta^2)^2}{k} \frac{J_1(k\Delta)}{k\Delta} + \frac{\partial J_1(k\Delta)}{\partial k\Delta} \frac{R(k)}{k} \right\} e^{-h\nu_\beta} \right] dk.$$

3. Case III ($\lambda = 90°, \delta = 45°$)

$$u_z = \frac{-U_0\,dS}{4\pi} \int_0^\infty [f_1(k) J_0(k\Delta) + \cos 2\phi f_2(k) J_2(k\Delta)] \frac{k\,dk}{R(k)},$$

$$u_\Delta = \frac{U_0\,dS}{4\pi} \int_0^\infty \frac{k^2}{R(k)} \left[f_3(k) J_1(k\Delta) + \cos 2\phi \left\{ f_4(k) \frac{\partial J_2(k\Delta)}{\partial k\Delta} + \frac{2R(k)}{\nu_\beta} e^{-h\nu_\beta} \frac{J_2(k\Delta)}{k\Delta} \right\} \right] dk, \tag{5.18}$$

$$u_\phi = \frac{-U_0\,dS}{4\pi} \sin 2\phi \int_0^\infty \frac{k^2}{R(k)} \left[2f_4(k) \frac{J_2(k\Delta)}{k\Delta} + \left\{ \frac{R(k)}{\nu_\beta} \frac{\partial J_2(k\Delta)}{\partial k\Delta} \right\} e^{-h\nu_\beta} \right] dk,$$

where

$$
\begin{aligned}
f_1(k) &= -(2k^2 - k_\beta^2)(3k^2 - 2k_\alpha^2)e^{-h\nu_\alpha} + 6k^2\nu_\alpha\nu_\beta e^{-h\nu_\beta},\\
f_2(k) &= 2k^2\nu_\alpha\nu_\beta e^{-h\nu_\beta} - k^2(2k^2 - k_\beta^2)e^{-h\nu_\alpha},\\
f_3(k) &= 2\nu_\beta(3k^2 - 2k_\alpha^2)e^{-h\nu_\alpha} - 3\nu_\beta(2k^2 - k_\beta^2)e^{-h\nu_\beta},\\
f_4(k) &= -2k^2\nu_\beta e^{-h\nu_\alpha} + \nu_\beta(2k^2 - k_\beta^2)e^{-h\nu_\beta}.
\end{aligned}
\tag{5.19}
$$

Because we have taken the z axis as pointing vertically upward, the angle ϕ is measured counterclockwise from the strike of the fault (Fig. 5.1).

In the case of a center of compression, we get [cf. Eqs. (5.7b) and (5.9)]

$$
\begin{aligned}
u_z &= \frac{-Mk_\beta^2}{2\pi(\lambda + 2\mu)} \int_0^\infty k \frac{2k^2 - k_\beta^2}{R(k)} e^{-\nu_\alpha h} J_0(k\Delta)dk,\\
u_\Delta &= \frac{-Mk_\beta^2}{\pi(\lambda + 2\mu)} \int_0^\infty \frac{k^2\nu_\beta}{R(k)} e^{-\nu_\alpha h} J_1(k\Delta)dk,\\
u_\phi &= 0.
\end{aligned}
\tag{5.20}
$$

The integral expressions in Eqs. (5.15)–(5.20) render the total spectral radiation field generated by a point dislocation or a center of compression. As such, these must yield all waves known to us already, including Rayleigh, P, and S waves. To demonstrate that this is indeed so, we shall next evaluate these integrals. Using the relation

$$2\frac{dJ_m(x)}{dx} = J_{m-1}(x) - J_{m+1}(x),$$

the integrals in Eqs. (5.15)–(5.20) can be reduced to the form

$$I_m = \int_0^\infty J_m(k\Delta)\frac{F(k)}{R(k)}dk \qquad (m = 0, 1, 2, 3) \tag{5.21}$$

where $F(k)$ is an even function of k when m is odd and an odd function of k when m is even. We substitute in the integrand of Eq. (5.21)

$$J_m(k\Delta) = \tfrac{1}{2}[H_m^{(2)}(k\Delta) + H_m^{(1)}(k\Delta)], \qquad H_m^{(1)}(-k\Delta) = (-1)^{m+1}H_m^{(2)}(k\Delta),$$

where $H_m^{(1)}(k\Delta)$ and $H_m^{(2)}(k\Delta)$ are the Hankel functions of order m of the first and second kind, respectively. With this provision the integral in Eq. (5.21) assumes the form

$$I_m = \frac{1}{2}\int_{-\infty}^{\infty} H_m^{(2)}(k\Delta)\frac{F(k)}{R(k)}dk. \tag{5.22}$$

The integral has four branch points at $k = \pm k_\alpha$ and $k = \pm k_\beta$ and poles at the roots of the equation $R(k) = 0$. Introducing the notation $k = \hat{\gamma}k_\beta$, the equation $R(k) = 0$ yields

$$(2\hat{\gamma}^2 - 1)^2 = 4\hat{\gamma}^2\sqrt{(\hat{\gamma}^2 - 1)}\sqrt{\hat{\gamma}^2 - \frac{\beta^2}{\alpha^2}}, \tag{5.23}$$

which is identical with the Rayleigh equation, Eq. (3.101).

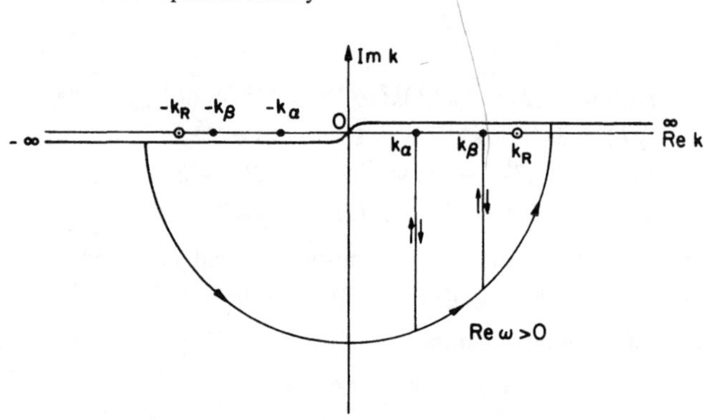

Figure 5.2. The contour of integration in the k plane for Re $\omega > 0$. k_R is the Rayleigh pole and $k_{\alpha,\beta}$ are branch points.

We evaluate I_m by employing Cauchy's residue theorem in the complex k plane. Deforming the integration contour as shown in Fig. 5.2, we end up with a single pole contribution at $k = k_R = \hat{\gamma} k_\beta$ and a contribution from the integration on both sides of the branch cuts through k_α and k_β. The pole contribution is

$$I_R = -\pi i H_m^{(2)}(k_R \Delta)\left[\frac{F(k)}{\partial R(k)/\partial k}\right]_{k=k_R}, \qquad \text{for} \quad \text{Re}\,\omega > 0. \tag{5.24}$$

For $\sigma = \frac{1}{4}$, $1/\hat{\gamma} = \sqrt{(2 - 2/\sqrt{3})}$ as obtained earlier in Eq. (3.104). Using the asymptotic approximation suitable for the far field

$$H_m^{(2)}(k_R \Delta) = \frac{\sqrt{2}}{(\pi k_R \Delta)^{1/2}} \exp\left[-i\left(k_R \Delta - \frac{\pi m}{2} - \frac{\pi}{4}\right)\right] + O\left[\frac{1}{(k_R \Delta)^{3/2}}\right] \tag{5.25}$$

and the result

$$\left.\frac{\partial R}{\partial k}\right|_{k=k_R} = k_\beta^3 G(\hat{\gamma}), \qquad G(\hat{\gamma}) = 2\,\frac{4\hat{\gamma}^2 - 1 - 8\hat{\gamma}^6(1 - \beta^2/\alpha^2)}{\hat{\gamma}(2\hat{\gamma}^2 - 1)^2}, \tag{5.26}$$

we can write the explicit form of the pole contribution of each of the field components. For example, the vertical motion for case I at the surface $z = 0$ is

$$u_z^R = \frac{U_0\, dS}{\sqrt{(2\pi\Delta)}} (\sin 2\phi) k_R^{3/2} \frac{\hat{\gamma}(2\hat{\gamma}^2 - 1)}{G(\hat{\gamma})}$$
$$\times \left[e^{-\gamma_\alpha k_R h} - \left(1 - \frac{c_R^2}{2\beta^2}\right) e^{-\gamma_\beta k_R h}\right] e^{i(\omega t - k_R \Delta) - (\pi i/4)}, \tag{5.27}$$

where $c_R = \omega/k_R$ is the Rayleigh-wave phase velocity and

$$\gamma_\alpha = \left(1 - \frac{c_R^2}{\alpha^2}\right)^{1/2}, \qquad \gamma_\beta = \left(1 - \frac{c_R^2}{\beta^2}\right)^{1/2}. \tag{5.28}$$

Similarly, the pole contribution to the vertical motion of case I for a surface source, at any depth z, obtained from Eq. (5.16) is

$$u_z^R = \frac{U_0\, dS}{\sqrt{(2\pi\Delta)}}(\sin 2\phi)k_R^{3/2}\frac{\hat{\gamma}(2\hat{\gamma}^2 - 1)}{G(\hat{\gamma})} \times \left[e^{\gamma_\beta k_R z} - \left(1 - \frac{c_R^2}{2\beta^2}\right)e^{\gamma_\alpha k_R z}\right]e^{i(\omega t - k_R\Delta) - (\pi i/4)}. \tag{5.29}$$

It may be noted that the dependence on the vertical coordinate z in Eq. (5.29) is the same as for free Rayleigh waves [Eq. (3.105)]. The pole at $k = k_R$ yields the surface Rayleigh waves and is therefore known as the *Rayleigh pole.* The differences between Eqs. (5.29) and (3.105) can be accounted for if we recede the point source to infinity in such a way that the radiations from it can be considered plane waves moving along the x axis. Note that our asymptotic approximation restricts the validity of Eqs. (5.27) and (5.29) to wavelengths that are much smaller than the epicentral distance, Δ. This approximation, however, is found to be sufficient for most seismologic applications. Note also that Rayleigh-wave amplitudes decay exponentially both with source depth and with the depth of the observation point relative to the free surface. The dependence of the Rayleigh-amplitudes on the azimuth relative to the fault's strike is known as the *radiation pattern* of the wave. Equation (5.27) predicts nodal lines at $\phi = 0, \pi/2, \pi$, and $3\pi/2$. Another important feature of Rayleigh waves is that their far-field spectral amplitudes over flat-earth models are inversly proportional to the square root of the distance. At this point we shall not treat the branch-line integrals. It will be shown later that they represent the P and SV body waves.

The same method that has led us to the derivation of Eqs. (5.16)–(5.20) can be used to derive integral expressions for displacements of more complicated half-space configurations. The next simple structure, which is very useful in simulating the earth's crust, is a single layer overlying a uniform half-space. The boundary conditions in this case yield nine equations in nine unknown amplitude coefficients. This set separates from the start into two independent sets: six equations for the coefficients of the coupled P–SV motion and three equations for the SH-type motion. The integral representations of the azimuthal component of the displacement at $z = 0$ for the SH motion are as follows:

1. Case I ($\lambda = 0°$, $\delta = 90°$)

$$u_\phi(h < H) = \frac{-U_0\, dS}{2\pi}\cos 2\phi\frac{\partial}{\partial\Delta}\int_0^\infty J_2(k\Delta)\frac{A^+(k, h, H)}{L(k, H)}\frac{k\, dk}{\nu_{1\beta}},$$
$$u_\phi(h > H) = \frac{-U_0\, dS}{\pi}\left(\frac{\mu_2}{\mu_1}\right)\cos 2\phi\frac{\partial}{\partial\Delta}\int_0^\infty J_2(k\Delta)e^{-\nu_{2\beta}(h-H)}\frac{k\, dk}{L(k, H)}; \tag{5.30}$$

2. Case II ($\lambda = 90°$, $\delta = 90°$)

$$u_\phi(h < H) = \frac{U_0\, dS}{2\pi}\cos\phi\frac{\partial}{\partial\Delta}\int_0^\infty J_1(k\Delta)\frac{A^-(k, h, H)}{L(k, H)}dk,$$
$$u_\phi(h > H) = \frac{-U_0\, dS}{\pi}\left(\frac{\mu_2}{\mu_1}\right)\cos\phi\frac{\partial}{\partial\Delta}\int_0^\infty J_1(k\Delta)e^{-\nu_{2\beta}(h-H)}\frac{\nu_{2\beta}\, dk}{L(k, H)}; \tag{5.31}$$

3. Case III ($\lambda = 90°$, $\delta = 45°$)

$$u_\phi(h < H) = \frac{-U_0\,dS}{4\pi}\sin 2\phi\,\frac{\partial}{\partial\Delta}\int_0^\infty J_2(k\Delta)\,\frac{A^+(k, h, H)}{L(k, H)}\,\frac{k\,dk}{\nu_{1\beta}},$$
$$u_\phi(h > H) = \frac{-U_0\,dS}{2\pi}\left(\frac{\mu_2}{\mu_1}\right)\sin 2\phi\,\frac{\partial}{\partial\Delta}\int_0^\infty J_2(k\Delta)e^{-\nu_{2\beta}(h-H)}\,\frac{k\,dk}{L(k, H)}; \tag{5.32}$$

where H is the layer thickness; μ_1 and μ_2 are the rigidities in the layer and in the half-space, respectively; β_1 and β_2 are the corresponding shear-wave velocities; and

$$\nu_{1\beta} = \left(k^2 - \frac{\omega^2}{\beta_1^2}\right)^{1/2}, \qquad \nu_{2\beta} = \left(k^2 - \frac{\omega^2}{\beta_2^2}\right)^{1/2},$$
$$L(k, H) = \left(\nu_{1\beta} + \frac{\mu_2}{\mu_1}\nu_{2\beta}\right)e^{H\nu_{1\beta}} - \left(\nu_{1\beta} - \frac{\mu_2}{\mu_1}\nu_{2\beta}\right)e^{-H\nu_{1\beta}},$$
$$A^+(k, h, H) = \left(\nu_{1\beta} + \frac{\mu_2}{\mu_1}\nu_{2\beta}\right)e^{(H-h)\nu_{1\beta}} + \left(\nu_{1\beta} - \frac{\mu_2}{\mu_1}\nu_{2\beta}\right)e^{-(H-h)\nu_{1\beta}}, \tag{5.33}$$
$$A^-(k, h, H) = -\left(\nu_{1\beta} + \frac{\mu_2}{\mu_1}\nu_{2\beta}\right)e^{(H-h)\nu_{1\beta}} + \left(\nu_{1\beta} - \frac{\mu_2}{\mu_1}\nu_{2\beta}\right)e^{-(H-h)\nu_{1\beta}}.$$

In all the three cases $u_z \equiv 0$ and the dependence of u_Δ on Δ in the far field is such that it is negligibly small as compared to u_ϕ.

We treat the *SH* integrals in Eqs. (5.30)–(5.32) as before. First, we note that the equation $L(k, H) = 0$ leads us to the dispersion relation for Love waves, derived earlier in Eq. (3.74), namely,

$$\tan[H\sqrt{(k_{\beta_1}^2 - k_L^2)}] = \frac{\mu_2}{\mu_1}\left[\frac{k_L^2 - k_{\beta_2}^2}{k_{\beta_1}^2 - k_L^2}\right]^{1/2}, \tag{5.34}$$

where

$$k_L = \frac{\omega}{c_L(\omega)}, \qquad k_{\beta_1} = \frac{\omega}{\beta_1}, \qquad k_{\beta_2} = \frac{\omega}{\beta_2}.$$

The condition $\beta_2 > c_L(\omega) > \beta_1$ (i.e., $k_{\beta_1} > k_L > k_{\beta_2}$) must be fulfilled in order to secure exponential decay with depth in the half-space and real roots of Eq. (5.34). Once again, the residues at the poles of the integrands in the *SH* displacement integrals will yield the contribution from Love waves. The remainder of the integrals can be interpreted as body-wave motion of the *SH* type. This is explained in detail in Chapter 7.

In Chapter 8 we shall treat the propagation of surface wave in a spherical-earth model of radius a. We shall then show rigorously that the dependence of the wave amplitude on the epicentral distance Δ (known as *geometric spread*) in the far field varies as $[a\sin(\Delta/a)]^{-1/2}$ and not as $\Delta^{-1/2}$, as is the case for a flat-earth model. The correction factor $[(\Delta/a)/\sin(\Delta/a)]^{1/2}$ results from *sphericity*. A simple geometric argument demonstrates this point (Fig. 5.3): Let the waves spread from a point source at O to an epicentral distance Δ. Because the initial energy spreads over a two-dimensional cap of circumference $2\pi a \sin(\Delta/a)$, the energy decay with distance is proportional to $[a \sin(\Delta/a)]^{-1}$. The amplitude decays, therefore, as $[a \sin(\Delta/a)]^{-1/2}$. In the case of a flat-earth model, the circumference of the circular region to which the energy spreads is $2\pi\Delta$ and, therefore, the amplitude decay is proportional to $\Delta^{-1/2}$. This agrees with Eqs. (5.27) and (5.29).

We now return to our integral expressions in Eqs. (5.15)–(5.20) and (5.30)–(5.32) and write the residue contribution of each of them. We introduce the notation

$$P_R = \frac{1}{4\pi a^2}(k_R a)^{3/2}\sqrt{(8\pi)}\left[e^{-\gamma_\alpha k_R h} - \left(1 - \frac{1}{2\hat{\gamma}^2}\right)e^{-\gamma_\beta k_R h}\right]\frac{\hat{\gamma}(2\hat{\gamma}^2 - 1)}{G(\hat{\gamma})},$$

$$Q_R = \frac{1}{4\pi a^2}(k_R a)^{3/2}\sqrt{(8\pi)}\,[e^{-\gamma_\alpha k_R h} - e^{-\gamma_\beta k_R h}]\frac{(2\hat{\gamma}^2 - 1)^3}{2G\hat{\gamma}^2\sqrt{(\hat{\gamma}^2 - 1)}},$$

$$S_R = \frac{1}{4\pi a^2}(k_R a)^{3/2}\sqrt{(8\pi)}\left[e^{-\gamma_\alpha k_R h} - \left(\frac{\hat{\gamma}^2 - 1/2}{\hat{\gamma}^2 - 2\beta^2/(3\alpha^2)}\right)e^{-\gamma_\beta k_R h}\right]$$
$$\times \frac{(2\hat{\gamma}^2 - 1)(3\hat{\gamma}^2 - 2\beta^2/\alpha^2)}{\hat{\gamma} G(\hat{\gamma})},$$

$$P_L(h < H) = \frac{1}{4\pi a^2}(k_L a)^{3/2}\sqrt{(8\pi)}\left[\frac{\sqrt{(k_L^2 - k_{\beta_2}^2)}\cos\{h\sqrt{(k_{\beta_1}^2 - k_L^2)}\}}{k_L \Phi}\right],$$

$$P_L(h > H) = \frac{1}{4\pi a^2}(k_L a)^{3/2}\sqrt{(8\pi)} \tag{5.35}$$
$$\times \left[\frac{(\mu_2/\mu_1)\sqrt{(k_L^2 - k_{\beta_2}^2)}\cos\{H\sqrt{(k_{\beta_1}^2 - k_L^2)}\}}{k_L \Phi}\right]e^{-(h-H)\sqrt{(k_L^2 - k_{\beta_2}^2)}},$$

$$Q_L(h < H) = \frac{1}{4\pi a^2}(k_L a)^{3/2}\sqrt{(8\pi)}$$
$$\times \left[\frac{\sqrt{(k_{\beta_1}^2 - k_L^2)}\sqrt{(k_L^2 - k_{\beta_2}^2)}\sin[h\sqrt{(k_{\beta_1}^2 - k_L^2)}]}{k_L^2 \Phi}\right],$$

$$Q_L(h > H) = \frac{1}{4\pi a^2}(k_L a)^{3/2}\sqrt{(8\pi)}$$
$$\times \left[\frac{\sqrt{(k_{\beta_1}^2 - k_L^2)}\sqrt{(k_L^2 - k_{\beta_2}^2)}\sin[H\sqrt{(k_{\beta_1}^2 - k_L^2)}]}{k_L^2 \Phi}\right]$$
$$\times e^{-(h-H)\sqrt{(k_L^2 - k_{\beta_2}^2)}},$$

$$\Phi = H\sqrt{(k_L^2 - k_{\beta_2}^2)} + \frac{(\mu_2/\mu_1)(k_{\beta_1}^2 - k_{\beta_2}^2)}{(k_{\beta_1}^2 - k_L^2) + (\mu_2/\mu_1)^2(k_L^2 - k_{\beta_2}^2)}.$$

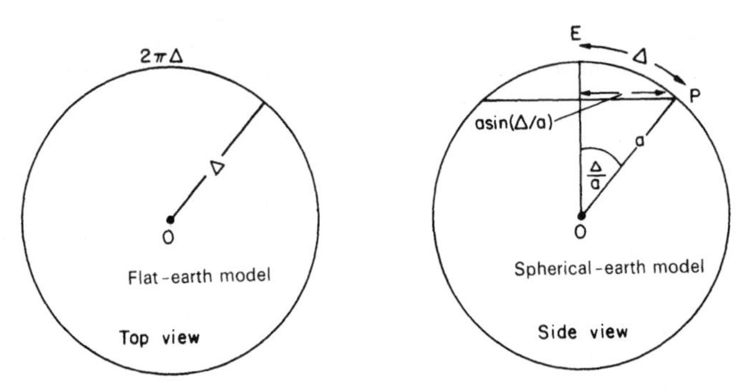

Figure 5.3. Geometric spreading in a flat-earth model and in a spherical-earth model.

With this notation, the vertical component of the far-field Rayleigh-wave displacement for case I, given in Eq. (5.27), may be written in the form

$$u_z^R = U_0\, dS \frac{e^{-\pi i/4}}{\sqrt{[\sin(\Delta/a)]}} P_R \sin 2\phi, \tag{5.36}$$

where the propagator $\exp[i(\omega t - k_R \Delta)]$ has been suppressed and the correction for the sphericity has been made. Similarly, the radial component of the displacement can be written in the form

$$u_\Delta^R = [\varepsilon(0) e^{\pi i/2}] u_z^R, \tag{5.37}$$

where $\varepsilon(0)$ is the Rayleigh-wave surface ellipticity, given in Eq. (3.109), which is independent of the source type or source depth. The azimuthal component of the far-field Rayleigh displacement is not important, because its amplitude is negligible in comparison with the amplitude of the vertical component. We have already mentioned that in the case of far-field Love-wave displacements for a two-layered earth model, the vertical component is identically zero and the radial component is negligible in comparison with the azimuthal component. Therefore, in the following we shall give only the vertical component of the far-field Rayleigh-wave displacement, u_z^R, at the free surface of a uniform half-space and the azimuthal component of the far-field Love-wave displacement, u_ϕ^L, in a two-layered earth model.

We derive first the displacement field for the three fundamental shear dislocations and then use Eq. (4.120) to get the field induced by an arbitrary shear dislocation. In this way, we obtain,

$$\begin{aligned} u_z^R &= [U_0\, dS] g(\omega) \frac{e^{-\gamma_R \Delta - \pi i/4}}{\sqrt{[\sin(\Delta/a)]}} [s_R S_R + p_R P_R + i q_R Q_R], \\ u_\phi^L &= [U_0\, dS] g(\omega) \frac{e^{-\gamma_L \Delta - 3\pi i/4}}{\sqrt{[\sin(\Delta/a)]}} [p_L P_L + i q_L Q_L], \end{aligned} \tag{5.38}$$

in which

$$\begin{aligned} s_R &= \sin\lambda\sin\delta\cos\delta, \\ q_R &= (\sin\lambda\cos 2\delta)\sin\phi + (\cos\lambda\cos\delta)\cos\phi, \\ p_R &= (\cos\lambda\sin\delta)\sin 2\phi - (\sin\lambda\sin\delta\cos\delta)\cos 2\phi, \\ q_L &= -(\cos\lambda\cos\delta)\sin\phi + (\sin\lambda\cos 2\delta)\cos\phi, \\ p_L &= (\sin\lambda\sin\delta\cos\delta)\sin 2\phi + (\cos\lambda\sin\delta)\cos 2\phi. \end{aligned} \tag{5.39}$$

In the case of a center of compression, Eq. (5.20) yields

$$u_z^R = \frac{M}{\lambda + 2\mu} g(\omega) \frac{e^{-\gamma_R\Delta - i\pi/4}}{\sqrt{[\sin(\Delta/a)]}} \left[-\frac{1}{4\pi a^2}(k_R a)^{3/2}\sqrt{(8\pi)}\,\frac{2\hat{\gamma}^2 - 1}{\hat{\gamma}G(\hat{\gamma})}\, e^{-\gamma_\alpha k_R h} \right], \tag{5.40}$$

taking the sphericity factor into account.

The following points about Eq. (5.38) are worth noting:

1. u_z^R represents the vertical (up) component of the Rayleigh-wave displacement and u_ϕ^L represents the azimuthal (counterclockwise) component of the Love-wave displacement.
2. The angle ϕ is the azimuth of the observer, measured counterclockwise from the strike of the fault.
3. $g(\omega)$ is the Fourier transform of the source time function. For a unit step source, $g(\omega) = 1/(i\omega)$.
4. The angles λ and δ are the strike and dip angles with the sign convention of Fig. 4.21.
5. The propagators $\exp[i(\omega t - k_R\Delta)]$ and $\exp[i(\omega t - k_L\Delta)]$ have been deleted in writing Eq. (5.38).
6. For Rayleigh waves, $u_\Delta^R = \varepsilon(0)e^{i\pi/2}u_z^R$, where $\varepsilon(0)$ is the surface ellipticity, given in Eq. (3.109). It is independent of the source type or source depth.
7. The factors $\exp(-\gamma_R\Delta)$ and $\exp(-\gamma_L\Delta)$ take care of the physical attenuation of the waves. The quantities $\gamma_R(\omega)$ and $\gamma_L(\omega)$ are the observed frequency-dependent *attenuation coefficients* for Rayleigh and Love waves, respectively.
8. The epicentral distance is usually measured in kilometers and the angle (Δ/a) is in radians. γ_R and γ_L are measured as per kilometer.
9. In the case of Rayleigh waves in a homogeneous half-space, there is only one pole at $k = k_R$. However, for Love waves in a two-layered half-space, the period equation (5.34) has an infinite number of roots at $k = k_L^{(n)}$, $n = 0, 1, 2, \ldots$, corresponding to the fundamental mode and overtones. P_L and Q_L are therefore functions of n as well.

Seismic stations are sometimes equipped with strain meters capable of recording ground strains. Because the stress vector across the free surface must vanish, and because Love waves are solenoidal and have no vertical component of motion, there are only four nonvanishing strain components at any field point on the free surface resulting from Love waves:

$$\varepsilon_{\Delta\Delta} = \frac{\partial u_\Delta}{\partial \Delta}, \qquad \varepsilon_{\phi\phi} = -\varepsilon_{\Delta\Delta}, \qquad \varepsilon_{\Delta\phi} = \varepsilon_{\phi\Delta} = \frac{1}{2}\left(\frac{\partial u_\phi}{\partial \Delta} - \frac{u_\phi}{\Delta} + \frac{1}{\Delta}\frac{\partial u_\Delta}{\partial \phi}\right). \tag{5.41}$$

In the far field, the only significant strain component is $\varepsilon_{\Delta\phi}$, as all other contributions are smaller by an order of magnitude in Δ. Therefore, from Eqs. (5.38) and (5.41)

$$\varepsilon_{\Delta\phi} \simeq \frac{1}{2}\frac{\partial u_\phi}{\partial \Delta} \simeq -\frac{i}{2} k_L u_\phi. \tag{5.42}$$

The extension at azimuth ϕ is accordingly $-ik_L u_\phi \sin\phi \cos\phi$ (see Sect. 1.2).

Similarly, the nonvanishing strain components on the free surface resulting from Rayleigh waves are, ε_{zz}, $\varepsilon_{\Delta\Delta}$, $\varepsilon_{\phi\phi}$ and $\varepsilon_{\Delta\phi} = \varepsilon_{\phi\Delta}$. They are connected through the relation

$$\varepsilon_{\Delta\Delta} + \varepsilon_{\phi\phi} = \text{areal strain} = -\frac{1-\sigma}{\sigma}\varepsilon_{zz} = \frac{1-\sigma}{1-2\sigma}\operatorname{div}\mathbf{u}. \tag{5.43}$$

The far-field approximation for the strain components then yields

$$\begin{aligned}
&\varepsilon_{\Delta\Delta} = -ik_R u_\Delta^R = k_R\,\varepsilon(0) u_z^R, \qquad \varepsilon_{\phi\phi} = 0, \qquad \varepsilon_{\Delta\phi} = 0,\\
&\varepsilon_{zz} = -k_R\,\varepsilon(0)\frac{\sigma}{1-\sigma}u_z^R,\\
&\operatorname{div}\mathbf{u} = k_R\,\varepsilon(0)\left(\frac{1-2\sigma}{1-\sigma}\right)u_z^R,
\end{aligned} \tag{5.44}$$

where $\varepsilon(0)$ is the surface ellipticity. The extension at azimuth ϕ is $k_R\,\varepsilon(0)u_z^R\cos^2\phi$.

EXAMPLE 5.1: Meteoritic Impact

Suppose that a meteorite, striking an atmosphereless planet at an angle θ with the vertical, has an effect that can be represented by a *point impulse* of magnitude K_0 dyne s. Assuming the medium to be modeled by a homogeneous elastic half-space, the vertical and horizontal surface displacements can be obtained from Eqs. (5.12) and (5.13) by inserting the values of i_m, etc., for the concentrated force obtainable from Table 4.4. We find

$$\begin{aligned}
u_z(\omega) &= \frac{K_0\cos\theta}{2\pi\mu}k_\beta^2\int_0^\infty J_0(k\Delta)\frac{\nu_\alpha k\,dk}{R(k)} - \frac{K_0\sin\theta\cos\phi}{2\pi\mu}\int_0^\infty J_1(k\Delta)\\
&\quad\times\frac{(2k^2-k_\beta^2)-2\nu_\alpha\nu_\beta}{R(k)}k^2\,dk,\\
u_\Delta(\omega) &= \frac{K_0\cos\theta}{2\pi\mu}\int_0^\infty J_1(k\Delta)\frac{(2k^2-k_\beta^2)-2\nu_\alpha\nu_\beta}{R(k)}k^2\,dk\\
&\quad-\frac{K_0\sin\theta\cos\phi}{2\pi\mu}k_\beta^2\int_0^\infty\left[J_0(k\Delta)-\frac{1}{k\Delta}J_1(k\Delta)\right]\frac{\nu_\beta k\,dk}{R(k)},\\
u_\phi(\omega) &= \frac{K_0\sin\theta\sin\phi}{2\pi\mu\Delta}k_\beta^2\int_0^\infty J_1(k\Delta)\frac{\nu_\beta\,dk}{R(k)},
\end{aligned} \tag{5.1.1}$$

where

$$R(k) = (2k^2-k_\beta^2)^2 - 4k^2\nu_\alpha\nu_\beta.$$

5.2. Generalization to a Vertically Inhomogeneous Half-Space

As in our previous treatment of the crustal response to plane body-waves, we cannot be satisfied with a model that is composed of a homogeneous half-space or a layer over a half-space, if we ultimately intend to apply our theoretical results to real-earth data. The earth's sphericity does not affect the calculated phase velocities in the period range 10–300 s by more than 0.5%, and a similar error is introduced in the calculation of the spectral amplitudes if we employ the geometric sphericity correction introduced earlier. However, using a homogeneous half-space model may introduce error in the calculated amplitudes as much as 50%. We must therefore generalize our previous results to take into account the vertical dependence of the structural parameters with depth. It will be assumed throughout our discussion that the seismic surface waves do not encounter horizontal discontinuities on their paths from the source to the recording station. By assuming different vertically inhomogeneous models we shall be able to model purely oceanic paths or purely continental paths. Mixed paths in principle can be treated only in an approximate way by the methods discussed in this section.

We shall describe two methods for the calculation of surface-wave amplitudes in vertically inhomogeneous media. The first method assumes that the structural parameters of the earth ρ, λ, and μ, are twice-differentiable functions of z. Our goal is to obtain explicit expressions for the *spectral transfer functions* of a vertically heterogeneous medium. We shall invoke the Rayleigh principle and express the results in terms of our previously derived plane-wave energy integrals. The theory will be presented in two stages. The *SH*- and Love-wave response will be obtained through the Green's dyadic and the Volterra relation. Then we shall give a different treatment for both Love and Rayleigh waves using the theory of inhomogeneous system of linear differential equations. In this way the reader is exposed to a variety of mathematical techniques that complement each other. We will continue to use a cylindrical coordinate system with the z axis drawn vertically upward.

5.2.1. Spectral *SH* Field

The Fourier-transformed equation of motion of the *SH* type in a vertically heterogeneous medium may be expressed in the form [see Eq. (3.79)]

$$\nabla^2 \mathbf{u} + \frac{\mu'}{\mu}\left[2\frac{\partial \mathbf{u}}{\partial z} + (\mathbf{e}_z \times \operatorname{curl} \mathbf{u})\right] + k_\beta^2 \mathbf{u} = 0, \tag{5.45}$$

where $\mathbf{u}$ is the spectral *SH*-displacement field, a prime denotes differentiation with respect to z, and $k_\beta = \omega/\beta(z)$. The substitution

$$\mathbf{u} = \operatorname{curl}(\mathbf{e}_z S) = \left(\mathbf{e}_\Delta \frac{1}{\Delta}\frac{\partial}{\partial \phi} - \mathbf{e}_\phi \frac{\partial}{\partial \Delta}\right) S, \tag{5.46}$$

leads to a scalar equation in the potential S

$$\nabla^2 S + \frac{\mu'}{\mu}\frac{\partial S}{\partial z} + k_\beta^2 S = 0. \tag{5.47}$$

Next, assuming

$$S(\Delta, \phi, z) = \int_0^\infty \sum_{m=-\infty}^{\infty} S_m\, kdk,$$

$$S_m = F(k, z) Y_m(k\Delta, \phi), \qquad Y_m(k\Delta, \phi) = J_m(k\Delta)e^{im\phi}, \tag{5.48}$$

we find that

$$\mathbf{u} = \int_0^\infty \sum_{m=-\infty}^{\infty} F(k, z)\mathbf{C}_m(k\Delta, \phi)k\, dk, \tag{5.49}$$

where

$$\mathbf{C}_m(k\Delta, \phi) = \left(\mathbf{e}_\Delta \frac{1}{k\Delta}\frac{\partial}{\partial\phi} - \mathbf{e}_\phi \frac{\partial}{\partial k\Delta}\right) Y_m(k\Delta, \phi). \tag{5.50}$$

Because

$$\nabla^2 S_m = \left(\frac{\partial^2}{\partial\Delta^2} + \frac{1}{\Delta}\frac{\partial}{\partial\Delta} + \frac{1}{\Delta^2}\frac{\partial^2}{\partial\phi^2} + \frac{\partial^2}{\partial z^2}\right) S_m = J_m(k\Delta)\left[\frac{d^2F}{dz^2} - k^2 F\right], \tag{5.51}$$

the partial differential equation (5.47) is reduced to an ordinary second-order equation in $F(k, z)$,

$$\frac{d^2F}{dz^2} + \frac{\mu'}{\mu}\frac{dF}{dz} + (k_\beta^2 - k^2)F = 0. \tag{5.52}$$

This equation may be written in the Sturm–Liouville form

$$\frac{d}{dz}\left(\mu \frac{dF}{dz}\right) - (\mu k^2 - \rho\omega^2)F = 0. \tag{5.53}$$

Introducing the functions

$y_1 = F$, (z factor of the displacement)

$y_2 = \mu \dfrac{dF}{dz}$, (z factor of the shear stress across a plane perpendicular to the z axis), (5.54)

we may replace Eq. (5.53) by a pair of coupled ordinary differential equations of the first order in the unknowns y_1 and y_2:

$$y_1' = \frac{1}{\mu} y_2,$$
$$y_2' = (\mu k^2 - \rho\omega^2) y_1. \tag{5.55}$$

When the medium is homogeneous, $\mu' = 0$ and Eq. (5.52) possesses the two independent solutions $F = e^{\nu_\beta z}, e^{-\nu_\beta z}$, where $\nu_\beta = \sqrt{(k^2 - k_\beta^2)}$. The displacements for this case are given by the familiar Hansen vectors $\mathbf{M}^+ = e^{\nu_\beta z}\mathbf{C}_m$, $\mathbf{M}^- = e^{-\nu_\beta z}\mathbf{C}_m$ (see Table 2.2). Therefore, we may generalize the concept of the Hansen vectors to inhomogeneous media and define

$$\mathbf{M}_m^+ = F^+(k, z)\mathbf{C}_m, \qquad \mathbf{M}_m^- = F^-(k, z)\mathbf{C}_m, \tag{5.56}$$

where $F^\pm(k, z) = y_1^\pm$ are the two solutions of Eq. (5.52) or the set (5.55), which reduce to $\exp(\pm\nu_\beta z)$ in the case of a homogeneous medium.

To continue the analogy between the two cases, we summon the *SH* part of the infinite-space Green's dyadic derived in Eq. (4.174) and write it in the form

$$\mathfrak{G}_\infty^L(\mathbf{z}|\mathbf{z}_0; \omega) = -\frac{1}{2\pi}\int_0^\infty \sum_{m=-\infty}^{\infty} \frac{1}{W}[\mathbf{M}_m^-(k, \mathbf{z}_>)\overset{*}{\mathbf{M}}{}_m^+(k, \mathbf{z}_<)]k\,dk, \tag{5.57}$$

where W is the Wronskian

$$W = y_1^+ y_2^- - y_1^- y_2^+, \tag{5.58}$$

(which is equal to $-2\mu\nu_\beta$ in the homogeneous case) and

$$\mathbf{z} = (\Delta, \phi, z), \qquad \mathbf{z}_0 = (\Delta_0, \phi_0, z_0), \qquad z_> = \max(z, z_0), \qquad z_< = \min(z, z_0).$$

It is clear that the form of the Green's dyadic for a vertically inhomogeneous medium is the same as in Eq. (5.57), with $\mathbf{M}_m$ from Eq. (5.56). We have already seen in Eq. (3.84) that W is independent of depth even for a vertically inhomogeneous medium. Seeking a Green's dyadic suitable for a half-space, we assume

$$\mathfrak{G}_H = \mathfrak{G}_\infty + \frac{1}{2\pi}\int_0^\infty \sum_{m=-\infty}^{\infty} \frac{1}{W(k)}\mathbf{M}_m^+(k, \mathbf{z})\mathbf{X}_m(k, \mathbf{z}_0)k\,dk, \tag{5.59}$$

where the superscript L has been deleted for convenience. $\mathbf{X}_m$ is an unknown source vector to be determined by the boundary condition

$$\mathbf{e}_z \cdot [\nabla\mathfrak{G}_H + (\nabla\mathfrak{G}_H)^{213}] = 0 \quad \text{at} \quad z = 0. \tag{5.60}$$

It is easily verified that

$$\mu\{\nabla\mathbf{M}_m(k, \mathbf{z}) + \mathbf{M}_m(k, \mathbf{z})\nabla\} \cdot \mathbf{e}_z = y_2(z)\mathbf{C}_m(k\Delta, \phi). \tag{5.61}$$

Hence, Eqs: (5.59)–(5.61) yield

$$\mathbf{X}_m(k, \mathbf{z}_0) = \frac{y_2^-(0)}{y_2^+(0)}\overset{*}{\mathbf{M}}{}_m^+(k, \mathbf{z}_0), \qquad y_2^\pm(z) = \mu\frac{dF^\pm}{dz}. \tag{5.62}$$

From Eqs. (5.57), (5.59), and (5.62), we have

$$\mathfrak{G}_H = -\frac{1}{2\pi}\int_0^\infty \sum_{m=-\infty}^{\infty} \frac{k\,dk}{W y_2^+(0)} \times \begin{cases} [y_2^+(0)\mathbf{M}_m^-(k, \mathbf{z}) - y_2^-(0)\mathbf{M}_m^+(k, \mathbf{z})]\overset{*}{\mathbf{M}}{}_m^+(k, \mathbf{z}_0), & z > z_0 \\ \mathbf{M}_m^+(k, \mathbf{z})[y_2^+(0)\overset{*}{\mathbf{M}}{}_m^-(k, \mathbf{z}_0) - y_2^-(0)\overset{*}{\mathbf{M}}{}_m^+(k, \mathbf{z}_0)], & z < z_0 \end{cases} \tag{5.63}$$

where

$$W(k) = y_1^+(0)y_2^-(0) - y_1^-(0)y_2^+(0). \tag{5.64}$$

Using Eqs. (5.56) and (5.64), the final form of the *spectral Green's dyadic* is

$$\mathfrak{G}_H(\mathbf{z}|\mathbf{z}_0) = -\frac{1}{2\pi}\int_0^\infty k\,dk \sum_{m=-\infty}^{\infty} \frac{\hat{y}_1(z)}{y_2^+(0)} \mathbf{C}_m(k\Delta, \phi)\overset{*}{\mathbf{C}}_m(k\Delta_0, \phi_0), \quad (5.65)$$

where

$$\hat{y}_1(z) = y_1^+(z_<)\left[\frac{y_2^+(0)y_1^-(z_>) - y_2^-(0)y_1^+(z_>)}{y_2^-(0)y_1^+(0) - y_2^+(0)y_1^-(0)}\right],$$

$$z_> = \max(z_0, z), \qquad z_< = \min(z_0, z), \quad (5.66)$$

$$\hat{y}_1(0) = -y_1^+(z_0).$$

In particular, at the free surface, the spectral Green's dyadic attains the simplified form

$$\mathfrak{G}_H(z=0) = \frac{1}{2\pi}\int_0^\infty k\,dk \sum_{m=-\infty}^{\infty} \left[\frac{y_1^+(z_0)}{y_2^+(0)}\right] \mathbf{C}_m(k\Delta, \phi)\overset{*}{\mathbf{C}}_m(k\Delta_0, \phi_0). \quad (5.67)$$

If $k = k_L$ is a root of the frequency equation, $y_2^+(k, 0) = 0$, we have

$$\hat{y}_1(k_L, z) = -\frac{y_1^+(k_L, z)y_1^+(k_L, z_0)}{y_1^+(k_L, 0)},$$

$$\hat{y}_1(k_L, 0) = -y_1^+(k_L, z_0). \quad (5.68)$$

Given the Green's dyadic, we can evaluate the displacement field resulting from a buried point dislocation at (Δ_0, ϕ_0, z_0), with the aid of the Volterra relation $[g(\omega) = 1]$

$$\mathbf{u} = \mu U_0\, dS[\mathbf{en}: \{\nabla_0 \mathfrak{G}_H + (\nabla_0 \mathfrak{G}_H)^{213}\}]. \quad (5.69)$$

We set a vector base $(\mathbf{e}_{\Delta_0}, \mathbf{e}_{\phi_0}, \mathbf{e}_{z_0})$ at the point source and assume, without loss of generality, that $\mathbf{n}$ is perpendicular to $\mathbf{e}_{\Delta_0}$, which is directed along the fault's strike. Therefore, according to Eq. (4.113) and Fig. 4.21, we have

$$\mathbf{e}_1^\circ = \mathbf{e}_{\Delta_0}, \qquad \mathbf{e}_2^\circ = \mathbf{e}_{\phi_0}, \qquad \mathbf{e}_3^\circ = \mathbf{e}_{z_0} = \mathbf{e}_z,$$

$$\mathbf{e} = \cos\lambda\,\mathbf{e}_{\Delta_0} + \sin\lambda\cos\delta\,\mathbf{e}_{\phi_0} + \sin\lambda\sin\delta\,\mathbf{e}_{z_0}, \quad (5.70)$$

$$\mathbf{n} = -\sin\delta\,\mathbf{e}_{\phi_0} + \cos\delta\,\mathbf{e}_{z_0}.$$

We shall next evaluate the displacement field resulting from the three fundamental shear dislocations in Table 4.3. Placing the source on the z axis at $z = z_0$ and taking the limit as $\Delta_0 \to 0$ and $\phi_0 \to 0$, we find

$$\mathbf{en}: \{\nabla_0[y_1(z_0)\overset{*}{\mathbf{C}}_m(k\Delta_0, \phi_0)] + [y_1(z_0)\overset{*}{\mathbf{C}}_m(k\Delta_0, \phi_0)]\nabla_0\}$$

$$= \frac{1}{2}ky_1(z_0)[\delta_{m,2} + \delta_{m,-2}], \qquad \text{for case I;}$$

$$= \frac{1}{2\mu_s}y_2(z_0)[\delta_{m,1} - \delta_{m,-1}], \qquad \text{for case II;}$$

$$= -\frac{1}{4}iky_1(z_0)[\delta_{m,2} - \delta_{m,-2}], \qquad \text{for case III,} \quad (5.71)$$

where μ_s is the value of μ at the source level. Equation (5.69) now yields the azimuthal displacement at the free surface for the three fundamental sources

$$
\begin{aligned}
u_\phi^{\mathrm{I}}(\omega) &= -\frac{\mu_s U_0\, dS}{2\pi}\cos 2\phi \frac{\partial}{\partial\Delta}\int_0^\infty \frac{y_1^+(z_0)}{y_2^+(0)} J_2(k\Delta)k\, dk,\\
u_\phi^{\mathrm{II}}(\omega) &= -\frac{U_0\, dS}{2\pi}\cos \phi \frac{\partial}{\partial\Delta}\int_0^\infty \frac{y_2^+(z_0)}{y_2^+(0)} J_1(k\Delta)dk, \qquad (5.72)\\
u_\phi^{\mathrm{III}}(\omega) &= -\frac{\mu_s U_0\, dS}{4\pi}\sin 2\phi \frac{\partial}{\partial\Delta}\int_0^\infty \frac{y_1^+(z_0)}{y_2^+(0)} J_2(k\Delta)k\, dk.
\end{aligned}
$$

As the inhomogeneous structure degenerates into a homogeneous half-space, $y_1^+(z)$ becomes $e^{\nu_\beta z}$ and the expressions given in Eq. (5.72) coincide with the *SH* part of the integrals in Eqs. (5.15)–(5.18), as they should.

Let us assume next that the vertical heterogeneity is such that Love waves exist. It means that the integrals in Eq. (5.72) may be considered a generalization of Eqs. (5.30)–(5.32). Then, the dispersion relation for Love waves reads

$$y_2^+(k_L, 0) = 0. \qquad (5.73)$$

Following our treatment in Section 5.1 and applying the residue theorem, we end up with *generalized amplitude transfer functions*

$$
\begin{aligned}
P_L &= \frac{1}{4\pi a^2}(k_L a)^{3/2}\sqrt{(8\pi)}\left[\frac{\mu_s y_1^+(z_0)}{(\partial/\partial k)y_2^+(0)}\right]_{k_L},\\
Q_L &= \left[\frac{y_2^+(z_0)}{\mu_s k y_1^+(z_0)}\right]_{k_L} P_L.
\end{aligned} \qquad (5.74)
$$

Because $y_2^+(k, z)$ is usually obtained in numerical form via a numerical integration of the equation pair (5.55), its derivative with respect to k_L is difficult to calculate. A simple transformation relieves us from this task and replaces the differentiation by integration with depth. Returning to our basic Eq. (5.52), we write it again for a particular value $k = k_L$

$$\frac{d^2F(k_L, z)}{dz^2} + \frac{\mu'}{\mu}\frac{dF(k_L, z)}{dz} + (k_\beta^2 - k_L^2)F(k_L, z) = 0. \qquad (5.75)$$

We multiply Eq. (5.52) by $F(k_L, z)$ and Eq. (5.75) by $F(k, z)$ and subtract. Simple manipulations then lead us to

$$\int_0^{-\infty} \mu(z)y_1^+(k, z)y_1^+(k_L, z)dz = \left[\frac{y_1^+(k_L, z)y_2^+(k, z) - y_1^+(k, z)y_2^+(k_L, z)}{k^2 - k_L^2}\right]_{z=0}^{z=-\infty}. \qquad (5.76)$$

Assuming that $y_1^+(k_L, -\infty) = y_2^+(k_L, -\infty) = 0$ and noting that $y_2^+(k_L, 0) = 0$, we get

$$-\int_0^{-\infty} \mu(z)y_1^+(k, z)y_1^+(k_L, z)dz = \frac{y_1^+(k_L, 0)y_2^+(k, 0)}{k^2 - k_L^2}. \qquad (5.77)$$

On taking the limit as $k \to k_L$ and noting that the medium occupies the region $z < 0$, we obtain

$$\frac{\partial}{\partial k_L} y_2^+(k_L, 0) = 2k_L y_1^+(k_L, 0) \int_{-\infty}^{\infty} \mu(z) \left[\frac{y_1^+(k_L, z)}{y_1^+(k_L, 0)}\right]^2 dz. \tag{5.78}$$

Equations (5.74) and (5.78) yield

$$P_L = \frac{1}{4\pi a^2} (k_L a)^{3/2} \sqrt{(8\pi)} \left[\frac{\mu_s \tilde{A}_L}{k_L}\right] \left[\frac{y_1^+(z_0)}{y_1^+(0)}\right], \tag{5.79}$$

$$Q_L = \left[\frac{1}{\mu_s k_L}\right] \left[\frac{y_2^+(z_0)}{y_1^+(z_0)}\right] P_L, \tag{5.80}$$

where

$$\frac{1}{\tilde{A}_L} = 2 \int_{-\infty}^{\infty} \mu(z) \left[\frac{y_1^+(z)}{y_1^+(0)}\right]^2 dz. \tag{5.81}$$

Note that the Love-wave field, which corresponds to the residues of the integrals in Eq. (5.72), is but part of the total seismic field. The *SH* body waves are contributed by the branch line integrals resulting from the branch points of the z eigenfunction (see Chap. 7).

5.2.2. Excitation of Love and Rayleigh Waves

Although the spectral Green's dyadic for the *SH* field can be derived with relative ease, the derivation of the *P–SV* counterpart by the same method is somewhat cumbersome. Because at present we are interested only in the medium response to surface waves, we in fact need only the Green's dyadic for Love and Rayleigh waves. In contradistinction to the spectral Green's dyadic, which governs the entire displacement field, we shall derive the *modal Green's dyadic* appropriate only for surface waves. We start with the general Fourier-transformed equation of motion resulting from a localized body force

$$\operatorname{div} \mathfrak{T}(\mathbf{u}) + \rho\omega^2 \mathbf{u} = -\mathbf{F}_0 g(\omega) \delta(\mathbf{z} - \mathbf{z}_0), \tag{5.82}$$

where $\mathbf{F}_0$ is a constant force and $g(\omega)$ is the Fourier transform of the force time dependence. This function shall be temporarily suppressed. This is equivalent to the assumption that the time dependence is that of a delta function. The solution of Eq. (5.82), with the appropriate boundary conditions, will yield the Green's dyadic because, by definition, the Green's dyadic is the medium's response to a localized single force.

We denote the *SH* field induced by the single force by $[y_1^L(z), y_2^L(z)]$ and the

P–SV field by $[y_1^R(z), y_2^R(z), y_3^R(z), y_4^R(z)]$. The solution of the homogeneous field equations can be represented in the form

$$\begin{aligned}\mathbf{u}(\mathbf{z}, \omega) &= \frac{1}{2\pi}\int_0^\infty k\,dk \sum_{m=-\infty}^{\infty} [y_1^R(z)\mathbf{P}_m(k\Delta, \phi) + y_3^R(z)\mathbf{B}_m(k\Delta, \phi) \\ &\quad + y_1^L(z)\mathbf{C}_m(k\Delta, \phi)], \\ \mathbf{e}_z \cdot \mathfrak{T}(\mathbf{u}) &= \frac{1}{2\pi}\int_0^\infty k\,dk \sum_{m=-\infty}^{\infty} [y_2^R(z)\mathbf{P}_m(k\Delta, \phi) + y_4^R(z)\mathbf{B}_m(k\Delta, \phi) \\ &\quad + y_2^L(z)\mathbf{C}_m(k\Delta, \phi)],\end{aligned} \tag{5.83}$$

where $\mathfrak{T}(\mathbf{u}) = \lambda\mathfrak{I}\,\text{div}\,\mathbf{u} + \mu(\nabla\mathbf{u} + \mathbf{u}\nabla)$. As we carry out the differentiation involved, we find that y_2^R and y_4^R are given in terms of the displacement amplitude functions y_1^R and y_3^R:

$$\begin{aligned} y_2^R &= (\lambda + 2\mu)\frac{dy_1^R}{dz} - k\lambda y_3^R \qquad \text{(normal stress amplitude)}, \\ y_4^R &= \mu\left[k y_1^R + \frac{dy_3^R}{dz}\right] \qquad \text{(tangential stress amplitude)}. \end{aligned} \tag{5.84}$$

Another pair of equations is derived by the substitution of the expansion of $\check{\mathbf{u}}$ into the equation of motion:

$$\begin{aligned} \frac{dy_2^R}{dz} &= -\rho\omega^2 y_1^R + k y_4^R, \\ \frac{dy_4^R}{dz} &= -\frac{k\lambda}{\lambda + 2\mu} y_2^R + \left[-\rho\omega^2 + 4k^2\frac{\mu(\lambda + \mu)}{\lambda + 2\mu}\right] y_3^R. \end{aligned} \tag{5.85}$$

We therefore note that the solution of the equation of motion, Eq. (5.82), with $\mathbf{F}_0 = 0$ is given by Eq. (5.83), provided (y_1^L, y_2^L) satisfy the set of Eqs. (5.55) and $(y_1^R, y_2^R, y_3^R, y_4^R)$ satisfy Eqs. (5.84) and (5.85), which may be expressed in the form of the set (3.123).

The representation of the source at $z = z_0$ is realized through the expansion

$$\begin{aligned}\mathbf{F}_0\delta(\mathbf{z} - \mathbf{z}_0) = \Bigg\{ \frac{1}{2\pi}\int_0^\infty \sum_{m=-\infty}^{\infty} & [f_2^R\mathbf{P}_m(k\Delta, \phi) + f_4^R\mathbf{B}_m(k\Delta, \phi) \\ & + f_2^L\mathbf{C}_m(k\Delta, \phi)]k\,dk\Bigg\}\delta(z - z_0).\end{aligned} \tag{5.86}$$

The orthogonality relations, Eq. (2.90), yield

$$\begin{aligned} f_2^R &= \mathbf{F}_0 \cdot \overset{*}{\mathbf{P}}_m(k\Delta_0, \phi_0), \\ f_4^R &= \mathbf{F}_0 \cdot \overset{*}{\mathbf{B}}_m(k\Delta_0, \phi_0), \\ f_2^L &= \mathbf{F}_0 \cdot \overset{*}{\mathbf{C}}_m(k\Delta_0, \phi_0). \end{aligned} \tag{5.87}$$

It is now easily seen that Eq. (5.83) represents the solution of the inhomogeneous Eq. (5.82) with $\mathbf{F}_0$ given by Eq. (5.86), provided the ys satisfy the inhomogeneous equation sets

$$\frac{d}{dz}\begin{bmatrix} y_1^L \\ y_2^L \end{bmatrix} = \begin{bmatrix} 0 & \mu^{-1} \\ \mu k^2 - \rho\omega^2 & 0 \end{bmatrix}\begin{bmatrix} y_1^L \\ y_2^L \end{bmatrix} - \begin{bmatrix} 0 \\ f_2^L \end{bmatrix}\delta(z - z_0), \tag{5.88}$$

$$\frac{d}{dz}\begin{bmatrix} y_1^R \\ y_2^R \\ y_3^R \\ y_4^R \end{bmatrix} = \begin{bmatrix} 0 & (\lambda + 2\mu)^{-1} & k\lambda(\lambda + 2\mu)^{-1} & 0 \\ -\rho\omega^2 & 0 & 0 & k \\ -k & 0 & 0 & \mu^{-1} \\ 0 & -k\lambda(\lambda + 2\mu)^{-1} & -\rho\omega^2 + 4\mu k^2(\lambda + \mu)/(\lambda + 2\mu) & 0 \end{bmatrix}$$

$$\times \begin{bmatrix} y_1^R \\ y_2^R \\ y_3^R \\ y_4^R \end{bmatrix} - \begin{bmatrix} 0 \\ f_2^R \\ 0 \\ f_4^R \end{bmatrix}\delta(z - z_0), \tag{5.89}$$

under the conditions

$$y_2^L = y_2^R = y_4^R = 0 \quad \text{at} \quad z = 0. \tag{5.90}$$

Both systems of equations have the general form

$$\frac{dy}{dz} = P(z)y - f\delta(z - z_0), \tag{5.91}$$

where $P(z)$ is a 2×2 or 4×4 matrix, whereas y and f are 2×1 or 4×1 column matrices. Let $\Omega(z, \zeta)$ be the matrizant of $P(z)$ and let the initial value of y be given at $z = z_1 < z_0$. Then the solution is [Eq. (3.213)]

$$\begin{aligned} y(z) &= \Omega(z, z_1)y(z_1) - \int_{z_1}^{z} \Omega(z, \zeta) f\delta(\zeta - z_0)d\zeta \\ &= \Omega(z, z_1)y(z_1) - \Omega(z, z_0) f H(z - z_0), \end{aligned} \tag{5.92}$$

where $H(z)$ is the unit-step function. Equation (5.92) simply restates the well-known theorem that the general solution of the inhomogeneous equation set is composed of the continuous solution $\Omega(z, z_1)$ of the homogeneous set and another solution satisfying the source condition. We note that

$$y(z_0 + 0) - y(z_0 - 0) = -f. \tag{5.93}$$

Therefore, we can either solve the inhomogeneous Eq. (5.91) or solve the homogeneous equation obtained from Eq. (5.91), by putting $f = 0$, and apply the source condition, Eq. (5.93). The latter method avoids the direct evaluation of the matrizant.

We shall next explain the procedure for *SH* waves. Let $y^+(z)$ denote the solution of (5.91) with $f = 0$, which is bounded at infinity, and $\tilde{y}(z)$ be another solution of the same equation. We put

$$y(z) = Ay^+(z) + \tilde{y}(z)H(z - z_0), \tag{5.94}$$

where A is a constant of integration. Equation (5.94) will satisfy the source condition, Eq. (5.93), and the boundary condition, Eq. (5.90), if

$$\tilde{y}(z_0) = -f, \qquad Ay_2^+(0) + \tilde{y}_2(0) = 0. \tag{5.95}$$

Equations (5.94) and (5.95) yield

$$y_j(z) = -\frac{\tilde{y}_2(0)}{y_2^+(0)} y_j^+(z) + \tilde{y}_j(z)H(z - z_0) \qquad (j = 1, 2). \tag{5.96}$$

The solution, Eq. (5.96), is now substituted into Eq. (5.83) and the residues are taken at the poles of the integrand according to the scheme outlined in Section 5.1. Using variants of Eqs. (5.24) and (5.78) and noting that the term $\tilde{y}_1(z)H(z - z_0)$ does not contribute anything to the residue, we find

$$\mathbf{u} = \frac{i}{4} \sum_{n=0}^{\infty} \sum_{m=-\infty}^{\infty} \frac{y_1^+(0)\tilde{y}_2(0)}{I_1^L} y_1^+(z)\left(\mathbf{e}_\Delta \frac{1}{k_n\Delta}\frac{\partial}{\partial\phi} - \mathbf{e}_\phi \frac{\partial}{\partial k_n\Delta}\right)H_m^{(2)}(k_n\Delta)e^{im\phi}$$

$$(\mathrm{Re}\,\omega > 0), \tag{5.97}$$

where

$$I_1^L = \int_{-\infty}^{\infty} \mu(z)[y_1^+(z)]^2\, dz, \tag{5.98}$$

and k_n are the roots of the equation $y_2^+(0) = 0$. Because $y_1^+(z)\tilde{y}_2(z) - \tilde{y}_1(z)y_2^+(z)$ is independent of z [Eq. (3.84)] and $\tilde{y}_1(z_0) = -f_1^L$, $\tilde{y}_2(z_0) = -f_2^L$ [Eq. (5.95)], we have

$$y_1^+(0)\tilde{y}_2(0) - \tilde{y}_1(0)y_2^+(0) = f_1^L y_2^+(z_0) - f_2^L y_1^+(z_0). \tag{5.99}$$

Noting that $y_2^+(0) = 0$ at $k = k_n$, Eqs. (5.87), (5.97), and (5.99) yield

$$\mathbf{u} = \frac{i}{4} \sum_{n=0}^{\infty} \sum_{m=-\infty}^{\infty} \frac{1}{I_1^L} [f_1^L y_2^+(z_0) - f_2^L y_1^+(z_0)] y_1^+(z)$$

$$\times \left(\mathbf{e}_\Delta \frac{1}{k_n\Delta}\frac{\partial}{\partial\phi} - \mathbf{e}_\phi \frac{\partial}{\partial k_n\Delta}\right)H_m^{(2)}(k_n\Delta)e^{im\phi}. \tag{5.100}$$

Equation (5.100) is the Love-wave displacement field for an arbitrary dipolar source. The source coefficients f_1^L and f_2^L are defined as [see Eq. (5.93)]

$$y_j^L(z_0 + 0) - y_j^L(z_0 - 0) = -f_j^L \qquad (j = 1, 2). \tag{5.101}$$

These represent jumps in the displacement and stress vectors across the plane $z = z_0$. These jumps were calculated in Section 4.5.4 for an arbitrary shear dislocation on the z axis. From Eqs. (4.182), (4.183), (5.100), and (5.101), we find $[g(\omega) = 1]$

$$\mathbf{u} = (U_0\, dS)\frac{i}{4} \sum_{n=0}^{\infty} \sum_{m=0}^{2} \sum_{\sigma=c,s} \frac{1}{I_1^L} [\mu_s k_n \bar{c}_m^\sigma y_1^+(z_0) - c_m^\sigma y_2^+(z_0)]$$

$$\times\, y_1^+(z)\left(\mathbf{e}_\Delta \frac{1}{k_n\Delta}\frac{\partial}{\partial\phi} - \mathbf{e}_\phi \frac{\partial}{\partial k_n\Delta}\right)H_m^{(2)}(k_n\Delta)(\cos m\phi, \sin m\phi), \tag{5.102}$$

where $\sigma = c(\cos m\phi)$ or $s(\sin m\phi)$ and c_m^σ, $\bar{c}_m^\sigma$ are given by Eq. (4.184). In the far field, we use Eq. (5.25) to obtain

$$u_\phi^L = -(U_0\, dS)\frac{i}{4}\sum_{n=0}^{\infty}\sum_{m=0}^{2}\sum_{\sigma=c,s}\frac{1}{I_1^L}[\mu_s k_n \bar{c}_m^\sigma y_1^+(z_0) - c_m^\sigma y_2^+(z_0)]$$
$$\times\, y_1^+(z)(\cos m\phi, \sin m\phi)\sqrt{(2/\pi k_n \Delta)}e^{i\pi(m/2-1/4)}, \tag{5.103}$$

where the propagator $\exp[i(\omega t - k_n\Delta)]$ is understood.

Inserting the values of c_m^σ and $\bar{c}_m^\sigma$ from Eq. (4.184) and suppressing the summation over n, Eq. (5.103) can be expressed in the form of Eq. (5.38), with

$$P_L = \frac{1}{4\pi a^2}(k_L a)^{3/2}\sqrt{(8\pi)}\left[\frac{\mu_s \tilde{A}_L}{k_L}\right]\left[\frac{y_1^+(z_0)}{y_1^+(0)}\right]\left[\frac{y_1^+(z)}{y_1^+(0)}\right],$$
$$Q_L = \left[\frac{y_2^+(z_0)}{\mu_s k_L y_1^+(z_0)}\right]P_L, \tag{5.104a}$$

where

$$\frac{1}{\tilde{A}_L} = 2\int_{-\infty}^{\infty}\mu(z)\left[\frac{y_1^+(z)}{y_1^+(0)}\right]^2 dz = 2c_L U_L\int_{-\infty}^{\infty}\rho(z)\left[\frac{y_1^+(z)}{y_1^+(0)}\right]^2 dz, \tag{5.104b}$$

[cf. Eq. (3.129)]. On taking $z = 0$, we get back Eq. (5.79).

In the case of a single force [Eq. (5.87)], $f_1^L = 0$, $f_2^L = \mathbf{F}_0 \cdot \overset{*}{\mathbf{C}}_m(k\Delta_0, \phi_0)$. Putting this into Eq. (5.100), we get the normal-mode Love-wave Green's dyadic for a half-space

$$\mathfrak{G}^L(\mathbf{z}|\mathbf{z}_0) = -\frac{i}{4}\sum_{n=0}^{\infty}\sum_{m=-\infty}^{\infty}\frac{y_1^+(z)y_1^+(z_0)}{I_1^L}\mathbf{C}_m^{(2)}(k_n\Delta, \phi)\overset{*}{\mathbf{C}}_m(k_n\Delta_0, \phi_0), \tag{5.105}$$

where $\mathbf{C}_m^{(2)}$ is obtained from $\mathbf{C}_m$ on replacing $J_m(k\Delta)$ by $H_m^{(2)}(k\Delta)$. With the aid of Eq. (3.129), I_1^L may be replaced by

$$c_L U_L I_0^L = c_L U_L\int_{-\infty}^{\infty}\rho(z)[y_1^+(z)]^2\, dz,$$

where U_L is the group velocity. It is instructive to compare the normal-mode Green's dyadic, Eq. (5.105), with the spectral Green's dyadic, Eq. (5.65).

The analysis of the Rayleigh-wave case is similar to that of Love waves. We find

$$\mathbf{u} = \frac{i}{4}\sum_{n=0}^{\infty}\sum_{m=-\infty}^{\infty}\frac{1}{c_R U_R I_0^R}[(f_1^R y_2^+ - f_2^R y_1^+) + (f_3^R y_4^+ - f_4^R y_3^+)]_{z_0}$$
$$\times\, \{y_1^+(z)\mathbf{P}_m^{(2)}(k_n\Delta, \phi) + y_3^+(z)\mathbf{B}_m^{(2)}(k_n\Delta, \phi)\} \qquad (\mathrm{Re}\,\omega > 0), \tag{5.106}$$

where $\mathbf{P}_m^{(2)}$ and $\mathbf{B}_m^{(2)}$ are obtained from $\mathbf{P}_m$ and $\mathbf{B}_m$ by replacing the Bessel function of the first kind by the Hankel function of the second kind, k_n are the roots of the Rayleigh characteristic equation, and

$$-f_j^R = y_j^R(z_0 + 0) - y_j^R(z_0 - 0) \qquad (j = 1, 2, 3, 4). \tag{5.107}$$

Moreover,

$$I_0^R = \int_{-\infty}^{\infty} \rho(z)[\{y_1^+(z)\}^2 + \{y_3^+(z)\}^2]dz.$$

Using Eqs. (4.182) and (4.183), Eq. (5.106) yields the following expression for the displacement field induced by an arbitrary shear dislocation on the z axis

$$\mathbf{u} = (U_0\, dS)\frac{i}{4}\sum_{n=0}^{\infty}\sum_{m=0}^{2}\sum_{\sigma=c,\,s}\frac{1}{c_R U_R I_0^R}[\mu_s k_n(\bar{p}_m^\sigma y_1^+ + \bar{b}_m^\sigma y_3^+) - (p_m^\sigma y_2^+ + b_m^\sigma y_4^+)]_{z_0}\{y_1^+(z)\mathbf{P}_m^{\sigma(2)}(k_n\Delta, \phi) + y_3^+(z)\mathbf{B}_m^{\sigma(2)}(k_n\Delta, \phi)\}. \quad (5.108)$$

p_m^σ, etc., are given in Eq. (4.184). Therefore, in the far field, we have

$$u_z^R = (U_0\, dS)\frac{i}{4}\sum_{n=0}^{\infty}\sum_{m=0}^{2}\sum_{\sigma=c,\,s}\frac{1}{c_R U_R I_0^R}[\mu_s k_n(\bar{p}_m^\sigma y_1^+ + \bar{b}_m^\sigma y_3^+) - (p_m^\sigma y_2^+ + b_m^\sigma y_4^+)]_{z_0} y_1^+(z)(\cos m\phi, \sin m\phi)\sqrt{(2/\pi k_n\Delta)}e^{i\pi(m/2+1/4)},$$

$$u_\Delta^R = \left[\frac{y_3^+(z)}{y_1^+(z)}e^{-i\pi/2}\right]u_z^R. \quad (5.109)$$

On inserting the values of the source coefficients from Eq. (4.184), making the sphericity correction, and expressing u_z^R in the form given in Eq. (5.38), we find

$$P_R = \frac{1}{4\pi a^2}(k_R a)^{3/2}\sqrt{(8\pi)}\left[\frac{\mu_s \tilde{A}_R}{k_R}\right]\left[\frac{y_3^+(z_0)}{y_1^+(0)}\right]\left[\frac{y_1^+(z)}{y_1^+(0)}\right],$$

$$Q_R = \frac{1}{\mu_s k_R}\left[\frac{y_4^+(z_0)}{y_3^+(z_0)}\right]P_R, \quad (5.110)$$

$$S_R = \left[\frac{1+\sigma_s}{1-\sigma_s} + \frac{1-2\sigma_s}{1-\sigma_s}\frac{1}{\mu_s k_R}\left\{\frac{y_2^+(z_0)}{y_3^+(z_0)}\right\}\right]P_R,$$

where σ_s is the Poisson ratio at source level and

$$\frac{1}{\tilde{A}_R} = 2c_R U_R \int_{-\infty}^{\infty}\rho(z)\left[\left\{\frac{y_1^+(z)}{y_1^+(0)}\right\}^2 + \left\{\frac{y_3^+(z)}{y_1^+(0)}\right\}^2\right]dz. \quad (5.111)$$

Returning to Eq. (5.106), we note that for a single force, $f_1^R = f_3^R = 0$ and f_2^R, f_4^R are given by Eq. (5.87). Therefore, the Rayleigh-wave normal-mode Green's dyadic for the half-space is

$$\mathfrak{G}(\mathbf{z}|\mathbf{z}_0) = -\frac{i}{4}\sum_{n=0}^{\infty}\sum_{m=-\infty}^{\infty}\frac{1}{c_R U_R I_0^R}\mathbf{Q}_m^{(2)}(k_n\Delta, \phi)\overset{*}{\mathbf{Q}}_m(k_n\Delta_0, \phi_0), \quad (5.112)$$

where

$$\mathbf{Q}_m(k\Delta, \phi) = y_1^+(z)\mathbf{P}_m(k\Delta, \phi) + y_3^+(z)\mathbf{B}_m(k\Delta, \phi). \quad (5.113)$$

It may be noted that $y_i^+(z)$ are the solutions of the system of homogeneous equations when the source is not present.

5.2.3. Numerical Procedure

The explicit expressions for the displacements and stresses of surface waves in a vertically heterogeneous half-space obtained here can be used as a tool for studying earthquake sources and earth structure provided we know how to evaluate numerically the amplitude functions $y_j^L(z)$ and $y_j^R(z)$, which are the solutions of the homogeneous sets of Eqs. (5.55), (5.84), and (5.85). This is achieved by means of a predictor–corrector integration scheme, using as initial values the known exact solution for a uniform elastic medium at a preassigned fiducial level $z = -H$. For SH motion, these are

$$y_1(z) = e^{\nu_\beta z}, \qquad \nu_\beta = \sqrt{(k^2 - k_\beta^2)}, \qquad k_\beta^2 = \omega^2 \frac{\rho_H}{\mu_H},$$
$$y_2(z) = \mu_H \nu_\beta e^{\nu_\beta z}, \qquad z \le -H, \tag{5.114}$$

where ρ_H and μ_H are the assumed density and rigidity, respectively, in the region $z < -H$. To calculate the phase velocity, we fix the value of ω, choose a trial value for k, and compute the initial values of y_1 and y_2 from Eq. (5.114). It is sufficient to choose H equal to twice the wavelength of the surface wave under consideration. The integration is then carried upward to $z = 0$. At this level the dispersion relation $y_2^+(k, 0) = 0$ is checked. The process is repeated for various values of k until a root of the dispersion relation is found. The phase velocity corresponding to a root thus found will then be computed from $c = \omega/k$. The group velocity can be found from Eq. (3.129).

The outlined procedure applies to P–SV (Rayleigh) waves too. In a uniform medium ($z < -H$), where λ, μ, ρ have constant values, Eqs. (5.84) and (5.85) have two sets of bounded independent solutions, corresponding to the P and SV constituents of the motion represented by the $\mathbf{L}_m$ and $\mathbf{N}_m$ vectors, respectively [Table 2.2, Eq. (5.4)]

$$\begin{aligned} y_{11} &= \nu_\alpha e^{\nu_\alpha z}, & y_{12} &= k e^{\nu_\beta z}, \\ y_{31} &= k e^{\nu_\alpha z}, & y_{32} &= \nu_\beta e^{\nu_\beta z}, \\ y_{21} &= \mu(2k^2 - k_\beta^2) e^{\nu_\alpha z}, & y_{22} &= 2k\mu\nu_\beta e^{\nu_\beta z}, \\ y_{41} &= 2k\mu\nu_\alpha e^{\nu_\alpha z}, & y_{42} &= \mu(2k^2 - k_\beta^2) e^{\nu_\beta z}, \qquad (z \le -H). \end{aligned} \tag{5.115}$$

We note that y_j^L and y_j^R are many-valued functions because of the radicals ν_α and ν_β in the initial values with which the integration is to be started. To be consistent with the condition that the displacement be finite at $z = -\infty$, we choose the sign of the radicals such that $\text{Re}(\nu_\alpha) > 0$, $\text{Re}(\nu_\beta) > 0$. When ω is real the Love-wave poles lie on the real axis in the k plane. When $\text{Re}\,\omega > 0$, $\text{Im}\,\omega < 0$, the poles are in the second and fourth quadrants in the complex k plane.

Once the roots of the dispersion relation are known, the eigenfunctions $y_j^L(z)$ and $y_j^R(z)$ can be evaluated directly by the predictor–corrector integration routine.

The set of differential equations (5.84) and (5.85) must be modified in the case of a liquid medium, e.g., when a solid half-space is overlain by a liquid layer. In the liquid, $y_4 = 0$ because the shear stress vanishes. Setting $\mu = 0$, $y_4 = 0$, the governing equations in the liquid layer become

$$\frac{dy_1}{dz} = \left(\frac{1}{\lambda} - \frac{k^2}{\omega^2\rho}\right)y_2, \qquad \frac{dy_2}{dz} = -\rho\omega^2 y_1, \qquad y_3 = -\frac{k}{\omega^2\rho}y_2. \tag{5.116}$$

The fourth equation, namely, $dy_3/dz = -ky_1$, is redundant. Because y_4 should vanish at the top of the half-space, y_1 and y_2 at the bottom of the liquid layer are determined except for a constant multiplier. The y_1 and y_2 thus determined are sufficient to specify the initial values at the bottom of the liquid layer, and the integration can be carried forward to the surface where $\Delta_R \equiv y_2(k, 0) = 0$ determines the dispersion relation.

In calculating the eigenvalues of the Rayleigh wave it often occurs that $\Delta_R(\omega, k) = 0$ even if a trial value of k is far from the roots of this equation. This is because individual terms in $\Delta_R(\omega, k)$ become large and, after subtraction, a large number of significant figures are lost. The difficulty is overcome by a simple artifice: We introduce six new variables in terms of the two independent solutions y_{ij} $(i = 1, 2, 3, 4; j = 1, 2)$ of the Rayleigh wave. Denoting $y_{ij}(z) = (ij)$, we define

$$\begin{aligned}
Y_1(z) &= (11)(32) - (12)(31),\\
Y_2(z) &= (21)(42) - (22)(41),\\
Y_3(z) &= (11)(22) - (12)(21),\\
Y_4(z) &= (11)(42) - (12)(41),\\
Y_5(z) &= (31)(22) - (32)(21),\\
Y_6(z) &= (31)(42) - (32)(41), \qquad Y_i(z = -\infty) = 0.
\end{aligned} \tag{5.117}$$

Because by Eq. (3.144) $\Delta_R(\omega, k) = Y_2(0)$, the value of $\Delta_R(\omega, k)$ will be obtained with the same accuracy as in Y_i. Differentiating $Y_6(z)$ and making use of Eqs. (5.84) and (5.85), we obtain $dY_6/dz = -dY_3/dz$, which integrates to $Y_6(z) = -Y_3(z) + \text{const}$. Then, because $Y_6(-\infty) = Y_3(-\infty) = 0$, we must have

$$Y_6(z) = -Y_3(z). \tag{5.118}$$

After the elimination of $Y_6(z)$, the new differential equations become

$$\begin{aligned}
\frac{dY_1}{dz} &= \frac{1}{\mu}Y_4 - \frac{1}{\lambda + 2\mu}Y_5,\\
\frac{dY_2}{dz} &= -\rho\omega^2 Y_4 + \left[\rho\omega^2 - 4k^2\frac{\mu(\lambda+\mu)}{\lambda + 2\mu}\right]Y_5,\\
\frac{dY_3}{dz} &= kY_4 + \frac{k\lambda}{\lambda + 2\mu}Y_5,\\
\frac{dY_4}{dz} &= \left[-\rho\omega^2 + 4k^2\frac{\mu(\lambda+\mu)}{\lambda + 2\mu}\right]Y_1 + \frac{1}{\lambda + 2\mu}Y_2 - \frac{2k\lambda}{\lambda + 2\mu}Y_3,\\
\frac{dY_5}{dz} &= \rho\omega^2 Y_1 - \frac{1}{\mu}Y_2 - 2kY_3.
\end{aligned} \tag{5.119}$$

The system of equations (5.119) should be integrated from $z = -H$ to $z = 0$ with the initial values

$$\begin{aligned} Y_1 &= \nu_\alpha \nu_\beta - k^2, \\ Y_2 &= \mu^2[(2k^2 - k_\beta^2)^2 - 4k^2\nu_\alpha\nu_\beta], \\ Y_3 &= \mu k[2\nu_\alpha\nu_\beta - (2k^2 - k_\beta^2)], \\ Y_4 &= -\nu_\alpha \rho\omega^2, \\ Y_5 &= \nu_\beta \rho\omega^2. \end{aligned} \tag{5.120}$$

The method just described not only improves the accuracy in calculating $\Delta_R(\omega, k)$ but also reduces the number of calculations, because there are only five unknown functions to be integrated compared with eight unknowns in the conventional method.

5.2.4. Matrix Method

We have so far discussed two basic numerical algorithms associated with surface waves: (1) the evaluation of the phase and group velocities of surface waves for a horizontally stratified medium by a matrix method, and (2) the evaluation of both dispersion and amplitude transfer functions by a numerical integration of the equations of motion. The matrix method can be extended to provide an efficient algorithm for the calculation of the amplitude transfer functions. An important feature of the matrix method is that it yields exact solutions in flat and homogeneous layers which approximate the earth's real structure. We are therefore faced with two alternatives: an exact solution to an approximate earth model or an approximate solution to an exact earth model. The two approaches should be compared on the basis of accuracy and speed of computation.

A closer look at Eqs. (5.104) and (5.110) reveals that the amplitude transfer functions for Love and Rayleigh waves are given in terms of the ratios

$$\begin{aligned} &\left[\frac{y_1^+(z)}{y_1^+(0)}\right]_{k_L}, \quad \left[\frac{y_1^+(z_0)}{y_1^+(0)}\right]_{k_L}, \quad \left[\frac{y_2^+(z_0)}{y_1^+(z_0)}\right]_{k_L}, \quad \left[\frac{y_1^+(z)}{y_1^+(0)}\right]_{k_R}, \\ &\left[\frac{y_3^+(z)}{y_1^+(0)}\right]_{k_R}, \quad \left[\frac{y_3^+(z_0)}{y_1^+(0)}\right]_{k_R}, \quad \left[\frac{y_2^+(z_0)}{y_3^+(z_0)}\right]_{k_R}, \quad \left[\frac{y_4^+(z_0)}{y_3^+(z_0)}\right]_{k_R}. \end{aligned} \tag{5.121}$$

Furthermore, all the *y*s appearing in expression (5.121) are solutions of the homogeneous equations of motion and are independent of the source. Consequently, these can be regarded as plane-wave amplitude functions and can be calculated by any suitable method. Later in this section we will explain how these can be found by the matrix method developed in Section 3.7. However, care is necessary in going from three-dimensional cylindrical waves to plane waves.

Using the asymptotic expansion (5.25) for the Hankel function, we find [see Eq. (5.108)]

$$y_1^+(z)\mathbf{P}_m^{(2)}(k\Delta, \phi) + y_3^+(z)\mathbf{B}_m^{(2)}(k\Delta, \phi)$$

$$= \sqrt{\left(\frac{2}{\pi k\Delta}\right)}[y_1^+(z)\mathbf{e}_z + y_3^+(z)e^{-\pi i/2}\mathbf{e}_\Delta]\exp\left[-i\left(k\Delta - \frac{\pi}{2}m - \frac{\pi}{4}\right) + im\phi\right]. \tag{5.122}$$

Therefore, the vertical and horizontal particle displacements in the far field are proportional to $y_1^+(z)$ and $y_3^+(z)\exp(-i\pi/2)$ and not just to y_1 and y_3. A similar argument will apply to stresses.

We use the notation of Section 3.7 *in toto*, taking the origin on the free surface and the z axis drawn vertically downward, into the medium. Because our aim is to calculate plane-wave amplitudes only, the direction of the x axis is of no consequence, as long as the system is right handed. Let a point source be situated on the z axis at a depth $z = h$ below the free surface in layer D_s ($z_{s-1} < h < z_s$). The source layer, D_s, is further divided into two layers, D_{s_1} and D_{s_2}, of identical elastic properties but of thicknesses $d_{s_1} = h - z_{s-1}$ and $d_{s_2} = z_s - h$, respectively (Fig. 5.4). Denoting the displacements in the directions of (x, y, z) by (u, v, w), noting that the ys are defined with the z axis vertically upward, and taking Eq. (5.122) into account, we have the following relations:

$$\begin{aligned} y_1^L(z) &= -v(z), & y_2^L(z) &= \tau_L(z), \\ y_1^R(z) &= -w(z), & y_2^R(z) &= \sigma_R(z), \\ y_3^R(z)e^{-i\pi/2} &= u(z), & y_4^R(z)e^{-i\pi/2} &= -\tau_R(z). \end{aligned} \tag{5.123}$$

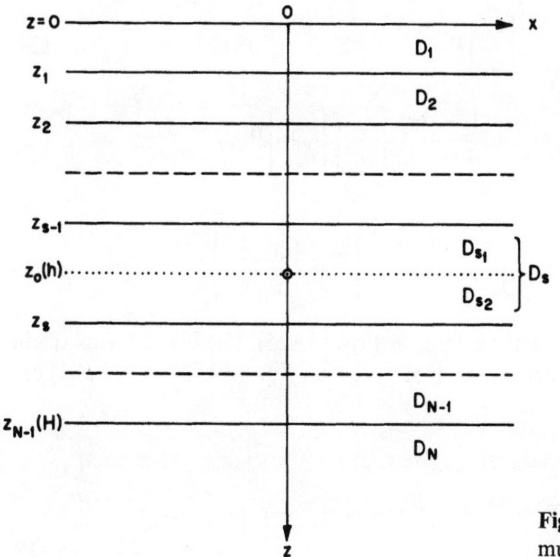

Figure 5.4. Point source in a multilayered half-space.

We introduce the notation

$$\frac{df}{dt} = \dot{f}, \qquad [f(z)]_{z=h} = f_s, \qquad [f(z)]_{z=0} = f_0,$$

where for a pure imaginary f,

$$f = i\bar{f}. \tag{5.124}$$

In Section 3.7, we noticed that it was advantageous to use velocities rather than displacements. Further, A_{mn}^L is real if $m + n =$ even and purely imaginary if $m + n =$ odd. Therefore, from Eq. (3.163) if $\dot{v}$ is real, τ_L is pure imaginary. With these considerations, Eq. (5.104a) may be written in the form

$$\begin{aligned} P_L &= \frac{1}{4\pi a^2}(k_L a)^{3/2}\sqrt{(8\pi)}\left[\frac{\mu_s \tilde{A}_L}{k_L}\right]\left[\frac{\dot{v}_s}{\dot{v}_0}\right]\left[\frac{\dot{v}(z)}{\dot{v}_0}\right], \\ Q_L &= \left[\frac{\bar{\tau}_{L_s}}{\mu_s}\right]\left[\frac{c_L}{\dot{v}_s}\right], \end{aligned} \tag{5.125}$$

where

$$\frac{1}{\tilde{A}_L} = 2c_L U_L \int_0^\infty \rho(z)\left[\frac{\dot{v}(z)}{\dot{v}_0}\right]^2 dz. \tag{5.126}$$

In the case of Rayleigh waves, if $\dot{w}$ is real, $\dot{u}$ and σ_R are pure imaginary, whereas τ_R is real. Therefore, Eqs. (5.110) become

$$\begin{aligned} P_R &= \frac{1}{4\pi a^2}(k_R a)^{3/2}\sqrt{(8\pi)}\left[\frac{\mu_s \tilde{A}_R}{k_R}\right]\left[\frac{\dot{\bar{u}}_s}{\dot{w}_0}\right]\left[\frac{\dot{w}(z)}{\dot{w}_0}\right], \\ Q_R &= -\left[\frac{\tau_{R_s}}{\mu_s}\right]\left[\frac{c_R}{\dot{\bar{u}}_s}\right]P_R, \\ S_R &= \left[\frac{1+\sigma_s}{1-\sigma_s} + \frac{1-2\sigma_s}{1-\sigma_s}\left\{\frac{\bar{\sigma}_{R_s}}{\mu_s}\right\}\left\{\frac{c_R}{\dot{\bar{u}}_s}\right\}\right]P_R, \end{aligned} \tag{5.127}$$

with

$$\frac{1}{\tilde{A}_R} = 2c_R U_R \int_0^\infty \rho(z)\left[\left\{\frac{\dot{w}(z)}{\dot{w}_0}\right\}^2 + \left\{\frac{\dot{\bar{u}}(z)}{\dot{w}_0}\right\}^2\right]dz. \tag{5.128}$$

We next express the ratios appearing in Eqs. (5.125)–(5.128) in terms of the layer matrices. In this connection, we first note that Eqs. (3.158) and (3.182) can be generalized to

$$\begin{aligned} [\dot{v}_n/c, \tau_n]_z &= a_n(z)[\dot{v}_{n-1}/c, \tau_{n-1}]_{z_{n-1}}, \qquad (z_{n-1} \le z \le z_n), \\ [\dot{u}_n/c, \dot{w}_n/c, \sigma_n, \tau_n]_z &= a_n(z)[\dot{u}_{n-1}/c, \dot{w}_{n-1}/c, \sigma_{n-1}, \tau_{n-1}]_{z_{n-1}} \end{aligned} \tag{5.129}$$

where $a_n(z)$ is obtained from a_n when $d_n = z_n - z_{n-1}$ is replaced by $d_n(z) = z - z_{n-1}$. The boundary conditions at the free surface now yield

$$
\begin{aligned}
&\frac{\dot{v}(z)}{\dot{v}_0} = [B_n(z)]_{11}, \qquad \frac{\tau_L(z)}{\dot{v}_0/c} = [B_n(z)]_{21}, \qquad (z_{n-1} \le z \le z_n),\\
&\frac{\dot{v}_s}{\dot{v}_0} = [B_s(h)]_{11}, \qquad \frac{\tau_{L_s}}{\dot{v}_0/c} = [B_s(h)]_{21}, \qquad (z_{s-1} < h < z_s),\\
&\frac{\dot{u}(z)}{\dot{w}_0} = [B_n(z)]_{11}\left(\frac{\dot{u}_0}{\dot{w}_0}\right) + [B_n(z)]_{12},\\
&\frac{\dot{w}(z)}{\dot{w}_0} = [B_n(z)]_{21}\left(\frac{\dot{u}_0}{\dot{w}_0}\right) + [B_n(z)]_{22},\\
&\frac{\sigma_R(z)}{\dot{w}_0/c} = [B_n(z)]_{31}\left(\frac{\dot{u}_0}{\dot{w}_0}\right) + [B_n(z)]_{32},\\
&\frac{\tau_R(z)}{\dot{w}_0/c} = [B_n(z)]_{41}\left(\frac{\dot{u}_0}{\dot{w}_0}\right) + [B_n(z)]_{42},\\
&\frac{\dot{u}_s}{\dot{w}_0} = [B_s(h)]_{11}\left(\frac{\dot{u}_0}{\dot{w}_0}\right) + [B_s(h)]_{12},\\
&\frac{\dot{w}_s}{\dot{w}_0} = [B_s(h)]_{21}\left(\frac{\dot{u}_0}{\dot{w}_0}\right) + [B_s(h)]_{22},\\
&\frac{\sigma_{R_s}}{\dot{w}_0/c} = [B_s(h)]_{31}\left(\frac{\dot{u}_0}{\dot{w}_0}\right) + [B_s(h)]_{32},\\
&\frac{\tau_{R_s}}{\dot{w}_0/c} = [B_s(h)]_{41}\left(\frac{\dot{u}_0}{\dot{w}_0}\right) + [B_s(h)]_{42},
\end{aligned}
\tag{5.130}
$$

where

$$
\begin{aligned}
B_n(z) &= a_n(z)a_{n-1}a_{n-2}\cdots a_1,\\
B_s(h) &= a_s(h)a_{s-1}a_{s-2}\cdots a_1,\\
a_s(h) &= a_{s_1}, \qquad a_n(z_n) = a_n.
\end{aligned}
\tag{5.131}
$$

The ratio $(\dot{u}_0/\dot{w}_0)$ is known from Eq. (3.199).

In Eqs. (5.130) and (5.131), $\dot{v}$, $\dot{w}$, and τ_R are real, whereas $\dot{u}$, τ_L, and σ_R are pure imaginary. Moreover, $[B_n(z)]_{pq}$ and $[B_s(h)]_{pq}$ are real if $p + q$ is even and pure imaginary if $p + q$ is odd. Keeping this in mind and using the notation of Eq. (5.124), all the relations in Eqs. (5.130) and (5.131) can be converted into equations involving real quantities only.

In terms of the present notation, Eqs. (5.109) read

$$
\frac{u_\Delta^R(z)}{u_z^R(z)} = -\frac{u(z)}{w(z)} = \varepsilon(z)e^{i\pi/2}, \qquad \varepsilon(z) = -\frac{\dot{\bar{u}}(z)}{\dot{w}(z)}, \tag{5.132}
$$

where $\varepsilon(z)$ is the Rayleigh-wave ellipticity. It may be noted that whereas u_Δ and u_z are the displacement components referred to a coordinate system with the z axis drawn vertically upward, u and w are the displacement components referred to a coordinate system with the z axis drawn vertically downward.

The symbols that appear in Eqs. (5.38), (5.125), and (5.127) may be grouped as follows:

1. Geometric parameters: Earth's radius (a), epicentral distance (Δ), source's depth (h), station azimuth (ϕ) with respect to a fault's strike, measured counterclockwise.
2. Source parameters: Slip (U_0), fault area (dS), slip angle (λ), dip angle (δ).
3. Wave parameters: Wave numbers (k_R, k_L), phase velocities (c_R, c_L), group velocities (U_R, U_L), attenuation coefficients (γ_R, γ_L).
4. Plane-wave functions: Surface particle velocities ($\dot{u}_0$, $\dot{v}_0$, $\dot{w}_0$), particle velocities at the source's depth ($\dot{u}_s$, $\dot{v}_s$, $\dot{w}_s$) and at depth z ($\dot{u}$, $\dot{v}$, $\dot{w}$), depth-dependent factors of the stresses (σ_{R_s}, τ_{R_s}, τ_{L_s}).

Table 5.2. Phase and Group Velocities (km/s) for Rayleigh Waves

T (s)	M_{11}		M_{21}		M_{12}	
	c	U	c	U	c	U
Continental						
300	5.083	3.849				
250	4.788	3.614				
200	4.482	3.562	6.482	5.351		
150	4.226	3.663	6.122	5.121		
100	4.053	3.807	5.610	4.534	6.479	5.378
50	3.948	3.830	4.846	4.214	5.442	4.379
Shield						
300	5.001	3.784				
250	4.718	3.625				
200	4.454	3.677	6.268	5.616		
150	4.259	3.851	6.017	5.093		
100	4.158	4.056	5.498	4.500	6.361	5.705
50	4.129	4.066	4.856	4.284	5.354	4.511
Oceanic						
300	5.075	3.832				
250	4.769	3.559				
200	4.457	3.547	6.277	5.575		
150	4.215	3.705	6.010	5.062		
100	4.075	3.909	5.489	4.477	6.377	5.806
50	4.036	4.041	4.751	4.121	5.327	4.467

5. Medium parameters: Poisson's ratio at the source's level (σ_s), rigidity at the source's level (μ_s).
6. Medium amplitude functions: $\tilde{A}_R$, $\tilde{A}_L$, independent of source type and depth.

Clearly, the spectral displacements are composite functions of the wave's period and mode, the source's depth and orientation, the station's azimuth, and the structural and dissipation parameters of the layered medium. The dependence of the displacements on the angle ϕ is known as the *radiation pattern* of the wave. Because both $u_z^R(\phi)$ and $u_\phi^L(\phi)$ are complex, there exist both amplitude and phase patterns.

We have completed the derivation of the *medium transfer functions* P_L, Q_L, P_R, Q_R, and S_R by which the ground motion is calculated for any given shear dislocation source. These entities are composite functions of the source's depth, mode, frequency and the earth's structure.

The structural parameters of three typical earth models, continental, shield, and oceanic, are given in Appendix L. Table 5.2 gives the phase and group velocities for Rayleigh waves for these models. The group velocity for the fundamental Rayleigh mode is also exhibited in Fig. 5.5. Table 5.3 gives the phase and group velocities of the fundamental Love mode and its first three overtones.

Table 5.3. Phase and Group Velocities (km/s) for Love Waves

T	L_0		L_1		L_2		L_3	
(s)	c	U	c	U	c	U	c	U
Continental								
300	5.168	4.269						
250	4.987	4.226						
200	4.811	4.210	6.621	5.534				
150	4.645	4.210	6.214	4.981				
100	4.489	4.199	5.574	4.396	6.512	5.383		
50	4.299	4.014	4.818	4.208	5.404	4.298	5.981	4.598
Shield								
300	5.189	4.408						
250	5.038	4.390						
200	4.894	4.398	6.410	6.035				
150	4.763	4.419	6.096	4.879				
100	4.646	4.434	5.495	4.444	6.399	5.661		
50	4.510	4.269	4.821	4.265	5.362	4.432	5.840	4.492
Oceanic								
300	5.168	4.309						
250	4.998	4.282						
200	4.837	4.289	6.418	6.227				
150	4.691	4.319	6.145	4.968				
100	4.569	4.362	5.529	4.411	6.417	6.078		
50	4.478	4.407	4.808	4.246	5.403	4.338	5.897	4.510

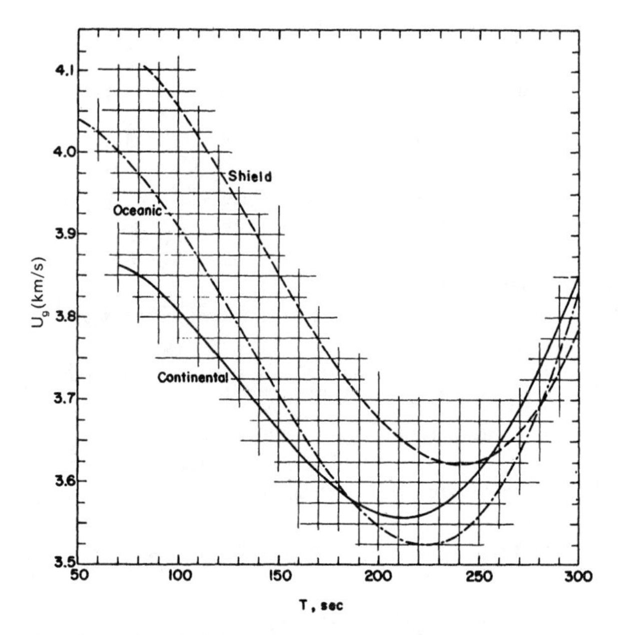

Figure 5.5. The effect of earth structural model on the Rayleigh-wave velocities (fundamental mode).

The variation of the five transfer functions with the source depth for $T = 50$, 100, 150, 200, 250, and 300 s is shown in Figs. 5.6–5.15 for the fundamental Love and Rayleigh modes for the three basic models of the earth. For comparison we also have included the Rayleigh excitation functions for a homogeneous half-space model in which $\sigma = \frac{1}{4}$, $\beta = 6.24$ km/s, and the Love excitation functions for a two-layered half-space in which $\alpha_1 = 6.30$ km/s, $\beta_1 = 3.64$ km/s, $\rho_1 = 2.87$ g/cm^3, $H = 40$ km, $\alpha_2 = 11.4$ km/s, $\beta_2 = 6.24$ km/s, $\rho_2 = 4.46$ g/cm^3. Amplitudes in all these figures are in micron-seconds per $(m \times km^2)$ (e.g., $U_0 = 1$ m, $dS = 1$ km^2). Note that all curves of P_R, S_R, and P_L have a marked discontinuity when the source's depth crosses the Mohorovičić (Moho) discontinuity. This is expected from the theory and is readily observed, for example, from Eq. (5.35). On substituting $h = H$, we find

$$
\begin{aligned}
P_L(H+) &= \frac{\mu_2}{\mu_1} P_L(H-), \\
Q_L(H+) &= Q_L(H-).
\end{aligned}
\tag{5.133}
$$

For a given structure $(\lambda_i, \mu_i, \rho_i)$, wave mode (n), source's depth (h), orientation (λ, δ), and frequency (ω), the radiation pattern at any azimuth (ϕ) and distance (Δ) is a complex number. The distributions of the amplitude and phase of this

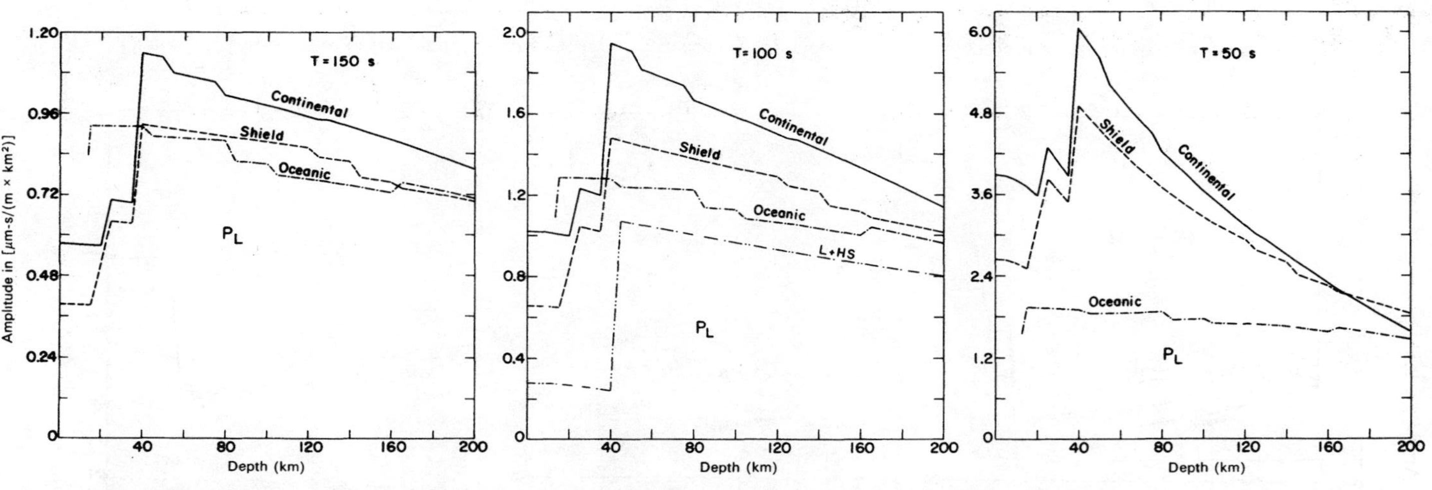

Figure 5.6. Dependence of the fundamental Love-mode displacement spectrum on source depth for a vertical strike-slip source. The curve marked $L + HS$ corresponds to the exact solution for a layer overlying a uniform half-space.

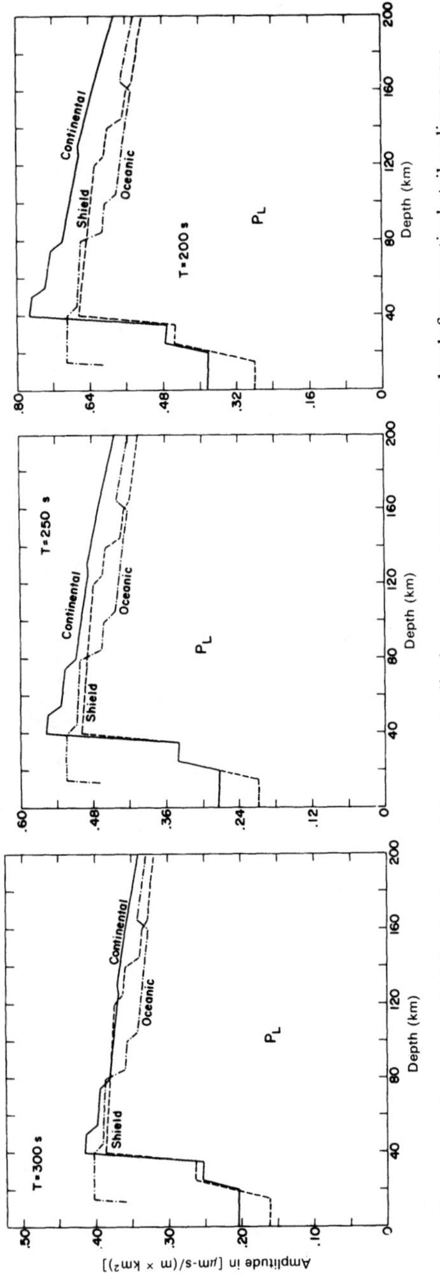

Figure 5.7. Dependence of the fundamental Love-mode displacement spectrum on source depth for a vertical strike-slip source.

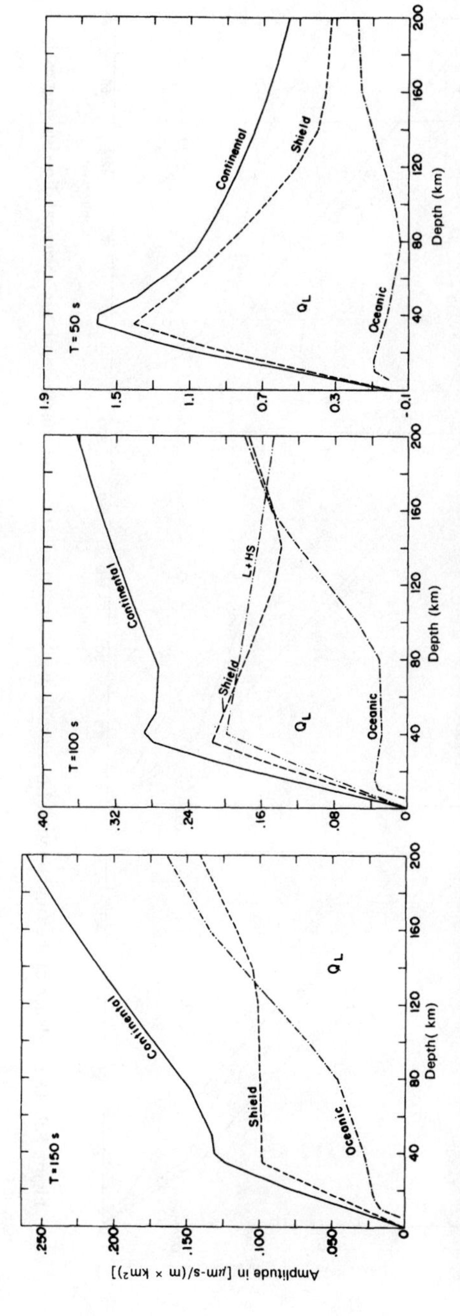

Figure 5.8. Dependence of the fundamental Love-mode displacement spectrum on source depth for a vertical dip-slip source. The curve marked $L + HS$ corresponds to the exact solution for a layer overlying a uniform half-space.

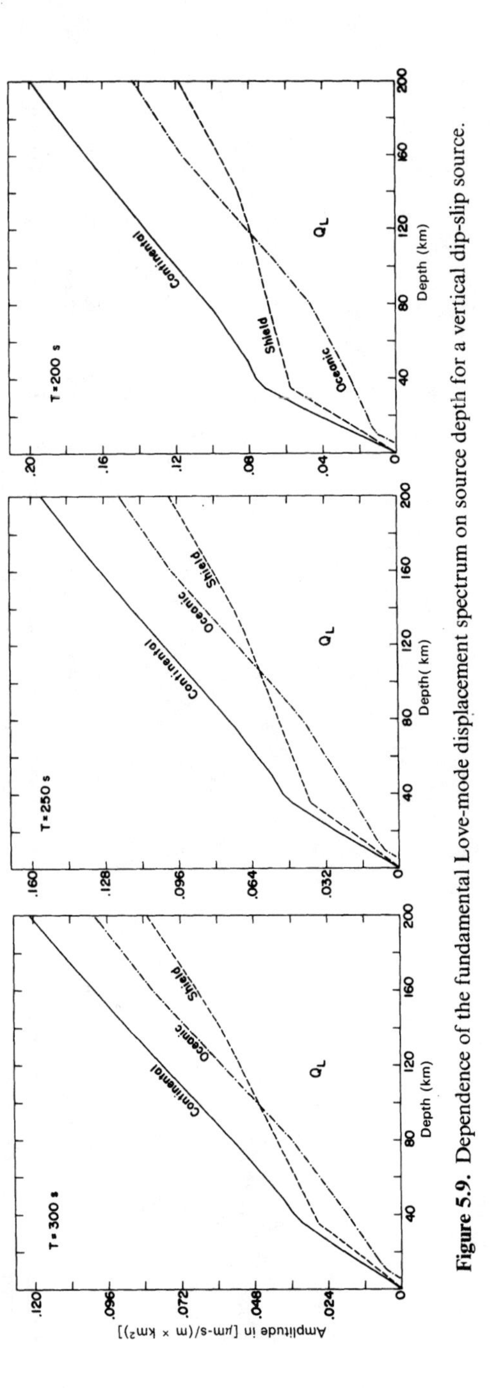

Figure 5.9. Dependence of the fundamental Love-mode displacement spectrum on source depth for a vertical dip-slip source.

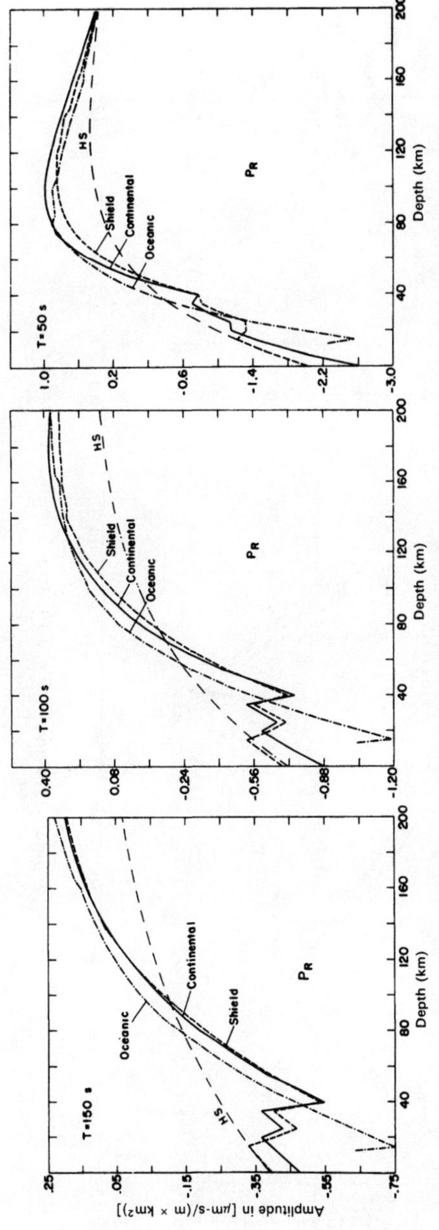

Figure 5.10. Dependence of the fundamental Rayleigh-mode displacement spectrum on source depth for a vertical strike-slip source for $T = 50$, 100, and 150 s. The curve marked *HS* corresponds to the exact solution for a homogeneous half-space.

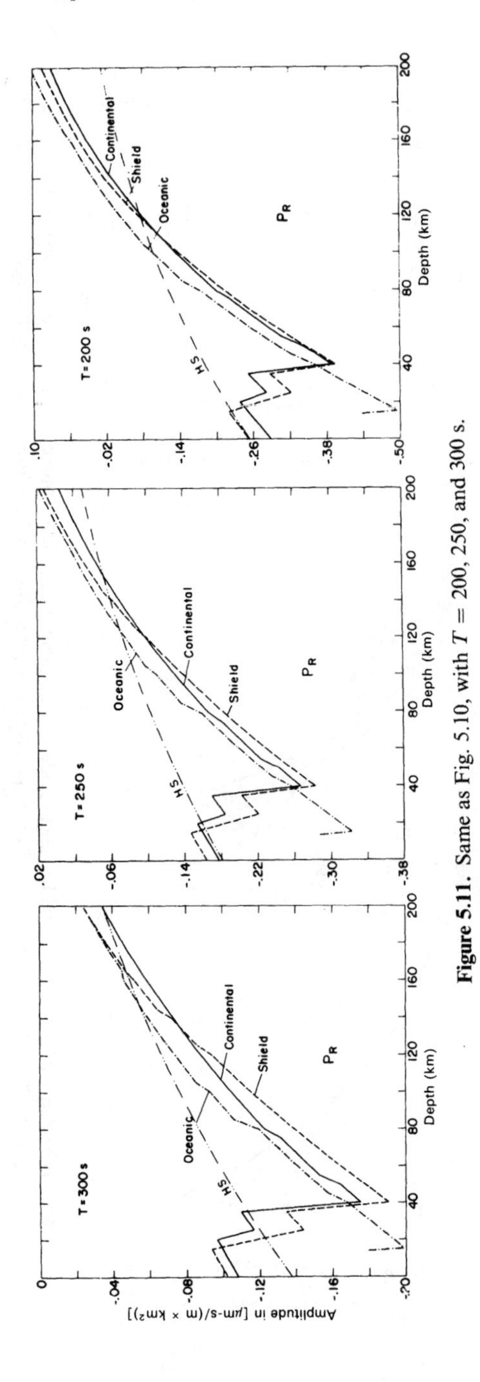

Figure 5.11. Same as Fig. 5.10, with $T = 200$, 250, and 300 s.

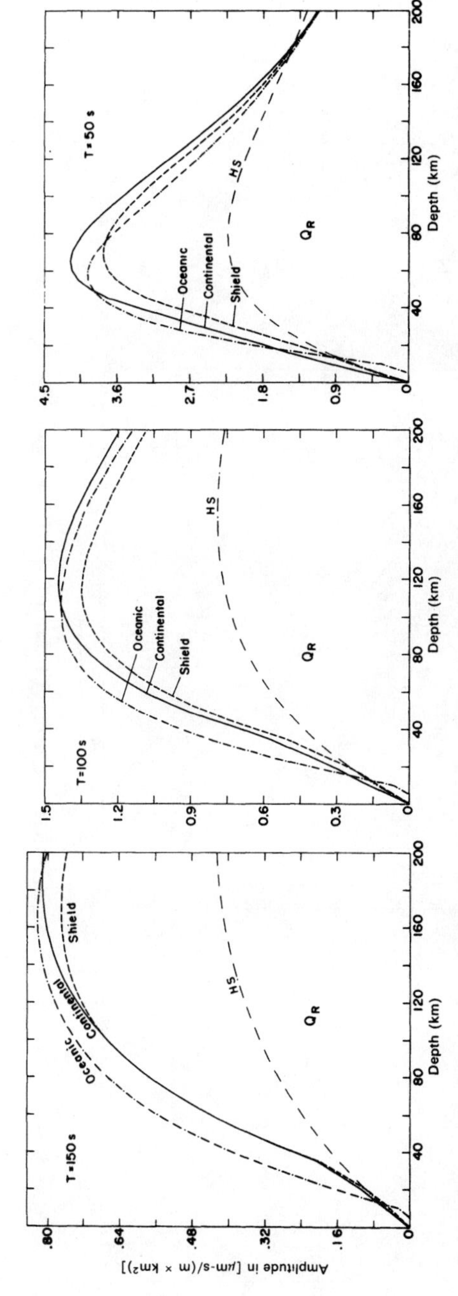

Figure 5.12. Dependence of the fundamental Rayleigh-mode displacement spectrum on source depth for a vertical dip-slip source for $T = 50,100$, and 150 s. The curve marked *HS* corresponds to the exact solution for a homogeneous half-space.

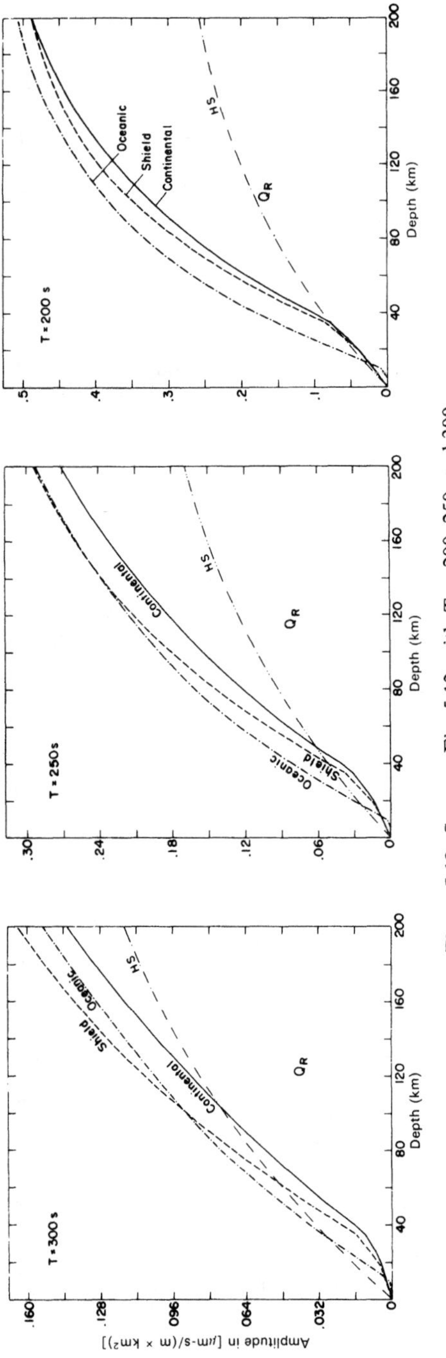

Figure 5.13. Same as Fig. 5.12, with $T = 200$, 250, and 300 s.

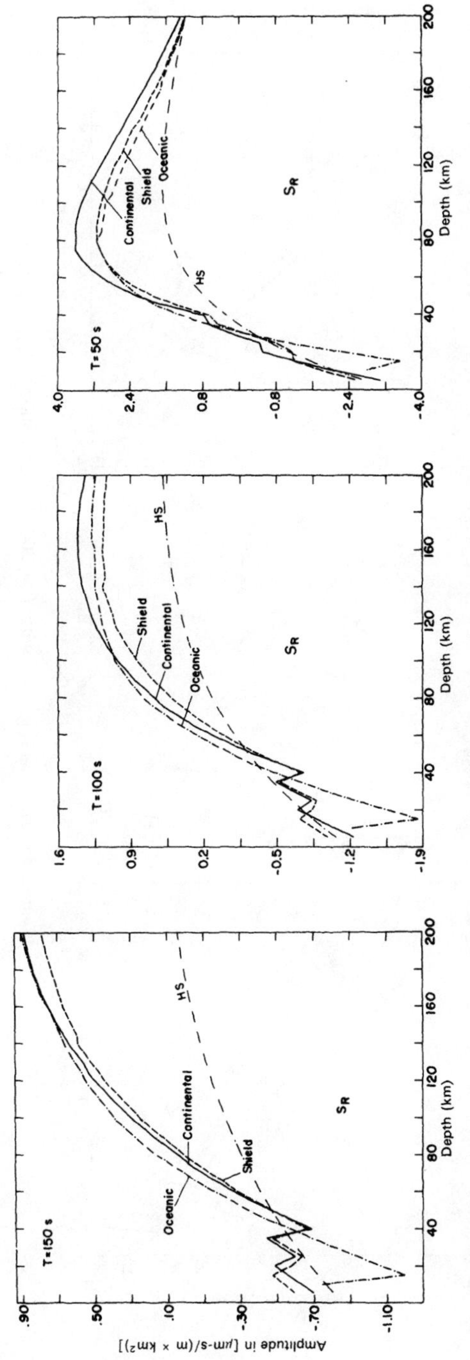

Figure 5.14. Dependence of the fundamental Rayleigh-mode displacement spectrum on source depth for $T = 50$, 100, and 150 s. The curve marked *HS* corresponds to the exact solution for a homogeneous half-space.

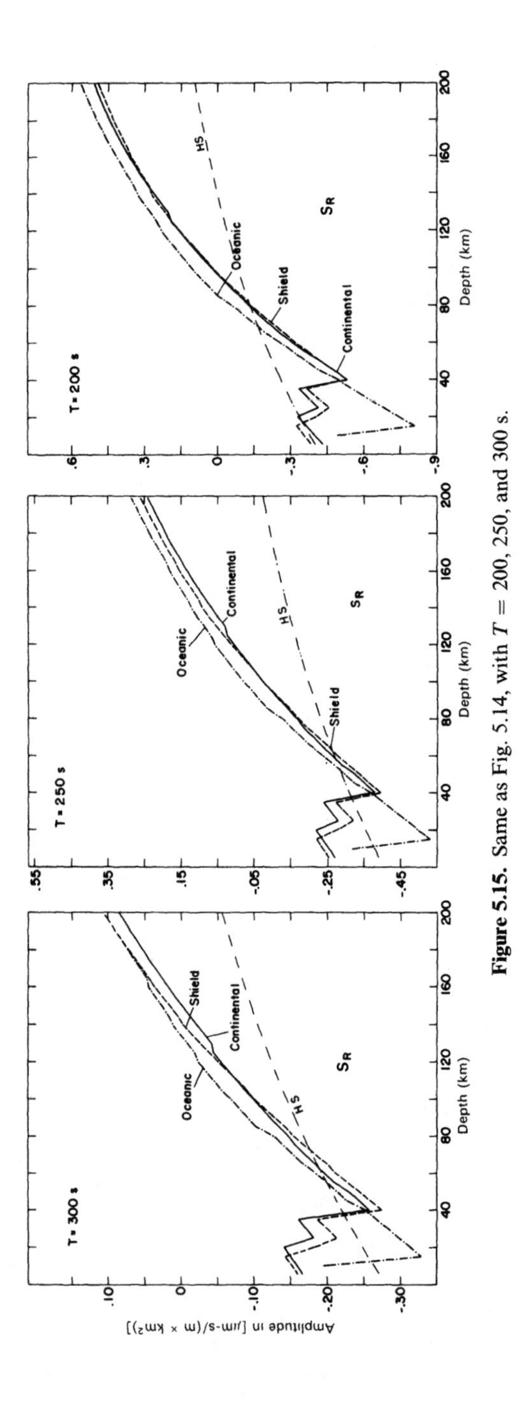

Figure 5.15. Same as Fig. 5.14, with $T = 200$, 250, and 300 s.

number with ϕ at a fixed distance are known as the *amplitude radiation pattern* and the *initial-phase pattern.* These are given explicitly as

$$\text{Initial spatial phase of } u_z^R = -\frac{\pi}{4} + \tan^{-1}\left[\frac{q_R Q_R}{s_R S_R + p_R P_R}\right],$$
$$\text{Initial spatial phase of } u_\phi^L = -\frac{3\pi}{4} + \tan^{-1}\left[\frac{q_L Q_L}{p_L P_L}\right]. \tag{5.134}$$

The radiation pattern function is a very useful tool for the determination and identification of seismic sources. However, it must be remembered that the inherent ambiguity of the fault plane and the auxiliary plane is also present in the radiation pattern. Certain symmetry relations, which are easily derived from Eq. (5.39), can be used to economize the calculation of the radiation patterns in the three-parameter space (ϕ, λ, δ). Let $U(\lambda, \delta; \phi) = \chi(\phi)e^{i\Lambda(\phi)}$ be the complex scalar radiation pattern function of either Rayleigh (u_z) or Love (u_ϕ) waves. Then

$$\begin{aligned}
U^L &= p_L P_L + iq_L Q_L,\\
U^R &= s_R S_R + p_R P_R + iq_R Q_R,\\
U(\lambda, \delta; \phi + \pi) &= U^*(\lambda, \delta; \phi),\\
U(\lambda + \pi, \delta; \phi) &= -U(\lambda, \delta; \phi),\\
U(-\delta) &= -U^*(\delta),\\
U^L(\pi - \lambda, \delta; \pi - \phi) &= -U^L(\lambda, \delta, \phi),\\
\chi(2\pi - \lambda, \delta; 2\pi - \phi) &= \chi(\lambda, \delta, \phi),
\end{aligned} \tag{5.135}$$

where an asterisk denotes a complex conjugate.

Figure 5.16 exhibits the meanings of the five amplitude transfer functions. For example, it shows that P_R is the magnitude of the vertical component of the Rayleigh-wave displacement for a strike-slip fault at an azimuth of 45° relative to the fault's strike. Figure 5.17 shows plots of calculated radiation patterns for horizontal components of the Rayleigh-wave displacement for two values of the source depth h. This clearly demonstrates the strong dependence of the radiation pattern on h. Similarly, Fig. 5.18 brings out the dependence of the radiation pattern on the chosen type of earth model. Figure 5.19 exhibits the radiation patterns for the first higher Rayleigh mode. The curves between the two concentric circles are the initial-phase patterns.

EXAMPLE 5.2: Rayleigh Waves from a Buried Explosive Source

We have shown in Section 4.6.1 that the spectral displacement field in an infinite elastic solid induced by a sudden application of a pressure p_0 to the wall of a spherical cavity of radius a is given by

$$\mathbf{u}(r; \omega) = \text{grad}\ \phi_s, \tag{5.2.1}$$

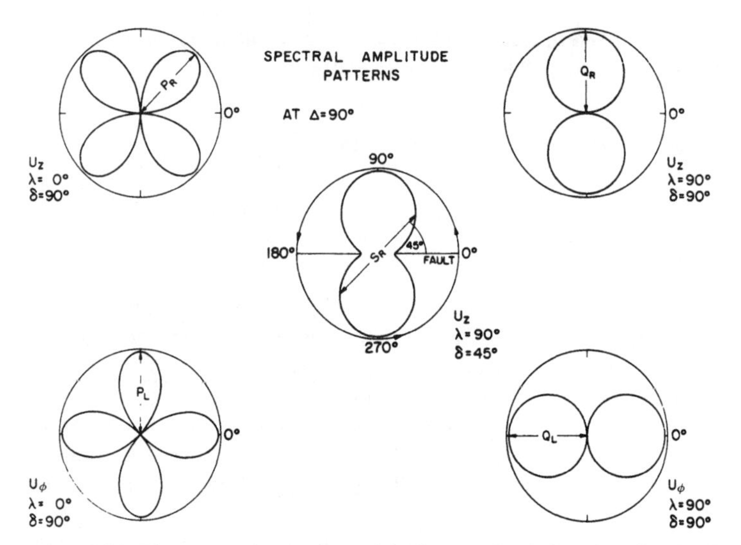

Figure 5.16. The geometric meanings of the five amplitude functions P_L, Q_L, P_R, Q_R, and S_R. Amplitudes are in arbitrary units.

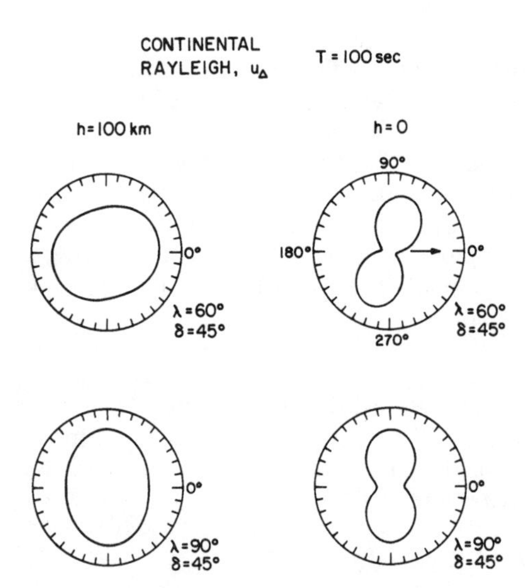

Figure 5.17. The effect of the source depth on the Rayleigh-wave radiation pattern. Amplitudes are in arbitrary units.

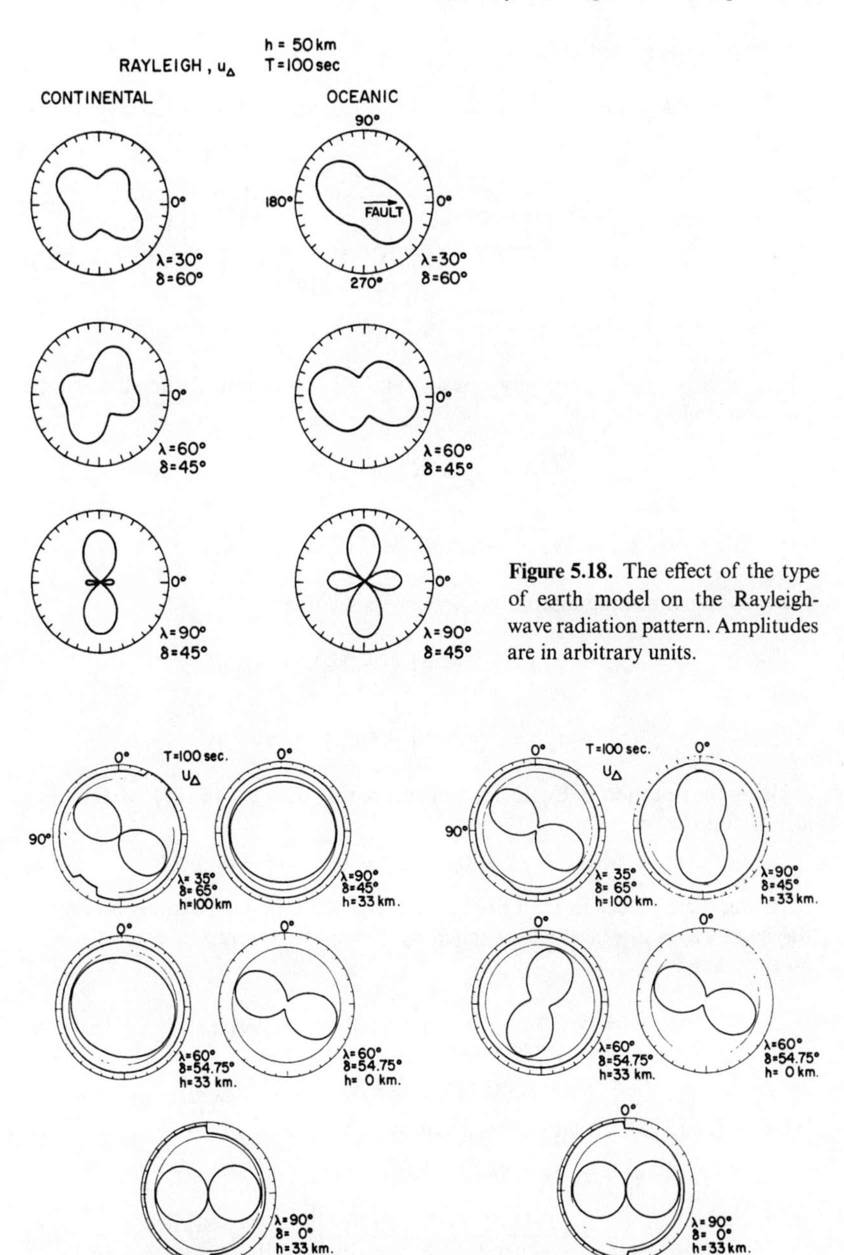

Figure 5.18. The effect of the type of earth model on the Rayleigh-wave radiation pattern. Amplitudes are in arbitrary units.

Figure 5.19. Radiation patterns for the first higher Rayleigh mode for two earth models. The amplitude patterns are given inside the inner circles and the phase patterns are given between the two concentric circles.

where ϕ_s is the potential function

$$\phi_s(r;\omega) = \gamma_0 \frac{e^{i(\omega t - k_\alpha r)}}{r} = \int_0^\infty \gamma_0 J_0(k\Delta) e^{-\nu_\alpha |z| + i\omega t} \frac{k\,dk}{\nu_\alpha} \tag{5.2.2}$$

and

$$\gamma_0 = -\frac{p_0 a^3}{4\mu} \frac{\exp\left[i\left\{k_\alpha a - \tan^{-1}\left[k_\alpha a\left(1 - \frac{k_\beta^2 a^2}{4}\right)^{-1}\right]\right\}\right]}{\left[\left(1 - \frac{k_\beta^2 a^2}{4}\right)^2 + k_\alpha^2 a^2\right]^{1/2}} g(\omega), \tag{5.2.3}$$

$$r^2 = \Delta^2 + z^2, \qquad \nu_\alpha = (k^2 - k_\alpha^2)^{1/2}.$$

The z axis is drawn vertically upward and the spectrum of the source-time function is $g(\omega)$. Because

$$u_\Delta = \frac{\partial\phi_s}{\partial\Delta}, \qquad u_z = \frac{\partial\phi_s}{\partial z},$$
$$\tau_{zz} = 2\mu\frac{\partial^2\phi_s}{\partial z^2} - \lambda k_\alpha^2\phi_s, \qquad \tau_{\Delta z} = 2\mu\frac{\partial^2\phi_s}{\partial\Delta\partial z}, \tag{5.2.4}$$

the only nonvanishing discontinuities across $z = 0$ are

$$(u_z^+ - u_z^-)_{z=0} = e^{i\omega t}\int_0^\infty \{-2\gamma_0\} J_0(k\Delta) k\,dk,$$
$$(\tau_{\Delta z}^+ - \tau_{\Delta z}^-)_{z=0} = e^{i\omega t}\int_0^\infty \{4\mu\gamma_0\} J_1(k\Delta) k^2\,dk. \tag{5.2.5}$$

Using the notation in Eq. (5.83) and comparing with Eq. (5.107), we find that $m = 0$ and

$$f_1^R = 4\pi\gamma_0, \qquad f_2^R = 0, \qquad f_3^R = 0, \qquad f_4^R = 8\pi\mu\gamma_0 k. \tag{5.2.6}$$

Inserting these values in Eq. (5.106), we derive the following expression for the Rayleigh waves displacements excited by an explosive source placed at $z = z_0$ on the z axis:

$$\mathbf{u} = i\pi\gamma_0 \sum_{n=0}^\infty \frac{1}{c_R U_R I_0^R}[y_2^+(z_0) - 2\mu_s k_n y_3^+(z_0)]$$
$$\times \{\mathbf{e}_z y_1^+(z) H_0^{(2)}(k_n\Delta) - \mathbf{e}_\Delta y_3^+(z) H_1^{(2)}(k_n\Delta)\}. \tag{5.2.7}$$

Using Eq. (5.25), the far field is given by

$$u_z(\Delta, z; k_R) = (\sqrt{8\pi}\,\gamma_0)(k_R\Delta)^{-1/2}\tilde{A}_R$$
$$\times\left[\frac{y_2^+(z_0)}{y_1^+(0)} - 2\mu_s k_R\frac{y_3^+(z_0)}{y_1^+(0)}\right]\left[\frac{y_1^+(z)}{y_1^+(0)}\right] e^{-i(k_R\Delta - 3\pi/4)}, \tag{5.2.8}$$

$$u_\Delta(\Delta, z; k_R) = \left[\frac{y_3^+(z)}{y_1^+(z)} e^{-i\pi/2}\right] u_z(\Delta, z; k_R),$$

where $\tilde{A}_R$ is defined in Eq. (5.111).

In terms of the notation introduced in Eqs. (5.123) and (5.124), Eq. (5.2.8) becomes

$$u_z = (\sqrt{8\pi\gamma_0})k_R^{1/2}\Delta^{-1/2}\tilde{A}_R\left[c_R\frac{\bar{\sigma}_{R_s}}{\dot{w}_0} - 2\mu_s\frac{\dot{u}_s}{\dot{w}_0}\right]\left[\frac{\dot{w}(z)}{\dot{w}_0}\right]e^{-i(k_R\Delta - 3\pi/4)}, \tag{5.2.9}$$

$$u_\Delta = [\varepsilon(z)e^{i\pi/2}]u_z,$$

where $\varepsilon(z)$ is the Rayleigh-wave ellipticity. All the parameters appearing in the expression (5.2.3) for the constant γ_0 are to be evaluated at the source level.

5.3. Surface Waves from a Finite Moving Source

5.3.1. Terrestrial Interferometry

Up to this point we have treated the seismic source as a point dislocation characterized by the tensor U_0**en**. It is well known that seismic waves are radiated from faults whose lengths can reach 800 km. Moreover, there is ample evidence to show that the seismic energy of earthquakes is released through a process of propagating rupture at subshear velocities. Because the periods of commonly recorded surface waves on long-period seismographs are in the range 50–300 s and because the corresponding wavelengths are in the range 200–1500 km, it is clear that both (fault length)/(wavelength) and (rupture time)/(wave period) are of the order unity. Under these conditions, the dislocation cannot be considered a point source and, consequently, the spectral displacement field must be evaluated by integrating over the fault area with proper time delays. Because most meaningful observations of long-period surface waves are made in the far field, we shall make corresponding approximations in our calculations.

The vertical component of the Rayleigh wave or the azimuthal component of the Love-wave spectral displacement at an epicentral distance, Δ, and azimuth, ϕ, from a point dislocation (λ, δ, h, $U_0\,dS$) can be expressed in the form [Eq. (5.38)]

$$u = U_0\,dS\,F(\lambda, \delta, h; \phi; \omega)\left\{\frac{e^{-\gamma\Delta + i(\omega t - k\Delta)}}{\sqrt{[\sin(\Delta/a)]}}\right\}. \tag{5.136}$$

The complex function F absorbs the overall dependence on source orientation (λ, δ), depth (h), azimuth (ϕ), frequency (ω), and structural parameters. The term in braces gives the dependence on the epicentral distance (Δ).

Consider a rectangular fault of length L and width W. Let the direction of rupture propagation be horizontal and coincident with the strike direction, $Q\xi$ (Fig. 5.20). We assume that the dislocation U_0 is constant over the entire fault. (A coordinate-dependent dislocation has been considered in Section 4.7.3 and the results have been given in Table 4.8.) Let $S(\xi, \eta)$ be any point on the

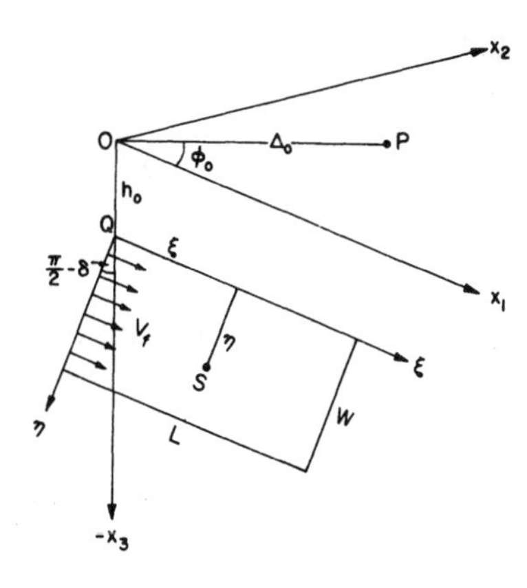

Figure 5.20. Geometry of a finite source.

fault. Then the coordinates of S in the (x_1, x_2, x_3) system are $(\xi, -\eta\cos\delta, -h_0 - \eta\sin\delta)$. Assuming $\Delta_0 \gg \xi, \eta$, we have

$$\begin{aligned}\Delta &= [(\Delta_0\cos\phi_0 - \xi)^2 + (\Delta_0\sin\phi_0 + \eta\cos\delta)^2]^{1/2}\\ &\simeq \Delta_0 - \xi\cos\phi_0 + \eta\cos\delta\sin\phi_0.\end{aligned} \tag{5.137}$$

Replacing dS in Eq. (5.136) by $d\xi d\eta$, integrating over ξ $(0 \le \xi \le L)$ and η $(0 \le \eta \le W)$, and introducing a time delay ξ/V_f, the far-field approximation yields for $g(\omega) = 1$

$$\begin{aligned}u = P_0 F(\lambda, \delta, h_0 + \tfrac{1}{2}W\sin\delta; \phi_0; \omega)&\left\{\frac{e^{-\gamma\Delta_0 + i(\omega t - k\Delta_0)}}{\sqrt{[\sin(\Delta_0/a)]}}\right\}\\ \times&\left\{\frac{1}{L}\int_0^L e^{-ik\xi(c/V_f - \cos\phi_0)}\,d\xi\right\},\end{aligned} \tag{5.138}$$

where $P_0 = U_0 LW$ is the source potency. It may be noted that Δ_0 is the horizontal distance of the observer from the point of rupture initiation, and ϕ_0 is the corresponding azimuth of the observer, relative to the strike.

The expression in the second pair of braces in Eq. (5.138) represents the effect of the finiteness of the fault and its motion. It has been shown in Section 4.7.3 that it is equal to

$$\frac{\sin X}{X}e^{-iX}$$

where

$$X = \frac{\omega L}{2V_f}\left(1 - \frac{V_f}{c}\cos\phi_0\right) = \frac{\pi L}{\Lambda}\left(\frac{c}{V_f} - \cos\phi_0\right) = \frac{\pi t_f}{T}\left(1 - \frac{V_f}{c}\cos\phi_0\right). \tag{5.139}$$

Above, Λ is the wavelength and $t_f = L/V_f$ is the rupture duration or the *rupture time.*

The radiation pattern in the far field depends strongly on the ratio L/Λ or t_f/T. Equation (5.139) represents a phenomenon reminiscent of the *Doppler effect* in electromagnetic theory with the complication that, in the seismic analog, the velocity c is frequency dependent. The factor $(\sin X/X)\exp(-iX)$ is a first-order correction for the source's finiteness and motion. The Doppler shift affects both the amplitude and the phase. An observer at the receding end will notice a *phase retardation* and a modulated spectral amplitude pattern with minima at various frequencies that depend upon the azimuth, ϕ_0; the rupture duration, L/V_f; and the *spectral Mach number*, $c(\omega)/V_f$. The minima are determined by the condition

$$\left(\frac{L}{V_f}\right)c(\omega) - L\cos\phi_0 = N\Lambda \qquad (N = 0, 1, 2, \ldots). \tag{5.140}$$

This is the condition of *destructive interference* at the point of observation. We therefore have a clear case of terrestrial interferometry. At a fixed azimuth, ϕ_0, the serial number of the "holes" in the spectrum resulting from a propagating rupture are determined by the equation

$$N = \frac{\nu L}{c(\nu)}\left[\frac{c(\nu)}{V_f} - \cos\phi_0\right] \tag{5.141}$$

where $\nu = \omega/2\pi$ is the frequency. However, for a fixed wavelength, Λ, the holes in the *radiation pattern* are given by

$$\phi_N = \arccos\left(\frac{c}{V_f} - \frac{N\Lambda}{L}\right). \tag{5.142}$$

Tables 5.4 and 5.5 give the finiteness factor $|\sin X/X|$ for two values of T (100 and 250 s) at a fixed $V_f = 3$ km/s as a function of ϕ_0 and t_f/T.

The influence of the finiteness factor $|\sin X/X|$ on the radiation pattern of Love waves is demonstrated in Fig. 5.21 for a source with rupture parameters similar to that of the great Alaskan earthquake of 28 March, 1964. It is obvious that the modulation of the pattern resulting from the finiteness factor is dominating. In Fig. 5.22, a comparison is made between the calculated and observed Love-wave radiation patterns for the Alaskan earthquake for four values of the source depth. Figure 5.23 compares the calculated and observed Rayleigh-wave radiation patterns for the Rat Island earthquake of 4 February, 1965.

5.3.2. Directivity

The directivity function has been derived in Section 4.7.4. In the case of surface waves the velocity α appearing in Eq. (4.260) is to be replaced by c_L for Love waves and c_R for Rayleigh waves. Fortunately for seismologists, the spherical shape of the earth provides a very simple way of applying the directivity function

Table 5.4. Absolute Values of the Finiteness Factor $|(\sin X)/X|$ [Eq. (5.139)] for Surface-Wave Spectra ($T = 250$ s)[a]

ϕ_0 (deg)	t_f/T 0.1	0.2	0.3	0.4	0.5	0.6	0.7	0.8	0.9	1.0	1.1	1.2	1.3	1.4	1.5	1.6	1.7	1.8	1.9	2.0
0	0.998	0.990	0.978	0.961	0.939	0.913	0.883	0.849	0.811	0.770	0.726	0.680	0.631	0.581	0.529	0.477	0.424	0.371	0.318	0.266
5	0.997	0.990	0.978	0.960	0.939	0.912	0.882	0.847	0.809	0.768	0.723	0.677	0.628	0.577	0.525	0.472	0.418	0.365	0.312	0.260
10	0.997	0.990	0.977	0.959	0.936	0.909	0.878	0.842	0.803	0.760	0.714	0.666	0.616	0.564	0.511	0.456	0.402	0.348	0.294	0.242
15	0.997	0.989	0.975	0.957	0.933	0.904	0.871	0.833	0.792	0.747	0.699	0.649	0.596	0.542	0.487	0.431	0.375	0.320	0.265	0.212
20	0.997	0.988	0.974	0.953	0.927	0.897	0.861	0.821	0.776	0.729	0.678	0.624	0.569	0.512	0.454	0.396	0.338	0.281	0.225	0.171
25	0.997	0.987	0.971	0.949	0.920	0.887	0.848	0.804	0.756	0.705	0.650	0.593	0.533	0.473	0.412	0.351	0.291	0.232	0.175	0.121
30	0.996	0.986	0.968	0.943	0.912	0.874	0.831	0.783	0.731	0.675	0.615	0.553	0.490	0.426	0.362	0.298	0.236	0.176	0.118	0.064
35	0.996	0.984	0.964	0.936	0.901	0.859	0.811	0.758	0.700	0.639	0.574	0.507	0.439	0.371	0.303	0.237	0.173	0.113	0.056	0.003
40	0.995	0.982	0.959	0.927	0.888	0.841	0.788	0.729	0.664	0.597	0.526	0.454	0.381	0.309	0.238	0.171	0.106	0.046	0.009	0.058
45	0.995	0.979	0.953	0.917	0.873	0.820	0.760	0.694	0.623	0.549	0.472	0.395	0.317	0.242	0.169	0.101	0.037	0.020	0.071	0.115
50	0.994	0.976	0.946	0.906	0.855	0.796	0.729	0.655	0.577	0.496	0.413	0.330	0.249	0.171	0.098	0.031	0.030	0.083	0.127	0.163
55	0.993	0.973	0.939	0.893	0.836	0.769	0.694	0.612	0.526	0.438	0.349	0.262	0.178	0.099	0.027	0.037	0.092	0.137	0.172	0.197
60	0.992	0.969	0.930	0.878	0.813	0.738	0.655	0.565	0.471	0.376	0.282	0.192	0.107	0.029	0.039	0.097	0.144	0.180	0.203	0.215
65	0.991	0.964	0.921	0.862	0.789	0.705	0.612	0.514	0.413	0.312	0.214	0.122	0.038	0.036	0.099	0.148	0.184	0.207	0.217	0.214
70	0.990	0.959	0.910	0.843	0.762	0.669	0.567	0.460	0.352	0.246	0.146	0.054	0.027	0.095	0.148	0.186	0.209	0.217	0.212	0.195
75	0.988	0.954	0.898	0.824	0.733	0.630	0.519	0.405	0.290	0.181	0.080	0.010	0.085	0.143	0.185	0.209	0.217	0.210	0.190	0.159

80	0.987	0.948	0.886	0.803	0.702	0.590	0.470	0.348	0.228	0.117	0.017	0.067	0.133	0.181	0.208	0.217	0.209	0.186	0.152	0.110
85	0.985	0.942	0.873	0.780	0.670	0.548	0.419	0.290	0.168	0.056	0.040	0.116	0.171	0.205	0.217	0.210	0.186	0.150	0.105	0.055
90	0.984	0.935	0.858	0.757	0.637	0.505	0.368	0.234	0.109	0.000	0.089	0.156	0.198	0.216	0.212	0.189	0.151	0.104	0.052	0.000
95	0.982	0.929	0.844	0.733	0.602	0.461	0.317	0.179	0.055	0.050	0.131	0.186	0.213	0.215	0.195	0.158	0.109	0.054	0.001	0.050
100	0.980	0.921	0.828	0.708	0.567	0.417	0.268	0.127	0.004	0.094	0.165	0.205	0.217	0.203	0.168	0.119	0.062	0.004	0.048	0.089
105	0.978	0.914	0.813	0.682	0.533	0.375	0.220	0.078	0.041	0.131	0.190	0.215	0.211	0.182	0.134	0.076	0.016	0.040	0.085	0.115
110	0.976	0.907	0.797	0.657	0.498	0.333	0.174	0.034	0.080	0.161	0.206	0.217	0.197	0.154	0.097	0.033	0.027	0.077	0.112	0.127
115	0.974	0.899	0.781	0.632	0.465	0.293	0.132	0.007	0.114	0.184	0.215	0.211	0.177	0.123	0.058	0.007	0.063	0.104	0.125	0.126
120	0.972	0.891	0.766	0.608	0.432	0.255	0.093	0.043	0.142	0.200	0.217	0.199	0.152	0.089	0.021	0.042	0.092	0.121	0.128	0.114
125	0.970	0.884	0.751	0.584	0.401	0.220	0.057	0.074	0.164	0.210	0.214	0.182	0.125	0.056	0.013	0.072	0.111	0.128	0.121	0.095
130	0.968	0.877	0.736	0.562	0.372	0.187	0.025	0.101	0.182	0.216	0.207	0.163	0.098	0.025	0.043	0.094	0.123	0.127	0.108	0.071
135	0.967	0.870	0.723	0.541	0.345	0.158	0.003	0.123	0.195	0.217	0.196	0.143	0.071	0.003	0.067	0.110	0.128	0.120	0.089	0.045
140	0.965	0.864	0.710	0.521	0.321	0.131	0.027	0.142	0.204	0.216	0.184	0.122	0.047	0.027	0.086	0.121	0.127	0.108	0.070	0.021
145	0.963	0.858	0.698	0.504	0.299	0.108	0.048	0.157	0.210	0.212	0.171	0.103	0.025	0.048	0.101	0.126	0.123	0.095	0.050	0.001
150	0.962	0.853	0.688	0.488	0.280	0.088	0.066	0.168	0.214	0.207	0.159	0.085	0.005	0.064	0.111	0.128	0.116	0.081	0.031	0.020
155	0.961	0.848	0.679	0.475	0.264	0.071	0.080	0.177	0.216	0.202	0.147	0.070	0.011	0.077	0.118	0.128	0.108	0.067	0.015	0.035
160	0.960	0.845	0.671	0.464	0.250	0.058	0.092	0.184	0.217	0.196	0.137	0.057	0.023	0.087	0.122	0.126	0.101	0.055	0.002	0.046
165	0.959	0.842	0.665	0.455	0.239	0.047	0.100	0.189	0.217	0.192	0.128	0.047	0.033	0.094	0.125	0.124	0.094	0.046	0.008	0.054
170	0.958	0.839	0.661	0.448	0.232	0.040	0.106	0.192	0.217	0.188	0.122	0.039	0.040	0.098	0.126	0.122	0.089	0.039	0.015	0.060
175	0.958	0.838	0.658	0.445	0.227	0.035	0.110	0.194	0.217	0.186	0.118	0.035	0.044	0.101	0.127	0.120	0.086	0.035	0.019	0.063
180	0.958	0.838	0.657	0.443	0.226	0.034	0.111	0.195	0.217	0.185	0.117	0.034	0.045	0.102	0.127	0.120	0.084	0.033	0.021	0.064

[a] The underlined values are zero-crossings of the finiteness factor.

Table 5.5. Absolute Values of the Finiteness Factor $|(\sin X)/X|$ for Surface-Wave Spectra ($T = 100$ s)[a]

ϕ_0	t_f/T																			
(deg)	0.1	0.2	0.3	0.4	0.5	0.6	0.7	0.8	0.9	1.0	1.1	1.2	1.3	1.4	1.5	1.6	1.7	1.8	1.9	2.0
0	0.998	0.994	0.987	0.976	0.963	0.947	0.928	0.907	0.883	0.856	0.828	0.797	0.765	0.730	0.694	0.657	0.619	0.579	0.539	0.498
5	0.998	0.994	0.986	0.976	0.962	0.946	0.927	0.905	0.881	0.854	0.825	0.794	0.761	0.726	0.690	0.652	0.613	0.573	0.532	0.491
10	0.998	0.994	0.986	0.974	0.960	0.943	0.923	0.900	0.875	0.847	0.816	0.784	0.749	0.713	0.675	0.636	0.595	0.554	0.512	0.469
15	0.998	0.993	0.984	0.972	0.957	0.938	0.916	0.892	0.864	0.834	0.801	0.766	0.730	0.691	0.650	0.609	0.566	0.552	0.478	0.433
20	0.998	0.992	0.983	0.969	0.952	0.931	0.907	0.880	0.849	0.816	0.780	0.742	0.701	0.659	0.615	0.570	0.524	0.477	0.430	0.383
25	0.998	0.991	0.980	0.965	0.945	0.922	0.895	0.864	0.829	0.792	0.752	0.709	0.664	0.618	0.570	0.520	0.470	0.420	0.370	0.320
30	0.997	0.990	0.977	0.959	0.937	0.910	0.878	0.843	0.804	0.762	0.716	0.668	0.618	0.566	0.513	0.459	0.405	0.351	0.297	0.245
35	0.997	0.988	0.973	0.952	0.926	0.895	0.858	0.818	0.773	0.724	0.673	0.618	0.562	0.505	0.446	0.388	0.329	0.272	0.216	0.161
40	0.996	0.986	0.968	0.944	0.913	0.877	0.834	0.787	0.735	0.680	0.621	0.560	0.497	0.434	0.370	0.307	0.245	0.185	0.128	0.073
45	0.996	0.983	0.962	0.934	0.898	0.855	0.806	0.751	0.691	0.628	0.562	0.494	0.424	0.355	0.287	0.220	0.156	0.095	0.039	0.013
50	0.995	0.980	0.956	0.922	0.880	0.829	0.772	0.709	0.641	0.570	0.495	0.420	0.345	0.271	0.199	0.130	0.066	0.007	0.046	0.092
55	0.994	0.976	0.948	0.908	0.858	0.800	0.734	0.662	0.585	0.505	0.423	0.341	0.260	0.183	0.110	0.042	0.019	0.073	0.119	0.156
60	0.993	0.972	0.938	0.892	0.835	0.767	0.692	0.610	0.523	0.435	0.346	0.258	0.174	0.096	0.024	0.040	0.095	0.140	0.174	0.199
65	0.992	0.968	0.928	0.874	0.808	0.730	0.645	0.553	0.457	0.361	0.266	0.175	0.090	0.013	0.054	0.111	0.155	0.188	0.208	0.217
70	0.990	0.962	0.916	0.854	0.778	0.690	0.594	0.492	0.388	0.285	0.186	0.093	0.010	0.061	0.120	0.165	0.196	0.213	0.217	0.209
75	0.989	0.956	0.904	0.832	0.746	0.647	0.540	0.429	0.317	0.209	0.108	0.017	0.061	0.124	0.171	0.201	0.216	0.215	0.201	0.176

80	0.987	0.950	0.890	0.809	0.711	0.601	0.484	0.364	0.246	0.135	0.034	0.052	0.121	0.171	0.203	0.217	0.213	0.195	0.164	0.125
85	0.986	0.943	0.874	0.783	0.675	0.554	0.426	0.298	0.176	0.064	0.032	0.110	0.167	0.202	0.217	0.212	0.190	0.156	0.112	0.063
90	0.984	0.935	0.858	0.757	0.637	0.505	0.368	0.234	0.109	0.000	0.089	0.156	0.198	0.216	0.212	0.189	0.151	0.104	0.052	0.000
95	0.982	0.928	0.842	0.729	0.597	0.455	0.310	0.172	0.047	0.057	0.137	0.189	0.214	0.214	0.192	0.153	0.102	0.047	0.008	0.056
100	0.979	0.919	0.824	0.701	0.558	0.405	0.254	0.113	0.009	0.105	0.173	0.209	0.216	0.198	0.159	0.107	0.049	0.009	0.059	0.098
105	0.977	0.911	0.806	0.671	0.518	0.357	0.200	0.059	0.058	0.145	0.198	0.217	0.206	0.171	0.119	0.058	0.003	0.057	0.098	0.122
110	0.975	0.902	0.788	0.642	0.478	0.309	0.149	0.009	0.101	0.175	0.212	0.214	0.186	0.136	0.074	0.009	0.049	0.094	0.121	0.128
115	0.973	0.893	0.770	0.614	0.440	0.264	0.102	0.035	0.136	0.197	0.217	0.202	0.159	0.097	0.030	0.034	0.085	0.118	0.128	0.118
120	0.970	0.885	0.752	0.585	0.403	0.222	0.059	0.073	0.163	0.210	0.214	0.183	0.127	0.058	0.011	0.070	0.110	0.128	0.122	0.096
125	0.968	0.876	0.734	0.558	0.368	0.182	0.020	0.105	0.184	0.216	0.205	0.160	0.094	0.020	0.047	0.097	0.124	0.126	0.105	0.067
130	0.966	0.868	0.717	0.532	0.335	0.146	0.014	0.131	0.199	0.217	0.191	0.134	0.061	0.014	0.076	0.115	0.128	0.115	0.081	0.035
135	0.964	0.860	0.701	0.508	0.305	0.114	0.043	0.153	0.209	0.213	0.175	0.108	0.030	0.043	0.097	0.125	0.124	0.098	0.055	0.004
140	0.962	0.852	0.686	0.486	0.277	0.085	0.068	0.170	0.215	0.206	0.157	0.083	0.003	0.067	0.112	0.128	0.115	0.078	0.029	0.022
145	0.960	0.845	0.672	0.466	0.253	0.060	0.089	0.183	0.217	0.197	0.139	0.059	0.021	0.085	0.122	0.126	0.102	0.058	0.005	0.044
150	0.958	0.839	0.660	0.448	0.231	0.039	0.107	0.193	0.217	0.188	0.122	0.039	0.040	0.099	0.127	0.121	0.088	0.038	0.016	0.060
155	0.957	0.834	0.650	0.432	0.213	0.021	0.121	0.200	0.216	0.178	0.106	0.021	0.056	0.109	0.128	0.114	0.075	0.021	0.032	0.072
160	0.956	0.829	0.641	0.420	0.198	0.007	0.132	0.205	0.214	0.170	0.093	0.007	0.068	0.115	0.128	0.107	0.062	0.007	0.045	0.080
165	0.955	0.826	0.634	0.410	0.186	0.004	0.140	0.208	0.211	0.162	0.083	0.004	0.077	0.120	0.127	0.101	0.052	0.004	0.054	0.084
170	0.954	0.823	0.629	0.402	0.178	0.012	0.145	0.210	0.209	0.157	0.075	0.012	0.083	0.122	0.125	0.096	0.045	0.012	0.060	0.087
175	0.953	0.821	0.626	0.398	0.173	0.017	0.148	0.211	0.208	0.154	0.070	0.017	0.086	0.124	0.124	0.092	0.040	0.017	0.063	0.088
180	0.953	0.821	0.625	0.397	0.171	0.018	0.149	0.211	0.208	0.152	0.069	0.018	0.087	0.124	0.124	0.091	0.039	0.018	0.064	0.089

[a] The underlined values are zero-crossings of the finiteness factor.

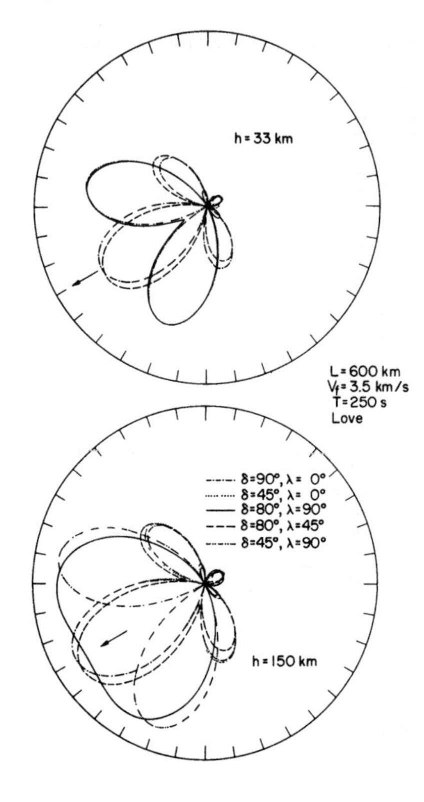

Figure 5.21. The effect of the source's finiteness and motion on the spectral amplitude radiation pattern. The arrows indicate the direction of the moving fault.

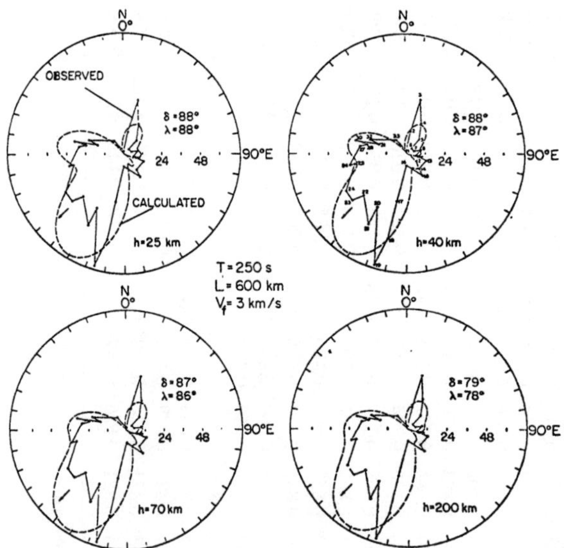

Figure 5.22. Comparison between the observed and calculated Love-wave radiation patterns for the Alaskan earthquake of 28 March, 1964. Arrow indicates the direction of the moving fault.

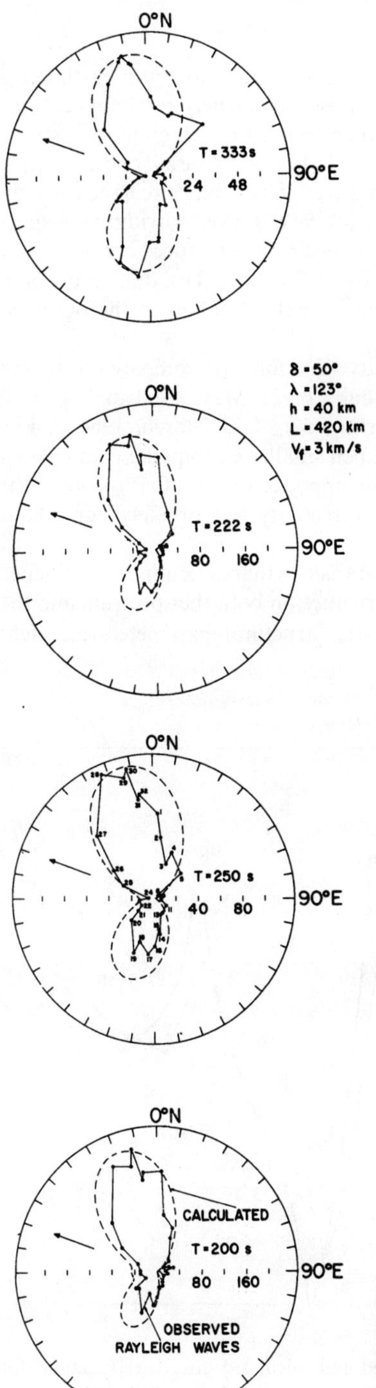

Figure 5.23. Comparison between the observed and calculated Rayleigh-wave radiation patterns for the Rat Island earthquake of 4 February, 1965. Arrow indicates the direction of the moving fault.

for the determination of the fault length and rupture velocity of earthquake faults. The method is useful especially in those cases where the faults are not accessible for field measurements, being under water or on high mountains, or when they do not break the surface at all. This is done with the aid of the multiple arrivals of long-period Love and Rayleigh waves, which may circle the earth up to 10 times or even more in the case of major earthquakes. Dividing the spectrum of an even-order wave, say R_{2n}, by the spectrum of its successor, R_{2n+1}, or its predecessor, R_{2n-1}, we get the directivity function. The derivation of L and V_f is then just a matter of fitting the theoretical values to the observed spectral ratios.

The calculated values of the directivity function are compared with the observed values for the Chilean earthquake of 22 May, 1960 in Fig. 5.24. The same is shown for the Alaskan earthquake of 28 March, 1964 in Fig. 5.25a. In Fig. 5.25b the directivity for a much smaller earthquake is calculated from the spectral ratio at two diametrically opposite stations that recorded the same wave. The best fit to the observed directivity was obtained for a fault length of 90 km and $V_f = 3$ km/s.

Tables 5.6a–5.6h contain the essential parameters that determine the influence of the earth's crustal and upper-mantle structure on both the spectrum and the real-earth seismograms of Rayleigh waves. Structural parameters for eight

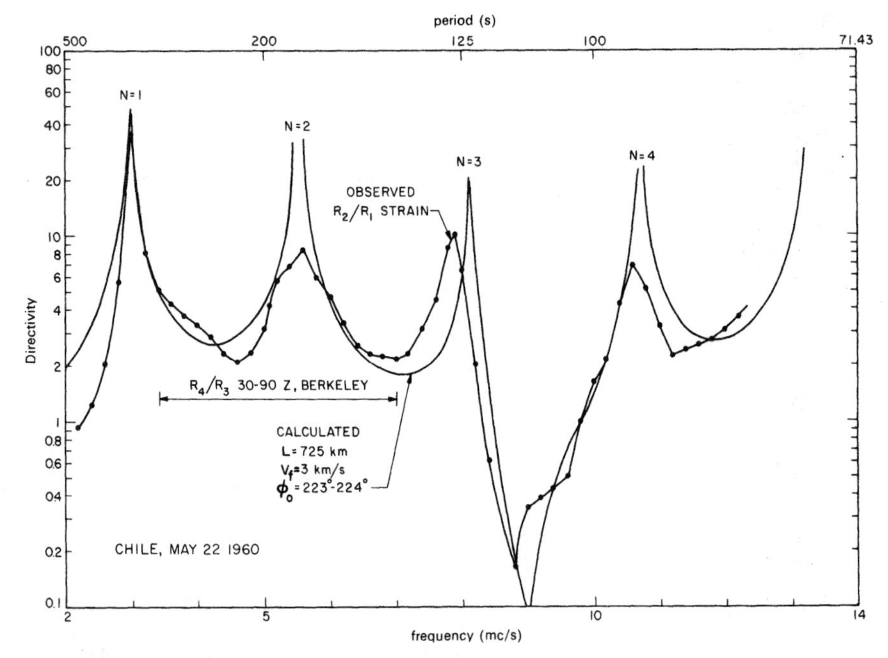

Figure 5.24. Comparison between the observed and calculated directivity functions for the Chilean earthquake of 22 May, 1960. Interference is observed up to the fourth order.

different geologic provinces are given with their corresponding amplitude functions. For periods in the range 14–100 s, the tables list the phase velocity c_R, the group velocity U_g, the surface ellipticity ε_0, the wavelength Λ, the medium spectral transfer functions, and the first and second derivatives of the group velocity. Calculations are made for the fundamental Rayleigh mode only for few common source depths. In Fig. 5.26 we show radiation patterns of Love and Rayleigh waves for an intermediate earthquake at $T = 20$ s. This pattern is mostly influenced by the crustal structure (which in this case is oceanic; Table 5.6c) and the rupture time, which is more than twice the wave's period. Figure 5.27 shows the extent to which the electronic computer can simulate a true earthquake record. A synthesized Rayleigh wavetrain is obtained by applying the Fourier transform to the spectral amplitudes in Eq. (5.38) with a step-function time dependence of the source. The response of the seismograph as well as the explicit dependence of γ_R on ω are taken into account in the calculated waveform. Figures 5.28 and 5.29 show seismogram traces of two of the greatest earthquakes that occured during the present century. In Fig. 5.28 we see the wavetrain of R_2 from the so-called "India Independence-Day earthquake" recorded on a strain meter [cf. Eq. (5.44)] in Southern California. The

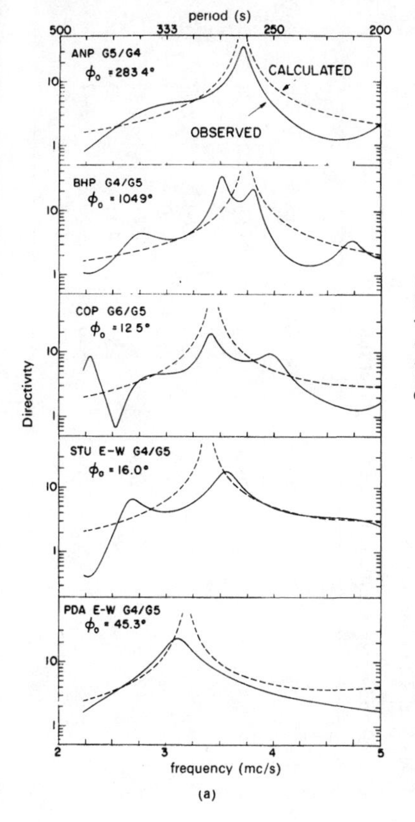

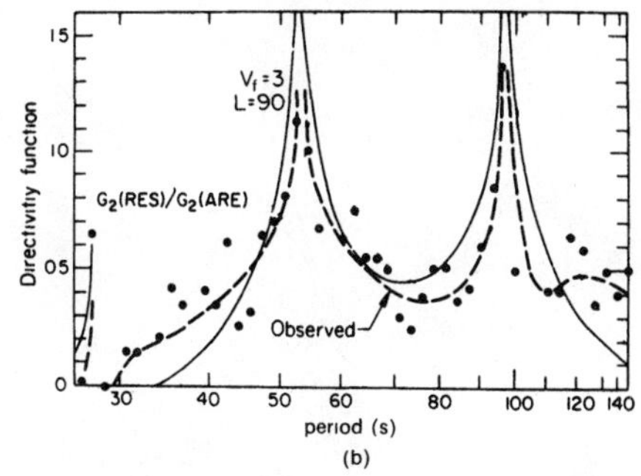

Figure 5.25. Comparison between the observed and calculated directivity functions. (a) Alaskan earthquake of 28 March, 1964. $L = 600$ km; $V_f = 3$ km/s; ANP = Anpu, Formosa (Taiwan); BHP = Balboa Heights (Panama); COP = Copenhagen (Denmark); STU = Stuttgart (Germany); PDA = Ponta Delgada (Azores); (b) Venezuela earthquake of 29 July, 1967. RES = Resolute Bay (Canada); ARE = Arequipa (Peru).

Table 5.6a. Continental U.S.A

Crustal and upper-mantle structure (H = layer thickness)

H (km)	α (km/s)	β (km/s)	ρ (g/cm^3)	μ, 10^{11} (dyne/cm^2)	λ, 10^{11} (dyne/cm^2)
28.0	6.15	3.55	2.74	3.453	3.457
12.0	6.70	3.80	3.00	4.332	4.803
13.0	7.96	4.60	3.37	7.131	7.091
25.0	7.85	4.50	3.39	6.864	7.161
50.0	7.85	4.41	3.42	6.651	7.772
75.0	7.85	4.41	3.45	6.710	7.841
50.0	8.20	4.50	3.47	7.027	9.279
∞	8.40	4.60	3.50	7.406	9.884

Spectral functions of Rayleigh waves R_{11}; $h = 15$ km

T (s)	c_R (km/s)	U_g (km/s)	ε_0	Λ (km)	P_R [μm/(m × km^2)]	Q_R [μm/(m × km^2)]	S_R [μm/(m × km^2)]	$10^2 \left\| \frac{dU_g}{dT} \right\|$ (km/s^2)	$10^3 \left\| \frac{d^2U_g}{dT^2} \right\|$ (km/s^3)
14	3.344	3.075	0.670	46.8	9.466	37.512	38.756	3.174	0.518
16	3.391	3.015	0.665	54.2	7.133	31.764	32.774	2.743	3.699
18	3.449	2.970	0.661	62.1	5.030	26.576	27.048	1.666	6.958
20	3.513	2.952	0.659	70.2	3.162	21.879	21.536	0.012	9.345
22	3.580	2.971	0.659	78.8	1.580	17.686	16.378	1.915	9.531
26	3.703	3.112	0.672	96.3	−0.547	11.101	8.115	4.742	3.683
28	3.753	3.212	0.683	105.0	−1.119	8.762	5.285	5.129	0.326
30	3.794	3.314	0.696	113.8	−1.454	6.975	3.241	4.954	1.871
34	3.853	3.492	0.724	131.0	−1.702	4.610	0.822	3.851	3.123
40	3.904	3.671	0.763	156.1	−1.654	2.770	−0.646	2.213	2.185
50	3.946	3.810	0.810	197.3	−1.390	1.470	−1.245	0.801	0.832
60	3.970	3.860	0.836	238.2	−1.157	0.911	−1.278	0.272	0.309
70	3.986	3.876	0.847	279.0	−0.977	0.615	−1.190	0.082	0.101
75	3.994	3.879	0.849	299.5	−0.903	0.516	−1.138	0.045	0.048
80	4.002	3.880	0.850	320.1	−0.835	0.438	−1.076	0.031	0.013
85	4.009	3.882	0.850	340.8	−0.775	0.375	−1.019	0.030	0.009
90	4.017	3.884	0.845	361.5	−0.720	0.323	−0.963	0.038	0.022
95	4.025	3.886	0.847	382.3	−0.671	0.280	−0.916	0.050	0.029
100	4.032	3.889	0.845	403.2	−0.626	0.244	−0.860	0.065	0.032

Table 5.6b. Canadian Shield

Crustal and upper-mantle structure

H (km)	α (km/s)	β (km/s)	ρ (g/cm^3)	μ, 10^{11} (dyne/cm^2)	λ, 10^{11} (dyne/cm^2)
6.0	5.64	3.47	2.70	3.251	2.087
10.5	6.15	3.64	2.80	3.710	3.171
18.7	6.60	3.85	2.85	4.224	3.966
80.0	8.10	4.72	3.30	7.352	6.948
100.0	8.20	4.54	3.44	7.090	8.950
100.0	8.30	4.51	3.53	7.180	9.958
80.0	8.70	4.76	3.60	8.157	10.935
∞	9.30	5.12	3.76	9.857	12.807

Spectral functions of Rayleigh waves R_{11}; $h = 15$ km

T (s)	c_R (km/s)	U_g (km/s)	ε_0	Λ (km)	P_R [μm/(m × km^2)]	Q_R [μm/(m × km^2)]	S_R [μm/(m × km^2)]	$10^2 \left\lvert \frac{dU_g}{dT} \right\rvert$ (km/s^2)	$10^3 \left\lvert \frac{d^2U_g}{dT^2} \right\rvert$ (km/s^3)
14	3.456	3.072	0.725	48.4	7.635	38.925	40.421	1.961	—
16	3.522	3.038	0.726	56.8	4.911	31.724	32.380	1.303	5.414
18	3.595	3.025	0.727	64.7	2.670	25.474	25.210	0.125	8.529
20	3.671	3.045	0.730	73.4	0.887	20.117	18.846	1.951	9.230
22	3.745	3.102	0.736	82.4	−0.434	15.666	13.438	3.638	7.264
26	3.869	3.288	0.756	100.6	−1.849	9.390	5.900	5.237	0.800
28	3.916	3.393	0.768	109.6	−2.120	7.344	3.594	5.170	1.343
30	3.955	3.493	0.781	118.6	−2.225	5.824	1.993	4.780	2.459
34	4.012	3.663	0.806	136.4	−2.178	3.838	1.538	3.682	2.804
40	4.061	3.838	0.837	162.4	−1.911	2.288	−0.923	2.247	1.942
50	4.099	3.987	0.868	204.9	−1.488	1.188	−1.296	0.914	0.865
60	4.115	4.044	0.880	246.9	−1.194	0.724	−1.253	0.309	0.411
70	4.126	4.059	0.881	288.9	−0.990	0.486	−1.141	0.009	0.213
75	4.131	4.057	0.879	309.8	−0.910	0.408	−1.082	0.083	0.156
80	4.136	4.051	0.876	330.9	−0.842	0.347	−1.027	0.149	0.113
85	4.142	4.042	0.872	352.1	−0.782	0.298	−0.974	0.197	0.079
90	4.148	4.031	0.868	373.3	−0.730	0.259	−0.924	0.230	0.052
95	4.155	4.019	0.863	394.7	−0.683	0.226	−0.878	0.250	0.029
100	4.163	4.007	0.858	416.3	−0.641	0.198	−0.835	0.259	0.009

Table 5.6c. Pacific

Crustal and upper-mantle structure

H (km)	α (km/s)	β (km/s)	ρ (g/cm^3)	μ, 10^{11} ($dyne/cm^2$)	λ, 10^{11} ($dyne/cm^2$)
5.0	1.52	0.0	1.03	0.0	0.238
1.0	2.10	1.00	2.10	0.210	0.506
5.5	6.70	3.80	3.00	4.332	4.803
48.5	8.17	4.70	3.34	7.378	7.538
160.0	8.17	4.50	3.44	6.966	9.030
80.0	8.49	4.60	3.53	7.469	10.505
100.0	8.81	4.80	3.60	8.294	11.353
∞	9.32	5.19	3.77	10.155	12.437

Spectral functions of Rayleigh waves R_{11}; $h = 15, 30, 50, 75$ km

T (s)	c_R (km/s)	U_g (km/s)	ε_0	Λ (km)	P_R [μm/(m × km²)]	Q_R [μm/(m × km²)]	S_R [μm/(m × km²)]
$h = 15$ km							
14					6.733	60.543	68.329
16					−4.209	28.860	17.481
18					−5.616	17.549	4.733
20					−5.703	12.230	0.021
22					−5.488	9.126	−2.178
26					−4.880	5.700	−3.891
28					−4.581	4.677	−4.185
30					−4.303	3.907	−4.309
34					−3.811	2.841	−4.293
40					−3.222	1.898	−4.009
50					−2.526	1.103	−3.424
60					−2.053	0.711	−2.916
70					−1.715	0.491	−2.508
75					−1.581	0.416	−2.337
80					−1.464	0.355	−2.183
85					−1.360	0.307	−2.045
90					−1.269	0.266	−1.919
95					−1.186	0.233	−1.806
100					−1.112	0.205	−1.702

Table 5.6c. (*Continued*)

T (s)	c_R (km/s)	U_g (km/s)	ε_0	Λ (km)	P_R [μm/(m × km^2)]	Q_R [μm/(m × km^2)]	S_R [μm/(m × km^2)]	$10^2 \left\| \frac{dU_g}{dT} \right\|$ (km/s^2)	$10^3 \left\| \frac{d^2U_g}{dT^2} \right\|$ (km/s^3)
h = 30 km									
14	3.461	1.313	0.397	48.4	14.732	39.744	52.455	72.331	502.944
16	3.878	2.801	0.647	62.0	8.924	31.432	34.648	51.275	265.779
18	3.993	3.445	0.741	71.9	5.337	23.460	22.833	19.272	83.160
20	4.044	3.712	0.780	80.9	3.376	18.570	16.282	9.142	29.432
22	4.072	3.850	0.798	89.6	2.138	15.197	12.075	5.171	13.256
26	4.100	3.985	0.807	106.6	0.707	10.802	7.036	2.166	4.151
28	4.108	4.021	0.806	115.0	0.277	9.297	5.448	1.498	2.671
30	4.113	4.047	0.804	120.3	−0.041	8.086	4.230	1.096	1.811
34	4.120	4.077	0.797	140.1	−0.459	6.274	2.528	0.531	0.926
40	4.126	4.096	0.785	165.0	−0.778	4.502	1.039	0.156	0.322
50	4.133	4.097	0.766	205.6	−0.956	2.833	−0.130	0.089	0.141
60	4.141	4.083	0.750	248.5	−0.968	1.927	−0.616	0.185	0.063
70	4.152	4.062	0.736	290.6	−0.926	1.383	−0.822	0.230	0.030
75	4.159	4.050	0.730	311.9	−0.898	1.189	−0.873	0.242	0.018
80	4.167	4.038	0.724	333.3	−0.867	1.031	−0.901	0.248	0.007
85	4.176	4.025	0.719	354.9	−0.836	0.900	−0.915	0.249	0.003
90	4.185	4.013	0.714	376.6	−0.805	0.792	−0.917	0.245	0.013
95	4.196	4.001	0.709	398.6	−0.774	0.700	−0.912	0.237	0.022
100	4.207	3.990	0.705	420.7	−0.744	0.621	−0.902	0.223	0.032

h = 50 km			
14	4.311	10.240	13.646
16	5.443	15.020	17.438
18	4.694	14.279	15.301
20	3.990	13.114	13.283
22	3.389	11.945	11.555
26	2.442	9.862	8.824
28	2.071	8.973	7.746
30	1.754	8.181	6.819
34	1.249	6.851	5.324
40	0.720	5.350	3.726
50	0.202	3.706	2.097
60	−0.076	2.690	1.171
h = 75 km			
14	0.632	1.419	1.920
16	1.763	4.481	5.359
18	2.087	5.694	6.372
20	2.174	6.230	6.682
22	2.157	6.433	6.685
26	1.982	6.331	6.268
28	1.865	6.152	5.966
30	1.742	5.931	5.642
34	1.498	5.440	4.987
40	1.167	4.703	4.085
50	0.743	3.659	2.898
60	0.453	3.873	2.059

Table 5.6d. Northern Europe

Crustal and upper-mantle structure

H (km)	α (km/s)	β (km/s)	ρ (g/cm^3)	μ, 10^{11} (dyne/cm^2)	λ, 10^{11} (dyne/cm^2)
8.0	5.60	3.20	2.70	2.765	2.938
12.0	6.20	3.60	2.75	3.564	3.443
20.0	6.40	3.75	2.85	4.008	3.658
60.0	8.00	4.55	3.30	6.832	7.456
13.0	7.95	4.50	3.40	6.885	7.719
218.0	8.43	4.60	3.50	7.406	10.061
4.5	8.78	4.70	3.60	7.952	11.847
97.5	8.95	4.80	3.64	8.387	12.384
∞	9.63	5.35	3.73	10.676	13.238

Spectral functions of Rayleigh waves R_{11}; $h = 15$ km

T (s)	c_R (km/s)	U_g (km/s)	ε_0	Λ (km)	P_R [μm/(m × km^2)]	Q_R [μm/(m × km^2)]	S_R [μm/(m × km^2)]	$10^2 \left\lvert \frac{dU_g}{dT} \right\rvert$ (km/s^2)	$10^3 \left\lvert \frac{d^2U_g}{dT^2} \right\rvert$ (km/s^3)
14	3.312	2.993	0.713	46.4	8.415	39.764	39.961	0.454	4.091
16	3.363	2.984	0.721	53.8	5.514	32.531	31.439	0.929	0.669
18	3.419	2.966	0.726	61.5	3.294	26.592	24.649	0.739	2.485
20	3.478	2.958	0.731	69.5	1.566	21.633	19.029	0.010	4.823
22	3.540	2.969	0.736	77.9	0.232	17.464	14.296	1.100	5.828
26	3.659	3.057	0.749	95.1	−1.441	11.149	7.165	3.151	3.697
28	3.712	3.126	0.758	103.9	−1.876	8.886	4.692	3.695	1.745
30	3.758	3.202	0.768	112.7	−2.120	7.118	2.845	3.870	0.096
34	3.384	3.352	0.789	130.3	−2.254	4.698	0.545	3.510	1.548
40	3.913	3.532	0.821	156.5	−2.082	2.744	−0.922	2.500	1.567
50	3.991	3.718	0.858	199.5	−1.648	1.353	−1.473	1.354	0.784
60	4.039	3.822	0.879	242.3	−1.302	0.780	−1.424	0.776	0.419
70	4.071	3.881	0.888	285.0	−1.054	0.497	−1.270	0.435	0.275
75	4.085	3.900	0.889	306.4	−0.958	0.408	−1.191	0.310	0.228
80	4.097	3.913	0.889	327.7	−0.876	0.340	−1.118	0.206	0.190
85	4.109	3.921	0.888	349.2	−0.806	0.287	−1.050	0.119	0.158
90	4.121	3.925	0.886	370.9	−0.745	0.245	−0.987	0.047	0.130
95	4.132	3.926	0.884	392.5	−0.692	0.211	−0.931	0.011	0.105
100	4.144	3.924	0.881	414.4	−0.645	0.184	−0.879	0.058	0.082

Table 5.6e. Siberian Platform

Crustal and upper-mantle structure

H (km)	α (km/s)	β (km/s)	ρ (g/cm^3)	μ, 10^{11} (dyne/cm^2)	λ, 10^{11} (dyne/cm^2)
20.0	6.15	3.55	2.82	3.554	3.558
20.0	6.58	3.80	2.93	4.231	4.224
13.0	8.14	4.70	3.30	7.290	7.286
25.0	7.85	4.50	3.39	6.865	7.161
50.0	7.85	4.41	3.42	6.651	7.772
75.0	8.00	4.41	3.45	6.710	8.661
50.0	8.20	4.50	3.47	7.027	9.279
100.0	8.40	4.60	3.50	7.406	9.884
100.0	9.00	4.95	3.60	8.821	11.518
100.0	9.63	5.31	3.89	10.968	14.138
100.0	10.17	5.63	4.13	13.091	16.534
100.0	10.59	5.92	4.33	15.175	18.210
∞	10.96	6.14	4.49	16.927	20.080

Spectral functions of Rayleigh waves R_{11}; $h = 30$ km

T (s)	c_R (km/s)	U_g (km/s)	ε_0	Λ (km)	P_R [μm/(m × km^2)]	Q_R [μm/(m × km^2)]	S_R [μm/(m × km^2)]	$10^2 \left\lvert \frac{dU_g}{dT} \right\rvert$ (km/s^2)	$10^3 \left\lvert \frac{d^2U_g}{dT^2} \right\rvert$ (km/s^3)
14	3.391	3.089	0.670	47.5	9.331	22.379	32.395	1.903	0.841
16	3.442	3.054	0.668	55.1	8.838	31.222	30.762	1.516	3.146
18	3.500	3.032	0.668	63.0	7.862	21.550	30.359	0.628	5.686
20	3.561	3.032	0.669	71.2	6.574	20.073	27.240	0.700	7.360
22	3.623	3.061	0.673	79.7	5.151	18.113	23.315	2.195	7.267
26	3.733	3.197	0.688	97.0	2.576	13.664	15.187	4.283	2.973
28	3.777	3.286	0.699	105.7	1.619	11.609	11.796	4.537	0.070
30	3.814	3.376	0.711	114.4	0.899	9.828	9.055	4.369	2.927
34	3.867	3.533	0.736	131.5	0.006	7.123	5.287	3.442	2.635
40	3.915	3.695	0.770	156.6	−0.563	4.658	2.363	2.027	1.936
50	3.955	3.824	0.810	197.7	−0.795	2.659	0.489	0.735	0.697
60	3.977	3.867	0.830	238.6	−0.794	1.715	−0.162	0.205	0.337
70	3.995	3.874	0.839	279.6	−0.743	1.191	−0.422	0.030	0.160
75	4.004	3.871	0.840	300.3	−0.713	1.012	−0.488	0.098	0.115
80	4.013	3.865	0.840	321.0	−0.683	0.869	−0.530	0.148	0.086
85	4.023	3.856	0.836	342.0	−0.653	0.752	−0.556	0.186	0.066
90	4.034	3.846	0.836	363.1	−0.625	0.656	−0.570	0.215	0.052
95	4.045	3.835	0.834	384.3	−0.598	0.576	−0.576	0.239	0.041
100	4.057	3.822	0.831	405.7	−0.572	0.508	−0.576	0.257	0.034

Table 5.6f. Central Asia

Crustal and upper-mantle structure

H (km)	α (km/s)	β (km/s)	ρ (g/cm^3)	μ, 10^{11} (dyne/cm^2)	λ, 10^{11} (dyne/cm^2)
23.0	5.60	3.40	2.65	3.063	2.184
23.0	6.50	3.74	2.85	3.986	4.068
∞	8.10	4.34	3.40	6.404	9.499

Spectral functions of Rayleigh waves R_{11}; $h = 30$ km

T (s)	c_R (km/s)	U_g (km/s)	ε_0	Λ (km)	P_R [μm/(m × km^2)]	Q_R [μm/(m × km^2)]	S_R [μm/(m × km^2)]	$10^2 \left\lvert \frac{dU_g}{dT} \right\rvert$ (km/s^2)	$10^3 \left\lvert \frac{d^2U_g}{dT^2} \right\rvert$ (km/s^3)
14	3.196	2.952	0.692	44.7	9.107	23.806	31.092	1.379	1.249
16	3.237	2.927	0.691	51.8	8.924	23.967	31.629	1.141	1.227
18	3.282	2.907	0.691	59.0	8.275	23.130	30.653	0.846	1.816
20	3.331	2.894	0.693	66.6	7.324	21.670	28.596	0.389	2.785
22	3.382	2.892	0.695	74.4	6.200	19.833	25.797	0.264	3.709
26	3.486	2.935	0.705	90.6	3.847	15.682	19.141	1.889	3.963
28	3.534	2.980	0.718	98.9	2.789	13.625	15.821	2.605	3.111
30	3.579	3.038	0.720	107.4	1.884	11.718	12.790	3.110	1.906
35	3.670	3.205	0.744	128.4	0.344	7.905	7.048	3.357	0.652
40	3.733	3.360	0.770	149.3	−0.405	5.420	3.631	2.772	1.440
50	3.807	3.569	0.813	190.3	−0.853	2.857	0.810	1.502	0.967
60	3.845	3.681	0.840	230.7	−0.872	1.722	−0.130	0.808	0.474
70	3.867	3.743	0.855	270.7	−0.802	1.136	−0.467	0.469	0.235
75	3.876	3.763	0.858	290.7	−0.760	0.946	−0.543	0.369	0.169
80	3.883	3.780	0.861	310.6	−0.718	0.798	−0.586	0.297	0.124
85	3.889	3.793	0.862	330.5	−0.678	0.680	−0.607	0.243	0.092
90	3.895	3.805	0.862	350.5	−0.640	0.585	−0.615	0.203	0.069
95	3.900	3.814	0.861	370.5	−0.604	0.508	−0.614	0.173	0.053
100	3.904	3.822	0.860	390.4	−0.571	0.444	−0.607	0.150	0.041

Table 3.6g. Africa

Crustal and upper-mantle structure

H (km)	α (km/s)	β (km/s)	ρ (g/cm³)	μ, 10^{11} (dyne/cm²)	λ, 10^{11} (dyne/cm²)
22.0	6.03	3.53	2.78	3.464	3.180
15.0	6.70	3.80	3.00	4.332	4.803
13.0	7.96	4.60	3.37	7.131	7.091
25.0	7.85	4.50	3.39	6.865	7.161
50.0	7.85	4.41	3.42	6.651	7.772
75.0	8.00	4.41	3.45	6.710	8.661
50.0	8.20	4.50	3.47	7.027	9.279
∞	8.40	4.60	3.50	7.406	9.884

Spectral functions of Rayleigh waves R_{11}; $h = 15$ km

T (s)	c_R (km/s)	U_g (km/s)	ε_0	Λ (km)	P_R [μm/(m × km²)]	Q_R [μm/(m × km²)]	S_R [μm/(m × km²)]	$10^2 \left\lvert \frac{dU_g}{dT} \right\rvert$ (km/s²)	$10^3 \left\lvert \frac{d^2U_g}{dT^2} \right\rvert$ (km/s³)
14	3.373	3.018	0.672	47.2	9.643	38.167	42.358	2.440	—
16	3.435	2.978	0.670	54.9	6.777	31.817	34.582	1.427	6.670
18	3.503	2.965	0.669	56.0	4.283	26.010	27.242	0.188	9.202
20	3.574	2.988	0.672	71.5	2.227	20.831	20.525	2.107	9.534
22	3.642	3.048	0.678	80.1	0.652	16.401	14.731	3.816	7.161
26	3.753	3.238	0.700	97.6	−1.139	10.000	6.620	5.192	0.063
28	3.794	3.340	0.714	106.2	−1.538	7.890	4.151	4.944	2.205
30	3.827	3.434	0.729	114.8	−1.742	6.317	2.443	4.395	3.119
34	3.873	3.584	0.758	131.7	−1.840	4.667	0.835	3.125	2.978
40	3.914	3.725	0.793	156.6	−1.717	2.621	−0.724	1.702	1.780
50	3.948	3.830	0.830	197.4	−1.419	1.422	−1.220	0.581	0.648
60	3.969	3.865	0.846	238.1	−1.176	0.887	−1.238	0.175	0.231

Table 5.6h. Middle East

Crustal and upper-mantle structure

H (km)	α (km/s)	β (km/s)	ρ (g/cm^3)	μ, 10^{11} (dyne/cm^2)	λ, 10^{11} (dyne/cm^2)
2.6	3.42	2.08	2.30	0.995	0.700
12.0	5.68	3.36	2.55	2.879	2.469
20.0	6.10	3.78	2.80	4.001	2.417
60.0	8.10	4.69	3.30	7.259	7.134
100.0	8.10	4.50	3.40	6.885	8.537
290.0	8.80	4.60	3.65	7.723	12.819
200.0	9.80	5.30	3.98	11.180	15.864
400.0	11.15	6.20	4.43	17.029	21.017
240.0	11.78	6.48	4.63	19.442	25.367
∞	12.02	6.62	4.71	20.641	26.768

Spectral functions of Rayleigh waves R_{11}; $h = 15$ km

T (s)	c_R (km/s)	U_g (km/s)	ε_0	$\gamma_R \times 10^4$ (km^{-1})	Λ (km)	P_R [μm/(m × km^2)]	Q_R [μm/(m × km^2)]	S_R [μm/(m × km^2)]	$10^2 \left\lvert \frac{dU_g}{dT} \right\rvert$ (km/s^2)	$10^3 \left\lvert \frac{d^2U_g}{dT^2} \right\rvert$ (km/s^3)
14	3.191	2.635	0.924	4.50	44.6	8.204	49.723	55.594	1.660	3.968
16	3.287	2.663	0.930	2.75	52.6	3.902	38.681	41.767	1.319	0.821
18	3.384	2.694	0.921	1.75	60.6	0.747	29.721	30.681	1.976	5.461
20	3.478	2.747	0.918	1.10	69.0	−1.500	22.524	21.682	3.333	7.562
22	3.567	2.828	0.919	0.41	77.7	−2.968	16.863	14.536	4.790	6.521
26	3.714	3.056	0.932	0.55	95.2	−4.131	9.391	5.305	6.178	0.207
28	3.770	3.179	0.942	0.60	104.0	−4.170	7.111	2.697	5.982	2.042
30	3.815	3.293	0.952	0.75	112.9	−4.040	5.480	0.980	5.438	3.235
34	3.880	3.483	0.971	1.05	130.8	−3.597	3.450	−0.839	4.032	3.468
40	3.939	3.669	0.989	1.50	157.8	−2.933	1.966	−1.736	2.298	2.264
50	3.990	3.814	0.996	1.30	203.6	−2.163	0.976	−1.870	0.839	0.851
60	4.020	3.867	0.987	1.10	249.2	−1.679	0.573	−1.676	0.314	0.259
70	4.045	3.889	0.970	0.85	283.1	−1.348	0.369	−1.453	0.140	0.090
75	4.056	3.895	0.960	0.80	304.2	−1.219	0.303	−1.349	0.105	0.051
80	4.067	3.899	0.951	0.75	325.3	−1.108	0.251	−1.253	0.082	0.034
85	4.078	3.903	0.941	0.68	346.6	−1.011	0.211	−1.164	0.064	0.030
90	4.089	3.906	0.932	0.63	368.0	−0.927	0.179	−1.084	0.046	0.030
95	4.099	3.907	0.923	0.58	389.4	−0.853	0.153	−1.011	0.025	0.030
100	4.110	3.908	0.914	0.53	411.0	−0.789	0.131	−0.945	0.024	0.048

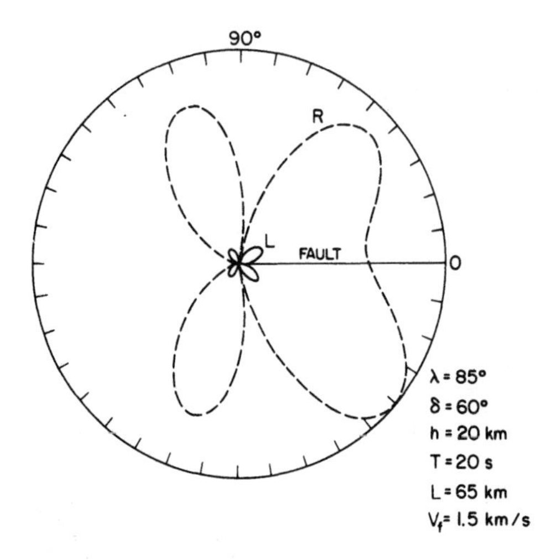

Figure 5.26. Radiation patterns of oceanic Rayleigh (R) and Love waves (L) from a moving source.

arrivals of the different periods that constitute the wave are indicated by the group-velocity scale in the figure. Note, in particular, the "Airy phase" at the wave's tail. Figure 5.29 is a "history-making" record on which the source's directivity was observed for the first time. The reader may note that the amplitude of G_4, for example, is much bigger than that of G_3, although the path length of G_4 exceeds that of G_3. Figure 5.30 shows two Rayleigh modes, R_{11} and R_{21}, from a deep-focus earthquake and Figs. 5.31 and 5.32 show crustal surface signals recorded at Eilat.

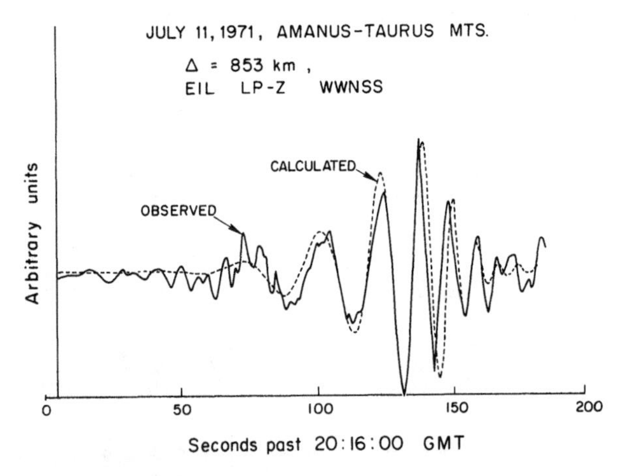

Figure 5.27. A computer vs. the earth: Simulation of source and structural conditions to match an observed vertical ground velocity recorded at Eilat (EIL) on a vertical Long-Period (LP-Z) standard seismograph. Path of wave is approximately along the Syrian–African rift valley.

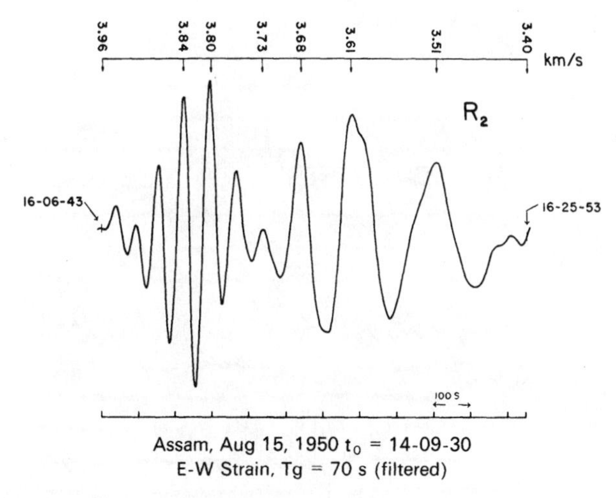

Figure 5.28. A Rayleigh wave traveling along the major arc from Assam to Pasadena, California.

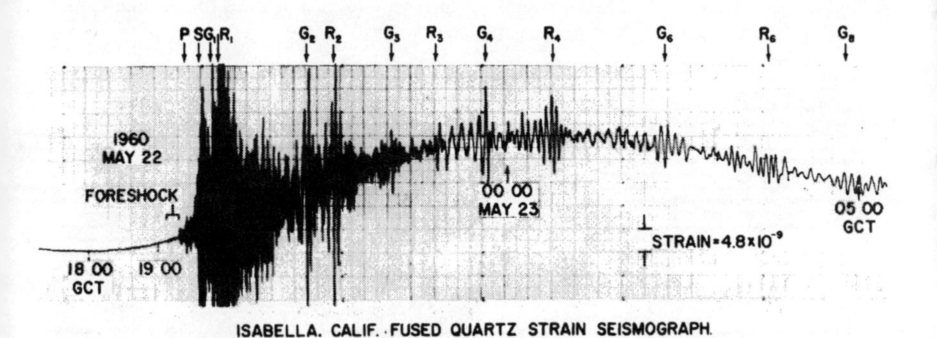

Figure 5.29. A gallery of long-period surface waves from the Chile earthquake of 22 May, 1960. The source's directivity can be observed through excessive amplitudes of even-order surface-wave arrivals.

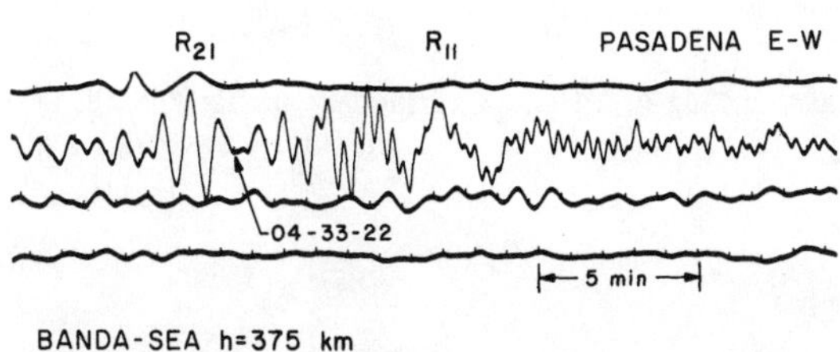

Figure 5.30. A higher mode Rayleigh wave arriving "on time" from a deep-focus earthquake.

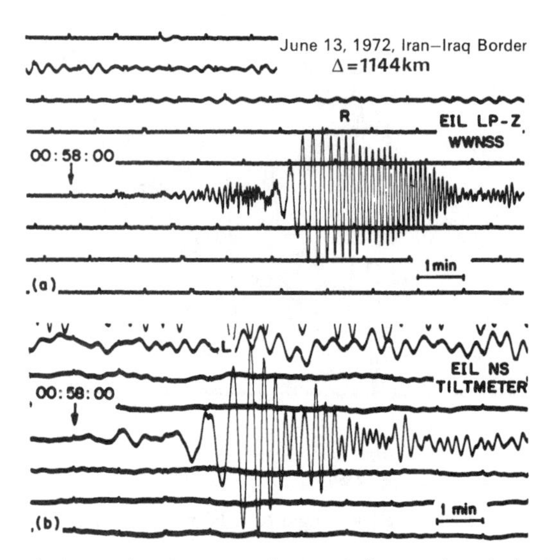

Figure 5.31. Typical crustal surface waves from a shallow earthquake in the middle east. (a) Worldwide standard long-period vertical ground motion at Eilat. (b) Ground motion at Eilat, recorded by a N-S mercury tiltmeter.

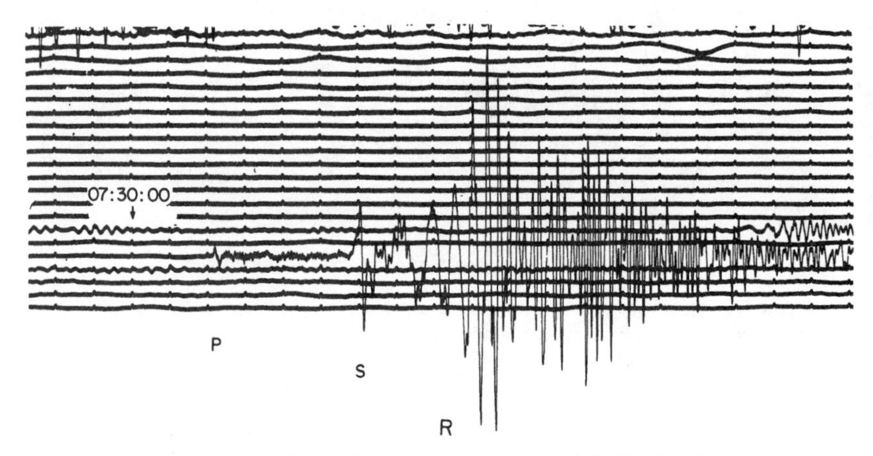

Figure 5.32. A recording of a shallow earthquake in Iran (58.2°E, 34.0°N) recorded at Eilat on 1 September, 1968. Crustal and mantle surface waves dominate the seismogram.

Bibliography

Aki K (1966) Generation and propagation of G waves from the Niigata earthquake of June 16, 1964. Bull Earthquake Res Inst (Tokyo) 44: 73–88.

Arkhangel'skaya VM (1964) A study of short-period seismic surface waves, II. Izv Geophys 9: 1334–1359. (English translation, pp. 807–821.)

Avetisyana RA, Yanovskaya TB (1973) Expansion of group velocities of Rayleigh waves in terms of spherical harmonics. Izv Phys Solid Earth 9: 706–710.

Backus GE (1962) The propagation of short elastic surface waves on a slowly rotating earth. Bull Seismol Soc Amer 52: 823–846.

Backus GE (1964) Geographical interpretation of measurements of average phase velocities of surface waves over great circular and great semicircular paths. Bull Seismol Soc Amer 54: 571–610.

Båth M, López Arroyo A (1962) Attenuation and dispersion of G-waves. Jour Geophys Res 67: 1933–1942.

Ben-Menahem A (1961) Radiation of seismic surface waves from finite moving sources. Bull Seismol Soc Amer 51: 401–435.

Ben-Menahem A (1971) The force system of the Chilean earthquake of 1960 May 22. Geophys Jour Roy Astron Soc (London) 25: 407–417.

Ben-Menahem A (1978) Source mechanism of the 1906 San-Francisco earthquake. Phys Earth Planet Int 17: 163–181.

Ben-Menahem A, Aboodi, E (1971) Tectonic patterns in the northern Red-Sea region. Jour Geophys Res 76: 2674–2689.

Ben-Menahem A, Aboodi E, Schild R (1974) The source of the great Assam earthquake—an interplate wedge motion. Phys Earth Planet Int 9: 265–289.

Ben-Menahem A, Harkrider DG (1964) Radiation patterns of seismic surface waves from buried dipolar point sources in a flat stratified earth. Jour Geophys Res 69: 2605–2620.

Ben-Menahem A, Rosenman M, Harkrider DG (1970) Fast evaluation of source parameters from isolated surface wave signals. Bull Seismol Soc Amer 60: 1337–1387.

Ben-Menahem A, Toksöz MN (1962) Source mechanism from spectra of long-period seismic surface waves. 1. The Mongolian earthquake of December 4, 1957. Jour Geophys Res 67: 1943–1955.

Ben-Menahem A, Toksöz, MN (1963a) Source mechanism from spectra of long-period seismic surface waves. 2. The Kamchatka earthquake of Nov. 4, 1952. Jour Geophys Res 68: 5207–5222.

Ben-Menahem A, Toksöz MN (1963b) Source mechanism from spectra of long-period seismic surface waves. 3. The Alaska earthquake of July 10, 1958. Bull Seismol Soc Amer 53: 905–919.

Brune JN (1961) Radiation pattern of Rayleigh waves from the southeast Alaska earthquake of July 10, 1958. Publ Dom Obs Ottawa 24: 373–383.

Brune JN, Dorman J (1963) Seismic waves and earth structure in the Canadian Shield. Bull Seismol Soc Amer 53: 167–209.

Gupta HK, Narain H (1967) Crustal structure in the Himalayan and Tibet plateau region from surface wave dispersion. Bull Seismol Soc Amer 57: 235–248.

Gutenberg B (1924) Dispersion und Extinction von seismischen Oberflächenwellen und der Aufbau der obersten Erdschichten. Phys Zeit 25: 377–381.

Gutenberg B (1932) Theorie der Erdbebenwellen. Handbuch der Geophysik, Vol 4, No 2, pp 1–298. Berlin, Verlag von Gebrüder Borntraeger.

Gutenberg B (1955) Channel waves in the earth's crust. Geophysics 20: 283–294.

Harkrider DG (1964) Surface waves in multilayered elastic media. I. Rayleigh and Love waves from buried sources in a multilayered elastic half-space. Bull Seismol Soc Amer 54: 627–679.

Harkrider DG (1970) Surface waves in multilayered elastic media. Part II. Higher mode spectra and spectral ratios from point sources in plane layered earth models. Bull Seismol Soc Amer 60: 1937–1987.

Keilis-Borok VI (1951) On the surface waves in a layer overlying a solid half-space (in Russian). Izv Akad Nauk, SSSR Ser Geograf I Geofiz 2: 17–39.

Knopoff L, Fonda AA (1975) Upper-mantle structure under the Arabian Peninsula. Tectonophysics 26: 121–134.

Kovach RL (1965) Seismic surface waves: some observations and recent developments. In: Ahrens LH, Press F, Runcorn SK, Urey HC (eds) Physics and Chemistry of the Earth, Vol 6, pp 251–314, Pergamon, New York.

Kovach RL (1978) Seismic surface waves and crustal and upper mantle structure. Rev Geophys Space Phys 16: 1–13.

Nakano H (1930) Love waves in cylindrical coordinates. Geophys Mag (Tokyo) 2: 37–51.

Rial JA (1978) The Caracas, Venezuela, earthquake of July 1967: A multiple-source event. Jour Geophys Res 83: 5405–5414.

Saito M, Takeuchi H (1966) Surface waves across the Pacific. Bull Seismol Soc Amer 56: 1067–1091.

Satô Y (1955) Analysis of dispersed surface waves by means of Fourier transform. Bull Earthquake Res Inst (Tokyo) 33: 33–48.

Satô Y (1958) Attenuation, dispersion, and the wave guide of the G wave. Bull Seismol Soc Amer 48: 231–251.

Scholte JGJ (1947). The range of existence of Rayleigh and Stoneley waves. Mon Not Roy Astron Soc (London) Geophys Suppl 5: 120–126.

Sezawa K (1927a) On the propagation of Rayleigh waves on plane and spherical surfaces. Bull Earthquake Res Inst (Tokyo) 2: 21–28.

Sezawa K (1927b) Dispersion of elastic waves propagated on the surface of stratified bodies and on curved surfaces. Bull Earthquake Res Inst (Tokyo) 3: 1–18.

Sezawa K (1935) Love waves generated from a source of a certain depth. Bull Earthquake Res Inst (Tokyo) 13: 1–17.

Sezawa K, Kanai K (1935) The M_2 seismic waves. Bull Earthquake Res Inst (Tokyo) 13: 471–475.

Sezewa K, Kanai K (1937) Relation between the thickness of a surface layer and the amplitude of dispersive Rayleigh waves. Bull Earthquake Res Inst (Tokyo) 15: 845–849.

Singh BM, Singh SJ, Chopra SD, Gogna ML (1976) On Love waves in a laterally and vertically heterogeneous layered media. Geophys Jour Roy Astron Soc (London) 45: 357–370.

Stoneley R (1924) Elastic waves at the surface of separation of two solids. Proc Roy Soç (London) A106: 416–428.

Stoneley R (1934) The transmission of Rayleigh waves in a heterogeneous medium. Mon Not Roy Astron Soc (London) Geophys Suppl 3: 222–232.

Takeuchi H, Dorman J, Saito M (1964) Partial derivatives of surface wave phase velocity with respect to physical parameter changes within the earth. Jour Geophys Res 69: 3429–3441.

Takeuchi H, Saito M (1972) Seismic surface waves. In: Bolt BA (ed) Methods of Computational Physics, Vol eleven, p 217–295. Academic Press, New York.

Warren DH, Healy JH (1973) Structure of the crust in the conterminous United States. Tectonophysics 20: 203–213.

Yanovskaya TB (1958) On the determination of the dynamic parameters of the focus hypocenter of an earthquake from records of surface waves. Izv Geophys Ser 289–301. (English translation, pp 161 167.)

CHAPTER 6

Normal-Mode Solution for Spherical Earth Models

. . . which shaketh the earth out of her place and the pillars thereof tremble.

(Job, 9;6)

6.1. Introduction

With the theory developed so far, we are now able to discuss the excitation of the free oscillations of the earth, which is one of the fundamental problems of seismology. Unless otherwise stated, the earth is assumed to be a radially heterogeneous, isotropic, gravitating sphere of radius a. The core of the earth is a concentric fluid sphere of radius b, and the region $b < r < a$, which we shall call the shell, is elastic.

In the theory of seismic waves, it is assumed that the energy is transmitted outward from the focus in the form of P, S, and surface waves. The emphasis is on the traveling waves, which affect only a relatively small part of the earth at any particular time. All such wave motions, however, can be looked at from a more general standpoint as belonging to some mode of vibration of the whole earth.

For small bodies the effect of mutual gravitational attraction of their particles is negligible compared to that of the elastic forces. For a sphere of the size and elasticity of the earth, however, the gravitational forces play an important role and must not be overlooked.

The vibrations of an elastic solid sphere were first considered by Poisson in 1829 (see Chapter 1). Lamb, in 1882, discussed the simpler modes of vibration of a uniform sphere and calculated the more important roots of the frequency equation. He classified the general types of vibrations of the sphere as of the "first class" and the "second class." In a vibration of the first class, the dilatation and the radial component of the displacement vanish everywhere, whereas in a vibration of the second class, the radial component of the curl of the displacement vanishes everywhere. Love presented the theory of oscillations of a uniform gravitating compressible sphere in 1911.

Following the Kamchatka earthquake of 4 November, 1952, Benioff (1958) recorded a ground motion of 58 min period with his strain seismograph. This stimulated a fundamental study of the free oscillations of the earth. The availability of high-speed electronic computers was fortunate at this stage.

The problem of the oscillations of an elastic sphere is so closely linked with the deformation of a sphere, either by surface tractions or by forces, that the two problems involve the same analysis. The static deformation is the limiting case of an oscillation in which the period is infinitely long. This will be reflected in the analysis that follows.

6.2. Oscillations of a Homogeneous Sphere

Let us first consider the earth as a homogeneous, nongravitating, elastic sphere. This relieves the free oscillation problem of its mathematical intricacies while retaining the important features.

In the case of a homogeneous, nongravitating medium, the Fourier-transformed equation of motion is Eq. (2.99)

$$\alpha^2 \operatorname{grad} \operatorname{div} \mathbf{u} - \beta^2 \operatorname{curl} \operatorname{curl} \mathbf{u} + \omega^2 \mathbf{u} = 0. \tag{6.1}$$

The displacement vector $\mathbf{u}$ satisfying the above equation has the general form

$$\mathbf{u} = \sum_{m,l} [\alpha_{ml} \mathbf{M}_{ml}^{+} + \beta_{ml} \mathbf{N}_{ml}^{+} + \gamma_{ml} \mathbf{L}_{ml}^{+}]. \tag{6.2}$$

We have selected only those eigenvectors that are bounded at the origin. Also, α_{ml}, etc., are arbitrary constants. Using Eq. (2.106) and Table 2.2, we get

$$\begin{aligned} \mathbf{e}_r \cdot \mathfrak{T}(\mathbf{u}) = \sum_{m,l} [&\hat{\alpha}_{ml}(r)\sqrt{\{l(l+1)\}}\mathbf{C}_{ml}(\theta, \phi) + \hat{\beta}_{ml}(r)\mathbf{P}_{ml}(\theta, \phi) \\ &+ \hat{\gamma}_{ml}(r)\sqrt{\{l(l+1)\}}\mathbf{B}_{ml}(\theta, \phi)], \end{aligned} \tag{6.3a}$$

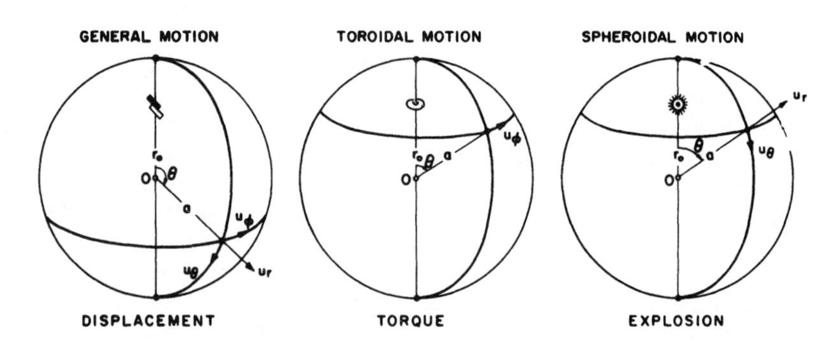

Figure 6.1. Various sources and the motion excited by them.

where

$$\begin{aligned}
\hat{\alpha}_{ml}(r) &= \mu k_\beta \eta F^+_{l,1}(\eta)\alpha_{ml}, \\
\hat{\beta}_{ml}(r) &= 2\mu[l(l+1)k_\beta F^+_{l,1}(\eta)\beta_{ml} + k_\alpha F^+_{l,3}(\xi)\gamma_{ml}], \\
\hat{\gamma}_{ml}(r) &= \mu[k_\beta F^+_{l,2}(\eta)\beta_{ml} + 2k_\alpha F^+_{l,1}(\xi)\gamma_{ml}], \\
\xi &= rk_\alpha, \qquad \eta = rk_\beta.
\end{aligned} \tag{6.3b}$$

The functions $F_{l,i}(z)$ are given by

$$\begin{aligned}
F^\pm_{l,1}(z) &= \frac{1}{z}\left(\frac{d}{dz} - \frac{1}{z}\right) f^\pm_l(z) = \frac{(l-1)}{z^2} f^\pm_l(z) - \frac{1}{z} f^\pm_{l+1}(z), \\
F^\pm_{l,2}(z) &= \left[\frac{2}{z^2}(l^2-1) - 1\right] f^\pm_l(z) + \frac{2}{z} f^\pm_{l+1}(z) \\
&= \frac{d^2 f^\pm_l}{dz^2} + \frac{(l-1)(l+2)}{z^2} f^\pm_l, \\
F^\pm_{l,3}(z) &= \left[\frac{1}{z^2} l(l-1) - \frac{1}{2}\left(\frac{\alpha}{\beta}\right)^2\right] f^\pm_l(z) + \frac{2}{z} f^\pm_{l+1}(z) = \frac{d^2}{dz^2} f^\pm_l - \frac{\lambda}{2\mu} f^\pm_l,
\end{aligned} \tag{6.4}$$

$f^+_l(z) = j_l(z)$; spherical Bessel function of the first kind,

$f^-_l(z) = h^{(2)}_l(z)$; spherical Hankel function of the second kind.

The boundary conditions are

$$\mathbf{e}_r \cdot \mathfrak{T}(\mathbf{u}) = 0 \quad \text{at} \quad r = a. \tag{6.5}$$

From Eq. (6.3b), we note that whereas $\hat{\alpha}_{ml}$ involves α_{ml} only, $\hat{\beta}_{ml}$ and $\hat{\gamma}_{ml}$ involve both β_{ml} and γ_{ml}, but not α_{ml}. Consequently, we may consider two types of oscillations, the first with $\beta_{ml} = \gamma_{ml} = 0$ and the second with $\alpha_{ml} = 0$. In the former case, **u** is made up of **M** vectors only, whereas in the latter case, **u** is made up of **N** and **L** vectors. The first type of oscillations is known as *toroidal*, *torsional*, or *oscillations of the first class.* The second type of oscillations is known as *spheroidal*, *poloidal*, or *oscillations of the second class.* Because

$$\operatorname{div} \mathbf{M} = 0, \qquad \mathbf{e}_r \cdot \mathbf{M} = 0, \tag{6.6}$$

we note that in the case of the toroidal oscillations, the dilatation and the radial component of the displacement vanish identically. Similarly, because

$$\mathbf{e}_r \cdot \operatorname{curl} \mathbf{N} = 0, \qquad \mathbf{e}_r \cdot \operatorname{curl} \mathbf{L} = 0, \tag{6.7}$$

the radial component of the curl of the displacement vanishes in the case of the spheroidal oscillations (Fig. 6.1).

6.2.1. Toroidal Oscillations

For the toroidal oscillations of a sphere, Eqs. (6.3) and (6.5) yield

$$F^{+}_{l,1}(ak_\beta) = 0.$$

Using Eq. (6.4), this may be written as

$$\chi \frac{d}{d\chi} j_l(\chi) = j_l(\chi), \tag{6.8}$$

or, as

$$\chi j_{l+1}(\chi) = (l-1) j_l(\chi), \tag{6.9}$$

where $\chi = ak_\beta$. This is the *frequency equation* or the *period equation* for the toroidal oscillations of a homogeneous sphere.

It is interesting to note that the frequency equation does not depend upon m. Because it is a transcendatal equation, for each l, Eq. (6.9) has an infinite number of roots. Let these roots be denoted by ${}_n\chi_l$; $n = 0, 1, 2, \ldots$. We may find the corresponding frequencies and periods through the relations

$${}_n\omega_l = \left(\frac{\beta}{a}\right) {}_n\chi_l, \qquad {}_nT_l = \frac{2\pi}{{}_n\omega_l}. \tag{6.10}$$

The ${}_n\omega_l$ are the *eigenfrequencies* and ${}_nT_l$ are the *eigenperiods*. From Eq. (6.2) the corresponding eigenfunctions are

$${}_n\mathbf{u}_{ml} = \alpha_{ml} j_l\left(\frac{r}{a}\, {}_n\chi_l\right) \sqrt{[l(l+1)]}\, \mathbf{C}_{ml}(\theta, \phi). \tag{6.11}$$

Figure 6.2 exhibits the toroidal oscillations of the first four orders for $m = 0$. The toroidal oscillations of order l are denoted by ${}_nT_l$; ${}_0T_l$ representing the fundamental mode and ${}_nT_l$ $(n = 1, 2, 3, \ldots)$ representing the nth overtone.

The eigenfunctions of the oscillations are governed by three numbers: l, m, and n. l is known as the *order of the oscillation* or the *colatitudinal mode*

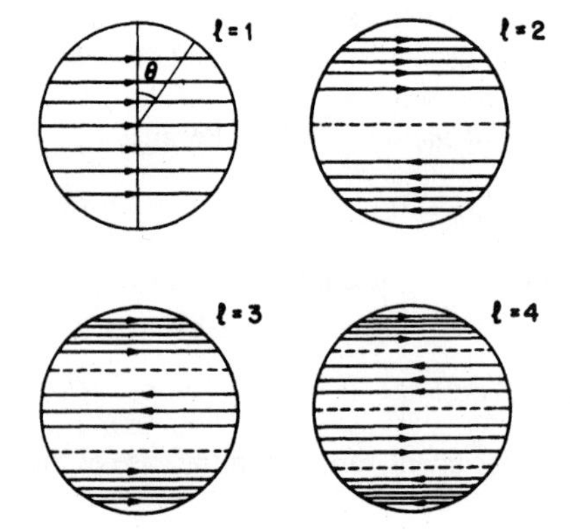

Figure 6.2. Toroidal oscillations of the first four orders. The broken lines represent nodal lines. Arrows show direction of particle motion.

number, m is known as the *azimuthal mode number*, and n is the *radial mode number*. The spherical harmonic

$$P_l^m(\cos\theta)(\cos m\phi, \sin m\phi)$$

from which the eigenfunction, Eq. (6.11), is derived has $(l - |m|)$ nodal latitude lines and $2|m|$ nodal longitude lines on the surface of the earth. The value of m depends upon the source exciting the oscillations and, for a nonrotating spherical-earth model, does not influence its periods of oscillations. This degeneracy is removed if the rotation of the earth is taken into account. For a given value of l, there exists a family of n concentric radial nodal surfaces, i.e., interior surfaces at which a particular eigenfunction vanishes. Some authors use $i = n + 1$ to denote the radial mode number. In that case $i = 1$ corresponds to the fundamental mode and $i = 2, 3$, etc., correspond to the first overtone, the second overtone, etc.

There are no toroidal oscillations corresponding to $l = 0$, because the corresponding displacement is identically zero. When $l = 1$, Eq. (6.11) gives

$${}_n\mathbf{u}_{m,1}^{c,s} = \alpha_{m,1} j_1\left(\frac{r}{a}\chi\right)\left(\mathbf{e}_\theta \frac{1}{\sin\theta}\frac{\partial}{\partial\phi} - \mathbf{e}_\phi \frac{\partial}{\partial\theta}\right) P_1^m(\cos\theta)(\cos m\phi, \sin m\phi),$$

where χ stands for ${}_n\chi_1$. Therefore, we have

$$\begin{aligned}
{}_n\mathbf{u}_{01}^c &= \alpha_{01} j_1\left(\frac{r}{a}\chi\right)\sin\theta\,\mathbf{e}_\phi = -\alpha_{01}\frac{1}{r} j_1\left(\frac{r}{a}\chi\right)(\mathbf{r}\times\mathbf{e}_3), \qquad {}_n\mathbf{u}_{01}^s = 0,\\
{}_n\mathbf{u}_{11}^c &= -\alpha_{11}\frac{1}{r} j_1\left(\frac{r}{a}\chi\right)(\mathbf{r}\times\mathbf{e}_1),\\
{}_n\mathbf{u}_{11}^s &= -\alpha_{11}\frac{1}{r} j_1\left(\frac{r}{a}\chi\right)(\mathbf{r}\times\mathbf{e}_2).
\end{aligned} \tag{6.12}$$

For $n > 0$ any spherical surface $r =$ const. oscillates as a rigid surface about the x_1 axis ($m = 1$, c), the x_2 axis ($m = 1$, s), or the x_3 axis ($m = 0$, c) with constant period $2\pi a/(\beta\chi)$ and with amplitude proportional to $j_1(\chi r/a)$. This type of motion is also known as *rotatory vibration*. The frequency equation for such vibrations is $j_2(\chi) = 0$. This yields

$$\frac{\tan\chi}{\chi} = \frac{3}{3 - \chi^2}. \tag{6.13}$$

The lowest roots of this equation are given, approximately, by

$$\chi = 5.763, 9.095, 12.322, 15.514, \ldots. \tag{6.14}$$

It may be noted that Eq. (6.14) lists the first four overtones. For $l = 1$, the fundamental mode does not exist. ${}_0T_1$ represents a *rigid rotation*. In Figure 6.3, we have drawn the graphs for $y = \tan\chi$ and $y = 3\chi/(3 - \chi^2)$. The latter curve (indicated as $l = 1$) does not intersect the branch of $y = \tan\chi$ through $\chi = \pi$ but does intersect every branch after that. The third curve in Fig. 6.3 is $y = \chi(\chi^2 - 12)/(5\chi^2 - 12)$, corresponding to $l = 2$. This curve intersects every branch of $y = \tan\chi$, including the one through the point $\chi = \pi$.

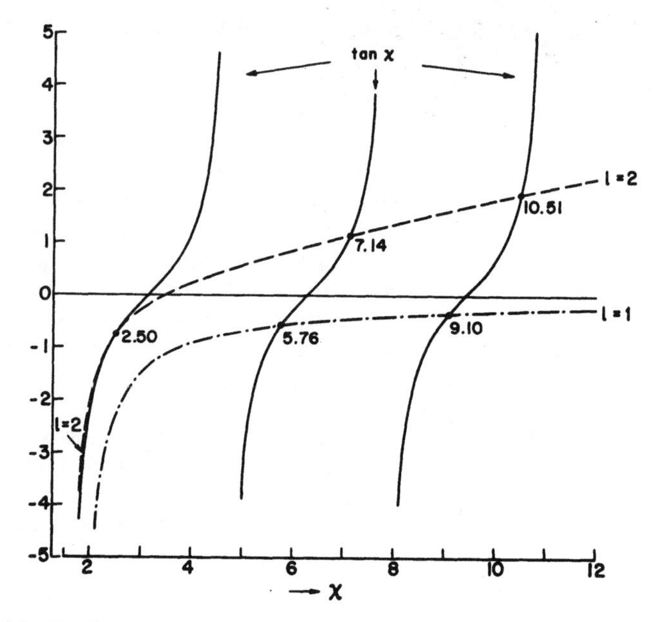

Figure 6.3. Graphs of $y = \tan \chi$ (continuous line), $y = 3\chi/(3 - \chi^2)(l = 1)$, and $y = \chi(\chi^2 - 12)/(5\chi^2 - 12)$ $(l = 2)$. The points of intersection at $\chi = 5.76, 9.10$, etc., correspond to the first-order toroidal oscillations. The points of intersection at $\chi = 2.50, 7.14$, etc., correspond to the second-order toroidal oscillations.

The nodal surfaces $r = R$ are given by $j_1((R/a)\chi) = 0$, i.e., by $\tan(R/a)\chi = (R/a)\chi$. The lowest roots of this equation are

$$\frac{R\chi}{a\pi} = 1.430, 2.459, 3.470, 4.477, \ldots.$$

On putting in the value of the dimensionless eigenfrequency χ from Eq. (6.14), we find that the first overtone has one nodal sphere, the second overtone has two nodal spheres, and so on. These have been given in Table 6.1 for $n = 1, 2, 3, 4$.

For $l = 2$, $m = 0$, we have

$$\mathbf{u} = \frac{3}{2} j_2\left(\frac{r}{a}\chi\right) \sin 2\theta \mathbf{e}_\phi. \tag{6.15}$$

A ring of particles symmetrical about the x_3 axis oscillates as a rigid circle, in its own plane, about that axis. The equatorial plane $x_3 = 0$ $(\theta = 90°)$ is a nodal plane, and the motion at equal distances on either side of this plane is equal and opposite (Fig. 6.2). It is known as a *torsional vibration.* The corresponding frequency equation is

$$\chi j_3(\chi) = j_2(\chi),$$

Table 6.1. Dimensionless Frequency (χ), Period (T), and Nodal Spheres (R_n) for $l = 1, 2$ for the Toroidal Oscillations of a Homogeneous Sphere

l	Frequency equation	n	χ	T[a] (s)	R_1/a	R_2/a	R_3/a	R_4/a
0	No motion							
1		0	[b]					
	$\frac{\tan \chi}{\chi} = \frac{3}{3 - \chi^2}$	1	5.763	1113	0.780			
		2	9.095	705	0.494	0.849		
		3	12.322	521	0.365	0.627	0.885	
		4	15.514	414	0.290	0.498	0.703	0.907
2		0	2.501	2565				
	$\frac{\tan \chi}{\chi} = \frac{\chi^2 - 12}{5\chi^2 - 12}$	1	7.136	899	0.808			
		2	10.514	610	0.548	0.865		
		3	13.771	466	0.419	0.660	0.895	
		4	16.983	378	0.339	0.535	0.725	0.914

[a] Assuming $a = 6371$ km, $\beta = 6.24$ km/s.
[b] For $l = 1$, the fundamental mode does not exist.

which may be put into the form

$$\frac{\tan \chi}{\chi} = \frac{\chi^2 - 12}{5\chi^2 - 12}. \tag{6.16}$$

The lowest roots of Eq. (6.16) are given, approximately, by

$$\chi = 2.501, 7.136, 10.514, 13.771, 16.983, \ldots. \tag{6.17}$$

For the first four eigenfrequencies of Eq. (6.17), the eigenfunctions $j_2(r\chi/a)$ are shown in Fig. 6.4 as functions of r/a. The eigenfunctions are so normalized

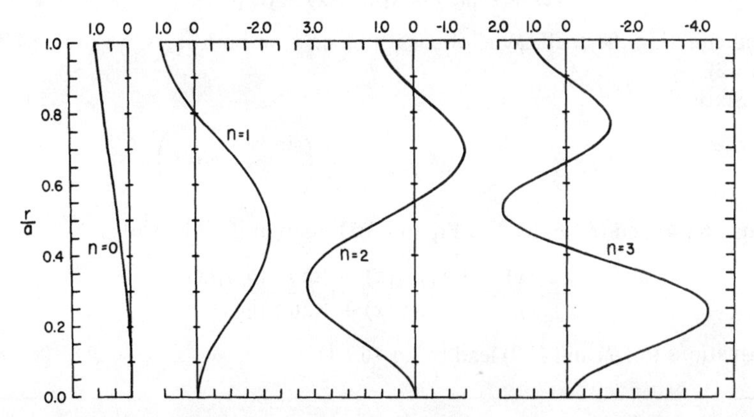

Figure 6.4. The eigenfunctions for the toroidal oscillations $l = 2$ of a homogeneous sphere corresponding to the fundamental tone and the first three overtones. The eigenfunctions are so normalized that each equals unity at the surface of the earth.

that each is equal to 1 at the surface of the sphere. The nodal spheres $r = R$ are given by $j_2(R\chi/a) = 0$. This yields

$$\frac{\tan x}{x} = \frac{3}{3 - x^2}, \qquad R = \left(\frac{x}{\chi}\right) a.$$

The roots of the above equation are given in Eq. (6.14) and χ is given by Eq. (6.17). The nodal spheres for $n = 1, 2, 3, 4$ are listed in Table 6.1.

EXAMPLE 6.1

Prove that the frequency equation for the toroidal oscillations of a homogeneous sphere may be put in the form

$$\frac{\tan \chi}{\chi} = -\frac{p_{l+1}(\chi) - (l - 1)p_l(\chi)}{q_{l+1}(\chi) - (l - 1)q_l(\chi)}, \tag{6.1.1}$$

where

$$p_{l+1}(x) = (2l + 1)p_l(x) - x^2 p_{l-1}(x),$$
$$q_{l+1}(x) = (2l + 1)q_l(x) - x^2 q_{l-1}(x), \tag{6.1.2}$$

and

$$p_0 = 0, \qquad p_1 = -1,$$
$$q_0 = 1, \qquad q_1 = 1. \tag{6.1.3}$$

Solution: Substituting

$$x^{l+1} j_l(x) = \{x \cos x\} p_l(x) + \{\sin x\} q_l(x), \tag{6.1.4}$$

in the recurrence relation

$$(2l + 1) j_l(x) = x[j_{l-1}(x) + j_{l+1}(x)], \tag{6.1.5}$$

and equating the coefficients of $\cos x$ and $\sin x$ on both sides, we obtain Eqs. (6.1.2).

Since

$$j_0(x) = \frac{\sin x}{x}, \qquad j_1(x) = \frac{1}{x}\left(\frac{\sin x}{x} - \cos x\right), \tag{6.1.6}$$

Eqs. (6.1.4) and (6.1.6) lead to Eq. (6.1.3). Equation (6.1.4) yields

$$\frac{j_{l+1}(x)}{j_l(x)} = \frac{1}{x} \frac{x p_{l+1}(x) + \{\tan x\} q_{l+1}(x)}{x p_l(x) + \{\tan x\} q_l(x)}. \tag{6.1.7}$$

Equations (6.1.7) and (6.9) lead to Eq. (6.1.1).

EXAMPLE 6.2

Find the frequency equation for the toroidal oscillations of an earth model consisting of a fluid core and homogeneous mantle.

Hint: Let

$$\mathbf{u} = \sum_{m,l} [c_1 j_l(\eta) + c_2 n_l(\eta)]\sqrt{[l(l+1)]}\mathbf{C}_{ml}(\theta, \phi), \qquad \eta = k_\beta r. \tag{6.2.1}$$

The boundary conditions are

$$\mathbf{e}_r \cdot \mathfrak{T}(\mathbf{u}) = 0 \quad \text{at} \quad r = a \quad \text{and} \quad r = b. \tag{6.2.2}$$

Inserting for **u** from Eq. (6.2.1) and eliminating c_1 and c_2, we get

$$\frac{(l-1)j_l(\chi) - \chi j_{l+1}(\chi)}{(l-1)n_l(\chi) - \chi n_{l+1}(\chi)} = \frac{(l-1)j_l(\Xi) - \Xi j_{l+1}(\Xi)}{(l-1)n_l(\Xi) - \Xi n_{l+1}(\Xi)}, \tag{6.2.3}$$

where $\chi = ak_\beta$, $\Xi = bk_\beta$.

6.2.2. Spheroidal Oscillations

Assuming a displacement field of the form in Eq. (6.2) with $\alpha_{ml} = 0$, we find from the boundary condition, Eq. (6.5), that

$$\begin{aligned} l(l+1)k_\beta F^+_{l,1}(\chi)\beta_{ml} + k_\alpha F^+_{l,3}(\zeta)\gamma_{ml} &= 0, \\ k_\beta F^+_{l,2}(\chi)\beta_{ml} + 2k_\alpha F^+_{l,1}(\zeta)\gamma_{ml} &= 0, \end{aligned} \tag{6.18}$$

where $\zeta = ak_\alpha$, $\chi = ak_\beta$. Eliminating β_{ml} and γ_{ml}, we get

$$F^+_{l,2}(\chi)F^+_{l,3}(\zeta) - 2l(l+1)F^+_{l,1}(\chi)F^+_{l,1}(\zeta) = 0. \tag{6.19}$$

This is the frequency equation for the spheroidal oscillations of a homogeneous sphere. Using Eqs. (6.4), Eq. (6.19) may be written in the following form, which is more suitable for numerical computation:

$$\begin{aligned} &2\left(\frac{\beta}{\alpha}\right)\left[\frac{1}{\chi} + \frac{(l-1)(l+2)}{\chi^2}\left\{\frac{j_{l+1}(\chi)}{j_l(\chi)} - \frac{l+1}{\chi}\right\}\right]\frac{j_{l+1}\{(\beta/\alpha)\chi\}}{j_l\{(\beta/\alpha)\chi\}} \\ &+ \left[-\frac{1}{2} + \frac{(l-1)(2l+1)}{\chi^2} + \frac{1}{\chi}\left\{1 - \frac{2l(l-1)(l+2)}{\chi^2}\right\}\frac{j_{l+1}(\chi)}{j_l(\chi)}\right] = 0. \end{aligned} \tag{6.20}$$

The ratio $j_{l+1}(x)/j_l(x)$ can be computed easily by using Eq. (6.1.7). It may be noted that, unlike Eq. (6.9) for the toroidal vibrations, Eq. (6.20) depends upon the velocity ratio β/α. Because Eq. (6.20) is independent of m, the eigenfrequencies are degenerate. For a given l, Eq. (6.20) has an infinite number of roots denoted by ${}_n\chi_l$; $n = 0, 1, 2, \ldots$. The spheroidal oscillations of order l are denoted by ${}_nS_l$; ${}_0S_l$ representing the fundamental mode and ${}_nS_l$ ($n = 1, 2, 3, \ldots$) representing the nth overtone.

When $l = 0$, it is better to start *ab initio*. We have

$$\begin{aligned} \mathbf{u} &= \gamma_{00}\mathbf{L}^+_{00} = -\gamma_{00}j_1(\xi)\mathbf{e}_r \qquad (\xi = rk_\alpha), \\ \mathbf{e}_r \cdot \mathfrak{T}(\mathbf{u}) &= 2\gamma_{00}\mu k_\alpha F^+_{0,3}(\xi)\mathbf{e}_r. \end{aligned} \tag{6.21}$$

Consequently, the frequency equation is

$$F^{+}_{0,3}(\zeta) = 0 \qquad (\zeta = ak_\alpha),$$

or, using Eq. (6.4),

$$j_1(\zeta) = \frac{1}{4}\left(\frac{\alpha}{\beta}\right)^2 \zeta j_0(\zeta).$$

This may be written in the form

$$\zeta \cot \zeta = 1 - \frac{1}{4}\left(\frac{\alpha}{\beta}\right)^2 \zeta^2. \tag{6.22}$$

For *Poisson's solid* ($\lambda = \mu$), the lowest roots of Eq. (6.22) are given by

$$\frac{\zeta}{\pi} = 0.816, 1.929, 2.936, 3.966, 4.973, \ldots. \tag{6.23}$$

From Eqs. (6.21) it is apparent that when $l = 0$, the displacement is in the radial direction. Therefore this mode of vibration is known as the *radial mode*.

When $l = 1$, Eq. (6.20) becomes

$$\frac{j_2(\chi)}{j_1(\chi)} + 2\frac{\beta}{\alpha}\frac{j_2\{(\beta/\alpha)\chi\}}{j_1\{(\beta/\alpha)\chi\}} = \frac{1}{2}\chi. \tag{6.24}$$

As in the toroidal case with $l = 1$, the fundamental mode does not exist. The roots of Eq. (6.24) corresponding to the first four overtones are ($\lambda = \mu$)

$$\chi = 3.424, 6.771, 7.444, 10.695. \tag{6.25}$$

The fundamental mode ${}_0S_1$ represents a rigid body translation, because the corresponding motion implies displacement of the center of the sphere. This can be seen as follows. Equations (6.21) show that for $l = 0$, $\mathbf{u} = 0$ at $r = 0$. For $l > 0$, it can be verified that, as $r \to 0$,

$$\begin{aligned}
\mathbf{M}^{+}_{ml} &\to 0 \qquad (l = 1, 2, \ldots),\\
\mathbf{N}^{+}_{ml}, \mathbf{L}^{+}_{ml} &\to 0 \qquad (l = 2, 3, \ldots),\\
\mathbf{N}^{+}_{(-1)1} &\to -\tfrac{1}{3}(\mathbf{e}_1 + i\mathbf{e}_2),\\
\mathbf{N}^{+}_{01} &\to \tfrac{2}{3}\mathbf{e}_3,\\
\mathbf{N}^{+}_{11} &\to \tfrac{2}{3}(\mathbf{e}_1 + i\mathbf{e}_2),\\
\mathbf{L}^{+}_{(-1)1} &\to -\tfrac{1}{6}(\mathbf{e}_1 + i\mathbf{e}_2),\\
\mathbf{L}^{+}_{01} &\to \tfrac{1}{3}\mathbf{e}_3,\\
\mathbf{L}^{+}_{11} &\to \tfrac{1}{3}(\mathbf{e}_1 + i\mathbf{e}_2).
\end{aligned} \tag{6.26}$$

Hence the motion at the center of the sphere is zero for all values of l except $l = 1$. For ${}_0S_1$, $\mathbf{N}^{+}_{01}$ and $\mathbf{L}^{+}_{01}$ represent the motion of the center of the sphere along the x_3 axis, and $\mathbf{N}^{+}_{(\pm 1)1}$ and $\mathbf{L}^{+}_{(\pm 1)1}$ represent this motion along the x_1 or x_2 axis, depending upon whether we consider the real part (i.e., $\mathbf{N}^c$, $\mathbf{L}^c$) or the imaginary part (i.e., $\mathbf{N}^s$, $\mathbf{L}^s$).

The vibrations of the second order ($l = 2$) are of particular interest. Taking $m = 0$, we have, from Eq. (6.2),

$$\mathbf{u} = \beta_{02}\mathbf{N}_{0,2}^{+} + \gamma_{02}\mathbf{L}_{0,2}^{+},$$

which yields

$$u_r = \tfrac{1}{4}U(r)(3\cos 2\theta + 1),$$
$$u_\theta = V(r)\sin 2\theta, \tag{6.27}$$

where

$$U(r) = 6\beta_{02}\frac{1}{\eta}j_2(\eta) + \gamma_{02}j_2'(\xi),$$
$$V(r) = -\frac{3}{2}\left[\beta_{02}\left\{j_2'(\eta) + \frac{1}{\eta}j_2(\eta)\right\} + \gamma_{02}\frac{1}{\xi}j_2(\xi)\right]. \tag{6.28}$$

The prime denotes differentiation with respect to the argument. The ratio β_{02}/γ_{02} is known from Eqs. (6.18). From Eqs. (6.27), it is easily seen that the oscillating sphere assumes alternatively the forms of a prolate and an oblate spheroid. This mode of oscillation is known as the *football mode.*

6.2.3. Spectral Green's Dyadic

We denote the Green's dyadic for an unbounded homogeneous medium by $\mathfrak{G}_\infty$. The eigenvector expansion of $\mathfrak{G}_\infty$ has been obtained in Chapter 4. From Eq. (4.172), we have

$$\mathfrak{G}_\infty(\mathbf{r}|\mathbf{r}_0) = -\frac{i}{\mu}k_\beta \sum_{l=1,1,0}^{\infty}\sum_{m=-l}^{l}\frac{1}{\Omega_{ml}}\left[\frac{1}{l(l+1)}\mathbf{M}_{ml}^{-}(k_\beta r_>)\overset{*}{\mathbf{M}}{}_{ml}^{+}(k_\beta r_<) + \frac{1}{l(l+1)}\mathbf{N}_{ml}^{-}(k_\beta r_>)\overset{*}{\mathbf{N}}{}_{ml}^{+}(k_\beta r_<) + \left(\frac{\beta}{\alpha}\right)^3 \mathbf{L}_{ml}^{-}(k_\alpha r_>)\overset{*}{\mathbf{L}}{}_{ml}^{+}(k_\alpha r_<)\right], \tag{6.29}$$

where $r_> = \max(r, r_0)$, $r_< = \min(r, r_0)$. Let the Green's dyadic for an elastic sphere of radius a that is centered at the origin be denoted by $\mathfrak{G}_a$. To obtain $\mathfrak{G}_a$, we must add to $\mathfrak{G}_\infty$ another dyadic $\hat{\mathfrak{G}}$ such that $\mathfrak{G}_\infty + \hat{\mathfrak{G}}$ satisfies the required boundary conditions at $r = a$. We assume

$$\mathfrak{G}_a = \mathfrak{G}_\infty - \frac{i}{\mu}k_\beta \sum_{l=1,1,0}^{\infty}\sum_{m=-l}^{l}\frac{1}{\Omega_{ml}}\left[\frac{1}{l(l+1)}\mathbf{M}_{ml}^{+}(k_\beta r)\mathbf{X}_{ml}(r_0) + \frac{1}{l(l+1)}\mathbf{N}_{ml}^{+}(k_\beta r)\mathbf{Y}_{ml}(r_0) + \left(\frac{\beta}{\alpha}\right)^3 \mathbf{L}_{ml}^{+}(k_\alpha r)\mathbf{Z}_{ml}(r_0)\right], \tag{6.30}$$

where $\mathbf{X}(r_0)$, $\mathbf{Y}(r_0)$, and $\mathbf{Z}(r_0)$ are unknown vectors of the $\mathbf{r}_0$ system. In the expansion for $\hat{\mathfrak{G}}$, we have taken only those eigenvectors of the $\mathbf{r}$ system that are finite at the origin. The boundary condition that requires the surface to be traction-free gives

$$\mathbf{e}_r \cdot \mathfrak{T}[\mathfrak{G}_a(\mathbf{r}|\mathbf{r}_0)] = 0, \qquad r = a, \tag{6.31}$$

where

$$\mathfrak{T}(\mathfrak{G}) = \lambda\mathfrak{I}\,\mathrm{div}\,\mathfrak{G} + \mu[\nabla\mathfrak{G} + (\nabla\mathfrak{G})^{213}]. \tag{6.32}$$

It may be noted that in the differentiations implied in $\mathfrak{T}(\mathfrak{G})$, the vector functions of the $\mathbf{r}_0$ system behave as constant vectors.

Inserting the value of $\mathfrak{G}_a$ into the boundary condition Eq. (6.31) and using Eq. (6.3), we obtain a dyadic equation. This equation must be satisfied for all θ and ϕ. Equating to zero the coefficient of $\mathbf{C}_{ml}(\theta, \phi)$, we get

$$\mathbf{X}_{ml}(r_0) = -\frac{F_{l,1}^-(\chi)}{F_{l,1}^+(\chi)}\overset{*}{\mathbf{M}}{}_{ml}^+(k_\beta r_0). \tag{6.33}$$

Similarly, on equating to zero the coefficients of $\mathbf{P}_{ml}(\theta, \phi)$ and $\mathbf{B}_{ml}(\theta, \phi)$, we obtain two simultaneous equations in $\mathbf{Y}_{ml}$ and $\mathbf{Z}_{ml}$. Solving these, we find

$$\begin{aligned}\mathbf{Y}_{ml}(r_0) &= \frac{\Delta_1}{\Delta_l^+}\overset{*}{\mathbf{N}}{}_{ml}^+(k_\beta r_0) + l(l+1)\frac{\Delta_2}{\Delta_l^+}\overset{*}{\mathbf{L}}{}_{ml}^+(k_\alpha r_0),\\ \mathbf{Z}_{ml}(r_0) &= \frac{\Delta_3}{\Delta_l^+}\overset{*}{\mathbf{N}}{}_{ml}^+(k_\beta r_0) + \frac{\Delta_4}{\Delta_l^+}\overset{*}{\mathbf{L}}{}_{ml}^+(k_\alpha r_0),\end{aligned} \tag{6.34}$$

where

$$\begin{aligned}\Delta_1 &= F_{l,2}^-(\chi)F_{l,3}^+(\zeta) - 2l(l+1)F_{l,1}^-(\chi)F_{l,1}^+(\zeta),\\ \Delta_2 &= 2\left(\frac{\beta}{\alpha}\right)^4[F_{l,1}^-(\zeta)F_{l,3}^+(\zeta) - F_{l,3}^-(\zeta)F_{l,1}^+(\zeta)] = -\frac{i}{\zeta\chi^4}[2(l+2)(l-1) - \chi^2],\\ \Delta_3 &= \left(\frac{\alpha}{\beta}\right)^4[F_{l,1}^-(\chi)F_{l,2}^+(\chi) - F_{l,1}^+(\chi)F_{l,2}^-(\chi)] = \left(\frac{\alpha}{\beta}\right)^3\Delta_2,\\ \Delta_4 &= F_{l,2}^+(\chi)F_{l,3}^-(\zeta) - 2l(l+1)F_{l,1}^-(\zeta)F_{l,1}^+(\chi),\end{aligned} \tag{6.35}$$

the functions $F_{l,1}^\pm(z)$, etc., are defined in Eq. (6.4) and

$$\Delta_l^+ = 2l(l+1)F_{l,1}^+(\zeta)F_{l,1}^+(\chi) - F_{l,2}^+(\chi)F_{l,3}^+(\zeta). \tag{6.36}$$

The expressions for Δ_2 and Δ_3 have been simplified with the help of the Wronskian relation for the spherical Bessel functions.

Inserting the values of **X**, **Y**, and **Z** from Eqs. (6.33) and (6.34) into Eq. (6.30) we find

$$\mathfrak{G}_a(\mathbf{r}|\mathbf{r}_0) = \mathfrak{G}_\infty - \frac{i}{\mu} k_\beta \sum_{l=1,1,1,0}^{\infty} \sum_{m=-l}^{l} \frac{1}{\Omega_{ml}} \left[-\frac{1}{l(l+1)} \frac{F_{l,1}^{-}(\chi)}{F_{l,1}^{+}(\chi)} \right.$$
$$\times \mathbf{M}_{ml}^{+}(k_\beta r)\overset{*}{\mathbf{M}}{}_{ml}^{+}(k_\beta r_0) + \frac{\Delta_2}{\Delta_l^{+}} \{\mathbf{L}_{ml}^{+}(k_\alpha r)\overset{*}{\mathbf{N}}{}_{ml}^{+}(k_\beta r_0)$$
$$+ \mathbf{N}_{ml}^{+}(k_\beta r)\overset{*}{\mathbf{L}}{}_{ml}^{+}(k_\alpha r_0)\} + \frac{1}{l(l+1)} \frac{\Delta_1}{\Delta_l^{+}} \mathbf{N}_{ml}^{+}(k_\beta r)\overset{*}{\mathbf{N}}{}_{ml}^{+}(k_\beta r_0)$$
$$\left. + \left(\frac{\beta}{\alpha}\right)^3 \frac{\Delta_4}{\Delta_l^{+}} \mathbf{L}_{ml}^{+}(k_\alpha r)\overset{*}{\mathbf{L}}{}_{ml}^{+}(k_\alpha r_0) \right]. \tag{6.37}$$

The expression for the Green's dyadic is simplified considerably at $r = a$. We obtain

$$\mathfrak{G}_a(\mathbf{a}|\mathbf{r}_0) = \frac{k_\beta}{\mu} \sum_{l=1,1,0}^{\infty} \frac{1}{l(l+1)} \sum_{m=-l}^{l} \frac{1}{\Omega_{ml}} \left[\frac{1}{\chi^3 F_{l,1}^{+}(\chi)} \sqrt{[l(l+1)]} \right.$$
$$\times \mathbf{C}_{ml}(\theta, \phi)\overset{*}{\mathbf{M}}{}_{ml}^{+}(k_\beta r_0) + \frac{1}{\chi^2 \Delta_l^{+}} \{l(l+1) F_{l,1}^{+}(\zeta)\mathbf{P}_{ml}(\theta, \phi)$$
$$- F_{l,3}^{+}(\zeta)\sqrt{[l(l+1)]}\mathbf{B}_{ml}(\theta, \phi)\}\overset{*}{\mathbf{N}}{}_{ml}^{+}(k_\beta r_0)$$
$$\left. + \frac{l(l+1)}{\zeta\chi\Delta_l^{+}} \{-\tfrac{1}{2}F_{l,2}^{+}(\chi)\mathbf{P}_{ml}(\theta, \phi) + F_{l,1}^{+}(\chi)\mathbf{B}_{ml}(\theta, \phi)\}\overset{*}{\mathbf{L}}{}_{ml}^{+}(k_\alpha r_0) \right]. \tag{6.38}$$

This expansion for the Green's dyadic for a sphere can be used to get the field induced by point sources placed inside the sphere. The procedure is explained in Section 6.3.

6.3. Oscillations of a Radially Inhomogeneous Self-Gravitating Earth Model

6.3.1. Equations of Motion

We take the center of the earth at the origin of a spherical coordinate system (r, θ, ϕ). Denoting the density by ρ, the acceleration of gravity by g, and Lamé's parameters by λ and μ, the Fourier-transformed equation of motion may be written in the form

$$\operatorname{div} \mathfrak{T} + \rho \mathbf{F} + \rho\omega^2 \mathbf{u} = 0, \tag{6.39}$$

where $\mathbf{F}$ is the body force per unit mass and ω is the angular frequency. We assume that in the undisturbed state the earth is in equilibrium under self-gravitation. The corresponding initial stress may be large. Consequently, when the earth is deformed by small disturbing forces from this initially stressed state, we cannot apply the ordinary equations of elasticity. In that case, the usual stress–strain relations have to be modified.

We make the following assumptions:

1. Small quantities of the second and higher orders are neglected.
2. The initial equilibrium stress, $\mathfrak{T}_i$, is given by a hydrostatic pressure p_0:

$$\mathfrak{T}_i = -p_0 \mathfrak{I},$$

$$\text{grad } p_0 = -g_0 \rho_0 \mathbf{e}_r, \tag{6.40}$$

   where the subscript zero refers to the equilibrium state.
3. The initial stress at a point which is at $(\mathbf{r})$ in the strained state is taken equal to the value of the initial stress at that point which is displaced to $(\mathbf{r})$ when the body is strained. This implies that a small element of the medium carries its initial stress with it when it moves from one place to another. This element acquires an additional stress depending upon the compression and distortion it suffers during the displacement. Hence $p_0 = p_0(\mathbf{r} - \mathbf{u})$.
4. In the strained state, the stress consists of two parts: the initial equilibrium stress $(\mathfrak{T}_i)$ and the additional elastic stress $(\mathfrak{T}_e)$. The additional stress is related to the strain, measured from the equilibrium state, by the usual stress–strain relations. Therefore, we have

$$\mathfrak{T} = \mathfrak{T}_i + \mathfrak{T}_e, \tag{6.41}$$

$$\mathfrak{T}_i = -p_0(\mathbf{r} - \mathbf{u})\mathfrak{I} \quad \text{[from assumption 3]}$$
$$= -(p_0 - \mathbf{u} \cdot \text{grad } p_0)\mathfrak{I} = -(p_0 + g_0 \rho_0 u_r)\mathfrak{I}, \tag{6.42}$$

$$\mathfrak{T}_e = \lambda \mathfrak{I} \,\text{div}\, \mathbf{u} + \mu(\nabla \mathbf{u} + \mathbf{u}\nabla). \tag{6.43}$$

5. The only body force is the force of gravity, which is derived from a gravitational potential Ψ:

$$\mathbf{F} = \text{grad } \Psi. \tag{6.44}$$

6. The values of ρ and Ψ in the strained state are equal to their equilibrium values ρ_0 and Ψ_0 plus small perturbations caused by the motion:

$$\rho = \rho_0 + \rho', \qquad \Psi = \Psi_0 + \psi. \tag{6.45}$$

7. The quantities λ, μ, ρ_0, g_0, and Ψ_0 are functions of r only.

The perturbation in the gravitational potential, ψ, results from: (1) a volume distribution of density ρ' in the region $0 < r < a$ caused by the perturbation in the density, and (2) a surface distribution of density $\rho_0 u_r$ on the surface $r = a$ caused by the wrinkling of the surface of the earth by the deformation.

Because div **u** measures the change in volume per unit volume, the equation of continuity yields

$$\begin{aligned}\rho(\mathbf{r}) &= \rho_0(\mathbf{r} - \mathbf{u})(1 - \operatorname{div}\mathbf{u}) = \rho_0(\mathbf{r}) - \mathbf{u}\cdot\operatorname{grad}\rho_0 - \rho_0\operatorname{div}\mathbf{u}\\ &= \rho_0(\mathbf{r}) - \operatorname{div}(\rho_0\mathbf{u}).\end{aligned} \tag{6.46}$$

Hence, we have

$$\rho' = \rho - \rho_0 = -\operatorname{div}(\rho_0\mathbf{u}). \tag{6.47}$$

From Eqs. (6.41)–(6.43), we find

$$\begin{aligned}\operatorname{div}\mathfrak{T} &= -\operatorname{grad}(p_0 + g_0\rho_0 u_r - \lambda\operatorname{div}\mathbf{u}) + \mu(\nabla^2\mathbf{u} + \operatorname{grad}\operatorname{div}\mathbf{u})\\ &\quad + \dot{\mu}\left(2\frac{\partial\mathbf{u}}{\partial r} + \mathbf{e}_r \times \operatorname{curl}\mathbf{u}\right)\\ &= g_0\rho_0\mathbf{e}_r - \rho_0\operatorname{grad}(g_0u_r) - \dot{\rho}_0 g_0 u_r\mathbf{e}_r + \operatorname{grad}(\lambda\operatorname{div}\mathbf{u})\\ &\quad + \mu(\nabla^2\mathbf{u} + \operatorname{grad}\operatorname{div}\mathbf{u}) + \dot{\mu}\left(2\frac{\partial\mathbf{u}}{\partial r} + \mathbf{e}_r \times \operatorname{curl}\mathbf{u}\right),\end{aligned} \tag{6.48}$$

where the dot signifies differentiation with respect to r. In deriving Eq. (6.48), we have used, besides Eqs. (6.40), the following identities:

$$\operatorname{div}(f\mathfrak{J}) = \operatorname{grad} f, \qquad \operatorname{div}(f\Phi) = f\operatorname{div}\Phi + \operatorname{grad} f\cdot\Phi$$

$$\mathbf{e}_r\cdot(\nabla\mathbf{u} + \mathbf{u}\nabla) = 2\frac{\partial\mathbf{u}}{\partial r} + \mathbf{e}_r \times \operatorname{curl}\mathbf{u}.$$

From Eqs. (6.44)–(6.48) and the relation grad $\Psi_0 = -g_0\mathbf{e}_r$, we have

$$\begin{aligned}\rho\mathbf{F} &= (\rho_0 - \dot{\rho}_0 u_r - \rho_0\operatorname{div}\mathbf{u})(\operatorname{grad}\psi - g_0\mathbf{e}_r)\\ &= \rho_0\operatorname{grad}\psi + g_0(\dot{\rho}_0 u_r + \rho_0\operatorname{div}\mathbf{u} - \rho_0)\mathbf{e}_r.\end{aligned} \tag{6.49}$$

Equations (6.39), (6.48), and (6.49) now give

$$\begin{aligned}&\mu(\nabla^2\mathbf{u} + \operatorname{grad}\operatorname{div}\mathbf{u}) + \frac{d\mu}{dr}\left(2\frac{\partial\mathbf{u}}{\partial r} + \mathbf{e}_r \times \operatorname{curl}\mathbf{u}\right) + \operatorname{grad}(\lambda\operatorname{div}\mathbf{u})\\ &\quad + \rho_0[\operatorname{grad}(\psi - g_0u_r) + g_0\mathbf{e}_r\operatorname{div}\mathbf{u}] + \omega^2\rho_0\mathbf{u} = 0.\end{aligned} \tag{6.50}$$

Equation (6.50) is the equation of motion of a radially heterogeneous, isotropic, prestressed, self-gravitating, elastic medium.

Because the gravitational potential, Ψ, satisfies the Poisson equation

$$\nabla^2\Psi = -4\pi G\rho,$$

the perturbation in the gravitational potential, ψ, also satisfies a similar equation:

$$\nabla^2\psi = -4\pi G\rho' = 4\pi G\operatorname{div}(\rho_0\mathbf{u}), \tag{6.51}$$

using Eq. (6.47). Here G denotes the universal constant of gravitation. Note that $g_0(r)$ can be determined from the density distribution $\rho_0(r)$ through the relation

$$g_0(r) = \frac{4\pi G}{r^2} \int_0^r \rho_0(x) x^2 \, dx. \tag{6.52}$$

Further, although ρ_0 and Ψ_0 are functions of r only, ρ' and ψ may be functions of θ and ϕ as well.

Equations (6.50) and (6.51) represent a differential system in the four variables: ψ and three components of the displacement. The coefficients that appear in this system are functions of r, namely, $\rho_0(r)$, $\lambda(r)$, $\mu(r)$, and $g_0(r)$.

6.3.2. Boundary Conditions

Equations (6.50) and (6.51) are to be solved under the following boundary conditions:

1. The solution is regular at the origin.
2. The stresses vanish at the *deformed* surface of the earth and are continuous at an internal deformed surface of discontinuity.
3. The displacements are continuous at an internal surface of discontinuity. However, at a liquid–solid interface, only the radial component of the displacement is continuous.
4. The gravitational potential and its normal derivative are continuous at the deformed surface of the earth and at an internal deformed surface of discontinuity.

Let $r = c$ be an internal surface of discontinuity which, in the strained state, takes the shape $r = c + u_r(\theta, \phi)$. Using Eqs. (6.40)–(6.43) and neglecting small quantities of the second order, we find

$$\mathfrak{T}(c + u_r) = \mathfrak{T}(c) + u_r \left[\frac{\partial \mathfrak{T}}{\partial r}\right]_c = \mathfrak{T}_e(c) - p_0(c)\mathfrak{I}. \tag{6.53}$$

The above result could have been anticipated: The additional elastic stress at c and at $c + u_r$ are equal to the first order in u_r. Moreover, as stated above, a small element of the medium carries its initial stress with it when it moves from one place to another. Combining these two observations, we get Eq. (6.53). We know that, in the unstrained state, the initial stress vanishes at the free surface of the earth and is continuous across an internal surface of discontinuity such as $r = c$. Boundary condition 2 with the help of Eq. (6.53) implies that the additional elastic stress vanishes at the undeformed surface of the earth and is continuous across an internal undeformed surface of discontinuity.

Boundary condition 4 yields

$$\Psi_1 = \Psi_2, \qquad \dot{\Psi}_1 = \dot{\Psi}_2; \qquad r = c + u_r, \tag{6.54}$$

where the subscripts 1 and 2 are used to distinguish quantities on the two sides of the surface of discontinuity. Using Eq. (6.45) and neglecting small quantities of the second order in u_r, we find

$$\psi_1 + \Psi_{01} + u_r \dot{\Psi}_{01} = \psi_2 + \Psi_{02} + u_r \dot{\Psi}_{02}; \qquad r = c, \tag{6.55}$$

$$\dot{\psi}_1 + \dot{\Psi}_{01} + u_r \ddot{\Psi}_{01} = \dot{\psi}_2 + \dot{\Psi}_{02} + u_r \ddot{\Psi}_{02}; \qquad r = c. \tag{6.56}$$

In the unstrained state, we have

$$\Psi_{01} = \Psi_{02}, \qquad \dot{\Psi}_{01} = \dot{\Psi}_{02}; \qquad r = c. \tag{6.57}$$

Making use of Eqs. (6.57) and the Poisson equation

$$\ddot{\Psi}_0 + \frac{2}{r}\dot{\Psi}_0 = -4\pi G \rho_0, \tag{6.58}$$

Eqs. (6.55) and (6.56) yield

$$\psi_1 = \psi_2, \qquad \dot{\psi}_1 - 4\pi G \rho_{01} u_r = \dot{\psi}_2 - 4\pi G \rho_{02} u_r; \quad r = c. \tag{6.59}$$

The corresponding equations at the free surface are

$$\psi = \psi_e, \qquad \dot{\psi} - 4\pi G \rho_0 u_r = \dot{\psi}_e; \qquad r = a, \tag{6.60}$$

where ψ_e denotes the value of ψ outside the earth.

It may be noted that, to the first order in u_r, the displacements at c and $c + u_r$ are equal. Therefore, boundary condition 3 may be applied at an internal undeformed surface of discontinuity.

EXAMPLE 6.3: Radial Oscillations of a Homogeneous Gravitating Sphere

If we put

$$\mathbf{u} = U(r)\mathbf{e}_r, \qquad \psi = \psi(r), \tag{6.3.1}$$

in Eqs. (6.50)–(6.52), we get

$$(\lambda + 2\mu)\left(\frac{d^2U}{dr^2} + \frac{2}{r}\frac{dU}{dr} - \frac{2}{r^2}U\right) + \rho_0\left(\frac{d\psi}{dr} - U\frac{dg_0}{dr} + \frac{2}{r}g_0 U\right) + \rho_0 \omega^2 U = 0, \tag{6.3.2}$$

$$\frac{1}{r^2}\frac{d}{dr}\left(r^2 \frac{d\psi}{dr}\right) = \frac{3A}{r^2}\frac{d}{dr}(r^2 U), \tag{6.3.3}$$

$$g_0 = Ar, \tag{6.3.4}$$

where

$$A = \tfrac{4}{3}\pi \rho_0 G. \tag{6.3.5}$$

Integrating Eq. (6.3.3) and remembering that ψ is finite at the origin, we get

$$\frac{d\psi}{dr} = 3AU. \tag{6.3.6}$$

Equations (6.3.2), (6.3.4), and (6.3.5) yield

$$\frac{d^2U}{dr^2} + \frac{2}{r}\frac{dU}{dr} + \left(k^2 - \frac{2}{r^2}\right)U = 0, \tag{6.3.7}$$

where

$$k^2 = \frac{\rho_0(\omega^2 + 4A)}{\lambda + 2\mu}. \tag{6.3.8}$$

Equation (6.3.7) is a differential equation satisfied by spherical Bessel functions of the first order. Because the solution has to be regular at the origin, we take

$$U = j_1(kr) = \frac{\sin kr}{k^2r^2} - \frac{\cos kr}{kr}, \tag{6.3.9}$$

where a constant factor with the dimensions of displacement has been suppressed. The boundary condition

$$\tau_{rr} = 0 \quad \text{at} \quad r = a \tag{6.3.10}$$

yields the frequency equation

$$\zeta \cot \zeta = 1 - \tfrac{1}{4}\left(\frac{\alpha}{\beta}\right)^2 \zeta^2, \qquad \zeta = ka. \tag{6.3.11}$$

Equation (6.3.11) is identical in form with the corresponding frequency equation for the case of radial oscillations of a homogeneous nongravitating sphere [cf. Eq. (6.22)] except for the fact that in the latter case the term $4A$ in Eq. (6.3.8) is missing. The fact that gravity appears only in the expression for k^2, and that too only in the combination of $(\omega^2 + 4A)$, exhibits the destabilizing effect of gravity. Writing Eq. (6.3.8) in the form

$$\omega^2 = \frac{\lambda + 2\mu}{\rho_0}k^2 - 4A, \tag{6.3.12}$$

and denoting the lowest root of Eq. (6.3.11) by ζ_1, it is apparent that ω^2 will become negative, or that *gravitational instability* will ensue, if the inequality

$$\frac{16\pi G\rho_0^2 a^2}{3(\lambda + 2\mu)} > \zeta_1^2 \tag{6.3.13}$$

is satisfied.

6.3.3. Toroidal Oscillations

As stated before, in purely toroidal oscillations, u_r and div $\mathbf{u}$ vanish identically. Equation (6.46) reveals that, in such a case, the density is unaffected. With no radial displacement of the boundaries and with the density remaining unaltered, there is no perturbation in the gravitational field. The motion is therefore controlled entirely by elastic forces. Because of the liquidity of the core, the motion is confined to the shell. The equation of motion, Eq. (6.50), reduces to

$$\mu \nabla^2 \mathbf{u} + \frac{d\mu}{dr}\left(2\frac{\partial \mathbf{u}}{\partial r} + \mathbf{e}_r \times \operatorname{curl} \mathbf{u}\right) + \omega^2 \rho_0 \mathbf{u} = 0. \tag{6.61}$$

In the case of a homogeneous nongravitating medium, we have seen that the vector **M** represents the toroidal motion and the vectors **N** and **L** represent the spheroidal motion. In going from a homogeneous nongravitating medium to a radially heterogeneous gravitating medium, only the r-dependent parts of the solutions of the equations of motion change. Consequently, for the toroidal oscillations of the real earth, we may assume

$$\mathbf{u} = \sum_{\sigma, m, l} y_1(r)\sqrt{[l(l+1)]}\,\mathbf{C}^{\sigma}_{ml}(\theta, \phi); \qquad \sigma = c, s. \tag{6.62}$$

Substituting this value of $\mathbf{u}$ into the equation of motion, Eq. (6.61), we find that $y_1(r)$ satisfies the equation

$$\mu\left(\frac{d^2 y_1}{dr^2} + \frac{2}{r}\frac{dy_1}{dr}\right) + \frac{d\mu}{dr}\left(\frac{dy_1}{dr} - \frac{y_1}{r}\right) + \left[\omega^2 \rho_0 - \frac{l(l+1)}{r^2}\mu\right] y_1 = 0. \tag{6.63}$$

On taking μ constant, the last equation reduces to the differential equation satisfied by the spherical Bessel functions, as it should. Equation (6.62) yields

$$\mathbf{e}_r \cdot \mathfrak{T}(\mathbf{u}) = \sum_{\sigma, m, l} y_2(r)\sqrt{[l(l+1)]}\,\mathbf{C}^{\sigma}_{ml}(\theta, \phi), \tag{6.64}$$

with

$$y_2 = \mu\left(\frac{dy_1}{dr} - \frac{y_1}{r}\right). \tag{6.65}$$

The boundary conditions are

$$y_2 = 0; \qquad r = a \quad \text{and} \quad r = b, \tag{6.66}$$

b denoting the radius of the core. At a solid–solid interface, both y_1 and y_2 must be continuous.

If Eq. (6.63) can be solved for a given dependence of μ and β on r, the period equation can then be obtained in closed form on making use of the boundary conditions in Eqs. (6.66). Table 6.2 gives $y_1(r)$ for some particular functional forms for μ and β. The constants were calculated by the method of least squares. Table 6.3 lists these constants for the first five models of Table 6.2 together with the periods of the toroidal oscillations of the earth calculated from the period equation.

Equation (6.63) is not suitable for numerical integration, because it becomes necessary to evaluate the derivative of the empirically determined quantity $\mu(r)$.

Table 6.2. Solution of Eq. (6.63) for Given $\mu(r)$ and $\beta(r)$

Model	$\mu(r)/\mu_0$	$\beta(r)/\beta_0$	$y_1(r)$	Remarks
I	1	1	$f_l(k_0 r)$ Spherical Bessel functions	$k_0 = \dfrac{\omega}{\beta_0}$
II	$\left(\dfrac{a}{r}\right)^p$	1	$r^{(p-1)/2} Z_\nu(k_0 r)$ Bessel functions	$\nu = \dfrac{1}{2}[(2l+1)^2 + p(p-6)]^{1/2}$
III	$\left(\dfrac{a}{r}\right)^p$	$\left(\dfrac{a}{r}\right)^s$	$r^{(p-1)/2} Z_\nu\left(\dfrac{\eta}{1+s} r^{s+1}\right)$	$\nu = \dfrac{1}{2(1+s)}[(2l+1)^2 + p(p-6)]^{1/2}$ $\eta = \dfrac{\omega}{\beta_0 a^s}$
IV	$\left(\dfrac{a}{r}\right)^2$	$1 - pr$	$\tau^{1/2} \begin{matrix}\cos\\ \sin\end{matrix} (s \ln \tau)$ for $l = 1$	$\tau = 1 - pr$ $s = \sqrt{\left(\left\lvert\dfrac{\omega^2}{p^2\beta_0^2} - \dfrac{1}{4}\right\rvert\right)}$
V	$\left(\dfrac{a}{r}\right)^2$	$(1 - pr)^{1/2}$	$\tau Z_1(\eta)$ for $l = 1$	$\tau = (1 - pr)^{1/2}$ $\eta = \dfrac{2\omega\tau}{p\beta_0}$
VI	$\dfrac{(1+\delta r)^2}{r^4}$	1	$\dfrac{r^{3/2}}{(1+\delta r)} Z_\nu(k_0 r)$	$\nu^2 = l(l+1) - \dfrac{7}{4}$ $k_0 = \dfrac{\omega}{\beta_0}$
VII	$\dfrac{1}{r^4} e^{2\delta r}$	1	$\dfrac{r^{3/2}}{e^{\delta r}} Z_\nu(\eta r)$	$\nu^2 = l(l+1) - \dfrac{7}{4}$ $\eta^2 = \dfrac{\omega^2}{\beta_0^2} - \delta^2$

Table 6.3. Structural Parameters for Five Analytical Models of the Earth's Mantle ($b \le r \le a$) and the Corresponding Periods of Toroidal Oscillations[a]

Model	Parameters: $\mu_0 (10^{12}\ \text{dyne/cm}^2)$	Parameters: β_0(km/s)	Parameters: Additional	l	Period equation[b]	Period, (min): $n = 0$	1	2	3
I	1.75	6.21	—	1	$\tan q = \dfrac{(7.882)q^3 + 9q}{(6.903)q^4 + (4.882)q^2 + q}, \quad q = k_0(a-b)$	—	14.1	7.5	5.1
				2	$\tan q = \dfrac{(34.517)q^5 + (51.058)q^3 + 144q}{(18.138)q^6 - (24.622)q^4 + (3.058)q^2 + 144}, \quad q = k_0(a-b)$	44.6	13.1	7.4	5.1
II	0.9	6.21	$p = 2$	1	$\tan q = \dfrac{q}{1 + (2.628)q^2}, \quad q = k_0(a-b)$	—	15.0	7.7	5.2
				2	$\dfrac{2.5J_0(x) + J_1(x)(x - 5/x)}{2.5N_0(x) + N_1(x)(x - 5/x)} =$ same expression with x replaced by y, $\quad x = k_0 a$, $y = k_0 b$	44.0	13.9	8.1	5.5
III	0.9	5.22	$s = 0.5$, $p = 2$	1	$\dfrac{2J_\nu(\theta_1) + \theta_1 A\{J_{\nu+1}(\theta_1) - J_{\nu-1}(\theta_1)\}}{2J_{-\nu}(\theta_1) - \theta_1 A\{J_{-\nu-1}(\theta_1) - J_{-\nu+1}(\theta_1)\}} =$ same expression with θ_1 replaced by θ_2	—	14.9		
	0.84	5	$s = 0.725$, $p = 2.576$	1	$\theta_1 = \dfrac{\eta a^{1+s}}{1+s}, \quad \theta_2 = \dfrac{\eta b^{1+s}}{1+s}, \quad A = \dfrac{2(1+s)}{3-p}$	—	15.0		
				2		44.8			
IV	0.9	11.31	$ap = 0.578$	1	$\tan\theta = \dfrac{(0.7)\theta}{\theta^2 + 0.773}, \quad s = (2.062)\theta$	—	14.8	7.7	5.2
V	0.9	10.09	$ap = 0.787$	1	$\dfrac{(0.542)J_1(\theta_1) + \theta_1 J_0(\theta_1)}{(0.542)N_1(\theta_1) + \theta_1 N_0(\theta_1)} = \dfrac{(2.667)J_1(\theta_2) + \theta_2 J_0(\theta_2)}{(2.667)N_1(\theta_2) + \theta_2 N_0(\theta_2)}$, $\quad \theta_2 = (1.63)\theta_1, \quad \theta_1 = \dfrac{2\omega}{p\beta_0}\sqrt{(1 - pa)}$	—	14.8		

[a] For a description of the models, see Table 6.2.
[b] $a = 6371$ km, $b = 3473$ km.

However, Eq. (6.63) is equivalent to the following system of two simultaneous linear differential equations of the first order in y_1 and y_2:

$$\frac{dy_1}{dr} = \frac{y_1}{r} + \frac{y_2}{\mu},$$
$$\frac{dy_2}{dr} = \left[\frac{l^2 + l - 2}{r^2}\mu - \omega^2\rho_0\right]y_1 - \frac{3}{r}y_2. \tag{6.67}$$

The set of equations (6.67) can be integrated numerically, subject to the boundary conditions in Eqs. (6.66). For a given earth model $[\mu(r), \rho_0(r)]$ and given l ($l = 1, 2, 3, \ldots$), there is a discrete set of values of ω for which the boundary conditions are satisfied. We denote this set by ${}_n\omega_l$ ($n = 0, 1, 2, \ldots$). Then ${}_n\omega_l$ are the *eigenfrequencies* and ${}_nT_l = 2\pi/{}_n\omega_l$ are the *eigenperiods* of the toroidal oscillations of the earth.

We have assumed above that y_1 and y_2 are continuous in the region $b < r < a$. However, this is not the case when we are considering the excitation of the free oscillations by internal sources. Let us suppose that a source is situated at the point $r = r_0$, $\theta = 0$. The source may be represented as discontinuity in y_1 and y_2 [Sect. 4.5.4]. Let

$$y_i(r_0 + 0) - y_i(r_0 - 0) = -f_i; \qquad i = 1, 2. \tag{6.68}$$

The functions f_i have been calculated in Section 4.5.4 for a dislocation source. Comparing Eq. (6.68) with Eqs. (4.179) and (4.180), we find

$$f_1 = \frac{-U_0\,dS}{4\pi r_0^2}(2l + 1)c^{\sigma}_{ml},$$
$$f_2 = \frac{-\mu(r_0)U_0\,dS}{4\pi r_0^3}(2l + 1)\bar{c}^{\sigma}_{ml}. \tag{6.69}$$

The above values of f_1 and f_2 correspond to a delta-function time dependence.

The determination of the eigenfrequencies does not require the presence of the source. Hence, for each ${}_n\omega_l$, a distribution of y_1 and y_2 is chosen throughout the shell that satisfies the conditions in Eqs. (6.66) and (6.68). These values of y_1 and y_2, when substituted in Eq. (6.62), give the spectral displacement throughout the earth.

6.3.3.1. Spectral Green's Dyadic. From Eq. (4.172), the toroidal Green's dyadic for a homogeneous medium is given by

$$\mathfrak{G}(\mathbf{r}\,|\,\mathbf{r}_0;\omega) = -\frac{ik_\beta}{\mu}\sum_{l=1}^{\infty}\sum_{m=-l}^{l}\frac{1}{\Omega_{ml}\,l(l+1)}\mathbf{M}^{-}_{ml}(r_>)\overset{*}{\mathbf{M}}{}^{+}_{ml}(r_<), \tag{6.70}$$

where

$$\mathbf{M}^{\pm}_{ml}(r) = y_1^{\pm}(r)\sqrt{[l(l+1)]}\,\mathbf{C}_{ml}(\theta,\phi),$$
$$y_1^{+}(r) = j_l(k_\beta r), \qquad y_1^{-}(r) = h_l^{(2)}(k_\beta r),$$
$$r_> = \max(r, r_0), \qquad r_< = \min(r, r_0),$$
$$\Omega_{ml} = \frac{4\pi}{2l+1}\cdot\frac{(l+m)!}{(l-m)!}. \tag{6.71}$$

We define

$$\begin{aligned} W &= y_1^+(r)\frac{d}{dr}y_1^-(r) - y_1^-(r)\frac{d}{dr}y_1^+(r) \\ &= \frac{1}{\mu}\left[y_1^+(r)y_2^-(r) - y_1^-(r)y_2^+(r)\right]. \end{aligned} \tag{6.72}$$

where y_1^+ and y_1^- are the two independent solutions of Eq. (6.63). Using the Wronksian relation for the spherical Bessel functions, we have

$$r^2 W(r) = \frac{1}{(ik_\beta)}. \tag{6.73}$$

Equation (6.70) can now be written as

$$\mathfrak{G}(\mathbf{r}\,|\,\mathbf{r}_0;\omega) = -\sum_{l=1}^{\infty}\sum_{m=-l}^{l}\frac{\mathbf{M}_{ml}^-(r_>)\overset{*}{\mathbf{M}}{}_{ml}^+(r_<)}{(r^2\mu W)l(l+1)\Omega_{ml}}. \tag{6.74}$$

For a radially inhomogeneous medium, the form of the Green's dyadic in Eq. (6.74) remains unaltered. However, in this general case, we do not have closed-form expressions for $y_1^+(r)$ and $y_1^-(r)$. From Eq. (6.63), we note that

$$\left(y_1^+\frac{d^2}{dr^2}y_1^- - y_1^-\frac{d^2}{dr^2}y_1^+\right) + \left(\frac{2}{r} + \frac{1}{\mu}\frac{d\mu}{dr}\right)\left(y_1^+\frac{d}{dr}y_1^- - y_1^-\frac{d}{dr}y_1^+\right) = 0,$$

i.e.,

$$r^2\mu\frac{d}{dr}\left(y_1^+\frac{d}{dr}y_1^- - y_1^-\frac{d}{dr}y_1^+\right) + \frac{d}{dr}(r^2\mu)\left(y_1^+\frac{d}{dr}y_1^- - y_1^-\frac{d}{dr}y_1^+\right) = 0.$$

Integrating, we get

$$r^2\mu W = \text{const.} = c \text{ (say)}. \tag{6.75}$$

Therefore, the toroidal Green's dyadic for a radially inhomogeneous medium of infinite extent is

$$\mathfrak{G}(\mathbf{r}\,|\,\mathbf{r}_0;\omega) = \frac{-1}{4\pi}\sum_{l=1}^{\infty}\sum_{m=-l}^{l}\frac{1}{c}\frac{2l+1}{l(l+1)}\frac{(l-m)!}{(l+m)!}\mathbf{M}_{ml}^-(r_>)\overset{*}{\mathbf{M}}{}_{ml}^+(r_<). \tag{6.76}$$

We denote by $\mathfrak{G}_a$ the Green's dyadic for a radially inhomogeneous, elastic sphere of radius a with zero surface tractions. To obtain $\mathfrak{G}_a$ we must add to $\mathfrak{G}$ a similar expression that is finite at the origin such that $\mathfrak{G}_a$ meets the boundary conditions of vanishing stresses across the surface $r = a$. We assume

$$\mathfrak{G}_a = \mathfrak{G} - \frac{1}{4\pi}\sum_{l=1}^{\infty}\sum_{m=-l}^{l}\frac{1}{c}\frac{2l+1}{l(l+1)}\frac{(l-m)!}{(l+m)!}\mathbf{M}_{ml}^+(r)\mathbf{X}_{ml}(r_0), \tag{6.77}$$

where $\mathbf{X}_{ml}(r_0)$ is an unknown vector. The boundary condition yields

$$\mathbf{e}_r\cdot[\nabla\mathfrak{G}_a + (\nabla\mathfrak{G}_a)^{213}] = 0, \quad \text{at} \quad r = a. \tag{6.78}$$

During differentiation implied by Eq. (6.78), the source vectors in $\mathfrak{G}_a$ behave as constants. Also, div $\mathfrak{G}_a = 0$. Solving for $\mathbf{X}$, we have

$$\mathbf{X}_{ml}(r_0) = -\frac{y_2^-(a)}{y_2^+(a)}\overset{*}{\mathbf{M}}{}^+_{ml}(r_0). \tag{6.79}$$

From Eqs. (6.76), (6.77), and (6.79), we get, for $r > r_0$,

$$\mathfrak{G}_a = -\frac{1}{4\pi}\sum_{l=1}^{\infty}\sum_{m=-l}^{l}\frac{1}{c}\frac{2l+1}{l(l+1)}\frac{(l-m)!}{(l+m)!}\{y_2^+(a)\mathbf{M}^-_{ml}(r) - y_2^-(a)\mathbf{M}^+_{ml}(r)\}$$
$$\times\frac{1}{y_2^+(a)}\overset{*}{\mathbf{M}}{}^+_{ml}(r_0), \tag{6.80}$$

and for $r < r_0$,

$$\mathfrak{G}_a = -\frac{1}{4\pi}\sum_{l=1}^{\infty}\sum_{m=-l}^{l}\frac{1}{c}\frac{2l+1}{l(l+1)}\frac{(l-m)!}{(l+m)!}\frac{1}{y_2^+(a)}\mathbf{M}^+_{ml}(r)\{y_2^+(a)\overset{*}{\mathbf{M}}{}^-_{ml}(r_0)$$
$$-y_2^-(a)\overset{*}{\mathbf{M}}{}^+_{ml}(r_0)\}. \tag{6.81}$$

Using Eqs. (6.72) and (6.75), the Green's dyadic at the free surface ($r = a$) can be reduced to the following simplified form

$$\mathfrak{G}_a(\mathbf{a}\,|\,\mathbf{r}_0;\omega) = \frac{1}{4\pi a^2}\sum_{l=1}^{\infty}\sum_{m=-l}^{l}(2l+1)\frac{(l-m)!}{(l+m)!}\frac{y_1^+(r_0)}{y_2^+(a)}\mathbf{C}_{ml}(\theta,\phi)\overset{*}{\mathbf{C}}_{ml}(\theta_0,\phi_0)$$
$$= \frac{1}{a^2}\sum_{l=1}^{\infty}\sum_{m=-l}^{l}\frac{1}{\Omega_{ml}}\frac{y_1^+(r_0)}{y_2^+(a)}\mathbf{C}_{ml}(\theta,\phi)\overset{*}{\mathbf{C}}_{ml}(\theta_0,\phi_0). \tag{6.82}$$

Equation (6.82) constitutes the spectral toroidal Green's dyadic for a radially heterogeneous earth model, assuming delta-function time dependence.

EXAMPLE 6.4: Torque in a Sphere

From Eq. (4.50), the displacement field resulting from a torque of moment M about the z axis (Fig. 6.5) is

$$\begin{aligned}\mathbf{u} &= M(\mathbf{e}_2\mathbf{e}_1 - \mathbf{e}_1\mathbf{e}_2) : \text{grad}_0\,\tilde{\mathfrak{G}}(\mathbf{r}\,|\,\mathbf{r}_0)\\ &= M(\mathbf{e}_3 \times \mathfrak{I}) : \text{grad}_0\,\tilde{\mathfrak{G}}\\ &= M\mathbf{e}_3\cdot(\mathfrak{I}\times\text{grad}_0\,\tilde{\mathfrak{G}})\\ &= M\mathbf{e}_3\cdot(\text{curl}_0\,\tilde{\mathfrak{G}}).\end{aligned} \tag{6.4.1}$$

With $\mathfrak{G}_a$, as in Eq. (6.82), we have

$$\mathbf{u} = \frac{M}{4\pi a^2}\sum_{l=1}^{\infty}\sum_{m=-l}^{l}(2l+1)\frac{(l-m)!}{(l+m)!}\frac{\mathbf{C}_{ml}(\theta,\phi)}{y_2^+(a)}\mathbf{e}_3$$
$$\cdot\left[\sqrt{\{l(l+1)\}}\frac{y_1^+(r_0)}{r_0}\mathbf{P}_{ml}(\theta_0,\phi_0) + \left(\frac{d}{dr_0}+\frac{1}{r_0}\right)y_1^+(r_0)\mathbf{B}_{ml}(\theta_0,\phi_0)\right]. \tag{6.4.2}$$

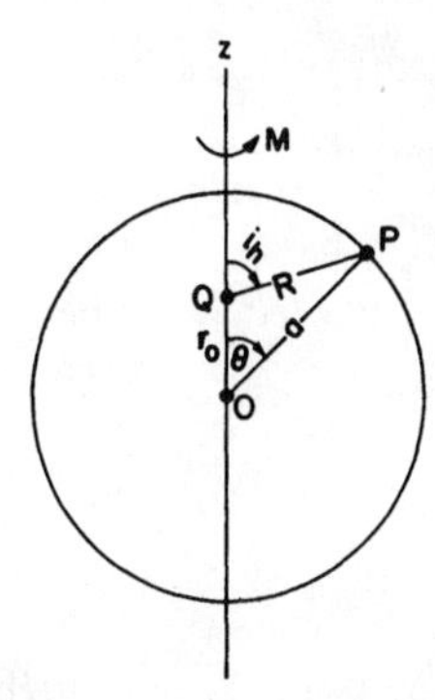

Figure 6.5. Torque about the z axis.

This result holds for a source at (r_0, θ_0, ϕ_0). For a source on the z axis, we take the limit as $\theta_0 \to 0$, $\phi_0 \to 0$. In this case

$$\mathbf{e}_3 \cdot \mathbf{P}_{ml}(\theta_0, \phi_0) = \delta_{m,0}, \qquad \mathbf{e}_3 \cdot \mathbf{B}_{ml}(\theta_0, \phi_0) = 0.$$

Therefore,

$$u_\phi(\omega) = \frac{-M}{4\pi a^2} \sum_{l=1}^{\infty} \frac{(2l+1)}{r_0} \frac{y_1^+(r_0)}{y_2^+(a)} \frac{\partial}{\partial \theta} P_l(\cos\theta), \qquad u_\theta = 0. \tag{6.4.3}$$

This is the spectral response for a delta-function time variation. Taking the inverse Fourier transform and assuming a unit step time-dependence, the residue theorem yields

$$u_\phi(t) = \frac{-M}{\pi a^2 r_0} \sum_{n=0}^{\infty} \sum_{l=1}^{\infty} \left(l + \frac{1}{2}\right) \left[\frac{y_1^+(r_0)}{\omega \dfrac{\partial}{\partial \omega} y_2^+(a)}\right]_{{}_n\omega_l} P_l^1(\cos\theta)(1 - \cos {}_n\omega_l t) H(t). \tag{6.4.4}$$

In particular, for a homogeneous sphere,

$$y_1^+(r_0) = j_l(k_\beta r_0),$$

$$\chi j_{l+1}(\chi) = (l-1) j_l(\chi), \qquad (\chi = a k_\beta).$$

The second relation is the period equation, Eq. (6.9). A further reduction then leads us to

$$u_\phi(t) = \frac{M}{\pi \mu a r_0} \sum_{n=0}^{\infty} \sum_{l=1}^{\infty} \left(l + \frac{1}{2}\right) P_l^1(\cos\theta) \left[\frac{j_l(k_\beta r_0)}{j_l(k_\beta a)} \cdot \frac{(1 - \cos\omega t)}{a^2 k_\beta^2 - (l-1)(l+2)}\right]_{{}_n\omega_l} H(t). \tag{6.4.5}$$

Taking the limit as $r_0 \to 0$ and noting that

$$\lim_{r_0 \to 0} \left[\frac{1}{r_0} j_l(k_\beta r_0)\right] = \frac{1}{3} k_\beta \delta_{l,1},$$

we get the following result for a source at the center of the sphere

$$u_\phi(t) = \frac{M}{2\pi\mu a^2} \sum_{n=0}^{\infty} \left[\frac{(1 - \cos \omega t)}{\sin \chi/\chi - \cos \chi} \right]_{n\omega_l} \sin \theta H(t). \tag{6.4.6}$$

The sphere then executes rotatory vibrations, subjected to the period equation

$$\frac{\tan \chi}{\chi} = \frac{3}{3 - \chi^2}.$$

Using this in Eq. (6.4.6), we get

$$u_\phi(t) = \frac{M}{2\pi\mu a^2} \sum_{n=0}^{\infty} \left[\frac{1}{\chi^2} (\chi^4 + 3\chi^2 + 9)^{1/2} (1 - \cos \omega t) \right]_{n\omega_l} \sin \theta H(t). \tag{6.4.7}$$

6.3.3.2. Normal-Mode Green's Dyadic. We denote the eigenvectors of the toroidal oscillations of the earth corresponding to the eigenfrequency ω_n by $\mathbf{M}_n$. These eigenvectors constitute a complete orthogonal basis. Let

$$\int_V \mathbf{M}_n(\mathbf{r}) \cdot \overset{*}{\mathbf{M}}_k(\mathbf{r}) \rho_0(r) dV = \delta_{nk} \Lambda_n, \tag{6.83}$$

where the integral extends over the entire volume of the earth, Λ_n is the normalizing factor, and the asterisk denotes a complex conjugate. We expand the toroidal displacement field in terms of the toroidal eigenvectors:

$$\mathbf{u}(\mathbf{r}) = \sum_{n=0}^{\infty} a_n \mathbf{M}_n(\mathbf{r}), \tag{6.84}$$

where, by Eq. (6.83),

$$a_n = \frac{1}{\Lambda_n} \int_V \mathbf{u} \cdot \overset{*}{\mathbf{M}}_n \rho_0 \, dV. \tag{6.85}$$

Substituting Eq. (6.84) into the Fourier-transformed equation of motion

$$\operatorname{div} \mathfrak{T} + \rho_0 \mathbf{F} + \rho_0 \omega^2 \mathbf{u} = 0, \tag{6.86}$$

where $\mathbf{F}$ represents the external body forces per unit mass, we get

$$\sum_p a_p \operatorname{div} \mathfrak{T}_p + \rho_0 \mathbf{F} + \rho_0 \omega^2 \sum_p a_p \mathbf{M}_p = 0 \tag{6.87}$$

in which

$$\mathfrak{T}_p = \mathfrak{T}(\mathbf{M}_p). \tag{6.88}$$

Taking the scalar product of Eq. (6.87) with $\overset{*}{\mathbf{M}}_n$, integrating over V, and using Eq. (6.83), we have

$$\int_V \sum_p a_p (\operatorname{div} \mathfrak{T}_p) \cdot \overset{*}{\mathbf{M}}_n \, dV + \int_V \overset{*}{\mathbf{M}}_n \cdot \mathbf{F} \rho_0 \, dV + \omega^2 \Lambda_n a_n = 0. \tag{6.89}$$

The eigenvector $\mathbf{M}_k$ satisfies Eq. (6.86) with $\mathbf{F} = 0$, i.e.,

$$\operatorname{div} \mathfrak{T}_k + \rho_0 \omega_k^2 \mathbf{M}_k = 0. \tag{6.90}$$

Taking the scalar product of Eq. (6.90) with $\overset{*}{\mathbf{M}}_n$, integrating over V, and using Eq. (6.83), it follows that

$$\int_V (\operatorname{div} \mathfrak{T}_k) \cdot \overset{*}{\mathbf{M}}_n \, dV + \omega_k^2 \delta_{nk} \Lambda_n = 0. \tag{6.91}$$

Using this result in Eq. (6.89), we find

$$a_n = \frac{1}{(\omega_n^2 - \omega^2)\Lambda_n} \int_V \overset{*}{\mathbf{M}}_n \cdot \mathbf{F} \rho_0 \, dV. \tag{6.92}$$

Equation (6.84) can now be written as

$$\mathbf{u}(\mathbf{r}) = \sum_{n=0}^{\infty} \frac{1}{(\omega_n^2 - \omega^2)\Lambda_n} \mathbf{M}_n(\mathbf{r}) \int_V \overset{*}{\mathbf{M}}_n \cdot \mathbf{F} \rho_0 \, dV. \tag{6.93}$$

This equation is of fundamental importance because it represents the field for a given distribution of body forces as a sum of normal modes of the earth.

If $\mathbf{F}$ is a unit point force localized at $\mathbf{r}_0$, we have

$$\rho_0 \mathbf{F} = \delta(\mathbf{r} - \mathbf{r}_0) \mathbf{a} g(\omega), \tag{6.94}$$

where $\delta(\mathbf{r} - \mathbf{r}_0)$ is the three-dimensional Dirac delta function and $\mathbf{a}$ is a unit vector in the direction of the force. Equations (6.93) and (6.94) yield

$$\mathbf{u}(\mathbf{r}) = \sum_n \frac{g(\omega)}{(\omega_n^2 - \omega^2)\Lambda_n} \mathbf{M}_n(\mathbf{r}) [\overset{*}{\mathbf{M}}_n(\mathbf{r}_0) \cdot \mathbf{a}]. \tag{6.95}$$

Therefore, the toroidal Green's dyadic is

$$\mathfrak{G}(\mathbf{r} | \mathbf{r}_0; \omega) = \sum_{n=0}^{\infty} \frac{\mathbf{M}_n(\mathbf{r}) \overset{*}{\mathbf{M}}_n(\mathbf{r}_0)}{(\omega_n^2 - \omega^2)\Lambda_n} g(\omega). \tag{6.96}$$

Executing the inverse Fourier transform and using Cauchy's residue theorem, we get the following expression for the Green's dyadic for *unit-step time function* $H(t)$

$$\mathfrak{G}(\mathbf{r} | \mathbf{r}_0; t) = \sum_{n=0}^{\infty} \mathbf{M}_n(\mathbf{r}) \overset{*}{\mathbf{M}}_n(\mathbf{r}_0) \frac{1 - \cos \omega_n t}{\omega_n^2 \Lambda_n} H(t). \tag{6.97}$$

It may be noted that the *static Green's dyadic*

$$\mathfrak{G}(\mathbf{r} | \mathbf{r}_0; st) = \sum_{n=0}^{\infty} \frac{1}{\omega_n^2 \Lambda_n} \mathbf{M}_n(\mathbf{r}) \overset{*}{\mathbf{M}}_n(\mathbf{r}_0) \tag{6.98}$$

which is obtained from Eq. (6.97) in the limit $t \to \infty$, can be expressed in terms of the normal modes of the earth.

Expressing $\mathbf{M}_n$ in terms of the spherical harmonics

$${}_n\mathbf{M}_{ml}(\mathbf{r}) = y_{1n}(r) \sqrt{[l(l+1)]} \, \mathbf{C}_{ml}(\theta, \phi), \tag{6.99}$$

we have, from Eq. (6.83),

$$ {}_n\Lambda_{ml} = l(l+1)\Omega_{ml} I_n^T, \tag{6.100} $$

where

$$ \Omega_{ml} = \frac{4\pi}{2l+1} \cdot \frac{(l+m)!}{(l-m)!}, \tag{6.101} $$

$$ I_n^T = \int_0^a y_{1n}^2 \rho_0 r^2 \, dr, \tag{6.102} $$

$$ 0 = \int_0^a y_{1n} y_{1k} \rho_0 r^2 \, dr \qquad (n \neq k). \tag{6.103} $$

It may be noted that ω_n, y_{1n} and I_n^T are functions of both n and l. However, they are independent of m. The Green's dyadic takes the form

$$ \mathfrak{G} = \sum_{l=1}^{\infty} \sum_{m=-l}^{l} \mathfrak{G}_{ml}, \tag{6.104} $$

where

$$ \mathfrak{G}_{ml}(\mathbf{r}\,|\,\mathbf{r}_0;\omega) = \frac{1}{l(l+1)\Omega_{ml}} \sum_{n=0}^{\infty} \frac{{}_n\mathbf{M}_{ml}(\mathbf{r})\,{}_n\overset{*}{\mathbf{M}}_{ml}(\mathbf{r}_0)}{(\omega_n^2 - \omega^2) I_n^T} \, g(\omega), \tag{6.105} $$

$$ \mathfrak{G}_{ml}(\mathbf{r}\,|\,\mathbf{r}_0;t) = \frac{1}{l(l+1)\Omega_{ml}} \sum_{n=0}^{\infty} {}_n\mathbf{M}_{ml}(\mathbf{r})\,{}_n\overset{*}{\mathbf{M}}_{ml}(\mathbf{r}_0) \, \frac{1-\cos\omega_n t}{\omega_n^2 I_n^T} \, H(t), \tag{6.106} $$

$$ \mathfrak{G}_{ml}(\mathbf{r}\,|\,\mathbf{r}_0;st) = \frac{1}{l(l+1)\Omega_{ml}} \sum_{n=0}^{\infty} \frac{1}{\omega_n^2 I_n^T} \, {}_n\mathbf{M}_{ml}(\mathbf{r})\,{}_n\overset{*}{\mathbf{M}}_{ml}(\mathbf{r}_0). \tag{6.107} $$

It is useful to note that $[\partial y_2(a)/\partial\omega]_{\omega_n}$ can be expressed in terms of I_n^T. Because y_{1n} satisfies Eq. (6.63) with ω replaced by ω_n, we have

$$ \mu\left(y_{1n}\frac{d^2 y_1}{dr^2} - y_1 \frac{d^2 y_{1n}}{dr^2}\right) + \left(\frac{d\mu}{dr} + \frac{2\mu}{r}\right)\left(y_{1n}\frac{dy_1}{dr} - y_1\frac{dy_{1n}}{dr}\right) $$
$$ + \rho_0(\omega^2 - \omega_n^2) y_{1n} y_1 = 0. \tag{6.108} $$

This may be written as

$$ \frac{d}{dr}\left[\mu r^2 \left(y_{1n}\frac{dy_1}{dr} - y_1\frac{dy_{1n}}{dr}\right)\right] = \rho_0 r^2 (\omega_n^2 - \omega^2) y_{1n} y_1. \tag{6.109} $$

Using Eq. (6.65), the last equation becomes

$$ \frac{d}{dr}\left[r^2 (y_{1n} y_2 - y_1 y_{2n})\right] = \rho_0 r^2 (\omega_n^2 - \omega^2) y_{1n} y_1. \tag{6.110} $$

On integration, this yields

$$ \int_0^a y_{1n} y_1 \rho_0 r^2 \, dr = \frac{1}{\omega_n^2 - \omega^2} \left[r^2 (y_{1n} y_2 - y_1 y_{2n})\right]\Big|_0^a = \frac{a^2 y_{1n}(a) y_2(a)}{\omega_n^2 - \omega^2}, \tag{6.111} $$

because $y_{2n}(a) = 0$. If $\omega = \omega_k$ $(n \neq k)$, we get

$$\int_0^a y_{1n} y_{1k} \rho_0 r^2 \, dr = 0, \tag{6.112}$$

because $y_{2k}(a) = 0$. This is the *orthogonality relation*, Eq. (6.103). Next, assuming that ω is not an eigenfrequency and taking the limit as $\omega \to \omega_n$; Eq. (6.111) yields

$$I_n^T = \int_0^a y_{1n}^2 \rho_0 r^2 \, dr = -\frac{a^2}{2}\left[\frac{y_1(a)}{\omega}\frac{\partial}{\partial\omega} y_2(a)\right]_{\omega_n}. \tag{6.113}$$

Hence, we have

$$\left[\frac{\partial}{\partial\omega} y_2(a)\right]_{\omega_n} = -\frac{2\omega_n I_n^T}{a^2 y_{1n}(a)}. \tag{6.114}$$

We see from Eq. (6.106) that the seismic response of a radially heterogeneous sphere can be represented as a Fourier series in time where the sum goes over all possible normal modes. The coefficients of $\cos \omega_n t$ in the expressions for the displacements are the *line spectra* of the time series.

To obtain these spectral amplitudes from earthquake seismograms, we must separate the sum into its spectral components. This is achieved in practice by applying the Fourier transform to the seismogram trace. Because of the earth's rotation, anelasticity, ellipticity, lateral heterogeneity, and the ever-present background noise, the lines are split and broadened in such a way that a spectral continuum is obtained. In this continuum, the spectral peaks corresponding to ${}_nT_l$ and ${}_nS_l$ are present.

From Eq. (6.105) we note that if the source is "driving" the sphere with a frequency that coincides with one of the eigenfrequencies, ${}_n\omega_l$, the Green's dyadic will become infinite; i.e., resonance will occur. In reality, however, spectral lines are of finite amplitude due to *physical attenuation.*

Equations (6.82), (6.105), and (6.106) are the representations of the same Green's dyadic. The form in Eq. (6.106) was obtained by taking the inverse Fourier transform of Eq. (6.105). Similarly, assuming unit-step function time dependence and taking the inverse Fourier transform of Eq. (6.82), we get Eq. (6.106).

In the case of a homogeneous sphere

$$y_{1n}(r) = j_l\left(\frac{r\omega_n}{\beta}\right) \tag{6.115}$$

and, therefore, Eq. (6.102) gives

$$I_n^T = \tfrac{1}{2}\rho_0 a^3 [j_l^2(\chi_n) - j_{l-1}(\chi_n) j_{l+1}(\chi_n)], \tag{6.116}$$

where $\chi_n = a\omega_n/\beta$. Noting that χ_n is a root of the frequency equation

$$\chi j_{l+1}(\chi) = (l-1) j_l(\chi), \tag{6.117}$$

and using the recurrence relation for $j_l(\chi)$, Eq. (6.116) may be reduced to the form

$$I_n^T = \frac{1}{2}\rho_0 a^3 j_l^2(\chi_n)\left[1 - \frac{(l-1)(l+2)}{\chi_n^2}\right]. \tag{6.118}$$

Substituting in Eq. (6.106), we get

$$\mathfrak{G}_{ml}(\mathbf{r}|\mathbf{r}_0; t) = \frac{2}{a\mu l(l+1)\Omega_{ml}} \sum_{n=0}^{\infty} \frac{{}_n\mathbf{M}_{ml}(\mathbf{r})_n\overset{*}{\mathbf{M}}_{ml}(\mathbf{r}_0)(1-\cos\omega_n t)H(t)}{j_l^2(\chi_n)[\chi_n^2 - (l-1)(l+2)]}. \tag{6.119}$$

Equation (6.119) can also be obtained by evaluating the inverse Fourier transform of the toroidal Green's dyadic for a homogeneous sphere given in Eq. (6.37).

6.3.3.3. Toroidal Field Resulting from a Shear Dislocation. Once the Green's dyadic is known, the Volterra relation gives directly the displacement field induced by a tangential dislocation. For a point source, the Volterra relation may be written as [cf. Eq. (4.107)]

$$\mathbf{u}(\mathbf{r}) = P_0\, \mathbf{en} : \mathfrak{T}_0(\tilde{\mathfrak{G}}), \tag{6.120}$$

where $P_0 = U_0\, dS$ is the source potency and

$$\mathfrak{T}_0(\mathbf{u}) = \lambda\mathfrak{I}\,\mathrm{div}_0\,\mathbf{u} + \mu(\nabla_0\mathbf{u} + \mathbf{u}\nabla_0).$$

Although we have derived the Volterra relation for a nongravitating medium, it can be shown that it remains valid even when the self-gravitation and initial hydrostatic stresses are taken into consideration.

It is easily seen that

$$\mathrm{div}_0\,\overset{*}{\mathbf{M}}_{ml}(\mathbf{r}_0) = 0,$$

$$\begin{aligned}
\mathrm{grad}_0\,\overset{*}{\mathbf{M}}_{ml}(\mathbf{r}_0) + \overset{*}{\mathbf{M}}_{ml}(\mathbf{r}_0)\mathrm{grad}_0 &= \left(\frac{d}{dr_0} - \frac{1}{r_0}\right) y_1(r_0)\sqrt{[l(l+1)]}\,(\mathbf{e}_{r_0}\overset{*}{\mathbf{C}}_{ml} + \overset{*}{\mathbf{C}}_{ml}\mathbf{e}_{r_0}) \\
&\quad + \frac{y_1(r_0)}{r_0}\left[(\mathbf{e}_{\theta_0}\mathbf{e}_{\theta_0} - \mathbf{e}_{\phi_0}\mathbf{e}_{\phi_0})\frac{2}{\sin\theta_0}\right. \\
&\quad \times \left(\frac{\partial}{\partial\theta_0} - \cot\theta_0\right)\frac{\partial\overset{*}{Y}_{ml}}{\partial\phi_0} \\
&\quad \left. - (\mathbf{e}_{\theta_0}\mathbf{e}_{\phi_0} + \mathbf{e}_{\phi_0}\mathbf{e}_{\theta_0})\left\{2\frac{\partial^2}{\partial\theta_0^2} + l(l+1)\right\}\overset{*}{Y}_{ml}\right],
\end{aligned} \tag{6.121}$$

where

$$\begin{aligned}
\overset{*}{\mathbf{M}}_{ml}(\mathbf{r}_0) &= y_1(r_0)\sqrt{[l(l+1)]}\,\overset{*}{\mathbf{C}}_{ml}(\theta_0, \phi_0), \\
\overset{*}{Y}_{ml} &= P_l^m(\cos\theta_0)\exp(-im\phi_0).
\end{aligned} \tag{6.122}$$

Clearly, **en** in Eq. (6.120) could be replaced by (**en** + **ne**)/2, because the operator $\mathfrak{T}_0$ is symmetric. Equations (4.115b), (6.104), (6.106), (6.120), and (6.121) yield the following expression for the displacement resulting from an arbitrary shear dislocation

$$\mathbf{u}(\mathbf{r}; t) = \sum_{l=1}^{\infty} \sum_{m=-l}^{l} \sum_{n=0}^{\infty} {}_n\mathbf{u}_{ml}(\mathbf{r}; t), \tag{6.123}$$

where

$$\begin{aligned}
{}_n\mathbf{u}_{ml}(\mathbf{r}; t) = {} & \frac{P_0(1 - \cos \omega_n t)H(t)}{l(l+1)\omega_n^2 I_n^T} \frac{1}{\Omega_{ml}} \left[\frac{\mu(r_0) y_{1n}(r_0)}{r_0} \left\{ p_1 \left(2 \frac{\partial^2}{\partial \theta_0^2} + l^2 + l \right) \right. \right. \\
& \left. - p_2 \frac{2}{\sin \theta_0} \left(\frac{\partial}{\partial \theta_0} - \cot \theta_0 \right) \frac{\partial}{\partial \phi_0} \right\} \overset{*}{Y}_{ml}(\theta_0, \phi_0) + y_{2n}(r_0) \\
& \left. \times \left(p_3 \frac{\partial}{\partial \theta_0} - p_4 \frac{1}{\sin \theta_0} \frac{\partial}{\partial \phi_0} \right) \overset{*}{Y}_{ml}(\theta_0, \phi_0) \right] \\
& \times [y_{1n}(r)][\sqrt{\{l(l+1)\}}\, \mathbf{C}_{ml}(\theta, \phi)],
\end{aligned} \tag{6.124}$$

$$\begin{aligned}
p_1 &= (\cos \lambda \sin \delta)\cos 2\hat{\phi} + (\sin \lambda \sin \delta \cos \delta)\sin 2\hat{\phi}, \\
p_2 &= (\cos \lambda \sin \delta)\sin 2\hat{\phi} - (\sin \lambda \sin \delta \cos \delta)\cos 2\hat{\phi}, \\
p_3 &= (\sin \lambda \cos 2\delta)\cos \hat{\phi} - (\cos \lambda \cos \delta)\sin \hat{\phi}, \\
p_4 &= (\sin \lambda \cos 2\delta)\sin \hat{\phi} + (\cos \lambda \cos \delta)\cos \hat{\phi}, \\
p_5 &= \sin \lambda \sin \delta \cos \delta.
\end{aligned} \tag{6.125}$$

Using the relation

$$P_l^{-m}(\cos \theta) = (-1)^m \frac{(l-m)!}{(l+m)!} P_l^m(\cos \theta),$$

it can be verified that

$${}_n\mathbf{u}_{(-m)l} = {}_n\overset{*}{\mathbf{u}}_{ml}, \tag{6.126}$$

showing thereby that the displacement $\mathbf{u}$ as expressed by Eq. (6.123) is indeed real.

The frequencies ${}_n\omega_l$ (written as ω_n in the above derivation) for a nonrotating, radially inhomogeneous, isotropic, spherical earth are degenerate, being independent of the azimuthal order number m. In Section 6.4, it is shown that the introduction of a slow angular rotation removes the degeneracy, each line ${}_n\omega_l$ being split into a multiplet of $(2l + 1)$ lines ${}_n\omega_l^m$. The effect of the ellipticity of the earth is to split each line into a multiplet of $(l + 1)$ lines. Though the expression in Eq. (6.124) has been derived for a nonrotating spherical earth, it may be taken as the zero-order approximation for a rotating elliptical earth and may be used to calculate the amplitudes of the split normal modes (l, n, m) of the earth. Given the geographic coordinates of the source (r_0, θ_0, ϕ_0) and the sensor (r, θ, ϕ), Eq. (6.124) is the final expression for the displacements of the toroidal oscillations of the earth excited by an arbitrary shear dislocation $(\lambda, \delta, \hat{\phi})$. Equation (6.126) shows that the spectral amplitudes of the pair of singlets of orders $\pm m$ are equal.

We must be extremely careful when comparing observed data with theoretical amplitudes. Significant differences exist between the spectral peaks of split and degenerate multiplets. Not only the locations of the peaks but also the amplitudes could be different depending upon the locations of the source and the receiver. The displacement of the multiplet for the degenerate case is obtained by the superposition of the displacements of the individual split singlets. We will get a large amplitude for the multiplet if the singlets interfere constructively. It is difficult to assess the extent of these differences until computations are actually carried out. Therefore, it is desirable to use the amplitudes of the split normal modes (l, n, m) rather than the amplitudes of the multiplets (l, n) obtained from such expressions as Eq. (6.124) by summing over m.

When theoretical seismograms are computed, we can ignore the splitting because we have to sum over m, n, and l anyway. In that case we make use of the addition theorem

$$P_l(\cos\varepsilon) = \sum_{m=-l}^{l} \frac{(l-m)!}{(l+m)!} Y_{ml}(\theta, \phi) \overset{*}{Y}_{ml}(\theta_0, \phi_0), \tag{6.127}$$

where

$$\cos\varepsilon = \cos\theta\cos\theta_0 + \sin\theta\sin\theta_0\cos(\phi - \phi_0). \tag{6.128}$$

Applying it to Eqs. (6.123) and (6.124), we obtain

$$\begin{aligned}\mathbf{u}(\mathbf{r}; t) = {} & \frac{P_0}{4\pi} \sum_{l=1}^{\infty} \frac{2l+1}{l(l+1)} \sum_{n=0}^{\infty} \frac{(1 - \cos\omega_n t)H(t)}{\omega_n^2 I_n^T} y_{1n}(r) \left(\mathbf{e}_\theta \frac{1}{\sin\theta} \frac{\partial}{\partial\phi} - \mathbf{e}_\phi \frac{\partial}{\partial\theta} \right) \\ & \times \left[\frac{\mu(r_0) y_{1n}(r_0)}{r_0} \left\{ p_1 \left(2\frac{\partial^2}{\partial\theta_0^2} + l^2 + l \right) - p_2 \frac{2}{\sin\theta_0} \left(\frac{\partial}{\partial\theta_0} - \cot\theta_0 \right) \right. \right. \\ & \left. \left. \times \frac{\partial}{\partial\phi_0} \right\} + y_{2n}(r_0) \left(p_3 \frac{\partial}{\partial\theta_0} - p_4 \frac{1}{\sin\theta_0} \frac{\partial}{\partial\phi_0} \right) \right] P_l(\cos\varepsilon). \end{aligned} \tag{6.129}$$

From Eq. (6.128), it is obvious that $(\varepsilon, \theta, \theta_0)$ are the three sides of the spherical triangle on the surface of the earth with the North Pole N, epicenter $B(a, \theta_0, \phi_0)$, and the point $C(a, \theta, \phi)$ as the three vertices (Fig. 6.6). If we denote the three angles by $A = \phi - \phi_0$, B, and C, respectively, the results in Eqs. (6.130) follow from the relations between the sides and the angles of a spherical triangle:

$$\begin{aligned} &\frac{\partial\varepsilon}{\partial\theta} = \cos C, \qquad \frac{\partial\varepsilon}{\partial\theta_0} = \cos B, \qquad \frac{1}{\sin\theta}\frac{\partial\varepsilon}{\partial\phi} = \sin C; \\ &\frac{\partial B}{\partial\theta} = \frac{\sin C}{\sin\varepsilon}, \qquad \frac{\partial B}{\partial\theta_0} = -\sin B\cot\varepsilon, \qquad \frac{1}{\sin\theta}\frac{\partial B}{\partial\phi} = -\frac{\cos C}{\sin\varepsilon}; \\ &\frac{\partial C}{\partial\theta} = -\sin C\cot\varepsilon, \qquad \frac{\partial C}{\partial\theta_0} = \frac{\sin B}{\sin\varepsilon}, \qquad \frac{1}{\sin\theta_0}\frac{\partial C}{\partial\phi} = -\frac{\cos B}{\sin\varepsilon}. \end{aligned} \tag{6.130}$$

Using these results and the recurrence relations for the Legendre functions, it follows that

$$\begin{aligned}
\frac{\partial}{\partial \theta_0} P_l(\cos \varepsilon) &= -\cos B P_l^1(\cos \varepsilon), \\
\frac{1}{\sin \theta_0} \frac{\partial}{\partial \phi_0} P_l(\cos \varepsilon) &= \sin B P_l^1(\cos \varepsilon), \\
\left[2 \frac{\partial^2}{\partial \theta_0^2} + l(l+1)\right] P_l(\cos \varepsilon) &= \cos 2B P_l^2(\cos \varepsilon), \\
\frac{1}{\sin \theta_0}\left(\frac{\partial^2}{\partial \theta_0 \partial \phi_0} - \cot \theta_0 \frac{\partial}{\partial \phi_0}\right) P_l(\cos \varepsilon) &= -\tfrac{1}{2} \sin 2B P_l^2(\cos \varepsilon).
\end{aligned} \tag{6.131}$$

With the aid of Eqs. (6.130) and (6.131), the expression in Eq. (6.129) can be put in the following convenient form:

$$\begin{aligned}
\mathbf{u}(\mathbf{r}; t) = \frac{P_0}{4\pi} \sum_{l=1}^{\infty} \frac{2l+1}{l(l+1)} \sum_{n=0}^{\infty} & \frac{(1 - \cos \omega_n t) H(t)}{\omega_n^2 I_n^T} y_{1n}(r) \left(\left\{ \frac{\mu(r_0) y_{1n}(r_0)}{r_0} \right. \right. \\
& \times \left[q_1 \sin C \frac{\partial}{\partial \varepsilon} P_l^2(\cos \varepsilon) - 2q_2 \frac{\cos C}{\sin \varepsilon} P_l^2(\cos \varepsilon) \right] \\
& \left. - y_{2n}(r_0) \left[q_3 \sin C \frac{\partial}{\partial \varepsilon} P_l^1(\cos \varepsilon) - q_4 \frac{\cos C}{\sin \varepsilon} P_l^1(\cos \varepsilon) \right] \right\} \mathbf{e}_\theta \\
& - \left\{ \frac{\mu(r_0) y_{1n}(r_0)}{r_0} \left[q_1 \cos C \frac{\partial}{\partial \varepsilon} P_l^2(\cos \varepsilon) + 2q_2 \frac{\sin C}{\sin \varepsilon} P_l^2(\cos \varepsilon) \right] \right. \\
& \left. \left. - y_{2n}(r_0) \left[q_3 \cos C \frac{\partial}{\partial \varepsilon} P_l^1(\cos \varepsilon) + q_4 \frac{\sin C}{\sin \varepsilon} P_l^1(\cos \varepsilon) \right] \right\} \mathbf{e}_\phi \right),
\end{aligned} \tag{6.132}$$

where q_i $(i = 1, 2, 3, 4)$ are obtained from p_i of Eq. (6.125) on replacing $\hat{\phi}$ by $\hat{\phi} - B$; i.e.,

$$q_i(\lambda, \delta, \hat{\phi}) = p_i(\lambda, \delta, \hat{\phi} - B) \qquad (i = 1, 2, 3, 4). \tag{6.133}$$

Equation (6.132) is our final expression for the toroidal displacement at (r, θ, ϕ) induced by an arbitrary shear dislocation placed at (r_0, θ_0, ϕ_0) within the earth. Given $(\theta, \phi; \theta_0, \phi_0)$, ε can be calculated from Eq. (6.128) and then B and C from the relations

$$\frac{\sin B}{\sin \theta} = \frac{\sin C}{\sin \theta_0} = \frac{\sin(\phi - \phi_0)}{\sin \varepsilon}. \tag{6.134}$$

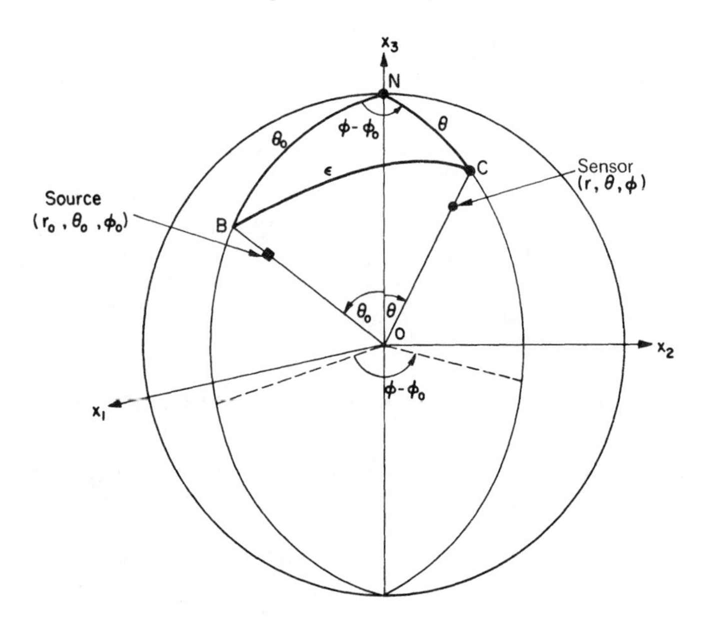

Figure 6.6. Location of the source (r_0, θ_0, ϕ_0) and the sensor (r, θ, ϕ) in the geographic system.

When the source lies on the x_3 axis, $\theta_0 = \phi_0 = 0$ and Eq. (6.132) can be simplified on putting $B = \pi - \phi$, $C = 0$, $\varepsilon = \theta$. We obtain

$$\mathbf{u}(\mathbf{r}; t) = \frac{P_0}{4\pi} \sum_{l=1}^{\infty} \frac{2l+1}{l(l+1)} \sum_{n=0}^{\infty} \frac{(1 - \cos \omega_n t)H(t)}{\omega_n^2 I_n^T} y_{1n}(r) \left[\left\{ y_{2n}(r_0) p_4^0 P_l^1(\cos\theta) - \frac{2\mu(r_0) y_{1n}(r_0)}{r_0} p_2^0 P_l^2(\cos\theta) \right\} \frac{\mathbf{e}_\theta}{\sin\theta} + \left\{ y_{2n}(r_0) p_3^0 \frac{\partial}{\partial\theta} P_l^1(\cos\theta) - \frac{\mu(r_0) y_{1n}(r_0)}{r_0} p_1^0 \frac{\partial}{\partial\theta} P_l^2(\cos\theta) \right\} \mathbf{e}_\phi \right], \tag{6.135}$$

where p_i^0 $(i = 1, 2, 3, 4)$ are obtained from p_i of Eqs. (6.125) on replacing $\hat{\phi}$ by ϕ^0, ϕ^0 $(= \phi + \hat{\phi} - \pi)$ being the azimuth of the sensor with respect to the fault's strike. For a vertical strike-slip source (case I), $p_1^0 = \cos 2\phi^0, p_2^0 = \sin 2\phi^0$, $p_3^0 = p_4^0 = p_5 = 0$; for a vertical dip-slip (case II), $p_3^0 = -\cos\phi^0$, $p_4^0 = -\sin\phi^0$, $p_1^0 = p_2^0 = p_5 = 0$; and, last, for case III, $p_1^0 = \frac{1}{2}\sin 2\phi^0$, $p_2^0 = -\frac{1}{2}\cos 2\phi^0$, $p_5 = \frac{1}{2}$, $p_3^0 = p_4^0 = 0$. It is interesting to compare the angular functions p_i^0 with the corresponding functions defined in Eq. (5.39) in connection with surface waves in a vertically heterogeneous half-space. We have

$$p_1^0 = p_L, \qquad p_2^0 = p_R, \qquad p_3^0 = q_L, \qquad p_4^0 = q_R, \qquad p_5 = s_R.$$

In the case of a homogeneous sphere, $y_{1n}(r) = j_l(r\omega_n/\beta)$, I_n^T is given by Eq. (6.118) and, the displacement is expressible in a closed form.

6.3.3.4. Spectral Field. Equation (6.106) gives the normal-mode toroidal Green's dyadic for a radially heterogeneous earth model. An application of the Volterra relation resulted in the expressions (6.132) and (6.135) for the displacement field in the time domain for an arbitrary shear dislocation. In Section 6.3.3.1 we have derived [Eq. (6.82)] the spectral Green's dyadic for the toroidal field. If we now apply the Volterra relation to this function, we will get the spectral toroidal displacement field induced by a shear dislocation situated in a radially heterogeneous earth model. Therefore, for a point shear dislocation (λ, δ) located at $(r = r_0, \theta = 0)$, the displacement field at the point (a, θ, ϕ) on the surface of the earth is found to be

$$\mathbf{u}(\mathbf{a}; \omega) = \frac{P_0}{4\pi a^2} \sum_{l=1}^{\infty} \frac{2l+1}{l(l+1)} \frac{1}{y_2(a)} \left[\left\{ y_2(r_0) p_4^0 P_l^1(\cos\theta) - \frac{2}{r_0} \mu(r_0) y_1(r_0) p_2^0 P_l^2(\cos\theta) \right\} \frac{\mathbf{e}_\theta}{\sin\theta} + \left\{ y_2(r_0) p_3^0 \frac{\partial}{\partial\theta} P_l^1(\cos\theta) - \frac{1}{r_0} \mu(r_0) y_1(r_0) p_1^0 \frac{\partial}{\partial\theta} P_l^2(\cos\theta) \right\} \mathbf{e}_\phi \right], \tag{6.136}$$

where the superscript (+) on the ys is understood. In the case of a homogeneous sphere,

$$y_1(r) = j_l(k_\beta r), \qquad y_2(r) = \mu\left(\frac{d}{dr} - \frac{1}{r}\right) j_l(k_\beta r),$$

and Eq. (6.136) can be written accordingly.

6.3.4. Spheroidal Oscillations

In the case of the spheroidal oscillations, we may assume

$$\mathbf{u} = \sum_{\sigma, m, l} [y_1(r)\mathbf{P}_{ml}^\sigma(\theta, \phi) + y_3(r)\sqrt{\{l(l+1)\}}\,\mathbf{B}_{ml}^\sigma(\theta, \phi)], \tag{6.137}$$

which yields

$$\begin{aligned} \operatorname{div} \mathbf{u} &= \sum_{\sigma, m, l} X(r) Y_{ml}^\sigma(\theta, \phi), \\ \operatorname{curl} \mathbf{u} &= \sum_{\sigma, m, l} Z(r)\sqrt{\{l(l+1)\}}\,\mathbf{C}_{ml}^\sigma(\theta, \phi); \end{aligned} \tag{6.138}$$

$$\begin{aligned} X &= \frac{dy_1}{dr} + \frac{2}{r} y_1 - \frac{l(l+1)}{r} y_3, \\ Z &= \frac{1}{r}(y_1 - y_3) - \frac{dy_3}{dr}. \end{aligned} \tag{6.139}$$

We further assume that

$$\psi = \sum_{\sigma, m, l} y_5(r) Y_{ml}^\sigma(\theta, \phi). \tag{6.140}$$

Equation (6.52) yields

$$\frac{dg_0}{dr} + \frac{2}{r}g_0 = 4\pi G\rho_0. \tag{6.141}$$

We now substitute the value of **u** from Eq. (6.137) into the equation of motion, Eq. (6.50), and use Eqs. (6.138)–(6.141). On equating to zero the coefficients of $\mathbf{P}_{ml}$ and $\mathbf{B}_{ml}$ in the equation thus obtained, we find

$$\begin{aligned}
&\mu\left[2\frac{dX}{dr} - \frac{l(l+1)}{r}Z\right] + \frac{d}{dr}(\lambda X) + 2\dot{\mu}\frac{dy_1}{dr} + \rho_0\left[\frac{dy_5}{dr} - 4\pi G\rho_0 y_1\right.\\
&\qquad \left. + g_0\left(X - \frac{dy_1}{dr} + \frac{2}{r}y_1\right) + \omega^2 y_1\right] = 0,\\
&(\lambda + 2\mu)\frac{X}{r} - \frac{d}{dr}(\mu Z) - \mu\frac{Z}{r} + 2\dot{\mu}\left(\frac{dy_3}{dr} + Z\right)\\
&\qquad + \rho_0\left[\frac{1}{r}(y_5 - g_0 y_1) + \omega^2 y_3\right] = 0.
\end{aligned} \tag{6.142}$$

Similarly, Eq. (6.51) yields

$$\frac{d^2y_5}{dr^2} + \frac{2}{r}\frac{dy_5}{dr} - \frac{l(l+1)}{r^2}y_5 = 4\pi G(\rho_0 X + \dot{\rho}_0 y_1). \tag{6.143}$$

Equations (6.43) and (6.137) give

$$\mathbf{e}_r \cdot \mathfrak{T}_e = \sum_{\sigma, m, l} [y_2(r)\mathbf{P}^{\sigma}_{ml}(\theta, \phi) + y_4(r)\sqrt{\{l(l+1)\}}\,\mathbf{B}^{\sigma}_{ml}(\theta, \phi)], \tag{6.144}$$

where

$$\begin{aligned}
y_2 &= \lambda X + 2\mu\frac{dy_1}{dr} = (\lambda + 2\mu)\frac{dy_1}{dr} + \frac{2\lambda}{r}y_1 - \lambda\frac{l(l+1)}{r}y_3,\\
y_4 &= \mu\left(Z + 2\frac{dy_3}{dr}\right) = \mu\left[\frac{1}{r}(y_1 - y_3) + \frac{dy_3}{dr}\right].
\end{aligned} \tag{6.145}$$

Further, because $\nabla^2\psi_e = 0$, one may take for the value of ψ outside the earth the expression

$$\psi_e = \sum_{\sigma, m, l}\psi_{eml}, \qquad \psi_{eml} \propto \frac{1}{r^{l+1}}Y^{\sigma}_{ml}(\theta, \phi). \tag{6.146}$$

Therefore,

$$\frac{\partial\psi_{eml}}{\partial r} = -\frac{l+1}{r}\psi_{eml}, \tag{6.147}$$

and the boundary condition, Eq. (6.60), becomes

$$\frac{dy_5}{dr} - 4\pi G\rho_0 y_1 = -\frac{l+1}{r}y_5; \qquad r = a, \tag{6.148}$$

which is independent of ψ_e.

We next introduce the notation

$$y_6 = \frac{dy_5}{dr} - 4\pi G\rho_0 y_1. \tag{6.149}$$

From Eqs. (6.144) and (6.148), the boundary conditions 2 and 4 of Section 6.3.2 reduce to

$$y_2 = 0, \qquad y_4 = 0, \qquad y_6 + \frac{l+1}{r} y_5 = 0; \qquad r = a. \tag{6.150}$$

At an internal surface of discontinuity, Eqs. (6.59), (6.137), (6.140), and (6.144) reveal that we must have the continuity of the six functions $y_1, \ldots, y_6$. However, if the surface of discontinuity is a solid–liquid interface, only y_1, y_2, y_4, y_5, and y_6 are continuous.

Equations (6.141)–(6.143) represent a differential system of the sixth order. In the present form, these equations are not suitable for a numerical solution, because we need to evaluate the derivatives of empirically determined quantities $\lambda(r)$, $\mu(r)$, $\rho_0(r)$. However, if we introduce $y_1, \ldots, y_6$ as new independent variables, Eqs. (6.141)–(6.143) reduce to six simultaneous, linear, differential equations of the first order which are free from the derivatives of λ, μ, and ρ_0. We find

$$\frac{dy_1}{dr} = -\frac{2\lambda}{(\lambda+2\mu)r} y_1 + \frac{1}{\lambda+2\mu} y_2 + \frac{l(l+1)\lambda}{(\lambda+2\mu)r} y_3, \tag{6.151}$$

$$\begin{aligned}\frac{dy_2}{dr} &= \left[-\omega^2\rho_0 - 4\frac{g_0\rho_0}{r} + \frac{4\mu(3\lambda+2\mu)}{(\lambda+2\mu)r^2}\right] y_1 - \frac{4\mu}{(\lambda+2\mu)r} y_2 \\ &\quad + \frac{l(l+1)}{r}\left[g_0\rho_0 - \frac{2\mu(3\lambda+2\mu)}{(\lambda+2\mu)r}\right] y_3 + \frac{l(l+1)}{r} y_4 - \rho_0 y_6,\end{aligned} \tag{6.152}$$

$$\frac{dy_3}{dr} = -\frac{1}{r} y_1 + \frac{1}{r} y_3 + \frac{1}{\mu} y_4, \tag{6.153}$$

$$\begin{aligned}\frac{dy_4}{dr} &= \left[\frac{g_0\rho_0}{r} - \frac{2\mu(3\lambda+2\mu)}{(\lambda+2\mu)r^2}\right] y_1 - \frac{\lambda}{(\lambda+2\mu)r} y_2 \\ &\quad + \left[-\omega^2\rho_0 + \{(2l^2+2l-1)\lambda + 2(l^2+l-1)\mu\}\frac{2\mu}{(\lambda+2\mu)r^2}\right] y_3 \\ &\quad - \frac{3}{r} y_4 - \frac{\rho_0}{r} y_5,\end{aligned} \tag{6.154}$$

$$\frac{dy_5}{dr} = 4\pi G\rho_0 y_1 + y_6, \tag{6.155}$$

$$\frac{dy_6}{dr} = -4\pi\frac{l(l+1)}{r} G\rho_0 y_3 + \frac{l(l+1)}{r^2} y_5 - \frac{2}{r} y_6. \tag{6.156}$$

This system is to be used in the shell. In the *liquid core*, we have

$$\mu = 0, \qquad y_2 = \lambda X, \qquad y_4 = 0. \tag{6.157}$$

Equation (6.154), then yields

$$y_3 = \frac{1}{\omega^2 r}\left(g_0 y_1 - \frac{1}{\rho_0} y_2 - y_5\right), \tag{6.158}$$

and y_1, y_2, y_5, y_6 satisfy the equations

$$\frac{dy_1}{dr} = -\frac{2}{r} y_1 + \frac{1}{\lambda} y_2 + \frac{l(l+1)}{r} y_3, \tag{6.159}$$

$$\frac{dy_2}{dr} = -\left(\omega^2 \rho_0 + \frac{4g_0\rho_0}{r}\right) y_1 + \frac{l(l+1)}{r} g_0 \rho_0 y_3 - \rho_0 y_6, \tag{6.160}$$

$$\frac{dy_5}{dr} = 4\pi G \rho_0 y_1 + y_6, \tag{6.161}$$

$$\frac{dy_6}{dr} = -4\pi \frac{l(l+1)}{r} G\rho_0 y_3 + \frac{l(l+1)}{r^2} y_5 - \frac{2}{r} y_6. \tag{6.162}$$

Having determined the eigenfrequencies for the spheroidal oscillations, we can now introduce the source. As in the case of the toroidal oscillations, let the source at the point $r = r_0, \theta = 0$ be represented as jumps in the displacements and stresses across the spherical surface through the source:

$$y_i(r_0 + 0) - y_i(r_0 - 0) = -f_i; \qquad i = 1, 2, 3, 4. \tag{6.163}$$

It may be noted that y_5 and y_6 are continuous across the surface $r = r_0$. Comparing Eq. (6.163) with Eqs. (4.179) and (4.180), we find

$$\begin{aligned} f_1 &= \frac{-U_0\,dS}{4\pi r_0^2}(2l+1)p^{\sigma}_{ml}, & f_3 &= \frac{-U_0\,dS}{4\pi r_0^2}(2l+1)b^{\sigma}_{ml}, \\ f_2 &= \frac{-\mu U_0\,dS}{4\pi r_0^3}(2l+1)\bar{p}^{\sigma}_{ml}, & f_4 &= \frac{-\mu U_0\,dS}{4\pi r_0^3}(2l+1)\bar{b}^{\sigma}_{ml}. \end{aligned} \tag{6.164}$$

One must solve Eqs. (6.151)–(6.156) subjected to the boundary conditions (6.150) and the source conditions (6.163).

EXAMPLE 6.5: Spheroidal Eigenfunctions for a Homogeneous Nongravitating Mantle Lying over a Liquid Core

In the solid mantle, the displacement vector is a linear combination of the following four Hansen vectors:

$$\mathbf{L}_{ml}^{\pm}(\xi) = \frac{d}{d\xi} f_l^{\pm}(\xi)\mathbf{P}_{ml}(\theta, \phi) + \frac{1}{\xi} f_l^{\pm}(\xi)[l(l+1)]^{1/2}\mathbf{B}_{ml}(\theta, \phi),$$

$$\mathbf{N}_{ml}^{\pm}(\eta) = \frac{1}{\eta} f_l^{\pm}(\eta) l(l+1)\mathbf{P}_{ml}(\theta, \phi) + \left(\frac{d}{d\eta} + \frac{1}{\eta}\right) f_l^{\pm}(\eta)[l(l+1)]^{1/2}\mathbf{B}_{ml}(\theta, \phi),$$

$$\xi = rk_\alpha, \qquad \eta = rk_\beta, \qquad f_l^{+}(x) = j_l(x), \qquad f_l^{-}(x) = h_l^{(2)}(x). \tag{6.5.1}$$

The corresponding stresses are given by [cf. Eq. (6.3a)]

$$\begin{aligned} \mathbf{e}_r \cdot \mathfrak{T}(\mathbf{L}) &= 2\mu k_\alpha[F_{l,3}(\xi)\mathbf{P}_{ml} + F_{l,1}(\xi)\{l(l+1)\}^{1/2}\mathbf{B}_{ml}], \\ \mathbf{e}_r \cdot \mathfrak{T}(\mathbf{N}) &= \mu k_\beta[2l(l+1)F_{l,1}(\eta)\mathbf{P}_{ml} + F_{l,2}(\eta)\{l(l+1)\}^{1/2}\mathbf{B}_{ml}], \end{aligned} \tag{6.5.2}$$

where $F_{l,1}$, etc., are defined in Eq. (6.4). Therefore, in the mantle, we can assume

$$\begin{bmatrix} y_1 \\ \dfrac{1}{2\mu} y_2 \\ y_3 \\ \dfrac{1}{\mu} y_4 \end{bmatrix} = \begin{bmatrix} \dfrac{d}{d\xi} f_l^{+}(\xi) & \dfrac{d}{d\xi} f_l^{-}(\xi) & \dfrac{N^2}{\eta} f_l^{+}(\eta) & \dfrac{N^2}{\eta} f_l^{-}(\eta) \\ k_\alpha F_{l,3}^{+}(\xi) & k_\alpha F_{l,3}^{-}(\xi) & k_\beta N^2 F_{l,1}^{+}(\eta) & k_\beta N^2 F_{l,1}^{-}(\eta) \\ \dfrac{1}{\xi} f_l^{+}(\xi) & \dfrac{1}{\xi} f_l^{-}(\xi) & \left(\dfrac{d}{d\eta} + \dfrac{1}{\eta}\right) f_l^{+}(\eta) & \left(\dfrac{d}{d\eta} + \dfrac{1}{\eta}\right) f_l^{-}(\eta) \\ 2k_\alpha F_{l,1}^{+}(\xi) & 2k_\alpha F_{l,1}^{-}(\xi) & k_\beta F_{l,2}^{+}(\eta) & k_\beta F_{l,2}^{-}(\eta) \end{bmatrix} \begin{bmatrix} A \\ B \\ C \\ D \end{bmatrix}, \tag{6.5.3}$$

where $N^2 = l(l+1)$ and A, B, C, D are arbitrary constants.

Equations (6.5.3) constitute the four independent solutions of the following set of four linear simultaneous differential equations:

$$\begin{bmatrix} \dot{y}_1 \\ \dot{y}_2 \\ \dot{y}_3 \\ \dot{y}_4 \end{bmatrix} =$$

$$\begin{bmatrix} \dfrac{-2\lambda}{(\lambda+2\mu)r} & \dfrac{1}{\lambda+2\mu} & \dfrac{N^2\lambda}{(\lambda+2\mu)r} & 0 \\ \left[-\omega^2\rho + \dfrac{4\mu(3\lambda+2\mu)}{(\lambda+2\mu)r^2}\right] & \dfrac{-4\mu}{(\lambda+2\mu)r} & -\dfrac{N^2}{r^2}\dfrac{2\mu(3\lambda+2\mu)}{\lambda+2\mu} & \dfrac{N^2}{r} \\ -\dfrac{1}{r} & 0 & \dfrac{1}{r} & \dfrac{1}{\mu} \\ \dfrac{-2\mu(3\lambda+2\mu)}{(\lambda+2\mu)r^2} & -\dfrac{\lambda}{(\lambda+2\mu)}\dfrac{1}{r} & M_1 & -\dfrac{3}{r} \end{bmatrix} \begin{bmatrix} y_1 \\ y_2 \\ y_3 \\ y_4 \end{bmatrix}, \tag{6.5.4}$$

where the dot signifies differentiation with respect to r and

$$M_1 = -\omega^2\rho + \frac{\lambda(2N^2 - 1) + 2\mu(N^2 - 1)}{(\lambda + 2\mu)r^2} 2\mu.$$

In the liquid core, Eq. (6.5.4) reduces to

$$\begin{bmatrix} \dot{y}_1^c \\ \dot{y}_2^c \end{bmatrix} = \begin{bmatrix} -\dfrac{2}{r} & \dfrac{1}{\lambda_c} - \dfrac{N^2}{\omega^2\rho_c r^2} \\ -\omega^2\rho_c & 0 \end{bmatrix} \begin{bmatrix} y_1^c \\ y_2^c \end{bmatrix}. \tag{6.5.5}$$

The corresponding eigenfunctions are

$$\begin{aligned} y_1^c &= \frac{1}{k_c}\frac{d}{dr} j_l(k_c r), \\ y_2^c &= -\lambda_c k_c j_l(k_c r), \\ y_3^c &= \frac{1}{k_c r} j_l(k_c r), \\ y_4^c &= 0, \end{aligned} \tag{6.5.6}$$

where $k_c = \omega/\alpha_c$, $\alpha_c^2 = \lambda_c/\rho_c$.

We may develop a matrix propagator for a layered sphere similar to the one used in Section 3.7.2 for a plane layered half-space. For this purpose, we use $h_l^{(1)}(x)$ in place of $j_l(x)$. Equation (6.5.3) is now written for the ith layer in the form

$$y_i = Y_i \Gamma_i, \tag{6.5.7}$$

where

$$\begin{aligned} y &= (y_1, y_3, y_2, y_4), \\ \Gamma &= (A', B', C', D'); \end{aligned} \tag{6.5.8}$$

$$\begin{aligned} Y_{11} &= \frac{1}{p}\frac{d}{dr} h_l^{(2)}(\xi), & Y_{12} &= \frac{l(l+1)}{s\eta} h_l^{(2)}(\eta), \\ Y_{21} &= \frac{1}{pr} h_l^{(2)}(\xi), & Y_{22} &= \frac{1}{s}\left(\frac{d}{d\eta} + \frac{1}{\eta}\right) h_l^{(2)}(\eta), \\ Y_{31} &= \frac{2}{p}\mu k_\alpha^2 F_{l,3}^{(2)}(\xi), & Y_{32} &= \frac{2}{s}\mu k_\beta l(l+1) F_{l,1}^{(2)}(\eta), \\ Y_{41} &= \frac{2}{p}\mu k_\alpha^2 F_{l,1}^{(2)}(\xi), & Y_{42} &= \frac{1}{s}\mu k_\beta F_{l,2}^{(2)}(\eta); \end{aligned} \tag{6.5.9}$$

$$p = \alpha(\rho k_\alpha \omega)^{1/2}, \qquad s = \beta k_\beta^{-1/2}\{\rho\omega l(l+1)\}^{1/2}. \tag{6.5.10}$$

The third and fourth columns of Y are obtained from the first and second columns, respectively, on replacing $h_l^{(2)}(x)$ by $h_l^{(1)}(x)$.

The condition of continuity of the components of stress and displacement vectors at the interface $r = a_i$ implies

$$y_{i-1}(a_i) = y_i(a_i). \tag{6.5.11}$$

Let the core–mantle boundary be at the Nth interface. Then, a repeated use of Eq. (6.5.7) yields

$$Y_N(a_N)\Gamma_N = E_{N-1}E_{N-2} \cdots E_1 Y_0(a_1)\Gamma_0, \tag{6.5.12}$$

where

$$E_i^{-1} = Y_i(a_i)Y_i^{-1}(a_{i+1}) \tag{6.5.13}$$

and

$$Y^{-1} = -\frac{i\omega r^2}{2}\begin{bmatrix} Y_{33} & l(l+1)Y_{43} & -Y_{13} & -l(l+1)Y_{23} \\ Y_{34} & l(l+1)Y_{44} & -Y_{14} & -l(l+1)Y_{24} \\ -Y_{31} & -l(l+1)Y_{41} & Y_{11} & l(l+1)Y_{21} \\ -Y_{32} & -l(l+1)Y_{42} & Y_{12} & l(l+1)Y_{22} \end{bmatrix}. \tag{6.5.14}$$

For the sake of numerical computation it is desirable to redefine the elements of the matrix Y such that

$$\bar{Y}_{j1} = \frac{pY_{j1}}{h_l^{(2)}(\xi)}, \quad \bar{Y}_{j2} = \frac{sY_{j2}}{h_l^{(2)}(\eta)}, \quad \bar{Y}_{j3} = \frac{pY_{j3}}{h_l^{(1)}(\xi)}, \quad \bar{Y}_{j4} = \frac{sY_{j4}}{h_l^{(1)}(\eta)},$$

$$j = 1, 2, 3, 4. \tag{6.5.15}$$

Then

$$E_i^{-1} = \bar{Y}_i(a_i)\Lambda_i \bar{Y}_i^{-1}(a_{i+1}), \tag{6.5.16}$$

where

$$\Lambda_i = \begin{bmatrix} \dfrac{h_l^{(2)}(k_\alpha a_i)}{h_l^{(2)}(k_\alpha a_{i+1})} & 0 & 0 & 0 \\ 0 & \dfrac{h_l^{(2)}(k_\beta a_i)}{h_l^{(2)}(k_\beta a_{i+1})} & 0 & 0 \\ 0 & 0 & \dfrac{h_l^{(1)}(k_\alpha a_i)}{h_l^{(1)}(k_\alpha a_{i+1})} & 0 \\ 0 & 0 & 0 & \dfrac{h_l^{(1)}(k_\beta a_i)}{h_l^{(1)}(k_\beta a_{i+1})} \end{bmatrix}. \tag{6.5.17}$$

With these modifications, Eq. (6.5.12) yields

$$\bar{\Gamma}_0 = [\bar{Y}_0^{-1}(a_1)E_1^{-1} \cdots E_{N-1}^{-1}\bar{Y}_N(a_N)]\bar{\Gamma}_N = R\bar{\Gamma}_N, \text{ say.} \tag{6.5.18}$$

The boundary conditions at the free surface and the core–mantle interface are introduced in the components of the vectors $\bar{\Gamma}_0$ and $\bar{\Gamma}_N$. Equation (6.5.18) then serves as a system of equations for the remaining unknown reflection coefficients.

A similar procedure exists for the toroidal (*SH*) field. Here we define the toroidal stress–displacement vector as

$$\hat{y} = (\hat{y}_1, \hat{y}_2) = \hat{Y}\Gamma,$$

where

$$\hat{Y}_{11} = \frac{1}{s} h_l^{(2)}(\eta), \qquad \hat{Y}_{21} = \frac{\mu}{s}\left(\frac{d}{dr} - \frac{1}{r}\right) h_l^{(2)}(\eta),$$

$$\hat{Y}_{12} = \frac{1}{s} h_l^{(1)}(\eta), \qquad \hat{Y}_{22} = \frac{\mu}{s}\left(\frac{d}{dr} - \frac{1}{r}\right) h_l^{(1)}(\eta), \tag{6.5.19}$$

$$\Gamma = \begin{bmatrix} A'' \\ B'' \end{bmatrix}.$$

Next, we show some useful orthogonality properties of the elements of the Y and $\hat{Y}$ matrices. We recall the expression (1.3.5) for the energy-flux density vector for time-harmonic displacements

$$\boldsymbol{\Sigma} = \frac{i\omega}{4}\{\mathbf{u}^* \cdot \mathfrak{T} - \mathbf{u} \cdot \mathfrak{T}^*\}$$

or

$$\Sigma_l = \frac{i\omega}{4}\sum_{j=1}^{3}(u_j^*\tau_{lj} - u_j\tau_{lj}^*), \tag{6.5.20}$$

where Σ_i is the energy-flux density in the x_i direction and the asterisks denote complex conjugates.

The radial energy-flux density is made up of expressions of the form

$$\frac{i\omega}{4}[(Y_{1j}Y_{3j}^* - Y_{1j}^*Y_{3j}) + l(l+1)(Y_{2j}Y_{4j}^* - Y_{2j}^*Y_{4j} + \hat{Y}_{1k}\hat{Y}_{2k}^* - \hat{Y}_{1k}^*\hat{Y}_{2k})],$$

$$j = 1, 2, 3, 4, \qquad k = 1, 2. \tag{6.5.21}$$

Using the Wronskian relation, Eq. (D.77),

$$W\{h_l^{(1)}(z), h_l^{(2)}(z)\} = -\frac{2i}{z^2},$$

together with the identities $h_l^{(1)*}(z) = h_l^{(2)}(z^*)$, $h_l^{(2)*}(z) = h_l^{(1)}(z^*)$, the expression in Eq. (6.5.21) reduces simply to $\pm(1/2r^2)$ for a given j or k. Therefore, the net flux through any spherical coordinate surface is constant for any frequency ω and order l. Defining a product between two spheroidal stress–displacement vectors $\mathbf{Y}_1(Y_{11}, Y_{21}, Y_{31}, Y_{41})$ and $\mathbf{Y}_2(Y_{12}, Y_{22}, Y_{32}, Y_{42})$ as

$$\mathbf{Y}_1 \times \mathbf{Y}_2 = (Y_{11}Y_{32}^* - Y_{12}^*Y_{31}) + l(l+1)(Y_{21}Y_{42}^* - Y_{22}^*Y_{41}), \tag{6.5.22}$$

we obtain the orthogonality relations

$$\mathbf{Y}_1 \times \mathbf{Y}_1 = \mathbf{Y}_2 \times \mathbf{Y}_2 = \frac{2i}{\omega r^2}, \qquad \mathbf{Y}_3 \times \mathbf{Y}_3 = \mathbf{Y}_4 \times \mathbf{Y}_4 = -\frac{2i}{\omega r^2},$$

$$\mathbf{Y}_i \times \mathbf{Y}_j = 0, \qquad i \neq j. \tag{6.5.23}$$

The connection of the vector product, Eq. (6.5.22), with the energy-flux density, Eq. (6.5.21), is obvious and the inverse matrix, Eq. (6.5.14), follows directly from the orthogonality condition, Eq. (6.5.23). These results are useful in the calculation of the reflection coefficients.

6.3.4.1. Spheroidal Green's Dyadic. When the eigenfrequencies of the spheroidal oscillations of the earth are denoted by ω_n and the corresponding eigenvectors by $\mathbf{Q}_n$ and the procedure explained in detail in the toroidal case (Sect. 6.3.3.2) is followed, it is clear that the spheroidal Green's dyadic has the same formal expression as the toroidal Green's dyadic except that $\mathbf{M}_n$ is to be replaced by $\mathbf{Q}_n$. Therefore, results corresponding to Eqs. (6.96)–(6.98) for the spheroidal case can be written immediately.

Expressing $\mathbf{Q}_n$ in terms of spherical harmonics

$$ {}_n\mathbf{Q}_{ml}(\mathbf{r}) = y_{1n}(r)\mathbf{P}_{ml}(\theta, \phi) + y_{3n}(r)\sqrt{\{l(l+1)\}}\,\mathbf{B}_{ml}(\theta, \phi), \qquad (6.165) $$

we have

$$ {}_n\Lambda_{ml} = \int_V {}_n\mathbf{Q}_{ml}(\mathbf{r}) \cdot {}_n\overset{*}{\mathbf{Q}}_{ml}(\mathbf{r})\rho_0(r)dV = \Omega_{ml} I_n^s, \qquad (6.166) $$

where

$$ I_n^s = \int_0^a [y_{1n}^2 + l(l+1)y_{3n}^2]\rho_0 r^2\, dr, \qquad (6.167) $$

$$ 0 = \int_0^a [y_{1n}y_{1k} + l(l+1)y_{3n}y_{3k}]\rho_0 r^2\, dr, \qquad (n \neq k). \qquad (6.168) $$

We now have $\mathfrak{G} = \sum_{l=0}^{\infty}\sum_{m=-l}^{l} \mathfrak{G}_{ml}$, where

$$ \mathfrak{G}_{ml}(\mathbf{r}\,|\,\mathbf{r}_0; \omega) = \frac{1}{\Omega_{ml}} \sum_{n=0}^{\infty} \frac{{}_n\mathbf{Q}_{ml}(\mathbf{r})\,{}_n\overset{*}{\mathbf{Q}}_{ml}(\mathbf{r}_0)}{(\omega_n^2 - \omega^2)I_n^s}\, g(\omega), \qquad (6.169) $$

$$ \mathfrak{G}_{ml}(\mathbf{r}\,|\,\mathbf{r}_0; t) = \frac{1}{\Omega_{ml}} \sum_{n=0}^{\infty} {}_n\mathbf{Q}_{ml}(\mathbf{r})\,{}_n\overset{*}{\mathbf{Q}}_{ml}(\mathbf{r}_0) \frac{1 - \cos\omega_n t}{\omega_n^2 I_n^s} H(t), \qquad (6.170) $$

$$ \mathfrak{G}_{ml}(\mathbf{r}\,|\,\mathbf{r}_0; st) = \frac{1}{\Omega_{ml}} \sum_{n=0}^{\infty} \frac{1}{\omega_n^2 I_n^s}\, {}_n\mathbf{Q}_{ml}(\mathbf{r})\,{}_n\overset{*}{\mathbf{Q}}_{ml}(\mathbf{r}_0). \qquad (6.171) $$

From Eqs. (6.151)–(6.156), it can be seen that

$$ \frac{d}{dr}\left[r^2(y_{1n}y_2 - y_1 y_{2n}) + l(l+1)r^2(y_{3n}y_4 - y_3 y_{4n}) + \frac{r^2}{4\pi G}(y_{5n}y_7 - y_5 y_{7n})\right] $$

$$ = (\omega_n^2 - \omega^2)\rho_0 r^2[y_{1n}y_1 + l(l+1)y_{3n}y_3], \qquad (6.172) $$

where

$$y_7 = y_6 + \frac{l+1}{r} y_5. \tag{6.173}$$

Because

$$y_{2n} = y_{4n} = y_{7n} = 0 \quad \text{at} \quad r = a, \tag{6.174}$$

Eq. (6.172), on integration, gives

$$\int_0^a [y_{1n}y_1 + l(l+1)y_{3n}y_3]\rho_0 r^2\, dr = \frac{a^2}{\omega_n^2 - \omega^2}\left[y_{1n}(a)y_2(a) + l(l+1)y_{3n}(a)y_4(a) + \frac{1}{4\pi G} y_{5n}(a)y_7(a)\right]. \tag{6.175}$$

If $\omega = \omega_k$ $(n \neq k)$, we have, on using relations similar to Eq. (6.174),

$$\int_0^a [y_{1n}y_{1k} + l(l+1)y_{3n}y_{3k}]\rho_0 r^2\, dr = 0, \tag{6.176}$$

which is the orthogonality relation, Eq. (6.168). However, if ω is not an eigenfrequency, we have, on taking the limit $\omega \to \omega_n$,

$$I_n^s = \int_0^a [y_{1n}^2 + l(l+1)y_{3n}^2]\rho_0 r^2\, dr$$

$$= -\frac{a^2}{2\omega_n}\left[y_1 \frac{\partial y_2}{\partial \omega} + l(l+1)y_3 \frac{\partial y_4}{\partial \omega} + \frac{1}{4\pi G} y_5 \frac{\partial y_7}{\partial \omega}\right]_{\omega=\omega_n,\, r=a}. \tag{6.177}$$

In a homogeneous nongravitating sphere, ${}_n\mathbf{Q}_{ml}$ is of the form [cf. Eq. (6.2)]

$${}_n\mathbf{Q}_{ml} = [\beta_{ml}\mathbf{N}_{ml}^{+}(k_\beta r) + \gamma_{ml}\mathbf{L}_{ml}^{+}(k_\alpha r)]_{\omega_n}. \tag{6.178}$$

From Eqs. (6.18), ${}_n\mathbf{Q}_{ml}$ will satisfy the boundary conditions at $r = a$ if

$$[2k_\alpha F_{l,1}^{+}(\zeta)\gamma_{ml} + k_\beta F_{l,2}^{+}(\chi)\beta_{ml}]_{\omega_n} = 0. \tag{6.179}$$

The other equation in Eqs. (6.18) will be identically satisfied because ω_n is a root of Eq. (6.19). On expressing $\mathbf{N}_{ml}$ and $\mathbf{L}_{ml}$ in terms of $\mathbf{P}_{ml}$ and $\mathbf{B}_{ml}$, we find from Eqs. (6.165), (6.178), and (6.179)

$$\begin{aligned} y_{1n}(r) &= \left[2\left(\frac{\beta}{\alpha}\right)^2 l(l+1)F_{l,1}^{+}(\zeta)\frac{1}{r}j_l(k_\beta r) - F_{l,2}^{+}(\chi)\frac{d}{dr}j_l(k_\alpha r)\right]_{\omega_n}, \\ y_{3n}(r) &= \left[2\left(\frac{\beta}{\alpha}\right)^2 F_{l,1}^{+}(\zeta)\left(\frac{d}{dr} + \frac{1}{r}\right)j_l(k_\beta r) - F_{l,2}^{+}(\chi)\frac{1}{r}j_l(k_\alpha r)\right]_{\omega_n}. \end{aligned} \tag{6.180}$$

We have omitted a common factor from y_{1n} and y_{3n}, because it is eliminated when ${}_n\mathbf{Q}_{ml}$ is substituted in the expression for $\mathfrak{G}_{ml}$.

The radial factors y_{2n} and y_{4n} can now be calculated from Eq. (6.145). The integral I_n^s then follows from the relation (6.177) with $y_5 = 0$.

6.3.4.2. Spheroidal Field Resulting from a Shear Dislocation. Because the procedure of calculating the field from the Green's dyadic using the Volterra relation has been explained in Section 6.3.3.3, we give here the final results without going into the details of their derivation. Assuming a *Heaviside unit step function*, we obtain the following expression for the displacement for an arbitrary shear dislocation $(\lambda, \delta, \hat{\phi})$:

$$\mathbf{u}(\mathbf{r}; t) = P_0 \sum_{l=0}^{\infty} \sum_{m=-l}^{l} \frac{1}{\Omega_{ml}} \sum_{n=0}^{\infty} \frac{(\cos \omega_n t - 1)H(t)}{\omega_n^2 I_n^s} \Bigg[\tfrac{1}{2} y_{8n}(r_0) \Bigg\{ p_2 \left(2 \frac{\partial^2}{\partial \theta_0^2} + l^2 + l \right)$$

$$+ 2p_1 \frac{1}{\sin \theta_0} \left(\frac{\partial}{\partial \theta_0} - \cot \theta_0 \right) \frac{\partial}{\partial \phi_0} \Bigg\} \overset{*}{Y}_{ml}(\theta_0, \phi_0) + y_{4n}(r_0) \left(p_4 \frac{\partial}{\partial \theta_0} \right.$$

$$\left. + p_3 \frac{1}{\sin \theta_0} \frac{\partial}{\partial \phi_0} \right) \overset{*}{Y}_{ml}(\theta_0, \phi_0) - \tfrac{1}{2} y_{9n}(r_0) p_5 \overset{*}{Y}_{ml}(\theta_0, \phi_0) \Bigg] [{}_n\mathbf{Q}_{ml}(\mathbf{r})], \tag{6.181}$$

where

$$y_{8n}(r_0) = 2\mu(r_0) \frac{y_{3n}(r_0)}{r_0},$$

$$y_{9n}(r_0) = 2\mu(r_0) \left[2 \left(\frac{d}{dr_0} - \frac{1}{r_0} \right) y_{1n}(r_0) + l(l+1) \frac{y_{3n}(r_0)}{r_0} \right], \tag{6.182}$$

and p_1 to p_5 are defined in Eqs. (6.125). Given the geographic coordinates of the source (r_0, θ_0, ϕ_0) and the sensor (r, θ, ϕ), Eq. (6.181) can be utilized to calculate the amplitudes of the split normal modes (l, n, m) of the spheroidal oscillations of the earth excited by an arbitrary shear dislocation $(\lambda, \delta, \hat{\phi})$. It is considered as the zero-order approximation of the corresponding result for a rotating elliptical earth.

Assuming that the splitting can be ignored, we use Eq. (6.127) to sum over m, obtaining

$$u_r(\mathbf{r}; t) = \frac{P_0}{8\pi} (-q_2 Q_1 + 2q_4 Q_6 + p_5 Q_4),$$

$$u_\theta(\mathbf{r}; t) = \frac{P_0}{8\pi} [2(q_1 Q_3 - q_3 Q_8) \sin C + (-q_2 Q_2 + 2q_4 Q_7 + p_5 Q_5) \cos C],$$

$$u_\phi(\mathbf{r}; t) = \frac{P_0}{8\pi} [(-q_2 Q_2 + 2q_4 Q_7 + p_5 Q_5) \sin C - 2(q_1 Q_3 - q_3 Q_8) \cos C], \tag{6.183}$$

where q_i ($i = 1, 2, 3, 4$) are obtained from p_i of Eqs. (6.125) on replacing $\hat{\phi}$ by $\hat{\phi} - B$ and

$$Q_1 = \sum_{l=2}^{\infty} \sum_{n=0}^{\infty} (2l+1) \left\{ \frac{(1 - \cos \omega t)H(t)}{\omega^2 I^s} y_8(r_0) y_1(r) \right\}_{\omega_n} P_l^2(\cos \varepsilon), \tag{6.184}$$

$$Q_2 = \sum_{l=2}^{\infty} \sum_{n=0}^{\infty} (2l+1) \left\{ \frac{(1 - \cos \omega t)H(t)}{\omega^2 I^s} y_8(r_0) y_3(r) \right\}_{\omega_n} \frac{\partial}{\partial \varepsilon} P_l^2(\cos \varepsilon), \tag{6.185}$$

$$Q_3 = \sum_{l=2}^{\infty} \sum_{n=0}^{\infty} (2l+1) \left\{ \frac{(1 - \cos \omega t)H(t)}{\omega^2 I^s} y_8(r_0) y_3(r) \right\}_{\omega_n} \frac{1}{\sin \varepsilon} P_l^2(\cos \varepsilon), \tag{6.186}$$

$$Q_4 = \sum_{l=0}^{\infty} \sum_{n=0}^{\infty} (2l+1) \left\{ \frac{(1 - \cos \omega t)H(t)}{\omega^2 I^s} y_9(r_0) y_1(r) \right\}_{\omega_n} P_l(\cos \varepsilon), \tag{6.187}$$

$$Q_5 = \sum_{l=0}^{\infty} \sum_{n=0}^{\infty} (2l+1) \left\{ \frac{(1 - \cos \omega t)H(t)}{\omega^2 I^s} y_9(r_0) y_3(r) \right\}_{\omega_n} \frac{\partial}{\partial \varepsilon} P_l(\cos \varepsilon), \tag{6.188}$$

$$Q_6 = \sum_{l=1}^{\infty} \sum_{n=0}^{\infty} (2l+1) \left\{ \frac{(1 - \cos \omega t)H(t)}{\omega^2 I^s} y_4(r_0) y_1(r) \right\}_{\omega_n} P_l^1(\cos \varepsilon), \tag{6.189}$$

$$Q_7 = \sum_{l=1}^{\infty} \sum_{n=0}^{\infty} (2l+1) \left\{ \frac{(1 - \cos \omega t)H(t)}{\omega^2 I^s} y_4(r_0) y_3(r) \right\}_{\omega_n} \frac{\partial}{\partial \varepsilon} P_l^1(\cos \varepsilon), \tag{6.190}$$

$$Q_8 = \sum_{l=1}^{\infty} \sum_{n=0}^{\infty} (2l+1) \left\{ \frac{(1 - \cos \omega t)H(t)}{\omega^2 I^s} y_4(r_0) y_3(r) \right\}_{\omega_n} \frac{1}{\sin \varepsilon} P_l^1(\cos \varepsilon), \tag{6.191}$$

$$Q_9 = \sum_{l=0}^{\infty} \sum_{n=0}^{\infty} (2l+1) \left\{ \frac{(1 - \cos \omega t)H(t)}{\omega^2 I^s} y_2(r_0) y_3(r) \right\}_{\omega_n} P_l(\cos \varepsilon). \tag{6.192}$$

When the source lies on the x_3 axis, $C = 0$, $\varepsilon = \theta$, and Eqs. (6.183) become

$$\begin{aligned} u_r(\mathbf{r}; t) &= \frac{P_0}{8\pi}(-p_2^0 Q_1 + 2p_4^0 Q_6 + p_5 Q_4), \\ u_\theta(\mathbf{r}; t) &= \frac{P_0}{8\pi}(-p_2^0 Q_2 + 2p_4^0 Q_7 + p_5 Q_5), \\ u_\phi(\mathbf{r}; t) &= \frac{P_0}{4\pi}(-p_1^0 Q_3 + p_3^0 Q_8), \end{aligned} \tag{6.193}$$

where Q_i ($i = 1, 2, \ldots, 8$) are given by Eqs. (6.184)–(6.191) with ε replaced by θ. Moreover, p_i^0 ($i = 1, 2, 3, 4$) are obtained from p_i of Eqs. (6.125) on replacing $\hat{\phi}$ by ϕ^0, where ϕ^0 is the azimuth of the sensor with respect to the fault's strike.

6.3.4.3. Spheroidal Field Resulting from a Center of Compression. The displacement field resulting from a center of compression placed at the point ($\mathbf{r}_0$) within the sphere is given by [cf. Eq. (4.46)]

$$\mathbf{u}(\mathbf{r}) = M \operatorname{div}_0 \tilde{\mathfrak{G}}(\mathbf{r}|\mathbf{r}_0). \tag{6.194}$$

Because

$$\operatorname{div}_0[y_1(r_0)\overset{*}{\mathbf{P}}_{ml}(\theta_0, \phi_0)] = \left(\frac{d}{dr_0} + \frac{2}{r_0}\right) y_1(r_0)\overset{*}{Y}_{ml}(\theta_0, \phi_0),$$

$$\operatorname{div}_0[y_3(r_0)\sqrt{\{l(l+1)\}}\,\overset{*}{\mathbf{B}}_{ml}(\theta_0, \phi_0)] = -l(l+1)\frac{1}{r_0}\, y_3(r_0)\overset{*}{Y}_{ml}(\theta_0, \phi_0), \quad (6.195)$$

Eqs. (6.170) and (6.194) yield

$$\mathbf{u}(\mathbf{r}; t) = M \sum_{l=0}^{\infty} \sum_{m=-l}^{l} \frac{1}{\Omega_{ml}} \sum_{n=0}^{\infty} \frac{(1 - \cos \omega_n t)H(t)}{\omega_n^2 I_n^s} \left[\left(\frac{d}{dr_0} + \frac{2}{r_0}\right) y_{1n}(r_0) - l(l+1)\frac{1}{r_0}\, y_{3n}(r_0)\right] [\overset{*}{Y}_{ml}(\theta_0, \phi_0)] [{}_n\mathbf{Q}_{ml}(\mathbf{r})]. \quad (6.196)$$

These are the displacement of the split normal mode (l, n, m) excited by a center of compression. If the splitting can be ignored, we use Eq. (6.127) to sum over m, obtaining

$$\mathbf{u}(\mathbf{r}; t) = \frac{M}{4\pi} \sum_{l=0}^{\infty} (2l+1) \sum_{n=0}^{\infty} \left\{\frac{(1 - \cos \omega t)H(t)}{\omega^2 I^s} \left[\left(\frac{d}{dr_0} + \frac{2}{r_0}\right) y_1(r_0) - \frac{l(l+1)}{r_0}\, y_3(r_0)\right] \left[y_1(r)\mathbf{e}_r + y_3(r)\left(\mathbf{e}_\theta \frac{\partial}{\partial\theta} + \mathbf{e}_\phi \frac{1}{\sin\theta}\frac{\partial}{\partial\phi}\right)\right] P_l(\cos \varepsilon)\right\}_{\omega_n}. \quad (6.197)$$

Using Eqs. (6.130), we get

$$\mathbf{u}(\mathbf{r}; t) = \frac{M}{4\pi} \sum_{l=0}^{\infty} (2l+1) \sum_{n=0}^{\infty} \left\{\frac{(1 - \cos \omega t)H(t)}{\omega^2 I^s} \left[\left(\frac{d}{dr_0} + \frac{2}{r_0}\right) y_1(r_0) - \frac{l(l+1)}{r_0}\, y_3(r_0)\right] \left[y_1(r)P_l(\cos \varepsilon)\mathbf{e}_r + y_3(r)\frac{\partial}{\partial\varepsilon} P_l(\cos \varepsilon)(\cos C\mathbf{e}_\theta + \sin C\mathbf{e}_\phi)\right]\right\}_{\omega_n}. \quad (6.198)$$

These are the displacement at the point (r, θ, ϕ) resulting from a center of compression placed at the point (r_0, θ_0, ϕ_0) within the earth. If the source lies on the x_3 axis, we obtain, on putting $\varepsilon = \theta$, $C = 0$,

$$\mathbf{u}(\mathbf{r}; t) = \frac{M}{4\pi} \sum_{l=0}^{\infty} (2l+1) \sum_{n=0}^{\infty} \left\{\frac{(1 - \cos \omega t)H(t)}{\omega^2 I^s} \left[\left(\frac{d}{dr_0} + \frac{2}{r_0}\right) y_1(r_0) - \frac{l(l+1)}{r_0}\, y_3(r_0)\right] \left[y_1(r)P_l(\cos \theta)\mathbf{e}_r + y_3(r)\frac{\partial}{\partial\theta} P_l(\cos \theta)\mathbf{e}_\theta\right]\right\}_{\omega_n}. \quad (6.199)$$

In the case of a homogeneous nongravitating sphere, $y_{1n}(r)$ and $y_{3n}(r)$ are given by Eqs. (6.180). On using these values, Eq. (6.198) simplifies to

$$\mathbf{u}(\mathbf{r}; t) = \frac{M}{4\pi\alpha^2} \sum_{l=0}^{\infty} (2l + 1) \sum_{n=0}^{\infty} \left\{ \frac{(1 - \cos \omega t)H(t)}{I^s} F_{l,2}^{+}(\chi) j_l(k_\alpha r_0) \times \left[y_1(r) P_l(\cos \varepsilon)\mathbf{e}_r + y_3(r) \frac{\partial}{\partial \varepsilon} P_l(\cos \varepsilon)(\cos C\mathbf{e}_\theta + \sin C\mathbf{e}_\phi) \right] \right\}_{\omega_n}. \tag{6.200}$$

A direct method of deriving Eq. (6.200) for $r = a$ without using the Green's dyadic is given in Example 6.6.

EXAMPLE 6.6.

Find the time-series solution for the spheroidal oscillations of a homogeneous sphere excited by a center of compression.

Solution: Let O be the center of a sphere of radius a and let the center of compression be placed at the point Q $(r = r_0, \theta = 0)$ inside the sphere. The displacement field resulting from this source in an unbounded medium may be expressed in the form (Table 4.4)

$$\mathbf{u}_0 = ik_\alpha g(\omega) A_0 \operatorname{grad} h_0^{(2)}(k_\alpha R), \qquad A_0 = \frac{M}{4\pi(\lambda + 2\mu)}. \tag{6.6.1}$$

However,

$$h_0^{(2)}(k_\alpha R) = \sum_{l=0}^{\infty} (2l + 1) j_l(k_\alpha r_0) h_l^{(2)}(k_\alpha r) P_l(\cos \theta), \qquad (r > r_0); \tag{6.6.2}$$

if $r < r_0$, r and r_0 are to be interchanged. Hence, we have

$$\mathbf{u}_0 = ik_\alpha^2 g(\omega) A_0 \sum_{0}^{\infty} (2l + 1) j_l(k_\alpha r_0) \mathbf{L}_{0,l}^{-}(k_\alpha r), \qquad (r > r_0). \tag{6.6.3}$$

When the medium is bounded by a sphere of radius a, we assume

$$\mathbf{u} = \mathbf{u}_0 - \sum_{l=0}^{\infty} (2l + 1)\{a_l \mathbf{N}_{0,l}^{+}(k_\beta r) + b_l \mathbf{L}_{0,l}^{+}(k_\alpha r)\}. \tag{6.6.4}$$

The boundary condition to be satisfied is

$$\mathbf{e}_r \cdot \mathfrak{T}(\mathbf{u}) = 0 \quad \text{at} \quad r = a. \tag{6.6.5}$$

Using the expressions for $\mathbf{e}_r \cdot \mathfrak{T}(\mathbf{N})$ and $\mathbf{e}_r \cdot \mathfrak{T}(\mathbf{L})$ computed in Eq. (6.3a), Eqs. (6.6.3)–(6.6.5) yield

$$a_l = \frac{-k_\alpha^2 g(\omega) A_0}{\Delta_l^{+} \chi \zeta^4} j_l(r_0 k_\alpha)[2(l - 1)(l + 2) - \chi^2], \tag{6.6.6}$$

$$b_l = ik_\alpha^2 g(\omega) \frac{A_0}{\Delta_l^{+}} j_l(r_0 k_\alpha)[2l(l + 1)F_{l,1}^{+}(\chi)F_{l,1}^{-}(\zeta) - F_{l,2}^{+}(\chi)F_{l,3}^{-}(\zeta)], \tag{6.6.7}$$

$$\Delta_l^{+} = 2l(l + 1)F_{l,1}^{+}(\chi)F_{l,1}^{+}(\zeta) - F_{l,2}^{+}(\chi)F_{l,3}^{+}(\zeta), \tag{6.6.8}$$

$$\zeta = ak_\alpha, \qquad \chi = ak_\beta, \tag{6.6.9}$$

where $F_{l,i}(x)$ are given by Eqs. (6.4).

At $r = a$, the Eq. (6.6.4) takes a somewhat simplified form

$$\mathbf{u}(\omega) = \frac{-g(\omega)A_0}{a^2}\left(\frac{\alpha}{\beta}\right)^2 \sum_0^{\infty} \frac{2l+1}{\Delta_l^+} j_l(r_0 k_\alpha)\left[F_{l,1}^+(\chi)\frac{dP_l(\cos\theta)}{d\theta}\mathbf{e}_\theta - \tfrac{1}{2}F_{l,2}^+(\chi)P_l(\cos\theta)\mathbf{e}_r\right]. \tag{6.6.10}$$

Assuming a unit-step time-variation so that $g(\omega) = 1/i\omega$, the displacement field in the time domain is given by

$$\mathbf{u}(t) = \frac{1}{2\pi i}\int_{-\infty}^{\infty} \mathbf{u}(\omega)e^{i\omega t}\frac{d\omega}{\omega}. \tag{6.6.11}$$

Because the roots of the frequency equation $\Delta_l^+(\omega) = 0$ are of the form $\pm\omega_n$, integration of Eq. (6.6.11) yields

$$\mathbf{u}(t) = \mathbf{u}(st) + \frac{A_0}{a^2}\left(\frac{\alpha}{\beta}\right)^2 \sum_{l=0}^{\infty}(2l+1)\sum_n\left[\frac{1}{D(\omega)} j_l(r_0 k_\alpha)\left\{F_{l,2}^+(\chi)P_l(\cos\theta)\mathbf{e}_r - 2F_{l,1}^+(\chi)\frac{dP_l(\cos\theta)}{d\theta}\mathbf{e}_\theta\right\}\cos\omega t H(t)\right]_{\omega_n}, \tag{6.6.12}$$

where $\mathbf{u}(st)$ is the contribution of the pole at the origin and represents the static response of the sphere. Further,

$$\mathbf{u}(st) = \lim_{\omega\to 0}\mathbf{u}(\omega),$$

and

$$\begin{aligned}
D(\omega_n) &= \left(\omega\frac{\partial\Delta_l^+}{\partial\omega}\right)_{\omega_n} \\
&= \left\{2 - \left(l + \frac{3}{2}\right)\left(\frac{\alpha}{\beta}\right)^2 - 2(l+2)(l^2-1)\chi^{-2}\right. \\
&\quad \left. + (l-1)(2l^2+4l+3)\zeta^{-2}\right\}j_l(\zeta)j_l(\chi) \\
&\quad + \left\{\frac{1}{2}\left(\frac{\alpha}{\beta}\right)\chi + (3l+1)\zeta^{-1}\right\}j_{l+1}(\zeta)j_l(\chi) + \left\{\frac{1}{2}\left(\frac{\alpha}{\beta}\right)^2\chi\right. \\
&\quad \left. - (2l^2-l-3)\left(\frac{\alpha}{\beta}\right)\zeta^{-1} + 2(l-1)(l+2)\chi^{-1}\right\}j_l(\zeta)j_{l+1}(\chi) \\
&\quad - 3\left(\frac{\alpha}{\beta}\right)j_{l+1}(\zeta)j_{l+1}(\chi); \qquad \omega = \omega_n.
\end{aligned} \tag{6.6.13}$$

In deriving Eq. (6.6.13), we have used the frequency equation $\Delta_l^+(\omega_n) = 0$.

EXAMPLE 6.7: Spheroidal Motion from a Buried Dislocation Source in a Homogeneous Sphere

Following the method of derivation given in Section 6.3.3.3 and using Eq. (6.38), we obtain the following expressions for the spectral field on the surface of a homogeneous sphere induced by a shear dislocation located at $r = r_0$, $\theta = 0$. The field at any point of the sphere can be calculated from Eq. (6.37) with $g(\omega) = 1$.

1. Vertical strike slip

$$u_r = \frac{P_0}{4\pi a^2} \sin 2\phi \sum_{l=2}^{\infty} (2l+1) \frac{D_1}{\Delta_R} P_l^2(\cos\theta),$$

$$u_\theta = \frac{P_0}{4\pi a^2} \sin 2\phi \sum_{l=2}^{\infty} \frac{2l+1}{l(l+1)} \frac{D_2}{\Delta_R} \frac{\partial}{\partial\theta} P_l^2(\cos\theta), \tag{6.7.1}$$

$$u_\phi = \frac{P_0}{2\pi a^2} \cos 2\phi \sum_{l=2}^{\infty} \frac{2l+1}{l(l+1)} \frac{D_2}{\Delta_R} \frac{P_l^2(\cos\theta)}{\sin\theta}.$$

2. Vertical dip slip

$$u_r = \frac{P_0}{4\pi a^2} \sin\phi \sum_{l=1}^{\infty} (2l+1) \frac{D_3}{\Delta_R} P_l^1(\cos\theta),$$

$$u_\theta = \frac{P_0}{4\pi a^2} \sin\phi \sum_{l=1}^{\infty} \frac{2l+1}{l(l+1)} \frac{D_4}{\Delta_R} \frac{\partial}{\partial\theta} P_l^1(\cos\theta), \tag{6.7.2}$$

$$u_\phi = \frac{P_0}{4\pi a^2} \cos\phi \sum_{l=1}^{\infty} \frac{2l+1}{l(l+1)} \frac{D_4}{\Delta_R} \frac{P_l^1(\cos\theta)}{\sin\theta}.$$

3. Case III (45° dip slip)

$$u_r = \frac{P_0}{8\pi a^2} \sum_{l=0}^{\infty} (2l+1) \frac{D_5}{\Delta_R} P_l(\cos\theta) - \frac{P_0}{8\pi a^2} \cos 2\phi \sum_{l=2}^{\infty} (2l+1) \frac{D_1}{\Delta_R} P_l^2(\cos\theta),$$

$$u_\theta = \frac{P_0}{4\pi a^2} \sum_{l=0}^{\infty} (2l+1) \frac{D_6}{\Delta_R} \frac{\partial}{\partial\theta} P_l(\cos\theta)$$

$$- \frac{P_0}{8\pi a^2} \cos 2\phi \sum_{l=2}^{\infty} \frac{2l+1}{l(l+1)} \frac{D_2}{\Delta_R} \frac{\partial}{\partial\theta} P_l^2(\cos\theta), \tag{6.7.3}$$

$$u_\phi = \frac{P_0}{4\pi a^2} \sin 2\phi \sum_{l=2}^{\infty} \frac{2l+1}{l(l+1)} \frac{D_2}{\Delta_R} \frac{P_l^2(\cos\theta)}{\sin\theta}.$$

We have used the following notation:

$$x = k_\alpha r_0, \qquad y = k_\beta r_0, \qquad \zeta = k_\alpha a, \qquad \chi = k_\beta a,$$

$$\begin{aligned}
D_1 &= \frac{1}{2x^2} j_l(x) F_{l,2}^+(\chi) - F_{l,1}^+(\zeta) F_{l,4}(y),\\
D_2 &= F_{l,3}^+(\zeta) F_{l,4}(y) - \frac{l(l+1)}{x^2} j_l(x) F_{l,1}^+(\chi),\\
D_3 &= F_{l,1}^+(x) F_{l,2}^+(\chi) - F_{l,1}^+(\zeta) F_{l,2}^+(y),\\
D_4 &= F_{l,2}^+(y) F_{l,3}^+(\zeta) - 2l(l+1) F_{l,1}^+(\chi) F_{l,1}^+(x),\\
D_5 &= 3l(l+1) F_{l,1}^+(\zeta) F_{l,1}^+(y) - \tfrac{1}{2} F_{l,2}^+(\chi) F_{l,5}(x),\\
D_6 &= F_{l,1}^+(\chi) F_{l,5}(x) - \tfrac{3}{2} F_{l,1}^+(y) F_{l,3}^+(\zeta),
\end{aligned} \tag{6.7.4}$$

$$F_{l,4}(z) = \left(\frac{1}{z}\frac{d}{dz} + \frac{1}{z^2}\right) j_l(z), \qquad F_{l,5}(z) = \left(3\frac{d^2}{dz^2} + 1\right) j_l(z),$$

$$f_l^+(z) = j_l(z), \qquad f_l^-(z) = h_l^{(2)}(z).$$

The functions $F_{l,1}$, $F_{l,2}$ and $F_{l,3}$ are defined in Eqs. (6.4) and $\Delta_R = \Delta_l^+$, in Eq. (6.36). Moreover, $P_0 = U_0\,dS$ denotes the source potency.

6.4. Effect of the Rotation of the Earth

So far, in our discussion of the oscillations of the earth, we have neglected the rotation of the earth. In this section, we consider the effect of the diurnal rotation of the earth on its eigenfrequencies by carrying out a first-order perturbation calculation. Let $\mathbf{\Omega}$ denote the uniform angular velocity of the earth about its center O. We assume that the observer is referred to a noninertial frame that, for all times, maintains a state of uniform rotation with angular velocity $\mathbf{\Omega}$. Let (x_1, x_2, x_3) be a Cartesian coordinate system in this uniformly rotating frame of reference, let the origin of the system coincide with the center of the earth, and let $\mathbf{e}_3$ be aligned along the axis of rotation, so that $\mathbf{\Omega} = \Omega\mathbf{e}_3$.

In the case of a nonrotating earth the equilibrium equations are (Sect. 6.3.1)

$$\begin{aligned}
\nabla p_0 &= \rho_0 \nabla \Psi_0,\\
\nabla^2 \Psi_0 &= -4\pi G \rho_0.
\end{aligned} \tag{6.201}$$

The field equations may be written in the form

$$\operatorname{div} \mathfrak{T}_e + \nabla(\rho_0 \mathbf{u} \cdot \nabla\Psi_0) - \operatorname{div}(\rho_0\mathbf{u})\nabla\Psi_0 + \rho_0\nabla\psi + \rho_0\omega^2\mathbf{u} = 0, \tag{6.202}$$

$$\nabla^2\psi = 4\pi G \operatorname{div}(\rho_0\mathbf{u}). \tag{6.203}$$

The rotation of the earth introduces two body forces. One of them is the *centrifugal force* that contributes an additional term in the gravitational potential,

$$\Phi_0 = \Psi_0 + \tfrac{1}{2}\Omega^2 r^2 \sin^2 \theta. \tag{6.204}$$

The other force introduced by the rotation is the *Coriolis force*, which, per unit mass, is given by

$$\mathbf{C} = 2\Omega \frac{\partial \mathbf{u}}{\partial t} \times \mathbf{e}_z. \tag{6.205}$$

When these two forces are taken into account, Eqs. (6.201) and (6.202) are modified as follows:

$$\nabla p_0 = \rho_0 \nabla \Phi_0, \tag{6.206}$$

$$\begin{aligned} \operatorname{div} \mathfrak{T}_e + \nabla[\rho_0 \mathbf{u} \cdot \nabla \Phi_0] - \operatorname{div}(\rho_0 \mathbf{u}) \nabla \Phi_0 + \rho_0 \nabla \psi + \rho_0 \omega^2 \mathbf{u} \\ + 2i\omega\rho_0 \Omega \mathbf{u} \times \mathbf{e}_z = 0. \end{aligned} \tag{6.207}$$

The entity Φ_0 is known as the *geopotential*. From Eq. (6.206) we note that

$$p_0 = p_0(\Phi_0), \qquad \rho_0 = \frac{dp_0}{d\Phi_0} = \rho_0(\Phi_0). \tag{6.208}$$

From Eqs. (6.201) and (6.204), we have

$$\nabla^2 \Phi_0 = -4\pi G \rho_0 + 2\Omega^2.$$

If we take $\rho_0 = 5$ g/cm^3, $G = 6.7 \times 10^{-8}$ cgs units, we get

$$\frac{2\Omega^2}{4\pi G \rho_0} \simeq \frac{1}{400}.$$

This shows that the effect of the centrifugal force is small. Remembering that Ψ_0 is a function of r only, it follows that the surfaces $\Phi_0 =$ const. are ellipsoids of revolution. However, it can be shown that these surfaces are nearly spherical throughout the volume of the earth, to within 1 part in 300. Therefore, in the following analysis we shall neglect the ellipticity of the geopotential surfaces and shall assume that p_0, ρ_0, and g_0 are functions of r only.

If the incremental Piola–Kirchoff stress tensor $\hat{\mathfrak{T}}$ is defined through the relation

$$\hat{\mathfrak{T}} = \mathfrak{T}_e - p_0(\mathfrak{I} \operatorname{div} \mathbf{u} - \mathbf{u}\nabla), \tag{6.209}$$

and Eqs. (6.206) and (6.208) are used, Eq. (6.207) may be written in the following compact form:

$$\operatorname{div} \hat{\mathfrak{T}} + \rho_0 \mathbf{u} \cdot \nabla\nabla\Phi_0 + \rho_0 \nabla\psi + \rho_0 \omega^2 \mathbf{u} + 2i\omega\rho_0 \Omega \mathbf{u} \times \mathbf{e}_z = 0. \tag{6.210}$$

Let us assume that the solution of Eqs. (6.203) and (6.210) with $\Omega = 0$ is $(\omega^0, \mathbf{u}^0, \psi^0)$ so that

$$\nabla^2 \psi^0 = 4\pi G \operatorname{div}(\rho_0 \mathbf{u}^0), \tag{6.211}$$

$$\operatorname{div} \hat{\mathfrak{T}}^0 + \rho_0 \mathbf{u}^0 \cdot \nabla\nabla\Psi_0 + \rho_0 \nabla\psi^0 + \rho_0 \omega^{0^2} \mathbf{u}^0 = 0, \tag{6.212}$$

where $\hat{\mathfrak{T}}^0 = \hat{\mathfrak{T}}(\mathbf{u}^0)$. We now carry out a first-order perturbation calculation, assuming

$$\omega = \omega^0 + \varepsilon\omega', \qquad \mathbf{u} = \mathbf{u}^0 + \varepsilon\mathbf{u}', \qquad \psi = \psi^0 + \varepsilon\psi', \qquad \varepsilon = \frac{\Omega}{\omega^0} \ll 1. \quad (6.213)$$

Substituting Eq. (6.213) in Eqs. (6.203) and (6.210), neglecting second and higher powers of ε, and using Eqs. (6.211) and (6.212), we find

$$\nabla^2\psi' = 4\pi G\,\mathrm{div}(\rho_0\mathbf{u}'), \quad (6.214)$$

$$\begin{aligned}\mathrm{div}\,\hat{\mathfrak{T}}' + \rho_0\mathbf{u}'\cdot\nabla\nabla\Psi_0 + \rho_0\nabla\psi' + \rho_0\omega^0(\omega^0\mathbf{u}' + 2\omega'\mathbf{u}^0)&\\ + 2i\omega^{0^2}\rho_0\mathbf{u}^0\times\mathbf{e}_z = 0.& \quad (6.215)\end{aligned}$$

Taking the scalar product of Eq. (6.212) with $\overset{*}{\mathbf{u}}'$, and that of Eq. (6.215) with $\overset{*}{\mathbf{u}}{}^0$ and subtracting, we get

$$\begin{aligned}\overset{*}{\mathbf{u}}'\cdot\mathrm{div}\,\hat{\mathfrak{T}}^0 - \overset{*}{\mathbf{u}}{}^0\cdot\mathrm{div}\,\hat{\mathfrak{T}}' + \rho_0(\overset{*}{\mathbf{u}}'\mathbf{u}^0 - \overset{*}{\mathbf{u}}{}^0\mathbf{u}')\colon\nabla\nabla\Psi_0 + \rho_0(\overset{*}{\mathbf{u}}'\cdot\nabla\psi^0 - \overset{*}{\mathbf{u}}{}^0\cdot\nabla\psi')&\\ + \rho_0\omega^{0^2}(\overset{*}{\mathbf{u}}'\cdot\mathbf{u}^0 - \overset{*}{\mathbf{u}}{}^0\cdot\mathbf{u}') = 2\rho_0\omega^0[\omega'\overset{*}{\mathbf{u}}{}^0\cdot\mathbf{u}^0 + i\omega^0(\overset{*}{\mathbf{u}}{}^0\times\mathbf{u}^0)\cdot\mathbf{e}_z],&\\ (6.216)&\end{aligned}$$

where an asterisk signifies the complex conjugate.

We express $\mathbf{u}$ and ψ in terms of spherical harmonics, and evaluate the volume integral

$$\int_0^a r^2\,dr\int_0^\pi \sin\theta\,d\theta\int_0^{2\pi}(\text{---})d\phi, \quad (6.217)$$

using Eqs. (6.211) and (6.214) together with the Gauss theorem. Applying the boundary conditions, and the orthogonality relations of the spherical harmonics, we find

$$\begin{aligned}\omega'\int_0^a \rho_0 r^2\,dr\int_0^\pi \sin\theta\,d\theta\int_0^{2\pi}\overset{*}{\mathbf{u}}{}^0\cdot\mathbf{u}^0\,d\phi&\\ + i\omega^0\int_0^a \rho_0 r^2\,dr\int_0^\pi \sin\theta\,d\theta\int_0^{2\pi}(\overset{*}{\mathbf{u}}{}^0\times\mathbf{u}^0)\cdot\mathbf{e}_z\,d\phi = 0.& \quad (6.218)\end{aligned}$$

This furnishes a relationship between ω', ω^0 and the solution for a nonrotating earth. In the case of the toroidal oscillations, we put

$$\mathbf{u}^0(\mathbf{r}) = \sum_{l=1}^{\infty}\sum_{m=-l}^{l} y_1^0(r)\sqrt{l(l+1)}\,\mathbf{C}_{ml}(\theta,\phi). \quad (6.219)$$

Using the orthogonality relation for $\mathbf{C}_{ml}(\theta,\phi)$ and the integral ($m \neq 0$)

$$\int_0^\pi P_l^m(\cos\theta)\frac{dP_l^m(\cos\theta)}{d\theta}\cos\theta\,d\theta = \frac{1}{2l+1}\cdot\frac{(l+m)!}{(l-m)!}, \quad (6.220)$$

we get

$$\omega' = \frac{m}{l(l+1)}\omega^0. \quad (6.221)$$

Equation (6.213) now yields

$$ {}_n\omega_l^m = {}_n\omega_l^0 + \frac{m}{l(l+1)}\Omega, \qquad -l \le m \le l. \tag{6.222}$$

For spheroidal oscillations, we may take

$$\mathbf{u}^0(\mathbf{r}) = \sum_{l=0}^{\infty}\sum_{m=-l}^{l}[y_1^0(r)\mathbf{P}_{ml}(\theta,\phi) + y_3^0(r)\sqrt{l(l+1)}\mathbf{B}_{ml}(\theta,\phi)]. \tag{6.223}$$

Substituting in Eq. (6.218), using the orthogonality relations of $\mathbf{P}_{ml}$, $\mathbf{B}_{ml}$ and the integral in Eq. (6.220) together with the relation

$$\int_0^\pi [P_l^m(\cos\theta)]^2 \sin\theta\, d\theta = \frac{2}{2l+1}\frac{(l+m)!}{(l-m)!}, \tag{6.224}$$

we get

$$\omega' = m\tau_l\omega^0,$$

where

$$\tau_l = \frac{\int_0^a (2y_1^0 + y_3^0)y_3^0\rho_0 r^2\, dr}{\int_0^a [y_1^{0^2} + l(l+1)y_3^{0^2}]\rho_0 r^2\, dr}. \tag{6.225}$$

Therefore, for the spheroidal case, we have

$$ {}_n\omega_l^m = {}_n\omega_l^0 + m\tau_l\Omega, \qquad -l \le m \le l. \tag{6.226}$$

We see that both for toroidal and spheroidal oscillations, the degenerate frequency ${}_n\omega_l$ is resolved by a slow rotation into $(2l+1)$ frequencies ${}_n\omega_l^m$ $(-l \le m \le l)$. For toroidal oscillations the splitting parameter $[l(l+1)]^{-1}$ does not depend upon the type of earth model. This splitting of the degenerate eigenfrequencies of the earth was observed for the first time in 1961, when the records of the great Chilean earthquake of 22 May, 1960 were analyzed.

6.5. Energy Integrals

Let

$$\mathbf{u}(\mathbf{r}, t) = \mathbf{q}(\mathbf{r})\cos\omega t. \tag{6.227}$$

Then, the kinetic energy at time t is given by

$$K(t) = \frac{1}{2}\omega^2 \sin^2\omega t \int_V \rho_0(\mathbf{q}\cdot\overset{*}{\mathbf{q}})dV. \tag{6.228}$$

The kinetic energy averaged over a cycle is

$$K = \frac{1}{T}\int_0^T K(t)dt = \frac{1}{4}\omega^2\int_V \rho_0(\mathbf{q}\cdot\overset{*}{\mathbf{q}})dV. \tag{6.229}$$

Substituting

$$\mathbf{q}(\mathbf{r}) = {}_nW_l(r)\sqrt{l(l+1)}\,\mathbf{C}_{ml}(\theta, \phi), \tag{6.230}$$

the average kinetic energy associated with the toroidal mode ${}_n\omega_l^m$ is found to be

$$K = \pi\frac{l(l+1)}{2l+1}\cdot\frac{(l+m)!}{(l-m)!}\cdot({}_n\omega_l^m)^2\,{}_nI_l^T. \tag{6.231}$$

In Eq. (6.231)

$${}_nI_l^T = \int_0^a \rho_0(r)_n W_l^2 r^2\,dr. \tag{6.232}$$

For the spheroidal mode ${}_n\omega_l^m$, we have

$$\mathbf{q}(\mathbf{r}) = {}_nU_l(r)\mathbf{P}_{ml}(\theta, \phi) + {}_nV_l(r)\sqrt{l(l+1)}\,\mathbf{B}_{ml}(\theta, \phi). \tag{6.233}$$

Therefore, the average kinetic energy is given by

$$K = \frac{\pi}{2l+1}\cdot\frac{(l+m)!}{(l-m)!}({}_n\omega_l^m)^2\,{}_nI_l^s, \tag{6.234}$$

where

$${}_nI_l^s = \int_0^a \rho_0(r)[{}_nU_l^2 + l(l+1)_nV_l^2]r^2\,dr. \tag{6.235}$$

The integrals ${}_nI_l^T$ and ${}_nI_l^s$ are known as *energy integrals.*

The strain or potential energy of deformation may be expressed as [Eq. (1.70)]

$$W(t) = \frac{1}{2}\cos^2\omega t\int_V [\lambda|\operatorname{div}\mathbf{q}|^2 + 2\mu\mathfrak{E}:\overset{*}{\mathfrak{E}}]dV, \tag{6.236}$$

where

$$2\mathfrak{E} = \nabla\mathbf{q} + \mathbf{q}\nabla. \tag{6.237}$$

Averaging over a cycle, we get

$$W = \frac{1}{T}\int_0^T W(t)dt = \frac{1}{4}\int_V [\lambda|\operatorname{div}\mathbf{q}|^2 + 2\mu\mathfrak{E}:\overset{*}{\mathfrak{E}}]dV. \tag{6.238}$$

Substituting for $\mathbf{q}$ from Eqs. (6.230) and (6.233) and using certain definite integrals involving associated Legendre functions, W can be calculated. For the toroidal oscillations, we find

$$W = \pi\frac{l(l+1)}{2l+1}\cdot\frac{(l+m)!}{(l-m)!}\int_0^a \mu(r)\left[r^2\left(\frac{d_nW_l}{dr}\right)^2 - 2r_nW_l\frac{d_nW_l}{dr} + (l^2+l-1)_nW_l^2\right]dr. \tag{6.239}$$

Similarly, the potential energy for the spheroidal oscillations is given by

$$W = \frac{\pi}{2l+1}\cdot\frac{(l+m)!}{(l-m)!}\int_0^a \left[(\lambda + 2\mu)\left\{r\frac{d_nU_l}{dr} + 2_nU_l - l(l+1)_nV_l\right\}^2\right.$$
$$+ l(l+1)\mu\left\{\left(r\frac{d_nV_l}{dr} - {}_nV_l + {}_nU_l\right)^2 + 4_nV_l\left({}_nU_l + r\frac{d_nU_l}{dr}\right) - 2_nV_l^2\right\}$$
$$\left. - 4\mu_nU_l\left({}_nU_l + 2r\frac{d_nU_l}{dr}\right)\right]dr. \tag{6.240}$$

Because the spheroidal oscillations perturb the gravitational field, the gravitational energy must also be taken into account. From Eq. (6.50), we note that gravity contributes a body force

$$\mathbf{F}_g = \rho_0[\operatorname{grad}(\psi - g_0 u_r) + g_0 \mathbf{e}_r \operatorname{div} \mathbf{u}] \tag{6.241}$$

per unit volume. We assume that

$$\psi(\mathbf{r}) = {}_nP_l(r)Y_{ml}(\theta, \phi)\cos {}_n\omega_l^m t. \tag{6.242}$$

Using Eqs. (6.141), (6.227), (6.233), (6.241), and (6.242) we get the following expression for the *gravitational energy* averaged over a cycle:

$$W' = \frac{1}{2T}\int_0^T \left[\int_V (-\mathbf{F}_g \cdot \overset{*}{\mathbf{u}})dV\right]dt$$
$$= \frac{\pi}{2l+1}\cdot\frac{(l+m)!}{(l-m)!}\int_0^a \left[4(\pi\rho_0 rG - g_0)_nU_l^2 - r_nU_l\frac{d_nP_l}{dr}\right.$$
$$\left. + l(l+1)_nV_l(2g_{0\,n}U_l - {}_nP_l)\right]\rho_0 r\,dr. \tag{6.243}$$

For the toroidal oscillations, the kinetic and potential energies, averaged over a cycle, are given by Eqs. (6.231) and (6.239), respectively. Equating the two energies, we get

$$\omega^2 I_0^T = I_1^T + l(l+1)I_2^T, \tag{6.244}$$

where

$$I_0^T = \int_0^a \rho W_l^2 r^2\,dr,$$
$$I_1^T = \int_0^a \mu\left\{-W_l^2 - 2rW_l\frac{dW_l}{dr} + r^2\left(\frac{dW_l}{dr}\right)^2\right\}dr, \tag{6.245}$$
$$I_2^T = \int_0^a \mu W_l^2\,dr.$$

On applying Rayleigh's principle to Eq. (6.244), we find

$$2\omega(\delta\omega)I_0^T + \omega^2(\delta I_0^T) = \delta I_1^T + l(l+1)(\delta I_2^T). \tag{6.246}$$

Equation (6.246) may be used to examine the variation in the eigenfrequency, $\delta\omega$, with the perturbation of a specific elastic parameter, keeping all the other parameters fixed. Therefore, if we perturb the density over an interval $r - \varepsilon \leq r \leq r + \varepsilon$, holding the rigidity fixed, we find that the change in the phase velocity[1] is given by

$$\frac{1}{c}\left(\frac{\partial c}{\partial \rho}\right)_{\mu} = -\frac{1}{2I_0^T}\int_{r-\varepsilon}^{r+\varepsilon} r^2 W_l^2\, dr \tag{6.247}$$

or

$$-\frac{\rho}{c}\left(\frac{\partial c}{\partial \rho}\right)_{\mu} = \frac{1}{2I_0^T}\int_{r-\varepsilon}^{r+\varepsilon} \rho r^2 W_l^2 dr. \tag{6.248}$$

The right-hand side of Eq. (6.248) is the ratio of the kinetic energy of the shell over which the density is perturbed to the total energy of the system. Similarly, the effect of the perturbation of the rigidity is

$$\frac{\mu}{c}\left(\frac{\partial c}{\partial \mu}\right)_{\rho} = \frac{1}{2\omega^2 I_0^T}\int_{r-\varepsilon}^{r+\varepsilon} \mu\left\{-W_l^2 - 2rW_l\frac{dW_l}{dr} + r^2\left(\frac{dW_l}{dr}\right)^2 - l(l+1)W_l^2\right\}dr, \tag{6.249}$$

which is the potential energy of the perturbed interval divided by the total energy.

Calculations show that the gravitational energy can be neglected for $l > 7$. Therefore, in the case of the spheroidal oscillations, we *neglect self-gravitation* and equate the kinetic and strain energies to find

$$\omega^2[I_1^s + l(l+1)I_2^s] = I_3^s + I_6^s + l(l+1)(I_7^s - 2I_4^s) + [l(l+1)]^2(I_5^s + 2I_8^s), \tag{6.250}$$

where

$$\begin{aligned}
I_1^s &= \int_0^a \rho U_l^2 r^2\, dr,\\
I_2^s &= \int_0^a \rho V_l^2 r^2\, dr,\\
I_3^s &= \int_0^a \lambda\left(2U_l + r\frac{dU_l}{dr}\right)^2 dr,\\
I_4^s &= \int_0^a \lambda\left(2U_l + r\frac{dU_l}{dr}\right)V_l\, dr,\\
I_5^s &= \int_0^a \lambda V_l^2\, dr,\\
I_6^s &= 2\int_0^a \mu\left\{2U_l^2 + r^2\left(\frac{dU_l}{dr}\right)^2\right\}dr,\\
I_7^s &= \int_0^a \mu\left\{U_l^2 - V_l^2 - 6U_lV_l + 2r(U_l - V_l)\frac{dV_l}{dr} + r^2\left(\frac{dV_l}{dr}\right)^2\right\}dr,\\
I_8^s &= \int_0^a \mu V_l^2\, dr.
\end{aligned} \tag{6.251}$$

[1] Surface waves in a spherical-earth model are discussed in Chapter 8, where the significance of the phase velocity is explained.

An application of the Rayleigh principle now yields

$$-\frac{\rho}{c}\left(\frac{\partial c}{\partial \rho}\right)_{\lambda,\mu} = \frac{\int_{r-\varepsilon}^{r+\varepsilon} \rho[U_l^2 + l(l+1)V_l^2]r^2\,dr}{[2\{I_1^s + l(l+1)I_2^s\}]}, \tag{6.252}$$

$$\frac{\mu}{c}\left(\frac{\partial c}{\partial \mu}\right)_{\rho,\lambda} = \int_{r-\varepsilon}^{r+\varepsilon} \mu\left[2\left\{2U_l^2 + r^2\left(\frac{dU_l}{dr}\right)^2\right\} + l(l+1)\left\{U_l^2 - V_l^2 - 6U_lV_l\right.\right.$$
$$\left.\left. + 2r(U_l - V_l)\frac{dV_l}{dr} + r^2\left(\frac{dV_l}{dr}\right)^2\right\} + 2\{l(l+1)\}^2V_l^2\right]dr$$
$$\bigg/[2\omega^2\{I_1^s + l(l+1)I_2^s\}], \tag{6.253}$$

$$\frac{\lambda}{c}\left(\frac{\partial c}{\partial \lambda}\right)_{\mu,\rho} = \int_{r-\varepsilon}^{r+\varepsilon} \lambda\left[\left\{2U_l + r\left(\frac{dU_l}{dr}\right)^2\right\}^2 - 2l(l+1)\left(2U_l + r\frac{dU_l}{dr}\right)V_l\right.$$
$$\left. + \{l(l+1)\}^2V_l^2\right]dr\bigg/[2\omega^2\{I_1^s + l(l+1)I_2^s\}]. \tag{6.254}$$

If α, β, ρ are taken as independent variables instead of λ, μ, ρ, we use the relations $\lambda = \rho(\alpha^2 - 2\beta^2)$, $\mu = \rho\beta^2$ to obtain

$$\left(\frac{\partial c}{\partial \alpha}\right)_{\beta,\rho} = \left(\frac{\partial c}{\partial \lambda}\right)\left(\frac{\partial \lambda}{\partial \alpha}\right) + \left(\frac{\partial c}{\partial \mu}\right)\left(\frac{\partial \mu}{\partial \alpha}\right) = 2\rho\alpha\left(\frac{\partial c}{\partial \lambda}\right)_{\mu,\rho},$$

$$\left(\frac{\partial c}{\partial \beta}\right)_{\rho,\alpha} = \left(\frac{\partial c}{\partial \lambda}\right)\left(\frac{\partial \lambda}{\partial \beta}\right) + \left(\frac{\partial c}{\partial \mu}\right)\left(\frac{\partial \mu}{\partial \beta}\right) = 2\rho\beta\left\{-2\left(\frac{\partial c}{\partial \lambda}\right)_{\mu,\rho} + \left(\frac{\partial c}{\partial \mu}\right)_{\rho,\lambda}\right\}, \tag{6.255}$$

$$\left(\frac{\partial c}{\partial \rho}\right)_{\alpha,\beta} = \left(\frac{\partial c}{\partial \rho}\right) + \left(\frac{\partial c}{\partial \lambda}\right)\left(\frac{\partial \lambda}{\partial \rho}\right) + \left(\frac{\partial c}{\partial \mu}\right)\left(\frac{\partial \mu}{\partial \rho}\right)$$
$$= \left(\frac{\partial c}{\partial \rho}\right)_{\lambda,\mu} + (\alpha^2 - 2\beta^2)\left(\frac{\partial c}{\partial \lambda}\right)_{\mu,\rho} + \beta^2\left(\frac{\partial c}{\partial \mu}\right)_{\rho,\lambda}.$$

The partial derivatives obtained in Eqs. (6.255) have an application in the calculation of the periods of free oscillation for realistic earth models. Even with a large computer numerical evaluation is tedious and time consuming. However, using partial derivatives we can assess which parameters are most important for any given mode and which parameters have little influence. The perturbation parameters can also be used in a matrix inversion scheme to modify a trial earth model in order to satisfy a given set of observed free-oscillation data.

It will be proved subsequently that for large values of l, $l(l+1) \simeq a^2k^2$, where k denotes the wave number for Love waves in the case of toroidal oscillations and for Rayleigh waves in the case of spheroidal oscillations. Using this approximation in Eqs. (6.244) and (6.250), keeping ρ, λ, and μ fixed, and applying Rayleigh's principle, we get expressions for the group velocities in terms of the energy integrals:

$$c_L U_L = c_L \left(\frac{d\omega}{dk}\right)_L = a^2 \left(\frac{I_2^T}{I_0^T}\right),$$

$$c_R U_R = \frac{a^2[I_7^s - 2I_4^s + 2l(l+1)(I_5^s + 2I_8^s) - \omega^2 I_2^s]}{[I_1^s + l(l+1)I_2^s]}, \tag{6.256}$$

$$c = \frac{a\omega}{(l+\frac{1}{2})}.$$

6.6. Source Effects

6.6.1. Effect of the Time Function

The spectral amplitudes of the earth's eigenvibrations naturally depend upon the source-time function. So far we have assumed a time dependence of a unit-step function. This yielded the factor [cf. Eq. (6.97)] $(1 - \cos \omega_n t)/\omega_n^2$ in all our summands. In general, if the time dependence is $g(t)$, the corresponding factor will be

$$\bar{g}(t) = \frac{1}{2\pi} \int_{-\infty}^{\infty} \frac{g(\omega)}{\omega_n^2 - \omega^2} e^{i\omega t} \, d\omega, \tag{6.257}$$

where $g(\omega)$ is the Fourier transform of $g(t)$:

$$g(\omega) = \int_{-\infty}^{\infty} g(t) e^{-i\omega t} \, dt. \tag{6.258}$$

Because $(\omega_n^2 - \omega^2)^{-1}$ is the Fourier transform of $\omega_n^{-1} \sin(\omega_n t) H(t)$, the convolution theorem yields

$$\bar{g}(t) = \frac{1}{\omega_n} \int_0^t \sin\{\omega_n T\} g(t - T) dT$$

$$= \frac{1}{\omega_n} \int_0^t \sin \omega_n (t - T) g(T) dT, \tag{6.259}$$

assuming $g(t) = 0$ for $t < 0$. Table 6.4 lists $\bar{g}(t)$ for six commonly used source-time functions.

6.6.2. Effect of the Finiteness of the Source

In the case of large earthquakes with fault lengths extending over several hundred kilometers, it is important to take into account the finiteness of the source and its motion when normal-mode amplitudes are calculated.

Table 6.4. Examples of the Source-Time Function $g(t)$ and Its Effect $\bar{g}(t)$ on the Normal-Mode Amplitudes

Name	$g(t)$	$g(\omega) = \int_{-\infty}^{\infty} g(t) e^{-i\omega t}\, dt$	$\bar{g}(t) = \frac{1}{2\pi}\int_{-\infty}^{\infty} \frac{g(\omega)}{\omega_n^2 - \omega^2} e^{i\omega t}\, d\omega$
Step function	$H(t)$	$\frac{1}{i\omega}$	$\frac{1 - \cos \omega_n t}{\omega_n^2}, \quad (t > 0)$
Gate function	$\frac{1}{2t_1}[H(t + t_1) - H(t - t_1)]$	$\frac{\sin \omega t_1}{\omega t_1}$	$\left(\frac{\sin \omega_n t_1}{\omega_n t_1}\right) \frac{1}{\omega_n} \sin \omega_n t$
Delta function	$\delta(t)$	1	$\frac{1}{\omega_n} \sin \omega_n t$
Exponential buildup	$H(t)(1 - e^{-t/\tau}), \quad \tau > 0$	$\frac{1}{i\omega(1 + i\omega\tau)}$	$\frac{1}{\omega_n^2}[1 - \cos \varepsilon \cos(\omega_n t - \varepsilon) - \sin^2 \varepsilon e^{-t/\tau}]$ $(t > 0, \tan \varepsilon = \omega_n \tau)$
Ramp function	$R(t, \tau) = \begin{cases} 0 & t < 0 \\ \frac{t}{\tau} & 0 \le t \le \tau \\ 1 & t > \tau \end{cases}$	$\frac{1}{i\omega} f(\omega) e^{-i\omega\tau/2}$ $\left[f(\omega) = \frac{\sin(\omega\tau/2)}{\omega\tau/2}\right]$	$\frac{1}{\omega_n^2}\left(\frac{t}{\tau}\right)\left(1 - \frac{\sin \omega_n t}{\omega_n t}\right), \quad (0 < t < \tau)$ $\frac{1}{\omega_n^2}\left[1 - f(\omega_n)\cos\left\{\omega_n\left(t - \frac{\tau}{2}\right)\right\}\right], \quad (t > \tau)$
Half-sine function	$H_s(t) = \begin{cases} 0 & t < 0 \\ \frac{1}{2}\left(1 - \cos \frac{\pi t}{\tau}\right) & 0 \le t \le \tau \\ 1 & t > \tau \end{cases}$	$\frac{1}{i\omega} f(\omega) e^{-i\omega\tau/2}$ $\left[f(\omega) = \frac{\cos(\omega\tau/2)}{1 - (\omega\tau/\pi)^2}\right]$	$\frac{1}{\omega_n^2}\left[1 - \frac{\cos^2(\omega_n t/2) - (\omega_n \tau/\pi)^2 \cos^2(\pi t/2\tau)}{1 - (\omega_n \tau/\pi)^2}\right], \quad (0 < t < \tau)$ $\frac{1}{\omega_n^2}\left[1 - f(\omega_n)\cos\left\{\omega_n\left(t - \frac{\tau}{2}\right)\right\}\right], \quad (t > \tau)$

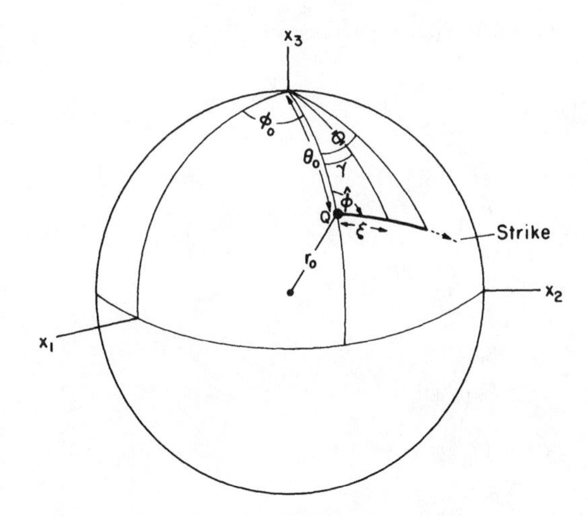

Figure 6.7. Geometry of a finite fault in a sphere.

Consider, for example, the toroidal Green's dyadic, Eq. (6.106), for a step source. It was shown in Section 6.6.1 that the Green's dyadic for a source time function $g(t)$,

$$\mathfrak{G}(\mathbf{r}\,|\,\mathbf{r}_0;t) = \frac{1}{2\pi}\int_{-\infty}^{\infty}\sum_{l,m,n}\frac{{}_n\mathbf{M}_{ml}(\mathbf{r})\,{}_n\overset{*}{\mathbf{M}}_{ml}(\mathbf{r}_0)g(\omega)e^{i\omega t}}{l(l+1)\Omega_{ml}(\omega_n^2-\omega^2)I_n^T}\,d\omega, \tag{6.260}$$

where $g(\omega)$ is the Fourier transform of $g(t)$. The displacement field of a shear dislocation can be obtained from Eq. (6.260) using the Volterra relation, Eq. (6.120). The effect of the finiteness of the source can be inferred from the Green's dyadic itself. Let the dislocation U_0 be constant along the fault. Introducing a time delay of (ξ/V_f), where V_f is the rupture velocity, and integrating along the strike of the fault (Fig. 6.7), we find

$$\mathfrak{G}(\mathbf{r}\,|\,\mathbf{r}_0;t) = \frac{1}{2\pi}\int_{-\infty}^{\infty}\sum_{l,m,n}\frac{{}_n\mathbf{M}_{ml}(\mathbf{r})g(\omega)}{l(l+1)\Omega_{ml}(\omega_n^2-\omega^2)I_n^T} \times\left\{\frac{1}{L}\int_0^L {}_n\overset{*}{\mathbf{M}}_{ml}(\hat{\mathbf{r}}_0)e^{i\omega(t-\xi/V_f)}\,d\xi\right\}d\omega, \tag{6.261}$$

where L is the fault length, $\hat{\mathbf{r}}_0 = [r_0, \theta(\xi), \phi_0+\gamma]$ and $\theta(0) = \theta_0$.

Suppose that the fault is confined between the azimuths $(\phi_0, \phi_0+\Phi)$, as shown in Fig. 6.7. Assuming that L is small in comparison with the source's polar distance and noting that

$${}_n\overset{*}{\mathbf{M}}_{ml}(\mathbf{r}_0) = y_{1n}(r_0)\left(\mathbf{e}_{\theta_0}\frac{1}{\sin\theta_0}\frac{\partial}{\partial\phi_0} - \mathbf{e}_{\phi_0}\frac{\partial}{\partial\theta_0}\right)P_l^m(\cos\theta_0)e^{-im\phi_0},$$

it is clear that the effect of the finiteness and the motion of the source in the frequency domain is determined by the factor

$$\frac{1}{L}\int_0^L e^{-im\gamma-i\omega\xi/V_f}\,d\xi. \tag{6.262}$$

Because $\xi \sin \hat{\phi} \simeq r_0 \gamma \sin \theta_0$, the above factor becomes

$$\frac{1}{\Phi} \int_0^{\Phi} \exp\left[-i\gamma\left(m + \frac{\omega r_0 \sin \theta_0}{V_f \sin \hat{\phi}}\right)\right] d\gamma = \frac{\sin X_m}{X_m} e^{-iX_m}, \tag{6.263}$$

where

$$X_m = \frac{1}{2}\Phi\left(m + \frac{\omega r_0 \sin \theta_0}{V_f \sin \hat{\phi}}\right) = \frac{L}{2V_f}\left(\omega + \frac{mV_f \sin \hat{\phi}}{r_0 \sin \theta_0}\right). \tag{6.264}$$

Defining

$$c_0 = \frac{\omega r_0}{(l + 1/2)},$$

Eq. (6.264) yields

$$X_m = \frac{\omega L}{2c_0}\left[\frac{c_0}{V_f} + \left(\frac{m}{l + 1/2}\right)\frac{\sin \hat{\phi}}{\sin \theta_0}\right] \quad (l = 1, 2, 3, \ldots, -l \leq m \leq l). \tag{6.265}$$

It is apparent from Eq. (6.263) that the ratio of the spectral displacements for $\pm m$ is

$$\begin{aligned} \frac{X_{-m}}{X_m}\left(\frac{\sin X_m}{\sin X_{-m}}\right) e^{-i(X_m - X_{-m})} &= \left[\frac{1 - (V_f/c_0)(m/(l + \frac{1}{2}))(\sin \hat{\phi}/\sin \theta_0)}{1 + (V_f/c_0)(m/(l + \frac{1}{2}))(\sin \hat{\phi}/\sin \theta_0)}\right] \\ &\times \left[\frac{\sin\{\omega L/(2V_f)[1 + (V_f/c_0)(m/(l + \frac{1}{2}))(\sin \hat{\phi}/\sin \theta_0)]\}}{\sin\{\omega L/(2V_f)[1 - (V_f/c_0)(m/(l + \frac{1}{2}))(\sin \hat{\phi}/\sin \theta_0)]\}}\right] \\ &\times \exp\left[-\frac{i\omega L}{c_0}\left(\frac{m}{l + 1/2}\right)\frac{\sin \hat{\phi}}{\sin \theta_0}\right]. \end{aligned} \tag{6.266}$$

This is similar to the directivity function, Eq. (4.260).

It is interesting to note that the finiteness factor, Eq. (6.263), depends upon the azimuthal order number, m. Therefore, if the normal modes of the earth are split because of rotation, ellipticity, or any other reason, the finiteness of the source will influence the amplitudes of different split normal modes differently. Another significant result is that the members of the pair ($\pm m$) of the split normal modes have different amplitudes, whereas the zero-order approximation in the case of a point source, yields the same amplitudes for ($\pm m$). Note also that the location of the source relative to the axis of rotation has a strong effect on the amplitudes of the split modes as given by the finiteness factor.

Because the finiteness factor depends upon m, we cannot use now the addition theorem, Eq. (6.127), to sum over m. If necessary, the summation over m must be performed numerically.

We have assumed that the dislocation is constant over the fault. However, if the dislocation varies along the fault and is of the type $U_0(\xi, t) = g(t)s(\xi)$, then the results in Section 4.7.3 and Table 4.8 can be used to calculate the finiteness factor. For example, if $s(\xi) = U \sin(\pi\xi/L)$, the factor $(\sin X/X)\exp(-iX)$ is replaced by

$$\frac{2}{\pi}\left(\frac{\cos X}{1 - 4X^2/\pi^2}\right)e^{-iX}.$$

From Eq. (6.264), it may be noted that

$$X_m = \frac{\tau}{2}\left(\omega + \frac{Lm \sin \hat{\phi}}{\tau r_0 \sin \theta_0}\right), \qquad \tau = \frac{L}{V_f}.$$

Therefore, keeping τ fixed, we find

$$\lim_{L \to 0}\left(\frac{\sin X_m}{X_m} e^{-iX_m}\right) = \frac{\sin(\omega\tau/2)}{\omega\tau/2} e^{-i\omega\tau/2}.$$

Therefore, the effect of a finite source with constant dislocation and unit-step time dependence is similar to the effect of a point source with a Ramp time function. The time of rupture plays the role of the rise time. Similarly,

$$\lim_{L \to 0}\left[\frac{2}{\pi}\left(\frac{\cos X_m}{1 - 4X_m^2/\pi^2}\right)e^{-iX_m}\right] = \frac{2}{\pi}\frac{\cos(\omega\tau/2)}{1 - (\omega\tau/\pi)^2} e^{-i\omega\tau/2}.$$

Comparing this equation with the ones in Table 6.4, we see that the effect of a finite source with a variable dislocation $s(\xi) = (\pi/2)\bar{U} \sin(\pi\xi/L)$ and unit-step time variation is similar to the effect of a point source with a half-sine time function. This shows that it may not always be possible to distinguish between the effect of the finiteness of the source and that of the source's time function.

6.6.3. Effect on the Energy

In the derivation of the expressions for the energy associated with various modes of oscillations, it was assumed in Section 6.5 that

$$\mathbf{u}(\mathbf{r}, t) = {}_nW_l(r)\sqrt{\{l(l+1)\}}\,\mathbf{C}_{ml}(\theta, \phi)\cos \omega t,$$

for toroidal oscillations and

$$\mathbf{u}(\mathbf{r}, t) = [{}_nU_l(r)\mathbf{P}_{ml}(\theta, \phi) + {}_nV_l(r)\sqrt{\{l(l+1)\}}\,\mathbf{B}_{ml}(\theta, \phi)]\cos \omega t,$$

for spheroidal oscillations. The radial functions ${}_nU_l(r)$, etc., obtained in Section 6.3 must be modified to take into account the finiteness of the source and its time function. From Eqs. (6.106) and (6.120), we note that

$${}_nW_l(r) = \frac{1}{l(l+1)\Omega_{ml}}[P_0 S^T(\lambda, \delta, \hat{\phi}; r_0, \theta_0, \phi_0)]\left[\frac{\sin X_m}{X_m}\right][\chi({}_n\omega_l)]\left[\frac{1}{I_n^T}\right] y_1(r), \tag{6.267}$$

where the expressions in the first three pairs of brackets are the effects of the source's location and orientation, the source's finiteness, and the source's time function, and the last pair of brackets contains the medium response. Furthermore, X_m is given by Eq. (6.264), and from Table 6.4,

$$\chi(\omega) = \begin{cases} -\omega^{-2} & \text{(step function)}, \\ -\omega^{-2}\left[\dfrac{\sin(\omega\tau/2)}{\omega\tau/2}\right] & \text{(Ramp function)}, \\ -\omega^{-2}\left[\dfrac{\cos(\omega\tau/2)}{1-(\omega\tau/\pi)^2}\right] & \text{(half-sine function)}. \end{cases} \tag{6.268}$$

In writing the above expressions for $\chi(\omega)$, we have ignored the *static and transient parts of* $\bar{g}(t)$ and considered only the *steady-state part*. The function S^T is given by

$$S^T = \mathbf{en} : \mathfrak{T}_0\{y_1(r_0)\sqrt{[l(l+1)]}\,\overset{*}{\mathbf{C}}_{ml}(\theta_0, \phi_0)\}. \tag{6.269}$$

Explicitly [see, e.g., Example 4.5 or Eq. (6.124)],

$$S^T = \left[\frac{\mu(r_0)y_1(r_0)}{r_0}\left\{p_1\left(2\frac{\partial^2}{\partial\theta_0^2} + l^2 + l\right) - p_2\frac{2}{\sin\theta_0}\left(\frac{\partial}{\partial\theta_0} - \cot\theta_0\right)\frac{\partial}{\partial\phi_0}\right\}\right.$$
$$\left. + y_2(r_0)\left(p_3\frac{\partial}{\partial\theta_0} - p_4\frac{1}{\sin\theta_0}\frac{\partial}{\partial\phi_0}\right)\right]\overset{*}{Y}_{ml}(\theta_0, \phi_0), \tag{6.270}$$

where the p_is are given by Eqs. (6.125). Finally,

$$I_n^T = \int_0^a \rho_0 r^2 y_1^2\, dr, \qquad P_0 = U_0 \times \text{source area}.$$

With the value of ${}_nW_l$ given in Eq. (6.267), the expression in Eq. (6.231) for the kinetic energy becomes

$$K = \frac{1}{16\pi}\cdot\frac{2l+1}{l(l+1)}\cdot\frac{(l-m)!}{(l+m)!}[P_0 S^T]^2\left[\frac{\sin X_m}{X_m}\right]^2[{}_n\omega_l\,\chi({}_n\omega_l)]^2\left[\frac{1}{I_n^T}\right]. \tag{6.271}$$

In the case of the spheroidal oscillations, we have

$$\begin{bmatrix} {}_nU_l(r) \\ {}_nV_l(r) \end{bmatrix} = \frac{1}{\Omega_{ml}}[P_0 S^s(\lambda, \delta, \hat{\phi}; r_0, \theta_0, \phi_0)]\left[\frac{\sin X_m}{X_m}\right][\chi({}_n\omega_l)]\left[\frac{1}{I_n^s}\right]\begin{bmatrix} y_1(r) \\ y_3(r) \end{bmatrix}, \tag{6.272}$$

where

$$I_n^s = \int_0^a \rho_0 r^2[y_1^2 + l(l+1)y_3^2]dr,$$
$$S^s = \mathbf{en} : \mathfrak{T}_0\{y_1(r_0)\overset{*}{\mathbf{P}}_{ml}(\theta_0, \phi_0) + y_3(r_0)\sqrt{[l(l+1)]}\,\overset{*}{\mathbf{B}}_{ml}(\theta_0, \phi_0)\}. \tag{6.273}$$

The explicit expression for S^s has been calculated in Example 4.5 [see also Eq. (6.181)]. Therefore, Eq. (6.234) yields

$$K = \frac{2l+1}{16\pi}\cdot\frac{(l-m)!}{(l+m)!}[P_0 S^s]^2\left[\frac{\sin X_m}{X_m}\right]^2[{}_n\omega_l\,\chi({}_n\omega_l)]^2\left[\frac{1}{I_n^s}\right]. \tag{6.274}$$

When the dislocation is not constant along the fault, the factor sin X/X should be replaced by a suitable factor to be determined from Table 4.8.

6.7. Numerical Procedures

6.7.1. Calculation of the Eigenfrequencies

The calculation of the free periods of oscillation of realistic earth models necessitates the use of computers for numerical integration. In this section we shall treat the numerical techniques that are applied to this problem and also illustrate how group velocities can be simultaneously calculated with the solution of the eigenvalues, thus obviating the need for subsequent numerical differentiation.

We introduce the dimensionless variables and parameters as follows:

$$\bar{r} = \frac{r}{a}, \qquad \bar{\tau} = \frac{b}{a}, \qquad \bar{\lambda} = \frac{\lambda}{\mu_r}, \qquad \bar{\mu} = \frac{\mu}{\mu_r}, \qquad \bar{\rho}_0 = \frac{\rho_0}{\rho_r},$$
$$\bar{\omega} = \frac{\omega}{\omega_r}, \qquad \bar{g}_0 = \frac{g_0}{g_r}, \qquad \bar{G} = \frac{G}{G_r}, \tag{6.275}$$

where

$$a = \text{radius of the earth},$$
$$b = \text{radius of the core},$$
$$\mu_r, \rho_r = \text{reference values of rigidity and density, respectively},$$
$$\omega_r = \left(\frac{1}{a}\right)\beta_r,$$
$$\beta_r = \left(\frac{\mu_r}{\rho_r}\right)^{1/2},$$
$$g_r = \frac{\mu_r}{a\rho_r},$$
$$G_r = \frac{\mu_r}{a^2\rho_r^2}. \tag{6.276}$$

We also define

$$\hat{y}_2 = \frac{a}{\mu_r} y_2 = \bar{\mu}\left(\frac{d}{d\bar{r}} - \frac{1}{\bar{r}}\right) y_1, \tag{6.277}$$

for toroidal oscillations and

$$\hat{y}_2 = \frac{a}{\mu_r} y_2 = (\bar{\lambda} + 2\bar{\mu})\frac{dy_1}{d\bar{r}} + \frac{\bar{\lambda}}{\bar{r}}[2y_1 - l(l+1)y_3],$$
$$\hat{y}_4 = \frac{a}{\mu_r} y_4 = \bar{\mu}\left(\frac{dy_3}{d\bar{r}} + \frac{y_1 - y_3}{\bar{r}}\right), \tag{6.278}$$

for spheroidal oscillations.

Equation (6.67) for the toroidal oscillations may now be written as

$$\frac{dy_1(\bar{r})}{d\bar{r}} = \frac{y_1(\bar{r})}{\bar{r}} + \frac{\hat{y}_2(\bar{r})}{\bar{\mu}},$$
$$-\frac{d\hat{y}_2(\bar{r})}{d\bar{r}} = \left[\bar{\omega}^2\bar{\rho}_0 - \frac{\bar{\mu}(l-1)(l+2)}{\bar{r}^2}\right]y_1(\bar{r}) + \frac{3}{\bar{r}}\hat{y}_2(\bar{r}). \tag{6.279}$$

These two differential equations are to be solved under the boundary conditions $\hat{y}_2(1) = \hat{y}_2(\bar{\tau}) = 0$. Numerical integration, using either a Runge–Kutta method or the Adams–Moulton predictor–corrector technique, is started at the core boundary with $\hat{y}_2 = 0$ at a trial frequency $\bar{\omega}$ and carried to the free surface. Newton's method is then used to find the necessary correction to the trial frequency so that $\hat{y}_2 = 0$ at the free surface. Once the correct eigenfrequency is found, the solution is normalized so that $y_1 = 1$ at the free surface.

Using relation (6.256), the phase and group velocities can be calculated from

$$\bar{c}_L = \bar{\omega}\left(l + \frac{1}{2}\right)^{-1}, \qquad \bar{U}_L = \frac{1}{\bar{c}_L}\left(\frac{\bar{I}_2^T}{\bar{I}^T}\right), \tag{6.280}$$

where

$$\bar{c}_L = \frac{c_L}{\beta_r}, \qquad \bar{U}_L = \frac{U_L}{\beta_r},$$
$$\bar{I}^T = \int_{\bar{\tau}}^{1} \bar{\rho}_0\left[\frac{y_1(\bar{r})}{y_1(1)}\right]^2 \bar{r}^2\, d\bar{r}, \qquad \bar{I}_2^T = \int_{\bar{\tau}}^{1} \bar{\mu}\left[\frac{y_1(\bar{r})}{y_1(1)}\right]^2 d\bar{r}. \tag{6.281}$$

However, we can recast the above equations in differential form

$$\frac{d}{d\bar{r}}\bar{I}^T = \bar{\rho}_0\bar{r}^2\left[\frac{y_1(\bar{r})}{y_1(1)}\right]^2,$$
$$\frac{d}{d\bar{r}}\bar{I}_2^T = \bar{\mu}\left[\frac{y_1(\bar{r})}{y_1(1)}\right]^2, \tag{6.282}$$

and solve these equations simultaneously with the two differential equations (6.279). This method of numerical integration allows group velocity to be calculated with the same precision that is used in solving for the eigenfrequency.

Equations corresponding to Eqs. (6.279) and (6.280) for the spheroidal case are obtained directly from Eqs. (6.151)–(6.162) and (6.256). If we consider vibrations of sufficiently high order $l \geq 18$, such that the motion is confined to the mantle, it is possible to integrate numerically the set of six linear differential equations for the spheroidal case using a fourth-order Runge–Kutta method. The procedure is as follows. The quantities y_1, y_3, and y_5 go to zero at the origin and, for higher order oscillations, are essentially zero within the core. At a certain depth in the mantle, $y_1 = y_3 = y_5$ are set equal to zero and three arbitrary starting values are given to y_2, y_4, and y_6. Three linearly independent sets of three values each are needed for three linearly independent solutions, and a linear combination of the values obtained after integration to the surface must satisfy the free-surface boundary conditions. That is,

$$\begin{pmatrix} y_2^{(1)} & y_2^{(2)} & y_2^{(3)} \\ y_4^{(1)} & y_4^{(2)} & y_4^{(3)} \\ y_6^{(1)} + \dfrac{l+1}{a} y_5^{(1)} & y_6^{(2)} + \dfrac{l+1}{a} y_5^{(2)} & y_6^{(3)} + \dfrac{l+1}{a} y_5^{(3)} \end{pmatrix} \begin{pmatrix} A \\ B \\ C \end{pmatrix} = \begin{pmatrix} 0 \\ 0 \\ 0 \end{pmatrix}. \tag{6.283}$$

Nontrivial solutions will be obtained only if the determinant of the 3×3 matrix vanishes. The value of the determinant is calculated for a trial frequency and the frequency is incremented until the determinant changes sign. Interpolation is finally used to find the frequency at which the determinant equals zero. When the eigenfrequency has been found, the three linear solutions may be calculated and the undetermined multipliers A, B, and C evaluated. y_1 is customarily set to unity at the surface for purposes of normalization

$$\begin{pmatrix} y_1^{(1)} & y_1^{(2)} & y_1^{(3)} \\ y_2^{(1)} & y_2^{(2)} & y_2^{(3)} \\ y_4^{(1)} & y_4^{(2)} & y_4^{(3)} \end{pmatrix} \begin{pmatrix} A \\ B \\ C \end{pmatrix} = \begin{pmatrix} 1 \\ 0 \\ 0 \end{pmatrix}. \tag{6.284}$$

For lower order vibrations that have motion in the core, the boundary conditions at the core–mantle interface must be considered. y_1, y_2, y_5, and y_6 are continuous across the core–mantle boundary and $y_4 = 0$ within the core and just at the boundary within the solid. y_3, which represents transverse motion, is zero within the core but discontinuous across the core–mantle boundary because slippage can occur at this boundary.

Numerical integration begins in the core with y_1, y_3, and y_5 set to zero with the additional constraint that y_4 also equals zero. Arbitrary starting values are assigned to y_2 and y_6 and a third starting value is assigned to y_3 at the base of the mantle. Three sets of starting values are therefore required, and a 3×3 determinant must be solved at each frequency step. Once the correct frequency is found, the phase and group velocities may be numerically calculated using Eq. (6.256).

Table 6.5 gives the calculated eigenperiods for the Jeffreys–Bullen A′ model of the earth for the fundamental mode and the first two overtones. For the sake

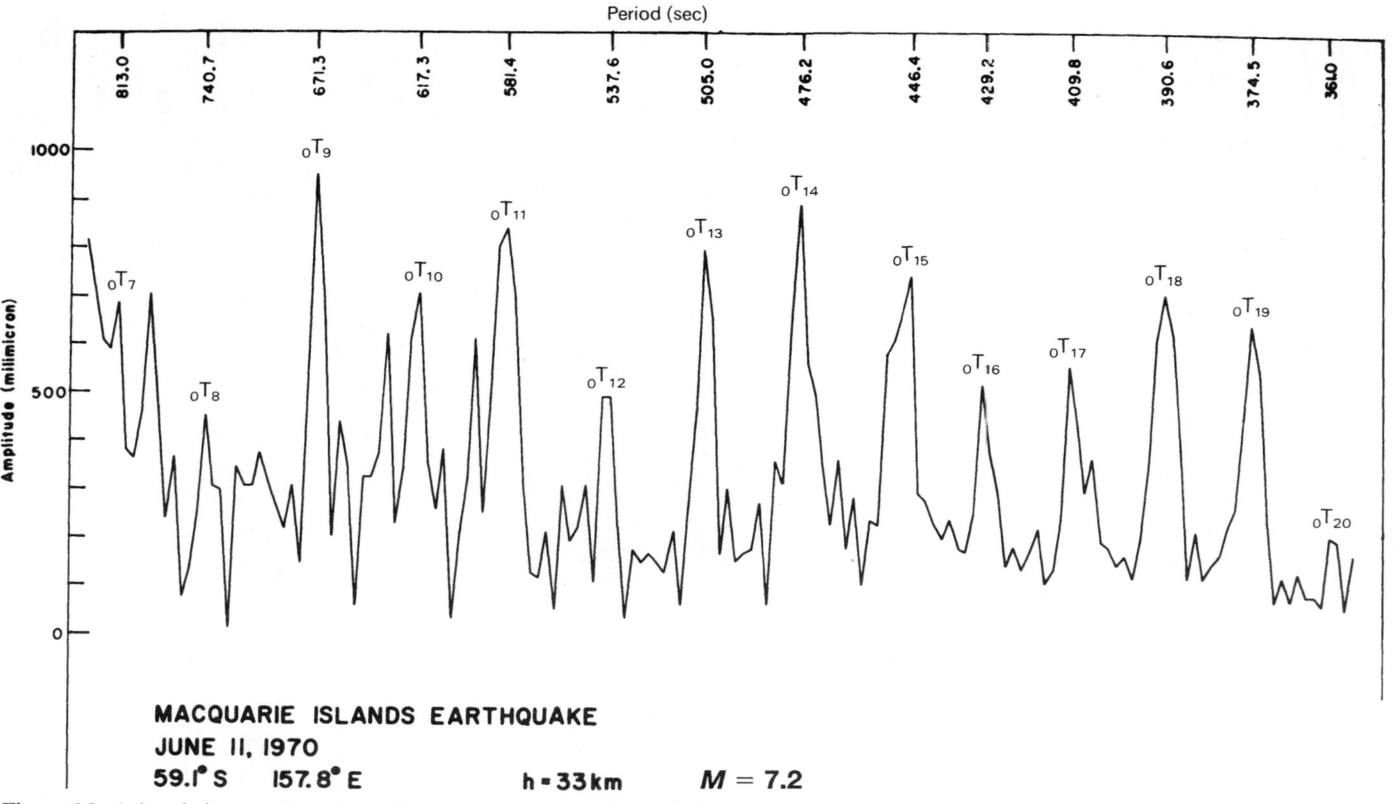

Figure 6.8. Azimuthal spectral trace amplitudes at Eilat. Average ground motion for $_0T_7$ is 2.5 μm, for $_0T_9$ 1.6 μm and for $_0T_{18}$ it is 0.7 μm. The periods, in seconds, for the fundamental toroidal oscillations of various orders are shown.

Table 6.5. Eigenperiods (s) for the Jeffreys–Bullen A' Model of the Earth[a]

	Toroidal			Spheroidal		
$l\backslash n$	0	1	2	0	1	2
1	[b]	805 (808)	452	[b]	2447 (2486)	1062
2	2610 (2579)	753	443	3206 (3232)	1458 (1469)	911
3	1690 (1707)	690	431	2116 (2134)	1056 (1064)	798
4	1293 (1306)	626	416	1531 (1546)	846	720
5	1067 (1076)	567	398	1179 (1189)	723	655
6	918	515	379	955	651	590
7	811	471	359	806	599	531
8	729	435	339	704	550	482
9	665	404	320	632	503	443
10	613	378	303	578	460	410
11	569	356	286	537	420	383
12	532	337	272	502	386	359
13	499	320	259	473	355	340
14	471	304	247	448	330	321
15	447	291	236	426	311	303

[a] Observed periods are shown in parentheses.
[b] Fundamental mode does not exist.

of comparison, observed periods are also included in this table. Figure 6.8 shows the amplitude spectrum for the Macquarie Island earthquake of 11 June, 1970. Figure 6.9 shows y_1 for $l = 2$ and $n = 0, 1, 2, 3$ for the toroidal oscillations of the Jeffreys–Bullen A′ model of the earth. Equations (6.248), (6.249), and (6.252)–(6.255) can be used to calculate the partial derivatives of the phase velocity with respect to the layer parameters. These partial derivatives are given in Tables 6.6–6.10 for the Gutenberg model for Love and Rayleigh waves.

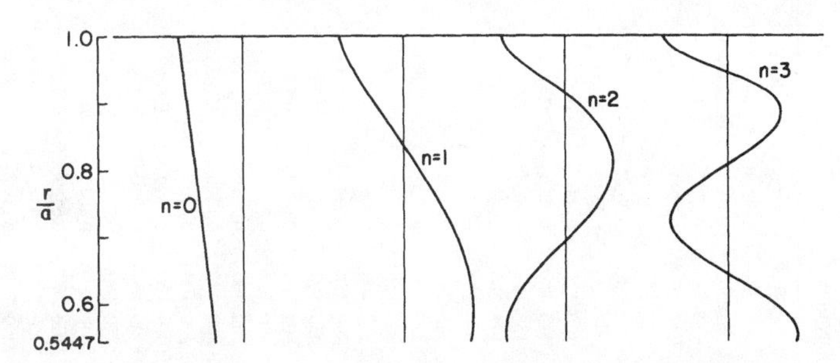

Figure 6.9. The eigenfunctions for the toroidal oscillations $l = 2$ of the Jeffreys–Bullen A′ model of the earth, corresponding to the fundamental mode and the first three overtones. The eigenfunctions are so normalized that each equals unity at the surface of the earth.

Table 6.6. Values of $(\partial c/\partial \beta)_\rho$ for Each Homogeneous Layer of the Gutenberg Model

	Love waves $(\partial c/\partial \beta)_\rho$											
Depth (km)	$l = 25$	$l = 30$	$l = 35$	$l = 40$	$l = 60$	$l = 80$	$l = 100$	$l = 120$	$l = 150$	$l = 200$	$l = 250$	$l = 500$
0–19	0.3528−1	0.4013−1	0.4456−1	0.4869−1	0.6399−1	0.7907−1	0.9578−1	0.1150	0.1500	0.2258	0.3150	0.6187
19–38	0.4098−1	0.4652−1	0.5153−1	0.5616−1	0.7276−1	0.8856−1	0.1051	0.1232	0.1540	0.2130	0.2699	0.3287
38–50	0.3431−1	0.3878−1	0.4276−1	0.4637−1	0.5863−1	0.6918−1	0.7905−1	0.8862−1	0.1024	0.1210	0.1283	0.6591−1
50–60	0.2831−1	0.3192−1	0.3511−1	0.3797−1	0.4743−1	0.5510−1	0.6180−1	0.6774−1	0.7513−1	0.8150−1	0.7789−1	0.2172−1
60–70	0.2783−1	0.3132−1	0.3437−1	0.3708−1	0.4582−1	0.5254−1	0.5800−1	0.6240−1	0.6692−1	0.6754−1	0.5902−1	0.9633−2
70–80	0.2729−1	0.3064−1	0.3354−1	0.3610−1	0.4412−1	0.4996−1	0.5434−1	0.5746−1	0.5972−1	0.5632−1	0.4518−1	0.4364−2
80–90	0.2680−1	0.3002−1	0.3277−1	0.3518−1	0.4253−1	0.4755−1	0.5098−1	0.5302−1	0.5348−1	0.4729−1	0.3494−1	0.2018−2
90–100	0.2631−1	0.2938−1	0.3199−1	0.3425−1	0.4092−1	0.4518−1	0.4775−1	0.4886−1	0.4789−1	0.3980−1	0.2717−1	0.9482−3
100–125	0.6407−1	0.7115−1	0.7705−1	0.8203−1	0.9588−1	0.1034	0.1066	0.1060	0.9892−1	0.7418−1	0.4450−1	0.7233−3
125–150	0.6203−1	0.6825−1	0.7323−1	0.7726−1	0.8711−1	0.9054−1	0.8965−1	0.8538−1	0.7400−1	0.4764−1	0.2358−1	0.1117−3
150–175	0.6005−1	0.6535−1	0.6937−1	0.7241−1	0.7829−1	0.7797−1	0.7385−1	0.6707−1	0.5371−1	0.2953−1	0.1204−1	0.1663−4
175–200	0.5802−1	0.6234−1	0.6535−1	0.6738−1	0.6942−1	0.6588−1	0.5935−1	0.5115−1	0.3761−1	0.1752−1	0.5855−2	0.2333−5
200–225	0.5569−1	0.5898−1	0.6096−1	0.6199−1	0.6048−1	0.5435−1	0.4632−1	0.3767−1	0.2524−1	0.9864−2	0.2684−2	0.3024−6
225–250	0.5308−1	0.5532−1	0.5628−1	0.5634−1	0.5172−1	0.4375−1	0.3506−1	0.2676−1	0.1622−1	0.5259−2	0.1155−2	0.3584−7
250–300	0.9757−1	0.9900−1	0.9810−1	0.9571−1	0.7945−1	0.6091−1	0.4424−1	0.3057−1	0.1589−1	0.3927−2	0.6427−3	0.4088−8
300–350	0.8511−1	0.8288−1	0.7886−1	0.7392−1	0.5245−1	0.3445−1	0.2142−1	0.1265−1	0.5159−2	0.8291−3	0.5030−3	
350–400	0.7186−1	0.6678−1	0.6068−1	0.5434−1	0.3223−1	0.1773−1	0.9237−2	0.4562−2	0.1417−2	0.1417−3	0.8797−5	
400–450	0.5911−1	0.5215−1	0.4502−1	0.3832−1	0.1861−1	0.8402−2	0.3593−2	0.1456−2	0.3346−3	0.1995−4	0.7168−6	
450–500	0.4778−1	0.3985−1	0.3254−1	0.2622−1	0.1024−1	0.3727−2	0.1285−2	0.4195−3	0.6961−4	0.2362−5	0.4511−7	
500–600	0.6749−1	0.5164−1	0.3874−1	0.2872−1	0.8131−2	0.2176−2	0.5594−3	0.1380−3	0.1544−4	0.2756−6		
600–700	0.3959−1	0.2662−1	0.1758−1	0.1148−1	0.1972−2	0.3229−3	0.5120−4	0.7858−5	0.4383−6			
700–800	0.2210−1	0.1294−1	0.7449−2	0.4245−2	0.4258−3	0.4110−4	0.3875−5	0.3578−6	0.9516−8			
800–900	0.1190−1	0.6025−2	0.3003−2	0.1482−2	0.8427−4	0.4654−5	0.2523−6	0.1367−7	0.1673−9			
900–1000	0.6202−2	0.2706−2	0.1162−2	0.4946−3	0.1560−4	0.4824−6	0.1457−7	0.4516−9	0.0			
1000–1200	0.4679−2	0.1663−2	0.5852−3	0.2055−3	0.3221−5	0.5466−7						
1200–1400	0.1080−2	0.2784−3	0.7131−4	0.1829−4	0.8557−7	0.4727−9						
1400–1600	0.2215−3	0.4063−4	0.7432−5	0.1368−5	0.1794−8	0.0						
1600–1800	0.4047−4	0.5180−5	0.6641−6	0.8613−7	0.0	0.0						
1800–2000	0.6569−5	0.5751−6	0.5070−7	0.4549−8	0.0	0.0						
2000–2200	0.9449−6	0.5543−7	0.3295−8	0.2009−9								
2200–2400	0.1192−6	0.4586−8	0.1801−9	0.0								
2400–2600	0.1305−7	0.3218−9	0.0	0.0								
2600–2800	0.1228−8	0.0	0.0	0.0								
2800–2898	0.0	0.0	0.0	0.0								

Table 6.7. Values of $(\partial c/\partial \rho)_\beta$ for Love Wave Eigenvalues, Gutenberg Model

	Love waves $(\partial c/\partial \rho)_\beta$											
Depth (km)	$l = 25$	$l = 30$	$l = 35$	$l = 40$	$l = 60$	$l = 80$	$l = 100$	$l = 120$	$l = 150$	$l = 200$	$l = 250$	$l = 500$
0–19	−0.2653−1	−0.2699−1	−0.2740−1	−0.2784−1	−0.3006−1	−0.3308−1	−0.3685−1	−0.4135−1	−0.4932−1	−0.6442−1	−0.7736−1	−0.7085−1
19–38	−0.2227−1	−0.2212−1	−0.2194−1	−0.2180−1	−0.2165−1	−0.2192−1	−0.2229−1	−0.2253−1	−0.2222−1	−0.1837−1	−0.9571−2	0.2929−1
38–50	−0.5739−2	−0.4561−2	−0.3490−2	−0.2524−2	0.5859−3	0.3062−2	0.5343−2	0.7642−2	0.1134−1	0.1811−1	0.2411−1	0.2058−1
50–60	−0.4846−2	−0.3893−2	−0.3030−2	−0.2256−2	0.1917−3	0.2063−2	0.3697−2	0.5251−2	0.7560−2	0.1120−1	0.1358−1	0.6503−2
60–70	−0.5069−2	−0.4164−2	−0.3347−2	−0.2617−2	−0.3453−3	0.1333−2	0.2740−2	0.4016−2	0.5791−2	0.8231−2	0.9311−2	0.2728−2
70–80	−0.5341−2	−0.4489−2	−0.3723−2	−0.3041−2	−0.9486−3	0.5538−3	0.1770−2	0.2831−2	0.4227−2	0.5915−2	0.6330−2	0.1159−2
80–90	−0.5500−2	−0.4687−2	−0.3959−2	−0.3314−2	−0.1357−2	0.1407−4	0.1093−2	0.2003−2	0.3140−2	0.4350−2	0.4412−2	0.5058−3
90–100	−0.5624−2	−0.4846−2	−0.4151−2	−0.3538−2	−0.1700−2	−0.4369−3	0.5341−3	0.1332−2	0.2290−2	0.3203−2	0.3102−2	0.2251−3
100–125	−0.1354−1	−0.1163−1	−0.9936−2	−0.8457−2	−0.4101−2	−0.1216−2	0.9072−3	0.2557−2	0.4358−2	0.5617−2	0.4829−2	0.1660−3
125–150	−0.1195−1	−0.9999−2	−0.8309−2	−0.6856−2	−0.2722−2	−0.1610−3	0.1565−2	0.2755−2	0.3789−2	0.3873−2	0.2652−2	0.2550−4
150–175	−0.9738−2	−0.7725−2	−0.6023−2	−0.4592−2	−0.7181−3	0.1445−2	0.2704−2	0.3392−2	0.3677−2	0.2862−2	0.1522−2	0.3935−5
175–200	−0.7396−2	−0.5352−2	−0.3676−2	−0.2306−2	0.1150−2	0.2794−2	0.3519−2	0.3699−2	0.3350−2	0.2029−2	0.8407−3	0.5770−6
200–225	−0.4879−2	−0.2843−2	−0.1238−2	0.2294−4	0.2891−2	0.3905−2	0.4054−2	0.3754−2	0.2911−2	0.1385−2	0.4466−3	0.7986−7
225–250	−0.2557−2	−0.6016−3	0.8668−3	0.1962−2	0.4090−2	0.4435−2	0.4063−2	0.3403−2	0.2299−2	0.8690−3	0.2192−3	0.1011−7
250–300	0.1192−2	0.4604−2	0.6911−2	0.8423−2	0.1012−1	0.8941−2	0.7042−2	0.5157−2	0.2877−2	0.7891−3	0.1428−3	0.1306−8
300–350	0.7818−2	0.1019−1	0.1139−1	0.1184−1	0.1022−1	0.7288−2	0.4757−2	0.2910−2	0.1240−2	0.2133−3	0.2342−4	
350–400	0.1203−1	0.1310−1	0.1310−1	0.1251−1	0.8432−2	0.4902−2	0.2638−2	0.1334−2	0.4266−3	0.4471−4	0.2913−5	
400–450	0.1370−1	0.1349−1	0.1248−1	0.1113−1	0.5950−2	0.2801−2	0.1228−2	0.5061−3	0.1189−3	0.7344−5	0.2765−6	
450–500	0.1320−1	0.1201−1	0.1037−1	0.8686−2	0.3677−2	0.1387−2	0.4886−3	0.1619−3	0.2735−4	0.9653−6	0.2165−7	
500–600	0.2181−1	0.1785−1	0.1398−1	0.1068−1	0.3216−2	0.8841−3	0.2309−3	0.5764−4	0.6543−5	0.1276−6		
600–700	0.1500−1	0.1063−1	0.7268−2	0.4862−2	0.8779−3	0.1470−3	0.2360−4	0.3656−5	0.2064−6			
700–800	0.9199−2	0.5630−2	0.3339−2	0.1943−2	0.2037−3	0.2005−4	0.1914−5	0.1781−6	0.4784−8			
800–900	0.5227−2	0.2753−2	0.1408−2	0.7085−3	0.4195−4	0.2360−5	0.1302−6	0.7090−8	0.0			
900–1000	0.2839−2	0.1282−2	0.5637−3	0.2440−3	0.7984−5	0.2510−6	0.8299−8	0.2533−9	0.0			
1000–1200	0.2215−2	0.8100−3	0.2908−3	0.1036−3	0.1674−5	0.2881−7						
1200–1400	0.5390−3	0.1422−3	0.3701−4	0.9602−5	0.4607−7	0.2575−9						
1400–1600	0.1155−3	0.2161−4	0.4005−5	0.7440−6	0.9966−9	0.0						
1600–1800	0.2176−4	0.2831−5	0.3670−6	0.4797−7	0.0	0.0						
1800–2000	0.3606−5	0.3201−6	0.2848−7	0.2574−8	0.0	0.0						
2000–2200	0.5269−6	0.3127−7	0.1873−8	0.1149−9								
2200–2400	0.6748−7	0.2621−8	0.1036−9	0.0								
2400–2600	0.7510−8	0.1866−9	0.0	0.0								
2600–2800	0.7219−9	0.0	0.0	0.0								
2800–2900	0.0	0.0	0.0	0.0								

Table 6.8. Values of $(\partial c/\partial \alpha)_{\rho,\beta}$ for Rayleigh Wave Eigenvalues, Gutenberg Model

	Rayleigh waves $(\partial c/\partial \alpha)_{\rho,\beta}$												
Depth (km)	$l = 25$	$l = 30$	$l = 35$	$l = 40$	$l = 60$	$l = 80$	$l = 100$	$l = 120$	$l = 150$	$l = 165$	$l = 200$	$l = 250$	$l = 500$
0–19	0.8657−2	0.4153−2	0.1216−1	0.1396−1	0.2117−1	0.2817−1	0.3481−1	0.4103−1	0.4956−1	0.5351−1	0.6209−1	0.7327−1	0.1018
19–38	0.9678−2	0.1130−1	0.1289−1	0.1444−1	0.2011−1	0.2500−1	0.2922−1	0.3284−1	0.3727−1	0.3910−1	0.4248−1	0.4534−1	0.2665−1
38–50	0.6477−2	0.7259−2	0.7969−2	0.8604−2	0.1047−1	0.1152−1	0.1204−1	0.1220−1	0.1199−1	0.1176−1	0.1100−1	0.9568−2	0.2171−2
50–60	0.5100−2	0.5599−2	0.6018−2	0.6362−2	0.7128−2	0.7248−2	0.7011−2	0.6584−2	0.5788−2	0.5371−2	0.4424−2	0.3224−2	0.3125−3
60–70	0.4783−2	0.5159−2	0.5445−2	0.5652−2	0.5889−2	0.5578−2	0.5028−2	0.4401−2	0.3484−2	0.3067−2	0.2237−2	0.1373−2	0.5978−4
70–80	0.4495−2	0.4771−2	0.4950−2	0.5050−2	0.4910−2	0.4342−2	0.3656−2	0.2986−2	0.2128−2	0.1777−2	0.1143−2	0.5854−3	0.1084−4
80–90	0.4260−2	0.4452−2	0.4545−2	0.4560−2	0.4149−2	0.3435−2	0.2705−2	0.2064−2	0.1325−2	0.1048−2	0.5926−3	0.2503−3	0.1748−5
90–100	0.4090−2	0.4215−2	0.4240−2	0.4189−2	0.3585−2	0.2793−2	0.2070−2	0.1486−2	0.8678−3	0.6541−3	0.3288−3	0.1162−3	0.2894−6
100–125	0.9894−2	0.9961−2	0.9773−2	0.9415−2	0.7292−2	0.5163−2	0.3486−2	0.2284−2	0.1166−2	0.8229−3	0.3551−3	0.1013−3	0.9281−7
125–150	0.9570−2	0.9328−2	0.8843−2	0.8226−2	0.5541−2	0.3425−2	0.2026−2	0.1165−2	0.4901−3	0.3140−3	0.1083−3	0.2231−4	0.2833−8
150–175	0.9036−2	0.8540−2	0.7837−2	0.7052−2	0.4164−2	0.2269−2	0.1190−2	0.6102−3	0.2183−3	0.1295−3	0.3778−4	0.6286−5	0.2883−9
175–200	0.8431−2	0.7737−2	0.6883−2	0.6001−2	0.3129−2	0.1516−2	0.7128−3	0.3303−3	0.1031−3	0.5760−4	0.1485−4	0.2144−5	0.0
200–225	0.7842−2	0.6992−2	0.6036−2	0.4105−2	0.2359−2	0.1021−2	0.4329−3	0.1826−3	0.5048−4	0.2674−4	0.6202−5	0.7880−6	0.0
225–250	0.7330−2	0.6365−2	0.5343−2	0.4393−2	0.1820−2	0.7132−3	0.2769−3	0.1084−3	0.2744−4	0.1405−4	0.3076−5	0.3697−6	0.0
250–300	0.1324−1	0.1104−1	0.8899−2	0.7019−2	0.2468−2	0.8271−3	0.2774−3	0.9476−4	0.1988−4	0.9329−5	0.1683−5	0.1567−6	0.0
300–350	0.1136−1	0.8974−2	0.6842−2	0.5106−2	0.1447−2	0.3967−3	0.1106−3	0.3185−4	0.5291−5	0.2228−5	0.3117−6	0.1990−7	
350–400	0.9549−2	0.7144−2	0.5155−2	0.3643−2	0.8371−3	0.1883−3	0.4350−4	0.1047−4	0.1340−5	0.5030−6	0.5174−7	0.2147−8	
400–450	0.7630−2	0.5370−2	0.3649−2	0.2432−2	0.4470−3	0.8152−4	0.1539−4	0.3040−5	0.2895−6	0.1045−6	0.7822−8	0.2681−9	
450–500	0.5754−2	0.3789−2	0.2414−2	0.1512−2	0.2211−3	0.3294−4	0.5174−5	0.8586−6	0.6344−7	0.3985−7	0.3083−8	0.2553−9	
500–600	0.7058−2	0.4139−2	0.2349−2	0.1310−2	0.1221−3	0.1195−4	0.1295−5	0.1587−6	0.8720−8	0.3373−7	0.2444−8		
600–700	0.3637−2	0.1840−2	0.9026−3	0.4364−3	0.2337−4	0.1351−5	0.8904−7	0.6811−8	0.1928−9				
700–800	0.1643−2	0.7068−3	0.2962−3	0.1228−3	0.3651−5	0.1207−6	0.4787−8	0.2196−9	0.0				
800–900	0.6556−3	0.2347−3	0.8221−4	0.2863−4	0.4417−6	0.7980−8	0.2740−9	0.0	0.0				
900–1000	0.2912−3	0.8836−4	0.2629−4	0.7785−5	0.6321−7	0.6050−9	0.2685−9	9.0	0.0				
1000–1200	0.1672−3	0.3911−4	0.9048−5	0.2107−5	0.7477−8	0.0							
1200–1400	0.3042−4	0.4966−5	0.8037−6	0.1316−6	0.1243−9	0.0							
1400–1600	0.4910−5	0.5506−6	0.6153−7	0.7007−8	0.0	0.0							
1600–1800	0.7176−6	0.5446−7	0.4132−8	0.3214−9	0.0	0.0							
1800–2000	0.8710−7	0.4369−8	0.2209−9	0.0	0.0	0.0							
2000–2200	0.1040−7	0.3362−9	0.0	0.0									
2200–2400	0.1183−8	0.0	0.0	0.0									
2400–2600	0.1457−9	0.0	0.0	0.0									
2600–2800	0.1078−9	0.0	0.0	0.0									
2800–2898	0.2089−9	0.0	0.0	0.0									

Table 6.9. Values of $(\partial c/\partial \beta)_{\rho,\alpha}$ for Rayleigh Wave Eigenvalues, Gutenberg Model

	Rayleigh waves $(\partial c/\partial \beta)_{\rho,\alpha}$												
Depth (km)	$l = 25$	$l = 30$	$l = 35$	$l = 40$	$l = 60$	$l = 80$	$l = 100$	$l = 120$	$l = 150$	$l = 165$	$l = 200$	$l = 250$	$l = 500$
0–19	0.9525−2	0.1149−1	0.1353−1	0.1558−1	0.2329−1	0.2992−1	0.3541−1	0.3984−1	0.4483−1	0.4677−1	0.5060−1	0.5687−1	0.1833
19–38	0.6130−2	0.6740−2	0.7339−2	0.7893−2	0.9600−2	0.1109−1	0.1325−1	0.1672−1	0.2522−1	0.3120−1	0.5023−1	0.9065−1	0.3491
38–50	0.4654−2	0.4808−2	0.4939−2	0.5035−2	0.5261−2	0.5978−2	0.7933−2	0.1144−1	0.1966−1	0.2498−1	0.3989−1	0.6462−1	0.9514−1
50–60	0.2619−2	0.2491−2	0.2413−2	0.2387−2	0.3051−2	0.5510−2	0.1006−1	0.1654−1	0.2893−1	0.3586−1	0.5255−1	0.7355−1	0.4817−1
60–70	0.1770−2	0.1553−2	0.1476−2	0.1546−2	0.3485−2	0.8284−2	0.1564−1	0.2479−1	0.4003−1	0.4763−1	0.6352−1	0.7801−1	0.2585−1
70–80	0.1112−2	0.9191−3	0.9725−3	0.1275−2	0.4973−2	0.1223−1	0.2193−1	0.3274−1	0.4856−1	0.5551−1	0.6773−1	0.7359−1	0.1273−1
80–90	0.6168−3	0.5420−3	0.8255−3	0.1457−2	0.7124−2	0.1657−1	0.2783−1	0.3916−1	0.5358−1	0.5899−1	0.6620−1	0.6408−1	0.5925−2
90–100	0.2398−3	0.3691−3	0.9681−3	0.2005−2	0.9725−2	0.2098−1	0.3304−1	0.4392−1	0.5569−1	0.5916−1	0.6132−1	0.5311−1	0.2668−2
100–125	−0.4087−3	0.1200−2	0.4381−2	0.8909−2	0.3600−1	0.6817−1	0.9673−1	0.1174	0.1317	0.1322	0.1206	0.8772−1	0.1797−2
125–150	−0.1341−3	0.3742−2	0.9501−2	0.1665−1	0.5115−1	0.8237−1	0.1024	0.1104	0.1051	0.9762−1	0.7470−1	0.4235−1	0.2073−3
150–175	0.2005−2	0.8225−2	0.1628−1	0.2540−1	0.6202−1	0.8640−1	0.9504−1	0.9165−1	0.7474−1	0.6442−1	0.4164−1	0.1858−1	0.2235−4
175–200	0.5280−2	0.1366−1	0.2350−1	0.3377−1	0.6815−1	0.8301−1	0.8125−1	0.7035−1	0.4931−1	0.3952−1	0.2165−1	0.7640−2	0.2279−5
200–225	0.9153−2	0.1932−1	0.3026−1	0.4085−1	0.6966−1	0.7466−1	0.6520−1	0.5076−1	0.3063−1	0.2284−1	0.1062−1	0.2964−2	0.2189−6
225–250	0.1338−1	0.2488−1	0.3628−1	0.4643−1	0.6778−1	0.6411−1	0.5000−1	0.3502−1	0.1818−1	0.1261−1	0.4972−2	0.1096−2	0.1975−7
250–300	0.3919−1	0.6392−1	0.8545−1	0.1020	0.1197	0.9453−1	0.6268−1	0.3775−1	0.1587−1	0.9962−2	0.3136−2	0.5072−3	0.1741−8
300–350	0.5274−1	0.7598−1	0.9236−1	0.1015	0.9016−1	0.5576−1	0.2945−1	0.1429−1	0.4408−2	0.2381−2	0.5316−3	0.5268−4	
350–400	0.6152−1	0.7994−1	0.8913−1	0.9060−1	0.6104−1	0.2944−1	0.1230−1	0.4768−2	0.1065−2	0.4913−3	0.7651−4	0.4534−5	
400–450	0.6483−1	0.7671−1	0.7875−1	0.7415−1	0.3781−1	0.1412−1	0.4627−2	0.1420−2	0.2261−3	0.8842−4	0.9405−5	0.3230−6	
450–500	0.6416−1	0.6941−1	0.6571−1	0.5730−1	0.2202−1	0.6316−2	0.1609−2	0.3869−3	0.4321−4	0.1458−4	0.1080−5	0.1876−7	
500–600	0.1173	0.1114	0.9365−1	0.7301−1	0.1864−1	0.3704−2	0.6741−3	0.1190−3	0.8740−5	0.2661−5	0.1700−6		
600–700	0.9373−1	0.7489−1	0.5345−1	0.3561−1	0.5021−2	0.5707−3	0.6097−4	0.6464−5	0.2297−6				
700–800	0.6827−1	0.4584−1	0.2770−1	0.1570−1	0.1196−2	0.7595−4	0.4647−5	0.2892−6	0.4818−8				
800–900	0.4693−1	0.2645−1	0.1348−1	0.6475−2	0.2614−3	0.9062−5	0.3100−6	0.1110−7	0.0				
900–1000	0.3130−1	0.1477−1	0.6338−2	0.2569−2	0.5399−4	0.1003−5	0.1833−7	0.3810−9	0.0				
1000–1200	0.3138−1	0.1158−1	0.3929−2	0.1274−2	0.1215−4	0.1177−6							
1200–1400	0.1025−1	0.2635−2	0.6265−3	0.1434−3	0.3663−6	0.1074−8							
1400–1600	0.2803−2	0.4951−3	0.8141−4	0.1298−4	0.8472−8	0.0							
1600–1800	0.6592−3	0.7883−4	0.8833−5	0.9668−6	0.1541−9	0.0							
1800–2000	0.1341−3	0.1068−4	0.8019−6	0.5934−7	0.0	0.0							
2000–2200	0.2382−4	0.1241−5	0.6140−7	0.3022−8									
2200–2400	0.3653−5	0.1220−6	0.3908−8	0.1257−9									
2400–2600	0.4800−6	0.1007−7	0.2046−9	0.0									
2600–2800	0.5395−7	0.6967−9	0.0	0.0									
2800–2898	0.3914−8	0.0	0.0	0.0									

Table 6.10. Values of $(\partial c/\partial \rho)_{\alpha,\beta}$ for Rayleigh Wave Eigenvalues, Gutenberg Model

	Rayleigh waves $(\partial c/\partial \rho)_{\alpha,\beta}$												
Depth (km)	$l = 25$	$l = 30$	$l = 35$	$l = 40$	$l = 60$	$l = 80$	$l = 100$	$l = 120$	$l = 150$	$l = 165$	$l = 200$	$l = 250$	$l = 500$
0–19	−0.2625−1	−0.2784−1	−0.2863−1	−0.2903−1	−0.2977−1	−0.3128−1	−0.3406−1	−0.3813−1	−0.4666−1	−0.5200−1	−0.6718−1	−0.9449−1	−0.1913
19–38	−0.2551−1	−0.2753−1	−0.2889−1	−0.2992−1	−0.3357−1	−0.3761−1	−0.4173−1	−0.4543−1	−0.4954−1	−0.5074−1	−0.5085−1	−0.4307−1	0.8432−1
38–50	−0.1302−1	−0.1433−1	−0.1531−1	−0.1615−1	−0.1955−1	−0.2316−1	−0.2632−1	−0.2852−1	−0.2957−1	−0.2002−1	−0.2500−1	−0.1334−1	0.3816−1
50–60	−0.1167−1	−0.1302−1	−0.1407−1	−0.1497−1	−0.1802−1	−0.2023−1	−0.2115−1	−0.2064−1	−0.1737−1	−0.1480−1	−0.7040−2	0.6144−2	0.2115−1
60–70	−0.1228−1	−0.1375−1	−0.1488−1	−0.1580−1	−0.1831−1	−0.1910−1	−0.1809−1	−0.1547−1	−0.9378−2	−0.5735−2	0.3285−2	0.1495−1	0.1161−1
70–80	−0.1273−1	−0.1424−1	−0.1533−1	−0.1615−1	−0.1767−1	−0.1681−1	−0.1396−1	−0.9718−2	−0.2109−2	0.1789−2	0.1010−1	0.1822−1	0.5725−2
80–90	−0.1302−1	−0.1447−1	−0.1545−1	−0.1610−1	−0.1638−1	−0.1390−1	−0.9592−2	−0.4376−2	0.3546−2	0.7102−2	0.1360−1	0.1791−1	0.2648−2
90–100	−0.1317−1	−0.1450−1	−0.1529−1	−0.1570−1	−0.1461−1	−0.1067−1	−0.5321−2	0.3007−3	0.7676−2	0.1055−1	0.1489−1	0.1595−1	0.1182−2
100–125	−0.3264−1	−0.3514−1	−0.3606−1	−0.3580−1	−0.2736−1	−0.1303−1	0.2048−2	0.1499−1	0.2807−1	0.3173−1	0.3422−1	0.2845−1	0.7932−3
125–150	−0.3092−1	−0.3190−1	−0.3107−1	−0.2896−1	−0.1402−1	0.2913−2	0.1631−1	0.2470−1	0.2907−1	0.2863−1	0.2396−1	0.1472−1	0.9155−4
150–175	−0.2809−1	−0.2737−1	−0.2482−1	−0.2109−1	−0.2312−1	0.1353−1	0.2265−1	0.2581−1	0.2372−1	0.2117−1	0.1447−1	0.6810−2	0.9970−5
175–200	−0.2460−1	−0.2220−1	−0.1812−1	−0.1317−1	0.6813−2	0.1929−1	0.2356−1	0.2258−1	0.1708−1	0.1401−1	0.7961−2	0.2916−2	0.1028−5
200–225	−0.2062−1	−0.1666−1	−0.1136−1	−0.5613−2	0.1344−1	0.2155−1	0.2160−1	0.1798−1	0.1142−1	0.8654−2	0.4126−2	0.1184−2	0.1009−6
225–250	−0.1630−1	−0.1098−1	−0.4820−2	0.1252−2	0.1774−1	0.2136−1	0.1826−1	0.1341−1	0.7216−2	0.5061−2	0.2031−2	0.4569−3	0.9315−8
250–300	−0.1944−1	−0.6009−2	0.7227−2	0.1861−1	0.4007−1	0.3621−1	0.2537−1	0.1571−1	0.6748−2	0.4260−2	0.1354−2	0.2217−3	0.8594−9
300–350	−0.3451−2	0.1130−1	0.2328−1	0.3160−1	0.3674−1	0.2447−1	0.1333−1	0.6570−2	0.2050−2	0.1111−2	0.2491−3	0.2487−4	
350–400	0.9783−2	0.2339−1	0.3212−1	0.3627−1	0.2850−1	0.1439−1	0.6127−2	0.2398−2	0.5389−3	0.2491−3	0.3885−4	0.2313−5	
400–450	0.1839−1	0.2907−1	0.3378−1	0.3404−1	0.1929−1	0.7441−2	0.2472−3	0.7632−3	0.1220−3	0.4777−4	0.5085−5	0.1760−6	
450–500	0.2231−1	0.2936−1	0.3047−1	0.2800−1	0.1171−1	0.3450−2	0.8889−3	0.2149−3	0.2408−4	0.8170−5	0.6061−6	0.1135−7	
500–600	0.4714−1	0.5118−1	0.4602−1	0.3733−1	0.1018−1	0.2064−2	0.3789−3	0.6717−4	0.4945−5	0.1571−5	0.9999−7		
600–700	0.4308−1	0.3773−1	0.2829−1	0.1941−1	0.2880−2	0.3328−3	0.3579−4	0.3807−5	0.1356−6				
700–800	0.3346−1	0.2418−1	0.1522−1	0.8853−2	0.7055−3	0.4546−4	0.2800−5	0.1747−6	0.2917−8				
800–900	0.2381−1	0.1428−1	0.7547−2	0.3709−2	0.1560−3	0.5488−5	0.1896−6	0.6800−8	0.0				
900–1000	0.1652−1	0.8197−2	0.3624−2	0.1498−2	0.3263−4	0.6138−6	0.1194−7	0.2444−9	0.0				
1000–1200	0.1703−1	0.6534−2	0.2271−2	0.7481−3	0.7352−5	0.7199−7							
1200–1400	0.5796−2	0.1532−2	0.3709−3	0.8592−4	0.2248−6	0.6650−9							
1400–1600	0.1632−2	0.2945−3	0.4913−4	0.7910−5	0.5264−8	0.0							
1600–1800	0.3904−3	0.4751−4	0.5386−5	0.5943−6	0.0	0.0							
1800–2000	0.8009−4	0.6472−5	0.4909−6	0.3658−7	0.0	0.0							
2000–2200	0.1431−4	0.7545−6	0.3765−7	0.1864−8									
2200–2400	0.2208−5	0.7448−7	0.2401−8	0.0									
2400–2600	0.2923−6	0.6182−8	0.1263−9	0.0									
2600–2800	0.3328−7	0.4302−9	0.0	0.0									
2800–2898	0.2688−8	0.0	0.0	0.0									

Having calculated these derivatives, we can calculate the change of c, say Δc, corresponding to arbitrary distribution of $\Delta\rho$, $\Delta\alpha$, and $\Delta\beta$ by the formula:

$$\Delta c = \int_0^a \left(\frac{\partial c}{\partial \rho} \Delta\rho + \frac{\partial c}{\partial \alpha} \Delta\alpha + \frac{\partial c}{\partial \beta} \Delta\beta \right) dr$$

6.7.2. Calculation of Spectral Line Amplitudes

We will first express the displacement components in terms of the dimensionless variables and parameters introduced in Section 6.7.1. In this connection, we further define

$$\bar{t} = \frac{r_0}{a}, \qquad \bar{\mu}_0 = \frac{\mu(r_0)}{\mu_r}, \qquad \bar{\lambda}_0 = \frac{\lambda(r_0)}{\lambda_r},$$

$$\bar{I}^T = \int_{\bar{t}}^1 \bar{\rho}_0 \bar{r}^2 \left[\frac{y_1(\bar{r})}{y_1(1)} \right]^2 d\bar{r},$$

$$G_1(\bar{t}) = \frac{\bar{\mu}_0}{\bar{t}} \frac{2l+1}{l(l+1)} \frac{1}{\bar{I}^T \bar{\omega}^2} \left[\frac{y_1(\bar{t})}{y_1(1)} \right], \tag{6.285}$$

$$G_2(\bar{t}) = \frac{2l+1}{l(l+1)} \frac{1}{\bar{I}^T \bar{\omega}^2} \left[\frac{\hat{y}_2(\bar{t})}{y_1(1)} \right];$$

$$\hat{y}_8(\bar{t}) = \frac{2\bar{\mu}_0 y_3(\bar{t})}{\bar{t}},$$

$$\hat{y}_9(\bar{t}) = \frac{2\bar{\mu}_0}{\bar{t}} \left(\frac{3\bar{\lambda}_0 + 2\bar{\mu}_0}{\bar{\lambda}_0 + 2\bar{\mu}_0} \right) [l(l+1) y_3(\bar{t}) - 2 y_1(\bar{t})] + \frac{4\bar{\mu}_0}{(\bar{\lambda}_0 + 2\bar{\mu}_0)} \hat{y}_2(\bar{t}),$$

$$F_1 = \frac{1}{2} \frac{2l+1}{\bar{\omega}^2 \bar{I}^s} \frac{\hat{y}_8(\bar{t})}{y_1(1)}, \qquad F_2 = F_1 \frac{y_3(\bar{t})}{y_1(1)},$$

$$F_3 = \frac{2l+1}{\bar{\omega}^2 \bar{I}^s} \frac{\hat{y}_4(\bar{t})}{y_1(1)}, \qquad F_4 = F_3 \frac{y_3(\bar{t})}{y_1(1)}, \tag{6.286}$$

$$F_5 = -\frac{1}{4} \frac{2l+1}{\bar{\omega}^2 \bar{I}^s} \frac{\hat{y}_9(\bar{t})}{y_1(1)}, \qquad F_6 = F_5 \frac{y_3(\bar{t})}{y_1(1)},$$

$$\bar{I}^s = \int_0^1 \bar{\rho}_0 \bar{r}^2 \left\{ \left[\frac{y_1(\bar{r})}{y_1(1)} \right]^2 + l(l+1) \left[\frac{y_3(\bar{r})}{y_1(1)} \right]^2 \right\} d\bar{r};$$

$$
\begin{aligned}
H_1 &= \frac{2P_l^2(\cos\theta)}{\sin\theta}, \qquad H_2 = \frac{\partial}{\partial\theta} P_l^2(\cos\theta),\\
H_3 &= \frac{P_l^1(\cos\theta)}{\sin\theta}, \qquad H_4 = \frac{\partial}{\partial\theta} P_l^1(\cos\theta),\\
H_5 &= \frac{2H_2 - H_1\cos\theta}{\sin\theta},\\
H_6 &= \frac{[8 - l(l+1)\sin^2\theta]H_1 - 4H_2\cos\theta}{2\sin\theta},\\
H_7 &= P_l^2(\cos\theta), \qquad H_8 = P_l^1(\cos\theta), \qquad H_9 = P_l(\cos\theta),\\
H_{10} &= \frac{[4 - l(l+1)\sin^2\theta]H_1 - 2H_2\cos\theta}{2\sin\theta} = \frac{\partial^2 P_l^2}{\partial\theta^2},\\
H_{11} &= \frac{[1 - l(l+1)\sin^2\theta]H_3 - H_4\cos\theta}{\sin\theta} = \frac{\partial^2 P_l^1}{\partial\theta^2},\\
H_{12} &= \frac{H_4\cos\theta - H_3}{\sin\theta}.
\end{aligned}
\tag{6.287}
$$

The displacement field induced by the toroidal oscillations excited by a shear dislocation with step-function time dependence and located at $r = r_0$, $\theta = 0$ is given by Eq. (6.135). In terms of the notation introduced above we have, at $r = a$,

$$
\begin{aligned}
u_\theta &= \Omega(G_1 H_1 p_2^0 - G_2 H_3 p_4^0),\\
u_\phi &= \Omega(G_1 H_2 p_1^0 - G_2 H_4 p_3^0),
\end{aligned}
\tag{6.288}
$$

where $\Omega = U_0\, dS/(4\pi a^2)$. The common time factor $\cos(\omega_n t)H(t)$ has been deleted for convenience. Summation over all permissible values of l and n is understood. The corresponding expressions for the strains are

$$
\begin{aligned}
\varepsilon_{\theta\theta} &= -\varepsilon_{\phi\phi} = \bar{\Omega}(G_1 H_5 p_2^0 - \tfrac{1}{2} G_2 H_1 p_4^0),\\
\varepsilon_{\theta\phi} &= \bar{\Omega}(G_1 H_6 p_1^0 + G_2 H_2 p_3^0),
\end{aligned}
\tag{6.289}
$$

where $\bar{\Omega} = \Omega/a$. The function G_1 corresponds to a vertical strike-slip source and G_2 to a vertical dip-slip source.

Similarly, starting from Eq. (6.193) for the spheroidal oscillations, we get the following expressions for the displacements and strains on the surface of the earth

$$
\begin{aligned}
u_r &= \Omega(F_1 H_7 p_2^0 - F_3 H_8 p_4^0 + 2F_5 H_9 p_5),\\
u_\theta &= \Omega(F_2 H_2 p_2^0 - F_4 H_4 p_4^0 - 2F_6 H_8 p_5),\\
u_\phi &= \Omega(F_2 H_1 p_1^0 - F_4 H_3 p_3^0),
\end{aligned}
$$

$$\varepsilon_{\theta\theta} = \bar{\Omega}[(F_1 H_7 + F_2 H_{10})p_2^0 - (F_3 H_8 + F_4 H_{11})p_4^0 + 2(F_5 H_9 - F_6 H_4)p_5],$$

$$\varepsilon_{\phi\phi} = \bar{\Omega}[\{F_1 H_7 + F_2(H_{10} - H_6)\}p_2^0 - (F_3 H_8 + F_4 H_{12})p_4^0 + 2\{F_5 H_9 - F_6(H_4 + H_7)\}p_5],$$

$$\varepsilon_{rr} = -\frac{\sigma}{1-\sigma}(\varepsilon_{\theta\theta} + \varepsilon_{\phi\phi}), \tag{6.290}$$

$$\varepsilon_{\theta\phi} = \bar{\Omega}(2F_2 H_5 p_1^0 + F_4 H_1 p_3^0).$$

The functions F_1 and F_2 correspond to the vertical and horizontal components of the displacement for a vertical strike-slip source and F_3 and F_4 correspond to the vertical and horizontal components of the displacement for a vertical dip-slip source.

In Eqs. (6.288)–(6.290), p_i^0 ($i = 1, 2, 3, 4$) and p_5 are functions of the slip angle (λ), the dip angle (δ), and the azimuth of the observer relative to the fault's strike (ϕ^0). The function p_5 is given in Eqs. (6.125) and p_i^0 can be obtained from the p_i of Eqs. (6.125) on replacing $\hat{\phi}$ by ϕ^0.

If the eigenfrequencies and eigenfunctions are known, the radial functions G_1 and G_2 for the toroidal case and F_i ($i = 1, 2, \ldots, 6$) for the spheroidal case can be computed without any difficulty.

In computing the associated Legendre polynomials, use is made of the following recursion formula for the Legendre polynomials

$$lP_l(\cos\theta) = (2l - 1)\cos\theta P_{l-1}(\cos\theta) - (l - 1)P_{l-2}(\cos\theta). \tag{6.291}$$

However, to control roundoff errors that may occur by the excessive use of this formula, its use is limited to 10 successive values of l. For the next two l values, the trigonometric series representation of the Legendre polynomials is used

$$P_l(\cos\theta) = \frac{(2l)!}{2^{2l}(l!)^2}\left[\cos l\theta + \frac{1}{1}\frac{l}{2l-1}\cos(l-2)\theta + \frac{1\cdot 3}{1\cdot 2}\frac{l(l-1)}{(2l-1)(2l-3)}\cos(l-4)\theta + \cdots + \cos(-l\theta)\right]. \tag{6.292}$$

Therefore, a proper computation of the Legendre polynomials for all values of l and θ is assured. To compute higher order polynomials and their derivatives the following relations are used:

$$P_l^1(\cos\theta) = \frac{-l}{\sin\theta}[\cos\theta P_l(\cos\theta) - P_{l-1}(\cos\theta)],$$

$$P_l^2(\cos\theta) = \left[-2l\frac{1+\cos 2\theta}{1-\cos 2\theta} - l(l+1)\right]P_l(\cos\theta) + 4l\frac{\cos\theta}{1-\cos 2\theta}P_{l-1}(\cos\theta),$$

$$\frac{\partial}{\partial\theta}P_l^n(\cos\theta) = -\frac{1}{\sin\theta}[(l+n)P_{l-1}^n(\cos\theta) - l\cos\theta P_l^n(\cos\theta)]. \tag{6.293}$$

These equations do not entail the compounding of roundoff errors because they are given directly in terms of previously computed Legendre polynomials. The colatitude functions H_1 to H_{12} can be calculated using the scheme for the evaluation of the Legendre functions and their derivatives.

The radial functions G_1, G_2, F_1 to F_6 for the fundamental mode are shown in Figs. 6.10 and 6.11 for the Jeffreys–Bullen A′ earth model. The effect of the source depth is demonstrated by calculating these functions for four values of h (550 km, 250 km, 100 km, and 30 km).

Given a structural model and a dislocation source of arbitrary orientation and depth, spheroidal and toroidal amplitudes are calculable for $l \leq 100$, $n \leq 4$ everywhere on the earth's surface. It has been shown that:

1. The combination of a realistic earth model with a dislocation source model produces theoretical displacements and strains that match the observations.
2. The theoretical calculations of the spectral line amplitudes can be used as an additional tool for the investigation of the earthquake source mechanism and the infrastructure of the interior of the earth.

The variation of the surface displacement and strain fields can be seen in Figs. 6.12–6.15. In our mapping of the surface of the earth on the plane of the paper, the entire sphere is mapped onto a tangent plane at the epicenter. Each point is moved in such a way that the epicentral distance and the azimuth are preserved. The angle ϕ^0 runs in a counterclockwise sense from the strike of the fault. The contour maps in these figures indicate the magnitude of the corresponding field component. Lines of equal magnitude are drawn as either solid

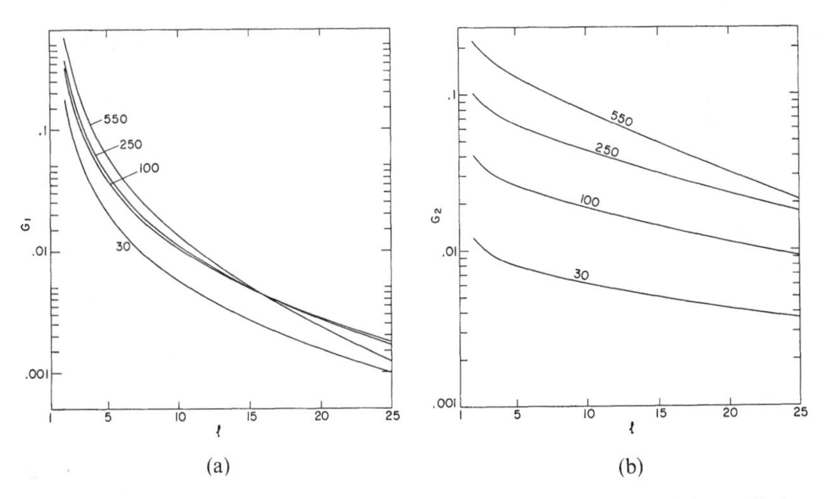

Figure 6.10. Dependence of the toroidal radial functions on the order of the oscillation for the Jeffreys–Bullen A′ model. (a) G_1 corresponds to the vertical strike-slip source and (b) G_2 to the vertical dip-slip source. The numbers labeling the curves give the source depths.

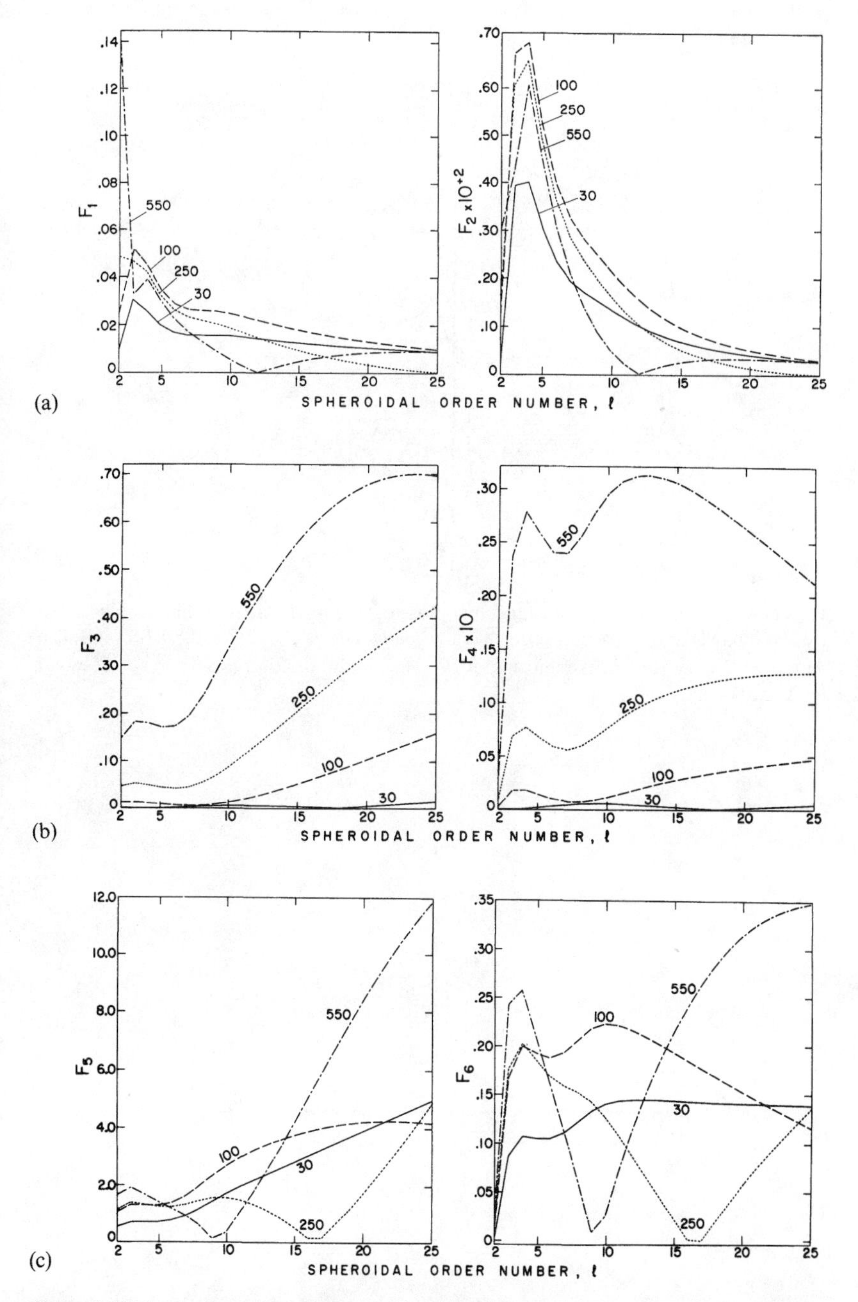

Figure 6.11. Dependence of the spheroidal radial functions on the order of the oscillation for the Jeffreys–Bullen A′ model. (a) F_1 and F_2 correspond to the vertical and horizontal components of the displacement for a vertical strike-slip source. (b) F_3 and F_4 are the corresponding functions for a vertical dip-slip source. (c) F_5 and F_6 are defined in Eq. (6.286). The numbers labeling the curves give the source depths.

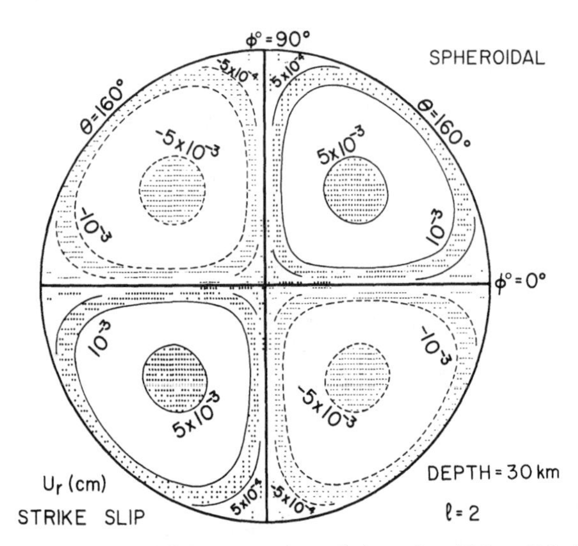

Figure 6.12. Azimuthal equidistant mapping of the spheroidal radial displacement caused by a vertical strike-slip source for the *football mode* ($l = 2$) for the Jeffreys–Bullen A′ model. Heavy solid lines indicate nodals. The amplitudes (in centimeters) correspond to $\Omega = 2000\ \mu$m.

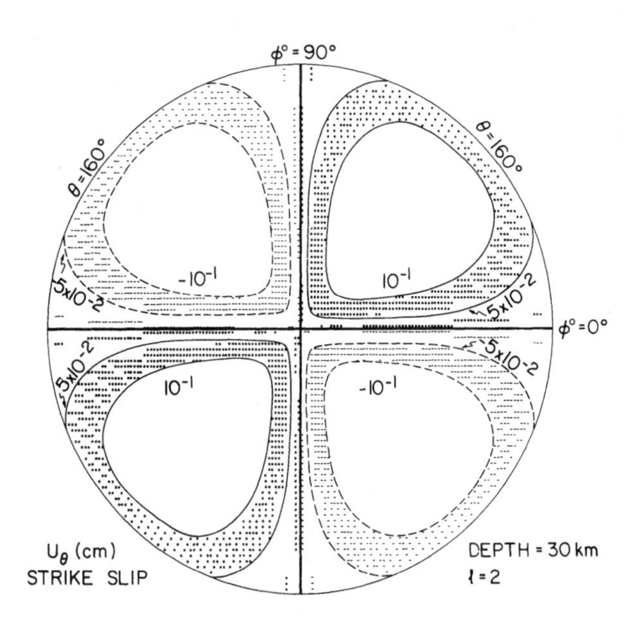

Figure 6.13. Same as Fig. 6.12 for the toroidal colatitudinal displacement.

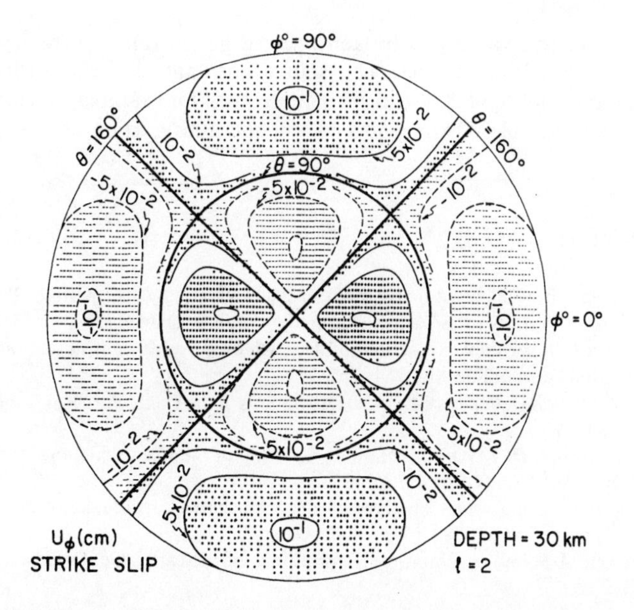

Figure 6.14. Same as Fig. 6.12 for the toroidal azimuthal displacement.

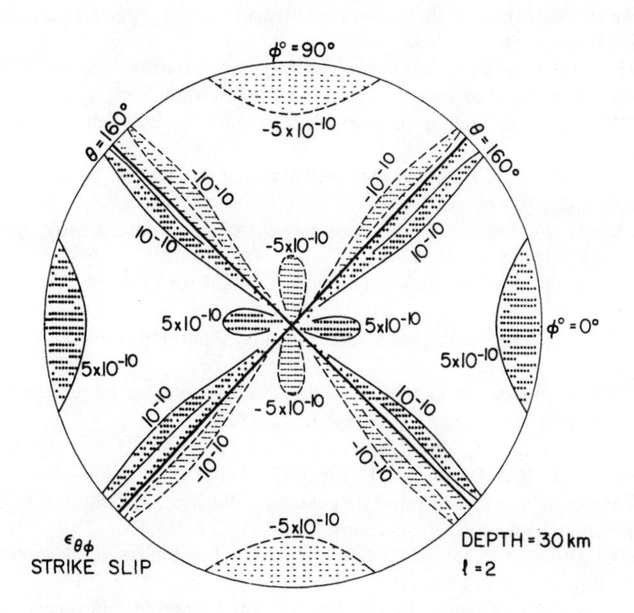

Figure 6.15. Same as Fig. 6.12 for the toroidal shearing strain.

thin lines (positive values) or broken lines (negative values). The nodals are exhibited as heavy solid lines. The source constant $\Omega = U_0\, dS/(4\pi a^2)$ was assigned a magnitude of 2000 μm. This represents, for example, an earthquake fault with a length of 600 km, width 80 km, and a dislocation of 20 m.

Bibliography

Alsop LE (1963) Free spheroidal vibrations of the earth at very long periods. Part I. Calculations of periods for several earth models. Bull Seismol Soc Amer 53: 483–502.

Alsop LE (1964) Spheroidal free periods of the earth observed at eight stations around the world. Bull Seismol Soc Amer 54: 755–776.

Alterman Z, Jarosch H, Pekeris CL (1959) Oscillations of the Earth. Proc Roy Soc (London) A252: 80–95.

Backus GE, Gilbert F (1961) The rotational splitting of the free oscillations of the earth. Proc Natl Acad Sci (USA) 47: 362–371.

Benioff H (1958) Long waves observed in the Kamchatka earthquake of Nov 4, 1952. Jour Geophys Res 63: 589–593.

Ben-Menahem A (1959) Free non-radial vibrations of the earth. Geofis Pura e Appl 43: 3–15.

Ben-Menahem A (1964) Spectral response of an elastic sphere to dipolar point-sources. Bull Seismol Soc Amer 54: 1323–1340.

Ben-Menahem A, Israel M, Levité U (1971) Theory and computation of amplitudes of terrestrial line spectra. Geophys Jour Roy Astron Soc (London) 25: 307–406.

Ben-Menahem A, Rosenman M, Israel M (1972) Source mechanism of the Alaskan earthquake of 1964 from amplitudes of free oscillations and surface waves. Phys Earth Planet Int 5: 1–29.

Bromwich TJ I'A (1898) On the influence of gravity on elastic waves and, in particular, on the vibrations of an elastic globe. Proc London Math Soc (Ser 1) 30: 98–120.

Derr JS (1969) Free oscillation observations through 1968. Bull Seismol Soc Amer 59: 2079–2099.

Gilbert F, MacDonald GJF (1960) Free oscillations of the earth. I. Toroidal oscillations. Jour Geophys Res 65: 675–693.

Hoskins LM (1920) The strain of a gravitating sphere of variable density and elasticity. Trans Amer Math Soc 21: 1–43.

Lamb H (1882) On the vibrations of an elastic sphere. Proc Lond Math Soc (Ser 1) 13: 189–212.

Love AEH (1911) Some Problems of Geodynamics. Cambridge University Press, Cambridge.

Luh PC (1974) Normal modes of a rotating, self-gravitating inhomogeneous earth. Geophys Jour Roy Astron Soc (London) 38: 187–224.

Madariaga R (1972) Toroidal free oscillations of the laterally heterogeneous earth. Geophys Jour Roy Astron Soc (London) 27: 81–100.

Ottelet I (1966) A variational principle governing the free oscillations of a gravitating elastic compressible planet. Astrophys Jour 143: 253–258.

Pekeris CL, Alterman Z, Jarosch H (1961) Rotational multiplets in the spectrum of the earth. Phys Rev 122: 1692–1700.

Rayleigh, Lord (1906) Scientific Papers, Vol 5, p 300. Cambridge University Press, Cambridge.

Saito M (1967) Excitation of free oscillations and surface waves by a point source in a vertically heterogeneous earth. Jour Geophys Res 72: 3689–3699.

Saito M (1971) Theory for the elastic-gravitational oscillation of a laterally heterogeneous earth. Jour Phys Earth 19: 259–270.

Satô Y, Usami T (1964) Propagation of spheroidal disturbances on an elastic sphere with a homogeneous mantle and core. Bull Earthquake Res Inst (Tokyo) 42: 407–425.

Satô Y, Usami T, Landisman M, Ewing M (1963) Basic study on the oscillation of a sphere. Part V. Propagation of torsional disturbances on a radially heterogeneous sphere. Case of a homogeneous mantle with a liquid core. Geophys Jour Roy Astron Soc. (London) 8: 44–63.

Singh SJ, Ben-Menahem A (1969a) Eigenvibrations of the earth excited by finite dislocations. I. Toroidal oscillations. Geophys Jour Roy Astron Soc (London) 17: 151–177.

Singh SJ, Ben-Menahem A (1969b) Eigenvibrations of the earth excited by finite dislocations. II. Spheroidal oscillations. Geophys Jour Roy Astron Soc (London) 17: 333–350.

Singh SJ, Ben-Menahem A (1969c) Deformation of a homogeneous gravitating sphere by internal dislocations. Pure Appl Geophys 76: 17–39.

Slichter LB (1961) The fundamental free mode of the earth's inner core. Proc Natl Acad Sci (USA) 47: 186–190.

Smith ML (1974) The scalar equations of infinitesimal elastic-gravitational motion for a rotating, slightly elliptical earth. Geophys Jour Roy Astron Soc (London) 37: 491–526.

Takeuchi H, Dorman J, Saito M (1964) Partial derivatives of surface wave phase velocity with respect to physical parameter changes within the earth. Jour Geophys Res 69: 3429–3441.

Usami T (1971) Effect of horizontal heterogeneity on the torsional oscillations of an elastic sphere. Jour Phys Earth 19: 175–180.

CHAPTER 7

Geometric Elastodynamics: Rays and Generalized Rays

Everything should be made as simple as possible, but not simpler.

(*Albert Einstein*)

7.1. Asymptotic Body Wave Theory

7.1.1. Decoupling the Equation of Motion for Radially Heterogeneous Media

The separation of dependent and independent variables is of fundamental importance in the analytic determination of the fields associated with the partial differential equations of mathematical physics. In particular, the determination of the displacement field at any point of a heterogeneous elastic medium is greatly simplified if the vector wave equation of elasticity can be split into three equations; corresponding to the P, SV, and SH waves of seismology. Although the SH motion can be separated from the P and SV motions for all radially heterogeneous isotropic media, the P and SV motions, in general, can at most be reduced to a system of two coupled scalar equations. The vector wave equation is called decoupled if these two scalar equations can be transformed into two uncoupled equations, one corresponding to P waves and the other to SV waves.

We consider a radially heterogeneous elastic medium for which $\rho = \rho(r)$, $\mu = \mu(r)$, $\lambda = \lambda(r)$. The equation of motion of such a medium may be written in the form [cf. Eq. (1.108)]

$$(\lambda + 2\mu)\text{grad div } \mathbf{u} - \mu \text{ curl curl } \mathbf{u} + (\lambda' \text{ div } \mathbf{u})\mathbf{e}_r + \mu'\left(2\frac{\partial \mathbf{u}}{\partial r} + \mathbf{e}_r \times \text{curl } \mathbf{u}\right)$$

$$= \rho\frac{\partial^2 \mathbf{u}}{\partial t^2}, \tag{7.1}$$

where a prime denotes differentiation with respect to r. Equation (7.1) is equivalent to three coupled scalar equations in the components of the displacement (u_r, u_θ, u_ϕ). For the following discussion it is better to take (u_r, S, Ω_r) as the three independent variables instead of (u_r, u_θ, u_ϕ), where

$$S = \operatorname{div}\mathbf{u}, \qquad \Omega_r = (\operatorname{curl}\mathbf{u})_r. \tag{7.2}$$

Replacing curl curl $\mathbf{u}$ in Eq. (7.1) by (grad div $\mathbf{u} - \nabla^2\mathbf{u}$) and taking its curl, the r component of the resulting equation gives

$$\nabla^2(r\Omega_r) + \frac{\mu'}{\mu}\left(\frac{\partial}{\partial r} - \frac{1}{r}\right)(r\Omega_r) = \frac{\rho}{\mu}\frac{\partial^2}{\partial t^2}(r\Omega_r). \tag{7.3}$$

The r component of Eq. (7.1) yields

$$\frac{\mu}{r}\nabla^2(ru_r) + 2\mu'\frac{\partial u_r}{\partial r} + (\lambda + \mu)\frac{\partial S}{\partial r} + \left(\lambda' - \frac{2\mu}{r}\right)S = \rho\frac{\partial^2 u_r}{\partial t^2}. \tag{7.4}$$

Equation (7.1) may be expressed in the form

$$\operatorname{grad}[(\lambda + 2\mu)S] - \operatorname{curl}(\mu\operatorname{curl}\mathbf{u}) + 2\mu'\left(\frac{\partial\mathbf{u}}{\partial r} - \mathbf{e}_r S + \mathbf{e}_r \times \operatorname{curl}\mathbf{u}\right) = \rho\frac{\partial^2\mathbf{u}}{\partial t^2}. \tag{7.5}$$

Taking the divergence of the last equation, we get

$$\nabla^2[(\lambda + 2\mu)S] - \frac{2}{r^3}\frac{\partial}{\partial r}(\mu' r^3 S) + 2r\left(\frac{\mu'}{r}\right)'\frac{\partial u_r}{\partial r} + \frac{2}{r}\mu'\nabla^2(ru_r) = \frac{\partial^2}{\partial t^2}(\rho' u_r + \rho S). \tag{7.6}$$

The equation of motion, Eq. (7.1), is equivalent to the three scalar Eqs. (7.3), (7.4), and (7.6). We note that Ω_r appears in Eq. (7.3) only, and that Eq. (7.3) is independent of u_r and S. Because for SH-type motion (toroidal motion), $u_r = S = 0$, we see that the SH-type motion is decoupled from the P–SV-type motion for all radially heterogeneous media and for all frequencies. Equation (7.3) therefore represents SH-type motion, whereas the P–SV motion is represented by the two coupled Eqs. (7.4) and (7.6). However, because the short-period body waves do appear as P and SV waves separately on a seismogram, the high frequency P motion can be expected to be decoupled from the SV motion. Let

$$\mathbf{u} = \mathbf{L} + \mathbf{N} + \mathbf{M}, \tag{7.7}$$

where $\mathbf{L}$, $\mathbf{N}$, and $\mathbf{M}$ correspond, respectively, to the P, SV, and SH waves. We assume

$$\mathbf{L} = \frac{1}{f_1}\operatorname{grad}(f_1\psi_1) = \operatorname{grad}\psi_1 + \mathbf{e}_r g_1\psi_1, \tag{7.8}$$

$$\mathbf{N} = \frac{1}{f_2}\operatorname{curl}\operatorname{curl}(\mathbf{r}f_2\psi_2) = \operatorname{curl}\operatorname{curl}(\mathbf{r}\psi_2) + g_2\mathbf{e}_r \times \operatorname{curl}(\mathbf{r}\psi_2), \tag{7.9}$$

$$\mathbf{M} = \operatorname{curl}(\mathbf{r}\psi_3), \tag{7.10}$$

where f_1 and f_2 are functions of r only and

$$g_1 = \frac{f'_1}{f_1}, \qquad g_2 = \frac{f'_2}{f_2}. \tag{7.11}$$

If we take f_1 and f_2 as constants, the representation in Eqs. (7.8)–(7.10) coincides with the corresponding representation used in the case of a homogeneous medium (Sect. 2.4).

On putting $\operatorname{div} \mathbf{u} = 0$ in Eq. (7.1), we find that the SH-motion is governed by the equation

$$\mu \nabla^2 \mathbf{u} + \mu' \left(2 \frac{\partial \mathbf{u}}{\partial r} + \mathbf{e}_r \times \operatorname{curl} \mathbf{u} \right) = \rho \frac{\partial^2 \mathbf{u}}{\partial t^2}. \tag{7.12}$$

From Eq. (7.10), we have

$$\mathbf{M} = \left(\mathbf{e}_\theta \frac{1}{\sin\theta} \frac{\partial}{\partial \phi} - \mathbf{e}_\phi \frac{\partial}{\partial \theta} \right) \psi_3,$$

$$\Omega_r = (\operatorname{curl} \mathbf{M})_r = -\frac{1}{r} B^2 \psi_3, \tag{7.13}$$

where B^2 denotes the *Beltrami operator*

$$B^2 \equiv \frac{1}{\sin\theta} \frac{\partial}{\partial\theta} \left(\sin\theta \frac{\partial}{\partial\theta} \right) + \frac{1}{\sin^2\theta} \frac{\partial^2}{\partial\phi^2}. \tag{7.14}$$

Substituting the value of Ω_r from Eqs. (7.13) into Eq. (7.3) and noting that ∇^2 and $\partial/\partial r$ commute with B^2, it is clear that Eq. (7.3) will be satisfied if ψ_3 satisfies the equation

$$\mu \nabla^2 \psi_3 + \mu' \left(\frac{\partial}{\partial r} - \frac{1}{r} \right) \psi_3 = \rho \frac{\partial^2 \psi_3}{\partial t^2}. \tag{7.15}$$

Defining

$$\Phi_3 = \sqrt{\mu}\, \psi_3, \tag{7.16}$$

we find from Eq. (7.15) that Φ_3 satisfies the equation

$$(\nabla^2 - b_3^2) \Phi_3 = \frac{\rho}{\mu} \frac{\partial^2 \Phi_3}{\partial t^2}, \tag{7.17}$$

where

$$b_3^2 = \frac{\mu''}{2\mu} - \frac{1}{4} \left(\frac{\mu'}{\mu} \right)^2 + \frac{2}{r} \frac{\mu'}{\mu}. \tag{7.18}$$

For a motion of the SV type ($\mathbf{u} = \mathbf{N}$), we see from Eq. (7.9) that $S = -g_2 u_r$. Putting this value in Eq. (7.4), we get

$$\mu\nabla^2(ru_r) + [-(\lambda+\mu)g_2 + 2\mu']\frac{\partial}{\partial r}(ru_r)$$
$$+\left[-(\lambda+\mu)g_2' + \left(\frac{\lambda+3\mu}{r} - \lambda'\right)g_2 - \frac{2}{r}\mu'\right](ru_r) = \rho\frac{\partial^2(ru_r)}{\partial t^2}. \quad (7.19)$$

Similarly, substituting $S = -g_2 u_r$ in Eq. (7.6), we get another equation in u_r. Comparing the coefficients of $\nabla^2 u_r$ and $\partial^2 u_r/\partial t^2$ in these two equations in u_r, we get the value of g_2:

$$g_2 = \frac{\mu}{\lambda+\mu}\left(\frac{2\mu'}{\mu} - \frac{\rho'}{\rho}\right). \quad (7.20)$$

Comparing the coefficients of the other terms in the two equations for u_r, we get conditions of decoupling of the P and SV motions. These are certain nonlinear relationships between ρ, λ, μ and their first and second derivatives. It can be shown that the known radial distributions of these parameters in the earth are consistent with the decoupling hypothesis and, therefore, these conditions need not concern us any further. Equation (7.9) shows that for SV-type motion,

$$ru_r = r(\mathbf{N})_r = -B^2\psi_2, \quad (7.21)$$

where, as in Eq. (7.14),

$$\nabla^2 \equiv \frac{\partial^2}{\partial r^2} + \frac{2}{r}\frac{\partial}{\partial r} + \frac{1}{r^2}B^2. \quad (7.22)$$

Because B^2 commutes with ∇^2, Eqs. (7.19)–(7.21) yield an equation in ψ_2:

$$\nabla^2\psi_2 + \frac{\rho'}{\rho}\frac{\partial\psi_2}{\partial r} + \frac{1}{\mu}\left\{-(\lambda+\mu)g_2' + \left(\frac{\lambda+3\mu}{r} - \lambda'\right)g_2 - \frac{2}{r}\mu'\right\}\psi_2 = \frac{\rho}{\mu}\frac{\partial^2\psi_2}{\partial t^2}. \quad (7.23)$$

On putting

$$\Phi_2 = \sqrt{\rho}\,\psi_2, \quad (7.24)$$

Eq. (7.23) transforms to

$$(\nabla^2 - b_2^2)\Phi_2 = \frac{\rho}{\mu}\frac{\partial^2\Phi_2}{\partial t^2}, \quad (7.25)$$

where

$$b_2^2 = \frac{\lambda+\mu}{\mu}g_2' - \frac{1}{\mu}\left(\frac{\lambda+3\mu}{r} - \lambda'\right)g_2 + \frac{1}{2}\frac{\rho''}{\rho} - \frac{1}{4}\left(\frac{\rho'}{\rho}\right)^2 + \frac{\rho'}{r\rho} + \frac{2\mu'}{r\mu}. \quad (7.26)$$

Similarly, starting with the displacement **L** for *P*-type motion, we derive the following two equations in ψ_1

$$(\lambda + 2\mu)\nabla^2\psi_1 + [(\lambda + \mu)g_1 + 2\mu']\frac{\partial\psi_1}{\partial r} + \left[(\lambda + \mu)g_1' + \left(2\frac{\lambda + 2\mu}{r} + \mu'\right)g_1 - \frac{2}{r}\mu'\right]\psi_1 = \rho\frac{\partial^2\psi_1}{\partial t^2}, \tag{7.27}$$

$$(\mu g_1 - 2\mu')\nabla^2\psi_1 + 2\left[\mu\left(\frac{g_1}{r}\right)' r + \frac{\mu'}{r} - \mu''\right]\frac{\partial\psi_1}{\partial r} + \left[\mu\left(\frac{g_1}{r}\right)'' r + 2\mu\left(\frac{g_1}{r}\right)' - \left(\mu'' + \frac{4\mu'}{r}\right)g_1 - \frac{2}{r}\left(\frac{\mu'}{r} - \mu''\right)\right]\psi_1 = \frac{\partial^2}{\partial t^2}[(\rho g_1 - \rho')\psi_1]. \tag{7.28}$$

Comparing the coefficients of $\nabla^2\psi_1$ and $\partial^2\psi_1/\partial t^2$ in Eqs. (7.27) and (7.28), we get the value of g_1:

$$g_1 = \frac{\lambda + 2\mu}{\lambda + \mu}\left(\frac{\rho'}{\rho} - \frac{2\mu'}{\lambda + 2\mu}\right). \tag{7.29}$$

Next, defining

$$\Phi_1 = \sqrt{\rho}\,\psi_1, \tag{7.30}$$

Eq. (7.27) becomes

$$(\nabla^2 - b_1^2)\Phi_1 = \frac{\rho}{\lambda + 2\mu}\frac{\partial^2\Phi_1}{\partial t^2}, \tag{7.31}$$

where

$$b_1^2 = -\frac{\lambda + \mu}{\lambda + 2\mu}g_1' - \left(\frac{2}{r} + \frac{\mu'}{\lambda + 2\mu}\right)g_1 + \frac{\rho''}{2\rho} - \frac{1}{4}\left(\frac{\rho'}{\rho}\right)^2 + \frac{\rho'}{r\rho} + \frac{2\mu'}{r(\lambda + 2\mu)}. \tag{7.32}$$

Equations (7.8)–(7.10), (7.16), (7.24), and (7.30) yield

$$\mathbf{L} = \text{grad}\left(\frac{\Phi_1}{\sqrt{\rho}}\right) + g_1\left(\frac{\Phi_1}{\sqrt{\rho}}\right)\mathbf{e}_r = \frac{1}{\sqrt{\rho}}[\text{grad}\,\Phi_1 - g_3\Phi_1\mathbf{e}_r], \tag{7.33}$$

$$\mathbf{N} = \text{curl curl}\left[\left(\frac{\Phi_2}{\sqrt{\rho}}\right)\mathbf{r}\right] - g_2\,\text{curl}\left[\left(\frac{\Phi_2}{\sqrt{\rho}}\right)\mathbf{r}\right] \times \mathbf{e}_r \tag{7.34}$$

$$= \frac{1}{\sqrt{\rho}}\left[\text{curl curl}(\mathbf{r}\Phi_2) + g_3\left(r\,\text{grad}\,\Phi_2 - \mathbf{r}\frac{\partial\Phi_2}{\partial r}\right)\right], \tag{7.35}$$

$$\mathbf{M} = \text{curl}\left[\left(\frac{\Phi_3}{\sqrt{\mu}}\right)\mathbf{r}\right] = \frac{1}{\sqrt{\mu}}(\text{grad}\,\Phi_3) \times \mathbf{r}, \tag{7.36}$$

where

$$2g_3 = g_2 - g_1. \tag{7.37}$$

Assuming a harmonic time dependence, Eqs. (7.17), (7.25), and (7.31) may be written as

$$(\nabla^2 + k_i^2 - b_i^2)\Phi_i = 0, \qquad (i = 1, 2, 3), \tag{7.38}$$

where

$$k_i^2 = \frac{\omega^2}{V_i^2}, \qquad V_1^2 = \frac{\lambda + 2\mu}{\rho}, \qquad V_2^2 = V_3^2 = \frac{\mu}{\rho}. \tag{7.39}$$

If the inhomogeneity is small over distances of the order of one wavelength, so that

$$\begin{gathered} \frac{\Lambda}{r} \ll 1, \qquad \Lambda\left|\frac{\mu'}{\mu}\right| \ll 1, \qquad \Lambda\left|\frac{\lambda'}{\lambda}\right| \ll 1, \qquad \Lambda\left|\frac{\rho'}{\rho}\right| \ll 1, \\ \Lambda\left|\frac{\mu''}{\mu}\right|^{1/2} \ll 1, \qquad \Lambda\left|\frac{\rho''}{\rho}\right|^{1/2} \ll 1, \end{gathered} \tag{7.40}$$

(Λ = wavelength, $r \sim a$ near the surface of the earth) then, from Eqs. (7.18), (7.26), and (7.32),

$$b_i^2 \ll k_i^2 \qquad (i = 1, 2, 3). \tag{7.41}$$

In the earth's mantle, for example, the average value of $|\mu'/\mu|$ is in the range (0.3×10^{-3} to 1.0×10^{-3}) per kilometer. Therefore, the corresponding wavelength must be much smaller than 1000 km. In cases when the rigidity changes are much more pronounced than the density changes, the condition $\Lambda|\mu'/\mu| \ll 1$ implies $\omega^{-1}|dV_2/dz| \ll 1$. A similar assumption is often made in the theory of electromagnetic waves.

If (7.41) is satisfied, Eq. (7.38) may be replaced by the Helmholtz equation with coordinate-dependent wave number:

$$(\nabla^2 + k_i^2)\Phi_i = 0, \qquad (i = 1, 2, 3). \tag{7.42}$$

In a homogeneous medium, P waves are associated with zero rotation and S waves with zero divergence. From Eqs. (7.8) and (7.9), we note that

$$\begin{aligned} \operatorname{curl} \mathbf{L} &= g_1 \mathbf{L} \times \mathbf{e}_r, \\ \operatorname{div} \mathbf{N} &= -g_2 \mathbf{N} \cdot \mathbf{e}_r. \end{aligned} \tag{7.43}$$

Hence, in the case of a heterogeneous medium, $\mathbf{L}$ is not always irrotational and $\mathbf{N}$ is not always solenoidal, although $\mathbf{M}$ is always solenoidal.

7.1.2. Displacement Potentials

Assuming the decoupling of the P and SV motions in the earth at sufficiently high frequencies, we wish now to derive simpler expressions for the displacements in terms of the potentials. First, let us summarize our former results:

$$\mathbf{L} = \frac{1}{f_1}\operatorname{grad}\left[\frac{f_1}{\sqrt{\rho}}\Phi_1\right] = \frac{1}{\sqrt{\rho}}[\operatorname{grad}\Phi_1 - g_3\Phi_1\mathbf{e}_r], \qquad P\text{ motion}$$

$$\begin{aligned}\mathbf{N} &= \frac{1}{f_2}\operatorname{curl}\operatorname{curl}\left[\frac{f_2}{\sqrt{\rho}}\mathbf{r}\Phi_2\right]\\ &= \frac{1}{\sqrt{\rho}}\left[\operatorname{curl}\operatorname{curl}(\mathbf{r}\Phi_2) + g_3\left(r\operatorname{grad}\Phi_2 - \mathbf{r}\frac{\partial\Phi_2}{\partial r}\right)\right], \qquad SV\text{ motion}\end{aligned} \tag{7.44}$$

$$\mathbf{M} = \operatorname{curl}\left[\mathbf{r}\frac{\Phi_3}{\sqrt{\mu}}\right] = \frac{1}{\sqrt{\mu}}\operatorname{curl}(\mathbf{r}\Phi_3), \qquad SH\text{ motion}$$

where, from Eqs. (7.11), (7.20), and (7.29),

$$\begin{aligned}f_1 &= \exp\left[\int\frac{\lambda+2\mu}{\lambda+\mu}\left(\frac{\rho'}{\rho} - \frac{2\mu'}{\lambda+2\mu}\right)dr\right],\\ f_2 &= \exp\left[\int\frac{\mu}{\lambda+\mu}\left(\frac{2\mu'}{\mu} - \frac{\rho'}{\rho}\right)dr\right],\\ 2g_3 &= \frac{4\mu'}{\lambda+\mu} - \frac{\lambda+3\mu}{\lambda+\mu}\left(\frac{\rho'}{\rho}\right),\end{aligned} \tag{7.45}$$

and

$$(\nabla^2 + k_i^2 - b_i^2)\Phi_i = 0, \qquad (i = 1, 2, 3). \tag{7.46}$$

Under the assumptions in (7.40), Eq. (7.46) reduces to

$$(\nabla^2 + k_i^2)\Phi_i = 0, \qquad (i = 1, 2, 3). \tag{7.47}$$

We will now simplify Eqs. (7.44) under the same assumptions.

To compare relative magnitudes of such quantities as Φ_1, $\partial\Phi_1/\partial r$, etc., we note that each spatial differentiation raises the frequency order by one, at most, i.e.,

$$\left(\frac{\partial}{\partial r}, \frac{1}{r}\frac{\partial}{\partial\theta}, \frac{1}{r}\frac{\partial}{\partial\phi}\right)\Phi = O\left(\frac{1}{\Lambda}\Phi\right) \quad \text{as} \quad \Lambda \to 0. \tag{7.48}$$

Moreover, from Eqs. (7.40) and (7.45), $\Lambda g_3 \ll 1$. Therefore, the expressions for **L** and **N** may be approximated as follows:

$$\mathbf{L} = \frac{1}{\sqrt{\rho}} \operatorname{grad} \Phi_1,$$

$$\mathbf{N} = \frac{1}{\sqrt{\rho}} \operatorname{curl} \operatorname{curl}(\mathbf{r}\Phi_2), \tag{7.49}$$

$$\mathbf{M} = \frac{1}{\sqrt{\mu}} \operatorname{curl}(\mathbf{r}\Phi_3),$$

where the expression for **M** does not involve any approximation and has been included for future reference.

7.1.3. The Eikonal Equation

Having reduced the Navier wave equation to three scalar Helmholtz equations, each with a variable wave number, we proceed to derive the laws governing the propagation of high-frequency P and S waves in boundless media under the conditions of weak inhomogeneity as stated in (7.40).

In unbounded homogeneous and isotropic elastic media, the concepts of *wavefront* and *direction of propagation* are strongly linked to the idea of a plane wave that has the property that its direction of propagation and amplitude are the same everywhere. The propagation of seismic waves, however, is determined by second-order partial differential equations with boundary conditions. If the geometry of the problem and the spatial variation of the material parameters λ, μ, and ρ are both complicated, there will be considerable mathematical difficulties in the solution of these boundary-value problems. There is one case in which we are able to furnish a relatively simple mathematical description of the physical field and this is precisely the case for which the P–SV decoupling occurs, namely, wave propagation in media in which the amplitude and the direction of propagation of the wave vary only slightly over distances of the order of a wavelength.

When these conditions prevail, certain useful concepts can be borrowed from plane-wave theory. The basic idea rests on the assumption that in relatively small spatial domains and in corresponding small time intervals, the scalar wave equation may render a solution of the form Ae^{-iQ}, which is similar to a plane-wave solution, $Be^{i(\omega t - \mathbf{k}\cdot\mathbf{r})}$. Further, A and Q are functions of the coordinates and time such that A is a slowly varying function, whereas the phase Q is almost linear. We assume that the generalized phase function Q can be developed into a Taylor series:

$$Q(\mathbf{r}_0 + \mathbf{r}, t_0 + t) = Q(\mathbf{r}_0, t_0) + \mathbf{r}\cdot(\nabla Q)_0 + t\left(\frac{\partial Q}{\partial t}\right)_0, \tag{7.50}$$

up to first-order terms in $|\mathbf{r}|$ and t. A comparison with the equivalent plane-wave representation at $(\mathbf{r}, t)$ yields

$$\omega = -\left(\frac{\partial Q}{\partial t}\right)_0, \qquad \mathbf{k} = (\nabla Q)_0, \tag{7.51}$$

where now ω and $\mathbf{k}$ are functions of $\mathbf{r}_0$ and t_0 and, therefore, are not identical with the frequency and wave number of a Fourier component. Because, for a plane wave in a nondissipative medium, $|\mathbf{k}|^2 = \omega^2/V^2$, where V is the intrinsic velocity of the wave, our analogy renders

$$(\nabla Q)^2 = \frac{1}{V^2}\left(\frac{\partial Q}{\partial t}\right)^2, \tag{7.52}$$

where $V(\mathbf{r})$ is a given function and represents the local velocity of wave propagation.

The scalar function $Q(\mathbf{r}, t)$ is known as the *eikonal*.[1] Equation (7.52) is the general equation satisfied by Q and is known as the *eikonal equation.* In seismologic applications, Q can be assumed to be of the form

$$Q(\mathbf{r}, t) = -\omega t + \omega\Psi(\mathbf{r}), \tag{7.53}$$

where $\Psi(\mathbf{r})$ is independent of the frequency. The eikonal equation then simplifies to

$$(\nabla\Psi)^2 = \frac{1}{V^2}, \tag{7.54}$$

where for P waves, $V^2 = (\lambda + 2\mu)/\rho$ and for S waves, $V^2 = \mu/\rho$.

A more rigorous derivation of Eq. (7.54) proceeds as follows. We suppress the subscripts of V and Φ and seek asymptotic solutions of [cf. Eq. (7.42)]

$$\nabla^2\Phi + \frac{\omega^2}{V^2(r)}\Phi = 0, \tag{7.55}$$

in the form

$$\Phi = e^{-i\omega\Psi}\sum_{n=0}^{\infty}(i\omega)^{-n}\xi_n, \tag{7.56}$$

where Ψ and ξ_n are frequency independent. Inserting this expansion in Eq. (7.55) and equating to zero the coefficients of each power of ω, it is seen that Ψ and ξ_n must satisfy the differential equations

$$|\text{grad}\,\Psi|^2 = \frac{1}{V^2}, \tag{7.57}$$

$$2(\text{grad}\,\Psi)\cdot\text{grad}\,\xi_n + \xi_n\nabla^2\Psi = \nabla^2\xi_{n-1}, \qquad (n \geq 0,\ \xi_{-1} = 0). \tag{7.58}$$

[1] Eikonal from Greek εἰκῶν, meaning image. The term was introduced in 1895 by H. Bruns.

We have set $\xi_{-1} = 0$ to permit the equation for $n = 0$ to be written in the same form as for $n \geq 1$. We shall be mainly interested in $\xi_0 = A$, for which Eq. (7.58) can be recast in the form

$$(\text{grad} \ln A) \cdot (\text{grad}\, \Psi) = -\frac{1}{2}\nabla^2\Psi. \tag{7.59}$$

Equation (7.57) is an inhomogeneous partial differential equation of the first order and second degree. Once Eq. (7.57) has been integrated, Eq. (7.59) yields the component of the gradient of (ln A) in the direction of grad Ψ. Equation (7.59) makes no statement about the gradient of (ln A) in directions perpendicular to grad Ψ. This equation permits discontinuities in A in these directions. Instead of the wave equation, which yields simultaneous information on both amplitude and phase, we now have two equations, one governing the phase and the other the amplitude, when the phase is known.

Let us first explore the properties of the Ψ function. The equation

$$\Psi(\mathbf{r}) = \text{const.} \tag{7.60}$$

describes a family of surfaces, known as *wave surfaces* or *wavefronts.* The phase of the wave is constant on a wavefront. On account of the eikonal equation, Eq. (7.57), V grad Ψ is a unit vector. If we denote this unit vector by $\mathbf{p}$, then $\mathbf{p}$ is along the normal to the surface in Eq. (7.60). Therefore,

$$\mathbf{p} = V \,\text{grad}\, \Psi = \frac{\text{grad}\, \Psi}{|\text{grad}\, \Psi|}. \tag{7.61}$$

Let us suppose that $\Psi(\mathbf{r})$ has continuous first-order partial derivatives in some region of space. This ensures that there exists a two-parameter family of orthogonal trajectories to the family of wavefronts, Eq. (7.60). This family of orthogonal trajectories to the wavefronts is known as *rays.* These are oriented space curves whose directions coincide everywhere with the corresponding directions of the vector grad Ψ.

The arc-length parameter along a ray will be denoted by s. If $\mathbf{r}(s)$ denotes the position vector of a point P on a ray, considered as a function of s, then $\mathbf{p} = d\mathbf{r}/ds$ and, therefore, from Eq. (7.61) the equation of the ray may be written as

$$\frac{d\mathbf{r}}{ds} = V \,\text{grad}\, \Psi. \tag{7.62a}$$

In spherical polar coordinates, this yields

$$\frac{dr}{\partial\Psi/\partial r} = \frac{r^2\, d\theta}{\partial\Psi/\partial\theta} = \frac{(r \sin\theta)^2\, d\phi}{\partial\Psi/\partial\phi} = \text{const.} \tag{7.62b}$$

Consider two neighboring wavefronts $\Psi = \text{const.}$ and $\Psi + \delta\Psi = \text{const.}$ Then

$$\frac{d\Psi}{ds} = \mathbf{p} \cdot \text{grad}\, \Psi = \frac{1}{V}. \tag{7.63}$$

Hence the distance δs between points on the two ends of a ray cutting the two wavefronts is directly proportional to the velocity, V. Writing $\delta s = V\delta t$, where t denotes time, Eq. (7.63) yields

$$\Psi - \Psi_0 = \int_{s_0}^{s} \frac{ds}{V} = t - t_0 \tag{7.64}$$

using Eq. (7.61). Hence Ψ measures, in fact, the travel-time along a ray.

Equation (7.64) gives an explicit expression for the calculation of the *travel time* along a given ray. Note that the wave-number vector $\mathbf{k}$ is related to $\mathbf{p}$ and Ψ through the relation

$$\mathbf{k} = \frac{\omega}{V}\mathbf{p} = \omega(\nabla\Psi). \tag{7.65}$$

Because V is assumed to be frequency independent, no dispersion takes place in our model.

Equation (7.62a) of a ray is in terms of Ψ. We next derive the equation that does not involve the phase factor directly. Defining *slowness*, $S = 1/V$, we find from Eq. (7.62a)

$$\frac{d}{ds}(S\mathbf{p}) = \frac{d}{ds}\left(S\frac{d\mathbf{r}}{ds}\right) = \frac{d}{ds}(\text{grad}\,\Psi) = \text{grad}\frac{d\Psi}{ds} = \text{grad}\,S, \tag{7.66}$$

using Eq. (7.63). Therefore, the equation of the ray becomes

$$S\frac{d^2\mathbf{r}}{ds^2} + \frac{dS}{ds}\frac{d\mathbf{r}}{ds} = \text{grad}\,S. \tag{7.67}$$

This is a second-order vector differential equation for rays in 3-space. The vector $S\mathbf{p}$ is called the *ray vector* and, because grad $\Psi = S\mathbf{p}$, it obeys the equation

$$\text{curl}(S\mathbf{p}) = 0. \tag{7.68}$$

In a homogeneous medium, $S = \text{const.}$ and the ray equation reduces to $d^2\mathbf{r}/ds^2 = 0$, with the solution $\mathbf{r} = \mathbf{b} + \mathbf{a}s$. This represents a straight line in the $\mathbf{a}$ direction, passing through $\mathbf{r} = \mathbf{b}$. Further, from Eq. (7.57) we obtain a particular solution

$$\Psi = \frac{1}{V}(ax + by + cz), \qquad a^2 + b^2 + c^2 = 1,$$

showing that the wavefronts are planes. Another solution is $\Psi = r/V$, which yields spherical wavefronts.

Equation (7.66) of the rays may be written in the form

$$S\frac{d\mathbf{p}}{ds} + \frac{dS}{ds}\mathbf{p} = \text{grad}\,S. \tag{7.69}$$

It is obvious that, in general, $\mathbf{p}$ will change its direction from point to point on the ray. The ray is, therefore, usually curved. If we construct the unit tangent vectors at two neighboring points on the ray, the *curvature* of the ray at P is defined as the limit of the angle between these two vectors divided by the corresponding arc length, δs, as $\delta s \to 0$.

Hence, if ρ denotes the *radius of curvature* of the ray,

$$\text{Curvature} = \frac{1}{\rho} = \frac{di}{ds}, \tag{7.70}$$

where i is the angle between the tangent to the ray at P and the radius vector (Fig. 7.1). From differential geometry, the *curvature vector* is

$$\mathbf{m} = \frac{d\mathbf{p}}{ds} = \frac{1}{\rho}\,\mathbf{v}, \qquad |\mathbf{m}| = \frac{1}{\rho}, \qquad \mathbf{m}\cdot\mathbf{p} = 0, \tag{7.71}$$

where $\mathbf{v}$ is the *unit principal normal.*

Equation (7.68) yields

$$S \operatorname{curl} \mathbf{p} + (\operatorname{grad} S) \times \mathbf{p} = 0. \tag{7.72}$$

Inserting the value of grad S from Eq. (7.69) and using Eq. (7.71), we find

$$\operatorname{curl} \mathbf{p} + \mathbf{m} \times \mathbf{p} = 0. \tag{7.73}$$

A vector multiplication with $\mathbf{p}$ results in

$$\mathbf{m} = (\operatorname{curl} \mathbf{p}) \times \mathbf{p} = \frac{1}{S}\mathbf{p} \times [(\operatorname{grad} S) \times \mathbf{p}] = \frac{1}{S}[\operatorname{grad} S - (\mathbf{p}\cdot\operatorname{grad} S)\mathbf{p}]$$

$$= \frac{1}{S}\left[\operatorname{grad} S - \frac{dS}{ds}\mathbf{p}\right]. \tag{7.74}$$

This relation shows that grad S lies in the plane of $\mathbf{p}$ and $\mathbf{m}$, also known as the *osculating plane* of the ray.

If we multiply Eq. (7.74) scalarly by $\mathbf{m}$ and use Eq. (7.71), we find

$$|\mathbf{m}| = \frac{1}{\rho} = \frac{1}{S}\,\mathbf{v}\cdot\operatorname{grad} S = -\mathbf{v}\cdot\operatorname{grad}(\ln V). \tag{7.75a}$$

Because ρ is always positive, this implies that the velocity decreases along the principal normal; i.e., the rays bend toward the region of decreasing velocity. From Eqs. (7.72) and (7.73), we have

$$\frac{1}{\rho} = |\mathbf{p} \times \mathbf{m}| = |\operatorname{curl} \mathbf{p}| = \frac{1}{S}|\mathbf{p} \times \operatorname{grad} S|. \tag{7.75b}$$

From Eq. (7.73) we notice that curl $\mathbf{p}$ is perpendicular to both $\mathbf{m}$ and $\mathbf{p}$. Hence curl $\mathbf{p}$ is in the direction of the normal in the osculating plane at P, known as the *binormal.* Therefore, we may apply the Serret–Frenet formulas and continue the discussion of rays as space curves in three dimensions. We shall not continue along this line, however, because in seismologic applications it is sufficient to assume that $S = S(r)$ and, as we shall see, this implies that the ray is always a plane curve with a *fixed binormal.*

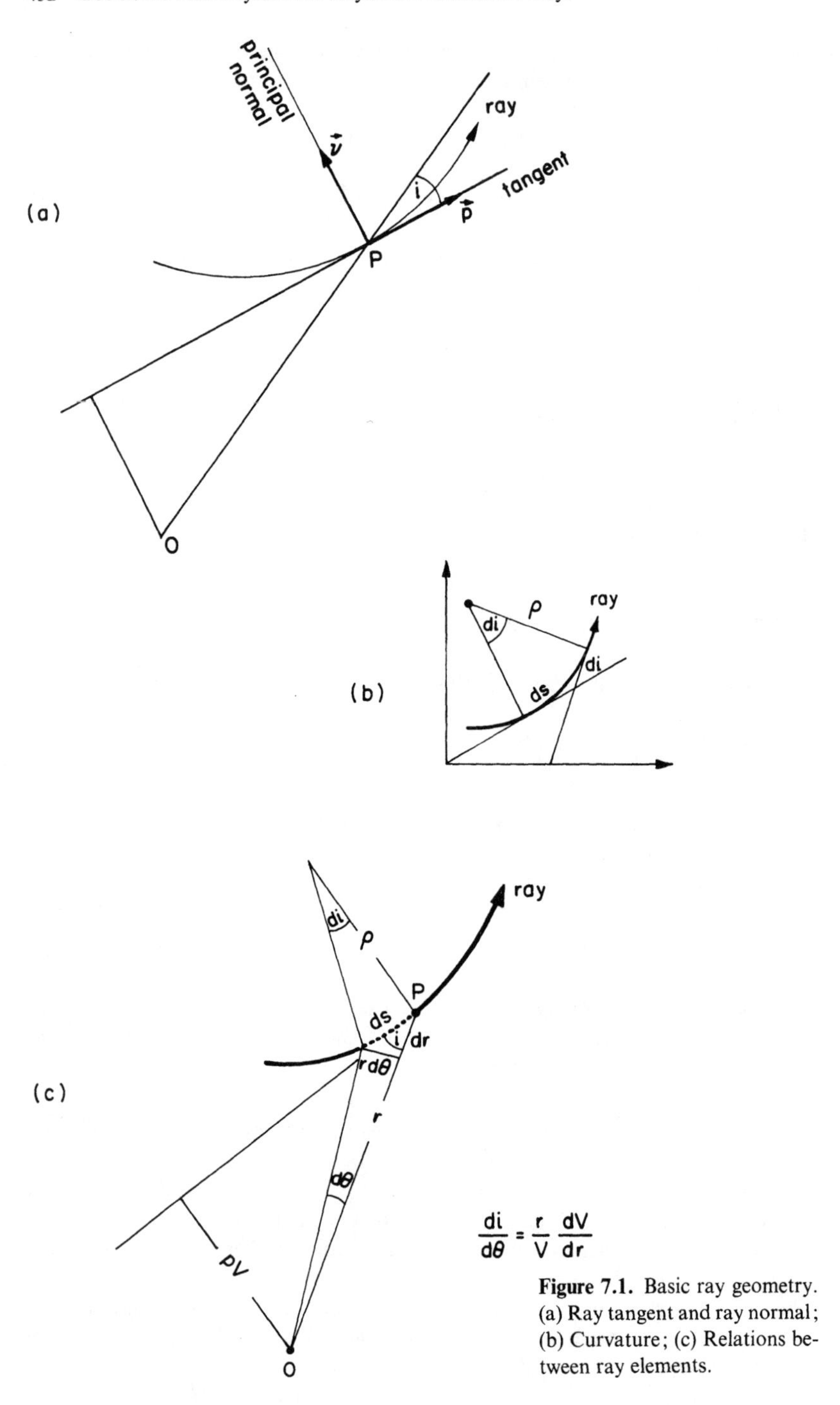

Figure 7.1. Basic ray geometry. (a) Ray tangent and ray normal; (b) Curvature; (c) Relations between ray elements.

7.1.3.1. Fermat's Principle. Let P and Q be any two points on a ray. We have seen in Eq. (7.64) that the travel-time along a ray is given by

$$t_{PQ} = \int_P^Q S\,ds = \int_P^Q \frac{1}{V}\,ds = \Psi_Q - \Psi_P. \tag{7.76}$$

Fermat's principle states that the ray-path between the points P and Q is also a path of stationary time between these points. In other words, the integral in Eq. (7.76) has a stationary value along the ray path. To prove it, we first notice that the variation of the travel-time is given by

$$\delta t_{PQ} = \int_P^Q [\delta S\,ds + S\delta(ds)]. \tag{7.77}$$

Furthermore,

$$\delta S = \delta\mathbf{r}\cdot\operatorname{grad} S, \qquad (ds)^2 = d\mathbf{r}\cdot d\mathbf{r}, \qquad ds\delta(ds) = d\mathbf{r}\cdot\delta(d\mathbf{r}), \quad \delta(ds) = \mathbf{p}\cdot d(\delta\mathbf{r}), \tag{7.78}$$

where $\mathbf{r} = \mathbf{r}(s)$, $\mathbf{p} = d\mathbf{r}/ds$. Integrating by parts and noting that $\delta\mathbf{r} = 0$ at P and Q, because P and Q are fixed points, we get

$$\int_P^Q S\delta(ds) = \int_P^Q S\mathbf{p}\cdot d(\delta\mathbf{r}) = -\int_P^Q \frac{d(S\mathbf{p})}{ds}\cdot\delta\mathbf{r}\,ds.$$

Equations (7.77) and (7.78) now yield

$$\delta t_{PQ} = \int_P^Q \left[\operatorname{grad} S - \frac{d}{ds}(S\mathbf{p})\right]\cdot\delta\mathbf{r}\,ds. \tag{7.79}$$

Therefore, using the equation of the ray in the form of Eq. (7.66), we find that $\delta t_{PQ} = 0$ for a ray; i.e., the time t_{PQ} is stationary along a ray.

We are assuming throughout this chapter that both Ψ and V are real. In absorbing media, the eikonal equation still holds, but Ψ and V are complex. Under these conditions Ψ can no longer be interpreted as time and Fermat's principle loses its simple physical meaning.

If the medium is homogeneous, it follows from Fermat's principle that the rays must be straight lines, because this minimizes the travel time. Let us now see what conclusions we can derive for rays crossing a plane discontinuity between two homogeneous media, M_1 and M_2. We assume that the ray travels from the point A in medium M_1 to the point B in medium M_2. If the velocities in the two media are v_1 and v_2, respectively, then the total time along the ray is (Fig. 7.2)

$$t(x) = \frac{1}{v_1}\sqrt{(a^2 + x^2)} + \frac{1}{v_2}\sqrt{[b^2 + (d - x)^2]},$$

where a is the distance of A from the plane of separation, b is the corresponding distance of the point B, and d is the horizontal distance between A and B. If $t(x)$ is extremal

$$t'(x) = \frac{1}{v_1}\frac{x}{\sqrt{(a^2 + x^2)}} - \frac{1}{v_2}\frac{(d - x)}{\sqrt{[b^2 + (d - x)^2]}} = \frac{\sin i_1}{v_1} - \frac{\sin i_2}{v_2} = 0.$$

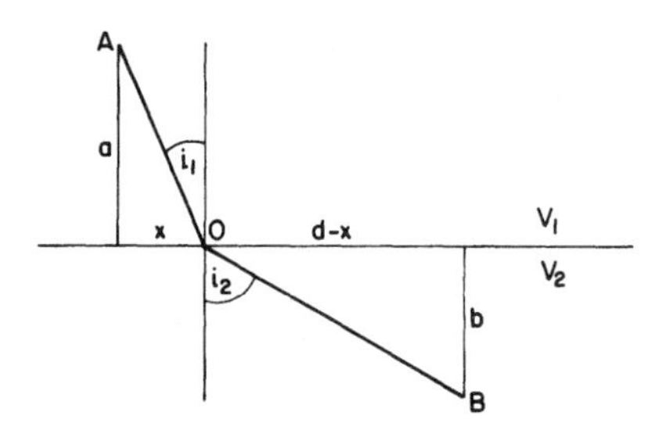

Figure 7.2. Refraction of a ray at a plane discontinuity.

Hence, at O

$$\frac{\sin i_1}{v_1} = \frac{\sin i_2}{v_2} \tag{7.80}$$

which is Snell's law at the refracting point O.

7.1.4. Rays in Spherically Symmetrical Media

From now on we shall suppose that the velocity V is a function of the distance, r, from a fixed point, O, only. Then $S = S(r)$ and Eq. (7.66) of the ray yields

$$\frac{d}{ds}[\mathbf{r} \times (S\mathbf{p})] = \frac{d\mathbf{r}}{ds} \times (S\mathbf{p}) + \mathbf{r} \times \frac{d}{ds}(S\mathbf{p})$$

$$= \mathbf{p} \times (S\mathbf{p}) + \mathbf{r} \times \operatorname{grad} S(r) = 0. \tag{7.81}$$

Therefore, $\mathbf{r} \times (S\mathbf{p})$ is constant along a ray and

$$Sr \sin i = \frac{r \sin i}{V} = p, \tag{7.82}$$

where p is constant along a ray. This is the general form of Snell's law for a spherically symmetrical medium. The quantity p is known as the *ray parameter*.

Equation (7.82) implies that a ray is a *plane curve* and, along each ray, the quantity ($Sr \sin i$) is preserved. If a particle were to move along the ray under the influence of a central force, then Eq. (7.82) would express the conservation of angular momentum of this particle. This analogy between ray theory and the motion of material particles is not accidental. Its origin lies in the fact that the eikonal equation is analogous to the Hamilton–Jacobi equation of classical mechanics.

The eikonal equation, Eq. (7.57), for the present case reduces to

$$\left(\frac{\partial \Psi}{\partial r}\right)^2 + \frac{1}{r^2}\left(\frac{\partial \Psi}{\partial \theta}\right)^2 = [S(r)]^2. \tag{7.83}$$

Assuming a separation of variables in the form

$$\Psi(r, \theta) = f(\theta) + g(r),$$

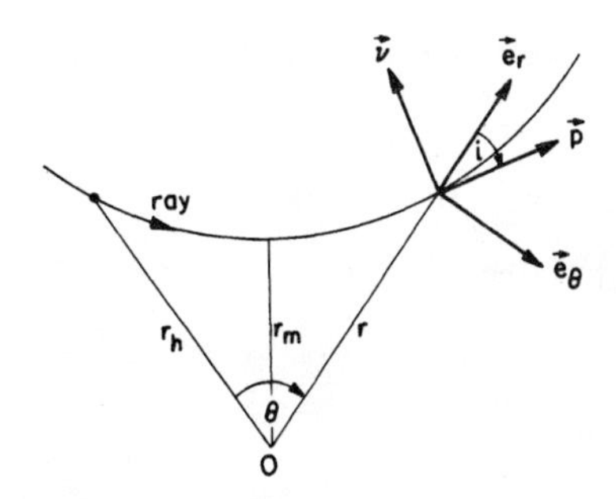

Figure 7.3. Ray elements in a spherically symmetric medium.

Eq. (7.83) yields the *complete solution*

$$\Psi = k\theta \pm \int_{r_h}^{r} \frac{dr}{r}\sqrt{(S^2r^2 - k^2)} \tag{7.84}$$[2]

where $\Psi = 0$ when $\theta = 0$ and $r = r_h$ and where k is an arbitrary constant that we determine in the following manner: From Fig. (7.3) we see that

$$\mathbf{p} = \mathbf{e}_\theta \sin i + \mathbf{e}_r \cos i. \tag{7.85}$$

However, from Eqs. (7.61) and (7.84),

$$\frac{\mathbf{p}}{V} = S\mathbf{p} = \text{grad}\,\Psi = \mathbf{e}_\theta \frac{k}{r} \pm \mathbf{e}_r \frac{1}{r}\sqrt{(r^2S^2 - k^2)}. \tag{7.86}$$

From Eqs. (7.85) and (7.86), we find

$$k = Sr \sin i = p, \tag{7.87}$$

which means that k in Eq. (7.84) is the ray parameter. Because Ψ is in fact the travel time along the ray, Eq. (7.84) may be written as

$$T = p\theta \pm \int_{r_h}^{r} (\eta^2 - p^2)^{1/2} \frac{dr}{r}, \tag{7.88}$$

where $\eta = Sr = r/V$. To derive the explicit equation of the ray, we differentiate Eq. (7.88) with respect to p, thus obtaining

$$\theta(r) = \pm p \int_{r_h}^{r} (\eta^2 - p^2)^{-1/2} \frac{dr}{r}. \tag{7.89}$$

Combining Eqs. (7.88) and (7.89), we have

$$T = \pm \int_{r_h}^{r} \eta^2(\eta^2 - p^2)^{-1/2} \frac{dr}{r}. \tag{7.90}$$

[2] The integration is along a ray and the + is for $dr > 0$ and the − for $dr < 0$. If dr changes sign, the interval (r_h, r) is split into two intervals: $(r_h, r_m; dr < 0)$ and $(r_m, r; dr > 0)$.

From Eq. (7.89), we also derive the following expression for the arc-length element along the ray

$$ds = [(dr)^2 + (r\,d\theta)^2]^{1/2} = \pm\eta(\eta^2 - p^2)^{-1/2}\,dr. \tag{7.91}$$

The curvature of the ray follows from Eq. (7.75b)

$$\begin{aligned}\frac{1}{\rho} &= \frac{1}{S}|\mathbf{p} \times \operatorname{grad} S(r)| = \frac{1}{S}\frac{dS}{dr}|\mathbf{p} \times \mathbf{e}_r| = -\frac{1}{V}\frac{dV}{dr}\sin i = -\frac{p}{r}\frac{dV}{dr} \\ &= \frac{p}{\eta}\left(\frac{1}{\eta}\frac{d\eta}{dr} - \frac{1}{r}\right).\end{aligned} \tag{7.92}$$

In particular, the radius of curvature at the lowest point of the ray, where $\sin i = \pi/2$, is

$$\rho_m = -\frac{V}{(dV/dr)}. \tag{7.93}$$

Consequently,

$$\frac{d\eta}{dr} = \frac{1}{V}\left(1 - \frac{r}{V}\frac{dV}{dr}\right) = \frac{1}{V}\left(1 + \frac{r}{\rho_m}\right). \tag{7.94}$$

If the downward (toward the center of the earth) curvature of the ray is taken as positive, Eq. (7.94) is modified to

$$\frac{d\eta}{dr} = \frac{1}{V}\left(1 - \frac{r}{\rho_m}\right).$$

If $d\eta/dr = 0$, $\rho_m = r$ and the ray is simply a circle of radius r with its center at O. However, if $d\eta/dr > 0$, $\rho_m > r$ and the ray bends upward. The ray therefore bends upward or downward according as $d\eta/dr > 0$ or <0, i.e., as $dV/dr <$ or $> V/r$. Because, except for a few limited regions in the earth, $dV/dr < 0$, the rays, in general, will bend upward.

7.1.5. Travel-Time Analysis

In seismology, observations are mostly made at seismograph stations on the earth's surface. Rays emitted from an earthquake source (*focus*), eventually reach the stations situated at various distances from the point of the earth's surface above the source (*epicenter*). The distance from the epicenter to the observing point is the *epicentral distance*. The geometric setup is depicted in Fig. 7.4. As long as we consider only a single ray, p is fixed and Eqs. (7.88)–(7.90) describe the equation of that particular ray and the travel time along it. We shall denote by Δ the total angle subtended by the ray at the earth's center.

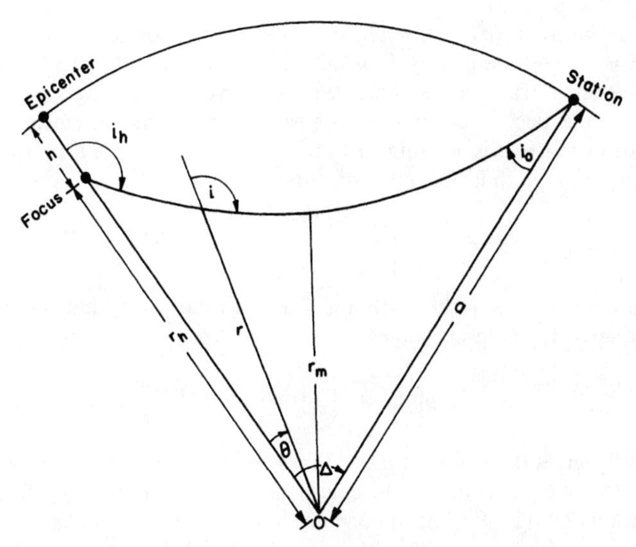

Figure 7.4. Parameters of a direct ray in the earth.

Suppose that we look at the totality of rays emerging from a single source in the earth. The coordinates of a point on any ray are (p, r), where p specifies the ray and r the moving coordinate along that ray. From Eq. (7.88), we have

$$\left(\frac{\partial T}{\partial \theta}\right)_r = p \tag{7.95}$$

which is *Benndorf's relation.* A simple geometric proof is supplied in Fig. 7.5. Consider two neighboring rays, FA and FB, subtending angles θ and $\theta + \delta\theta$ at the center of the earth. Let the corresponding travel times be T and $T + \delta T$. Then, it is clear from the figure that

$$\delta T = \frac{BC}{V(r)} = \frac{r\delta\theta \sin i}{V} = p\delta\theta.$$

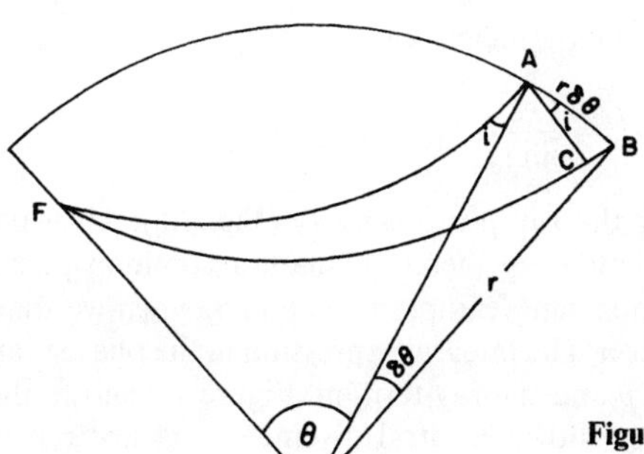

Figure 7.5. The geometry of Benndorf's relation.

Hence, in the limit, $(\partial T/\partial\theta)_r = p$. For points on the surface of the earth, this derivative may be denoted by $dT/d\Delta$. It can be measured directly by observing the difference in travel times of a given ray at two neighboring stations on the earth's surface. Because seismology is based chiefly on measurements made on the earth's surface, it is advantageous to write Eq. (7.88) in a form that relates to the entire ray path from source to station:

$$T = p\Delta + \left[\int_{r_m}^{r_h} + \int_{r_m}^{a}\right](\eta^2 - p^2)^{1/2}\frac{dr}{r}, \tag{7.96}$$

where a is the radius of the earth and T is the total travel time along a given ray. Similarly, Eq. (7.89) becomes

$$\Delta(p) = p\left[\int_{r_m}^{r_h} + \int_{r_m}^{a}\right](\eta^2 - p^2)^{-1/2}\frac{dr}{r}. \tag{7.97}$$

Table 7.1 gives the observed travel times and the values of the ray parameter $p = dT/d\Delta$ for P waves at various epicentral distances from 0 to 100°.

Equation (7.96) is of prime importance in seismology as it links the observed travel-times of body-waves with the intrinsic P and S velocity distributions in the earth. Most of our present knowledge about the earth's interior has come from the study of seismic body waves. These studies have been largely confined to the travel times of first arriving waves. Recently surface waves have supplied an independent view of the structure of the upper mantle that is in substantial agreement with the gross features of some of the standard body-wave models. With the improved coverage of the earth's surface with standardized instruments and arrays, the use of accurately timed and located large explosions as seismic sources, and the increased accuracy and convenience of data processing with large digital computers, there is an increased emphasis on the use of body waves for detailed study of the earth's deep interior.

The phase of a ray solution is given by the factor $e^{i\omega(t-T)}$, where T is the travel time along the ray. Inserting the explicit expression for T from Eq. (7.96), the phase angle becomes

$$\omega\left[t - \frac{a\Delta}{V_a} - \left\{\int_{r_m}^{r_h} + \int_{r_m}^{a}\right\}(\eta^2 - p^2)^{1/2}\frac{dr}{r}\right],$$

where $a\Delta$ is the epicentral distance in kilometers and

$$V_a = \frac{V_0}{\sin i_0}$$

is the *apparent surface velocity* or the *ray phase velocity* (Fig. 7.6). All wave frequencies that travel along the same ray therefore share a common phase velocity. We shall return to this important feature in Chapter 8, when we shall link it with the normal-mode solution. The integral expression in the phase is an additional phase that depends on p and the ray termini. Figure 7.7 shows the variation of the ray phase velocity with epicentral distance for a few representative phases.

Table 7.1. Observed Times of P Waves, Surface Focus

	$\Delta^{\circ a}$	Time (s)	$dT/d\Delta$ (s/deg)	r_m (km)		$\Delta^{\circ a}$	Time (s)	$dT/d\Delta$ (s/deg)	r_m (km)
P_g	0.0	0.0000	18.5326	6371.0		4.0	62.2906	13.7038	6326.5
	0.1	1.8533	18.5326	6371.0		4.1	63.6609	13.7026	6326.3
	0.2	3.7065	18.5325	6371.0		4.2	65.0311	13.7014	6326.0
	0.3	5.5598	18.5325	6371.0		4.3	66.4012	13.7001	6325.7
	0.4	7.4130	18.5325	6371.0		4.4	67.7712	13.6988	6325.4
	0.5	9.2663	18.5324	6371.0		4.5	69.1410	13.6975	6325.1
	0.6	11.1195	18.5323	6370.9		4.6	70.5107	13.6961	6324.7
	0.7	12.9727	18.5322	6370.9		4.7	71.8802	13.6947	6324.4
	0.8	14.8259	18.5321	6370.9		4.8	73.2496	13.6933	6324.1
	0.9	16.6791	18.5320	6370.8		4.9	74.6188	13.6918	6323.7
P_g	1.0	18.5323	18.5319	6370.[illegible]		5.0	75.9880	13.6903	6323.4
P^*	1.1	20.3788	16.4344	6355.9		5.5	82.8312	13.6825	6321.5
	1.2	22.0222	16.4343	6355.9		6.0	89.6703	13.6738	6319.4
	1.3	23.6656	16.4342	6355.9		6.5	96.5049	13.6645	6317.1
	1.4	25.3090	16.4341	6355.8		7.0	103.3346	13.6543	6314.6
	1.5	26.9525	16.4340	6355.8		7.5	110.1591	13.6435	6311.8
	1.6	28.5958	16.4339	6355.7		8.0	116.9779	13.6318	6308.9
	1.7	30.2392	16.4337	6355.7		8.5	123.7908	13.6195	6305.8
P^*	1.8	31.8826	16.4336	6355.6		9.0	130.5973	13.6063	6302.5
P_n	1.9	33.4908	13.7224	6330.5		9.5	137.3970	13.5925	6298.9
	2.0	34.8630	13.7218	6330.4		10.0	144.1896	13.5778	6295.2
	2.1	36.2351	13.7212	6330.3		10.5	150.9747	13.5625	6291.2
	2.2	37.6072	13.7206	6330.1		11.0	157.7519	13.5463	6287.0
	2.3	38.9792	13.7199	6330.0		11.5	164.5209	13.5295	6282.7
	2.4	40.3512	13.7192	6329.9		12.0	171.2813	13.5118	6278.1
	2.5	41.7231	13.7185	6329.7		12.5	178.0326	13.4935	6273.3
	2.6	43.0949	13.7177	6329.6	P_n	13.0	184.7746	13.4743	6268.3
	2.7	44.4666	13.7169	6329.4		13.5	191.4964	13.4206	6255.9
	2.8	45.8383	13.7161	6329.2		14.0	198.1926	13.3619	6243.6
	2.9	47.2098	13.7152	6329.1		14.5	204.8555	13.2862	6229.0
	3.0	48.5813	13.7143	6328.9		15.0	211.4756	13.1906	6211.9
	3.1	49.9527	13.7134	6328.7		15.5	218.0429	13.0758	6192.9
	3.2	51.3240	13.7125	6328.5		16.0	224.5485	12.9444	6172.4
	3.3	52.6952	13.7115	6328.3		16.5	230.9845	12.7960	6150.4
	3.4	54.0663	13.7105	6328.0		17.0	237.3414	12.6284	6126.7
	3.5	55.4373	13.7095	6327.8		17.5	243.6096	12.4411	6101.1
	3.6	56.8082	13.7084	6327.6		18.0	249.7793	12.2341	6073.8
	3.7	58.1789	13.7073	6327.3		18.5	255.8408	12.0099	6045.0
	3.8	59.5496	13.7062	6327.1		19.0	261.7872	11.7738	6015.2
	3.9	60.9202	13.7050	6326.8		19.5	267.6136	11.5314	5984.9

(continued)

Table 7.1. (cont.)

$\Delta^{\circ a}$	Time (s)	$dT/d\Delta$ (s/deg)	r_m (km)	$\Delta^{\circ a}$	Time (s)	$dT/d\Delta$ (s/deg)	r_m (km)
20.0	273.3185	11.2895	5954.8	40.0	455.7020	8.3024	5410.6
20.5	278.9036	11.0504	5924.9	40.5	459.8449	8.2690	5398.6
21.0	284.3693	10.8115	5894.7	41.0	463.9710	8.2354	5386.5
21.5	289.7160	10.5786	5865.0	41.5	468.0802	8.2013	5374.1
22.0	294.9501	10.3618	5836.8	42.0	472.1723	8.1672	5361.6
22.5	300.0806	10.1612	5810.0	42.5	476.2473	8.1329	5348.9
23.0	305.1134	9.9709	5784.0	43.0	480.3051	8.0981	5335.9
23.5	310.0533	9.7913	5758.8	43.5	484.3454	8.0630	5322.7
24.0	314.9070	9.6258	5734.9	44.0	488.3680	8.0274	5309.3
24.5	319.6818	9.4767	5712.6	44.5	492.3728	7.9916	5295.7
25.0	324.3869	9.3478	5692.7	45.0	496.3596	7.9557	5281.9
25.5	329.0331	9.2395	5675.1	45.5	500.3285	7.9196	5268.0
26.0	333.6295	9.1490	5659.8	46.0	504.2791	7.8828	5253.8
26.5	338.1848	9.0748	5646.6	46.5	508.2111	7.8452	5239.1
27.0	342.7068	9.0154	5635.4	47.0	512.1242	7.8068	5224.1
27.5	347.2025	8.9704	5626.3	47.5	516.0178	7.7678	5208.8
28.0	351.6796	8.9409	5619.8	48.0	519.8920	7.7290	5193.5
28.5	356.1456	8.9240	5615.7	48.5	523.7469	7.6907	5178.3
29.0	360.6048	8.9139	5612.8	49.0	527.5828	7.6531	5163.2
29.5	365.0596	8.9045	5609.8	49.5	531.4001	7.6163	5148.4
30.0	369.5086	8.8894	5605.1	50.0	535.1992	7.5800	5133.6
30.5	373.9477	8.8671	5598.3	50.5	538.9802	7.5440	5118.8
31.0	378.3751	8.8424	5590.8	51.0	542.7433	7.5084	5104.1
31.5	382.7900	8.8171	5583.1	51.5	546.4887	7.4731	5089.3
32.0	387.1923	8.7929	5575.7	52.0	550.2164	7.4379	5074.5
32.5	391.5831	8.7699	5568.4	52.5	553.9266	7.4028	5059.6
33.0	395.9621	8.7453	5560.5	53.0	557.6192	7.3675	5044.6
33.5	400.3281	8.7188	5551.9	53.5	561.2941	7.3319	5029.3
34.0	404.6807	8.6913	5543.0	54.0	564.9510	7.2958	5013.7
34.5	409.0193	8.6629	5533.7	54.5	568.5899	7.2597	4998.0
35.0	413.3435	8.6340	5524.1	55.0	572.2107	7.2238	4982.3
35.5	417.6532	8.6046	5514.3	55.5	575.8137	7.1879	4966.5
36.0	421.9479	8.5739	5504.0	56.0	579.3986	7.1517	4950.4
36.5	426.2269	8.5417	5493.1	56.5	582.9653	7.1150	4934.1
37.0	430.4894	8.5081	5481.8	57.0	586.5135	7.0777	4917.4
37.5	434.7347	8.4733	5470.0	57.5	590.0430	7.0403	4900.7
38.0	438.9626	8.4382	5458.0	58.0	593.5538	7.0031	4883.9
38.5	443.1730	8.4035	5446.1	58.5	597.0462	6.9666	4867.3
39.0	447.3662	8.3693	5434.2	59.0	600.5205	6.9307	4850.9
39.5	451.5425	8.3357	5422.4	59.5	603.9770	6.8955	4834.6

Table 7.1. (cont.)

$\Delta^{\circ a}$	Time (s)	$dT/d\Delta$ (s/deg)	r_m (km)	$\Delta^{\circ a}$	Time (s)	$dT/d\Delta$ (s/deg)	r_m (km)
60.0	607.4162	6.8613	4818.7	80.0	730.6349	5.4035	4066.3
60.5	610.8385	6.8281	4803.2	80.5	733.3270	5.3649	4044.9
61.0	614.2444	6.7957	4787.8	81.0	735.9998	5.3263	4023.5
61.5	617.6343	6.7639	4772.7	81.5	738.6533	5.2875	4001.8
62.0	621.0084	6.7326	4757.6	82.0	741.2871	5.2477	3979.6
62.5	624.3668	6.7011	4742.3	82.5	743.9007	5.2060	3956.2
63.0	627.7094	6.6689	4726.6	83.0	746.4926	5.1609	3930.9
63.5	631.0356	6.6359	4710.4	83.5	749.0611	5.1133	3904.4
64.0	634.3452	6.6024	4693.9	84.0	751.6058	5.0657	3877.8
64.5	637.6379	6.5685	4677.1	84.5	754.1271	5.0198	3852.1
65.0	640.9137	6.5345	4660.2	85.0	756.6260	4.9764	3827.6
65.5	644.1724	6.5005	4643.3	85.5	759.1042	4.9369	3805.2
66.0	647.4142	6.4667	4626.3	86.0	761.5636	4.9016	3785.0
66.5	650.6392	6.4334	4609.5	86.5	764.0064	4.8698	3766.6
67.0	653.8477	6.4005	4592.8	87.0	766.4338	4.8399	3749.0
67.5	657.0398	6.3675	4576.0	87.5	768.8465	4.8114	3732.1
68.0	660.2151	6.3336	4558.6	88.0	771.2455	4.7847	3716.1
68.5	663.3731	6.2985	4540.5	88.5	773.6315	4.7598	3701.0
69.0	666.5134	6.2627	4522.1	89.0	776.0056	4.7368	3686.8
69.5	669.6355	6.2253	4502.8	89.5	778.3687	4.7162	3673.9
70.0	672.7383	6.1851	4482.1	90.0	780.7222	4.6982	3662.4
70.5	675.8202	6.1424	4460.2	90.5	783.0673	4.6824	3652.2
71.0	678.8805	6.0991	4438.0	91.0	785.4049	4.6681	3642.7
71.5	681.9193	6.0560	4415.9	91.5	787.7356	4.6547	3633.6
72.0	684.9366	6.0134	4393.9	92.0	790.0597	4.6418	3624.8
72.5	687.9329	5.9723	4372.7	92.5	792.3774	4.6293	3616.1
73.0	690.9092	5.9332	4352.4	93.0	794.6891	4.6177	3607.8
73.5	693.8665	5.8960	4332.9	93.5	796.9953	4.6073	3600.3
74.0	696.8054	5.8599	4313.9	94.0	799.2966	4.5982	3593.6
74.5	699.7264	5.8244	4295.1	94.5	801.5937	4.5904	3587.7
75.0	702.6299	5.7895	4276.4	95.0	803.8872	4.5839	3582.6
75.5	705.5519	5.7545	4257.6	95.5	806.1777	4.5784	3578.2
76.0	708.3843	5.7189	4238.3	96.0	808.4658	4.5739	3574.4
76.5	711.2346	5.6821	4218.4	96.5	810.7518	4.5702	3571.2
77.0	714.0661	5.6433	4197.3	97.0	813.0361	4.5673	3568.5
77.5	716.8776	5.6030	4175.5	97.5	815.3192	4.5653	3566.6
78.0	719.6690	5.5628	4153.6	98.0	817.6016	4.5645	3565.7
78.5	722.4405	5.5230	4131.9	98.5	819.8838	4.5644	3565.6
79.0	725.1920	5.4829	4110.0	99.0	822.1660	4.5643	3565.4
79.5	727.9234	5.4428	4088.0	99.5	824.4481	4.5643	3565.4
				100.0	826.7303	4.5643	3565.4

[a] P_q, P^* and P_n are defined in Section 7.4.4.1 and shown in Fig. 7.39.

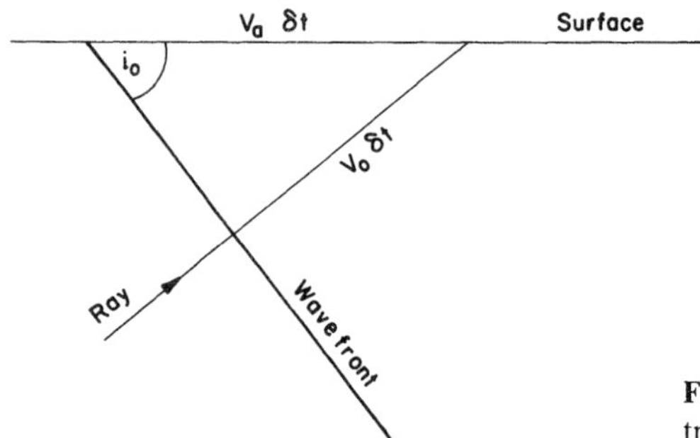

Figure 7.6. Apparent (V_a) and intrinsic (V_0) velocities.

Consider the problem of the travel-time perturbation. In a spherically symmetrical earth, the velocity distribution $V(r)$ is perturbed by a small amount $\delta V(r)$. It is necessary to evaluate the corresponding perturbation in the travel time $(\delta T)_\Delta$ at a fixed epicentral distance, Δ. Because the end points of the ray are fixed,

$$T + \delta T = \int_P^Q \frac{ds}{V(s) + \delta V(s)}, \tag{7.98}$$

along the new perturbed ray, which is the Fermat extremal-time path for the new distribution $V + \delta V$. We then have

$$(\delta T)_\Delta = \int_P^Q \frac{ds}{V + \delta V} - \int_P^Q \frac{ds}{V} \simeq - \int_P^Q \left(\frac{\delta V}{V^2}\right) ds = - \int_P^Q \left(\frac{\delta V}{V}\right) dT, \tag{7.99}$$

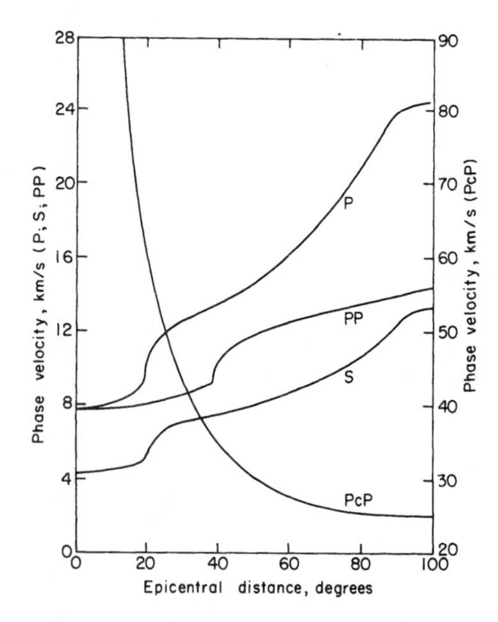

Figure 7.7. Variation of the phase velocity with epicentral distance for P, S, PP, and PcP rays.

where now the integration is along the original ray. In practice it is preferable to integrate over r. To this end, we use Eq. (7.91) and obtain for a surface focus

$$(\delta T)_\Delta = -2\int_{r_m}^{a} \frac{\delta V(r)}{V^3} \frac{r\,dr}{\sqrt{(r^2/V^2 - p^2)}}. \tag{7.100}$$

Still another useful form, which we shall use later, is to convert Eq. (7.100) into an integral over Δ. Using Eq. (7.89), we get

$$(\delta T)_\Delta = -\frac{1}{p}\int_0^{\Delta} r^2 \frac{\delta V}{V^3}\,d\theta. \tag{7.101}$$

7.1.5.1. Inversion of Travel-Time Data. Travel-time data of seismic rays from natural earthquakes and anthropogenic explosions have been accumulating since the turn of the century. Average travel-time tables, $T(\Delta)$, for many seismic rays in the earth are now in common use for various seismologic research projects. One of the prime uses of these data, however, is the determination of the variation of the velocity–depth functions $\alpha(r)$ and $\beta(r)$ from the surface measurement of $T(\Delta)$. This was probably the first inverse problem encountered in geophysics and its mathematical solution was found by Herglotz in 1907. Clearly, the determination of $T(\Delta)$ from a given distribution $V(r)$ is unique. The uniqueness of the inverse process is not so obvious. Consider a source on the earth's surface. Then, by Eq. (7.97),

$$\Delta(p) = 2p\int_{r_m}^{a} \frac{d(\ln r)}{\sqrt{(r^2/V^2 - p^2)}}. \tag{7.102}$$

When we introduce

$$y = \left(\frac{r}{aV}\right)^2, \tag{7.103}$$

Eq. (7.102) becomes

$$\frac{1}{2}V_a\Delta = \int_{y_m}^{y_0} \frac{d/dy(\ln r)}{\sqrt{(y - y_m)}}\,dy, \tag{7.104}$$

$$y_0 = \frac{1}{V_0^2}, \qquad y_m = \left(\frac{r_m}{aV_m}\right)^2 = \frac{1}{V_a^2}, \qquad V_a = \frac{a}{p}.$$

We know that the solution of the *Abel integral equation*

$$f(x) = \int_x^b \frac{u(y)dy}{(y - x)^k}, \qquad (0 < k < 1) \tag{7.105}$$

is

$$u(y) = -\frac{\sin \pi k}{\pi}\frac{d}{dy}\int_y^b \frac{f(x)dx}{(x - y)^{1-k}}. \tag{7.106}$$

The necessary and sufficient conditions that Eq. (7.105) will have a continuous solution, Eq. (7.106), in the interval $y \le x \le b$ are:

1. $f(x)$ must be continuous in the interval except for a finite number of discontinuities.
2. $f(b) = 0$.

Therefore, the solution of the integral equation (7.104) is

$$\frac{d}{dy} \ln r(y) = \frac{-1}{2\pi} \frac{d}{dy} \int_y^{y_0} V_a \Delta (y_m - y)^{-1/2} \, dy_m. \tag{7.107}$$

Integrating with respect to y, we get

$$\ln\left(\frac{r}{a}\right) = \frac{-1}{2\pi} \int_y^{y_0} V_a \Delta (y_m - y)^{-1/2} \, dy_m. \tag{7.108}$$

Suppose that we wish to calculate the velocity at the level, $r = r_1$. We then write Eq. (7.108) as

$$\ln\left(\frac{r_1}{a}\right) = \frac{-1}{2\pi} \int_{y_1}^{y_0} V_a \Delta (y_m - y_1)^{-1/2} \, dy_m, \qquad y_1 = \left(\frac{r_1}{aV_1}\right)^2. \tag{7.109}$$

It may be noted that whereas in the original Eq. (7.102), the integration is along a given ray (p fixed) that bottoms at $r = r_m$, the inverse relation, Eq. (7.109), entails integration over a *family of rays*, having $y_1 \le y_m \le y_0$, i.e., $r_1 \le r_m \le a$ (Fig. 7.8). In terms of the epicentral distance, we have $\Delta_1 \le \Delta \le 0$, where Δ_1 is the epicentral distance for a ray that bottoms at $r = r_1$.

Putting $y_m = p^2/a^2$ in Eq. (7.109), we get

$$\ln\left(\frac{r_1}{a}\right) = \frac{-1}{\pi} \int_{p_1}^{p_0} \Delta (p^2 - p_1^2)^{-1/2} \, dp,$$

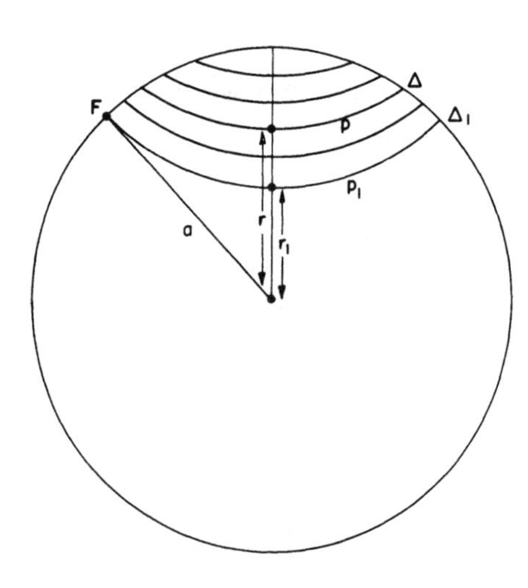

Figure 7.8. Geometrical interpretation of the integration over a family of rays (Eq. 7.109).

where $p_1 = r_1/V_1$, $p_0 = a/V_0$. Integrating by parts and noting that $\Delta = 0$ when $p = p_0$, we obtain

$$\ln\left(\frac{r_1}{a}\right) = \frac{-1}{\pi}\int_0^{\Delta_1} \mathrm{ch}^{-1}\left(\frac{p}{p_1}\right) d\Delta, \tag{7.110}$$

which may be written as

$$V(r_1) = \frac{a}{p(\Delta_1)} \exp\left[\frac{-1}{\pi}\int_0^{\Delta_1} \mathrm{ch}^{-1}\left\{\frac{p(\Delta)}{p(\Delta_1)}\right\} d\Delta\right]. \tag{7.111}$$

Therefore, given the numbers $p(\Delta) = dT/d\Delta$ for a sufficiently dense grid of points in the interval $0 \leq \Delta \leq \Delta_1$ we are able to calculate, in principle, the velocity V_1 at $r = r_1$, corresponding to the lowest point of the ray that reaches $\Delta = \Delta_1$. The method is valid under the assumptions that the rays do not intersect and that the ratio (V/r) in the earth grows monotonically with depth, i.e., $-d/dr(V/r) = r^{-1}(V/r - dV/dr) > 0$, because otherwise no ray will have its deepest point at a given r.

As long as $V(r)$ is continuous, V slowly increases as r decreases and dV/dr changes slowly, the travel time $T(\Delta)$ is a single-valued monotonically increasing function of Δ and the slope of the travel-time curve, $dT/d\Delta$, at each point determines the value of the ray parameter, p, there. This is the usual case for most regions of the mantle of the earth and in limited regions of its deep interior. However, the above conditions on V are not always satisfied. One or more of the following anomalies in $V(r)$ are common:

1. Rapid increase in velocity with depth.
2. Discontinuous increase in velocity with depth.
3. Decrease in velocity with depth.

Case 1 is depicted in Fig. 7.9a. If $V/(-dV/dr)$ decreases with depth, then from Eq. (7.93) it follows that a ray going down at a steeper angle will have a smaller radius of curvature at the lowest point and, consequently, will appear closer to the epicenter (Fig. 7.9b). Equation (7.111) still holds, for to each value of p there corresponds a single Δ. However, difficulties in observing $dT/d\Delta$ render the inversion method ineffective.

Introducing the notation

$$\xi = 2\Big/\left(1 - \frac{r}{V}\frac{dV}{dr}\right), \tag{7.112}$$

integrating Eq. (7.102) by parts, and then differentiating with respect to p, we get

$$\frac{d\Delta}{dp} = -\xi_0\left(\frac{a^2}{V_0^2} - p^2\right)^{-1/2} + \int_{r_m}^{a}\left(\frac{r^2}{V^2} - p^2\right)^{-1/2}\left(\frac{d\xi}{dr}\right) dr, \tag{7.113}$$

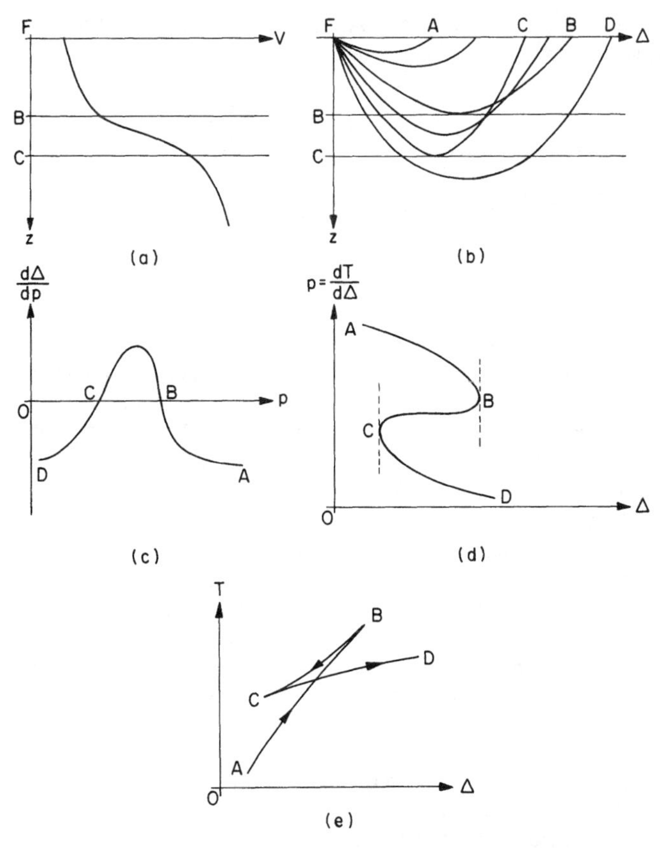

Figure 7.9. (a) Rapid increase of velocity with depth in the region BC. (b) Effect on the points of emergence of rays on the surface. (c) $d\Delta/dp$ may change sign. (d) Variation of p with Δ. (e) Triplication of the travel-time curve caused by the rapid increase of velocity with depth.

where ξ_0 is the value of ξ at the surface. The sign of $d\Delta/dp$ will depend upon the value of $d\xi/dr$. The first term on the right-hand side of Eq. (7.113) is negative. Therefore, if $V(r)$ is such that

$$\frac{d\xi}{dr} = 2\frac{d}{dr}\left(\frac{r}{V}\frac{dV}{dr}\right)\left(1 - \frac{r}{V}\frac{dV}{dr}\right)^{-2} \tag{7.114}$$

is small, $d\Delta/dp$ will be negative. However, if $d\xi/dr$ is positive and sufficiently large, this will make $d\Delta/dp > 0$ (Fig. 7.9c) for a range of values of p. This results in the triplication of the travel-time curve (Fig. 7.9e). If $d\xi/dr$, although positive, is not sufficiently large to make $d\Delta/dp$ positive, no triplication of the travel-time curve occurs although this curve will have a considerable curvature in the vicinity of the appropriate value of Δ. The points B and C on the (T, Δ) curve are known as *cusps*. At a cusp, $d\Delta/dp = 0$, and we shall see later, in Section 7.5, that the ray amplitude at such a point is relatively high. The travel-time curve is triple-valued between B and C and reverses itself at B and again at C. The segments meeting at the cusps are mutually tangential. The segment BC is referred to as a *reverse segment* of the travel-time curve.

In the second case, there is a discontinuity surface across which V increases by a finite amount from above to below, the behavior on both sides of the surface being ordinary. Because of rays that are totally reflected upward at the discontinuity, there is a gap in the values of p for the rays reaching the surface. These rays are not included in Eq. (7.111).

The third case has an important application in seismology as it can serve as a model for a *low-velocity layer* that is believed to exist below the earth's crust. The situation is shown in Fig. 7.10. Up to the point B, rays emerge in a normal manner and Eq. (7.111) can be used. At steeper takeoff angles, rays are affected by the low-velocity layer such that they emerge beyond D. The zone $\Delta_B \leq \Delta \leq \Delta_D$ is a *shadow zone* for real rays and ray theory breaks down there. The diffracted field in the zone must then be determined by a direct solution of the wave equation itself. When this is done, it is found that the amplitudes of both P and S waves inside the shadow zone in the vicinity of B decay exponentially at a rate proportional to a one-third power of the frequency and a two-thirds power of the velocity gradient near the surface [see Eq. (7.9.23)]. At the outer boundary of the shadow zone, a focusing of the amplitudes will occur because of the cusp at D (Fig. 7.10e). In general, this focusing is a characteristic of a

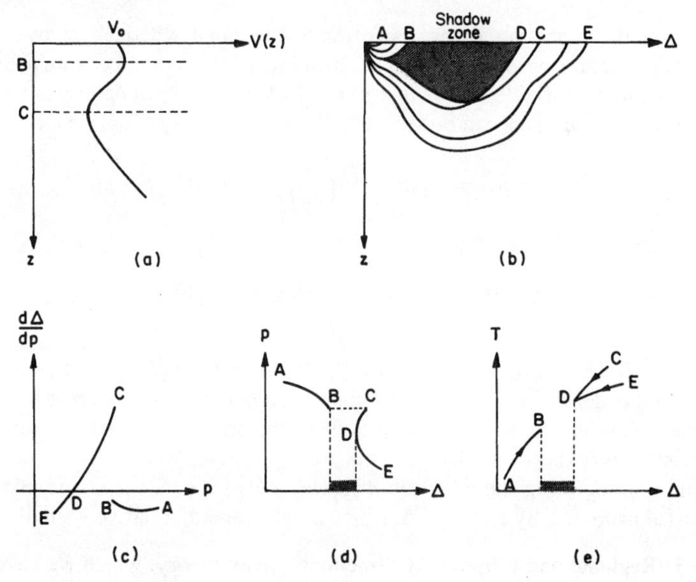

Figure 7.10. (a) Low-velocity channel. (b) Shadow zone. (c), (d), (e) The discontinuities in the $d\Delta/dp$–p, p–Δ, and T–Δ curves.

velocity structure with a negative gradient section. The generated shadow zone is followed by a focusing point at its outer boundary. The diffracted field in the shadow zone will be treated separately in Chapter 8.

7.1.5.2. Unfolding of the Travel-Time Curve. We have seen above that a rapid increase in velocity with depth may lead to a triplication in the T–Δ curve. In such a situation, the time intervals between successive arrivals at a given station may be very short and, therefore, we may have difficulty in choosing the appropriate branch of the travel-time curve. To overcome this difficulty, we look for an independent variable such that the data become a single-valued function. This would lead to the "unfolding" of any multiplicities that might exist in the travel-time curve.

We notice that both T and Δ are single-valued functions of p. Therefore, the combination $\tau = T - p\Delta$ is also a single-valued function of p. From Eq. (7.96), τ can be expressed in the form

$$\tau = T - p\Delta = \left[\int_{r_m}^{r_h} + \int_{r_m}^{a}\right](\eta^2 - p^2)^{1/2}\frac{dr}{r}. \tag{7.115}$$

Because $p = dT/d\Delta$, we have $d\tau/dp = -\Delta(p)$. Therefore, τ is a monotonically decreasing, single-valued function of p leading to the "unfolding" of the travel-time curve. Note that the $(\tau$–$p)$ curve may have discontinuities in the values of τ because of low-velocity zones, but the p values will be continuous.

Equations (7.100) and (7.115) yield

$$(\delta\tau)_p = (\delta T)_\Delta.$$

Therefore, the inversion theory is formally the same whether we use $T(\Delta)$ data or $\tau(p)$ data. Another advantage of using $\tau(p)$ data is that, because p is now the independent variable, errors in $\tau(p)$ results from errors of δp in p are of the second order in δp. This can be seen as follows: If $p_1 = p_0 + \delta p$, we have

$$\tau(p_1) = \tau(p_0) + \delta p\left(\frac{d\tau}{dp}\right)_0 + O(\delta p)^2.$$

Because $d\tau/dp = -\Delta$, we find

$$\begin{aligned}\tau(p_1) &= \tau(p_0) - (p_1 - p_0)\Delta(p_0) + O(\delta p)^2 \\ &= T(p_0) - p_1\Delta(p_0) + O(\delta p)^2.\end{aligned}$$

Therefore, if we approximate $\tau(p_1)$ by $\hat{\tau}(p_1) = T(p_0) - p_1\Delta(p_0)$, the error is of the second order in $\delta p = p_1 - p_0$. In this way, a major source of error has been reduced. Now the error in $\tau(p)$ results from the errors in T and Δ. However, if we have an experimentally determined $\Delta(p)$ curve, we can construct $\tau(p)$ by simply integrating the equation $d\tau/dp = -\Delta(p)$. Then, the errors in τ will be caused primarily by errors in Δ, which are, in general, relatively small.

7.1.5.3. Regional and Ellipticity Corrections. Ray theory, which we have developed so far, applies to an ideal earth in which the surfaces $V(r) =$ const. are exact spheres for all $0 < r < a$. This does not happen in the case of the real earth.

Nevertheless, it is convenient to define a standard of comparison as an earth model in which each surface of equal P or S velocity is spherical and encloses the same volume as in the real earth. This is known as the *mean sphere*. Standard travel-time tables in the earth, such as the *Jeffreys–Bullen* (*J–B*) *tables* are related to the mean sphere and use geographic rather than spherical polar coordinates. (The geographic latitude of a place is the angle between the normal to the level surface at the place and the equatorial plane.) These tables are primarily based on global observation of P and S phases. These data are then inverted on the assumption that (V/r) is monotonically increasing with depth and travel times of additional common seismic rays are calculated on the basis of the derived $\alpha(r)$ and $\beta(r)$.

Because the actual earth deviates from the mean spherical model, certain corrections need to be applied to the observations before comparison can be made with J–B tables. Chief among these is the ellipticity correction. The true shape is given to sufficient accuracy (for seismologic purposes) by an *ellipsoid of revolution* with semiaxes

$$a = 6378.388 \text{ km}, \qquad b = 6356.912 \text{ km}$$

and a surface ellipticity

$$e_0 = \frac{a - b}{a} = \frac{1}{297.0}.$$

With these figures, the ellipticity correction may reach the maximum of 2 s for direct P waves and 3 s for direct S waves. To evaluate the expression for this correction, we assume that the surfaces of equal velocity are given by

$$r' = r + \delta r(r, \theta'),$$

$$\delta r = re(r)\left(\frac{1}{3} - \cos^2 \theta'\right),$$

where (r', θ', ϕ') are the spherical polar coordinates, called *geocentric coordinates*, with the equator as the xy plane, and r is the radius of the mean sphere.

The geometric setup is shown in Fig. 7.11. The change δV corresponding to δr is obtained by the expansion of $V(r')$ by a Taylor series about r:

$$V(r') = V(r + \delta r) \simeq V(r) + \delta r \frac{dV}{dr}.$$

Therefore,

$$\delta V = V(r) - V(r') = -\delta r \frac{dV}{dr} = -re(r)\left(\frac{1}{3} - \cos^2 \theta'\right)\frac{dV}{dr}. \tag{7.116}$$

Substituting into the general expression for δT obtained in Eq. (7.101), we get

$$(\delta T)_1 = \frac{1}{p}\int_0^{\Delta} \frac{r^3}{V^3} e(r)\left(\frac{1}{3} - \cos^2 \theta'\right)\frac{dV}{dr}\, d\theta, \tag{7.117}$$

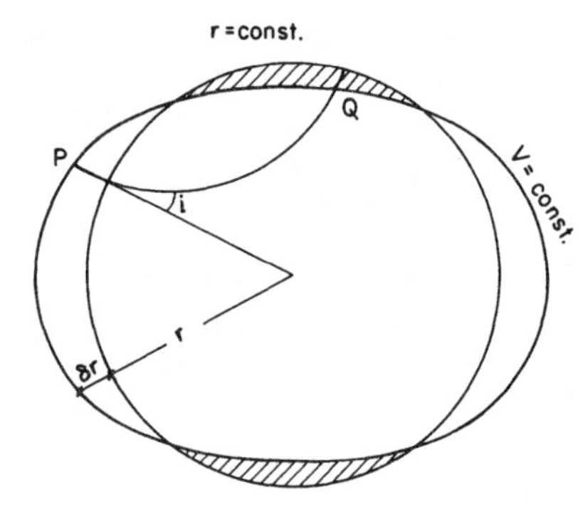

Figure 7.11. Ellipticity correction.

which comprises the first part of the ellipticity correction. It may be noted that the angle θ is the angle shown in Fig. 7.4, and the angle $\theta'(\theta)$ is the geocentric colatitude of the corresponding point on the ray.

The second part results from the displacement of the ray-ends by (δr), with no change in the internal distribution of the velocity. It is simply equal to $(\delta T)_2 = [(\delta r/V_0)\cos i_0]_{0,\Delta}$, calculated at both ends of the ray. This is clearly seen in Fig. 7.11, where at one end the effect has a positive sign and at the other end a negative sign. The total ellipticity correction, to be subtracted from the observed times to get times over the mean sphere, is then

$$\delta T = e_0\left[\left(\frac{1}{3} - \cos^2\theta'\right)\left(\frac{a^2}{V_0^2} - p^2\right)^{1/2}\right]_{0,\Delta} + (\delta T)_1. \qquad (7.118)^3$$

In addition to ellipticity, which, to first order, is independent of the azimuth ϕ, there are *regional corrections*, which are caused by horizontal inhomogeneities in the earth's crust and upper mantle. If the δV caused by these regional deviations from the mean sphere are known, the corresponding travel-time corrections are again calculated with the aid of Eq. (7.101).

7.2. Ray-Amplitude Theory

7.2.1. Intrinsic Coordinate System

So far we have dealt only with the travel times of rays via the eikonal equation, Eq. (7.57). Knowing Ψ (or t) on each ray, we may proceed to solve Eq. (7.58) for the ray amplitudes. To this end, we set up a spherical coordinate system

[3] The symbol $[\ \ ]_{0,\Delta}$ means that the quantities within the brackets are to be evaluated at $\theta = 0$ (P in Fig. 7.11) and $\theta = \Delta$ (Q in Fig. 7.11). The two values are then to be added.

(r, θ, ϕ) with its origin at the earth's center and its polar axis through the focus. Simultaneously, we define another coordinate system (t, i_h, ϕ) with its origin at the focus and call it the *intrinsic coordinate system* (Fig. 7.12). Here t denotes the travel time along the ray, i_h the takeoff angle, and ϕ the usual azimuthal angle.

The surfaces $t =$ const. represent wavefronts and the curves $i_h =$ const., $\phi =$ const. are the orthogonal trajectories to the wavefronts (rays). The ranges of variation of t, i_h, and ϕ are

$$t \geq 0, \qquad 0 \leq i_h \leq \pi, \qquad 0 \leq \phi \leq 2\pi. \tag{7.119}$$

Let $(\mathbf{e}_t, \mathbf{e}_{i_h}, \mathbf{e}_\phi)$ denote the moving trihedral, so that $\mathbf{e}_t$ is the unit vector in the sense of t increasing, $\mathbf{e}_{i_h}$ in the sense of i_h increasing, and $\mathbf{e}_\phi$ in the sense of ϕ increasing. If the angle between the radius vector and the tangent to the ray at any point $P(t, i_h, \phi)$ is denoted by i, then Snell's law may be stated as

$$p = \frac{r \sin i}{V(r)} = \frac{r_h \sin i_h}{V(r_h)}, \tag{7.120}$$

where p is the ray parameter and r_h is the radius at the focus.

It is clear from Fig. 7.12 that

$$\mathbf{e}_t = \mathbf{e}_r \cos i + \mathbf{e}_\theta \sin i,$$

$$\mathbf{e}_{i_h} = -\mathbf{e}_r \sin i + \mathbf{e}_\theta \cos i.$$

We then have [cf. Eqs. (7.88) and (7.90)]

$$t = \int_{r_h}^{r} \frac{dr}{V \cos i} = \pm \int_{r_h}^{r} \frac{r\,dr}{V^2(r^2/V^2 - p^2)^{1/2}} = p\theta \pm \int_{r_h}^{r} \left(\frac{r^2}{V^2} - p^2\right)^{1/2} \frac{dr}{r}. \tag{7.121}$$

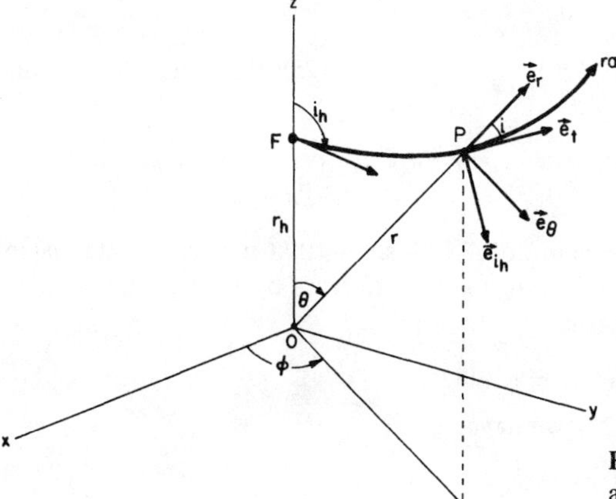

Figure 7.12. Intrinsic coordinates.

Simple differentiation yields

$$\left(\frac{\partial r}{\partial t}\right)_{i_h} = V \cos i, \qquad \left(\frac{\partial \theta}{\partial t}\right)_{i_h} = \frac{V \sin i}{r},$$
$$\left(\frac{\partial t}{\partial r}\right)_{\theta} = \frac{\cos i}{V}, \qquad \left(\frac{\partial t}{\partial \theta}\right)_{r} = p, \tag{7.122}$$
$$\left(\frac{\partial \theta}{\partial i_h}\right)_{t} = pI_h \cos^2 i \cot i_h, \qquad \left(\frac{\partial r}{\partial i_h}\right)_{t} = -p^2 I_h V \cos i \cot i_h,$$

where

$$I_h(r, i_h) = \left(\frac{\partial p}{\partial \theta}\right)_r^{-1} = \pm \int_{r_h}^{r} (r^2 - V^2 p^2)^{-(3/2)} V r \, dr. \tag{7.123}$$

Next, using the relation [Eq. (7.89)]

$$\theta = \pm p \int_{r_h}^{r} \frac{dr}{r(r^2/V^2 - p^2)^{1/2}},$$
$$\tan i = r\left(\frac{\partial \theta}{\partial r}\right)_{i_h}, \tag{7.124}$$

we get

$$\left(\frac{\partial i_h}{\partial r}\right)_{\theta} = -\frac{V \tan i_h}{r^2 I_h \cos i}, \qquad \left(\frac{\partial i_h}{\partial \theta}\right)_{r} = \frac{\tan i_h}{pI_h}. \tag{7.125}$$

The orthogonality of the coordinate system (t, i_h, ϕ) can be proved easily. We have

$$dr = \left(\frac{\partial r}{\partial t}\right)_{i_h} dt + \left(\frac{\partial r}{\partial i_h}\right)_{t} di_h,$$
$$d\theta = \left(\frac{\partial \theta}{\partial t}\right)_{i_h} dt + \left(\frac{\partial \theta}{\partial i_h}\right)_{t} di_h, \tag{7.126}$$
$$d\phi = d\phi.$$

Consequently, using Eqs. (7.120), (7.122), and (7.125), the square of the line element is given by

$$(dr)^2 + (r\, d\theta)^2 + (r \sin \theta\, d\phi)^2 = V^2 (dt)^2 + (prI_h \cos i \cot i_h)^2 (di_h)^2 + (r \sin \theta)^2 (d\phi)^2. \tag{7.127}$$

The absence of the cross-terms in Eq. (7.127) shows that the coordinate system (t, i_h, ϕ) is orthogonal. If we denote the scale factors by h_1, h_2, and h_3, then the square of the line element is

$$h_1^2 (dt)^2 + h_2^2 (di_h)^2 + h_3^2 (d\phi)^2. \tag{7.128}$$

Comparing it with Eq. (7.127), we have

$$h_1 = V, \qquad h_2 = |prI_h \cos i \cot i_h|, \qquad h_3 = r \sin \theta. \tag{7.129}$$

We now return to Eq. (7.58), which, using Eq. (7.61), is transformed to

$$\frac{\partial \xi_n}{\partial t} + \left(\frac{1}{2} V^2 \nabla^2 t\right) \xi_n = \frac{1}{2} V^2 \nabla^2 \xi_{n-1}. \tag{7.130}$$

This is a first-order differential equation of the type

$$\frac{dy}{dx} + a(x) y = b(x)$$

and can be solved by Lagrange's method of "variation of the parameters." Its solution is

$$\xi_n(t) = Q(t/t_0)\left[\xi_n(t_0) + \frac{1}{2}\int_{t_0}^{t} \frac{V^2 \nabla^2 \xi_{n-1}}{Q(t/t_0)}\, dt\right], \tag{7.131}$$

where

$$Q(t/t_0) = \frac{q(t)}{q(t_0)} = \exp\left[-\frac{1}{2}\int_{t_0}^{t} V^2 (\nabla^2 t) dt\right]. \tag{7.132}$$

This is a recursive equation for the determination of the ξs. Putting $\xi_{-1} = 0$, we have

$$\xi_0(t) = Q(t/t_0)\xi_0(t_0). \tag{7.133a}$$

Inserting $\xi_0(t)$ into Eq. (7.131), we get

$$\xi_1(t) = Q(t/t_0)\left[\xi_1(t_0) + \frac{1}{2}\int_{t_0}^{t} \frac{V^2 \nabla^2 \{Q(t/t_0)\xi_0(t_0)\}}{Q(t/t_0)}\, dt\right]. \tag{7.133b}$$

In principle, this iteration process can yield explicit expression for any $\xi_n(t)$. It may be noted that if ξ_0 is defined to be identically zero, Eq. (7.133a) becomes an equation in ξ_1 and then Eq. (7.133b) yields ξ_2.

Now, because

$$\nabla^2 t = \frac{1}{h_1 h_2 h_3}\frac{\partial}{\partial t}\left(\frac{h_2 h_3}{h_1}\right) = \frac{1}{V^2}\frac{\partial}{\partial t}\ln\left(\frac{h_2 h_3}{h_1}\right), \tag{7.134}$$

Eqs. (7.120), (7.125), and (7.129) enable us to express $q(t)$ as

$$q(t) = \exp\left(-\frac{1}{2}\int^{t} V^2 \nabla^2 t\, dt\right)$$

$$= \left(\frac{h_1}{h_2 h_3}\right)^{1/2} \tag{7.135a}$$

$$= \frac{1}{r}\left|\frac{V \tan i_h}{p I_h \sin\theta \cos i}\right|^{1/2} \tag{7.135b}$$

$$= \frac{1}{r}\left|\frac{V}{\sin\theta \cos i}\left(\frac{\partial i_h}{\partial \theta}\right)_r\right|^{1/2}. \tag{7.135c}$$

It is therefore obvious that

$$q(t) = \left(\frac{h_1}{h_2 h_3}\right)^{1/2} \propto \left(\frac{V}{w}\right)^{1/2}, \tag{7.136}$$

where w is an element of area of the (i_h, ϕ) surface; i.e., it is the area of a cross section of a tube of rays $[(i_h, \phi), (i_h + di_h, \phi + d\phi)]$. From Eqs. (7.132) and (7.136) we find

$$Q(t/t_0) = \left(\frac{V w_0}{V_0 w}\right)^{1/2}. \tag{7.137}$$

From Eq. (1.124), we note that under conditions of equilibrium and for $\mathbf{F} = 0$ the energy-flux density $\boldsymbol{\Sigma}$ obeys the equation

$$\operatorname{div} \boldsymbol{\Sigma} = 0. \tag{7.138}$$

Because energy flows along the rays, we can put

$$\boldsymbol{\Sigma} = \mathbf{p} A$$

where

$$A = |\boldsymbol{\Sigma}|.$$

Using the relation $\mathbf{p} = V \operatorname{grad} \Psi$, we have

$$\begin{aligned} \operatorname{div} \boldsymbol{\Sigma} &= \operatorname{div}(A\mathbf{p}) = \operatorname{div}(AV\nabla\Psi) = AV\nabla^2\Psi + \nabla\Psi \cdot \nabla(AV) \\ &= AV\nabla^2\Psi + \frac{1}{V}\mathbf{p} \cdot \nabla(AV) = AV\nabla^2\Psi + \frac{1}{V}\frac{\partial}{\partial s}(AV). \end{aligned} \tag{7.139}$$

Equations (7.64) and (7.138) therefore yield

$$\frac{1}{AV}\frac{\partial}{\partial s}(AV) = -V\nabla^2 t.$$

Integrating, we find

$$\left(\frac{VA}{V_0 A_0}\right) = \exp\left[-\int_{s0}^{s} V\nabla^2 t \, ds\right] = \exp\left[-\int_{t0}^{t} V^2\nabla^2 t \, dt\right]. \tag{7.140}$$

Since A is the energy flowing per unit area per unit time, it transpires that $A \propto 1/w$, where w is the area of a cross section of a tube of rays. Therefore, Eq. (7.140) may be written as

$$\frac{V w_0}{V_0 w} = \exp\left[-\int_{t0}^{t} V^2\nabla^2 t \, dt\right], \tag{7.141}$$

which coincides with the corresponding result obtained from Eqs. (7.132) and (7.137).

EXAMPLE 7.1: Rays in a Homogeneous Sphere

We shall derive here the travel-time and amplitude equations of rays in a homogeneous sphere. We already know that in a homogeneous medium, the rays are straight lines. From the geometry of the triangle OFA (Fig. 7.13)

$$\begin{aligned} i_h &= i_0 + \Delta = i + \theta, \\ di &= -d\theta \quad \text{(for a fixed ray)}, \\ D(r) &= r\cos i - r_h \cos i_h, \\ pV_0 &= a\sin i_0 = r_h \sin i_h = r\sin i. \end{aligned} \tag{7.1.1}$$

Therefore, the travel time for the direct ray is $T = D(a)/V_0$. If the source is on the surface, it is easily seen that

$$T = \frac{2a}{V_0}\sin\frac{\Delta}{2} = \frac{D}{V_0}, \qquad p = \frac{dT}{d\Delta} = \frac{a}{V_0}\cos\frac{\Delta}{2}, \qquad \frac{dp}{d\Delta} = -\frac{D}{4V_0}. \tag{7.1.2}$$

We can, of course, derive the above equations directly from the general results. From Eq. (7.123), we find

$$I_h(r, i_h) = \frac{V_0}{r_h \cos i_h} - \frac{V_0}{r\cos i} = \frac{V_0 D(r)}{r r_h \cos i \cos i_h}. \tag{7.1.3}$$

Therefore,

$$h_1 = V_0, \qquad h_2 = D(r), \qquad h_3 = D\sin i_h. \tag{7.1.4}$$

Consequently, by Eq. (7.135a)

$$q(t) = \frac{1}{D}\left(\frac{V_0}{\sin i_h}\right)^{1/2}, \qquad Q(t/t_0) = \frac{D_0}{D} = \frac{t_0}{t}. \tag{7.1.5}$$

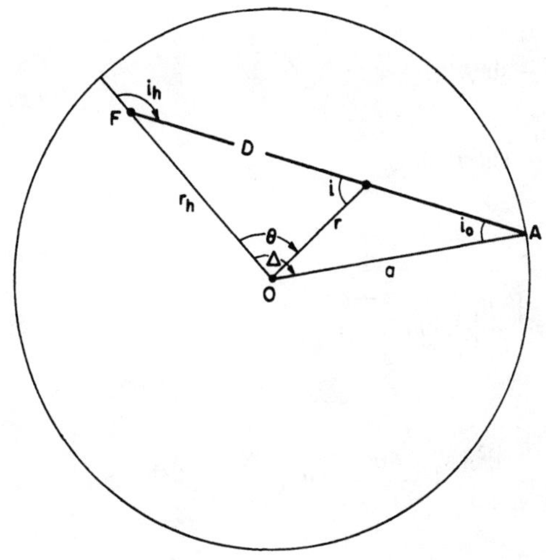

Figure 7.13. Ray geometry in a homogeneous medium.

The first term in the ray amplitude expansion therefore becomes,

$$\xi_0(t) = \xi_0(t_0)\left(\frac{D_0}{D}\right),$$

where $\xi_0(t_0)D_0$ is a source constant and cannot be specified unless the source is explicitly given. The total zero-order field potential is then [cf. Eq. (7.56)]

$$\Phi = D_0\xi_0(t_0)\left(\frac{e^{-ik_0D}}{D}\right), \qquad k_0 = \frac{\omega}{V_0}. \tag{7.1.6}$$

Note that (e^{-ik_0D}/D) is an exact particular solution of the wave equation, $\nabla^2\Phi + k_0^2\Phi = 0$. Comparing Eqs. (7.1.6) and (7.49) with the exact solution for an explosion source in a homogeneous medium [Eq. (4.138)], we get (for delta-function time dependence)

$$\xi_0(t_0) = \left[\frac{-M}{4\pi(\lambda + 2\mu)^{1/2}V_0^{3/2}}\right]t_0^{-1}, \qquad \xi_n(t_0) = 0, \qquad n \geq 1.$$

In a similar fashion, by comparing the solution with the exact solution for a dislocation source in an infinite medium, we can show that three terms of the eikonal expansion are needed and $\xi_n(t_0) = 0$ for $n \geq 3$.

EXAMPLE 7.2: The Law $V(r) = a - br^2$

First, we solve the scalar wave equation

$$\nabla^2\Phi + \frac{\omega^2}{V^2}\Phi = 0, \tag{7.2.1}$$

in intrinsic coordinates. Using Eq. (7.129), we derive the explicit expressions for the scale factors

$$\begin{aligned} h_1 &= a - br^2, \\ h_2 &= \frac{a - br^2}{\gamma}\,\mathrm{sh}(\gamma t), \qquad \gamma = 2\sqrt{(ab)}, \\ h_3 &= \frac{a - br^2}{\gamma}\,\mathrm{sh}(\gamma t)\sin i_h. \end{aligned} \tag{7.2.2}$$

Applying the transformation

$$\Phi = \left[\frac{h_1}{h_2 h_3}\sin i_h\right]^{1/2} S, \tag{7.2.3}$$

Eq. (7.2.1) becomes

$$\frac{\partial^2 S}{\partial t^2} + \frac{4ab}{\mathrm{sh}^2(\gamma t)}\frac{\partial^2 S}{\partial i_h^2} + \frac{4ab}{\mathrm{sh}^2(\gamma t)\sin^2 i_h}\frac{\partial^2 S}{\partial \phi^2} + \frac{4ab}{\mathrm{sh}^2(\gamma t)}\cot i_h\frac{\partial S}{\partial i_h} + (\omega^2 - ab)S = 0. \tag{7.2.4}$$

Although this equation can be solved exactly, we are interested here in solutions for long times ($\gamma t \gg 1$) and high frequencies. Then Eq. (7.2.4) reduces to

$$\frac{\partial^2 S}{\partial t^2} + \omega^2 S = 0.$$

The solutions are $S = e^{\pm i\omega t}$, where t is travel time along a ray. The expression for Φ is then

$$\Phi = \frac{\gamma}{\operatorname{sh} \gamma t} \frac{1}{\sqrt{(a - br^2)}} e^{-i\omega t}. \tag{7.2.5}$$

Equating the expression in Eq. (7.2.2) for h_3 to $r \sin \theta$ and making minor modifications, we get

$$t(r, \theta) = \frac{1}{\sqrt{(ab)}} \operatorname{sh}^{-1} \left[\frac{(r - r_h)^2 + 4rr_h \sin^2(\theta/2)}{(a - br^2)(a - br_h^2)} ab \right]^{1/2}. \tag{7.2.6}$$

When both source and sensor are on the surface, ($r = r_h$, $\theta = \Delta$), Eq. (7.2.6) reduces to

$$T = \frac{1}{\sqrt{(ab)}} \operatorname{sh}^{-1} \left[\operatorname{sh} \beta \sin \frac{\Delta}{2} \right], \tag{7.2.7}$$

where

$$\operatorname{ch} \beta = \frac{2a}{V_0} - 1, \qquad V_0 = a - br_h^2.$$

Using Eq. (7.92), we note that the radius of curvature of any ray in this medium is $\rho = 1/2bp = \text{const}$. The rays are therefore circular arcs of radii $(2bp)^{-1}$. Consequently, the wavefronts are spherical surfaces. Also, for a surface focus,

$$\Delta = 2 \tan^{-1} \left[\frac{\cot i_0}{\operatorname{ch} \beta} \right]. \tag{7.2.8}$$

The velocity model $V = a - br^2$ was fitted to the Jeffreys–Bullen shear velocity profile in the earth's mantle. The results are given in Table 7.2.

EXAMPLE 7.3: The Law $V = Ar^b$; $b < 1$

From Eq. (7.92)

$$\rho = -\frac{r^{2-b}}{Abp}. \tag{7.3.1}$$

For a surface focus

$$T = \frac{2a^{1-b}}{A(1-b)} \sin \left[\frac{1}{2}(1-b)\Delta \right] = \frac{2}{1-b} \left[\left(\frac{a^{1-b}}{A} \right)^2 - p^2 \right]^{1/2},$$

$$\Delta = \frac{2}{1-b} \cos^{-1} \left[\frac{Ap}{a^{1-b}} \right], \quad \frac{dT}{d\Delta} = \frac{a}{V_0} \cos \left[\frac{1}{2}(1-b)\Delta \right], \quad \frac{d^2T}{d\Delta^2} = -\left(\frac{1-b}{2} \right)^2 T. \tag{7.3.2}$$

Table 7.2. Least-Squares Fit of $V = a - br^2$ to J–B Shear Velocity Distribution in the Earth's Mantle.

r (km)	$V(r)$ (km/s)	$a - br^2$ (km/s)[a]	$V(r) - (a - br^2)$[a]
6338	4.35	4.715	$-3.65E-1$
6270	4.45	4.804	$-3.54E-1$
6170	4.60	4.935	$-3.35E-1$
6070	4.76	5.063	$-3.03E-1$
5970	4.94	5.189	$-2.49E-1$
5957	4.96	5.205	$-2.45E-1$
5870	5.32	5.313	$6.69E-3$
5770	5.66	5.435	$2.24E-1$
5670	5.93	5.554	$3.75E-1$
5570	6.13	5.672	$4.57E-1$
5470	6.27	5.788	$4.81E-1$
5370	6.36	5.901	$4.58E-1$
5170	6.50	6.122	$3.77E-1$
4970	6.62	6.334	$2.85E-1$
4770	6.73	6.538	$1.91E-1$
4570	6.83	6.734	$9.58E-2$
4370	6.92	6.921	$-1.32E-3$
4170	7.02	7.100	$-8.01E-2$
3970	7.12	7.270	$-1.50E-1$
3770	7.21	7.432	$-2.22E-1$
3570	7.30	7.586	$-2.86E-1$
3470	7.30	7.660	$-3.60E-1$

[a] $a = 8.9205$ km/s, $b = 1.0469 \times 10^{-7}$/(km.s)

This model provides a close approximation to actual velocity variation over particular ranges in the earth. In practice, the mantle is divided into a number of layers where in each layer the constants A and b are adjusted to fit the travel times of a given earth model. Then, the fit can be improved by perturbing the parameters in each layer. From Eq. (7.3.2), we note that

$$\frac{\partial}{\partial A}(T - p\Delta) = -\frac{T}{A},$$
$$\frac{\partial}{\partial a}(T - p\Delta) = \frac{1-b}{a}T, \qquad \frac{\partial}{\partial b}(T - p\Delta) = \frac{T - p\Delta}{1-b} - T\ln a. \tag{7.3.3}$$

The total change in the travel time is given by

$$\delta T(A, b, a) = \frac{\partial T}{\partial A}\delta A + \frac{\partial T}{\partial b}\delta b + \frac{\partial T}{\partial a}\delta a. \tag{7.3.4}$$

7.2.2. The Divergence Coefficient

In the following, we continue to retain only the lowest order term in the eikonal expansion. This corresponds to the geometric optics field of the theory of high-frequency electromagnetic waves in the visible spectrum. In seismology, we follow suit and name it the *geometric elastodynamic approximation*, or GEA.

Consider a unit *focal sphere* about F. The velocity inside this sphere is considered as constant and equal to V_h. Let t_0 be the time that a ray takes to travel from the focus to a point on the surface of the unit focal sphere. As noted in Example 7.1, rays inside the focal sphere are straight lines. Moreover, for points on the unit focal sphere,

$$h_1 = V_h, \qquad h_2 = 1, \qquad h_3 = \sin i_h,$$
$$q(t_0) = \left(\frac{V_h}{\sin i_h}\right)^{1/2}. \tag{7.142}$$

Equation (7.135c) then yields

$$Q(t/t_0) = \frac{q(t)}{q(t_0)} = \frac{1}{r}\left|\frac{V(t)}{V(t_0)} \cdot \frac{\sin i_h}{\sin\theta\cos i}\left(\frac{\partial i_h}{\partial\theta}\right)_r\right|^{1/2}. \tag{7.143}$$

The displacements are obtained by inserting the expansion (7.56) for the potentials into Eq. (7.49). Keeping only terms with the highest powers in ω in each expression and remembering that these terms are obtained by differentiating the factor $\exp(-i\omega t_j)$, we obtain

$$\mathbf{u}_j = \mathbf{e}_j\chi_j e^{-i\omega t_j}, \tag{7.144}$$

where

$$\mathbf{u}_1 = \mathbf{L}, \qquad \mathbf{u}_2 = \mathbf{N}, \qquad \mathbf{u}_3 = \mathbf{M},$$
$$\mathbf{e}_1 = \mathbf{e}_t, \qquad \mathbf{e}_2 = \mathbf{e}_{i_h}, \qquad \mathbf{e}_3 = \mathbf{e}_\phi, \tag{7.145}$$

$$\chi_1 = -\frac{1}{V_1\sqrt{\rho}}\xi_{1,1}, \qquad \chi_2 = \frac{p_2}{V_2\sqrt{\rho}}\xi_{2,2},$$

$$\chi_3 = \frac{p_3}{\sqrt{\mu}}\xi_{1,3}, \qquad \xi_{0,1} = \xi_{0,2} = \xi_{1,2} = \xi_{0,3} = 0, \tag{7.146}$$

$$p_2 = p_3 = \frac{r\sin i}{V_2}.$$

Using Eq. (7.133a), the χs may be expressed as

$$\chi_j(t) = \chi_j(t_0)G_j(t/t_0), \qquad (j = 1, 2, 3) \tag{7.147}$$

where

$$G_j(t/t_0) = \frac{V_j(t_0)}{V_j(t)}\left[\frac{\rho(t_0)}{\rho(t)}\right]^{1/2} Q_j(t/t_0). \tag{7.148}$$

Equations (7.145) and (7.147) yield

$$\mathbf{u}_j(t) = |\mathbf{u}_j(t_0)| G_j(t/t_0)\mathbf{e}_j \exp(-i\omega t_j). \tag{7.149}$$

The function $G(t/t_0)$ is known as a *divergence coefficient* or a divergence factor. Equations (7.137) and (7.148) show that it may be expressed as

$$G_j(t/t_0) = \left[\frac{V_j(t_0)\rho(t_0)w(t_0)}{V_j(t)\rho(t)w(t)}\right]^{1/2}, \tag{7.150}$$

where ρ is the density. Alternatively, from Eqs. (7.143) and (7.148), we have

$$\begin{aligned} G_j(t/t_0) &= \frac{1}{r}\left|\frac{V_j(t_0)\rho(t_0)}{V_j(t)\rho(t)} \cdot \frac{\sin i_h \tan i_h}{pI_h \sin\theta \cos i}\right|^{1/2} \\ &= \frac{1}{r}\left|\frac{V_j(t_0)\rho(t_0)}{V_j(t)\rho(t)} \cdot \frac{\sin i_h}{\sin\theta\cos i} \cdot \left(\frac{\partial i_h}{\partial\theta}\right)_r\right|^{1/2}. \end{aligned} \tag{7.151}$$

Equation (7.149) shows that for the GEA, $\mathbf{u}_1$ is along $\mathbf{e}_1$, $\mathbf{u}_2$ is along $\mathbf{e}_2$, and $\mathbf{u}_3$ is along $\mathbf{e}_3$. In other words, P waves are purely longitudinal; SV waves are purely transverse, with motion in the plane of propagation (plane of the ray); and SH waves are purely transverse, with motion perpendicular to the plane of propagation. Combining the two S waves, we get [cf. Eq. (7.163)]

$$\begin{gathered} \mathbf{u}_2 + \mathbf{u}_3 = G_2(t/t_0)A_0(\mathbf{e}_2 \cos\varepsilon_0 + \mathbf{e}_3 \sin\varepsilon_0)\exp(-i\omega t_2), \\ A_0^2 = |\mathbf{u}_2(t_0)|^2 + |\mathbf{u}_3(t_0)|^2, \qquad \tan\varepsilon_0 = \frac{|\mathbf{u}_3(t_0)|}{|\mathbf{u}_2(t_0)|}. \end{gathered} \tag{7.151a}$$

Equation (7.151a) reveals that ε_0, the *angle of polarization of* S waves, remains constant along a ray in a spherically symmetrical medium. In a more general media, it can be demonstrated that in the GEA, the direction of motion of a transverse wave rotates about the ray as the wave progresses.

Let the function $F(i_h, \phi)$ indicate the amplitude distribution on the unit focal sphere. The curved rays are supposed to pick up (on this sphere) the corresponding amplitudes as given by $F(i_h, \phi)$ and carry them along the rays. We have just proved (Eq. 7.149) that along the ray, P is purely longitudinal, SV is purely transverse in the plane of the ray, and SH is purely transverse, with motion perpendicular to the plane of the ray, while the angle of polarization remains fixed along the ray. Because the far-field signal is not disturbed along the ray, F retains its initial form and the sign of the first motion is also preserved. Therefore, the nodal lines on the focal sphere will be mapped by the rays onto nodal lines on the surface of the earth.

The factor $[w(t_0)/w(t)]^{1/2}$ appearing in the expression (7.150) for the divergence coefficient is known as *geometric spreading*, because it measures the spreading of a tube of rays. This can be easily calculated from geometric considerations. If t_0 refers to the unit focal sphere and t to that of the surface of the earth, then it is obvious from Fig. 7.14 that

$$\frac{w(t_0)}{w(t)} = \frac{(\delta i_h)(\sin i_h \delta\phi)}{[(a\delta\Delta)(a \sin \Delta\delta\phi)]\cos i_0}$$

$$= \frac{\sin i_h}{a^2 \sin \Delta \cos i_0}\left(\frac{\partial i_h}{\partial \theta}\right)_a, \tag{7.152}$$

in the limit.

Because

$$p = \left(\frac{\partial T}{\partial \theta}\right)_r = \frac{r_h}{V_h} \sin i_h, \qquad \left(\frac{\partial i_h}{\partial \theta}\right)_r = \frac{V_h}{r_h}\frac{1}{\cos i_h}\left(\frac{\partial^2 T}{\partial \theta^2}\right)_r, \tag{7.153}$$

Eq. (7.151) may be expressed as

$$G = \frac{V(r_h)}{r}\left|\frac{\rho(r_h)}{\rho(r)} \cdot \frac{1}{r_h V}\frac{\tan i_h}{\sin \theta \cos i}\left(\frac{\partial^2 T}{\partial \theta^2}\right)_r\right|^{1/2}, \tag{7.154}$$

where $\partial^2 T/\partial\theta^2$ is measured in seconds per square radian. The divergence coefficient of Eq. (7.154) may be recast into the following form, which is more useful for numerical evaluation:

$$G = \frac{1}{r}\frac{\eta}{\eta_h}\left[\frac{r_h \rho_h}{r\rho}\frac{p}{\sin\theta}\left|\frac{\partial^2 T}{\partial\theta^2}\right|_r (\eta^2 - p^2)^{-1/2}(\eta_h^2 - p^2)^{-1/2}\right]^{1/2}, \tag{7.155}$$

where $p = (\partial T/\partial \theta)_r$ is the ray parameter and $\eta = r/V$.

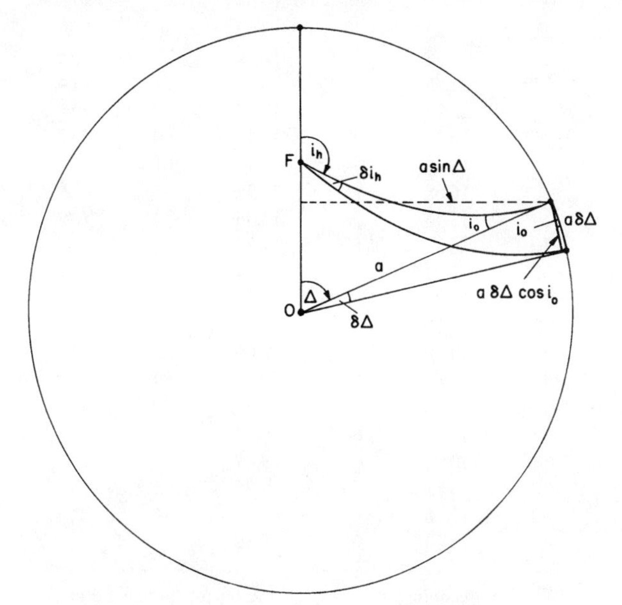

Figure 7.14. Divergence coefficient and geometric spreading.

The values of p and $(\partial^2 T/\partial\theta^2)_r$ can be calculated as functions of θ by using a cubic spline interpolation of the travel-time tables. This method gives the first and second derivatives, which are free from discontinuities at the tabulated points. For multibranch phases, we can still use cubic splines as long as we continue considering a single specified branch, treating each branch separately.

For core phases, $\sin i_h \to 0$ as $\sin\Delta \to 0$. For phases that do not reach the core, $\sin i_h \to 0$ as $\Delta \to 0$ ($r_h \neq a$). When $r_h = a$, $\Delta \to 0$ implies that the point of observation approaches the focus. Therefore, from Eq. (7.155), we conclude that for phases which do not reach the core, $G \to \infty$ as $\Delta \to \pi$. When $G \to \infty$, one can expect large amplitudes. When considering the points $\Delta = 0$ and $\Delta = \pi$, we must integrate over ϕ to get the resultant field. In the neighborhood of $\Delta = \pi$, the same phase arrives almost simultaneously following two different paths corresponding to the azimuthal angles ϕ_h and $\phi_h + \pi$.

Table 7.3. Divergence Coefficient $G(10^{-10}/\text{cm})$ for P Waves[a]

$\Delta°$	G	$\Delta°$	G	$\Delta°$	G
15	53.70	42	13.53	69	9.02
16	49.55	43	13.24	70	8.91
17	45.80	44	12.97	71	8.80
18	42.42	45	12.71	72	8.70
19	39.30	46	12.46	73	8.60
20	36.37	47	12.22	74	8.48
21	33.68	48	12.09	75	8.36
22	31.29	49	12.12	76	8.26
23	28.56	50	12.14	77	8.15
24	25.85	51	12.14	78	8.04
25	23.17	52	12.14	79	7.94
26	20.81	53	12.14	80	7.82
27	18.75	54	12.10	81	7.71
28	16.68	55	11.76	82	7.60
29	15.54	56	11.45	83	7.49
30	16.36	57	11.14	84	7.39
31	17.03	58	10.84	85	7.29
32	17.26	59	10.55	86	6.90
33	15.70	60	10.27	87	6.39
34	14.10	61	10.09	88	5.86
35	12.45	62	9.94	89	5.30
36	11.90	63	9.78	90	4.69
37	11.39	64	9.64	91	4.00
38	10.91	65	9.49	92	3.52
39	11.81	66	9.35	93	3.24
40	12.76	67	9.23	94	2.94
41	13.54	68	9.12		

[a] The recording station is on the top of the mantle and $h = 413$ km.

Table 7.3 gives the divergence coefficient for P waves calculated from the Jeffreys–Bullen travel-time tables. The depth of the focus is 413 km and the depth of the receiver 33 km. Table 7.4 shows the effect of the source depth on the divergence coefficient for S waves. The depths, in kilometers, below the free surface corresponding to the 14 h values in Table 7.4 are 0, 33, 96, 160, 223, 287, 350, 413, 477, 540, 603, 667, 730, and 794.

Table 7.4. Divergence Coefficient $G(10^{-11}/\text{cm})$ for S Waves[a]

	$\theta°$									
h	3	6	9	12	15	18	21	24	27	30
Surf	847	557	449	400	355	299[a]	337	199	122	79
0.00	2751	1354	959	754	611	481	476	275	170	110
0.01	3030	1175	1005	782	618	489	449	261	170	112
0.02	3081	1553	1074	817	629		428	245	169	113
0.03	2881	1638	1181	826	643	621	410	245	167	114
0.04	2682	1627	1118	835		586	392	245	163	114
0.05	2495	1613	1165			560	373	245	161	113
0.06	2324	1573	896		720	537	347	242	162	111
0.07	2333	1711	1445	1088	790	569	368	259	177	124
0.08	2285	1761	1607	0	834	579	392	265	188	160
0.09	2221	1761	1493	0	867	566	408	275	192	203
0.10	2136	1726	1296	0	836	583	407	286	205	251
0.11	2040	1680	1328	973	742	567	398	281	277	302
0.12	1940	1623	1295	957	219	504	384	330	350	329

	$\theta°$									
h	33	36	39	42	45	48	51	54	57	60
Surf	55	59	68	65	61	56	52	50	50	51
0.00	76	86	96	92	85	78	73	70	70	72
0.01	75	94	102	97	89	82	76	73	75	75
0.02	78	104	107	102	94	86	81	77	79	79
0.03	90	114	114	107	98	91	85	82	85	83
0.04	102	125	120	113	103	95	90	87	90	88
0.05	116	134	128	118	108	100	95	93	96	92
0.06	131	142	136	124	114	105	100	99	102	97
0.07	160	164	155	141	130	120	115	115	116	110
0.08	193	187	174	158	145	135	129	130	130	123
0.09	224	210	193	175	160	150	143	146	143	135
0.10	251	234	211	191	175	164	159	162	155	147
0.11	277	253	227	206	188	178	175	176	166	158
0.12	300	268	240	218	202	190	190	187	175	167

(continued)

Table 7.4. (cont.)

h	$\theta°$									
	63	66	69	72	75	78	81	84	87	90
Surf	49	47	47	48	44	42	43	44	44	42
0.00	68	66	65	66	62	59	60	61	60	58
0.01	72	70	69	70	64	62	63	64	63	60
0.02	76	73	73	74	67	66	67	68	67	62
0.03	80	77	77	77	70	69	70	71	70	64
0.04	84	81	82	80	73	73	74	75	73	66
0.05	88	86	87	84	77	77	79	79	77	68
0.06	93	91	92	88	81	82	83	83	81	71
0.07	106	103	104	98	91	93	94	94	91	78
0.08	118	116	116	108	102	103	104	104	101	84
0.09	130	128	129	118	113	114	115	113	110	89
0.10	141	140	138	126	123	124	124	122	117	93
0.11	152	151	146	133	132	133	133	130	122	96
0.12	161	161	153	139	140	141	140	136	125	97

[a] In this table and Tables 7.11–7.18, the end of the early travel-time branch (caused by the "20° discontinuity") is indicated by an underscore.

7.2.3. Finiteness Correction for GEA Amplitudes

In Section 4.7 we obtained the basic expressions for the effect of the source's finiteness and motion upon the far-field displacements. It was concluded that the spectral amplitudes were modulated by a finiteness factor ($\sin X/X$), in which $X = (\omega L/2V_f)(1 - (V_f/V_h)\cos\Theta)$, where V_h is the intrinsic wave velocity at the source, $\cos\Theta = \sin i_h \cos\phi_h$, L is the source length, and V_f is the velocity of rupture. We note that X can be recast in the form

$$X = \frac{\omega L}{2V_a}\left(\frac{V_a}{V_f} - \cos\phi_h\right), \qquad V_a = \frac{V_h}{\sin i_h} \simeq \frac{V_0}{\sin i_0}, \tag{7.156}$$

assuming that $r_h/a \simeq 1$. Therefore the role of the phase velocity in the surface-wave case [Eq. (5.139)] is taken now by the apparent velocity, V_a.

7.2.4. Ray-Theoretic Spectral Displacements of P and S Waves

The far-field displacements in an infinite elastic solid from a point dislocation were derived in Eq. (4.192) in the time domain. The Fourier transform yields corresponding spectral displacements. The finiteness correction was added in Eq. (4.255). If we examine again the physical conditions under which these expressions are valid, we realize that the far-field approximation in the frequency domain requires that $(\omega R/V) \gg 1$, which is certainly true for high frequencies provided R is sufficiently large. It therefore transpires that if we consider Eq. (4.192) to be valid on the surface of a focal sphere of radius R_0 such that $(\omega/V)R_0 \gg 1$, we can embed this focal sphere in a radially heterogeneous sphere and use Eq. (4.192) or its Fourier transform for the direct P and S arrivals on the earth's surface provided we do the following:

1. Replace $1/R$ in Eq. (4.192) by the divergence coefficient G.
2. Account for physical absorption.
3. Account for the crustal structure at the receiving end. If the source is shallow, correction must be made also for the crustal structure at the source.

For example, the spectral displacement at the base of the crust for P waves from a deep source of potency P_0 is

$$u_R(\omega) = \frac{P_0}{12\pi\beta_h}[i\omega g(\omega)]\left(\frac{\beta}{\alpha}\right)_h^3 F(\lambda, \delta; i_h, \phi_h)G_\alpha \frac{\sin X_\alpha}{X_\alpha} e^{-\omega t^*(\Delta, h)} e^{i\omega(t - T_\alpha) - iX_\alpha}. \tag{7.157}$$

Similar expressions are obtained from Eq. (4.192) for the direct arrivals of SV and SH types. In Eq. (7.157),

$$\begin{aligned} P_0 &= U_0\, dS, \\ g(\omega) &= \text{source spectrum}, \\ X_\alpha &= \frac{\omega L}{2\alpha_a}\left(\frac{\alpha_a}{V_f} - \cos\phi_h\right), \\ \alpha_a &= \frac{\alpha_0}{\sin i_0}, \\ \exp[-\omega t^*(\Delta, h)] &= \text{attenuation factor (see Sect. 10.4.2)}, \\ h &= \text{source depth}, \\ \Delta &= \text{epicentral distance}. \end{aligned} \tag{7.158}$$

The function $F(\lambda, \delta; i_h, \phi_h)$ is given in Eq. (4.154).

In practice, seismologists use Eq. (7.157) to extract the source parameters $P_0, \lambda, \delta, \phi_h, L, V_f, g(\omega)$ from observed amplitude spectra of P waves at a large number of seismic stations. The analysis is carried out by a method known as *amplitude equalization*. Granted that h, i_h, Δ, ϕ, and G_α are known for each ray, and assuming also a fair knowledge of the crustal structure beneath each station, the method consists of compensating the observed spectrums for instrumental and propagation effects. The resulting function is then matched against theoretical models.

This equalization procedure is conceptually analogous to an inverse filtering process through a series of three linear filters: the recording instrument, the earth's crust, and the earth's interior. If the transfer function of each of these is known, we can, in principle, remove its effects and plot the resulting equalized spectrums, at any given frequency, either as a function of i_h for fixed azimuth ϕ_h (*vertical radiation pattern*) or as a function of ϕ_h for a fixed i_h (*horizontal radiation pattern*). Some theoretical patterns are shown in Figs. 7.15 and 7.16 for P, SH, and SV waves. From Eqs. (4.154) and (4.192), the vertical patterns are given as

$$F_V(i_h) = c_0 + c_1 \sin i_h + d_1 \cos i_h + c_2 \sin 2i_h + d_2 \cos 2i_h, \quad (7.159)$$

where the coefficients c_i and d_i are given in Table 7.5. Likewise, the horizontal radiation patterns may be expressed in the form

$$F_H(\phi_h) = a_0 + a_1 \cos \phi_h + b_1 \sin \phi_h + a_2 \cos 2\phi_h + b_2 \sin 2\phi_h, \quad (7.160)$$

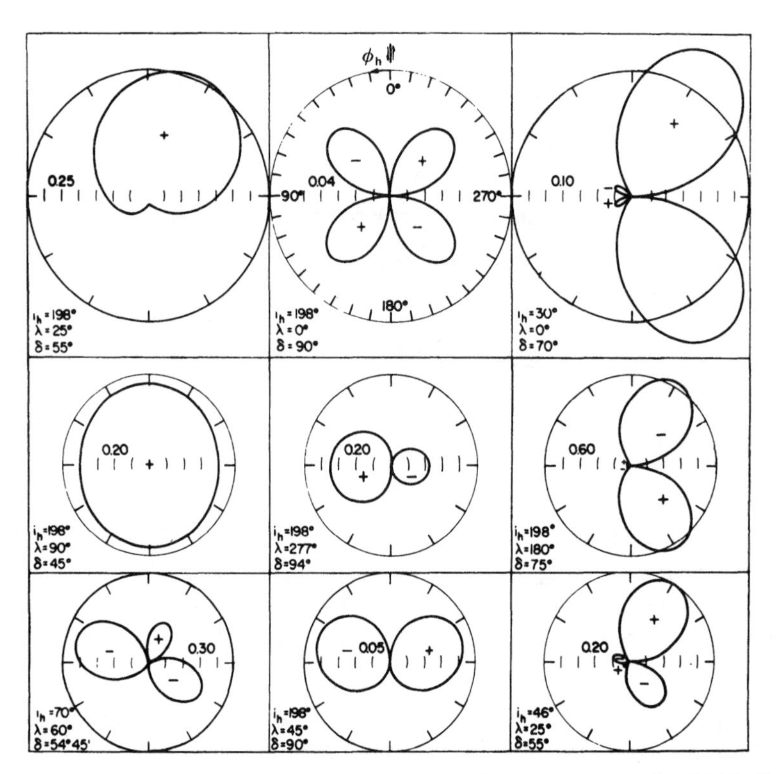

Figure 7.15. Far-field horizontal radiation patterns of P waves from a shear-fault dislocation. Positive and negative signs indicate compression (outward motion) and dilatation, respectively. Numbers on the radial scale indicate amplitude of radiation pattern in dimensionless units.

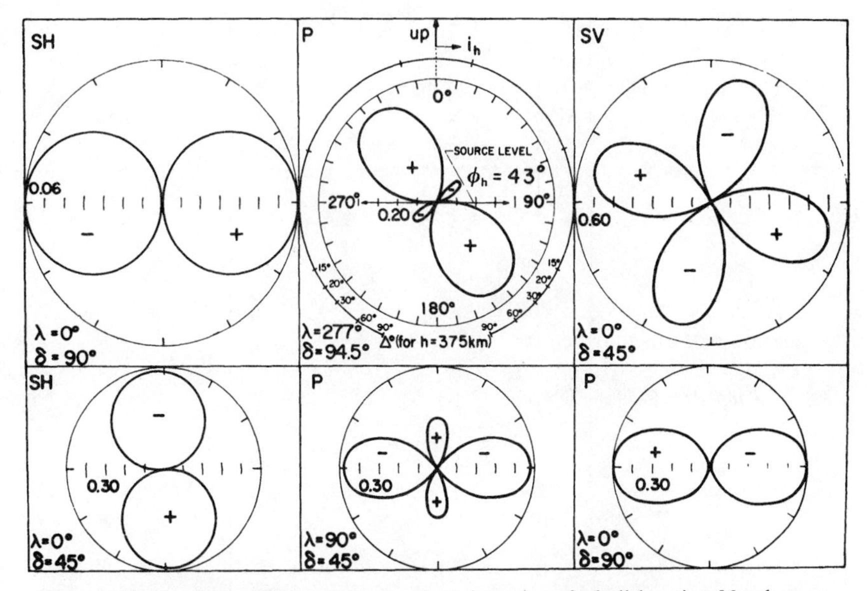

Figure 7.16. Far-field vertical radiation patterns for a shear-fault dislocation. Numbers on the radial scale are as in Figure 7.15.

where the coefficients a_i, b_i are given in Table 7.6. In Table 7.5

$$
\begin{aligned}
p_1 &= (\sin\lambda\sin\delta\cos\delta)\sin 2\phi_h + (\cos\lambda\sin\delta)\cos 2\phi_h,\\
p_2 &= (\cos\lambda\sin\delta)\sin 2\phi_h - (\sin\lambda\sin\delta\cos\delta)\cos 2\phi_h,\\
p_3 &= -(\cos\lambda\cos\delta)\sin\phi_h + (\sin\lambda\cos 2\delta)\cos\phi_h,\\
p_4 &= (\sin\lambda\cos 2\delta)\sin\phi_h + (\cos\lambda\cos\delta)\cos\phi_h,\\
p_5 &= \sin\lambda\sin\delta\cos\delta,\\
4p_6 &= 4s_0 + 2 - \sin^2\delta, \qquad s_0 = \frac{\sigma_h}{1-2\sigma_h}, \qquad \sigma_h = \text{Poisson's ratio.}
\end{aligned}
\tag{7.161}
$$

Note that the amplitude patterns obey the following symmetry relations:

$$
\begin{aligned}
F(\phi_h, i_h) &= F(\phi_h, i_h \pm \pi) && P, SV\\
&= -F(\phi_h, i_h \pm \pi) && SH\\
&= F(\phi_h + \pi, 2\pi - i_h) && P\\
&= -F(\phi_h + \pi, 2\pi - i_h) && SV, SH\\
F(\lambda + \pi) &= -F(\lambda).
\end{aligned}
\tag{7.162}
$$

The *polarization angle* is defined as [cf. Eq. (7.151a)]

$$\varepsilon_0 = \arg[u_{SV} + iu_{SH}]. \tag{7.163}$$

Figures 7.17–7.20 show comparisons between the calculated and observed horizontal radiation patterns for four deep earthquakes.

Table 7.5. Vertical Radiation Patterns for Body Waves

	c_0	c_1	d_1	c_2	d_2
P waves					
Shear Fault	$\frac{3}{2}(p_5 - p_2)$	0	0	$3p_4$	$\frac{3}{2}(3p_5 + p_2)$
Tensile fault	$s_0 + \frac{1}{2} - \frac{1}{4}\sin^2\delta(1 + \cos 2\phi_h)$	0	0	$-\frac{1}{2}\sin 2\delta \sin\phi_h$	$\frac{1}{4}[2 + \sin^2\delta(\cos 2\phi_h - 3)]$
SH waves					
Shear fault	0	$-6p_1$	$6p_3$	0	0
Tensile fault	0	$-\frac{1}{2}\sin^2\delta \sin 2\phi_h$	$\frac{1}{2}\sin 2\delta \cos\phi_h$	0	0
SV waves					
Shear fault	0	0	0	$-3(3p_5 + p_2)$	$6p_4$
Tensile fault	0	0	0	$\frac{1}{4}[(1 - 3\cos^2\delta) - \sin^2\delta\cos 2\phi_h]$	$-\frac{1}{2}\sin\phi_h \sin 2\delta$

Table 7.6. Horizontal Radiation Patterns for Body Waves

	a_0	a_1	b_1	a_2	b_2
P waves					
Shear fault	$\frac{3}{2}\sin\lambda\sin 2\delta\cdot(3\cos^2 i_h - 1)$	$3\cos\lambda\cos\delta\sin 2i_h$	$3\sin\lambda\cos 2\delta\sin 2i_h$	$\frac{3}{2}\sin\lambda\sin 2\delta\sin^2 i_h$	$-3\cos\lambda\sin\delta\sin^2 i_h$
Tensile fault	$p_6 + \frac{1}{4}(2 - 3\sin^2\delta)\cdot\cos 2i_h$	0	$-\frac{1}{2}\sin 2\delta\sin 2i_h$	$\frac{1}{4}\sin^2\delta(\cos 2i_h - 1)$	0
SH waves					
Shear fault	0	$6\sin\lambda\cos 2\delta\cos i_h$	$-6\cos\lambda\cos\delta\cos i_h$	$-6\cos\lambda\sin\delta\sin i_h$	$-3\sin\lambda\sin 2\delta\sin i_h$
Tensile fault	0	$-\frac{1}{2}\sin 2\delta\cos i_h$	0	0	$\frac{1}{2}\sin^2\delta\sin i_h$
SV waves					
Shear fault	$-\frac{9}{2}\sin\lambda\sin 2\delta\cdot\sin 2i_h$	$6\cos\lambda\cos\delta\cos 2i_h$	$6\sin\lambda\cos 2\delta\cos 2i_h$	$\frac{3}{2}\sin\lambda\sin 2\delta\sin 2i_h$	$-3\cos\lambda\sin\delta\sin 2i_h$
Tensile fault	$\frac{1}{4}\sin 2i_h\cdot(1 - 3\cos^2\delta)$	0	$-\frac{1}{2}\cos 2i_h\sin 2\delta$	$-\frac{1}{4}\sin 2i_h\sin^2\delta$	0

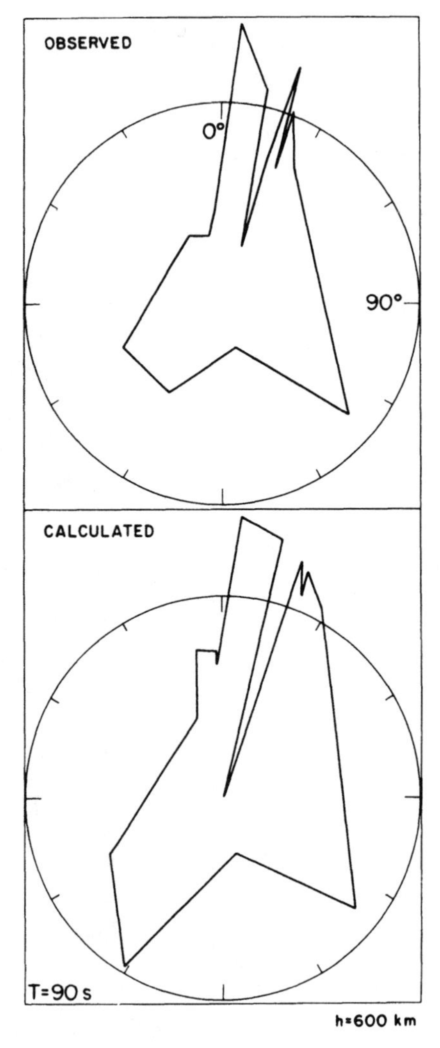

Figure 7.17. A comparison between the observed and calculated horizontal radiation patterns for November 9, 1963 West Brazil earthquake.

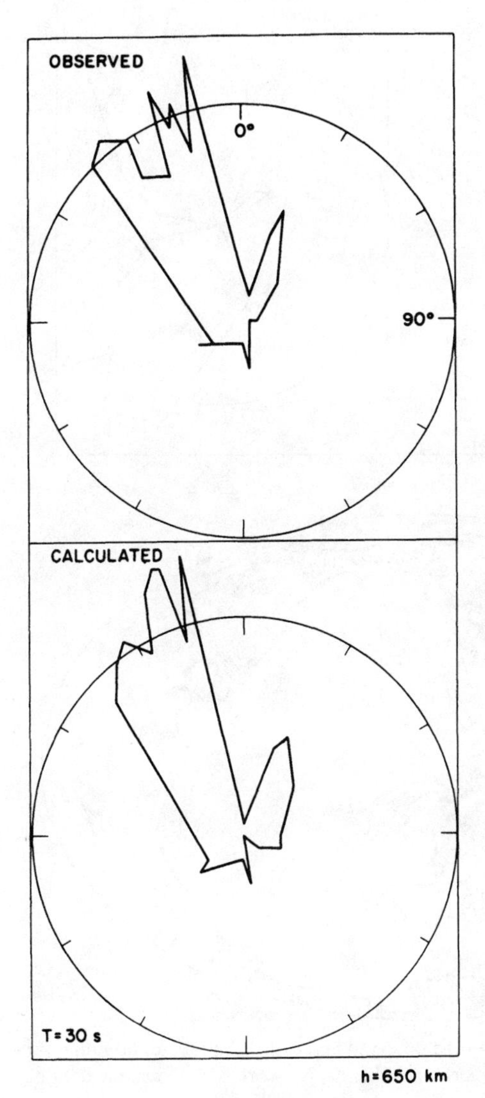

Figure 7.18. A comparison between the observed and calculated horizontal radiation patterns for December 15, 1963 Java Sea earthquake.

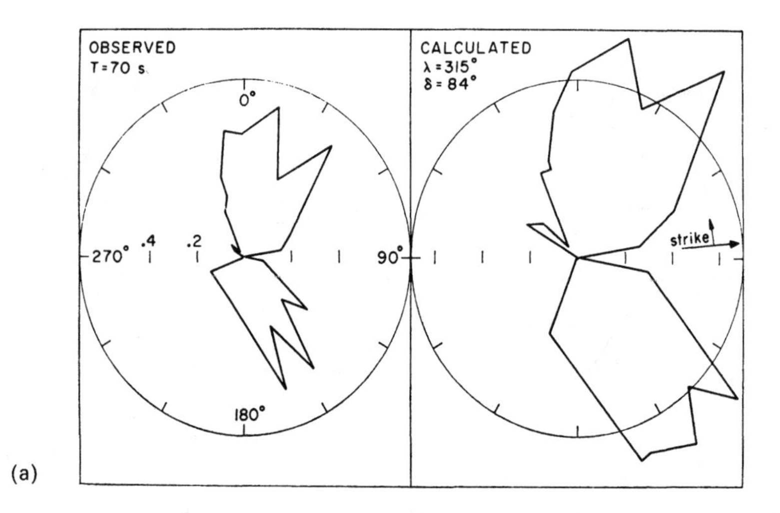

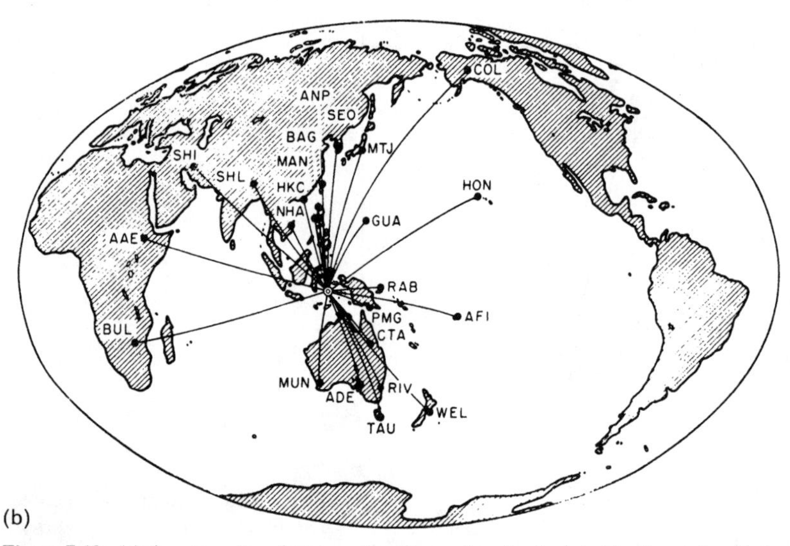

Figure 7.19. (a) A comparison between the observed and calculated horizontal radiation patterns for 21 March, 1964 Banda-Sea earthquake. (b) Network of stations, the data of which were used in the above analysis.

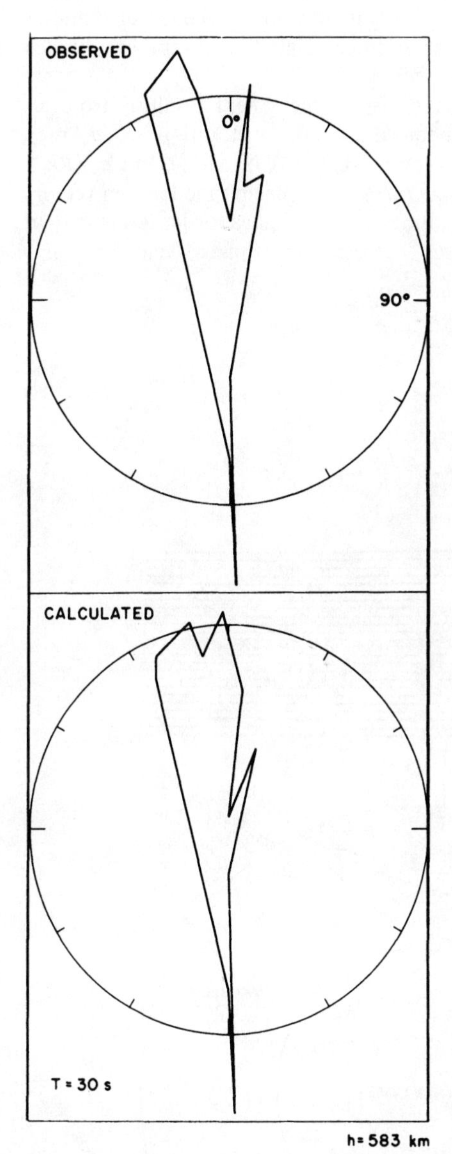

Figure 7.20. A comparison between the observed and calculated horizontal radiation patterns for the November 3 1965 Peru–Brazil earthquake.

7.2.5. Generalization to Reflected and Transmitted Waves

So far we have treated only the theory of the direct P and S arrivals. Figure 7.21 shows some important reflected phases in addition to the direct P and S. We now consider modifications necessary to calculate the spectral amplitudes of these and other derived phases.

We have seen in Chapter 4 that the displacement field resulting from an arbitrary source placed in a homogeneous unbounded medium $(\rho_h, \lambda_h, \mu_h)$ may be expressed in terms of the Hansen vectors $\mathbf{M}_{ml}(Rk_\beta)$, $\mathbf{N}_{ml}(Rk_\beta)$, and $\mathbf{L}_{ml}(Rk_\alpha)$. *Assuming a Heaviside unit function time variation*, replacing the Hansen vectors by their explicit expressions, and retaining only the highest order terms in ω in the eikonal-type expansion, we can see that, apart from phase factors,

$$
\begin{aligned}
u_P &= u_R = \frac{P_0}{R} F_1, \\
u_{SV} &= u_{i_h} = \frac{P_0}{R} F_2, \\
u_{SH} &= u_{\phi_h} = \frac{P_0}{R} F_3.
\end{aligned}
\tag{7.164}
$$

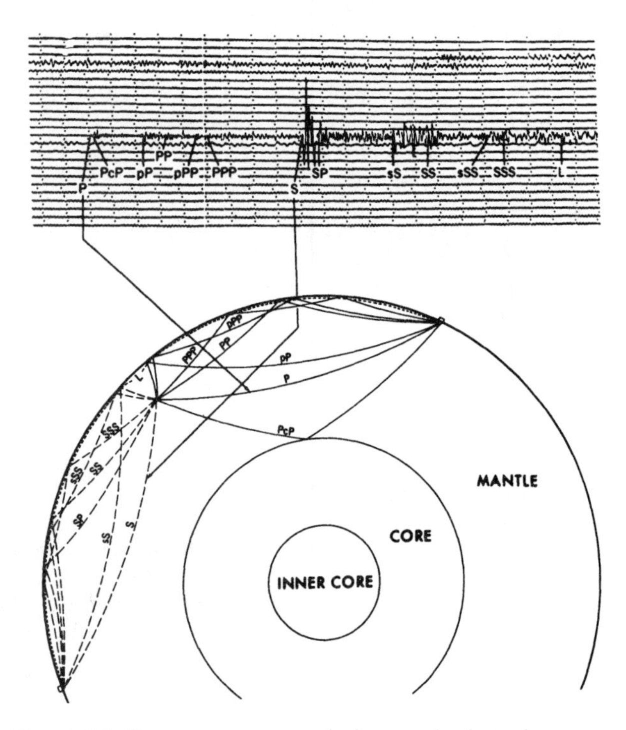

Figure 7.21. Some common ray arrivals appearing in a seismogram.

In Eqs. (7.164) P_0 is the source potency and the functions F_1, F_2, and F_3 give the P, SV, and SH radiation patterns of the source. Therefore, in the case of a shear dislocation, we have (Sect. 4.5)

$$P_0 = U_0 S,$$

$$F_1 = \frac{1}{12\pi\beta_h}\left(\frac{\beta_h}{\alpha_h}\right)^3 [6p_5 P_2(\cos i_h) + 2p_4 P_2^1(\cos i_h) - p_2 P_2^2(\cos i_h)],$$

$$F_2 = \frac{1}{24\pi\beta_h}\left[6p_5 \frac{\partial}{\partial i_h} P_2(\cos i_h) + 2p_4 \frac{\partial}{\partial i_h} P_2^1(\cos i_h) - p_2 \frac{\partial}{\partial i_h} P_2^2(\cos i_h)\right], \quad (7.165)$$

$$F_3 = \frac{1}{12\pi\beta_h} \cdot \frac{1}{\sin i_h} [p_3 P_2^1(\cos i_h) - p_1 P_2^2(\cos i_h)],$$

where U_0 is the magnitude of the dislocation, S is the fault area, and p_i are defined in Eqs. (7.161).

Equations (7.164) hold for a homogeneous medium in which rays are straight lines. We assume that the earth is a radially heterogeneous, isotropic, elastic sphere with the possibility of a finite number of concentric spherical discontinuities across which the density ρ and the wave velocities α and β may have finite jumps in their values. Because of the radial heterogeneity, the rays become curved. A tube of rays of the P or SV type, striking at a discontinuity, gives rise in general to four derived tubes. If dissipation is neglected, the energy emitted by the source, within a small tube of rays, does not leave the tube until the tube meets a discontinuity. In the steady state, therefore, the rate of flow of the energy across a cross section of the tube is constant along its length. However, the rate of flow of energy across a cross section of a derived tube is, in general, different from the corresponding quantity for the incident tube. The amplitude of the displacement at the receiver is influenced by the heterogeneity of the medium as well as by each discontinuity that the tube encounters on its way from the source to the receiver.

Figure 7.22a shows a small tube of rays $[(i_h, \phi_h), (i_h + di_h, \phi_h + d\phi_h)]$ of the ScP type which strikes the mantle–core boundary at A and reaches the surface of the earth at $P(a, \theta, \phi_h)$. Let the area of the cross section of the tube just before it strikes the mantle–core boundary (where it is of the SV type) be w_1. Let the corresponding area just after the tube strikes the boundary (where it is of the P type) be w_2 and let it be w_0 at the free surface. We consider a unit-focal sphere about Q. The medium within this sphere is assumed to be homogeneous, and, therefore, rays within this sphere are straight lines. Figure 7.22b shows the portion QP_H of the tube of rays within the focal sphere. We denote the cross-sectional area of the tube at P_H by w_H. Let E denote the energy density that flows across a cross section of a tube per unit time. Assuming that there is no loss of energy because of dissipation, we have the following relations:

$$E_H w_H = E_1 w_1,$$

$$E_2 w_2 = E_0 w_0, \quad (7.166)$$

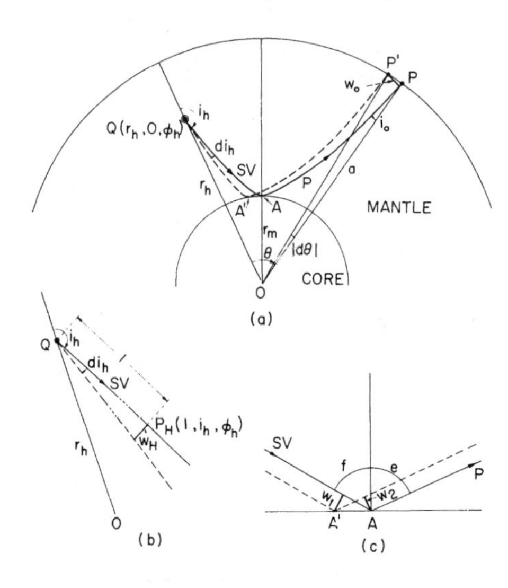

Figure 7.22. (a) A tube of rays of the *ScP* type. (b) The corresponding tube in a homogeneous medium. (c) Approximation at the mantle–core boundary.

where the subscripts of E and w have similar meanings. Hence

$$\frac{E_0}{E_H} = \frac{E_0}{E_2}\frac{E_2}{E_1}\frac{E_1}{E_H} = \frac{E_2}{E_1}\frac{w_2}{w_1}\frac{w_H}{w_0}. \tag{7.167}$$

It is seen from Fig. 7.22 that [see also Eq. (7.152)]

$$\frac{w_H}{w_0} = \left(\frac{1}{a}\right)^2 \frac{\sin i_h}{\sin\theta\cos i_0}\left|\frac{di_h}{d\theta}\right|, \tag{7.168}$$

where i_0 is the angle of incidence for *ScP* at the free surface or the Moho, as the case may be. The modulus in Eq. (7.168) is required for cases in which di_h and $d\theta$ are of opposite signs.

The value of w_2/w_1 depends upon the condition at the mantle–core boundary. In reality, we have a curved wavefront incident upon a curved surface. However, if the distance QA and the radius of the core are large compared to wavelengths of interest, we may regard the situation at the mantle–core boundary as that of a plane wavefront incident upon a plane surface. With this approximation (Fig. 7.22c), we have

$$\frac{w_2}{w_1} = \frac{\cos e}{\cos f}, \tag{7.169}$$

where e and f are the angles which the P and S waves, respectively, make with the normal. Equations (7.167)–(7.169) then yield

$$\frac{E_0}{E_H} = \frac{E_2\cos e}{E_1\cos f}\left(\frac{1}{a}\right)^2 \frac{\sin i_h}{\sin\theta\cos i_0}\left|\frac{di_h}{d\theta}\right|. \tag{7.170}$$

It may be noted that, whereas E_1 is the amount of energy that flows across a unit area of cross section of the incident SV tube in a unit time, $E_1 \cos f$ is the amount of energy that is transported to a unit area of the mantle–core boundary in a unit time by the incident SV wave. Similarly, $E_2 \cos e$ is the rate at which energy is being removed by the reflected P wave per unit area of the reflecting surface per unit time.

We know that the mean value of the energy density E over a cycle is given by the relation [Eq. (2.117)]

$$E = \frac{1}{2}\rho\,(\text{velocity})(\text{ang. frequency})^2(\text{amplitude})^2. \tag{7.171}$$

From Eqs. (7.170) and (7.171), we find that the ratio of the amplitude at the free surface to the corresponding amplitude on the unit-focal sphere is given by

$$\frac{(\text{amp})_0}{(\text{amp})_H} = ZG, \tag{7.172}$$

where $(\text{amp})_H$ is given by Eqs. (7.164) with R = 1 and

$$Z = \left[\frac{E_2 \cos e}{E_1 \cos f}\right]^{1/2}, \tag{7.173}$$

$$G = \frac{1}{a}\left[\frac{\rho_h V_h}{\rho_0 V_0}\frac{\sin i_h}{\sin\theta \cos i_0}\left|\frac{di_h}{d\theta}\right|\right]^{1/2}. \tag{7.174}$$

In the Eqs. (7.172)–(7.174) the subscript 0 refers to the free surface (or the Moho) and the subscript h to the source. In addition, for ScP, we have $V_h = \beta(r_h)$, $V_0 = \alpha(a)$. It is assumed that the values of the density and of the velocities at the source level are equal to the corresponding quantities for the homogeneous case.

Using Eqs. (7.164) and (7.172), we get the following expression for the amplitude of ScP at the free surface

$$u_{ScP} = P_0 Z G F_2. \tag{7.175}$$

The derivation of Eq. (7.175) is quite general and can be applied to any phase. Therefore, the expression for the displacement at the free surface corresponding to an arbitrary phase may be written in the form

$$u = P_0 Z G F, \tag{7.176}$$

where P_0 is the source potency.

For a signal arriving after a number of reflections and/or transmissions, Z is a product of quantities of the type in Eq. (7.173). We call Z the reflection coefficient, although, in general it is a product of several reflection and transmission coefficients:

$$Z = \prod_j Z_j. \tag{7.177}$$

We have defined the reflection coefficient Z_j as the square root of the ratio of the energy of the derived wave per unit area of the boundary to the corresponding energy of the incident wave. We can find the relationship between the reflection coefficient and the ratio of the amplitude of the displacement of the derived wave to the amplitude of the displacement of the incident wave as follows. Let E_1 be the amount of energy that flows in a unit time across a unit area of cross section of the incident tube of rays, A_1 be the corresponding amplitude of the displacement, V_1 be the velocity, i_1 be the angle with the normal, and ρ_1 be the density of the medium of incidence. The corresponding quantities for the derived (reflected or transmitted) wave are denoted by E_2, A_2, V_2, i_2, and ρ_2, respectively. Then, from Eqs. (7.171) and (7.173) and Snell's law $\sin e_2/V_2 = \sin e_1/V_1$, we have

$$Z = \left[\frac{E_2 \cos e_2}{E_1 \cos e_1}\right]^{1/2} = \left|\frac{A_2}{A_1}\right| \left(\frac{\rho_2 \sin 2e_2}{\rho_1 \sin 2e_1}\right)^{1/2}. \tag{7.178}$$

In the case of reflection, $\rho_2 = \rho_1$. Further, if the reflected wave is of the same type (P or S) as the incident wave, then $e_2 = e_1$. Table 7.7 gives the reflection

Table 7.7. Relationship Between the Reflection/Transmission Coefficients and the Displacement Amplitude Ratios

Reflection (free surface)	Reflection (mantle–core boundary)
$PP(1) = \left\lvert\frac{A^0}{A}\right\rvert$	$PP(11) = \left\lvert\frac{A_1}{A}\right\rvert$
$PS(1) = \left\lvert\frac{B^0}{A}\right\rvert \left(\frac{\sin 2f}{\sin 2e}\right)^{1/2}$	$PS(11) = \left\lvert\frac{B_1}{A}\right\rvert \left(\frac{\sin 2f}{\sin 2e}\right)^{1/2}$
$SP(1) = \left\lvert\frac{A^0}{B}\right\rvert \left(\frac{\sin 2e}{\sin 2f}\right)^{1/2}$	$SP(11) = \left\lvert\frac{A_1}{B}\right\rvert \left(\frac{\sin 2e}{\sin 2f}\right)^{1/2}$
$SS(1) = \left\lvert\frac{B^0}{B}\right\rvert$	$SS(11) = \left\lvert\frac{B_1}{B}\right\rvert$

Transmission (mantle–core boundary)	Transmission (core–mantle boundary)
$PP(12) = \left\lvert\frac{A'}{A}\right\rvert \left(\varepsilon \frac{\sin 2e'}{\sin 2e}\right)^{1/2}$	$PP(21) = \left\lvert\frac{A''}{A}\right\rvert \left(\frac{1}{\varepsilon} \frac{\sin 2e}{\sin 2e'}\right)^{1/2}$
$SP(12) = \left\lvert\frac{A'}{B}\right\rvert \left(\varepsilon \frac{\sin 2e'}{\sin 2f}\right)^{1/2}$	$PS(21) = \left\lvert\frac{B''}{A}\right\rvert \left(\frac{1}{\varepsilon} \frac{\sin 2f}{\sin 2e'}\right)^{1/2}$

Reflection (core–mantle boundary)
$PP(22) = \left\lvert\frac{A_2}{A}\right\rvert$

coefficients for a single reflection/transmission in terms of the amplitude ratios. In this table, ε *denotes the ratio of the density of the core to the density of the mantle* at the mantle–core boundary. Other symbols are defined in Table 7.8. The amplitude ratios are given in Table 7.9. These ratios have been derived in Chapter 3, although the notation is slightly different in some cases. Table 7.10 contains the reflection coefficients Z for some important phases. In expression (7.174) for G, V_h is either $\alpha(r_h)$ or $\beta(r_h)$, depending upon whether the phase begins as a P wave or as an S wave. Similarly, V_0 is $\alpha(a)$ or $\beta(a)$, depending upon whether the phase arrives as a P wave or as an S wave. The function F depends upon the source and the phase. In the case of a shear dislocation, F is F_1, F_2, or F_3, according to whether the signal leaves the source as a P, SV, or SH wave. In general, F can be obtained by solving the corresponding canonical problem for a homogeneous unbounded medium.

7.2.5.1. Computation of the Reflection Coefficient. The reflection coefficient Z for a given phase is the product of all the reflection and transmission coefficients pertaining to that phase. The individual reflection and transmission coefficients are computed from the corresponding theoretical expressions. We have defined the reflection and transmission coefficients as the square root of the energy ratios and not as the amplitude ratios. Consequently, the energy consideration requires that the sum of the squares of the reflection and transmission coefficients associated with a wave incident at a boundary must be equal to unity. For example, in the case of an SV wave incident at the mantle–core boundary, we have the energy equation

$$[SP(11)]^2 + [SS(11)]^2 + [SP(12)]^2 = 1. \qquad (7.179)$$

Table 7.8. Nomenclature of Velocities, Angles, and Amplitudes for Reflection/Transmission of Plane Waves

	Velocity	Angle with normal	Amplitude
Free surface			
Incident P	α	e	A
Incident S	β	f	B
Reflected P	α	e	A°
Reflected S	β	f	B°
Mantle–core boundary			
Incident P	α	e	A
Incident S	β	f	B
Reflected P	α	e	A_1
Reflected S	β	f	B_1
Transmitted P	α'	e'	A'
Core–mantle boundary			
Incident P	α'	e'	A
Reflected P	α'	e'	A_2
Transmitted P	α	e	A''
Transmitted S	β	f	B''

Table 7.9. Amplitude Ratios for Reflection/Transmission of Plane Waves ($\varepsilon = \rho_c/\rho_m$)

Free surface

Incident P	Incident SV
$\dfrac{A^0}{A} = D_1^{-1}\left[\left(\dfrac{\beta}{\alpha}\right)^2 \sin 2e \sin 2f - \cos^2 2f\right]$	$\dfrac{A^0}{B} = D_1^{-1}\left[\left(\dfrac{\beta}{\alpha}\right)\sin 4f\right]$
$\dfrac{B^0}{A} = D_1^{-1}\left[-2\left(\dfrac{\beta}{\alpha}\right)\sin 2e \cos 2f\right]$	$\dfrac{B^0}{B} = \dfrac{A^0}{A}$

$$D_1 = \left(\frac{\beta}{\alpha}\right)^2 \sin 2e \sin 2f + \cos^2 2f$$

Mantle–core boundary

Incident P	Incident SV
$\dfrac{A_1}{A} = D_{12}^{-1}\left[\left(\dfrac{\beta}{\alpha}\right)^2 \sin 2e \sin 2f \cos e' + \varepsilon\left(\dfrac{\alpha'}{\alpha}\right)\cos e - \cos^2 2f \cos e'\right]$	$\dfrac{A_1}{B} = D_{12}^{-1}\left[\left(\dfrac{\beta}{\alpha}\right)\sin 4f \cos e'\right]$
$\dfrac{B_1}{A} = D_{12}^{-1}\left[-2\left(\dfrac{\beta}{\alpha}\right)\sin 2e \cos 2f \cos e'\right]$	$\dfrac{B_1}{B} = D_{12}^{-1}\left[\left(\dfrac{\beta}{\alpha}\right)^2 \sin 2e \sin 2f \cos e' - \varepsilon\left(\dfrac{\alpha'}{\alpha}\right)\cos e - \cos^2 2f \cos e'\right]$
$\dfrac{A'}{A} = D_{12}^{-1}[2 \cos e \cos 2f]$	$\dfrac{A'}{B} = D_{12}^{-1}\left[-2\left(\dfrac{\beta}{\alpha}\right)\cos e \sin 2f\right]$

$$D_{12} = \left(\frac{\beta}{\alpha}\right)^2 \sin 2e \sin 2f \cos e' + \varepsilon\left(\frac{\alpha'}{\alpha}\right)\cos e + \cos^2 2f \cos e'$$

Core–mantle boundary

$$\frac{A_2}{A} = D_{12}^{-1}\left[\left(\frac{\beta}{\alpha}\right)^2 \sin 2e \sin 2f \cos e' - \varepsilon\left(\frac{\alpha'}{\alpha}\right)\cos e + \cos^2 2f \cos e'\right]$$

$$\frac{A''}{A} = D_{12}^{-1}\left[2\varepsilon\left(\frac{\alpha'}{\alpha}\right)\cos 2f \cos e'\right]$$

$$\frac{B''}{A} = D_{12}^{-1}\left[-4\varepsilon\left(\frac{\alpha'}{\alpha}\right)\cos e \sin f \cos e'\right]$$

Table 7.10. Reflection Coefficient (Z) for Some Common Waves

Wave	Z	Wave	Z
P	1	SPP	$SP(1) \cdot PP(1)$
S	1	SPS	$SP(1) \cdot PS(1)$
SH	1	SSP	$SS(1) \cdot SP(1)$
PP	$PP(1)$	SSS	$SS(1) \cdot SS(1)$
PS	$PS(1)$	$SSSH$	1
SP	$SP(1)$	PcP	$PP(11)$
SS	$SS(1)$	PcS	$PS(11)$
SSH	1	ScP	$SP(11)$
PPP	$PP(1) \cdot PP(1)$	ScS	$SS(11)$
PPS	$PP(1) \cdot PS(1)$	PKP	$PP(12) \cdot PP(21)$
PSP	$PS(1) \cdot SP(1)$	PKS	$PP(12) \cdot PS(21)$
PSS	$PS(1) \cdot SS(1)$	SKP	$SP(12) \cdot PP(21)$
		SKS	$SP(12) \cdot PS(21)$

This equation serves as a good check on the computation of the reflection and transmission coefficients.

For a given phase, the angle of incidence at a boundary can be calculated from Snell's law. Again taking ScP as our example, we have

$$p = \frac{dT}{d\Delta} = \frac{r_h \sin i_h}{\beta_h} = \frac{r_m \sin f}{\beta_m} = \frac{r_m \sin e}{\alpha_m} = \frac{a \sin i_0}{\alpha_0}, \tag{7.180}$$

where T is the travel time in seconds, Δ is the epicentral distance in radians, r is the distance from the center of the earth in kilometers, and α, β are the wave velocities in kilometers per second. The subscripts h refer to the focus, m to the mantle at the mantle–core boundary, and 0 to the observer. The ray parameter p can be computed from the travel-time tables by using cubic splines. Knowing p for a given Δ, we can calculate i_h, e, f, and i_0 from Eq. (7.180). The angle of incidence, i_0, at the base of the crust for some important phases is given in Tables 7.11–7.18. These tables are based on the Jeffreys–Bullen travel-time tables. Further, the depths (in kilometers) of focus corresponding to the 14 values of h given in Table 7.11 are 0, 33, 96, 160, 223, 287, 350, 413, 477, 540, 603, 667, 730, and 794.

In the case of a P wave incident at the free surface, the angle e is given and the angle f can be found from Snell's law. Because $\alpha > \beta$, f is always real. When the incident wave is of the SV type, however, we notice that $\sin e > 1$ for $f > f_1$, where f_1 [$=\arcsin(\beta/\alpha)$] is the critical angle. Therefore, for $f > f_1$, the angle e is imaginary. We then define

$$\cos e = -i(\sin^2 e - 1)^{1/2}, \qquad (f > f_1),$$

and, therefore,

$$\frac{B^0}{B} = -\frac{2i(\beta/\alpha)^2 \sin e(\sin^2 e - 1)^{1/2} \sin 2f + \cos^2 2f}{2i(\beta/\alpha)^2 \sin e(\sin^2 e - 1)^{1/2} \sin 2f - \cos^2 2f}. \tag{7.181}$$

Table 7.11. Angle of Incidence (i_0) in Degrees at the Base of the Crust

	P					
	$\theta°$					
h	3	6	9	12	15	18
Surface	83.9	79.7	75.1	70.2	65.2	59.9
0.00	82.9	79.0	74.4	69.4	64.3	58.9
0.01	72.3	73.8	70.9	67.0	62.5	57.7
0.02	61.8	68.3	67.3	64.3	60.3	
0.03	52.7	63.0	63.7	61.5	58.2	46.4
0.04	45.2	57.9	60.1	58.8		45.5
0.05	39.0	53.1	56.6			44.6
0.06	33.9	48.6	52.9		46.4	43.7
0.07	29.7	44.3	49.0		45.2	42.8
0.08	26.1	40.2	45.3	45.4	43.7	41.8
0.09	23.1	36.7	42.2	43.2	42.3	40.7
0.10	20.5	33.5	39.4	41.1	40.8	39.7
0.11	18.4	30.8	36.9	39.2	39.3	38.7
0.12	16.6	28.3	34.6	37.3	38.0	37.8

	$\theta°$									
h	21	24	27	30	33	36	39	42	45	48
Surface	45.5	42.3	39.9	38.3	37.3	36.3	35.5	34.8	34.0	32.9
0.12	37.3	36.7	35.9	35.3	34.8	34.1	33.2	32.4	31.5	30.6

	$\theta°$									
h	51	54	57	60	63	66	69	72	75	78
Surface	32.0	30.9	29.7	28.6	27.5	26.5	25.5	24.5	23.5	22.6
0.12	29.5	28.5	27.6	26.7	25.8	24.9	24.0	23.1	22.2	21.3

	$\theta°$								
h	81	84	87	90	93	96	99	102	105
Surface	21.6	20.7	19.8	19.1	18.7	18.5	18.4	18.2	17.5
0.12	20.4	19.6	19.0	18.6	18.5	18.4	18.1		

Table 7.12. Angle of Incidence (i_0) in Degrees at the Base of the Crust

	PP								
	h					h			
$\theta°$	Surface	0.04	0.08	0.12	$\theta°$	Surface	0.04	0.08	0.12
6	83.9				111	30.3	30.0	29.6	29.1
9	81.9				114	29.7	29.4	29.0	28.6
12	79.7				117	29.1	28.8	28.5	28.0
15	77.5				120	28.6	28.3	28.0	27.6
18	75.1				123	28.0	27.8	27.5	27.1
21	72.7				126	27.5	27.3	27.0	26.6
24	70.2				129	27.0	26.8	26.5	26.1
27	67.7	60.1			132	26.5	26.3	26.0	25.6
30	65.2	59.1			135	26.0	25.8	25.5	25.2
33	62.6	57.4	45.4		138	25.5	25.3	25.0	24.7
36	59.9		44.4		141	25.0	24.8	24.6	24.2
39	57.2	45.5	43.2		144	24.5	24.3	24.1	23.8
42	45.5	43.9	42.0		147	24.0	23.8	23.6	23.3
45	43.8	42.5	40.9		150	23.5	23.3	23.1	22.8
48	42.3	41.2	39.9	37.9	153	23.0	22.9	22.6	22.4
51	41.0	40.1	39.1	37.6	156	22.6	22.4	22.2	21.9
54	39.9	39.1	38.4	37.3	159	22.1	21.9	21.7	21.4
57	39.0	38.4	37.9	37.0	162	21.6	21.4	21.2	21.0
60	38.3	37.9	37.5	36.5	165	21.1	21.0	20.8	20.5
63	37.8	37.5	37.0	36.1	168	20.7	20.5	20.3	20.1
66	37.3	37.0	36.5	35.7	171	20.2	20.1	19.9	19.7
69	36.8	36.4	36.0	35.4	174	19.8	19.6	19.5	19.3
72	36.3	35.9	35.6	35.1	177	19.4	19.3	19.2	19.0
75	35.8	35.6	35.3	34.8	180	19.1	19.0	18.9	18.8
78	35.5	35.2	35.0	34.5	183	18.8	18.8	18.7	18.7
81	35.2	35.0	34.7	34.0	186	18.7	18.6	18.6	18.6
84	34.8	34.6	34.2	33.6	189	18.6	18.5	18.5	18.5
87	34.5	34.1	33.7	33.1	192	18.5	18.5	18.5	18.4
90	34.0	33.6	33.2	32.6	195	18.4	18.4	18.4	18.4
93	33.4	33.1	32.8	32.2	198	18.4	18.4	18.4	18.3
96	32.9	32.6	32.3	31.7	201	18.3	18.3	18.2	18.1
99	32.4	32.2	31.8	31.2	204	18.2	18.1	18.0	17.9
102	32.0	31.7	31.3	30.7	207	17.9	17.8	17.7	
105	31.4	31.1	30.7	30.2	210	17.5			
108	30.9	30.6	30.2	29.6					

Table 7.13. Angle of Incidence (i_0) in Degrees at the Base of the Crust

PS				
	h			
$\theta°$	Surface	0.04	0.08	0.12
48	34.0			
51	33.8			
54	33.5			
57	33.2			
60	32.8			
63	32.3			
66	31.9			
69	31.4			
72	31.0			
75	30.4	29.0		
78	29.8	28.6		
81	29.3	28.2		
84	28.7	27.7		
87	28.1	27.1		
90	27.5	25.4		
93	26.8	24.9	23.6	
96	25.0	24.3	23.3	
99	24.4	23.8	22.9	
102	23.8	23.2	22.4	
105	23.2	22.7	21.9	
108	22.6	22.1	21.4	20.2
111	22.0	21.5	21.0	20.1
114	21.5	21.0	20.6	19.9
117	20.9	20.6	20.2	19.7
120	20.5	20.2	20.0	19.5
123	20.2	20.0	19.7	19.3
126	19.9	19.7	19.5	19.2
129	19.6	19.5	19.3	19.1
132	19.4	19.3	19.2	19.0
135	19.2	19.2	19.1	19.0
138	19.1	19.1	19.0	
141	19.1	19.0	18.9	
144	19.0	18.9		
147	18.9			

PSS				
	h			
$\theta°$	Surface	0.04	0.08	0.12
93	34.0			
96	33.9			
99	33.7			
102	33.5			
105	33.2			
108	33.0			
111	32.7			
114	32.5			
117	32.2			
120	31.9			
123	31.6			
126	31.3			
129	31.0			
132	30.7			
135	30.4			
138	30.0	29.1		
141	29.7	28.8		
144	29.3	28.6		
147	29.0	28.3		
150	28.6	28.0		
153	28.3	27.6		
156	27.9	27.3		
159	27.5	27.0		
162	27.2	26.6		
165	26.8	25.5		
168	25.6	25.2		
171	25.3	24.9		
174	24.9	24.5	23.7	
177	24.6	24.2	23.5	
180	24.2	23.8	23.2	
183	23.9	23.5	22.9	
186	23.5	23.1	22.6	
189	23.1	22.7	22.2	
192	22.7	22.4	21.9	
195	22.3	22.0	21.6	
198	22.0	21.6	21.2	
201	21.6	21.3	20.9	20.2
204	21.2	21.0	20.6	20.0
207	20.9	20.6	20.4	19.9
210	20.6	20.4	20.1	19.7
213	20.3	20.1	19.9	19.6
216	20.1	19.9	19.8	19.5

Table 7.13. (cont.)

PS					PSS				
	h					h			
$\theta°$	Surface	0.04	0.08	0.12	$\theta°$	Surface	0.04	0.08	0.12
					219	19.9	19.8	19.6	19.3
					222	19.7	19.6	19.4	19.2
					225	19.5	19.4	19.3	19.2
					228	19.4	19.3	19.2	19.1
					231	19.3	19.2	19.2	19.1
					234	19.2	19.2	19.1	19.1
					237	19.2	19.1	19.1	19.0
					240	19.1	19.1	19.0	19.0
					243	19.1	19.0	19.0	18.9
					246	19.0	19.0	19.0	
					249	19.0	18.9	18.9	
					252	18.9			

Table 7.14. Angle of Incidence (i_0) in Degrees at the Base of the Crust

PPP									
	h					h			
$\theta°$	Surface	0.04	0.08	0.12	$\theta°$	Surface	0.04	0.08	0.12
9	83.9				63	45.5	44.4	42.9	
12	82.6				66	44.4	43.4	42.1	
15	81.2				69	43.3	42.4	41.3	
18	79.7				72	42.3	41.5	40.6	
21	78.2				75	41.4	40.7	39.9	
24	76.7				78	40.6	40.0	39.3	38.0
27	75.1				81	39.9	39.4	38.8	37.8
30	73.5				84	39.2	38.8	38.4	37.6
33	71.9				87	38.7	38.4	38.0	37.3
36	70.2				90	38.3	38.0	37.7	37.1
39	68.6				93	37.9	37.7	37.4	36.8
42	66.9				96	37.6	37.4	37.1	36.4
45	65.2	60.1			99	37.3	37.1	36.8	36.1
48	63.5	59.3			102	37.0	36.8	36.4	35.9
51	61.7	58.1			105	36.6	36.4	36.1	35.6
54	59.9		45.4		108	36.3	36.0	35.8	35.4
57	58.1	46.7	44.7		111	35.9	35.8	35.6	35.2
60	46.8	45.5	43.8		114	35.7	35.5	35.3	35.0

(continued)

Table 7.14. (cont.)

PPP									
	h					h			
$\theta°$	Surface	0.04	0.08	0.12	$\theta°$	Surface	0.04	0.08	0.12
117	35.5	35.3	35.1	34.8	216	24.5	24.4	24.2	24.0
120	35.3	35.1	35.0	34.6	219	24.2	24.1	23.9	23.7
123	35.1	34.9	34.7	34.3	222	23.9	23.7	23.6	23.4
126	34.8	34.7	34.5	34.0	225	23.5	23.4	23.3	23.1
129	34.6	34.4	34.1	33.7	228	23.2	23.1	22.9	22.7
132	34.3	34.1	33.8	33.4	231	22.9	22.8	22.6	22.4
135	34.0	33.7	33.5	33.1	234	22.6	22.4	22.3	22.1
138	33.6	33.4	33.1	32.7	237	22.2	22.1	22.0	21.8
141	33.3	33.1	32.8	32.4	240	21.9	21.8	21.7	21.5
144	32.9	32.7	32.5	32.1	243	21.6	21.5	21.4	21.2
147	32.6	32.4	32.2	31.8	246	21.3	21.2	21.1	20.9
150	32.3	32.1	31.8	31.5	249	21.0	20.9	20.7	20.6
153	32.0	31.8	31.5	31.1	252	20.7	20.6	20.4	20.3
156	31.6	31.4	31.2	30.8	255	20.4	20.3	20.1	20.0
159	31.3	31.0	30.8	30.4	258	20.1	20.0	19.9	19.7
162	30.9	30.7	30.4	30.0	261	19.8	19.7	19.6	19.4
165	30.5	30.3	30.0	29.6	264	19.5	19.4	19.3	19.2
168	30.1	29.9	29.6	29.3	267	19.3	19.2	19.1	19.0
171	29.7	29.5	29.3	28.9	270	19.1	19.0	18.9	18.9
174	29.3	29.1	28.9	28.6	273	18.9	18.9	18.8	18.8
177	28.9	28.7	28.5	28.2	276	18.8	18.7	18.7	18.7
180	28.6	28.4	28.2	27.9	279	18.7	18.7	18.6	18.6
183	28.2	28.0	27.8	27.5	282	18.6	18.6	18.6	18.5
186	27.9	27.7	27.5	27.2	285	18.5	18.5	18.5	18.5
189	27.5	27.3	27.2	26.9	288	18.5	18.5	18.5	18.4
192	27.2	27.0	26.8	26.6	291	18.5	18.4	18.4	18.4
195	26.8	26.7	26.5	26.2	294	18.4	18.4	18.4	18.4
198	26.5	26.3	26.2	25.9	297	18.4	18.4	18.4	18.4
201	26.2	26.0	25.8	25.6	300	18.4	18.3	18.3	18.3
204	25.8	25.7	25.5	25.3	303	18.3	18.3	18.2	18.2
207	25.5	25.4	25.2	25.0	306	18.2	18.1	18.1	18.0
210	25.2	25.0	24.9	24.6	309	18.0	17.9	17.9	17.8
213	24.8	24.7	24.5	24.3	312	17.8	17.7	17.6	

Table 7.15. Angle of Incidence (i_0) in Degrees at the Base of the Crust

S

	$\theta°$					
h	3	6	9	12	15	18
Surface	82.8	79.5	76.0	71.9	67.3	62.7
0.00	82.4	78.9	75.4	71.4	66.8	62.2
0.01	72.4	74.3	72.1	68.6	64.6	60.6
0.02	61.9	69.0	68.4	65.9	62.5	
0.03	52.9	63.7	64.8	63.2	60.5	50.1
0.04	45.4	58.6	61.3	60.6		48.1
0.05	39.3	53.8	57.8			46.3
0.06	34.2	49.3	54.0		49.2	44.7
0.07	30.0	44.9	50.1	49.9	46.9	43.3
0.08	26.4	41.0	46.5	47.0	44.8	42.0
0.09	23.4	37.4	43.3	44.3	42.9	40.9
0.10	20.8	34.1	40.3	41.9	41.2	39.9
0.11	18.7	31.3	37.6	39.8	39.7	38.9
0.12	16.8	28.7	35.2	37.8	38.4	38.1

	$\theta°$									
h	21	24	27	30	33	36	39	42	45	48
Surface	49.1	42.5	39.7	38.3	37.6	37.1	36.2	35.3	34.3	33.5
0.12	37.6	37.2	36.7	36.0	35.2	34.5	33.7	33.0	32.3	31.6

	$\theta°$									
h	51	54	57	60	63	66	69	72	75	78
Surface	32.7	31.9	31.1	30.3	29.4	28.5	27.6	26.7	25.7	24.8
0.12	30.9	30.1	29.3	28.5	27.7	26.9	26.0	25.1	24.3	23.4

	$\theta°$								
h	81	84	87	90	93	96	99	102	105
Surface	23.9	22.9	21.9	20.8	19.9	19.4	19.2	19.1	19.0
0.12	22.5	21.5	20.5	19.8	19.3	19.2	19.1	19.0	

Table 7.16. Angle of Incidence (i_0) in Degrees at the Base of the Crust

			SS						
		h					*h*		
$\theta°$	Surface	0.04	0.08	0.12	$\theta°$	Surface	0.04	0.08	0.12
3	84.0				111	31.5	31.3	31.0	30.6
6	82.8				114	31.1	30.9	30.6	30.2
9	81.2				117	30.7	30.5	30.2	29.8
12	79.5				120	30.3	30.0	29.7	29.3
15	77.7				123	29.8	29.6	29.3	28.9
18	76.0				126	29.4	29.1	28.9	28.5
21	74.0				129	28.9	28.7	28.5	28.1
24	71.9				132	28.5	28.3	28.0	27.7
27	69.7				135	28.1	27.8	27.6	27.2
30	67.3	61.0			138	27.6	27.4	27.1	26.8
33	65.0	59.9	47.2		141	27.2	26.9	26.7	26.3
36	62.7		45.9		144	26.7	26.4	26.2	25.8
39	60.5	48.5	44.0		147	26.2	26.0	25.7	25.4
42	49.1	45.2	42.2		150	25.7	25.5	25.3	25.0
45	45.2	42.7	40.8	38.4	153	25.3	25.1	24.9	24.6
48	42.5	41.0	39.8	38.1	156	24.8	24.6	24.4	24.1
51	40.8	39.9	39.0	37.8	159	24.4	24.2	24.0	23.7
54	39.7	39.0	38.4	37.6	162	23.9	23.7	23.5	23.2
57	38.8	38.4	38.0	37.4	165	23.4	23.2	23.0	22.7
60	38.3	38.0	37.6	37.2	168	22.9	22.7	22.5	22.2
63	37.9	37.6	37.4	36.9	171	22.4	22.2	22.0	21.7
66	37.6	37.4	37.2	36.5	174	21.9	21.7	21.4	21.1
69	37.3	37.1	36.8	36.1	177	21.3	21.1	20.9	20.6
72	37.1	36.8	36.4	35.7	180	20.8	20.6	20.4	20.2
75	36.7	36.4	36.0	35.3	183	20.3	20.1	20.0	19.8
78	36.2	35.9	35.5	34.8	186	19.9	19.8	19.7	19.6
81	35.7	35.4	35.0	34.4	189	19.6	19.5	19.4	19.4
84	35.3	34.9	34.6	34.0	192	19.4	19.3	19.3	19.2
87	34.8	34.5	34.1	33.6	195	19.2	19.2	19.2	19.2
90	34.3	34.0	33.7	33.2	198	19.2	19.2	19.1	19.1
93	33.9	33.6	33.3	32.8	201	19.1	19.1	19.1	19.1
96	33.5	33.2	32.9	32.5	204	19.1	19.1	19.0	19.0
99	33.1	32.8	32.5	32.1	207	19.0	19.0	19.0	19.0
102	32.7	32.4	32.2	31.8	210	19.0	19.0	18.9	18.9
105	32.3	32.1	31.8	31.4	213	18.9	18.9		
108	31.9	31.7	31.4	31.0					

Table 7.17. Angle of Incidence (i_0) in Degrees at the Base of the Crust

	SP					*SPP*			
		h					*h*		
$\theta°$	Surface	0.04	0.08	0.12	$\theta°$	Surface	0.04	0.08	0.12
42			86.6	83.0	45		86.1	85.4	83.4
45		86.0	83.9	80.4	48	85.8	85.1	83.9	81.8
48	85.0	83.3	81.2	77.9	51	84.6	83.6	82.3	80.2
51	82.3	80.6	78.6	75.6	54	83.0	81.9	80.6	78.5
54	79.6	77.9	76.1	73.4	57	81.3	80.2	79.0	77.0
57	77.0	75.5	73.8	71.5	60	79.6	78.6	77.3	75.4
60	74.6	73.2	71.7	69.6	63	77.9	76.9	75.8	74.0
63	72.4	71.1	69.8	67.7	66	76.3	75.4	74.2	72.5
66	70.3	69.2	67.8	65.9	69	74.8	73.8	72.7	71.1
69	68.4	67.2	65.9	64.0	72	73.2	72.3	71.3	69.7
72	66.4	65.3	64.0	62.2	75	71.7	70.9	69.9	68.4
75	64.4	63.3	62.1	60.5	78	70.3	69.5	68.5	67.1
78	62.5	61.4	60.3	58.8	81	68.9	68.1	67.1	65.7
81	60.6	59.6	58.6	57.2	84	67.5	66.7	65.7	64.3
84	58.8	57.9	56.9	55.8	87	66.1	65.3	64.3	63.0
87	57.0	56.2	55.2	53.9	90	64.6	63.8	62.9	61.6
90	55.3	54.5	53.5	49.3	93	63.2	62.4	61.6	60.3
93	53.5	49.6	48.9	47.9	96	61.8	61.1	60.2	59.1
96	48.8	48.2	47.5	46.6	99	60.4	59.7	58.9	57.8
99	47.4	46.8	46.1	45.3	102	59.1	58.4	57.6	56.5
102	46.0	45.4	44.8	44.0	105	57.7	57.1	56.3	55.3
105	44.6	44.1	43.5	42.7	108	56.4	55.8	55.0	49.4
108	43.2	42.7	42.2	41.5	111	55.1	54.5	49.0	48.3
111	41.9	41.4	40.9	40.3	114	49.0	48.5	48.0	47.3
114	40.7	40.2	39.8	39.3	117	47.9	47.5	47.0	46.3
117	39.5	39.2	38.8	38.4	120	46.9	46.5	46.0	45.4
120	38.6	38.3	38.1	37.7	123	45.9	45.5	45.0	44.4
123	37.9	37.7	37.5	37.2	126	44.9	44.5	44.0	43.5
126	37.3	37.1	36.9	36.7	129	43.9	43.5	43.1	42.6
129	36.8	36.6	36.4	36.2	132	42.9	42.6	42.2	41.7
132	36.3	36.1	36.0	35.9	135	42.0	41.7	41.4	40.9
135	35.9	35.9	35.8	35.7	138	41.2	40.9	40.6	40.1
138	35.7	35.7	35.6	35.6	141	40.4	40.1	39.8	39.5
141	35.6	35.5	35.5	35.4	144	39.7	39.4	39.2	38.9
144	35.4	35.4	35.3		147	39.0	38.8	38.6	38.4
147	35.2				150	38.5	38.3	38.1	37.9
					153	38.0	37.9	37.8	37.6
					156	37.7	37.5	37.4	37.3
					159	37.3	37.2	37.1	36.9
					162	37.0	36.9	36.7	36.5
					165	36.6	36.5	36.4	36.2
					168	36.3	36.2	36.1	36.0
					171	36.0	35.9	35.9	35.8
					174	35.8	35.8	35.7	35.7
					177	35.7	35.6	35.6	35.5
					180	35.6	35.5	35.3	35.3
					183	35.4	35.4	35.3	35.3
					186	35.3	35.2		

Table 7.18. Angle of Incidence (i_0) in Degrees at the Base of the Crust

SSS									
	h					h			
$\theta°$	Surface	0.04	0.08	0.12	$\theta°$	Surface	0.04	0.08	0.12
3	84.3				129	34.9	34.7	34.5	34.1
6	83.7				132	34.6	34.4	34.2	33.8
9	82.8				135	34.3	34.1	33.9	33.6
12	81.7				138	34.0	33.8	33.6	33.3
15	80.6				141	33.7	33.6	33.4	33.0
18	79.5				144	33.5	33.3	33.1	32.8
21	78.3				147	33.2	33.0	32.8	32.5
24	77.2				150	32.9	32.8	32.6	32.3
27	76.0				153	32.7	32.5	32.3	32.1
30	74.7				156	32.4	32.3	32.1	31.8
33	73.3				159	32.2	32.0	31.8	31.6
36	71.9				162	31.9	31.8	31.6	31.3
39	70.4				165	31.7	31.5	31.3	31.1
42	68.9				168	31.4	31.3	31.1	30.8
45	67.3				171	31.1	31.0	30.8	30.5
48	65.7	61.2			174	30.9	30.7	30.5	30.2
51	64.2	60.5			177	30.6	30.4	30.2	29.9
54	62.7	59.5			180	30.3	30.1	29.9	29.6
57	61.2	51.4	46.4		183	30.0	29.8	29.6	29.4
60	59.8	48.6	45.0		186	29.7	29.5	29.3	29.1
63	49.1	46.3	43.6		189	29.4	29.2	29.0	28.8
66	46.3	44.3	42.3		192	29.1	28.9	28.8	28.5
69	44.2	42.6	41.2		195	28.8	28.6	28.5	28.2
72	42.5	41.4	40.4		198	28.5	28.4	28.2	27.9
75	41.3	40.5	39.8	38.3	201	28.2	28.1	27.9	27.7
78	40.4	39.8	39.2	38.2	204	27.9	27.8	27.6	27.4
81	39.7	39.2	38.7	38.0	207	27.6	27.5	27.3	27.1
84	39.1	38.7	38.4	37.8	210	27.3	27.2	27.0	26.7
87	38.6	38.4	38.1	37.6	213	27.0	26.8	26.7	26.4
90	38.3	38.1	37.8	37.4	216	26.7	26.5	26.3	26.1
93	38.0	37.8	37.6	37.3	219	26.3	26.2	26.0	25.8
96	37.7	37.6	37.4	37.1	222	26.0	25.9	25.7	25.5
99	37.6	37.4	37.3	36.9	225	25.7	25.6	25.4	25.2
102	37.4	37.3	37.1	36.7	228	25.4	25.3	25.1	24.9
105	37.3	37.1	36.9	36.4	231	25.1	25.0	24.9	24.7
108	37.1	36.9	36.6	36.1	234	24.8	24.7	24.6	24.4
111	36.8	36.6	36.3	35.8	237	24.5	24.4	24.3	24.1
114	36.5	36.3	36.0	35.6	240	24.2	24.1	24.0	23.8
117	36.2	36.0	35.7	35.3	243	23.9	23.8	23.6	23.4
120	35.9	35.7	35.4	35.0	246	23.6	23.5	23.3	23.1
123	35.6	35.4	35.1	34.7	249	23.3	23.1	23.0	22.8
126	35.3	35.0	34.8	34.4	252	22.9	22.8	22.6	22.4

Table 7.18. (cont.)

	SSS								
	h					*h*			
$\theta°$	Surface	0.04	0.08	0.12	$\theta°$	Surface	0.04	0.08	0.12
255	22.6	22.4	22.3	22.1	291	19.3	19.3	19.2	19.2
258	22.2	22.1	21.9	21.7	294	19.2	19.2	19.2	19.2
261	21.9	21.7	21.6	21.4	297	19.2	19.2	19.2	19.1
264	21.5	21.4	21.2	21.0	300	19.1	19.1	19.1	19.1
267	21.1	21.0	20.9	20.7	303	19.1	19.1	19.1	19.1
270	20.8	20.6	20.5	20.4	306	19.1	19.1	19.1	19.0
273	20.4	20.3	20.2	20.1	309	19.0	19.0	19.0	19.0
276	20.1	20.0	29.0	19.8	312	19.0	19.0	19.0	19.0
279	19.9	19.8	19.7	19.6	315	19.0	19.0	19.0	18.9
282	19.7	19.6	19.6	19.5	318	18.9	18.9	18.9	
285	19.5	19.5	19.4	19.4	321	18.9			
288	19.4	19.3	19.3	19.3					

Obviously, $|B^0/B| = 1$, and $SS(1) = 1$. The energy equation will be satisfied only if we take $SP(1) = 0$. Therefore, for $f > f_1$, the reflected P wave carries no energy; the whole energy goes into the reflected SV wave.

The discussion at the mantle–core boundary is similar. For a P wave incident at this boundary (assuming $\alpha > \alpha'$), there is no critical angle. For an incident SV wave (assuming $\alpha > \alpha' > \beta$), there are two critical angles, f_1 and f_2, corresponding to the reflected and the transmitted P waves, respectively. We define

$$\begin{aligned} f_1 &= \arcsin\left(\frac{\beta}{\alpha}\right), \qquad f_2 = \arcsin\left(\frac{\beta}{\alpha'}\right), \\ \cos e &= -i(\sin^2 e - 1)^{1/2}, \qquad \text{for} \quad f > f_1, \\ \cos e' &= -i(\sin^2 e' - 1)^{1/2}, \qquad \text{for} \quad f > f_2. \end{aligned} \tag{7.182}$$

The choice of the sign in Eqs. (7.181) and (7.182) is made in such a manner that the disturbance tends to zero away from the boundary.

When f is greater than both the critical angles, $SS(11) = 1$. In other words, the whole energy goes into the reflected SV wave, the reflected P and transmitted P waves being physically insignificant. When f is equal to either of the two critical angles, we have $SP(11) = 0$, $SP(12) = 0$, and $SS(11) = 1$.

Important points concerning the computation of reflection and transmission coefficients may be summarized as follows:

1. Only the absolute values of the amplitude ratios should be used. When the angle of incidence becomes greater than a critical angle, the amplitude ratios become complex. In that case, the corresponding moduli should be used.

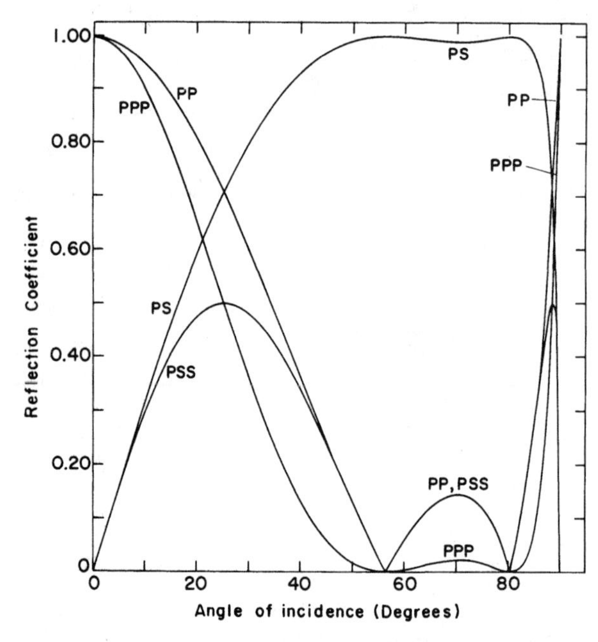

Figure 7.23. The reflection coefficients of PP, PS, PPP, and PSS waves ($\beta/\alpha = 0.587$).

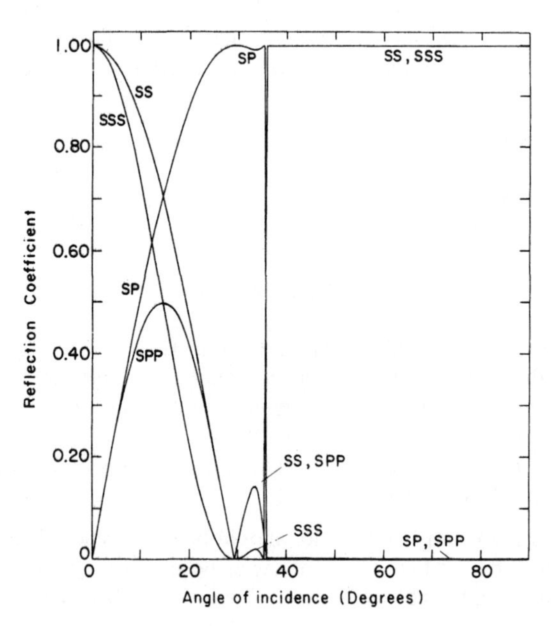

Figure 7.24. The reflection coefficients of SP, SS, SPP and SSS waves ($\beta/\alpha = 0.587$). The critical angle is at 36.0°.

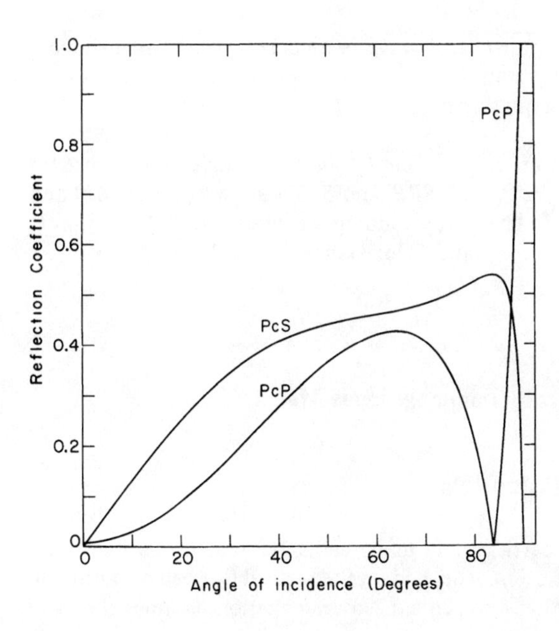

Figure 7.25. The reflection coefficients of *PcP* and *PcS* waves.

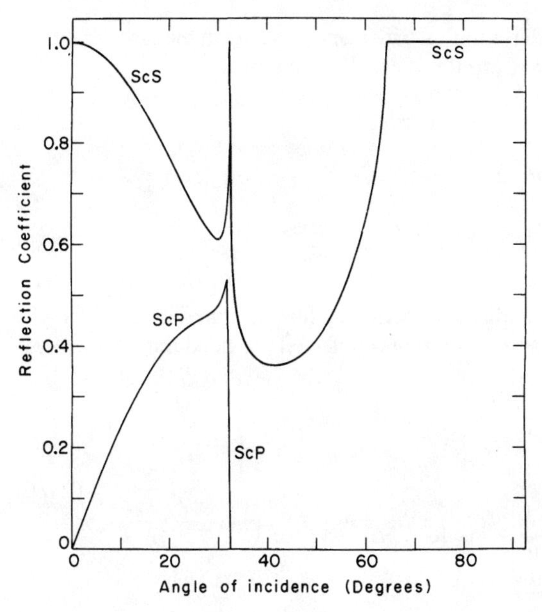

Figure 7.26. The reflection coefficients of *ScP* and *ScS* waves.

2. When a reflected or a transmitted wave is associated with an imaginary angle, it carries no energy and the corresponding reflection or transmission coefficient should be taken as zero.

The variation of the reflection coefficient (Z) with the angle of incidence for the waves *PP*, *PS*, *SP*, *SS*, *PPP*, *PSS*, *SPP*, and *SSS* is shown in Figs. 7.23 and 7.24 assuming $\beta/\alpha = 0.587$. The corresponding variation for *PcP*, *PcS*, *ScP*, and *ScS* is shown in Figs. 7.25 and 7.26, assuming $\varepsilon = 1.714$, $\beta/\alpha = 0.535$, $\alpha'/\alpha = 0.594$.

7.3. Ray Theory in Vertically Inhomogeneous Media

7.3.1. Displacement Potentials

Wave propagation in the earth can, in many cases, be treated in a half-space model where the parameters λ, μ, ρ depend on the vertical Cartesian coordinate z only. With this assumption, the general Navier equation assumes the form [cf. Eq. (1.107)]

$$\mu\nabla^2\mathbf{u} + (\lambda + \mu)\text{grad div }\mathbf{u} + \mathbf{e}_z\lambda' \text{ div }\mathbf{u} + \mu'\left(2\frac{\partial\mathbf{u}}{\partial z} + \mathbf{e}_z \times \text{curl }\mathbf{u}\right) = \rho\frac{\partial^2\mathbf{u}}{\partial t^2}, \tag{7.183}$$

where the prime denotes differentiation with respect to z. As in the case of radially heterogeneous media, the *SH* motion is exactly decoupled,

$$\mathbf{M} = \text{curl}(\mathbf{e}_z\psi_3), \tag{7.184}$$

where ψ_3 satisfies the equation

$$\mu\nabla^2\psi_3 + \mu'\frac{\partial\psi_3}{\partial z} = \rho\frac{\partial^2\psi_3}{\partial t^2}. \tag{7.185}$$

Proceeding along the lines indicated previously for the spherical case, a high-frequency decoupling of P and SV motions is effected provided that (Λ = wavelength)

$$\left|\Lambda\frac{\lambda'}{\lambda}\right| \ll 1, \quad \left|\Lambda\frac{\mu'}{\mu}\right| \ll 1, \quad \left|\Lambda\frac{\rho'}{\rho}\right| \ll 1,$$
$$\Lambda\left|\frac{\mu''}{\mu}\right|^{1/2} \ll 1, \quad \Lambda\left|\frac{\rho''}{\rho}\right|^{1/2} \ll 1. \tag{7.186}$$

We can then show that the vectors **L**, **N**, and **M**, representing P, SV, and SH motions, respectively, are:

$$\mathbf{L} = \frac{1}{\sqrt{\rho}} \operatorname{grad} \Phi_1,$$
$$\mathbf{N} = \frac{1}{\sqrt{\rho}} \operatorname{curl} \operatorname{curl}(\mathbf{e}_z \Phi_2), \qquad (7.187)$$
$$\mathbf{M} = \frac{1}{\sqrt{\mu}} \operatorname{curl}(\mathbf{e}_z \Phi_3),$$

where

$$\nabla^2 \Phi_i + \frac{\omega^2}{v_i^2(z)} \Phi_i = 0 \qquad (i = 1, 2, 3),$$
$$v_1^2 = (\lambda + 2\mu)/\rho, \qquad v_2^2 = v_3^2 = \mu/\rho \qquad (7.188)$$

7.3.2. Earth-Flattening Transformation

In various problems in seismology, especially when dealing with wave propagation over distances such that the earth's curvature can be neglected or at least taken into account by simple corrections, it is advantageous to develop the fundamental equations of ray theory in cylindrical coordinates. It is possible to set a simple correspondence that enables us to translate our previous results into the half-space "language." To this end, we write down, side by side, the eikonal equations in both systems under the condition of azimuthal symmetry:

$$\left(\frac{\partial t}{\partial x}\right)^2 + \left(\frac{\partial t}{\partial z}\right)^2 = \frac{1}{v^2(z)}, \qquad \text{(cylindrical)} \qquad (7.189)$$

$$\frac{1}{a^2}\left(\frac{\partial t}{\partial \theta}\right)^2 + \frac{r^2}{a^2}\left(\frac{\partial t}{\partial r}\right)^2 = \left(\frac{r}{a}\right)^2 \frac{1}{V^2(r)}, \qquad \text{(spherical)} \qquad (7.190)$$

where a is the radius of the earth and x denotes the cylindrical radial coordinate. A term by term correspondence then yields the relations

$$\frac{\partial}{\partial x} = \frac{1}{a}\frac{\partial}{\partial \theta} \Rightarrow x = a\theta,$$
$$\frac{\partial}{\partial z} = -\frac{r}{a}\frac{\partial}{\partial r} \Rightarrow r = ae^{-z/a} \quad \text{or} \quad z = a \ln\left(\frac{a}{r}\right), \qquad (7.191)$$
$$\frac{1}{v(z)} = \frac{r}{aV(r)} \Rightarrow v(z) = \frac{a}{r} V(r).$$

The introduction of a is on dimensional considerations, so that $z = 0$ corresponds to $r = a$.

Therefore, by the change of variables:

$$r = ae^{-z/a}, \qquad \theta = \frac{x}{a}, \qquad V(r) = e^{-z/a}v(z), \tag{7.192}$$

the eikonal equation in spherical coordinates becomes the corresponding equation in cylindrical coordinates. Then, apart from the vicinity of singularities that might be introduced by the change of variables, the two equations are mathematically equivalent. Equations (7.192) are known as the *earth-flattening transformation.*

If $z/a \ll 1$, we have $r \simeq a - z$ from Eq. (7.191) and, therefore,

$$v(z) = \left(1 + \frac{z}{a}\right)V(a - z). \tag{7.193}$$

This is the *earth-flattening approximation.* The interpretation of this equation is as follows: As far as the travel times are concerned, spherical curvature is accounted for by imposing a constant velocity gradient upon the existing velocity distribution. In particular, a homogeneous sphere is transformed into a half-space with a linear velocity profile. This transformation is, of course, subject to the condition $z/a \ll 1$; i.e., the ray does not penetrate deeply into the medium. Figure 7.27 exhibits this interesting duality and interchangeability of the curvature of the rays and the bounding surface.

Table 7.19 summarizes the sphere–half-space correspondence for the exact earth-flattening transformation. In this table the half-space parameters j, X, t, z_m correspond to the sphere parameters i, Δ, T, r_m (Fig. 7.28).

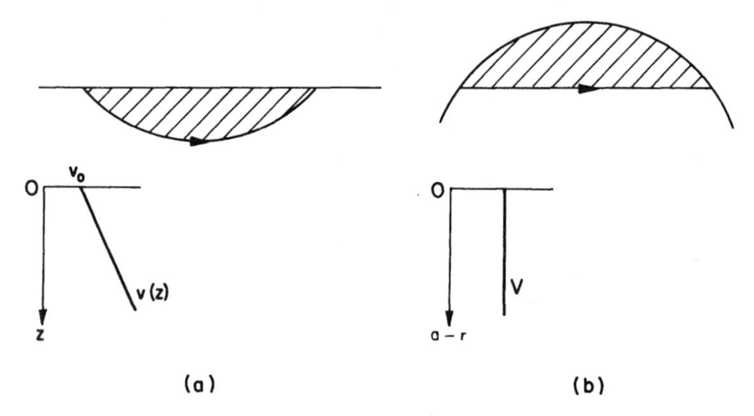

Figure 7.27. (a) Curved rays in a half-space with constant velocity gradient. (b) Straight rays in a homogeneous sphere.

Table 7.19. Earth-Flattening Transformation under GEA

Sphere		Half-space
$a\theta$		x
r		$ae^{-z/a}$
$a \ln\left(\frac{a}{r}\right)$		z
$\frac{a}{r} V(r)$		$v(z)$
$\frac{\partial}{\partial r}$		$-\frac{a}{r}\frac{\partial}{\partial z}$
$a^{-1}\frac{\partial}{\partial \theta}$		$\frac{\partial}{\partial x}$
$\frac{dV(r)}{dr}$		$\frac{1}{a} v - \frac{dv}{dz}$
$r\frac{d^2V}{dr^2}$		$a\frac{d^2v}{dz^2} - \frac{dv}{dz}$
$\frac{r \sin i}{V(r)} = \frac{r_m}{V_m} = p = \frac{dT}{d\theta}$	(Snell's law)	$\frac{\sin j}{v(z)} = \frac{1}{v_m} = p = \frac{dt}{dx}$
$\Delta = 2p\int_{r_m}^{a} \frac{dr}{r\sqrt{(r^2/V^2 - p^2)}}$		$X = 2p\int_{0}^{z_m} \frac{dz}{\sqrt{(1/v^2 - p^2)}}$
$= 2\int_{r_m}^{a} \tan i \frac{dr}{r}$		$= 2\int_{0}^{z_m} (\tan j) dz$
$= \frac{2}{p}\int_{i_0}^{\pi/2} \frac{\sin i \, di}{(V/r - dV/dr)}$		$= \frac{2}{p}\int_{j_0}^{\pi/2} \frac{\sin j \, dj}{dv/dz}$
$T = p\theta \pm \int_{r_h}^{r} \sqrt{\left(\frac{r^2}{V^2} - p^2\right)}\frac{dr}{r}$		$t = px \pm \int_{z_h}^{z} \sqrt{\left[\frac{1}{v^2(z)} - p^2\right]} dz$
$= \pm \int_{r_h}^{r} \frac{r}{V^2}\frac{dr}{\sqrt{(r^2/V^2 - p^2)}}$		$= \pm \int_{z_h}^{z} \frac{dz}{v^2\sqrt{(1/v^2 - p^2)}}$
$-\frac{p}{r}\frac{dV}{dr}$	(ray curvature)	$p\frac{dv}{dz}$

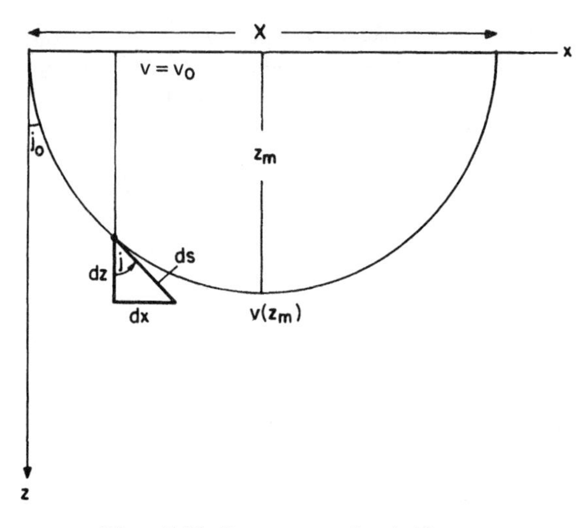

Figure 7.28. Ray geometry in a half-space.

EXAMPLE 7.4: The law $v(z) = v_0 \operatorname{ch}[\pi(z/a)]$

In this case

$$t = 2\int_0^{z_m} \frac{dz}{v^2(1/v^2 - p^2)^{1/2}} = 2\int_{v_0}^{v_m} \frac{(dz/dv)dv}{v(1 - v^2/v_m^2)^{1/2}}$$

$$= \frac{2a}{\pi}\int_{v_0}^{v_m} \frac{dv}{v(1 - v^2/v_m^2)^{1/2}(v^2 - v_0^2)^{1/2}}$$

$$= \frac{a}{\pi v_m}\int_{u_0}^{1} \frac{du}{u(1 - u)^{1/2}(u - u_0)^{1/2}},$$

where

$$u = \left(\frac{v}{v_m}\right)^2, \qquad u_0 = \left(\frac{v_0}{v_m}\right)^2.$$

This yields, on integration,

$$t = \frac{a}{v_0}.$$

The travel-time is therefore a fixed number for all the rays. By calculating the range, it can be easily seen that the rays are brought to a common focus, at a distance a, so that the travel-time curve degenerates into a single point, $X = a$, $t = a/v_0$.

EXAMPLE 7.5: Travel time for a transition zone

At certain depths in the earth there is rapid change in the intrinsic seismic velocities. Although for convenience in computation we usually represent such a case as a material discontinuity in the form of an interface across which there is a jump in rigidity and density, it is useful sometimes to model it as a transition layer with gradually changing material properties. A function of this type is (Fig. 7.29)

$$v_n(z) = \frac{v_0 + v_1 e^{n(z-h)}}{1 + e^{n(z-h)}}, \qquad (z, h \geq 0).$$

This describes a velocity–depth profile such that

$$\lim_{n\to\infty} v_n(z) = \begin{cases} v_0, & z < h \\ \frac{1}{2}(v_0 + v_1), & z = h \\ v_1, & z > h. \end{cases}$$

For fixed n,

$$v_n(\infty) = v_1, \qquad v_n(h) = \frac{1}{2}(v_1 + v_0), \qquad v_n(0) = v_0 \frac{1 + (v_1/v_0)e^{-nh}}{1 + e^{-nh}}.$$

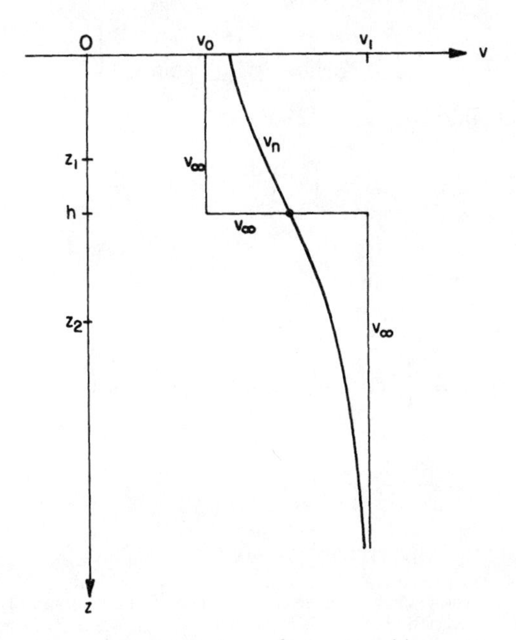

Figure 7.29. A velocity profile in a transition zone.

If $v_1 > v_0$, $v_n(0) > v_0$ and if $v_1 < v_0$, then $v_n(0) < v_0$. The travel time of a ray crossing the transition zone at normal incidence from z_1 to z_2 is ($z_1 < h < z_2$)

$$t_n = \int_{z_1}^{z_2} \frac{dz}{v_n(z)} = \frac{z_2 - z_1}{v_0} - \frac{1}{n}\left(\frac{1}{v_0} - \frac{1}{v_1}\right)\ln\frac{v_0 + v_1 e^{n(z_2-h)}}{v_0 + v_1 e^{-n(h-z_1)}},$$

$$t = \lim_{n\to\infty} t_n = \frac{h - z_1}{v_0} + \frac{z_2 - h}{v_1}.$$

EXAMPLE 7.6[4]: The Law $V(z) = V_0(1 + \gamma z)$

This distribution is the half-space analog to the distribution $V = a - br^2$ in a sphere considered in Example 7.2, where we have proved that the rays are circular arcs. In the present case, too, we note that the ray curvature is

$$\frac{dj}{ds} = p\frac{dV}{dz} = pV_0\gamma = \text{const.} = \gamma \sin j_0 = \frac{\sin j_h}{h + 1/\gamma},$$

implying thereby that the rays are circular arcs of radii $R = (h + 1/\gamma)/\sin j_h$. The centers of all the circles lie on the line $z = -1/\gamma$, $1/\gamma$ units above the free surface (Fig. 7.30).

The travel-time along a ray from a source at depth h to a point at depth z is

$$t = \int_0^s \frac{ds}{V} = \int_{j_h}^{j} \frac{1}{V}\frac{ds}{dj}\,dj = \int_{j_h}^{j} \frac{dj}{V\gamma \sin j_0}$$

$$= \int_{j_h}^{j} \frac{dj}{\gamma V_0 \sin j} = \frac{1}{\gamma V_0}\ln\left[\frac{\tan j/2}{\tan j_h/2}\right]. \tag{7.6.1}$$

The horizontal range is

$$x = \int dx = \int_{j_h}^{j} \frac{dx}{ds}\frac{ds}{dj}\,dj = \int_{j_h}^{j} \frac{\sin j}{\gamma \sin j_0}\,dj$$

$$= \frac{1}{\gamma \sin j_0}(\cos j_h - \cos j). \tag{7.6.2}$$

From Snell's law

$$z = \frac{1}{\gamma V_0}(V - V_0) = \frac{1}{\gamma \sin j_0}(\sin j - \sin j_0). \tag{7.6.3}$$

Therefore,

$$z - h = \frac{1}{\gamma \sin j_0}(\sin j - \sin j_h). \tag{7.6.4}$$

[4] From here to the end of the chapter we use $V(z)$ to signify a general velocity in the half-space since the symbol v is reserved for slowness from section 7.6.4 onwards. The symbols α and β are used specifically for longitudinal and shear velocities respectively.

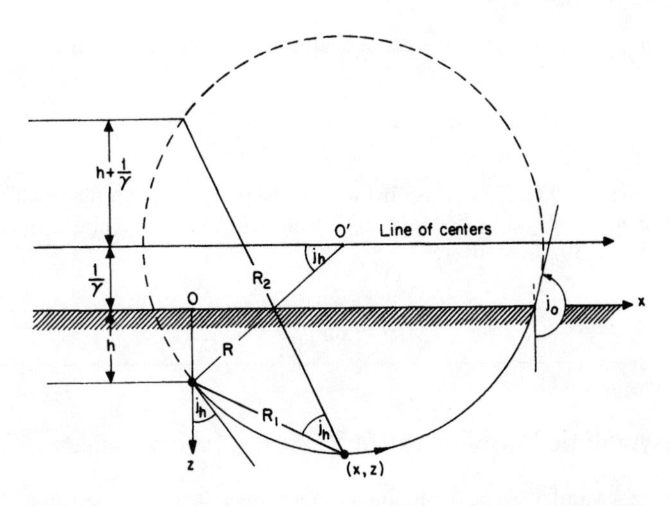

Figure 7.30. A circular ray in a half-space with a constant velocity gradient.

Equations (7.6.2) and (7.6.4) yield

$$R_1 = [x^2 + (z - h)^2]^{1/2} = \frac{2}{\gamma \sin j_0} \sin \frac{j - j_h}{2},$$
$$R_2 = \left[x^2 + \left(z + h + \frac{2}{\gamma}\right)^2\right]^{1/2} = \frac{2}{\gamma \sin j_0} \sin \frac{j + j_h}{2}. \tag{7.6.5}$$

Equations (7.6.1) and (7.6.5) now give

$$t = \frac{2}{\gamma V_0} \operatorname{th}^{-1}\left[\frac{\tan j/2 - \tan j_h/2}{\tan j/2 + \tan j_h/2}\right] = \frac{2}{\gamma V_0} \operatorname{th}^{-1}\left(\frac{R_1}{R_2}\right). \tag{7.6.6}$$

From Eq. (7.6.2) and Snell's law, we have

$$x = \frac{V_h}{\gamma V_0 \sin j_h}(\cos j_h - \cos j) = \frac{V_h}{\gamma V_0 \sin j_h}\left[\cos j_h - \left(1 - \frac{V^2}{V_h^2}\sin^2 j_h\right)^{1/2}\right].$$

It can now be seen that

$$\left(\frac{\partial x}{\partial j_h}\right)_z = \frac{x}{\cos j \sin j_h}. \tag{7.6.7}$$

For a half-space the divergence coefficient, Eq. (7.151), is replaced by

$$G(z/z_0) = \left|\frac{V_h \rho_h}{V\rho} \cdot \frac{\sin j_h}{x \cos j} \cdot \left(\frac{\partial j_h}{\partial x}\right)_z\right|^{1/2}. \tag{7.6.8}$$

Thus, using Eq. (7.6.7), the divergence coefficient for a constant velocity gradient is

$$G = \left(\frac{V_h \rho_h}{V \rho}\right)^{1/2} \frac{\sin j_h}{x}. \tag{7.6.9}$$

It may be noted that R_1 is the distance of the point (x, z) on the ray from the source $(0, h)$ and R_2 is its distance to the image of the source above the line of centers (Fig. 7.30). Moreover,

$$R_1 R_2 = \frac{2x}{\gamma \sin j_0} = 2xR. \tag{7.6.10}$$

7.4. Asymptotic Wave-Theory in Vertically Inhomogeneous Media

In Chapters 4 and 5, we have obtained exact explicit integral expressions for the field of point dislocations. Even if we could evaluate these integrals over the wave number k, there still would remain an integration over the frequency ω in order to obtain the field in the time domain. We shall learn how to do this in Section 7.6; in the meantime, however, we would like to get additional information out of these integrals without having to evaluate them exactly.

Part of the total field is already known to us at this point. In Chapter 5 we have seen that the surface-wave field is obtained as the residues at the poles of the integrand. We shall next show that reflected and transmitted P and S waves whose wavelengths are small compared to the distance from the source are obtained by an ingenious approximation, known as the *method of steepest descents* [cf. App. E]. The idea behind this method is to deform the path of integration in the complex plane in such a way that the main contribution to the integral is determined by a small neighborhood of a point, known as a *saddle point*. It turns out that the saddle point is given by $k = \omega p$, where p is the ray parameter. The steepest descents method serves, therefore, as a mathematical tool to sort out the rays out of the entire field. Those parts of the field that result neither from real poles nor from saddle points must be evaluated by other methods. Chief among them are the *head waves*, which usually travel along interfaces. These are contributed by *branch points* of the integrand. A Taylor expansion of the integrand around these points will yield a fair approximation of their behavior. We shall demonstrate this in a few examples.

After concluding this brief survey of approximate methods of evaluating the k integrals, we shall turn to exact evaluation of the time-domain field, otherwise known as *theoretical seismograms*. It can be obtained in two alternative ways:

1. By converting the k integral into an infinite series of so-called *generalized rays* and effecting the ω integration term by term.
2. By evaluating the integral as the sum of residues at all real and complex poles of the integrand, thus obtaining an infinite sum of normal modes.

7.4.1. *SH* Plane Waves: Internal Reflection

If the heterogeneity is weak so that the conditions in relations (7.186) are satisfied, the *SH* field is given by [cf. Eq. (7.187)]

$$\mathbf{M} = \frac{1}{\sqrt{\mu}}\,\mathrm{curl}(\mathbf{e}_z\Phi_3) = -\frac{\mathbf{e}_y}{\sqrt{\mu}}\frac{\partial\Phi_3}{\partial x}, \tag{7.194}$$

assuming the motion to be independent of the y coordinate. The potential Φ_3 satisfies the equation

$$\frac{\partial^2\Phi_3}{\partial x^2} + \frac{\partial^2\Phi_3}{\partial z^2} = \frac{1}{V^2}\frac{\partial^2\Phi_3}{\partial t^2}, \tag{7.195}$$

where $V = V_3(z)$ is the S-wave velocity. Hence, we may take

$$u_y(z) = \frac{1}{\sqrt{\mu}}F(z)\cos k(ct - x), \tag{7.196}$$

where F is an integral of the equation

$$\frac{d^2F}{dz^2} + k^2\left(\frac{c^2}{V^2} - 1\right)F = 0. \tag{7.197}$$

Let us take the origin at the level where $V = c$ and the z axis in the direction of V increasing (Fig. 7.31). Then the Wentzel–Kramers–Brillouin–Jeffreys (WKBJ) solution (see App. H) of Eq. (7.197) is

$$F(z) = |q|^{-1/4} \times \begin{cases} A\sin kL + B\cos kL, & (V < c) \\ Ce^{-kM}, & (V > c) \end{cases} \tag{7.198}$$

$$L = \int_z^0 \sqrt{\left(\frac{c^2}{V^2} - 1\right)}dz, \qquad M = \int_0^z \sqrt{\left(1 - \frac{c^2}{V^2}\right)}dz, \qquad q = \frac{c^2}{V^2} - 1.$$

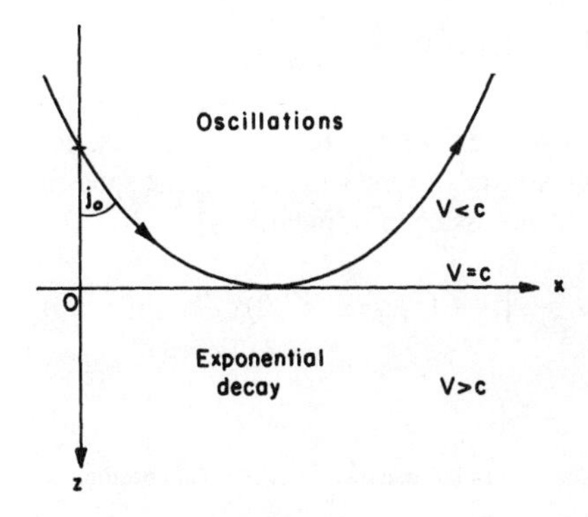

Figure 7.31. Internal reflection of *SH* plane waves.

Considering the level $z=0$ (at which $V = c$) as a virtual reflecting surface, we match the displacements (F values) and stresses (dF/dz values) on both sides of $z = 0$, thus obtaining $A = B = C$. Hence, the solution for $z < 0$ is

$$u_y(z<0) = Aq^{-1/4}\left(\frac{2}{\mu}\right)^{1/2} \cos k(ct - x)\sin\left(kL + \frac{\pi}{4}\right)$$

$$= \frac{Aq^{-1/4}}{\sqrt{(2\mu)}}\left[\sin\left\{k(ct - x + L) + \frac{\pi}{4}\right\} - \sin\left\{k(ct - x - L) - \frac{\pi}{4}\right\}\right]. \tag{7.199}$$

In a homogeneous medium $V = V_0$,

$$L = -z\sqrt{\left(\frac{c^2}{V_0^2} - 1\right)}, \qquad M = z\sqrt{\left(1 - \frac{c^2}{V_0^2}\right)}, \qquad \sin j_0 = \frac{V_0}{c},$$

and $\pm\sin\{k(ct - x \mp z\cot j_0) \pm (\pi/4)\}$ describe downgoing and upgoing plane waves. In a nonhomogeneous medium the dependence on z is more complicated but the motion is again composed of a superposition of downgoing and upgoing waves such that there is a phase advance of $\pi/2$ in the upgoing wave relative to the downgoing wave, independent of the frequency. As we have already seen in Section 3.2.3, this causes a distortion of the entire pulse in the time domain. If the downgoing wave, for example, is a unit-step function with c independent of frequency, i.e., $H(ct - x - z\cot j_0)$, the upgoing wave will be the allied function

$$\frac{1}{\pi}\int_0^\infty \cos k(ct - x + z\cot j_0)\frac{dk}{k}.$$

Therefore, the high-frequency plane wave is characterised by the transfer of energy along a specified path together with a *total reflection* at its lowest point.

7.4.2. *SH* Waves from a Point Dislocation

Consider a vertically heterogeneous half-space. Taking the positive z direction into the medium and a point dislocation source of type I at $z = z_0$, the spectral *SH* displacement field is [Eq. (5.72) with necessary modifications]

$$u_\phi^I(\omega) = \frac{\mu_s U_0\, dS e^{i\omega t}}{4\pi}\cos 2\phi \frac{\partial}{\partial \Delta}\int_{-\infty}^{\infty} G(z|z_0)H_2^{(2)}(k\Delta)k\, dk, \tag{7.200}$$ [5]

[5] From here to the end of the chapter Δ denotes the radial distance in cylindrical coordinates.

where

$$G(z|z_0) = \frac{y_1^-(z_>)}{W}\left[y_1^+(z_<) - \frac{y_2^+(0)}{y_2^-(0)}\,y_1^-(z_<)\right], \tag{7.201}$$

$$z_> = \max(z, z_0), \qquad z_< = \min(z, z_0), \qquad z, z_0 \geq 0,$$

$$W = y_2^+(0)y_1^-(0) - y_2^-(0)y_1^+(0).$$

For simplicity we take $z_0 = 0$ (surface source), which yields

$$G(z|0) = \frac{-y_1^-(z)}{y_2^-(0)}. \tag{7.202}$$

The function $y_1(z)$ satisfies Eq. (5.52)

$$\frac{d^2y_1}{dz^2} + \frac{1}{\mu}\frac{d\mu}{dz}\frac{dy_1}{dz} + \omega^2\left(\frac{1}{V^2} - \frac{1}{c^2}\right)y_1 = 0, \qquad (\omega = ck). \tag{7.203}$$

For weak heterogeneity, this may be replaced by

$$\frac{d^2y_1}{dz^2} + \omega^2\left[\frac{1}{V^2(z)} - \frac{1}{c^2}\right]y_1 = 0. \tag{7.204}$$

At high frequencies, the WKBJ solution applies (App. H) and we can replace the function $H_2^{(2)}(k\Delta)$ in the integrand of Eq. (7.200) by its asymptotic value:

$$H_2^{(2)}(k\Delta) = \sqrt{\left(\frac{2}{\pi k\Delta}\right)}e^{-i(k\Delta - 5\pi/4)}. \tag{7.205}$$

Away from the turning point $V(z) = c$, we have

$$\begin{aligned} y_1(z) &\simeq Q^{-1/4}\exp\left(\pm i\pi/4\right)\exp\left(\pm i\omega\int_{z_m}^{z}\sqrt{Q}\,dz\right), \\ y_2(z) = \mu\frac{dy_1}{dz} &\simeq \pm i\omega\mu Q^{1/4}\exp\left(\pm i\pi/4\right)\exp\left(\pm i\omega\int_{z_m}^{z}\sqrt{Q}\,dz\right), \end{aligned} \tag{7.206}$$

where the upper sign is for the upgoing waves, the lower sign for the downgoing waves, and

$$Q = \frac{1}{V^2} - \frac{1}{c^2}. \tag{7.207}$$

Hence, for $V < c$,[6]

$$G(z|0) = \frac{\exp\left[-i\omega\left(\pm\int_0^{z_m} + \int_z^{z_m}\right)\sqrt{Q}\,dz \pm i\pi/4 - i\pi/4\right]}{\omega\mu_s Q^{1/4}(0)Q^{1/4}(z)}. \tag{7.208}$$

[6] This assumption is justified because the saddle point of the integral obeys this inequality except at one point on the ray.

Making the substitutions

$$p = \frac{1}{c}, \qquad k = \omega p, \qquad \Phi = p\Delta \mp \int_0^z \sqrt{\left(\frac{1}{V^2} - p^2\right)} dz, \tag{7.209}$$

and using Eqs. (7.205) and (7.208), Eq. (7.200) assumes the form

$$u_\phi^I(\omega) = -\omega^{1/2} \frac{U_0 \, dS e^{i\omega t}}{4\pi} \sqrt{\left(\frac{2}{\pi\Delta}\right)} e^{\pm i\pi/4} \cos 2\phi$$

$$\times \frac{\partial}{\partial\Delta} \int_{-\infty}^{\infty} \frac{p^{1/2}}{[(1/V_0^2 - p^2)(1/V^2 - p^2)]^{1/4}} e^{-i\omega\Phi(p)} \, dp. \tag{7.210}$$

The saddle points are the roots p_0 of $\Phi'(p) = 0$. This equation takes the form

$$\Delta = \mp p_0 \int_0^z \frac{dz}{\sqrt{(1/V^2 - p_0^2)}}. \tag{7.211}$$

However, Eq. (7.211) is exactly the range equation of a ray (Table 7.19). Therefore, at the saddle point

$$p = \text{ray parameter}.$$

This means that the value of the integral at high frequencies is the amplitude of the seismic ray; i.e., *the saddle point yields, in fact, Snell's law.* The amplitude of the ray is obtained via the *stationary-phase approximation* (Appendix E)

$$\int_{-\infty}^{\infty} A(p) e^{-i\omega\Phi(p)} \, dp \simeq A(p_0) \left[\frac{2\pi}{\omega|\Phi''(p_0)|}\right]^{1/2}$$

$$\times \exp\left[-i\omega\Phi(p_0) - \left(\frac{\pi i}{4}\right) \operatorname{sgn} \Phi''(p_0)\right], \tag{7.212}$$

where

$$\Phi''(p_0) = -\frac{\partial\Delta}{\partial p} = -\frac{\partial\Delta}{\partial j_h} \bigg/ \frac{\partial p}{\partial j_h} = -\frac{\partial\Delta}{\partial j_h} \frac{V_h}{\cos j_h}. \tag{7.213}$$

Because for a fixed z, x decreases as j_h increases ($\partial\Delta/\partial j_h < 0$), the integral in Eq. (7.210) is given by (for $z = 0$)

$$u_\phi^I(\omega) = -\omega \frac{U_0(\omega) dS}{2\pi V_0} (\cos 2\phi \sin j_0) \left[\frac{\tan j_0}{\Delta} \left|\frac{\partial j_0}{\partial \Delta}\right|\right]^{1/2} e^{i\omega(t-T)}, \tag{7.214}$$

where T is the travel time.

The square-root factor is none other than the geometric spreading in a half-space. The factor ($\cos 2\phi \sin j_0$) is the radiation pattern for SH waves from a strike-slip source and fits with our former results. We therefore see that the WKBJ solutions yield results that are identical with those obtained from the GEA.

7.4.3. Point Source in a Layered Half-Space

7.4.3.1. Source in the Layer. A dislocation source of type I is buried at depth h in a layer of thickness $H(>h)$ overlying a uniform half-space. The rigidity and shear-wave velocity in the layer are μ_1 and β_1, respectively, and μ_2, β_2 in the half-space. When the z axis is taken as going into the medium, the spectral displacements on the free surface at $z = 0$ are [Eqs. (5.30)–(5.33) with necessary modifications]

$$u_{\phi}^{I}(\omega) = \frac{U_0(\omega)dS}{4\pi} \cos 2\phi \frac{\partial}{\partial \Delta} \int_{-\infty}^{\infty} \frac{A(k, h, H)}{L(k, H)} H_2^{(2)}(k\Delta) \frac{k\,dk}{\nu_1}, \qquad (7.215)^7$$

where

$$A = (\lambda \nu_1 + \nu_2)e^{(H-h)\nu_1} + (\lambda \nu_1 - \nu_2)e^{-(H-h)\nu_1},$$

$$L = (\lambda \nu_1 + \nu_2)e^{H\nu_1} - (\lambda \nu_1 - \nu_2)e^{-H\nu_1}, \qquad (7.216)$$

$$\lambda = \frac{\mu_1}{\mu_2} < 1, \qquad \nu_1 = \sqrt{(k^2 - k_{\beta_1}^2)}, \qquad \nu_2 = \sqrt{(k^2 - k_{\beta_2}^2)}, \qquad k_\beta = \frac{\omega}{\beta}.$$

We expand A/L into a convergent infinite series of exponentials:

$$\frac{A}{L} = \sum_{n=0}^{\infty} \Lambda^n e^{-\nu_1(2nH+h)} + \sum_{n=0}^{\infty} \Lambda^{n+1} e^{-\nu_1[2(n+1)H - h]}, \qquad (7.217)$$

where

$$\Lambda = \frac{\lambda \nu_1 - \nu_2}{\lambda \nu_1 + \nu_2} = \exp\left[-2\text{th}^{-1}\left(\frac{\mu_2}{\mu_1}\frac{\nu_2}{\nu_1}\right)\right] = -\exp\left[-2\text{th}^{-1}\left(\frac{\mu_1}{\mu_2}\frac{\nu_1}{\nu_2}\right)\right]. \qquad (7.218)$$

Substituting

$$k = k_{\beta_1}\sin f_1 = k_{\beta_2}\sin f_2, \qquad \nu_1 = ik\cot f_1, \qquad \nu_2 = ik\cot f_2, \qquad (7.219)$$

in Eq. (7.218), we find

$$\Lambda = \frac{\sin 2f_1 - (\rho_2/\rho_1)\sin 2f_2}{\sin 2f_1 + (\rho_2/\rho_1)\sin 2f_2}. \qquad (7.220)$$

Equation (7.220) coincides with the reflection coefficient derived in Eq. (3.41). Therefore, Λ represents the reflection coefficient for *SH* waves incident from the medium (μ_1, β_1) at the interface between two half-spaces.

Using Eqs. (7.217) and (7.218) and the asymptotic expansion of the Hankel function, Eq. (7.215) yields

$$u_{\phi}^{I}(\omega) = \frac{U_0(\omega)dS}{4\pi} \cos 2\phi \left(\frac{2}{\pi}\right)^{1/2} e^{i5\pi/4} \frac{\partial}{\partial \Delta}\left[\Delta^{-1/2} \sum_{n=0}^{\infty} (I_n^+ + I_n^-)\right], \qquad (7.221)$$

[7] In section 7.4.3 λ is *not* the Lamé parameter and Λ is *not* wavelength.

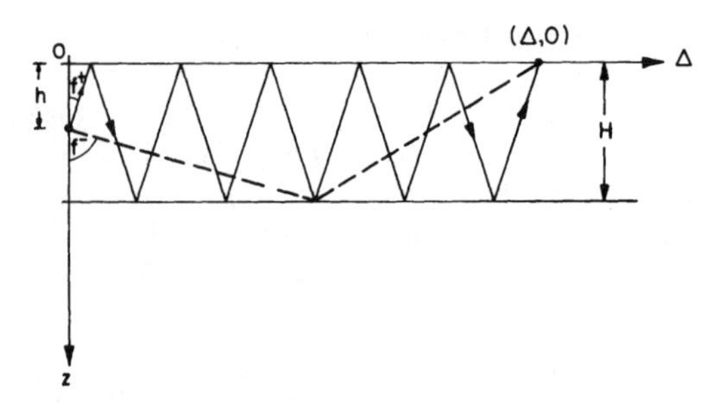

Figure 7.32. Multiple reflection of *SH* waves in a layer.

where

$$I_n^{\pm} = \int_{-\infty}^{\infty} e^{-i\Phi_n^{\pm}(k)} \frac{k^{1/2}\,dk}{\nu_1},$$

$$\Phi_n^{+}(k) = k\Delta - i\nu_1(2nH + h) - i2n\,\mathrm{th}^{-1}\frac{\mu_2\nu_2}{\mu_1\nu_1}, \tag{7.222}$$

$$\Phi_n^{-}(k) = k\Delta - i\nu_1[2(n+1)H - h] - i2(n+1)\,\mathrm{th}^{-1}\frac{\mu_2\nu_2}{\mu_1\nu_1}.$$

The integrals $I_n^{\pm}$ can be evaluated by the saddle-point approximation. Except at the grazing incidence and near the critical angle, ν_2/ν_1 is a slowly varying function of k and, therefore, its derivative contributes little to the location of the saddle points. If we ignore its contribution, the saddle points are such that,

$$\begin{aligned} \Delta &= (2nH + h)\tan f^{+} \quad \text{for} \quad I_n^{+}, \\ \Delta &= [2(n+1)H - h]\tan f^{-} \quad \text{for} \quad I_n^{-}, \end{aligned} \tag{7.223}$$

writing $f^{\pm}$ for the angle f_1 defined in Eq. (7.219).

The integrals $I_n^{\pm}$ have a simple physical interpretation. I_n^{+} represents a pulse that is upgoing at the source and undergoes n reflections at the interface and n reflections at the free surface. In contrast, I_n^{-} corresponds to the pulse that is downgoing at the source and undergoes $(n+1)$ reflections at the interface and n reflections at the free surface. Figure 7.32 shows I_5^{+} (continuous line) and I_0^{-} (broken line).

We give now a method of evaluating the integrals $I_n^{\pm}$ approximately. Let the frequency ω be complex with $\mathrm{Im}\,\omega < 0$ and let k be a complex variable. Then the integrand in I_n is a four-valued function of k.[8] The branch points of the integrand are given by

$$\nu_1 = 0, \qquad \nu_2 = 0, \quad \text{i.e.,} \quad k = \pm k_{\beta_1}, \pm k_{\beta_2}.$$

[8] It may be noted that the integrand in (7.215) is an even function of ν_1. Therefore, $\nu_1 = 0$ does not give relevant branch points of the integrand in (7.215). However, the integrand in I_n is not even in ν_1.

To make the integrand single valued, we confine ourselves to the top Riemann sheet, on which Re $\nu_1, \nu_2 \geq 0$. Let Re $\omega > 0$ and

$$k = \xi + i\eta, \qquad \omega = c - is, \qquad (c, s > 0). \tag{7.224}$$

Then the branch points lie in the second and fourth quadrants of the k plane. The branch lines along which the four Riemann sheets coalesce are given by Re $\nu_1 = 0$, Re $\nu_2 = 0$. Let us consider first Re $\nu_1 = 0$. This implies that

$$\nu_1^2 = k^2 - \frac{\omega^2}{\beta_1^2} = \xi^2 - \eta^2 - \frac{(c^2 - s^2)}{\beta_1^2} + 2i\left(\xi\eta + \frac{cs}{\beta_1^2}\right)$$

is real and negative. Hence the equations of the branch lines are

$$\xi\eta = -\frac{cs}{\beta_1^2}, \qquad \xi^2 - \eta^2 < \frac{(c^2 - s^2)}{\beta_1^2}. \tag{7.225}$$

These represent arcs of hyperbolas in the second and fourth quadrants having the axes as asymptotes. The arcs start at $\pm k_{\beta_1}$ and proceed to $\mp i\infty$ (Fig. 7.33a). Similarly, branch lines corresponding to ν_2 start at $\pm k_{\beta_2}$ and proceed to $\mp i\infty$.

We may write $I_n^{\pm}$ of Eq. (7.222) in the form

$$I_n = \left(-\int_0^{-\infty} + \int_0^{\infty}\right) e^{-i\Phi_n} \frac{k^{1/2}\, dk}{\nu_1}. \tag{7.226}$$

In the first integral on the right-hand side of the above equation, we deform the path of integration from the negative real axis into (Fig. 7.34a)

1. The negative imaginary axis, followed by
2. The arc AB of the infinite circle in the third quadrant.

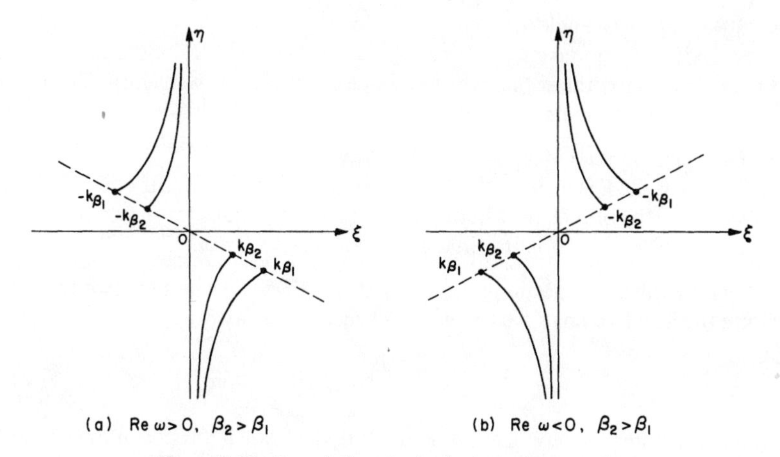

Figure 7.33. Branch lines in the k plane (Im $\omega < 0$).

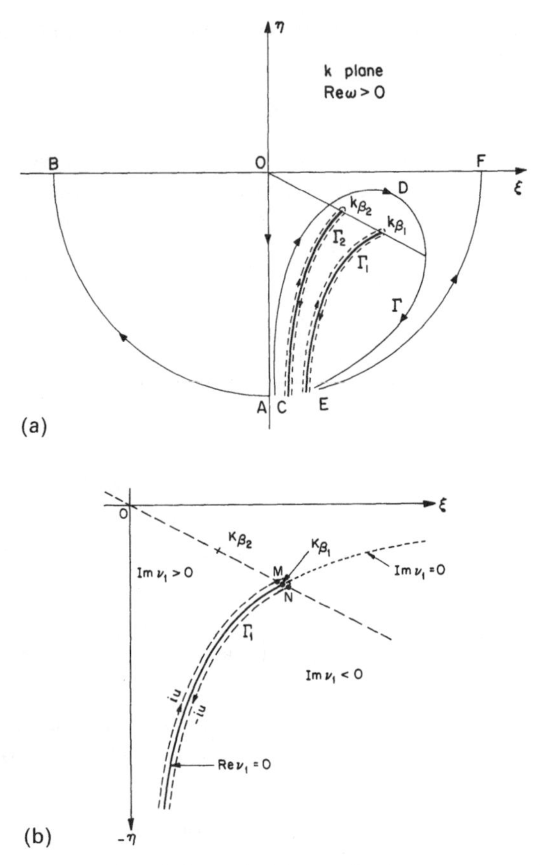

Figure 7.34. (a) Deformation of the path of integration in the k plane. (b) The loop Γ_1 around the branch line Re $\nu_1 = 0$.

In the second integral on the right-hand side of Eq. (7.226), we deform the path of integration from the positive real axis into

1. The negative imaginary axis, followed by
2. The loop CDE which starts at $-i\infty$ in the fourth quadrant and, after surrounding the branch lines in the fourth quadrant, returns to $-i\infty$ and then to
3. The arc EF of the infinite circle in the fourth quadrant.

Because the integrals along the arcs AB and EF vanish and the two integrals along the negative imaginary axis cancel each other, we are left with

$$I_n = \int_\Gamma e^{-i\Phi_n} \frac{k^{1/2}\, dk}{\nu_1}, \qquad (\text{Re } \omega > 0), \tag{7.227}$$

where Γ denotes the loop CDE. If Re $\omega < 0$, the branch lines lie in the first and third quadrants (Fig. 7.33b) and the integral will be along a loop in the first quadrant. However, it is, in general, not necessary to consider the case Re $\omega < 0$ separately. Therefore, we continue to assume Re $\omega > 0$.

The loop Γ is equivalent to two loops, Γ_1 and Γ_2, around the branch lines Re $\nu_1 = 0$ and Re $\nu_2 = 0$, respectively. If the integrand has poles, their contribution must be taken into account when we deform the path of integration. The loops $\Gamma_{1,2}$ start at $-i\infty$ on the left banks of the respective branch lines and, after surrounding the branch lines, return to $-i\infty$ on the right banks. We suppose that these loops are drawn indefinitely close to the branch lines so that for the purpose of evaluating the integrals along them, they may be supposed to coincide with the two sides of the respective branch lines.

Consider the loop Γ_1 first. The radical ν_1 is purely imaginary on both banks of the cut Re $\nu_1 = 0$. We have to determine its sign. Let the line of branch points cut the contour Γ_1 at M on the left bank and at N on the right bank (Fig. 7.34b). Let $k = qk_{\beta_1}$ at M or N, so that $q < 1$ for M and $q > 1$ for N. Noting that $\omega = c - is$ with c and s real and positive, and Re $\nu_1 \geq 0$, we see that, for M, $\nu_1 = ((s + ic)/\beta_1)\sqrt{(1 - q^2)}$ and for N, $\nu_1 = ((c - is)/\beta_1)\sqrt{(q^2 - 1)}$. Therefore Im $\nu_1 > 0$ for M and Im $\nu_1 < 0$ for N. In the right half of the k plane, Im $\nu_1 = 0$ only on the dotted line shown in Fig. 7.34b, and so Im ν_1 can change sign only on crossing this line. Hence we conclude that Im ν_1 is positive on the left bank of the cut Re $\nu_1 = 0$ and negative on the right bank. We write $\nu_1 = iu$ on the left bank and $\nu_1 = -iu$ on the right bank, where u is real and positive and varies from 0 to ∞ as we move away from k_{β_1} along either bank of the cut. Then

$$k^2 - k_{\beta_1}^2 = \nu_1^2 = -u^2,$$

and so,

$$k\, dk = -u\, du.$$

If Δ is large, the factor $e^{-i\Delta k}$ in the integrand, and so the integrand itself, decays very rapidly as we move away from the branch point k_{β_1} along either bank of the cut. This is because in that case $|\text{Im } k|$ increases when Im k is negative. Therefore, the major contribution to the value of the integral comes from the neighborhood of k_{β_1}, i.e., from small values of u. Hence, for the evaluation along Γ_1, we make the approximations

$$\begin{aligned} k &= k_{\beta_1} - \frac{u^2}{2k_{\beta_1}} \text{ in the exponential,} \\ &= k_{\beta_1} \text{ elsewhere,} \end{aligned} \tag{7.228}$$

$$\nu_2 = (k_{\beta_1}^2 - k_{\beta_2}^2)^{1/2} = k_{12} = \frac{\omega}{c_{12}}, \qquad c_{12}^{-2} = \beta_1^{-2} - \beta_2^{-2},$$

$$\Lambda = -e^{-2\,\text{th}^{-1}(\lambda\nu_1/\nu_2)} = -e^{-2\,\text{th}^{-1}(\pm iu\lambda/k_{12})} \simeq -e^{\mp 2iu\lambda/k_{12}}, \qquad \left(\lambda = \frac{\mu_1}{\mu_2}\right).$$

The upper sign is for the left bank and the lower sign for the right bank.

When these approximations are made, the integrals reduce to forms that can be evaluated exactly by using the following known results:

$$\int_0^\infty e^{imu^2} \cos nu \, du = \frac{1}{2}\left(\frac{\pi i}{m}\right)^{1/2} e^{-in^2/4m},$$

$$\int_0^\infty e^{imu^2} u \sin nu \, du = \frac{in}{4m}\left(\frac{\pi i}{m}\right)^{1/2} e^{-in^2/4m}, \tag{7.229}$$

$$\int_0^\infty e^{imu^2} u^2 \cos nu \, du = \left(\frac{i}{4m} + \frac{n^2}{8m^2}\right)\left(\frac{\pi i}{m}\right)^{1/2} e^{-in^2/4m}.$$

More results can be obtained by differentiating under the sign of integration. These integrals are valid for $\operatorname{Im} m > 0$.

With this preparation, we are now in a position to evaluate I_n along Γ_1. Denoting the contribution of Γ_1 with a suffix 1, we get

$$I_{n,1}^+(\omega) = \int_{\Gamma_1} \exp[-ik\Delta - \nu_1(2nH + h)]\Lambda^n \frac{k^{1/2}\, dk}{\nu_1}$$

$$= -2ik_{\beta_1}^{-1/2}(-1)^n e^{-i\Delta k_{\beta_1}} \int_0^\infty e^{i\Delta u^2/(2k_{\beta_1})} \cos\left(2nH + h + \frac{2n\lambda}{k_{12}}\right) u \, du$$

$$= (-1)^n\left(\frac{2\pi}{i\Delta}\right)^{1/2} \exp\left[-\frac{i}{\beta_1}\left\{\omega\left[\Delta + \frac{1}{2\Delta}(2nH + h)^2\right]\right.\right.$$

$$\left.\left. + \frac{2n\lambda}{\Delta} c_{12}(2nH + h) + \frac{2n^2\lambda^2 c_{12}^2}{\omega\Delta}\right\}\right]. \tag{7.230}$$

The inverse Fourier transform of $I_{n,1}^+(\omega)$ can be calculated by using the integral

$$\frac{1}{2\pi} \int_{-\infty}^{\infty} \exp\left[i\left(x\omega + \frac{y}{\omega}\right)\right] \frac{d\omega}{(i\omega)^{n+1}} = J_n[2(xy)^{1/2}]\left(\frac{x}{y}\right)^{n/2} H(x), \tag{7.231}$$

where $n > -1$ and the contour of integration is drawn just below the real axis in order to avoid the singularity at the origin.

When Eq. (7.230) is multiplied by $1/(i\omega)$, corresponding to unit-step function time dependence, and the inverse Fourier transform is found, it is apparent from Eq. (7.231) that the displacement will be identically zero up to time

$$T_{n,1}^+ = \frac{1}{\beta_1}\left[\Delta + \frac{1}{2\Delta}(2nH + h)^2\right] \simeq \frac{1}{\beta_1}[\Delta^2 + (2nH + h)^2]^{1/2},$$

for large Δ. If we consider a wave that is upgoing at the source and reaches the observer on the free surface after n reflections at the interface, then $T_{n,1}^+$ is exactly equal to its travel time. Figure 7.35 shows the situation when $n = 2$. It is also apparent from Eqs. (7.221) and (7.230) that the displacement decays with epicentral distance as Δ^{-2}.

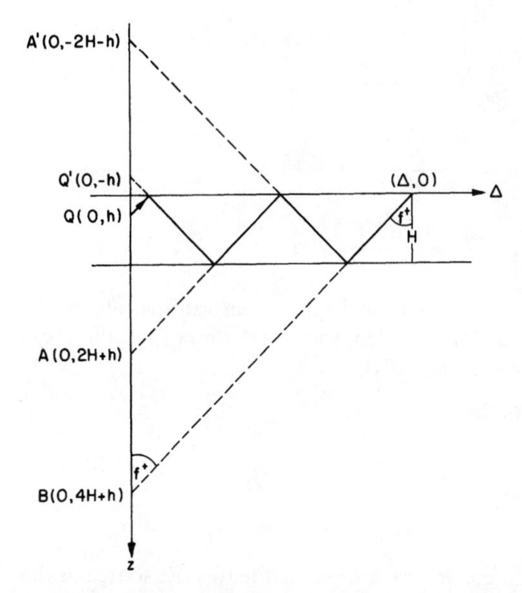

Figure 7.35. Successive reflections and images.

Similarly, evaluating I_n^- along Γ_1 we find that the corresponding displacement is zero up to the time

$$T_{n,1}^- = \frac{1}{\beta_1}\left[\Delta + \frac{1}{2\Delta}\{2(n+1)H - h\}^2\right].$$

Evaluating I_n^- along the loop Γ_2, we write $\nu_2 = \pm iu$, the upper sign for the left bank of the cut and the lower sign for the right bank. Then, $k\,dk = -u\,du$, and we have the approximations

$$\begin{aligned} k &= k_{\beta_2} - \frac{u^2}{2k_{\beta_2}} \text{ in the exponential,} \\ &= k_{\beta_2} \text{ elsewhere,} \\ \nu_1 &= i\left(k_{12} + \frac{u^2}{2k_{12}}\right) \text{ in the exponential,} \\ &= ik_{12} \text{ elsewhere,} \\ \Lambda &= e^{-2\,\mathrm{th}^{-1}(\mu_2\nu_2/\mu_1\nu_1)} = e^{-2\,\mathrm{th}^{-1}(\pm iu/\lambda ik_{12})} \simeq e^{\mp 2u/\lambda k_{12}}, \end{aligned} \tag{7.232}$$

where k_{12} is defined in Eq. (7.228). The result of integration along Γ_2 is then

$$\begin{aligned} I_{n,2}^+(\omega) &= \frac{2}{k_{\beta_2}^{1/2}k_{12}}\, e^{-i\Delta k_{\beta_2} - ik_{12}(2nH+h)} \int_0^\infty \exp\left[\frac{iu^2}{2}\left(\frac{\Delta}{k_{\beta_2}} - \frac{2nH+h}{k_{12}}\right)\right] \\ &\quad \times \sin\left(\frac{2nui}{\lambda k_{12}}\right) u\,du \\ &= -\frac{2n}{\lambda\omega}(2\pi i)^{1/2}\frac{c_{12}^2}{\beta_2}\Delta_{n^+}^{-3/2}\exp(-i\omega T_{n,2}^+)\exp\left(\frac{i}{\omega}d_{n^+}\right), \end{aligned} \tag{7.233}$$

where

$$d_{n^+} = \frac{2c_{12}^2 n^2}{\beta_2 \lambda^2} \Delta_{n^+}^{-1},$$

$$\Delta_{n^+} = \Delta - \frac{c_{12}}{\beta_2}(2nH + h), \tag{7.234}$$

$$T_{n,2}^+ = \frac{\Delta}{\beta_2} + \frac{2nH + h}{c_{12}}.$$

Appealing to Eq. (7.231), we note that the displacement vanishes identically up to time $T_{n,2}^+$. Equations (7.221) and (7.233) show that, for large Δ, the decay with distance is as Δ^{-3}. If $f_c = \arcsin(\beta_1/\beta_2)$,

$$T_{n,2}^+ = \frac{\Delta}{\beta_2} + \frac{(2nH + h)\cos f_c}{\beta_1}$$

$$= \frac{2nH + h}{\beta_1 \cos f_c} + \frac{\Delta - (2nH + h)\tan f_c}{\beta_2} = \frac{2nH + h}{\beta_1 \cos f_c} + \frac{\Delta_{n^+}}{\beta_2}. \tag{7.235}$$

This shows that $T_{n,2}^+$ is the time taken by a wave that leaves the source in the upward direction at an angle f_c with the vertical and reaches the observer after n reflections at the free surface. It may be noted that the angle f_c is the critical angle for reflection at the interface, and before reaching the observer, the wave travels a distance Δ_{n^+} parallel to the interface but below it. Such waves are known as *head waves, conical waves,* or *refraction arrivals* and are of paramount importance in geophysical prospecting. Figure 7.36b shows the head wave $T_{1,2}^+$.

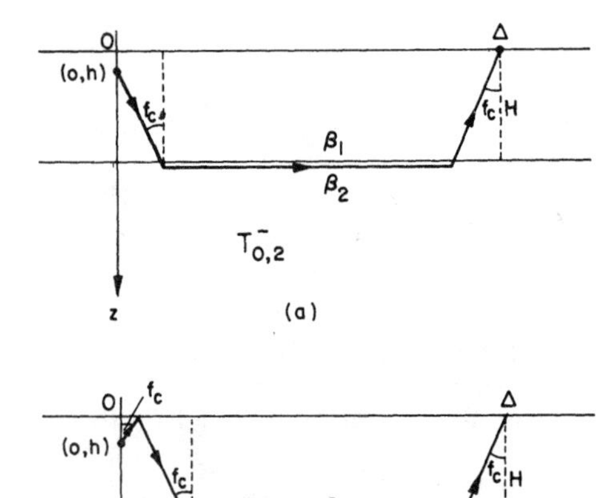

Figure 7.36. (a) The head wave $T_{0,2}^-$. (b) The head wave $T_{1,2}^+$. $f_c = \arcsin(\beta_1/\beta_2)$ is the critical angle.

Similarly, the arrival time for the head wave, which is downgoing at the source and reaches the observer after n reflections at the free surface is

$$T_{n,2}^{-} = \frac{2(n+1)H - h}{\beta_1 \cos f_c} + \frac{\Delta - [2(n+1)H - h]\tan f_c}{\beta_2}. \quad (7.236)$$

Because for large epicentral distances the arrival time of the direct SH wave for $\Delta \gg h$ is about Δ/β_1, while for the head wave $T_{0,2}^{-}$ (Fig. 7.36a) it is about Δ/β_2, the head wave $T_{0,2}^{-}$ will arrive before the direct wave $T_{0,1}^{+}$.

From Eqs. (7.235) and (7.236) it is clear that the conditions for the existence of the two head waves are, respectively,

$$\begin{aligned} \Delta_{n^+} &= \Delta - (2nH + h)\tan f_c > 0 \qquad (n = 1, 2, 3, \ldots), \\ \Delta_{n^-} &= \Delta - [2(n+1)H - h]\tan f_c > 0 \qquad (n = 0, 1, 2, \ldots). \end{aligned} \quad (7.237)$$

For a unit-step time function, $U_0(\omega) = U_0/(i\omega)$. Performing the inverse Fourier transform and using Eqs. (7.221), (7.231), and (7.233), we find

$$\begin{aligned} [u_\phi^I(t)]_{n^+} = &- \frac{U_0\, dS}{\pi}\left(\frac{\beta_2}{2}\right)^{1/2} \tan f_c \cos 2\phi \frac{\partial}{\partial \Delta}\left[\Delta_{n^+}^{-1}\left(\frac{\tau_{n^+}}{\Delta}\right)^{1/2}\right. \\ &\left. \times J_1\left\{2\sqrt{2}\, n \frac{\mu_2}{\mu_1}(\tan f_c)\left(\frac{\tau_{n^+}}{\tau_{n^+}^{*}}\right)^{1/2}\right\} H(\tau_{n^+})\right], \end{aligned}$$

where (7.238)

$$\tau_{n^+}^{*} = \frac{\Delta_{n^+}}{\beta_2},$$

$$\tau_{n^+} = t - T_{n,2}^{+} = t - \left(\frac{2nH + h}{\beta_1 \cos f_c} + \frac{\Delta_{n^+}}{\beta_2}\right).$$

When τ_{n^+} is small,

$$[u_\phi^I(t)]_{n^+} = \frac{U_0\, dS}{\pi}\left(n\frac{\mu_2}{\mu_1}\right)\tan^2 f_c \cos 2\phi \left[\frac{1}{\Delta^{1/2}\Delta_{n^+}^{3/2}} H(\tau_{n^+})\right]. \quad (7.239)$$

This shows that the time-domain amplitude of the initial motion decreases with distance as Δ^{-2}. For large epicentral distances, when $\Delta_{n^+} \simeq \Delta$, we have

$$\begin{aligned} [u_\phi^I(t)]_{n^+} &= \frac{U_0\, dS}{2\pi}\left(\frac{1}{2\beta_2}\right)^{1/2} \tan f_c \cos 2\phi \Delta^{-3/2}\left(t - \frac{\Delta}{\beta_2}\right)^{-1/2} \\ &\quad \times J_1\left\{2\sqrt{2}\, n\frac{\mu_2}{\mu_1}(\tan f_c)\left(\frac{t\beta_2}{\Delta} - 1\right)^{1/2}\right\} H\left(t - \frac{\Delta}{\beta_2}\right) \\ &= \frac{U_0\, dS}{2\pi}\left(n\frac{\mu_2}{\mu_1}\right)\tan^2 f_c \cos 2\phi \Delta^{-2}\left[\frac{2J_1(x)}{x}\right] H\left(t - \frac{\Delta}{\beta_2}\right), \end{aligned} \quad (7.240)$$

where

$$x = 2\sqrt{2}n\frac{\mu_2}{\mu_1}\left(\frac{t\beta_2}{\Delta} - 1\right)^{1/2} \tan f_c. \tag{7.241}$$

The first two zeros of the function $J_1(x)/x$ are at $x = 3.84$, 7.02, approximately (Fig. 7.37). This determines the period of the observed initial cycle at (Δ, t).

7.4.3.2. Source in the Half-Space. Consider next the situation when the source lies below the interface, at $(0, h)$ $h > H$. From Eq. (5.30), we have

$$\begin{aligned} u_\phi^I &= -\frac{U_0\,dS}{2\pi}\frac{\mu_2}{\mu_1}\cos 2\phi\,\frac{\partial}{\partial\Delta}\int_{-\infty}^{\infty} H_2^{(2)}(k\Delta)e^{-\nu_2(h-H)} \\ &\quad\times \frac{k\,dk}{(\lambda\nu_1 + \nu_2)e^{H\nu_1} - (\lambda\nu_1 - \nu_2)e^{-H\nu_1}} \\ &= -\frac{U_0\,dS}{2\pi}\frac{\mu_2}{\mu_1}\cos 2\phi\,\frac{\partial}{\partial\Delta}\sum_{n=0}^{\infty}\int_{-\infty}^{\infty}\Lambda^n H_2^{(2)}(k\Delta)e^{-\nu_2(h-H)-(2n+1)H\nu_1}\frac{k\,dk}{\lambda\nu_1 + \nu_2}, \end{aligned} \tag{7.242}$$

with the notation of Eqs. (7.216)–(7.218). Replacing the Hankel function by its asymptotic expansion, we find

$$\begin{aligned} u_\phi^I(\omega) &= -\frac{U_0(\omega)dS}{2\pi\lambda}\left(\frac{2}{\pi}\right)^{1/2} e^{i5\pi/4}\cos 2\phi\,\frac{\partial}{\partial\Delta}\sum_{n=0}^{\infty} \\ &\quad\times\left[\Delta^{-1/2}\int_{-\infty}^{\infty}\Lambda^n e^{-ik\Delta-\nu_2(h-H)-(2n+1)H\nu_1}\frac{k^{1/2}\,dk}{\lambda\nu_1 + \nu_2}\right]. \end{aligned} \tag{7.243}$$

The use of the method of stationary phase yields the equation for the saddle point:

$$\Delta = (h - H)\frac{ik}{\nu_2} + (2n + 1)H\frac{ik}{\nu_1}. \tag{7.244}$$

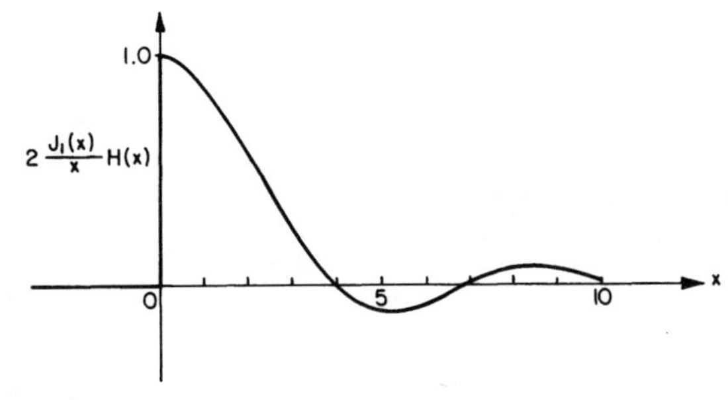

Figure 7.37. Amplitude variation with time at a fixed receiver for an *SH* source in the layer.

Equations (7.219) and (7.244) yield the ray-path equation

$$\Delta = (h - H)\tan f_2 + (2n + 1)H \tan f_1. \tag{7.245}$$

The path for $n = 0$ is shown in Fig. 7.38a. For higher values of n the ray is reflected $2n$ times inside the layer before it reaches the observer at $(\Delta, 0)$.

The evaluation along the branch line $\operatorname{Re} \nu_2 = 0$ yields

$$\begin{aligned} u^I_\phi(\omega) = {} & \frac{U_0(\omega)dS}{2\pi\lambda^2}\left(\frac{c_{12}}{\beta_2}\right)\left[(h - H) + \frac{(2n+1)c_{12}}{i\omega\lambda}\right]\cos 2\phi \\ & \times \frac{\partial}{\partial\Delta}\sum_{n=0}^{\infty}\left[\Delta^{-1/2}\Delta_n^{-3/2}\exp\left\{-(h - H)\frac{(2n+1)c_{12}}{\lambda\Delta_n\beta_2}\right\}\right. \\ & \left.\times \exp\left(-i\omega T_{n,2} + \frac{i}{\omega}d_n\right)\right], \end{aligned} \tag{7.246}$$

where

$$\begin{aligned} T_{n,2} &= \frac{\Delta}{\beta_2} + \frac{(2n+1)H}{c_{12}} + \frac{(h - H)^2}{2\beta_2\Delta_n}, \\ \Delta_n &= \Delta - (2n+1)H\frac{c_{12}}{\beta_2} = \Delta - (2n+1)H \tan f_c, \\ d_n &= \frac{(2n+1)^2 c_{12}^2}{2\lambda^2\beta_2\Delta_n}. \end{aligned} \tag{7.247}$$

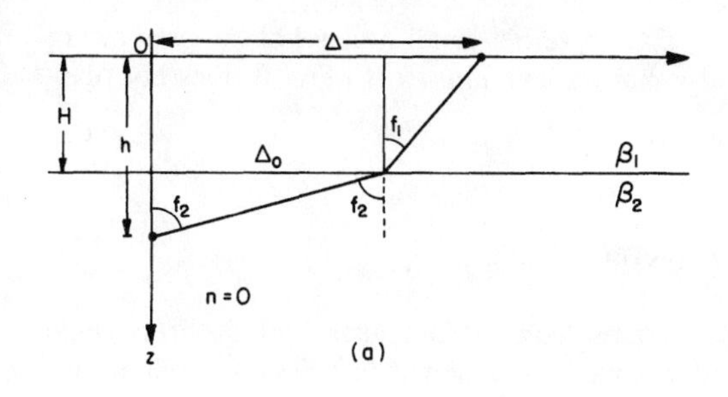

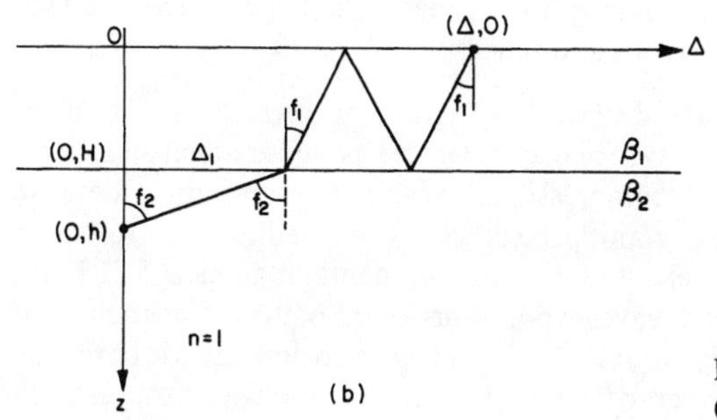

Figure 7.38. (a) Direct ray-path. (b) Multireflected ray-path.

Assuming a unit-step function time dependence $[U_0(\omega) = U_0/(i\omega)]$ and finding the inverse Fourier transform with the help of Eq. (7.231), we get

$$u_\phi^I(t) = \frac{U_0\, dS}{2\pi}\left(\frac{\mu_2}{\mu_1}\right)^2 \tan f_c \cos 2\phi \frac{\partial}{\partial \Delta}\sum_{n=0}^{\infty}$$
$$\times \left[\Delta^{-1/2}\Delta_n^{-3/2}\exp\left\{-(h-H)\frac{(2n+1)\tan f_c}{\Delta_n}\left(\frac{\mu_2}{\mu_1}\right)\right\}\right.$$
$$\left.\times \{(h-H)J_0(x_n) + (2\beta_2\tau_n\Delta_n)^{1/2}J_1(x_n)\}H(\tau_n)\right], \tag{7.248}$$

where

$$\tau_n = t - T_{n,2},$$
$$x_n = (2n+1)\left(\frac{\mu_2}{\mu_1}\right)\tan f_c\left(\frac{2\beta_2\tau_n}{\Delta_n}\right)^{1/2} \tag{7.249}$$

As in Eq. (7.235), we can write

$$T_{n,2} = \frac{(2n+1)H}{\beta_1 \cos f_c} + \frac{\Delta_n}{\beta_2} + \frac{(h-H)^2}{2\beta_2\Delta_n} \simeq \frac{(2n+1)H}{\beta_1\cos f_c} + \frac{[\Delta_n^2 + (h-H)^2]^{1/2}}{\beta_2}. \tag{7.250}$$

Therefore, $T_{n,2}$ is approximately equal to the travel-time of a pulse reflected up and down n times in the layer, Δ_n being the horizontal distance traveled in the lower medium (Fig. 7.38b).

The surface motion in Eq. (7.248) is rather small because of the presence of the negative exponential, and is known as a *diffracted wave*. It does not obey the least-time principle.

7.4.4. Rays in a Three-Layered Crust

Observed short-period seismograms from earthquakes and anthropogenic explosions exhibit at short and intermediate ranges ($\Delta < 800$ km) certain prominent signals whose arrival times are consistent with a model of a crust that is composed of three layers overlying a homogeneous half-space. These divide into two groups: refraction arrivals and reflections.

7.4.4.1. Refraction Arrivals (Head Waves). Refraction arrivals are branch-line contributions. A pair of waves for each interface, compressional and shear, is shown in Fig. 7.39. The first compressional wave is P_g (subscript g, because of its horizontal path in the granitic layer below the sediments). Its shear counterpart is S_g. A typical velocity for P_g in continental regions is 5.71 km/s. Similarly, P^* and S^* are head waves whose horizontal paths are at the top of the basaltic layer. The pair P_n, S_n travels as head waves below the Mohorovičić (Moho) discontinuity at an average depth of 35 km in continental zones. Each

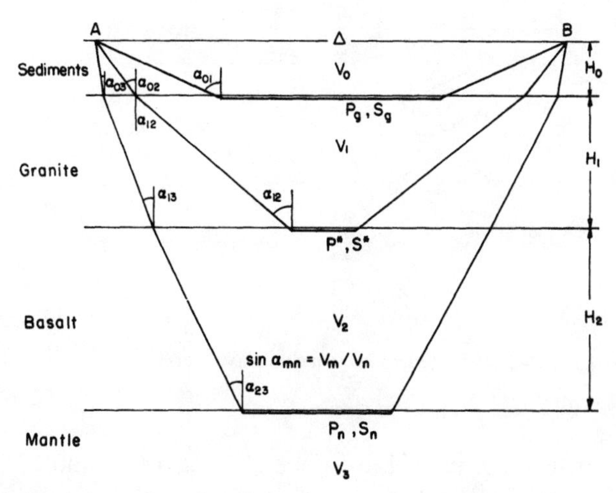

Figure 7.39. Head waves in a three-layered crust over a homogeneous mantle, $\alpha_{mn} = \arcsin(V_m/V_n)$ are critical angles.

ray obeys Snell's law, with a critical angle that depends on the velocity contrast of the two sides of the interface. The travel time equations are

$$T_g = \frac{\Delta}{V_1} + 2H_0\left(\frac{V_1}{V_0} - \frac{V_0}{V_1}\right)(V_1^2 - V_0^2)^{-1/2},$$

$$T^* = \frac{\Delta}{V_2} + 2H_0\left(\frac{V_2}{V_0} - \frac{V_0}{V_2}\right)(V_2^2 - V_0^2)^{-1/2} + 2H_1\left(\frac{V_2}{V_1} - \frac{V_1}{V_2}\right)(V_2^2 - V_1^2)^{-1/2},$$

$$T_n = \frac{\Delta}{V_3} + 2H_0\left(\frac{V_3}{V_0} - \frac{V_0}{V_3}\right)(V_3^2 - V_0^2)^{-1/2} + 2H_1\left(\frac{V_3}{V_1} - \frac{V_1}{V_3}\right)(V_3^2 - V_1^2)^{-1/2}$$
$$+ 2H_2\left(\frac{V_3}{V_2} - \frac{V_2}{V_3}\right)(V_3^2 - V_2^2)^{-1/2}. \tag{7.251}$$

7.4.4.2. Reflections. The reflections from the first, second, and third discontinuities are named P_0P, P_IP, and P_MP for compressional waves and S_0S, S_IS, and S_MS for shear waves. The dependence of its travel-time on the epicentral distance Δ is given implicitly by the pair of parametric equations (Fig. 7.40).

$$\Delta = 2\sum_{i=0}^{n} H_i \tan\theta_{in}, \qquad (n = 0, 1, 2),$$
$$T = 2\sum_{i=0}^{n} \frac{H_i}{V_i\cos\theta_{in}}. \tag{7.252}$$

Denoting by p_n the ray-parameter of the n-th ray, Snell's law yields

$$p_n = \frac{\sin\theta_{in}}{V_i}, \qquad (i \le n, n = 0, 1, 2).$$

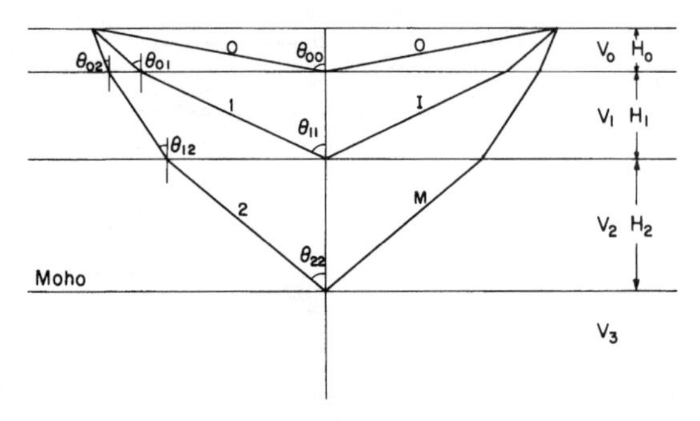

Figure 7.40. Reflected waves in a three-layered crust.

7.4.5. Point-Source in a Half-Space Having a Constant Velocity Gradient

Solving Eq. (7.203) with

$$V(z) = V_0(1 + \gamma z), \qquad \mu = \mu_0(1 + \gamma z)^2, \qquad \rho = \text{const.}$$

we obtain

$$\begin{aligned} y_1^+(z) &= (1 + \gamma z)^{-1/2} I_{iv}\left[\frac{k}{\gamma}(1 + \gamma z)\right], \\ y_1^-(z) &= (1 + \gamma z)^{-1/2} K_{iv}\left[\frac{k}{\gamma}(1 + \gamma z)\right], \\ y_2(z) &= \mu \frac{dy_1(z)}{dz}, \\ W &= \gamma\mu_s, \qquad v = \sqrt{\left[\left(\frac{\omega}{\gamma V_0}\right)^2 - \frac{1}{4}\right]}. \end{aligned} \tag{7.253}$$

Equation (7.201) now becomes

$$\begin{aligned} G(z\,|\,z_0) = \frac{1}{\gamma\mu_s[(1 + \gamma z)(1 + \gamma z_0)]^{1/2}} K_{iv}\left[\frac{k}{\gamma}(1 + \gamma z_>)\right]\bigg\{ I_{iv}\left[\frac{k}{\gamma}(1 + \gamma z_<)\right] \\ - K_{iv}\left[\frac{k}{\gamma}(1 + \gamma z_<)\right] E\left(\frac{k}{\gamma}\right)\bigg\}, \end{aligned} \tag{7.254}$$

where

$$E\left(\frac{k}{\gamma}\right) = \frac{I_{iv}(k/\gamma) - (2k/\gamma) I'_{iv}(k/\gamma)}{K_{iv}(k/\gamma) - (2k/\gamma) K'_{iv}(k/\gamma)}.$$

The Bateman integral is

$$\begin{aligned}\int_0^\infty K_{i\nu}(ak)I_{i\nu}(bk)J_0(\Delta k)k\,dk &= \frac{1}{R_1R_2}\left(\frac{R_2-R_1}{R_2+R_1}\right)^{i\nu} \\ &= \frac{1}{R_1R_2}\exp\left[-2i\nu\,\mathrm{th}^{-1}\left(\frac{R_1}{R_2}\right)\right] \\ &= \frac{1}{R_1R_2}\exp\left[-i\tau\left(\omega^2-\frac{1}{4}\gamma^2V_0^2\right)^{1/2}\right], \end{aligned} \tag{7.255}$$

where

$$\begin{gathered}\operatorname{Im}\nu<1, \qquad \operatorname{Re}a>|\operatorname{Re}b|, \\ R_1=[\Delta^2+(a-b)^2]^{1/2}, \qquad R_2=[\Delta^2+(a+b)^2]^{1/2}, \\ \tau=\frac{2}{\gamma V_0}\mathrm{th}^{-1}\left(\frac{R_1}{R_2}\right).\end{gathered} \tag{7.256}$$

The function $G(z|z_0)$ in Eq. (7.254) is the Green's function for the medium. For an axially symmetrical source, the displacement field is given by

$$u_\phi(\omega)=A_0\frac{\partial}{\partial\Delta}\int_0^\infty G(z|z_0)J_0(k\Delta)k\,dk=u_1+u_2, \tag{7.257}$$

where u_1 corresponds to the first term inside the braces on the right-hand side of Eq. (7.254) and u_2 corresponds to the second term. Using Eq. (7.255), we find

$$\begin{aligned}u_1(\omega) &= \frac{A_0}{\gamma\mu_s(1+\gamma z)^{1/2}(1+\gamma h)^{1/2}}\frac{\partial}{\partial\Delta}\int_0^\infty K_{i\nu}\left[\frac{k}{\gamma}(1+\gamma z_>)\right] \\ &\quad\times I_{i\nu}\left[\frac{k}{\gamma}(1+\gamma z_<)\right]J_0(k\Delta)k\,dk \\ &= \frac{A_0}{\gamma\mu_s(1+\gamma z)^{1/2}(1+\gamma h)^{1/2}}\frac{\partial}{\partial\Delta}\left[\frac{1}{R_1R_2}\exp\left\{-i\tau\left(\omega^2-\frac{1}{4}\gamma^2V_0^2\right)^{1/2}\right\}\right],\end{aligned} \tag{7.258}$$

where

$$\begin{gathered}R_1=[\Delta^2+(z-h)^2]^{1/2}, \qquad R_2=\left[\Delta^2+\left(z+h+\frac{2}{\gamma}\right)^2\right]^{1/2}, \\ \tau=\frac{2}{\gamma V_0}\mathrm{th}^{-1}\left(\frac{R_1}{R_2}\right), \qquad h=z_0.\end{gathered} \tag{7.259}$$

Assuming a delta-function source in time and evaluating the inverse Fourier transform, we get

$$\begin{aligned}u_1(t) &= \frac{A_0}{\gamma\mu_s(1+\gamma z)^{1/2}(1+\gamma h)^{1/2}}\frac{\partial}{\partial\Delta}\left[\frac{1}{R_1R_2}\left\{\delta(t-\tau)\right.\right. \\ &\quad\left.\left.-\,\mathrm{th}^{-1}\left(\frac{R_1}{R_2}\right)(t^2-\tau^2)^{-1/2}J_1\left(\frac{\gamma V_0}{2}(t^2-\tau^2)^{1/2}\right)H(t-\tau)\right\}\right].\end{aligned} \tag{7.260}$$

In Example 7.6 we have seen that τ represents the travel-time along the ray from the point $(0, h)$ to the point (Δ, z). Therefore $u_1(t)$ represents the direct SH arrival.

The second part of u_ϕ, namely, u_2, represents the field induced by the presence of the boundary. The surface-wave part of it can be evaluated as a residue contribution and the reflected waves are determined by the method of stationary phase.

7.5. Breakdown of the GEA: Caustics

We have seen earlier that cusps in travel-time curves are associated with abnormally large amplitudes resulting from the relatively large values of $d^2T/d\Delta^2$ at these points. We also remember that the validity of ray theory depends on the fulfilment of the conditions in Eqs. (7.186). These conditions are indeed violated in regions where strong variation of $V(r)$ occurs, and ray theory cannot account for diffraction and focusing phenomena. However, certain approximate methods can be applied successfully to evaluate the seismic field in these cases. An important example is furnished by the caustics formed by seismic waves in the earth.

The name *caustic*[9] is given in optics to the envelope of rays after reflection or refraction, associated with exceptional brightness. The same applies to seismic rays associated with exceptionally large amplitudes. A noted case is the caustic that occurs at epicentral distances of about 142°. This caustic is a consequence of the refraction of waves that cross the mantle–core boundary. Caustics are also associated with the cusps B and C in Fig. 7.9e. Figure 7.41 shows two simple examples of caustics. Figure 7.41a shows the formation of a caustic as an envelope of SH rays reflected from the free surface of an elastic half-space in which the velocity is a linear function of depth. A detailed study of this case is given in Section 7.5.3. Figure 7.41b describes the formation of a caustic near the point of total reflection of a spherical SH wave at the boundary between two elastic half-spaces.

7.5.1. The Geometry of Caustics

The totality of rays in a half-space $(z > 0)$ can be written in the form

$$F(x, z, p) = 0, \tag{7.261}$$

where p is the ray parameter. For each value of p, Eq. (7.261) defines a curve in the x–z plane, namely, a ray. Assigning to p all permissible values we obtain a family of curves. Let us assume that there exists a curve, $z = f(x)$, every point of which is a point of contact with a curve of the family. Then we call this curve the

[9] From the Greek: καυστικοφ, meaning "burning" or "that which burns."

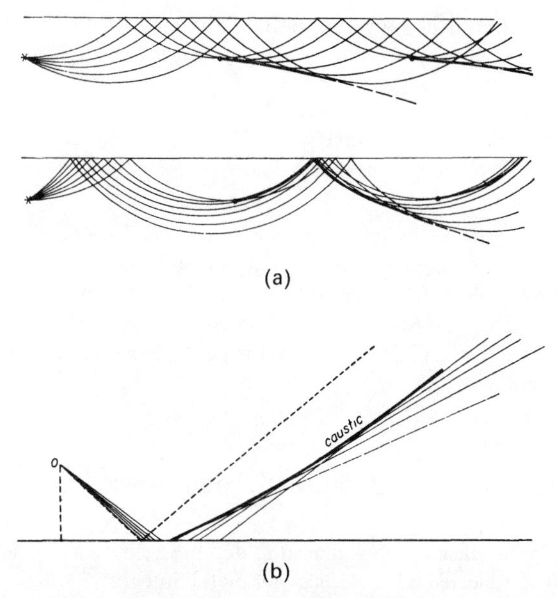

Figure 7.41. (a) Formation of a caustic as an envelope of *SH* rays reflected from the free surface of a half-space in which the velocity varies linearly with depth. (b) Formation of a caustic near the point of total reflection of spherical waves.

envelope of the given family. Clearly, the tangent at each point of the envelope is also a tangent to a curve of the family which passes through that point.

From Eq. (7.261), the slope of a given ray (p fixed) can be determined from the equation

$$\frac{\partial F}{\partial x}\,dx + \frac{\partial F}{\partial z}\,dz = 0. \tag{7.262}$$

Let $M(x, z)$ be a point on the envelope. This point also lies on a ray of the family, the ray parameter for which can, in principle, be found from Eq. (7.261) as $p = p(x, z)$. Therefore, the equation of the envelope can be written as

$$F[x, z, p(x, z)] = 0, \tag{7.263}$$

where p is no longer a constant. Hence, the slope of the envelope is given by

$$\frac{\partial F}{\partial x}\,dx + \frac{\partial F}{\partial z}\,dz + \frac{\partial F}{\partial p}\,dp = 0. \tag{7.264}$$

The slope, dz/dx, of the tangent to the envelope must, by definition, be the same as that of the tangent to the curve of the family that passes through $M(x, z)$. Hence, from Eqs. (7.262) and (7.264), we find

$$\frac{\partial F}{\partial p} = 0. \tag{7.265}$$

The equation of the envelope of the family is obtained by eliminating p from the two equations

$$F(x, z, p) = 0, \tag{7.266}$$

$$\frac{\partial F(x, z, p)}{\partial p} = 0. \tag{7.267}$$

Consider now the locus of points for which $\partial F/\partial x = 0$, $\partial F/\partial z = 0$. Such points are known as *singular points* of F, and they lie on a curve, $z = g(x)$. It is clear from Eqs. (7.262) and (7.264) that the coordinates of the singular points will also satisfy Eqs. (7.266) and (7.267). Therefore, after we obtain the curve that satisfies Eqs. (7.266) and (7.267), we must further find whether it is an envelope or a locus of singular points.

7.5.2. Caustic Formed Because of Total Reflection

A strike-slip point dislocation is placed at depth h below a solid–solid interface (Fig. 7.42). The reflected SH field is given by the integral

$$u_{\phi,\,\mathrm{Refl}} = \frac{U_0\, dS}{4\pi} \cos 2\phi \int_{-\infty}^{\infty} \frac{\nu_1 - \eta\nu_2}{\nu_1 + \eta\nu_2}$$

$$\times \left[2\,\frac{H_2^{(2)}(k\Delta)}{\Delta} - kH_1^{(2)}(k\Delta) \right] e^{-\nu_1(h+z)}\,\frac{k\,dk}{\nu_1}, \tag{7.268}$$

where

$$\nu_1 = \sqrt{(k^2 - k_{\beta_1}^2)}, \qquad \nu_2 = \sqrt{(k^2 - k_{\beta_2}^2)}, \qquad \eta = \frac{\mu_2}{\mu_1}.$$

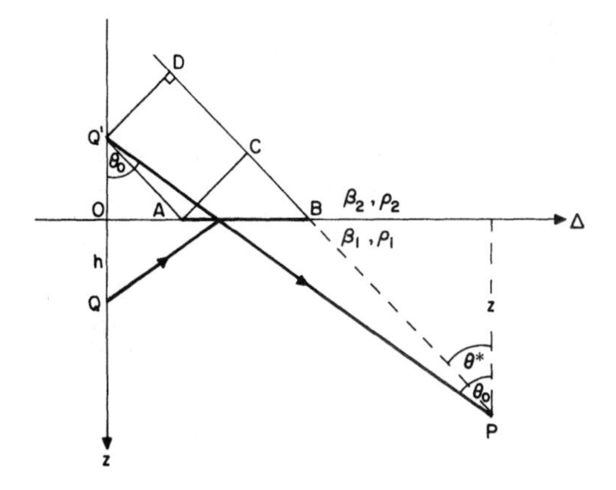

Figure 7.42. The ray displacement caused by the sphericity of the wavefront.

This result is derived easily by using the source representation given in Table 4.4 and solving the boundary-value problem. Alternatively, it can be inferred from Eq. (7.215) by simply taking

$$\frac{A}{L} = -\frac{\nu_1 - \eta\nu_2}{\nu_1 + \eta\nu_2} e^{-\nu_1(z+h)},$$

the negative sign resulting from the difference in the direction of the z axis. When we apply the usual asymptotic formulas for $H_1^{(2)}(k\Delta)$ and $H_2^{(2)}(k\Delta)$ in the far field ($k\Delta \gg 1$), the integral assumes the form

$$u_{\phi,\,\mathrm{Refl}} = \sqrt{\left(\frac{2}{\pi\Delta}\right)} \frac{U_0\, dS}{4\pi} \cos 2\phi e^{\pi i/4} \int_{-\infty}^{\infty} \frac{k^{3/2}}{\nu_1} \frac{\nu_1 - \eta\nu_2}{\nu_1 + \eta\nu_2} e^{-ik\Delta - \nu_1(h+z)}\, dk. \tag{7.269}$$

Note that the reflection coefficient $(\nu_1 - \eta\nu_2)/(\nu_1 + \eta\nu_2)$ is, for a fixed value of k, none other than the plane *SH*-wave reflection coefficient [Eq. (7.220)]. As in Eq. (7.219), it is convenient to introduce a change of variable that will bring out the latent geometric features of the problem:

$$\begin{aligned} k &= k_{\beta_1} \sin\theta, & \nu_1 &= ik_{\beta_1} \cos\theta, \\ \Delta &= R_1 \sin\theta_0, & z + h &= R_1 \cos\theta_0, \end{aligned} \tag{7.270}$$

where θ is complex. Equation (7.269) then becomes

$$u_{\phi,\,\mathrm{Refl}}(\omega) = -\sqrt{\left(\frac{1}{2\pi\Delta}\right)} \frac{U_0\, dS}{2\pi} \cos 2\phi k_{\beta_1}^{3/2} e^{-\pi i/4} \times \int_{\pi/2 - i\infty}^{3\pi/2 + i\infty} (\sin\theta)^{3/2} e^{-ik_{\beta_1} R_1 \chi(\theta)}\, d\theta, \tag{7.271}$$

with

$$\begin{aligned} \chi(\theta) &= \cos(\theta - \theta_0) + \frac{g(\theta)}{k_{\beta_1} R_1}, \\ g(\theta) &= -2 \tan^{-1}\left\{\eta \frac{\sqrt{[\sin^2\theta - (\beta_1/\beta_2)^2]}}{\cos\theta}\right\}. \end{aligned} \tag{7.272}$$

The contour of integration in Eq. (7.271) starts at $\pi/2 - i\infty$, crosses the real axis at $(\pi, 0)$, and comes to an end at $3\pi/2 + i\infty$. It can be replaced by an equivalent contour. The saddle point, $\theta = \theta^*$, is given by $\chi'(\theta^*) = 0$, i.e., by

$$\sin(\theta^* - \theta_0) = \frac{g'(\theta^*)}{k_{\beta_1} R_1}. \tag{7.273}$$

Because $k_{\beta_1} R_1$ is large, a rough approximation is $\theta^* \simeq \theta_0$. Substituting this into the right-hand side of Eq. (7.273), we get the next approximation

$$\theta^* = \theta_0 + \sin^{-1}\left[\frac{g'(\theta_0)}{k_{\beta_1} R_1}\right]. \tag{7.274}$$

Evaluating the integral in Eq. (7.271) by the method of stationary phase, we obtain

$$u_{\phi,\mathrm{Refl}}(\omega) = \frac{U_0\, dS}{2\pi}(k_{\beta_1})\cos 2\phi(\sin\theta^*)^{3/2}[R_1\Delta|\chi''(\theta^*)|]^{-1/2}$$
$$\times \exp\left\{i\omega\left[t - \left(\frac{R_1}{\beta_1}\right)\chi(\theta^*)\right] + \frac{i\pi}{4}[3 - \operatorname{sgn}\chi''(\theta^*)]\right\}. \quad (7.275)$$

The fact that θ^* is in general not equal to θ_0 means that because of the sphericity of the wavefronts, the time of arrival of the wave to the observation point at P will not be simply R_1/β_1. Because θ^* is frequency dependent, a small dispersion will develop as if β_1 were virtually frequency dependent. Moreover, since $\theta_0 = \tan^{-1}[\Delta/(z + h)]$, the ray will arrive at the receiver at an angle $\theta^* \neq \theta_0$, causing the ray to be displaced by a certain amount. From Fig. 7.42, the amount of the displacement is

$$AB = AC\sec\theta^* = R_1\sin(\theta_0 - \theta^*)\sec\theta^*$$
$$= -\frac{\sec\theta^*}{k_{\beta_1}}g'(\theta^*), \quad (7.276)$$

using Eq. (7.273).

Because this effect is of the order of $(1/\omega)$, it is usually ignored in seismic ray theory. Furthermore, even at moderate frequencies, the function $g(\theta)$ is slowly varying except near $\theta_0 = \pi/2$ and the critical angle $\theta_c = \sin^{-1}(\beta_1/\beta_2)$. Clearly, the focusing effect at the caustic is caused by the ray displacement mentioned above. We therefore expect $\chi''(\theta^*) = 0$ near the critical angle. Let us then derive the explicit equations for the caustic. Suppose that $\chi''(\theta^*) = 0$ in addition to $\chi'(\theta^*) = 0$. At this point ray theory breaks down. According to Eq. (7.272) the caustic is given by the equations

$$k_{\beta_1}R_1\sin(\theta^* - \theta_0) = g'(\theta^*), \qquad k_{\beta_1}R_1\cos(\theta^* - \theta_0) = g''(\theta^*). \quad (7.277)$$

At angles θ close to $\theta_c = \sin^{-1}(\beta_1/\beta_2)$, $g(\theta)$ in Eq. (7.272) is approximately given by

$$g(\theta) \simeq -a\sqrt{(\theta - \theta_c)}, \qquad a = \frac{2\mu_2}{\mu_1}\left(\frac{2\beta_1/\beta_2}{1 - (\beta_1/\beta_2)^2}\right)^{1/2}. \quad (7.278)$$

Because $(\theta^* - \theta_0)$ is small, Eqs. (7.277) and (7.278) yield

$$\theta_0 = \theta_c + \frac{3}{4}\left(\frac{2a}{k_{\beta_1}R_1}\right)^{2/3}. \quad (7.279)$$

This is the equation of the caustic in polar coordinates (θ_0, R_1) centered at the image point Q' (Fig. 7.42). Figure 7.41b shows the formation of the caustic as a result of ray displacement. Note that as long as $\theta_0 \neq \theta_c$, the reflected ray arrives separately from the head wave, which is a refraction arrival at the critical angle. As $\theta_0 \to \theta_c$, the saddle point, which is responsible for the reflected wave, approaches the branch point of the integrand in Eq. (7.268) and the two waves coalesce.

The field *on* the caustic is obtained by expanding $\chi(\theta)$ about θ^* and keeping terms up to third order

$$\chi(\theta) = \chi(\theta^*) + (\theta - \theta^*)\chi'(\theta^*) + \frac{1}{2}(\theta - \theta^*)^2\chi''(\theta^*) + \frac{1}{6}(\theta - \theta^*)^3\chi'''(\theta^*). \tag{7.280}$$

Because on the caustic $\chi'(\theta^*) = \chi''(\theta^*) = 0$ and because $(\sin\theta)^{3/2}$ in the integrand of Eq. (7.271) is a slowly varying function near θ^*, the integral to be evaluated is

$$\int_{\pi/2 - i\infty}^{3\pi/2 + i\infty} e^{-ik_{\beta_1}R_1\chi(\theta)}\,d\theta \simeq e^{-ik_{\beta_1}R_1\chi(\theta^*)}\int_{\pi/2 - i\infty}^{3\pi/2 + i\infty} e^{-(i/6)k_{\beta_1}R_1(\theta - \theta^*)^3\chi'''(\theta^*)}\,d\theta. \tag{7.281}$$

Introducing the new integration variable

$$u = \tau(\theta - \theta^*)e^{i\pi/6},$$
$$\tau = 2^{-1/3}[\chi'''(\theta^*)R_1k_{\beta_1}]^{1/3}, \tag{7.282}$$

the integral reduces to

$$\frac{1}{\tau}e^{-\pi i/6 - ik_{\beta_1}R_1\chi(\theta^*)}\int_{-\infty\exp(2\pi i/3)}^{\infty\exp(2\pi i/3)} e^{-u^3/3}\,du = \frac{2\pi}{\tau}Ai(0)e^{i\pi/3 - ik_{\beta_1}R_1\chi(\theta^*)}, \tag{7.283}$$

where $Ai(z)$ is the Airy function (Appendix G). For $k_{\beta_1}R_1 \gg 1$, on the caustic, we may put $\theta^* \simeq \theta_0$, $\chi(\theta^*) \simeq 1$, $|\chi'''(\theta^*)| \simeq |\chi'''(\theta_0)| \simeq 1$ in the amplitude factors only. Equation (7.271) then yields

$$u_{\phi,\,\text{Caustic}}(\omega) = -[U_0\,dS]\cos 2\phi\left[\frac{2^{1/3}Ai(0)}{\sqrt{(2\pi)}}\right]\left[\frac{(k_{\beta_1})^{7/6}}{\Delta^{1/2}R_1^{1/3}}\right](\sin\theta_0)^{3/2}$$
$$\times\, e^{i[\omega t - k_{\beta_1}R_1\chi(\theta^*) + \pi/12]}. \tag{7.284}$$

7.5.3. The Field Near a Caustic

Consider a point source at depth $z = h$ in a weak inhomogeneous elastic half-space $z > 0$. Whether the inhomogeneity varies continuously with z or is simulated by a stack of horizontally stratified layers, the far-field always reduces to an integral of the type

$$I = \int_{-\infty}^{\infty} e^{-i\Phi(\Delta, z, h; k)}\,dk, \qquad \Phi(\Delta, z, h; k) = k\Delta + f(z, h; k). \tag{7.285}$$

In this expression, slowly varying factors have already been taken out of the integral sign and the Hankel function has been approximated by its asymptotic expression. At a given point of observation (Δ, z), the saddle points, k_s, are determined from

$$\frac{\partial\Phi}{\partial k} = 0 \quad\text{or}\quad -\frac{\partial f(z, h; k)}{\partial k} = \Delta. \tag{7.286}$$

The equation can also be written as

$$\Delta = \Delta(z, h; k_s). \tag{7.287}$$

It represents a ray that starts from $z = h$ and passes through the given point (Δ, z). Expanding Φ about $k = k_s$, we have

$$\Phi(\Delta, z, h; k) = \Phi(\Delta, z, h; k_s) + (k - k_s)\Phi' + \frac{1}{2}(k - k_s)^2\Phi'' + \frac{1}{6}(k - k_s)^3\Phi'''. \tag{7.288}$$

Let us examine three distinct cases:

1. $\Phi'(\Delta, z, h; k_s) = 0, \Phi''(\Delta, z, h; k_s) \neq 0, |\Phi'''| \ll |\Phi''|$. The point (Δ, z) is a regular point on the ray and the integral has the well-known approximation

$$I = \sqrt{\left[\frac{2\pi}{i\Phi''(k_s)}\right]} e^{-i\Phi(k_s)}. \tag{7.289}$$

There are two saddle points at $\pm k_s$ and we take the sum of their contributions.

2. $\Phi'(\Delta, z, h; k_s) = 0, \Phi''(\Delta, z, h; k_s) = 0, \Phi''' \neq 0$. The point (Δ, z) is on a caustic given by the simultaneous equations

$$\begin{gathered}\frac{\partial \Phi}{\partial k} = 0 \Rightarrow \Delta = \Delta(z, h; k_s), \\ \left.\frac{\partial^2 \Phi}{\partial k^2}\right|_{k=k_s} = 0 \Rightarrow \left.\frac{\partial^2 f(z, h; k)}{\partial k^2}\right|_{k=k_s} = 0 \Rightarrow \left.\frac{\partial \Delta(z, h; k)}{\partial k}\right|_{k=k_s} = 0.\end{gathered} \tag{7.290}$$

Then, by our previous results

$$I = \frac{2\pi Ai(0)}{[-\frac{1}{2}\Phi'''(k_s)]^{1/3}} e^{-i\Phi(k_s)}. \tag{7.291}$$

3. $\Phi'(\Delta, z, h; k_s) \neq 0, \Phi''(\Delta, z, h; k_s) = 0, \Phi''' \neq 0$. Let the point of observation be not on a caustic but close to it. Then if (Δ, z) is such that $\Phi''(k_s) = 0$, we must not have $\Phi'(k_s) = 0$, because (Δ, z) is not on the caustic. Accordingly

$$\begin{gathered}I = \frac{2\pi Ai(\sigma)}{[-\frac{1}{2}\Phi'''(k_s)]^{1/3}} e^{-i\Phi(k_s)}, \\ \sigma = \frac{\Phi'(k_s)}{[-\frac{1}{2}\Phi'''(k_s)]^{1/3}}.\end{gathered} \tag{7.292}$$

The interpretation of $\Phi(k_s)$ and its derivatives in terms of ray parameters is as follows: We have shown in Table 7.19 that for a continuously heterogeneous medium

$$\begin{gathered}k_s = p\omega = \frac{\omega}{V(h)} \sin j_h, \\ \Phi(\Delta, z, h; k_s) = k_s\Delta \pm \int_h^z \sqrt{\left(\frac{\omega^2}{V^2} - k_s^2\right)} dz.\end{gathered} \tag{7.293}$$

Hence

$$\left.\frac{\partial\Phi}{\partial k}\right|_{k_s} = \Delta \mp k_s \int_h^z \frac{dz}{\sqrt{(\omega^2/V^2 - k_s^2)}} = 0,$$

$$\Phi''(k_s) = -\frac{\partial\Delta}{\partial k_s} = -\frac{\partial\Delta}{\partial j_h}\bigg/\frac{\partial k_s}{\partial j_h} = -\frac{\partial\Delta}{\partial j_h}\frac{V_h}{\omega\cos j_h},$$

$$\Phi'''(k_s) = -\frac{\partial^2\Delta}{\partial k_s^2} = -\frac{1}{(\partial k_s/\partial j_h)}\frac{\partial}{\partial j_h}\left(\frac{\partial\Delta}{\partial k_s}\right) \tag{7.294}$$

$$= \frac{V_h^2}{\omega^2\cos^2 j_h}\left[\frac{\partial\Delta}{\partial j_h}\tan j_h + \frac{\partial^2\Delta}{\partial j_h^2}\right].$$

At a caustic, $\Phi''(k_s) = 0$ and, therefore,

$$\Phi'''(k_s) = \frac{V_h^2}{\omega^2\cos^2 j_h}\frac{\partial^2\Delta}{\partial j_h^2}. \tag{7.295}$$

Consider, for example, the case of a linear velocity profile

$$V(z) = V_0(1 + \gamma z).$$

From the triangle $P_1D_1Q_1$ (Fig. 7.43)

$$\zeta^2 + [\Delta - (R^2 - \zeta_h^2)^{1/2}]^2 = R^2,$$

where (7.296)

$$\zeta = z + \frac{1}{\gamma}, \qquad \zeta_h = h + \frac{1}{\gamma}.$$

Equation (7.296) is the equation of the ray leaving the source Q, with $j_h < \pi/2$. Similarly, the equation for the nth surface reflected ray is

$$\zeta^2 + \left[\Delta \mp (R^2 - \zeta_h^2)^{1/2} - 2n\left(R^2 - \frac{1}{\gamma^2}\right)^{1/2}\right]^2 = R^2. \tag{7.297}$$

Because the ray parameter is $1/V_0\gamma R$ and γV_0 is fixed, we can take R to be the ray parameter. The minus sign applies to the rays originally downgoing from the source and the plus sign to the rays originally upgoing. Differentiating with respect to R, we get

$$\left[\Delta \mp \sqrt{(R^2 - \zeta_h^2)} - 2n\sqrt{\left(R^2 - \frac{1}{\gamma^2}\right)}\right]\left[\frac{\mp 1}{\sqrt{(R^2 - \zeta_h^2)}} - \frac{2n}{\sqrt{(R^2 - 1/\gamma^2)}}\right] = 1. \tag{7.298}$$

The envelope is obtained by eliminating R between Eqs. (7.297) and (7.298). For given n and γ, we choose a certain value of R. Then the value of Δ is calculated from Eq. (7.298) and the corresponding value of z from Eq. (7.297). For each given value of R, Eqs. (7.297) and (7.298) yield the corresponding points (Δ, z). The locus of these points eventually constitutes the caustic. The envelope will

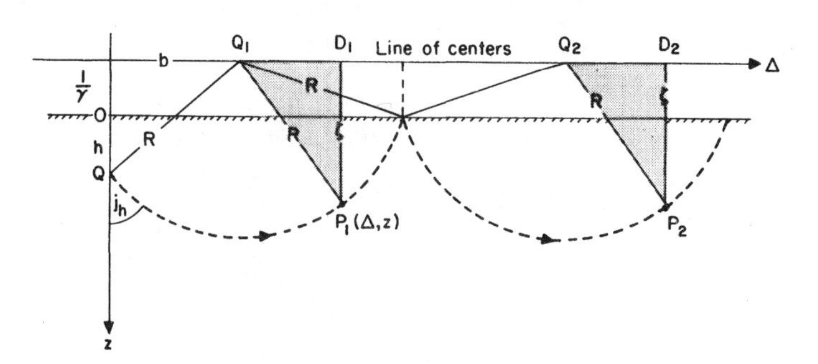

Figure 7.43. Circular rays in a medium with a constant velocity gradient.

have a single branch for upgoing rays and two branches for downgoing rays (Fig. 7.41a). For a surface source $h = 0$, the two caustics coalesce into the hyperbola

$$z^2 = \frac{\Delta^2}{4n(n+1)} + \frac{1}{\gamma^2}. \tag{7.299}$$

7.6. Theoretical Seismograms

7.6.1. Introduction

Seismologic data, obtained in the form of seismograms, can be processed and interpreted in either the frequency domain or the time domain. The first step in frequency-domain analysis is usually that of the application of the Fourier transform to the raw data. Time domain studies do not need this first step. However, the mathematics involved in such analysis is more complicated. As a result, frequency-domain techniques are more frequently used by seismologists.

The basic idea of using synthetic seismograms is to compare computed seismograms directly with observed ones. The theoretical models or assumptions involved in the synthesis of seismograms can be changed until agreement with observations is achieved.

We shall first derive an exact analytic solution in closed form for the surface displacements resulting from a surface point dislocation in a homogeneous elastic half-space. We shall then derive the exact integral representation for the displacement field everywhere in a homogeneous half-space resulting from a point dislocation at arbitrary depth. This integral is then evaluated numerically. It is known as *Lamb's problem*. We shall also derive the exact integral representation for the displacement field induced by a point dislocation on either side of a plane interface between two half-spaces. These results can be extended to a multilayered half-space.

7.6.2. Lamb's Problem

Let us have a second look at the expressions for the displacements at any point of a uniform half-space caused by a point source. Taking the origin at the free surface, the z axis vertically upward, and a source on the z axis at a depth h below the free surface, the total field at a depth d below the free surface is given by [cf. Eqs. (5.12)–(5.14)]

$$\mathbf{u}(\omega) = \sum_m \int_0^\infty \mathbf{u}_m(k, \omega) k\, dk, \tag{7.300}$$

where

$$\begin{aligned}
\mathbf{u}_m(k, \omega) &= U_P(k, \omega)\mathbf{P}_m(k) + U_B(k, \omega)\mathbf{B}_m(k) + U_C(k, \omega)\mathbf{C}_m(k),\\
U_P(k, \omega) &= -\nu_\alpha i_m\{\varepsilon^{m+1} e^{-\nu_\alpha|d-h|} + R^z_{PP} e^{-\nu_\alpha(h+d)} + R^z_{PS} e^{-\nu_\alpha h - \nu_\beta d}\}\\
&\quad + k j_m\{\varepsilon^{m+1} e^{-\nu_\beta|d-h|} + R^z_{SS} e^{-\nu_\beta(h+d)} + R^z_{SP} e^{-\nu_\beta h - \nu_\alpha d}\},\\
U_B(k, \omega) &= k i_m\{\varepsilon^m e^{-\nu_\alpha|d-h|} + R^\Delta_{PP} e^{-\nu_\alpha(h+d)} + R^\Delta_{PS} e^{-\nu_\alpha h - \nu_\beta d}\}\\
&\quad - \nu_\beta j_m\{\varepsilon^m e^{-\nu_\beta|d-h|} + R^\Delta_{SS} e^{-\nu_\beta(h+d)} + R^\Delta_{SP} e^{-\nu_\beta h - \nu_\alpha d}\},\\
U_C(k, \omega) &= k_\beta k_m\{\varepsilon^m e^{-\nu_\beta|d-h|} + R^H_{SS} e^{-\nu_\beta(d+h)}\},\\
R_{PP} &= -R^z_{PP} = R^\Delta_{PP}, \qquad R_{SS} = R^z_{SS} = -R^\Delta_{SS}, \qquad R^H_{SS} = 1,\\
R^z_{PS} &= -\left(\frac{\sin f}{\cos e}\right) R_{PS} = -\frac{ik}{\nu_\alpha}\left(\frac{\beta}{\alpha}\right) R_{PS},\\
R^\Delta_{PS} &= -\left(\frac{\cos f}{\sin e}\right) R_{PS} = \frac{i\nu_\beta}{k}\left(\frac{\beta}{\alpha}\right) R_{PS},\\
R^z_{SP} &= \left(\frac{\cos e}{\sin f}\right) R_{SP} = -\frac{i\nu_\alpha}{k}\left(\frac{\alpha}{\beta}\right) R_{SP},\\
R^\Delta_{SP} &= \left(\frac{\sin e}{\cos f}\right) R_{SP} = \frac{ik}{\nu_\beta}\left(\frac{\alpha}{\beta}\right) R_{SP}.
\end{aligned} \tag{7.301}$$

Here e and f are the angles that the P and SV waves, respectively, make with the normal to the boundary. The vector cylindrical harmonics $\mathbf{P}_m$, $\mathbf{B}_m$, and $\mathbf{C}_m$ are as defined in Eq. (2.87) with $\exp(im\phi)$ replaced by $\cos m\phi$ or $\sin m\phi$, as the case may be. The algebraic symbol of the reflection coefficient for SH waves, R^H_{SS}, although equal to unity, has been retained for future generalization.

Introducing the notation

$$\begin{aligned}
s &= i\omega, \qquad k = su,\\
a &= \left(u^2 + \frac{1}{\alpha^2}\right)^{1/2}, \qquad b = \left(u^2 + \frac{1}{\beta^2}\right)^{1/2},\\
\Omega &= u^2 + \frac{1}{2\beta^2},\\
\Delta_R &= \Omega^2 - u^2 ab,
\end{aligned} \tag{7.302}$$

we find

$$\begin{aligned}
R_{PP}^{z} &= \frac{(\Omega^2 + u^2 ab)}{\Delta_R} = -R_{SS}^{z}, \qquad R_{SS}^{H} = 1, \\
R_{PS}^{z} &= -\frac{2u^2\Omega}{\Delta_R}, \qquad R_{SP}^{z} = \frac{2ab\Omega}{\Delta_R}, \\
R_{PP}^{\Delta} &= -R_{SS}^{\Delta} = -R_{PP}^{z}, \\
R_{PS}^{\Delta} &= R_{SP}^{z}, \qquad R_{SP}^{\Delta} = R_{PS}^{z}.
\end{aligned} \tag{7.303}$$

The integral in Eq. (7.300) assumes the form

$$\mathbf{u}(s) = \sum_m s^2 \int_0^\infty \mathbf{u}_m(u, s) u \, du, \tag{7.304}$$

where

$$\begin{aligned}
\mathbf{u}_m(u, s) &= U_P(u, s)\mathbf{P}_m(su) + U_B(u, s)\mathbf{B}_m(su) + U_C(u, s)\mathbf{C}_m(su), \\
U_P(u, s) &= -a\hat{\imath}_m\{\varepsilon^{m+1}e^{-sa|d-h|} + R_{PP}^{z}e^{-sa(h+d)} + R_{PS}^{z}e^{-s(ah+bd)}\} \\
&\quad + u\hat{\jmath}_m\{\varepsilon^{m+1}e^{-sb|d-h|} + R_{SS}^{z}e^{-sb(h+d)} + R_{SP}^{z}e^{-s(bh+ad)}\}, \\
U_B(u, s) &= u\hat{\imath}_m\{\varepsilon^{m}e^{-sa|d-h|} + R_{PP}^{\Delta}e^{-sa(h+d)} + R_{PS}^{\Delta}e^{-s(ah+bd)}\} \\
&\quad - b\hat{\jmath}_m\{\varepsilon^{m}e^{-sb|d-h|} + R_{SS}^{\Delta}e^{-sb(h+d)} + R_{SP}^{\Delta}e^{-s(bh+ad)}\}, \\
U_C(u, s) &= \hat{k}_m\{\varepsilon^{m}e^{-sb|d-h|} + R_{SS}^{H}e^{-sb(h+d)}\}.
\end{aligned} \tag{7.305}$$

The source coefficients $\hat{\imath}_m$, $\hat{\jmath}_m$, and $\hat{k}_m$ are given in Table 7.20.

Table 7.20. Canonical Source Coefficients

Source	m	$\hat{\imath}_m$		$\hat{\jmath}_m$		$\hat{k}_m$		Common factor
Vertical strike slip	2	$\beta^2 \frac{u^2}{a}$	sin	$\beta^2 u$	sin	$\frac{u}{b}$	cos	
Vertical dip slip	1	$2\beta^2 u$	sin	$2\beta^2 \frac{\Omega}{b}$	sin	1	cos	$\frac{U_0(s)dS}{4\pi}$
Case III	0	$-\frac{1}{2}\beta^2\left(\frac{3u^2 + 2/\alpha^2}{a}\right)$	cos	$-\frac{3}{2}\beta^2 u$	cos	0	—	
	2	$-\frac{1}{2}\beta^2 \frac{u^2}{a}$	cos	$-\frac{1}{2}\beta^2 u$	cos	$\frac{1}{2}\frac{u}{b}$	sin	
Center of compression	0	$-\frac{3}{16\pi\mu}\frac{E_0(s)}{a}$	cos	0	—	0	—	1

Individual terms in Eqs. (7.305) correspond to incident and reflected waves. Our main interest is in reflected waves. We shall use the vertical component of the reflected *PS* wave, namely,

$$[-a\hat{\imath}_m][R^z_{PS}][e^{-s(ah+bd)}]$$

for the sake of illustration. Each reflected wave is the product of three factors. The first is the source factor, $[-a\hat{\imath}_m]$ in our example. The source factor depends upon four things: the nature of the source through $(\hat{\imath}_m, \hat{\jmath}_m, \hat{k}_m)$; the component of the field under consideration; whether the wave is *P*, *SV*, or *SH* at the source; and, last, whether the wave is upgoing or downgoing at the source.

The second factor, $[R^z_{PS}]$ in our example, is the *generalized reflection coefficient.*[10] The significance of the adjective "generalized" will become clear as we proceed further. The coefficient R_{PS}, as defined in Section 5.1, is the ratio of the displacement amplitude of the reflected *S* wave to the amplitude of the incident *P* wave. In contrast, R^z_{PS} is the corresponding ratio for the vertical component of the displacement. The relationship

$$R^z_{PS} = -\left(\frac{\sin f}{\cos e}\right) R_{PS}$$

follows directly from Eqs. (3.1) and (3.2). We therefore have two sets of generalized reflection coefficients for the displacement, one for the horizontal component and one for the vertical.

The third factor in a reflected wave, which is $\exp[-s(ah + bd)]$ in the case of reflected *PS* wave, is the path effect. Note that, in our example, h is the vertical distance covered by the wave with velocity α and d is the corresponding distance covered with velocity β.

We now proceed to find the amplitude of an individual wave induced by a source in a multilayered half-space. Consider first two half-spaces in welded contact. Let the source be in the lower half-space ($z < 0$) and let the upper half-space ($z > 0$) be characterized by appending a prime to its parameters. The displacement field in the lower half-space is still given by Eqs. (7.304) and (7.305). However, R^z_{PP}, etc., now denote the generalized reflection coefficients for reflection at the interface. These can be obtained from the results of Section 3.3. In terms of the present notation, we have

$$R^z_{PP} = -R^\Delta_{PP} = \frac{Y_1}{Y}, \qquad R^z_{SS} = -R^\Delta_{SS} = -\frac{Y_3}{Y},$$

$$R^z_{PS} = R^\Delta_{SP} = -u^2\left(\frac{Y_2}{Y}\right), \qquad R^z_{SP} = R^\Delta_{PS} = ab\left(\frac{Y_2}{Y}\right), \tag{7.306}$$

$$R^H_{SS} = \frac{\mu b - \mu' b'}{\mu b + \mu' b'},$$

[10] Generalized reflection and transmission coefficients used here correspond to the amplitude ratios of Sections 3.3 and 7.2.5. The reflection and transmission coefficients as defined in Section 7.2.5 have, in general, an additional factor (Table 7.7).

where

$$\begin{bmatrix} Y \\ Y_1 \\ Y_3 \end{bmatrix} = u^2(u^2 + q + q')^2 \pm u^2 aa'bb' \mp ab(u^2 + q')^2$$
$$- a'b'(u^2 + q)^2 \pm qq'(ab' \pm a'b),$$

$$Y_2 = 2(u^2 + q + q')(u^2 + q') - 2a'b'(u^2 + q),$$

$$a' = \left(u^2 + \frac{1}{\alpha'^2}\right)^{1/2}, \qquad b' = \left(u^2 + \frac{1}{\beta'^2}\right)^{1/2},$$

$$\Omega' = u^2 + \frac{1}{2\beta'^2},$$

$$\frac{1}{q} = 2\beta^2\left(1 - \frac{\mu'}{\mu}\right), \qquad \frac{1}{q'} = 2\beta'^2\left(1 - \frac{\mu}{\mu'}\right).$$

The displacement field at a point in the upper half-space which is at a distance d' from the interface is given by Eq. (7.304) with

$$\begin{aligned} U_P &= -a\hat{\imath}_m\{T^z_{PP} e^{-s(ah + a'd')} + T^z_{PS} e^{-s(ah + b'd')}\} \\ &\quad + u\hat{\jmath}_m\{T^z_{SS} e^{-s(bh + b'd')} + T^z_{SP} e^{-s(bh + a'd')}\}, \\ U_B &= u\hat{\imath}_m\{T^\Delta_{PP} e^{-s(ah + a'd')} + T^\Delta_{PS} e^{-s(ah + b'd')}\} \\ &\quad - b\hat{\jmath}_m\{T^\Delta_{SS} e^{-s(bh + b'd')} + T^\Delta_{SP} e^{-s(bh + a'd')}\}, \\ U_C &= \hat{k}_m\{T^H_{SS} e^{-s(bh + b'd')}\}. \end{aligned} \tag{7.307}$$

The transmission coefficients, T^z_{PP}, etc., are:

$$\begin{aligned} T^z_{PP} &= a'\left(\frac{Y_4}{Y}\right), & T^\Delta_{PP} &= a\left(\frac{Y_4}{Y}\right), \\ T^z_{SS} &= b\left(\frac{Y_6}{Y}\right), & T^\Delta_{SS} &= b'\left(\frac{Y_6}{Y}\right), \\ T^z_{PS} &= u^2\left(\frac{Y_5}{Y}\right), & T^\Delta_{PS} &= ab'\left(\frac{Y_5}{Y}\right), \\ T^z_{SP} &= a'b\left(\frac{Y_7}{Y}\right), & T^\Delta_{SP} &= u^2\left(\frac{Y_7}{Y}\right), \\ T^H_{SS} &= \frac{2\mu b}{\mu b + \mu' b'}, \end{aligned} \tag{7.308}$$

where

$$Y_4 = 2q[b(u^2 + q') - b'(u^2 + q)],$$
$$Y_5 = 2q[u^2 + q + q' - a'b],$$
$$Y_6 = 2q[a(u^2 + q') - a'(u^2 + q)],$$
$$Y_7 = 2q[u^2 + q + q' - ab'].$$

When the upper medium is a fluid, Eqs. (7.306) and (7.308) must be replaced by

$$\begin{aligned}
&T_{SS} = T_{PS} = 0,\\
&R^z_{PP} = -R^\Delta_{PP} = \frac{\Lambda_1}{\Lambda}, \qquad R^z_{SS} = -R^\Delta_{SS} = -\frac{\Lambda_3}{\Lambda},\\
&R^z_{PS} = R^\Delta_{SP} = -2\frac{u^2 a'\Omega}{\Lambda}, \qquad R^z_{SP} = R^\Delta_{PS} = \frac{2aa'b\Omega}{\Lambda}, \qquad R^H_{SS} = 1,\\
&T^z_{PP} = \frac{a'\Omega}{\beta^2\Lambda}, \qquad T^\Delta_{PP} = \frac{a\Omega}{\beta^2\Lambda},\\
&T^z_{SP} = \frac{aa'b}{\beta^2\Lambda}, \qquad T^\Delta_{SP} = \frac{au^2}{\beta^2\Lambda},
\end{aligned} \tag{7.309}$$

where

$$\begin{bmatrix}\Lambda\\ \Lambda_1\\ \Lambda_3\end{bmatrix} = a'\Omega^2 \mp aa'bu^2 \underset{+}{\pm} \underset{+}{} \frac{a\rho'}{4\mu\beta^2}.$$

It may be noted that Eqs. (7.309) give the coefficients when the waves are incident from the solid (unprimed parameters) on a solid–liquid interface. The corresponding results for waves incident from the liquid (unprimed parameters) on a liquid–solid boundary are

$$\begin{aligned}
&R_{PS} = 0,\\
&R^z_{PP} = -R^\Delta_{PP} = \frac{\Gamma_1}{\Gamma}, \qquad T^z_{PS} = \frac{\rho a' u^2}{\mu'\Gamma}, \qquad T^\Delta_{PS} = \frac{\rho a a' b'}{\mu'\Gamma},\\
&T^z_{PP} = -\frac{a'\Omega'\rho}{\mu'\Gamma}, \qquad T^\Delta_{PP} = -\frac{a\Omega'\rho}{\mu'\Gamma},
\end{aligned} \tag{7.310}$$

$$\begin{bmatrix}\Gamma\\ \Gamma_1\end{bmatrix} = a\Omega'^2 + aa'b'u^2 \mp \frac{a'\rho}{4\mu'\beta'^2}.$$

It may be emphasized here that since the z axis is drawn vertically upward, Eqs. (7.305) and (7.307) yield the reflected and transmitted fields, respectively, for waves that are upgoing at the source. In other words, the interface is above the source. When the interface is below the source so that we consider the reflected and transmitted field generated by waves that are downgoing at the source, then $\varepsilon = -1$ at the interface. Equation (5.8) then tells us that the expressions for U_P, U_B, and U_C must be multiplied by $(-)^{m+1}$, $(-)^m$, and $(-)^m$, respectively.

So far we have derived an integral representation of the displacement field induced by a point source in the presence of a single interface. This method can now be applied to any number of interfaces. The reflected–transmitted field from one interface acts as the incident field for the next interface. After we choose a particular type of wave at the source (P, SV, or SH) and the component of the field of interest (vertical, radial, or azimuthal), we follow it to the first boundary it encounters. The incident field at the boundary is then calculated. This is then multiplied by the proper reflection–transmission coefficient to get the derived field at the first interface. It now acts as the incident field for the second interface and the whole process is repeated.

This procedure is useful only if we are interested in a particular wave, not in the whole field. In Section 7.4.3, we have noticed that if, in the exact integral representation of the total field induced by a point source in a layer over a half-space, the denominator is expanded in negative powers of the exponentials, each individual term in the expansion corresponds to a reflected wave. The procedure explained above is equivalent to picking one wave out of this expansion. The advantage is that we can manage here without knowing the whole expansion. This is very convenient because when the number of interfaces is more than two, the solution of the boundary value problem becomes exceedingly laborious. Even the expression for the denominator, which is simply the surface-wave determinant, becomes unwieldy.

A word of caution before we proceed further. The integral representation corresponding to a given wave, say a reflected PS, is not simply the reflected PS body wave of the GEA, but it contains, in addition to this reflected wave, information about associated least-time arrivals, such as head waves, and also non-least-time arrivals. All those arrivals which are P before incidence and S after reflection are obtainable from the integral corresponding to the reflected PS. It is important not to confuse the generalized coefficients with the plane-wave coefficients. In plane-wave theory, the expressions for the derived waves are obtained by multiplying the incident wave by the plane-wave coefficients. In the case of waves generated by a point source (curved wavefronts), the generalized coefficients do not multiply the source term itself but, instead, multiply the integrand in the integral representation of the source. Therefore, we first represent the source as a sum of plane waves and then deal with them as in plane-wave theory.

7.6.3. Ray Generation

It is clear from Eq. (7.304) and the expressions for the generalized reflection and transmission coefficients that the displacement of a given wave at any point of a multilayered half-space can be expressed in terms of integrals of the type

$$I = s^n \int_0^{\infty} J_m(su\Delta)u^{2l+m+1}f(u^2)e^{-sg(u^2)}\,du, \tag{7.311}$$

where

$$g(u^2) = \sum_j h_j\left(u^2 + \frac{1}{c_j^2}\right)^{1/2}, \qquad (c = \alpha, \beta)$$

$$f(u^2) = D(u^2)S_{\mu\nu}(u^2).$$

Moreover, $D(u^2)$ is a source factor, $S_{\mu\nu}$ is a product of the generalized reflection–transmission coefficients, c_j is the wave velocity of P or S waves in the jth layer, and h_j is the projection of the wave path on the z axis in that layer.

The reliability of synthetic seismograms depends strongly on the family of rays that is selected from the total number of rays. Automatic generation of generalized rays is incorporated into the program for constructing synthetic seismograms. A short discussion of the method follows.

We define two generalized rays as *kinematically equivalent* when their $g(u^2)$ functions are identical. Similarly, two generalized rays are *dynamically equivalent* when their $f(u^2)$ functions are identical. In Fig. 7.44a, all of the kinematic analogs and their partition into six different dynamic analogs are shown for a representative case. Figure 7.44b is an example of the detailed identification of phases in a synthetic seismogram. The corresponding phases are given in Table 7.21. In this table, for example, $P3$ indicates that the ray has traveled layer 3 as a P wave. Computational experience has taught us several rules of thumb that reduce the work considerably. Experience in reading observed seismograms leads frequently to the same rules. These are:

1. The most important rays (rays with large-amplitude contributions) are the rays with a minimal (or near-minimal) number of reflections and therefore a maximal (or near-maximal) number of transmissions.
2. When some kind of wave guide exists in the model, rays that propagate mainly in this wave guide may be of paramount importance, although they violate rule 1.
3. Converted rays may contribute when the conversion takes place at a reflection but are usually negligible when converted at transmission.

Assuming these rules, a simple scheme for generating multiconverted rays from unconverted ones may be used. The only ray–interface interactions eligible for conversion are, according to rule 3, those occurring as reflections at certain interfaces. For the sake of brevity, we call these "eligible points." Suppose an unconverted ray has n eligible points, and suppose further that we are interested in rays that have at most k conversions. Then, there exists a one-to-one correspondence between all multiconverted rays generated from the given uncoverted ray (having exactly k conversions) and all compositions of $n + 1$ having exactly $k + 1$ nonzero terms. Therefore, to a composition $n_1 + n_2 + \cdots + n_{k+1} = n + 1$, there corresponds the following multiconverted ray: Its first

Table 7.21. Nomenclature of Generalized Rays in a Layered Half-space

No.	Phase	Remarks
1	*P3P4P4P3P2P1*	*P* Moho head wave
2	*P3P3P2P1*	*P* head wave
3	*P3P2P1*	Direct wave
4	*P3P3P2P1*	Reflected wave
5	*P3P3P3P2P1*	
6	*P3P2P2P3P3P2P1*	
7	*P3P4P4P3P2P2P3P4P4P3P2P1*	
8	*P3P4P4P3P2P1P1P2P3P4P4P3P2P1*	
9	*S3S2S1P1P2P3P4P4P3P2P1*	
10	*S3S2S1P1P2P3P3P2P1*	
11	*S3S2S1P1P2P2P1*	
12	*P3P4P4P3P2P1S1S2S3S4P4P3P2P1*	
13	*S3S4S4S3S2S1P1P2P3P3P2P1*	
14	*P3P2P1S1S2S3S4S4S3S2S1*	
15	*S3S4S4S3S2S1*	*S* Moho head wave
16	*S3S4S4S3S2S1*	
17	*S3S3S2S1*	*S* head wave
18	*S3S4S4S4S4S3S2S1*	
19	*S3S2S1*	Direct wave
20	*S3S3S2S1*	
21	*S3S3S3S2S1*	
22	*S3S4S5S5S4S3S2S1*	
23	*S3S3S2S1S1S1*	
24	*S3S3S3S3S2S1*	
25	$(S3S4S4S3)_2\,S2S1$	
26	*S3S2S2S3S3S2S1*	
27	*S3S2S2S3S4S4S3S2S1*	
28	*S3S2S1S1S2S3S4S4S3S2S1*	
29	*S3S2S1S1S2S3S4S4S4S4S3S2S1*	
30	*S3S2S1S1S2S3S3S2S1*	
31	*S3S4S4S3S2S2S3S4S4S3S2S1*	
32	*S3S2S1S1S2S3S4S4S3S2S1S1S1*	
33	*S3S2S1S1S2S3S3S2S1S1S1*	
34	$S3S2S1(S1S2S2S1)_2$	
35	*S3S4S4S3S2S1S1S2S3S4S4S3S2S1*	
36	*S3S3S4S4S3S2S1S1S2S3S4S4S3S2S1*	
37	*S3S2S1S1S2S3S4S4S3S2S1S1S2S2S1*	
38	*S3S3S4S4S3S2S1S1S2S3S4S4S3S2S1S1S1*	
39	*S3S2S2S3S4S4S3S2S1S1S2S3S4S4S3S2S1*	
40	*S3S2S1S1S2S3S4S4S3S2S1S1S2S3S3S2S1*	
41	$S3S2S1(S1S2S3S4S4S3S2S1)_2$	
42	$S3S2S1(S1S2S3S4S4S3S2S1)_2S1S1$	
43	*S3S4S4S3S2S1S1S2S3S4S4S3S2S1S1S2S3S3S2S1*	
44	$S3S2S1(S1S2S3S4S4S3S2S1)_2S1S2S2S1$	

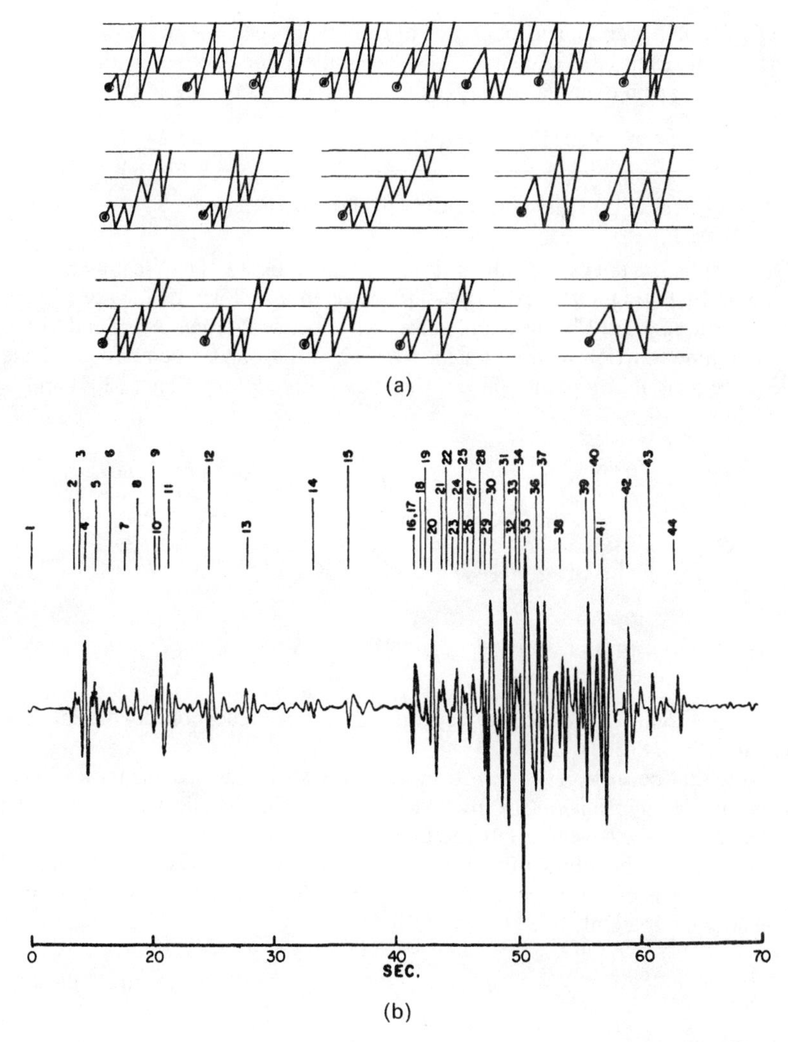

Figure 7.44. Generalized rays. (a) Eighteen kinematic analogs divided into six different groups of dynamic analogs. (b) A synthetic seismogram. Numbers on lines refer to Table 7.21.

conversion occurs at the n_1th "eligible point," its second conversion occurs at the $(n_1 + n_2)$th "eligible point," and so on, up to the $(n_1 + n_2 + \cdots + n_k)$th "eligible point." We adopt, arbitrarily, the convention that all rays begin as P rays. Later, the interchange of P and S can be simply carried out to yield an appropriate ray.

Inasmuch as compositions are easily constructed, this correspondence yields at once a practical way of generating all multiconverted rays, which have

n eligible points and k conversions. The number of such distinct rays, dynamic analogy disregarded, is of course $\binom{n}{k}$. No general method to compute the number of kinematic and dynamic analogs of multiconverted rays is known.

7.6.4. Path Integrals and Their Interpretation

Our next task is to evaluate integrals of the type in Eq. (7.311). We illustrate the method by taking a vertical strike-slip source in a uniform half-space ($z < 0$) as our example. Let h be the depth of the source below the free surface and let the source-time function be a unit step, $H(t)$. Then the vertical component of the displacement at any point $(\Delta, \phi, -d)$ follows directly from Eq. (7.304) and is given by

$$u_z(s) = \frac{U_0\, dS}{4\pi}\beta^2(\sin 2\phi)s\Big\{-\varepsilon\int_0^\infty J_2(su\Delta)u^3[e^{-sa|h-d|} - e^{-sb|h-d|}]du$$
$$-\int_0^\infty J_2(su\Delta)u^3[R^z_{PP}e^{-sa(h+d)} - R^z_{SS}e^{-sb(h+d)}$$
$$+ R^z_{PS}e^{-s(ah+bd)} - R^z_{SP}e^{-s(bh+ad)}]du\Big\}, \tag{7.312}$$

where $\varepsilon = \operatorname{sgn}(h - d)$ and the generalized vertical reflection coefficients are given by Eqs. (7.303).

Note that because of the transformation $s = i\omega$, u_z has become the *Laplace-transformed ground motion*. In order to obtain the surface displacements in the time domain, we have to apply the Laplace inversion integral.

We shall see next how this routine can be avoided as a result of the special analytic forms of the integrands. This will be done in a number of steps. First, consider the integral

$$I = s^n\int_0^\infty J_m(su\Delta)u^{2l+m+1}f(u^2)e^{-sg(u^2)}\,du, \tag{7.313a}$$

where

$$g(u^2) = \sum_j h_j\left(u^2 + \frac{1}{c_j^2}\right)^{1/2}, \qquad h_j = h, d, \qquad c_j = \alpha, \beta.$$

The total field, as given in Eq. (7.312), is composed of six partial fields as shown in Eq. (7.313a). The first two integrals correspond to the direct arrivals of P and S (Fig. 7.45). The next two result from the arrivals of the reflected waves PP and SS and the last two correspond to the converted reflections PS and SP. We shall call each of these six partial fields a *generalized ray*. We make the change of variable

$$u = -iv$$

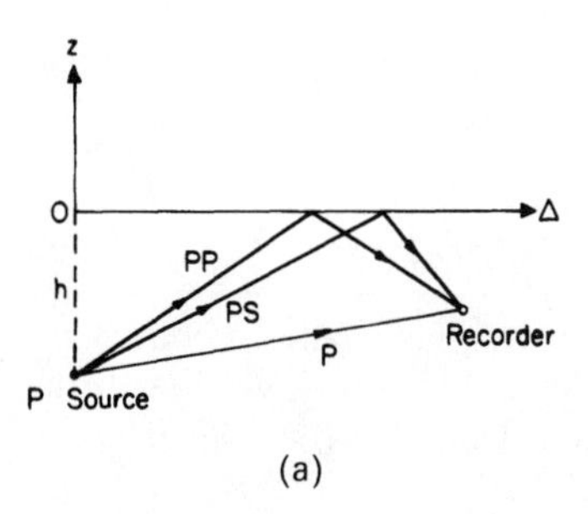

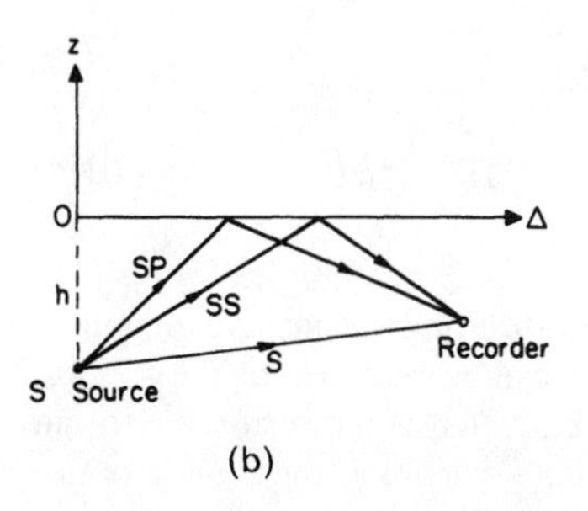

Figure 7.45. Six partial fields that comprise all direct and reflected rays emerging from a buried source and received at a buried recorder. (a) *PP*, *PS*, and *P* rays; (b) *SP*, *SS*, and *S* rays.

and use the relations

$$J_m(su\Delta) = \operatorname{Im}\{iH_m^{(2)}(su\Delta)\},$$

$$iH_m^{(2)}(z) = -\left(\frac{2}{\pi}\right)e^{i\pi m/2}K_m(iz), \qquad -\pi < \arg z \leq \frac{\pi}{2},$$

where K_m is the modified Bessel function of the second kind. The integral $I(s)$ then assumes the form

$$I(s) = (-)^l \frac{2s^n}{\pi} \operatorname{Im}\left[\int_0^{i\infty} K_m(sv\Delta)f(-v^2)v^{2l+m+1}e^{-sg(-v^2)}\,dv\right]. \quad (7.313b)$$

As our next step we invoke the integral representation

$$K_m(\hat{a}s)e^{\hat{a}s} = \frac{\Gamma(\frac{1}{2})}{\Gamma(m+\frac{1}{2})}\left(\frac{s}{2\hat{a}}\right)^m \int_0^{\infty} e^{-sy}(y^2 + 2\hat{a}y)^{m-1/2}\,dy,$$

$$(\operatorname{Re} m > -\tfrac{1}{2}, \qquad |\arg \hat{a}| < \pi).$$

When we change the variable to $y = t - \hat{b} - \hat{a}$, this becomes

$$K_m(\hat{a}s)e^{-\hat{b}s} = \frac{\Gamma(\frac{1}{2})}{\Gamma(m+\frac{1}{2})}\left(\frac{s}{2\hat{a}}\right)^m \int_0^{\infty} e^{-st}[(t-\hat{b}-\hat{a})(t-\hat{b}+\hat{a})]^{m-1/2}$$
$$\times H(t-\hat{b}-\hat{a})dt. \quad (7.314)$$

We now substitute

$$\hat{a} = v\Delta, \qquad \hat{b} = g(-v^2),$$

$$\tau(v) = \hat{a} + \hat{b} = v\Delta + \sum_j h_j\left(\frac{1}{c_j^2} - v^2\right)^{1/2}, \tag{7.315}$$

with the requirement that τ be real and positive. This secures a convenient contour of integration in the complex v plane and also provides τ with the meaning of real time. When we substitute from Eqs. (7.314) and (7.315), Eq. (7.313b) becomes

$$I(s) = (-)^l \frac{2s^n}{\pi(2m-1)!!}\left(\frac{s}{\Delta}\right)^m \operatorname{Im}\int_0^{i\infty} v^{2l+1} f(-v^2)dv$$

$$\times \int_0^{\infty} e^{-st}[(t-\tau)(t-\tau+2v\Delta)]^{m-1/2} H(t-\tau)dt. \tag{7.316}$$

Note that for real values of v, $\tau(v)$ is real for $v \leq 1/\alpha$ and complex for $v > 1/\alpha$. Equation (7.316) indicates that we have to integrate along the positive imaginary axis in the complex v plane. This particular path can be modified, by the application of Cauchy's theorem, into an equivalent path that is more suitable to our needs. Clearly, the singularities of $f(-v^2)$ and $g(-v^2)$ are at the branch points $v = \pm 1/\alpha, \pm 1/\beta$. Hence, the cuts $(-\infty, -1/\alpha)$ and $(1/\alpha, \infty)$ along the real v axis and just below it will secure the singlevaluedness of the functions $f(-v^2)$ and $g(-v^2)$.

Each of the radicals that appears in the expressions for $g(-v^2)$ and $f(-v^2)$ has two branches given by $\operatorname{Re}(1/c_j^2 - v^2)^{1/2} \gtrless 0$. The convergence of the integral in Eq. (7.313b) requires that we choose the branch that secures the boundness of $e^{-sg(-v^2)}$. It follows from this convention that the radical denotes the positive square root when v is real, positive, and less than $1/c_j$. This convention is continued when v leaves the real axis. Therefore, we may write the conditions of convergence in the form

$$\operatorname{Re}\left(\frac{1}{c_j^2} - v^2\right)^{1/2} \geq 0, \qquad \operatorname{Im} v = 0,$$

$$\operatorname{sgn}\left[\operatorname{Im}\left(\frac{1}{c_j^2} - v^2\right)^{1/2}\right] = -\operatorname{sgn}\operatorname{Im}(v^2), \qquad \operatorname{Im} v \neq 0.$$

In addition to the branch points, the integrand has two poles at $v = \pm 1/c_R$, which are the roots of the Rayleigh equation $\Delta_R = 0$.

We now apply Cauchy's integral theorem to the region bounded by the positive imaginary axis, the circular arc γ_4 in the first quadrant and the path $\Gamma(\gamma_1 + \gamma_2 + \gamma_3)$, which is obtained by our physical postulation that the variable τ remains always a real quantity (Fig. 7.46a). Mathematically, Γ is determined by the condition

$$\operatorname{Im}\tau = \Delta\{\operatorname{Im} v\} + \sum_j h_j \operatorname{Im}\left(\frac{1}{c_j^2} - v^2\right)^{1/2} = 0.$$

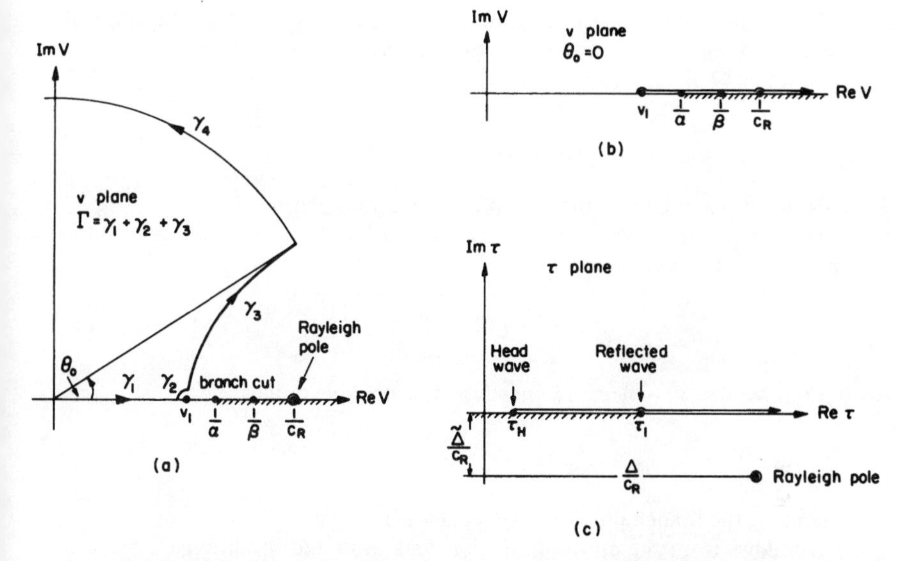

Figure 7.46. Transformation of integration contours: (a) Bromwich contour transformed into an equivalent contour in the complex v plane. (b) A special case of (a) for $\theta_0 = 0$. (c) Final mapping onto the τ plane.

The curve Γ has been chosen such that no singularity of the integrand is inside the said region. Special care has to be exercised to handle singularities that may lie on Γ.

Let $\partial\tau/\partial v = 0$ at $v = v_1$. In the neighborhood of this point we have, by Taylor's expansion,

$$\tau = \tau_1 + \frac{1}{2}\left(\frac{\partial^2\tau}{\partial v^2}\right)_{v_1}(v - v_1)^2 + \cdots,$$

where $\tau_1 = \tau(v_1)$ and

$$\frac{\partial^2\tau}{\partial v^2} = -\sum \frac{h_j}{c_j^2}\left(\frac{1}{c_j^2} - v^2\right)^{-3/2}.$$

In the interval $0 < v < 1/\alpha$, $\partial^2\tau/\partial v^2$ is real and negative. Hence, $\partial\tau/\partial v$ is a monotonically decreasing function there. However, because $(\partial\tau/\partial v)_{v=0} = \Delta > 0$ and $(\partial\tau/\partial v)_{v=1/\alpha} = -\infty$, $\partial\tau/\partial v$ has exactly one zero at $v = v_1$ in the interval $0 < v < 1/\alpha$. Neglecting third- and higher order terms in $(v - v_1)$, we may write

$$\tau - \tau_1 = -\frac{1}{2}A(v - v_1)^2, \qquad \left[A = -\left(\frac{\partial^2\tau}{\partial v^2}\right)_{v_1} > 0\right]. \tag{7.317a}$$

This equation cannot be satisfied for $\tau > \tau_1$ unless v is complex. Therefore, at $v = v_1$ the contour Γ in the v plane leaves the real axis (Fig. 7.46a).

Equation (7.317a) yields

$$\frac{\partial v}{\partial \tau} = -\frac{1}{A(v - v_1)} = \frac{1}{\sqrt{(2A)}}(\tau_1 - \tau)^{-1/2}. \tag{7.317b}$$

This shows that τ_1 is a branch point of $v(\tau)$ in the τ plane.

There may arise cases in which a branch point of $f(-v^2)$ lies on γ_1. Consider, for example, the wave associated with the phase function

$$g(-v^2) = (h + d)\left(\frac{1}{\beta^2} - v^2\right)^{1/2}.$$

In this case $0 < v_1 < 1/\beta$ and, therefore, it is possible to have

$$0 < \frac{1}{\alpha} < v_1 < \frac{1}{\beta}.$$

Therefore, the branch point $v = 1/\alpha$ of $f(-v^2)$ lies on γ_1.

It follows from the above discussion that while the condition $\operatorname{Im} \tau = 0$ is common to all points on Γ, the condition $\operatorname{Im} v = 0$ holds only for $v < v_1$. Beyond $v = v_1$, the condition $\operatorname{Im} \tau = 0$ cannot be fulfilled any longer for real v values alone and v must become complex in order to render τ real along the rest of Γ. In order to fix our ideas we specify the integration path appropriate for the reflected P wave whose phase function is

$$g(-v^2) = \hat{z}\sqrt{\left(\frac{1}{\alpha^2} - v^2\right)}, \qquad \hat{z} = h + d.$$

Denoting $v = \alpha^{-1}\cos(\theta_0 + i\theta_1)$, the condition $\operatorname{Im} \tau = 0$ can be written as $\hat{z}/\Delta = \tan\theta_0$. Therefore, if $\theta_1 = 0$,

$$v = v_1 = \frac{1}{\alpha}\cos\theta_0 = \frac{\Delta}{\alpha R}, \qquad R^2 = \Delta^2 + \hat{z}^2. \tag{7.318a}$$

Hence, by Eq. (7.315),

$$\tau_1 = \tau(v_1) = \frac{R}{\alpha}. \tag{7.318b}$$

It can be shown that $\partial\tau/\partial v$ has no zeros on γ_3. The path Γ makes a small detour to avoid the singular point at $v = v_1$ where $\partial\tau/\partial v = 0$, denoted by γ_2 in Fig. 7.46a. The relation (7.318b) is recognized as the travel time of the reflected P ray. In fact, the corresponding relation $v_1 = (1/\alpha)\sin(\pi/2 - \theta_0)$ is none other than Snell's law, with v_1 in the role of the ray parameter. Note again that τ_1 is a branch point of $v(\tau)$ in the τ plane. In fact, it follows from Eqs. (7.315) that in the present case

$$v = \tau\frac{\Delta}{R^2} \pm \frac{\hat{z}}{R^2}(\tau_1^2 - \tau^2)^{1/2}. \tag{7.319}$$

For values of v such that $|v| \gg 1/\alpha$, we have

$$\text{Im } v = \frac{\hat{z}}{\Delta}(\text{Re } v) = \tan\theta_0(\text{Re } v).$$

Therefore, the curve $\text{Im } \tau = 0$ is asymptotic to a line through the origin, making an angle θ_0 with the real axis. Because the integral along γ_4 vanishes as $|v| \to \infty$, the integration path along the positive imaginary axis is replaced by the path Γ. Equation (7.316) now reads

$$I(s) = (-)^l \frac{2s^n}{\pi(2m-1)!!}\left(\frac{s}{\Delta}\right)^m \text{Im P} \int_0^\infty e^{-st}\,dt$$
$$\times \left\{\int_\Gamma v^{2l+1} f(-v^2)[(t-\tau)(t-\tau+2\Delta v)]^{m-1/2} H(t-\tau)\,dv\right\},$$

where the change of the order of integration is permissible on account of uniform convergence. The symbol P denotes the principal value in the Cauchy sense. Its need arises only when $\theta_0 = 0$. We shall henceforth suppress this symbol in our formulas and restore it only when necessary. Changing variables from v to τ, we get

$$I(s) = (-)^l \frac{2s^n}{\pi(2m-1)!!}\left(\frac{s}{\Delta}\right)^m \int_0^\infty e^{-st}\,dt$$
$$\times \left\{\text{Im} \int_{\tau_H}^t v^{2l+1} f(-v^2)[(t-\tau)(t-\tau+2v\Delta)]^{m-1/2} H(t-\tau)\frac{\partial v}{\partial \tau}\,d\tau\right\}. \tag{7.320}$$

Because τ is real, the integral over the complex path Γ is transformed into the real and positive interval (τ_H, t). The upper limit is t because of the presence of $H(t-\tau)$ in the integrand, whereas the lower limit $\tau_H < t$ corresponds to a certain threshold value of τ, prior to which the integrand is real. The Rayleigh pole is mapped into

$$\tau_R = \frac{\Delta}{c_R} - i\hat{z}\sqrt{\left(\frac{1}{c_R^2} - \frac{1}{\alpha^2}\right)} = \frac{1}{c_R}(\Delta - i\tilde{\Delta}).$$

It lies *below* the integration path.

The case $\hat{z} = 0$ requires special attention. Here Γ lies on the real axis in the v plane, from $v = 0$ to infinity, and must encounter the Rayleigh pole (Fig. 7.46b). The corresponding integral over Γ will be written as the principal value of an integral over the real axis plus a contribution from a small semicircular detour above the pole. The transformation of the integral to the τ plane proceeds as in the case $\hat{z} \neq 0$.

From Eq. (7.319), $\tau = \pm\tau_1$ are the branch points on the real axis of the τ plane. We make a branch cut in this plane, connecting the two branch points along $\text{Im } \tau = 0$. For $\tau < \tau_1$, $\text{Im } v = 0$ and from Eq. (7.319) there are two real values of v for each real τ. Of these, only the smaller lies on Γ. For $\tau > \tau_1$, v on Γ

becomes complex. Here, there are two complex values of v for each real τ. In order to stay on Γ, we must choose that root for which Im v is algebraically bigger. Therefore, the overall choice that is concordant with Γ is

$$v = \tau\frac{\Delta}{R^2} - \frac{\hat{z}}{R^2}\left[\frac{R^2}{\alpha^2} - \tau^2\right]^{1/2}, \qquad \tau < \frac{R}{\alpha};$$
$$v = \tau\frac{\Delta}{R^2} + \frac{i\hat{z}}{R^2}\left[\tau^2 - \frac{R^2}{\alpha^2}\right]^{1/2}, \qquad \tau > \frac{R}{\alpha}. \tag{7.321}$$

The path of integration in the τ plane is along the real axis (above the cut), except near τ_R, where it has a small detour above the real axis. If we take this segment as an arc of radius δ, the integrand behaves on it as $\delta^{-1/2}$, at most, whereas the length of the path is of the order of δ. Therefore, the integral along this arc tends to zero as $\delta \to 0$. The integral in Eq. (7.320) should be written as the sum of two improper integrals $\int_{\tau_H}^{\tau_1} + \int_{\tau_1}^{t}$ where the singularity at τ_1 is integrable and poses no difficulties.

It is essential to remember that although the formal expression for the generalized ray integral in Eq. (7.320) is the same for all rays, the entities τ_H, $f(-v^2)$, and $v(\tau)$ will vary from ray to ray. Therefore, the results obtained so far for the reflected P wave will apply, *mutatis mutandis*, to all the individual integrals in Eq. (7.312).

It now remains for us to discuss the behavior of $\partial v/\partial\tau$ in the τ plane. Because the only zero value of $\partial\tau/\partial v$ is located at $v = v_1$, the sole singularity of $\partial v/\partial\tau$ will be at that point. It is avoided by a small detour above the real axis as explained above.

We are now ready to effect the inversion of $I(s)$ by inspection. Invoking the following theorem from the theory of the Laplace transform

$$\int_0^\infty e^{-st}\left\{\frac{d^N}{dt^N}F(t)\right\}dt = s^N\int_0^\infty e^{-st}F(t)dt - \sum_{k=1}^{N} s^{N-k}F^{(k-1)}(+0),$$

we notice that the right-hand side of Eq. (7.320) has the form of the Laplace transform of a known function. This function is

$$(-1)^l\frac{2}{\pi}\frac{d^n}{dt^n}Q_m(t),$$

where

$$(2m-1)!!Q_m(t) = \text{Im}\left\{\frac{1}{\Delta^m}\frac{d^m}{dt^m}\int_{\tau_H}^{t} H(t-\tau)v^{2l+1}f(-v^2)\right.$$
$$\left.\times\,[(t-\tau)(t-\tau+2v\Delta)]^{m-1/2}\frac{\partial v}{\partial\tau}\,d\tau\right\}. \tag{7.322}$$

Hence, from Eq. (7.312),

$$u_z(t) = \frac{U_0\,dS\beta^2}{2\pi^2}(-)^l\sin 2\phi\,\frac{d}{dt}Q_2(t). \tag{7.323}$$

It is understood that in Eq. (7.323) we have to sum the contributions from the separate six integrals that appear in Eq. (7.312).

Because the integrand vanishes at the upper limit of the integral for $m > 0$, the m-fold differentiation is allowed to operate directly on the integrand. Furthermore, no contribution to the integral arises from the derivatives of $H(t - \tau)$. For $m = 0, 1, 2$, therefore, we obtain,

$$Q_0(t) = H(t - \tau_H)\mathrm{Im} \int_{\tau_H}^{t} v^{2l+1} f(-v^2) \frac{\partial v}{\partial \tau} \frac{d\tau}{[(t - \tau)(t - \tau + 2v\Delta)]^{1/2}},$$

$$Q_1(t) = \frac{1}{\Delta} H(t - \tau_H)\mathrm{Im} \int_{\tau_H}^{t} v^{2l+1} f(-v^2) \frac{\partial v}{\partial \tau} \frac{(t - \tau + v\Delta)}{[(t - \tau)(t - \tau + 2v\Delta)]^{1/2}} d\tau,$$

$$Q_2(t) = \frac{1}{\Delta^2} H(t - \tau_H)\mathrm{Im} \int_{\tau_H}^{t} v^{2l+1} f(-v^2) \frac{\partial v}{\partial \tau} \frac{[2(t - \tau)^2 + 4(t - \tau)v\Delta + v^2\Delta^2]}{[(t - \tau)(t - \tau + 2v\Delta)]^{1/2}} d\tau. \quad (7.324)$$

By induction

$$Q_m(t) = H(t - \tau_H)\mathrm{Im} \int_{\tau_H}^{t} v^{2l+m+1} f(-v^2) \frac{\partial v}{\partial \tau} \frac{T_m[1 + (t - \tau)/v\Delta]}{[(t - \tau)(t - \tau + 2v\Delta)]^{1/2}} d\tau, \quad (7.325)$$

where T_m are the Chebyshev polynomials of the first kind.

The final expression for the displacements in Eq. (7.323) is intentionally written with the operator d/dt outside Q_2 because $u(t)$ will eventually be convolved with a seismograph response function and with a function describing the real time dependence of the source where these functions are known explicitly. Integration of the convolution integral by parts will cause the operator d/dt to operate on explicitly known functions. Because this operation may be done analytically, it can save a great deal of computational work.

We shall now apply the general expressions given in Eqs. (7.324) to the special case in which both $d = 0$ and $h = 0$. Equation (7.312) then reduces to the simpler form

$$u_z(s) = \frac{U_0\, dS}{4\pi} (\sin 2\phi) s \int_0^{\infty} J_2(su\Delta) u^3 \left[\frac{ab - \Omega}{\Delta_R(u^2)}\right] du. \quad (7.326)$$

The final form of the τ integral for this case ($m = 2$, $l = 0$, $\tau = v\Delta$), is

$$u_z(t) = \frac{U_0\, dS}{2\pi^2\Delta^4} \sin 2\phi H\left(t - \frac{\Delta}{\alpha}\right) \frac{d}{dt} \mathrm{Im}\, \mathrm{P} \int_{\Delta/\alpha}^{t} \tau$$
$$\times \frac{\sqrt{(1/\alpha^2 - v^2)}\sqrt{(1/\beta^2 - v^2)} - (1/2\beta^2 - v^2)}{(1/2\beta^2 - v^2)^2 + v^2\sqrt{(1/\alpha^2 - v^2)}\sqrt{(1/\beta^2 - v^2)}} \frac{(2t^2 - \tau^2)}{\sqrt{(t^2 - \tau^2)}} d\tau. \quad (7.327)$$

If we introduce the dimensionless times $\beta t/\Delta = \theta$, $\beta\tau/\Delta = y$ and the parameter $\eta = \beta/\alpha < 1$, Eq. (7.327) can be recast in the form

$$u_z(\theta) = \frac{U_0\, dS}{2(\pi\Delta)^2} \sin 2\phi H(\theta - \eta) \frac{d}{d\theta} \mathrm{P} \int_\eta^\theta \frac{\mathrm{Im}\, f(y, \eta)}{\sqrt{(\theta^2 - y^2)}} (2\theta^2 - y^2) y\, dy, \quad (7.328)$$

where

$$f(y, \eta) = -\frac{(\frac{1}{2} - y^2) - \sqrt{(1 - y^2)}\sqrt{(\eta^2 - y^2)}}{(\frac{1}{2} - y^2)^2 + y^2\sqrt{(1 - y^2)}\sqrt{(\eta^2 - y^2)}}, \quad (7.329)$$

and the derivative $d/d\theta$ is taken at a fixed Δ.

An examination of $f(y, \eta)$ shows that

$$\mathrm{Im}\, f(y, \eta) = \begin{cases} \dfrac{4(1 - 2y^2)\sqrt{(y^2 - \eta^2)}\sqrt{(1 - y^2)}}{1 - 8y^2 + 8(3 - 2\eta^2)y^4 - 16(1 - \eta^2)y^6}, & \eta < y < 1, \\ 0 & y < \eta \text{ or } y > 1. \end{cases} \quad (7.330)$$

Hence,

$$u_z(\theta) = \begin{cases} 0, & \theta < \eta, \ \left(t < \dfrac{\Delta}{\alpha}\right) \\ \dfrac{U_0\, dS}{2(\pi\Delta)^2} \sin 2\phi \dfrac{d}{d\theta} \mathrm{P} \displaystyle\int_\eta^\theta \frac{\mathrm{Im}\, f(y, \eta)}{\sqrt{(\theta^2 - y^2)}} (2\theta^2 - y^2) y\, dy, & \\ & \eta < \theta < 1, \ \left(\dfrac{\Delta}{\alpha} < t < \dfrac{\Delta}{\beta}\right) \\ \dfrac{U_0\, dS}{2(\pi\Delta)^2} \sin 2\phi \dfrac{d}{d\theta} \left[\mathrm{P} \displaystyle\int_\eta^1 \frac{\mathrm{Im}\, f(y, \eta)}{\sqrt{(\theta^2 - y^2)}} (2\theta^2 - y^2) y\, dy \right. & \\ \qquad \left. - \pi A \dfrac{H(\theta - \hat{\gamma})}{\sqrt{(\theta^2 - \hat{\gamma}^2)}} (2\theta^2 - \hat{\gamma}^2)\right], & \theta > 1, \ \left(t > \dfrac{\Delta}{\beta}\right), \end{cases} \quad (7.331)$$

where

$$A = \frac{\frac{1}{2}(2\hat{\gamma}^2 - 1)^3}{4\hat{\gamma}^2 - 1 - 8\hat{\gamma}^6(1 - \eta^2)}.$$

The extra term for the case $\theta > 1$ arises from the residue of the Rayleigh pole at $y = \hat{\gamma}$. The same result can be obtained independently as follows. Starting from Eq. (5.24), assuming a unit-step time variation, and taking the inverse Fourier transform, we find (for $h = 0$)

$$u_z(t) = A \frac{U_0\, dS}{8\pi c_R^2} \sin 2\phi \int_{-\infty}^{\infty} H_2^{(2)}\left(\frac{\omega}{c_R} \Delta\right) e^{i\omega t} \omega\, d\omega, \quad (7.332)$$

where the integral over ω is a "hook integral." Using the result

$$\int_{-\infty}^{\infty} H_2^{(2)}\left(\frac{\omega\Delta}{c_R}\right) e^{i\omega t}\omega\, d\omega = -4\frac{d}{dt}\left[\left\{2\left(\frac{c_R t}{\Delta}\right)^2 - 1\right\}\frac{H(t-\Delta/c_R)}{\sqrt{(t^2-\Delta^2/c_R^2)}}\right], \quad (7.333)$$

we obtain

$$u_z(t) = -\frac{U_0\, dSA}{2\pi c_R^2}\sin 2\phi\frac{d}{dt}\left[\left\{2\left(\frac{c_R t}{\Delta}\right)^2 - 1\right\}\frac{H(t-\Delta/c_R)}{\sqrt{(t^2-\Delta^2/c_R^2)}}\right], \quad (7.334)$$

which is identical to the Rayleigh-pole contribution in Eqs. (7.331). The integrals in Eqs. (7.331) are evaluated by decomposing their integrands into partial fractions. It can be shown that the denominator of Im $f(y, \eta)$ [Eq. (7.330)] has six positive real zeros $\pm y_i$ $(i = 1, 2, 3)$ such that

$$0 < y_1 < y_2 < \eta < 1 < y_3, \quad (7.335)$$

provided that the Poisson ratio $\sigma < \sigma_0 = 0.2631$ (or, equivalently, $0.3215 < \eta^2 < 1$). Substituting

$$\begin{aligned} y^2 &= \eta^2 + (\theta^2 - \eta^2)\sin^2\varepsilon, && (\eta < \theta < 1)\\ y^2 &= \eta^2 + (1 - \eta^2)\sin^2\varepsilon, && (\theta > 1) \end{aligned} \quad (7.336a)$$

in Eqs. (7.331), we obtain

$$u_z(\theta) = \begin{cases} 0, & \theta < \eta \\ \dfrac{2}{\sqrt{(1-\eta^2)}}\dfrac{U_0\, dS}{(\pi\Delta)^2}\sin 2\phi\dfrac{\partial}{\partial\theta}\left[\dfrac{1}{8(1-\eta^2)}\right. \\ \quad\times\left\{\dfrac{2\eta^4 - 3\eta^2 + 2}{2(1-\eta^2)} - 2\theta^2\right\}K(k) - \dfrac{1}{8}E(k) \\ \quad\left. + \sum_{i=1}^{3} Y_i(y_i^2 - 2\theta^2)(2y_i^2 - 1)(y_i^2 - 1)\Pi(n_i k^2, k)\right], & \eta < \theta < 1 \\ \dfrac{2}{\sqrt{(\theta^2-\eta^2)}}\dfrac{U_0\, dS}{(\pi\Delta)^2}\sin 2\phi\dfrac{\partial}{\partial\theta}\left[\dfrac{1}{8(1-\eta^2)}\right. \\ \quad\times\left\{\dfrac{\eta^2(2\eta^2-1)}{2(1-\eta^2)} - \theta^2\right\}K\left(\dfrac{1}{k}\right) - \dfrac{k^2}{8}E\left(\dfrac{1}{k}\right) \\ \quad\left. + \sum_{i=1}^{3} Y_i(y_i^2 - 2\theta^2)(2y_i^2 - 1)(y_i^2 - 1)\Pi\left(n_i, \dfrac{1}{k}\right)\right] \\ \quad - A\dfrac{U_0\, dS}{2\pi\Delta^2}\sin 2\phi\dfrac{\partial}{\partial\theta}\left[\dfrac{H(\theta - \hat{\gamma})}{\sqrt{(\theta^2 - \hat{\gamma}^2)}}(2\theta^2 - \hat{\gamma}^2)\right], & \theta > 1 \end{cases} \quad (7.336b)$$

where $K(k)$, $E(k)$, and $\Pi(n, k)$ are, respectively, the complete elliptic integrals of the first, second, and third kind

$$K(k) = \int_0^{\pi/2} \frac{d\varepsilon}{\sqrt{(1 - k^2 \sin^2 \varepsilon)}}, \qquad E(k) = \int_0^{\pi/2} \sqrt{(1 - k^2 \sin^2 \varepsilon)}\, d\varepsilon,$$
$$\Pi(n, k) = \int_0^{\pi/2} \frac{d\varepsilon}{(1 + n \sin^2 \varepsilon)\sqrt{(1 - k^2 \sin^2 \varepsilon)}}. \tag{7.337}$$

Further,

$$k^2(\theta) = \frac{\theta^2 - \eta^2}{1 - \eta^2}, \qquad n_i = \frac{1 - \eta^2}{\eta^2 - y_i^2}, \qquad n_1, n_2 > 0, \qquad n_3 < 0,$$
$$Y_1 = \frac{y_0}{(y_1^2 - y_2^2)(y_1^2 - y_3^2)}, \qquad y_0 = -\frac{1}{16(1 - \eta^2)}, \tag{7.338}$$
$$Y_2 = \frac{y_0}{(y_2^2 - y_3^2)(y_2^2 - y_1^2)}, \qquad Y_3 = \frac{y_0}{(y_3^2 - y_1^2)(y_3^2 - y_2^2)}.$$

The following relations are useful

$$\frac{\partial E(k)}{\partial \theta} = \frac{\theta}{\theta^2 - \eta^2}[E(k) - K(k)],$$
$$\frac{\partial K(k)}{\partial \theta} = \frac{\theta}{\theta^2 - \eta^2}[\Pi(-k^2, k) - K(k)]. \tag{7.339}$$

For Poisson materials ($\sigma = \frac{1}{4}$), we find

$$\eta^2 = \frac{1}{3}, \qquad y_1^2 = \frac{1}{4}, \qquad y_2^2 = \frac{3 - \sqrt{3}}{4}, \qquad y_3^2 = \frac{3 + \sqrt{3}}{4}, \qquad A = \frac{1}{4}. \tag{7.340}$$

We have thus obtained an explicit exact expression for the vertical surface displacement caused by a point dislocation at the surface. These results for $h = 0$ and $z = 0$ can be extended to obtain the three components of the ground motion arising from the three fundamental sources.

If the source is buried at depth $h \neq 0$, a solution cannot be obtained in simple form and the analytical solution will not go beyond that given in relations (7.325). Some efficient numerical method of integration is then needed to evaluate the displacements. The integrals to be considered in this case are obtained directly from Eq. (7.312) by simply setting $d = 0$ therein. The general form of the transformed displacements, apart from a suppressed constant, can still be written in the form of Eq. (7.313a), where now

$$f(u^2) = D(u^2)R_{\mu\nu}(u^2). \tag{7.341}$$

In Eq. (7.341), $R_{\mu\nu}$ is a reflection coefficient (R_{PP}, R_{SS}, etc.) and $D(u^2)$ is a source term. The corresponding integral on the real τ axis will assume the form given in Eq. (7.320), where, for a P arrival,

$$\tau = v\Delta + h\sqrt{\left(\frac{1}{\alpha^2} - v^2\right)}. \tag{7.342}$$

From Eqs. (7.321) and (7.342), we find

$$\frac{dv}{d\tau} = \sqrt{\left(\frac{1/\alpha^2 - v^2}{R^2/\alpha^2 - \tau^2}\right)}, \qquad R^2 = \Delta^2 + h^2. \tag{7.343}$$

For $h = 0$ we have $\tau = v\Delta$ and $dv/d\tau = 1/\Delta$, as it should be.

For the purpose of illustration we restrict our attention to the surface displacements ($z = 0$) generated by a buried vertical strike-slip dislocation. Setting $d = 0$ in Eq. (7.312), we have

$$\begin{aligned} u_z &= u_{zP} + u_{zS}, \\ u_{zP} &= -\frac{U_0\, dS}{4\pi}(\sin 2\phi)s\int_0^\infty J_2(su\Delta)\frac{\Omega}{\Delta_R}u^3 e^{-sah}\, du, \\ u_{zS} &= \frac{U_0\, dS}{4\pi}(\sin 2\phi)s\int_0^\infty J_2(su\Delta)\frac{ab}{\Delta_R}u^3 e^{-sbh}\, du. \end{aligned} \tag{7.344}$$

From Eq. (7.322), the corresponding form of u_{zP} in the time domain is

$$\begin{aligned} u_{zP}(t) &= \frac{-U_0\, dS}{2\pi^2}\sin 2\phi H(t - \tau_p) \\ &\times \operatorname{Re}\frac{d}{dt}\int_{\tau_p}^{t}\left[\frac{\sqrt{(1/\alpha^2 - v^2)}(1/2\beta^2 - v^2)v^3}{(1/2\beta^2 - v^2)^2 + v^2\sqrt{(1/\alpha^2 - v^2)}\sqrt{(1/\beta^2 - v^2)}}\right] \\ &\times \frac{T_2[1 + (t - \tau)/v\Delta]\, d\tau}{\sqrt{[(\tau + \tau_p)(t - \tau + 2v\Delta)]}\sqrt{[(\tau - \tau_p)(t - \tau)]}}. \end{aligned} \tag{7.345}$$

The integrand in Eq. (7.345) has singular points at both end points of the integration interval. These singularities, however, are integrable. Changing the variable via the relation

$$q = \sin^{-1}\sqrt{\left(\frac{t - \tau}{t - \tau_p}\right)}, \qquad \tau_p = \frac{R}{\alpha}, \tag{7.346}$$

we have

$$\begin{aligned} u_{zP}(t) &= \frac{-U_0\, dS}{(\pi\Delta)^2}\sin 2\phi H(t - \tau_p) \\ &\times \operatorname{Re}\frac{d}{dt}\int_0^{\pi/2}\left[\frac{\sqrt{(1/\alpha^2 - v^2)}(1/2\beta^2 - v^2)v}{(1/2\beta^2 - v^2)^2 + v^2\sqrt{(1/\alpha^2 - v^2)}\sqrt{(1/\beta^2 - v^2)}}\right] \\ &\times \left[\frac{v^2\Delta^2 + 2(t - \tau)(t - \tau + 2v\Delta)}{\sqrt{[(\tau + \tau_p)(t - \tau + 2v\Delta)]}}\right]dq, \end{aligned} \tag{7.347}$$

where

$$\tau(q) = t - (t - \tau_p)\sin^2 q,$$

$$v(\tau) = \tau\frac{\Delta}{R^2} + i\frac{h}{R^2}\sqrt{\left(\tau^2 - \frac{R^2}{\alpha^2}\right)}, \qquad \tau \geq \frac{r}{\alpha}. \tag{7.348}$$

Note that the integrand is a complex function, but the path of integration and τ are real.

A convenient way of rendering a physical interpretation to integrals of the form of Eq. (7.347) arises from the recognition that this integral is a slowly varying function along Γ (Fig. 7.46) except near $v = v_1$ and the branch point at $v = 1/\alpha$. This is the reason that the main contribution comes from the vicinity of these points.

The integration path for u_{zP} when $z = 0$ is shown in Fig. 7.47a. The integration starts at $v = v_1$ with the arrival of the P wave, which is here the sum of P, PP, and PS. As we continue to move along γ_3 away from v_1, we come under the influence of the singularities at $v = 1/\alpha$ and $v = 1/\beta$. We substitute $v = x + iy$ ($x > 0$, $y > 0$) in Eq. (7.342) and get

$$\tau = \left(x\Delta + h\frac{xy}{\eta}\right) + i(y\Delta - h\eta), \tag{7.349}$$

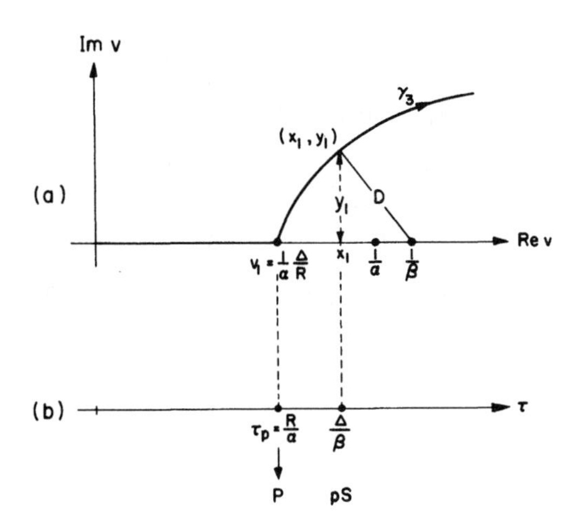

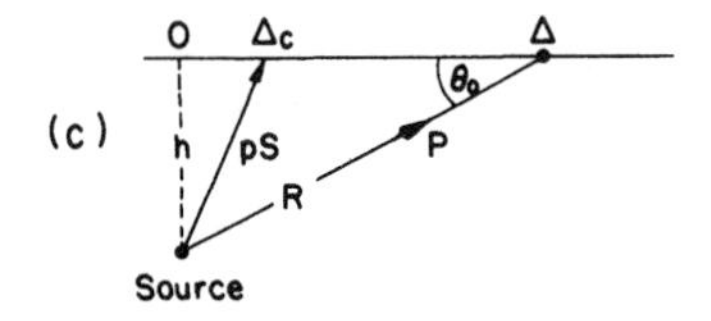

Figure 7.47. (a) Integration path for u_{zP} when $z = 0$. (b) P and pS arrival times. (c) pS (secondary P) wave vs. the direct P wave.

where

$$2\eta^2 = [a^2 + 4x^2y^2]^{1/2} - a, \qquad a = \frac{1}{\alpha^2} - x^2 + y^2 < 0. \tag{7.350}$$

The condition Im $\tau = 0$ then implies

$$\eta = y\frac{\Delta}{h}, \qquad \tau = x\Delta\left(1 + \frac{h^2}{\Delta^2}\right) = x\frac{R^2}{\Delta}. \tag{7.351}$$

The equation of the curve γ_3 in the (x, y) plane is

$$x^2 - \left(\frac{\Delta}{h}y\right)^2 = \left(\frac{\Delta}{\alpha R}\right)^2, \qquad R^2 = \Delta^2 + h^2. \tag{7.352}$$

It is to be expected that the largest contribution to u_{zP} comes from that part of γ_3 which is nearest to a branch point. Let $(x_0, 0)$ be a branch point. Then, the shortest distance D (Fig. 7.47a) from $(x_0, 0)$ to a point (x_1, y_1) on γ_3 is given by the equation

$$x_1 - x_0 + y_1 y_1' = 0.$$

However,

$$x_1^2 - \left(\frac{\Delta}{h}\right)^2 y_1^2 = \left(\frac{\Delta}{\alpha R}\right)^2.$$

This yields

$$x_1 = \left(\frac{\Delta^2}{R^2}\right)x_0, \qquad y_1 = \frac{h}{R}\left(\frac{x_0^2\Delta^2}{R^2} - \frac{1}{\alpha^2}\right)^{1/2}, \qquad D = \frac{h}{R}(x_0^2 - \alpha^{-2})^{1/2}. \tag{7.353}$$

Therefore, y_1 will be real only if

$$\frac{\Delta^2}{R^2} > \frac{1}{(x_0\alpha)^2},$$

i.e.,

$$\Delta > \frac{h}{(x_0^2\alpha^2 - 1)^{1/2}}.$$

This condition is not satisfied by the branch point $(1/\alpha, 0)$. For the branch point $(1/\beta, 0)$, we must have

$$\Delta > \frac{h}{(\alpha^2/\beta^2 - 1)^{1/2}} = \Delta_c.$$

The associated travel time is

$$\tau = x_1\frac{R^2}{\Delta} = \frac{\Delta}{\beta}. \tag{7.354}$$

Hence, at epicentral distances such that $\Delta > \Delta_c$, the major contribution from the branch point at $v = 1/\beta$ arrives at time $\tau = \Delta/\beta$. Because

$$\frac{\Delta}{\beta} = \frac{(h^2 + \Delta_c^2)^{1/2}}{\alpha} + \frac{\Delta - \Delta_c}{\beta},$$

the time $\tau = \Delta/\beta$ is equivalent to the travel time of a P wave from the source up to the surface at $\Delta = \Delta_c$, followed by motion along the surface with the S-wave velocity. This peculiar wave is known as a *secondary P wave* and is designated as pS (Fig. 7.47b). It is basically a diffracted wave with the following characteristics:

1. Exists only for $\Delta > \Delta_c$.
2. No sharp commencement.
3. Arrival at non-least-time, $\tau = \Delta/\beta$.
4. Amplitude diminishes with increase of D and, therefore, with the increase of $h/R\beta$.

For a Poisson solid, $\Delta_c = h/\sqrt{2}$. At $h = 0$, both source and receiver are on the surface and $\tau = \Delta/\beta$ becomes a least-time path.

The treatment of the S part of the field is more complicated. Here, the existence and times of arrival of the various phases are governed by the equations

$$\tau = \Delta v + h\sqrt{\left(\frac{1}{\beta^2} - v^2\right)},$$

$$v(\tau) = \begin{cases} \tau\dfrac{\Delta}{R^2} - \dfrac{h}{R^2}\sqrt{\left(\dfrac{R^2}{\beta^2} - \tau^2\right)}, & \tau < \dfrac{R}{\beta}; \\[2ex] \tau\dfrac{\Delta}{R^2} + \dfrac{ih}{R^2}\sqrt{\left(\tau^2 - \dfrac{R^2}{\beta^2}\right)}, & \tau > \dfrac{R}{\beta}. \end{cases} \tag{7.355}$$

The surface displacement is given by

$$u_{zS}(t) = \frac{U_0\,dS}{2(\pi\Delta)^2}\sin 2\phi H(t - \tau_H)\operatorname{Im}\frac{d}{dt}\int_{\tau_H}^{t}\left[\frac{v(1/\beta^2 - v^2)\sqrt{(1/\alpha^2 - v^2)}}{\Delta_R}\right]$$
$$\times\left[\frac{v^2\Delta^2 + 2(t - \tau)(t - \tau + 2v\Delta)}{\sqrt{[(\tau + \tau_s)(t - \tau + 2v\Delta)]}}\right]\frac{d\tau}{\sqrt{[(t - \tau)(\tau_s - \tau)]}}. \tag{7.356}$$

Two distinct cases arise:

1. $\Delta < \Delta_c = h/(\alpha^2/\beta^2 - 1)^{1/2}$. The inequality $\Delta < h/\sqrt{(\alpha^2/\beta^2 - 1)}$ implies that

$$v_1 = \frac{1}{\beta}\frac{\Delta}{R} < \frac{1}{\alpha} < \frac{1}{\beta}. \tag{7.357}$$

The branch points $(1/\alpha, 0)$ and $(1/\beta, 0)$ yield no contribution. Using again the substitution $\sin q = \sqrt{[(t-\tau)/(t-\tau_s)]}$, the surface displacement, Eq. (7.356), becomes

$$u_{zS}(t) = \frac{U_0\,dS}{(\pi\Delta)^2}\sin 2\phi H(t-\tau_s)$$

$$\times \operatorname{Re}\frac{d}{dt}\int_0^{\pi/2}\left[\frac{(1/\beta^2 - v^2)\sqrt{(1/\alpha^2 - v^2)}v}{(1/2\beta^2 - v^2)^2 + v^2\sqrt{(1/\alpha^2 - v^2)}\sqrt{(1/\beta^2 - v^2)}}\right]$$

$$\times \left[\frac{v^2\Delta^2 + 2(t-\tau)(t-\tau+2v\Delta)}{\sqrt{[(\tau+\tau_s)(t-\tau+2v\Delta)]}}\right]dq, \qquad \left(\tau_s = \frac{R}{\beta}\right). \tag{7.358}$$

It is convenient to write Eqs. (7.347) and (7.358) in a unified form where the source terms are grouped together in a single symbol B_j

$$u_{z,j}(t) = \frac{U_0\,dS}{\pi^2}H(t-\tau_j)\operatorname{Re}\frac{d}{dt}\int_0^{\pi/2}\frac{B_j\,dq}{\Delta_R\sqrt{[(\tau+\tau_j)(t-\tau+2v\Delta)]}},$$

$$j = P, S. \tag{7.359}$$

The coefficient B_j is given in Table 7.22 for the three fundamental seismic sources and Δ_R is given by

$$\Delta_R(-v^2) = \left(\frac{1}{2\beta^2} - v^2\right)^2 + v^2\left(\frac{1}{\alpha^2} - v^2\right)^{1/2}\left(\frac{1}{\beta^2} - v^2\right)^{1/2}.$$

It may be noted that B_p is related to $\hat{\imath}_m$ and B_s is related to $\hat{\jmath}_m$ of Table 7.20.

2. $\Delta > \Delta_c$. The location of the singular point $v = v_1$ in this case is such that (Fig. 7.48a)

$$\frac{1}{\alpha} < v_1 < \frac{1}{\beta}. \tag{7.360}$$

Table 7.22. Source Term, B_j, for Vertical Surface Displacements from a Dislocation Source Buried in a Half-Space[a]

	Vertical strike slip	Vertical dip slip	Case III
P field (B_p)	$-\frac{1}{\Delta^2}va\Omega Q\sin 2\phi$	$-\frac{2}{\Delta}va^2\Omega T\sin\phi$	$\frac{3}{2}va\Omega\left(\frac{2}{3\alpha^2} - v^2\right) + \frac{1}{2\Delta^2}va\Omega Q\cos 2\phi$
S field (B_s)	$\frac{1}{\Delta^2}vab^2Q\sin 2\phi$	$\frac{2}{\Delta}vab\Omega T\sin\phi$	$\frac{3}{2}v^3ab^2 - \frac{1}{2\Delta^2}vab^2Q\cos 2\phi$

[a] $a = (1/\alpha^2 - v^2)^{1/2}$, $b = (1/\beta^2 - v^2)^{1/2}$, $\Omega = (1/2\beta^2) - v^2$, $\Delta_R = \Omega^2 + v^2ab$, $T = t - \tau + v\Delta$, and $Q = 2(t-\tau)(t-\tau+2v\Delta) + v^2\Delta^2$.

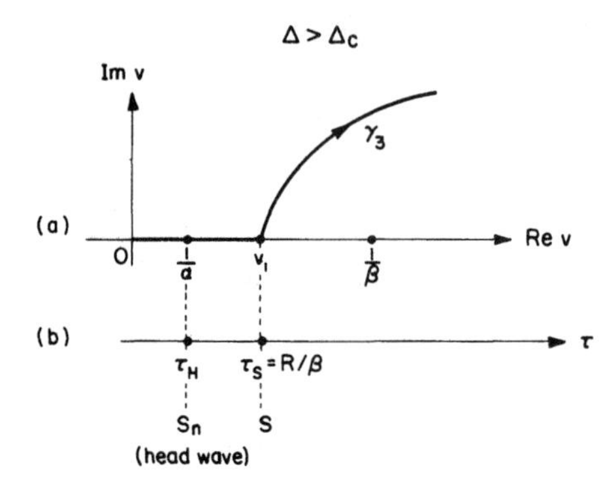

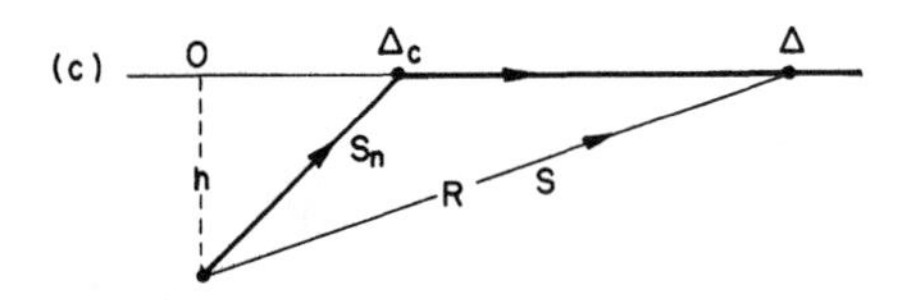

Figure 7.48. S head waves. (a) Integration path for u_{zS} when $z = 0$; (b) S_n and S arrival times; (c) Ray paths of S_n and direct S waves.

The travel time corresponding to the branch point at $v = 1/\alpha$ is

$$\tau_H = \frac{\Delta}{\alpha} + \frac{h}{\alpha}\sqrt{\left(\frac{\alpha^2}{\beta^2} - 1\right)}. \tag{7.361}$$

Simple algebraic operations reveal that $\tau_H < \tau_s$ and that τ_H is the time required for an emergent S wave to hit the surface at $\Delta = \Delta_c$ and continue from that point onward with the P wave velocity (Fig. 7.48b). A wave of this kind is known as a head wave. It is a least-time arrival and is denoted as S_n or the SP head wave.

The explicit expressions for the ground displacement in this case will depend on whether t is smaller, greater, or equal to τ_s. Starting with Eq. (7.356), we derive the following results.

a. $t < \tau_s$. The only singularity at $\tau = t$ can be eliminated by the change of variable $q = \sin^{-1}\sqrt{(\tau/t)}$, yielding

$$u_{zS}(t) = \frac{U_0\, dS}{\pi^2} H(t - \tau_H)\,\mathrm{Im}\,\frac{d}{dt}\int_{q_c}^{\pi/2} \frac{\sqrt{\tau}B_s\, dq}{\Delta_R\sqrt{[(\tau + \tau_s)(\tau_s - \tau)(t - \tau + 2v\Delta)]}}, \tag{7.362}$$

where $q_c = \sin^{-1}\sqrt{(\tau_H/t)}$.

b. $t > \tau_s$. The integration interval (τ_H, t) is broken into two subintervals (τ_H, τ_s) and (τ_s, t). The first integral, after being subjected to a change of variable $q = \sin^{-1}\sqrt{(\tau/\tau_s)}$, assumes the form

$$u_{zS}(t) = \frac{U_0\, dS}{\pi^2} \operatorname{Im} \frac{d}{dt} \int_{q_d}^{\pi/2} \frac{\sqrt{\tau} B_s\, dq}{\Delta_R \sqrt{[(\tau + \tau_s)(t - \tau)(t - \tau + 2v\Delta)]}},$$
$$q_d = \sin^{-1}\sqrt{\left(\frac{\tau_H}{\tau_s}\right)}. \tag{7.363}$$

The second integral is identical with the integral of case 1.

c. $t = \tau_s = R/\beta$. The integral can be evaluated approximately. In the vicinity of τ_s the dominant term becomes

$$u_{zS}(t) = \frac{-U_0\, dS}{4\pi^2} \operatorname{Im}\left\{\frac{B_s}{\Delta_R \sqrt{(v_1 \tau_s \Delta)}}\right\} \ln|\tau_s - t| \tag{7.364}$$

and the arrival of the S phase is marked by a logarithmic infinity. This term is absent for $h = 0$.

7.6.4.1. The Rayleigh Wave in the Time Domain. We know [Eq. (7.334)] that the vertical Rayleigh displacement resulting from a surface dislocation of the strike-slip type and a step-function time dependence is

$$u_z(t) = -\frac{U_0\, dS}{2\pi c_R^2} A \sin 2\phi \frac{d}{dt}\left\{\left[2\left(\frac{c_R t}{\Delta}\right)^2 - 1\right] \frac{H(t - \Delta/c_R)}{\sqrt{(t^2 - \Delta^2/c_R^2)}}\right\}. \tag{7.365}$$

Close to the arrival time of this wave $t \simeq \Delta/c_R$, we may write

$$\sqrt{\left(t^2 - \frac{\Delta^2}{c_R^2}\right)} \propto \sqrt{\Delta}\sqrt{(|\Delta - c_R t|)}. \tag{7.366}$$

Therefore, in the neighborhood of the Rayleigh wavefront, the wave amplitude varies with range as $\Delta^{-1/2}$, as a cylindrical wave should. In the limit $t \to \infty$, the Rayleigh wave amplitude merges into the steady-state solution

$$u_z(\infty) = -\frac{U_0\, dS}{\pi \Delta^2} A \sin 2\phi, \tag{7.367}$$

as can be easily verified by carrying out the differentiation in Eq. (7.365). It can be shown that at the time of arrival of the Rayleigh wave, the horizontal displacement is marked by an infinite discontinuity followed immediately by a steady state.

The exact form of $u_z(t)$ for source depths that differ from zero also can be evaluated in closed form. The result, however, is a rather lengthy algebraic expression. To simplify the analysis without losing too much accuracy, we use the asymptotic form of the Hankel function $H_2^{(2)}(x) = \sqrt{(2/\pi x)}e^{-ix + 5\pi i/4}$

which is accurate, in magnitude, to within 2% already at $x = 2\pi$. Therefore, for $\Delta \gg$ (Rayleigh wavelength), we may use Eq. (5.27) and find the time-domain solution by taking its inverse Fourier transform. For this purpose, we need Euler's formula

$$\int_0^\infty \omega^{n-1} e^{-p\omega} \cos(\omega\tau + \delta) d\omega = \Gamma(n)(p^2 + \tau^2)^{-n/2} \cos(n\chi + \delta), \quad (7.368)$$

where $\tan \chi = \tau/p$. Therefore, for a unit-step source, we find

$$u_z(t) = \frac{U_0\, dS}{2\pi} \sin 2\phi (2\hat{\gamma} h^3 \Delta)^{-1/2} \cdot \frac{2\hat{\gamma}^2 - 1}{G(\hat{\gamma})} \left[\cos^{3/2} \chi_1 \cos\left(\frac{3}{2}\chi_1 + \frac{5\pi}{4}\right) - \left(1 - \frac{c_R^2}{2\beta^2}\right) \cos^{3/2} \chi_2 \cos\left(\frac{3}{2}\chi_2 + \frac{5\pi}{4}\right)\right], \quad (7.369)$$

where

$$\tan \chi_1 = \frac{c_R t - \Delta}{h\gamma_\alpha}, \qquad \tan \chi_2 = \frac{c_R t - \Delta}{h\gamma_\beta},$$

$$\gamma_\alpha = \left(1 - \frac{c_R^2}{\alpha^2}\right)^{1/2}, \qquad \gamma_\beta = \left(1 - \frac{c_R^2}{\beta^2}\right)^{1/2},$$

$$\hat{\gamma} = \frac{\beta}{c_R}.$$

The function $G(\hat{\gamma})$ is defined in Eq. (5.26).

The horizontal component of the displacement is obtained through Eq. (5.37). We find

$$u_\Delta(t) = \frac{U_0\, dS}{2\pi} \sin 2\phi (2\hat{\gamma} h^3 \Delta)^{-1/2} \cdot \frac{2\hat{\gamma}^2 - 1}{G(\hat{\gamma})} \cdot \varepsilon(0) \left[-\cos^{3/2} \chi_1 \sin\left(\frac{3}{2}\chi_1 + \frac{5\pi}{4}\right) + \left(1 - \frac{c_R^2}{2\beta^2}\right) \cos^{3/2} \chi_2 \sin\left(\frac{3}{2}\chi_2 + \frac{5\pi}{4}\right)\right], \quad (7.370)$$

where

$$\varepsilon(0) = \frac{(1 - c_R^2/2\beta^2)}{(1 - c_R^2/\alpha^2)^{1/2}},$$

is the Rayleigh-wave surface ellipticity [Eq. (3.109)]. For a Poisson solid

$$\lambda = \mu, \qquad \varepsilon(0) = 0.6812,$$

$$\frac{1}{\hat{\gamma}} = 0.9195, \qquad \gamma_\alpha = 0.8475, \qquad \gamma_\beta = 0.3933.$$

The amplitude dependence upon source-depth is as $h^{-3/2}$. As $h \to 0$,

$$\frac{\cos \chi_1}{h} \to \frac{\gamma_\alpha}{c_R t - \Delta},$$

$$\frac{\cos \chi_2}{h} \to \frac{\gamma_\beta}{c_R t - \Delta},$$

and, therefore, $u_z(t)$ tends to an expression that is proportional to $\Delta^{-1/2}(c_R t - \Delta)^{-3/2}$, as it should.

7.6.4.2. The Horizontal Ground Motion. So far we have treated mainly the vertical ground motion. When the procedure is repeated for the horizontal displacement, it is found that the horizontal field is composed of two terms, one of which contains the derivative $(d/dt)Q_m$ and the other Q_m itself. It may be seen that the term with $(d/dt)Q_m$ dominates for large values of Δ. Hence, the term with $(d/dt)Q_m$ is identified as the far-field contribution, whereas the term with Q_m is essentially the near-field contribution.

When suitable values of the source term B_j are substituted from Tables 7.22 and 7.23, Table 7.24 gives the vertical as well as horizontal surface displacements. The derivative outside the sign of integration is not present in the case of horizontal displacements in the near field.

7.6.4.3. Convolution of Time Functions. Let us denote the ground displacement caused by the action of a point source having a unit-step time-dependence by $u(t)$. For a source with time dependence $M(t)$, the ground displacement $G(t)$ will be

$$G(t) = M(+0)u(t) + \int_{+0}^{t} M'(\tau)u(t-\tau)d\tau, \qquad (7.371)$$

where a prime denotes time derivation. We assume that the time history of the source has no discontinuities; i.e., $M(+0) = 0$. Therefore

$$G(t) = \int_{+0}^{t} M'(\tau)u(t-\tau)d\tau = M'(t) * u(t), \qquad (7.372)$$

where the asterisk denotes convolution.

Consider now a seismograph with response $I(t)$ to a $\delta(t)$ input ground displacement. The seismograph trace for a $G(t)$ ground displacement will then be

$$G(t) * I(t) = M'(t) * u(t) * I(t). \qquad (7.373)$$

Assuming [see Eq. (7.323)]

$$u(t) = \frac{d}{dt} Q(t), \qquad (7.374)$$

Eq. (7.373) becomes

$$\begin{aligned} M'(t) * Q'(t) * I(t) &= Q'(t) * [I(t) * M'(t)] \\ &= Q(t) * [I(t) * M'(t)]'. \end{aligned} \qquad (7.375)$$

The advantage of the last form in Eqs. (7.375) is obvious. $Q(t)$ is a numerically given function, which has finite jumps and, in certain cases, logarithmic singularities. The numerical differentiation of such a function is of somewhat doubtful numerical value. In contrast, $I(t)$ and $M(t)$ are given in analytical form, and its exact differentiation can be easily carried out. Furthermore, $Q(t)$ is ray dependent and its differentiation must be carried out anew for each ray. $I(t)$, $M(t)$ are ray independent and the function $[I(t) * M'(t)]'$ need be computed only once. Usually,

Table 7.23a. Source Term, B_j, for Horizontal Far-Field Surface Displacements from a Dislocation Source Buried in a Half-space[a]

	Vertical strike slip		Vertical dip slip		Case III	
	Radial	Azimuthal	Radial	Azimuthal	Radial	Azimuthal
P field (B_p)	$\frac{-1}{\Delta} v^3 abT \sin 2\phi$	0	$-2v^3a^2b \sin \phi$	0	$\frac{3}{2\Delta} v\left(\frac{2}{3\alpha^2} - v^2\right)abT + \frac{1}{2\Delta} v^3 abT \cos 2\phi$	0
S field (B_s)	$\frac{-1}{\Delta} vb^2\Omega T \sin 2\phi$	$-\frac{2}{\Delta} v\Delta_R T \cos 2\phi$	$-2vb\Omega^2 \sin \phi$	$-2vb\Delta_R \cos \phi$	$-\frac{1}{2\Delta} vb^2\Omega T(3 - \cos 2\phi)$	$-\frac{1}{\Delta} v\Delta_R T \sin 2\phi$

[a] For notation, see Table 7.22.

Table 7.23b. Source Term, B_j, for Horizontal Near-Field Surface Displacements from a Dislocation Source Buried in a Half-space[a]

	Vertical strike slip		Vertical dip slip		Case III	
	Radial	Azimuthal	Radial	Azimuthal	Radial	Azimuthal
P field (B_p)	$\frac{-2}{\Delta^3} vabQ \sin 2\phi$	$\frac{2}{\Delta^3} vabQ \cos 2\phi$	$-\frac{2}{\Delta^2} va^2bT \sin \phi$	$\frac{2}{\Delta^2} va^2bT \cos \phi$	$\frac{1}{\Delta^3} vabQ \cos 2\phi$	$\frac{1}{\Delta^3} vabQ \sin 2\phi$
S field (B_s)	$\frac{-2}{\Delta^3} v(\Omega - 2ab)Q \times \sin 2\phi$	$\frac{2}{\Delta^3} v(\Omega - 2ab)Q \times \cos 2\phi$	$\frac{2}{\Delta^2} vab^2T \sin \phi$	$\frac{-2}{\Delta^2} vab^2T \cos \phi$	$\frac{1}{\Delta^3} v(\Omega - 2ab)Q \times \cos 2\phi$	$\frac{1}{\Delta^3} v(\Omega - 2ab)Q \times \sin 2\phi$

[a] For notation, see Table 7.22.

Table 7.24. Theoretical Seismograms for a Buried Dislocation in a Homogeneous Half-Space

Field type	Displacement
P field	$\dfrac{U_0\,dS}{\pi^2} H(t-\tau_p)\mathrm{Re}\dfrac{d}{dt}\displaystyle\int_0^{\pi/2} \dfrac{B_p\,dq}{\Delta_R[(\tau_p+\tau)(t-\tau+2v\Delta)]^{1/2}}$
S field $\Delta < \Delta_c$	$\dfrac{U_0\,dS}{\pi^2} H(t-\tau_s)\mathrm{Re}\dfrac{d}{dt}\displaystyle\int_0^{\pi/2} \dfrac{B_s\,dq}{\Delta_R[(\tau_s+\tau)(t-\tau+2v\Delta)]^{1/2}}$
$\Delta > \Delta_c,\ \tau_s > t$	$\dfrac{U_0\,dS}{\pi^2} H(t-\tau_H)\mathrm{Im}\dfrac{d}{dt}\displaystyle\int_{q_c}^{\pi/2} \dfrac{B_s\sqrt{\tau}\,dq}{\Delta_R[(\tau_s-\tau)(\tau_s+\tau)(t-\tau+2v\Delta)]^{1/2}}$
$\Delta > \Delta_c,\ \tau_s < t$ (first integral only)	$\dfrac{U_0\,dS}{\pi^2} H(t-\tau_H)\mathrm{Im}\dfrac{d}{dt}\displaystyle\int_{q_d}^{\pi/2} \dfrac{B_s\sqrt{\tau}\,dq}{\Delta_R[(\tau_s+\tau)(t-\tau)(t-\tau+2v\Delta)]^{1/2}}$

$M(t)$ is approximated by a convenient analytic function having the properties of a realistic source time-function.

Theoretical seismograms for the vertical, radial, and azimuthal components of ground displacements generated by various shear dislocations are shown in Figs. 7.49–7.56. The seismograms were computed for a source with potency

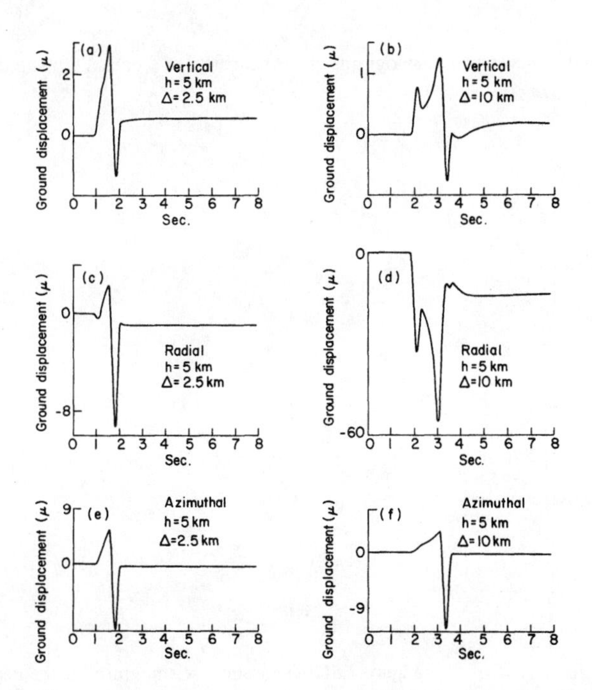

Figure 7.49a-f. Theoretical seismograms at the near field of a buried strike-slip dislocation.

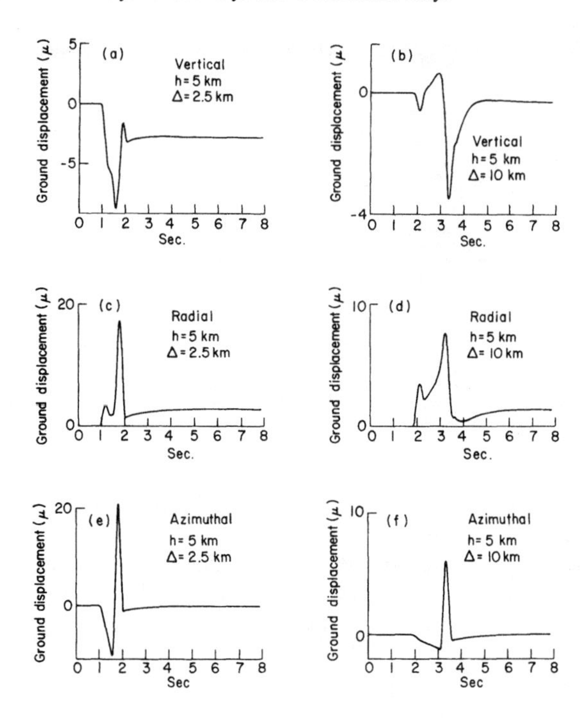

Figure 7.50a-f. Theoretical seismograms at the near field of a buried dip-slip dislocation.

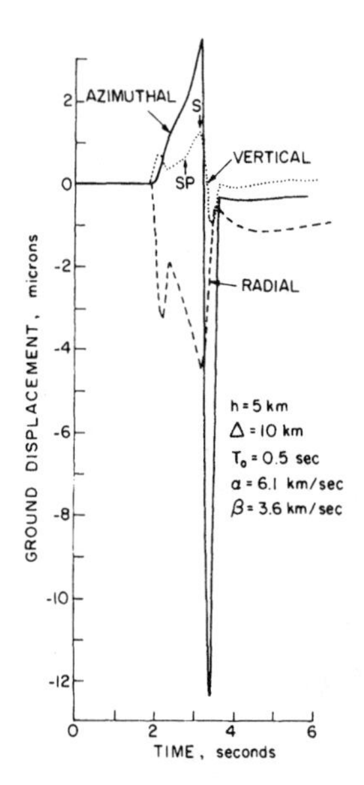

Figure 7.51. Theoretical seismograms at the near field of a strike-slip shear dislocation.

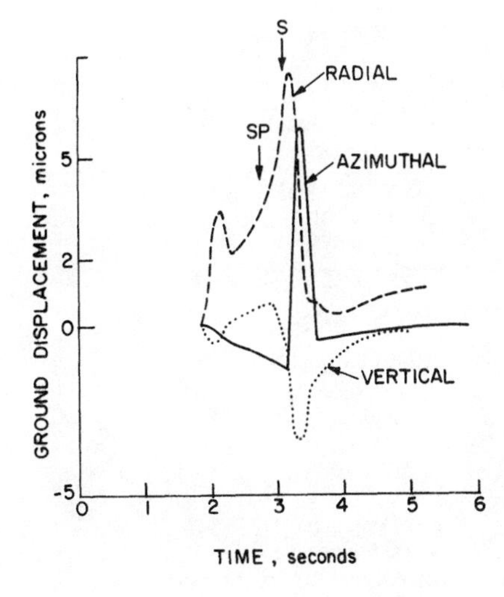

Figure 7.52. Theoretical seismograms at the near field of a dip-slip shear dislocation.

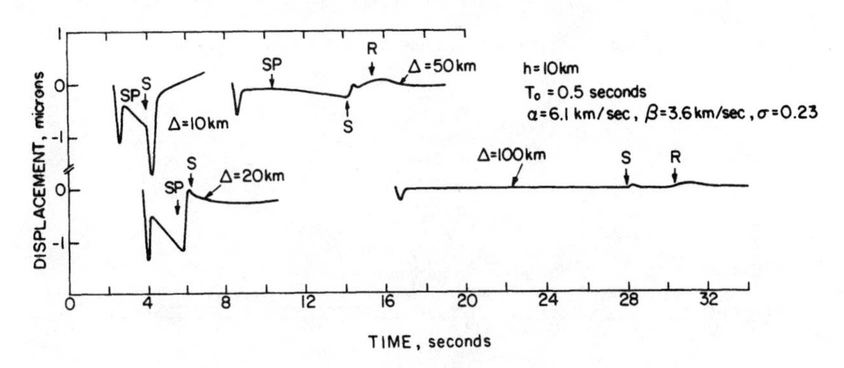

Figure 7.53. Radial displacement of a strike-slip source.

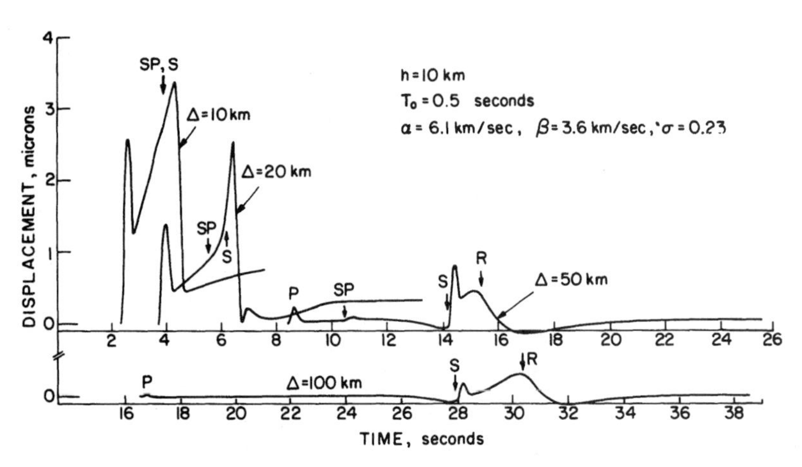

Figure 7.54. Radial displacement of a dip-slip source.

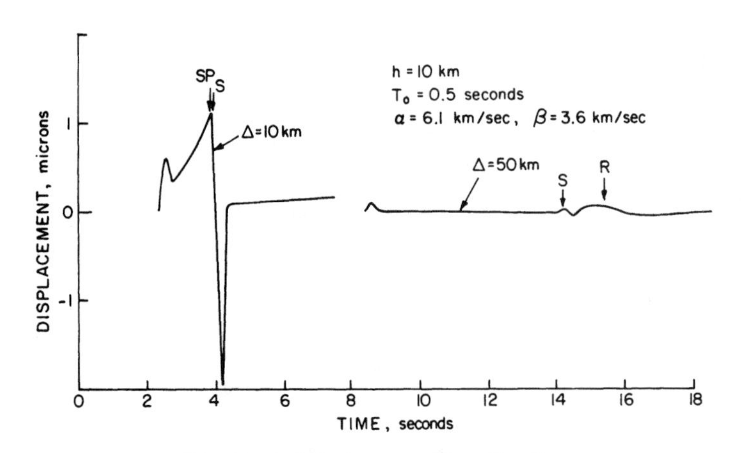

Figure 7.55. Vertical displacement of a strike-slip source.

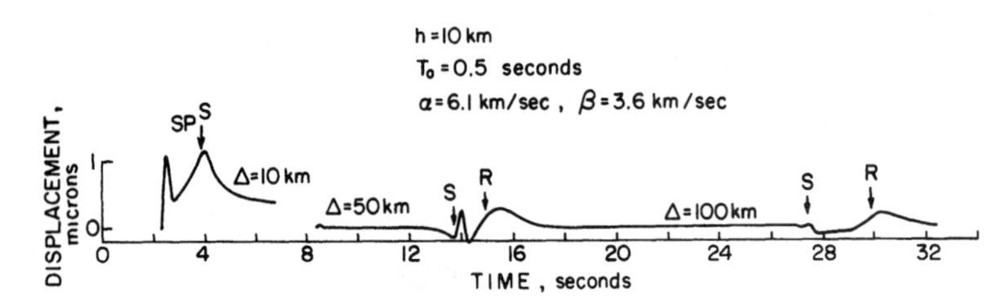

Figure 7.56. Vertical displacement of a dip-slip source.

$U_0\,dS = 10^{-6}\text{m} \times \text{km}^2$ and a source-time function of the form of a modulated ramp:

$$M(t) = \begin{cases} 0, & t < 0 \\ \dfrac{t}{T_0}\left(1 - \dfrac{\sin \omega_0 t}{\omega_0 t}\right), & 0 \le t \le T_0 \\ 1, & t > T_0 \qquad \left(\omega_0 = \dfrac{2\pi}{T_0}\right). \end{cases} \tag{7.376}$$

This time function is essentially a unit step with rounded shoulders. It was chosen to simplify the numerical procedure of time differentiation. It has the advantage of having continuous first and second derivatives.

The P and S wave arrivals can easily be recognized on the seismograms, whereas the SP head wave can only be identified by a change of slope. No identifiable Rayleigh wave exists at the ranges considered. As $t \to \infty$, all the displacement components tend to a limit, known as the *residual deformation.* At short distances this is achieved very soon after the arrival of the S wave.

EXAMPLE 7.7: The Exact Field at $\Delta = 0$

An explosion takes place at a depth h below the free surface of a homogeneous elastic half-space ($z \le 0$). The displacements in the time domain resulting from a unit-step function source are given by [cf. Eqs. (5.7), (5.9), and (5.12)]

$$u_z(0, z, t) = \frac{-S_0}{2\pi} \int_{-\infty}^{\infty} \frac{e^{i\omega t}}{i\omega}\,d\omega \int_0^{\infty} \left\{ -\varepsilon e^{-\nu_\alpha |z+h|} - \frac{R^+(k)}{R(k)} e^{-\nu_\alpha (h-z)} + \frac{L(k)}{R(k)} e^{-\nu_\alpha h + \nu_\beta z} \right\} k\,dk,$$

$$u_\Delta = u_\phi = 0, \qquad S_0 = \frac{3E_0}{16\pi\mu}, \qquad \varepsilon = \operatorname{sgn}(h+z), \qquad L(k) = 4k^2(2k^2 - k_\beta^2). \tag{7.7.1}$$

Each of the three integrals will be dealt with separately. The first represents the direct P wave. Because

$$-\varepsilon \int_0^{\infty} e^{-\nu_\alpha |z+h|} k\,dk = \frac{\partial}{\partial z} \int_0^{\infty} e^{-\nu_\alpha |z+h|} \frac{k\,dk}{\nu_\alpha} = \frac{\partial}{\partial z} \frac{e^{-ik_\alpha |z+h|}}{|z+h|},$$

we have

$$u_{z,1} = \frac{-S_0}{2\pi} \int_{-\infty}^{\infty} \frac{e^{i\omega t}}{i\omega} \frac{\partial}{\partial z} \frac{e^{-ik_\alpha |z+h|}}{|z+h|}\,d\omega = -S_0 \frac{\partial}{\partial z} \frac{H(t - (|z+h|/\alpha))}{|z+h|}$$

$$= S_0 \varepsilon \left[\frac{1}{\alpha} \frac{\delta(t - (|z+h|/\alpha))}{|z+h|} + \frac{H(t - (|z+h|/\alpha))}{|z+h|^2} \right]. \tag{7.7.2}$$

In the second integral (representing the reflected P wave) we change variable to $k = \omega p$ and then to $q = (h - z)\sqrt{(1/\alpha^2 - p^2)}$. This yields

$$
\begin{aligned}
u_{z,2} &= -\frac{S_0}{2\pi} \int_{-\infty}^{\infty} \frac{e^{i\omega t}}{i\omega} (i\omega)^2 \, d\omega \int_0^{\infty} \frac{R^+(p)}{R(p)} e^{-i\omega\sqrt{(1/\alpha^2 - p^2)}(h-z)} p \, dp \\
&= \frac{\partial}{\partial t}\left(\frac{S_0}{2\pi}\right) \int_{-\infty}^{\infty} e^{i\omega(t-q)} \, d\omega \int_{(h-z)/\alpha}^{0, +i\infty} \frac{1}{(h-z)} \left\{ \frac{R^+(p)}{R(p)} \sqrt{\left(\frac{1}{\alpha^2} - p^2\right)} \right\} dq \\
&= S_0 \frac{\partial}{\partial t} \int_{(h-z)/\alpha}^{0, +i\infty} \frac{\delta(t-q)}{h-z} \left\{ \frac{R^+(p)}{R(p)} \sqrt{\left(\frac{1}{\alpha^2} - p^2\right)} \right\} dq \\
&= S_0 \frac{\partial}{\partial t} \left[\frac{H[t - (h-z)/\alpha]}{h-z} \left\{ \frac{R^+(p)}{R(p)} \sqrt{\left(\frac{1}{\alpha^2} - p^2\right)} \right\}_{p=p(t)} \right],
\end{aligned}
\tag{7.7.3}
$$

where

$$
t = (h - z)\sqrt{\left(\frac{1}{\alpha^2} - p^2\right)}.
$$

In a similar manner, the substitutions $k = \omega p$ and $q = h\sqrt{(1/\alpha^2 - p^2)} - z\sqrt{(1/\beta^2 - p^2)}$ yield the reflected S field

$$
\begin{aligned}
u_{z,3} = -S_0 \frac{\partial}{\partial t} \Bigg[H\left(t - \frac{h}{\alpha} + \frac{z}{\beta}\right) \\
\times \left\{ \frac{L(p)}{R(p)} \frac{1}{h((1/\alpha^2) - p^2)^{-1/2} - z((1/\beta^2) - p^2)^{-1/2}} \right\}_{p=p(t)} \Bigg],
\end{aligned}
\tag{7.7.4}
$$

where p is given implicitly by

$$
t = h\sqrt{\left(\frac{1}{\alpha^2} - p^2\right)} - z\sqrt{\left(\frac{1}{\beta^2} - p^2\right)}.
$$

7.6.4.4. Multilayered Media. The evaluation of the integral in Eq. (7.311) for the multilayered case is more involved. There will obviously be more branch points than in the case of the homogeneous half-space and it is therefore convenient to introduce the following notation:

$$
\frac{1}{c_{\max}} = \min\left\{\frac{1}{c_j}\right\},
\tag{7.377}
$$

where c_j is a branch point of $g(-v^2)$, and

$$
\frac{1}{c_H} = \min\left\{\frac{1}{{}_f c_j}\right\} \leq \frac{1}{c_{\max}},
\tag{7.378}
$$

where ${}_f c_j$ is a branch point of $f(-v^2)$. The cuts $(-\infty, -1/c_H)$, $(1/c_H, \infty)$ along the real v axis and just below it will secure the singlevaluedness of the functions f and g. At this point the discussion follows closely the one just given for the homogeneous half-space. Let us first examine the significance of the points τ_H and τ_1 from a purely geometric viewpoint. From Eqs. (7.315)

$$\tau_H = \tau\left(\frac{1}{c_H}\right) = \frac{\Delta}{c_H} + \sum_j h_j\left[\frac{1}{c_j^2} - \frac{1}{c_H^2}\right]^{1/2} = \frac{\Delta}{c_H} + \sum_j \frac{h_j}{\cos\theta_j}\left[\frac{1}{c_j} - \frac{\sin\theta_j}{c_H}\right]$$
$$= \left[\frac{\Delta}{c_H} - \sum_j \frac{l_j \sin\theta_j}{c_H}\right] + \sum_j \frac{l_j}{c_j} = \frac{\Delta_H}{c_H} + \sum_j \frac{l_j}{c_j}, \tag{7.379}$$

where $\theta_j = \sin^{-1}(c_j/c_H) = \cos^{-1}(h_j/l_j)$, according to Snell's law, is the angle that the ray makes with the normal to the $(j-1)$ interface, h_j is the thickness of the jth layer and Δ_H is the ray path along the interface (Fig. 7.57). This ray corresponds to what is known as a head wave. Likewise, each branch point encountered prior to v_1 is associated with a head wave along a certain interface. The geometric reflected ray is associated with v_1, which is the root of $\partial\tau/\partial v = 0$. Indeed, differentiation of Eqs. (7.315) with respect to v yields

$$\Delta = v_1 \sum_j h_j\left[\frac{1}{c_j^2} - v_1^2\right]^{-1/2}. \tag{7.380}$$

Geometric considerations yield (Fig. 7.57),

$$\frac{\sin\phi_j}{c_j} = p \quad \text{(Snell's law)},$$

$$\Delta = \sum_j h_j \tan\phi_j = \sum_j h_j \frac{pc_j}{[1 - p^2c_j^2]^{1/2}} = p\sum_j h_j\left[\frac{1}{c_j^2} - p^2\right]^{-1/2}. \tag{7.381}$$

Hence by Eqs. (7.380) and (7.381)

$$v = \frac{\sin\phi_j}{c_j} = p = \text{ray parameter}. \tag{7.382}$$

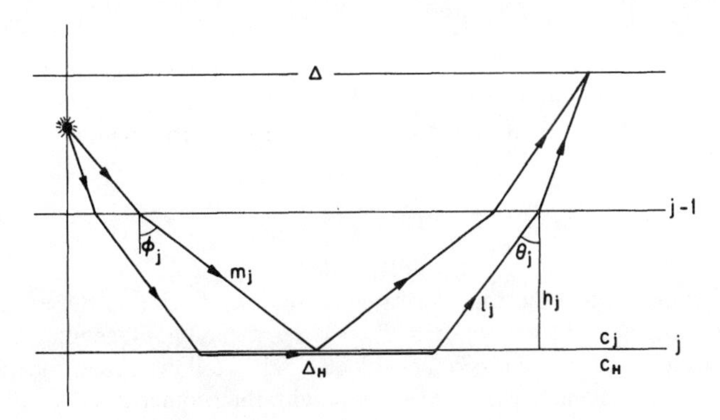

Figure 7.57. Ray-paths of reflected vs. head waves.

The travel time corresponding to v_1 is, therefore,

$$\begin{aligned}\tau(v_1) &= \Delta v_1 + \sum_j h_j\left(\frac{1}{c_j^2} - v_1^2\right)^{1/2} = \Delta v_1 + \sum_j \frac{h_j}{c_j}\cos\phi_j \\ &= \Delta v_1 + \sum_j \frac{h_j}{c_j\cos\phi_j}(1 - \sin^2\phi_j) \\ &= \left[\Delta v_1 - \sum_j \frac{\sin\phi_j}{c_j}(h_j\tan\phi_j)\right] + \sum_j \frac{h_j}{c_j\cos\phi_j} \\ &= \sum_j \frac{m_j}{c_j} = \tau_{\text{Reflected}} = \tau_1,\end{aligned} \tag{7.383}$$

which is recognized as the travel time of the geometric reflected ray (Fig. 7.57). A similar analysis holds for the transmitted ray. Therefore, as t changes from τ_H onward, the integral in Eq. (7.325) will contain the contributions from the head and reflected (transmitted) waves. We shall use the term "geometric wave" for brevity, meaning either a reflected or a transmitted wave.

Next, we recast Eq. (7.325) in the form

$$\begin{aligned}Q_m(t) &= H(t-\tau_1)\operatorname{Im} S^0_{\mu\nu}\int_{\tau_1}^{t} v^{2l+m+1}D(-v^2)\frac{\partial v}{\partial\tau}\frac{T_m(\zeta)}{[(t-\tau)(t-\tau+2v\Delta)]^{1/2}}\,d\tau \\ &\quad + H(t-\tau_1)\operatorname{Im}\int_{\tau_H}^{\tau_1} v^{2l+m+1}D(-v^2)S_{\mu\nu}\frac{\partial v}{\partial\tau} \\ &\quad \times \frac{T_m(\zeta)}{[(t-\tau)(t-\tau+2v\Delta)]^{1/2}}\,d\tau \\ &\quad + H(t-\tau_1)\operatorname{Im}\int_{\tau_1}^{t} v^{2l+m+1}D(-v^2)[S_{\mu\nu} - S^0_{\mu\nu}]\frac{\partial v}{\partial\tau} \\ &\quad \times \frac{T_m(\zeta)}{[(t-\tau)(t-\tau+2v\Delta)]^{1/2}}\,d\tau \\ &= I_1 + I_2 + I_3,\end{aligned} \tag{7.384}$$

where $\zeta = (1/v\Delta)(t-\tau) + 1$. The function $S_{\mu\nu}$ is a product of the generalized reflection or transmission coefficients, where S stands for R or T and the indices μ and ν stand for P or S. The function $S^0_{\mu\nu}$ is the corresponding coefficient at $v = v_1$:

$$S_{\mu\nu}(v_1) = S^0_{\mu\nu}. \tag{7.385}$$

Note that, for a given wave, $S^0_{\mu\nu}$ is independent of ω and k.

It is important to note that the same v_1 that renders $\tau(v_1)$ the meaning of geometric ray travel-time renders $S_{\mu\nu}(v_1)$ the meaning of geometric plane-wave reflection or transmission coefficient. This follows from the physical meaning of v_1 and enables us to identify I_1 with the geometric wave.

We now discuss each of the terms I_1, I_2, I_3 appearing in Eq. (7.384). I_1 is simply the geometric ray contribution, beginning at τ_1.

The main contribution to I_2 is the sum of all possible head waves that arrive before the reflected pulse. In the v plane, the path of integration lies along the positive real axis, above the branch cut. Whenever the path passes near a branch point, the integral has a contribution that, by its τ, is identified with a head wave. In many cases of interest, there is no branch point before v_1. In these cases, I_2 is identified with the P head wave.

The integral I_3 has no such simple physical meaning. Along its path of integration in the v plane, no branch point is encountered. Moreover, for a buried source, there is only one apparent singularity of the integrand, namely, at τ_1. However, because $S_{\mu\nu}(v_1) = S^0_{\mu\nu}$, and $\tau_1 = \tau(v_1)$, the contribution of I_3 to the reflected wave is negligible compared with that of I_1. Because I_3 contains no head-wave arrival prior to τ_1, we conclude that I_3 has no geometric ray interpretation. In fact, I_3 yields contributions of non-least-time arrivals. Such arrivals are not contained in either I_1 or I_2. The secondary P wave encountered in the case of a half-space is an example of the contribution of I_3.

The theory of generalized rays gives a mathematical expression that can be numerically evaluated for the ground displacement inside any layer of the medium caused by a single generalized ray that is emitted by a source with unit-step time dependence. Observed seismograms, however, consist of the instrument response to the surface motion caused by all rays arriving to the observation point. These rays come from a source that has a highly complicated time history and are attenuated on their way to the receiver. To take these phenomena into account, one must devise a suitable method of generating, enumerating, and classifying all rays. The effect of the source-time function and the instrument response also should be taken into account.

When only a few isolated observed arrivals are under investigation, we model every observed phase by all the generalized rays in a layered structure that meet the following conditions: (1) The generalized ray must arrive inside the time interval wherein the observed phase arrives. (2) Both the generalized ray and the observed phase must have undergone the same number of reflections. The bending of observed phases at their deepest point of penetration of the earth is counted as a reflection. (3) The number of reflections at the free surface must be the same for both the theoretical generalized ray and the observed phase.

The horizontal displacements from multipolar sources in the presence of a single plane boundary consist of P, SV, and SH contributions. Even when the horizontal displacement is written in terms of components in radial and azimuthal directions, the existence of contributions of all three field components persists. This seems contrary to the generally accepted notion that the azimuthal field is purely SH, whereas the radial field is purely of the SV and P type. However, it can be shown that, for large epicentral distances, the SV and SH displacements are perpendicular to each other such that the azimuthal displacement is purely SH, whereas the radial displacement is purely SV. This result leads to significant reduction in the calculation of horizontal components.

7.6.4.5. The Major Contributions Approximation. Although the decomposition in Eq. (7.384) renders physical meaning to the various terms, it is not useful for numerical work. We therefore introduce another decomposition, which is more suitable for computations.

Let Λ be the set of the branch points of $f(-v^2)$ and let Ω be a subset of Λ. For the present discussion, it is necessary to write $f(-v^2)$ in the form

$$f(-v^2) = D(-v^2)\prod_j {}_jS_{\mu\nu}. \tag{7.386}$$

Let J be the set of indices over which the multiplication in Eq. (7.386) is performed. We decompose J into two distinct subsets, J_1 and J_2, such that $J_1 \cap J_2 = \phi$ and $J_1 \cup J_2 = J$, and such that for $j \in J_1, {}_jS_{\mu\nu}(v)$ has at least one element of Ω as a branch point, whereas for $j \in J_2, {}_jS_{\mu\nu}(v)$ has no element of Ω as a branch point (each of the sets may be empty). We may write

$$\prod_J {}_jS_{\mu\nu}(v) = \prod_{J_1} {}_jS_{\mu\nu}(v)\prod_{J_2} {}_jS_{\mu\nu}(v). \tag{7.387}$$

Equation (7.384) therefore may be recast in the form

$$\begin{aligned} Q_m(t) &= I_2 + H(t-\tau_1)\mathrm{Im}\int_{\tau_1}^{t} A(v)\prod_{J_1} {}_jS_{\mu\nu}(v)\prod_{J_2} {}_jS_{\mu\nu}(v_1)\,d\tau \\ &\quad + H(t-\tau_1)\mathrm{Im}\int_{\tau_1}^{t} A(v)\prod_{J_1} {}_jS_{\mu\nu}(v)\left[\prod_{J_2} {}_jS_{\mu\nu}(v) - \prod_{J_2} {}_jS_{\mu\nu}(v_1)\right]d\tau \\ &= I_2 + I_1^{\Omega} + I_3^{\Omega}, \end{aligned} \tag{7.388}$$

where

$$A(v) = D(-v^2)v^{2l+m+1}\frac{\partial v}{\partial \tau}\frac{T_m(\zeta)}{[(t-\tau)(t-\tau+2v\Delta)]^{1/2}}.$$

This decomposition is exact. However, I_1^{Ω} does not retain the meaning of a geometric ray contribution. The decomposition in Eq. (7.388) is more general, in the sense that when Ω degenerates and is empty ($\Omega = \phi$), Eq. (7.388) degenerates into Eq. (7.384).

I_2 retains its form and meaning and I_3^{Ω} contains mainly non-least-time contributions, inasmuch as at $\tau = \tau_1$ ($v = v_1$) the integrand of I_3^{Ω} vanishes. This means that the contribution of I_3^{Ω} to the reflected wave will, in general, be small compared with that of I_1^{Ω}. Obviously I_3^{Ω} does not contribute to the head wave. I_1^{Ω}, in contrast, contains contributions from the geometric reflected ray as well as contributions from those branch points that are elements of Ω. It will be shown how a judicious choice of the elements of Ω may yield an approximation to $Q_m(t)$. This method is acceptable even when geometric ray theory is unacceptable.

To this end, we discuss shortly the limitations of geometric ray theory from the point of view of our method. It is well known that ray theory is unapplicable at ranges near the critical distance. Synthetic seismograms computed by geometric ray methods are incomplete near the critical distance, and other methods must be used to circumvent the difficulty.

Moreover, given a range that is large enough and a model containing thin enough layers, a ray may be found whose critical distance is near the given range. For this reason all synthetic seismograms built by ray methods alone are in error. It is true that in many cases the amplitude of this ray is quite negligible, but a different method must be used to show that such is indeed the case. At the critical distance the travel times of the head wave and the reflected wave are equal and the two waves interfere with each other, yielding a wave that cannot be described by the first approximation of the ray method. In our formulation, the interference of the two waves at the critical distance is equivalent to saying that v_1 equals $v_H = 1/c_H$. Because v_1 is a monotone increasing function of Δ, the statement that v_1 is near v_H is equivalent to the statement that Δ is near the critical distance. For v_1 near enough to v_H, ray theory is therefore unapplicable. The term "enough" is just a qualitative description. A quantitative measure is obtained by solving $v_1 = v(\tau_1)$, $v_H = v(\tau_H)$ and finding the appropriate δv interval from the $\delta\tau$ interval. This, in turn, depends both on the source's time dependence and on the time constants of the recording instrument.

Another handicap of ray theory may arise for large Δ even when curvature corrections are made, because v_1 then moves near to $1/c_{\max}$. This may be seen from the relations

$$v_1 < \frac{1}{c_{\max}}, \qquad \lim_{\Delta\to\infty} v_1 = \frac{1}{c_{\max}}. \tag{7.389}$$

Disregarding the interference resulting from the proximity of v_1 to $1/c_{\max}$ may lead to serious errors. Further, other branch points of the integrand in Eq. (7.384), related to other layers, may be near v_1 and so contribute to the wave.

When interference does occur, the effect may be even more pronounced. In the geometric ray method, the relative magnitudes of the different contributions are not known and only the contribution of v_1 is retained. The use of the saddle point approximation to evaluate integrals of the form in Eq. (7.311) may be in error, unless attention is paid to branch points near the saddle point.

A practical criterion for determining the applicability of ray theory is obtained by computing v_1 and the elements of Ω. If v_1 is far enough from any λ ($\lambda \in \Lambda$), ray theory is applicable. In that case, we use the decomposition given in Eq. (7.384) and neglect I_3 in the computation. If, however, v_1 is too close to λ ($\lambda \in \Lambda$), we insert λ into Ω and use the decomposition, Eq. (7.388), neglecting I_3^{Ω} with the same results as above.

The essence of the method outlined above is to incorporate into the calculations only the contributions of those singular and branch points which are expected to substantially affect the amplitudes and waveforms in the seismogram. Hence we term this method the major contributions approximation.

An important feature of this procedure is that the omission of I_3^{Ω} does not change the form of I_1^{Ω}. Other approximations (e.g., the initial motion approximation) may be applied to the simplified integral instead of the whole integral.

7.6.5. Applications

Once a satisfactory method of generating theoretical seismograms is available, one may study the source mechanism of seismic events by comparing the observed seismograms with their theoretical counterparts. Whenever agreement between both sets of seismograms is acceptable, the parameters of the calculated seismograms are necessarily regarded as representative of the parameters of the real event.

In matching the observed and synthetic seismograms, we usually have several degrees of freedom: the upper mantle structure and the source parameters. The constraints, however, are also numerous. We require the agreement of dominant body-wave peak amplitudes, the arrival times of these peaks, and their pulse shapes. A solution, if found, satisfying all these constraints is acceptable for our purposes, although not necessarily unique.

The following procedure is quite satisfactory for most purposes. As a starting point, a coarse upper-mantle model is assumed. We then compute the fundamental seismograms resulting from strike-slip, dip-slip, and Case III sources, for a given station. The computer combines these fundamental seismograms by means of loops over different values of λ, δ, and azimuth, thus generating compound theoretical seismograms. The observed and theoretical seismograms are then compared, and a search is made for λ, δ, and azimuth that render a best fit to the observed waveforms.

A comparison between the observed and calculated seismograms for the seismic event associated with the Siberian meteor of 30 June, 1908 is shown in Fig. 7.58. Similarly, Fig. 7.59 compares the observed with the calculated seismogram of the Nile earthquake of 29 April, 1974. Figure 7.60 shows the observed

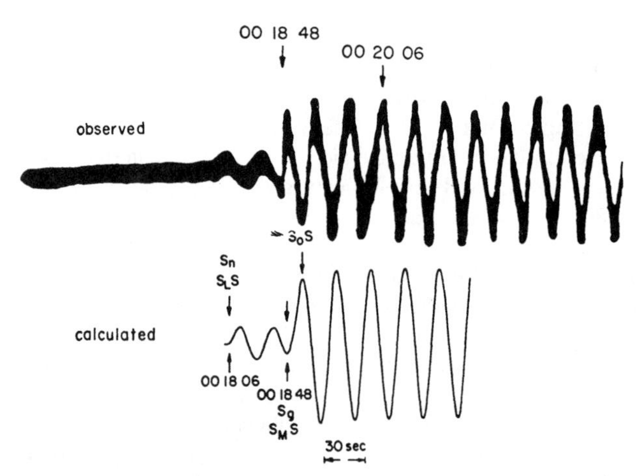

Figure 7.58. Observed vs. calculated seismograms of the *SH* ground motion at Irkutsk resulting from a ballistic wave accompanying the Siberian cometary explosion on 30 June, 1908, 00 14 28 GMT.

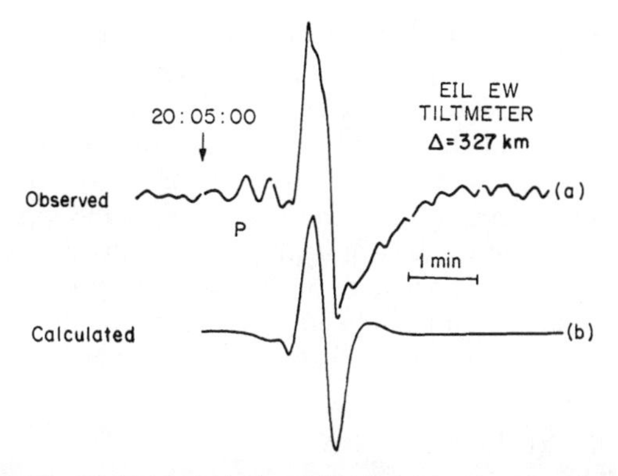

Figure 7.59. Observed (a) vs. calculated (b) tiltmeter seismograms at Eilat, Israel from an earthquake in the Nile Delta in Egypt, 29 April, 1974.

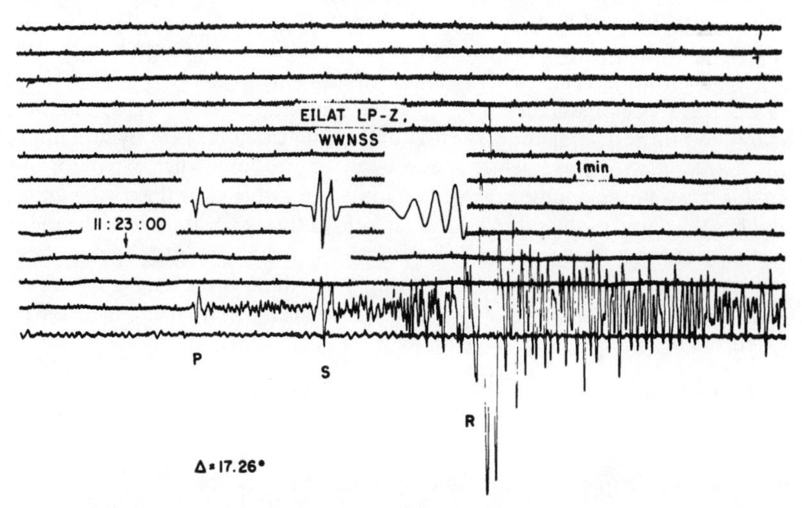

Figure 7.60. Calculated vs. observed *P*, *S*, and Rayleigh waves from an Iranian earthquake recorded at Eilat, Israel, 23 February, 1970.

seismogram of an Iranian earthquake. Because the source parameters of this shock together with the crustal and upper-mantle structure along the path to Eilat were known, theoretical seismograms of the *P* and *S* arrivals were calculated. The long-period portion of the Rayleigh wave was also calculated.

7.7. Spectral Asymptotic Approximations

The spectral amplitudes of body waves from point sources have been discussed twice in this chapter so far. The far-field spectral displacements of *P*, *SV*, and *SH* waves from deep sources were dealt with in Section 7.2.4. The radiation patterns of the source, the attenuation, the sphericity, and the finiteness effects were included. The crustal structure at the receiver was taken locally into account on the assumption that the incoming waves were plane. The crustal structure at the source was ignored and only direct arrivals were considered. This theory was generalized to core and surface reflections in Section 7.2.5. However, the assumption of an isolated signal was not removed and hence the applicability of this theory is limited to sources that are sufficiently removed from the free surface of the earth. In this section we conclude the treatment of the spectral amplitudes of body waves by taking into account the effect of the free surface on the radiation patterns of body waves, both at the source and at the station.

7.7.1. Shallow and Surface Sources

The study of surface and shallow sources became important in connection with nuclear explosion seismology. In Section 7.2.4, we have studied the radiation patterns of body waves from sources in infinite media. This assumption is justified only when the source is deep and the direct wave can be separated from the surface reflections. It is useful to use generalized ray theory (GRT) to obtain approximations for body-wave amplitudes valid for surface and shallow sources. The method consists of forming the total field by adding the direct and reflected wave fields. The resultant expression is then expanded in a Taylor series in the nondimensional quantity (kh), where k is the wave number and h is the depth of the source. The first term in the expansion yields the radiation pattern for surface sources. The second term in the series may be considered a correction to be added to the first term, for small nonvanishing h, and so on.

Consider, for example, the vertical displacement component of a buried strike-slip source. From Eq. (7.312), we have

$$\begin{aligned} u_{z(d>h)} = {} & J_2(su\Delta)u^3 s \sin 2\phi\{[e^{-sa(d-h)} - R^z_{PP}e^{-sa(d+h)} \\ & + R^z_{SP}e^{-s(ad+bh)}] - [e^{-sb(d-h)} - R^z_{SS}e^{-sb(d+h)} + R^z_{PS}e^{-s(bd+ah)}]\}, \end{aligned} \tag{7.390}$$

where we have suppressed, for the time being, the operator

$$\frac{\beta^2 U_0\, dS}{4\pi}\int_0^\infty \cdots du.$$

Expanding in sah or sbh, we have

$$\begin{aligned} u_{z(d>h)} = {} & J_2(su\Delta)u^3 s \sin 2\phi\left\{e^{-sad}\left[\left\{1 + sah + \frac{1}{2}(sah)^2 + \cdots\right\}\right.\right. \\ & - R^z_{PP}\left\{1 - sah + \frac{1}{2}(sah)^2 - \cdots\right\} + \left.R^z_{SP}\left\{1 - sbh + \frac{1}{2}(sbh)^2 - \cdots\right\}\right] \\ & - e^{-sbd}\left[\left\{1 + sbh + \frac{1}{2}(sbh)^2 + \cdots\right\} - R^z_{SS}\left\{1 - sbh + \frac{1}{2}(sbh)^2 - \cdots\right\}\right. \\ & \left.\left. + R^z_{PS}\left\{1 - sah + \frac{1}{2}(sah)^2 - \cdots\right\}\right]\right\}, \end{aligned} \tag{7.391}$$

which can be rearranged as

$$u_{z(d>h)} = J_2(su\Delta)u^3 s \sin 2\phi\Big\{e^{-sad}\Big[(1 - R^z_{PP} + R^z_{SP}) + sah\Big(1 + R^z_{PP} - \frac{b}{a} R^z_{SP}\Big)$$

$$+ \frac{1}{2}(sah)^2\Big(1 - R^z_{PP} + \frac{b^2}{a^2} R^z_{SP}\Big) + \cdots\Big]$$

$$- e^{-sbd}\Big[(1 - R^z_{SS} + R^z_{PS}) + sbh\Big(1 + R^z_{SS} - \frac{a}{b} R^z_{PS}\Big)$$

$$+ \frac{1}{2}(sbh)^2\Big(1 - R^z_{SS} + \frac{a^2}{b^2} R^z_{PS}\Big) + \cdots\Big]\Big\}. \qquad (7.392)$$

The P field is given by the terms containing e^{-sad}, whereas the S field is given by the terms containing e^{-sbd}.

Considering now only the P field and reinstating the operator $\int_0^\infty \cdots du$, we have, for the limiting case $h \to 0$ (zero approximation for small h),

$$U^0_{zP} = s \sin 2\phi \int_0^\infty J_2(su\Delta)u^3 e^{-sad}(1 - R^z_{PP} + R^z_{SP})du. \qquad (7.393)$$

A first-order approximation for small h is

$$U^1_{zP} = U^0_{zP} + s^2(\sin 2\phi)h \int_0^\infty J_2(su\Delta)u^3 e^{-sad} a\Big(1 + R^z_{PP} - \frac{b}{a} R^z_{SP}\Big)du, \qquad (7.394)$$

and a second-order approximation is

$$U^2_{zP} = U^1_{zP} + \frac{1}{2} s^3(\sin 2\phi)h^2 \int_0^\infty J_2(su\Delta)u^3 e^{-sad} a^2\Big(1 - R^z_{PP} - \frac{b^2}{a^2} R^z_{SP}\Big)du. \qquad (7.395)$$

Higher order corrections may be obtained along the same lines.

A comparison of Eq. (7.393) with the expressions for the displacement generated by the direct field of a buried strike-slip source from Eq. (7.390) reveals that the P surface source term introduced by the free surface is

$$D(u^2) = 1 - R^z_{PP} + R^z_{SP} = \frac{1}{\beta^2}\frac{ab}{\Delta_R}. \qquad (7.396)$$

This result can be generalized immediately. By putting $h = 0$ in Eqs. (7.305) it becomes clear that the zero-order approximation of the effect of the free surface may be expressed in terms of a source term. This essentially takes account of the surface reflections. We therefore find that the surface source term introduced by the free surface is given by the following equations:

$$\text{Vertical } P: \quad D(u^2) = 1 + (-1)^{m+1}\left[R_{PP}^{z} - \frac{u\hat{\jmath}_m}{a\hat{\imath}_m} R_{SP}^{z}\right];$$

$$\text{Vertical } SV: \quad D(u^2) = 1 + (-1)^{m+1}\left[R_{SS}^{z} - \frac{a\hat{\imath}_m}{u\hat{\jmath}_m} R_{PS}^{z}\right];$$

$$\text{Horizontal } P: \quad D(u^2) = 1 + (-1)^{m}\left[R_{PP}^{\Delta} - \frac{b\hat{\jmath}_m}{u\hat{\imath}_m} R_{SP}^{\Delta}\right]; \tag{7.397}$$

$$\text{Horizontal } SV: \quad D(u^2) = 1 + (-1)^{m}\left[R_{SS}^{\Delta} - \frac{u\hat{\imath}_m}{b\hat{\jmath}_m} R_{PS}^{\Delta}\right];$$

$$\text{Horizontal } SH: \quad D(u^2) = 1 + (-1)^{m} R_{SS}^{H}.$$

The expressions for R_{PP}^{z}, etc., are given in Eqs. (7.303) and $\hat{\imath}_m$, etc., in Table 7.20.

In obtaining Eqs. (7.397) we have assumed that, for a given m, either both $\hat{\imath}_m$ and $\hat{\jmath}_m$ are zero or both are nonzero. If this is not true, these equations must be modified. For example, in the case of a center of compression, $\hat{\imath}_0 \neq 0$ and $\hat{\jmath}_0 = 0$. When $h = 0$, Eqs. (7.305) yield

$$U_P = a\hat{\imath}_0\{(1 - R_{PP}^{z})e^{-sad} - R_{PS}^{z} e^{-sbd}\},$$

$$U_B = u\hat{\imath}_0\{(1 + R_{PP}^{\Delta})e^{-sad} + R_{PS}^{\Delta}\, e^{-sbd}\}.$$

These relations show that even though a center of compression in an infinite medium does not generate SV waves, the same source on the free surface of a uniform half-space does yield SV waves. The source term $D(u^2)$ introduced by the free surface is $(1 - R_{PP}^{z})$ for P waves. Because SV waves are absent before the introduction of the boundary, we cannot specify $D(u^2)$ for SV waves. The SV field can be obtained by recognizing that for $d = 0$, the ratio of the integrands of the vertical SV field to the corresponding P field is

$$\frac{-R_{PS}^{z}}{1 - R_{PP}^{z}} = -\frac{\Omega}{ab}.$$

Similarly, the ratio of the integrands of the horizontal SV field to the corresponding P field is

$$\frac{R_{PS}^{\Delta}}{1 + R_{PP}^{\Delta}} = -\frac{\Omega}{u^2}.$$

The explicit expressions for $D(u^2)$ are given in Table 7.25. These are valid for both horizontal and vertical components of the displacement. We note that the field induced by a surface vertical dip-slip source is identically zero.

Table 7.25. Surface Source-Factor, $D(u^2)$, Resulting from the Free Surface, for a Buried Point Source in a Homogeneous Half-space[a]

Source	m	P	SV	SH	Common factor
Vertical strike slip	2	ab	Ω	2	$\frac{1}{\beta^2\Delta_R}$ (it is 1 for SH)
Vertical dip slip	1	0	0	0	
Case III	0	$abu^2\left(\frac{3-4\beta^2/\alpha^2}{3u^2+2/\alpha^2}\right)$	$\left(1-\frac{4\beta^2}{3\alpha^2}\right)\Omega$	2	
	2	ab	Ω	2	
Center of compression	0	$-2u^2ab$	not defined	—	$\frac{1}{\Delta_R}$

[a] $a = (u^2 + 1/\alpha^2)^{1/2}$, $b = (u^2 + 1/\beta^2)^{1/2}$, $\Omega = u^2 + 1/(2\beta^2)$, $\Delta_R = \Omega^2 - u^2ab$.

Therefore, for such a source at a shallow depth, we cannot use the zero-order approximation ($h = 0$) and we must consider the first-order approximation corresponding to Eq. (7.394). In the case of a shallow vertical dip-slip source, the displacement field is proportional to h. In order to determine the radiation patterns, we must find the limit of the displacements divided by h as $h \to 0$. We make the usual change of variables $u = -iv$ and recall that, for geometric waves, $v = v_1$ = ray parameter $= \sin e/\alpha_h = \sin f/\beta_h$, where e and f are the angles that the P and S waves, respectively, make with the normal to the free surface and α_h and β_h are the corresponding wave velocities at the source. In applications, however, it is more convenient to express the results in terms of the takeoff angle i_h (Fig. 7.4). For a surface source, this angle is obtuse. Therefore, for P waves, putting $e = \pi - i_h$, we get

$$u^2 = -\frac{\sin^2 i_h}{\alpha_h^2}, \qquad a = \frac{1}{\alpha_h}|\cos i_h|, \qquad b = \frac{1}{\beta_h}(1 - \gamma_h^2 \sin^2 i_h)^{1/2},$$

$$\Omega = \frac{1}{2\beta_h^2}(1 - 2\gamma_h^2 \sin^2 i_h), \qquad \Delta_R = \frac{1}{4\alpha_h^4}\Delta_P^h, \tag{7.398}$$

where

$$\gamma_h = \frac{\beta_h}{\alpha_h},$$

$$\Delta_P^h = (\gamma_h^{-2} - 2\sin^2 i_h)^2 + 4\sin^2 i_h |\cos i_h| (\gamma_h^{-2} - \sin^2 i_h)^{1/2}.$$

In the case of SV waves, in contrast, we denote the takeoff angle by j_h, put $f = \pi - j_h$, and get

$$u^2 = -\frac{\sin^2 j_h}{\beta_h^2}, \qquad b = \frac{1}{\beta_h}|\cos j_h|,$$

$$a = \begin{cases} \dfrac{1}{\beta_h}(\gamma_h^2 - \sin^2 j_h)^{1/2}; & \sin j_h < \gamma_h \\[2ex] -\dfrac{i}{\beta_h}(\sin^2 j_h - \gamma_h^2)^{1/2}; & \sin j_h > \gamma_h, \end{cases}$$

$$\Omega = \frac{1}{2\beta_h^2}\cos 2j_h, \tag{7.399}$$

$$\Delta_R = \begin{cases} \dfrac{1}{4\beta_h^4}\Delta_S^h; & \sin j_h < \gamma_h \\[2ex] \dfrac{1}{4\beta_h^4}[\cos^2 2j_h - 4i\sin^2 j_h|\cos j_h|(\sin^2 j_h - \gamma_h^2)^{1/2}]; & \sin j_h > \gamma_h, \end{cases}$$

where

$$\Delta_S^h = \cos^2 2j_h + 4\sin^2 j_h|\cos j_h|(\gamma_h^2 - \sin^2 j_h)^{1/2}.$$

Equations (7.399) are valid for $\text{Im } s > 0$ ($\text{Re } \omega > 0$). The results for $\text{Im } s < 0$ are obtained by changing i to $-i$.

Equations (7.165) give the far-field radiation patterns in an infinite medium. Using Table 7.25 and Eqs. (7.398) and (7.399), the corresponding patterns for the case when the source lies on the free surface are given by

$$F_1^0 = \frac{-1}{\pi\beta_h\Delta_P^h}[(3 - 4\gamma_h^2)p_5 + p_2]\sin^2 i_h|\cos i_h|(1 - \gamma_h^2\sin^2 i_h)^{1/2},$$

$$F_2^0 = \begin{cases} \dfrac{1}{2\pi\beta_h\Delta_S^h}[(3 - 4\gamma_h^2)p_5 + p_2]\sin j_h|\cos j_h|\cos 2j_h; & \sin j_h < \gamma_h \\[2ex] \dfrac{1}{2\pi\beta_h\bar{\Delta}_S^h}[(3 - 4\gamma_h^2)p_5 + p_2]\sin j_h|\cos j_h|\cos 2j_h; & \sin j_h > \gamma_h, \end{cases} \tag{7.400}$$

$$F_3^0 = -\frac{1}{2\pi\beta_h}p_1\sin j_h,$$

where

$$\bar{\Delta}_S^h = [\cos^4 2j_h + 16\sin^4 j_h\cos^2 j_h(\sin^2 j_h - \gamma_h^2)]^{1/2}.$$

Note that for $\sin j_h > \gamma_h$, Δ_R is complex. Therefore, the expression for F_2^0 for $\sin j_h > \gamma_h$ has an additional phase factor, which has been suppressed in Eqs. (7.400).

Reinstating the phase factors in Eqs. (7.400) and finding the inverse Fourier transform, we get the following *time-domain expressions*

$$F_1^0 = \frac{-1}{\pi\beta_h\Delta_P^h}[(3-4\gamma_h^2)p_5 + p_2]\sin^2 i_h|\cos i_h|(1-\gamma_h^2\sin^2 i_h)^{1/2}\delta\left(t-\frac{R}{\alpha_h}\right),$$

$$F_2^0 = \frac{1}{2\pi\beta_h\Delta_S^h}[(3-4\gamma_h^2)p_5 + p_2]\sin j_h|\cos j_h|\cos 2j_h\delta\left(t-\frac{R}{\beta_h}\right), \qquad \sin j_h < \gamma_h$$

$$= \frac{1}{2\pi\beta_h(\bar{\Delta}_S^h)^2}[(3-4\gamma_h^2)p_5 + p_2]\sin j_h|\cos j_h|\cos 2j_h\left[\cos^2 2j_h\delta\left(t-\frac{R}{\beta_h}\right)\right.$$

$$\left. + 4\sin^2 j_h|\cos j_h|(\sin^2 j_h - \gamma_h^2)^{1/2}\hat{\delta}\left(t-\frac{R}{\beta_h}\right)\right]; \qquad \sin j_h > \gamma_h,$$

$$F_3^0 = -\frac{1}{2\pi\beta_h}p_1\sin j_h\delta\left(t-\frac{R}{\beta_h}\right), \tag{7.401}$$

where $\hat{\delta}(t-R/\beta_h)$ is the function allied to $\delta(t-R/\beta_h)$. To get the displacements, we must multiply the above expressions by $U_0\,dS/R$.

Equations (7.400) and (7.401) give the radiation patterns of P (F_1^0), SV (F_2^0), and SH (F_3^0) waves excited by an arbitrary shear dislocation on the surface. The functions p_1, p_2, and p_5 are given in Eqs. (7.161). For a vertical strike-slip source, $p_1 = \cos 2\phi_h$, $p_2 = \sin 2\phi_h$, and $p_5 = 0$. For a dip-slip source on a 45° dipping fault (case III), $p_1 = \frac{1}{2}\sin 2\phi_h$, $p_2 = -\frac{1}{2}\cos 2\phi_h$, and $p_5 = \frac{1}{2}$. The field induced by a vertical dip-slip source vanishes identically. It may be noted that in the notation used in Eqs. (7.164) and (7.165), along the ray from the source, F_1^0 is the displacement in the line of sight, F_2^0 is the displacement perpendicular to F_1^0 and points downward (in the sense of i_h increasing) in a vertical plane, and F_3^0 points to the observer's left. Equations (7.400) assume a unit-step time dependence.

A center of compression in an infinite homogeneous medium generates only P waves. The far spectral field can be obtained from Eq. (4.138). We have

$$u_P = u_R = \frac{P_0}{R}F_1,$$

where the phase factor has been withheld and

$$P_0 = \frac{E_0}{\mu_h} = \frac{4}{3\mu_h}\left(\frac{\beta}{\alpha}\right)_h^2 M, \qquad F_1 = \frac{3}{16\pi\alpha_h}. \tag{7.402}$$

From Table 7.25 and Eqs. (7.398), the radiation pattern functions, of a source situated on the free surface, are given by

$$F_1^0 = \frac{3}{2\pi\beta_h \Delta_P^h} \sin^2 i_h |\cos i_h| (1 - \gamma_h^2 \sin^2 i_h)^{1/2},$$

$$F_2^0 = \frac{3}{4\pi\beta_h} \begin{cases} -\dfrac{1}{\Delta_S^h} \sin j_h \cos 2j_h (\gamma_h^2 - \sin^2 j_h)^{1/2}; & \sin j_h < \gamma_h \\ \dfrac{1}{\overline{\Delta_S^h}} \sin j_h \cos 2j_h (\sin^2 j_h - \gamma_h^2)^{1/2}; & \sin j_h > \gamma_h, \end{cases} \quad (7.403)$$

$$F_3^0 = 0.$$

The function F_2^0 is obtained by using the result [Sec. 3.2.2.] that the ratio of the total surface SV displacement to the P displacement is $-(\alpha/\beta)_h \cos 2j_h / |\sin 2j_h|$. Snell's law $(\sin i_h)/\alpha_h = (\sin j_h)/\beta_h$ lead us to the result $\Delta_S^h/\Delta_P^H = \gamma_h^4$.

7.7.2. Effect of the Boundary at the Station

If the observations are made on the surface, as is the case in practice, we must take into consideration the effect of the free surface at the station. The theory has been developed in Section 3.2. The final expressions for the displacements which take into account the free boundary at the source as well as at the station are given in Eqs. (7.404)–(7.407). In these equations i_0 and j_0 denote the angles of incidence, and α_0 and β_0 the intrinsic velocities of P and S waves, respectively, at the free surface. The suffixes z and Δ designate, respectively, the vertical (up) and horizontal (away from the source) components of the displacement.

P waves:

$$u_{zP} = \frac{2A}{\gamma_0^4 \Delta_P^0} \cos i_0 (1 - 2\gamma_0^2 \sin^2 i_0),$$

$$u_{\Delta P} = \frac{2A}{\gamma_0^3 \Delta_P^0} \sin 2i_0 (1 - \gamma_0^2 \sin^2 i_0)^{1/2}, \quad (7.404)$$

$$\frac{u_{zP}}{u_{\Delta P}} = \frac{1 - 2\gamma_0^2 \sin^2 i_0}{2\gamma_0 \sin i_0 (1 - \gamma_0^2 \sin^2 i_0)^{1/2}},$$

where

$$\gamma_0 = \frac{\beta_0}{\alpha_0}, \qquad A = \frac{F_1^0 P_0}{R}, \qquad B = \frac{F_2^0 P_0}{R},$$

$$\Delta_P^0 = (\gamma_0^{-2} - 2\sin^2 i_0)^2 + 4\sin^2 i_0 \cos i_0 (\gamma_0^{-2} - \sin^2 i_0)^{1/2}.$$

SV waves:

$$u_{zS} = \begin{cases} -\dfrac{2B}{\Delta_S^0} \sin 2j_0(\gamma_0^2 - \sin^2 j_0)^{1/2}; & \sin j_0 < \gamma_0 \\ -\dfrac{2B}{\bar{\Delta}_S^0} \sin 2j_0(\sin^2 j_0 - \gamma_0^2)^{1/2}; & \sin j_0 > \gamma_0, \end{cases}$$

$$u_{\Delta S} = \begin{cases} \dfrac{2B}{\Delta_S^0} \cos j_0 \cos 2j_0; & \sin j_0 < \gamma_0 \\ \dfrac{2B}{\bar{\Delta}_S^0} \cos j_0 \cos 2j_0; & \sin j_0 > \gamma_0, \end{cases} \tag{7.405}$$

$$\frac{u_{zS}}{u_{\Delta S}} = -\tan j_0 \tan 2j_0 \left(\frac{\gamma_0^2}{\sin^2 j_0} - 1 \right)^{1/2}; \qquad \sin j_0 < \gamma_0,$$

where

$$\Delta_S^0 = \cos^2 2j_0 + 4 \sin^2 j_0 \cos j_0(\gamma_0^2 - \sin^2 j_0)^{1/2},$$

$$\bar{\Delta}_S^0 = [\cos^4 2j_0 + 16 \sin^4 j_0 \cos^2 j_0(\sin^2 j_0 - \gamma_0^2)]^{1/2}. \tag{7.406}$$

SH waves:

$$u_{SH} = \frac{2F_3^0 P_0}{R}. \tag{7.407}$$

7.8. Initial Motion

As early as 1917, seismologists had noticed the existence of certain regularities in the distribution of the polarities of the initial *P* wave motion as observed on mechanical seismographs. In 1923, they investigated the initial motion patterns resulting from the classical Stokes–Love point force model and certain combinations of such forces in the far field of an infinite medium. It was concluded that the initial motion studies might prove useful in identifying the equivalent force system at the earthquake focus. Using these results as an Archimedian pivot, seismologists were next confronted with two basic questions: First, what force system should be chosen to model best the earthquake sources, and second, how should the first motion observations be inverted and interpreted?

Of the various source equivalents suggested, two were considered to be more "real" than the others. These are the couple and the double couple. A couple is obtained by two antiparallel forces fixed end to end with a small separation between them. The couple is then realized by shrinking the lateral separation to zero while increasing the force magnitude to infinity in such a way that their product, known as the moment, remains finite. A double couple is obtained by an orthogonal superposition of two couples such that the combined moment vanishes.

The first question was answered by seismologists in two different ways: Seismologists in California, motivated by the surface features of the San Andreas fault, considered the earthquake sources as a release of seismic energy through a movement on a planar fault. Intuitively, they chose the single couple as the appropriate force equivalent to the fault. Japanese seismologists, in contrast, viewed the source as a point where pressure and tension acted simultaneously at right angles. This can be shown to be equivalent to a double couple at the far field (Sect. 4.2).

A novel idea resulted in a working method that gave a simple answer to the second question. According to this method, the observed initial motion at a given station (compression or dilatation) is plotted on a plane with the proper azimuth but at a distance that is obtained by the stereographic image of the extended position of the station on an equatorial plane. From the position of the nodal lines (curves separating regions of compression from regions of dilatation), on the projection plane, it is possible, in principle, to deduce two orientations at the source, one of which coincides with the actual motion vector. However, the distribution of the initial motions in the far field cannot distinguish between a single-couple source and a double-couple source and a unique determination of the true motion vector can be made by determining the nodal lines for S waves from the corresponding initial S motions. We have shown in Section 4.5.5 that the radiation pattern of P waves in the far field of an infinite homogeneous medium is given by the function $F(\lambda, \delta; i_h, \phi_h)$. Let us imagine a hypothetical unit sphere surrounding a point dislocation. We could consider F to be the amplitude distribution of the source over this sphere, with nodal lines given by the equations

$$(\mathbf{e} \cdot \mathbf{e}_R) = 0, \qquad (\mathbf{n} \cdot \mathbf{e}_R) = 0. \tag{7.408}$$

We have shown earlier that seismic rays in media with weak heterogeneity conserve the polarity of the first motion together with the angle of polarization. Therefore, rays leaving the focal sphere in a given direction (i_h, ϕ_h) will pick the polarity information $\operatorname{sgn}[F(i_h, \phi_h)]$ and carry it unchanged to the surface of the earth, provided the source is not too close to the free surface. The nodal lines on the focal sphere will therefore be "transplanted" by the rays as new nodal lines on the earth's surface. The deduction of the orientation (λ, δ) of the point dislocation from the observed distribution of polarities is the objective of the initial motion method.

From Eqs. (4.113) and (4.152) we find that the condition $(\mathbf{n} \cdot \mathbf{e}_R) = 0$ is equivalent to

$$\cot i_h - \sin \phi_h \tan \delta = 0,$$

and the condition $(\mathbf{e} \cdot \mathbf{e}_R) = 0$ is equivalent to

$$\cot i_h \sin \delta + \cos \delta \sin \phi_h + \cot \lambda \cos \phi_h = 0.$$

Define a vector (x, y) on an equatorial plane in the source system such that its polar representation is given by (ρ, ϕ_h), where $\rho = \cot i_h$. The equations of the nodal lines on this plane then become

$$x^2 + y^2 - y \tan \delta = 0, \tag{7.409}$$

$$x^2 + y^2 + x\left(\frac{\cot \lambda}{\sin \delta}\right) + y \cot \delta = 0. \tag{7.410}$$

The first line is a circle centered at $(0, \frac{1}{2}\tan\delta)$ with diameter $\tan\delta$ (Fig. 7.61). The second nodal line is another circle centered at $[-\frac{1}{2}(\cot\lambda|\sin\delta), -\frac{1}{2}\cot\delta]$ with diameter $\tan\delta'$, where $\cos\delta' = \sin\lambda\sin\delta$ [see Sec. 4.4.3].

The results are utilized in the following way: The observed polarity of the P wave first motion at a given station is marked on the equatorial plane with the proper azimuth of the station, at a distance that is obtained by a stereographic image of the extended position on that plane ($\rho = \cot i_h$). The procedure is repeated for each station of the network. The resulting nodal lines are the above-mentioned circles that separate regions of compression (+) from regions of dilatation (−). By comparing the observed results with the above theory, it is possible, in principle, to deduce the orientation of the fault plane and the auxiliary plane. An unavoidable ambiguity arises from the symmetry of F in **e** and **n** and cannot be removed. Additional information is then needed to decide which of the two is the fault plane on which the dislocation has taken place. Note that the theoretical interpretation of the initial motion polarity distribution rests on the assumption that the initial motion is contributed by the high-frequency far-field portion of the spectrum. We shall show that the initial motion in the time domain renders the same results.

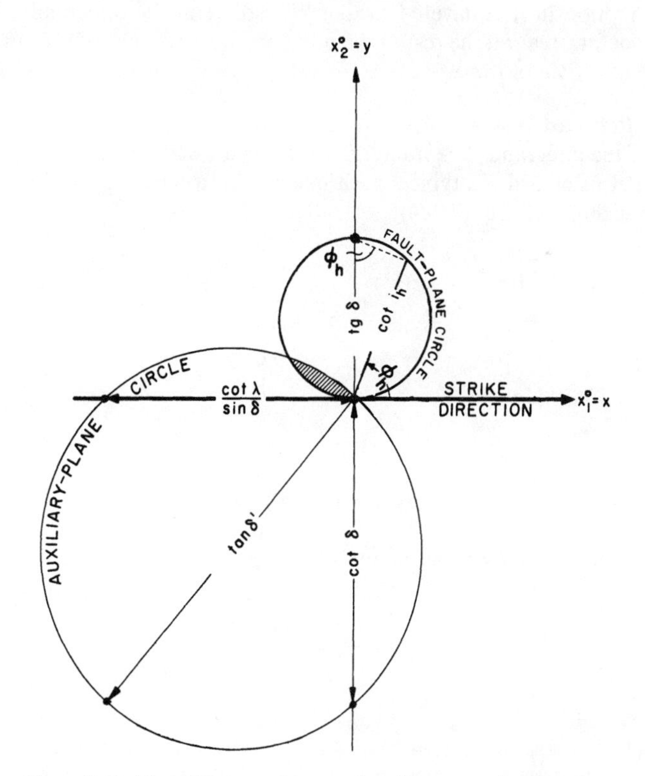

Figure 7.61. Nodal lines on the extended-distance projection-plane.

So far we have obtained expressions for the spectral amplitudes of body waves. It is desirable, however, to develop methods by which the initial portion of the seismogram can be compared with simple theoretical formulas in order to facilitate data analysis from a network of stations. There are three methods by which this goal can be achieved:

1. Initial motion approximation using the exact GRT integrals.
2. Inverse Fourier transform of the results in the frequency domain. Because these results pertain to the far-field high-frequency approximation, their equivalent in the time domain is the initial motion.
3. Application of the method of steepest descents together with a Tauberian theorem.

7.8.1. Initial Motion from Generalized Ray Theory

A method will now be developed that avoids the complete evaluation of the Q_m integrals and renders closed-form expressions for the initial portions of the ground motion in multilayered media. We isolate in the integrand a rapidly varying factor, treating the rest of the integrand as constant, to be evaluated at the time of arrival of the wave.

7.8.1.1. Reflected Waves. We assume that v_1 is well isolated from the branch points of the integrand; i.e., no interference occurs between head and reflected waves. Let us consider a typical path integral for a reflected wave in a multilayered medium [cf. Eq. (7.384)]

$$I(t) = (-)^l \frac{2}{\pi} \operatorname{Im} \frac{d}{dt} \int_{\tau_1}^{t} vE(v^2) \frac{\partial v}{\partial \tau} \Pi S^0_{\mu\nu} \frac{H(t - \tau_1)T_m(\zeta)d\tau}{\sqrt{[(t - \tau)(t - \tau + 2v\Delta)]}}, \quad (7.411)$$

where

$$\zeta = 1 + \frac{t - \tau}{v\Delta}, \quad (7.412)$$

$$E(v^2) = v^{2l+m}D(-v^2). \quad (7.413)$$

We now limit ourselves to initial motions only; i.e.,

$$\varepsilon = \frac{t - \tau}{v\Delta} \ll 1. \quad (7.414)$$

Because

$$T_0(1 + \varepsilon) = 1, \qquad T_1(1 + \varepsilon) = 1 + \varepsilon, \qquad T_2(1 + \varepsilon) = 1 + 4\varepsilon + 2\varepsilon^2, \quad (7.415)$$

we can make the approximation

$$T_m\left(1 + \frac{t - \tau}{v\Delta}\right) = 1 + O(\varepsilon), \tag{7.416}$$

for initial motions. Equation (7.416) is exact for $m = 0$.

For $m > 0$, it assumes

$$\frac{t - \tau}{v\Delta} \ll 2^{-m}. \tag{7.417}$$

A typical case of short-period motion at short distances is $v = \frac{1}{8}$ s/km, $\Delta \simeq$ 200 km, $t - \tau \simeq 1$ s. The inaccuracy thus introduced (for a strike-slip case, $m = 2$) may be as large as 15%. At shorter distances or for longer periods, the error becomes intolerable.

The denominator of the integrand in Eq. (7.411) may be approximated as follows:

$$\frac{1}{[(t - \tau)(t - \tau + 2v\Delta)]^{1/2}} \simeq \frac{1}{[(t - \tau)2v\Delta]^{1/2}}. \tag{7.418}$$

Considering only the rapidly varying factor and treating the rest of the integrand as constant, to be evaluated at $v = v_1$, the integral in Eq. (7.411) transforms into a convolution integral

$$I(t) = \frac{2(-1)^l}{\pi\sqrt{(2\Delta)}} \operatorname{Im}\left[\frac{d}{dt}\left\{H(t - \tau_1)\frac{1}{\sqrt{t}} * \left(\frac{\partial v}{\partial \tau}\right)_t\right\} v_1^{1/2} E(v_1^2) \Pi S_{\mu\nu}^0\right]. \tag{7.419}$$

As in Eq. (7.317b), it can be shown that

$$\frac{\partial v}{\partial \tau} = \frac{1}{\sqrt{[2(\partial^2\tau/\partial v^2)_{v_1}]}} \frac{1}{\sqrt{(\tau - \tau_1)}}. \tag{7.420}$$

Inserting this value in Eq. (7.419), we get

$$I(t) = \frac{(-1)^l}{\pi} \operatorname{Im}\left[\frac{d}{dt}\left\{H(t - \tau_1)\frac{1}{\sqrt{t}} * \frac{1}{\sqrt{(t - \tau_1)}}\right\} E(v_1^2) \left(\frac{v}{\Delta(\partial^2\tau/\partial v^2)}\right)_{v_1}^{1/2} \Pi S_{\mu\nu}^0\right]. \tag{7.421}$$

We now use two results. The first is

$$\frac{1}{\sqrt{t}} * \frac{1}{\sqrt{(t - \tau_1)}} = \int_{\tau_1}^{t} \frac{1}{\sqrt{(t - \tau)}\sqrt{(\tau - \tau_1)}}\, d\tau = \pi \tag{7.422}$$

where we have neglected $\int_{\tau_H}^{\tau_1} [1/\sqrt{(t - \tau)}\sqrt{(\tau - \tau_1)}]d\tau$, assuming no arrival prior to the arrival of the reflected wave. Next, because of the relation

$$v_1 = \frac{\sin\theta_s}{c_s} = \frac{\sin\theta_j}{c_j}, \tag{7.423}$$

where s denotes the layer containing the source and θ the angle with the vertical, we have

$$\left.\frac{\partial^2\tau}{\partial v^2}\right|_{v_1} = -\sum_j \frac{h_j}{c_j^2}\left(\frac{1}{c_j^2} - v_1^2\right)^{-3/2} = -\sum_j \frac{h_j c_j}{\cos^3\theta_j},$$

$$\left[\frac{v}{\Delta(\partial^2\tau/\partial v^2)}\right]_{v_1}^{1/2} = -i\left[\Delta\left(\frac{c_s}{\sin\theta_s}\right)\sum_j \frac{h_j c_j}{\cos^3\theta_j}\right]^{-1/2} = -iL^{-1}\left(\frac{\cos\theta_s}{c_s}\right), \quad (7.424)$$

where L is the *geometric spreading factor*

$$L = \left[\frac{\Delta\cos^2\theta_s}{c_s\ \sin\theta_s}\sum_j \frac{h_j c_j}{\cos^3\theta_j}\right]^{1/2}. \quad (7.425)$$

For a single interface, the reflected and incident waves are of the same type, and we have

$$L = \left[\frac{\Delta(h_1 + h_2)}{\sin\theta_s\cos\theta_s}\right]^{1/2} = [\Delta^2 + (h_1 + h_2)^2]^{1/2}.$$

Clearly, in this particular case, the geometric spreading factor is the distance traveled by the wave. Therefore, we get altogether for the reflected pulse

$$I(t) = -\frac{(-1)^l}{L}\left(\frac{\cos\theta_s}{c_s}\right)\mathrm{Re}\{E(v_1^2)\Pi S_{\mu\nu}^0\}\delta(t - \tau_1), \quad (7.426)$$

where $\Pi S_{\mu\nu}^0$ is exactly the geometric reflection coefficient.

It may be noted that the derivation of Eq. (7.426) assumes that there are no arrivals before the reflected wave. If this condition is not met (this happens when there is a head wave preceding the reflected wave), Eq. (7.422) is replaced by

$$\frac{1}{\sqrt{t}} * \frac{1}{\sqrt{(t-\tau_1)}} = \pi + \int_{\tau_H}^{\tau_1} \frac{d\tau}{\sqrt{(t-\tau)}\sqrt{(\tau-\tau_1)}}$$
$$= \pi - i\ln\frac{|t-\tau_1|}{\tau_H^*}, \quad (7.427)$$

where

$$\tau_H^* = 2\tau_H - t - \tau_1 + 2[\tau_H(\tau_H - t - \tau_1) + t\tau_1]^{1/2}.$$

Equation (7.426) can be modified accordingly. The displacement of the reflected wave will now contain both the delta function and its allied function.

7.8.1.2. Head Waves. For the pure head wave, we have [Eq. (7.384)]

$$I(t) = \frac{2}{\pi}(-1)^l\ \mathrm{Im}\frac{d}{dt}\int_{\tau_H}^{t} vE(v^2)\frac{\partial v}{\partial\tau}\Pi S_{\mu\nu}\frac{H(t - t_H)T_m(\zeta)d\zeta}{[(t-\tau)(t-\tau+2v\Delta)]^{1/2}}. \quad (7.428)$$

This equation assumes that v_H is well separated from other singular points of the integrand. Limiting ourselves to initial motions only, we find

$$I(t) = \frac{2}{\pi}(-1)^l \frac{1}{\sqrt{(2\Delta)}} \operatorname{Im}\left[E(v_H^2)v_H^{1/2}\left(\frac{\partial v}{\partial \tau}\right)_{v_H} \Pi_0(S_{\mu\nu})_{v_H}\frac{d}{dt}\right.$$
$$\left.\times\left\{H(t-\tau_H)\frac{1}{\sqrt{t}} * (\Pi_H S_{\mu\nu})_t\right\}\right], \tag{7.429}$$

where $\Pi S_{\mu\nu} = \Pi_0 S_{\mu\nu}\Pi_H S_{\mu\nu}$ and $\Pi_0 S_{\mu\nu}$ includes all the reflection–transmission coefficients that do not depend on v_H.

Consider the case of a single reflection from a high-velocity layer. We then use Eq. (7.306) to recast $\Pi_H S_{\mu\nu}$ in the form [cf. Eq. (7.8.3)]

$$\Pi_H S_{\mu\nu} = \frac{x(v) + \eta y(v)}{x_1(v) + \eta y_1(v)}, \qquad \eta^2 = v_H^2 - v^2. \tag{7.430}$$

Near v_H

$$\Pi_H S_{\mu\nu} = \frac{x}{x_1} + \eta\frac{x_1 y - x y_1}{x_1^2}, \qquad v(\tau) = v(\tau_H) + (\tau - \tau_H)\left(\frac{\partial v}{\partial \tau}\right)_{\tau_H},$$
$$\eta^2 = (v_H + v)(v_H - v) \simeq -2v_H\frac{\partial v}{\partial \tau}(\tau - \tau_H). \tag{7.431}$$

Further,

$$\tau = v\Delta + \sum_j h_j\left(\frac{1}{c_j^2} - v^2\right)^{1/2},$$
$$\left(\frac{\partial \tau}{\partial v}\right)_{v_H} = \Delta - \sum_j \frac{h_j v_H}{((1/c_j^2) - v_H^2)^{1/2}} = \Delta_0,$$

where Δ_0 is the length of the head wave-path in the refracting layer. Substituting in Eq. (7.429), and using Eq. (7.422), we get

$$I(t) = (-1)^l \operatorname{Re}\left[\frac{v_H E(v_H^2)\Pi_0(S_{\mu\nu})_{v_H}}{\Delta^{1/2}\Delta_0^{3/2}}\left(\frac{x_1 y - x y_1}{x_1^2}\right)\right]H(t - \tau_H). \tag{7.432}$$

Remembering that Eq. (7.411) corresponds to an $H(t)$ source, Eqs. (7.426) and (7.432) reveal that whereas the initial motion of the head wave has the time variation of the source itself, the time dependence of the initial motion of the geometric wave is the derivative of the source-time function.

In the event of another phase arriving before the head wave under consideration, the displacement for the head wave will contain a logarithmic singularity [see Eq. (7.427)]. Therefore, the S head wave will contain a logarithmic singularity on account of the P head wave, which arrives before it.

EXAMPLE 7.8: First Motions for Reflected and Head P Waves

Consider two homogeneous half-spaces, $(H; \alpha, \beta)$ and $(H'; \alpha', \beta')$, in contact. Let there be a point source of the vertical strike-slip type in the upper medium (H) at a height h above the interface and a receiver in the same medium at a height d. The integral for the vertical (up) displacement corresponding to the reflected P wave may be written in the form (Sect. 7.6.2)

$$u_z(s) = -\beta^2 \frac{U_0\,dS}{4\pi}(\sin 2\phi)s\int_0^\infty R_{PP}^z e^{-sa(h+d)}J_2(k\Delta)u^3\,du, \tag{7.8.1}$$

where the generalized reflection coefficient, R_{PP}^z, is given in Eqs. (7.306). Assuming $\alpha' > \alpha$, and comparing Eq. (7.8.1) with Eq. (7.311), we find

$$\begin{gathered} n = 1, \qquad m = 2, \qquad l = 0, \qquad g(u^2) = a(h + d), \\ E(v^2) = v^2, \qquad \Pi S_{\mu\nu} = f(u^2) = R_{PP}^z, \\ v_1 = \frac{1}{\alpha}\sin\theta_s, \qquad v_H = \frac{1}{\alpha'}, \qquad \sin\theta_c = \frac{\alpha}{\alpha'}, \\ \Delta_0 = \Delta - (h + d)\tan\theta_c, \qquad \Delta = (h + d)\tan\theta_s. \end{gathered} \tag{7.8.2}$$

Moreover, from Eqs. (7.306), we have

$$R_{PP}^z = \frac{x(v) + a'y(v)}{x_1(v) + a'y_1(v)}, \tag{7.8.3}$$

where

$$\begin{bmatrix} x(v) \\ x_1(v) \end{bmatrix} = -v^2(q + q' - v^2)^2 \pm ab(q' - v^2)^2 \mp qq'ab',$$

$$\begin{bmatrix} y(v) \\ y_1(v) \end{bmatrix} = \pm v^2abb' - b'(q - v^2)^2 + qq'b.$$

Equations (7.426) and (7.427) now yield the first motion for the reflected P wave:

$$u_z(t) = \left(\frac{\beta}{\alpha}\right)^2 \frac{U_0\,dS}{4\pi\alpha R}\sin 2\phi\cos\theta_s\sin^2\theta_s\left[\mathrm{Re}\{(R_{PP}^z)_{v=v_1}\}\delta\left(t - \frac{R}{\alpha}\right)\right.$$
$$\left. - \mathrm{Im}\{(R_{PP}^z)_{v=v_1}\}\hat{\delta}\left(t - \frac{R}{\alpha}\right)\right],$$

where

$$\hat{\delta}\left(t - \frac{R}{\alpha}\right) = \frac{-1}{\pi(t - R/\alpha)}, \qquad R = [\Delta^2 + (h + d)^2]^{1/2}. \tag{7.8.4}$$

Similarly, Eq. (7.432) yields the first motion for the P head wave:

$$u_z(t) = -\left(\frac{\beta}{\alpha'}\right)^2 \frac{U_0\,dS}{4\pi\alpha'}\sin 2\phi\frac{1}{\Delta^{1/2}\Delta_0^{3/2}}\mathrm{Re}\left(\frac{x_1y - xy_1}{x_1^2}\right)_{v=1/\alpha'} H(t - \tau_H),$$

where

$$\tau_H = \frac{\Delta}{\alpha'} + (h + d)\left(\frac{1}{\alpha^2} - \frac{1}{\alpha'^2}\right)^{1/2}. \tag{7.8.5}$$

7.8.2. Initial Motion by Means of a Tauberian Theorem

In this section, we show how initial motions can be inferred by using the saddle point method in conjunction with a Tauberian theorem.

The displacement field at a point on the free surface of a uniform half-space can be obtained from Eqs. (7.304) and (7.305) on putting $d = 0$. If we take the z axis as vertically upward and a point source of the strike-slip type on the z axis at a depth h below the free surface, the vertical component of the displacement is given by [see also Eq. (7.312)]

$$u_z(s) = u_{zP}(s) + u_{zS}(s),$$

$$u_{zP}(s) = -\frac{U_0\, dS}{8\pi}\beta^2(\sin 2\phi)s\int_{-\infty}^{\infty}(1 + R^z_{PP} + R^z_{PS})e^{-sah}H_2^{(2)}(su\Delta)u^3\, du,$$

$$u_{zS}(s) = \frac{U_0\, dS}{8\pi}\beta^2(\sin 2\phi)s\int_{-\infty}^{\infty}(1 + R^z_{SS} + R^z_{SP})e^{-sbh}H_2^{(2)}(su\Delta)u^3\, du; \tag{7.433}$$

where R^z_{PP}, etc., are the generalized vertical reflection coefficients given in Eqs. (7.303) and a unit-step time variation is assumed.

Replacing the Hankel function by the first term of its asymptotic expansion, we find

$$u_{zP}(s) = -\frac{U_0\, dS}{4\pi}\beta^2\sin 2\phi\left(\frac{s}{2\pi\Delta}\right)^{1/2}\int_{-\infty}^{\infty}(1 + R^z_{PP} + R^z_{PS})e^{-s(ah+iu\Delta)}(iu)^{5/2}\, du. \tag{7.434}$$

Assuming s to be large, we evaluate the integral in Eq. (7.434) by the method of steepest descents. The saddle point in the complex u plane is $u = u_s$, where

$$\frac{d}{du}(ah + iu\Delta) = 0, \qquad a = \left(u^2 + \frac{1}{\alpha^2}\right)^{1/2}. \tag{7.435}$$

This yields

$$u_s = -i\frac{\sin e}{\alpha}, \qquad \tan e = \frac{\Delta}{h}. \tag{7.436}$$

The path of steepest descents goes through the saddle point on the imaginary axis in the u plane. There are branch points in the u plane at $\pm i/\alpha$, $\pm i/\beta$. The integrand is made singlevalued by drawing cuts from i/α to $i\infty$ and from $-i/\alpha$ to $-i\infty$ along the imaginary axis (Fig. 7.62). The saddle point u_s lies between the two nearest branch points, $\pm i/\alpha$. Therefore, the steepest descent integral is not significantly perturbed by the presence of the branch points and the path of integration does not intersect the branch cuts. Hence, we may approximate the integral over u in Eq. (7.434) by the integral through the saddle point. This process yields

$$u_{zP}(s) = -\frac{U_0\, dS}{4\pi R}\left(\frac{\beta^2}{\alpha^3}\right)\sin 2\phi\cos e\sin^2 e(1 + R^z_{PP} + R^z_{PS})_{u_s}e^{-s\tau_P},$$

$$\tau_P = \frac{1}{\alpha}(h\cos e + \Delta\sin e) = \frac{R}{\alpha}. \tag{7.437}$$

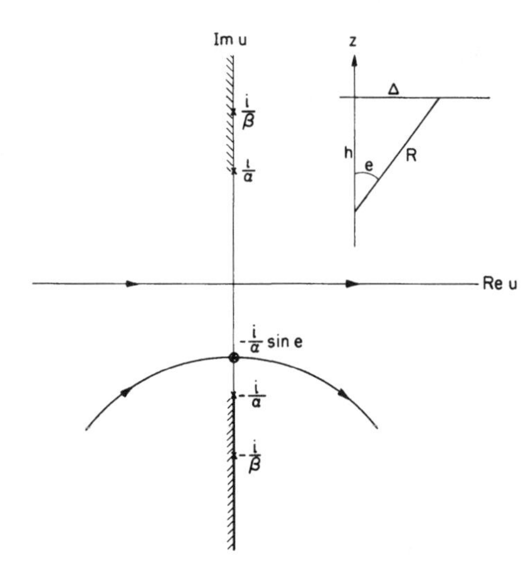

Figure 7.62. Steepest descents path through the saddle point in the complex u plane, yielding initial motion approximation for u_{zP}.

It may be noted that the values of the generalized reflection coefficients at the saddle point are simply the plane-wave reflection coefficients.

The inverse Laplace transform of Eq. (7.437) yields

$$\begin{aligned} u_{zP}(t) &= -\frac{U_0\,dS}{4\pi R}\left(\frac{\beta^2}{\alpha^3}\right)\sin 2\phi \cos e \sin^2 e(1 + R^z_{\dot{P}P} + R^z_{\dot{P}S})\delta\left(t - \frac{R}{\alpha}\right) \\ &= -\frac{U_0\,dS}{2\pi R\alpha\Delta_P}\left(\frac{\alpha}{\beta}\right)^2 \sin 2\phi \cos e \sin^2 e(1 - 2\gamma^2 \sin^2 e)\delta\left(t - \frac{R}{\alpha}\right), \end{aligned} \tag{7.438}$$

where $\gamma = \beta/\alpha$ and

$$\Delta_P = (\gamma^{-2} - 2\sin^2 e)^2 + 4\sin^2 e \cos e(\gamma^{-2} - \sin^2 e)^{1/2}.$$

A Tauberian theorem of the Laplace transform states that if $f(s)$ is the Laplace transform of $F(t)$, then

$$\lim_{t\to+0} F(t) = \lim_{s\to\infty}[sf(s)]. \tag{7.439}$$

Remembering that we indeed assumed s to be large when evaluating the integral in Eq. (7.434) at the saddle point, it is clear that Eq. (7.438) renders the initial motion.

7.8.2.1. *SV* Motion. We treat the integral for u_{zS} in the same way. Now the saddle point is located at

$$u_s = -i\frac{\sin f}{\beta}, \qquad \tan f = \frac{\Delta}{h}.$$

Here we have denoted the angle $\tan^{-1}(\Delta/h)$ by f for clarity, although it is essentially the angle that has been denoted as e in the previous case. If $\sin f < \beta/\alpha$, then the location of the saddle point is as depicted in Fig. 7.62. The evaluation of the integral at the saddle point yields

$$u_{zS}(t) = \frac{U_0\, dS}{4\pi\beta R} \sin 2\phi \cos f \sin^2 f(1 + R_{SS}^z + R_{SP}^z)\delta\left(t - \frac{R}{\beta}\right)$$

$$= \frac{U_0\, dS}{4\pi\beta\Delta_S} \sin 2\phi \sin^2 2f(\gamma^2 - \sin^2 f)^{1/2}\delta\left(t - \frac{R}{\beta}\right),$$

where $\sin f < \gamma$ and

$$\Delta_S = \cos^2 2f + 4\sin^2 f \cos f(\gamma^2 - \sin^2 f)^{1/2}. \tag{7.440}$$

When $\sin f \geq \beta/\alpha$, the situation is more complex. In this case, the saddle point lies on the branch cut. The contour of integration is then deformed from the real axis to a contour that has two straight parts from the saddle point to the branch point $(-i/\alpha)$ and back, and a curved part (the tip) around the branch point $(-i/\alpha)$, as shown in Fig. 7.63. Significant contribution to the integral will be obtained from the tip of the contour near the branch point $(-i/\alpha)$, in addition to the contribution from the saddle point. The latter contribution is

$$u_{zS}(s) = \frac{U_0\, dS}{4\pi\beta R} \sin 2\phi \cos f \sin^2 f(1 + R_{SS}^z + R_{SP}^z)_{u_s} e^{-sR/\beta}. \tag{7.441}$$

However, there is a significant difference between this expression and the one obtained for $\sin f < \beta/\alpha$. In the present case the values of R_{SS}^z and R_{SP}^z at

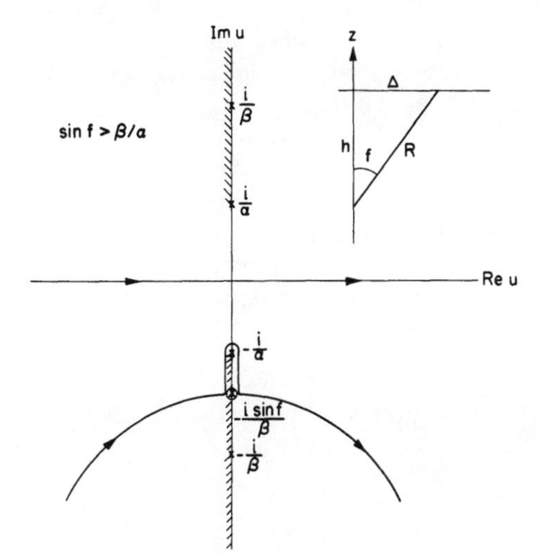

Figure 7.63. Initial motion approximation for u_{zS} beyond the critical angle.

$u = u_s = -i \sin f/\beta$ are complex, whereas for $\sin f < \beta/\alpha$ they were real. In fact, from Eqs. (7.303), we now have

$$1 + R_{SS}^z + R_{SP}^z = \frac{4\cos f(\sin^2 f - \gamma^2)^{1/2}}{\bar{\Delta}_S^2} \times [4\sin^2 f \cos f(\sin^2 f - \gamma^2)^{1/2} - i\cos^2 2f], \quad (7.442)$$

where

$$\bar{\Delta}_S^2 = \cos^4 2f + 16\sin^4 f \cos^2 f(\sin^2 f - \gamma^2).$$

Inserting in Eq. (7.441) and taking the inverse transform, we find

$$u_{zS}(t) = \frac{U_0\, dS}{4\pi\beta R\bar{\Delta}_S^2} \sin 2\phi \sin^2 2f(\sin^2 f - \gamma^2)^{1/2} \times \left[4\sin^2 f\cos f(\sin^2 f - \gamma^2)^{1/2}\delta\left(t - \frac{R}{\beta}\right) - \cos^2 2f\hat{\delta}\left(t - \frac{R}{\beta}\right)\right], \quad \sin f > \gamma, \quad (7.443)$$

where $\hat{\delta}(t - R/\beta)$ is the allied function corresponding to the delta function $\delta(t - R/\beta)$ and is given by (see Sect. 3.2.3)

$$\hat{\delta}\left(t - \frac{R}{\beta}\right) = -\frac{1}{\pi(t - R/\beta)}. \quad (7.444)$$

The contribution of the branch point $-i/\alpha$ to the integral for u_{zS} is simply the SP head wave.

It may be noted that whereas the source time-function is a unit step $H(t)$, the initial motion for P and S waves has the time dependence $\delta(t - R/v)$, when $\sin f < \beta/\alpha$. However, when $\sin f > \beta/\alpha$, the initial motion for SV waves is a combination of the delta function $\delta(t - R/\beta)$ and its allied function $\hat{\delta}(t - R/\beta)$. In general, if $H(t)$ is replaced by $g(t)$, then $\delta(t)$ must be replaced by dg/dt, and $\hat{\delta}(t)$ by the function allied to dg/dt. Here $v = \alpha$ or β.

It is instructive to note that Eqs. (7.438), (7.440), and (7.443) could be obtained directly from Eq. (7.165), taking into consideration the effect of the boundary at the recorder by the method of Section 7.7.2, and then applying the inverse transform. This procedure can be used without difficulty for an arbitrary source and for all the components of the displacement. Therefore, the initial motion of the displacements resulting from a deep source in a *uniform half-space* are:

P waves:

$$u_z = \frac{P_0}{R} F_1\left(\frac{2}{\gamma^4\Delta_P}\right)\cos e(1 - 2\gamma^2\sin^2 e)\delta\left(t - \frac{R}{\alpha}\right),$$

$$u_\Delta = \frac{P_0}{R} F_1\left(\frac{2}{\gamma^3\Delta_P}\right)\sin 2e(1 - \gamma^2\sin^2 e)^{1/2}\delta\left(t - \frac{R}{\alpha}\right), \qquad u_\phi = 0. \quad (7.445)$$

SV waves:

$$
\begin{aligned}
u_z &= \frac{P_0}{R} F_2\left(-\frac{2}{\Delta_S}\right)\sin 2f(\gamma^2 - \sin^2 f)^{1/2}\delta\left(t - \frac{R}{\beta}\right); \qquad \sin f < \gamma \\
&= \frac{P_0}{R} F_2\left(-\frac{2}{\bar{\Delta}_S^2}\right)\sin 2f(\sin^2 f - \gamma^2)^{1/2}\left[4\sin^2 f \cos f(\sin^2 f - \gamma^2)^{1/2}\right. \\
&\quad \left. \times\, \delta\left(t - \frac{R}{\beta}\right) - \cos^2 2f\hat{\delta}\left(t - \frac{R}{\beta}\right)\right]; \qquad \sin f > \gamma, \\
u_\Delta &= \frac{P_0}{R} F_2\left(\frac{2}{\Delta_S}\right)\cos f \cos 2f\delta\left(t - \frac{R}{\beta}\right); \qquad \sin f < \gamma \\
&= \frac{P_0}{R} F_2\left(\frac{2}{\bar{\Delta}_S^2}\right)\cos f \cos 2f\left[\cos^2 2f\delta\left(t - \frac{R}{\beta}\right) + 4\sin^2 f \cos f(\sin^2 f - \gamma^2)^{1/2}\right. \\
&\quad \left. \times\, \hat{\delta}\left(t - \frac{R}{\beta}\right)\right]; \qquad \sin f > \gamma, \\
u_\phi &= 0.
\end{aligned}
\tag{7.446}
$$

SH waves:

$$
\begin{aligned}
u_z &= u_\Delta = 0, \\
u_\phi &= 2\frac{P_0}{R} F_3\delta\left(t - \frac{R}{\beta}\right).
\end{aligned}
\tag{7.447}
$$

In Eqs. (7.445)–(7.447), $\gamma = \beta/\alpha$, Δ_P, Δ_S and $\bar{\Delta}_S$ are given by Eqs. (7.438), (7.440), and (7.442), respectively, and the source-time function is $H(t)$. The functions F_i $(i = 1, 2, 3)$ for an arbitrary shear dislocation are given by Eq. (7.165), wherein $i_h = e$ for F_1, $i_h = f$ for F_2 and F_3, $\alpha_h = \alpha$, $\beta_h = \beta$, and P_0 is the source potency. The z axis is pointing vertically upward.

EXAMPLE 7.9: The Effect of Small Velocity Gradients on Head-Wave Amplitudes

So far, we have studied head waves formed at the plane boundary of two homogeneous media. In the earth, especially at the crust–mantle boundary, head waves are known to propagate in the presence of the earth's curvature and possible physical gradients of the constitutive parameters ρ, μ, α, and β. In the following discussion we shall show that the presence of such small velocity gradients in the refracting medium as well as a slight curvature of the refracting boundary may affect the amplitudes of the head waves. We have shown in Section 7.3.2 that because of the "earth-flattening transformation" the curvature can be substituted, to first order, by a constant positive gradient in the intrinsic wave velocity. Let us therefore study, through a simple model, the effect of such gradients on the formation and propagation of head waves.

We adopt the model of Example 7.8 except that the lower half-space ($z < 0$) is now assumed to be inhomogeneous with

$$\alpha(z \leq 0) = \frac{\alpha_0}{\sqrt{(n_0^2 + \gamma z)}}, \qquad \beta(z \leq 0) = \frac{\beta_0}{\sqrt{(n_0^2 + \gamma z)}}, \tag{7.9.1}$$

where $(\alpha_0, \beta_0, \rho_0)$ are the fixed parameters of the upper homogeneous half-space ($z > 0$). Just below the discontinuity at $z = 0$ we have

$$\alpha(0^-) = \frac{\alpha_0}{n_0}, \qquad \beta(0^-) = \frac{\beta_0}{n_0}, \qquad n_0 < 1. \tag{7.9.2}$$

As in Example 7.8, the source is at $(0, z_h)$ and the receiver is at (Δ, z), both in the upper half-space, with the z axis pointing upward and $z = 0$ at the discontinuity (Fig. 7.64).

The effect of the boundary curvature maps into an *effective positive velocity gradient* according to the earth-flattening transformation, Eq. (7.192). The overall longitudinal velocity profile for $|\gamma z/n_0^2| \ll 1$ and $|z/a| \ll 1$, is

$$\hat{\alpha}(z \leq 0) \simeq \alpha(z)\left(1 - \frac{z}{a}\right) \simeq \frac{\alpha_0}{n_0}\left[1 - \frac{\hat{\gamma}}{2n_0^2} z\right] \simeq \frac{\alpha_0}{\sqrt{(n_0^2 + \hat{\gamma} z)}}, \tag{7.9.3}$$

where the *total gradient* is determined by

$$\hat{\gamma} = \frac{2n_0^2}{a} + \gamma, \tag{7.9.4}$$

a being the earth's radius. The constant-gradient approximation of $\alpha(z)$ is sufficient for the purposes of this example because the head wave is most affected by the immediate vicinity below the boundary, which is of the order of a wavelength or so. There is a critical value $\gamma_c = -2n_0^2/a$ that is equal and opposite to the effective curvature gradient. At $\gamma = \gamma_c$ we fall back on the classical head wave for flat homogeneous layers treated in Example 7.8. The cases $\hat{\gamma} > 0$ and $\hat{\gamma} < 0$ will now be treated in some detail. To this end, we may still use the formal integral representation given in Eq. (7.8.1) provided we evaluate anew the reflection coefficient R_{PP}^z. Changing back to the frequency domain, we substitute there $s = i\omega$, $d = z$, $h = z_h$, $k = su$ and consider the range of integration along the entire real k axis. This leads us to the generalized *Sommerfeld–integral* [cf. Eq. (2.154)] for the vertical spectral displacement of the reflected P wave in the upper medium $z > 0$ [$\nu = \sqrt{(k^2 - k_{\alpha_0}^2)}$, $\Delta = a\theta$]

$$\{u_z(\Delta, z > 0, z_h; \omega)\}_{PP} = -i\frac{\beta_0^2}{\omega^3}\frac{U_0(\omega)dS}{8\pi}\sin 2\phi\sqrt{\left(\frac{\theta}{\sin\theta}\right)}\int_{-\infty}^{\infty} H_2^{(2)}(k\Delta)\{\hat{R}_{PP}^z\}$$
$$\times\, e^{-\nu(z+z_h)}k^3\, dk. \tag{7.9.5}$$

Here, $\hat{R}_{PP}^z$ is the new *generalized reflection coefficient* appropriate for the lower inhomogeneous medium. For a given value of $k = k_\alpha \sin\vartheta$, it is the reflection coefficient of a *compressional plane wave* impinging in the upper half-space on the boundary at an angle ϑ with the normal. Its explicit form is derived as

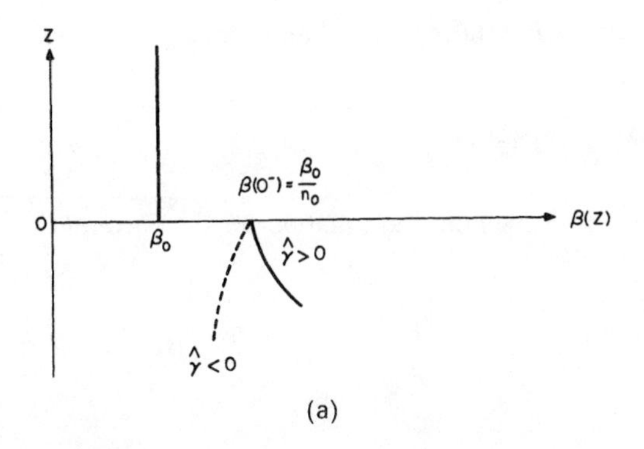

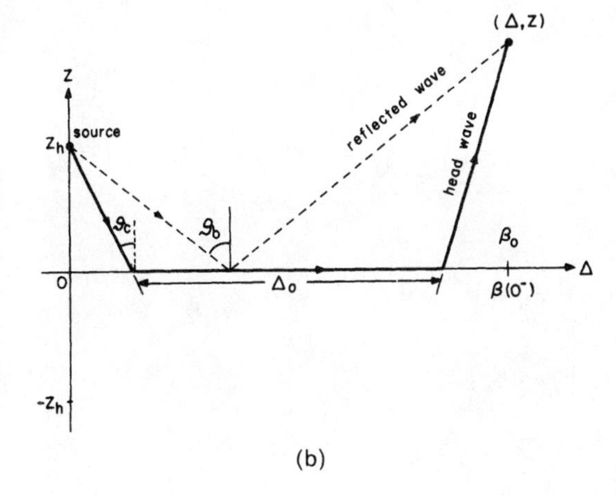

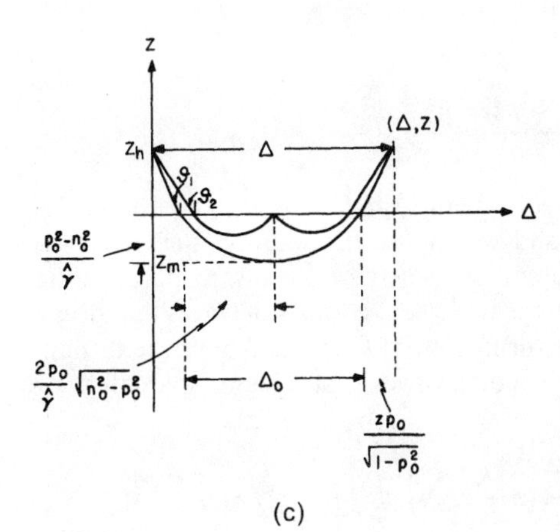

Figure 7.64. (a) Shear velocity β, as a function of depth. (b) Geometry of head-wave and reflected wave at a discontinuity between two elastic half-spaces. (c) The effect of a positive gradient in the lower medium on transmitted waves near the critical angle.

follows: The vertical eigenfunctions $F(z)$ in the upper half-space are solutions of the equations

$$\frac{d^2}{dz^2} F_i(z) - [k^2 - k_i^2] F_i(z) = 0, \qquad (i = \alpha_0, \beta_0), \tag{7.9.6}$$

namely, $e^{-\nu z}$ and $e^{-\nu' z}$. In the lower medium the equation for the decoupled P wave is

$$\frac{d^2 F_\alpha}{dz^2} + \left[\frac{\omega^2}{\hat{\alpha}^2(z)} - k^2\right] F = 0,$$

where, by Eq. (7.9.3),

$$\frac{1}{\hat{\alpha}^2} = \frac{1}{\alpha^2(0^-)}\left[1 + \frac{\hat{\gamma}}{n_0^2} z\right].$$

Defining

$$k_\alpha = \frac{\omega}{\alpha(0^-)} = \frac{\omega}{\alpha_0} n_0 = k_{\alpha 0} n_0, \qquad k_{\alpha 0} = \frac{\omega}{\alpha_0}, \qquad n_0 < 1, \tag{7.9.7}$$

the equation becomes

$$\frac{d^2 F_\alpha}{dz^2} + [M + Nz] F_\alpha = 0, \qquad (z < 0)$$

$$M = k_\alpha^2 - k^2, \qquad N = \left(\frac{k_\alpha^2}{n_0^2}\right)\hat{\gamma} = \left(\frac{\omega^2}{\alpha_0^2}\right)\hat{\gamma} = k_{\alpha 0}^2 \hat{\gamma}. \tag{7.9.8}$$

Denoting

$$k = k_{\alpha 0} p_\alpha, \qquad p_\alpha = \sin \vartheta_\alpha, \tag{7.9.9}$$

we observe that Eq. (7.9.8) has a *turning point* at

$$z_{m\alpha} = \frac{p_\alpha^2 - n_0^2}{\hat{\gamma}} = \frac{k^2 - k_\alpha^2}{\hat{\gamma} k_{\alpha 0}^2}. \tag{7.9.10}$$

Two independent solutions of Eq. (7.9.8) are the Airy functions Ai and Bi [cf. Eq. (G.10)]. To suit the problem at hand we choose the solution such that, for $\hat{\gamma} < 0$ and $z \to -\infty$, the wave field consists solely of downward propagating waves, whereas, for $\hat{\gamma} > 0$, $z \to -\infty$, the wave field consists of waves that decay exponentially with depth below the turning point ($z < z_m$). Above the turning point ($z > z_m$) both upgoing and downgoing waves exist. The suitable solutions are therefore

$$F_\alpha(z \le 0) = \begin{cases} Ai\left[-\dfrac{|\hat{\gamma}|}{s_\alpha^2}(z - z_{m\alpha}) e^{-2\pi i/3}\right], & \hat{\gamma} < 0 \\[2ex] Ai\left[-\dfrac{\hat{\gamma}}{s_\alpha^2}(z - z_{m\alpha})\right], & \hat{\gamma} > 0 \end{cases} \tag{7.9.11}$$

where

$$s_\alpha = \left(\frac{|\hat{\gamma}|}{k_{\alpha 0}}\right)^{1/3}.$$

Similar expressions are derived for $F_\beta(z \le 0)$. The longitudinal field, transmitted to the lower medium, is then given by the generalized ray

$$\{u_z(\Delta, z < 0, z_h; \omega)\}_P = -i\frac{\beta_0^2}{\omega^3}\frac{U_0(\omega)dS}{8\pi}\sin 2\phi\sqrt{\left(\frac{\theta}{\sin\theta}\right)}\int_{-\infty}^{\infty} H_2^{(2)}(k\Delta)$$
$$\times \{\hat{T}_{PP}^z\}F_\alpha(z)k^3\,dk. \tag{7.9.12}$$

The explicit expressions for $\hat{R}_{PP}^z$ and $\hat{T}_{PP}^z$ are the plane-wave reflection and transmission coefficients at $z = 0$. They can be found by solving an ancilliary problem in which a plane wave of unit amplitude impinges on the boundary from above at an incidence angle ϑ. However, we may bypass this exercise by using the expressions in Sect. 7.6.2 as they stand provided we set the correspondence

$$sa' = \nu \Rightarrow -\left.\frac{(d/dz)Ai[-\zeta_\alpha(z)]}{Ai[-\zeta_\alpha(z)]}\right|_{z=0} = -\left(\frac{\zeta_{\alpha 0}}{z_{m\alpha}}\right)\frac{Ai'[-\zeta_{\alpha 0}]}{Ai[-\zeta_{\alpha 0}]},$$

$$\zeta_\alpha(z) = \frac{\hat{\gamma}}{s_\alpha^2}(z - z_{m\alpha})(\operatorname{sgn}\hat{\gamma})^{-2/3}, \tag{7.9.13}$$

$$\zeta_\alpha(0) = \zeta_{\alpha 0} = \frac{(n_0^2 - p_\alpha^2)}{s_\alpha^2}(\operatorname{sgn}\hat{\gamma})^{-2/3},$$

$$sb' = \nu' \Rightarrow -\left.\frac{(d/dz)Ai[-\zeta_\beta(z)]}{Ai[-\zeta_\beta(z)]}\right|_{z=0} = -\left(\frac{\zeta_{\beta 0}}{z_{m\beta}}\right)\frac{Ai'[-\zeta_{\beta 0}]}{Ai[-\zeta_{\beta 0}]}, \tag{7.9.14}$$

$$\bar{q} = \frac{1}{2\beta_0^2[1 - \mu(0^-)/\mu_0]}, \qquad \bar{q}' = \frac{1}{2\beta^2(0^-)[1 - \mu_0/\mu(0^-)]}, \tag{7.9.15}$$

where $\bar{q}$ in Eq. (7.9.15) replaces q of Eq. (7.306). Note that when $\hat{\gamma} \to 0$ it follows from Eq. (7.9.8) that $F_\alpha \to e^{-z\sqrt{(k^2 - k_\alpha^2)}}$, as it should. Also, by using Eq. (G.10), we find

$$\frac{Ai'(z)}{Ai(z)} \simeq -\sqrt{z}, \qquad |z| \gg 1, \qquad |\arg z| < \pi. \tag{7.9.16}$$

Substituting it in Eq. (7.9.13), we obtain

$$\nu \Rightarrow \sqrt{(k^2 - k_\alpha^2)}, \qquad \nu' \Rightarrow \sqrt{(k^2 - k_\beta^2)}.$$

In a similar way we derive from Eqs. (7.305), for $z > 0$,

$$\{u_\phi\}_{SSH} = -\frac{U_0(\omega)dS}{8\pi}\cos 2\phi\sqrt{\left(\frac{\theta}{\sin\theta}\right)}\frac{\partial}{\partial\Delta}\int_{-\infty}^{\infty} H_2^{(2)}(k\Delta)\{\hat{R}_{SSH}\}$$
$$\times e^{-(z+z_h)\sqrt{(k^2 - k_{\beta 0}^2)}}\frac{k\,dk}{\sqrt{(k^2 - k_{\beta 0}^2)}} \tag{7.9.17}$$

and, for $z < 0$,

$$\{u_\phi\}_{SSH} = -\frac{U_0(\omega)dS}{8\pi}\cos 2\phi\sqrt{\left(\frac{\theta}{\sin\theta}\right)}\frac{\partial}{\partial\Delta}\int_{-\infty}^{\infty} H_2^{(2)}(k\Delta)\{\hat{T}_{SSH}\}F_\beta(z)$$

$$\times\frac{k\,dk}{\sqrt{(k^2 - k_{\beta_0}^2)}},$$

where

$$\hat{R}_{SSH} = \frac{imq - s_\beta X}{imq + s_\beta X},\qquad \hat{T}_{SSH} = \frac{2imq}{imq + s_\beta X},\qquad m = \frac{\mu_0}{\mu(0^-)},\qquad q = (1 - p^2)^{1/2}, \tag{7.9.18}$$

$$k = k_{\beta 0}p,\qquad X = \left[\frac{Ai'(-\zeta_{\beta 0})}{Ai(-\zeta_{\beta 0})}\right],\qquad \zeta_{\beta 0} = \zeta_0. \tag{7.9.19}$$

From this point on we restrict our discussion to SH waves only. Consequently, all values of n_0, $\hat{\gamma}$, $Ai(z)$, s etc. will refer exclusively to shear waves. For $\hat{\gamma} > 0$ and $k < k_{\beta 0}(\vartheta_\beta$ real, $v = 1/\beta(0^-)$, $v_H = 1/\beta_0)$ X is real and the modulus of the plane-wave reflection coefficient is unity. Consequently, all the energy entering the lower medium is eventually reflected back into the upper homogeneous half-space and the reflection coefficient can be expressed as $e^{2i\tan^{-1}\chi}$, where $\chi = (s_\beta/mq)X$.

We define the *critical velocity gradient* Γ_c as the velocity gradient corresponding to the critical value $\gamma_c = -2n_0^2/a$. For shear waves we have, from Eq. (7.9.3),

$$\Gamma_c = \frac{d\beta(z)}{d(-z)} = \frac{\beta_0\gamma}{2n_0^3}\bigg|_{\gamma=\gamma_c} = -\frac{\beta(0^-)}{a}. \tag{7.9.20}$$

The critical velocity gradient corresponds to the case in which the curvature of the ray at its turning point matches the curvature of the earth [cf. Table 7.19]. Using the critical gradient as a reference, the cases $\hat{\gamma} < 0$ and $\hat{\gamma} > 0$ can be referred to as having *subcritical* and *supercritical* gradients, respectively. Because of curvature, a homogeneous lower half-space ($\Gamma_c = 0$) has an effective supercritical gradient. In the following, we shall evaluate the head-wave contribution of the integral in Eq. (7.9.17) by means of the saddle-point method. Figure 7.65 shows a schematic representation of the effects of curvature and velocity gradient on waves near the critical angle of refraction. Pure head waves are just a particular case of a group of generalized rays that includes *interference head waves*, or diving waves, ($\hat{\gamma} > 0$, $\Gamma_c \geq 0$), *pure head waves* [$\hat{\gamma} = 0$, $\Gamma_c = -\beta(0^-)/a$], and *diffraction head waves* [$\hat{\gamma} < 0$, $\Gamma_c < -\beta(0^-)/a$].

We return to Eq. (7.9.17) and replace $H_2^{(2)}(k\Delta)$ by its asymptotic form $\sqrt{(2/\pi k\Delta)}\,e^{-i[k\Delta + 3\pi/4]}$ (App. D). We then change the variable of integration

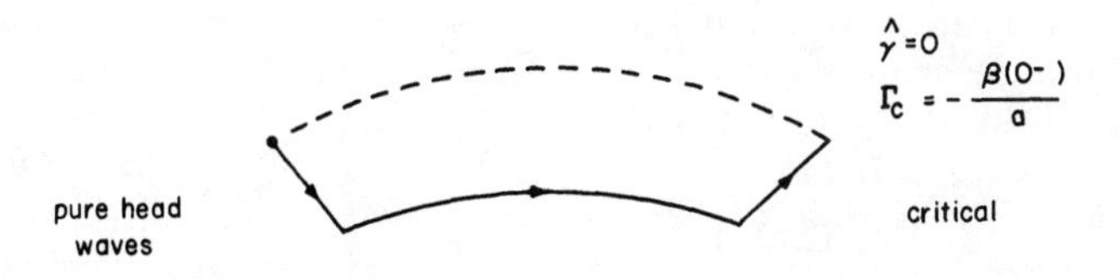

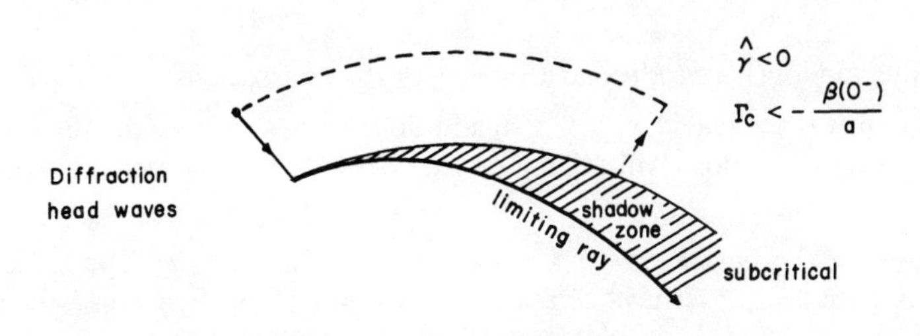

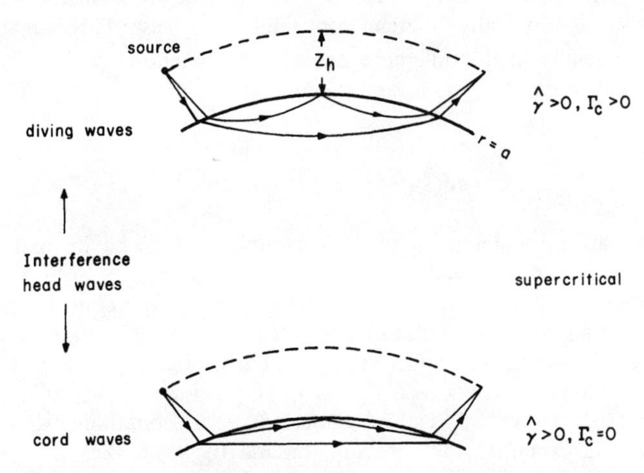

Figure 7.65. Schematic representation of effects of curvature and velocity gradients on near-critical waves.

from the horizontal wave number k to $p = p_\beta$ through $k = k_0 p$. Equation (7.9.17) then becomes

$$\{u_\phi\}_{SSH} = -k_0^{3/2} \frac{U_0(\omega)dS}{4\pi\sqrt{(2\pi\Delta^*)}} \cos 2\phi e^{\pi i/4} \int_{-\infty}^{\infty} \{\hat{R}_{SSH}\} e^{-ik_0[p\Delta + q(z+z_h)]} \frac{p^{3/2}\, dp}{q}, \tag{7.9.21}$$

where

$$\sqrt{(k^2 - k_0^2)} = iqk_0, \qquad k_0 = \frac{\omega}{\beta_0}, \qquad \Delta^* = a \sin\theta, \qquad \Delta = a\theta.$$

The integrand has *branch points* at $p = \pm 1$ $(q = 0)$ associated with the radical q and at $p = 0$ associated with $\sqrt{p}$. In addition, we have to consider the *poles* of $\{\hat{R}_{SSH}\}$ whose locations will depend on the sign of $\hat{\gamma}$. We consider three cases.

1. *The critical gradient* $[\hat{\gamma} = 0,\ \Gamma_c = -\beta(0^-)/a,\ \gamma_c = -2n_0^2/a]$. The reflection coefficient $\{\hat{R}_{SSH}\}$ reduces to the form for a lower homogeneous half-space [cf. Eqs. (7.306)]. The saddle-point contribution then renders the reflected wave, whereas the pure head wave arises from the contribution of the branch point at $p = n_0$ associated with the radical $\sqrt{(n_0^2 - p^2)}$ in $\{R_{SSH}\}$, as we have stated in Sections 7.4–7.8.
2. *The subcritical gradient* [diffracted head wave, $\hat{\gamma} < 0$, $\Gamma_c < -\beta(0^-)/a$]. No matter how small $\hat{\gamma}$ is, provided it is not identically zero, the reflection coefficient is physically different from the limiting case. This is manifested mathematically in the difference between the function $\sqrt{(n_0^2 - p^2)}$ which has a branch point at $p = n_0$ and the function

$$\frac{Ai'[-((n_0^2 - p^2)/s_\beta^2)(\operatorname{sgn}\hat{\gamma})^{-2/3}]}{Ai[-((n_0^2 - p^2)/s_\beta^2)(\operatorname{sgn}\hat{\gamma})^{-2/3}]}$$

which is an entire function in the finite complex p plane, having real poles at the roots d_j of $Ai(-d) = 0$.

Because we know that for $\hat{\gamma} = 0$, the singularity that contributes to the head waves is at $p = n_0$, we shall assume that for $0 < |\hat{\gamma}| \ll 1$ also, the head-wave contribution comes from the neighborhood of $p = n_0$. With this in mind, we look for roots of $s_\beta X + imq = 0$ [cf. Eq. (7.9.18)] in the region $|p - n_0| \ll 1$. However, in this region, the term containing $Ai(-\zeta_0)$ in $\{\hat{R}_{SSH}\}$ will dominate the denominator and the roots of $s_\beta X + imq = 0$ will be near the zeros of $Ai(-\zeta_0)$. Denoting the real positive zeros of $Ai(-d)$ by d_j and noticing that [cf. Eq. (7.9.13)]

$$-\zeta_0 \simeq -\frac{2n_0}{s_\beta^2}(n_0 - p)e^{-2\pi i/3},$$

the roots of $s_\beta X + imq = 0$ are approximately at

$$p_j \simeq n_0 + \frac{s_\beta^2 d_j}{2n_0} e^{-\pi i/3}, \tag{7.9.22}$$

where $d_j \simeq [(3\pi/8)(4j-1)]^{2/3}$ as shown in Appendix H. The steepest descents path and head-wave poles in the complex p plane are shown in Fig. 7.66a. The contributions of the poles decrease exponentially with their distance from the real p axis such that the dominant contribution comes from the vicinity of the critical point $p_0 = n_0$ (critical angle of refraction). It is

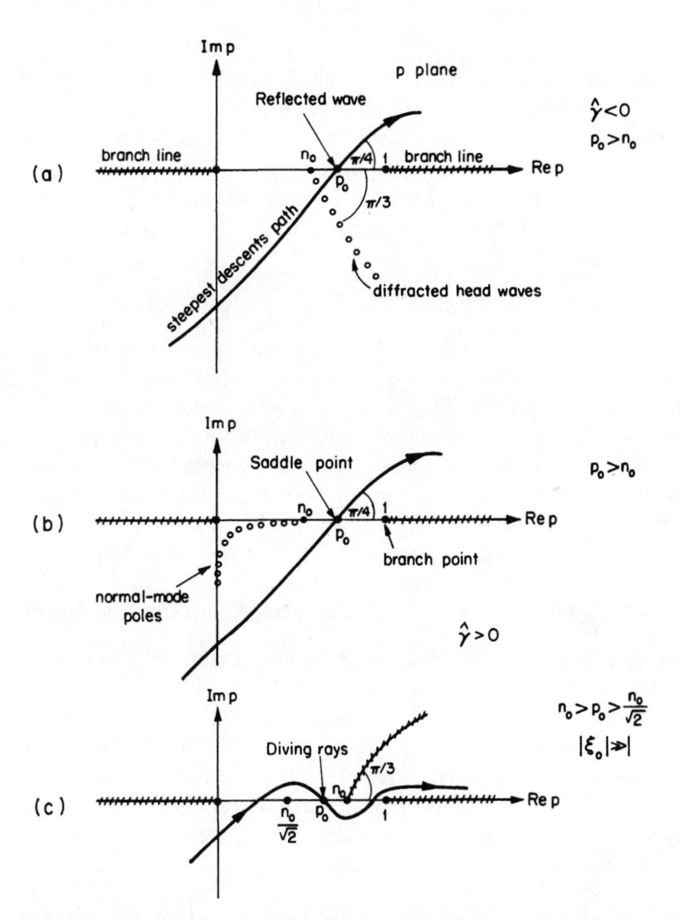

Figure 7.66. Steepest descents path for critically refracted waves in the complex p plane: (a) Negative gradient: the saddle point yields the reflected wave and the poles render diffracted head waves. (b, c) Positive gradient: the poles' contribution represents head waves at short ranges (b) and interference head-waves (c) at the far field. The saddle point yields diving waves (c).

clear from Fig. 7.66a that, when $p_0 < n_0$, the only contribution comes from the saddle point p_0, i.e., the reflected wave. When $p_0 > n_0$, however, the steepest descents path through p_0 passes through the line of poles associated with $\{\hat{R}_{SSH}\}$ and the two contributions may interfere. Applying the residue theorem to those poles in the region $|p - n_0| \ll 1$ at large distances from the source, the dominant contribution comes from the pole nearest the real p axis. The analysis then shows that the diffraction head-wave spectral amplitude decays as

$$\exp\left\{-\frac{\sqrt{3}}{4}\frac{d_1\Delta_0}{n_0\beta_0^{1/3}}|\hat{\gamma}|^{2/3}\omega^{1/3}\right\}, \tag{7.9.23}$$

where Δ_0 is defined in Eq. (7.8.2) (see also Fig. 7.64).

3. *Supercritical gradient* (Interference head waves, $\hat{\gamma} > 0$, $\Gamma_c \geq 0$). The reflection coefficient has a line of poles that lies just below the real p axis in the interval $0 < p < n_0$, as shown in Fig. 7.66b. As before, the term containing $Ai(-\zeta_0)$ dominates in the denominator of the reflection coefficient, but it is necessary to carry an extra term in the expansion about $p = n_0$ to ensure that the residue series converges. Therefore, the poles are at

$$p_j \simeq \left(n_0 - \frac{s_\beta^2}{2n_0}d_j\right) - i\frac{s_\beta}{\sqrt{(1-n_0^2)}}\left(1 + \frac{s_\beta^4}{n_0^3}d_j\right). \tag{7.9.24}$$

These poles describe normal modes propagating in the wave guide formed between the discontinuity at $z = 0$ and the continuous increase of velocity with depth. The poles in the vicinity of $p = n_0$ are associated with normal modes that have phase velocities near the pure head-wave phase velocity. The poles closer to the origin describe deeply penetrating normal modes with high phase velocities and low group velocities. Evaluation of the residue contribution shows that only at relatively short distances, such that $\Delta_0 \ll 2n_0/(k_0\hat{\gamma}^2)^{1/3}$, does the sum of the normal modes yield a pure head wave plus a correction term of the order $k_0\Delta_0 s_\beta^2/2n_0$. With $\hat{\gamma} \simeq 10^{-3}$ km^{-1}, $\beta_0 = 3$ km/s, $\beta(0^-) = 6$ km/s, and $\omega = 5$ Hz, Δ_0 must be less than 50 km.

To obtain the field at large distances from the source, it is convenient to expand $\{\hat{R}_{SSH}\}$ into an infinite series, the nth term of which can be identified as the reflection coefficient for the generalized ray bottoming n times in the heterogeneous medium and reflecting $(n - 1)$ times at the boundary $z = 0$. The expansion is accomplished by replacing the standing wave-form of the Airy function $Ai[-(n_0^2 - p^2)/s_\beta^2]$ with a combination of the Airy functions $Ai[-(n_0^2 - p^2)s_\beta^{-2}(e^{2\pi i/3})]$, and $Ai[-(n_0^2 - p^2)s_\beta^{-2}(e^{-2\pi i/3})]$, representing upgoing and downgoing traveling waves, respectively

$$Ai\left[-\frac{(n_0^2 - p^2)}{s_\beta^2}\right] = e^{-\pi i/3}Ai\left[-\frac{(n_0^2 - p^2)}{s_\beta^2}e^{2\pi i/3}\right] + e^{\pi i/3}Ai\left[-\frac{(n_0^2 - p^2)}{s_\beta^2}e^{-2\pi i/3}\right]. \tag{7.9.25}$$

Defining the ratios

$$A^{(1)} = \frac{Ai'[-\zeta_0 e^{-2\pi i/3}]}{Ai[-\zeta_0 e^{-2\pi i/3}]}, \qquad A^{(2)} = \frac{Ai'[-\zeta_0 e^{2\pi i/3}]}{Ai[-\zeta_0 e^{2\pi i/3}]}, \qquad A = \frac{Ai[-\zeta_0 e^{-2\pi i/3}]}{Ai[-\zeta_0 e^{2\pi i/3}]} \tag{7.9.26}$$

we can express X [Eq. (7.9.19)] as

$$\frac{\varepsilon A^{(2)} + AA^{(1)}}{1 + \varepsilon A},$$

where $\varepsilon = e^{2\pi i/3}$ and the prime means derivative with respect to the argument. It then transpires that

$$\begin{aligned}
\hat{R}_{SSH} &= \frac{[iqm - s_\beta A^{(2)}\varepsilon] + A[iqm\varepsilon - s_\beta A^{(1)}]}{[iqm + s_\beta A^{(2)}\varepsilon] + A[iqm\varepsilon + s_\beta A^{(1)}]} \\
&= \frac{iqm - s_\beta A^{(2)}\varepsilon}{iqm + s_\beta A^{(2)}\varepsilon}\left[\frac{1 + \varepsilon A(iqm - s_\beta A^{(1)}\varepsilon^{-1})/(iqm - s_\beta A^{(2)}\varepsilon)}{1 + \varepsilon A(iqm + s_\beta A^{(1)}\varepsilon^{-1})/(iqm + s_\beta A^{(2)}\varepsilon)}\right] \\
&= \frac{iqm - s_\beta A^{(2)}\varepsilon}{iqm + s_\beta A^{(2)}\varepsilon}\left[1 + \varepsilon A \frac{2iqms_\beta[A^{(2)}\varepsilon - A^{(1)}\varepsilon^{-1}]}{[iqm - s_\beta A^{(2)}\varepsilon][iqm + s_\beta A^{(2)}\varepsilon]}\right. \\
&\quad \times \left.\frac{1}{1 + \varepsilon A(iqm + s_\beta A^{(1)}\varepsilon^{-1})/(iqm + s_\beta A^{(2)}\varepsilon)}\right] \\
&= Y_0 + \sum_{n=1}^{\infty} Y_n,
\end{aligned} \tag{7.9.27}$$

where

$$Y_0 = \frac{iqm - s_\beta e^{2\pi i/3} A^{(2)}}{iqm + s_\beta e^{2\pi i/3} A^{(2)}}, \tag{7.9.28}$$

$$Y_n = (-)^{n-1}\varepsilon^n(2iqms_\beta)A^n[A^{(2)}\varepsilon - A^{(1)}\varepsilon^{-1}]\frac{[iqm + s_\beta A^{(1)}\varepsilon^{-1}]^{n-1}}{[iqm + s_\beta A^{(2)}\varepsilon]^{n+1}}. \tag{7.9.29}$$

When substituted into the integral in Eq. (7.9.21), this series representation of the reflection coefficient provides a complete description of the reflected field in terms of an infinite number of generalized rays. The integral with Y_0 describes the reflected *SSH* wave from the discontinuity at $z = 0$. The rest describe the infinite number of "diving waves" reflected by the positive velocity gradient in $z < 0$.

The singularities associated with each term in the series involve a *string of poles* at approximately the n roots of $Ai[-\zeta_0(p)e^{2\pi i/3}] = 0$, namely at $p_j \simeq n_0 + (s_\beta^2/2n_0)d_j e^{\pi i/3}$. They extend from $p = n_0$ into the upper half of the complex p plane at an angle of $(\pi/3)$ (Fig. 7.66c). The singularities are simple poles for Y_0 and poles of order $(n+1)$ for the nth term under the summation sign of Eq. (7.9.27).

An approximate expression for the field at large distances can be obtained by replacing the Airy-function ratios by their asymptotic forms for $|\zeta_0| \gg 1$. Using Eq. (7.9.16), we have

$$A^{(1)} \to i\frac{(n_0^2 - p^2)^{1/2}}{s_\beta}e^{-\pi i/3}, \qquad A^{(2)} \to -i\frac{(n_0^2 - p^2)^{1/2}}{s_\beta}e^{\pi i/3},$$
$$A \to ie^{-(4k_0 i/3\hat{\gamma})(n_0^2 - p^2)^{3/2} + \pi i/3}. \tag{7.9.30}$$

Substituting these expressions and Eq. (7.9.27) into Eq. (7.9.21), we obtain

$$\{u_\phi\}_{SSH} = k_0^{3/2}\frac{U_0(\omega)dS}{4\pi\sqrt{(2\pi\Delta^*)}}\cos 2\phi e^{\pi i/4}\left[\int_{-\infty}^{\infty} e^{-ik_0 g_0(p)}V_0\frac{p^{3/2}}{q}dp\right.$$
$$\left.+\sum_{n=1}^{\infty} i^n\int_{-\infty}^{\infty} e^{-ik_0 g_n(p)}V_0^{n-1}(1 - V_0^2)\frac{p^{3/2}}{q}dp\right], \tag{7.9.31}$$

where

$$V_0 = \frac{qm - \sqrt{(n_0^2 - p^2)}}{qm + \sqrt{(n_0^2 - p^2)}},$$
$$g_n = p\Delta + q(z + z_h) + \frac{4n}{3\hat{\gamma}}(n_0^2 - p^2)^{3/2}. \tag{7.9.32}$$

Each integral has a pair of branch points at $p = \pm n_0$ associated with the radical $(n_0^2 - p^2)^{3/2}$ in the exponential function in addition to branch points at $p = \pm 1$ and $p = 0$ described earlier. These new branch cuts are the asymptotic equivalents of the line of poles associated with the exact series representation of $\{\hat{R}_{SSH}\}$.

To ensure that the integrals in Eq. (7.9.31) are convergent along the original integration path $-\infty < p < \infty$, these new branch cuts are chosen to follow the lines defined by the poles $p_j = n_0 + (s_\beta^2/2n_0)d_j e^{\pi i/3}$ (Fig. 7.66c).

The saddle point for the nth term in Eq. (7.9.31) is given by $(d/dp)g_n(p) = 0$ or the p root of

$$\Delta - \frac{p}{\sqrt{(1 - p^2)}}(z + z_h) - \frac{4n}{\hat{\gamma}}p\sqrt{(n_0^2 - p^2)} = 0. \tag{7.9.33}$$

Equation (7.9.33) is solved by $p = \sin\vartheta_n$ where ϑ_n is the angle of incidence of the diving ray entering the lower medium and bottoming n times beneath the interface $z = 0$ before arriving at the receiver at (Δ, z) [see Fig. 7.64c]. The steepest descents path for this integrand approaches the saddle point at an angle of $3\pi/4$ for $n_0/\sqrt{2} < p_n < n_0$ and $-3\pi/4$ for $p_n < n_0/\sqrt{2}$, as shown in

Fig. 7.66c. As no singularities are crossed in deforming the contour, the entire contribution for each term in the series comes from the immediate vicinity of the respective saddle points.

Evaluating Eq. (7.9.31) by the saddle-point method for $n_0/\sqrt{2} < p_n < n_0$, we find for the nth diving wave

$$\{u_\phi\}_{SSH} \simeq \left[\frac{k_0 U_0(\omega) dS}{\sqrt{(\Delta^*/\hat{\gamma})}4\pi}\right] \cos 2\phi \left[\frac{V_0^{n-1}(1 - V_0^2)}{q_n}\right]$$

$$\times \left[\frac{p_n^3 q_n^3 (n_0^2 - p_n^2)^{1/2}}{4nq_n^3(n_0^2 - 2p_n^2) + \hat{\gamma}(n_0^2 - p_n^2)^{1/2}(z + z_h)}\right]^{1/2} e^{i\omega(t-\tau) + i\phi_0}, \quad (7.9.34)$$

$$V_0 = V_0(p_n), \qquad q_n = \sqrt{(1 - p_n^2)}, \qquad p_n = \sin \vartheta_n,$$

$$\phi_0 = \frac{\pi}{2}(n - 1), \qquad \tau = \frac{1}{\beta_0} g(p_n).$$

For $z_h = 0$ and angles of incidence $p_n = n_0/\sqrt{2}$, the amplitude term in Eqs. (7.9.34) becomes infinite at $z = 0$. This is because of the formation of a caustic (Fig. 7.41) by diving waves that bottom at depths greater than $|z_m| = n_0/2\hat{\gamma}$ [cf. Eq. (7.9.10)]. For $p_n < n_0/\sqrt{2}$, the result is the same but with $\phi_0 = -\pi n/2$. The difference is caused by the $(-\pi/2)$ phase-shift acquired by the waves as they pass through the caustic. The phase of the nth diving wave differs from the geometric ray theory by $(\pi/2)(n - 1)$, which is attributed to a phase of $\{\pi(n - 1)\}$ acquired by $(n - 1)$ *internal reflections* at $z = 0$ and a phase of $\{-(\pi/2)(n - 1)\}$ associated with $(n - 1)$ *internal caustics* formed by the multiply reflected waves (Fig. 7.41). No phase-shift is associated with the first diving wave corresponding to $n = 1$.

The caustic for waves with angles of incidence corresponding to $p_n \leq n_0/\sqrt{2}$ is a result of the particular velocity distribution chosen in Eq. (7.9.1). Such a caustic will not occur in media with a constant positive gradient or a positive gradient that decreases monotonically with depth. Because we are concerned here with the effect of small gradients on waves near the critical angle of incidence, we will limit our consideration to diving waves with $n_0/\sqrt{2} < p_n < n_0$. These waves are not involved with the caustic formed by the more deeply penetrating diving waves and they provide a reliable analog of the family of cord waves in the earth reflected internally for moderate epicentral distances $\Delta < 30°$ (Fig. 7.65).

We must also remember that the result in Eq. (7.9.34) is limited by two approximations: The first is the assumption that $|\zeta_0| \gg 1$, made in the process of simplification of the reflection coefficient. The second approximation comes from retaining only the first term in the steepest descents evaluation of the integral (geometric ray theory approximation). The first assumption restricts the depth of penetration of a diving ray below the boundary $z = 0$ to $z > z_{\text{min}}$ where $z_{\text{min}} \simeq 3(\hat{\gamma} k_0^2)^{-1/3}$.

We therefore see that for distances such that $\Delta_0 > (\frac{4}{3})p_0/(k_0\hat{\gamma}^2)^{1/3}$, the reflected field can be described as a sum of diving waves, provided that the bottom of the ray is in the range $3/(\hat{\gamma}k_0^2)^{1/3} < |z_m| < n_0/2\hat{\gamma}$ and the ray does not touch the caustic defined by $p_n = n_0/\sqrt{2}$.

At sufficiently large distances, the direct diving wave will arrive first and will be followed at successively later times by waves making an increasing number of internal reflections at $z = 0$. The last group of diving waves to arrive will be those that propagate just below the boundary, making a great number of internal reflections. Their travel time will be essentially that of the pure head wave but their amplitude, decreasing as $n^{-1/2}$, will be small relative to that of the first few diving waves.

7.9 Normal-Mode versus Ray Solutions for Vertically Inhomogeneous Media

7.9.1 Normal-Mode Solution for a Vertically Inhomogeneous Half-space

The normal-mode Love-wave Green's dyadic for a vertically inhomogeneous half-space is given by Eq. (5.105). The propagation of a transient pulse with a source spectrum $S(\omega)$ is, therefore, controlled by the function

$$\mathfrak{G}(\mathbf{z}|\mathbf{z}_0;t) = (\mathrm{curl})(\mathrm{curl})_0[\mathbf{e}_z\mathbf{e}_{z_0}g(\mathbf{z}|\mathbf{z}_0;t)], \tag{7.448}$$

where

$$\begin{aligned} g(\mathbf{z}|\mathbf{z}_0;t) &= \frac{1}{4\pi}\,\mathrm{Re}\int_0^\infty e^{i\omega t - i\pi/2}S(\omega)\sum_{n=0}^\infty\sum_{m=-\infty}^\infty\left[\frac{y_1^+(z)y_1^+(z_0)}{I_1^L}\right] \\ &\quad\times H_m^{(2)}(k_n\Delta)J_m(k_n\Delta_0)e^{im(\phi-\phi_0)}\,d\omega \\ &= \frac{1}{4\pi}\,\mathrm{Re}\cdot\int_0^\infty e^{i\omega t - i\pi/2}S(\omega)\sum_{n=0}^\infty\left[\frac{y_1^+(z)y_1^+(z_0)}{I_1^L}\right]H_0^{(2)}(k_nD)\,d\omega, \end{aligned} \tag{7.449}$$

$$D^2 = \Delta^2 + \Delta_0^2 - 2\Delta\Delta_0\cos(\phi - \phi_0).$$

We approximate the Hankel function by its asymptotic expansion and obtain the far field

$$g(\mathbf{z}|\mathbf{z}_0;t) = \frac{1}{4\pi}\left(\frac{2}{\pi D}\right)^{1/2}\mathrm{Re}\sum_{n=0}^\infty\int_0^\infty S(\omega)k_n^{-1/2}\left[\frac{y_1^+(z)y_1^+(z_0)}{I_1^L}\right]e^{i(\omega t - k_nD - \pi/4)}\,d\omega. \tag{7.450}$$

The major contribution to the integral in Eq. (7.450) comes from the point of stationary phase $\omega = \omega_0$ given by

$$\frac{\partial}{\partial\omega}[\omega t - k_n(\omega)D] = 0, \qquad \text{i.e.,} \quad U_g^n(\omega_0) = \frac{D}{t}, \tag{7.451}$$

where U_g^n denotes the group velocity of the nth mode. Therefore [cf. Sect. 3.6.3]

$$g(\mathbf{z}|\mathbf{z}_0; t) = \frac{1}{2\pi D}\operatorname{Re}\sum_{n=0}^{\infty}\left[\left\{\frac{y_1^+(z)y_1^+(z_0)}{I_1^L}\right\}S(\omega)U_g^n\left\{k_n\left|\frac{dU_g^n}{d\omega}\right|\right\}^{-1/2}\right]_{\omega_0}$$
$$\times \exp\left\{i\left[\omega_0 t - k_n(\omega_0)D - \frac{\pi}{4} - \frac{\pi}{4}\operatorname{sgn} k_n''(\omega_0)\right]\right\}. \tag{7.452}$$

If $(dU_g/d\omega)_{\omega_0} = 0$, Eq. (7.452) should be modified as in Eq. (3.96). The solution, Eq. (7.452), is known as the normal-mode solution.

In the derivation of Eq. (5.105), on which Eq. (7.452) is based, it was tacitly assumed that the only singularities of the integrand in the k plane were poles. However, if the integrand has branch points as well, we shall get additional contribution from these singularities. We have seen already that the branch points give rise to head waves whose amplitude decrease with epicentral distance as $1/\Delta^2$, except near cutoff frequencies of the wave guide, where multiple reflections interfere constructively to yield $1/\Delta$ dependence. Because the normal modes behave as $1/\Delta$, these will predominate at large epicentral distances.

7.9.2. Transformation of the Normal-Mode Solution in Flat-Earth[11] Models into the Ray Solution by the Poisson Transformation

The displacement field of a point source is expressible either as a sum of rays or as a sum of normal modes. Rays can be considered as waves emanating from a set of images. In the immediate vicinity of the source, the number of rays necessary to determine the field is small because the nearby images dominate the field. However, the number of modes required to describe the field at close ranges is large. At great distances, the situation is reversed, the normal-mode solution (e.g., Love and Rayleigh waves in solid-earth configurations) being as a rule more rapidly convergent than the ray solution. In the intermediate ranges, both the number of rays required and the normal modes are large. Such is also the case in spherical-earth models at distances beyond 30°.

It is therefore tempting to find the mathematical relationship between the ray and normal-mode representations. We shall now derive it for a simple case where both solutions are exact, namely, in a case in which there are no branch-line integrals. Because the problem is basically a mathematical one, let

[11] The relationship between the ray and normal-mode solutions in a sphere will be dealt with in Chapter 8.

us skip the physical details of the problem. It is sufficient to know that the field potential resulting from a point source at $z = h$ in a *plate* bounded by two perfectly reflecting boundaries at $z = 0$ and $z = H$ has the integral representation

$$\Psi = 2\int_0^\infty J_0(k\Delta)\left\{\sin \zeta z_< \frac{\sin \zeta(H - z_>)}{\sin \zeta H}\right\}\frac{k\,dk}{\zeta},$$

$$z_< = \min(z, h), \qquad z_> = \max(z, h), \tag{7.453}$$

$$\zeta = \sqrt{(k_0^2 - k^2)} = -i\sqrt{(k^2 - k_0^2)}, \qquad 0 < z, h < H.$$

The ray solution can be written down directly because the reflection coefficient at both boundaries is -1. The effect of the boundaries can be exactly simulated by replacing them with a system of images strung along a vertical line through the source. These consist first of the dipole formed by the source and its negative image at $z = -h$, plus other dipoles spaced at a distance $2H$ apart (Fig. 7.35), with neighboring dipoles having opposite polarity. Mathematically, we make the following expansion of the integral for $0 < z < h$

$$2\sin \zeta z \frac{\sin \zeta(H - h)}{\sin \zeta H} = \frac{1}{i}[1 + e^{-2i\zeta H} + e^{-4i\zeta H} + \cdots]$$
$$\times [e^{i\zeta(z-h)} + e^{-i\zeta(z-h+2H)} - e^{-i\zeta(z+h)} - e^{i\zeta(z+h-2H)}]. \tag{7.454}$$

Using the Sommerfeld integral [Eq. (2.153)]

$$\int_0^\infty J_0(k\Delta)e^{-i\zeta|z|}\frac{k\,dk}{i\zeta} = \frac{e^{-ik_0R}}{R} = g(R^2) = g(\Delta^2 + z^2) \tag{7.455}$$

we obtain

$$\Psi(h > z) = g[\Delta^2 + (h - z)^2] - g[\Delta^2 + (h + z)^2]$$
$$+ \sum_{n=1}^{\infty}\left\{\begin{array}{l} g[\Delta^2 + (h - z + 2nH)^2] + g[\Delta^2 + (z - h + 2nH)^2] \\ -g[\Delta^2 + (z + h + 2nH)^2] - g[\Delta^2 + (-z - h + 2nH)^2]. \end{array}\right. \tag{7.456}$$

An identical result is obtained for the case $z > h$. The first two terms represent the direct ray and the ray reflected from the boundary $z = 0$. Each quadruplet of terms in the foregoing sum represents the four rays that undergo n reflections at the plane $z = H$. The normal-mode solution is obtained by evaluating the integral in Eqs. (7.453) in terms of the residues of the integrand. This is accomplished by first making the transformation

$$\Psi = \int_{-\infty}^{\infty} H_0^{(2)}(k\Delta)\left\{\sin \zeta z_< \frac{\sin \zeta(H - z_>)}{\sin \zeta H}\right\}\frac{k\,dk}{\zeta}. \tag{7.457}$$

The residues at the roots $k = k_n$ of $\sin \zeta H = 0$ then yield

$$\Psi = -\frac{2\pi i}{H}\sum_{n=1}^{\infty} H_0^{(2)}(k_n\Delta)\sin\left(\frac{\pi n h}{H}\right)\sin\left(\frac{\pi n z}{H}\right), \tag{7.458}$$

where

$$k_n = \sqrt{\left[k_0^2 - \left(\frac{\pi n}{H}\right)^2\right]}; \qquad k_0 H > \pi n$$

$$k_n = -i\sqrt{\left[\left(\frac{\pi n}{H}\right)^2 - k_0^2\right]}; \qquad k_0 H < \pi n.$$

At long ranges where $H_0^{(2)}(k_n\Delta) \simeq \sqrt{(2/\pi k_n\Delta)}e^{i(\pi/4 - k_n\Delta)}$, and whenever $k_0 < \pi n/H$, the factor $e^{-ik_n\Delta}$ gives an exponential attenuation with range

$$\exp\left\{-\pi n\left(\frac{\Delta}{H}\right)\sqrt{\left[1 - \left(\frac{2H}{n\Lambda}\right)^2\right]}\right\} \qquad \left(\Lambda = \frac{2\pi}{k_0}\right).$$

Therefore when $\Lambda > 2H$, all modes suffer exponential attenuation with range. For $H < \Lambda < 2H$, the first mode is free, but all higher modes are attenuated, and so on. For moderate and large ranges in terms of H, the number n of modes required in the representation in Eq. (7.458) is less than $(2H/\Lambda)$. The normal-mode solution has poor convergence when the wavelength is a small fraction of the thickness H of the plate.

Let us now see how the normal-mode solution, Eq. (7.458), is related to the ray solution, Eq. (7.456). If

$$F(x) = \sqrt{\left(\frac{2}{\pi}\right)}\int_0^{\infty} f(t)\cos xt\, dt, \tag{7.459}$$

then it can be shown by an application of the basic properties of the Fourier integral that

$$\beta^{1/2}\left[\frac{1}{2}F(0) + \sum_{n=1}^{\infty} F(n\beta)\right] = \alpha^{1/2}\left[\frac{1}{2}f(0) + \sum_{n=1}^{\infty} f(n\alpha)\right], \qquad (\alpha\beta = 2\pi), \tag{7.460}$$

which is known as the *Poisson transformation*. For $\alpha = 1$, $\beta = 2\pi$, this yields

$$\sqrt{(2\pi)}\left[\frac{1}{2}F(0) + \sum_{n=1}^{\infty} F(2\pi n)\right] = \frac{1}{2}f(0) + \sum_{n=1}^{\infty} f(n). \tag{7.461}$$

In our case

$$f(n) = -i\left(\frac{2\pi}{H}\right)H_0^{(2)}\left[\Delta\sqrt{\left(k_0^2 - \frac{\pi^2 n^2}{H^2}\right)}\right]\sin\left(\pi n\frac{h}{H}\right)\sin\left(\pi n\frac{z}{H}\right). \tag{7.462}$$

With $u = \pi t/H$, $v = xH/\pi$, we get from Eqs. (7.459) and (7.462)

$$F(x) = -2i\sqrt{\left(\frac{2}{\pi}\right)}\int_0^\infty H_0^{(2)}[\Delta\sqrt{(k_0^2 - u^2)}]\sin(uh)\sin(uz)\cos(uv)du$$

$$= -\frac{i}{\sqrt{(2\pi)}}\int_0^\infty H_0^{(2)}[\Delta\sqrt{(k_0^2 - u^2)}][\cos u(v + z - h)$$

$$+ \cos u(v - z + h) - \cos u(v + z + h) - \cos u(v - z - h)]du. \tag{7.463}$$

Using the integral

$$\int_0^\infty H_0^{(2)}[\Delta\sqrt{(k_0^2 - x^2)}]\cos(zx)dx = \frac{ie^{-ik_0\sqrt{(\Delta^2 + z^2)}}}{\sqrt{(\Delta^2 + z^2)}}, \tag{7.464}$$

Eq. (7.463) gives

$$F(x) = \frac{1}{\sqrt{(2\pi)}}\begin{cases} g[\Delta^2 + (v + z - h)^2] + g[\Delta^2 + (v - z + h)^2] \\ -g[\Delta^2 + (v + z + h)^2] - g[\Delta^2 + (v - z - h)^2], \end{cases}$$

$$\frac{1}{2}F(0) = \frac{1}{\sqrt{(2\pi)}}\{g[\Delta^2 + (z - h)^2] - g[\Delta^2 + (z + h)^2]\}. \tag{7.465}$$

Inserting Eqs. (7.465) in Eq. (7.461), we get the ray solution, Eq. (7.456). Therefore, for the wave propagation problem considered, the ray solution and normal-mode solution stand in the relationship of a Poisson transform.

7.9.3. *SH* Pulse in a Layered Half-space

The problem of an *SH* pulse in a layered half-space has been discussed in Section 7.4.3. The expression (7.215) for the displacements at the free surface may be written in the form

$$u_\phi^I(\omega) = \frac{U_0(\omega)dS}{2\pi}\cos 2\phi\sum_{n=0}^\infty(I_1 + I_2), \tag{7.466}$$

where

$$I_1 = \frac{\partial}{\partial\Delta}\int_0^\infty J_2(k\Delta)\Lambda^n e^{-\nu_1(h + 2nH)}\frac{k\,dk}{\nu_1},$$

$$I_2 = \frac{\partial}{\partial\Delta}\int_0^\infty J_2(k\Delta)\Lambda^{n+1}e^{-\nu_1[(2n+1)H + (H - h)]}\frac{k\,dk}{\nu_1}. \tag{7.467}$$

Let us first evaluate I_1. Putting

$$s = i\omega, \qquad k = su, \tag{7.468}$$

we get

$$I_1 = s\frac{\partial}{\partial\Delta}\int_0^{\infty} J_2(su\Delta)e^{-sg_1(u^2)}f_1(u^2)u\,du, \tag{7.469}$$

with

$$g_1(u^2) = (h + 2nH)\sqrt{\left(u^2 + \frac{1}{\beta_1^2}\right)},$$

$$f_1(u^2) = \frac{1}{\sqrt{(u^2 + 1/\beta_1^2)}}\left[\frac{\lambda\sqrt{(u^2 + 1/\beta_1^2)} - \sqrt{(u^2 + 1/\beta_2^2)}}{\lambda\sqrt{(u^2 + 1/\beta_1^2)} + \sqrt{(u^2 + 1/\beta_2^2)}}\right]^n \tag{7.470}$$

$$= \frac{1}{\sqrt{(u^2 + 1/\beta_1^2)}}F_n^{(1)}(u^2).$$

Following the development in Section 7.6.4, we now put

$$u = -iv, \qquad J_m(su\Delta) = \mathrm{Im}\{iH_m^{(2)}(su\Delta)\} = \mathrm{Im}\{iH_m^{(2)}(-isv\Delta)\}$$

$$= -\frac{2}{\pi}\,\mathrm{Im}\{e^{\pi im/2}K_m(sv\Delta)\}, \tag{7.471}$$

$$\frac{\partial}{\partial\Delta}J_2(su\Delta) = \frac{2}{\Delta}J_0(su\Delta) + \left(su - \frac{4}{su\Delta^2}\right)J_1(su\Delta).$$

Equation (7.469) then becomes

$$I_1 = I_{11} + I_{12} + I_{13}, \tag{7.472}$$

where

$$I_{11}(s) = \frac{8}{\pi\Delta^2}\,\mathrm{Im}\int_0^{i\infty} K_1(sv\Delta)e^{-sg_1(-v^2)}F_n^{(1)}(-v^2)\frac{dv}{\sqrt{(1/\beta_1^2 - v^2)}},$$

$$I_{12}(s) = \frac{4s}{\pi\Delta}\,\mathrm{Im}\int_0^{i\infty} K_0(sv\Delta)e^{-sg_1(-v^2)}F_n^{(1)}(-v^2)\frac{v\,dv}{\sqrt{(1/\beta_1^2 - v^2)}}, \tag{7.473}$$

$$I_{13}(s) = \frac{2s^2}{\pi}\,\mathrm{Im}\int_0^{i\infty} K_1(sv\Delta)e^{-sg_1(-v^2)}F_n^{(1)}(-v^2)\frac{v^2\,dv}{\sqrt{(1/\beta_1^2 - v^2)}}.$$

We next use the identities

$$K_0(sa)e^{-sb} = \int_0^{\infty} e^{-st}[(t - b - a)(t - b + a)]^{-1/2}H(t - b - a)dt,$$

$$K_1(sa)e^{-sb} = \frac{s}{a}\int_0^{\infty} e^{-st}[(t - b - a)(t - b + a)]^{1/2}H(t - b - a)dt, \tag{7.474}$$

and denote

$$\tau_1 = \Delta v + (h + 2nH)\sqrt{\left(\frac{1}{\beta_1^2} - v^2\right)}, \tag{7.475}$$

where τ_1 is forced, by the choice of the integration contour, to be always real and positive.

Let the source vary with time as $\Theta(t)$. Proceeding as in Section 7.6.4, we find

$$u_\phi^I(t) = u_1(t) + u_2(t),$$

where

$$u_1(t) = \frac{U_0\,dS}{\pi^2}\cos 2\phi \sum_{n=0}^{\infty}(I_{11} + I_{12} + I_{13})$$

$$I_{11} = Q_{11}^{(1)} * \Theta(t),$$

$$I_{12} = \left[\frac{d}{dt}Q_{12}^{(1)}\right] * \Theta(t) = Q_{12}^{(1)} * \frac{d}{dt}\Theta(t),$$

$$I_{13} = \left[\frac{d^2}{dt^2}Q_{13}^{(1)}\right] * \Theta(t) = Q_{13}^{(1)} * \frac{d^2}{dt^2}\Theta(t), \tag{7.476}$$

$$Q_{11}^{(1)}(t) = \frac{4}{\Delta^2}H(t - \tau_1)\,\mathrm{Im}\int_{\tau_1}^{t}\frac{F_n^{(1)}(-v^2)}{\sqrt{(1/\beta_1^2 - v^2)}}\,\frac{T_1[1 + (t - \tau)/v\Delta]}{\sqrt{[(t - \tau)(t - \tau + 2v\Delta)]}}\left(\frac{\partial v}{\partial\tau}\right)d\tau,$$

$$Q_{12}^{(1)}(t) = \frac{2}{\Delta}H(t - \tau_1)\,\mathrm{Im}\int_{\tau_1}^{t}\frac{vF_n^{(1)}(-v^2)}{\sqrt{(1/\beta_1^2 - v^2)}}\left(\frac{\partial v}{\partial\tau}\right)\frac{d\tau}{\sqrt{[(t - \tau)(t - \tau + 2v\Delta)]}},$$

$$Q_{13}^{(1)}(t) = H(t - \tau_1)\,\mathrm{Im}\int_{\tau_1}^{t}\frac{v^2F_n^{(1)}(-v^2)}{\sqrt{(1/\beta_1^2 - v^2)}}\left(\frac{\partial v}{\partial\tau}\right)\frac{T_1[1 + (t - \tau)/v\Delta]}{\sqrt{[(t - \tau)(t - \tau + 2v\Delta)]}}\,d\tau.$$

Similar expressions can be obtained for $u_2(t)$. The asterisk stands for the convolution integral and

$$v(\tau) = \tau\frac{\Delta}{\Delta^2 + (h + 2nH)^2} - \frac{h + 2nH}{\Delta^2 + (h + 2nH)^2}\left[\frac{\Delta^2 + (h + 2nH)^2}{\beta_1^2} - \tau^2\right]^{1/2},$$

$$\frac{\partial v}{\partial\tau} = \frac{\Delta}{\Delta^2 + (h + 2nH)^2} + \frac{h + 2nH}{\Delta^2 + (h + 2nH)^2} \times \frac{\tau}{[(\Delta^2 + (h + 2nH)^2)/\beta_1^2 - \tau^2]^{1/2}}. \tag{7.477}$$

7.9.3.1. Numerical Results. In order to calculate the ensuing displacement field, the expressions in Eqs. (7.476) and (7.477) are converted into a non-dimensional form. To simplify the calculation while still retaining the salient features of the theoretical seismogram, we have chosen a somewhat simpler source. If a torque of moment M_0 is placed in the layer at $z = H/2$, it is found that the spectral surface displacements are given by

$$u_\phi = \frac{M_0(\omega)}{4\pi\mu_1}\int_0^\infty J_1(k\Delta)\frac{A}{L}\frac{k^2\,dk}{v_1}$$
$$= \frac{M_0(\omega)}{4\pi\mu_1}\int_0^\infty J_1(k\Delta)\frac{k^2}{v_1}\left\{\sum_{n=0}^{\infty} e^{-(n+(1/2))v_1 H}\Lambda^{[(n+1)/2]}\right\}dk, \tag{7.478}$$

where A/L is given in Eq. (7.216) and $[(n+1)/2] = (n+1)/2$ if n is odd and $= n/2$ if n is even.

We now introduce the nondimensional quantities

$$v = \frac{1}{\beta_1}x, \qquad \eta = \frac{\beta_1}{\beta_2}, \qquad \rho = \frac{\Delta}{H}, \qquad \bar{\tau} = \frac{\beta_1}{H}\tau, \qquad \bar{t} = \frac{\beta_1}{H}t, \qquad m = n + \frac{1}{2}$$
$$\bar{\tau}_1 = \rho x + m\sqrt{(1-x^2)}, \qquad \bar{\tau}_0 = \sqrt{(m^2+\rho^2)}, \qquad \bar{\tau}_H = \eta\rho + m\sqrt{(1-\eta^2)},$$
$$x = \bar{\tau}\frac{\rho}{\bar{\tau}_0^2} - \frac{m}{\bar{\tau}_0^2}\sqrt{(\bar{\tau}_0^2 - \bar{\tau}^2)}, \tag{7.479}$$
$$\frac{\partial x}{\partial \bar{\tau}} = \frac{\rho}{\bar{\tau}_0^2} + \frac{\bar{\tau}}{\sqrt{(\bar{\tau}_0^2-\bar{\tau}^2)}}\frac{m}{\bar{\tau}_0^2}.$$

Here $\bar{\tau}_1$ corresponds to the time of arrival of the reflected wave, whereas $\bar{\tau}_H$ corresponds to the time of arrival of the head wave. The displacements are then given by

$$u_\phi(t) = \frac{M_0}{2\pi^2\mu_1}\sum_{n=0}^{\infty}\left[Q_n(t) * \frac{d^2\Theta}{dt^2}\right] = \frac{M_0}{2\pi^2\mu_1}\sum_{n=0}^{\infty}\left[\left\{\int_0^t Q_n(t)dt\right\} * \frac{d^3\Theta}{dt^3}\right], \tag{7.480}$$
$$Q_n(t) = H(t-\tau_H)\,\mathrm{Im}\int_{\tau_H}^{t}\frac{v^2\Lambda^{[(n+1)/2]}(-v^2)}{\sqrt{(1/\beta_1^2 - v^2)}}\frac{T_1[1+(t-\tau)/v\Delta]}{\sqrt{[(t-\tau)(t-\tau+2v\Delta)]}}\frac{\partial v}{\partial \tau}\,d\tau.$$

Writing in the dimensionless form, we have

$$Q_n(\bar{t}) = \frac{1}{\beta_1 H}H(\bar{t}-\bar{\tau}_H)\,\mathrm{Im}\int_{\bar{\tau}_H}^{\bar{t}}\frac{x^2\Lambda^{[(n+1)/2]}(-x^2)}{\sqrt{(1-x^2)}}\frac{T_1[1+(\bar{t}-\bar{\tau})/x\rho]}{\sqrt{[(\bar{t}-\bar{\tau})(\bar{t}-\bar{\tau}+2x\rho)]}}$$
$$\times\frac{\partial x}{\partial \bar{\tau}}\,d\bar{\tau},$$
$$u_\phi(\bar{t}) = \frac{M_0}{2\pi^2\mu_1 H^2}\sum_{n=0}^{\infty}\left[U_n(\bar{t}) * \frac{d^3\Theta(\bar{t})}{d\bar{t}^3}\right], \tag{7.481}$$
$$U_n(\bar{t}) = \int_0^{\bar{t}} Q_n(\bar{t})d\bar{t}.$$

7.9.3.2. Specification of the Time Function. Time functions of earthquake sources are best approximated by a Heaviside step function. However, the application of $H(t)$ in the above integrals leads to infinite displacements at the time of arrival of each reflected wave that is preceeded by a refracted wave. In order to suppress these singularities, the time variation of the applied source is taken to be a step function with rounded shoulders. One of the many ways in which this can be achieved is to choose

$$\begin{aligned}\Theta(t) &= 0; && t < 0\\ &= \frac{1}{2}\left(\frac{t}{T}\right)^2; && 0 < t < T\\ &= \frac{1}{2}\left(\frac{t}{T}\right)^2 - \left(1 - \frac{t}{T}\right)^2; && T < t < 2T\\ &= 1; && t > 2T.\end{aligned} \quad (7.482)^{12}$$

Therefore

$$T^2 \frac{d^3}{dt^3} \Theta = \delta(t) - 2\delta(t - T) + \delta(t - 2T).$$

Hence, the convolution integral disappears and we get

$$u_\phi(\bar{t}) = \frac{M_0}{2\pi^2 \mu_1 H^2} \sum_{n=0}^{\infty} \left[\frac{U_n(\bar{t}) - 2U_n(\bar{t} - d) + U_n(\bar{t} - 2d)}{d^2}\right], \quad \left(d = \frac{\beta_1 T}{H}\right). \tag{7.483}$$

Figure 7.67 shows the theoretical seismograms for the case $\Delta = 100H$ which in the earth would correspond to a continental epicentral distance of about 3500 km. Because $H/\beta_1 \simeq 3.5$ s, the *source's rise time*, $2T$, ranges from 0.35 to 3 s. The influence of the rise time on the relative amplitudes of the Love-wave modes is very pronounced. As T increases the spectral content is gradually shifted toward the longer periods, and it seems as if the whole seismogram is being filtered. The dominant period of the longer Love waves depends on d or Δ/H and is equal to about 2.5 H/β_1.

7.9.3.3. The Normal Mode Solution. The theoretical seismograms shown in Fig. 7.67 have been obtained by the exact ray theory for large ranges, where the normal-mode theory should also give a good approximation. In our case the roots of the denominator in Eq. (7.478) occur on the real axis between $k = k_{\beta_2}$ and k_{β_1} and are associated with the Love waves. With $x_n = H\sqrt{(k_{\beta_1}^2 - k_n^2)}$, the period equation is

$$x_n \tan x_n = \frac{\mu_2}{\mu_1} \sqrt{[H^2(k_{\beta_1}^2 - k_{\beta_2}^2) - x_n^2]}. \tag{7.484}$$

[12] In this context T is not travel-time but the rise-time parameter.

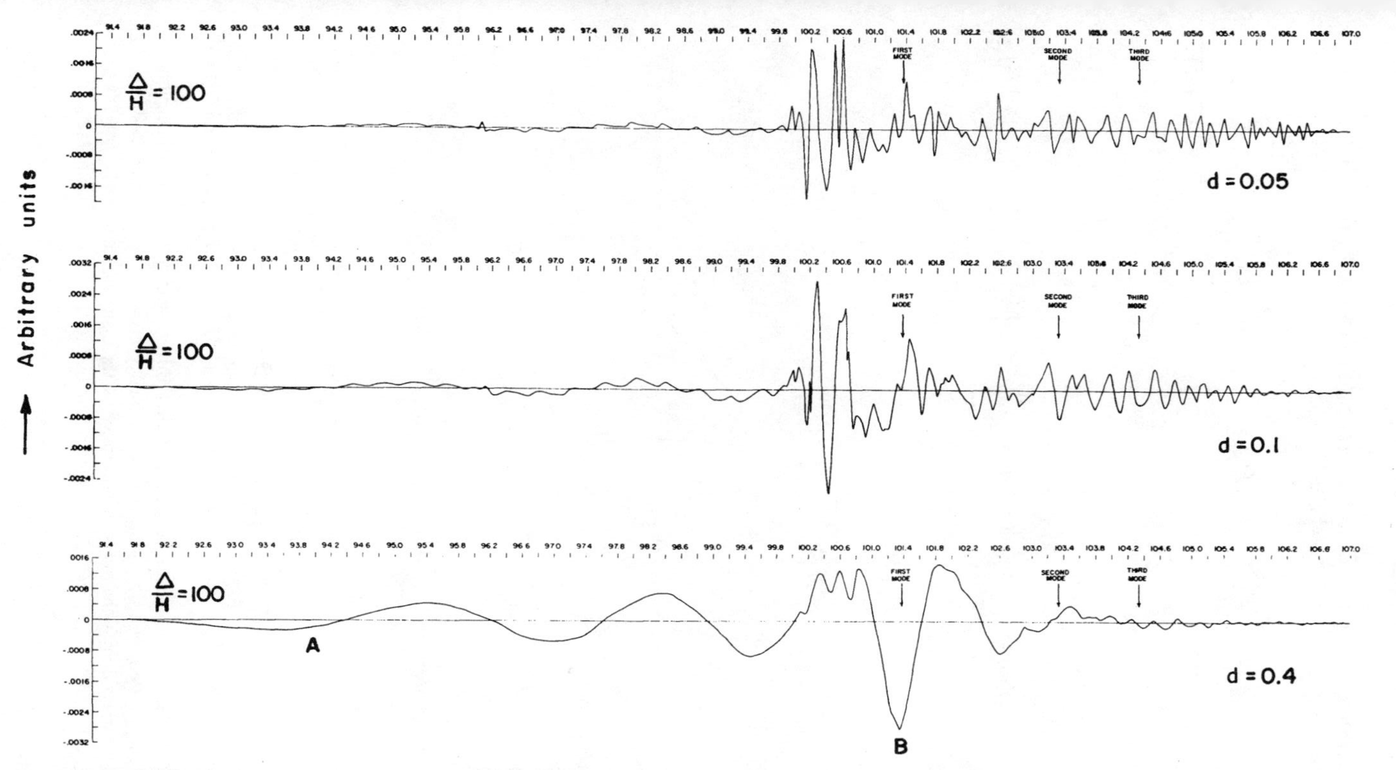

Figure 7.67. Far-field seismograms consisting of horizontal surface ground motion caused by the application of an *SH*-torque pulse in a layered elastic half-space. Source is in midlayer. Displacements are mainly Love waves. The long-period end of the fundamental mode (A) arrives first. It is followed by the Airy phase of that mode (B). Higher modes are smaller in amplitude. Head waves and reflected body waves contributed insignificantly to these seismograms. $\mu_2/\mu_1 = 2$, $\rho_2/\rho_1 = 1.65$, $\beta_2/\beta_1 = 1.1$.

The spectrum of the nth mode at large range Δ (obtained by the usual method of residues) is $q_n e^{i[\omega t - k_n\Delta + \pi/4]}$, where

$$q_n = \frac{M_0(\omega)}{4\mu_1 H}\sqrt{\left(\frac{2k_n}{\pi\Delta}\right)}F(x_n)\cos\left(x_n\frac{h}{H}\right)\cos\left(x_n\frac{z}{H}\right),$$

$$F(x_n) = \frac{x_n}{(\mu_2/\mu_1)^2\cos x_n\cot x_n + \sin x_n\cos x_n + x_n}. \tag{7.485}$$

The spectrum of the time function $\Theta(t)$ is

$$g(\omega) = \int_{-\infty}^{\infty}\Theta(t)e^{-i\omega t}\,dt = \frac{4i}{\omega^3T^2}\sin^2\left(\frac{\omega T}{2}\right)e^{-i\omega T}. \tag{7.486}$$

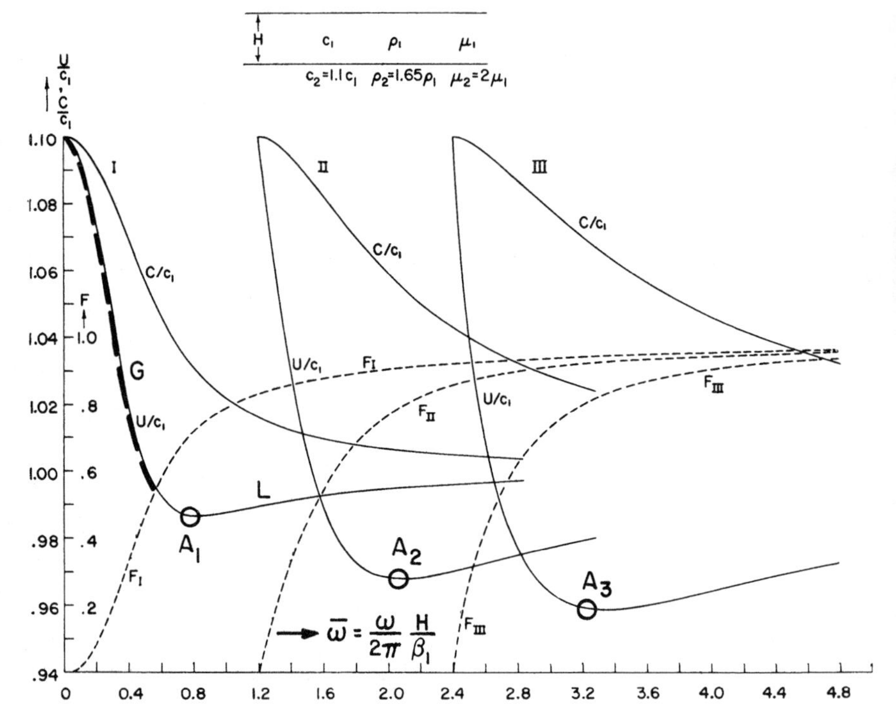

Figure 7.68. Group and phase velocities of first three shear modes. Nondimensional velocities are plotted against nondimensional frequency. $c_1 = \beta_1$, and A indicates the Airy phase. G (dashed line) is the long-period branch of the dispersion curve whereas L is the short-period branch. The F curves give spectral content of theoretical seismograms in the far field.

The stationary phase approximation of the nth-mode time series is evaluated through Eq. (3.96):

$$\{u_\phi(t)\}_n = \frac{1}{\pi} \operatorname{Re} \int_0^\infty g(\omega) q_n(\omega) e^{i(\omega t - k_n\Delta + \pi/4)}\, d\omega$$

$$= \frac{2}{\sqrt{2\pi\Delta}} \operatorname{Re}\left(g(\omega_0) q_n(\omega_0) \frac{U_g}{\sqrt{(|dU_g/d\omega|_{\omega_0})}} \right.$$

$$\left. \times \exp\left\{ i \left[\omega_0 t - k_n(\omega_0)\Delta + \frac{\pi}{4} - \frac{\pi}{4} \operatorname{sgn} k_n''(\omega_0) \right] \right\} \right). \qquad (7.487)$$

As in Eq. (7.452), ω_0 is the root of the equation

$$\frac{t}{\Delta} = \frac{dk_n}{d\omega} = \frac{1}{U_g(\omega)}.$$

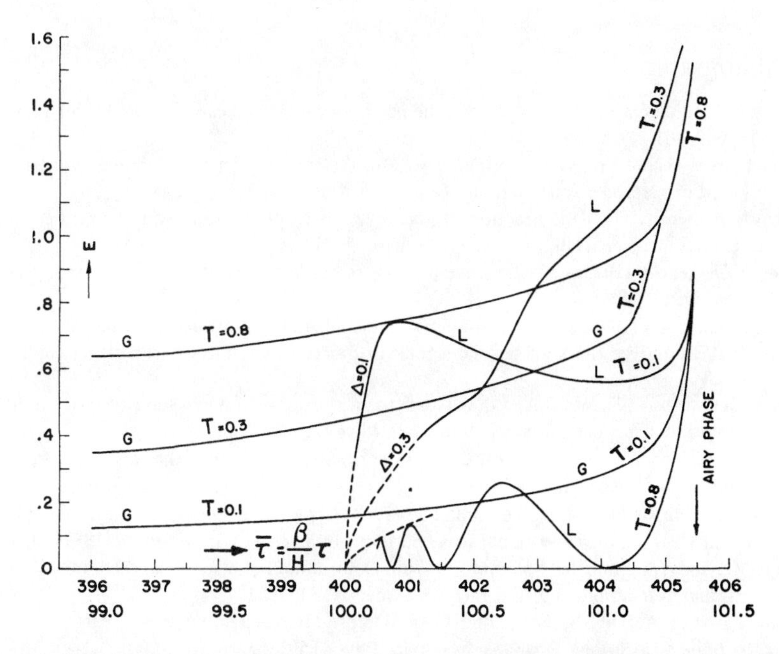

Figure 7.69. Time-domain amplitudes of the two dispersion branches of the first Love mode. G denotes the amplitudes of the long-period branch and L denotes amplitudes of the short-period branch. The dashed curves are the envelopes of the maxima. Amplitude curves are given for three different values of the source rise-time parameter T.

The period equation [cf. Eq. (3.78)] can be recast in the form $\bar{\omega} = f(c)$, when $\bar{\omega}$ is a nondimensional frequency

$$\bar{\omega} = \frac{\omega H}{2\pi\beta_1} = \frac{1}{2\pi}\frac{(c/\beta_1)}{\sqrt{[(c/\beta_1)^2 - 1]}}\tan^{-1}\left\{\sqrt{\left[\frac{(\beta_2/\beta_1)^2 - (c/\beta_1)^2}{(c/\beta_1)^2 - 1}\right]}\frac{\mu_2}{\mu_1}\frac{\beta_1}{\beta_2}\right\}. \tag{7.488}$$

From this formula, analytic expressions for $U_g = c^2/[c - \omega(dc/d\omega)]$ and $dU_g/d\omega$ as functions of ω and c, can be derived. The overall amplitude dependence of $\{u_\phi(t)\}_n$ for the case at hand $z = 0$, $h/H = \frac{1}{2}$, is given by the excitation function

$$E_n = F(x_n)\cos\left(\frac{x_n}{2}\right)\frac{\sin^2\{\pi\bar{\omega}\beta_1 T/H\}}{\bar{\omega}^2 T}\frac{1}{\sqrt{[\bar{\omega}(c/\beta_1)|dU_g/d\bar{\omega}|]}}. \tag{7.489}$$

The group-velocity and phase-velocity calculations are shown in Fig. 7.68. This figure also shows the function $F(x_n)$ of Eq. (7.485). The excitation function $E(x_n)$ is shown in Fig. 7.69. These calculations confirm that the identification of the prominent arrivals in the theoretical seismograms shown in Fig. 7.67 with the *SH* surface waves in the far field is correct. The arrival times of the dominant periods and the relative amplitudes of the long and shorter waves yields a good agreement between exact ray theory and normal-mode theory.

Bibliography

Airy GB (1838) On intensity of light in the neighbourhood of a caustic. Camb Phil Trans 6: 379–401.

Bateman H (1928) Transverse seismic waves on the surface of a semi-infinite solid composed of heterogeneous material. Bull Amer Math Soc 34: 343–348.

Ben-Menahem A (1960) Diffraction of elastic waves from a surface-source in a heterogeneous medium. Bull Seism Soc Amer 50: 15–33.

Ben-Menahem A (1975a) Elastic wave-motion across a vertical discontinuity. Jour Eng Math 9: 145–158.

Ben-Menahem A (1975b) Source parameters of the Siberian explosion of June 30, 1908 from analysis and synthesis of seismic signals at four stations. Phys Earth Planet Int 11: 1-35.

Ben-Menahem A, Cisternas A (1963) The dynamic response of an elastic half-space to an explosion in a spherical cavity. Jour Math Phys 42: 112–125.

Ben-Menahem A, Gillon A (1970) Crustal deformation by earthquakes and explosions. Bull Seismol Soc Amer 60: 193–215.

Ben-Menahem A, Jarosch H, Rosenman M (1968) Large scale processing of seismic data in search of regional and global stress patterns. Bull Seismol Soc Amer 58: 1899–1932.

Ben-Menahem A, Singh SJ, Solomon F (1969) Deformation of a spherical earth model by internal dislocations. Bull Seismol Soc Amer 59: 813–853.

Ben-Menahem A, Singh SJ, Solomon F (1970) Deformation of a homogeneous earth model by finite dislocations. Rev Geophys Space Phys 8: 591–632.

Ben-Menahem A, Smith SW, Teng TL (1965) A procedure for source studies from spectrums of long-period seismic body waves. Bull Seismol Soc Amer 55: 203–235.

Ben-Menahem A, Weinstein M (1970) The *P–SV* decoupling condition and its bearing on the structure of the earth. Geophys Jour Roy Astron Soc (London) 21: 131–135.

Born M, Wolf E (1964) Principles of Optics (2nd rev ed) Pergamon, New York, 808 pp.

Brekhovskikh LM (1948) Distribution of sound in a liquid layer with a constant velocity gradient (in Russian). Dokl Akad Nauk SSSR 62: 469–471.

Bruns H (1895) Das Eikonal. Abh Kgl Sächs Ges Wiss Math-Phys Kl 21: 370–391.

Bullen KE (1961) Seismic ray theory. Geophys Jour Roy Astron Soc (London) 4: 93–105.

Burridge R, Lapwood ER, Knopoff L (1964) First motions from seismic sources near a free surface. Bull Seismol Soc Amer 54: 1889–1913.

Byerly P (1926) The Montana Earthquake of June 28, 1925. Bull Seismol Soc Amer 16: 209–265.

Cagniard L (1939) Réflection et réfraction des ondes séismiques progressive. Gauthier-Villars and Cie, Paris.

Chao CC, Bleich HH, Sackman J (1961) Surface waves in an elastic half-space. Jour Appl Mech 28: 300–301.

Clebsch A (1863) Über die Reflection an einer Kugelfläche. Jour Math 61: 195–262.

Dix CH (1961) The seismic head pulse, reflection and pseudo-reflection pulses. Jour Geophys Res 66: 2945–2951.

Dziewonski AM, Gilbert F (1976) The effect of small, aspherical perturbations on travel times and re-examination of the corrections for ellipticity. Geophys Jour Roy Astron Soc (London) 44: 7–17.

Epstein PS (1930) Geometrical optics in absorbing media. Proc Natl Acad Sci (US) 16: 37–45.

Gerjuoy E (1953) Total reflection of waves from a point source. Comm Pure Appl Math 6: 73–91.

Gerver M, Markushevich V (1966) Determination of a seismic wave velocity from the travel-time curve. Geophys Jour Roy Astron Soc (London) 11: 165–173.

Helmberger DV (1968) The crust–mantle transition in the Bering-Sea. Bull Seismol Soc Amer 58: 179–214.

Herglotz G (1907) Über das Benndorfsche Problem der Fortpflanzungsgeschwindigkeit der Erdbenstrahlen. Phys Zeit 8: 145–147.

Herrin E (1968) Seismological Tables for *P* phases. Bull Seismol Soc Amer 58: 1193–1241.

Hill DP (1973) Critically refracted waves in a spherically symmetric radially heterogeneous earth model. Geophys Jour Roy Astron Soc (London) 34: 149–177.

Honda H (1962) Earthquake mechanism and seismic waves. Jour Phys Earth 10: 1–97.

Hook JF (1962) Contributions to a theory of separability of the vector wave equation of elasticity for inhomogeneous media. Jour Acoust Soc Amer 34: 946–953.

Hron F (1972) Numerical methods of ray generation in multilayered media. In: Bolt BA (ed.) Methods in Computational Physics 12: 1–34, Academic Press, New York.

Ingram RE, Hodgson JH (1956) Phase change of *PP* and *pP* on reflection at a free surface. Bull Seismol Soc Amer 46: 203–213.

Israel M, Ben-Menahem A (1974) Residual displacements and strains due to faulting in real earth models. Phys Earth Planet Int 8: 23–45.

Israel M, Vered M (1977) Near-field source parameters by finite-source theoretical seismograms. Bull Seismol Soc Amer 67: 631–640.

Jeffreys H (1931) The formation of Love waves (Querwellen) in a two-layer crust. Gerl Beit Geophys 30: 336–350.

Jeffreys H, Bullen KE (1940) Seismological Tables. British Association for the Advancement of Science, London.

Johnson LE, Gilbert F (1972) Inversion and inference for teleseismic ray data. In: Bolt BA (ed) Methods in Computational Physics 12: 231–266, Academic Press, New York.

Karal FC Jr, Keller JB (1959) Elastic wave propagation in homogeneous and inhomogeneous media. Jour Acoust Soc Amer 31: 694–705.

Kawasumi H (1937) A historical sketch of the development of knowledge concerning the initial motion of an earthquake. Publ Bureau Central Seismol Intern Ser A Travaux Sci 15: 258–330.

Kline M, Kay IW (1965) Electromagnetic Theory and Geometrical Optics. Interscience, New York.

Knopoff L, Gilbert F (1959) First motion methods in theoretical seismology. Jour Acoust Soc Amer 31: 1161–1168.

Knott CG (1919) The propagation of earthquake waves through the earth and connected problems. Proc Roy Soc Edinburgh 39: 157–208.

Lamb H (1904) On the propagation of tremors over the surface of an elastic solid. Phil Trans Roy Soc (London) A203: 1–42.

Lapwood ER (1949) The disturbance due to a line source in a semi-infinite elastic medium. Phil Trans Roy Soc (London) A242: 63–100.

Mohorovičić A (1910) Das Beben vom 8, Okt., 1909. Jharb Meteorol Obs Zagreb (Agram) 9: Part IV, Zagreb.

Mooney HM (1974) Some numerical solutions for Lamb's problem. Bull Seismol Soc Amer 64: 473–491.

Nakamura ST (1922) On the distribution of the first movement of the earthquake. Jour Meteorol Soc Japan 2: 1–10.

Oldham RD (1900) On the propagation of earthquake motion to great distances. Phil Trans Roy Soc (London) A194: 135–174.

Pekeris CL (1950) Ray theory vs. normal mode theory in wave propagation problems. Proc Symp Appl Math 2: 71–75.

Pekeris CL, Alterman Z, Abramovici F (1963) Propagation of an *SH*-torque pulse in a layered solid. Bull Seismol Soc Amer 53: 39–57.

Pridmore-Brown DC, Ingrad U (1955) Sound propagation into the shadow zone in a temperature-stratified atmosphere above a plane boundary. Jour Acoust Soc Amer 27: 36–42.

Rabinowitz P (1970) Gaussian integration of functions with branch point singularities. Intern Jour Comp Math 2: 297–306.

Shimshoni M, Ben-Menahem A (1970) Computation of the divergence coefficient for seismic phases. Geophys Jour Roy Astron Soc (London) 21: 285–294.

Singh SJ, Ben-Menahem A (1969a) Displacement and strain fields due to faulting in a sphere. Phys Earth Planet Int 2: 77–87.

Singh SJ, Ben-Menahem A (1969b) Decoupling of the vector wave equation of elasticity for radially heterogeneous media. Jour Acoust Soc Amer 46: 655–660.

Singh SJ, Ben-Menahem A (1969c) Asymptotic theory of body waves in a radially heterogeneous earth. Bull Seismol Soc Amer 59: 2039–2059.

Singh SJ, Ben-Menahem A, Shimshoni M (1972) Theoretical amplitudes of body waves from a dislocation source in the earth. I. Core reflections. Phys Earth Planet Int 5: 231–263.

Slichter LB (1932) The theory of the interpretation of seismic travel-time curves in horizontal structures. Physics 3: 273–295.

Smirnov V (1933) Sur l'application de la méthode nouvelle à l'étude les vibrations élastiques dans l'espace à symétrie axiale. Publ Inst Seismol Acad Sci URSS No. 29.

Smirnov V, Sobolev S (1932, 1933) Sur une méthode nouvelle dans le probléme plan des vibrations élastiques. Publ Inst Séismol Acad Sci URSS Nos. 20, 29.

Sommerfeld A, Runge J (1911) Anwendung der Vektorrechnung auf die Grundlagen der Geometrischen Optik. Ann Phys 35: 277–298.

Spencer TW (1960) The method of generalized reflection and transmission coefficients. Geophys 25: 624–641.
Tolstoy I, Usdin E (1953) Dispersive properties of stratified elastic and liquid media: A ray theory. Geophys 18: 844–870.
Vered M, Ben-Menahem A (1974) Application of synthetic seismograms to the study of low-magnitude earthquakes and crustal structure in the northern Red-Sea region. Bull Seismol Soc Amer 64: 1221–1237.
Vered M, Ben-Menahem A (1976) Generalized multipolar ray theory for surface and shallow sources. Geophys Jour Roy Astron Soc (London) 45: 195–198.
Vered M, Ben-Menahem A, Aboodi E (1975) Computer generated *P* and *S* waveforms from an earthquake source. Pure Appl Geophys 113: 651–659.
Werth GC, Herbst RF, Springer DL (1962) Amplitudes of seismic arrivals from the M discontinuity. Jour Geophys Res 67: 1587–1610.
Wiechert E, Geiger L (1910) Bestimmung des Weges der Erdbebenwellen im Erdinnern. Phys Zeit 11: 294–311.

CHAPTER 8

Asymptotic Theory of the Earth's Normal Modes

After having spent years trying to be accurate, we must spend as many more in discovering when and how to be inaccurate.

(Ambrose Bierce)

8.1. Jeans' Formula

We have seen that the complete seismic field induced in a radially heterogeneous sphere can be expressed as an infinite sum of standing waves, namely the normal modes. However, we know from seismogram analysis that most of the recorded earth motion can be explained in terms of propagating waves. There must exist a link, therefore, between these two seemingly different aspects of wave motion.

Consider a general term of the normal-mode solution. The factor of this term that depends upon the time and the colatitudinal angle is

$$e^{i_n\omega_l t} P_l^m(\cos\theta). \tag{8.1}$$

Replacing $P_l^m(\cos\theta)$ by its asymptotic approximation for large values of l (App. H), this factor becomes

$$(-l)^m\left(\frac{1}{2\pi l\sin\theta}\right)^{1/2}\left[\exp\left[i\left\{{}_n\omega_l t-\left(l+\frac{1}{2}\right)\theta+\frac{\pi}{4}-\frac{m\pi}{2}\right\}\right]\right.$$
$$\left.+\exp\left[i\left\{{}_n\omega_l t+\left(l+\frac{1}{2}\right)\theta-\frac{\pi}{4}+\frac{m\pi}{2}\right\}\right]\right]. \tag{8.2}$$

The first term in expression (8.2) describes a wave motion of frequency ${}_n\omega_l$ diverging from the pole, $\theta = 0$, whereas the second term describes a wave motion diverging from the antipode, $\theta = \pi$. A diverging wave in circular cylinder coordinates may be expressed as

$$\exp\{i({}_n\omega_l t - {}_nk_l\Delta)\} = \exp\left\{i\,{}_n\omega_l\left[t-\frac{\Delta}{c({}_n\omega_l)}\right]\right\}, \tag{8.3}$$

where ${}_nk_l$ is the wave number corresponding to the phase velocity $c({}_n\omega_l)$. Putting $\Delta = a\theta$[1] and comparing expressions (8.2) and (8.3), we get *Jeans' formula*

$$l + \frac{1}{2} = {}_nk_l a = \frac{{}_n\omega_l a}{c({}_n\omega_l)}. \tag{8.4}$$

The relation in Eq. (8.4) means that if l is large, every mode of oscillation can be interpreted as a propagating wave whose phase velocity depends upon l and n according to the following equations:

$$ {}_nc_l = \frac{{}_n\omega_l a}{l + \frac{1}{2}}, \qquad {}_nT_l = \frac{2\pi a}{(l + \frac{1}{2}){}_nc_l}, \qquad {}_n\Lambda_l = \frac{2\pi a}{l + \frac{1}{2}}, \tag{8.5}$$

where ${}_nT_l$ is the period of the oscillation and ${}_n\Lambda_l$ is the associated wavelength. The relations in Eqs. (8.5) are found to yield a good approximation for $l \geq 7$.

Jeans' formula is used mainly in the study of seismic surface waves, where it associates l with long-period Fourier components, but it is valid for all seismic waves, including body waves. Knowing l and ${}_nT_l$, $c({}_nT_l)$ can be easily calculated through Jeans' formula to yield surface-wave dispersion curves in the period range where the formula is applicable. The group velocity can then be found from the relation

$$U_g = \left(\frac{d\omega}{dk}\right)_{{}_n\omega_l} = a\,\frac{d\,{}_n\omega_l}{dl}. \tag{8.6}$$

The derivative with respect to l is obtained, in practice, by finite differencing. Equation (8.6) is not valid for small values of l.

Table 8.1 compares the phase velocities of the fundamental Rayleigh mode (obtained from the fundamental spheroidal periods through Jeans' formula) with standard values. Table 8.2 gives the calculated periods of the toroidal

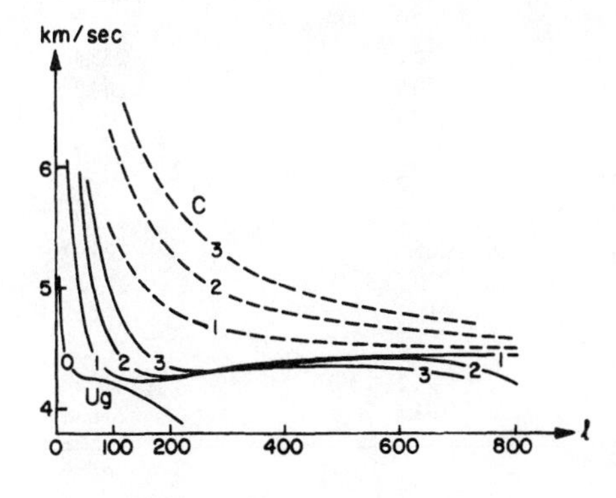

Figure 8.1. Calculated toroidal phase velocities (dashed lines) and group velocities (solid lines) as functions of l for normal modes $n = 0,1,2,3$ in a Gutenberg–Bullard I (App. L) earth model.

[1] This notation holds up to Sec. 8.7.3.

Table 8.1. Spheroidal Free Periods (in seconds) for the Chilean Earthquake of 22 May 1960

l	Agra	Chester	Hallett	Hong Kong	Kyoto	Los Angeles	Lwiro	Mt. Tsukuba	Palisades
2		3264			3204	3233			3252
3		2124			2148	2133			2149
4		1530			1530	1551			1550
5		1176			1194	1190			1197
6		956			962	964			966
7		806			812	805	811	809	
8	710	707			707	707	709	706	704
9	637	632			634	634	635	631	634
10	583	576	579		581	581	582	578	572
11	538	536		539	536	536	535	532	533
12	504.5	501		502.5	503.1	502.1	502.6		501.4
13	475.6	472.2	473.4	474.5		472.9	473.6	474.5	471.7
14	450.0	448.2	448.0	449.4		448.1	449.0	448.4	447.6
15	428.5	424.2	426.3	427.2		426.1	426.7	427.0	426.6
16	409.2	405.0	406.9			406.8	408.2	406.7	406.7
17	390.1	390.0	388.8			389.3	390.0		390.0
18	375.4	375.0	373.7			373.9	374.7		374.1
19	361.0	360.0	360.5	361.6		360.1	360.5	360.9	
20	349.2	346.2	347.4	348.3		346.7			346.9
21	337.2	336.0	335.6	336.8		336.5		335.8	335.2
22	326.1	325.2	324.8	326.0		325.4		325.6	325.1
23	316.5	315.0	315.4	316.2		315.3	315.7		315.2
24		306.0	306.1	306.9		306.2	305.6	306.7	306.6
25		297.6	297.4	298.4		297.5	298.5		297.4
26		290.4	290.1	290.5		289.7			289.4
27		282.6	282.2			282.2	280.7		281.4
28		275.4	275.0			275.1	276.3		274.4
29		268.2	268.4	269.1		268.6			267.7
30		261.6		262.6		262.0	261.8		
31		255.0	256.0	256.7		256.2	257.1		255.4
32				251.0		250.0			249.2
33		245.4	245.2	245.5		245.3			
34		239.4	239.6	240.2		239.9			238.7
35			234.7	235.3		235.2			
36				230.4		230.2			230.7
37		225.0				225.3			226.9
38						220.9			219.2
39						216.7			214.9
40									211.5
41						208.5			207.3
42						204.3			204.3

Paris	Pasadena	Suva	Tiefen-ort	Trieste	Uppsala	Average phase velocity (km/s)	Difference between observed and standard (km/s)
	3234			3186		4.959	0.008
	2133			2172		5.336	−0.014
1552	1548		1548	1554		5.756	0.015
1194	1188		1182	1188	1191	6.121	0.032
965	960			960		6.402	0.053
	810				809	6.598	0.093
707	709	708	708	708	706	6.657	0.100
635	634	636				6.643	0.123
	580	580			579	6.582	0.152
	539	536			538	6.491	0.174
	502.2	502.3			501.5	6.374	0.172
	472.8				473.0	6.262	0.162
	448.2	447.6			446.8	6.157	0.163
	426.0					6.054	0.157
	406.8	406.7				5.960	0.152
	388.8	388.3			388.7	5.874	0.150
	373.8	373.5			372.6	5.784	0.137
	360.6	362.2			359.4	5.691	0.118
	346.8	348.3			346.9	5.620	0.117
	335.4	336.0			334.8	5.542	0.104
	323.4	325.0				5.470	0.095
	315.6	314.9				5.398	0.082
	306.0	305.9			305.2	5.337	0.076
	297.6	297.2			296.9	5.274	0.066
	289.8	289.1				5.211	0.053
	281.4	281.8				5.166	0.056
	275.4	274.7				5.103	0.039
	268.2	267.9				5.057	0.035
	262.2	261.5				5.010	0.028
	256.2	255.5				4.963	0.021
	250.8	249.7				4.923	0.017
	244.8	244.5				4.874	0.003
	240.0	239.2				4.843	0.005
	235.2	233.9				4.801	−0.006
	229.2	228.8				4.771	−0.006
	223.8	224.7				4.741	−0.007
	219.6					4.728	0.006
		216.1				4.693	−0.002

Table 8.2. Calculated Toroidal Eigenperiods and Velocities of the Gutenberg–Bullard I Earth Model

$n = 0$				$n = 1$				$n = 2$			
l	${}_nT_l$	c	U_g	l	${}_nT_l$	c	U_g	l	${}_nT_l$	c	U_g
2	2658.90	6.021		1	818.80	32.588		1	452.28	58.996	
3	1722.90	6.637		2	764.34	20.945		2	443.40	36.106	
4	1320.60	6.735	6.690	3	698.64	16.368	5.542	3	430.83	26.543	3.059
5	1090.00	6.676	6.152	4	631.96	14.074	6.481	4	415.30	21.417	3.886
6	938.03	6.564	5.768	5	570.69	12.751	7.055	5	397.60	18.302	4.685
7	828.81	6.439	5.492	6	517.53	11.898	7.306	6	378.55	16.266	5.443
8	745.65	6.315	5.288	7	472.73	11.289	7.316	7	358.87	14.870	6.135
9	679.72	6.198	5.131	8	435.42	10.814	7.178	8	339.29	13.878	6.720
10	625.87	6.090	5.007	9	404.29	10.421	6.977	9	320.40	13.149	7.157
11	580.86	5.992	4.906	10	378.05	10.083	6.772	10	302.70	12.593	7.422
12	542.57	5.901	4.823	11	355.61	9.787	6.594	11	286.48	12.148	7.520
13	509.51	5.819	4.753	12	336.14	9.525	6.448	12	271.86	11.778	7.486
14	480.61	5.743	4.694	13	319.03	9.293	6.330	13	258.80	11.456	7.367
15	455.10	5.674	4.644	14	303.82	9.085	6.231	14	247.15	11.168	7.209
16	432.37	5.610	4.602	15	290.19	8.898	6.146	15	236.73	10.908	7.049
17	411.97	5.552	4.566	16	277.89	8.729	6.069	16	227.35	10.669	6.905
18	393.54	5.497	4.534	17	266.71	8.575	5.998	17	218.85	10.451	6.783
19	376.80	5.447	4.508	18	256.52	8.434	5.930	18	211.08	10.250	6.683
20	361.49	5.401	4.485	19	247.18	8.304	5.864	19	203.94	10.064	6.599
21	347.45	5.358	4.465	20	238.59	8.183	5.800	20	197.34	9.894	6.529
22	334.51	5.318	4.447	21	230.65	8.071	5.738	21	191.21	9.736	6.469
23	322.54	5.280	4.432	22	223.31	7.966	5.676	22	185.50	9.589	6.416
24	311.44	5.245	4.419	23	216.49	7.867	5.617	23	180.16	9.453	6.369
25	301.10	5.213	4.408	24	210.14	7.774		24	175.16	9.327	
26	291.44	5.182	4.398	25	204.21	7.686		25	170.46	9.208	

l	$_nT_l$	c	U_g
27	282.41	5.154	4.389
28	273.93	5.127	4.381
29	265.96	5.101	4.375
30	258.45	5.077	4.369
31	251.37	5.055	4.363
32	244.67	5.033	4.359
33	238.32	5.013	4.354
34	232.30	4.994	4.351
35	226.58	4.976	4.347
36	221.14	4.959	4.344
37	215.96	4.942	4.342
38	211.02	4.927	4.339
39	206.30	4.912	4.337
40	201.79	4.897	4.335
41	197.47	4.884	4.333
42	193.34	4.871	4.331
43	189.38	4.858	4.330
44	185.58	4.847	4.328
45	181.93	4.835	4.327
46	178.42	4.824	4.326
47	175.04	4.814	4.325
48	171.79	4.804	4.324
49	168.67	4.794	4.323
50	165.65	4.785	4.322
51	162.74	4.776	
52	159.93	4.767	

$n = 3$				$n = 4$			
l	$_nT_l$	c	U_g	l	$_nT_l$	c	U_g
1	309.18	86.301		1	232.30	114.865	
2	306.33	52.262		2	231.10	69.275	
3	302.17	37.844	2.091	3	229.34	49.862	1.549
4	296.84	29.963	2.668	4	227.04	39.174	1.980
5	290.48	25.052	3.233	5	224.25	32.450	2.404
6	283.26	21.738	3.786	6	221.02	27.860	2.821
7	275.36	19.380	4.329	7	217.39	24.549	3.230
8	266.91	17.641	4.864	8	213.41	22.064	3.631
9	258.09	16.324	5.391	9	209.14	20.145	4.028
10	249.02	15.307	5.906	10	204.62	18.629	4.420
11	239.83	14.511	6.394	11	199.90	17.410	4.812
12	230.68	13.880	6.834	12	195.02	16.418	5.206
13	221.71	13.372	7.194	13	190.02	15.602	5.603
14	213.06	12.955	7.447	14	184.93	14.926	6.002
15	204.85	12.605	7.581	15	179.78	14.363	6.396
16	197.18	12.302	7.601	16	174.61	13.892	6.769
17	190.09	12.032	7.534	17	169.48	13.495	7.099
18	183.57	11.785	7.413	18	164.44	13.156	7.359
19	177.59	11.557	7.272	19	159.56	12.864	7.531
20	172.09	11.345	7.133	20	154.88	12.606	7.607
21	167.02	11.146	7.010	21	150.45	12.373	7.598
22	162.30	10.960	6.905	22	146.29	12.160	7.525
23	157.91	10.785	6.818	23	142.40	11.960	7.417
24	153.80	10.622		24	138.77	11.772	
25	149.93	10.469		25	135.37	11.595	

oscillations of the Gutenberg–Bullard I (G–BI) model of the earth together with the phase and group velocities obtained from Jeans' formula for the fundamental mode and the first four overtones. Similarly, Table 8.3 gives the calculated periods, phase velocities, and group velocities for the spheroidal oscillations of the same model. Figure 8.1 shows the calculated toroidal phase and group velocities vs. l.

Table 8.3. Calculated Spheroidal Eigenperiods and Velocities of the Gutenberg–Bullard I Earth Model

$n = 0$			
l	${}_nS_l$	c	U_g
2	3233.60	4.951	
3	2137.60	5.350	
4	1549.50	5.740	7.436
5	1195.00	6.089	7.792
6	969.83	6.349	7.708
7	820.37	6.505	7.273
8	718.19	6.556	6.582
9	646.21	6.520	5.864
10	592.88	6.429	5.325
11	550.89	6.318	4.978
12	516.25	6.202	4.844
13	486.00	6.100	4.699
14	460.51	5.994	4.513
15	437.83	5.898	4.464
16	417.62	5.808	4.384
17	399.50	5.725	4.309
18	383.16	5.646	4.242
19	368.31	5.573	4.178
20	354.77	5.503	4.113
21	342.38	5.437	4.066
22	330.92	5.375	4.014
23	320.37	5.316	3.967
24	310.54	5.261	3.933
25	301.40	5.208	3.892
26	292.85	5.157	3.868
27	284.82	5.110	3.831
28	277.29	5.065	3.813
29	270.15	5.022	3.798
30	263.44	4.981	3.766
31	257.07	4.943	3.756
32	251.02	4.906	3.750
33	245.27	4.871	3.720
34	239.83	4.837	3.713
35	234.58	4.806	3.739
36	229.56	4.777	3.706
37	224.81	4.748	3.710
38	220.19	4.721	3.714
39	215.81	4.695	3.679
40	211.61	4.670	3.684
41	207.57	4.646	3.685
42	203.67	4.624	3.682
43	199.92	4.602	3.732
44	196.23	4.583	3.747
45	192.73	4.564	3.651
46	189.43	4.544	3.658
47	186.16	4.526	3.718
48	183.00	4.510	3.719
49	179.94	4.494	3.728
50	176.97	4.479	3.748
51	174.08	4.464	
52	171.38	4.449	

$n = 1$			
l	${}_nS_l$	c	U_g
1	2508.40	10.637	
2	1489.50	10.748	
3	1079.30	10.595	9.759
4	865.19	10.280	8.504
5	741.75	9.811	6.706
6	668.97	9.204	5.401
7	614.26	8.688	5.490
8	563.35	8.358	6.311
9	514.33	8.191	
10	468.85	8.130	

8.2. Watson's Transformation of the Spectral Field

Let the spectral field in a radially heterogeneous medium be given in the form of an infinite series of the type

$$\sum_{l=m}^{\infty} (2l + 1) f_l P_l^m(\cos\theta), \tag{8.7}$$

where $P_l^m(\cos\theta)$ is the associated Legendre function. It has been found that, in general, the numerical evaluation of this series is hampered by its poor convergence. We will now show that this series can be converted into a rapidly converging series with the help of Cauchy's residue theorem.

Consider the complex integral

$$\int_{C_1} f_{s-1/2} P_{s-1/2}^m(-\cos\theta) \frac{s\,ds}{i\cos s\pi}, \tag{8.8}$$

taken along the contour C_1 of Fig. 8.2a. This contour starts at $+\infty$ and returns to $+\infty$ after enclosing the zeros $s = l + \frac{1}{2}$ $(l = 0, 1, 2, \ldots)$ of $\cos s\pi$ in a clockwise sense. These zeros are denoted by crosses on the Re s axis in Fig. 8.2a. In the integral (8.8),

$$\begin{aligned} P_s^m(x) &= (1 - x^2)^{m/2} \frac{d^m}{dx^m} P_s(x), \qquad (m \geq 0) \\ P_s(x) &= {}_2F_1\left(-s, s+1, 1; \frac{1-x}{2}\right), \\ P_s^{-m}(x) &= (-1)^m \frac{\Gamma(s - m + 1)}{\Gamma(s + m + 1)} P_s^m(x), \end{aligned} \tag{8.9}$$

where ${}_2F_1(a, b, c; x)$ is the hypergeometric function. The above definition holds for arbitrary s. When s is an integer, $P_s(x)$ is identical with the Legendre polynomial.

The residue at the pole $s = l + \frac{1}{2}$ of the integrand in the integral (8.8) equals

$$-\frac{(-1)^m}{2\pi i}(2l + 1) f_l P_l^m(\cos\theta). \tag{8.10}$$

In deriving expression (8.10), we have used the relationship

$$P_l^m(-\cos\theta) = (-1)^{l-m} P_l^m(\cos\theta), \tag{8.11}$$

valid for integral values of l and m.

Let us assume that $f_{s-1/2}$ has real poles within C_1 which are denoted by s_j $(j = 0, 1, 2, \ldots)$. A typical pole of this type is denoted by the dot on the Re s axis in Fig. 8.2a. Let $f_{s-1/2} = g(s)/\Delta(s)$ and assume that s_j are simple poles. The residue of the integrand at the pole s_j is equal to

$$P_{s_j-1/2}^m(-\cos\theta)\left[\frac{g(s)}{\partial\Delta(s)/\partial s}\right]_{s_j} \frac{s_j}{i\cos s_j\pi}. \tag{8.12}$$

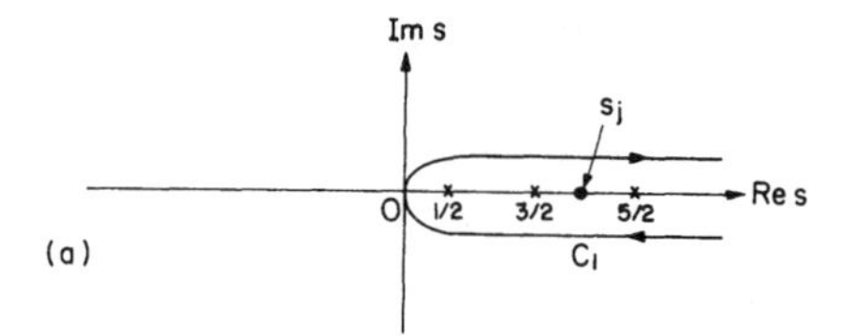

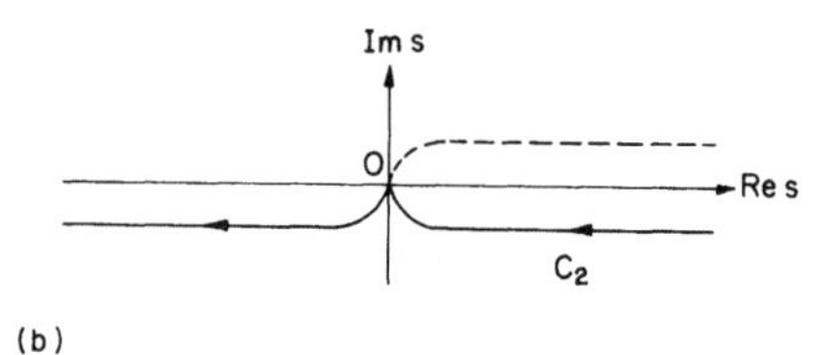

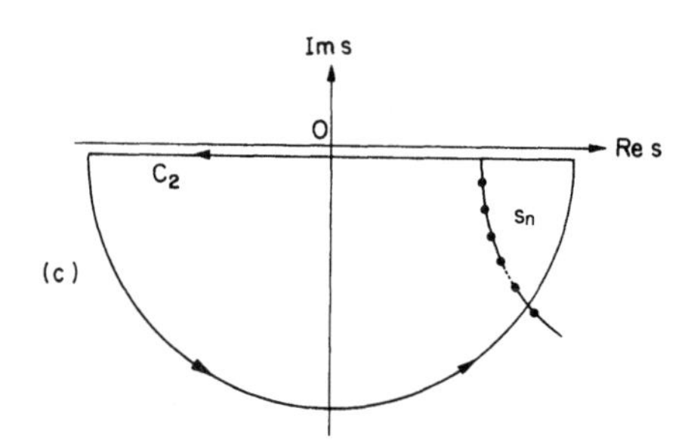

Figure 8.2. Contours in the s plane associated with the Watson transformation.

Applying Cauchy's residue theorem to integral (8.8), we get

$$(-1)^m \sum_{l=m}^{\infty} (2l+1) f_l P_l^m(\cos\theta) - 2\pi \sum_{j=0}^{\infty} P_{s_j-1/2}^m(-\cos\theta)\left[\frac{g(s)}{\partial\Delta(s)/\partial s}\right]_{s_j} \frac{s_j}{\cos s_j\pi}$$
$$= \int_{C_1} f_{s-1/2} P_{s-1/2}^m(-\cos\theta)\,\frac{s\,ds}{i\cos s\pi}.$$

This may be written in the form

$$\sum_{l=m}^{\infty} (2l+1) f_l P_l^m(\cos\theta)$$
$$= (-1)^m \pi \sum_{j=0}^{\infty} \frac{(2\nu_j+1)}{\cos(\nu_j+\frac{1}{2})\pi}\left[\frac{g(s)}{\partial\Delta(s)/\partial s}\right]_{s=\nu_j+1/2} P_{\nu_j}^m(-\cos\theta)$$
$$+ (-1)^m \int_{C_1} f_{s-1/2} P_{s-1/2}^m(-\cos\theta)\,\frac{s\,ds}{i\cos s\pi}, \tag{8.13}$$

where $\nu_j = s_j - \frac{1}{2}$. We have assumed above that the zeros of $\{\cos s\pi\}$ and $\Delta(s)$ do not overlap. This is true as long as ω does not coincide with any of the eigenfrequencies ${}_n\omega_l$. We will assume that this condition is satisfied.

We assume that $f_{s-1/2}$ is an even function of s. Because $P^m_{s-1/2} = P^m_{-s-1/2}$, the integrand in Eq. (8.13) is an odd function of s. When s is replaced by $-s$ in the integral along the upper portion of the contour C_1, the original contour is transformed into one below the entire real axis. This contour is denoted by C_2 in Fig. 8.2b.

When the contour C_2 is closed with the help of a semicircle of large radius in the lower half of the s plane (Fig. 8.2c), it can be seen that

$$\begin{aligned}
&\sum_{l=m}^{\infty}(2l+1)f_l P_l^m(\cos\theta)\\
&\quad= (-1)^m\pi\sum_{j=0}^{\infty}\frac{(2\nu_j+1)}{\cos(\nu_j+\frac{1}{2})\pi}\left[\frac{g(s)}{\partial\Delta(s)/\partial s}\right]_{s=\nu_j+1/2} P^m_{\nu_j}(-\cos\theta)\\
&\quad+ (-1)^m\pi\sum_{n=0}^{\infty}\frac{(2\nu_n+1)}{\cos(\nu_n+\frac{1}{2})\pi}\left[\frac{g(s)}{\partial\Delta(s)/\partial s}\right]_{s=\nu_n+1/2} P^m_{\nu_n}(-\cos\theta),
\end{aligned}\tag{8.14}$$

where $s = s_n = \nu_n + \frac{1}{2}$ are the complex poles of $f_{s-1/2}$ in the lower half of the s plane. It may be noted that the first sum on the right-hand side of Eq. (8.14) originates in the real poles of $f_{s-1/2}$ and represents surface waves. In contrast, the second sum stems from the complex poles of $f_{s-1/2}$ in the lower half of the s plane and corresponds to diffracted waves. In deriving Eq. (8.14), it has been assumed that the integral along the infinite semicircle in the lower half of the s plane vanishes.

We therefore see that by means of the Watson transformation the spectral response of a radially heterogeneous medium given in expression (8.7) can be split into two parts. The first part is again an infinite sum representing surface waves that arise from the real poles of $f_{s-1/2}$. The second part is in the form of a contour integral along C_1, which can be replaced by C_2 if $f_{s-1/2}$ is an even function of s. The integral along C_2, when evaluated in terms of residues at the complex poles of $f_{s-1/2}$ in the lower half of the s plane, gives rise to diffracted waves. However, if we evaluate the integral along C_2 by the saddle-point method, we shall get body waves. Therefore, the Watson transformation not only improves the convergence of the infinite sum but also splits the spectral field into physically meaningful components. A modification of the Watson transformation is sometimes useful. If $(s + m)$ is not an integer, we use the representation

$$P_s^m(-\cos\theta) = e^{-i\pi(s+m)}P_s^m(\cos\theta) + i\sin(s+m)\pi\left[P_s^m(\cos\theta) + \frac{2i}{\pi}Q_s^m(\cos\theta)\right].\tag{8.14a}$$

Assuming that $f_{s-1/2}$ is an even function in s, the substitution of $P_s^m(-\cos\theta)$ from Eq. (8.14a) into the contour integral in Eq. (8.13) yields

$$\begin{aligned}
\frac{(-)^m}{i}\int_{C_1} sf_{s-1/2}P^m_{s-1/2}(-\cos\theta)\sec(\pi s)\,ds &= \int_{C_2} sf_{s-1/2}P^m_{s-1/2}(\cos\theta)e^{-\pi is}\sec(\pi s)\,ds\\
&- 2\int_{C_2} sf_{s-1/2}E^{(1)}_{s-1/2,\,m}(\cos\theta)\,ds,
\end{aligned}\tag{8.14b}$$

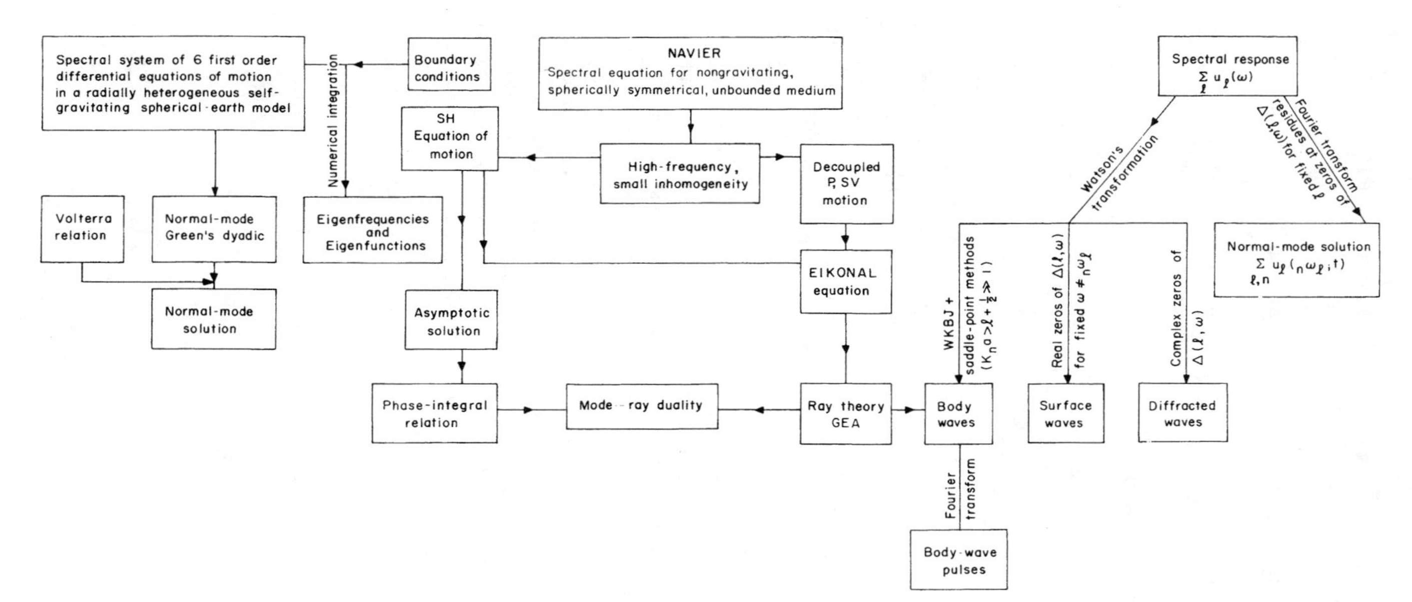

Figure 8.3. Flow-chart showing transformations, approximations, and interrelations of the various seismic fields induced in a spherical-earth model by seismic sources.

where $E^{(1,2)}_{s-1/2,m}$ are defined in Appendix H. The first integral on the right-hand side of Eq. (8.14b) can be evaluated by closing the contour in the lower half-plane and calculating the residues at the poles of the integrand. The other integral renders propagating body waves as we shall show in Section 8.7. Note that because $E^{(1,2)}_{s-1/2,m}$ have poles at the negative half integers, the Watson poles of the integrand have been transferred to the negative s axis.

The results obtained by the saddle-point method are easily interpretable in terms of geometric elastodynamics but are of use in the "lit" area only; in the geometric seismic shadow zone the results obtained by Watson's residue method are usable, but these results are untractable outside this area. Therefore, the two methods yield complementary formulas that together enable us to calculate the field at every point of the medium. Because the diffracted field is in fact the contribution of poles, it is obtained from the infinite sum representing the "secondary" field; the infinite sum representing the "primary" field does not contribute anything to the diffracted field. Here the primary field means the field of the source in an infinite medium, whereas the secondary field is the effect of the boundaries. The results may be summarized in the form of a flow-chart given in Fig. 8.3.

8.3. Surface Waves on a Sphere

Consider first the surface waves. We recall that these waves were introduced for models with plane-parallel boundaries. However, Love and Rayleigh surface waves with periods up to 300 s are commonly observed in earthquake seismograms. The phase and group velocities of surface waves in the period range $10\ s < T < 300$ s, computed by means of Jeans' formula, agree within 3% with the results obtained through the half-space matrix methods. The real earth, however, differs from a vertically heterogeneous half-space model in three important respects: sphericity, gravity, and liquid core. Numerical integration of the equations of motion of a gravitating-earth model has shown that in the period range 50 s $< T <$ 300 s, the analysis can be considerably simplified by using a flat-earth model with a suitable earth-flattening approximation. For $T <$ 50 s, the flat-earth model yields a good approximation.

At the other spectral end, propagating waves in the period range 300–600 s are sometimes observed in seismograms of major earthquakes. The corresponding normal modes have significant amplitudes not only in the crust and mantle but in the core as well. Their motion is governed by the elastic restoring forces as well as by gravitational forces. In the range 300 s $< T <$ 400 s, the normal-mode amplitudes in the core become negligibly small so that it is sufficient to carry out the integration from the base of the mantle to the surface. In this restricted normal-mode method, we assume that all the components of the displacement and stress vanish at the core–mantle boundary. The ranges of applicability of the various methods are shown in Table 8.4.

Table 8.4. Ranges of Applicability of Various Methods for the Calculation of the Earth's Normal Modes

Range of period T (s)	Range of l	Method	Core included	Gravity included
$3220 > T > 400$	$17 > l > 2$	Complete normal mode	Yes	Yes
$400 > T > 300$	$25 > l > 17$	Restricted normal mode	No	Yes
$300 > T > 50$	$200 > l > 25$	Earth-flattening approximation	No	No
$50 > T$	$l > 200$	Flat earth	No	No

For Im $\nu < 0$, we have the expansion

$$\frac{1}{\cos(\nu + \frac{1}{2})\pi} = 2 \sum_{N=0}^{\infty} (-1)^N e^{-(2N+1)(\nu+1/2)\pi i}. \tag{8.15}$$

However, the asymptotic expansion of the Legendre function

$$\pi(-1)^m P_\nu^m[\cos(\pi - \theta)] \sim \nu^m \left(\frac{\pi}{2\nu \sin \theta}\right)^{1/2} \{e^{i[(\nu+1/2)(\pi-\theta)+m\pi/2-\pi/4]} + e^{-i[(\nu+1/2)(\pi-\theta)+m\pi/2-\pi/4]}\}, \tag{8.16}$$

is valid for large ν, and for θ sufficiently removed from the pole and the antipode. Let $\Delta = a\theta$ be the epicentral distance and let $k_j a = \nu_j + \frac{1}{2}$. We then have

$$\begin{aligned} e^{i\omega t}(-1)^m \pi \sum_{j=0}^{\infty} \frac{(2\nu_j + 1)}{\cos(\nu_j + \frac{1}{2})\pi} \left[\frac{g(s)}{\partial\Delta(s)/\partial s}\right]_{s=\nu_j+1/2} P_{\nu_j}^m(-\cos\theta) \\ = \left(\frac{8\pi a}{\Delta}\right)^{1/2} \left(\frac{\theta}{\sin\theta}\right)^{1/2} \sum_{j=0}^{\infty} ak_j \left(ak_j - \frac{1}{2}\right)^{m-1/2} \\ \times \left[\frac{g(s)}{\partial\Delta(s)/\partial s}\right]_{s=ak_j} \sum_{N=0}^{\infty} (e^{iS_1} + e^{iS_2}), \end{aligned} \tag{8.17}$$

where

$$\begin{aligned} S_1 &= \omega t - k_j(\Delta + 2N\pi a) + N\pi + \frac{m\pi}{2} - \frac{\pi}{4}, \\ S_2 &= \omega t - k_j[(2\pi a - \Delta) + 2N\pi a] + N\pi - \frac{m\pi}{2} + \frac{\pi}{4}. \end{aligned} \tag{8.18}$$

Equation (8.17) has the following features:

1. The amplitudes of surface waves vary as $\Delta^{-1/2}$, modulated by the *sphericity correction factor* $(\theta/\sin\theta)^{1/2}$.

2. S_1 is the phase of the wave that travels along a minor arc ($N = 0$) from the source to the observer and continues on its way to complete one revolution ($N = 1$), two revolutions ($N = 2$), etc. The corresponding waves are known in seismology as G_1, G_3, G_5, etc., in the case of toroidal motion and as R_1, R_3, R_5, etc., in the case of spheroidal motion. In contrast, S_2 is the phase of a wave that leaves the source in the opposite direction and reaches the observer via the major arc. These waves are known as G_2, G_4, G_6, etc., and R_2, R_4, R_6, etc., as the case may be.
3. The parameter k_j plays the role of the wave number and is equal to $\omega/c(\omega)$, where c is the phase velocity. The relation $k_j a = \nu_j + \frac{1}{2}$ is similar in form to Jeans' formula. However, whereas Jeans' formula is valid for all normal modes of large integral order, the present relation is valid only for surface waves and the phase velocity defined by it is that of surface waves only. Both relations are not exact.
4. The phases S_1 and S_2 include a phase advance of π for every complete revolution around the earth. This means that there is a phase advance of $\pi/2$ for each polar or antipodal crossing. This is known as the *polar phase shift* of surface waves on a sphere. This shift is independent of the frequency. In the time domain, the signal is transformed to its allied function on each polar passage. Figure 8.4 shows clearly the necessity of introducing a polar phase shift into the asymptotic representation of the Legendre functions. This figure compares $P_7(\cos\theta)$ with a sinusoidal wave of wavelength equal to the asymptotic wavelength of $P_7(\cos\theta)$. The amplitude and the phase of the sinusoidal wave are matched with those of the spherical harmonic wave at 90°. It is clear that if the sinusoidal wave is extended through 180° with the same wavelength it will fail to match the spherical harmonic $P_7(\cos\theta)$ in the region $180° + \delta$ to $360° - \delta$ unless the sinusoidal wave is advanced by $\Lambda/4$ in phase. Here δ is of the order of $\Lambda/4$. If the angular range in Fig. 8.4 is extended to the right, an additional $\Lambda/4$ phase advance is required at each crossing of the pole or antipode. When m is different from zero, there is an initial source phase that is $m\pi/2$ for S_1 and $-m\pi/2$ for S_2.

Consider, for example, the spectral toroidal field induced by a strike-slip dislocation in a radially heterogeneous sphere. The azimuthal component of the surface displacement is given by [cf. Eq. (6.136) with $r_0 \to r_h$]

$$u_\phi = -\frac{U_0\,dS}{4\pi a^2}\,e^{i\omega t}\cos 2\phi\,\frac{\partial}{\partial\theta}\sum_{l=2}^{\infty}\frac{2l+1}{l(l+1)}\left[\frac{y_1(r_h;\,l)/r_h}{y_2(a;\,l)/\mu(r_h)}\right]P_l^2(\cos\theta).$$

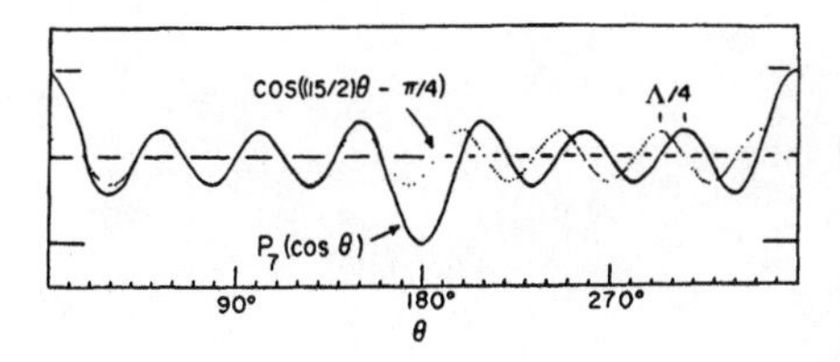

Figure 8.4. A comparison of P_7 (cos θ) with its asymptotic representation.

Proceeding as above, we get the following expression for the Love-wave displacement from Eq. (8.17) for $v_j \gg 1$

$$u_\phi = \frac{U_0\,dS}{4\pi a^2}\cos 2\phi\left(\frac{8\pi}{\sin\theta}\right)^{1/2}(ak_j)^{3/2}\left[\frac{a\mu(r_h)}{r_h}\frac{y_1(r_h;ak_j)}{(\partial/\partial k_j)y_2(a;ak_j)}\right]\sum_{N=0}^{\infty}(e^{iS_1}+e^{iS_2}),$$

where

$$S_1 = \omega t - k_j(\Delta + 2N\pi a) + N\pi - \frac{3\pi}{4},$$

$$S_2 = \omega t - k_j[(2\pi a - \Delta) + 2N\pi a] + N\pi + \frac{3\pi}{4}.$$

If we replace a/r_h by 1, take $N = 0$, and consider surface waves corresponding to S_1, we get

$$u_\phi = \frac{U_0\,dS}{4\pi a^2}\cos 2\phi(8\pi)^{1/2}\left(\sin\frac{\Delta}{a}\right)^{-1/2}(ak_j)^{3/2}\left[\frac{\mu(r_h)y_1(r_h;ak_j)}{(\partial/\partial k_j)y_2(a;ak_j)}\right]e^{i(\omega t - k_j\Delta - 3\pi/4)}.$$

This equation coincides with the corresponding result obtained from Eqs. (5.38) and (5.74) for the Love-wave spectral displacements in a vertically heterogeneous half-space.

8.3.1. Surface Waves on a Rotating Earth

The normal-mode solution of a nonrotating spherically symmetrical earth is degenerate, because the eigenfrequencies of such a model are independent of the azimuthal order number m. In this case m is determined by the source of excitation; e.g., $m = 0$, 1, 2 for a shear dislocation. However, this degeneracy is removed if the rotation of the earth is taken into account. Then m assumes all integral values such that $-l \le m \le l$.

It is shown in Appendix H that the spherical harmonic $y_{ml}(\theta, \phi)$ has the asymptotic expansion

$$y_{ml}(\theta,\phi) \sim \frac{e^{i\Phi^+} + e^{i\Phi^-}}{2\pi(\sin^2\theta - \sin^2\theta_0)^{1/4}}, \tag{8.19}$$

where both m and l are large and

$$\Phi^{\pm} = \pm\int_{\theta_0}^{\theta} p_\theta\,d\theta + m\phi \mp \frac{\pi}{4}, \qquad p_\theta = \frac{m}{\sin\theta\sin\theta_0}(\sin^2\theta - \sin^2\theta_0)^{1/2},$$
$$\theta_0 = \sin^{-1}\left(\frac{m}{l+\frac{1}{2}}\right). \tag{8.20}$$

Here $y_{ml}(\theta, \phi)$ is a fully normalized spherical harmonic:

$$y_{ml}(\theta,\phi) = (-1)^m\left[\frac{2l+1}{4\pi}\cdot\frac{(l-m)!}{(l+m)!}\right]^{1/2} Y_{ml}(\theta,\phi),$$
$$Y_{ml}(\theta,\phi) = P_l^m(\cos\theta)e^{im\phi}. \tag{8.21}$$

In the equatorial region $\theta_0 < \theta < \pi - \theta_0$, y_{ml} in (8.19) describes an oscillating function, whereas in the polar regions, $\theta < \theta_0$, $\theta > \pi - \theta_0$, it decays exponentially as the point of observation approaches the poles from $\theta = \theta_0$ or $\pi - \theta_0$, as the case may be. Introducing the factor $\exp(i\omega t)$, it is easily seen that y_{ml} in (8.19) represents a wave motion in the counterclockwise direction whose ray trajectories are great circle paths similar to the one shown in Fig. 8.5. The first exponential (Φ^+) describes an ascending wave in the front hemisphere, whereas the second exponential (Φ^-) corresponds to a descending wave in the back hemisphere. The axis of the great circle path forms an angle $\Theta = \pi/2 - \theta_0$ with the vertical axis, so that each pair (m, l) determines an orientation of a great circle. Given (m, l), the corresponding great circle is determined by a vector of length $l + \frac{1}{2}$ along the axis of that great circle and the projection m along the vertical coordinate axis (say, the axis of rotation of the earth). Because θ_0 is a space angle common to all great circles that "precess" around the vertical axis, each $y_{ml}(\theta, \phi)$ with m and l fixed defines a family of great circles with a common θ_0.

We have seen in Section 6.4 that the eigenfrequencies of a rotating earth may be expressed in the form

$$ {}_n\omega_l^m = {}_n\omega_l^0 + m\tau_l\Omega, \tag{8.22}$$

where ${}_n\omega_l^0$ correspond to a nonrotating earth, Ω is the uniform angular velocity of the earth, and $\tau_l = 1/l(l+1)$ for toroidal oscillations. For spheroidal oscillations, τ_l is given by Eq. (6.225). Using Eqs. (8.20), Eq. (8.22) may be written as

$$ {}_n\omega_l^m = {}_n\omega_l^0 + \left(l + \frac{1}{2}\right)\tau_l\Omega \sin\theta_0. \tag{8.23}$$

The corresponding result for the phase velocity is

$$ c({}_n\omega_l^m) = c({}_n\omega_l^0) + a\tau_l\Omega \sin\theta_0, \tag{8.24}$$

where we have used Jeans' formula $l + \frac{1}{2} = {}_nk_l a$.

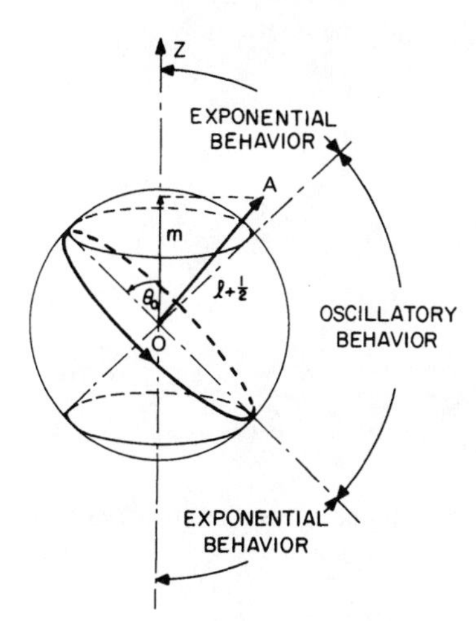

Figure 8.5. The great circle paths of surface waves associated with $y_{ml}(\theta, \phi)$. The asymptotic expression of this function represents a wave propagating around the sphere such that a vector of length $l + \frac{1}{2}$ along the axis OA has a projection of length m along the z axis.

8.3.2. Sphericity Corrections for Half-Space Models

In Chapter 3, we have shown how to calculate the phase and group velocities for surface waves in multilayered flat-earth models. Because the data used for the structural parameters of these models are derived from spherical-earth data, it is logical that we should either use spherical-earth models or, at least, introduce a sphericity correction when using flat-earth models. Because the first approach requires elaborate calculations, we shall demonstrate how a sphericity correction can be applied in the simpler case of Love waves.

The radial factor for the toroidal oscillations of a radially heterogeneous earth satisfies the equation [Eq. (6.63)]

$$\frac{d}{dr}\left[\mu\left(\frac{dV}{dr} - \frac{V}{r}\right)\right] + \frac{3\mu}{r}\frac{dV}{dr} + \left[\omega^2\rho - \frac{l(l+1)+1}{r^2}\mu\right]V = 0. \quad (8.25)$$

The transformation [cf. Eq. (7.191)]

$$V = rF, \qquad z = a\ln\left(\frac{a}{r}\right),$$

changes Eq. (8.25) into

$$\frac{d}{dz}\left[\mu\left(\frac{r}{a}\right)^3\frac{dF}{dz}\right] + \left[\omega^2\rho\left(\frac{r}{a}\right)^5 - \mu\left(\frac{r}{a}\right)^3\left\{\frac{(l-1)(l+2)}{a^2}\right\}\right]F = 0. \quad (8.26)$$

From Eq. (3.81), we know that Love-wave propagation in a vertically heterogeneous earth, $\mu = \mu(z)$, $\rho = \rho(z)$, is governed by the equation

$$\frac{d}{dz}\left[\mu\frac{dF}{dz}\right] + (\omega^2\rho - \mu k^2)F = 0. \quad (8.27)$$

A direct comparison of Eqs. (8.26) and (8.27) yields the relations

$$\mu_f(z) = \left(\frac{r}{a}\right)^3\mu_s(r), \qquad \rho_f(z) = \left(\frac{r}{a}\right)^5\rho_s(r),$$
$$(k_f a)^2 = (l-1)(l+2). \quad (8.28)$$

where z is the depth below the free surface. Therefore,

$$\beta_f = \left(\frac{\mu_f}{\rho_f}\right)^{1/2} = \left(\frac{\mu_s}{\rho_s}\right)^{1/2}\left(\frac{a}{r}\right) = \beta_s\left(\frac{a}{r}\right) = \beta_s e^{z/a} \simeq \beta_s\left(1 + \frac{z}{a}\right). \quad (8.29)$$

Here, the subscript f refers to the flat-earth model and the subscript s refers to the spherical-earth model.

We have seen before that for surface waves on a sphere, $k_s a = l + \frac{1}{2}$, approximately. Therefore, as far as surface waves are concerned

$$(k_f a)^2 = (l-1)(l+2) = l(l+1) - 2 = \left(k_s a - \frac{1}{2}\right)\left(k_s a + \frac{1}{2}\right) - 2$$
$$= (k_s a)^2 - \frac{9}{4},$$

i.e.,

$$k_s = \left(k_f^2 + \frac{9}{4a^2}\right)^{1/2}. \qquad (8.30)$$

In practice, Eqs. (8.28) and (8.30) are applied as follows: A spherical, layered-earth model is chosen. In each layer of this model $\mu_s(r)$ and $\rho_s(r)$ are known. Positions of the interfaces in the corresponding flat-earth model and the distributions $\mu_f(z)$ and $\rho_f(z)$ are then determined from Eq. (8.28). The Love-wave dispersion $k_f(\omega)$ in the resulting flat-earth model is determined by the method described in Section 3.7. Equation (8.30) then yields the Love-wave dispersion $k_s(\omega)$ for the spherical model. Because the group velocity is equal to $d\omega/dk$, we have

$$U_s = U_f\left[1 + \frac{9}{4(k_f a)^2}\right]^{1/2}. \qquad (8.31)$$

8.4. Mode-Ray Duality

The normal-mode solution represents the entire seismic field, which is given in the form of a double-infinite sum over the indices l and n. This representation is convenient for cases in which we wish to interpret the observed amplitude of a particular spectral line (l, n), obtained by the Fourier analysis of the entire seismogram of a given earthquake. Most of the conspicuous signals observed in a common seismogram are time limited and frequency limited with a threshold at $T < 600$ s. These constitute partial sums of the normal-mode solution. In order to be able to interpret various seismic phases in terms of normal modes, certain transformations must be applied either to the normal-mode solution or to the differential equations themselves.

8.4.1. Asymptotic Solution of the Equation of Motion

The toroidal oscillations of a radially heterogeneous earth model are governed by the equation [cf. Eq. (6.63)]

$$\mu\left(\frac{d^2V}{dr^2} + \frac{2}{r}\frac{dV}{dr}\right) + \frac{d\mu}{dr}\left(\frac{dV}{dr} - \frac{V}{r}\right) + \left[{}_n\omega_l^2\rho - \frac{l(l+1)\mu}{r^2}\right]V = 0. \quad (8.32)$$

Writing

$$y_1 = V, \qquad y_2 = \mu\left(\frac{dV}{dr} - \frac{V}{r}\right), \tag{8.33}$$

Eq. (8.32) may be transformed to

$$\frac{dy_1}{dr} = \frac{y_1}{r} + \frac{y_2}{\mu}, \tag{8.34}$$

$$\frac{dy_2}{dr} = \left[(l-1)(l+2)\frac{\mu}{r^2} - {}_n\omega_l^2\rho\right]y_1 - \frac{3y_2}{r}. \tag{8.35}$$

Equations (8.34) and (8.35) are to be solved subject to the boundary conditions

$$y_2 = 0 \quad \text{at} \quad r = a \quad \text{and at} \quad r = b, \tag{8.36}$$

where a is the radius of the earth and b is the radius of its core.

Putting

$$V(r) = y_1 = rK(r), \qquad y_2 = r^{-3}L(r), \qquad h_l^2 = (l-1)(l+2) = \left(l + \frac{1}{2}\right)^2 - \frac{9}{4}, \tag{8.37}$$

Eq. (8.32) takes the Sturm–Liouville form

$$\frac{d}{dr}\left(\mu r^4 \frac{dK}{dr}\right) + ({}_n\omega_l^2\rho r^4 - h_l^2\mu r^2)K = 0. \tag{8.38}$$

Or, equivalently,

$$\frac{dK}{dr} = \frac{L}{\mu r^4}, \qquad \frac{dL}{dr} = (h_l^2\mu r^2 - {}_n\omega_l^2\rho r^4)K. \tag{8.39}$$

It is convenient to introduce nondimensional quantities as follows:

$$\begin{gathered} x = \frac{r}{a}, \qquad \rho(r) = \rho_0\,\varepsilon(x), \qquad \mu(r) = \mu_0\,\tau(x), \\ \beta(r) = \beta_0\,c(x), \qquad c^2(x) = \tau(x)/\varepsilon(x), \\ L(r) = a^3\mu_0 M(x), \qquad {}_ns_l = \frac{{}_n\omega_l a}{\beta_0}, \qquad \beta_0^2 = \frac{\mu_0}{\rho_0}. \end{gathered} \tag{8.40}$$

Equations (8.38) and (8.39) now become

$$\frac{d}{dx}\left(\tau x^4 \frac{dK}{dx}\right) + ({}_ns_l^2\varepsilon x^4 - h_l^2\tau x^2)K = 0, \tag{8.41}$$

$$\frac{dK}{dx} = \frac{M}{\tau x^4}, \qquad \frac{dM}{dx} = (h_l^2\tau x^2 - {}_ns_l^2\varepsilon x^4)K. \tag{8.42}$$

The boundary conditions reduce to

$$M = 0 \quad \text{at} \quad x = \frac{b}{a} \quad \text{and at} \quad x = 1. \tag{8.43}$$

Let

$${}_np_l = \frac{h_l}{{}_ns_l}, \qquad y(x) = \int_{x_m}^{x} \frac{dx}{x^4 \tau(x)}, \qquad x_m = {}_np_l c(x_m). \tag{8.44}$$

With this change of variable, Eq. (8.41) becomes

$$\frac{d^2K}{dy^2} \pm Q^2 K = 0, \tag{8.45}$$

with the + for $x/c > {}_np_l$ and the − for $x/c < {}_np_l$ and

$$Q = {}_ns_l \varepsilon c x^3 \begin{cases} (x^2 - {}_np_l^2 c^2)^{1/2}; & \dfrac{x}{c} > {}_np_l, \quad (8.46) \\[2ex] ({}_np_l^2 c^2 - x^2)^{1/2}; & \dfrac{x}{c} < {}_np_l. \quad (8.47) \end{cases}$$

We make still another change of variable

$$z = \int_{y_m}^{y} Q\, dy = \int_{x_m}^{x} Q \frac{dy}{dx}\, dx = {}_ns_l \int_{x_m}^{x} \frac{1}{xc} (x^2 - {}_np_l^2 c^2)^{1/2}\, dx, \tag{8.48}$$

thereby transforming Eq. (8.45) to

$$\frac{d^2G(z)}{dz^2} + \left(1 + \frac{5}{36z^2}\right) G(z) = \frac{AG}{{}_ns_l^{4/3} z^{2/3}} + O(G\, {}_ns_l^{-2}). \tag{8.49}$$

with

$$G(z) = K(x)\sqrt{Q(x)}. \tag{8.50}$$

Here, A is a constant independent of ${}_ns_l$. Assuming ${}_ns_l \gg 1$, we neglect the right-hand side of Eq. (8.49) and write its solution in the form

$$G = \sqrt{z}[J_{1/3}(z) + J_{-1/3}(z)]; \qquad \frac{x}{c} > {}_np_l, \tag{8.51}$$

$$G = \frac{\sqrt{(3v)}}{\pi} K_{1/3}(v); \qquad \frac{x}{c} < {}_np_l, \tag{8.52}$$

where

$$v = {}_ns_l \int_{x}^{x_m} \frac{1}{xc} ({}_np_l^2 c^2 - x^2)^{1/2}\, dx. \tag{8.53}$$

Therefore, the solution has an oscillatory character for $x > x_m$ and decays exponentially for $x < x_m$.

We must now examine how the boundary conditions, Eqs. (8.43), can be satisfied by the approximate solution. Because for large v, G in Eq. (8.52) behaves as $(3/2\pi)^{1/2}e^{-v}$, the boundary conditions, Eqs. (8.43), are satisfied at $x = b/a$. At the free surface we must satisfy the boundary condition

$$\frac{1}{K}\frac{dK}{dx} = \frac{1}{G}\frac{dG}{dz}\,{}_ns_l(1 - {}_np_l^2)^{1/2} - \frac{1}{2Q}\frac{dQ}{dx} = 0, \qquad \text{at} \quad x = 1. \tag{8.54}$$

Because ${}_ns_l$ is assumed to be large, the roots of Eq. (8.54) are close to the roots of the equation

$$\frac{dG}{dz} = \frac{d}{dz}\{\sqrt{z}[J_{1/3}(z) + J_{-1/3}(z)]\} = 0. \tag{8.55}$$

The first seven roots of Eq. (8.55) are 0.880167, 3.945062, 7.078484, 10.217004, 13.356982, 16.497518, and 19.638495.

On the assumption that the rigidity and shear-wave velocity are constant near the surface, we have from Eq. (8.46)

$$\frac{1}{Q}\frac{dQ}{dx} = 3 + \frac{1}{1 - {}_np_l^2} \qquad \text{at} \quad x = 1. \tag{8.56}$$

In the neighborhood of the roots z_n of Eq. (8.55), we have

$$\frac{d}{dz}\left(\frac{1}{G}\frac{dG}{dz}\right) = \frac{1}{G}\frac{d^2G}{dz^2} - \left(\frac{1}{G}\frac{dG}{dz}\right)^2 \simeq \frac{1}{G}\frac{d^2G}{dz^2},$$

and, therefore, from Eq. (8.49),

$$\frac{1}{G}\frac{dG}{dz} \simeq -\left(1 + \frac{5}{36z_n^2}\right)(z - z_n). \tag{8.57}$$ [2]

Substituting Eqs. (8.56) and (8.57) in Eq. (8.54) and using Eq. (8.48), we get

$${}_nz_l = {}_ns_l\int_{x_m}^{1}\frac{1}{xc}(x^2 - {}_np_l^2c^2)^{1/2}\,dx = z_n - {}_n\delta_l, \tag{8.58}$$

where

$${}_n\delta_l = \frac{[3(1 - {}_np_l^2)^{-1/2} + (1 - {}_np_l^2)^{-3/2}]}{2\,{}_ns_l\left(1 + \dfrac{5}{36z_n^2}\right)}. \tag{8.59}$$

Using Eqs. (8.40), Eq. (8.58) becomes

$${}_n\omega_l\int_{r_m}^{a}\left[\frac{r^2}{\beta^2(r)} - \frac{(l-1)(l+2)}{{}_n\omega_l^2}\right]^{1/2}\frac{dr}{r} = z_n - {}_n\delta_l, \tag{8.60}$$

[2] We use the symbol $\sim$ for asymptotic approximations and the symbol $\simeq$ for other approximations.

where $r_m = ax_m = \beta(r_m)h_l/{}_n\omega_l = \beta(r_m)[(l-1)(l+2)]^{1/2}{}_n\omega_l^{-1}$. This equation is known as the *phase-integral relation.*

Equation (8.58) serves as an approximate period equation for the evaluation of the frequency parameter, ${}_ns_l$, because ${}_np_l = [(l-1)(l+2)]^{1/2}{}_ns_l^{-1}$ and z_n is a numerical constant for each mode. An important feature of Eq. (8.58) is that the density distribution does not enter into it at all. Therefore, the observed periods of the free toroidal oscillations provide no information on the density distribution. Equation (8.58) has been used to determine the approximate periods of the free toroidal oscillations T_l^{appr} for the Jeffreys–Bullen A' (J–B A') model of the earth. These values together with the exact values are given in Table 8.5. It is noticed that for $l = 10$, $T_l^{\text{exact}}/T_l^{\text{appr}}$ is close to 1 within 1%.

The phase-integral relation can be simplified if, from Eq. (8.48), we note that $z \gg 1$ follows from the assumption ${}_ns_l \gg 1$. In that case, Eq. (8.49) reduces to $d^2G/dz^2 + G = 0$ with a solution $G \propto \sin[\pi/4 + z]$. Then, the boundary condition $(1/G)(dG/dz) = 0$ renders the relation $\cot(\pi/4 + z) = 0$. Consequently

$$z_n \simeq \frac{\pi}{4} + \pi n, \qquad n = 0, 1, 2, 3, \ldots.$$

A comparison between the true z_n and this approximation is shown in Table 8.6. The phase integral relation now becomes

$$\int_{r_m}^{a} \frac{dr}{r}\sqrt{\left[\frac{r^2}{\beta^2(r)} - \frac{(l-1)(l+2)}{{}_n\omega_l^2}\right]} = \frac{\pi(n+\frac{1}{4})}{{}_n\omega_l}. \tag{8.60a}$$

8.4.2. Asymptotic Distribution of Eigenfrequencies

We derive next an explicit expression for the asymptotic dependence of ${}_n\omega_l$ on the radial mode number n for large values of n. To this end, we assume that μ and $d\mu/dr$ are continuous throughout the shell (b, a) and apply the *Liouville transformation*

$$V = \frac{rZ}{M}, \qquad t = \int_b^r \frac{dr}{\beta(r)}, \qquad M(r) = r^2\sqrt{(\beta\rho)},$$

to the governing equation, Eq. (8.32). This yields

$$\frac{d^2Z}{dt^2} + [{}_n\omega_l^2 - q(t)]Z = 0, \tag{8.61}$$

where

$$q(t) = \frac{1}{M}\frac{d^2M}{dt^2} + \frac{(l-1)(l+2)}{r^2}\beta^2.$$

Table 8.5. Exact vs. Approximate Eigenperiods for the J–B A' Earth Model[a]

l	${}_0s_l$	${}_0p_l$	${}_0d_l$ (km)	${}_0z_l$	$z_0 - {}_0\delta_l$	${}_0T_l^{\text{exact}}$ (s)	${}_0T_l^{\text{appr.}}$ (s)	${}_0T_l^{\text{exact}}/{}_0T_l^{\text{appr.}}$
2	4.3185	0.4631	1054	0.2947	0.4068	2610.70	2247.70	1.1615
3	6.6729	0.4739	978	0.4288	0.5706	1689.60	1529.90	1.1043
4	8.7152	0.4868	897	0.5226	0.6401	1293.60	1211.40	1.0679
5	10.5672	0.5008	828	0.5901	0.6792	1066.90	1019.70	1.0463
6	12.2851	0.5148	768	0.6403	0.7046	917.72	889.98	1.0312
7	13.9062	0.5284	717	0.6782	0.7226	810.74	794.87	1.0200
8	15.4557	0.5413	676	0.7089	0.7361	729.46	721.00	1.0117
9	16.9506	0.5534	641	0.7334	0.7468	665.13	661.57	1.0054
10	18.4026	0.5647	609	0.7533	0.7554	612.65	612.14	1.0008
11	19.8198	0.5753	584	0.7706	0.7626	568.84	570.50	0.9971
12	21.2084	0.5851	563	0.7849	0.7687	531.60	534.57	0.9944
13	22.5732	0.5944	543	0.7965	0.7740	499.46	503.16	0.9926
14	23.9178	0.6030	525	0.8061	0.7786	471.38	475.45	0.9914
15	25.2456	0.6111	508	0.8141	0.7827	446.59	450.87	0.9905
16	26.5589	0.6187	494	0.8213	0.7863	424.50	428.85	0.9899
17	27.8600	0.6258	482	0.8272	0.7896	404.68	408.92	0.9896
18	29.1506	0.6325	471	0.8318	0.7925	386.76	390.78	0.9897
19	30.4323	0.6389	461	0.8352	0.7952	370.47	374.20	0.9900
20	31.7063	0.6448	451	0.8375	0.7977	355.59	358.97	0.9906
21	32.9736	0.6504	442	0.8387	0.8000	341.92	344.93	0.9913
22	34.2353	0.6558	434	0.8388	0.8022	329.32	331.94	0.9921
23	35.4919	0.6608	426	0.8379	0.8042	317.66	319.86	0.9931
24	36.7442	0.6655	418	0.8358	0.8060	306.83	308.58	0.9943
25	37.9927	0.6700	410	0.8318	0.8078	296.75	298.03	0.9957
26	39.2379	0.6743	398	0.8266	0.8094	287.33	288.20	0.9970
27	40.4801	0.6783	387	0.8217	0.8110	278.52	279.03	0.9982
28	41.7197	0.6822	376	0.8167	0.8125	270.24	270.43	0.9993
29	42.9570	0.6858	366	0.8118	0.8139	262.46	262.37	1.0003
30	44.1922	0.6893	357	0.8070	0.8152	255.12	254.78	1.0013
31	45.4254	0.6927	348	0.8022	0.8165	248.19	247.63	1.0023
32	46.6570	0.6958	339	0.7976	0.8177	241.64	240.89	1.0031
33	47.8870	0.6989	331	0.7930	0.8188	235.44	235.51	1.0040
34	49.1155	0.7018	324	0.7885	0.8199	229.55	228.47	1.0047
35	50.3427	0.7045	316	0.7841	0.8210	223.95	222.74	1.0055
36	51.5686	0.7072	309	0.7797	0.8220	218.63	217.29	1.0062
37	52.7934	0.7097	302	0.7755	0.8230	213.56	212.11	1.0068
38	54.0170	0.7122	296	0.7712	0.8239	208.72	207.18	1.0074
39	55.2396	0.7146	289	0.7670	0.8248	204.10	202.48	1.0080
40	56.4612	0.7168	283	0.7629	0.8257	199.68	197.99	1.0086
41	57.6818	0.7190	277	0.7589	0.8265	195.46	193.70	1.0091
42	58.9015	0.7211	271	0.7551	0.8274	191.41	189.59	1.0096
43	60.1204	0.7231	266	0.7513	0.8281	187.53	185.66	1.0101
44	61.3383	0.7251	261	0.7477	0.8289	183.81	181.89	1.0105
45	62.5554	0.7270	255	0.7441	0.8296	180.23	178.28	1.0110
46	63.7716	0.7288	250	0.7406	0.8303	176.79	174.81	1.0114
47	64.9870	0.7306	246	0.7372	0.8310	173.49	171.47	1.0118
48	66.2016	0.7323	241	0.7339	0.8317	170.30	168.26	1.0121
49	67.4154	0.7339	237	0.7306	0.8323	167.24	165.18	1.0125
50	68.6284	0.7355	232	0.7274	0.8329	164.28	162.20	1.0128

[a] ${}_0s_l = 2\pi a/{}_0T_l\beta_0$, ${}_0d_l = a - r_m$, $z_0 = 0.880167$, $n = 0$

Table 8.6. Roots z_n of

$$\frac{d}{dz}[\sqrt{z}\{J_{1/3}(z) + J_{-1/3}(z)\}] = 0$$

and Their Asymptotic Approximations

n	z_n	$\pi(n + \frac{1}{4})$
0	0.880167	0.785398
1	3.945062	3.926991
2	7.078484	7.068583
3	10.217004	10.210176
4	13.356982	13.351768
5	16.497578	16.493361
6	19.638495	19.634954

The boundary conditions in Eqs. (8.36) transform to

$$\begin{aligned} \frac{dZ}{dt} - h_1 Z = 0 \quad \text{at} \quad t = 0, \\ \frac{dZ}{dt} - h_2 Z = 0 \quad \text{at} \quad t = \gamma, \end{aligned} \tag{8.62}$$

where

$$h_1 = \left(\frac{1}{M}\frac{dM}{dt}\right)_{r=b}, \qquad h_2 = \left(\frac{1}{M}\frac{dM}{dt}\right)_{r=a}, \qquad \gamma = \int_b^a \frac{dr}{\beta}.$$

If $|q(t)| \ll {}_n\omega_l^2$ in $0 < t < \gamma$, Eq. (8.61) takes the form

$$\frac{d^2Z}{dt^2} + {}_n\omega_l^2 Z = 0.$$

The general solution of this equation is

$$Z(t) = A \sin {}_n\omega_l t + B \cos {}_n\omega_l t.$$

From Eqs. (8.62), we derive the period equation

$$\tan({}_n\omega_l \gamma) = \frac{{}_n\omega_l(h_1 - h_2)}{{}_n\omega_l^2 + h_1 h_2}.$$

Therefore,

$${}_n\omega_l \gamma = \pi n + \tan^{-1}\left[\frac{{}_n\omega_l(h_1 - h_2)}{{}_n\omega_l^2 + h_1 h_2}\right] \simeq \pi n + \frac{h_1 - h_2}{{}_n\omega_l}, \tag{8.63}$$

for large ${}_n\omega_l$. As a first approximation, ${}_n\omega_l = \pi n/\gamma$. Putting this value in the right-hand side of Eq. (8.63), we get

$${}_n\omega_l \simeq \frac{\pi n}{\gamma} + \frac{h_1 - h_2}{\pi n}.$$

To obtain a better approximation, we write $Q(t) = {}_n\omega_l^2 - q(t)$ and introduce the substitution

$$\cot \phi = Q^{-1/2} \frac{1}{z} \frac{dz}{dt} \tag{8.64}$$

To determine the distribution of the eigenvalues of Eq. (8.61), we need only the differential equation for the phase ϕ of Z. This equation is obtained from Eq. (8.61) in the form:

$$\phi' = ({}_n\omega_l^2 - q)^{1/2} - \frac{q'}{4({}_n\omega_l^2 - q)} \sin 2\phi, \tag{8.65}$$

prime indicating derivative with respect to t.

The boundary conditions become

$$Q^{1/2}(0) \cot \phi = h_1 \quad \text{at} \quad t = 0 \; (r = b), \tag{8.66a}$$

$$Q^{1/2}(\gamma) \cot \phi = h_2 \quad \text{at} \quad t = \gamma \; (r = a). \tag{8.66b}$$

We choose a solution $\phi(t, {}_n\omega_l)$ of Eq. (8.65) satisfying Eq. (8.66a). If ${}_n\omega_l$ is an eigenvalue of this system,

$$\tan \phi(\gamma, {}_n\omega_l) = \frac{Q^{1/2}(\gamma)}{h_2},$$

and $\phi(\gamma, {}_n\omega_l)$ must exceed $\phi(0, {}_n\omega_l)$ by approximately $n\pi$.

For large ${}_n\omega_l$, we write Eq. (8.65) in the form

$$\phi' = {}_n\omega_l - \frac{q(t)}{2\,{}_n\omega_l} + O\left(\frac{1}{{}_n\omega_l^2}\right).$$

Integrating, we get

$$\phi(t, {}_n\omega_l) = \phi(0, {}_n\omega_l) + {}_n\omega_l t - \frac{1}{2\,{}_n\omega_l} \int_0^t q(\tau) d\tau + O\left(\frac{1}{{}_n\omega_l^2}\right),$$

so that

$$\phi(\gamma, {}_n\omega_l) = \phi(0, {}_n\omega_l) + {}_n\omega_l \gamma - \frac{1}{2\,{}_n\omega_l} \int_0^\gamma q(\tau) d\tau + O\left(\frac{1}{{}_n\omega_l^2}\right). \tag{8.67}$$

Because $h_1 Q^{-1/2}(0)$ and $h_2 Q^{-1/2}(\gamma)$ are of order $1/{}_n\omega_l$, we can approximate Eqs. (8.66) by

$$\phi(0, {}_n\omega_l) = \frac{\pi}{2} - \frac{h_1}{{}_n\omega_l}, \qquad \phi(\gamma, {}_n\omega_l) = n\pi + \frac{\pi}{2} - \frac{h_2}{{}_n\omega_l}.$$

Therefore, from Eq. (8.67), we have approximately

$${}_n\omega_l \gamma = n\pi + \frac{h_1 - h_2}{{}_n\omega_l} + \frac{1}{2\,{}_n\omega_l} \int_0^\gamma q(\tau) d\tau.$$

The zero-order solution of this equation is ${}_n\omega_l = n\pi/\gamma$. Inserting it into the right-hand side, we get the first-order approximation

$$ {}_n\omega_l = \frac{n\pi}{\gamma} + \frac{1}{n\pi}\left[h_1 - h_2 + \frac{1}{2}\int_0^\gamma q(\tau)d\tau\right]. \tag{8.68}$$

We consider next the case of a homogeneous shell of external radius a surrounding a liquid core of radius b. The frequency equation for the toroidal oscillations of this model has been derived in Example 6.2 and may be put in the form

$$ \mathscr{J}_l(k_\beta a)\mathscr{N}_l(k_\beta b) - \mathscr{J}_l(k_\beta b)\mathscr{N}_l(k_\beta a) = 0, \tag{8.69}$$

where $k_\beta = \omega/\beta_0$ and

$$ \mathscr{J}_l(z) = \frac{1}{z}j_l'(z) - \frac{1}{z^2}j_l(z), \qquad \mathscr{N}_l(z) = \frac{1}{z}n_l'(z) - \frac{1}{z^2}n_l(z). $$

We shall next obtain asymptotic forms of the frequency equation, Eq. (8.69), under various conditions.

8.4.2.1. $k_\beta b$ Large Compared with l. For z real and large and l not large, we have the following asymptotic approximations for the spherical Bessel functions

$$ j_l(z) \sim \frac{1}{z}\left\{\cos\left(z - \frac{l+1}{2}\pi\right) - \frac{l(l+1)}{2z}\sin\left(z - \frac{l+1}{2}\pi\right)\right\}, $$

$$ n_l(z) \sim \frac{1}{z}\left\{\sin\left(z - \frac{l+1}{2}\pi\right) + \frac{l(l+1)}{2z}\cos\left(z - \frac{l+1}{2}\pi\right)\right\}. $$

These yield

$$ \mathscr{J}_l(z) \sim \frac{1}{z^2}\left\{\cos\left(z - \frac{1}{2}l\pi\right) - \frac{l(l+1)+4}{2z}\sin\left(z - \frac{1}{2}l\pi\right)\right\}, $$

$$ \mathscr{N}_l(z) \sim \frac{1}{z^2}\left\{\sin\left(z - \frac{1}{2}l\pi\right) + \frac{l(l+1)+4}{2z}\cos\left(z - \frac{1}{2}l\pi\right)\right\}. $$

With these approximations, the frequency equation, Eq. (8.69), becomes

$$ \sin(Z_a - Z_b) + \frac{l(l+1)+4}{2k_\beta}\left(\frac{1}{a} - \frac{1}{b}\right)\cos(Z_a - Z_b) + O\left[\frac{\{l(l+1)+4\}^2}{k_\beta^2 ab}\right] = 0, $$

where

$$ Z_a = k_\beta a - \frac{1}{2}l\pi, \qquad Z_b = k_\beta b - \frac{1}{2}l\pi. $$

Hence, we have

$$ \tan k_\beta(a-b) = \frac{l(l+1)+4}{2k_\beta}\left(\frac{1}{b} - \frac{1}{a}\right) + O\left[\frac{\{l(l+1)+4\}^2}{k_\beta^2 ab}\right]. \tag{8.70}$$

The zero-order approximation to solutions of Eq. (8.70), obtained by neglecting the whole of the right-hand side, is $k_\beta(a - b) = n\pi$, where n is an integer. Hence

$$_n\omega_l = \frac{n\pi\beta_0}{(a - b)}. \tag{8.71}$$

Equation (8.71) states that, for oscillations corresponding to a fixed l, the time taken by a wave to travel along a radius with speed β_0 from the surface to the core boundary and back is n times the period $_nT_l = 2\pi/_n\omega_l$ of the nth radial overtone. Here we have a short-wave approximation, in which the period may be calculated by constructive interference, neglecting the curvature of the surfaces of constant radius. That is why l does not appear in Eq. (8.71).

To obtain the first-order approximation, we put $k_\beta(a - b) = n\pi$ in the right-hand side of Eq. (8.70). This yields

$$k_\beta(a - b) = n\pi + \tan^{-1}\left[\frac{l(l + 1) + 4}{2n\pi ab}(a - b)^2\right] + O\left(\frac{l^4}{n^2}\right),$$

i.e.,

$$_n\omega_l \simeq n\pi\left(\frac{\beta_0}{a - b}\right)\left[1 + \frac{l(l + 1) + 4}{2n^2\pi^2 ab}(a - b)^2\right]. \tag{8.72}$$

It is interesting to note that relation (8.72) can also be obtained from Eq. (8.68) by taking $\beta = $ const. This assumption implies

$$\gamma = \frac{(a - b)}{\beta_0}, \qquad t = \frac{(r - b)}{\beta_0}, \qquad q(t) = \frac{l(l + 1)\beta_0^2}{r^2}, \qquad h_1 = \frac{2\beta_0}{b}, \qquad h_2 = \frac{2\beta_0}{a}.$$

8.4.2.2. $k_\beta b$ Large, l Large, $k_\beta b > l + \frac{1}{2}$. The approximation in Eq. (8.72) breaks down when $l^2/(k_\beta b)$ is not small. If both z and l are large, and $z > l + \frac{1}{2}$, we use the asymptotic formulas

$$\begin{aligned} j_l(z) &\sim \frac{1}{z\sqrt{(\sin\alpha)}}\cos\eta(z), & n_l(z) &\sim \frac{1}{z\sqrt{(\sin\alpha)}}\sin\eta(z), \\ j_l'(z) &\sim \frac{-\sqrt{(\sin\alpha)}}{z}\sin\eta(z), & n_l'(z) &\sim \frac{\sqrt{(\sin\alpha)}}{z}\cos\eta(z), \end{aligned} \tag{8.73}$$

where

$$\cos\alpha = \frac{(l + \frac{1}{2})}{z}, \qquad \eta(z) = \left(l + \frac{1}{2}\right)(\tan\alpha - \alpha) - \frac{\pi}{4}.$$

This implies

$$\mathscr{J}_l(z) \sim -\frac{1}{z^2}\sqrt{[\sin\alpha]}\sin\eta(z), \qquad \mathscr{N}_l(z) \sim \frac{1}{z^2}\sqrt{[\sin\alpha]}\cos\eta(z).$$

The frequency equation, Eq. (8.69), is now asymptotically given by

$$\sin\{\eta(k_\beta a) - \eta(k_\beta b)\} = 0.$$

The zero-order approximation is

$$\eta(k_\beta a) - \eta(k_\beta b) = n\pi,$$

or,

$$(l + \tfrac{1}{2})(\tan\alpha_a - \alpha_a - \tan\alpha_b + \alpha_b) = n\pi, \tag{8.74}$$

where

$$\cos\alpha_a = \frac{(l + \frac{1}{2})}{(k_\beta a)}, \qquad \cos\alpha_b = \frac{(l + \frac{1}{2})}{(k_\beta b)}.$$

8.4.2.3. $k_\beta b$ **Large,** l **Large,** $k_\beta a > l + \frac{1}{2} > k_\beta b$. When both z and l are large, and $z < l + \frac{1}{2}$, we have the asymptotic formulas

$$\begin{aligned} j_l(z) &\sim \frac{1}{2z\sqrt{(\text{sh}\, B)}} \exp\left[-\left(l + \frac{1}{2}\right)(B - \text{th}\, B)\right], \\ n_l(z) &\sim \frac{-1}{z\sqrt{(\text{sh}\, B)}} \exp\left[\left(l + \frac{1}{2}\right)(B - \text{th}\, B)\right], \end{aligned} \tag{8.75}$$

where

$$\text{ch}\, B = \frac{(l + \frac{1}{2})}{z}.$$

This yields

$$\begin{aligned} \mathscr{J}_l(z) &\sim \frac{\sqrt{(\text{sh}\, B)}}{2z^2} \exp\left[-\left(l + \frac{1}{2}\right)(B - \text{th}\, B)\right], \\ \mathscr{N}_l(z) &\sim \frac{\sqrt{(\text{sh}\, B)}}{z^2} \exp\left[\left(l + \frac{1}{2}\right)(B - \text{th}\, B)\right]. \end{aligned} \tag{8.76}$$

Because $B - \text{th}\, B > 0$, the frequency equation, Eq. (8.69), reduces to $\sin\eta(k_\beta a) = 0$, approximately, with zero-order solution $\eta(k_\beta a) = n\pi$. This leads to

$$\left(l + \frac{1}{2}\right)(\tan\alpha_a - \alpha_a) = n\pi. \tag{8.77}$$

8.4.3. Correspondence of Rays and Normal Modes

When a localized source excites elastic waves in a nonrotating earth, the resulting motion can be represented as a double sum of standing waves, the summation being over the colatitudinal mode number (l) and the radial mode number (n).

The travel time along a ray in a radially heterogeneous earth is given by [cf. Eq. (7.88)]

$$T = p\theta + 2\int_{r_m}^{a}\left(\frac{r^2}{v^2} - p^2\right)^{1/2}\frac{dr}{r}, \tag{8.78}$$

where r_m refers to the lowest point on the ray, $v(r)$ is the radial velocity distribution and p is the ray parameter. Moreover, Eq. (8.78) assumes surface focus and θ is the angular epicentral distance. Snell's law then yields

$$p = \frac{dT}{d\theta} = \frac{r\sin i}{v} = \frac{r_m}{v_m} = \frac{a\sin i_0}{v_0} = \frac{a}{c_a}, \tag{8.79}$$

where $c_a = v_0/\sin i_0$ is the apparent velocity. Combining it with Jeans' formula, Eq. (8.4), we get

$$p = \frac{a\sin i_0}{v_0} = \frac{a}{c} = \frac{l+\frac{1}{2}}{{}_n\omega_l}, \tag{8.80}$$

where $c = c_a = c({}_n\omega_l)$. Because c is constant along a ray, the ray can be considered as the locus of all normal modes that travel with the same phase velocity.

Clearly, the conditions under which Eq. (8.80) is valid include the condition under which Jeans' formula is valid, namely, $l \gg m$. Because by Eq. (8.80) $l + \frac{1}{2} = ({}_n\omega_l a/v_0)\sin i_0$, the ray–mode correspondence is subject to the condition

$$\frac{{}_n\omega_l a}{v_0} > l + \frac{1}{2} \gg m.$$

Combining Eqs. (8.60a), (8.78), and (8.79), we obtain

$$t_s - \Delta\frac{dt_s}{d\Delta} = \left(n + \frac{1}{4}\right){}_nT_l, \tag{8.81}$$

where t_s is the travel time of the direct S wave along its ray, ${}_nT_l$ is a toroidal eigenperiod, and Δ is the epicentral distance.

Consider, next, the spherical eigenfunction in a homogeneous sphere

$$\Phi_{ml} = h_l^{(2)}({}_nk_l a)P_l^m(\cos\theta)e^{i{}_n\omega_l t}, \qquad {}_nk_l = \frac{{}_n\omega_l}{v_0},$$

which is the building block of the normal-mode solution. To interpret it in terms of traveling waves, we write

$$2\,\Phi_{ml} = \Phi_{ml}^{(1)} + \Phi_{ml}^{(2)},$$

where

$$\Phi_{ml}^{(1,\,2)} = h_l^{(2)}({}_nk_l a)\left[P_l^m(\cos\theta) \pm \frac{2i}{\pi}\,Q_l^m(\cos\theta)\right]e^{i{}_n\omega_l t}. \tag{8.82}$$

We employ the asymptotic approximations (App. H)

$$h_l^{(2)}({}_nk_l a) \sim \frac{1}{{}_nk_l a(\cos\tau)^{1/2}} \exp\left[-i\left[{}_nk_l a\left\{\cos\tau - \left(\frac{\pi}{2} - \tau\right)\sin\tau\right\} - \frac{\pi}{4}\right]\right], \tag{8.83}$$

$$P_l^m(\cos\theta) + \frac{2i}{\pi} Q_l^m(\cos\theta) \sim \left(\frac{2}{\pi\sin\theta}\right)^{1/2} l^{m-1/2} \exp\left[-i\left[\left(l + \frac{1}{2}\right)\theta - \frac{m\pi}{2} - \frac{\pi}{4}\right]\right],$$

valid for ${}_nk_l a > l + \frac{1}{2} \gg m > 1$, $\varepsilon < \theta < \pi - \varepsilon$, $\sin\tau = (l + \frac{1}{2})/{}_nk_l a$. With these approximations

$$\Phi_{ml}^{(1)} = \left(\frac{2}{\pi\sin\theta\cos\tau}\right)^{1/2} \frac{l^{m-1/2}}{{}_nk_l a} \exp\left[i\left[{}_n\omega_l t - {}_nk_l D + \frac{(m+1)\pi}{2}\right]\right], \tag{8.84}$$

where

$$D = a[\cos\tau - (\alpha - \theta)\sin\tau], \qquad \alpha = \frac{\pi}{2} - \tau.$$

We interpret the phase in Eq. (8.84) as follows: Consider a sphere centered at O (Fig. 8.6) with radius

$$r_m = \frac{(l + \frac{1}{2})}{{}_nk_l} = a\sin\tau. \tag{8.85}$$

It is then clear from the geometry of the figure that

$$\begin{aligned} SP &= QP - QS \\ &\simeq QP - QS' = QP - (\alpha - \theta)r_m = a[\cos\tau - (\alpha - \theta)\sin\tau]. \end{aligned}$$

Therefore, the quantity

$$D = a[\cos\tau - (\alpha - \theta)\sin\tau] \tag{8.86}$$

is approximately equal to the ray path from the source S to the station P provided that the angle τ is interpreted as the angle of incidence i_0 of the ray at

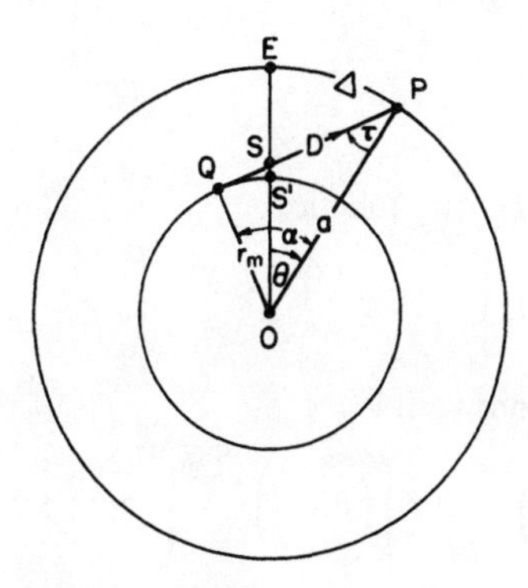

Figure 8.6. Geometric interpretation of the saddle-point approximation for a spherical wave.

the surface. In general, given l and ${}_nk_l = {}_n\omega_l/v_0$, we face three possibilities:

1. $l + \frac{1}{2} < {}_nk_l a \Rightarrow r_m < a$: The rays are real.
2. $l + \frac{1}{2} > {}_nk_l a \Rightarrow r_m > a$: No real ray exists.
3. $l + \frac{1}{2} = {}_nk_l a \Rightarrow r_m = a$: The surface of the sphere coincides with the ray envelope (caustic).

EXAMPLE 8.1: Rays versus normal modes in a homogeneous sphere

The normal-mode solution for toroidal motion caused by a buried strike-slip dislocation in a homogeneous sphere is included in Eq. (6.135) and has the explicit exact form

$$\mathbf{u}(a, r_h; t) = -\frac{U_0\,dS}{2\pi a r_h} H(t) \sum_{l=2}^{\infty} \frac{2l+1}{l(l+1)} \sum_{n=0}^{\infty} \frac{j_l(k_n r_h)[1 - \cos({}_n\omega_l t)]}{j_l(k_n a)[(l-1)(l+2) - k_n^2 a^2]} \times \sqrt{[l(l+1)]}\,\mathbf{C}^c_{l,2}(\theta, \phi). \qquad (8.1.1)$$

Here

$$k_n = \frac{{}_n\omega_l}{\beta_0}, \qquad {}_nz_l = k_n a, \qquad {}_nz_l\, j_{l+1}({}_nz_l) = (l-1) j_l({}_nz_l). \qquad (8.1.2)$$

Because $j_l(z) = \sqrt{[\pi/2z]}J_{l+1/2}(z)$, the period equation can also be written in the form

$$J'_{l+1/2}(z) = \frac{3}{2z} J_{l+1/2}(z). \qquad (8.1.3)$$

For large values of z, the approximate period equation is $J'_{l+1/2}(z) = 0$. Employing the approximation (App. H)

$$J'_\nu(z) = -\left(\frac{2}{\nu}\right)^{2/3} Ai'\left[-2^{1/3}\left(\frac{z-\nu}{\nu^{1/3}}\right)\right],$$

the roots of the period equation for large values of l are given by

$${}_nz_\nu = \nu - (2^{-1/3}\bar{q}_n)\nu^{1/3} + O(\nu^{-1/3}), \qquad (8.1.4)$$

where

$$Ai'(\bar{q}_n) = 0.$$

The values of $\bar{q}_n$ are given in Table 8.7. For the Airy function,

$$-\bar{q}_n = \left[\left(\frac{3\pi}{8}\right)(4n+1)\right]^{2/3}$$

is a good approximation already for $n = 1$. Hence with $\nu = l + \frac{1}{2}$

$${}_nz_l = \left(l + \frac{1}{2}\right) + \frac{1}{2}\left[3\pi\left(n + \frac{1}{4}\right)\right]^{2/3}\left(l + \frac{1}{2}\right)^{1/3} + O\left[\left(l + \frac{1}{2}\right)^{-1/3}\right]. \qquad (8.1.5)$$

Table 8.7. Zeros of the Airy Function and Its Derivative

n	$-q_n$	$[(3\pi/8)(4n+3)]^{2/3}$	$-\bar{q}_n$	$[(3\pi/8)(4n+1)]^{2/3}$
0	2.338107	2.320251	1.018792	1.115460
1	4.087949	4.081810	3.248197	3.261625
2	5.520559	5.517164	4.820099	4.826316
3	6.786708	6.784454	6.163307	6.167128
4	7.944133	7.942486	7.372177	7.374853
5	9.022650	9.021373	8.488486	8.490506
6	10.040174	10.039142	9.535449	9.537050
7	11.008524	11.007665	10.527660	10.528974
8	11.936015	11.935284	11.475056	11.476163
9	12.828776	12.828145	12.384788	12.385738

By refining this process we are able to obtain additional terms in the expansion. In particular

$$\begin{aligned}
{}_0z_l &= \left(l+\frac{1}{2}\right) + 0.8086\left(l+\frac{1}{2}\right)^{1/3} - 0.8550\left(l+\frac{1}{2}\right)^{-1/3} + O\left[\left(l+\frac{1}{2}\right)^{-1}\right],\\
{}_1z_l &= \left(l+\frac{1}{2}\right) + 2.5781\left(l+\frac{1}{2}\right)^{1/3} + 1.6645\left(l+\frac{1}{2}\right)^{-1/3} + O\left[\left(l+\frac{1}{2}\right)^{-1}\right],\\
{}_2z_l &= \left(l+\frac{1}{2}\right) + 3.8257\left(l+\frac{1}{2}\right)^{1/3} + 4.1690\left(l+\frac{1}{2}\right)^{-1/3} + O\left[\left(l+\frac{1}{2}\right)^{-1}\right].
\end{aligned} \tag{8.1.6}$$

Clearly, for $l \gg 1$, ${}_nz_l \simeq l + \frac{1}{2}$ and

$${}_nT_l \simeq \frac{2\pi a}{(l+\frac{1}{2})\beta_0}. \tag{8.1.7}$$

${}_nT_l$ can be obtained also from Jeans' formula in the limit $c({}_n\omega_l) \to \beta_0$ for large values of l.

The former result can be derived in another way with the use of the phase-integral relation of Eq. (8.60a). Taking there $\beta(r) = \beta_0 = \text{const.}$, replacing $(l-1)(l+2)$ by $(l+\frac{1}{2})^2$, and evaluating the integral, we obtain

$$\left(l+\frac{1}{2}\right)(\tan\tau_n - \tau_n) = \pi\left(n+\frac{1}{4}\right), \tag{8.1.8}$$

where

$${}_nz_l = \left(l+\frac{1}{2}\right)\sec\tau_n.$$

When we make use of the known expansions

$$\sec \tau_n = 1 + \frac{1}{2}\tau_n^2 + \frac{5}{24}\tau_n^4 + \cdots, \tag{8.1.9}$$

$$\tan \tau_n - \tau_n = \frac{1}{3}\tau_n^3 + O(\tau_n^5), \tag{8.1.10}$$

Eq. (8.1.8) yields for $|\tau_n| < \pi/2$

$${}_nz_l = \left(l + \frac{1}{2}\right) + \frac{1}{2}\left[3\pi\left(n + \frac{1}{4}\right)\right]^{2/3}\left(l + \frac{1}{2}\right)^{1/3} + O\left[\left(l + \frac{1}{2}\right)^{-1/3}\right] \tag{8.1.11}$$

in accord with Eq. (8.1.5).

Because $\gamma = a/\beta_0$ is the travel-time along the radius of the sphere, we can write the explicit dependence of ${}_n\omega_l$ on both l and n

$${}_n\omega_l = \frac{l + \frac{1}{2}}{\gamma}\left[1 + \frac{1}{2}\left[3\pi\left(n + \frac{1}{4}\right)\right]^{2/3}\left(l + \frac{1}{2}\right)^{-2/3} + O\left\{\left(\frac{n}{l}\right)^{4/3}\right\}\right], \tag{8.1.11a}$$

where

$$\frac{l}{n} > \frac{24}{\pi^2} = 2.43\cdots.$$

In the parameter range ${}_nz_l > n > l$ we must consider angles τ_n close to $\pi/2$ where $\tan \tau_n \gg \tau_n$. There Eqs. (8.1.2) and (8.1.8) yield

$${}_n\omega_l \simeq \frac{\pi(n + \frac{1}{4})}{\gamma}\left[1 + \frac{1}{2}\left\{\frac{l + \frac{1}{2}}{\pi(n + \frac{1}{4})}\right\}^2\right] + O(n^{-3}). \tag{8.1.12}$$

We wish next to explore the relations between the period equation and travel times of *SH* waves in a homogeneous sphere. To this end we appeal to the asymptotic approximation

$$j_l(z) = \frac{\sin[(l + \frac{1}{2})(\tan \delta - \delta) + \pi/4]}{z\sqrt{[\sin \delta]}}, \tag{8.1.13}$$

with

$$\cos \delta = \frac{l + \frac{1}{2}}{z}, \qquad z > \left(l + \frac{1}{2}\right) \gg 1.$$

We substitute $j_l(z)$ from Eq. (8.1.13) into the period equation, Eq. (8.1.2), and use the duality relation (8.80), $z/(l + \frac{1}{2}) = c/\beta_0$. We then obtain

$$\sin\left[\left(l + \frac{1}{2}\right)\left\{\sqrt{\left[\frac{c^2}{\beta_0^2} - 1\right]} - \tan^{-1}\sqrt{\left[\frac{c^2}{\beta_0^2} - 1\right]}\right\} - \pi\phi_0\right] = 0, \tag{8.1.14}$$

where

$$\frac{{}_n\omega_l a}{\beta_0} > l + \frac{1}{2} \gg 1, \qquad {}_n\omega_l = \frac{c}{a}\left(l + \frac{1}{2}\right),$$

$$\pi\phi_0 = \tan^{-1}\left\{\frac{D \sin \Omega}{1 - D \cos \Omega}\right\} - \frac{\pi}{4},$$

$$D = \frac{l + \frac{1}{2}}{l - 1} \frac{\tan(\theta/2)}{\cos(\theta/2)\tan(\theta'/2)}, \tag{8.1.15}$$

$$\Omega = \left(l + \frac{1}{2}\right)\left[\tan\frac{\theta'}{2} - \tan\frac{\theta}{2} - \frac{\theta'}{2} + \frac{\theta}{2}\right] - \frac{\theta'}{2},$$

$$\tan\frac{\theta}{2} = \left[\frac{c^2}{\beta_0^2} - 1\right]^{1/2}, \qquad \tan\frac{\theta'}{2} = \left[\frac{c^2}{\beta_0^2} - \left\{\frac{l + 3/2}{l + 1/2}\right\}^2\right]^{1/2}.$$

In the limit $l \to \infty$

$$\theta' = \theta, \qquad \Omega = -\frac{\theta}{2}, \qquad D = \frac{1}{\cos(\theta/2)}, \qquad {}_n\omega_l \to \infty$$

and consequently

$$\phi_0 \to \frac{1}{4}. \tag{8.1.16}$$

We next recast Eq. (8.1.14) as

$$\sin\left[\frac{\pi}{{}_nT_l}\left\{\frac{2a}{c}\tan\frac{\theta}{2} - \frac{a\theta}{c}\right\} - \pi\phi_0\right] = 0 \tag{8.1.17}$$

and recall [Eq. (7.1.2)] that the travel-time along a direct ray in a homogeneous sphere is given by [cf. Ex. 7.1]

$$t = \frac{2a}{\beta_0}\sin\frac{\theta}{2}, \qquad \frac{dt}{d\theta} = \frac{a}{\beta_0}\cos\frac{\theta}{2}. \tag{8.1.18}$$[3]

Hence, with $a\theta = \Delta$, Eqs. (8.79), (8.1.17) and (8.1.18) render the period equation in its new form

$$\sin\left[\frac{\pi}{{}_nT_l}\left\{t - \frac{\Delta}{{}_nc_l}\right\} - \pi\phi_0\right] = 0 \tag{8.1.19}$$

or

$$t = \frac{\Delta}{{}_nc_l} + (n + \phi_0)\,{}_nT_l. \tag{8.1.20}$$

[3] We use t here for travel-time along the ray, instead of T, which we used in Chapter 7, in order to avoid confusion with the eigenperiod symbol ${}_nT_l$.

In the limit $l \to \infty$, $\phi_0 \to \frac{1}{4}$ in the direct wave and Eq. (8.1.20) coincides with Eq. (8.81). Equation (8.1.20) can also be put in the form

$$ {}_nT_l = \frac{t - \Delta(dt/d\Delta)}{n + \phi_0}, \qquad \phi_0 = \phi_0({}_nT_l, c), \qquad c = {}_nc_l. \tag{8.1.21}$$

Here, ϕ_0 is the interference phase shift (in circle units) resulting from the reflection of the wave from the concave free surface. Alternatively, we can regard Eq. (8.1.21) as a definition of the parameter ϕ_0, which relates the normal-mode periods to the travel-time curves. This equation was derived from the phase interference requirement for a steady state. For a homogeneous sphere, the parameter ϕ_0 can be calculated from the analytical expressions for the travel-time and the ray parameter as a function of epicentral distance.

The same analysis can be carried out for P and SV waves in a homogeneous sphere. No new idea is involved but the analysis is more complicated because of the conversion of P and SV waves at the free surface. We shall give here only a brief survey of the mode–ray correspondence: Consider P and S rays, in a homogeneous sphere, that share a common phase velocity c. Invoking Eq. (8.80), we have

$$c = \frac{\alpha_0}{\sin i_0} = \frac{\beta_0}{\sin j_0} = \frac{a\,{}_n\omega_l}{l + \frac{1}{2}}. \tag{8.1.22}$$

Hence

$$\sin i_0 = \frac{l + \frac{1}{2}}{(a\,{}_n\omega_l/\alpha_0)}, \qquad \sin j_0 = \frac{l + \frac{1}{2}}{(a\,{}_n\omega_l/\beta_0)}, \tag{8.1.23}$$

where i_0 and j_0 are the P and S angles of incidence at the surface of the sphere.

The following cases may arise:

1. $c < \beta$. From Eq. (8.1.22) we deduce that i_0 and j_0 are complex angles. Therefore, in the ordinary sense, there is no body wave, that has a phase velocity in this range. Because complex angles of propagation are associated with Rayleigh waves (Chap. 3) the fundamental Rayleigh mode will satisfy this condition at high frequencies.
2. $\beta < c < \alpha$. This condition is equivalent to

$$\sin^{-1}\left(\frac{\beta}{\alpha}\right) < j_0 < \frac{\pi}{2}.$$

   Therefore, S waves are always totally reflected and ordinary P waves do not exist.
3. $c > \alpha$. Both P and S waves exist and are reflected and converted into each other.

8.4.4. Saddle-Point Approximation of the Watson Integral

Consider the spectral response of a radially heterogeneous earth to a buried dislocation source of the vertical strike-slip type located at $r = r_h, \theta = 0$. From Eq. (6.136), the azimuthal component of the surface displacement is given by

$$u_\phi = -\frac{U_0\, dS}{4\pi a^2} \cos 2\phi e^{i\omega t} \frac{\partial}{\partial\theta} \sum_{l=2}^{\infty} (2l+1) f_l P_l^2(\cos\theta), \tag{8.87}$$

where

$$f_l = \frac{1}{l(l+1)} \left[\frac{y_1(l; r_h)/r_h}{y_2(l; a)/\mu_h}\right].$$

The radial function y_1 satisfies Eq. (6.63) and $y_2 = \mu(dy_1/dr - y_1/r)$. Substituting

$$y_1 = \left(\frac{\mu_h}{\mu}\right)^{1/2} \frac{F_l(r)}{r}, \qquad \mu_h = \mu(r_h), \tag{8.88}$$

in Eq. (6.63) and assuming $\mu'/\mu \ll 1$, $\mu''/\mu \ll 1$, we obtain for $F_l(r)$ the equation

$$\frac{d^2 F_l(r)}{dr^2} + \omega^2 \left[\frac{1}{v^2} - \frac{l(l+1)}{\omega^2 r^2}\right] F_l(r) = 0. \tag{8.89}$$

The WKBJ solution (App. H) of Eq. (8.89) is

$$F_l(r) = \exp\left(\mp\frac{\pi i}{4}\right)\left[\frac{1}{v^2} - \frac{(l+\frac{1}{2})^2}{\omega^2 r^2}\right]^{-1/4} \exp\left[\pm i\omega \int_{r_m}^{r} \left[1 - \frac{v^2(l+\frac{1}{2})^2}{\omega^2 r^2}\right]^{1/2} \frac{dr}{v}\right], \quad r > r_m \tag{8.90}$$

$$F_l(r) = \left[\frac{(l+\frac{1}{2})^2}{\omega^2 r^2} - \frac{1}{v^2}\right]^{-1/4} \exp\left[-\omega \int_{r}^{r_m} \left[\frac{v^2(l+\frac{1}{2})^2}{\omega^2 r^2} - 1\right]^{1/2} \frac{dr}{v}\right], \qquad r < r_m \tag{8.91}$$

where r_m, the transition point of the differential equation (8.89), is given by

$$r_m = \left(l + \frac{1}{2}\right)\frac{v(r_m)}{\omega}. \tag{8.92}$$

The upper sign is used for the downgoing waves and the lower sign for the upgoing waves. Because

$$\frac{1}{\mu} y_2(r) \simeq \frac{1}{r}\left[\frac{\mu_h}{\mu}\right]^{1/2} \frac{dF_l}{dr} \tag{8.93}$$

we find

$$f_l(a, r_h; \omega) \simeq \frac{1}{l(l+1)(k_0 a)} \left[\frac{\mu_h}{\mu_0}\right]^{1/2} \left[\frac{a}{r_h}\right]^{3/2} \left[1 - \frac{(l+\frac{1}{2})^2}{k_0^2 a^2}\right]^{-1/4}$$

$$\times \left[\frac{v_0^2 r_h^2}{v_h^2 a^2} - \frac{(l+\frac{1}{2})^2}{k_0^2 a^2}\right]^{-1/4} e^{-i\omega\tau - \frac{\pi i}{4} \mp \frac{\pi i}{4}}, \tag{8.94}$$

where $v_0 = v(a)$, $\mu_0 = \mu(a)$, $k_0 = \omega/v_0$, and

$$\tau = \left[\mp \int_{r_m}^{r_h} + \int_{r_m}^{a}\right] \left\{\frac{r^2}{v^2} - \frac{(l+\frac{1}{2})^2}{\omega^2}\right\}^{1/2} \frac{dr}{r}. \tag{8.95}$$

We note that for values of l and ω such that $(l + \frac{1}{2}) > k_0 a$, f_l decays exponentially with l. In this case, the Legendre sum converges rapidly and the modal solution, Eq. (8.87), needs no further transformation. However, if $(l + \frac{1}{2}) < k_0 a$, the terms in the sum in Eq. (8.87) are periodic for $r > r_m$ and the convergence is poor. One then transforms Eq. (8.87) with the help of the Watson transformation. When the rays are real, therefore, the convergence of the Legendre sum is poor and we use the Watson transformation to improve its convergence. When no real rays exist, however, the Legendre sum converges rapidly and the modal solution needs no further transformation. We will show next how to obtain rays by a saddle point approximation of the transformed Legendre sum.

Applying the Watson transformation to the sum in Eq. (8.87) and ignoring the real poles of the integrand which give rise to surface waves, we have, as in Eq. (8.13),

$$\sum_{l=2}^{\infty} (2l+1) f_l P_l^2(\cos\theta) = \int_{C_2} f_{s-1/2} P_{s-1/2}^2(-\cos\theta) \frac{s\,ds}{i\cos s\pi}, \tag{8.96}$$

where C_2 is the contour shown in Fig. 8.2. In deriving Eq. (8.96) we have used the fact that $f_{-s-1/2} = f_{s-1/2}$.

Using Eqs. (8.15) and (8.16), we have

$$\sec\left(\nu + \frac{1}{2}\right)\pi \frac{\partial}{\partial\theta} P_\nu^2(-\cos\theta) = i\left(\frac{2}{\pi\sin\theta}\right)^{1/2} \nu^{5/2} \sum_{N=0}^{\infty} (-1)^N (-e^{-i\Lambda_1} + e^{-i\Lambda_2}), \tag{8.97}$$

where $\operatorname{Im}\nu < 0$, $\varepsilon < \theta < \pi - \varepsilon$, $\varepsilon > 0$, and

$$\Lambda_1 = \left(\nu + \frac{1}{2}\right)(\theta + 2\pi N) - \frac{3\pi}{4},$$

$$\Lambda_2 = \left(\nu + \frac{1}{2}\right)[(2\pi - \theta) + 2\pi N] + \frac{3\pi}{4}.$$

Combining the results of Eqs. (8.87), (8.94), (8.95), and (8.97), we obtain for the direct ray ($N = 0$ with Λ_1 term only)

$$u_\phi(a, r_h; \omega) = -\frac{U_0\, dS}{2\pi r_h^2}\left[\frac{\mu_h}{\mu_0}\right]^{1/2} \frac{\cos 2\phi}{(k_0 a)[2\pi \sin\theta]^{1/2}}$$
$$\times \int_{-\infty}^{\infty} \frac{\exp[i[\omega(t-\tau) - (\nu + \frac{1}{2})\theta] + \frac{\pi i}{2} \mp \frac{\pi i}{4}]}{[1 - \nu(\nu+1)/(k_0 a)^2]^{1/4}[(v_0/v_h)^2 - \nu(\nu+1)/(k_0 r_h)^2]^{1/4}}$$
$$\times\, \nu^{3/2}\, d\nu.$$

Introducing the new variable $i_0 = i_0(\nu)$,

$$\sqrt{[\nu(\nu+1)]} \simeq \nu + \frac{1}{2} = (k_0 a)\sin i_0 = (k_0 r_h)\frac{v_0}{v_h}\sin i_h,$$

the above equation assumes the form

$$u_\phi = -\frac{U_0\, dS}{2\pi r_h^2}\left[\frac{\mu_h\, v_h}{\mu_0\, v_0}\right]^{1/2} \frac{\cos 2\phi}{(k_0 a)[2\pi \sin\theta]^{1/2}}$$
$$\times \int_{-\infty}^{\infty} \frac{\exp\{i[\omega(t-\tau) - (\nu + \frac{1}{2})\theta] + \frac{\pi i}{2} \mp \frac{\pi i}{4}\}}{[\cos i_0(\nu) \cos i_h(\nu)]^{1/2}}\, \nu^{3/2}\, d\nu. \tag{8.98}$$

To evaluate the integral by the method of stationary phase for the large parameter ω, we use the result (App. E)

$$\int_{-\infty}^{\infty} e^{-i\chi(\nu)} A(\nu) d\nu = \left[\frac{2\pi}{|\chi''(\nu_0)|}\right]^{1/2} A(\nu_0)\exp\left[-i\chi(\nu_0) - \frac{\pi i}{4}\,\mathrm{sgn}\,\chi''(\nu_0)\right], \tag{8.99}$$

where the saddle point ν_0 is the root of the equation $\chi'(\nu) = 0$, i.e.,

$$\frac{\partial}{\partial \nu}\left[\left(\nu + \frac{1}{2}\right)\theta + \omega\tau(\nu)\right] = 0.$$

Inserting the explicit form of $\tau(\nu)$ [Eq. (8.95)], we find that the roots $\nu = \nu_0$ are given by the integral relation

$$\theta(\nu_0) = p\left\{\int_{r_m}^{r_h} + \int_{r_m}^{a}\right\} \frac{dr}{r\sqrt{[r^2/v^2 - p^2]}}, \tag{8.100}$$

where

$$p = \frac{\nu_0 + \frac{1}{2}}{\omega}.$$

Now, comparing Eq. (8.100) with Eq. (7.97), we see that p is identical with the ray parameter and, by Eq. (8.79), the saddle point is at

$$\nu_0 \simeq (k_0 a)\sin i_0.$$

With this result at hand, we return to Eqs. (8.98) and (8.99). Because θ does not depend upon ν but does depend on ν_0 on a given ray,

$$\left.\frac{\partial^2\chi}{\partial\nu^2}\right|_{\nu=\nu_0} = \frac{\partial^2}{\partial\nu^2}\left[\left(\nu+\frac{1}{2}\right)\theta + \omega\tau\right]_{\nu=\nu_0} = \omega\left.\frac{\partial^2\tau}{\partial\nu^2}\right|_{\nu=\nu_0}$$

$$= -\frac{\partial\theta(\nu_0)}{\partial\nu_0} = -\frac{\partial\theta}{\partial i_h}\frac{\partial i_h}{\partial\nu_0} = \frac{-1}{k_h r_h \cos i_h(\partial i_h/\partial\theta)}.$$

Hence

$$|\chi''(\nu_0)|^{-1/2} = \left[k_h r_h \cos i_h \left|\frac{\partial i_h}{\partial\theta}\right|\right]^{1/2}.$$

The sign of $\chi''(\nu_0)$ will depend on the wave in question. For a downgoing ray, $\cos i_h$ is negative and $\partial i_h/\partial\theta > 0$, so that $\chi''(\nu_0) > 0$. For an upgoing ray, $\chi''(\nu_0) < 0$. Therefore, the spectral displacements at the surface are found to be

$$u_\phi(a, r_h; \omega) = \omega \exp\left[-\frac{\pi i}{4} \mp \frac{\pi i}{4}\right]\frac{U_0(\omega)dS}{2\pi v_h}(\cos 2\phi \sin i_h)$$
$$\times\left\{\frac{1}{a}\left[\frac{\rho_h v_h}{\rho_0 v_0}\frac{\sin i_h}{\sin\theta\cos i_0}\left|\frac{\partial i_h}{\partial\theta}\right|\right]^{1/2}\right\}e^{i\omega(t-T_s)}. \tag{8.101}$$

The expression in the braces coincides with the *divergence coefficient* given in Eq. (7.174) and ($\cos 2\phi \sin i_h$) is the appropriate radiation pattern given in Eq. (7.165). The result in Eq. (8.101) is similar to the half-space analog given in Eq. (7.214). Comparing Eq. (8.101) with the corresponding expression for a homogeneous infinite medium [Eq. (7.164)] we see that while the divergence coefficient is introduced by the heterogeneity of the medium, the displacements are doubled because of the presence of the free surface.

If the point of observation is near a caustic, the approximation, Eq. (8.99), is no longer valid. To derive a suitable expression for u_ϕ in the neighborhood of a caustic, we follow the treatment given in Section 7.5. Starting with the ν-dependent phase of the integrand in Eq. (8.98)

$$\chi = \left(\nu+\frac{1}{2}\right)\theta + \omega\left[\int_{r_m}^{r_h} + \int_{r_m}^{a}\right]\left\{\left(\frac{r^2}{v^2} - \left[\frac{\nu+\frac{1}{2}}{\omega}\right]^2\right)^{1/2}\frac{dr}{r}\right\}$$

we have

$$\frac{\partial\chi}{\partial\nu} = \theta - \frac{\nu+\frac{1}{2}}{\omega}\left[\int_{r_m}^{r_h} + \int_{r_m}^{a}\right]\left\{\left(\frac{r^2}{v^2} - \left[\frac{\nu+\frac{1}{2}}{\omega}\right]^2\right)^{-1/2}\frac{dr}{r}\right\}.$$

The saddle point ν_0 is the solution of $\partial\chi/\partial\nu = 0$, that is, of $\theta(\nu_0) = \theta$, with $\theta(\nu_0)$ given by Eq. (8.100). If the observation point is on the caustic, we must have, in addition to Eq. (8.100), the condition

$$\frac{\partial^2\chi}{\partial\nu^2} = -\frac{\partial\theta(\nu_0)}{\partial\nu_0} = 0.$$

In the *neighborhood* of the caustic, however, the point of observation does not lie on a ray and ν_0 is not a saddle point. Let θ_{ca} be a point on the caustic closest to our point of observation θ, and let ν_{ca} be a ray that touches the caustic at $\theta = \theta_{ca}$. We may then expand $\chi(\nu)$ about ν_{ca} in a Taylor series

$$\chi(\nu) = \chi(\nu_{ca}) + (\theta - \theta_{ca})(\nu - \nu_{ca}) - \left(\frac{\partial^2\theta}{\partial\nu^2}\right)_{ca} \frac{(\nu - \nu_{ca})^3}{6} + \cdots.$$

The integral in Eq. (8.99) then takes the form

$$J = \int_{-\infty}^{\infty} e^{-i\chi(\nu)} A(\nu) d\nu$$

$$= A(\nu_{ca}) e^{-i\chi(\nu_{ca})} \int_{-\infty}^{\infty} \exp\left[-i\left[(\theta - \theta_{ca})(\nu - \nu_{ca}) - \left(\frac{\partial^2\theta}{\partial\nu^2}\right)_{ca} \frac{(\nu - \nu_{ca})^3}{6}\right]\right] d\nu.$$

Putting

$$x = \frac{2}{3}\left[2\Big/\left(\frac{\partial^2\theta}{\partial\nu^2}\right)_{ca}\right]^{1/2} (\theta - \theta_{ca})^{3/2}, \qquad s = \left[\frac{1}{6}\left(\frac{\partial^2\theta}{\partial\nu^2}\right)_{ca}\right]^{1/3} (\nu - \nu_{ca}),$$

$$J = A(\nu_{ca}) e^{-i\chi(\nu_{ca})} I$$

we obtain

$$I = \left[6\Big/\left(\frac{\partial^2\theta}{\partial\nu^2}\right)_{ca}\right]^{1/3} \int_{-\infty}^{\infty} e^{i[s^3 - 3(x/2)^{2/3}s]}\, ds$$

$$= 2\pi\left[2\Big/\left(\frac{\partial^2\theta}{\partial\nu^2}\right)_{ca}\right]^{1/3} Ai\left[-2^{1/3}(\theta - \theta_{ca})\Big/\left(\frac{\partial^2\theta}{\partial\nu^2}\right)_{ca}^{1/3}\right]. \tag{8.101a}$$

In particular, on the caustic $\theta = \theta_{ca}$, we have

$$I_0 = 2\pi\left[2\Big/\left(\frac{\partial^2\theta}{\partial\nu^2}\right)_{ca}\right]^{1/3} Ai(0). \tag{8.101b}$$

Whenever θ is close to θ_{ca}, we may obtain a sufficiently good approximation by developing $Ai(-z)$, $z = (\theta - \theta_{ca})[2/(\partial^2\theta/\partial\nu^2)_{ca}]^{1/3}$ in a power series about θ_{ca}. Specifically

$$Ai(0 - z) = Ai(0) - \frac{z}{1!} Ai'(0) + \frac{z^2}{2!} Ai''(0) - \frac{z^3}{3!} Ai'''(0) + \frac{z^4}{4!} Ai^{\text{IV}}(0) + \cdots.$$

However, because $Ai''(z) = zAi(z)$, we have

$$Ai(0 - z) = Ai(0)\left[1 - \left\{\frac{Ai'(0)}{Ai(0)}\right\} z - \frac{1}{6} z^3 + \frac{1}{12}\left\{\frac{Ai'(0)}{Ai(0)}\right\} z^4 + \cdots\right],$$

where

$$\frac{Ai'(0)}{Ai(0)} = -3^{1/3}\frac{\Gamma(\frac{2}{3})}{\Gamma(\frac{1}{3})} = -0.72901,$$

$$Ai(0) = 0.3550280, \qquad Ai'(0) = -0.2588194.$$

The result

$$Ai(-z) \simeq Ai(0)[1 + (0.72901)z - (0.16667)z^3 - (0.060750)z^4] \quad (8.101c)$$

is valid on both sides ($\theta > \theta_{ca}$ and $\theta < \theta_{ca}$) of the caustic. We note that the *maximum intensity* occurs at $z \simeq 1.2075$, which is not on the caustic but on its convex side, ($\theta > \theta_{ca}$).

8.5. Ray Analysis in a Homogeneous Sphere

We know already that displacement field induced in a solid sphere of radius a can be represented either as a doubly infinite sum over all normal modes or else as a Fourier integral over the spectral displacements. In thc particular case of a homogeneous sphere $y_1(l, r) = j_l(r\omega/v_0)$. Therefore, the azimuthal component of the surface displacement caused by a vertical strike-slip source at $r = r_h$, $\theta = 0$ with a step-function time dependence becomes [cf. Eqs. (8.1.1) and (6.136)]

$$u_\phi(a, r_h; t) = \frac{U_0\, dS}{2\pi a r_h} H(t)\cos 2\phi \frac{\partial}{\partial\theta} \sum_{l=2}^{\infty} \sum_{n} (2l+1) F_l P_l^2(\cos\theta), \quad (8.102)$$

$$F_l = \frac{-1}{l(l+1)} \frac{j_l(k_n r_h)}{j_l(k_n a)} \frac{1 - \cos({}_n\omega_l t)}{(k_n a)^2 - (l-1)(l+2)},$$

or, alternatively,

$$u_\phi(a, r_h; t) = \frac{1}{2\pi} \int_{-\infty}^{\infty} \frac{e^{i\omega t}}{i\omega}\, d\omega \left\{ \frac{U_0\, dS}{4\pi a r_h} \cos 2\phi \frac{\partial}{\partial\theta} \sum_{l=2}^{\infty} (2l+1) f_l P_l^2(\cos\theta) \right\},$$

$$f_l = \frac{1}{l(l+1)} \left[\frac{j_l(k_0 r_h)}{j_l(k_0 a) - (k_0 a) j_l'(k_0 a)} \right], \qquad k_0 = \frac{\omega}{v_0}. \quad (8.103)$$

Prior to the evaluation of these sums, we examine the asymptotic behavior of the functions that appear in the summand. The spherical Bessel functions have the known asymptotic representations

$$j_l(z) = \begin{cases} \dfrac{\cos[(l+\frac{1}{2})(\tan\alpha - \alpha) - \pi/4]}{z\sqrt{[\sin\alpha]}}, & z > l + \dfrac{1}{2} \gg 1, \quad \dfrac{(l+\frac{1}{2})}{z} = \cos\alpha \\[2ex] \dfrac{e^{-(l+1/2)(\beta - \operatorname{th}\beta)}}{2z\sqrt{[\operatorname{sh}\beta]}}, & l + \dfrac{1}{2} > z \gg 1, \quad \dfrac{(l+\frac{1}{2})}{z} = \operatorname{ch}\beta. \end{cases} \quad (8.104)^4$$

We note that the sum in Eqs. (8.102) will converge rapidly if $l \gg k_n a$, because then

$$\frac{j_l(k_n r_h)}{j_l(k_n a)} \simeq \left(\frac{a}{r_h}\right) \left[\frac{\operatorname{sh}\beta_a}{\operatorname{sh}\beta_h}\right]^{1/2} \exp\left[-\left(l + \frac{1}{2}\right)[\beta_h - \operatorname{th}\beta_h - \beta_a + \operatorname{th}\beta_a]\right].$$

[4] In Sec. 8.5 (Sec. 8.6.6), α and β denote angles; the shear wave velocity is denoted by v_0 (v).

This ratio decays exponentially with l because $\beta - \text{th}\,\beta > 0$ and $\beta_h > \beta_a$. In fact, when $r_h \sim a$, $l + \frac{1}{2} > k_n a \gg 1$

$$\frac{j_l(k_n r_h)}{j_l(k_n a)} \simeq \exp\left[-k_n(a - r_h)\left[\left\{\frac{(l + \frac{1}{2})}{k_n a}\right\}^2 - 1\right]^{1/2}\right].$$

When $k_n a > l + \frac{1}{2}$, however, the terms in the sum are periodic in l and convergence is hampered. The evaluation of the sum by means of the Watson transformation followed by saddle-point integration offers a faster convergence. Each saddle-point integral then "selects" the mode contributions that correspond to its ray event.

Figure 8.7 shows the region of convergence of the mode series. The above arguments hold true also for the Fourier integral representation of Eq. (8.103) if we replace $k_n = {}_n\omega_l/v_0$ by $k_0 = \omega/v_0$. We shall next be concerned with the approximate evaluation of this sum in the *wave zone* $k_0 r_h > l + \frac{1}{2} \gg 1$. Using the asymptotic approximation for $j_l(z)$ given in Eq. (8.73) we obtain

$$f_l \simeq -\frac{j_l(k_0 r_h)}{l(l+1)(k_0 a)j_l'(k_0 a)}$$

$$= \frac{1}{l(l+1)}\left(\frac{\cos\eta_h}{\sin\eta_a}\right)\frac{1}{(k_0 r_h)[\{1 - ((l+\frac{1}{2})/k_0 r_h)^2\}\{1 - ((l+\frac{1}{2})/k_0 a)^2\}]^{1/4}},$$

where

$$\eta_h = \left[(k_0 r_h)^2 - \left(l + \frac{1}{2}\right)^2\right]^{1/2} - \left(l + \frac{1}{2}\right)\cos^{-1}\left(\frac{l + \frac{1}{2}}{k_0 r_h}\right) - \frac{\pi}{4},$$
$$\eta_a = \left[(k_0 a)^2 - \left(l + \frac{1}{2}\right)^2\right]^{1/2} - \left(l + \frac{1}{2}\right)\cos^{-1}\left(\frac{l + \frac{1}{2}}{k_0 a}\right) - \frac{\pi}{4}. \tag{8.105}$$

Because $j_{-s-1/2}(z) = (-)^s j_{s-1/2}(z)$, it follows from Eq. (8.103) that $f_{-s-1/2} = f_{s-1/2}$ and Eq. (8.96) can be directly applied. Furthermore,

$$\frac{\cos\eta_h}{\sin\eta_a} = i\{e^{i(\eta_h - \eta_a)} + e^{-i(\eta_h + \eta_a)}\}\sum_{j=0}^{\infty} e^{-2ij\eta_a}. \tag{8.106}$$

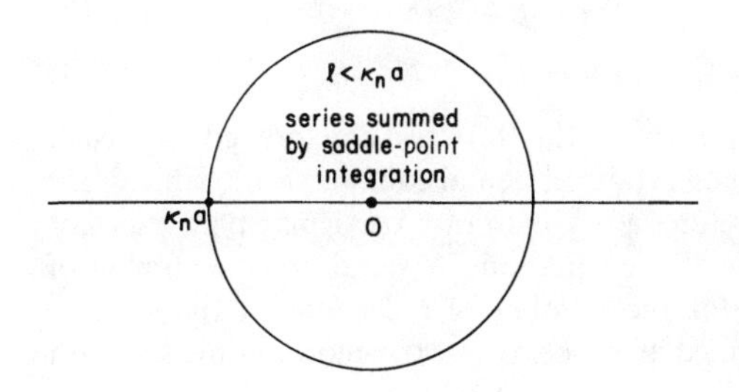

Figure 8.7. Region of convergence of the mode series and region of effect of the saddle-point integration.

If we write $\nu = s - \frac{1}{2}$, Eqs. (8.103)–(8.106) yield

$$u_\phi = \frac{U_0\,dS}{2\pi a r_h}\frac{\cos 2\phi}{(2\pi\sin\theta)^{1/2}}\int_{-\infty}^{\infty}\frac{\nu^{3/2}(k_0 r_h)^{-1}\,d\nu}{[\{1-((\nu+\frac{1}{2})/k_0 r_h)^2\}\{1-((\nu+\frac{1}{2})/k_0 a)^2\}]^{1/4}}$$
$$\times\sum_{N,j}(-1)^{N+1}(e^{-i\chi_1}+e^{-i\chi_2}-e^{-i\chi_3}-e^{-i\chi_4}), \tag{8.106a}$$

where

$$\chi_{1,2} = \left(\nu+\frac{1}{2}\right)(\theta+2\pi N)\mp\eta_h+(2j+1)\eta_a-\frac{\pi}{4},$$

$$\chi_{3,4} = \left(\nu+\frac{1}{2}\right)[(2\pi-\theta)+2\pi N]\pm\eta_h+(2j+1)\eta_a+\frac{5\pi}{4},$$

and $\eta_{h,a}$ are given by Eq. (8.105) with l replaced by ν. We evaluate the integral over ν by the method of stationary phase. We first observe that

$$\frac{\partial\eta_h}{\partial\nu}\simeq-\cos^{-1}\left(\frac{\nu+\frac{1}{2}}{k_0 r_h}\right),\qquad \frac{\partial^2\eta_h}{\partial\nu^2}\simeq\frac{1}{k_0 r_h[1-((\nu+\frac{1}{2})/k_0 r_h)^2]^{1/2}}, \tag{8.107}$$

together with similar expressions for the derivatives of η_a. Therefore,

$$\frac{\partial^2\chi_1}{\partial\nu^2}=\frac{1}{k_0 r_h}\left\{\frac{r_h}{a}\frac{(2j+1)}{[1-((\nu+\frac{1}{2})/k_0 a)^2]^{1/2}}-\frac{1}{[1-((\nu+\frac{1}{2})/k_0 r_h)^2]^{1/2}}\right\}. \tag{8.108}$$

The equation giving the saddle point $\nu=\nu_0$ for the χ_1 integral is $\partial\chi_1/\partial\nu=0$, i.e.,

$$\theta+2\pi N+\cos^{-1}\left(\frac{\nu_0+\frac{1}{2}}{k_0 r_h}\right)-(2j+1)\cos^{-1}\left(\frac{\nu_0+\frac{1}{2}}{k_0 a}\right)=0. \tag{8.109}$$

If we put

$$\nu_0+\frac{1}{2}=k_0 a\cos\alpha=k_0 r_h\cos\beta, \tag{8.110}$$

we get

$$\theta+2\pi N+\beta-(2j+1)\alpha=0,$$

i.e.,

$$\theta=(2j+1)\alpha-\beta-2\pi N. \tag{8.111}$$

Likewise, for χ_2, χ_3, and χ_4, we get, respectively,

$$\theta=(2j+1)\alpha+\beta-2\pi N, \tag{8.112}$$
$$2\pi-\theta=(2j+1)\alpha+\beta-2\pi N, \tag{8.113}$$
$$2\pi-\theta=(2j+1)\alpha-\beta-2\pi N. \tag{8.114}$$

We can express α in terms of β through Eq. (8.110). Therefore, given θ, each of the Eqs. (8.111)–(8.114) becomes a transcendental equation for β, which may not posses a real solution. Therefore, a ray path of a particular type may not necessarily exist. Note that in the above equations, N stands for the number of complete circuits and j stands for the number of reflections at the surface. Moreover, 2α is the angle subtended at the center by the chord on the ray-path between two reflections, and β is the angle subtended at the center by the projection of the radius vector to the source on the first partial chord of the ray-path. Equation (8.111) and, consequently, χ_1 correspond to a ray that is upgoing at the source. In contrast, Eq. (8.112) correspond to a ray that is downgoing at the source. χ_3 and χ_4 represent rays that traverse the path in the counterclockwise sense (Fig. 8.8a). Figure 8.8b shows a χ_1 ray for $j = 1, N = 0$ and Fig. 8.8c shows a χ_2 ray for $j = 1, N = 0$.

From Eqs. (8.105) and (8.110), we have, for $l = \nu_0$,

$$\begin{aligned}\eta_h &= k_0 r_h \sin\beta - \left(\nu_0 + \frac{1}{2}\right)\beta - \frac{\pi}{4},\\ \eta_a &= k_0 a \sin\alpha - \left(\nu_0 + \frac{1}{2}\right)\alpha - \frac{\pi}{4}.\end{aligned} \tag{8.115}$$

Equations (8.106a), (8.110), (8.111), and (8.115) now yield

$$\chi_1 = \omega\tau_1 - (2j+1)\frac{\pi}{4}, \tag{8.116}$$

where

$$\tau_1 = \frac{D_1}{v_0}, \qquad D_1 = (2j+1)a\sin\alpha - r_h \sin\beta. \tag{8.117}$$

Similarly,

$$\begin{aligned}\chi_2 &= \omega\tau_2 - (2j+3)\frac{\pi}{4}, \qquad \chi_3 = \omega\tau_3 - (2j-3)\frac{\pi}{4},\\ \chi_4 &= \omega\tau_4 - (2j-5)\frac{\pi}{4},\end{aligned} \tag{8.118}$$

where

$$\begin{aligned}\tau_2 &= \frac{D_2}{v_0}, \qquad \tau_3 = \frac{D_3}{v_0}, \qquad \tau_4 = \frac{D_4}{v_0},\\ D_2 &= (2j+1)a\sin\alpha + r_h\sin\beta, \qquad D_3 = D_2, \qquad D_4 = D_1.\end{aligned} \tag{8.119}$$

Therefore, τ_i ($i = 1, 2, 3, 4$) gives the time along the ray from the source to the surface observer and D_i is the corresponding length of path.

When Eq. (8.106a) is evaluated at the saddle point, ν_0, the displacement of a χ_1 ray for a given j and N and for a unit-step source is found to be

$$\begin{aligned}[u_\phi(t)]_{1jN} &= \frac{U_0\, dS}{2\pi v_0}(-1)^{N+1}\cos\beta\cos 2\phi\\ &\times \left[\frac{\cos\beta}{a\sin\theta\,|a\sin\alpha - (2j+1)r_h\sin\beta|}\right]^{1/2}\\ &\times \left\{\frac{1}{2\pi i}\int_{-\infty}^{\infty} e^{i[\omega(t-\tau_1)+\nu_1]}\,d\omega\right\},\end{aligned} \tag{8.120}$$

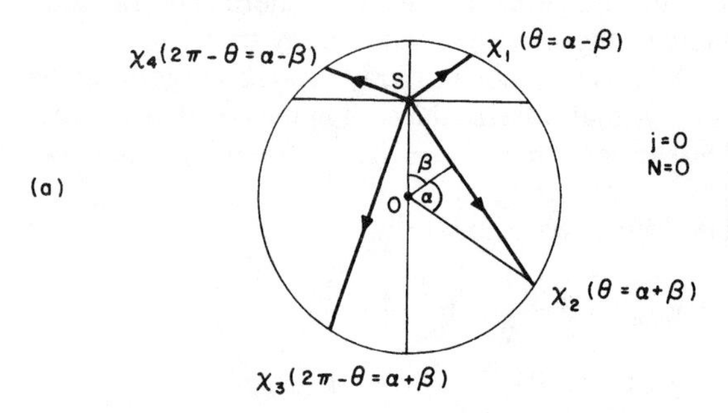

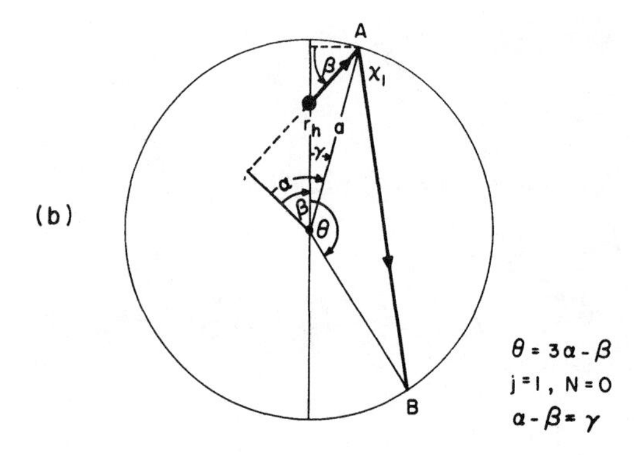

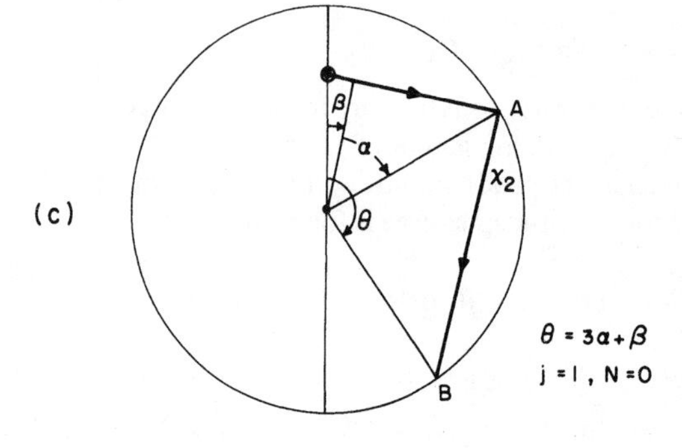

Figure 8.8. Geometric interpretation of the saddle-point approximation yielding direct rays (a) and once reflected rays (b, c) in a homogeneous sphere.

where θ is given by Eq. (8.111) and

$$\nu_1 = \frac{\pi}{4}[(2j+1) - \operatorname{sgn} \chi_1''(\nu_0)].$$

Similarly, for χ_2, χ_3, and χ_4 rays, we find

$$\begin{aligned}
[u_\phi(t)]_{2jN} &= \frac{U_0\, dS}{2\pi v_0}(-1)^{N+1} \cos\beta \cos 2\phi \\
&\quad \times \left[\frac{\cos\beta}{a \sin\theta\{a\sin\alpha + (2j+1)r_h \sin\beta\}}\right]^{1/2} \\
&\quad \times \left\{\frac{1}{2\pi i}\int_{-\infty}^{\infty} e^{i[\omega(t-\tau_2)+\nu_2]}\, d\omega\right\}, \\
[u_\phi(t)]_{3jN} &= -\frac{U_0\, dS}{2\pi v_0}(-1)^{N+1} \cos\beta \cos 2\phi \\
&\quad \times \left[\frac{\cos\beta}{a \sin\theta\{a\sin\alpha + (2j+1)r_h \sin\beta\}}\right]^{1/2} \\
&\quad \times \left\{\frac{1}{2\pi i}\int_{-\infty}^{\infty} e^{i[\omega(t-\tau_3)+\nu_3]}\, d\omega\right\}, \\
[u_\phi(t)]_{4jN} &= -\frac{U_0\, dS}{2\pi v_0}(-1)^{N+1} \cos\beta \cos 2\phi \\
&\quad \times \left[\frac{\cos\beta}{a \sin\theta|a\sin\alpha - (2j+1)r_h \sin\beta|}\right]^{1/2} \\
&\quad \times \left\{\frac{1}{2\pi i}\int_{-\infty}^{\infty} e^{i[\omega(t-\tau_4)+\nu_4]}\, d\omega\right\},
\end{aligned} \tag{8.121}$$

where

$$\nu_2 = \frac{\pi}{4}[(2j+3) - \operatorname{sgn} \chi_2''(\nu_0)],$$

$$\nu_3 = \frac{\pi}{4}[(2j-3) - \operatorname{sgn} \chi_3''(\nu_0)],$$

$$\nu_4 = \frac{\pi}{4}[(2j-5) - \operatorname{sgn} \chi_4''(\nu_0)].$$

Direct Rays (Fig. 8.8a). For the direct rays ($j = 0,\ N = 0$), Eqs. (8.120) and (8.121) yield

$$[u_\phi(t)]_{100} = -\frac{U_0\, dS}{2\pi v_0} \cos\beta \cos 2\phi \left(\frac{\cos\beta}{aD_1 \sin\theta}\right)^{1/2} \delta\left(t - \frac{D_1}{v_0}\right), \qquad (\theta = \alpha - \beta)$$

$$[u_\phi(t)]_{200} = -\frac{U_0\, dS}{2\pi v_0} \cos\beta \cos 2\phi \left(\frac{\cos\beta}{aD_2 \sin\theta}\right)^{1/2} \delta\left(t - \frac{D_2}{v_0}\right), \qquad (\theta = \alpha + \beta)$$

$$[u_\phi(t)]_{300} = -\frac{U_0\, dS}{2\pi v_0} \cos\beta \cos 2\phi \left[\frac{\cos\beta}{aD_3 \sin(\theta - \pi)}\right]^{1/2} \delta\left(t - \frac{D_3}{v_0}\right), \tag{8.122}$$

$$(2\pi - \theta = \alpha + \beta)$$

$$[u_\phi(t)]_{400} = -\frac{U_0\, dS}{2\pi v_0} \cos\beta \cos 2\phi \left[\frac{\cos\beta}{aD_4 \sin(\theta - \pi)}\right]^{1/2} \delta\left(t - \frac{D_4}{v_0}\right),$$

$$(2\pi - \theta = \alpha - \beta)$$

where D_i are given by Eqs. (8.117) and (8.119) with $j = 0$.

8.5.1. Equation of the *SS* Caustic

For once-reflected rays ($j = 1$) corresponding to χ_1 or χ_2 we have $\cos\beta = \cos(3\alpha - \theta)$. Using Eq. (8.110), this yields

$$\cos\alpha = h\cos(3\alpha - \theta), \qquad h = \frac{r_h}{a}. \tag{8.123}$$

For once-reflected rays corresponding to χ_3 or χ_4,

$$\cos\alpha = h\cos(3\alpha + \theta). \tag{8.124}$$

The equation of a once-reflected ray corresponding to χ_1 may be written in the form (Fig. 8.8b)

$$\theta = (\alpha - \beta) + \alpha \pm \cos^{-1}\left(\frac{a}{r}\cos\alpha\right),$$

or

$$\cos(2\alpha - \beta - \theta) = \frac{a}{r}\cos\alpha. \tag{8.125}$$

Putting $\alpha - \beta = \gamma$, the polar equation of the ray becomes

$$\frac{1}{r} = A\cos(2\gamma - \theta) - B\sin(2\gamma - \theta), \tag{8.126}$$

where

$$A = \frac{\cos\beta}{a\cos\alpha}, \qquad B = \frac{\sin\beta}{a\cos\alpha}. \tag{8.127}$$

From Eq. (8.110) and Fig. (8.8b), we find

$$A = \frac{1}{r_h}, \qquad B = \frac{1}{r_h} \tan \beta = \frac{a \cos \gamma - r_h}{a r_h \sin \gamma}. \tag{8.128}$$

With these values of A and B, Eq. (8.126) reduces to

$$\frac{\sin \gamma}{r} = \frac{\sin(\theta - \gamma)}{r_h} + \frac{\sin(2\gamma - \theta)}{a}. \tag{8.129}$$

Putting $2\gamma - \theta = 2\Phi$, Eq. (8.129) becomes

$$\frac{1}{a} \sin 2\Phi = \frac{1}{r} \sin\left(\Phi + \frac{\theta}{2}\right) + \frac{1}{r_h} \sin\left(\Phi - \frac{\theta}{2}\right), \tag{8.130}$$

or, equivalently,

$$\frac{2}{a} = \left(\frac{1}{r} - \frac{1}{r_h}\right) \frac{\sin(\theta/2)}{\sin \Phi} + \left(\frac{1}{r} + \frac{1}{r_h}\right) \frac{\cos(\theta/2)}{\cos \Phi}. \tag{8.131}$$

Equation (8.131) is of the form

$$f(r, \theta, \Phi) \equiv \frac{P}{\sin \Phi} + \frac{Q}{\cos \Phi} - 1 = 0. \tag{8.132}$$

The equation of the caustic, which is the envelope of the family of rays, is obtained by eliminating Φ between $f = 0$ and $\partial f / \partial \Phi = 0$. This yields

$$\frac{P}{\sin^3 \Phi} = \frac{Q}{\cos^3 \Phi} = \lambda \quad \text{(say)}. \tag{8.133}$$

Equation (8.132) shows that $\lambda = 1$. Therefore,

$$P^{2/3} + Q^{2/3} = 1. \tag{8.134}$$

Substituting the expressions for P and Q, the equation of the caustic is found to be

$$\left[\left(\frac{1}{r} - \frac{1}{r_h}\right) \sin \frac{\theta}{2}\right]^{2/3} + \left[\left(\frac{1}{r} + \frac{1}{r_h}\right) \cos \frac{\theta}{2}\right]^{2/3} = \left(\frac{2}{a}\right)^{2/3}. \tag{8.135}$$

If we begin with rays of the type χ_2, χ_3, or χ_4, we end up with the same equation for its caustics.

Consider the once-reflected rays, Eq. (8.123), once again. If we consider α as a parameter and eliminate it between Eq. (8.123) and the equation

$$\sin \alpha = 3h \sin(3\alpha - \theta), \tag{8.136}$$

we will get the value θ_0 for θ at which the caustic meets the surface of the sphere. Therefore, for any given h, $\theta_0(h)$ is determined from

$$\cos \alpha = h \cos (3\alpha - \theta_0), \tag{8.137}$$

$$\frac{1}{3} \sin \alpha = h \sin(3\alpha - \theta_0). \tag{8.138}$$

Squaring and adding, we get

$$\cos^2 \alpha = \frac{(9h^2 - 1)}{8}. \tag{8.139}$$

Equation (8.137) now gives

$$\theta_0 = 3 \cos^{-1}\left(\frac{9h^2 - 1}{8}\right)^{1/2} - \cos^{-1}\left(\frac{9h^2 - 1}{8h^2}\right)^{1/2}, \tag{8.140}$$

or,

$$\cos \theta_0 = \frac{(27h^4 - 18h^2 - 1)}{8h}. \tag{8.141}$$

Table 8.8 gives θ_0 for various values of h, assuming $a = 6371$ km. Figure 8.9b shows θ_0 as a function of h according to Eq. (8.140).

Table 8.8. The Dependence of the Co-latitude of the *SS* Caustic-Crossing on the Source Depth

Source depth (km)	h	$\theta_0(°)$
0	1	0
64	0.99	22.93
159	0.975	36.26
319	0.95	51.35
637	0.90	72.70
3185	0.50	162.40
4247	$\frac{1}{3}$	180

8.5.2. Pulse-Shapes of Once-Reflected Rays

Given θ, Eq. (8.123) becomes an equation for α. We will have real rays if this equation renders positive roots such that $0 < \alpha < \pi/2$. Consider the graphs for $\cos \alpha$ and $\{h \cos(3\alpha - \theta)\}$. As θ increases, the curve for $\{h \cos(3\alpha - \theta)\}$ moves to the right (Fig. 8.9a) across the curve for $\cos \alpha$. Consider the critical case when the graphs for $\cos \alpha$ and $\{h \cos(3\alpha - \theta)\}$ touch each other (i.e., have a common ordinate and a common tangent). In this case

$$\cos \alpha = h \cos(3\alpha - \theta), \qquad \sin \alpha = 3h \sin(3\alpha - \theta).$$

These equations are identical with Eqs. (8.137) and (8.138), respectively. Therefore, there is no positive root α of Eq. (8.123) in $(0, \pi/2)$ if $\theta < \theta_0$, where θ_0 is the point at which the caustic cuts the sphere. Therefore, no ray of the type χ_1 or χ_2 arrives in the angle range $0 < \theta < \theta_0$. Between $\theta = \theta_0$ and $\theta = \pi$, there are two positive roots of Eq. (8.123), as shown in Fig. 8.9a. The greater of

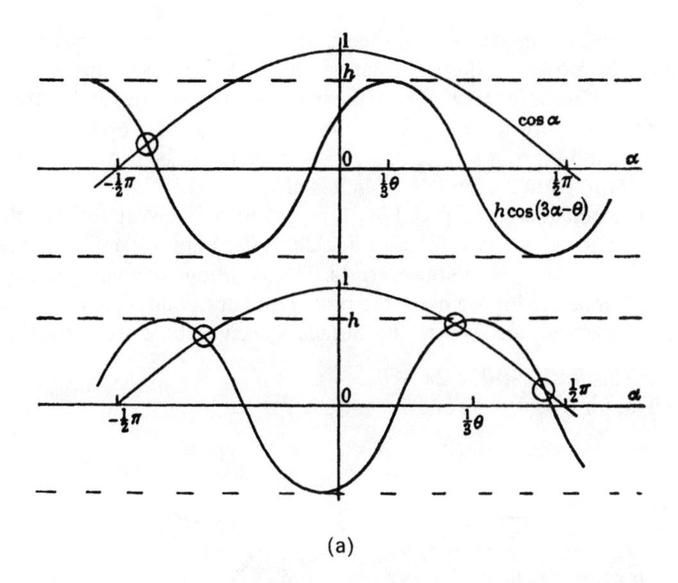

(a)

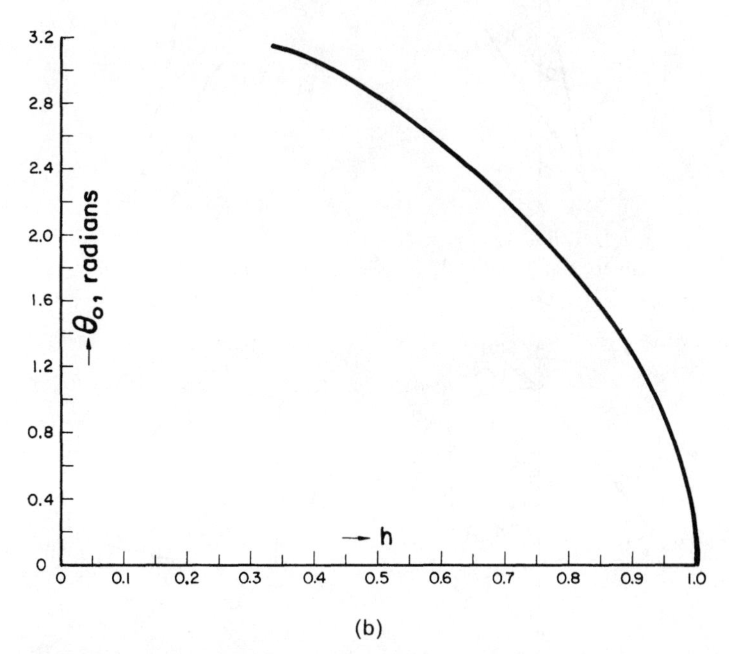

(b)

Figure 8.9. (a) Solution of $\cos\alpha = h\cos(3\alpha - \theta)$. (b) The critical angular distance θ_0 as a function of h [cf. Eq. (8.140)].

these roots corresponds to χ_1 and the smaller to χ_2. As θ increases, the curve of $h\cos(3\alpha - \theta)$ moves to the right across the curve of $\cos\alpha$, and when θ lies between π and 2π, only the smaller of the two positive roots lies within the region $(0, \frac{1}{2}\pi)$ of α. Therefore, for $0 < \theta < \theta_0$, neither χ_1 nor χ_2 arrives; for $\theta_0 < \theta < \pi$, both arrive; and for $\pi < \theta < 2\pi$, only χ_2 arrives. It can be shown that when χ_1 exists, the expression $3r_h \sin\beta - a\sin\alpha$ is always positive.

Similarly, considering Eq. (8.124) for χ_3 and χ_4 rays, it follows that for $0 < \theta < \pi$, only χ_3 arrives; for $\pi < \theta < 2\pi - \theta_0$, both χ_3 and χ_4 arrive; and when $2\pi - \theta_0 < \theta < 2\pi$, neither arrives. The situation is depicted in Fig. 8.10: On a great circle section we mark the pole A and antipode A', and D, D' are the intersections of the caustic with the surface. Given θ, there are altogether:

Three arrivals for $\theta_0 < \theta < 2\pi - \theta_0$,
One arrival for $0 < \theta < \theta_0$ and $2\pi - \theta_0 < \theta < 2\pi$,

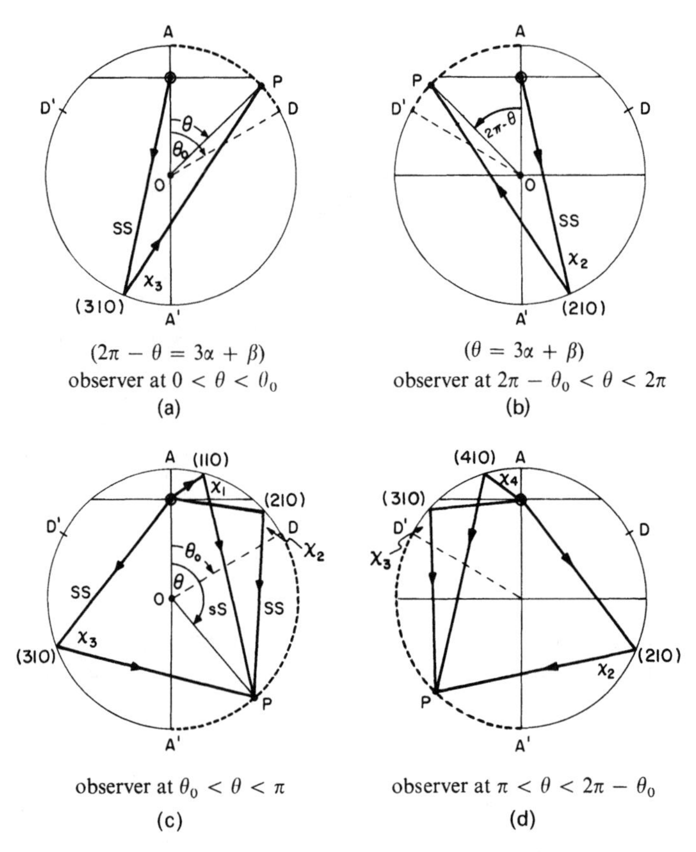

$(2\pi - \theta = 3\alpha + \beta)$
observer at $0 < \theta < \theta_0$
(a)

$(\theta = 3\alpha + \beta)$
observer at $2\pi - \theta_0 < \theta < 2\pi$
(b)

observer at $\theta_0 < \theta < \pi$
(c)

observer at $\pi < \theta < 2\pi - \theta_0$
(d)

Figure 8.10. Source-depth and the location of the point of observation on the surface determine the number of rays that may reach the observer at P.

where $\theta_0 = \theta_0(h)$ depends on the source depth according to Eq. (8.141). Indeed, we see in Fig. 8.10 that if P is in $D'AD$, there is only one arrival from a reflection on the opposite hemisphere. If P is in $DA'D'$, there are two reflections from the same side and one from the opposite side.

We can also put it in another way: When the source is below a certain depth ($h < \frac{1}{3}$), only one once-reflected ray arrives at all distances ($0 \leq \theta \leq \pi$). Moreover, when $h > \frac{1}{3}$, there arrive three different once-reflected rays at any distance $\theta > \theta_0(h)$ and for $\theta < \theta_0$, only one once-reflected ray arrives. At the surface, $h = 1$, $\theta_0 = 0$, and three once-reflected rays arrive at all θ.

We have shown in Eq. (8.101b) that the amplitude at a point where the caustic intersects the surface ($\theta = \theta_0$) is proportional to

$$\left(\frac{\partial^2 \theta}{\partial v^2}\right)_{ca}^{-1/3}.$$

Figure 8.11 shows SH caustics for six different source-depths. Figure 8.12 shows ray paths with different j and N values for various observation points on the surface.

Let us determine the stationary values of $\tau_1 = (3a \sin \alpha - r_h \sin \beta)/v_0$ under the condition $\beta = 3\alpha - \theta$ (θ being fixed). $d\tau_1/d\alpha = 0$ gives: $a \cos \alpha = r_h \cos \beta$, and

$$\frac{d^2 \tau_1}{d\alpha^2} = \frac{3}{v_0}(-a \sin \alpha + 3r_h \sin \beta),$$

which is positive in the region of existence of χ_1 rays. Therefore, χ_1 yields a minimum-time path. However, considering $\tau_2 = (3a \sin \alpha + r_h \sin \beta)/v_0$, $\beta = \theta - 3\alpha$, we find: $a \cos \alpha = r_h \cos \beta$ for stationary values of τ_2, and

$$\frac{d^2 \tau_2}{d\alpha^2} = \frac{3}{v_0}(-a \sin \alpha - 3r_h \sin \beta) < 0.$$

It means that χ_2 corresponds to a maximum-time path. Here, "maximum" is used in reference to the family of paths confined to the diametral plane through the source and observer. If paths that diverge from this plane are admitted, the path of χ_2 will no longer be a true maximum-time path.

For $0 < \theta < \theta_0$, the amplitude of the once-reflected arrival ($j = 1$, $N = 0$) obtained from Eq. (8.121) is (Fig. 8.10a)

$$[u_\phi(t)]_{310} = \frac{U_0\, dS}{2\pi v_0} \cos \beta \cos 2\phi \left[\frac{\cos \beta}{a \sin \theta (3r_h \sin \beta + a \sin^{\cdot} \alpha)}\right]^{1/2}$$

$$\times\, \delta\left[t - \frac{1}{v_0}(3a \sin \alpha + r_h \sin \beta)\right], \qquad (2\pi - \theta = 3\alpha + \beta),$$

$$(8.142)$$

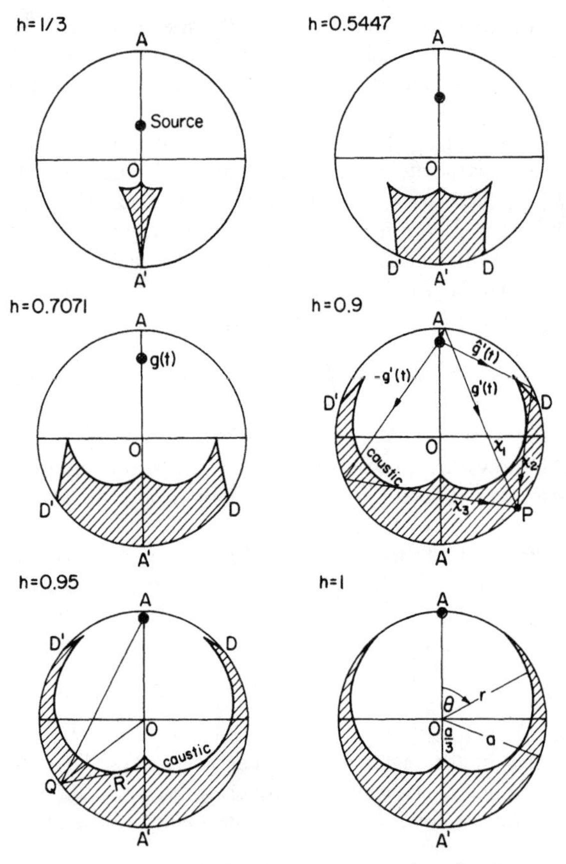

Figure 8.11. *SH* caustics in a homogeneous sphere for six source depths. A ray from the source is reflected at Q ($h = 0.95$ in the figure), such that the angles AQO and OQR are equal. The envelope of the line QR generates the caustic.

where α is a root of Eq. (8.124). For $\theta_0 < \theta < \pi$, there are three once-reflected arrivals. The amplitude of one is given by Eq. (8.142) and the amplitudes of the other two arrivals are (Fig. 8.10c)

$$
\begin{aligned}
[u_\phi(t)]_{110} &= -\frac{U_0\,dS}{2\pi v_0}\cos\beta\cos 2\phi\left[\frac{\cos\beta}{a\sin\theta(3r_h\sin\beta - a\sin\alpha)}\right]^{1/2} \\
&\quad\times\delta\left[t - \frac{1}{v_0}(3a\sin\alpha - r_h\sin\beta)\right], \qquad (\theta = 3\alpha - \beta) \\
[u_\phi(t)]_{210} &= -\frac{U_0\,dS}{2\pi v_0}\cos\beta\cos 2\phi\left[\frac{\cos\beta}{a\sin\theta(3r_h\sin\beta + a\sin\alpha)}\right]^{1/2} \\
&\quad\times\hat{\delta}\left[t - \frac{1}{v_0}(3a\sin\alpha + r_h\sin\beta)\right], \qquad (\theta = 3\alpha + \beta).
\end{aligned}
\tag{8.143}
$$

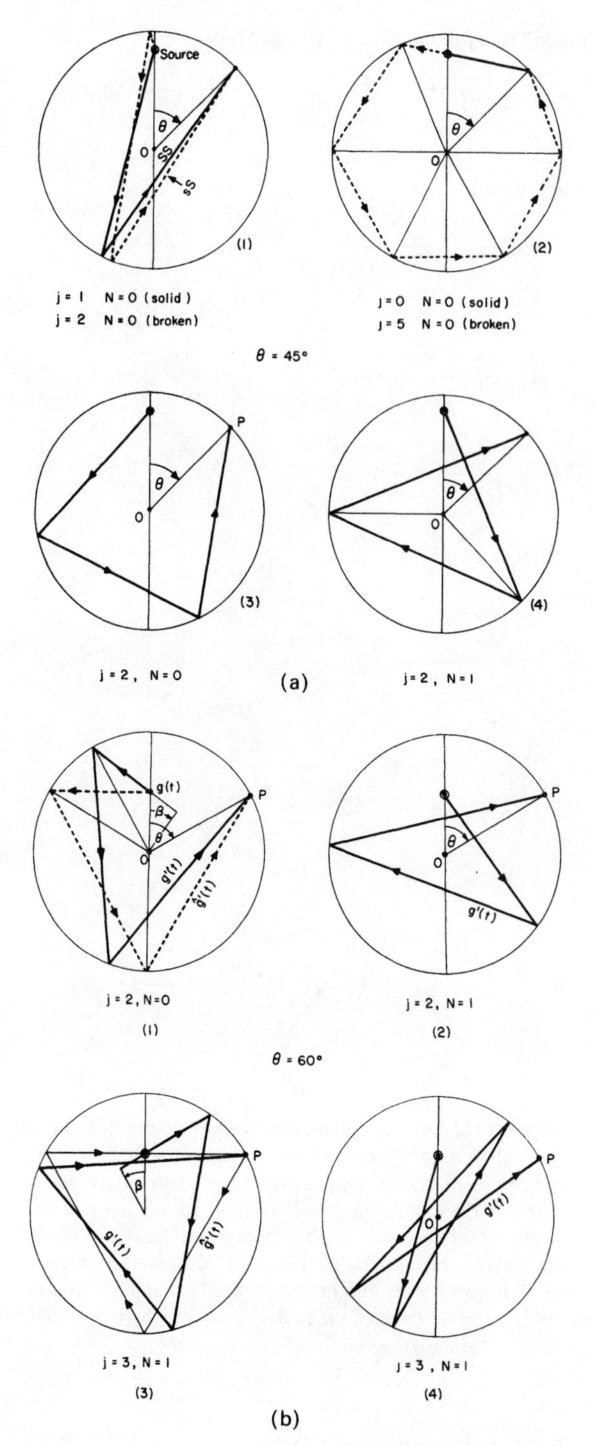

[**Figure 8.12** continues overleaf]

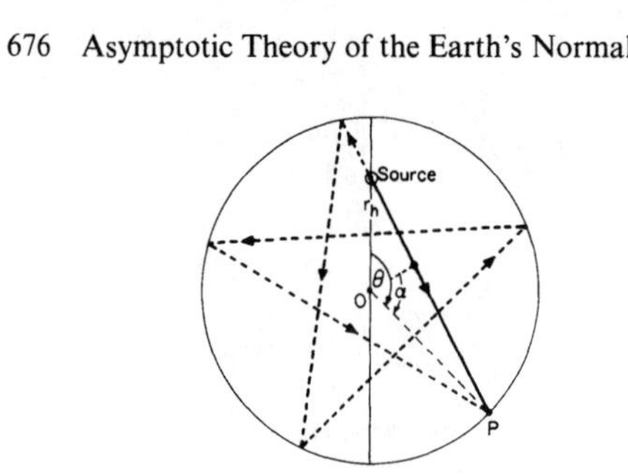

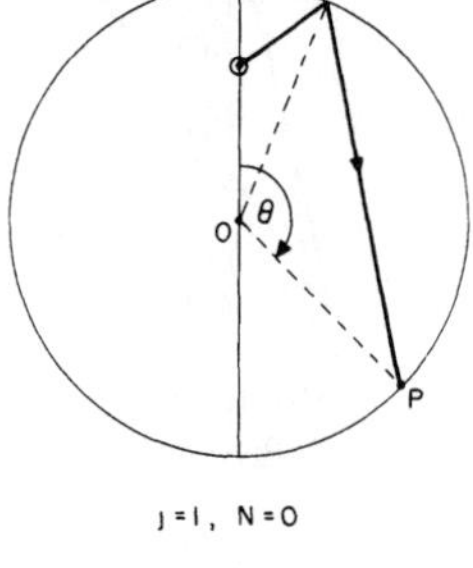

j = 0, N = 0 (solid)
j = 4, N = 2 (broken)

(1)

θ = 135°

j = 1, N = 0

(2)

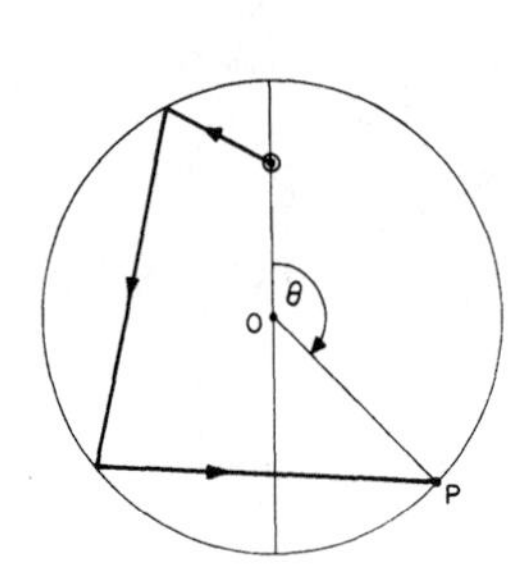

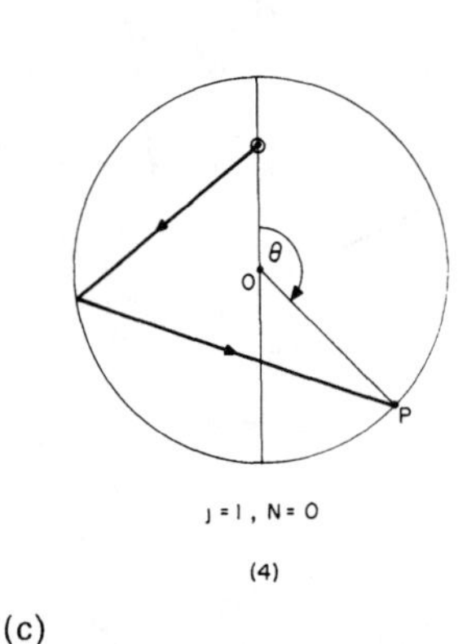

j = 2, N = 0

(3)

j = 1, N = 0

(4)

(c)

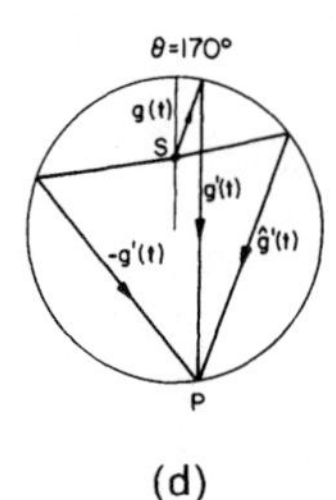

(d)

Figure 8.12. *SH* 'ray-billiards' for various source depths and four observation points on the surface of a homogeneous sphere. In all examples $g(t)$ marks a ray whose time function is that of the source, whereas $\hat{g}(t)$ is a ray whose time function is the allied function of $g(t)$. In example (3), for $\theta = 60°$, the ray is thrice reflected before it reaches the observer at P. It goes on for another two reflections and reaches the observer again through the source. (a) $\theta = 45°$, (b) $\theta = 60°$, (c) $\theta = 135°$, (d) $\theta = 170°$.

Here α satisfies Eq. (8.123) and $\hat{\delta}(t)$ denotes the allied function corresponding to $\delta(t)$. If the source-time function is $g(t)$ instead of $H(t)$, $\delta(t)$ should be replaced by $g'(t)$ and $\hat{\delta}(t)$ by the function allied to $g'(t)$. It may be noted that the shape of the pulse will depend both upon the source and the type of waves under consideration. We can give a physical explanation of the difference between the shapes of the three once-reflected pulses (see Fig. 8.11, $h = 0.9$): First, the ray χ_1 is not tangent to the caustic inside the sphere and there is no change of phase associated with the caustic. In contrast, the ray χ_2 touches the caustic within the sphere, adjacent rays cross, and each component of the spectrum changes phase by $\pi/2$. With χ_3 the story is yet different. Here, rays cross simultaneously in two planes: the great circle section plane and a plane perpendicular to it in which the rays are adjacent in azimuth. As each crossing changes the phase by $\pi/2$, there is a total phase change of each Fourier component by π. This causes a change of sign relative to χ_1.

EXAMPLE 8.2: Half-space displacements as a limiting case of displacements in a sphere

Consider a point source at $r = r_h$, $\theta = 0$ in a sphere of radius a centered at the origin, and a station on the surface of the sphere. As $a \to \infty$ and $r_h \to \infty$ in such a way that $a - r_h = h$ remains fixed, the sphere is transformed into a half-space. The components (u_r, u_θ, u_ϕ) of the displacement in a sphere transform into the components (u_z, u_Δ, u_ϕ) of the half-space, where z is taken vertically upward through the source. We can then set the correspondence

$$a \to \infty, \qquad r_h \to \infty, \qquad a - r_h \to h, \qquad \theta \to 0, \qquad a\theta \to \Delta. \tag{8.2.1}$$

In addition, using Jeans' formula, we have

$$l \to \infty, \qquad \frac{l}{a} \to k, \qquad \left(\frac{r_h}{a}\right)^l = \left(1 - \frac{h}{a}\right)^l = \left(1 - \frac{h}{a}\right)^{(a/h)\cdot(lh/a)} \to e^{-kh}. \tag{8.2.2}$$

These transformations are meaningful only when infinite sums are transformed into integrals for the half-space according to the scheme

$$\frac{1}{a^2}\sum_l l\{\cdots\} \to \int_0^\infty \{\cdots\}k\,dk. \tag{8.2.3}$$

For the Legendre functions, we use a result valid for large values of ν,

$$P_{\nu-1/2}(\cos\theta) \sim \left(\frac{\theta}{\sin\theta}\right)^{1/2} J_0(\nu\theta). \tag{8.2.4}$$

Hence

$$P_l(\cos\theta) \to J_0(k\Delta). \tag{8.2.5}$$

Using the relation

$$P_l^m(\cos\theta) = \sin^m\theta\left(-\frac{1}{\sin\theta}\frac{\partial}{\partial\theta}\right)^m P_l(\cos\theta),$$

we have

$$\frac{1}{l^m} P_l^m(\cos\theta) \to \Delta^m\left(-\frac{1}{\Delta}\frac{\partial}{\partial k\Delta}\right)^m J_0(k\Delta). \tag{8.2.6}$$

The radial functions are transformed with the help of the Debye approximation, Eq. (8.75). This yields

$$\frac{j_l(r_h k_0)}{j_l(ak_0)} \to e^{-h(k^2-k_0^2)^{1/2}}, \qquad \frac{j_l'(ak_0)}{j_l(ak_0)} \to \frac{1}{k_0}(k^2-k_0^2)^{1/2},$$
$$\frac{j_j''(ak_0)}{j_l(ak_0)} \to \frac{1}{k_0^2}(k^2-k_0^2). \tag{8.2.7}$$

When the radius of the sphere is large compared with the wavelength, it is expected that the effect of the curvature of the surface may be neglected. In particular, under the transformation expressed by Eqs. (8.2.1)–(8.2.7), the field in the sphere is transformed into the corresponding field in the half-space. We will show this with the help of an example.

Consider the radial component of the displacement resulting from a center of compression placed inside a sphere [cf. Eq. (6.6.10)]. We recast it in the form

$$u_r = \frac{M}{8\pi\mu a^2}\sum_{l=2}^{\infty}(2l+1)R_{PP}^{-}\left[\frac{j_l(x)}{j_l(\zeta)}\right]P_l(\cos\theta), \tag{8.2.8}$$

where

$$\zeta = ak_\alpha, \qquad \chi = ak_\beta, \qquad x = r_h k_\alpha,$$

$$R_{PP}^{-} = \frac{[j_l''(\chi)]/[j_l(\chi)] + [(l-1)(l+2)]/\chi^2}{\dfrac{2l(l+1)}{\zeta\chi}\left[\dfrac{j_l'(\zeta)}{j_l(\zeta)} - \dfrac{1}{\zeta}\right]\left[\dfrac{j_l'(\chi)}{j_l(\chi)} - \dfrac{1}{\chi}\right] - \left[\dfrac{j_l''(\chi)}{j_l(\chi)} + \dfrac{(l-1)(l+2)}{\chi^2}\right]\left[\dfrac{j_l''(\zeta)}{j_l(\zeta)} - \dfrac{\lambda}{2\mu}\right]}. \tag{8.2.9}$$

Under the above transformation

$$\frac{j_l(x)}{j_l(\zeta)} \to e^{-h(k^2-k_\alpha^2)^{1/2}}, \qquad R_{PP}^{-} \to R_{PP}^{HS}$$

with

$$R_{PP}^{HS} = \frac{-2k_\alpha^2(2k^2-k_\beta^2)}{(2k^2-k_\beta^2)^2 - 4k^2(k^2-k_\alpha^2)^{1/2}(k^2-k_\beta^2)^{1/2}}. \tag{8.2.10}$$

Therefore,

$$u_r \to -\frac{M}{4\pi\mu}\int_0^\infty R_{PP}^{HS} e^{-h(k^2-k_\alpha^2)^{1/2}} J_0(k\Delta)k\,dk. \tag{8.2.11}$$

This coincides with the expression in Eq. (5.20) for the vertical component of the displacement induced by a center of compression in a half-space.

8.6. *SH*-Field Analysis in a Uniform Shell Overlying a Fluid Core

We next treat *SH*-wave propagation in the simplest real-earth model: an elastic layer with the constitutive parameters of the earth's average mantle overlying a fluid core. Here, in addition to surface waves and rays, we shall meet also a diffracted displacement field in the geometric shadow zone where no real rays are supposed to exist.

The geometric configuration is shown in Fig. 8.13. According to Eq. (6.135), the surface azimuthal toroidal displacement resulting from a vertical strike-slip point-dislocation is given explicitly by the normal-mode solution

$$u_\phi(a, r_h; t) = -\frac{\mu_h U_0\, dS}{4\pi r_h}\cos 2\phi H(t)\frac{\partial}{\partial\theta}\sum_{l=2}^{\infty}\sum_{n}\frac{2l+1}{l(l+1)} \times \frac{y_1(a)y_1(r_h)[1-\cos(\omega_n t)]}{\omega_n^2 I_n} P_l^2(\cos\theta), \tag{8.144}$$

where

$$I_n = \int_b^a \rho(r)y_1^2(r)r^2\, dr, \tag{8.145}$$

b, a are respective radii of the core and the sphere, and

$$y_1(r) = (k_n a)j_l(k_n r)\frac{d}{dx}\left[\frac{n_l(x)}{x}\right]_{x=k_n a} - (k_n b)n_l(k_n r)\frac{d}{dx}\left[\frac{j_l(x)}{x}\right]_{x=k_n b}, \quad k_n = \frac{\omega_n}{v}. \tag{8.146}$$

Using the indefinite integral

$$\int x^2 z_l^2(x)dx = \frac{1}{2}x^3[z_l^2(x) - z_{l+1}(x)z_{l-1}(x)], \qquad z_l = \text{spherical Bessel function}$$

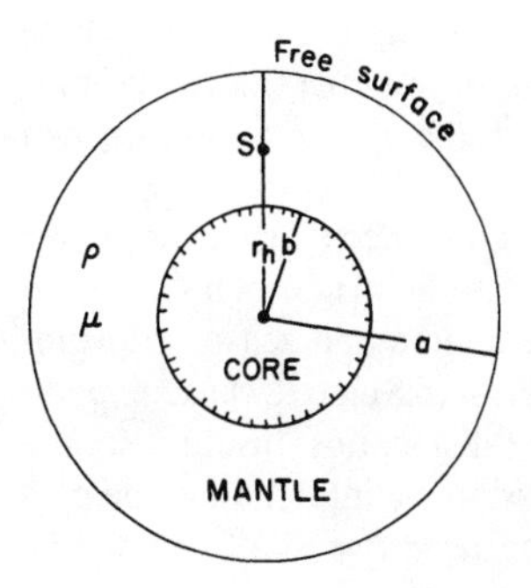

Figure 8.13. An *SH* source at $(r_h, 0, 0)$ in a uniform layer of thickness $(a - b)$ overlying a fluid core of radius b.

the integral in Eq. (8.145) yields a generalization of Eq. (6.118)

$$I_n = \frac{1}{2}\rho_0 a^3\left[1 - \frac{(l-1)(l+2)}{(k_n a)^2}\right] y_1^2(a) - \frac{1}{2}\rho_0 b^3\left[1 - \frac{(l-1)(l+2)}{(k_n b)^2}\right] y_1^2(b). \tag{8.147}$$

The period equation, Eq. (8.69), was used in the derivation of Eq. (8.147). The displacements are then written as

$$u_\phi(a, r_h; t) = \frac{U_0\, dS}{2\pi a r_h} \cos 2\phi H(t) \frac{\partial}{\partial \theta} \sum_{l=2}^{\infty} \sum_n (2l+1) f_l P_l^2(\cos\theta), \tag{8.148}$$

$$f_l = -\frac{1}{l(l+1)}\left[\frac{y_1(r_h)}{y_1(a)}\right]\left[\frac{1-\cos(\omega_n t)}{k_n^2 a^2 - (l-1)(l+2)}\right]\left[\frac{1}{1-X}\right],$$
$$X = \frac{b}{a}\left[\frac{y_1(b)}{y_1(a)}\right]^2\left[\frac{(k_n b)^2 - (l-1)(l+2)}{(k_n a)^2 - (l-1)(l+2)}\right]. \tag{8.149}$$

When $b \to 0$, $X \to 0$, and we fall back on Eq. (8.102).

If the source is an *SH* torque with a time function $g(t)$ (Table 6.4), Eq. (6.4.5) yields

$$u_\phi^{(\text{torque})}(a, r_h; t) = \frac{(M_0/\mu)}{2\pi a r_h} \sum_{l=1}^{\infty} \sum_n (2l+1) P_l^1(\cos\theta)$$
$$\times \left[\frac{\omega_n^2 \bar{g}(t)}{(k_n a)^2 - (l-1)(l+2)}\right]\left[\frac{y_1(r_h)}{y_1(a)}\right]\left[\frac{1}{1-X}\right]. \tag{8.150}$$

The dimensions of (M_0/μ) are that of volume.

8.6.1. The Diffracted Field in the Shadow Zone

In Section 8.6.2 we shall treat the exact numerical calculation of Eq. (8.150). In doing so we shall meet a type of seismic field hitherto undiscussed, namely, waves diffracted by the mantle–core boundary. In order to study the nature of these waves, we prefer to isolate this phenomenon by first setting up a simpler boundary-value problem. Because this medium is infinite the spectral solution will be used.

Let, therefore, a dislocation source be placed at a distance r_h from the center of a spherical cavity of radius b, which coincides with the origin of a Cartesian coordinate system. In this system $\mathbf{r}_0$ and $\mathbf{r}$ are the source and sensor coordinates, respectively (Fig. 8.14). We have derived in Eqs. (6.81)–(6.82) the explicit form of the Green's dyadic $\mathfrak{G}_a(\mathbf{r}/\mathbf{r}_0; \omega)$ for a point dislocation inside a sphere with a spectral time-function $F(\omega)$. If in this expression we interchange the plus and minus signs everywhere except in the source term, we shall obtain the

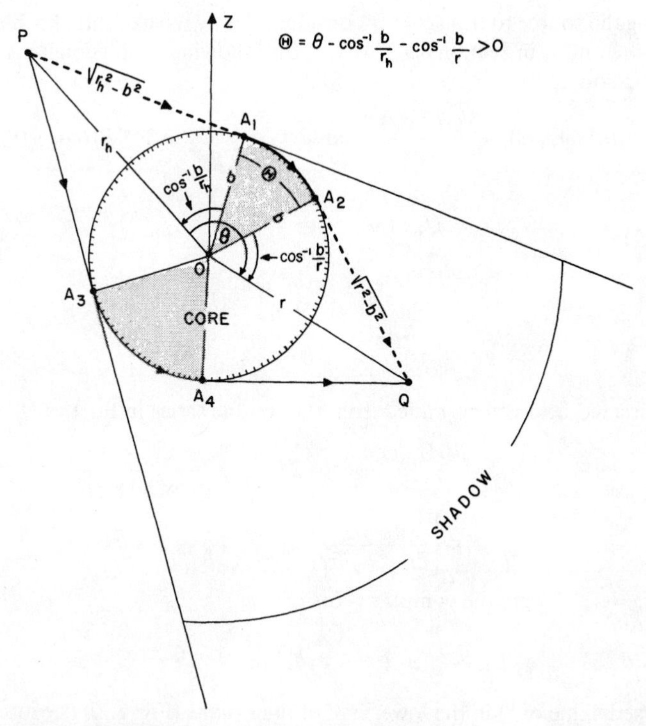

Figure 8.14. Diffraction arrivals in the shadow zone of a cavity in an infinite elastic medium.

Green's dyadic $\hat{\mathfrak{G}}_a$ for the cavity problem. Explicitly, for a homogeneous medium, $[\mathbf{r}_0 \equiv (r_h, \theta_0, \phi_0)]$

$$\hat{\mathfrak{G}}_b(\mathbf{r}/\mathbf{r}_0; \omega) = -\frac{ik_\beta}{\mu} F(\omega) \sum_{l=1}^{\infty} \sum_{m=-l}^{l} \frac{\mathbf{C}_{ml}(\theta, \phi)\overset{*}{\mathbf{C}}_{ml}(\theta_0, \phi_0)}{\Omega_{ml}} \Phi_l(r, r_h; \omega), \quad (8.151)$$

where

$$\Phi_l(r, r_h; \omega) = j_l(kr_<)h_l^{(2)}(kr_>) - h_l^{(2)}(kr_h)h_l^{(2)}(kr) \frac{j_l(kb) - (kb)j_l'(kb)}{h_l^{(2)}(kb) - (kb)h_l^{(2)\prime}(kb)} \quad (8.152)$$

$$= \frac{1}{2} h_l^{(1)}(kr_<)h_l^{(2)}(kr_>) - \frac{1}{2} h_l^{(2)}(kr_h)h_l^{(2)}(kr)$$

$$\times \frac{h_l^{(1)}(kb) - (kb)h_l^{(1)\prime}(kb)}{h_l^{(2)}(kb) - (kb)h_l^{(2)\prime}(kb)} = \frac{\Delta_1}{\Delta_l}, \quad (8.153)$$

$$r_> = \max(r, r_h), \qquad r_< = \min(r, r_h), \qquad k = k_\beta.$$

Taking the source to be at $\theta = 0$, considering only the azimuthal component, and proceeding as in Section 6.3.3, we find the following expression for a vertical strike-slip source

$$u_\phi(r, r_h; \omega) = \frac{ik_\beta U_0(\omega)dS}{4\pi r_h} \cos 2\phi \frac{\partial}{\partial \theta} \sum_{l=2}^{\infty} (2l + 1) f_l P_l^2(\cos \theta), \quad (8.154)$$

where

$$U_0(\omega) = U_0 F(\omega),$$
$$\Delta_l = h_l^{(2)}(z) - z \frac{\partial}{\partial z} h_l^{(2)}(z), \qquad z = k_\beta b, \quad (8.155)$$
$$g_l = \frac{\Delta_1}{l(l + 1)}, \qquad f_l = \frac{g_l}{\Delta_l}, \qquad k_\beta = \frac{\omega}{v}.$$

The diffracted field is determined from the residue series in Eq. (8.14)

$$\{u_\phi\}_{\text{diff}} = \frac{ik_\beta U_0(\omega)dS}{4r_h} \cos 2\phi \frac{\partial}{\partial \theta} \sum_{n=0}^{\infty} \frac{2v_n + 1}{\cos(v_n + \frac{1}{2})\pi} \times \left[\frac{g_{s-1/2}}{\partial \Delta_{s-1/2}/\partial s}\right]_{s=v_n+1/2} P_{v_n}^2(-\cos \theta), \quad (8.156)$$

where $s_n = v_n + \frac{1}{2}$ are the complex roots of

$$h_{s-1/2}^{(2)}(z) - z \frac{\partial}{\partial z} h_{s-1/2}^{(2)}(z) = 0 \quad (8.157)$$

(for a fixed value of z) in the lower half of the s plane (Fig. 8.2). Because

$$h_{-s-1/2}^{(1,2)}(z) = e^{\pm i\pi s} h_{s-1/2}^{(1,2)}(z),$$

the function $f_{s-1/2}$ is even in s.

We will next give a physical interpretation to the sum in Eq. (8.156). Because all subsequent results will depend on the roots of Eq. (8.157), we must first derive the roots of this equation. We know already that the expression $\{P_v(-\cos \theta)/\cos \pi(v + \frac{1}{2})\}$ yields wave-type motion only for sufficiently large values of $|v|$. We also restrict our attention to values of the argument $z = x$ that are real and positive, because we ignore for the moment the absorption of seismic waves in the earth. Furthermore, we shall be interested in the seismic field at distances that are much greater than the radiation's wavelength, i.e., $x \gg 1$. Under these conditions, the period equation can be approximated by the simpler equation

$$\frac{\partial}{\partial x} h_{s-1/2}^{(2)}(x) = 0. \quad (8.158)$$

Keeping x fixed, we seek the roots of Eq. (8.158) with respect to order, namely, the roots $s_n(x)$ under the conditions $|s_n| \gg 1$, $x \gg 1$. Using Eq. (H.43), the roots of Eq. (8.158) are given in the form

$$\text{Re}\, s_n = x - \left(\frac{x}{16}\right)^{1/3} \bar{q}_n - \frac{1}{120}\left(\bar{q}_n^2 - \frac{21}{\bar{q}_n}\right)\left(\frac{2}{x}\right)^{1/3} + O(x^{-1}),$$
$$\text{Im}\, s_n = \frac{\sqrt{3}}{2}\left\{\left(\frac{x}{2}\right)^{1/3} \bar{q}_n - \frac{1}{60}\left(\bar{q}_n^2 - \frac{21}{\bar{q}_n}\right)\left(\frac{2}{x}\right)^{1/3}\right\} + O(x^{-1}), \quad (8.159)$$

where

$$Ai'(\bar{q}_n) = 0.$$

Table 8.7 lists the first 10 zeros of the Airy function and its derivative. The error involved in the approximate zeros for $n > 1$ is less than 0.5%.

To first approximation, $-\bar{q}_n = [(3\pi/8)(4n + 1)]^{2/3}$, $n = 0, 1, 2, \ldots$ and we therefore write,

$$s_n = k_\beta b\left[1 + \frac{1}{2}(4n + 1)^{2/3}\left(\frac{3\pi}{4k_\beta b}\right)^{2/3} e^{-\pi i/3}\right], \qquad n = 0, 1, 2, \ldots$$

$$\text{Re } s_n = k_\beta b + \frac{1}{4}\left[\frac{3\pi}{4}(4n + 1)\right]^{2/3}(k_\beta b)^{1/3} > 0, \tag{8.160}$$

$$\text{Im } s_n = -\left(\frac{\sqrt{3}}{4}\right)\left[\frac{3\pi}{4}(4n + 1)\right]^{2/3}(k_\beta b)^{1/3} < 0.$$

Having the explicit expressions of the roots, we proceed to substitute in Eq. (8.156) the proper asymptotic expansions of the radial and azimuthal eigenfunctions for $x > |s_n| \gg 1$. First, because $\text{Im } s_n < 0$ and $|\text{Im } \nu_n|$ is increasing with n, it is sufficient to consider only the first term in the expansion of

$$\left[\cos \pi\left(\nu_n + \frac{1}{2}\right)\right]^{-1}$$

in Eq. (8.15). Therefore

$$\frac{P^2_{\nu_n}(-\cos\theta)}{\cos\pi(\nu_n + \frac{1}{2})} \sim \frac{2\nu_n^{3/2}e^{\pi i/4}}{\sqrt{[2\pi\sin\theta]}}[e^{-i(\nu_n + 1/2)\theta} + ie^{-i(\nu_n + 1/2)(2\pi - \theta)}] + O(\nu_n^{-3/2}). \tag{8.161}$$

The approximation of the radial functions is made by means of the asymptotic approximation given in Appendix H. When this is applied to the spherical Hankel functions that appear in Eq. (8.156), we find

$$h^{(2)}_{\nu_n}(k_\beta r_h)h^{(2)}_{\nu_n}(k_\beta r) \simeq \frac{e^{i\Lambda + i\pi/2}}{(k_\beta r)(k_\beta r_h)[1 - (b/r)^2]^{1/4}[1 - (b/r_h)^2]^{1/4}},$$

$$\Lambda = s_n\left[\cos^{-1}\frac{b}{r} + \cos^{-1}\frac{b}{r_h}\right] - k_\beta[\sqrt{(r^2 - b^2)} + \sqrt{(r_h^2 - b^2)}], \tag{8.162}$$

$$\frac{h^{(1)}_{\nu_n}(k_\beta b) - (k_\beta b)(\partial/\partial(k_\beta b))h^{(1)}_{\nu_n}(k_\beta b)}{[(\partial/\partial\nu)[h^{(2)}_\nu(k_\beta b) - (k_\beta b)(\partial/\partial(k_\beta b))h^{(2)}_\nu(k_\beta b)]]_{\nu_n}} \simeq \frac{-i}{\alpha_n}$$

Here,

$$s_n = \nu_n + \frac{1}{2} = (k_\beta b)\cos\alpha_n \simeq k_\beta b\left[1 - \frac{1}{2}\alpha_n^2\right] \tag{8.163}$$

which together with Eq. (8.160) yields the explicit approximation for α_n

$$\alpha_n \simeq \left[(4n+1)^{1/3}\left(\frac{3\pi}{4}\right)^{1/3} e^{\pi i/3}\right](k_\beta b)^{-1/3}. \tag{8.164}$$

Because the term $\frac{1}{2}h_l^{(1)}(k_\beta r_<)h_l^{(2)}(k_\beta r_>)$ in Eq. (8.153) does not contribute to the residues, the insertion of Eqs. (8.161)–(8.164) into Eq. (8.156) results in the approximate displacements

$$\{u_\phi\}_{\text{diff}} = \frac{(k_\beta b)^{5/6} b U_0(\omega) dS}{2rr_h^2(1-b^2/r^2)^{1/4}(1-b^2/r_h^2)^{1/4}}\left[\frac{4}{3\pi}\right]^{1/3}\frac{\cos 2\phi}{\sqrt{[2\pi\sin\theta]}}$$
$$\times\, e^{i(\omega t - k_\beta D_1 - \phi_0)}\sum_n (4n+1)^{-1/3}\{e^{-is_n\Theta} + e^{-i(s_n\Theta' + \pi/2)}\},$$

$$\phi_0 = \frac{13}{12}\pi, \qquad D_1 = \sqrt{(r^2 - b^2)} + \sqrt{(r_h^2 - b^2)}, \tag{8.165}$$

$$\Theta = \theta - \cos^{-1}\frac{b}{r} - \cos^{-1}\frac{b}{r_h},$$

$$\Theta' = (2\pi - \theta) - \cos^{-1}\frac{b}{r} - \cos^{-1}\frac{b}{r_h}.$$

Owing to the exponential factor $e^{-|\text{Im}\, s_n|\Theta}$ that occurs in Eq. (8.165), only the zeros s_n nearest to the real axis in the s plane carry most of the contribution to u_ϕ.

The interpretation of the sum in Eq. (8.165) is facilitated with the aid of Fig. 8.14. The fundamental features of the solution are:

1. Two separate waves travel from the source at $(r_h, 0, 0)$ to the sensor at (r, θ, ϕ). Because $-is_n = -|\text{Im}\, s_n| - i|\text{Re}\, s_n|$, the amplitude and phase of the first group of waves, for a given value of n, have the form

$$e^{-|\text{Im}\, s_n|\Theta} e^{i[\omega t - kD_1 - |\text{Re}\, s_n|\Theta]}. \tag{8.166}$$

   If we imagine a string that is stretched between the source and the sensor such that it is constrained to be tangent to the core, then D_1 is that part of its length that does not touch the core and Θ is the angle subtended by the touching arc (Fig. 8.14). It therefore transpires that $D_1/v + (1/\omega)|\text{Re}\, s_n|\Theta$ is the total travel time of a harmonic component. The second wave travels along the opposite side of the core. The nontangent portions of its route are the same as the first wave but the angle subtended by the core is different.
2. The summation over n can be carried out only if the terms decrease with n. This takes place only if $\Theta > 0$, which restricts the usefulness of the Watson transformation to the shadow region. However, in this region, the convergence is very fast owing to the factor $e^{-n^{2/3}}$. Note that the direct wave and the once-reflected wave from the core (body waves) do not appear in our final results because they have been eliminated *a priori* by the residue process and the requirement that the Bessel functions be approximated in the transition region $s \sim x$. We shall see in Section 8.6.2 that in order to derive these body

waves, we must approximate the value of the integrals by the saddle point method. The results obtained in this way are of use only in the "lit" area. In the seismic shadow area the results obtained by Watson's residue method are usable. However, these results are untractable outside this area. The two methods therefore yield complementary formulas that together enable us to calculate the seismic field everywhere.

3. The wave amplitude decreases exponentially with the distance traveled along the core boundary area. The damping constant is proportional to $\omega^{1/3}$.
4. If Re s_n were equal exactly to $(k_\beta b)$, no dispersion would take place. However, because (Re $s_n - k_\beta b) \propto n^{2/3}\omega^{1/3}$, the velocity of propagation is slightly increasing with ω and n, resulting in dispersion of the wave trains.
5. The factor $1/\sqrt{(\sin\theta)}$ is the effect of the geometric spreading. We have already met this effect in the case of surface waves.
6. In a radially heterogeneous earth model and in the presence of a source that is not necessarily of the *SH* type, it can be shown that Eq. (8.160) is modified to

$$s_n = kb\left[1 + \frac{1}{2}(4n + c)^{2/3}\left(1 - \frac{bv'(b)}{v(b)}\right)\left(\frac{3\pi}{4kb}\right)^{2/3} e^{-\pi i/3}\right], \tag{8.167}$$

where $-1 \le c \le 1$ depends on the boundary conditions at the core–mantle interface and $v(r)$ is the intrinsic velocity profile in the earth.

8.6.2. The Direct and Reflected Fields in the "Lit" Zone

We have just noticed that $\mathfrak{G}_\infty$ does not contribute to the field at the observation point in the shadow zone. The displacement field resulting from this part of the total Green's dyadic represents the field at Q in the absence of the core. In the context of our problem it has separate physical meaning only at field points that are sufficiently away from the core boundary, such that most of the arriving signal is unperturbed by the diffracted and reflected waves from the core. However, if the point of observation is in the "lit" zone but too close to the boundary, a superposition of these three waves will take place close to the arrival of the direct wave and a direct evaluation of the infinite sum must be made. We begin with the evaluation of the direct field. According to Eq. (8.96),

$$\begin{aligned} u_\phi &= \frac{ik_\beta U_0(\omega)dS}{8\pi r_h}\cos 2\phi \frac{\partial}{\partial\theta}\sum_{l=2}^{\infty}\frac{2l+1}{l(l+1)} h_l^{(1)}(k_\beta r_<)h_l^{(2)}(k_\beta r_>)P_l^2(\cos\theta) \\ &= -\frac{k_\beta U_0(\omega)dS}{16\pi r_h}\cos 2\phi\frac{\partial}{\partial\theta}\int_{-\infty}^{\infty}\frac{2\nu+1}{\nu(\nu+1)} h_\nu^{(1)}(k_\beta r_<)h_\nu^{(2)}(k_\beta r_>) \\ &\quad\times\left\{\frac{P_\nu^2(-\cos\theta)}{\cos\pi(\nu+\frac{1}{2})}\right\}d\nu. \end{aligned} \tag{8.168}$$

In Eq. (8.168) the sum has been transformed to a Watson integral with an integrand that is an odd function in $\nu + \frac{1}{2}$.

We next insert in Eq. (8.168) the representations

$$P_\nu^2(-\cos\theta) = \left(\frac{\partial^2}{\partial\theta^2} - \cot\theta\frac{\partial}{\partial\theta}\right)P_\nu(-\cos\theta),$$

$$\left[\cos\pi\left(\nu+\frac{1}{2}\right)\right]^{-1} = 2\sum_{N=0}^{\infty}(-)^N e^{-(2N+1)(\nu+1/2)\pi i},$$

$$h_\nu^{(1)}(z) = \frac{1}{\sqrt{(2\pi z)}}\int_{-\eta+i\infty}^{\eta-i\infty} e^{i[z\cos\tau_1+(\nu+1/2)(\tau_1-\pi/2)]}\,d\tau_1, \tag{8.169}$$

$$h_\nu^{(2)}(z) = \frac{-1}{\sqrt{(2\pi z)}}\int_{(\pi-\eta)+i\infty}^{(\eta-\pi)-i\infty} e^{-i[z\cos\tau_2+(\nu+1/2)(\tau_2-\pi/2)]}\,d\tau_2, \tag{8.170}$$

$$(-\arg z < \eta < \pi - \arg z)$$

$$P_\nu(-\cos\theta) = \frac{1}{2\pi}\int_{-\pi}^{\pi} e^{\nu\ln[\cos(\theta-\pi)-i\sin(\theta-\pi)\cos\tau_3]}\,d\tau_3. \tag{8.171}$$

The resulting fourfold integral is

$$u_\phi = C_0\frac{\partial}{\partial\theta}\left(\frac{\partial^2}{\partial\theta^2} - \cot\theta\frac{\partial}{\partial\theta}\right)\int_{-\infty}^{\infty} d\nu\int_{-\pi}^{\pi} d\tau_3\iint\frac{2\nu+1}{\nu(\nu+1)}e^{i\Psi}\,d\tau_1\,d\tau_2, \tag{8.172}$$

where the exponent reads

$$\Psi = -\pi\nu(2N+1) - i\nu\ln[\cos(\theta-\pi) - i\sin(\theta-\pi)\cos\tau_3]$$
$$+ k_\beta[r_<\cos\tau_1 - r_>\cos\tau_2] + \left(\nu+\frac{1}{2}\right)(\tau_1-\tau_2) - \frac{\pi}{2}$$

and

$$C_0 = \frac{1}{32\pi^3}\frac{U_0\,dS\cos 2\phi}{(r_>r_<)^{1/2}r_h}.$$

The saddle points are determined from

$$\frac{\partial\Psi}{\partial\nu} = \frac{\partial\Psi}{\partial\tau_1} = \frac{\partial\Psi}{\partial\tau_2} = \frac{\partial\Psi}{\partial\tau_3} = 0. \tag{8.173}$$

With the particular choice $N = 0$, these points are determined by the equations:

$$\nu + \frac{1}{2} = k_\beta r_<\sin\tau_1 = k_\beta r_>\sin\tau_2,$$

$$\tau_3 = M\pi, \qquad M = 0, 1, 2, \ldots, \qquad \theta = \tau_1 - \tau_2\ (M\ \text{even}), \tag{8.174}$$
$$2\pi - \theta = \tau_1 - \tau_2\ (M\ \text{odd}).$$

Of these, only a single saddle point lies on the path of integration. Its coordinates in the parameter space are given by $\tau_3 = 0$, $0 < \tau_1 < \pi$, $0 < \tau_2 < \pi$ and the explicit forms of τ_1, τ_2, ν as functions of $r_<, r_>$, and θ are obtained from Eq. (8.174). The geometric configuration is shown in Fig. 8.15. The total contribution of this point is

$$\frac{2\nu_o + 1}{\nu_o(\nu_o + 1)} e^{i\Psi_o} \frac{[i\sqrt{(2\pi)}]^4}{\sqrt{|H_s|}}. \tag{8.175}$$

H_s is a symmetrical determinant of order 4, known as the *Hessian*, whose elements are the values of $\partial^2\Psi/\partial z_i \partial z_j$ at the saddle point, $z_i = (\tau_1, \tau_2, \tau_3, \tau_4)$, $\tau_4 = \nu$

$$H_s = \begin{Vmatrix} -k_\beta r_< \cos\tau_1 & 0 & 0 & 1 \\ 0 & k_\beta r_> \cos\tau_2 & 0 & -1 \\ 0 & 0 & \nu_o \sin\theta e^{i\theta} & 0 \\ 1 & -1 & 0 & 0 \end{Vmatrix}$$

$$= \nu_o k_\beta \sin\theta e^{i\theta}(r_< \cos\tau_1 - r_> \cos\tau_2), \tag{8.176}$$

$$\Psi_o = k_\beta(r_< \cos\tau_1 - r_> \cos\tau_2) + \frac{1}{2}\theta - \frac{\pi}{2}.$$

It is clear from Fig. 8.15 that $D = r_> \cos\tau_2 - r_< \cos\tau_1$ is equal to the distance between the source and the sensor. The substitution of Eq. (8.175) into Eq. (8.172) yields in the far field

$$u_\phi^{\text{Direct}} = \frac{-ik_\beta U_0(\omega)dS}{4\pi} \cos 2\phi \sin\tau_2 \frac{e^{-ik_\beta D}}{D} \tag{8.176a}$$

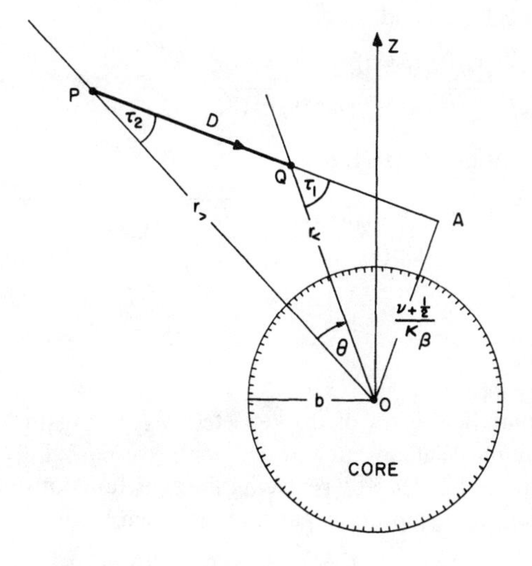

Figure 8.15. Geometry of the direct ray in the lit zone of an infinite medium surrounding a spherical cavity.

which agrees with Eq. (7.164) for $\lambda = 0$, $\delta = \pi/2$. An examination of terms with $N \neq 0$ shows that the corresponding path of integration does not cross the saddle point of the exponent and they are therefore omitted.

8.6.3. Spherical Reflection Coefficients and Mode–Ray Correspondence of the *ScS* Wave

In Section 7.2 we have presented the far-field high-frequency approximation of the spectral field of body waves. The reflection coefficients at the earth's surface and the mantle–core boundary were those of plane waves incident on a planar interface and as such were frequency independent. The effect of the source was manifested in the radiation-pattern function $F(\lambda, \delta, \phi_h, i_h)$ and the effect of the radially heterogeneous medium was carried through the divergence coefficients, which were also assumed to be frequency independent. The only place where the earth's sphericity entered was in the geometric spreading factor of the divergence coefficient.

The ensuing expressions for the amplitudes of the various rays therefore rendered only the amplitudes of the initial motions. However, if we wish to obtain further information concerning the time dependence of the entire signal, we must take into consideration the curvature of the discontinuities in the earth. To this end we return to Eq. (8.154) and recast it in the form

$$u_\phi(\omega) = \frac{ik_\beta U_0(\omega)dS}{8\pi r_h} \cos 2\phi e^{i\omega t} \sum_{l=2}^{\infty} \frac{2l+1}{l(l+1)} \times \left[h_l^{(2)}(k_\beta r_h) h_l^{(2)}(k_\beta r) \frac{\partial}{\partial\theta} P_l^2(\cos\theta)\right]\left[\frac{h_l^{(1)}(k_\beta b)}{h_l^{(2)}(k_\beta b)}\right] R_c, \tag{8.177}$$

where the source term has been neglected and

$$-R_c = \frac{1 - (k_\beta b)\{[h_l'^{(1)}(k_\beta b)]/[h_l^{(1)}(k_\beta b)]\}}{1 - (k_\beta b)\{[h_l'^{(2)}(k_\beta b)]/[h_l^{(2)}(k_\beta b)]\}}. \tag{8.178}$$

According to the Debye approximation (App. H)

$$\frac{h_l'^{(1)}(k_\beta b)}{h_l^{(1)}(k_\beta b)} \to +i\frac{\sqrt{(k_\beta^2 - k^2)}}{k_\beta}, \qquad \frac{h_l'^{(2)}(k_\beta b)}{h_l^{(2)}(k_\beta b)} \to -i\frac{\sqrt{(k_\beta^2 - k^2)}}{k_\beta},$$

$$kb = l + \frac{1}{2}, \qquad k_\beta b > l + \frac{1}{2} \gg 1$$

and $R_c \to 1$, in the high-frequency limit.

To obtain the approximate analytical form of the reflected ray, we again replace the infinite sum over l by the Watson integral. However, we pass the evaluation of a Hessian determinant of order 5 by replacing the eigenfunctions in the integrand by their corresponding asymptotic expansions and then evaluate the contribution of the saddle point in a single variable. Applying the results given in Appendix H, we find

$$h_\nu^{(1,2)}(x) = \frac{1}{x\sqrt{(\sin\delta)}}e^{\pm iz}\left[1 + O\left(\frac{1}{\nu}\right)\right],$$

$$\cos\delta = \frac{\nu + \frac{1}{2}}{x}, \qquad z = \left(\nu + \frac{1}{2}\right)\int_1^{x/(\nu+1/2)}\left[\sqrt{(y^2 - 1)}\frac{dy}{y}\right] - \frac{\pi}{4}, \tag{8.179}$$

$$\frac{h_l^{(1)}(k_\beta b)}{h_l^{(2)}(k_\beta b)} = \exp\left\{2i\left(\nu + \frac{1}{2}\right)\int_1^{(k_\beta b)/(\nu+1/2)}\left[\sqrt{(y^2 - 1)}\frac{dy}{y}\right] - \frac{\pi i}{2}\right\}\left[1 + O\left(\frac{1}{\nu}\right)\right], \tag{8.180}$$

$$h_\nu^{(2)}(k_\beta r_h)h_\nu^{(2)}(kr)$$

$$= \frac{\exp\left\{\frac{\pi i}{2} - i\left(\nu + \frac{1}{2}\right)\left[\int_1^{k_\beta r/(\nu+1/2)}\sqrt{(y^2 - 1)}\frac{dy}{y} + \int_1^{k_\beta r_h/(\nu+1/2)}\sqrt{(y^2 - 1)}\frac{dy}{y}\right]\right\}}{k_\beta^2 r r_h\sqrt{(\sin\delta\sin\delta_0)}}$$

$$\times\left[1 + O\left(\frac{1}{\nu}\right)\right], \tag{8.181}$$

$$\frac{(\partial/\partial\theta)P_\nu^2(-\cos\theta)}{\cos\pi(\nu + \frac{1}{2})} = -i\frac{\sqrt{2}\nu^{5/2}}{\sqrt{(\pi\sin\theta)}}e^{-i\theta(\nu+1/2)+3\pi i/4}\left[1 + O\left(\frac{1}{\nu}\right)\right].$$

Applying the Watson transformation and inserting these asymptotic approximations in the integrand, we get

$$u_\phi(r, r_h; \omega) = \frac{iU_0(\omega)dS}{4\pi k_\beta r_h^2 r}\frac{\cos 2\phi}{\sqrt{(2\pi\sin\theta)}}\int_{-\infty}^{\infty}\frac{R_c(\nu)\nu^{3/2}}{\sqrt{(\sin\delta\sin\delta_0)}}e^{-i\Psi(\nu)}d\nu, \tag{8.182}$$

$$\cos\delta = \frac{\nu + \frac{1}{2}}{k_\beta r}, \qquad \cos\delta_0 = \frac{\nu + \frac{1}{2}}{k_\beta r_h}, \tag{8.183}$$

$$\Psi(\nu) = \left(\nu + \frac{1}{2}\right)\left\{-2\int_1^{k_\beta b/(\nu+1/2)}\sqrt{(y^2 - 1)}\frac{dy}{y} + \int_1^{k_\beta r/(\nu+1/2)}\sqrt{(y^2 - 1)}\frac{dy}{y}\right.$$

$$\left. + \int_1^{k_\beta r_h/(\nu+1/2)}\sqrt{(y^2 - 1)}\frac{dy}{y} + \theta\right\} - \frac{3\pi}{4}. \tag{8.184}$$

We apply again the stationary-phase approximation

$$\int_{-\infty}^{\infty}A(\nu)e^{-i\Psi(\nu)}\,d\nu = A(\nu_0)\left[\frac{2\pi}{|\Psi''(\nu_0)|}\right]^{1/2}e^{-i\Psi(\nu_0)-\pi i/4\,\mathrm{sgn}\,\Psi''(\nu_0)}$$

and introduce the variables τ, τ_b, and τ_h such that

$$\nu_o + \frac{1}{2} = k_\beta b\sin\tau_b = k_\beta r\sin\tau = k_\beta r_h\sin\tau_h. \tag{8.185}$$

We then find that the saddle-point condition $\Psi'(\nu_0) = 0$ has the geometric interpretation (Fig. 8.16)

$$\theta = 2\tau_b - \tau - \tau_h. \tag{8.186}$$

Simple algebraic manipulations then lead to

$$\Psi''(\nu_0) = -\frac{1}{k_\beta}\frac{D_1 r\cos\tau + D_2 r_h \cos\tau_h}{b r r_h \cos\tau\cos\tau_h\cos\tau_b} < 0, \tag{8.187}$$

$$D_1 = r_h\cos\tau_h - b\cos\tau_b, \qquad D_2 = r\cos\tau - b\cos\tau_b.$$

The reflected displacement assumes the form

$$u_\phi = \left[-i\frac{k_\beta U_0(\omega)dS}{4\pi}\right][G_{ScS}][R_c][\sin\tau_h\cos 2\phi]e^{i[\omega t - k_\beta(D_1+D_2)]}, \tag{8.188}$$

where G_{ScS} is the *divergence coefficient*

$$G_{ScS} = \frac{b}{\sqrt{(rr_h)}}\left[\frac{\cos\tau_b}{\cos\tau_h}\right]^{1/2}\frac{1}{\left[\dfrac{b\sin\theta}{\sin\tau_h}\right]^{1/2}\left(D_1\dfrac{r\cos\tau}{r_h\cos\tau_h} + D_2\right)^{1/2}}. \tag{8.189}$$

Note that $(D_1 + D_2)$ is the path of the ScS wave from the source to the sensor. Figure 8.17 shows all the relevant singularities in the complex $[(\nu + \frac{1}{2})/k_\beta b]$ plane that contribute to the diffracted, direct, and reflected waves. The complex

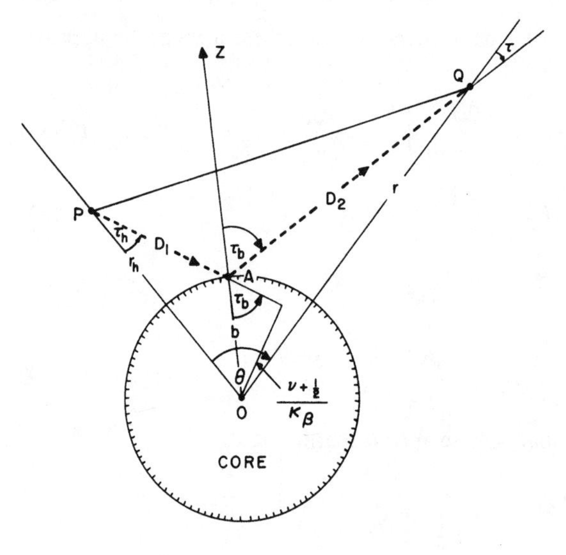

Figure 8.16. The geometry of the ScS wave.

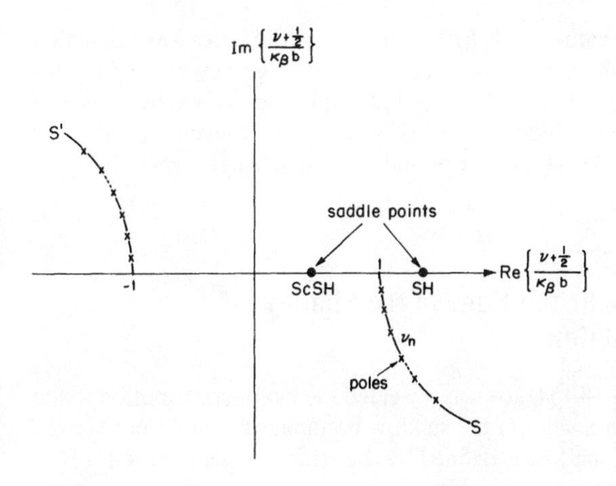

Figure 8.17. Singularities in the complex plane that make major contributions to the direct, reflected, and diffracted waves in the shell.

poles whose residues contributed to the diffracted waves in the shadow zone are located on S and S'. Of these, only those on S are relevant. The saddle point of the direct wave is located on the real axis at the root $\nu = \nu_1(\theta)$ of the equation [cf. Eq. (8.174), minus sign for obtuse τ_1]

$$\cos\theta = \frac{(\nu + \frac{1}{2})^2}{(k_\beta r)(k_\beta r_h)} \pm \left[1 - \left(\frac{\nu + \frac{1}{2}}{k_\beta r}\right)^2\right]^{1/2}\left[1 - \left(\frac{\nu + \frac{1}{2}}{k_\beta r_h}\right)^2\right]^{1/2}. \quad (8.190)$$

We therefore have derived approximate expressions for the azimuthal displacement field induced by a strike-slip point source in both the lit and the shadow zones. It is of interest to compare the general features of solutions with the direct evaluation of the sum $\sum_{l=2}^{\infty}(2l+1)f_l P_l(\cos\theta)$ given in Eq. (8.154). This is shown in Fig. 8.18 as a function of θ for certain fixed values of the parameters r_h/r, b/r and $k_\beta b$. This figure shows the spectral amplitude $\{k_\beta|S(\omega)|\}$ in

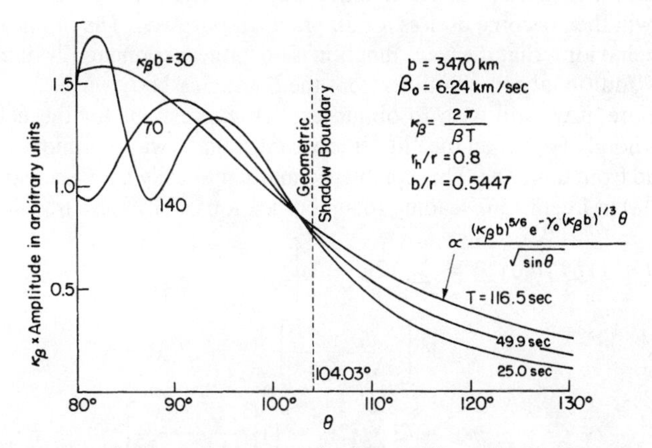

Figure 8.18. Spectral wave amplitudes outside and inside the geometric shadow boundary. γ_0 is a constant independent of frequency.

arbitrary units for three values of $(k_\beta b)$. Summation of the series was calculated until convergence was obtained. It was not necessary to go beyond $l = 150$ for any value of $k_\beta b$. As seen in Fig. 8.18, computed amplitudes for smaller values of θ show undulation when $k_\beta b$ is large. This is caused by the superposition of direct waves, reflected waves at the core boundary, and diffracted waves.

8.6.4. Amplitudes Near the Edge of the Shadow: Fresnel Diffraction

The exact solution in Eq. (8.154) so far has yielded the major contribution to the displacement field on both sides of the shadow boundary but sufficiently away from it. The edge of the shadow is defined by the critical angle [Eq. (8.165)],

$$\theta_s = \cos^{-1}\frac{b}{r} + \cos^{-1}\frac{b}{r_h}. \tag{8.191}$$

For $r = r_h = a$ (the earth's radius), $\theta_s = 113.99°$, whereas for $r = a$, $a - r_h = 77$ (km), $\theta_s = 113.54°$. The amplitude inside the shadow zone is then governed by an infinite sum, the first term of which is proportional to [Eq. (8.165)]

$$\exp\left[-(\theta - \theta_s)(k_\beta b)^{1/3}\frac{\sqrt{3}}{4}\left(\frac{3\pi}{4}\right)^{2/3}\right].$$

The first few terms of the sum will give a significant contribution if

$$\theta - \theta_s \gg (k_\beta b)^{-1/3}. \tag{8.192}$$

For $\theta < \theta_s$, as we have already noted in Section 8.6.1, the sum pertaining to the first term in the expansion of $[\cos \pi(\nu_n + \frac{1}{2})]^{-1}$ will increase exponentially with n, although higher terms will still converge. The residue series, as given in Eq. (8.165), will then become useless for all practical purposes. The physical reason for this behavior is that the wave function is no longer exponentially damped but contains additional contributions from the S and ScS body waves.

Therefore, if we still wish to obtain explicit expressions for the field in the neighborhood of θ_s but on the "lit" side, we must find a way to remove the body-wave field from this zone. This is achieved mathematically by the substitution of Eq. (8.14a) in Eq. (8.156), leading to a modification of Watson's transformation

$$\begin{aligned}\frac{1}{\pi}\sum_{l=2}^{\infty}(2l+1)f_l P_l^2(\cos\theta) &= \sum_{n=0}^{\infty}(2\nu_n+1)\left[\frac{e^{-\pi i\nu_n}}{\cos\pi(\nu_n+\frac{1}{2})}\right] \\ &\quad\times\left[\frac{g_{s-1/2}}{\partial\Delta_{s-1/2}/\partial s}\right]_{s=\nu_n+1/2} P_{\nu_n}^2(\cos\theta) \\ &\quad - 2i\sum_{n=0}^{\infty}(2\nu_n+1)\left[\frac{g_{s-1/2}}{\partial\Delta_{s-1/2}/\partial s}\right]_{s=\nu_n+1/2} E_{\nu_n,2}^{(1)}(\cos\theta).\end{aligned} \tag{8.193}$$

The second sum can be retransformed into an integral that, on application of the method of steepest descents, yields only body waves because it no longer has poles at the positive half-integers. The first sum, however, is a residue series that renders the displacements in the "lit" zone near the shadow boundary. The evaluation of this sum by the method outlined in Section 8.6.1, provides a continuation of Eq. (8.165) into the "lit" zone. The convergence of this sum requires the condition $\theta_s - \theta \gg (k_\beta b)^{-1/3}$.

Let us now consider the behavior of the displacement field in the neighborhood of the geometric shadow boundary $\theta \simeq \theta_s$. This is already included in the body wave field that in itself is a part of the complete displacement field given by the Legendre sum representation, Eq. (8.154). The body-wave field, as we know, is composed of a direct wave and a core reflection, contributed by two saddle points on the real ν axis [cf. Eqs. (8.174) and (8.185)]. The *ScS* saddle point is closer to the origin than that of direct wave. As θ approaches θ_s, the angle τ_b (Fig. 8.16) tends to $\pi/2$. Consequently, the two saddle points merge into a single point at $\nu + \frac{1}{2} = k_\beta b$, close to the first pole on S (Fig. 8.17). In that case

$$\theta_s = \cos^{-1}\left\{\frac{b^2}{rr_h} - \left[\left(1 - \frac{b^2}{r^2}\right)\left(1 - \frac{b^2}{r_h^2}\right)\right]^{1/2}\right\} = \cos^{-1}\frac{b}{r} + \cos^{-1}\frac{b}{r_h}. \quad (8.194)$$

However, at $\theta = \theta_s$, the divergence coefficient and the reflection coefficient of *ScS* are both zero [cf. Eqs. (8.178) and (8.189)] implying that most of the contribution to the field near $\theta = \theta_s$ comes from the saddle point of the direct wave. Because the wave is grazing the core en route to the surface, its amplitude is no longer given by Eq. (8.176a). We will show that its amplitude is given by the expression

$$u_\phi \simeq u_\phi^{\text{direct}}\left\{\frac{1}{2} - \frac{1}{\sqrt{2}}\left[\exp\left(\frac{\pi i}{4}\right)\right]\int_0^X e^{-(\pi i/2)\tau^2}\, d\tau\right\}, \quad (8.195)$$

where

$$X = \left[\frac{k_\beta D}{4\pi}\right]^{1/2}(\theta - \theta_s). \quad (8.196)$$

The correction factor, recognized as the *Fresnel integral*, has the value 0.5 at $\theta = \theta_s$ and zero at $|X| = \infty$ ($\omega \to \infty, \theta \neq \theta_s$). To prove it, we start again from the Legendre sum in Eq. (8.168). As in Eq. (8.14b), we employ the modified Watson transform

$$\sum_{l=2}^{\infty}(2l + 1)f_l P_l^2(\cos\theta) = -2\int_{C_2} f_{\nu-1/2}E^{(1)}_{\nu-1/2,\,2}(\cos\theta)\nu\, d\nu, \quad (8.197)$$

where $f_{\nu-1/2}$ is even in ν. Inserting the asymptotic approximation $E^{(1)}_{\nu-1/2,\,2} \sim \nu^{3/2}(2\pi\sin\theta)^{-1/2}e^{-i[\nu\theta - 5\pi/4]}$, [Eq. (H.60)], Eq. (8.168) yields, for $r \geq r_h$,

$$u_\phi = \frac{k_\beta U_0(\omega)dS}{2(2\pi)^{3/2}r_h}\frac{\cos 2\phi}{\sqrt{(\sin\theta)}}\left[\exp\left(\frac{\pi i}{4}\right)\right]\int_{\nu_1(\theta_s)=k_\beta b}^{\infty} h^{(2)}_{\nu-1/2}(k_\beta r_h)h^{(2)}_{\nu-1/2}(k_\beta r)e^{-i\nu\theta}\nu^{7/2}\, d\nu. \quad (8.198)$$

The direction of the original contour of integration, C_2, is reversed and a minus sign is placed in front of the integral. The new contour is then drawn just below the poles of $E^{(1)}_{\nu-1/2,\,2}$ at the negative half-integers and continues to go through the saddle point at $\nu + \frac{1}{2} = k_\beta b$ on the positive ν axis. Because the integrand is an even function of ν, the integral is folded over the positive ν axis. As the contribution of the ScS wave is very small in the neighborhood of θ_s, the value of the integral in Eq. (8.198) from $\nu = 0$ up to $\nu_1(\theta)$ can be neglected as compared to the remainder and we may put the lower limit of integration at $\nu_1(\theta_s) = k_\beta b$.

Consider the case $r = r_h$. We introduce a change of variable $\nu = k_\beta r \cos(\sigma/2)$. The geometric interpretation of σ is shown later in Fig. 8.34. Consequently, the lower limit of the integral in Eq. (8.198) is placed at θ_s. We next replace the spherical Hankel functions by their Debye approximation, Eq. (H.22)

$$e^{-i\nu\theta}[h^{(2)}_{\nu-1/2}(k_\beta r)]^2 \sim (k_\beta r)^{-2}\left(\sin\frac{\sigma}{2}\right)^{-1} \times \exp\left\{-2ik_\beta r\left[\sin\frac{\sigma}{2} - \cos\frac{\sigma}{2}\left(\frac{\sigma}{2} - \frac{\theta}{2}\right)\right] + \frac{\pi i}{2}\right\}. \quad (8.199)$$

At $\theta \sim \theta_s$, the main contribution comes from the neighborhood of the lower limit of integration, so that we may expand the integrand about $\sigma/2 = \theta/2$

$$\sin\frac{\sigma}{2} = \sin\frac{\theta}{2} + \left(\frac{\sigma}{2} - \frac{\theta}{2}\right)\cos\frac{\theta}{2} - \frac{1}{2}\left(\frac{\sigma}{2} - \frac{\theta}{2}\right)^2 \sin\frac{\theta}{2} + \cdots,$$

$$\cos\frac{\sigma}{2} = \cos\frac{\theta}{2} - \left(\frac{\sigma}{2} - \frac{\theta}{2}\right)\sin\frac{\theta}{2} - \frac{1}{2}\left(\frac{\sigma}{2} - \frac{\theta}{2}\right)^2 \cos\frac{\theta}{2} + \cdots,$$

and extend the corresponding range of integration to infinity. Taking note that the source–sensor distance D at grazing incidence is equal to $2r\sin(\theta/2)$ [cf. Example 7.1], the result, Eq. (8.195), follows.

When $r \neq r_h$, we change variables to $\sigma = \cos^{-1}(\nu/k_\beta r) + \cos^{-1}(\nu/k_\beta r_h)$ and although the analysis is somewhat more complicated, the result is the same.

The condition for the validity of this approximation is that the higher order terms in the expansion of the exponent in Eq. (8.199) shall be negligible in the relevant portion of the domain of integration. This leads to the *transition-zone conditions*

$$|\theta - \theta_s| \ll (k_\beta b)^{-1/3}, \qquad k_\beta D \gg (k_\beta b)^{2/3}. \quad (8.200)$$

The contribution of the ScS wave in the neighborhood of θ_s can be worked out in a similar way. Because the WKBJ approximation breaks down at angles of grazing incidence at the core–mantle boundary, the Airy function approximations to $h^{(1,2)}_{\nu-1/2}$ must be employed. When this is done, it is found that the contribution of the ScS wave to the diffracted field is of order $(\theta - \theta_s)(k_\beta b)^{1/3}$ and its neglect is justified.

Therefore, if conditions (8.200) are satisfied, the transition from "light" to shadow is described by an angular Fresnel diffraction pattern very similar to the classical one for an opaque screen (See Fig. 8.18). The effects of curvature of this screen come in through small correction terms of order $(\theta - \theta_s)(k_\beta b)^{1/3}$. We shall show later, in Section 8.7.4, that the above result can be generalized to a radially inhomogeneous earth model for which high-frequency separability of P and SV exists. In that case Eq. (8.195) still remains valid, provided [cf. Eq. (8.324)]

$$X = (\theta - \theta_s)\left(-\frac{\omega}{\pi}\frac{dp}{d\theta}\right)^{1/2}. \tag{8.201}$$

8.6.4.1. Mode–Ray Correspondence. We wish to examine the mode–ray correspondence for an *SH* core reflection (known as the *ScSH* wave). Here the travel time and its derivative are given exactly by the equations

$$\begin{aligned} t &= \frac{2a}{v}\left[1 + \xi^2 - 2\xi\cos\left(\frac{\theta}{2}\right)\right]^{1/2}, \qquad \xi = \frac{b}{a}, \\ \frac{v}{b}\frac{dt}{d\theta} &= \frac{\sin(\theta/2)}{[1 + \xi^2 - 2\xi\cos(\theta/2)]^{1/2}}, \end{aligned} \tag{8.202}$$

where b is the radius of the core. We take

$$\begin{aligned} v &= 6.24 \text{ km/s (average shear velocity in the mantle)}, \\ \xi &= 3473/6371 = 0.5447, \\ a &= 6371 \text{ km}. \end{aligned}$$

Table 8.9 lists numerical results for the simplest earth model consisting of a homogeneous mantle overlying a fluid core. The toroidal eigenperiods are obtained by solving the period equation in Example 6.2. The phase velocity is then calculated with the help of Jeans' formula. The time $\delta t = t - \Delta(dt/d\Delta) = (n + \frac{1}{4})_n T_l$ appropriate for the direct S wave is also given.

Table 8.10 exhibits the mode–ray correspondence for the *ScSH* wave in a homogeneous mantle. The eigenperiods are obtained by solving the period equation. The phase velocity derived through Jeans' formula agrees quite well with $a(dt/d\theta)^{-1}$. Appealing to Eq. (8.1.16), we note that for *ScSH* waves $\phi_0 \to 0$ as ${}_nT_l \to 0$, because $\phi_0 \to \frac{1}{4}$ for the surface reflection and $\phi_0 \to -\frac{1}{4}$ for the reflection from the mantle–core boundary. As can be seen from Table 8.10, ϕ_0 is indeed very small when ${}_nT_l$ is small. For a given θ, the value of $t - \Delta\, dt/d\Delta$ is obtained from Eq. (8.202). After n is chosen, the mode number l is selected in such a manner that

$$\left| {}_nc_l - \frac{a}{dt/d\theta} \right|$$

is minimum. We note that at different points on the surface of the sphere, different modes are needed to build the *ScSH* wave at that point.

Figure 8.19 exhibits the dependence of the phase velocity ${}_nc_l$ on ${}_nT_l$ for eight radial overtones of the toroidal oscillations of the Gutenberg–Bullard I earth model.

Table 8.9. Some Toroidal Mode Parameters of a Homogeneous Mantle Overlying a Fluid Core[a]

	n											
	1				3				5			
l	${}_nT_l$, exact (s)	${}_nc_l$ (km/s)	${}_nk_l a$	δt (s)	${}_nT_l$, exact (s)	${}_nc_l$ (km/s)	${}_nk_l a$	δt (s)	${}_nT_l$, exact (s)	${}_nc_l$ (km/s)	${}_nk_l a$	δt (s)
100	56.8	7.00	112.9	71.0	51.4	7.75	124.7	167.0				
75	73.9	7.17	86.8	92.4	65.6	8.08	97.8	213.2				
50	106.1	7.47	60.5	132.6	91.0	8.71	70.4	295.7	81.2	9.76	79.0	426.3
25	190.7	8.23	33.6	238.4	152.3	10.31	42.1	495.0	129.3	12.15	49.6	678.8
10	383.7	9.94	16.7	479.6	252.1	15.12	25.4	819.3	171.2	22.25	37.5	899.3
8	448.3	10.51	14.3	560.3	268.0	17.58	23.9	871.0	175.8	26.79	36.5	922.9
6	536.8	11.47	11.9	671.0	282.5	21.80	22.7	918.4	179.6	34.28	35.7	942.9
4	654.0	13.60	9.8	817.7	294.8	30.17	21.7	958.1	182.5	48.71	35.1	958.6
2	787.0	20.34	8.1	984.1	303.4	52.77	21.1	986.0	184.5	86.78	34.8	968.6
1	842.2	31.68	7.6	1053.1	306.0	87.20	20.9	994.5	185.1	144.15	34.7	971.8

[a] $v = 6.24$ km/s, $a = 6371$ km, $b = 3473$ km, $2\pi a = 40024$ km, $\delta t = (n + \frac{1}{4})_n T_l$, ${}_nc_l = 2\pi a/(l + \frac{1}{2})_n T_l$; ${}_nk_l = 2\pi a/v_n T_l$, above the line in the table body ${}_nk_l a > l > n$, below the line ${}_nk_l a > n \geq l$ and $2\gamma = 2(a - b)/v = 929.5s$.

Table 8.10. Mode–Ray Correspondence for the *ScSH* Wave in a Homogeneous Average Mantle Overlying a Liquid Core

$\theta = 15°$, $t - \Delta(dt/d\Delta) = 909.4$ s						$\theta = 30°$, $t - \Delta(dt/d\Delta) = 857.0$ s						$\theta = 45°$, $t - \Delta(dt/d\Delta) = 789.6$ s						$\theta = 60°$, $t - \Delta(dt/d\Delta) = 723.5$ s					
n	l	${}_nc_l$ (km/s)	$\frac{a}{dt/d\theta}$ (km/s)	${}_nT_l$ (s)	ϕ_0	n	l	${}_nc_l$ (km/s)	$\frac{a}{dt/d\theta}$ (km/s)	${}_nT_l$ (s)	ϕ_0	n	l	${}_nc_l$ (km/s)	$\frac{a}{dt/d\theta}$ (km/s)	${}_nT_l$ (s)	ϕ_0	n	l	${}_nc_l$[a] (km/s)	$\frac{a}{dt/d\theta}$ (km/s)	${}_nT_l$ (s)	ϕ_0
1	1	31.68	40.84	842.2	0.08	1	2	20.34	21.88	787.0	0.09	1	3	15.86	16.13	720.9	0.10	1	4	13.60	13.61	654.0	0.11
2	2	36.09	40.84	443.6	0.05	2	4	21.32	21.88	417.1	0.05	2	6	16.07	16.13	383.1	0.06	2	8	13.57	13.61	347.0	0.08
3	3	38.17	40.84	299.6	0.03	3	6	21.80	21.88	282.5	0.03	3	9	16.20	16.13	260.0	0.04	3	12	13.58	13.61	235.8	0.07
4	4	39.73	40.84	225.9	0.03	4	8	22.08	21.88	213.3	0.02	4	12	16.29	16.13	196.6	0.02	4	16	13.59	13.61	178.4	0.05
5	5	40.17	40.84	181.2	0.02	5	10	22.26	21.88	171.2	0.01	5	15	16.35	16.13	157.9	0.00	5	20	13.61	13.61	143.5	0.04
$l = n$						$l = 2n$						$l = 3n$						$l = 4n$					

[a] ${}_nc_l$ shows some scattering because it is evaluated by Jeans' formula.

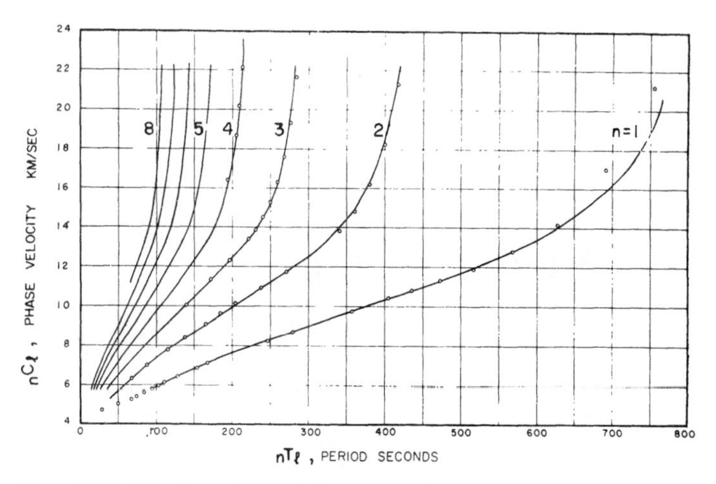

Figure 8.19. The dependence of the phase velocity ${}_nc_l$ for eight radial overtones for the Gutenberg–Bullard I earth model (App. L) (open circles). Solid curves are computed from Eqs. (8.1.21) and (8.1.22) with $\phi_0 = \frac{1}{4}$ by using travel-time curves for the Gutenberg–Bullard I model.

8.6.5. Numerical Normal-Mode Solution

In the preceding section we have explained the salient features of the diffracted field that arises from the presence of the material discontinuity at the core. We are now ready to explore the complete field induced in an earth model composed of a uniform shell over a liquid core. At this time we do not employ any asymptotic approximation but effect a straightforward numerical evaluation of the normal mode sum for a surface *SH* torque. The analytical expression is given in Eq. (8.150) and the numerical results are shown in Fig. 8.20. Displacements are given at various points on the surface by summing the contributions from 360 normal modes for $0 \leq n \leq 5$, $1 \leq l \leq 60$.

Two types of signals stand out:

1. Surface waves, similar to those observed from earthquakes, are caused primarily by the fundamental radial mode ($n = 0$).
2. Body waves, primarily related to the higher radial modes ($n > 0$). These include, in particular, the core reflections *ScS*, which may be multiple-reflected from the free surface.

For the sake of computation, the time dependence of the source was taken to be the symmetric "gate function" $f(t) = (1/2t_1)[H(t + t_1) - H(t - t_1)]$ whose Fourier transform is the real function

$$F(\omega) = \frac{1}{2t_1} \int_{-t_1}^{t_1} e^{i\omega t}\, dt = \frac{\sin \omega t_1}{\omega t_1}.$$

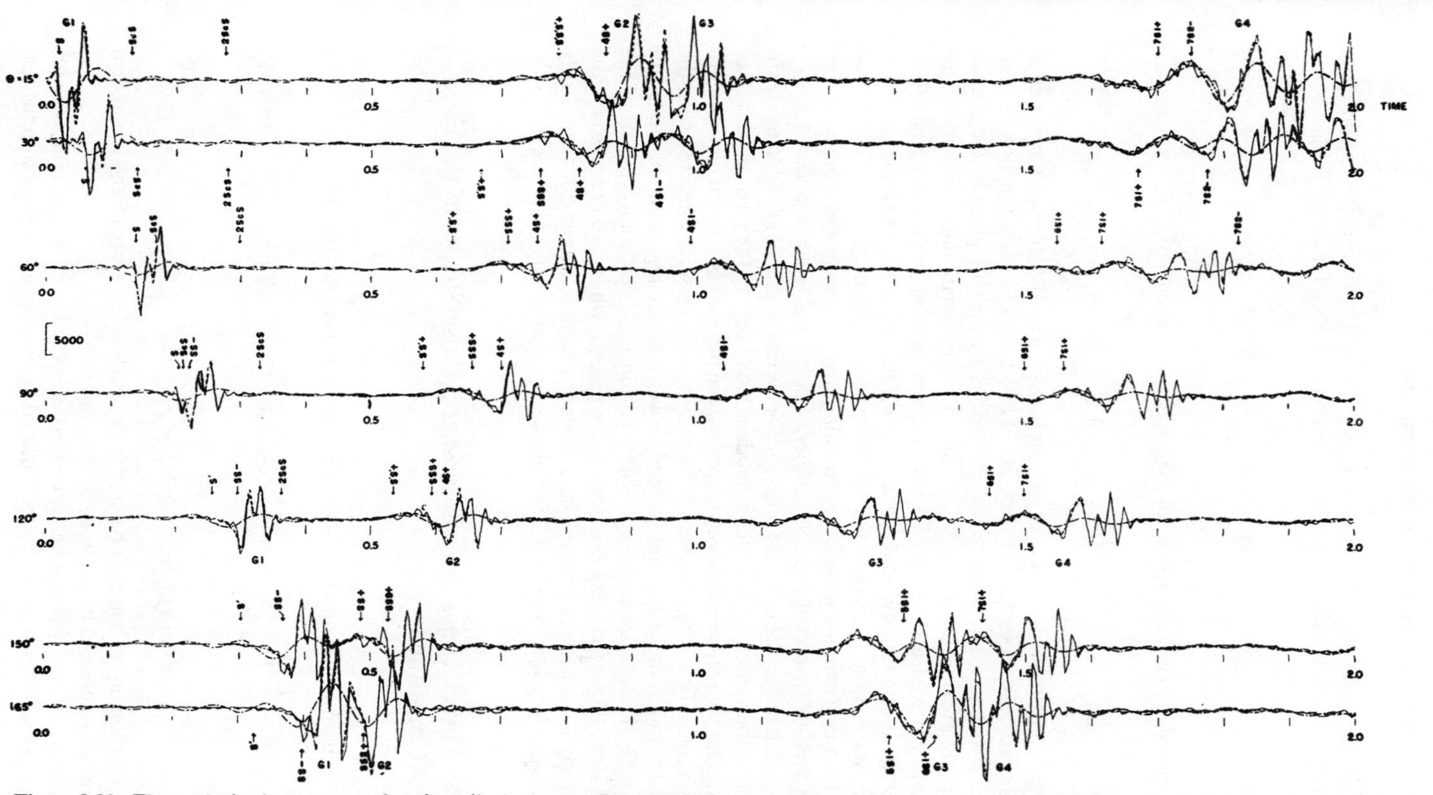

Figure 8.20. Theoretical seismograms of surface displacements [Eq. (8.150)] induced by an areal torque around the pole calculated at various epicentral distances. Arrows show the arrival times of various phases calculated by ray theory. Solid line, ${}_5u_{60} = \sum_{n=0}^{5} \sum_{l=1}^{60}$; broken line, ${}_0u_{60}$; chain line, ${}_0u_{10}$. $\theta_1 = 0.02$ rad, $\theta_2 = 0.04$ rad, $t_1 = 0.015$ units [cf. Eqs. (8.203) and (8.204)]. Time measured in units of $2\pi a/v$.

Clearly, in the limit $t_1 \to 0$, $F(\omega) = 1, f(t) = \delta(t)$. The time unit was chosen as $2\pi a/v$, where v is the shear velocity in the shell. We then find from Table 6.4 that

$$\omega_n^2 \bar{g}(t) = \left(\frac{\sin \omega_n t_1}{t_1}\right) \sin \omega_n t. \tag{8.203}$$

The spatial dependence $\Phi(\theta, \phi)$ of the source was taken as

$$\Phi(\theta, \phi) = \begin{cases} 1, & \theta_1 < \theta < \theta_2 \\ 0, & \theta < \theta_1, \theta > \theta_2. \end{cases} \tag{8.204}$$

Figure 8.20 shows the results of the summation. The displacements are given at the angular distances 15°, 30°, 60°, 90°, 120°, 150°, 165° as a function of non-dimensional time. Expected arrivals of various S waves are given by the arrows. Ray paths are shown in Fig. 8.21. All ray paths are straight lines and from simple geometry it is clear that the S shadow zone starts at the critical angle $\theta_s =$ 113.987°. In the shadow zone, where $\theta > \theta_s$, the diffracted wave supplants the direct wave. Long surface waves are observed in each curve of Fig. 8.20, at first by a fairly simple waveform consisting of a few peaks and troughs. Later, the waves are dispersed and the number of peaks and troughs increases.

It is seen in Fig. 8.20 that the difference between the sum of all radial modes to the contribution of the first radial mode is small during the passage of surface waves, and the main discrepancy between these two curves exists in the phases of body waves. This suggests that the surface wave is closely connected with the fundamental radial mode and the body wave with the higher radial modes. In Fig. 8.22, the fundamental mode contribution has been subtracted from the total sum, thus enhancing the body-wave phases, which are obscured by the relatively large surface waves. The ScS wave therefore stands out remarkably well. Other repeatedly-reflected ScS phases are also clearly observed.

8.6.6. Travel Times, Amplitudes, and Pulse Shapes of the Field Constituents

As in the simple case of a homogeneous sphere, the displacements are expressible as a Fourier transform of the spectral displacements [Eq. (8.87)]

$$u_\phi(a, r_h; t) = \frac{1}{2\pi} \int_{-\infty}^{\infty} \frac{e^{i\omega t}}{i\omega}\, d\omega \left\{ \frac{U_0\, dS}{4\pi a r_h} \cos 2\phi \sum_{l=2}^{\infty} (2l+1) f_l(a, r_h; \omega) \frac{\partial}{\partial\theta} P_l^2(\cos\theta) \right\},$$

$$f_l = -\frac{1}{l(l+1)} \left[\frac{y_1(r_h)}{a\{dy_1(a)/da\} - y_1(a)} \right]. \tag{8.205}$$

This representation can also be used independently to evaluate the theoretical seismograms created by a source in the spherical layer.

Because $y_1(a)$, $y_1(r_h)$ can be approximated as in the case of a homogeneous sphere, the whole procedure can be repeated, with, of course, a more complicated

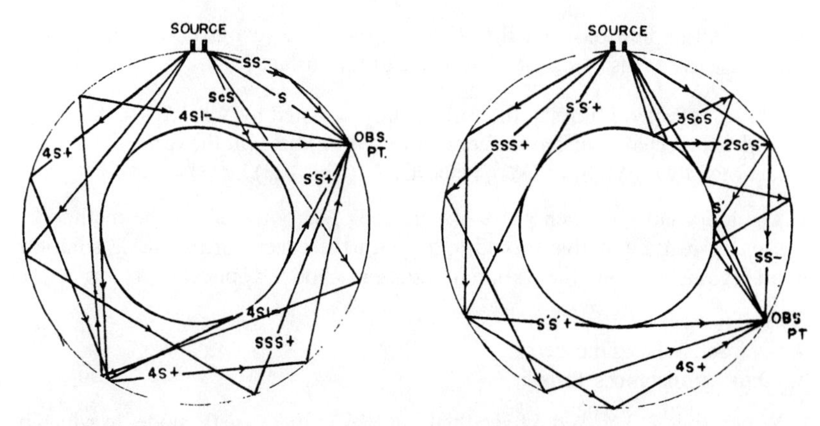

Figure 8.21. Some ray paths in the shell. The notation $nSm^{\mp}$ means that the S wave was reflected $(n - 1)$ times at the free surface and arrived at a point with an epicentral distance $\theta + 2m\pi\ (nSm^{-})$ or $-\theta + 2(m + 1)\pi\ (nSm^{+})$, after having traveled along the minor $(-)$ or major $(+)$ arc.

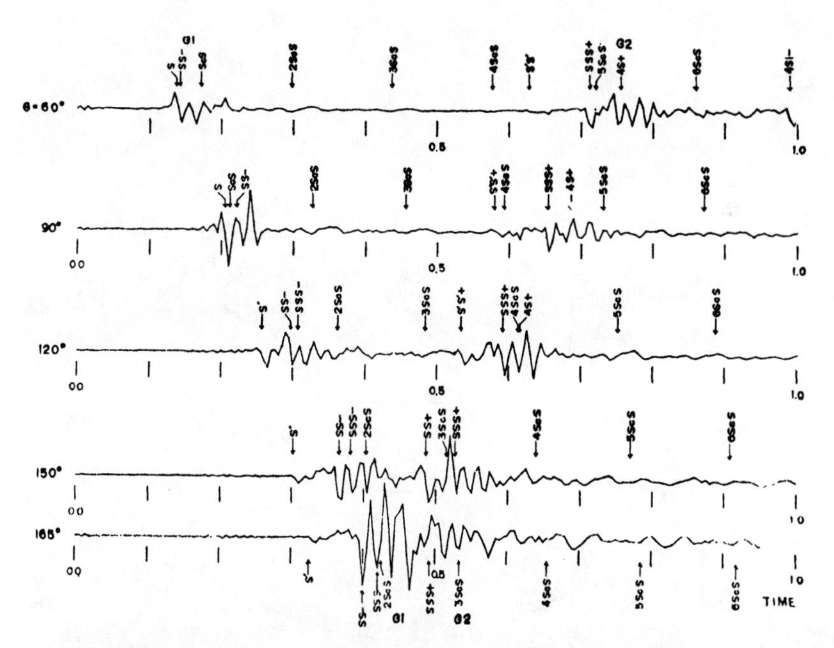

Figure 8.22. Displacements obtained by subtracting the contribution of the fundamental mode from the total displacements $[{}_5u_{60} - {}_0u_{60}]$ making the body waves more prominent on the record.

algebra but no new basic results. We shall, therefore, just give the final results: There are five types of signals that arrive at the surface:

1. Pulses reflected j times at the surface only, denoted by S_jS, $[S_1S \equiv SS]$.
2. Pulses reflected j times at the core and $j - 1$ or j times at the surface. They are denoted by $(ScS)_j$ or $S(ScS)_j$, respectively (Fig. 8.23).

The amplitudes of each pulse in item 1 or 2 is obtained by the method of *stationary phase* and this should agree with the exact normal-mode solution in all regions in which the stationary phase method is applicable.

3. Surface waves.
4. Diffractions at the core.
5. Diffractions at the surface.

Wave types 2, 3, 4, and 5 have their analogs in a flat-earth model in which a single layer overlies a homogeneous half-space (see Chap. 7). Type 4 is essentially a head wave at the core–mantle boundary and type 5 corresponds to the non-least-time arrivals in a homogeneous half-space.

Rays for which $\alpha < \cos^{-1}(b/a)$ are not reflected at the core, but only at the free surface. Snell's law reads: $a \cos \alpha = r_h \cos \beta = b \cos \gamma$ [see footnote, p. 662]

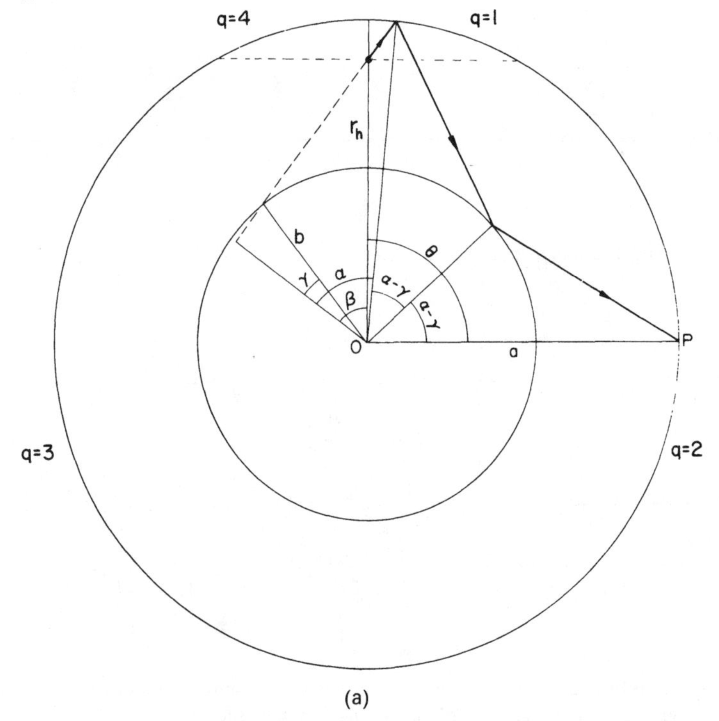

Figure 8.23 (see legend on facing page).

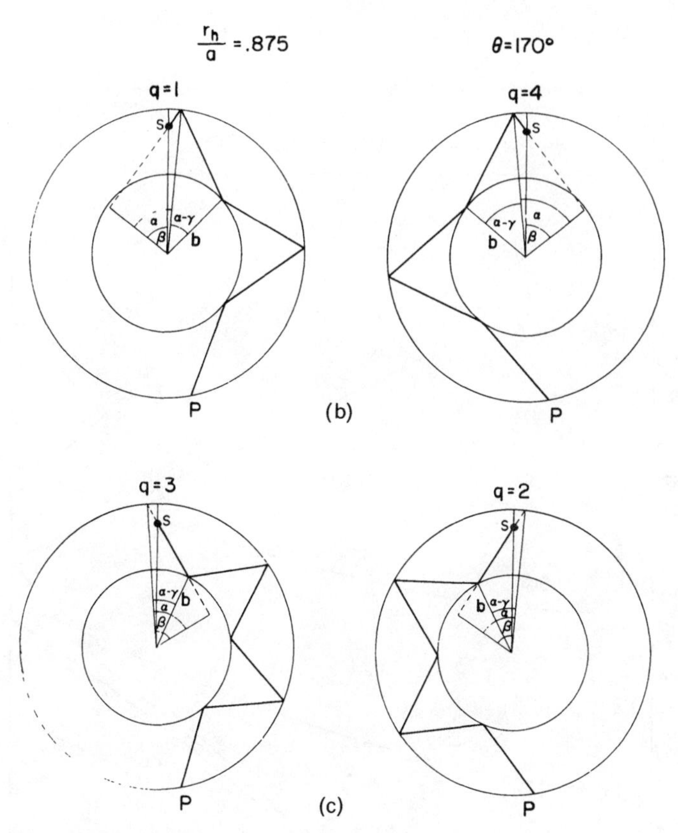

Figure 8.23. Geometry of rays corresponding to pulses reflected at the core and the surface. (a) $S(ScS)$ (b) $S(ScS)_2$ (c) $(ScS)_3$.

8.6.6.1. Features of the $S(ScS)_j$ Wave. Travel-time:

$$t = \frac{1}{v}[(2j + 1)a \sin \alpha - 2jb \sin \gamma - r_h \sin \beta]. \tag{8.206}$$

Divergence coefficient:

$$\left[\frac{\cos \alpha}{Dr_h \sin \theta}\right]^{1/2},$$
$$D = \left[(2j + 1)r_h \sin \beta - 2j\left(\frac{r_h a}{b}\right)\frac{\sin \beta \sin \alpha}{\sin \gamma} - a \sin \alpha\right]. \tag{8.207}$$

Saddle point condition:

$$\left.\begin{array}{ll}\text{Clockwise} & \theta + 2N\pi \\ \text{Counterclockwise} & (2\pi - \theta) + 2N\pi\end{array}\right\} = (2j + 1)\alpha - \beta - 2j\gamma. \tag{8.208}$$

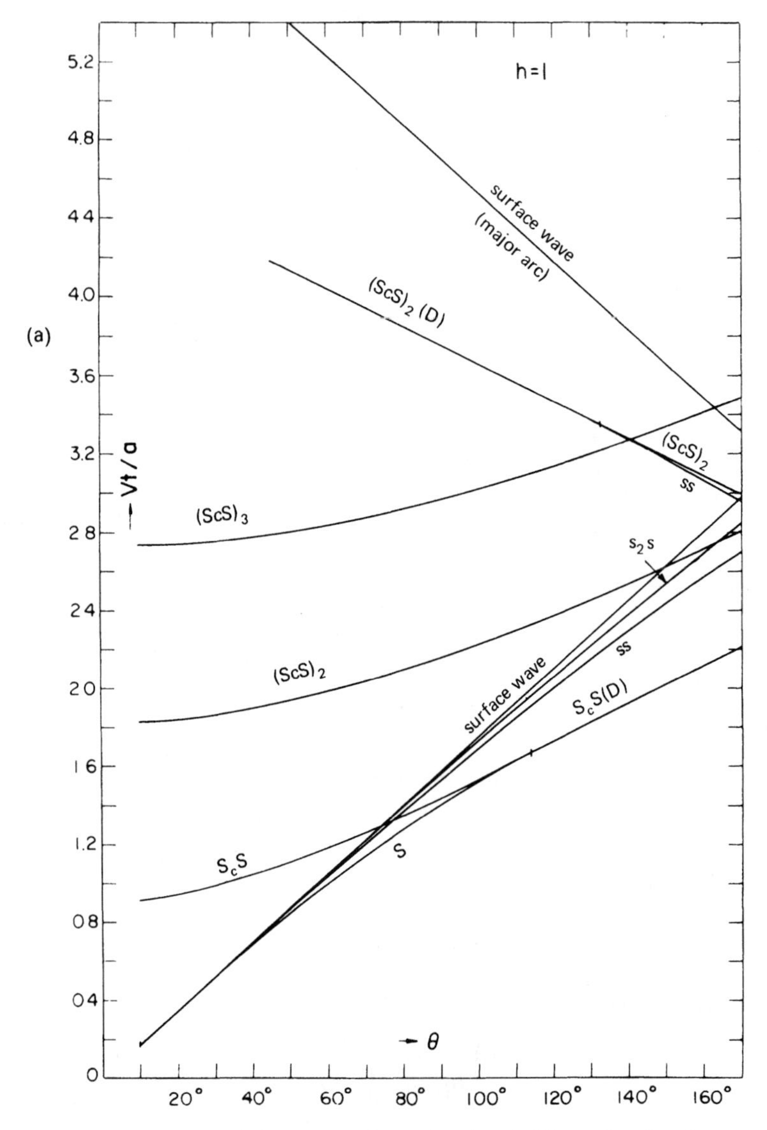

Figure 8.24. Travel-time curves for reflected and diffracted pulses from a point source in the shell: (a) Source at the surface.

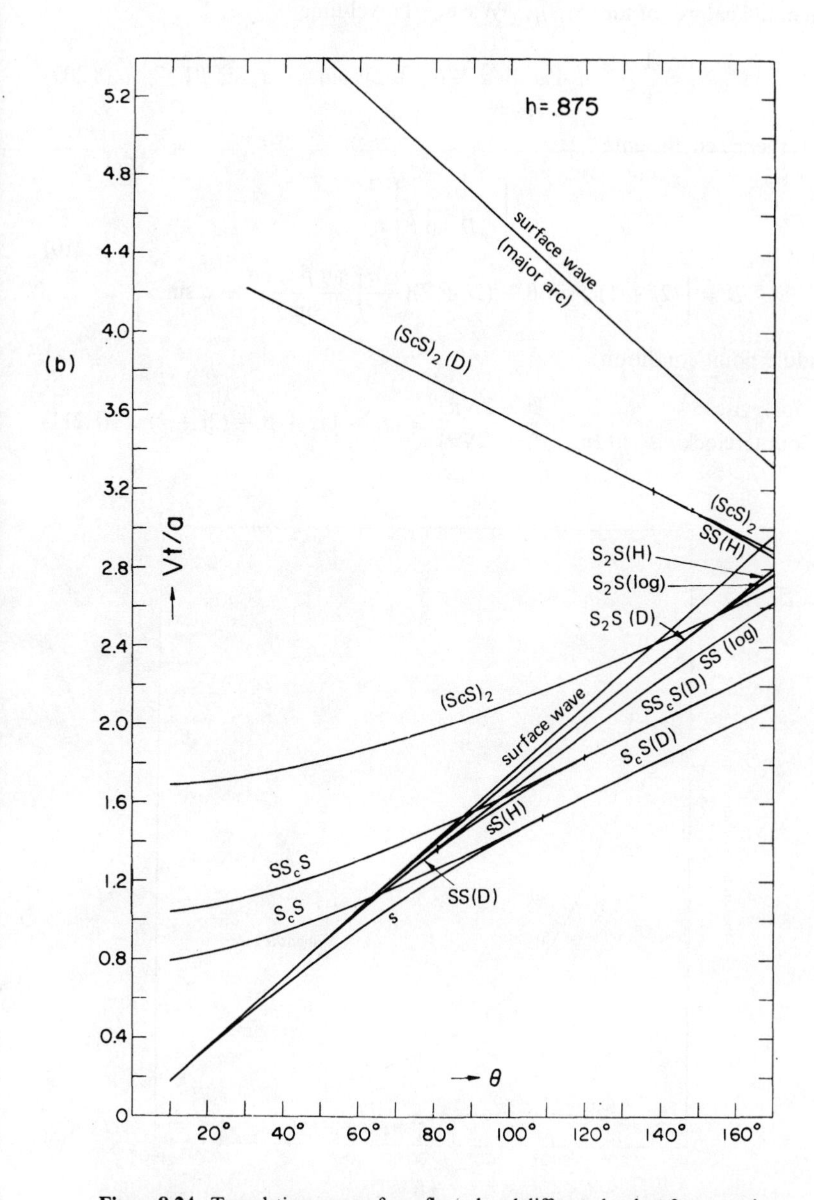

Figure 8.24. Travel-time curves for reflected and diffracted pulses from a point source in the shell: (b) Buried source at $r_h/a = 0.875$. (H) stands for $g'(t)$, (log) stands for $\hat{g}'(t)$, and (D) is core diffraction. Time is in units of (a/v).

8.6.6.2. Features of the $(ScS)_{j+1}$ Wave. Travel-time:

$$t = \frac{1}{v}\left[(2j+1)a\sin\alpha - (2j+2)b\sin\gamma + r_h\sin\beta\right]. \tag{8.209}$$

Divergence coefficient:

$$\left[\frac{\cos\alpha}{r_h D\sin\theta}\right]^{1/2},$$
$$D = \left[(2j+1)r_h\sin\beta - (2j+2)\left(\frac{r_h a}{b}\right)\frac{\sin\beta\sin\alpha}{\sin\gamma} + a\sin\alpha\right]. \tag{8.210}$$

Saddle point condition:

$$\begin{matrix}\text{Clockwise} & \theta + 2N\pi \\ \text{Counterclockwise} & (2\pi-\theta) + 2N\pi\end{matrix}\Bigg\} = (2j+1)\alpha + \beta - (2j+2)\gamma. \tag{8.211}$$

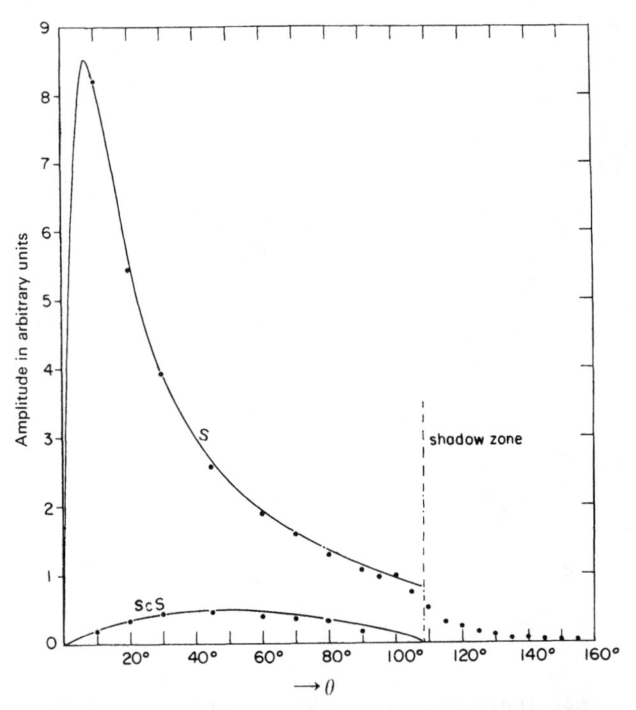

Figure 8.25. Variation with epicentral distance of amplitudes (in arbitrary units) of the direct S and the ScS pulses as obtained by steepest descents analysis (initial motion). The circles indicate amplitudes read from theoretical seismograms. Torque source is at $h = 0.875$.

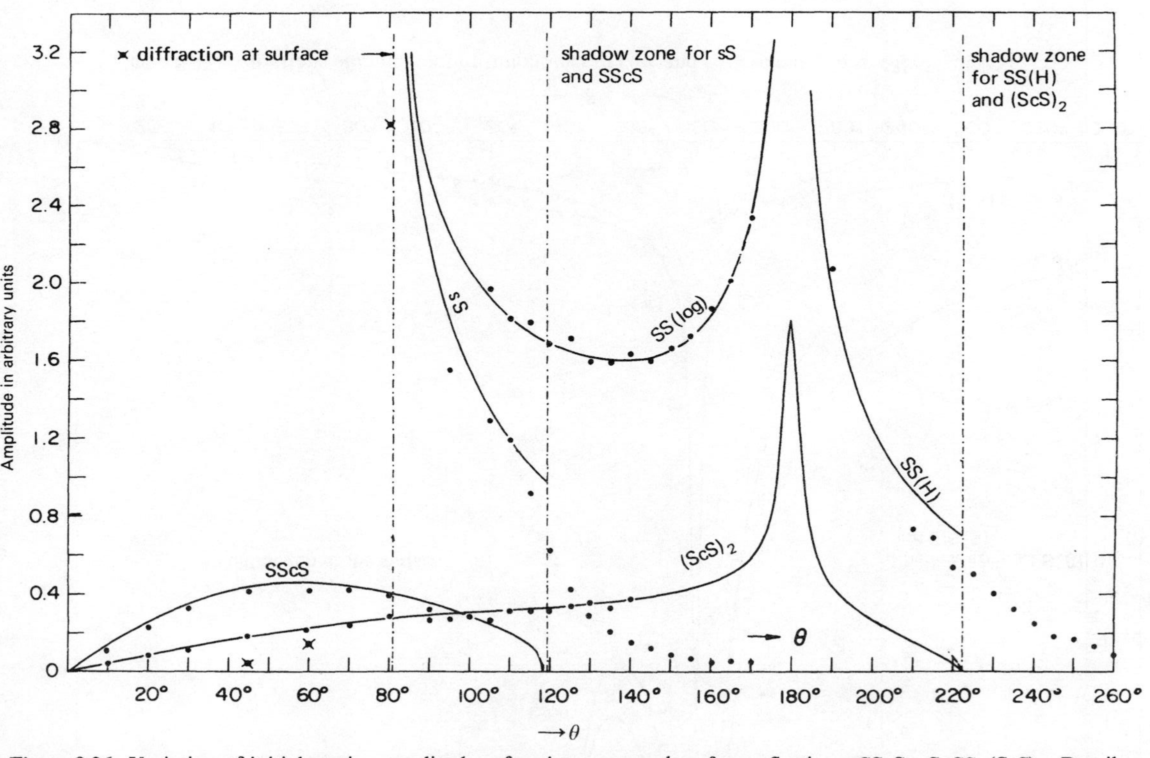

Figure 8.26. Variation of initial motion amplitudes of various core and surface reflections: *SScS*, *sS*, *SS*, $(ScS)_2$. Details as in Fig. 8.25 and notations as in Fig. 8.24.

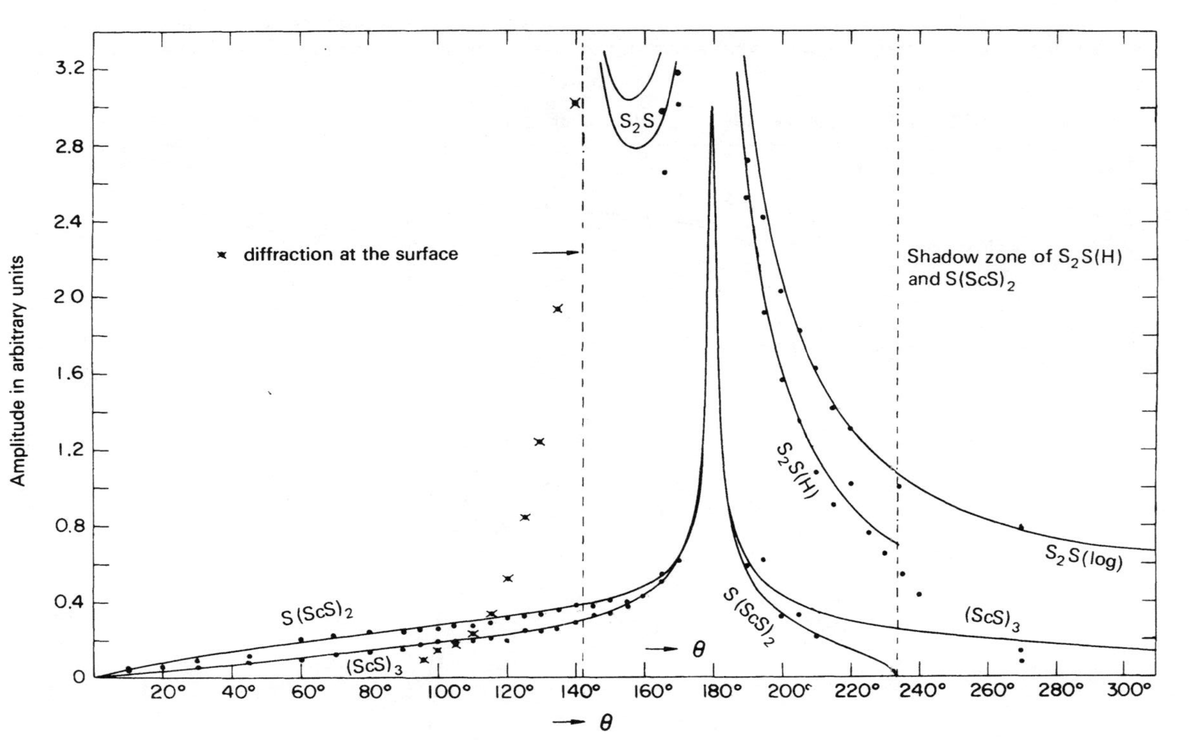

Figure 8.27. Variations of initial motion amplitudes of core and surface reflections: $S(ScS)_2$, $(ScS)_3$, S_2S.

The direct S pulse arrives at distances $0 \leq \theta \leq \theta_s$ where θ_s marks the beginning of the shadow zone

$$\theta_s = \cos^{-1}\frac{b}{a} + \cos^{-1}\frac{b}{r_h}.$$

The ScS wave arrives up to the same distance. For $r_h = a$ (surface source), $\theta_s = 2\cos^{-1}(b/a) = 113.987°$ in the earth. The travel times of the body waves are shown in Fig. 8.24.

8.6.6.3. Diffraction at the Core. The diffracted wave has been treated in Section 8.6.1. Its amplitude and travel times are shown in Figs. 8.24 and 8.25. These pulses are denoted as $ScS(D)$, $SS(D)$, etc. Note that the pulses SS, $(ScS)_2$, which travel in the counterclockwise direction, have a common shadow at 222.4°.

8.6.6.4. Diffractions at the Surface. According to GEA, the rays reflected once at the surface are disallowed at $\theta < 81°$ and the amplitude curves for Ss and SS do not start before 81°. However, diffracted pulses $SS(D)$ do arrive at $\theta < 81°$. These amplitudes can be read from theoretical seismograms. They are shown by dots in Figs. 8.26 and 8.27. Other surface reflections, such as $S_jS(D)$, may have large amplitudes. Figure 8.24b shows travel-time curves from a deep source. The continuation of the curves beyond 180° is possible because arrivals along the major arc may occur.

Seismologic observations show that in the case of deep-focus earthquakes the once-reflected waves sS and SS (also pP and PP) differ in shape on arrival at the surface. We have treated this problem in detail in Section 8.5 when dealing with the homogeneous sphere. A similar phenomenon takes place in the present case.

8.7. Generalized Rays in Spherical-Earth Models[5]

In this section we intend to apply the concept of a generalized ray to wave propagation in a sphere. As in the parallel case of a half-space, we begin with the interaction of a spherical elastic wave with a single spherical interface.

8.7.1. Fields of Multipolar Sources in the Presence of a Single Interface

In the following we shall formulate the problem in terms of the Laplace-transformed displacements in order to secure convergence of the expansion of the solution in infinite series. Later, when dealing with a particular generalized ray, it will be more convenient to return to the frequency domain.

[5] From Section 8.7 to the end of Chap. 8 we denote the longitudinal and shear velocities by α and β respectively. A general velocity is denoted by v.

Consider an elastic nongravitating sphere composed of $(N-1)$ concentric layers enclosing a homogeneous core. We introduce a right-handed spherical coordinate system $(\mathbf{e}_r, \mathbf{e}_\theta, \mathbf{e}_\phi)$ with its origin at the sphere's center and a source at $r = r_h$. The Laplace-transformed field in each homogeneous stratum is of the form

$$\bar{\mathbf{u}}_n(s) = \text{grad}\, \bar{\phi}_n + \text{curl}(\bar{\psi}_n \mathbf{r}) + \text{curl curl}(\bar{\chi}_n \mathbf{r}), \tag{8.212}$$

$$\mathbf{e}_r \cdot \bar{\mathfrak{T}}_n(s) = \lambda_n \mathbf{e}_r \,\text{div}\, \bar{\mathbf{u}}_n + \mu_n \mathbf{e}_r \cdot [\nabla \bar{\mathbf{u}}_n + \bar{\mathbf{u}}_n \nabla], \qquad n = 1, 2, 3, \ldots, N \tag{8.213}$$

where $\bar{\mathbf{u}}_n$ is the transformed displacement vector in the nth layer; $\bar{\mathfrak{T}}_n$ is the corresponding transformed stress tensor; λ_n, μ_n are the Lamé parameters; and $\bar{\phi}$, $\bar{\psi}$, $\bar{\chi}$ satisfy the transformed scalar wave equations

$$\nabla^2 \bar{\phi}_n - \frac{s^2}{\alpha_n^2} \bar{\phi}_n = 0, \qquad \nabla^2 \bar{\psi}_n - \frac{s^2}{\beta_n^2} \bar{\psi}_n = 0, \qquad \nabla^2 \bar{\chi}_n - \frac{s^2}{\beta_n^2} \bar{\chi}_n = 0 \tag{8.214}$$

with

$$\bar{\phi}_n = \int_0^\infty e^{-st} \phi_n(t)\, dt,$$

etc. The longitudinal and shear-wave velocities in the nth layer are denoted by α_n and β_n, respectively.

We introduce the spherical vector harmonics as in Eq. (2.71)

$$\mathbf{P}_{m,l} = \mathbf{e}_r Y_{m,l}, \qquad \sqrt{[l(l+1)]}\mathbf{B}_{m,l} = \left\{\mathbf{e}_\theta \frac{\partial}{\partial\theta} + \mathbf{e}_\phi \frac{1}{\sin\theta} \frac{\partial}{\partial\phi}\right\} Y_{m,l},$$
$$Y_{m,l} = P_l^m(\cos\theta) e^{im\phi}, \ \sqrt{[l(l+1)]}\mathbf{C}_{m,l} = \left\{\mathbf{e}_\theta \frac{1}{\sin\theta} \frac{\partial}{\partial\phi} - \mathbf{e}_\phi \frac{\partial}{\partial\theta}\right\} Y_{m,l}, \tag{8.215}$$

and the transformed eigenfunctions

$$\bar{\phi}^+_{m,l}(s) = Y_{m,l}(\theta, \phi) i_l\left(\frac{s}{\alpha} r\right); \qquad \bar{\phi}^-_{m,l}(s) = Y_{m,l}(\theta, \phi) k_l\left(\frac{s}{\alpha} r\right), \tag{8.216}$$

etc., where $P_l^m(\cos\theta)$ is the associated Legendre polynomial of the first kind and $i_l(x)$, $k_l(x)$ are the modified spherical Bessel functions of the first and second kinds, respectively. Equations (8.212) and (8.213) then assume the form

$$\bar{\mathbf{u}}_n(s) = \sum_{m,l} \bar{\mathbf{u}}_{mln}(s), \qquad \bar{\mathfrak{T}}_n \cdot \mathbf{e}_r = \sum_{m,l} \bar{\mathbf{T}}_{mln}(s), \tag{8.217}$$

where

$$\bar{\mathbf{u}}_{mln}(s) = \bar{x}_{mln} \mathbf{P}_{m,l} + \bar{y}_{mln} \sqrt{[l(l+1)]}\mathbf{B}_{m,l} + \bar{z}_{mln} \sqrt{[l(l+1)]}\mathbf{C}_{m,l},$$
$$\bar{\mathbf{T}}_{mln}(s) = \bar{X}_{mln} \mathbf{P}_{m,l} + \bar{Y}_{mln} \sqrt{[l(l+1)]}\mathbf{B}_{m,l} + \bar{Z}_{mln} \sqrt{[l(l+1)]}\mathbf{C}_{m,l}; \tag{8.218}$$

$$\bar{x}_{mln} = [A'_{mln}k'_l(\hat{a}_n) + A''_{mln}i'_l(\hat{a}_n)] + \frac{l(l+1)}{\hat{b}_n}[B'_{mln}k_l(\hat{b}_n) + B''_{mln}i_l(\hat{b}_n)],$$

$$\bar{y}_{mln} = \frac{1}{\hat{a}_n}[A'_{mln}k_l(\hat{a}_n) + A''_{mln}i_l(\hat{a}_n)] + \left[B'_{mln}\left\{k'_l(\hat{b}_n) + \frac{1}{\hat{b}_n}k_l(\hat{b}_n)\right\}\right. \tag{8.219}$$

$$\left. + B''_{mln}\left\{i'_l(\hat{b}_n) + \frac{1}{\hat{b}_n}i_l(\hat{b}_n)\right\}\right],$$

$$\bar{z}_{mln} = C'_{mln}k_l(\hat{b}_n) + C''_{mln}i_l(\hat{b}_n);$$

$$\bar{X}_{mln} = 2\frac{s}{\alpha_n}\mu_n[A'_{mln}F^-_{l_2}(\hat{a}_n) + A''_{mln}F^+_{l_2}(\hat{a}_n)]$$

$$+ 2\frac{s}{\beta_n}\mu_n l(l+1)[B'_{mln}F^-_{l_1}(\hat{b}_n) + B''_{mln}F^+_{l_1}(\hat{b}_n)],$$

$$\bar{Y}_{mln} = 2\frac{s}{\alpha_n}\mu_n[A'_{mln}F^-_{l_1}(\hat{a}_n) + A''_{mln}F^+_{l_1}(\hat{a}_n)] + \frac{s}{\beta_n}\mu_n[B'_{mln}F^-_{l_3}(\hat{b}_n) \tag{8.220}$$

$$+ B''_{mln}F^+_{l_3}(\hat{b}_n)],$$

$$\bar{Z}_{mln} = \hat{b}_n\frac{s}{\beta_n}\mu_n[C'_{mln}F^-_{l_1}(\hat{b}_n) + C''_{mln}F^+_{l_1}(\hat{b}_n)], \qquad \hat{a}_n = \frac{s}{\alpha_n}r, \qquad \hat{b}_n = \frac{s}{\beta_n}r,$$

and A'_{mln}, A''_{mln}, B'_{mln}, B''_{mln}, C'_{mln}, C''_{mln} are the six coefficients of each layer, to be determined from the boundary and source conditions. A prime denotes differentiation with respect to the argument and

$$F^{\pm}_{l_1}(\eta) = \frac{1}{\eta}\left[\frac{d}{d\eta} - \frac{1}{\eta}\right]f^{\pm}_l(\eta),$$

$$F^{\pm}_{l_2}(\eta) = \left[\frac{d^2}{d\eta^2} + \frac{\lambda}{2\mu}\right]f^{\pm}_l(\eta),$$

$$F^{\pm}_{l_3}(\eta) = \left[\frac{d^2}{d\eta^2} + \frac{(l-1)(l+2)}{\eta^2}\right]f^{\pm}_l(\eta),$$

$$F^{\pm}_{l_4}(\eta) = \left(\frac{d}{d\eta} + \frac{1}{\eta}\right)f^{\pm}_l(\eta), \tag{8.221}$$

$$F^{\pm}_{l_5}(\eta) = \frac{1}{\eta}f^{\pm}_l(\eta),$$

$$F^{\pm}_{l_6}(\eta) = \frac{d}{d\eta}f^{\pm}_l(\eta),$$

$$F^{\pm}_{l_7}(\eta) = \frac{d^2}{d\eta^2}f^{\pm}_l(\eta).$$

In Eq. (8.221), $f^+_l(\eta) = i_l(\eta)$, $f^-_l(\eta) = k_l(\eta)$, and $\eta = \hat{a}_n$ or $\hat{b}_n$.

Next, it is necessary to develop the source's displacement and stress fields into series of vector spherical harmonics. We therefore write, similar to Eqs. (8.217) and (8.218),

$$\bar{\mathbf{u}}^{(0)} = \sum_{m,l} \bar{\mathbf{u}}_{m,l}^{(0)}(s), \qquad \mathbf{e}_r \cdot \mathfrak{T}^{(0)}(s) = \sum_{m,l} \mathbf{T}_{m,l}^{(0)}(s), \tag{8.222}$$

where the summation over m extends over all values appropriate to a given multipolar point-source and the said expansion has the form

$$\begin{aligned} \bar{\mathbf{u}}_{m,l}^{(0)}(s) &= \bar{x}_{m,l}^{(0)}\mathbf{P}_{m,l} + \bar{y}_{m,l}^{(0)}\sqrt{[l(l+1)]}\mathbf{B}_{m,l} + \bar{z}_{m,l}^{(0)}\sqrt{[l(l+1)]}\mathbf{C}_{m,l}, \\ \bar{\mathbf{T}}_{m,l}^{(0)}(s) &= \bar{X}_{m,l}^{(0)}\mathbf{P}_{m,l} + \bar{Y}_{m,l}^{(0)}\sqrt{[l(l+1)]}\mathbf{B}_{m,l} + \bar{Z}_{m,l}^{(0)}\sqrt{[l(l+1)]}\mathbf{C}_{m,l}. \end{aligned} \tag{8.223}$$

We shall derive the unknown coefficients in Eqs. (8.223) for two sources under whose action the sphere remains in equilibrium. The first source is that of an explosion at $(r_h, 0, 0)$. It is represented in terms of the spherical eigenvector $\bar{\mathbf{L}}_{0,0}^{-}$ (Table 4.4)

$$\bar{\mathbf{u}}^{(0)} = -\frac{im_0(s^2/\alpha^2)}{4\pi(\lambda+2\mu)}\bar{\mathbf{L}}_{0,0}^{-}\left(\frac{s}{\alpha}R\right), \qquad m_0 = m_0(s), \tag{8.224}$$

where R is the distance from the source to the sensor. The transformation to a coordinate system having its origin at the sphere's center is given by (App. J)

$$\begin{aligned} \bar{\mathbf{L}}_{0,0}^{-}\left(\frac{s}{\alpha}R\right) = \frac{2}{\pi}\sum_{l=0}^{\infty}(2l+1)f_l^{\varepsilon}\left(\frac{s}{\alpha}r_h\right)\Bigg[F_{l_6}^{-\varepsilon}\left(\frac{s}{\alpha}r\right)\mathbf{P}_{0,l} \\ + \sqrt{[l(l+1)]}\mathbf{B}_{0,l}F_{l_5}^{-\varepsilon}\left(\frac{s}{\alpha}r\right)\Bigg], \end{aligned} \tag{8.225}$$

$$\varepsilon = \operatorname{sgn}(r - r_h).$$

The second source is a tangential dislocation at $(r_h, 0, 0)$ for which we have (Table 4.4)

$$\bar{\mathbf{u}}^{(0)} = -i\frac{U_0\,dS}{24\pi}\left(\frac{s^2}{\beta^2}\right) \times \begin{cases} -2i\left(\dfrac{\beta}{\alpha}\right)^4\bar{\mathbf{L}}_{2,2}^{-}\left(\dfrac{s}{\alpha}R\right) - i\bar{\mathbf{N}}_{2,2}^{-}\left(\dfrac{s}{\beta}R\right) & \text{case I} \quad (\lambda = 0, \delta = 90°) \\ -4i\left(\dfrac{\beta}{\alpha}\right)^4\bar{\mathbf{L}}_{1,2}^{-}\left(\dfrac{s}{\alpha}R\right) - 2i\bar{\mathbf{N}}_{1,2}^{-}\left(\dfrac{s}{\beta}R\right) & \text{case II} \quad (\lambda = 90°, \delta = 90°) \\ \left.\begin{array}{l} -2\left(\dfrac{\beta}{\alpha}\right)^4\left[3\bar{\mathbf{L}}_{0,2}^{-}\left(\dfrac{s}{\alpha}R\right) + \dfrac{1}{2}\bar{\mathbf{L}}_{2,2}^{-}\left(\dfrac{s}{\alpha}R\right)\right] \\ -3\bar{\mathbf{N}}_{0,2}^{-}\left(\dfrac{s}{\beta}R\right) - \dfrac{1}{2}\bar{\mathbf{N}}_{2,2}^{-}\left(\dfrac{s}{\beta}R\right) \end{array}\right\} & \text{case III} \ (\lambda = 90°, \delta = 45°) \end{cases} \tag{8.226}$$

where $U_0 = U_0(s)$,

$$\bar{\mathbf{N}}_{0,2}^{-}\left(\frac{s}{\beta}R\right) = \sum_{l=0}^{\infty} 3H_{l,0}^{\varepsilon}(\hat{b}_h)[l(l+1)F_{l_5}^{-\varepsilon}(\hat{b})\mathbf{P}_{0,l} + F_{l_4}^{-\varepsilon}(\hat{b})\sqrt{\{l(l+1)\}}\mathbf{B}_{0,l}],$$

$$\bar{\mathbf{N}}_{1,2}^{-}\left(\frac{s}{\beta}R\right) = \sum_{l=1}^{\infty} 3H_{l,1}^{\varepsilon}(\hat{b}_h)[l(l+1)F_{l_5}^{-\varepsilon}(\hat{b})\mathbf{P}_{1,l} + F_{l_4}^{-\varepsilon}(\hat{b})\sqrt{\{l(l+1)\}}\mathbf{B}_{1,l}]$$

$$- 3\hat{b}_h \sum_{l=1}^{\infty} \frac{2l+1}{l(l+1)} E_{l,1}^{\varepsilon}(\hat{b}_h) f_l^{-\varepsilon}(\hat{b})\sqrt{\{l(l+1)\}}\mathbf{C}_{1,l}, \tag{8.227}$$

$$\bar{\mathbf{N}}_{2,2}^{-}\left(\frac{s}{\beta}R\right) = \sum_{l=2}^{\infty} 3H_{l,2}^{\varepsilon}(\hat{b}_h)[l(l+1)F_{l_5}^{-\varepsilon}(\hat{b})\mathbf{P}_{2,l} + F_{l_4}^{-\varepsilon}(\hat{b})\sqrt{\{l(l+1)\}}\mathbf{B}_{2,l}]$$

$$- 6\hat{b}_h \sum_{l=2}^{\infty} \frac{2l+1}{l(l+1)} E_{l,2}^{\varepsilon}(\hat{b}_h) f_l^{-\varepsilon}(\hat{b})\sqrt{\{l(l+1)\}}\mathbf{C}_{2,l};$$

$$\bar{\mathbf{L}}_{0,2}^{-}\left(\frac{s}{\alpha}R\right) = \sum_{l=0}^{\infty} 3(2l+1)E_{l,0}^{\varepsilon}(\hat{a}_h)[F_{l_6}^{-\varepsilon}(\hat{a})\mathbf{P}_{0,l} + F_{l_5}^{-\varepsilon}(\hat{a})\sqrt{\{l(l+1)\}}\mathbf{B}_{0,l}],$$

$$\bar{\mathbf{L}}_{1,2}^{-}\left(\frac{s}{\alpha}R\right) = \sum_{l=1}^{\infty} 3(2l+1)E_{l,1}^{\varepsilon}(\hat{a}_h)[F_{l_6}^{-\varepsilon}(\hat{a})\mathbf{P}_{1,l} + F_{l_5}^{-\varepsilon}(\hat{a})\sqrt{\{l(l+1)\}}\mathbf{B}_{1,l}],$$

$$\bar{\mathbf{L}}_{2,2}^{-}\left(\frac{s}{\alpha}R\right) = \sum_{l=2}^{\infty} 3(2l+1)E_{l,2}^{\varepsilon}(\hat{a}_h)[F_{l_6}^{-\varepsilon}(\hat{a})\mathbf{P}_{2,l} + F_{l_5}^{-\varepsilon}(\hat{a})\sqrt{\{l(l+1)\}}\mathbf{B}_{2,l}]; \tag{8.228}$$

$$E_{l,0}^{\varepsilon}(\eta) = \frac{2i}{\pi}\left\{\frac{1}{2}l(l-1)\frac{f_l^{\varepsilon}(\eta)}{\eta^2} - \varepsilon\frac{f_{l+1}^{\varepsilon}(\eta)}{\eta} + \frac{f_l^{\varepsilon}(\eta)}{3}\right\},$$

$$E_{l,1}^{\varepsilon}(\eta) = \frac{2i}{\pi}\left\{(l-1)\frac{f_l^{\varepsilon}(\eta)}{\eta^2} + \varepsilon\frac{f_{l+1}^{\varepsilon}(\eta)}{\eta}\right\}, \tag{8.229}$$

$$E_{l,2}^{\varepsilon}(\eta) = \frac{2i}{\pi}\cdot\frac{f_l^{\varepsilon}(\eta)}{\eta^2}, \qquad \hat{a}_h = \frac{s}{\alpha}r_h, \qquad \hat{b}_h = \frac{s}{\beta}r_h;$$

$$H_{l,0}^{\varepsilon} = (2l+1)E_{l,0}^{\varepsilon}(\eta) + i\eta E_{l-1,0}^{\varepsilon}(\eta) + i\eta E_{l+1,0}^{\varepsilon}(\eta),$$

$$H_{l,1}^{\varepsilon} = (2l+1)E_{l,1}^{\varepsilon}(\eta) + i\eta\frac{l-1}{l}E_{l-1,1}^{\varepsilon}(\eta) + i\eta\frac{l+2}{l+1}E_{l+1,1}^{\varepsilon}(\eta), \tag{8.230}$$

$$H_{l,2}^{\varepsilon} = (2l+1)E_{l,2}^{\varepsilon}(\eta) + i\eta\frac{l-2}{l}E_{l-1,2}^{\varepsilon}(\eta) + i\eta\frac{l+3}{l+1}E_{l+1,2}^{\varepsilon}(\eta).$$

We therefore obtain

$$\bar{x}^{(0)}_{m,l} = (2l+1)I^{\varepsilon}_{m,l}\left(\frac{s}{\alpha}r_h\right)F^{-\varepsilon}_{l_6}\left(\frac{s}{\alpha}r\right) + l(l+1)J^{\varepsilon}_{m,l}\left(\frac{s}{\beta}r_h\right)F^{-\varepsilon}_{l_5}\left(\frac{s}{\beta}r\right),$$

$$\bar{y}^{(0)}_{m,l} = (2l+1)I^{\varepsilon}_{m,l}\left(\frac{s}{\alpha}r_h\right)F^{-\varepsilon}_{l_5}\left(\frac{s}{\alpha}r\right) + J^{\varepsilon}_{m,l}\left(\frac{s}{\beta}r_h\right)F^{-\varepsilon}_{l_4}\left(\frac{s}{\beta}r\right), \tag{8.231}$$

$$\bar{z}^{(0)}_{m,l} = \frac{2l+1}{l(l+1)}K^{\varepsilon}_{m,l}\left(\frac{s}{\beta}r_h\right)f^{-\varepsilon}_{l}\left(\frac{s}{\beta}r\right),$$

where

$$I^{\varepsilon}_{m,l}(\eta) = \begin{cases} \dfrac{-im_0 s^2}{4\pi\alpha^2(\lambda+2\mu)}\cdot\dfrac{2}{\pi}f^{\varepsilon}_l(\eta) & \text{explosion} \quad (m=0) \\ -\dfrac{s^2U_0\,dS}{4\pi\beta^2}\left(\dfrac{\beta}{\alpha}\right)^4 E^{\varepsilon}_{l,2}(\eta) & \text{case I} \quad (m=2) \\ -\dfrac{s^2U_0\,dS}{2\pi\beta^2}\left(\dfrac{\beta}{\alpha}\right)^4 E^{\varepsilon}_{l,1}(\eta) & \text{case II} \quad (m=1) \\ \dfrac{is^2U_0\,dS}{8\pi\beta^2}\left(\dfrac{\beta}{\alpha}\right)^4 [6E^{\varepsilon}_{l,0}(\eta) + E^{\varepsilon}_{l,2}(\eta)] & \text{case III} \quad (m=0,2), \end{cases} \tag{8.232}$$

$$J^{\varepsilon}_{m,l}(\eta) = \begin{cases} 0 & \text{explosion} \\ -\dfrac{s^2U_0\,dS}{8\pi\beta^2}H^{\varepsilon}_{l,2}(\eta) & \text{case I} \\ -\dfrac{s^2U_0\,dS}{4\pi\beta^2}H^{\varepsilon}_{l,1}(\eta) & \text{case II} \\ i\dfrac{s^2U_0\,dS}{8\pi\beta^2}\left[3H^{\varepsilon}_{l,0}(\eta) + \dfrac{1}{2}H^{\varepsilon}_{l,2}(\eta)\right] & \text{case III,} \end{cases} \tag{8.233}$$

$$K^{\varepsilon}_{m,l}(\eta) = \begin{cases} 0 & \text{explosion} \\ \dfrac{s^2U_0\,dS}{4\pi\beta^2}\eta E^{\varepsilon}_{l,2}(\eta) & \text{case I} \\ \dfrac{s^2U_0\,dS}{4\pi\beta^2}\eta E^{\varepsilon}_{l,1}(\eta) & \text{case II} \\ -\dfrac{is^2U_0\,dS}{8\pi\beta^2}\eta E^{\varepsilon}_{l,2}(\eta) & \text{case III,} \end{cases} \tag{8.234}$$

$$m_0 = m_0(s), \qquad U_0 = U_0(s).$$

Note that summation corresponding to each member of Eqs. (8.223) starts at $l = m$. The appropriate m value is given in Eqs. (8.232), (8.233), and (8.234).

The stress coefficients of the source are obtained from Eqs. (8.231) with the aid of the equations

$$\bar{X}^{(0)}_{m,l} = \left[(\lambda + 2\mu)\frac{\partial}{\partial r} + \frac{2\lambda}{r}\right]\bar{x}^{(0)}_{m,l} - \frac{\lambda}{r}\, l(l+1)\bar{y}^{(0)}_{m,l},$$

$$\bar{Y}^{(0)}_{m,l} = \frac{\mu}{r}\bar{x}^{(0)}_{m,l} + \mu\left[\frac{\partial}{\partial r} - \frac{1}{r}\right]\bar{y}^{(0)}_{m,l}, \tag{8.235}$$

$$\bar{Z}^{(0)}_{m,l} = \mu\left[\frac{\partial}{\partial r} - \frac{1}{r}\right]\bar{z}^{(0)}_{m,l}.$$

Explicitly, with $\hat{a} = (s/\alpha)r$, $\hat{b} = (s/\beta)r$,

$$\bar{X}^{(0)}_{m,l} = 2\mu\frac{s}{\alpha}(2l+1)I^{\varepsilon}_{m,l}(\hat{a}_h)F^{-\varepsilon}_{l_2}(\hat{a}) + 2\mu\frac{s}{\beta}\, l(l+1)J^{\varepsilon}_{m,l}(\hat{b}_h)F^{-\varepsilon}_{l_1}(\hat{b}),$$

$$\bar{Y}^{(0)}_{m,l} = 2\mu\frac{s}{\alpha}(2l+1)I^{\varepsilon}_{m,l}(\hat{a}_h)F^{-\varepsilon}_{l_1}(\hat{a}) + \mu\frac{s}{\beta}J^{\varepsilon}_{m,l}(\hat{b}_h)F^{-\varepsilon}_{l_3}(\hat{b}), \tag{8.236}$$

$$\bar{Z}^{(0)}_{m,l} = \frac{s}{\beta}\mu\frac{2l+1}{l(l+1)}K^{\varepsilon}_{m,l}(\hat{b}_h)\hat{b}F^{-\varepsilon}_{l_1}(\hat{b}).$$

The parameters α, β, λ, μ in Eqs. (8.224)–(8.236) are taken at the source level. The source's time dependence is embedded in $U_0(s)$ and $m_0(s)$, which are characteristic of the dislocation and the explosion sources. We adopt the convention that upon the transformation of Eq. (8.217) and (8.222) back to the time domain, $e^{im\phi}$ is replaced by $\cos m\phi$ and $(-i)e^{im\phi}$ by $\sin m\phi$.

Consider next a configuration that consists of two elastic media separated by a spherical boundary at $r = r_{q-1}$ and a point source at level $r = r_h < r_{q-1}$ (Fig. 8.28). The boundary conditions at this interface are

$$\begin{aligned} {}_{q-1}\bar{x}_{m,l} &= {}_q\bar{x}_{m,l} + \bar{x}^{(0)}_{m,l}; & {}_{q-1}\bar{X}_{m,l} &= {}_q\bar{X}_{m,l} + \bar{X}^{(0)}_{m,l},\\ {}_{q-1}\bar{y}_{m,l} &= {}_q\bar{y}_{m,l} + \bar{y}^{(0)}_{m,l}; & {}_{q-1}\bar{Y}_{m,l} &= {}_q\bar{Y}_{m,l} + \bar{Y}^{(0)}_{m,l},\\ {}_{q-1}\bar{z}_{m,l} &= {}_q\bar{z}_{m,l} + \bar{z}^{(0)}_{m,l}; & {}_{q-1}\bar{Z}_{m,l} &= {}_q\bar{Z}_{m,l} + \bar{Z}^{(0)}_{m,l}. \end{aligned} \tag{8.237}$$

The displacement and stress fields above the interface are equal to $\bar{\mathbf{u}}_{q-1}$ and $\bar{\mathfrak{T}}_{q-1}$, respectively. Below the interface they are equal to $\bar{\mathbf{u}}^{(0)} + \bar{\mathbf{u}}_q$ and $\bar{\mathfrak{T}}^{(0)} + \bar{\mathfrak{T}}_q$, respectively. The undetermined coefficients are ${}_{q-1}A'_{m,l}$, ${}_{q-1}B'_{m,l}$, ${}_{q-1}C'_{m,l}$ above the interface and ${}_qA''_{m,l}$, ${}_qB''_{m,l}$, ${}_qC''_{m,l}$ below it.

Similarly for the case where the source is above the interface ($r_h > r_{q-1}$), the boundary conditions at the interface are

$$\begin{aligned} {}_{q-1}\bar{x}_{m,l} + \bar{x}^{(0)}_{m,l} &= {}_q\bar{x}_{m,l}; & {}_{q-1}\bar{X}_{m,l} + \bar{X}^{(0)}_{m,l} &= {}_q\bar{X}_{m,l},\\ {}_{q-1}\bar{y}_{m,l} + \bar{y}^{(0)}_{m,l} &= {}_q\bar{y}_{m,l}; & {}_{q-1}\bar{Y}_{m,l} + \bar{Y}^{(0)}_{m,l} &= {}_q\bar{Y}_{m,l},\\ {}_{q-1}\bar{z}_{m,l} + \bar{z}^{(0)}_{m,l} &= {}_q\bar{z}_{m,l}; & {}_{q-1}\bar{Z}_{m,l} + \bar{Z}^{(0)}_{m,l} &= {}_q\bar{Z}_{m,l}. \end{aligned} \tag{8.238}$$

Substituting from Eqs. (8.219), (8.220), and (8.231) into Eqs. (8.237) and (8.238), we solve the six equations for the above six coefficients. This system of equations

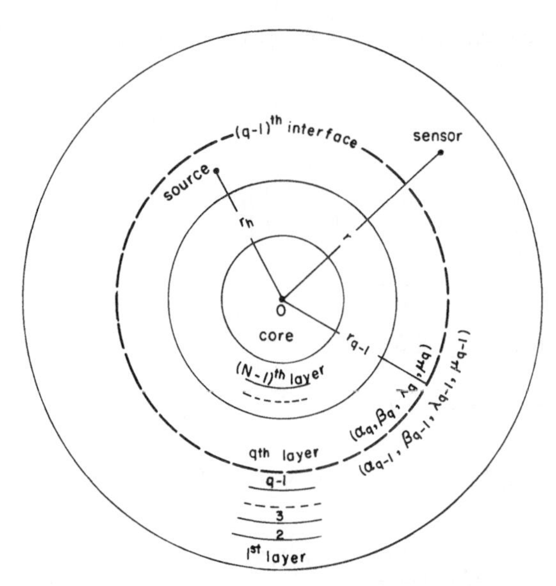

Figure 8.28. Layers and interfaces in a sphere composed of N-1 solid layers enclosing a core.

splits into two independent sets: four equations for the P–SV waves and two equations for the SH waves. The solutions, written in matrix form, are:

$$\delta = \operatorname{sgn}(r_{q-1} - r_h) = +1 \quad \text{(source below the interface)}$$

$$\begin{bmatrix} {}_{q-1}A'_{m,l} \\ {}_{q-1}B'_{m,l} \\ {}_{q}A''_{m,l} \\ {}_{q}B''_{m,l} \end{bmatrix} = (2l+1) I^{+}_{m,l}\left(\frac{s}{\alpha_h} r_h\right) \begin{bmatrix} T^{+}_{PP} \\ T^{+}_{PS} \\ R^{-}_{PP} \\ R^{-}_{PS} \end{bmatrix} + J^{+}_{m,l}\left(\frac{s}{\beta_h} r_h\right) \begin{bmatrix} T^{+}_{SP} \\ T^{+}_{SSV} \\ R^{-}_{SP} \\ R^{-}_{SSV} \end{bmatrix}, \tag{8.239}$$

$$\begin{bmatrix} {}_{q}C''_{m,l} \\ {}_{q-1}C'_{m,l} \end{bmatrix} = \frac{2l+1}{l(l+1)} K^{+}_{m,l}\left(\frac{s}{\beta_h} r_h\right) \begin{bmatrix} R^{-}_{SSH} \\ T^{+}_{SSH} \end{bmatrix}; \tag{8.240}$$

$$\delta = \operatorname{sgn}(r_{q-1} - r_h) = -1 \quad \text{(source above interface)}$$

$$\begin{bmatrix} {}_{q-1}A'_{m,l} \\ {}_{q-1}B'_{m,l} \\ {}_{q}A''_{m,l} \\ {}_{q}B''_{m,l} \end{bmatrix} = (2l+1) I^{-}_{m,l}\left(\frac{s}{\alpha_h} r_h\right) \begin{bmatrix} R^{+}_{PP} \\ R^{+}_{PS} \\ T^{-}_{PP} \\ T^{-}_{PS} \end{bmatrix} + J^{-}_{m,l}\left(\frac{s}{\beta_h} r_h\right) \begin{bmatrix} R^{+}_{SP} \\ R^{+}_{SSV} \\ T^{-}_{SP} \\ T^{-}_{SSV} \end{bmatrix}, \tag{8.241}$$

$$\begin{bmatrix} {}_{q}C''_{m,l} \\ {}_{q-1}C'_{m,l} \end{bmatrix} = \frac{2l+1}{l(l+1)} K^{-}_{m,l}\left(\frac{s}{\beta_h} r_h\right) \begin{bmatrix} T^{-}_{SSH} \\ R^{+}_{SSH} \end{bmatrix}. \tag{8.242}$$

The sign appearing as a superscript in R, T indicates the position of the source. For example, R^{+} is a "reflection up," that is, the source is above the interface, while T^+ is a "transmission up," the source being below the interface. In Eqs. (8.240) and (8.242) the R/T coefficients are:

$$T^{+}_{SSH} = \frac{(\hat{b}_q(\mu_q/\beta_q)F^{-}_{l_1}(\hat{b}_q)f_l^{+}(\hat{b}_q) - \hat{b}_q(\mu_q/\beta_q)F^{+}_{l_1}(\hat{b}_q)f_l^{-}(\hat{b}_q)}{\Delta_H},$$

$$R^{-\delta}_{SSH} = \frac{\hat{b}_q(\mu_q/\beta_q)F^{-\delta}_{l_1}(\hat{b}_q)f_l^{-\delta}(\hat{b}_{q-1}) - \hat{b}_{q-1}(\mu_{q-1}/\beta_{q-1})F^{-\delta}_{l_1}(\hat{b}_{q-1})f_l^{-\delta}(\hat{b}_q)}{\Delta_H},$$

where (8.243)

$$\hat{b}_q = (s/\beta_q)r_{q-1}, \qquad \hat{b}_{q-1} = (s/\beta_{q-1})r_{q-1},$$

$$\Delta_H = \hat{b}_{q-1}\frac{\mu_{q-1}}{\beta_{q-1}}F^{-}_{l_1}(\hat{b}_{q-1})f_l^{+}(\hat{b}_q) - \hat{b}_q\frac{\mu_q}{\beta_q}F^{+}_{l_1}(\hat{b}_q)f_l^{-}(\hat{b}_{q-1}).$$

In T^{-}_{SSH} we replace q by $q-1$ in the numerator. The explicit expressions for the P–SV field are algebraically more involved. The following procedure, however, saves some labor. We define

$$G^{\pm}_{l_i}\begin{Bmatrix}\hat{a}_r\\ \hat{b}_r\end{Bmatrix} = s\cdot\mu_r\begin{Bmatrix}\dfrac{2}{\alpha_r}\\ \dfrac{2l(l+1)}{\beta_r}\end{Bmatrix}F^{\pm}_{l_i}\begin{Bmatrix}\hat{a}_r\\ \hat{b}_r\end{Bmatrix}, \qquad i = 1, 2$$

$$G^{\pm}_{l_3}\begin{Bmatrix}\hat{a}_r\\ \hat{b}_r\end{Bmatrix} = s\cdot\mu_r\begin{Bmatrix}\dfrac{1}{\alpha_r}\\ \dfrac{1}{\beta_r}\end{Bmatrix}F^{\pm}_{l_3}\begin{Bmatrix}\hat{a}_r\\ \hat{b}_r\end{Bmatrix},$$

$$G^{\pm}_{l_i}\begin{Bmatrix}\hat{a}_r\\ \hat{b}_r\end{Bmatrix} = F^{\pm}_{l_i}\begin{Bmatrix}\hat{a}_r\\ \hat{b}_r\end{Bmatrix}, \qquad i = 4, 6$$

$$G^{\pm}_{l_5}\begin{Bmatrix}\hat{a}_r\\ \hat{b}_r\end{Bmatrix} = \begin{Bmatrix}1\\ l(l+1)\end{Bmatrix}F^{\pm}_{l_5}\begin{Bmatrix}\hat{a}_r\\ \hat{b}_r\end{Bmatrix}, \qquad (8.244)$$

and

$$\mathbf{U}_{\{\substack{1\\2}\}}(\hat{a}_r) = \begin{bmatrix}G^{\pm}_{l_6}(\hat{a}_r)\\ G^{\pm}_{l_5}(\hat{a}_r)\\ G^{\pm}_{l_2}(\hat{a}_r)\\ G^{\pm}_{l_1}(\hat{a}_r)\end{bmatrix}, \qquad \mathbf{V}_{\{\substack{1\\2}\}}(\hat{b}_r) = \begin{bmatrix}G^{\pm}_{l_5}(\hat{b}_r)\\ G^{\pm}_{l_4}(\hat{b}_r)\\ G^{\pm}_{l_1}(\hat{b}_r)\\ G^{\pm}_{l_3}(\hat{b}_r)\end{bmatrix}, \qquad (8.245)$$

where in the definitions of the $G_l^{\pm}$ we now have $f_l^{+}(\eta) = k_l^{(1)}(\eta)$, $f_l^{-}(\eta) = k_l^{(2)}(\eta)$ [Eq. (D.113)]. Incorporating the relation

$$i_l(z) = \frac{1}{2}\{k_l^{(1)}(z) - (-1)^l k_l^{(2)}(z)\}, \qquad k_l(z) = \frac{\pi}{2}k_l^{(2)}(z),$$

the R/T coefficients for a source below the interface become

$$T_{PP}^{+} = \frac{1}{\Delta} \|\mathbf{U}_2(\hat{a}_q); \mathbf{U}_1(\hat{a}_q) - (-1)^l \mathbf{U}_2(\hat{a}_q); \mathbf{V}_2(\hat{b}_p); \mathbf{V}_1(\hat{b}_q) - (-1)^l \mathbf{V}_2(\hat{b}_q)\|,$$

$$R_{PP}^{-} = -\frac{2}{\Delta} \|\mathbf{U}_2(\hat{a}_p); \mathbf{U}_2(\hat{a}_q); \mathbf{V}_2(\hat{b}_p); \mathbf{V}_1(\hat{b}_q) - (-1)^l \mathbf{V}_2(\hat{b}_q)\|, \ (p = q - 1)$$

$$T_{PS}^{+} = \frac{1}{\Delta} \|\mathbf{U}_2(\hat{a}_p); \mathbf{U}_1(\hat{a}_q) - (-1)^l \mathbf{U}_2(\hat{a}_q); \mathbf{U}_2(\hat{a}_q); \mathbf{V}_1(\hat{b}_q) - (-1)^l \mathbf{V}_2(\hat{b}_q)\|,$$

$$R_{PS}^{-} = -\frac{2}{\Delta} \|\mathbf{U}_2(\hat{a}_p); \mathbf{U}_1(\hat{a}_q) - (-1)^l \mathbf{U}_2(\hat{a}_q); \mathbf{V}_2(\hat{b}_p); \mathbf{U}_2(\hat{a}_q)\|, \tag{8.246}$$

$$T_{SP}^{+} = \frac{1}{\Delta} \|\mathbf{V}_2(\hat{b}_q); \mathbf{U}_1(\hat{a}_q) - (-1)^l \mathbf{U}_2(\hat{a}_q); \mathbf{V}_2(\hat{b}_p); \mathbf{V}_1(\hat{b}_q) - (-1)^l \mathbf{V}_2(\hat{b}_q)\|,$$

$$R_{SP}^{-} = -\frac{2}{\Delta} \|\mathbf{U}_2(\hat{a}_p); \mathbf{V}_2(\hat{b}_q); \mathbf{V}_2(\hat{b}_p); \mathbf{V}_1(\hat{b}_q) - (-1)^l \mathbf{V}_2(\hat{b}_q)\|,$$

$$T_{SSV}^{+} = \frac{1}{\Delta} \|\mathbf{U}_2(\hat{a}_p); \mathbf{U}_1(\hat{a}_q) - (-1)^l \mathbf{U}_2(\hat{a}_q); \mathbf{V}_2(\hat{b}_q); \mathbf{V}_1(\hat{b}_q) - (-1)^l \mathbf{V}_2(\hat{b}_q)\|,$$

$$R_{SSV}^{-} = -\frac{2}{\Delta} \|\mathbf{U}_2(\hat{a}_p); \mathbf{U}_1(\hat{a}_q) - (-1)^l \mathbf{U}_2(\hat{a}_q); \mathbf{V}_2(\hat{b}_p); \mathbf{V}_2(\hat{b}_q)\|,$$

$$\Delta = \|\mathbf{U}_2(\hat{a}_p); \mathbf{U}_1(\hat{a}_q) - (-)^l \mathbf{U}_2(\hat{a}_q); \mathbf{V}_2(\hat{b}_p); \mathbf{V}_1(\hat{b}_q) - (-)^l \mathbf{V}_2(\hat{b}_q)\|.$$

8.7.1.1. Generalized R/T Coefficients. When the source is above the interface, $\mathbf{U}_2$ and $\mathbf{V}_2$, wherever they appear alone in Eq. (8.246), should be replaced by $\mathbf{U}_1$ and $\mathbf{V}_1$ respectively. The sign of R/T changes accordingly. From Eqs. (8.218), (8.236), (8.238), (8.239)–(8.242) we obtain the total displacement field above and below the interface, where the expressions for $I_{m,l}$, $J_{m,l}$, and $K_{m,l}$ are to be taken from Eqs. (8.232)–(8.234):

$$\begin{aligned}(u_\phi)_{\text{incident}} &= ime^{im\phi} \sum_{l=m}^{\infty} (2l+1) I_{m,l}^{\varepsilon}(\hat{a}_h) F_{l_5}^{-\varepsilon}(\hat{a}) \frac{P_l^m(\cos\theta)}{\sin\theta} \\ &+ ime^{im\phi} \sum_{l=m}^{\infty} J_{m,l}^{\varepsilon}(\hat{b}_h) F_{l_4}^{-\varepsilon}(\hat{b}) \frac{P_l^m(\cos\theta)}{\sin\theta} \\ &- e^{im\phi} \sum_{l=m}^{\infty} \frac{2l+1}{l(l+1)} K_{m,l}^{\varepsilon}(\hat{b}_h) f_l^{-\varepsilon}(\hat{b}) \frac{\partial}{\partial\theta} P_l^m(\cos\theta),\end{aligned} \tag{8.247}$$

$$\begin{aligned}(u_\phi)_{\text{reflected}} &= ime^{im\phi} \sum_{l=m}^{\infty} [(2l+1) I_{m,l}^{\delta}(\hat{a}_h) R_{PP}^{-\delta} + J_{m,l}^{\delta}(\hat{b}_h) R_{SP}^{-\delta}] F_{l_5}^{\delta}(\hat{a}) \frac{P_l^m(\cos\theta)}{\sin\theta} \\ &+ ime^{im\phi} \sum_{l=m}^{\infty} [(2l+1) I_{m,l}^{\delta}(\hat{a}_h) R_{PS}^{-\delta} + J_{m,l}^{\delta}(\hat{b}_h) R_{SSV}^{-\delta}] F_{l_4}^{\delta}(\hat{b}) \frac{P_l^m(\cos\theta)}{\sin\theta} \\ &- e^{im\phi} \sum_{l=m}^{\infty} \frac{2l+1}{l(l+1)} K_{m,l}^{\delta}(\hat{b}_h) R_{SSH}^{-\delta} f_l^{\delta}(\hat{b}) \frac{\partial}{\partial\theta} P_l^m(\cos\theta),\end{aligned} \tag{8.248}$$

$$(u_\phi)_{\text{transmitted}} = ime^{im\phi} \sum_{l=m}^{\infty} [(2l+1)I_{m,l}^{\delta}(\hat{a}_h)T_{PP}^{\delta} + J_{m,l}^{\delta}(\hat{b}_h)T_{SP}^{\delta}]F_{l_5}^{-\delta}(\hat{b}) \frac{P_l^m(\cos\theta)}{\sin\theta}$$

$$+ ime^{im\phi} \sum_{l=m}^{\infty} [(2l+1)I_{m,l}^{\delta}(\hat{a}_h)T_{PS}^{\delta} + J_{m,l}^{\delta}(\hat{b}_h)T_{SSV}^{\delta}]F_{l_4}^{-\delta}(\hat{b}) \frac{P_l^m(\cos\theta)}{\sin\theta}$$

$$- e^{im\phi} \sum_{l=m}^{\infty} \frac{2l+1}{l(l+1)} K_{m,l}^{\delta}(\hat{b}_h)T_{SSH}^{\delta} f_l^{-\delta}(\hat{b}) \frac{\partial}{\partial\theta} P_l^m(\cos\theta), \tag{8.249}$$

$$\varepsilon = \operatorname{sgn}(r - r_h), \qquad \delta = \operatorname{sgn}(r_{q-1} - r_h);$$

$$(u_r)_{\text{incident}} = e^{im\phi} \sum_{l=m}^{\infty} (2l+1)I_{m,l}^{\varepsilon}(\hat{a}_h)F_{l_6}^{-\varepsilon}(\hat{a})P_l^m(\cos\theta)$$

$$+ e^{im\phi} \sum_{l=m}^{\infty} J_{m,l}^{\varepsilon}(\hat{b}_h)l(l+1)F_{l_5}^{-\varepsilon}(\hat{b})P_l^m(\cos\theta), \tag{8.250}$$

$$(u_r)_{\text{reflected}} = e^{im\phi} \sum_{l=m}^{\infty} [(2l+1)I_{m,l}^{\delta}(\hat{a}_h)R_{PP}^{-\delta} + J_{m,l}^{\delta}(\hat{b}_h)R_{SP}^{-\delta}]F_{l_6}^{\delta}(\hat{a})P_l^m(\cos\theta)$$

$$+ e^{im\phi} \sum_{l=m}^{\infty} [(2l+1)I_{m,l}^{\delta}(\hat{a}_h)R_{PS}^{-\delta} + J_{m,l}^{\delta}(\hat{b}_h)R_{SSV}^{-\delta}]$$

$$\times\, l(l+1)F_{l_5}^{\delta}(\hat{b})P_l^m(\cos\theta), \tag{8.251}$$

$$(u_r)_{\text{transmitted}} = e^{im\phi} \sum_{l=m}^{\infty} [(2l+1)I_{m,l}^{\delta}(\hat{a}_h)T_{PP}^{\delta} + J_{m,l}^{\delta}(\hat{b}_h)T_{SP}^{\delta}]F_{l_6}^{-\delta}P_l^m(\cos\theta)$$

$$+ e^{im\phi} \sum_{l=m}^{\infty} [(2l+1)I_{m,l}^{\delta}(\hat{a}_h)T_{PS}^{\delta} + J_{m,l}^{\delta}(\hat{b}_h)T_{SSV}^{\delta}]$$

$$\times\, l(l+1)F_{l_5}^{-\delta}(\hat{b})P_l^m(\cos\theta); \tag{8.252}$$

$$(u_\theta)_{\text{incident}} = e^{im\phi} \sum_{l=m}^{\infty} (2l+1)I_{m,l}^{\varepsilon}(\hat{a}_h)F_{l_5}^{-\varepsilon}(\hat{a}) \frac{\partial}{\partial\theta} P_l^m(\cos\theta)$$

$$+ e^{im\phi} \sum_{l=m}^{\infty} J_{m,l}^{\varepsilon}(\hat{b}_h)F_{l_4}^{-\varepsilon}(\hat{b}) \frac{\partial}{\partial\theta} P_l^m(\cos\theta)$$

$$+ ime^{im\phi} \sum_{l=m}^{\infty} \frac{2l+1}{l(l+1)} K_{m,l}^{\varepsilon}(\hat{b}_h) f_l^{-\varepsilon}(\hat{b}) \frac{P_l^m(\cos\theta)}{\sin\theta}, \tag{8.253}$$

$$(u_\theta)_{\text{reflected}} = e^{im\phi} \sum_{l=m}^{\infty} [(2l+1)I_{m,l}^{\delta}(\hat{a}_h)R_{PP}^{-\delta} + J_{m,l}^{\delta}(\hat{b}_h)R_{SP}^{-\delta}]F_{l_5}^{\delta}(\hat{a}) \frac{\partial}{\partial\theta} P_l^m(\cos\theta)$$

$$+ e^{im\phi} \sum_{l=m}^{\infty} [(2l+1)I_{m,l}^{\delta}(\hat{a}_h)R_{PS}^{-\delta} + J_{m,l}^{\delta}(\hat{b}_h)R_{SSV}^{-\delta}]F_{l_4}^{\delta}(\hat{b}) \frac{\partial}{\partial\theta} P_l^m(\cos\theta)$$

$$+ ime^{im\phi} \sum_{l=m}^{\infty} \frac{2l+1}{l(l+1)} K_{m,l}^{\delta}(\hat{b}_h)R_{SSH}^{-\delta} f_l^{\delta}(\hat{b}) \frac{P_l^m(\cos\theta)}{\sin\theta}, \tag{8.254}$$

$$(u_\theta)_{\text{transmitted}} = e^{im\phi} \sum_{l=m}^{\infty} [(2l+1)I_{m,l}^{\delta}(\hat{a}_h)T_{PP}^{\delta} + J_{m,l}^{\delta}(\hat{b}_h)T_{SP}^{\delta}]F_{l_5}^{-\delta}(\hat{a}) \frac{\partial}{\partial\theta} P_l^m(\cos\theta)$$

$$+ e^{im\phi} \sum_{l=m}^{\infty} [(2l+1)I_{m,l}^{\delta}(\hat{a}_h)T_{PS}^{\delta} + J_{m,l}^{\delta}(\hat{b}_h)T_{SSV}^{\delta}]$$

$$\times F_{l_4}^{-\delta}(\hat{b}) \frac{\partial}{\partial\theta} P_l^m(\cos\theta)$$

$$+ ime^{im\phi} \sum_{l=m}^{\infty} \frac{2l+1}{l(l+1)} K_{m,l}^{\delta}(\hat{b}_h)T_{SSH}^{\delta} f_l^{-\delta}(\hat{b}) \frac{P_l^m(\cos\theta)}{\sin\theta}. \tag{8.255}$$

The parameters α and β in $\hat{a}$ and $\hat{b}$, respectively, are to be taken as the velocities corresponding to the distance r.

Equations (8.247)–(8.255) are valid when the source is either above or below the interface. For example, if the source is above the interface, we have $\hat{b} = (s/\beta_{q-1})r$ for the reflected S wave and $\hat{a} = (s/\alpha_q)r$ for the transmitted P wave. For $r = r_{q-1}$ we have to take the velocity of the layer from which we are approaching the interface. Equations (8.247)–(8.255) show clearly how the reflected–transmitted field is built from the incident one. Disregarding for the moment the summation over l, every term in these equations is a product of three factors: The first is an *amplitude factor* of the form $M_{m,l}^{\delta} S_{\sigma\eta}$, where $M = I$, J or K, $S = R$, T or 1, and σ, η are P or S. The second term contains the *angular dependence*, and the third term is a *phase factor* depending on r through the variable $(s/\alpha)r$ for a P field or through the variable $(s/\beta)r$ for an S field. Upon reflection or transmission through an interface, the resultant displacement differs from the displacements of the incident field as follows:

The angular dependence does not change. The phase factor retains its functional form, when conversion does not occur, with the possible change of ε to $\pm\delta$ and $\hat{a}_j/\hat{b}_j$ to $\hat{a}_k/\hat{b}_k$. When conversion occurs, the phase factor must be appropriately changed. The amplitude factor changes by a factor of the form $S_{\sigma\eta}$, which depends only on the type of the reflection and/or transmission at that interface (and of course on the radius r_q and on the properties of the material on both sides of the interface). Apart from these changes, the form of the expressions for the displacements is maintained. Any component of the resultant field may now be regarded as a new incident field, and similar changes will have to be inserted in its formal form to yield the resultant displacements upon interaction with another interface.

Consider, for example, u_θ. This component of the incident field [Eq. (8.253)] has three terms, which by virtue of the amplitude term I, J, or K and the dependence on $(s/\alpha)r$ or $(s/\beta)r$ may be identified as the contributions of the P, SV, and SH fields. Consider now the five terms of Eq. (8.254). The first of these is identified with the P incident–P reflected field. This term clearly results from the P incident field because it contains the I amplitude factor. It results from P to P reflection as evidenced by both the R_{PP} term and the dependence on $(s/\alpha)r$. Note that this term has exactly the form of the incident P field. Likewise, the second term in Eq. (8.254) is the S incident–P reflected contribution to u_θ. It results from the incident S field, because it contains the J amplitude factor. It is multiplied by the R_{SP} reflection coefficient to yield the S to P reflected field, and its phase factor behaves as a P wave by virtue of the dependence on $(s/\alpha)r$. Likewise, the third term of Eq. (8.254) is the P to S reflected field and the fourth term is the SV to SV reflected field. The fifth term is the SH to SH reflected field. The same procedure holds for the transmitted field, where the only change is that the transmission coefficients replace the reflection coefficients. Note that in each case the reflected–transmitted component has exactly the same form as the incident one. The same method is valid when we wish to propagate a part of the resultant field across an interface.

8.7.2. The "Rainbow Expansion"

Expressions (8.247)–(8.255) give a complete description of the seismic field on either side of the discontinuity. Aside from the direct (incident) field, the functions $R_{PP}^{\pm}$, $R_{SP}^{\pm}$, $R_{PS}^{\pm}$, and $R_{SSV}^{\pm}$ control the P–SV field on the source's side, whereas $T_{PP}^{\pm}$, $T_{SP}^{\pm}$, $T_{PS}^{\pm}$, and $T_{SSV}^{\pm}$ control the P–SV field on the opposite side of the source. Neither of these coefficients is associated with a particular ray-path but each includes the sum of many generalized rays. In order to derive all generalized rays from a given R/T coefficient, a power series expansion must be made. Each term in this expansion represents a particular ray-path that can be associated with a seismic ray. This procedure is known as the "rainbow expansion" because it was first used to interpret the rainbow amplitude patterns formed by superposition of multiple reflections of sunlight in spherical raindrops.

A few words about the nomenclature are relevant here: The symbol $R_{PP}^{\pm}$, for example, does not mean that it is composed of a product of true reflection coefficients alone. Consider a source in the shell above the core. When making the "rainbow expansion" of this coefficient we will find that the first term corresponds to the well-known seismic ray PcP, which is simply the reflected P wave from the core. Another term in the expansion is a wave that is transmitted to the core as a P wave and then emerges back into the shell as a P wave, known in seismology as the PKP wave. The contribution of R_{PP}^{+} in this case is found to be $\{{}^*T_{PP}^{-}\}\{{}^*T_{PP}^{+}\}$, which is a product of two true *transmission coefficients.*

The algebraic steps are as follows: Let

$$^*\Delta = \|\mathbf{U}_2(\hat{a}_{q-1});\, \mathbf{U}_1(\hat{a}_q);\, \mathbf{V}_2(\hat{b}_{q-1});\, \mathbf{V}_1(\hat{b}_q)\|,$$

$$^*T_{PP}^{+} = \frac{1}{^*\Delta}\|\mathbf{U}_2(\hat{a}_q);\, \mathbf{U}_1(\hat{a}_q);\, \mathbf{V}_2(\hat{b}_{q-1});\, \mathbf{V}_1(\hat{b}_q)\|,$$

$$^*R_{PP}^{-} = -\frac{1}{^*\Delta}\|\mathbf{U}_2(\hat{a}_{q-1});\, \mathbf{U}_2(\hat{a}_q);\, \mathbf{V}_2(\hat{b}_{q-1});\, \mathbf{V}_1(\hat{b}_q)\|,$$

$$
\begin{aligned}
{}^*T^+_{PS} &= \frac{1}{{}^*\Delta}\,\|\mathbf{U}_2(\hat{a}_{q-1});\,\mathbf{U}_1(\hat{a}_q);\,\mathbf{U}_2(\hat{a}_q);\,\mathbf{V}_1(\hat{b}_q)\|,\\
{}^*R^-_{PS} &= -\frac{1}{{}^*\Delta}\,\|\mathbf{U}_2(\hat{a}_{q-1});\,\mathbf{U}_1(\hat{a}_q);\,\mathbf{V}_2(\hat{b}_{q-1});\,\mathbf{U}_2(\hat{a}_q)\|,\\
{}^*T^+_{SP} &= \frac{1}{{}^*\Delta}\,\|\mathbf{V}_2(\hat{b}_q);\,\mathbf{U}_1(\hat{a}_q);\,\mathbf{V}_2(\hat{b}_{q-1});\,\mathbf{V}_1(\hat{b}_q)\|,\\
{}^*R^-_{SP} &= -\frac{1}{{}^*\Delta}\,\|\mathbf{U}_2(\hat{a}_{q-1});\,\mathbf{V}_2(\hat{b}_q);\,\mathbf{V}_2(\hat{b}_{q-1});\,\mathbf{V}_1(\hat{b}_q)\|,\\
{}^*T^+_{SSV} &= \frac{1}{{}^*\Delta}\,\|\mathbf{U}_2(\hat{a}_{q-1});\,\mathbf{U}_1(\hat{a}_q);\,\mathbf{V}_2(\hat{b}_q);\,\mathbf{V}_1(\hat{b}_q)\|,\\
{}^*R^-_{SSV} &= -\frac{1}{{}^*\Delta}\,\|\mathbf{U}_2(\hat{a}_{q-1});\,\mathbf{U}_1(\hat{a}_q);\,\mathbf{V}_2(\hat{b}_{q-1});\,\mathbf{V}_2(\hat{b}_q)\|.
\end{aligned}
\tag{8.256}
$$

Then, because

$$
\begin{aligned}
\mathbf{U}_2(\hat{a}_q) &= \{{}^*T^+_{PP}\}\mathbf{U}_2(\hat{a}_{q-1}) - \{{}^*R^-_{PP}\}\mathbf{U}_1(\hat{a}_q)\\
&\quad + \{{}^*T^+_{PS}\}\mathbf{V}_2(\hat{b}_{q-1}) - \{{}^*R^-_{PS}\}\mathbf{V}_1(\hat{b}_q),\\
\mathbf{V}_2(\hat{b}_q) &= \{{}^*T^+_{SP}\}\mathbf{U}_2(\hat{a}_{q-1}) - \{{}^*R^-_{SP}\}\mathbf{U}_1(\hat{a}_q)\\
&\quad + \{{}^*T^+_{SSV}\}\mathbf{V}_2(\hat{b}_{q-1}) - \{{}^*R^-_{SSV}\}\mathbf{V}_1(\hat{b}_q),
\end{aligned}
\tag{8.257}
$$

Eqs. (8.246) and (8.256) lead us to the result

$$
T^+_{PP} = \frac{\left\|\begin{matrix} \{{}^*T^+_{PP}\} & -(-1)^l\{{}^*T^+_{PP}\} & -(-1)^l\{{}^*T^+_{SP}\}\\ -\{{}^*R^-_{PP}\} & 1+(-1)^l\{{}^*R^-_{PP}\} & (-1)^l\{{}^*R^-_{SP}\}\\ -\{{}^*R^-_{PS}\} & (-1)^l\{{}^*R^-_{PS}\} & 1+(-1)^l\{{}^*R^-_{SSV}\}\end{matrix}\right\|}{\left\|\begin{matrix} 1+(-1)^l\{{}^*R^-_{PP}\} & (-1)^l\{{}^*R^-_{SP}\}\\ (-1)^l\{{}^*R^-_{PS}\} & 1+(-1)^l\{{}^*R^-_{SSV}\}\end{matrix}\right\|}.
\tag{8.258}
$$

We may use the expansions

$$
\begin{aligned}
\frac{Z}{(1-X)(1-W)-YZ} &= \frac{Z}{(1-X)(1-W)}\sum_{k=0}^{\infty}\left[\frac{YZ}{(1-X)(1-W)}\right]^k\\
&= Z\sum_{n=0}^{\infty}X^n\sum_{m=0}^{\infty}W^m\sum_{k=0}^{\infty}Y^kZ^k\left\{\sum_{j=0}^{\infty}X^j\sum_{r=0}^{\infty}W^r\right\}^k,
\end{aligned}
\tag{8.259}
$$

$$
\frac{1-X}{(1-X)(1-W)-YZ} = \sum_{n=0}^{\infty}W^n\sum_{m=0}^{\infty}Y^mZ^m\left\{\sum_{k=0}^{\infty}\sum_{j=0}^{\infty}W^kX^j\right\}^m,
\tag{8.260}
$$

to obtain

$$T_{PP}^{+} = \{{}^*T_{PP}^{+}\} + \{{}^*R_{PP}^{-}\}\{{}^*R_{PS}^{-}\}\{{}^*T_{SP}^{+}\} \sum_{n=0}^{\infty} \sum_{m=0}^{\infty} (-1)^{ln+lm+n+m}$$

$$\times \{{}^*R_{PP}^{-}\}^n\{{}^*R_{SSV}^{-}\}^m \sum_{k=0}^{\infty} \{{}^*R_{SP}^{-}\}^k\{{}^*R_{PS}^{-}\}^k$$

$$\times \left[\sum_{j=0}^{\infty} \sum_{r=0}^{\infty} (-1)^{jl+rl+j+r}\{{}^*R_{PP}^{-}\}^j\{{}^*R_{SSV}^{-}\}^r\right]^k$$

$$-(-1)^l\{{}^*R_{PP}^{-}\}\{{}^*T_{PP}^{+}\} \sum_{n=0}^{\infty} (-1)^{ln+n}\{{}^*R_{PP}^{-}\}^n \sum_{m=0}^{\infty} \{{}^*R_{PS}^{-}\}^m\{{}^*R_{SP}^{-}\}^m$$

$$\times \left[\sum_{k=0}^{\infty} \sum_{j=0}^{\infty} (-1)^{kl+jl+k+j}\{{}^*R_{PP}^{-}\}^k\{{}^*R_{SSV}^{-}\}^j\right]^m$$

$$+ \{{}^*R_{PS}^{-}\}\{{}^*T_{PP}^{+}\}\{{}^*R_{SP}^{-}\} \sum_{n=0}^{\infty} \sum_{m=0}^{\infty} (-1)^{ln+lm+n+m}\{{}^*R_{PP}^{-}\}^n\{{}^*R_{SSV}^{-}\}^m$$

$$\times \sum_{k=0}^{\infty} \{{}^*R_{SP}^{-}\}^k\{{}^*R_{PS}^{-}\}^k\left[\sum_{j=0}^{\infty} \sum_{r=0}^{\infty} (-1)^{jl+rl+j+r}\{{}^*R_{PP}^{-}\}^j\{{}^*R_{SSV}^{-}\}^r\right]^k$$

$$-(-1)^l\{{}^*R_{PS}^{-}\}\{{}^*T_{SP}^{+}\} \sum_{n=0}^{\infty} (-1)^{ln+n}\{{}^*R_{SSV}^{-}\}^n \sum_{m=0}^{\infty} \{{}^*R_{SP}^{-}\}^m\{{}^*R_{PS}^{-}\}^m$$

$$\times \left[\sum_{k=0}^{\infty} \sum_{j=0}^{\infty} (-1)^{kl+jl+k+j}\{{}^*R_{PP}^{-}\}^k\{{}^*R_{SSV}^{-}\}^j\right]^m. \tag{8.261}$$

Similar expansions are obtained for the other reflection and transmission coefficients of Eq. (8.256). In order to explore the physical meaning of this expansion, we take, for example, in the fourth term $n = 1$, $m = 2$, $k = 1$, and $j = r = 0$ to obtain

$$(-1)^{l+1}\{{}^*R_{PS}^{-}\}^2\{{}^*R_{SP}^{-}\}^2\{{}^*R_{PP}^{-}\}\{{}^*R_{SSV}^{-}\}^2\{{}^*T_{PP}^{+}\}.$$

This is seen to correspond to the ray in Fig. 8.29. It should be mentioned that such a path is not unique, because the terms of the "rainbow expansion" determine only the number of reflections occurring in the internal surface of the earth. Note that the R/T coefficients in Eq. (8.256) are represented by the ratio of two functional determinants of order 4. The complexity of this function reflects the nature of the interaction of a spherical wavefront with a spherical boundary.

So far we have applied the "rainbow expansion" to waves interacting with a single discontinuity, such as the core–mantle boundary or the free surface. The "rainbow expansion," however, is not limited to a single interface. If this expansion is applied to the analytical solution of a point source in a multi-layered sphere, the Legendre sum for each generalized ray will include a product of transmission and reflection coefficients at the layers relevant to the path of that ray. In this sense the situation is very similar to the half-space analog that we have discussed in detail in Chapter 7.

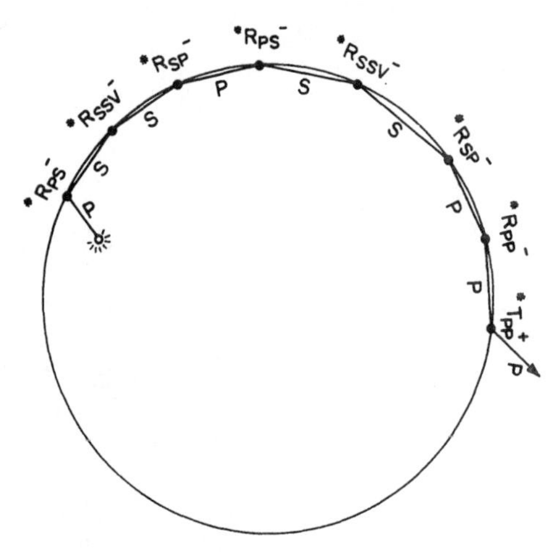

Figure 8.29. A generalized ray originating from a source below the interface and recorded above the interface. It represents a single term in a "rainbow expansion."

8.7.3. Generalized Rays in a Uniform Shell Overlying a Liquid Core: *PcP*, *PcS*, *ScP*, *ScS*, *pP*, *PP*, and *PKP*

8.7.3.1. *P* and *SV* Core Reflections. The *ScSH* wave has been studied thoroughly in Section 8.6.3. We shall treat now the other core reflections, known as *PcP*, *PcS*, *ScP*, and *ScSV*, that arise from the reflection of *P* and *S* waves at the core–mantle boundary. Consider a vertical strike-slip point dislocation in a shell overlying a liquid core. Disregarding the presence of the earth's free surface, we use Eq. (8.251) to evaluate the reflected radial displacement. The total reflected P field is

$$(u_r)_P = e^{i2\phi} \sum_{l=2}^{\infty} (2l + 1) I^{-}_{2,l}(\hat{a}_h) F^{-}_{l_6}(\hat{a}) P_l^2(\cos\theta)[R^{+}_{PP}], \tag{8.262}$$

where $r_h > r_{q-1}$ implies $\delta = -1$. Therefore, with the use of Eqs. (8.221), (8.229), (8.232) and a passage to the frequency domain, we find

$$(u_r)_P = \frac{iU_0\, dS}{8\pi r_h^2}\left(\frac{\beta_h}{\alpha_h}\right)^2 \sin 2\phi \sum_{l=2}^{\infty} (2l + 1) P_l^2(\cos\theta) h_l^{(2)}(k_\alpha r_h) h_l'^{(2)}(k_\alpha r)[R^{+}_{PP}], \tag{8.263}$$

where R^{+}_{PP} is given by an equation similar to Eq. (8.256). The results for a liquid core are obtained from Eqs. (8.247)–(8.255) by taking the limit of vanishing rigidity in the layer below the source. The part of R^{+}_{PP} that contributes to the *PcP* wave, say, is then found from the "rainbow expansion" of this coefficient. In the same way we derive the reflected *S* field. There are four basic partial fields that are of special interest to seismology. For a vertical strike-slip source they are given explicitly by their Legendre sums

$$u_r(PcP) = \frac{iU_0\,dS}{8\pi r_h^2}\left(\frac{\beta_h}{\alpha_h}\right)^2 \sin 2\phi \sum_{l=2}^{\infty}(2l+1)P_l^2(\cos\theta)h_l^{(2)}(k_\alpha r_h)h_l'^{(2)}(k_\alpha r)$$
$$\times\left[\frac{h_l^{(1)}(k_\alpha b)}{h_l^{(2)}(k_\alpha b)}\right](LL), \tag{8.264}$$

$$u_r(PcS) = \frac{iU_0\,dS}{8\pi r_h^2}\left(\frac{\beta_h}{\alpha_h}\right)^2 \sin 2\phi \sum_{l=2}^{\infty}(2l+1)P_l^2(\cos\theta)h_l^{(2)}(k_\alpha r_h)h_l^{(2)}(k_\beta r)$$
$$\times\left[\frac{l(l+1)}{k_\beta r}\right]\left[\frac{h_l^{(1)}(k_\alpha b)}{h_l^{(2)}(k_\beta b)}\right](LT), \tag{8.265}$$

$$u_r(ScP) = \frac{iU_0\,dS}{8\pi r_h^2}\sin 2\phi \sum_{l=2}^{\infty}(2l+1)P_l^2(\cos\theta)h_l^{(2)}(k_\beta r_h)h_l'^{(2)}(k_\alpha r)$$
$$\times\Gamma_l\left[\frac{h_l^{(1)}(k_\beta b)}{h_l^{(2)}(k_\alpha b)}\right](TL), \tag{8.266}$$

$$u_r(ScSV) = \frac{iU_0\,dS}{8\pi r_h^2}\sin 2\phi \sum_{l=2}^{\infty}(2l+1)P_l^2(\cos\theta)h_l^{(2)}(k_\beta r_h)h_l^{(2)}(k_\beta r)$$
$$\times\left[\frac{l(l+1)}{k_\beta r}\right]\Gamma_l\left[\frac{h_l^{(1)}(k_\beta b)}{h_l^{(2)}(k_\beta b)}\right](TT), \tag{8.267}$$

$$\Gamma_l = \frac{1}{2}\left[1 + \frac{(1-(2/l))h_{l-1}^{(2)}(k_\beta r_h) + (1+2/(l+1))h_{l+1}^{(2)}(k_\beta r_h)}{h_{l-1}^{(2)}(k_\beta r_h) + h_{l+1}^{(2)}(k_\beta r_h)}\right]. \tag{8.268}$$

The symbols (LL), (LT), (TL), and (TT) are *spherical amplitude ratios* [Eq. (8. 276), L = longitudinal, T = transverse]. For large values of l and (kr) (such that the periods of P and SV waves do not exceed 40 and 75 s, respectively) they are not very different from the amplitude ratios of plane waves interacting with a flat boundary as given in Table 7.9.

The approximate evaluation of the Legendre sums in Eqs. (8.264)–(8.268) by the Watson transformation and the saddle-point method proceeds along the same lines as in Section 8.6.3. The divergence coefficients of PcP rays have the same form as that of the $ScSH$ rays given in Eq. (8.189). The divergence coefficients of the PcS rays are found to be (Fig. 8.30)

$$G_{PcS} = \frac{b}{r_h}\left[\frac{\cos e}{\cos\tau_h}\right]^{1/2}$$
$$\times\frac{(\tan e/\tan f)}{\{(b\sin\theta/\sin\tau)[D_1(\tan e/\tan f)(r/r_h)(\cos\tau/\cos\tau_h) + D_2]\}^{1/2}}. \tag{8.269}$$

Similar expressions can be obtained for G_{ScP} and G_{ScSV}. In the shadow zone, amplitudes of diffracted PcP rays follow decay laws similar to those of $ScSH$ rays.

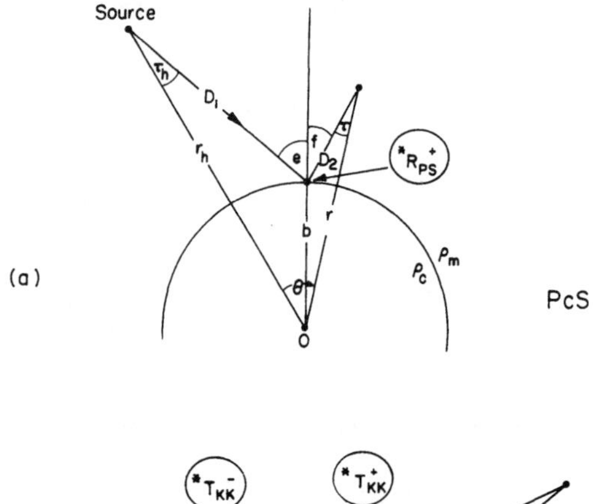

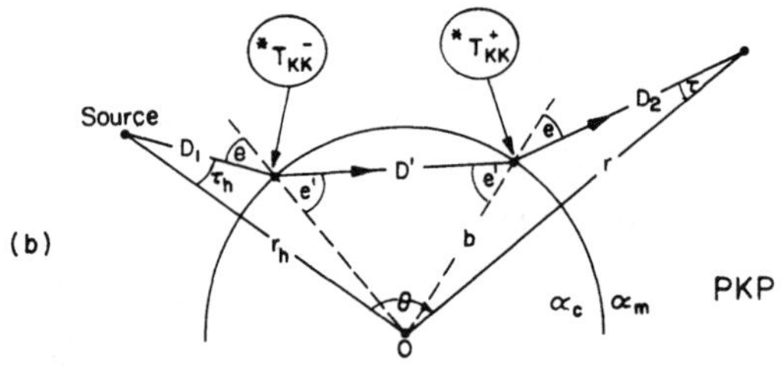

Figure 8.30. Generalized rays representing two well-known seismic waves that interact with the earth's core. (a) *PcS*, (b) *PKP*.

In contrast to the divergence factor for the *PcP* wave, G_{PcS} depends not only on the curvature of the reflecting surface but also on the value of α/β. If the curvature tends to zero, the limiting value is

$$G_{PcS} = \frac{(\tan e/\tan f)}{\{((\alpha_h/\beta_h)D_1 + D_2)((\alpha_h/\beta_h)D_1(\cos^2 f/\cos^2 e) + D_2)\}^{1/2}}$$

$$> \frac{1}{D_1 + (\alpha_h/\beta_h)D_2}. \quad (8.270)$$

Because the wave potential decays with distance as $(D_1 + (\alpha_h/\beta_h)D_2)^{-1}$, the wave will tend to *converge* in the far field. The diverging effect of the curvature is of importance only for small values of the angle of incidence.

8.7.3.2. Surface Reflections: The *pP* and *PP* Waves. In the foregoing we investigated the effect of the earth's core on elastic waves emitted from a point source outside this core. In this section we shall consider the effect of the other boundary of the mantle, namely, the surface of the earth. To this end, we consider the motion inside a homogeneous sphere with radius a, surrounded by vacuum. We consider again the radial spectral displacement caused by a vertical strike-slip point source at $(\theta = 0, r = r_h)$ and employ Eq. (8.251) to give us the reflected field from the free surface. In contradistinction to the *PcP* wave, we have here $r = a > r_h$, $q = 1$, $\delta = +1$, $\varepsilon = +1$, and hence

$$(u_r)_P = e^{2i\phi} \sum_{l=2}^{\infty} (2l+1) I^+_{2,l}(\hat{a}_h) F^+_{l_6}(\hat{a}) P_l^2(\cos\theta)[R^-_{PP}]. \tag{8.271}$$

Applying the "rainbow expansion" to $[R^-_{PP}]$ via Eq. (8.256), we find the contribution of this coefficient to a once-reflected wave from the free surface to be $\{{}^*R^-_{PP}\}$. The explicit Legendre sum that represents the *PP* wave is then

$$(u_r)_{PP} = \frac{iU_0\,dS}{8\pi r_h^2}\left(\frac{\beta}{\alpha}\right)_h^2 \sin 2\phi \sum_{l=2}^{\infty} (2l+1) P_l^2(\cos\theta) h_l^{(1)}(k_\alpha r_h) h_l'^{(1)}(k_\alpha a) \times \left[\frac{h_l^{(2)}(k_\alpha a)}{h_l^{(1)}(k_\alpha a)}\right](ll). \tag{8.272}$$

In the high-frequency limit, (ll) goes into the flat free-surface amplitude-ratio [see Eq. (3.10)]

$$(ll) \simeq \frac{\sin 2e \sin 2f - (\alpha/\beta)^2 \cos^2 f}{\sin 2e \sin 2f + (\alpha/\beta)^2 \cos^2 f}. \tag{8.273}$$

Clearly

$$\frac{h_l^{(2)}(k_\alpha a)}{h_l^{(1)}(k_\alpha a)}(ll) = {}^*R^-_{PP}. \tag{8.274}$$

The approximate evaluation of the Legendre sum in Eq. (8.272) proceeds as before. However in the present case, the saddle-point equations yield two solutions. The wave corresponding to the smaller of these roots is commonly indicated by *pP* and that to the larger one by *PP*. Details are not given here because the situation is quite similar to the *sS* and *SS* waves discussed in Section 8.5. Here also, the existence of two roots is connected with the appearance of a caustic which intersects the surface of the sphere.

Near the antipode, a focusing of the *PP* wave occurs, similar to that of the *PKP* wave. Figure 8.31 shows some common seismic rays and their associated reflection coefficients.

8.7.3.3. The *PKP* Wave. The symbol *PKP* means that wave enters the liquid core as a *P* wave, travels in it with longitudinal velocity, and emerges as a *P* wave in the solid shell. The radial component of the displacement is obtained via Eq. (8.251) from the second term in the "rainbow expansion" of $\{R^+_{PP}\}$, namely, $\{{}^*T^-_{PK}\}\{{}^*T^+_{KP}\}$. The explicit Legendre sum associated with this wave for the radial displacement of a vertical strike-slip source at $r = r_h$, $\theta = 0$ is then

$$u_r(PKP) = \frac{iU_0(\omega)dS}{8\pi r_h^2}\left(\frac{\beta}{\alpha}\right)_h^2 \sin 2\phi \sum_{l=2}^{\infty} (2l+1) P_l^2(\cos\theta) h_l^{(2)}(k_\alpha r_h) h_l'^{(2)}(k_\alpha r) \times \left[\frac{h_l^{(1)}(k_\alpha b)}{h_l^{(1)}(k_c b)}\right]\left[\frac{h_l^{(2)}(k_c b)}{h_l^{(2)}(k_\alpha b)}\right](LL')(L'L), \tag{8.275}$$

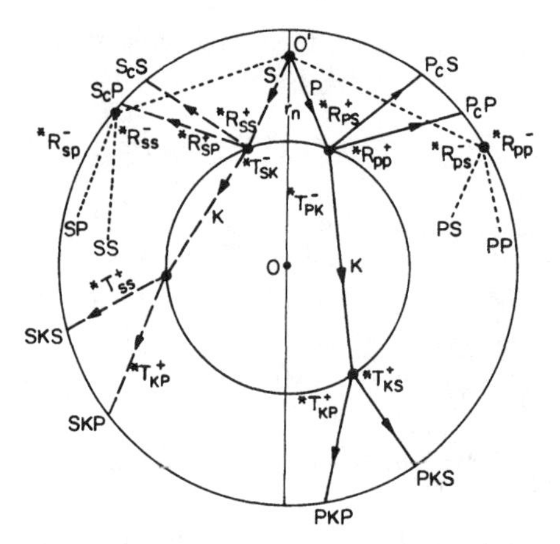

Figure 8.31. Principal rays observed in seismograms and their associated R/T coefficients.

where (LL') and $(L'L)$ [like (LL), (TL), (LT), and (TT) introduced earlier] go in the high-frequency limit into the flat layer amplitude ratios. They relate to the "rainbow expansion" coefficients in the following way

$$\frac{h_l^{(1)}(k_\alpha b)}{h_l^{(2)}(k_\alpha b)}(LL) = {}^*R_{PP}^+, \qquad \frac{h_l^{(1)}(k_\alpha b)}{h_l^{(2)}(k_\beta b)}(LT) = {}^*R_{PS}^+,$$

$$\frac{h_l^{(1)}(k_\beta b)}{h_l^{(2)}(k_\alpha b)}(TL) = {}^*R_{SP}^+, \qquad \frac{h_l^{(1)}(k_\beta b)}{h_l^{(2)}(k_\beta b)}(TT) = {}^*R_{SSV}^+, \tag{8.276}$$

$$\frac{h_l^{(1)}(k_\alpha b)}{h_l^{(1)}(k_c b)}(LL') = {}^*T_{PK}^-, \qquad \frac{h_l^{(2)}(k_c b)}{h_l^{(2)}(k_\alpha b)}(L'L) = {}^*T_{KP}^+.$$

The approximate evaluation of the sum in Eq. (8.275) is now accomplished by the method outlined in Section 8.6.2. We replace the Legendre sum in Eq. (8.275) by its equivalent integral and find that the Hessian determinant is

$$H_s = -\nu_0 \sin\theta e^{i\theta}(k_\alpha^4 k_c^2)(rr_h b^4)\cos\tau_h \cos\tau \cos^2 e \cos^2 e'$$

$$\times\left[\frac{2}{k_\alpha b\cos e} - \frac{1}{k_\alpha r_h \cos\tau_h} - \frac{1}{k_\alpha r\cos\tau} - \frac{2}{k_c b\cos e'}\right], \tag{8.277}$$

where τ_h, τ, e, and e' are angles shown in Fig. 8.30 and k_c refers to the value of k in the core.

The saddle point ν_0 is determined from the equations

$$\nu_0 + \frac{1}{2} = k_\alpha r_h \sin\tau_h = k_c b \sin e' = k_\alpha r \sin\tau = k_\alpha b \sin e \tag{8.278}$$

and is given by

$$\pi = \theta + \tau_h + \tau + 2e' - 2e \tag{8.279}$$

or

$$\theta = \pi - \sin^{-1}\frac{\nu_0 + \frac{1}{2}}{k_\alpha r_h} - \sin^{-1}\frac{\nu_0 + \frac{1}{2}}{k_\alpha r} - 2\sin^{-1}\frac{\nu_0 + \frac{1}{2}}{k_c b} + 2\sin^{-1}\frac{\nu_0 + \frac{1}{2}}{k_\alpha b}$$

$$= f(\nu_0). \tag{8.280}$$

After some reduction, we find (Fig. 8.30)

$$u_r(PKP) = \left\{\frac{-i\omega U_0(\omega)dS}{4\pi\alpha_h}\left(\frac{\beta}{\alpha}\right)_h^2 \sin 2\phi\right\}\Phi(PKP),$$

$$\Phi(PKP) = (LL')(L'L)G_{PKP}e^{i[\omega t - \{k_\alpha(D_1 + D_2) + k_c D'\}]}\sin^2\tau_h\cos\tau, \tag{8.281}$$

with the divergence coefficient

$$G_{PKP} = \frac{b/rr_h}{\left[\sin\theta\dfrac{\cos\tau_h\cos\tau}{\sin e\cos e}\left\{\dfrac{D_1}{r_h\cos\tau_h} + \dfrac{D_2}{r\cos\tau} - \left(\dfrac{\alpha_c}{\alpha}\right)\dfrac{D'}{b\cos e'}\right\}\right]^{1/2}} \tag{8.282}$$

and the transmission coefficient at high frequencies (Tables 7.7–7.9)

$$(LL')(L'L) \simeq \frac{4((\rho_c/\rho)(\alpha_c/\alpha)(\cos e/\cos e'))\cos^2 2f}{[((\rho_c/\rho)(\alpha_c/\alpha)(\cos e/\cos e')) + \cos^2 2f + (\beta^2/\alpha^2)\sin 2e\sin 2f]^2}. \tag{8.283}$$

Now, because

$$\frac{\partial\theta}{\partial\nu_0} = \frac{2}{k_\alpha b\cos e} - \frac{1}{k_\alpha r_h\cos\tau_h} - \frac{1}{k_\alpha r\cos\tau} - \frac{2}{k_c b\cos e'} \tag{8.284}$$

the Hessian determinant vanishes for a certain value of θ and Eq. (8.281) will cease to be valid there. The two equations $\theta = f(\nu)$ and $\partial\theta/\partial\nu = 0$ determine the *PKP caustic*. It is known from earthquake seismology that the intersection of the earth's surface with the *PKP* caustic occurs at about $\theta = 143°$.

To complete the calculation of the *PKP* wave amplitude for points *near* the caustic, we use the results obtained earlier in Section 8.4.4. In order to simplify the expression for Φ as much as possible, we apply the method used by us in Section 8.6.3 and replace the eigenfunctions appearing in the integrand of the Watson integral by their corresponding asymptotic approximations. In this way we obtain

$$\Phi(PKP) = \frac{1}{k_\alpha^3 r_h^3 r}\int_{-\infty}^{\infty}\frac{(\cos\tau)\nu^{5/2}\,e^{-i\phi(\nu) - \pi i/4}}{[2\pi\sin\theta\cos\tau_h\cos\tau]^{1/2}}(LL')(L'L)d\nu \tag{8.285}$$

with

$$\phi = \left(\nu + \frac{1}{2}\right)[\theta - f(\nu)] + k_\alpha r_h \cos\tau_h + k_\alpha r \cos\tau + 2k_c b \cos e' - 2k_\alpha b \cos e.$$

At the caustic, $\theta = \theta_{ca} = f(\nu)$ and, therefore,

$$\phi_{ca} = k_\alpha r_h \cos\tau_h + k_\alpha r \cos\tau + 2k_c b \cos e' - 2k_\alpha b \cos e.$$

For a point *near* the caustic, we have by Eq. (8.101a)

$$\Phi(PKP) = \left(\frac{\sin^2\tau_h}{k_\alpha r_h r}\right)\left(\frac{\cos\tau}{\cos\tau_h}\right)^{1/2}\left\{\frac{(LL')(L'L)}{\sqrt{(2\pi\sin\theta)}}\right\}\left[\frac{6}{(\partial^2\theta/\partial\nu^2)_{ca}}\right]^{1/3}$$
$$\times\left[\frac{2\pi}{3^{1/3}}Ai(-z)\right]e^{i[\omega t - \phi_{ca}] - \pi i/4}. \tag{8.285a}$$

Then, denoting the value of Φ on the caustic ($z = 0$) by Φ_{ca} we use Eq. (8.101c) to get

$$\Phi(PKP) \simeq \Phi_{ca}[1 + (0.72901)z - (0.16667)z^3 - (0.060750)z^4],$$
$$z = (\theta - \theta_{ca})\left[\frac{2}{(\partial^2\theta/\partial\nu^2)_{ca}}\right]^{1/3}. \tag{8.286}$$

On the earth's surface ($r = a, \tau = i_0$), we have by Snell's law $\nu + \frac{1}{2} = (k_\alpha a)\sin i_0$. Therefore

$$\frac{\partial^2\theta}{\partial\nu^2} = \frac{1}{(k_\alpha a)^2}\frac{\partial^2\theta}{\partial(\sin i_0)^2} = \left(\frac{1}{2\pi a}\right)^2\left[\Lambda^2\frac{\partial^2\theta^2}{\partial(\sin i_0)^2}\right]. \tag{8.287}$$

Thus, the quantity z, which governs the amplitude of the *PKP* wave near the caustic, is given by

$$z = \frac{25.7(\theta - \theta_{ca})}{\left[\Lambda^2\dfrac{\partial^2\theta}{\partial(\sin i_0)^2}\right]^{1/3}}, \tag{8.288}$$

where the distance $(\theta - \theta_{ca})$ is measured in degrees and the wavelength Λ in kilometers. As

$$\frac{\partial^2\theta}{\partial(\sin i_0)^2} = \frac{2}{\sin^2 i_0}[\tan^3 e - \tan^3 e' - \tan^3 i_0], \tag{8.289}$$

the numerical value of z can easily be found. Taking $e = 46°$, $e' = 26°$, and $i_0 = 23°$, which correspond to the ray of minimum deviation $\theta_{ca} = 143°$, we obtain

$$z = 11.23(\theta - \theta_{ca})\Lambda^{-2/3}. \tag{8.290}$$

For points of observation *near* the caustic, we use the asymptotic approximations of the Airy function valid for large $|z|$ [App. G],

$$Ai(-z) \sim \begin{cases} \dfrac{1}{\sqrt{\pi}} z^{-1/4} \sin\left[\dfrac{2}{3} z^{3/2} + \dfrac{\pi}{4}\right], & z > 0, (\theta > \theta_{ca}) \\ \dfrac{1}{2\sqrt{\pi}} |z|^{-1/4} e^{-(2/3)|z|^{3/2}}, & z < 0, (\theta < \theta_{ca}). \end{cases} \tag{8.291}$$

On the concave side of the caustic ($\theta < \theta_{ca}$) the amplitudes decay exponentially into the shadow zone. The exponential decrease, as obtained by a combined use of Eqs. (8.290) and (8.291), is

$$e^{-25 \cdot 1(\theta - \theta_{ca})^{3/2} \Lambda^{-1}}.$$

As stated in Section 8.4.4, the maximum intensity for a given wavelength is not at $\theta = \theta_{ca}$ but at a value of $\theta > \theta_{ca}$ such that $z \simeq 1.21$. The *PKP* field on the convex side of the caustic ($\theta > \theta_{ca}$) is governed by the function

$$\sin\left[\frac{2}{3} z^{3/2} + \frac{\pi}{4}\right] e^{i[\omega t - \phi] - \pi i/4}$$

which represents the result of the interference of two waves, the travel-times of which can be written as

$$T - T_{ca} = A(\theta - \theta_{ca}) \pm B(\theta - \theta_{ca})^{3/2}. \tag{8.292}$$

Here A and B are frequency-dependent constants and the subscript *ca* refers to values at the caustic.

Near the antipode of an earthquake epicenter ($\theta = 180°$), the divergence coefficient G_{PKP} becomes singular because the Laplace asymptotic approximation that we have employed for $P_l^2(\cos\theta)$ is not valid there. For θ close to 180° and large order, we must use the Hilb–Szegö approximation [Eqs. (H.61)–(H.67)]

$$P^m_{\nu-1/2}(-\cos\theta) \simeq \nu^m \left[\frac{(\pi - \theta)}{\sin(\pi - \theta)}\right]^{1/2} J_m[\nu(\pi - \theta)][1 + O(\nu^{-3/2})], \tag{8.293}$$

$$\nu \simeq k_\alpha a \sin\tau.$$

Because $J_m(0) = \delta_{0m}$ and because $J_0(x)$ attains its maximal value of unity at $x = 0$, a focusing of seismic waves at the epicenter's antipode will occur only for sources that are capable of generating fields with azimuthally symmetrical components, ($m = 0$). According to Eqs. (8.232)–(8.234) and (8.247)–(8.255), this may occur either for an explosion or case III of a displacement dislocation or else for a vertical single force. The physical reason for this focusing phenomenon is that energy flows to the antipode from all azimuths rather than from a single azimuth, as is the case when θ is sufficiently removed from the antipode.

8.7.4. Generalized Rays in Radially Heterogeneous Earth Models and the Fresnel Diffraction Integral

Hitherto we have treated generalized rays in a simple earth model composed of a homogeneous shell overlying a liquid core. We have neglected the frequency dependence of the reflection and transmission coefficients and have ruled out all those phenomena with which the saddle-point method is unable to cope. These restrictions exist even in a multilayered-earth model, which is inadequate to describe certain aspects of full-wave theory, such as the core-grazing P rays. However, we may employ the concept of a generalized ray in radially heterogeneous earth models with continuous velocity profiles in the mantle (crust excluded) provided the decoupling conditions (7.40) hold and provided we abandon the exact inversion of the Laplace integrals (Sect. 7.6.4).

The main departure from our previous formal results [Eqs. (8.247)–(8.255)] will be the appearance of different radial eigenfunctions instead of the spherical Bessel functions and the need to evaluate integrals in the complex p (ray parameter) plane. We shall now examine this method in some detail. The spherical Bessel functions that appear in the Fourier-transform equivalents of $I^{\varepsilon}_{m,l}(\hat{a}_h)$ and $F^{-\varepsilon}_{l_6}(\hat{a})$, etc., are known to be the solutions of the differential equation (D.60) in the modified form

$$\frac{d^2}{dr^2}(rg_l^{(1,2)}) + \omega^2\left[\frac{1}{v^2} - \frac{l(l+1)}{\omega^2 r^2}\right](rg_l^{(1,2)}) = 0. \tag{8.294}$$

When $v = \text{const.}$ and $k = \omega/v$, the solutions are $g_l^{(1)} = h_l^{(1)}(kr)$, $g_l^{(2)} = h_l^{(2)}(kr)$. However, if $v = v(r)$, Eqs. (8.247)–(8.255) will still hold under decoupling conditions, provided we replace $h_l^{(1,2)}(kr)$ by the solutions of Eq. (8.294). Also note that if we put $l = \nu - \frac{1}{2}$ in Eq. (8.294), its basic solutions $g^{(1)}_{\nu-1/2}$ and $g^{(2)}_{\nu-1/2}$ remain invariant to a change of sign in ν, up to constants that may depend on ν but not on $v(r)$. These constants are then determined from the homogeneous case ($v = \text{const.}$), yielding

$$g^{(1)}_{-\nu-1/2}(r) = e^{i\nu\pi}g^{(1)}_{\nu-1/2}(r), \qquad g^{(2)}_{-\nu-1/2}(r) = e^{-i\nu\pi}g^{(2)}_{\nu-1/2}(r) \tag{8.295}$$

or

$$[g^{(1)}_{-\nu-1/2}(r) + g^{(2)}_{-\nu-1/2}(r)] = [g^{(1)}_{\nu-1/2}(r) + g^{(2)}_{\nu-1/2}(r)]e^{i\nu\pi} - 2i(\sin\nu\pi)g^{(2)}_{\nu-1/2}(r). \tag{8.296}$$

We therefore see that

$$\{g^{(1)}_{\nu-1/2}(r_1)g^{(2)}_{\nu-1/2}(r_2)\}$$

is even in ν, whereas

$$\{g^{(2)}_{-\nu-1/2}(r_1)g^{(2)}_{-\nu-1/2}(r_2)\}e^{i\nu\pi} = \{g^{(2)}_{\nu-1/2}(r_1)g^{(2)}_{\nu-1/2}(r_2)\}e^{-i\nu\pi}. \tag{8.297}$$

To be specific we use the WKBJ solutions [Eqs. (H.1)–(H.10) and (8.88)–(8.90)] with the correspondence [see Eqs. (8.90)–(8.91)]

$$h_l^{(1,2)}(kr) \Rightarrow \frac{vf_0}{\omega r}\left[1 - \frac{v^2(l+\frac{1}{2})^2}{\omega^2 r^2}\right]^{-1/4} \times \exp\left\{\pm i\omega \int_{r_m}^{r} \frac{dr}{v(r)}\left[1 - \frac{v^2(l+\frac{1}{2})^2}{\omega^2 r^2}\right]^{1/2} \mp \frac{\pi i}{4}\right\} = g_l^{(1,2)}(r), \tag{8.298}$$

$$\frac{d}{dr} h_l^{(1,2)}(kr) \Rightarrow \frac{f_0}{r}\left[1 - \frac{v^2(l+\frac{1}{2})^2}{\omega^2 r^2}\right]^{1/4} \times \exp\left\{\pm i\omega \int_{r_m}^{r} \frac{dr}{v(r)}\left[1 - \frac{v^2(l+\frac{1}{2})^2}{\omega^2 r^2}\right]^{1/2} \pm \frac{\pi i}{4}\right\} = \frac{\partial}{\partial r} g_l^{(1,2)}(r),$$

$$f_0 = \left(\frac{v_h}{v}\right)^{1/2}, \qquad \omega r > (l+\tfrac{1}{2})v(r). \tag{8.299}$$

The parameters in this correspondence have been adjusted in such a way that for a homogeneous medium (v = const. = ω/k), the WKBJ solution does coincide with the Debye approximation of the spherical Hankel functions in the wave zone $kr > (l + \frac{1}{2})$ [Eqs. (H.22)–(H.23)]. The factor

$$f_0 = \left(\frac{v_h}{v}\right)^{1/2}$$

is equal to unity in a homogeneous medium and also in the case where the source and the receiver are on the same level. It is derived by comparing the asymptotic value of a "rainbow expansion" with its equivalent ray-theoretical form.

The above correspondence is valid only if the differential equation for the radial eigenfunction has no turning points. If there is a single turning point of the first order [App. H] the correct approximation for the spherical Hankel functions, valid on both sides of the transition point, is represented with the aid of the Airy function and given explicitly in Eqs. (H.17a) and (H.17b). The corresponding WKBJ solutions are then adjusted to yield

$$h_l^{(1,2)}(kr) \Rightarrow \frac{2f_0\sqrt{\pi}e^{\mp \pi i/3}v^{1/2}}{(\omega^{5/6})r}\left[\frac{\Phi}{Q}\right]^{1/4} Ai\{-\omega^{2/3}\Phi e^{\pm 2\pi i/3}\} = g_l^{(1,2)}(r), \tag{8.300}$$

$$\frac{d}{dr} h_l^{(1,2)}(kr) \Rightarrow \frac{-2f_0\sqrt{\pi}e^{\pm \pi i/3}v^{1/2}}{(\omega^{1/6})r}\left[\frac{Q}{\Phi}\right]^{1/4} Ai'\{-\omega^{2/3}\Phi e^{\pm 2\pi i/3}\} = \frac{\partial g_l^{(1,2)}(r)}{\partial r},$$

$$Q = \frac{1}{v^2}\left[1 - \frac{v^2(l+\frac{1}{2})^2}{r^2\omega^2}\right], \qquad \Phi = \left[\frac{3}{2}\int_{r_m}^{r} \frac{dr}{v}\left\{1 - \frac{v^2(l+\frac{1}{2})^2}{\omega^2 r^2}\right\}^{1/2}\right]^{2/3}. \tag{8.301}$$

However, we know from ray theory, Eq. (7.88), and ray–mode duality, Eq. (8.80), that if ${}_n\omega_l$ is sufficiently large (as is indeed assumed in the WKBJ approximation) the relations

$$l + \frac{1}{2} = p\omega, \qquad \int_{r_m}^{r} \frac{dr}{v}\left[1 - \frac{v^2(l+\frac{1}{2})^2}{\omega^2 r^2}\right]^{1/2} = T - p\Delta \tag{8.302}$$[6]

can be assumed to hold for any spectral component of the body-wave spectra because for each ω there is an eigenvalue ${}_n\omega_l$ sufficiently close to it. Hence, to the order of accuracy of the WKBJ approximation,

$$\Phi \simeq \left[\frac{3}{2}(T - p\Delta)\right]^{2/3}, \tag{8.303}$$

where T is the travel time along the ray and p is the ray parameter. The eigenfunctions $g^{(1,2)}(r)$ can alternatively be expressed in terms of Hankel functions of order $\frac{1}{3}$. Using Eq. (H.24) in Eq. (8.300), we obtain the *Langer approximation*

$$g^{(1,2)}(r) \simeq \frac{e^{\pm \pi i/6} f_0}{\omega r}\left[\frac{\pi v \omega \xi}{2q}\right]^{1/2} H_{1/3}^{(1,2)}(\omega\xi), \tag{8.304}$$

$$\frac{\partial}{\partial r} g^{(1,2)}(r) \simeq \frac{e^{\pm 5\pi i/6} f_0}{r}\left[\frac{\pi v q \omega \xi}{2}\right]^{1/2} H_{2/3}^{(1,2)}(\omega\xi), \tag{8.305}$$

where

$$q = Q^{1/2} = \left[\frac{1}{v^2} - \frac{p^2}{r^2}\right]^{1/2}, \qquad \xi = \int_{r_m}^{r} q\, dr = T - p\Delta. \tag{8.306}$$

We note that $g^{(1)}$ corresponds to downgoing waves and $g^{(2)}$ to upgoing waves at radii above the turning point. The Wronskian of $g^{(1)}$ and $g^{(2)}$ is simply

$$g^{(1)}\frac{\partial g^{(2)}}{\partial r} - g^{(2)}\frac{\partial g^{(1)}}{\partial r} = -\frac{2iv_h}{\omega r^2}. \tag{8.307}$$

At worst, the approximation given in Eqs. (8.301)–(8.305) can introduce errors of order $[v(r)/\omega r]^{2/3}$, and this remains below 5% for almost all body waves with periods less than 40 s. Where this error is unacceptable, it is possible to develop explicitly the next term in the asymptotic series in the Langer approximation [App. H]. The integral for ξ can be made a sum, if the interpolation law $v(r) = Ar^b$ (Example 7.3) is used between radii at which v is specified.

The approximations given in Eqs. (8.304) and (8.305) are not valid for media in which more than one turning point r_m exists for each p. One then replaces the Hankel function by the solutions of the adequate comparison equation [App. H]. When r is near a turning point r_m, we expand $Q(r)$ in Taylor series about r_m

$$Q(r) \simeq Q(r_m) + (r - r_m)\left(\frac{\partial Q}{\partial r}\right)_m = (r - r_m)\left\{\frac{2}{r_m v_m^2}\left[1 - \frac{r_m}{v_m}\left(\frac{dv}{dr}\right)_m\right]\right\}. \tag{8.308}$$

[6] In Secs. 8.7.4–8.7.7, Δ denotes the angle subtended at the center of the earth.

Denoting

$$\frac{r_m}{v_m}\left(\frac{dv}{dr}\right)_m = b_m, \qquad \eta = \frac{2(1-b_m)}{pv_m^3}, \qquad F = rg, \tag{8.309}$$

Eq. (8.294) becomes

$$\frac{d^2F}{dr^2} + \omega^2\eta(r - r_m)F = 0. \tag{8.310}$$

Its solutions, representing outgoing and incoming waves, are known to be [Eq. (G.8)]

$$Ai[-\eta^{1/3}\omega^{2/3}(r - r_m)e^{\pm 2\pi i/3}].$$

Hence

$$g^{(1,2)} \propto \frac{1}{r} Ai\left\{\left[\frac{2(1-b_m)}{p}\right]^{1/3} \omega^{2/3}\left(p - \frac{r}{v_m}\right)e^{\pm 2\pi i/3}\right\}. \tag{8.311}$$

We also get this result directly from Eq. (8.300) after verifying that $\Phi = \eta^{1/3}(r - r_m)$ and $[\Phi/Q]^{1/4} = \eta^{-1/6}$. Hence the full expressions at $r \sim r_m$ are

$$g_l^{(1,2)}(r) \simeq \frac{2\sqrt{\pi}e^{\mp \pi i/3}\sqrt{(vv_m)}f_0}{(\omega^{5/6})r}\left[\frac{2(1-b_m)}{p}\right]^{-1/6}$$
$$\times\ Ai\left\{\left[\frac{2(1-b_m)}{p}\right]^{1/3} \omega^{2/3}\left(p - \frac{r}{v_m}\right)e^{\pm 2\pi i/3}\right\}, \tag{8.312}$$

$$\frac{\partial g_l^{(1,2)}(r)}{\partial r} \simeq \frac{-2\sqrt{\pi}e^{\pm(\pi i/3)}f_0}{(\omega^{1/6})r}\left(\frac{v}{v_m}\right)^{1/2}\left[\frac{2(1-b_m)}{p}\right]^{1/6}$$
$$\times\ Ai'\left\{\left[\frac{2(1-b_m)}{p}\right]^{1/3} \omega^{2/3}\left(p - \frac{r}{v_m}\right)e^{\pm 2\pi i/3}\right\}. \tag{8.313}$$

Having derived the explicit expressions for the radial eigenfunctions appropriate for a radially heterogeneous earth model, the recipe for any generalized ray is as follows:

1. We choose a particular generalized ray from the "rainbow expansion" of the homogeneous sphere solution and replace the functions $h_l^{(1,2)}(kr)$ in the Legendre sum by their counterparts $g_l^{(1,2)}(r)$, and the functions

$$\frac{d}{d(kr)}\,h_l^{(1,2)}(kr) \quad \text{by} \quad \left\{\frac{v}{\omega}\frac{d}{dr}\,g_l^{(1,2)}(r)\right\}.$$

2. Using Eq. (8.14b), we replace the Legendre sum by its equivalent Watson integral according to the relation

$$\sum_{l=m}^{\infty}(2l+1)f_l P_l^m(\cos\theta) = -2\int_{C_2} f_{\nu-1/2}(r,\omega)E^{(1)}_{\nu-1/2,m}(\cos\theta)\nu\, d\nu, \quad (8.314a)$$

where $f_{\nu-1/2}$ is assumed to be an even function in ν.

3. We invoke the fundamental relation of mode–ray duality [cf. Eq. (8.80)] in the form of a variable-change $\nu = p\omega$ where p is now a complex ray parameter. The Watson-integral assumes the form

$$2\omega^2\int_{\Gamma} f_{p\omega-1/2}(r,p,\omega)E^{(1)}_{p\omega-1/2,m}(\cos\theta)p\, dp. \quad (8.314b)$$

The significance of this procedure is illustrated in the following example, in which we consider a P wave emerging from a strike-slip source at $\theta = 0$, $r = r_h$ in a homogeneous sphere. Its radial component, before it strikes any discontinuity, is worked out from Eq. (8.250). Substituting therein $s = i\omega$ and using Eq. (D.96), we obtain the Fourier-transformed displacements ($r > r_h$)

$$(u_r)_P = \frac{iU_0(\omega)dS}{4\pi r_h^2}\left(\frac{\beta}{\alpha}\right)_h^2 \sin 2\phi \sum_{l=2}^{\infty}(2l+1)P_l^2(\cos\theta)j_l(k_\alpha r_h)h_l'^{(2)}(k_\alpha r). \quad (8.315)$$

Recapitulating the three steps indicated above, we find the radial component of the direct P wave from the same source in a radially inhomogeneous earth model, for $r > r_h$, to be

$$(u_r)_P = \frac{i\alpha_h\omega U_0(\omega)dS}{4\pi r_h^2}\left(\frac{\beta}{\alpha}\right)_h^2 \sin 2\phi \int_{\Gamma} g^{(1,2)}_{p\omega-1/2}(r_h)g'^{(2)}_{p\omega-1/2}(r)E^{(1)}_{p\omega-1/2,2}(\cos\Delta_0)p\, dp, \quad (8.316)$$

where we currently denote $\theta = \Delta_0$. In this expression, $g_l^{(1)}(r_h)$ is used when the ray departs upward, whereas $g_l^{(2)}(r_h)$ is used when the ray departs downward.

The contour Γ is shown in Fig. 8.32. We have already noted in Eq. (8.14b) that the Watson poles have been moved to the negative real axis. It is then convenient to get rid of the minus sign in Eq. (8.14b) by reversing the direction of the equivalent path C_2'. The result is the path Γ_1 adequate for a direct ray with no turning point. When the ray has a turning point (departs downward) the function $f_{\nu-1/2}$ in Eq. (8.14b), namely, $g_l^{(2)}(r_h)g_l'^{(2)}(r)$, is not even in ν and consequently, the result, Eq. (8.14b), is not valid, as it stands. Simple transformations incorporating Eq. (8.295), however, show that Eq. (8.314b) still holds provided the path of integration is taken as in Fig. 8.32d.

We are next concerned with the evaluation of the integral in Eq. (8.316). Although integrals such as this can be evaluated numerically, it is sufficient for many practical purposes in seismology, especially at body-wave periods of less than 40 s, to replace the eigenfunctions $g^{(1,2)}_{p\omega-1/2}$ and $E^{(1)}_{p\omega-1/2,2}$ by their asymptotic approximations. Therefore, inserting

$$E^{(1)}_{p\omega-1/2,2}(\cos\Delta_0) \sim (2\pi\sin\Delta_0)^{-1/2}p^{3/2}\omega^{3/2}e^{-ip\omega\Delta_0+5\pi i/4}$$

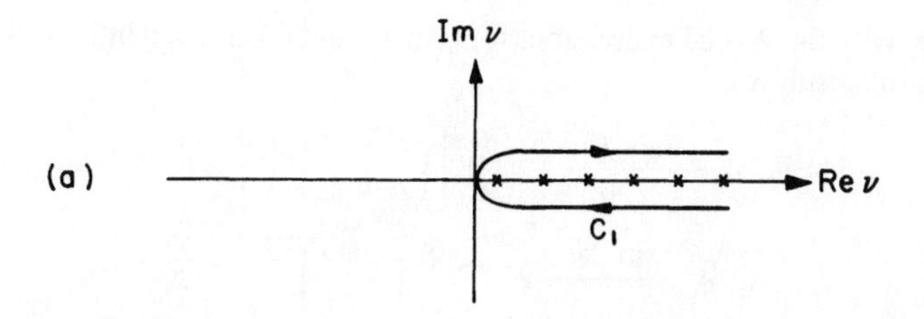

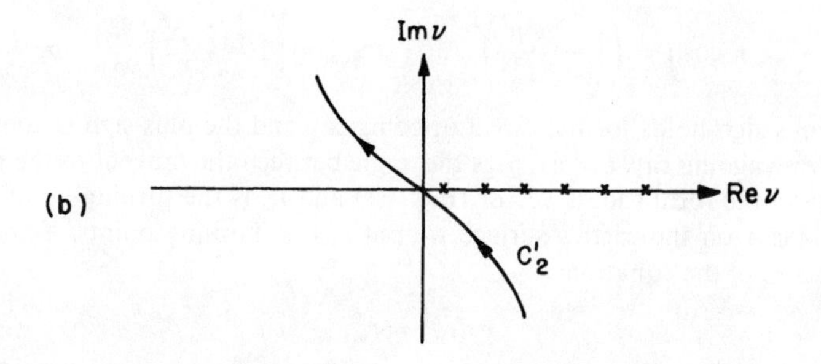

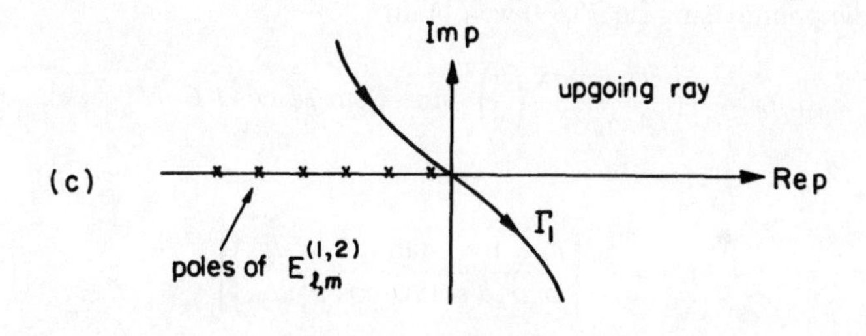

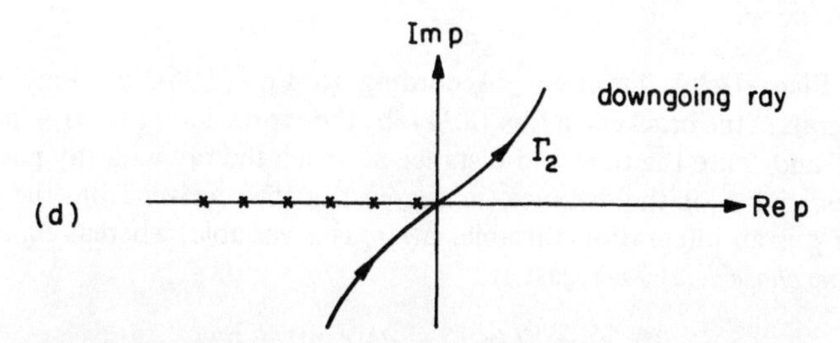

Figure 8.32. Integration paths in the complex ray-parameter plane used to evaluate generalized rays. (a) Watson's contour, (b)–(d) modified Watson contours.

together with the WKBJ expressions for $g_l^{(1,2)}(r_h)g_l'^{(2)}(r)$ as given in Eqs. (8.298)–(8.299), the result is

$$(u_r)_P = -\frac{i\omega U_0(\omega)dS}{4\pi}\left(\frac{\beta}{\alpha}\right)_h^2\left[\left(\frac{\omega}{2\pi}\right)^{1/2}\frac{\alpha_h^2 f_0}{rr_h^3}\right]$$
$$\times\frac{\sin 2\phi}{\sqrt{(\sin\Delta_0)}}e^{\mp\pi i/4}\int_\Gamma\left(\frac{\cos i}{\cos i_h}\right)^{1/2}e^{-i\omega J}p^{5/2}\,dp,$$
$$J(r,p) = p\Delta_0 + \left[\int_{r_m}^{r}\left(\frac{\cos i}{\alpha}\right)dr \mp \int_{r_m}^{r_h}\left(\frac{\cos i}{\alpha}\right)dr\right], \tag{8.317}$$
$$\sin i = \frac{p\alpha}{r},\qquad \cos i = \left(1-\frac{\alpha^2p^2}{r^2}\right)^{1/2},\qquad \cos i_h = \left(1-\frac{\alpha_h^2p^2}{r_h^2}\right)^{1/2},\qquad \alpha_h = \alpha(r_h).$$

The minus sign holds for the direct upgoing ray and the plus sign is adequate for the downgoing ray. $i = i(r, p)$ is the angle between the tangent to the ray at level r and the local radius vector (Fig. 7.1) and r_m is the turning-point level. For a sensor on the earth's surface we put $r = a$. Turning points, if any, are the roots r_m of the equation

$$p\alpha(r_m) = r_m.$$

Because the integrand has only a single saddle point on the real p axis, the contour Γ is drawn parallel to the p axis, just below it (for Re $p < 0$) and through the saddle point. Using Eq. (8.99) we obtain

$$(u_r)_P = \frac{-i\omega U_0(\omega)dS}{4\pi\alpha_h}\left(\frac{\beta}{\alpha}\right)_h^2\sin^2 i_h\sin 2\phi\cos i\{G_P e^{-i\omega T}\}$$

where

$$G_P = \frac{\alpha_h}{r}\left[\frac{\rho_h}{\rho}\cdot\frac{1}{r_h\alpha}\frac{\tan i_h}{\sin\Delta_0\cos i}\left|\frac{dp}{d\Delta_0}\right|\right]^{1/2} \tag{8.318}$$

is the divergence coefficient derived in Eq. (7.154). T is the ray travel time and the extra factor ($\cos i$) arises from resolving the longitudinal motion in the radial direction.

8.7.4.1. Phase-Delay Function. According to Eq. (7.96), we may replace the integrals in the brackets in Eqs. (8.317) by the expression $\{T(r, p) - p\Delta(r, p)\}$ where T and Δ are the time and distance at which the ray with ray parameter p arrives, and Δ_0 is the distance (in degrees) of the receiver from the source. Because p is an integration variable, $\Delta(r,p)$ is a variable, whereas Δ_0 is fixed. Hence the *phase delay* J is recast as

$$J(r, p) = \{T(r, p) - p\Delta(r, p)\} + p\Delta_0. \tag{8.319}$$

According to Eqs. (7.95) and (7.96), $\partial T/\partial p = p(\partial\Delta/\partial p)$. The saddle points occur at values of p_0 such that $\partial J/\partial p = 0$, or at $\Delta(r, p_0) = \Delta_0$, i.e., at just the ray parameters for which there is a ray between source and sensor. Near such a saddle point, p_0, a Taylor series expansion gives

$$J(r, p) = T(r, p_0) + \frac{1}{2}(p - p_0)^2\left(-\frac{\partial\Delta}{\partial p}\right)_{p=p_0}. \tag{8.320}$$

Equation (8.320) has some interesting consequences: First, consider the simplest case where there is just one solution to the saddle-point equation, i.e., a single ray between source and sensor. If $(d/dr)(v/r) < 0$ (i.e., the rays do not intersect), then $\partial i_h/\partial\Delta > 0$ for both upgoing and downgoing rays from the source, whereas $\cos i_h > 0$ for upgoing rays and <0 for downgoing rays. Consequently, $[-\partial\Delta/\partial p] > 0$ for a ray with a turning point and <0 for a ray without it. Second, consider a downgoing ray (i.e., the ray has a turning point) with ray parameter p. We effect a small perturbation in the vertical plane containing source and sensor such that the ray parameter $p + \delta p$. The perturbed ray then passes through a point somewhat displaced radially from the original turning point. Then, putting $p - p_0 = \delta p$ in Eq. (8.320), we find that the corresponding perturbation in the travel time is

$$\delta T = J(r, p) - T(r, p_0) = -\frac{1}{2}(\delta p)^2\left(\frac{d^2T}{d\Delta^2}\right)^{-1}. \tag{8.321}$$

However, $\delta T > 0$ means that the travel time along the ray is a local minimum. Hence, whenever $d^2T/d\Delta^2 < 0$ for a ray with a turning point, the ray path has a *minimum-time* property, whereas $d^2T/d\Delta^2 > 0$ for a ray with a turning point will indicate a *maximum-time* property. Normally, as can be clearly seen from observed travel-time curves of seismic waves, $d^2T/d\Delta^2$ is negative. Core reflections usually have $d^2T/d\Delta^2 > 0$ but because their rays do not have turning points, they still have the minimum-time property.

If the equation $\Delta(r, p) = \Delta_0$ has more than one solution in p, the contour Γ_2 in the complex p plane will be deformed to cross each saddle point by the steepest descent path. This corresponds to ray arrivals to the sensor over different paths. Each ray has its own ray parameter and the arrivals correspond to separate branches on the T–Δ curve as we have shown in Figs. 7.9 and 7.10 and also discussed in detail in Section 7.1.5.

Near each saddle point, the phase factor $J(r, p)$ is given by Eq. (8.320). It becomes clear from this form that the sign of $d^2T/d\Delta^2$ will influence the orientation of the path of steepest descent. Following our discussion in Section E.3, we know that such a path has the property that

$$e^{-i\omega J} = e^{-i\omega T(p_0)}e^{-A^2},$$

where A is real. It follows from Eq. (8.320) that

$$p - p_0 = |A|\left[\frac{2}{\omega}\left|\frac{d^2T}{d\Delta^2}\right|\right]^{1/2}\exp\left[-\frac{\pi i}{4}\operatorname{sgn}\left(\frac{d^2T}{d\Delta^2}\right)\right]. \tag{8.322}$$

Therefore, the path of integration makes an angle $+\pi/4$ with the real p axis when $d^2T/d\Delta^2 < 0$ and an angle of $-\pi/4$ when $d^2T/d\Delta^2 > 0$.

Figure 8.33a shows a T–Δ curve with a triplication, and Fig. 8.33b gives the integration path. For each successive arrival, $d^2T/d\Delta^2$ changes its sign. The steepest descent contribution from each saddle point then has a frequency-independent factor of either $e^{\pi i/4}$ or $e^{-\pi i/4}$, giving rise to a $\pi/2$ phase shift between successive saddle points.

8.7.4.2. Fresnel Diffraction Integral at a Shadow Boundary. We shall prove Eq. (8.201) and thus furnish a useful tool for a study of the influence of the earth's core on the observed seismic amplitudes in the vicinity of the shadow edge. To this end we start from the integral in Eq. (8.198), in which the spherical Hankel functions are replaced by the WKBJ approximations of $g_l^{(1,2)}(r)$ as given in Eq. (8.298). Substituting $v = p\omega$, the integral becomes

$$\int_{p_1(\theta_s)}^{\infty} \frac{p^{7/2}}{(\cos i \cos i_h)^{1/2}} e^{-i\omega J}\, dp. \tag{8.323}$$

Let p_0 be the parameter of a ray emerging at θ close to the shadow edge at θ_s. Expanding J about θ_s and using Eq. (8.320) with Δ replaced by θ, we have

$$e^{-i\omega J} = e^{-i\omega T(p_0)} \exp\left[i \frac{\omega}{2} \frac{(p - p_0)^2}{\partial p/\partial \theta}\right].$$

The integral in Eq. (8.323) then leads to the Fresnel integral $\int_0^X e^{-(\pi i/2)\tau^2}\, d\tau$ where

$$X = \frac{\omega[p_0 - p_1(\theta_s)]}{(-\pi\omega(\partial p/\partial\theta))^{1/2}}. \tag{8.324}$$

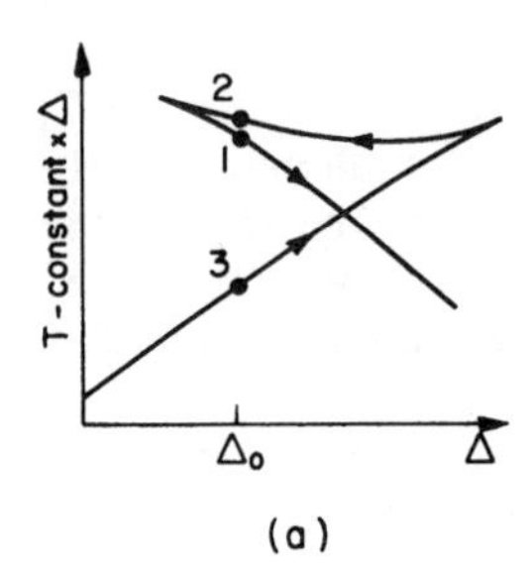

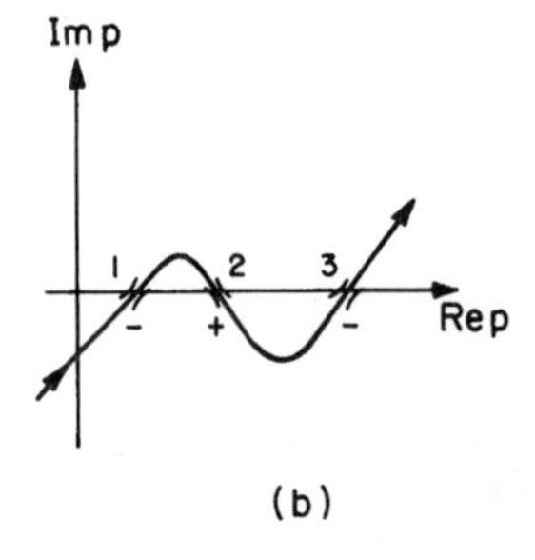

Figure 8.33. Travel-time curve of a generalized ray showing a triplication and its associated integration path in the complex p plane. (a) Three arrivals at $\Delta = \Delta_0$. The first and third fall on forward branches and the second on a receding branch. (b) Saddle points corresponding to the three arrivals. Signs are those of $d^2T/d\Delta^2$ at the saddle points.

Table 8.11. Corrections to Ray Theory, for 2-s P Waves, in a Model with Geometric Shadow Boundary at 97°

					Fresnel factor	
$\theta°$	p (s/rad)	$-\partial p/\partial\theta$ (s/rad²)	Time between PcP and P (s)	$X(\theta)$	Amplitude	Phase (°)
90	266.4	154	0.65	−0.969	1.12	9.51
91	263.8	143	0.40	−0.791	1.01	13.58
92	261.5	124	0.25	−0.643	0.91	15.12
93	259.5	107	0.15	−0.498	0.80	15.41
94	257.8	86	0.07	−0.371	0.71	13.86
95	256.4	75	0.03	−0.231	0.63	10.77
96	255.2	65	0.01	−0.107	0.57	6.01
97	254.4	50	0.0	0.0	0.50	0.0

Expanding

$$p(\theta) = p_1(\theta_s) + (\theta - \theta_s)\frac{\partial p}{\partial \theta} + \cdots$$

we arrive at Eq. (8.201). In a homogeneous mantle (Example 7.1), $-\partial p/\partial\theta = D/4v_0$ and Eq. (8.201) leads directly to Eq. (8.196), as it should. The effect of the earth's core on the grazing P wave is shown in Fig. 8.34.

The *Fresnel factor*

$$F(X) = \frac{1}{2} - \frac{1}{\sqrt{2}}\exp\left[\frac{\pi i}{4}\right]\int_0^X \exp\left[-\left(\frac{\pi i}{2}\right)\tau^2\right]d\tau \tag{8.325}$$

is a correction to ray theory for the effect of the core on the P wave at finite frequencies. Table 8.11 gives calculated values of the amplitude and phase of $F(X)$ for 2-s P-waves in an inhomogeneous earth model (App. L). For a 2-s wave, the shadow edge is shifted to $\theta = 97°$ [see Eq. (8.101c) and our discussion following Eq. (8.352) in Section 8.7.6].

8.7.5. Airy Theory of the Spherical Reflection–Transmission Coefficients

The integral in the complex p plane associated with a generalized ray may have (up to a function of ω) the general form

$$\int_\Gamma g^{(1,2)}(r_h)g'^{(2)}(r)\left\{\prod_{j,k} R_j T_k\right\}E^{(1)}(\cos\Delta_0)p\,dp, \tag{8.326}$$

where the expression in the braces stands for the product of all the reflection and transmission coefficients needed to specify the ray in question. After the

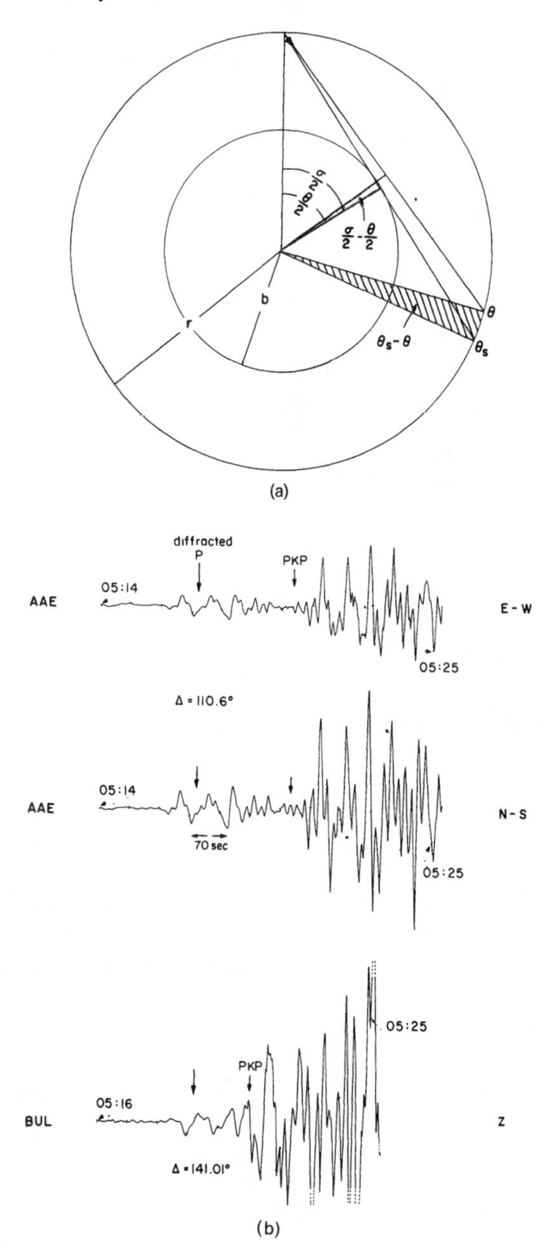

Figure 8.34. (a) Fresnel diffraction at the boundary of the core's shadow. (b) Seismogram showing diffracted P waves, Rat Island, 4 February, 1965 recorded at Addis Ababa (AAE) and Bulawayo (BUL).

application of the WKBJ approximation to the functions $E^{(1)}$ and $g_l^{(1,2)}$, one is usually left with an expression containing $\{\prod_{j,k} R_j T_k\}$ evaluated at the relevant saddle points. This expression can be approximated on four levels, depending on the particular ray at hand and the desired accuracy:

1. At high frequencies (large ω), where a discontinuity is not near a turning point of P and S rays, or if one of these rays does not intersect a discontinuity at a near-grazing angle, we may safely assume $|\omega\xi| \gg 1$ [Eq. (8.306)] and use plane-wave reflection–transmission (R/T) coefficients for plane boundaries. $\prod_{j,k}$ is then frequency independent and p dependent.
2. Away from turning points and angles of grazing incidence (high ξ) but at periods not exceeding 40 s or so, sphericity must be taken into account. The R/T coefficients are then expressed in terms of the spherical Hankel functions $h^{(1,2)}((\omega/v_0)r)$ where v_0 is a constant velocity. The R/T coefficients then become frequency dependent.
3. Same conditions as in case 2, taking into account also small gradients near the discontinuity. Spherical Hankel functions are replaced by their WKBJ counterparts [Eq. (8.298)].
4. Reflection–transmission coefficients are expressed in terms of Airy-function ratios, which incorporate all corrections required for the curvature of the boundary, frequency dependence, and earth structure near the boundary. Periods are limited to about 40 s for P waves and a single turning point is assumed.

As we already know [Sects. 3.2–3.5, 5.1, 7.2.5], the R/T coefficients are products of amplitude ratios and another factor that, for homogeneous plane layers, is given by Eq. (7.178) and for homogeneous spherical layers is a ratio of spherical Hankel functions as in Eq. (8.276). Expressed in terms of the generalized radial eigenfunctions, it can have any one of the 16 forms, for any given interface $r = r_0$,

$$\frac{g^{(1,2)}(v_j, r_0)}{g^{(1,2)}(v_i, r_0)} = e^{i(0,\,\pm\pi/3)}\left[\frac{\xi_j}{\xi_i}\frac{q_i}{q_j}\right]^{1/2}\frac{H_{1/3}^{(1,2)}(\omega\xi_j)}{H_{1/3}^{(1,2)}(\omega\xi_i)} \qquad i, j = \alpha, \beta.$$

To obtain the *corrected amplitude ratios* for any particular generalized ray we employ an artifice already used by us in Example 7.9 [Eqs. (7.9.14)–(7.9.16)]: By Snell's law, $\cos i = (1 - p^2v^2/r^2)^{1/2}$, and from Eqs. (8.304)–(8.305)

$$\frac{\partial g^{(1,2)}/\partial r}{g^{(1,2)}} = \left[\frac{\pm i\omega}{v}\right]\left\{e^{\pm\pi i/6}\,\frac{H_{2/3}^{(1,2)}(\omega\xi)}{H_{1/3}^{(1,2)}(\omega\xi)}\right\}\left[1 - \frac{p^2v^2}{r^2}\right]^{1/2}. \tag{8.327}$$

Denoting the expression in the braces by $\Xi^{(1,2)}$ we use the asymptotic series of the Hankel functions (App. H) to show that

$$\lim_{|\omega\xi|\to\infty} \Xi^{(1,2)} = 1.$$

We may then set the correspondence

$$\cos i \Rightarrow \frac{v}{\pm i\omega}\frac{(\partial/\partial r)g_l^{(1,2)}(r)}{g_l^{(1,2)}(r)} = \Xi^{(1,2)}\left[1 - \frac{p^2v^2}{r^2}\right]^{1/2} = C^{(1,2)}(r, p, \omega). \quad (8.328)$$

For real values of p, $C^{(1)}$ and $C^{(2)}$ are complex conjugate. If $p^2v^2/r^2 > 1$, then the branches chosen to make $g^{(1)}$ and $g^{(2)}$ analytic in p are such that

$$C^{(1,2)} \sim \pm i\left[\frac{p^2v^2}{r^2} - 1\right]^{1/2} \quad \text{as} \quad \omega \to \infty. \quad (8.329)$$

It may be shown that $C^{(1)}$ and $C^{(2)}$ are analytic functions of p, with singularities consisting of a string of poles that have properties similar to branch cuts. The poles of $C^{(1)}$ are at the zeros of $H_{1/3}^{(1)}(\omega\xi)$, namely at the zeros of

$$Ai[-\omega^{2/3}\Phi e^{2\pi i/3}],$$

and the poles of $C^{(2)}$ are at the zeros of $Ai[-\omega^{2/3}\Phi e^{-2\pi i/3}]$. In regions far from a turning point ($|\omega\xi|$ large), the coefficient of $(1 - p^2v^2/r^2)^{1/2}$ in Eq. (8.328) is practically unity and so $C^{(1)} = C^{(2)} = \cos i$. However, in the region of the turning point where $(1 - p^2v^2/r^2)^{1/2}$ and $|\omega\xi|$ are near zero, $C^{(j)}$ remains finite and depends on frequency, curvature, and earth structure near the discontinuity. $C^{(1)}$ may be referred to as the downgoing generalized cosine (because it is defined by the downgoing radial wave function) and $C^{(2)}$ is the upgoing generalized cosine.

To assess the significance of differences between values of $\cos i$ and $C^{(1)}$, we consider an example of an internally reflected core wave. The plane-wave reflection coefficient $*R_{\bar{K}K}$ for P waves within the core, incident upon the core–mantle boundary from below, will now be written in the symbolic form

$$*R_{\bar{K}K} = R_{\bar{K}K}T_{\bar{K}K}, \quad (8.330)$$

where

$$T_{\bar{K}K} = \frac{g^{(2)}(\alpha_c, r_{cm})}{g^{(1)}(\alpha_c, r_{cm})} = e^{-\pi i/3}\left[\frac{H_{1/3}^{(2)}(\omega\xi_c)}{H_{1/3}^{(1)}(\omega\xi_c)}\right], \quad (8.331)$$

$$\xi_c = \int_{r_m}^{r_c}\left[\frac{1}{\alpha^2} - \frac{p^2}{r^2}\right]^{1/2} dr.$$

The dependence of $|*R_{\bar{K}K}|$ on i_c is shown in Fig. 8.35a, and

$$R_{\bar{K}K} = \frac{-\rho_c\alpha_c\cos i_m + \rho_m\cos i_c\{\alpha_m\cos^2 2j_m + 4p^2\beta_m^3r_{cm}^{-2}\cos j_m\cos i_m\}}{\rho_c\alpha_c\cos i_m + \rho_m\cos i_c\{\alpha_m\cos^2 2j_m + 4p^2\beta_m^3r_{cm}^{-2}\cos j_m\cos i_m\}} \quad (8.332)$$

as in Table 7.9. In deriving Eq. (8.332) we have introduced the notation $\varepsilon = \rho_c/\rho_m$, $\alpha' = \alpha_c$, $\rho' = \rho_c$, $e = i_m$, $f = j_m$, $e' = i_c$ and used Snell's law: $\sin i_m = (\alpha_m/r_{cm})p$, $\sin j_m = (\beta_m/r_{cm})p$. The subscripts c and m, respectively, relate to the core and the mantle, $r = r_{cm} = 3473$ km (Fig. 8.35b). The core-grazing parameter at r_{cm} for the J–B A′ earth model (App. L) is $p = 3473/13.64 = 254.6$ s/rad.

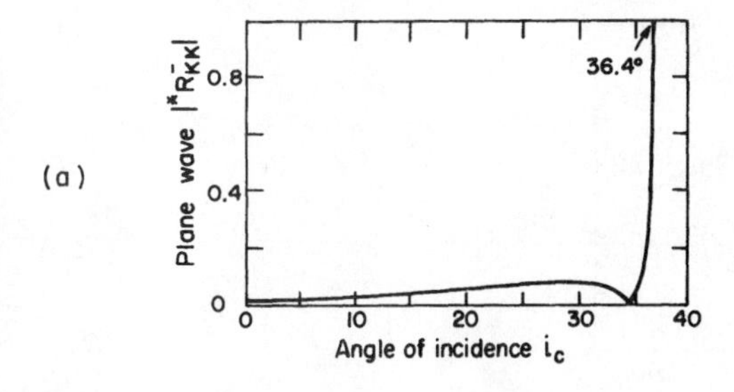

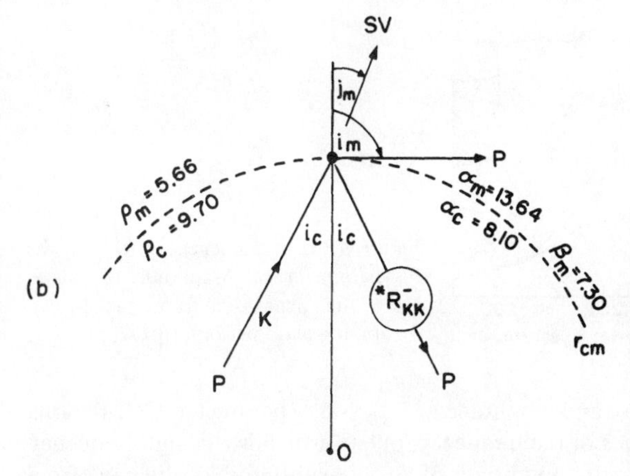

Figure 8.35. Reflection coefficient for a P wave within the core incident at the core–mantle boundary. (a) Reflection coefficient as a function of angle of incidence in the core. (b) Geometry of a core–mantle reflection.

Consider the range $248 \lesssim p \le 262$ s/rad. The incident P wave in the core will be partly reflected at an angle i_c and partly transmitted into the mantle as an SV wave with angle of emergence j_m and a P wave with angle of emergence i_m. Because of Snell's law (Fig. 8.35b)

$$\frac{p}{r_{cm}} = \frac{\sin i_m}{\alpha_m} = \frac{\sin j_m}{\beta_m} = \frac{\sin i_c}{\alpha_c}. \tag{8.333}$$

As i_c is increased, i_m ($>j_m$) will increase until the transmitted P will eventually graze the core. Then $i_m = 90°, i_c = \sin^{-1}[\alpha_c/\alpha_m] = \sin^{-1}[8.10/13.64] = 36.4°$, and $j_m = \sin^{-1}[\beta_m/\alpha_m] = \sin^{-1}[7.30/13.64] = 32.3°$. It can also be seen that i_c and j_m are nowhere near 90° for the entire range $248 \le p \le 262$ s/rad and so $\cos i_c$ and $\cos j_m$ can be retained in Eq. (8.332) and there is no need to replace them by the generalized cosines. As a rule, *cosines of angles that are not close to* 90° *should be left unchanged.* In our case, only $\cos i_m$ should be replaced. To determine where it should be replaced by $C^{(1)}$ and where by $C^{(2)}$, we note that the problem has no downgoing P wave in the mantle, so $C^{(1)}$ does not appear and we have

$$R^{-}_{KK} = \frac{-\rho_c \alpha_c C^{(2)} + \rho_m \cos i_c\{\alpha_m \cos^2 2j_m + 4p^2 \beta_m^3 r_{cm}^{-2} \cos j_m C^{(2)}\}}{\rho_c \alpha_c C^{(2)} + \rho_m \cos i_c\{\alpha_m \cos^2 2j_m + 4p^2 \beta_m^3 r_{cm}^{-2} \cos j_m C^{(2)}\}}. \tag{8.334}$$

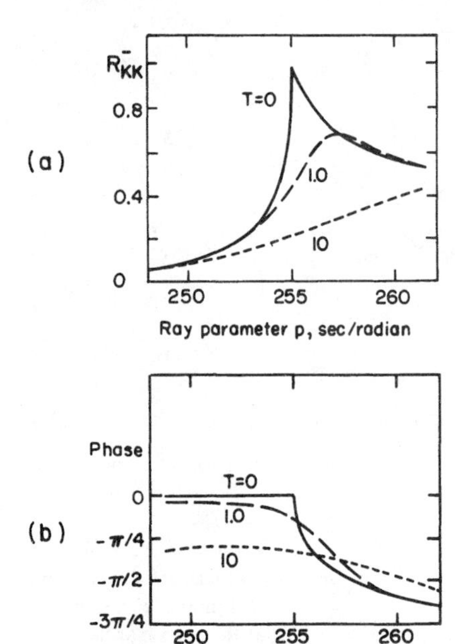

Figure 8.36. Dependence of R^-_{KK} on the ray parameter. (a) Amplitude, (b) phase. Note the departure from plane-wave theory at a wave period of 10 s.

Fig. 8.36 shows the phase and amplitude of R^-_{KK} vs. ray parameter. Calculations are made for three values of the frequency, one of which is the high-frequency limit. Even at $T = 1$ s, the amplitude of R^-_{KK} is substantially different from the plane-wave value throughout the range $254 < p < 257$ s/rad. It is within this range that core phases such as *P4KP* are observed and in this case of multiple internal reflections, R^-_{KK} is raised to the third power. At $T = 10$ s, R^-_{KK} is poorly represented by the plane-wave formula.

8.7.6. Amplitude Theory of Isolated Body Waves in the Real Earth: *PKKP* and *PKJKP*

Given a dislocation source at depth and a structural earth model, we may write down at once the integral representation of any seismic ray of interest. Numerical integration will then yield the wave spectrum at any desired epicentral distance. We give here two examples.

8.7.6.1. The Generalized Ray Integral for the Once Internally Reflected "Rainbow" Wave, *PKKP*. The integral representation for any generalized ray other than the direct wave will differ from Eq. (8.316) only in the presence of the appropriate reflection and transmission coefficients in its integrand. Consider, for example, the *PKKP* wave (Fig. 8.37), which is a *P* wave refracted into the core, reflected back at the core–mantle boundary, and refracted out of the core, into the mantle again. The ray has two turning points in the core. It is a wave of

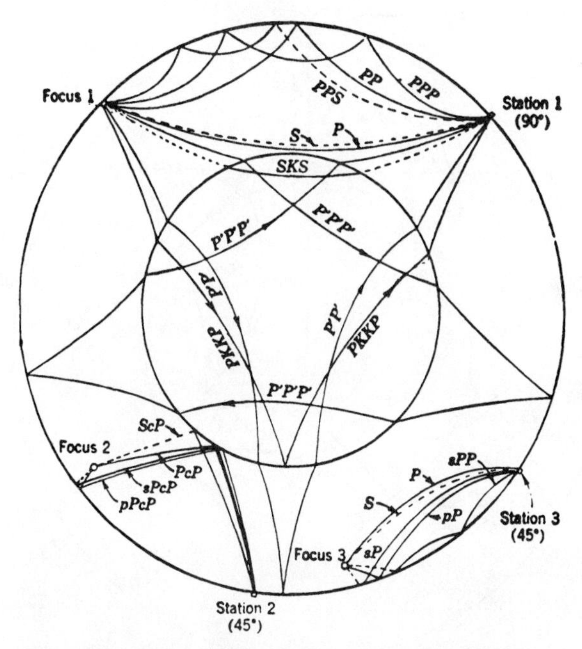

Figure 8.37. Ray-paths of various seismic waves in the earth. The wave *PKPPKPPKP* is denoted *P′P′P′* for brevity.

considerable theoretical interest, because its study may reveal physical conditions within the core and also afford a check on the core's dimensions.

This ray configuration occurs also for light rays refracted and reflected by a sphere of water in air. Enhanced amplitudes near the caustic of once internally reflected rays are responsible for the primary rainbow. Secondary and tertiary rainbows are occasionally observed in the sky at directions associated with caustics in *PKKKP* and *PKKKKP*. Seismologists have reported observations of *PmKP* up to $m = 7$. The caustic for $m = 2$ occurs at distances near 241° (119°), for $m = 4$ at about 415°, and for $m = 7$ at 668°.

The geometry of the *PKKP* wave is depicted in Fig. 8.38. Leaving a surface source at an angle e with the normal, the ray is refracted into the core. Its angle of incidence with the core (i_m) is decreased to i_c because of the velocity drop as it crosses the mantle–core boundary. The ranges of variation of e, i_m, and i_c are determined from Snell's law along the ray

$$p = \frac{a \sin e}{\alpha_0} = \frac{r_{cm} \sin i_m}{\alpha_m} = \frac{r_{cm} \sin i_c}{\alpha_c} = \frac{r \sin i}{\alpha} \text{ (in the core)}, \qquad (8.335)$$

where

$$a = 6371 \text{ km}, \qquad \alpha_0 = 6 \text{ km/s},$$
$$r_{cm} = 3473 \text{ km}, \qquad \alpha_m = 13.64 \text{ km/s}, \qquad \alpha_c = 8.10 \text{ km/s}.$$
$$r_{ic} = 1250 \text{ km}, \qquad \alpha_{ic}^{+} = 9.40 \text{ km/s},$$

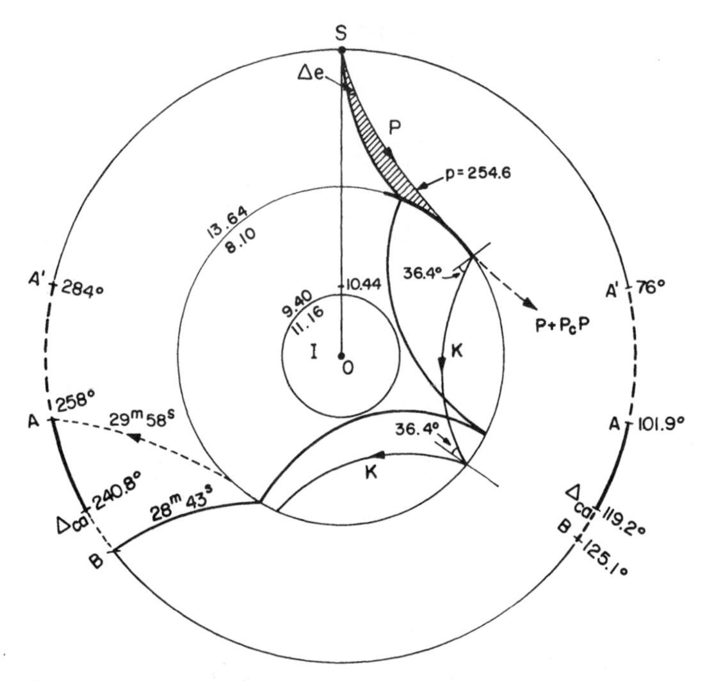

Figure 8.38. Ray-paths of *PKKP* waves. Numbers at the boundaries are for longitudinal velocities. A, A', Δ_{ca}, and B refer to Figure 8.39. Travel times for two particular rays are indicated.

The lower limit for e is determined by the requirement that the *PKKP* ray graze the inner core boundary. This will occur for $p = r_{ic}/\alpha_{ic}^{+} = 133.2$ s/rad. The corresponding e value is 7.2°. The upper limit for e is determined by the core–mantle grazing incidence cutoff at $p = r_{cm}/\alpha_m = 254.6$ s/rad. The corresponding value of e is 13.9°. The ranges of i_m and i_c are then determined from Eq. (8.335) as

$$18.1° \leq i_c \leq 36.4°, \qquad 31.5° \leq i_m \leq 90°.$$

Figure 8.38 shows the ray-path of the *PKKP* wave in the earth and Fig. 8.39 exhibits its observed travel-times.

There are two branches with a cusp at about 125.1°. This is caused by velocity gradients on both sides of the core–mantle boundary and the lower part of the liquid core. As a result there exists a caustic that cuts the earth's surface at about $\Delta = 119.2°$ (240.8°). For a symmetrical surface source (e.g., an explosion or a single vertical force), *PKKP* amplitudes are observable inside a belt that ranges from about 76° to about 125°. Large-amplitude arrivals in the range $76° < \Delta < 101.9°$ (corresponding to the segment AA' in the later travel-time branch of Fig. 8.39) cannot be accounted for by plane-wave theory. These distances are beyond the cutoff dictated by P waves bottoming at the core–mantle boundary while plane-wave theory predicts zero amplitude at $i_m = 90°$. We shall soon show that these waves are predicted by the spherical theory

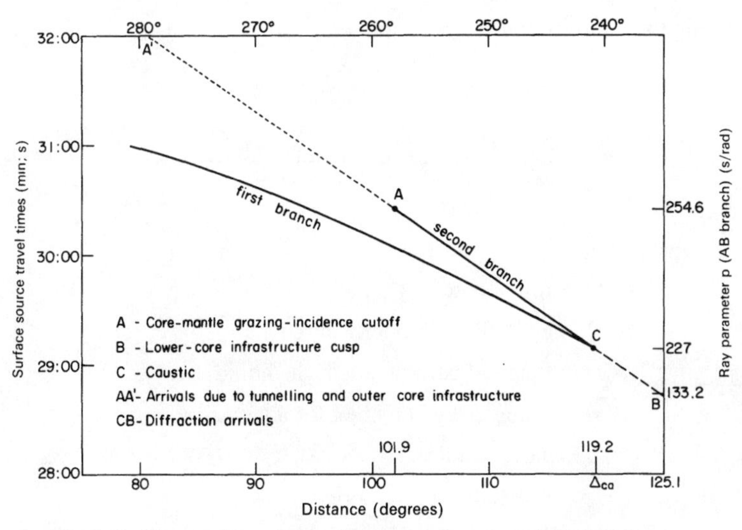

Fig. 8.39. Observed travel-times for the two branches of the *PKKP* wave.

through the analytic properties of the Airy functions. First, however, let us see how far can we go on the strength of plane-wave theory alone.

To this end we write the explicit form of the displacement integral [Eq. (8.316)]

$$(u_r)_{PKKP} = \frac{i\alpha_h \omega U_0(\omega) dS}{4\pi r_h^2}\left(\frac{\beta_h}{\alpha_h}\right)^2 \sin 2\phi \int_\Gamma g^{(2)}_{p\omega-1/2}(r_h) g'^{(2)}_{p\omega-1/2}(r)$$
$$\times \{{}^*T^-_{PK}\}\{{}^*R^-_{KK}\}\{{}^*T^+_{KP}\} E^{(1)}_{p\omega-1/2,2}(\cos\Delta_0) p\, dp. \tag{8.336}$$

Substituting $E^{(1)}$ and the reflection coefficients by their asymptotic approximations, we obtain

$$(u_r)_{PKKP} = \frac{i\omega^{5/2}\alpha_h U_0(\omega) dS}{4\pi(2\pi)^{1/2} r_h^2} e^{5\pi i/4}\left(\frac{\beta}{\alpha}\right)_h^2 \frac{\sin 2\phi}{(\sin\Delta_0)^{1/2}} \int_\Gamma p^{5/2} F_{PKKP}(r, r_h, \omega, p)$$
$$\times \{T^-_{PK} R^-_{KK} T^+_{KP}\} e^{-i\omega p \Delta_0}\, dp, \tag{8.337}$$

where

$$F_{PKKP} = g'^{(2)}(\alpha, r) g^{(2)}(\alpha_h, r_h)\left[\frac{g^{(1)}(\alpha_m, r_{cm})}{g^{(2)}(\alpha_m, r_{cm})}\right]\left[\frac{g^{(2)}(\alpha_c, r_{cm})}{g^{(1)}(\alpha_c, r_{cm})}\right]^2. \tag{8.338}$$

Introducing the WKBJ asymptotic expressions for the radial eigenfunctions in the integrand from Eq. (8.298), namely,

$$g^{(1,2)}(\alpha, r) \sim \frac{\alpha f_0}{\omega r\sqrt{(\cos i)}} \exp\left(\pm i\omega \int_{r_{\min}}^{r} \frac{\cos i}{\alpha}\, dr \mp \frac{\pi i}{4}\right), \tag{8.339}$$

$$g'^{(2)}(\alpha, r) \sim \frac{f_0}{r}\sqrt{(\cos i)} \exp\left(-i\omega \int_{r_{\min}}^{r} \frac{\cos i}{\alpha}\, dr - \frac{\pi i}{4}\right), \tag{8.340}$$

we obtain for a surface source and a surface sensor $\alpha = \alpha_h = \alpha_0, r = r_h = a$

$$e^{-i\omega p\Delta_0} F_{PKKP} \sim \frac{i\alpha_0}{\omega a^2} e^{-i\omega J(p)},$$

$$J(p) = 2\int_{r_{cm}}^{a} \left(\frac{\cos i}{\alpha}\right) dr + 4\int_{r_{min}}^{r_{cm}} \left(\frac{\cos i'}{\alpha'}\right) dr + p\Delta_0, \tag{8.341}$$

where

$$\begin{aligned}
m &= \text{bottom of mantle} \\
c &= \text{top of core} \\
\alpha(r) &= \text{longitudinal velocity in the mantle} \\
\alpha'(r) &= \text{longitudinal velocity in the liquid core} \\
J(p) &= \text{"time-delay" function for a surface source} \\
a &= \text{earth's radius} \\
r_{cm} &= b = \text{radius of the core} \\
r_{min} &= \text{radius of turning point in the liquid core} \\
&= p\alpha'(r_{min}) > 1250 \text{ km} \\
\cos i &= \left(1 - \frac{\alpha^2 p^2}{r^2}\right)^{1/2}, \qquad \cos i' = \left(1 - \frac{\alpha'^2 p^2}{r^2}\right)^{1/2}.
\end{aligned} \tag{8.342}$$

The displacement integral then assumes the form

$$(u_r)_{PKKP} = \frac{\omega^{3/2}\beta_0^2 U_0(\omega) dS}{4\pi(2\pi)^{1/2} a^4} \frac{\sin 2\phi}{(\sin\Delta_0)^{1/2}} e^{\pi i/4} \int_\Gamma p^{5/2}\{T_{PK}^- R_{KK}^- T_{KP}^+\} e^{-i\omega J(p)}\, dp. \tag{8.343}$$

The plane-wave reflection coefficients (Tables 7.7–7.9) are obtained from the WKBJ approximations of the R/T coefficients. Omitting the phase factors, these coefficients are

$$|{}^*R_{KK}^-| = \left|\left[1 - \frac{2\rho_c\alpha_c \cos i_m}{D}\right]\right|, \tag{8.344}$$

$$|{}^*T_{PK}^-| = \left|\left[\frac{2\rho_m\alpha_m \cos i_m \cos 2j_m}{D}\right]\right| \left[\frac{\rho_c}{\rho_m}\left|\frac{\sin 2i_c}{\sin 2i_m}\right|\right]^{1/2}, \tag{8.345}$$

$$|{}^*T_{KP}^+| = \left|\left[\frac{2\rho_c\alpha_c \cos i_c \cos 2j_m}{D}\right]\right| \left[\frac{\rho_m}{\rho_c}\left|\frac{\sin 2i_m}{\sin 2i_c}\right|\right]^{1/2}, \tag{8.346}$$

$$D = \cos i_m[\rho_c\alpha_c + 4p^2\beta_m^3 r_{cm}^{-2}\rho_m \cos j_m \cos i_c] + \rho_m\alpha_m \cos^2 2j_m \cos i_c, \tag{8.347}$$

$$\begin{aligned}
\cos i_m &= \left(1 - \frac{p^2\alpha_m^2}{r_{cm}^2}\right)^{1/2}, \qquad \cos i_c = \left(1 - \frac{p^2\alpha_c^2}{r_{cm}^2}\right)^{1/2}, \\
\cos j_m &= \left(1 - \frac{p^2\beta_m^2}{r_{cm}^2}\right)^{1/2}, \qquad \cos 2j_m = 1 - \frac{2p^2\beta_m^2}{r_{cm}^2}.
\end{aligned} \tag{8.348}$$

At grazing incidence, $i_m = 90°$. Consequently, plane-wave theory predicts that at this angle,

$$|{}^*T_{PK}^-| = 0, \qquad |{}^*R_{KK}^-| = 1, \qquad |{}^*T_{KP}^+| = 0$$

and no energy is transmitted into the core. On the other hand, P-wave energy that is already present in the core cannot leak out efficiently at all angles of incidence i_c, in particular at $i_c = 36.4°$ ($i_m = 90°$) where $|{}^*R_{KK}^-| = 1$ (see Figs. 8.35 and 8.40). We shall soon see that this is not the case in the earth. The saddle points of the integral in Eq. (8.343) are given by the p solutions of the equation

$$\Delta_0 = 2p \int_{r_{cm}}^{a} \frac{dr}{r\sqrt{(r^2/\alpha^2 - p^2)}} + 4p \int_{r_{\min}}^{r_{cm}} \frac{dr}{r\sqrt{(r^2/\alpha'^2 - p^2)}} = \Delta(p) \quad (8.349)$$

and the caustic distance $\Delta = \Delta_{ca}$ is obtained from the condition $\partial\Delta/\partial p = 0$. Given $\alpha(r)$ and $\alpha'(r)$, the two equations can be solved numerically. It is found that there is a caustic at $\Delta_{ca} = 119.2°$ with ray parameter $p_{ca} = 227$ s/rad. The saddle points are located as follows (Fig. 8.41): For $\Delta_0 < \Delta_{ca}$ (assuming $\Delta_0 < 180°$) the equation $\Delta_0 = \Delta(p)$ has two real solutions, p_1 and p_2, as in Fig. 8.41. The figure also shows a steepest descent path Γ crossing each saddle. The corresponding values of the reflection coefficient will differ on the two branches. It turns out that the coefficient $|{}^*R_{KK}^-|$ is much smaller for values of p on the second branch ($p > p_{ca}$), going through zero near $p = 240$ s/rad. Therefore, the contributions from the two branches do not interfere strongly. For $\Delta_0 > \Delta_{ca}$ the equation $\Delta_0 = \Delta(p)$ has two complex conjugate roots, giving saddles as in Fig. 8.41. In this case the steepest descent path is taken across one saddle only. The arrivals corresponding to this contribution are in the range $\Delta_{ca} < \Delta_0 < 125.1°$, extending from the caustic to the cusp B in Fig. 8.39. Because $\text{Im}\, p_1 < 0$, the displacement will be proportional to $e^{-\omega\Delta_0|\text{Im}\, p_1|}$ for Δ_0 values that are sufficiently removed from the caustic ($\partial^2\Delta/\partial p^2 \neq 0$). Integrand singularities nearest $p = p_{ca}$ occur near $p_j = 254.6$ s/rad (A in Fig. 8.39) and are associated with rays in the mantle incident at near-grazing angles ($i_m \sim 90°$) on the core. This singularity, shown in Fig. 8.41, is well away from the path Γ in Fig. 8.41a and Fig. 8.41b. Numerical

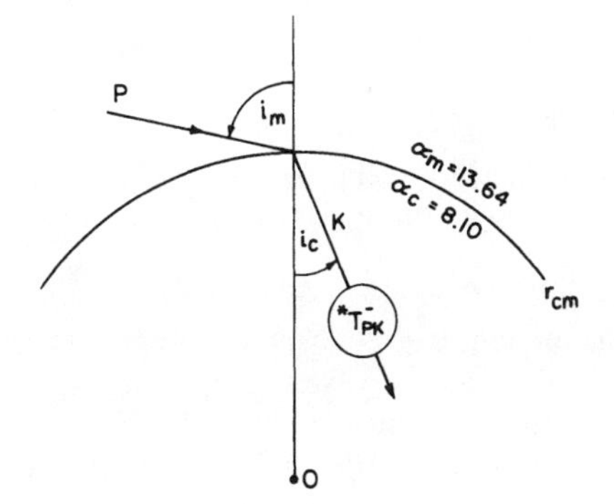

Figure 8.40. P wave transmitted into the core. Velocities are in kilometers per second.

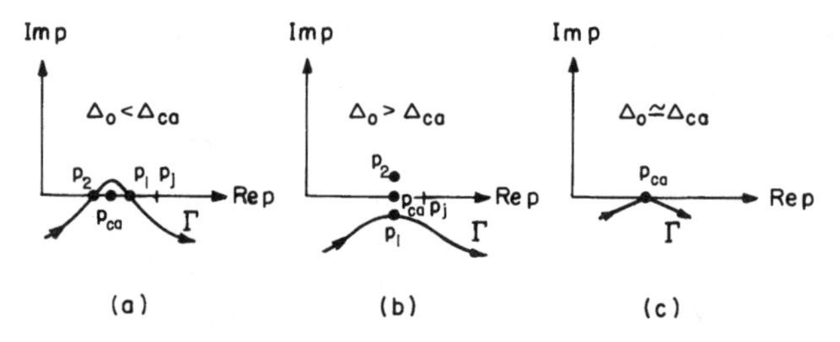

Figure 8.41. Paths of integration and disposition of singularities in the complex ray-parameter plane associated with the *PKKP* wave, at three typical epicentral distances.

evaluation of Eq. (8.343) can be achieved for all Δ_0 near Δ_{ca} with the path Γ of Fig. 8.41.

At epicentral distances close to Δ_{ca} as well as at Δ_{ca} itself, we apply the method used in Section 8.7.3.3 for the *PKP* caustic. We expand $J(\Delta_0, p)$ about (Δ_{ca}, p_{ca}) in the form

$$J(\Delta_0, p) = T_{ca} + (\Delta_0 - \Delta_{ca})p - \frac{1}{6}\left(\frac{\partial^2 \Delta}{\partial p^2}\right)_{p=p_{ca}} (p - p_{ca})^3 \tag{8.350}$$

correct to third order in the small quantities $\Delta_0 - \Delta_{ca}, p - p_{ca}$. Here

$$T_{ca} = T(p_{ca}) = J(\Delta_{ca}, p_{ca}) = \text{travel-time to distance } \Delta_{ca}.$$

Substituting Eq. (8.350) in Eq. (8.343) and evaluating all but the phase at $p = p_{ca}$, we obtain

$$(u_r)_{PKKP} = \left[\frac{\omega^{3/2}\beta_0^2 U_0(\omega)dS}{4\pi(2\pi)^{1/2}a^4}\right]\left[\frac{\sin 2\phi}{(\sin \Delta_0)^{1/2}}\right][e^{\pi i/4}][p^{5/2}T^-_{PK}R^-_{KK}T^+_{KP}]_{p=p_{ca}}\{I\}, \tag{8.351}$$

$$I = 2\pi\left[\frac{2}{\omega(\partial^2\Delta/\partial p^2)_{p_{ca}}}\right]^{1/3} e^{-i\omega[T_{ca}+(\Delta_0-\Delta_{ca})p_{ca}]} Ai\left\{2^{1/3}\omega^{2/3}\left(\frac{\partial^2\Delta}{\partial p^2}\right)^{-1/3}(\Delta_{ca} - \Delta_0)\right\} \tag{8.352}$$

as in Eq. (8.101a). The curvature factor $(\partial^2\Delta/\partial p^2)_{p=p_{ca}}$ has here the numerical value of 2.12×10^{-4} rad/s^2. Equation (8.352) is valid for $\Delta_0 > 180°$. For $\Delta_0 < 180°$ the sign of $(\Delta_{ca} - \Delta_0)$ in the argument of the Airy function must be reversed.

We have already noted in Eq. (8.101c) that the peak displacement is not at $\Delta = \Delta_{ca}$ but is instead displaced by the amount $0.96\omega^{-2/3}(\partial^2\Delta/\partial p^2)^{1/3}$. In the present case the shift is $(0.96\tau^{2/3})$ degrees, where τ is the wave's period in seconds. Therefore the transition zone between the "illuminated" and the shadow regions broadens, (together with a *shadow-boundary shift*) as the frequency decreases. At high frequencies all shifts approach zero and the shadow zone boundary approaches its geometric limit.

Finally, we examine those *PKKP* rays for which i_m is close to 90°. Here the WKBJ formulas break down because the energy transmission of the *P* wave from the mantle to the core is made at the turning point of the *P* wave as it grazes the core. Therefore, we must use here the Airy function representation of the radial eigenfunctions. Applying our former results of Eqs. (8.300)–(8.301), we find that the proper amplitude ratios are

$$\begin{aligned} T_{PK}^{-} &= \rho_m \alpha_m \cos 2j_m \frac{[C^{(1)} + C^{(2)}]}{D}, \\ R_{KK}^{-} &= 1 - \frac{2\rho_c \alpha_c C^{(1)}}{D}, \\ T_{KP}^{+} &= \frac{2\rho_c \alpha_c \cos i_c \cos 2j_m}{D}, \end{aligned} \tag{8.353}$$

where

$$D = C^{(1)}[\rho_c \alpha_c + 4p^2 \beta_m^3 r_{cm}^{-2} \rho_m \cos j_m \cos i_c] + \rho_m \alpha_m \cos^2 2j_m \cos i_c, \tag{8.354}$$

$$\begin{aligned} C^{(1,2)} &= \pm i e^{\pm 2\pi i/3} \left[\frac{2(1-b_m)}{\omega p}\right]^{1/3} \frac{Ai'[Z_m e^{\pm 2\pi i/3}]}{Ai[Z_m e^{\pm 2\pi i/3}]} \\ &= e^{\pm \pi i/6} \frac{H_{2/3}^{(1,2)}(\omega\xi)}{H_{1/3}^{(1,2)}(\omega\xi)} \left[1 - \frac{p^2 \alpha_m^2}{r_m^2}\right]^{1/2}, \end{aligned} \tag{8.355}$$

$$Z_m = \left(p - \frac{r_m}{\alpha_m}\right)\left[\frac{2\omega^2(1-b_m)}{p}\right]^{1/3}, \qquad \xi = \int_{r_m}^{r} \left[\frac{1}{v^2} - \frac{p^2}{r^2}\right]^{1/2} dr. \tag{8.356}$$

As p decreases below the core-grazing value r_m/α_m ($Z_m < 0$), $C^{(1)}$ and $C^{(2)}$ both become approximately equal to $\cos i_m$. However, as p increases above r_m/α_m ($Z_m > 0$) one finds

$$C^{(1,2)} \sim \pm i \left[\frac{p^2 \alpha_m^2}{r_m^2} - 1\right]^{1/2} \quad \text{and} \quad T_{PK}^{-} \to 0.$$

For intermediate values of p, i.e., for *P* waves that have turning points just above the core, the wave function $\{C^{(1)} + C^{(2)}\}$ in T_{PK} effects an exponential decay of the wave amplitude between the turning point and the core–mantle interface.

Therefore, wave energy leaks down into the core. The effect is known as *tunnelling*. This is shown for *P4KP* in Fig. 8.42, where the amplitude of the reflection coefficient is shown as a function of ray parameter. Plane-wave theory predicts a cutoff at $p = 254.6$ s/rad because the ray just grazes the core. The Airy function method permits energy at finite frequencies to cross into the core, even from mantle ray-paths that do not intersect the core. Figure 8.43 shows

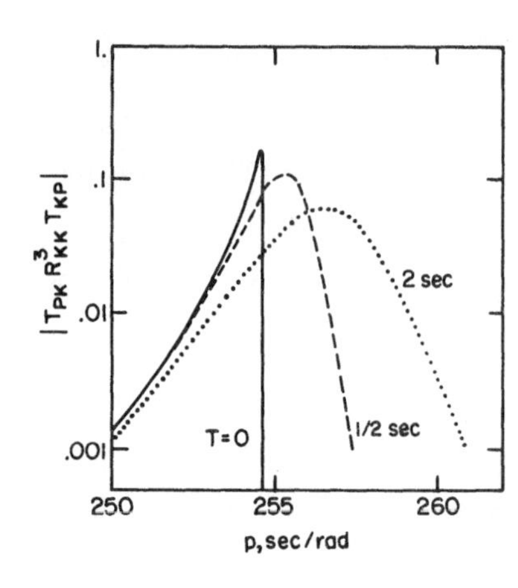

Figure 8.42. Total transmission coefficient for *P4KP* as a function of ray parameter. Plane-wave theory predicts a cutoff at $p = 254.6$ rad/s, when the mantle ray just grazes the core. The Airy method permits energy at finite frequencies to cross into the core, even from mantle ray-paths that bottom above the core.

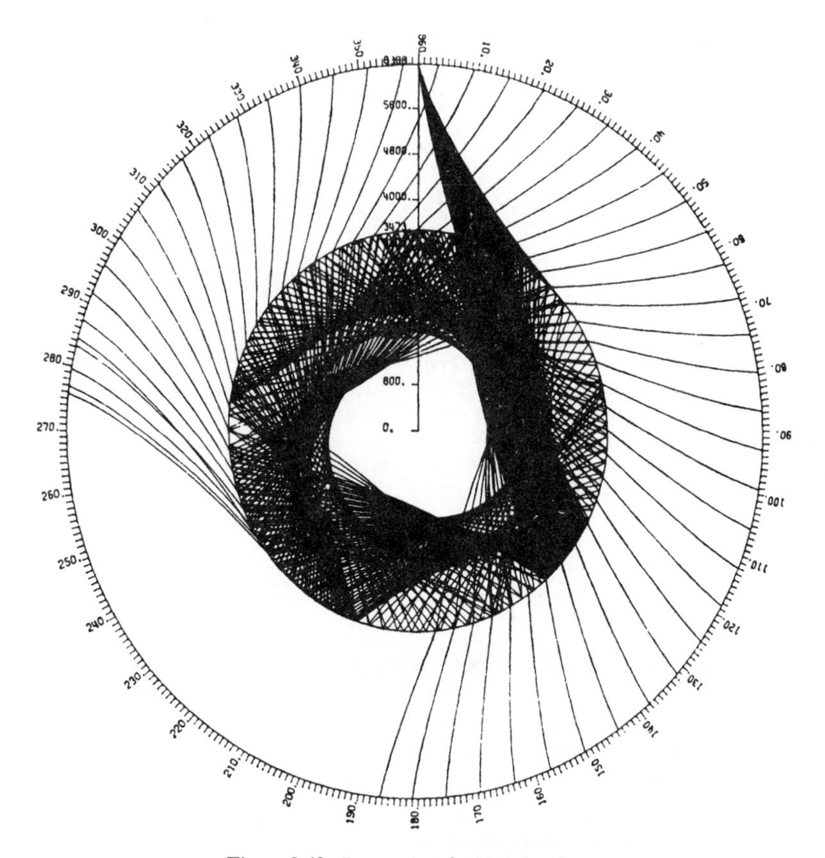

Figure 8.43. Ray-paths of *P7KP* in the earth.

the "rainbow wave" *P7KP*, having six internal reflections from the mantle–core boundary into the outer core.

8.7.6.2. The *PKJKP* Wave. Assume that the inner core of the earth is an elastic solid such that the Navier equations apply to it. Assume also that it can accommodate the boundary conditions applied to liquid and elastic media in contact. Then there is a *theoretical possibility* that a *P* wave is refracted at its boundary into an *S* wave (dashed line in Fig. 8.44). It continues to traverse the outer core as a *P* wave, reaching the earth's surface as a wave labeled *PKJKP*. For a strike-slip source, the radial surface displacement is

$$(u_r)_{PKJKP} = \frac{i\alpha_h \omega^{5/2} U_0(\omega) dS}{4\pi(2\pi)^{1/2} r_h^2} \left(\frac{\beta_h}{\alpha_h}\right)^2 \frac{\sin 2\phi}{(\sin \Delta_0)^{1/2}} e^{5\pi i/4} \int_\Gamma p^{5/2} F_{PKJKP}(a, r_h, \omega, p)$$

$$\times \{T_{PK}^- T_{KJ}^- T_{JK}^+ T_{KP}^+\} e^{-i\omega p \Delta_0} \, dp, \qquad (8.357)$$

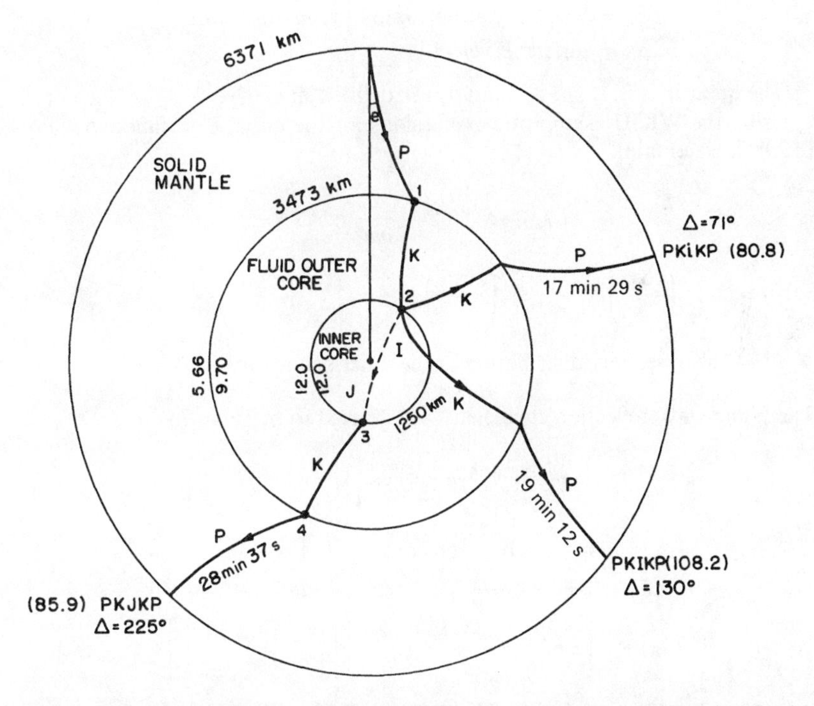

Figure 8.44. Three different interactions with the inner core: External reflection (*PKiKP*), *P* transmission (*PKIKP*), and *S* transmission (*PKJKP*). Travel-times, (minutes and seconds), epicentral distances, (degrees), and ray parameters (s/rad), are indicated for each of the rays.

where

$$F_{PKJKP} = g'^{(2)}(\alpha_0, a)g^{(2)}(\alpha_h, r_h)\left[\frac{g^{(1)}(\alpha_m, b)}{g^{(1)}(\alpha_c, b)}\right]\left[\frac{g^{(1)}(\alpha_c^-, d)}{g^{(1)}(\beta_{ic}, d)}\right]$$
$$\times \left[\frac{g^{(2)}(\beta_{ic}, d)}{g^{(2)}(\alpha_c^-, d)}\right]\left[\frac{g^{(2)}(\alpha_c, b)}{g^{(2)}(\alpha_m, b)}\right] \tag{8.358}$$

ic = inner core
c = core
m = mantle
d = radius of the inner core
α_m = P velocity at bottom of mantle
α_c = P velocity at top of outer core
α_c^- = P velocity at bottom of outer core
β_{ic} = S velocity at top of inner core ($\simeq 3.6$ km/s)
α_{ic} = P velocity at top of inner core
$\beta_{\min}$ = S velocity at bottom of ray in inner core
α_0 = surface P velocity.

The quantities $i_c, i_c^-, \rho_c^-, j_{ic}$ and ρ_{ic} are defined similarly.

Using the WKBJ asymptotic expressions for the radial eigenfunctions [Eq. (8.298)], we obtain

$$F_{PKJKP}e^{-i\omega p\Delta_0} \sim \frac{\alpha_0}{\omega a^2}e^{-i\omega J(p)},$$

$$J(p) = 2\int_b^a\left(\frac{\cos i}{\alpha}\right)dr + 2\int_d^b\left(\frac{\cos i'}{\alpha'}\right)dr + 2\int_{r_{\min}}^d\left(\frac{\cos j''}{\beta''}\right)dr + p\Delta_0, \tag{8.359}$$

β'' = shear velocity distribution in the inner solid core.

The plane-wave reflection coefficients are derived from Table 7.9

$$|{*T_{PK}^-}| = \left|\left[\frac{2\rho_m\alpha_m \cos i_m \cos 2j_m}{D}\right]\right|\left[\frac{\rho_c}{\rho_m}\left|\frac{\sin 2i_c}{\sin 2i_m}\right|\right]^{1/2}, \tag{8.360}$$

$$|{*T_{KJ}^-}| = \left|\left[\frac{-4\rho_c^-\alpha_c^- \cos i_c^- \cos i_{ic}\sin j_{ic}}{\hat{D}}\right]\right|\left[\frac{\rho_{ic}}{\rho_c^-}\left|\frac{\sin 2j_{ic}}{\sin 2i_c^-}\right|\right]^{1/2}, \tag{8.361}$$

$$|{*T_{JK}^+}| = \left|\left[\frac{-2\rho_{ic}\beta_{ic}\cos i_{ic}\sin 2j_{ic}}{\hat{D}}\right]\right|\left[\frac{\rho_c^-}{\rho_{ic}}\left|\frac{\sin 2i_c^-}{\sin 2j_{ic}}\right|\right]^{1/2}, \tag{8.362}$$

$$|{*T_{KP}^+}| = \left|\left[\frac{2\rho_c\alpha_c \cos i_c \cos 2j_m}{D}\right]\right|\left[\frac{\rho_m}{\rho_c}\left|\frac{\sin 2i_m}{\sin 2i_c}\right|\right]^{1/2}, \tag{8.363}$$

$$\hat{D} = \cos\ i_{ic}[\rho_c^-\alpha_c^- + 4p^2\beta_{ic}^3 d^{-2}\rho_{ic}\cos j_{ic}\cos i_c^-] + \rho_{ic}\alpha_{ic}\cos^2 2j_{ic}\cos i_c^-, \tag{8.364}$$

and then expressed in terms of p via Snell's law:

$$p = \frac{a \sin e}{\alpha_0} = \frac{b \sin i_m}{\alpha_m} = \frac{b \sin i_c}{\alpha_c} = \frac{d \sin i_c^-}{\alpha_c^-} = \frac{d \sin i_{ic}}{\alpha_{ic}} = \frac{d \sin j_{ic}}{\beta_{ic}}$$

$$= \frac{r_{\min}}{\beta_{\min}}. \tag{8.365}$$

If $\alpha(r)$, $\alpha'(r)$ and $\beta''(r)$ are known, the relation

$$\Delta_0 = 2p \int_b^a \frac{dr}{r}\left(\frac{r^2}{\alpha^2} - p^2\right)^{-1/2} + 2p \int_d^b \frac{dr}{r}\left(\frac{r^2}{\alpha'^2} - p^2\right)^{-1/2}$$

$$+ 2p \int_{r_{\min}}^d \frac{dr}{r}\left(\frac{r^2}{\beta''^2} - p^2\right)^{-1/2} \tag{8.366}$$

will render $\Delta_0 = \Delta_0(p)$ and consequently $\Delta_0 = \Delta_0(e)$. Figure 8.44 shows the waves *PKiKP* and *PKIKP*. This last wave is shown also in Fig. 8.45.

8.7.7. Pulse-Shapes of Seismic Body Waves

A dislocation point source with time function $g(t)$ sends various body wave pulses through the earth and these are recorded on seismographs on the earth's surface. It is of interest to relate the pulse shapes of these waves to the original source-time function.

We have shown in Eq. (4.191) that the far field in a homogeneous infinite medium behaves as $g'(t)$. Then, in Section 8.5, when dealing with the propagation of *SH* waves in a homogeneous sphere, we met rays with two types of time dependence: $g'(t)$ and $\hat{g}'(t)$ [allied function of $g'(t)$]. The first type was associated with minimum travel-time. The second type had a non-minimum travel-time path because it touched a caustic. We know from Example 7.9 that crossing a caustic introduces a phase advance of $\pi/2$ at all frequencies. Therefore, crossing a caustic in the plane of the ray virtually changes the sign of $d^2T/d\Delta^2$ for that ray and transforms the shape of the body-wave signal into that of a non-minimum-time ray. If the shape of the pulse was $g'(t)$ prior to its encounter with the caustic, it will change to the Hilbert transform of this function (Sect. 3.2.3 and App. F). Triplications in the travel-time curve (Sect. 7.1.5) indicates the existence of three arrivals (Fig. 8.33); two are minimum-time rays and the third, with the reverse branch, is a non-minimum-time ray.

Reflections beyond the critical angle at discontinuities can also cause a constant phase shift that eventually distorts the signal's shape [Sect. 3.2.3]. The radiation pattern at the source introduces a change of polarity according to the sign of the function $F(\lambda, \delta, i_h, \phi_h)$ given in Eq. (4.154). Frequency-dependent phase shifts are caused by the ray's interaction with discontinuities chiefly because of the frequency dependence of reflection and transmission coefficients. If the frequency dependence of the R/T coefficients is neglected, theoretical

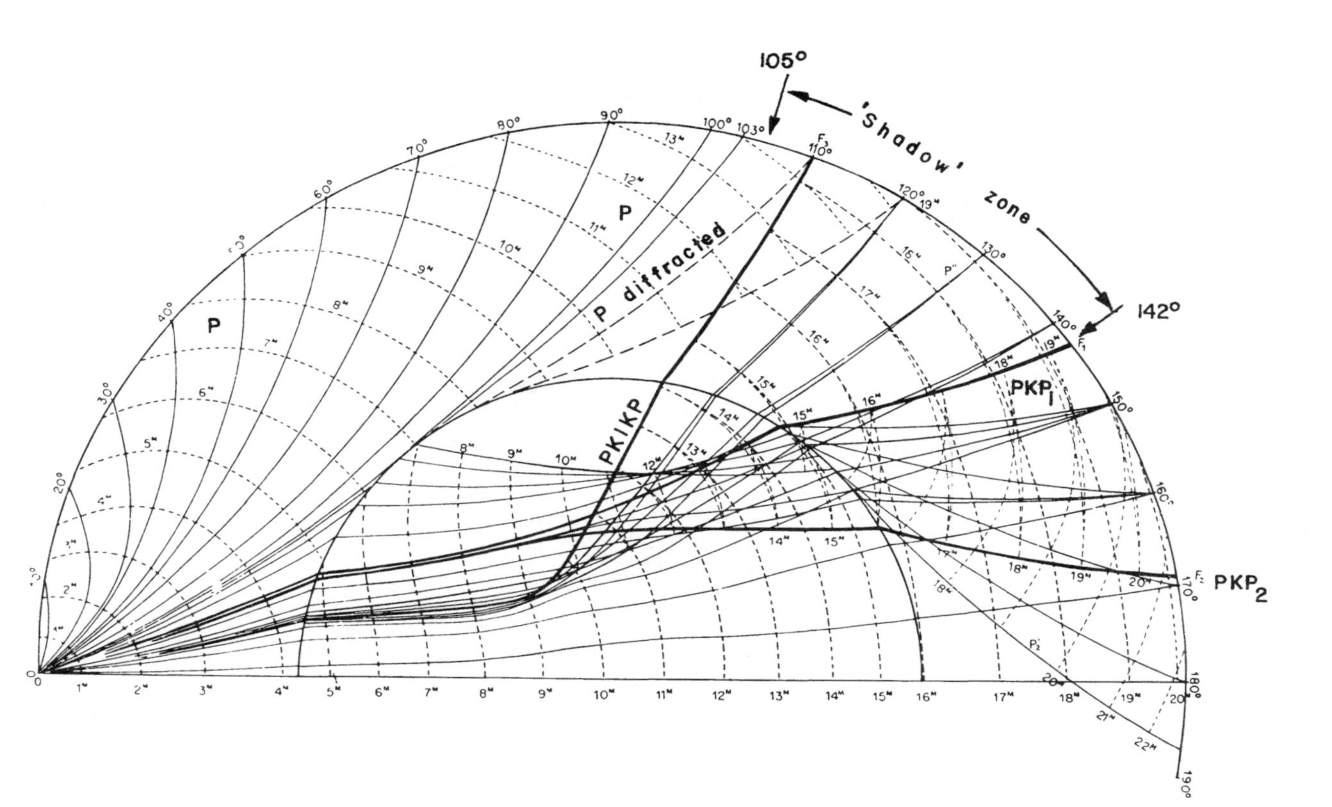

Figure 8.45. Prominent wave arrivals at $\Delta_0 \leq 190°$. Wavefronts and travel-times are shown.

seismograms can easily be calculated, because the pulse shape of a generalized ray can be worked out directly by applying the Fourier transform to the displacement p integral. As far as the frequency dependence is concerned, the displacement spectrum of any generalized ray originating from a dislocation source can be represented by the integral

$$u(\Delta_0, \omega) = A_0 \omega^{3/2} U_0(\omega) \int_\Gamma f(p, \Delta_0) e^{-i\omega J(p, \Delta_0)}\, dp, \qquad (8.367)$$

where f and J are independent of the frequency and A_0 absorbs all the factors that depend neither on ω nor on p. The theoretical seismogram then assumes the form

$$u(\Delta_0, t) = \frac{A_0}{\pi} \int_0^\infty \omega^{3/2} U_0(\omega) \mathrm{Re}\left\{ \int_\Gamma f e^{i\omega(t-J)}\, dp \right\} d\omega. \qquad (8.368)$$

8.7.8. Earth-Flattening Transformation of Wave Amplitudes

Although the earth-flattening transformation was derived in Section 7.3.2 for the purpose of estimating the effect of the earth's curvature on travel-times of seismic waves, it can also be used to calculate first-order corrections of the planar reflection coefficients caused by the sphericity of material discontinuities. We present for simplicity the case of an SH wave reflected from a spherical surface of discontinuity.

Consider the Legendre sum representation of the reflected SH wave from a spherical surface of discontinuity in an infinite homogeneous medium as given by Eq. (8.248) for a vertical strike-slip source. The source and the receiver are above the interface. The constitutive parameters on the source side are μ_1, ρ_1, β_1 and below the interface they are μ_2, ρ_2, β_2. The corresponding Watson integral over the ray parameter p, for $r > r_h$, is

$$(u_\phi)_{\text{reflected}} = \frac{ik_\beta U_0(\omega) dS}{2\pi r_h} \cos 2\phi \frac{\partial}{\partial\theta} \int_{\Gamma_2} h^{(2)}_{p\omega-1/2}(k_{\beta_1} r_h) h^{(2)}_{p\omega-1/2}(k_{\beta_1} r)$$
$$\times \{{}^*R^+_{SSH}\} E^{(1)}_{p\omega-1/2,\,2}(\cos\theta) \frac{dp}{p}, \qquad (8.369)$$

where

$$\{{}^*R^+_{SSH}\} = \left[\frac{h^{(1)}_{p\omega-1/2}(k_{\beta_1} b)}{h^{(2)}_{p\omega-1/2}(k_{\beta_1} b)}\right]$$
$$\times \left[\frac{\left\{1 - (k_{\beta_1} b)\dfrac{h_l'^{(1)}(k_{\beta_1} b)}{h_l^{(1)}(k_{\beta_1} b)}\right\} + \dfrac{\mu_2}{\mu_1}\left\{(k_{\beta_2} b)\dfrac{h_l'^{(2)}(k_{\beta_2} b)}{h_l^{(1)}(k_{\beta_2} b)} - \dfrac{h_l^{(2)}(k_{\beta_2} b)}{h_l^{(1)}(k_{\beta_2} b)}\right\}}{\left\{1 - (k_{\beta_1} b)\dfrac{h_l'^{(2)}(k_{\beta_1} b)}{h_l^{(2)}(k_{\beta_1} b)}\right\} + \dfrac{\mu_2}{\mu_1}\left\{(k_{\beta_2} b)\dfrac{h_l'^{(1)}(k_{\beta_2} b)}{h_l^{(1)}(k_{\beta_2} b)} - 1\right\}}\right]_{l=p\omega-1/2}$$
$$(8.370)$$

When $\mu_2 \to 0$ we fall back on Eq. (8.177).

We now apply the exact earth-flattening transformation to the radial part of the integrand. We already know that the Debye asymptotic approximation yields for $k_\beta b > \nu \gg 1$, $kb = \nu = p\omega$

$$\frac{h'^{(1)}_{\nu-1/2}(k_{\beta_1}b)}{h^{(1)}_{\nu-1/2}(k_{\beta_1}b)} \sim +\frac{i}{k_{\beta_1}}\eta, \qquad \frac{h'^{(2)}_{\nu-1/2}(k_{\beta_1}b)}{h^{(2)}_{\nu-1/2}(k_{\beta_1}b)} \sim -\frac{i}{k_{\beta_1}}\eta, \qquad \eta = [k^2_{\beta_1} - k^2]^{1/2}, \tag{8.371}$$

where k is the horizontal wave number and η is the vertical wave number. Also, as in Eqs. (8.180) and (8.181),

$$\left[\frac{h^{(1)}_{\nu-1/2}(k_{\beta_1}b)}{h^{(2)}_{\nu-1/2}(k_{\beta_1}b)}\right][h^{(2)}_{\nu-1/2}(k_{\beta_1}r_h)h^{(2)}_{\nu-1/2}(k_{\beta_1}r)]$$
$$\simeq \frac{e^{-i\chi}}{k_{\beta_1}rr_h\left(k^2_{\beta_1} - k^2\dfrac{b^2}{r^2}\right)^{1/4}\left(k^2_{\beta_1} - k^2\dfrac{b^2}{r_h^2}\right)^{1/4}}, \tag{8.372}$$

$$\chi = \eta[(r-b) + (r_h - b)] + kb\left[\cos^{-1}\frac{k}{k_{\beta_1}} - \cos^{-1}\frac{kb}{k_{\beta_1}r}\right]$$
$$+ kb\left[\cos^{-1}\frac{k}{k_{\beta_1}} - \cos^{-1}\frac{kb}{k_{\beta_1}r_h}\right]. \tag{8.373}$$

We now let $b \to \infty$ in such a way that $(r - b)$ and $(r_h - b)$ remain fixed and z is directed in the positive radial direction [see Example 8.2]. We then employ the earth-flattening approximation [Eqs. (7.192) and (7.193), Table 7.19]

$$r = be^{z/b} = b\left[1 + \frac{z}{b} + \frac{z^2}{2b^2} + \cdots\right],$$
$$r_h = be^{z_h/b} = b\left[1 + \frac{z_h}{b} + \frac{z_h^2}{2b^2} + \cdots\right], \tag{8.374}$$

and expand the second and third terms of χ in Eq. (8.373) and the denominator in Eq. (8.372) in a Taylor series up to the order $(z/b)^3$ according to the approximation

$$f(x) - f(\varepsilon x) \simeq xf'(x)(1-\varepsilon) + \frac{1}{2}x^2 f''(x)(1-\varepsilon)^2,$$
$$\varepsilon = \frac{b}{r}, \qquad 1 - \varepsilon = \frac{z}{b} - \frac{z^2}{2b^2} + O\left(\frac{z^3}{b^3}\right), \qquad (1-\varepsilon)^2 = \frac{z^2}{b^2} + O\left(\frac{z^3}{b^3}\right). \tag{8.375}$$

Then, with Eq. (H.67) and the relations

$$b\theta \to \Delta, \qquad p\,dp \to \frac{b^2}{\omega^2}k\,dk, \tag{8.376}$$ [7]

[7] In Eqs. (8.376)–(8.392), the symbol Δ denotes the radial cylindrical coordinate and v is the wave slowness.

we are able to replace Eq. (8.369) by its half-space equivalent

$$(u_\phi)_{\text{reflected}} = \frac{U_0(\omega)dS}{4\pi} \cos 2\phi \left[\frac{\theta}{\sin\theta}\right]^{1/2} \times \frac{\partial}{\partial\Delta} \int_{-\infty}^{\infty} H_2^{(2)}(k\Delta)\{R_c\}\, e^{-i\eta(z+z_h)(1+\delta_1)} \frac{k\,dk}{\eta(1+\delta_2)}, \tag{8.377}$$

where R_c is the reflection coefficient and

$$\delta_1 = \frac{k^2}{\eta^2 b}\frac{z^2 + z_h^2}{z + z_h} + O\left(\frac{z^2}{b^2}\right), \qquad \delta_2 = \frac{k^2 + 2\eta^2}{\eta^2}\frac{z + z_h}{b} + O\left(\frac{z^2}{b^2}\right). \tag{8.378}$$

If δ_1 and δ_2 are neglected, the integral reduces to the Sommerfeld integral [see Eq. (7.215)] for plane boundaries. If terms of $O(z^2/b^2)$ are neglected, the δ_i become the first-order corrections for boundary curvature corresponding to a linear approximation to the earth-flattening approximation. This gives a valid description of the reflected wave field in a spherical boundary under the following conditions.

1. $|z/b| \ll 1$. Curvature effects are compensated only in a vicinity above the boundary $r = b$.
2. $|kb| \gg 1$. Solution is limited to high frequencies. Wavelength is much less than the radius b.
3. $|\eta b| \gg 1$. Near-normal and near-grazing incidence are both avoided.
4. Solutions are valid at angular distances greater than several wavelengths from the source if $H_2^{(2)}(k\Delta)$ is replaced by its asymptotic approximation.

The factors $(1 + \delta_i)$ correspond to corrections to the angle of incidence because of curvature above the boundary $r = b$. The effects of curvature below the boundary are contained in the reflection coefficient R_c. For moderate angles of incidence and for $|z_h/b| \ll 1$, these factors can be replaced by unity.

The foregoing results are therefore useful for body waves impinging on major discontinuities in the crust and upper mantle of the earth near critical angles of incidence. For the outer layers of the earth, the spherical reflection coefficient $R_c(k)$ can be approximated by a reflection coefficient appropriate for a plane boundary at $r = b$ but with original velocity distribution, $\beta(z)$, modified according to Eq. (7.193).

The earth-flattening approximation may be made to enter implicitly into the solution by transforming the Watson integral into the form used by us to represent a generalized ray in a layered half-space [cf. Eq. (7.313b)]. This method is now described again for the reflected *SH* pulse from a spherical cavity in an infinite elastic solid. We construct the *Laplace-transformed* solution from Eq. (8.154) by replacing $i\omega$ in it by s. Eq. (8.154) then reads

$$\bar{u}_\phi(r|r_h; s) = \frac{sU_0(s)dS}{4\pi r_h \beta} \cos 2\phi \frac{\partial}{\partial\theta} \sum_{l=2}^{\infty} (2l+1) f_l P_l^2(\cos\theta), \tag{8.379}$$

$$f_l = \frac{-(2/\pi)}{l(l+1)}\left[i_l(\hat{k}r_<)k_l(\hat{k}r_>) - k_l(\hat{k}r_h)k_l(\hat{k}r)\frac{i_l(\hat{k}b) - (\hat{k}b)i_l'(\hat{k}b)}{k_l(\hat{k}b) - (\hat{k}b)k_l'(\hat{k}b)}\right]. \tag{8.380}$$

The Laplace-transformed ScS displacements are then obtained in a straightforward manner

$$\bar{u}_{ScS}(r\,|\,r_h\,;\,s) = \frac{sU_0(s)dS}{2\pi^2 r_h \beta}\cos 2\phi\,\frac{\partial}{\partial\theta}\sum_{l=2}^{\infty}\frac{2l+1}{l(l+1)}\hat{k}_l(\hat{k}r_h)k_l(\hat{k}r)$$

$$\times\, P_l^2(\cos\theta)\left[\frac{i_l(\hat{k}b)}{k_l(\hat{k}b)}\right]R_c, \tag{8.381}$$

where

$$R_c(\hat{k}b) = \frac{1-(\hat{k}b)[i_l'(\hat{k}b)/i_l(\hat{k}b)]}{1-(\hat{k}b)[k_l'(\hat{k}b)/k_l(\hat{k}b)]}, \qquad \hat{k} = \frac{s}{\beta_0}. \tag{8.382}$$

We next apply the approximations (Apps. D and H) valid for x, $\lambda \gg 1$, $\mathrm{Re}(x^2+\lambda^2)^{1/2} \gg 0$

$$k_{\lambda-1/2}(x) = \frac{\pi}{2\sqrt{x}}\frac{e^{-\xi}}{[\lambda^2+x^2]^{1/4}}\left[1+O\left(\frac{1}{(\lambda^2+x^2)^{1/2}}\right)\right],$$

$$i_{\lambda-1/2}(x) = \frac{1}{2\sqrt{x}}\frac{e^{\xi}}{[\lambda^2+x^2]^{1/4}}\left[1+O\left(\frac{1}{(\lambda^2+x^2)^{1/2}}\right)\right], \tag{8.383}$$

$$\xi = [\lambda^2+x^2]^{1/2} - \lambda\,\mathrm{sh}^{-1}\left(\frac{\lambda}{x}\right) = \int^x [x^2+\lambda^2]^{1/2}\frac{dx}{x}, \qquad 0 < \left|\arg\frac{x}{\lambda}\right| < \frac{\pi}{2}$$

$$\frac{i_{\lambda-1/2}(x)}{k_{\lambda-1/2}(x)} \simeq \frac{1}{\pi}e^{2\xi}, \qquad \frac{xi'_{\lambda-1/2}(x)}{i_{\lambda-1/2}(x)} \simeq [\lambda^2+x^2]^{1/2},$$

$$\frac{xk'_{\lambda-1/2}(x)}{k_{\lambda-1/2}(x)} \simeq -[\lambda^2+x^2]^{1/2}, \qquad R_c \simeq \frac{1-[\lambda^2+\hat{k}^2b^2]^{1/2}}{1+[\lambda^2+\hat{k}^2b^2]^{1/2}}.$$

Next, we let Eq. (8.381) undergo the Watson transformation and replace the functions in the integrand by their approximations according to Eqs. (8.383). The resulting integral representation for the ScS wave is

$$\bar{u}(r\,|\,r_h\,;\,s) = i\,\frac{U_0(s)dS}{8\pi r_h(r_h r)^{1/2}}\cos 2\phi\,\frac{\partial}{\partial\theta}\int_{-\infty}^{\infty}\frac{\sec(\pi\lambda)P^2_{\lambda-1/2}(-\cos\theta)e^{-\Phi(\lambda)}}{\lambda(\lambda^2+\hat{k}^2r_h^2)^{1/4}(\lambda^2+\hat{k}^2r^2)^{1/4}}R_c(\lambda)d\lambda,$$

$$\Phi(\lambda) = \xi(\hat{k}r_h) + \xi(\hat{k}r) - 2\xi(\hat{k}b). \tag{8.384}$$

Through the substitution $\lambda = iv$, Eq. (8.384) is modified into

$$\bar{u}(r\,|\,r_h\,;\,s) = \frac{U_0(s)dS}{4\pi r_h(rr_h)^{1/2}}\cos 2\phi$$

$$\times\,\frac{\partial}{\partial\theta}\left[\frac{1}{2i}\left\{\int_0^{i\infty} - \int_0^{-i\infty}\right\}\frac{\mathrm{sech}(\pi v)P^2_{iv-1/2}(-\cos\theta)e^{-\Phi(iv)}}{[\hat{k}^2r_h^2-v^2]^{1/4}[\hat{k}^2r^2-v^2]^{1/4}}R_c(iv)\frac{dv}{v}\right]. \tag{8.385}$$

However, the integrand is analytic in ν and therefore

$$\frac{1}{2i}\left\{\int_0^{i\infty} - \int_0^{-i\infty}\right\} = \text{Im}\int_0^{i\infty},$$

$$\bar{u}(r\,|\,r_h; s) = \frac{U_0(s)dS}{4\pi r_h(rr_h)^{1/2}}\cos 2\phi$$

$$\times \frac{\partial}{\partial\theta}\,\text{Im}\int_0^{i\infty}\frac{\text{sech}(\pi\nu)P^2_{i\nu-1/2}(-\cos\theta)e^{-\Phi(i\nu)}}{[\hat{k}^2 r_h^2 - \nu^2]^{1/4}[\hat{k}^2 r^2 - \nu^2]^{1/4}}R_c\frac{d\nu}{\nu}. \quad (8.386)$$

Applying Eqs. (8.14a), (H.67), and (H.57), we have

$$P^2_{i\nu-1/2}(-\cos\theta) \sim ie^{\pi\nu}\left[P^2_{i\nu-1/2}(\cos\theta) - \frac{2i}{\pi}Q^2_{i\nu-1/2}(\cos\theta)\right] + O(e^{-2\pi\nu}),$$

$$P^2_{i\nu-1/2}(\cos\theta) - \frac{2i}{\pi}Q^2_{i\nu-1/2}(\cos\theta) \sim -\nu^2\left(\frac{\theta}{\sin\theta}\right)^{1/2}H_2^{(1)}(i\nu\theta)[1 + O(\nu^{-1})].$$

$$(8.387)$$

Moreover, because

$$\text{sech}(\pi\nu) \sim 2e^{-\pi\nu}[1 + O(e^{-2\pi\nu})]$$

we arrive at the asymptotic approximation valid for large values of ν,

$$\text{sech}(\pi\nu)P^2_{i\nu-1/2}(-\cos\theta) \sim \frac{4}{\pi}\nu^2\left[\frac{\theta}{\sin\theta}\right]^{1/2}K_2(\nu\theta). \quad (8.388)$$

With the final change of variable $\nu = sv$, the transformed displacements assume the form

$$\bar{u}_{ScS}(r\,|\,r_h; s) = \frac{sU_0(s)dS}{\pi^2 r_h(rr_h)^{1/2}}\cos 2\phi\left[\frac{\theta}{\sin\theta}\right]^{1/2}$$

$$\times \frac{\partial}{\partial\theta}\,\text{Im}\int_0^{i\infty}K_2(sv\theta)f(-v^2)\,e^{-\Psi(s,v)}v\,dv,$$

$$f(-v^2) = \left[\frac{r_h^2}{\beta^2} - v^2\right]^{-1/4}\left[\frac{r^2}{\beta_0^2} - v^2\right]^{-1/4}, \quad (8.389)$$

$$\Psi(s, v) = \left\{\int_{r=b}^{r_h} + \int_{r=b}^{r}\right\}[x^2 - \nu^2]^{1/2}\frac{dx}{x} - \ln R_c$$

$$\sim s\left[\left\{\int_b^{r_h} + \int_b^{r}\right\}\left\{\frac{1}{\beta_0^2} - \frac{v^2}{r^2}\right\}^{1/2}dr\right] - \ln(-1),$$

because, for large s,

$$\ln R_c \sim \ln\frac{1 - s[(b^2/\beta_0^2) - v^2]^{1/2}}{1 + s[(b^2/\beta_0^2) - v^2]^{1/2}} \sim \ln(-1). \quad (8.390)$$

Let us now compare the integral in Eq. (8.389) with its half-space equivalent in Eq. (7.313b). The two integrands have the same structure with two exceptions: θ in Eq. (8.389) vs. Δ in Eq. (7.313b) and the radical $(1/\beta_0^2 - v^2/r^2)^{1/2}$ in Eq.

(8.389) vs. the half-space expression $(1/\beta_0^2 - v^2)^{1/2}$. We remember, however, that Snell's law in terms of the half-space slowness variable v reads [Eq. (7.382)] $v = (\sin i)/\beta_0 = p$ = ray parameter, whereas in spherical coordinates we have

$$v = \frac{r}{\beta_0} \sin i = p = \text{ray parameter.}$$

Therefore, the variable v/r in a sphere corresponds to v in a half-space and $sv\theta = (s(v/r))(r\theta) = s(v/r)\Delta$ will correspond to the $(sv\Delta)$ in Eq. (7.313b).

In view of this analogy we may proceed from Eq. (8.389), as we did in Eq. (7.313b), and develop the theory of generalized rays in a layered sphere parallel to the theory of generalized rays in a layered half-space, as developed in Section 7.6. We must note, however, that the inversion of the Laplace integral as given for plane layers in Eq. (7.320) is valid for spherical layers only if the frequency-independent reflection and transmission coefficients for a plane interface are used. This evidently limits the validity of the method to wavelengths that are small compared with some characteristic distance in the earth. We take this distance to be such that the corresponding function $\sin\theta/\theta \le 0.998$, i.e., $\Lambda \le 600$ km. If we take $c = 8$ km/s for S waves we have $T \le 75$ s. For P waves we take $c = 15$ km/s and find $T \le 40$ s. So P and S pulses with these limiting periods could reasonably be expected to be well represented by generalized ray theory. Therefore, at those frequencies where ray theory renders a sufficient approximation to normal-mode theory, the concept of a generalized ray can be extended to spherical-earth models simulated by a stack of homogeneous layers.

Consider for example the Laplace-transformed surface displacements for the ray-path shown in Fig. 8.46, in which the factor in front of the integral

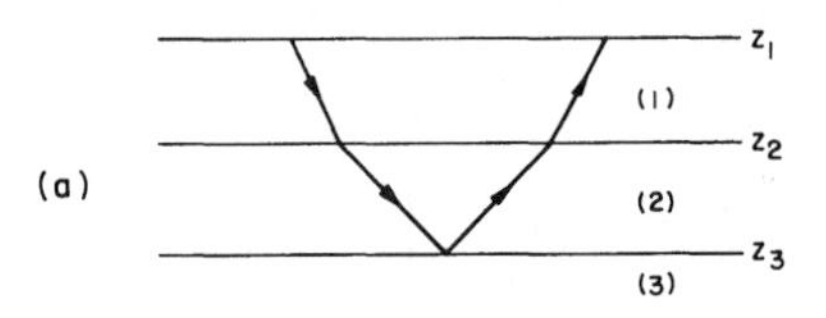

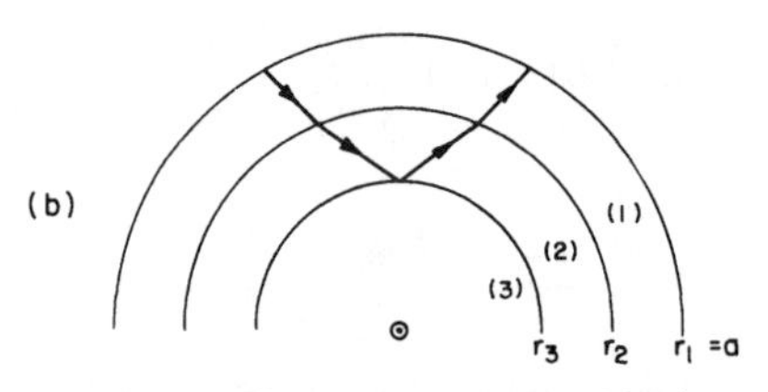

Figure 8.46. Earth - flattening approximation for generalized rays. (a) Flat layers; (b) spherical layers.

has been suppressed. It has a form similar to Eq. (7.313b),

$$\frac{1}{\pi i}\left(\frac{\theta}{\sin\theta}\right)^{1/2}\int_{-i\infty}^{i\infty} K_m(sv\theta)F\left(\frac{v}{a}\right)T_{12}\left(\frac{v}{r_2}\right)R_{23}\left(\frac{v}{r_3}\right)T_{21}\left(\frac{v}{r_2}\right)e^{-s\Psi(v)} \times v^{2l+2m+1}f(-v^2)dv, \quad (8.391)$$

where

$$\Psi(v) = 2\int_{r_3}^{r_2}\left(\frac{1}{\beta_2^2}-\frac{v^2}{r^2}\right)^{1/2} dr + 2\int_{r_2}^{a}\left(\frac{1}{\beta_1^2}-\frac{v^2}{r^2}\right)^{1/2} dr. \quad (8.392)$$

T, R are the plane transmission and reflection coefficients, respectively, at the internal interfaces, and $F(v/a)$ is the combined source and sensor term.

Bibliography

Alsop E (1964) Spheroidal free periods of the earth observed at eight stations around the world. Bull. Seism. Soc. Am 54: 755–776.

Alterman Z, Jarosch H, Pekeris CL (1961) Propagation of Rayleigh waves in the earth. Geophys Jour Roy Astron Soc (London) 4: 219–241.

Alterman Z, Kornfeld P (1966) Effect of a fluid core on propagation of an *SH*-torque pulse from a point-source in a sphere. Geophys 31: 741–763.

Ansell JH (1978) On the scattering of *SH* waves from a point source by a sphere. Geophys Jour Roy Astron Soc (London) 54: 349–387.

Ben-Menahem A (1962) Radiation of seismic body waves from a finite moving source in the earth. Jour Geophys Res 67: 345–350.

Ben-Menahem A (1964) Mode-ray duality. Bull Seismol Soc Amer 54: 1315–1321.

Benndorf H (1905) Über die Art der Fortpflanzung der Erdbebenwellen in Erdinnern. Mitt Erdbebenkomm Wien 29: 1407.

Biswas NN, Knopoff L (1970) Exact earth-flattening calculations for Love waves. Bull Seismol Soc Amer 60: 1123–1137.

Bremmer H (1949) Terrestrial Radio Waves. Elsevier, New York, 343 pp.

Bromwich TJ I'A (1904) The caustic by reflection, of a circle. Amer Jour Math 26: 33–44.

Brune JN (1964) Travel times, body waves, and normal modes of the earth. Bull Seismol Soc Amer 54: 2099–2128.

Brune JN (1966) *P* and *S* wave travel times and spheroidal normal modes of a homogeneous sphere. Jour Geophys Res 71: 2959–2965.

Brune JN, Nafe JE, Alsop LE (1961) The polar phase shift of surface waves on a sphere. Bull Seismol Soc Amer 51: 247–257.

Cayley A (1857) A memoir upon caustics. Phil Trans Roy Soc (London) A147: 273–312.

Chapman CH (1969) Seismic wave diffraction theory. Ph.D. Thesis, Cambridge University.

Duwalo G, Jacobs JA (1959) Effects of a liquid core on the propagation of seismic waves. Canad Jour Phys 37: 109–128.

Engdahl ER (1968) Seismic waves within Earth's outer core: multiple reflection. Science 161: 263–264.

Engdahl ER, Flinn EA, Romney CR (1970) Seismic waves reflected from the Earth's inner core. Nature 228: 852–853.

Franz W (1954) Über die Greenschen Funktionen des Zylinders und der Kugel. Zeit Naturforschung 9a: 705–716.

Gans R (1915) Fortpflanzung des Lichtes durch ein inhomogenes Medium. Ann Phys 47: 709–736.
Gilbert F (1976) The representation of seismic displacements in terms of travelling waves Geophys Jour Roy Astron Soc (London) 44: 275–280.
Gilbert F, Helmberger DV (1972) Generalized ray theory for a layered sphere. Geophys Jour Roy Astron Soc (London) 27: 57–80.
Gilbert F, MacDonald GJF (1960) Free oscillations of the earth. I. Toroidal oscillations. Jour Geophys Res 65: 675–693.
Gutenberg B (1951) *PKKP*, *P'P'* and the Earth's core. Trans Amer Geophys Union 32: 373–390.
Herrin E (1968) Seismological tables for *P* phases. Bull Seismol Soc Amer 58: 1193–1241.
Imai I (1954) Die Beugung Electromagnetischer Wellen an einen Kreiszylinder. Zeit Phys 137: 31–48.
Jacobs JA (1975) The Earth's Core. Academic Press, London, 253 pp.
Jeans JH (1923) The propagation of earthquake waves. Proc Roy Soc (London) A102: 554–574.
Jeffreys H, Lapwood ER (1957) The reflection of a pulse within a sphere. Proc Roy Soc (London) A241: 455–479.
Jobert N (1960) Calcul de la dispersion des ondes de Love de grand periode a la surface de la terre. Ann Geophys 16: 393–413.
Keller JB, Rubinow SI, Goldstein M (1963) Zeros of Hankel functions and poles of scattering amplitudes. Jour Math Phys 4: 829–832.
Kennett BLN (1972) Seismic wave scattering by obstacles on interfaces. Geophys Jour Roy Astron Soc (London) 28: 249–266.
Landisman M, Satô Y, Nafe J (1965) Free vibrations of the earth and the properties of its deep interior regions. Part I. Density. Geophys Jour Roy Astron Soc (London) 9: 439–502.
Langer RE (1931) On the asymptotic solutions of ordinary differential equations, with an application to the Bessel functions of large order. Trans Amer Math Soc 33: 23–64.
Langer RE (1937) On the connection formulas and the solutions of the wave equation. Phys Rev 51: 669–676.
Luneburg RK (1964) Mathematical Theory of Optics. University of California Press, Berkeley.
Nussenzveig HM (1965) High-frequency scattering by an impenetrable sphere. Ann Phys 34: 23–95.
Odaka T (1978) Derivation of asymptotic frequency equations in terms of ray and normal mode theory and some related problems. Jour Phys Earth 26: 105–121.
Pekeris CL (1965) Asymptotic theory of the free torsional oscillations of the earth. Proc Natl Acad Sci (US) 53: 1253–1261.
Richards PG (1973) Calculation of body waves, for caustics and tunnelling in core phases. Geophys Jour Roy Astron Soc (London) 35: 243–264.
Richards PG (1976) On the adequacy of plane-wave reflection/transmission coefficients in the analysis of seismic body waves. Bull Seismol Soc Amer 66: 701–717.
Sato R (1969) Body wave amplitude near shadow boundary: *SH* wave. Jour Phys Earth 17: 1–12.
Sato R, Lapwood ER (1977) The asymptotic distribution of torsional eigenfrequencies of a spherical shell. I. Jour Phys Earth 25: 257–282.
Satô Y (1961) Normal mode interpretation of the sound propagation in whispering galleries. Nature 189: 475–476.

Satö Y, Usami T, Landisman M, Ewing M (1963) Basic study of the oscillations of a sphere, Part V. Propagation of Torsional disturbances on a radially heterogeneous sphere. Case of a homogeneous mantle with a liquid core. Geophys Jour Roy Astron Soc (London) 8: 44–63.

Satô Y, Usami T, Landisman M (1968) Theoretical seismograms of torsional disturbances excited at a focus within a heterogeneous spherical earth—case of a Gutenberg–Bullen A′ earth model. Bull Seismol Soc Amer 58: 133–170.

Scholte JGJ (1956) On Seismic Waves in a Spherical Earth. Publ. 65, Roy. Nederland Meteorological Institute. The Hague, 55 pp.

Shimamura H, Sato R (1965) Model experiments on body waves—travel times, amplitudes, wave-forms and attentuation. Jour Phys Earth 13: 10–33.

Shimshoni M, Ben-Menahem A (1970) Computation of the divergence coefficient for seismic phases. Geophys Jour Roy Astron Soc (London) 21: 285–294.

Shimshoni M, Sylman Y, Ben-Menahem A (1973) Theoretical amplitudes of body waves from a dislocation source in the earth. II. Core phases. Phys Earth Planet Int 7: 59–91.

Sills LB (1978) Scattering of horizontally polarized shear waves by surface irregularities. Geophys Jour Roy Astron Soc (London) 54: 319–348.

Singh SJ, Ben-Menahem A (1969a). Decoupling of the vector wave equation of elasticity for radially heterogeneous media. Jour Acoust Soc Amer 46: 655–660.

Singh SJ, Ben-Menahem A (1969b) Asymptotic theory of body waves in a radially heterogeneous earth. Bull Seismol Soc Amer 59: 2039–2059.

Singh SJ, Ben-Menahem A, Shimshoni M (1972) Theoretical amplitudes of body waves from a dislocation source in the earth. I. Core reflections. Phys Earth Planet Int 5: 231–263.

Sommerfeld A (1964) Optics. Academic Press, New York, 383 pp.

Sommerfeld A (1964) Partial Differential Equations in Physics. Academic Press, New York, 335 pp.

Sylman Y, Shimshoni M, Ben-Menahem A (1974) Theoretical amplitudes of body waves from a dislocation source in the earth. III. *P*, *S* and surface reflections. Phys Earth Planet Int 8: 130–147.

Van der Pol B, Bremmer H (1937) The diffraction of electromagnetic waves from an electrical point-source round a finitely conducting sphere, with application to radiotelegraph, and the theory of the rainbow. Phil Mag (Ser 7) 24: 141–176, 825–864.

Vered M, Sylman Y, Ben-Menahem A (1974) Generalized reflection and transmission coefficients for seismic sources in a multi-layered spherical earth model. Pure Appl Geophys 112: 821–835.

Watson GN (1918) The diffraction of electric waves by the earth. Proc Roy Soc (London) A95: 83–99.

Woodhouse JH (1972) Diffraction by anomalous regions in the earth's mantle. Geophys Jour Roy Astron Soc (London) 32: 295–324.

Yanovskaya TB (1958) The dispersion of Rayleigh waves in a spherical layer. Izv Geophys Ser 801–817 (English translation, 461–467).

Zoeppritz K, Geiger L, Gutenberg B (1912) Über Erdbebenwellen. V Nachr Kgl Ges Wiss Göttingen Math-Phys Kl, 121–206.

CHAPTER 9

Atmospheric and Water Waves and Companion Seismic Phenomena

He that calleth for the waters of the sea, and poureth them out upon the face of the earth.

(*Amos*, 9;6)

9.1. The Navier–Stokes Equation

9.1.1. Introduction

Elastic waves, traveling in the solid body of the earth, sometimes convert part of their energy to other waves that propagate in the water or in the air. The opposite also happens. Large explosions in the atmosphere or in the ocean produce waves that travel through the solid earth. For example, a volcanic eruption in the ocean will simultaneously generate air, sea, and earth waves. Submarine earthquakes generate *gravity sea waves* (*tsunamis*) that carry the destructive powers of the source to distant shores, thousands of kilometers away. Also, seismic waves from major earthquakes are known to cause huge waves (*seiches*) in lakes and canals several thousand kilometers away. We therefore cannot fully explore the nature of earthquake waves without having some knowledge about wave phenomena in fluids.

A fluid is a material that can usually sustain only small shearing stresses and so offers little resistance to change of shape. Where this assumption does not hold, shear forces must be considered. They are called friction or viscous stresses and depend on the velocity gradient in the flowing fluid normal to the direction of flow. Such stresses appear, for example, when water flows in pipes and channels or when a solid body moves through a liquid. Fluids in which friction must be taken into account are called *viscous*.

Under usual conditions, water undergoes negligible density changes even under high pressure. Fluids that behave in this way can be considered *incompressible*. Such a material is a true liquid and has practically a constant volume and density. Fluids that are free from friction and are incompressible and homogeneous are called *ideal fluids*. Gases, such as air, have low resistance to change of volume and are *compressible*.

9.1.2. Stress Tensor in Fluids

In an inviscid fluid, the stress on any plane is normal to that plane and is independent of the orientation of the plane. Therefore, remembering that a negative sign indicates a compression, the stress dyadic is

$$\mathfrak{T} = -p\mathfrak{J}. \tag{9.1}$$

The scalar p is the magnitude of the compressive normal stress and is known as the *hydrostatic pressure* or simply *pressure*.

Because in fluids the displacements in general are large, there is not much physical sense in associating the fluid particles with a displacement field. Instead, we introduce here a velocity field $\mathbf{v} = \partial\mathbf{u}/\partial t$. Consequently, for a fluid, we define the *rate of strain dyadic*

$$\mathfrak{U} = \frac{\partial\mathfrak{E}}{\partial t} = \frac{1}{2}(\nabla\mathbf{v} + \mathbf{v}\nabla), \tag{9.2}$$

so that

$$\text{trace } \mathfrak{U} = \text{div } \mathbf{v}.$$

We turn now to viscous fluids and define a *Newtonian fluid* as a viscous fluid for which the shearing stress is linearly related to its rate of deformation, i.e.,

$$\mathfrak{T} = -p\mathfrak{J} + \mathbf{D} : \mathfrak{U}. \tag{9.3}$$

$\mathbf{D}$ is a tensor whose elements are the viscosity coefficients of the fluid. The term $(-p\mathfrak{J})$ represents the state of stress possible in a fluid at rest ($\mathfrak{U} = 0$). We assume that the elements of $\mathbf{D}$ may depend on the temperature, but not on the stress or the rate of deformation. If the fluid is isotropic, D_{ijkl} is an isotropic tensor of the fourth order. Moreover, because $\mathfrak{T}$ and $\mathfrak{U}$ are symmetrical, D_{ijkl} can be expressed in terms of two independent constants a and b [see Eq. (1.67)]

$$D_{ijkl} = a\delta_{ij}\delta_{kl} + b(\delta_{ik}\delta_{jl} + \delta_{il}\delta_{jk}). \tag{9.4}$$

A substitution of this relation in Eq. (9.3) yields the *Navier–Poisson law* for isotropic linear viscous fluids

$$\mathfrak{T} = -p\mathfrak{J} + a\mathfrak{J}\,\text{div } \mathbf{v} + 2b\mathfrak{U}. \tag{9.5}$$

The identification of the parameters a and b with intrinsic physical properties of the viscous fluid is done in the following way. The effect of any general distortion of a fluid element can be broken into two contributions; that arising from the change in the geometric shape of the element at a constant volume (*shear distortion*) and that arising from change of volume when the geometric shape remains similar (*dilatation*). The first kind is governed by a pure shearing strain rate with no change of volume (i.e., with zero trace), namely,

$$\mathfrak{T}_1 = 2\eta\left(\mathfrak{U} - \frac{1}{3}\mathfrak{I}\,\mathrm{div}\,\mathbf{v}\right),$$

where η is the coefficient of shear friction (viscosity). When the fluid is expanding without shear ($\mathfrak{U} = 0$), there arise frictional effects from pure expansion and the pressure is affected by the rate of expansion. The resulting stress dyadic is

$$\mathfrak{T}_2 = \bar{\lambda}\mathfrak{I}\,\mathrm{div}\,\mathbf{v}.$$

The total stress dyadic is

$$\begin{aligned}\mathfrak{T} &= (-p + \bar{\lambda}\,\mathrm{div}\,\mathbf{v})\mathfrak{I} + 2\eta\left(\mathfrak{U} - \frac{1}{3}\mathfrak{I}\,\mathrm{div}\,\mathbf{v}\right)\\ &= -p\mathfrak{I} + \left(\bar{\lambda} - \frac{2}{3}\eta\right)\mathfrak{I}\,\mathrm{div}\,\mathbf{v} + 2\eta\mathfrak{U}.\end{aligned} \tag{9.6}$$

Therefore,

$$a = \bar{\lambda} - \frac{2}{3}\eta, \qquad b = \eta. \tag{9.7}$$

The coefficient $\bar{\lambda}$ is known as the *bulk viscosity*, and η is the *shear viscosity*. Note that the similarity between the viscous and the elastic stress tensors is only in their forms. There is, however, a fundamental difference in their content: Whereas the elastic stresses are conservative, the viscous stresses are dissipative, being proportional to velocity gradients.

A fluid for which $\bar{\lambda} = 0$ (i.e., the viscosity is controlled by one material constant, η) is called *Stokes' fluid*. If $\eta = 0$ as well, we get back the constitutive equation for an inviscid fluid, $\mathfrak{T} = -p\mathfrak{I}$.

The interpretation of η will become clear through a simple example. Consider the velocity field in a Cartesian coordinate system

$$v_1 = f(x_2), \qquad v_2 = 0, \qquad v_3 = 0. \tag{9.8}$$

For this flow, the only nonzero element of $\mathfrak{U}$ is $U_{12} = \frac{1}{2}(df/dx_2)$, so that,

$$\tau_{11} = \tau_{22} = \tau_{33} = -p, \qquad \tau_{13} = \tau_{23} = 0 \quad \text{and} \quad \tau_{12} = \eta\left(\frac{dv_1}{dx_2}\right).$$

Therefore,

$$\eta = \frac{\tau_{12}}{(dv_1/dx_2)}, \tag{9.9}$$

and the coefficient of the shear viscosity can be determined from the ratio of the tangential stress to the velocity gradient. The ratio $\nu = \eta/\rho$ is called the *kinematic viscosity*.

The unit of η in the metric system is *poise.* It is the shearing stress needed to maintain a fixed velocity gradient of 1 cm/s per centimeter. The shear viscosity of water, for example, at 20°C is 1.005×10^{-2} gm/cm · s and that of air at the same temperature is 1.808×10^{-4} gm/cm · s.

9.1.3. Equations of Motion for Isothermal Flows

Euler's equation of motion, Eq. (1.98), may be written in the form

$$\rho \frac{D\mathbf{v}}{Dt} = \rho\mathbf{F} + \operatorname{div} \mathfrak{T}, \tag{9.10}$$

where $\mathbf{F}$ is the external body force per unit mass. We define the momentum density, $\mathbf{P} = \rho\mathbf{v}$, and the *momentum flux dyadic*, $\mathfrak{J} = \rho\mathbf{v}\mathbf{v}$. Using the tensor identity

$$\operatorname{div}(\rho\mathbf{v}\mathbf{v}) = \mathbf{v} \operatorname{div}(\rho\mathbf{v}) + \rho(\mathbf{v} \cdot \operatorname{grad} \mathbf{v}),$$

together with the equation of continuity, Eq. (1.96), and the relation (App. C)

$$\frac{D\mathbf{v}}{Dt} = \frac{\partial \mathbf{v}}{\partial t} + \mathbf{v} \cdot \operatorname{grad} \mathbf{v}, \tag{9.11}$$

we can rewrite the Eulerian equation of motion for fluids in the form

$$\frac{\partial \mathbf{P}}{\partial t} + \operatorname{div}(\mathfrak{J} - \mathfrak{T}) = \rho\mathbf{F}. \tag{9.12}$$

Inserting the expression in Eq. (9.6) for the stress dyadic into Eq. (9.10) and noting that, for constant $\bar{\lambda}$ and η,

$$\operatorname{div}\left\{\mathfrak{J}\left[-p + \left(\bar{\lambda} - \frac{2}{3}\eta\right)\operatorname{div} \mathbf{v}\right]\right\} = -\operatorname{grad} p - \left(\frac{2}{3}\eta - \bar{\lambda}\right)\operatorname{grad} \operatorname{div} \mathbf{v},$$

$$\operatorname{div}(\nabla\mathbf{v} + \mathbf{v}\nabla) = \operatorname{div} \operatorname{grad} \mathbf{v} + \operatorname{grad} \operatorname{div} \mathbf{v},$$

we obtain the *Navier–Stokes* equation,

$$\rho \frac{D\mathbf{v}}{Dt} = \rho\mathbf{F} - \operatorname{grad} p + \eta\nabla^2\mathbf{v} + \left(\bar{\lambda} + \frac{1}{3}\eta\right)\operatorname{grad} \operatorname{div} \mathbf{v}. \tag{9.13}$$

In addition, we have, from the equation of continuity [Eq. (1.96)],

$$\frac{D\rho}{Dt} + \rho \operatorname{div} \mathbf{v} = \frac{\partial \rho}{\partial t} + \operatorname{div}(\rho\mathbf{v}) = 0. \tag{9.14}$$

If $D\rho/Dt = 0$, i.e., if the density remains constant as the fluid element moves about, the fluid is incompressible and the equation of continuity becomes $\operatorname{div} \mathbf{v} = 0$.

Boundary conditions over a rigid boundary are such that the normal velocity vanishes for an ideal fluid and $\mathbf{v} = 0$ for a viscous fluid. Note that the Navier–Stokes equation is nonlinear in $\mathbf{v}$ because of the acceleration term [Eq. (9.11)].

9.1.4. Dynamic Similarity and the Reynolds Number

Consider a homogeneous incompressible fluid. Let V and L be characteristic velocity and length, respectively, pertaining to the special physical problem under consideration. Introducing the dimensionless variables

$$x_i' = \frac{x_i}{L}, \qquad v_i' = \frac{v_i}{V}, \qquad p' = \frac{p}{\rho V^2}, \qquad t' = \frac{Vt}{L}$$

and the parameter $R = VL\rho/\eta$, the force-free Navier–Stokes equation assumes the dimensionless form,

$$\frac{D\mathbf{v}'}{Dt'} = -\text{grad}' \, p' + \frac{1}{R} \nabla'^2 \mathbf{v}', \qquad \text{div}\, \mathbf{v}' = 0, \qquad \nabla' \equiv \mathbf{e}_i \frac{\partial}{\partial x_i'}. \tag{9.15}$$

Equation (9.15) shows that flows about geometrically similar bodies at the same values of R are completely similar in the sense that the functions $p'(\mathbf{r}', t')$ and $\mathbf{v}'(\mathbf{r}', t')$ are the same for the various flows. The parameter R is called the *Reynolds number* and it governs the *dynamic similarity*. Because $R = \rho V^2/(\eta V/L)$, R expresses the ratio of the inertial force to the viscous force. One can show that Reynolds' number is the only parameter characterizing the isothermal flow if the velocity of flow is much less than the velocity of sound in the fluid.

9.1.5. Linearization of the Equations of Motion

For sufficiently small velocities, the term $(\mathbf{v} \cdot \text{grad}\ \mathbf{v})$ is small compared to $\partial \mathbf{v}/\partial t$ and, therefore, the Navier–Stokes equation, Eq. (9.13), can be linearized as follows

$$\rho \frac{\partial \mathbf{v}}{\partial t} = -\text{grad}(\rho\chi + p) + \eta \nabla^2 \mathbf{v} + \left(\bar{\lambda} + \frac{1}{3}\eta\right) \text{grad div}\, \mathbf{v}. \tag{9.16}$$

Here, it is assumed that, $\mathbf{F} = -\text{grad}\ \chi$; i.e., the external forces are *conservative* and $\rho = \text{const}$. Under conditions of incompressibility ($\text{div}\ \mathbf{v} = 0$), Eq. (9.16) reduces to

$$\eta \nabla^2 \mathbf{v} - \rho \left(\frac{\partial \mathbf{v}}{\partial t}\right) = \text{grad}(p + \rho\chi). \tag{9.17}$$

If the motion is steady (i.e., $\partial \mathbf{v}/\partial t = 0$) and there are no external forces, the viscous forces dominate the inertial forces and the equations further reduce to

$$\text{grad}\ p = \eta \nabla^2 \mathbf{v}, \qquad \text{div}\, \mathbf{v} = 0, \tag{9.18}$$

which describe the *Stokes' flow*. These are also known as Stokes equations or *creep equations*. The equations are solved under the condition that $\mathbf{v}$ vanishes at the surface of the containing boundary.

9.2. Sound Waves

Let the external field of force be conservative and the motion *irrotational* so that $\mathbf{F} = -\operatorname{grad} \chi$ and $\mathbf{v} = \operatorname{grad} \psi$. Then, because $\nabla^2 \mathbf{v} = \operatorname{grad} \operatorname{div} \mathbf{v} - \operatorname{curl} \operatorname{curl} \mathbf{v}$, the linearized equation of motion, Eq. (9.16), takes the form

$$\operatorname{grad}\left[-(p + \rho\chi) + \left(\frac{4}{3}\eta + \bar{\lambda}\right)\nabla^2\psi\right] = \rho \operatorname{grad} \frac{\partial \psi}{\partial t}. \tag{9.19}$$

Differentiating this equation with respect to t and noting that $\rho\chi$, η, and $\bar{\lambda}$ are time independent, we get

$$\operatorname{grad}\left[-\frac{\partial p}{\partial t} + \left(\frac{4}{3}\eta + \bar{\lambda}\right)\nabla^2\left(\frac{\partial \psi}{\partial t}\right)\right] = \frac{\partial \rho}{\partial t} \operatorname{grad} \frac{\partial \psi}{\partial t} + \rho \operatorname{grad} \frac{\partial^2 \psi}{\partial t^2}. \tag{9.20}$$

The right-hand side of the above equation equals

$$\operatorname{grad}\left(\rho \frac{\partial^2 \psi}{\partial t^2}\right) + \left[\frac{\partial \rho}{\partial t} \operatorname{grad} \frac{\partial \psi}{\partial t} - \left(\frac{\partial^2 \psi}{\partial t^2}\right) \operatorname{grad} \rho\right].$$

Because each term in the brackets is a product of two terms of the first order and therefore small relative to $\operatorname{grad}(\rho\, \partial^2\psi/\partial t^2)$, it can be neglected. Equation (9.20) can then be integrated to yield

$$\frac{\partial^2 \psi}{\partial t^2} = -\frac{(\partial p/\partial t)}{\rho} + \left[\frac{4\eta + 3\bar{\lambda}}{3\rho}\right]\nabla^2\left(\frac{\partial \psi}{\partial t}\right). \tag{9.21}$$

In order to turn this into an equation in ψ alone, we must express $\partial p/\partial t$ in terms of ψ. In the case of fluids, the rate of change of pressure is proportional to the divergence of the velocity [compare with Eq. (1.79) for solids]. Hence we have

$$\frac{\partial p}{\partial t} + \kappa \operatorname{div} \mathbf{v} = 0,$$

i.e.,

$$-\frac{(\partial p/\partial t)}{\rho} = c^2 \nabla^2 \psi, \qquad c^2 = \frac{\kappa}{\rho}, \tag{9.22}$$

where c has the dimensions of velocity. The incompressibility κ is assumed to be constant over a large range of pressures.

Combining Eqs. (9.21) and (9.22), we obtain *Stokes' wave equation*

$$\frac{\partial^2 \psi}{\partial t^2} = c^2 \nabla^2 \psi + \left[\frac{4\eta + 3\bar{\lambda}}{3\rho}\right]\nabla^2\left(\frac{\partial \psi}{\partial t}\right). \tag{9.23}$$

Note that because

$$\rho \frac{\partial p}{\partial \rho} = \kappa,$$

we have

$$\frac{\partial p}{\partial \rho} = \frac{\kappa}{\rho} = c^2. \tag{9.24}$$

Equation (9.23) governs the oscillatory motion of small amplitudes in a compressible viscous fluid. It is known as *sound waves*, or *acoustic waves*. These waves are *attenuated* on account of the viscosity.

The effect of the attenuation can be manifested in two ways. Applying the Fourier transform to Eq. (9.23) we obtain the Helmholtz equation, $\nabla^2\psi + \hat{k}^2\psi = 0$, in which $\hat{k}^2$ is complex

$$\hat{k}^2 = \frac{\omega^2}{c_a^2}, \qquad c_a^2 = c^2 + i\omega\left[\frac{4\eta + 3\bar{\lambda}}{3\rho}\right]. \tag{9.25}$$

We shall show in Chapter 10 that this represents a wave whose amplitude decays as it travels in space. However, if we assume that $\psi(\mathbf{r}, t)$ can be represented as a product, $\psi = \psi_1(\mathbf{r})\psi_2(t)$, the space part is found to satisfy the equation, $\nabla^2\psi_1 + K^2\psi_1 = 0$, whereas the temporal part obeys the equation of a damped linear oscillator

$$\frac{d^2\psi_2}{dt^2} + \left[\frac{4\eta + 3\bar{\lambda}}{3\rho}\right]K^2\left(\frac{d\psi_2}{dt}\right) + c^2K^2\psi_2 = 0. \tag{9.26}$$

If $\bar{\lambda} = \eta = 0$, we fall back on the standard wave equation $\partial^2\psi/\partial t^2 = c^2\nabla^2\psi$. We shall now discuss this simpler case and study the nature of sound waves from a different angle.

Let us consider a fluid in equilibrium. The changes of pressure, density, and velocity induced by the acoustic fields are considered as small perturbations of the corresponding equilibrium values. Therefore, we make the *sound-wave approximation*

$$\rho = \rho_0 + \rho', \qquad p = p_0 + p', \qquad \mathbf{v} = \operatorname{grad}\psi, \tag{9.27}$$

where ρ_0 and p_0 are the distributions of the density and pressure, respectively, in the fluid prior to the passage of the sound wave and $\rho' \ll \rho_0$, $p' \ll p_0$. The linearized equations of continuity and motion now become ($\bar{\lambda} = \eta = 0$)

$$\frac{\partial \rho'}{\partial t} + \operatorname{div}(\rho_0\mathbf{v}) = 0, \qquad \frac{\partial \mathbf{v}}{\partial t} + \frac{1}{\rho_0}\operatorname{grad} p' = 0. \tag{9.28}$$

These are four equations in the five unknowns $\mathbf{v}$, p', and ρ'. We assume, therefore, a fifth relation, known as the *equation of state*, $p = p(\rho)$. Applying the Taylor expansion at constant *entropy*, S, we get

$$p = p_0 + p' = f(\rho_0 + \rho') = f(\rho_0) + \rho'\left(\frac{\partial p_0}{\partial \rho_0}\right)_S + \text{higher order terms}. \tag{9.29}$$

Consequently, we derive the relations

$$p' = c^2\rho', \qquad c^2 = \left(\frac{\partial p_0}{\partial \rho_0}\right)_S. \tag{9.30}$$

Because, to first order,

$$\frac{1}{\rho_0}\operatorname{grad} p' = \operatorname{grad}\left(\frac{1}{\rho_0} p'\right),$$

Eq. (9.28) together with the relation $\mathbf{v} = \operatorname{grad}\psi$ yield

$$p' = -\rho_0 \frac{\partial \psi}{\partial t}. \tag{9.31}$$

Combining Eqs. (9.28), (9.30), and (9.31), we derive

$$\nabla^2 \psi = \frac{1}{c^2}\frac{\partial^2 \psi}{\partial t^2}, \qquad \nabla^2 p' = \frac{1}{c^2}\frac{\partial^2 p'}{\partial t^2}. \tag{9.32}$$

We mention at this point that the reduction of Eqs. (9.28) to the wave equation, Eq. (9.32), is not feasible for arbitrary dependence of ρ_0 on the coordinates. It can be shown that the heterogeneity of ρ_0 is limited by the condition

$$\frac{|\nabla \rho_0|}{\rho_0} L \ll 1, \tag{9.33}$$

where L is the smallest wavelength of the acoustic signal that propagates through the fluid.

The relation $c^2 = \partial p_0/\partial \rho_0$ is a linearized version of the more general *adiabatic energy equation*

$$\frac{Dp_0}{Dt} = c^2 \frac{D\rho_0}{Dt}. \tag{9.34}$$

It may be noted that $\mathbf{v}$ and ρ' also satisfy the wave equation, Eq. (9.32), provided that condition (9.33) holds.

The velocity of sound $c = (\partial p_0/\partial \rho_0)^{1/2}$ depends on the relationship between the pressure and the density. If the flow is adiabatic, $p_0 \propto \rho_0^\gamma$ and $c = \sqrt{(\gamma p_0/\rho_0)}$, where $\gamma = c_p/c_v$ is the ratio of the specific heats for a perfect gas. Under isothermal conditions, $c = (p_0/\rho_0)^{1/2}$. In all other cases, it is necessary to introduce the temperature explicitly as a variable.

The *Lagrangian density* for sound waves in an inviscid fluid is defined by [compare with Eq. (1.125)]

$$\mathscr{L} = \frac{1}{2}\rho_0 \mathbf{v}^2 - \frac{1}{2\kappa} p'^2. \tag{9.35}$$

Inserting

$$\mathbf{v} = \operatorname{grad}\psi, \qquad p' = -\rho_0 \frac{\partial \psi}{\partial t}, \qquad c^2 = \frac{\kappa}{\rho_0},$$

we obtain

$$\mathscr{L} = \frac{1}{2}\rho_0\left[|\operatorname{grad}\psi|^2 - \frac{1}{c^2}\left(\frac{\partial \psi}{\partial t}\right)^2\right] = \frac{1}{2}\rho_0 |\operatorname{grad}_4 \psi|^2, \tag{9.36}$$

where grad_4 is the operator $(\partial/\partial x_1, \partial/\partial x_2, \partial/\partial x_3, \partial/\partial ict)$.

9.3. Gravity Waves in Liquids

9.3.1. General Considerations

We shall derive the basic theory of wave motion in liquids in the presence of gravity. Wave motion results from the action of gravity, which tends to restore the undisturbed state, energy being transferred by the waves from regions of relative excess to those of relative deficiency. Because liquids have no intrinsic rigidity, they can transmit only longitudinal wave motion. Shear waves cannot be maintained without strong damping.

We shall assume in all cases that the motion is irrotational ($\mathbf{v} = \text{grad}\, \psi$) and that the liquid is inviscid and incompressible. The motion of the liquid is then governed by the Laplace equation in the *velocity potential*,

$$\nabla^2 \psi = 0, \tag{9.37}$$

obtained from the equation of continuity $\text{div}\, \mathbf{v} = 0$. To derive the boundary conditions, we start again with the equation of motion [Eq. (9.13) with $\bar{\lambda} = \eta = 0$]

$$\frac{\partial \mathbf{v}}{\partial t} + \mathbf{v} \cdot \text{grad}\, \mathbf{v} = \mathbf{F} - \frac{1}{\rho} \text{grad}\, p. \tag{9.38}$$

Substituting $\mathbf{v} = \text{grad}\, \psi$, $\mathbf{F} = -\text{grad}\, \chi$, and using the identity $\mathbf{v} \cdot \text{grad}\, \mathbf{v} = \text{grad}\, \frac{1}{2}\mathbf{v}^2 - (\mathbf{v} \times \text{curl}\, \mathbf{v})$, we obtain the *Bernoulli equation* for a homogeneous liquid ($\rho = \text{const.}$)

$$\frac{\partial \psi}{\partial t} + \frac{1}{2}\mathbf{v}^2 + \chi + \frac{p}{\rho} = C(t). \tag{9.39}$$

In deriving Eq. (9.39), we have assumed that ρ is constant. If the external force is due to gravity only and if we choose the gravity field in the z direction, where z is directed upward, then $\chi = gz$. Also, neglecting $\frac{1}{2}\mathbf{v}^2$ and absorbing $C(t)$ in $\partial \psi / \partial t$, Eq. (9.39) may be written as

$$\frac{p}{\rho} + \frac{\partial \psi}{\partial t} + gz = \text{const.} \tag{9.40}$$

The conditions under which $\mathbf{v}^2$ is neglected can be derived. Let (Λ, T, a) be the wavelength, period, and amplitude of the gravity wave. Then the velocity $|\mathbf{v}|$ is of the order (a/T), the time derivative of the velocity is of the order (a/T^2), and its space derivative is of the order $a/(T\Lambda)$. The quantity $(\mathbf{v} \cdot \text{grad}\, \mathbf{v})$ can be neglected in comparison with $\partial \mathbf{v}/\partial t$ if the condition $a \ll \Lambda$ is satisfied.

At the liquid–air boundary, the equation of continuity is replaced by a special surface condition. Let $F(x, y, z, t) = 0$ be the equation of the boundary surface. For this surface to be a boundary of the fluid, the velocity of a particle lying in it, relative to the surface, must be wholly tangential (or zero), because otherwise, we should have a finite flow of liquid across it. To find the mathematical expression for this condition, we consider the state of the surface at time $t + \delta t$. Any change in the shape of the surface can be expressed in terms of the velocity of its elements. Let $\mathbf{v}_s$ be the velocity of this motion and let $\mathbf{n}(\mathbf{r}, t)$ be the unit normal to F. Then, the equation of the surface at time $t + \delta t$ will be $F(\mathbf{r} + \mathbf{v}_s \delta t, t + \delta t) = 0$. Subtracting the corresponding values of F at times $t + \delta t$ and t and dividing the result by δt, we obtain, in the limit $\delta t \to 0$;

$$\frac{\partial F}{\partial t} + \mathbf{v}_s \cdot \nabla F = 0.$$

However, because

$$\mathbf{n} = \pm \frac{\nabla F}{|\nabla F|}, \qquad (\mathbf{v} - \mathbf{v}_s) \cdot \mathbf{n} = 0,$$

we have on $F(\mathbf{r}, t) = 0$

$$\frac{\partial F}{\partial t} + \mathbf{v} \cdot \operatorname{grad} F = 0. \tag{9.41}$$

Consider a liquid body of unlimited extent in the x and y directions but of uniform depth, H. We take the origin of the coordinate system at the undisturbed level and let the z axis be drawn vertically upward (Fig. 9.1). The elevation of the surface relative to the undisturbed level is denoted by $\zeta(x, y, t)$. Then, by Eq. (9.40), we have at the surface,

$$p_0 = -\rho g \zeta - \rho \left(\frac{\partial \psi}{\partial t} \right)_{z=\zeta}. \qquad (9.42)^1$$

Note that, instead of the potential ψ, we can use the potential $\psi' = \psi + (p_0/\rho)t$ because $\mathbf{v} = \operatorname{grad} \psi = \operatorname{grad} \psi'$. Therefore, omitting the prime over ψ, we have

$$\zeta = -\left[\frac{1}{g} \frac{\partial \psi}{\partial t} \right]_{z=\zeta} = -\left[\frac{1}{g} \frac{\partial \psi}{\partial t} \right]_{z=0}, \tag{9.43}$$

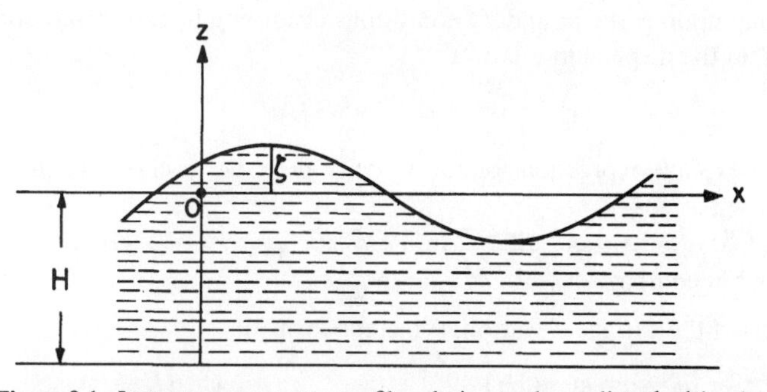

Figure 9.1. Instantaneous water profile relative to the undisturbed level.

[1] In the case of gravity waves in liquids (Section 9.3), p_0 denotes the value of the liquid pressure at the surface. In contrast, in the case of sound waves (Section 9.2) and acoustic-gravity waves (Sections 9.4–9.7), p_0 denotes the value of the pressure in the equilibrium state.

where the error in placing the boundary condition at $z = 0$ instead of $z = \zeta$ is of the order of terms already neglected. Because the equation of the boundary is $z = \zeta$, we put $F = z - \zeta$ in Eq. (9.41) and neglect small quantities of the second order. The result is,

$$\frac{\partial \zeta}{\partial t} = \left[\frac{\partial \psi}{\partial z}\right]_{z=0}. \tag{9.44}$$

Eliminating ζ between Eqs. (9.43) and (9.44), we obtain the *kinematic boundary condition* at the free surface

$$\frac{\partial^2 \psi}{\partial t^2} + g\frac{\partial \psi}{\partial z} = 0 \qquad \text{at} \quad z = 0. \tag{9.45}$$

Finally, there is the bottom condition, which states that there is no vertical motion at the bottom. If there is no source of disturbance at this level, the condition reads

$$\left.\frac{\partial \psi}{\partial z}\right|_{z=-H} = 0. \tag{9.46}$$

Let us assume that the motion is two-dimensional and is confined in planes parallel to the xz plane.

A solution of the Laplace equation, $\nabla^2 \psi = 0$, which at the same time represents a progressive plane wave, can be written as

$$\psi(x, z, t) = [A \operatorname{sh} kz + B \operatorname{ch} kz]\cos(\omega t - kx), \tag{9.47}$$

where t plays the role of a parameter. Consequently, by Eq. (9.43),

$$\zeta(x, t) = \zeta_{\max} \sin(\omega t - kx), \qquad \zeta_{\max} = \frac{kc}{g} B, \qquad \omega = ck. \tag{9.48}$$

Imposing upon ψ the boundary conditions according to Eqs. (9.45) and (9.46), we obtain the dispersion relation

$$\omega^2 = (gk)\operatorname{th}(kH), \tag{9.49}$$

and the explicit expression for the velocity potential everywhere in the fluid,

$$\psi(x, z, t) = B\frac{\operatorname{ch} k(z + H)}{\operatorname{ch} kH}\cos(\omega t - kx), \qquad 0 > z > -H. \tag{9.50}$$

In terms of the surface elevation, ψ has the form

$$\psi(x, z, t) = \frac{g\zeta_{\max}}{kc}\frac{\operatorname{ch} k(z + H)}{\operatorname{ch} kH}\cos(\omega t - kx) = \psi_{\max}\frac{\operatorname{ch} k(H + z)}{\operatorname{ch} kH}\cos(\omega t - kx), \tag{9.51}$$

$$\psi_{\max} = \frac{g\zeta_{\max}}{kc}.$$

It is seen that ζ satisfies the one-dimensional wave equation:

$$\frac{\partial^2 \zeta}{\partial x^2} = \frac{1}{c^2} \frac{\partial^2 \zeta}{\partial t^2}. \tag{9.52}$$

For a fixed z, ψ also satisfies the same equation, although for a fixed t, it satisfied the Laplace equation.

Let us next examine the dependence of the wave motion on the z coordinate in the fluid. Because $(\partial\psi/\partial x, \partial\psi/\partial z)$ are components of the particle velocity, we may define a displacement vector (X, Z), which measures the particle displacement from the undisturbed position resulting from the wave motion:

$$\begin{aligned} \frac{\partial X}{\partial t} &= \frac{\partial \psi}{\partial x} = \frac{g\zeta_{\max}}{c} \frac{\operatorname{ch} k(z+H)}{\operatorname{ch} kH} \sin(\omega t - kx), \\ \frac{\partial Z}{\partial t} &= \frac{\partial \psi}{\partial z} = \frac{g\zeta_{\max}}{c} \frac{\operatorname{sh} k(z+H)}{\operatorname{ch} kH} \cos(\omega t - kx). \end{aligned} \tag{9.53}$$

Integrating with respect to time and eliminating t between the two ensuing expressions, we get

$$\frac{X^2}{\alpha^2} + \frac{Z^2}{\beta^2} = 1, \tag{9.54}$$

where

$$\alpha = \frac{g\zeta_{\max}}{kc^2} \frac{\operatorname{ch} k(z+H)}{\operatorname{ch} kH}, \qquad \beta = \frac{g\zeta_{\max}}{kc^2} \frac{\operatorname{sh} k(z+H)}{\operatorname{ch} kH}. \tag{9.55}$$

Clearly, the path of a particle is an ellipse with semiaxes α and β. At the bottom of the liquid ($z = -H$) the ellipse degenerates into a straight line. Consequently, particles at the bottom move to and fro only and do not rise and fall. Particles below a crest or a trough are all moving horizontally in the same vertical line. In particular, the particle at a crest is moving forward, whereas at a trough it is moving backward. Any undisturbed vertical liquid plane is bent by the passage of a wave, most at the surface and progressively less until the bottom is reached.

The pressure at any point in the liquid caused by progressive surface waves is determined from the Bernoulli equation, Eq. (9.40), and Eq. (9.43)

$$\left[p + \rho g z + \rho \frac{\partial \psi}{\partial t}\right]_z = \left[p + \rho g z + \rho \frac{\partial \psi}{\partial t}\right]_{z=\zeta} = p_0, \tag{9.56}$$

where $p = p_0$ at the surface. However, from Eqs. (9.48) and (9.51)

$$\frac{\partial \psi}{\partial t} = -g\zeta(x, t) \frac{\operatorname{ch} k(z+H)}{\operatorname{ch} kH}. \tag{9.57}$$

Hence

$$p(z, t) = p_0 + g\rho\left[\zeta(x, t) \frac{\operatorname{ch} k(z+H)}{\operatorname{ch} kH} - z\right]. \tag{9.58}$$

Dispersion is determined by Eq. (9.49). Using the relation $k = \omega/c$, we find that the phase velocity c is given by the equation

$$\frac{c}{c_0} = \sqrt{\left[\frac{\text{th}\, kH}{kH}\right]} \leq 1, \qquad c_0 = \sqrt{(gH)}. \tag{9.59}$$

The group velocity, in contrast, is derived through the differential expression $U_g = d\omega/dk$ and is found to be

$$\frac{U_g}{c} = \frac{1}{2}\left[1 + \frac{2kH}{\text{sh}(2kH)}\right] \leq 1. \tag{9.60}$$

The expression for c is independent of the source of excitation of the fluid. The dependence of c on wavelength is shown in Fig. 9.2.

Two limiting cases are of particular interest:

1. *The short-wave limit* ($kH \to \infty$ or $\zeta \ll \Lambda \ll H$): Because for large values of kH, $\text{th}(kH) \simeq 1$, we have $c^2 \simeq g/k = g\Lambda/2\pi$ or $c \simeq gT/2\pi$, and $U_g \simeq c/2$. As the wave profile moves, individual waves appear at the back, overtake the group with relative velocity $c/2$, and then disappear at the front of the profile. This may occur to ocean waves, caused by wind and storm disturbances. From Eq. (9.55), we have

$$\frac{\beta}{\alpha} = \text{th}\left[\frac{2\pi}{\Lambda}(z + H)\right] \simeq 1. \tag{9.61}$$

Therefore, the particle motion is circular, and the radius of the circle decreases with depth until it ultimately vanishes. From Eq. (9.51)

$$\psi \simeq \psi_{\max} e^{kz} \cos(\omega t - kx).$$

The disturbance in the liquid becomes negligibly small at a depth of about one half of a wavelength. That is why these disturbances are called *surface waves.* A submarine submerged at a depth of half a wavelength would hardly notice the motion of surface waves created by storms.

The pressure at any point in the liquid is obtained from Eq. (9.58). Under the condition $\Lambda \ll H$, we have $[\text{ch}\, k(z + H)/\text{ch}\, kH] \simeq e^{kz}$. Taking the time average of Eq. (9.58) over one period we obtain, with $\langle \zeta \rangle_T = 0$,

$$\langle p(z, t) \rangle_T = \langle p_0 \rangle_T - \rho g z. \tag{9.62}$$

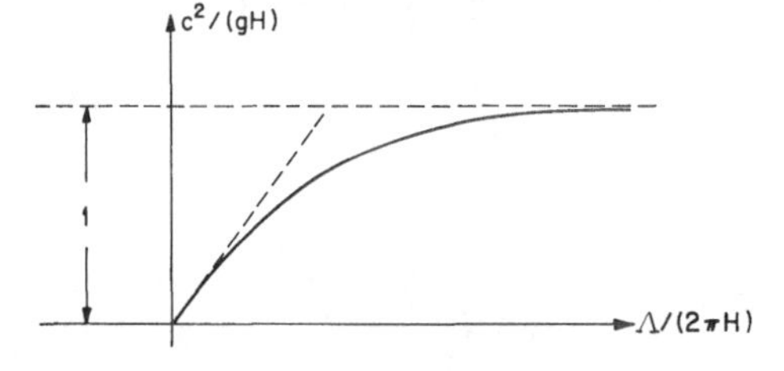

Figure 9.2. Dependence of the phase velocity (c) of water waves on wavelength (Λ). $c \propto \sqrt{\Lambda}$ for surface waves and $c = \sqrt{gH}$ for long waves.

Hence the average dynamic pressure at any fluid particle position, z, is equal to the hydrostatic pressure at the mean position of that particle.

2. *The long-wave limit* ($kH \to 0$, $\zeta \ll H \ll \Lambda$): In order to derive the field equations under these conditions, we rewrite the equations of continuity and motion in component form. For the sake of clarity, consider first a two-dimensional velocity field $\mathbf{v} = (u, 0, w)$. We then have

$$\frac{\partial u}{\partial x} + \frac{\partial w}{\partial z} = 0; \tag{9.63}$$

$$\begin{aligned} a_x &= \frac{\partial u}{\partial t} + u\frac{\partial u}{\partial x} + w\frac{\partial u}{\partial z} = -\frac{1}{\rho}\frac{\partial p}{\partial x}, \\ a_z &= \frac{\partial w}{\partial t} + u\frac{\partial w}{\partial x} + w\frac{\partial w}{\partial z} = -\frac{1}{\rho}\frac{\partial p}{\partial z} - g; \end{aligned} \tag{9.64}$$

with the provision that u, w are much smaller than c. We denote

$$\theta = \frac{H}{\Lambda} \ll 1.$$

Using the symbol $\sim$ to relate quantities of the same order of magnitude, we write

$$\Delta x \sim \Lambda, \qquad \Delta z \sim H, \qquad \Delta u \sim u, \qquad \Delta w \sim w, \qquad \Delta t \sim \frac{\Lambda}{c}. \tag{9.65}$$

From Eq. (9.63) and (9.65), we see that

$$\begin{gathered} \frac{\partial u}{\partial x} \sim \frac{u}{\Lambda}, \qquad \frac{\partial w}{\partial z} \sim \frac{w}{H} = \frac{w}{\Lambda\theta}, \qquad \frac{u}{\Lambda} \sim \frac{\partial u}{\partial x} = -\frac{\partial w}{\partial z} \sim \frac{w}{H}, \qquad \frac{\partial u}{\partial t} \sim \frac{uc}{\Lambda}, \\ u\frac{\partial u}{\partial x} \sim \frac{u^2}{\Lambda}, \qquad w\frac{\partial u}{\partial z} \sim \frac{u^2}{\Lambda}, \qquad \frac{\partial w}{\partial t} \sim \theta\frac{uc}{\Lambda}, \qquad u\frac{\partial w}{\partial x} \sim \frac{\theta u^2}{\Lambda}, \qquad w\frac{\partial w}{\partial z} \sim \frac{\theta u^2}{\Lambda}. \end{gathered} \tag{9.66}$$

Because $u \ll c$ and $\theta \ll 1$, these relations indicate that

$$a_x \gg a_z, \qquad u \gg w.$$

Hence, using Eqs. (9.66), the Euler equations (9.64) reduce to the simpler form

$$a_x = \frac{\partial u}{\partial t} = -\frac{1}{\rho}\frac{\partial p}{\partial x}, \qquad \frac{1}{\rho}\frac{\partial p}{\partial z} = -g. \tag{9.67}$$

Because the pressure at the surface ($z = \zeta$) must be p_0, the second equation yields

$$p = p_0 + g\rho(\zeta - z). \tag{9.68}$$

The pressure at depth z is therefore the hydrostatic pressure there. Substituting this expression in the first equation in Eqs. (9.67), we obtain

$$\frac{\partial u}{\partial t} = -g\frac{\partial \zeta}{\partial x} - \frac{1}{\rho}\frac{\partial p_0}{\partial x}. \tag{9.69}$$

In the case of a three-dimensional velocity field, $\mathbf{v} = (u, v, w)$, we have an additional equation

$$\frac{\partial v}{\partial t} = -g\frac{\partial \zeta}{\partial y} - \frac{1}{\rho}\frac{\partial p_0}{\partial y}. \tag{9.70}$$

Because $\zeta = \zeta(x, y, t)$, the horizontal acceleration is independent of z. We need now a third equation to determine the three unknowns u, v, and ζ. We shall use the law of conservation of mass under conditions of incompressibility, i.e., a continuity equation that suits the needs of our special problem. The matter that in a unit time enters into a column which stands on the elementary rectangle $(\delta x \delta y)$, must wholly manifest itself as a change in the fluid level, because the fluid is incompressible. Therefore, to the first order,

$$-\frac{\partial}{\partial x}(\rho u H \delta y)\delta x - \frac{\partial}{\partial y}(\rho v H \delta x)\delta y = \frac{\partial}{\partial t}(\rho \zeta \delta x \delta y),$$

or

$$\frac{\partial \zeta}{\partial t} = -H\left[\frac{\partial u}{\partial x} + \frac{\partial v}{\partial y}\right]. \tag{9.71}$$

Let (F_x, F_y, F_z) be the components of a small disturbing force per unit mass such that $F_z \ll g$. Then, Eq. (9.68) remains unaltered and Eq. (9.38) shows that Eqs. (9.69) and (9.70) will be modified to

$$\frac{\partial u}{\partial t} = F_x - g\frac{\partial \zeta}{\partial x} - \frac{1}{\rho}\frac{\partial p_0}{\partial x}, \qquad \frac{\partial v}{\partial t} = F_y - g\frac{\partial \zeta}{\partial y} - \frac{1}{\rho}\frac{\partial p_0}{\partial y}. \tag{9.72}$$

These equations show that the horizontal component of the disturbing force is alone important. Eliminating u and v from Eqs. (9.71) and (9.72), we get

$$\frac{\partial^2 \zeta}{\partial t^2} = gH\nabla_1^2\zeta - H \operatorname{div}_1 \mathbf{F} + \frac{H}{\rho}\nabla_1^2 p_0, \tag{9.73}$$

where

$$\nabla_1^2 = \frac{\partial^2}{\partial x^2} + \frac{\partial^2}{\partial y^2}, \qquad \operatorname{div}_1 \mathbf{F} = \frac{\partial F_x}{\partial x} + \frac{\partial F_y}{\partial y}. \tag{9.74}$$

If $\mathbf{F}$ is taken to be the periodic tide-generating force of the moon, the solution of Eq. (9.73) will represent the well-known effect of the tidal waves of the ocean.

We shall sum up the main differences between the two extreme cases of gravity waves:

1. *Surface waves.* The wave profile moves with the liquid; i.e., the velocities of wave and liquid are the same at the liquid's surface. The wave amplitude decreases rapidly with depth.
2. *Long waves.* The profile does not, in general, move with the liquid and the motion is not necessarily irrotational. The vertical acceleration may be neglected and the horizontal acceleration is the same at all depths. The pressure at depth is the hydrostatic pressure.

9.3.2. Long Waves in Canals

9.3.2.1. One-Dimensional Wave Propagation. Consider the case of waves traveling along a straight canal of depth H having a horizontal bed and parallel vertical sides. Let the x axis be taken along the length of the canal, the z axis vertically upward, and the origin at the undisturbed level of the liquid. We suppose that the motion is two dimensional, so that $\partial/\partial y \equiv 0$. Equation (9.73) now becomes

$$\frac{\partial^2 \zeta}{\partial t^2} = gH \frac{\partial^2 \zeta}{\partial x^2} - H \frac{\partial F_x}{\partial x} + \frac{H}{\rho} \frac{\partial^2 p_0}{\partial x^2}. \tag{9.75}$$

Similarly, the equation of continuity, Eq. (9.71), reduces to

$$\frac{\partial \zeta}{\partial t} = -H \frac{\partial u}{\partial x}. \tag{9.76}$$

It is convenient to introduce a new variable Θ as follows:

$$u = \frac{\partial \Theta}{\partial t} \quad \text{or} \quad \Theta = \int^t u \, dt. \tag{9.77}$$

Clearly, Θ is the time integral of the velocity past the plane $x = \text{const.}$ up to the time t. In the case of small motions, it will be equal to the displacement of the particles which originally occupied the plane $x = \text{const.}$ or which actually occupy this plane at time t.

From Eqs. (9.76) and (9.77), we have

$$\zeta = -H \frac{\partial \Theta}{\partial x}. \tag{9.78}$$

Substituting the values of u and ζ from Eqs. (9.77) and (9.78) into the first equation in Eqs. (9.72), we get

$$\frac{\partial^2 \Theta}{\partial t^2} = F_x + gH \frac{\partial^2 \Theta}{\partial x^2} - \frac{1}{\rho} \frac{\partial p_0}{\partial x}. \tag{9.79}$$

In the case of a straight canal of arbitrary cross section, H should be replaced by S/b, where S is the sectional area of the undisturbed fluid and b is the breadth at the free surface. This can be easily seen by deriving the equation of continuity for this case.

Two particular cases will be considered here.

9.3.2.1.1. TRAVELING PRESSURE WAVE ON THE FREE SURFACE. A traveling pressure wave on the free surface is a good model of a meteorologic front. Let

$$\frac{1}{\rho} p_0(x, t) = f(Vt - x). \tag{9.80}$$

If we assume $F_x = 0$, Eqs. (9.75) and (9.80) yield

$$\frac{\partial^2 \zeta}{\partial t^2} = gH \frac{\partial^2 \zeta}{\partial x^2} + Hf''(Vt - x). \tag{9.81}$$

Assuming a solution of the form

$$\zeta = Af(Vt - x),$$

we get

$$A = \frac{H}{V^2 - gH}.$$

This yields,

$$\zeta = p_0 \frac{H}{\rho(V^2 - gH)}. \tag{9.82}$$

Consequently, the surface depression will be in phase with the pressure, or 180° out of phase, according to $V \lessgtr \sqrt{(gH)}$.

9.3.2.1.2. TRAVELING DISTURBANCE ON THE OCEAN BOTTOM. Let a seismic disturbance $\zeta_0 = f(Vt - x)$ travel along the bottom of the canal. In this case, the equation of continuity, Eq. (9.76), is modified to

$$\frac{\partial}{\partial t}(\zeta - \zeta_0) = -H \frac{\partial u}{\partial x}. \tag{9.83}$$

Proceeding as before, we get instead of Eq. (9.75),

$$\frac{\partial^2}{\partial t^2}(\zeta - \zeta_0) = gH \frac{\partial^2 \zeta}{\partial x^2}, \tag{9.84}$$

assuming $F_x = p_0 = 0$. Hence ζ is of the form

$$\zeta = Af(Vt - x).$$

Inserting this value of ζ in Eq. (9.84), we get

$$\zeta = \zeta_0 \frac{V^2}{V^2 - gH}. \tag{9.85}$$

This is the amplitude of the long water wave at the surface induced by the seismic disturbance at the bottom. The expression

$$\frac{V^2}{V^2 - gH} \tag{9.86}$$

is the dynamic amplification factor of the water wave-guide. The crests and troughs at the free surface and the bottom correspond or are opposite according as $V \gtrless \sqrt{(gH)}$. Maximum energy coupling occurs at $V = \sqrt{(gH)}$.

9.3.2.2. Two-Dimensional Wave Propagation. Consider the problem of long waves on a plane sheet of water of unlimited extent and uniform depth, H. We will suppose that the disturbance is caused by a variable pressure $p_0(x, y, t)$ applied to the surface. We have seen in Eq. (9.42) that at the surface,

$$g\zeta = -\frac{\partial\psi}{\partial t} - \frac{p_0}{\rho}. \tag{9.87}$$

Eliminating ζ between Eqs. (9.71) and (9.87) and remembering that $(u, v) = (\partial\psi/\partial x, \partial\psi/\partial y)$, we get

$$\nabla_1^2\psi = \frac{1}{c_0^2}\frac{\partial^2\psi}{\partial t^2} + \frac{1}{\rho c_0^2}\frac{\partial p_0}{\partial t}. \tag{9.88}$$

Equation (9.88) could be replaced by the homogeneous equation

$$\nabla_1^2\psi = \frac{1}{c_0^2}\frac{\partial^2\psi}{\partial t^2}, \tag{9.89}$$

together with an appropriate boundary condition to account for the deleted term. Let us suppose that $p_0(x, y, t)$ is of appreciable magnitude over an area, S, whose linear dimensions are small compared with the length of the waves generated by it but large as compared with H. We multiply both sides of Eq. (9.88) with $dS = dx\,dy$ and integrate over the area in question,

$$\int_S \nabla_1^2\psi\, dS = \frac{1}{c_0^2}\frac{\partial^2}{\partial t^2}\int_S \psi\, dS + \frac{1}{\rho c_0^2}\frac{\partial}{\partial t}\int_S p_0\, dS. \tag{9.90}$$

Using Gauss' theorem, we have

$$\int_S \nabla_1^2\psi\, dS = \int_C \frac{\partial\psi}{\partial n}\, ds,$$

where ds is an element of the boundary C of the area S, and $\mathbf{n}$ is the horizontal normal to ds, drawn outward. Further,

$$\frac{1}{\rho c_0^2}\frac{\partial}{\partial t}\int_S p_0\, dS = \frac{1}{\rho g H}\frac{\partial P_0}{\partial t} = f(t), \quad \text{(say)},$$

where, $P_0(t) = \int_S p_0\, dS$, is the total disturbing force. Noting that the first integral on the right-hand side of Eq. (9.90) goes to zero as $S \to 0$, we find

$$f(t) = \int_C \frac{\partial\psi}{\partial n}\, ds. \tag{9.91}$$

Turning to two-dimensional polar coordinates, we assume symmetry about the origin. Then, Eqs. (9.89) and (9.91) become

$$\frac{\partial^2 \psi}{\partial \Delta^2} + \frac{1}{\Delta}\frac{\partial \psi}{\partial \Delta} = \frac{1}{c_0^2}\frac{\partial^2 \psi}{\partial t^2}, \tag{9.92}$$

$$f(t) = \lim_{\Delta \to 0} \int_C \frac{\partial \psi}{\partial \Delta} \Delta \, d\phi = \lim_{\Delta \to 0} \left(2\pi\Delta \frac{\partial \psi}{\partial \Delta}\right), \tag{9.93}$$

where $f(t)$, the strength of the source, is given *a priori.*

9.3.2.3. Basins with Variable Depth. If the fluid is of variable depth, $H = H(x, y)$, the equation of continuity, Eq. (9.71), is replaced by

$$\frac{\partial \zeta}{\partial t} = -\left[\frac{\partial}{\partial x}(uH) + \frac{\partial}{\partial y}(vH)\right]. \tag{9.94}$$

Eliminating u and v with the aid of Eqs. (9.72), where $\mathbf{F} = 0$ and $p_0 = $ constant, we find

$$\frac{\partial^2 \zeta}{\partial t^2} = g\left[\frac{\partial}{\partial x}\left(H\frac{\partial \zeta}{\partial x}\right) + \frac{\partial}{\partial y}\left(H\frac{\partial \zeta}{\partial y}\right)\right]. \tag{9.95}$$

In polar coordinates, Eq. (9.95) becomes

$$\frac{\partial^2 \zeta}{\partial t^2} = gH\left(\frac{\partial^2 \zeta}{\partial \Delta^2} + \frac{1}{\Delta}\frac{\partial \zeta}{\partial \Delta} + \frac{1}{\Delta^2}\frac{\partial^2 \zeta}{\partial \phi^2}\right) + g\left(\frac{\partial \zeta}{\partial \Delta}\frac{\partial H}{\partial \Delta} + \frac{1}{\Delta^2}\frac{\partial \zeta}{\partial \phi}\frac{\partial H}{\partial \phi}\right). \tag{9.96}$$

Assuming, for example, $\zeta(\Delta, \phi, t) = f(\Delta)\cos m\phi \cos(\omega t + \theta_0)$, $H = H(\Delta)$, we have

$$H\left(\frac{d^2 f}{d\Delta^2} + \frac{1}{\Delta}\frac{df}{d\Delta} - \frac{m^2}{\Delta^2}f\right) + \frac{dH}{d\Delta}\frac{df}{d\Delta} + \frac{\omega^2}{g}f = 0. \tag{9.97}$$

For a *circular basin* of radius a, which shelves gradually from the center to the edge according to the law

$$H = H_0\left(1 - \frac{\Delta^2}{a^2}\right), \tag{9.98}$$

Eq. (9.97) yields

$$\left(1 - \frac{\Delta^2}{a^2}\right)\left(\frac{d^2 f}{d\Delta^2} + \frac{1}{\Delta}\frac{df}{d\Delta} - \frac{m^2}{\Delta^2}f\right) - \frac{2\Delta}{a^2}\frac{df}{d\Delta} + \frac{\omega^2}{gH_0}f = 0. \tag{9.99}$$

The exact solution of Eq. (9.99) is

$$f(\Delta) = A_m\left(\frac{\Delta}{a}\right)^m {}_2F_1\left(\alpha, \beta, \alpha + \beta, \frac{\Delta^2}{a^2}\right), \tag{9.100}$$

where

$$\alpha = \frac{m + n}{2}, \qquad \beta = 1 + \frac{m - n}{2}, \qquad n(n - 2) = m^2 + \frac{\omega^2 a^2}{gH_0}. \tag{9.101}$$

The hypergeometric series for ${}_2F_1$ will not converge for $\Delta = a$ unless it terminates. This can happen only when n is an integer of the form $m + 2j$. Hence, $\alpha = m + j$, $\beta = 1 - j$, and

$$2j(j-1) + m(2j-1) = \frac{\omega^2 a^2}{2gH_0}, \tag{9.102}$$

$$\zeta = A_m \left(\frac{\Delta}{a}\right)^m {}_2F_1\left(m + j, 1 - j, m + 1, \frac{\Delta^2}{a^2}\right) \cos m\phi \cos(\omega t + \theta_0)$$

$$= A_m \left(\frac{\Delta}{a}\right)^m \left[\frac{(j-1)!m!}{(m+j-1)!}\right] P_{j-1}^{(m,0)}(\cos\chi) \cos m\phi \cos(\omega t + \theta_0), \tag{9.103}$$

where $\sin(\chi/2) = \Delta/a$ and $P_j^{(m,k)}$ denote the *Jacobi polynomials*. Equation (9.102) is the frequency equation. We consider two modes of vibration.

9.3.2.3.1. SYMMETRICAL MODES ($m = 0, j > 1$). In this case

$$\zeta = A_0 P_{j-1}(\cos\chi)\cos(\omega t + \theta_0). \tag{9.104}$$

There are $(j - 1)$ concentric nodal circles of radii lying between 0 and a. These radii are determined by the zeros of the Legendre polynomial. The eigenfrequencies are determined from Eq. (9.102)

$$\frac{\omega a}{\sqrt{(gH_0)}} = 2\sqrt{[j(j-1)]} = 2.828, 4.899, 6.928, 8.944, \ldots. \tag{9.105}$$

The gravest symmetrical mode ($j = 2$) has a nodal circle of radius $a/\sqrt{2}$.

9.3.2.3.2. SLOWEST UNSYMMETRICAL MODES. For a given m, the value of j that makes the period largest is $j = 1$. This yields

$$\omega^2 = \frac{2mgH_0}{a^2}, \qquad \zeta = A_m \left(\frac{\Delta}{a}\right)^m \cos m\phi \cos(\omega t + \theta_0). \tag{9.106}$$

However, if $m = 1$ and j is arbitrary, Eq. (9.102) yields

$$\frac{\omega^2 a^2}{gH_0} = 2(2j^2 - 1),$$

i.e.,

$$\frac{\omega a}{\sqrt{(gH_0)}} = 1.414, 3.742, 5.831, 7.874, \ldots. \tag{9.107}$$

Clearly, if a source is located arbitrarily in the basin and with an arbitrary time-function, we must sum over all modes and integrate over the frequency.

9.3.2.4. Seiches Induced by Earthquakes. Forced or free oscillations of long water waves in an enclosed or semienclosed basin are called *seiches*. Among the causes of seiches are the horizontal Love and Rayleigh oscillations caused by major (even distant) earthquakes, passage of meteorologic pressure fronts, impact of wind gusts, flood discharge, direct air–water coupling of atmospheric disturbance, etc. Coastal seiches may be energized by long-period waves from the open ocean. These seiches may be forced or free. As long as the long-period wave energy is impressed on the mouth of a bay or the entrance of a harbor basin, the oscillations will be forced. When this energy source is withdrawn, the seiche will be free and will be sustained as long as the remaining energy can hold against the dissipative tendencies of bottom friction, internal viscosity, and turbulence of flushing at the basin mouth.

Seiches in bodies of water remote from the epicenter have been associated with a number of large earthquakes. In these cases, the natural period of a basin must be very close to the periods of the earthquake waves and the energy of the earthquake must be above a certain threshold value.

Assuming a velocity-dependent *dissipative force* and an excitation in the form of a surface-wave horizontal acceleration $F_x(x, t)$, Eq. (9.79) becomes

$$\frac{\partial^2 \Theta}{\partial t^2} + k\frac{\partial \Theta}{\partial t} - gH\frac{\partial^2 \Theta}{\partial x^2} = F_x(x, t), \qquad (k > 0). \tag{9.108}$$

The boundary conditions are

$$\Theta(0, t) = \Theta(L, t) = 0; \qquad \Theta(x, 0) = \frac{\partial \Theta}{\partial t}(x, 0) = 0, \qquad 0 \leq L \leq x; \tag{9.109}$$

where L is the channel's width.

The solution of the problem proceeds in the following steps:

1. $\sin(n\pi x/L)$ is a complete set of eigenfunctions on $0 \leq x \leq L$. Therefore, Θ can be expanded into a Fourier *sine series*:

$$\Theta(x, t) = \sum_{n=1}^{\infty} A_n(t)\sin\left(\frac{n\pi x}{L}\right). \tag{9.110}$$

2. We assume that the change in $F_x(x, t)$ is small over L so that $F_x(x, t) \simeq F(t)$. Using the sum

$$\frac{4}{\pi}\sum_{n=0}^{\infty}\frac{\sin[(2n+1)(\pi x/L)]}{2n+1} = \begin{cases} 1, & 0 < x/L < 1 \\ 0, & x = 0, L \end{cases} \tag{9.111}$$

we may write

$$F(t) = \frac{4F(t)}{\pi}\sum_{n=0}^{\infty}\frac{1}{2n+1}\sin\left[(2n+1)\frac{\pi x}{L}\right]. \tag{9.112}$$

Substituting the values of Θ and F from Eqs. (9.110) and (9.112) into Eq. (9.108) and equating the coefficients of $\sin[(2n + 1)(\pi x/L)]$, we obtain a differential equation for the unknown coefficients

$$\frac{d^2}{dt^2}A_{2n+1}(t) + k\frac{d}{dt}A_{2n+1}(t) + \frac{(2n+1)^2\pi^2 c_0^2}{L^2}A_{2n+1}(t) = \frac{4}{\pi(2n+1)}F(t), \tag{9.113}$$

where $c_0 = \sqrt{(gH)}$ is the velocity of free gravity waves in water of depth much less than the water wavelength.

3. An explicit solution of Eq. (9.113) is (assuming $k \ll 1$)

$$A_{2n+1}(t) = \frac{-4}{\pi(2n+1)} \int_0^t F(\tau)e^{-(1/2)k(t-\tau)} \left\{\frac{\sin[\beta(t-\tau)]}{\beta}\right\} d\tau,$$

$$A_{2n}(t) = 0, \qquad \beta = \frac{\pi c_0}{L}(2n+1). \tag{9.114}$$

Therefore, from Eqs. (9.78) and (9.110), the surface elevation is

$$\zeta(x,t) = -H\frac{\partial\Theta}{\partial x} = \frac{4}{\pi}\sqrt{\left(\frac{H}{g}\right)} \sum_{n=0}^{\infty} \left\{\frac{\cos[(2n+1)(\pi x/L)]}{2n+1}\right\}$$

$$\times \int_0^t F(\tau)e^{-(1/2)k(t-\tau)} \sin[\beta(t-\tau)]d\tau. \tag{9.115}$$[2]

This expression yields the seiche waveform. Its amplitude is proportional to $\sqrt{H}$; i.e., deeper channels have higher seiches.

Note that we can decompose $\zeta(x, t)$ into a sum of two progressive waves:

$$\zeta(x,t) = \frac{1}{2}\sqrt{\left(\frac{H}{g}\right)} \int_0^t F(\tau)e^{-(1/2)k(t-\tau)}$$

$$\times \frac{4}{\pi}\sum_{n=0}^{\infty}\left[\frac{\sin(2n+1)\theta^+}{(2n+1)} - \frac{\sin(2n+1)\theta^-}{(2n+1)}\right] d\tau, \tag{9.116}$$

$$\theta^{\pm} = \frac{\pi}{L}[x \pm c_0(t-\tau)].$$

The damping constant, k, incorporates the effects of both imperfect reflection coefficient at the channel edges and the bottom drag-force damping. It may depend on the wavelength such that higher seiche harmonics will be preferentially attenuated. The existence of standing-wave modes depends upon the basin's shape. The higher modes, with their shorter wavelengths, will be affected more by detail in the channel structure, which will tend to damp them out faster because irregular structures will absorb, diffuse, and randomize the otherwise harmonic oscillations. A value of $k = 4 \times 10^{-3}$/s is consistent with most data.

The forcing function $F(t)$ is the horizontal acceleration of Love or Rayleigh waves at the canal. It depends on the source parameters of the earthquake and the coordinates of the canal relative to the source.

Note that the seiche equation in ζ is [Eq. (9.75)],

$$\frac{\partial^2\zeta}{\partial t^2} + k\frac{\partial\zeta}{\partial t} - c_0^2\frac{\partial^2\zeta}{\partial x^2} = -H\frac{\partial F_x(x,t)}{\partial x}. \tag{9.117}$$

[2] Note that $H(t)$ represents the Heaviside unit-step function [Eq. (F.43)], whereas the parameter H denotes constant water depth.

9.3.2.5. Seiches Resulting from a Traveling Pressure Wave. If we introduce a velocity-dependent dissipative force and assume $F_x = 0$, Eq. (9.75) becomes

$$\frac{\partial^2 \zeta}{\partial t^2} + k\frac{\partial \zeta}{\partial t} - c_0^2 \frac{\partial^2 \zeta}{\partial x^2} = \frac{H}{\rho}\frac{\partial^2 p_0}{\partial x^2}. \tag{9.118}$$

Assuming

$$p_0 = P_0 \exp\left[i\omega_0\left(t - \frac{x}{V}\right)\right], \qquad \zeta = \zeta_0 \exp\left[i\omega_0\left(t - \frac{x}{V}\right)\right],$$

we find

$$|\zeta_0(\omega_0)| = \frac{HP_0/(\rho V^2)}{\sqrt{[(1 - c_0^2/V^2)^2 + (k/\omega_0)^2]}}. \tag{9.119}$$

9.3.3. Tsunamis

Consider again an infinite homogeneous incompressible ocean of uniform depth H. At time $t = 0$, a limited region of the sea floor is elevated instantaneously with a velocity that has the time dependence of the Dirac delta function as a result of a submarine earthquake. This disturbance generates a surface wave that travels outward in all directions and carries part of the earthquake energy across the ocean. Whenever these waves reach the shallow waters of a continental coast, they may cause great damage and destruction. These waves are known as *tsunamis*. Because tsunamis are essentially long gravity waves, the theory of Section 9.3.1 applies and the equations governing the phenomenon are

$$\nabla^2\psi = 0, \qquad \zeta(x, y, t) = -\left[\frac{1}{g}\frac{\partial \psi}{\partial t}\right]_{z=0}, \qquad \left[\frac{\partial^2 \psi}{\partial t^2} + g\frac{\partial \psi}{\partial z}\right]_{z=0} = 0. \tag{9.120}$$

In order to evaluate the tsunami wave-amplitude, we must specify the source of the excitation. This we write as

$$\left[\frac{\partial \psi}{\partial z}\right]_{z=-H} = u_0(x, y)\delta(t), \tag{9.121}$$

where $u_0(x, y)$ is the bottom elevation. Note that the time dependence of this elevation is a step function in displacement and therefore a delta function in velocity.

The retention of the term $\frac{1}{2}\mathbf{v}^2 = \frac{1}{2}(\nabla\psi)^2$ in the Bernoulli equation and the corresponding second-order term in the kinematic boundary condition gives rise to a wave of finite amplitude known as a *solitary wave*. This wave consists of a single elevation, which may travel for a considerable distance with little change of profile. However, we will not be considering solitary waves here.

In order to arrive at a suitable representation of the source, we require the *Fourier–Bessel theorem*:

$$F(\Delta) = \int_0^\infty J_m(k\Delta)k\, dk \int_0^\infty F(\Delta_0)J_m(k\Delta_0)\Delta_0\, d\Delta_0, \tag{9.122}$$

the *Fourier series* representation:

$$F(\phi) = \frac{1}{2\pi} \sum_{m=-\infty}^{\infty} \int_0^{2\pi} e^{im(\phi-\phi_0)}F(\phi_0)d\phi_0, \tag{9.123}$$

the *addition theorem* (Fig. 9.3):

$$\begin{aligned} J_0(kR) &= \sum_{m=-\infty}^{\infty} J_m(k\Delta)J_m(k\Delta_0)e^{im(\phi-\phi_0)}, \\ R^2 &= \Delta^2 + \Delta_0^2 - 2\Delta\Delta_0\cos(\phi - \phi_0), \end{aligned} \tag{9.124}$$

and the *Neumann integral theorem*

$$F(\Delta, \phi) = \frac{1}{2\pi} \int_0^\infty k\, dk \int_0^\infty \Delta_0\, d\Delta_0 \int_0^{2\pi} F(\Delta_0, \phi_0)J_0(kR)d\phi_0. \tag{9.125}$$

We use Eq. (9.125) to represent the source excitation function

$$u_0(\Delta, \phi)\delta(t) = \frac{1}{(2\pi)^2} \int_{-\infty}^{\infty} e^{i\omega t}\, d\omega \int_0^\infty k\, dk \int_0^\infty \Delta_0\, d\Delta_0 \int_0^{2\pi} u_0(\Delta_0, \phi_0)J_0(kR)d\phi_0. \tag{9.126}$$

The general solution of the Laplace equation in cylindrical coordinates can be forced into a similar representation

$$\begin{aligned} \psi(\Delta, \phi, z, t) &= \frac{1}{2\pi} \int_{-\infty}^{\infty} e^{i\omega t}\, d\omega \int_0^\infty k\, dk \int_0^\infty \Delta_0\, d\Delta_0 \\ &\quad \times \int_0^{2\pi} [A(\Delta_0, \phi_0)\text{ch } kz + B(\Delta_0, \phi_0)\text{sh } kz]J_0(kR)d\phi_0. \end{aligned} \tag{9.127}$$

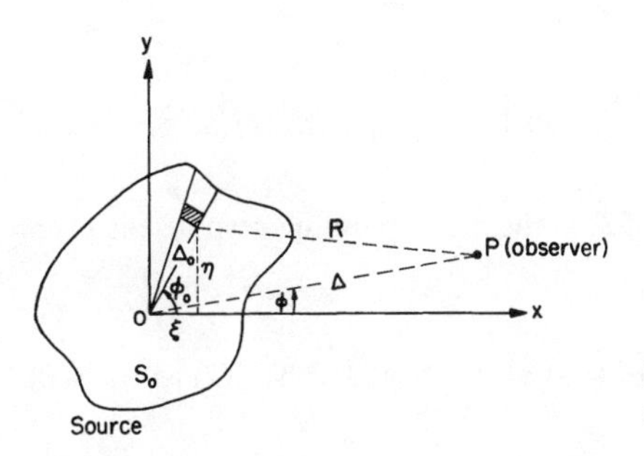

Figure 9.3. The source–observer geometry.

The surface elevation at $z = 0$ therefore becomes [cf. Eq. (9.120)]

$$\zeta(x, y, t) = \frac{-1}{2\pi g} \int_{-\infty}^{\infty} i\omega e^{i\omega t}\, d\omega \int_0^{\infty} k\, dk \int_0^{\infty} \Delta_0\, d\Delta_0 \int_0^{2\pi} A(\Delta_0, \phi_0) J_0(kR) d\phi_0.$$

The application of the boundary and source conditions yields two simultaneous equations for A and B. Substituting the value of A in the last equation, we obtain

$$\zeta(x, y, t) = \frac{-1}{(2\pi)^2} \int_0^{\infty} \frac{k\, dk}{\operatorname{ch} kH} \int_{-\infty}^{\infty} \frac{i\omega e^{i\omega t}\, d\omega}{\omega^2 - gk \operatorname{th} kH} \times \left[\int_0^{\infty} \int_0^{2\pi} u_0(\Delta_0, \phi_0) J_0(kR) \Delta_0\, d\phi_0\, d\Delta_0 \right]. \tag{9.128}$$

Inserting for $J_0\,(kR)$ from Eq. (9.124), we get

$$\zeta(x, y, t) = \frac{-1}{(2\pi)^2} \int_0^{\infty} \frac{k\, dk}{\operatorname{ch} kH} \int_{-\infty}^{\infty} \frac{i\omega e^{i\omega t}\, d\omega}{\omega^2 - gk \operatorname{th} kH} \left[\sum_{m=0}^{\infty} \varepsilon_m J_m(k\Delta) I_m(k) \right], \tag{9.129}$$

where $\varepsilon_m = 1 (m = 0);\ 2 (m = 1, 2, \ldots)$ and

$$I_m(k) = \int_0^{\infty} \int_0^{2\pi} u_0(\Delta_0, \phi_0) J_m(k\Delta_0) \cos m(\phi - \phi_0) \Delta_0\, d\phi_0\, d\Delta_0. \tag{9.130}$$

The evaluation of the integral over ω is done by the method of residues. The poles of the integrand are the roots $\omega = \omega(k)$ of the previously derived dispersion relation $\omega^2 = gk \operatorname{th} kH$. These roots are at $\omega = \pm\, \omega_1$, where $\omega_1 = \sqrt{(gk \operatorname{th} kH)}$. To ensure that the disturbance vanishes for $t < 0$, we assume that the path of integration in the ω plane passes below the poles. Evaluating the residues at the poles, we find

$$\frac{1}{2\pi} \int_{-\infty}^{\infty} \frac{i\omega e^{i\omega t}\, d\omega}{(\omega - \omega_1)(\omega + \omega_1)} = -H(t) \cos \omega_1 t.$$

The surface elevation function is therefore

$$\zeta(x, y, t) = \frac{H(t)}{2\pi} \int_0^{\infty} \frac{\cos \omega_1 t}{\operatorname{ch} kH} k\, dk \left[\sum_{m=0}^{\infty} \varepsilon_m J_m(k\Delta) I_m(k) \right]. \tag{9.131}$$

If the time-function is $f(t)$ instead of $\delta(t)$, the corresponding response is given by the convolution integral

$$\zeta(x, y, t) = \frac{1}{2\pi} \int_0^{\infty} \frac{k\, dk}{\operatorname{ch} kH} \int_0^{t} f(\tau) \cos \omega_1(t - \tau) d\tau \left[\sum_{m=0}^{\infty} \varepsilon_m J_m(k\Delta) I_m(k) \right], \quad (t > 0).$$

When the source is axially symmetrical,

$$u_0(\Delta_0, \phi_0) = u_0(\Delta_0), \qquad I_m = 0 \quad \text{for} \quad m > 0,$$

$$I_0 = 2\pi \int_0^\infty u_0(\Delta_0) J_0(k\Delta_0)\Delta_0 \, d\Delta_0,$$

and Eq. (9.131) becomes

$$\zeta(x, y, t) = \frac{H(t)}{2\pi} \int_0^\infty \frac{\cos \omega_1 t}{\operatorname{ch} kH} J_0(k\Delta) I_0(k) \, k \, dk. \tag{9.132}$$

Returning to Eq. (9.128), we assume that $u_0(\Delta_0, \phi_0)$ vanishes outside a small region such that $\Delta_0 \ll \Delta$. Denoting

$$Q_0 = \int_0^\infty \int_0^{2\pi} u_0(\Delta_0, \phi_0)\Delta_0 \, d\phi_0 \, d\Delta_0,$$

Equation (9.131) then reduces to

$$\zeta(x, y, t) = \frac{Q_0 H(t)}{2\pi} \int_0^\infty \frac{\cos \omega_1 t}{\operatorname{ch} kH} J_0(k\Delta) k \, dk. \tag{9.133}$$

The quantity Q_0 is of the dimensions $[L^3]$ and represents the strength of the source. From the limiting process that leads to the definition of Q_0 it is clear that, in the case of a displacement dislocation, Q_0 is proportional to the product of the projection of the fault area (S_0) over the horizontal plane and the vertical component of the dislocation. Therefore, in the case of a tensile dislocation, $Q_0 = A_0 U_0 S_0 \cos^2 \delta$, where U_0 is the magnitude of the dislocation, δ is the dip of the fault, and A_0 is a dimensionless source constant. In the case of a shear dislocation (Fig. 4.21), the vertical component of U_0 is $U_0 \sin \lambda \sin \delta$, and therefore, $Q_0 = A_0 U_0 S_0 \sin \lambda \sin \delta \cos \delta$, where λ is the slip angle. The same conclusion is arrived at if we evaluate the epicentral static displacements caused by a shear dislocation in a uniform half-space, which also shows that A_0 is inversely proportional to the square of the source depth. Therefore, only shallow earthquakes will be able to generate tsunamis of appreciable amplitude. Moreover, the most "efficient" shear dislocation in exciting tsunami waves has $\lambda = 90°$, $\delta = 45°$, which is a dip slip on a 45° dipping fault, labeled as source III in Chapter 4.

9.3.3.1. Dispersion. The observed group velocity of the head of the tsunami wave in midocean is about 235 m/s at periods around 600 s. The corresponding wavelength, ca. 140 km, is much greater than the average ocean depth ($H \simeq 5.5$ km). The long-wave approximation, therefore, applies to tsunami waves. Indeed, the calculated value of $\sqrt{(gH)}$ yields a velocity of 232 m/s. The phase and group velocities of the tsunami's head are therefore almost equal and the wave attains the maximum velocity that a gravity wave may have for any given depth of the fluid (Fig. 9.4).

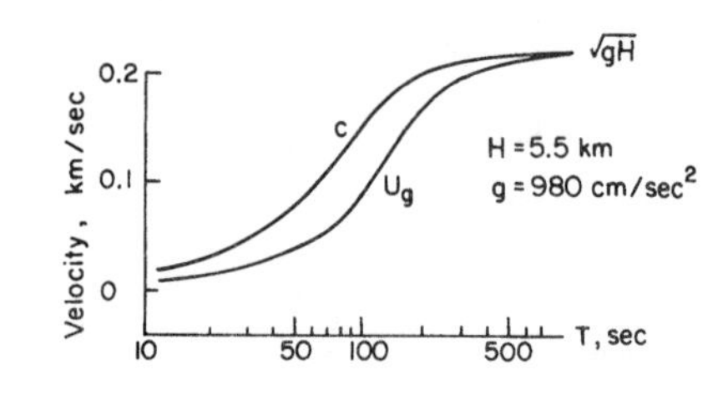

Figure 9.4. Dispersion curves for tsunami waves. (c = phase velocity, U_g = group velocity).

In general, we may approximate the dispersion relation for long waves such that the evaluation of the k integral becomes analytically feasible. We, therefore, set $kH \ll 1$, $k\Delta \gg 1$. Consequently,

$$\omega_1 = k\sqrt{(gH)}\left(\frac{\text{th } kH}{kH}\right)^{1/2} = k\sqrt{(gH)}\left[1 - \frac{1}{6}k^2H^2 + O(k^4H^4)\right],$$

$$J_0(k\Delta) = \sqrt{\left(\frac{2}{\pi k\Delta}\right)}\cos\left(k\Delta - \frac{\pi}{4}\right).$$

Equation (9.133) now becomes

$$\zeta(x, y, t) = \frac{Q_0 H(t)}{2\pi}\int_0^\infty \left(\frac{k}{2\pi\Delta}\right)^{1/2}\left[\cos\left(k\Delta - \frac{\pi}{4} + \omega_1 t\right) + \cos\left(k\Delta - \frac{\pi}{4} - \omega_1 t\right)\right]\frac{dk}{\text{ch } kH}.$$

Considering only waves diverging from the source and replacing ω_1 by its approximate value, we get

$$\zeta(x, y, t) = \frac{Q_0 H(t)}{2\pi}\int_0^\infty \left(\frac{k}{2\pi\Delta}\right)^{1/2}\cos\left[(\Delta - c_0 t)k + \frac{1}{6}c_0 t H^2 k^3 - \frac{\pi}{4}\right]\frac{dk}{\text{ch } kH}, \tag{9.134}$$

where $c_0 = \sqrt{(gH)}$. If we substitute $s = (c_0 t H^2/2)^{1/3}k$, Eq. (9.134) reduces to

$$\zeta(x, y, t) = \frac{Q_0}{2H(\pi^3 c_0 t\Delta)^{1/2}}\int_0^\infty \cos\left(ps + \frac{1}{3}s^3 - \frac{\pi}{4}\right)\frac{\sqrt{s}\, ds}{\text{ch } as}, \qquad (t > 0) \tag{9.135}$$

where

$$p = \frac{\Delta - c_0 t}{(c_0 H^2 t/2)^{1/3}}, \qquad a = \left(\frac{2H}{c_0 t}\right)^{1/3}. \tag{9.136}$$

Equation (9.135) implies that, in the two-dimensional case, the surface elevation decays with distance as Δ^{-1}. Assuming ch $as \simeq 1$ and using the definite integral

$$\int_0^\infty x^{k-1}e^{i\mu x^\alpha}\,dx = \frac{\Gamma(k/\alpha)}{\alpha\mu^{(k/\alpha)}}e^{k\pi i/2\alpha}, \tag{9.137}$$

the particular case $p = 0$ lends itself to the explicit result,

$$\zeta\left(\Delta, t = \frac{\Delta}{c_0}\right) = \frac{(0.09)Q_0}{H\Delta}. \tag{9.138}$$

By applying the method of stationary phase, it can be shown that, near $t = \Delta/c_0$, the amplitude of the converging wave is much smaller than the amplitude of the diverging wave.

9.3.3.2. Extended Source. Let us suppose that the source extends to infinity in the y direction and u_0 is independent of y. Then, if L is the source width, we have [cf. Eq. (9.128)],

$$\begin{aligned}\int_0^\infty \int_0^{2\pi} & u_0(\Delta_0, \phi_0) J_0(kR)\Delta_0 \, d\phi_0 \, d\Delta_0 \\ &= \int_{-L/2}^{L/2} u_0(\xi) d\xi \int_{-\infty}^{\infty} J_0[k\{(x-\xi)^2 + (y-\eta)^2\}^{1/2}] d\eta \\ &= \int_{-L/2}^{L/2} u_0(\xi) \left[\frac{2}{k} \cos k(x-\xi)\right] d\xi. \qquad (9.139)\end{aligned}$$

Equation (9.131) and the long-wave approximation ($kH \ll 1$) now yield

$$\begin{aligned}\zeta(x,t) &= \frac{H(t)}{\pi} \int_{-L/2}^{L/2} u_0(\xi) d\xi \int_0^\infty \frac{\cos \omega_1 t \cos k(x-\xi)}{\operatorname{ch} kH} dk, \\ \omega_1 &= k\sqrt{(gH)}\left(1 - \frac{1}{6} k^2 H^2\right).\end{aligned} \qquad (9.140)$$

Assuming that $u_0 = U_0$ is constant over the source width L, and that $x \gg \xi$, we have

$$\begin{aligned}\zeta(x,t) &\simeq \frac{1}{\pi} U_0 L H(t) \int_0^\infty \cos \omega_1 t \cos kx \, dk \\ &= \frac{1}{\pi} U_0 L H(t) \int_0^\infty \frac{1}{2} [\cos(\omega_1 t - kx) + \cos(\omega_1 t + kx)] dk.\end{aligned}$$

Considering only the disturbance moving in the positive x direction, inserting the values of ω_1 from Eq. (9.140), and making the change of variable, $s = (c_0 t H^2/2)^{1/3} k$, we get

$$\begin{aligned}\zeta(x,t) \simeq \frac{U_0 L}{2\pi} \int_0^\infty \cos(\omega_1 t - kx) dk &= \frac{U_0 L}{2\pi}\left(\frac{2}{c_0 t H^2}\right)^{1/3} \int_0^\infty \cos\left(ps + \frac{1}{3}s^3\right) ds \\ &= \frac{1}{2} U_0 L \left(\frac{2}{c_0 t H^2}\right)^{1/3} Ai(p), \qquad (t > 0)\end{aligned} \qquad (9.141)$$

where $p = (x - c_0 t)/(\frac{1}{2} c_0 t H^2)^{1/3}$ and $Ai(p)$ is the Airy function. Because $Ai(0) = \Gamma(\frac{1}{3})/(2\pi \cdot 3^{1/6}) \simeq 0.355\ldots$, the dependence of the far-field elevation on the distance x becomes

$$\zeta\left(x, t = \frac{x}{c_0}\right) \simeq \frac{0.45(U_0 L/2)}{H^{2/3} x^{1/3}}. \qquad (9.142)$$

The Airy function $Ai(p)$ has the general form shown in Fig. G.1 (App. G). The asymptotic behavior of the Airy function is given by

$$Ai(p) \sim \begin{cases} \dfrac{1}{2\sqrt{\pi}} p^{-1/4} e^{-(2/3)p^{3/2}}, & p \to +\infty \\ \dfrac{1}{\sqrt{\pi}} |p|^{-1/4} \sin\left[\dfrac{2}{3}|p|^{3/2} + \dfrac{\pi}{4}\right], & p \to -\infty. \end{cases} \tag{9.143}$$

It is clear that $\zeta(x, t)$ decays exponentially ahead of $x = c_0 t$ and becomes oscillatory behind. Exactly at $x = c_0 t$, $\zeta \propto t^{-1/3}$. The width of the transition region is proportional to $(\frac{1}{2}c_0 t H^2)^{1/3}$. Away from the transition region, as $(x - c_0 t)/(\frac{1}{2}c_0 t H^2)^{1/3} \to -\infty$,

$$\zeta(x, t) \to \frac{U_0 L}{2\sqrt{\pi}} \frac{\sin[\frac{2}{3}(c_0 t - x)^{3/2}/(\frac{1}{2}c_0 t H^2)^{1/2} + \pi/4]}{[\frac{1}{2}c_0 t H^2 (c_0 t - x)]^{1/4}}, \tag{9.144}$$

which is shown to agree with the stationary-phase approximation.

The disturbance moving in the negative x direction can be considered similarly. It is found that near $t = x/c_0$ the amplitude of the wave moving in the negative x direction is very small compared with the amplitude of the wave moving in the positive x direction.

When this theory is applied to long shear dislocations, it is clear that U_0 in the above equations should be replaced by the vertical component of the dislocation, $U_0 \sin \lambda \sin \delta$, and L by the projection of the width of the fault on the horizontal plane, $L \cos \delta$.

9.4. Acoustic–Gravity Waves in the Atmosphere

9.4.1. Basic Equations[3]

We shall confine our discussion to the oscillations of a nonhomogeneous atmosphere resting on a flat ground when disturbed from its static equilibrium by a source such as an explosion or an earthquake. The following assumptions will be made: (1) The atmosphere consists of an ideal fluid (gas) and the sound propagation in it is adiabatic. (2) The deviations from equilibrium are small such that the sound-wave approximation holds. (3) The constitutive parameters depend on the vertical coordinate alone. If the positive direction of the z axis is taken upward, then the pressure distribution, $p_0(z)$, the density distribution, $\rho_0(z)$, in the equilibrium state, the acceleration of gravity, g, and the temperature distribution, $T(z)$, are related through the equations

$$\frac{dp_0}{dz} = -g\rho_0, \qquad p_0 = \mathrm{K}\rho_0 T. \tag{9.145}$$

[3] In this section T denotes the temperature on the Kelvin scale.

Their solutions are

$$p_0(z) = p_0(z_0)\exp\left(-\frac{1}{\mathrm{K}}\int_{z_0}^{z}\frac{g}{T}\,dz\right), \qquad \rho_0(z) = \rho_0(z_0)\frac{T_0}{T}\exp\left(-\frac{1}{\mathrm{K}}\int_{z_0}^{z}\frac{g}{T}\,dz\right),$$
$$T_0 = T(z_0). \quad (9.146)$$

Here T is the temperature on the Kelvin scale and K is the universal gas constant per unit molecular weight. Upon this equilibrium state, we superimpose small oscillations that are governed by the Euler equations of hydrodynamics for a gravitating and temperature-stratified compressible atmosphere:

$$\rho_{\text{tot}}\frac{D\mathbf{v}}{Dt} = -\operatorname{grad} p_{\text{tot}} - g\rho_{\text{tot}}\mathbf{e}_z \qquad \text{(conservation of momentum)}, \quad (9.147)$$

$$\frac{D\rho_{\text{tot}}}{Dt} + \operatorname{div}(\rho_{\text{tot}}\mathbf{v}) = 0 \qquad \text{(conservation of mass)}, \quad (9.148)$$

$$\frac{Dp_{\text{tot}}}{Dt} = \gamma\frac{p_{\text{tot}}}{\rho_{\text{tot}}}\frac{D\rho_{\text{tot}}}{Dt} \qquad \text{(conservation of energy)}. \quad (9.149)$$

In these equations

$$p_{\text{tot}} = p_0 + p, \qquad \rho_{\text{tot}} = \rho_0 + \rho, \qquad \frac{D}{Dt} \equiv \frac{\partial}{\partial t} + \mathbf{v}\cdot\nabla, \quad (9.150)$$

where ρ and p are the perturbations in the density and pressure, respectively, $\mathbf{v}$ is the velocity, and γ is the ratio of the specific heats. The pressure perturbation, p, is also known as *overpressure.*

If $p/p_0 \ll 1$, $\rho/\rho_0 \ll 1$ and $|\mathbf{v}/\alpha| \ll 1$, where $\alpha^2 = \gamma(p_0/\rho_0) = \gamma \mathrm{K}T$, the above equations simplify to the linear set

$$\rho_0\frac{\partial\mathbf{v}}{\partial t} = -\operatorname{grad} p - g\rho\mathbf{e}_z, \qquad \frac{\partial\rho}{\partial t} + \operatorname{div}(\rho_0\mathbf{v}) = 0,$$
$$\left(\frac{\partial p}{\partial t} - \alpha^2\frac{\partial\rho}{\partial t}\right) + \mathbf{v}\cdot(\operatorname{grad} p_0 - \alpha^2\operatorname{grad}\rho_0) = 0. \quad (9.151)$$

All in all, we have five scalar equations for the determination of the five variables $\mathbf{v}(u, v, w)$, p and ρ:

$$\rho_0\frac{\partial u}{\partial t} = -\frac{\partial p}{\partial x}, \qquad \rho_0 = \rho_0(z)$$
$$\rho_0\frac{\partial v}{\partial t} = -\frac{\partial p}{\partial y},$$
$$\rho_0\frac{\partial w}{\partial t} = -\frac{\partial p}{\partial z} - g\rho, \qquad \rho = \rho(z, t) \quad (9.152)$$
$$\frac{\partial\rho}{\partial t} + w\frac{\partial\rho_0}{\partial z} = -\rho_0\chi,$$
$$\frac{\partial p}{\partial t} - g\rho_0 w = -\gamma p_0\chi,$$

where $\chi = (\partial u/\partial x) + (\partial v/\partial y) + (\partial w/\partial z)$ is the time derivative of the volume dilatation. From Eqs. (9.145) and (9.152), we find

$$\frac{1}{\rho_0}\frac{d\rho_0}{dz} = -\frac{1}{\alpha^2}\left(\gamma g + \frac{d\alpha^2}{dz}\right). \tag{9.153}$$

Eliminating p and ρ from Eqs. (9.152), we are left with three equations in three unknown functions (u, v, w):

$$\begin{aligned}
\frac{\partial^2 u}{\partial t^2} &= \frac{\partial}{\partial x}(\alpha^2\chi - gw),\\
\frac{\partial^2 v}{\partial t^2} &= \frac{\partial}{\partial y}(\alpha^2\chi - gw),\\
\frac{\partial^2 w}{\partial t^2} &= \frac{\partial}{\partial z}(\alpha^2\chi - gw) - \left[\frac{d\alpha^2}{dz} + (\gamma - 1)g\right]\chi.
\end{aligned} \tag{9.154}$$

Taking the derivatives of the Eqs. (9.154) with respect to x, y, and z, respectively, adding, and neglecting the derivatives of g, we find

$$\frac{\partial^2\chi}{\partial t^2} = (\alpha^2\nabla^2\chi - g\nabla^2 w) - \left[\frac{d\alpha^2}{-dz} + (\gamma - 1)g\right]\frac{\partial\chi}{\partial z}. \tag{9.155}$$

Because

$$\frac{\partial^2}{\partial t^2}\boldsymbol{\Omega} = \operatorname{curl}\left(\frac{\partial^2 u}{\partial t^2}, \frac{\partial^2 v}{\partial t^2}, \frac{\partial^2 w}{\partial t^2}\right) = \left(\frac{\partial\ddot{w}}{\partial y} - \frac{\partial\ddot{v}}{\partial z}, \frac{\partial\ddot{u}}{\partial z} - \frac{\partial\ddot{w}}{\partial x}, \frac{\partial\ddot{v}}{\partial x} - \frac{\partial\ddot{u}}{\partial y}\right),$$

the components of the second time derivative of the vorticity, $\boldsymbol{\Omega} = \operatorname{curl}(u, v, w)$, found from Eq. (9.154), are

$$\begin{aligned}
\frac{\partial^2\Omega_x}{\partial t^2} &= -\left[\frac{d\alpha^2}{dz} + (\gamma - 1)g\right]\frac{\partial\chi}{\partial y},\\
\frac{\partial^2\Omega_y}{\partial t^2} &= \left[\frac{d\alpha^2}{dz} + (\gamma - 1)g\right]\frac{\partial\chi}{\partial x}, \qquad \frac{\partial^2\Omega_z}{\partial t^2} = 0.
\end{aligned} \tag{9.156}$$

Now, differentiating Eq. (9.155) twice with respect to time and noting that

$$\begin{aligned}
\frac{\partial^2}{\partial t^2}\left(\nabla^2 w - \frac{\partial\chi}{\partial z}\right) &= \frac{\partial}{\partial y}\left(\frac{\partial^2\Omega_x}{\partial t^2}\right) - \frac{\partial}{\partial x}\left(\frac{\partial^2\Omega_y}{\partial t^2}\right)\\
&= -\left[\frac{d\alpha^2}{dz} + (\gamma - 1)g\right]\left(\frac{\partial^2\chi}{\partial x^2} + \frac{\partial^2\chi}{\partial y^2}\right),
\end{aligned} \tag{9.157}$$

we obtain, to terms of the first order, a fourth-order partial differential equation for χ

$$\frac{\partial^4\chi}{\partial t^4} = \alpha^2\nabla^2\frac{\partial^2\chi}{\partial t^2} - \left(\frac{d\alpha^2}{dz} + \gamma g\right)\frac{\partial^3\chi}{\partial t^2\,\partial z} + g\left[\frac{d\alpha^2}{-dz} + (\gamma - 1)g\right]\nabla_1^2\chi. \tag{9.158}$$

As in the case of gravity waves in fluids (Sect. 9.3.3), we represent χ in the form of the Neumann integral:

$$\chi(\Delta, \phi, z; t) = \frac{1}{(2\pi)^2} \int_{-\infty}^{\infty} e^{i\omega t}\, d\omega \int_0^{\infty} k\, dk \int_0^{\infty} \Delta_0\, d\Delta_0 \int_0^{2\pi} A(\Delta_0, \phi_0; k, \omega) \times \bar{\chi}(z; k, \omega) J_0(kR) d\phi_0, \tag{9.159}$$

where (Fig. 9.3)

$$R^2 = \Delta^2 + \Delta_0^2 - 2\Delta\Delta_0 \cos(\phi - \phi_0).$$

The function A will be determined from the boundary and source conditions. Similarly, we put

$$w(\Delta, \phi, z; t) = \frac{1}{(2\pi)^2} \int_{-\infty}^{\infty} e^{i\omega t}\, d\omega \int_0^{\infty} k\, dk \int_0^{\infty} \Delta_0\, d\Delta_0 \int_0^{2\pi} A(\Delta_0, \phi_0; k, \omega) \times \bar{w}(z; k, \omega) J_0(kR) d\phi_0, \tag{9.160}$$

$$p(\Delta, \phi, z; t) = \frac{1}{(2\pi)^2} \int_{-\infty}^{\infty} e^{i\omega t}\, d\omega \int_0^{\infty} k\, dk \int_0^{\infty} \Delta_0\, d\Delta_0 \int_0^{2\pi} A(\Delta_0, \phi_0; k, \omega) \times \bar{p}(z; k, \omega) J_0(kR) d\phi_0.$$

From Eqs. (9.158) and (9.159), we find that $\bar{\chi}$ satisfies the equation

$$\alpha^2 \frac{d^2\bar{\chi}}{dz^2} - \left(\frac{d\alpha^2}{dz} + \gamma g\right)\frac{d\bar{\chi}}{dz} + \left[\frac{gk^2}{\omega^2}\frac{d\alpha^2}{dz} + \omega^2 - k^2\alpha^2\left(1 - \frac{\omega_B^2}{\omega^2}\right)\right]\bar{\chi} = 0, \tag{9.161}$$

where

$$\omega_B = \frac{g}{\alpha}(\gamma - 1)^{1/2}$$

is the *Brunt frequency*.

The entity ω_B has an interesting physical interpretation. Consider a bit of fluid to be enclosed in a small balloon-like membrane such that the pressures inside and outside are always the same but adiabatic changes of density can occur unimpeded. The balloon, initially at level z, is displaced to level $z + \xi$, with a resulting change in its pressure, $\delta p = -\rho_0 g \xi$. The density inside the balloon has changed adiabatically by the amount $(\delta\rho_0)_{\text{inside}} = (\delta p)/\alpha^2 = -(\xi\rho_0 g/\alpha^2)$. The density outside the balloon, however, is affected by a change in the elevation: $(\delta\rho_0)_{\text{outside}} = \xi(d\rho_0/dz)$. The balloon therefore experiences a buoyant force per unit volume, equal to

$$g[(\delta\rho_0)_{\text{outside}} - (\delta\rho_0)_{\text{inside}}] = g\xi\left[\frac{d\rho_0}{dz} + \frac{\rho_0 g}{\alpha^2}\right]. \tag{9.162}$$

If we define

$$\omega_B^2 = -g\left[\frac{1}{\rho_0}\frac{d\rho_0}{dz} + \frac{g}{\alpha^2}\right], \tag{9.163}$$

the equation of motion of the balloon is

$$\frac{d^2\xi}{dt^2} + \omega_B^2\xi = 0. \tag{9.164}$$

The balloon will therefore oscillate with the frequency ω_B provided ω_B is real. If ω_B^2 is negative, the balloon will be unstable and depart widely from its initial position. In our atmosphere, $(1/\rho_0)d\rho_0/dz = -\gamma g/\alpha^2$, and therefore $\omega_B^2 = (\gamma - 1)g^2/\alpha^2$ is positive.

From Eqs. (9.152) and (9.154), we can easily derive the following relations among the entities $\bar{\chi}$, $\bar{w}$, and $\bar{p}$:

$$\bar{w} = (g^2k^2 - \omega^4)^{-1}\left[\omega^2\alpha^2\frac{d\bar{\chi}}{dz} + g(k^2\alpha^2 - \gamma\omega^2)\bar{\chi}\right] = \frac{i}{\omega\rho_0 h}\cdot\left[\frac{d\bar{p}}{dz} + \frac{g}{\alpha^2}\bar{p}\right],$$

$$h = 1 - \left(\frac{\omega_B}{\omega}\right)^2, \tag{9.165}$$

$$\bar{p} = i\omega\rho_0(g^2k^2 - \omega^4)^{-1}\left[-g\alpha^2\frac{d\bar{\chi}}{dz} + (\gamma g^2 - \omega^2\alpha^2)\bar{\chi}\right] = \frac{i\rho_0}{\omega}(\alpha^2\bar{\chi} - g\bar{w}).$$

Now, because $\bar{w}$ is the vertical velocity of the harmonic vibration of a fluid particle, the expression $\bar{w}/(i\omega)$ is to first order equal to the particle vertical displacement from its static equilibrium position. The pressure associated with this motion is $\delta\bar{p} = -\rho_0 g(-i\bar{w}/\omega)$. Therefore, the *parcel pressure*, $\bar{P}$, is given by the sum of the pressure perturbation, $\bar{p}$, and the additional pressure from the change of elevation of the test particle, i.e.,

$$\bar{P} = \bar{p} + \delta\bar{p} = i\left(\frac{\alpha^2\rho_0}{\omega}\right)\bar{\chi}, \qquad \bar{p} = \bar{P} - i\frac{g\rho_0}{\omega}\bar{w}. \tag{9.166}$$

Let us consider now the boundary and source conditions. The ground at $z = 0$ is usually assumed to be rigid, so that $\bar{w}(0) = 0$ outside the source region (if the source is at the ground level). The pressure variation at the ground is then given by Eq. (9.165):

$$\frac{\bar{p}(0; k, \omega)}{p_0(0)} = i\frac{\gamma}{\omega}\bar{\chi}(0; k, \omega), \tag{9.167}$$

where $p_0(0)$ is the undisturbed atmospheric pressure at $z = 0$. The dispersion relation, $k = k(\omega)$, is determined by the vanishing of $\bar{w}$ at $z = 0$, namely,

$$\left[\omega^2\alpha^2\frac{d\bar{\chi}}{dz} + g(k^2\alpha^2 - \gamma\omega^2)\bar{\chi}\right]_{z=0} = 0, \tag{9.168}$$

where $\bar{\chi}$ is a solution of Eq. (9.161). Clearly, Eq. (9.161) differs considerably from the Helmholtz equation that governs sound waves. This describes a motion that is strongly affected and modified by the presence of the earth gravity field. The ensuing waves are known as *acoustic–gravity waves.* However, in order to bring out the contrast with the laws of propagation of acoustic waves ($g \to 0$), where the motion is irrotational (Sect. 9.2), we put

$$u = \frac{\partial \psi}{\partial x}, \qquad v = \frac{\partial \psi}{\partial y}, \qquad p = -\rho_0 \frac{\partial \psi}{\partial t}. \tag{9.169}$$

We then find, from Eqs. (9.152) and (9.154),

$$g\bar{w} - \alpha^2 \bar{\chi} = \omega^2 \bar{\psi},$$

$$\bar{w} = \frac{\partial \bar{\psi}}{\partial z} - \left[(\gamma - 1)g + \frac{d\alpha^2}{dz}\right] \frac{\left(\omega^2 \bar{\psi} - g\dfrac{\partial \bar{\psi}}{\partial z}\right)}{\left[-g\dfrac{d\alpha^2}{dz} + \omega^2 \alpha^2 - (\gamma - 1)g^2\right]}. \tag{9.170}$$

Hence $\mathbf{v} = \text{grad}\,\psi$ only if $(\gamma - 1)g + d\alpha^2/dz = 0$. This shows that the motion of the acoustic–gravity waves is rotational unless $\alpha = \text{const.}$, $\gamma = 1$, or $dT/dz = -(\gamma - 1)g/\gamma K$. The first case is that of uniform equilibrium temperature and the second occurs when the actual temperature gradient is equal to the dry-adiabatic lapse rate. In either of these special cases, the velocity potential satisfies the equation

$$\frac{\partial^2 \psi}{\partial t^2} = \alpha^2 \nabla^2 \psi - g\frac{\partial \psi}{\partial z}.$$

This condition, however, does not occur in reality, because the normal temperature gradient in the atmosphere is only about 0.6 of the adiabatic.

9.4.2. Air Waves Excited by Earthquakes

Let us now introduce a source of excitation in the form of a given upward particle velocity w_s over the ground $z = 0$ [cf. Eq. (9.126)]:

$$\begin{aligned} w_s(\Delta, \phi; t) &= \frac{\partial}{\partial t}[u_0(\Delta, \phi)H(t)] = u_0(\Delta, \phi)\delta(t) \\ &= \frac{1}{(2\pi)^2}\int_{-\infty}^{\infty} e^{i\omega t}\, d\omega \int_0^{\infty} k\, dk \int_0^{\infty} \Delta_0\, d\Delta_0 \int_0^{2\pi} u_0(\Delta_0, \phi_0) J_0(kR) d\phi_0. \end{aligned} \tag{9.171}$$

Comparing this with Eq. (9.160) and using Eq. (9.165) we obtain

$$A(\Delta_0, \phi_0; k, \omega) = \frac{(g^2k^2 - \omega^4)u_0(\Delta_0, \phi_0)}{\left[\omega^2\alpha^2 \dfrac{d\bar{\chi}}{dz} + g(k^2\alpha^2 - \gamma\omega^2)\bar{\chi}\right]_{z=0}}. \tag{9.172}$$

The overpressure at an observation point on the ground (source level, $z = 0$) at (Δ, ϕ) is then given by Eqs. (9.160) and (9.167):

$$\frac{p(\Delta, \phi, 0; t)}{p_0(0)} = \frac{-\gamma}{(2\pi)^2} \int_{-\infty}^{\infty} \frac{e^{i\omega t}}{i\omega} d\omega \int_0^{\infty} \frac{(g^2k^2 - \omega^4)\bar{\chi}(0; k, \omega)k\, dk}{\left[\omega^2\alpha^2 \dfrac{d\bar{\chi}}{dz} + g(k^2\alpha^2 - \gamma\omega^2)\bar{\chi}\right]_{z=0}}$$
$$\times \left[\int_0^{\infty} \int_0^{2\pi} u_0(\Delta_0, \phi_0) J_0(kR)\Delta_0\, d\phi_0\, d\Delta_0\right]. \tag{9.173}$$

The source integral can be dealt with as in Eq. (9.129). In the case of an axially symmetrical point source, it is simply equal to $[Q_0 J_0(k\Delta)]$, where Q_0 gives the source strength and is of the dimensions $[L^3]$. For a shear dislocation, $Q_0 = A_0 U_0 S_0 \sin\lambda \sin\delta \cos\delta$, where U_0 is the magnitude of the dislocation, λ and δ the slip and dip angles, respectively, of the fault, and S_0 is the fault area. Moreover, A_0 is a dimensionless source constant that varies inversely as the square of the source depth.

A prerequisite for the evaluation of the overpressure integral is the solution of the second-order differential equation, Eq. (9.161), for $\bar{\chi}(z; k, \omega)$. In order to solve it, we must specify the vertical variation of the temperature. The evaluation of the integral in Eq. (9.173) leads to good agreement between theory and observations. For example, the Alaska earthquake of 28 March, 1964 generated an atmospheric wave that was well recorded on various microbarographs along the Pacific coast. At Berkeley ($\Delta = 3130$ km), a signal having a peak amplitude of 40 μbar at a period of about 600 s was recorded.

9.4.3. Pressure-Induced Surface Waves

We examine next a physical process that is the reverse of the previous case. There we studied the generation of air waves by a vertical ground motion. Here we study the excitation of elastic surface waves by infrasonic air waves. This coupling is possible only if the earth's surface is considered not as a rigid boundary but as an elastic interface.

Consider an elastic half-space with the z axis normal to its boundary and drawn vertically upward. A plane sound wave of unit amplitude strikes the half-space at an angle e with the normal. The problem has been discussed in detail in Section 3.5. Let the total velocity in the air be given by

$$\mathbf{v} = [(\mathbf{e}_x \sin e - \mathbf{e}_z \cos e)e^{i\omega z \cos e/\alpha} + A_1(\mathbf{e}_x \sin e + \mathbf{e}_z \cos e)e^{-i\omega z \cos e/\alpha}]e^{i\omega(t - x/\hat{c})}. \tag{9.174}$$

Similarly, let the total velocity in the elastic half-space be

$$\mathbf{v}' = [A'(\mathbf{e}_x \sin e' - \mathbf{e}_z \cos e')e^{i\omega z \cos e'/\alpha'} + B'(\mathbf{e}_x \cos f' + \mathbf{e}_z \sin f')e^{i\omega z \cos f'/\beta'}]e^{i\omega(t - x/\hat{c})}. \tag{9.175}$$

In these equations

$$
\begin{aligned}
e &= \text{angle of incidence of sound waves,}\\
e', f' &= \text{angles of transmission for } P \text{ and } SV \text{ waves,}\\
A_1 &= \text{amplitude of the reflected sound wave,}\\
A', B' &= \text{amplitudes of the transmitted } P \text{ and } SV \text{ waves,}\\
\alpha &= \text{sound-wave velocity in the air,}\\
\alpha', \beta' &= \text{velocities of } P \text{ and } SV \text{ waves in the solid,}
\end{aligned}
$$

$$
\hat{c} = \frac{\alpha}{\sin e} = \frac{\alpha'}{\sin e'} = \frac{\beta'}{\sin f'}. \tag{9.176}
$$

The amplitude ratio can be calculated from Eqs. (3.64). We find

$$
A_1 = \frac{m_1 - m_2}{m_1 + m_2}, \qquad A' = \frac{m_3}{m_1 + m_2}, \qquad B' = \frac{m_4}{m_1 + m_2}, \tag{9.177}
$$

where

$$
\begin{aligned}
m_1 &= \cos e \left[\left(\frac{\beta'}{\alpha'} \right)^2 \sin 2e' \sin 2f' + \cos^2 2f' \right],\\
m_2 &= \left(\frac{\rho\alpha}{\rho'\alpha'} \right) \cos e',\\
m_3 &= 2 \left(\frac{\rho\alpha}{\rho'\alpha'} \right) \cos 2f' \cos e,\\
m_4 &= -4 \left(\frac{\rho\alpha}{\rho'\alpha'} \right) \cos e' \sin f' \cos e,\\
\rho &= \text{density of the air,}\\
\rho' &= \text{density of the solid.}
\end{aligned} \tag{9.178}
$$

From Eqs. (9.175) and (9.177), the total interface velocity in the solid is given by

$$
\begin{aligned}
v_x^0 &= 2 \left(\frac{\rho\alpha}{\rho'\alpha'} \right) \frac{\cos e \sin(e' - 2f')}{m_1 + m_2} e^{i\omega(t - x/\hat{c})},\\
v_z^0 &= \frac{-2m_2 \cos e}{m_1 + m_2} e^{i\omega(t - x/\hat{c})}.
\end{aligned} \tag{9.179}
$$

Similarly, from Eq. (9.174), the total interface pressure in the air can be calculated:

$$
\frac{\partial p}{\partial t} = -\lambda \operatorname{div} \mathbf{v}, \qquad p_0 = \frac{i\lambda}{\omega} \operatorname{div} \mathbf{v} = P_0 e^{i\omega(t - x/\hat{c})}, \qquad P_0 = \frac{2\rho\alpha m_1}{m_1 + m_2}. \tag{9.180}
$$

From Eqs. (9.179) and (9.180), we have

$$v_x^0 = P_0 \frac{\cos e \sin(e' - 2f')}{\rho'\alpha' m_1} e^{i\omega(t - x/\hat{c})},$$
$$v_z^0 = -P_0 \frac{m_2 \cos e}{\rho\alpha m_1} e^{i\omega(t - x/\hat{c})}, \qquad \frac{v_z^0}{v_x^0} = -\frac{\rho'\alpha' m_2}{\rho\alpha \sin(e' - 2f')}. \tag{9.181}$$

It has been established experimentally that sound waves arriving from a distant source in the atmosphere hit the ground at an angle e that is close to 90° (grazing incidence). Because of this and the large velocity contrast, $\sin e' \gg 1$, $\sin f' \gg 1$. Therefore, $\cos e'$ and $\cos f'$ will be purely imaginary. The velocity $\mathbf{v}'$ in the solid will tend to zero as $|z| \to \infty$, only if we choose

$$\cos e' = -i(\sin^2 e' - 1)^{1/2},$$
$$\cos f' = -i(\sin^2 f' - 1)^{1/2}. \tag{9.182}$$

Using these values in Eq. (9.175), we see that $\mathbf{v}'$ represents surface waves.

For grazing incidence, we have, from Eq. (9.176),

$$\eta = \frac{1}{\sin f'} = \frac{\alpha}{\beta' \sin e} \ll 1, \qquad \zeta = \frac{1}{\sin e'} = \frac{\alpha}{\alpha' \sin e} \ll 1. \tag{9.183}$$

Therefore,

$$\cos f' = -\frac{i}{\eta}\left[1 - \frac{1}{2}\eta^2 - \frac{1}{8}\eta^4 + O(\eta^6)\right],$$
$$\cos e' = -\frac{i}{\zeta}\left[1 - \frac{1}{2}\zeta^2 - \frac{1}{8}\zeta^4 + O(\zeta^6)\right]. \tag{9.184}$$

From Eqs. (9.181)–(9.184), we get

$$\sin(e' - 2f') = -\frac{1}{\zeta}\left(\frac{\beta'}{\alpha'}\right)^2\left[1 + \frac{1}{4}\zeta^2 + \frac{1}{4}\left(\frac{\alpha'}{\beta'}\right)^4\zeta^2 - \frac{1}{2}\left(\frac{\alpha'}{\beta'}\right)^2\zeta^2 + O(\zeta^4)\right],$$
$$m_1 = 2\frac{\cos e}{\eta^2}\left(\frac{\beta'}{\alpha'}\right)^2\left[1 - \left(\frac{\alpha'}{\beta'}\right)^2 + \frac{1}{4}\zeta^2 + \frac{3}{4}\left(\frac{\alpha'}{\beta'}\right)^4\zeta^2 - \frac{1}{2}\left(\frac{\alpha'}{\beta'}\right)^2\zeta^2 + O(\zeta^4)\right],$$
$$m_2 = \frac{\rho\alpha}{\rho'\alpha'}\left(-\frac{i}{\zeta}\right)\left[1 - \frac{1}{2}\zeta^2 - \frac{1}{8}\zeta^4 + O(\zeta^6)\right]. \tag{9.185}$$

Substituting in Eq. (9.181), we finally have

$$v_x^0 = \frac{P_0 \hat{c}}{2(\lambda' + \mu')} e^{i\omega(t - x/\hat{c})},$$
$$v_z^0 = P_0 \frac{\hat{c} e^{-i\pi/2}}{2(\lambda' + \mu')}\left(\frac{\lambda' + 2\mu'}{\mu'}\right) e^{i\omega(t - x/\hat{c})}, \tag{9.186}$$
$$\frac{v_z^0}{v_x^0} = \left(\frac{\lambda' + 2\mu'}{\mu'}\right) e^{-i\pi/2}.$$

In terms of the displacements, Eq. (9.186) reads

$$u_x = \frac{-i\hat{c}P_0}{2\omega(\lambda' + \mu')} e^{i\omega(t - x/\hat{c})},$$
$$u_z = \frac{-\hat{c}P_0}{2\omega(\lambda' + \mu')} \left(\frac{\lambda' + 2\mu'}{\mu'}\right) e^{i\omega(t - x/\hat{c})}, \tag{9.187}$$
$$\frac{u_z}{u_x} = \left(\frac{\lambda' + 2\mu'}{\mu'}\right) e^{-i\pi/2},$$

where $\hat{c} = \alpha/\sin e$ is the horizontal phase velocity (apparent velocity).

For plane waves whose normal makes an angle ε with the north, we replace x in the previous equations by $(x' \sin \varepsilon + y' \cos \varepsilon)$, where x' is in the east–west direction and y' is in the north-south direction. Consequently, the horizontal ground motion at the recording site is given, for $\lambda' = \mu'$, by

$$u_{\mathrm{EW}} = \frac{-iP_0\hat{c} \sin \varepsilon}{4\omega\mu'} \exp(i\chi_0),$$
$$u_{\mathrm{NS}} = \frac{-iP_0\hat{c} \cos \varepsilon}{4\omega\mu'} \exp(i\chi_0), \qquad \chi_0 = \omega\left(t - \frac{x}{\hat{c}}\right). \tag{9.188}$$

Ground tilt is obtained by differentiating the vertical displacement with respect to x' and y', respectively:

$$(\mathrm{tilt})_{\mathrm{EW}} = \frac{\partial u_z}{\partial x'} = \frac{3iP_0 \sin \varepsilon}{4\mu'} \exp(i\chi_0),$$
$$(\mathrm{tilt})_{\mathrm{NS}} = \frac{\partial u_z}{\partial y'} = \frac{3iP_0 \cos \varepsilon}{4\mu'} \exp(i\chi_0).$$

Clearly, both the displacements and the tilts have the characteristics of a surface wave that propagates in the x direction with velocity $\hat{c}$. Note that P_0 depends on the source of the sound waves as well as on their path in the atmosphere and is therefore a complex function of ω.

The amplitude factors in Eq. (9.188) contain two intrinsic physical quantities: the phase velocity $\hat{c}$, which is characteristic of the fluid, and the rigidity μ' of the elastic solid. Because the surface-wave amplitude decays with depth as $\exp[-2\pi|z|/\hat{c}T]$, wave periods from 200 to 300 s will sample an effective depth of about 10–15 km. The value of μ' appropriate for the above equations is accordingly the mean rigidity of the upper crust of the earth.

When $\rho' \to \infty$, we fall back on the limiting case of a rigid boundary, for which $A_1 = 1$, $A' = B' = 0$, and $v_z^0 = 0$. We claim, therefore, that the pressure-induced surface waves consist of a second-order amplitude effect.

Figure 9.5 shows the east–west mercury tiltmeter record at Eilat, Israel, of 14 October, 1970. The signal at the bottom arrived from Lop Nor, China

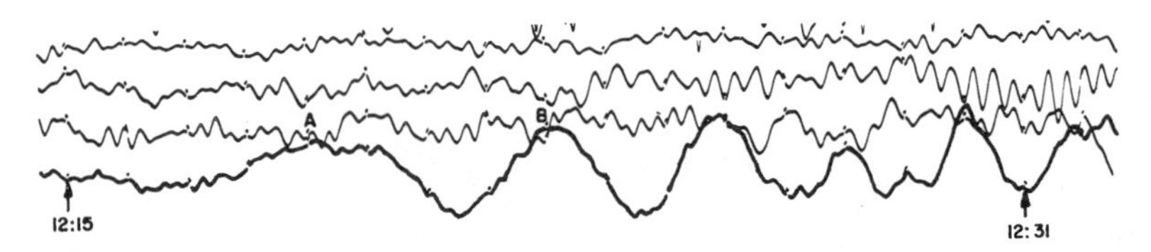

Figure 9.5. East–West mercury tiltmeter record at Eilat, Israel, showing pressure-induced surface waves from atmospheric nuclear explosion on 14 October, 1970 above Lop Nor, China.

(40.9°N, 89.4°E), at $t_0 = 07^h29^m58.6^s$. Because the tiltmeter motion is not affected by long-period changes in air density that induce buoyancy forces on ordinary vertical pendulums, the observed signal is from pressure-induced surface waves. At $T = 240$ s (AB in Fig. 9.5) a tilt of 3×10^{-10} rad and a displacement of 3 μm were observed.

9.5. Waves Generated by Atmospheric Explosions

The worldwide pressure signals that resulted from the volcanic explosion in Krakatoa (27 August, 1883) and the comet explosion over central Siberia (30 June, 1908) initiated interest in the problem of pulse propagation in the atmosphere. These studies were prompted by the desire to account for the velocity of the pulse, to explain its peculiar signature, and to provide information concerning the constitution of the atmosphere. Interest in the problem was renewed with the detonation of nuclear devices over various test sites in the world since 1952. These explosions excited long atmospheric waves that were recorded on a worldwide network of sensitive barographs. In the meantime, the structure of the atmosphere became sufficiently well known from rocket soundings and satellite observations so that major emphasis shifted to explaining the significant features observed on the barograms by means of calculations performed with the aid of high-speed digital computers. Similar methods were used to analyze waveforms of Rayleigh waves excited by underground nuclear explosions and atmospheric explosions.

We shall now apply our former results to the study of the effects of the source and the structural models upon the resulting waveforms. We have formerly established the basic dynamic equations for the free (sourceless) acoustic wave motion in a continuous atmospheric model in which the ambient density, pressure, and temperature are functions of the altitude, z. We shall represent the complex vertical temperature structure of the atmosphere by a large number of horizontally stratified isothermal layers and place a point source in one of them. The strength of the source in the frequency domain is scaled so that the pressure variations with time of the direct wave near the source is in agreement with the observations near actual nuclear detonations.

9.5.1. Dispersion

Let us consider first the problem of dispersion alone. We assume symmetry about the z axis and express χ, w, and p in the form

$$\begin{bmatrix} \chi(\Delta, z; t) \\ w(\Delta, z; t) \\ p(\Delta, z; t) \end{bmatrix} = \frac{1}{2\pi} \int_{-\infty}^{\infty} e^{i\omega t}\, d\omega \int_0^{\infty} \begin{bmatrix} \bar{\chi}(k, z; \omega) \\ \bar{w}(k, z; \omega) \\ \bar{p}(k, z; \omega) \end{bmatrix} J_0(k\Delta)k\, dk. \tag{9.189}$$

However, in the following discussion, we write χ, w, p instead of $\bar{\chi}$, $\bar{w}$, $\bar{p}$, respectively. From Eq. (9.161), the equation for χ for a constant-velocity layer in a horizontally stratified atmosphere (Fig. 9.6) is

$$\alpha_m^2 \chi_m'' - \gamma g_m \chi_m' + \left[\omega^2 - k^2\alpha_m^2\left(1 - \frac{\omega_{Bm}^2}{\omega^2}\right)\right]\chi_m = 0, \qquad \omega_{Bm} = \frac{g_m}{\alpha}\sqrt{(\gamma - 1)},$$
$$z_m > z > z_{m-1} \tag{9.190}$$

where prime signifies differentiation with respect to z, which is taken positive in the vertical (up) direction. The frequency ω_B is known as the *Brunt resonant frequency*. It is the period of the free oscillation of an individual particle, caused by difference between the ambient temperature gradient and the dry-adiabatic lapse time.

Once χ_m is known, the vertical velocity $w_m(z)$ and the overpressure $p_m(z)$ are derived via the expressions [Eq. (9.165)]

$$w_m(z) = \frac{\omega^2\alpha_m^2}{\delta_m}\chi_m' + \frac{g_m}{\delta_m}(k^2\alpha_m^2 - \gamma\omega^2)\chi_m, \qquad \delta_m = g_m^2 k^2 - \omega^4,$$

$$p_m(z) = i\frac{\omega}{\delta_m}\rho_m^0(z)[-g_m\alpha_m^2\chi_m' + (\gamma g_m^2 - \omega^2\alpha_m^2)\chi_m] \tag{9.191}$$

$$= \frac{i\rho_m^0(z)}{\omega}[-g_m w_m + \alpha_m^2\chi_m],$$

where α_m is the layer intrinsic sound velocity, g_m is the gravitational constant, k is the wave number in the horizontal direction, $\gamma = c_p/c_v$ is the ratio of the specific heats and ρ_m^0 is the value of ρ_0 in layer m.

Equation (9.190) is also satisfied by $d\chi_m/dz$ and, therefore, by w_m. However, because of the relation $d\rho_0/dz = -\gamma g\rho_0/\alpha^2$, the function $p_m(z)$ satisfies the following equation

$$\alpha_m^2 p_m'' + \gamma g_m p_m' + \left[\omega^2 - k^2\alpha_m^2\left(1 - \frac{\omega_{Bm}^2}{\omega^2}\right)\right]p_m = 0. \tag{9.192}$$

In the absence of any disturbance, the equilibrium state in each layer is determined by the equations

$$\frac{dp_0}{dz} = -g\rho_0, \qquad \alpha^2 = \gamma\frac{p_0}{\rho_0},$$

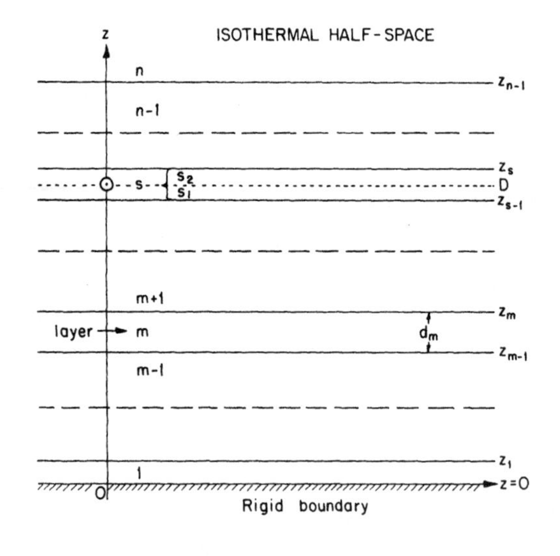

Figure 9.6. Stratification of the atmosphere over a flat-earth model. The configuration consists of $(n-1)$ layers underlying an isothermal half-space.

where p_0 is the static pressure. These equations yield the density distribution in the mth layer $z_{m-1} < z < z_m$

$$\rho_m^0(z) = \rho_m^0(z_{m-1})e^{-2\Gamma_m(z-z_{m-1})}, \qquad \Gamma_m = \frac{\gamma g_m}{2\alpha_m^2}. \tag{9.193}$$

Boundary conditions are:

1. Continuity of the vertical particle velocity and the parcel pressure across each boundary that separates two layers.
2. At $z = 0$, layer 1 is in contact with a flat rigid boundary where the vertical particle velocity vanishes and the parcel pressure is equal to the atmospheric static pressure.
3. The atmosphere is bounded from above at a finite height by an isothermal half-space (nth layer) such that the motion in the half-space decays exponentially with z. This requirement guarantees that the kinetic energy integrated over a column of atmosphere will be finite.

From Eqs. (9.166) and (9.191), the parcel pressure in the mth layer is given by

$$P_m(z) = i\frac{\rho_m^0(z)}{\omega}\alpha_m^2\chi_m = p_m(z) + \frac{i\rho_m^0 g_m}{\omega}w_m. \tag{9.194}$$

At the layer interface we require

$$P_{m-1}(z_{m-1}) = P_m(z_{m-1}), \qquad w_{m-1}(z_{m-1}) = w_m(z_{m-1}). \tag{9.195}$$

The general solution of Eq. (9.190) is given by

$$\begin{aligned} \chi_m(z;\omega) &= [\Delta_m' e^{z(\Gamma_m - \nu_m)} + \Delta_m'' e^{z(\Gamma_m + \nu_m)}], \qquad 1 \le m \le n-1, \\ \chi_n(z;\omega) &= \Delta_n' e^{z(\Gamma_n - \nu_n)}, \end{aligned} \tag{9.196}$$

where Δ'_m, Δ''_m are constants to be determined from the boundary conditions and

$$\gamma \geq 1, \qquad \nu_m = \left[\Gamma_m^2 - \frac{\omega^2}{\alpha_m^2} + k^2\left(1 - \frac{\omega_B^2}{\omega^2}\right)\right]^{1/2}, \qquad \omega_B^2 = \frac{g_m^2}{\alpha_m^2}(\gamma - 1). \tag{9.197}$$

Introducing the notation $\beta_m^2 = (4(\gamma - 1)/\gamma^2)\alpha_m^2$, $\omega = kc$, we can eliminate ω and rewrite the solution in terms of the wave number k and the wave phase velocity, $c(k)$:

$$\nu_m(k, c) = \left[k^2\left(1 - \frac{c^2}{\alpha_m^2}\right) - \frac{\omega_B^2}{c^2}\left(1 - \frac{c^2}{\beta_m^2}\right)\right]^{1/2}. \tag{9.198}$$

Substituting Eq. (9.196) into Eqs. (9.191) and (9.194), evaluating at z_m and z_{m-1}, and eliminating the constants Δ'_m and Δ''_m, we obtain the following matrix relation

$$\begin{bmatrix} w_m(z_m) \\ P_m(z_m) \end{bmatrix} = \begin{bmatrix} (a_m)_{11} & (a_m)_{12} \\ (a_m)_{21} & (a_m)_{22} \end{bmatrix} \begin{bmatrix} w_m(z_{m-1}) \\ P_m(z_{m-1}) \end{bmatrix}, \qquad 1 \leq m \leq n - 1 \tag{9.199}$$

where

$$\begin{aligned}
(a_m)_{11} &= \left\{\mathrm{ch}(\nu_m d_m) + \frac{g_m}{\alpha_m^2}\left(\frac{\alpha_m^2}{c^2} - \frac{\gamma}{2}\right)\frac{\mathrm{sh}(\nu_m d_m)}{\nu_m}\right\}e^{\Gamma_m d_m}, \\
(a_m)_{12} &= i\frac{(kc)^3}{\rho_m^0 \alpha_m^4 \delta_m}\left\{g_m^2\left(\frac{\alpha_m^2}{c^2} - \frac{\gamma}{2}\right)^2 - \alpha_m^4 \nu_m^2\right\}\frac{\mathrm{sh}(\nu_m d_m)}{\nu_m}, \\
(a_m)_{21} &= i\frac{\rho_m^0 \delta_m}{(kc)^3}\frac{\mathrm{sh}(\nu_m d_m)}{\nu_m}, \\
(a_m)_{22} &= \left\{\mathrm{ch}(\nu_m d_m) - \frac{g_m}{\alpha_m^2}\left(\frac{\alpha_m^2}{c^2} - \frac{\gamma}{2}\right)\frac{\mathrm{sh}(\nu_m d_m)}{\nu_m}\right\}e^{-\Gamma_m d_m}, \\
d_m &= z_m - z_{m-1}.
\end{aligned} \tag{9.200}$$

It is seen that the determinant of a_m is equal to unity. Further, the matrix element $(a_m)_{jk}$ is real or imaginary, for (c, k) real, depending on whether $j + k$ is even or odd, respectively. The elements of a matrix resulting from the matrix multiplication of any number of layer matrices will be real or imaginary in the same manner as the individual elements of the matrices.

By a repeated use of Eqs. (9.195) and (9.199), we are able to write the matrix relation

$$\begin{bmatrix} w_{n-1}(z_{n-1}) \\ P_{n-1}(z_{n-1}) \end{bmatrix} = a_{n-1}a_{n-2}\cdots a_1 \begin{bmatrix} w_1(0) \\ P_1(0) \end{bmatrix} = A\begin{bmatrix} 0 \\ p_0 \end{bmatrix}, \tag{9.201}$$

where $A = a_{n-1}a_{n-2}\cdots a_1$. However, from Eqs. (9.191) and (9.196), we derive the expressions for the pressure and vertical velocity at the interface $z = z_{n-1}$:

$$\begin{bmatrix} w_{n-1}(z_{n-1}) \\ P_{n-1}(z_{n-1}) \end{bmatrix} = \begin{bmatrix} w_n(z_{n-1}) \\ P_n(z_{n-1}) \end{bmatrix} = \Delta'_n \begin{bmatrix} b_{1n} \\ b_{2n} \end{bmatrix}, \tag{9.202}$$

where

$$b_{1n} = \frac{(kc)^2}{\delta_n} e^{z_{n-1}(\Gamma_n - \nu_n)} \left\{ g_n \left(\frac{\alpha_n^2}{c^2} - \frac{\gamma}{2} \right) - \alpha_n^2 \nu_n \right\},$$

$$b_{2n} = i e^{z_{n-1}(\Gamma_n - \nu_n)} \left\{ \frac{\rho_n^0(z_{n-1})\alpha_n^2}{kc} \right\}. \tag{9.203}$$

Substituting Eq. (9.202) into Eq. (9.201) and eliminating Δ_n' from the two resulting linear equations, we obtain the period equation for the n-layered fluid half-space

$$F(k, c) \equiv \bar{A}_{12} + A_{22} \left[\frac{(kc)^3}{\rho_n^0(z_{n-1})\alpha_n^2 \delta_n} \left\{ g_n \left(\frac{\alpha_n^2}{c^2} - \frac{\gamma}{2} \right) - \alpha_n^2 \nu_n \right\} \right] = 0. \tag{9.204}$$

Here A_{ij} are the elements of A, while $\bar{A}_{12} = -iA_{12}$ and A_{22} is real. Because imaginary values of ν_n define vertically propagating waves, we must limit ν_n in the half-space to real values. This leads to a cutoff region in the (c, k) plane defined by $\nu_n = 0$. The boundary of this region is obtained in terms of c and the period T

$$\frac{T(c, \alpha_n)}{T_{Bn}} = \left[\frac{c^2/\alpha_n^2 - 1}{c^2/\beta_n^2 - 1} \right]^{1/2}, \tag{9.205}$$

where $T_{Bn} = 2\pi\alpha_n/(g_n\sqrt{(\gamma - 1)})$ is the Brunt resonant period of the half-space.

Clearly there is no cutoff region for either $\beta_n \leq c \leq \alpha_n$ or $[(2/\gamma)\sqrt{(\gamma - 1)}]T_{Bn} \leq T \leq T_{Bn}$. However, for

$$\begin{aligned} c &\to \infty, & T &\to \left[\frac{2}{\gamma} \sqrt{(\gamma - 1)} \right] T_{Bn}; \\ c &= \alpha_n, & T &= 0; \\ c &\to \beta_n, & T &\to \infty; \\ c &= 0, & T &= T_{Bn}. \end{aligned} \tag{9.206}$$

The general computational procedure to find the zeros of $F(k, c)$ for a given c and a given structure of the atmosphere is as follows: A trial value for k is specified and the product of the layer matrices calculated. The value of $F(k)$ is then evaluated. The process is repeated for $F(k + \Delta k)$, etc., until a root is bracketed by a change of sign of F. Linear interpolation and extrapolation are then repeatedly used to find small F values until ks of different F sign are within the prescribed precision interval. The resulting interpolated value of k and the corresponding prescribed value of c constitute a point (c, k) on the dispersion curve for the preselected mode. As c changes, the curve $c(k)$ is obtained.

Figure 9.7 shows Air Research and Development Command (ARDC, U.S. Air Force) models of the atmosphere, exhibiting the dependence of the temperature on height. The effects of geographic and seasonal variations are also

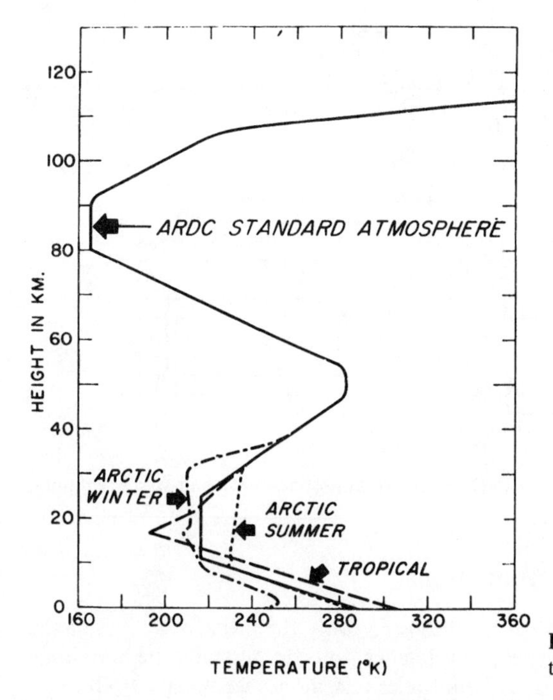

Figure 9.7. Standard and extreme atmospheric models.

exhibited. Dispersion curves for this model are shown in Fig. 9.8. They are separated into modes S_0, S_1, GR_0, and GR_1. The S modes are the first two of an infinite set, which also exist (although differently shaped) in an atmosphere without gravity. They are consequently named "acoustic modes." Their high-frequency limit is the sound velocity in the upper channel (at ca. 85 km; see Fig. 9.7). The modes GR_0 and GR_1 are not present for the nongravitating model and represent the increasing effect of gravity on the longer periods. Indeed, as the wavelength increases, so do the changes of density in the vertical direction. These changes are affected in turn by buoyancy forces caused by gravity. Higher modes of the types S_n and GR_n exist but are not discussed here. The dispersion curves shown in Fig. 9.8 have the following features:

1. The first-arriving waves correspond to the region of flat group-velocity curves for the S_0 and GR modes. These are transients containing periods of 3–10 min that arrive at times corresponding to propagation velocities in the range 290–325 m/s, depending on the season, wind regime, and geographic features of the path.
2. In general, the pressure associated with the S_n acoustic mode has n nodes in the z direction, whereas the vertical particle velocity has $n + 1$ nodes. For the GR_n modes, pressure has $n + 1$ nodes and the vertical velocity has n nodes. Comparison with experimental data is shown in Fig. 9.9.

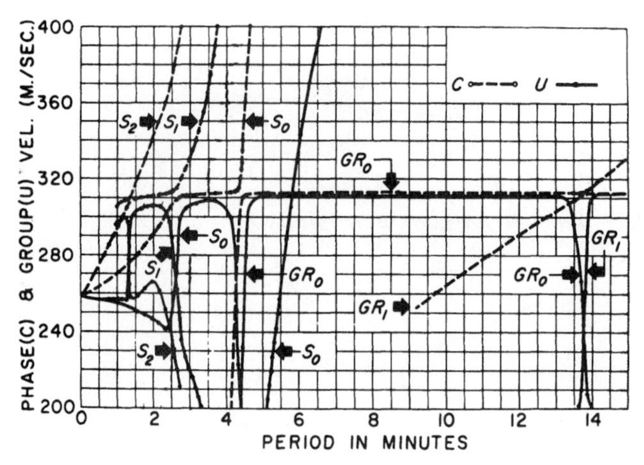

Figure 9.8. Dispersion curves for ARDC standard atmosphere with half-space beginning at 220 km.

9.5.2. Point-Source Formulations

In the preceding section we have discussed only the order of arrival of the various modes and their respective velocities. In order to study the amplitude distribution among the various frequencies and modes we must introduce an explosive source and specify its time and space behavior.

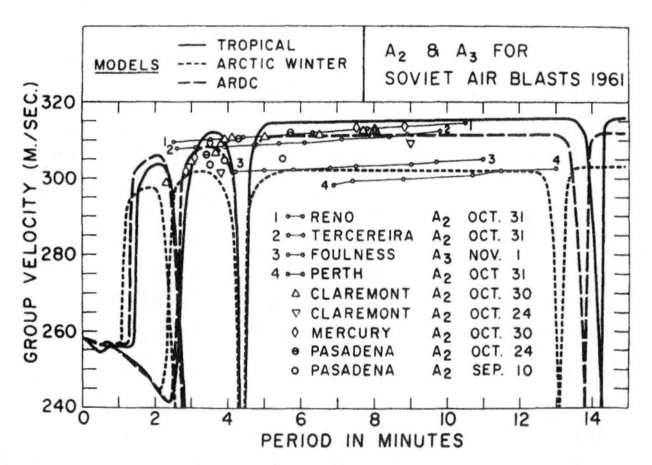

Figure 9.9. Comparison of observed and theoretical curves for A_2 and A_3 waves from Novaya Zemlya explosions. A_2 is an air wave traveling from the source along the major arc, whereas A_3 is the one traveling as A_1 along the minor arc and then completing one revolution around the earth. In contradistinction to A_1 observations, the scatter found for A_2 and A_3 waves is reduced because wind and seasonal and geographic effects average out for these longer paths.

We consider first the general case in which the source is situated in a gravitating isothermal layer and has both mass and energy injection components. The equations of conservation of mass, momentum, and energy in the linear region may be written in the form

$$\frac{D\rho}{Dt} + \operatorname{div}(\rho_0 \mathbf{v}) = F_M(\mathbf{r}, t), \tag{9.207}$$

$$\frac{D\mathbf{v}}{Dt} = -\frac{1}{\rho}\operatorname{grad} p - \mathbf{e}_z g, \tag{9.208}$$

$$\frac{Dp}{Dt} - \alpha^2 \frac{D\rho}{Dt} = F_E(\mathbf{r}, t), \tag{9.209}$$

where F_M and F_E are, respectively, the mass and energy injection source functions. Applying the relation $p_0 = (\alpha^2/\gamma)\rho_0$, we have

$$\begin{aligned}\frac{Dp}{Dt} - \alpha^2 \frac{D\rho}{Dt} &= \frac{\partial p}{\partial t} - \alpha^2 \frac{\partial \rho}{\partial t} + \mathbf{v}\cdot(\operatorname{grad} p_0 - \alpha^2 \operatorname{grad} \rho_0)\\ &= \frac{\partial p}{\partial t} - \alpha^2 \frac{\partial \rho}{\partial t} + (\mathbf{v}\cdot\mathbf{e}_z)\frac{\rho_0}{g}\alpha^2\omega_B^2.\end{aligned}$$

A linearization of Eqs. (9.207)–(9.209) yields

$$\frac{\partial \rho}{\partial t} + \operatorname{div}(\rho_0 \mathbf{v}) = F_M(\mathbf{r}, t), \tag{9.210}$$

$$\rho_0 \frac{\partial \mathbf{v}}{\partial t} = -\operatorname{grad} p - \mathbf{e}_z g\rho, \tag{9.211}$$

$$\left(\frac{1}{\alpha^2}\frac{\partial p}{\partial t} - \frac{\partial \rho}{\partial t}\right) + (\mathbf{v}\cdot\mathbf{e}_z)\omega_B^2 \frac{\rho_0}{g} = \frac{1}{\alpha^2} F_E(\mathbf{r}, t). \tag{9.212}$$

It may be noted that

$$\rho = \rho(z; t), \qquad \rho_0 = \rho_0(z), \qquad p = p(\Delta, z; t), \qquad \mathbf{v} = \mathbf{v}(\Delta, z; t). \tag{9.213}$$

Taking the time derivative of Eq. (9.210) and substituting therein the expression for $\partial\mathbf{v}/\partial t$ from Eq. (9.211), we get

$$\left(\frac{\partial^2}{\partial t^2} - g\frac{\partial}{\partial z}\right)\rho = \dot{F}_M + \nabla^2 p, \qquad (\dot{\ }) \equiv \frac{\partial}{\partial t}. \tag{9.214}$$

Likewise, taking the time derivative of Eq. (9.212) and substituting therein the expression for $\partial\mathbf{v}/\partial t$ from Eq. (9.211), we have

$$\left(\frac{\partial^2}{\partial t^2} + \omega_B^2\right)\rho = \frac{1}{\alpha^2}\frac{\partial^2 p}{\partial t^2} - \frac{\omega_B^2}{g}\frac{\partial p}{\partial z} - \frac{1}{\alpha^2}\dot{F}_E. \tag{9.215}$$

Because the operators $((\partial^2/\partial t^2) - g(\partial/\partial z))$ and $((\partial^2/\partial t^2) + \omega_B^2)$ commute, the variable ρ can be eliminated between Eqs. (9.214) and (9.215) to obtain a fourth-order partial differential equation in the overpressure p,

$$\left(\frac{\partial^2}{\partial t^2} + \omega_B^2\right)(\nabla^2 p + \dot{F}_M) = \left(\frac{\partial^2}{\partial t^2} - g\frac{\partial}{\partial z}\right)\left(\frac{1}{\alpha^2}\frac{\partial^2 p}{\partial t^2} - \frac{\omega_B^2}{g}\frac{\partial p}{\partial z} - \frac{1}{\alpha^2}\dot{F}_E\right). \quad (9.216)$$

Introducing the variable $q = \rho_0^{-1/2}p$ and using the relations

$$\frac{d\rho_0}{dz} = -\frac{\gamma g}{\alpha^2}\rho_0, \qquad \frac{d^2\rho_0}{dz^2} = \left(\frac{\gamma g}{\alpha^2}\right)^2\rho_0,$$

$$\rho_0^{-1/2}\nabla p = \nabla q - \mathbf{e}_z\frac{\omega_A}{\alpha}q, \qquad \rho_0^{-1/2}\frac{\partial p}{\partial z} = \frac{\partial q}{\partial z} - \frac{\omega_A}{\alpha}q, \quad (9.217)$$

$$\rho_0^{-1/2}\nabla^2 p = \nabla^2 q - \frac{2\omega_A}{\alpha}\frac{\partial q}{\partial z} + \left(\frac{\omega_A}{\alpha}\right)^2 q, \qquad \omega_A = \frac{\gamma g}{2\alpha},$$

we obtain the final equation

$$\left[\left(\frac{\partial^2}{\partial t^2} + \omega_B^2\right)\nabla_1^2 + \frac{\partial^2}{\partial t^2}\left(\frac{\partial^2}{\partial z^2} - \frac{1}{\alpha^2}\left\{\frac{\partial^2}{\partial t^2} + \omega_A^2\right\}\right)\right]q$$

$$= -\left[\left(\frac{\partial^2}{\partial t^2} + \omega_B^2\right)\frac{\dot{F}_M}{\sqrt{\rho_0}} + \frac{1}{\alpha^2}\left\{\frac{\partial^2}{\partial t^2} - g\frac{\partial}{\partial z} + \frac{\gamma\omega_B^2}{2(\gamma - 1)}\right\}\left(\frac{\dot{F}_E}{\sqrt{\rho_0}}\right)\right]. \quad (9.218)$$

We note at once that for $g = 0$, F_M is equivalent to $(1/\alpha^2)F_E$. We next consider the two sources corresponding to the mass and energy injections separately.

9.5.2.1. Mass-Injection Source. At the explosion point $\mathbf{r} = \mathbf{r}_0[\Delta = 0, z = D]$, let $f_M(t)$ units of mass be expelled per unit time. Then

$$F_M(\mathbf{r}, t) = f_M(t)\delta(\mathbf{r} - \mathbf{r}_0), \qquad F_E(\mathbf{r}, t) = 0. \quad (9.219)$$

Because the sound-wave approximation certainly does not hold in the immediate vicinity of the source, we consider here an equivalent acoustic source which, if placed at $R = 0$ ($R = |\mathbf{r} - \mathbf{r}_0|$), would generate the same overpressure on a spherical surface at a distance $R = R_0$ as is produced by the explosion itself. At this distance, known as the "fireball radius," the overpressure and velocity are sufficiently small to be considered as acoustic fields governed by the linearized equations [Fig. (9.10a)]. The function $f_M(t)$, is, therefore, chosen such that the linearized equations predict as accurately as possible the known properties of the acoustic field in the vicinity of the source.

Substituting $p = -\rho_0(\partial\psi/\partial t)$, $\rho = -(\rho_0/\alpha^2)(\partial\psi/\partial t)$, $\mathbf{v} = \operatorname{grad}\psi$ (Sect. 9.2) into Eq. (9.210), assuming that ρ_0 is constant in the "nonacoustic" source region, we obtain the inhomogeneous wave equations

$$\nabla^2\psi - \frac{1}{\alpha^2}\frac{\partial^2\psi}{\partial t^2} = \frac{1}{\rho_0}f_M(t)\delta(\mathbf{r} - \mathbf{r}_0),$$

$$\nabla^2 p - \frac{1}{\alpha^2}\frac{\partial^2 p}{\partial t^2} = -\dot{f}_M(t)\delta(\mathbf{r} - \mathbf{r}_0), \quad (9.220)$$

valid for a homogeneous nongravitating medium with $\gamma = 1$. With $R^2 = \Delta^2 + (z - D)^2$, the solution to the pressure equation is known to be

$$p(R; t) = \frac{1}{4\pi R} \frac{\partial}{\partial t} f_M\left(t - \frac{R}{\alpha}\right). \tag{9.221}$$

If the source is at ground level, the factor $1/4\pi$ in Eq. (9.221) is replaced by $1/2\pi$. It has been established from empirical data that the shape of the pressure pulse at $R \geq R_0$, where R_0 is the limit of the linear region, can be approximated by the function

$$p(R, t) = p_{as} \frac{R_0}{R}\left(1 - \frac{t - R/\alpha}{\tau}\right) \exp\left[-\frac{(t - R/\alpha)}{\tau}\right] H\left(t - \frac{R}{\alpha}\right), \quad (D > R \geq R_0) \tag{9.222}$$

where p_{as} is the pressure at $R = R_0$. The variation of the pressure $p(R_0, t)$ with time is shown in Fig. 9.10b. The duration of the positive phase is τ and t_0 is the time required for the disturbance to reach the distance R_0. Comparing Eq. (9.221) with Eq. (9.222) we get the explicit form of $f_M(t)$,

$$f_M(t) = (4\pi p_{as} R_0)(t - t_0) \exp\left[-\frac{t - t_0}{\tau}\right] H(t - t_0), \qquad t_0 = \frac{R_0}{\alpha}. \tag{9.223}$$

Defining the total ejected mass

$$M_0 = \int_{t_0}^{\infty} dt \int_S \rho_0(\mathbf{v} \cdot \mathbf{n}) dS = \int_{t_0}^{\infty} f_M(t) dt = 4\pi p_{as} R_0 \tau^2, \tag{9.224}$$

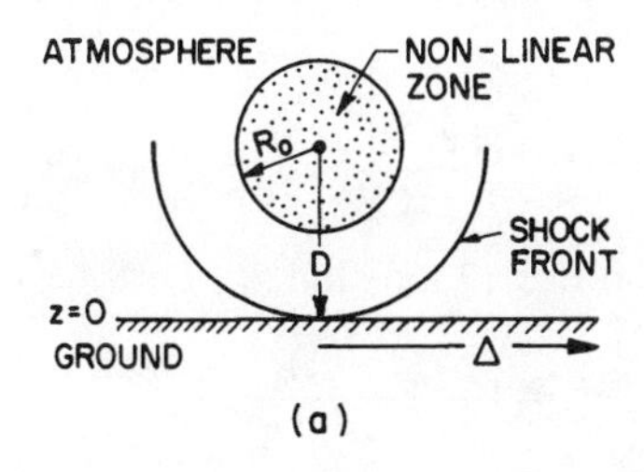

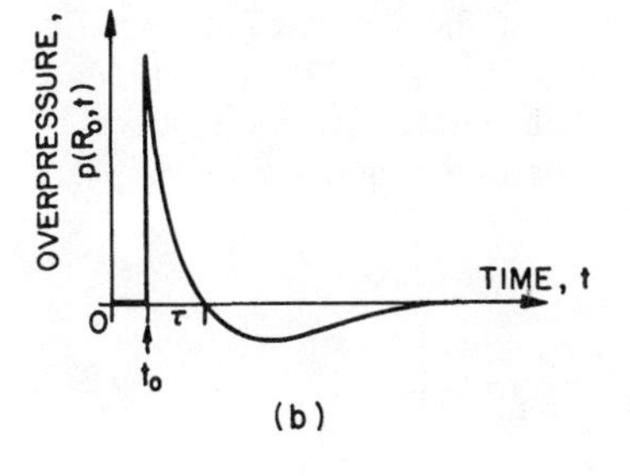

Figure 9.10. (a) The geometry of an atmospheric nuclear explosion. (b) Variation of overpressure with time for an atmospheric nuclear explosion. t_0 is the arrival time of the pressure pulse.

where S is the surface of a sphere of radius R_0, we obtain

$$f_M(t) = M_0 G(t), \qquad G(t) = \frac{t - t_0}{\tau^2} \exp\left[-\frac{t - t_0}{\tau}\right] H(t - t_0), \qquad \frac{M_0}{4\pi R_0 \tau^2} = p_{as}. \tag{9.225}$$

The equivalent volume of gas involved in the explosion is

$$V_0 = \frac{M_0}{\rho_0} = \frac{4\pi p_{as} R_0 \tau^2}{\rho_0}. \tag{9.226}$$

The pressure pulse at $R = R_0$ is then

$$p(t; R_0) = \frac{M_0}{4\pi R} \frac{\partial}{\partial t} G(t) = p_{as} \frac{\partial}{\partial t} \left\{ (t - t_0) \exp\left(-\frac{t - t_0}{\tau}\right) \right\}, \quad (t > t_0). \tag{9.227}$$

Applying the Fourier transform to Eq. (9.216) and using Eq. (9.219), we have a second-order equation for the overpressure in the linear acoustic region

$$\frac{1}{\Delta} \frac{\partial}{\partial \Delta} \left(\Delta \frac{\partial p}{\partial \Delta} \right) + \frac{1}{h} \left(\frac{\partial^2 p}{\partial z^2} + \frac{\gamma g}{\alpha^2} \frac{\partial p}{\partial z} + \frac{\omega^2}{\alpha^2} p \right) = S_M(\omega) \cdot \frac{\delta^+(\Delta)}{2\pi\Delta} \delta(z - D), \tag{9.228}$$

where

$$S_M(\omega) = -\int_{-\infty}^{\infty} \dot{f}_M(t) e^{-i\omega t}\, dt = -i\omega m(\omega), \qquad h = 1 - \left(\frac{\omega_B}{\omega}\right)^2, \tag{9.229}$$

and $m(\omega)$ is the Fourier transform of $f_M(t)$. The factor $(2\pi)^{-1}$ on the right-hand side of Eq. (9.228) is introduced by the integration over the azimuthal angle, ϕ, assuming the source to be axially symmetrical. A particular solution of Eq. (9.228) is

$$p(\Delta, z; \omega) = -\frac{1}{4\pi} S_M(\omega) \sqrt{h}\, e^{-\Gamma(z - D)} \frac{e^{-K_s R^*}}{R^*}. \tag{9.230}$$

Here, in accord with our previous definitions in Eqs. (9.165) and (9.193).

$$K_s^2 = \frac{\Gamma^2 \alpha^2 - \omega^2}{h\alpha^2}, \qquad \Gamma = \frac{\gamma g}{2\alpha^2}, \qquad h = 1 - \frac{(\gamma - 1) g^2}{\omega^2 \alpha^2},$$
$$R^{*2} = \Delta^2 + h(z - D)^2. \tag{9.231}$$

It is easy to verify by direct substitution that the Green's function, Eq. (9.230), is indeed a solution of the homogeneous Eq. (9.228) for all (Δ, z) except at the point $(0, D)$. Also, if we put $g = 0$, the Green's function reduces to the well-known point-source representation for outgoing spherical pressure waves in an acoustic medium, in which $\Gamma = 0$ and $h = 1$.

From Eqs. (9.223) and (9.229), we have

$$S_M(\omega) = 4\pi p_{as} R_0 \tau^2 \left[\frac{-i\omega e^{-i\omega t_0}}{(1 + i\omega\tau)^2} \right]. \tag{9.232}$$

Substituting this value of $S_M(\omega)$ in Eq. (9.230), we get

$$p(\Delta, z; \omega) = p_{as} R_0 h^{1/2} \tau^2 \left[\frac{i\omega e^{-i\omega t_0}}{(1 + i\omega\tau)^2}\right]\left[\frac{e^{-\Gamma(z-D) - K_s R^*}}{R^*}\right]. \tag{9.233}$$

Expression (9.223) for the mass ejected by the source and the corresponding expression (9.232) for $S_M(\omega)$ are derived under the assumption that the source region is homogeneous and nongravitating, with $\gamma = 1$. Let us next assume that $S_M(\omega)$ is not given by Eq. (9.229) but instead is unknown. From Eq. (9.222), we have the empirical result

$$p(R_0, \omega) = p_{as} \tau^2 \left[\frac{i\omega e^{-i\omega t_0}}{(1 + i\omega\tau)^2}\right]. \tag{9.234}$$

We equate this value with the overpressure obtained from Eq. (9.230) at $\Delta = R_0$, $z = D$, obtaining

$$S_M(\omega) = -4\pi p_{as} \tau^2 R_0 h^{-1/2} \left[\frac{i\omega e^{-i\omega t_0}}{(1 + i\omega\tau)^2}\right] e^{K_s R_0}. \tag{9.235}$$

Substituting this value of $S_M(\omega)$ in Eq. (9.230), we have

$$p(\Delta, z; \omega) = p_{as} \tau^2 R_0 \left[\frac{i\omega e^{-i\omega t_0}}{(1 + i\omega\tau)^2}\right]\left[\frac{1}{R^*} e^{-\Gamma(z-D) - K_s(R^* - R_0)}\right]. \tag{9.236}$$

Using the Sommerfeld integral [Eq. (2.153)], we find

$$\frac{e^{-K_s R^*}}{R^*} = \int_0^\infty J_0(k\Delta) e^{-\hat{\nu} h^{1/2}|z-D|} \frac{k\,dk}{\hat{\nu}}, \tag{9.237}$$

where $\hat{\nu} = (k^2 + K_s^2)^{1/2}$. Eq. (9.237), can be recast in the form

$$\frac{e^{-K_s R^*}}{R^*} = h^{1/2} \int_0^\infty J_0(k\Delta) e^{-\nu|z-D|} \frac{k\,dk}{\nu}, \tag{9.238}$$

where

$$\nu = h^{1/2}\hat{\nu} = \left[\Gamma^2 - \frac{\omega^2}{\alpha^2} + k^2\left(1 - \frac{\omega_B^2}{\omega^2}\right)\right]^{1/2}.$$

Equations (9.236) and (9.238) now yield

$$p(\Delta, z; \omega) = p_{as} \tau^2 R_0 h^{1/2} e^{-\Gamma(z-D) + K_s R_0} \left[\frac{i\omega e^{-i\omega t_0}}{(1 + i\omega\tau)^2}\right] \times \int_0^\infty J_0(k\Delta) e^{-\nu|z-D|} \frac{k\,dk}{\nu}. \tag{9.239}$$

It is clear that p is continuous across the plane $z = D$, whereas $\partial p/\partial z$ is discontinuous. Using the relations [Eq. (9.165)]

$$w = \frac{i}{\omega\rho_0 h}\left(\frac{\partial p}{\partial z} + \frac{g}{\alpha^2} p\right), \qquad P = \left(1 - \frac{g^2}{h\omega^2\alpha^2}\right)p - \frac{g}{h\omega^2}\frac{\partial p}{\partial z}, \tag{9.240}$$

we get the following representation of the source in terms of the jumps in w and P across $z = D$:

$$\delta w = w(z = D + 0) - w(z = D - 0) = \frac{-2iS_{m0}}{\omega \rho_0 h} \int_0^\infty J_0(k\Delta) k \, dk,$$

$$\delta P = P(z = D + 0) - P(z = D - 0) = \frac{ig\rho_0}{\omega} \delta w, \tag{9.241}$$

where P is the parcel pressure and

$$S_{m0} = p_{as} \tau^2 R_0 h^{1/2} e^{K_s R_0} \left[\frac{i\omega e^{-i\omega t_0}}{(1 + i\omega\tau)^2} \right]. \tag{9.242}$$

9.5.2.2. Energy-Injection Source. The quantities M_0 and V_0 are rather ineffective representations of a nuclear explosion source in the atmosphere because the equivalent volume or mass ejection of nuclear explosions cannot be estimated directly. It is more convenient, instead, to define a parameter E_c as that part of the explosive energy that produces the compressive effect. Under average conditions, E_c will be approximately 50% of the total yield energy Y.

Let us again imagine the source to consist of an initially isothermal sphere in an unbounded isothermal atmosphere with negligible gravity. The initial sphere contains a fluid at very high temperature and pressure.

We put

$$F_M(\mathbf{r}, t) = 0, \qquad F_E(\mathbf{r}, t) = f_E(t)\delta(\mathbf{r} - \mathbf{r}_0), \tag{9.243}$$

where $f_E(t)$ is of the dimensions of energy per unit time and is related to the total yield.

Applying the Fourier transform to Eq. (9.216) and using Eq. (9.243), we get

$$\frac{1}{\Delta}\frac{\partial}{\partial \Delta}\left(\Delta \frac{\partial p}{\partial \Delta}\right) + \frac{1}{h}\left[\frac{\partial^2 p}{\partial z^2} + \frac{\gamma g}{\alpha^2}\frac{\partial p}{\partial z} + \frac{\omega^2}{\alpha^2} p\right] = \frac{S_E(\omega)}{\alpha^2 h}\left(1 + \frac{g}{\omega^2}\frac{\partial}{\partial z}\right) \times \frac{1}{2\pi\Delta}\delta^+(\Delta)\delta(z - D), \tag{9.244}$$

where

$$S_E(\omega) = -\int_{-\infty}^{\infty} \dot{f}_E(t) e^{-i\omega t} \, dt = -i\omega e(\omega), \tag{9.245}$$

and $e(\omega)$ is the Fourier transform of $f_E(t)$. The Green's function of Eq. (9.244), in the absence of the operator $(g/\omega^2)(\partial/\partial z)$ on the right-hand side, is simply

$$G(\mathbf{r} - \mathbf{r}_0; \omega) = -\frac{S_E(\omega)}{4\pi\alpha^2\sqrt{h}} e^{-\Gamma(z-D)} \frac{e^{-K_s R^*}}{R^*}$$

$$= -\frac{S_E(\omega)}{4\pi\alpha^2} e^{-\Gamma(z-D)} \int_0^\infty e^{-\nu|z-D|} J_0(k\Delta) \frac{k\,dk}{\nu},$$

using Eq. (9.238). Therefore, a particular solution of Eq. (9.244) is

$$p(\Delta, z; \omega) = \left(1 + \frac{g}{\omega^2}\frac{\partial}{\partial z}\right)G = -\frac{S_E(\omega)}{4\pi\alpha^2}e^{-\Gamma(z-D)}$$

$$\times \int_0^\infty \left(1 - \frac{g\Gamma}{\omega^2} - \varepsilon\frac{g\nu}{\omega^2}\right)e^{-\nu|z-D|}J_0(k\Delta)\frac{k\,dk}{\nu}, \qquad (9.246)$$

where

$$\begin{aligned} \varepsilon &= 1, \qquad z > D \\ &= -1, \qquad z < D. \end{aligned}$$

It has been found that

$$f_E(t) = c_0 EG(t),$$

where $G(t)$ is given by Eq. (9.225) and c_0 is a dimensionless constant. The value of c_0 for nuclear explosions in the atmosphere is in the range 0.016–0.033 and

$$E \text{ (in ergs)} = (4.2 \times 10^{19}) \times \text{Yield in kilotons of TNT.}$$

Equation (9.246) then implies

$$\delta p(D) = p(z = D + 0) - p(z = D - 0) = \left(\frac{2g}{\omega^2}\right)S_{e0}\int_0^\infty J_0(k\Delta)k\,dk,$$

$$\delta\left[\frac{\partial p(D)}{\partial z}\right] = 2\left(1 - \frac{2g\Gamma}{\omega^2}\right)S_{e0}\int_0^\infty J_0(k\Delta)k\,dk, \qquad (9.247)$$

where

$$S_{e0} = \frac{S_E(\omega)}{4\pi\alpha^2}. \qquad (9.248)$$

From Eqs. (9.240) and (9.247), we have the source representation:

$$\begin{aligned} \delta w &= \frac{2iS_{e0}}{\omega\rho_0}\int_0^\infty J_0(k\Delta)k\,dk, \\ \delta P &= 0. \end{aligned} \qquad (9.249)$$

9.5.3. Amplitude Theory

We now extend our matrix formulation of Section 9.5.1 to include an energy-injection source. We label quantities corresponding to the source layer (Fig. 9.6) with the subscript *s*. The source is represented in terms of jumps in the vertical component of the velocity and the parcel pressure across the plane

$z = D$, where D is the height of the source above the ground. For an axially symmetrical point source, we have

$$w_s(z = D + 0) - w_s(z = D - 0) = \frac{1}{2\pi} \int_{-\infty}^{\infty} e^{i\omega t} \, d\omega \int_0^{\infty} \delta w_s(k, \omega) J_0(k\Delta) k \, dk,$$

$$P_s(z = D + 0) - P_s(z = D - 0) = \frac{1}{2\pi} \int_{-\infty}^{\infty} e^{i\omega t} \, d\omega \int_0^{\infty} \delta P_s(k, \omega) J_0(k\Delta) k \, dk, \tag{9.250}$$

where δw_s and δP_s are assumed to be known.

We split the source layer $s(z_{s-1} \leq z \leq z_s)$ into two layers, s_1 $(z_{s-1} \leq z \leq D)$ and s_2 $(D \leq z \leq z_s)$ of identical properties. Equations (9.195) and (9.199) now yield

$$\begin{bmatrix} w_{n-1}(z_{n-1}) \\ P_{n-1}(z_{n-1}) \end{bmatrix} = a_{n-1} a_{n-2} \cdots a_{s_2} \begin{bmatrix} w_{s_2}(D) \\ P_{s_2}(D) \end{bmatrix}, \tag{9.251}$$

$$\begin{bmatrix} w_{s_1}(D) \\ P_{s_1}(D) \end{bmatrix} = a_{s_1} a_{s-1} \cdots a_1 \begin{bmatrix} w_1(0) \\ P_1(0) \end{bmatrix}. \tag{9.252}$$

Using the convention set in Eq. (9.189), Eq. (9.250) may be written as

$$\begin{bmatrix} \delta w_s \\ \delta P_s \end{bmatrix} = \begin{bmatrix} w_{s_2}(D) \\ P_{s_2}(D) \end{bmatrix} - \begin{bmatrix} w_{s_1}(D) \\ P_{s_1}(D) \end{bmatrix}. \tag{9.253}$$

Combining Eqs. (9.251)–(9.253) and using Eq. (9.202) and the relations $a_{s_2} a_{s_1} = a_s$, $w_1(0) = 0$, $P_1(0) = p_1(0)$, we have

$$\Delta'_n \begin{bmatrix} b_{1n} \\ b_{2n} \end{bmatrix} = \begin{bmatrix} w_{n-1}(z_{n-1}) \\ P_{n-1}(z_{n-1}) \end{bmatrix} = A \begin{bmatrix} 0 \\ p_1(0) \end{bmatrix} + B \begin{bmatrix} \delta w_s \\ \delta P_s \end{bmatrix}, \tag{9.254}$$

where

$$A = a_{n-1} a_{n-2} \cdots a_1, \qquad B = a_{n-1} a_{n-2} \cdots a_{s_2}. \tag{9.255}$$

Eliminating Δ'_n from Eqs. (9.254), we find

$$p_1(0) = \frac{F_w}{F_A} \delta w_s + \frac{F_P}{F_A} \delta P_s, \tag{9.256}$$

where

$$F_A = A_{12} - \frac{b_{1n}}{b_{2n}} A_{22},$$

$$F_w = -B_{11} + \frac{b_{1n}}{b_{2n}} B_{21}, \tag{9.257}$$

$$F_P = -B_{12} + \frac{b_{1n}}{b_{2n}} B_{22}.$$

Substituting the value of $p_1(0)$ from Eqs. (9.256) into Eq. (9.189), we find

$$p_1(\Delta, 0; \omega) = \int_0^\infty (F_w \delta w_s + F_P \delta P_s) J_0(k\Delta) \frac{k\,dk}{F_A}. \tag{9.258}$$

It may be noted that $F_A(k, c) = 0$ is the frequency equation derived earlier in Eq. (9.204). Let the roots of this equation be denoted by $k_j(\omega)$. The integral in Eq. (9.258) is of the same form as the integral encountered in the case of seismic surface waves and can be evaluated similarly. Therefore, Eqs. (5.21), (5.24), and (5.25) show that, for Re $\omega > 0$, the ground pressure for the jth mode is given by

$$p_1(\Delta, 0; k_j) = -\pi i \left[k(F_w \delta w_s + F_P \delta P_s) \left(\frac{\partial F_A}{\partial k} \right)^{-1} H_0^{(2)}(k\Delta) \right]_{k_j}. \tag{9.259}$$

In the case of an energy-injection source, Eqs. (9.249) and (9.250) show that

$$\delta w_s(k, \omega) = \frac{2iS_{e0}}{\omega \rho_s^0}, \qquad \delta P_s(k, \omega) = 0. \tag{9.260}$$

Therefore, Eq. (9.259) becomes

$$p_1(\Delta, 0; k_j) = -\pi i \left[kF_w \delta w_s \left(\frac{\partial F_A}{\partial k} \right)^{-1} H_0^{(2)}(k\Delta) \right]_{k_j}. \tag{9.261}$$

We define

$$C = a_{s_1} a_{s-1} \cdots a_1, \tag{9.262}$$

so that $A = BC$. Noting that for $k = k_j$, $F_A = 0$ and that the determinant of the matrix B is unity (because the determinant of a_m is unity), we have, for $k = k_j$,

$$F_w = \frac{1}{A_{22}} (B_{21} A_{12} - B_{11} A_{22}) = -\frac{C_{22}}{A_{22}}. \tag{9.263}$$

Hence, we get

$$p_1(\Delta, 0; k_j) = \pi i \left[k \delta w_s \left(\frac{C_{22}}{A_{22}} \right) \left(\frac{\partial F_A}{\partial k} \right)^{-1} H_0^{(2)}(k\Delta) \right]. \tag{9.264}$$

Inserting the value of δw_s, we have, for nuclear explosions in the atmosphere,

$$\begin{aligned} p_1(\Delta, 0; k_j) &= \frac{ic_0 E}{2\rho_s^0 \alpha_s^2} \left[\frac{k e^{-i\omega t_0}}{(1 + i\omega\tau)^2} \left(\frac{C_{22}}{A_{22}} \right) \left(\frac{\partial F_A}{\partial k} \right)^{-1} H_0^{(2)}(k\Delta) \right]_{k_j}, \\ &= \frac{c_0 E}{\rho_s^0 \alpha_s^2} (2\pi\Delta)^{-1/2} \left[\frac{k^{1/2} e^{-i\omega t_0}}{(1 + i\omega\tau)^2} \left(\frac{C_{22}}{A_{22}} \right) \left(\frac{\partial F_A}{\partial k} \right)^{-1} e^{-i(k\Delta - 3\pi/4)} \right]_{k_j}, \end{aligned} \tag{9.265}$$

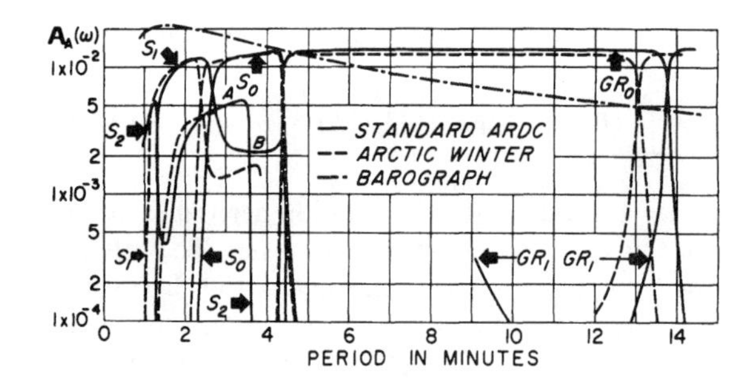

Figure 9.11. Spectral amplitude for the standard and arctic winter atmospheres. The spectral amplitude of the barograph is superimposed. In this figure, $A_A(\omega) =$

$$\left\{\left(\frac{C_{22}}{A_{22}}\right)\left(\frac{\partial F_A}{\partial k}\right)^{-1}\right\}_{k_1}.$$

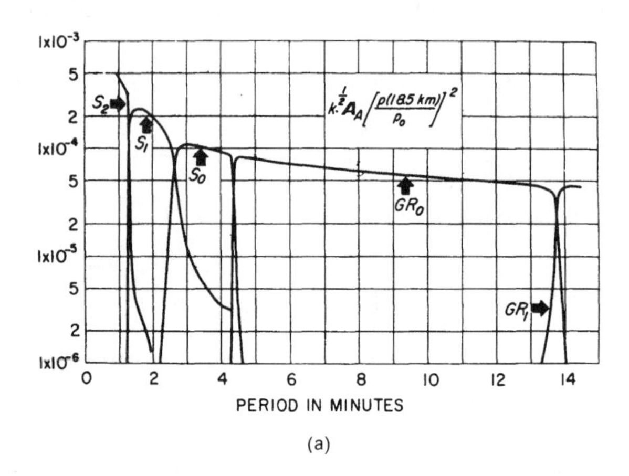

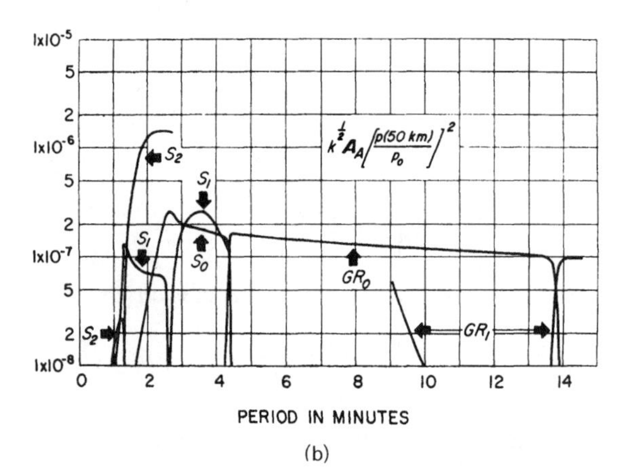

Figure 9.12. Spectral amplitudes for $S_{0,1,2}$ and $GR_{0,1}$ modes for a source and sensor at an altitude of (a) 18.5 km (b) 50 km.

valid for large values[4] of $k_j\Delta$. As in the case of seismic surface waves, the ratio (C_{22}/A_{22}) can be expressed in terms of plane-wave amplitudes. In fact, when we put $w(0) = \delta w_s = \delta P_s = 0$, Eqs. (9.251) and (9.252) yield

$$\frac{C_{22}}{A_{22}} = \frac{P_s(D)}{P_{n-1}(z_{n-1})}. \tag{9.265a}$$

The evaluation of $p_1(\Delta, 0)$ in Eq. (9.265) is done as follows: First we find the root k_j that satisfies $F_A = 0$ for a given input phase velocity, c_j. Once the root is determined, the values of A and $U_g = d\omega/dk = -(\partial F_A/\partial k)_\omega/(\partial F_A/\partial\omega)_k$ are computed. To calculate $(\partial F_A/\partial k)_\omega$, we note that F_A is essentially a product of 2×2 matrices and the derivative of each of them follows the rule $[(\partial a_m/\partial k)_\omega]_{lq} = [\partial(a_m)_{lq}/\partial k]_\omega$. However, because $A_m = a_m A_{m-1}$, we have $(\partial A_m/\partial k)_\omega = (\partial a_m/\partial k)_\omega A_{m-1} + a_m(\partial A_{m-1}/\partial k)_\omega$. In each layer we can use analytic expression for $\partial(a_m)_{lq}/\partial k$ and then, starting from the surface layer, we can calculate the matrix $(\partial A/\partial k)_\omega$ and hence $(\partial F_A/\partial k)_\omega$. Amplitude response curves are shown in Figs. 9.11 and 9.12.

Figure 9.13 shows a microbarogram recorded at Eilat (29.7°N, 34.9°E) on 14 October, 1970, from an atmospheric nuclear explosion at 40.9°N, 89.4°E of yield ~10 MT. Its waves originated at 07 h 29 min 58.6 s UT. The first peak-to-trough amplitude on this record corresponds to a pressure change of about 250 μb at a period of 285 s. It traveled 5024 km with a group velocity of 290 m/s and constituted the main arrival of the first gravity mode, GR_0. It was followed by the fundamental acoustic mode, S_0, at a period of 240 s.

Figure 9.14 shows a composite drawing of a microbarogram recorded at six English stations (centered at 51°30′N, 0°20′W) on 30 June, 1908, from the comet explosion at 60°55′N, 101°57′E (Central Siberia) at 00 h 14 m 28 s. Waves traveled about 5760 km. The yield of the explosion is estimated at 12½ MT. The maximal pressure change is about 200 μb at a period of 300 s. It traveled with a group velocity of 323 m/sec and corresponded to the first gravity mode GR_0. Velocity may have been increased by a few percent because of winds in the middle atmosphere and arctic summer conditions.

The paths of the acoustic–gravity waves generated by these two explosions in the lower atmosphere are shown in Fig. 9.15.

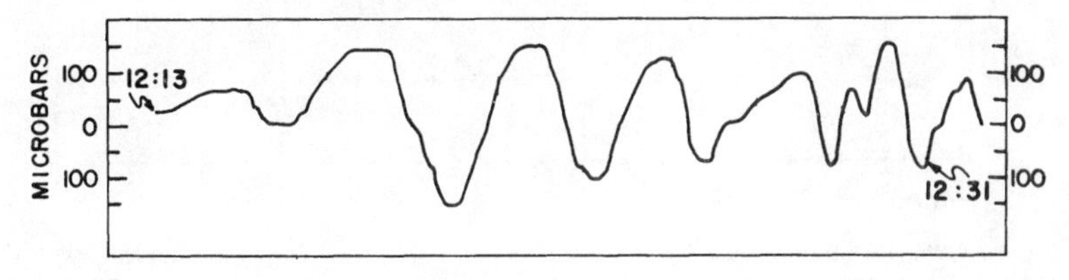

Figure 9.13. A microbarogram recorded at Eilat from a Lop Nor atmospheric nuclear explosion on 14 October, 1970 of yield ~10 MT.

[4] In realistic models of the atmosphere, we must introduce a sphericity-correction factor $(\Delta/a)^{1/2}[\sin(\Delta/a)]^{-1/2}$.

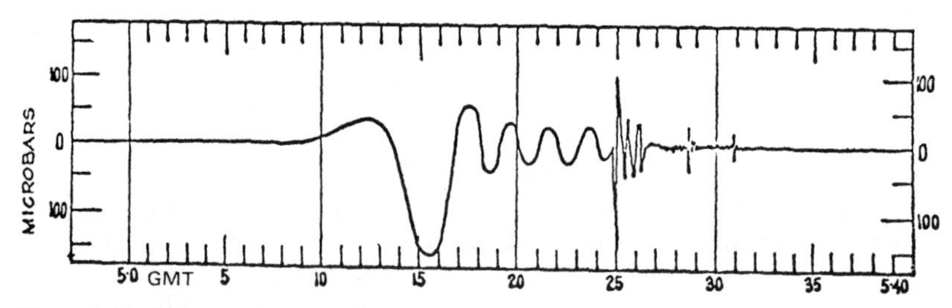

Figure 9.14. A composite drawing of a microbarogram recorded at six English stations on 30 June, 1908 from the comet explosion in Central Siberia.

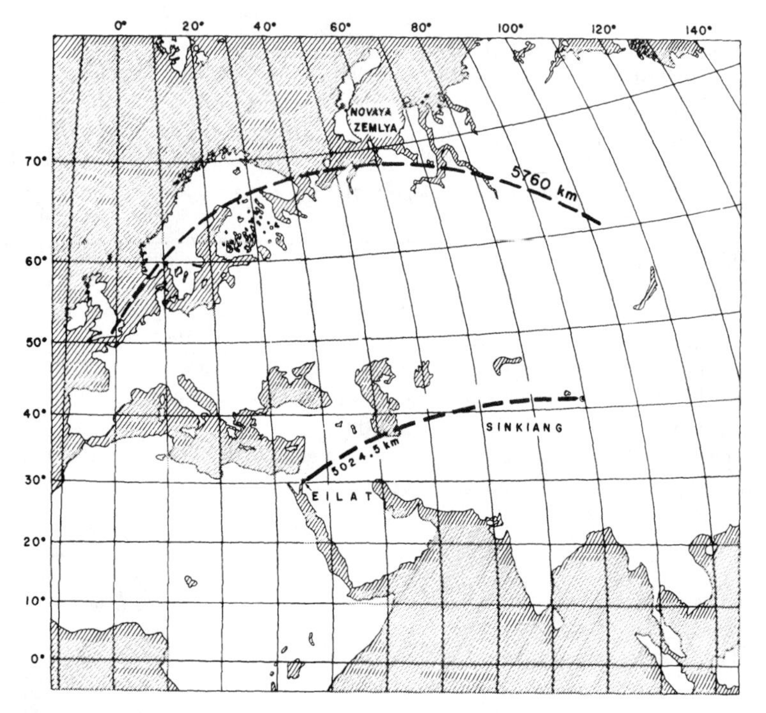

Figure 9.15. Paths of acoustic-gravity waves generated by a Lop Nor nuclear explosion (14 October, 1970; lower trace) and the Siberian comet explosion (30 June, 1908; upper trace) in the lower atmosphere.

9.6. Coupled Air–Sea Waves

It is well known that certain violent atmospheric explosions are able to excite sea waves that are observable at large distances from the explosion source. Following the Krakatoa explosion of 27 August, 1883, arrival times of major tide-gauge disturbances could be correlated with the arrival time of the first or second air waves. It seems plausible that the energy in the sea waves was transferred to the water from the air waves by *resonant coupling* when this disturbance approached the station from the ocean side. It could not be identified with free wave motion in the ocean over all water paths. The following theory will show that internal ocean gravity waves, with phase and group velocities of atmospheric acoustic–gravity waves, can be excited by a transfer of energy from the atmosphere to the ocean.

The linearized equations governing the overpressure in sea water differ somewhat from their atmospheric counterparts. The difference arises in the application of the energy equation [Eq. (9.151)]

$$\left(\frac{1}{\alpha^2}\frac{\partial p}{\partial t} - \frac{\partial \rho}{\partial t}\right) + (\mathbf{v}\cdot\mathbf{e}_z)\left(-\frac{d\rho_0}{dz} + \frac{1}{\alpha^2}\frac{dp_0}{dz}\right) = 0, \tag{9.266}$$

to a liquid in which $dp_0/dz = -g\rho_0$. Invoking our former general definition [Eq. (9.163)] of the Brunt frequency, $\omega_B^2 = -g((1/\rho_0)(d\rho_0/dz) + (g/\alpha^2))$, the overpressure equation for a homogeneous liquid layer reads

$$\frac{d^2p}{dz^2} + \left(\frac{\omega_B^2}{g} + \frac{g}{\alpha^2}\right)\frac{dp}{dz} + \left(\frac{\omega^2}{\alpha^2} - hk^2\right)p = 0, \qquad h = 1 - \frac{\omega_B^2}{\omega^2}. \tag{9.267}$$

If we substitute $\omega_B^2 = (\gamma - 1)g^2/\alpha^2$ in Eq. (9.267), it reduces to the corresponding equation for an atmospheric layer [Eq. (9.192)].

The value of ω_B for sea water varies with depth from 1.0×10^{-3}/s to 8×10^{-3}/s. With this provision, we may extend the acoustic–gravity theory and the numerical matrix techniques of calculating dispersion and source excitation to a system in which the ocean is the lower part of the layered waveguide. Therefore, with the rigid surface $z = 0$ now at the ocean bottom, we take the input parameters for each ocean layer to be the layer thickness, acoustic velocity, density, and Brunt frequency.

Figures 9.16–9.22 show the results for a model used to explain observed sea waves from the Krakatoa explosion. The model used in the calculations consisted of a standard ARDC atmosphere underlain by a constant-density, constant-velocity ocean. The atmosphere was terminated by a free surface [$P_{n-1}(z_{n-1}) = 0$, $F_A = A_{22}$, Eq. (9.201)] and the ocean bottom was assumed to be perfectly rigid. The explosive source was placed at the surface and its time variation was taken as a delta function or as a single cycle of a sine wave with period ranging from 10 to 120 min.

The phase- and group-velocity curves for the atmosphere alone and the ocean alone are shown in Figs. 9.16 and 9.17, respectively. The atmospheric curves show the acoustic (S) modes and the gravity (GR) modes. The flat

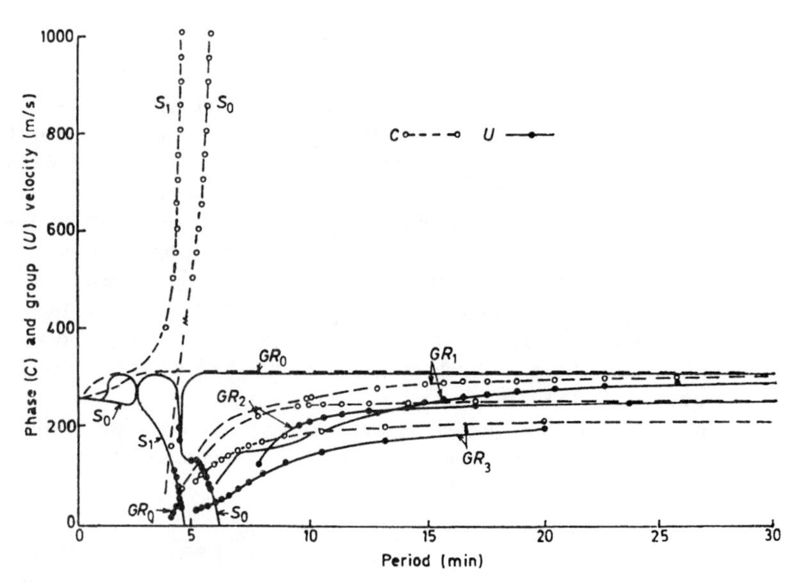

Figure 9.16. Dispersion curves for an ARDC standard atmosphere with free surface at 220 km and terminated by a rigid boundary at the bottom.

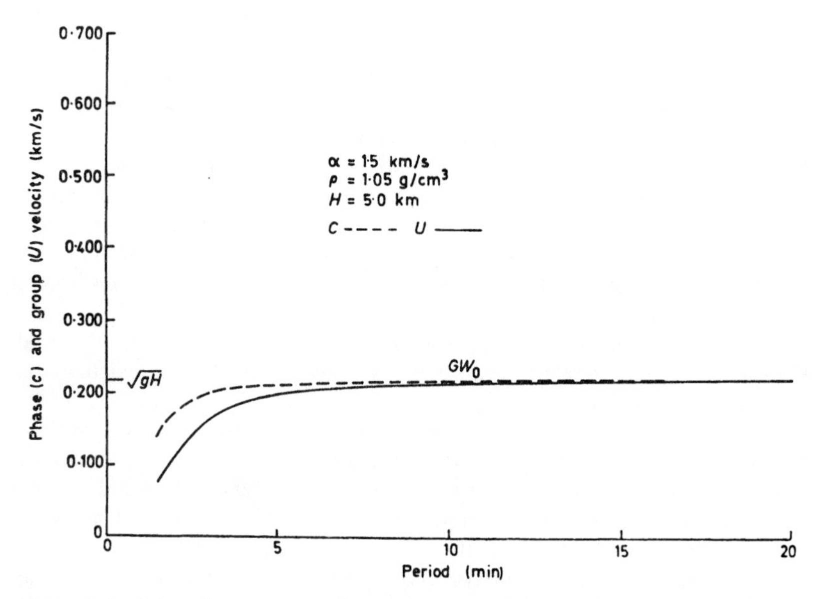

Figure 9.17. Dispersion curves for the fundamental gravity mode in a constant-velocity, constant-density ocean of 5-km depth, terminated by a rigid boundary at the bottom.

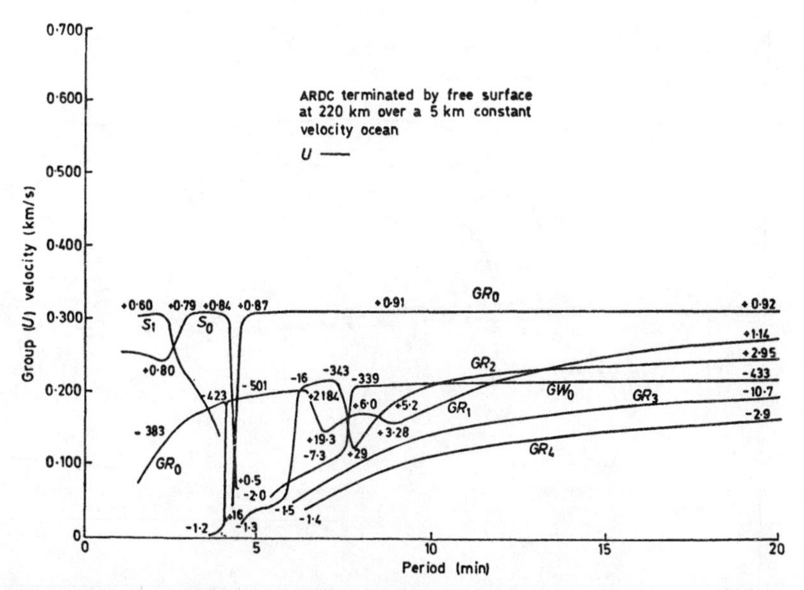

Figure 9.18. Group-velocity dispersion curves of the air–ocean system. The numbers shown are the dynamic ratios at their corresponding periods.

segments of group velocities of S_1, S_0, and GR_0 are nearly connected to form a common group-velocity curve, which accounts for the main features of the Krakatoa air wave. Several interesting features emerge when the two systems are coupled as shown in Fig. 9.18. The characteristics of the individual systems are retained in pseudodispersion curves consisting of segments of the continuous model curves of the coupled system. Therefore, the GW_0 mode of the separate

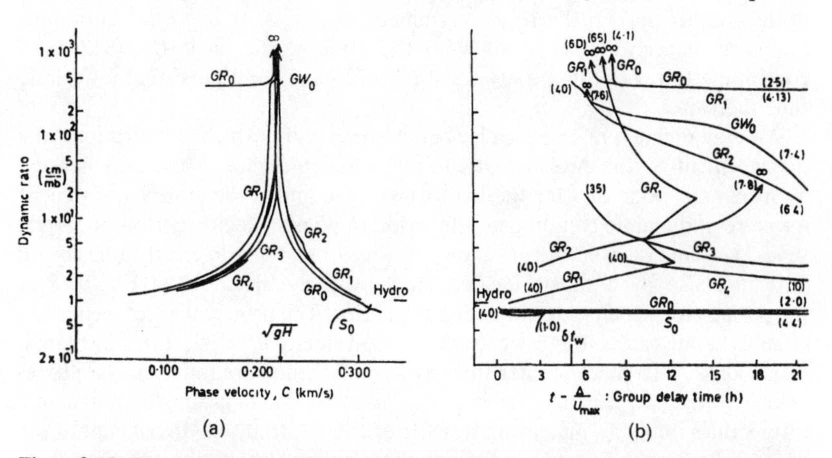

Figure 9.19. Dynamic ratio as a function of phase velocity (a) and group delay with respect to the first arrival at San Francisco (b). $U_{\max} = 312$ m/s and $\Delta = 14{,}053$ km. The numbers in parentheses indicate the period ranges of the curves.

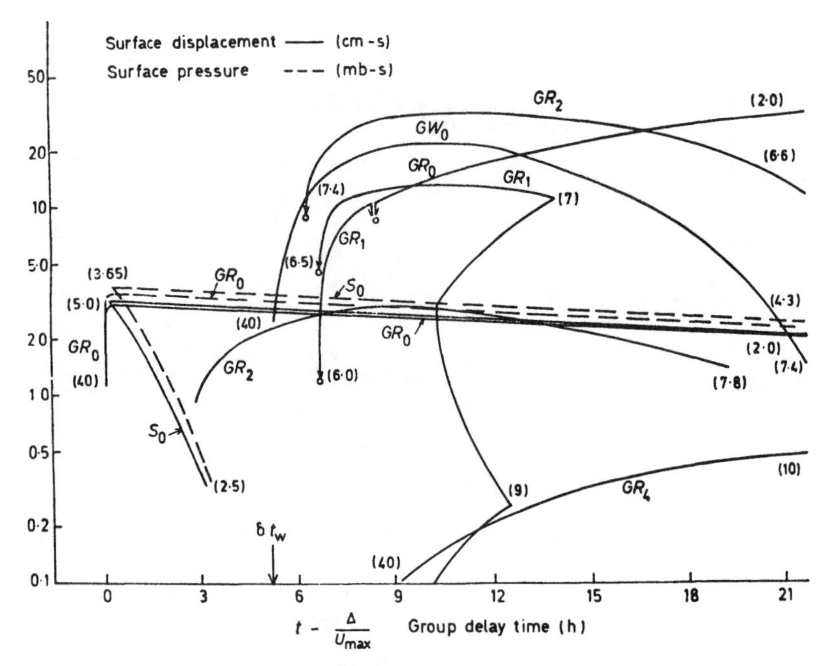

Figure 9.20. Spectra of sea-level displacements and pressures from a surface explosion with delta-function time variation. Abscissa is group delay with respect to the first arrival in San Francisco. The numbers in parentheses are the periods of the waves at the times indicated.

ocean system is duplicated by joining segments of GW_0, GR_2, GR_1, and GR_0 of the coupled system. These dispersion curves lead us to expect a continuous sequence of arrivals, beginning with the atmospheric pulse GR (312 m/s), continuing to the ocean gravity wave GW (220 m/s), and persisting for a long time thereafter.

We may define a quantity called the *dynamic ratio*, which is the ratio of the displacement to the pressure at the air–ocean interface. This ratio has the approximate value of 1 for the hydrostatic case and is therefore a convenient measure of dynamic coupling as a function of phase velocity, period or arrival time. Dynamic ratio curves are shown in Fig. 9.19a, the numerical values being indicated also by numbers located on the dispersion curves of Fig. 9.18 at the periods for which they have been calculated. The principal effect seen is the large dynamic ratios found where the phase velocity is near $\sqrt{(gH)}$. Dynamic ratios at least 10 times greater than hydrostatic value are found in the phase-velocity range 195–230 m/s. When plotted as a function of group delay, dynamic ratio values imply an increasing transfer of energy from the atmosphere to the ocean following the arrival of the initial air pulse. At a station as far away as San Francisco (Fig. 9.19b), waves with dynamic ratios greater than 10 begin to arrive about 6 h after the initial air pulse.

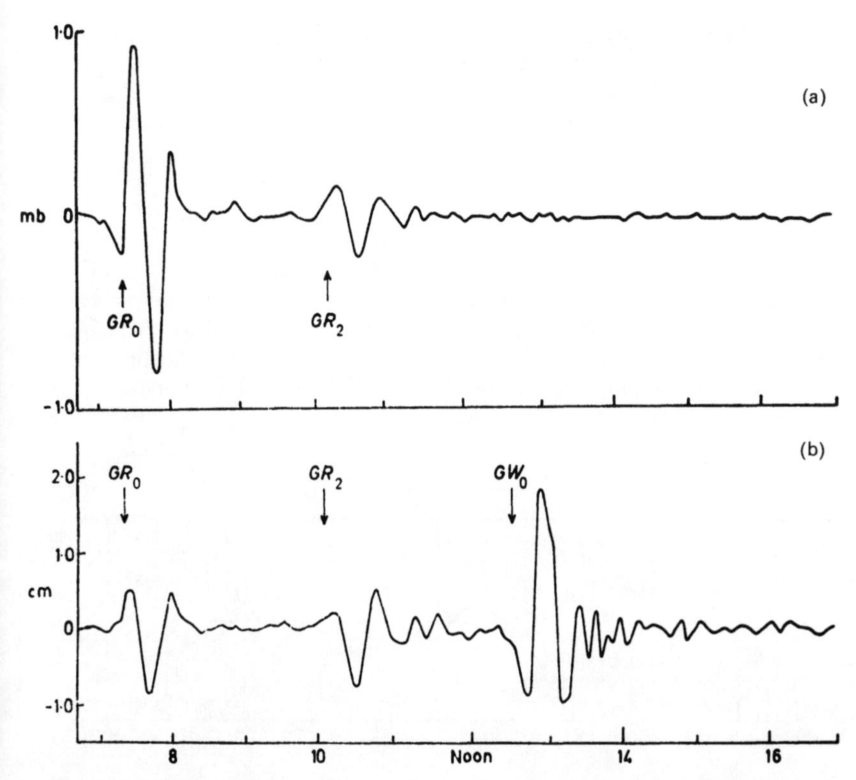

Figure 9.21. Synthetic barogram (a) and marigram (b) for San Francisco. Source time-function is a single-cycle sine wave of 40-min. period. Time is local time, 27 August, 1883.

Dynamic ratios and dispersion indicate the displacement-pressure values at the air–sea interface and the phase and group characteristics of all possible modes. It remains to discuss which portions of the different modes are excited by a specific source. The relative excitations of several modes are shown in Fig. 9.20 for the case of a surface source with a delta-function time variation. The abscissa is time following the first arrival for a station distance of 14 053 km (San Francisco). It is seen that the first arrival is an atmospheric pulse and a sea wave with dynamic ratio close to the hydrostatic value of 1. Contributions from GR_0 and S_0 make up this signal. The oscillations continue at about the same level until the time $\delta t_w = \Delta/\sqrt{(gH)} - \Delta/U_{\max}$ (where Δ is the distance and $U_{\max}$ is 312 m/s), when the water waves increase several times in amplitude. The numbers in parentheses indicate periods of the waves at the times indicated.

Finally, a synthetic barogram and a synthetic marigram are shown in Fig. 9.21 for a surface source with time variation of a single-cycle sine wave of period 40 min. Instrumental response was assumed to be flat and the distance corresponds to that of San Francisco. The main effects, already discussed, appear in the records. The initial pulse is a transient and shows as a disturbance in the atmosphere and ocean. This is the main Krakatoa pulse, which was observed at many places over the world. It is formed almost entirely from the GR_0 mode. The second event also shows on the barogram and marigram and is a contribution from the mode GR_2. Finally, the largest disturbance of sea level occurs with the arrival of the GW_0 mode with hardly any pressure pulse accompanying it. However, the sea wave is excited by atmospheric waves with the same phase velocity, although they are not apparent on the record because of the large dynamic ratio.

The response of the barographs and tide gauges that recorded the Krakatoa signals in 1883 are not known to us. There is reason to believe that nonlinear responses may be involved, because the instruments were driven at higher frequencies than anticipated in their design. Furthermore, the tide gauges may have been responding to resonance in the harbor excited by the sea waves. For these reasons one can only make a qualitative comparison of the theory with the observations.

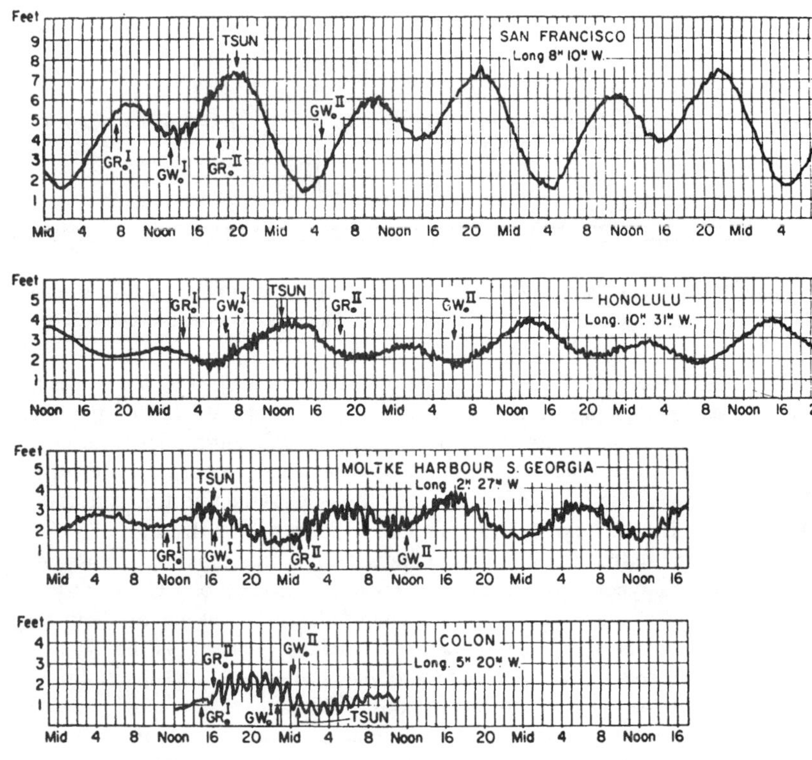

Figure 9.22. Marigrams for San Francisco, Honolulu, South Georgia, and Colon. Arrows indicate theoretical arrival times. Roman superscripts indicate minor (I) and major (II) great circle paths. Abscissa is local civil time beginning 27 August, 1883, except Honolulu, which begins 26 August.

A selection of marigrams is shown in Fig. 9.22. The arrows indicate the theoretical times of the modes GR_0 and GW_0. Also indicated are the theoretical arrival times of tsunamis, which follow least-time paths around obstacles formed by continents, archipelagos, etc. Except for South Georgia, the tide gauge disturbance occurred much too early to be explained as a tsunami, the discrepancy being of the order of 10 h. Moreover, the paths of Colon, Honolulu, San Francisco, and the English Channel ports require such circuitous routes and passage across barriers as to eliminate the possibility of tsunami action. Excitation by the atmospheric pulses is almost a necessity.

It seems reasonable to conclude that the main Krakatoa atmospheric pulse corresponds to propagation in the GR_0 mode of the atmosphere. The corresponding sea disturbance is essentially hydrostatic. The main sea waves are in the GW_0 mode and propagate along great circle paths with phase velocities near $\sqrt{(gH)}$. These are excited by atmospheric waves with phase velocities near $\sqrt{(gH)}$. Intervening land barriers are jumped by the air waves, which can reexcite the sea wave if a sufficiently long fetch is available.

9.7. Rayleigh Waves from Atmospheric Explosions

We assume a system that consists of a layered atmosphere overlying a layered ocean. The ocean, in turn, rests on top of a layered elastic solid. The source is in the layer s in the atmosphere and the entire structure is bounded above the atmosphere by a free surface and below in the solid by a homogeneous half-space. The positive z axis is directed upward into the atmosphere and the origin is taken at the solid–fluid interface. To avoid confusion, we denote the vertical particle velocity in the fluid by v, whereas the vertical displacement is denoted by w. The matrix relation for the fluid (air–sea) are those derived already in Section 9.5.3:

$$\begin{bmatrix} v_{s_2}(D) \\ P_{s_2}(D) \end{bmatrix} = \begin{bmatrix} v_{s_1}(D) \\ P_{s_1}(D) \end{bmatrix} + \begin{bmatrix} \delta v_s \\ \delta P_s \end{bmatrix}, \tag{9.268}$$

$$\begin{bmatrix} v_{n-1}(z_{n-1}) \\ P_{n-1}(z_{n-1}) \end{bmatrix} = B \begin{bmatrix} v_{s_2}(D) \\ P_{s_2}(D) \end{bmatrix}, \tag{9.269}$$

$$\begin{bmatrix} v_{s_1}(D) \\ P_{s_1}(D) \end{bmatrix} = C \begin{bmatrix} v_0 \\ P_0 \end{bmatrix}, \tag{9.270}$$

with $v_0 = v_1(0)$, $P_0 = P_1(0)$. In the present case, $v_0 \neq 0$. At the free surface $(z = z_{n-1})$ we must have

$$\begin{bmatrix} v_{n-1}(z_{n-1}) \\ P_{n-1}(z_{n-1}) \end{bmatrix} = \begin{bmatrix} v_n \\ 0 \end{bmatrix}. \tag{9.271}$$

From Eqs. (9.268)–(9.271), we derive the matrix relation

$$\begin{bmatrix} v_n \\ 0 \end{bmatrix} = A \begin{bmatrix} v_0 \\ P_0 \end{bmatrix} + B \begin{bmatrix} \delta v_s \\ \delta P_s \end{bmatrix}. \tag{9.272}$$

For an energy-injection source, $\delta P_s = 0$ [Eq. (9.249)]. Therefore, Eq. (9.272) admits the result

$$v_0 = \frac{-B_{21}\delta v_s}{[A_{21} + (P_0/v_0)A_{22}]}. \tag{9.273}$$

The boundary conditions at the solid–fluid interface require that, at $z = 0$,

1. The vertical component of the velocity is continuous
2. The normal stress is continuous (9.274)
3. The tangential stress in the solid must vanish

Because the z axis is now drawn vertically upward and because the displacement and stress fields in the solid half-space must tend to zero as $|z| \to \infty$, Eq. (3.189) gives

$$[\Omega_\alpha^+, \Omega_\alpha^+, \Omega_\beta^+, \Omega_\beta^+] = J[\dot{u}_0/c, -\dot{w}_0/c, \sigma_0, -\tau_0]. \tag{9.275}$$

Using the boundary conditions in the list (9.274), Eq. (9.275) yields

$$\frac{P_0}{v_0/c} = \frac{(J_{12} - J_{22})(J_{31} - J_{41}) - (J_{32} - J_{42})(J_{11} - J_{21})}{(J_{13} - J_{23})(J_{31} - J_{41}) - (J_{33} - J_{43})(J_{11} - J_{21})} = \frac{KN - LM}{GN - LH}, \tag{9.276}$$

where K, L, M, and N are defined in terms of the elements of the solid layer product matrix, A^R, [cf. Eq. (3.200) with A_{ij} replaced by A_{ij}^R] and

$$G = \gamma_n \eta_{\alpha n} A_{13}^R + (\gamma_n - 1)A_{23}^R - \frac{\eta_{\alpha n} A_{33}^R}{(\rho_n c^2)} + \frac{A_{43}^R}{(\rho_n c^2)},$$

$$H = -(\gamma_n - 1)A_{13}^R + \gamma_n \eta_{\beta n} A_{23}^R + \frac{A_{33}^R}{(\rho_n c^2)} + \frac{\eta_{\beta n} A_{43}^R}{(\rho_n c^2)}. \tag{9.277}$$

It is useful to note that

$$\frac{G}{L} = \frac{J_{13} - J_{23}}{J_{11} - J_{21}}, \quad \frac{H}{N} = \frac{J_{33} - J_{43}}{J_{31} - J_{41}}, \quad \frac{K}{L} = \frac{J_{12} - J_{22}}{J_{11} - J_{21}},$$

$$\frac{M}{N} = \frac{J_{32} - J_{42}}{J_{31} - J_{41}}. \tag{9.278}$$

In the case when the fluid is absent ($P_0 = 0$), Eq. (9.276) leads us to the known Rayleigh-wave frequency equation for a layered solid, viz., $KN - LM = 0$.

Equations (9.273) and (9.276) now render

$$v_0 = -c\frac{B_{21}}{A_{22}}(GN - LH)\frac{\delta v_s}{F_e(k, \omega)}, \tag{9.279}$$

where

$$F_e(k, \omega) = KN - LM + c\left(\frac{A_{21}}{A_{22}}\right)(GN - LH). \tag{9.280}$$

If we introduce the suppressed Δ dependence [Eq. (9.189)], we obtain

$$v_0(0, \Delta; \omega) = -\int_0^\infty c\delta v_s\left(\frac{B_{21}}{A_{22}}\right)(GN - LH)J_0(k\Delta)\frac{k\,dk}{F_e(k, \omega)}. \quad (9.281a)$$

Denoting the roots of the equation $F_e(k, \omega) = 0$ by $k_j(\omega)$ and evaluating the residues, we find, for Re $\omega > 0$,

$$v_0(0, \Delta; k_j) = \pi i\left[\omega\delta v_s\left(\frac{B_{21}}{A_{22}}\right)(GN - LH)\left(\frac{\partial F_e}{\partial k}\right)^{-1}H_0^{(2)}(k\Delta)\right]_{k_j}. \quad (9.281b)$$

We next substitute the value of δv_s for an energy-injection source. Eqs. (9.281) then yield for $k_j\Delta \gg 1$,

$$\begin{aligned} v_0(0, \Delta; k_j) &= \frac{ic_0 E}{2\rho_s^0\alpha_s^2}\left[\frac{\omega e^{-i\omega t_0}}{(1 + i\omega\tau)^2}\left(\frac{B_{21}}{A_{22}}\right)(GN - LH)\left(\frac{\partial F_e}{\partial k}\right)^{-1}H_0^{(2)}(k\Delta)\right]_{k_j} \\ &= \frac{c_0 E}{\rho_s^0\alpha_s^2}(2\pi\Delta)^{-1/2} \\ &\quad\times\left[\frac{ck^{1/2}e^{-i\omega t_0}}{(1 + i\omega\tau)^2}\left(\frac{B_{21}}{A_{22}}\right)(GN - LH)\left(\frac{\partial F_e}{\partial k}\right)^{-1}e^{-i(k\Delta - 3\pi/4)}\right]_{k_j}. \end{aligned} \quad (9.282)$$

The ratio (B_{21}/A_{22}) can be expressed in terms of plane-wave amplitudes. Therefore, if we put $P_{n-1}(z_{n-1}) = \delta v_s = \delta P_s = 0$, Eqs. (9.269) and (9.272) yield

$$\frac{B_{21}}{A_{22}} = \frac{-P_s(D)}{v_0}.$$

In deriving the above relation, we have used the fact that the determinant of the matrix a_m (and therefore those of A and B) is unity. The numerical evaluation of $v_0(k_j)$ is similar to the scheme described in Sect. 9.5.3. Once a root of $F_e(k, \omega) = 0$ for a fixed ω is found, the group velocity, $(GN - LH)$, $(\partial F_e/\partial k)$, and B_{21}/A_{22} can be calculated.

The group velocity U_g and $\partial F_e/\partial k$ are obtained by computing the analytic derivatives with respect to ω and k of the individual layer matrices and then using the chain rule. For a nongravitating isothermal ocean of depth $\hat{H}$ terminated at sea level by a free surface, A_{21} and A_{22} are given by

$$A_{21} = -ic\rho_0\frac{\text{sh}[k\hat{H}\sqrt{1 - c^2/\alpha^2}]}{\sqrt{1 - c^2/\alpha^2}}, \qquad A_{22} = \text{ch}\left[k\hat{H}\sqrt{1 - \frac{c^2}{\alpha^2}}\right]. \quad (9.283)$$

This follows from Eq. (9.200) on putting $g = 0$, $d = \hat{H}$. The accompanying period equation appropriate for oceanic Rayleigh waves is derived from Eq. (9.280),

$$[NK - LM] - i\frac{c^2\rho_0}{\sqrt{(1 - c^2/\alpha^2)}}[GN - LH]\text{th}\left[k\hat{H}\sqrt{1 - \frac{c^2}{\alpha^2}}\right] = 0. \quad (9.284)$$

The period equation, $F_e = 0$, determines the dispersion relation for all surface waves in the coupled system. The roots correspond not only to Rayleigh waves in the earth but also the acoustic–gravity waves in the atmosphere–ocean system. The Rayleigh waves are obtained by choosing the trial wave-number k so that the nearest root is a Rayleigh-wave mode. The free-surface boundary condition at the top of the atmosphere was chosen in order to obtain real roots for the Rayleigh waves. The termination of the top of the atmosphere by an isothermal half-space causes Rayleigh roots to be complex owing to the radiation of energy out of the atmospheric waveguide.

The effect of the atmospheric wave guide on the dispersion of Rayleigh waves is very small. Therefore, a reasonable estimate of $[B_{21}/A_{22}]$ can be obtained by using as input to an eigenfunction program the values of the Rayleigh root calculated from $\Delta_R = 0$. The medium response function A^R for four earth structures is shown in Fig. 9.23. These results and others indicate that:

1. For most source altitudes, the energy coupling from the atmosphere into Rayleigh waves is more efficient for the continental-earth structure than for the oceanic structure.
2. As far as Rayleigh-wave excitation is concerned, the mass-injection and energy-injection sources are equivalent.
3. For source altitudes 0–10 km, the Rayleigh-wave excitation is independent of source type.
4. The part of the total yield energy converted to Rayleigh waves varies from 3×10^{-7} to 3×10^{-5}, depending on the yield, height of burst, period of wave, and crustal structure of the wave-path.

The evaluation of the inverse Fourier transform

$$\frac{1}{2\pi}\int_{-\infty}^{\infty} \{v_0(\omega)\}_j e^{i\omega t}\, d\omega,$$

yields the time function of the expected Rayleigh-wave signal from an atmospheric explosion. If this integral is evaluated by the method of stationary phase and physical attenuation is included, the peak-to-trough vertical ground displacement amplitude for the fundamental Rayleigh mode at time $t = \Delta/U_g$ and stationary period $T = T_0$, (Δ, epicentral distance in degrees and U_g, the group velocity in km/s) is found to abide by the theoretical–empirical law, valid for source heights 2–10 km,

$$|u_z(\text{in } \mu\text{m})| = \left[\frac{0.2Y}{1 + (Y/Y_0)^{2/3}}\right]\left[\frac{e^{-\gamma_R \Delta}}{\sqrt{(\Delta \sin \Delta)}}\right]\left[\frac{(10^{15}A^R)k_R^{-1/2}}{(T/U_g)\sqrt{(|dU_g/dT|)}}\right]_{T=T_0}, \tag{9.285}$$ [5]

where

$\gamma_R(T)$ = attenuation coefficient, per kilometer,
k_R = Rayleigh wave number, per kilometer,
Y = yield energy, in megatons (MT) of TNT,
A^R = solid earth response function, in microns per dyne,
Y_0 = a reference yield of the order of 500 MT for sea-level explosions.

[5] The constant 0.2 in Eq. (9.285) is of dimensions dyn · km^{-1} · sec. The entity Y is dimensionless.

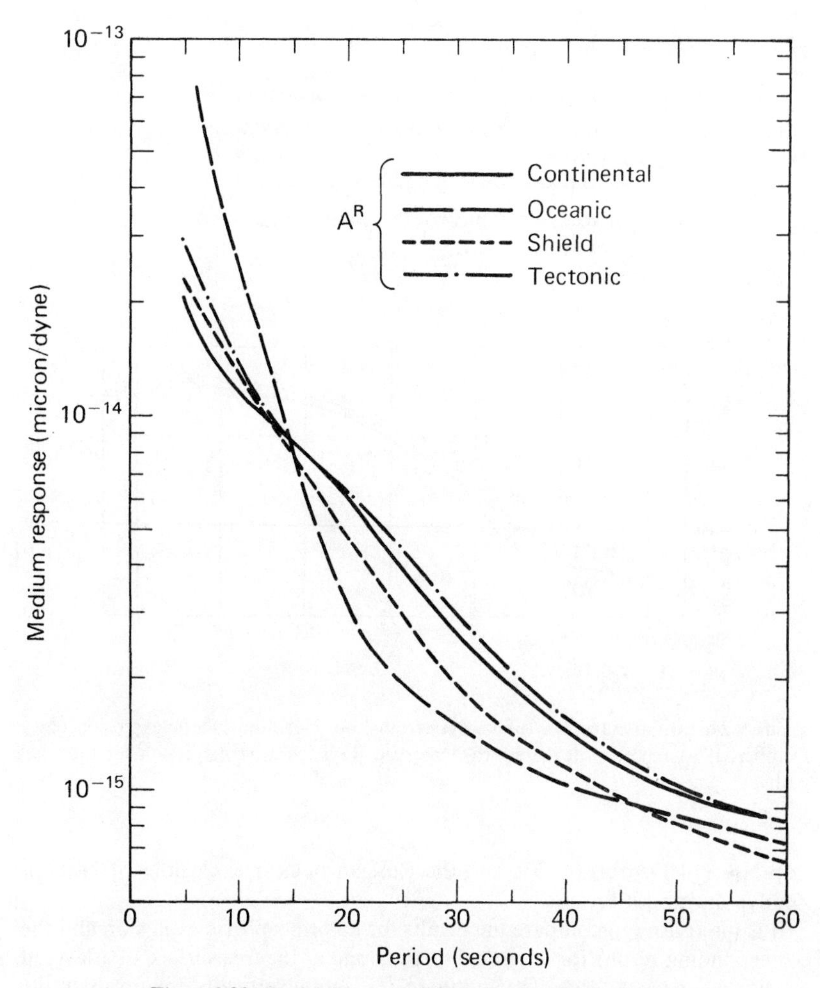

Figure 9.23. Rayleigh-wave solid-earth response functions.

Typical values at $T_0 = 20$ s, for the path Uppsala (Sweden)–Novaya Zemlya, are

$$10^{15}A^R = 6\ \mu\text{m/dyne}, \qquad c = 3.4\ \text{km/s}, \qquad U_g = 2.78\ \text{km/s},$$

$$\left(\frac{T}{U_g}\right)\sqrt{\left|\frac{dU_g}{dT}\right|} = 1.135\ \text{km}^{-1/2}\ \text{s}, \qquad \gamma_R = 1.0 \times 10^{-4}\ \text{km}^{-1},$$

$$k_R^{-1/2} = 3.28\ \text{km}^{1/2}. \tag{9.286}$$

Some observed results are shown in Figs. 9.24 and 9.25. Figure 9.25 refers to the comet explosion in Siberia (30 June, 1908), the Chinese nuclear explosion at

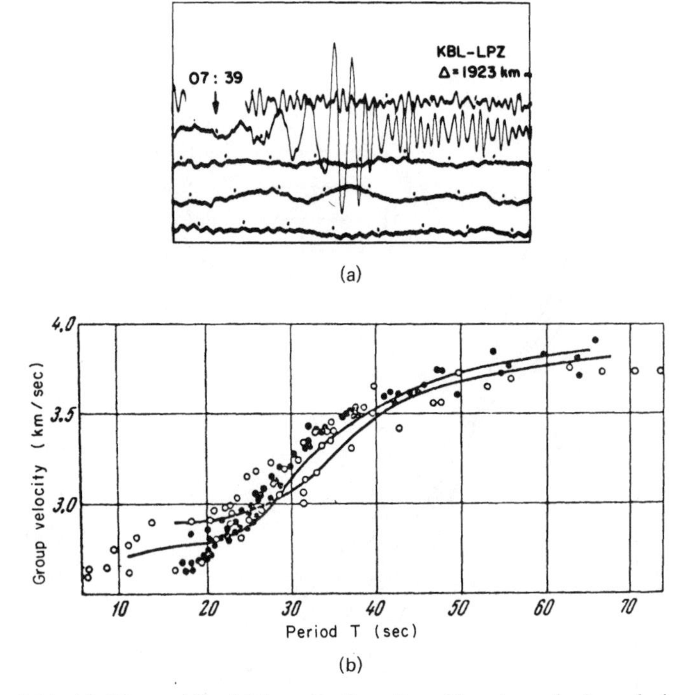

Figure 9.24. (a) Observed Rayleigh waves from Lop Nor atmospheric explosion of 14 October, 1970, recorded at Kabul. (b) Observed Rayleigh-wave dispersion for Eurasian paths.

Lop-Nor (14 October, 1970), and the Russian nuclear explosions at Novaya Zemlya during 1961.

It is interesting to compare the results for an atmospheric explosion and the corresponding results for a vertical force acting at the free surface of a layered elastic half-space. A force of magnitude $f_0(\omega)$ acting vertically downward at the origin has the representation[6] (see App. F)

$$\tau_{zz} = \frac{f_0(\omega)}{2\pi} \int_0^{\infty} J_0(k\Delta)k\,dk = \frac{f_0(\omega)}{2\pi} \frac{\delta^+(\Delta)}{\Delta}. \tag{9.287}$$

Therefore, τ_{zz} vanishes for all but infinitesimal values of Δ, where it becomes infinite in such a way that its integral over the whole plane is finite:

$$2\pi \int_0^{\infty} \tau_{zz}\Delta\,d\Delta = f_0. \tag{9.288}$$

[6] This can be obtained by the method of Section 4.5.4.

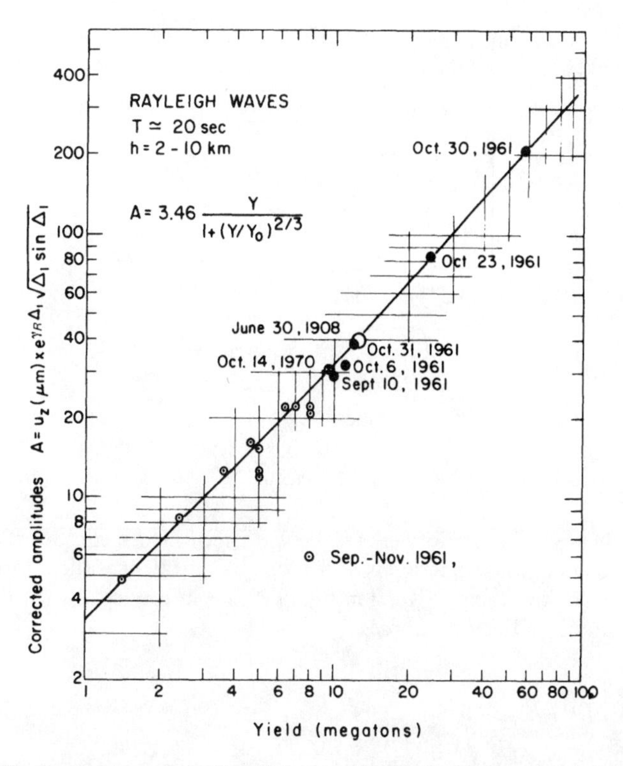

Figure 9.25. Dependence of the Rayleigh-wave amplitude at 20 s on the yield (Y) of atmospheric explosions.

Putting $P_0 = f_0(\omega)/2\pi$ in Eq. (9.276) and introducing the suppressed Δ-dependence according to Eq. (9.189), we get the following expression for the vertical velocity at any point of the free surface of a layered elastic half-space from a concentrated force f_0 acting vertically downward at the origin:

$$v_0(0, \Delta; \omega) = \frac{1}{2\pi} \int_0^\infty c f_0(\omega) \frac{GN - LH}{KN - LM} J_0(k\Delta) k \, dk. \tag{9.289}$$

As noted before, $F_e(k, \omega)$ in Eq. (9.281a) can be approximated by the Rayleigh-wave function, $KN - LM$, for a layered elastic half-space. Inserting the value of S_{e0} for an energy-injection source from Eqs. (9.245), (9.248) and (9.260), we see that Eq. (9.281a) will coincide with Eq. (9.289) if

$$f_0(\omega) = -\frac{1}{\rho_0 \alpha^2} \left(\frac{B_{21}}{A_{22}} \right) e(\omega), \tag{9.290}$$

where $e(\omega)$ is the Fourier transform of $f_E(t)$ [Eq. (9.245)]. If the atmosphere is considered as a nongravitating, isothermal half-space and if D denotes the source height above the fluid–solid interface, Eqs. (9.200) and (9.255) yield

$$\frac{B_{21}}{A_{22}} = -\frac{i\omega\rho_0}{\nu}e^{-\nu D},$$
$$\nu = (k^2 - k_\alpha^2)^{1/2}, \qquad k_\alpha = \frac{\omega}{\alpha}, \tag{9.291}$$

where α is the constant sound-wave velocity in the atmosphere. Because $k_j/k_\alpha \ll 1$, $\nu \simeq ik_\alpha$, Eq. (9.291) yields

$$\left(\frac{B_{21}}{A_{22}}\right)_{k_j} \simeq -\rho_0\alpha e^{-ik_\alpha D}. \tag{9.292}$$

With this value, Eq. (9.290) becomes

$$f_0(\omega) = \frac{1}{\alpha}e(\omega)e^{-ik_\alpha D}. \tag{9.293}$$

Therefore, as far as the Rayleigh-wave vertical displacements are concerned, we can model an atmospheric explosion by an equivalent vertical force at the epicenter. The factor $\exp(-ik_\alpha D)$ in Eq. (9.293) is introduced by the difference in the locations of the two sources.

Bibliography

Ben-Menahem A (1972) Mercury tiltmeter as an infrasonic detector: Theory, observations, and applications. Jour Geophys Res 77: 818–825.

Ben-Menahem A, Rosenman M (1972) Amplitude patterns of tsunami waves from submarine earthquakes. Jour Geophys Res 77: 3097–3128.

Ben-Menahem A, Vered M (1975) Modeling of atmospheric nuclear explosions over a mountainous region by vertical and horizontal single forces. Bull Seismol Soc Amer 65: 971–980.

Brunt D (1927) The period of simple vertical oscillations in the atmosphere. Quart Jour Roy Meteorol Soc (London) 53: 30–32.

Donn WL, Ewing M (1962) Atmospheric waves from nuclear explosions. Part II. The Soviet test of 30 Oct 1961. Jour Atmos Sci 19: 264–273.

Donn WL, Posmentier ES (1964) Ground-coupled air waves from the great Alaskan earthquake. Jour Geophys Res 69: 5357–5361.

George, TM (ed) (1968) Acoustic-Gravity Waves in the Atmosphere. ESSA-ARPA Symposium Proceedings, Boulder, Colorado, 15–17 July, 1968. U.S. Government Printing Office, Washington, D.C.

Harkrider DG (1964) Theoretical and observed acoustic-gravity waves from explosive sources in the atmosphere. Jour Geophys Res 69: 5295–5321.

Harkrider DG, Newton CA, Flinn EA (1974) Theoretical effect of yield and burst-height of atmospheric explosions on Rayleigh wave amplitudes. Geophys Jour Roy Astron Soc (London) 36: 191–225.

Harkrider DG, Press F (1967) The Krakatoa air-sea waves: an example of pulse propagation in coupled systems. Geophys Jour Roy Astron Soc (London) 13: 149–159.

Helmholtz HLF von (1889) Über atmosphärische Bewegung. Sitzungsberichte Preuss Akad Wiss.

Johnson NK (1929) Atmospheric oscillations shown by the microbarograph. Quart Jour Roy Meteorol Soc (London) 55: 19–30.

Kajiura K (1963) The leading wave of a tsunami. Bull Earthquake Res Inst (Tokyo) 41: 535–571.

Lamb H (1945) Hydrodynamics. Dover, New York.

McGarr A (1965) Excitation of seiches in channels by seismic waves. Jour Geophys Res 70: 847–854.

Momoi T (1963) Diffraction of tsunami invading a semi-circular peninsula. Bull Earthquake Res Inst (Tokyo) 41: 589–594.

Omer GC, Hall HH (1949) The scattering of a tsunami by a cylindrical island. Bull Seismol Soc Amer 39: 257–260.

Pekeris CL (1948) The propagation of a pulse in the atmosphere. II. Phys Rev 73: 145–154.

Pierce AD (1965) Propagation of acoustic-gravity waves in a temperature and wind stratified atmosphere. Jour Acoust Soc Amer 37: 218–227.

Press F, Harkrider DG (1962) Propagation of acoustic-gravity waves in the atmosphere. Jour Geophys Res 67: 3889–3908.

Rodean HC (1971) Nuclear-explosion Seismology. U.S. Atomic Energy Commission, Division of Technical Information, Lawrence Livermore Laboratory, University of California, 156 pp.

Row RV (1967) Acoustic-gravity waves in the upper atmosphere due to a nuclear detonation and an earthquake. Jour Geophys Res 72: 1599–1610.

Scholte JGJ (1949) On true and pseudo-Rayleigh waves. Proc Koninkl Ned Akad Wetenschap Amst 52: 652–653.

Strick E (1959) The pseudo-Rayleigh wave. Phil Trans Roy Soc (London) A251: 488–523.

Van Hulsteyn DB (1965a) The atmospheric pressure wave generated by a nuclear explosion. Part I. Jour Geophys Res. 70: 257–269.

Van Hulsteyn DB (1965b) The atmospheric pressure wave generated by a nuclear explosion. Part II. Jour Geophys Res 70: 271–278.

Weston VH (1961) The pressure pulse produced by a large explosion in the atmosphere. I. Can Jour Phys 39: 993–1009.

Weston VH (1962) The pressure pulse produced by a large explosion in the atmosphere. II. Can Jour Phys 40: 431–445.

CHAPTER 10

Seismic Wave Motion in Anelastic Media

Men argue, Nature acts.
(*Voltaire*)

10.1. The Specific Dissipation Parameter

10.1.1. Damped Linear Oscillator

Consider a mechanical linear oscillator with a single degree of freedom driven by an external time-dependent acceleration $f(t)$. The time behavior of this system is governed by the equation

$$\ddot{x} + \frac{1}{Q}\omega_0 \dot{x} + \omega_0^2 x = f(t), \qquad \left(Q \gg 1,\ \dot{x} = \frac{dx}{dt}\right) \tag{10.1}$$

where Q is a dimensionless parameter to be studied later. Mathematically, the solution of Eq. (10.1), under the initial conditions $x(0) = x_0$, $\dot{x}(0) = \dot{x}_0$ can formally be considered as the sum $x(t) = x_1(t) + x_2(t)$, where

$$\ddot{x}_1(t) + \frac{1}{Q}\omega_0 \dot{x}_1(t) + \omega_0^2 x_1(t) = 0, \qquad x_1(0) = x_0, \qquad \dot{x}_1(0) = \dot{x}_0,$$

$$\ddot{x}_2(t) + \frac{1}{Q}\omega_0 \dot{x}_2(t) + \omega_0^2 x_2(t) = f(t), \qquad x_2(0) = 0, \qquad \dot{x}_2(0) = 0. \tag{10.2}$$

In physical terms, we say that the total response of the system to the input stimulus $f(t)$ is composed of two components:

1. *Source-free motion* (also called transient, natural, homogeneous, complementary). Explicitly,

$$x_1(t) = Ae^{-\omega_0 t/2Q}\sin[\omega t + \chi_0], \qquad \omega = \omega_0\left(1 - \frac{1}{4Q^2}\right)^{1/2}, \tag{10.3}$$

$$A = \left[x_0^2 + \left\{\frac{\dot{x}_0 + (1/2Q)\omega_0 x_0}{\omega}\right\}^2\right]^{1/2}, \qquad \sin\chi_0 = \frac{x_0}{A}, \qquad T_0 = \frac{2\pi}{\omega_0}. \tag{10.4}$$

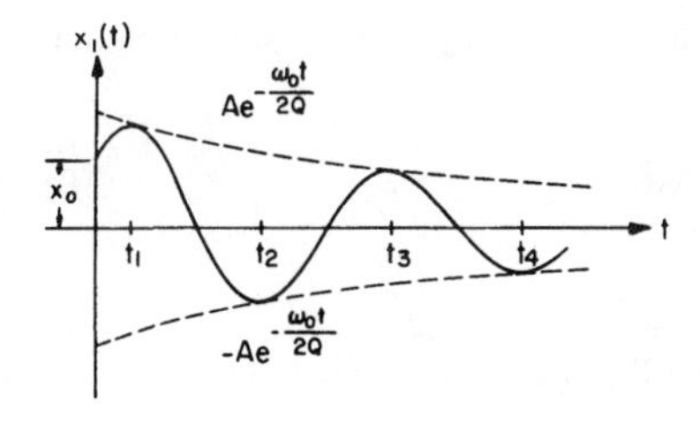

Figure 10.1. Transient solution of the equation of a linear oscillator, describing a free, underdamped motion.

The motion depends on the intrinsic parameters of the system and the *initial energy* left in the system in the form of a stressed spring, charged capacitor, etc. At $t = 0$, this energy is let free with the result that the ensuing oscillations die out as work is done against frictional forces.

The oscillator executes *periodic zero crossings* with period $T = T_0/[1 - 1/4Q^2]^{1/2} > T_0$ and an exponentially decaying amplitude (Fig. 10.1). Whereas T_0 is known as the *natural* (or free) period of vibration (i.e., for a system that obeys $\ddot{x} + \omega_0^2 x = 0$), we call T the *damped period.* For small damping we have approximately $T \simeq T_0$. It is easily verified that the sequence of minima and maxima of $x_1(t)$ is given by ($Q \gg 1$)

$$t_n = \frac{1}{\omega_0}[\tan^{-1}\{2Q\} - \chi_0 + \pi n], \qquad n = 1, 2, \ldots. \tag{10.5}$$

Because the points of contact of $x_1(t)$ with $\{\pm Ae^{-\omega_0 t/2Q}\}$ are given by

$$\hat{t}_n = \frac{1}{\omega_0}\left[-\chi_0 + \pi\left(n + \frac{1}{2}\right)\right], \tag{10.6}$$

the lag of t_n behind $\hat{t}_n$ will be

$$\delta = \frac{1}{\omega_0}\tan^{-1}\left\{\frac{1}{2Q}\right\}. \tag{10.7}$$

However, the values of $x_1(t)$ at the corresponding arguments t_n and $\hat{t}_n$ are assumed to be close and we may therefore measure the amplitude decay via the *damping ratio* $\Delta = |x_1(t_n)/x_1(t_{n+2})| \simeq e^{\omega_0 T/2Q}$, $T = t_{n+2} - t_n$. We define the *logarithmic decrement*

$$\ln\Delta = \ln\left|\frac{x_1(t_n)}{x_1(t_{n+2})}\right| = \frac{\pi}{Q}, \tag{10.8}$$

where the dimensionless parameter Q is known as the *quality factor* of the system. In addition, the decay can be measured by the time it takes the amplitude to drop to $1/e$ of its initial value, known as the *relaxation time*, $\tau_0 = 2Q/\omega_0$. In terms of τ_0, the equation of motion is $\ddot{x}_1 + (2/\tau_0)\dot{x}_1 + \omega_0^2 x_1 = 0$. For a lightly damped oscillator, we have

$$\delta = \frac{1}{2Q\omega_0}, \qquad \omega = \omega_0\left(1 - \frac{1}{4Q^2}\right)^{1/2} \simeq \omega_0\left(1 - \frac{1}{8Q^2}\right). \tag{10.9}$$

2. *Forced motion* (also called steady state, driven, inhomogeneous, particular). We have

$$x_2(t) = \int_0^t f(\tau)G(t - \tau)d\tau, \tag{10.10}$$

$$G(t) = \frac{1}{\omega_0(1 - h^2)^{1/2}} \sin[\omega_0(1 - h^2)^{1/2} t] e^{-\omega_0 t/2Q},$$

$$h = \frac{1}{2Q}. \tag{10.11}$$

The motion is determined both by the system constants and the driving force. The name "steady state" stems from the feature that part of the motion in Eq. (10.10) is maintained undamped long after the transients have died out. We say "part," because $x_2(t)$ may include transients of its own that result not from the initial conditions but from discontinuities in $f(t)$ at $t = 0$ or/and any other specified time. Two examples follow:

$f_1(t) = a_0 H(t)$

$$x_2(t) = \frac{a_0}{\omega_0^2}\left[1 - \frac{\sin\{\omega_0(1 - h^2)^{1/2} t + \theta_0\}}{\sin\theta_0} e^{-\omega_0 t/2Q}\right] H(t),$$

$$\sin\theta_0 = (1 - h^2)^{1/2}. \tag{10.12}$$

$f_2(t) = a_0 H(t) \sin \omega t$

$$x_2(t) = \frac{a_0}{D}\left[\sin(\omega t - \chi_1) + \frac{\omega}{\omega_0}\frac{\sin\{\omega_0(1 - h^2)^{1/2} t + \chi_2\}}{(1 - h^2)^{1/2}} e^{-\omega_0 t/2Q}\right] H(t),$$

$$D = [(\omega_0^2 - \omega^2)^2 + (\omega_0 \omega Q^{-1})^2]^{1/2}, \tag{10.13}$$

$$\tan\chi_1 = \frac{\omega_0 \omega Q^{-1}}{\omega_0^2 - \omega^2}, \qquad \tan\chi_2 = \frac{\omega_0^2(1 - h^2)^{1/2} Q^{-1}}{\omega^2 - \omega_0^2(1 - 2h^2)}.$$

The response to $f_1(t)$, known as the *step response* of the system, is shown in Fig. 10.2a.

Time response is specified with the aid of a number of useful parameters that tell us how fast the system moves to follow the input, how oscillatory it is, or how long does it take to practically reach the final value:

1. *Rise time*: the time required for the response to reach the steady-state level for the first time. For $Q \gg 1$, it is

$$t_r = \frac{\pi - \tan^{-1}[2Q]}{\omega_0}. \tag{10.14}$$

2. *Peak time*: the time required for the response to reach its peak. Differentiating Eq. (10.12) with respect to time and equating to zero, we get, for $Q \gg 1$,

$$t_p = \frac{\pi}{\omega_0(1 - h^2)^{1/2}}. \tag{10.15}$$

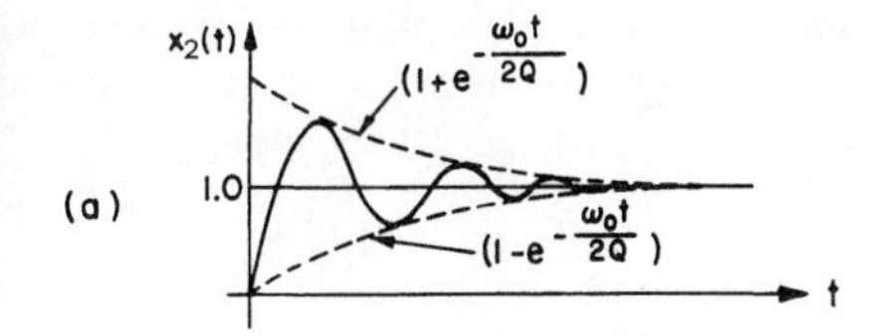

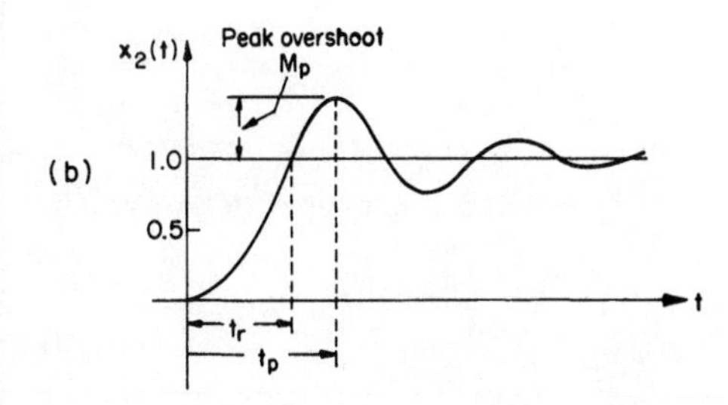

Figure 10.2. Step response of an underdamped linear oscillator with $a_0 = \omega_0^2$ [Eq. (10.12)]. (a) Envelopes of minima and maxima. (b) Overshoot parameters.

3. *Peak overshoot*: the fractional rise of the response above the steady-state level. For $Q \gg 1$, it is

$$M_p = \frac{x_2(t_p)}{x_2(\infty)} - 1 = e^{-\pi/2Q}. \tag{10.16}$$

These quantities are shown in Fig. 10.2b.

To obtain the *steady-state solution*, free of all transients, we apply the Fourier transform to Eq. (10.1). Defining

$$X(\omega) = \int_{-\infty}^{\infty} x(t)e^{-i\omega t}\,dt, \qquad F(\omega) = \int_{-\infty}^{\infty} f(t)e^{-i\omega t}\,dt$$

and applying the initial conditions

$$x_0 = 0, \qquad \dot{x}_0 = 0,$$

we derive the relation $X(\omega) = M(\omega)F(\omega)$ where $M(\omega)$ is the *response function*, (App. K)

$$M(\omega) = \left[(\omega_0^2 - \omega^2) + i\frac{1}{Q}\omega\omega_0\right]^{-1} = \frac{1}{\omega_0^2}\frac{\exp\left\{-i\tan^{-1}\left[\dfrac{\omega/\omega_0}{1-(\omega/\omega_0)^2}\dfrac{1}{Q}\right]\right\}}{[(1-\omega^2/\omega_0^2)^2 + (\omega/\omega_0 Q)^2]^{1/2}}. \tag{10.17}$$

Therefore, the displacement lags behind the driving force by the angle

$$\tan^{-1}\left[\frac{\dfrac{1}{Q}\dfrac{\omega}{\omega_0}}{(1-\omega^2/\omega_0^2)}\right]. \tag{10.17a}$$

If we watch the change in $M(\omega)$ as ω increases from zero toward ω_0, we shall note a critical change at $\omega = \omega_0$. This effect of synchronization of the forcing term and the response is known as *resonance.* At resonance, the response

$$M(\omega_0) = \frac{Q}{\omega_0^2} e^{-\pi i/2} \tag{10.18}$$

grows without limit as $Q \to \infty$.

The *peak response* occurs at a frequency such that $(d/du)[(1-u^2)^2 + (2hu)^2]^{-1/2} = 0$, where $u = \omega/\omega_0$. This frequency is given by

$$\omega = \omega_p = \omega_0\left(1 - \frac{1}{2Q^2}\right)^{1/2}. \tag{10.19}$$

It does not occur at either the natural or the damped frequency of the system. The value of $M(\omega)$ at peak response is

$$M(\omega_p) \simeq \frac{Q}{\omega_0^2} e^{-i\tan^{-1}[2Q]}. \tag{10.20}$$

The important properties of the response function $M(\omega)$ are as follows (Fig. 10.3):

1. The peak response exists only for $Q > 1/\sqrt{2}$. At $Q = 1/\sqrt{2}$, $\omega_p = 0$ and the response curve is parallel to the ω axis at $\omega = 0$. As $1/Q$ decreases, the peak moves closer to the resonance point $\omega = \omega_0$. The peak response decreases monotonically with the increase of the resonant frequency.

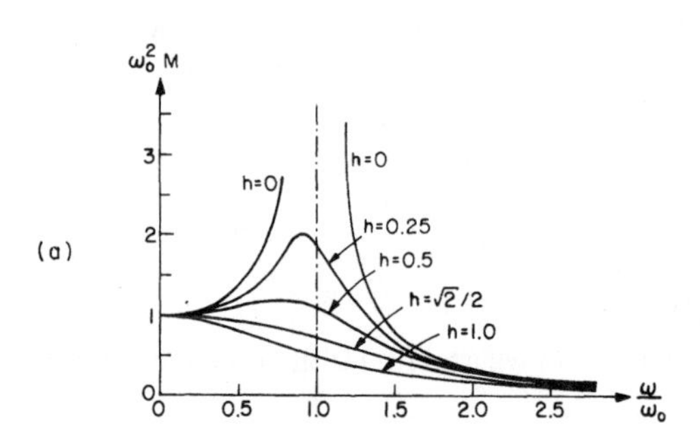

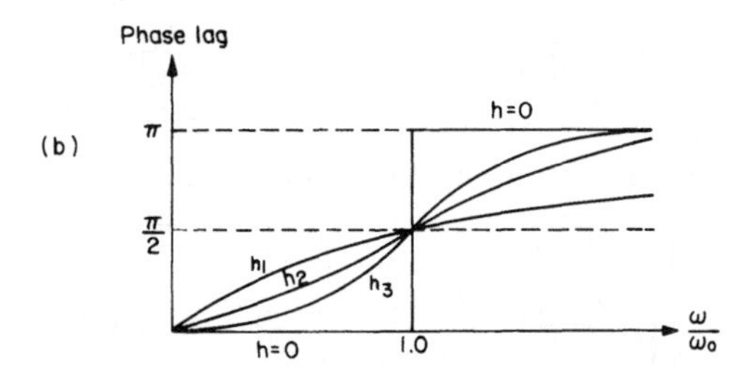

Figure 10.3. Steady-state response of an underdamped linear system. Resonance occurs at $\omega = \omega_0$. At $h = 0.7071$ the amplitude response is more flat whereas the phase response is more linear than at other h values. $h_1 = 0.5$, $h_2 = \sqrt{2}/2$, $h_3 = 0.25$. (a) Amplitude response for various values of the damping parameter h. (b) Phase response.

2. It follows from expression (10.17a) that at resonance the displacement lags by $\pi/2$ relative to $f(t)$. At lower frequencies, the lag is less than $\pi/2$, whereas above resonance the lag is greater than $\pi/2$. At very high frequencies, the inertial term dominates the motion and the driving force is practically in phase with the inertial term, or out of phase with the displacement by almost π. Since the derivative of the phase function ϕ is

$$\frac{d\phi}{d\omega} = \left(\frac{2h}{\omega_0}\right) \frac{(1 + u^2)}{(1 - u^2)^2 + 4h^2u^2}, \tag{10.21}$$

we find that $\phi'(0) = \phi'(1) = \sqrt{2}/\omega_0$ at $h = \sqrt{2}/2$. The phase is then nearly a linear function of frequency in the region where the amplitude response is flat.

10.1.2. Diffusion of Seismic Energy

At epicentral distances below 100 km, the dominant wavelengths of seismic signals interact strongly with inhomogeneities in the earth and the propagation of elastic waves may assume the character of diffusion. The gross features of the recorded ground motion can be attributed to numerous reflections and refractions. Each reflection and refraction at an interface gives rise to new compressional and shear waves. The number of possible ray paths in the heterogeneous earth increases rapidly such that the overall description of the phenomenon approaches that of a diffusion process. Truly, one of the major properties of wave propagation is the directed flow of energy in the direction of propagation. Strong scattering, however, will destroy the directional property of seismic waves. It is assumed instead that the energy-flux density [Eq. (1.122)] is in the direction of the gradient of the energy and proportional to that gradient. Let $H(\mathbf{r}, t; \omega)$ be the seismic energy per unit volume within a unit frequency band around ω. Then, we put $\mathbf{F} = 0$ in Eqs. (1.119)–(1.124), and assume that under the conditions stated above

$$\left\langle \mathfrak{T} \cdot \frac{\partial \mathbf{u}}{\partial t} \right\rangle_{\text{average}} = \frac{1}{2} D \,\text{grad}\left[\mathfrak{T} : \mathfrak{E} + \rho \left(\frac{\partial \mathbf{u}}{\partial t} \right)^2 \right],$$

where D is the diffusivity. Equation (1.124) then turns into the diffusion equation

$$\frac{\partial H}{\partial t} = D\nabla^2 H. \tag{10.21a}$$

Adding terms to account for linear dissipation and the presence of a point source at $\mathbf{r} = \mathbf{r}_0$, it becomes

$$\frac{\partial H}{\partial t} = D\nabla^2 H - \frac{\omega}{Q} H - H_0 \delta(\mathbf{r} - \mathbf{r}_0)\delta(t - t_0). \tag{10.22}$$

The term $\{-(\omega/Q)H\}$ represents a process through which seismic energy is transformed into heat. Q is the *specific dissipation parameter* of the medium and H_0 is the seismic energy generated by the source within a unit frequency band around ω. Note that Eq. (10.22) is valid only at sufficiently high frequencies. For that reason, ω must enter explicitly as a parameter into the differential equation, along with its dual, the time t. Although P, S, Love, and Rayleigh waves have different seismic velocities and scattering properties (and hence different diffusivities), it may be assumed that a radiative equilibrium between the four types depends only on the properties of the medium and is rapidly set up. A better model, of course, will take into account a diffusivity, D_H say, caused by horizontally propagating waves (Love, Rayleigh, and the horizontal component of body waves) and a vertical diffusivity, D_V, for the vertical components of body waves. Deleting the source term in Eq. (10.22) we obtain the anisotropic diffusion equation

$$\frac{\partial H}{\partial t} = D_H\left[\frac{\partial^2 H}{\partial x^2} + \frac{\partial^2 H}{\partial y^2}\right] + D_V\frac{\partial^2 H}{\partial z^2} - \frac{\omega}{Q}H. \tag{10.23}$$

A solution of Eq. (10.22) for body-wave scattering in a boundless medium is

$$H(\mathbf{r}, t; \omega) = \frac{H_0}{[4\pi D(t - t_0)]^{3/2}}\exp\left\{-\left[\frac{|\mathbf{r} - \mathbf{r}_0|^2}{4D(t - t_0)} + \frac{\omega}{Q}(t - t_0)\right]\right\}. \tag{10.24}$$

To obtain the total energy, we integrate Eq. (10.24) over the frequency with a possible dependence of both H_0 and D on ω.

10.1.3. Limitations of the Group-Velocity Concept in Absorbing Media

Consider the one-dimensional wave equation with a dissipation term

$$\frac{1}{c^2}\frac{\partial^2 u}{\partial t^2} + 2b\frac{\partial u}{\partial t} + au - \frac{\partial^2 u}{\partial x^2} = 0, \qquad b > 0. \tag{10.25}$$

Looking for a steady-state solution in the form of propagating waves with phase velocity $v(\omega)$ and spatial attenuation $\gamma(\omega)$, we propose the solution

$$u(x, t) = \mathrm{Im}[Ae^{i\omega t - x(\gamma + i\omega/v)}] = Ae^{-\gamma x}\sin\omega\left(t - \frac{x}{v}\right), \tag{10.26}$$

where A, $v(\omega)$, and $\gamma(\omega)$ are real and positive. We substitute Eq. (10.26) into Eq. (10.25) and obtain

$$\gamma^2 = \omega^2\left[\frac{1}{v^2} - \frac{1}{c^2}\right] + a, \quad \gamma = bv.$$

Solving for $\gamma(\omega)$ and $v(\omega)$, we find

$$v(\omega) = \frac{c}{\xi\sqrt{\chi}}, \qquad \gamma(\omega) = \frac{\omega}{2Qv} = \frac{1}{\xi}\frac{\omega\sqrt{\chi}}{2cQ_0}, \qquad Q(\omega) = Q_0\xi^2, \tag{10.27}$$

$$\xi = \left[\frac{1}{2} + \frac{1}{2}(1 + Q_0^{-2})^{1/2}\right]^{1/2}, \qquad \xi > 0, \qquad \chi = 1 - \frac{ac^2}{\omega^2}, \qquad Q_0 = \frac{\chi\omega}{2bc^2}.$$

To ensure $\gamma(\omega) > 0$ and $v(\omega) > 0$, we must require that $\chi > 0$ and $\xi > 0$. For $a \neq 0$, this condition puts a lower limit on the frequency, namely, $\omega > c\sqrt{a}$. As $\omega \to \infty$, we find

$$\chi \to 1, \qquad \xi \to 1, \qquad v(\omega) \to c, \qquad \gamma \to bc,$$

and

$$\lim_{\omega \to \infty} \left[\omega^2\left(\frac{1}{v^2} - \frac{1}{c^2}\right)\right] = b^2c^2 - a.$$

With the aid of Eq. (3.85), a formal expression for the group velocity is obtained

$$U_g = v\left[\frac{2\xi^2 - 1}{(\xi^2 - 1) + (1/\chi)}\right]. \tag{10.28}$$

It is easy to show that for $a = 0$, $U_g > c$ occurs if $\omega > bc^2/(2 + \sqrt{5})^{1/2}$. As $\omega \to \infty$, $U_g \to v \to c$.

Note that if the wave propagator is written as $\exp[i\omega(t - x/(c/n))]$ where $n(\omega)$ has the meaning of a complex *refraction index*, then $n = (c/v)(1 - i\gamma(v/\omega))$ or $n^2(\omega) = 1 - (c^2/\omega^2)(a + 2ib\omega)$. This example serves to show that only for large values of the dissipation parameter Q_0 ($\xi \simeq 1$) does U_g retain its physical meaning as the velocity of energy flow in the medium.

10.1.4. The Physical Meanings of Q

Apart from the relation in Eq. (10.8), Q is also associated with the "sharpness" of the response curve: If we look for points on this curve where the response $|M(\omega)|$ is half of its peak value, Eq. (10.17) yields, for $Q \gg 1$,

$$\frac{Q}{2\omega_0^2} \simeq \frac{1}{|\omega_0^2 - \omega^2|}. \tag{10.29}$$

However, if ω is close to ω_0, $\omega_0^2 - \omega^2 \simeq 2\omega_0(\omega_0 - \omega)$ and hence

$$\frac{1}{Q} = \frac{\Delta\omega}{2\omega_0}, \tag{10.30}$$

where $\Delta\omega$ is the *width* of the response curve at half its peak value. The *magnification* at resonance is Q/ω_0^2. Equation (10.30) changes to

$$\frac{1}{Q} = \frac{\Delta\omega}{\omega_0} \tag{10.31}$$

when *power response* is considered instead of displacement response. Another interpretation of Q is obtained as follows: Suppose that an initial energy E_0 is sent through a damped linear system governed by the equation

$$\ddot{x} + \frac{\omega_0}{Q}\dot{x} + \omega_0^2 x = 0. \tag{10.32}$$

Multiplying both sides by $\dot{x}$ we get

$$\frac{d}{dt}\left[\frac{1}{2}\dot{x}^2 + \frac{1}{2}\omega_0^2 x^2\right] = -\left(\frac{\omega_0}{Q}\right)\dot{x}^2 = -P,$$

where P is the power loss. Therefore, the rate of loss of the initial energy because of work done against the frictional forces is $(\omega_0/Q)\dot{x}^2$. We define the time averages over one cycle

$$\langle x \rangle = \frac{1}{T}\int_0^T x(t)dt,$$

$$\langle B \rangle = \frac{1}{2}\langle \dot{x}^2 \rangle + \frac{1}{2}\omega_0^2 \langle x^2 \rangle.$$

The explicit solution of Eq. (10.32) is

$$x(t) = Ae^{-\omega_0 t/2Q}\sin(\omega_0 t + \chi).$$

Working out the averages

$$\langle x^2 \rangle = \frac{A^2}{2} e^{-\omega_0 T/Q}, \qquad \langle B \rangle = \frac{A^2}{2}\omega_0^2 e^{-\omega_0 T/Q},$$

$$\langle \dot{x}^2 \rangle = \frac{A^2}{2}\omega_0^2 e^{-\omega_0 T/Q}, \qquad \langle P \rangle = \frac{\omega_0}{Q}\frac{A^2}{2}\omega_0^2 e^{-\omega_0 T/Q},$$

we find

$$\frac{\langle B \rangle}{T\langle P \rangle} = \frac{Q}{2\pi},$$

or,

$$\frac{1}{Q} = \frac{1}{2\pi}\,\frac{\text{energy dissipated in one cycle}}{\text{peak energy stored during the cycle}}. \tag{10.33}$$

10.2. Linear Viscoelastic Solid

10.2.1. Anelasticity: Creep, Recovery, Relaxation, and Flow

Material media with a constitutive relation, such as given by Eq. (1.65), do not have internal energy losses. Ideal materials of this kind do not exist in nature, let alone in the solid earth, where the attenuation of seismic energy is an established fact. This deviation of material behavior from that of pure elastic substances is known as anelasticity.

Because the observed attenuation of seismic waves in the earth is an important source of information regarding the composition, state, and temperature of the deep interior, it is necessary to introduce damping into the elastic constitutive equations.

Anelastic damping usually depends, in a rather complicated manner, upon temperature, frequency and type of vibration. Because of the large number of physical mechanisms that contribute to this phenomenon, it is not possible to represent them all by a single modification of the constitutive equations.

The simplest model which is the basis for further generalization is based upon superposition of two mechanisms of resistance to deformation: *linear elasticity* and *Stokes' viscosity.* A material of this kind is called viscoelastic. A material is said to be *linearly viscoelastic* if stress components are linearly related to strain components at a given time and the principle of linear superposition holds. The second requirement states that the strain output from a combination of two arbitrary but different stress inputs applied at different times equals the sum of the strain outputs resulting from these stresses, each acting separately.

In this section it will be assumed that earth materials display linear anelesticity over the ranges of stress, strain, and time associated with the passage of seismic waves in the earth, with the possible exclusion of the close vicinity of earthquake faults. There are various strain responses to *constant stress*: Elastic strain is instantaneous and reversible but not necessarily linear. The limiting stress, above which the behavior is no longer elastic, is called the *elastic limit.* Beyond the elastic limit the strain does not disappear after removal of the stress.

Viscoelastic behavior introduces a number of new concepts that have been hitherto absent in the elastic regime. *Creep* is the slow continuous deformation of a material under constant stress (Fig. 10.4). As the stress is removed, the strain gradually diminishes through a process known as *recovery.* If the recovery is complete, we speak of *elastic creep.* This will occur for stresses less than the strength of the material. *Elastic flow* refers to a situation where the recovery is partial. If there is no recovery at all, there is *flow.* Flow in which the strain rate is linear with the stress is *viscous flow.* Flow in which the strain rate is nonlinear with the stress is a *plastic flow.* A creep that occurs in an increasing rate may terminate in *rupture.*

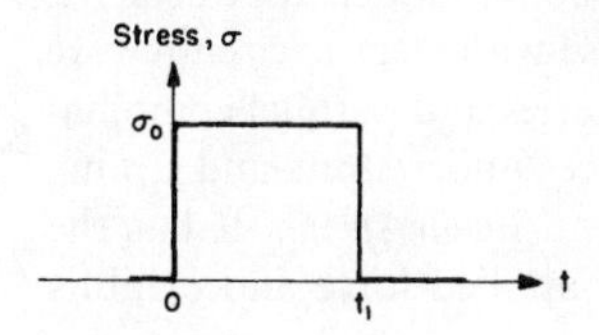

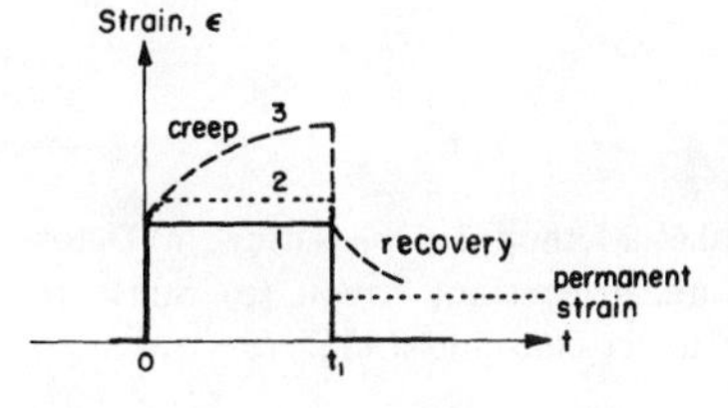

Figure 10.4. Strain behavior under constant stress for various substances: (1) Instantaneous elastic response. (2) Plastic response resulting in a permanent residual strain. (3) Viscoelastic creep and recovery.

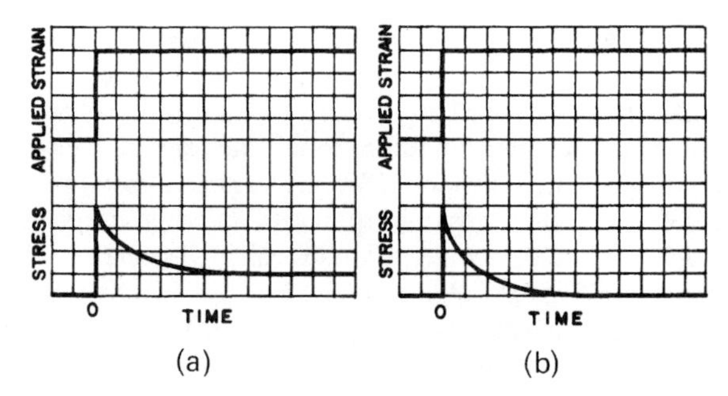

Figure 10.5. Stress relaxation at constant strain for a substance having characteristic of (a) elastic flow; (b) elastic creep.

In viscoelastic materials subjected to a constant strain, the stress will gradually diminish in a process known as *relaxation* (Fig. 10.5). The creep and relaxation behaviors of materials are tested in the following manner: To subject the material to creep we apply to it a constant stress and measure the deformation (extension, torsion, etc.) as a function of time (Fig. 10.6a). In a relaxation test, a constant deformation is imposed (say, extension) and the stress required to keep the deformation is measured as a function of time. Figure 10.6 shows the experimental arrangements needed for these tests. In Fig. 10.6a a fixed load P is applied to a specimen, and its relative decrease in length $(\Delta l/l_0)$ is measured as a function of time. In Fig. 10.6b, in contrast, the initial deformation $(\Delta l_0/l_0)$ is kept constant and the load is measured by means of calibrated springs.

10.2.2. Mechanical Viscoelastic Analogs

We wish next to construct the simplest one-dimensional models that bear the characteristics of viscoelastic materials. As is the case with electric circuitry, we shall use basic lumped elements which, in various series and parallel combinations, will simulate the behavior of various substances under stress and strain.

The simplest representative of linear elasticity is a *linear spring*. It has the property that the extension is proportional to the applied force and exhibits

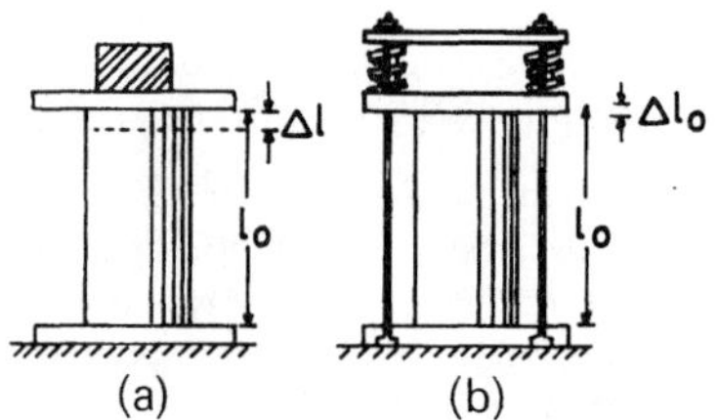

Figure 10.6. Methods for measuring: (a) Deformation under constant stress. (b) Stress relaxation under constant strain.

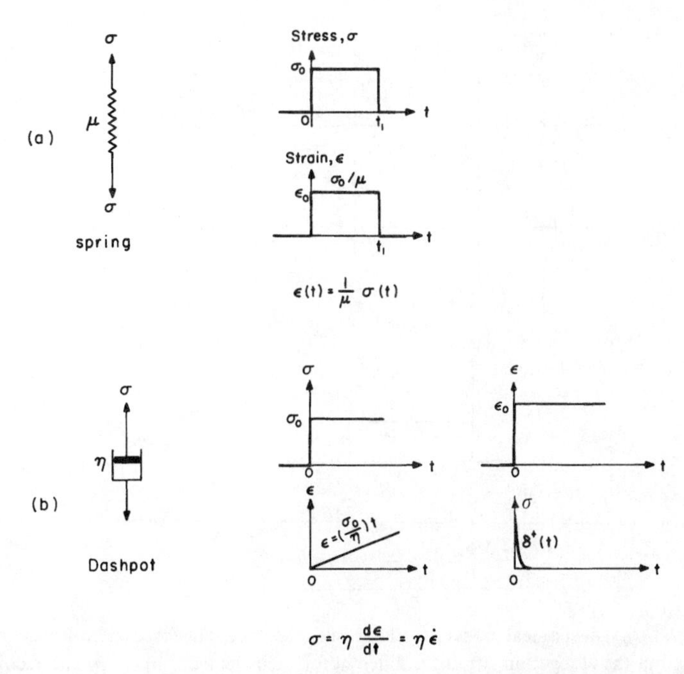

Figure 10.7. The basic mechanical elements and their strain response under constant stress and vice versa. (a) Spring analog of a perfectly elastic body. (b) Dashpot analog of a perfectly viscous fluid.

instantaneous elasticity and instantaneous recovery. It is shown in Fig. 10.7a. Time plays no role in determining the extension of the spring, because the application of the force leads to immediate extension. Conversely, if the force is removed, the spring reverts instantaneously to its unloaded state. This behavior is also characteristic of elastic solids: Stress depends only on the deformation and not at all on the rate of strain. Moreover, there is a fiducial state to which the elastic solid reverts when stress is removed. Note that the applied force serves as an analog of shear stress as well as tensile stress. This analog, like the others that we shall shortly describe, models the behavior of the body only at a given point. When the body is homogeneous, it may serve to model also a finite mass. Although the spring is a one-dimensional element, μ will denote here stress (force/area) and not stiffness (force/length).

The mechanical model that is analogous to a viscous fluid is the *linear dashpot element* shown in Fig. 10.7b. It is used to represent Stokes' linear viscosity. Indeed, this element has the property that its rate of extension (velocity of piston) is proportional to the applied force. Therefore, when subjected to a step of constant stress, it will deform continuously at a constant rate. If we substitute "rate of shearing strain" for "rate of extension" and "shearing stress" for "applied force," the dashpot complies with the basic property of a viscous fluid.

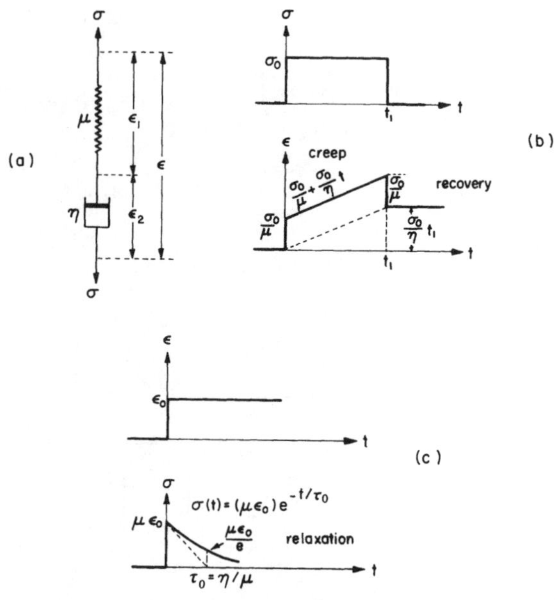

Figure 10.8. (a) Mechanical model for a Maxwell substance. The force on both elements is the same but the elongation (strain) is different for each element. (b) Creep and recovery. (c) Stress relaxation under constant strain. During the relaxation time τ_0, the stress falls to e^{-1} of its initial value.

However, when a step of constant strain is imposed on the dashpot, the behavior of the stress is indicated by a delta function in time. Because infinite stress is not realistic, it is impossible to impose an instantaneous finite deformation upon the dashpot. From these two basic elements we shall now compose models for two basic viscoelastic materials. The energy stored in the spring, i.e., the potential strain energy is $\sigma^2/2\mu$, whereas the rate of energy dissipation in the dashpot is $(1/\eta)\sigma^2$. In a complex spring–dashpot system we must sum over all springs and dashpots.

10.2.2.1. Maxwell Model. The mechanical analog for the Maxwell model is a linear spring connected in *series* with a dashpot. The force acting on both elements is always the same but the elongation distributes itself as a function of time, depending on the load. The model is shown in Fig. 10.8. In this material, a given stress σ generates at the same time two types of strains: an *elastic strain* $\varepsilon_1 = (1/\mu)\sigma$ and a *strain rate* $\dot{\varepsilon}_2 = (1/\eta)\sigma$, also proportional to the same stress.[1]

The *total rate of strain* is therefore

$$\dot{\varepsilon} = \dot{\varepsilon}_1 + \dot{\varepsilon}_2 = \frac{1}{\mu}\dot{\sigma} + \frac{1}{\eta}\sigma. \tag{10.34}$$

[1] Here μ denotes the elastic modulus of the spring.

Equation (10.34) can be viewed as a differential equation for σ, given ε and the constitutive parameters μ and η.

Defining the *relaxation time*

$$\tau_0 = \frac{\eta}{\mu},$$

the integration of Eq. (10.34) yields

$$\sigma = \mu e^{-t/\tau_0}\left[\varepsilon(0) + \int_0^t \dot{\varepsilon}(t)e^{t/\tau_0}\,dt\right]. \tag{10.35}$$

Integration by parts renders an alternative expression

$$\frac{\sigma}{\mu} = \varepsilon - \frac{1}{\tau_0}\int_0^t \varepsilon(\theta)e^{-(t-\theta)/\tau_0}\,d\theta. \tag{10.36}$$

The strain as a function of stress is given by a direct integration of Eq. (10.34)

$$\varepsilon(t) = \frac{1}{\mu}\sigma(t) + \frac{1}{\eta}\int_0^t \sigma(\theta)d\theta. \tag{10.37}$$

We examine a number of typical cases:

1. A step of stress, $\sigma = \sigma_0 H(t)$. Directly from Eq. (10.37)

$$\varepsilon(t) = \frac{\sigma_0}{\mu}\left[1 + \frac{t}{\tau_0}\right]H(t), \qquad \sigma_0 = \sigma(0). \tag{10.38}$$

2. A stress "boxcar," $\sigma = \sigma_0[H(t) - H(t - t_1)]$. The resulting strain is

$$\varepsilon(t) = \begin{cases} \dfrac{\sigma_0}{\mu}\left[1 + \dfrac{t}{\tau_0}\right]H(t), & t < t_1 \\[2ex] \dfrac{\sigma_0}{\eta}t_1, & t > t_1. \end{cases} \tag{10.39}$$

Note that case 2 includes case 1 as a special case ($t_1 \to \infty$) and is shown in Fig. 10.8b. There is a linear *creep* up to $t = t_1$ followed by a *permanent residual strain* when the stress is removed.

3. Harmonic stress cycle, $\sigma = \sigma_0 \sin \omega_0 t$. Equation (10.37) yields

$$\begin{aligned} \varepsilon(t) &= \frac{\sigma_0}{\mu}\left[\sin \omega_0 t - \frac{1}{Q}\cos \omega_0 t + \frac{1}{Q}\right] \\ &= \frac{\sigma_0}{\mu}\left\{\left[1 + \frac{1}{Q^2}\right]^{1/2} \sin[\omega_0 t - \chi_0] + \frac{1}{Q}\right\}, \end{aligned} \tag{10.40}$$

$$\tan \chi_0 = \frac{1}{Q}, \qquad Q = \omega_0 \tau_0 = \omega_0 \frac{\eta}{\mu}.$$

The strain lags behind the activating stress by the angle $\tan^{-1} 1/Q \simeq 1/Q$ when $Q \gg 1$. Note that the algebraic curve relating ε to σ, after time has been eliminated, is a canonical ellipse whose axes have been rotated by an angle χ_0 relative to the Cartesian $\varepsilon - \sigma$ axes.

4. A strain step, $\varepsilon = \varepsilon_0 H(t)$. Equation (10.36) renders

$$\sigma(t) = \mu\varepsilon_0 e^{-t/\tau_0}. \tag{10.41}$$

Equation (10.41) describes the *stress relaxation* phenomenon under constant strain. This phenomenon is shown in Fig. 10.8. The *rate of stress* at $t = 0$ is $\dot{\sigma}(0) = -\sigma_0(\mu/\eta) = -(\sigma_0/\tau_0)$. The intercept of the tangent at $t = 0$ with the time axis is τ_0. Therefore, when the model is subjected to a sudden extension that is then maintained, both elements experience the same force at all times, but the extension is initially only in the spring because a finite force cannot cause a finite extension of the dashpot in infinitesimal time. As time proceeds, the spring contracts, pulling out the dashpot to maintain a constant elongation of the entire system. When the spring has contracted to its equilibrium position, the internal deformation process stops and the force required to maintain the elongation has relaxed completely.

10.2.2.2. Kelvin–Voigt Model. In this model the spring and the dashpot are connected in parallel (Fig. 10.9a). If a constant force is applied to the model and maintained thereafter (say, by attaching a weight at $t = 0$), this stress will produce no immediate displacement because the dashpot, arranged in parallel with the spring, will not move instantaneously. Instead, a deformation will gradually build up while the spring takes an increasingly greater share of the load. A maximum value of the extension will eventually be reached, which is determined by the magnitude of the applied force. As the stress is suddenly dropped to zero, the strain again lags behind because of the exponential relaxation of the dashpot. No permanent deformation remains after full recovery.

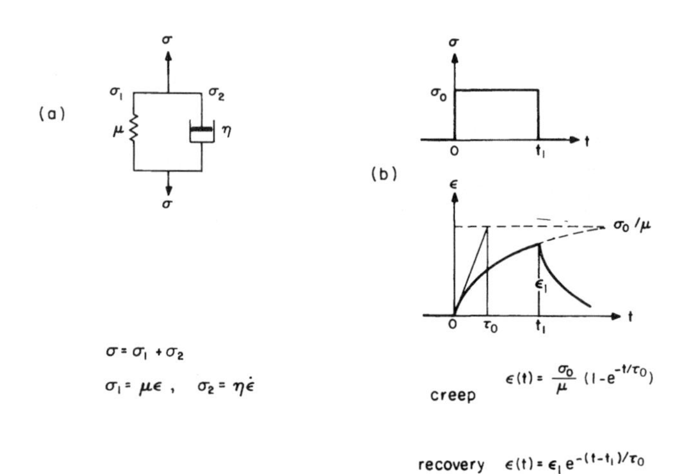

Figure 10.9. (a) Mechanical model for a Kelvin–Voigt substance. The strain is the same on both elements but the forces are different. (b) Creep and recovery.

To put these ideas in mathematical terms, we first suppose that the system is displaced by a time-dependent displacement $x(t)$, common to both elements. Different forces will develop in each "arm" of the system. The elastic "arm," by Hooke's law, will maintain a force $f_e = kx$, where k is the stiffness of the spring. The viscous element will maintain a viscous force $f_v = k'(\partial x/\partial t)$. The total force is

$$F = f_e + f_v = kx + k' \frac{\partial x}{\partial t}. \qquad (10.42)$$

To go one step further, we substitute the applied force with stress and the spring displacement with strain. Then, the total stress is composed of an elastic stress $\sigma_1 = \mu\varepsilon$ and a viscous stress $\sigma_2 = \eta\dot{\varepsilon}$. Together

$$\sigma = \sigma_1 + \sigma_2 = \mu\varepsilon + \eta\dot{\varepsilon}. \qquad (10.43)$$

Whereas, in the Maxwell model the force is the same on both elements, here the force is different but the strain is the same on both "arms" at all times. If $\sigma(t)$ is given, the strain is obtained by solving the differential equation (10.43) for $\varepsilon(t)$. The result is

$$\varepsilon(t) = \varepsilon(0)e^{-t/\tau_0} + \frac{1}{\eta}\int_0^t \sigma(\theta)e^{-(t-\theta)/\tau_0}\, d\theta. \qquad (10.44)$$

We consider two examples.

10.2.2.2.1. CREEP UNDER CONSTANT STRESS: $\sigma = \sigma_0 H(t)$, $\varepsilon(0) = 0$. Equation (10.44) yields

$$\varepsilon(t) = \frac{\sigma_0}{\mu}[1 - e^{-t/\tau_0}]. \qquad (10.45)$$

As shown in Fig. 10.9b, $\varepsilon(t)$ increases with decreasing rate and approaches asymptotically the value σ_0/μ. As stated already above, the stress is first carried entirely by the viscous element η. Under the stress, the viscous element elongates, transferring a greater and greater portion of the stress to the elastic element. Finally, the entire stress is carried by the elastic element. The initial strain rate is $\dot{\varepsilon}(0) = \sigma_0/\eta$. If the strain were to increase at this rate it would reach the asymptotic value at time τ_0.

If the stress is removed at time $t = t_1$, a *recovery* will take place. We put $\sigma(t) = \sigma_0[H(t) - H(t - t_1)]$ in Eq. (10.44) and obtain, for $t > t_1$,

$$\varepsilon(t) = \frac{\sigma_0}{\mu} e^{-t/\tau_0}(e^{t_1/\tau_0} - 1) = \varepsilon_1 e^{-(t-t_1)/\tau_0}. \qquad (10.46)$$

The situation is displayed in Fig. 10.9b.

10.2.2.2.2. RELAXATION UNDER CONSTANT STRAIN: $\varepsilon = \varepsilon_0 H(t)$. From Eq. (10.43)

$$\sigma(t) = \varepsilon_0[\mu H(t) + \eta\delta(t)]. \qquad (10.47)$$

Hence, the Kelvin–Voigt model does not show a time-dependent relaxation. The presence of the delta function is explained by the fact that an infinite stress

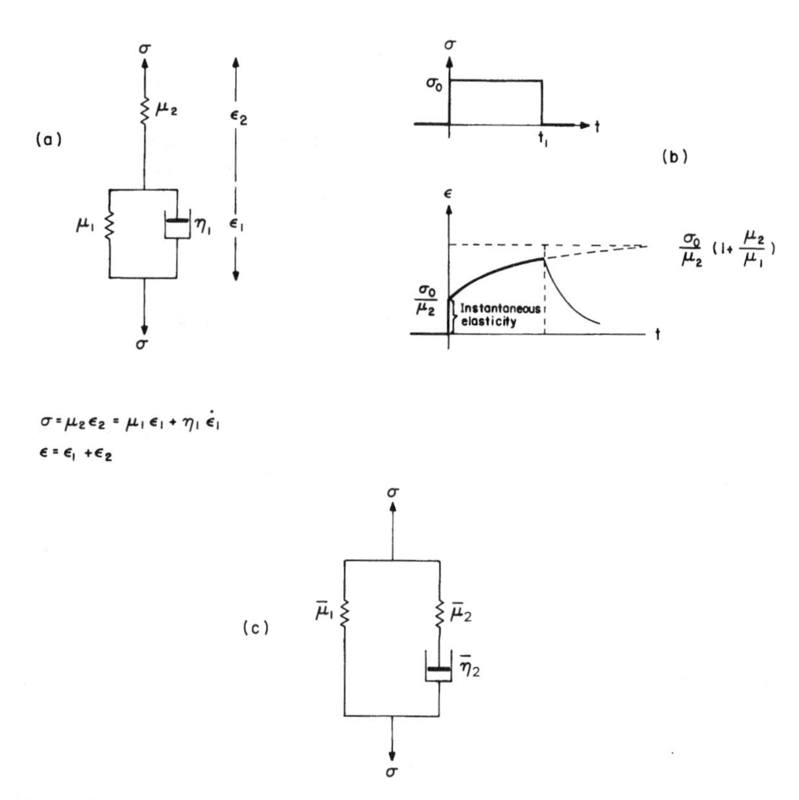

Figure 10.10. (a) Mechanical model for a standard linear solid. (b) Creep and recovery. (c) Nonuniqueness of the analog representation.

is needed in order to accommodate an abrupt change of strain in the dashpot. At $t > 0$, the entire stress is carried by the spring.

Neither the Maxwell nor the Kelvin–Voigt model is sufficient to account for the behavior of most viscoelastic materials. A more satisfactory model for the behavior of the earth's crust and mantle under stresses and strains associated with seismic vibrations, is presented next.

10.2.2.3. Standard Linear Solid (also called a three-element solid). When a stress is suddenly applied to a Kelvin–Voigt solid there is no instantaneous or initial strain. The strain gradually approaches an asymptotic value. Rocks, however, do usually build up an instantaneous strain upon sudden application of stress. Therefore, a more appropriate mechanical model is a Voigt element (μ_1, η_1) connected in series with a spring μ_2 (Fig. 10.10a). The material has an instantaneous elasticity $\varepsilon(0^+) = \sigma_0/\mu_2$ and also an asymptotic elastic behavior for $\sigma = \sigma_0 H(t)$ of the amount $\varepsilon(\infty) = \sigma_0(1/\mu_1 + 1/\mu_2)$. The equations that govern this combination are

$$\sigma = \mu_1\varepsilon_1 + \eta_1\dot{\varepsilon}_1 = \mu_2\varepsilon_2, \qquad \varepsilon = \varepsilon_1 + \varepsilon_2. \tag{10.48}$$

Eliminating ε_1 and ε_2 from these equations, we get

$$\sigma + \tau_\sigma \dot{\sigma} = M_R[\varepsilon + \tau_\varepsilon \dot{\varepsilon}], \tag{10.49}$$

where

$$\tau_\sigma = \frac{\eta_1}{\mu_1 + \mu_2}, \qquad \tau_\varepsilon = \frac{\eta_1}{\mu_1} > \tau_\sigma, \qquad M_R = \frac{\mu_1 \mu_2}{\mu_1 + \mu_2}.$$

Simple manipulations lead from Eq. (10.49) to the solution

$$\varepsilon(t) = \frac{1}{M_R}\left[\sigma(t) - \left(1 - \frac{\tau_\sigma}{\tau_\varepsilon}\right)\int_0^t e^{-(t-\theta)/\tau_\varepsilon}\dot{\sigma}(\theta)d\theta\right]. \tag{10.50}$$

As $\mu_2 \to \infty$ the standard linear solid reduces to the Kelvin–Voigt solid. The material exhibits both creep and relaxation. Setting $\sigma = \sigma_0 H(t)$ in Eq. (10.50), we get

$$\begin{aligned}\varepsilon(t) &= \frac{\sigma_0}{M_R}\left[1 - \left(1 - \frac{\tau_\sigma}{\tau_\varepsilon}\right)e^{-t/\tau_\varepsilon}\right]H(t) \\ &= \frac{\sigma_0}{\mu_2}\left[1 + \left(\frac{\tau_\varepsilon}{\tau_\sigma} - 1\right)\{1 - e^{-t/\tau_\varepsilon}\}\right]H(t),\end{aligned} \tag{10.51}$$

with the limiting values

$$\varepsilon(\infty) = \frac{\sigma_0}{M_R}, \qquad \varepsilon(0) = \frac{\sigma_0}{\mu_2}. \tag{10.51a}$$

The dependence of $\varepsilon(t)$ on time is shown in Fig. 10.10b. Note that the final value of the stress/strain ratio is M_R. This quantity is known as the *relaxed elastic modulus.* The initial value of the stress/strain ratio is μ_2 and is known as the *unrelaxed elastic modulus.* The solution of Eq. (10.49) for $\varepsilon = \varepsilon_0 H(t)$ is obtained from Eq. (10.51) by exchanging the roles of τ_σ and τ_ε and substituting M_R^{-1} for M_R. The stress relaxation with time is then

$$\sigma(t) = M_R \varepsilon_0\left[1 + \left(\frac{\tau_\varepsilon}{\tau_\sigma} - 1\right)e^{-t/\tau_\sigma}\right]H(t). \tag{10.52}$$

Consequently

$$\sigma(0) = \mu_2 \varepsilon_0, \qquad \sigma(\infty) = M_R \varepsilon_0.$$

τ_σ is known as the *stress relaxation time under constant strain,* and τ_ε is the *strain relaxation time under constant stress.*

10.2.2.3.1. HARMONIC STRESS CYCLE: $\sigma = Se^{i\omega t}$, $\varepsilon = Ee^{i\omega t}$. From the basic Eq. (10.49)

$$S[1 + i\omega\tau_\sigma] = M_R E[1 + i\omega\tau_\varepsilon]. \tag{10.53}$$

Therefore

$$\frac{S}{E} = M_R \frac{1 + i\omega\tau_\varepsilon}{1 + i\omega\tau_\sigma} = \mu_2 \frac{\mu_1 + i\omega\eta_1}{(\mu_1 + \mu_2) + i\omega\eta_1} = Ke^{i\delta}, \tag{10.54}$$

where

$$\tan\delta = \frac{\omega(\tau_\varepsilon - \tau_\sigma)}{1 + \omega^2\tau_\varepsilon\tau_\sigma} = \frac{1}{Q}. \tag{10.55}$$

Here again, δ measures the lag of the strain behind the stress and provides a measure of damping.

Note that the representation of the standard linear solid by mechanical analogs is not unique. Suppose that we consider the model shown in Fig. 10.10c. The associated stress–strain relation is

$$\dot{\sigma} + \frac{\bar{\mu}_2}{\bar{\eta}_2}\sigma = (\bar{\mu}_1 + \bar{\mu}_2)\dot{\varepsilon} + \frac{\bar{\mu}_1\bar{\mu}_2}{\bar{\eta}_2}\varepsilon$$

or

$$\sigma + \tau'_\sigma\dot{\sigma} = M'_R[\varepsilon + \tau'_\varepsilon\dot{\varepsilon}], \tag{10.56}$$

where

$$M'_R = \bar{\mu}_1, \qquad \tau'_\sigma = \frac{\bar{\eta}_2}{\bar{\mu}_2}, \qquad \tau'_\varepsilon = \frac{\bar{\eta}_2}{\bar{\mu}_2}\left(1 + \frac{\bar{\mu}_2}{\bar{\mu}_1}\right). \tag{10.57}$$

We find that the two sets of constants for the two analogs can be adjusted such that both models represent the same standard linear solid.

In the last three sections (10.2.2.1–10.2.2.3) we have evaluated the strain response of certain simple linear materials to a step of stress $\sigma_0 H(t)$. All results can be written as

$$\varepsilon(t) = \sigma_0\phi_c(t)H(t), \tag{10.58}$$

where $\phi_c(t)$ is called the *creep compliance.* It is the creep strain per unit applied stress. Creep compliance is a material property. The values of $\phi_c(t)$ for some materials are given below

$$\begin{array}{lll}
\text{Elastic solid} & \dfrac{1}{\mu} & \\[2ex]
\text{Maxwell substance} & \dfrac{1}{\mu} + \dfrac{1}{\eta}t & \\[2ex]
\text{Kelvin–Voigt substance} & \dfrac{1}{\mu}[1 - e^{-t/\tau_0}], & \tau_0 = \dfrac{\eta}{\mu} \\[2ex]
\text{Standard linear solid} & \dfrac{1}{M_R}\left[1 - \left(1 - \dfrac{\tau_\sigma}{\tau_\varepsilon}\right)e^{-t/\tau_\varepsilon}\right]. &
\end{array} \tag{10.58a}$$

In contrast, the stress per unit applied strain is called the *relaxation modulus* $\psi_c(t)$,

$$\sigma(t) = \varepsilon_0\psi_c(t)H(t), \qquad \varepsilon(t) = \varepsilon_0 H(t). \tag{10.59}$$

It is useful for further applications to separate from $\phi_c(t)$ the part $(\varepsilon_\infty \sigma_0^{-1})$ which expresses the instantaneous component of strain. Therefore, we put

$$\phi_c(t) = \varepsilon_\infty \sigma_0^{-1} + \phi(t), \qquad \phi(0) = 0, \tag{10.60}$$

[2]

where $\phi(t)$ is the *creep function* and ε_∞ is the instantaneous strain produced by the application of a unit stress. Likewise, we write

$$\psi_c(t) = Y_0 + \psi(t), \qquad \psi(\infty) = 0, \qquad Y_\infty = Y_0 + \psi(0), \tag{10.61}$$

where $\psi(t)$ is the *relaxation function,* Y_0 is the static stress produced by the application of a unit strain, and Y_∞ is the instantaneous stress produced by the application of a unit strain. For a Maxwell element, $Y_0 = 0$.

10.2.3. Generalized Linear Solid in One Dimension

Neither the Maxwell model nor the Kelvin–Voigt model accurately represent the behavior of most viscoelastic materials. The Kelvin–Voigt model cannot accommodate an abrupt change in strain and does not show residual strain after unloading, and the Maxwell model has no creep features. Both models have finite initial strain rates, $\dot{\varepsilon}(0)$, whereas the apparent initial strain rate for most materials is very rapid. In addition, the models discussed cannot describe the behavior of many viscoelastic materials over a wide range of variables, especially for both large and small values of time.

Furthermore, real materials behave as though they have several relaxation times. To deal with this situation, a finite number of Kelvin–Voigt and/or Maxwell elements can be connected in series or parallel. First we notice that the combination of N Maxwell elements in series yields the constitutive equation

$$\dot{\varepsilon} = \dot{\sigma}\left[\sum_{i=1}^{N} \frac{1}{\mu_i}\right] + \sigma\left[\sum_{i=1}^{N} \frac{1}{\eta_i}\right].$$

The result is therefore a new Maxwell element with parameters

$$\mu = \left\{\sum_{i=1}^{N} \frac{1}{\mu_i}\right\}^{-1}, \qquad \eta = \left\{\sum_{i=1}^{N} \frac{1}{\eta_i}\right\}^{-1}.$$

A similar result occurs with N parallel Kelvin–Voigt models

$$\sigma = \varepsilon \sum_{i=1}^{N} \mu_i + \dot{\varepsilon} \sum_{i=1}^{N} \eta_i.$$

[2] In Eqs. (10.51)–(10.52) we have used $\varepsilon(\infty)$ and $\sigma(\infty)$ to indicate the value of these time functions at $t = \infty$. The symbols ε_0 and σ_0 indicate initially prescribed values of the respective entities. However, ε_∞ and Y_∞ are the high frequency limits of these entities which we discuss later in Section 10.3.1. In the present case $\varepsilon(0) = \varepsilon_\infty$ and $Y(\infty) = Y_0$.

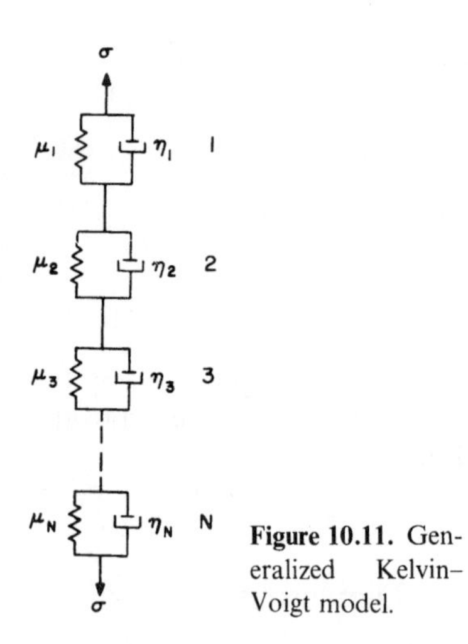

Figure 10.11. Generalized Kelvin–Voigt model.

If, however, we combine Kelvin–Voigt elements in series (Fig. 10.11), the stress σ on each unit is the same. If the applied stress is a step of stress $\sigma = \sigma_0 H(t)$, the total creep strain of the system is the sum of the creep strains of the individual Kelvin–Voigt elements

$$\varepsilon(t) = \sigma_0 \sum_{r=1}^{N} \frac{1}{\mu_r} [1 - e^{-t/\tau_r}] H(t), \tag{10.62}$$

where $\tau_r = \eta_r/\mu_r$ is the rth *retardation time*. The reciprocal rigidity is called *compliance* and is denoted by $J_r = 1/\mu_r$.

As the number of elements in the generalized model increases indefinitely in such a way that $J_r \to 0$ for all r and $\sum J_r \to$ constant > 0, the above equation becomes

$$\varepsilon(t) = H(t)\sigma_0 \int_0^\infty j(\tau)[1 - e^{-t/\tau}] d\tau. \tag{10.63}$$

The function $j(\tau)$ is known as the *retardation spectrum*. It is a distribution function of retardation times. Comparing Eq. (10.63) with Eq. (10.58) and noting that $\varepsilon(0) = 0$, we find the integral representation of the *creep function*:

$$\phi(t) = \int_0^\infty j(\tau)[1 - e^{-t/\tau}] d\tau \tag{10.64}$$

and, therefore,

$$\frac{d\phi(t)}{dt} = \int_0^\infty \left[\frac{1}{\tau} j(\tau)\right] e^{-t/\tau} \, d\tau. \tag{10.64a}$$

Note that because the creep function is assumed to be continuous, uniformly increasing function, Eq. (10.64) can be used as a definition of $j(\tau)$ without the need of the limiting process that led to Eq. (10.63).

If in Eq. (10.64) we put $z = 1/\tau$ and denote $j(\tau) = G(z)$, we have

$$\phi(t) = \int_0^\infty z^{-2} G(z)\{1 - e^{-tz}\} dz, \tag{10.65}$$

$$\dot{\phi}(t) = \int_0^\infty z^{-1} G(z) e^{-tz}\, dz. \tag{10.66}$$

Consequently,

$$z^{-1} G(z) = \mathscr{L}^{-1}\{\dot{\phi}(t)\}, \tag{10.67}$$

where $\mathscr{L}^{-1}$ is the inverse Laplace transform operator and t is considered the Laplace transform variable instead of the usual s.

In the same way we may show that a series connection of a large number of Maxwell elements yields the relaxation stress

$$\sigma(t) = H(t)\varepsilon_0 \int_0^\infty y(\tau) e^{-t/\tau}\, d\tau, \tag{10.67a}$$

where $y(\tau)$ is the *relaxation spectrum*. Comparison of Eqs. (10.59) and (10.67a) yields (with $Y_0 = 0$)

$$\psi(t) = \int_0^\infty y(\tau) e^{-t/\tau}\, d\tau. \tag{10.68}$$

10.2.3.1. Fourier Transforms of Creep and Relaxation Functions. Let the stress–strain relation be generalized into the form

$$P\sigma(t) = S\varepsilon(t), \tag{10.69}$$

where

$$P = p_0 + p_1 \frac{\partial}{\partial t} + \cdots + p_L \frac{\partial^L}{\partial t^L},$$

$$S = s_0 + s_1 \frac{\partial}{\partial t} + \cdots + s_M \frac{\partial^M}{\partial t^M},$$

and

$$\sigma = \dot{\sigma} = \ddot{\sigma} = \cdots = \sigma^{(L-1)} = 0 \qquad \text{at } t = \pm\infty,$$

$$\varepsilon = \dot{\varepsilon} = \ddot{\varepsilon} = \cdots = \varepsilon^{(M-1)} = 0 \qquad \text{at } t = \pm\infty.$$

We apply the Fourier transform to Eq. (10.69) with the initial conditions stated above and define

$$\hat{P}(i\omega) = \sum_{l=0}^{L} p_l (i\omega)^l,$$

$$\hat{S}(i\omega) = \sum_{m=0}^{M} s_m (i\omega)^m.$$

Assuming, without loss of generality, that $p_0 = 1$, we find

$$\hat{\varepsilon}(\omega) = J^*(i\omega)\hat{\sigma}(\omega) = \frac{\hat{P}(i\omega)}{\hat{S}(i\omega)}\hat{\sigma}(\omega)$$

$$= \frac{1 + p_1(i\omega) + p_2(i\omega)^2 + \cdots + p_L(i\omega)^L}{s_0 + s_1(i\omega) + s_2(i\omega)^2 + \cdots + s_M(i\omega)^M}\hat{\sigma}(\omega), \tag{10.70}$$

where $J^*(i\omega)$ is the *complex creep compliance*, $\hat{\varepsilon}(\omega) = \int_{-\infty}^{\infty} \varepsilon(t)e^{-i\omega t}\,dt$ and $\hat{\sigma}(\omega) = \int_{-\infty}^{\infty} \sigma(t)e^{-i\omega t}\,dt$. Expanding Eq. (10.70) into partial fractions

$$J^*(i\omega) = \sum_{r=1}^{M} \frac{\alpha_r}{\beta_r + i\omega}, \tag{10.71}$$

we substitute Eq. (10.71) back in Eq. (10.70) and apply the inverse Fourier transform to this equation. With the aid of the convolution theorem, we obtain

$$\varepsilon(t) = \int_{-\infty}^{t} \sigma(\tau)\bar{A}(t-\tau)d\tau = \int_{-\infty}^{t} \sigma(\tau)\sum_{r=1}^{M} \alpha_r e^{-\beta_r(t-\tau)}\,d\tau, \tag{10.72}$$

where

$$\bar{A}(t) = \frac{1}{2\pi}\int_{-\infty}^{\infty} J^*(i\omega)e^{i\omega t}\,d\omega. \tag{10.73}$$

The concepts of compliance and relaxation, as defined in Eqs. (10.58) and (10.59), can also be used in cases where the applied stresses and strains are periodic. Let $\sigma = \sigma_0 e^{i\omega t}$ and let the resulting strain be $\varepsilon = \varepsilon^* e^{i\omega t}$, where $\varepsilon^* = \varepsilon_0 e^{-i\delta}$. Inserting these expressions into Eq. (10.69), we obtain

$$\sigma_0[1 + p_1(i\omega) + p_2(i\omega)^2 + \cdots + p_L(i\omega)^L] = \varepsilon^*[s_0 + s_1(i\omega) + \cdots + s_M(i\omega)^M]$$

or

$$\frac{\varepsilon^*}{\sigma_0} = J^*(i\omega) = \frac{1 + p_1(i\omega) + \cdots + p_L(i\omega)^L}{s_0 + s_1(i\omega) + \cdots + s_M(i\omega)^M} = \hat{J}_1(\omega) - i\hat{J}_2(\omega). \tag{10.74}$$

Clearly

$$|J^*(i\omega)| = \frac{\varepsilon_0}{\sigma_0}, \qquad \tan\delta = \frac{\hat{J}_2}{\hat{J}_1}. \tag{10.75}$$

Similarly, the stress response $\sigma = \sigma^* e^{i\omega t} = \sigma_0 e^{i(\omega t + \delta)}$ to the input strain $\varepsilon = \varepsilon_0 e^{i\omega t}$ leads to the relation

$$\frac{\sigma^*}{\varepsilon_0} = Y^*(i\omega) = \frac{s_0 + s_1(i\omega) + \cdots + s_M(i\omega)^M}{1 + p_1(i\omega) + \cdots + p_L(i\omega)^L} = \hat{Y}_1(\omega) + i\hat{Y}_2(\omega), \tag{10.76}$$

where $Y^*(i\omega)$ is the *complex relaxation modulus* for which

$$\frac{\sigma_0}{\varepsilon_0} = |Y^*(i\omega)|, \qquad \tan\delta = \frac{\hat{Y}_2}{\hat{Y}_1}. \tag{10.77}$$

It then follows from Eq. (10.75) that for the same viscoelastic substance

$$J^*(i\omega)Y^*(i\omega) = 1, \qquad |J^*(i\omega)| = \frac{1}{|Y^*(i\omega)|}, \qquad \tan\delta = \frac{\hat{Y}_2}{\hat{Y}_1} = \frac{\hat{J}_2}{\hat{J}_1}. \tag{10.78}$$

Applying the Fourier transform to the relations

$$\sigma(t) = \sigma_0 H(t), \qquad \varepsilon(t) = \sigma_0 \phi_c(t) H(t),$$

we obtain

$$\hat{\sigma}(\omega) = \frac{\sigma_0}{i\omega}, \qquad \hat{\varepsilon}(\omega) = \sigma_0 \int_0^\infty \phi_c(t) e^{-i\omega t}\, dt.$$

Considering Eqs. (10.58) and (10.60), we find that the decomposition of $J^*(i\omega)$ into real and imaginary parts reveals that

$$\hat{J}_1(\omega) = \varepsilon_\infty \sigma_0^{-1} + J_1(\omega), \qquad \hat{J}_2(\omega) = J_2(\omega), \tag{10.79}$$

where

$$\begin{aligned} J(i\omega) &= J_1(\omega) - iJ_2(\omega) \\ &= i\omega \int_0^\infty \phi(t) e^{-i\omega t}\, dt = \int_0^\infty \dot{\phi}(t) e^{-i\omega t}\, dt \end{aligned} \tag{10.80}$$

is the part of the complex compliance associated with the phenomenon of creep. Likewise, if

$$\hat{Y}_1(\omega) = Y_0 + Y_1(\omega), \qquad \hat{Y}_2(\omega) = Y_2(\omega) \tag{10.81}$$

then

$$Y(i\omega) = Y_1(\omega) + iY_2(\omega) \tag{10.82}$$

is the part of the relaxation modulus associated with the phenomenon of relaxation. Clearly,

$$Y(i\omega) = i\omega \int_0^\infty \psi(t) e^{-i\omega t}\, dt = \psi(0) + \int_0^\infty \dot{\psi}(t) e^{-i\omega t}\, dt. \tag{10.83}$$

An important property of Eq. (10.78) is that it allows establishment of relations between the coefficients of Stokes' viscosity, the instantaneous and static elastic moduli, and the constants $\phi(\infty)$ and $\psi(0)$. To obtain these relations we first combine Eqs. (10.79) and (10.80), yielding

$$J^*(i\omega) = \varepsilon_\infty \sigma_0^{-1} + \int_0^\infty \dot{\phi}(t) e^{-i\omega t}\, dt. \tag{10.84}$$

Likewise, from Eqs. (10.82) and (10.61), we find

$$Y^*(i\omega) = Y_\infty + \int_0^\infty \dot{\psi}(t) e^{-i\omega t}\, dt. \tag{10.85}$$

The substitution of Eqs. (10.84) and (10.85) in Eq. (10.78) yields, for $\omega = \infty$,

$$Y_\infty \varepsilon_\infty \sigma_0^{-1} = 1 \tag{10.86}$$

and, for $\omega \to 0$,

$$[\varepsilon_\infty \sigma_0^{-1} + \phi(\infty)] Y_0 = 1. \tag{10.87}$$

Substituting $\varepsilon_\infty \sigma_0^{-1} = 1/Y_\infty = 1/\{Y_0 + \psi(0)\}$, Eq. (10.87) yields

$$\phi(\infty) = \frac{\psi(0)}{Y_0[Y_0 + \psi(0)]}. \tag{10.88}$$

In general,

$$\frac{(-1)}{Y_\infty} \int_0^\infty \dot{\psi}(t) e^{-i\omega t}\, dt = \frac{(\sigma_0/\varepsilon_\infty) \int_0^\infty \dot{\phi}(t) e^{-i\omega t}\, dt}{1 + (\sigma_0/\varepsilon_\infty) \int_0^\infty \dot{\phi}(t) e^{-i\omega t}\, dt}, \tag{10.89}$$

where Eqs. (10.78) and (10.86) have been used.

Other useful relations that follow readily from Eqs. (10.80) and (10.83) are

$$\begin{aligned} J_1(\omega) &= \omega \int_0^\infty \phi(t) \sin \omega t\, dt, \qquad J_2(\omega) = -\omega \int_0^\infty \phi(t) \cos \omega t\, dt, \\ Y_1(\omega) &= \omega \int_0^\infty \psi(t) \sin \omega t\, dt, \qquad Y_2(\omega) = \omega \int_0^\infty \psi(t) \cos \omega t\, dt. \end{aligned} \tag{10.90}$$

The inverse Fourier transform leads to the reciprocal relations

$$\begin{aligned} \phi(t) &= \frac{2}{\pi} \int_0^\infty \frac{J_1(\omega)}{\omega} \sin \omega t\, d\omega = -\frac{2}{\pi} \int_0^\infty \frac{J_2(\omega)}{\omega} \cos \omega t\, d\omega, \\ \psi(t) &= \frac{2}{\pi} \int_0^\infty \frac{Y_1(\omega)}{\omega} \sin \omega t\, d\omega = \frac{2}{\pi} \int_0^\infty \frac{Y_2(\omega)}{\omega} \cos \omega t\, d\omega. \end{aligned} \tag{10.91}$$

A comparison of Eq. (10.73) with Eq. (10.84) renders

$$\bar{A}(t) = \varepsilon_\infty \sigma_0^{-1} \delta(t) + A(t), \tag{10.92}$$

where

$$A(t) = \frac{\partial \phi}{\partial t}$$

is the *rate of creep*.

10.2.3.2. Energy Loss under a Loading Cycle. In Eq. (10.33) the dimensionless dissipation parameter Q has been defined in terms of the relative energy loss per cycle of a damped harmonic oscillator. Then, in Eq. (10.40), we derived the response of a Maxwell substance to a harmonic stress cycle. Let us now generalize that result to any linear viscoelastic solid. The energy loss per cycle is

$$\Delta W = \int_0^T \sigma(t) \left(\frac{\partial \varepsilon}{\partial t}\right) dt. \tag{10.93}$$

Substituting $\sigma = \sigma_0 \sin \omega t$, $\dot{\varepsilon} = \varepsilon_0 \omega \cos(\omega t - \delta)$, where δ is given by Eq. (10.75), we obtain via Eq. (10.77)

$$\Delta W = \pi \sigma_0 \varepsilon_0 \sin \delta = \pi \varepsilon_0^2 \hat{Y}_2. \tag{10.94}$$

The maximum average energy W that the substance can store in one cycle is obtained when the stress and strain are in phase ($\delta = 0$)

$$W = \frac{1}{T} \int_0^T \sigma \varepsilon \, dt = \frac{1}{T} \int_0^T \sigma_0 \varepsilon_0 \sin^2 \omega t \, dt = \frac{1}{2} \sigma_0 \varepsilon_0.$$

Therefore,

$$\frac{1}{Q} = \frac{1}{2\pi} \frac{\Delta W}{W} = \sin \delta = \frac{\hat{J}_2/\hat{J}_1}{[1 + (\hat{J}_2/\hat{J}_1)^2]^{1/2}}. \tag{10.95}$$

10.2.4. Boltzmann Superposition Principle in One Dimension

Consider the stress–strain relation in one dimension. If we apply at time τ a stress $\sigma(\tau)$ to a viscoelastic material and maintain it for a time $\delta\tau$, the contribution of this stress to the residual strain at a later time t will be given by $\sigma(\tau)M(t - \tau)\delta\tau$, where $M(t)$ is a *memory function* characteristic of the medium. The *total* contribution to strain at any time t is given by the sum of all these infinitesimal contributions, namely

$$\varepsilon(t) = \int_{-\infty}^{t} \sigma(\tau) M(t - \tau) d\tau, \tag{10.96}$$

where $M \equiv \bar{A}$ according to Eq. (10.72). We shall henceforth assume that all stresses and strains are causal and therefore take the lower integration limit at $t = 0$.

Equation (10.96) can be derived in a different way: A constant stress σ_1 is applied at time $t = \tau_1$, namely, $\sigma(t) = \sigma_1 H(t - \tau_1)$. According to Eq. (10.58), the strain is $\varepsilon(t) = \sigma_1 \phi_c(t - \tau_1) H(t - \tau_1)$. A series of N stress steps $\Delta\sigma_n = \sigma_{n+1} - \sigma_n$ added consecutively at times $\tau_N > \tau_{N-1} > \cdots > \tau_{n+1} > \tau_n > \cdots > \tau_1$ will then induce the total strain

$$\varepsilon(t) = \sum_{n=1}^{N} \varepsilon_n(t - \tau_n) = \sum_{n=1}^{N} \Delta\sigma_n \phi_c(t - \tau_n) H(t - \tau_n)$$

$$\Rightarrow \int_0^t \phi_c(t - \tau) d\sigma(\tau) = \int_0^t \phi_c(t - \tau) \left[\frac{d\sigma(\tau)}{d\tau}\right] d\tau. \tag{10.97}$$

The use of Eq. (10.60) then leads to

$$\varepsilon(t) = \int_0^t \frac{d\sigma(\tau)}{d\tau} \left[\varepsilon_\infty \sigma_0^{-1} + \phi(t - \tau)\right] d\tau. \tag{10.98}$$

An alternative form of Eq. (10.98) is obtained by employing integration by parts. The result is

$$\varepsilon(t) = \varepsilon_\infty \frac{\sigma(t)}{\sigma_0} - \int_{0^-}^{t} \sigma(\tau) \frac{\partial \phi(t-\tau)}{\partial \tau} d\tau$$

$$= \varepsilon_\infty \frac{\sigma(t)}{\sigma_0} + \int_{0}^{t} \sigma(\tau) A(t-\tau) d\tau \tag{10.99}$$

in accord with Eqs. (10.92) and (10.96).

The same arguments hold when a strain step is applied. The stress relaxation under arbitrarily prescribed strain $\varepsilon(t)$ is

$$\sigma(t) = \int_0^t \frac{d\varepsilon(\tau)}{d\tau} [Y_0 + \psi(t-\tau)] d\tau$$

$$= Y_0 \varepsilon(t) + \int_0^t \frac{d\varepsilon(\tau)}{d\tau} \psi(t-\tau) d\tau. \tag{10.100}$$

Using Eq. (10.61) this becomes

$$\sigma(t) = Y_\infty \varepsilon(t) - \int_0^t \varepsilon(\tau) \frac{\partial \psi(t-\tau)}{\partial \tau} d\tau. \tag{10.101}$$

Figure 10.12 shows a schematic diagram of all anelastic models treated so far.

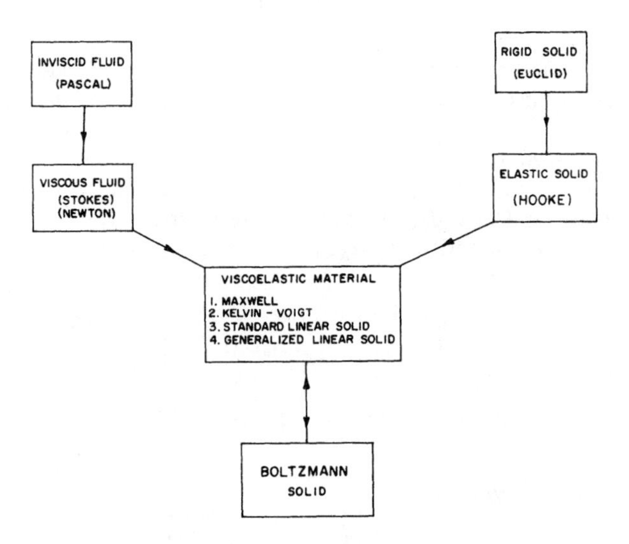

Figure 10.12. Summary of anelastic models of common use in seismology.

EXAMPLE 10.1: Scott–Blair Stress–Strain Law

Consider a creep function of the form

$$\phi(t) = \frac{t^{\nu}}{K\Gamma(1+\nu)} H(t), \qquad 0 < \nu \leq 1, \qquad K > 0. \tag{10.1.1}$$

If $\nu = 0$, this is an elastic solid. If $\nu = 1$, it is a viscous fluid. The case $\nu = 0$ will be excluded in this example. To show that $\phi(t)$ is a creep function of a linear viscoelastic substance it is sufficient to show that there exists a corresponding retardation spectrum that is nonnegative. Comparing Eq. (10.64a) with the derivative of Eq. (10.1.1), we obtain

$$\frac{t^{\nu-1}}{K\Gamma(\nu)} H(t) = \int_0^{\infty} \left[\frac{1}{p} j\left(\frac{1}{p}\right)\right] e^{-pt}\, dp, \tag{10.1.2}$$

where we have substituted $p = 1/\tau$. The evaluation of the inverse Laplace transform yields

$$j\left(\frac{1}{p}\right) = \frac{p^{1-\nu}}{K\Gamma(1-\nu)\Gamma(\nu)}. \tag{10.1.3}$$

Clearly $j(1/p) > 0$ for $0 < p < \infty$ and, therefore, Eq. (10.1.1) is a creep function of a linear viscoelastic substance.

Going back to Eq. (10.99) we substitute therein the expression for $\phi(t)$, obtaining

$$\varepsilon(t) = \frac{1}{K\Gamma(\nu)} \int_{-\infty}^{t} \sigma(\tau)(t-\tau)^{\nu-1}\, d\tau. \tag{10.1.4}$$

This convolution integral can be inverted to yield the stress–strain law

$$\sigma(t) = \frac{K}{\Gamma(1-\nu)} \int_{-\infty}^{t} (t-\tau)^{-\nu} \left[\frac{d\varepsilon(\tau)}{d\tau}\right] d\tau \tag{10.1.5}$$

which, for integral values $\nu = n$, reduces to the form

$$\sigma(t) = K \frac{d^n}{dt^n} \varepsilon(t) \tag{10.1.6}$$

known as the Scott–Blair stress–strain law.

From Eqs. (10.1.5) and (10.100) we find the relaxation function

$$\psi(t) = \frac{K}{\Gamma(1-\nu)} t^{-\nu} H(t). \tag{10.1.7}$$

The complex creep compliance and relaxation modulus are obtained from the Fourier transforms of $\phi(t)$ and $\psi(t)$, respectively

$$J(i\omega) = i\omega \int_0^{\infty} \phi(t) e^{-i\omega t}\, dt = \frac{1}{K} (i\omega)^{-\nu}, \tag{10.1.8}$$

$$Y(i\omega) = i\omega \int_0^{\infty} \psi(t) e^{-i\omega t}\, dt = K(i\omega)^{\nu}. \tag{10.1.9}$$

10.2.5. Linear Anelasticity in Three Dimensions

So far we have treated viscoelasticity in a single dimension only. To account for the damping of seismic waves in the earth, a more general treatment is needed. Starting from the generalized Hooke's law [cf. Eq. (1.65)] and recalling the viscosity tensor associated with a Newtonian fluid [cf. Eq. (9.3)], we write the stress–strain relations for a Kelvin–Voigt solid in the form[3]

$$\sigma_{ij} = C_{ijlm}\varepsilon_{lm} + D_{ijlm}\frac{\partial \varepsilon_{lm}}{\partial t}. \tag{10.102}$$

Symbolically

$$\mathfrak{T} = \mathbf{C} : \mathfrak{E} + \mathbf{D} : \frac{\partial \mathfrak{E}}{\partial t} = \left(\mathbf{C} + \mathbf{D}\frac{\partial}{\partial t}\right) : \mathfrak{E}. \tag{10.103}$$

In isotropic materials, the viscosity tensor, like the tensor of elastic moduli, C_{ijkl}, will have only 21 nonvanishing components, of which only two are independent [cf. Eq. (1.67)]. The viscous behavior of the substance is defined by the Stokes constants of viscosity $\bar{\lambda}$ and η [Eq. (9.7)], which we shall rename here λ' and μ'. The overall viscoelasticity of the material is governed by the four functions λ, μ, λ', and μ'. In the light of Eq. (10.103) we may formally draw an analogy with the theory of pure elasticity provided we set the correspondence

$$\lambda \to \lambda + \lambda'\frac{\partial}{\partial t}, \qquad \mu \to \mu + \mu'\frac{\partial}{\partial t}.$$

The stress–strain relations for a Kelvin–Voigt body could then be written as

$$\mathfrak{T}(\mathbf{r}, t) = \left(\lambda + \lambda'\frac{\partial}{\partial t}\right)\mathfrak{I}\,\mathrm{div}\,\mathbf{u}(\mathbf{r}, t) + 2\left(\mu + \mu'\frac{\partial}{\partial t}\right)\mathfrak{E}(\mathbf{r}, t). \tag{10.104}$$

Consequently, the corresponding equation of motion for a homogeneous material obtained from Eq. (1.112) is

$$\frac{\partial^2 \mathbf{u}}{\partial t^2} = \mathbf{F} + \alpha^2\left(1 + \tau_1\frac{\partial}{\partial t}\right)\mathrm{grad}\,\mathrm{div}\,\mathbf{u} - \beta^2\left(1 + \tau_2\frac{\partial}{\partial t}\right)\mathrm{curl}\,\mathrm{curl}\,\mathbf{u}, \tag{10.105}$$

where

$$\tau_1 = \frac{\lambda' + 2\mu'}{\lambda + 2\mu}, \qquad \tau_2 = \frac{\mu'}{\mu}.$$

Note that the Kelvin–Voigt solid includes the following particular cases:

$\lambda' = \mu' = 0,\ \lambda = \mu = \infty$	rigid body
$\lambda\,\mathrm{div}\,\mathbf{u} = -p,\ \mu = \mu' = \lambda' = 0$	ideal fluid
$\mu' = \lambda' = 0$	elastic solid
$\lambda\,\mathrm{div}\,\mathbf{u} = -p,\ \mu = 0,\ \mu' = \eta,\ \lambda' = \bar{\lambda} - \frac{2}{3}\eta$	Newtonian fluid
$\lambda\,\mathrm{div}\,\mathbf{u} = -p,\ \mu = 0,\ \mu' = \eta,\ \lambda' = -\frac{2}{3}\eta$	Stokes' fluid.

[3] In Chapter 10, σ_{ij} denote the components of the stress tensor because τ is used for a time variable. The Poisson ratio is denoted by $\bar{\sigma}$.

Consider next a three-dimensional Maxwell solid. A generalization of Eq. (10.34), with the exclusion of volumetric strain, is

$$\frac{1}{2\mu}\dot{\sigma}_{ij} + \frac{1}{2\eta}\sigma_{ij} = \dot{\varepsilon}_{ij}, \qquad (i \neq j, \quad i, j = 1, 2, 3) \tag{10.106}$$

where μ is the rigidity and η is the shear viscosity. Under conditions of constant strain, $\sigma_{ij}(t) = \sigma_{ij}(0)e^{-t/\tau_0}$ where $\tau_0 = \eta/\mu$ is the stress relaxation time. The solution of Eq. (10.106) for an arbitrary assigned shear stress is

$$\varepsilon_{ij}(t) = \left[\varepsilon_{ij}(0) - \frac{1}{2\mu}\sigma_{ij}(0)\right] + \frac{1}{2\mu}\sigma_{ij}(t) + \frac{1}{2\mu\tau_0}\int_0^t \sigma_{ij}(\theta)d\theta. \tag{10.107}$$

When $t < \tau_0$, the contribution of the integral is small compared with $\sigma_{ij}(t)$ and Hooke's law prevails; i.e., the body behaves as if elastic. When $t > \tau_0$, there may be finite strain associated with infinitely small stress and the body will behave as a viscous fluid. In particular, under constant effective stress, the strain is $\varepsilon_{ij}(t) = \varepsilon_{ij}(0) + [(1/2\eta)\sigma_{ij}(0)]t$ and the body will flow.

The equation of motion for a plane shear wave propagating in the y direction and having a displacement vector in the x direction, $u(y, t)\mathbf{e}_x$, is

$$\rho\frac{\partial^2 u}{\partial t^2} = \frac{\partial \sigma_{12}}{\partial y}, \qquad \varepsilon_{12} = \frac{1}{2}\frac{\partial u}{\partial y}.$$

Combining this with Eq. (10.106), we get

$$\frac{1}{\beta^2}\frac{\partial^2 u}{\partial t^2} = \tau_0\frac{\partial}{\partial t}\left[\frac{\partial^2 u}{\partial y^2} - \frac{1}{\beta^2}\frac{\partial^2 u}{\partial t^2}\right]. \tag{10.108}$$

A further generalization of Eqs. (10.100) and (10.102), which constitutes the most general law of linear viscoelasticity for isotropic materials, is the tensor form of the *Boltzmann superposition principle*

$$\sigma_{ij}(t) = \int_{-\infty}^{t} \Psi_{ijlm}(t - \tau)\left[\frac{d\varepsilon_{lm}(\tau)}{d\tau}\right]d\tau, \tag{10.109}$$

where σ, ε, and Ψ may depend on the coordinates. Here Ψ_{ijlm} is the fourth-order *relaxation tensor*. Elasticity and linear anelacticity are special cases of this generalized formulation. Therefore, the particular choice [cf. Eqs. (1.67) and (9.4)]

$$\Psi_{ijlm} = \Psi_{ijlm}^{(0)} = C_{ijlm}H(t) + D_{ijlm}\delta(t) \tag{10.110}$$

leads us back to Eq. (10.102).

In order to construct the relaxation tensor for a general isotropic anelastic substance, we shall assume, to begin with, that the material is characterized by two relaxation functions: one for shear and the other for bulk compression. We have shown in Eq. (1.83) that the stress–strain relations in terms of the isotropic parts of the stress and strain tensors and their deviators are

$$\sigma_{mm}(t) = 3\kappa\varepsilon_{mm}, \qquad \text{(summation over } m)$$
$$\sigma_{ij}(t) = 2\mu\varepsilon_{ij}, \qquad i \neq j.$$

The corresponding Boltzmann superposition integrals

$$\begin{aligned}\sigma_{mm}(\mathbf{r}, t) &= \int_0^t \psi_\kappa(t-\tau)\left[\frac{\partial \varepsilon_{mm}(\mathbf{r}, \tau)}{\partial \tau}\right]d\tau,\\ \sigma_{ij}(\mathbf{r}, t) &= \int_0^t \psi_\beta(t-\tau)\left[\frac{\partial \varepsilon_{ij}(\mathbf{r}, \tau)}{\partial \tau}\right]d\tau, \qquad i \neq j\end{aligned} \tag{10.111}$$

can be reformulated in the single equation

$$\sigma_{ij}(\mathbf{r}, t) = \int_0^t \psi_\beta(t-\tau)\dot{\varepsilon}_{ij}(\mathbf{r}, \tau)d\tau + \frac{1}{3}\delta_{ij}\int_0^t [\psi_\kappa(t-\tau) - \psi_\beta(t-\tau)]\dot{\varepsilon}_{mm}(\tau)d\tau. \tag{10.112}$$

Equation (10.111) reduces to the elastic case [cf Eq. (1.83)] whenever we take $\psi_\kappa(t) = 3\kappa H(t)$ and $\psi_\beta(t) = 2\mu H(t)$. However, Eq. (10.111) is a special case of Eq. (10.109), with

$$\Psi_{ijlm}(t) = \frac{1}{3}[\psi_\kappa(t) - \psi_\beta(t)]\delta_{ij}\delta_{lm} + \frac{1}{2}\psi_\beta(t)[\delta_{il}\delta_{jm} + \delta_{im}\delta_{jl}]. \tag{10.113}$$

From Eq. (10.113) it follows that [cf. Eq. (1.67)]

$$\psi_\beta = 2\Psi_{1212}, \qquad \psi_\kappa = 2\Psi_{1212} + 3\Psi_{1122}. \tag{10.114}$$

Integration by parts of Eqs. (10.111) yields

$$\begin{aligned}\sigma_{mm}(\mathbf{r}, t) &= 3\kappa_\infty\varepsilon_{mm}(\mathbf{r}, t) + \int_0^t \varepsilon_{mm}(\mathbf{r}, t-\tau)\dot{\psi}_\kappa(\tau)d\tau,\\ \sigma_{ij}(\mathbf{r}, t) &= 2\mu_\infty\varepsilon_{ij}(\mathbf{r}, t) + \int_0^t \varepsilon_{ij}(\mathbf{r}, t-\tau)\dot{\psi}_\beta(\tau)d\tau, \qquad i \neq j\end{aligned} \tag{10.115}$$

where

$$\psi_\kappa(t=0) = \psi_\kappa(\omega=\infty) = 3\kappa_\infty, \qquad \psi_\beta(0) = 2\mu_\infty.$$

Equations (10.115) generalize Eq. (10.100) to three dimensions. Note the appearance of the two time functions of which $\psi_\kappa(t)$ is the *bulk relaxation function* and $\psi_\beta(t)$ is the *shear relaxation function*. The relaxation function in Eq. (10.100) is unspecified and may stand for either of them depending on the physical significance of Y_0.

Strains can be expressed in terms of stresses by means of Eq. (1.72). The inverse relations of Eqs. (10.115) are

$$\begin{aligned}\varepsilon_{mm}(\mathbf{r}, t) &= \frac{1}{3\kappa_\infty}\sigma_{mm}(\mathbf{r}, t) + \int_0^t \sigma_{mm}(\mathbf{r}, t-\tau)\dot{\phi}_\kappa(\tau)d\tau,\\ \varepsilon_{ij}(\mathbf{r}, t) &= \frac{1}{2\mu_\infty}\sigma_{ij}(\mathbf{r}, t) + \int_0^t \sigma_{ij}(\mathbf{r}, t-\tau)\dot{\phi}_\beta(\tau)d\tau, \qquad i \neq j\end{aligned} \tag{10.116}$$

where ϕ_β and ϕ_κ are two causal material creep functions and Eqs. (10.116) generalize Eq. (10.99) to 3-space. Note that the anelastic stress–strain laws in Eqs. (10.115) and (10.116) can also be condensed into the two equations:

$$\sigma_{ij}(\mathbf{r}, t) = [2\mu_\infty \varepsilon_{ij}(\mathbf{r}, t) + \lambda_\infty \delta_{ij}\varepsilon_{mm}(\mathbf{r}, t)] + \int_0^t \dot{\psi}_\beta(\tau)\varepsilon_{ij}(\mathbf{r}, t-\tau)d\tau$$

$$+ \frac{1}{3}\delta_{ij}\int_0^t [\dot{\psi}_\kappa(\tau) - \dot{\psi}_\beta(\tau)]\varepsilon_{mm}(\mathbf{r}, t-\tau)d\tau, \tag{10.117}$$

$$\varepsilon_{ij}(\mathbf{r}, t) = \left[\frac{1}{2\mu_\infty}\sigma_{ij}(\mathbf{r}, t) - \frac{\lambda_\infty}{6\kappa_\infty \mu_\infty}\delta_{ij}\sigma_{mm}(\mathbf{r}, t)\right] + \int_0^t \dot{\phi}_\beta(\tau)\sigma_{ij}(\mathbf{r}, t-\tau)d\tau$$

$$+ \frac{1}{3}\delta_{ij}\int_0^t [\dot{\phi}_\kappa(\tau) - \dot{\phi}_\beta(\tau)]\sigma_{mm}(\mathbf{r}, t-\tau)d\tau, \tag{10.118}$$

where

$$\lambda_\infty = \frac{1}{3}[3\kappa_\infty - 2\mu_\infty].$$

Returning to the general form in Eq. (10.109), we assume that $\sigma_{ij} = \varepsilon_{ij} = 0$ for $t < 0$ and that there is a finite strain jump $\varepsilon_{kl}(0^+)$ at $t = 0$. Equation (10.109) then reads

$$\sigma_{ij}(\mathbf{r}, t) = \varepsilon_{lm}(\mathbf{r}, 0^+)\Psi_{ijlm}(t) + \int_0^t \Psi_{ijlm}(t-\tau)\left[\frac{\partial \varepsilon_{lm}(\mathbf{r}, \tau)}{\partial \tau}\right]d\tau. \tag{10.119}$$

When $\partial\varepsilon_{ij}/\partial t$ and $\partial\Psi_{ijlm}/\partial t$ exist and are continuous in $0 \le t \le \infty$, integration by parts of Eq. (10.109) yields

$$\sigma_{ij}(t) = \Psi_{ijlm}(0)\varepsilon_{lm}(\mathbf{r}, t) + \int_0^t \varepsilon_{lm}(\mathbf{r}, t-\tau)\left[\frac{\partial \Psi_{ijlm}(\tau)}{\partial \tau}\right]d\tau$$

$$= \frac{\partial}{\partial t}\int_0^t \varepsilon_{lm}(\mathbf{r}, t-\tau)\Psi_{ijlm}(\tau)d\tau. \tag{10.120}$$

It can be shown that if $\Psi_{ijkl}(\mathbf{r}, t)$ is twice differentiable and if $\Psi_{ijkl}(\mathbf{r}, 0) \neq 0$, there exists the following relation inverse to Eq. (10.109)

$$\varepsilon_{ij}(\mathbf{r}, t) = \int_{-\infty}^t \Phi_{ijlm}(t-\tau)\left[\frac{\partial \sigma_{lm}(\mathbf{r}, \tau)}{\partial \tau}\right]d\tau \tag{10.121}$$

in which Φ_{ijlm} represents a fourth-order creep tensor. Again, if the motion starts at $t = 0$ such that $\sigma_{ij}(0^+)$ is the stress jump at $t = 0$, we may write

$$\varepsilon_{ij}(\mathbf{r}, t) = \sigma_{lm}(\mathbf{r}, 0^+)\Phi_{ijlm}(t) + \int_0^t \Phi_{ijlm}(t-\tau)\left[\frac{\partial \sigma_{lm}(\mathbf{r}, \tau)}{\partial \tau}\right]d\tau$$

$$= \Phi_{ijlm}(0)\sigma_{lm}(\mathbf{r}, t) + \int_0^t \sigma_{lm}(\mathbf{r}, \tau)\left[\frac{\partial \Phi_{ijlm}(t-\tau)}{\partial (t-\tau)}\right]d\tau. \tag{10.122}$$

Note that Eqs. (10.120) and (10.122) have the form of a *Duhamel integral* [cf. Eq. (K.4)]

$$g(t) = f(0^+)h(t) + \int_0^t h(t-\tau)f'(\tau)d\tau$$

in which the creep and relaxation tensors are in the role of the system's step response.

10.3. Pulse Propagation in Unbounded Anelastic Media

10.3.1. Generalized Navier Equation and Complex Propagation Functions

We combine the Cauchy equation of motion [Eq. (1.100)] with the Boltzmann superposition law. To this end we write the strains in terms of the displacements and apply the divergence operator to Eq. (10.117). Remembering that ψ_κ and ψ_β are functions of time only, we obtain, in the absence of body forces,

$$\frac{\partial^2 \mathbf{u}(\mathbf{r},t)}{\partial t^2} = \alpha_\infty^2 \operatorname{grad}\operatorname{div}\left\{\mathbf{u}(\mathbf{r},t) + \int_0^t \mathbf{u}(\mathbf{r},t-\tau)\left[\frac{\dot\kappa(\tau)+\frac{4}{3}\dot\mu(\tau)}{\kappa_\infty+\frac{4}{3}\mu_\infty}\right]d\tau\right\}$$
$$-\beta_\infty^2 \operatorname{curl}\operatorname{curl}\left\{\mathbf{u}(\mathbf{r},t) + \int_0^t \mathbf{u}(\mathbf{r},t-\tau)\left[\frac{\dot\mu(\tau)}{\mu_\infty}\right]d\tau\right\}, \tag{10.123}$$

where

$$\psi_\beta(t) = 2\mu(t), \qquad \psi_\beta(0) = 2\mu_\infty, \qquad \beta_\infty^2 = \frac{\mu_\infty}{\rho},$$

$$\psi_\kappa(t) = 3\kappa(t), \qquad \psi_\kappa(0) = 3\kappa_\infty, \qquad \alpha_\infty^2 = \frac{\kappa_\infty + \frac{4}{3}\mu_\infty}{\rho}.$$

Because the bulk of information about the viscoelastic behavior of the earth is obtained through frequency-domain data, it is more convenient to work with the spectrum of the displacement field. Therefore, applying the Fourier transform to Eq. (10.123), we obtain

$$\hat\alpha^2 \operatorname{grad}\operatorname{div}\mathbf{U} - \hat\beta^2 \operatorname{curl}\operatorname{curl}\mathbf{U} + \omega^2\mathbf{U} = 0, \tag{10.124}$$

where

$$\mathbf{U}(\mathbf{r},\omega) = \int_{-\infty}^{\infty} \mathbf{u}(\mathbf{r},t)e^{-i\omega t}\,dt, \qquad \lim_{t\to\infty}\frac{\partial \mathbf{u}}{\partial t} = 0,$$

$$\hat\alpha^2(\omega) = \frac{1}{\rho}\left[\left(\kappa_\infty + \frac{4}{3}\mu_\infty\right) + \int_0^\infty \left\{\dot\kappa(t) + \frac{4}{3}\dot\mu(t)\right\}e^{-i\omega t}\,dt\right], \tag{10.125}$$

$$\hat\beta^2(\omega) = \frac{1}{\rho}\left[\mu_\infty + \int_0^\infty \dot\mu(t)e^{-i\omega t}\,dt\right].$$

Although earth materials tend to behave elastically at both very high and very low frequencies, the response at high frequencies represents the instantaneous response and therefore corresponds to the more common interpretation of elastic behavior. We shall therefore require in this chapter that all the frequency-dependent constitutive parameters of the viscoelastic solid tend to their elastic reference values in the high-frequency limit, $\omega \to \infty$. Hence

$$\psi_\kappa(0) = 3\kappa_\infty, \qquad \psi_\beta(0) = 2\mu_\infty, \tag{10.126}$$

where κ_∞ is the elastic bulk modulus and μ_∞ is the elastic shear modulus. The wave velocities at infinite frequency are denoted here by α_∞ and β_∞.

Let us define, for real values of ρ and ω, the *complex elastic moduli*, *complex wave velocities*, and *complex wave numbers*:

$$\begin{aligned}
\hat{\mu}(\omega) &= \mu_\infty + \int_0^\infty \dot{\mu}(t)e^{-i\omega t}\,dt = \mu + i\mu^*,\\
\hat{\kappa}(\omega) &= \kappa_\infty + \int_0^\infty \dot{\kappa}(t)e^{-i\omega t}\,dt = \kappa + i\kappa^*,\\
\hat{\alpha}(\omega) &= \left[\frac{\hat{\kappa}(\omega) + \frac{4}{3}\hat{\mu}(\omega)}{\rho}\right]^{1/2} = \alpha + i\alpha^*,\\
\hat{\beta}(\omega) &= \left[\frac{\hat{\mu}(\omega)}{\rho}\right]^{1/2} = \beta + i\beta^*.
\end{aligned} \tag{10.127}$$

The function $\hat{\mu}(\omega)$ is known as the *dynamic shear modulus*. The dynamic wave numbers are

$$\begin{aligned}
\hat{k}_\alpha &= \frac{\omega}{\hat{\alpha}(\omega)} = k_\alpha - ik_\alpha^*, &\qquad \hat{k}_\beta &= \frac{\omega}{\hat{\beta}(\omega)} = k_\beta - ik_\beta^*,\\
k_\alpha^* &= \gamma_\alpha(\omega) > 0, & k_\beta^* &= \gamma_\beta(\omega) > 0,\\
k_\alpha &= \frac{\omega}{c_\alpha(\omega)}, & k_\beta &= \frac{\omega}{c_\beta(\omega)},
\end{aligned} \tag{10.128}$$

where

$$\begin{aligned}
c_\alpha(\omega) &= \frac{1}{\mathrm{Re}[\rho/\{\hat{\kappa}(\omega) + \frac{4}{3}\hat{\mu}(\omega)\}]^{1/2}}, &\qquad \gamma_\alpha(\omega) &= -\omega\,\mathrm{Im}\left[\frac{\rho}{\hat{\kappa}(\omega) + \frac{4}{3}\hat{\mu}(\omega)}\right]^{1/2},\\
c_\beta(\omega) &= \frac{1}{\mathrm{Re}[\rho/\hat{\mu}(\omega)]^{1/2}}, & \gamma_\beta(\omega) &= -\omega\,\mathrm{Im}\left[\frac{\rho}{\hat{\mu}(\omega)}\right]^{1/2}.
\end{aligned} \tag{10.129}$$

If we define the specific dissipation parameters [cf. Section 10.1]

$$\frac{1}{Q_\alpha} = \frac{\kappa^* + (4/3)\mu^*}{\kappa + (4/3)\mu}, \qquad \frac{1}{Q_\beta} = \frac{\mu^*}{\mu} \tag{10.130}$$

and put

$$\hat{\mu}(\omega) = \mu\left[1 + \frac{i}{Q_\beta}\right] = \mu\left[1 + \frac{1}{Q_\beta^2}\right]^{1/2} \exp\left(i \tan^{-1}\frac{1}{Q_\beta}\right),$$

we obtain from Eq. (10.129)

$$c_\beta(\omega) = \left[\frac{\mu(\omega)}{\rho}\right]^{1/2}\left[\left(1 + \frac{1}{Q_\beta^2}\right)^{3/2} - \left(1 + \frac{1}{Q_\beta^2}\right)\right]^{1/2} Q_\beta\sqrt{2}, \tag{10.131}$$

$$\gamma_\beta(\omega) = \omega\left[\frac{\rho}{\mu(\omega)}\right]^{1/2}\left[\frac{(1 + 1/Q_\beta^2)^{-1/2} - (1 + 1/Q_\beta^2)^{-1}}{2}\right]^{1/2}, \tag{10.132}$$

$$\frac{\gamma_\beta(\omega)c_\beta(\omega)}{Q_\beta(\omega)} = \left(1 + \frac{1}{Q_\beta^2}\right)^{1/2} - 1, \qquad \frac{1}{Q_\beta} = \frac{2(\gamma_\beta c_\beta/\omega)}{1 - (\gamma_\beta c_\beta/\omega)^2}. \tag{10.133}$$

There are similar expressions for $c_\alpha(\omega)$, $\gamma_\alpha(\omega)$ and $\hat{k}_\alpha(\omega)$. For large values of Q_β, Eq. (10.133) yields the approximation

$$\frac{1}{Q_\beta(\omega)} \simeq \frac{2\gamma_\beta(\omega)c_\beta(\omega)}{\omega}. \tag{10.134}$$

All starred quantities and their counterparts are real. We also define the *complex indices of refraction*

$$\hat{n}_\alpha(\omega) = \frac{\hat{k}_\alpha(\omega)}{k_\alpha(\infty)} = \frac{\alpha_\infty}{\hat{\alpha}(\omega)}, \qquad \hat{n}_\beta(\omega) = \frac{\hat{k}_\beta(\omega)}{k_\beta(\infty)} = \frac{\beta_\infty}{\hat{\beta}(\omega)}, \tag{10.135}$$

where

$$k_\alpha(\infty) = \frac{\omega}{\alpha_\infty}, \qquad k_\beta(\infty) = \frac{\omega}{\beta_\infty} \tag{10.136}$$

are the elastic wave numbers at infinite frequency. Clearly,

$$\begin{aligned} \hat{n}_\alpha(\omega) &= \frac{\alpha_\infty}{c_\alpha(\omega)} - i\frac{\alpha_\infty}{\omega}\gamma_\alpha(\omega), \\ \hat{n}_\beta(\omega) &= \frac{\beta_\infty}{c_\beta(\omega)} - i\frac{\beta_\infty}{\omega}\gamma_\beta(\omega). \end{aligned} \tag{10.137}$$

The indices of refraction can also be written as

$$\hat{n}_\beta(\omega) = \left[\frac{\mu_\infty}{\hat{\mu}(\omega)}\right]^{1/2} = \left[1 - \int_0^t B_\beta(t)e^{-i\omega t}\,dt\right]^{-1/2}, \tag{10.138}$$

$$\hat{n}_\alpha(\omega) = \left[\frac{\kappa_\infty + \frac{4}{3}\mu_\infty}{\hat{\kappa}(\omega) + \frac{4}{3}\hat{\mu}(\omega)}\right]^{1/2} = \left[1 - \int_0^t B_\alpha(t)e^{-i\omega t}\,dt\right]^{-1/2}, \tag{10.139}$$

where

$$B_\beta(t) = -\frac{\dot{\mu}(t)}{\mu_\infty} > 0, \qquad B_\alpha(t) = -\frac{\dot{\kappa}(t) + \frac{4}{3}\dot{\mu}(t)}{\kappa_\infty + \frac{4}{3}\mu_\infty} > 0. \tag{10.140}$$

Let $B(t)$, $Q(\omega)$, $\hat{n}(\omega)$, $\gamma(\omega)$ and $c(\omega)$ apply to both P and S waves. We define

$$\int_0^t B(t)e^{-i\omega t}\,dt = M(\omega) - iH(\omega). \tag{10.141}$$

Then by Eqs. (10.127) and (10.130), we have

$$\frac{1}{Q(\omega)} = \frac{H(\omega)}{1 - M(\omega)} = \frac{\int_0^\infty B(t)\sin\omega t\,dt}{1 - \int_0^\infty B(t)\cos\omega t\,dt}, \tag{10.142}$$

Similarly, by Eqs. (10.137) and (10.138)

$$\hat{n}(\omega) = [1 - M(\omega)]^{-1/2}\left[1 + \frac{i}{Q}\right]^{-1/2} \tag{10.143}$$

At high frequencies, both $H(\omega)$ and $M(\omega)$ tend to zero and $1/Q(\omega)$ is very small. The expansions of $c(\omega)$ and $\gamma(\omega)$ in Eqs. (10.131) and (10.132) then yield

$$\gamma(\omega) \simeq \frac{\omega}{2c_\infty}H(\omega), \qquad c(\omega) \simeq c_\infty\left[1 + \frac{3}{8Q^2}\right], \tag{10.144}$$

$$\hat{n}(\omega) \simeq 1 - \frac{i}{2Q}, \qquad \frac{1}{Q(\omega)} \simeq \frac{2\gamma(\omega)c_\infty}{\omega} \ll 1. \tag{10.145}$$

where c_∞ denotes either α_∞ or β_∞.

10.3.1.1. The Correspondence Principle. We note that the Fourier-transformed Navier equation for an anelastic medium, Eq. (10.124), is similar in form to the corresponding equation of linear elasticity theory, except that the Lamé constants are replaced by complex moduli. Therefore, any formal solution of Navier's equation in the theory of linear elasticity offers a corresponding solution for a linear viscoelastic body if the elastic moduli that occur in the elastic solution are replaced by the corresponding complex moduli. This is known as the *correspondence principle.*

In Example 10.2 we show how this principle is used in the solution of a special class of problems, known as *quasistatic problems.* Let an anelastic material be subjected to the action of slowly-varying forces such that the inertial terms in the equations of motion are negligible compared to the other terms. The Navier equation is then time dependent only through its creep and relaxation functions. The mathematical problem is equivalent to solving the static Navier equation with frequency-dependent elastic constants. After the solution has been expressed in terms, say, of the force spectrum $f(\omega)$ $\hat{\mu}(\omega)$, and $\hat{\kappa}(\omega)$, the resulting expression is subjected to an inverse transform (Fourier or Laplace) leading eventually to explicit expressions for $\mathbf{u}(\mathbf{r}, t)$. This approach is valid also in those cases where the acting forces are impulsive, provided we are interested only in the long-term behavior of the system, after all the transients have died out.

EXAMPLE 10.2: The Boussinesq Problem for an Anelastic Medium

Let the xy plane coincide with the free surface of an *elastic* half-space and let the z axis, along which acts a force $\mathbf{e}_z(F_0/\rho)$, be directed into the half-space. The point of application of the force is taken as the origin of the coordinates (Fig. 10.13). From Eq. (4.3), we obtain the *static Navier equation* in the limit $\omega \to 0$,

$$\nabla^2 \mathbf{u} + \frac{1}{1 - 2\bar{\sigma}} \operatorname{grad} \operatorname{div} \mathbf{u} = -\mathbf{e}_z \frac{F_0}{\mu} \delta(\mathbf{r} - \mathbf{r}_0). \tag{10.2.1}$$

The solution of Eq. (10.2.1) is composed of a particular solution of the inhomogeneous equation and a solution of the homogeneous equation. The particular solution $\mathbf{u}_0$ is readily found from Eq. (4.27) by choosing $g(t) = H(t)$ and taking the limit as $t \to \infty$. One then arrives at the solution of the dyadic equation ($\bar{\sigma}$ = Poisson's ratio)

$$\nabla^2 \mathfrak{G}_\infty + \frac{1}{1 - 2\bar{\sigma}} \operatorname{grad} \operatorname{div} \mathfrak{G}_\infty = -\frac{1}{\mu} \mathfrak{I} \delta(\mathbf{r} - \mathbf{r}_0) \tag{10.2.2}$$

known as the *Somigliana tensor*

$$\mathfrak{G}_\infty = \frac{1}{4\pi\mu} \left[\frac{\mathfrak{I}}{R} - \frac{1}{4(1 - \bar{\sigma})} \left\{ \frac{\mathfrak{I} - \mathbf{e}_R \mathbf{e}_R}{R} \right\} \right]. \tag{10.2.3}$$

In terms of $\mathfrak{G}_\infty$, the particular solution is

$$\mathbf{u}_0 = F_0(\mathbf{e}_z \cdot \mathfrak{G}_\infty). \tag{10.2.4}$$

The substitution $\mathbf{u} = \operatorname{grad} B_0$ in the homogeneous equation corresponding to Eq. (10.2.1), reveals at once that B_0 is an arbitrary harmonic function. We choose $B_0 = \ln(r + z)$ and subject the complete solution

$$\mathbf{u} = A\mathbf{u}_0 + B \operatorname{grad}[\ln(r + z)], \qquad (r_0 = 0) \tag{10.2.5}$$

to the boundary conditions

$$\sigma_{xz} = \sigma_{yz} = \sigma_{zz} = 0 \quad \text{at} \quad z = 0.$$

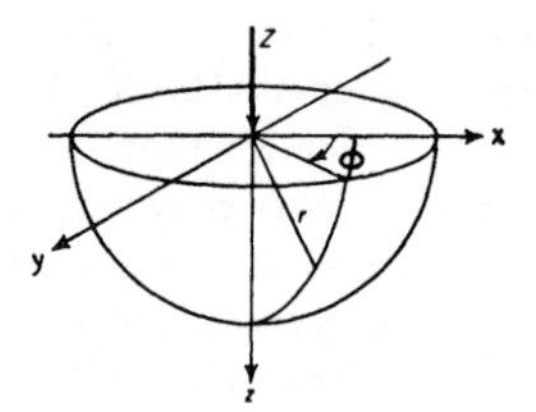

Figure 10.13. Boussinesq problem for an anelastic medium.

The expressions for the displacements and stresses are then found to be

$$u_x = \frac{F_0}{4\pi\mu}\left[\frac{z}{r^3} - \frac{1-2\bar{\sigma}}{r(r+z)}\right]x,$$

$$u_y = \frac{F_0}{4\pi\mu}\left[\frac{z}{r^3} - \frac{1-2\bar{\sigma}}{r(r+z)}\right]y, \tag{10.2.6}$$

$$u_z = \frac{F_0}{4\pi\mu}\left[\frac{z^2}{r^3} + \frac{2(1-\bar{\sigma})}{r}\right],$$

$$\sigma_{\Delta z} = \frac{-3F_0}{2\pi}\frac{\Delta z^2}{r^5}, \qquad \sigma_{zz} = \frac{-3F_0}{2\pi}\frac{z^3}{r^5},$$

$$\sigma_{\Delta\Delta} = \frac{F_0}{2\pi}\left[-\frac{3\Delta^2 z}{r^5} + \frac{1-2\bar{\sigma}}{r(r+z)}\right], \tag{10.2.7}$$

$$\sigma_{\phi\phi} = \frac{(1-2\bar{\sigma})}{2\pi}F_0\left[\frac{z}{r^3} - \frac{1}{r(r+z)}\right], \qquad \sigma_{\Delta\phi} = \sigma_{z\phi} = 0.$$

At this point we restore the frequency dependence of F_0 and $\bar{\sigma}$, where ω is now in the role of a parameter. Invoking the *correspondence principle*, we write $\sigma_{\Delta\Delta}$, say, in the form

$$\hat{\sigma}_{\Delta\Delta}(\Delta, z; \omega) = \frac{F_0(\omega)}{2\pi}\left[\frac{1-2\hat{\bar{\sigma}}(\omega)}{r(r+z)} - \frac{3\Delta^2 z}{r^5}\right], \tag{10.2.8}$$

where

$$\hat{\bar{\sigma}}(\omega) = \frac{\hat{\lambda}(\omega)}{2[\hat{\lambda}(\omega) + \hat{\mu}(\omega)]}. \tag{10.2.9}$$

An inverse Fourier transform yields

$$\sigma_{\Delta\Delta}(\Delta, z, t) = \frac{1}{2\pi}\left[\frac{G(t)}{r(r+z)} - \frac{3\Delta^2 z}{r^5}g(t)\right], \tag{10.2.10}$$

where

$$g(t) = \frac{1}{2\pi}\int_{-\infty}^{\infty} F_0(\omega)e^{i\omega t}\,d\omega, \qquad \xi(t) = \frac{1}{2\pi}\int_{-\infty}^{\infty}[1 - 2\hat{\bar{\sigma}}(\omega)]e^{i\omega t}\,d\omega, \tag{10.2.11}$$

$$G(t) = \frac{1}{2\pi}\int_{-\infty}^{\infty} F_0(\omega)[1 - 2\hat{\bar{\sigma}}(\omega)]e^{i\omega t}\,d\omega = \int_0^t g(t-\tau)\xi(\tau)d\tau.$$

If, for example, the half-space behaves as a Kelvin–Voigt substance under deviatoric stress and is elastic for dilatational stress with $g(t) = g_0 H(t)$, the three-dimensional generalization of Eq. (10.43) is

$$\sigma_{ij} = \left(2\mu + 2\eta\frac{\partial}{\partial t}\right)\varepsilon_{ij}, \qquad i \neq j$$

$$\sigma_{ii} = 3\kappa\varepsilon_{ii}. \tag{10.2.12}$$

Consequently

$$1 - 2\hat{\bar{\sigma}} = \frac{3\hat{\mu}}{3\hat{\kappa} + \hat{\mu}}, \qquad \hat{\sigma}_{ij} = (2\mu + 2i\omega\eta)\hat{\varepsilon}_{ij} = 2\hat{\mu}\hat{\varepsilon}_{ij},$$
$$\hat{\mu} = \mu + i\omega\eta, \qquad \hat{\kappa} = \kappa. \tag{10.2.13}$$

Hence, the frequency-dependence of $(1 - 2\hat{\bar{\sigma}})$ is

$$1 - 2\hat{\bar{\sigma}}(\omega) = \frac{3(\mu + i\omega\eta)}{3\kappa + \mu + i\omega\eta}. \tag{10.2.14}$$

It now follows from Eqs. (10.2.11) and (10.2.14) that

$$G(t) = \frac{g_0}{2\pi} \int_{-\infty}^{\infty} \frac{1}{i\omega} \frac{3(\mu + i\omega\eta)}{3\kappa + \mu + i\omega\eta} e^{i\omega t}\, d\omega$$
$$= \frac{3g_0}{3\kappa + \mu}[\mu + 3\kappa e^{-\{(3\kappa+\mu)/\eta\}t}]H(t). \tag{10.2.15}$$

10.3.2. General Vector Plane Waves and Energy Dissipation in Low-Loss Media: The Quality Factors $Q_\alpha(\omega)$ and $Q_\beta(\omega)$

In Section 10.3.1 we met the complex wave velocities $\hat{\alpha}(\omega)$, $\hat{\beta}(\omega)$; their associated complex scalar wave numbers $\hat{k}_\alpha(\omega)$, $\hat{k}_\beta(\omega)$; and the complex structural parameters $\hat{\mu}(\omega)$, $\hat{\lambda}(\omega)$, $\hat{\kappa}(\omega)$. We wish next to derive the complex vector solutions of the transformed equation, Eq. (10.124). We know already from our discussion in Section 2.4 that if ψ_α, ψ_β are solutions of the Helmholtz equation

$$\nabla^2\psi + \hat{k}^2(\omega)\psi = 0, \qquad \hat{k}(\omega) = \hat{k}_\alpha(\omega) \quad \text{or} \quad \hat{k}_\beta(\omega), \tag{10.146}$$

then the vectors $\nabla\psi_\alpha$, $\text{curl}(\mathbf{b}\psi_\beta)$, and curl $\text{curl}(\mathbf{b}\psi_\beta)$ are three independent vector solutions of Eq. (10.124)

A general solution of Eq. (10.146) representing plane waves is of the form

$$\psi = Be^{-i\mathbf{K}\cdot\mathbf{r}}, \tag{10.147}$$

where $\mathbf{K}$ is the bivector

$$\mathbf{K} = \mathbf{P} - i\mathbf{A}. \tag{10.148}$$

The phase velocity of the wave is $\omega/|\mathbf{P}|$. The vectors $\mathbf{P}$ and $\mathbf{A}$ are not necessarily parallel. Denoting the angle between them by χ and substituting Eq. (10.147) into Eq. (10.146), we obtain

$$[\hat{k}(\omega)]^2 = (k - ik^*)^2 = |\mathbf{K}|^2 = |\mathbf{P}|^2 - |\mathbf{A}|^2 - 2i|\mathbf{P}||\mathbf{A}|\cos\chi. \tag{10.149}$$

The vector $\mathbf{P}$ is normal to the planes of constant phase defined by $\mathbf{P}\cdot\mathbf{r} = \text{const.}$ The vector $\mathbf{A}$ is normal to the planes of constant amplitude defined by $\mathbf{A}\cdot\mathbf{r} =$

const. If $\chi = 0$, the wave is called homogeneous. If $\chi \neq 0$ the wave is inhomogeneous. Clearly,

$$\mathrm{Re}(\hat{k}^2) = k^2 - k^{*2} = |\mathbf{P}|^2 - |\mathbf{A}|^2, \tag{10.150}$$

$$\mathrm{Im}(\hat{k}^2) = -2kk^* = -2|\mathbf{P}||\mathbf{A}|\cos\chi, \tag{10.151}$$

$$\frac{\mathrm{Im}(\hat{k}^2)}{\mathrm{Re}(\hat{k}^2)} = -\frac{2(k^*/k)}{1-(k^*/k)^2} = -\frac{2|\mathbf{A}|/|\mathbf{P}|}{1-|\mathbf{A}|^2/|\mathbf{P}|^2}\cos\chi = -\frac{1}{Q}\ \text{(say)}. \tag{10.152}$$

The physical requirement that the amplitude of the wave is bounded in the direction of propagation leads to the condition $0 \leq \chi \leq \pi/2$.

For elastic materials, $\mathrm{Im}(\hat{k}^2) = 0$ and Eq. (10.151) implies that either $\mathbf{A} = \mathbf{0}$ (unattenuated homogeneous plane wave) or $\chi = \pi/2$ [e.g., a surface wave in multilayered media; cf. Eq. (3.68)]. The velocity of planes of constant phase of the general plane wave in Eq. (10.147) is given as

$$\mathbf{v} = \frac{\omega}{|\mathbf{P}|^2}\mathbf{P}. \tag{10.153}$$

Equations (10.150) and (10.151) may be solved simultaneously to yield $|\mathbf{P}|$ and $|\mathbf{A}|$ in terms of $\mathrm{Im}(\hat{k}^2)$, $\mathrm{Re}(\hat{k}^2)$, and χ

$$|\mathbf{P}| = \left[\frac{1}{2}\mathrm{Re}(\hat{k}^2)\right]^{1/2}\left[\left(1+\frac{1}{Q^2\cos^2\chi}\right)^{1/2}+1\right]^{1/2}, \tag{10.154}$$

$$|\mathbf{A}| = \left[\frac{1}{2}\mathrm{Re}(\hat{k}^2)\right]^{1/2}\left[\left(1+\frac{1}{Q^2\cos^2\chi}\right)^{1/2}-1\right]^{1/2}. \tag{10.155}$$

For any medium in which $\hat{k}^2(\omega)$ and χ are given, Eqs. (10.154) and (10.155) serve to construct the solving potential ψ. The particle motion is then found by evaluating the Hansen vectors **L**, **M**, **N** associated with the potentials ψ_α and ψ_β. It will be shown later in this section that particle motions for both *P* and *S* waves are elliptical. For a homogeneous *P* wave, the elliptical particle motion degenerates to motion along a straight line parallel to the direction of propagation. For a homogeneous *S* wave, the elliptical motion degenerates into motion along a straight line normal to the direction of propagation.

10.3.2.1. Energy of Plane Viscoelastic Waves. In the absence of body forces, the energy equation (1.118) becomes

$$\frac{\partial}{\partial t}\left[\int_V \tfrac{1}{2}(\mathfrak{T}:\nabla\mathbf{u} + \rho\dot{\mathbf{u}}^2)\,dV\right] = \int_S (\mathfrak{T}\cdot\dot{\mathbf{u}})\cdot d\mathbf{S}, \tag{10.156}$$

where

$$\mathfrak{T} = \lambda\theta\mathfrak{I} + \mu(\nabla\mathbf{u} + \mathbf{u}\nabla), \qquad \theta = \operatorname{div}\mathbf{u}.$$

When the displacement vector is complex, we must use its real part in Eq. (10.156).

Assuming sinusoidal time-dependence, the stress-strain relation (10.117) for a viscoelastic medium may be written in the form

$$\mathfrak{T} + i\mathfrak{T}^* = \hat{\lambda}(\omega)\Theta\mathfrak{I} + \hat{\mu}(\omega)(\nabla\mathbf{U} + \mathbf{U}\nabla), \tag{10.157}$$

where $\mathbf{U}(\mathbf{r},t) = \mathbf{U}_0(r)\exp(i\omega t) = \mathbf{u} + i\mathbf{u}^*$, $\Theta = \operatorname{div}\mathbf{U}$, $\hat{\lambda} = \lambda + i\lambda^* = \hat{k} - (2/3)\hat{\mu}$, $\hat{\mu} = \mu + i\mu^*$, $\hat{k} = k + ik^*$, and $\hat{\mu}$ and $\hat{k}$ are given in Eq. (10.127). Equating the real and imaginary parts on the two sides of Eq. (10.157), and noting that, on account of the sinusoidal time-dependence, $\dot{\mathbf{u}} = -\omega\mathbf{u}^*$, we find

$$\mathfrak{T} = \lambda\theta\mathfrak{I} + \mu(\nabla\mathbf{u} + \mathbf{u}\nabla) + \frac{1}{\omega}\left[\lambda^*\dot{\theta}\mathfrak{I} + \mu^*(\nabla\dot{\mathbf{u}} + \dot{\mathbf{u}}\nabla)\right]. \tag{10.158}$$

Equations (10.156) and (10.158) yield

$$\frac{\partial}{\partial t}\int_V H\,dV + \int_V D\,dV = -\int_S \Sigma\cdot d\mathbf{S}, \tag{10.159}$$

where $\Sigma = \Sigma_1 + \Sigma_2$, and

$$H = \frac{1}{2}\left[\rho\dot{u}^2 + \lambda\theta^2 + \mu\nabla\mathbf{u}:(\nabla\mathbf{u} + \mathbf{u}\nabla)\right], \tag{10.160}$$

$$D = \frac{1}{\omega}\left[\lambda^*\dot{\theta}^2 + \mu^*\nabla\dot{\mathbf{u}}:(\nabla\dot{\mathbf{u}} + \dot{\mathbf{u}}\nabla)\right], \tag{10.161}$$

$$\Sigma_1 = -\left[\lambda\theta\dot{\mathbf{u}} + \mu\dot{\mathbf{u}}\cdot(\nabla\mathbf{u} + \mathbf{u}\nabla)\right], \tag{10.162}$$

$$\Sigma_2 = -\frac{1}{\omega}\left[\lambda^*\dot{\theta}\dot{\mathbf{u}} + \mu^*\dot{\mathbf{u}}\cdot(\nabla\dot{\mathbf{u}} + \dot{\mathbf{u}}\nabla)\right]. \tag{10.163}$$

In these equations, H is the total mechanical energy per unit volume (Hamiltonian density) and D is the rate of mechanical energy dissipation per unit volume. The energy-flux density Σ is split into two parts: Σ_1 represents the rate of work exchange between the material inside S and outside it, and Σ_2 is the rate at which energy is convected across the surface S. The second law of thermodynamics requires that the total amount of dissipated energy increase with time.

Expressing D as the sum of its dilatational and deviatoric parts as we did in Eq. (1.84) for W, it can be shown that D is nonnegative if

$$\mu^*(\omega) \geq 0, \qquad k^*(\omega) \geq 0. \tag{10.164}$$

Note that the dissipation function D is quadratic in the components of the rate of strain dyadic.

10.3.2.2. Dissipation Parameters. The explicit expressions for **u** are needed to evaluate the expressions for H and D in Eqs. (10.160) and (10.161). Starting from the explicit P-wave potential, Eq. (10.147) becomes

$$\psi_\alpha = B_\alpha e^{-\mathbf{A}_\alpha \cdot \mathbf{r} + i(\omega t - \mathbf{P}_\alpha \cdot \mathbf{r})}, \tag{10.165}$$

and the corresponding real displacement field is

$$\begin{aligned} \mathbf{u}_P &= \mathrm{Re}\{\mathbf{U}_P\} = \mathrm{Re}\{\nabla\psi_\alpha\} = e^{-\mathbf{A}_\alpha \cdot \mathbf{r}} \mathrm{Re}\left\{-i\mathbf{K}_\alpha B_\alpha e^{i(\omega t - \mathbf{P}_\alpha \cdot \mathbf{r})}\right\} \\ &= e^{-\mathbf{A}_\alpha \cdot \mathbf{r}} |B_\alpha \hat{k}_\alpha(\omega)| \mathrm{Re}\left\{\left[\frac{\mathbf{P}_\alpha - i\mathbf{A}_\alpha}{\hat{k}_\alpha(\omega)}\right] \exp\left[i\left\{\omega t - \mathbf{P}_\alpha \cdot \mathbf{r} + \arg(\hat{k}_\alpha B_\alpha) - \frac{\pi}{2}\right\}\right]\right\}. \end{aligned} \tag{10.166}$$

Defining

$$\begin{aligned} \zeta(t) &= \omega t - \mathbf{P}_\alpha \cdot \mathbf{r} + \arg(\hat{k}_\alpha B_\alpha) - \frac{\pi}{2}, \\ W_\alpha &= |B_\alpha \hat{k}_\alpha(\omega)| e^{-\mathbf{A}_\alpha \cdot \mathbf{r}}, \\ \hat{k}_\alpha(\omega) &= k_\alpha - i k_\alpha^*, \\ \mathbf{e}_1 &= \frac{k_\alpha \mathbf{P}_\alpha + k_\alpha^* \mathbf{A}_\alpha}{|\hat{k}_\alpha^2|}, \qquad \mathbf{e}_2 = \frac{-k_\alpha^* \mathbf{P}_\alpha + k_\alpha \mathbf{A}_\alpha}{|\hat{k}_\alpha^2|}, \end{aligned} \tag{10.167}$$

Eq. (10.166) bears the compact form

$$\mathbf{u}_P = W_\alpha[\mathbf{e}_1 \cos\zeta(t) + \mathbf{e}_2 \sin\zeta(t)]. \tag{10.168}$$

It then follows from Eqs. (10.150) and (10.151) that

$$\mathbf{e}_1 \cdot \mathbf{e}_2 = 0, \qquad \mathbf{e}_1 \cdot \mathbf{e}_1 - \mathbf{e}_2 \cdot \mathbf{e}_2 = 1. \tag{10.169}$$

Equation (10.168) means that the particle motion for a P wave is elliptical in the plane of $\mathbf{P}_\alpha$ and $\mathbf{A}_\alpha$ with a major axis $W_\alpha|\mathbf{e}_1|$, minor axis $W_\alpha|\mathbf{e}_2|$, and eccentricity $1/|\mathbf{e}_1|$. The direction of rotation is from $\mathbf{P}_\alpha$ to $\mathbf{A}_\alpha$. The angle between the major axis of the ellipse and the direction of propagation is given by

$$\cos\eta_\alpha = \frac{\mathbf{e}_1 \cdot \mathbf{P}_\alpha}{|\mathbf{e}_1||\mathbf{P}_\alpha|} = \frac{k_\alpha}{|\hat{k}_\alpha|}\left[1 + \frac{(k_\alpha^*)^2}{|\mathbf{P}_\alpha|^2}\right]^{1/2}. \tag{10.170}$$

The S-wave solution is

$$\mathbf{u}_S = \mathrm{Re}\{\mathrm{curl}(\mathbf{b}\psi_\beta)\} \tag{10.171}$$

with the provision

$$\mathrm{div}(\mathbf{b}\psi_\beta) = -[\mathbf{A}_\beta + i\mathbf{P}_\beta] \cdot \mathbf{b}\psi_\beta = 0, \tag{10.172}$$

where **b** is a real unit vector, normal to both $\mathbf{A}_\beta$ and $\mathbf{P}_\beta$. It then follows from Eq. (10.171) that

$$\mathbf{u}_S = W_\beta[\mathbf{g}_1 \cos\xi(t) + \mathbf{g}_2 \sin\xi(t)], \tag{10.173}$$

where

$$\xi(t) = \omega t - \mathbf{P}_\beta \cdot \mathbf{r} + \arg(\hat{k}_\beta B_\beta) + \frac{\pi}{2},$$

$$W_\beta = |B_\beta \hat{k}_\beta(\omega)| e^{-\mathbf{A}_\beta \cdot \mathbf{r}}, \tag{10.174}$$

$$\hat{k}_\beta(\omega) = k_\beta - ik_\beta^*,$$

$$\mathbf{g}_1 = \left[\frac{k_\beta \mathbf{P}_\beta + k_\beta^* \mathbf{A}_\beta}{|\hat{k}_\beta|^2}\right] \times \mathbf{b}, \qquad \mathbf{g}_2 = \left[\frac{-k_\beta^* \mathbf{P}_\beta + k_\beta \mathbf{A}_\beta}{|\hat{k}_\beta|^2}\right] \times \mathbf{b}, \tag{10.175}$$

$$\mathbf{g}_1 \cdot \mathbf{g}_2 = 0, \qquad \mathbf{g}_1 \cdot \mathbf{g}_1 - \mathbf{g}_2 \cdot \mathbf{g}_2 = 1.$$

Hence, the particle motion for an *SV* wave is also elliptical in the plane of $\mathbf{P}_\beta$ and $\mathbf{A}_\beta$. The ellipse has a major axis $W_\beta |\mathbf{g}_1|$, minor axis $W_\beta |\mathbf{g}_2|$, and eccentricity $1/|\mathbf{g}_1|$. The direction of rotation is from $\mathbf{P}_\beta$ to $\mathbf{A}_\beta$. The angle η_β between the vector $\mathbf{P}_\beta \times \mathbf{b}$ and the major axis $\mathbf{g}_1$ is given by

$$\cos \eta_\beta = \frac{(\mathbf{P}_\beta \times \mathbf{b}) \cdot \mathbf{g}_1}{|\mathbf{P}_\beta \times \mathbf{b}||\mathbf{g}_1|} = \frac{k_\beta}{|\hat{k}_\beta|}\left[1 + \frac{(k_\beta^*)^2}{|\mathbf{P}_\beta|^2}\right]^{1/2}. \tag{10.176}$$

For homogeneous waves ($\chi_\alpha = 0, \chi_\beta = 0$)

$$\begin{aligned} |\mathbf{P}_\alpha| &= k_\alpha, & |\mathbf{A}_\alpha| &= k_\alpha^*, & |\mathbf{e}_2| &= 0, & \eta_\alpha &= 0, \\ |\mathbf{P}_\beta| &= k_\beta, & |\mathbf{A}_\beta| &= k_\beta^*, & |\mathbf{g}_2| &= 0, & \eta_\beta &= 0, \end{aligned} \tag{10.177}$$

and the ellipses degenerate into straight lines.

Having determined the *P*- and *SV*-wave particle motion, we substitute from Eqs. (10.168) and (10.173) into the expressions for *H*, *D*, and $\mathbf{\Sigma}$ given in Eqs. (10.160)–(10.163). The time parameter is eliminated by averaging over a time interval equal to one period $T = 2\pi/\omega$. The analysis eventually yields the following exact formulas for either *P*- or *SV* waves

$$\langle H_1 \rangle = \langle \text{kinetic energy density} \rangle = \frac{1}{4}\rho\omega^2 |B|^2 e^{-2\mathbf{A}\cdot\mathbf{r}}[|\mathbf{P}|^2 + |\mathbf{A}|^2], \tag{10.178}$$

$$\langle H_2 \rangle = \langle \text{potential energy density} \rangle$$

$$= \frac{1}{4}|B|^2 e^{-2\mathbf{A}\cdot\mathbf{r}}[\rho\omega^2(|\mathbf{P}|^2 - |\mathbf{A}|^2) + 8\mu|\mathbf{P} \times \mathbf{A}|^2], \tag{10.179}$$

$$\langle H \rangle = \langle H_1 + H_2 \rangle = |B|^2 e^{-2\mathbf{A}\cdot\mathbf{r}}\left[\frac{1}{2}\rho\omega^2|\mathbf{P}|^2 + 2\mu|\mathbf{P} \times \mathbf{A}|^2\right], \tag{10.180}$$

$$\langle D \rangle = \omega |B|^2 e^{-2\mathbf{A}\cdot\mathbf{r}}[\rho\omega^2(\mathbf{P}\cdot\mathbf{A}) + 4\mu^*|\mathbf{P} \times \mathbf{A}|^2], \tag{10.181}$$

$$\langle \mathbf{\Sigma} \rangle = \omega |B|^2 e^{-2\mathbf{A}\cdot\mathbf{r}}\left[\frac{1}{2}\rho\omega^2\mathbf{P} + 2(\mathbf{P} \times \mathbf{A}) \times (\mu^*\mathbf{P} - \mu\mathbf{A})\right]. \tag{10.182}$$

These expressions admit the simple relations

$$\langle H \rangle = \frac{1}{\omega}\mathbf{P}\cdot\langle \mathbf{\Sigma} \rangle, \qquad \langle D \rangle = 2\mathbf{A}\cdot\langle \mathbf{\Sigma} \rangle. \tag{10.183}$$

This means that the mean energy density is determined by the component of the mean energy-flux density in the direction of propagation and the mean rate of energy dissipation density depends on the component of the mean energy flux in the direction of maximum attenuation.

A useful concept derived from these expressions is the *energy velocity* (either P or S)

$$\mathbf{V}_E = \frac{\langle \mathbf{\Sigma} \rangle}{\langle H \rangle} = \frac{\omega \langle \mathbf{\Sigma} \rangle}{\mathbf{P} \cdot \langle \mathbf{\Sigma} \rangle}. \tag{10.184}$$

For homogeneous waves (either P or S), $\mathbf{P} \times \mathbf{A} = 0$, and therefore

$$\begin{aligned}
\langle \mathbf{\Sigma} \rangle &= \left\{ \frac{1}{2} \rho \omega^3 |B|^2 e^{-2\mathbf{A}\cdot\mathbf{r}} \right\} \mathbf{P}, \\
\mathbf{V}_E &= \frac{\omega}{|\mathbf{P}|^2} \mathbf{P} \qquad \text{[cf. Eq. (10.153)]}, \\
\langle D \rangle &= \rho \omega^3 |B|^2 e^{-2\mathbf{A}\cdot\mathbf{r}} |\mathbf{P}| |\mathbf{A}|, \\
\langle H \rangle &= \frac{1}{2} \rho \omega^2 |B|^2 e^{-2\mathbf{A}\cdot\mathbf{r}} |\mathbf{P}|^2, \\
\langle 2H_2 \rangle &= \frac{1}{2} \rho \omega^2 |B|^2 e^{-2\mathbf{A}\cdot\mathbf{r}} (|\mathbf{P}|^2 - |\mathbf{A}^2|), \\
\frac{1}{\omega} \frac{\langle D \rangle}{\langle 2H \rangle} &= \frac{|\mathbf{A}|}{|\mathbf{P}|}.
\end{aligned} \tag{10.185}$$

For an elastic solid

$$\langle \mathbf{\Sigma} \rangle = \left\{ \frac{1}{2} \omega |B|^2 e^{-2\mathbf{A}\cdot\mathbf{r}} [\rho \omega^2 + 4\mu |\mathbf{A}|^2] \right\} \mathbf{P}. \tag{10.186}$$

Hence, for homogeneous waves in either elastic or anelastic solids and for inhomogeneous waves in elastic solids, the direction of maximum energy flow is the same as the direction in which planes of constant phase propagate. However, for inhomogeneous waves in anelastic solids the two directions do not necessarily coincide.

We now return to our basic definition of $1/Q$ in Eq. (10.33) and apply it to the present case. Clearly, the peak energy density during the cycle is equal to the maximum value of the potential energy density, i.e., $\max(H_2)$. The average dissipated energy per unit volume per unit time during a cycle is $\langle D \rangle$. The total dissipated energy during a cycle is $T\langle D \rangle$ per unit volume.

Hence, the desired expression for Q^{-1} is

$$\frac{1}{Q} = \frac{1}{2\pi} \frac{T\langle D \rangle}{\max(H_2)} = \frac{[\rho \omega^2 (\mathbf{P} \cdot \mathbf{A}) + 4\mu^* |(\mathbf{P} \times \mathbf{A})|^2]}{[\frac{1}{2} \rho \omega^2 (|\mathbf{P}|^2 - |\mathbf{A}|^2) + 2\mu |(\mathbf{P} \times \mathbf{A})|^2]}. \tag{10.187}$$

For the sake of seismic wave propagation in the earth's crust and mantle it is reasonable to assume that the wave is homogeneous and that

$$|\mathbf{A}|^2 \ll |\mathbf{P}|^2. \tag{10.188}$$

Then, using Eq. (10.177), Eq. (10.187) reduces to

$$\frac{1}{Q_\alpha} \simeq 2\frac{|\mathbf{A}_\alpha|}{|\mathbf{P}_\alpha|} = \frac{2k_\alpha^*}{k_\alpha}, \tag{10.189}$$

$$\frac{1}{Q_\beta} \simeq 2\frac{|\mathbf{A}_\beta|}{|\mathbf{P}_\beta|} = \frac{2k_\beta^*}{k_\beta}. \tag{10.190}$$

However, because of the relations (10.127)–(10.132), the smallness of k_β^*/k_β implies also the smallness of the ratios μ^*/μ and β^*/β. Expanding $\hat{\beta}(\omega)$ in power series[4] of (μ^*/μ), we have

$$\hat{\beta}(\omega) = \left(\frac{\mu}{\rho}\right)^{1/2}\left[1 + i\frac{\mu^*}{2\mu}\right] = \beta\left[1 + i\frac{\mu^*}{2\mu}\right].$$

Consequently,

$$\hat{k}_\beta(\omega) = \frac{\omega}{\hat{\beta}(\omega)} = \frac{\omega}{\beta}\left[1 - i\frac{\mu^*}{2\mu}\right] = k_\beta - ik_\beta^* = \frac{\omega}{c_\beta(\omega)} - i\gamma_\beta(\omega).$$

Therefore, to the same order of smallness

$$\frac{1}{Q_\beta} = \frac{2k_\beta^*}{k_\beta} = \frac{\mu^*}{\mu} = \frac{2\beta^*}{\beta}, \tag{10.191}$$

$$\beta(\omega) = c_\beta(\omega) = \left(\frac{\mu}{\rho}\right)^{1/2}, \qquad \beta^* = \left(\frac{\beta}{2\mu}\right)\mu^*, \qquad \gamma_\beta(\omega) = \frac{\omega\mu^*}{2\beta\mu}, \tag{10.192}$$

$$\hat{\beta}(\omega) = \beta(\omega)\left[1 + \frac{i}{2Q_\beta(\omega)}\right], \qquad \hat{k}_\beta(\omega) = k_\beta\left[1 - \frac{i}{2Q_\beta(\omega)}\right]. \tag{10.193}$$

In the same way, we can show that

$$\frac{1}{Q_\alpha} = \frac{2k_\alpha^*}{k_\alpha} = \frac{\kappa^* + \frac{4}{3}\mu^*}{\kappa + \frac{4}{3}\mu} = \frac{2\alpha^*}{\alpha}, \tag{10.194}$$

$$\alpha(\omega) = c_\alpha(\omega) = \left(\frac{\kappa + \frac{4}{3}\mu}{\rho}\right)^{1/2}, \qquad \gamma_\alpha(\omega) = \frac{\omega}{2\alpha}\frac{\kappa^* + \frac{4}{3}\mu^*}{\kappa + \frac{4}{3}\mu}, \tag{10.195}$$

$$\hat{\alpha}(\omega) = \alpha(\omega)\left[1 + \frac{i}{2Q_\alpha(\omega)}\right], \qquad \hat{k}_\alpha(\omega) = k_\alpha\left[1 - \frac{i}{2Q_\alpha(\omega)}\right], \tag{10.196}$$

$$\frac{Q_\beta}{Q_\alpha} = \left[\frac{\beta(\omega)}{\alpha(\omega)}\right]^2\left[\frac{4}{3} + \frac{\kappa^*}{\mu^*}\right] = \left(\frac{\beta_\infty}{\alpha_\infty}\right)^2\left[\frac{4}{3} + \frac{\kappa^*}{\mu^*}\right]. \tag{10.197}$$

It is sometimes convenient to define a *bulk dissipation parameter*

$$\frac{1}{Q_\kappa} = \frac{\kappa^*}{\kappa}. \tag{10.198}$$

[4] A more exact procedure would be that of expanding the anelastic state in a Taylor series about the elastic ($\omega = \infty$) state, i.e., $\hat{\mu}(\omega) = \mu_\infty + \delta\mu$, etc., where $\delta\mu = (\mu - \mu_\infty) + i\mu^*$. However, because $\mu^* \ll \mu_\infty$, the two approximations yield the same result, as far as terms of order $1/Q$ are concerned.

This parameter is linked to the longitudinal and shear dissipation parameters by the relations

$$\frac{1}{Q_\alpha} = \frac{L}{Q_\beta} + \frac{1 - L}{Q_\kappa}, \qquad L = \frac{4}{3}\left[\frac{\beta(\omega)}{\alpha(\omega)}\right]^2, \tag{10.199}$$

$$\frac{1 + \frac{4}{3}(\mu/\kappa)}{Q_\alpha} = \frac{1}{Q_\kappa} + \frac{4}{3}\left(\frac{\mu}{\kappa}\right)\frac{1}{Q_\beta}. \tag{10.200}$$

Observations made on the relative damping of P and S waves in the earth point to the fact that P waves are in general less attenuated than S waves at the same frequency over the same path. This observation can be explained by the assumption that bulk losses are smaller than shear losses in the earth's mantle. It then follows from Eqs. (10.192) and (10.195) that

$$\frac{\gamma_\beta}{\gamma_\alpha} = \begin{cases} \dfrac{3}{4}\left(\dfrac{\alpha}{\beta}\right)^3 & \text{if} \quad \kappa^* = 0 \quad \text{(upper bound)} \\ \dfrac{3}{7}\left(\dfrac{\alpha}{\beta}\right)^3 & \text{if} \quad \kappa^* = \mu^* \quad \text{(lower bound).} \end{cases} \tag{10.201}$$

Therefore, if all losses result from shear, Q_α is about 2.4–2.6 times the value of Q_β for velocity ratios that occur in the earth's mantle. Because α/β in the mantle is about 1.84, shear waves at a given period will attenuate about $\frac{3}{4}(1.84)^3 \simeq 4.6$ times faster along the ray than a P wave having the same period. By requiring the density to be real, we ignore the possibility of losses from *imperfect inertia.* If this is not the case, we assume $0 < \rho^*/\rho \ll 1$ and then

$$\frac{\mu + i\mu^*}{\rho + i\rho^*} \simeq \frac{\mu}{\rho}\left[1 + i\left(\frac{\mu^*}{\mu} - \frac{\rho^*}{\rho}\right)\right].$$

Consequently,

$$\frac{2k_\beta^*}{k_\beta} = \frac{\mu^*}{\mu} - \frac{\rho^*}{\rho}. \tag{10.202}$$

Likewise

$$\frac{2k_\alpha^*}{k_\alpha} = \frac{\kappa^* + \frac{4}{3}\mu^*}{\kappa + \frac{4}{3}\mu} - \frac{\rho^*}{\rho}. \tag{10.203}$$

The last two equations can also be written in the form

$$\frac{1}{Q_\beta} = \frac{1}{Q_\mu} - \frac{1}{Q_\rho}, \qquad \frac{1}{Q_\alpha} = \frac{L}{Q_\mu} + \frac{1 - L}{Q_\kappa} - \frac{1}{Q_\rho}, \tag{10.204}$$

where

$$\frac{1}{Q_\rho} = \frac{\rho^*}{\rho}, \qquad \frac{1}{Q_\mu} = \frac{\mu^*}{\mu}. \tag{10.205}$$

Equations (10.204) represent the total damping in terms of imperfections in rigidity, bulk modulus, and density.

EXAMPLE 10.3: Plane Waves in Uniaxially Deformed Media: Longitudinal Strain

In a state of *elastic longitudinal strain* [cf. Eq. (1.145)], the nonzero components of elastic stress, strain, and displacements are

$$\varepsilon_{11} = \frac{\partial u_1}{\partial x_1}, \qquad \sigma_{11} = (\lambda + 2\mu)\frac{\partial u_1}{\partial x_1}, \qquad \sigma_{22} = \sigma_{33} = \lambda\frac{\partial u_1}{\partial x_1},$$
$$\mathbf{u} = u_1(x_1, t)\mathbf{e}_1, \qquad \operatorname{div} \mathfrak{T}_e = (\lambda + 2\mu)\frac{\partial^2 u_1}{\partial x_1^2}\,\mathbf{e}_1. \tag{10.3.1}$$

Denoting $\sigma_{11} = \sigma$, $\varepsilon_{11} = \varepsilon$, we have from Eq. (10.101), for the *anelastic* case,

$$\operatorname{div} \mathfrak{T} = \left[(\lambda_\infty + 2\mu_\infty)\frac{\partial^2 u_1}{\partial x_1^2} + \int_0^t \frac{\partial^2 u_1(x_1, \tau)}{\partial x_1^2}\,\dot{\psi}(t - \tau)d\tau\right]\mathbf{e}_1.$$

The Cauchy equation [cf. Eq. (1.100)] in the absence of body forces is

$$\rho\frac{\partial^2 \mathbf{u}}{\partial t^2} = \operatorname{div} \mathfrak{T}. \tag{10.3.2}$$

Therefore

$$\frac{1}{\alpha_\infty^2}\frac{\partial^2 u_1}{\partial t^2} = \frac{\partial^2 u_1}{\partial x_1^2} + \frac{1}{\lambda_\infty + 2\mu_\infty}\int_0^t \frac{\partial^2 u_1(x_1, \tau)}{\partial x_1^2}\,\dot{\psi}(t - \tau)d\tau, \tag{10.3.3}$$

where $\alpha_\infty^2 = (\lambda_\infty + 2\mu_\infty)/\rho$. The application of the Fourier transform to Eq. (10.3.3) leads us to the ordinary differential equation

$$\frac{d^2 U}{dx_1^2} + \frac{\omega^2}{\alpha_\infty^2}\hat{n}_\alpha^2 U = 0, \tag{10.3.4}$$

where

$$U(x_1, \omega) = \int_{-\infty}^{\infty} u_1(x_1, t)e^{-i\omega t}\,dt,$$

$$\hat{n}_\alpha(\omega) = \left[1 + \frac{1}{\lambda_\infty + 2\mu_\infty}\int_0^t \dot{\psi}(t)e^{-i\omega t}\,dt\right]^{-1/2}.$$

EXAMPLE 10.4: Plane Waves in Uniaxially Deformed Media: Longitudinal Stress

Consider next a state of *elastic longitudinal stress* [cf. Eqs. (1.147)–(1.148)], in which the elastic deformation tensors are

$$\mathfrak{T} = \sigma_{11}\mathbf{e}_1\mathbf{e}_1, \qquad \mathfrak{E} = \frac{\sigma_{11}}{2\mu}\left[\mathbf{e}_1\mathbf{e}_1 - \frac{\lambda}{3\lambda + 2\mu}\mathfrak{I}\right]. \tag{10.4.1}$$

We seek a plane-wave solution such that the displacement depends only on x_1 and t and denote: $\sigma = \sigma_{11}, \varepsilon = \varepsilon_{11} = \partial u_1/\partial x_1 = \sigma_{11}/Y_\infty$. The anelastic deformation is then given by the superposition law, Eq. (10.99),

$$\varepsilon_{11}(t) = \frac{1}{Y_\infty}\sigma_{11}(t) + \int_0^t \sigma_{11}(\tau)\dot{\phi}(t-\tau)d\tau. \tag{10.4.2}$$

Differentiating this equation with respect to x_1 and using Eq. (10.3.2), we have

$$\frac{\partial^2 u_1}{\partial x_1^2} = \frac{1}{c_\infty^2}\left[\frac{\partial^2 u_1}{\partial t^2} + Y_\infty \int_0^t \frac{\partial^2 u_1(\tau)}{\partial \tau^2}\dot{\phi}(t-\tau)d\tau\right], \tag{10.4.3}$$

where

$$Y_\infty = \frac{\mu_\infty}{\lambda_\infty + \mu_\infty}(3\lambda_\infty + 2\mu_\infty), \qquad c_\infty^2 = \frac{Y_\infty}{\rho}, \qquad \dot{\phi}(t) > 0. \tag{10.4.4}$$

Taking the Fourier transform of Eq. (10.4.3), we obtain

$$\frac{d^2U}{dx_1^2} + \frac{\omega^2}{c_\infty^2}\hat{n}^2 U = 0, \tag{10.4.5}$$

where

$$\hat{n}(\omega) = \left[1 + Y_\infty \int_0^\infty \dot{\phi}(t)e^{-i\omega t}dt\right]^{1/2}, \qquad U(x_1, \omega) = \int_{-\infty}^\infty u_1(x_1, t)e^{-i\omega t}dt. \tag{10.4.6}$$

10.3.3. Causality and Dispersion Relations

In this section we shall derive certain relations between the real and imaginary parts of the complex propagation functions. These relations follow from the principle of causality and predict a causal dispersion resulting from anelasticity regardless of the physical absorption mechanism. The principle of causality, in its strict sense, simply states that in a physical system, no output can occur before the input. Formulated in terms of signals that carry information across the medium, it says that a signal originating at $x = 0$, $t = 0$ cannot arrive at the distance $x = ct$ before time t, where c is some finite limiting velocity characteristic of the medium. Now, the theory of viscoelasticity is embodied in the Boltzmann equation. In one dimension [cf. Eq. (10.99)]

$$\varepsilon(t) = \int_0^t \sigma(\tau)\dot{\phi}(t-\tau)d\tau + \varepsilon_\infty\frac{\sigma(t)}{\sigma_0}, \qquad (10.206)^5$$

[5] Substances for which $\varepsilon_\infty = \varepsilon(0) = 0$ need special treatment. For such substances, Eq. (10.206) becomes $\varepsilon(t) = \int_0^t \sigma(\tau)\dot{\phi}(t-\tau)d\tau$. Thus in the case of a Kelvin–Voigt solid we obtain [cf. Eqs. (10.45), (10.58a), and (10.60)]

$$\varepsilon(t) = \frac{\sigma_0}{\mu}(1 - e^{-t/\tau_0}), \qquad \phi(t) = \frac{1}{\mu}(1 - e^{-t/\tau_0}),$$

$$\dot{\phi}(t) = \frac{1}{\eta}e^{-t/\tau_0}, \qquad S(\omega) = (1 + i\omega\tau_0)^{-1},$$

$$\hat{n}(\omega) = [S(\omega)]^{1/2} = (1 + i\omega\tau_0)^{-1/2}.$$

where $\phi(t)$, the creep function, is determined by the mechanism of dissipation in the medium. Because this equation assumes that the strain at time t, $\varepsilon(t)$, is caused by a linear superposition of the total stress history, $\sigma(t)$, up to time t, it must incorporate both the property of linearity and the causality principle. We have already mentioned in Section 10.2.5 that Boltzmann's equation has the form of Duhamel's integral in which $\sigma(t)$ is the input, $\varepsilon(t)$ is the output, and $\dot{\phi}(t)$ has the role of the *impulse response.*

Therefore, the causality principle, inherent in Boltzmann's equation, is carried further into the anelastic Navier equation of motion that has been derived from it. We have shown in Example 10.4 that a longitudinal plane wave traveling in a medium whose viscoelastic behavior is governed by Eq. (10.206), satisfies the equation

$$\frac{d^2U}{dx^2} + \frac{\omega^2}{c_\infty^2}\hat{n}^2 U = 0. \tag{10.207}$$

The refraction index $\hat{n}(\omega)$ is related to the phase velocity, attenuation coefficient, and rate of creep functions via Eqs. (10.136) and (10.4.6)

$$\hat{n}(\omega) = [1 + S(\omega)]^{1/2} = [1 + R(\omega)]^{1/2}\left[1 - \frac{i}{Q(\omega)}\right]^{1/2} = \frac{c_\infty}{c(\omega)} - i\frac{c_\infty}{\omega}\gamma(\omega),$$

$$S(\omega) = Y_\infty \int_0^\infty \dot{\phi}(t)e^{-i\omega t}\,dt = R(\omega) - iX(\omega); \tag{10.208}$$

$$\frac{1}{Q(\omega)} = \frac{X(\omega)}{1 + R(\omega)} = \frac{\int_0^\infty \dot{\phi}(t)\sin\omega t\,dt}{Y_\infty^{-1} + \int_0^\infty \dot{\phi}(t)\cos\omega t\,dt}. \tag{10.209}$$

The attenuation function $\gamma(\omega)$ [and through it, the velocity function $c(\omega)$[6]] are determined by the physical mechanism of attenuation.

Given the initial condition

$$u(0, t) = f(t) = \frac{u_0}{2\pi}\int_{-\infty}^{\infty} F(\omega)e^{i\omega t}\,d\omega, \tag{10.210}$$

the solution of Eq. (10.207) is represented by the Fourier integral [cf. Eq. (K.5)]

$$u(x, t) = \frac{u_0}{2\pi}\int_{-\infty}^{\infty} F(\omega)\exp\left\{i\omega\left[t - \frac{x}{c_\infty}\hat{n}(\omega)\right]\right\}d\omega. \tag{10.211}$$

Equation (10.211) states that a plane wave traveling in a linear viscoelastic medium can be represented as a superposition of plane waves of all frequencies.

Given $f(t)$, $\phi(t)$, and c_∞, we may, in principle, evaluate the shape of the pulse $u(x, t)$ for all values of x and t. In addition, we may calculate the functions $c(\omega)$, $\gamma(\omega)$, and $Q(\omega)$ and compare them with observations in order to obtain

[6] In this context the medium is assumed to be homogeneous and the frequency dependence of the velocity results totally from anelasticity. The explicit relations between $c(\omega)$ and $\gamma(\omega)$ are the main concern of this section.

the best estimate for $\phi(t)$ in the earth. However, before we apply the theory to earth data, we shall be concerned with certain interrelations among the spectral functions that follow from the principle of causality. Because most of the results are proved in Appendix K, we shall list here their important features only:

1. If $\dot{\phi}(t)$ is real, causal, and devoid of impulses, the relations given in Eqs. (K.5)–(K.21) hold. In particular

$$Q(-\omega) = -Q(\omega), \qquad \lim_{\omega \to \infty} i\omega S(\omega) = Y_\infty \dot{\phi}(0). \tag{10.212}$$

2. Let $S(z)$ be a *bounded analytic function* of the complex variable $z = \omega - i\Omega$ in the lower half of the z plane and let $[1 + S(z)]$ have no zeros in the lower half of the z plane such that $\hat{n}(z)$ is also analytic in this region. Then, the real and imaginary parts of $\hat{n}(z)$ in the limit $\Omega \to 0$ are Hilbert transforms of each other. These are the so-called *dispersion relations*[7] [Eqs. (K.26) and (K.30)] which bear the explicit form

$$\frac{c_\infty}{c(\omega)} = 1 + \frac{2c_\infty}{\pi} \mathrm{P} \int_0^\infty \frac{\gamma(x)}{x^2 - \omega^2}\, dx = 1 + \frac{c_\infty}{\omega\pi} \int_0^\infty \frac{d}{dx}[\gamma(x)] \ln \left| \frac{x + \omega}{x - \omega} \right| dx \geq 0, \tag{10.213}$$

$$\gamma(\omega) = \frac{-2\omega^2}{\pi} \mathrm{P} \int_0^\infty \frac{1}{c(x)} \frac{dx}{x^2 - \omega^2} \geq 0. \tag{10.214}$$

From Eqs. (10.213) and (10.214) it follows that

$$\lim_{\omega \to \infty} c(\omega) = c_\infty; \qquad c(-\omega) = c(\omega), \qquad \gamma(-\omega) = \gamma(\omega). \tag{10.215}$$

These results are in accord with our previous notion that c_∞ is the wave velocity in our *reference inviscid medium.*

3. If $S(z)$ and $F(z)$ obey the condition given in the preceding section and if, in addition, $\lim \operatorname{Im}\{\hat{n}(z)\} = 0$, $\lim \operatorname{Re}\{\hat{n}(z)\} = 1$, as $|z| \to \infty$, then $u(0, t < 0) = 0$ implies $u(x, t < x/c_\infty) = 0$. The proof of this statement is given in Example 10.5. It means that even though Eq. (10.213) does not prohibit $c(\omega) > c_\infty$ for some frequencies, the signal velocity (i.e., the velocity of propagation of energy and information) in the viscoelastic medium can at most be c_∞. As we have stated before, this statement is equivalent to the principle of causality.
4. It can be shown that if

$$\int_0^\infty \frac{\gamma(\omega)}{1 + \omega^2}\, d\omega \tag{10.216}$$

[7] If $[1 + S(z)]$ vanishes at some $z = z_0$ in the lower half-plane then, because $u(x, t)$ is real, we must have $S^*(z_0) = S(-z_0^*)$. Consequently, $[1 + S(z)]$ will vanish also at $z = -z_0^*$. Because both z_0 and $(-z_0)^*$ lie in the lower half-plane, we must draw a *branch cut* from z_0 to $-z_0^*$ in order to render $\hat{n}(z)$ analytic in the lower half-plane. The dispersion relations will be modified accordingly.

converges, one can find a function $c(\omega)$ such that $u(x, t)$, as represented in Eq. (10.211), is causal (the Payley–Wiener theorem). The above integral converges if

$$\gamma(\omega) = O(\omega^m), \qquad 0 < m < 1. \tag{10.217}$$

This condition then defines a class of attenuation functions that can characterize the absorption of elastic waves in real media regardless of the absorption mechanism. Experiments with the attenuation of seismic waves in solids show that $\gamma(\omega)$ is almost linear over a very wide range of frequencies. The result in Eq. (10.217) states that a *linear dependence of $\gamma(\omega)$ over all frequencies is not compatible with causality.*

In Examples 10.6, 10.7, and 10.8, we use the dispersion relations to evaluate velocity dispersion and Q functions for a number of trial attenuation functions that are consistent with seismic observations.

EXAMPLE 10.5: The Principle of Limiting Velocity

Because $u(0, t < 0) = 0$ we may invert the Fourier transform in Eq. (10.210) and write

$$F(\omega) = \int_0^\infty u(0, t)e^{-i\omega t}\, dt. \tag{10.5.1}$$

We continue $F(\omega)$ into the complex $z = \omega - i\Omega$ plane through the definition

$$F(z) = \int_0^\infty u(0, t)e^{-izt}\, dt, \tag{10.5.2}$$

where $F(z)$ is analytic for $\Omega > 0$ and $|F(z)| \to 0$ as $|z| \to \infty$ on the infinite arc. We also assume that $\hat{n}(z)$ is analytic in $\Omega > 0$ and that

$$\operatorname{Im}\{\hat{n}(z)\} \to 0, \qquad \operatorname{Re}\{\hat{n}(z)\} \to 1, \qquad |z| \to \infty. \tag{10.5.3}$$

Because both $F(z)$ and $\hat{n}(z)$ are analytic in the lower half-plane, we must have

$$\int_\Gamma F(z)\exp\left\{iz\left[t - \frac{\hat{n}(z)}{c_\infty}x\right]\right\}dz = 0; \tag{10.5.4}$$

where the contour Γ is shown in Fig. 10.14. Hence

$$\int_{-R}^{R} F(\omega)\exp\left\{i\omega\left[t - \frac{\hat{n}(\omega)}{c_\infty}x\right]\right\}d\omega + \int_{S_R} F(z)\exp\left\{iz\left[t - \frac{\hat{n}(z)}{c_\infty}x\right]\right\}dz = 0, \tag{10.5.5}$$

$$z = R\cos\theta - iR\sin\theta, \qquad 0 \le \theta \le \pi.$$

Denoting the integral over S_R by I_R and using relations (10.5.3), we have

$$|I_R| \le 2R\int_0^{\pi/2} |F(z)|\exp\left[-R\sin\theta\left(\frac{x}{c_\infty} - t\right)\right]d\theta$$

$$\le 2R\int_0^{\pi/2} |F(z)|\exp\left[-\frac{2\theta}{\pi}R\left(\frac{x}{c_\infty} - t\right)\right]d\theta$$

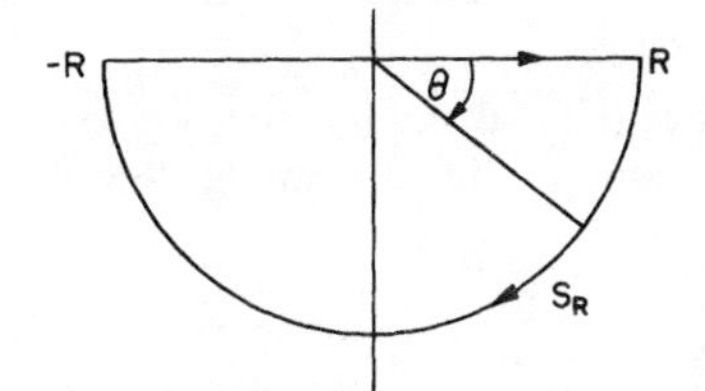

Figure 10.14. Integration contour in the complex $z = \omega - i\Omega$ plane used to demonstrate the principle of limiting velocity.

because $\sin\theta \le 2\theta/\pi$ for $0 \le \theta \le \pi/2$. Also, because $|F(z)| \to 0$ on S_R as $|z| \to \infty$, we may write $|F(z)| \le a/R^\lambda$, ($\lambda > 0$). Then

$$|I_R| \le 2\frac{a}{R^{\lambda-1}}\int_0^{\pi/2} \exp\left[-\frac{2\theta R}{\pi}\left(\frac{x}{c_\infty} - t\right)\right] d\theta$$

$$= \frac{\pi a}{R^\lambda}\left\{\frac{1 - \exp[-R(x/c_\infty - t)]}{(x/c_\infty - t)}\right\}. \tag{10.5.6}$$

As $R \to \infty$ while $x > c_\infty t$, I_R will vanish and by Eqs. (10.5.5) and (10.211) we shall have $u(x, t < x/c_\infty) = 0$. Therefore

$$u(0, t < 0) = 0 \quad \text{implies} \quad u\left(x, t < \frac{x}{c_\infty}\right) = 0. \tag{10.5.7}$$

EXAMPLE 10.6: Causal Dispersion in Solids with Power Law Attenuation

We consider a medium in which

$$\gamma(\omega) = m|\omega|^{1-\nu}, \qquad 0 < \nu < 1. \tag{10.6.1}$$

Using Eq. (K.24) and Table F.1, the causal velocity dispersion is

$$\frac{c_\infty}{c(\omega)} = 1 + \frac{mc_\infty}{\pi} \mathrm{P}\int_{-\infty}^{\infty} \operatorname{sgn} x \frac{|x|^{-\nu}\,dx}{x - \omega} = 1 + c_\infty m|\omega|^{-\nu}\tan\frac{\pi(1-\nu)}{2}. \tag{10.6.2}$$

Consequently, for $\omega > 0$

$$Q(\omega) = \frac{\omega}{2\gamma(\omega)c(\omega)} = \frac{1}{2}\tan\frac{\pi(1-\nu)}{2} + \frac{1}{2c_\infty m}\omega^\nu, \tag{10.6.3}$$

$$c(\omega) = c_\infty\left[1 + \frac{c_\infty m}{|\omega|^\nu}\tan\frac{\pi(1-\nu)}{2}\right]^{-1}. \tag{10.6.4}$$

The impulse response of our solid is [cf. Eq. (10.211)]

$$u(x, t > 0) = \frac{u_0}{\pi}\int_0^\infty e^{-mx\omega^{1-\nu}} \cos\left[\omega\left(t - \frac{x}{c_\infty}\right) - mx\omega^{1-\nu}\tan\frac{\pi(1-\nu)}{2}\right] d\omega. \tag{10.6.5}$$

EXAMPLE 10.7: "Constant-Q" Solid

We apply the dispersion relation, Eq. (K.29), to the real and imaginary parts of $\hat{\mu}(\omega) = \mu + i\mu^*$. So, if $\hat{\mu}(z)$ is a bounded analytic function in the lower $z = \omega - i\Omega$ plane,

$$\mu^*(\omega) = \frac{1}{\pi}\int_0^\infty \left[\frac{d}{dx}\mu(x)\right]\ln\left|\frac{x+\omega}{x-\omega}\right| dx$$
$$\simeq \frac{\pi}{2}\left[\frac{d\mu}{d\omega}\right]\omega \quad \text{[as in Eq. (K.39)]}. \tag{10.7.1}$$

Let us further assume that in our solid

$$\mu^* = \frac{1}{Q_0}\mu \quad \text{for all} \quad \omega, \tag{10.7.2}$$

where $Q_0 \gg 1$ is independent of frequency. Then, it follows from Eqs. (10.7.1) and (10.7.2) that

$$\mu(\omega) = \mu(\omega_r)\exp\left[\left(\frac{2}{\pi Q_0}\right)\ln\left(\frac{\omega}{\omega_r}\right)\right] \simeq \mu(\omega_r)\left[1 + \frac{2}{\pi Q_0}\ln\frac{\omega}{\omega_r}\right], \tag{10.7.3}$$

where $\omega_r > \omega$ is some fixed reference frequency and Q_0 is large. Because $c(\omega) = [\mu(\omega)/\rho]^{1/2}$, Eq. (10.7.3) yields

$$c(\omega) \simeq c(\omega_r)\left[1 - \frac{1}{\pi Q_0}\ln\frac{\omega_r}{\omega}\right], \tag{10.7.4}$$

$$\gamma(\omega) = \gamma(\omega_r)\left[\frac{\omega}{\omega_r}\right]^{1-1/\pi Q_0} \simeq \left[\frac{\gamma(\omega_r)}{\omega_r}\right]\omega, \tag{10.7.5}$$

$$\frac{1}{Q(\omega)} = \frac{1}{Q(\omega_r)}\left[1 - \frac{1}{\pi Q_0}\ln\frac{\omega_r}{\omega}\right]. \tag{10.7.6}$$

EXAMPLE 10.8: A "Boxcar-Q" Solid

Denote the real and imaginary parts of the complex refraction index by n and $-n^*$, respectively,

$$\hat{n}(\omega) = \frac{c_\infty}{c(\omega)} - i\frac{c_\infty}{\omega}\gamma(\omega) = n - in^*. \tag{10.8.1}$$

To first order in Q^{-1},

$$\gamma(\omega) = \frac{\omega}{2c_\infty Q(\omega)}. \tag{10.8.2}$$

Accordingly,

$$n = \frac{c_\infty}{c(\omega)}, \qquad n^* = \frac{1}{2Q(\omega)}. \tag{10.8.3}$$

Now, consider an anelastic medium in which

$$\frac{1}{Q(\omega)} = \begin{cases} \dfrac{1}{Q_m} \operatorname{sgn} \omega, & \omega_1 \leq |\omega| \leq \omega_2 \\ 0, & \text{otherwise} \end{cases} \tag{10.8.4}$$

and Q_m is independent of frequency. Assume that $\hat{n}(\omega)$ qualifies analytically for the dispersion relation, Eq. (K.24), namely,

$$n(\omega) = \frac{c_\infty}{c(\omega)} = 1 + \frac{1}{\pi} \mathrm{P} \int_{-\infty}^{\infty} \frac{n^*(x)}{x - \omega} dx = 1 - \frac{1}{2\pi} \mathrm{P} \int_{-\infty}^{\infty} \frac{1}{Q(x)} \frac{dx}{\omega - x}. \tag{10.8.5}$$

Carrying out the integration according to Eq. (10.8.4), we find that in the interval $\omega_1 \ll \omega \ll \omega_2$

$$c(\omega) = c_\infty \left[1 - \frac{1}{2\pi Q_m} \ln \left|\frac{\omega^2 - \omega_1^2}{\omega_2^2 - \omega^2}\right|\right]^{-1} \simeq c_\infty \left[1 - \frac{1}{\pi Q_m} \ln \frac{\omega_2}{\omega}\right] \leq c_\infty. \tag{10.8.6}$$

10.3.4. Signal Distortion in Low-Loss Solids

The equation of motion of shear plane waves in a Kelvin–Voigt solid is obtained from Eq. (10.105) through the substitution $\mathbf{u} = u_\beta(x, t)\mathbf{e}_y$. Because div $\mathbf{u} = 0$ and curl curl $\mathbf{u} = -(\partial^2 u_\beta/\partial x^2)\mathbf{e}_y$, the resulting equation is

$$\frac{1}{\beta^2} \frac{\partial^2 u_\beta}{\partial t^2} = \frac{\partial^2}{\partial x^2} \left(u_\beta + \frac{\mu'}{\rho\beta^2} \frac{\partial u_\beta}{\partial t}\right). \tag{10.218}$$[8]

This equation bears similarity to Stokes' equation in one dimension for damped sound waves [cf. Eq. (9.23)]. In Eq. (10.218) we transform the dependent and independent variables in the following manner:

$$\omega_0 = \frac{\mu}{\mu'}, \qquad T = \omega_0 t, \qquad X = \frac{\omega_0}{\beta} x. \tag{10.219}$$

The ensuing nondimensionalized equation is

$$\frac{\partial^2}{\partial X^2} \left(u_\beta + \frac{\partial u_\beta}{\partial T}\right) = \frac{\partial^2 u_\beta}{\partial T^2}. \tag{10.220}$$

[8] $\beta = (\mu/\rho)^{1/2}$ in Sections 10.3.4 and 10.3.5 plays the role of $\beta_\infty = (\mu_\infty/\rho)^{1/2}$ of Section 10.3.1 and should not be confused with $\beta(\omega)$ of Eq. (10.127).

We shall show a number of different ways to solve this equation. First, the substitution of the integral representation

$$u_\beta = 2\int_0^\infty e^{-Xf(\lambda)}\cos[Xg(\lambda) - \lambda T]d\lambda \equiv 2\,\mathrm{Re}\int_0^\infty e^{-i\lambda h(\lambda)}\,d\lambda$$

into Eq. (10.220) yields the expressions

$$f(\lambda) = \lambda(1+\lambda^2)^{-1/4}\sin\left(\frac{1}{2}\tan^{-1}\lambda\right),$$

$$g(\lambda) = \lambda(1+\lambda^2)^{-1/4}\cos\left(\frac{1}{2}\tan^{-1}\lambda\right),$$

$$h(\lambda) = \frac{X}{(1+i\lambda)^{1/2}} - T.$$

A second method is by means of the rapidly converging infinite sum

$$u_\beta = u_0\sum_{n=0}^{\infty}\left(\frac{2}{X}\right)^{(n+2)/2}\Psi_n(\xi), \qquad \xi = \left(\frac{2}{X}\right)^{1/2}(T - X) = \left[\frac{2\beta\omega_0}{x}\right]^{1/2}\left(t - \frac{x}{\beta}\right). \tag{10.221}$$

The substitution from Eqs. (10.221) into Eq. (10.220) yields a difference differential equation for $\Psi_n(\xi)$

$$\begin{aligned}[16\Psi_n''' + 8\xi\Psi_n'' + 8(n+3)\Psi_n'] &+ [8\xi\Psi_{n-1}''' + (\xi^2 + 8n + 24)\Psi_{n-1}'' \\ &+ (2n+5)\xi\Psi_{n-1}' + (n+1)(n+3)\Psi_{n-1}] \\ &+ [\xi^2\Psi_{n-2}''' + (2n+5)\xi\Psi_{n-2}'' + (n+1)(n+3)\Psi_{n-2}'] = 0, \\ \Psi_{-m} &= 0 \qquad (m = 1, 2, 3, \ldots),\end{aligned} \tag{10.222}$$

where a prime means a differentiation with respect to ξ. Using the known integrals

$$\begin{aligned}\int_0^\infty y^{2n}e^{-y^2}\cos y\xi\,dy &= (-)^n\sqrt{\pi}e^{-\xi^2/4}2^{-(2n+1)}H_{2n}\left(\frac{\xi}{2}\right), \\ \int_0^\infty y^{2n+1}e^{-y^2}\sin y\xi\,dy &= (-)^n\sqrt{\pi}e^{-\xi^2/4}2^{-(2n+2)}H_{2n+1}\left(\frac{\xi}{2}\right), \qquad (\xi > 0)\end{aligned} \tag{10.223}$$

in which H_n is the *Hermite polynomial of degree n*, we find that Eq. (10.222) is solved by the functions

$$\Psi_n(\xi) = \frac{\sqrt{\pi}}{2}e^{-\xi^2/4}f_n(\xi), \tag{10.224}$$

where $f_n(\xi)$ are polynomials of degree $3n$ which are linear combinations of the Hermite polynomials. For example,

$$f_0 = -G_1, \qquad f_1 = \frac{3}{4}G_4, \qquad f_2 = -\left(\frac{5}{8}G_5 + \frac{9}{32}G_7\right), \quad \text{etc.,}$$

$$G_n(\xi) = \left(-\frac{1}{2}\right)^n H_n\left(\frac{\xi}{2}\right).$$

The displacement is given by the sum

$$u_\beta(x,t) = \frac{\sqrt{\pi}}{4} u_0 \exp\left[-\frac{\beta\mu}{2\mu' x}\left(t - \frac{x}{\beta}\right)^2\right] \sum_{n=0}^{\infty} \left(\frac{2}{X}\right)^{(n+2)/2} \left(\frac{3}{32}\right)^n \frac{1}{n!} S_n(\xi), \tag{10.225}$$

where the first three polynomials are

$$S_0 = \xi,$$
$$S_1 = \xi^4 - 12\xi^2 + 12,$$
$$S_2 = \xi^7 - \frac{298}{9}\xi^5 + \frac{2180}{9}\xi^3 - \frac{920}{3}\xi.$$

Higher polynomials can be derived without difficulty. Because of the presence of the factor $(\frac{3}{32})^n(1/n!)$, the convergence of the sum is very rapid at almost all values of x and t. The solution of Eq. (10.220) represents a waveform that simultaneously decays in amplitude and broadens as x is increased. Writing ξ in the form

$$\xi = \frac{t - x/\beta}{(x/2\omega_0\beta)^{1/2}}, \tag{10.226}$$

we see that the broadening is proportional to a *scale time* $(x/2\omega_0\beta)^{1/2}$.

A third method of solving Eq. (10.220), most commonly used in linear wave equations, is the substitution of a spectral plane wave solution,

$$u_\beta = e^{-\gamma x + i\omega(t - x/c)},$$

where γ and c are real functions of ω. The results are the same as in Eqs. (10.131) and (10.132), with $Q^{-1} = \omega\mu'/\mu$ and $[\mu(\omega)/\rho]^{1/2} = \beta$. Denoting $\tau_0 = \mu'/\mu = 1/\omega_0$, we obtain the explicit expressions for a Kelvin–Voigt solid

$$\gamma(\omega) = \frac{\omega}{\beta}\left[\frac{(1+\omega^2\tau_0^2)^{1/2} - 1}{2(1+\omega^2\tau_0^2)}\right]^{1/2},$$
$$c(\omega) = \beta\left[\frac{2(1+\omega^2\tau_0^2)[(1+\omega^2\tau_0^2)^{1/2} - 1]}{\omega^2\tau_0^2}\right]^{1/2}. \tag{10.227}$$

With the aid of the definition

$$\frac{\omega}{\omega_0} = \tan 2\theta,$$

Eqs. (10.227) are recast as

$$\gamma(\omega) = \frac{\omega_0}{\beta}\tan 2\theta \sin\theta(\cos 2\theta)^{1/2},$$
$$\frac{\omega}{c(\omega)} = \frac{\omega_0}{\beta}\tan 2\theta \cos\theta(\cos 2\theta)^{1/2}, \tag{10.228}$$
$$\frac{c(\omega)\gamma(\omega)}{\omega} = \tan\theta.$$

When $\omega/\omega_0 \ll 1$ it follows from Eqs. (10.227) that

$$\gamma(\omega) = \frac{\omega_0}{2\beta}\left(\frac{\omega}{\omega_0}\right)^2, \qquad c(\omega) = \beta\left[1 + \frac{3}{8}\left(\frac{\omega}{\omega_0}\right)^2 - \frac{17}{128}\left(\frac{\omega}{\omega_0}\right)^4 + \cdots\right] \simeq \beta,$$

$$Q^{-1}(\omega) = \frac{2c(\omega)\gamma(\omega)}{\omega} = \frac{\omega}{\omega_0}. \tag{10.229}$$

Therefore, for small values of the viscosity μ', the attenuation coefficient depends on the square of the frequency, whereas the dispersion resulting from attenuation is a second-order effect. This is indeed predicted by the dispersion relation in Eq. (10.213).

At the other extreme where $\omega/\omega_0 \gg 1$

$$\gamma(\omega) = \frac{\omega_0}{2\beta}\left(\frac{2\omega}{\omega_0}\right)^{1/2}, \qquad c(\omega) = \beta\left(\frac{2\omega}{\omega_0}\right)^{1/2}, \qquad Q^{-1}(\omega) = 2. \tag{10.230}$$

At $\omega = \omega_0$, known as the *corner frequency*, there is a change in the behavior of the attenuation and the onset of great dispersion. Writing Stokes' equation in the form

$$\frac{1}{\beta^2}\frac{\partial^2 u}{\partial t^2} = \frac{\partial^2}{\partial x^2}\left(u + \frac{1}{\omega_0}\frac{\partial u}{\partial t}\right) \tag{10.231}$$

we note that as the corner frequency moves toward higher values, the range over which viscous behavior prevails is increased.

To obtain the complex propagation functions for a Maxwell solid we solve Eq. (10.108) with $u(y, t) = e^{i[\omega t - \hat{k}_\beta(\omega)y]}$. The dispersion equation is

$$[\hat{k}_\beta(\omega)]^2 = \frac{\omega^2}{\beta^2} - \frac{i\omega}{\tau_0\beta^2} = \left[\frac{\omega}{c(\omega)} - i\gamma(\omega)\right]^2. \tag{10.232}$$

Solving for $c(\omega)$ and $\gamma(\omega)$, we find

$$c(\omega) = \beta\left[\frac{2}{(1 + 1/\omega^2\tau_0^2)^{1/2} + 1}\right]^{1/2}, \qquad c(\infty) = \beta, \tag{10.233}$$

$$\gamma(\omega) = \frac{\omega}{\beta}\left[\frac{(1 + 1/\omega^2\tau_0^2)^{1/2} - 1}{2}\right]^{1/2}, \qquad \gamma(\infty) = \frac{1}{2\beta\tau_0}, \qquad Q^{-1}(\infty) = 0.$$

For $\omega \ll 1/\tau_0$

$$\gamma(\omega) \simeq \left[\frac{\rho}{2\eta}\right]^{1/2}\omega^{1/2}, \qquad c(\omega) \simeq \left[\frac{2\eta}{\rho}\right]^{1/2}\omega^{1/2}, \qquad Q^{-1}(\omega) = 2. \tag{10.234}$$

The propagation of a plane-wave pulse in a viscoelastic solid with the initial condition $u(0, t) = u_0 H(t)$ is given by the integral representation [cf. Eq. (10.211)],

$$u(x, t) = \frac{u_0}{2\pi}\oint_{-\infty}^{\infty}\exp\left\{i\omega\left[t - \frac{x}{\beta}\hat{n}(\omega)\right]\right\}\frac{d\omega}{i\omega}, \tag{10.235}$$

where $\hat{n}(\omega) = (\beta/c) - i(\beta/\omega)\gamma = (\beta/c(\omega))[1 - (i/2Q(\omega))]$ and $c(\omega)$ and $Q(\omega)$ are given in Eqs. (10.131)–(10.133). In general, $\hat{n}(\omega)$ is not analytic in the lower complex ω plane. Consequently, $u(x, t)$ will not be causal, although the source function $u(0, t)$ is. The exact evaluation of the integral in Eq. (10.235) is usually complicated.

In earth materials, however, $Q(\omega)$ is almost constant over a wide frequency range such that the approximation

$$\hat{n}(\omega) \simeq 1 - \frac{i}{2Q_0} \operatorname{sgn} \omega, \qquad \omega_1 \leq |\omega| \leq \omega_2 \tag{10.236}$$

is valid. There seems to be a problem in the neighborhood of $\omega = 0$, where $(1/\omega)$ is large, but because the stationary point of the exponent is excluded from this range, the approximation in Eq. (10.236) still renders the major contribution. Consequently, for any "constant-Q" solid in the sense of Eq. (10.236) [see Eq. (K.8)]

$$\begin{aligned} u(x, t) &\simeq \frac{u_0}{2\pi} \int_{-\infty}^{\infty} \exp\left[i\omega\left(t - \frac{x}{\beta}\right) - \frac{x|\omega|}{2Q_0\beta}\right] \frac{d\omega}{i\omega} \\ &= u_0 \left\{ \frac{1}{2} + \frac{1}{\pi} \int_0^{\infty} \sin \omega \left(t - \frac{x}{\beta}\right) \exp\left(-\frac{\omega x}{2Q_0\beta}\right) \frac{d\omega}{\omega} \right\} \\ &= u_0 \left\{ \frac{1}{2} + \frac{1}{\pi} \tan^{-1}\left[\frac{t - x/\beta}{x/2Q_0\beta}\right] \right\}. \end{aligned} \tag{10.237}$$

This represents a blunted step function that does not vanish for $t < x/\beta$. As $Q_0 \to \infty$, $u(x, t) \to H(t - x/\beta)$ as it should, and causality is restored.

If we evaluate the integral in Eq. (10.235) with the aid of the *stationary-phase method* (App. E), we obtain a similar approximation

$$u(x, t) = u_0 \left\{ \frac{1}{2} + \frac{1}{2} \operatorname{Erf}\left[\frac{t - x/\beta}{(x/2Q_0\beta)^{1/2}}\right] \right\}, \tag{10.238}$$

where

$$\operatorname{Erf} y = \frac{2}{\sqrt{\pi}} \int_0^y e^{-\xi^2}\, d\xi. \tag{10.239}$$

The displacement begins a little prior to the time x/β and increases continuously. It tends to unity in the limit $t \to \infty$ for any finite x.

10.3.5. Signal Distortion in Media with Given Creep Laws

We have established earlier the connection between the dynamic elastic moduli and their corresponding creep and relaxation time functions. These time functions were extrapolated from laboratory experiments on rock specimens. The time spans in these experiments are usually of the order of 10^2 days, which is much less than the stress relaxation time in the earth as a whole ($\eta/\mu \simeq 10^{10}$ s).

However, the periods of seismic waves in the earth range from 1 to 10^3 s. In spite of the gross incomensurability of these time intervals, some empirical creep laws seem to be compatible with observations of the attenuation of seismic body waves in the earth.

To fix our ideas we return to Boltzmann's law, in which the shear components of the strain tensor are related to the corresponding stress components by means of the rate of creep function [cf. Eq. (10.116)]

$$\varepsilon_{ij}(t) = \frac{1}{2\mu_\infty}\sigma_{ij}(t) + \int_0^t \sigma_{ij}(t-\tau)\dot{\phi}_\beta(\tau)d\tau, \qquad i \neq j.$$

To simplify the forthcoming analysis, we consider only one component, say $i = 1, j = 2$. We denote $\varepsilon_{12} = \varepsilon(t)$ and choose $\sigma_{12} = \sigma_0 H(t)$. It is also convenient to nondimensionalize the creep function. In Eq. (10.116), $\phi_\beta(t)$ has the dimensions of inverse stress. Defining the *normalized creep function* by means of the relation

$$\phi_\beta(t) = \frac{1}{2\mu_\infty}\underline{\phi}(t), \qquad \underline{\phi}(0) = 0, \tag{10.239a}$$

we have

$$\varepsilon(t) = \frac{\sigma_0}{2\mu_\infty}[1 + \underline{\phi}(t)], \qquad t > 0. \tag{10.240}$$[9]

Given $\underline{\phi}(t)$, the complex index of refraction $\hat{n}(\omega)$ and the dissipation parameter $Q_\beta(\omega)$ are related through equations of the type of Eqs. (10.208) and (10.209)

$$S(\omega) = \int_0^\infty \dot{\underline{\phi}}(t)e^{-i\omega t}\,dt = R(\omega) - iX(\omega),$$

$$\frac{1}{Q_\beta(\omega)} = \frac{\int_0^\infty \dot{\underline{\phi}}(t)\sin\omega t\,dt}{1 + \int_0^\infty \dot{\underline{\phi}}(t)\cos\omega t\,dt} = \frac{X(\omega)}{1 + R(\omega)}, \tag{10.241a}$$

$$\hat{n}(\omega) = \left[1 + \int_0^\infty \dot{\underline{\phi}}(t)e^{-i\omega t}\,dt\right]^{1/2} = \frac{\beta}{c(\omega)} - i\frac{\beta}{\omega}\gamma(\omega).$$

By Eq. (K.17) $\lim_{\omega\to\infty} i\omega S(\omega) = \dot{\underline{\phi}}(0^+)$, $S(\omega = 0) = \underline{\phi}(\infty)$ and

$$c(\omega) = \sqrt{2}\beta X^{-1}[\{(1 + R)^2 + X^2\}^{1/2} - (1 + R)]^{1/2}, \tag{10.241b}$$[10]

$$\gamma(\omega) = \left(\frac{\omega}{\sqrt{2}\beta}\right)[\{(1 + R)^2 + X^2\}^{1/2} - (1 + R)]^{1/2}.$$

[9] Eqs. (10.240)–(10.241) are not valid for the Kelvin–Voigt solid [cf. footnote on page 887 and Eqs. (10.208) and (10.227)]. Also note that the normalizing elastic parameter is chosen to suit the particular creep function under discussion.

[10] The expressions for $c(\omega)$, $\gamma(\omega)$, $Q(\omega)$, and $\hat{n}(\omega)$ in Eqs. (10.241 a, b) seem to differ from the corresponding equations in Eqs. (10.130)–(10.143). The reason for this difference is that, whereas in Section 10.3.1 $Q(\omega)$ is expressed in terms of the relaxation function $\psi(t)$, in Section 10.3.5 it is given in terms of the creep function $\phi(t)$. In low-loss materials, both representations are identical as far as terms of order Q^{-1} are concerned.

Through Eq. (10.80) we define the *normalized relaxation function* $\underline{\psi}(t)$. It is related to the normalized creep function via the relation

$$\int_0^\infty \dot{\underline{\psi}}(t)e^{-i\omega t}\,dt = \frac{\int_0^\infty \dot{\underline{\phi}}(t)e^{-i\omega t}\,dt}{1 + \int_0^\infty \dot{\underline{\phi}}(t)e^{-i\omega t}\,dt}. \tag{10.242}$$

In media with small anelasticity the denominator on the right-hand side of Eq. (10.242) is close to unity and $\underline{\phi}(t)$ and $\underline{\psi}(t)$ differ only by a constant.

Note that the application of the Fourier transform to Eq. (10.64a) yields, together with Eq. (10.241a)

$$S(\omega) = \int_0^\infty \frac{j(\tau)}{1 + i\omega\tau}\,d\tau = \int_0^\infty \frac{z^{-1}G(z)}{z + i\omega}\,dz.$$

Hence

$$\begin{aligned} R(\omega) &= \int_0^\infty \frac{G(z)dz}{z^2 + \omega^2} = \int_0^\infty \frac{j(\tau)d\tau}{1 + \omega^2\tau^2}, \\ X(\omega) &= \omega\int_0^\infty \frac{z^{-1}G(z)dz}{z^2 + \omega^2} = \omega\int_0^\infty \frac{\tau j(\tau)d\tau}{1 + \omega^2\tau^2}. \end{aligned} \tag{10.243}$$

A creep function that gives a fair description of the strain in rock specimens under long continued stress, is the *Trouton–Rankine* logarithmic function

$$\underline{\phi}(t) = q\ln\left(1 + \frac{t}{t_0}\right). \tag{10.244}$$

For $t/t_0 \ll 1$

$$\varepsilon(t) \simeq \varepsilon(0)\left[1 + q\frac{t}{t_0}\right], \qquad \frac{\partial\varepsilon}{\partial t} \simeq \frac{q}{t_0}\,\varepsilon(0),$$

implying that for relatively short time intervals, creep is linear in time with a constant rate of creep. However, although this creep function leads to a finite initial rate of creep, it predicts that primary creep grows indefinitely as $\ln(t)$. This is in contrast to satellite observations of the gravitational potential, which show that stress differences of the order of 100 bars must exist in the earth's upper mantle on the geologic time scale. We shall nevertheless examine the compatibility of Eq. (10.244) with the observed attenuation of seismic waves. For the evaluation of the integrals in Eq. (10.241a), we shall need the results

$$\begin{aligned} \frac{1}{t_0}\int_0^\infty \frac{\sin\omega t\,dt}{1 + t/t_0} &= \left[\frac{\pi}{2} - Si(\omega t_0)\right]\cos(\omega t_0) + Ci(\omega t_0)\sin(\omega t_0), \\ \frac{1}{t_0}\int_0^\infty \frac{\cos\omega t\,dt}{1 + t/t_0} &= \left[\frac{\pi}{2} - Si(\omega t_0)\right]\sin(\omega t_0) - Ci(\omega t_0)\cos(\omega t_0), \end{aligned} \tag{10.245}$$

where

$$Si(x) = \int_0^x \frac{\sin\xi}{\xi}\,d\xi = \sum_{n=0}^{\infty}(-)^n \frac{x^{2n+1}}{(2n+1)(2n+1)!},$$

$$Ci(x) = -\int_x^{\infty} \frac{\cos\xi}{\xi}\,d\xi = \gamma + \ln x + \sum_{n=1}^{\infty}(-)^n \frac{x^{2n}}{2n(2n)!}, \tag{10.246}$$

$$\gamma = 0.577215\ldots \quad \text{(Euler's constant)}.$$

In particular, for small values of (ωt_0),

$$Si(\omega t_0) \simeq \omega t_0, \qquad Ci(\omega t_0) \simeq \ln\left(\frac{\omega}{\omega_0}\right),$$

where

$$t_0 e^{\gamma} = [1.78107]t_0 = \frac{1}{\omega_0}.$$

Assuming that

$$\omega t_0 \ll 1, \qquad (\omega t_0)\ln(\omega t_0) \ll 1, \qquad q\ln(\omega t_0) \ll 1, \qquad Q_0 = \frac{2}{\pi q} \gg 1,$$

we substitute power series expansions of the sine, cosine, sine integral, and cosine integral functions appearing in Eqs. (10.245). The resulting approximations, valid for $\omega \ll \omega_0$, are

$$\gamma_\beta(\omega) \simeq \frac{\omega}{2Q_0\beta}\left[1 - \frac{1}{\pi Q_0}\ln\left(\frac{\omega_0}{\omega}\right)\right],$$

$$c_\beta(\omega) \simeq \beta\left[1 - \frac{1}{\pi Q_0}\ln\left(\frac{\omega_0}{\omega}\right)\right], \tag{10.247}$$

$$\frac{1}{Q_\beta(\omega)} \simeq \frac{1}{Q_0}\left[1 - \frac{2}{\pi Q_0}\ln\left(\frac{\omega_0}{\omega}\right)\right], \qquad \hat{n}_\beta(\omega) \simeq 1 - \frac{i}{2Q_\beta(\omega)}.$$

There is a striking similarity between Eqs. (10.247) and those obtained on the strength of the dispersion relations for a "boxcar-Q" solid (Example 10.8). This similarity is by no means accidental and stems from the fact that Boltzmann's superposition law already incorporates the principle of causality.

We have pointed out in Example 10.7 that the assumption that Q is frequency independent over the entire spectrum is incompatible with the principles of linearity and causality that underline the theory of linear viscoelasticity. Yet, observations of attenuation of seismic waves in the earth and in laboratory rock specimens indicate that Q is almost constant over a wide spectral range. The exact expression for Q^{-1}, obtained for the Trouton–Rankine creep function, is

$$\frac{1}{Q_\beta(\omega)} = \frac{1}{Q_0}\left[\frac{[1 - (2/\pi)Si(\omega t_0)]\cos(\omega t_0) + (2/\pi)Ci(\omega t_0)\sin(\omega t_0)}{1 + (1/Q_0)\{[1 - (2/\pi)Si(\omega t_0)]\sin(\omega t_0) - (2/\pi)Ci(\omega t_0)\cos(\omega t_0)\}}\right]. \tag{10.248}$$

Evaluation of the right-hand side of Eq. (10.248) reveals that $1/Q_\beta(\omega)$ is characterized by an extended horizontal plateau in the range

$$10^{-10} < \omega t_0 < 10^{-3}.$$

Within this range, the approximations in Eqs. (10.247) hold, and the pulse distortion in this medium is given by Eq. (10.237).

10.3.5.1. Jeffreys' Creep Law. The Trouton–Rankine creep function cannot furnish a simultaneous account for seismic and nonseismic dissipation phenomena in the earth. We know, for example, that the damping of the 14-month precessional motion of the earth's instantaneous axis of rotation about its axis of figure (known as the "Chandler wobble") is associated with a Q value of about 30. If this Q value is used in Eq. (10.237) for an S wave that has emerged at 80° after traveling for some 1200 s, the wave's pulse shape will have the form

$$\left\{\frac{1}{2} + \frac{1}{\pi}\tan^{-1}\frac{(t - x/\beta)}{(x/2Q\beta)}\right\}.$$

This means that even at $t - x/\beta = 30$ s its amplitude would still reach 0.8 of its maximal value. This spreading is definitely not observed in earthquake seismograms, where the peak amplitude is reached within 2–3 s. The reason is that Q_β for S waves in the earth is not 30 but is higher by one order of magnitude (see Table 10.4).

We must therefore modify the Trouton–Rankine law such that it can accommodate these variations in Q. A time function that seems to answer this need is

$$\underline{\phi}(t) = \frac{q}{\nu}\left[\left(1 + \frac{t}{t_0}\right)^{\nu} - 1\right], \qquad 0 \le \nu < 1 \tag{10.249}$$

where q, ν, and t_0 are constants. This function has the following properties:

1. In the limit $\nu \to 0$, $\qquad \underline{\phi}(t) = q\ln\left(1 + \dfrac{t}{t_0}\right)$ (Trouton–Rankine law)

2. For $\dfrac{t}{t_0} \ll 1$ (or $\nu = 1$), $\qquad \underline{\phi}(t) \simeq q\dfrac{t}{t_0}, \qquad \underline{\phi}(0) = 0$

3. For $\dfrac{t}{t_0} \gg 1, \qquad \nu \neq 0, \qquad \underline{\phi}(t) \simeq \dfrac{q}{\nu}\left(\dfrac{t}{t_0}\right)^{\nu}.$

To evaluate the propagation functions $\gamma(\omega)$, $c(\omega)$, and $\hat{n}(\omega)$, we shall need the sine and cosine transforms of the rate of creep function $\dot{\underline{\phi}}(t) = (q/t_0)(1 + t/t_0)^{\nu-1}$. They are

$$R(\omega) = \int_0^\infty \dot{\underline{\phi}}(t)\cos\omega t\,dt = q|\omega t_0|^{1/2-\nu}(1 - \nu)S_{\nu-3/2,\,1/2}(|\omega t_0|), \tag{10.250}$$

$$X(\omega) = \int_0^\infty \dot{\underline{\phi}}(t)\sin\omega t\,dt = q|\omega t_0|^{1/2-\nu}\operatorname{sgn}(\omega t_0)S_{\nu-1/2,\,1/2}(|\omega t_0|), \tag{10.251}$$

where $S_{\mu,1/2}(z)$ is the *Lommel function*,

$$S_{\mu,1/2}(z) = \frac{z^{\mu+1}}{(\mu+\frac{1}{2})(\mu+\frac{3}{2})}\,{}_1F_2\left(1, \frac{\mu+\frac{5}{2}}{2}, \frac{\mu+\frac{7}{2}}{2}; -\frac{z^2}{4}\right)$$
$$+ \left(\frac{2}{\pi}\right)^{1/2} 2^{\mu-1}\Gamma\left(\frac{\mu+\frac{1}{2}}{2}\right)\Gamma\left(\frac{\mu+\frac{3}{2}}{2}\right)\cos\left[\frac{\mu-\frac{1}{2}}{2}\pi - z\right]z^{-1/2}. \quad (10.252)$$

For values of ω such that $|z| \ll 1$ (i.e., $\omega t_0 \ll 1$) and $\nu > 0$, we have

$$z^{1/2}S_{\nu-3/2,1/2}(z) \simeq \frac{z^\nu}{\nu(\nu-1)} - \Gamma(\nu-1)\cos\left(\frac{\pi\nu}{2}\right),$$
$$z^{1/2}S_{\nu-1/2,1/2}(z) \simeq \frac{z^{\nu+1}}{\nu(\nu+1)} + \Gamma(\nu)\sin\left(\frac{\pi\nu}{2}\right), \quad (10.253)$$

and, therefore,

$$R(\omega) \simeq q\Gamma(\nu)\cos\left(\frac{\pi\nu}{2}\right)|\omega t_0|^{-\nu},$$
$$X(\omega) \simeq q\Gamma(\nu)\sin\left(\frac{\pi\nu}{2}\right)(\operatorname{sgn}\omega)|\omega t_0|^{-\nu}. \quad (10.254)$$

The same results are obtained if we assume *a priori* that $t/t_0 \gg 1$ and apply the sine and cosine transforms to the approximated creep function $\underline{\phi}(t) = (q/\nu)(t/t_0)^\nu$. For real-earth materials, typical values of the parameters are

$$\nu \simeq 0.2, \qquad t_0 \simeq 1.6 \times 10^{-4}\text{ s}, \qquad q \simeq 10^{-3} \text{ to } 10^{-2}. \quad (10.255)$$

Because $\Gamma(\nu) \simeq 4.5$, $\sin(\pi\nu/2) \simeq 0.3$, and $\cos(\pi\nu/2) \simeq 0.95$, we obtain

$$X \simeq \frac{T^{1/5}}{200}, \qquad R \simeq \frac{T^{1/5}}{65}.$$

Therefore, in the period range $10^{-2} < T < 10^6$ s, (ωt_0), $X(\omega)$, and $R(\omega)$ are all smaller than unity and we are led to the expressions

$$\frac{1}{Q(\omega)} \simeq \frac{X}{1+R} \simeq \frac{q\Gamma(\nu)\sin(\pi\nu/2)}{(|\omega t_0|)^\nu}, \quad (10.256)$$

$$\gamma(\omega) \simeq \frac{\omega}{2Q\beta} = m|\omega|^{1-\nu}, \qquad m = \frac{q\Gamma(\nu)\sin(\pi\nu/2)}{2\beta t_0^\nu}, \quad \text{[cf. Eq. (10.6.1)]} \quad (10.257)$$

$$\hat{n}(\omega) \simeq 1 - \frac{i}{2Q}\operatorname{sgn}\omega. \quad (10.258)$$

At low frequencies such that $|R(\omega)| \gg 1$ and $|\omega t_0| \ll 1$,

$$\frac{1}{Q} \simeq \tan\frac{\pi\nu}{2} = \text{constant}.$$

10.3.5.2. Pulse Propagation in a Jeffreys Solid. We consider the case where both $R(\omega)$ and $X(\omega)$ are small. Substituting then the expression of $\hat{n}(\omega)$ into the exponent of the integral in Eq. (10.235), we obtain

$$u(x, t) = \frac{u_0}{2\pi} \int_{-\infty}^{\infty} \exp\left\{i\left[\omega\tau + b\varepsilon|\omega|^{1-\nu}\left(-\cos\frac{\pi\nu}{2} + i\varepsilon\sin\frac{\pi\nu}{2}\right)\right]\right\}\frac{d\omega}{i\omega}, \tag{10.259}$$

where

$$\varepsilon = \operatorname{sgn}\omega, \qquad \tau = t - \frac{x}{\beta}, \qquad b = \left(\frac{q\Gamma(\nu)}{2\beta t_0^{\nu}}\right)x.$$

Note that the integration extends over the entire frequency range whereas the expression for $\hat{n}(\omega)$ is valid only at sufficiently high frequencies. However, the errors caused by this discrepancy are negligible because the main contribution to the integral arises from stationary points of the exponent, which are located away from $\omega = 0$. The substitution

$$\omega = BV, \qquad B = b^{1/\nu}(1-\nu)^{1/\nu}\tau^{-1/\nu}$$

transforms Eq. (10.259) into

$$u(x, t) = \frac{u_0}{2\pi} \int_{-\infty}^{\infty} \frac{A(V)}{iV} e^{ipV}\, dV = u_0 H\left(t - \frac{x}{\beta}\right)\left[\frac{1}{2} + \frac{1}{2\pi}\, \mathrm{P}\int_{-\infty}^{\infty} \frac{A(V)}{iV} e^{ipV}\, dV\right], \tag{10.260}$$

where

$$p = B\tau, \qquad A(V) = \exp\left[ip\varepsilon\frac{|V|^{1-\nu}}{1-\nu}\left(-\cos\frac{\pi\nu}{2} + i\varepsilon\sin\frac{\pi\nu}{2}\right)\right]$$

and Eq. (K.8) has been used with $A(0) = 1$ for $\nu < 1$. Consequently

$$u(x, t) = u_0 H\left(t - \frac{x}{\beta}\right)\left\{\frac{1}{2} + \frac{1}{\pi}\int_0^{\infty} \sin\left[pV - \frac{p\cos(\pi\nu/2)}{1-\nu}V^{1-\nu}\right]\right.$$
$$\left. \times \exp\left[-\frac{p\sin(\pi\nu/2)}{1-\nu}V^{1-\nu}\right]\frac{dV}{V}\right\}. \tag{10.261}$$

The second transformation

$$W = p\frac{V^{1-\nu}}{1-\nu} \tag{10.262}$$

changes Eq. (10.261) into

$$u(x, t) = u_0 H\left(t - \frac{x}{\beta}\right)\left\{\frac{1}{2} + \frac{1}{\pi(1-\nu)}\int_0^{\infty} \exp\left[-W\sin\left(\frac{\pi\nu}{2}\right)\right]\right.$$
$$\left. \times \sin\left(W\left[\left(\frac{W}{p}\right)^{\nu/(1-\nu)}(1-\nu)^{1/(1-\nu)} - \cos\frac{\pi\nu}{2}\right]\right)\frac{dW}{W}\right\} \tag{10.263}$$

which is convenient for numerical evaluation because the integrand is bounded near $W = 0$. The expression for $Q^{-1}(\omega)$ as given in Eq. (10.256) depends on two parameters, ν and $\{qt_0^{-\nu}\}$. The latter appears also in the theory of damping of the Chandler wobble and can be given in terms of ν only. We know that the Q value for an S wave that traversed the entire mantle is about 400 (see Table 10.4). These two independent measurements enable us to assign to ν the range of values 0.14–0.21.

EXAMPLE 10.9: Broadening of a Wavelet

A seismic signal broadens during its propagation in an anelastic solid. This can be looked upon as a continuous process of filtering in which waves of shorter wavelength are being removed faster from the wave profile than the longer waves. Thus, the original source-time function becomes relatively richer in longer periods and consequently the profile broadens with time. A simple time function for which a width is defined can be represented by the one-sided Fourier integral

$$f(t) = u(0, t) = \frac{2u_0}{\pi}\int_0^\infty \frac{1 - \cos a\omega}{\omega}\sin \omega t \, d\omega = \begin{cases} +u_0, & 0 < t < a \\ -u_0, & -a < t < 0 \\ 0, & \text{otherwise.} \end{cases} \tag{10.9.1}$$

A plane wave in an elastic medium resulting from the above source function at $x = 0$, will have the integral form

$$u(x, t) = \frac{2u_0}{\pi}\int_0^\infty \frac{1 - \cos a\omega}{\omega}\sin \omega\left(t - \frac{x}{\beta}\right) d\omega. \tag{10.9.2}$$

In a low-loss medium in which the attenuation is of the type given in Eq. (10.257) and a is small so that $(1 - \cos a\omega)/\omega \simeq (\omega/2)a^2$, the integral representation of the pulse shape is

$$u(x, t) = \frac{a^2 u_0}{\pi}\int_0^\infty \omega \sin \omega\left(t - \frac{x}{\beta}\right) e^{-mx\omega^{1-\nu}} \, d\omega. \tag{10.9.3}$$

Let

$$\omega = \frac{\xi}{(mx)^\theta}, \qquad z = \frac{t - x/\beta}{(mx)^\theta}, \qquad \theta = \frac{1}{1 - \nu}. \tag{10.9.4}$$

Then

$$\frac{u(x, t)}{u_0} = \frac{a^2}{\pi(mx)^{2\theta}}\int_0^\infty \xi e^{-\xi^{(1/\theta)}} \sin(\xi z) d\xi. \tag{10.9.5}$$

Expanding the sine function in a power series we have,

$$\begin{aligned} \frac{u(x, t)}{u_0} &= \frac{a^2}{\pi(mx)^{2\theta}} \sum_{n=0}^{\infty} \frac{(-)^n z^{2n+1}}{(2n + 1)!}\int_0^\infty \xi^{2n+2} e^{-\xi^{(1/\theta)}} d\xi \\ &= \frac{a^2\theta}{\pi(mx)^{2\theta}} \sum_{n=0}^{\infty} (-)^n z^{2n+1} \frac{\Gamma[(2n + 3)\theta]}{\Gamma(2n + 2)}. \end{aligned} \tag{10.9.6}$$

When $\theta^{-1} = 1, 2$ and ∞, the integral in Eq. (10.9.5) can be evaluated in terms of elementary functions. For $\theta = 1$, for example, ($\nu = 0$, Trouton–Rankine law)

$$\frac{u(x,t)}{u_0} = \frac{2a^2}{\pi} \frac{mx(t - x/\beta)}{[m^2x^2 + (t - x/\beta)^2]^2}, \qquad m = \frac{1}{2Q_0\beta}. \tag{10.9.7}$$

This expression is related to the second derivative of the solution given in Eq. (10.237), as it should be. In the general expression given in Eq. (10.9.6), anelasticity introduces an amplitude decay with distance as $x^{-2\theta}$ (a diverging spherical wave will decay as $x^{-(1+2\theta)}$). The coefficient θ measures the steepness of the cutoff of the medium absorption spectrum. The dependence of the amplitude on z [Eq. (10.9.4)] causes the waveform to broaden as x increases and this broadening is proportional to the θth power of the distance and hence of the travel time of the center of the disturbance.

EXAMPLE 10.10: Modifications of the Creep Law

Although the creep function in Eq. (10.249) leads to a finite initial rate of creep, it still predicts an infinite strain for constant applied stress. It therefore represents a physically unrealizable linear system. The associated distribution of relaxation times is derived from the integral

$$\int_0^\infty \frac{e^{-t_0 z} - e^{-(t+t_0)z}}{z^{1+\nu}}\, dz = \Gamma(1-\nu) t_0^\nu \frac{(1 + t/t_0)^\nu - 1}{\nu}. \tag{10.10.1}$$

However, because $\int_0^\infty j(\tau)d\tau = \infty$ [cf. Eq. (10.64)], both the creep function and the distribution of relaxation times are nonnormalizable. It is, however, possible to modify the definition of the creep function so that the resulting function represents a physically realizable linear system while the essentials of the response of the system remain unaltered in regions where the original function shows good agreement with experiments. One way to achieve this is through the *modified rate of creep*

$$\dot{\underline{\phi}}(t) = \left(\frac{q}{t_0}\right)\left[1 + \frac{t}{t_0}\right]^{\nu-1} e^{-t/t_\infty}. \tag{10.10.2}$$

The corresponding creep function is

$$\underline{\phi}(t) = qa^{-\nu}e^a \int_a^{a(1+t/t_0)} e^{-\xi}\xi^{\nu-1}\, d\xi, \qquad a = \frac{t_0}{t_\infty}. \tag{10.10.3}$$

The associated distribution of relaxation times is found from Eq. (10.67)

$$j(\tau) = b\tau^{\nu-1}\left(1 - \frac{\tau}{t_\infty}\right)^{-\nu} e^{-t_0/\tau}, \qquad \tau < t_\infty \tag{10.10.4}$$

where

$$b = qe^a[t_0^\nu\Gamma(1-\nu)]^{-1}.$$

When $a = t_0/t_\infty$ is much smaller than unity, $\underline{\phi}(t)$ and $j(\tau)$ will show the Jeffreys' response for long times. By taking t_∞ sufficiently large, the range of agreement can be extended indefinitely and the final strain will be finite. Employing the integral definitions of the incomplete gamma function

$$\Gamma(\nu, z) = z^{\nu} e^{-z} \int_0^{\infty} \frac{e^{-zu}\, du}{(1+u)^{1-\nu}} = \int_z^{\infty e^{i\delta}} \xi^{\nu-1} e^{-\xi}\, d\xi, \qquad (10.10.5)$$

$$\operatorname{Re} z > 0, \qquad \delta \text{ real}, \qquad |\delta| = \frac{\pi}{2}, \qquad 0 < \operatorname{Re} \nu < 1,$$

the complex propagation functions for the present case are given as

$$\int_0^{\infty} \underline{\phi}(t) e^{-i\omega t}\, dt = q \int_0^{\infty} \frac{e^{-zu}}{(1+u)^{1-\nu}}\, du, \qquad (z = a + i\omega t_0)$$

$$= \frac{q a^{-\nu}}{[1 + (\omega t_{\infty})^2]^{\nu/2}} e^{-i\nu \operatorname{tg}^{-1}(\omega t_{\infty})} \left\{ e^z \int_z^{\infty e^{i\delta}} \xi^{\nu-1} e^{-\xi}\, d\xi \right\}. \qquad (10.10.6)$$

Therefore

$$R(\omega) = q \int_0^{\infty} \frac{\cos(\omega t_0 u)}{(1+u)^{1-\nu}} e^{-au}\, du, \qquad (10.10.7)$$

$$X(\omega) = q \int_0^{\infty} \frac{\sin(\omega t_0 u)}{(1+u)^{1-\nu}} e^{-au}\, du. \qquad (10.10.8)$$

For geophysical applications it is justified to assume that

$$\frac{t_0}{t_{\infty}} \ll 1, \qquad \omega t_0 \ll 1, \qquad q \ll 1. \qquad (10.10.9)$$

The expression in the braces in Eq. (10.10.6) is then simplified to $\Gamma(\nu)$ and the overall result is

$$\int_0^{\infty} \underline{\phi}(t) e^{-i\omega t}\, dt \simeq \frac{q \Gamma(\nu) a^{-\nu}}{[1 + (\omega t_{\infty})^2]^{\nu/2}} e^{-i\nu \operatorname{tg}^{-1}(\omega t_{\infty})}. \qquad (10.10.10)$$

Note that the assumption $\omega t_0 \ll 1$ is equivalent to the corresponding time-domain restriction

$$t \gg t_0. \qquad (10.10.11)$$

Indeed, if we approximate the rate of creep by the expression $\dot{\underline{\phi}}(t) \simeq (q/t_0)(t/t_0)^{\nu-1} e^{-t/t_{\infty}}$, Eq. (10.10.10) will ensue. The use of Eq. (10.10.10) then leads to the approximate propagation functions, valid under the assumptions, $0 < t_0/t_{\infty} \ll 1$, $\omega t_0 \ll 1$, and $q \ll 1$

$$X(\omega) = \frac{1}{Q_0} \frac{a^{-\nu}}{[1 + x^2]^{\nu/2}} \left[\Gamma(\nu + 1) \frac{\sin(\nu \operatorname{tg}^{-1} x)}{(\pi\nu/2)} \right], \qquad (10.10.12)$$

$$R(\omega) = \frac{2\Gamma(\nu)}{\pi Q_0 a^{\nu}} \frac{\cos(\nu \operatorname{tg}^{-1} x)}{(1 + x^2)^{\nu/2}}, \qquad (10.10.13)$$

$$a = \frac{t_0}{t_{\infty}}, \qquad x = \omega t_{\infty}.$$

The rate of creep function in Eq. (10.10.2) is by no means unique. Another useful creep function is derived by the following reasoning: We recognize that the lack of convergence of $\int_0^\infty j(\tau)d\tau$, where

$$j(\tau) = q[\Gamma(1-\nu)t_0^\nu]^{-1}\tau^{\nu-1}e^{-t_0/\tau}, \quad \text{(Table 10.1)} \tag{10.10.14}$$

arises from the slow decay of the density of relaxation times for large τ. Normalization can be achieved if an increase in the decay rate of $j(\tau)$ for large τ is made to exceed a threshold value that we call t_∞. Such an increase will result in *a smaller density of relaxation times for* $\tau > t_\infty$ than that given by Eq. (10.10.14). Therefore, convergence is secured if we take, for example,

$$j(\tau) = \frac{q}{t_0^\nu \Gamma(1-\nu)} \times \begin{cases} \dfrac{1}{\tau^{1-\nu}} e^{-t_0/\tau}, & 0 \le \tau \le t_\infty \\ \left(\dfrac{t_\infty}{\tau}\right)^2 \dfrac{1}{t_\infty^{1-\nu}} e^{-t_0/t_\infty}, & \tau \ge t_\infty. \end{cases} \tag{10.10.15}$$

For $\tau \ge t_\infty$, the slope of $j(\tau)$ on a log–log plot is (-2) instead of the value $(\nu - 1)$ which follows from Eq. (10.10.15). Any value less than (-1) could have been used to ensure the convergence of $\int_0^\infty j(\tau)d\tau$. The creep-rate and creep functions associated with the new $j(\tau)$ are found to be

$$\dot{\underline{\phi}}(t) = \frac{q}{t_0\Gamma(1-\nu)}\left[\frac{\Gamma(1-\nu, a+aT)}{(1+T)^{1-\nu}} + a^{1-\nu}e^{-a}\frac{1-(1+aT)e^{-aT}}{(aT)^2}\right],$$

$$T = \frac{t}{t_0}, \qquad a = \frac{t_0}{t_\infty}, \tag{10.10.16}$$

$$\underline{\phi}(t) = \frac{q}{\Gamma(1-\nu)}\left[\Gamma(-\nu, a) - (1+T)^\nu\Gamma(-\nu, a+aT) - a^{-\nu}e^{-a} \times \left\{\frac{1-e^{-aT}}{aT} - 1\right\}\right]. \tag{10.10.17}$$

From these expressions it follows that

$$\underline{\phi}(0) = 0, \qquad \underline{\phi}(\infty) = \frac{q}{\Gamma(1-\nu)}[\Gamma(-\nu, a) + a^{-\nu}e^{-a}]. \tag{10.10.18}$$

In particular, for $a \ll 1$, a series expansion of the incomplete gamma function leads to

$$\underline{\phi}(\infty) = \begin{cases} q(1-\gamma-\ln a), & \nu = 0 \\ \dfrac{q}{\Gamma(1-\nu)}\left[a^{-\nu}\left(1+\dfrac{1}{\nu}+\dfrac{a}{1-\nu}\right) - \dfrac{1}{\nu}\Gamma(1-\nu)\right], & 0 < \nu < 1 \end{cases} \tag{10.10.19}$$

where γ is Euler's constant. Therefore, the final strain is finite and positive for $a > 0$. The functions $R(\omega)$ and $X(\omega)$ corresponding to Eq. (10.10.16), are

$$R(\omega) = \frac{2}{\pi Q_0 \Gamma(1-\nu)} \left[\frac{1 - |\omega t_0/a| \{\operatorname{tg}^{-1}|a/\omega t_0|\}}{a^{\nu} e^{a}} + \int_a^{\infty} \frac{z^{1-\nu} e^{-z}}{z^2 + \omega^2 t_0^2}\, dz \right], \tag{10.10.20}$$

$$X(\omega) = \frac{(\omega t_0)}{\pi \Gamma(1-\nu) Q_0} \left[\frac{\ln(1 + a^2/\omega^2 t_0^2)}{a^{1+\nu} e^{a}} + 2\int_a^{\infty} \frac{z^{-\nu} e^{-z}}{z^2 + \omega^2 t_0^2}\, dz \right]. \tag{10.10.21}$$

10.3.6. Macroscopic Models of Anelasticity: Standard Linear Solid with a Continuous Relaxation

In this section we construct a physical model of an anelastic solid in which Q is constant over a finite range. It is based on the standard linear solid that we have described in Section 10.2.2.3. From Eqs. (10.50)–(10.52), (10.58), (10.60), and (10.206) it follows that the normalized creep function for the standard linear solid is given by

$$\underline{\phi}(t) = \underline{\phi}(\infty)(1 - e^{-t/\tau_\varepsilon}),$$

$$\underline{\phi}(\infty) = \frac{\tau_\varepsilon}{\tau_\sigma} - 1, \qquad \varepsilon(t) = \frac{\sigma_0}{\mu_2}[1 + \underline{\phi}(t)]. \tag{10.264}$$[11]

Because

$$S(\omega) = \int_0^{\infty} \dot{\underline{\phi}}(t) e^{-i\omega t}\, dt = \frac{\tau_\varepsilon/\tau_\sigma - 1}{1 + i\omega\tau_\varepsilon}, \tag{10.265}$$

the complex propagation functions are

$$\hat{n}(\omega) = \left[\frac{\tau_\varepsilon}{\tau_\sigma}\right]^{1/2} \left[\frac{1 + i\omega\tau_\sigma}{1 + i\omega\tau_\varepsilon}\right]^{1/2}, \tag{10.266}$$

$$\hat{k}(\omega) = \frac{\omega}{v_e}\left[\frac{1 + i\omega\tau_\sigma}{1 + i\omega\tau_\varepsilon}\right]^{1/2}, \qquad v_e = \left[\frac{M_R}{\rho}\right]^{1/2}. \tag{10.267}$$

[11] Note that in the limit $\mu_2 \to \infty$, the standard linear solid degenerates into the Kelvin–Voigt solid, $\varepsilon(t) \to (\sigma_0/\mu_1)(1 - e^{-t/\tau_1})$, in conformity with Eq. (10.45). The corresponding limits of $\underline{\phi}$, $S(\omega)$, and $\hat{n}(\omega)$ do not exist. However,

$$(\tau_\sigma/\tau_\varepsilon)\underline{\phi}(t) \to 1 - e^{-t/\tau_\varepsilon}, \qquad (\tau_\sigma/\tau_\varepsilon)S(\omega) \to (1 + i\omega\tau_\varepsilon)^{-1},$$

and

$$(\tau_\sigma/\tau_\varepsilon)^{1/2}\hat{n}(\omega) \to (1 + i\omega\tau_\varepsilon)^{-1/2}.$$

From Eqs. (10.267) we obtain

$$c(\omega) = v_e \left[\frac{2[1 + \omega^2\tau_0^2(\tau_\varepsilon/\tau_\sigma)]}{[(1 + \omega^2\tau_0^2)^2 + 4\omega^2\tau_0^2 Q_0^{-2}]^{1/2} + (1 + \omega^2\tau_0^2)} \right]^{1/2}, \tag{10.268}$$

$$\gamma(\omega) = \frac{\omega}{v_e} \left[\frac{[(1 + \omega^2\tau_0^2)^2 + 4\omega^2\tau_0^2 Q_0^{-2}]^{1/2} - (1 + \omega^2\tau_0^2)}{2[1 + \omega^2\tau_0^2(\tau_\varepsilon/\tau_\sigma)]} \right]^{1/2}, \tag{10.269}$$

$$\frac{2\gamma(\omega)c(\omega)}{\omega} = Q_0 \frac{[(1 + \omega^2\tau_0^2)^2 + 4\omega^2\tau_0^2 Q_0^{-2}]^{1/2} - (1 + \omega^2\tau_0^2)}{\omega\tau_0}, \tag{10.270}$$

where

$$\frac{1}{Q_0} = \frac{\tau_\varepsilon - \tau_\sigma}{2\tau_0}, \qquad \tau_0 = (\tau_\varepsilon \tau_\sigma)^{1/2}. \tag{10.271}$$

Then, by Eq. (10.143)

$$\frac{1}{Q(\omega)} = \frac{1}{Q_0} \left[\frac{2\omega\tau_0}{1 + \omega^2\tau_0^2} \right] \tag{10.272}$$

which agrees with Eq. (10.55). The curve $Q(\omega)$ has its peak at $\omega_0 = 1/\tau_0$ and the value of $1/Q$ at the peak is

$$\frac{1}{Q(\omega_0)} = \frac{1}{Q_0}. \tag{10.273}$$

The same result is derived from the equation of motion. Starting from Eq. (10.49), we differentiate it with respect to the coordinate y and substitute $\varepsilon = \varepsilon_{12} = \frac{1}{2}\partial u/\partial y$, $\partial\sigma_{12}/\partial y = \rho(\partial^2 u/\partial t^2)$. The elimination of ε_{12} and σ_{12} then yields

$$\frac{1}{v_e^2} \left[\frac{\partial^2 u}{\partial t^2} + \tau_\sigma \frac{\partial^3 u}{\partial t^3} \right] = \frac{\partial^2 u}{\partial y^2} + \tau_\varepsilon \frac{\partial^3 u}{\partial t\, \partial y^2}. \tag{10.274}$$

Assuming a plane-wave solution $u = u_0 e^{i[\omega t - \hat{k}(\omega)y]}$, Eq. (10.267) is obtained and the rest follows. The branch points of $\hat{k}(\omega)$ are in the upper half of the ω plane at

$$\omega_1 = \frac{i}{\tau_\sigma}, \qquad \omega_2 = \frac{i}{\tau_\varepsilon}. \tag{10.275}$$

Consequently, $\hat{k}(\omega)$ is an analytic function in the lower ω plane and causality is maintained.

10.3.6.1. Continuous Relaxation Model. The frequency variation of $1/Q$ for the standard linear solid is shown in Fig. 10.15a. It has a single peak at $\omega_0 = 1/\tau_0$. As it stands it cannot serve as basis for the attenuation mechanism in the earth because the observed Q is flat over a wide frequency range. However, if we superimpose many such relaxtion mechanisms, each one shifted somewhat relative to the other, we can form a flat dependence of $1/Q$ over an arbitrary finite frequency range. The mathematical basis for this idea has been discussed

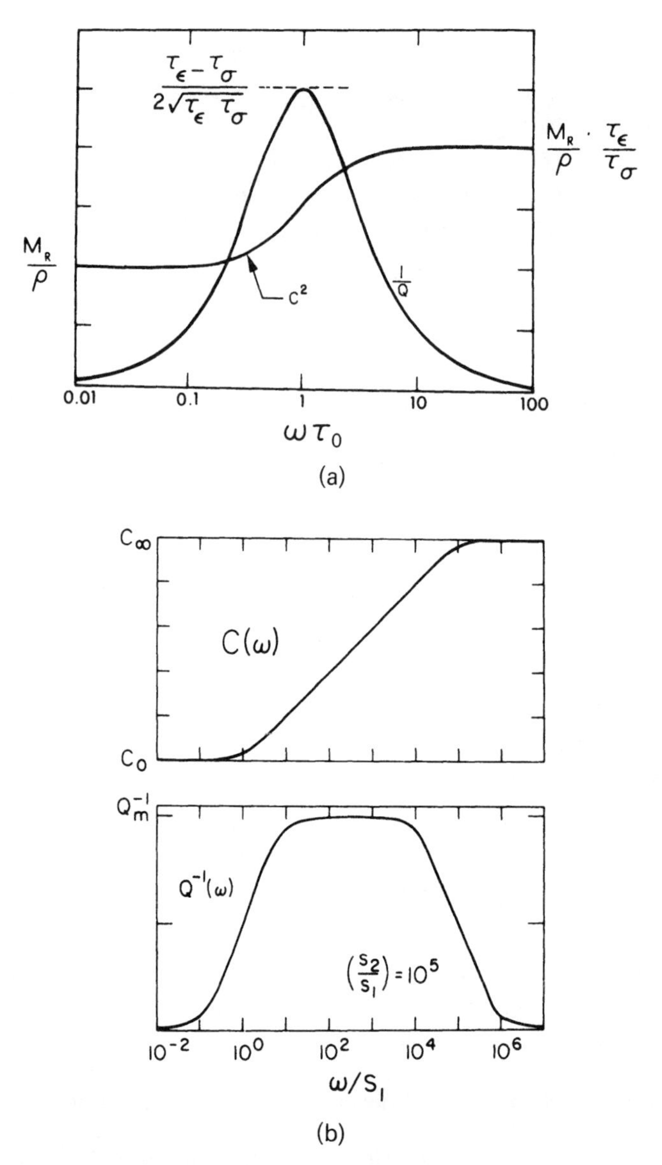

Figure 10.15. Generation of a band-limited constant Q model by a continuous superposition of single relaxation mechanisms. (a) Phase velocity and specific dissipation parameter for a standard linear solid. (b) Phase velocity and specific dissipation parameter for a model with continuous band-limited relaxation times.

in Section 10.2.3: We effect a linear superposition of creep elements, each of which bears the characteristics of a standard linear solid with a weighting function $j(\tau_\varepsilon)$. The addition of these elements is subjected to the restriction that

$$\frac{\tau_\varepsilon}{\tau_\sigma} - 1 = \underline{\phi}(\infty) \ll 1. \tag{10.276}$$

Then, following Eq. (10.64), the total creep function reads

$$\underline{\phi}(t) = \int_0^\infty j(\tau_\varepsilon)[1 - e^{-t/\tau_\varepsilon}]d\tau_\varepsilon. \tag{10.277}$$

We adopt *Becker's distribution function*

$$j(\tau) = \begin{cases} A/\tau, & \tau_2 < \tau < \tau_1 \\ 0, & \text{otherwise.} \end{cases} \tag{10.278}$$

A is a constant to be determined later. The total creep is then

$$\underline{\phi}(t) = A\left[\ln\frac{s_2}{s_1} - \int_{s_1}^{s_2} e^{-st}\frac{ds}{s}\right], \qquad s_1 = \frac{1}{\tau_1}, \qquad s_2 = \frac{1}{\tau_2}$$

or

$$\underline{\phi}(t) = A\left[\ln\frac{s_2}{s_1} + \{Ei(-s_1 t) - Ei(-s_2 t)\}\right], \tag{10.279}$$

where, for $x > 0$, $-\pi < \arg x < \pi$

$$Ei(-x) = -\int_x^\infty \frac{e^{-t}}{t}\,dt = 0.5772156\cdots + \ln x$$

$$- e^{-x}\sum_{n=1}^{\infty}\left(1 + \frac{1}{2} + \cdots + \frac{1}{n}\right)\frac{x^n}{n!}.$$

In particular

$$\underline{\phi}(\infty) = A\ln\frac{s_2}{s_1}, \qquad \underline{\phi}(\text{small } t) \simeq A(s_2 - s_1)t. \tag{10.280}$$

As we compare the creep laws of Eqs. (10.244) and (10.279) we note that the behavior of these functions is very similar except that in Eq. (10.279) there is no singularity at $t = \infty$ and an appropriate *static limit* is reached. It is not surprising that we get this result because both functions are associated with a constant Q over a wide frequency range.

We have already seen in Eq. (10.242) that in media where the anelasticity is small, the creep and relaxation functions differ by a constant. Here

$$\underline{\psi}(t) = A[Ei(-s_1 t) - Ei(-s_2 t)]. \tag{10.281}$$

In order to determine the constant A we evaluate the integral

$$\int_0^\infty \underline{\dot{\phi}}(t)e^{-i\omega t}\,dt = A\int_0^\infty \frac{e^{-s_1 t} - e^{-s_2 t}}{t}e^{-i\omega t}\,dt$$

Table 10.1. Anelasticity Models and Their Creep and Propagation Functions[a]

Solid				
Type	τ_0	$\hat{n}(\omega)$	Creep function $\underline{\phi}(t)$	Distribution of relaxation times $j(\tau)$
1 Maxwell	$\dfrac{\eta}{\mu}$	$\left(1 + \dfrac{1}{i\omega\tau_0}\right)^{1/2}$	$1 + \dfrac{t}{\tau_0}$	—
2 Kelvin–Voigt	$\dfrac{\eta}{\mu}$	$(1 + i\omega\tau_0)^{-1/2}$	$1 - e^{-t/\tau_0}$	$\delta(\tau - \tau_0)$
3 Standard linear	$[\tau_\varepsilon \tau_\sigma]^{1/2}$	$\left[\dfrac{\tau_\varepsilon}{\tau_\sigma} \dfrac{1 + i\omega\tau_\sigma}{1 + i\omega\tau_\varepsilon}\right]^{1/2}$	$\left(\dfrac{\tau_\varepsilon}{\tau_\sigma} - 1\right)(1 - e^{-t/\tau_\varepsilon})$	$\left(\dfrac{\tau_\varepsilon}{\tau_\sigma} - 1\right)\delta(\tau - \tau_\varepsilon)$
4 Continuous relaxation $(\omega_1 \le \omega \le \omega_2)$	$\dfrac{1}{\omega_2}$	$\left[\dfrac{2}{\pi Q_m} \ln \dfrac{\omega_2 + i\omega}{\omega_1 + i\omega} + 1\right]^{1/2}$	$\dfrac{2}{\pi Q_m}\left[\ln \dfrac{\omega_2}{\omega_1} - \displaystyle\int_{\omega_1}^{\omega_2} e^{-st} \dfrac{ds}{s}\right]$	$\begin{cases} \dfrac{2}{\pi Q_m \tau}, & \tau_2 < \tau < \tau_1 \\ 0, & \text{otherwise} \end{cases}$
5 Jeffreys $\nu \ne 0$	t_0	From Eqs. (10.250) and (10.251)	$\dfrac{2}{\pi Q_0 \nu}\left[\left(1 + \dfrac{t}{t_0}\right)^{\nu} - 1\right]$	$\dfrac{2}{\pi Q_0} \dfrac{B(\nu)}{\tau^{1-\nu}} e^{-t_0/\tau}$
$\nu = 0$	$(1.78107)t_0$	$1 - \dfrac{1}{\pi Q_0} \ln(i\omega t_0)$	$\dfrac{2}{\pi Q_0} \ln\left(1 + \dfrac{t}{t_0}\right)$	$\dfrac{2}{\pi Q_0} \dfrac{1}{\tau} e^{-t_0/\tau}$
6 Modified Jeffreys	t_∞		$\dfrac{2}{\pi Q_0} \dfrac{e^{t_0/t_\infty}}{(t_0/t_\infty)^{\nu}} \displaystyle\int_{t_0/t_\infty}^{t_0/t_\infty(1 + t/t_0)} e^{-\xi} \xi^{\nu - 1}\, d\xi$	$\begin{cases} \dfrac{b}{\tau^{1-\nu}} e^{-t_0/\tau}\left(1 - \dfrac{\tau}{t_\infty}\right)^{-\nu}, & \tau < t_\infty \\ 0, & \tau > t_\infty \end{cases}$

Table 10.1. (Continued)

Solid	Normalized attenuation coefficient $\tau_0 c_\infty \gamma(x)$, $x = \omega\tau_0$				Normalized phase velocity $c(x)/c_\infty$			
		$x \ll 1$	$x = 1$	$x \gg 1$		$x \ll 1$	$x = 1$	$x \gg 1$
1 Maxwell	$\frac{x}{\sqrt{2}}\left[\left(1+\frac{1}{x^2}\right)^{1/2} - 1\right]^{1/2}$	$\left(\frac{x}{2}\right)^{1/2}$	$0.455\ldots$	$\frac{1}{2}$	$x\sqrt{2}\left[\left(1+\frac{1}{x^2}\right)^{1/2} - 1\right]^{1/2}$	$[2x]^{1/2}$	$0.910\ldots$	1
2 Kelvin–Voigt	$x\left[\frac{(1+x^2)^{1/2} - 1}{2(1+x^2)}\right]^{1/2}$	$\frac{1}{2}x^2$	$0.321\ldots$	$\left(\frac{x}{2}\right)^{1/2}$	$\frac{\sqrt{2}}{x}[(1+x^2)(\sqrt{1+x^2} - 1)]^{1/2}$	1	$1.287\ldots$	$\sqrt{2x}$
3 Standard linear	$\frac{x}{\sqrt{2}}(1+x^2)^{1/2}\left\{\frac{\left[1+\frac{4x^2Q_0^{-2}}{(1+x^2)^2}\right]^{1/2} - 1}{\left(1+x^2\frac{\tau_\varepsilon}{\tau_\sigma}\right)}\right\}^{1/2}$	$\frac{1}{Q_0}x^2$	$\frac{1}{2Q_0}$	$\frac{1}{Q_0}\left[\frac{\tau_\sigma}{\tau_\varepsilon}\right]^{1/2}$	$\left\{\frac{2\left(1+x^2\frac{\tau_\varepsilon}{\tau_\sigma}\right)\Big/(1+x^2)}{\left[1+\frac{4x^2Q_0^{-2}}{(1+x^2)^2}\right]^{1/2} + 1}\right\}^{1/2}$	1	$1 - \frac{1}{8Q_0^2}$	$\left(\frac{\tau_\varepsilon}{\tau_\sigma}\right)^{1/2}$
	$Q_0^{-1} = \frac{(\tau_\varepsilon - \tau_\sigma)}{2\tau_0}$							
4 Continuous relaxation ($\omega_1 \le \omega \le \omega_2$)	$\frac{x}{2Q_m}$				$1 + \frac{1}{\pi Q_m}\ln x$		1	
5 Jeffreys $\nu \ne 0$	$\frac{\lvert x\rvert^{1-\nu}}{2Q_0}D(\nu)$				$\frac{1}{1+\Gamma(\nu)\frac{x^{-\nu}}{\pi Q_0}\cos\frac{\pi\nu}{2}}$, $\nu > 0$			
$\nu = 0$	$\frac{\lvert x\rvert}{2Q_0}\left[1+\frac{1}{\pi Q_0}\ln x\right]$				$1 + \frac{1}{\pi Q_0}\ln x$			

[a] $b = B(\nu)\frac{2}{\pi Q_0}e^{t_0/t_\infty}$, $B(\nu) = [\Gamma(1-\nu)t_0^\nu]^{-1}$, $D(\nu) = \Gamma(\nu+1)\frac{\sin \pi\nu/2}{\pi\nu/2}$.

by using the known quadratures

$$\int_0^\infty \frac{e^{-s_1 t} - e^{-s_2 t}}{t} \sin \omega t \, dt = \tan^{-1} \frac{(s_2 - s_1)\omega}{\omega^2 + s_1 s_2}, \tag{10.282}$$

$$\int_0^\infty \frac{e^{-s_1 t} - e^{-s_2 t}}{t} \cos \omega t \, dt = \frac{1}{2} \ln \frac{\omega^2 + s_2^2}{\omega^2 + s_1^2}. \tag{10.283}$$

If we choose small s_1 and large s_2 so that

$$s_1 \ll \omega \ll s_2, \tag{10.284}$$

the integral in Eq. (10.282) will reduce to $\pi/2$, yielding [cf. Eq. (10.241a)]

$$\frac{1}{Q(\omega)} \simeq A \tan^{-1} \frac{(s_2 - s_1)\omega}{\omega^2 + s_1 s_2} \simeq \frac{A\pi}{2} = \frac{1}{Q_m}, \tag{10.285}$$

where Q_m^{-1} is defined as the constant value of Q^{-1} in the interval defined in Eq. (10.284). Hence

$$A = \frac{2}{\pi Q_m}, \qquad \underline{\phi}(\infty) = \frac{2}{\pi Q_m} \ln \frac{s_2}{s_1}. \tag{10.286}$$

The dispersion of the phase velocity is then obtained from Eqs. (10.208) and (10.283) on the assumption that A is small. Therefore, equating the real and imaginary parts in the two expressions for $\hat{n}(\omega)$

$$\hat{n}(\omega) \simeq 1 + \frac{1}{2} \int_0^\infty \dot{\underline{\phi}}(t) e^{-i\omega t} \, dt = \frac{c_\infty}{c(\omega)} - i \frac{c_\infty}{\omega} \gamma(\omega),$$

we obtain,

$$c(\omega) = c_\infty \left[1 + \frac{1}{2\pi Q_m} \ln \frac{\omega^2 + s_2^2}{\omega^2 + s_1^2} \right]^{-1}. \tag{10.287}$$

Clearly

$$c(\infty) = c_\infty = c(0) \left[1 + \frac{1}{\pi Q_m} \ln \frac{s_2}{s_1} \right], \quad \frac{1}{\pi Q_m} = \frac{[c(\infty)/c(0)] - 1}{\ln(s_2/s_1)},$$

$$\underline{\phi}(\infty) = 2 \left[\frac{c(\infty)}{c(0)} - 1 \right]. \tag{10.288}$$

In the region where Q is constant, the phase-velocity dispersion is approximated by the expression

$$c(\omega) \simeq c_\infty \left[1 - \frac{1}{\pi Q_m} \ln \frac{s_2}{\omega} \right] \tag{10.289}$$

which is in agreement with our former results obtained by invoking causality [Eq. (10.7.4)] and the application of the Trouton–Rankine creep law [Eq.

(10.247)]. Note that Eq. (10.289) can also be written in a form that is free from s_2, namely

$$c(\omega) = c(\omega_r)\left[1 - \frac{1}{\pi Q_m} \ln \frac{\omega_r}{\omega}\right], \tag{10.290}$$

where ω_r is an arbitrary frequency in the band where Q is constant.

Above we have examined three alternative approaches that eventually led us to the same results:

1. We have invoked the dispersion relations under the condition that Q is constant over a finite frequency range that excludes the origin $\omega = 0$ and the point at infinity $\omega = \infty$.
2. We have invoked the Boltzmann superposition law with logarithmic creep function, or a modified power law. Causality does not enter explicitly. This theory does not pass in the limit to the static case and predicts that primary creep will grow indefinitely.
3. Starting from the standard linear solid we have constructed a continuous relaxation model with Becker's distribution function $j(\tau) = A/\tau$ over a finite interval of strain relaxation times. The resulting model has a constant Q over a finite frequency range and exhibits velocity dispersion. A static limit is predicted.

Figure 10.15a shows the curves for $c^2(\omega)$ and $Q^{-1}(\omega)$ for a standard linear solid with a single relaxation mechanism (Table 10.1). Figure 10.15b shows curves for $c(\omega)$ and $Q^{-1}(\omega)$ for a viscoelastic solid with a continuous relaxation mechanism over a limited band of relaxation times. This results in a band-limited "constant-Q" model.

Table 10.1 compares the creep and propagation functions for the viscoelastic solids treated in this chapter.

10.4. Attenuation of Seismic Waves in the Earth

10.4.1. Introduction

The foregoing theory will next be applied to phenomena associated with attenuation of seismic energy in the earth. The seismic *absorption band* is that spectral window through which seismic waves with measurable amplitudes reach the global network of seismometers. In practice, this band extends from a fraction of a second (say 0.1 s) to the gravest mode of the free oscillations of the earth, which has a period of about 1 h. Although the creep laws discussed in Section 10.3 are believed to apply over a wider spectral range, especially at the long-period end, observational evidence for or against these laws outside the seismic band will not be presented here. Let us therefore summarize the basic anelastic properties of a low-loss linear solid:

1. The dissipation function is a quadratic function of the strain components and is always positive definite.

2. The causality principle is satisfied and, as a result, the complex wave number is an analytic function of the frequency in the lower half of the ω plane. This property is not just a mathematical abstraction but has a definite physical meaning. Because the medium is governed by a linear equation, a signal traveling in it must propagate with a finite velocity and any point in the medium will remain at rest until reached by the signal. Then, the analyticity of $\hat{k}(\omega)$ follows mathematically from causality and, as a result, the attenuation leads to causal phase shifts, which introduce dispersion.
3. In the limit $\omega \to \infty$ ($t \to 0$), the anelastic equations of state pass to the elastic equation.
4. The viscoelastic parameters of the equation of state must be related to the internal structure of the medium on the molecular scale because internal friction has its origin in the microstructure of matter.

10.4.2. Attenuation of Body Waves in the Earth

In Eq. (7.157) we gave the structure of the spectrum of a direct body wave in a real-earth model. In Chapter 8 we generalized this result to various reflected and transmitted rays in the mantle and core. An attenuation factor $e^{-\omega t^*(\Delta, h)}$ was defined there without specifying its origin.

Let $\gamma(\omega, s)$ be the attenuation coefficient at point s on the ray, s being the arc-length parameter. The total attenuation along the ray is then given by the factor $\exp\{-\int_{\text{ray}} \gamma(\omega, s)|ds|\}$. We know, however, that over the frequency band of observable body waves ($T = 0.1$ s–100 s), Q changes so slightly that it can be regarded as frequency independent. We shall denote it by $Q_m(s)$. Then, in the earth's absorption band

$$\gamma(\omega, s) \simeq \frac{\omega}{2Q_m(s)v_m(s)}, \qquad \omega_1 < \omega < \omega_2$$

where v_m is the intrinsic velocity. Inserting the value of $|ds| = \eta\, dr/[\eta^2 - p^2]^{1/2}$ from Eq. (7.91), the function $t^*(\Delta, h)$ assumes the explicit form

$$t^*(\Delta, h) = \frac{1}{2}\int_{r(\text{source})}^{r(\text{observer})} \frac{dr}{Q_m(r)v_m(r)[1 - p^2v_m^2/r^2]^{1/2}}. \tag{10.291}$$

If we replace $Q_m(r)$ by a constant average value Q_{av}, then $t^*(\Delta, h) = \tau/2Q_{\text{av}}$, where τ is the total travel-time along the ray. The attenuation factor then reduces to $e^{-\omega\tau/2Q_{\text{av}}}$.

For generalized rays the attenuation is introduced through a convolution of each generalized ray time-function with the inverse Fourier transform of $e^{-|\omega|t^*(\Delta, h)}$, which is

$$\frac{1}{\pi}\left\{\frac{t^*}{t^2 + t^{*2}}\right\}.$$

The causal dispersion will introduce an additional phase shift in Eq. (7.157). This phase shift is

$$\exp\left\{-i\omega \int_{\text{ray}} \left[\frac{1}{v(s,\omega)} - \frac{1}{v(s,\infty)}\right] ds\right\}.$$

Substituting from Eq. (10.287)

$$v^{-1}(s,\omega) = v_0^{-1}(s,\infty)\left[1 + \frac{1}{2\pi Q_m(s)} \ln \frac{\omega^2 + s_2^2}{\omega^2 + s_1^2}\right],$$

the body-wave propagator assumes the form

$$\exp\left[i\omega\left(t - \frac{1}{\pi} t^* \ln \frac{\omega^2 + s_2^2}{\omega^2 + s_1^2}\right)\right]. \tag{10.292}$$

Under condition (10.284), we have

$$\begin{aligned} \text{Phase delay} &= \frac{2t^*}{\pi} \ln\left(\frac{s_2}{\omega}\right), \\ \text{Group delay} &= \frac{2t^*}{\pi}\left[\ln\left(\frac{s_2}{\omega}\right) - 1\right]. \end{aligned} \tag{10.293}$$

The group delay at T relative to T' is $(2t^*/\pi)\ln(T/T')$. Observations of the attenuation of body waves in the earth indicate an average Q value of about 300 for the whole mantle. The shear-wave group delay at $T = 3$ s, say, relative to $T' = 1$ s is $\tau/\pi Q_{av} \simeq \tau/1000$ s. For an S wave at 80°, $\tau \simeq 22$ m, and the relative group delay amounts to about 1 s.

10.4.3. Observed Attenuation of Dispersed Surface Waves

The attenuation of propagating seismic waves is apparent in the records of many earthquakes and anthropogenic explosions. A finite amount of energy is imparted by the seismic source to the earth, and this energy, whether in the form of propagating waves or as normal modes, slowly decays with time and distance until the signal dies into the noise. Were it otherwise, every signal would have rung forever and seismology would have been an impossible science.

If the earth were a strongly absorbing medium, however, sharp beginnings of seismic reflections and transmissions would have been impossible, and the theoretical interpretation of seismic waves would have become extremely difficult. Seismology, in such a case, would have been ruined all the same. We therefore owe seismology to the simple fact that the dissipation of seismic energy in our planet is just right, small but not zero. A quantitative measure of this attenuation in the case of long-period surface waves is, in principle, rather simple: We consider two stations $P_1(\Delta_1)$ and $P_2(\Delta_2 > \Delta_1)$, located on a great circle path that goes through the source. Here Δ_1 and Δ_2 are the epicentral distances at these points. If the exponential decay law is valid, the spectral displacement at a distance Δ from the source will attenuate physically as $e^{-\gamma\Delta}$, where $\gamma = \gamma(\omega)$ is the *attenuation coefficient* measured per kilometer. [This factor appears in the expression given in Eq. (5.38) for the spectral displacements of Rayleigh and Love waves.] It will be denoted as γ_R for Rayleigh waves and γ_L for Love waves. Therefore, the spectral *amplitude ratio* of the surface displacements at P_1 and P_2 is

$$\frac{u(\omega, \Delta_1)}{u(\omega, \Delta_2)} = \left[\frac{\sin(\Delta_2/a)}{\sin(\Delta_1/a)}\right]^{1/2} e^{\gamma(\omega)[\Delta_2 - \Delta_1]}.$$

Hence

$$\gamma(\omega) = \frac{\ln[A(\omega, \Delta_1)/A(\omega, \Delta_2)]}{\Delta_2 - \Delta_1}, \tag{10.294}$$

where

$$A(\omega, \Delta) = u(\omega, \Delta)\left[\sin\left(\frac{\Delta}{a}\right)\right]^{1/2}.$$

In particular, if measurements are made at a single station from multiple arrivals of the same wave [e.g., G_2 and G_4], the difference $(\Delta_2 - \Delta_1)$ will be approximately equal to the length of the great circle that passes through the source and station. Whenever γ is evaluated directly from a seismogram by a stationary-phase approximation, Eq. (3.96), the dispersion effects must be compensated. We then use the expression

$$\gamma(\omega) = \frac{\ln[A(\omega_0, \Delta_1)/A(\omega_0, \Delta_2)] - \alpha(\Delta_2/\Delta_1)}{\Delta_2 - \Delta_1}, \tag{10.295}$$

where α is equal to $\frac{1}{3}$ or $\frac{1}{2}$, depending on whether the angular frequency ω_0 is at an Airy phase or not. Figure 10.16 shows the effect of attenuation on a Love wave that reappeared at a station after completing an additional revolution around the earth. It is obvious from the figure that attenuation increases with frequency because G_2 seems as if it has gone through a low-pass filter. Figure 10.17 shows the same effect for Rayleigh waves at about 20 s.

We have seen earlier that in an infinite medium the dimensionless dissipation parameter Q is related to $\gamma(\omega)$ via the relation $Q = \omega/(2\gamma c)$, where c is the frequency-independent wave velocity. We shall next show that in the case of surface waves this velocity is the wave group velocity.

Consider a seismogram of a surface wave (the fundamental mode, say) at a large distance. It may be represented by the Fourier integral

$$F(t, \Delta) = \frac{L_0}{[\sin(\Delta/a)]^{1/2}} \operatorname{Re} \int_0^\infty A(\omega)\exp\left[-\frac{\omega t}{2Q} + i(\omega t - k\Delta + \phi_0)\right]dk, \tag{10.296}$$

where Δ/a is the given epicentral distance in radians, $\Lambda = 2\pi/k$ is the wavelength, L_0 is a source constant, ϕ_0 is the source initial phase, and Q is the dimensionless dissipation parameter. A dispersion relation $\omega = \omega(k)$ is understood.

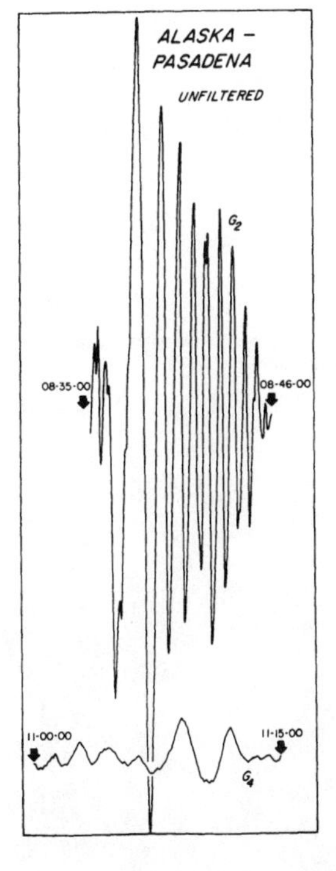

Figure 10.16. Attenuation of long-period Love waves in the earth: A G_2-wave from the Alaskan earthquake of 10 July, 1958, was recorded at Pasadena, California. It reappeared at the recording station after completing an additional revolution around the earth, losing in the process its high frequencies.

The saddle point is given by

$$i\left(t\frac{d\omega}{dk} - \Delta\right) - \frac{d}{dk}\left(\frac{\omega t}{2Q}\right) = 0. \tag{10.297}$$

If Q is large enough for the variation of $(\omega t/2Q)$ to be much less than that of ωt within a wavelength, the saddle point will be determined by the simpler equation $t\, d\omega/dk - \Delta = 0$, whence $d\omega/dk = \Delta/t = U$, the group velocity. Let the solution of this equation be k_0, ω_0. If Δ and t are known for a given point in the seismogram at a given location, the relation $\Delta/t = U(\omega_0)$ renders ω_0 when U is known by using standard group-velocity curves.

The second derivative of the index in Eq. (10.296) is $\{it(d^2\omega/dk^2)\}$ and the approximation near k_0 contains

$$\exp i\left[\omega t - k\Delta + \phi_0 + \frac{1}{2}t\omega_0''(k - k_0)^2\right]. \tag{10.298}$$

The steepest descents method then yields, according to the sign of ω_0'' (App. E),

$$F(t, \Delta) = \frac{L_0}{[\sin(\Delta/a)]^{1/2}} A(\omega_0)\exp\left[-\frac{\omega_0 t}{2Q}\right]\left(\frac{2\pi}{t|\omega_0''|}\right)^{1/2} \times \cos\left[\omega_0 t - k_0\Delta + \phi_0 + \frac{\pi}{4}\operatorname{sgn}\omega_0''\right], \tag{10.299}$$

$$k_0 = k_0(t, \Delta), \qquad \omega_0 = \omega_0(k_0)$$

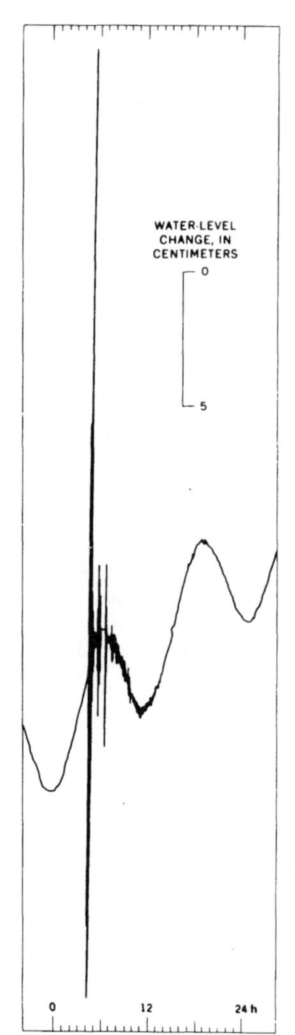

Figure 10.17. Attenuation of Rayleigh waves at 20–22 s in the earth: Water-level fluctuation in a deep well at Heibaart, Belgium, resulting from the Alaskan earthquake of 28 March, 1964. Observe the decay of the envelopes of R_1, R_2, R_3, R_4, R_5, and R_6. Note that the amplitude ratio of R_1/R_3 at $T = 22$ s is 5.2 yields an attenuation coefficient of $0.41 \times 10^{-4}\,\text{km}^{-1}$. After traveling for about 10 h, most of the Rayleigh waves had disappeared into the noise.

which is valid for large distances, for large values of Q, and away from extrema of the group velocity. There is no loss of generality in this last restriction because the attenuation is not affected.

Consider next the phase angle $\theta = \omega_0 t - k_0 \Delta$ in Eq. (10.299). Because of $d\omega_0/dk_0 = \Delta/t$, we have

$$\frac{\partial \theta}{\partial \Delta} = t \frac{d\omega_0}{dk_0} \frac{\partial k_0}{\partial \Delta} - k_0 - \Delta \frac{\partial k_0}{\partial \Delta} = -k_0,$$

$$\frac{\partial \theta}{\partial t} = \omega_0 + t \frac{d\omega_0}{dk_0} \frac{\partial k_0}{\partial t} - \Delta \frac{\partial k_0}{\partial t} = \omega_0.$$

Consequently, for values of Δ near Ut, the motion is approximately harmonic with wavelength $2\pi/k_0$, and for values of t near Δ/U it is approximately harmonic with period $2\pi/\omega_0$.

The function $F(t, \Delta)$ represents an individual harmonic component that passes the coordinate Δ at time t with a local phase velocity $c = c(\Delta, t) = \omega_0/k_0$. However, if we fix the value of Δ and consider a surface-wave seismogram in the finite time interval $t_1 \leq t \leq t_2$, a correspondence can be set up between the values of t in this interval and the values of $\omega_0(t)$ in the corresponding interval

$$\omega_1(t_1) \leq \omega_0(t) \leq \omega_2(t_2).$$

In this sense, Eq. (10.299) is in effect another spectral representation of $F(t, \Delta)$ with $t = t(\omega_0)$. Putting $T = 2\pi/\omega_0$ we substitute $t = \Delta/U$ into $\exp\{-\omega_0 t/2Q\}$ with the result that each spectral component of $F(t, \Delta)$ attenuates as $\exp\{-\pi\Delta/QUT\}$. Combining this with the equivalent spatial decay law $\exp\{-\gamma\Delta\}$, we find

$$Q = \frac{\pi}{\gamma UT}, \tag{10.300}$$

where $\gamma(\omega)$ like U is measurable from seismograms.

Table 10.2 demonstrates the evaluation of Q for Love and Rayleigh waves from measurements of $\gamma(\omega)$ and the group velocities over an all-oceanic path. Analysis was made with the waves G_2, G_3 and R_1, R_3 that originated at the Kurile Islands (44.7°N, 150.7°E) on 20 October, 1963, 00 h 53 min 07 s UT, and were recorded at Scott Base, Antarctica. Group velocities were calculated by fitting a polynomial to the phase velocity data and then differentiating this polynomial analytically according to $U_g^{-1} = c^{-1}[1 + (T/c)(dc/dT)]$. Table 10.3 shows similar calculations for shorter periods of Love and Rayleigh waves that traversed the oceanic path from Toledo, Spain, to Trinidad. Table 10.4 summarizes observations of Q values in the seismic absorption band 1 s to 1 h. Figures 10.18a, b show the dependence of the attenuation coefficients of Love and Rayleigh waves upon the period in the range $T = 10$–300 s.

Several features can be seen in the data:

1. Q decreases monotonically from 300 s to a broad minimum centered near 60 s. Even a 60-s wave is sampling the upper mantle.
2. Q for Rayleigh waves reaches a maximum value at a period of $\sim$23 s and decreases at shorter periods.
3. Q for Love waves rises abruptly at periods less than 40 s. This means that Q is significantly higher in the crust as compared to the upper mantle.
4. Q for Rayleigh waves is somewhat greater than Q for Love waves.

Measurements of the attenuation of Rayleigh waves over predominantly oceanic and continental paths reveal that the anelastic properties of the structure sampled by waves in the 20–50 s period range are statistically different. In particular, lower values of Q are observed over oceanic paths and the trend to lower values of Q occurs at shorter periods for oceanic paths. This, in turn,

Table 10.2. Measured Attenuation and Dispersion of Surface Waves along an Oceanic Path Between Scott Base, Antarctica, and the Kurile Islands

Period,	Fundamental Love mode (L_0)				Fundamental Rayleigh mode (R_{11})			
T (s)	$\gamma_L \times 10^4$ (km^{-1})	c_L (km/s)	U_L (km/s)	$Q_L = \pi/\gamma_L U_L T$	$\gamma_R \times 10^4$ (km^{-1})	c_R (km/s)	U_R (km/s)	$Q_R = \pi/\gamma_R U_R T$
277.8	0.187	5.155	4.440	136		5.133	3.636	
270.3	0.194	5.132	4.180			5.077	3.598	
263.1	0.201	5.109	4.400			5.022	3.568	
256.4	0.207	5.088	4.386			4.966	3.543	
250.0	0.214	5.068	4.376			4.913	3.526	
243.9	0.221	5.049	4.369	133		4.866	3.519	
238.1	0.227	5.031	4.364			4.825	3.520	
232.6	0.234	5.014	4.362			4.786	3.526	
227.3	0.241	4.998	4.363		0.195	4.751	3.535	200
222.2	0.247	4.982	4.366		0.198	4.718	3.545	
217.4	0.253	4.968	4.369	131	0.203	4.689	3.556	
212.8	0.258	4.954	4.373		0.210	4.661	3.566	
208.3	0.264	4.941	4.377		0.217	4.636	3.577	194
204.1	0.270	4.928	4.381		0.225	4.609	3.584	
200.0	0.275	4.916	4.385		0.232	4.576	3.586	
196.1	0.279	4.904	4.389		0.238	4.541	3.586	
192.3	0.284	4.893	4.394	131	0.243	4.517	3.598	187
188.7	0.288	4.883	4.398		0.249	4.500	3.616	
185.2	0.292	4.873	4.400		0.256	4.482	3.632	
181.8	0.295	4.863	4.403		0.263	4.466	3.646	
178.6	0.298	4.854	4.405		0.271	4.451	3.658	
175.4	0.301	4.845	4.407	135	0.279	4.435	3.665	175
172.4		4.837	4.408		0.287	4.418	3.670	
169.5		4.829	4.407		0.297	4.398	3.670	
166.7		4.821	4.405		0.308	4.380	3.670	
163.9		4.814	4.402		0.323	4.366	3.673	
161.3		4.807	4.397		0.339	4.355	3.675	156

Table 10.3. Propagation Parameters, Toledo–Trinidad

		Rayleigh waves, R_{11}					Love waves, L_0					
$\omega/2\pi$ (mc/s)	T (s)	Λ (km)	c_R (km/s)	U_R (km/s)	$\gamma_R \times 10^4$ (km^{-1})	Q_R	Λ (km)	c_L (km/s)	U_L (km/s)	$\gamma_L \times 10^4$ (km^{-1})	Q_L	Q_R/Q_L
8.0	125.0	586	4.250	3.70	0.45	151	586	4.690	4.35	0.58	109	1.38
8.8	113.6	477	4.196	3.72	0.47	158	531	4.673	4.35	0.58	109	1.45
9.6	104.2	435	4.173	3.77	0.50	160	486	4.662	4.35	0.61	114	1.40
10.4	96.2	401	4.167	3.82	0.56	152	447	4.652	4.35	0.62	121	1.25
11.2	89.3	371	4.159	3.86	0.63	145	414	4.640	4.35	0.63	128	1.13
12.0	83.3	347	4.167	3.90	0.70	138	386	4.627	4.35	0.67	129	1.07
12.8	78.1	326	4.169	3.98	0.75	134	360	4.613	4.35	0.74	125	1.07
13.6	73.5	306	4.160	4.04	0.80	132	338	4.603	4.34	0.88	112	1.18
14.4	69.4	290	4.170	4.08	0.85	130	319	4.595	4.34	0.98	106	1.23
15.2	65.8	274	4.163	4.10	0.92	127	302	4.588	4.33	1.09	101	1.26
16.0	62.5	260	4.159	4.09	1.00	123	286	4.577	4.33	1.18	100	1.23
16.8	59.5	247	4.155	4.07	1.09	119	271	4.558	4.32	1.23	100	1.19
17.6	56.8	236	4.150	4.06	1.15	118	258	4.539	4.31	1.27	101	1.17
18.4	54.3	225	4.145	4.05	1.20	119	246	4.522	4.29	1.30	103	1.15
19.2	52.1	216	4.141	4.04	1.23	121	235	4.519	4.28	1.31	107	1.13
20.0	50.0	207	4.136	4.04	1.24	125	225	4.498	4.27	1.32	111	1.13

Table 10.4. Observed Q Values of Seismic Wave Motion in the Earth

Free oscillations						Surface waves (fundamental mode)				Body waves (0.5–70 s)	
Spheroidal			Toroidal			Rayleigh, R_{11}		Love, L_0			
$_0S_l$	T (s)	Q_S	$_0T_l$	T (s)	Q_T	T (s)	Q_R	T (s)	Q_L	P	S
$_0S_0$	1227.6	4110				300	228	300	140		
$_0S_1$						290		290			
$_0S_2$	3234.1	581	$_0T_2$	2630.8	308	280		280			
$_0S_3$	2134.7	460	$_0T_3$	1702.5	290	270		270			
$_0S_4$	1546.9	411	$_0T_4$	1303.7	290	260	206	260			
$_0S_5$	1190.8	352	$_0T_5$	1075.5	280	250	196	250	131		
$_0S_6$	962.8	343	$_0T_6$	925.5	280	240		240			
$_0S_7$	812.1	373	$_0T_7$	818.0		230	186	230			
$_0S_8$	708.2	357	$_0T_8$	736.4	217	220		220			
$_0S_9$	634.2	320	$_0T_9$	671.8		210		210			
$_0S_{10}$	580.5	329	$_0T_{10}$	619.1	228	200	176	200	125		
$_0S_{11}$			$_0T_{11}$	575.1		190		190			
$_0S_{12}$	502.6	280	$_0T_{12}$	537.6	185	180		180		Upper mantle Q_α = 100– 500	Upper mantle Q_β = 110– 350
$_0S_{13}$	473.7	290	$_0T_{13}$	506.1	165	170	166	170		Lower mantle Q_α = 1600–4000	Lower mantle Q_β = 500–2000
$_0S_{14}$	448.1	272	$_0T_{14}$	477.0	161	160	160	160			
$_0S_{15}$		290	$_0T_{15}$	452.0	157	150	162	150	118		
$_0S_{16}$	407.4	259	$_0T_{16}$	429.7		140		140			
$_0S_{17}$	389.6		$_0T_{17}$	409.6		130		130			
$_0S_{18}$	374.1		$_0T_{18}$	391.5		120	151	120			
$_0S_{19}$	360.5		$_0T_{19}$	375.0		110		110			
$_0S_{20}$	347.6	210	$_0T_{20}$	359.9		100	145	100	108		
$_0S_{21}$	335.9	222	$_0T_{21}$			90		90			
$_0S_{22}$	325.2	200	$_0T_{22}$		114	80		80	110		
$_0S_{23}$	315.3	198	$_0T_{23}$		123	70	130	70			
$_0S_{24}$	306.1	197	$_0T_{24}$			60		60	100		
$_0S_{25}$	297.2	200	$_0T_{25}$	300.1	115	50	125	50			

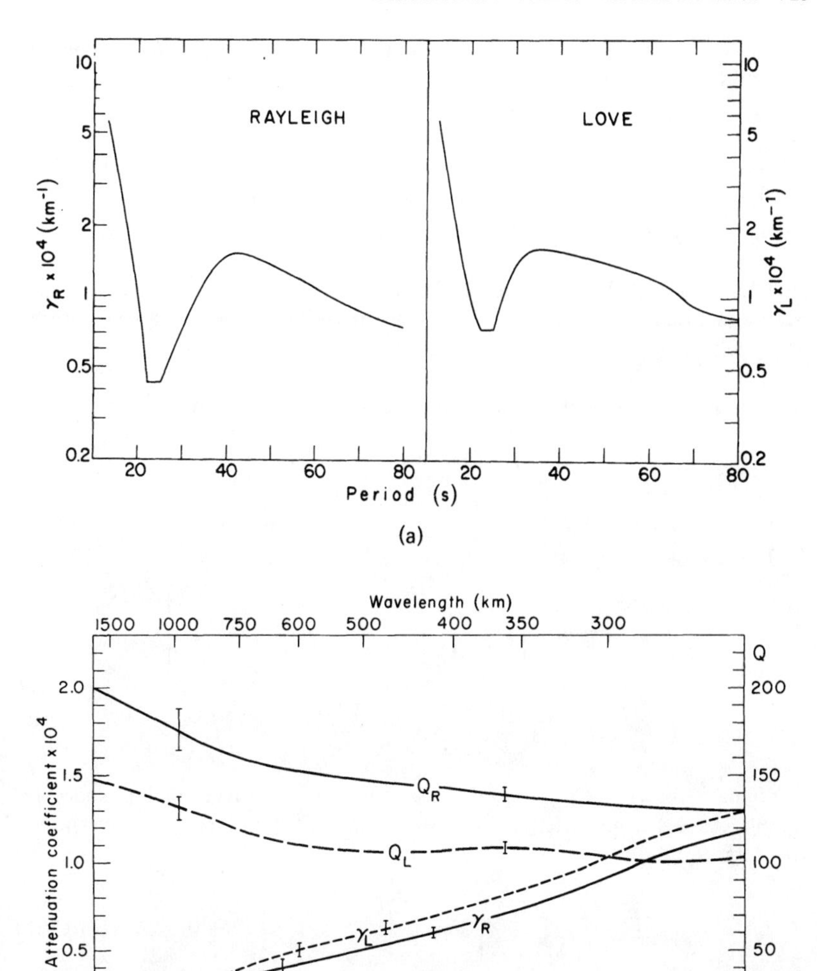

Figure 10.18. Summary of observed attenuation of Love and Rayleigh waves in the earth as measured from earthquake recordings. (a) Crustal and subcrustal waves. (b) Waves in the upper mantle of the earth.

implies that increased dissipation occurs at shallower depths under the oceans than under the continents.

10.4.4. Attenuation of Surface Waves in a Multilayered Flat-Earth Model

10.4.4.1. Rayleigh Waves in an Anelastic Half-Space. Whenever dissipation occurs in an anelastic solid, we have to distinguish between two types of body-wave velocities:

1. The intrinsic elastic velocities in the absence of dissipation

$$\alpha_\infty = \left[\frac{\lambda + 2\mu}{\rho}\right]^{1/2}, \qquad \beta_\infty = \left[\frac{\mu}{\rho}\right]^{1/2}.$$

2. The complex velocities

$$\hat{\alpha} = \alpha(\omega)\left[1 + \frac{i}{2Q_\alpha}\right], \qquad \hat{\beta} = \beta(\omega)\left[1 + \frac{i}{2Q_\beta}\right],$$

$$\lim_{\omega\to\infty} \hat{\alpha} = \alpha_\infty, \qquad \lim_{\omega\to\infty} \hat{\beta} = \beta_\infty.$$

We have shown in Chapter 3 that Rayleigh waves in an elastic half-space are neither dispersed nor attenuated with distance. Their velocity c is determined from the equation

$$\left(2 - \frac{c^2}{\beta_\infty^2}\right)^2 - 4\left[1 - \frac{c^2}{\alpha_\infty^2}\right]^{1/2}\left[1 - \frac{c^2}{\beta_\infty^2}\right]^{1/2} = 0. \tag{10.301}$$

Suppose that the half-space is weakly anelastic such that

$$\hat{\alpha} = \alpha_\infty\left[1 + \frac{i}{2Q_\alpha}\right], \qquad \hat{\beta} = \beta_\infty\left[1 + \frac{i}{2Q_\beta}\right] \tag{10.302}$$

in accord with Eqs. (10.193) and (10.196), except that here we assume initially that all entities in Eq. (10.302) are frequency independent and that $Q_\alpha \gg 1$, $Q_\beta \gg 1$. This is physically reasonable because Rayleigh waves in a perfect elastic and homogeneous half-space are nondispersive and because the anelasticity is taken to be weak. We may therefore put

$$\hat{c} = c\left[1 + \frac{i}{2Q_R}\right], \qquad Q_R \gg 1 \tag{10.303}$$

where Q_R is to be expressed in terms of the known Q_α and Q_β. Using Eqs. (10.302) and (10.303) we substitute $\hat{\alpha}$, $\hat{\beta}$, and $\hat{c}$ for the corresponding entities α_∞, β_∞, c in Eq. (10.301). The Rayleigh equation becomes

$$\left\{2 - \left(\frac{c}{\beta_\infty}\right)^2 \left[\frac{1 + i/2Q_R}{1 + i/2Q_\beta}\right]^2\right\}^4 = 16\left\{1 - \frac{c^2}{\beta_\infty^2}\left[\frac{1 + i/2Q_R}{1 + i/2Q_\beta}\right]^2\right\}$$
$$\times \left\{1 - \frac{c^2}{\alpha_\infty^2}\left[\frac{1 + i/2Q_R}{1 + i/2Q_\alpha}\right]^2\right\}. \quad (10.304)$$

We expand each side of Eq. (10.304) in powers of $1/Q_\alpha$, $1/Q_\beta$, and $1/Q_R$, retaining only terms up to the first power of these entities. We then separate it into real and imaginary parts. The real part of the equation yields back Eq. (10.301), whereas the imaginary part renders, after some algebraic steps,

$$\frac{1}{Q_R} = \frac{m}{Q_\alpha} + \frac{1 - m}{Q_\beta}, \quad (10.305)$$

where

$$m = \frac{1}{1 - \theta},$$
$$\theta = \frac{(\alpha_\infty^2/\beta_\infty^2)(1 - c^2/\alpha_\infty^2)(2 - 3c^2/\beta_\infty^2)}{(2 - c^2/\beta_\infty^2)(1 - c^2/\beta_\infty^2)}. \quad (10.306)$$

Therefore, Q_R^{-1} is interpreted as a weighted average value of Q_α^{-1} and Q_β^{-1}. Clearly, m is a function of Poisson's ratio only. In particular, for $\bar{\sigma} = \frac{1}{4}$,

$$\frac{1}{Q_R} = \frac{0.134}{Q_\alpha} + \frac{0.866}{Q_\beta}. \quad (10.307)$$

Other values of m are given in Table 10.5. The Rayleigh wave spectral displacement will still be given by Eq. (3.105), except that the factor

$$\{e^{-\omega\Delta/2Q_R C}\}$$

will multiply the propagator $e^{i(\omega t - kx)}$. When one or more of the $1/Q_i$s are appreciably greater than 0.1, the complex analog of Eq. (10.301) must be solved using the exact forms

$$\hat{\alpha} = \alpha_\infty\left[1 + \frac{i}{Q_\alpha}\right]^{1/2}, \qquad \hat{\beta} = \beta_\infty\left[1 + \frac{i}{Q_\beta}\right]^{1/2}, \qquad \hat{c} = c\left[1 + \frac{i}{Q_R}\right]^{1/2}. \quad (10.308)$$

It is then necessary to solve the resulting cubic equation with complex coefficients. Note that for $\beta_\infty/\alpha_\infty \geq 0.567008$ ($\bar{\sigma} \leq 0.24328$) there are three real roots of Eq. (10.301) rather than the single root that appears for smaller ratios of $\beta_\infty/\alpha_\infty$.

Table 10.5. Dissipation Parameters of Rayleigh Waves in a Uniform Half-space

$\dfrac{\beta_\infty}{\alpha_\infty}$	$\bar{\sigma} = \dfrac{1 - 2\beta_\infty^2/\alpha_\infty^2}{2 - 2\beta_\infty^2/\alpha_\infty^2}$	$\dfrac{c}{\beta_\infty}$	m
0	0.5	0.95531	$0\ (Q_R = Q_\beta)$
0.05	0.49874	0.95516	3.2411×10^{-4}
0.10	0.49494	0.95469	1.3233×10^{-3}
0.15	0.48849	0.95389	3.0823×10^{-3}
0.20	0.47916	0.95271	5.7570×10^{-3}
0.25	0.46666	0.95112	9.6002×10^{-3}
0.30	0.45054	0.94903	1.5008×10^{-2}
0.35	0.43019	0.94633	2.2597×10^{-2}
0.40	0.40476	0.94286	3.3347×10^{-2}
0.45	0.37304	0.93837	4.8851×10^{-2}
0.50	0.33333	0.93253	7.1789×10^{-2}
0.55	0.28315	0.92476	0.10685
0.60	0.21875	0.91419	0.16257
0.65	0.13419	0.89937	0.25512
0.7071	0.0	0.87785	0.41586

However, it is only the smallest of the three roots that is of physical significance for Rayleigh-wave propagation.

10.4.4.2. Love Waves in a Two-Layered Anelastic Medium. The mathematical procedure of the previous section is now applied to the case of Love waves in a homogeneous layer with an anelastic parameter $Q_{\beta 1}$ overlying a homogeneous half-space with the anelastic parameter $Q_{\beta 2}$. When the result of Eq. (3.74) is combined with Eqs. (10.191)–(10.196), the appropriate dispersion equation for a viscoelastic layer overlying a viscoelastic half-space will be

$$\tan\left\{\hat{k}_L H\left[\frac{\hat{c}^2}{\hat{\beta}_1^2} - 1\right]^{1/2}\right\} = \frac{\hat{\mu}_2}{\hat{\mu}_1}\left[\frac{1 - (\hat{c}/\hat{\beta}_2)^2}{(\hat{c}/\hat{\beta}_1)^2 - 1}\right]^{1/2} \tag{10.309}$$

with

$$\hat{\beta}_1^2 = \beta_1^2\left(1 + \frac{i}{Q_{\beta 1}}\right), \quad \hat{\beta}_2^2 = \beta_2^2\left(1 + \frac{i}{Q_{\beta 2}}\right), \quad \hat{k}_L = k_L\left(1 - \frac{i}{2Q_L}\right),$$

$$\hat{\mu}_1 = \mu_1\left(1 + \frac{i}{Q_{\beta 1}}\right), \quad \hat{\mu}_2 = \mu_2\left(1 + \frac{i}{Q_{\beta 2}}\right), \quad \hat{c} = c\left(1 + \frac{i}{2Q_L}\right),$$

$$Q_{\beta 1} \gg 1, \qquad Q_{\beta 2} \gg 1, \qquad Q_L \gg 1. \tag{10.310}$$

We insert these entities into Eq. (10.309), use the addition law for the tangent function, and then expand the radicals in power series of $Q_{\beta 1}^{-1}$, $Q_{\beta 2}^{-1}$, and Q_L^{-1}. The real part of the ensuing equation gives the usual period equation of Love waves in which β_1 and β_2 are slightly frequency dependent on account of the medium's anelasticity. The imaginary part yields the new result,

$$\frac{1}{Q_L} = \frac{A}{Q_{\beta 1}} + \frac{B}{Q_{\beta 2}}, \tag{10.311}$$

where

$$A = \frac{p_1}{q_L}, \qquad B = \frac{p_2}{q_L},$$

$$q_L = \frac{kH}{\eta_2}\left[1 + \left(\frac{\mu_2\eta_2}{\mu_1\eta_1}\right)^2\right] + \frac{\mu_2}{\mu_1}\left(\frac{1}{\eta_1^2} + \frac{1}{\eta_2^2}\right),$$

$$p_1 = \frac{kH}{\eta_2}\left(\frac{c}{\beta_1}\right)^2\left[1 + \left(\frac{\mu_2\eta_2}{\mu_1\eta_1}\right)^2\right] + \frac{\mu_2}{\mu_1}\left(\frac{1}{\eta_1^2} - 1\right), \tag{10.312}$$

$$p_2 = \frac{\mu_2}{\mu_1}\left(\frac{1}{\eta_2^2} + 1\right),$$

$$\eta_2 = \left[1 - \left(\frac{c}{\beta_2}\right)^2\right]^{1/2}, \qquad \eta_1 = \left[\left(\frac{c}{\beta_1}\right)^2 - 1\right]^{1/2}.$$

In this example we have, for the first time, made the assumption that Q_L depends not only upon the frequency, but also on the depth, via $Q_{\beta 1}$ and $Q_{\beta 2}$. The reader can easily verify the interesting relations:

$$A = \frac{\beta_1}{c}\left(\frac{\partial c}{\partial \beta_1}\right), \qquad B = \frac{\beta_2}{c}\left(\frac{\partial c}{\partial \beta_2}\right). \tag{10.313}$$

10.4.4.3. A Variational Formulation for Love Waves in Multilayered Low-Loss Solids. The results of the two previous examples can be generalized to a multilayered half-space or a half-space with a continuous velocity distribution by the application of the Rayleigh principle and the assumption that the attenuation is small. Using some of our earlier results from Chapter 3, we shall derive a relation like those given in Eqs. (10.305) and (10.311). It relates the observed $Q_R(\omega)$, $Q_L(\omega)$ at the surface $z = 0$ to the discrete or continuous intrinsic distributions of $Q_\alpha(z, \omega)$ and $Q_\beta(z, \omega)$ at depth. Although we restrict our present discussion to cylindrical coordinates, the same method can be used for a spherical-earth model when the energy integrals in spherical coordinates are used.

In Section 3.6.6 we have presented a variational formulation for Love waves in a multilayered solid. The use of Rayleigh's principle for conservative systems requires the equating of the time averages of kinetic and potential energies [cf. Sect. 3.6.6]. This, in turn, results in the vanishing of the time-averaged Lagrangian for the system. When the system is dissipative, the total mechanical energy is not conserved and Rayleigh's principle cannot be applied in its usual form.

We shall next introduce a new formalism that will enable us to carry on calculations for dissipative systems as though they were conservative systems. This can be done by writing a Lagrangian for two systems operating simultaneously.

One of them is the system at hand, which loses energy because of friction. The other is a "mirror image" system with negative friction into which goes the energy that is drained from the dissipative system, such that the total energy is conserved.

As an example, we consider Love waves in a layered, vertically inhomogeneous, attenuating half-space. According to Eq. (10.157), the equation of motion in a homogeneous layer is

$$\mu \nabla^2 u + \frac{\mu^*}{\omega} \nabla^2 \dot{u} = \rho \ddot{u}, \tag{10.314}$$

where u is the transverse particle displacement. The solution of Eq. (10.314) was given in Eq. (10.173) and we write it as

$$u = V(z) e^{-k^* x} \cos(\omega t - kx). \tag{10.315}$$

The equation of the mirror image system is

$$\mu \nabla^2 u^+ - \frac{\mu^*}{\omega} \nabla^2 \dot{u}^+ = \rho \ddot{u}^+ \tag{10.316}$$

with the corresponding solution

$$u^+ = V^+(z) e^{k^* x} \cos(\omega t - kx). \tag{10.317}$$

Let us assume that the Lagrangian density of the combined system exists. Then, according to Eq. (1.130) and because $\mathscr{L}$ must depend also on u itself, the Euler–Lagrange equations become

$$\frac{\partial}{\partial t} \frac{\partial \mathscr{L}}{\partial \dot{u}} + \frac{\partial}{\partial x} \frac{\partial \mathscr{L}}{\partial(\partial u/\partial x)} + \frac{\partial}{\partial z} \frac{\partial \mathscr{L}}{\partial(\partial u/\partial z)} = \frac{\partial \mathscr{L}}{\partial u}, \tag{10.318}$$

$$\frac{\partial}{\partial t} \frac{\partial \mathscr{L}}{\partial \dot{u}^+} + \frac{\partial}{\partial x} \frac{\partial \mathscr{L}}{\partial(\partial u^+/\partial x)} + \frac{\partial}{\partial z} \frac{\partial \mathscr{L}}{\partial(\partial u^+/\partial z)} = \frac{\partial \mathscr{L}}{\partial u^+}. \tag{10.319}$$

A function $\mathscr{L}$ that satisfies both Eq. (10.318) and Eq. (10.319) exists in the explicit form

$$\mathscr{L} = \rho \dot{u} \dot{u}^+ - \frac{1}{2} \frac{\mu^*}{\omega} \nabla^2 [u \dot{u}^+ - \dot{u} u^+] - \mu \left[\frac{\partial u}{\partial x} \frac{\partial u^+}{\partial x} + \frac{\partial u}{\partial z} \frac{\partial u^+}{\partial z} \right]. \tag{10.320}$$

We now substitute Eqs. (10.315) and (10.317) into Eq. (10.320) and then integrate both sides over space and time (one cycle). The result is an explicit form for the Lagrangian of the combined system (compare with Eq. (3.237))

$$\mathsf{L} = \omega^2 I_0^L - (k^2 - k^{*2}) I_1^L - I_2^L \tag{10.321}$$

with the energy integrals

$$I_0^L = \int_0^\infty \rho V V^+ \, dz, \quad I_1^L = \int_0^\infty \mu V V^+ \, dz, \quad I_2^L = \int_0^\infty \mu \left[\frac{dV}{dz} \frac{dV^+}{dz} \right] dz. \tag{10.322}$$

The first term on the right-hand side of Eq. (10.321) is interpreted as the time average of the kinetic energy and the remaining two terms as the time average of the potential energy. This allows the application of the *virial theorem*, which we have already stated and used in Section 3.6.6

Consequently, $\mathsf{L} \equiv 0$. The resulting equation

$$\omega^2 I_0^L = (k^2 - k^{*2}) I_1^L + I_2^L, \tag{10.323}$$

is a generalization of Eq. (3.127) for anelastic media. As in Section 3.6.6, we continue with the application of the Rayleigh principle to Eq. (10.323), and obtain

$$c_L U_L = \left[1 - \frac{k^*}{k}\left(\frac{\delta k^*}{\delta k}\right)\right]\frac{I_1^L}{I_0^L}. \tag{10.324}$$

Using the relations $k^* \simeq \omega/2U_L Q_L$ [Eq. (10.300)], $k = \omega/c_L$, and

$$\frac{\delta k^*}{\delta k} \simeq \frac{\partial k^*/\partial\omega}{\partial k/\partial\omega} = U_L \frac{\partial}{\partial\omega}\left(\frac{\omega}{2U_L Q_L}\right) \simeq \frac{1}{2Q_L} + O\left(\frac{1}{Q_L^2}\right), \tag{10.325}$$

Eq. (10.324) yields, approximately,

$$c_L U_L = \left[1 - \frac{c_L}{U_L}\frac{1}{4Q_L^2}\right]\frac{I_0^L}{I_1^L}. \tag{10.326}$$

We therefore see that the correction term to the elastic case is of the order of $1/Q_L^2$ and Eq. (3.127) will apply also to low-loss media provided we keep terms up to first order in Q_L^{-1}. It can be shown that the same statement is valid for Rayleigh waves.

10.4.4.4. Perturbation of the Energy Integrals in Anelastic Solids. We have just shown that the basic idea associated with the Rayleigh principle and the consequent derivation of the Love and Rayleigh integrals remain valid for complex values of the parameters. For Love waves these include k and μ. For Rayleigh waves λ is added. It also applies to complex values of the inertia [Eqs. (10.202)–(10.205)] and complex values of the frequency. This latter case calls for some explanation: So far, ω has been held real as befits the mathematical description of a propagating wave. The earth, however, is capable of sustaining standing waves in the form of normal modes (free oscillations). The time dependence of the displacements in this case is $u(t) \propto e^{i\hat{\omega}t}$, where $\hat{\omega} = \omega + i\omega^*$. In this process the anelasticity will be manifested in the ratio of the strain energy dissipated per cycle to the maximum strain energy stored during this cycle. The value of Q^{-1} associated with this process is $2\omega^*/\omega$ if $\omega^*/\omega \ll 1$.

Let us therefore explore the immediate mathematical consequences of this generalization. For the sake of simplicity we shall treat in detail only the case of Love waves. The subscript L will be deleted from all parameters and variables except Q. We substitute in the energy integrals [given in Eqs. (3.127) and (3.140)] the values

$$\begin{aligned} \hat{k} &= k - ik^*, \\ \hat{\omega} &= \omega + i\omega^*, \\ \hat{\mu} &= \mu + i\mu^*, \\ \hat{\lambda} &= \lambda + i\lambda^*, \\ \hat{\rho} &= \rho + i\rho^*, \end{aligned} \tag{10.327}$$

instead of the real quantities k, ω, μ, λ, ρ. Starting from Eq. (3.127) we obtain

$$(\omega + i\omega^*)^2 \int_0^\infty (\rho + i\rho^*)V^2\, dz = (k - ik^*)^2 \int_0^\infty (\mu + i\mu^*)V^2\, dz + \int_0^\infty (\mu + i\mu^*)\left(\frac{dV}{dz}\right)^2 dz. \tag{10.328}$$

Our next step is to assume that the imaginary parts of all the parameters are small perturbations. This assumption is based on the physical evidence that in the earth the attenuation is small. Expanding the products in Eq. (10.328) and equating the real and imaginary parts on both sides, the real part gives back the energy equation for the limiting elastic medium, whereas the imaginary part yields the new result:

$$\omega\omega^* \int_0^\infty \rho V^2\, dz + \frac{1}{2}\omega^2 \int_0^\infty \left(\frac{\rho^*}{\rho}\right)\rho V^2\, dz = -kk^* \int_0^\infty \mu V^2\, dz + \frac{1}{2}\int_0^\infty \left(\frac{\mu^*}{\mu}\right)\mu\left[\left(\frac{dV}{dz}\right)^2 + k^2V^2\right]dz. \tag{10.329}$$

Introducing the notation

$$\frac{1}{2}\mu\left[\left(\frac{dV}{dz}\right)^2 + k^2V^2\right] = \text{potential-energy density} = \theta(z, \omega),$$

$$\frac{\omega^2}{2}\rho V^2 = \text{kinetic-energy density} = T(z),$$

$$\omega^2 \int_0^\infty \rho V^2\, dz = \text{total wave energy} = E,$$

Eq. (10.329) reads

$$\int_0^\infty \frac{\mu^*}{\mu}\theta(z, \omega)dz - \int_0^\infty \frac{\rho^*}{\rho} T(z)dz = \omega\omega^* \int_0^\infty \rho V^2\, dz + kk^* \int_0^\infty \mu V^2\, dz. \tag{10.330}$$

However, from Eq. (3.129), to first order in Q_L^{-1} [cf. Eq. (10.326)],

$$k^2 \int_0^\infty \mu V^2\, dz = \frac{U}{c}\omega^2 \int_0^\infty \rho V^2\, dz. \tag{10.331}$$

Substituting this relation in Eq. (10.330), we have

$$\frac{\int_0^\infty (\mu^*/\mu)\theta(z, \omega)dz - \int_0^\infty (\rho^*/\rho)T(z)dz}{\omega^2 \int_0^\infty \rho V^2\, dz} = \frac{\omega^*}{\omega} + \frac{U}{c}\frac{k^*}{k}, \tag{10.332}$$

or

$$\int_0^\infty \frac{\mu^*}{\mu}\left[\frac{\theta(z, \omega)}{E}\right] dz - \int_0^\infty \frac{\rho^*}{\rho}\left[\frac{T(z)}{E}\right]dz = \frac{\omega^*}{\omega} + \left(\frac{U}{c}\right)\frac{k^*}{k}. \tag{10.333}$$

The interpretation of Eq. (10.333) is based on our previous results: For traveling Love waves in a fully inertial medium

$$\rho^* = 0, \qquad \omega^* = 0, \qquad \frac{\mu^*}{\mu} = \frac{1}{Q_\beta(z, \omega)}, \qquad k^* = \gamma(\omega), \qquad k = \frac{\omega}{c(\omega)}.$$

Then, by Eq. (10.300),

$$\frac{2U}{c}\frac{k^*}{k} = \frac{1}{Q_L(\omega)}.$$

Substituting these expressions in Eq. (10.333), the *apparent attentuation* of propagating Love waves as measured on the surface is related to the intrinsic attenuation in the interior of the medium through the expression

$$\frac{1}{Q_L(\omega)} = \frac{\int_0^\infty (1/Q_\beta(z))\theta(z, \omega)dz}{\int_0^\infty \theta(z, \omega)dz}. \tag{10.334}$$

Note that Q_L^{-1} for the entire medium is interpreted as a weighted average over the depth coordinate. We are assuming that Q_β is a function of z only for the entire Love-wave frequency band. This is certainly not true for all types of earth waves. Equation (10.334) shows three important facts:

1. The observed specific attenuation function Q_L on the earth's surface is a weighted average of the intrinsic dissipation–depth function. The weighting function is the fractional potential-energy density at each depth.
2. Given $Q_L(\omega)$, Eq. (10.334) can be considered as an integral equation for $Q_\beta(z)$. Although a unique inversion is not possible, it is nevertheless possible to approximate the data with several models.
3. If the intrinsic dissipation factor Q_β is independent of depth, the apparent dissipation factor Q_L is constant and equal to Q_β. Whenever $Q_\beta = Q_\beta(z)$, even if it is frequency independent, Q_L must depend on the frequency through $V(z, \omega)$.

10.4.4.5. The Inversion of Surface-Wave "Q" Observations. Equation (10.334) is not convenient for numerical computation because it requires the knowledge of the explicit amplitudes V. We wish to replace it by an equivalent expression in which V does not appear. We apply again our perturbation scheme and replace the continuous half-space model by a stack of N discrete homogeneous layers. In each layer μ, ρ are constants but V is variable such that V and dV/dz are continuous at the interfaces. Taking the z axis downward and denoting the y component of displacement in the sth layer by V_s (it is proportional to $e^{i\omega t - ikx}$) the differential equation is

$$\frac{d}{dz}\left[\mu_s \frac{dV_s}{dz}\right] - \mu_s k^2 V_s = -\omega^2 \rho_s V_s.$$

Multiplying both sides by V_s and integrating with respect to z from the upper boundary of the sth layer, z_{s-1}, to the lower boundary, z_s, one obtains

$$\left[\mu_s\left(\frac{dV_s}{dz}\right)V_s\right]_{z_{s-1}}^{z_s} - \int_{z_{s-1}}^{z_s} \mu_s\left(\frac{dV_s}{dz}\right)^2 dz - k^2\int_{z_{s-1}}^{z_s} \mu_s V_s^2\,dz = -\omega^2\int_{z_{s-1}}^{z_s} \rho_s V_s^2\,dz.$$

Because V_s and $\mu_s(dV_s/dz)$ are continuous at the boundaries, this can be written as

$$\sum_{s=1}^{N}\int_{z_{s-1}}^{z_s} \mu_s\left[\left(\frac{dV_s}{dz}\right)^2 + k^2V_s^2\right]dz = \omega^2\sum_{s=1}^{N}\int_{z_{s-1}}^{z_s} \rho_s V_s^2\,dz, \tag{10.335}$$

where the left-hand side is equal to $2\sum\theta_s$, and the right-hand side to $2\sum T_s$. Substituting $\mu_s + i\mu_s^*$, $\rho_s + i\rho_s^*$, $\omega + i\omega^*$, and $k - ik^*$ in place of μ_s, ρ_s, ω, and k in Eq. (10.335) and taking the imaginary parts, we find

$$\sum_{s=1}^{N}\int_{z_{s-1}}^{z_s}\left(\frac{\mu_s^*}{\mu_s}\right)\mu_s\left[\left(\frac{dV_s}{dz}\right)^2 + k^2V_s^2\right]dz - \frac{2k^*}{k}\sum_{s=1}^{N}\int_{z_{s-1}}^{z_s} k^2\mu_s V_s^2\,dz$$

$$= \frac{2\omega^*}{\omega}\omega^2\sum_{s=1}^{N}\int_{z_{s-1}}^{z_s}\rho_s V_s^2\,dz + \omega^2\sum_{s=1}^{N}\int_{z_{s-1}}^{z_s}\left(\frac{\rho_s^*}{\rho_s}\right)\rho_s V_s^2\,dz. \tag{10.336}$$

Partial differentiation of both sides of Eq. (10.335) with respect to ω yields

$$\frac{\omega}{k}\frac{\partial k}{\partial\omega} = \frac{c}{U} = \frac{[\omega^2\sum_{s=1}^{N}\int_{z_{s-1}}^{z_s}\rho_s V_s^2\,dz]}{[k^2\sum_{s=1}^{N}\int_{z_{s-1}}^{z_s}\mu_s V_s^2\,dz]}. \tag{10.337}$$

Hence, Eq. (10.336) can be reduced to

$$\sum_{s=1}^{N}\left(\frac{\mu_s^*}{\mu_s}\right)\theta_s - \sum_{s=1}^{N}\left(\frac{\rho_s^*}{\rho_s}\right)T_s = E\left[\frac{\omega^*}{\omega} + \frac{k^*}{k}\frac{U}{c}\right] \tag{10.338}$$

which is equivalent to Eq. (10.333). We now write Eq. (10.334) in the discrete form

$$\frac{1}{Q_L(\omega)} = \sum_{s=1}^{N}\frac{\theta_s}{\theta}\frac{1}{Q_{\beta,s}}, \tag{10.339}$$

where

$$\theta_s = \frac{1}{2}\int_{z_{s-1}}^{z_s}\mu\left[k^2V^2 + \left(\frac{dV}{dz}\right)^2\right]dz, \qquad \frac{1}{Q_{\beta,s}} = \left(\frac{\mu^*}{\mu}\right)_s,$$

$$\theta = \frac{1}{2}\int_0^\infty\mu\left[k^2V^2 + \left(\frac{dV}{dz}\right)^2\right]dz = \frac{1}{2}\omega^2\int_0^\infty\rho V^2\,dz = \frac{c}{2U}k^2\int_0^\infty\mu V^2\,dz. \tag{10.340}$$

However, in Eq. (3.233) we have proved that

$$\left(\frac{\partial c}{\partial\mu_s}\right)_{k,\rho} = \frac{U}{k^2}\frac{\int_{z_{s-1}}^{z_s}[k^2V^2 + (dV/dz)^2]dz}{2\int_0^\infty\mu V^2\,dz}. \tag{10.341}$$

Because μ_s is constant in the layer, we may multiply each side by μ_s and put $\mu_s = \mu$ inside the integral. Hence

$$2\mu_s\left(\frac{\partial c_L}{\partial \mu_s}\right)_{k,\rho} = \frac{U}{k^2}\frac{\int_{z_{s-1}}^{z_s}\mu[k^2V^2 + (dV/dz)^2]dz}{(U/c_Lk^2)\int_0^\infty \mu[k^2V^2 + (dV/dz)^2]dz} = c_L\frac{\theta_s}{\theta}$$

or

$$\frac{\theta_s}{\theta} = \frac{2\mu_s}{c_L}\left(\frac{\partial c_L}{\partial \mu_s}\right)_{k,\rho} = \frac{\beta_s}{c_L}\left(\frac{\partial c_L}{\partial \beta_s}\right)_{k,\rho}.$$

Therefore, from Eq. (10.339),

$$\frac{1}{Q_L(\omega)} = \sum_{s=1}^{N}\left(\frac{\beta_s}{c_L}\frac{\partial c_L}{\partial \beta_s}\right)_{k,\rho}\frac{1}{Q_{\beta,s}} = \sum_{s=1}^{N}\left(\frac{\partial Q_L^{-1}}{\partial Q_{\beta,s}^{-1}}\right)\frac{1}{Q_{\beta,s}}. \tag{10.342}$$

In a similar way, using Eqs. (3.234)–(3.235), we get

$$\begin{aligned}\frac{1}{Q_R(\omega)} &= \sum_{s=1}^{N}\left(\frac{\alpha_s}{c_R}\frac{\partial c_R}{\partial \alpha_s}\right)_{k,\rho,\beta}\frac{1}{Q_{\alpha,s}} + \sum_{s=1}^{N}\left(\frac{\beta_s}{c_R}\frac{\partial c_R}{\partial \beta_s}\right)_{k,\rho,\alpha}\frac{1}{Q_{\beta,s}}\\ &= \sum_{s=1}^{N}\left(\frac{\partial Q_R^{-1}}{\partial Q_{\alpha,s}^{-1}}\right)\frac{1}{Q_{\alpha,s}} + \sum_{s=1}^{N}\left(\frac{\partial Q_R^{-1}}{\partial Q_{\beta,s}^{-1}}\right)\frac{1}{Q_{\beta,s}}.\end{aligned} \tag{10.343}$$

The subscript s is the layer index; the subscripts R, L, α, and β associated with Q identify the wave type. Other subscripts refer to quantities being held constant.

Equations (10.342) and (10.343) can be used to invert the observed surface-wave attenuation, $Q_L(\omega)$ and $Q_R(\omega)$, for the purpose of deriving the dependence of Q_α and Q_β on depth. A satisfactory Q distribution for the earth must explain the observed frequency dependence of both Love and Rayleigh waves and be consistent with body-wave attenuation observations and laboratory results. Figure 10.19 shows the theoretical curves that have been fitted to the observed Q_R and Q_L given in Tables 10.2–10.4. The solid curve for Rayleigh waves was derived for a theoretical model in which $Q_\alpha = 2.25Q_\beta$. This relation is derived from the equation

$$\frac{Q_\beta}{Q_\alpha} = \left(\frac{\beta}{\alpha}\right)^2\left(\frac{\kappa^*}{\mu^*} + \frac{4}{3}\right)$$

by inserting therein the values

$$\kappa^* = 0, \qquad \left(\frac{\beta}{\alpha}\right)^2 = \frac{1}{3}; \tag{10.344}$$

i.e., there are no losses in pure compression in a Poisson's solid. If

$$\kappa^* = \frac{1}{3}\mu^*, \qquad \left(\frac{\beta}{\alpha}\right)^2 = \frac{1}{3} \tag{10.345}$$

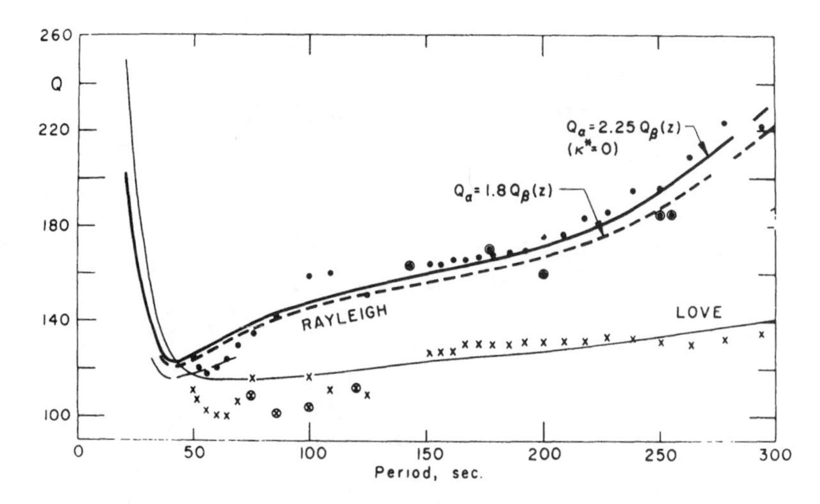

Figure 10.19. From the observed $Q_R(T)$ and $Q_L(T)$ of surface waves, we are able to deduce the depth profiles of the intrinsic shear and longitudinal dissipation parameters $Q_\alpha(z)$ and $Q_\beta(z)$ given in Table 10.6. The better the fit of the observed Q_L and Q_R (crosses and circles) to the corresponding calculated values (lines), the greater the fidelity of the resulting Q_α and Q_β models.

we get $Q_\alpha = 1.8Q_\beta$, which fits the broken line in Fig. 10.19. Table 10.6 lists the resulting distribution of Q_α and Q_β with depth for the case $Q_\alpha = 2.25Q_\beta$.

An alternative method of deriving Eqs. (10.342) and (10.343) is to consider $i\lambda^*$, $i\mu^*$, $i\alpha^*$, $i\beta^*$, $i\omega^*$, and ik^* as small perturbations over the purely elastic configuration and then expand $\omega(\alpha_s + i\alpha_s^*, \beta_s + i\beta_s^*) = \omega + i\omega^*$ in a power series about $\omega(\alpha_s, \beta_s)$. Therefore, for a fixed k, we have

$$\omega^* = \sum_{s=1}^{N} \left(\frac{\partial \omega}{\partial \alpha_s}\right)_{k,\beta} \alpha_s^* + \sum_{s=1}^{N} \left(\frac{\partial \omega}{\partial \beta_s}\right)_{k,\alpha} \beta_s^* + \cdots. \tag{10.346}$$

Using the relations

$$\alpha^* = \frac{\alpha}{2Q_\alpha}, \qquad \beta^* = \frac{\beta}{2Q_\beta}, \qquad \omega^* = \frac{\omega}{2Q}, \tag{10.347}$$

we obtain

$$\frac{1}{Q_{\text{toroidal}}} = \sum_{s=1}^{N} \frac{\beta_s}{\omega_T} \left(\frac{\partial \omega_T}{\partial \beta_s}\right)_k \frac{1}{Q_{\beta,s}}, \tag{10.348}$$

$$\frac{1}{Q_{\text{spheroidal}}} = \sum_{s=1}^{N} \frac{\alpha_s}{\omega_S} \left(\frac{\partial \omega_S}{\partial \alpha_s}\right)_{k,\beta} \frac{1}{Q_{\alpha,s}} + \sum_{s=1}^{N} \frac{\beta_s}{\omega_S} \left(\frac{\partial \omega_S}{\partial \beta_s}\right)_{k,\alpha} \frac{1}{Q_{\beta,s}}. \tag{10.349}$$

On putting $\omega = ck$, Eqs. (10.348) and (10.349) reduce to Eqs. (10.342) and (10.343), respectively.

Table 10.6. A Model for the Distribution of Q_α and Q_β with Depth[a] in the Earth

d (km)	Q_β	Q_α
38	450	1012.5
22	60	135.0
10	80	180.0
55	100	225.0
375	150	337.5
100	180	405.0
100	250	562.5
100	450	1012.5
100	500	1125.0
100	600	1350.0

[a] d = layer thickness, $Q_\alpha/Q_\beta = 2.25$.

We note that it is not possible to derive an equation like Eq. (10.334) for $1/Q_R$. Here we find that

$$\frac{1}{Q_R} = \frac{1}{E}\sum_{s=1}^{N}\int_{z_{s-1}}^{z_s}\left[\frac{\lambda_s + 2\mu_s}{Q_{\alpha,s}}\left\{\left(\frac{dW_s}{dz}\right)^2 + k^2U_s^2\right\} + \frac{\mu_s}{Q_{\beta,s}}\left\{\left(\frac{dU_s}{dz}\right)^2 + k^2W_s^2\right\}\right]dz \tag{10.350}$$

while the potential energy in the sth layer is

$$\theta_s = \frac{1}{2}\int_{z_{s-1}}^{z_s}\left[(\lambda_s + 2\mu_s)\left\{\left(\frac{dW_s}{dz}\right)^2 + k^2U_s^2\right\} + \mu_s\left\{\left(\frac{dU_s}{dz}\right)^2 + k^2W_s^2\right\}\right]dz. \tag{10.351}$$

Therefore, unless

$$Q_{\alpha,s} = Q_{\beta,s}$$

for all layers, a form similar to Eq. (10.334) cannot be obtained.

10.4.4.6. Causal Dispersion. We consider an N-layered spherical-earth model. In each layer the dispersion relation for body waves is given by Eq. (10.290), namely,

$$\begin{aligned}\alpha_s(\omega) &= \alpha_s(\omega_r)\left[1 - \frac{1}{\pi Q_{\alpha,s}}\ln\left(\frac{\omega_r}{\omega}\right)\right],\\ \beta_s(\omega) &= \beta_s(\omega_r)\left[1 - \frac{1}{\pi Q_{\beta,s}}\ln\left(\frac{\omega_r}{\omega}\right)\right],\end{aligned} \tag{10.352}$$

where

$$\begin{aligned} \alpha &= P\text{-wave phase velocity} \\ \beta &= S\text{-wave phase velocity} \\ \omega &= \text{angular frequency of body wave} \\ Q_{\alpha,s}, Q_{\beta,s} &= \text{body-wave dissipation parameters} \\ s &= \text{layer index} \\ \omega_r &= \text{reference angular frequency.} \end{aligned}$$

Note that causal phase shifts make the body-wave phase velocity frequency dependent and ruin the concept of frequency-independent intrinsic velocity, which is appropriate only for purely elastic media.

Assume that an earth model has been derived from the inversion of travel-time data of short-period body waves at a frequency $\omega = \omega_r$. Let $c_L(\omega_r)$ be the Love-wave phase velocity calculated by using elastic moduli at $\omega = \omega_r$. If we want to evaluate the Love phase-velocity $c_L(\omega)$ at other frequencies for the same elastic structure, we must take into account the changes of $\alpha_s(\omega)$ and $\beta_s(\omega)$ in each layer owing to causality. The corresponding change in $c_L(\omega)$ to first order in $1/Q_{\beta,s}$ is

$$\begin{aligned} \Delta c_L &= \sum_{s=1}^{N} \frac{\partial c_L}{\partial \beta_s} (\Delta\beta)_s = \sum_{s=1}^{N} \frac{\partial c_L}{\partial \beta_s} \left[\frac{\beta_s}{\pi Q_{\beta,s}} \ln\left(\frac{\omega}{\omega_r}\right) \right] \\ &= \frac{c_L(\omega)}{\pi} \ln\left(\frac{\omega}{\omega_r}\right) \sum_{s=1}^{N} \frac{\beta_s}{c_L} \left(\frac{\partial c_L}{\partial \beta}\right)_s \frac{1}{Q_{\beta,s}} = \frac{c_L(\omega)}{\pi Q_L(\omega)} \ln\left(\frac{\omega}{\omega_r}\right) \end{aligned}$$

or

$$\frac{\Delta c_L}{c_L} = \frac{1}{\pi Q_L} \ln\left(\frac{\omega}{\omega_r}\right) = -\frac{1}{\pi Q_L} \ln\left(\frac{T}{T_r}\right). \tag{10.353}$$

Similarly, for Rayleigh waves,

$$\frac{\Delta c_R}{c_R} = \frac{1}{\pi Q_R} \ln\left(\frac{\omega}{\omega_r}\right). \tag{10.354}$$

Here $Q_L(\omega)$, $Q_R(\omega)$ are the quality factors of Love and Rayleigh waves. Equations (10.353) and (10.354) can therefore be written in a form similar to Eq. (10.352)

$$c_m(\omega) = c_m(\omega_r)\left[1 - \frac{1}{\pi Q_m} \ln\left(\frac{\omega_r}{\omega}\right)\right], \qquad m = L, R$$

or

$$c_m(T) = c_m(T_r)\left[1 - \frac{1}{\pi Q_m} \ln\left(\frac{T}{T_r}\right)\right]. \tag{10.355}$$

Here again, $c(\omega_r)$ is understood as the phase velocity at ω, calculated by using elastic moduli at $\omega = \omega_r$. Note that since the dispersion relations are derived by

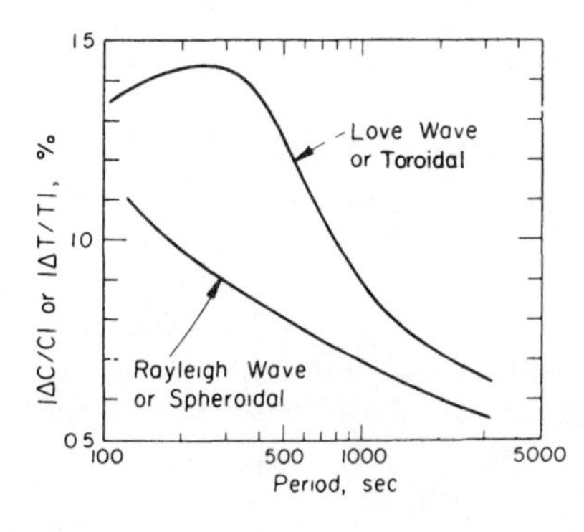

Figure 10.20. Rayleigh and Love fractional phase-velocity shifts, resulting from anelasticity.

integration over the entire frequency axis, Equation (10.355) is affected by values of Q outside the seismic frequency band which are mostly unknown.

The above corrections amount to 0.5–1.4% (Fig. 10.20). Equation (10.355) means that anelasticity diminishes the phase velocities relative to an elastic earth such that the longer periods are slowed more than the shorter ones.

10.4.5. Attenuation of the Earth's Free Oscillations

In an elastic earth, the time dependence of a free oscillation is like $e^{i\omega t}$, where $\omega = {}_n\omega_l$ is the frequency of a certain spectral line. We have shown earlier that anelasticity in this case is mathematically equivalent to the exchange of ω with $\hat{\omega} = \omega + i\omega^*$ such that $\omega^* = \omega/2Q$. Therefore, the overall time dependence of the oscillation is $e^{i\omega t}e^{-\omega t/2Q}$.

Consider a fixed point on earth where the observation is made. At this point we can write

$$\xi(t) = \xi(t_0)e^{-(\pi/QT)(t-t_0)} = \xi_{\max}\, e^{-t/\tau_0}, \qquad \tau_0 = \frac{QT}{\pi} \tag{10.356}$$

where $\xi(t)$ is the amplitude at time t of an oscillation of period T and t_0 is an arbitrary fiducial time. The quality factor Q can be derived from observations in two ways: First, from the "half-power width" of well-resolved spectral lines [cf. Eq. (10.31)] and second, from amplitude decay in time of a given spectral peak at a given location. To this end we define the mean amplitude over a time interval $t_b - t_a$

$$\bar{\xi}_{ab} = \frac{1}{t_b - t_a}\int_{t_a}^{t_b} \xi(t)dt = \tau_0\xi_{\max}\frac{e^{-t_a/\tau_0} - e^{-t_b/\tau_0}}{(t_b - t_a)}. \tag{10.357}$$

If we repeat this operation for another time interval, $t_d - t_c = t_b - t_a$, we have

$$\frac{\bar{\xi}_{ab}}{\bar{\xi}_{cd}} = \frac{e^{-t_a/\tau_0} - e^{-t_b/\tau_0}}{e^{-t_c/\tau_0} - e^{-t_d/\tau_0}} = \exp\left[\frac{(t_c - t_a)}{\tau_0}\right]$$

and hence

$$Q = \frac{\pi}{T} \frac{(t_c - t_a)}{\ln(\bar{\xi}_{ab}/\bar{\xi}_{cd})}. \tag{10.358}$$

The entities $\bar{\xi}_{ab}$ and $\bar{\xi}_{cd}$ are the spectral amplitudes obtained by subjecting the earthquake record to a Fourier analysis. $\bar{\xi}_{ab}$ is the Fourier transform over the time interval from t_a to t_b and $\bar{\xi}_{cd}$ is the Fourier transform over the time interval from t_c to t_d. $t_c - t_a$ is the time between the beginnings of the two intervals. To secure complete resolution of the spectral lines, one must choose the analyzed time intervals such that

$$t_d - t_c = t_b - t_a > \frac{1}{\tau_l F} > T,$$

because the entire pattern of motion corresponding to a given order rotates relative to the earth's surface with a period equal to $1/\tau_l F$, where F is the rotational frequency of the earth and τ_l is the splitting parameter (cf. Sect. 6.4).

Figure 10.21 shows the amplitude decay of the spectral peaks of ${}_0T_5$, from which Q values have been calculated according to Eq. (10.358). Note that the derivation of Q from the standing-wave data is essentially similar to the derivation of Q from the propagating-wave data, except that in the former case the comparison is made in time and in the latter it is made in space.

10.4.5.1. Causal Effects on the Free Periods. As in the case of surface waves, the anelasticity of the earth has a double effect on the free oscillation time-series and spectrum. The prime effect is the decay of the amplitudes according to Eq. (10.356). The second effect, namely, a causal period lengthening, is by no

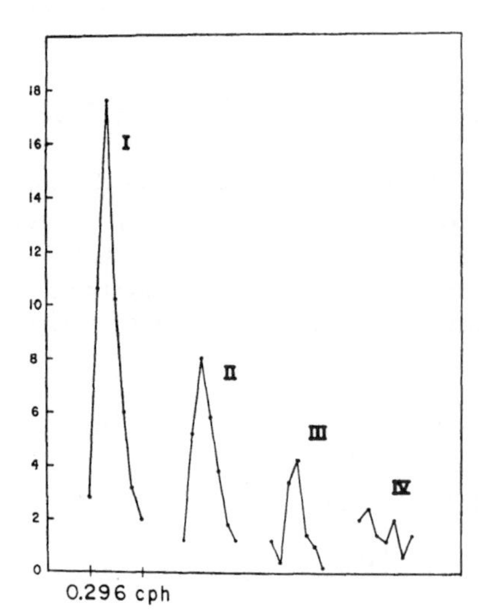

Figure 10.21. Relative amplitude decay with time of the spectral peak of the mode ${}_0T_5$ from the Chile earthquake of 22 May, 1960. Analysis was made from a strain meter seismogram in four consecutive 20.35-h intervals. The frequency interval between the end points of the peak is indicated at the bottom of the figure.

means negligible. To derive an analytic expression for this phenomenon, we refer to Jeans' formula [Eq. (8.4)]

$$cT = \frac{2\pi a}{l + \frac{1}{2}} \qquad \text{(fixed for a given mode)}$$

which upon differentiation yields

$$\frac{\Delta c}{c} = -\frac{\Delta T}{T} = \frac{\Delta \omega}{\omega}. \tag{10.359}$$

Eq. (10.355) and Eq. (10.359) then give

$$\Delta({}_nT_l) = \frac{{}_nT_l}{\pi Q} \ln\left(\frac{{}_nT_l}{T_r}\right) \tag{10.360}$$

or

$${}_nT_l = {}_nT_l(T_r)\left[1 + \frac{1}{\pi Q} \ln\left(\frac{{}_nT_l}{T_r}\right)\right]. \tag{10.361}$$

If $T_r = 1$ s, a period lengthening of 22 s is predicted for ${}_0T_2$ ($Q \simeq 300$).

The meaning of Eq. (10.361) is this: Suppose that we start from an earth model based upon travel time data of short-period body waves at T_r and compute theoretical free periods for this model. The observed free periods are longer than those calculated on the basis of the elastic model and must therefore be corrected according to Eq. (10.360) and reduced by $|\Delta T|$. The reduced periods are then inverted to yield an elastic model that fits the body-wave data. It is agreed to take $T_r = 1$ s and define a 1-s *elastic-earth model.* In short, the presence of anelasticity in the earth causes free oscillation periods to be shifted to periods longer than the periods that would be observed in a perfectly elastic earth.

The attenuation of seismic body waves and the decay of free oscillations certainly emphasize that the earth is not a perfectly elastic body. One implication of this observation is that the elastic moduli and seismic velocities must depend on frequency. The effect of dispersion over the entire seismic frequency band can amount to 1% and is a nonnegligible effect when earth models based on free oscillation and long-period surface-wave data are to be compared with earth models inferred from body-wave travel-times.

If the dispersion caused by anelasticity is neglected there is an observed discrepancy between earth models based on body-wave, surface-wave, and free oscillation data in that the inversion of body-wave data gives somewhat higher velocities. However, if prior to inversion the observed data are corrected to the same reference frequency, the discrepancy between free oscillation and body-wave results is removed. The effect of energy dissipation on wave motion in the earth is summarized in Table 10.7.

Table 10.7. The Effect of Dissipation on Seismic Wave Motion in the Earth

	Body waves	Surface waves	Free oscillations
Spectral attenuation factor	$e^{-2\pi t^*/T}$ $t^* = \frac{1}{2}\int_{\text{ray}} \frac{ds}{Q_m(s)v_m(s)} \to \frac{\tau}{2Q_{\text{av}}}$ τ = travel time	$e^{-\pi\Delta/QUT}$	$e^{-\pi t/Q_n T_l}$
Causal dispersion	Relative phase and group delay at $T > T_r$ $\Delta\tau = \frac{\tau}{\pi Q_{\text{av}}} \ln\left(\frac{T}{T_r}\right)$ Delay increases with both travel time and period	$c(T > T_r) = c(T_r)\left[1 - \frac{1}{\pi Q} \ln\left(\frac{T}{T_r}\right)\right]$ T_r = reference period Anelasticity diminishes velocities relative to elastic-earth model; longer periods are slowed more	$\Delta({}_nT_l) = \frac{{}_nT_l}{\pi Q} \ln\left(\frac{{}_nT_l}{T_r}\right)$ Anelasticity shifts forward spectral lines relative to elastic-earth model; shift increases with period

Bibliography

Alsop LE, Sutton GH, Ewing M (1961) Measurement of Q for very long period free oscillations. Jour Geophys Res 66: 2911–2915.

Anderson DL, Ben-Menahem A, Archambeau CB (1965). Attenuation of Seismic Energy in the Upper Mantle. Jour Geophys Res 70: 1441–1448.

Becker R (1925) Elastische Nachwirkung und Plastizität. Zeit Phys 33: 185–213.

Ben-Menahem A (1965) Observed attenuation and Q values of seismic surface waves in the upper mantle. Jour Geophys Res 70: 4641–4651.

Ben-Menahem A, Jeffreys H (1971) The saddle-point approximation for damped surface waves. Geophys Jour Roy Astron Soc (London) 24: 1–2.

Boltzmann L (1876) Zur Theorie der elastischen Nachwirkung. Ann Phys Chem (Poggendorff) Ergänzungsband 7: 624–654.

Borcherdt RD (1973) Energy and plane waves in linear viscoelastic media. Jour Geophys Res 78: 2442–2453.

Brune JN (1962) Attenuation of dispersed wave trains. Bull Seismol Soc Amer 52: 109–112.

Canas JA, Mitchell B (1978) Lateral variations of surface-wave anelastic attenuation across the Pacific. Bull Seismol Soc Amer 68: 1637–1650.

Collins F, Lee CC (1956) Seismic wave attenuation characteristics from pulse experiments. Geophysics 21: 16–40.

Findley WN, Lai JS, Onaran K (1976) Creep and Relaxation of Nonlinear Viscoelastic Materials. North-Holland, Amsterdam, 367 pp.

Flügge W (1975) Viscoelasticity, 2nd edn. Springer-Verlag, Berlin, 194 pp.

Futterman WI (1962) Dispersive body waves. Jour Geophys Res 67: 5279–5291.

Gross B (1953) Mathematical structure of the Theories of Viscoelasticity. Publ Inst Natl Tech Brazil. Hermann and Co, Paris, 74 pp.

Gurtin ME, Sternberg E (1962) On the linear theory of viscoelasticity. Arch Ration Mech Anal 11: 291–356.

Gutenberg B (1932) Theorie der Erdbebenwellen. Handbuch Geophys 4(2): 255–258.

Jeffreys H (1976) The Earth, 6th edn. Cambridge University Press, Cambridge.

Jeffreys H, Crampin S (1970) On the modified Lomnitz law of damping. Mon Not Roy Astron Soc (London) 147: 295–301.

Kanamori H (1977) Importance of physical dispersion in surface wave and free oscillation problems: Review. Rev Geophys Space Phys. 15: 105–112.

Kelvin, Lord (William Thomson) (1890) Collected Works, Vol 3. Cambridge University Press, Cambridge, pp 1–112.

Knopoff L (1956) The seismic pulse in materials possessing solid friction. I. Plane waves. Bull Seismol Soc Amer 46: 175–184.

Knopoff L (1964) Q. Rev Geophys 2: 625–660.

Kogan S Ya (1966) A brief review of seismic wave absorption theories, I, II (English translation). Bull Acad Sci USSR, Earth Phys 11: 671–683.

Kolsky H (1956) The propagation of stress pulses in viscoelastic solids. Phil Mag (Ser 8) 1: 693–710.

Liu HP, Kanamori H (1976) Velocity dispersion due to anelasticity: Implications for seismology and mantle composition. Geophys Jour Roy Astron Soc (London) 47: 41–58.

Macdonald JR (1959) Rayleigh-wave dissipation function in low-loss media. Geophys Jour Roy Astron Soc (London) 2: 132–135.

Macdonald JR (1961) Theory and application of a superposition model of internal friction and creep. Jour Appl Phys 32: 2385–2398.

Marshall PD, Carpenter EW (1966). Estimates of Q for Rayleigh waves. Geophys Jour Roy Astron Soc (London) 10: 549–550.

Maxwell JC (1866) Collected Works, Vol 2. Dover, New York, pp 26–78.

Meyer OE (1874a) Zur Theorie der inner Reibung. Jour Reine Angew Math 78: 130–135.

Meyer OE (1874b) Theorie der elastischen Nachwirkung. Ann Phys 227: 108–119.

Mitchell B (1973) Radiation and attenuation of Rayleigh waves from the south-eastern Missouri earthquake of Oct 21, 1965. Jour Geophys Res 78: 886–899.

Ness NF, Harrison JC, Slichter LB (1961) Observations of the free oscillations of the earth. Jour Geophys Res 66: 621–629.

Newlands M (1954) Lamb's problem with internal dissipation, I. Jour Acoust Soc Amer 26: 434–448.

Nuttli OW (1973) Seismic wave attenuation and magnitude relations for eastern North America. Jour Geophys Res 78: 876–885.

Phillips P (1905) The slow stretch in India rubber, glass, and metal wires when subjected to a constant pull. Phil Mag (Ser 6) 9: 513–531.

Ricker N (1940) Wavelet functions and their polynomials. Geophysics 9: 314–323.

Sailor RV, Dziewonski AM (1978) Measurements and interpretation of normal mode attenuation. Geophys Jour Roy Astron Soc (London) 53: 559–581.

Sezawa K (1927) On the decay of waves in viscoelastic solid bodies. Bull Earthquake Res Inst (Tokyo) 3: 43–54.

Sezawa K, Kanai K (1938) Damping of periodic viscoelastic wave with increase in focal distance. Bull Earthquake Res Inst (Tokyo) 16: 491–503.

Silva W (1976) A variational formulation for Love waves in a layered anelastic solid. Geophys Jour Roy Astron Soc (London) 45: 445–450.

Singh SJ, Rosenman M (1973) Quasi-static strains and tilts due to faulting in a visco-elastic half-space. Bull Seismol Soc Amer 63: 1737–1752.

Singh SJ, Rosenman M (1974) Quasi-static deformation of a viscoelastic half-space by shear dislocations. Phys Earth Planet Int 8: 87–101.

Sterling A, Smets E (1971) Study of earth tides, earthquakes and terrestrial spectroscopy by analysis of the level fluctuations in a borehole at Heibaart (Belgium). Geophys Jour Roy Astron Soc (London) 23: 225–242.

Stokes, GG (1849) On the theories of the internal friction of fluids in motion and of the equilibrium and motion of elastic solids. Trans Camb Phil Soc 8: 287–319.

Strick E (1967) The determination of Q, dynamic viscosity and transient creep curves from wave propagation measurements. Geophys Jour Roy Astron Soc (London) 13: 197–218.

Trouton FT, Rankine AO (1904) On the stretching and torsion of lead wire beyond the elastic limit. Phil Mag (Ser 6) 8: 538–556.

Tryggvason E (1965) Dissipation of Rayleigh wave energy. Jour Geophys Res 70: 1449–1455.

Voigt W (1892) Über innere Reibung fester Körper, insbesondere der Metalle. Ann Phys 47: 671–693.

Wesley JP (1965) Diffusion of seismic energy in the near range. Jour Geophys Res 70: 5099–5106.

Zener C (1948) Elasticity and Anelasticity of Metals. University of Chicago Press, Chicago.

Appendices

. . . Vether it's worth while goin' through so much, to learn so little, as the charity-boy said ven he got to the end of the alphabet, is a matter o' taste . . .

(Charles Dickens, The Pickwick Papers, Chapter 27)

APPENDIX A

Algebra and Calculus of Dyadics

A.1. Algebra of Dyadics

Consider a right Cartesian system O, with unit vectors $\mathbf{e}_1$, $\mathbf{e}_2$, and $\mathbf{e}_3$ along the respective axes x_1, x_2, and x_3. Let (x_1, x_2, x_3) and (ξ_1, ξ_2, ξ_3) represent the vectors $\mathbf{r}$ and $\boldsymbol{\xi}$ in this system and let T_{ij} represent a 3×3 matrix such that

$$\begin{bmatrix} \xi_1 \\ \xi_2 \\ \xi_3 \end{bmatrix} = \begin{bmatrix} T_{11} & T_{12} & T_{13} \\ T_{21} & T_{22} & T_{23} \\ T_{31} & T_{32} & T_{33} \end{bmatrix} \begin{bmatrix} x_1 \\ x_2 \\ x_3 \end{bmatrix}. \tag{A.1}$$

Equation (A.1) is equivalent to the set

$$\begin{aligned} \xi_1 &= T_{11}x_1 + T_{12}x_2 + T_{13}x_3, \\ \xi_2 &= T_{21}x_1 + T_{22}x_2 + T_{23}x_3, \\ \xi_3 &= T_{31}x_1 + T_{32}x_2 + T_{33}x_3. \end{aligned} \tag{A.2}$$

It therefore appears that the matrix T_{ij} acts as a linear operator transforming the vector $\mathbf{r}$ into another vector $\boldsymbol{\xi}$ in the same system.

Suppose next that we wish to write Eq. (A.1) in an alternative form that involves the vectors $\mathbf{r}$ and $\boldsymbol{\xi}$ rather than their components. To this end, we put

$$\begin{aligned} \boldsymbol{\xi} &= \xi_1\mathbf{e}_1 + \xi_2\mathbf{e}_2 + \xi_3\mathbf{e}_3, \\ \mathbf{r} &= x_1\mathbf{e}_1 + x_2\mathbf{e}_2 + x_3\mathbf{e}_3, \end{aligned} \tag{A.3}$$

and define the vectors

$$\begin{aligned} \mathbf{T}_1 &= T_{11}\mathbf{e}_1 + T_{12}\mathbf{e}_2 + T_{13}\mathbf{e}_3, \\ \mathbf{T}_2 &= T_{21}\mathbf{e}_1 + T_{22}\mathbf{e}_2 + T_{23}\mathbf{e}_3, \\ \mathbf{T}_3 &= T_{31}\mathbf{e}_1 + T_{32}\mathbf{e}_2 + T_{33}\mathbf{e}_3. \end{aligned} \tag{A.4}$$

It then follows from Eqs. (A.2)–(A.4) that

$$\boldsymbol{\xi} = \mathbf{e}_1(\mathbf{T}_1 \cdot \mathbf{r}) + \mathbf{e}_2(\mathbf{T}_2 \cdot \mathbf{r}) + \mathbf{e}_3(\mathbf{T}_3 \cdot \mathbf{r}), \tag{A.5}$$

where a dot denotes a scalar (dot) product.

Alternatively, Eq. (A.5) can be written in the symbolic form

$$\boldsymbol{\xi} = (\mathbf{e}_1 \mathbf{T}_1 + \mathbf{e}_2 \mathbf{T}_2 + \mathbf{e}_3 \mathbf{T}_3) \cdot \mathbf{r}, \tag{A.6}$$

or, simply

$$\boldsymbol{\xi} = \mathfrak{T} \cdot \mathbf{r}, \tag{A.7}$$

where $\mathfrak{T} = \mathbf{e}_1 \mathbf{T}_1 + \mathbf{e}_2 \mathbf{T}_2 + \mathbf{e}_3 \mathbf{T}_3$ is a new entity. Substituting for $\mathbf{T}_1, \mathbf{T}_2$, and $\mathbf{T}_3$ from Eq. (A.4) we get an explicit expression for $\mathfrak{T}$ in terms of the unit vectors

$$\begin{aligned} \mathfrak{T} &= T_{11}\mathbf{e}_1\mathbf{e}_1 + T_{12}\mathbf{e}_1\mathbf{e}_2 + T_{13}\mathbf{e}_1\mathbf{e}_3 + T_{21}\mathbf{e}_2\mathbf{e}_1 + T_{22}\mathbf{e}_2\mathbf{e}_2 + T_{23}\mathbf{e}_2\mathbf{e}_3 \\ &\quad + T_{31}\mathbf{e}_3\mathbf{e}_1 + T_{32}\mathbf{e}_3\mathbf{e}_2 + T_{33}\mathbf{e}_3\mathbf{e}_3 \\ &= \sum_{i,j=1}^{3} T_{ij}\mathbf{e}_i\mathbf{e}_j. \end{aligned} \tag{A.8}$$

We may write $T_{ij}\mathbf{e}_i\mathbf{e}_j = \mathbf{e}_i\mathbf{e}_j T_{ij}$ but the order of the unit vectors cannot be changed. Clearly, Eqs. (A.1) and (A.7) differ in form but are equivalent in their content. The entity $\mathfrak{T}$ is known as a dyadic and each of its elements a *dyad*. T_{ij} are the *components* of the dyadic. The matrix T_{ij} is known as the *matrix of the dyadic* $\mathfrak{T}$ and is denoted by $[\mathfrak{T}]$. The representation of $\mathfrak{T}$ in Eq. (A.8) is known as the *nonion* form of the dyadic.

The most general dyad is the juxtaposition of any two vectors **a** and **u** defined as a particular case of Eq. (A.8) with $T_{ij} = a_i u_j$. It is written as **au** and known sometimes as the *algebraic product* of **a** and **u** in contradistinction to the *scalar product* $\mathbf{a} \cdot \mathbf{u}$ and the *vector product* $\mathbf{a} \times \mathbf{u}$. We have

$$\begin{aligned} \mathbf{au} &= a_1 u_1 \mathbf{e}_1\mathbf{e}_1 + a_1 u_2 \mathbf{e}_1\mathbf{e}_2 + a_1 u_3 \mathbf{e}_1\mathbf{e}_3 + a_2 u_1 \mathbf{e}_2\mathbf{e}_1 + a_2 u_2 \mathbf{e}_2\mathbf{e}_2 + a_2 u_3 \mathbf{e}_2\mathbf{e}_3 \\ &\quad + a_3 u_1 \mathbf{e}_3\mathbf{e}_1 + a_3 u_2 \mathbf{e}_3\mathbf{e}_2 + a_3 u_3 \mathbf{e}_3\mathbf{e}_3. \end{aligned} \tag{A.9}$$

The first vector in a dyad is called the *antecedent*; and the second vector, the *consequent*.

It can be shown that any dyadic $\mathfrak{T}$ can be reduced to the form

$$\mathfrak{T} = \mathbf{au} + \mathbf{bv} + \mathbf{cw}, \tag{A.10}$$

in which either $(\mathbf{a}, \mathbf{b}, \mathbf{c})$ or $(\mathbf{u}, \mathbf{v}, \mathbf{w})$, but not both, are arbitrarily chosen noncoplanar vectors. A dyadic $\mathfrak{T}$ is said to be *complete* if and only if for each vector $\boldsymbol{\xi}$ there exists a vector **r** such that Eq. (A.7) is satisfied. A necessary and sufficient condition that $\mathfrak{T}$ be complete is that in Eq. (A.10) $(\mathbf{a}, \mathbf{b}, \mathbf{c})$ and $(\mathbf{u}, \mathbf{v}, \mathbf{w})$ are two sets of noncoplanar vectors; i.e., $[\mathbf{abc}][\mathbf{uvw}] \neq 0$.[1] This is equivalent to the statement that the matrix T_{ij} is nonsingular. A complete dyadic cannot be reduced to the sum of fewer than three dyads.

The complete dyadic $\mathfrak{I} = \mathbf{e}_1\mathbf{e}_1 + \mathbf{e}_2\mathbf{e}_2 + \mathbf{e}_3\mathbf{e}_3$ transforms every vector **a** into itself. It is therefore called the *unit dyadic* or the *idemfactor*. Therefore,

$$\mathfrak{I} \cdot \mathbf{a} = \mathbf{a} \cdot \mathfrak{I} = \mathbf{a}. \tag{A.11}$$

[1] $[\mathbf{abc}] = \mathbf{a} \cdot (\mathbf{b} \times \mathbf{c}) = (\mathbf{a} \times \mathbf{b}) \cdot \mathbf{c}$.

A *vector (cross) product of a dyad* **au** with a vector **b** is defined as

$$\mathbf{au} \times \mathbf{b} = \mathbf{a}(\mathbf{u} \times \mathbf{b}). \tag{A.12}$$

This leads to the interesting result

$$\mathfrak{J} \times \mathbf{a} = \mathbf{a} \times \mathfrak{J} = a_1(\mathbf{e}_3\mathbf{e}_2 - \mathbf{e}_2\mathbf{e}_3) + a_2(\mathbf{e}_1\mathbf{e}_3 - \mathbf{e}_3\mathbf{e}_1) + a_3(\mathbf{e}_2\mathbf{e}_1 - \mathbf{e}_1\mathbf{e}_2). \tag{A.13}$$

It follows from Eqs. (A.12) and (A.13) that for any dyadic $\mathfrak{T}$

$$\begin{aligned}
(\mathfrak{T} \times \mathbf{a}) \cdot \mathbf{b} &= \mathfrak{T} \cdot (\mathbf{a} \times \mathbf{b}), \qquad \mathfrak{T} \times \mathbf{a} = \mathfrak{T} \cdot (\mathfrak{J} \times \mathbf{a}), \\
\mathbf{b} \cdot (\mathfrak{T} \times \mathbf{a}) &= (\mathbf{b} \cdot \mathfrak{T}) \times \mathbf{a}, \qquad \mathbf{a} \times \mathfrak{T} = (\mathfrak{J} \times \mathbf{a}) \cdot \mathfrak{T}, \\
\mathbf{a} \times (\mathfrak{T} \times \mathbf{b}) &= (\mathbf{a} \times \mathfrak{T}) \times \mathbf{b} = \mathbf{a} \times \mathfrak{T} \times \mathbf{b}, \\
\mathbf{b} \times (\mathbf{a} \times \mathfrak{T}) &= \mathbf{a}(\mathbf{b} \cdot \mathfrak{T}) - (\mathbf{a} \cdot \mathbf{b})\mathfrak{T}, \\
(\mathfrak{J} \times \mathbf{a}) \cdot (\mathfrak{J} \times \mathbf{b}) &= \mathbf{ba} - \mathfrak{J}(\mathbf{a} \cdot \mathbf{b}).
\end{aligned} \tag{A.14}$$

Two dyadics $\mathfrak{T}$ and $\mathfrak{E}$ are equal when either $\mathbf{r} \cdot \mathfrak{T} = \mathbf{r} \cdot \mathfrak{E}$ or $\mathfrak{T} \cdot \mathbf{r} = \mathfrak{E} \cdot \mathbf{r}$ for every vector **r**. The *zero dyadic* $\mathfrak{O}$ is such that $\mathfrak{O} \cdot \mathbf{r} = \mathbf{0}$ or $\mathbf{r} \cdot \mathfrak{O} = \mathbf{0}$ for every **r**.

It we reverse the order of the vectors in each of the dyads of a dyadic, we get the *conjugate* of the dyadic. Denoting the conjugate of $\mathfrak{T}$ by $\tilde{\mathfrak{T}}$, it is evident that

$$\mathfrak{T} \cdot \mathbf{a} = \mathbf{a} \cdot \tilde{\mathfrak{T}}. \tag{A.15}$$

The conjugate is also known as the *transpose*. A dyadic is *symmetrical if* $\mathfrak{T} = \tilde{\mathfrak{T}}$, it is *antisymmetrical* if $\mathfrak{T} = -\tilde{\mathfrak{T}}$. Every dyadic can be expressed in just one way as the sum of a symmetrical and an antisymmctrical dyadic

$$\mathfrak{T} = \frac{1}{2}(\mathfrak{T} + \tilde{\mathfrak{T}}) + \frac{1}{2}(\mathfrak{T} - \tilde{\mathfrak{T}}),$$

where $\frac{1}{2}(\mathfrak{T} + \tilde{\mathfrak{T}})$ is symmetrical and $\frac{1}{2}(\mathfrak{T} - \tilde{\mathfrak{T}})$ is antisymmetrical. The general antisymmetrical matrix must therefore be of the form

$$\begin{bmatrix} 0 & -a_3 & a_2 \\ a_3 & 0 & -a_1 \\ -a_2 & a_1 & 0 \end{bmatrix}.$$

According to Eq. (A.13), this matrix represents the dyadic $(\mathfrak{J} \times \mathbf{a})$.

Every dyadic $\mathfrak{T}$ has a scalar and a vector associated with it, which we denote by T and $\langle\mathfrak{T}\rangle$, respectively. Using the representation (A.10), we define

$$\begin{aligned}
T &= (\mathbf{a} \cdot \mathbf{u}) + (\mathbf{b} \cdot \mathbf{v}) + (\mathbf{c} \cdot \mathbf{w}), \\
\langle\mathfrak{T}\rangle &= (\mathbf{a} \times \mathbf{u}) + (\mathbf{b} \times \mathbf{v}) + (\mathbf{c} \times \mathbf{w}).
\end{aligned} \tag{A.16}$$

In terms of the elements of $\mathfrak{T}$, they are

$$\begin{aligned}
T &= T_{11} + T_{22} + T_{33}, \\
\langle\mathfrak{T}\rangle &= (T_{23} - T_{32})\mathbf{e}_1 + (T_{31} - T_{13})\mathbf{e}_2 + (T_{12} - T_{21})\mathbf{e}_3.
\end{aligned} \tag{A.17}$$

T is also known as the *trace of* $\mathfrak{T}$.

EXAMPLE A.1: Algebra of Rotations

In a fixed Cartesian system $O(x, y, z)$ space is rotated by an angle ω about the z axis. This rotation carries the position vector $\mathbf{r}$ into a new position $\mathbf{r}'$ relative to O. From geometric considerations

$$\begin{bmatrix} x' \\ y' \\ z' \end{bmatrix} = \begin{bmatrix} \cos\omega & -\sin\omega & 0 \\ \sin\omega & \cos\omega & 0 \\ 0 & 0 & 1 \end{bmatrix} \begin{bmatrix} x \\ y \\ z \end{bmatrix}. \tag{A.1.1}$$

The dyadic form of this matrix relation is

$$\mathbf{r}' = \mathfrak{R} \cdot \mathbf{r},$$

where

$$\mathfrak{R}(\mathbf{e}_z, \omega) = \cos\omega(\mathbf{e}_x\mathbf{e}_x + \mathbf{e}_y\mathbf{e}_y) + \sin\omega(\mathbf{e}_y\mathbf{e}_x - \mathbf{e}_x\mathbf{e}_y) + \mathbf{e}_z\mathbf{e}_z. \tag{A.1.2}$$

For an arbitrary axis of rotation with a unit vector $\mathbf{e}$,

$$\mathfrak{R}(\boldsymbol{\omega}) = \mathfrak{R}(\mathbf{e}, \omega) = \mathfrak{I}\cos\omega + (1 - \cos\omega)\mathbf{ee} + (\mathfrak{I} \times \mathbf{e})\sin\omega. \tag{A.1.3}$$

Therefore, the dyadic $\mathfrak{R}(\mathbf{e}, \omega)$, known sometimes as a *versor*, describes a rotation of space about $\mathbf{e}$ by an angle ω. The matrix of $\mathfrak{R}$ is orthogonal.

The applications of two or more rotations in succession is given by the product of the corresponding dyadics in the given order. Therefore, if $\mathfrak{R}(\boldsymbol{\omega}_1)$ is followed by $\mathfrak{R}(\boldsymbol{\omega}_2)$, and finally by $\mathfrak{R}(\boldsymbol{\omega}_3)$, the resulting rotation will be

$$\mathfrak{R}_{\text{total}} = \mathfrak{R}(\boldsymbol{\omega}_3) \cdot \mathfrak{R}(\boldsymbol{\omega}_2) \cdot \mathfrak{R}(\boldsymbol{\omega}_1). \tag{A.1.4}$$

When $\delta\omega$ is small such that second-order infinitesimals may be neglected, we put $\cos\delta\omega \simeq 1$, $\sin\delta\omega \simeq \delta\omega$, $\delta\boldsymbol{\omega} = (\delta\omega)\mathbf{e}$. The ensuing infinitesimal rotation is given by

$$\mathfrak{R}(\mathbf{e}, \delta\omega) = \mathfrak{I} + (\mathfrak{I} \times \delta\boldsymbol{\omega}) + O[(\delta\omega)^2]. \tag{A.1.5}$$

Let the whole space be subjected to two consecutive infinitesimal rotations $\delta\boldsymbol{\omega}_1 = \mathbf{e}_1\delta\omega_1$ and $\delta\boldsymbol{\omega}_2 = \mathbf{e}_2\delta\omega_2$. The combined rotation, to first order in $(\delta\omega)$, is given by the scalar product

$$\mathfrak{R}_2 \cdot \mathfrak{R}_1 = \mathfrak{I} + [\mathfrak{I} \times (\delta\boldsymbol{\omega}_1 + \delta\boldsymbol{\omega}_2)] + O[(\delta\omega_1)(\delta\omega_2)].$$

Therefore, two infinitesimal rotations combine according to the law of vector addition. Finite rotations do not. In addition to the representation of a rotation by the vector $\boldsymbol{\omega} = \mathbf{e}\omega$, it can also be represented by the three *Euler angles* (α, β, γ):

$$\mathfrak{R}(\mathbf{e}, \omega) = \mathfrak{R}_3(\mathbf{e}_z, \alpha) \cdot \mathfrak{R}_2(\mathbf{e}_y, \beta) \cdot \mathfrak{R}_1(\mathbf{e}_z, \gamma), \tag{A.1.6}$$

where

$$\mathfrak{R}_3 = \begin{bmatrix} \cos\alpha & -\sin\alpha & 0 \\ \sin\alpha & \cos\alpha & 0 \\ 0 & 0 & 1 \end{bmatrix}; \quad \mathfrak{R}_2 = \begin{bmatrix} \cos\beta & 0 & \sin\beta \\ 0 & 1 & 0 \\ -\sin\beta & 0 & \cos\beta \end{bmatrix}; \tag{A.1.7}$$

$$\mathfrak{R}_1 = \begin{bmatrix} \cos\gamma & -\sin\gamma & 0 \\ \sin\gamma & \cos\gamma & 0 \\ 0 & 0 & 1 \end{bmatrix}.$$

The dyadic $\mathfrak{R}_1$ represents the *rotation of space* (known also as *active rotation* in contradistinction to a rotation of the axes, which is called *passive rotation*) relative to the fixed axes (x, y, z) about the z axis by an angle γ. This is followed by a rotation β about the new y axis and terminated by a rotation α about the new z axis.

Suppose that $\mathfrak{R}(\boldsymbol{\omega})$ is followed by an infinitesimal rotation $\mathfrak{R}(d\boldsymbol{\omega})$ with respect to the fixed axes. Let, as in Eq. (A.1.6).

$$\mathfrak{R}(d\boldsymbol{\omega}) \cdot \mathfrak{R}(\boldsymbol{\omega}) = \mathfrak{R}_3(\mathbf{e}_z, \alpha + d\alpha) \cdot \mathfrak{R}_2(\mathbf{e}_y, \beta + d\beta) \cdot \mathfrak{R}_1(\mathbf{e}_z, \gamma + d\gamma). \quad \text{(A.1.8)}$$

Because

$$d\boldsymbol{\omega} = (d\omega_x)\mathbf{e}_x + (d\omega_y)\mathbf{e}_y + (d\omega_z)\mathbf{e}_z \quad \text{(A.1.9)}$$

the corresponding first-order changes in the Euler angles are given by

$$d\boldsymbol{\omega} = (d\alpha)\mathbf{e}_z + \mathfrak{R}_3(\mathbf{e}_z, \alpha) \cdot \{d\beta \mathbf{e}_y\} + \mathfrak{R}_3(\mathbf{e}_z, \alpha) \cdot \mathfrak{R}_2(\mathbf{e}_y, \beta) \cdot \{d\gamma \mathbf{e}_z\}, \quad \text{(A.1.10)}$$

where the form of each term in Eq. (A.1.10) reflects the order of rotations; i.e., the rotation $d\gamma$ is followed by the rotations β and α, the rotation $d\beta$ is followed only by the rotation α, and the rotation $d\alpha$ is final. Substituting from Eqs. (A.1.6) and (A.1.7), we have

$$\begin{aligned} \mathfrak{R}_3(\mathbf{e}_z, \alpha) \cdot \mathbf{e}_y &= -\sin\alpha \mathbf{e}_x + \cos\alpha \mathbf{e}_y, \\ \mathfrak{R}_3(\mathbf{e}_z, \alpha) \cdot \mathfrak{R}_2(\mathbf{e}_y, \beta) \cdot \mathbf{e}_z &= (\cos\alpha \sin\beta)\mathbf{e}_x + (\sin\alpha\sin\beta)\mathbf{e}_y + \cos\beta \mathbf{e}_z. \end{aligned} \quad \text{(A.1.11)}$$

Eqs. (A.1.9)–(A.1.11) yield

$$\begin{bmatrix} d\alpha \\ d\beta \\ d\gamma \end{bmatrix} = \begin{bmatrix} -\cot\beta\cos\alpha & -\cot\beta\sin\alpha & 1 \\ -\sin\alpha & \cos\alpha & 0 \\ \dfrac{\cos\alpha}{\sin\beta} & \dfrac{\sin\alpha}{\sin\beta} & 0 \end{bmatrix} \begin{bmatrix} d\omega_x \\ d\omega_y \\ d\omega_z \end{bmatrix}. \quad \text{(A.1.12)}$$

These are the relations between the differentials of the two representations of infinitesimal rotations.

A.2. Calculus of Dyadics

Let $f(\mathbf{r}) = f(x_1, x_2, x_3)$ be a scalar point function differentiable in a region R in 3-space. The *gradient* of f at a point P with position vector $\mathbf{r} = x_1\mathbf{e}_1 + x_2\mathbf{e}_2 + x_3\mathbf{e}_3$ is defined as a vector

$$\operatorname{grad} f = \nabla f = \mathbf{e}_1 \frac{\partial f}{\partial x_1} + \mathbf{e}_2 \frac{\partial f}{\partial x_2} + \mathbf{e}_3 \frac{\partial f}{\partial x_3}. \quad \text{(A.18)}$$

Let $\mathbf{e}$ be a unit vector with direction cosines (α, β, γ). We define the *directional derivative* of f at P in the direction $\mathbf{e}$ as

$$\frac{df}{ds} = \lim_{s \to 0} \frac{f(\mathbf{r} + s\mathbf{e}) - f(\mathbf{r})}{s}, \quad \text{(A.19)}$$

where $P_1(\mathbf{r} + s\mathbf{e})$ is a point near $P(\mathbf{r})$. Because $\mathbf{e} = \mathbf{e}_1 \cos \alpha + \mathbf{e}_2 \cos \beta + \mathbf{e}_3 \cos \gamma$, it is obvious that $df/ds = \mathbf{e} \cdot \nabla f$, which is the rate of change of f in the $\mathbf{e}$ direction. Note that df/ds is by definition independent of the choice of the coordinate system. So is grad f. If $f(u, v, w)$ is a function of (u, v, w), which themselves are functions of (x_1, x_2, x_3), then

$$\mathbf{e} \cdot \nabla f = \frac{df}{ds} = \frac{\partial f}{\partial u}\frac{du}{ds} + \frac{\partial f}{\partial v}\frac{dv}{ds} + \frac{\partial f}{\partial w}\frac{dw}{ds} = \mathbf{e} \cdot \left(\nabla u \frac{\partial f}{\partial u} + \nabla v \frac{\partial f}{\partial v} + \nabla w \frac{\partial f}{\partial w}\right).$$

Because this relation holds for every $\mathbf{e}$, we have

$$\nabla f = \nabla u \frac{\partial f}{\partial u} + \nabla v \frac{\partial f}{\partial v} + \nabla w \frac{\partial f}{\partial w}. \tag{A.20}$$

Consider next a vector point function in R. It is said to be continuous at P if, as $\mathbf{r}_1$ approaches $\mathbf{r}$ in any manner, $\lim \mathbf{f}(\mathbf{r}_1) = \mathbf{f}(\mathbf{r})$. In rectangular coordinates we represent $\mathbf{f}(\mathbf{r})$ in the form $\mathbf{f}(\mathbf{r}) = f_1\mathbf{e}_1 + f_2\mathbf{e}_2 + f_3\mathbf{e}_3$. The continuity of $f_i(x_1, x_2, x_3)$ follows from the continuity of $\mathbf{f}(\mathbf{r})$. We define the directional derivative of a vector point function $\mathbf{f}(\mathbf{r})$ at P in the $\mathbf{e}$ direction as

$$\frac{d\mathbf{f}}{ds} = \lim_{s \to 0} \frac{\mathbf{f}(\mathbf{r} + s\mathbf{e}) - \mathbf{f}(\mathbf{r})}{s}. \tag{A.21}$$

Formally,

$$\frac{d\mathbf{f}}{ds} = \frac{\partial \mathbf{f}}{\partial x_1}\frac{dx_1}{ds} + \frac{\partial \mathbf{f}}{\partial x_2}\frac{dx_2}{ds} + \frac{\partial \mathbf{f}}{\partial x_3}\frac{dx_3}{ds} = \frac{\partial \mathbf{f}}{\partial x_1}(\mathbf{e} \cdot \mathbf{e}_1) + \frac{\partial \mathbf{f}}{\partial x_2}(\mathbf{e} \cdot \mathbf{e}_2) + \frac{\partial \mathbf{f}}{\partial x_3}(\mathbf{e} \cdot \mathbf{e}_3)$$

$$= \mathbf{e} \cdot \left(\mathbf{e}_1 \frac{\partial \mathbf{f}}{\partial x_1} + \mathbf{e}_2 \frac{\partial \mathbf{f}}{\partial x_2} + \mathbf{e}_3 \frac{\partial \mathbf{f}}{\partial x_3}\right), \tag{A.22}$$

where $\partial \mathbf{f}/\partial x_i$ are defined as in Eq. (A.21). Analogous to Eq. (A.18) we define the dyadic in the parenthesis as the gradient of $\mathbf{f}(\mathbf{r})$, i.e.,

$$\nabla \mathbf{f} = \mathbf{e}_1 \frac{\partial \mathbf{f}}{\partial x_1} + \mathbf{e}_2 \frac{\partial \mathbf{f}}{\partial x_2} + \mathbf{e}_3 \frac{\partial \mathbf{f}}{\partial x_3}, \tag{A.23}$$

and hence,

$$\frac{d\mathbf{f}}{ds} = \mathbf{e} \cdot \nabla \mathbf{f}. \tag{A.24}$$

Inserting $\mathbf{f}(\mathbf{r}) = f_1\mathbf{e}_1 + f_2\mathbf{e}_2 + f_3\mathbf{e}_3$ into Eq. (A.23), the matrix of the dyadic grad $\mathbf{f}$ is seen to be

$$\begin{bmatrix} \dfrac{\partial f_1}{\partial x_1} & \dfrac{\partial f_2}{\partial x_1} & \dfrac{\partial f_3}{\partial x_1} \\ \dfrac{\partial f_1}{\partial x_2} & \dfrac{\partial f_2}{\partial x_2} & \dfrac{\partial f_3}{\partial x_2} \\ \dfrac{\partial f_1}{\partial x_3} & \dfrac{\partial f_2}{\partial x_3} & \dfrac{\partial f_3}{\partial x_3} \end{bmatrix}, \tag{A.25}$$

with the general element $a_{ij} = \partial f_j/\partial x_i$. The scalar of the dyadic grad **f** is div **f**, and its vector is curl **f**. The particular case $\mathbf{f}(\mathbf{r}) = \mathbf{r}$ yields

$$\nabla \mathbf{r} = \mathfrak{I}. \tag{A.26}$$

It can be shown that the vector of a dyadic $\mathfrak{T}$ may be expressed in the form

$$\langle \mathfrak{T} \rangle = \operatorname{div}(\mathfrak{T} \times \mathbf{r}) - (\operatorname{div} \mathfrak{T}) \times \mathbf{r}. \tag{A.27}$$

If we choose $\mathbf{f}(\mathbf{r}) = \operatorname{grad} \chi$, where $\chi(x_1, x_2, x_3)$ is a scalar point function, the resulting dyadic is grad grad χ with the general element $a_{ij} = \partial^2\chi/\partial x_i \partial x_j = a_{ji}$. The scalar of this symmetrical dyadic is $\nabla^2\chi$ and its vector is the zero vector. The transpose of $\nabla \mathbf{f}$ is denoted by $\mathbf{f}\nabla$ and its general element is $a_{ij} = \partial f_i/\partial x_j$. We may write

$$\mathbf{f}\nabla = \frac{\partial \mathbf{f}}{\partial x_1}\mathbf{e}_1 + \frac{\partial \mathbf{f}}{\partial x_2}\mathbf{e}_2 + \frac{\partial \mathbf{f}}{\partial x_3}\mathbf{e}_3. \tag{A.28}$$

When **f** is a function of the variables (u, v, w), which are themselves functions of (x, y, z), we have

$$\mathbf{e} \cdot \nabla \mathbf{f} = \frac{d\mathbf{f}}{ds} = \frac{\partial \mathbf{f}}{\partial u}\frac{du}{ds} + \frac{\partial \mathbf{f}}{\partial v}\frac{dv}{ds} + \frac{\partial \mathbf{f}}{\partial w}\frac{dw}{ds} = \mathbf{e} \cdot \left(\nabla u \frac{\partial \mathbf{f}}{\partial u} + \nabla v \frac{\partial \mathbf{f}}{\partial v} + \nabla w \frac{\partial \mathbf{f}}{\partial w}\right)$$

for any **e**. Hence,

$$\nabla \mathbf{f} = \nabla u \frac{\partial \mathbf{f}}{\partial u} + \nabla v \frac{\partial \mathbf{f}}{\partial v} + \nabla w \frac{\partial \mathbf{f}}{\partial w}. \tag{A.29}$$

In particular, for $\mathbf{f} = \mathbf{r}$,

$$\nabla \mathbf{r} = \mathfrak{I} = \nabla u \frac{\partial \mathbf{r}}{\partial u} + \nabla v \frac{\partial \mathbf{r}}{\partial v} + \nabla w \frac{\partial \mathbf{r}}{\partial w}. \tag{A.30}$$

It is sometimes useful to introduce the notation

$$\operatorname{grad} \phi = \frac{\partial \phi}{\partial \mathbf{r}} \tag{A.31}$$

through which we stress the fact that ϕ is changed because of a change in the position vector $d\mathbf{r}$. However, $\partial\phi/\partial\mathbf{r}$ cannot be regarded as the limit of a quotient of two small quantities. Consequently, we may write symbolically

$$\nabla = \frac{\partial}{\partial \mathbf{r}} = \mathbf{e}_x \frac{\partial}{\partial x} + \mathbf{e}_y \frac{\partial}{\partial y} + \mathbf{e}_z \frac{\partial}{\partial z}. \tag{A.32}$$

This result is capable of a simple generalization. Let

$$\mathbf{q} = \mathbf{e}_x u + \mathbf{e}_y v + \mathbf{e}_z w$$

where u, v and w are functions of the coordinates. Then

$$\frac{\partial}{\partial \mathbf{q}} = \mathbf{e}_x \frac{\partial}{\partial u} + \mathbf{e}_y \frac{\partial}{\partial v} + \mathbf{e}_z \frac{\partial}{\partial w}. \tag{A.33}$$

A.3. Identities

Here u, v, w are scalars; $\mathbf{f}$, $\mathbf{g}$, $\mathbf{h}$, $\mathbf{q}$ are vectors; and $\mathfrak{A}$, $\mathfrak{B}$, $\mathfrak{C}$ are dyadics. The Cartesian coordinates are denoted either by (x, y, z) or by (x_1, x_2, x_3), with unit vectors $(\mathbf{e}_x, \mathbf{e}_y, \mathbf{e}_z)$ or $(\mathbf{e}_1, \mathbf{e}_2, \mathbf{e}_3)$, respectively. The spherical coordinates are denoted by (r, θ, ϕ), with unit vectors $(\mathbf{e}_r, \mathbf{e}_\theta, \mathbf{e}_\phi)$. Similarly, the circular cylinder coordinates (Δ, ϕ, z) have unit vectors $(\mathbf{e}_\Delta, \mathbf{e}_\phi, \mathbf{e}_z)$. The idemfactor is denoted by $\mathfrak{I}$. The subscripts i, j, k may take the values 1, 2, 3, and the summation convention is used.

$$\operatorname{grad}(uv) = (\operatorname{grad} u)v + u \operatorname{grad} v \tag{A.34}$$

$$\operatorname{grad}(u\mathbf{f}) = (\operatorname{grad} u)\mathbf{f} + u \operatorname{grad} \mathbf{f} \tag{A.35}$$

$$\operatorname{grad}(\mathbf{fg}) = (\operatorname{grad} \mathbf{f})\mathbf{g} + (\mathbf{f} \operatorname{grad} \mathbf{g})^{213} \tag{A.36}$$

$$\operatorname{grad}(u\mathfrak{A}) = (\operatorname{grad} u)\mathfrak{A} + u \operatorname{grad} \mathfrak{A} \tag{A.37}$$

$$\operatorname{grad}(u\mathfrak{I}) = (\operatorname{grad} u)\mathfrak{I} \tag{A.38}$$

$$\begin{aligned} \operatorname{grad}(\mathbf{f} \cdot \mathbf{g}) &= (\operatorname{grad} \mathbf{f}) \cdot \mathbf{g} + (\operatorname{grad} \mathbf{g}) \cdot \mathbf{f} \\ &= \mathbf{g} \cdot \operatorname{grad} \mathbf{f} + \mathbf{g} \times \operatorname{curl} \mathbf{f} + \mathbf{f} \cdot \operatorname{grad} \mathbf{g} + \mathbf{f} \times \operatorname{curl} \mathbf{g} \end{aligned} \tag{A.39}$$

$$\operatorname{grad}(\mathfrak{A} \cdot \mathbf{f}) = (\operatorname{grad} \mathfrak{A}) \cdot \mathbf{f} + (\operatorname{grad} \mathbf{f}) \cdot \tilde{\mathfrak{A}} \tag{A.40}$$

$$\operatorname{grad}(\mathbf{f} \cdot \mathfrak{A}) = (\operatorname{grad} \mathbf{f}) \cdot \mathfrak{A} + \mathbf{f} \cdot (\operatorname{grad} \mathfrak{A})^{213} \tag{A.41}$$

$$\operatorname{grad}(\mathbf{f} \times \mathbf{g}) = (\operatorname{grad} \mathbf{f}) \times \mathbf{g} - (\operatorname{grad} \mathbf{g}) \times \mathbf{f} \tag{A.42}$$

$$\begin{aligned} \operatorname{grad}(\mathbf{f} \times \mathfrak{I}) &= \operatorname{grad}(\mathfrak{I} \times \mathbf{f}) \\ &= (\operatorname{grad} \mathbf{f}) \times \mathfrak{I}; \qquad \nabla\mathbf{f} - \mathbf{f}\nabla = -\mathfrak{I} \times \operatorname{curl} \mathbf{f} \end{aligned} \tag{A.43}$$

$$\operatorname{div}(u\mathbf{f}) = u \operatorname{div} \mathbf{f} + (\operatorname{grad} u) \cdot \mathbf{f} \tag{A.44}$$

$$\operatorname{div}(\mathbf{fg}) = (\operatorname{div} \mathbf{f})\mathbf{g} + \mathbf{f} \cdot \operatorname{grad} \mathbf{g} \tag{A.45}$$

$$\operatorname{div}(u\mathfrak{A}) = u \operatorname{div} \mathfrak{A} + (\operatorname{grad} u) \cdot \mathfrak{A} \tag{A.46}$$

$$\operatorname{div}(u\mathfrak{I}) = \operatorname{grad} u \tag{A.47}$$

$$\operatorname{div}(\mathfrak{A}\mathbf{f}) = (\operatorname{div} \mathfrak{A})\mathbf{f} + \tilde{\mathfrak{A}} \cdot \operatorname{grad} \mathbf{f} \tag{A.48}$$

$$\operatorname{div}(\mathfrak{I}\mathbf{f}) = \operatorname{grad} \mathbf{f} \tag{A.49}$$

$$\operatorname{div}(\mathbf{f}\mathfrak{A}) = (\operatorname{div} \mathbf{f})\mathfrak{A} + \mathbf{f} \cdot \operatorname{grad} \mathfrak{A} \tag{A.50}$$

$$\operatorname{div}(\mathbf{f}\mathfrak{I}) = (\operatorname{div} \mathbf{f})\mathfrak{I} \tag{A.51}$$

$$\operatorname{div}(\mathbf{f} \cdot \mathfrak{A}) = \operatorname{grad} \mathbf{f} : \mathfrak{A} + \mathbf{f} \cdot \operatorname{div} \tilde{\mathfrak{A}} \tag{A.52}$$ [2]

$$\operatorname{div}(\mathfrak{A} \cdot \mathbf{f}) = (\operatorname{div} \mathfrak{A}) \cdot \mathbf{f} + \tilde{\mathfrak{A}} : \operatorname{grad} \mathbf{f} \tag{A.53}$$

[2] We define

$$(\mathbf{fg}) : (\mathbf{hq}) = (\mathbf{f} \cdot \mathbf{q})(\mathbf{g} \cdot \mathbf{h}), \qquad (\mathbf{fg}) \overset{\cdot}{\times} (\mathbf{hq}) = (\mathbf{f} \times \mathbf{q})(\mathbf{g} \cdot \mathbf{h}),$$

$$(\mathbf{fg}) \overset{\times}{\times} (\mathbf{hq}) = (\mathbf{f} \times \mathbf{q})(\mathbf{g} \times \mathbf{h}). \qquad (\mathbf{fg}) \overset{\times}{\cdot} (\mathbf{hq}) = (\mathbf{f} \cdot \mathbf{q})(\mathbf{g} \times \mathbf{h})$$

$$\operatorname{div}(\mathbf{f} \times \mathbf{g}) = (\operatorname{curl} \mathbf{f}) \cdot \mathbf{g} - \mathbf{f} \cdot \operatorname{curl} \mathbf{g} \tag{A.54}$$

$$\operatorname{div}(\mathbf{f} \times \mathfrak{A}) = (\operatorname{curl} \mathbf{f}) \cdot \mathfrak{A} - \mathbf{f} \cdot \operatorname{curl} \mathfrak{A} \tag{A.55}$$

$$\operatorname{div}(\mathfrak{A} \times \mathbf{f}) = (\operatorname{div} \mathfrak{A}) \times \mathbf{f} + \tilde{\mathfrak{A}} \underset{\times}{\cdot} \operatorname{grad} \mathbf{f} \tag{A.56}$$

$$\operatorname{div}(\mathfrak{I} \times \mathbf{f}) = \operatorname{curl} \mathbf{f} \tag{A.57}$$

$$\operatorname{div} \operatorname{grad} \mathbf{f} = \nabla^2 \mathbf{f} \tag{A.58}$$

$$\operatorname{div}(\operatorname{grad} \mathbf{f})^{\sim} = \operatorname{grad} \operatorname{div} \mathbf{f} \tag{A.59}$$

$$\operatorname{curl}(u\mathbf{f}) = u \operatorname{curl} \mathbf{f} + (\operatorname{grad} u) \times \mathbf{f} \tag{A.60}$$

$$\operatorname{curl}(\mathbf{fg}) = (\operatorname{curl} \mathbf{f})\mathbf{g} - \mathbf{f} \times \operatorname{grad} \mathbf{g} \tag{A.61}$$

$$\operatorname{curl}(u\mathfrak{A}) = u \operatorname{curl} \mathfrak{A} + (\operatorname{grad} u) \times \mathfrak{A} \tag{A.62}$$

$$\operatorname{curl}(u\mathfrak{I}) = (\operatorname{grad} u) \times \mathfrak{I} \tag{A.63}$$

$$\operatorname{curl}(u\mathfrak{I} \times \mathfrak{I}) = \mathfrak{I} \operatorname{grad} u - (\mathfrak{I} \operatorname{grad} u)^{132} \tag{A.64}$$

$$\begin{aligned} \operatorname{curl}(\mathbf{f} \times \mathbf{g}) &= \operatorname{div}(\mathbf{gf} - \mathbf{fg}) \\ &= \mathbf{f} \operatorname{div} \mathbf{g} - \mathbf{g} \operatorname{div} \mathbf{f} + \mathbf{g} \cdot \operatorname{grad} \mathbf{f} - \mathbf{f} \cdot \operatorname{grad} \mathbf{g} \end{aligned} \tag{A.65}$$

$$\operatorname{curl}(\mathfrak{A} \times \mathbf{f}) = (\operatorname{curl} \mathfrak{A}) \times \mathbf{f} - (\operatorname{grad} \mathbf{f}) \underset{\times}{\overset{\times}{}} \tilde{\mathfrak{A}} \tag{A.66}$$

$$\operatorname{curl}(\mathbf{f} \times \mathfrak{I}) = \operatorname{curl}(\mathfrak{I} \times \mathbf{f}) = (\operatorname{grad} \mathbf{f})^{\sim} - \mathfrak{I} \operatorname{div} \mathbf{f} \tag{A.67}$$

$$\operatorname{curl} \operatorname{grad} \mathbf{f} = 0 \tag{A.68}$$

$$\operatorname{curl}(\operatorname{grad} \mathbf{f})^{\sim} = (\operatorname{grad} \operatorname{curl} \mathbf{f})^{\sim} \tag{A.69}$$ [3]

$$\operatorname{curl} \operatorname{curl} \mathbf{f} = \operatorname{grad} \operatorname{div} \mathbf{f} - \nabla^2 \mathbf{f} \tag{A.70}$$

$$\mathfrak{I} : \operatorname{grad} \mathbf{f} = \operatorname{div} \mathbf{f} \tag{A.71}$$

$$\mathfrak{I} \underset{\times}{\cdot} \operatorname{grad} \mathbf{f} = \operatorname{curl} \mathbf{f} \tag{A.72}$$

$$\mathfrak{I} \underset{\times}{\overset{\times}{}} \operatorname{grad} \mathbf{f} = \mathfrak{I} \operatorname{div} \mathbf{f} - \operatorname{grad} \mathbf{f} \tag{A.73}$$

$$\nabla^2 \mathbf{f} = \mathbf{e}_i \nabla^2 f_i \text{ (Cartesian system only)} \tag{A.74}$$

$$\nabla^2 \mathfrak{A} = \mathbf{e}_i (\nabla^2 A_{ij}) \mathbf{e}_j \text{ (Cartesian system only)} \tag{A.75}$$

$$\nabla^2(uv) = (\nabla^2 u)v + u\nabla^2 v + 2(\operatorname{grad} u) \cdot (\operatorname{grad} v) \tag{A.76}$$

$$\nabla^2(u\mathbf{f}) = (\nabla^2 u)\mathbf{f} + u\nabla^2 \mathbf{f} + 2(\operatorname{grad} u) \cdot (\operatorname{grad} \mathbf{f}) \tag{A.77}$$

$$\nabla^2(\mathbf{fg}) = (\nabla^2 \mathbf{f})\mathbf{g} + \mathbf{f}\nabla^2 \mathbf{g} + 2(\operatorname{grad} \mathbf{f})^{\sim} \cdot (\operatorname{grad} \mathbf{g}) \tag{A.78}$$

$$\nabla^2(u\mathfrak{A}) = (\nabla^2 u)\mathfrak{A} + u\nabla^2 \mathfrak{A} + 2(\operatorname{grad} u) \cdot (\operatorname{grad} \mathfrak{A}) \tag{A.79}$$

$$\nabla^2(\mathbf{f} \cdot \mathbf{g}) = (\nabla^2 \mathbf{f}) \cdot \mathbf{g} + \mathbf{f} \cdot \nabla^2 \mathbf{g} + 2(\operatorname{grad} \mathbf{f})^{\sim} : (\operatorname{grad} \mathbf{g}) \tag{A.80}$$

[3] The symbols $(\ )^{\sim}$, $(\ )^{213}$, and $(\ )^{132}$ denote, respectively, the transpose of a dyadic, the left transpose of a triadic, and the right transpose of a triadic; e.g. $(\mathbf{ab})^{\sim} = \mathbf{ba}$, $(\mathbf{abc})^{213} = \mathbf{bac}$, $(\mathbf{abc})^{132} = \mathbf{acb}$.

$$\nabla^2(\mathbf{f} \cdot \mathfrak{A}) = (\nabla^2\mathbf{f}) \cdot \mathfrak{A} + \mathbf{f} \cdot \nabla^2\mathfrak{A} + 2(\text{grad } \mathbf{f})^{\sim} : (\text{grad } \mathfrak{A}) \tag{A.81}$$

$$\nabla^2(\mathfrak{A} \cdot \mathbf{f}) = (\nabla^2\mathfrak{A}) \cdot \mathbf{f} + \mathfrak{A} \cdot \nabla^2\mathbf{f} + 2(\text{grad } \mathfrak{A})^{312} : (\text{grad } \mathbf{f}) \tag{A.82}$$

$$\nabla^2(\mathbf{f} \times \mathfrak{I}) = \nabla^2(\mathfrak{I} \times \mathbf{f}) = (\nabla^2\mathbf{f}) \times \mathfrak{I} \tag{A.83}$$

$$\nabla^2 \text{ grad } u = \text{grad } \nabla^2 u \tag{A.84}$$

$$\nabla^2 \text{ curl } \mathbf{f} = \text{curl } \nabla^2\mathbf{f} \tag{A.85}$$

$$\nabla^2 \text{ curl curl } \mathbf{f} = \text{curl curl } \nabla^2\mathbf{f} \tag{A.86}$$

$$\text{curl curl}(u\mathfrak{I}) = \text{curl}[(\text{grad } u) \times \mathfrak{I}] = \text{grad grad } u - \mathfrak{I}\nabla^2 u. \tag{A.87}$$

A.4. Cartesian Coordinates (x, y, z)

$$\text{grad}(\mathbf{e}_z \cdot \mathbf{f}) = \frac{\partial \mathbf{f}}{\partial z} + \mathbf{e}_z \times \text{curl } \mathbf{f} \tag{A.88}$$

$$\text{div}(\mathbf{e}_z \times \mathbf{f}) = -\mathbf{e}_z \cdot \text{curl } \mathbf{f} \tag{A.89}$$

$$\text{curl}(\mathbf{e}_z u) = (\text{grad } u) \times \mathbf{e}_z \tag{A.90}$$

$$\text{curl}(\mathbf{e}_z \times \mathbf{f}) = -\frac{\partial \mathbf{f}}{\partial z} + \mathbf{e}_z \text{ div } \mathbf{f} \tag{A.91}$$

$$\begin{aligned} \nabla^2(\mathbf{e}_z u) &= \mathbf{e}_z \nabla^2 u \\ \nabla^2(\mathbf{e}_z \times \mathbf{f}) &= \mathbf{e}_z \times \nabla^2\mathbf{f} \end{aligned} \tag{A.92}$$

$$\begin{aligned} \frac{1}{2}(\nabla\mathbf{f} + \mathbf{f}\nabla) = {} & \mathbf{e}_1\mathbf{e}_1 \frac{\partial f_1}{\partial x_1} + \mathbf{e}_2\mathbf{e}_2 \frac{\partial f_2}{\partial x_2} + \mathbf{e}_3\mathbf{e}_3 \frac{\partial f_3}{\partial x_3} \\ & + (\mathbf{e}_1\mathbf{e}_2 + \mathbf{e}_2\mathbf{e}_1)\frac{1}{2}\left(\frac{\partial f_1}{\partial x_2} + \frac{\partial f_2}{\partial x_1}\right) \\ & + (\mathbf{e}_2\mathbf{e}_3 + \mathbf{e}_3\mathbf{e}_2)\frac{1}{2}\left(\frac{\partial f_2}{\partial x_3} + \frac{\partial f_3}{\partial x_2}\right) \\ & + (\mathbf{e}_3\mathbf{e}_1 + \mathbf{e}_1\mathbf{e}_3)\frac{1}{2}\left(\frac{\partial f_3}{\partial x_1} + \frac{\partial f_1}{\partial x_3}\right). \end{aligned} \tag{A.93}$$

A.5. Circular Cylinder Coordinates (Δ, ϕ, z) (Fig. A.1)

$$\Delta^2 = x^2 + y^2; \qquad \tan\phi = \frac{y}{x}; \qquad x = \Delta\cos\phi; \qquad y = \Delta\sin\phi \tag{A.94}$$

$$ds^2 = d\Delta^2 + (\Delta d\phi)^2 + dz^2 \tag{A.95}$$

$$\Delta\frac{\partial}{\partial\Delta} = x\frac{\partial}{\partial x} + y\frac{\partial}{\partial y}; \qquad \frac{\partial}{\partial\phi} = x\frac{\partial}{\partial y} - y\frac{\partial}{\partial x} \tag{A.96}$$

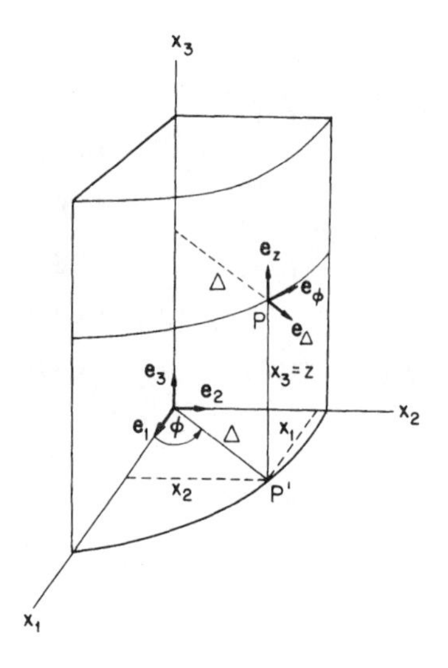

Figure A.1. Circular cylinder coordinates.

$$
\begin{aligned}
\mathbf{e}_\Delta &= \mathbf{e}_x \cos\phi + \mathbf{e}_y \sin\phi; \qquad \mathbf{e}_\phi = -\mathbf{e}_x \sin\phi + \mathbf{e}_y \cos\phi \\
\mathbf{e}_x &= \mathbf{e}_\Delta \cos\phi - \mathbf{e}_\phi \sin\phi; \qquad \mathbf{e}_y = \mathbf{e}_\Delta \sin\phi + \mathbf{e}_\phi \cos\phi
\end{aligned}
\tag{A.97}
$$

$$
\frac{\partial \mathbf{e}_\Delta}{\partial \phi} = \mathbf{e}_\phi; \qquad \frac{\partial \mathbf{e}_\phi}{\partial \phi} = -\mathbf{e}_\Delta;
\tag{A.98}
$$

$$
\frac{\partial \mathbf{e}_\Delta}{\partial \Delta} = \frac{\partial \mathbf{e}_\Delta}{\partial z} = \frac{\partial \mathbf{e}_\phi}{\partial \Delta} = \frac{\partial \mathbf{e}_\phi}{\partial z} = \frac{\partial \mathbf{e}_z}{\partial \Delta} = \frac{\partial \mathbf{e}_z}{\partial \phi} = \frac{\partial \mathbf{e}_z}{\partial z} = 0
\tag{A.99}
$$

$$
\operatorname{grad} u = \mathbf{e}_\Delta \frac{\partial u}{\partial \Delta} + \mathbf{e}_\phi \frac{1}{\Delta}\frac{\partial u}{\partial \phi} + \mathbf{e}_z \frac{\partial u}{\partial z}
\tag{A.100}
$$

$$
\operatorname{div} \mathbf{f} = \frac{\partial f_\Delta}{\partial \Delta} + \frac{f_\Delta}{\Delta} + \frac{1}{\Delta}\frac{\partial f_\phi}{\partial \phi} + \frac{\partial f_z}{\partial z}
\tag{A.101}
$$

$$
\begin{aligned}
\operatorname{curl} \mathbf{f} &= \mathbf{e}_\Delta \left(\frac{1}{\Delta}\frac{\partial f_z}{\partial \phi} - \frac{\partial f_\phi}{\partial z} \right) + \mathbf{e}_\phi \left(\frac{\partial f_\Delta}{\partial z} - \frac{\partial f_z}{\partial \Delta} \right) \\
&\quad + \mathbf{e}_z \left(\frac{\partial f_\phi}{\partial \Delta} + \frac{f_\phi}{\Delta} - \frac{1}{\Delta}\frac{\partial f_\Delta}{\partial \phi} \right)
\end{aligned}
\tag{A.102}
$$

$$
\nabla^2 u = \frac{\partial^2 u}{\partial \Delta^2} + \frac{1}{\Delta}\frac{\partial u}{\partial \Delta} + \frac{1}{\Delta^2}\frac{\partial^2 u}{\partial \phi^2} + \frac{\partial^2 u}{\partial z^2}
\tag{A.103}
$$

$$
\begin{aligned}
\nabla^2 \mathbf{f} &= \mathbf{e}_\Delta \left(\nabla^2 f_\Delta - \frac{f_\Delta}{\Delta^2} - \frac{2}{\Delta^2}\frac{\partial f_\phi}{\partial \phi} \right) + \mathbf{e}_\phi \left(\nabla^2 f_\phi - \frac{f_\phi}{\Delta^2} + \frac{2}{\Delta^2}\frac{\partial f_\Delta}{\partial \phi} \right) \\
&\quad + \mathbf{e}_z \nabla^2 f_z
\end{aligned}
\tag{A.104}
$$

$$\begin{aligned}\frac{1}{2}(\nabla\mathbf{f} + \mathbf{f}\nabla) = \mathbf{e}_\Delta\mathbf{e}_\Delta \frac{\partial f_\Delta}{\partial \Delta} + \mathbf{e}_\phi\mathbf{e}_\phi \frac{1}{\Delta}\left(\frac{\partial f_\phi}{\partial \phi} + f_\Delta\right) + \mathbf{e}_z\mathbf{e}_z \frac{\partial f_z}{\partial z} \\ + (\mathbf{e}_\Delta\mathbf{e}_\phi + \mathbf{e}_\phi\mathbf{e}_\Delta)\frac{1}{2}\left(\frac{\partial f_\phi}{\partial \Delta} - \frac{f_\phi}{\Delta} + \frac{1}{\Delta}\frac{\partial f_\Delta}{\partial \phi}\right) \\ + (\mathbf{e}_\Delta\mathbf{e}_z + \mathbf{e}_z\mathbf{e}_\Delta)\frac{1}{2}\left(\frac{\partial f_z}{\partial \Delta} + \frac{\partial f_\Delta}{\partial z}\right) \\ + (\mathbf{e}_\phi\mathbf{e}_z + \mathbf{e}_z\mathbf{e}_\phi)\frac{1}{2}\left(\frac{\partial f_\phi}{\partial z} + \frac{1}{\Delta}\frac{\partial f_z}{\partial \phi}\right).\end{aligned} \tag{A.105}$$

A.6. Spherical Coordinates (r, θ, ϕ) (Fig. A.2)

$$r^2 = x^2 + y^2 + z^2; \qquad \tan\theta = \frac{(x^2 + y^2)^{1/2}}{z}; \qquad \tan\phi = \frac{y}{x} \tag{A.106}$$

$$x = r\sin\theta\cos\phi; \qquad y = r\sin\theta\sin\phi; \qquad z = r\cos\theta \tag{A.107}$$

$$ds^2 = dr^2 + (r\,d\theta)^2 + (r\sin\theta\,d\phi)^2 \tag{A.108}$$

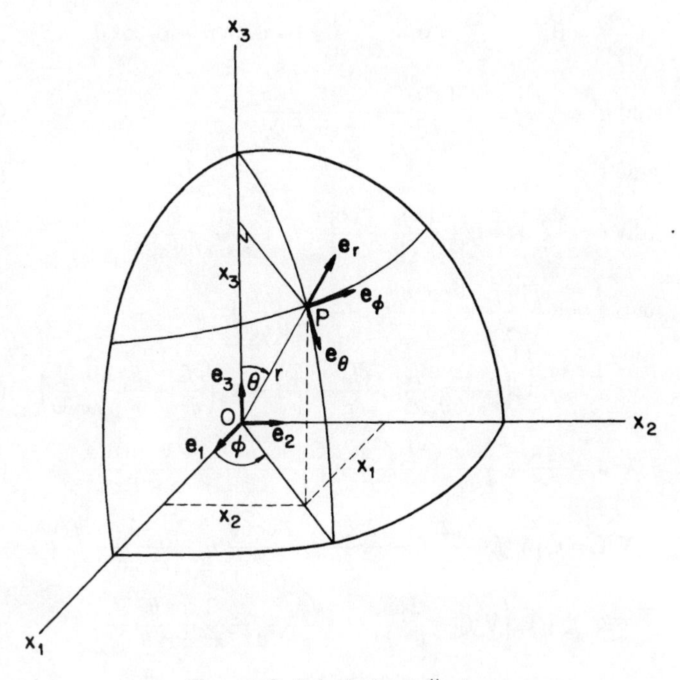

Figure A.2. Spherical coordinates.

$$r\frac{\partial}{\partial r} = x\frac{\partial}{\partial x} + y\frac{\partial}{\partial y} + z\frac{\partial}{\partial z} \tag{A.109}$$

$$\frac{\partial}{\partial \theta} = \frac{xz}{(x^2+y^2)^{1/2}}\frac{\partial}{\partial x} + \frac{yz}{(x^2+y^2)^{1/2}}\frac{\partial}{\partial y} - (x^2+y^2)^{1/2}\frac{\partial}{\partial z} \tag{A.110}$$

$$\frac{\partial}{\partial \phi} = x\frac{\partial}{\partial y} - y\frac{\partial}{\partial x} \tag{A.111}$$

$$\mathbf{e}_r = \mathbf{e}_x \sin\theta\cos\phi + \mathbf{e}_y \sin\theta\sin\phi + \mathbf{e}_z\cos\theta \tag{A.112}$$

$$\mathbf{e}_\theta = \mathbf{e}_x \cos\theta\cos\phi + \mathbf{e}_y \cos\theta\sin\phi - \mathbf{e}_z\sin\theta \tag{A.113}$$

$$\mathbf{e}_\phi = -\mathbf{e}_x\sin\phi + \mathbf{e}_y\cos\phi \tag{A.114}$$

$$\mathbf{e}_x = \mathbf{e}_r\sin\theta\cos\phi + \mathbf{e}_\theta\cos\theta\cos\phi - \mathbf{e}_\phi\sin\phi \tag{A.115}$$

$$\mathbf{e}_y = \mathbf{e}_r\sin\theta\sin\phi + \mathbf{e}_\theta\cos\theta\sin\phi + \mathbf{e}_\phi\cos\phi \tag{A.116}$$

$$\mathbf{e}_z = \mathbf{e}_r\cos\theta - \mathbf{e}_\theta\sin\theta \tag{A.117}$$

$$\frac{\partial \mathbf{e}_r}{\partial r} = 0; \qquad \frac{\partial \mathbf{e}_r}{\partial \theta} = \mathbf{e}_\theta; \qquad \frac{\partial \mathbf{e}_r}{\partial \phi} = \sin\theta\,\mathbf{e}_\phi; \tag{A.118}$$

$$\frac{\partial \mathbf{e}_\theta}{\partial r} = 0; \qquad \frac{\partial \mathbf{e}_\theta}{\partial \theta} = -\mathbf{e}_r; \qquad \frac{\partial \mathbf{e}_\theta}{\partial \phi} = \cos\theta\,\mathbf{e}_\phi; \tag{A.119}$$

$$\frac{\partial \mathbf{e}_\phi}{\partial r} = 0; \qquad \frac{\partial \mathbf{e}_\phi}{\partial \theta} = 0; \qquad \frac{\partial \mathbf{e}_\phi}{\partial \phi} = -\mathbf{e}_r\sin\theta - \mathbf{e}_\theta\cos\theta \tag{A.120}$$

$$\operatorname{grad} u = \mathbf{e}_r\frac{\partial u}{\partial r} + \mathbf{e}_\theta\frac{1}{r}\frac{\partial u}{\partial \theta} + \mathbf{e}_\phi\frac{1}{r\sin\theta}\frac{\partial u}{\partial \phi} \tag{A.121}$$

$$\operatorname{grad}\mathbf{r} = \mathfrak{I} \tag{A.122}$$

$$\operatorname{div}\mathbf{f} = \frac{\partial f_r}{\partial r} + \frac{2}{r}f_r + \frac{1}{r}\frac{\partial f_\theta}{\partial \theta} + \frac{\cot\theta}{r}f_\theta + \frac{1}{r\sin\theta}\frac{\partial f_\phi}{\partial \phi} \tag{A.123}$$

$$\begin{aligned}\operatorname{curl}\mathbf{f} = \mathbf{e}_r&\left(\frac{1}{r}\frac{\partial f_\phi}{\partial \theta} + \frac{\cot\theta}{r}f_\phi - \frac{1}{r\sin\theta}\frac{\partial f_\theta}{\partial \phi}\right)\\ &+ \mathbf{e}_\theta\left(\frac{1}{r\sin\theta}\frac{\partial f_r}{\partial \phi} - \frac{\partial f_\phi}{\partial r} - \frac{f_\phi}{r}\right) + \mathbf{e}_\phi\left(\frac{\partial f_\theta}{\partial r} + \frac{f_\theta}{r} - \frac{1}{r}\frac{\partial f_r}{\partial \theta}\right)\end{aligned} \tag{A.124}$$

$$\nabla^2 u = \frac{\partial^2 u}{\partial r^2} + \frac{2}{r}\frac{\partial u}{\partial r} + \frac{1}{r^2}\frac{\partial^2 u}{\partial \theta^2} + \frac{\cot\theta}{r^2}\frac{\partial u}{\partial \theta} + \frac{1}{r^2\sin^2\theta}\frac{\partial^2 u}{\partial \phi^2} \tag{A.125}$$

$$\begin{aligned}\nabla^2\mathbf{f} = \mathbf{e}_r&\left(\nabla^2 f_r - \frac{2}{r^2}f_r - \frac{2}{r^2}\frac{\partial f_\theta}{\partial \theta} - \frac{2\cot\theta}{r^2}f_\theta - \frac{2}{r^2\sin\theta}\frac{\partial f_\phi}{\partial \phi}\right)\\ &+ \mathbf{e}_\theta\left(\nabla^2 f_\theta + \frac{2}{r^2}\frac{\partial f_r}{\partial \theta} - \frac{f_\theta}{r^2\sin^2\theta} - \frac{2\cos\theta}{r^2\sin^2\theta}\frac{\partial f_\phi}{\partial \phi}\right)\\ &+ \mathbf{e}_\phi\left(\nabla^2 f_\phi + \frac{2}{r^2\sin\theta}\frac{\partial f_r}{\partial \phi} - \frac{f_\phi}{r^2\sin^2\theta} + \frac{2\cos\theta}{r^2\sin^2\theta}\frac{\partial f_\theta}{\partial \phi}\right)\end{aligned} \tag{A.126}$$

$$\frac{1}{2}(\nabla\mathbf{f} + \mathbf{f}\nabla) = \mathbf{e}_r\mathbf{e}_r\frac{\partial f_r}{\partial r} + \mathbf{e}_\theta\mathbf{e}_\theta\frac{1}{r}\left(\frac{\partial f_\theta}{\partial\theta} + f_r\right)$$

$$+ \mathbf{e}_\phi\mathbf{e}_\phi\frac{1}{r}\left(\frac{1}{\sin\theta}\frac{\partial f_\phi}{\partial\phi} + f_r + \cot\theta\, f_\theta\right)$$

$$+ (\mathbf{e}_r\mathbf{e}_\theta + \mathbf{e}_\theta\mathbf{e}_r)\frac{1}{2}\left(\frac{1}{r}\frac{\partial f_r}{\partial\theta} + \frac{\partial f_\theta}{\partial r} - \frac{f_\theta}{r}\right)$$

$$+ (\mathbf{e}_r\mathbf{e}_\phi + \mathbf{e}_\phi\mathbf{e}_r)\frac{1}{2}\left(\frac{1}{r\sin\theta}\frac{\partial f_r}{\partial\phi} + \frac{\partial f_\phi}{\partial r} - \frac{f_\phi}{r}\right)$$

$$+ (\mathbf{e}_\theta\mathbf{e}_\phi + \mathbf{e}_\phi\mathbf{e}_\theta)\frac{1}{2r}\left(\frac{1}{\sin\theta}\frac{\partial f_\theta}{\partial\phi} + \frac{\partial f_\phi}{\partial\theta} - \cot\theta\, f_\phi\right). \qquad \text{(A.127)}$$

A.7. Source–Observer System

We use the notation

$$R = |\mathbf{r} - \mathbf{r}_0| = [(x - x_0)^2 + (y - y_0)^2 + (z - z_0)^2]^{1/2} \qquad \text{(A.128)}$$

$$R\mathbf{e}_R = \mathbf{R} = \mathbf{r} - \mathbf{r}_0 = \mathbf{e}_x(x - x_0) + \mathbf{e}_y(y - y_0) + \mathbf{e}_z(z - z_0) \qquad \text{(A.129)}$$

$$\operatorname{grad}_0 u = \mathbf{e}_x\frac{\partial u}{\partial x_0} + \mathbf{e}_y\frac{\partial u}{\partial y_0} + \mathbf{e}_z\frac{\partial u}{\partial z_0}. \qquad \text{(A.130)}$$

Then,

$$\operatorname{grad} F(R) = \mathbf{e}_R F'(R) = -\operatorname{grad}_0 F(R) \qquad \text{(A.131)}$$

$$\operatorname{grad} \mathbf{R} = \mathfrak{I} \qquad \text{(A.132)}$$

$$\operatorname{grad}[\mathbf{e}_R F(R)] = \mathbf{e}_R\mathbf{e}_R\left(F' - \frac{F}{R}\right) + \mathfrak{I}\frac{F}{R} \qquad \text{(A.133)}$$

$$\operatorname{div}[\mathbf{e}_R F(R)] = F'(R) + \frac{2}{R}F(R) = -\operatorname{div}_0[\mathbf{e}_R F(R)] \qquad \text{(A.134)}$$

$$\operatorname{div}[\mathfrak{I}F(R)] = F'(R)\mathbf{e}_R \qquad \text{(A.135)}$$

$$\operatorname{grad}\operatorname{div}[\mathfrak{I}F(R)] = \left(F'' - \frac{F'}{R}\right)\mathbf{e}_R\mathbf{e}_R + \frac{F'}{R}\mathfrak{I} \qquad \text{(A.136)}$$

$$\operatorname{div}[\mathbf{e}_R\mathbf{e}_R F(R)] = \mathbf{e}_R\left(F' + \frac{2}{R}F\right) \qquad \text{(A.137)}$$

$$\operatorname{curl}[\mathbf{e}_R F(R)] = 0 \qquad \text{(A.138)}$$

$$\nabla^2 F(R) = F''(R) + \frac{2}{R}F'(R) = \nabla_0^2 F(R) \qquad \text{(A.139)}$$

$$\nabla^2\mathbf{R} = 0 \qquad \text{(A.140)}$$

$$\operatorname{grad} F\left(t - \frac{R}{\alpha}\right) = -\frac{\mathbf{e}_R}{\alpha}F'\left(t - \frac{R}{\alpha}\right), \qquad \text{(A.141)}$$

where the prime denotes differentiation with respect to the argument.

$$\operatorname{grad}\left[\frac{1}{R} F\left(t-\frac{R}{\alpha}\right)\right] = -\mathbf{e}_R\left[\frac{1}{\alpha R} F'\left(t-\frac{R}{\alpha}\right) + \frac{1}{R^2} F\left(t-\frac{R}{\alpha}\right)\right] \tag{A.142}$$

$$\nabla^2 F\left(t-\frac{R}{\alpha}\right) = \frac{1}{\alpha^2} F''\left(t-\frac{R}{\alpha}\right) - \frac{2}{\alpha R} F'\left(t-\frac{R}{\alpha}\right) \tag{A.143}$$

$$\nabla^2\left[\frac{1}{R} F\left(t-\frac{R}{\alpha}\right)\right] = \frac{1}{\alpha^2 R} F''\left(t-\frac{R}{\alpha}\right) \quad (R \neq 0). \tag{A.144}$$

If $\mathbf{a}$ is a constant vector and $R \neq 0$,

$$\operatorname{grad}(\mathbf{R}\cdot\mathbf{a}) = \mathbf{a} \tag{A.145}$$

$$\operatorname{div}(R\mathbf{a}) = \mathbf{e}_R\cdot\mathbf{a} \tag{A.146}$$

$$\operatorname{curl}(R\mathbf{a}) = \mathbf{e}_R \times \mathbf{a} \tag{A.147}$$

$$\operatorname{curl}(\mathbf{a}\times\mathbf{R}) = 2\mathbf{a} \tag{A.148}$$

$$\operatorname{grad}\operatorname{div}\frac{\mathfrak{I}}{R} = \operatorname{grad}\operatorname{grad}\frac{1}{R} = \frac{1}{R^3}(3\mathbf{e}_R\mathbf{e}_R - \mathfrak{I}) \tag{A.149}$$

$$\operatorname{grad}\operatorname{div}\frac{\mathbf{RR}}{R^3} = \frac{1}{R^3}(\mathfrak{I} - 3\mathbf{e}_R\mathbf{e}_R) \tag{A.150}$$

$$\operatorname{div}(\mathfrak{I}\mathbf{e}_R) = \operatorname{grad}\mathbf{e}_R = \frac{1}{R}(\mathfrak{I} - \mathbf{e}_R\mathbf{e}_R) \tag{A.151}$$

$$\operatorname{div}(\mathbf{e}_R\mathfrak{I}) = \frac{2}{R}\mathfrak{I} \tag{A.152}$$

$$\operatorname{div}(\mathbf{e}_R\mathbf{e}_R\mathbf{e}_R) = \frac{2}{R}\mathbf{e}_R\mathbf{e}_R. \tag{A.153}$$

APPENDIX B

Orthogonal Curvilinear Coordinates

The simplest coordinate system is the rectangular Cartesian frame in which the three unit vectors $\mathbf{e}_i$ have a fixed orientation in space. These coordinates, however, are suitable only for the solution of boundary-value problems involving bodies of rectangular shape. In geophysics we often need to treat finite configurations with spherical, spheroidal, or ellipsoidal boundaries, and infinite domains with axial or spherical symmetry. We must therefore learn how to write the components of vectors and tensors in a wider class of coordinates. To gain insight into the principle of this generalization we recapitulate the well-known relations between the rectangular and spherical coordinates

$$
\begin{aligned}
r &= (x_1^2 + x_2^2 + x_3^2)^{1/2},\\
\theta &= \cos^{-1}\left\{\frac{x_3}{(x_1^2 + x_2^2 + x_3^2)^{1/2}}\right\},\\
\phi &= \cos^{-1}\left\{\frac{x_1}{(x_1^2 + x_2^2)^{1/2}}\right\},\\
x_1 &= r \sin\theta \cos\phi,\\
x_2 &= r \sin\theta \sin\phi,\\
x_3 &= r \cos\theta.
\end{aligned}
\tag{B.1}
$$

Clearly the *coordinate surfaces* $r = r_0$, $\theta = \theta_0$, and $\phi = \phi_0$ consist of a sphere, a cone, and a plane, respectively. Each pair of coordinate surfaces intersect at a *coordinate line*, along which a single coordinate varies. The three unit vectors, tangent to the coordinate lines at a given point and oriented in the directions of increasing θ, ϕ, and r, are denoted by $\mathbf{e}_\theta$, $\mathbf{e}_\phi$, and $\mathbf{e}_r$, respectively. These form a right-hand orthogonal system. In contradistinction to the Cartesian frame, the coordinate lines are space curves.

Because

$$
\mathbf{r} = x_1\mathbf{e}_1 + x_2\mathbf{e}_2 + x_3\mathbf{e}_3 = r(\sin\theta\cos\phi\mathbf{e}_1 + \sin\theta\sin\phi\mathbf{e}_2 + \cos\theta\mathbf{e}_3) = r\mathbf{e}_r,
\tag{B.2}
$$

we can easily establish the relations

$$
\frac{\partial\mathbf{r}}{\partial r} = \mathbf{e}_r, \qquad \frac{\partial\mathbf{r}}{\partial\theta} = r\mathbf{e}_\theta, \qquad \frac{\partial\mathbf{r}}{\partial\phi} = r\sin\theta\mathbf{e}_\phi.
\tag{B.3}
$$

The unit vectors are therefore proportional to the derivatives of the position vector with respect to the coordinates (r, θ, ϕ).

Consider now the general coordinates (q_1, q_2, q_3) of a point P in 3-space which are related to the Cartesian coordinates (x_1, x_2, x_3) of P via the relations

$$x_1 = x_1(q_1, q_2, q_3), \qquad x_2 = x_2(q_1, q_2, q_3), \qquad x_3 = x_3(q_1, q_2, q_3), \quad \text{(B.4)}$$

or, in the concise forms

$$\mathbf{r} = \mathbf{r}(q_1, q_2, q_3), \qquad x_i = x_i(q_1, q_2, q_3), \quad i = 1, 2, 3. \quad \text{(B.5)}$$

The tangent direction along a coordinate line is given by the expression $\partial\mathbf{r}/\partial q_\alpha$ $(\alpha = 1, 2, 3)$, and the corresponding unit vector will therefore be

$$\mathbf{e}_\alpha = \frac{1}{h_\alpha}\frac{\partial \mathbf{r}}{\partial q_\alpha}, \qquad h_\alpha = \left|\frac{\partial \mathbf{r}}{\partial q_\alpha}\right| = \left(\frac{\partial \mathbf{r}}{\partial q_\alpha}\cdot\frac{\partial \mathbf{r}}{\partial q_\alpha}\right)^{1/2}. \quad \text{(B.6)}$$

In an orthonormal right-hand base, the following relations hold

$$\mathbf{e}_\alpha \cdot \mathbf{e}_\beta = \delta_{\alpha\beta}, \qquad \mathbf{e}_\alpha \times \mathbf{e}_\beta = \mathbf{e}_\gamma, \quad \text{(B.7)}$$

where $\delta_{\alpha\beta}$ is the *Kronecker delta* and α, β, and γ are in cyclic order (123, 231, 312). Consideration will henceforth be limited to orthogonal systems.

The functions h_α are called *scale factors*. Because, by Eqs. (B.6)

$$d\mathbf{r} = \sum_{\alpha=1}^{3} \frac{\partial \mathbf{r}}{\partial q_\alpha}\, dq_\alpha = \sum_{\alpha=1}^{3} h_\alpha \mathbf{e}_\alpha\, dq_\alpha, \quad \text{(B.8)}$$

the square of the arc element, ds, will have the form

$$ds^2 = d\mathbf{r}\cdot d\mathbf{r} = h_1^2(dq_1)^2 + h_2^2(dq_2)^2 + h_3^2(dq_3)^2. \quad \text{(B.9)}$$

In practice, Eq. (B.9) serves as the simplest way to derive explicit expressions for the scale factors because this equation can be derived by application of the Pythagorean rule to a space element.

We shall next evaluate the derivatives $\partial\mathbf{e}_\alpha/\partial q_\beta$. Because this expression is a vector quantity, it can be represented as a linear combination of the base vectors $\mathbf{e}_\gamma$, i.e.,

$$\frac{\partial \mathbf{e}_\alpha}{\partial q_\beta} = c_1(\alpha, \beta)\mathbf{e}_1 + c_2(\alpha, \beta)\mathbf{e}_2 + c_3(\alpha, \beta)\mathbf{e}_3 = \sum_{\gamma=1}^{3} c_\gamma(\alpha, \beta)\mathbf{e}_\gamma. \quad \text{(B.10)}$$

It is convenient to introduce the notation

$$c_\gamma(\alpha, \beta) = ({}_\alpha{}^\gamma{}_\beta), \quad \text{(B.11)}$$

known as the *Christoffel symbol of the second kind.* Its meaning is simply the γ component of the derivative of the α-unit vector along the β coordinate. Each symbol is a function of the scale factors only. To establish the explicit form of this functional dependence, we start by substituting in the relation $\mathbf{e}_\gamma \cdot (\partial \mathbf{e}_\alpha / \partial q_\beta) = ({}_\alpha{}^\gamma{}_\beta)$ the explicit expression for

$$\frac{\partial \mathbf{e}_\alpha}{\partial q_\beta} = \frac{\partial}{\partial q_\beta}\left(\frac{1}{h_\alpha}\frac{\partial \mathbf{r}}{\partial q_\alpha}\right). \tag{B.12}$$

Using the equation

$$\mathbf{e}_\gamma \cdot \mathbf{e}_\alpha = \delta_{\gamma\alpha} = \frac{1}{h_\gamma h_\alpha}\frac{\partial \mathbf{r}}{\partial q_\gamma} \cdot \frac{\partial \mathbf{r}}{\partial q_\alpha},$$

it is easy to show that

$$\begin{aligned} &({}_\alpha{}^\gamma{}_\beta) = 0 \quad \text{for} \quad \gamma \neq \beta \neq \alpha, \\ &({}_\alpha{}^\alpha{}_\beta) = 0, \qquad ({}_\alpha{}^\alpha{}_\alpha) = 0, \\ &({}_\alpha{}^\gamma{}_\alpha) = -\frac{1}{h_\gamma}\frac{\partial h_\alpha}{\partial q_\gamma}, \qquad ({}_\alpha{}^\gamma{}_\gamma) = \frac{1}{h_\alpha}\frac{\partial h_\gamma}{\partial q_\alpha}. \end{aligned} \tag{B.13}$$

The expression

$$\frac{\partial \mathbf{e}_\alpha}{\partial q_\beta} = \sum_{\gamma=1}^{3} ({}_\alpha{}^\gamma{}_\beta)\mathbf{e}_\gamma, \tag{B.14}$$

has therefore been fully defined in terms of the scale factors.

We now proceed to derive the basic differential expression for the gradient operator. Let $\phi(q_1, q_2, q_3)$ be a scalar function. Then

$$\nabla\phi = f_1 \mathbf{e}_1 + f_2 \mathbf{e}_2 + f_3 \mathbf{e}_3, \tag{B.15}$$

where $f_i(q_1, q_2, q_3)$ are the components of $\nabla\phi$. However, the differential $d\phi$ can be written in two ways

$$d\phi = \frac{\partial\phi}{\partial q_1} dq_1 + \frac{\partial\phi}{\partial q_2} dq_2 + \frac{\partial\phi}{\partial q_3} dq_3$$

and

$$d\phi = \nabla\phi \cdot d\mathbf{r} = h_1 f_1 \, dq_1 + h_2 f_2 \, dq_2 + h_3 f_3 \, dq_3.$$

Hence, we have the formal identity

$$\nabla \equiv \frac{\mathbf{e}_1}{h_1}\frac{\partial}{\partial q_1} + \frac{\mathbf{e}_2}{h_2}\frac{\partial}{\partial q_2} + \frac{\mathbf{e}_3}{h_3}\frac{\partial}{\partial q_3} = \sum_{\alpha=1}^{3} \frac{\mathbf{e}_\alpha}{h_\alpha}\frac{\partial}{\partial q_\alpha}. \tag{B.16}$$

In particular,

$$\nabla q_\alpha = \frac{\mathbf{e}_\alpha}{h_\alpha}.$$

If the operator ∇ is applied to the vector $\mathbf{u} = u_1 \mathbf{e}_1 + u_2 \mathbf{e}_2 + u_3 \mathbf{e}_3$, the result is the dyadic

$$(\nabla \mathbf{u})_{\alpha\beta} = \frac{1}{h_\alpha}\frac{\partial u_\beta}{\partial q_\alpha} + \sum_{\gamma=1}^{3} \frac{u_\gamma}{h_\alpha} ({}_\gamma{}^\beta{}_\alpha). \tag{B.17}$$

Using Eq. (B.16), we arrive at the expressions for the divergence and curl of a vector

$$
\begin{aligned}
\operatorname{div} \mathbf{u} &= \nabla \cdot \sum_{\alpha=1}^{3} u_\alpha \mathbf{e}_\alpha = \sum_{\alpha=1}^{3} \sum_{\beta=1}^{3} \left[\frac{1}{h_\beta} \mathbf{e}_\beta \cdot \frac{\partial}{\partial q_\beta} (u_\alpha \mathbf{e}_\alpha) \right] \\
&= \sum_{\alpha=1}^{3} \sum_{\beta=1}^{3} \left[\frac{1}{h_\alpha} \frac{\partial u_\alpha}{\partial q_\alpha} + \frac{u_\alpha}{h_\beta} \left({}^{\ \beta}_{\alpha\ \beta}\right) \right] \\
&= \frac{1}{h_1 h_2 h_3} \left[\frac{\partial}{\partial q_1} (h_2 h_3 u_1) + \frac{\partial}{\partial q_2} (h_3 h_1 u_2) + \frac{\partial}{\partial q_3} (h_1 h_2 u_3) \right],
\end{aligned}
\tag{B.18}
$$

$$
\operatorname{curl} \mathbf{u} = \nabla \times \sum_{\alpha=1}^{3} u_\alpha \mathbf{e}_\alpha = \nabla \times \sum_{\alpha=1}^{3} (u_\alpha h_\alpha) \nabla q_\alpha .
$$

However, because

$$
\nabla \times (\lambda \mathbf{a}) = (\nabla \lambda) \times \mathbf{a} + \lambda \operatorname{curl} \mathbf{a}, \qquad \operatorname{curl} \operatorname{grad} \equiv 0,
$$

we have

$$
\begin{aligned}
\operatorname{curl} \mathbf{u} &= \sum_{\alpha=1}^{3} \nabla(u_\alpha h_\alpha) \times \nabla q_\alpha = \sum_{\alpha=1}^{3} \left[\nabla(u_\alpha h_\alpha) \times \frac{\mathbf{e}_\alpha}{h_\alpha} \right] \\
&= \sum_{\gamma=1}^{3} \mathbf{e}_\gamma \left\{ \sum_{\alpha=1}^{3} \sum_{\beta=1}^{3} \frac{\varepsilon_{\beta\alpha\gamma}}{h_\alpha h_\beta} \frac{\partial}{\partial q_\beta} (u_\alpha h_\alpha) \right\},
\end{aligned}
\tag{B.19}
$$

where ε_{ijk} is the Levi–Civita symbol

$$
\varepsilon_{ijk} = \begin{cases} 1 & \text{if} \quad ijk \text{ are in cyclic order} \\ -1 & \text{if} \quad ijk \text{ are in anticyclic order} \\ 0 & \text{if} \quad \text{any two indices are equal.} \end{cases}
\tag{B.20}
$$

Therefore,

$$
\begin{aligned}
(\operatorname{curl} \mathbf{u})_1 &= \frac{1}{h_2 h_3} \left[\frac{\partial(u_3 h_3)}{\partial q_2} - \frac{\partial(u_2 h_2)}{\partial q_3} \right], \\
(\operatorname{curl} \mathbf{u})_2 &= \frac{1}{h_3 h_1} \left[\frac{\partial(u_1 h_1)}{\partial q_3} - \frac{\partial(u_3 h_3)}{\partial q_1} \right], \\
(\operatorname{curl} \mathbf{u})_3 &= \frac{1}{h_1 h_2} \left[\frac{\partial(u_2 h_2)}{\partial q_1} - \frac{\partial(u_1 h_1)}{\partial q_2} \right].
\end{aligned}
\tag{B.21}
$$

APPENDIX C

The Material Derivative

Consider a continuum of matter in a state of motion with density $\rho = \rho(\mathbf{r}, t)$. The total mass in a given volume V is $m = \int_V \rho \, dV$. An infinitesimal volume of mass $\rho \, dV$ will be referred to as a particle. Given a fixed frame of reference in space, the state of each particle in time and space is characterized by its velocity vector $\mathbf{v}(t)$. Let $A(\mathbf{r}, t)$ denote a continuously differentiable scalar, vector, or tensor field associated with the particle, such as its mass, momentum, or stress. Then, the *total change* of A along the path of the particle over a time interval dt is

$$DA = A(\mathbf{r} + \mathbf{v}\, dt, t + dt) - A(\mathbf{r}, t).$$

Rearranging this difference in the form

$$[A(\mathbf{r}, t + dt) - A(\mathbf{r}, t)] + [A(\mathbf{r} + \mathbf{v}\, dt, t + dt) - A(\mathbf{r}, t + dt)],$$

we find

$$\frac{DA}{Dt} = \lim_{dt \to 0} \left[\frac{A(\mathbf{r}, t + dt) - A(\mathbf{r}, t)}{dt} \right] + \lim_{dt \to 0} \left[\frac{A(\mathbf{r} + \mathbf{v}\, dt, t + dt) - A(\mathbf{r}, t + dt)}{dt} \right]. \tag{C.1}$$

Expanding to first order in dt,

$$A(\mathbf{r} + \mathbf{v}\, dt, t) = A(\mathbf{r}, t) + dt(\mathbf{v} \cdot \nabla A).$$

Therefore, we obtain the explicit expression for the *material derivative*

$$\underset{\text{Material derivative}}{\frac{DA}{Dt}} = \underset{\text{Local derivative}}{\frac{\partial A}{\partial t}} + \underset{\text{Convective term}}{(\mathbf{v} \cdot \nabla A)} . \tag{C.2}$$

The physical interpretation of Eq. (C.2) is this: if we measure the change of A over the time interval dt by making observations at a fixed point in space, we find that the rate of change is $\partial A / \partial t$. However, if we make this measurement while "riding" a moving particle, we have to add to the local derivative the spatial change of A along the particle's trajectory during dt.

The material derivative of a volume integral

$$I(t) = \int_{V(t)} A(\mathbf{r}, t) dV,$$

defined over a spatial domain V with boundaries occupied by a given set of particles, is of special importance. We have

$$\frac{DI}{Dt} = \lim_{dt\to 0}\left[\frac{\int_{V+\Delta V} A(\mathbf{r}, t+dt)dV - \int_V A(\mathbf{r}, t)dV}{dt}\right]$$

$$= \lim_{dt\to 0}\left[\int_V \left\{\frac{A(\mathbf{r}, t+dt) - A(\mathbf{r}, t)}{dt}\right\} dV + \frac{1}{dt}\int_{\Delta V} A(\mathbf{r}, t+dt)dV\right]. \qquad \text{(C.3)}$$

However, during the time interval dt, the additional volume dV is swept by particles occupying an element of area dS such that $dV = (\mathbf{v}\cdot\mathbf{n})dS\,dt$, where $\mathbf{n}$ is a unit outward-drawn normal. Hence, to first order in dt,

$$\int_{\Delta V} A(\mathbf{r}, t+dt)dV = dt\int_S A(\mathbf{r}, t)(\mathbf{v}\cdot\mathbf{n})dS = dt\int_V \operatorname{div}(\mathbf{v}A)dV$$

$$= dt\int_V [\mathbf{v}\cdot\nabla A + A(\operatorname{div}\mathbf{v})]dV. \qquad \text{(C.4)}$$

Combining Eqs. (C.3) and (C.4), we have *Reynolds' transport theorem*

$$\frac{D}{Dt}\left[\int_{V(t)} A(\mathbf{r}, t)dV\right] = \int_{V(t)}\left[\frac{DA}{Dt} + A(\operatorname{div}\mathbf{v})\right]dV. \qquad \text{(C.5)}$$

Therefore, in general, spatial integration and material derivation do not commute.

APPENDIX D

Bessel and Legendre Functions

D.1. Bessel Functions

D.1.1. Differential Equation

The Bessel functions $Z_m(x)$ are the solutions of the Bessel differential equation

$$\frac{d^2Z}{dx^2} + \frac{1}{x}\frac{dZ}{dx} + \left(1 - \frac{m^2}{x^2}\right)Z = 0. \tag{D.1}$$

The solution, which is bounded at $x = 0$, is known as the Bessel function of the first kind of order m and is denoted by $J_m(x)$. It is given by

$$\begin{aligned} J_m(x) &= \sum_{n=0}^{\infty} \frac{(-1)^n}{n!(m+n)!}\left(\frac{x}{2}\right)^{m+2n} \\ &= \frac{1}{m!}\left(\frac{x}{2}\right)^m\left[1 - \frac{(x/2)^2}{1!(m+1)} + \frac{(x/2)^4}{2!(m+1)(m+2)} - \cdots\right]. \end{aligned} \tag{D.2}$$

The second independent solution of Eq. (D.1) is the Bessel function of the second kind or the Neumann function, $N_m(x)$. It is sometimes denoted by $Y_m(x)$. The Bessel function of the second kind is singular at the origin. For m that is not an integer

$$N_m(x) = \frac{(\cos m\pi)J_m(x) - J_{-m}(x)}{\sin m\pi}. \tag{D.3}$$

When $m = p$ is an integer,

$$N_p(x) = \lim_{m \to p} N_m(x).$$

The Bessel functions of the third kind, or Hankel functions, are defined in terms of $J_m(x)$ and $N_m(x)$:

$$H_m^{(1)}(x) = J_m(x) + iN_m(x), \tag{D.4}$$

$$H_m^{(2)}(x) = J_m(x) - iN_m(x). \tag{D.5}$$

Any two out of the four functions $J_m(x)$, $N_m(x)$, $H_m^{(1)}(x)$, and $H_m^{(2)}(x)$ constitute two independent solutions of the Bessel equation. In the following, $Z_m(x)$

will stand for any one of the four functions $J_m(x)$, $N_m(x)$, $H_m^{(1)}(x)$ and $H_m^{(2)}(x)$ or a linear combination of these functions with coefficients independent of x and m.

D.1.2. General Properties

$$\begin{aligned}
Z_{-m}(x) &= (-1)^m Z_m(x) \quad (m \text{ integer}) \\
J_m(-x) &= e^{m\pi i} J_m(x) \\
N_m(-x) &= e^{-m\pi i} N_m(x) + 2i\cos(m\pi) J_m(x) \\
H_m^{(1)}(xe^{\pi i}) &= -e^{-m\pi i} H_m^{(2)}(x) \\
H_m^{(2)}(xe^{-\pi i}) &= -e^{m\pi i} H_m^{(1)}(x)
\end{aligned} \tag{D.6}$$

$$J_m(x) = \frac{1}{\pi}\int_0^{\pi} \cos(m\theta - x\sin\theta)\,d\theta \quad (m \text{ integer}) \tag{D.7}$$

$$\cos m\phi J_m(x) = \frac{i^{-m}}{2\pi}\int_0^{2\pi} e^{ix\cos(\theta-\phi)} \cos m\theta\, d\theta \quad (m \text{ integer}) \tag{D.8}$$

$$e^{(x/2)(t-k^2/t)} = \sum_{m=-\infty}^{\infty} \left(\frac{t}{k}\right)^m J_m(kx) \tag{D.9}$$

$$e^{ix\sin\phi} = \sum_{m=-\infty}^{\infty} e^{im\phi} J_m(x) \tag{D.10}$$

$$J_m(x) = (-x)^m \left(\frac{1}{x}\frac{d}{dx}\right)^m J_0(x) \quad (m \text{ integer}). \tag{D.11}$$

D.1.3. Wronskian Relations

The Wronskian of a pair of solutions F_1, F_2 of a linear, homogeneous, second-order differential equation is defined as

$$W(F_1, F_2) = F_1 F_2' - F_1' F_2, \tag{D.12}$$

where the prime denotes differentiation with respect to the independent variable. We have

$$\begin{aligned}
W(J_m, N_m) &= J_m(x) N_m'(x) - J_m'(x) N_m(x) \\
&= J_m(x) N_{m-1}(x) - J_{m-1}(x) N_m(x) = \frac{2}{\pi x}
\end{aligned} \tag{D.13}$$

$$W[J_m, H_m^{(1,2)}] = \pm\frac{2i}{x\pi} \tag{D.14}$$

$$J_m(x) N_m''(x) - J_m''(x) N_m(x) = -\frac{2}{\pi x^2} \tag{D.15}$$

$$J_m'(x) N_m''(x) - J_m''(x) N_m'(x) = \frac{2}{\pi x}\left(1 - \frac{m^2}{x^2}\right). \tag{D.16}$$

In formulas (D.13), (D.15), and (D.16), J_m, N_m may be replaced by $H_m^{(1)}$, $H_m^{(2)}$, respectively, if the expressions in the right-hand side are multiplied by $-2i$.

D.1.4. Addition Theorems

If

$$D^2 = \Delta^2 + \Delta_0^2 - 2\Delta\Delta_0 \cos(\phi - \phi_0); \qquad 0 < \Delta < \Delta_0, \tag{D.17}$$

we have

$$Z_0(kD) = \sum_{m=-\infty}^{\infty} e^{im(\phi-\phi_0)} J_m(k\Delta) Z_m(k\Delta_0). \tag{D.18}$$

If

$$R^2 = D^2 + (z - z_0)^2 = \Delta^2 + \Delta_0^2 - 2\Delta\Delta_0 \cos(\phi - \phi_0) + (z - z_0)^2, \tag{D.19}$$

then

$$\frac{1}{R} = \sum_{m=-\infty}^{\infty} e^{im(\phi-\phi_0)} \int_0^{\infty} J_m(k\Delta) J_m(k\Delta_0) e^{-k|z-z_0|}\, dk, \tag{D.20}$$

$$\frac{e^{-ik_0R}}{R} = \sum_{m=-\infty}^{\infty} e^{im(\phi-\phi_0)} \int_0^{\infty} J_m(k\Delta) J_m(k\Delta_0) e^{-\nu|z-z_0|} \frac{k\, dk}{\nu}, \tag{D.21}$$

where

$$\nu = (k^2 - k_0^2)^{1/2}. \tag{D.22}$$

D.1.5. Recurrence and Differential Relations

$$\frac{2m}{x} Z_m(x) = Z_{m-1}(x) + Z_{m+1}(x) \tag{D.23}$$

$$\frac{d}{dx} Z_m(x) = \frac{1}{2}[Z_{m-1}(x) - Z_{m+1}(x)] \tag{D.24}$$

$$= Z_{m-1}(x) - \frac{m}{x} Z_m(x) \tag{D.25}$$

$$= -Z_{m+1}(x) + \frac{m}{x} Z_m(x) \tag{D.26}$$

In particular,

$$\frac{d}{dx} Z_0(x) = -Z_1(x).$$

$$\frac{d}{dx}[Z_m(\alpha x)] = -\alpha Z_{m+1}(\alpha x) + \frac{m}{x} Z_m(\alpha x)$$

D.1.6. Orthogonality

$$\int_0^\infty J_m(kx)J_m(k'x)x\,dx = \frac{\delta(k-k')}{\sqrt{kk'}} \tag{D.27}$$

$$\int_0^{2\pi}\int_0^\infty Y_m(k\Delta,\phi)\overset{*}{Y}_{m'}(k'\Delta,\phi)\Delta\,d\Delta\,d\phi = \frac{2\pi}{\sqrt{kk'}}\delta_{mm'}\delta(k-k'), \tag{D.28}$$

where

$$Y_m(k\Delta,\phi) = J_m(k\Delta)e^{im\phi}.$$

$$\int_0^\infty J_m(kx)k\,dk\int_0^\infty F(y)J_m(ky)y\,dy = F(x), \quad \text{Fourier–Bessel integral.} \tag{D.29}$$

D.1.7. Asymptotic Expansions

If $|x| \gg |m|$, $|x| \gg 1$, we have

$$J_m(x) \sim \left(\frac{2}{\pi x}\right)^{1/2}\cos\left(x - \frac{\pi}{4} - \frac{m\pi}{2}\right) \tag{D.30}$$

$$N_m(x) \sim \left(\frac{2}{\pi x}\right)^{1/2}\sin\left(x - \frac{\pi}{4} - \frac{m\pi}{2}\right) \tag{D.31}$$

$$H_m^{(1)}(x) \sim \left(\frac{2}{\pi x}\right)^{1/2} e^{ix - i\pi(m/2+1/4)} \tag{D.32}$$

$$H_m^{(2)}(x) \sim \left(\frac{2}{\pi x}\right)^{1/2} e^{-ix + i\pi(m/2+1/4)}. \tag{D.33}$$

If m is large,

$$J_m(x) \sim \left(\frac{1}{2\pi m}\right)^{1/2}\left(\frac{ex}{2m}\right)^m \tag{D.34}$$

$$N_m(x) \sim -\left(\frac{2}{\pi m}\right)^{1/2}\left(\frac{2m}{ex}\right)^m. \tag{D.35}$$

D.1.8. Expansions Near the Origin

If $|x| \ll 1$, $m \geq 1$, we have (m integer)

$$J_m(x) \simeq \frac{x^m}{2^m m!}; \qquad J_0(x) \simeq 1 - \frac{x^2}{4} \tag{D.36}$$

$$N_m(x) \simeq -\frac{2^m(m-1)!}{\pi x^m}; \qquad N_0(x) \simeq \frac{2}{\pi}\ln\frac{x}{2}. \tag{D.37}$$

D.2. Modified Bessel Functions

D.2.1. Definition

Bessel functions of purely imaginary argument are known as modified Bessel functions. These are defined as follows:

$$I_m(x) = e^{-m\pi i/2} J_m(ix); \qquad -\pi < \arg x < \frac{\pi}{2} \tag{D.38}$$

$$K_m(x) = \frac{\pi}{2} e^{(m+1)\pi i/2}[J_m(ix) + iN_m(ix)]; \qquad -\pi < \arg x < \frac{\pi}{2}. \tag{D.39}$$

The function K_m is also known as Macdonald's function.

D.2.2. General Properties

The functions $I_m(x)$ and $K_m(x)$ satisfy the differential equation

$$\frac{d^2Z}{dx^2} + \frac{1}{x}\frac{dZ}{dx} - \left(1 + \frac{m^2}{x^2}\right)Z = 0 \tag{D.40}$$

$$I_m(x) = \sum_{n=0}^{\infty} \frac{1}{n!(m+n)!}\left(\frac{x}{2}\right)^{m+2n} \tag{D.41}$$

$$K_{-m}(x) = K_m(x) = \frac{\pi}{2} e^{(m+1)\pi i/2} H_m^{(1)}(xe^{\pi i/2}); \qquad -\pi < \arg x < \frac{\pi}{2} \tag{D.42}$$

$$I_{-m}(x) = I_m(x) \quad (m \text{ integer}) \tag{D.43}$$

$$\begin{aligned} I_m(-x) &= e^{m\pi i} I_m(x) \\ K_m(-x) &= e^{-m\pi i} K_m(x) - \pi i I_m(x). \end{aligned} \tag{D.44}$$

D.2.3. Wronskian Relations

$$\begin{aligned} I_m(x)\frac{d}{dx}K_m(x) - \frac{d}{dx}I_m(x)K_m(x) &= -I_m(x)K_{m+1}(x) - I_{m+1}(x)K_m(x) \\ &= -\frac{1}{x}. \end{aligned} \tag{D.45}$$

D.2.4. Addition Theorems

With the definition in Eq. (D.17) for D, we have

$$I_0(kD) = \sum_{m=-\infty}^{\infty} (-1)^m e^{im(\phi-\phi_0)} I_m(k\Delta) I_m(k\Delta_0) \tag{D.46}$$

$$K_0(kD) = \sum_{m=-\infty}^{\infty} e^{im(\phi-\phi_0)} I_m(k\Delta) K_m(k\Delta_0). \tag{D.47}$$

D.2.5. Recurrence and Differential Relations

$$\frac{2m}{x} I_m(x) = I_{m-1}(x) - I_{m+1}(x) \tag{D.48}$$

$$2\frac{d}{dx} I_m(x) = I_{m-1}(x) + I_{m+1}(x) \tag{D.49}$$

$$-\frac{2m}{x} K_m(x) = K_{m-1}(x) - K_{m+1}(x) \tag{D.50}$$

$$-2\frac{d}{dx} K_m(x) = K_{m-1}(x) + K_{m+1}(x). \tag{D.51}$$

In particular,

$$\frac{d}{dx} I_0(x) = I_1(x), \qquad \frac{d}{dx} K_0(x) = -K_1(x). \tag{D.52}$$

D.2.6. Asymptotic Expansions

If $|x| \gg |m|$, $|x| \gg 1$, we have

$$I_m(x) \sim \frac{e^x}{(2\pi x)^{1/2}} \tag{D.53}$$

$$K_m(x) \sim \left(\frac{\pi}{2x}\right)^{1/2} e^{-x}. \tag{D.54}$$

D.2.7. Expansions Near the Origin

If $|x| \ll 1$, $m \geq 1$, we have (m integer)

$$I_m(x) \simeq \frac{x^m}{2^m m!}; \qquad I_0(x) \simeq 1 + \frac{x^2}{4}, \tag{D.55}$$

$$K_m(x) \simeq \frac{2^{m-1}(m-1)!}{x^m}; \qquad K_0(x) \simeq \log\frac{2}{x}. \tag{D.56}$$

D.3. Spherical Bessel Functions

D.3.1. Definition

Spherical Bessel functions of the first and second kinds and the spherical Hankel functions of the first and second kinds are defined by (l integer)

$$j_l(x) = \left(\frac{\pi}{2x}\right)^{1/2} J_{l+1/2}(x), \tag{D.57}$$

$$n_l(x) = \left(\frac{\pi}{2x}\right)^{1/2} N_{l+1/2}(x), \tag{D.58}$$

$$h_l^{(1,2)}(x) = \left(\frac{\pi}{2x}\right)^{1/2} H_{l+1/2}^{(1,2)}(x). \tag{D.59}$$

These functions satisfy the differential equation

$$\frac{d^2f}{dx^2} + \frac{2}{x}\frac{df}{dx} + \left[1 - \frac{l(l+1)}{x^2}\right] f = 0. \tag{D.60}$$

Out of the four functions, the spherical Bessel function of the first kind is bounded at the origin and the spherical Hankel function of the second kind represents outgoing waves [see asymptotic expansion of $h_l^{(2)}(x)$]. Any two of the four functions $j_l(x)$, $n_l(x)$, $h_l^{(1)}(x)$, and $h_l^{(2)}(x)$ constitute two independent solutions of the differential equation (D.60). In the following $f_l(x)$ will stand for any of the four functions $j_l(x)$, $n_l(x)$, $h_l^{(1)}(x)$, and $h_l^{(2)}(x)$ or a linear combination of these functions with coefficients independent of x and l.

D.3.2. General Properties

$$h_l^{(1)}(x) = j_l(x) + i n_l(x) \tag{D.61}$$

$$h_l^{(2)}(x) = j_l(x) - i n_l(x) \tag{D.62}$$

$$j_{-l-1}(x) = (-1)^{l+1} n_l(x); \qquad n_{-l-1}(x) = (-1)^l j_l(x) \tag{D.63}$$

$$h_{-l-1}^{(1)}(x) = i(-1)^l h_l^{(1)}(x); \qquad h_{-l-1}^{(2)}(x) = -i(-1)^l h_l^{(2)}(x) \tag{D.64}$$

$$j_0(x) = \frac{\sin x}{x}; \qquad j_1(x) = \frac{1}{x}\left(\frac{\sin x}{x} - \cos x\right) \tag{D.65}$$

$$j_2(x) = \left(\frac{3}{x^2} - 1\right)\frac{\sin x}{x} - \frac{3}{x^2}\cos x \tag{D.66}$$

$$j_l(x) = (-x)^l \left(\frac{1}{x}\frac{d}{dx}\right)^l \left(\frac{\sin x}{x}\right) \tag{D.67}$$

$$n_0(x) = -\frac{\cos x}{x}; \qquad n_1(x) = -\frac{1}{x}\left(\frac{\cos x}{x} + \sin x\right) \tag{D.68}$$

$$n_2(x) = -\left(\frac{3}{x^2} - 1\right)\frac{\cos x}{x} - \frac{3}{x^2}\sin x \tag{D.69}$$

$$n_l(x) = -(-x)^l\left(\frac{1}{x}\frac{d}{dx}\right)^l\left(\frac{\cos x}{x}\right) \tag{D.70}$$

$$h_0^{(1)}(x) = \frac{e^{ix}}{ix}; \qquad h_1^{(1)}(x) = -\left(1 - \frac{1}{ix}\right)\frac{e^{ix}}{x} = \sum_0^\infty i^{n+1}x^{n-2}\frac{n-1}{n!} \tag{D.71}$$

$$h_2^{(1)}(x) = -\left(\frac{1}{i} + \frac{3}{x} - \frac{3}{ix^2}\right)\frac{e^{ix}}{x}; \qquad h_3^{(1)}(x) = \left(1 - \frac{6}{ix} - \frac{15}{x^2} + \frac{15}{ix^3}\right)\frac{e^{ix}}{x} \tag{D.72}$$

$$h_l^{(1)}(x) = -i(-x)^l\left(\frac{1}{x}\frac{d}{dx}\right)^l\left(\frac{e^{ix}}{x}\right) \tag{D.73}$$

For $h_l^{(2)}(x)$, change i to $-i$ in Eqs. (D.71)–(D.73).

$$e^{\pm ikr\cos\theta} = \sum_{l=0}^\infty (\pm i)^l(2l+1)P_l(\cos\theta)j_l(kr) \tag{D.74}$$

D.3.3. Wronskian Relations

[For definition, see Eq. (D.12)]

$$W(j_l, n_l) = \frac{1}{x^2} \tag{D.75}$$

$$W[j_l, h_l^{(1,2)}] = \pm\frac{i}{x^2} \tag{D.76}$$

$$W[h_l^{(1)}, h_l^{(2)}] = -\frac{2i}{x^2} \tag{D.77}$$

D.3.4. Addition Theorems

If

$$R^2 = r^2 + r_0^2 - 2rr_0\cos\gamma, \tag{D.78}$$

we have

$$f_0(k_0R) = \sum_{l=0}^\infty (2l+1)P_l(\cos\gamma)j_l(k_0r)f_l(k_0r_0); \qquad 0 < r < r_0. \tag{D.79}$$

In particular, using Eq. (D.71)

$$\frac{e^{\pm ik_0R}}{R} = \pm ik_0\sum_{l=0}^\infty (2l+1)P_l(\cos\gamma)j_l(k_0r)h_l^{(1,2)}(k_0r_0); \qquad 0 < r < r_0. \tag{D.80}$$

D.3.5. Recurrence and Differential Relations

$$\frac{2l+1}{x} f_l(x) = f_{l-1}(x) + f_{l+1}(x) \tag{D.81}$$

$$\frac{d}{dx} f_l(x) = \frac{l}{2l+1} f_{l-1}(x) - \frac{l+1}{2l+1} f_{l+1}(x) \tag{D.82}$$

$$= \frac{l}{x} f_l(x) - f_{l+1}(x) \tag{D.83}$$

$$= -\frac{l+1}{x} f_l(x) + f_{l-1}(x) \tag{D.84}$$

In particular,

$$\frac{d}{dx} f_0(x) = -f_1(x). \tag{D.85}$$

D.3.6. Asymptotic Expansions

If $|x| \gg 1$,

$$j_l(x) \sim \frac{1}{x} \cos\left[x - (l+1)\frac{\pi}{2}\right] \tag{D.86}$$

$$n_l(x) \sim \frac{1}{x} \sin\left[x - (l+1)\frac{\pi}{2}\right] \tag{D.87}$$

$$h_l^{(1)}(x) \sim \frac{1}{x} (-i)^{l+1} e^{ix} \tag{D.88}$$

$$h_l^{(2)}(x) \sim \frac{1}{x} (i)^{l+1} e^{-ix}. \tag{D.89}$$

D.3.7. Expansions Near the Origin

If $|x| \ll 1$,

$$j_l(x) = 2^l \sum_{n=0}^{\infty} \frac{(-1)^n (l+n)!}{n!(2l+2n+1)!} x^{l+2n} \tag{D.90}$$

$$= \frac{1}{(2l+1)!!}\left[x^l - \frac{1}{2\cdot(2l+3)} x^{l+2} + \frac{1}{2.4\cdot(2l+3)(2l+5)} x^{l+4} + O(x^{l+6})\right],$$

where

$$(2l+1)!! = (2l+1)(2l-1)(2l-3)\cdots 3.1$$

$$\begin{aligned} n_l(x) &= -\frac{1}{2^l}\sum_{n=0}^{\infty}\frac{\Gamma(2l-2n+1)}{\Gamma(n+1)\Gamma(l-n+1)}x^{2n-l-1} \\ &= -(2l-1)!!\left[x^{-l-1}+\frac{1}{2\cdot(2l-1)}x^{-l+1}\right. \\ &\quad \left. + \frac{1}{2.4\cdot(2l-1)(2l-3)}x^{-l+3}+O(x^{-l+5})\right]. \end{aligned} \tag{D.91}$$

D.4. Modified Spherical Bessel Functions

D.4.1. Definition (l integer)

$$i_l(x) = \left(\frac{\pi}{2x}\right)^{1/2} I_{l+1/2}(x) \tag{D.92}$$

$$k_l(x) = \left(\frac{\pi}{2x}\right)^{1/2} K_{l+1/2}(x) \tag{D.93}$$

These functions satisfy the differential equation

$$\frac{d^2f}{dx^2}+\frac{2}{x}\frac{df}{dx}-\left[1+\frac{l(l+1)}{x^2}\right]f = 0. \tag{D.94}$$

D.4.2. General Properties

$$i_l(x) = i^{-l}j_l(ix), \qquad -\pi < \arg x < \frac{\pi}{2} \tag{D.95}$$

$$\begin{aligned} k_l(x) &\doteq -\frac{\pi}{2}i^l h_l^{(1)}(ix); \qquad -\pi < \arg x < \frac{\pi}{2} \\ k_{-l-1}(x) &= k_l(x) \end{aligned} \tag{D.96}$$

$$i_l(x) = \frac{1}{2x}\left[(-1)^{l+1}e^{-x}Q_l\left(\frac{1}{x}\right)+e^{x}Q_l\left(-\frac{1}{x}\right)\right] \tag{D.97}$$

$$k_l(x) = \frac{\pi}{2x}e^{-x}Q_l\left(\frac{1}{x}\right). \tag{D.98}$$

Here $Q_l(u)$ is a polynomial of degree l in u:

$$Q_l(u) = \sum_{p=0}^{l}\frac{(l+p)!}{p!(l-p)!}\left(\frac{u}{2}\right)^p, \qquad Q_{-l-1}(u) = Q_l(u). \tag{D.99}$$

In particular,

$$
\begin{aligned}
Q_0(u) &= 1 \\
Q_1(u) &= 1 + u \\
Q_2(u) &= 1 + 3u + 3u^2 \\
Q_3(u) &= 1 + 6u + 15u^2 + 15u^3
\end{aligned}
\tag{D.100}
$$

D.4.3. Recurrence and Differential Relations

$$(2l + 1)\frac{1}{x}\, i_l(x) = i_{l-1}(x) - i_{l+1}(x) \tag{D.101}$$

$$(2l + 1)\frac{d}{dx}\, i_l(x) = l i_{l-1}(x) + (l + 1) i_{l+1}(x) \tag{D.102}$$

$$(2l + 1)\frac{1}{x}\, k_l(x) = -k_{l-1}(x) + k_{l+1}(x) \tag{D.103}$$

$$(2l + 1)\frac{d}{dx}\, k_l(x) = -l k_{l-1}(x) - (l + 1) k_{l+1}(x) \tag{D.104}$$

D.4.4. Wronskian Relation

$$W(i_l, k_l) = -\frac{\pi}{2x^2} \tag{D.105}$$

D.4.5. Addition Theorem

$$\frac{e^{-\lambda R}}{R} = \frac{2\lambda}{\pi}\sum_{l=0}^{\infty}(2l + 1)k_l(\lambda r_0) i_l(\lambda r) P_l(\cos\gamma), \tag{D.106}$$

where

$$R^2 = r^2 + r_0^2 - 2rr_0\cos\gamma; \qquad 0 < r < r_0.$$

D.4.6. Ancillary Relations

Let

$$K_\nu^{(1)}(z) = e^{\nu\pi i/2} H_\nu^{(1)}(z e^{-\pi i/2}), \tag{D.107}$$

$$K_\nu^{(2)}(z) = -i e^{-\nu\pi i/2} H_\nu^{(2)}(z e^{-\pi i/2}). \tag{D.108}$$

From the known relations

$$H^{(1)}_{-\nu}(z) = e^{\nu\pi i}H^{(1)}_\nu(z), \tag{D.109}$$

$$H^{(2)}_{-\nu}(z) = e^{-\nu\pi i}H^{(2)}_\nu(z), \tag{D.110}$$

we note that the functions $K^{(1)}_\nu$, $K^{(2)}_\nu$ are even in ν. We further have

$$\begin{aligned} I_\nu(z) &= e^{\nu\pi i/2}J_\nu(ze^{-i\pi/2}) \\ &= \frac{1}{2}\,e^{\nu\pi i/2}\{H^{(1)}_\nu(ze^{-i\pi/2}) + H^{(2)}_\nu(ze^{-i\pi/2})\} \\ &= \frac{1}{2}\,\{K^{(1)}_\nu(z) + ie^{\nu\pi i}K^{(2)}_\nu(z)\} \end{aligned} \tag{D.111}$$

and the Wronskians

$$\begin{aligned} W\{K^{(1)}_\nu(z), K^{(2)}_\nu(z)\} &= -\frac{4}{\pi z}, \\ W\{I_\nu(z), K^{(1)}_\nu(z)\} &= \frac{2i}{\pi z}\,e^{\nu\pi i}, \\ W\{I_\nu(z), K^{(2)}_\nu(z)\} &= -\frac{2}{\pi z}. \end{aligned} \tag{D.112}$$

Similarly, for the corresponding spherical functions

$$\begin{aligned} i_l(z) &= \sqrt{\left(\frac{\pi}{2z}\right)}\,I_{l+1/2}(z), \\ k^{(1)}_l(z) &= \sqrt{\left(\frac{\pi}{2z}\right)}\,K^{(1)}_{l+1/2}(z), \\ k^{(2)}_l(z) &= \sqrt{\left(\frac{\pi}{2z}\right)}\,K^{(2)}_{l+1/2}(z), \end{aligned} \tag{D.113}$$

we obtain

$$\begin{aligned} W\{k^{(1)}_l(z), k^{(2)}_l(z)\} &= -\frac{2}{z^2}, \\ W\{i_l(z), k^{(1)}_l(z)\} &= \frac{1}{z^2}\,(-1)^{l+1}, \\ W\{i_l(z), k^{(2)}_l(z)\} &= -\frac{1}{z^2}. \end{aligned} \tag{D.114}$$

D.5. Legendre Functions

D.5.1. Legendre Equation

$$(1 - x^2)\frac{d^2y}{dx^2} - 2x\frac{dy}{dx} + \left[l(l+1) - \frac{m^2}{1-x^2}\right]y = 0 \qquad \text{(D.115)}$$

Putting $x = \cos\theta$, we get

$$\frac{d^2y}{d\theta^2} + \cot\theta\frac{dy}{d\theta} + \left[l(l+1) - \frac{m^2}{\sin^2\theta}\right]y = 0. \qquad \text{(D.116)}$$

We shall assume that l is a positive integer and m is a positive or a negative integer. Further, $-1 \leq x \leq 1$, i.e., $0 \leq \theta \leq \pi$. Two linearly independent solutions of Eq. (D.115) are denoted by $P_l^m(x)$ and $Q_l^m(x)$ and are called the associated Legendre functions of the first and the second kinds of degree l and order m. When $m = 0$, $P_l^m(x)$ is a polynomial of degree l in x:

$$P_l(x) = \frac{(2l)!}{2^l(l!)^2}\left[x^l - \frac{l(l-1)}{2(2l-1)}x^{l-2} + \frac{l(l-1)(l-2)(l-3)}{2.4\cdot(2l-1)(2l-3)}x^{l-4} - \cdots\right] \qquad \text{(D.117)}$$

$$= \frac{(2l)!}{2^{2l}(l!)^2}\left[\cos l\theta + \frac{1}{1}\frac{l}{2l-1}\cos(l-2)\theta + \frac{1\cdot 3}{1\cdot 2}\frac{l(l-1)}{(2l-1)(2l-3)}\right.$$
$$\times \cos(l-4)\theta + \frac{1\cdot 3\cdot 5}{1\cdot 2\cdot 3}\frac{l(l-1)(l-2)}{(2l-1)(2l-3)(2l-5)}$$
$$\left.\times \cos(l-6)\theta + \cdots + \cos(-l)\theta\right]. \qquad \text{(D.118)}$$

D.5.2. General Properties

$$P_l(x) = \frac{1}{2^l l!}\frac{d^l}{dx^l}[(x^2-1)^l]; \quad \text{Rodrigues' formula} \qquad \text{(D.119)}$$

$$P_l^m(x) = (1-x^2)^{m/2}\frac{d^m}{dx^m}P_l(x) = \frac{1}{2^l}\frac{1}{l!}(1-x^2)^{m/2}\frac{d^{l+m}}{dx^{l+m}}(x^2-1)^l; \qquad m \geq 0 \qquad \text{(D.120)[1]}$$

$$P_l^{-m}(x) = (-1)^m\frac{(l-m)!}{(l+m)!}P_l^m(x) \qquad \text{(D.121)}$$

[1] This is Ferrer's definition. Hobson defines the Legendre function of the first kind as $(-1)^m P_l^m(x)$ and denotes it by $P_l^m(x)$ itself. Ferrer's definition yields $P_1^1(\cos\theta) = \sin\theta$, whereas with Hobson's definition, $P_1^1(\cos\theta) = -\sin\theta$.

$$P_l^m(x) = 0; \qquad |m| > l \tag{D.122}$$

$$P_{-l-1}^m(x) = P_l^m(x) \tag{D.123}$$

$$P_l^m(-x) = (-1)^{l-m} P_l^m(x) \tag{D.124}$$

$$P_{2l+m+1}^m(0) = 0; \qquad P_{2l+m}^m(0) = \frac{(-1)^l (2l+2m)!}{2^{2l+m} l!\,(l+m)!} \tag{D.125}$$

$$P_m^m(\cos\theta) = (2m-1)!!\,(\sin\theta)^m \tag{D.126}$$

$$P_l(1) = 1; \qquad P_l(-1) = (-1)^l; \qquad P_l^m(\pm 1) = 0 \quad \text{if} \quad m \neq 0 \tag{D.127}$$

$$P_0(x) = 1 \tag{D.128}$$

$$P_1(x) = x = \cos\theta \tag{D.129}$$

$$P_1^1(x) = (1-x^2)^{1/2} = \sin\theta \tag{D.130}$$

$$P_2(x) = \frac{1}{2}(3x^2 - 1) = \frac{1}{4}(3\cos 2\theta + 1) \tag{D.131}$$

$$P_2^1(x) = 3x(1-x^2)^{1/2} = \frac{3}{2}\sin 2\theta \tag{D.132}$$

$$P_2^2(x) = 3(1-x^2) = \frac{3}{2}(1 - \cos 2\theta) \tag{D.133}$$

D.5.3. Recurrence and Differential Relations

$$(l-m)P_l^m(x) = (2l-1)xP_{l-1}^m(x) - (l+m-1)P_{l-2}^m(x) \tag{D.134}$$

$$= (l+m)xP_{l-1}^m(x) - (1-x^2)^{1/2} P_{l-1}^{m+1}(x) \tag{D.135}$$

$$= lxP_{l-1}^m(x) - (1-x^2)\frac{d}{dx} P_{l-1}^m(x) \tag{D.136}$$

$$(l+m+1)P_l^m(x) = (2l+3)xP_{l+1}^m(x) - (l-m+2)P_{l+2}^m(x) \tag{D.137}$$

$$= (l-m+1)xP_{l+1}^m(x) + (1-x^2)^{1/2} P_{l+1}^{m+1}(x) \tag{D.138}$$

$$= (l+1)xP_{l+1}^m(x) + (1-x^2)\frac{d}{dx} P_{l+1}^m(x) \tag{D.139}$$

$$(1-x^2)^{1/2} P_l^m(x) = 2(m-1)xP_l^{m-1}(x) - (l+m-1)(l-m+2)(1-x^2)^{1/2} P_l^{m-2}(x) \tag{D.140}$$

$$= (l+m)xP_l^{m-1}(x) - (l-m+2)P_{l+1}^{m-1}(x) \tag{D.141}$$

$$= (l+m-1)P_{l-1}^{m-1}(x) - (l-m+1)xP_l^{m-1}(x) \tag{D.142}$$

$$= (m-1)xP_l^{m-1}(x) + (1-x^2)\frac{d}{dx} P_l^{m-1}(x) \tag{D.143}$$

$$(l + m + 1)(l - m)(1 - x^2)^{1/2} P_l^m(x)$$
$$= 2(m + 1)x P_l^{m+1}(x) - (1 - x^2)^{1/2} P_l^{m+2}(x) \tag{D.144}$$
$$= (m + 1)x P_l^{m+1}(x) - (1 - x^2)\frac{d}{dx} P_l^{m+1}(x) \tag{D.145}$$

$$(1 - x^2)\frac{d}{dx} P_l^m(x) = mx P_l^m(x) - (l + m)(l - m + 1)(1 - x^2)^{1/2} P_l^{m-1}(x) \tag{D.146}$$
$$= (1 - x^2)^{1/2} P_l^{m+1}(x) - mx P_l^m(x) \tag{D.147}$$
$$= (l + m) P_{l-1}^m(x) - lx P_l^m(x) \tag{D.148}$$
$$= (l + 1)x P_l^m(x) - (l - m + 1) P_{l+1}^m(x) \tag{D.149}$$

D.5.4. Integral Relations

$$\int_0^\pi P_{2l}(\cos\theta)d\theta = \pi\left[\frac{(2l)!}{2^{2l}(l!)^2}\right]^2$$
$$\int_0^\pi P_{2l+1}(\cos\theta)\cos\theta\, d\theta = \pi\frac{(2l)!(2l+2)!}{[2^{2l+1}l!(l+1)!]^2} \tag{D.150}$$

$$\int_0^\pi P_l^m(\cos\theta)P_{l'}^m(\cos\theta)\sin\theta\, d\theta = \frac{1}{2\pi}\Omega_{ml}\delta_{ll'}; \qquad \text{(orthogonality relation)} \tag{D.151}$$

where

$$\Omega_{ml} = \frac{4\pi}{2l+1}\cdot\frac{(l+m)!}{(l-m)!} \tag{D.152}$$

$$\int_0^\pi P_l^m(\cos\theta)P_l^{m'}(\cos\theta)\frac{d\theta}{\sin\theta} = \frac{1}{m}\frac{(l+m)!}{(l-m)!}\delta_{mm'} \tag{D.153}$$

$$\int_0^\pi \left(\frac{dP_l^m}{d\theta}\frac{dP_{l'}^m}{d\theta} + \frac{m^2}{\sin^2\theta}P_l^m P_{l'}^m\right)\sin\theta\, d\theta = \frac{l(l+1)}{2\pi}\Omega_{ml}\delta_{ll'} \tag{D.154}$$

D.6. Surface Spherical Harmonics

Let $\mathbf{r}(1, \theta, \phi)$ denote the tip of the position vector on a unit sphere. The function

$$Y_{ml}(\theta, \phi) = P_l^m(\cos\theta)e^{im\phi} \tag{D.155}$$

is known as the surface spherical harmonic of order m and degree l. The surface spherical harmonics form a complete orthogonal set of functions on the surface of the sphere. Their orthogonality relation is

$$\int_0^{2\pi}\int_0^\pi Y_{ml}\overset{*}{Y}_{m'l'}\sin\theta\, d\theta\, d\phi = \Omega_{ml}\delta_{mm'}\delta_{ll'}, \tag{D.156}$$

where Ω_{ml} is given in Eq. (D.152). It is sometimes convenient to split the surface harmonics into their real and imaginary parts and denote each part with a special symbol

$$\operatorname{Re}\{Y_{ml}\} = Y^c_{ml} = P^m_l(\cos\theta)\cos m\phi; \qquad \operatorname{Im}\{Y_{ml}\} = Y^s_{ml} = P^m_l(\cos\theta)\sin m\phi. \tag{D.157}$$

The expression

$$Y_l(\theta, \phi) = \sum_{m=0}^{l} (a_m Y^c_{ml} + b_m Y^s_{ml}) \tag{D.158}$$

is called a surface harmonic of degree l.

The polynomial $P_l(x)$ has l distinct zeros in the interval $(-1, +1)$ arranged symmetrically about $x = 0$ so that $P_l(\cos\theta)$ has l zeros between 0 and π, arranged symmetrically about $\theta = \pi/2$. Accordingly, on a sphere with origin at the center, the function $P_l(\cos\theta)$ vanishes on l parallels of latitude that are symmetrically situated with respect to the equator in the northern and southern hemispheres. The sphere is thus divided into $(l + 1)$ zones and the functions $P_l(\cos\theta)$ are called *zonal harmonics*. The function

$$P^{(m)}_l(x) = \frac{d^m}{dx^m} P_l(x) = \frac{1}{2^l l!} \frac{d^{m+l}}{dx^{m+l}} (x^2 - 1)^l \tag{D.159}$$

has $(l - m)$ zeros in the interval $(-1, +1)$ and none elsewhere and therefore vanishes along $(l - m)$ parallels of latitude. The function

$$P^m_l(x) = (1 - x^2)^{m/2} P^{(m)}_l(x) \tag{D.160}$$

has an additional set of zeros of order $m/2$ at each pole. For $l > m > 0$, the spherical harmonic

$$(a_m \cos m\phi + b_m \sin m\phi) P^m_l(\cos\theta) \tag{D.161}$$

vanishes along $2m$ lines of longitude determined by the roots of the equation

$$a_m \cos m\phi + b_m \sin m\phi = 0.$$

The angle between two successive nodal meridians is π/m. Consequently, the sphere is divided into a net of $2m(l - m + 1)$ units, or tesserae. The corresponding functions in Eq. (D.161) are called *tesseral harmonics*.

A special case arises for $l = m$. The corresponding spherical harmonic

$$(2m - 1)!!(a_m \cos m\phi + b_m \sin m\phi)\sin^m\theta \tag{D.162}$$

vanishes at $\theta = 0, \pi$ and on m great circles through these points. The sphere therefore is divided into $2m$ sectors. The corresponding functions in Eq. (D.162) are called *sectoral harmonics.*

The usefulness of the spherical harmonics in the solution of problems of mathematical physics stems mainly from their simple orthogonality relation

$$\int_0^{2\pi} \int_0^{\pi} Y^\sigma_{ml}(\theta, \phi) Y^{\sigma'}_{m'l'}(\theta, \phi) \sin\theta \, d\theta \, d\phi = \frac{1}{\varepsilon_m} \Omega_{ml} \delta_{mm'} \delta_{ll'} \delta_{\sigma\sigma'}, \tag{D.163}$$

where $\sigma, \sigma' = c$ or s and $\varepsilon_m = 1$ $(m = 0)$, $2 (m > 0)$.

Let $\mathbf{r}(1, \theta, \phi)$ and $\mathbf{r}'(1, \theta_0, \phi_0)$ be two points on the unit sphere. The angle γ between their position vectors is such that

$$\cos\gamma = \cos\theta\cos\theta_0 + \sin\theta\sin\theta_0\cos(\phi - \phi_0). \tag{D.164}$$

Assuming an expansion in terms of $Y_{ml}(\theta, \phi)$ and using Eq. (D.163), it is easily seen that

$$\begin{aligned} P_l(\cos\gamma) &= \sum_{m=-l}^{l} \frac{(l-m)!}{(l+m)!} Y_{ml}(\theta, \phi)\overset{*}{Y}_{ml}(\theta_0, \phi_0) \qquad \text{(D.165)} \\ &= P_l(\cos\theta)P_l(\cos\theta_0) + 2\sum_{m=1}^{l} \frac{(l-m)!}{(l+m)!} \\ &\quad \times P_l^m(\cos\theta)P_l^m(\cos\theta_0)\cos m(\phi - \phi_0). \end{aligned}$$

This is the addition theorem for zonal harmonics.

We may apply the same procedure to an arbitrary function $f(\theta, \phi)$, which together with its first and second derivatives are continuous functions of θ and ϕ in the range $0 < \theta < \pi$, $0 < \phi < 2\pi$. It is found that

$$f(\theta, \phi) = \frac{1}{4\pi}\sum_{l=0}^{\infty}(2l + 1)\int_0^{2\pi}\int_0^{\pi} f(\theta_0, \phi_0)P_l(\cos\gamma)\sin\theta_0\, d\theta_0\, d\phi_0. \tag{D.166}$$

APPENDIX E

Asymptotic Evaluation of Special Integrals

E.1. The Method of Laplace

This is applicable to integrals of the form

$$L(s) = \int_0^\infty g(t)e^{-sf(t)}\,dt, \tag{E.1}$$

where $f(t)$, $g(t)$ are real and continuous functions and s is a large positive constant. It is assumed that $f'(t)$ and $f''(t)$ are continuous and that $f(t)$ has a single stationary point $t = t_0$ inside the interval of integration, i.e.,

$$f'(t_0) = 0, \qquad f''(t_0) \neq 0. \tag{E.2}$$

Because s is large, the main contribution to the integral comes from the neighborhood of t_0. Setting $f(t) = \xi$, Eq. (E.1) takes the form

$$\int e^{-s\xi}m(\xi)d\xi,$$

where $m(\xi) = g(t)/f'(t)$. If $m(\xi)$ is nonsingular [i.e., if $f(t)$ has no stationary points], the integral can be shown to have an asymptotic expansion, which is obtained by term by term integration of the series expansion of $m(\xi)$. However, $m(\xi)$ becomes singular at a stationary point of $f(t)$ and the contribution to the integral is then explicitly derivable from a simple expression that we shall now obtain. Because the influence of t_0 is critical, it is sufficient to consider only the immediate neighborhood of t_0. Expanding $f(t)$ in a Taylor series, we have

$$\begin{aligned} f(t) &= f(t_0) + (t - t_0)f'(t_0) + \frac{1}{2}(t - t_0)^2 f''(t_0) + \cdots \\ &= f(t_0) + p^2, \end{aligned} \tag{E.3}$$

where

$$p \simeq (t - t_0)\left[\frac{f''(t_0)}{2}\right]^{1/2} \quad \text{for} \quad f''(t_0) \neq 0. \tag{E.4}$$

Then

$$L(s) = e^{-sf(t_0)} \int g(t)\left[\frac{2}{f''(t_0)}\right]^{1/2} e^{-sp^2}\,dp, \tag{E.5}$$

where $t = t(p)$. Assuming that $g(t)$ varies slowly, we replace $g(t)$ in Eq. (E.5) by $g(t_0)$ and get

$$L(s) = 2e^{-sf(t_0)} \frac{g(t_0)}{\sqrt{[2f''(t_0)]}} \int_{-p_0}^{\infty} e^{-sp^2}\,dp, \tag{E.6}$$

with

$$p_0 = t_0\left[\frac{1}{2}f''(t_0)\right]^{1/2} \neq 0.$$

However,

$$\int_{-p_0}^{\infty} e^{-sp^2}\,dp = \sqrt{\left(\frac{\pi}{s}\right)} - \frac{e^{-p_0^2 s}}{2p_0 s} + \cdots \sim \sqrt{\left(\frac{\pi}{s}\right)}.$$

Hence, from Eq. (E.6), we have *Laplace's formula*

$$\int_0^{\infty} e^{-sf(t)}g(t)dt \sim \left[\frac{2\pi}{sf''(t_0)}\right]^{1/2} g(t_0)e^{-sf(t_0)}. \tag{E.7}$$

E.2. Kelvin's Method of Stationary Phase

This is applicable to integrals of the form

$$K(\lambda) = \int_{-\infty}^{\infty} g(\omega)e^{i\lambda f(\omega)}\,d\omega, \tag{E.8}$$

where $f(\omega)$, $g(\omega)$ are real functions and λ is a large, positive constant. When λ is large, the exponential function $e^{i\lambda f(\omega)}$ will, in general, oscillate very rapidly. If g changes slowly, the rapid phase change will sample successive positive and negative values of $g(\omega)$ such that their sum at each value of ω will tend to zero. In total, the integrand will vanish except around points $\omega = \omega_0$ where $f(\omega)$ is stationary, i.e., $f'(\omega_0) = 0$. The quantitative treatment is as follows: Consider first an interval (a, b) in which $f(\omega)$ has no stationary points. Substituting $u = f(\omega)$, the integral in Eq. (E.8) becomes

$$K_{ab}(\lambda) = \int_{f(a)}^{f(b)} e^{i\lambda u}\left[\frac{g(\omega)}{f'(\omega)}\right]du, \qquad \omega = \omega(u). \tag{E.9}$$

Assume that $f(\omega)$ is monotonically increasing on (a, b) and $\omega(u)$ exists. Integrating Eq. (E.9) by parts, we easily can show that $K_{ab}(\lambda) = O(1/\lambda)$ and this is also true for the sum of all segments that do not contain ω_0.

We now proceed to show that the integral over an interval containing a stationary point is $O(1/\sqrt{\lambda})$. Indeed, expanding both $g(\omega)$ and $f(\omega)$ in their

Taylor series about ω_0 and substituting $\omega - \omega_0 = \xi/\sqrt{\lambda}$, we have, from Eq. (E.8)

$$K(\lambda) = \frac{e^{i\lambda f(\omega_0)}}{\sqrt{\lambda}} g(\omega_0) \int_{-\infty}^{\infty} e^{(1/2)i\xi^2 f''(\omega_0)} \times \left[1 + \frac{g'(\omega_0)}{g(\omega_0)} \frac{\xi}{\sqrt{\lambda}} + \frac{if'''(\omega_0)}{6\sqrt{\lambda}} \xi^3 + O\left(\frac{1}{\lambda}\right)\right] d\xi, \tag{E.10}$$

where we have brought down all the terms from the exponent except $f(\omega_0)$ and $f''(\omega_0)$. Keeping only the leading term in λ and using the Fresnel integral

$$\int_{-\infty}^{\infty} e^{(1/2)if''(\omega_0)\xi^2}\, d\xi = \left[\frac{2\pi}{-if''(\omega_0)}\right]^{1/2} = \left[\frac{2\pi}{|f''(\omega_0)|}\right]^{1/2} e^{(\pi i/4)\operatorname{sgn} f''(\omega_0)}, \tag{E.11}$$

we arrive at *Kelvin's formula*

$$\int_{-\infty}^{\infty} g(\omega) e^{i\lambda f(\omega)}\, d\omega \sim \left[\frac{2\pi}{\lambda |f''(\omega_0)|}\right]^{1/2} g(\omega_0) e^{i\lambda f(\omega_0) + (\pi i/4)\operatorname{sgn} f''(\omega_0)}. \tag{E.12}$$

If there is more than one stationary point, the corresponding sum must be taken.

E.3. The Riemann–Debye Method of Steepest Descents or the "Saddle-Point Method"

This method is basically an extension of Laplace's method to integrals in the complex plane. In the present case, the integral to be asymptotically approximated is given as a contour integral

$$I(\lambda) = \int_C e^{\lambda f(z)} g(z) dz, \qquad (\lambda > 0) \tag{E.13}$$

where $f(z)$, $g(z)$ are analytic functions of z and λ is large. We note first that the integrand is largest where $\text{Re}[f(z)]$ is maximum. Because of the exponential dependence, it might be hoped that a greater part of the integral would come from the neighborhood of this maximum. This, however, is not necessarily so because $\text{Re}[f(z)]$ may not fall rapidly enough from the said maximum and because the imaginary part of $f(z)$ may be rapidly varying, causing rapid oscillations in the integrand that may diminish its value near the point in question.

How, then, should we select a new contour for which these difficulties do not arise? In principle, we should choose it so that near the maximum of $\text{Re}[f(z)]$, $\text{Re}[f(z)]$ decreases as rapidly as possible while, at the same time, $\text{Im}[f(z)]$ varies as little as possible. Fortunately, it is possible to satisfy these conditions simultaneously by virtue of the following theorems in the theory of the functions of a complex variable:

1. If $f(z) = u(x, y) + iv(x, y)$, the curves u = const. and v = const. are orthogonal and u varies most rapidly along v = const. A path on which u = const.

is said to be a level curve of $f(z)$. A path on which $v = \text{const.}$ is a path of *steepest descent.*

2. Neither u nor v can have a maximum or a minimum at any point at which $f(z)$ is analytic.

Now, at points where $f'(z)$ is zero, both u and v will have extreme values. However, by the second theorem above, this extremum can be neither a maximum nor a minimum. It must then be a saddle point such that the function increases for some displacements from the extremum and decreases for others. Using the substitution $w = f(z)$, Eq. (E.13) becomes

$$I(\lambda) = \int_C \left[\frac{g(z)}{f'(z)}\right] e^{\lambda w}\, dw, \qquad z = z(w). \tag{E.14}$$

The contour is deformed in the w plane in order to obtain the most rapid decrease in the absolute value of the integrand with increasing distance from the saddle point. Therefore, the method of steepest descents consists essentially of choosing a path of integration with a particular geometric property. To find the path, we consider the directional derivative in the θ direction given by

$$\frac{\partial}{\partial s} = \cos\theta \frac{\partial}{\partial x} + \sin\theta \frac{\partial}{\partial y},$$

where s is the arc length. The necessary condition that u increases or decreases most rapidly in the direction θ is simply $\partial u/\partial\theta = 0$. In terms of the Cartesian derivatives, it means

$$-\sin\theta \frac{\partial u}{\partial x} + \cos\theta \frac{\partial u}{\partial y} = 0.$$

On using the Cauchy–Riemann equations, we get

$$\frac{\partial v}{\partial s} = 0.$$

If $\partial u/\partial s < 0 (> 0)$ we say that it is a path of steepest descent (ascent) of $f(z)$.

In short: The paths of steepest descent away from the saddle point are simultaneously paths of constant phase, because the gradient vector of a harmonic function $u = \text{Re}[f(z)]$ is tangent to the level curves of the conjugate function $v = \text{Im}[f(z)]$. Note that the slope of the level lines is $(dy/dx)_u = -(u_x/u_y)$ and the slope of the steepest descent paths is $(dy/dx)_v = -(v_x/v_y) = (u_y/u_x) = -1/(dy/dx)_u$. The situation is shown in Fig. E.1. The level curve through the saddle point separates the nearby part of P into valleys below the saddle point and hills above the saddle point. The curves $v = \text{const.}$ are the orthogonal trajectories to the level curves. The projection of the steepest descent curves on the xy plane is shown in Fig. E.1(b).

Having deformed the contour according to the above specifications, we now expand $f(z)$ near z_0 as a convergent power series

$$f(z) = f(z_0) + a_2(z - z_0)^2 + \cdots, \qquad a_2 = \frac{1}{2} f''(z_0). \tag{E.15}$$

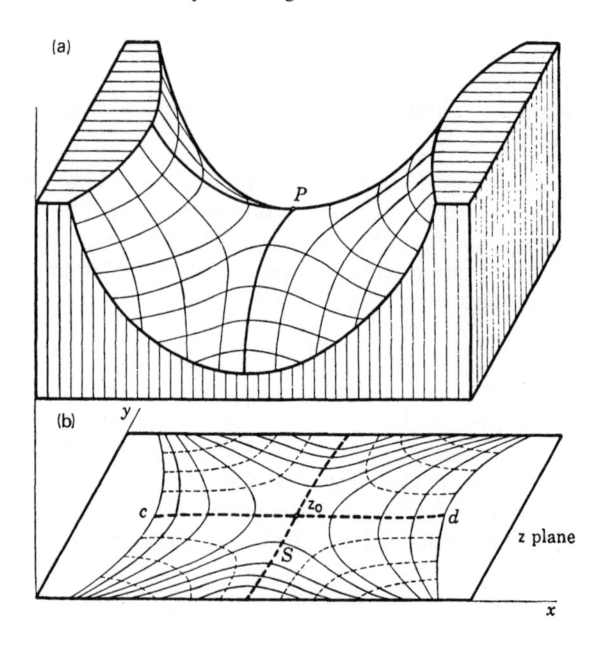

Figure E.1. (a) A saddle point of $w = f(z) = u(x, y) + iv(x, y)$ at P. (b) The curves $u =$ const. (level curves) and $v =$ const. (steepest descents curves) in the (x, y) plane. The projection of P is at z_0.

We choose the path of integration near z_0 to be a straight line on which the second term in the series in Eq. (E.15) is real and negative. This direction is called the critical direction. It is the tangent at z_0 to the two directions of steepest descent from z_0. It can be shown that, in the neighborhood of z_0,

$$g(z) = g(z_0) + O\left(\frac{1}{\lambda^{\varepsilon}}\right), \tag{E.16}$$

$$e^{\lambda f(z)} = e^{\lambda[f(z_0) + a_2(z - z_0)^2]}\left[1 + O\left(\frac{1}{\lambda^{3\varepsilon - 1}}\right)\right],$$

where $\frac{1}{3} < \varepsilon < \frac{1}{2}$. The contribution from the neighborhood $|z - z_0| \leq 1/\lambda^{\varepsilon}$ of the saddle point z_0 is, therefore,

$$I(\lambda) = g(z_0)e^{\lambda f(z_0)} \int e^{\lambda a_2(z - z_0)^2}\, dz\left[1 + O\left(\frac{1}{\lambda^{3\varepsilon - 1}}\right)\right]. \tag{E.17}$$

We now write $a_2 = r_0 e^{i\alpha}$, $z = z_0 + re^{i\theta}$, $r_0 > 0$. Then,

$$a_2(z - z_0)^2 = r_0 r^2 e^{i(\alpha + 2\theta)}, \tag{E.18}$$

which is real and negative when $\theta = \pm\pi/2 - \alpha/2$. This gives two opposite directions of steepest descent from z_0 (Fig. E.1). Choosing the upper sign, we can show that

$$I(\lambda) = g(z_0)e^{\lambda f(z_0)}\left[\frac{-2}{\lambda f''(z_0)}\right]^{1/2} \int_{-\infty}^{\infty} e^{-u^2}\, du\left[1 + O\left(\frac{1}{\lambda^{3\varepsilon-1}}\right)\right], \qquad u^2 = \lambda r_0 r^2. \tag{E.19}$$

Finally, we have

$$I(\lambda) \sim g(z_0)\left[\frac{-2\pi}{\lambda f''(z_0)}\right]^{1/2} e^{\lambda f(z_0)}. \tag{E.20}$$

This approximation holds when there is only one saddle point that is a simple zero of $f'(z)$. If there are several saddle points at the same height, the approximation will consist of the sum of their contributions. If the highest saddle point is a zero of higher order, relation (E.20) must be modified.

Sometimes it is simpler not to take the path at the saddle point in the critical direction, but to take it in a direction making an angle less than 45° with the critical direction. This makes $a_2(z - z_0)^2$ complex with a negative real part and there will be a slight change in the analysis. However, if we take a direction making an angle of 45° with the critical direction, the real part of $a_2(z - z_0)^2$ is zero and we fall back on the method of stationary phase. Therefore, in general

$$I(\lambda) \sim g(z_0)\left[\frac{2\pi}{\lambda |f''(z_0)|}\right]^{1/2} e^{\lambda f(z_0) + i\theta\, \mathrm{sgn}[f''(z_0)]}. \tag{E.21}$$

APPENDIX F

Generalized Functions

Well-behaved functions are, in general, not suitable to model physical situations involving discontinuities, jumps, or singularities. It is desirable, therefore, to generalize the function concept so as to include a whole class of mathematical entities that have the above properties. These are known as *generalized functions* and can be obtained from ordinary functions by limiting processes.

F.1. The Dirac Delta Function

Consider the Cauchy function

$$\delta(x, \alpha) = \frac{\alpha}{\pi(1 + \alpha^2 x^2)}, \tag{F.1}$$

where $\alpha > 0$ is a parameter. For a fixed value of α, the bell-shaped curve for $\delta(x, \alpha)$ has its maximum at $x = 0$ and $\delta(0, \alpha) = \alpha/\pi$. As α increases, the corresponding curve becomes narrower and taller (Fig. F.1). Finally, in the limit as α tends to infinity, the curve reduces to a spike at $x = 0$, because

$$\lim_{\alpha \to \infty} \delta(x, \alpha) = \begin{cases} 0 & (x \neq 0), \\ \infty & (x = 0). \end{cases} \tag{F.2}$$

The area under the curve is equal to

$$\int_{-\infty}^{\infty} \delta(x, \alpha) dx = \int_{-\infty}^{\infty} \frac{\alpha\, dx}{\pi(1 + \alpha^2 x^2)} = 1, \tag{F.3}$$

which is independent of α. Therefore, in the limit, as α tends to infinity, $\delta(x, \alpha)$ becomes zero everywhere except at the origin, where it tends to infinity in such a way that the area under the curve remains unity. We write

$$\lim_{\alpha \to \infty} \delta(x, \alpha) = \delta(x). \tag{F.4}$$

The limit is known as the *Dirac delta function*. This calls for some clarifications and precautions, as follows:

1. The properties of the delta function are independent of the particular form of the auxiliary function $\delta(x, \alpha)$. Simple alternatives are

$$\frac{\alpha}{2} J_\nu(\alpha |x|), \qquad \frac{\alpha}{\sqrt{\pi}} e^{-\alpha^2 x^2}, \qquad \frac{\sin \alpha x}{\pi x}. \tag{F.5}$$

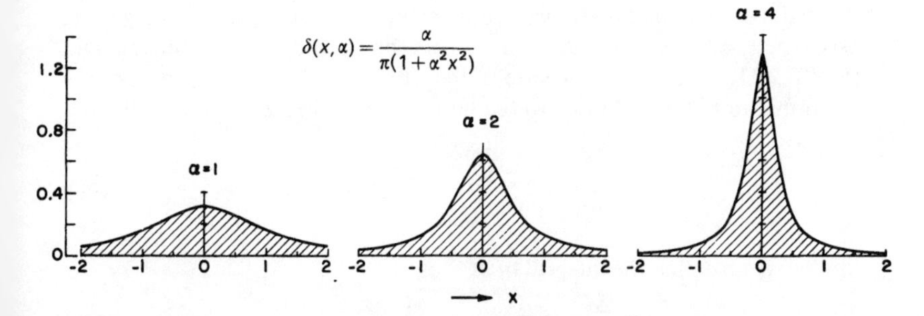

Figure F.1. The Dirac delta function as a limit of the Cauchy family of functions.

2. The auxiliary function $\delta(x, \alpha)$ is continuous, whereas the delta function itself is discontinuous. Therefore, the convergence of $\delta(x, \alpha)$ to $\delta(x)$ is not uniform and the meaning of the derivative of $\delta(x)$ does not follow automatically from the definition of $\delta(x)$. Instead, we must *define* the derivative of the limit to be equal to the limit of the derivative

$$\frac{d}{dx}\,\delta(x) = \lim_{\alpha\to\infty} \frac{d}{dx}\,\delta(x, \alpha). \tag{F.6}$$

Substituting for $\delta(x, \alpha)$ from Eq. (F.1), it is seen that (Fig. F.2)

$$\frac{d}{dx}\,\delta(x) = \lim_{\alpha\to\infty}\left[\frac{-2\alpha^3 x}{\pi(1+\alpha^2x^2)^2}\right]. \tag{F.7}$$

The function inside the brackets posseses a maximum at $x = -1/(\alpha\sqrt{3})$ and a minimum at $x = 1/(\alpha\sqrt{3})$. As α tends to infinity these extrema converge toward

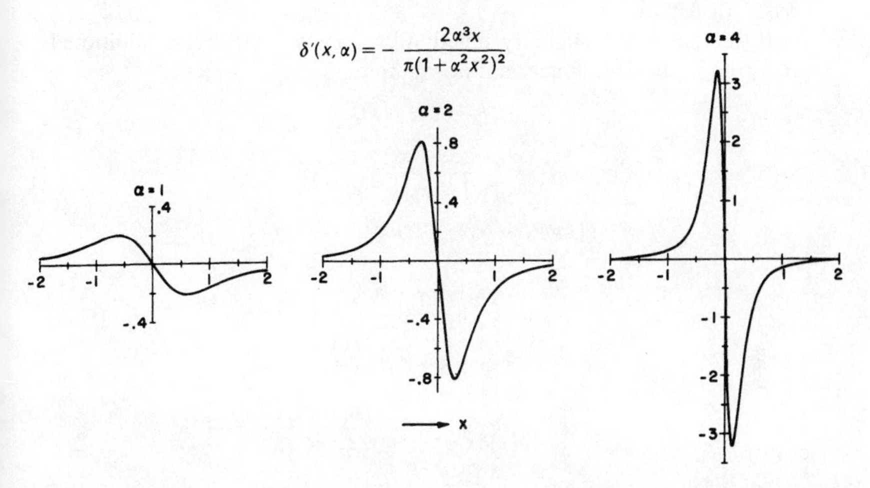

Figure F.2. The derivative of the delta function as a limit of the derivative of the Cauchy family of functions.

the ordinate axis in such a way as to make the function zero everywhere except at $x = 0$ where it tends to $+\infty$ from the left and to $-\infty$ from the right (Fig. F.2). In contrast to the case of the simpler limit of $\delta(x, \alpha)$, we are not permitted to put $x = 0$ prior to the limiting operation $\alpha \to \infty$. Hence,

$$\frac{d}{dx}\delta(x) = \begin{cases} \infty & x = -0 \\ 0 & x \neq 0 \\ -\infty & x = +0. \end{cases} \tag{F.8}$$

Similarly, we define the integral of $\delta(x)$ as

$$\int_a^b \delta(x)dx = \lim_{\alpha\to\infty} \int_a^b \delta(x, \alpha)dx. \tag{F.9}$$

3. The axis of symmetry of $\delta(x, \alpha)$ can be shifted away from $x = 0$ and be placed at an arbitrary point $x = x_0$. In that case

$$\lim_{\alpha\to\infty} \delta(x - x_0, \alpha) = \delta(x - x_0). \tag{F.10}$$

4. Because $\delta(x - x_0)$ vanishes for $x \neq x_0$, one may write symbolically

$$\begin{aligned} \int_a^b \delta(x - x_0)f(x)dx &= f(x_0)\int_a^b \delta(x - x_0)dx \\ &= f(x_0); \qquad a < x_0 < b \\ &= 0; \qquad x_0 < a \quad \text{or} \quad x_0 > b \\ &= \frac{1}{2}f(x_0); \qquad x_0 = a \quad \text{or} \quad x_0 = b \end{aligned} \tag{F.11}$$

which is known as the *sifting property* of the delta function. It is sometimes used as its definition.

5. With the aid of the auxiliary function $\delta(x, \alpha)$, we can derive additional properties of the delta function. For example,

$$\begin{gathered} \delta(-x) = \delta(x), \\ \delta[f(x)] = \sum_k \frac{\delta(x - x_k)}{|f'(x_k)|}; \qquad f(x_k) = 0, \\ f(x)\delta(x - x_0) = f(x_0)\delta(x - x_0), \\ \delta(ax) = \frac{1}{|a|}\delta(x), \\ \delta^{(n)}(x) = (-)^n n!\frac{\delta(x)}{x^n}, \\ \int_{-\infty}^0 \delta(x)dx = \int_0^\infty \delta(x)dx = \frac{1}{2}, \\ \int_a^b f(x)\delta^{(n)}(x - x_0)dx = (-)^n f^{(n)}(x_0), \qquad a < x_0 < b. \end{gathered} \tag{F.12}$$

Equations (F.2) and (F.3) define the one-dimensional delta function. Similarly, we can define the three-dimensional delta function as

$$\delta(\mathbf{r}-\mathbf{r}_0)=\begin{cases}0 & (\mathbf{r}\neq\mathbf{r}_0),\\ \infty & (\mathbf{r}=\mathbf{r}_0),\end{cases}\qquad \int_V \delta(\mathbf{r}-\mathbf{r}_0)dV(\mathbf{r})=1, \tag{F.13}$$

where the point $\mathbf{r}_0$ lies within the volume V. For any continuous function $f(\mathbf{r})$, the sifting property is

$$\int_V \delta(\mathbf{r}-\mathbf{r}_0)f(\mathbf{r})dV=f(\mathbf{r}_0), \tag{F.14}$$

$\mathbf{r}_0$ being within V. Note that $\nabla^2(1/|\mathbf{r}-\mathbf{r}_0|)=-4\pi\delta(\mathbf{r}-\mathbf{r}_0)$.

In orthogonal curvilinear coordinates (q_1, q_2, q_3), we require

$$\iiint \delta(q_1-q_1^0)\delta(q_2-q_2^0)\delta(q_3-q_3^0)dq_1\,dq_2\,dq_3=1. \tag{F.15}$$

Because $dV=h_1h_2h_3\,dq_1\,dq_2\,dq_3$, we have, from Eqs. (F.13) and (F.15),

$$\delta(\mathbf{r}-\mathbf{r}_0)=\frac{1}{h_1h_2h_3}\delta(q_1-q_1^0)\delta(q_2-q_2^0)\delta(q_3-q_3^0). \tag{F.16}$$

In particular,

$$\begin{aligned}\delta(\mathbf{r}-\mathbf{r}_0)&=\delta(x-x_0)\delta(y-y_0)\delta(z-z_0) && \text{Cartesian coordinates;}\\ &=\frac{1}{\Delta}\delta(\Delta-\Delta_0)\delta(\phi-\phi_0)\delta(z-z_0) && \text{circular cylinder coordinates;}\\ &=\frac{1}{r^2\sin\theta}\delta(r-r_0)\delta(\theta-\theta_0)\delta(\phi-\phi_0), && \text{spherical coordinates.}\end{aligned} \tag{F.17}$$

F.1.1. Expansions of the Delta Function

We first show that

$$\frac{1}{\Delta}\delta(\Delta-\Delta_0)\delta(\phi-\phi_0)=\frac{1}{2\pi}\int_0^\infty \sum_{m=-\infty}^{\infty} Y_m(k\Delta,\phi)\overset{*}{Y}_m(k\Delta_0,\phi_0)k\,dk, \tag{F.18}$$

where

$$Y_m(k\Delta,\phi)=J_m(k\Delta)e^{im\phi}.$$

Let

$$\frac{1}{\Delta}\delta(\Delta-\Delta_0)\delta(\phi-\phi_0)=\int_0^\infty \sum_{m=-\infty}^{\infty} D_m(k)Y_m(k\Delta,\phi)k\,dk. \tag{F.19}$$

Multiplying both sides with $\overset{*}{Y}_{m'}(k'\Delta,\phi)$ and integrating over the plane $z=0$, we obtain

$$\overset{*}{Y}_{m'}(k'\Delta_0,\phi_0)=\int_0^\infty \sum_{m=-\infty}^{\infty} D_m(k)\left[\int_0^{2\pi}\int_0^\infty Y_m(k\Delta,\phi)\overset{*}{Y}_{m'}(k'\Delta,\phi)\Delta\,d\Delta\,d\phi\right]k\,dk,$$

where the sifting property of the delta function has been used. The orthogonality relation, Eq. (D.28), now yields

$$\overset{*}{Y}_{m'}(k'\Delta_0, \phi_0) = \int_0^\infty \sum_{m=-\infty}^{\infty} D_m(k)\left[\frac{2\pi}{k}\,\delta_{mm'}\,\delta(k - k')\right]k\,dk,$$

i.e.,

$$D_{m'}(k') = \frac{1}{2\pi}\,\overset{*}{Y}_{m'}(k'\Delta_0, \phi_0).$$

Using this value in Eq. (F.19), we obtain Eq. (F.18).

Recalling the addition theorem

$$J_0(kD) = \sum_{m=-\infty}^{\infty} Y_m(k\Delta, \phi)\overset{*}{Y}_m(k\Delta_0, \phi_0),$$

where (F.20)

$$D^2 = \Delta^2 + \Delta_0^2 - 2\Delta\Delta_0\cos(\phi - \phi_0),$$

Eq. (F.18) reduces to

$$\frac{1}{\Delta}\,\delta(\Delta - \Delta_0)\delta(\phi - \phi_0) = \frac{1}{2\pi}\int_0^\infty J_0(kD)k\,dk,$$

$$\Delta > 0, \qquad \Delta_0 > 0, \qquad 0 < \phi_0 < 2\pi. \tag{F.21}$$

Similarly, assuming

$$\frac{1}{\Delta}\,\delta(\Delta - \Delta_0)\delta(\phi - \phi_0)\mathfrak{I} = \int_0^\infty \sum_{m=-\infty}^{\infty} [\mathbf{P}_m(k\Delta, \phi)\mathbf{D}_1 + \mathbf{B}_m(k\Delta, \phi)\mathbf{D}_2 + \mathbf{C}_m(k\Delta, \phi)\mathbf{D}_3]k\,dk, \tag{F.22}$$

and using the orthogonality relations (2.88)–(2.90), we derive

$$\frac{1}{\Delta}\,\delta(\Delta - \Delta_0)\delta(\phi - \phi_0)\mathfrak{I} = \frac{1}{2\pi}\int_0^\infty \sum_{m=-\infty}^{\infty} [\mathbf{P}_m(k\Delta, \phi)\overset{*}{\mathbf{P}}_m(k\Delta_0, \phi_0) + \mathbf{B}_m(k\Delta, \phi)\overset{*}{\mathbf{B}}_m(k\Delta_0, \phi_0) + \mathbf{C}_m(k\Delta, \phi)\overset{*}{\mathbf{C}}_m(k\Delta_0, \phi_0)]k\,dk. \tag{F.23}$$

Expansions corresponding to Eq. (F.18) and (F.23), in spherical coordinates, are:

$$\frac{\delta(\theta - \theta_0)\delta(\phi - \phi_0)}{\sin\theta} = \sum_{l=0}^{\infty}\sum_{m=-l}^{l}\frac{1}{\Omega_{ml}}\,Y_{ml}(\theta, \phi)\overset{*}{Y}_{ml}(\theta_0, \phi_0), \tag{F.24}$$

$$\frac{\delta(\theta - \theta_0)\delta(\phi - \phi_0)}{\sin\theta}\mathfrak{I} = \sum_{l=0,1,1}^{\infty}\sum_{m=-l}^{l}\frac{1}{\Omega_{ml}}\,[\mathbf{P}_{ml}(\theta, \phi)\overset{*}{\mathbf{P}}_{ml}(\theta_0, \phi_0) + \mathbf{B}_{ml}(\theta, \phi)\overset{*}{\mathbf{B}}_{ml}(\theta_0, \phi_0) + \mathbf{C}_{ml}(\theta, \phi)\overset{*}{\mathbf{C}}_{ml}(\theta_0, \phi)], \tag{F.25}$$

where

$$0 < \theta_0 < \pi, \qquad 0 < \phi_0 < 2\pi, \qquad \Omega_{ml} = \frac{4\pi}{2l+1} \cdot \frac{(l+m)!}{(l-m)!}.$$

We next use the addition theorem

$$P_l(\cos \varepsilon) = \sum_{m=-l}^{l} \frac{(l-m)!}{(l+m)!} Y_{ml}(\theta, \phi) \overset{*}{Y}_{ml}(\theta_0, \phi_0), \tag{F.26}$$

where

$$\cos \varepsilon = \cos \theta \cos \theta_0 + \sin \theta \sin \theta_0 \cos(\phi - \phi_0), \tag{F.27}$$

to reduce Eq. (F.24) to the form

$$\frac{\delta(\theta - \theta_0)\delta(\phi - \phi_0)}{\sin \theta} = \frac{1}{4\pi} \sum_{l=0}^{\infty} (2l+1) P_l(\cos \varepsilon). \tag{F.28}$$

Various interesting results can be derived from Eq. (F.25). Using the relations $\mathfrak{I} \cdot \mathbf{e}_{r_0} = \mathbf{e}_{r_0}$, $\mathbf{B}_{ml}(\theta_0, \phi_0) \cdot \mathbf{e}_{r_0} = \mathbf{C}_{ml}(\theta_0, \phi_0) \cdot \mathbf{e}_{r_0} = 0$, we obtain

$$\begin{aligned} \frac{\delta(\theta - \theta_0)\delta(\phi - \phi_0)}{\sin \theta} \mathbf{e}_{r_0} &= \sum_{l=0}^{\infty} \sum_{m=-l}^{l} \frac{1}{\Omega_{ml}} \mathbf{P}_{ml}(\theta, \phi) \overset{*}{Y}_{ml}(\theta_0, \phi_0) \\ &= \frac{1}{4\pi} \sum_{l=0}^{\infty} (2l+1) P_l(\cos \varepsilon) \mathbf{e}_r, \end{aligned} \tag{F.29}$$

using Eq. (F.26). Similarly,

$$\begin{aligned} \frac{\delta(\theta - \theta_0)\delta(\phi - \phi_0)}{\sin \theta} \mathbf{e}_{\theta_0} &= \sum_{l=1}^{\infty} \sum_{m=-l}^{l} \frac{1}{\sqrt{[l(l+1)]}\Omega_{ml}} \left[\mathbf{B}_{ml}(\theta, \phi) \frac{\partial}{\partial \theta_0} \overset{*}{Y}_{ml}(\theta_0, \phi_0) \right. \\ &\qquad \left. - \mathbf{C}_{ml}(\theta, \phi) \frac{1}{\sin \theta_0} \frac{\partial}{\partial \phi_0} \overset{*}{Y}_{ml}(\theta_0, \phi_0) \right] \\ &= \frac{1}{4\pi} \sum_{l=1}^{\infty} \frac{2l+1}{l(l+1)} \left[\left(\mathbf{e}_\theta \frac{\partial}{\partial \theta} + \mathbf{e}_\phi \frac{1}{\sin \theta} \frac{\partial}{\partial \phi} \right) \frac{\partial}{\partial \theta_0} \right. \\ &\qquad \left. + \left(\mathbf{e}_\theta \frac{1}{\sin \theta} \frac{\partial}{\partial \phi} - \mathbf{e}_\phi \frac{\partial}{\partial \theta} \right) \frac{1}{\sin \theta_0} \frac{\partial}{\partial \phi_0} \right] P_l(\cos \varepsilon), \end{aligned} \tag{F.30}$$

and

$$\begin{aligned} \frac{\delta(\theta - \theta_0)\delta(\phi - \phi_0)}{\sin \theta} \mathbf{e}_{\phi_0} &= \sum_{l=1}^{\infty} \sum_{m=-l}^{l} \frac{1}{\sqrt{[l(l+1)]}\Omega_{ml}} \left[\mathbf{B}_{ml}(\theta, \phi) \frac{1}{\sin \theta_0} \right. \\ &\qquad \left. \times \frac{\partial}{\partial \phi_0} \overset{*}{Y}_{ml}(\theta_0, \phi_0) - \mathbf{C}_{ml}(\theta, \phi) \frac{\partial}{\partial \theta_0} \overset{*}{Y}_{ml}(\theta_0, \phi_0) \right] \\ &= \frac{1}{4\pi} \sum_{l=1}^{\infty} \frac{2l+1}{l(l+1)} \left[\left(\mathbf{e}_\theta \frac{\partial}{\partial \theta} + \mathbf{e}_\phi \frac{1}{\sin \theta} \frac{\partial}{\partial \phi} \right) \frac{1}{\sin \theta_0} \frac{\partial}{\partial \phi_0} \right. \\ &\qquad \left. - \left(\mathbf{e}_\theta \frac{1}{\sin \theta} \frac{\partial}{\partial \phi} - \mathbf{e}_\phi \frac{\partial}{\partial \theta} \right) \frac{\partial}{\partial \theta_0} \right] P_l(\cos \varepsilon). \end{aligned} \tag{F.31}$$

Relation (F.27) reveals that ε, θ, and θ_0 may be taken as the sides of a spherical triangle as shown in Fig. 6.6. We denote the angles of this triangle by $A = \phi - \phi_0$, B, and C, respectively. Then using Eqs. (6.130), Eqs. (F.30) and (F.31) can be reduced to

$$\frac{\delta(\theta-\theta_0)\delta(\phi-\phi_0)}{\sin\theta}\mathbf{e}_{\theta_0} = \frac{1}{4\pi}\sum_{l=1}^{\infty}\frac{2l+1}{l(l+1)}\left[-\left(\mathbf{e}_\theta\frac{\partial}{\partial\theta}+\mathbf{e}_\phi\frac{1}{\sin\theta}\frac{\partial}{\partial\phi}\right)\right.$$
$$\left.\times\left\{\cos BP_l^1(\cos\varepsilon)\right\}+\left(\mathbf{e}_\theta\frac{1}{\sin\theta}\frac{\partial}{\partial\phi}-\mathbf{e}_\phi\frac{\partial}{\partial\theta}\right)\left\{\sin BP_l^1(\cos\varepsilon)\right\}\right] \quad \text{(F.32)}$$

and

$$\frac{\delta(\theta-\theta_0)\delta(\phi-\phi_0)}{\sin\theta}\mathbf{e}_{\phi_0} = \frac{1}{4\pi}\sum_{l=1}^{\infty}\frac{2l+1}{l(l+1)}\left[\left(\mathbf{e}_\theta\frac{\partial}{\partial\theta}+\mathbf{e}_\phi\frac{1}{\sin\theta}\frac{\partial}{\partial\phi}\right)\right.$$
$$\left.\times\left\{\sin BP_l^1(\cos\varepsilon)\right\}+\left(\mathbf{e}_\theta\frac{1}{\sin\theta}\frac{\partial}{\partial\phi}-\mathbf{e}_\phi\frac{\partial}{\partial\theta}\right)\left\{\cos BP_l^1(\cos\varepsilon)\right\}\right]. \quad \text{(F.33)}$$

The expansions obtained so far are valid for arbitrary values of θ_0 and ϕ_0. From Fig. 6.6, we note that $\varepsilon \to \theta$ as $\theta_0 \to 0$. Equations (F.28) and (F.29), therefore, yield

$$\frac{\delta(\theta)\delta(\phi-\phi_0)}{\sin\theta} = \frac{1}{4\pi}\sum_{l=0}^{\infty}(2l+1)P_l(\cos\theta), \quad \text{(F.34)}$$

$$\frac{\delta(\theta)\delta(\phi-\phi_0)}{\sin\theta}\mathbf{e}_{r_0} = \frac{1}{4\pi}\sum_{l=0}^{\infty}(2l+1)\mathbf{P}_{0,l}(\theta,\phi). \quad \text{(F.35)}$$

Integrating Eqs. (F.34) and (F.35) with respect to ϕ over $(0, 2\pi)$, we get

$$2\frac{\delta(\theta)}{\sin\theta} = \sum_{l=0}^{\infty}(2l+1)P_l(\cos\theta) = 2\delta(1-\cos\theta), \quad \text{(F.36)}$$

$$2\frac{\delta(\theta)}{\sin\theta}\mathbf{e}_{r_0} = \sum_{l=0}^{\infty}(2l+1)\mathbf{P}_{0,l}(\theta,\phi). \quad \text{(F.37)}$$

Further, as $\theta_0 \to 0$, $\phi_0 \to 0$, we note that $\varepsilon \to \theta$ and $B \to \pi - \phi$. Hence Eqs. (F.32) and (F.33) become

$$\frac{\delta(\theta)\delta(\phi)}{\sin\theta}\mathbf{e}_{\theta_0} = \frac{1}{4\pi}\sum_{l=1}^{\infty}\frac{2l+1}{\sqrt{[l(l+1)]}}[\mathbf{B}^c_{1,l}(\theta,\phi)+\mathbf{C}^s_{1,l}(\theta,\phi)], \quad \text{(F.38)}$$

$$\frac{\delta(\theta)\delta(\phi)}{\sin\theta}\mathbf{e}_{\phi_0} = \frac{1}{4\pi}\sum_{l=1}^{\infty}\frac{2l+1}{\sqrt{[l(l+1)]}}[\mathbf{B}^s_{1,l}(\theta,\phi)-\mathbf{C}^c_{1,l}(\theta,\phi)]. \quad \text{(F.39)}$$

F.2. Discontinuous Functions Related to the Delta Function

F.2.1. The Unit-Step Function

Consider the auxiliary function

$$U(x, \alpha) = \frac{1}{2} + \frac{1}{\pi} \tan^{-1}(\alpha x). \tag{F.40}$$

We have (Fig. F.3)

$$U(x) = \lim_{\alpha \to \infty} U(x, \alpha) = \begin{cases} 0 & (x < 0), \\ \frac{1}{2} & (x = 0), \\ 1 & (x > 0). \end{cases}$$

The function $U(x)$ is known as the *unit-step function.* Because, from Eq. (F.40),

$$\frac{d}{dx} U(x, \alpha) = \frac{\alpha}{\pi(1 + \alpha^2 x^2)} = \delta(x, \alpha),$$

we have

$$\delta(x) = \lim_{\alpha \to \infty} \delta(x, \alpha) = \lim_{\alpha \to \infty} \frac{d}{dx} U(x, \alpha). \tag{F.41}$$

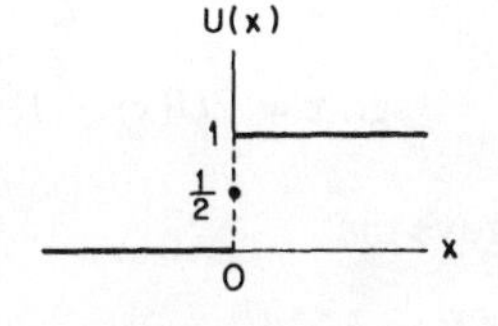

(a) Unit-step function

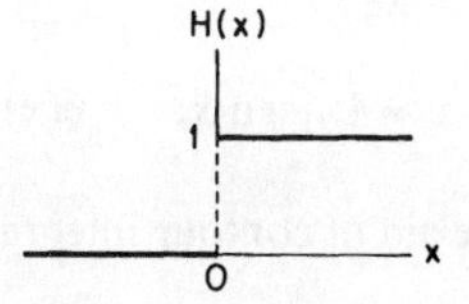

(b) Heaviside unit-step function

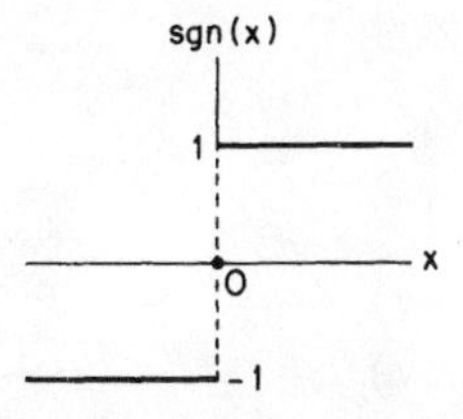

(c) The signum function

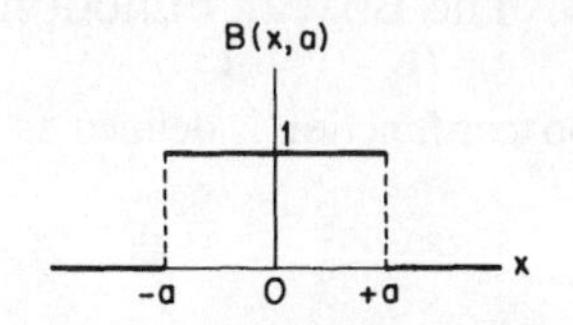

(d) The boxcar function

Figure F.3. The unit-step function and related functions.

If we define

$$\frac{d}{dx} U(x) = \lim_{\alpha \to \infty} \frac{d}{dx} U(x, \alpha),$$

we have the symbolic relation

$$\delta(x) = \frac{d}{dx} U(x). \tag{F.42}$$

F.2.2. The Heaviside Unit-Step Function

It is defined as

$$H(x) = \begin{cases} 0 & (x < 0), \\ 1 & (x > 0). \end{cases} \tag{F.43}$$

The value of the function at $x = 0$ is left undefined.

F.2.3. The Signum Function

The signum function is defined as

$$\operatorname{sgn} x = \lim_{\alpha \to \infty} \left[\frac{2}{\pi} \tan^{-1}(\alpha x) \right] = \begin{cases} -1 & (x < 0), \\ 0 & (x = 0), \\ 1 & (x > 0). \end{cases} \tag{F.44}$$

Clearly,

$$x = |x| \operatorname{sgn} x, \qquad \delta(x) = \frac{1}{2} \frac{d}{dx} \operatorname{sgn} x, \qquad \operatorname{sgn} x = 2U(x) - 1. \tag{F.45}$$

With the aid of contour integration, we can prove that

$$\operatorname{sgn} x = \frac{1}{\pi} \int_{-\infty}^{\infty} \frac{\sin sx}{s} \, ds. \tag{F.46}$$

F.2.4. The Boxcar Function

The boxcar function is defined as

$$B(x, a) = \begin{cases} 0 & (|x| > a), \\ 1 & (|x| < a). \end{cases} \tag{F.47}$$

Obviously,

$$B(x, a) = H(x + a) - H(x - a).$$

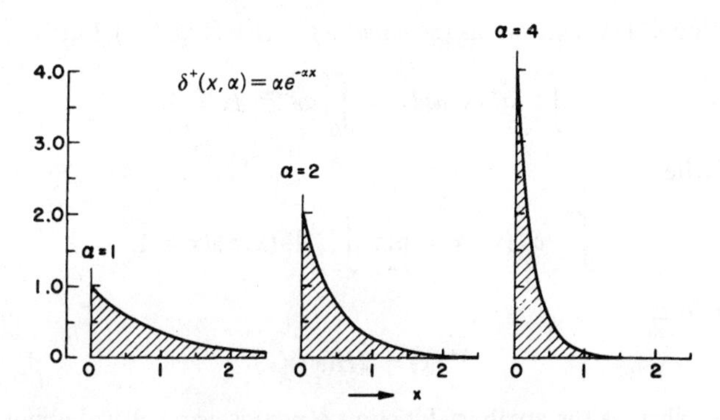

Figure F.4. The impulse function.

F.2.5. The Unit Gate Function

We define

$$G_a(x) = \begin{cases} 0, & x < -\dfrac{a}{2}, \quad x > \dfrac{a}{2} \\ \dfrac{1}{2}, & x = \pm\dfrac{a}{2} \\ 1, & -\dfrac{a}{2} < x < \dfrac{a}{2} \end{cases} \tag{F.48}$$

or, alternatively,

$$G_a(x) = \left[U\left(x + \frac{a}{2}\right) - U\left(x - \frac{a}{2}\right)\right].$$

We also can show that

$$G_a(x) = \frac{2}{\pi}\int_0^{\infty} \sin\left(\frac{\omega a}{2}\right)\cos(\omega x)\frac{d\omega}{\omega} = \frac{1}{\pi}\int_{-\infty}^{\infty} e^{i\omega x}\frac{\sin(\omega a/2)}{\omega}\,d\omega. \tag{F.49}$$

F.2.6. The Impulse Function

Consider the auxiliary function

$$\delta^+(x,\alpha) = \begin{cases} \alpha e^{-\alpha x} & (x \geq 0), \\ 0 & (x < 0), \end{cases} \tag{F.50}$$

where $\alpha > 0$ is a parameter. We have (Fig. F.4)

$$\delta^+(x) = \lim_{\alpha\to\infty} \delta^+(x,\alpha) = \begin{cases} 0 & (x \neq 0), \\ \infty & (x = 0). \end{cases} \tag{F.51}$$

The function $\delta^+(x)$ is known as the *impulse function* (Fig. F.4). Further, because

$$\int_{-\infty}^{\infty} \delta^+(x, \alpha)dx = \int_0^{\infty} \alpha e^{-\alpha x}\, dx = 1,$$

we may write

$$\int_{-\infty}^{\infty} \delta^+(x)dx = \lim_{\alpha\to\infty} \int_{-\infty}^{\infty} \delta^+(x, \alpha)dx = 1,$$

and, therefore,

$$\delta^+(x) = 2H(x)\delta(x). \tag{F.52}$$

Note that whereas the auxiliary function $\delta(x, \alpha)$ is symmetrical about the line $x = 0$, $\delta^+(x, \alpha)$ is zero for $x < 0$. Because of this property, $\delta^+(x)$ is sometimes more useful in physical applications than $\delta(x)$.

F.3. The Allied Function

We start from the Fourier spectral representation of $f(t)$

$$f(t) = \frac{1}{2\pi}\int_{-\infty}^{\infty} F(\omega)e^{i\omega t}\, d\omega = \frac{1}{\pi}\int_0^{\infty} d\omega \int_{-\infty}^{\infty} f(\tau)\cos\omega(\tau - t)d\tau \tag{F.53}$$

and define an associated function $f_\varepsilon(t)$ as

$$f_\varepsilon(t) = \frac{1}{2\pi}\int_{-\infty}^{\infty} F(\omega)e^{[i\omega t + 2i\varepsilon\,\mathrm{sgn}\,\omega]}\, d\omega. \tag{F.54}$$

Note that the spectral amplitudes of $f(t)$ and $f_\varepsilon(t)$ are the same. Their spectra differ, however, by a phase function that is equal to 2ε for positive frequencies and to (-2ε) for negative frequencies.

We next rewrite $f_\varepsilon(t)$ in the form

$$f_\varepsilon(t) = \frac{1}{2\pi}\int_0^{\infty} F(\omega)e^{i(\omega t + 2\varepsilon)}\, d\omega + \frac{1}{2\pi}\int_0^{\infty} F(-\omega)e^{-i(\omega t + 2\varepsilon)}\, d\omega \tag{F.55}$$

and substitute therein

$$F(\omega) = \int_{-\infty}^{\infty} f(\tau)e^{-i\omega\tau}\, d\tau.$$

After some simple algebraic steps, we arrive at the relation

$$f_\varepsilon(t) = (\cos 2\varepsilon)f(t) + (\sin 2\varepsilon)\hat{f}(t) = \mathrm{Re}\{e^{2i\varepsilon}(f - i\hat{f})\}, \tag{F.56}$$

where

$$\hat{f}(t) = \frac{1}{\pi}\int_0^{\infty} d\omega \int_{-\infty}^{\infty} f(\tau)\sin\omega(\tau - t)d\tau \tag{F.57}$$

is known as the *allied function* of $f(t)$. The structure of $\hat{f}(t)$ is similar to that of $f(t)$ except that the sine function replaces the cosine function inside the double integral. Nevertheless, $\hat{f}(t)$ is a new integral transform of $f(t)$. In fact, $\hat{f}(t)$ is the *Hilbert transform* of $f(t)$. To prove this we define

$$\hat{f}(t, A) = \frac{1}{\pi}\int_0^A d\omega \int_{-\infty}^{\infty} f(\tau)\sin\omega(\tau - t)d\tau = \frac{1}{\pi}\int_{-\infty}^{\infty} f(\tau)d\tau \int_0^A \sin\omega(\tau - t)d\omega. \tag{F.58}$$

On the assumptions that $f(t)$ is of bounded variation, $|f(t)|$ is integrable from $-\infty$ to $+\infty$ and $f(t)$ satisfies the Lipschitz condition[1] for all values of t in the interval, we have

$$\hat{f}(t) = \lim_{A\to\infty}\left\{\frac{1}{\pi}\int_{-\infty}^{\infty} f(\tau)\frac{1 - \cos A(\tau - t)}{\tau - t}d\tau\right\} = \frac{1}{\pi}\,\mathrm{P}\int_{-\infty}^{\infty}\frac{f(\tau)}{\tau - t}d\tau. \tag{F.59}$$

This transform represents $\hat{f}(t)$ as the convolution of $f(t)$ with $(-1/\pi t)$ and is known as the Hilbert transform of $f(t)$. The inverse transform is given by

$$f(t) = -\frac{1}{\pi}\,\mathrm{P}\int_{-\infty}^{\infty}\frac{\hat{f}(\tau)d\tau}{\tau - t}. \tag{F.60}$$

For example, the allied function to $H(t - t_0)$ is

$$\frac{1}{\pi}\,\mathrm{P}\int_{t_0}^{\infty}\frac{d\tau}{\tau - t} = -\frac{1}{\pi}\ln|t - t_0|. \tag{F.61}$$

Likewise, the allied function of $\delta(t)$ is

$$\hat{\delta}(t) = -\frac{1}{\pi}\int_0^{\infty}\sin\omega t\,d\omega = \begin{cases}-1/\pi t, & t \neq 0\\ 0, & t = 0.\end{cases} \tag{F.62}$$

In conclusion

$$\begin{aligned} f(t) &= \frac{1}{2\pi}\int_{-\infty}^{\infty} F(\omega)e^{i\omega t}\,d\omega = -\frac{1}{\pi}\,\mathrm{P}\int_{-\infty}^{\infty}\frac{\hat{f}(\tau)d\tau}{\tau - t},\\ \hat{f}(t) &= \frac{1}{2\pi}\int_{-\infty}^{\infty}\hat{F}(\omega)e^{i\omega t}\,d\omega = \frac{1}{2\pi}\int_{-\infty}^{\infty} F(\omega)e^{[i\omega t + (\pi i/2)\operatorname{sgn}\omega]}\,d\omega \\ &= \frac{1}{\pi}\,\mathrm{P}\int_{-\infty}^{\infty}\frac{f(\tau)d\tau}{\tau - t}. \end{aligned} \tag{F.63}$$

Observe that because

$$\begin{aligned} f(t) - i\hat{f}(t) &= \frac{1}{2\pi}\int_{-\infty}^{\infty} F(\omega)e^{i\omega t}[1 - ie^{(\pi i/2)\operatorname{sgn}\omega}]d\omega \\ &= \frac{1}{\pi}\int_0^{\infty} F(\omega)e^{i\omega t}\,d\omega, \end{aligned} \tag{F.64}$$

[1] $|f(\tau) - f(t)| \leq M|\tau - t|^{\alpha}$, $0 < \alpha \leq 1$ for some M and α.

we obtain the useful relation for any real $f(t)$

$$\frac{1}{\pi}\int_0^\infty F(\omega)e^{i\omega t}\,d\omega = f(t) - \frac{i}{\pi}\,\mathrm{P}\int_{-\infty}^{\infty}\frac{f(\tau)d\tau}{\tau - t}.$$

Therefore,

$$\begin{aligned} f(t) &= \mathrm{Re}\left\{\frac{1}{\pi}\int_0^\infty F(\omega)e^{i\omega t}\,d\omega\right\},\\ \hat{f}(t) &= \mathrm{Im}\left\{-\frac{1}{\pi}\int_0^\infty F(\omega)e^{i\omega t}\,d\omega\right\}. \end{aligned} \tag{F.65}$$

It is sometimes convenient to split $F(\omega)$ into its real and imaginary parts, $(1/\pi)F(\omega) = a(\omega) - ib(\omega)$. Equations (F.65) will then yield

$$\begin{aligned} f(t) &= \int_0^\infty [a(\omega)\cos\omega t + b(\omega)\sin\omega t]d\omega,\\ \hat{f}(t) &= -\int_0^\infty [a(\omega)\sin\omega t - b(\omega)\cos\omega t]d\omega, \end{aligned} \tag{F.66}$$

where

$$a(\omega) = \frac{1}{\pi}\int_{-\infty}^{\infty} f(\tau)\cos\omega\tau\,d\tau, \qquad b(\omega) = \frac{1}{\pi}\int_{-\infty}^{\infty} f(\tau)\sin\omega\tau\,d\tau.$$

Table F.1 gives some useful time functions and their allied functions.

Table F.1. Some Allied Functions (Hilbert Transforms)[a]

$f(t)$	$\hat{f}(t) = \frac{1}{\pi}\,\mathrm{P}\int_{-\infty}^{\infty}\frac{f(x)}{x-t}\,dx$
$\delta(t)$	$-\frac{1}{\pi t}\quad [\hat{f}(0) = 0]$
$H(t - t_0)$	$-\frac{1}{\pi}\ln\lvert t - t_0\rvert$
$H(t-b) - H(t-a)$	$\frac{1}{\pi}\ln\left\lvert\frac{t-a}{t-b}\right\rvert$
$\frac{1}{t}H(t-a),\quad a > 0$	$\frac{1}{\pi t}\ln\left\lvert\frac{a}{t-a}\right\rvert\quad (t \neq a, t \neq 0)$
$\frac{1}{t^2}H(t-a),\quad a > 0$	$\frac{1}{\pi t^2}\ln\left\lvert\frac{a}{t-a}\right\rvert - \frac{1}{\pi a t}\quad (t \neq a, t \neq 0)$
$t^{\nu-1}H(t),\quad 0 < \mathrm{Re}\,\nu < 1$	$\begin{cases}\frac{(-t)^{\nu-1}}{\sin\pi\nu}, & -\infty < t < 0\\ [-\mathrm{ctg}\,\pi\nu]t^{\nu-1}, & 0 < t < \infty\end{cases}$
$\sin(a\sqrt{t})H(t),\quad a > 0$	$\begin{cases}e^{-a\sqrt{\lvert t\rvert}}, & -\infty < t < 0\\ \cos(a\sqrt{t}), & 0 < t < +\infty\end{cases}$

$\{H(t+1) - H(t-1)\}P_n(t), \quad n = 0, 1, 2, \ldots$	$-\frac{2}{\pi} Q_n(t) \quad (t \neq 1, -1)$
$\cos \omega t, \quad \omega > 0$	$-\sin \omega t$
$\sin \omega t, \quad \omega > 0$	$\cos \omega t$
$\frac{\sin \omega t}{\omega t}, \quad \omega > 0$	$-\frac{\sin^2(\omega t/2)}{\omega t/2}$
$\frac{t}{a^2 + t^2}, \quad \operatorname{Re} a > 0$	$\frac{a}{a^2 + t^2}$
$\frac{a}{a^2 + t^2}, \quad \operatorname{Re} a > 0$	$-\frac{t}{a^2 + t^2}$
$\lvert t\rvert^{\nu/2} J_\nu(a\lvert t\rvert^{1/2}), \quad a > 0, \quad -1 < \operatorname{Re} \nu < \frac{3}{2}$	$-\operatorname{sgn} t \lvert t\rvert^{\nu/2}\left[\frac{2}{\pi} K_\nu(a\lvert t\rvert^{1/2}) + N_\nu(a\lvert t\rvert^{1/2})\right]$
$\lvert t\rvert^\nu J_\nu(\lvert t\rvert) \operatorname{sgn} t, \quad -\frac{1}{2} < \operatorname{Re} \nu < \frac{3}{2}$	$-\lvert t\rvert^\nu N_\nu(\lvert t\rvert)$
$t^{\nu/2} J_\nu(a\sqrt{t}) H(t), \quad a > 0, \quad -1 < \operatorname{Re} \nu < \frac{3}{2}$	$\begin{cases} \frac{2}{\pi}(-t)^{\nu/2} K_\nu\{a\sqrt{(-t)}\}, & -\infty < t < 0 \\ -t^{\nu/2} N_\nu(a\sqrt{t}), & 0 < t < \infty \end{cases}$
$e^{-at} I_0(at) H(t), \quad a > 0$	$\frac{1}{\pi} e^{-at} K_0(a\lvert t\rvert)$
$e^{-b\lvert t\rvert}, \quad b > 0$	$\begin{cases} \frac{1}{\pi} \operatorname{sgn} t[e^{b\lvert t\rvert} Ei(-b\lvert t\rvert) \\ -e^{-b\lvert t\rvert} \bar{E}i(b\lvert t\rvert)] \end{cases}$
$\operatorname{sgn} t \lvert t\rvert^{s-1}, \quad 0 < \operatorname{Re} s < 1$	$\tan\left(\frac{\pi s}{2}\right)\lvert t\rvert^{s-1}$
$ci(a\lvert t\rvert), \quad a > 0$	$si(a\lvert t\rvert) \operatorname{sgn} t$
$si(a\lvert t\rvert) \operatorname{sgn} t, \quad a > 0$	$-ci(a\lvert t\rvert)$
$\frac{T_n(t)}{\sqrt{(1 - t^2)}} B(t)$	$U_{n-1}(t), \quad -1 < t < 1$
$\sqrt{(1 - t^2)} U_n(t) B(t)$	$-T_{n+1}(t), \quad -1 < t < 1$
$(1 - t)^\alpha (1 + t)^\beta P_n^{(\alpha, \beta)}(t) B(t), \quad \operatorname{Re} \alpha > -1, \ \operatorname{Re} \beta > -1$	$\frac{-2}{\pi}(1 + t)^\beta Q_n^{(\alpha, \beta)}(t) \times \begin{cases} (1 - t)^\alpha, & -1 < t < 1 \\ (t - 1)^\alpha, & \text{otherwise} \end{cases}$
$\sin(at) J_n(bt), \quad a > b > 0$	$\cos(at) J_n(bt)$
$\cos(at) J_n(bt), \quad a > b > 0$	$-\sin(at) J_n(bt)$

[a] $B(t) = 1$, if $-1 < t < 1$; $B(t) = 0$ otherwise. $T_n(t)$, $U_n(t)$ are the Chebyshev polynomials of the first and second kinds, respectively. $si(t) = Si(t) - \pi/2$, $ci(t) = Ci(t)$ [see Eq. (10.246)]. $Ei(-x)$ is defined in Eq. (10.279). $\bar{E}i(x) = \frac{1}{2}[Ei(x + i0) + Ei(x - i0)]$, $x > 0$.

F.3.1. Properties of $f(t)$ and $\hat{f}(t)$

Note that

$$\int_{-\infty}^{\infty} f(t)\hat{f}(t)dt = \int_{-\infty}^{\infty} f(t)\left[\frac{1}{\pi} \mathrm{P} \int_{-\infty}^{\infty} \frac{f(\tau)}{\tau - t} d\tau\right]dt$$

$$= \int_{-\infty}^{\infty} f(\tau)\left[\frac{1}{\pi} \mathrm{P} \int_{-\infty}^{\infty} \frac{f(t)}{\tau - t} dt\right]d\tau \quad \text{(changing orders of integration)}$$

$$= \int_{-\infty}^{\infty} f(\tau)\left[\frac{1}{\pi} \mathrm{P} \int_{-\infty}^{\infty} \frac{f(t)}{t - \tau} dt\right]d\tau \quad \text{(interchanging } t \text{ and } \tau \text{ in the first equation)}$$

$$= 0 \quad \text{(compatibility condition of the second and third equations).} \tag{F.67}$$

Therefore, $f(t)$ and $\hat{f}(t)$ are "orthogonal."

We also can prove easily that each function pair $f(t)$ and $\hat{f}(t)$ has the same energy

$$\int_{-\infty}^{\infty} f^2(t)dt = \int_{-\infty}^{\infty} \hat{f}^2(t)dt \tag{F.68}$$

provided that these integrals exist and are finite. A third important property of the pair is as follows: If $f(t)$ is band limited $[F(\omega) = 0$ for $|\omega| > \omega_c]$ then $\hat{f}(t)$ is also limited to the same band. The reader may also prove without difficulty that:

1. $$\hat{f}(t) = \frac{1}{\pi}\int_0^{\infty} \frac{f(t+x) - f(t-x)}{x} dx,$$

$$f(t) = -\frac{1}{\pi}\int_0^{\infty} \frac{\hat{f}(t+x) - \hat{f}(t-x)}{x} dx. \tag{F.69}$$

2. If $f(t)$ is a square integrable in $(-\infty, \infty)$,

$$\hat{f}(t) = -\frac{1}{\pi}\frac{d}{dt}\int_{-\infty}^{\infty} f(x)\ln\left|1 - \frac{t}{x}\right|dx,$$

$$f(t) = \frac{1}{\pi}\frac{d}{dt}\int_{-\infty}^{\infty} \hat{f}(x)\ln\left|1 - \frac{t}{x}\right|dx. \tag{F.70}$$

F.3.2. Analytic Signals

From Eq. (F.63) it follows that

$$f(t) - i\hat{f}(t) = \frac{1}{\pi i} \mathrm{P} \int_{-\infty}^{\infty} \frac{f(\tau) - i\hat{f}(\tau)}{\tau - t} d\tau. \tag{F.71}$$

If we define the function

$$A(t) = f(t) - i\hat{f}(t) \tag{F.72}$$

we have the relation

$$A(t) = \frac{1}{\pi i} \, \mathrm{P} \int_{-\infty}^{\infty} \frac{A(\tau) d\tau}{\tau - t}. \tag{F.73}$$

We continue the function $A(t)$ into the complex plane through the following extensions of Eq. (F.66):

$$\begin{aligned} X(t, \theta) &= \int_0^\infty \{a(\omega)\cos \omega t + b(\omega)\sin \omega t\} e^{-\theta\omega} \, d\omega \to f(t) \quad \text{as} \quad \theta \to 0, \\ Y(t, \theta) &= \int_0^\infty \{a(\omega)\sin \omega t - b(\omega)\cos \omega t\} e^{-\theta\omega} \, d\omega \to -\hat{f}(t) \quad \text{as} \quad \theta \to 0. \end{aligned} \tag{F.74}$$

Defining

$$z = t + i\theta, \qquad A(z) = X + iY, \tag{F.75}$$

we combine Eqs. (F.74) into the single equation

$$\begin{aligned} A(z) &= \int_0^\infty [a(\omega) - ib(\omega)] e^{iz\omega} \, d\omega \\ &= \frac{1}{\pi} \int_0^\infty F(\omega) e^{iz\omega} \, d\omega = -\frac{i}{\pi} \int_0^\infty \hat{F}(\omega) e^{iz\omega} \, d\omega. \end{aligned} \tag{F.76}$$

Define

$$H(z, \omega) = e^{iz\omega} H(\omega), \qquad H(z, -\omega) = e^{-iz\omega} H(-\omega),$$

where $H(\omega)$ is the unit step function. Then, applying Parseval's formula for the functions $F(\omega)$ and $H(z, \omega)$ in Eq. (F.76), we get

$$A(z) = \frac{1}{\pi} \int_{-\infty}^{\infty} F(\omega) H(z, \omega) d\omega = 2 \int_{-\infty}^{\infty} f(\tau) h(z, \tau) d\tau, \tag{F.77}$$

where

$$\begin{aligned} h(z, \tau) &= \frac{1}{2\pi} \int_{-\infty}^{\infty} H(z, -\omega) e^{i\omega\tau} \, d\omega \\ &= \frac{1}{2\pi i(\tau - z)}, \qquad (\theta > 0) \end{aligned} \tag{F.78}$$

because $(\tau - z)$ is complex and never vanishes for $\theta \neq 0$. Consequently,

$$A(z) = \frac{1}{\pi i} \int_{-\infty}^{\infty} \frac{f(\tau)}{\tau - z} \, d\tau, \qquad (\theta > 0). \tag{F.79}$$

Separating the real and imaginary parts of Eq. (F.79), we obtain the Poisson integral formulas

$$X(t, \theta) = \frac{\theta}{\pi} \int_{-\infty}^{\infty} \frac{f(\tau) d\tau}{(\tau - t)^2 + \theta^2} = \frac{\theta}{\pi} \int_{-\infty}^{\infty} \frac{X(\tau, 0) d\tau}{(\tau - t)^2 + \theta^2}, \tag{F.80}$$

$$Y(t, \theta) = -\frac{1}{\pi} \int_{-\infty}^{\infty} \frac{(\tau - t) f(\tau) d\tau}{(\tau - t)^2 + \theta^2}. \tag{F.81}$$

Through these relations we may continue $A(\tau)$ from Eq. (F.73) into the complex z-plane for $\theta = \text{Im } z > 0$.

In the limit $\theta \to 0$ we find that $\theta/[(\tau - t)^2 + \theta^2] \to \pi\delta(\tau - t)$ and hence $X(t, 0) \to f(t)$, $Y(t, 0) \to -(1/\pi)\int_{-\infty}^{\infty} f(\tau)d\tau/(\tau - t) = -\hat{f}(t)$, as it should.

Now, because the Fourier transform of $\hat{f}(t)$ is $\hat{F}(\omega)$, we may appeal to Parseval's formula once again to derive the relation

$$iA(z) = \frac{1}{\pi}\int_0^{\infty} \hat{F}(\omega)e^{iz\omega}\,d\omega = \frac{1}{\pi}\int_{-\infty}^{\infty} \hat{F}(\omega)H(z, \omega)d\omega$$

$$= 2\int_{-\infty}^{\infty} \hat{f}(\tau)h(z, \tau)d\tau = \frac{1}{\pi i}\int_{-\infty}^{\infty} \frac{\hat{f}(\tau)}{\tau - z}\,d\tau,$$

that is,

$$A(z) = -\frac{1}{\pi}\int_{-\infty}^{\infty} \frac{\hat{f}(\tau)d\tau}{\tau - z}, \qquad (\theta > 0) \tag{F.82}$$

$$X(t, \theta) = -\frac{1}{\pi}\int_{-\infty}^{\infty} \frac{(\tau - t)\hat{f}(\tau)d\tau}{(\tau - t)^2 + \theta^2}, \tag{F.83}$$

$$Y(t, \theta) = -\frac{\theta}{\pi}\int_{-\infty}^{\infty} \frac{\hat{f}(\tau)d\tau}{(\tau - t)^2 + \theta^2}. \tag{F.84}$$

Combining Eqs. (F.79) and (F.82) we have, for real τ

$$A(z) = \frac{1}{2\pi i}\int_{-\infty}^{\infty} \frac{A(\tau)}{\tau - z}\,d\tau, \qquad \theta > 0. \tag{F.85}$$

Therefore, when $A(z)$ is an analytic function, regular for $\theta > 0$ and $\int_{-\infty}^{\infty} |A(z)|^2\,dt$ is bounded and exists for every $\theta > 0$, we may continue $f(t) - i\hat{f}(t)$ into the complex plane $z = t + i\theta$. The result of this continuation is known as the *analytic signal.*

Simple manipulations with Eqs. (F.79)–(F.85) yield

$$X(t, \theta) = \frac{1}{\pi}\int_{-\infty}^{\infty} Y(\tau, 0)\frac{(\tau - t)d\tau}{(\tau - t)^2 + \theta^2}, \tag{F.86}$$

$$Y(t, \theta) = -\frac{1}{\pi}\int_{-\infty}^{\infty} X(\tau, 0)\frac{(\tau - t)d\tau}{(\tau - t)^2 + \theta^2}, \tag{F.87}$$

which coincide, respectively, with Eqs. (F.60) and (F.63) for $\theta \to 0$. Also

$$A(z) = \frac{1}{\pi i}\int_{-\infty}^{\infty} \frac{A(\tau)(\tau - t)}{(\tau - t)^2 + \theta^2}\,d\tau. \tag{F.88}$$

APPENDIX G

The Airy Integral

Consider the Fourier integral $(1/2\pi)\int_{-\infty}^{\infty} e^{i(\omega t - k\Delta)}G(\omega)d\omega$ and expand $k(\omega)$ and $G(\omega)$ in a Taylor series in $\omega - \omega_0 = u$ about ω_0, where $(d^2k/d\omega^2)_{\omega_0} = 0$,

$$G(\omega) = G(\omega_0) + O(u),$$

$$k(\omega) = k(\omega_0) + u\left.\frac{dk}{d\omega}\right|_{\omega_0} + \frac{1}{2}u^2\left.\frac{d^2k}{d\omega^2}\right|_{\omega_0} + \frac{1}{6}u^3\left.\frac{d^3k}{d\omega^3}\right|_{\omega_0} + O(u^4)$$

$$= k(\omega_0) + \frac{u}{U_0} + \frac{1}{6}u^3k^{(3)}(\omega_0). \tag{G.1}$$

Now, with $G(\omega_0) \neq 0$, $u \ll \omega_0 \neq 0$, $k(\omega_0) = k_0$, the Fourier integral takes the form

$$\frac{1}{2\pi}\int_{-\infty}^{\infty} e^{i(\omega t - k\Delta)}G(\omega)d\omega = \frac{1}{2\pi}G(\omega_0)e^{i(\omega_0 t - k_0\Delta)}\int_{-\infty}^{\infty} e^{i[u(t-\Delta/U_0) - u^3\Delta k_0^{(3)}/6]}\,du. \tag{G.2}$$

We next make the change of variable

$$s = \mp\left\{\frac{1}{2}\Delta|k_0^{(3)}|\right\}^{1/3}u, \tag{G.3}$$

where the upper sign is taken when $k_0^{(3)} > 0$, and the lower sign when $k_0^{(3)} < 0$. This yields

$$\frac{1}{2\pi}\int_{-\infty}^{\infty} e^{i(\omega t - k\Delta)}G(\omega)d\omega = G(\omega_0)e^{i(\omega_0 t - k_0\Delta)}\frac{Ai(z)}{\{\frac{1}{2}\Delta|k_0^{(3)}|\}^{1/3}}, \tag{G.4}$$

where

$$z = \pm\frac{(\Delta/U_0 - t)}{\{\frac{1}{2}\Delta|k_0^{(3)}|\}^{1/3}} \tag{G.5}$$

and where $Ai(z)$ is the *Airy function*, defined by

$$Ai(z) = \frac{1}{\pi}\int_0^{\infty}\cos\left(zs + \frac{1}{3}s^3\right)ds = \frac{1}{2\pi}\int_{-\infty}^{\infty} e^{i(zs + s^3/3)}\,ds.$$

Applying the relations

$$\frac{dk}{d\omega} = \frac{1}{U}, \qquad \frac{d^2k}{d\omega^2} = \frac{1}{2\pi}\left(\frac{T}{U}\right)^2 \frac{dU}{dT}, \qquad \frac{d^3k}{d\omega^3} = -\frac{T^4}{4\pi^2U^2}\frac{d^2U}{dT^2},$$

where the last result assumes $dU/dT = 0$, we obtain

$$\frac{1}{2\pi}\int_{-\infty}^{\infty} e^{i(\omega t - k\Delta)}G(\omega)d\omega = G(\omega_0)e^{i(\omega_0 t - k_0\Delta)}\frac{2\pi^{2/3}Ai(z)}{\{\Delta(T^4/U^2)|d^2U/dT^2|\}_0^{1/3}},$$

$$z = \frac{\pm(\Delta/U_0 - t)}{\{\Delta(T^4/U^2)|d^2U/dT^2|\}_0^{1/3}}(2\pi^{2/3}). \tag{G.6}$$

From the graph of $Ai(z)$ in Fig. G.1, it is seen that at the head of the disturbance there is a slow rise to a maximum somewhat after the instant given by $\Delta = U_0 t$. This is followed by a succession of maxima and minima of decreasing amplitude and period.

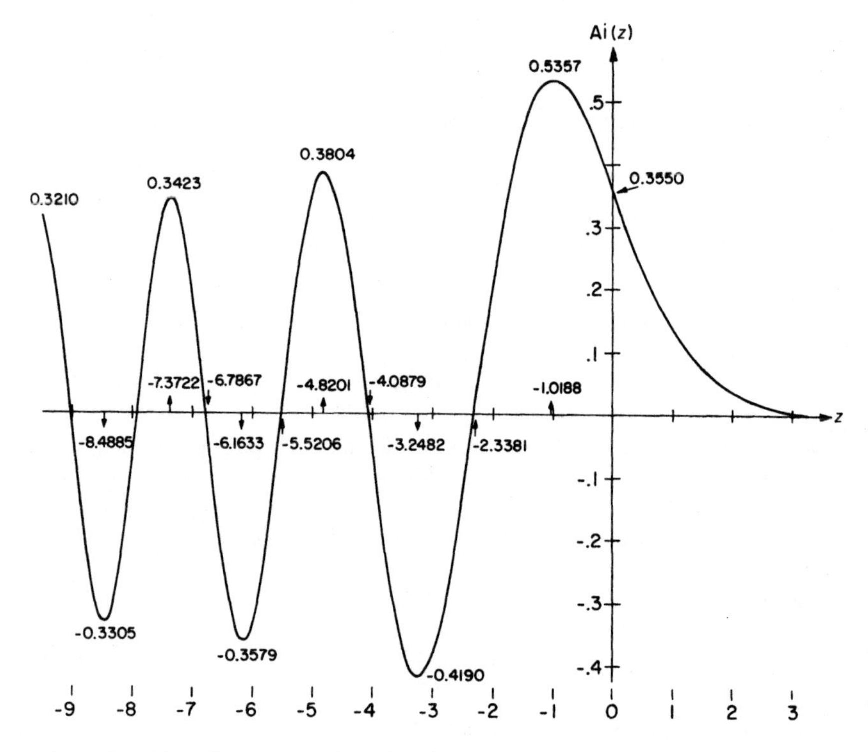

Figure G.1. The Airy function $Ai(z)$. The abscissas and ordinates of zeros, extrema, and intercept are indicated by arrows. Ordinates of extrema are given (e.g. 0.3210, −0.3305, etc.).

It may be noted that if $k(-\omega) = -k(\omega)$, $U(-\omega) = U(\omega)$, and if $d^2k/d\omega^2$ vanishes for $\omega = \omega_0$ then it vanishes for $\omega = -\omega_0$ too. If the contributions of $\pm\omega_0$ are added under the assumption $G(-\omega) = G(\omega)$, we shall get

$$\frac{1}{2\pi}\int_{-\infty}^{\infty} e^{i(\omega t - k\Delta)}G(\omega)d\omega = G(\omega_0)\cos(\omega_0 t - k_0\Delta)\frac{4\pi^{2/3}Ai(z)}{\{\Delta(T^4/U^2)|d^2U/dT^2|\}_0^{1/3}}. \tag{G.7}$$

It is easily seen that $Ai(z)$ satisfies the differential equation

$$\frac{d^2f(z)}{dz^2} - zf(z) = 0. \tag{G.8}$$

Pairs of linearly independent solutions of Eq. (G.8) are

$$\{Ai(z), Bi(z)\}, \qquad \{Ai(z), Ai(ze^{2\pi i/3})\}, \qquad \{Ai(z), Ai(ze^{-2\pi i/3})\},$$

where

$$\begin{aligned} Bi(z) &= \frac{1}{\pi}\int_0^{\infty}\left\{e^{zs-(1/3)s^3} + \sin\left(zs + \frac{1}{3}s^3\right)\right\}ds \\ &= e^{\pi i/6}Ai(ze^{2\pi i/3}) + e^{-\pi i/6}Ai(ze^{-2\pi i/3}). \end{aligned} \tag{G.9}$$

The salient features of the Airy functions are ($\zeta = \frac{2}{3}z^{3/2}$):

$$\int_{-\infty}^{\infty} Ai(z)dz = 1,$$

$$Ai(-z) = \frac{1}{3}\sqrt{z}[J_{1/3}(\zeta) + J_{-1/3}(\zeta)],$$

$$Ai(z) = \frac{1}{3}\sqrt{z}[I_{-1/3}(\zeta) - I_{1/3}(\zeta)] = \frac{1}{\pi}\left(\frac{z}{3}\right)^{1/2}K_{1/3}(\zeta),$$

$$Ai(0) = \frac{1}{\pi}\int_0^{\infty}\cos\left(\frac{x^3}{3}\right)dx = \frac{3^{-2/3}}{\Gamma(\frac{2}{3})} \simeq 0.355028,$$

$$Ai(z) = \begin{cases} \dfrac{1}{2\sqrt{\pi}}z^{-1/4}e^{-(2/3)z^{3/2}}[1 + O(z^{-3/2})], & |z| \to +\infty, \quad -\pi < \arg z < \pi \\ \dfrac{1}{\sqrt{\pi}}|z|^{-1/4}\left[\sin\left\{\dfrac{2}{3}|z|^{3/2} + \dfrac{\pi}{4}\right\} + O(z^{-3/2})\right], & z \to -\infty \end{cases}$$

$$e^{\mp\pi i/3}Ai(ze^{\pm 2\pi i/3}) = \frac{1}{2}[Ai(z) \mp iBi(z)],$$

$$Ai(-ze^{2\pi i/3}) = \frac{1}{2}e^{\pi i/2}\sqrt{\left(\frac{z}{3}\right)}H_{1/3}^{(1)}\left(\frac{2}{3}z^{3/2}\right),$$

$$Ai(-ze^{-2\pi i/3}) = \frac{1}{2}e^{-\pi i/2}\sqrt{\left(\frac{z}{3}\right)}H_{1/3}^{(2)}\left(\frac{2}{3}z^{3/2}\right). \tag{G.10}$$

The case $\omega_0 = 0$ requires a special treatment. When $G(\omega)$ is an even function of ω, Eq. (G.4) reads

$$\frac{1}{2\pi}\int_{-\infty}^{\infty} G(\omega)e^{i(\omega t - k\Delta)}\,d\omega = \frac{G(0)Ai(z)}{|\frac{1}{2}\Delta k^{(3)}(0)|^{1/3}}. \tag{G.11}$$

However, when $G(\omega)$ is an odd function of ω,

$$\frac{1}{2\pi}\int_{-\infty}^{\infty} G(\omega)e^{i(\omega t - k\Delta)}\,d\omega = \frac{iG(0)Gi(z)}{|\frac{1}{2}\Delta k^{(3)}(0)|^{1/3}}, \tag{G.12}$$

where

$$Gi(z) = \frac{1}{\pi}\int_{0}^{\infty} \sin\left(zs + \frac{1}{3}s^3\right)ds. \tag{G.13}$$

Integrals of the type

$$\frac{1}{2\pi}\int_{-\infty}^{\infty} F(k)e^{i(\omega t - k\Delta)}\,dk$$

are reduced to integrals over the frequency simply by substituting $dk = d\omega/U$.

APPENDIX H

Asymptotic Solutions of Second-Order Linear Differential Equations

Usually, asymptotic expansions of the Bessel functions are based on saddle-point approximations of their integral representations. In some cases, however, it is more convenient to obtain a direct expansion of the function from the differential equation that it obeys without the need to know its explicit integral representation.

Consider the equation

$$\frac{d^2W}{dz^2} + \nu^2 Q(z, \nu)W = 0, \tag{H.1}$$

where the parameter ν is taken to be large and positive and where $Q(z, \nu)$ tends to a limit as $\nu \to \infty$ for fixed z. Let us change the variables from (z, W) to (u, X) through the relations

$$z = z(u), \qquad W = X\left(\frac{dz}{du}\right)^{1/2}. \tag{H.2}$$

Equation (H.1) then transforms into the equation

$$\frac{d^2X}{du^2} + (\nu^2 Q z'^2)X = \left[\frac{3}{4}\left(\frac{z''}{z'}\right)^2 - \frac{1}{2}\frac{z'''}{z'}\right]X, \qquad \left(' = \frac{d}{du}\right). \tag{H.3}$$

We now choose z such that

$$\nu^2 Q\left(\frac{dz}{du}\right)^2 = -u, \tag{H.4}$$

yielding thereby

$$u = (\pm i)^{2/3}\nu^{2/3}\Phi(z), \tag{H.5}$$

where

$$\Phi(z) = \left[\frac{3}{2}\int_{z_0}^{z}\sqrt{Q}\,dz\right]^{2/3}.$$

Equation (H.3) therefore assumes the form

$$\frac{d^2X}{du^2} - uX = \frac{1}{\nu^{4/3}} r_1(u)X, \tag{H.6}$$

where

$$r_1(u) = -\frac{1}{2}\nu^{4/3}\{z, u\}$$

and

$$\{z, u\} = \frac{z'''}{z'} - \frac{3}{2}\left(\frac{z''}{z'}\right)^2$$

is the *Schwarzian derivative* of z with respect to u. Because, from Eqs. (H.4) and (H.5),

$$\frac{dz}{du} \propto \nu^{-2/3}, \qquad \frac{d^2z}{du^2} \propto \nu^{-4/3}, \qquad \frac{d^3z}{du^3} \propto \nu^{-2},$$

it follows that $r_1(u)$ is independent of ν. Neglecting the right-hand side of Eq. (H.6) for large ν, we note that the solutions of this equation tend to those of the Airy equation

$$\frac{d^2X}{du^2} - uX = 0. \tag{H.7}$$

Using Eqs. (H.2) and (H.4), we write the uniformly asymptotic solution to Eq. (H.1) as

$$W = \nu^{-1/3}\left[\frac{\Phi}{Q}\right]^{1/4}[c_1 Ai\{-e^{2\pi i/3}\nu^{2/3}\Phi\} + c_2 Ai\{-e^{-2\pi i/3}\nu^{2/3}\Phi\}], \tag{H.8}$$

where c_1, c_2 are constants. By *uniformly asymptotic* we mean that the error is at most $(1/\nu)$ times the main approximation and that it is maintained throughout the domain. Equation (H.7) is known as the *comparison equation* and its solutions are the "vehicle" for the solution of the original differential equation. The behavior of the solution of Eq. (H.1) is much more involved if Q has a zero at $z = c$ (say) in the complex plane. Such a point is called a *transition point* or a *turning point* of the differential equation. This point separates an interval in which the solutions are monotonic from one in which they are *oscillatory*, and the transition from one kind of behavior to the other takes place in the immediate vicinity of c. A similar phenomenon occurs whenever Q has a singularity in the interval of interest. A turning point is said to be of the nth order if Q and all its derivatives up to order $(n - 1)$ vanish at $z = c$.

H.1. Turning Points and Comparison Equation: WKBJ

The choice of the comparison equation is dictated by the number and order of the turning points of the differential equation. If we wish to construct a uniform asymptotic solution in a given domain in the z plane, then the solutions of the comparison equation must reflect this property. If this is not the case, different analytic solutions will be needed in each region between the turning points and a *connecting formula* will be needed in the *transition region*. Let us now demonstrate these ideas for some simple cases.

H.1.1. No Turning Points

In this case we assume

$$\nu^2 Q\left(\frac{dz}{du}\right)^2 = -1,$$

yielding the relation

$$u = \pm i\nu \int_{z_0}^{z} \sqrt{Q}\, dz. \tag{H.9}$$

Equation (H.3) is then reduced to

$$\frac{d^2X}{du^2} - X = \frac{1}{\nu^2} r_0(z)X,$$

where

$$r_0(z) = -\frac{1}{2}\nu^2\{z, u\} = \frac{1}{4Q^2}\left[\frac{5Q'^2}{4Q} - Q''\right].$$

Therefore, if Q is not too near zero, the condition

$$\frac{r_0(z)}{\nu^2} \ll 1$$

will secure the finiteness of the approximate solutions. We then have

$$X = c_1 e^u + c_2 e^{-u}.$$

Equations (H.2) and (H.9) now yield

$$W = \nu^{-1/2}Q^{-1/4}\left[c_1 \exp\left(i\nu \int_{z_0}^{z} \sqrt{Q}\, dz\right) + c_2 \exp\left(-i\nu \int_{z_0}^{z} \sqrt{Q}\, dz\right)\right]. \tag{H.10}$$

This is known as the Wentzel–Kramers–Brillouin–Jeffreys (abbreviated as WKBJ solution). Generally, the integral in Eq. (H.10) contains both real and imaginary parts so that both phase and amplitude of W vary with z. Taken together, the integral forms a complex phase factor known as the *phase integral.*

H.1.2. A Single Transition Point of the First Order

Suppose that $Q(z, \nu) = 0$ at some $z = z_0$ such that $Q'(z_0) = \xi_0 > 0$. If z is real, the two exponents in Eq. (H.10) will be purely imaginary for $z > z_0$ and real for $z < z_0$. The solution therefore will be oscillatory in character for $z > z_0$

and exponential in character for $z < z_0$. The transition between the two kinds of behavior must take place in a region containing z_0. A particular asymptotic solution of Eq. (H.1) will be represented by quite differently behaving functions on either side of z_0 and it will be necessary to find a way to connect one side of z_0 with the other. The provision of such connection formulas is part of the WKBJ method. The lines separating the solution in the different sectors are known as *Stokes' lines* if z is complex. The explicit form of the connecting formula and the Stokes' lines are not given here because they can be avoided altogether by using the uniform solution, Eq. (H.8). Because the comparison equation already takes care of the transition point, this single asymptotic formula is valid at the transition point as well as on its two sides. Equation (H.8) ceases to be valid in the neighborhood of a second zero of $Q(z, v)$. There, the solutions of the comparison equation are the parabolic cylinder functions, which can also be expressed in terms of the Whittaker functions.

A single turning point of the nth order is treated with the solutions of

$$\frac{d^2X}{du^2} - u^nX = 0.$$

Uniform asymptotic expansion can be obtained for certain types of third- and fourth-order equations by using suitable comparison equations.

H.2. Application to Bessel Functions

Asymptotic approximations of the Bessel functions that we shall be using can be grouped as follows:

1. Expansions of the *Hankel type* for large argument and fixed order
$$H_\nu^{(1,2)}(z) = \left[\frac{2}{\pi z}\right]^{1/2} \exp\left[\pm i\left\{z - \left(\frac{\pi}{2}\right)\nu - \left(\frac{\pi}{4}\right)\right\}\right]\left[1 + O\left(\frac{1}{z}\right)\right].$$
2. Expansions of the *Debye type* for large argument and large order such that ν/z is fixed but $|\nu|$ is not close to z.
3. Expansions in the neighborhood of the transition zone $|\nu| \sim z$. These are of two kinds: Second-order saddle-point approximation (known sometimes as the *tangent approximation*) and third-order saddle-point approximation (known also as the *Watson approximation*).
4. Expansions of the *Langer type*, which represent uniform asymptotic approximation expressed by the Airy functions.

We shall give a brief summary of the features of these approximations. The functions

$$W = \sqrt{y}\,H_\nu^{(1,2)}(\nu y), \qquad 0 < y < \infty, \tag{H.11}$$

satisfy the differential equation

$$\frac{d^2W}{dy^2} + \left[\nu^2\left(1 - \frac{1}{y^2}\right) + \frac{1}{4y^2}\right]W = 0. \tag{H.12}$$

For large values of ν, we set $Q = 1 - 1/y^2$. Then, by Eq. (H.5)

$$\frac{2}{3}[-\Phi(y)]^{3/2} = \int_y^1 \left(\frac{1}{t^2} - 1\right)^{1/2} dt = -\sqrt{(1-y^2)} + \ln\frac{1+\sqrt{(1-y^2)}}{y}, \qquad (0 < y \leq 1) \quad \text{(H.13)}$$

$$\frac{2}{3}[\Phi(y)]^{3/2} = \int_1^y \left(1 - \frac{1}{t^2}\right)^{1/2} dt = \sqrt{(y^2-1)} - \cos^{-1}\left(\frac{1}{y}\right), \qquad (1 \leq y < \infty). \quad \text{(H.14)}$$

Using Eq. (H.8) in the region $1 \leq y < \infty$ and denoting $\nu y = x(\geq \nu)$, we obtain

$$H_\nu^{(1,2)}(x) = \left(\frac{\Phi}{x^2/\nu^2 - 1}\right)^{1/4} \nu^{-1/3}[c_1 Ai(-\nu^{2/3}\Phi e^{2\pi i/3}) + c_2 Ai(-\nu^{2/3}\Phi e^{-2\pi i/3})]. \quad \text{(H.15)}$$

To obtain the values of the numerical constants c_1 and c_2, we compare these approximations with some known approximations of $H_\nu^{(1,2)}(x)$. We know, for example, that for $x \gg 1$, $x \gg \nu$

$$H_\nu^{(1,2)}(x) = \sqrt{\left(\frac{2}{\pi x}\right)}\exp\left[\pm i\left(x - \frac{\pi\nu}{2} - \frac{\pi}{4}\right)\right]\left[1 + O\left(\frac{1}{x}\right)\right],$$

$$Ai(-ze^{\pm 2\pi i/3}) = \frac{e^{\pm \pi i/3}}{2\sqrt{\pi}\, z^{1/4}} \exp\left[\pm i\left(\frac{2}{3}z^{3/2} - \frac{\pi}{4}\right)\right], \quad \text{(see Appendix G)} \quad \text{(H.16)}$$

$$\Phi = \left(\frac{3}{2}\right)^{2/3}\left(\frac{x}{\nu} - \frac{\pi}{2}\right)^{2/3}, \quad \text{(from Eq. (H.14)]}$$

$$z = \nu^{2/3}\Phi = \left(\frac{3}{2}\right)^{2/3}\left(x - \frac{\pi\nu}{2}\right)^{2/3}.$$

The approximations given in the above equation are known as the *Hankel approximations* and are valid for large argument and fixed order. Equations (H.15) and (H.16) yield

$$c_1 = 2\sqrt{2}\,e^{-i\pi/3}, \qquad c_2 = 0 \quad \text{for} \quad H_\nu^{(1)}(x);$$

$$c_1 = 0, \qquad c_2 = 2\sqrt{2}\,e^{i\pi/3} \quad \text{for} \quad H_\nu^{(2)}(x).$$

Therefore, we finally have, for x and ν large and $x \geq \nu$,

$$H_\nu^{(1,2)}(x) = 2\sqrt{2}\,\nu^{-1/3}e^{\mp \pi i/3}\left(\frac{\Phi}{x^2/\nu^2 - 1}\right)^{1/4} Ai(-\nu^{2/3}\Phi e^{\pm 2\pi i/3}), \quad \text{(H.17a)}$$

where

$$\frac{2}{3}\Phi^{3/2} = \left(\frac{x^2}{\nu^2} - 1\right)^{1/2} - \cos^{-1}\left(\frac{\nu}{x}\right).$$

This is known as the *Langer approximation*. For the derivative, we have

$$x \frac{d}{dx} H_\nu^{(1,2)}(x) = 2\sqrt{2}\,\nu^{1/3} e^{\pm 4\pi i/3} \left(\frac{x^2/\nu^2 - 1}{\Phi}\right)^{1/4} Ai'(-\nu^{2/3} e^{\pm 2\pi i/3} \Phi). \quad \text{(H.17b)}$$

From Eq. (H.17a), we obtain the asymptotic expansion for $J_\nu(x)$:

$$\begin{aligned} J_\nu(x) &= \frac{1}{2}[H_\nu^{(1)}(x) + H_\nu^{(2)}(x)] \\ &= \left[\frac{4\Phi}{x^2/\nu^2 - 1}\right]^{1/4} \nu^{-1/3} \{e^{-\pi i/3} Ai(-\nu^{2/3} \Phi e^{2\pi i/3}) + e^{\pi i/3} Ai(-\nu^{2/3} \Phi e^{-2\pi i/3})\} \\ &= \left[\frac{4\Phi}{x^2/\nu^2 - 1}\right]^{1/4} \nu^{-1/3} Ai(-\nu^{2/3} \Phi), \end{aligned} \quad \text{(H.17c)}$$

valid for large x and ν with $x \geq \nu$.

When $\nu \sim x$, we put $\nu/x = \cos\alpha$ and find

$$\frac{2}{3}\Phi^{3/2} = \tan\alpha - \alpha \simeq \frac{1}{3}\alpha^3,$$

$$\Phi \simeq 2^{-2/3}\alpha^2 \simeq 2^{1/3}(1 - \cos\alpha) = \frac{2^{1/3}(x - \nu)}{x} \simeq \frac{2^{1/3}(x - \nu)}{\nu}, \quad \text{(H.17d)}$$

$$\frac{\Phi}{x^2/\nu^2 - 1} = 2^{-2/3}.$$

Equations (H.17a, b) then yield

$$H_\nu^{(1,2)}(x) = 2e^{\mp \pi i/3}\left(\frac{2}{\nu}\right)^{1/3} Ai\left[e^{\pm 2\pi i/3}\left(\frac{2}{\nu}\right)^{1/3}(\nu - x)\right], \quad \text{(H.18a)}$$

$$\frac{d}{dx} H_\nu^{(1,2)}(x) = 2e^{\pm 4\pi i/3}\left(\frac{2}{\nu}\right)^{2/3} Ai'\left[e^{\pm 2\pi i/3}\left(\frac{2}{\nu}\right)^{1/3}(\nu - x)\right]. \quad \text{(H.18b)}$$

We next consider the equation

$$\frac{d^2 W}{dz^2} + \nu^2(e^{2z} - 1)W = 0, \quad \text{(H.19)}$$

the exact solutions of which are $H_\nu^{(1,2)}(\nu e^z)$. Introducing the notation

$$e^z = \frac{1}{\cos\alpha}, \quad \text{(H.20)}$$

we find

$$\int_0^z \sqrt{Q}\, dz = \int_0^z \sqrt{(e^{2z} - 1)}\, dz = \sqrt{(e^{2z} - 1)} - \tan^{-1}\sqrt{(e^{2z} - 1)} = \tan\alpha - \alpha,$$

$$\Phi = \left[\frac{3}{2}(\tan\alpha - \alpha)\right]^{2/3}. \quad \text{(H.21)}$$

Therefore, by Eq. (H.8), the uniform asymptotic expansion of $H_\nu^{(1,2)}(x)$ for large order ($\nu \gg 1$) can be expressed in terms of

$$Ai\left[-\nu^{2/3}e^{2\pi i/3}\left\{\frac{3}{2}(\tan\alpha - \alpha)\right\}^{2/3}\right].$$

Using the asymptotic expansion of the Airy function and evaluating the constants c_1 and c_2, as in the previous example, we obtain

$$H_\nu^{(1,2)}\left(\frac{\nu}{\cos\alpha}\right) = \left(\frac{2}{\pi\nu\tan\alpha}\right)^{1/2} \exp\left\{\pm i\left[\nu(\tan\alpha - \alpha) - \frac{\pi}{4}\right]\right\}.$$

If $x = \nu/\cos\alpha > \nu$,

$$H_\nu^{(1,2)}(x) = \left(\frac{2}{\pi x\sin\alpha}\right)^{1/2} \exp\left\{\pm i\left[x(\sin\alpha - \alpha\cos\alpha) - \frac{\pi}{4}\right]\right\}. \tag{H.22}$$

This is known as the *Debye approximation* which breaks down for $x \sim \nu$. The corresponding result for the derivative is

$$\frac{d}{dx}H_\nu^{(1,2)}(x) = \left(\frac{2\sin\alpha}{\pi x}\right)^{1/2} \exp\left\{\pm i\left[x(\sin\alpha - \alpha\cos\alpha) + \frac{\pi}{4}\right]\right\}. \tag{H.23}$$

In the limit as $\alpha \to \pi/2$, Eq. (H.22) yields the Hankel approximation.

H.2.1. Watson's Approximation

The Airy function is related to the Hankel functions of order $\frac{1}{3}$ through the equation

$$Ai[-ze^{\pm 2\pi i/3}] = \frac{1}{2}e^{\pm\pi i/2}\sqrt{\left(\frac{z}{3}\right)}H_{1/3}^{(1,2)}\left(\frac{2}{3}z^{3/2}\right). \tag{H.24}$$

Equations (H.17a) and (H.24) yield

$$H_\nu^{(1,2)}(x) = \sqrt{\left(\frac{2}{3}\right)}e^{\pm\pi i/6}\left(\frac{x^2}{\nu^2} - 1\right)^{-1/4} \Phi^{3/4}H_{1/3}^{(1,2)}\left(\frac{2}{3}\nu\Phi^{3/2}\right). \tag{H.25}$$

In the right-hand side of Eq. (H.25), we use the asymptotic expansion of $H_{1/3}^{(1,2)}$ for large argument given in Eq. (H.16). We obtain

$$H_\nu^{(1,2)}(x) = \sqrt{\left(\frac{2}{\pi\nu}\right)}e^{\pm\pi i/6}\left(\frac{x^2}{\nu^2} - 1\right)^{-1/4} \exp\left[\pm i\left(\frac{2}{3}\nu\Phi^{3/2} - \frac{5\pi}{12}\right)\right]. \tag{H.26}$$

Using the identity

$$\int\frac{(t^2-1)^{1/2}}{t}\,dt = \frac{1}{3}(t^2-1)^{3/2} - \int\frac{(t^2-1)^{3/2}}{t}\,dt,$$

in Eq. (H.14), we have

$$\frac{2}{3}[\Phi(y)]^{3/2} = \frac{1}{3}(y^2-1)^{3/2} - \Psi(y), \tag{H.27}$$

where $y = x/\nu$ and

$$\Psi(y) = \int_1^y \frac{(t^2 - 1)^{3/2}}{t}\,dt = \frac{1}{3}\tan^3\alpha - \tan\alpha + \alpha, \qquad \cos\alpha = \frac{\nu}{x}.$$

Using Eq. (H.27) in Eq. (H.26) and appealing to the asymptotic formula of $H_{1/3}^{(1,2)}$ for large argument once again, we get the *Watson approximation*

$$H_\nu^{(1,2)}(x) = \left[\frac{1}{3}\left(\frac{x^2}{\nu^2} - 1\right)\right]^{1/2} e^{\pm i(\pi/6 - \nu\Psi)} H_{1/3}^{(1,2)}\left[\frac{\nu}{3}\left(\frac{x^2}{\nu^2} - 1\right)^{3/2}\right]. \quad \text{(H.28)}$$

It is valid for x and ν large and $x \geq \nu$, where

$$-\frac{\pi}{2} < \arg\left(\frac{x^2}{\nu^2} - 1\right)^{1/2} < \frac{\pi}{2}.$$

In passing, we mention another approximation of the Hankel function, known as the *tangent approximation*:

$$H_\nu^{(1)}(x) = 2i\left(\frac{2}{\pi x \sin\alpha}\right)^{1/2} \sin\left[x(\sin\alpha - \alpha\cos\alpha) - \frac{\pi}{4}\right]\left[1 + O\left(\frac{1}{x}\right)\right],$$
$$H_\nu^{(2)}(x) = 2\left(\frac{2}{\pi x \sin\alpha}\right)^{1/2} \sin\left[x(\sin\alpha - \alpha\cos\alpha) + \frac{\pi}{4}\right]\left[1 + O\left(\frac{1}{x}\right)\right], \quad \text{(H.29)}$$

valid for

$$\left|\frac{x - \nu}{\nu^{1/3}}\right| = O(1), \qquad x > \nu, \qquad \cos\alpha = \frac{\nu}{x}.$$

Both the Watson and tangent approximations are valid in the transition zone, where $\nu \sim x$. Out of these two, the Watson approximation is the more exact one. This applies particularly to $x = \nu$, where the tangent approximation even becomes infinite, whereas the Watson approximation remains finite. In the neighborhood of $\nu = x$, for large values of ν, both approximations chiefly depend upon the ratio

$$\left|\frac{\nu - x}{x^{1/3}}\right|.$$

Assuming

$$\tau x^{-2/3} = \frac{x - \nu}{x} \ll 1,$$

the tangent approximation yields

$$H_{x-\tau x^{1/3}}^{(1)}(x) = \left(\frac{8}{\pi}\right)^{1/2} e^{i\pi/2} x^{-1/3}(2\tau)^{-1/4} \sin\left[\frac{1}{3}(2\tau)^{3/2} - \frac{\pi}{4}\right], \quad \text{(H.30a)}$$

$$H_{x-\tau x^{1/3}}^{(2)}(x) = \left(\frac{8}{\pi}\right)^{1/2} x^{-1/3}(2\tau)^{-1/4} \sin\left[\frac{1}{3}(2\tau)^{3/2} + \frac{\pi}{4}\right]. \quad \text{(H.30b)}$$

Similarly, the Watson approximation becomes

$$H^{(1,2)}_{x-\tau x^{1/3}}(x) = e^{\pm i\pi/6} x^{-1/3}\left(\frac{2\tau}{3}\right)^{1/2} H^{(1,2)}_{1/3}\left[\frac{1}{3}(2\tau)^{3/2}\right]. \tag{H.30c}$$

Asymptotic approximations of the spherical Hankel functions follow from their definition

$$h^{(1,2)}_{\nu-1/2}(x) = \left[\frac{\pi}{2x}\right]^{1/2} H^{(1,2)}_{\nu}(x).$$

The Debye approximation given in Eq. (H.22) can also be obtained in another way by approximating the Sommerfeld integral representation

$$H^{(1)}_{\nu}(x) = \frac{1}{\pi}\int_S \exp\left[ix\cos\theta + i\nu\left(\theta - \frac{\pi}{2}\right)\right]d\theta$$

(S goes from $a_1 + i\infty$ to $b_1 - i\infty$, $-\pi < a_1 < 0$, $0 < b_1 < \pi$) by the saddle-point method. There are two saddle points of unequal altitude, one of which yields the dominating contribution given in Eq. (H.22). In the limit $\alpha \to \pi/2$, the Debye approximation goes into the ordinary Hankel asymptotic formula, Eq. (H.16).

H.2.2. Roots of $H^{(1)}_{\nu}(x) = 0$ and Related Equations

We consider now the roots $\nu_n(x)$, $n = 1, 2, 3 \cdots$ of the transcendental equation

$$H^{(1)}_{\nu}(x) = 0 \tag{H.31}$$

for fixed real values of x, assuming $|\nu|$ and x large. From Eq. (H.22), it would seem that no roots of Eq. (H.31) could exist even for complex values of ν, because the exponential function in Eq. (H.22) vanishes for no finite values of the exponent, as long as x is real. However, as $|\nu|$ draws closer to x, i.e., α comes closer to zero, the contributions from the two saddle points tend to become equal. Then, the representation in Eq. (H.22) ceases to be valid because it is based on taking the contribution from one saddle point only. At $\alpha = 0$, the two saddle points are of equal altitude, and if the required path of integration is made to lead over both passes, then the sum of the two exponential expressions yields a trigonometric function that makes the existence of the roots possible. We then obtain the tangent approximation, the result of which is quoted in Eq. (H.29). The roots of Eq. (H.31) are therefore obtained from the subsidiary equation

$$x(\sin\alpha - \alpha\cos\alpha) - \frac{\pi}{4} = -n\pi, \qquad n = 1, 2, 3, \ldots. \tag{H.32}$$

Here, the choice of the negative sign on the right side follows from the particular path of integration used in the evaluation of $H^{(1)}_{\nu}(x)$. For small α, Eq. (H.32) renders $\frac{1}{3}x\alpha^3 = -(4n - 1)\pi/4$. Making the correct choice of the cube root of unity, we find $\alpha = x^{-1/3}[(3\pi/4)(4n - 1)]^{1/3}e^{-\pi i/3}$. Substituting for α from $\nu = x \cos \alpha \simeq x(1 - \alpha^2/2)$, we finally arrive at the *Van der Pol formula,*

$$\nu_n(x) = x + \frac{1}{2}x^{1/3}\left[\frac{3\pi}{4}(4n - 1)\right]^{2/3} e^{\pi i/3} + O(x^{-1/3}), \tag{H.33}$$

$$x \gg |\nu| > 0.$$

The same result is obtainable from Eq. (H.30a) by putting $\nu_n = x - x^{1/3}\tau_n$ and evaluating τ_n from the condition $\sin[\pi/4 - \frac{1}{3}(2\tau_n)^{3/2}] = 0$. It yields,

$$\tau_n = -2^{-1/3}q_n e^{2\pi i/3}, \tag{H.34}$$

where

$$q_n = -\left[\frac{3\pi}{8}(4n - 1)\right]^{2/3}$$

are large zeros of the Airy function, i.e.,

$$Ai(q_n) = 0, \qquad n \gg 1. \tag{H.35}$$

It is then clear from Eq. (H.18a) that q_n is related to the roots of $H_\nu^{(1)}(x)$ through the relation

$$q_n = e^{2\pi i/3}(2/x)^{1/3}(\nu_n - x)$$

or

$$\nu_n = x + [q_n 2^{-1/3} e^{-2\pi i/3}]x^{1/3} + O(x^{-1/3}). \tag{H.36}$$

Equation (H.36) is more general than Eq. (H.33) in the sense that q_n in Eq. (H.36) are not necessarily large zeros of $Ai(z)$.

Higher terms in Eq. (H.36) are obtained in the following way: We write

$$q_n = e^{\pi i/3}(x/2)^{2/3}\eta(\nu_n), \tag{H.37}$$

where

$$\eta = [3(\sin \alpha - \alpha \cos \alpha)]^{2/3} = \alpha^2\left[1 - \frac{1}{15}\alpha^2\right] + O(\alpha^6), \tag{H.38}$$

$$\frac{\nu_n}{x} = \cos \alpha = 1 - \frac{1}{2}\alpha^2 + \frac{1}{24}\alpha^4 + O(\alpha^6). \tag{H.39}$$

Solving Eq. (H.38) for α^2 in terms of η, we obtain

$$\alpha^2 = \eta + \frac{1}{15}\eta^2 + O(\eta^3). \tag{H.40}$$

The above value of α^2 is substituted into Eq. (H.39). Then, inserting the value of η from Eq. (H.37) into the resulting equation, we get

$$\nu_n = x + q_n[2^{-1/3}e^{-2\pi i/3}]x^{1/3} - \frac{1}{60}q_n^2[2^{1/3}e^{-\pi i/3}]x^{-1/3} + O\left(\frac{1}{x}\right). \tag{H.41}$$

Likewise, the roots of $(d/dx)H_\nu^{(2)}(x) = 0$ are derived from Eq. (H.18b)

$$\nu_n = x + \bar{q}_n[2^{-1/3}e^{2\pi i/3}]x^{1/3} - 2^{1/3}e^{\pi i/3}\left[\frac{\bar{q}_n^2}{60} - \frac{1}{10\bar{q}_n}\right]x^{-1/3}, \qquad \text{(H.42)}$$

where $\bar{q}_n$ are the zeros of $Ai'(z)$.

In a similar way, we find that the roots of

$$\frac{d}{dx}\left[\sqrt{\left(\frac{\pi}{2}\right)}\frac{H_\nu^{(2)}(x)}{\sqrt{x}}\right] = \frac{d}{dx}h_{\nu-1/2}^{(2)}(x) = 0$$

are

$$\nu_n = x + \bar{q}_n[2^{-1/3}e^{2\pi i/3}]x^{1/3} - 2^{1/3}e^{\pi i/3}\left[\frac{\bar{q}_n^2}{60} - \frac{7}{20\bar{q}_n}\right]x^{-1/3}, \qquad \text{(H.43)}$$

whereas the roots of

$$\frac{d}{dx}\left[\sqrt{\left(\frac{\pi}{2}\right)}\sqrt{x}\,H_\nu^{(2)}(x)\right] = \frac{d}{dx}[xh_{\nu-1/2}^{(2)}(x)] = 0$$

are

$$\nu_n = x + \bar{q}_n[2^{-1/3}e^{2\pi i/3}]x^{1/3} - 2^{1/3}e^{\pi i/3}\left[\frac{\bar{q}_n^2}{60} + \frac{3}{20\bar{q}_n}\right]x^{-1/3}. \qquad \text{(H.44)}$$

The roots of $H_\nu^{(2)}(x) = 0$ are the complex conjugate of the roots of $H_\nu^{(1)}(x) = 0$.

H.3. Application to Legendre Functions

The defining equation of the Legendre polynomial $u = P_l(\cos\theta)$ is

$$\frac{1}{\sin\theta}\frac{d}{d\theta}\left[\sin\theta\frac{du}{d\theta}\right] + l(l+1)u = 0, \qquad \text{(H.45)}$$

where l is an integer. Substituting $z = u\sqrt{(\sin\theta)}$, the equation becomes

$$\frac{d^2z}{d\theta^2} + \left(l + \frac{1}{2}\right)^2 z = -\frac{z}{4\sin^2\theta}. \qquad \text{(H.46)}$$

Regarding the right-hand side of Eq. (H.46) as a known function, we find that its general solution is

$$z(\theta) = P_l(0)\cos\left[\left(l+\frac{1}{2}\right)\left(\frac{\pi}{2}-\theta\right)\right] + \frac{P_l'(0)}{l+\frac{1}{2}}\sin\left[\left(l+\frac{1}{2}\right)\left(\frac{\pi}{2}-\theta\right)\right]$$
$$+ \frac{1}{4(l+\frac{1}{2})}\int_\theta^{\pi/2} z(t)\sin\left[\left(l+\frac{1}{2}\right)(\theta - t)\right]\frac{dt}{\sin^2 t}, \qquad \text{(H.47)}$$

where

$$z\left(\frac{\pi}{2}\right) = P_l(0) = \begin{cases} (-)^{l/2} \dfrac{\Gamma(l/2 + \frac{1}{2})}{\sqrt{\pi}\,\Gamma(l/2 + 1)} = (-)^{l/2} A_l, & l \text{ even} \\ 0, & l \text{ odd}, \end{cases}$$

$$z'\left(\frac{\pi}{2}\right) = -P_l'(0) = \begin{cases} (-)^{(l-1)/2} \dfrac{2\Gamma(l/2 + 1)}{\sqrt{\pi}\,\Gamma(l/2 + \frac{1}{2})} = (-)^{(l+1)/2}(l + \frac{1}{2})B_l, & l \text{ odd} \\ 0, & l \text{ even}. \end{cases} \tag{H.48}$$

It follows that Eq. (H.47) can be written in the form

$$z(\theta) = \alpha_l \left\{ \sin\left[\left(l + \frac{1}{2}\right)\theta + \frac{\pi}{4}\right] + r_l(\theta) \right\}, \tag{H.49}$$

where α_l denotes A_l or B_l of Eq. (H.48) depending on whether l is even or odd, and

$$r_l(\theta) = \frac{1}{4(l + \frac{1}{2})\alpha_l} \int_\theta^{\pi/2} z(t) \sin\left[\left(l + \frac{1}{2}\right)(\theta - t)\right] \frac{dt}{\sin^2 t}. \tag{H.50}$$

If θ is confined to the interval $\varepsilon \le \theta \le \pi - \varepsilon$ where ε is a positive fixed number, it can be easily shown that $r_l(\theta) = O(1/l)$ uniformly in the interval $[\varepsilon, \pi - \varepsilon]$. Therefore, Eq. (H.49) leads to the asymptotic formula valid for $\varepsilon \le \theta \le \pi - \varepsilon$,

$$z(\theta) = \alpha_l \sin\left[\left(l + \frac{1}{2}\right)\theta + \frac{\pi}{4}\right]\left[1 + O\left(\frac{1}{l}\right)\right]. \tag{H.51}$$

Making some simple calculations based on Stirling's formula, we find that

$$\alpha_l = \left[\frac{2}{\pi l}\right]^{1/2} + O\left(\frac{1}{l^{3/2}}\right).$$

By the combined use of Eq. (H.51) and $u = z/(\sin\theta)^{1/2}$, we obtain the *Laplace formula*

$$P_l(\cos\theta) = \left[\frac{2}{\pi l \sin\theta}\right]^{1/2} \sin\left[\left(l + \frac{1}{2}\right)\theta + \frac{\pi}{4}\right]\left[1 + O\left(\frac{1}{l}\right)\right]. \tag{H.52}$$

Likewise, it can be shown that

$$Q_l(\cos\theta) = \left[\frac{\pi}{2l \sin\theta}\right]^{1/2} \cos\left[\left(l + \frac{1}{2}\right)\theta + \frac{\pi}{4}\right]\left[1 + O\left(\frac{1}{l}\right)\right]. \tag{H.53}$$

A better approximation is

$$P_l(\cos\theta) = \left[\frac{2}{\pi l \sin\theta}\right]^{1/2}\left[\left(1 - \frac{1}{4l}\right)\sin\phi - \frac{1}{8l}\cot\theta\cos\phi\right]\left[1 + O\left(\frac{1}{l^2}\right)\right],$$

$$Q_l(\cos\theta) = \left[\frac{\pi}{2l \sin\theta}\right]^{1/2}\left[\left(1 - \frac{1}{4l}\right)\cos\phi + \frac{1}{8l}\cot\theta\sin\phi\right]\left[1 + O\left(\frac{1}{l^2}\right)\right],$$

$$\phi = \left(l + \frac{1}{2}\right)\theta + \frac{\pi}{4}. \tag{H.54}$$

The former expressions can be generalized to the associated Legendre polynomials, $l \geq m$; l, m integers

$$l^{-m}P_l^m(\cos\theta) = (-)^m\left[\frac{2}{\pi l \sin\theta}\right]^{1/2} \sin\left[\left(l+\frac{1}{2}\right)\theta + \frac{\pi}{4} + \frac{m\pi}{2}\right]\left[1 + O\left(\frac{1}{l}\right)\right], \tag{H.55}$$

$$l^{-m}Q_l^m(\cos\theta) = (-)^m\left[\frac{\pi}{2l \sin\theta}\right]^{1/2} \cos\left[\left(l+\frac{1}{2}\right)\theta + \frac{\pi}{4} + \frac{m\pi}{2}\right]\left[1 + O\left(\frac{1}{l}\right)\right]. \tag{H.56}$$

We define two *traveling-wave functions*

$$E_{lm}^{(1,2)}(\cos\theta) = \frac{1}{2}\left[P_l^m(\cos\theta) \pm \frac{2i}{\pi} Q_l^m(\cos\theta)\right]. \tag{H.57}$$

In terms of these functions

$$P_l^m(\cos\theta) = E_{lm}^{(1)}(\cos\theta) + E_{lm}^{(2)}(\cos\theta), \tag{H.58}$$

$$\frac{2i}{\pi} Q_l^m(\cos\theta) = E_{lm}^{(1)}(\cos\theta) - E_{lm}^{(2)}(\cos\theta). \tag{H.59}$$

The asymptotic expressions for $E_{lm}^{(1,2)}(\cos\theta)$ can be written in terms of exponential functions

$$l^{-m}E_{lm}^{(1,2)}(\cos\theta) = \frac{1}{[2\pi l \sin\theta]^{1/2}} \exp\left\{\mp i\left[\left(l+\frac{1}{2}\right)\theta - \frac{\pi}{4} - \frac{m\pi}{2}\right]\right\}\left[1 + O\left(\frac{1}{l}\right)\right], \tag{H.60}$$

$$\sin\theta \gg \frac{1}{l}, \qquad |l| \gg m, \qquad |l|\varepsilon \gg 1, \qquad \varepsilon \leq \theta \leq \pi - \varepsilon, \qquad \varepsilon > 0.$$

Equations (H.55)–(H.60) are valid also for complex values of l and m provided that

$$|l| > |m|, \qquad |\arg l| < \pi, \qquad |l|\sin\theta \gg 1.$$

Another useful approximation stems from *Hilb's relation*

$$P_{\nu-1/2}(\cos\theta) = \left[\frac{\theta}{\sin\theta}\right]^{1/2} J_0(\nu\theta) + O\left(\frac{1}{\nu^{3/2}}\right), \qquad \theta \leq \frac{\pi}{2}, \tag{H.61}$$

which reduces to Eq. (H.52) for $l = \nu - \frac{1}{2}$ and $|\nu|\theta \gg 1$ and remains valid for $\theta \to 0$. Likewise, we quote the *Szegö relation*

$$Q_{\nu-1/2}(\cos\theta) = -\frac{\pi}{2}\left[\frac{\theta}{\sin\theta}\right]^{1/2} N_0(\nu\theta)\left[1 + O\left(\frac{1}{\nu}\right)\right]. \tag{H.62}$$

Hence

$$E_{\nu-1/2,0}^{(1,2)}(\cos\theta) = \frac{1}{2}\left[\frac{\theta}{\sin\theta}\right]^{1/2} H_0^{(2,1)}(\nu\theta) + O\left(\frac{1}{\nu^{3/2}}\right), \qquad \theta \leq \frac{\pi}{2}. \tag{H.63}$$

Generalization to $m \neq 0$ can be made by using the relation

$$P_l^m(\cos\theta) = \sin^m\theta\left[-\frac{1}{\sin\theta}\frac{\partial}{\partial\theta}\right]^m P_l(\cos\theta) \tag{H.64}$$

and keeping terms of the highest order in l. For example,

$$\begin{aligned} P^1_{\nu-1/2}(\cos\theta) &= -\frac{\partial}{\partial\theta}\left\{\left[\frac{\theta}{\sin\theta}\right]^{1/2} J_0(\nu\theta)\right\} \\ &= \left\{-\frac{\partial}{\partial\theta}\left[\frac{\theta}{\sin\theta}\right]^{1/2}\right\} J_0(\nu\theta) + \nu\left[\frac{\theta}{\sin\theta}\right]^{1/2} J_1(\nu\theta) \end{aligned} \tag{H.65}$$

yields

$$\nu^{-1}P^1_{\nu-1/2}(\cos\theta) = \left[\frac{\theta}{\sin\theta}\right]^{1/2} J_1(\nu\theta) + O\left(\frac{1}{\nu^{3/2}}\right). \tag{H.66}$$

In general

$$\nu^{-m}E^{(1,2)}_{\nu-1/2,m}(\cos\theta) = \frac{1}{2}\left[\frac{\theta}{\sin\theta}\right]^{1/2} H_m^{(2,1)}(\nu\theta) + O\left(\frac{1}{\nu^{3/2}}\right). \tag{H.67}$$

H.3.1. Asymptotic Expansion of $y_{ml}(\theta, \phi)$ When Both m and l Are Large

We begin with the Legendre equation

$$(1 - x^2)\frac{d^2y}{dx^2} - 2x\frac{dy}{dx} + \left[l(l+1) - \frac{m^2}{1-x^2}\right]y = 0 \tag{H.68}$$

and make the substitution

$$W = y\sqrt{(1-x^2)}, \qquad -1 < x < 1.$$

The equation in W is

$$\frac{d^2W}{dx^2} + \nu^2 Q(x, \nu)W = 0, \tag{H.69}$$

where

$$\begin{gathered} \varepsilon^2 = \frac{1}{l(l+1)} \ll 1, \qquad \nu = \frac{1}{\varepsilon} \gg 1, \qquad \sin\theta_0 = \frac{m}{\sqrt{[l(l+1)]}} = m\varepsilon, \\ Q = \frac{a^2 + \varepsilon^2 - x^2}{(1-x^2)^2}, \qquad a = \cos\theta_0 = \left[1 - \frac{m^2}{l(l+1)}\right]^{1/2}. \end{gathered} \tag{H.70}$$

The WKBJ solution yields the expression

$$W = \begin{cases} A\dfrac{(1-x^2)^{1/2}}{(a^2+\varepsilon^2-x^2)^{1/4}}\exp\left[\pm i\nu\displaystyle\int^x \frac{(a^2+\varepsilon^2-x^2)^{1/2}}{1-x^2}\,dx\right], & a^2+\varepsilon^2 > x^2 \\ B\dfrac{(1-x^2)^{1/2}}{(a^2+\varepsilon^2-x^2)^{1/4}}\exp\left[\pm \nu\displaystyle\int^x \frac{(x^2-\varepsilon^2-a^2)^{1/2}}{1-x^2}\,dx\right], & a^2+\varepsilon^2 < x^2. \end{cases} \tag{H.71}$$

Therefore, in the oscillatory region, with $a^2 + \varepsilon^2 \simeq a^2$ and

$$1 - m^2/l(l+1) > x^2 = \cos^2\theta$$

we have the following approximation for the fully normalized spherical harmonics defined in Eq. (8.21):

$$\begin{aligned} y_{ml}(\theta, \phi) &= \frac{W}{(1-x^2)^{1/2}} e^{im\phi} = \frac{A_1}{(a^2 - x^2)^{1/4}} \exp\left[iv \int^x \frac{(a^2-x^2)^{1/2}}{1-x^2}\, dx + im\phi\right] \\ &+ \frac{A_2}{(a^2-x^2)^{1/4}} \exp\left[-iv \int^x \frac{(a^2-x^2)^{1/2}}{1-x^2}\, dx + im\phi\right]. \end{aligned} \tag{H.72}$$

Written in a slightly modified form, with the proper normalizing factors $A_1 = (1/2\pi)\exp(\pi i/4)$, $A_2 = (1/2\pi)\exp(-\pi i/4)$ [which can be obtained by putting $m = 0$ and comparing with Eq. (H.52)], it becomes

$$y_{ml}(\theta, \phi) = \frac{1}{2\pi}\left[\frac{l + \frac{1}{2}}{p_\theta \sin\theta}\right]^{1/2} [e^{i\Phi^+} + e^{i\Phi^-}] + O\left(\frac{1}{l^2}\right) \tag{H.73}$$

where

$$p_\theta = \left[\left(l + \frac{1}{2}\right)^2 - \frac{m^2}{\sin^2\theta}\right]^{1/2},$$

$$m = \left(l + \frac{1}{2}\right)\sin\theta_0, \qquad \Phi^\pm = \pm \int_{\theta_0}^{\theta} p_\theta\, d\theta + m\phi \mp \frac{\pi}{4}. \tag{H.74}$$

Equation (H.73) can be written in yet another form

$$y_{ml}(\theta, \phi) = \frac{e^{i\Phi^+} + e^{i\Phi^-}}{2\pi(\sin^2\theta - \sin^2\theta_0)^{1/4}} + O\left(\frac{1}{l^2}\right). \tag{H.75}$$

Noting that

$$\begin{aligned} I &= \int_{\theta_0}^{\theta} \frac{1}{\sin\theta} [\sin^2\theta - \sin^2\theta_0]^{1/2} d\theta \\ &= \int_{\theta_0}^{\theta} \frac{\sin\theta\, d\theta}{[\sin^2\theta - \sin^2\theta_0]^{1/2}} - \sin^2\theta_0 \int_{\theta_0}^{\theta} \frac{d\theta}{\sin\theta[\sin^2\theta - \sin^2\theta_0]^{1/2}}, \end{aligned}$$

$$\int_{\theta_0}^{\theta} \frac{\sin\theta\, d\theta}{[\sin^2\theta - \sin^2\theta_0]^{1/2}} = \frac{\pi}{2} - \sin^{-1}\left(\frac{\cos\theta}{\cos\theta_0}\right),$$

$$\int_{\theta_0}^{\theta} \frac{d\theta}{\sin\theta[\sin^2\theta - \sin^2\theta_0]^{1/2}} = -\frac{1}{\sin\theta_0}\tan^{-1}\frac{\cos\theta\sin\theta_0}{[\sin^2\theta - \sin^2\theta_0]^{1/2}} + \frac{1}{\sin\theta_0}\frac{\pi}{2},$$

$$\sin\theta_0 = \frac{m}{l + \frac{1}{2}} \ll 1, \qquad \cos\theta_0 = \left[1 - \frac{m^2}{(l + \frac{1}{2})^2}\right]^{1/2} \simeq 1,$$

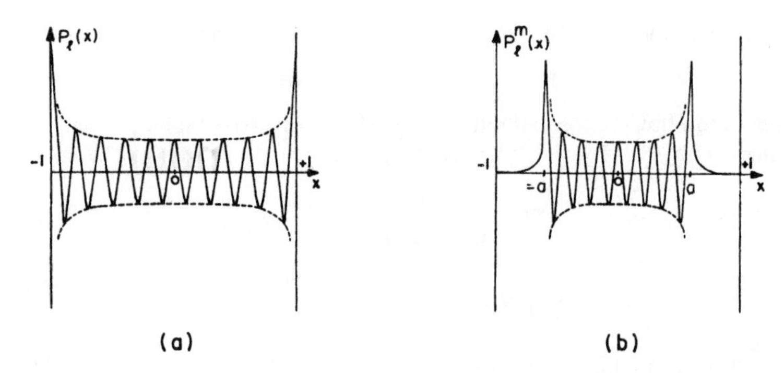

Figure H.1. (a) Oscillatory behavior of the Legendre polynomial $P_l(x)$ in the interval $|x| \leq 1$ for large values of l. (b) Oscillatory behavior of the associated Legendre polynomial $P_l^m(x)$ in the interval $|x| < a = \cos\theta_0$ and exponential behavior for $|x| > a$.

we find

$$
\begin{aligned}
i\int_{\theta_0}^{\theta} p_\theta\, d\theta &= i\left(l+\frac{1}{2}\right)\int_{\theta_0}^{\theta}\frac{1}{\sin\theta}[\sin^2\theta - \sin^2\theta_0]^{1/2} d\theta \\
&= i\left(l+\frac{1}{2}\right)\left[\frac{\pi}{2} - \sin^{-1}\left(\frac{\cos\theta}{\cos\theta_0}\right) + \frac{m}{l+\frac{1}{2}}\right. \\
&\quad \left.\times \tan^{-1}\frac{\cos\theta\sin\theta_0}{[\sin^2\theta - \sin^2\theta_0]^{1/2}} - \frac{\pi}{2}\frac{m}{l+\frac{1}{2}}\right] \\
&\simeq i\left(l+\frac{1}{2}\right)\left[\theta + \frac{\pi}{2}\frac{m}{l+\frac{1}{2}}\right].
\end{aligned}
$$

Now, because

$$
y_{ml}(\theta,\phi) = (-)^m\left[\frac{2l+1}{4\pi}\frac{(l-m)!}{(l+m)!}\right]^{1/2} P_l^m(\cos\theta)e^{im\phi}
$$

we get

$$
P_l^m(\cos\theta) = \frac{1}{\pi}\frac{(-)^m}{\sqrt{(\sin\theta)}}\sin\left[\left(l+\frac{1}{2}\right)\theta + \frac{\pi}{4} + \frac{m\pi}{2}\right]\left[\frac{2\pi}{l+\frac{1}{2}}\frac{(l+m)!}{(l-m)!}\right]^{1/2},
$$

$$
\left[\frac{(l-m)!}{(l+m)!}\right]^{1/2} P_l^m(\cos\theta) = (-)^m\left[\frac{2}{\pi l\sin\theta}\right]^{1/2}\sin\left[\left(l+\frac{1}{2}\right)\theta + \frac{\pi}{4} + \frac{m\pi}{2}\right]. \tag{H.76}
$$

However, for $l \gg m$,

$$
\left[\frac{(l-m)!}{(l+m)!}\right]^{1/2} \simeq l^{-m},
$$

and Eq. (H.55) is obtained. The asymptotic behavior of $P_l^m(\cos\theta)$ is shown in Fig. H.1b.

APPENDIX I

Generalized Spherical Harmonics

I.1. Spherical Harmonics

Let a scalar function $\psi(\mathbf{r})$ be defined in 3-space. Consider a point whose position vector is $\mathbf{r}$ and let it be translated to a new point $\mathbf{r} + d\mathbf{r}$. The change in ψ will be $d\psi = \psi(\mathbf{r} + d\mathbf{r}) - \psi(\mathbf{r}) = d\mathbf{r} \cdot \nabla\psi$. We write

$$\frac{d\psi}{d\mathbf{r}} = \nabla\psi \quad \text{or} \quad \frac{d}{d\mathbf{r}} = \nabla. \tag{I.1}$$

Analogously, we may effect an infinitesimal rotation $d\boldsymbol{\omega}$ such that the point $\mathbf{r}$ goes to $\mathbf{r}' = \mathbf{r} + (d\boldsymbol{\omega} \times \mathbf{r})$. The corresponding change in ψ is

$$d\psi = d\mathbf{r} \cdot \nabla\psi = (d\boldsymbol{\omega} \times \mathbf{r}) \cdot \nabla\psi = d\boldsymbol{\omega} \cdot (\mathbf{r} \times \nabla\psi). \tag{I.2}$$

Written symbolically it reads

$$\frac{d\psi}{d\boldsymbol{\omega}} = \mathbf{r} \times \nabla\psi \quad \text{or} \quad \frac{d}{d\boldsymbol{\omega}} = \mathbf{r} \times \nabla. \tag{I.3}$$

The operator $d/d\boldsymbol{\omega}$ is denoted here by $\mathbf{B}$. Its components in Cartesian and spherical coordinates are

$$\begin{aligned}
B_x &= y\frac{\partial}{\partial z} - z\frac{\partial}{\partial y} = -\left[\sin\phi\frac{\partial}{\partial\theta} + \cot\theta\cos\phi\frac{\partial}{\partial\phi}\right], \\
B_y &= z\frac{\partial}{\partial x} - x\frac{\partial}{\partial z} = \left[\cos\phi\frac{\partial}{\partial\theta} - \cot\theta\sin\phi\frac{\partial}{\partial\phi}\right], \\
B_z &= x\frac{\partial}{\partial y} - y\frac{\partial}{\partial x} = \frac{\partial}{\partial\phi}, \\
\mathbf{B} &= \mathbf{e}_\phi\frac{\partial}{\partial\theta} - \mathbf{e}_\theta\frac{1}{\sin\theta}\frac{\partial}{\partial\phi}.
\end{aligned} \tag{I.4}$$

The square of the operator $\mathbf{B}$ is known as the *Beltrami operator* [cf. Eq. (7.14)]

$$(\mathbf{B} \cdot \mathbf{B}) = B^2 = \frac{1}{\sin\theta}\frac{\partial}{\partial\theta}\left(\sin\theta\frac{\partial}{\partial\theta}\right) + \frac{1}{\sin^2\theta}\frac{\partial^2}{\partial\phi^2}. \tag{I.5}$$

Furthermore, Eq. (2.71) yields

$$\sqrt{[l(l+1)]}\,\mathbf{C}_{ml}(\theta, \phi) = -\mathbf{B}\{Y_{ml}\} = -\mathbf{B}\{P_l^m(\cos\theta)e^{im\phi}\}. \qquad \text{(I.6)}^{(1)}$$

Next, we wish to show that the infinitesimal rotation operator has a rather simple effect on the spherical harmonics. Indeed, using the recurrence relations

$$\begin{aligned} 2m\cot\theta P_l^m(\cos\theta) &= P_l^{m+1}(\cos\theta) + (l+m)(l-m+1)P_l^{m-1}(\cos\theta), \\ 2\frac{\partial}{\partial\theta}P_l^m(\cos\theta) &= -P_l^{m+1}(\cos\theta) + (l+m)(l-m+1)P_l^{m-1}(\cos\theta), \end{aligned} \qquad \text{(I.7)}$$

Eqs. (I.4) yield

$$\begin{aligned} B_z\{Y_{ml}\} &= imY_{ml}, \\ (B_x + iB_y)\{Y_{ml}\} &= -iY_{(m+1)l}, \\ (B_x - iB_y)\{Y_{ml}\} &= -i(l-m+1)(l+m)Y_{(m-1)l}, \\ B^2\{Y_{ml}\} + l(l+1)Y_{ml} &= 0. \end{aligned} \qquad \text{(I.8)}$$

Consequently, an infinitesimal rotation of Y_{ml} about the z axis simply changes the phase of Y_{ml} and a rotation about the x or y axis brings in only the two neighboring functions $Y_{(m+1)l}$ and $Y_{(m-1)l}$. We note that the spherical harmonics are eigenfunctions of the operator B^2.

I.2. Generalized Spherical Harmonics

We shall next evaluate the differential operators corresponding to an infinitesimal rotation by the three Euler angles. Consider the rotation vector $\boldsymbol{\omega}$ with respect to the fixed system (x, y, z). Then

$$\boldsymbol{\omega} = \mathbf{e}_x\omega_x + \mathbf{e}_y\omega_y + \mathbf{e}_z\omega_z$$

and

$$\mathbf{B} = \frac{d}{d\boldsymbol{\omega}} = \mathbf{e}_x\frac{\partial}{\partial\omega_x} + \mathbf{e}_y\frac{\partial}{\partial\omega_y} + \mathbf{e}_z\frac{\partial}{\partial\omega_z}.$$

Clearly, $B_x = \partial/\partial\omega_x$, $B_y = \partial/\partial\omega_y$, $B_z = \partial/\partial\omega_z$. Introducing the three Euler angles (α, β, γ), we may write

$$\omega_x = \omega_x(\alpha, \beta, \gamma), \qquad \omega_y = \omega_y(\alpha, \beta, \gamma), \qquad \omega_z = \omega_z(\alpha, \beta, \gamma).$$

Therefore,

$$\begin{aligned} B_x &= \frac{\partial}{\partial\omega_x} = \frac{\partial\alpha}{\partial\omega_x}\frac{\partial}{\partial\alpha} + \frac{\partial\beta}{\partial\omega_x}\frac{\partial}{\partial\beta} + \frac{\partial\gamma}{\partial\omega_x}\frac{\partial}{\partial\gamma}, \\ B_y &= \frac{\partial}{\partial\omega_y} = \frac{\partial\alpha}{\partial\omega_y}\frac{\partial}{\partial\alpha} + \frac{\partial\beta}{\partial\omega_y}\frac{\partial}{\partial\beta} + \frac{\partial\gamma}{\partial\omega_y}\frac{\partial}{\partial\gamma}, \\ B_z &= \frac{\partial}{\partial\omega_z} = \frac{\partial\alpha}{\partial\omega_z}\frac{\partial}{\partial\alpha} + \frac{\partial\beta}{\partial\omega_z}\frac{\partial}{\partial\beta} + \frac{\partial\gamma}{\partial\omega_z}\frac{\partial}{\partial\gamma}. \end{aligned} \qquad \text{(I.9)}$$

[1] The infinitesimal rotation operator **B** should not be confused with the vector spherical harmonic $\mathbf{B}_{ml}(\theta, \phi)$ defined in Eq. (2.71).

However, from Eq. (A.1.12) we know that

$$\frac{\partial \alpha}{\partial \omega_x} = -\cot \beta \cos \alpha, \qquad \frac{\partial \alpha}{\partial \omega_y} = -\cot \beta \sin \alpha, \qquad \frac{\partial \alpha}{\partial \omega_z} = 1,$$

$$\frac{\partial \beta}{\partial \omega_x} = -\sin \alpha, \qquad \frac{\partial \beta}{\partial \omega_y} = \cos \alpha, \qquad \frac{\partial \beta}{\partial \omega_z} = 0, \tag{I.10}$$

$$\frac{\partial \gamma}{\partial \omega_x} = \frac{\cos \alpha}{\cos \beta}, \qquad \frac{\partial \gamma}{\partial \omega_y} = \frac{\sin \alpha}{\sin \beta}, \qquad \frac{\partial \gamma}{\partial \omega_z} = 0.$$

Substituting in Eqs. (I.9), we have, after denoting $\beta = \theta$, $\alpha = \phi$,

$$B_x = -\left[\sin \phi \frac{\partial}{\partial \theta} + \cos \phi \cot \theta \frac{\partial}{\partial \phi} - \frac{\cos \phi}{\sin \theta} \frac{\partial}{\partial \gamma}\right],$$

$$B_y = \left[\cos \phi \frac{\partial}{\partial \theta} - \sin \phi \cot \theta \frac{\partial}{\partial \phi} + \frac{\sin \phi}{\sin \theta} \frac{\partial}{\partial \gamma}\right], \tag{I.11}$$

$$B_z = \frac{\partial}{\partial \phi},$$

$$\mathbf{B} = \mathbf{e}_\phi \frac{\partial}{\partial \theta} - \mathbf{e}_\theta \frac{1}{\sin \theta} \frac{\partial}{\partial \phi} + \mathbf{e}_\Delta \frac{1}{\sin \theta} \frac{\partial}{\partial \gamma}, \tag{I.12}$$

$$\mathbf{e}_\Delta = \mathbf{e}_r \sin \theta + \mathbf{e}_\theta \cos \theta = \mathbf{e}_x \cos \phi + \mathbf{e}_y \sin \phi.$$

Consequently,

$$B^2 = \left[\frac{1}{\sin \theta} \frac{\partial}{\partial \theta}\left(\sin \theta \frac{\partial}{\partial \theta}\right) + \frac{1}{\sin^2 \theta} \frac{\partial^2}{\partial \phi^2}\right] + \left[\frac{1}{\sin^2 \theta} \frac{\partial^2}{\partial \gamma^2} - \frac{2 \cos \theta}{\sin^2 \theta} \frac{\partial^2}{\partial \phi \, \partial \gamma}\right]. \tag{I.13}$$

If the function on which B^2 is operating, say Y, is independent of γ, the solution of the equation

$$B^2 Y + l(l+1) Y = 0 \tag{I.14}$$

renders the well-known spherical harmonics $Y_{ml}(\theta, \phi)$. However, if there exists an *a-priori* dependence on ϕ, θ, and γ, we must then seek a *generalized spherical harmonic*, say $T_l^{mn}(\phi, \theta, \gamma)$, such that Eq. (I.14) is satisfied with B^2 given by Eq. (I.13).

This leads us to the partial differential equation

$$\frac{\partial^2 T_l^{mn}}{\partial \theta^2} + \cot \theta \frac{\partial T_l^{mn}}{\partial \theta} + \frac{1}{\sin^2 \theta}\left[\frac{\partial^2 T_l^{mn}}{\partial \phi^2} - 2 \cos \theta \frac{\partial^2 T_l^{mn}}{\partial \phi \, \partial \gamma} + \frac{\partial^2 T_l^{mn}}{\partial \gamma^2}\right]$$
$$+ l(l+1) T_l^{mn} = 0. \tag{I.15}$$

If we substitute

$$T_l^{mn}(\phi, \theta, \gamma) = e^{im\phi + in\gamma} P_l^{mn}(\cos \theta), \tag{I.16}$$

the equation for $P_l^{mn}(\cos\theta)$ becomes

$$\frac{d^2 P_l^{mn}}{d\theta^2} + \cot\theta \frac{dP_l^{mn}}{d\theta} + \left[l(l+1) - \frac{n^2 - 2mn\cos\theta + m^2}{\sin^2\theta}\right] P_l^{mn} = 0. \quad \text{(I.17)}$$

Here m and n must be integers to secure the single-valuedness of T_l^{mn}. Note that as far as Eq. (I.17) is concerned, P_l^{mn} is symmetrical with respect to n and m. The functions $P_l^{mn}(\mu)$ are called *ultraspherical functions.*

The substitution

$$P_l^{mn}(\mu) = F(l, m, n)(1-\mu)^{|m-n|/2}(1+\mu)^{(m+n)/2} V(\mu) \quad \text{(I.18)}$$

transforms Eq. (I.17) into the differential equation

$$(1-\mu^2)V'' + [\beta - \alpha - (\alpha + \beta + 2)\mu]V' + s(\alpha + \beta + s + 1)V = 0,$$

which is the standard equation for the *Jacobi polynomials*

$$\begin{gathered} V(\mu) = P_s^{(\alpha,\beta)}(\mu), \qquad m \geq 0, \qquad n \geq 0, \\ \mu = \cos\theta, \qquad \alpha = |m-n|, \qquad \beta = m + n, \\ s = l - \frac{1}{2}(m+n) - \frac{1}{2}|m-n| = l - \max(m, n) \end{gathered} \quad \text{(I.19)}$$

(α, β should not be confused with the Euler angles). The normalization factor

$$F(l, m, n) = \frac{1}{2^{\max(m,n)}} \left\{ \frac{[l + \max(m,n)]!\,[l - \max(m,n)]!}{[l + \min(m,n)]!\,[l - \min(m,n)]!} \right\}^{1/2} \quad \text{(I.20)}$$

guarantees the symmetry properties

$$\begin{aligned} P_l^{mn}(\mu) &= (-)^{m+n} P_l^{nm}(\mu) = P_l^{(-n)(-m)}(\mu), \\ P_l^{mn}[\cos(\pi - \theta)] &= (-)^{l-m} P_l^{m(-n)}(\cos\theta), \end{aligned} \quad \text{(I.21)}$$

and the orthogonality relations

$$\int_0^\pi P_l^{mn}(\cos\theta) P_{l'}^{mn}(\cos\theta) \sin\theta \, d\theta = \frac{2}{2l+1} \delta_{ll'}, \quad \text{(I.22)}$$

$$\int_0^{2\pi} \int_0^\pi \int_0^{2\pi} \overset{*}{T}{}_l^{mn}(\phi, \theta, \gamma) T_{l'}^{m'n'}(\phi, \theta, \gamma) d\phi \sin\theta \, d\theta \, d\gamma = \frac{8\pi^2}{2l+1} \delta_{ll'} \delta_{mm'} \delta_{nn'}. \quad \text{(I.23)}$$

From Eqs. (I.18), (I.19), and (I.20) and on account of the properties of the Jacobi polynomials, we derive an expression from which the ultraspherical functions can be constructed explicitly

$$P_l^{mn}(\mu) = \frac{(-)^s}{2^l s!} \left\{ \frac{[l + \max(m,n)]!\,[l - \max(m,n)]!}{[l + \min(m,n)]!\,[l - \min(m,n)]!} \right\}^{1/2} \times \frac{(d^s/d\mu^s)[(1-\mu)^{s+\alpha}(1+\mu)^{s+\beta}]}{(1-\mu)^{\alpha/2}(1+\mu)^{\beta/2}}. \quad \text{(I.24)}$$

For $n = 0$, the combination of Eqs. (I.24) and (D.119)–(D.121) yields

$$P_l^{m0}(\mu) = \left[\frac{(l-m)!}{(l+m)!}\right]^{1/2} P_l^m(\mu)$$

with $P_l^m(\mu)$ as in Eq. (D.120). However, if we redefine

$$P_{ml}(\mu) = \left[\frac{2l+1}{4\pi}\frac{(l-m)!}{(l+m)!}\right]^{1/2} P_l^m(\mu),$$
$$Y_l^m(\theta, \phi) = \left[\frac{(l-m)!}{(l+m)!}\right]^{1/2} e^{im\phi} P_l^m(\cos\theta) = \left[\frac{(l-m)!}{(l+m)!}\right]^{1/2} Y_{ml}(\theta, \phi), \tag{I.24a}$$

we have

$$P_l^{m0}(\mu) = \left[\frac{4\pi}{2l+1}\right]^{1/2} P_{ml}(\mu),$$

$$T_l^{m0}(\phi, \theta, \gamma) = Y_l^m(\theta, \phi) = \left[\frac{4\pi}{2l+1}\right]^{1/2} e^{im\phi} P_{ml}(\mu), \qquad Y_l^{-m} = (-)^m \overset{*}{Y}{}_l^m. \tag{I.25}$$

Another useful representation of $P_l^{mn}(\mu)$ is derived by means of a series representation of the Jacobi polynomials

$$P_s^{(\alpha,\beta)}(\mu) = 2^{-s}\sum_{r=0}^{s}\binom{s+\alpha}{r}\binom{s+\beta}{s-r}(\mu+1)^r(\mu-1)^{s-r}.$$

Equations (I.18)–(I.20) then yield,

$$P_l^{mn}(\mu) = 2^{-l}\left\{\frac{[l+\max(m,n)]!\,[l-\max(m,n)]!}{[l+\min(m,n)]!\,[l-\min(m,n)]!}\right\}^{1/2}$$
$$\times \sum_{r=0}^{l-\max(m,n)} \binom{|n-m| + l - \max(m,n)}{r}\binom{m+n+l-\max(m,n)}{l-\max(m,n)-r}$$
$$\times (1+\mu)^A (1-\mu)^B (-)^{l-\max(m,n)-r}; \tag{I.26}$$

$$A = r + \frac{1}{2}(m+n), \qquad B = l - \frac{1}{2}(m+n) - r.$$

With the aid of Eq. (I.26) we prove, after some lengthy but straightforward algebra,

$$\int_0^{2\pi}\int_0^{\pi}\int_0^{2\pi} T_{l_1}^{m_1 n_1}(\phi,\theta,\gamma) T_{l_2}^{m_2 n_2}(\phi,\theta,\gamma) T_{l_3}^{m_3 n_3}(\phi,\theta,\gamma) d\phi \sin\theta\, d\theta\, d\gamma$$
$$= 8\pi^2 \begin{pmatrix} l_1 & l_2 & l_3 \\ n_1 & n_2 & n_3 \end{pmatrix}\begin{pmatrix} l_1 & l_2 & l_3 \\ m_1 & m_2 & m_3 \end{pmatrix}. \tag{I.27}$$

Each of the two symbols on the right-hand side of Eq. (I.27) is known as *Wigner's symbol.* Its explicit form is

$$\begin{pmatrix} l_1 & l_2 & l_3 \\ m_1 & m_2 & m_3 \end{pmatrix} = c \sum_N \frac{(-)^K}{N!}$$
$$\times \left[\frac{(l_1+m_1)!(l_1-m_1)!(l_2+m_2)!(l_2-m_2)!(l_3+m_3)!(l_3-m_3)!}{(l_1+l_2-l_3-N)!(l_1-m_1-N)!(l_2+m_2-N)! \times (l_3-l_2+m_1+N)!(l_3-l_1-m_2+N)!} \right],$$

$$c = \left[\frac{(2p-2l_1)!(2p-2l_2)!(2p-2l_3)!}{(2p+1)!} \right]^{1/2} \delta_{m_1+m_2,\,-m_3}, \tag{I.28}$$
$$2p = l_1 + l_2 + l_3,$$
$$K = l_1 - l_2 - m_3 + N.$$

The summation is over positive values of N such that the arguments in the denominator are nonnegative. Therefore, the sum contains only a finite number of terms.

Although it is hard to see in this form, the numerical value of the symbol is unchanged upon an even permutation of the columns, whereas an odd permutation is equivalent to multiplication by $(-)^{l_1+l_2+l_3}$. The Wigner symbol is different from zero only if $m_1 + m_2 + m_3 = 0$ and l_1, l_2, l_3 are the sides of a triangle, i.e., $|l_\beta - l_\gamma| \le l_\alpha \le l_\beta + l_\gamma$. In particular,

$$\begin{pmatrix} l_1 & l_2 & l_3 \\ 0 & 0 & 0 \end{pmatrix} = \begin{cases} \dfrac{(-)^p p!}{(p-l_1)!(p-l_2)!(p-l_3)!} \left[\dfrac{(2p-2l_1)!(2p-2l_2)!(2p-2l_3)!}{(2p+1)!} \right]^{1/2}, & \text{if } 2p \text{ is even} \\ 0, & \text{if } 2p \text{ is odd} \end{cases} \tag{I.28a}$$

$$\begin{pmatrix} l_1 & l_2 & 0 \\ m_1 & m_2 & 0 \end{pmatrix} = \frac{(-)^{l_1-m_1}}{\sqrt{(2l_1+1)}} \delta_{l_1 l_2} \delta_{m_1(-m_2)},$$
$$\begin{pmatrix} l_1 & l_2 & l_3 \\ -m_1 & -m_2 & -m_3 \end{pmatrix} = (-)^{l_1+l_2+l_3} \begin{pmatrix} l_1 & l_2 & l_3 \\ m_1 & m_2 & m_3 \end{pmatrix}. \tag{I.28b}$$

Special cases of Eq. (I.27) are, for $n_i = 0$,

$$\int_0^{2\pi}\int_0^{\pi} Y_{l_1}^{m_1}(\theta,\phi) Y_{l_2}^{m_2}(\theta,\phi) Y_{l_3}^{m_3}(\theta,\phi) \sin\theta \, d\theta \, d\phi = 4\pi \begin{pmatrix} l_1 & l_2 & l_3 \\ 0 & 0 & 0 \end{pmatrix} \begin{pmatrix} l_1 & l_2 & l_3 \\ m_1 & m_2 & m_3 \end{pmatrix} \tag{I.28c}$$

and, for $n_i = m_i = 0$,

$$\int_0^{\pi} P_{l_1}(\cos\theta) P_{l_2}(\cos\theta) P_{l_3}(\cos\theta) \sin\theta \, d\theta = 2 \begin{pmatrix} l_1 & l_2 & l_3 \\ 0 & 0 & 0 \end{pmatrix}^2, \tag{I.28d}$$

where $P_l(\cos\theta)$ is the Legendre polynomial.

I.2.1. Recurrence Relations

Recurrence relations exist among the generalized spherical harmonics. Some are obtained by a factorization of Eq. (I.17)

$$\left[\frac{\partial}{\partial\theta} + \frac{(m+1)\cos\theta - n}{\sin\theta}\right]\left[\frac{\partial}{\partial\theta} + \frac{n - m\cos\theta}{\sin\theta}\right] P_l^{mn} = [m(m+1) - l(l+1)]P_l^{mn}. \tag{I.29}$$

Knowing the recurrence relations in the case $n = 0$ [App. D], we generalize by assuming

$$\left[\frac{\partial}{\partial\theta} + \frac{n - m\cos\theta}{\sin\theta}\right] P_l^{mn} = -\Gamma_{m+1} P_l^{(m+1)n}, \tag{I.30}$$

$$\left[\frac{\partial}{\partial\theta} + \frac{m\cos\theta - n}{\sin\theta}\right] P_l^{mn} = \Gamma_m P_l^{(m-1)n}. \tag{I.31}$$

Substituting Eq. (I.30) into Eq. (I.31) and using Eq. (I.29), we obtain

$$\Gamma_m = \sqrt{(l+m)(l-m+1)}.$$

Subtracting Eq. (I.30) from Eq. (I.31), we form the new recurrence relation

$$\Gamma_{m+1} P_l^{(m+1)n} = 2\frac{m\cos\theta - n}{\sin\theta} P_l^{mn} - \Gamma_m P_l^{(m-1)n}. \tag{I.32}$$

Interchanging the roles of n and m in Eqs. (I.30)–(I.32) and using Eq. (I.21), we find

$$\Gamma_{n+1} P_l^{m(n+1)} + \Gamma_n P_l^{m(n-1)} = -2\frac{n\cos\theta - m}{\sin\theta} P_l^{mn}, \tag{I.33}$$

$$\Gamma_n P_l^{m(n-1)} - \Gamma_{n+1} P_l^{m(n+1)} = -2\frac{\partial}{\partial\theta} P_l^{mn}, \tag{I.34}$$

$$\Gamma_n = \sqrt{(l+n)(l-n+1)}.$$

From Eq. (I.24),

$$P_1^{mn}(\cos\theta) = \begin{bmatrix} \frac{1}{2}(1+\cos\theta) & -\frac{1}{\sqrt{2}}\sin\theta & \frac{1}{2}(1-\cos\theta) \\ \frac{1}{\sqrt{2}}\sin\theta & \cos\theta & -\frac{1}{\sqrt{2}}\sin\theta \\ \frac{1}{2}(1-\cos\theta) & \frac{1}{\sqrt{2}}\sin\theta & \frac{1}{2}(1+\cos\theta) \end{bmatrix} \begin{matrix} n=+1 \\ \\ n=0 \\ \\ n=-1. \end{matrix} \tag{I.35}$$

$$m=+1 \qquad m=0 \qquad m=-1$$

I.2.2. Addition Theorems

Because both n and m sweep the integer range from $-l$ to $+l$, there are $(2l+1)^2$ elements of $P_l^{mn}(\cos\theta)$ for each value of l. These can be put in the form of a square matrix of $(2l+1)$ rows. Such a matrix can be considered as a representation of the rotation $\mathfrak{R}$ in the sense of our previous definition of the finite rotation $\mathfrak{R}(\phi, \theta, \gamma)$ (cf. Example A.1). Because the effect of two successive independent rotations $\mathfrak{R}_1 = (\phi_1, \theta_1, \gamma_1)$ and $\mathfrak{R}_2 = (\phi_2, \theta_2, \gamma_2)$, in that order, is given by the product of their corresponding matrices, the combined rotation $\mathfrak{R}_2 \cdot \mathfrak{R}_1$ will have the representation of the matrix product of the individual representations,

$$T_l^{mn}(\mathfrak{R}_2 \cdot \mathfrak{R}_1) = \sum_{k=-l}^{l} T_l^{mk}(\mathfrak{R}_2) T_l^{kn}(\mathfrak{R}_1). \tag{I.36}$$

Equation (I.36) should be interpreted in the following sense: The (mn)th element of the matrix of the combined rotation $(\mathfrak{R}_2 \cdot \mathfrak{R}_1)$ is equal to the (mn)th element of the product of the matrix representation of $\mathfrak{R}_2$ and $\mathfrak{R}_1$. Equation (I.36) is known as the *addition formula* for the generalized spherical harmonics. To simplify Eq. (I.36) we assume, without loss of generality, that

$$\mathfrak{R}_1 = (\phi_1, \theta_1, 0), \qquad \mathfrak{R}_2 = (0, \theta_2, \gamma_2).$$

Then, because the rotation γ_2 follows the rotation ϕ_1 about the same axis, we may denote

$$\Lambda = \phi_1 + \gamma_2,$$

and get

$$e^{im\phi + in\gamma} P_l^{mn}(\theta) = \sum_{k=-l}^{l} e^{ik\Lambda} P_l^{mk}(\theta_2) P_l^{kn}(\theta_1), \tag{I.37}$$

where ϕ, θ, γ are yet unknown functions of $\Lambda, \theta_1, \theta_2$. To find these functions it is sufficient to choose some particular matrix elements on both sides. Taking

$$l = 1, \qquad (m, n) = (0, 0), (1, 0), (-1, 0)$$

and equating the real and imaginary parts in the resulting three equations we find, with the aid of Eq. (I.35),

$$\begin{aligned}
\cos\theta &= \cos\theta_1 \cos\theta_2 - \sin\theta_1 \sin\theta_2 \cos\Lambda, \\
\tan\phi &= \frac{\sin\Lambda \sin\theta_1}{\cos\theta_1 \sin\theta_2 + \sin\theta_1 \cos\theta_2 \cos\Lambda}, \\
\tan\gamma &= \frac{\sin\Lambda \sin\theta_2}{\cos\theta_2 \sin\theta_1 + \sin\theta_2 \cos\theta_1 \cos\Lambda}.
\end{aligned} \tag{I.38}$$

Three special cases are of particular interest:

1. $m = n = 0$

$$P_l(\cos\theta) = \sum_{k=-l}^{l} (-)^k \frac{(l-k)!}{(l+k)!} Y_{kl}(\theta_1, \phi_1) Y_{kl}(\theta_2, \gamma_2)$$

$$= \sum_{k=-l}^{l} (-)^k Y_l^k(\theta_2, \gamma_2) Y_l^k(\theta_1, \phi_1) \tag{I.39}$$

which coincides with the addition theorem, Eq. (D.165), for the Legendre polynomials on putting $\gamma_2 = \pi - \phi_2$.

2. $n = 0$

$$Y_l^m(\theta, \phi) = \sum_{k=-l}^{l} (-)^{k+m} \underset{\text{vector}}{Y_l^k(\theta_1, \phi_1)} \underset{\text{matrix}}{T_l^{km}(0, \theta_2, \gamma_2)}. \tag{I.40}$$

This is the transformation law of the spherical harmonics under the rotation of space as seen by an observer in a fixed frame. Equation (I.40) tells us that under the transformation

$$\mathbf{r}'(\theta, \phi) = \mathfrak{R}(0, \theta_2, \gamma_2) \cdot \mathbf{r}(\theta_1, \phi_1)$$

the value of Y_l^m at the new point (θ, ϕ) on the unit sphere (for fixed values of m and l) is given by a linear combination of all its orders from $m = -l$ to $m = +l$ at the former point (θ_1, ϕ_1). The coefficients of this transformation are the generalized spherical harmonics of the rotation $(0, \theta_2, \gamma_2)$.

3. $\Lambda = 0\ (\theta = \theta_1 + \theta_2, \phi = \gamma = 0)$

$$P_l^{mn}[\cos(\theta_1 + \theta_2)] = \sum_{k=-l}^{l} P_l^{mk}(\cos\theta_2) P_l^{kn}(\cos\theta_1). \tag{I.41}$$

In particular, for $n = 0$, we obtain an addition theorem for the associated Legendre polynomials

$$P_{ml}[\cos(\theta_1 + \theta_2)] = \sum_{k=-l}^{l} (-)^{m+k} P_{kl}(\cos\theta_1) P_l^{km}(\cos\theta_2). \tag{I.42}$$

I.2.3. Expansions

Assume the expansion

$$T_{l_1}^{m_1 n_1}(\mathfrak{R}) T_{l_2}^{m_2 n_2}(\mathfrak{R}) = \sum_{l, m, n} (2l+1) K \overset{*}{T}{}_l^{mn}(\mathfrak{R}). \tag{I.43}$$

Multiply each side by $T_{l_3}^{m_3 n_3}(\mathfrak{R})$ and integrate over α, β, γ on a unit sphere using Eqs. (I.23) and (I.27). The immediate result is

$$K = \begin{pmatrix} l_1 & l_2 & l \\ m_1 & m_2 & m \end{pmatrix} \begin{pmatrix} l_1 & l_2 & l \\ n_1 & n_2 & n \end{pmatrix}.$$

Because $m = -(m_1 + m_2)$, $n = -(n_1 + n_2)$, there remains in Eq. (I.43) a summation over l only. The sum over l is finite in the integer range

$$|l_1 - l_2| \leq l \leq l_1 + l_2.$$

In particular,

$$\overset{*}{Y}{}_{l_1}^{m_1}(\theta, \phi)\overset{*}{Y}{}_{l_2}^{m_2}(\theta, \phi) = \sum_l (2l + 1)\begin{pmatrix} l_1 & l_2 & l \\ m_1 & m_2 & m \end{pmatrix}\begin{pmatrix} l_1 & l_2 & l \\ 0 & 0 & 0 \end{pmatrix} Y_l^m(\theta, \phi),$$

$$m = -(m_1 + m_2) \quad \text{(I.44)}$$

$$P_{l_1}^{m_1}(\cos\theta)P_{l_2}^{m_2}(\cos\theta) = \sum_l (2l + 1)\begin{pmatrix} l_1 & l_2 & l \\ m_1 & m_2 & -(m_1 + m_2) \end{pmatrix}\begin{pmatrix} l_1 & l_2 & l \\ 0 & 0 & 0 \end{pmatrix}$$
$$\times (-)^{m_1 + m_2}\left[\frac{(l - m_1 - m_2)!(l_1 + m_1)!(l_2 + m_2)!}{(l + m_1 + m_2)!(l_1 - m_1)!(l_2 - m_2)!}\right]^{1/2}$$
$$\times P_l^{m_1 + m_2}(\cos\theta).$$

(I.45)

APPENDIX J

Transformation of Wave Functions under Translation and Rotation of the Coordinate Axes

J.1. Scalar Wave Functions under Translation of the Axes

In a (r, θ, ϕ) coordinate system [$O(x, y, z)$ in Fig. J.1], let the point O' at (r_0, θ_0, ϕ_0) be taken as the origin of a second coordinate system [$O'(x', y', z')$ in Fig. J.1] and oriented so that a rigid translation by a vector $\mathbf{r}_0$ takes the system O to O'. We consider a fixed point Q with coordinates (r, θ, ϕ) in O and (R, i_h, ϕ_h) in O'. The angle between $\mathbf{r}$ and $\mathbf{r}_0$ is denoted by χ (Fig. J.1). The axes (x', y', z') are parallel to the axes (x, y, z), respectively. Our aim is to express the spherical eigenfunctions $f_l(k_c r)Y_l^m(\theta, \phi)$ in terms of $f_l(k_c R)Y_l^m(i_h, \phi_h)$. Here $f_l(\xi)$ is either the spherical Bessel function $j_l(\xi)$ or the spherical Hankel function of the first or second kind $h_l^{(1,2)}(\xi)$ and Y_l^m is defined in Eq. (I.24a).

To begin with, we appeal to Eqs. (2.142) and (2.143), which we combine in the single expression

$$4\pi i^{-l} f_l(k_c r)Y_l^m(\theta, \phi) = \varepsilon \iint e^{-i\mathbf{k}\cdot\mathbf{r}} Y_l^m(e, \lambda) \sin e \, de \, d\lambda, \tag{J.1}$$

where $\varepsilon = 2$ for $f_l = h_l^{(1,2)}$ and $\varepsilon = 1$ for $f_l = j_l$. We also recast Eq. (2.141) in the form

$$e^{-i\mathbf{k}\cdot\mathbf{r}} = \sum_{\nu=0}^{\infty} \sum_{\mu=-\nu}^{\nu} (2\nu + 1) i^{-\nu} j_\nu(k_c r) Y_\nu^\mu(\theta, \phi) \overset{*}{Y}{}_\nu^\mu(e, \lambda). \tag{J.2}$$

Multiplying both sides of Eq. (J.2) by $Y_l^m(e, \lambda)$ and integrating over e, λ we get the following expansion

$$\iint Y_l^m(e, \lambda) e^{-i\mathbf{k}\cdot\mathbf{r}} \sin e \, de \, d\lambda = \sum_{\nu=0}^{\infty} \sum_{\mu=-\nu}^{\nu} (2\nu + 1) i^{-\nu} j_\nu(k_c r) Y_\nu^\mu(\theta, \phi)$$
$$\times \iint \overset{*}{Y}{}_\nu^\mu(e, \lambda) Y_l^m(e, \lambda) \sin e \, de \, d\lambda. \tag{J.3}$$

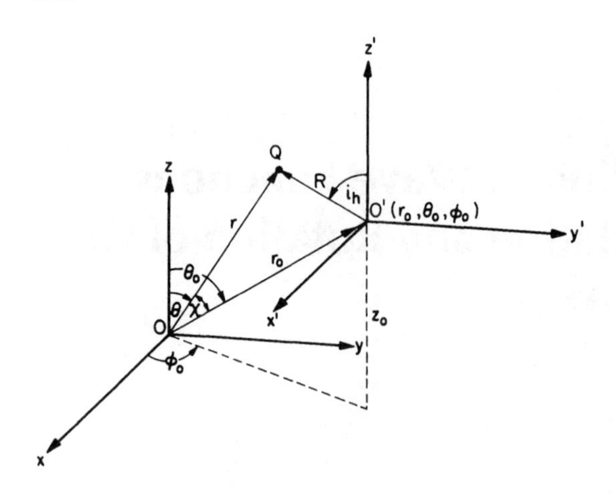

Figure J.1. General translation of the Cartesian axes.

Next, because $\mathbf{r} = \mathbf{r}_0 + \mathbf{R}$ (Fig. J.1), we have

$$e^{-i\mathbf{k}\cdot\mathbf{r}} = e^{-i\mathbf{k}\cdot\mathbf{R}}e^{-i\mathbf{k}\cdot\mathbf{r}_0}. \tag{J.4}$$

Let us assume that $R > r_0$. Then, by substituting Eq. (J.4) into Eq. (J.1), using Eq. (J.2) to expand $e^{-i\mathbf{k}\cdot\mathbf{r}_0}$, and interchanging the order of summation and integration, we obtain

$$4\pi i^{-l} f_l(k_c r) Y_l^m(\theta, \phi) = \varepsilon \sum_{\nu=0}^{\infty} \sum_{\mu=-\nu}^{\nu} (2\nu + 1) i^{-\nu} j_\nu(k_c r) Y_\nu^\mu(\theta_0, \phi_0)$$
$$\times \iint e^{-i\mathbf{k}\cdot\mathbf{R}} \overset{*}{Y}{}_\nu^\mu(e, \lambda) Y_l^m(e, \lambda) \sin e \, de \, d\lambda. \tag{J.5}$$

From Eqs. (I.25) and (I.44), we find

$$(-)^m Y_l^m(e, \lambda) \overset{*}{Y}{}_\nu^\mu(e, \lambda) = \sum_{q=|l-\nu|}^{l+\nu} (2q + 1) Y_q^p(e, \lambda) \begin{pmatrix} l & \nu & q \\ -m & \mu & p \end{pmatrix} \begin{pmatrix} l & \nu & q \\ 0 & 0 & 0 \end{pmatrix},$$
$$p = m - \mu. \tag{J.6}$$

Using Eq. (J.6) in Eq. (J.5) and noting from Eq. (J.1) that

$$\varepsilon \iint e^{-i\mathbf{k}\cdot\mathbf{R}} Y_q^p(e, \lambda) \sin e \, de \, d\lambda = 4\pi i^{-q} f_q(k_c R) Y_q^p(i_h, \phi_h), \tag{J.7}$$

we get

$$f_l(k_c r) Y_l^m(\theta, \phi) = \sum_{\nu=0}^{\infty} \sum_{\mu=-\nu}^{\nu} \sum_{q=|l-\nu|}^{l+\nu} A(l, m|\nu, \mu|q) \{j_\nu(k_c r_0) Y_\nu^\mu(\theta_0, \phi_0)\}$$
$$\times \{f_q(k_c R) Y_q^p(i_h, \phi_h)\}, \tag{J.8}$$

where

$$A(l, m|\nu, \mu|q) = (-)^m i^{l-\nu-q}(2\nu + 1)(2q + 1)$$
$$\times \begin{pmatrix} l & \nu & q \\ m & -\mu & \mu - m \end{pmatrix} \begin{pmatrix} l & \nu & q \\ 0 & 0 & 0 \end{pmatrix}. \tag{J.9}$$

Obviously, $A(l, m|\nu, \mu|q)$ is nonzero only when $l + \nu + q =$ even. The result in Eq. (J.8) is valid for $r_0 < R$. If the converse is true, $\mathbf{r}_0$ and $\mathbf{R}$ must be interchanged. Written in a form valid for both cases it becomes

$$f_l(k_c r) Y_l^m(\theta, \phi) = \sum_{\nu=0}^{\infty} \sum_{\mu=-\nu}^{\nu} \sum_{q=|l-\nu|}^{l+\nu} A(l, m|\nu, \mu|q) \{j_\nu(k_c r_<) Y_\nu^\mu(\theta_<, \phi_<)\}$$
$$\times \{f_q(k_c r_>) Y_q^{m-\mu}(\theta_>, \phi_>)\}, \tag{J.10}$$

where

$$(r_<, \theta_<, \phi_<) = \mathbf{r}_0, \qquad (r_>, \theta_>, \phi_>) = \mathbf{R} \quad \text{if} \quad R > r_0,$$
$$(r_<, \theta_<, \phi_<) = \mathbf{R}, \qquad (r_>, \theta_>, \phi_>) = \mathbf{r}_0 \quad \text{if} \quad R < r_0.$$

If $O(r, \theta, \phi)$ has its origin at the center of the earth and $O'(R, i_h, \phi_h)$ has its origin at the source, it is sometimes necessary to express the wave function relative to the source in terms of the wave functions relative to the earth's center. Instead of Eq. (J.5) we shall then write, for the case $r > r_0$,

$$4\pi i^{-l} f_l(k_c R) Y_l^m(i_h, \phi_h) = \varepsilon \sum_{\nu=0}^{\infty} \sum_{\mu=-\nu}^{\nu} (2\nu + 1) i^\nu j_\nu(k_c r_0) Y_\nu^\mu(\theta_0, \phi_0)$$
$$\times \iint e^{-i\mathbf{k}\cdot\mathbf{r}} \overset{*}{Y}{}_\nu^\mu(e, \lambda) Y_l^m(e, \lambda) \sin e \, de \, d\lambda.$$

Applying Eq. (J.6) together with Eq. (J.1), we obtain

$$f_l(k_c R) Y_l^m(i_h, \phi_h) = \sum_{\nu=0}^{\infty} \sum_{\mu=-\nu}^{\nu} \sum_{q=|l-\nu|}^{l+\nu} \tau^q (-\tau)^\nu A(l, m|\nu, \mu|q)$$
$$\times \{j_\nu(k_c r_<) Y_\nu^\mu(\theta_<, \phi_<)\} \{f_q(k_c r_>) Y_q^{m-\mu}(\theta_>, \phi_>)\},$$
$$\tau = \operatorname{sgn}(r - r_0) \tag{J.11}$$

which is the expected result because the coordinates of O relative to O' are $(r_0, \pi - \theta_0, \pi + \phi_0)$ and because

$$Y_\nu^\mu(\pi - \theta_0, \pi + \phi_0) = (-)^\nu Y_\nu^\mu(\theta_0, \phi_0).$$

J.1.1. Special Case: Translation Along the z Axis

A common case in seismology is the translation along the z axis of the initial coordinate system. Then $\theta_0 = 0$, and only terms corresponding to $\mu = 0$ remain. For $r > r_h$, we have

$$f_l(k_c R) Y_l^m(i_h, \phi_h) = \sum_{\nu=0}^{\infty} \sum_{q=|l-\nu|}^{l+\nu} (-)^\nu A(l, m|\nu, 0|q) j_\nu(k_c r_h) f_q(k_c r) Y_q^m(\theta, \phi). \tag{J.12}$$

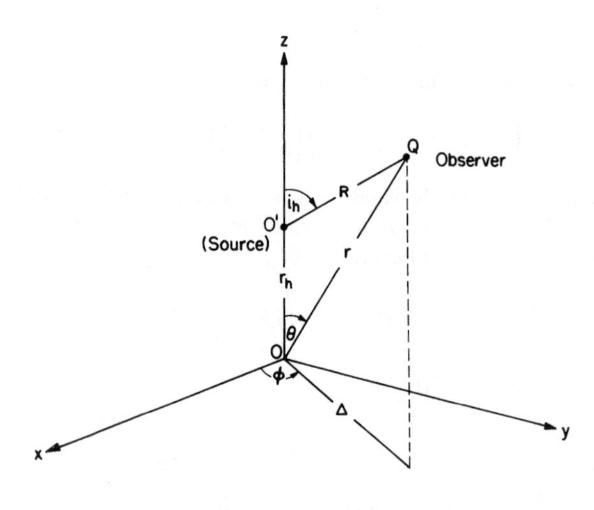

Figure J.2. Translation along the z axis.

The coefficient A vanishes unless $l + v + q$ is even and $m \leq q$. Because the inner sum is finite, the order of the sums can be reversed. Moreover, the sets $v = (0, \infty)$, $q = (|l - v|, l + v)$ and $q = (0, \infty)$, $v = (|q - l|, q + l)$ yield the same terms in Eq. (J.12) but in a different order. Reordering the terms, we have the alternative representation,

$$f_l(k_c R) Y_l^m(i_h, \phi_h) = \sum_{q=0}^{\infty} \sum_{v=|q-l|}^{q+l} (-)^v A(l, m | v, 0 | q) j_v(k_c r_h) f_q(k_c r) Y_q^m(\theta, \phi) \quad \text{(J.13)}$$

where the source is at $O'(r_h, 0, 0)$ as in Fig. J.2.

We shall next transform Eq. (J.13) into another form in which the coefficient A does not appear. To this end we apply some results derived in Chapter 2. We have seen in Eq. (2.6.10) that

$$j_v(k_c r_h) = i^{-v} P_v\left(\frac{\partial}{\partial i k_c r_h}\right) j_0(k_c r_h).$$

With this, Eq. (J.13) reads

$$f_l(k_c R) Y_l^m(i_h, \phi_h) = \sum_{q=0}^{\infty} \left\{ f_q(k_c r) Y_q^m(\theta, \phi) \sum_{v=|q-l|}^{q+l} (-)^v A(l, m | v, 0 | q) \times i^{-v} P_v\left(\frac{\partial}{\partial i k_c r_h}\right) j_0(k_c r_h) \right\}.$$

Now, Eqs. (I.28), (I.44), and (J.9) yield

$$\sum_{v=|q-l|}^{q+l} (-)^v A(l, m | v, 0 | q) i^{-v} P_v\left(\frac{\partial}{\partial i k_c r_h}\right) = (-)^m i^{l-q} (2q + 1) \times P_q^{-m}\left(\frac{\partial}{\partial i k_c r_h}\right) P_l^m\left(\frac{\partial}{\partial i k_c r_h}\right).$$

Replacing l by $l + m$ and changing the index of summation from q to $n + m$, we obtain

$$f_{m+l}(k_c R)P_{m+l}^m(\cos i_h)e^{im\phi_h} = \sum_{n=0}^{\infty} i^{l-n}(-)^m(2n + 2m + 1)$$

$$\times \{f_{m+n}(k_c r)P_{m+n}^m(\cos\theta)e^{im\phi}\}$$

$$\times P_{m+l}^m\left(\frac{\partial}{\partial ik_c r_h}\right)P_{m+n}^{-m}\left(\frac{\partial}{\partial ik_c r_h}\right)j_o(k_c r_h).$$

However,

$$P_{m+l}^m(-ix) = (2m-1)!!(1+x^2)^{m/2}i^{-l}F_{m+l}^m(x),$$

where $F_{m+l}^m(x)$ are polynomials that obey the recurrence relation

$$(2m + 2l + 1)xF_{m+l}^m(x) = (l+1)F_{m+l+1}^m(x) - (2m + l)F_{m+l-1}^m(x),$$

$$F_m^m(x) = 1, \qquad F_{m+1}^m(x) = (2m+1)x.$$

These are connected with the Gegenbauer polynomials through the relation $F_{m+l}^m(x) = i^l C_l^{m+1/2}(-ix)$. Because the operators $(1 + x^2)^m$ and $F_{m+n}^m(x)$ commute and because

$$\left(1 + \frac{\partial^2}{\partial\xi^2}\right)^m f_0(\xi) = m!2^m\left\{\frac{f_m(\xi)}{\xi^m}\right\},$$

[hint: use the differential equation of $f_m(\xi)$]

$$F_{m+n}^m\left(\frac{\partial}{\partial\xi}\right)\frac{f_m(\xi)}{\xi^m} = (-)^n\frac{(2m+n)!}{(2m)!n!}\frac{f_{m+n}(\xi)}{\xi^m}, \qquad \text{[from Eq. (2.6.23)]}$$

it transpires that

$$\left[1 + \frac{\partial^2}{\partial(k_c r_h)^2}\right]^m F_{m+n}^m\left[\frac{\partial}{\partial k_c r_h}\right]f_0(k_c r_h) = (-)^n\frac{(2m+n)!}{n!(2m-1)!!}\frac{f_{m+n}(k_c r_h)}{(k_c r_h)^m}. \qquad \text{(J.14)}$$

Consequently,

$$f_{m+l}(k_c R)P_{m+l}^m(\cos i_h)e^{im\phi_h} = \sum_{n=0}^{\infty}\{g_{m+n}(k_c r)P_{m+n}^m(\cos\theta)e^{im\phi}\}E_{m,m+l}^{m+n}(k_c r_h), \qquad \text{(J.15)}$$

where

$$E_{m,m+l}^{m+n}(x) = (2m-1)!!(2m + 2n + 1)F_{m+l}^m\left(\frac{\partial}{\partial x}\right)\left\{\frac{z_{m+n}(x)}{x^m}\right\}, \qquad \text{(J.16)}$$

$$z_l(x) = f_l(x), \qquad g_l(x) = j_l(x); \qquad r < r_h$$

$$z_l(x) = j_l(x), \qquad g_l(x) = f_l(x); \qquad r > r_h.$$

In particular,

$$\begin{aligned} E_{22}^{n+2}(x) &= 3(2n+5)\frac{z_{n+2}(x)}{x^2}, \\ E_{12}^{n+1}(x) &= 3(2n+3)\left[n\frac{z_{n+1}(x)}{x^2} - \frac{z_{n+2}(x)}{x}\right], \\ E_{02}^{n}(x) &= 3(2n+1)\left[\frac{n(n-1)}{2}\frac{z_n(x)}{x^2} + \frac{z_{n+1}(x)}{x} - \frac{z_n(x)}{3}\right]. \end{aligned} \tag{J.17}$$

J.2. Scalar and Vector Wave Functions under Rotation of the Axes

Because the radial part of the wave function, $f_l(k_c r)$, is invariant under rotation (active or passive), the transformation law of $\chi_{ml}^{\pm}(\mathbf{r}) = f_l^{\pm}(k_c r)Y_l^m(\theta, \phi)$ follows that of $Y_l^m(\theta, \phi)$, which we have already derived in Eq. (I.40). Because $\mathbf{M}_l^m(\mathbf{r}) = (\nabla \chi_{ml}) \times \mathbf{r}$ and because ∇ is defined independently of the coordinate system, the transformation of $\mathbf{M}_l^m(\mathbf{r})$ under rotation will follow the same rule as that of $Y_l^m(\theta, \phi)$. Similar arguments apply to the vector fields $\mathbf{L}_l^m$ and $\mathbf{N}_l^m$.

J.3. Translational Addition Theorems for Spherical Vector Wave Functions

We restate the transformation law in Eq. (J.11) for the scalar potential under translation

$$\Phi_{ml}^{\pm}(\mathbf{R}) = \sum_{\nu=0}^{\infty} \sum_{\mu=-\nu}^{\nu} \sum_{q=|l-\nu|}^{l+\nu} \tau^q(-\tau)^\nu B(l, m|\nu, \mu|q)\Phi_{\mu\nu}^{+}(\mathbf{r}_<)\Phi_{(m-\mu)q}^{\pm}(\mathbf{r}_>), \tag{J.18}$$

where

$$\begin{aligned} \mathbf{r}_< &= \mathbf{r}_0, \quad \mathbf{r}_> = \mathbf{r} \quad \text{if} \quad r_0 < r, \\ \mathbf{r}_< &= \mathbf{r}, \quad \mathbf{r}_> = \mathbf{r}_0 \quad \text{if} \quad r < r_0, \end{aligned}$$

$$\Phi_{ml}(\mathbf{r}) = f_l(k_c r)P_l^m(\cos\theta)e^{im\phi},$$

$$B(l, m|\nu, \mu|q) = \left[\frac{(l+m)!(\nu-\mu)!(q-m+\mu)!}{(l-m)!(\nu+\mu)!(q+m-\mu)!}\right]^{1/2} A(l, m|\nu, \mu|q),$$

$$\tau = \operatorname{sgn}(r - r_0)$$

and A is given explicitly in Eq. (J.9).

Because $\mathbf{L}_{ml}^{\pm} = (1/k_c)\nabla\Phi_{ml}^{\pm}$ and because the gradient operator is invariant to a transformation of the coordinate system, we can immediately write, for $r > r_0$,

$$\mathbf{L}_{ml}^{\pm}(\mathbf{R}) = \sum_{\nu=0}^{\infty} \sum_{\mu=-\nu}^{\nu} \sum_{q=|l-\nu|}^{l+\nu} (-)^\nu B(l, m|\nu, \mu|q)\Phi_{\mu\nu}^{+}(\mathbf{r}_0)\mathbf{L}_{(m-\mu)q}^{\pm}(\mathbf{r}). \tag{J.19}$$

The transformation laws of $\mathbf{M}_{ml}^{\pm}$ and $\mathbf{N}_{ml}^{\pm}$ are more involved. Because $\mathbf{R} = \mathbf{r} - \mathbf{r}_0$, we have, for $r > r_0$,

$$\begin{aligned} \mathbf{M}_{ml}^{\pm}(\mathbf{R}) = \operatorname{curl}(\mathbf{R}\Phi_{ml}^{\pm}) &= \sum_{\nu=0}^{\infty} \sum_{\mu=-\nu}^{\nu} \sum_{q=|l-\nu|}^{l+\nu} (-)^\nu B\Phi_{\mu\nu}^{+}(\mathbf{r}_0)\mathbf{M}_{(m-\mu)q}^{\pm}(\mathbf{r}) \\ &\quad - \sum_{\nu=0}^{\infty} \sum_{\mu=-\nu}^{\nu} \sum_{q=|l-\nu|}^{l+\nu} (-)^\nu B\Phi_{\mu\nu}^{+}(\mathbf{r}_0)\operatorname{curl}_r[\mathbf{r}_0\Phi_{(m-\mu)q}^{\pm}(\mathbf{r})]. \end{aligned} \tag{J.20}$$

In terms of the Cartesian unit vectors of O (Fig. J.1)

$$\mathbf{r}_0 = r_0[\mathbf{e}_x \sin\theta_0 \cos\phi_0 + \mathbf{e}_y \sin\theta_0 \sin\phi_0 + \mathbf{e}_z \cos\theta_0]. \tag{J.21}$$

We have, however, shown in Eq. (2.7.10) that

$$\frac{1}{k_c}\operatorname{curl}(\mathbf{e}_z \Phi_{ml}) = \frac{1}{2l+1}\left[\frac{l+m}{l}\mathbf{M}_{m(l-1)}(\mathbf{r}) + \frac{l-m+1}{l+1}\mathbf{M}_{m(l+1)}(\mathbf{r})\right] + \frac{im}{l(l+1)}\mathbf{N}_{ml}(\mathbf{r}), \qquad l \geq 1,$$

$$\frac{1}{k_c}\operatorname{curl}(\mathbf{e}_z \Phi_{00}) = \mathbf{M}_{01}. \tag{J.22}$$

Following the method outlined in Example 2.7, we can prove that, for $l = m = 0$,

$$\frac{1}{k_c}\operatorname{curl}(\mathbf{e}_x \Phi_{00}) = \frac{1}{2}\mathbf{M}_{11}(\mathbf{r}) - \mathbf{M}_{-1,1}(\mathbf{r}),$$

$$\frac{1}{k_c}\operatorname{curl}(\mathbf{e}_y \Phi_{00}) = -i\left[\frac{1}{2}\mathbf{M}_{11}(\mathbf{r}) + \mathbf{M}_{-1,1}(\mathbf{r})\right],$$

whereas for $l \geq 1$,

$$\begin{aligned}\frac{1}{k_c}\operatorname{curl}(\mathbf{e}_x \Phi_{ml}) = &-\frac{i}{2}\frac{1}{l(l+1)}\left[\mathbf{N}_{(m+1)l}(\mathbf{r}) + (l+m)(l-m+1)\mathbf{N}_{(m-1)l}(\mathbf{r})\right] \\ &+ \frac{1}{2(2l+1)}\left[-\frac{1}{l}\mathbf{M}_{(m+1)(l-1)}(\mathbf{r}) + \frac{1}{l+1}\mathbf{M}_{(m+1)(l+1)}(\mathbf{r})\right. \\ &+ \frac{(l+m)(l+m-1)}{l}\mathbf{M}_{(m-1)(l-1)}(\mathbf{r}) \\ &\left.- \frac{(l-m+1)(l-m+2)}{l+1}\mathbf{M}_{(m-1)(l+1)}(\mathbf{r})\right],\end{aligned} \tag{J.23}$$

$$\begin{aligned}\frac{1}{k_c}\operatorname{curl}(\mathbf{e}_y \Phi_{ml}) = &-\frac{1}{2}\frac{1}{l(l+1)}\left[\mathbf{N}_{(m+1)l}(\mathbf{r}) - (l+m)(l-m+1)\mathbf{N}_{(m-1)l}(\mathbf{r})\right] \\ &+ \frac{i}{2(2l+1)}\left[\frac{1}{l}\mathbf{M}_{(m+1)(l-1)}(\mathbf{r}) - \frac{1}{l+1}\mathbf{M}_{(m+1)(l+1)}(\mathbf{r})\right. \\ &+ \frac{(l+m)(l+m-1)}{l}\mathbf{M}_{(m-1)(l-1)}(\mathbf{r}) \\ &\left.- \frac{(l-m+1)(l-m+2)}{l+1}\mathbf{M}_{(m-1)(l+1)}(\mathbf{r})\right].\end{aligned} \tag{J.24}$$

From Eqs. (J.20)–(J.24) and after some manipulations, the addition theorem for $\mathbf{M}_{ml}$ under rigid translation of the coordinates for $r > r_0$ assumes the form

$$\mathbf{M}_{ml}^{\pm}(\mathbf{R}) = \sum_{\nu=0}^{\infty} \sum_{\mu=-\nu}^{\nu} \{C_{\mu\nu}^{ml}\mathbf{M}_{\mu\nu}^{\pm}(\mathbf{r}) + D_{\mu\nu}^{ml}\mathbf{N}_{\mu\nu}^{\pm}(\mathbf{r})\}. \tag{J.25}$$

Because the curl operator is independent of the coordinate system, the application of this operator to Eq. (J.25) yields

$$\mathbf{N}_{ml}^{\pm}(\mathbf{R}) = \sum_{\nu=0}^{\infty} \sum_{\mu=-\nu}^{\nu} \{C_{\mu\nu}^{ml}\mathbf{N}_{\mu\nu}^{\pm}(\mathbf{r}) + D_{\mu\nu}^{ml}\mathbf{M}_{\mu\nu}^{\pm}(\mathbf{r})\}, \tag{J.26}$$

with

$$\begin{aligned} C_{\mu\nu}^{ml} &= \sum_{q=|l-\nu|}^{l+\nu} (-)^q B(l, m|\nu, \mu|q) a(l, \nu, q) \Phi_{(m-\mu)q}^{+}(\mathbf{r}_0), \\ D_{\mu\nu}^{ml} &= \sum_{q=|l-\nu|}^{l+\nu} (-)^q G(l, m|\nu, \mu|q) b(l, \nu, q) \Phi_{(m-\mu)q}^{+}(\mathbf{r}_0), \end{aligned} \tag{J.27}$$

where B is defined in Eq. (J.18) and

$$\begin{aligned} a &= 1 + \frac{(\nu+1)(l-\nu+q+1)(l+\nu-q) - \nu(\nu-l+q+1)(l+\nu+q+2)}{2\nu(\nu+1)(2\nu+1)}, \\ b &= \frac{[(l+\nu+q+1)(\nu-l+q)(l-\nu+q)(l+\nu-q+1)]^{1/2}}{2\nu(\nu+1)}, \\ G &= (-)^m i^{l-\nu-q}(2\nu+1)(2q+1)\left[\frac{(l+m)!(\nu-\mu)!(q-m+\mu)!}{(l-m)!(\nu+\mu)!(q+m-\mu)!}\right]^{1/2} \\ &\quad \times \begin{pmatrix} l & \nu & q \\ m & -\mu & \mu-m \end{pmatrix} \begin{pmatrix} l & \nu & q-1 \\ 0 & 0 & 0 \end{pmatrix}. \end{aligned} \tag{J.28}$$

The addition theorems (J.19), (J.25), and (J.26) are valid when $r > r_0$. For $r < r_0$, one must replace $j_n(k_c r_0)$ by $f_n^{\pm}(k_c r_0)$ and $f_n^{\pm}(k_c r)$ by $j_n(k_c r)$ in the right-hand sides of these relations. The same applies to Eqs. (J.29) and (J.31) below.

J.3.1. Translation Along the z Axis

For most seismic source studies in a nonrotating earth model it is sufficient to consider translation along the z axis. Starting, therefore, from Eq. (J.15) we find, for $r > r_h$,

$$\mathbf{L}_{ml}^{\pm}(\mathbf{R}) = \sum_{n=m}^{\infty} E_{ml}^{n,+}(k_c r_h)\mathbf{L}_{mn}^{\pm}(\mathbf{r}). \tag{J.29}$$

For dipolar seismic sources (ml) can take up the values (0, 0), (0, 1), (1, 1), (0, 2), (1, 2), and (2, 2) only. The corresponding E^n_{ml} are found from Eq. (J.16)

$$
\begin{aligned}
E^{n,\pm}_{m,m}(x) &= (2m-1)!!(2n+1)\frac{f^{\pm}_n(x)}{x^m},\\
E^{n,\pm}_{m,m+1}(x) &= (2m+1)!!(2n+1)\left[(n-m)\frac{f^{\pm}_n(x)}{x^{m+1}} - \frac{f^{\pm}_{n+1}(x)}{x^m}\right],\\
E^{n,\pm}_{m,m+2}(x) &= (2m+3)!!(2n+1)\left[\frac{1}{2}(n-m)(n-m-1)\frac{f^{\pm}_n(x)}{x^{m+2}}\right.\\
&\quad \left. + (m+1)\frac{f^{\pm}_{n+1}(x)}{x^{m+1}} - \frac{m+1}{2m+3}\frac{f^{\pm}_n(x)}{x^m}\right].
\end{aligned}
\tag{J.30}
$$

Next, because $\mathbf{R} = \mathbf{r} - r_h\mathbf{e}_z$, Eqs. (J.15) and (J.22) yield at once the required transformations $(r > r_h)$

$$
\begin{aligned}
\mathbf{M}^{\pm}_{ml}(\mathbf{R}) &= \sum_{n=m}^{\infty} [C^n_{ml}(k_c r_h)\mathbf{M}^{\pm}_{mn}(\mathbf{r}) + D^n_{ml}(k_c r_h)\mathbf{N}'^{\pm}_{mn}(\mathbf{r})],\\
\mathbf{N}^{\pm}_{ml}(\mathbf{R}) &= \sum_{n=m}^{\infty} [C^n_{ml}(k_c r_h)\mathbf{N}^{\pm}_{mn}(\mathbf{r}) + D^n_{ml}(k_c r_h)\mathbf{M}'^{\pm}_{mn}(\mathbf{r})],
\end{aligned}
\tag{J.31}
$$

$$
\begin{aligned}
C^n_{ml} &= E^{n,+}_{ml}(k_c r_h) - \frac{(n-m)k_c r_h}{n(2n-1)} E^{n-1,+}_{ml}(k_c r_h)\\
&\quad - \frac{(n+m+1)k_c r_h}{(n+1)(2n+3)} E^{n+1,+}_{ml}(k_c r_h),
\end{aligned}
\tag{J.32}
$$

$$
D^n_{ml} = \frac{-k_c r_h}{n(n+1)} E^{n,+}_{ml}(k_c r_h),
$$

$$
\mathbf{M}'_{mn}(\mathbf{r}) = \operatorname{curl}\left[\mathbf{r}\frac{\partial}{\partial\phi}\Phi_{mn}(k_c r)\right], \qquad \mathbf{N}'_{mn}(r) = \frac{1}{k_c}\operatorname{curl}\operatorname{curl}\left[\mathbf{r}\frac{\partial}{\partial\phi}\Phi_{mn}(k_c r)\right].
\tag{J.33}
$$

APPENDIX K

The Mathematics of Causality

K.1. Causal Systems

A time function is called a *causal function* if it is equal to zero for negative values of its argument. The input–output relations of a linear system are given by the known convolution integral

$$g(t) = \int_{-\infty}^{\infty} f(\tau)a(t-\tau)d\tau = \int_{-\infty}^{\infty} f(t-\tau)a(\tau)d\tau, \tag{K.1}$$

where $f(t)$ is the input, $g(t)$ is the output and $a(t)$ is the system's *impulse response*. If $a(t)$ is an arbitrary function, the causality of the output $g(t)$ is not guaranteed by the causality of the input $f(t)$. Those linear systems for which the causality of $f(t)$ implies the causality of $g(t)$ are called *causal systems*. Because, in nature, effects can never precede their causes, linear passive physical systems are causal. Given that the system is causal, Eq. (K.1) implies that $a(t)$ must be causal. This condition is sometimes called the condition of *physical realizability* (a misnomer, because systems may be impossible to construct because of other reasons). There are useful linear systems that are not physically realizable. Such are, for example, some filters that must not act in real time and, therefore, operate not only on the "past history" of the signal but also on its "future."

In general, there are four possible cases:

$$\begin{aligned} g(t) &= \int_{-\infty}^{\infty} f(\tau)a(t-\tau)d\tau \quad \text{general } f(t) \text{ and } a(t) \\ &= \int_{0}^{t} f(\tau)a(t-\tau)d\tau \quad \text{causal } f(t) \text{ and } a(t) \\ &= \int_{0}^{\infty} f(\tau)a(t-\tau)d\tau \quad \text{causal } f(t) \text{ and general } a(t) \\ &= \int_{-\infty}^{t} f(\tau)a(t-\tau)d\tau \quad \text{general } f(t) \text{ and causal } a(t). \end{aligned} \tag{K.2}$$

Because, in a causal system, $a(t-\tau) = 0$ for $t < \tau$, the convolution integral for causal inputs will assume the form

$$g(t) = \int_{0}^{t} f(\tau)a(t-\tau)d\tau = \int_{0}^{t} f(t-\tau)a(\tau)d\tau. \tag{K.3}$$

Next we put $a(t-\tau) = (d/dt)h(t-\tau) = -(d/d\tau)h(t-\tau)$ in Eq. (K.3) and integrate by parts

$$g(t) = -\int_0^t f(\tau)\frac{d}{d\tau}h(t-\tau)d\tau = -f(\tau)h(t-\tau)\Big|_0^t + \int_0^t f'(\tau)h(t-\tau)d\tau.$$

Assuming $h(0) = 0$, we find

$$g(t) = f(0^+)h(t) + \int_0^t f'(\tau)h(t-\tau)d\tau \tag{K.4}$$

which is known as the *superposition* or *Duhamel's integral.* The function $h(t)$ is known as the *step response* of the system. In Eq. (K.4), $f(0^+)$ is the jump value of $f(t)$ at $t = 0$.

The response of a linear system to the harmonic excitation $e^{i\omega t}$ is known as the *system's function* or *transfer function* $S(\omega)$. If $F(\omega)$ is the Fourier transform of the input function $f(t)$, the output can be represented by the Fourier integral

$$g(t) = \frac{1}{2\pi}\,\mathrm{P}\int_{-\infty}^{\infty} F(\omega)S(\omega)e^{i\omega t}\,d\omega, \tag{K.5}$$

where P means the principal value of the integral. If we choose, as a special case, $f(t) = \delta(t)$, we get $F(\omega) = 1$ and Eq. (K.5) then yields

$$a(t) = \frac{1}{2\pi}\,\mathrm{P}\int_{-\infty}^{\infty} S(\omega)e^{i\omega t}\,d\omega, \tag{K.6}$$

which is a relation between the system's function and the system's impulse response. Clearly,

$$S(\omega) = \int_0^{\infty} a(t)e^{-i\omega t}\,dt. \tag{K.7}$$

Moreover, using Eq. (K.5) with $f(t) = H(t)$, we find a relation between the system's function and the system's *step response*,

$$\begin{aligned} h(t) &= \frac{1}{2\pi}\int_C \frac{S(\omega)}{i\omega}e^{i\omega t}\,d\omega \\ &= \frac{1}{2}S(0) + \frac{1}{2\pi}\,\mathrm{P}\int_{-\infty}^{\infty}\frac{S(\omega)}{i\omega}e^{i\omega t}\,d\omega \\ &= \frac{1}{2\pi}\,\mathrm{P}\int_{-\infty}^{\infty}\left[\pi S(0)\delta(\omega) + \left(\frac{1}{i\omega}\right)S(\omega)\right]e^{i\omega t}\,d\omega, \end{aligned} \tag{K.8}$$

where C is drawn just below the real axis to secure the vanishing of $h(t)$ for $t < 0$. It is assumed that $S(\omega) \to 0$ as $|\omega| \to \infty$ on the infinite circle. A schematic diagram for a linear system response to standard inputs is shown in Fig. K.1.

Steady-state response

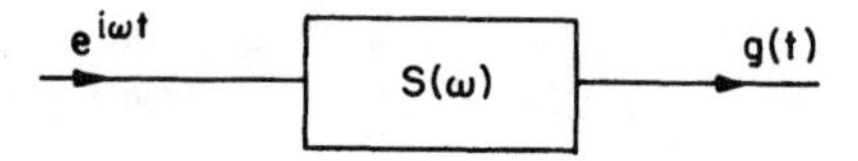

Impulse response

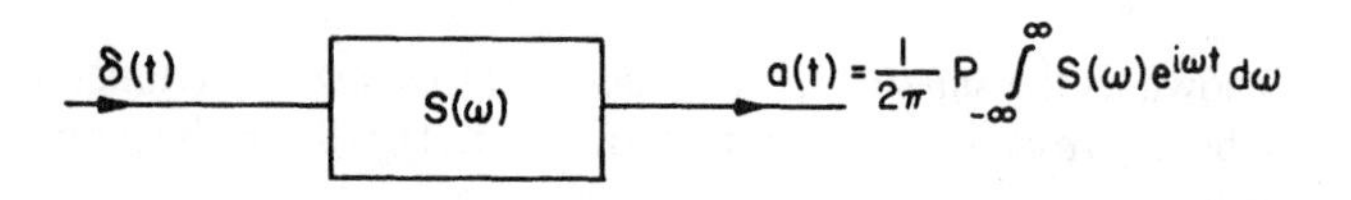

Step response

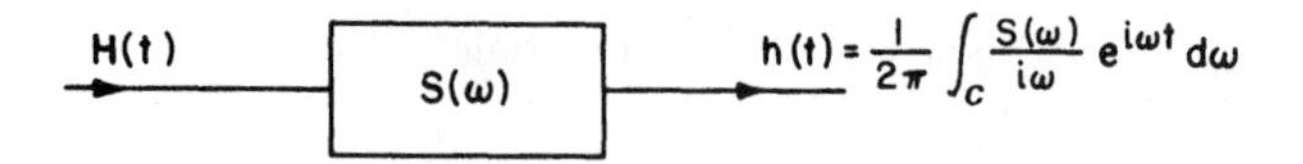

Figure K.1. Response of a linear system to standard inputs.

K.2. Integral Theorems for Causal Systems

Further insight into the general behavior of linear systems is gained via certain integral relations that link the impulse response $a(t)$ and its spectral image. The results will be primarily based on the Fourier integral theorem. In addition, it shall be assumed that $a(t)$ is real, causal, and devoid of impulses. Denoting the real and imaginary parts of $S(\omega)$ by $R(\omega)$ and $X(\omega)$, respectively, we easily verify the following simple results:

1. Because $a(t)$ is real we must have $R(-\omega) = R(\omega)$, $X(-\omega) = -X(\omega)$. Also, if $a(t) = a_e(t) + a_o(t)$ (e = even, o = odd), the relation

$$R(\omega) + iX(\omega) = \int_{-\infty}^{\infty} [a_e(t) + a_o(t)] e^{-i\omega t} \, dt \tag{K.9}$$

implies

$$R(\omega) = 2 \int_0^{\infty} a_e(t) \cos \omega t \, dt = \int_{-\infty}^{\infty} a_e(t) e^{-i\omega t} \, dt, \tag{K.10}$$

$$iX(\omega) = -2i \int_0^{\infty} a_o(t) \sin \omega t \, dt = \int_{-\infty}^{\infty} a_o(t) e^{-i\omega t} \, dt. \tag{K.11}$$

Their inverses are

$$a_e(t) = \frac{1}{2\pi}\int_{-\infty}^{\infty} R(\omega)e^{i\omega t}\,d\omega = \frac{1}{\pi}\int_{0}^{\infty} R(\omega)\cos\omega t\,d\omega, \tag{K.12}$$

$$a_o(t) = \frac{i}{2\pi}\int_{-\infty}^{\infty} X(\omega)e^{i\omega t}\,d\omega = -\frac{1}{\pi}\int_{0}^{\infty} X(\omega)\sin\omega t\,d\omega. \tag{K.13}$$

Causality, however, implies $a(-t) = a_e(-t) + a_o(-t) = a_e(t) - a_o(t) = 0$. Therefore, for $t > 0$

$$a(t) = 2a_e(t) = 2a_o(t) = \frac{2}{\pi}\int_{0}^{\infty} R(\omega)\cos\omega t\,d\omega = -\frac{2}{\pi}\int_{0}^{\infty} X(\omega)\sin\omega t\,d\omega. \tag{K.14}$$

2. It is known that at a point where $a(t)$ is discontinuous, the Fourier inversion formula yields the average value

$$\lim_{\lambda\to\infty}\frac{1}{2\pi}\int_{-\lambda}^{\lambda} S(\omega)e^{i\omega x}\,d\omega = \frac{1}{2}[a(x+0) + a(x-0)],$$

where $a(t)$ is square integrable and of bounded variation in the neighborhood of $t = x$. Applying this theorem to $a(t)$ at $t = 0$, it is found that

$$\frac{1}{2}a(0^+) = \frac{1}{2\pi}\int_{-\infty}^{\infty} S(\omega)d\omega = \frac{1}{2\pi}\int_{-\infty}^{\infty} R(\omega)d\omega, \tag{K.15}$$

because $a(0^-) = 0$ (causality) and $X(\omega)$ is odd (reality). The causality of $a(t)$ can be used again to our advantage. Integrating by parts we obtain

$$\int_{0}^{\infty} e^{-i\omega t}\left[\frac{da(t)}{dt}\right]dt = [e^{-i\omega t}a(t)]_{0^+}^{\infty} + i\omega\int_{0}^{\infty} e^{-i\omega t}a(t)dt = i\omega S(\omega) - a(0^+). \tag{K.16}$$

If we invoke the Riemann–Lebesgue lemma, Eqs. (K.15) and (K.16) yield

$$\lim_{\omega\to\infty}[i\omega S(\omega)] = a(0^+) = \frac{1}{\pi}\int_{-\infty}^{\infty} R(\omega)d\omega. \tag{K.17}$$

3. Applying the known theorem of Plancherel

$$\int_{-\infty}^{\infty} |a(t)|^2\,dt = \frac{1}{2\pi}\int_{-\infty}^{\infty} |S(\omega)|^2\,d\omega \tag{K.18}$$

to our time functions and their spectrums, we obtain

$$\int_{-\infty}^{\infty} a_e^2(t)dt = \frac{1}{2\pi}\int_{-\infty}^{\infty} R^2(\omega)d\omega, \qquad \int_{-\infty}^{\infty} a_o^2(t)dt = \frac{1}{2\pi}\int_{-\infty}^{\infty} X^2(\omega)d\omega. \tag{K.19}$$

The causality of $a(t)$ now yields

$$\int_0^\infty a^2(t)dt = \frac{2}{\pi}\int_0^\infty R^2(\omega)d\omega = \frac{2}{\pi}\int_0^\infty X^2(\omega)d\omega. \tag{K.20}$$

4. Suppose that we form a new complex function $Z(\omega)$ such that Re $Z(\omega) = |S(\omega)|^2$. Then, by Eq. (K.17),

$$\lim_{\omega\to\infty}\{i\omega Z(\omega)\} = \frac{1}{\pi}\int_{-\infty}^{\infty}\{\text{Re } Z(\omega)\}d\omega = \frac{1}{\pi}\int_{-\infty}^{\infty}|S(\omega)|^2\, d\omega. \tag{K.21}$$

K.3. Analyticity and Causality

In the preceding section we have assumed that the system's impulse response $a(t)$ is real and causal (i.e., physical realizability) and have derived from this assumption a number of relations between $a(t)$ and $S(\omega)$. Therefore, the causality of $a(t)$ implies certain analytic properties of the real and imaginary parts of $S(\omega)$.

We shall now be concerned with the reverse situation. We begin with the assumption that the transfer function $S(\omega)$ is analytic and bounded in the lower half of the complex ω plane and derive certain integral relations between the real and imaginary parts of $S(\omega)$. These relations, known as *dispersion relations*, then guarantee the causality of $a(t)$.

Consider a closed contour in the lower $z = \omega - i\Omega$ plane (Fig. K.2), where $S(z)$ is analytic. Because a function of a complex variable that is not constant or zero must have at least one singularity, the singularities of $S(z)$ must be located in the upper half-plane. We shall assume that $S(z)$ has no poles on the real axis. This last requirement is not mandatory and is made only for the sake of simplicity. According to Cauchy's integral theorem,

$$\int_C \frac{S(z)}{z-\omega_0}\, dz = 0,$$

where ω_0 is an arbitrary point on the real axis and C consists of the segments $(-R_0, \omega_0 - r)$, $(\omega_0 + r, R_0)$ of the real axis, a large semicircle c_1 of radius R_0

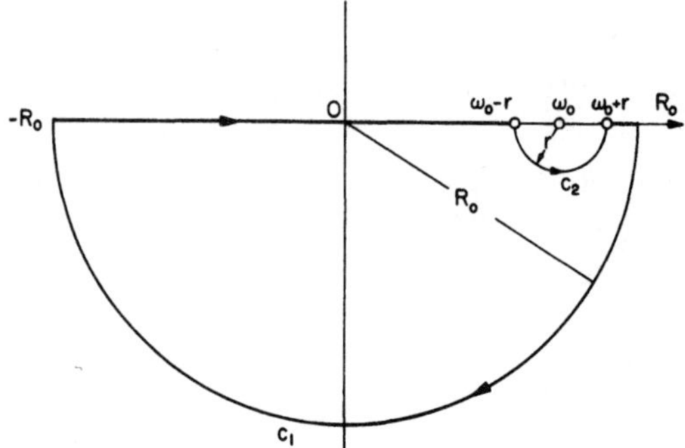

Figure K.2. Integration contour for the derivation of the dispersion relations.

and a small semicircular indentation c_2 at $z = \omega_0$ on the real axis. The integral over c_1 is given by

$$\lim_{R_0\to\infty}\int_{c_1}\frac{S(R_0e^{i\theta})}{R_0e^{i\theta}-\omega_0}iR_0e^{i\theta}\,d\theta = iS(\infty)\int_0^{-\pi}d\theta = -\pi iS(\infty). \quad \text{(K.22)}$$

As $r\to 0$ and $R_0\to\infty$, the contribution of the segments tends to the principal value of the integral along the real axis and the contribution of the indentation is $\pi iS(\omega_0)$. Therefore,

$$0 = -\pi iS(\infty) + \lim_{\substack{r\to 0\\ R_0\to\infty}}\left[\int_{-R_0}^{\omega_0-r}\frac{S(x)}{x-\omega_0}dx + \int_{\omega_0+r}^{R_0}\frac{S(x)}{x-\omega_0}dx\right] + \pi iS(\omega_0),$$

where the boundedness of $S(z)$ secures the convergence of the integral on the infinite arc. Equating the real and imaginary parts of both sides and replacing ω_0 by ω, we get

$$X(\omega) = \frac{1}{\pi}\,\mathrm{P}\int_{-\infty}^{\infty}\frac{R(x)}{x-\omega}dx + X(\infty), \quad \text{(K.23)}$$

$$R(\omega) = R(\infty) - \frac{1}{\pi}\,\mathrm{P}\int_{-\infty}^{\infty}\frac{X(x)}{x-\omega}dx, \quad \text{(K.24)}$$

where the principal value is needed at ω and ∞. Thus, by making a few reasonable assumptions about a linear system, we have found interrelations between the real and imaginary parts of the physical entity $S(\omega)$. These are known as *Kramers–Kronig dispersion relations.*

Next we obtain alternative forms for Eq. (K.23) and (K.24) in order to throw more light on these relationships and also to bring out certain points that are not immediately apparent.

Making use of the result

$$\mathrm{P}\int_{-\infty}^{\infty}\frac{dx}{x-\omega} = \lim_{\substack{r\to 0\\ R_0\to\infty}}\left[\int_{-R_0}^{\omega-r}\frac{dx}{x-\omega} + \int_{\omega+r}^{R_0}\frac{dx}{x-\omega}\right]$$

$$= \lim_{\substack{r\to 0\\ R_0\to\infty}}\left[\ln\left|\frac{R_0-\omega}{r}\right| - \ln\left|\frac{R_0+\omega}{r}\right|\right] = 0,$$

we obtain from Eqs. (K.23) and (K.24)

$$X(\omega) = X(\infty) + \frac{1}{\pi}\int_{-\infty}^{\infty}\frac{R(x)-R(\omega)}{x-\omega}dx, \quad \text{(K.25)}$$

$$X(\omega) = \frac{2\omega}{\pi}\int_0^{\infty}\frac{R(x)-R(\omega)}{x^2-\omega^2}dx, \quad \text{(K.26)}$$

$$R(\omega) = R(\infty) - \frac{1}{\pi}\int_{-\infty}^{\infty}\frac{X(x)-X(\omega)}{x-\omega}dx, \quad \text{(K.27)}$$

$$R(\omega) = R(\infty) - \frac{2}{\pi}\int_0^{\infty}\frac{xX(x)-\omega X(\omega)}{x^2-\omega^2}dx. \quad \text{(K.28)}$$

In contradistinction to Eqs. (K.23) and (K.24), there is no singularity at $x = \omega$. Integrating by parts, we get from Eqs. (K.26) and (K.28)

$$X(\omega) = \frac{1}{\pi}\int_0^\infty \left[\frac{dR(x)}{dx}\right]\ln\left|\frac{x+\omega}{x-\omega}\right| dx, \tag{K.29}$$

$$R(\omega) - R(\infty) = -\frac{1}{\omega\pi}\int_0^\infty \frac{d}{dx}[xX(x)]\ln\left|\frac{x+\omega}{x-\omega}\right| dx. \tag{K.30}$$

We notice that rapid changes in attenuation cause large phase shifts.

In cases where $R(\infty)$ is not known, but $R(\omega_0)$ for some ω_0 can be determined, the value of $R(\infty)$ can be eliminated and one obtains the *once-subtracted dispersion relations*

$$R(\omega) = R(\omega_0) - \frac{1}{\pi}(\omega - \omega_0)\,\mathrm{P}\int_{-\infty}^{\infty} \frac{X(x)dx}{(x-\omega)(x-\omega_0)}, \tag{K.31}$$

$$X(\omega) = X(\omega_0) + \frac{1}{\pi}(\omega - \omega_0)\,\mathrm{P}\int_{-\infty}^{\infty} \frac{R(x)dx}{(x-\omega)(x-\omega_0)}. \tag{K.32}$$

Note that because $\mathrm{P}\int_{-\infty}^{\infty} dx/(x-\omega) = 0$, both functions $R(\omega)$ and $X(\omega)$ can be determined from each other up to an *additive constant*, which can always be determined from the given conditions of the problem at hand. Apart from that, the analyticity and boundedness of $S(\omega)$ in the lower ω half-plane guarantees a *one-to-one relationship* (uniqueness) between the real and imaginary parts of $S(\omega)$. It can be demonstrated by simple examples that if $S(\omega)$ has, say, poles in the lower ω half-plane, then for each given $R(\omega)$, one can find more than one imaginary part $X(\omega)$ such that the dispersion relations are obeyed.

The causality of $a(t)$ follows directly from Eq. (K.24). Assuming $R(\infty) = 0$, we write

$$R(\omega) = -\frac{1}{\pi}\,\mathrm{P}\int_{-\infty}^{\infty} \frac{X(x)}{x-\omega}\,dx.$$

Using the *frequency convolution theorem*, we have from Eqs. (K.12) and (K.13)

$$a_e(t) = \frac{1}{2\pi}\int_{-\infty}^{\infty} R(\omega)e^{i\omega t}\,d\omega = a_o(t)\mathrm{sgn}\,t.$$

Therefore,

$$a_e(t) = a_o(t), \qquad t > 0;$$

$$a_e(t) = -a_o(t), \qquad t < 0.$$

Consequently, $a(t) = a_o(t) + a_e(t)$ is a causal function.

Taking into account both the causality of $a(t)$ and the analyticity of $S(\omega) = R(\omega) + iX(\omega)$, we easily derive the integral relations

$$R(\omega) = \int_0^\infty a(t)\cos \omega t \, dt, \qquad X(\omega) = -\int_0^\infty a(t)\sin \omega t \, dt,$$

$$\begin{aligned} a(t) &= \frac{1}{\pi}\int_0^\infty [R \cos \omega t - X \sin \omega t] d\omega \\ &= \frac{2}{\pi}\int_0^\infty R(\omega)\cos \omega t \, d\omega = -\frac{2}{\pi}\int_0^\infty X(\omega)\sin \omega t \, d\omega, \qquad (t > 0) \end{aligned} \quad \text{(K.33)}$$

$$h(t) = \frac{2}{\pi}\int_0^\infty \frac{R(\omega)}{\omega} \sin \omega t \, d\omega = S(0) + \frac{2}{\pi}\int_0^\infty \frac{X(\omega)}{\omega} \cos \omega t \, d\omega, \quad \text{(K.34)}$$

$$\frac{R(\omega)}{\omega} = \int_0^\infty h(t)\sin \omega t \, dt,$$

$$\pi S(0)\delta(\omega) + \frac{X(\omega)}{\omega} = \int_0^\infty h(t)\cos \omega t \, dt. \quad \text{(K.35)}$$

K.4. Minimum-Phase Systems

If we wish to examine the relations between the amplitude and the phase of a transfer function, it is necessary to use the polar representation of the transfer function

$$S(\omega) = |S(\omega)| e^{i\phi(\omega)}, \qquad |S(\omega)|^2 = S(\omega)S^*(\omega). \quad \text{(K.36)}$$

Because $S(\omega) = \int_{-\infty}^{\infty} a(t)e^{-i\omega t} \, dt$, the reality of $a(t)$ implies the symmetry conditions

$$S(-\omega) = S^*(\omega), \qquad \phi(-\omega) = -\phi(\omega), \qquad \omega \text{ real}$$

$$S(-z^*) = S^*(z), \qquad z \text{ complex}.$$

Moreover, because $S^2(\omega) = S(\omega)S^*(\omega)e^{2i\phi(\omega)}$, we have

$$\phi(\omega) = \frac{1}{2i} \ln \frac{S(\omega)}{S(-\omega)}, \qquad \omega \text{ real}.$$

Taking the logarithm of both sides of Eq. (K.36), we get

$$\ln S(\omega) = \ln|S(\omega)| + i\phi(\omega).$$

If $a(t)$ is causal, we can apply the dispersion relations to the function pair $\ln|S(\omega)|$ and $\phi(\omega)$, provided $\ln S(\omega)$ is analytic and bounded in the lower z plane. The regularity of $S(\omega)$ itself in the said region is not sufficient because $\ln|S(\omega)|$ is not analytic at the zeros of $S(\omega)$. If $S(\omega)$ does not have zeros in the lower z plane (including the real axis), in addition to its being analytic and bounded there, then $\ln S(\omega)$ is analytic and bounded in the lower z plane and

$\ln|S(\omega)|$ and $\phi(\omega)$ are Hilbert transforms of each other just as $R(\omega)$ and $X(\omega)$ are in Eqs. (K.23) and (K.24),

$$\ln|S(\omega)| = -\frac{1}{\pi} \mathrm{P} \int_{-\infty}^{\infty} \frac{\phi(x)}{x - \omega} dx,$$
$$\phi(\omega) = \frac{1}{\pi} \mathrm{P} \int_{-\infty}^{\infty} \frac{\ln|S(x)|}{x - \omega} dx. \tag{K.37}$$

Systems that satisfy this condition are called *minimum-phase systems.* They are characterized by the property that of all possible systems with the same logarithmic amplitude, they provide the smallest phase changes. However, if $S(z)$ has zeros in the lower z half-plane, $\phi(\omega)$ will not be uniquely determined from $|S(\omega)|$ and we shall have a class of causal time functions with the same amplitudes but different phases for any given frequency. Consider, for example, the causal time functions

$$a_1(t) = (3e^{-2t} - 2e^{-t})H(t),$$
$$a_2(t) = e^{-2t}H(t),$$

whose transfer functions are

$$S_1(\omega) = \frac{1}{2 + i\omega} \frac{i\omega - 1}{i\omega + 1}, \qquad S_2(\omega) = \frac{1}{2 + i\omega}.$$

Both transfer functions have the same amplitude but differ in their phase functions. The reason is that $S_1(z)$ has a zero in the lower z half-plane at $z = -i$ that is a singularity of $\ln S_1(z)$. Therefore $S_2(\omega)$ *is a minimum-phase function,* whereas $S_1(\omega)$ is not.

Another time function,

$$a_3(t) = \delta(t) - 2\alpha e^{-\alpha t}H(t),$$

is associated with the "all-pass" transfer function

$$S_3(\omega) = \frac{i\omega - \alpha}{i\omega + \alpha} = \exp\left[-2i \operatorname{tg}^{-1}\left(\frac{\omega}{\alpha}\right)\right] = \exp\left[i \operatorname{tg}^{-1}\left[\frac{2\omega\alpha}{(\omega^2 - \alpha^2)}\right]\right]$$

and constitutes the simplest type of a non-minimum-phase system. We note that the minimum-phase system of this class is obtained for $\alpha = 0$.

If $S(\omega)$ is not a minimum-phase function, it can be written as a product of a minimum-phase function and another function of unit amplitude and negative phase (all pass).

In minimum phase systems the dispersion relations can sometimes be simplified in the following manner: Combining Eqs. (K.29) and (K.37), we write [with $|S(\omega)| = A(\omega)$]

$$\phi(\omega) = \frac{1}{\pi} \int_0^{\infty} \frac{d \ln A(x)}{dx} \ln\left|\frac{x + \omega}{x - \omega}\right| dx = \frac{1}{\pi} \int_0^{\infty} \frac{d \ln A(x)}{d \ln x} \ln\left|\frac{x + \omega}{x - \omega}\right| \frac{dx}{x}. \tag{K.38}$$

The logarithmic factor of the integrand becomes infinite at $x = \omega$ but tends to zero if $x \to 0$ or $x \to \infty$. In contrast, the logarithmic derivative of the amplitude usually changes between narrow limits. Therefore, the first factor can be approximated to be equal to the value attained at $x = \omega$:

$$\phi(\omega) \simeq \frac{1}{\pi} \frac{d \ln A(\omega)}{d \ln \omega} \left[\int_0^\omega \ln\left(\frac{\omega + x}{\omega - x}\right) \frac{dx}{x} + \int_\omega^\infty \ln\left(\frac{x + \omega}{x - \omega}\right) \frac{dx}{x} \right].$$

Because

$$\int_0^\omega \ln\left(\frac{\omega + x}{\omega - x}\right) \frac{dx}{x} = \frac{\pi^2}{4},$$

$$\int_\omega^\infty \ln\left(\frac{x + \omega}{x - \omega}\right) \frac{dx}{x} = \int_0^{1/\omega} \ln\left(\frac{1/\omega + y}{1/\omega - y}\right) \frac{dy}{y} = \frac{\pi^2}{4}, \qquad \left(x = \frac{1}{y}\right),$$

we obtain the approximation

$$\phi(\omega) \simeq \frac{\pi}{2} \frac{d \ln A(\omega)}{d \ln \omega}$$

or

$$\frac{\phi(\omega)}{\omega} \simeq \frac{\pi}{2} \frac{1}{A} \frac{dA}{d\omega}. \tag{K.39}$$

The effectiveness of this approximation is reflected in a simple example:

$$S(\omega) = \frac{1}{1 + i\omega}, \qquad A(\omega) = \frac{1}{\sqrt{(1 + \omega^2)}}, \qquad \phi(\omega) = -\mathrm{tg}^{-1}\omega.$$

This leads to the approximation

$$\tan^{-1} \omega = \frac{\pi}{2} \frac{\omega^2}{1 + \omega^2}.$$

APPENDIX L

Models of the Earth and the Atmosphere

Table L.1. Structural Constants of the Jeffreys–Bullen A' (J–B A') and the Gutenberg–Bullard I (G–B I) Models of the Earth[a]

Radius	Shear velocity (km/s)		Compressional velocity (km/s)		Density (g/cm^3)		μ (10^{11} dyne/cm^2)		$\kappa = \lambda + 2\mu/3$ (10^{11} dyne/cm^2)		g (cm/s^2)	
(km)	J–B A′	G–B I	J–B A′	G–B I	J–B A′	G–B I	J–B A′	G–B I	J–B A′	G–B I	J–B A′	G–B I
6370.0	3.55	3.55	6.30	6.30	2.84	2.84	3.58	3.58	6.50	6.50	982.7	981.6
6338.0	3.55	3.55	6.30	6.30	2.84	2.84	3.58	3.58	6.50	6.50	985.0	983.9
6338.0	4.35	4.65	7.75	8.16	3.32	3.67	6.28	7.94	11.56	13.86	985.0	983.9
6310.0	—	4.60	—	8.15	—	3.70	—	7.83	—	14.14	—	984.0
6270.0	4.45	4.40	7.95	8.00	3.38	3.74	6.69	7.24	12.44	14.28	987.1	984.0
6220.0	—	4.35	—	7.85	—	3.79	—	7.17	—	13.79	—	984.0
6170.0	4.60	4.40	8.26	8.05	3.47	3.83	7.34	7.41	13.88	14.93	990.2	983.9
6070.0	4.76	4.60	8.58	8.50	3.55	3.92	8.04	8.29	15.41	17.26	993.2	983.6
5970.0	4.94	—	8.93	—	3.63	—	8.86	—	17.14	—	996.2	—
5957.0	4.96	5.00	8.97	9.06	3.64	4.01	8.96	10.04	17.34	19.57	996.6	983.0
5870.0	5.32	5.30	9.66	9.60	3.87	4.08	10.95	11.46	21.51	22.32	998.5	982.4
5770.0	5.66	5.60	10.24	10.10	4.10	4.15	13.13	13.01	25.48	24.98	999.5	981.6
5670.0	5.93	5.90	10.67	10.50	4.30	4.21	15.12	14.66	28.79	26.88	999.2	980.9
5570.0	6.13	6.15	11.01	10.90	4.46	4.27	16.76	16.16	31.72	29.21	998.0	980.3
5470.0	6.27	6.30	11.25	11.30	4.57	4.33	17.97	17.21	33.88	32.41	996.3	979.7
5370.0	6.36	6.35	11.43	11.40	4.65	4.40	18.81	17.74	35.67	33.53	994.4	979.2
5170.0	6.50	6.50	11.71	11.80	4.77	4.50	20.15	19.01	38.54	37.31	990.8	978.9
4970.0	6.62	6.60	11.99	12.05	4.89	4.62	21.43	20.12	41.73	40.25	987.9	979.8
4770.0	6.73	6.75	12.26	12.30	5.00	4.72	22.65	21.51	44.96	42.73	986.1	982.1
4570.0	6.83	6.85	12.53	12.55	5.11	4.82	23.84	22.62	48.44	45.76	985.8	986.4
4370.0	6.92	6.95	12.79	12.80	5.22	4.92	25.00	23.76	52.06	48.92	987.5	993.3

Table L.1. (Continued)

Radius	Shear velocity (km/s)		Compressional velocity (km/s)		Density (g/cm^3)		μ (10^{11} $dyne/cm^2$)		$\kappa = \lambda + 2\mu/3$ (10^{11} $dyne/cm^2$)		g (cm/s^2)	
(km)	J–B A′	G–B I	J–B A′	G–B I	J–B A′	G–B I	J–B A′	G–B I	J–B A′	G–B I	J–B A′	G–B I
4170.0	7.02	7.00	13.03	13.00	5.32	5.02	26.22	24.60	55.37	52.04	991.8	1003.5
3970.0	7.12	7.10	13.27	13.20	5.42	5.11	27.48	25.76	58.81	54.69	999.7	1018.0
3770.0	7.21	7.20	13.50	13.45	5.52	5.21	28.70	27.01	62.34	58.24	1011.9	1037.7
3570.0	7.30	7.25	13.64	13.70	5.61	5.30	29.90	27.86	64.51	62.33	1029.9	1064.1
3490.0	—	7.20	—	13.70	—	5.35	—	27.73	—	63.43	—	1076.9
3470.0	7.30	7.20	13.64	13.65	5.66	5.36	30.16	27.79	65.09	62.82	1041.5	1080.3
3470.0			8.10	8.04	9.70	10.06			63.64	65.03	1041.5	1080.3
3123.0			8.53	8.44	10.21	10.60			74.29	75.51	963.2	999.0
2776.0			9.03	8.90	10.62	11.06			86.60	87.61	876.9	908.6
2429.0			9.44	9.31	11.02	11.46			98.20	99.33	783.9	810.6
2082.0			9.78	9.63	11.38	11.79			108.85	109.34	684.2	706.0
1735.0			10.10	9.88	11.70	12.06			119.35	117.72	578.4	596.0
1388.0			10.44	10.08	11.90	12.28			129.70	124.77	467.9	481.8
1318.6			—	10.11	—	12.32			—	125.93	—	458.4
1297.8			—	10.11	—	12.32			—	125.98	—	451.4
1283.9			—	11.17	—	12.33			—	127.58	—	446.7
1250.0			9.40	—	12.00	—			106.03	—	422.9	—
1250.0			11.16	—	12.00	—			149.45	—	422.9	—

1249.2	—	10.48	—	12.35	—	135.64	—	435.0
1214.5	—	10.76	—	12.36	—	143.16	—	423.2
1179.8	—	10.93	—	12.38	—	147.90	—	411.4
1145.1	—	11.04	—	12.40	—	151.13	—	399.6
1125.0	11.19	—	12.06	—	151.01	—	381.5	—
1110.4	—	11.09	—	12.41	—	152.63	—	387.8
1075.7	—	11.12	—	12.42	—	153.64	—	375.9
1041.0	—	11.13	—	12.44	—	154.10	—	364.0
1000.0	11.21	—	12.10	—	152.05	—	339.8	—
875.0	11.23	—	12.13	—	152.97	—	297.9	—
867.5	—	11.15	—	12.50	—	155.40	—	304.1
750.0	11.25	—	12.17	—	154.03	—	255.8	—
694.0	—	11.17	—	12.54	—	156.46	—	243.9
625.0	11.27	—	12.20	—	154.96	—	213.5	—
520.5	—	11.17	—	12.58	—	157.02	—	183.3
500.0	11.28	—	12.22	—	155.49	—	171.0	—
375.0	11.29	—	12.24	—	156.08	—	128.4	—
347.0	—	11.16	—	12.61	—	157.05	—	122.3
250.0	11.30	—	12.26	—	156.61	—	85.7	—
173.5	—	11.15	—	12.62	—	156.89	—	61.2
0.0	11.31	11.15	12.30	12.63	157.34	157.02	0.0	0.0

[a] A vacancy (indicated by "—") means that the value of the corresponding parameter at the assigned depth has not been uniquely determined. It can be estimated by interpolation.

Table L.2. Structural Constants for a Continental-Earth Model

Depth (km)	Layer thickness (km)	α (km/s)	β (km/s)	ρ (g/cm^3)	μ ($dyne/cm^2 \times 10^{12}$)	λ ($dyne/cm^2 \times 10^{12}$)
11.0	22	6.03	3.53	2.78	0.35	0.32
29.5	15	6.70	3.80	3.00	0.43	0.48
43.5	13	7.96	4.60	3.37	0.71	0.71
62.5	25	7.85	4.50	3.39	0.69	0.72
100.0	50	7.85	4.41	3.42	0.67	0.78
162.5	75	8.00	4.41	3.45	0.67	0.87
225.0	50	8.20	4.50	3.47	0.70	0.93
300.0	100	8.40	4.60	3.50	0.74	0.99
400.0	100	9.00	4.95	3.63	0.89	1.16
500.0	100	9.63	5.31	3.89	1.10	1.41
600.0	100	10.17	5.63	4.13	1.31	1.65
700.0	100	10.59	5.92	4.33	1.51	1.82
800.0	100	10.96	6.14	4.49	1.69	2.00
900.0	100	11.28	6.29	4.60	1.82	2.21
1025.0	150	11.46	6.38	4.69	1.91	2.34
1200.0	200	11.76	6.50	4.80	2.03	2.58
1400.0	200	12.02	6.61	4.91	2.15	2.80
1600.0	200	12.28	6.74	5.03	2.29	3.02
1800.0	200	12.54	6.85	5.13	2.41	3.25
2000.0	200	12.80	6.96	5.24	2.54	3.51
2200.0	200	13.02	7.00	5.34	2.62	3.82
2400.0	200	13.24	7.10	5.44	2.74	4.05
	∞	13.48	7.20	5.54	2.87	4.32

Table L.3. Structural Constants for an Oceanic-Earth Model

Depth (km)	Layer thickness (km)	α (km/s)	β (km/s)	ρ (g/cm^3)	μ (dyne/cm^2 $\times$ 10^{12})	λ (dyne/cm^2 $\times$ 10^{12})
15.5	9	8.11	4.61	3.40	0.72	0.79
22.0	5	8.12	4.61	3.40	0.72	0.80
32.5	15	8.12	4.61	3.40	0.72	0.80
50.0	20	8.01	4.56	3.37	0.70	0.76
70.0	20	7.95	4.56	3.37	0.70	0.73
90.0	20	7.71	4.40	3.37	0.65	0.70
110.0	20	7.68	4.34	3.33	0.63	0.71
130.0	20	7.78	4.34	3.33	0.63	0.76
150.0	20	7.85	4.34	3.33	0.63	0.80
170.0	20	8.10	4.45	3.33	0.66	0.87
220.0	80	8.12	4.45	3.33	0.66	0.88
270.0	20	8.12	4.45	3.35	0.66	0.88
290.0	20	8.12	4.45	3.36	0.67	0.88
310.0	20	8.12	4.45	3.37	0.67	0.89
330.0	20	8.12	4.45	3.38	0.67	0.89
250.0	20	8.24	4.50	3.39	0.69	0.93
365.0	10	8.30	4.53	3.44	0.71	0.96
380.0	20	8.36	4.56	3.50	0.73	0.99
402.5	25	8.75	4.80	3.68	0.85	1.13
425.0	20	9.15	5.04	3.88	0.99	1.28
440.0	10	9.43	5.22	3.90	1.06	1.35
455.0	20	9.76	5.40	3.92	1.14	1.45
477.5	25	9.77	5.40	3.93	1.15	1.46
502.5	25	9.78	5.40	3.95	1.15	1.47
527.5	25	9.78	5.40	3.96	1.15	1.48
552.5	25	9.78	5.40	3.99	1.16	1.49
577.5	25	9.79	5.40	4.02	1.17	1.51
602.5	25	9.79	5.40	4.06	1.18	1.52
627.5	25	9.80	5.40	4.09	1.19	1.54
652.5	25	9.80	5.40	4.12	1.20	1.55
677.5	25	10.16	5.60	4.17	1.31	1.69
702.5	25	10.49	5.80	4.21	1.42	1.80
727.5	25	10.82	6.10	4.26	1.58	1.81
752.5	25	11.12	6.20	4.30	1.65	2.01
777.7	25	11.14	6.21	4.48	1.72	2.10
802.5	25	11.15	6.21	4.63	1.79	2.19
827.5	25	11.17	6.22	4.80	1.85	2.27
852.5	25	11.18	6.23	4.94	1.92	2.34
877.5	25	11.22	6.25	4.94	1.93	2.37
902.5	25	11.27	6.28	4.95	1.95	2.38
927.5	25	11.31	6.30	4.95	1.96	2.41
952.5	25	11.35	6.32	4.95	1.98	2.42
977.5	25	11.39	6.34	4.95	1.99	2.45
1002.5	25	11.43	6.36	4.95	2.00	2.47
1027.5	25	11.48	6.38	4.96	2.01	2.50
1052.5	25	11.52	6.39	4.96	2.02	2.53
1077.5	25	11.56	6.41	4.96	2.03	2.56
	∞	11.60	6.42	4.96	2.05	2.58

Table L.4. Structural Constants for a Shield-Earth Model

Depth (km)	Layer thickness (km)	α (km/s)	β (km/s)	ρ (g/cm^3)	μ (dyne/cm^2 $\times$ 10^{12})	λ (dyne/cm^2 $\times$ 10^{12})
19.0	5	6.40	3.70	3.08	0.42	0.42
27.8	13.5	6.70	3.92	3.42	0.53	0.48
37.5	5	8.15	4.75	3.42	0.77	0.73
50.0	20	8.16	4.75	3.42	0.77	0.73
70.0	20	8.21	4.75	3.42	0.77	0.76
90.0	20	8.26	4.75	3.42	0.77	0.79
110.0	20	8.32	4.75	3.42	0.77	0.82
130.0	20	8.30	4.70	3.40	0.75	0.84
150.0	20	8.28	4.58	3.40	0.71	0.91
170.0	20	8.28	4.54	3.40	0.70	0.93
190.0	20	8.28	4.54	3.41	0.70	0.93
210.0	20	8.28	4.54	3.42	0.70	0.94
280.0	120	8.28	4.54	3.45	0.71	0.95
350.0	20	8.31	4.54	3.45	0.71	0.96
365.0	10	8.51	4.64	3.45	0.74	1.01
380.0	20	8.70	4.75	3.45	0.78	1.05
402.5	25	8.74	4.75	3.66	0.83	1.14
524.0	20	8.76	4.75	3.88	0.88	1.23
440.0	10	9.04	5.00	3.90	0.98	1.24
455.0	20	9.49	5.25	3.92	1.08	1.37
477.5	25	9.50	5.25	3.93	1.09	1.38
502.5	25	9.52	5.26	3.95	1.09	1.39
527.5	25	9.53	5.26	3.96	1.10	1.40
552.5	25	9.58	5.29	3.99	1.11	1.43
577.5	25	9.63	5.31	4.02	1.14	1.46
602.5	25	9.68	5.34	4.06	1.16	1.49
627.5	25	9.74	5.37	4.09	1.18	1.52
652.5	25	9.78	5.39	4.12	1.20	1.55
677.5	25	10.01	5.52	4.17	1.27	1.64
702.5	25	10.18	5.63	4.21	1.34	1.69
727.5	25	10.19	5.75	4.26	1.41	1.74
752.5	25	10.49	5.85	4.30	1.47	1.79
777.5	25	10.68	5.95	4.48	1.58	1.93
802.5	25	10.85	6.04	4.63	1.69	2.07
827.5	25	11.03	6.14	4.80	1.81	2.21
852.5	25	11.18	6.23	4.94	1.92	2.34
877.5	25	11.22	6.25	4.94	1.93	2.37
902.5	25	11.27	6.28	4.95	1.95	2.38
927.5	25	11.31	6.30	4.95	1.96	2.41
952.5	25	11.35	6.32	4.95	1.98	2.42
977.5	25	11.39	6.34	4.95	1.99	2.45
1002.5	25	11.43	6.36	4.95	2.00	2.47
1027.5	25	11.48	6.38	4.96	2.01	2.50
1052.5	25	11.52	6.39	4.96	2.02	2.53
1077.5	25	11.56	6.41	4.96	2.03	2.56
	∞	11.60	6.42	4.96	2.05	2.58

Table L.5. ARDC Standard Model of the Atmosphere

Altitude (km)	Temperature (°K)	Pressure (mbar)	Acceleration of gravity (m/s^2)	Density (kg/m^3)	Sound velocity (m/s)	Kinematic viscosity (m^2/s)
0	288.16	1.01325 + 3	9.8067	1.2250	340.29	1.4607 − 5
1	281.66	8.9876 + 2	9.8036	1.1117	336.43	1.5813
2	275.16	7.9501	9.8005	1.0066	332.52	1.7148
3	268.67	7.0121	9.7974	9.0926 − 1	328.58	1.8629
4	262.18	6.1660	9.7943	8.1935	324.59	2.0275
5	255.69	5.4048	9.7912	7.3643	320.54	2.2111
6	249.20	4.7217	9.7882	6.6011	316.45	2.4162
7	242.71	4.1105	9.7851	5.9002	312.30	2.6462
8	236.23	3.5651	9.7820	5.2578	308.10	2.9046
9	229.74	3.0800	9.7789	4.6706	303.85	3.1958
10	223.26	2.6500	9.7759	4.1351	299.53	3.5253
11	216.78	2.2700	9.7728	3.6480	295.15	3.8990
12	216.66	1.9399	9.7696	3.1194	295.07	4.5576
13	216.66	1.6579	9.7667	2.6659	295.07	5.3327
14	216.66	1.4170	9.7636	2.2785	295.07	6.2394
15	216.66	1.2112	9.7605	1.9475	295.07	7.2998
16	216.66	1.0353	9.7575	1.6647	295.07	8.5401
17	216.66	8.8496 + 1	9.7544	1.4230	295.07	9.9906
18	216.66	7.5652	9.7513	1.2165	295.07	1.1687 − 4
19	216.66	6.4674	9.7483	1.0399	295.07	1.3671
20	216.66	5.5293	9.7452	8.8909 − 2	295.07	1.5990
21	216.66	4.7274	9.7422	7.6015	295.07	1.8702
22	216.66	4.0420	9.7391	6.4995	295.07	2.1874
23	216.66	3.4562	9.7361	5.5575	295.07	2.5581
24	216.66	2.9554	9.7330	4.7522	295.07	2.9916
25	216.66	2.5273	9.7300	4.0639	295.07	3.4983
26	219.34	2.1632	9.7269	3.4359	296.89	4.1805
28	225.29	1.5949	9.7208	2.4663	300.89	5.9550
30	231.24	1.1855	9.7147	1.7861	304.83	8.4018
32	237.18	8.8802 + 0	9.7087	1.3044	308.73	1.1747 − 3
34	243.12	6.7007	9.7026	9.6020 − 3	312.57	1.6282
36	249.05	5.0914	9.6965	7.1221	316.36	2.2384
38	254.98	3.8944	9.6904	5.3210	320.10	3.0534
40	260.91	2.9977	9.6844	4.0028	323.80	4.1342
42	266.83	2.3215	9.6783	3.0310	327.46	5.5581
44	272.75	1.8082	9.6723	2.3096	331.07	7.4218
46	278.67	1.4162	9.6662	1.7704	334.64	9.8465
48	282.66	1.1147	9.6602	1.3739	337.03	1.2830 − 2
50	282.66	8.7858 − 1	9.6542	1.0829	337.03	1.6279
52	282.66	6.9256	9.6481	8.5360 − 4	337.03	2.0651
54	280.21	5.4586	9.6421	6.7867	335.56	2.5798
56	271.36	4.2786	9.6361	5.4931	330.22	3.1079
58	262.52	3.3273	9.6301	4.4156	324.80	3.7661
60	253.68	2.5657	9.6241	3.5235	319.29	4.5922
62	244.86	1.9607	9.6181	2.7897	313.68	5.6367
64	236.03	1.4838	9.6121	2.1901	307.98	6.9682

Table L.5. (Continued)

Altitude (km)	Temperature (°K)	Pressure (mbar)	Acceleration of gravity (m/s^2)	Density (kg/m^3)	Sound velocity (m/s)	Kinematic viscosity (m^2/s)
66	227.21	1.1113 − 1	9.6061	1.7039 − 4	302.17	8.6802 − 2
68	218.40	8.2298 − 2	9.6001	1.3128	296.25	1.0902 − 1
70	209.59	6.0209	9.5942	1.0008	290.22	1.3815
72	200.79	4.3470	9.5882	7.5424 − 5	284.06	1.7674
74	191.99	3.0937	9.5822	5.6137	277.76	2.2849
76	183.2	2.167	9.576	4.122	271.3	2.988
78	174.4	1.492	9.570	2.981	264.7	3.956
80	165.7	1.008	9.564	2.120	258.0	5.311
82	165.7	6.744 − 3	9.558	1.418	258.0	7.940
84	165.7	4.512	9.553	9.489 − 6	258.0	1.187 + 0
86	165.7	3.020	9.547	6.350	258.0	1.773
88	165.7	2.021	9.541	4.251	258.0	2.649
90	165.7	1.353	9.535	2.846	258.0	3.957
92	168.4	9.074 − 4	9.529	1.877		
94	176.1	6.172	9.523	1.221		
96	183.7	4.270	9.517	8.087 − 7		
98	191.4	3.000	9.511	5.452		
100	199.0	2.138	9.505	3.734		
102	206.6	1.544	9.499	2.596		
104	214.2	1.128	9.493	1.829		
106	221.8	8.341 − 5	9.488	1.305		
108	248.4	6.274	9.482	8.759 − 8		
110	286.7	4.906	9.476	5.930		
112	325.0	3.957	9.470	4.218		
114	363.1	3.270	9.464	3.117		
116	401.2	2.755	9.458	2.375		
118	439.1	2.358	9.452	1.856		
120	477.0	2.044	9.447	1.480		
125	571.3	1.497	9.432	9.032 − 9		
130	664.9	1.151	9.417	5.953		
135	757.8	9.167 − 6	9.403	4.150		
140	849.9	7.502	9.389	3.020		
145	941.0	6.272	9.374	2.274		
150	1031.0	5.334	9.360	1.759		
155	1120.0	4.602	9.345	1.392		
160	1207.0	4.018	9.331	1.123		
165	1285.0	3.544	9.317	9.256 − 10		
170	1323.0	3.145	9.302	7.932		
175	1359.0	2.803	9.288	6.841		
180	1371.0	2.505	9.274	6.015		
185	1381.0	2.243	9.260	5.300		
190	1389.0	2.013	9.246	4.680		
195	1397.0	1.809	9.231	4.142		
200	1404.0	1.629	9.217	3.673		

Table L.5. (Continued)

Altitude (km)	Temperature (°K)	Pressure (mbar)	Acceleration of gravity (m/s^2)	Density (kg/m^3)	Sound velocity (m/s)	Kinematic viscosity (m^2/s)
205	1411.0	1.470 − 6	9.203	3.264 − 10		
210	1414.0	1.328	9.189	2.914		
215	1414.0	1.201	9.175	2.610		
220	1414.0	1.088	9.161	2.339		
225	1414.0	9.863 − 7	9.148	2.100		
230	1415.0	8.953	9.134	1.887		
240	1415.0	7.401	9.106	1.530		
250	1415.0	6.143	9.078	1.246		
260	1416.0	5.120	9.051	1.019		
270	1417.0	4.284	9.024	8.375 − 11		
280	1418.0	3.598	8.996	6.909		
290	1420.0	3.033	8.969	5.722		
300	1423.0	2.565	8.942	4.757		

Table L.6. Structural Constants for the Gutenberg Model of the Earth

Depth (km)	ρ (g/cm^3)	α (km/s)	β (km/s)	g (cm/s^2)
0–19	2.74	6.14	3.55	982
19–38	3.00	6.58	3.80	983
38–50	3.32	8.20	4.65	984
50–60	3.34	8.17	4.62	985
60–70	3.35	8.14	4.57	985
70–80	3.36	8.10	4.51	986
80–90	3.37	8.07	4.46	986
90–100	3.38	8.02	4.41	986
100–125	3.39	7.93	4.37	986
125–150	3.41	7.85	4.35	987
150–175	3.43	7.89	4.36	988
175–200	3.46	7.98	4.38	989
200–225	3.48	8.10	4.42	989
225–250	3.50	8.21	4.46	990
250–300	3.53	8.38	4.54	991
300–350	3.58	8.62	4.68	992
350–400	3.62	8.87	4.85	993
400–450	3.69	9.15	5.04	995
450–500	3.82	9.45	5.21	996
500–600	4.01	9.88	5.45	997
600–700	4.21	10.30	5.76	998
700–800	4.40	10.71	6.03	998
800–900	4.56	11.10	6.23	997
900–1000	4.63	11.35	6.32	995
1000–1200	4.74	11.60	6.42	993
1200–1400	4.85	11.93	6.55	990
1400–1600	4.96	12.17	6.69	986
1600–1800	5.07	12.43	6.80	983
1800–2000	5.19	12.67	6.90	982
2000–2200	5.29	12.90	6.97	981
2200–2400	5.39	13.10	7.05	984
2400–2600	5.49	13.32	7.15	989
2600–2800	5.59	13.59	7.23	997
2800–2898	5.69	13.70	7.20	1011

Bibliography

Abramowitz M, Stegun IA (eds) (1965) Handbook of Mathematical Functions. Dover, New York.

Barut AO, Wilson R (1976) Some new identities of Clebsch–Gordan coefficients and representation functions of SO(2, 1) and SO(4). Jour Math Phys 17: 900–915.

Bateman H, Archibald RC (1944) A guide to tables of Bessel functions. Math Tables Aids Comp 1: 205–208.

Ben-Menahem A (1962) An operational representation of the addition theorems for spherical waves. Jour Math and Phys 41: 201–204.

Ben-Menahem A (1966) Summation of certain Legendre series and related difference equations. Jour Math and Phys 45: 224–228.

Ben-Menahem A (1975) Properties and applications of a certain operator associated with the Kontorovich–Lebedev transform. Glasgow Math Jour 16: 109–122.

Biedenharn LC, Van Dam H (eds) (1965) Quantum Theory of Angular Momentum, A Collection of Reprints and Original Papers. Academic Press, New York, 332 pp.

Brand L (1947) Vector and Tensor Analysis. John Wiley, New York.

Brussaard PJ, Tolhoek HA (1957) Classical limits of Clebsch–Gordan coefficients, Racah coefficients and $D^l_{mn}(\phi, \theta, \psi)$-functions. Physica 23: 955–971.

Cruzan OR (1962) Translational addition theorems for spherical vector wave functions. Quart Appl Math 20: 33–40.

Drew TB (1961) Handbook of Vector and Polyadic Analysis. Reinhold Publishing Co, New York.

Edelstein LA (1963) On the one centre expansion of scattered distorted spherical and hyperspherical waves. Proc Camb Phil Soc 59: 185–196.

Edmonds AR (1957) Angular Momentum in Quantum Mechanics. Princeton University Press, Princeton, NJ, 146 pp.

Erdélyi A (1937) Zur Theorie der Kugelwellen. Physica 4: 107–120.

Erdélyi A (ed) (1954) Tables of Integral Transforms, Vols I and II. Bateman Manuscript Project, McGraw-Hill, New York.

Erdélyi A (1960) Asymptotic solutions of differential equations with transition points or singularities. Jour Math Phys 1: 16–26.

Gelfand IM, Minlos RA, Shapiro ZY (1963) Representations of the Rotation and Lorentz Groups and Their Applications. Pergamon, New York.

Gibbs JW (1960) Vector Analysis. Dover, New York.

Heading J (1962) An Introduction to Phase-Integral Methods. Methuen, London.

Heine HE (1878) Handbuch der Kugelfunktionen. G Reimer, Berlin.

Hilb E (1919) Über die Laplacesche Reihe. Math Zeit 5: 17–25.

Jones DS (1966) Generalized Functions. McGraw-Hill, London, 482 pp.

Laplace PS (1823) Mécanique Céleste, Vol 5, Book 11, Suppl 1. Hillad, Gray, Little and Witkins, Boston.

Lighthill MJ (1962) Fourier Analysis and Generalised Functions. Cambridge University Press, 79 pp.

Luré AI (1964) Three-dimensional Problems of the Theory of Elasticity. Interscience, New York.

Magnus W, Oberhettinger F, Soni RP (1966) Special Functions of Mathematical Physics, 3rd edn. Springer-Verlag, New York.

Maxwell JC (1873) Treatise on Electricity and Magnetism. Clarendon Press, Oxford.

Mirsky L (1955) An Introduction to Linear Algebra. Clarendon Press, Oxford.

Rayleigh, Lord (Strutt JW) (1872) On the vibrations of a gas contained within a rigid spherical envelope. Proc Lond Math Soc (Ser 1) 4: 93–103.

Robin L (1958) Function Spheriques de Legendre et Function Spheroidales, Vol II. Gauthier-Villars, Paris, 384 pp.

Stein S (1961) Addition theorems for spherical wave functions. Quart Appl Math 19: 15–24.

Szegö G (1934) Über einige Asymptotische Entwicklungen der Legendreschen Funktionen. Proc Lond Math Soc (Ser 2) 36: 427–450.

Titchmarsh EC (1967) Introduction to the Theory of Fourier Integrals. Clarendon Press, Oxford.

Van der Pol B (1936) A generalization of Maxwell's definition of solid harmonics to waves in n dimensions. Physica 3: 393–397.

Van der Pol B, Bremmer H (1959) Operational Calculus. Cambridge University Press, Cambridge.

Wason HR, Singh SJ (1971) Transformation of earthquake displacement field for spherical earth. Bull Seismol Soc Amer 61: 289–295.

Watson GN (1966) A Treatise on the Theory of Bessel Functions. Cambridge University Press, Cambridge, 804 pp.

List of Symbols

It is not feasible to give a complete list of symbols used in the book. The following list includes those used repeatedly. The number(s) following commas is (are) the page number(s) on which the symbol is discussed.

Latin and German Alphabets

a	Radius of a cavity, 221; mean radius of the earth, 266; $= (u^2 + 1/\alpha^2)^{1/2}$, 531.
$a(t)$	Impulse response of a linear system, 1046.
a_n	Layer matrix, 128, 132, 809.
a_x, a_y, a_z	Components of acceleration, 781.
$\mathbf{a}$	$d\mathbf{r}/ds$, 39.
$\mathbf{a}_x, \mathbf{a}_y, \mathbf{a}_z$	Unit vectors, 105.
$\mathbf{a}_1, \mathbf{a}_2, \mathbf{a}_3$	Unit vectors along x_1, x_2, x_3, respectively, 89; Eigenvectors, 5, 169.
A	$= [A_{ij}]$, acoustic medium response function, 809; work, 17.
A, B, C	Angles in a spherical triangle, 368.
A_A	, 822.
$A(t)$	Rate of creep $(\partial\phi/\partial t)$, 864.
$A(l, m \mid v, \mu \mid q)$	Expansion coefficients, 1039.
$Ai(z)$	Airy function, 1007.
A^L	2×2 matrix used in the theory of SH (Love) wave propagation in a multilayered solid, 129.
A_{ij}^L	Elements of A^L, 129.
A^R	4×4 matrix used in the theory of P and SV (Rayleigh) wave propagation in a multilayered solid, 136.
A_{ij}^R	Elements of A^R, 134.
$\tilde{A}_L$	Love-wave medium response function, 276, 286.
$\tilde{A}_R$	Rayleigh-wave medium response function, 281, 286.
$\mathbf{A}, \mathbf{A}_\alpha, \mathbf{A}_\beta$	Vector normal to the planes of constant amplitude, 878.
A	, 197.

b	Mean radius of the core, 337; $=(u^2 + 1/\beta^2)^{1/2}$, 531.
$\mathbf{b}$	"Null vector", 183.
B	$=[B_{ij}]$, acoustic medium response function, 820.
B^2	Beltrami operator, 422.
$B(l, m \mid \nu, \mu \mid q)$	Expansion coefficients, 1042.
$B(t)$	Boxcar function, 1003.
$B(x, a)$	Boxcar function, 998.
$B(m, n)$	Beta function, 238.
$Bi(z)$	Airy function, 1009.
$\mathbf{B}$	Rotation operator, 1028.
$\mathbf{B}_m$	Vector cylindrical harmonic, 60.
$\mathbf{B}_{ml}$	Vector spherical harmonic, 57.
$\mathfrak{B}$	Wave stress-energy tensor, 29.
B	, 197.
c	Intrinsic wave velocity, 44; phase velocity, 96; sound wave velocity, 773.
c_0	$=(gH)^{1/2}$, 780.
c_∞	$=c(\omega = \infty)$, 889.
$ci(t)$	$=Ci(t)$, 1003.
c_L	Love-wave phase velocity, 122.
c_R	Rayleigh-wave phase velocity, 114, 264.
${}_nc_l$	$=c({}_n\omega_l)$, 623.
c_p, c_v	Specific heats, 775.
$c_\alpha(\omega), c_\beta(\omega)$	, 873.
C	$=[C_{ij}]$, acoustic medium response function, 821.
C_1, C_2	Paths of integration in the complex s plane, 630.
$C^{(1,2)}$	Generalized cosine, 744.
$Ci(x)$	Cosine integral, 248, 900.
$C_n^\nu(x)$	Gegenbauer polynomial, 82.
$C_{ijkl}(=\mathsf{C})$	Tensor of elastic moduli, 18, 19, 868.
$\mathbf{C}_m$	Vector cylindrical harmonic, 60.
$\mathbf{C}_{ml}$	Vector spherical harmonic, 57.
C	, 197.
d	Thickness of layer, 126, 808.
ds	Arc-element, 962.
dS	Point-source fault area, 178; surface element, 2.
$d\mathbf{S}$	Vector element of surface area ($\mathbf{n}dS$), 4.
D	Diffusivity, 845; rate of mechanical energy dissipation per unit volume, 880; 651.
D_H	Horizontal diffusivity, 846.
D_V	Vertical diffusivity, 846.
D_α	Directivity, 244.
$\mathfrak{D}^L$	$=[D_{ij}]$, Lagrange deformation tensor, 39.
D	$=[D_{ijkl}]$, viscosity tensor, 769, 868.

e	$= 2.71828\ldots$, base of natural logarithm, 31; angle with the normal for P waves, 89; elongation, 39.
$e(r)$	Earth's ellipticity at level r, 449.
e_0	Surface ellipticity of the earth, 449.
$\mathbf{e}$	Unit-slip vector, 179; unit vector, 949.
$\mathbf{en}$	Moment tensor for a shear dislocation, normalized, 179.
$\mathbf{e}_1, \mathbf{e}_2, \mathbf{e}_3$	Unit vectors along x_1, x_2, x_3 (or q_1, q_2, q_3), respectively, 2, 962.
$\mathbf{e}_1^\circ, \mathbf{e}_2^\circ, \mathbf{e}_3^\circ$	Unit vectors in the source system, 182.
$\mathbf{e}_r, \mathbf{e}_\theta, \mathbf{e}_\phi$	Unit vectors in a spherical coordinate system, 953.
$\mathbf{e}_\Delta, \mathbf{e}_\phi, \mathbf{e}_z$	Unit vectors in a cylindrical coordinate system, 953.
$\mathbf{e}_R, \mathbf{e}_{i_h}, \mathbf{e}_{\phi_h}$	Unit vectors in a spherical coordinate system with origin at the source, 200.
$E(k)$	Complete elliptic integral of the second kind, 550.
$E(k, \phi)$	Incomplete elliptic integral of the second kind, 550.
$Ei(-x)$	Exponential integral, 911.
$Erf(z)$	Error function, 897.
E_0	, 223, 259.
E_n	4×4 matrix used in the theory of P and SV waves propagating in a multilayered medium, 132.
$E_{lm}^{(1,2)}(\cos\theta)$	Traveling-wave functions, 1023.
$E_{m,l}^{n}(x)$	, 1041.
$\mathfrak{E}(\mathbf{u})$	$= \frac{1}{2}(\nabla\mathbf{u} + \mathbf{u}\nabla)$, strain dyadic, 10.

f	Angle with the normal for S waves, 89.
$f(r_0 + 0)$	Value of f just above r_0, 374.
$f(r_0 - 0)$	Value of f just below r_0, 374.
$\hat{f}(t)$	Allied function of $f(t)$, 1001.
$f(st)$	$\lim_{\omega \to 0} f(\omega)$, 363, 385.
$f_l(x)$	Spherical Bessel function, 973.
$f_l^+(x)$	$= j_l(x)$, spherical Bessel function of the first kind, 59; $= i_l(x)$, modified spherical Bessel function, 711.
$f_l^-(x)$	$= h_l^{(2)}(x)$, spherical Hankel function of the second kind, 59; $= k_l(x)$, spherical Macdonald function, 711.
$F(\lambda, \delta; i_h, \phi_h)$	Radiation pattern of a shear dislocation, 200.
F_0	Magnitude of single force, 152.
$\dot{F}$	dF/dt, 26, 813.
F_A	, 820.
F_e	, 832.
$F(x_n)$	, 616.
$F_{l,i}(x)$	, 222, 339.
${}_2F_1(a, b, c; z)$	Hypergeometric function, 629.
$F_l^m(x)$	, 1041.
$\mathbf{F}$	Body force per unit mass, 4.

g	Acceleration of gravity, 349, 776, 798.
$g(t)$	Source time-function, 152.
$\hat{g}(t)$	Allied function of $g(t)$, 676.
$g(\omega)$	Fourier transform of source time-function, 152, 395.
$g(\mathbf{z}\mid\mathbf{z}_0; t)$	Scalar Green's function, 606.
g_0	Acceleration of gravity in the equilibrium state, 350.
$\bar{g}$	, 395.
$g_l^{(1,2)}$	, 732.
G	Constant of gravitation, 352, 388.
G, H, L, K, M, N	, 136, 832.
G, G_α, G_β	Divergence coefficient, 459.
$G(z\mid z_0)$	Scalar Green's function, 505.
$G(\hat{\gamma})$	, 264.
$G_a(x)$	Unit-gate function, 999.
G_i	Love-wave multiple arrivals, 635.
$G_i^j(P, Q)$	$= G_{ij}(P, Q)$, components of $\mathfrak{G}$: displacement in the x_i direction due to a single unit force in the x_j direction, 163.
$Gi(z)$	, 1010.
GR_i	Acoustic-gravity modes, 811.
GW_i	Oceanic-gravity modes, 827.
$\mathfrak{G}$	Green's dyadic, 153, 174.
$\mathfrak{G}_a$	Green's dyadic for a sphere of radius a, 347.
$\mathfrak{G}_H$	Green's dyadic for a half-space, 373.
$\mathfrak{G}_\infty$	Green's dyadic for an unbounded medium, 174.
$\tilde{\mathfrak{G}}$	Transpose of $\mathfrak{G}$, 163.
h	Source-depth, 258, 285; 816; $= r_h/a$, 668, 670.
$h(t)$	Step-response of a linear system, 1047.
h_1, h_2, h_3	Scale factors, 962.
$h_l^{(1,2)}(x)$	Spherical Hankel functions of the first and second kinds, 51, 973.
H	Hamiltonian density, 26; layer thickness, 106; water depth, 777; 832.
$H(t)$	Heaviside unit-step function, 98, 998.
$H_n(x)$	Hermite polynomial, 894.
$H_m^{(1,2)}(x)$	Hankel functions of the first and second kinds, 967.
H	$\equiv H_{kl}^{ij}$, (ij) stress component resulting from a (kl) dipolar source, 199, 202.
i	Angle between the ray and radius vector, 431; $= (-1)^{1/2}$, 31.
i_0	Angle of incidence, 481.
i_h	Takeoff angle, 451.
i_m, j_m, k_m	Source coefficients, 260.
$i_l(x)$	Modified spherical Bessel function, 976.
$\mathbf{i}_1, \mathbf{i}_2, \mathbf{i}_3$	Unit vectors in the fault coordinate system, 188.
I	Unit matrix, 138.

$I_m(x)$ Modified Bessel function, 109, 971.
I_h , 452.
I_n^S Spheroidal energy integral, 379, 391.
I_n^T Toroidal energy integral, 364, 391.
I_i^L Love-wave energy integrals, 121, 279.
I_i^R Rayleigh-wave energy integrals, 123, 281.
$\mathfrak{I}$ Idemfactor (unit dyadic), 6, 947.

j Angle between the ray and the vertical, 496.
$j(t)$ Retardation spectrum, 860.
$j_l(x)$ Spherical Bessel function of the first kind, 51, 973.
J $= [J_{ij}]$, 134.
$J(t)$ Compliance, 860.
$J^*(i\omega)$; $J(i\omega)$ Complex creep compliance, 862, 863.
J_1, J_2, J_3 Invariants of a dyadic, 42.
$J_m(x)$ Bessel function of the first kind, 50, 967.
$\mathfrak{J}$ Momentum flux dyadic, 771.

k Separation variable, 49; wave number, 105.
k_0 $= \omega/v_0$, 662.
k_c, k_α, k_β $= \omega/c, \omega/\alpha, \omega/\beta$, wave numbers, 46, 63.
k_f Wave number in a flat-earth model, 638.
k_s Wave number in a spherical-earth model, 639.
$k_l(x)$ Spherical Macdonald function, 976.
$k_l^{(1,2)}(z)$, 978.
k_L Wave number of Love waves, 122.
k_R Wave number of Rayleigh waves, 124.
k_n $= {}_n\omega_l/\beta_0$, 652.
k_x, k_y, k_z Cartesian components of the propagation vector, 48.
${}_nk_l$ $= {}_n\omega_l/c({}_n\omega_l)$, 623.
$\hat{k}_{\alpha,\beta}$ $= k_{\alpha,\beta} - ik^*_{\alpha,\beta}$, complex wave number, 873, 931.
$\mathbf{k}$ Propagation vector, 46.
K , 136; universal gas constant per unit molecular weight, 797.
$K(k)$ Complete elliptic integral of the first kind, 550.
$\mathbf{K}$ Anelastic wave-number bivector, 878.
$K_m(z)$ Macdonald function, 109, 971.
K_s , 816.
$K_\nu^{(1,2)}(z)$, 977.
$\mathfrak{K}$ Matter stress-energy tensor, 29.

l Colatitudinal mode number, 340.
l, m, n Direction cosines, 11.
L Fault-length, 229; 136.
$L(k, H)$ Love-wave determinant, 266.
$\mathscr{L}$ Lagrangian density, 27, 31, 775.
L Lagrangian function, 143, 930.

L_n Love modes, 289.
$\mathbf{L}, \mathbf{M}, \mathbf{N}$ Hansen eigenvectors, 56.
$\mathbf{L}^c, \mathbf{M}^c, \mathbf{N}^c$ Hansen eigenvectors with cos $m\phi$, 201, 202.
$\mathbf{L}^s, \mathbf{M}^s, \mathbf{N}^s$ Hansen eigenvectors with sin $m\phi$, 201, 202.
$\mathbf{L}_m$ Cylindrical eigenvector, 62.
$\mathbf{L}_m^\sigma$, 202.
$\mathbf{L}_{ml}$ Spherical eigenvector, 59.
$\mathbf{L}_{ml}^{c,s}$, 201.

m Azimuthal mode number, 340; mass, 965.
$\mathbf{m}$ Curvature vector, 431.
M Source moment, 163; 136.
$M(t)$ Memory function, 865.
M_R Relaxed elastic modulus, 857.
M_i Eigenvalues of $\mathfrak{M}$, 169.
M_{ij} Components of the source moment tensor, 168; Rayleigh modes, 118.
$\mathbf{M}$ Total moment of momentum, 22.
$\mathbf{M}_m$ Cylindrical eigenvector, 62.
$\mathbf{M}_m^\sigma$, 202.
$\mathbf{M}_{ml}$ Spherical eigenvector, 59.
$\mathfrak{M}$ Source moment tensor, 168, 186.

n Radial mode number, 340; integer, 507.
$\hat{n}(\omega)$ $= n - in^*$, complex refraction index, 888, 892.
$n_l(x)$ Spherical Bessel function of the second kind, 345, 973.
$\hat{n}_\alpha(\omega)$ $= \alpha_\infty/\hat{\alpha}(\omega)$, complex refraction index for P waves, 874.
$\hat{n}_\beta(\omega)$ $= \beta_\infty/\hat{\beta}(\omega)$, complex refraction index for S waves, 874.
$\mathbf{n}$ Unit normal, 1, 172.
N Normal stress, 5; positive integer, 125; 136; $= [l(l+1)]^{1/2}$, 375.
$N_m(x)$ Bessel function of the second kind, 967.
$\mathbf{N}_m$ Cylindrical eigenvector, 62.
$\mathbf{N}_m^\sigma$, 202.
$\mathbf{N}_{ml}$ Spherical eigenvector, 59.

$\mathbf{O}$ Zero vector, 948.
$\mathfrak{O}$ Zero dyadic, 948.

$\bar{p}$, 799.
p Isotropic part of the stress tensor, 7; hydrostatic pressure, 769, 774; overpressure, 797; ray parameter, 434; plunge angle, 185.
$\bar{P}$ Parcel pressure, 800.
p_0 Hydrostatic pressure at the surface; 779; hydrostatic pressure in the equilibrium state, 350, 774, 796.
p' Perturbation in p_0, 774.

$p_1, \ldots, p_5$	Functions of λ, δ and $\hat{\phi}$, 187.
p_{as}	Air pressure at border of linear zone $R = R_0$, 816.
p_L, q_L, p_R, q_R, s_R	Functions of λ, δ and ϕ, 269.
$\mathbf{p}$	$= l\mathbf{e}_x + m\mathbf{e}_y + n\mathbf{e}_z$, unit vector in the direction of propagation, 45; $= d\mathbf{r}/ds$, 429; unit source vector, 193.
P	Longitudinal wave, 69.
P_0	Source potency, 231.
P_g, P^*, P_n	Longitudinal head waves in the earth's crust, 518.
$P_l(x)$	Legendre polynomial, 979.
$P_l^m(x)$	Associated Legendre function of the first kind, 979.
$P_l^{(m)}(x)$	, 982.
$P_{ml}(\mu)$	, 1031.
$P_l^{mn}(\mu)$	Ultraspherical function, 1030.
$P_l^{(\alpha,\beta)}(\mu)$	Jacobi polynomial, 1030.
P_L, Q_L	Love-wave amplitude transfer functions, 267, 275, 280, 286.
P_R, Q_R, S_R	Rayleigh-wave amplitude transfer functions, 267, 281, 286.
$\mathbf{P}$	Canonical momentum density, 26.
$\mathbf{P}(t)$	Total linear momentum, 22.
$\mathbf{P}, \mathbf{B}, \mathbf{C}$	Vector surface harmonics, 57.
$\mathbf{P}, \mathbf{P}_\alpha, \mathbf{P}_\beta$	Vector normal to the planes of constant phase, 878.
$\mathbf{P}_m$	Vector cylindrical harmonic, 60.
$\mathbf{P}_m^c, \mathbf{B}_m^c, \mathbf{C}_m^c$	Vector cylindrical harmonics with cos $m\phi$,61.
$\mathbf{P}_m^s, \mathbf{B}_m^s, \mathbf{C}_m^s$	Vector cylindrical harmonics with sin $m\phi$, 61.
$\mathbf{P}_{ml}$	Vector spherical harmonic, 57.
q	, 503.
q_1, q_2, q_3	Orthogonal curvilinear coordinates, 13, 962.
$q(t)$	, 453; 643.
q_n	Zeros of the Airy function $Ai(z)$, 653, 1020.
$\bar{q}_n$	Zeros of $Ai'(z)$, 653, 1021.
$\mathbf{q}(\mathbf{r})$	Eigenvectors, 391.
Q	Specific dissipation parameter, 841, 847, 853, 858, 865, 873, 883; eikonal, 428; 641; 646; 733.
$Q(t/t_0)$	, 453.
Q_L	Love-wave dissipation parameter, 928.
Q_R	Rayleigh-wave dissipation parameter, 927.
Q_α	Longitudinal-wave dissipation parameter, 884.
Q_β	Shear-wave dissipation parameter, 884.
Q_k	Bulk dissipation parameter, 884.
Q_μ	Shear dissipation parameter, 885.
Q_ρ	Inertial dissipation parameter, 885.
$Q_m(t)$	, 547.
$Q_l^m(x)$	Associated Legendre function of the second kind, 631.
$\mathbf{Q}$	Wave momentum density, 29; source vector, 167, 170.
${}_n\mathbf{Q}_{ml}$	, 379.
$\mathbf{Q}_m$	, 281.

r, θ, ϕ	Spherical coordinates, 14, 957.
r_0, θ_0, ϕ_0	Coordinates of a point source, 181.
$r_>$	$= \max(r, r_0)$, 347.
$r_<$	$= \min(r, r_0)$, 347.
r_h	$= r_0$, radius at the source, 451.
r_m	Radius at the lowest point of the ray, 657.
$\mathbf{r}$	$= r\mathbf{e}_r = x_1\mathbf{e}_1 + x_2\mathbf{e}_2 + x_3\mathbf{e}_3$, radius vector, 4, 961.
$\mathbf{r}_0$	Radius vector to a point source, 152.
R	Reynolds number, 772; $= \lvert\mathbf{r} - \mathbf{r}_0\rvert$, 152, 959.
R^*	, 816.
$R(k)$	Rayleigh-wave determinant, 261.
$R(\omega)$	Real part of $S(\omega)$, 1048.
R_i	Rayleigh-wave multiple arrivals, 314, 635.
$R_{PS}^{\pm}$	Amplitude ratios for incident P reflected S (+ refers to reflected up and − refers to reflected down), 716.
$R_{PS}^{\Delta, z}$	Amplitude ratios for incident P reflected S (Δ refers to the horizontal component and z refers to the vertical component of the displacement), 533.
$\mathbf{R}$	$= \mathbf{r} - \mathbf{r}_0$, 152, 959.
$\mathfrak{R}$	Rotation (spin) dyadic, 10.
s	Laplace transform variable, 531; arc-length parameter, 429; $= l + \frac{1}{2}$, 629.
sgn x	Signum function, 98, 998.
$si(t)$	$= Si(t) - \pi/2$, 1003.
s_j	Real poles of $f_{s-1/2}$, 629.
S	Shear wave, 69; shearing stress, 6; surface, 2.
SH	Horizontally polarized S wave, 69.
SV	Vertically polarized S wave, 69.
$S(\omega)$	Transfer function of a linear system, 1047.
S_g, S^*, S_n	Shear head-waves in the earth's crust, 518.
S_i	Pure acoustic modes, 811.
$Si(x)$	Sine integral, 248.
${}_nS_l$	Spheroidal oscillations, 345.
$S_{\nu,\mu}(x)$	Lommel function, 902.
t	Time, 22; trend of motion, 185.
t^*	, 465, 916.
t_f	Rupture time, 306.
t_{d_α}	Duration of P signal, 231.
t_{d_β}	Duration of S signal, 231.
t_α, t_β	, 231.
$\mathbf{t}$	Unit source vector, 193.
T	Period, 47; temperature on the Kelvin scale, 797; travel-time, 435; rise time, 241, 614.

$T_n(\mu)$	Chebyshev polynomial of the first kind, 82.
${}_nT_l$	$= 2\pi/{}_n\omega_l$, eigenperiod, 340, 623; toroidal oscillations, 340.
$T_k^{ij}(= T_{ij}^k)$	, 175, 197, 206, 219, the x_k component of displacement at P resulting from a dipolar source at Q: double-couple in the x_ix_j plane if $i \neq j$, dipole in the x_i direction plus a center of compression if $i = j$.
T_{Bn}	Brunt resonant period, 810.
$T_l^{mn}(\phi, \theta, \gamma)$	Generalized spherical harmonic, 1029.
$T_{PS}^{\pm}$	Amplitude ratios for incident P transmitted S (+ refers to transmitted up and − refers to transmitted down), 716.
$\mathbf{T}(\mathbf{n})$	Stress vector (traction) across a plane with normal $\mathbf{n}$, 1.
$\mathfrak{T}$	Stress dyadic, 3.
$\mathfrak{T}_0$	Prestress dyadic, 229.
$\mathfrak{T}(\mathbf{u})$	$= \lambda\mathfrak{T}$ div $\mathbf{u} + \mu(\nabla\mathbf{u} + \mathbf{u}\nabla)$, for isotropic media, 19.
T_α, T_β	, 197.
u	$= k/s$, 531.
u, v, w	Components of velocity, 782; amplitudes of plane waves at depth in a multilayered media, 127, 131; components of the displacement vector, 39.
$\dot{u}_s, \dot{v}_s, \dot{w}_s$	Velocities of plane waves at source level, 286, 288.
$\dot{u}_0, \dot{v}_0, \dot{w}_0$	Velocities of plane waves at the free surface, 129, 134.
$u_{k,j}$	Partial derivative of the x_k component of $\mathbf{u}$ with respect to the x_j coordinate, 27.
$\mathbf{u}$	Displacement vector, 9.
$\mathbf{u}_{\mathrm{I}}, \mathbf{u}_{\mathrm{II}}, \mathbf{u}_{\mathrm{III}}$	Displacement fields for the three fundamental shear dislocations, 188.
U	Rayleigh-wave horizontal displacement at depth (y_3) in a sourceless elastic half-space with a continuous vertical inhomogeneity, 115; total mechanical energy, 26.
$U(x)$	Unit-step function, 997.
$U_0(\omega)$	$= U_0 g(\omega)$, 179.
$\bar{U}, U_0$	Dislocation, 179, 230.
U_f	Group velocity in a flat-earth model, 639.
U_g	Group velocity, 110.
U_L	Love-wave group velocity, 122.
U_R	Rayleigh-wave group velocity, 124.
U_s	Group velocity in a spherical-earth model, 639.
$U_n(\mu)$	Chebyshev polynomial of the second kind, 1003.
${}_nU_l$	$= y_{ln}$, spheroidal radial function, 391.
$\mathfrak{U} = \dfrac{\partial \mathfrak{G}}{\partial t}$	Strain-rate dyadic, 769.
v	$= iu$, 540; intrinsic wave velocity, 495; volume, 42.
v_0, V_0	Fixed values of the wave velocity or its values at a fixed level, 438, 455, 477, 499, 519, 650, 658.

v_e	, 908.
$v_n(z)$	, 499.
$\mathbf{v}$	Velocity vector, 769.
V	Intrinsic wave velocity, 428; Love-wave displacement (y_1), 107; volume, 4.
V_1, V_2, V_3	Intrinsic wave velocities, 425.
V_a	Apparent velocity, 438.
V_f	Rupture velocity, 229.
${}_nV_l$	$= y_{3n}$, spheroidal radial function, 391.
$\mathbf{V}_E$	Velocity of energy transport, 122, 883.
w	$= q_1 + iq_2$, 52; vertical component of velocity, 781.
$\overline{w}$	, 799.
W	Fault-width, 229; Rayleigh-wave vertical displacement (y_1), 115; strain energy density, 18.
$W(F_1, F_2)$	$= F_1F_2' - F_1'F_2$, Wronskian, 273, 968.
${}_nW_l$	$= y_{ln}$, toroidal radial function, 391.
$W_{l,m}(z)$	Whittaker function, 109.
$\mathfrak{W}$	$= [W_{ij}]$, wave force dyadic, 29.
x	$= \cos\theta$, 979.
x_1, x_2, x_3	Cartesian coordinates (x, y, z), 953.
X	Finiteness parameter, 306.
$X(\omega)$	Imaginary part of $S(\omega)$, 1048.
X_i	$= x_i - y_i$, 154.
X_α, X_β	, 237.
$y(t)$	Relaxation spectrum, 861.
y_1, y_2	Vertical functions for Love waves, 109, 272; radial functions for toroidal oscillations, 355.
$y_1, \ldots, y_4$	Vertical functions of displacements and stresses for Rayleigh waves in an inhomogeneous half-space, 120.
$y_1, \ldots, y_6$	Radial functions of displacements, stresses and gravitational potential for spheroidal oscillations in a real earth model, 371–373.
$y_{ml}(\theta, \phi)$	Normalized spherical harmonic, 636.
Y	Young's modulus, 20.
Y_0	$= Y(t = \infty)$, 859
Y_∞	$= Y(\omega = \infty)$, 859, 887.
Y_α	, 243.
$Y^*(i\omega), Y(i\omega)$	Complex relaxation modulus, 862; 863.
$Y_m(k\Delta, \phi)$	$= J_m(k\Delta)e^{im\phi}$, cylindrical harmonic, 50.
$Y_m^{c,s}(k\Delta, \phi)$	$= J_m(k\Delta)^{\cos}_{\sin} m\phi$, cylindrical harmonic, 50.
$Y_{ml}(\theta, \phi)$	$= P_l^m(\cos\theta)e^{im\phi}$, spherical harmonic, 51.
$Y_{ml}^{c,s}(\theta, \phi)$	$= P_l^m(\cos\theta)^{\cos}_{\sin} m\phi$, spherical harmonic, 51.
$Y_l^m(\theta, \phi)$	, 1031.

z Vertical coordinate, 955.
z_0 Source-depth, 274.
$z_>$ $= \max(z, z_0)$, 273.
$z_<$ $= \min(z, z_0)$, 273.
$z_l(x)$ Spherical Bessel function, 51.
z_m Depth of the lowest point of a ray, 497.
${}_nz_l$ $= k_n a$, 652.
Z Reflection coefficient, 477.
$Z_m(z)$ Bessel function, 49, 967.

Greek Alphabet

α Compressional wave velocity, 25; sound wave velocity, 797.
α, β Angles, 662.
α, β, γ Euler angles, 949; direction cosines, 5.
$\hat{\alpha}$ $= \alpha + i\alpha^*$, complex compressional wave velocity, 873.
α_∞ $= \alpha(\omega = \infty)$, 873, 926.

β Shear wave velocity, 25.
$\hat{\beta}$ $= \beta + i\beta^*$, complex shear wave velocity, 873.
β_0 Fixed value of the shear velocity or its value at a fixed level, 640, 648, 652.
β_∞ $= \beta(\omega = \infty)$, 873, 926.

γ Attenuation coefficient, 888; ratio of specific heats (c_p/c_v), 775; 0.577215 . . . Euler's constant, 248; angle, 397, 703; 645; velocity gradient, 594.
$\gamma_{\alpha\beta}$, 40.
$\hat{\gamma}$ $= k_R/k_\beta = \beta/c_R$, 264.
γ_L, γ_R Love and Rayleigh wave attenuation coefficients, 269.
γ_n $= 2\beta_n^2/c^2$, 131.
$\gamma_1, \gamma_2, \gamma_3$ Integration paths in the complex v plane, 543.
γ_α $= (1 - c^2/\alpha^2)^{1/2}$, 105; P wave attenuation coefficient, 873.
γ_β $= (1 - c^2/\beta^2)^{1/2}$, 105; S wave attenuation coefficient, 873.
Γ $= \gamma g/(2\alpha^2)$, 816; path of integration in the complex plane, 543.

$\Gamma(z)$ Gamma function, 629.
$\Gamma(v, z)$ Incomplete gamma function, 907.
$\Gamma_i^{jk}(P, Q)$ $(\partial/\partial y_k)G_j^i(P, Q)$, third order tensor: x_i component of the displacement per unit moment at P resulting from a dipole in the x_j direction at $Q(j = k)$ or from a single couple at $Q(j \neq k)$ whose forces are parallel to the x_j direction and whose arm is in the x_k direction, 164.
$\mathbf{\Gamma}$ Total applied torque, 22.

δ	Dip angle, 183; lag of strain behind stress, 841, 858, 862.
δ', λ'	Dip and slip angles of an auxiliary plane, 190.
$\delta(x)$	Dirac delta function, 990.
$\hat{\delta}(x)$	Allied function of $\delta(x)$, 1001.
$\delta^+(x)$	Unit impulse function, 999.
$\delta(\mathbf{r} - \mathbf{r}_0)$	Three-dimensional delta function, 993.
δ_{kl}	$= 0(k \neq l), 1(k = l)$, Kronecker delta, 18.
δF	Variation in F.
Δ	Epicentral distance in units of length, 266; epicentral distance in radians, 436, 734; radial distance in cylindrical coordinates, 14, 504, 760, 955.
Δ_R	Rayleigh wave determinant, 117, 124, 283, 531.
Δ_l	, 222.
Δ_l^+	, 348.
ε	Epicentral distance, 368; extension (elongation) in the direction (l, m, n), 11; strain, 852; $= \operatorname{sgn}(r - r_0)$, 214; $= \operatorname{sgn}(z - z_0)$, 76; small quantity, 13, 163; angle, 370, 549.
$\varepsilon(z)$	Rayleigh-wave ellipticity, 268, 287.
ε_0	Surface ellipticity (Rayleigh wave), 315; initial strain, 859.
ε_∞	$= \varepsilon(\omega = \infty)$, 859.
ε_i	Eigenvalues of the strain dyadic, 12.
ε_m	$= 1(m = 0), 2(m > 0)$, Neumann factor, 59.
ε_{ijk}	Levi-Civita symbol, 964.
$\varepsilon_\alpha, \varepsilon_\beta$	Spectral energy density for P and S waves, 247.
$\varepsilon_{\alpha\beta}$	Strain components, 11, 13, 14.
$\boldsymbol{\varepsilon}$	Polarization vector, 86.
ζ	$= ak_\alpha$, 222, 678; 804.
$\zeta(t)$	, 881.
$\zeta(x, y, t)$	Elevation of water above undisturbed level, 777.
η	$= \eta_2$, 229; $= rk_\beta$, 339; $= r/V$, 435; coefficient of shear viscosity, 770; μ_2/μ_1, 524; 553; 804.
η_1, η_2, η_3	Coordinates in the fault system, 188.
η_a, η_h	, 663.
η_α	$= (c^2/\alpha^2 - 1)^{1/2}$, 105.
η_β	$= (c^2/\beta^2 - 1)^{1/2}$, 105.
$\eta_{\alpha n}, \eta_{\beta n}$	, 131.
θ	Polar angle, 957; dilatation, 879.
θ_s	Beginning of shadow zone, 692.
Θ	, 230, 783.
κ	Bulk modulus, 20.
$\hat{\kappa}$	$= \kappa + i\kappa^*$, complex bulk modulus, 873.
$\kappa(t)$	, 872.
κ_∞	$= \kappa(\omega = \infty)$, elastic bulk modulus, 873.

λ Lamé parameter, 18; slip angle, 183.
$\bar{\lambda}$ $= \lambda'$, coefficient of bulk viscosity, 770, 868.
$\hat{\lambda}$ $= \lambda + i\lambda^*$, complex Lamé parameter, 931.
λ_∞ $= \lambda(\omega = \infty)$, 871.
Λ Wavelength, 47; reflection coefficient for *SH* waves, 507.
$\mathbf{\Lambda}$ Total applied force, 22.
$\mathbf{\Lambda}^i_{ml}$ Vector spherical harmonics, 59.

μ Rigidity, 18; $= \cos\theta$, 81.
$\mu(t)$, 872.
$\hat{\mu}$ $= \mu + i\mu^*$, complex rigidity, 873.
μ' $= \eta$, coefficient of shear viscosity, 868.
μ_∞ Elastic rigidity, 873.
μ_s Rigidity at the source level, 274.

ν Frequency $(\omega/2\pi)$, 247.
ν_j $= s_j - \frac{1}{2}$, 630.
ν_c $= (k^2 - k_c^2)^{1/2}$, 49.
ν_α $= (k^2 - k_\alpha^2)^{1/2}$, 258.
ν_β $= (k^2 - k_\beta^2)^{1/2}$, 258.
$\boldsymbol{\nu}$ Unit vector, 163; unit principal normal, 431.

ξ $= \eta_1$, 229; $= rk_\alpha$, 339; coordinate along strike of fault, 397; 445; b/a, 695; 894.
ξ_n , 428.
$\Pi(n, k)$ Complete elliptic integral of the third kind, 550.
$\mathbf{\Pi}$ Unit tetradic, 19.
$\mathbf{\Pi}'$, 19.

ρ Density, 4, 771, 774; perturbation in density, 797; radius of curvature, 431.
$\hat{\rho}$ $= \rho + i\rho^*$, complex density, 931.
ρ' Perturbation in ρ_0, 774.
ρ_0 Density in the equilibrium state, 350, 774, 796.

σ c or s, 50; Poisson's ratio, 20; stress, 131, 852; angle, 742.
$\bar{\sigma}$ Poisson's ratio, 868.
σ_0 Initial stress, 859.
σ_s Poisson's ratio at the source level, 281.
σ_{ij} Components of stress, 868.
$\mathbf{\Sigma}$ Energy-flux density, 26.

τ Rise time, 232; shearing stress, 127, 131; $T - p\Delta$, 448; integration variable, 741; angle, 651.
τ_0 Relaxation time, 841, 853.
τ_1, τ_2, τ_3 Principal stresses, 5.

τ_i Eigenvalues of stress dyadic, 5.
$\tau_p, \tau_s, \tau_H, \tau_R$ Arrival times, 545, 561, 567.
τ_l Splitting parameter, 390.
τ_ε Strain relaxation time, 857.
τ_σ Stress relaxation time, 857.
τ_{ij} Components of stress, 3.

ϕ Azimuth, 955, 957.
$\hat{\phi}$ Strike azimuth, 182.
$\phi(t)$ Creep function, 859.
$\underline{\phi}(t)$ Normalized creep function, 898.
$\phi_c(t)$ Creep compliance, 858.
$\phi_\beta(t)$ Shear creep function, 870.
$\phi_\kappa(t)$ Bulk creep function, 870.
Φ_{ijkl} Creep tensor, 871.
$\Phi_m^{\pm}(k_c\Delta)$ $= e^{\pm\nu_c z} Y_m(k\Delta, \phi)$, cylindrical wave-functions, 50.
$\Phi_m^{\sigma,\pm}(k_c\Delta)$ $= e^{\pm\nu_c z} Y_m^{\sigma}(k\Delta, \phi)$, cylindrical wave-functions, 50.
$\Phi_{ml}^{\pm}(k_c r)$ $= f_l^{\pm}(k_c r) Y_{ml}(\theta, \phi)$, spherical wave-functions, 51.
$\Phi_{ml}^{c,s}(k_c r)$ $= f_l(k_c r) Y_{ml}^{c,s}(\theta, \phi)$, spherical wave-functions, 51.

χ $= ak_\beta$, 222; force potential, 772; time derivative of dilatation, 798; dimensionless frequency, 340.
$\bar{\chi}$, 799.

ψ Perturbation of the gravitational potential, 350; velocity potential, 773.
$\psi(t)$ Relaxation function, 859.
$\underline{\psi}(t)$ Normalized relaxation function, 899.
$\psi_c(t)$ Relaxation modulus, 858.
$\psi_\beta(t)$ Shear relaxation function, 870.
$\psi_\kappa(t)$ Bulk relaxation function, 870.
$\boldsymbol{\psi}$ Deformation dyadic, 9.
Ψ Gravitational potential, 350.
Ψ_{ijkl} Relaxation tensor, 869.

ω Angular frequency $(2\pi/T)$, 31, 47.
$\hat{\omega}$ $= \omega + i\omega^*$, complex angular frequency, 931.
ω_B Brunt frequency, 799.
ω_n $= {}_n\omega_l$, eigenfrequency, 679.
${}_n\omega_l$ Eigenfrequency $(2\pi/{}_nT_l)$, 340.
${}_n\omega_l^m$ Eigenfrequencies of the split normal modes, 390.
$\boldsymbol{\omega}$ $= \frac{1}{2}\operatorname{curl} \mathbf{u}$, 41.
Ω Angular velocity of the earth, 387; $= u^2 + 1/(2\beta^2)$, 531; matrizant, 138; 890.
Ω_{ml} $= \dfrac{4\pi}{2l+1} \cdot \dfrac{(l+m)!}{(l-m)!}$, 57.
$\boldsymbol{\Omega}$ Vorticity vector, 798.

Special Symbols

$\times$	Vector multiplication, 948.
$\cdot$	Scalar multiplication (dot product), 946.
$:$	Double dot product, 953.
$\overset{\times}{\cdot}$	Cross-dot product, 953.
$\underset{\times}{\cdot}$	Dot-cross product, 953.
$\overset{\times}{\times}$	Double-cross product, 953.
$*$	Complex conjugate, 31; convolution, 559.
$\sim$	Denotes asymptotic approximation, 642; denotes conjugate (transpose) of a dyadic, 948.
$\simeq$	Denotes approximation in general, 642.
$\gg$	Large compared to, 545.
$\ll$	Small compared to, 10.
$\sum$	Summation, 29.
$\prod$	Product, 570.
$[\ ]$	Matrix, 4.
$\lvert\ \rvert$	Modulus, 988; magnitude, 6, 41, 71.
$\lVert\ \rVert$	Determinant, 7, 687.
$-\!\!\!\!\!\int$	Path of integration with a semicircular indentation, 512.
$(\partial f/\partial x)_y$	Partial derivative of f with respect to x when y is kept fixed, 142.
$F'(z)$	dF/dz, 68.
$\dot{F}$	dF/dt, 26, 813.
∇	Gradient operator, 952.
$\nabla\mathbf{f}$	Gradient of $\mathbf{f}$, 951.
$\mathbf{f}\nabla$	Conjugate of $\nabla\mathbf{f}$, 952.
∇_0	Gradient operator in the $\mathbf{r}_0$ system, 163, 959.
∇^2	Laplacian operator, 954.
∇_1^2	Two-dimensional Laplacian operator, 35.
$\dfrac{D}{Dt}$	$\dfrac{\partial}{\partial t} + (\mathbf{v}\cdot\nabla)$, material derivative, 22, 965.
$n!$	$= n(n-1)\cdots 3\cdot 2\cdot 1$, factorial function, 57.
$(2n-1)!!$	$= (2n-1)(2n-3)\cdots 5\cdot 3\cdot 1$, 1041.
$\left({}_{j}{}^{i}{}_{k}\right)$	Christoffel symbol of the second kind, 962.
$\dbinom{m}{n}$	$= \dfrac{m(m-1)(m-2)\cdots(m-n+1)}{n!}$, 540.
$\begin{pmatrix} l_1 & l_2 & l_3 \\ m_1 & m_2 & m_3 \end{pmatrix}$	Wigner's symbol, 1032.
$\hat{f}(t)$	Hilbert transform of $f(t)$, 98, 1000; see also Eq. (4.22), 155.
$O[f(z)]$	Order of magnitude of $f(z)$, 976, 986, 988.
$\langle f(t)\rangle$	Average of $f(t)$ over a cycle, 31.
$\langle\mathfrak{T}\rangle$	Vector of the dyadic $\mathfrak{T}$, 948.
$[\mathfrak{T}]$	Matrix of the dyadic $\mathfrak{T}$, 948.
$(\mathbf{abc})^{213}$	Left transpose of a triadic, 954.
$(\mathbf{abc})^{132}$	Right transpose of a triadic, 954.

List of Abbreviations

ARDC	Air Research and Development Command
arg	Argument
ch	Hyperbolic cosine
cos	Circular cosine
cth	Hyperbolic cotangent
div	Divergence
EW	East-West
G-B	Gutenberg-Bullard
GEA	Geometric Elastodynamic Approximation
GMT	Greenwich Mean Time
grad	Gradient
GRT	Generalized Ray Theory
Im	Imaginary part
J-B	Jeffreys-Bullen
L	Love-wave
ln	Natural logarithm
LPZ	Long-Period Vertical Seismogram
max	Maximum
min	Minimum
Moho	Mohorovičić discontinuity
MT	Megaton
NS	North-South
P	Primary (Longitudinal)
P	Principal-value
R	Rayleigh-wave
Re	Real part
R/T	Reflection/Transmission
S	Secondary (shear)
sgn	Sign
sh	Hyperbolic sine
SH	Horizontally polarized shear
sin	Circular sine
SV	Vertically polarized shear
tg	Circular tangent
th	Hyperbolic tangent
UT	Universal Time
WKBJ	Wentzel-Kramers-Brillouin-Jeffreys
WWNSS	Worldwide Network of Standardized Stations

Subject Index

A CATALOG OF SELECTED

DOVER BOOKS

IN SCIENCE AND MATHEMATICS

A CATALOG OF SELECTED

DOVER BOOKS

IN SCIENCE AND MATHEMATICS

QUALITATIVE THEORY OF DIFFERENTIAL EQUATIONS, V.V. Nemytskii and V.V. Stepanov. Classic graduate-level text by two prominent Soviet mathematicians covers classical differential equations as well as topological dynamics and ergodic theory. Bibliographies. 523pp. 5⅜ x 8½. 65954-2 Pa. $14.95

MATRICES AND LINEAR ALGEBRA, Hans Schneider and George Phillip Barker. Basic textbook covers theory of matrices and its applications to systems of linear equations and related topics such as determinants, eigenvalues and differential equations. Numerous exercises. 432pp. 5⅜ x 8½. 66014-1 Pa. $12.95

QUANTUM THEORY, David Bohm. This advanced undergraduate-level text presents the quantum theory in terms of qualitative and imaginative concepts, followed by specific applications worked out in mathematical detail. Preface. Index. 655pp. 5⅜ x 8½. 65969-0 Pa. $15.95

ATOMIC PHYSICS (8th edition), Max Born. Nobel laureate's lucid treatment of kinetic theory of gases, elementary particles, nuclear atom, wave-corpuscles, atomic structure and spectral lines, much more. Over 40 appendices, bibliography. 495pp. 5⅜ x 8½. 65984-4 Pa. $13.95

ELECTRONIC STRUCTURE AND THE PROPERTIES OF SOLIDS: The Physics of the Chemical Bond, Walter A. Harrison. Innovative text offers basic understanding of the electronic structure of covalent and ionic solids, simple metals, transition metals and their compounds. Problems. 1980 edition. 582pp. 6⅛ x 9¼. 66021-4 Pa. $19.95

BOUNDARY VALUE PROBLEMS OF HEAT CONDUCTION, M. Necati Özisik. Systematic, comprehensive treatment of modern mathematical methods of solving problems in heat conduction and diffusion. Numerous examples and problems. Selected references. Appendices. 505pp. 5⅜ x 8½. 65990-9 Pa. $12.95

A SHORT HISTORY OF CHEMISTRY (3rd edition), J.R. Partington. Classic exposition explores origins of chemistry, alchemy, early medical chemistry, nature of atmosphere, theory of valency, laws and structure of atomic theory, much more. 428pp. 5⅜ x 8½. (Available in U.S. only) 65977-1 Pa. $12.95

A HISTORY OF ASTRONOMY, A. Pannekoek. Well-balanced, carefully reasoned study covers such topics as Ptolemaic theory, work of Copernicus, Kepler, Newton, Eddington's work on stars, much more. Illustrated. References. 521pp. 5⅜ x 8½. 65994-1 Pa. $15.95

PRINCIPLES OF METEOROLOGICAL ANALYSIS, Walter J. Saucier. Highly respected, abundantly illustrated classic reviews atmospheric variables, hydrostatics, static stability, various analyses (scalar, cross-section, isobaric, isentropic, more). For intermediate meteorology students. 454pp. 6⅛ x 9¼. 65979-8 Pa. $14.95

ORDINARY DIFFERENTIAL EQUATIONS, Morris Tenenbaum and Harry Pollard. Exhaustive survey of ordinary differential equations for undergraduates in mathematics, engineering, science. Thorough analysis of theorems. Diagrams. Bibliography. Index. 818pp. 5⅜ x 8½. 64940-7 Pa. $19.95

STATISTICAL MECHANICS: Principles and Applications, Terrell L. Hill. Standard text covers fundamentals of statistical mechanics, applications to fluctuation theory, imperfect gases, distribution functions, more. 448pp. 5⅜ x 8½. 65390-0 Pa. $14.95

ORDINARY DIFFERENTIAL EQUATIONS AND STABILITY THEORY: An Introduction, David A. Sánchez. Brief, modern treatment. Linear equation, stability theory for autonomous and nonautonomous systems, etc. 164pp. 5⅜ x 8¼. 63828-6 Pa. $6.95

THIRTY YEARS THAT SHOOK PHYSICS: The Story of Quantum Theory, George Gamow. Lucid, accessible introduction to influential theory of energy and matter. Careful explanations of Dirac's anti-particles, Bohr's model of the atom, much more. 12 plates. Numerous drawings. 240pp. 5⅜ x 8½. 24895-X Pa. $7.95

THEORY OF MATRICES, Sam Perlis. Outstanding text covering rank, nonsingularity and inverses in connection with the development of canonical matrices under the relation of equivalence, and without the intervention of determinants. Includes exercises. 237pp. 5⅜ x 8½. 66810-X Pa. $8.95

GREAT EXPERIMENTS IN PHYSICS: Firsthand Accounts from Galileo to Einstein, edited by Morris H. Shamos. 25 crucial discoveries: Newton's laws of motion, Chadwick's study of the neutron, Hertz on electromagnetic waves, more. Original accounts clearly annotated. 370pp. 5⅜ x 8½. 25346-5 Pa. $11.95

INTRODUCTION TO PARTIAL DIFFERENTIAL EQUATIONS WITH APPLICATIONS, E.C. Zachmanoglou and Dale W. Thoe. Essentials of partial differential equations applied to common problems in engineering and the physical sciences. Problems and answers. 416pp. 5⅜ x 8½. 65251-3 Pa. $11.95

BURNHAM'S CELESTIAL HANDBOOK, Robert Burnham, Jr. Thorough guide to the stars beyond our solar system. Exhaustive treatment. Alphabetical by constellation: Andromeda to Cetus in Vol. 1; Chamaeleon to Orion in Vol. 2; and Pavo to Vulpecula in Vol. 3. Hundreds of illustrations. Index in Vol. 3. 2,000pp. 6⅛ x 9¼. 23567-X, 23568-8, 23673-0 Pa., Three-vol. set $46.85

CHEMICAL MAGIC, Leonard A. Ford. Second Edition, Revised by E. Winston Grundmeier. Over 100 unusual stunts demonstrating cold fire, dust explosions, much more. Text explains scientific principles and stresses safety precautions. 128pp. 5⅜ x 8½. 67628-5 Pa. $5.95

AMATEUR ASTRONOMER'S HANDBOOK, J.B. Sidgwick. Timeless, comprehensive coverage of telescopes, mirrors, lenses, mountings, telescope drives, micrometers, spectroscopes, more. 189 illustrations. 576pp. 5⅜ x 8¼. (Available in U.S. only) 24034-7 Pa. $13.95

GEOMETRY OF COMPLEX NUMBERS, Hans Schwerdtfeger. Illuminating, widely praised book on analytic geometry of circles, the Moebius transformation, and two-dimensional non-Euclidean geometries. 200pp. 5⅝ x 8¼. 63830-8 Pa. $8.95

MECHANICS, J.P. Den Hartog. A classic introductory text or refresher. Hundreds of applications and design problems illuminate fundamentals of trusses, loaded beams and cables, etc. 334 answered problems. 462pp. 5⅜ x 8½. 60754-2 Pa. $12.95

TOPOLOGY, John G. Hocking and Gail S. Young. Superb one-year course in classical topology. Topological spaces and functions, point-set topology, much more. Examples and problems. Bibliography. Index. 384pp. 5⅝ x 8¼. 65676-4 Pa. $11.95

STRENGTH OF MATERIALS, J.P. Den Hartog. Full, clear treatment of basic material (tension, torsion, bending, etc.) plus advanced material on engineering methods, applications. 350 answered problems. 323pp. 5⅜ x 8½. 60755-0 Pa. $10.95

ELEMENTARY CONCEPTS OF TOPOLOGY, Paul Alexandroff. Elegant, intuitive approach to topology from set-theoretic topology to Betti groups; how concepts of topology are useful in math and physics. 25 figures. 57pp. 5⅜ x 8½.
60747-X Pa. $4.95

ADVANCED STRENGTH OF MATERIALS, J.P. Den Hartog. Superbly written advanced text covers torsion, rotating disks, membrane stresses in shells, much more. Many problems and answers. 388pp. 5⅜ x 8½. 65407-9 Pa. $11.95

COMPUTABILITY AND UNSOLVABILITY, Martin Davis. Classic graduate-level introduction to theory of computability, usually referred to as theory of recurrent functions. New preface and appendix. 288pp. 5⅜ x 8½. 61471-9 Pa. $8.95

GENERAL CHEMISTRY, Linus Pauling. Revised 3rd edition of classic first-year text by Nobel laureate. Atomic and molecular structure, quantum mechanics, statistical mechanics, thermodynamics correlated with descriptive chemistry. Problems. 992pp. 5⅜ x 8½. 65622-5 Pa. $19.95

AN INTRODUCTION TO MATRICES, SETS AND GROUPS FOR SCIENCE STUDENTS, G. Stephenson. Concise, readable text introduces sets, groups, and most importantly, matrices to undergraduate students of physics, chemistry, and engineering. Problems. 164pp. 5⅜ x 8½. 65077-4 Pa. $7.95

THE HISTORICAL BACKGROUND OF CHEMISTRY, Henry M. Leicester. Evolution of ideas, not individual biography. Concentrates on formulation of a coherent set of chemical laws. 260pp. 5⅜ x 8½. 61053-5 Pa. $8.95

THE PHILOSOPHY OF MATHEMATICS: An Introductory Essay, Stephan Körner. Surveys the views of Plato, Aristotle, Leibniz & Kant concerning propositions and theories of applied and pure mathematics. Introduction. Two appendices. Index. 198pp. 5⅜ x 8½. 25048-2 Pa. $8.95

THE DEVELOPMENT OF MODERN CHEMISTRY, Aaron J. Ihde. Authoritative history of chemistry from ancient Greek theory to 20th-century innovation. Covers major chemists and their discoveries. 209 illustrations. 14 tables. Bibliographies. Indices. Appendices. 851pp. 5⅜ x 8½. 64235-6 Pa. $18.95